Developments in Mathematics

Volume 76

Aims and Scope

The **Developments in Mathematics** (DEVM) book series is devoted to publishing well-written monographs within the broad spectrum of pure and applied mathematics. Ideally, each book should be self-contained and fairly comprehensive in treating a particular subject. Topics in the forefront of mathematical research that present new results and/or a unique and engaging approach with a potential relationship to other fields are most welcome. High-quality edited volumes conveying current state-of-the-art research will occasionally also be considered for publication. The DEVM series appeals to a variety of audiences including researchers, postdocs, and advanced graduate students.

Dorina Mitrea · Irina Mitrea · Marius Mitrea

Geometric Harmonic Analysis V

Fredholm Theory and Finer Estimates for Integral Operators, with Applications to Boundary Problems

Springer

Dorina Mitrea
Department of Mathematics
Baylor University
Waco, TX, USA

Irina Mitrea
Department of Mathematics
Temple University
Philadelphia, PA, USA

Marius Mitrea
Department of Mathematics
Baylor University
Waco, TX, USA

ISSN 1389-2177 ISSN 2197-795X (electronic)
Developments in Mathematics
ISBN 978-3-031-31560-2 ISBN 978-3-031-31561-9 (eBook)
https://doi.org/10.1007/978-3-031-31561-9

Mathematics Subject Classification: 32A, 26B20, 31B, 35J, 42B

This Springer imprint is published by the registered company Springer Nature Switzerland AG
The registered company address is: Gewerbestrasse 11, 6330 Cham, Switzerland

Dedicated with love to our parents

Description of Volume V

The current work is part of a series, comprised of five volumes, [129], [130], [131], [132], [133]. In broad terms, the principal aim is to develop tools in Real and Harmonic Analysis, of geometric measure theoretic flavor, capable of treating a broad spectrum of boundary value problems formulated in rather general geometric and analytic settings.

In Volume I ([129]) we establish a sharp version of Divergence Theorem (aka Fundamental Theorem of Calculus) which allows for an inclusive class of vector fields whose boundary trace is only assumed to exist in a nontangential pointwise sense.

Volume II ([130]) is concerned with function spaces measuring size and/or smoothness, such as Hardy spaces, Besov spaces, Triebel-Lizorkin spaces, Sobolev spaces, Morrey spaces, Morrey-Campanato spaces, spaces of functions of Bounded Mean Oscillations, etc., in general geometric settings.Work here also highlights the close interplay between differentiability properties of functions and singular integral operators.

The topic of singular integral operators is properly considered in Volume III ([131]), where we develop a versatile Calderón-Zygmund theory for singular integral operators of convolution type (and with variable coefficient kernels) on uniformly rectifiable sets in the Euclidean ambient, and the setting of Riemannian manifolds. Applications to scattering by rough obstacles are also discussed in this volume.

In Volume IV ([132]) we focus on singular integral operators of boundary layer type which enjoy more specialized properties (compared with generic, garden variety singular integral operators treated earlier in Volume III). Applications to Complex Analysis in several variables are subsequently presented, starting from the realizations that many natural integral operators in this setting, such as the Bochner-Martinelli operator, are actual particular cases of double layer potential operators associated with the complex Laplacian.

The ultimate goal in (the current) Volume V is to prove well-posedness and Fredholm solvability results concerning boundary value problems for elliptic second-order homogeneous constant (complex) coefficient systems, and domains of a rather general geometric nature. The formulation of the boundary value problems treated here is optimal from a multitude of points of view, having to do with:

(1) Geometry: as we identify the most natural generalization of the class of bounded $\mathscr{C}^1$ domains from the perspective of geometric measure theory (through the consideration of infinitesimally flat AR domains and other related classes of sets quantifying flatness in Chapter 3);
(2) Functional Analysis: through the consideration of a large variety of function spaces, without restrictions on the indices involved and including spaces whichare not locally convex, as boundary data;
(3) Topology: as no explicit topological assumptions are made on the underlying set; this being said, the sort of flatness assumptions we make (see item (1) above) does have a number of topological implications.
(4) Partial Differential Equations: through the consideration of strongly elliptic second-order homogeneous systems with constant (complex) coefficients, without any symmetry or additional positivity assumptions made on the associated bilinear form.

In relation to item (4), we wish to note that the Fredholm solvability for the Dirichlet Problem may fail in the absence of Legendre-Hadamard strong ellipticity, even when the underlying domain is smooth. To substantiate this claim, work in the two-dimensional setting and consider the second-order, homogeneous, real constant coefficient, weakly elliptic 2×2 system

$$\mathbb{L}_B := \frac{1}{4}\begin{pmatrix} \partial_x^2 - \partial_y^2 & -2\partial_x\partial_y \\ 2\partial_x\partial_y & \partial_x^2 - \partial_y^2 \end{pmatrix}. \tag{0.0.1}$$

This is closely related to Bitsadze's (scalar) operator

$$L_B := \partial_{\bar{z}}^2 = \frac{1}{4}\big(\partial_x^2 - \partial_y^2 + 2\mathrm{i}\partial_x\partial_y\big), \tag{0.0.2}$$

where $\partial_{\bar{z}} := \frac{1}{2}\big(\partial_x - \frac{1}{\mathrm{i}}\partial_y\big)$ is the Cauchy-Riemann operator in the plane $\mathbb{R}^2 \equiv \mathbb{C}$. Specifically, for any complex-valued function u of class $\mathscr{C}^2$ we have

$$\mathbb{L}_B(\operatorname{Re} u, \operatorname{Im} u) = \big(\operatorname{Re}\big(\partial_{\bar{z}}^2 u\big), \operatorname{Im}\big(\partial_{\bar{z}}^2 u\big)\big). \tag{0.0.3}$$

Hence, if

$$\mathbb{D} := \{z \in \mathbb{C} : |z| < 1\} \tag{0.0.4}$$

denotes the unit disk in the plane, then $\{u_k\}_{k\in\mathbb{N}} \subseteq \big[\mathscr{C}^\infty\big(\overline{\mathbb{D}}\big)\big]^2$ with

$$u_k(z) := \Big(\mathrm{Re}\big[(1-|z|^2)z^k\big],\ \mathrm{Im}\big[(1-|z|^2)z^k\big]\Big),\ z \in \overline{\mathbb{D}},\ k \in \mathbb{N}_0, \tag{0.0.5}$$

is an infinite family of linearly independent null-solutions for the classical Dirichlet Problem for the homogeneous, second-order, constant (real) coefficient system $\mathbb{L}_B$ in the (smooth, bounded) domain $\Omega := \mathbb{D}$. While the system $\mathbb{L}_B$ is weakly elliptic, it fails to satisfy the Legendre-Hadamard strong ellipticity condition.

A higher-dimensional version of this counterexample to the Fredholm solvability of the Dirichlet Problem for a weakly elliptic second-order homogeneous systems with constant (complex) coefficients in the upper half-space $\mathbb{R}^n_+$, with $n \geq 2$ arbitrary, is discussed at length in Chapter 2. In this scenario, the role of $\mathbb{L}_B$ is played by the weakly elliptic, homogeneous, second-order, constant (real) coefficient $n \times n$ system

$$L_D := \Delta - 2\nabla\mathrm{div}. \tag{0.0.6}$$

Our work in the present volume aligns itself with the program stemming from A.P. Calderón's 1978 ICM plenary address in which he advocates the use of layer potentials "*for much more general elliptic systems* [*than the Laplacian*]" – see [28, p. 90], and may be regarded as a far-reaching, optimal extension, of the pioneering work of E.B. Fabes, M. Jodeit, and N.M. Rivière in [52] (where layer potential methods have been first used to solve boundary value problems for the Laplacian in bounded $\mathscr{C}^1$ domains). In this endeavor, we have been also motivated by the problem posed by A.P. Calderón on [28, p. 95], asking to identify the function spaces on which singular integral operators (of boundary layer type) are well defined and continuous. This is relevant since, as Calderón mentions, "*A clarification of this question would be very important in the study of boundary value problems for elliptic equations* [*in rough domains*]. *The methods employed so far seem to be insufficient for the treatment of these problems.*" In this vein, our work in this volume is also in line with the issue raised as an open problem by C. Kenig in [93, Problem 3.2.2, pp. 116-117], where he asked whether operators of layer potential type may be inverted on appropriate Lebesgue and Sobolev spaces in suitable subclasses on NTA domains with compact Ahlfors regular boundaries.

In the general geometric/functional analytic setting we presently consider, methods based on positivity, maximum principles, or harmonic measures are hopelessly inadequate. Throughout, we employ boundary layer potentials and (in the case of domains with compact boundaries) rely on Fredholm theory. Even though there are infinitely many double layer potential operators associated with a given weakly elliptic second-order system, most (typically, all except possibly one) fail to be sufficiently in tune with the geometry of the underlying domain (more specifically, fail to be sensitive to flatness). It is only for what we call "distinguished" coefficient tensors that the corresponding boundary-to-domain double layer operators respond well to the (global/infinitesimal) flatness of the "surface" on which they are defined.

We shall often work in Chapter 8 under the hypothesis that the elliptic second order system possesses such a "distinguished" coefficient tensor. However, this is by necessity, since its absence may cause the boundary value problem for the system in question to fail to be Fredholm solvable. For example, this is the case for the

Dirichlet Problem for the system $\mathbb{L}_B$ from (0.0.1) in the unit disk in the plane, and for the Dirichlet Problem for the system L_D from (0.0.6) in the upper half-space $\mathbb{R}^n_+$, with $n \geq 2$ arbitrary (again, see the discussion in Chapter 2).

It is in the treatment of boundary value problems (in Chapter 8) where everything comes together. In Volumes I-IV we have already developed a great deal of machinery which is essential in this endeavor though, to accomplish this task, a number of outstanding issues remain. These are dealt with elsewhere in the present volume. Below, we briefly elaborate on the contents of this volume.

The notion of distinguished coefficient tensor (already referred to above) is introduced and studied in Chapter 1 where, among other things, issues such as equivalent definitions, existence, uniqueness, stability under transposition are discussed. Here we also provide relevant examples of weakly elliptic homogeneous constant (complex) coefficient second-order systems possessing distinguished coefficient tensors, including certain Lamé-like systems, and the entire class of *scalar* weakly elliptic homogeneous constant (complex) coefficient second-order operators in $\mathbb{R}^n$ with $n \geq 3$.

As noted earlier, the principal task in Chapter 2 is to produce concrete examples, in all space dimensions, of weakly elliptic homogeneous constant (complex) coefficient second-order systems with the property that their associated L^p Dirichlet Problems in the upper half-space fail to be Fredholm solvable. This is a delicate task, particularly in view of the fact that the L^p Dirichlet Problem for second-order, homogeneous, constant (complex) coefficient systems satisfying the *strong* Legendre-Hadamard ellipticity condition is actually well posed in the upper half-space (see, e.g., the discussion in [114]). The manner in which this ties up with our earlier work in Chapter 1 is that we look for such pathological weakly elliptic systems in the class of those which fail to possess a distinguished coefficient tensor.

The main issue addressed in Chapter 3 is how to quantify global and infinitesimal flatness in classes of Euclidean sets of locally finite perimeter which may otherwise lack structural qualities which have traditionally been used to describe regularity (like being locally the upper-graphs of functions). The philosophy emerging from the body of results developed here (and subsequently reinforced by work in future chapters) is that measuring the mean oscillations of the geometric measure theoretic outward unit normal (either globally, or at an infinitesimal scale) is an effective way of quantifying the flatness of a large category of sets which is congruous with the functional analytic nature of singular integral operators of layer potential type and, ultimately, with the solvability properties of the boundary value problems formulated on such sets.

The goal in Chapter 4 is to estimate singular integral operators of "chord-dot-normal" type (of the sort introduced earlier in [132, §5.2]), associated with Ahlfors regular domains with unbounded boundaries, in a manner which makes it clear how the flatness of said domains (measured as indicated above) affects the size of these operators on various function spaces. Singular integral operators of "chord-dot-normal" type are sensitive to flatness, in the sense that they become identically zero when the underlying domain is a half-space in $\mathbb{R}^n$. By way of contrast, such a feature is not enjoyed by the Cauchy singular integral operator (to give just one example). Indeed, when considered in relation to the upper half-plane, the principal value Cauchy integral operator becomes the Hilbert transform on the real line. What makes the difference is the algebraic structure of the integral kernel, specifically the presence, or absence, of the inner product between the geometric measure theoretic outward unit normal ν with the "chord" $x - y$. In fact, this is not an accident, since here we show (see Theorem 4.1.10) that this is the only algebraic template (for SIO's whose integral kernel depends "linearly" on ν) guaranteeing "sensitivity to flatness" as interpreted above. The manner in which this work ties up with boundary layer potentials is as follows. For a given weakly elliptic, homogeneous, constant (complex) coefficient, second-order system L in $\mathbb{R}^n$ which has a distinguished coefficient tensor A, we may then conclude that the corresponding double layer potentials K_A, $K_A^{\#}$ (of the sort discussed at length in Volume IV; cf. [132]), associated with A in a UR domain Ω are of "chord-dot-normal" type. As such, the present results apply and prove that K_A, $K_A^{\#}$ have small operator norms if the BMO-seminorm of the De Giorgi-Federer outward unit normal ν to Ω is small enough. In turn, this permits us to invert $\pm\frac{1}{2}I + K_A$ and $\pm\frac{1}{2}I + K_A^{\#}$ (which are relevant in the context of boundary value problems for the system L) using a Neumann series argument. This program works well in domains with *unbounded boundaries*, since the demand on the smallness of $\|\nu\|_{\mathrm{BMO}}$ automatically implies that $\partial\Omega$ is unbounded.

In Chapter 5 we pursue similar aims, now working in UR domains $\Omega \subseteq \mathbb{R}^n$ with *compact boundaries*. In such a setting, it is the task of estimating the distance to the space of compact operators (aka the *essential norm*) for a singular integral operator of "chord-dot-normal" type on $\partial\Omega$ that becomes the focal point of our work (in place of the operator norm, as was the case before, in Chapter 4). Also, in place of the BMO-seminorm of the De Giorgi-Federer outward unit normal ν for the domain Ω, this essential norm must now be controlled in terms of the distance (measured in the John-Nirenberg space BMO) of the normal vector ν to the Sarason space VMO, of functions of vanishing mean oscillations on $\partial\Omega$. This turns out to be a vastly intricate affair, but the rewards are very palpable. Indeed, the results in Chapters 4-5 have transformational impact: what used to be a rather monotone, featureless landscape of mostly generic integral operators of Calderón-Zygmund type on uniformly rectifiable sets is now populated by classes of SIO's with distinct, remarkable characteristics, well-suited for dealing with boundary value problems (as we treat in Chapter 8).

In Chapter 6, in honor of the pioneering work of J. Radon and T. Carleman to singular integral operators in potential theory, we coin the term *Radon-Carleman Problem* as a broad label for questions having to do with computing and/or estimating the essential norm and/or Fredholm radius of singular integral operators of double layer type, associated with elliptic operators/systems, on function spaces naturally intervening in the formulation of boundary value problems for said partial differential operators. In this chapter we review the genesis of this topic and produce new results, culminating with Theorem 6.3.3 which, in essence, is our principal contribution to the Radon-Carleman Problem.

Chapter 7 may be subsumed into a broad, central theme in Geometric Measure Theory, namely understanding how geometric qualities translate into analytic properties. Specifically, here we are concerned with the correlation between the infinitesimal flatness of a given set of locally finite perimeter, quantified in terms of the mean oscillations of its geometric measure theoretic outward unit normal, and the functional analytic nature of singular integral operators defined on the boundary of the set in question. Succinctly put, "good geometry" plus "chord-dot-normal" structure ensure favorable functional analytic properties for double layer potentials which, in turn, permit one to solve boundary value problems via layer potential operators. We pursue this principle in its natural progression: in this chapter we establish Fredholm and invertibility properties of boundary layer potentials on compact surfaces, then in §8.1 we use these results to treat boundary value problems in domains with compact boundaries. In a larger picture, this body of work points to the fact that: *the category of δ-oscillating* AR *domains, with $\delta \in (0, 1)$ suitably small (introduced in Chapter* 3) *is a most natural geometric environment where Fredholm theory, as originally envisioned by Ivar Fredholm, becomes applicable to boundary layer potential operators, thus making it possible to solve boundary value problems via layer potential methods for weakly elliptic systems.*

Lastly, in Chapter 8 we formulate and solve boundary value problems for elliptic second-order systems, involving a large variety of boundary conditions and function spaces (as boundary data), in geometric setting displaying new levels of generality and inclusiveness, compared with previouswork. For example,we are able to successfully implement the method of boundary layer potentials for the Dirichlet/ Neumann Problems with data in Muckenhouptweighted Lebesgue spaces, as well as Hardy, Sobolev, BMO, VMO, Hölder, Morrey, Besov, Triebel-Lizorkin spaces, for elliptic second-order systems with complex coefficients in domains that are allowed to develop certain spiral singularities on their boundaries. In this chapter we also take the first steps in the direction of combining geometric measure theory with scattering theory, by solving the basic boundary value problems in acoustic theory in novel geometric settings. Some of the virtues of the work undertaken here have already been heralded in the first part of the current narrative, so we shall not repeat them. Instead, we wish to close by once more pondering on the fundamental role played by the brand of Divergence Theorem developed in Volume I ([129]) throughout this journey, from its origins all the way to the work in the present chapter.

One representative chain of implications goes as follows: The Divergence Theorem from [129, Chapter 1] is the key ingredient in the proof of the Green representation formulas deduced in [131, §1.5]. These integral representation formulas, together with nontangential trace and jump-formulas from [131, §2.5], are then used to produce (composition) operator identities of the sort described in [132, Theorem 1.5.1, item (*xiii*)] for boundary layer potentials. In concert with the Fredholmness results obtained in Chapter 7 for $\pm\frac{1}{2}I + K$ and $\pm\frac{1}{2}I + K^{\#}$, such operator identities then permit us to conclude that the (boundary-to-boundary) single layer operator S associated with an elliptic second-order system which possesses a distinguished coefficient tensor is itself Fredholm with index zero (see Corollary 7.1.6). Granted the injectivity results for S from [132, §1.7], we then arrive at the conclusion that S actually an invertible operator. For details, see Theorem 7.1.9, Theorem 7.1.10, Theorem 7.1.13, Theorem 7.1.14, Theorem 7.1.15, Theorem 7.1.17, Theorem 7.1.18, Theorem 7.1.19, Theorem 7.1.20, Theorem 7.1.34, Theorem 7.1.35, Theorem 7.1.39, and Theorem 7.1.40 in this regard.

Ultimately, this allows us to prove solvability results for the Dirichlet and Regularity Problems for the system in question using a single layer representation for the solution. This blueprint may be successfully implemented for a large variety of function spaces and, as opposed to an approach which favors inverting the double layer potential operator ($\pm\frac{1}{2}I + K$ and $\pm\frac{1}{2}I + K^{\#}$ to be exact), this has the distinct virtue of not requiring any additional topological properties for the underlying domain (inverting double layer operators very much depends on the topology of the domain; see Theorem 7.1.11).

Acknowledgments: The authors gratefully acknowledge partial support from the Simons Foundation (through grants # 426669, # 958374, # 318658, # 616050, # 637481), as well as NSF (grant # 1900938). Portions of this work have been completed at Baylor University in Waco, Temple University in Philadelphia, the Institute for Advanced Study in Princeton, MSRI in Berkeley, and the American Institute of Mathematics in San Jose. We wish to thank these institutions for their generous hospitality. Last, but not least, we are grateful to Michael E. Taylor for gently yet persistently encouraging us over the years to complete this project.

Contents

Check for updates

Chapter 1
Distinguished Coefficient Tensors

One of the main goals in the current volume is to employ boundary layer potential operators associated with a given second-order weakly elliptic system L and a given uniformly rectifiable domain $\Omega \subseteq \mathbb{R}^n$ for the purpose of solving boundary value problems for L in Ω. Any such system L may be expressed in infinitely many equivalent algebraic formats, each of which employs a specific coefficient tensor. In turn, each coefficient tensor gives rise to a distinct double layer potential operator. While all these integral operators share many common properties (such as nontangential maximal function estimates, boundedness properties, jump-relations, etc.), some more specialized functional analytic features are heavily dependent on the nature of the coefficient tensor involved.

We are particularly interested in those double layer potential operators which are of "chord-dot-normal" type (see [132, Theorem 5.2.2]), in case they exist. In particular, the boundary-to-boundary versions of such double layer potential operators vanish whenever the underlying domain is a half-space. Coefficient tensors giving rise to such double layers of "chord-dot-normal" type will be referred to as *distinguished coefficient tensors*.

To further motivate and elaborate on this aspect, consider the task of solving (the interior and exterior) L^2 Dirichlet Problem for a given second-order, weakly elliptic, homogeneous, constant (complex) coefficient, $M \times M$ system L in $\mathbb{R}^n$ in some UR domain $\Omega \subseteq \mathbb{R}^n$ using a double layer potential operator. There are infinitely many choices of a coefficient tensors $A = \big(a_{rs}^{\alpha\beta}\big)_{\substack{1\leq r,s\leq n\\ 1\leq\alpha,\beta\leq M}}$ with complex entries, with the property that the $M \times M$ given system may be expressed as

$$L = \big(a_{rs}^{\alpha\beta}\partial_r\partial_s\big)_{1\leq\alpha,\beta\leq M} \tag{1.0.1}$$

with the summation convention over repeated indices in effect. Associated with Ω and each such coefficient tensor A we may then consider a boundary-to-boundary double layer potential operator K_A, as in (A.0.116). In view of the format of the jump-formula for the corresponding boundary-to-domain double layer operator $\mathcal{D}_A$ (cf. [132, Theorem 1.5.1, item *(iv)*]), we are motivated to attempt to invert $\pm\frac{1}{2}I + K_A$ on the space $\big[L^2(\partial\Omega,\sigma)\big]^M$.

It turns out that the latter issue is drastically affected by the choice of the coefficient tensor A. As an example, consider the case when $L := \Delta$ is the Laplacian in $\mathbb{R}^2$.

D. Mitrea et al., *Geometric Harmonic Analysis V*, Developments in Mathematics 76,
https://doi.org/10.1007/978-3-031-31561-9_1

Then $n = 2$ and $M = 1$. In this scalar case, we agree to drop the Greek superscripts labeling the entries of the coefficient tensor A used to write the operator $L = \Delta$ explicitly. Hence, we shall consider writings $\Delta = a_{jk}\partial_j\partial_k$ corresponding to various choices of the matrix $A = (a_{jk})_{1\leq j,k\leq 2} \in \mathbb{C}^{2\times 2}$. Two such natural choices are

$$A_0 := \begin{pmatrix} 1 & 0 \\ 0 & 1 \end{pmatrix}, \qquad A_1 := \begin{pmatrix} 1 & \mathrm{i} \\ -\mathrm{i} & 1 \end{pmatrix}, \tag{1.0.2}$$

corresponding to which the recipe given in (A.0.116) yields singular integral operators acting on each function $f \in L^2(\partial\Omega, \sigma)$ according to

$$K_{A_0} f(x) = \lim_{\varepsilon\to 0^+} \frac{1}{2\pi} \int_{\partial\Omega\setminus\overline{B(x,\varepsilon)}} \frac{\langle \nu(y), y - x\rangle}{|x - y|^2} f(y)\,\mathrm{d}\sigma(y) \tag{1.0.3}$$

for σ-a.e. $x \in \partial\Omega$, i.e., the classical (two-dimensional) harmonic boundary-to-boundary double layer potential operator and, under the natural identification $\mathbb{R}^2 \equiv \mathbb{C}$,

$$K_{A_1} f(z) = \lim_{\varepsilon\to 0^+} \frac{1}{2\pi\mathrm{i}} \int_{\partial\Omega\setminus\overline{B(z,\varepsilon)}} \frac{f(\zeta)}{\zeta - z}\,\mathrm{d}\zeta \quad \text{for } \sigma\text{-a.e. } z \in \partial\Omega, \tag{1.0.4}$$

i.e., the boundary-to-boundary Cauchy integral operator, respectively.

Let us now further specialize matters to the case when $\Omega := \mathbb{R}^2_+$, the upper half-plane. From the very algebraic make up of the classical harmonic boundary-to-boundary double layer operator in (1.0.3) we see that in this case $K_{A_0} = 0$. Hence, on the one hand, $\pm\frac{1}{2}I + K_{A_0} = \pm\frac{1}{2}I$ are trivially invertible on $L^2(\partial\Omega, \sigma)$ in such a scenario. On the other hand, the operator K_{A_1} from (1.0.4) becomes (under the natural identification $\partial\Omega \equiv \mathbb{R}$)

$$K_{A_1} f(x) = \lim_{\varepsilon\to 0^+} \frac{1}{2\pi\mathrm{i}} \int_{\mathbb{R}\setminus[x-\varepsilon,x+\varepsilon]} \frac{f(y)}{y - x}\,\mathrm{d}y \quad \text{for } \mathcal{L}^1\text{-a.e. } x \in \mathbb{R}, \tag{1.0.5}$$

i.e., $K_{A_1} = (i/2)H$ where

$$Hf(x) := \lim_{\varepsilon\to 0^+} \frac{1}{\pi} \int_{\substack{y\in\mathbb{R}\\ |x-y|>\varepsilon}} \frac{f(y)}{x - y}\,\mathrm{d}y \quad \text{for } \mathcal{L}^1\text{-a.e. } x \in \mathbb{R}, \tag{1.0.6}$$

is the classical Hilbert transform on the real line. In particular, the classical formula $H^2 = -I$ (on the space of square-integrable functions on the real line) implies $\big(K_{A_1}\big)^2 = 4^{-1}I$. In turn, this goes to show that in this case we have

$$\big(\tfrac{1}{2}I + K_{A_1}\big)\big(-\tfrac{1}{2}I + K_{A_1}\big) = 0 \text{ on } L^2(\partial\Omega, \sigma), \tag{1.0.7}$$

which precludes the operators $\pm\frac{1}{2}I + K_{A_1}$ from being invertible on $L^2(\partial\Omega, \sigma)$ in such a scenario. This discussion makes it clear that the choice of the coefficient tensor strongly influences the functional analytic properties of the corresponding boundary-to-boundary double layer potential operator.

The above comments bring up the question of determining *which of the many coefficient tensors A that may be used in the representation of the given system L as in* (1.0.1) *give rise to double layer potential operators K_A (via the blueprint* (A.0.116)) *that actually vanish identically whenever the underlying domain Ω is a half-space in $\mathbb{R}^n$*. We shall refer to any such coefficient tensor as being "distinguished." In this chapter our main goal is to identify such "distinguished" coefficient tensors (if they exist) among the infinite library of coefficient tensors of the given system. This task is of a purely algebraic nature, and the work here paves the way for employing boundary layer potentials in the treatment of boundary value problems in domains with unbounded boundaries later on, in §8.4.

The existence of "distinguished" coefficient tensors is also fundamentally relevant in implementing Fredholm theory in the treatment of boundary value problems via boundary layer potentials later on, in §8.1. Indeed, a boundary-to-boundary double layer potential operator, associated with a coefficient tensor used to represent a given weakly elliptic homogeneous constant coefficient second-order system, becomes compact (on, say, Lebesgue spaces) on the boundary of any smooth bounded domain if and only if the coefficient tensor in question is distinguished. See Theorem 5.3.2 in this regard.

It is actually possible to directly indicate how the choice of a coefficient tensor affects the Fredholm properties of boundary layer potential operators considered on domains Ω with compact boundaries. Specifically, in such a setting one is interested in determining whether $\pm\frac{1}{2}I + K_A$ are Fredholm operators on, say, the space $\big[L^2(\partial\Omega,\sigma)\big]^M$. To illustrate how this issue is decisively influenced by the choice of the coefficient tensor A used to expressed the given system L as in (1.0.1), consider (as before) the case when $L := \Delta$ is the Laplacian in $\mathbb{R}^2$ (so $n = 2$ and $M = 1$). This time, take $\Omega := \mathbb{D}$, the unit disk in the plane. Writing $\Delta = a_{jk}\partial_j\partial_k$ corresponding to the two choices of the matrix $A = (a_{jk})_{1\leq j,k\leq 2} \in \mathbb{C}^{2\times 2}$ given in (1.0.2) now leads (in place of (1.0.3)-(1.0.4)) to the singular integral operators acting on each function $f \in L^2(\partial\Omega,\sigma)$ according to

$$K_{A_0}f = \frac{1}{2}\fint_{S^1} f\,\mathrm{d}\mathcal{H}^1, \tag{1.0.8}$$

and, under the natural identification $\mathbb{R}^2 \equiv \mathbb{C}$,

$$K_{A_1}f(z) = \lim_{\varepsilon\to 0^+}\frac{1}{2\pi\mathrm{i}}\int_{S^1\setminus\overline{B(z,\varepsilon)}}\frac{f(\zeta)}{\zeta - z}\,\mathrm{d}\zeta \ \text{ for } \mathcal{H}^1\text{-a.e. } z \in S^1, \tag{1.0.9}$$

i.e., the classical Cauchy singular integral operator on the unit circle, respectively. Then K_{A_0} is a finite-rank operator (since its range consists of constant functions) and continuous, hence compact on $L^2(\partial\Omega,\sigma)$. Thus, on the one hand, $\pm\frac{1}{2}I + K_{A_0}$ are Fredholm operators on $L^2(\partial\Omega,\sigma)$. On the other hand, since the composition of the classical Cauchy singular integral operator on the unit circle with itself is known to be $(1/4)I$ on $L^2(\partial\Omega,\sigma)$, we see that (1.0.7) remains true in the present scenario, and this precludes the operators $\pm\frac{1}{2}I + K_{A_1}$ from being Fredholm on $L^2(\partial\Omega, \sigma)$ in the current setting.

Finally, we wish to note that, through the consideration of Clifford algebras as a substitute for the field of complex numbers, all the above considerations may be extended to the higher-dimensional setting in a fairly straightforward fashion.

The existence and uniqueness of a distinguished coefficient tensor are rather delicate issues. To briefly elaborate on this topic, suppose L is a given second-order, homogeneous, constant complex coefficient, $M \times M$ weakly elliptic system in $\mathbb{R}^n$, where $n \geq 2$. Denote by $\mathfrak{A}_L$ the entire library of coefficient tensors that can be used to write L, and use the symbol $\mathfrak{A}_L^{\rm dis}$ for the subclass of $\mathfrak{A}_L$ consisting of distinguished coefficient tensors.

The scalar case (i.e., when $M = 1$) is basically fully understood. Specifically, it turns out that

$$\text{if } n \geq 3 \text{ then for each weakly elliptic, scalar, homogeneous, second-order operator } L = {\rm div}A\nabla \text{ in } \mathbb{R}^n \text{ with constant complex coefficients, the class } \mathfrak{A}_L^{\rm dis} \text{ consists precisely of one matrix, namely } {\rm sym}\, A := (A + A^\top)/2, \tag{1.0.10}$$

while

$$\mathfrak{A}_L^{\rm dis} = \big\{{\rm sym}\, A\big\} \text{ for every strongly elliptic constant complex coefficient operator } L = {\rm div}A\nabla \text{ in } \mathbb{R}^n \text{ with } n \geq 2. \tag{1.0.11}$$

The situation for genuine systems (i.e., when $M \geq 2$) is vastly more intricate. For one thing, it turns out that there exist second-order, weakly elliptic, constant real coefficient, symmetric systems L with the property that $\mathfrak{A}_L^{\rm dis} = \varnothing$ (see Proposition 1.6.8). In the case of the complex Lamé system with Lamé moduli $\mu, \lambda \in \mathbb{C}$ in the regime guaranteeing weak ellipticity, i.e., for

$$L_{\mu,\lambda} := \mu\Delta + (\mu + \lambda)\nabla{\rm div} \text{ with } \mu \neq 0 \text{ and } 2\mu + \lambda \neq 0, \tag{1.0.12}$$

we have (see the discussion in Remark 1.6.10)

$$\mathfrak{A}_{L_{\mu,\lambda}}^{\rm dis} \neq \varnothing \Longleftrightarrow 3\mu + \lambda \neq 0. \tag{1.0.13}$$

As for generic weakly elliptic systems L, in Theorem 1.5.5 we show that

$$\text{if } \mathfrak{A}_L^{\rm dis} \neq \varnothing \text{ and } \mathfrak{A}_{L^\top}^{\rm dis} \neq \varnothing \text{ then in fact } \mathfrak{A}_L^{\rm dis} = \{A\} \text{ and } \mathfrak{A}_{L^\top}^{\rm dis} = \{A^\top\} \text{ for some } A \in \mathfrak{A}_L \text{ (hence both } \mathfrak{A}_L^{\rm dis} \text{ and } \mathfrak{A}_{L^\top}^{\rm dis} \text{ are singletons).} \tag{1.0.14}$$

If actually L satisfies the Legendre-Hadamard (strong) ellipticity condition (cf. [131, (1.3.4) in Definition 1.3.2]), then

$$\begin{aligned} &\text{either } \mathfrak{A}_L^{\rm dis} = \varnothing \text{ and } \mathfrak{A}_{L^\top}^{\rm dis} = \varnothing, \\ &\text{or } \mathfrak{A}_L^{\rm dis} = \{A\} \text{ and } \mathfrak{A}_{L^\top}^{\rm dis} = \{A^\top\} \text{ for some } A \in \mathfrak{A}_L. \end{aligned} \tag{1.0.15}$$

See Theorem 1.5.2. Lastly, we wish to mention that the issue of existence of a distinguished coefficient tensor for a given weakly elliptic system L is linked to the existence of Poisson kernels for L in arbitrary half-spaces in the Euclidean ambient and their compatibility with one another. This is discussed in Theorem 1.6.4.

In a broader historical perspective, the quest for finding a distinguished coefficient tensor was first carried out by Erik Ivar Fredholm in his fundamental 1906 paper on the Dirichlet Problem in elasticity theory (cf. [64]). Following his earlier breakthrough in employing boundary layer potentials to solve the Dirichlet Problem for the Laplacian in a smooth set with compact boundary (cf. [61], [62]), Fredholm sought to replicate his success to the case of the three dimensional Lamé system of elasticity. However, in stark contrast to the case of the classical (boundary-to-boundary) harmonic double layer operator, the "natural" double layer operator for the Lamé system, corresponding to the so-called stress conormal operator, fails to be weakly singular on smooth compact surfaces[1]. In a surprising twist, Fredholm went on to remedy this by constructing a *different* elastic double layer operator (something based on a conormal derivative for the Lamé system which eventually became known as "pseudo-stress"; cf. [131, (1.7.52)]) which turned out to be weakly singular on smooth compact surfaces. This is fortunate since from (1.0.13)-(1.0.14) we now know that there is at most one such double layer for the Lamé system. In contrast to the *ad hoc* methods of Fredholm, our goal in this chapter is to develop a systematic approach to the task of existence and uniqueness of distinguished coefficient tensors for general weakly elliptic second-order systems.

The contents of this chapter are as follows. In §1.1 we discuss algebraic generalities regarding various natural classes of coefficient tensors. In §1.2 we formally introduce the notion of distinguished coefficient tensor, by adopting a point of view void of any geometric considerations (see Definition 1.2.2). In §1.3 we then elaborate on the significance of distinguished coefficient tensors vis-à-vis to the nature of the double layer potential operators associated with them. In §1.4 we address the following question: is the quality of being a distinguished coefficient tensor stable under transposition? More specifically, if A is a distinguished coefficient tensor for a given weakly elliptic system L, is it true that $A^\top$ is also a distinguished coefficient tensor for $L^\top$? Finally, the issue of uniqueness of a distinguished coefficient tensor is considered in §1.5, while in §1.6 we concern ourselves with the issue of existence of a distinguished coefficient tensor.

1.1 Some Useful Classes of Coefficient Tensors

Fix two background integers $n, M \in \mathbb{N}$. Throughout, denote by $\mathfrak{A}(n, M)$ the collection of blocks of the form $A = \left(a_{rs}^{\alpha\beta}\right)_{\substack{1\leq r,s\leq n\\ 1\leq\alpha,\beta\leq M}}$ with each $a_{rs}^{\alpha\beta}$ a complex number.

[1] a critique of erroneous attempts to employ Fredholm theory for singular integral operators arising in elasticity which fail to be weakly singular is given by Hermann Weyl in [205]

Elements in $\mathfrak{A}(n, M)$ are referred to as `coefficient tensors with complex entries` (of type (n, M)).

Equality of such coefficient tensors is understood entry-by-entry. Hence, the collection of coefficient tensors with constant complex entries is given by

$$\mathfrak{A}(n, M) := \Big\{A = \big(a_{jk}^{\alpha\beta}\big)_{\substack{1\leq\alpha,\beta\leq M\\ 1\leq j,k\leq n}} : \text{ each } a_{jk}^{\alpha\beta} \text{ belongs to } \mathbb{C}\Big\}, \tag{1.1.1}$$

Adopting natural operations (i.e., componentwise addition and multiplication by scalars), this becomes a finite dimensional vector space (over $\mathbb{C}$). We shall endow $\mathfrak{A}(n, M)$ with the norm

$$\|A\| := \sum_{\substack{1\leq\alpha,\beta\leq M\\ 1\leq j,k\leq n}} \big|a_{jk}^{\alpha\beta}\big| \text{ for each } A = \big(a_{jk}^{\alpha\beta}\big)_{\substack{1\leq\alpha,\beta\leq M\\ 1\leq j,k\leq n}} \in \mathfrak{A}(n, M). \tag{1.1.2}$$

In fact, we can turn $\mathfrak{A}(n, M)$ into an algebra (which is typically non-commutative) by defining the multiplication law

$$\begin{gathered} A \boxdot B := \big(a_{rs}^{\alpha\beta} b_{sk}^{\beta\gamma}\big)_{\substack{1\leq r,k\leq n\\ 1\leq\alpha,\gamma\leq M}} \text{ for all} \\ A = \big(a_{rs}^{\alpha\beta}\big)_{\substack{1\leq r,s\leq n\\ 1\leq\alpha,\beta\leq M}} \in \mathfrak{A}(n, M) \text{ and } B = \big(b_{sk}^{\beta\gamma}\big)_{\substack{1\leq s,k\leq n\\ 1\leq\beta,\gamma\leq M}} \in \mathfrak{A}(n, M). \end{gathered} \tag{1.1.3}$$

Above and elsewhere, we adopt the standard convention of summation over repeated indices. Since

$$\|A \boxdot B\| \leq \|A\|\|B\| \text{ for all } A, B \in \mathfrak{A}(n, M) \tag{1.1.4}$$

it follows that $\mathfrak{A}(n, M)$ is in fact a Banach algebra.

Corresponding to $M = 1$, there is a canonical identification of $\mathfrak{A}(n, 1)$ with $n \times n$ complex matrices, i.e.,

$$\mathfrak{A}(n, 1) \ni A = \big(a_{rs}^{11}\big)_{1\leq r,s\leq n} =: \big(a_{rs}\big)_{1\leq r,s\leq n} \in \mathbb{C}^{n\times n}. \tag{1.1.5}$$

Given a coefficient tensor $A = \big(a_{rs}^{\alpha\beta}\big)_{\substack{1\leq r,s\leq n\\ 1\leq\alpha,\beta\leq M}} \in \mathfrak{A}(n, M)$, we let

$$A^{\top} := \big(a_{sr}^{\beta\alpha}\big)_{\substack{1\leq r,s\leq n\\ 1\leq\alpha,\beta\leq M}} \text{ and } A^{c} := \big(\overline{a_{rs}^{\alpha\beta}}\big)_{\substack{1\leq r,s\leq n\\ 1\leq\alpha,\beta\leq M}} \tag{1.1.6}$$

denote its (global) transpose, and its complex conjugate, respectively (compare with [131, (1.7.2)]). Then the assignment $\mathfrak{A}(n, M) \ni A \mapsto A^{\top} \in \mathfrak{A}(n, M)$ is an isometry. Call A `symmetric` if $A = A^{\top}$, and call A `real` if $A = A^{c}$. Clearly, for every $A, B \in \mathfrak{A}(n, M)$ we have

$$(A \boxdot B)^{\top} = B^{\top} \boxdot A^{\top} \text{ and } (A \boxdot B)^{c} = A^{c} \boxdot B^{c}. \tag{1.1.7}$$

Define the transpose of $A = \big(a_{rs}^{\alpha\beta}\big)_{\substack{1\leq r,s\leq n\\ 1\leq\alpha,\beta\leq M}} \in \mathfrak{A}(n, M)$ in the lower indices as

$$A^t := \left(a_{sr}^{\alpha\beta}\right)_{\substack{1\le r,s\le n\\ 1\le\alpha,\beta\le M}}. \tag{1.1.8}$$

Of course, in the case $M = 1$ the two brands of transposition (global, and in the lower indices) defined in (1.1.6) and (1.1.8), respectively, are compatible. That is, $A^\top = A^t$ for all $A \in \mathfrak{A}(n, 1)$. Going further, for any coefficient tensor with complex entries $A = \left(a_{rs}^{\alpha\beta}\right)_{\substack{1\le r,s\le n\\ 1\le\alpha,\beta\le M}}$ of type (n, M) we define the symmetric part of A (in the lower indices) as

$$\operatorname{sym} A := \tfrac{1}{2}\left(A + A^t\right) = \left(\tfrac{1}{2}\left(a_{rs}^{\alpha\beta} + a_{sr}^{\alpha\beta}\right)\right)_{\substack{1\le r,s\le n\\ 1\le\alpha,\beta\le M}}, \tag{1.1.9}$$

and the antisymmetric part of A (in the lower indices) as

$$\operatorname{ant} A := A - \operatorname{sym} A = \tfrac{1}{2}\left(A - A^t\right) = \left(\tfrac{1}{2}\left(a_{rs}^{\alpha\beta} - a_{sr}^{\alpha\beta}\right)\right)_{\substack{1\le r,s\le n\\ 1\le\alpha,\beta\le M}}. \tag{1.1.10}$$

Hence, for every $A, B \in \mathfrak{A}(n, M)$ and $\lambda \in \mathbb{C}$ we have

$$\begin{gathered}\operatorname{sym}(\operatorname{sym} A) = \operatorname{sym} A, \quad \operatorname{sym}(\lambda A) = \lambda \operatorname{sym} A,\\ \text{and } \operatorname{sym}(A + B) = \operatorname{sym} A + \operatorname{sym} B,\end{gathered} \tag{1.1.11}$$

with similar properties satisfied by the antisymmetric part. Call A `symmetric in the lower indices` whenever $\operatorname{sym} A = A$, i.e., when $A^t = A$, and call A `antisymmetric in the lower indices` provided $\operatorname{ant} A = A$, i.e., if $A^t = -A$. Then it is clear from definitions that

for each $A \in \mathfrak{A}(n, M)$, the coefficient tensor $\operatorname{sym} A$ is symmetric in the lower indices, while $\operatorname{ant} A$ is antisymmetric in the lower indices. (1.1.12)

Let us introduce

$$\mathfrak{A}^{\text{ant}}(n, M) := \Big\{A = \left(a_{jk}^{\alpha\beta}\right)_{\substack{1\le\alpha,\beta\le M\\ 1\le j,k\le n}} \in \mathfrak{A}(n, M) : a_{jk}^{\alpha\beta} = -a_{kj}^{\alpha\beta} \text{ for } 1 \le j,k \le n \text{ and } 1 \le \alpha,\beta \le M\Big\}, \tag{1.1.13}$$

i.e., the collection of all coefficient tensors which are antisymmetric in the lower indices. In particular, $\mathfrak{A}^{\text{ant}}(n, M)$ is a closed linear subspace of $\mathfrak{A}(n, M)$. On the collection of all coefficient tensors with complex entries of type (n, M), we may then define the following equivalence relation:

$$A \sim B \overset{\text{def}}{\Longleftrightarrow} A - B \in \mathfrak{A}^{\text{ant}}(n, M). \tag{1.1.14}$$

Then $A \sim \operatorname{sym} A$ for all $A \in \mathfrak{A}(n, M)$, and for every $A, B \in \mathfrak{A}(n, M)$ we have

$$A \sim B \Longleftrightarrow \operatorname{sym} A = \operatorname{sym} B. \tag{1.1.15}$$

Next, define the `Hermitian adjoint` of $A = \left(a_{rs}^{\alpha\beta}\right)_{\substack{1\le r,s\le n\\ 1\le\alpha,\beta\le M}} \in \mathfrak{A}(n, M)$ as

$$A^* := (A^\top)^c = (A^c)^\top = \left(\overline{a^{\beta\alpha}_{sr}}\right)_{\substack{1\le r,s\le n\\ 1\le\alpha,\beta\le M}}. \tag{1.1.16}$$

Also, call a coefficient tensor $A = \left(a^{\alpha\beta}_{rs}\right)_{\substack{1\le r,s\le n\\ 1\le\alpha,\beta\le M}}$ `complex symmetric` (or Hermitian self-adjoint) if $A^* = A$, i.e.,

$$a^{\alpha\beta}_{rs} = \overline{a^{\beta\alpha}_{sr}}, \qquad \forall \alpha,\beta,r,s. \tag{1.1.17}$$

Clearly, for each $A \in \mathfrak{A}(n,M)$ we have

$$A \text{ complex symmetric} \iff A^\top \text{ complex symmetric}. \tag{1.1.18}$$

Given a coefficient tensor with complex entries $A = \left(a^{\alpha\beta}_{rs}\right)_{\substack{1\le r,s\le n\\ 1\le\alpha,\beta\le M}}$, define the bilinear form $\langle A\cdot,\cdot\rangle$ by setting

$$\langle A\zeta,\eta\rangle := a^{\alpha\beta}_{rs}\zeta^\beta_s\eta^\alpha_r, \quad \forall \zeta := (\zeta^\beta_s)_{\beta,s} \in \mathbb{C}^{n\times M}, \quad \forall \eta := (\eta^\alpha_r)_{\alpha,r} \in \mathbb{C}^{n\times M}. \tag{1.1.19}$$

In this regard, it is useful to observe that

$$A \text{ complex symmetric} \Longrightarrow \begin{cases} \langle A\zeta,\overline{\eta}\rangle = \langle \zeta,\overline{A\eta}\rangle \text{ for every } \zeta,\eta \in \mathbb{C}^{n\times M}, \\ \langle A\zeta,\overline{\zeta}\rangle \in \mathbb{R} \text{ for every } \zeta \in \mathbb{C}^{n\times M}. \end{cases} \tag{1.1.20}$$

Definition 1.1.1 *Call* $A = \left(a^{\alpha\beta}_{rs}\right)_{\substack{1\le r,s\le n\\ 1\le\alpha,\beta\le M}} \in \mathfrak{A}(n,M)$ `positive definite` *provided there exists some real number* $\kappa > 0$ *such that*

$$\operatorname{Re}\langle A\zeta,\overline{\zeta}\rangle = \operatorname{Re}\left(a^{\alpha\beta}_{rs}\zeta^\beta_s\overline{\zeta^\alpha_r}\right) \ge \kappa|\zeta|^2, \quad \forall \zeta = (\zeta^\alpha_r)_{\substack{1\le r\le n\\ 1\le\alpha\le M}} \in \mathbb{C}^{n\times M}. \tag{1.1.21}$$

Also, call $A = \left(a^{\alpha\beta}_{rs}\right)_{\substack{1\le r,s\le n\\ 1\le\alpha,\beta\le M}} \in \mathfrak{A}(n,M)$ `positive semi-definite` *if*

$$\operatorname{Re}\langle A\zeta,\overline{\zeta}\rangle = \operatorname{Re}\left(a^{\alpha\beta}_{rs}\zeta^\beta_s\overline{\zeta^\alpha_r}\right) \ge 0, \quad \forall \zeta = (\zeta^\alpha_r)_{\substack{1\le r\le n\\ 1\le\alpha\le M}} \in \mathbb{C}^{n\times M}. \tag{1.1.22}$$

In the class of coefficient tensors with complex entries of type (n,M),

being positive definite (or positive semi-definite) is preserved under global transposition, complex conjugation, and taking Hermitian adjoints. (1.1.23)

Definition 1.1.2 *Let* $A = \left(a^{\alpha\beta}_{rs}\right)_{\substack{1\le r,s\le n\\ 1\le\alpha,\beta\le M}}$ *be a coefficient tensor with complex entries of type* (n,M)*. Say that* A *is* `weakly elliptic` *provided*

$$\det\left[\left(a^{\alpha\beta}_{rs}\xi_r\xi_s\right)_{1\le\alpha,\beta\le M}\right] \ne 0 \text{ for each } \xi = (\xi_r)_{1\le r\le n} \in \mathbb{R}^n \setminus \{0\}. \tag{1.1.24}$$

Also, recall from [131, Definition 1.3.2] *that A is* **Legendre-Hadamard elliptic** *provided there exists $\kappa > 0$ such that the following condition is satisfied:*

$$\begin{aligned} &\mathrm{Re}\left[a_{rs}^{\alpha\beta}\xi_r\xi_s\overline{\eta_\alpha}\eta_\beta\right] \geq \kappa|\xi|^2|\eta|^2 \text{ for all} \\ &\xi = (\xi_r)_{1\leq r\leq n} \in \mathbb{R}^n \text{ and } \eta = (\eta_\alpha)_{1\leq\alpha\leq M} \in \mathbb{C}^M. \end{aligned} \tag{1.1.25}$$

Finally, define

$$\begin{aligned} \mathfrak{A}_{\mathrm{WE}}(n, M) &:= \left\{A \in \mathfrak{A}(n, M) : \ A \text{ is weakly elliptic}\right\}, \\ \mathfrak{A}_{\mathrm{LH}}(n, M) &:= \left\{A \in \mathfrak{A}(n, M) : \ A \text{ is Legendre-Hadamard elliptic}\right\}. \end{aligned} \tag{1.1.26}$$

We remark that $\mathfrak{A}_{\mathrm{WE}}(n, M)$ is an open subset of $\mathfrak{A}(n, M)$. Since for any real matrix $C \in \mathbb{R}^{M\times M}$ with the property that there exists some $c > 0$ such that $\langle C\theta, \theta\rangle \geq c|\theta|^2$ for each $\theta \in \mathbb{R}^M$ we have

$$\begin{aligned} \mathrm{Re}\,\langle C\zeta, \overline{\zeta}\rangle &= \big\langle C(\mathrm{Re}\,\zeta), (\mathrm{Re}\,\zeta)\big\rangle + \big\langle C(\mathrm{Im}\,\zeta), (\mathrm{Im}\,\zeta)\big\rangle \\ &\geq c|\mathrm{Re}\,\zeta|^2 + c|\mathrm{Im}\,\zeta|^2 = c|\zeta|^2 \ \text{ for each } \zeta \in \mathbb{C}^M, \end{aligned} \tag{1.1.27}$$

it follows that whenever the coefficient tensor $A = \big(a_{rs}^{\alpha\beta}\big)_{\substack{1\leq r,s\leq n\\ 1\leq\alpha,\beta\leq M}}$ has real entries it suffices to test (1.1.25) only for real vectors $\eta = (\eta_\alpha)_{1\leq\alpha\leq M} \in \mathbb{R}^M$.

It is also relevant to note that the positive definiteness condition (1.1.21) implies the Legendre-Hadamard ellipticity condition (1.1.25). This is readily seen by observing that, given any vectors $\xi = (\xi_r)_r \in \mathbb{R}^n$ and $\eta = (\eta_\alpha)_\alpha \in \mathbb{C}^M$, if we consider $\zeta = (\zeta_r^\alpha)_{r,\alpha} \in \mathbb{C}^{n\times M}$ with components $\zeta_r^\alpha := \xi_r\eta_\alpha$ then $|\zeta| = |\xi||\eta|$. In turn, Legendre-Hadamard ellipticity implies weak ellipticity. In summary, for every $A \in \mathfrak{A}(n, M)$ we have:

$$\begin{aligned} A \text{ is positive definite } &\Longrightarrow A \text{ is Legendre-Hadamard elliptic} \\ &\Longrightarrow A \text{ is weakly elliptic.} \end{aligned} \tag{1.1.28}$$

A key feature of a Legendre-Hadamard elliptic coefficient tensor $A \in \mathfrak{A}(n, M)$, which may be easily justified based on repeated applications of Plancherel's theorem and (1.1.25), is the fact that there exists $c = c(A, n) \in (0, \infty)$ with the property that

$$\mathrm{Re}\int_{\mathbb{R}^n}\big\langle A\nabla u, \overline{\nabla u}\big\rangle\,d\mathcal{L}^n \geq c\int_{\mathbb{R}^n}|\nabla u|^2\,d\mathcal{L}^n, \qquad \forall u \in \big[W^{1,2}(\mathbb{R}^n)\big]^M. \tag{1.1.29}$$

It is also clear from definitions that, in the class of coefficient tensors with complex entries of type (n, M),

being weakly elliptic (or Legendre-Hadamard elliptic) is preserved under global transposition, complex conjugation, and taking Hermitian adjoints. (1.1.30)

Finally, in terms of the equivalence relation introduced in (1.1.14) we have if $A_1, A_2 \in \mathfrak{A}(n, M)$ are such that $A_1 - A_2$ is antisymmetric in the lower indices, then A_1 is weakly elliptic (respectively, Legendre-Hadamard elliptic) if and only if A_2 is weakly elliptic (respectively, Legendre-Hadamard elliptic), (1.1.31)

and we may be rephrase (1.1.31) as follows:

if the coefficient tensors $A_1, A_2 \in \mathfrak{A}(n, M)$ satisfy $A_1 \sim A_2$, then A_1 is weakly elliptic (respectively, Legendre-Hadamard elliptic) if and only if A_2 is weakly elliptic (respectively, Legendre-Hadamard elliptic). (1.1.32)

We next turn our attention to operators. Fix $n \in \mathbb{N}$ with $n \geq 2$ along with $M \in \mathbb{N}$, and denote by $\mathfrak{L}$ the collection of all homogeneous constant complex coefficient second-order $M \times M$ systems L in $\mathbb{R}^n$. Hence, any element L in $\mathfrak{L}$ may be written as a matrix of differential operators of the form $L = \left(a_{jk}^{\alpha\beta} \partial_j \partial_k\right)_{1 \leq \alpha,\beta \leq M}$ for some complex numbers $a_{jk}^{\alpha\beta}$ (here and elsewhere, we shall use the usual convention of summation over repeated indices). In particular, the action of L on any given vector-valued distribution $u = (u_\beta)_{1 \leq \beta \leq M}$ may be described as

$$Lu = \left(a_{jk}^{\alpha\beta} \partial_j \partial_k u_\beta\right)_{1 \leq \alpha \leq M}. \tag{1.1.33}$$

We shall denote by $L^\top := \left(a_{kj}^{\beta\alpha} \partial_j \partial_k\right)_{1 \leq \alpha,\beta \leq M}$ the (real) transpose of the operator L, by $L^c := \left(\overline{a_{kj}^{\beta\alpha}} \partial_j \partial_k\right)_{1 \leq \alpha,\beta \leq M}$ the complex conjugate of L, and by $L^* := (L^c)^\top$ the Hermitian adjoint of L. We also define the characteristic matrix of L as

$$L(\xi) := -\left[\left(a_{jk}^{\alpha\beta} \xi_j \xi_k\right)_{1 \leq \alpha,\beta \leq M}\right] \quad \text{for each } \xi = (\xi_i)_{1 \leq i \leq n} \in \mathbb{R}^n, \tag{1.1.34}$$

and introduce

$$\mathfrak{L}_{\mathrm{WE}} := \left\{L \in \mathfrak{L} : \det[L(\xi)] \neq 0 \text{ for each } \xi \in \mathbb{R}^n \setminus \{0\}\right\}. \tag{1.1.35}$$

We shall refer to a system $L \in \mathfrak{L}$ as being `weakly elliptic` if actually $L \in \mathfrak{L}_{\mathrm{WE}}$.

For each coefficient tensor $A = \left(a_{jk}^{\alpha\beta}\right)_{\substack{1 \leq \alpha,\beta \leq M \\ 1 \leq j,k \leq n}} \in \mathfrak{A}(n, M)$ we agree to associate the second-order $M \times M$ system $L_A \in \mathfrak{L}$ according to

$$L_A := \left(a_{jk}^{\alpha\beta} \partial_j \partial_k\right)_{1 \leq \alpha,\beta \leq M}. \tag{1.1.36}$$

Then the map

$$\mathfrak{A}(n, M) \ni A \longmapsto L_A \in \mathfrak{L} \tag{1.1.37}$$

is linear and surjective, though it fails to be injective. Specifically, for any two coefficient tensors $A, \widetilde{A} \in \mathfrak{A}(n, M)$ we have

$$L_A = L_{\widetilde{A}} \Longleftrightarrow A - \widetilde{A} \in \mathfrak{A}^{\text{ant}}(n, M). \tag{1.1.38}$$

Define the "library" of (admissible) coefficient tensors of the system L as

$$\mathfrak{A}_L := \{A \in \mathfrak{A}(n, M) : L = L_A\} \text{ for each } L \in \mathfrak{L}, \tag{1.1.39}$$

and for each $L \in \mathfrak{L}$ set (with the distance considered in the normed vector space $\mathfrak{A}$)

$$\|L\| := \text{dist}\big(A, \mathfrak{A}^{\text{ant}}(n, M)\big) \text{ for each/some } A \in \mathfrak{A}_L. \tag{1.1.40}$$

Then $\mathfrak{L} \ni L \mapsto \|L\|$ is an unambiguously defined norm on the vector space $\mathfrak{L}$. In the topology induced by this norm, $\mathfrak{L}_{\text{WE}}$ from (1.1.35) is an open subset of $\mathfrak{L}$, the mapping (1.1.37) is continuous, and $\mathfrak{L} \ni L \mapsto L^{\top} \in \mathfrak{L}$ is an isometry.

Finally, we note that $\mathfrak{A}_{\text{WE}}(n, M)$ (the collection of all weakly elliptic coefficient tensors of type (n, M)) may be described as

$$\mathfrak{A}_{\text{WE}}(n, M) = \{A \in \mathfrak{A}(n, M) : L_A \in \mathfrak{L}_{\text{WE}}\}. \tag{1.1.41}$$

1.2 What Constitutes a Distinguished Coefficient Tensor

To each weakly elliptic system L we may canonically associate a fundamental solution E as in [131, Theorem 1.4.2]. Having fixed a UR domain, this is then used to create a variety of double layer potential operators K_A, in relation to each choice of a coefficient tensor $A \in \mathfrak{A}_L$. While any such double layer K_A has a rich Calderón-Zygmund theory (as discussed in, e.g., [132, Theorem 1.5.1]), seeking more specialized properties requires placing additional demands on the coefficient tensor A. The following result describes said demands phrased in a multitude of equivalent forms.

Proposition 1.2.1 *Let L be a homogeneous, second-order, constant complex coefficient, weakly elliptic $M \times M$ system in $\mathbb{R}^n$, and consider the inverse of the characteristic matrix of L, i.e., introduce the matrix-valued function defined as*

$$\mathcal{E}(\xi) := \big(\mathcal{E}_{\gamma\beta}(\xi)\big)_{1 \le \gamma,\beta \le M} := \big[L(\xi)\big]^{-1} \in \mathbb{C}^{M \times M} \textit{ for each } \xi \in \mathbb{R}^n \setminus \{0\} \tag{1.2.1}$$

(recall that the characteristic matrix $L(\xi) \in \mathbb{C}^{M \times M}$, for $\xi \in \mathbb{R}^n$, of the system L has been defined in (1.1.34)*). Also, denote by $E = \big(E_{\alpha\beta}\big)_{1 \le \alpha,\beta \le M}$ the fundamental solution associated with the given system L as in* [131, Theorem 1.4.2].

Then for each coefficient tensor $A = \big(a_{rs}^{\alpha\beta}\big)_{\substack{1 \le \alpha,\beta \le M \\ 1 \le r,s \le n}} \in \mathfrak{A}_L$ (cf. (1.1.39)*) the following conditions are equivalent:*

(a) For each $s, s' \in \{1, \dots, n\}$ and each $\alpha, \gamma \in \{1, \dots, M\}$ there holds

$$\big(x_{s'} a_{rs}^{\beta\alpha} - x_s a_{rs'}^{\beta\alpha}\big)(\partial_r E_{\gamma\beta})(x) = 0 \textit{ for all } x = (x_j)_{1 \le j \le n} \in \mathbb{R}^n \setminus \{0\}. \tag{1.2.2}$$

(b) For each $s, s' \in \{1, \dots, n\}$ and each $\alpha, \gamma \in \{1, \dots, M\}$ there holds

$$\big(x_{s'}a^{\beta\alpha}_{rs} - x_s a^{\beta\alpha}_{rs'}\big)(\partial_r E_{\gamma\beta})(x) = 0 \ \ in \ \ \mathcal{S}'(\mathbb{R}^n). \tag{1.2.3}$$

(c) For each $s, s' \in \{1, \dots, n\}$ *and each* $\alpha, \gamma \in \{1, \dots, M\}$ *one has*

$$\Big[a^{\beta\alpha}_{rs}\partial_{\xi_{s'}} - a^{\beta\alpha}_{rs'}\partial_{\xi_s}\Big]\big[\xi_r \mathcal{E}_{\gamma\beta}(\xi)\big] = 0 \ \ in \ \ \mathcal{S}'(\mathbb{R}^n). \tag{1.2.4}$$

(d) For each $s, s' \in \{1, \dots, n\}$ *and each* $\alpha, \gamma \in \{1, \dots, M\}$ *one has*

$$\big(a^{\beta\alpha}_{s's} - a^{\beta\alpha}_{ss'} + \xi_r a^{\beta\alpha}_{rs}\partial_{\xi_{s'}} - \xi_r a^{\beta\alpha}_{rs'}\partial_{\xi_s}\big)\mathcal{E}_{\gamma\beta}(\xi) = 0 \ for\ all \ \ \xi \in \mathbb{R}^n \setminus \{0\} \tag{1.2.5}$$

and also

$$\int_{S^1} \big(a^{\beta\alpha}_{rs}\xi_{s'} - a^{\beta\alpha}_{rs'}\xi_s\big)\xi_r\, \mathcal{E}_{\gamma\beta}(\xi)\, d\mathcal{H}^1(\xi) = 0 \ \ if \ n = 2. \tag{1.2.6}$$

(e) One has

$$\begin{gathered}\xi_r\xi_j\Big[a^{\beta\alpha}_{rs'}\big(a^{\lambda\mu}_{sj} + a^{\lambda\mu}_{js}\big) - a^{\beta\alpha}_{rs}\big(a^{\lambda\mu}_{s'j} + a^{\lambda\mu}_{js'}\big)\Big]\mathcal{E}_{\mu\beta}(\xi) + a^{\lambda\alpha}_{ss'} - a^{\lambda\alpha}_{s's} = 0 \\ for\ all\ \xi \in S^{n-1},\ all\ s, s' \in \{1, \dots, n\},\ and\ all\ \alpha, \lambda \in \{1, \dots, M\},\end{gathered} \tag{1.2.7}$$

with the cancellation condition

$$\begin{gathered}\int_{S^1} \Big(a^{\beta\alpha}_{rs}\xi_{s'} - a^{\beta\alpha}_{rs'}\xi_s\Big)\xi_r \mathcal{E}_{\lambda\beta}(\xi)\, d\mathcal{H}^1(\xi) = 0 \\ for\ all\ s, s' \in \{1, \dots, n\}\ and\ \alpha, \lambda \in \{1, \dots, M\},\end{gathered} \tag{1.2.8}$$

additionally imposed in the case when $n = 2$.

(f) For each $\xi \in S^{n-1}$ *and each* $\alpha, \lambda \in \{1, \dots, M\}$,

$$\begin{gathered}the\ expression\ \ \big(a^{\lambda\mu}_{sj} + a^{\lambda\mu}_{js}\big)\mathcal{E}_{\mu\beta}(\xi)\xi_j\xi_r a^{\beta\alpha}_{rs'} - a^{\lambda\alpha}_{s's} \\ is\ symmetric\ in\ the\ indices\ s, s' \in \{1, \dots, n\},\end{gathered} \tag{1.2.9}$$

with the condition that for each $\alpha, \lambda \in \{1, \dots, M\}$*, the expression*

$$\int_{S^1} a^{\beta\alpha}_{rs}\xi_{s'}\xi_r\mathcal{E}_{\lambda\beta}(\xi)\, d\mathcal{H}^1(\xi) \ \ is\ symmetric\ in\ s, s' \in \{1, 2\}, \tag{1.2.10}$$

also imposed in the case when $n = 2$.

(g) There exists a matrix-valued function

$$k = \big\{k_{\gamma\alpha}\big\}_{1 \le \gamma, \alpha \le M} : \mathbb{R}^n \setminus \{0\} \longrightarrow \mathbb{C}^{M\times M} \tag{1.2.11}$$

with the property that for each $\gamma, \alpha \in \{1, \dots, M\}$ *and* $s \in \{1, \dots, n\}$ *one has*

$$a^{\beta\alpha}_{rs}(\partial_r E_{\gamma\beta})(x) = x_s k_{\gamma\alpha}(x) \ \ for\ all \ \ x \in \mathbb{R}^n \setminus \{0\}. \tag{1.2.12}$$

(h) For each $\alpha, \gamma \in \{1, \dots, M\}$ *and each* $\xi = (\xi_s)_{1 \leq s \leq n} \in \mathbb{R}^n$ *one has*

$$a^{\beta\alpha}_{rs} \xi_s (\partial_r E_{\gamma\beta})(x) = 0 \text{ for each } x \in \langle \xi \rangle^{\top} \setminus \{0\}. \tag{1.2.13}$$

In relation to item *(g)* above it is worth noting that for each $x_* \in \mathbb{R}^n \setminus \{0\}$ we may find an open neighborhood O of x_* and some $s \in \{1, \dots, n\}$ with the property that $x_s \neq 0$ for each $x \in O$. From this, (1.2.12), and [131, Theorem 1.4.2] it follows that

the entries of the matrix-valued function k from (1.2.11) belong to $\mathscr{C}^\infty(\mathbb{R}^n \setminus \{0\})$, are even, and positive homogeneous of degree $-n$. (1.2.14)

In addition, from (1.2.12) and [131, (1.4.25), (1.4.26)] we see that

$$\int_{S^{n-1}} k \, d\mathcal{H}^{n-1} = I_{M \times M} \tag{1.2.15}$$

and in fact, for any vector $\omega \in S^{n-1}$,

$$\int\limits_{\xi \in S^{n-1},\ \langle \xi, \omega \rangle > 0} k \, d\mathcal{H}^{n-1} = \tfrac{1}{2} I_{M \times M}. \tag{1.2.16}$$

We now turn to the proof of Proposition 1.2.1.

Proof of Proposition 1.2.1 To justify the implication *(g)* $\Rightarrow$ *(a)*, fix $s, s' \in \{1, \dots, n\}$ together with $\alpha, \gamma \in \{1, \dots, M\}$. We may then use (1.2.12) to compute

$$\big(x_{s'} a^{\beta\alpha}_{rs} - x_s a^{\beta\alpha}_{rs'}\big)(\partial_r E_{\gamma\beta})(x) = x_{s'} x_s k_{\alpha\gamma}(x) - x_s x_{s'} k_{\alpha\gamma}(x) = 0, \tag{1.2.17}$$

for all $x = (x_i)_{1 \leq i \leq n} \in \mathbb{R}^n \setminus \{0\}$, and (1.2.2) follows. Consider now the implication *(a)* $\Rightarrow$ *(g)*. Assume that (1.2.2) holds for each $s, s' \in \{1, \dots, n\}$ and $\alpha, \gamma \in \{1, \dots, M\}$, the goal being to construct a family of functions $\{k_{\alpha\gamma}\}_{1 \leq \alpha, \gamma \leq M}$ as in item *(g)*. To this end, fix some arbitrary indices $\alpha, \gamma \in \{1, \dots, M\}$ and suppose that a point $x^* = (x_1^*, \dots, x_n^*) \in \mathbb{R}^n \setminus \{0\}$ has been given. Then there exists an index $s \in \{1, \dots, n\}$ such that $x_s \neq 0$ for every point $x = (x_1, \dots, x_n)$ in some open neighborhood U of x^*. For every $x \in U$ we then define

$$k_{\alpha\gamma}(x) := \sum_{r=1}^{n} \sum_{\beta=1}^{M} a^{\beta\alpha}_{rs} \frac{(\partial_r E_{\gamma\beta})(x)}{x_s}. \tag{1.2.18}$$

The reader is alerted to the fact that, in (1.2.18), *no summation on the index s takes place*. So far, formula (1.2.18) yields a well-defined, $\mathscr{C}^\infty$ function, in the neighborhood U of the (arbitrarily chosen) point $x^* \in \mathbb{R}^n \setminus \{0\}$. The crux of the matter is that (1.2.2) ensures that the right-hand side of (1.2.18) is independent of the choice of the index $s \in \{1, \dots, n\}$ with the property that $x_s \neq 0$ for every $x \in U$. As such, these local definitions are compatible and, collectively, yield a well-defined function $k_{\alpha\gamma} \in \mathscr{C}^\infty\big(\mathbb{R}^n \setminus \{0\}\big)$. Let us set

$$\mathbb{R}^n_* := \{(x_1, \dots, x_n) : x_s \in \mathbb{R} \setminus \{0\} \text{ for every } s \in \{1, \dots, n\}\}. \tag{1.2.19}$$

By design,

$$x_s k_{\alpha\gamma}(x) = a_{rs}^{\beta\alpha}(\partial_r E_{\gamma\beta})(x) \text{ for each } s \in \{1, \dots, n\} \text{ and each } x \in \mathbb{R}^n_*. \tag{1.2.20}$$

From (1.2.20) it is clear that the identity in (1.2.12) is satisfied if we restrict x to $\mathbb{R}^n_*$. Note that the functions in both the left-hand side and the right-hand side of the identity in (1.2.12) are continuous (in fact smooth) in $\mathbb{R}^n \setminus \{0\}$. Since $\mathbb{R}^n_*$ is dense in $\mathbb{R}^n$, we conclude that (1.2.12) holds as stated. This finishes the proof of the implication *(a)*$\Rightarrow$*(g)*. Hence, so far, conditions *(a)* and *(g)* are equivalent.

Next, we make two observations regarding the fundamental solution E. First, if $\mathcal{E}(\xi)$ is as in (1.2.1), then $\mathcal{E}(\xi) = O(|\xi|^{-2})$. Hence, for $1 \leq j \leq n$ the function $\xi_j \mathcal{E}(\xi)$ belongs to $\big[L^1_{\rm loc}(\mathbb{R}^n, \mathcal{L}^n)\big]^M$ for $n \geq 2$ and and is bounded at infinity. Thus,

$$\xi_j \mathcal{E}(\xi) \in \big[\mathcal{S}'(\mathbb{R}^n)\big]^M \text{ for } n \geq 2 \text{ and } 1 \leq j \leq n. \tag{1.2.21}$$

Second, since $L_A E = \delta I_{M\times M}$ in $\big[\mathcal{S}'(\mathbb{R}^n)\big]^M$, where δ is the Dirac distribution with mass at the origin in $\mathbb{R}^n$, for every $r \in \{1, \dots, n\}$ we may write

$$L_A \partial_r E = \partial_r L_A E = (\partial_r \delta) I_{M\times M} \text{ in } \big[\mathcal{S}'(\mathbb{R}^n)\big]^M. \tag{1.2.22}$$

Taking the Fourier transform of both sides of (1.2.22) then leads to

$$L(\xi)\widehat{\partial_r E} = i\xi_r I_{M\times M} \text{ in } \big[\mathcal{S}'(\mathbb{R}^n)\big]^M, \text{ for every } r \in \{1, \dots, n\}. \tag{1.2.23}$$

Thus, thanks to (1.2.21) and the fact that E is a tempered distribution in $\mathbb{R}^n$,

$$u_r := \widehat{\partial_r E} - i\xi_r \mathcal{E}(\xi) \in \big[\mathcal{S}'(\mathbb{R}^n)\big]^M \text{ for every } r \in \{1, \dots, n\}. \tag{1.2.24}$$

In particular, from (1.2.24), (1.2.23), and (1.2.1), for every $r \in \{1, \dots, n\}$ we obtain

$$L(\xi)u_r = L(\xi)\widehat{\partial_r E} - i\xi_r L(\xi)\mathcal{E}(\xi) = i\xi_r I_{M\times M} - i\xi_r I_{M\times M} = 0 \text{ in } \big[\mathcal{S}'(\mathbb{R}^n)\big]^M.$$

Consequently, if $\varphi \in \big[\mathcal{S}(\mathbb{R}^n)\big]^M$ has $\operatorname{supp}\varphi \cap \{0\} = \varnothing$, this implies

$${}_{[\mathcal{S}'(\mathbb{R}^n)]^M}\big\langle u_r, \varphi\big\rangle_{[\mathcal{S}(\mathbb{R}^n)]^M} = {}_{[\mathcal{S}'(\mathbb{R}^n)]^M}\Big\langle L(\xi)u_r, \big([L(\xi)]^{-1}\big)^\top \varphi\Big\rangle_{[\mathcal{S}(\mathbb{R}^n)]^M} = 0$$

for every $r \in \{1, \dots, n\}$. This shows that $\operatorname{supp} u_r \subseteq \{0\}$ for every $r \in \{1, \dots, n\}$. In addition, from (1.2.24), [129, (4.5.42)], (1.2.1), and [131, Theorem 1.4.2] we see that u_r is homogeneous of degree -1 for every r. As such, [129, Proposition 4.5.4] applies and gives that $u_r = 0$ in $\big[\mathcal{S}'(\mathbb{R}^n)\big]^M$ for every $r \in \{1, \dots, n\}$ or, equivalently,

$$\widehat{\partial_r E} = i\xi_r \mathcal{E}(\xi) \text{ in } \big[\mathcal{S}'(\mathbb{R}^n)\big]^M, \text{ for every } r \in \{1, \dots, n\}. \tag{1.2.25}$$

(Parenthetically, we note that this also may be seen from [131, (1.4.31)] and (1.2.1).)

Now suppose *(a)* holds. Fix $s, s' \in \{1,\dots,n\}$ and $\alpha, \gamma \in \{1,\dots,M\}$ arbitrary. The expression $\big(x_{s'}a^{\beta\alpha}_{rs} - x_s a^{\beta\alpha}_{rs'}\big)(\partial_r E_{\gamma\beta})(x)$ is a tempered distribution which is positive homogeneous of degree $2-n$ in $\mathbb{R}^n$ (cf. [131, Theorem 1.4.2]). Then [129, Proposition 4.5.4] applies and gives that said distribution is zero, hence *(b)* holds. Thus, *(a)* $\Rightarrow$ *(b)* and since the converse is clear we conclude that actually *(a)* $\Leftrightarrow$ *(b)*.

Suppose next that *(b)* holds. Again, fix $s, s' \in \{1,\dots,n\}$ and $\alpha, \gamma \in \{1,\dots,M\}$ arbitrary. Taking Fourier transforms in (1.2.3) we conclude (using standard Fourier calculus and (1.2.25)) that $a^{\beta\alpha}_{rs}\partial_{\xi_{s'}}\big(\xi_r \mathcal{E}_{\gamma\beta}(\xi)\big) - a^{\beta\alpha}_{rs'}\partial_{\xi_s}\big(\xi_r \mathcal{E}_{\gamma\beta}(\xi)\big) = 0$ in $\mathcal{S}'(\mathbb{R}^n)$. This proves the implication *(b)* $\Rightarrow$ *(c)*. Conversely, if *(c)* holds then, as mentioned earlier, the Fourier transform of the tempered distribution $\big(x_{s'}a^{\beta\alpha}_{rs} - x_s a^{\beta\alpha}_{rs'}\big)(\partial_r E_{\gamma\beta})(x)$ is zero for each pair of indices $s, s' \in \{1,\dots,n\}$ and $\alpha, \gamma \in \{1,\dots,M\}$, so (1.2.3) is true. This proves that *(c)* $\Rightarrow$ *(b)* so, ultimately, *(b)* $\Leftrightarrow$ *(c)*.

Moving on, we make a few observations of a general nature. Specifically, having fixed two pairs of indices, $s, s' \in \{1,\dots,n\}$ and $\alpha, \gamma \in \{1,\dots,M\}$, for each Schwartz function $\varphi \in \mathcal{S}(\mathbb{R}^n)$ we may use (1.2.21) to write

$$\begin{aligned}
&{}_{\mathcal{S}'(\mathbb{R}^n)}\Big\langle \big(a^{\beta\alpha}_{rs}\partial_{\xi_{s'}} - a^{\beta\alpha}_{rs'}\partial_{\xi_s}\big)\big(\xi_r \mathcal{E}_{\gamma\beta}(\xi)\big), \varphi(\xi)\Big\rangle_{\mathcal{S}(\mathbb{R}^n)} \qquad (1.2.26)\\
&\quad = -{}_{\mathcal{S}'(\mathbb{R}^n)}\Big\langle \xi_r \mathcal{E}_{\gamma\beta}(\xi), \big(a^{\beta\alpha}_{rs}\partial_{\xi_{s'}} - a^{\beta\alpha}_{rs'}\partial_{\xi_s}\big)\varphi(\xi)\Big\rangle_{\mathcal{S}(\mathbb{R}^n)}\\
&\quad = -\int_{\mathbb{R}^n} \xi_r \mathcal{E}_{\gamma\beta}(\xi)\big(a^{\beta\alpha}_{rs}\partial_{\xi_{s'}} - a^{\beta\alpha}_{rs'}\partial_{\xi_s}\big)\varphi(\xi)\,d\xi\\
&\quad = -\lim_{\varepsilon\to 0^+}\int_{|\xi|>\varepsilon} \xi_r \mathcal{E}_{\gamma\beta}(\xi)\big(a^{\beta\alpha}_{rs}\partial_{\xi_{s'}} - a^{\beta\alpha}_{rs'}\partial_{\xi_s}\big)\varphi(\xi)\,d\xi\\
&\quad = \lim_{\varepsilon\to 0^+}\int_{|\xi|>\varepsilon} \big(a^{\beta\alpha}_{rs}\partial_{\xi_{s'}} - a^{\beta\alpha}_{rs'}\partial_{\xi_s}\big)\big(\xi_r \mathcal{E}_{\gamma\beta}(\xi)\big)\varphi(\xi)\,d\xi\\
&\qquad + \lim_{\varepsilon\to 0^+}\frac{1}{\varepsilon}\int_{|\xi|=\varepsilon} \big(a^{\beta\alpha}_{rs}\xi_{s'} - a^{\beta\alpha}_{rs'}\xi_s\big)\xi_r \mathcal{E}_{\gamma\beta}(\xi)[\varphi(\xi)-\varphi(0)]\,d\mathcal{H}^{n-1}(\xi)\\
&\qquad + \varphi(0)\lim_{\varepsilon\to 0^+}\frac{1}{\varepsilon}\int_{|\xi|=\varepsilon} \big(a^{\beta\alpha}_{rs}\xi_{s'} - a^{\beta\alpha}_{rs'}\xi_s\big)\xi_r \mathcal{E}_{\gamma\beta}(\xi)\,d\mathcal{H}^{n-1}(\xi) =: \mathrm{I} + \mathrm{II} + \mathrm{III}.
\end{aligned}$$

Note that

$$\mathrm{I} = \lim_{\varepsilon\to 0^+}\int_{|\xi|>\varepsilon} \varphi(\xi)\Big(a^{\beta\alpha}_{s's} - a^{\beta\alpha}_{ss'} + \xi_r a^{\beta\alpha}_{rs}\partial_{\xi_{s'}} - \xi_r a^{\beta\alpha}_{rs'}\partial_{\xi_s}\Big)\mathcal{E}_{\gamma\beta}(\xi)\,d\xi. \qquad (1.2.27)$$

Next, since

$$\big(a^{\beta\alpha}_{rs}\xi_{s'} - a^{\beta\alpha}_{rs'}\xi_s\big)\xi_r \mathcal{E}_{\gamma\beta}(\xi)[\varphi(\xi)-\varphi(0)] = O(\varepsilon) \text{ when } |\xi| = \varepsilon, \qquad (1.2.28)$$

we have $\mathrm{II} = 0$. Finally, a change of variables gives

$$\mathrm{III} = \varphi(0)\bigg(\int_{S^{n-1}} \Big(a^{\beta\alpha}_{rs}\xi_{s'} - a^{\beta\alpha}_{rs'}\xi_s\Big)\xi_r \mathcal{E}_{\gamma\beta}(\xi)\,d\mathcal{H}^{n-1}(\xi)\bigg)\Big(\lim_{\varepsilon\to 0^+}\varepsilon^{n-2}\Big). \qquad (1.2.29)$$

As a consequence,

$$\mathrm{III} = 0 \text{ whenever } n \geq 3. \tag{1.2.30}$$

Returning to the mainstream discussion, consider the equivalence *(c)* ⇔ *(d)*. It is immediate that if *(c)* holds, then (1.2.5) also holds. In order to prove the additional condition formulated in (1.2.6) in the two-dimensional setting, fix $s, s' \in \{1, \dots, n\}$ together with $\alpha, \gamma \in \{1, \dots, M\}$ and observe that in the case when $n = 2$ formula (1.2.26) reduces to

$$\begin{aligned}
&{}_{\mathcal{S}'(\mathbb{R}^2)}\Big\langle \big(a^{\beta\alpha}_{rs}\partial_{\xi_{s'}} - a^{\beta\alpha}_{rs'}\partial_{\xi_s}\big)\big(\xi_r \mathcal{E}_{\gamma\beta}(\xi)\big), \varphi(\xi)\Big\rangle_{\mathcal{S}(\mathbb{R}^2)} \\
&\quad = \lim_{\varepsilon\to 0^+}\int_{|\xi|>\varepsilon} \varphi(\xi)\Big(a^{\beta\alpha}_{s's} - a^{\beta\alpha}_{ss'} + \xi_r a^{\beta\alpha}_{rs}\partial_{\xi_{s'}} - \xi_r a^{\beta\alpha}_{rs'}\partial_{\xi_s}\Big)\mathcal{E}_{\gamma\beta}(\xi)\,\mathrm{d}\xi \\
&\qquad + \varphi(0)\int_{S^1}\Big(a^{\beta\alpha}_{rs}\xi_{s'} - a^{\beta\alpha}_{rs'}\xi_s\Big)\xi_r \mathcal{E}_{\gamma\beta}(\xi)\,\mathrm{d}\mathcal{H}^1(\xi).
\end{aligned} \tag{1.2.31}$$

Under the current assumptions, the term in the left-hand side of (1.2.31) is zero, and so is the term under the first integral in the right-hand side of (1.2.31). From these and the fact that (1.2.31) holds for arbitrary test functions $\varphi \in \mathcal{S}(\mathbb{R}^2)$, we conclude that the additional condition formulated in (1.2.6) is satisfied in the two-dimensional setting. This completes the proof of the implication *(c)* ⇒ *(d)*.

Conversely, suppose that *(d)* is satisfied. Again, fix arbitrary $s, s' \in \{1, \dots, n\}$ and $\alpha, \gamma \in \{1, \dots, M\}$. Then $\mathrm{I} = \mathrm{II} = \mathrm{III} = 0$ so (1.2.26) proves that (1.2.4) holds. This completes the proof of *(d)* ⇒ *(c)*. Consider next the equivalence *(d)* ⇔ *(e)*. The starting point is the observation that

$$\begin{aligned}
&(\partial_s \mathcal{E}_{\gamma\beta})(\xi) = \xi_j\big(a^{\lambda\mu}_{sj} + a^{\lambda\mu}_{js}\big)\mathcal{E}_{\gamma\lambda}(\xi)\mathcal{E}_{\mu\beta}(\xi) \\
&\text{for all } \xi \in \mathbb{R}^n\setminus\{0\}, \text{ all } s \in \{1, \dots, n\}, \text{ and all } \gamma, \beta \in \{1, \dots, M\},
\end{aligned} \tag{1.2.32}$$

Indeed, $L(\xi)\mathcal{E}(\xi) = I_{M\times M}$ for each $\xi \in \mathbb{R}^n \setminus \{0\}$, hence

$$\partial_{\xi_s}\big[L(\xi)\mathcal{E}(\xi)\big] = 0 \text{ for each } s \in \{1, \dots, n\} \text{ and } \xi \in \mathbb{R}^n \setminus \{0\}. \tag{1.2.33}$$

This further entails that, for each $s \in \{1, \dots, n\}$ and $\xi \in \mathbb{R}^n \setminus \{0\}$,

$$\partial_s \mathcal{E}(\xi) = -\mathcal{E}(\xi)\big(\partial_s[L(\xi)]\big)\mathcal{E}(\xi). \tag{1.2.34}$$

From this, (1.2.32) now follows upon observing that for each $s \in \{1, \dots, n\}$ and each $\xi \in \mathbb{R}^n \setminus \{0\}$ we have

$$\partial_s[L(\xi)] = -\big(\xi_j a^{\lambda\mu}_{sj} + \xi_k a^{\lambda\mu}_{ks}\big)_{1\leq\lambda,\mu\leq M} = -\big(\xi_j a^{\lambda\mu}_{sj} + \xi_j a^{\lambda\mu}_{js}\big)_{1\leq\lambda,\mu\leq M}. \tag{1.2.35}$$

The relevance of (1.2.32) is that in the case when $n \geq 3$ it allows us to equivalently recast condition (1.2.5) as

$$\xi_r\xi_j\big(a^{\beta\alpha}_{rs'}a^{\lambda\mu}_{sj}+a^{\beta\alpha}_{rs'}a^{\lambda\mu}_{js}-a^{\beta\alpha}_{rs}a^{\lambda\mu}_{s'j}-a^{\beta\alpha}_{rs}a^{\lambda\mu}_{js'}\big)\mathcal{E}_{\gamma\lambda}(\xi)\mathcal{E}_{\mu\beta}(\xi) \qquad (1.2.36)$$
$$-\big(a^{\beta\alpha}_{s's}-a^{\beta\alpha}_{ss'}\big)\mathcal{E}_{\gamma\beta}(\xi)=0 \text{ in } \mathbb{R}^n\setminus\{0\},$$

for all $s,s'\in\{1,\dots,n\}$ and $\alpha,\gamma\in\{1,\dots,M\}$. In addition, relabeling indices as to write

$$\big(a^{\beta\alpha}_{s's}-a^{\beta\alpha}_{ss'}\big)\mathcal{E}_{\gamma\beta}(\xi)=\big(a^{\lambda\alpha}_{s's}-a^{\lambda\alpha}_{ss'}\big)\mathcal{E}_{\gamma\lambda}(\xi) \qquad (1.2.37)$$
$$\text{for all } s,s'\in\{1,\dots,n\} \text{ and } \alpha,\gamma\in\{1,\dots,M\},$$

and keeping in mind that $\big(\mathcal{E}_{\gamma\lambda}(\xi)\big)_{1\leq\gamma,\lambda\leq M}$ is an invertible matrix, it follows that (1.2.36) is further equivalent to

$$\xi_r\xi_j\big(a^{\beta\alpha}_{rs'}a^{\lambda\mu}_{sj}+a^{\beta\alpha}_{rs'}a^{\lambda\mu}_{js}-a^{\beta\alpha}_{rs}a^{\lambda\mu}_{s'j}-a^{\beta\alpha}_{rs}a^{\lambda\mu}_{js'}\big)\mathcal{E}_{\mu\beta}(\xi)+a^{\lambda\alpha}_{ss'}-a^{\lambda\alpha}_{s's}=0 \qquad (1.2.38)$$
$$\text{for all } \xi\in\mathbb{R}^n\setminus\{0\}, \text{ all } s,s'\in\{1,\dots,n\}, \text{ and all } \alpha,\lambda\in\{1,\dots,M\}.$$

The latter may be equivalently expressed as in (1.2.7), establishing *(d)*$\Leftrightarrow$*(e)*. Since conditions (1.2.7), (1.2.8) may be further recast as the demands formulated in (1.2.9) and (1.2.10), respectively, we also have *(e)*$\Leftrightarrow$*(f)*.

Next, it is clear that *(g)*$\Rightarrow$*(h)*. Finally, to see that *(h)*$\Rightarrow$*(a)*, having fixed $s,s'\in\{1,\dots,n\}$ and $\alpha,\gamma\in\{1,\dots,M\}$ write (1.2.13) for $x\in\mathbb{R}^n\setminus\{0\}$ and $\xi:=x_{s'}\mathbf{e}_s-x_s\mathbf{e}_{s'}\in\mathbb{R}^n$, which satisfy $\langle x,\xi\rangle=0$. This yields (1.2.2). The proof of Proposition 1.2.1 is therefore complete. □

The stage has now been set for us to make the following definition, which plays a fundamental role in subsequent considerations.

Definition 1.2.2 *Given a second-order, weakly elliptic, homogeneous, $M\times M$ system L in $\mathbb{R}^n$, with constant complex coefficients, call*

$$A=\big(a^{\alpha\beta}_{rs}\big)_{\substack{1\leq r,s\leq n\\ 1\leq\alpha,\beta\leq M}}\in\mathfrak{A}_L \qquad (1.2.39)$$

a `distinguished coefficient tensor` *for the system L provided any of the condition (a)-(h) in Proposition 1.2.1 holds. Also, denote by $\mathfrak{A}^{\rm dis}_L$ the family of such distinguished coefficient tensors for L, say,*

$$\mathfrak{A}^{\rm dis}_L:=\Big\{A=\big(a^{\alpha\beta}_{rs}\big)_{\substack{1\leq r,s\leq n\\ 1\leq\alpha,\beta\leq M}}\in\mathfrak{A}_L:\ \textit{conditions } (1.2.7)\text{-}(1.2.8)$$
$$\textit{hold for each } s,s'\in\{1,\dots,n\} \textit{ and } \alpha,\lambda\in\{1,\dots,M\}\Big\}. \qquad (1.2.40)$$

Finally, introduce the class of weakly elliptic systems which posses a distinguished coefficient tensor, by setting

$$\mathfrak{L}^{\rm dis}:=\big\{L\in\mathfrak{L}_{\rm WE}:\ \mathfrak{A}^{\rm dis}_L\neq\varnothing\big\}. \qquad (1.2.41)$$

As we shall see in due time, not every weakly elliptic possesses a distinguished coefficient tensor, so $\mathfrak{L}^{\rm dis}$ is a proper subclass of $\mathfrak{L}_{\rm WE}$. For now we wish to note that, staring from definitions, and using (1.2.2) together with the last formula in

[131, (1.4.32)], it is clear that for each second-order, weakly elliptic, homogeneous, constant complex coefficient, $M \times M$ system L in $\mathbb{R}^n$, and each coefficient tensor $A = \big(a^{\alpha\beta}_{jk}\big)_{\substack{1\leq\alpha,\beta\leq M\\ 1\leq j,k\leq n}} \in \mathfrak{A}_L$ the following properties are valid:

$$A \in \mathfrak{A}^{\rm dis}_L \Longleftrightarrow A^c \in \mathfrak{A}^{\rm dis}_{L^c}, \tag{1.2.42}$$

and, for each $\lambda \in \mathbb{C}\setminus\{0\}$,

$$A \in \mathfrak{A}^{\rm dis}_L \Longleftrightarrow \lambda A \in \mathfrak{A}^{\rm dis}_{\lambda L}. \tag{1.2.43}$$

More generally, we have the following result:

Proposition 1.2.3 *Fix $n, M \in \mathbb{N}$ with $n \geq 2$, and let $A = \big(a^{\alpha\beta}_{jk}\big)_{\substack{1\leq\alpha,\beta\leq M\\ 1\leq j,k\leq n}} \in \mathfrak{A}(n, M)$ be an arbitrary coefficient tensor. For any matrix $C = (c_{\beta\gamma})_{1\leq\beta,\gamma\leq M} \in \mathbb{C}^{M\times M}$ define*

$$A \diamond C := \big(a^{\alpha\beta}_{jk}c_{\beta\gamma}\big)_{\substack{1\leq\alpha,\gamma\leq M\\ 1\leq j,k\leq n}} \in \mathfrak{A}(n, M), \tag{1.2.44}$$

and for any matrix $C = (c_{\gamma\alpha})_{1\leq\gamma,\alpha\leq M} \in \mathbb{C}^{M\times M}$ define

$$C \diamond A := \big(c_{\gamma\alpha}a^{\alpha\beta}_{jk}\big)_{\substack{1\leq\gamma,\beta\leq M\\ 1\leq j,k\leq n}} \in \mathfrak{A}(n, M). \tag{1.2.45}$$

Then, if L is a second-order, weakly elliptic, homogeneous, constant complex coefficient, $M \times M$ system in $\mathbb{R}^n$, and $A \in \mathfrak{A}_L$, then for any invertible matrix $C \in \mathbb{C}^{M\times M}$ one has $A \diamond C \in \mathfrak{A}_{LC}$ (with the system LC interpreted in the sense of multiplication of $M \times M$ matrices) and

$$A \in \mathfrak{A}^{\rm dis}_L \Longleftrightarrow A \diamond C \in \mathfrak{A}^{\rm dis}_{LC}. \tag{1.2.46}$$

Likewise, any invertible matrix $C \in \mathbb{C}^{M\times M}$ one has $C \diamond A \in \mathfrak{A}_{CL}$ (with the system CL interpreted in the sense of multiplication of $M \times M$ matrices) and

$$A \in \mathfrak{A}^{\rm dis}_L \Longleftrightarrow C \diamond A \in \mathfrak{A}^{\rm dis}_{CL}. \tag{1.2.47}$$

Proof All claims follow from definitions, (1.2.2), and [131, (1.4.34)]. An alternative proof is to combine [132, (1.5.268)] with the equivalence *(i)*$\Leftrightarrow$*(ii)* in Proposition 1.3.2, stated a little later. □

We close by augmenting the above proposition with a result to the effect that the property of being a distinguished coefficient tensor is stable under conjugation by invertible matrices, in the specific manner indicated below.

Proposition 1.2.4 *Fix two integers $n, M \in \mathbb{N}$, with $n \geq 2$, and suppose L is a second-order, homogeneous, complex constant coefficient, $M \times M$ system in $\mathbb{R}^n$. Also, pick an invertible matrix $W \in \mathbb{R}^{n\times n}$. Then for any coefficient tensor $A \in \mathfrak{A}_L$ one has*

$$A \in \mathfrak{A}^{\rm dis}_L \Longleftrightarrow W^\top \circ A \circ W \in \mathfrak{A}^{\rm dis}_{L\circ W} \tag{1.2.48}$$

(*with notation introduced in* [131, (1.4.56)-(1.4.58)]).

Proof Write explicitly $W = (w_{ij})_{1\le i,j\le n} \in \mathbb{R}^{n\times n}$ and $A = \big(a_{ij}^{\alpha\beta}\big)_{\substack{1\le i,j\le n\\ 1\le\alpha,\beta\le M}} \in \mathfrak{A}_L$. In addition, define $\widetilde{L} := L\circ W$ (which is known to be a weakly elliptic system, by [131, Proposition 1.4.3]), and introduce

$$\widetilde{A} := W^{\top}\circ A\circ W = \big(a_{ij}^{\alpha\beta} w_{ir} w_{js}\big)_{\substack{1\le r,s\le n\\ 1\le\alpha,\beta\le M}} = \big(\widetilde{a}_{rs}^{\alpha\beta}\big)_{\substack{1\le r,s\le n\\ 1\le\alpha,\beta\le M}} \in \mathfrak{A}_{\widetilde{L}}. \tag{1.2.49}$$

Finally, bring in the matrix-valued fundamental solutions $E := E_L$ and $\widetilde{E} := E_{\widetilde{L}}$ associated with the weakly elliptic systems $L, \widetilde{L}$ as in [131, Theorem 1.4.2].

To proceed, suppose $A \in \mathfrak{A}_L^{\rm dis}$. Recall from item *(g)* in Proposition 1.2.1 that this membership entails the existence of a matrix-valued function $k = \big\{k_{\gamma\alpha}\big\}_{1\le\gamma,\alpha\le M}$ as in (1.2.11)-(1.2.12). Pick $\gamma,\alpha \in \{1,\dots,M\}$ and $s \in \{1,\dots,n\}$. Then for each $x \in \mathbb{R}^n \setminus \{0\}$ we may compute

$$\begin{aligned}
\widetilde{a}_{rs}^{\beta\alpha}(\partial_r\widetilde{E}_{\gamma\beta})(x) &= |\det W|^{-1} a_{ij}^{\beta\alpha} w_{ir} w_{js}\partial_r\Big[E_{\gamma\beta}\big((W^{-1})^{\top}x\big)\Big]\\
&= |\det W|^{-1} a_{ij}^{\beta\alpha} w_{ir} w_{js}\big((W^{-1})^{\top}\big)_{\ell r}\big(\partial_\ell E_{\gamma\beta}\big)\big((W^{-1})^{\top}x\big)\\
&= |\det W|^{-1} a_{ij}^{\beta\alpha} w_{js}\delta_{i\ell}\big(\partial_\ell E_{\gamma\beta}\big)\big((W^{-1})^{\top}x\big)\\
&= |\det W|^{-1} a_{ij}^{\beta\alpha} w_{js}\big(\partial_i E_{\gamma\beta}\big)\big((W^{-1})^{\top}x\big)\\
&= |\det W|^{-1} w_{js}\big((W^{-1})^{\top}x\big)_j\, k_{\gamma\alpha}\big((W^{-1})^{\top}x\big)\\
&= |\det W|^{-1} x_s\, k_{\gamma\alpha}\big((W^{-1})^{\top}x\big),
\end{aligned} \tag{1.2.50}$$

with the first equality provided by (1.2.49) and [131, (1.4.64), (1.4.65)], the second equality implied by Chain Rule, the penultimate equality a consequence of (1.2.12), and all others matrix formalism. If we now define

$$\widetilde{k} := \Big\{|\det W|^{-1} k_{\gamma\alpha}\circ (W^{-1})^{\top}\Big\}_{1\le\gamma,\alpha\le M} : \mathbb{R}^n\setminus\{0\} \longrightarrow \mathbb{C}^{M\times M} \tag{1.2.51}$$

then (1.2.50) shows that the property in item *(g)* of Proposition 1.2.1 is satisfied by $\widetilde{A}$, $\widetilde{E}$, and $\widetilde{k}$. According to Definition 1.2.2, we therefore have $\widetilde{A} \in \mathfrak{A}_{\widetilde{L}}^{\rm dis}$. This finishes the proof of the right-pointing implication in (1.2.48). Lastly, the left-pointing implication in (1.2.48) follows from what we have just proved.

In closing, we wish to note that an alternative proof is to combine [132, (1.5.264)] with the equivalence *(i)* ⇔ *(ii)* in Proposition 1.3.2, stated a little later. □

1.3 The Significance of Distinguished Coefficient Tensors

The relevance of the distinguished coefficient tensors is most apparent from Proposition 1.3.2 proved a little later below. As a preamble, we first establish the following useful result.

Lemma 1.3.1 *Pick some $n, N \in \mathbb{N}$ and consider a vector-valued function*

$$\vec{k} \in \left[\mathscr{C}^N(\mathbb{R}^n \setminus \{0\})\right]^n, \text{ odd, positive homogeneous of degree } 1-n. \tag{1.3.1}$$

Then there exists a scalar-valued function $k \in \mathscr{C}^N(\mathbb{R}^n \setminus \{0\})$ which is even and positive homogeneous of degree $-n$ with the property that

$$\vec{k}(x) = x\,k(x) \text{ for each } x \in \mathbb{R}^n \setminus \{0\} \tag{1.3.2}$$

if and only if for each hyper-plane $H = \langle\omega\rangle^{\perp} \subseteq \mathbb{R}^n$, with $\omega \in S^{n-1}$, and each function $\phi \in \mathscr{C}_c^{\infty}(\mathbb{R}^n)$ one has, at $\mathcal{H}^{n-1}$-a.e. $x \in H$,

$$\lim_{\varepsilon \to 0^+} \int_{y \in H,\ |x-y|>\varepsilon} \big\langle \omega, \vec{k}(x-y)\big\rangle \phi(y)\, d\mathcal{H}^{n-1}(y) = 0. \tag{1.3.3}$$

Parenthetically we wish to note that, whenever (1.3.2) holds, Euler's formula for positive homogeneous functions implies that $\vec{k}$ is divergence-free in $\mathbb{R}^n \setminus \{0\}$.

Proof of Lemma 1.3.1 For starters, we wish to note that, thanks to the assumptions in (1.3.1), the limit in (1.3.3) exists at each point $x \in H$ for every $\phi \in \mathscr{C}_c^{\infty}(\mathbb{R}^n)$. To start in earnest, observe that if (1.3.2) holds then the claim made in relation to (1.3.3) is true simply by virtue of the fact that $\langle \omega, x-y\rangle = 0$ whenever $\omega \in S^{n-1}$ and $x, y \in H = \langle\omega\rangle^{\perp}$.

Assume next that (1.3.3) holds for each hyper-plane $H = \langle\omega\rangle^{\perp} \subseteq \mathbb{R}^n$, with $\omega \in S^{n-1}$, and each function $\phi \in \mathscr{C}_c^{\infty}(\mathbb{R}^n)$. Pick an arbitrary $\omega \in S^{n-1}$, define $H := \langle\omega\rangle^{\perp} \subseteq \mathbb{R}^n$, and select an arbitrary $\phi \in \mathscr{C}_c^{\infty}(\mathbb{R}^n)$. Then (1.3.3) implies

$$\int_H \big\langle \omega, \vec{k}(x-y)\big\rangle \phi(y)\, d\mathcal{H}^{n-1}(y) = 0 \text{ for each } x \in H \setminus \operatorname{supp}\phi. \tag{1.3.4}$$

Granted this, [129, Corollary 3.7.3] applied with $s := n-1$, an arbitrary open set $O \subseteq \mathbb{R}^n$, $X := H \cap O$, and $f := \langle \omega, \vec{k}(x-\cdot)\rangle$ for some arbitrary fixed $x \in H \setminus O$, then guarantees that $\langle \omega, \vec{k}(x-\cdot)\rangle = 0$ for $\mathcal{H}^{n-1}$-a.e. point in $H \cap O$. In view of the arbitrariness of the open set $O \subseteq \mathbb{R}^n$ and the continuity of $\vec{k}$ in $\mathbb{R}^n \setminus \{0\}$, this forces

$$\big\langle \omega, \vec{k}(x-y)\big\rangle = 0 \text{ for each } \omega \in S^{n-1} \text{ and each } x, y \in \langle\omega\rangle^{\perp} \text{ with } x \neq y. \tag{1.3.5}$$

Specializing this to the case when $y = 0$ and observing that having $x \in \langle\omega\rangle^{\perp}$ is equivalent to having $\omega \in \langle x\rangle^{\perp}$, we arrive at

$$\langle \omega, \vec{k}(x) \rangle = 0 \text{ whenever } x \neq 0 \text{ and } \omega \in \langle x \rangle^{\perp}. \tag{1.3.6}$$

This is the same as saying that for each vector $x \in \mathbb{R}^n \setminus \{0\}$ the vector $\vec{k}(x)$ is a scalar multiple of x. Thus, there exists a scalar function k defined in $\mathbb{R}^n \setminus \{0\}$ such that (1.3.2). A posteriori, the latter identity together with (1.3.1) imply that said function is even, positive homogeneous of degree $-n$, and of class $\mathscr{C}^N$ in $\mathbb{R}^n \setminus \{0\}$. □

Here is the result advertised a little earlier, which establishes a fundamental link between the quality of being distinguished (in the sense of Definition 1.2.2) for a given coefficient tensor and the nature of the double layer potential operator associated with said coefficient tensor. To put it succinctly, each system L has an infinite "library" of coefficient tensors $\mathfrak{A}_L$ (that may be used to write L; cf. (1.1.39)) and, among this library, some coefficient tensors $A \in \mathfrak{A}_L$ may yield double layer potential operators K_A which are decisively "better" in terms of recognizing flatness (for the underlying "surface" on which they are defined).

In a nutshell, distinguished coefficient tensors are precisely those giving rise to double layer potential operators of "chord-dot-normal" type (see [132, Theorem 5.2.2]). In particular, the boundary-to-boundary versions of double layer potential operators associated with distinguished coefficient tensors vanish whenever the underlying domain is a half-space.

Proposition 1.3.2 *Let L be a homogeneous, second-order, constant complex coefficient, weakly elliptic $M \times M$ system in $\mathbb{R}^n$, and suppose $A \in \mathfrak{A}_L$. Then the following statements are equivalent.*

(i) The coefficient tensor A belongs to $\mathfrak{A}_L^{\text{dis}}$.

(ii) Whenever Ω is a half-space in $\mathbb{R}^n$, the boundary-to-boundary double layer potential K associated with A and Ω as in (A.0.116) *(i.e., K is regarded as a mapping sending functions $f \in \Big[L^1\Big(\partial_*\Omega, \frac{\sigma(x)}{1+|x|^{n-1}}\Big)\Big]^M$ into σ-measurable functions on $\partial\Omega$ according to* (A.0.116)*) is actually the zero operator.*

(iii) Whenever Ω is a half-space in $\mathbb{R}^n$ with the property that $0 \in \partial\Omega$, the modified boundary-to-boundary double layer operator K_{mod} associated as in (A.0.117) *with the set Ω and the given coefficient tensor A is actually the zero operator.*

(iii') Whenever Ω is a half-space in $\mathbb{R}^n$, the modified boundary-to-boundary double layer operator K_{mod} associated as in (A.0.117) *with the set Ω and the given coefficient tensor A maps functions from $\big[\mathscr{C}_c^\infty(\partial\Omega)\big]^M$ into constants in $\mathbb{C}^M$.*

(iv) Whenever Ω is a half-space in $\mathbb{R}^n$, the boundary-to-boundary "transpose" double layer potential $K^\#$ associated with A and Ω as in (A.0.118) *(i.e., $K^\#$ is regarded as a mapping sending functions $f \in \Big[L^1\Big(\partial\Omega, \frac{\sigma(x)}{1+|x|^{n-1}}\Big)\Big]^M$ into σ-measurable functions on $\partial\Omega$ according to* (A.0.118)*) is the zero operator.*

(v) There exists a matrix-valued function $k \in \big[\mathscr{C}^\infty(\mathbb{R}^n \setminus \{0\})\big]^{M\times M}$ which is even, positive homogeneous of degree $-n$, and with the property that for any Lebesgue

measurable set $\Omega \subseteq \mathbb{R}^n$ *of locally finite perimeter the (matrix-valued) integral kernel of the double layer potential operator* K *associated with* A *and* Ω *as in* (A.0.116) *has the form*

$$\langle \nu(y), x - y\rangle k(x - y) \text{ for each } x \in \partial\Omega \text{ and } \mathcal{H}^{n-1}\text{-a.e. } y \in \partial_*\Omega, \tag{1.3.7}$$

where ν *is the geometric measure theoretic outward unit normal to* Ω.

(vi) There exists a matrix-valued function $k^\# \in \big[\mathscr{C}^\infty(\mathbb{R}^n \setminus \{0\})\big]^{M\times M}$ *which is even, positive homogeneous of degree* $-n$, *and with the property that, for each Lebesgue measurable set* $\Omega \subseteq \mathbb{R}^n$ *such that* $\partial\Omega$ *is countably rectifiable of dimension* $n-1$, *the (matrix-valued) integral kernel of the "transpose" double layer potential operator* $K^\#$ *associated with* A *and* Ω *as in* (A.0.118) *has the form*

$$\langle \nu(x), y - x\rangle k^\#(x - y) \text{ for } \mathcal{H}^{n-1}\text{-a.e. } x \in \partial_*\Omega \text{ and each } y \in \partial\Omega, \tag{1.3.8}$$

where ν *is the geometric measure theoretic outward unit normal to* Ω.

Moreover, whenever either (hence all) of the above conditions materializes, the matrices $k, k^\#$ *in items (v), (vi) above are related to each other via* $k^\# = k^\top$, *where the superscript* $\top$ *indicates transposition.*

Proof Assume the coefficient tensor $A = \big(a_{rs}^{\alpha\beta}\big)_{\substack{1\le r,s\le n\\ 1\le\alpha,\beta\le M}}$ belongs to $\mathfrak{A}_L^{\rm dis}$. Then, according to item *(g)* in Proposition 1.2.1 and the observation made in (1.2.14), there exists a matrix-valued function

$$k = \{k_{\gamma\alpha}\}_{1\le\gamma,\alpha\le M} : \mathbb{R}^n \setminus \{0\} \longrightarrow \mathbb{C}^{M\times M}, \text{ whose entries belong to } \mathscr{C}^\infty(\mathbb{R}^n \setminus \{0\}), \text{ are even, and positive homogeneous of degree } -n, \tag{1.3.9}$$

with the property that for each $\gamma, \alpha \in \{1, \dots, M\}$ and $s \in \{1, \dots, n\}$ we have

$$a_{rs}^{\beta\alpha}(\partial_r E_{\gamma\beta})(x) = x_s k_{\gamma\alpha}(x) \text{ for all } x \in \mathbb{R}^n \setminus \{0\}, \tag{1.3.10}$$

where $E = (E_{\alpha\beta})_{1\le\alpha,\beta\le M}$ is the matrix-valued fundamental solution associated with the given system L as in [131, Theorem 1.4.2]. Consider next a Lebesgue measurable set $\Omega \subseteq \mathbb{R}^n$ of locally finite perimeter and denote by $\nu = (\nu_1, \dots, \nu_n)$ the geometric measure theoretic outward unit normal to Ω. From (A.0.116) and (1.3.10) we then see that the integral kernel of the double layer potential operator K associated with A and Ω as in (A.0.116) has the form

$$\begin{aligned}\Big(-\nu_s(y)a_{rs}^{\beta\alpha}(\partial_r E_{\gamma\beta})(x-y)\Big)_{1\le\gamma,\alpha\le M} &= -\Big(\nu_s(y)(x_s - y_s)k_{\gamma\alpha}(x-y)\Big)_{1\le\gamma,\alpha\le M}\\ &= -\langle \nu(y), x-y\rangle k(x-y)\end{aligned} \tag{1.3.11}$$

for each $x \in \partial\Omega$ and $\mathcal{H}^{n-1}$-a.e. $y \in \partial_*\Omega$. This establishes the claim in (1.3.7) (with k replaced by $-k$), hence we have *(i)* $\Rightarrow$ *(v)*. Clearly, *(v)* $\Rightarrow$ *(ii)* since the inner product in (1.3.7) vanishes identically whenever Ω is a half-space in $\mathbb{R}^n$.

To prove that *(ii)* ⇒ *(i)*, assume that the boundary-to-boundary double layer potential K associated with A and Ω as in (A.0.116) is actually the zero operator whenever Ω is a half-space in $\mathbb{R}^n$. From this, (A.0.116), and Lemma 1.3.1 applied to each $\vec{k}_{\gamma\alpha} := \big(a_{rs}^{\beta\alpha}\partial_r E_{\gamma\beta}\big)_{1\leq s\leq n}$ with $\alpha,\gamma \in \{1,\dots,M\}$ we then conclude that for each $\alpha,\gamma \in \{1,\dots,M\}$ there exists a scalar-valued function $k_{\gamma\alpha} \in \mathscr{C}^\infty(\mathbb{R}^n \setminus \{0\})$ which is even and positive homogeneous of degree $-n$ with the property that

$$\Big(a_{rs}^{\beta\alpha}(\partial_r E_{\gamma\beta})(x)\Big)_{1\leq s\leq n} = x\, k_{\gamma\alpha}(x) \text{ for each } x \in \mathbb{R}^n \setminus \{0\}. \tag{1.3.12}$$

In turn, this proves that the property in item *(g)* of Proposition 1.2.1 holds, hence the coefficient tensor A belongs to $\mathfrak{A}_L^{\rm dis}$ according to Definition 1.2.2. This finishes the proof of the implication *(ii)* ⇒ *(i)* hence, at this stage, we have *(i)* ⇔ *(ii)* ⇔ *(v)*.

Next, assume the coefficient tensor $A = \big(a_{rs}^{\alpha\beta}\big)_{\substack{1\leq r,s\leq n\\ 1\leq\alpha,\beta\leq M}}$ belongs to $\mathfrak{A}_L^{\rm dis}$; in particular, there exists a matrix-valued function as in (1.3.9)-(1.3.10). Also, suppose $\Omega \subseteq \mathbb{R}^n$ is a Lebesgue measurable set with the property that $\partial\Omega$ is countably rectifiable of dimension $n-1$. Then the integral kernel of the "transpose" double layer potential operator $K^\#$ associated with A and Ω as in (A.0.118) has the form

$$\begin{aligned}\Big(\nu_s(x)a_{rs}^{\beta\alpha}(\partial_r E_{\gamma\beta})(x-y)\Big)_{1\leq\alpha,\gamma\leq M} &= \Big(\nu_s(x)(x_s-y_s)k_{\gamma\alpha}(x-y)\Big)_{1\leq\alpha,\gamma\leq M}\\ &= \langle\nu(x), x-y\rangle k^\top(x-y)\end{aligned} \tag{1.3.13}$$

for $\mathcal{H}^{n-1}$-a.e. $x \in \partial_*\Omega$ and each $y \in \partial\Omega$. This shows that *(i)* ⇒ *(vi)*, with the matrix-valued functions $k, k^\#$ related to each other via $k^\# = k^\top$. Also, *(vi)* ⇒ *(iv)* since the inner product in (1.3.8) vanishes identically whenever Ω is a half-space in $\mathbb{R}^n$. That *(iv)* ⇒ *(ii)* is a consequence of the duality result in [132, Theorem 1.5.1, item *(iii)*].

Assume next that $K_{\rm mod}$ is the zero operator whenever Ω is a half-space in $\mathbb{R}^n$ with the property that $0 \in \partial\Omega$. Fix $\omega \in S^{n-1}$ and consider $\Omega := \{x \in \mathbb{R}^n : x\cdot\omega < 0\}$. Then, as is apparent from (A.0.117) and the fact that $k_\varepsilon^{(r\gamma\beta)} := (\partial_r E_{\gamma\beta})\cdot \mathbf{1}_{\mathbb{R}^n\setminus\overline{B(0,\varepsilon)}}$ for each $\varepsilon > 0$, this forces

$$\Bigg(\omega_s a_{rs}^{\beta\alpha}\Big\{(\partial_r E_{\gamma\beta})(x-y) - (\partial_r E_{\gamma\beta})(-y)\cdot\mathbf{1}_{\mathbb{R}^n\setminus\overline{B(0,1)}}(-y)\Big\}\Bigg)_{1\leq\gamma,\alpha\leq M} = 0 \quad \text{for all } x,y \in \langle\omega\rangle^\perp \text{ with } x\neq y. \tag{1.3.14}$$

If for each fixed $y \in \langle\omega\rangle^\perp$ we send x to infinity from within $\langle\omega\rangle^\perp$ we arrive at the conclusion that (cf. [131, Theorem 1.4.2])

$$\Big(\omega_s a_{rs}^{\beta\alpha}(\partial_r E_{\gamma\beta})(-y)\cdot\mathbf{1}_{\mathbb{R}^n\setminus\overline{B(0,1)}}(-y)\Big)_{1\leq\gamma,\alpha\leq M} = 0 \text{ for all } y \in \langle\omega\rangle^\perp. \tag{1.3.15}$$

Using this back into (1.3.14) we obtain

$$\Big(\omega_s a_{rs}^{\beta\alpha}(\partial_r E_{\gamma\beta})(x-y)\Big)_{1\le\gamma,\alpha\le M} = 0 \text{ for all } x, y \in \langle\omega\rangle^\perp \text{ with } x \neq y. \tag{1.3.16}$$

Hence, K is the zero operator whenever $\Omega := \{x \in \mathbb{R}^n : x \cdot \omega < 0\}$ with $\omega \in S^{n-1}$. Ultimately, this proves that *(iii)* $\Rightarrow$ *(ii)*.

To show that *(i)* $\Rightarrow$ *(iii)*, assume the coefficient tensor $A = \big(a_{rs}^{\alpha\beta}\big)_{\substack{1\le r,s\le n\\ 1\le\alpha,\beta\le M}}$ is in $\mathfrak{A}_L^{\rm dis}$. Then there exists a matrix-valued function $k = \{k_{\gamma\alpha}\}_{1\le\gamma,\alpha\le M}$ as in (1.3.9) such that the identity in (1.3.10) holds for each $\gamma, \alpha \in \{1,\dots,M\}$ and $s \in \{1,\dots,n\}$. In turn, this implies that (1.3.14) holds for each $\omega \in S^{n-1}$ since for all $x, y \in \langle\omega\rangle^\perp$ with $x \neq y$ and all $\gamma, \alpha \in \{1,\dots,M\}$ we have

$$\begin{aligned} &\omega_s a_{rs}^{\beta\alpha}\Big\{(\partial_r E_{\gamma\beta})(x-y) - (\partial_r E_{\gamma\beta})(-y)\cdot \mathbf{1}_{\mathbb{R}^n\setminus\overline{B(0,1)}}(-y)\Big\} \\ &\quad = \omega_s\Big\{(x_s-y_s)k_{\gamma\alpha}(x-y) - (-y_s)k_{\gamma\alpha}(-y)\cdot\mathbf{1}_{\mathbb{R}^n\setminus\overline{B(0,1)}}(-y)\Big\} = 0, \end{aligned} \tag{1.3.17}$$

given that $\omega_s(x_s - y_s) = \langle\omega, x-y\rangle = 0$ and $\omega_s(-y_s) = \langle\omega, -y\rangle = 0$. Consequently, $K_{\rm mod}$ is the zero operator whenever $\Omega := \{x \in \mathbb{R}^n : x\cdot\omega < 0\}$ with $\omega \in S^{n-1}$, as the left-most expression in (1.3.17) is a typical entry in the matrix-valued integral kernel of said operator in this setting.

Let us now show that we have the equivalence *(iii')* $\Leftrightarrow$ *(ii)*. To deal with the left-to-right implication, pick a point $x_0 \in \mathbb{R}^n$ and fix a vector $\omega \in S^{n-1}$. Then the set $\Omega := \{x \in \mathbb{R}^n : (x-x_0)\cdot\omega < 0\}$ is a half-space in $\mathbb{R}^n$ with boundary $\partial\Omega = x_0 + \langle\omega\rangle^\perp$. Choose an arbitrary $f = (f_\alpha)_{1\le\alpha\le M} \in \big[\mathscr{C}_c^\infty(\partial\Omega)\big]^M$. The current working hypothesis implies that there exists $c_f \in \mathbb{C}^M$ such that $(K_{\rm mod} f)(x) = c_f$ for $\mathcal{H}^{n-1}$-a.e. point $x \in \partial\Omega$. In particular,

$$\begin{aligned} c_f &= \lim_{\partial\Omega\ni x\to\infty}(K_{\rm mod} f)(x) \\ &= -\Bigg(\int_{y\in x_0+\langle\omega\rangle^\perp} \omega_s a_{rs}^{\beta\alpha}(\partial_r E_{\gamma\beta})(-y)\cdot\mathbf{1}_{\mathbb{R}^n\setminus\overline{B(0,1)}}(-y) f_\alpha(y)\,{\rm d}\mathcal{H}^{n-1}(y)\Bigg)_{1\le\gamma\le M}. \end{aligned} \tag{1.3.18}$$

thanks to (A.0.117) and [131, Theorem 1.4.2]. Together with (A.0.117) and (A.0.116) this further implies that $0 = K_{\rm mod} f - c_f = Kf$ for each $f \in \big[\mathscr{C}_c^\infty(\partial\Omega)\big]^M$. In turn, this forces $\Big(\omega_s a_{rs}^{\beta\alpha}(\partial_r E_{\gamma\beta})(x-y)\Big)_{1\le\gamma,\alpha\le M} = 0$ for all $x, y \in x_0 + \langle\omega\rangle^\perp$ with $x \neq y$. Thus, K is the zero operator, ultimately, proving that *(iii')* $\Rightarrow$ *(ii)*.

There remains to show that *(ii)* $\Rightarrow$ *(iii')*. To this end, assume Ω is a half-space in $\mathbb{R}^n$. Observe from definitions that for each $f \in \big[\mathscr{C}_c^\infty(\partial\Omega)\big]^M$ the difference $K_{\rm mod} f - Kf$ is a constant in $\mathbb{C}^M$. Hence, if K (sending functions $f \in \Big[L^1\Big(\partial_*\Omega, \frac{\sigma(x)}{1+|x|^{n-1}}\Big)\Big]^M$ into $\mathcal{H}^{n-1}$-measurable functions on $\partial\Omega$ according to (A.0.116)) is actually the zero operator then the modified boundary-to-boundary double layer operator $K_{\rm mod}$ maps each $\big[\mathscr{C}_c^\infty(\partial\Omega)\big]^M$ function into a constant in $\mathbb{C}^M$. This establishes *(ii)* $\Rightarrow$ *(iii')*. □

1.4 Behavior of Distinguished Coefficient Tensors Under Transposition

In the class of systems satisfying the strong Legendre-Hadamard ellipticity condition, it turns out that the quality of possessing a distinguished coefficient tensor is stable under transposition. In fact, the following more precise result holds:

Proposition 1.4.1 *Let $n, M \in \mathbb{N}$ with $n \geq 2$ and consider a homogeneous, complex constant coefficient, second-order $M \times M$ system L in $\mathbb{R}^n$ satisfying the Legendre-Hadamard (strong) ellipticity condition* [131, Definition 1.3.2]. *Then for each coefficient tensor $A \in \mathfrak{A}_L$ one has*

$$A \in \mathfrak{A}_L^{\text{dis}} \textit{ if and only if } A^\top \in \mathfrak{A}_{L^\top}^{\text{dis}}. \tag{1.4.1}$$

Proof Pick an arbitrary coefficient tensor $A \in \mathfrak{A}_L$. For a given UR domain $\Omega \subseteq \mathbb{R}^n$, let K_{mod} be the modified boundary-to-boundary double layer potential associated with A and Ω as in (A.0.117). Also, let $K^{\#}_{A^\top}$ be the "transpose" double layer potential associated with $A^\top$ and Ω as in (A.0.118). Finally, recall that S_{mod} is the modified version of the boundary-to-boundary single layer operator, introduced in (A.0.232). From [132, Theorem 1.8.26, item *(1)*] we know that for each $f \in \big[L^p(\partial\Omega, \sigma)\big]^M$, where $\sigma := \mathcal{H}^{n-1}\lfloor\partial\Omega$ and $p \in (1, \infty)$, there exists c_f, which is the nontangential trace on $\partial\Omega$ of some $\mathbb{C}^M$-valued locally constant function in Ω, such that

$$S_{\text{mod}}\big(K^{\#}_{A^\top} f\big) = K_{\text{mod}}\big(S_{\text{mod}} f\big) + c_f \text{ at } \sigma\text{-a.e. point on } \partial\Omega. \tag{1.4.2}$$

Let us now assume that $A = \big(a_{rs}^{\alpha\beta}\big)_{\substack{1\leq r,s\leq n\\ 1\leq\alpha,\beta\leq M}}$ belongs to $\mathfrak{A}_L^{\text{dis}}$, and that Ω is a half-space in $\mathbb{R}^n$; in particular, $\partial\Omega$ is an $(n-1)$-dimensional plane. Then, as a consequence of (A.0.116) and the implication *(i)*$\Rightarrow$*(ii)* in Proposition 1.3.2, we conclude that the integral kernel of the boundary-to-boundary double layer operator K (associated with A and Ω) is zero, i.e., we have

$$\begin{gathered} \nu_s(y) a_{rs}^{\beta\alpha}(\partial_r E_{\gamma\beta})(x-y) = 0 \text{ for any pair of indices} \\ \gamma, \alpha \in \{1, \dots, M\} \text{ and any two distinct points } x, y \in \partial\Omega. \end{gathered} \tag{1.4.3}$$

Fix now an arbitrary function $f \in \big[L^p(\partial\Omega, \sigma)\big]^M$ and define $g := S_{\text{mod}} f$. Then $g \in \Big[L^1\big(\partial\Omega, \frac{\sigma(x)}{1+|x|^n}\big)\Big]^M$ by [132, (1.5.76)], and we claim that

$$K_{\text{mod}}\, g \text{ is a constant function on } \partial\Omega. \tag{1.4.4}$$

Indeed, as seen from the definition of K_{mod} in (A.0.117), for any $x_0, x_1 \in \partial\Omega$ (save for a set of surface measure zero) we may express $\big(K_{\text{mod}} g\big)(x_0) - \big(K_{\text{mod}} g\big)(x_1)$ as

$$\left(-\lim_{\varepsilon\to 0^+}\int_{\partial\Omega}\nu_s(y)a^{\beta\alpha}_{rs}\left\{k^{(r\gamma\beta)}_{\varepsilon}(x_0-y)-k^{(r\gamma\beta)}_{\varepsilon}(x_1-y)\right\}g_\alpha(y)\,\mathrm{d}\sigma(y)\right)_{1\leq\gamma\leq M}$$
$$\text{where } k^{(r\gamma\beta)}_{\varepsilon}:=(\partial_r E_{\gamma\beta})\cdot\mathbf{1}_{\mathbb{R}^n\setminus\overline{B(0,\varepsilon)}} \text{ for each } \varepsilon>0. \tag{1.4.5}$$

In turn, from (1.4.3) and (1.4.5) we then see that $\big(K_{\text{mod}}g\big)(x_0)-\big(K_{\text{mod}}g\big)(x_1)=0$, from which the claim made in (1.4.4) follows. We may rephrase (1.4.4) as

$$K_{\text{mod}}\big(S_{\text{mod}}f\big) \text{ is constant on } \partial\Omega \text{ for each } f\in\big[L^p(\partial\Omega,\sigma)\big]^M. \tag{1.4.6}$$

In concert with (1.4.2) this implies that $S_{\text{mod}}\big(K^{\#}_{A^\top}f\big)$ is constant on $\partial\Omega$ for each $f\in\big[L^p(\partial\Omega,\sigma)\big]^M$. We may now invoke [132, Proposition 1.8.11] (with $w\equiv 1$) to conclude that $K^{\#}_{A^\top}f=0$ for each $f\in\big[L^p(\partial\Omega,\sigma)\big]^M$. The latter readily yields that the integral kernel of $K^{\#}_{A^\top}$ is zero, hence $K^{\#}_{A^\top}$ regarded as a mapping associated with $A^\top$ and Ω as in (A.0.118) is the zero operator. With this in hand, the implication *(iv)*$\Rightarrow$*(i)* in Proposition 1.3.2 (with A replaced by $A^\top$ and L replaced by $L^\top$) then proves that $A^\top\in\mathfrak{A}_{L^\top}$ satisfies $A^\top\in\mathfrak{A}^{\text{dis}}_{L^\top}$. This completes the proof of the right-pointing implication in (1.4.1). Finally, the opposite implication follows from what we just proved applied to $L^\top$. □

We next elaborate on connections with Poisson kernels, which we introduce by singling out the most essential features which have the potential of identifying these objects uniquely.

Definition 1.4.2 *Let L be a weakly elliptic $M\times M$ homogeneous constant complex coefficient second-order system in $\mathbb{R}^n$. A* `Poisson kernel` *for L in $\mathbb{R}^n_+$ is a matrix-valued function*

$$P^L=\big(P^L_{\alpha\beta}\big)_{1\leq\alpha,\beta\leq M}:\mathbb{R}^{n-1}\longrightarrow\mathbb{C}^{M\times M} \tag{1.4.7}$$

satisfying the following properties:

(a) There exists some constant $C\in(0,\infty)$ such that

$$|P^L(x')|\leq\frac{C}{(1+|x'|^2)^{\frac{n}{2}}} \text{ for each } x'\in\mathbb{R}^{n-1}. \tag{1.4.8}$$

(b) The function P^L is Lebesgue measurable and

$$\int_{\mathbb{R}^{n-1}}P^L(x')\,\mathrm{d}x'=I_{M\times M}, \tag{1.4.9}$$

where $I_{M\times M}$ denotes the $M\times M$ identity matrix.

(c) If one sets

$$K^L(x',t):=P^L_t(x')=t^{1-n}P^L(x'/t) \quad \text{for each } x'\in\mathbb{R}^{n-1} \text{ and } t>0, \tag{1.4.10}$$

then the $\mathbb{C}^{M\times M}$-valued function K^L satisfies (with L acting on the columns of K^L in the sense of distributions)

$$LK^L = 0 \cdot I_{M\times M} \ \textit{in} \ \big[\mathcal{D}'(\mathbb{R}^n_+)\big]^{M\times M}. \tag{1.4.11}$$

From (1.4.8) we see that there exists some constant $C \in (0,\infty)$ with the property that we have $|K^L(x',t)| \le Ct/|(x',t)|^n$ for each $(x',t) \in \mathbb{R}^n_+$. In particular, K^L belongs to $\big[L^1_{\rm loc}(\mathbb{R}^n_+, \mathcal{L}^n)\big]^{M\times M}$, so the demand in (1.4.11) is meaningful. In fact, via elliptic regularity (cf. [127]), from (1.4.11) we conclude that $K^L \in \big[\mathscr{C}^\infty(\mathbb{R}^n_+)\big]^{M\times M}$ which, in view of (1.4.10), further entails

$$P^L \in \big[\mathscr{C}^\infty(\mathbb{R}^{n-1})\big]^{M\times M}. \tag{1.4.12}$$

It is also instructive to note that any Poisson kernel P^L has the property that

$$P^L_t \longrightarrow \delta_{\mathbb{R}^{n-1}} \cdot I_{M\times M} \ \text{in} \ \big[\mathcal{D}'(\mathbb{R}^{n-1})\big]^{M\times M} \ \text{as} \ t \to 0^+, \tag{1.4.13}$$

where $\delta_{\mathbb{R}^{n-1}}$ stands for Dirac's distribution in $\mathbb{R}^{n-1}$ (see [127, Exercise 2.26, p. 29]), hence the kernel function K^L associated with P^L as in (1.4.10) solves the boundary value problem

$$\begin{cases} K^L \in \big[\mathscr{C}^\infty(\mathbb{R}^n_+)\big]^{M\times M}, \\ LK^L = 0 \cdot I_{M\times M} \ \text{in} \ \mathbb{R}^n_+, \\ K^L\big|_{\partial\mathbb{R}^n_+} = \delta_{\mathbb{R}^{n-1}} \cdot I_{M\times M} \ \text{on} \ \mathbb{R}^{n-1}, \end{cases} \tag{1.4.14}$$

with the last property understood as $\lim\limits_{t\to 0^+} K^L(x',t) = \delta_{\mathbb{R}^{n-1}}(x') \cdot I_{M\times M}$ in the space $\big[\mathcal{D}'(\mathbb{R}^{n-1})\big]^{M\times M}$ for all x' in $\mathbb{R}^{n-1}$. We next record the following basic result.

Theorem 1.4.3 *Let L be an $M \times M$ homogeneous constant complex coefficient second-order system in $\mathbb{R}^n$ which satisfies the Legendre-Hadamard (strong) ellipticity condition* [131, Definition 1.3.2]. *Then L has a unique Poisson kernel P^L (in the sense of Definition 1.4.2).*

Concerning Theorem 1.4.3, we note that the existence part follows from the classical work of S. Agmon, A. Douglis, and L. Nirenberg in [4]-[5] (cf. also [109], [186], [195], [196] and the discussion in [102, p. 24]). The uniqueness property has been proved in [114]. We wish to augment Theorem 1.4.3 with the following result which identifies yet another scenario when a Poisson kernel exists.

Proposition 1.4.4 *Let L be an $M \times M$ homogeneous constant complex coefficient second-order system in $\mathbb{R}^n$ which is weakly elliptic, and denote by $E = \big(E_{\gamma\beta}\big)_{1\le\gamma,\beta\le M}$ the matrix-valued fundamental solution associated with L as in* [131, Theorem 1.4.2]. *Assume that the class of* `distinguished coefficient tensors adapted to` $\mathbb{R}^n_+$ *for L, defined as*

$$\mathfrak{A}_L^{\rm dis}(\mathbb{R}^n_+) := \Big\{ A = \big(a_{rs}^{\alpha\beta}\big)_{\substack{1\le r,s\le n\\ 1\le\alpha,\beta\le M}} \in \mathfrak{A}_L : \textit{there exists a matrix-valued function}$$
$$k = \big(k_{\gamma\alpha}\big)_{1\le\gamma,\alpha\le M} \in \big[\mathscr{C}^\infty(\mathbb{R}^n\setminus\{0\})\big]^{M\times M} \ \textit{satisfying}$$
$$a_{rn}^{\beta\alpha}(\partial_r E_{\gamma\beta})(x',x_n) = x_n\, k_{\gamma\alpha}(x',x_n) \ \textit{for each pair} \tag{1.4.15}$$
$$\gamma,\alpha\in\{1,\dots,M\} \ \textit{and all}\ (x',x_n)\in\mathbb{R}^n\setminus\{0\}\Big\},$$

is a nonempty set. Fix some coefficient tensor $A = \big(a_{rs}^{\alpha\beta}\big)_{\substack{1\le r,s\le n\\ 1\le\alpha,\beta\le M}} \in \mathfrak{A}_L^{\rm dis}(\mathbb{R}^n_+)$ *and bring in the function* $k \in \big[\mathscr{C}^\infty(\mathbb{R}^n\setminus\{0\})\big]^{M\times M}$ *associated with* A *as in* (1.4.15). *Then the matrix-valued function*

$$P : \mathbb{R}^{n-1} \longrightarrow \mathbb{C}^{M\times M} \ \textit{defined by}$$
$$P(x') := 2k(x',1) \ \textit{for each}\ x'\in\mathbb{R}^{n-1} \tag{1.4.16}$$

is a Poisson kernel for L *in* $\mathbb{R}^n_+$ *(in the sense of Definition 1.4.2).*

Moreover, the aforementioned Poisson kernel interfaces tightly with the boundary-to-domain double layer potential operator $\mathcal{D}$ *associated as in* (A.0.72) *with the given coefficient tensor* A *and the set* $\Omega := \mathbb{R}^n_+$, *and its modified version* $\mathcal{D}_{\rm mod}$ *associated with the coefficient tensor* A *and the set* $\Omega := \mathbb{R}^n_+$ *as in* (A.0.73). *Specifically, the Poisson kernel for* L *in* $\mathbb{R}^n_+$ *from* (1.4.16) *has the property that for each function*

$$f \in \Big[L^1\Big(\mathbb{R}^{n-1}, \frac{dx'}{1+|x'|^{n-1}}\Big)\Big]^M \tag{1.4.17}$$

one has

$$(P_t * f)(x') = 2(\mathcal{D}f)(x',t) \ \textit{for each}\ (x',t)\in\mathbb{R}^n_+, \tag{1.4.18}$$

and for each function

$$f \in \Big[L^1\Big(\mathbb{R}^{n-1}, \frac{dx'}{1+|x'|^{n}}\Big)\Big]^M \tag{1.4.19}$$

one has

$$(P_t * f)(x') = 2\big(\mathcal{D}_{\rm mod} f\big)(x',t) \ \textit{for each}\ (x',t)\in\mathbb{R}^n_+. \tag{1.4.20}$$

Finally, in view of item (g) of Proposition 1.2.1, all the above results are true if in place of $A \in \mathfrak{A}_L^{\rm dis}(\mathbb{R}^n_+)$ *one works under the stronger assumption* $\mathfrak{A}_L^{\rm dis} \neq \varnothing$ *and* $A \in \mathfrak{A}_L^{\rm dis}$.

It is apparent from definitions that for any weakly elliptic $M\times M$ homogeneous constant complex coefficient second-order system L in $\mathbb{R}^n$ one has

$$\mathfrak{A}_L^{\rm dis} \subseteq \mathfrak{A}_L^{\rm dis}(\mathbb{R}^n_+). \tag{1.4.21}$$

Proof of Proposition 1.4.4 We first observe that that the matrix-valued function

$$
\begin{array}{c} k = (k_{\gamma\alpha})_{1\leq\gamma,\alpha\leq M} \in \big[\mathscr{C}^\infty(\mathbb{R}^n \setminus \{0\})\big]^{M\times M} \\ \text{is even and positive homogeneous of degree } -n. \end{array} \tag{1.4.22}
$$

Indeed, from [131, Theorem 1.4.2] we know that ∇E is odd and positive homogeneous of degree $1-n$. In concert with the formula in (1.4.15), this shows that the function k is even and positive homogeneous of degree $-n$ when restricted to $\mathbb{R}^n \setminus \{(x', 0) : x' \in \mathbb{R}^{n-1}\}$. By density, these properties then extend to $\mathbb{R}^n \setminus \{0\}$, proving (1.4.22).

Let us prove that the matrix-valued function P defined in (1.4.16) is a Poisson kernel for the system L in $\mathbb{R}^n_+$ in the sense of Definition 1.4.2. By the homogeneity property enjoyed by k, for each $x \in \mathbb{R}^n \setminus \{0\}$ we may write

$$
|k(x)| = \Big|k\Big(|x|\frac{x}{|x|}\Big)\Big| = |x|^{-n}\Big|k\Big(\frac{x}{|x|}\Big)\Big| \leq \Big(\sup_{S^{n-1}} |k|\Big)|x|^{-n}. \tag{1.4.23}
$$

Thus, for each $x' \in \mathbb{R}^{n-1}$, we obtain

$$
|P(x')| = 2|k(x', 1)| \leq 2\Big(\sup_{S^{n-1}} |k|\Big)|(x', 1)|^{-n} = \frac{2\Big(\sup_{S^{n-1}} |k|\Big)}{(1+|x'|^2)^{\frac{n}{2}}}, \tag{1.4.24}
$$

which shows that property (a) in Definition 1.4.2 holds with $C := 2\Big(\sup_{S^{n-1}} |k|\Big)$.

To proceed, for each $t \in (0, \infty)$ define $P_t := t^{1-n}P(\cdot/t)$ in $\mathbb{R}^{n-1}$. Also, select an arbitrary

$$
f = (f_\alpha)_{1\leq\alpha\leq M} \in \Big[L^1\Big(\mathbb{R}^{n-1}, \frac{dy'}{1+|y'|^{n-1}}\Big)\Big]^M \tag{1.4.25}
$$

and consider the vector-valued function u defined at each point $(x', t) \in \mathbb{R}^n_+$ by the formula

$$
\begin{aligned} u(x', t) := (P_t * f)(x') &= \int_{\mathbb{R}^{n-1}} t^{1-n}\, P\Big(\frac{x'-y'}{t}\Big)\, f(y')\, dy' \\ &= \Big(\int_{\mathbb{R}^{n-1}} 2t\, k_{\gamma\alpha}(x'-y', t)\, f_\alpha(y')\, dy'\Big)_{1\leq\gamma\leq M} \\ &= \Big(\int_{\mathbb{R}^{n-1}} 2a^{\beta\alpha}_{rn}(\partial_r E_{\gamma\beta})(x'-y', t)\, f_\alpha(y')\, dy'\Big)_{1\leq\gamma\leq M} \end{aligned} \tag{1.4.26}
$$

where the third equality is based on (1.4.16) and (1.4.22), while the last equality uses the formula in (1.4.15). For each $\gamma, \alpha \in \{1, \ldots, M\}$ consider

$$
\theta_{\gamma\alpha}(x) := 2a^{\beta\alpha}_{rn}(\partial_r E_{\gamma\beta})(x), \qquad \forall x \in \mathbb{R}^n \setminus \{0\}, \tag{1.4.27}
$$

and associate with each such kernel the boundary-to-domain integral operator $\mathcal{T}_{\gamma\alpha}$ acting on each function $\phi \in L^1\big(\mathbb{R}^{n-1}, \frac{dy'}{1+|y'|^{n-1}}\big)$ according to

$$(\mathcal{T}_{\gamma\alpha}\phi)(x) := \int_{\mathbb{R}^{n-1}} \theta_{\gamma\alpha}(x'-y',x_n)\,\phi(y')\,\mathrm{d}y' \text{ for each } x=(x',x_n)\in\mathbb{R}^n_+. \tag{1.4.28}$$

In term of these operators, formula (1.4.26) may be simply re-written as

$$u = \big(\mathcal{T}_{\gamma\alpha}f_\alpha\big)_{1\leq\gamma\leq M} \text{ in } \mathbb{R}^n_+. \tag{1.4.29}$$

Fix $\kappa > 0$, and select an arbitrary $\gamma \in \{1,\dots,M\}$. Since the kernels (1.4.27) are smooth, odd, and positive homogeneous of degree $1-n$ (cf. [131, (2.3.3)]), as a special case of the jump-formula in [131, Theorem 2.5.1] (used here with $\Omega := \mathbb{R}^n_+$, hence $\nu = -\mathbf{e}_n$) we see that at $\mathcal{L}^{n-1}$-a.e. point $x' \in \mathbb{R}^{n-1}$ we have

$$\Big(u\Big|^{\kappa-\text{n.t.}}_{\partial\mathbb{R}^n_+}\Big)_\gamma(x') = \frac{1}{2\sqrt{-1}}\widehat{\theta_{\gamma\alpha}}(-\mathbf{e}_n)f_\alpha(x') + \lim_{\varepsilon\to0^+}\int\limits_{\substack{y'\in\mathbb{R}^{n-1}\\ |x'-y'|>\varepsilon}} \theta_{\gamma\alpha}(x'-y',0)f_\alpha(y')\,\mathrm{d}y'. \tag{1.4.30}$$

Also, as seen from (1.4.27) and the identity in (1.4.15), we have $\theta_{\gamma\alpha}(x'-y',0)=0$, so the above formula implies that for each $\gamma\in\{1,\dots,M\}$ there holds

$$\Big(u\Big|^{\kappa-\text{n.t.}}_{\partial\mathbb{R}^n_+}\Big)_\gamma(x') = \frac{1}{2\sqrt{-1}}\widehat{\theta_{\gamma\alpha}}(-\mathbf{e}_n)f_\alpha(x') \tag{1.4.31}$$

at $\mathcal{L}^{n-1}$-a.e. point $x'\in\mathbb{R}^{n-1}$. Moreover, if we abbreviate

$$B := -L(-\mathbf{e}_n) = \big(a^{\alpha\beta}_{nn}\big)_{1\leq\alpha,\beta\leq M} \in \mathbb{C}^{M\times M} \tag{1.4.32}$$

then from (1.4.27) and [131, (1.4.30)] we deduce that

$$\widehat{\theta_{\gamma\alpha}}(-\mathbf{e}_n) = 2\sqrt{-1}B_{\beta\alpha}\big(B^{-1}\big)_{\gamma\beta} = 2\sqrt{-1}\delta_{\gamma\alpha}. \tag{1.4.33}$$

In concert, (1.4.31)-(1.4.33) ultimately imply $u\big|^{\kappa-\text{n.t.}}_{\partial\mathbb{R}^n_+} = f$ at $\mathcal{L}^{n-1}$-a.e. point in $\mathbb{R}^{n-1}$. In light of (1.4.26) this shows that

$$\lim_{t\to0^+}(P_t * f)(x') = f(x') \text{ for } \mathcal{L}^{n-1}\text{-a.e. } x'\in\mathbb{R}^{n-1}. \tag{1.4.34}$$

Fix $j\in\{1,\dots,M\}$, define $\mathbf{e}_j := (\delta_{j\ell})_{1\leq\ell\leq M}\in\mathbb{C}^M$, and consider the function $f := \mathbf{1}_{B_{n-1}(0',1)}\mathbf{e}_j$, where $B_{n-1}(0',1)$ is the unit ball in $\mathbb{R}^{n-1}$ centered at the origin. Then there exists a point $x'\in B_{n-1}(0',1)$ at which (1.4.34) holds. For such a point x' we therefore have

$$\begin{aligned}\mathbf{e}_j &= f(x') = \lim_{t\to0^+}(P_t*f)(x') = \lim_{t\to0^+}\int_{\mathbb{R}^{n-1}} P(z')f(x'-tz')\,\mathrm{d}z'\\ &= \Big(\int_{\mathbb{R}^{n-1}} P(z')\,\mathrm{d}z'\Big)f(x') = \Big(\int_{\mathbb{R}^{n-1}} P(z')\,\mathrm{d}z'\Big)\mathbf{e}_j,\end{aligned} \tag{1.4.35}$$

where the third equality in (1.4.35) is obtained via a change of variables, while the penultimate equality follows by applying Lebesgue's Dominated Convergence Theorem. Thus, $\int_{\mathbb{R}^{n-1}} P(z')\,dz'$ is an $M \times M$ matrix whose action preserves $\mathbf{e}_j$. Since $j \in \{1, \dots, M\}$ was arbitrary, this readily implies that (*b*) in Definition 1.4.2 holds.

Parenthetically, we wish to point out that, in the special case when $\mathfrak{A}_L^{\text{dis}} \neq \varnothing$ and $A \in \mathfrak{A}_L^{\text{dis}}$, an alternative proof of the normalization condition in item (*b*) of Definition 1.4.2 may be given using (1.4.16), (1.6.42), and (1.6.12).

As far as property (*c*) in Definition 1.4.2 is concerned, observe that, thanks to the definition of P (cf. (1.4.16)), the fact that k is positive homogeneous of degree $-n$, and the formula in (1.4.15), the entries in the matrix $K = \big(K_{\gamma\alpha}\big)_{1 \leq \gamma, \alpha \leq M}$ associated with P as in (1.4.10) are presently given by

$$K_{\gamma\alpha}(x', t) = \big(P_t(x')\big)_{\gamma\alpha} = 2t\, k_{\gamma\alpha}(x', t) = 2a_{rn}^{\beta\alpha}(\partial_r E_{\gamma\beta})(x', t), \tag{1.4.36}$$

for each $\gamma, \alpha \in \{1, \dots, M\}$ and $(x', t) \in \mathbb{R}^n_+$. If we express L as

$$L = A_{rs}\partial_r\partial_s \text{ with } A_{rs} := \big(a_{rs}^{\alpha\beta}\big)_{1 \leq \alpha, \beta \leq M} \in \mathbb{C}^{M \times M} \text{ for } 1 \leq r, s \leq n, \tag{1.4.37}$$

then we may recast (1.4.36) simply as the following equality of matrix-valued functions: $K = 2(\partial_r E)A_{rn}$ in $\mathbb{R}^n_+$. Bearing in mind that $LE = \delta \cdot I_{M \times M}$ in the sense of distributions in $\mathbb{R}^n$ (cf. part (2) in [131, Theorem 1.4.2]), we therefore see that $LK = 2\partial_r(LE)A_{rn} = 0$ in $\mathbb{R}^n_+$. Hence, property (*c*) in Definition 1.4.2 holds as well.

Going further, fix an arbitrary function f as in (1.4.17) and denote by $(f_\alpha)_{1 \leq \alpha \leq M}$ its scalar components. Then based on (1.4.16), (1.4.22), and the formula in (1.4.15) we may write

$$\begin{aligned}
(P_t * f)(x') &= \int_{\mathbb{R}^{n-1}} P_t(x' - y') f(y')\,dy' \\
&= 2\bigg(\int_{\mathbb{R}^{n-1}} t k_{\gamma\alpha}(x' - y', t) f_\alpha(y')\,dy'\bigg)_{1 \leq \gamma \leq M} \\
&= 2\bigg(\int_{\mathbb{R}^{n-1}} a_{rn}^{\beta\alpha}(\partial_r E_{\gamma\beta})(x' - y', t) f_\alpha(y')\,dy'\bigg)_{1 \leq \gamma \leq M} = 2(\mathcal{D} f)(x', t)
\end{aligned} \tag{1.4.38}$$

for each $(x', t) \in \mathbb{R}^n_+$, bearing in mind that, in this setting, $\nu_s = -\delta_{sn}$ for each $s \in \{1, \dots, n\}$. This establishes (1.4.18).

Next, recall from (A.0.73) that the boundary-to-domain modified double layer potential operator $\mathcal{D}_{\text{mod}}$ associated with the coefficient tensor A and the set $\Omega := \mathbb{R}^n_+$ acts on any function $f = (f_\alpha)_{1 \leq \alpha \leq M}$ as in (1.4.19) according to (again, keeping in mind that $\nu_s = -\delta_{sn}$ for each $s \in \{1, \dots, n\}$ in the current setting)

$$(\mathcal{D}_{\rm mod} f)(x',t) = \bigg(\int_{\mathbb{R}^{n-1}} \{a^{\beta\alpha}_{rn}(\partial_r E_{\gamma\beta})(x'-y',t) \tag{1.4.39}$$
$$- a^{\beta\alpha}_{rn}(\partial_r E_{\gamma\beta})(0'-y',0)\cdot \mathbf{1}_{\mathbb{R}^{n-1}\setminus\overline{B_{n-1}(0',1)}}(y')\} f_\alpha(y')\,\mathrm{d}y'\bigg)_{1\leq\gamma\leq M}$$

at each point (x',t) belonging to $\mathbb{R}^n_+$. In view of the fact that the formula in (1.4.15) implies $a^{\beta\alpha}_{rn}(\partial_r E_{\gamma\beta})(0'-y',0) = 0$ for each $y' \in \mathbb{R}^{n-1}\setminus\{0'\}$, the same type of argument as in (1.4.38) presently yields (1.4.20). The proof of Proposition 1.4.4 is therefore complete. □

Property (c) in Definition 1.4.2 links the Poisson kernel to partial differential equations, while properties (a)-(b) in Definition 1.4.2 make the Poisson kernel a good approximation to the identity. Regrading the latter, here is a general result of purely real-variable nature, proved in [114, Lemma 3.3].

Lemma 1.4.5 *Let $P = (P_{\alpha\beta})_{1\leq\alpha,\beta\leq M} : \mathbb{R}^{n-1} \to \mathbb{C}^{M\times M}$ be a Lebesgue measurable function satisfying, for some $c \in (0,\infty)$,*

$$|P(x')| \leq \frac{c}{(1+|x'|^2)^{\frac{n}{2}}} \quad \textit{for each } x' \in \mathbb{R}^{n-1}, \tag{1.4.40}$$

and set $P_t(x') := t^{1-n}P(x'/t)$ for each $x' \in \mathbb{R}^{n-1}$ and $t \in (0,\infty)$. Then, for each $t \in (0,\infty)$ fixed, the operator

$$\Big[L^1\Big(\mathbb{R}^{n-1}, \frac{\mathrm{d}x'}{1+|x'|^n}\Big)\Big]^M \ni f \mapsto P_t * f \in \Big[L^1\Big(\mathbb{R}^{n-1}, \frac{\mathrm{d}x'}{1+|x'|^n}\Big)\Big]^M \tag{1.4.41}$$

is well-defined, linear and bounded, with operator norm controlled by $C(t+1)$. Moreover, for every $\kappa > 0$ there exists a finite constant $C_\kappa > 0$ with the property that for each $x' \in \mathbb{R}^{n-1}$,

$$\sup_{|x'-y'|<\kappa t} \big|(P_t * f)(y')\big| \leq C_\kappa\, \mathcal{M}f(x'), \quad \forall f \in \Big[L^1\Big(\mathbb{R}^{n-1}, \frac{\mathrm{d}x'}{1+|x'|^n}\Big)\Big]^M, \tag{1.4.42}$$

where $\mathcal{M}$ is the Hardy-Littlewood maximal operator on $\mathbb{R}^{n-1}$. Finally, given any function

$$f = (f_\beta)_{1\leq\beta\leq M} \in \Big[L^1\Big(\mathbb{R}^{n-1}, \frac{\mathrm{d}x'}{1+|x'|^n}\Big)\Big]^M \subset \big[L^1_{\rm loc}(\mathbb{R}^{n-1}, \mathcal{L}^{n-1})\big]^M, \tag{1.4.43}$$

at every Lebesgue point $x'_0 \in \mathbb{R}^{n-1}$ of f there holds

$$\lim_{\substack{(x',t)\to(x'_0,0)\\ |x'-x'_0|<\kappa t}} (P_t * f)(x') = \Big(\int_{\mathbb{R}^{n-1}} P(x')\,\mathrm{d}x'\Big) f(x'_0), \tag{1.4.44}$$

and the function

$$\mathbb{R}^n_+ \ni (x', t) \mapsto (P_t * f)(x') \in \mathbb{C}^M \text{ is locally integrable in } \mathbb{R}^n_+. \tag{1.4.45}$$

Thanks to such features, Poisson kernels may be used to solve Dirichlet boundary value problems in the upper half-space in a variety of function spaces on which the Hardy-Littlewood operator is bounded, and in [114] this program has been carried out in the class of homogeneous, constant (complex) coefficient, second-order elliptic systems, satisfying the Legendre-Hadamard (strong) ellipticity condition. In this class, the existence of a Poisson kernel in the sense of Definition 1.4.2 is guaranteed by Theorem 1.4.3 (which also tells us that such a Poisson kernel is unique). Here we wish to note that similar solvability results may be obtained if the system in question is only assumed to be weakly elliptic, provided said system possesses a distinguished coefficient tensor adapted to $\mathbb{R}^n_+$. In such a scenario, Proposition 1.4.4 takes the role of Theorem 1.4.3 in ensuring the existence of a Poisson kernel in the sense of Definition 1.4.2. Remarkably, it turns out that the special format of the particular Poisson kernel constructed in (1.4.16) is well-suited for dealing with the Regularity Problem in $\mathbb{R}^n_+$ as well, an aspect on which we elaborate below.

Proposition 1.4.6 *Let L be an $M \times M$ homogeneous constant complex coefficient second-order system in $\mathbb{R}^n$ which is weakly elliptic and satisfies $\mathfrak{A}^{\mathrm{dis}}_L(\mathbb{R}^n_+) \neq \varnothing$ (cf.* (1.4.15)*). Also, fix an aperture parameter $\kappa > 0$ along with an integrability exponent $p \in (1, \infty)$. Then the following statements are true.*

(i) The Inhomogeneous Regularity Problem for L in $\mathbb{R}^n_+$, i.e.,

$$\begin{cases} u \in \big[\mathscr{C}^\infty(\mathbb{R}^n_+)\big]^M, \quad Lu = 0 \text{ in } \mathbb{R}^n_+, \\ \mathcal{N}_\kappa u,\ \mathcal{N}_\kappa(\nabla u) \in L^p(\mathbb{R}^{n-1}, \mathcal{L}^{n-1}), \\ u\big|^{\kappa-\mathrm{n.t.}}_{\partial\mathbb{R}^n_+} = f \text{ at } \mathcal{L}^{n-1}\text{-a.e. point in } \mathbb{R}^{n-1}, \end{cases} \tag{1.4.46}$$

has a solution for each boundary datum $f \in \big[L^p_1(\mathbb{R}^{n-1}, \mathcal{L}^{n-1})\big]^M$.
More specifically, pick a coefficient tensor $A \in \mathfrak{A}^{\mathrm{dis}}_L(\mathbb{R}^n_+)$ and bring in the matrix-valued function $k : \mathbb{R}^n \setminus \{0\} \to \mathbb{C}^{M\times M}$ associated with A as in (1.4.15)*. Use this to define the matrix-valued function $P : \mathbb{R}^{n-1} \to \mathbb{C}^{M\times M}$ by setting $P(x') := 2k(x', 1)$ for each $x' \in \mathbb{R}^{n-1}$ (which, according to Proposition 1.4.4, is a Poisson kernel for L in $\mathbb{R}^n_+$ in the sense of Definition 1.4.2). Finally, for each $t \in (0, \infty)$, define $P_t := t^{1-n}P(\cdot/t)$ in $\mathbb{R}^{n-1}$. Then the function defined as*

$$u(x', t) := (P_t * f)(x') \text{ for each } (x', t) \in \mathbb{R}^{n-1} \times (0, \infty) = \mathbb{R}^n_+ \tag{1.4.47}$$

is a solution of the Inhomogeneous Regularity Problem (1.4.46)*. Furthermore, in the case when $\mathfrak{A}^{\mathrm{dis}}_{L^\top}(\mathbb{R}^n_+) \neq \varnothing$ there is at most one solution to the Inhomogeneous Regularity Problem* (1.4.46)*. As a consequence,*

$$\text{the Inhomogeneous Regularity Problem (1.4.46) has a unique solution for each boundary datum } f \in \big[L^p_1(\mathbb{R}^{n-1}, \mathcal{L}^{n-1})\big]^M \text{ whenever } \mathfrak{A}^{\mathrm{dis}}_L(\mathbb{R}^n_+) \neq \varnothing \text{ and } \mathfrak{A}^{\mathrm{dis}}_{L^\top}(\mathbb{R}^n_+) \neq \varnothing. \tag{1.4.48}$$

(ii) Analogous results are valid for the Dirichlet Problem for L in $\mathbb{R}^n_+$. *Specifically, the function u defined as in* (1.4.47) *in relation to any given boundary datum* $f \in \big[L^p(\mathbb{R}^{n-1}, \mathcal{L}^{n-1})\big]^M$ *solves the boundary value problem*

$$\begin{cases} u \in \big[\mathscr{C}^\infty(\mathbb{R}^n_+)\big]^M, \quad Lu = 0 \text{ in } \mathbb{R}^n_+, \\ \mathcal{N}_\kappa u \in L^p(\mathbb{R}^{n-1}, \mathcal{L}^{n-1}), \\ u\big|^{\kappa-\mathrm{n.t.}}_{\partial\mathbb{R}^n_+} = f \text{ at } \mathcal{L}^{n-1}\text{-a.e. point in } \mathbb{R}^{n-1}, \end{cases} \tag{1.4.49}$$

and this solution is unique under the additional assumption that $\mathfrak{A}^{\mathrm{dis}}_{L^\top}(\mathbb{R}^n_+) \neq \varnothing$. *Therefore,*

$$\text{the Dirichlet Problem (1.4.49) has a unique solution for each boundary datum } f \in \big[L^p(\mathbb{R}^{n-1}, \mathcal{L}^{n-1})\big]^M \text{ if } \mathfrak{A}^{\mathrm{dis}}_{L}(\mathbb{R}^n_+) \neq \varnothing \text{ and } \mathfrak{A}^{\mathrm{dis}}_{L^\top}(\mathbb{R}^n_+) \neq \varnothing. \tag{1.4.50}$$

(iii) Similar results are valid for the Homogeneous Regularity Problem for L in $\mathbb{R}^n_+$. *Concretely, for each boundary datum* $f \in \big[\dot{L}^p_1(\mathbb{R}^{n-1}, \mathcal{L}^{n-1})\big]^M$ *the function u defined as in* (1.4.47) *in relation to this f solves the boundary value problem*

$$\begin{cases} u \in \big[\mathscr{C}^\infty(\mathbb{R}^n_+)\big]^M, \quad Lu = 0 \text{ in } \mathbb{R}^n_+, \\ \mathcal{N}_\kappa(\nabla u) \in L^p(\mathbb{R}^{n-1}, \mathcal{L}^{n-1}), \\ u\big|^{\kappa-\mathrm{n.t.}}_{\partial\mathbb{R}^n_+} = f \text{ at } \mathcal{L}^{n-1}\text{-a.e. point in } \mathbb{R}^{n-1}, \end{cases} \tag{1.4.51}$$

and this solution is unique under the additional assumption that $\mathfrak{A}^{\mathrm{dis}}_{L^\top}(\mathbb{R}^n_+) \neq \varnothing$. *Altogether,*

$$\text{the Homogeneous Regularity Problem (1.4.51) has a unique solution for each boundary datum } f \in \big[\dot{L}^p_1(\mathbb{R}^{n-1}, \mathcal{L}^{n-1})\big]^M \text{ if } \mathfrak{A}^{\mathrm{dis}}_{L}(\mathbb{R}^n_+) \neq \varnothing \text{ and } \mathfrak{A}^{\mathrm{dis}}_{L^\top}(\mathbb{R}^n_+) \neq \varnothing. \tag{1.4.52}$$

Proof Pick a coefficient tensor $A \in \mathfrak{A}^{\mathrm{dis}}_{L}(\mathbb{R}^n_+)$. From (1.4.18), [132, (1.3.24), (1.5.8)], and the approximation to the identity property for the Poisson kernel (cf. (1.4.44)) we then conclude that u defined as in (1.4.47) solves the Inhomogeneous Regularity Problem (1.4.46) if $f \in \big[L^p_1(\mathbb{R}^{n-1}, \mathcal{L}^{n-1})\big]^M$, and u solves the Dirichlet Problem (1.4.49) if $f \in \big[L^p(\mathbb{R}^{n-1}, \mathcal{L}^{n-1})\big]^M$. Also, from (1.4.20) and [132, (1.8.142)] we see that u defined as in (1.4.47) solves the Homogeneous Regularity Problem (1.4.51) if $f \in \big[\dot{L}^p_1(\mathbb{R}^{n-1}, \mathcal{L}^{n-1})\big]^M$.

Henceforth make the additional assumption that $\mathfrak{A}^{\mathrm{dis}}_{L^\top}(\mathbb{R}^n_+) \neq \varnothing$. Then from what we have proved already to this point it follows that the Inhomogeneous Regularity Problem for the transpose system $L^\top$ in the upper half-space $\mathbb{R}^n_+$ is always solvable for boundary data in $\big[L^{p'}_1(\mathbb{R}^{n-1}, \mathcal{L}^{n-1})\big]^M$, where the exponent $p' \in (1, \infty)$ is such that $1/p + 1/p' = 1$. Granted this, [131, Theorem 4.3.1] guarantees that there exists at most one solution for the Dirichlet Problem (1.4.49) for any boundary datum in $\big[L^p(\mathbb{R}^{n-1},\mathcal{L}^{n-1})\big]^M$. As a consequence, the Inhomogeneous Regularity Problem (1.4.46) also has at most one solution for any boundary datum in $\big[L^p_1(\mathbb{R}^{n-1},\mathcal{L}^{n-1})\big]^M$. There remains to prove uniqueness for Homogeneous Regularity Problem (1.4.51). Hence, assume

$$\begin{cases} u \in \big[\mathscr{C}^\infty(\mathbb{R}^n_+)\big]^M, \quad Lu = 0 \text{ in } \mathbb{R}^n_+, \\ \mathcal{N}_\kappa(\nabla u) \in L^p(\mathbb{R}^{n-1},\mathcal{L}^{n-1}), \\ u\big|^{\kappa-\text{n.t.}}_{\partial\mathbb{R}^n_+} = 0 \text{ at } \mathcal{L}^{n-1}\text{-a.e. point in } \mathbb{R}^{n-1}, \end{cases} \tag{1.4.53}$$

with the goal of proving that $u = 0$ in $\mathbb{R}^n_+$. To this end, pick a coefficient tensor

$$A \in \mathfrak{A}_L \text{ such that } A^\top \in \mathfrak{A}^{\text{dis}}_{L^\top}(\mathbb{R}^n_+). \tag{1.4.54}$$

Then the integral representation formula [132, Theorem 1.8.19, (1.8.200)] presently reduces to

$$u = -\mathscr{S}_{\text{mod}}\big(\partial^A_\nu u\big) + c \text{ in } \mathbb{R}^n_+ \tag{1.4.55}$$

for some constant $c \in \mathbb{C}^M$. Here $\mathscr{S}_{\text{mod}}$ is the modified single layer potential operator associated with the system L and the set $\Omega := \mathbb{R}^n_+$ as in (A.0.229), and $\partial^A_\nu u$ is the conormal derivative of u, associated with the coefficient tensor A and the set $\Omega := \mathbb{R}^n_+$ as in (A.0.203). Apply ∂^A_ν to both sides of (1.4.55). On account of the jump-formula [132, (1.5.58)] we then obtain

$$\partial^A_\nu u = -\partial^A_\nu \mathscr{S}_{\text{mod}}\big(\partial^A_\nu u\big) = -\big(-\tfrac{1}{2}I + K^\#_{A^\top}\big)\big(\partial^A_\nu u\big), \tag{1.4.56}$$

where I is the identity operator, and $K^\#_{A^\top}$ is the operator associated as in (A.0.118) with the coefficient tensor $A^\top$ and the set $\Omega := \mathbb{R}^n_+$. Since (1.4.54) and the equivalence *(i)* $\Leftrightarrow$ *(iv)* in Proposition 1.3.2 presently imply that actually $K^\#_{A^\top} = 0$, we conclude from (1.4.56) that $\partial^A_\nu u = 0$. Using this back in (1.4.55) proves that u is a constant in $\mathbb{R}^n_+$ which, in light of the last condition in (1.4.53), forces $u \equiv 0$ in $\mathbb{R}^n_+$. □

In addition to the Poisson integral representation formula (1.4.47), solutions to the Regularity Problems discussed above may also be expressed in terms of single layers. Somewhat more specifically, if both the given weakly elliptic system and its the transpose have distinguished coefficient tensors adapted to $\mathbb{R}^n_+$ then any function satisfying the first three conditions in the formulation of the Homogeneous Regularity Problem (1.4.51) may be represented as a (modified) single layer. Here is a precise statement to this effect.

Corollary 1.4.7 *Assume L is a weakly elliptic, homogeneous, constant (complex) coefficient, second-order $M \times M$ system in $\mathbb{R}^n$ for which $\mathfrak{A}^{\text{dis}}_L(\mathbb{R}^n_+) \neq \varnothing$ and $\mathfrak{A}^{\text{dis}}_{L^\top}(\mathbb{R}^n_+) \neq \varnothing$ (cf. (1.4.15)). Choose a coefficient tensor $A \in \mathfrak{A}^{\text{dis}}_L(\mathbb{R}^n_+)$ and pick an aperture parameter $\kappa > 0$ together with an exponent $p \in (1,\infty)$. Then for any*

$$u \in \big[\mathscr{C}^\infty(\mathbb{R}^n_+)\big]^M, \quad Lu = 0 \text{ in } \mathbb{R}^n_+, \quad \mathcal{N}_\kappa(\nabla u) \in L^p(\mathbb{R}^{n-1},\mathcal{L}^{n-1}), \tag{1.4.57}$$

there exists a constant $c \in \mathbb{C}^M$ such that

$$u = -2\mathscr{S}_{\rm mod}\big(\partial_\nu^A u\big) + c \ \text{ in } \mathbb{R}^n_+ \tag{1.4.58}$$

where $\mathscr{S}_{\rm mod}$ is the modified single layer potential operator associated with the system L and the set $\Omega := \mathbb{R}^n_+$ as in (A.0.229)*, and where $\partial_\nu^A u$ is the conormal derivative of u, associated with the coefficient tensor A and the set $\Omega := \mathbb{R}^n_+$ as in* (A.0.203).

Proof From [132, Theorem 1.8.19] we know that for any function u as in (1.4.57) the nontangential boundary trace $u\big|_{\partial\mathbb{R}^n_+}^{\kappa-\text{n.t.}}$ is a well-defined function in the space $\big[\dot{L}^p_1(\mathbb{R}^{n-1}, \mathcal{L}^{n-1})\big]^M$, the conormal derivative $\partial_\nu^A u$ is a well-defined function in the space $\big[L^p(\mathbb{R}^{n-1}, \mathcal{L}^{n-1})\big]^M$, and there exists some $c \in \mathbb{C}^M$ such that

$$u = \mathcal{D}_{\rm mod}\Big(u\big|_{\partial\mathbb{R}^n_+}^{\kappa-\text{n.t.}}\Big) - \mathscr{S}_{\rm mod}\big(\partial_\nu^A u\big) + c \ \text{ in } \mathbb{R}^n_+ \tag{1.4.59}$$

where $\mathcal{D}_{\rm mod}$ is the modified boundary-to-domain double layer potential operator associated with the coefficient tensor A and the set $\Omega := \mathbb{R}^n_+$ as in (A.0.73). Taking nontangential boundary traces in (1.4.59) then yields, on account of (1.4.20) and [132, (1.5.80)], $u\big|_{\partial\mathbb{R}^n_+}^{\kappa-\text{n.t.}} = \frac{1}{2}\Big(u\big|_{\partial\mathbb{R}^n_+}^{\kappa-\text{n.t.}}\Big) - S_{\rm mod}\big(\partial_\nu^A u\big) + c$ at $\mathcal{L}^{n-1}$-a.e. point in $\mathbb{R}^{n-1}$. Hence, $u\big|_{\partial\mathbb{R}^n_+}^{\kappa-\text{n.t.}} = -2S_{\rm mod}\big(\partial_\nu^A u\big) + c$ at $\mathcal{L}^{n-1}$-a.e. point in $\mathbb{R}^{n-1}$. Then the functions u and $-2\mathscr{S}_{\rm mod}\big(\partial_\nu^A u\big) + c$ solve the same Homogeneous Regularity Problem, formulated as in (1.4.51) for the boundary datum $f := u\big|_{\partial\mathbb{R}^n_+}^{\kappa-\text{n.t.}} \in \big[\dot{L}^p_1(\mathbb{R}^{n-1}, \mathcal{L}^{n-1})\big]^M$. In view of (1.4.52), this then implies that (1.4.58) holds. □

The goal of our next proposition is to address the issue of uniqueness of the Poisson kernel for weakly elliptic systems whose transpose has a distinguished coefficient tensor adapted to $\mathbb{R}^n_+$.

Proposition 1.4.8 *Let L be a weakly elliptic, homogeneous, constant complex coefficient, second-order $M \times M$ system in $\mathbb{R}^n$ with the property that $\mathfrak{A}^{\rm dis}_{L^\top}(\mathbb{R}^n_+) \neq \varnothing$ (cf.* (1.4.15)*). Then there exists at most one Poisson kernel P^L for L in $\mathbb{R}^n_+$ in the sense of Definition 1.4.2.*

Proof Let $P^{(1)}$ and $P^{(2)}$ be two Poisson kernels for L in $\mathbb{R}^n_+$ in the sense of Definition 1.4.2. Fix an arbitrary $f \in \big[\mathscr{C}^\infty_c(\mathbb{R}^{n-1})\big]^M$ and for each $(x',t) \in \mathbb{R}^n_+$ define $u_1(x',t) := (P^{(1)}_t * f)(x')$ and $u_2(x',t) := (P^{(2)}_t * f)(x')$. Then from Definition 1.4.2 and Lemma 1.4.5 we see that, for any given exponent $p \in (1,\infty)$ and aperture parameter $\kappa \in (0,\infty)$, both u_1 and u_2 solve the Dirichlet Problem for L in $\mathbb{R}^n_+$ with datum f, regarded as a function in $\big[L^p(\mathbb{R}^{n-1}, \mathcal{L}^{n-1})\big]^M$, i.e., for $j = 1, 2$ we have

$$\begin{cases} u_j \in \big[\mathscr{C}^\infty(\mathbb{R}^n_+)\big]^M, \quad Lu_j = 0 \ \text{ in } \mathbb{R}^n_+, \\ \mathcal{N}_\kappa u_j \in L^p(\mathbb{R}^{n-1}, \mathcal{L}^{n-1}), \\ u_j\big|_{\partial\mathbb{R}^n_+}^{\kappa-\text{n.t.}} = f \ \text{ at } \mathcal{L}^{n-1}\text{-a.e. point in } \mathbb{R}^{n-1}. \end{cases} \tag{1.4.60}$$

Since we are presently assuming $\mathfrak{A}^{\mathrm{dis}}_{L^\top}(\mathbb{R}^n_+) \neq \varnothing$, the Regularity Problem for the system $L^\top$ in the upper half-space $\mathbb{R}^n_+$ is always solvable (as noted in Proposition 1.4.6). Granted this, [131, Theorem 4.3.1] guarantees that $u_1 = u_2$ in $\mathbb{R}^n_+$, which further translates into $\big(P^{(1)}_t * f\big)(x') = \big(P^{(2)}_t * f\big)(x')$ for all $(x', t) \in \mathbb{R}^n_+$ and all $f \in \big[\mathscr{C}^\infty_c(\mathbb{R}^{n-1})\big]^M$. In turn, this yields $P^{(1)} = P^{(2)}$ at $\mathcal{L}^{n-1}$-a.e. point in $\mathbb{R}^{n-1}$, hence everywhere by the continuity of $P^{(1)}$ and $P^{(2)}$ (cf. (1.4.12)). □

Putting together Proposition 1.4.4 with Proposition 1.4.8 yields the following remarkable corollary.

Corollary 1.4.9 *Let L be a weakly elliptic, homogeneous, constant complex coefficient, second-order $M \times M$ system in $\mathbb{R}^n$ with the property that $\mathfrak{A}^{\mathrm{dis}}_{L}(\mathbb{R}^n_+) \neq \varnothing$ and $\mathfrak{A}^{\mathrm{dis}}_{L^\top}(\mathbb{R}^n_+) \neq \varnothing$ (cf. (1.4.15)). Then there exists a unique Poisson kernel P^L for L in $\mathbb{R}^n_+$ in the sense of Definition 1.4.2.*

Proof Indeed, existence is ensured by Proposition 1.4.4, while uniqueness follows from Proposition 1.4.8. □

With a proof anticipating results obtained later, we can complement Proposition 1.4.4, Proposition 1.4.8, and Corollary 1.4.9 by the following nonexistence result:

Proposition 1.4.10 *For each $n \in \mathbb{N}$ with $n \geq 2$ there exists a weakly elliptic, homogeneous, constant (real) coefficient, second-order system in $\mathbb{R}^n$ which does not possess a Poisson kernel in $\mathbb{R}^n_+$ in the sense of Definition 1.4.2.*

Proof This is the case for the $n \times n$ system $L_D := \Delta - 2\nabla\mathrm{div}$ in $\mathbb{R}^n$. Indeed, this is a weakly elliptic, homogeneous, constant (real) coefficient, second-order system, but the failure of the L^p Dirichlet Problem with $p \in (1, \infty)$ for L_D in $\mathbb{R}^n_+$ to be solvable for arbitrary boundary data in $\big[L^p(\mathbb{R}^{n-1}, \mathcal{L}^{n-1})\big]^n$ (noted in Theorem 8.6.1) implies that there cannot possibly be a Poisson kernel for L_D in $\mathbb{R}^n_+$ in the sense of Definition 1.4.2 since otherwise, by Proposition 1.4.6, said problem would always have a solution. □

1.5 The Issue of Uniqueness of a Distinguished Coefficient Tensor

The point made in our first result is that any system satisfying the Legendre-Hadamard strong ellipticity condition can have at most one distinguished coefficient tensor.

Proposition 1.5.1 *Let $M, n \in \mathbb{N}$ with $n \geq 2$ and consider a homogeneous, second-order, constant complex coefficient, $M \times M$ system L in $\mathbb{R}^n$ which satisfies the Legendre-Hadamard (strong) ellipticity condition* [131, Definition 1.3.2]. *Then $\#\mathfrak{A}^{\mathrm{dis}}_L \leq 1$. In other words, if $M, n \in \mathbb{N}$ with $n \geq 2$ and L is a homogeneous, second-order, constant complex coefficient, $M \times M$ system in $\mathbb{R}^n$ satisfying the Legendre-Hadamard (strong) ellipticity condition, then $\mathfrak{A}^{\mathrm{dis}}_L$ is either empty or a singleton.*

Proof In the case $\mathfrak{A}_L^{\rm dis} = \varnothing$ there is nothing to prove. Assume next that there exists some $A = \big(a_{jk}^{\alpha\beta}\big)_{\substack{1\leq\alpha,\beta\leq M\\ 1\leq j,k\leq n}} \in \mathfrak{A}_L^{\rm dis}$. The goal is to show that this coefficient tensor is uniquely determined by L. To this end, bring in the function $k : \mathbb{R}^n \setminus \{0\} \longrightarrow \mathbb{C}^{M\times M}$ from (1.2.11), associated with A as in item *(g)* of Proposition 1.2.1. Then from Proposition 1.4.4 and Theorem 1.4.3 we conclude that the matrix-valued function P defined as in (1.4.16) is the unique Poisson kernel for L in $\mathbb{R}^n_+$ in the sense of Definition 1.4.2. Given that, as seen from (1.4.16), we have $k(x',1) = 2^{-1}P(x')$ for each $x' \in \mathbb{R}^{n-1}$, it follows that the assignment $\mathbb{R}^{n-1} \ni x' \mapsto k(x',1) \in \mathbb{C}^{M\times M}$ is uniquely determined by L (in the sense that it is independent of the choice of $A \in \mathfrak{A}_L^{\rm dis}$). Since k is even and positive homogeneous of degree $-n$ in $\mathbb{R}^n \setminus \{0\}$ (cf. (1.2.14)), we then deduce that the assignment

$$\mathbb{R}^n \setminus \big(\mathbb{R}^{n-1} \times \{0\}\big) \ni x \longmapsto k(x) \in \mathbb{C}^{M\times M} \tag{1.5.1}$$

is also uniquely determined by L (in the same sense as before). Based on this and the fact that the entries in k are continuous in $\mathbb{R}^n \setminus \{0\}$ (again, see (1.2.14)) we conclude (via density) that the assignment $\mathbb{R}^n \ni x \longmapsto k(x) \in \mathbb{C}^{M\times M}$ is once more uniquely determined by L. From this and (1.2.12) we then see that for each index $s \in \{1,\dots,n\}$ the assignment

$$\mathbb{R}^n \setminus \{0\} \ni x \longmapsto \Big(a_{rs}^{\beta\alpha}(\partial_r E_{\gamma\beta})(x)\Big)_{1\leq\gamma,\alpha\leq M} \in \mathbb{C}^{M\times M} \tag{1.5.2}$$

is uniquely determined by L. Fix $s \in \{1,\dots,n\}$ and regard $\big(a_{rs}^{\beta\alpha}\partial_r E_{\gamma\beta}\big)_{1\leq\gamma,\alpha\leq M}$ as a matrix whose entries are locally integrable functions in $\mathbb{R}^n$ (with respect to $\mathcal{L}^n$) which induce tempered distributions in $\mathbb{R}^n$ via ordinary integration against Schwartz functions (cf. [131, Theorem 1.4.2]). From this and the fact that (1.5.2) is uniquely determined by L we then conclude that the $M \times M$ matrix of tempered distributions

$$\big(a_{rs}^{\beta\alpha}\partial_r E_{\gamma\beta}\big)_{1\leq\gamma,\alpha\leq M} \in \big[\mathcal{S}'(\mathbb{R}^n)\big]^{M\times M} \tag{1.5.3}$$

is uniquely determined by L. Taking the Fourier transform then restricting to $\mathbb{R}^n \setminus \{0\}$ gives (on account of [131, (1.4.30)]) that for each $s \in \{1,\dots,n\}$ the assignment

$$\mathbb{R}^n \setminus \{0\} \ni \xi \mapsto \Big(a_{rs}^{\beta\alpha}\xi_r\big(L(\xi)^{-1}\big)_{\gamma\beta}\Big)_{1\leq\gamma,\alpha\leq M} \in \mathbb{C}^{M\times M} \tag{1.5.4}$$

is uniquely determined by L. To proceed, for each pair of indices $s, r \in \{1,\dots,n\}$ define the matrix $A_{rs} := \big(a_{rs}^{\alpha\beta}\big)_{1\leq\alpha,\beta\leq M} \in \mathbb{C}^{M\times M}$. We may then recast what we have just proved as the statement that for each $r \in \{1,\dots,n\}$ the assignment

$$\mathbb{R}^n \setminus \{0\} \ni \xi \longmapsto L(\xi)^{-1}\xi_r A_{rs} \in \mathbb{C}^{M\times M} \tag{1.5.5}$$

is uniquely determined by L. Multiplying by the characteristic matrix $L(\xi)$ then proves that for each index $r \in \{1,\dots,n\}$ the matrix-valued assignment

$$\mathbb{R}^n \setminus \{0\} \ni \xi \longmapsto \xi_r A_{rs} \in \mathbb{C}^{M\times M} \tag{1.5.6}$$

is uniquely determined by L. This readily implies that the family of matrices A_{rs} with $1 \le r, s \le n$ is uniquely determined by L, hence so is the original coefficient tensor $A = \big(a_{jk}^{\alpha\beta}\big)_{\substack{1\le\alpha,\beta\le M\\ 1\le j,k\le n}} \in \mathfrak{A}_L^{\rm dis}$. This ultimately shows that the cardinality of $\mathfrak{A}_L^{\rm dis}$ is at most one, finishing the proof of the proposition. □

We are now in a position to formulate and prove the following basic theorem:

Theorem 1.5.2 *Let $M, n \in \mathbb{N}$ with $n \ge 2$ and consider a homogeneous, second-order, constant complex coefficient, $M \times M$ system L in $\mathbb{R}^n$ which satisfies the Legendre-Hadamard (strong) ellipticity condition* [131, Definition 1.3.2]. *Then either*

$$\mathfrak{A}_L^{\rm dis} = \varnothing \ \textit{and}\ \mathfrak{A}_{L^\top}^{\rm dis} = \varnothing, \tag{1.5.7}$$

or

$$\textit{there exists some } A \in \mathfrak{A}_L \textit{ such that } \mathfrak{A}_L^{\rm dis} = \{A\} \textit{ and } \mathfrak{A}_{L^\top}^{\rm dis} = \{A^\top\}. \tag{1.5.8}$$

As a corollary, if $M, n \in \mathbb{N}$ with $n \ge 2$ and L is a homogeneous, second-order, constant complex coefficient, $M\times M$ system in $\mathbb{R}^n$ satisfying the Legendre-Hadamard (strong) ellipticity condition, then $\mathfrak{A}_L^{\rm dis}$ is either empty or a singleton.

Proof This readily follows from Proposition 1.5.1 and Proposition 1.4.1. □

Turning our attention to systems that are merely weakly elliptic, we first note the following result:

Proposition 1.5.3 *Let $M, n \in \mathbb{N}$ with $n \ge 2$ and consider a weakly elliptic, homogeneous, second-order, constant complex coefficient, $M \times M$ system L in $\mathbb{R}^n$. Then*

$$\mathfrak{A}_{L^\top}^{\rm dis} \neq \varnothing \Longrightarrow A^\top \in \mathfrak{A}_{L^\top}^{\rm dis} \textit{ for each } A \in \mathfrak{A}_L^{\rm dis}. \tag{1.5.9}$$

As a consequence,

$$\begin{aligned} &\textit{if both } \mathfrak{A}_L^{\rm dis} \neq \varnothing \textit{ and } \mathfrak{A}_{L^\top}^{\rm dis} \neq \varnothing \textit{ then for each } A \in \mathfrak{A}_L \\ &\quad \textit{one has } A \in \mathfrak{A}_L^{\rm dis} \textit{ if and only if } A^\top \in \mathfrak{A}_{L^\top}^{\rm dis}. \end{aligned} \tag{1.5.10}$$

Equivalently,

$$\textit{if both } \mathfrak{A}_L^{\rm dis} \neq \varnothing \textit{ and } \mathfrak{A}_{L^\top}^{\rm dis} \neq \varnothing \textit{ then } \mathfrak{A}_{L^\top}^{\rm dis} = \big(\mathfrak{A}_L^{\rm dis}\big)^\top := \big\{A^\top : A \in \mathfrak{A}_L^{\rm dis}\big\}. \tag{1.5.11}$$

Proof Assume $\mathfrak{A}_{L^\top}^{\rm dis} \neq \varnothing$ and pick an arbitrary coefficient tensor $A \in \mathfrak{A}_L$. For any given UR domain $\Omega \subseteq \mathbb{R}^n$, let $K_{\rm mod}$ be the modified boundary-to-boundary double layer potential associated with A and Ω as in (A.0.117). Also, let $K^{\#}_{A^\top}$ be the "transpose" double layer potential associated with $A^\top$ and Ω as in (A.0.118). Finally, recall that $S_{\rm mod}$ is the modified version of the boundary-to-boundary single layer operator, introduced in (A.0.232). From [132, Theorem 1.8.26, item *(1)*] we know that for each $f \in \big[L^p(\partial\Omega,\sigma)\big]^M$, where $\sigma := \mathcal{H}^{n-1}\lfloor\partial\Omega$ and $p \in (1,\infty)$, there exists c_f, which is the nontangential trace on $\partial\Omega$ of some $\mathbb{C}^M$-valued locally constant function in Ω, with the property that

$$S_{\text{mod}}\big(K^{\#}_{A^\top} f\big) = K_{\text{mod}}\big(S_{\text{mod}} f\big) + c_f \ \text{ at } \sigma\text{-a.e. point on } \partial\Omega. \tag{1.5.12}$$

Let us now assume that $A \in \mathfrak{A}_L^{\text{dis}}$ and that Ω is a half-space in $\mathbb{R}^n$. Then from (1.5.12) and (1.4.6) we conclude that

$$S_{\text{mod}}\big(K^{\#}_{A^\top} f\big) \text{ is constant on } \partial\Omega \text{ for each } f \in \big[L^p(\partial\Omega,\sigma)\big]^M. \tag{1.5.13}$$

Since we are assuming $\mathfrak{A}_{L^\top}^{\text{dis}} \neq \varnothing$, it is possible to select a coefficient tensor $\widetilde{A} \in \mathfrak{A}_L$ with the property that $\widetilde{A}^\top \in \mathfrak{A}_{L^\top}^{\text{dis}}$. Denote by $\widetilde{\mathcal{D}}_{\text{mod}}$ the modified version of the double layer operator associated with $\widetilde{A}$ and Ω as in (A.0.73). Then from (1.5.13) and [132, (1.8.145)-(1.8.146)] we see that, on the one hand,

$$(\partial_\nu^{\widetilde{A}} \widetilde{\mathcal{D}}_{\text{mod}}) S_{\text{mod}}\big(K^{\#}_{A^\top} f\big) = 0 \ \text{ for each } f \in \big[L^p(\partial\Omega,\sigma)\big]^M. \tag{1.5.14}$$

On the other hand, [132, (1.8.251)] implies

$$\begin{aligned}(\partial_\nu^{\widetilde{A}} \widetilde{\mathcal{D}}_{\text{mod}}) S_{\text{mod}}\big(K^{\#}_{A^\top} f\big) &= \big(\tfrac{1}{2}I + K^{\#}_{\widetilde{A}^\top}\big)\big(-\tfrac{1}{2}I + K^{\#}_{\widetilde{A}^\top}\big)\big(K^{\#}_{A^\top} f\big) \\ &= -\tfrac{1}{4}K^{\#}_{A^\top} f \ \text{ for each } f \in \big[L^p(\partial\Omega,\sigma)\big]^M,\end{aligned} \tag{1.5.15}$$

where the last equality is a consequence of the fact that, since $\widetilde{A}^\top \in \mathfrak{A}_{L^\top}^{\text{dis}}$, the operator $K^{\#}_{\widetilde{A}^\top}$ vanishes identically (see the implication *(i)*$\Rightarrow$*(iv)* in Proposition 1.3.2). Together, (1.5.14) and (1.5.15) imply $K^{\#}_{A^\top} f = 0$ for each $f \in \big[L^p(\partial\Omega,\sigma)\big]^M$. In turn, this forces the integral kernel of $K^{\#}_{A^\top}$ to be zero, hence $K^{\#}_{A^\top}$ regarded as a mapping associated with $A^\top$ and Ω as in (A.0.118) is the zero operator. With this in hand, the implication *(iv)*$\Rightarrow$*(i)* in Proposition 1.3.2 (with A replaced by $A^\top$ and L replaced by $L^\top$) then proves that $A^\top \in \mathfrak{A}_{L^\top}$ satisfies $A^\top \in \mathfrak{A}_{L^\top}^{\text{dis}}$. This completes the proof of (1.5.9). Finally, (1.5.10) is an immediate consequence of (1.5.9). □

We continue by establishing the uniqueness result in the proposition below.

Proposition 1.5.4 *Let $M, n \in \mathbb{N}$ with $n \geq 2$. Consider a weakly elliptic, homogeneous, second-order, constant complex coefficient, $M \times M$ system L in $\mathbb{R}^n$. Then*

$$\mathfrak{A}_{L^\top}^{\text{dis}} \neq \varnothing \Longrightarrow \#\mathfrak{A}_L^{\text{dis}} \leq 1. \tag{1.5.16}$$

Proof Work under the assumption that $\mathfrak{A}_{L^\top}^{\text{dis}} \neq \varnothing$. If $\mathfrak{A}_L^{\text{dis}} = \varnothing$ there is nothing to prove, so assume there exists $A \in \mathfrak{A}_L^{\text{dis}}$. The goal is to show that this is uniquely determined by L. To this end, bring in the function $k : \mathbb{R}^n \setminus \{0\} \to \mathbb{C}^{M\times M}$ from (1.2.11), associated with A as in item *(g)* of Proposition 1.2.1. Then from Proposition 1.4.4 and Proposition 1.4.8 (whose applicability is guaranteed by our assumption that $\mathfrak{A}_{L^\top}^{\text{dis}} \neq \varnothing$) we conclude that the matrix-valued function P defined as in (1.4.16) is the unique Poisson kernel for L in $\mathbb{R}^n_+$ in the sense of Definition 1.4.2. Given that, as seen from (1.4.16), we have $k(x', 1) = 2^{-1}P(x')$ at each point $x' \in \mathbb{R}^{n-1}$, it follows that the assignment $\mathbb{R}^{n-1} \ni x' \mapsto k(x', 1) \in \mathbb{C}^{M\times M}$ is uniquely determined by L (in the sense that it is independent of the choice of $A \in \mathfrak{A}_L^{\text{dis}}$).

From this point on, the same type of argument as in the end-game of the proof of Proposition 1.5.1 applies and gives that the coefficient tensor A is uniquely determined by L. Hence, that the cardinality of $\mathfrak{A}_L^{\text{dis}}$ is at most one, as wanted. □

The following is a companion result to Theorem 1.5.2, dealing with the class of weakly elliptic systems.

Theorem 1.5.5 *Let $M, n \in \mathbb{N}$ with $n \geq 2$. Consider a weakly elliptic, homogeneous, second-order, constant complex coefficient, $M \times M$ system L in $\mathbb{R}^n$ with the property that $\mathfrak{A}_L^{\text{dis}} \neq \emptyset$ and $\mathfrak{A}_{L^\top}^{\text{dis}} \neq \emptyset$. Then both $\mathfrak{A}_L^{\text{dis}}$ and $\mathfrak{A}_{L^\top}^{\text{dis}}$ are singletons. In fact, $\mathfrak{A}_L^{\text{dis}} = \{A\}$ and $\mathfrak{A}_{L^\top}^{\text{dis}} = \{A^\top\}$ for some $A \in \mathfrak{A}_L$.*

In particular, if $M, n \in \mathbb{N}$ with $n \geq 2$ and L is a symmetric, weakly elliptic, homogeneous, second-order, constant complex coefficient, $M \times M$ system in $\mathbb{R}^n$, then $\mathfrak{A}_L^{\text{dis}}$ is either empty or a singleton and, in the latter case, one has $\mathfrak{A}_L^{\text{dis}} = \{A\}$ for some $A \in \mathfrak{A}_L$ satisfying $A^\top = A$.

Proof This is an immediate consequence of Proposition 1.5.4 and Proposition 1.5.3. □

1.6 The Issue of Existence of a Distinguished Coefficient Tensor

We begin by extending the scope of Definition 1.4.2. Specifically, associated with a given second-order, homogeneous, constant complex coefficient, weakly elliptic system we introduce the notion of Poisson kernel (and kernel function) in arbitrary half-spaces in the Euclidean ambient. As we will see later (in Theorem 1.6.4), their existence and compatibility with one another is strongly linked to the property that the system in question has a distinguished coefficient tensor.

Definition 1.6.1 *Fix $M \in \mathbb{N}$ and let L be an $M \times M$ second-order, homogeneous, constant complex coefficient, weakly elliptic system in $\mathbb{R}^n$. Also, fix $h \in S^{n-1}$ and introduce*

$$H := \big\{x \in \mathbb{R}^n : x \cdot h = 0\big\}, \quad H^+ := \big\{x \in \mathbb{R}^n : x \cdot h > 0\big\} = \langle h \rangle^\perp, \tag{1.6.1}$$

along with the orthogonal projection of $\mathbb{R}^n$ onto H:

$$\pi_H : \mathbb{R}^n \longrightarrow H, \quad \pi_H(x) := x - (x \cdot h)h \text{ for all } x \in \mathbb{R}^n. \tag{1.6.2}$$

(1) Call a function $P : H \longrightarrow \mathbb{C}^{M\times M}$ a `Poisson kernel for` *L* `in` *H^+ provided P satisfies the following conditions:*

(a) there exists some $C \in (0, \infty)$ such that $|P(x)| \leq \dfrac{C}{(1 + |x|^2)^{n/2}}$ for all $x \in H$;

(b) P is measurable and satisfies $\int_H P \, d\mathcal{H}^{n-1} = I_{M\times M}$, where $I_{M\times M}$ denotes the $M \times M$-identity matrix;
(c) one has $LK = 0$ in $\big[\mathcal{D}'(H^+)\big]^{M\times M}$ if

$$K(x) := (x \cdot h)^{1-n} P\Big(\frac{\pi_H(x)}{x \cdot h}\Big) \text{ for each } x \in H^+. \tag{1.6.3}$$

(2) Call $K : H^+ \to \mathbb{C}^{M\times M}$ a `kernel function for` *L* `in` *H^+ provided K satisfies the following conditions:*

(i) $K \in \big[\mathscr{C}^\infty(H^+)\big]^{M\times M}$ and $LK = 0 \cdot I_{M\times M}$ in H^+;
(ii) K is positive homogeneous of degree $1 - n$;
(iii) there exists some $C \in (0, \infty)$ such that $|K(x)| \leq C\frac{x \cdot h}{|x|^n}$ for all $x \in H^+$;
(iv) $\int_H K(x + h) \, d\mathcal{H}^{n-1}(x) = I_{M\times M}$.

Related to the notions just defined, a couple of remarks are in order.

Remark 1.6.2 *Retain the context of Definition 1.6.1.*

(1) If P is a Poisson kernel for L in H^+ then the function K associated with P according to (1.6.3) satisfies

$$|K(x)| \leq C(x \cdot h)^{1-n}\Big(1 + \Big|\frac{\pi_H(x)}{x \cdot h}\Big|^2\Big)^{-n/2} = C\frac{x \cdot h}{|x|^n} \text{ for all } x \in H^+. \tag{1.6.4}$$

Hence $K \in \big[L^1_{\rm loc}(H^+, \mathcal{L}^n)\big]^{M\times M}$.
(2) If P is a Poisson kernel for L in H^+ then the function K associated with P according to (1.6.3) is a kernel function for L in H^+. Indeed, that this K satisfies (i) follows from (c) in Definition 1.6.1 and elliptic regularity, while (1.6.4) ensures that (iii) is satisfied. Also, the definition of K and the fact that the function $H^+ \ni x \mapsto \pi_H(x)/(x \cdot h)$ is positive homogeneous of degree 0 imply that K is positive homogeneous of degree $1-n$. It remains to show that K satisfies the integral formula in (iv). The latter is seen by writing

$$\begin{aligned} \int_H K(x + h) \, d\mathcal{H}^{n-1}(x) &= \int_H [(x + h) \cdot h] \, P\Big(\frac{\pi_H(x + h)}{(x + h) \cdot h}\Big) \, d\mathcal{H}^{n-1}(x) \\ &= \int_H P(x) \, d\mathcal{H}^{n-1}(x) = I_{M\times M}, \end{aligned} \tag{1.6.5}$$

where for the second equality in (1.6.5) we observed that since $h \in S^{n-1}$ we have

$$(x + h) \cdot h = 1 \text{ and } \pi_H(x + h) = x \text{ for all } x \in H = \langle h\rangle^\perp. \tag{1.6.6}$$

(3) If K is a kernel function for L in H^+ then the mapping $P : H \to \mathbb{C}^{M\times M}$ defined by $P(x) := K(x + h)$ for all $x \in H$ is a Poisson kernel for L in H^+. Indeed,

$h \in S^{n-1}$ implies $|x+h|^n = (1+|x|^2)^{n/2}$ for all $x \in H$, so property (iii) for K and the first identity in (1.6.6) yield

$$|P(x)| = |K(x+h)| \le C\frac{(x+h)\cdot h}{|x+h|^n} = \frac{C}{(1+|x|^2)^{n/2}} \text{ for all } x \in H. \tag{1.6.7}$$

This shows that P satisfies condition (a) in Definition 1.6.1, while condition (b) in Definition 1.6.1 is immediate from property (iv) satisfied by K. Finally, bearing in mind that K is positive homogeneous of degree $1-n$, we may write

$$\begin{aligned}(x\cdot h)^{1-n}P\Big(\frac{\pi_H(x)}{x\cdot h}\Big) &= (x\cdot h)^{1-n}K\Big(\frac{\pi_H(x)}{x\cdot h}+h\Big)\\ &= K\big(\pi_H(x)+(x\cdot h)h\big) = K(x),\end{aligned} \tag{1.6.8}$$

for all $x \in H^+$. Together with property (i) for K, this shows that P satisfies (c) in Definition 1.6.1.

In what follows we will make use of the following change of variables result.

Proposition 1.6.3 *Fix an integer $n \ge 2$. Then for any vector $h \in S^{n-1}$ and any continuous function $f : \{\omega \in S^{n-1} : \omega\cdot h \ge 0\} \to \mathbb{C}$ one has*

$$\int_{x\in\langle h\rangle^\perp} f\Big(\frac{x+h}{|x+h|}\Big)\frac{1}{|x+h|^n}\,d\mathcal{H}^{n-1}(x) = \int_{\omega\in S^{n-1},\,\omega\cdot h>0} f(\omega)\,d\mathcal{H}^{n-1}(\omega). \tag{1.6.9}$$

As a corollary, for any $h \in S^{n-1}$ and any $f \in L^1(S^{n-1},\mathcal{H}^{n-1})$ one has

$$\int_{S^{n-1}} f\,d\mathcal{H}^{n-1} = 2\int_{x\in\langle h\rangle^\perp} f_{\rm even}\Big(\frac{x+h}{|x+h|}\Big)\frac{1}{|x+h|^n}\,d\mathcal{H}^{n-1}(x), \tag{1.6.10}$$

where $f_{\rm even}$ is the even part of f, defined as

$$f_{\rm even}(\omega) := \tfrac{1}{2}\big(f(\omega)+f(-\omega)\big) \text{ for } \mathcal{H}^{n-1}\text{-a.e. } \omega \in S^{n-1}. \tag{1.6.11}$$

Proof We shall focus on (1.6.9), as (1.6.10) is a consequence of (1.6.9), (1.6.11), and a standard density result. Using the rotation invariance of the integrals in (1.6.9), maters are reduced to proving (1.6.9) in the case when $h = \mathbf{e}_n$ which reads

$$\int_{x'\in\mathbb{R}^{n-1}} f\Big(\frac{(x',1)}{|(x',1)|}\Big)\frac{1}{|(x',1)|^n}\,dx' = \int_{\omega\in S^{n-1}_+} f(\omega)\,d\mathcal{H}^{n-1}(\omega), \tag{1.6.12}$$

where $S^{n-1}_+ := S^{n-1}\cap\mathbb{R}^n_+$ and $f \in \mathscr{C}^0(\overline{S^{n-1}_+})$. The current assumptions on f ensure that both integrals in (1.6.12) are well-defined.

To proceed with the proof of (1.6.12), consider $P : \mathbb{R}^{n-1} \to \mathbb{R}^n$ defined by

$$P(x') := \frac{(x',1)}{|(x',1)|} \text{ for each } x' \in \mathbb{R}^{n-1}. \tag{1.6.13}$$

We shall eventually show that this a global parametrization of the surface S^{n-1}_{+}. For starters, it is immediate from the above definition that $P \in \mathscr{C}^{\infty}(\mathbb{R}^{n-1})$ and the image of P is contained in S^{n-1}_{+}. In addition, we claim that P is a bijection, its Jacobian matrix DP has $\operatorname{rank}[(DP)(x')] = n-1$ for all $x' \in \mathbb{R}^{n-1}$, and the norm of the vector $\partial_1 P \times \cdots \times \partial_{n-1} P$ is given by

$$\big|(\partial_1 P)(x') \times \cdots \times (\partial_{n-1} P)(x')\big| = \frac{1}{(1+|x'|^2)^{\frac{n}{2}}} \quad \text{for all } x' \in \mathbb{R}^{n-1}. \tag{1.6.14}$$

Here and in what follows, if $v_1 = (v_{11}, \dots, v_{1n}), \dots, v_{n-1} = (v_{n-1\,1}, \dots, v_{n-1\,n})$ are $n-1$ vectors in $\mathbb{R}^n$, their cross product is defined as

$$v_1 \times v_2 \times \cdots \times v_{n-1} := \det \begin{pmatrix} v_{11} & v_{12} & \dots & v_{1n} \\ v_{21} & v_{22} & \dots & v_{2n} \\ \vdots & \vdots & \dots & \vdots \\ v_{n-1\,1} & v_{n-1\,2} & \dots & v_{n-1\,n} \\ \mathbf{e}_1 & \mathbf{e}_2 & \dots & \mathbf{e}_n \end{pmatrix} \tag{1.6.15}$$

where the determinant is understood as computed by formally expanding it with respect to the last row, the result being a vector in $\mathbb{R}^n$. More precisely,

$$v_1 \times \ldots \times v_{n-1} \text{ is defined as} \tag{1.6.16}$$

$$\sum_{j=1}^{n} (-1)^{j+n} \det \begin{pmatrix} v_{11} & \dots & v_{j-1} & v_{j+1} & \dots & v_{1n} \\ \vdots & \dots & \vdots & \vdots & \dots & \vdots \\ v_{n-1\,1} & \dots & v_{n-1\,j-1} & v_{n-1\,j+1} & \dots & v_{n-1\,n} \end{pmatrix} \mathbf{e}_j .$$

Once all the above claims are established, formula (1.6.12) follows via the change of variables $\omega = P(x')$ for $x' \in \mathbb{R}^{n-1}$, which allows us to write

$$\begin{aligned} \int_{S^{n-1}_{+}} f(\omega)\, d\mathcal{H}^{n-1}(\omega) &= \int_{\mathbb{R}^{n-1}} f(P(x'))\, |(\partial_1 P)(x') \times \cdots \times (\partial_{n-1} P)(x')|\, dx' \\ &= \int_{\mathbb{R}^{n-1}} f\left(\frac{(x',1)}{|(x',1)|}\right) \frac{1}{(1+|x'|^2)^{\frac{n}{2}}}\, dx'. \end{aligned} \tag{1.6.17}$$

Regarding the proof of the earlier claims, first observe that the injectivity of P is immediate from its definition. To prove that the image of P is precisely S^{n-1}_{+}, pick an arbitrary $\omega = (\omega', \omega_n) := (\omega_1, \dots, \omega_n) \in S^{n-1}_{+}$. Since $|(\omega'/\omega_n, 1)| = 1/\omega_n$ we have $P(\frac{1}{\omega_n}\omega') = \omega$, from which the desired conclusion follows.

Next, we focus on proving (1.6.14). As is apparent from (1.6.16), the norm of the cross product of the vectors $\partial_1 P, \dots, \partial_{n-1} P$ may be expressed as

$$|\partial_1 P \times \cdots \times \partial_{n-1} P| = \sqrt{\sum_{i=1}^{n} (\det A_i)^2} \tag{1.6.18}$$

where each A_i is the $(n-1)\times(n-1)$ matrix obtained by removing row $i \in \{1, \dots, n\}$ in the Jacobian matrix DP, that is

$$A_i = (\partial_k P_j)_{\substack{1\le j\le n, j\neq i\\ 1\le k\le n-1}} \quad \text{for each } i \in \{1, \dots, n\}, \tag{1.6.19}$$

if $P = (P_1, \dots, P_n)$. Fix $x' \in \mathbb{R}^{n-1}$ and to simplify notation introduce

$$a := (1 + |x'|^2)^{1/2}. \tag{1.6.20}$$

Then (1.6.13) implies

$$(\partial_k P)(x') = \left(\left(\frac{\delta_{jk}}{a} - \frac{x_j x_k}{a^3} \right)_{1\le j\le n-1}, \frac{-x_k}{(1 + |x'|^2)^{3/2}} \right) \tag{1.6.21}$$

for each $k \in \{1, \dots, n-1\}$, thus

$$(DP)(x') = \begin{pmatrix} \frac{1}{a} - \frac{x_1^2}{a^3} & -\frac{x_1x_2}{a^3} & -\frac{x_1x_3}{a^3} & \cdots & -\frac{x_1x_{n-1}}{a^3} \\ -\frac{x_2x_1}{a^3} & \frac{1}{a} - \frac{x_2^2}{a^3} & -\frac{x_2x_3}{a^3} & \cdots & -\frac{x_2x_{n-1}}{a^3} \\ -\frac{x_3x_1}{a^3} & -\frac{x_3x_2}{a^3} & \frac{1}{a} - \frac{x_3^2}{a^3} & \cdots & -\frac{x_3x_{n-1}}{a^3} \\ \vdots & \vdots & \vdots & \ddots & \vdots \\ -\frac{x_{n-1}x_1}{a^3} & -\frac{x_{n-1}x_2}{a^3} & -\frac{x_{n-1}x_3}{a^3} & \cdots & \frac{1}{a} - \frac{x_{n-1}^2}{a^3} \\ -\frac{x_1}{a^3} & -\frac{x_2}{a^3} & -\frac{x_3}{a^3} & \cdots & -\frac{x_{n-1}}{a^3} \end{pmatrix}. \tag{1.6.22}$$

In order to make use of (1.6.18), we need to compute $\det A_i$ for each $i \in \{1, \dots, n\}$. Corresponding to $i = 1$, we have

$$A_1 = \begin{pmatrix} -\frac{x_2x_1}{a^3} & \frac{1}{a} - \frac{x_2^2}{a^3} & -\frac{x_2x_3}{a^3} & \cdots & -\frac{x_2x_{n-1}}{a^3} \\ -\frac{x_3x_1}{a^3} & -\frac{x_3x_2}{a^3} & \frac{1}{a} - \frac{x_3^2}{a^3} & \cdots & -\frac{x_3x_{n-1}}{a^3} \\ \vdots & \vdots & \vdots & \ddots & \vdots \\ -\frac{x_{n-1}x_1}{a^3} & -\frac{x_{n-1}x_2}{a^3} & -\frac{x_{n-1}x_3}{a^3} & \cdots & \frac{1}{a} - \frac{x_{n-1}^2}{a^3} \\ -\frac{x_1}{a^3} & -\frac{x_2}{a^3} & -\frac{x_3}{a^3} & \cdots & -\frac{x_{n-1}}{a^3} \end{pmatrix}. \tag{1.6.23}$$

For each $2 \le j \le n-1$, multiply the last row of A_1 by $-x_j$ and then add the result to the $(j-1)$-th row of A_1 to obtain the row-equivalent matrix

$$\widetilde{A}_1 := \begin{pmatrix} 0 & \frac{1}{a} & 0 & \cdots & 0 \\ 0 & 0 & \frac{1}{a} & \cdots & 0 \\ \vdots & \vdots & \vdots & \ddots & \vdots \\ 0 & 0 & 0 & \cdots & \frac{1}{a} \\ -\frac{x_1}{a^3} & -\frac{x_2}{a^3} & -\frac{x_3}{a^3} & \cdots & -\frac{x_{n-1}}{a^3} \end{pmatrix}. \tag{1.6.24}$$

This yields

$$|\det A_1| = |\det \widetilde{A}_1| = \frac{|x_1|}{a^3} \cdot \left(\frac{1}{a}\right)^{n-2} = \frac{|x_1|}{a^{n+1}}. \tag{1.6.25}$$

Performing a similar computation, we obtain that

$$|\det A_i| = \frac{|x_i|}{a^{n+1}} \quad \text{for each } i \in \{1, \ldots, n-1\}. \tag{1.6.26}$$

There remains to treat the case when $i = n$. By removing the last row of DP in (1.6.22) we arrive at the $(n-1) \times (n-1)$ matrix

$$A_n = \frac{1}{a} I_{(n-1)\times(n-1)} - \frac{1}{a^3} x' \otimes x' = \frac{1}{a}\left[I_{(n-1)\times(n-1)} - \left(\frac{x'}{a}\right) \otimes \left(\frac{x'}{a}\right)\right], \tag{1.6.27}$$

where $x' \otimes x'$ denotes the tensor product of x' with itself in $\mathbb{R}^{n-1}$ (cf. (A.0.9)). Recall Sylvester's determinant theorem, stating that for any $M \times N$ matrix A and any $N \times M$ matrix B we have

$$\det(I_{M\times M} - AB) = \det(I_{N\times N} - BA). \tag{1.6.28}$$

Then

$$\det A_n = a^{1-n} \det\left[I_{(n-1)\times(n-1)} - \left(\frac{x'}{a}\right) \otimes \left(\frac{x'}{a}\right)\right] \tag{1.6.29}$$

$$= a^{1-n}\left(1 - \frac{|x'|^2}{a^2}\right) = \frac{1}{a^{n+1}}, \tag{1.6.30}$$

where the second equality above is a consequence of (1.6.28) used with the $(n-1)\times 1$ matrix $A := (x_1/a, \ldots, x_{n-1}/a)^\top$ and the $1 \times (n-1)$ matrix $B := (x_1/a, \ldots, x_{n-1}/a)$, while the last equality is based on (1.6.20).

In concert, (1.6.18), (1.6.26), (1.6.29), and (1.6.20) yield

$$\left|(\partial_1 P)(x') \times \cdots \times (\partial_{n-1} P)(x')\right| = \sqrt{\sum_{i=1}^{n-1} \frac{x_i^2}{a^{2(n+1)}} + \frac{1}{a^{2(n+1)}}}$$

$$= \frac{1}{a^{n+1}} \sqrt{|x'|^2 + 1} = \frac{1}{a^n} = \frac{1}{(1 + |x'|^2)^{\frac{n}{2}}} \tag{1.6.31}$$

and the proof of (1.6.14) is finished.

Finally, the claim that $\operatorname{rank}[(DP)(x')] = n - 1$ is a consequence of (1.6.14) since having proved that $|\partial_1 P \times \cdots \times \partial_{n-1} P| \neq 0$ ensures $\partial_1 P, \ldots, \partial_{n-1} P$ are linearly independent vectors in $\mathbb{R}^n$. □

Our next theorem contains equivalent conditions ensuring that a given weakly elliptic system L possesses a distinguished coefficient tensor. It turns out that having $\mathfrak{A}_L^{\text{dis}} \neq \varnothing$ hinges on the existence of Poisson kernels for L in arbitrary half-spaces of the Euclidean ambient and their compatibility with one another.

Theorem 1.6.4 *Let $M, n \in \mathbb{N}$ with $n \geq 2$ and suppose L is an $M \times M$ second-order, homogeneous, constant complex coefficient, weakly elliptic system in $\mathbb{R}^n$. Then the following are equivalent.*

(1) The system L possesses a distinguished coefficient tensor, i.e., $\mathfrak{A}_L^{\text{dis}} \neq \varnothing$.

(2) There exists a matrix-valued function $k \in \left[\mathscr{C}^\infty(\mathbb{R}^n \setminus \{0\})\right]^{M \times M}$ which is even, positive homogeneous of degree $-n$ and satisfies

$$\int_{S^{n-1}} k(\omega)\, d\mathcal{H}^{n-1}(\omega) = I_{M \times M} \tag{1.6.32}$$

(where $I_{M\times M}$ is the $M \times M$ identity matrix), as well as

$$L\big(x_s k(x)\big) = 0 \cdot I_{M \times M} \text{ in } \mathbb{R}^n \setminus \{0\} \text{ for each } s \in \{1, \ldots, n\}, \tag{1.6.33}$$

with the operator L acting on the columns of $x_s k(x)$.

(3) There exists an even function $\mathbb{H} : S^{n-1} \longrightarrow \mathbb{C}^{M \times M}$ with the property that for each $s \in \{1, \ldots, n\}$ it follows that

$$K^{(s)}(x) := \frac{x \cdot \mathbf{e}_s}{|x|^n}\, \mathbb{H}\Big(\frac{x}{|x|}\Big) \text{ for each } x = (x_1, \ldots, x_n) \in \mathbb{R}^n \text{ with } x_s > 0 \tag{1.6.34}$$

is a kernel function for L in the half-space $\{x \in \mathbb{R}^n : x \cdot \mathbf{e}_s > 0\}$.

(4) There exists a family of kernel functions $K^{(s)}$ for L in $\{x \in \mathbb{R}^n : x \cdot \mathbf{e}_s > 0\}$, for $s \in \{1, \ldots, n\}$, such that the even extensions of $K^{(s)}(x)/(x \cdot \mathbf{e}_s)$ to the set $\mathbb{R}^n \setminus \{x \in \mathbb{R}^n : x \cdot \mathbf{e}_s = 0\}$ agree with one another (on the common domain) for any two values of $s \in \{1, \ldots, n\}$.

(5) There exists an even function $\mathbb{H} : S^{n-1} \longrightarrow \mathbb{C}^{M \times M}$ such that for every $h \in S^{n-1}$ the expression

$$K^{(h)}(x) := \frac{x \cdot h}{|x|^n}\, \mathbb{H}\Big(\frac{x}{|x|}\Big) \text{ for each } x \in \mathbb{R}^n \text{ with } x \cdot h > 0, \tag{1.6.35}$$

is a kernel function for L in the half-space $\{x \in \mathbb{R}^n : x \cdot h > 0\}$.

(6) There exists an even function $k : \mathbb{R}^n \setminus \{0\} \longrightarrow \mathbb{C}^{M \times M}$ *that is positive homogeneous of degree* $-n$ *and such that for each* $s \in \{1, \ldots, n\}$ *it follows that*

$$P^{(s)}(x) := 2k(x + \mathbf{e}_s) \textit{ for all } x = (x_1, \ldots, x_n) \in \mathbb{R}^n \textit{ with } x_s = 0, \tag{1.6.36}$$

is a Poisson kernel for L in the half-space $\{x \in \mathbb{R}^n : x \cdot \mathbf{e}_s > 0\}$.

(7) There exists an even function $k : \mathbb{R}^n \setminus \{0\} \longrightarrow \mathbb{C}^{M \times M}$ *that is positive homogeneous of degree* $-n$ *and such that for each* $h \in S^{n-1}$ *it follows that*

$$P^{(h)}(x) := 2k(x + h) \textit{ for all } x \in \mathbb{R}^n \textit{ with } x \cdot h = 0, \tag{1.6.37}$$

is a Poisson kernel for L in $\{x \in \mathbb{R}^n : x \cdot h > 0\}$.

Moreover, if $\#\mathfrak{A}_L^{\mathrm{dis}} = 1$ *then a function k satisfying the conditions in (2) is unique.*

We observe that the requirement in *(2)* for k to be even may be dropped since if one has a function k satisfying all the conditions listed in *(2)* except being even, then $\frac{1}{2}(k(x) - k(-x))$ satisfies all the conditions listed in *(2)*, including being even.

Proof To begin, let $E = (E_{\alpha\beta})_{1 \le \alpha,\beta \le M}$ be the fundamental solution associated with the given system L as in [131, Theorem 1.4.2]. Also, in what follows, for each $A = \big(a_{rs}^{\alpha\beta}\big)_{\substack{1 \le r,s \le n \\ 1 \le \alpha,\beta \le M}} \in \mathfrak{A}_L$ we abbreviate

$$A_{rs} := \big(a_{rs}^{\alpha\beta}\big)_{1 \le \alpha,\beta \le M} \text{ for each } r, s \in \{1, \ldots, n\}. \tag{1.6.38}$$

First we prove *(1)*⇒*(2)*. Suppose *(1)* is true and pick

$$A = \big(a_{rs}^{\alpha\beta}\big)_{\substack{1 \le r,s \le n \\ 1 \le \alpha,\beta \le M}} \in \mathfrak{A}_L^{\mathrm{dis}}. \tag{1.6.39}$$

Since A is a distinguished coefficient tensor there exists a matrix-valued function (cf. Definition 1.2.2)

$$k = \{k_{\gamma\alpha}\}_{1 \le \gamma,\alpha \le M} \in \big[\mathscr{C}^\infty(\mathbb{R}^n \setminus \{0\})\big]^{M \times M} \tag{1.6.40}$$

which is even, positive homogeneous of degree $-n$, and such that

$$\begin{aligned} &a_{rs}^{\beta\alpha}(\partial_r E_{\gamma\beta})(x) = x_s k_{\gamma\alpha}(x) \text{ for all } x \in \mathbb{R}^n \setminus \{0\}, \\ &\text{for each } \gamma, \alpha \in \{1, \ldots, M\} \text{ and } s \in \{1, \ldots, n\}. \end{aligned} \tag{1.6.41}$$

This allows us to write

$$\begin{aligned} \int_{S^{n-1}} k(\omega)\, d\mathcal{H}^{n-1}(\omega) &= \Big(\int_{S^{n-1}} \omega_s \omega_s k_{\gamma\alpha}(\omega)\, d\mathcal{H}^{n-1}(\omega)\Big)_{1 \le \gamma,\alpha \le M} \\ &= \Big(\int_{S^{n-1}} \omega_s a_{rs}^{\beta\alpha}(\partial_r E_{\gamma\beta})(\omega)\, d\mathcal{H}^{n-1}(\omega)\Big)_{1 \le \gamma,\alpha \le M} = I_{M \times M}, \end{aligned} \tag{1.6.42}$$

where the last equality in (1.6.42) is a consequence of [131, (1.4.25)] (cf. also (1.2.15)). In addition, starting with (1.6.41), for each $s \in \{1, \dots, n\}$ we may compute (recall the abbreviation in (1.6.38))

$$\begin{aligned} L\big(x_s k(x)\big) &= L\big(x_s k_{\gamma\alpha}(x)\big)_{1\leq\gamma,\alpha\leq M} = L\big(a^{\beta\alpha}_{rs}(\partial_r E_{\gamma\beta})(x)\big)_{1\leq\gamma,\alpha\leq M} \\ &= L\Big(\partial_r E(x) A_{rs}\Big) = \partial_r (LE)(x) A_{rs} = 0 \cdot I_{M\times M} \text{ pointwise in } \mathbb{R}^n \setminus \{0\}. \end{aligned} \tag{1.6.43}$$

This completes the proof of the implication *(1)* $\Rightarrow$ *(2)*.

Next, we focus on proving *(2)* $\Rightarrow$ *(1)*. Let k be a matrix-valued function as in *(2)* and fix some index $s \in \{1, \dots, n\}$. Then the function $\mathbb{R}^n \ni x \longmapsto x_s k(x) \in \mathbb{C}^{M\times M}$ is defined $\mathcal{L}^n$-a.e. in $\mathbb{R}^n$, is locally integrable, and positive homogeneous of degree $1-n$. In particular, it gives rise to a tempered distribution, via integration against Schwartz functions in $\mathbb{R}^n$, which is positive homogeneous of degree $1 - n$. Consequently,

$$\text{the tempered distribution } L(x_s k(x)) \text{ is positive homogeneous of degree } -1-n. \tag{1.6.44}$$

Also, as a consequence of (1.6.33), the support of the matrix-valued distribution $L(x_s k(x))$ is contained in the set $\{0\}$. Hence, each of its entries is of the form $\sum_{|\alpha|\leq N} c_\alpha \partial^\alpha \delta$ for some $N \in \mathbb{N}_0$ and $c_\alpha \in \mathbb{C}$ for each $\alpha \in \mathbb{N}_0^n$ with $|\alpha| \leq N$ (see, e.g., [127, Exercise 2.75, p. 47]). Since the Dirac Distribution δ is positive homogeneous of degree $-n$, the only terms in the latter sums that are positive homogeneous of degree $-1 - n$ are those corresponding to derivatives of order one of δ. Recalling now (1.6.44), we may conclude that there exists a family of constant matrices $B_{rs} \in \mathbb{C}^{M\times M}$, indexed by $r, s \in \{1, \dots, n\}$, such that

$$L(x_s k(x)) = B_{rs}\partial_r \delta \text{ in } \mathcal{S}'(\mathbb{R}^n) \text{ for each } s \in \{1, \dots, n\}. \tag{1.6.45}$$

In addition, [129, (4.5.47)] and (1.6.32) imply

$$\partial_s L(x_s k(x)) = L\Big(\partial_s\big(x_s k(x)\big)\Big) = L\big(I_{M\times M}\delta\big) \text{ in } \mathcal{S}'(\mathbb{R}^n), \tag{1.6.46}$$

which in view of (1.6.45) yields

$$L\big(I_{M\times M}\delta\big) = B_{rs}\partial_r\partial_s\delta \text{ in } \mathcal{S}'(\mathbb{R}^n). \tag{1.6.47}$$

After an application of the Fourier transform, (1.6.47) becomes

$$L(\xi) = -B_{rs}\xi_r\xi_s \text{ in } \mathcal{S}'(\mathbb{R}^n), \tag{1.6.48}$$

with the notation $\xi = (\xi_1, \dots, \xi_n) \in \mathbb{R}^n$. If

$$B_{rs} = \big(b^{\alpha\beta}_{rs}\big)_{1\leq\alpha,\beta\leq M} \text{ for each } r, s \in \{1, \dots, n\}, \tag{1.6.49}$$

and we define the matrix $B := \big(b^{\alpha\beta}_{rs}\big)_{\substack{1\leq r,s\leq n\\ 1\leq\alpha,\beta\leq M}}$, then from (1.6.48) we may infer that

$$B \in \mathfrak{A}_L. \tag{1.6.50}$$

Also, applying the Fourier transform to both sides in (1.6.45) yields

$$L(\xi)\widehat{x_s k(x)}(\xi) = \mathrm{i}B_{rs}\xi_r \text{ in } \mathcal{S}'(\mathbb{R}^n) \text{ for each } s \in \{1, \dots, n\}. \tag{1.6.51}$$

Consequently, for each $s \in \{1, \dots, n\}$ we have

$$\big(\widehat{x_s k(x)}\big|_{\mathbb{R}^n \setminus \{0\}}\big)(\xi) = \mathrm{i}L(\xi)^{-1}B_{rs}\xi_r \text{ for all } \xi \in \mathbb{R}^n \setminus \{0\}. \tag{1.6.52}$$

Now observe that the tempered distribution $\widehat{x_s k(x)}$ is positive homogeneous of degree $-n - (1 - n) = -1$ (being the Fourier transform of a tempered distribution that is positive homogeneous of degree $1-n$). We may therefore apply [129, Lemma 4.5.5], with $f(\xi) := \mathrm{i}L(\xi)^{-1}B_{rs}\xi_r$ for all $\xi \in \mathbb{R}^n \setminus \{0\}$ and $u := \widehat{x_s k(x)}$, to arrive at

$$\widehat{x_s k(x)} = \mathrm{i}L(\xi)^{-1}B_{rs}\xi_r \text{ in } \mathcal{S}'(\mathbb{R}^n) \text{ for each } s \in \{1, \dots, n\}. \tag{1.6.53}$$

Given that, as seen from [131, (1.4.31)] in [131, Theorem 1.4.2],

$$\widehat{\partial_r E} = \mathrm{i}\xi_r \left[L(\xi)\right]^{-1} \text{ in } \left[\mathcal{S}'(\mathbb{R}^n)\right]^{M\times M} \text{ for each } s \in \{1, \dots, n\}, \tag{1.6.54}$$

the right-hand side of the equality in (1.6.53) is precisely the Fourier transform of $(\partial_r E)(x)B_{rs}$. As such, for each $s \in \{1, \dots, n\}$ we obtain $x_s k(x) = (\partial_r E)(x)B_{rs}$ in $\mathcal{S}'(\mathbb{R}^n)$, which further implies $x_s k(x) = (\partial_r E)(x)B_{rs}$ in a pointwise sense for each $x \in \mathbb{R}^n \setminus \{0\}$. Componentwise, this matrix equality becomes (recall (1.6.49))

$$x_s k_{\gamma\alpha}(x) = (\partial_r E_{\gamma\beta})(x)b^{\beta\alpha}_{rs} \text{ for all } x \in \mathbb{R}^n \setminus \{0\}, \tag{1.6.55}$$

and every $\alpha, \beta \in \{1, \dots, M\}$ and every $s \in \{1, \dots, n\}$. Bearing in mind (1.6.50), from (1.6.55) and Definition 1.2.2 we may now conclude that $B \in \mathfrak{A}^{\mathrm{dis}}_L$. The proof of *(2)*$\Rightarrow$*(1)* is therefore finished.

Focusing next on *(1)*$\Rightarrow$*(3)*, pick $A = \big(a^{\alpha\beta}_{rs}\big)_{\substack{1\le r,s\le n\\ 1\le\alpha,\beta\le M}} \in \mathfrak{A}^{\mathrm{dis}}_L$. As above, by the definition of a distinguished coefficient tensor, there exists k satisfying (1.6.40)-(1.6.41) that is also even and positive homogeneous of degree $-n$. Fix such a k and note that, as seen from the proof *(1)*$\Rightarrow$*(2)*, it follows that k also satisfies (1.6.32)-(1.6.33). Thus, if we now set

$$\mathbb{H} : S^{n-1} \longrightarrow \mathbb{C}^{M\times M}, \quad \mathbb{H} := 2k\big|_{S^{n-1}}, \tag{1.6.56}$$

then $\mathbb{H}$ is even and for each $s \in \{1, \dots, n\}$ and $x = (x_1, \dots, x_n) \in \mathbb{R}^n \setminus \{0\}$ we have

$$\frac{x_s}{|x|^n}\,\mathbb{H}\Big(\frac{x}{|x|}\Big) = 2x_s k\Big(\frac{x}{|x|}\cdot|x|\Big) = 2x_s k(x) = 2\Big(a^{\beta\alpha}_{rs}(\partial_r E_{\gamma\beta})(x)\Big)_{1\le\gamma,\alpha\le M}. \tag{1.6.57}$$

To proceed, fix $s \in \{1, \dots, n\}$, denote

$$H_s := \{x \in \mathbb{R}^n : x \cdot \mathbf{e}_s = 0\}, \quad H_s^+ := \{x \in \mathbb{R}^n : x \cdot \mathbf{e}_s > 0\}, \tag{1.6.58}$$

and define

$$K^{(s)}(x) := \frac{x \cdot \mathbf{e}_s}{|x|^n} \mathbb{H}\Big(\frac{x}{|x|}\Big) \text{ for all } x \in H_s^+. \tag{1.6.59}$$

We claim that $K^{(s)}$ is a kernel function for L in the half-space H_s^+ corresponding to $h := \mathbf{e}_s$. Indeed, from (1.6.57) (recall also the abbreviation made in (1.6.38)) we have $K^{(s)}(x) = 2\partial_r E(x) A_{rs}$ for all $x \in H_s^+$, which, in light of the properties of E, implies $K^{(s)} \in \big[\mathscr{C}^\infty(H_s^+)\big]^{M\times M}$, the entries of K are positive homogeneous of degree $1-n$, and $LK^{(s)} = 0$ in H_s^+ (as proved in (1.6.43)). In addition,

$$|K^{(s)}(x)| \leq \big(\sup_{S^{n-1}} |\mathbb{H}|\big) \frac{x \cdot \mathbf{e}_s}{|x|^n} \text{ for all } x \in H_s^+, \tag{1.6.60}$$

and,

$$\begin{aligned}
\int_{H_s} K^{(s)}(x+\mathbf{e}_s)\, d\mathcal{H}^{n-1}(x) &= \int_{H_s} \frac{1}{|x+\mathbf{e}_s|^n} \mathbb{H}\Big(\frac{x+\mathbf{e}_s}{|x+\mathbf{e}_s|}\Big)\, d\mathcal{H}^{n-1}(x) \\
&= \int\limits_{\omega\in S^{n-1},\, \omega\cdot\mathbf{e}_s>0} \mathbb{H}(\omega)\, d\mathcal{H}^{n-1}(\omega) = 2 \int\limits_{\omega\in S^{n-1},\, \omega\cdot\mathbf{e}_s>0} k(\omega)\, d\mathcal{H}^{n-1}(\omega) \\
&= \int_{S^{n-1}} k(\omega)\, d\mathcal{H}^{n-1}(\omega) = I_{M\times M}.
\end{aligned} \tag{1.6.61}$$

The first equality in (1.6.61) is a consequence of the fact that $(x+\mathbf{e}_s)\cdot\mathbf{e}_s = 1$ for all $x \in H_s$, the second equality is an application of (1.6.9) in Proposition 1.6.3, the fourth uses the fact that k is even, while the last one is property (1.6.33). We have just checked that each $K^{(s)}$ satisfies properties *(i)*- *(iv)* in Definition 1.6.1. This finishes the proof of *(1)*$\Rightarrow$*(3)*.

To prove *(3)*$\Rightarrow$*(2)*, start with $\mathbb{H}$ and $K^{(s)}$, $s \in \{1,\dots,n\}$, as in *(3)* then define

$$k(x) := \frac{1}{2|x|^n} \mathbb{H}\Big(\frac{x}{|x|}\Big) \text{ for all } x \in \mathbb{R}^n \setminus \{0\}. \tag{1.6.62}$$

By design, k is positive homogeneous of degree $-n$ and even (recall that, by assumption, H is even). Fix $s \in \{1,\dots,n\}$ and employ the notation from (1.6.58). Since $K^{(s)}$ is a kernel function for L in the half-space H_s^+ and (1.6.34) holds, we have $\mathbb{H} \in \big[\mathscr{C}^\infty(S^{n-1})\big]^{M\times M}$, thus

$$k \in \big[\mathscr{C}^\infty(\mathbb{R}^n \setminus \{0\})\big]^{M\times M}. \tag{1.6.63}$$

Moreover, (1.6.62) and (1.6.34) imply $x_s k(x) = \frac{1}{2}K^{(s)}(x)$ for all $x \in H_s^+$. Recalling that $LK^{(s)} = 0$ in H_s^+, the latter implies $L(x_s k(x)) = 0$ for all $x \in H_s^+$. In light of the fact that the mapping $x \mapsto x_s k(x)$ is an odd function in $\mathbb{R}^n \setminus \{0\}$ and L is a homogeneous system, this further entails

$$L(x_s k(x)) = 0 \text{ for all } x \in \mathbb{R}^n \setminus H_s. \tag{1.6.64}$$

Now (1.6.33) follows from (1.6.64) by density since $L(x_s k(x))$ belongs to the space $[\mathscr{C}^\infty(\mathbb{R}^n \setminus \{0\})]^{M\times M}$ (the latter a consequence of (1.6.63)).

There remains to observe that (1.6.62), $\mathbb{H}$ being even, the change of variables formula (1.6.9) (with $h := \mathbf{e}_s$), (1.6.34) (note that if $x \in H_s$ then $x + \mathbf{e}_s \in H_s^+$), and property *(iv)* in Definition 1.6.1 (for $K := K^{(s)}$ and $h := \mathbf{e}_s$) allow us to compute

$$\begin{aligned}\int_{S^{n-1}} k(\omega)\,d\mathcal{H}^{n-1}(\omega) &= \frac{1}{2}\int_{S^{n-1}} \mathbb{H}(\omega)\,d\mathcal{H}^{n-1}(\omega) = \int\limits_{\omega\in S^{n-1},\,\omega\cdot\mathbf{e}_s>0} \mathbb{H}(\omega)\,d\mathcal{H}^{n-1}(\omega)\\ &= \int\limits_{H_s} \frac{1}{|x+\mathbf{e}_s|^n}\,\mathbb{H}\Big(\frac{x+\mathbf{e}_s}{|x+\mathbf{e}_s|}\Big)\,d\mathcal{H}^{n-1}(x) = \int\limits_{H_s} K^{(s)}(x+\mathbf{e}_s)\,d\mathcal{H}^{n-1}(x) = I_{M\times M}.\end{aligned} \tag{1.6.65}$$

This shows that k defined in (1.6.62) satisfies all properties listed in *(2)*.

Moving on, whenever *(3)* holds, the given kernel functions $K^{(s)}$, $s \in \{1,\dots,n\}$, and even function $\mathbb{H}$ satisfy (1.6.34), hence, for each $s \in \{1,\dots,n\}$, the even extension of the function $K^{(s)}(x)/(x\cdot\mathbf{e}_s)$ to $\mathbb{R}^n \setminus \{x \in \mathbb{R}^n : x\cdot\mathbf{e}_s = 0\}$ is equal to $\frac{1}{|x|^n}\mathbb{H}\big(\frac{x}{|x|}\big)$ there. The fact that any two such extensions agree with one another on the common domain is now immediate. This shows *(3)*$\Rightarrow$*(4)*.

Conversely, assume there exists a family $K^{(s)}$, $s \in \{1,\dots,n\}$, of kernel functions for L as in *(4)*. For each $s \in \{1,\dots,n\}$, we denote by $\mathrm{Ext}^{(s)}$ the even extension of $K^{(s)}(x)/(x\cdot\mathbf{e}_s)$ to the set $\mathbb{R}^n \setminus \{x \in \mathbb{R}^n : x\cdot\mathbf{e}_s = 0\}$. Then we may define a function $\mathbb{H} : S^{n-1} \to \mathbb{C}^{M\times M}$ as follows. For each $\omega = (\omega_1,\dots,\omega_n) \in S^{n-1}$ arbitrary, pick $s_0 \in \{1,\dots,n\}$ such that $\omega\cdot\mathbf{e}_{s_0} \neq 0$, and set

$$\mathbb{H}(\omega) := \mathrm{Ext}^{(s_0)}(\omega). \tag{1.6.66}$$

Thanks to the compatibility conditions satisfied by the extensions $\{\mathrm{Ext}^{(s)}\}_{1\leq s\leq n}$ we have that H is unambiguously defined. In addition, from (1.6.66) it follows that we have $\mathbb{H}(\omega) = \frac{K^{(s)}(\omega)}{\omega\cdot\mathbf{e}_s}$ whenever $\omega \in S^{n-1}$ satisfies $\omega\cdot\mathbf{e}_s > 0$, for some $s \in \{1,\dots,n\}$. Hence, if $x \in \mathbb{R}^n$ is such that $x\cdot\mathbf{e}_s > 0$ then

$$\mathbb{H}\Big(\frac{x}{|x|}\Big) = \frac{K^{(s)}\big(\frac{x_s}{|x|}\big)}{\frac{x}{|x|}} = \frac{|x|^{n-1}K^{(s)}(x)}{|x|^{-1}x_s} = \frac{|x|^n}{x_s}K^{(s)}(x), \tag{1.6.67}$$

where the second equality in (1.6.67) used the positive homogeneity of order $1-n$ of $K^{(s)}$. As such, (1.6.34) holds and we have shown *(4)* $\Longrightarrow$ *(3)*.

To prove the implication *(5)*$\Rightarrow$*(3)*, take $K^{(s)} := K^{(\mathbf{e}_s)}$ for each $s \in \{1,\dots,n\}$.

To see why *(1)*$\Rightarrow$*(5)* is true, pick some A as in (1.6.39) and, from what we proved earlier, we know that there exists k satisfying (1.6.40)-(1.6.41), which also has the properties singled out in *(2)*. Let $\mathbb{H}$ be the even function defined in (1.6.56) for this k, pick $h \in S^{n-1}$ arbitrary, and recall the notation from (1.6.1). If we now consider the function $K^{(h)}$ defined according to (1.6.35), then

$$K^{(h)}(x) = \frac{x_s h_s}{|x|^n}\,\mathbb{H}\Big(\frac{x}{|x|}\Big) = 2h_s x_s k\Big(\frac{x}{|x|}\cdot|x|\Big) = 2h_s x_s k(x)$$
$$= \Big(2h_s a_{rs}^{\beta\alpha}(\partial_r E_{\gamma\beta})(x)\Big)_{1\leq\gamma,\alpha\leq M} \tag{1.6.68}$$

for all $x \in H^+$. Having established (1.6.68), the fact that $K^{(h)}$ is a kernel function for L in the half-space H^+ is seen by suitably adjusting the reasoning in the proof of the fact that $K^{(s)}$ from (1.6.59) is a kernel function for L in the half-space H_s^+. Specifically, with the abbreviation in (1.6.38), the formula in (1.6.68) becomes $K^{(h)}(x) = 2h_s\partial_r E(x)A_{rs}$ for all $x \in H^+$. The latter implies that $K^{(h)} \in \big[\mathscr{C}^\infty(H_s^+)\big]^{M\times M}$, the function $K^{(h)}$ is positive homogeneous of degree $1-n$, the differential equation $LK^{(h)} = 2h_s\partial_r(LE)(x)A_{rs} = 0$ holds in H^+, we have the estimate $|K^{(h)}(x)| \leq \big(\sup_{S^{n-1}}|\mathbb{H}|\big)\frac{x\cdot h}{|x|^n}$ for all $x \in H^+$, and

$$\int_H K^{(h)}(x+h)\,d\mathcal{H}^{n-1}(x) = \int_H \frac{(x+h)\cdot h}{|x+h|^n}\,\mathbb{H}\Big(\frac{x+h}{|x+h|}\Big)\,d\mathcal{H}^{n-1}(x)$$
$$= \int_{\omega\in S^{n-1},\,\omega\cdot h>0} \mathbb{H}(\omega)\,d\mathcal{H}^{n-1}(\omega) = 2\int_{\omega\in S^{n-1},\,\omega\cdot h>0} k(\omega)\,d\mathcal{H}^{n-1}(\omega)$$
$$= \int_{S^{n-1}} k(\omega)\,d\mathcal{H}^{n-1}(\omega) = I_{M\times M}. \tag{1.6.69}$$

The second equality in (1.6.69) is a consequence of the identity $(x+h)\cdot h = |h| = 1$ for all $h \in H$ and formula (1.6.9), for the fourth equality we used that k is even, while the last one is simply (1.6.33). In conclusion, $K^{(h)}$ is a kernel function for L in the half-space H^+ and *(1)* $\Rightarrow$ *(5)* is proved.

Moving on, assume *(2)* holds with the goal of showing that *(6)* follows. For k as in *(2)* and $s \in \{1,\dots,n\}$ arbitrary, let $P^{(s)}$ be the function defined in (1.6.36) and recall the notation in (1.6.58). Then $P^{(s)} \in \big[\mathscr{C}^\infty(H_s)\big]^{M\times M}$ and since k is positive homogeneous of degree $-n$ and $|x+\mathbf{e}_s|^2 = |x|^2+1$ for every $x \in H_s$, we have

$$P^{(s)}(x) = 2k\Big(\frac{x+\mathbf{e}_s}{|x+\mathbf{e}_s|}\Big)\cdot\frac{1}{|x+\mathbf{e}_s|^n} = 2k\Big(\frac{x+\mathbf{e}_s}{|x+\mathbf{e}_s|}\Big)\cdot\frac{1}{(|x|^2+1)^{n/2}} \tag{1.6.70}$$

for all $x \in H_s$, hence $|P^{(s)}(x)| \leq 2\Big(\sup_{S^{n-1}}|k|\Big)\frac{1}{(|x|^2+1)^{n/2}}$ for all $x \in H_s$. Also, (1.6.70), the change of variables formula (1.6.9), the fact that k is even, and identity (1.6.32), allow us to write

$$\int_{H_s} P^{(s)}(x)\,d\mathcal{H}^{n-1}(x) = 2\int_{H_s}\frac{1}{|x+\mathbf{e}_s|^n}\,k\Big(\frac{x+\mathbf{e}_s}{|x+\mathbf{e}_s|}\Big)\,d\mathcal{H}^{n-1}(x) \tag{1.6.71}$$
$$= 2\int_{\omega\in S^{n-1},\,\omega\cdot\mathbf{e}_s>0} k(\omega)\,d\mathcal{H}^{n-1}(\omega) = \int_{S^{n-1}} k(\omega)\,d\mathcal{H}^{n-1}(\omega) = I_{M\times M}.$$

So far we have seen that $P^{(s)}$ has properties *(a)* and *(b)* in Definition 1.6.1 corresponding to $h := \mathbf{e}_s$. To see that *(c)* in Definition 1.6.1 is also satisfied consider the function $K^{(s)}(x) := (x \cdot \mathbf{e}_s)^{1-n} P^{(s)}\big(\frac{\pi_{H_s}(x)}{x \cdot \mathbf{e}_s}\big)$ for each $x \in H_s^+$. Then the positive homogeneity of degree $-n$ of k and (1.6.36) imply

$$K^{(s)}(x) = 2x_s^{1-n} k\Big(\frac{x - x_s \mathbf{e}_s}{x_s} + \mathbf{e}_s\Big) = 2x_s k(x), \quad \forall x = (x_1, \dots, x_n) \in H_s^+. \tag{1.6.72}$$

In turn, (1.6.72) and (1.6.33) yield $LK^{(s)} = 0$ in $\big[\mathcal{D}'(H_s^+)\big]^{M \times M}$. In summary, $P^{(s)}$ is a Poisson kernel for L in H_s^+, so the proof of *(2)⇒(6)* is complete.

To justify *(6)⇒(3)*, start with k as in *(6)* and define $\mathbb{H} : S^{n-1} \to \mathbb{C}^{M \times M}$ by setting $\mathbb{H} := 2k\big|_{S^{n-1}}$. We claim that $\mathbb{H}$ satisfies all properties listed in *(3)*. Clearly, $\mathbb{H}$ is even. Fix $s \in \{1, \dots, n\}$ and observe that under the current assumptions the equalities in (1.6.57) hold. Consequently, if $K^{(s)}(x) := \frac{x \cdot \mathbf{e}_s}{|x|^n} \mathbb{H}\big(\frac{x}{|x|}\big)$ for each $x \in H_s^+$ (again employing the notation in (1.6.58)) then, for all $x = (x_1, \dots, x_n) \in H_s^+$,

$$K^{(s)}(x) = 2x_s k(x) = 2x_s^{1-n} k\Big(\frac{x - x_s \mathbf{e}_s}{x_s} + \mathbf{e}_s\Big) = (x \cdot \mathbf{e}_s)^{1-n} P^{(s)}\Big(\frac{\pi_{H_s}(x)}{x \cdot \mathbf{e}_s}\Big). \tag{1.6.73}$$

Since, by assumption, $P^{(s)}$ is a Poisson kernel for L in H_s^+, it follows that $K^{(s)}$ is a kernel function for L in H_s^+. Thus, *(3)* holds.

The implication *(2)⇒(7)* may be obtained with a reasoning similar to the one above for *(2)⇒(6)*. Specifically, let k be as in *(2)*, pick $h \in S^{n-1}$ arbitrary, and consider H and H^+ as in (1.6.1). If $P^{(h)}$ is the function defined in (1.6.37), then $P^{(h)} \in \big[\mathscr{C}^\infty(H)\big]^{M \times M}$ and may be bounded, for all $x \in H$, by

$$|P^{(h)}(x)| = 2\Big|k\Big(\frac{x + h}{|x + h|}\Big)\Big| \cdot \frac{1}{(|x|^2 + 1)^{n/2}} \leq 2\Big(\sup_{S^{n-1}} |k|\Big) \frac{1}{(|x|^2 + 1)^{n/2}} \tag{1.6.74}$$

since k is positive homogeneous of degree $-n$ and $|x + h|^2 = |x|^2 + 1$ for all $x \in H$. The computation in (1.6.71) written with h in place of $\mathbf{e}_s$ (and, consequently, with H in place of H_s) now gives

$$\begin{aligned} \int_H P^{(h)}(x + h)\, d\mathcal{H}^{n-1}(x) &= 2 \int_H \frac{1}{|x + h|^n}\, k\Big(\frac{x + h}{|x + h|}\Big)\, d\mathcal{H}^{n-1}(x) \\ &= 2 \int\limits_{\omega \in S^{n-1},\ \omega \cdot h > 0} k(\omega)\, d\mathcal{H}^{n-1}(\omega) = \int\limits_{S^{n-1}} k(\omega)\, d\mathcal{H}^{n-1}(\omega) = I_{M \times M}. \end{aligned} \tag{1.6.75}$$

Finally, let us set $K^{(h)}(x) := (x \cdot h)^{1-n} P^{(h)}\big(\frac{\pi_H x}{x \cdot h}\big)$ for all $x \in H^+$ and observe that

$$K^{(h)}(x) = 2(x \cdot h)^{1-n} k\Big(\frac{\pi_H x}{x \cdot h} + h\Big) = 2(x \cdot h) k(x) = 2\frac{x \cdot h}{|x|^n}\, k\Big(\frac{x}{|x|}\Big) = \frac{x \cdot h}{|x|^n}\, \mathbb{H}\Big(\frac{x}{|x|}\Big) \tag{1.6.76}$$

for all $x \in H^+$, where $\mathbb{H} := 2k\big|_{S^{n-1}}$. Hence, $K^{(h)}$ is the function from (1.6.68) which, in the proof of *(1)⇒(5)*, has been shown to be a kernel function for L in the half-space H^+. In summary, $P^{(h)}$ is a Poisson kernel for L in H^+.

The last step in the proof of the theorem is showing *(7)⇒(5)*. Let k be as in *(7)* and define the even function $\mathbb{H} := 2k\big|_{S^{n-1}}$. Pick $h \in S^{n-1}$, recall the notation in (1.6.1), and note that, by assumption, $P^{(h)}$ as in (1.6.37) is a Poisson kernel for L in H^+. Then for all $x \in H^+$ we have

$$\begin{aligned}\frac{x\cdot h}{|x|^n}\,\mathbb{H}\Big(\frac{x}{|x|}\Big) &= 2\frac{x\cdot h}{|x|^n}\,k\Big(\frac{x}{|x|}\Big) = 2(x\cdot h)k(x) \\ &= 2(x\cdot h)^{1-n}k\Big(\frac{\pi_H x}{x\cdot h}+h\Big) = 2(x\cdot h)^{1-n}P^{(h)}\Big(\frac{\pi_H x}{x\cdot h}\Big).\end{aligned} \tag{1.6.77}$$

In particular, (1.6.77) shows that $K^{(h)}(x) := \frac{x\cdot h}{|x|^n}\mathbb{H}\Big(\frac{x}{|x|}\Big)$ for each $x \in H$, is the Poisson kernel function canonically associated with $P^{(h)}$ in H^+ as in Definition 1.6.3 (recall Remark 1.6.2). This proves *(5)*.

Finally, if we assume that $\#\mathfrak{A}_L^{\rm dis} = 1$ then a function k satisfying the conditions in item *(2)* of Theorem 1.6.4 exists. In addition, following the proof of *(2)⇒(1)*, we observe that by multiplying the identity in (1.6.55) by x_s, then summing up over $s \in \{1,\dots,n\}$ and, finally, dividing by $|x|^2$, we arrive at the conclusion that $k_{\gamma\alpha}(x) = \frac{x_s}{|x|^2}(\partial_r E_{\gamma\beta})(x)b_{rs}^{\beta\alpha}$ for all $x \in \mathbb{R}^n \setminus \{0\}$ and every $\alpha, \beta \in \{1,\dots,M\}$. These equalities determine k uniquely since $b_{rs}^{\beta\alpha}$'s are the entries of the unique distinguished coefficient tensor for L. This finishes the proof of Theorem 1.6.4. □

We single out the following useful consequence of Theorem 1.6.4.

Corollary 1.6.5 *Let $M, n \in \mathbb{N}$ with $n \geq 2$ and suppose L is an $M \times M$ second-order, homogeneous, constant complex coefficient, weakly elliptic system in $\mathbb{R}^n$. Then the following are equivalent.*

(1) The system L possesses a distinguished coefficient tensor, i.e., $\mathfrak{A}_L^{\rm dis} \neq \varnothing$.

(2) There exists an even function $\mathbb{H} \in \big[\mathscr{C}^\infty(S^{n-1})\big]^{M\times M}$ such that

$$\int_{S^{n-1}} \mathbb{H}\,d\mathcal{H}^{n-1} \text{ is an invertible } M\times M \text{ matrix, and} \tag{1.6.78}$$

$$L\bigg(\frac{x\cdot h}{|x|^n}\,\mathbb{H}\Big(\frac{x}{|x|}\Big)\bigg) = 0\cdot I_{M\times M} \text{ in } \mathbb{R}^n\setminus\{0\}, \text{ for each } h \in S^{n-1}. \tag{1.6.79}$$

Proof If *(1)* holds, let k be as in item *(2)* of Theorem 1.6.4 and define the function $\mathbb{H} : S^{n-1} \to \mathbb{C}^{M\times M}$ by setting $\mathbb{H} := k\big|_{S^{n-1}}$. Then (1.6.32) implies (1.6.78). In addition, *(5)* in Theorem 1.6.4 also holds with the same $\mathbb{H}$. In particular, for each $h \in S^{n-1}$, it follows that $K^{(h)}$ from (1.6.35) is a kernel function for the operator L in the half-space $\{x \in \mathbb{R}^n : x\cdot h > 0\}$ thus the PDE in (1.6.79) is satisfied.

Conversely, if $\mathbb{H}$ is a matrix-valued function as in *(2)*, set

$$k(x) := \frac{1}{|x|^n}\,\mathbb{H}\Big(\frac{x}{|x|}\Big)\Big(\int_{S^{n-1}}\mathbb{H}\,d\mathcal{H}^{n-1}\Big)^{-1} \text{ for all } x \in \mathbb{R}^n\setminus\{0\}. \tag{1.6.80}$$

Then from (1.6.78) and (1.6.80) we see that k is a well-defined matrix-valued function which is even, positive homogeneous of degree $-n$, belongs to $\big[\mathscr{C}^\infty(\mathbb{R}^n\setminus\{0\})\big]^{M\times M}$, and satisfies (1.6.32). Also, (1.6.79) written for $h := \mathbf{e}_s$, with $s \in \{1,\dots,n\}$, yields (1.6.33). Consequently, this k satisfies *(2)* in Theorem 1.6.4, which in turn proves the claim in the present item *(1)*. □

An immediate consequence of Corollary 1.6.5 is that, whenever L_1, L_2 are two $M \times M$ second-order, homogeneous, constant complex coefficient, weakly elliptic systems in $\mathbb{R}^n$,

$$\text{if } \mathfrak{A}^{\rm dis}_{L_1} \neq \varnothing \text{ and } \mathfrak{A}^{\rm dis}_{L_2} \neq \varnothing \text{ then } \mathfrak{A}^{\rm dis}_{\mathbf{L}} \neq \varnothing, \tag{1.6.81}$$

where

$$\mathbf{L} := \begin{pmatrix} L_1 & 0 \\ 0 & L_2 \end{pmatrix}, \quad \text{or } \mathbf{L} := \begin{pmatrix} 0 & L_1 \\ L_2 & 0 \end{pmatrix}. \tag{1.6.82}$$

Theorem 1.6.4 also has the following noteworthy consequence:

Corollary 1.6.6 *Fix $M, n \in \mathbb{N}$ with $n \geq 2$. Let L be an $M \times M$ second-order, homogeneous, constant complex coefficient, weakly elliptic system in $\mathbb{R}^n$. Assume that there exists a matrix-valued function $k_* \in \big[\mathscr{C}^\infty(\mathbb{R}^n\setminus\{0\})\big]^{M\times M}$ which is positive homogeneous of degree $-n$, is not identically zero, and satisfies*

$$\int_{S^{n-1}} k_* \, d\mathcal{H}^{n-1} = 0 \cdot I_{M\times M} \tag{1.6.83}$$

(where $I_{M\times M}$ is the $M \times M$ identity matrix), as well as

$$L\big(x_s k_*(x)\big) = 0 \cdot I_{M\times M} \text{ in } \mathbb{R}^n \setminus \{0\} \text{ for each } s \in \{1,\dots,n\}. \tag{1.6.84}$$

Then either $\mathfrak{A}^{\rm dis}_{L} = \varnothing$, or $\mathfrak{A}^{\rm dis}_{L^\top} = \varnothing$.

Proof Seeking a contradiction, assume that both $\mathfrak{A}^{\rm dis}_{L} \neq \varnothing$ and $\mathfrak{A}^{\rm dis}_{L^\top} \neq \varnothing$. Then Theorem 1.5.5 ensures that $\mathfrak{A}^{\rm dis}_{L}$ is a singleton. Granted this, the last part of Theorem 1.6.4 guarantees that there is only one function k as in item *(2)* of said theorem. However, $k + k_*$ is also as in item *(2)* of Theorem 1.6.4 and does not coincide with k. This contradiction finishes the proof. □

Consider next the following complexified version of the Lamé system (originally arising in the study of linear elasticity), defined for any two parameters $\mu, \lambda \in \mathbb{C}$ (referred to as Lamé moduli) as

$$L_{\mu,\lambda} := \mu\Delta + (\mu + \lambda)\nabla\mathrm{div}, \tag{1.6.85}$$

acting on vector fields defined in open subsets of $\mathbb{R}^n$, with the Laplacian applied componentwise. Hence, $L_{\mu,\lambda} = (L_{\mu,\lambda})^\top$, and from [131, (1.3.9)] we know that

$$\begin{array}{l}\text{the complex Lamé system (1.6.85) is weakly}\\ \text{elliptic if and only if } \mu \neq 0 \text{ and } 2\mu + \lambda \neq 0.\end{array} \tag{1.6.86}$$

We may express the complex Lamé system L as in (1.1.33) (with $M := n$) using a variety of coefficient tensors. In particular, it has been noted in [132, (1.4.169)] that we may employ the one-parameter family

$$A(\zeta) = \big(a^{\alpha\beta}_{jk}(\zeta)\big)_{\substack{1\leq\alpha,\beta\leq n\\ 1\leq j,k\leq n}} \text{ defined for each } \zeta \in \mathbb{C} \text{ according to}$$
$$a^{\alpha\beta}_{jk}(\zeta) := \mu\delta_{jk}\delta_{\alpha\beta} + (\mu+\lambda-\zeta)\delta_{j\alpha}\delta_{k\beta} + \zeta\delta_{j\beta}\delta_{k\alpha}, \quad 1 \leq j,k,\alpha,\beta \leq n. \tag{1.6.87}$$

In other words, for each vector field $u = (u_\beta)_{1\leq\beta\leq n} \in \big[\mathcal{D}'(\Omega)\big]^n$ with Ω open subset of $\mathbb{R}^n$, and each parameter $\zeta \in \mathbb{C}$, the Lamé system (1.6.85) satisfies

$$L_{\mu,\lambda}u = \Big(a^{\alpha\beta}_{jk}(\zeta)\partial_j\partial_k u_\beta\Big)_{1\leq\alpha\leq n} \quad \text{in } \big[\mathcal{D}'(\Omega)\big]^n. \tag{1.6.88}$$

Fix now $\mu, \lambda \in \mathbb{C}$ with $\mu \neq 0$ and $2\mu + \lambda \neq 0$, and pick an arbitrary $\zeta \in \mathbb{C}$. Recall that the integral kernel of the principal-value double layer potential operator $K_{A(\zeta)}$ associated with the Lamé system $L_{\mu,\lambda}$ (cf. (1.6.85)) written as in (1.6.88) for the coefficient tensor $A(\zeta)$ from [132, (1.4.169)] in any Lebesgue measurable set $\Omega \subseteq \mathbb{R}^n$ of locally finite perimeter has been identified in [132, (1.4.173)] (with the constants $C_1(\zeta), C_2(\zeta)$ as in [132, (1.4.172)]). From this and the equivalence *(i)* $\Leftrightarrow$ *(v)* in Proposition 1.3.2 we then see that for each $\zeta \in \mathbb{C}$ we have

$$A(\zeta) \in \mathfrak{A}^{\rm dis}_{L_{\mu,\lambda}} \Longleftrightarrow C_2(\zeta) = 0 \Longleftrightarrow 3\mu + \lambda \neq 0 \text{ and } \zeta = \frac{\mu(\mu+\lambda)}{3\mu+\lambda}. \tag{1.6.89}$$

Moreover, corresponding to this value of ζ, the entries in $A(\zeta)$ become

$$a^{\alpha\beta}_{rs} = \mu\delta_{rs}\delta_{\alpha\beta} + \frac{(\lambda+\mu)(2\mu+\lambda)}{3\mu+\lambda}\delta_{r\alpha}\delta_{s\beta} + \frac{\mu(\lambda+\mu)}{3\mu+\lambda}\delta_{r\beta}\delta_{s\alpha} \tag{1.6.90}$$

for $\alpha, \beta, r, s, \alpha, \beta \in \{1, \dots, n\}$. This ultimately shows that

whenever the Lamé moduli $\mu, \lambda \in \mathbb{C}$ are such that $\mu \neq 0$, $2\mu + \lambda \neq 0$, and $3\mu + \lambda \neq 0$, the Lamé operator $L_{\mu,\lambda}$ defined as in (1.6.88) has the property that $\mathfrak{A}^{\rm dis}_{L_{\mu,\lambda}} = \mathfrak{A}^{\rm dis}_{L^\top_{\mu,\lambda}} \neq \emptyset$. (1.6.91)

An alternative, direct proof of (1.6.91) is presented in the next proposition.

Proposition 1.6.7 *Fix $n \in \mathbb{N}$ with $n \geq 2$ and let $\mu, \lambda \in \mathbb{C}$ be such that $\mu \neq 0$ and $2\mu + \lambda \neq 0$. Consider the even matrix-valued function $\mathbb{H} : S^{n-1} \to \mathbb{C}^{n\times n}$ defined by*

$$\mathbb{H}(\omega) := \frac{1}{\omega_{n-1}}\big(2\mu I_{n\times n} + n(\mu+\lambda)\,\omega\otimes\omega\big) \text{ for all } \omega \in S^{n-1}. \tag{1.6.92}$$

Then

$$\int_{S^{n-1}} \mathbb{H}(\omega)\,d\mathcal{H}^{n-1}(\omega) = (3\mu+\lambda)I_{n\times n} \tag{1.6.93}$$

and, for each $h \in S^{n-1}$,

$$L_{\mu,\lambda}\left(\frac{x\cdot h}{|x|^n}\,\mathbb{H}\Big(\frac{x}{|x|}\Big)\right)=0\ \ in\ \mathbb{R}^n\setminus\{0\},\tag{1.6.94}$$

where $L_{\mu,\lambda}$ is the Lamé system from (1.6.85).

In particular,

$$\mathfrak{A}^{\text{dis}}_{L_{\mu,\lambda}}\neq\varnothing\iff 3\mu+\lambda\neq 0.\tag{1.6.95}$$

Proof Denote $\omega=(\omega_1,\dots,\omega_n)\in S^{n-1}$. Starting with (1.6.92) we compute

$$\begin{aligned}
&\int_{S^{n-1}}\mathbb{H}(\omega)\,d\mathcal{H}^{n-1}(\omega)\\
&=\frac{1}{\omega_{n-1}}2\mu\,\omega_{n-1}\,I_{n\times n}+\frac{n(\mu+\lambda)}{\omega_{n-1}}\left(\int_{S^{n-1}}\omega_j\omega_k\,d\mathcal{H}^{n-1}(\omega)\right)_{1\leq j,k\leq n}\\
&=2\mu\,I_{n\times n}+\frac{n(\mu+\lambda)}{\omega_{n-1}}\left(\frac{\pi^{\frac{n}{2}}}{\Gamma(\frac{n}{2}+1)}\delta_{jk}\right)_{1\leq j,k\leq n}=(3\mu+\lambda)I_{n\times n},
\end{aligned}\tag{1.6.96}$$

where the second equality in (1.6.96) is a consequence of

$$\int_{S^{n-1}}\omega_j\omega_k\,d\mathcal{H}^{n-1}(\omega)=\frac{\pi^{\frac{n}{2}}}{\Gamma(\frac{n}{2}+1)}\delta_{jk}\ \text{ for } j,k\in\{1,\dots,n\},\tag{1.6.97}$$

(see, e.g., [127, Proposition 14.69, p. 581]) and the last equality uses the formula $\frac{n\pi^{\frac{n}{2}}}{\Gamma(\frac{n}{2}+1)}=\omega_{n-1}$. This proves (1.6.93). Next, define the matrix-valued function k_* by

$$k_*(x):=\frac{1}{|x|^n}\,\mathbb{H}\Big(\frac{x}{|x|}\Big)=2\mu\frac{1}{\omega_{n-1}}\frac{1}{|x|^n}I_{n\times n}+(\mu+\lambda)\frac{n}{\omega_{n-1}}\frac{x\otimes x}{|x|^{n+2}}\tag{1.6.98}$$

for all $x\in\mathbb{R}^n\setminus\{0\}$. Also, recall the standard fundamental solution for the Laplacian in $\mathbb{R}^n$, defined at each point $x\in\mathbb{R}^n\setminus\{0\}$ according to

$$E_\Delta(x):=\begin{cases}\dfrac{1}{(2-n)\omega_{n-1}}\dfrac{1}{|x|^{n-2}} & \text{if } n\geq 3,\\[2mm] \dfrac{1}{2\pi}\ln|x| & \text{if } n=2.\end{cases}\tag{1.6.99}$$

Since for each $x\in\mathbb{R}^n\setminus\{0\}$ the Hessian matrix of E_Δ is

$$\big((\partial_i\partial_jE_\Delta)(x)\big)_{1\leq i,j\leq n}=\left(\frac{\delta_{ij}}{\omega_{n-1}}\frac{1}{|x|^n}-\frac{n}{\omega_{n-1}}\frac{x_ix_j}{|x|^{n+2}}\right)_{1\leq i,j\leq n}\tag{1.6.100}$$

we have

$$k_*(x)=-(\mu+\lambda)\big((\partial_i\partial_jE_\Delta)(x)\big)_{1\leq i,j\leq n}+(3\mu+\lambda)\frac{1}{\omega_{n-1}}\frac{1}{|x|^n}I_{n\times n}\tag{1.6.101}$$

for all $x\in\mathbb{R}^n\setminus\{0\}$. Fix $s\in\{1,\dots,n\}$ and use (1.6.101) to write

$$x_s k_*(x) = -(\mu+\lambda)\big(x_s(\partial_i\partial_j E_\Delta)(x)\big)_{1\le i,j\le n} + (3\mu+\lambda)(\partial_s E_\Delta)(x) I_{n\times n} \tag{1.6.102}$$

for all $x \in \mathbb{R}^n \setminus \{0\}$. Then, for each $x \in \mathbb{R}^n \setminus \{0\}$, we have

$$\begin{aligned} &L_{\mu,\lambda}\big(x_s(\nabla\partial_j E_\Delta)(x)\big)_{1\le j\le n} \\ &\quad = \mu\Delta\big(x_s(\nabla\partial_j E_\Delta)(x)\big)_{1\le j\le n} + (\lambda+\mu)\nabla\mathrm{div}\big(x_s(\nabla\partial_j E_\Delta)(x)\big)_{1\le j\le n}. \end{aligned} \tag{1.6.103}$$

Having fixed an arbitrary index $j \in \{1,\dots,n\}$, at each $x \in \mathbb{R}^n \setminus \{0\}$ we compute

$$\begin{aligned} \Delta\big(x_s(\nabla\partial_j E_\Delta)(x)\big) &= (\Delta x_s)(\nabla\partial_j E_\Delta)(x) + 2\big((\nabla x_s)\cdot\nabla(\partial_i\partial_j E_\Delta)(x)\big)_{1\le i\le n} \\ &\quad + x_s\Delta(\nabla\partial_j E_\Delta)(x) = 2\big((\partial_s\partial_i\partial_j E_\Delta)(x)\big)_{1\le i\le n} = 2\nabla\big((\partial_s\partial_j E_\Delta)(x)\big), \end{aligned} \tag{1.6.104}$$

and

$$\nabla\mathrm{div}\big(x_s(\nabla\partial_j E_\Delta)(x)\big) = \nabla\Big(\mathbf{e}_s\cdot(\nabla\partial_j E_\Delta)(x) + x_s\Delta(\partial_j E_\Delta)(x)\Big) = \nabla\big((\partial_s\partial_j E_\Delta)(x)\big). \tag{1.6.105}$$

Collectively, (1.6.103)-(1.6.105) prove that

$$L_{\mu,\lambda}\big(x_s(\nabla\partial_j E_\Delta)(x)\big)_{1\le j\le n} = (3\mu+\lambda)\nabla\big((\partial_s\partial_j E_\Delta)(x)\big)_{1\le j\le n} \tag{1.6.106}$$

in $\mathbb{R}^n \setminus \{0\}$. Also,

$$\begin{aligned} L_{\mu,\lambda}\big((\partial_s E_\Delta)(x) I_{n\times n}\big) &= \mu\Delta\big((\partial_s E_\Delta)(x)\big) I_{n\times n} + (\lambda+\mu)\nabla\mathrm{div}\big((\partial_s E_\Delta)(x) I_{n\times n}\big) \\ &= 0\cdot I_{n\times n} + (\lambda+\mu)\nabla\big((\partial_s\partial_j E_\Delta)(x)\big)_{1\le j\le n} \text{ in } \mathbb{R}^n \setminus \{0\}. \end{aligned} \tag{1.6.107}$$

Combining (1.6.98), (1.6.102), (1.6.106), and (1.6.107) we arrive at

$$\begin{aligned} L_{\mu,\lambda}\left(\frac{x_s}{|x|^n}\mathbb{H}\Big(\frac{x}{|x|}\Big)\right) &= L_{\mu,\lambda}\big(x_s k_*(x)\big) = -(\mu+\lambda)L_{\mu,\lambda}\big(x_s(\partial_i\partial_j E_\Delta)(x)\big)_{1\le i,j\le n} \\ &\quad + (3\mu+\lambda)L_{\mu,\lambda}(\partial_s E_\Delta)(x) I_{n\times n} = -(\mu+\lambda)(3\mu+\lambda)\nabla\big((\partial_s\partial_j E_\Delta)(x)\big)_{1\le j\le n} \\ &\quad + (3\mu+\lambda)(\lambda+\mu)\nabla\big((\partial_s\partial_j E_\Delta)(x)\big)_{1\le j\le n} = 0\cdot I_{n\times n} \end{aligned} \tag{1.6.108}$$

in $\mathbb{R}^n \setminus \{0\}$. From this (1.6.94) follows by linearity since $s \in \{1,\dots,n\}$ is arbitrary.

The right-to-left implication in (1.6.95) is a consequence of (1.6.93), (1.6.94), and Corollary 1.6.5. For the opposite implication, note that k_* is positive homogeneous of degree $-n$ in $\mathbb{R}^n \setminus \{0\}$, is not identically zero, and if $3\mu+\lambda = 0$ then k_* satisfies the hypothesis of Corollary 1.6.6 for $L := L_{\mu,\lambda}$. Hence, taking also into account the fact that $L_{\mu,\lambda}$ is symmetric, by Corollary 1.6.6 we obtain $\mathfrak{A}^{\mathrm{dis}}_{L_{\mu,\lambda}} = \varnothing$. □

The above result brings into focus the special case corresponding to $\mu = 1$ and $\lambda = -3$ in (1.6.85). For these values of the Lamé moduli we shall use the notation

$$L_D := \Delta - 2\nabla\mathrm{div}. \tag{1.6.109}$$

We shall study this system systematically in §2. For now, we note that the weakly elliptic system L_D does not have a distinguished coefficient tensor.

Proposition 1.6.8 *For each $n \in \mathbb{N}$ with $n \geq 2$, the $n \times n$ system L_D in $\mathbb{R}^n$ from* (1.6.109) *is weakly elliptic, second-order, homogeneous, constant real coefficient, symmetric, and has the property that $\mathfrak{A}^{\rm dis}_{L_D} = \mathfrak{A}^{\rm dis}_{L_D^\top} = \varnothing$.*

Proof That L_D is a second-order, homogeneous, constant real coefficient, symmetric $n \times n$ system in $\mathbb{R}^n$ is clear from (1.6.109). Also, from (1.6.86) we see that L_D in $\mathbb{R}^n$ is weakly elliptic. The fact that $\mathfrak{A}^{\rm dis}_{L_D} = \mathfrak{A}^{\rm dis}_{L_D^\top} = \varnothing$ follows from (1.6.95). □

In turn, we may use Proposition 1.6.8 to prove the following result for the system

$$\mathbb{L}_B := \frac{1}{4}\begin{pmatrix} \partial_x^2 - \partial_y^2 & -2\partial_x\partial_y \\ 2\partial_x\partial_y & \partial_x^2 - \partial_y^2 \end{pmatrix}. \tag{1.6.110}$$

Corollary 1.6.9 *For the second-order, homogeneous, real constant coefficient, weakly elliptic 2×2 system*

$$\mathbb{L}_B := \frac{1}{4}\begin{pmatrix} \partial_x^2 - \partial_y^2 & -2\partial_x\partial_y \\ 2\partial_x\partial_y & \partial_x^2 - \partial_y^2 \end{pmatrix} \tag{1.6.111}$$

in $\mathbb{R}^2$ one has

$$\mathfrak{A}^{\rm dis}_{\mathbb{L}_B} = \varnothing \text{ and } \mathfrak{A}^{\rm dis}_{\mathbb{L}_B^\top} = \varnothing. \tag{1.6.112}$$

Proof Bring in the two-dimensional version of the special system $L_D := \Delta - 2\nabla\mathrm{div}$ from (1.6.109), i.e.,

$$L_D = \begin{pmatrix} \partial_y^2 - \partial_x^2 & -2\partial_x\partial_y \\ -2\partial_x\partial_y & \partial_x^2 - \partial_y^2 \end{pmatrix}. \tag{1.6.113}$$

If we now let $V := \begin{pmatrix} -1 & 0 \\ 0 & 1 \end{pmatrix}$, then $V^\top = V = V^{-1}$ and we may write $\mathbb{L}_B = \frac{1}{4}L_D V$ and $\mathbb{L}_B^\top = \frac{1}{4}V L_D$. These together with (1.2.46) and (1.2.47) show that

$$A \in \mathfrak{A}^{\rm dis}_{L_D} \iff A \diamond V \in \mathfrak{A}^{\rm dis}_{\mathbb{L}_B}, \tag{1.6.114}$$

$$A \in \mathfrak{A}^{\rm dis}_{L_D} \iff V \diamond A \in \mathfrak{A}^{\rm dis}_{\mathbb{L}_B^\top}. \tag{1.6.115}$$

Since we have proved that $\mathfrak{A}^{\rm dis}_{L_D} = \varnothing$ (cf. Proposition 1.6.8), the equivalences in (1.6.114)-(1.6.115) imply (1.6.112). □

We continue by making the following observation in relation to the Lamé system.

Remark 1.6.10 *Consider the complex Lamé system $L_{\mu,\lambda}$, defined earlier in* (1.6.85), *in the regime $\mu, \lambda \in \mathbb{C}$ with $\mu \neq 0$ and $2\mu + \lambda \neq 0$. From* (1.6.86) *we know that this is equivalent with the weak ellipticity of $L_{\mu,\lambda}$. Hence, this is the range in which we may consider the issue of whether $L_{\mu,\lambda}$ possesses distinguished coefficient tensors. In* *this regard, we note that* (1.6.91) *and Theorem 1.5.5 imply that* $\mathfrak{A}^{\rm dis}_{L_{\mu,\lambda}}$ *is a singleton when* $3\mu + \lambda \neq 0$. *In addition, if* $3\mu + \lambda = 0$ *then* $L_{\mu,\lambda} = \mu L_D$, *so Proposition 1.6.8 implies that* $\mathfrak{A}^{\rm dis}_{L_{\mu,\lambda}}$ *is empty. Collectively, these observations prove that*

$$\begin{gathered}\text{given any } \mu, \lambda \in \mathbb{C} \text{ with } \mu \neq 0 \text{ and } 2\mu + \lambda \neq 0, \text{ then } \mathfrak{A}^{\rm dis}_{L_{\mu,\lambda}} \neq \varnothing \text{ if} \\ \text{and only if } 3\mu + \lambda \neq 0 \text{ if and only if } \mathfrak{A}^{\rm dis}_{L_{\mu,\lambda}} \text{ is a singleton (namely the} \\ \text{coefficient tensor } A(\zeta) \text{ described in (1.6.87), corresponding to the} \\ \text{choice } \zeta = \tfrac{\mu(\mu+\lambda)}{3\mu+\lambda}).\end{gathered} \tag{1.6.116}$$

Intriguingly, $\mu \neq 0$, $2\mu + \lambda \neq 0$, $3\mu + \lambda \neq 0$ *are precisely the conditions making the Dirichlet Problem for the Lamé system* $L_{\mu,\lambda}$ *regular elliptic, in the sense of Shapiro-Lopatinskiĭ* ([109], [187]; *see, e.g., the discussion in* [98, Example 1, pp. 117–118] *and* [207, pp. 633-634]).

As an illustration of these considerations, let us focus on the family of homogenous second-order $n \times n$ *systems* $L_\alpha := \Delta - \alpha\nabla\mathrm{div}$ *indexed by* $\alpha \in \mathbb{C}$ *(which contains the operator* $L_D := \Delta - 2\nabla\mathrm{div}$*, discussed earlier). This fits into the larger family of Lamé-like systems (for the choice* $\mu := 1$ *and* $\lambda := -1-\alpha$*), so* [131, Proposition 1.3.3] *applies and guarantees that*

$$L_\alpha \text{ is weakly elliptic if and only if } \alpha \neq 1. \tag{1.6.117}$$

In the regime $\alpha \in \mathbb{C} \setminus \{1\}$, *from* (1.6.116) *we then conclude that*

$$\mathfrak{A}^{\rm dis}_{L_\alpha} \neq \varnothing \text{ if and only } \alpha \neq 2. \tag{1.6.118}$$

This sheds further light on the special role played by the coefficient 2 *in the definition of* L_D *given in* (1.6.109).

It is also of interest to observe that *there exist (weakly elliptic) coefficient tensors which are distinguished but not Legendre-Hadamard (strongly) elliptic*. In this vein, it is relevant to note that if Δ stands for the Laplacian in $\mathbb{R}^n$ then

$$L := \begin{pmatrix} -\Delta & 0 \\ 0 & \Delta \end{pmatrix} \tag{1.6.119}$$

is a homogeneous, constant coefficient, weakly elliptic, second-order, 2×2 system in $\mathbb{R}^n$ with the property that $\mathfrak{A}^{\rm dis}_{L} \neq \varnothing$ (as seen from (1.6.81)-(1.6.82)), and yet

$$\begin{gathered}\text{for any } \theta \in [0, 2\pi) \text{ the system } e^{i\theta} L \text{ does not satisfy the} \\ \text{Legendre-Hadamard (strong) ellipticity condition.}\end{gathered} \tag{1.6.120}$$

As regards scalar operators, let us first recall that a second-order, homogeneous, constant complex coefficient, scalar differential operator L in $\mathbb{R}^n$ is said to be `strongly elliptic` provided

$$L = a_{jk}\partial_j\partial_k \text{ with } a_{jk} \in \mathbb{C} \text{ for } j, k \in \{1, \dots, n\} \tag{1.6.121}$$

having the property that there exists a constant $c \in (0, \infty)$ such that

$$-\mathrm{Re}\big[L(\xi)\big] = \mathrm{Re}\Big[\sum_{j,k=1}^{n} a_{jk}\xi_j\xi_k\Big] \geq c|\xi|^2, \quad \forall \xi = (\xi_1, \dots, \xi_n) \in \mathbb{R}^n. \tag{1.6.122}$$

Note that this is what the Legendre-Hadamard (strong) ellipticity condition [131, (1.3.4) in Definition 1.3.2] becomes in the scalar case (i.e., for $M = 1$). In relation to this class of operators, we have the following result.

Proposition 1.6.11 *For every scalar, homogeneous, second-order, constant complex coefficient, strongly elliptic (in the sense of (1.6.122)) operator L in $\mathbb{R}^n$, it follows that $\mathfrak{A}_L^{\mathrm{dis}}$ is a singleton if $n \geq 2$. In fact,*

$$\mathfrak{A}_L^{\mathrm{dis}} = \big\{\mathrm{sym}\, A\big\} \text{ for every strongly elliptic constant complex coefficient operator } L = \mathrm{div} A \nabla \text{ in } \mathbb{R}^n \text{ with } n \geq 2. \tag{1.6.123}$$

Proof Let L be as in (1.6.121). Introduce $A := (a_{jk})_{1\leq j,k\leq n} \in \mathbb{C}^{n\times n}$ and define

$$(\widetilde{a}_{jk})_{1\leq j,k\leq n} := \mathrm{sym}\, A := \frac{A + A^{\top}}{2}, \quad (b_{jk})_{1\leq j,k\leq n} := (\mathrm{sym}\, A)^{-1}. \tag{1.6.124}$$

In particular,

$$L = L_{\mathrm{sym}\, A} := \widetilde{a}_{jk}\partial_j\partial_k, \tag{1.6.125}$$

i.e., the coefficient matrix sym A may be used to represent the given differential operator L. As noted in [131, Proposition 1.4.17], the fundamental solution E canonically associated with the operator L as in [131, Theorem 1.4.2] may be explicitly identified as the function $E \in L^1_{\mathrm{loc}}(\mathbb{R}^n, \mathcal{L}^n)$ given at each point $x \in \mathbb{R}^n \setminus \{0\}$ by

$$E(x) = \begin{cases} -\dfrac{1}{(n-2)\,\omega_{n-1}\sqrt{\det(\mathrm{sym}\, A)}}\langle(\mathrm{sym}\, A)^{-1}x, x\rangle^{-\frac{n-2}{2}} & \text{if } n \geq 3, \\ \dfrac{1}{4\pi\sqrt{\det(\mathrm{sym}\, A)}}\log(\langle(\mathrm{sym}\, A)^{-1}x, x\rangle) + c_A & \text{if } n = 2, \end{cases} \tag{1.6.126}$$

where log denotes the principal branch of the complex logarithm (defined for complex numbers $z \in \mathbb{C} \setminus (-\infty, 0]$ so that $z^a = e^{a \log z}$ for each $a \in \mathbb{R}$), and c_A is a complex constant which depends solely on A. Since both sym A and $(\mathrm{sym}\, A)^{-1}$ are symmetric matrices, for each index $j \in \{1, \dots, n\}$ and each point $x = (x_i)_{1\leq i\leq n} \in \mathbb{R}^n \setminus \{0\}$ we therefore have (in all dimensions $n \geq 2$)

$$\begin{aligned}(\partial_j E)(x) &= \frac{\langle(\mathrm{sym}\, A)^{-1}x, x\rangle^{-\frac{n}{2}}(\delta_{rj}b_{rs}x_s + \delta_{sj}b_{rs}x_r)}{2\omega_{n-1}\sqrt{\det(\mathrm{sym}\, A)}} \\ &= \frac{\langle(\mathrm{sym}\, A)^{-1}x, x\rangle^{-\frac{n}{2}} b_{rj}x_r}{\omega_{n-1}\sqrt{\det(\mathrm{sym}\, A)}}. \end{aligned} \tag{1.6.127}$$

As a consequence of this and (1.6.124), for each $s \in \{1, \dots, n\}$ we then have

$$(\operatorname{sym} A)_{js}(\partial_j E)(x) = \frac{\langle(\operatorname{sym} A)^{-1}x, x\rangle^{-\frac{n}{2}} x_s}{\omega_{n-1}\sqrt{\det(\operatorname{sym} A)}}, \tag{1.6.128}$$

at each point $x = (x_i)_{1\leq i\leq n} \in \mathbb{R}^n \setminus \{0\}$. To proceed, set

$$k(x) := \frac{\langle(\operatorname{sym} A)^{-1}x, x\rangle^{-\frac{n}{2}}}{\omega_{n-1}\sqrt{\det(\operatorname{sym} A)}}, \qquad \forall x \in \mathbb{R}^n \setminus \{0\}, \tag{1.6.129}$$

and observe that k is well defined, smooth, even, and positive homogeneous of degree $-n$ in $\mathbb{R}^n \setminus \{0\}$. In addition, thanks to [127, (7.12.47), p. 314], we have

$$\int_{S^{n-1}} k(\omega)\,\mathrm{d}\mathcal{H}^{n-1}(\omega) = 1, \tag{1.6.130}$$

while (1.6.128) implies (bearing in mind that $LE = 0$ in $\mathbb{R}^n \setminus \{0\}$) that

$$L\big(x_s k(x)\big) = 0 \text{ in } \mathbb{R}^n \setminus \{0\} \text{ for each } s \in \{1, \dots, n\}. \tag{1.6.131}$$

As such, the demands in (1.6.32)-(1.6.33) are also satisfied. In view of these properties and the equivalence *(1)*$\Leftrightarrow$*(2)* in Theorem 1.6.4, we then conclude that the scalar differential operator L possesses a distinguished coefficient tensor. Lastly, since $L^\top = L$ (as may be seen from (1.6.125)), Theorem 1.5.2 implies that L actually has a unique distinguished coefficient tensor. The justification of the first claim in the statement is therefore complete.

Lastly, based on (1.6.125), item *(a)* in Proposition 1.2.1, and (1.6.127) it may be checked without difficulty that $\operatorname{sym} A \in \mathfrak{A}_L^{\mathrm{dis}}$ for every strongly elliptic constant complex coefficient operator $L = \operatorname{div} A\nabla$ in $\mathbb{R}^n$. In concert with what we have proved already, this establishes (1.6.123). □

Moving on, we wish to show that, for scalar operators in dimensions $n \geq 3$, weak ellipticity itself guarantees the existence of a unique distinguished coefficient tensor.

Proposition 1.6.12 *If $n \geq 3$ then for each weakly elliptic, scalar, homogeneous, second-order operator $L = \operatorname{div} A\nabla$ in $\mathbb{R}^n$ with constant complex coefficients, the class $\mathfrak{A}_L^{\mathrm{dis}}$ consists precisely of one matrix, namely* $\operatorname{sym} A := (A + A^\top)/2$.

Proof Suppose $n \geq 3$, and consider an arbitrary second-order, homogeneous, constant complex coefficient, scalar differential operator $L = a_{jk}\partial_j\partial_k$ in $\mathbb{R}^n$, with $a_{jk} \in \mathbb{C}$ for $j, k \in \{1, \dots, n\}$, which, in contrast to Proposition 1.6.11, now is only assumed to be weakly elliptic, i.e., $\sum_{j,k=1}^{n} a_{jk}\xi_j\xi_k \neq 0$ for all $\xi = (\xi_1, \dots, \xi_n) \in \mathbb{R}^n \setminus \{0\}$. Introduce $A := (a_{jk})_{1\leq j,k\leq n} \in \mathbb{C}^{n\times n}$. It has been shown in [131, Proposition 1.4.39] (here is where $n \geq 3$ is used) that

> there exists an angle $\theta \in [0, 2\pi)$ such that if we set $A_\theta := e^{i\theta} A$ then the matrix $\operatorname{sym} A_\theta := (A_\theta + A_\theta^\top)/2 \in \mathbb{C}^{n\times n}$ is strongly elliptic, in the sense that there exists some $c \in (0, \infty)$ such that $\operatorname{Re}\langle(\operatorname{sym} A_\theta)\xi, \xi\rangle \geq c|\xi|^2$ for each $\xi \in \mathbb{R}^n$ (cf. (1.6.122)). (1.6.132)

Based on (1.2.47) with $C := e^{i\theta}$, (1.6.125) for A_θ in place of A, and (1.6.123) for A_θ in place of A (whose applicability uses (1.6.132)), we may then write

$$e^{i\theta} \cdot \mathfrak{A}_L^{\rm dis} = \mathfrak{A}_{e^{i\theta}L}^{\rm dis} = \mathfrak{A}_{L_{A_\theta}}^{\rm dis} = \mathfrak{A}_{L_{\mathrm{sym}A_\theta}}^{\rm dis} = \{\mathrm{sym}\, A_\theta\} = e^{i\theta} \cdot \{\mathrm{sym}\, A\}, \tag{1.6.133}$$

from which the desired conclusion follows. □

Here is another perspective on the scalar case. Since $M = 1$ the Greek indices become irrelevant (as they are all equal to 1). We agree to drop them and simply write $A = (a_{jk})_{1\le j,k\le n}$ for a given weakly elliptic coefficient tensor in the scalar case. With this convention in mind, in place of (1.2.1) we now have

$$\mathcal{E}(\xi) = -\frac{1}{\langle A\xi, \xi\rangle}, \qquad \text{for all } \xi \in \mathbb{R}^n \setminus \{0\}, \tag{1.6.134}$$

and the expression in (1.2.9) becomes

$$\big(a_{sj} + a_{js}\big)\frac{\xi_j \xi_r}{\langle A\xi, \xi\rangle} a_{rs'} + a_{s's} = \left((A + A^\top)\frac{\xi \otimes \xi}{\langle A\xi, \xi\rangle} A + A^\top \right)_{ss'} \tag{1.6.135}$$

for each $s, s' \in \{1, \dots, n\}$. Thus, its symmetry in the indices $s, s' \in \{1, \dots, n\}$ amounts to having

$$(A + A^\top)\frac{\xi \otimes \xi}{\langle A\xi, \xi\rangle} A + A^\top \text{ symmetric matrix, for each } \xi \in S^{n-1}. \tag{1.6.136}$$

Since $A^\top(\xi \otimes \xi)A$ is always symmetric, (1.6.136) shows that for scalar weakly elliptic operators the symmetry condition formulated in (1.2.9) may be equivalently rephrased as the demand that the matrix

$$A(\xi \otimes \xi)A + \langle A\xi, \xi\rangle A^\top \text{ is symmetric for each } \xi \in \mathbb{R}^n. \tag{1.6.137}$$

Given that (1.6.137) is satisfied with A replaced by $\mathrm{sym}\, A := (A+A^\top)/2 \in \mathbb{C}^{n\times n}$, and bearing in mind that $\mathrm{sym}\, A \in \mathfrak{A}_L$, we conclude from Definition 1.2.2 that actually $\mathrm{sym}\, A \in \mathfrak{A}_L^{\rm dis}$. Granted these properties, Theorem 1.5.5 applies (bearing in mind that we presently have $L = L^\top$) and gives another proof of Proposition 1.6.12.

The two-dimensional version of Proposition 1.6.12 is discussed below.

Proposition 1.6.13 *Consider a scalar homogeneous constant (complex) coefficient second-order operator in* $\mathbb{R}^2$,

$$L = \sum_{j,k=1}^{2} a_{jk}\partial_j\partial_k, \tag{1.6.138}$$

assumed to be weakly elliptic. Introduce $A := (a_{jk})_{1\le j,k\le 2} \in \mathbb{C}^{2\times 2}$ *and define* $\mathrm{sym}\, A := \frac{1}{2}(A + A^\top)$. *Then, with the classes of matrices* M_2^0, $M_2^\pm$ *define as in* [131, (1.4.187)-(1.4.189)], *one has*

$$\mathfrak{A}_L^{\mathrm{dis}} = \varnothing \iff \operatorname{sym} A \in M_2^{\pm}, \tag{1.6.139}$$

$$\mathfrak{A}_L^{\mathrm{dis}} = \{\operatorname{sym} A\} \iff \operatorname{sym} A \in M_2^0. \tag{1.6.140}$$

As a corollary, in the class of scalar operators with real coefficients the following result holds (compare with Proposition 1.6.11):

$$\text{for any weakly elliptic operator } L = \operatorname{div} A\nabla \text{ with } A \in \mathbb{R}^{2\times 2} \text{ one has } \mathfrak{A}_L^{\mathrm{dis}} = \{\operatorname{sym} A\}, \text{ and either } L \text{ or } -L \text{ is Legendre-Hadamard elliptic.} \tag{1.6.141}$$

Proof The fact that L is weakly elliptic implies (see [131, (1.4.178)] and [131, (1.4.186)]) that

$$\operatorname{sym} A \in M_2 = M_2^0 \sqcup M_2^+ \sqcup M_2^-. \tag{1.6.142}$$

The structural result in [131, Lemma 1.4.18] also tells us that we may express

$$\operatorname{sym} A = a\begin{pmatrix} 1 & -\frac{\lambda_1+\lambda_2}{2} \\ -\frac{\lambda_1+\lambda_2}{2} & \lambda_1\lambda_2 \end{pmatrix} \text{ for some } a \in \mathbb{C}\setminus\{0\} \text{ and } \lambda_1, \lambda_2 \in \mathbb{C}\setminus\mathbb{R}. \tag{1.6.143}$$

Then any matrix $B \in \mathfrak{A}_L$ is of the form

$$B = a\begin{pmatrix} 1 & -\frac{\lambda_1+\lambda_2}{2} + z \\ -\frac{\lambda_1+\lambda_2}{2} - z & \lambda_1\lambda_2 \end{pmatrix} \text{ for some } z \in \mathbb{C}. \tag{1.6.144}$$

To proceed, fix B as above and assume first that $\operatorname{sym} A \in M_2^0$. In view of [131, Proposition 1.4.29] this implies that canonical fundamental solution of L constructed in [131, Theorem 1.4.2] is the function defined at each $x = (x_1, x_2) \in \mathbb{R}^2 \setminus \{0\}$ as

$$E(x) = C_1 \cdot \log\big[a(x_2 + \lambda_1 x_1)(x_2 + \lambda_2 x_1)\big] + C_2, \tag{1.6.145}$$

where $C_1 \in \mathbb{C}\setminus\{0\}$ and $C_2 \in \mathbb{C}$ are constants, while log is the logarithm branch from [131, Lemma 1.4.27]. In particular, for each $x = (x_1, x_2) \in \mathbb{R}^2 \setminus \{0\}$ we have

$$(\partial_1 E)(x) = C_1 \frac{(\lambda_1 + \lambda_2)x_2 + 2\lambda_1\lambda_2 x_1}{(x_2 + \lambda_1 x_1)(x_2 + \lambda_2 x_1)}, \tag{1.6.146}$$

$$(\partial_2 E)(x) = C_1 \frac{2x_2 + (\lambda_1 + \lambda_2)x_1}{(x_2 + \lambda_1 x_1)(x_2 + \lambda_2 x_1)}. \tag{1.6.147}$$

According to Definition 1.2.2, in order see whether $B \in \mathfrak{A}_L^{\mathrm{dis}}$ it suffices to check that the condition formulated in item *(a)* of Proposition 1.2.1 is presently satisfied (with $n = 2$ and $M = 1$). Specifically, we need to show that each $s, s' \in \{1, 2\}$ and each $x = (x_1, x_2) \in \mathbb{R}^2 \setminus \{0\}$ we have

$$\big(x_{s'} b_{rs} - x_s b_{rs'}\big)(\partial_r E)(x) = 0, \tag{1.6.148}$$

where $(b_{jk})_{1\leq j,k\leq 2}$ are the entries of B. Since (1.6.148) is trivially true when $s = s'$, matters are reduced to considering the cases when $s = 1$ and $s' = 2$, and when $s = 2$ and $s' = 1$, respectively. We shall explicitly deal only with the first case, since the second one is handled similarly. Concretely, in view of (1.6.144) and (1.6.146)-(1.6.147), we have

$$\begin{aligned}
&\big(x_2 b_{r1} - x_1 b_{r2}\big)(\partial_r E)(x)\\
&\qquad = \big(x_2 b_{11} - x_1 b_{12}\big)(\partial_1 E)(x) + \big(x_2 b_{21} - x_1 b_{22}\big)(\partial_2 E)(x)\\
&\qquad = a\,C_1\Big[x_2 + x_1\Big(\frac{\lambda_1+\lambda_2}{2} - z\Big)\Big]\frac{(\lambda_1+\lambda_2)x_2 + 2\lambda_1\lambda_2 x_1}{(x_2+\lambda_1 x_1)(x_2+\lambda_2 x_1)}\\
&\qquad\quad - a\,C_1\Big[x_2\Big(\frac{\lambda_1+\lambda_2}{2} + z\Big) + x_1\lambda_1\lambda_2\Big]\frac{2x_2 + (\lambda_1+\lambda_2)x_1}{(x_2+\lambda_1 x_1)(x_2+\lambda_2 x_1)}\\
&\qquad = -a\,C_1\,z\frac{2\lambda_1\lambda_2 x_1^2 + (2+\lambda_1+\lambda_2)x_1x_2 + 2x_2^2}{(x_2+\lambda_1 x_1)(x_2+\lambda_2 x_1)}. \qquad (1.6.149)
\end{aligned}$$

The last expression above is zero for each $x = (x_1, x_2) \in \mathbb{R}^2 \setminus \{0\}$ if and only if $z = 0$. Bearing in mind that B simply becomes $\operatorname{sym} A$ when $z = 0$, this ultimately proves

$$\operatorname{sym} A \in M_2^0 \Longrightarrow \mathfrak{A}_L^{\rm dis} = \{\operatorname{sym} A\}. \qquad (1.6.150)$$

Consider next the case when $\operatorname{sym} A \in M_2^{\pm}$ and $\lambda_1 \neq \lambda_2$. From [131, Proposition 1.4.32] we know that the canonical fundamental solution of L constructed in [131, Theorem 1.4.2] presently takes the form

$$E(x) = C_1 \cdot \log\Big[a\frac{x_2+\lambda_1 x_1}{x_2+\lambda_2 x_1}\Big] + C_2, \quad \text{for all } x = (x_1, x_2) \in \mathbb{R}^2 \setminus \{0\}, \qquad (1.6.151)$$

where $C_1 \in \mathbb{C} \setminus \{0\}$ and $C_2 \in \mathbb{C}$ are constants, while log denotes the logarithm branch from [131, Lemma 1.4.30]. For each $x = (x_1, x_2) \in \mathbb{R}^2 \setminus \{0\}$ we now have

$$(\partial_1 E)(x) = C_1\frac{(\lambda_1-\lambda_2)x_2}{(x_2+\lambda_1 x_1)(x_2+\lambda_2 x_1)}, \qquad (1.6.152)$$

$$(\partial_2 E)(x) = C_1\frac{(\lambda_2-\lambda_1)x_1}{(x_2+\lambda_1 x_1)(x_2+\lambda_2 x_1)}. \qquad (1.6.153)$$

In light of Definition 1.2.2, verifying the membership of B to $\mathfrak{A}_L^{\rm dis}$ is equivalent to checking that the condition formulated in item *(a)* of Proposition 1.2.1 is presently satisfied (with $n = 2$ and $M = 1$). Hence, we need to check whether (1.6.148) holds for each $s, s' \in \{1, 2\}$ and each $x = (x_1, x_2) \in \mathbb{R}^2 \setminus \{0\}$ where, as before, $(b_{jk})_{1\leq j,k\leq 2}$ are the entries of B, but where now the partial derivatives $\partial_r E$ (with $r \in \{1, 2\}$) are as in (1.6.152)-(1.6.153). Let us consider in detail the case when $s = 1$ and $s' = 2$, i.e., see if it is possible to select $z \in \mathbb{C}$ such that

$$\begin{aligned}
0 &= \big(x_2 b_{r1} - x_1 b_{r2}\big)(\partial_r E)(x) \\
&= \big(x_2 b_{11} - x_1 b_{12}\big)(\partial_1 E)(x) + \big(x_2 b_{21} - x_1 b_{22}\big)(\partial_2 E)(x) \\
&= a\, C_1 \Big[x_2 + x_1\Big(\frac{\lambda_1+\lambda_2}{2} - z\Big)\Big] \frac{(\lambda_1-\lambda_2)x_2}{(x_2+\lambda_1 x_1)(x_2+\lambda_2 x_1)} \\
&\quad - a\, C_1 \Big[x_2\Big(\frac{\lambda_1+\lambda_2}{2} + z\Big) + x_1\lambda_1\lambda_2\Big] \frac{(\lambda_2-\lambda_1)x_1}{(x_2+\lambda_1 x_1)(x_2+\lambda_2 x_1)} \\
&= a\, C_1\, (\lambda_1-\lambda_2) \frac{\lambda_1\lambda_2 x_1^2 + (\lambda_1+\lambda_2)x_1x_2 + x_2^2}{(x_2+\lambda_1 x_1)(x_2+\lambda_2 x_1)}
\end{aligned} \tag{1.6.154}$$

for each $x = (x_1, x_2) \in \mathbb{R}^2 \setminus \{0\}$. However, this is an impossibility, hence

$$\mathfrak{A}_L^{\rm dis} = \varnothing \text{ whenever } \operatorname{sym} A \in M_2^{\pm} \text{ with } \lambda_1 \neq \lambda_2. \tag{1.6.155}$$

Finally, let us analyze the case when $\operatorname{sym} A \in M_2^{\pm}$ has $\lambda_1 = \lambda_2$. Abbreviate $\lambda := \lambda_1 = \lambda_2$. Then [131, Proposition 1.4.34] shows that the canonical fundamental solution of L constructed in [131, Theorem 1.4.2] presently takes the form

$$E(x) = C_1 \frac{x_1}{x_2 + \lambda x_1} + C_2, \quad \text{for all } x = (x_1, x_2) \in \mathbb{R}^2 \setminus \{0\}, \tag{1.6.156}$$

where $C_1 \in \mathbb{C} \setminus \{0\}$ and $C_2 \in \mathbb{C}$. Then for each $x = (x_1, x_2) \in \mathbb{R}^2 \setminus \{0\}$ we now have

$$(\partial_1 E)(x) = \frac{C_1 x_2}{(x_2+\lambda x_1)^2} \text{ and } (\partial_2 E)(x) = \frac{-C_1 x_1}{(x_2+\lambda x_1)^2}. \tag{1.6.157}$$

Also, the matrix B from (1.6.144) currently becomes $B = a \begin{pmatrix} 1 & -\lambda + z \\ -\lambda - z & \lambda^2 \end{pmatrix}$ with $z \in \mathbb{C}$. By Definition 1.2.2, if $B \in \mathfrak{A}_L^{\rm dis}$ then the condition formulated in item *(a)* of Proposition 1.2.1 holds when written for the entries $(b_{jk})_{1\le j,k\le 2}$ of B. Let us focus on the special case when $s = 1$ and $s' = 2$, and see if it is possible to select $z \in \mathbb{C}$ such that, with $\partial_r E$ as in (1.6.157) (with $r \in \{1, 2\}$), we have

$$\begin{aligned}
0 &= \big(x_2 b_{r1} - x_1 b_{r2}\big)(\partial_r E)(x) \\
&= \big(x_2 b_{11} - x_1 b_{12}\big)(\partial_1 E)(x) + \big(x_2 b_{21} - x_1 b_{22}\big)(\partial_2 E)(x) \\
&= a\, C_1 \big[x_2 + x_1(\lambda - z)\big] \frac{x_2}{(x_2+\lambda x_1)^2} - a\, C_1 \big[x_2(\lambda+z) + x_1\lambda^2\big] \frac{x_1}{(x_2+\lambda x_1)^2} \\
&= a\, C_1 \frac{-\lambda^2 x_1^2 - 2z x_1 x_2 + x_2^2}{(x_2+\lambda x_1)^2} \quad \text{for each } x = (x_1, x_2) \in \mathbb{R}^2 \setminus \{0\}.
\end{aligned} \tag{1.6.158}$$

Since this never materializes, the conclusion is that

$$\mathfrak{A}_L^{\rm dis} = \varnothing \text{ whenever } \operatorname{sym} A \in M_2^{\pm} \text{ with } \lambda_1 = \lambda_2. \tag{1.6.159}$$

At this stage, the claims made in (1.6.139)-(1.6.140) are clear from (1.6.150), (1.6.155), and (1.6.159), bearing in mind (1.6.150).

Let us now justify (1.6.141). Since we are presently assuming $A \in \mathbb{R}^{2\times 2}$, from (1.6.143) we see that $a \in \mathbb{R}$ and, further, that $\lambda_1 + \lambda_2$ belongs to $\mathbb{R}$. As such, $\operatorname{Im}\lambda_1 = -\operatorname{Im}\lambda_2$ which goes to show that $\operatorname{sym} A \in M_2^0$. Granted this, (1.6.140) applies and gives the first claim in (1.6.141). To deal with the second claim in (1.6.141) note that $S^1 \ni \xi \mapsto L(\xi) \in \mathbb{R}$ is a well-defined continuous mapping, which therefore maps the compact connected set S^1 into a compact interval which (thanks to weak ellipticity) does not contain zero. Thus, either $L(\xi) > 0$ for each $\xi \in S^1$, or $L(\xi) < 0$ for each $\xi \in S^1$. Ultimately, this shows that either L, or $-L$, is Legendre-Hadamard elliptic. □

As an illustration, we may use Proposition 1.6.13 to conclude that Bitsadze's operator $\partial_{\bar z}^2$ lacks a distinguished coefficient tensor.

Example 1.6.14 *Work in the two-dimensional setting and bring back Bitsadze's operator $L_B = \partial_{\bar z}^2$, where $\partial_{\bar z} := \frac{1}{2}\big(\partial_x - \frac{1}{\mathrm{i}}\partial_y\big)$ is the Cauchy-Riemann operator in the plane. This is a weakly elliptic scalar operator which may be written as in* (1.6.138) *for the symmetric 2×2 matrix $A = (a_{jk})_{1\le j,k\le 2}$ given by*

$$A = \frac{1}{4}\begin{pmatrix} 1 & \mathrm{i} \\ \mathrm{i} & -1 \end{pmatrix}. \tag{1.6.160}$$

Since this fits into the template [131, (1.4.179)] *with $a := 1/4 \in \mathbb{C}\setminus\{0\}$ and $\lambda_1 = \lambda_2 = -\mathrm{i} \in \mathbb{C}\setminus\mathbb{R}$, we conclude from* [131, (1.4.189)] *that* $\operatorname{sym} A = A \in M_2^-$. *Granted this, we conclude from* (1.6.139) *that*

$$\mathfrak{A}_{L_B}^{\mathrm{dis}} = \varnothing. \tag{1.6.161}$$

Bitsadze's operator $\partial_{\bar z}^2$ is a special case of the family of operators $L_\lambda := \partial_{\bar z}^2 - \lambda^2\partial_z^2$ indexed by $\lambda \in \mathbb{C}$ (corresponding to $\lambda = 0$). The point made in our next example is that all weakly elliptic operators in this family turn out not to possess a distinguished coefficient tensor.

Example 1.6.15 *Continue working in the two-dimensional setting and for each $\lambda \in \mathbb{C}$ define $L_\lambda := \partial_{\bar z}^2 - \lambda^2\partial_z^2$, where $\partial_{\bar z} := \frac{1}{2}\big(\partial_x - \frac{1}{\mathrm{i}}\partial_y\big)$ and $\partial_z := \frac{1}{2}\big(\partial_x + \frac{1}{\mathrm{i}}\partial_y\big)$ are, respectively, the Cauchy-Riemann operator in the plane and its complex conjugate. Since for each $\lambda \in \mathbb{C}$ we may express L_λ as*

$$L_\lambda = \operatorname{div} A \nabla \ \text{ where } \ A := \frac{1}{4}\begin{pmatrix} 1-\lambda^2 & 2\mathrm{i}(1+\lambda^2) \\ 2\mathrm{i}(1+\lambda^2) & -(1-\lambda^2) \end{pmatrix}, \tag{1.6.162}$$

from [131, Example 1.4.25] *we then see (bearing in mind* [131, (1.4.178)]*) that, for any given $\lambda \in \mathbb{C}$,*

$$L_\lambda \text{ is weakly elliptic if and only if } \lambda \in \mathbb{C}\setminus\{\pm\mathrm{i}\} \text{ and } \frac{\lambda^2+1}{\lambda^2-1} \in \mathbb{C}\setminus \mathrm{i}\mathbb{R}, \tag{1.6.163}$$

In addition, from [131, (1.4.239)] *we deduce that*

$$\begin{aligned} &\text{whenever } \lambda \in \mathbb{C} \setminus \{\pm \mathrm{i}\} \text{ and } \tfrac{\lambda^2+1}{\lambda^2-1} \in \mathbb{C} \setminus \mathrm{i}\mathbb{R} \text{ then} \\ &A \in M_2^+ \text{ if } \mathrm{Im}\Big((-\mathrm{i})\tfrac{\lambda^2+1}{\lambda^2-1}\Big) < 0, \text{ and } A \in M_2^- \text{ if } \mathrm{Im}\Big((-\mathrm{i})\tfrac{\lambda^2+1}{\lambda^2-1}\Big) > 0. \end{aligned} \tag{1.6.164}$$

In particular, from (1.6.163), (1.6.164), *and Proposition 1.6.13 we conclude that*

$$\text{if } L_\lambda \text{ is weakly elliptic then } A \in M_2^{\pm}, \text{ hence } \mathfrak{A}_{L_\lambda}^{\mathrm{dis}} = \varnothing. \tag{1.6.165}$$

In other words, L_λ does not possess a distinguished coefficient tensor in the regime where this operator is weakly elliptic, i.e.,

$$\mathfrak{A}_{L_\lambda}^{\mathrm{dis}} = \varnothing \text{ for each } \lambda \in \mathbb{C} \setminus \{\pm \mathrm{i}\} \text{ with } \frac{\lambda^2+1}{\lambda^2-1} \in \mathbb{C} \setminus \mathrm{i}\mathbb{R}. \tag{1.6.166}$$

Here is another significant example where Proposition 1.6.13 plays a role.

Example 1.6.16 *Fix $n \in \mathbb{N}$ with $n \geq 2$ and consider the scalar, homogeneous, second-order differential operator*

$$L := \partial_1^2 + \cdots + \partial_{n-1}^2 + \mathrm{i}\partial_n^2 \tag{1.6.167}$$

in $\mathbb{R}^n$. Note that $L(\xi) = -(\xi_1^2 + \cdots + \xi_{n-1}^2 + \mathrm{i}\xi_n^2)$ for each $\xi = (\xi_1, \dots, \xi_n) \in \mathbb{R}^n$. In particular, $L(\xi) \neq 0$ for each $\xi \in \mathbb{R}^n \setminus \{0\}$ so L is weakly elliptic. However, L is not strongly elliptic since (1.6.122) *presently fails. Nonetheless, Proposition 1.6.12 applies and gives that if $n \geq 3$ then $\mathfrak{A}_L^{\mathrm{dis}}$ consists precisely of one matrix, namely the symmetric diagonal matrix*

$$A := \begin{pmatrix} 1 & 0 & 0 & \cdots & 0 & 0 \\ 0 & 1 & 0 & \cdots & 0 & 0 \\ \vdots & \vdots & \vdots & \ddots & \vdots & \vdots \\ 0 & 0 & 0 & \cdots & 1 & 0 \\ 0 & 0 & 0 & \cdots & 0 & \mathrm{i} \end{pmatrix} \in \mathbb{C}^{n \times n}. \tag{1.6.168}$$

In the case when $n = 2$, the above matrix becomes

$$A = \begin{pmatrix} 1 & 0 \\ 0 & \mathrm{i} \end{pmatrix} \in \mathbb{C}^{2\times 2} \tag{1.6.169}$$

which may be written as in [131, (1.4.179)] *for the choice*

$$a := 1, \quad \lambda_1 := (1-\mathrm{i})/\sqrt{2}, \quad \lambda_2 := (-1+\mathrm{i})/\sqrt{2}. \tag{1.6.170}$$

Hence, $\mathrm{Im}\,\lambda_1 \cdot \mathrm{Im}\,\lambda_2 < 0$ *which in view of* [131, (1.4.187)] *puts A in M_2^0. As such, Proposition 1.6.13 applies and* (1.6.140) *shows that if $n = 2$ then $\mathfrak{A}_L^{\mathrm{dis}}$ consists precisely of one matrix, namely* (1.6.169).

In summary, the differential operator L defined in (1.6.167) *is weakly elliptic and $\mathfrak{A}_L^{\rm dis}$ consists precisely of one matrix, given in* (1.6.168), *in all dimensions $n \geq 2$.*

Next, the reader is reminded that the notion of mildly elliptic matrix has been introduced in [131, Definition 1.4.10] (see also [131, (1.4.141)]). From Proposition 1.6.13 and [131, Lemma 1.4.21] we then see that:

$$\begin{array}{l}\text{for each weakly elliptic, scalar, homogeneous, second-order operator}\\ L = \operatorname{div} A\nabla \text{ in } \mathbb{R}^2 \text{ with constant complex coefficients, one has } \mathfrak{A}_L^{\rm dis} \neq \varnothing\\ \text{if and only if } A \text{ is mildly elliptic, in which case } \mathfrak{A}_L^{\rm dis} \text{ consists precisely}\\ \text{of one matrix, namely } \operatorname{sym} A := (A + A^\top)/2.\end{array} \tag{1.6.171}$$

We conclude with a remark paving the way for future work.

Remark 1.6.17 *Whenever L is a second-order, homogeneous, constant* (*complex*) *coefficient, weakly elliptic $M \times M$ system in $\mathbb{R}^n$ with the property that $\mathfrak{A}_L^{\rm dis} \neq \varnothing$, it follows from Proposition 1.4.6 that both the L^p Dirichlet Problem and the L_1^p Regularity Problem for the system L in the upper half-space $\mathbb{R}^n_+$ are solvable for each $p \in (1, \infty)$. As a consequence, for any second-order, homogeneous, constant* (*complex*) *coefficient, weakly elliptic system L for which either of the aforementioned boundary value problems fails to be solvable we necessarily have $\mathfrak{A}_L^{\rm dis} = \varnothing$.*

For example, Remark 1.6.17 and the results in [132, §8.1] imply that for second-order, homogeneous, complex constant coefficient, 2×2 weakly elliptic system

$$\mathbf{L}_B := \begin{pmatrix} \partial_{\bar z}^2 & 0 \\ 0 & \partial_{\bar z}^2 \end{pmatrix} \tag{1.6.172}$$

we have

$$\mathfrak{A}_{\mathbf{L}_B}^{\rm dis} = \varnothing. \tag{1.6.173}$$

Chapter 2
Failure of Fredholm Solvability for Weakly Elliptic Systems

It is known (see the discussion in [114]) that the L^p Dirichlet Problem for constant coefficient second-order systems satisfying the Legendre-Hadamard (strong) ellipticity condition is well posed in the upper half-space. We have already seen in [132, Chapter 8] that this may not be the case in the class of weakly elliptic scalar operators in the complex plane. As we shall see in this chapter, counterexamples exist for weak elliptic systems in all space dimensions. In fact, the failure of the corresponding L^p Dirichlet Problems to be well posed is at a fundamental level, in the sense that as we shall see momentarily, there exist weakly elliptic systems in $\mathbb{R}^n$ with $n \geq 2$ for which the L^p Dirichlet Problem is not even Fredholm solvable. The manner in which this ties up with our earlier work in Chapter 1 is that we shall look for such a pathological weakly elliptic system in the class of those which fail to possess a distinguished coefficient tensor.

2.1 Nontangential Boundary Traces in the Upper Half-Space

We start with a lemma that may be regarded as an embodiment of the heuristic principle that tangential derivatives commute with nontangential boundary traces.

Lemma 2.1.1 *Fix an aperture parameter $\kappa > 0$ and select a truncation parameter $\rho \in (0, \infty)$. Suppose $\omega \in W^{1,1}_{\rm loc}(\mathbb{R}^n_+)$ has the property that $\mathcal{N}^{\rho}_{\kappa}\omega$ and $\mathcal{N}^{\rho}_{\kappa}(\nabla\omega)$ belong to $L^1_{\rm loc}(\mathbb{R}^{n-1}, \mathcal{L}^{n-1})$ and $\omega\big|^{\kappa-{\rm n.t.}}_{\partial\mathbb{R}^n_+}$ exists $\mathcal{L}^{n-1}$-a.e. in $\mathbb{R}^{n-1} \equiv \partial\mathbb{R}^n_+$. If also $(\partial_j\omega)\big|^{\kappa-{\rm n.t.}}_{\partial\mathbb{R}^n_+}$ exists $\mathcal{L}^{n-1}$-a.e. in $\mathbb{R}^{n-1}$ for some $j \in \{1, \dots, n-1\}$, then*

$$\partial_j\big(\omega\big|^{\kappa-{\rm n.t.}}_{\partial\mathbb{R}^n_+}\big) = (\partial_j\omega)\big|^{\kappa-{\rm n.t.}}_{\partial\mathbb{R}^n_+} \ \text{ in } \mathcal{D}'(\mathbb{R}^{n-1}). \tag{2.1.1}$$

Thus, if $(\nabla\omega)\big|^{\kappa-{\rm n.t.}}_{\partial\mathbb{R}^n_+}$ exists $\mathcal{L}^{n-1}$-a.e. in $\mathbb{R}^{n-1}$, then $\omega\big|^{\kappa-{\rm n.t.}}_{\partial\mathbb{R}^n_+} \in L^1_{1,{\rm loc}}(\mathbb{R}^{n-1}, \mathcal{L}^{n-1})$.

Proof This is established reasoning as in the proof of [130, Proposition 11.3.2], using the Divergence Theorem in $\mathbb{R}^n_+$ from [129, Corollary 1.2.2]. □

Corollary 2.1.2 *Fix an exponent $p \in [1, \infty)$ along with an aperture parameter $\kappa > 0$ and a truncation parameter $\rho \in (0, \infty)$. Suppose $\omega \in W^{1,1}_{\rm loc}(\mathbb{R}^n_+)$ is such that*

D. Mitrea et al., *Geometric Harmonic Analysis V*, Developments in Mathematics 76,
https://doi.org/10.1007/978-3-031-31561-9_2

$$\mathcal{N}_\kappa^\rho \omega \text{ and } \mathcal{N}_\kappa^\rho(\nabla\omega) \text{ belong to } L^p_{\mathrm{loc}}(\mathbb{R}^{n-1}, \mathcal{L}^{n-1}) \tag{2.1.2}$$

and the boundary traces $\omega\big|_{\partial\mathbb{R}^n_+}^{\kappa-\mathrm{n.t.}}$ *and* $(\nabla\omega)\big|_{\partial\mathbb{R}^n_+}^{\kappa-\mathrm{n.t.}}$ *exist* $\mathcal{L}^{n-1}$*-a.e. in* $\mathbb{R}^{n-1}$*. Then*

$$\omega\big|_{\partial\mathbb{R}^n_+}^{\kappa-\mathrm{n.t.}} \in L^p_{1,\,\mathrm{loc}}(\mathbb{R}^{n-1}, \mathcal{L}^{n-1}). \tag{2.1.3}$$

Also, if in place of (2.1.2) *one assumes that* $\mathcal{N}_\kappa^\rho\omega$ *and* $\mathcal{N}_\kappa^\rho(\nabla\omega)$ *belong to* $L^p(\mathbb{R}^{n-1}, \mathcal{L}^{n-1})$*, then in place of* (2.1.3) *one obtains* $\omega\big|_{\partial\mathbb{R}^n_+}^{\kappa-\mathrm{n.t.}} \in L^p_1(\mathbb{R}^{n-1}, \mathcal{L}^{n-1})$.

Proof This is a consequence of Lemma 2.1.1 and [129, (8.9.8)]. □

The next corollary amounts to a higher-order version of Lemma 2.1.1.

Corollary 2.1.3 *Fix* $p \in (1,\infty)$, $\kappa > 0$, *and* $k \in \mathbb{N}_0$. *Suppose* $\omega \in W^{k,1}_{\mathrm{loc}}(\mathbb{R}^n_+)$ *is such that, for every* $\alpha \in \mathbb{N}_0^n$ *with* $|\alpha| \le k$,

$$\begin{aligned} &\mathcal{N}_\kappa(\partial^\alpha\omega) \text{ belongs to } L^p(\mathbb{R}^{n-1}, \mathcal{L}^{n-1}) \text{ and} \\ &\text{the trace } (\partial^\alpha\omega)\big|_{\partial\mathbb{R}^n_+}^{\kappa-\mathrm{n.t.}} \text{ exists at } \mathcal{L}^{n-1}\text{-a.e. point in } \mathbb{R}^{n-1}. \end{aligned} \tag{2.1.4}$$

Then $\omega\big|_{\partial\mathbb{R}^n_+}^{\kappa-\mathrm{n.t.}} \in L^p_k(\mathbb{R}^{n-1}, \mathcal{L}^{n-1})$ *and for every* $\beta \in \mathbb{N}_0^{n-1}$ *with* $|\beta| \le k$ *one has*

$$\partial^\beta\big(\omega\big|_{\partial\mathbb{R}^n_+}^{\kappa-\mathrm{n.t.}}\big) = (\partial^\beta\omega)\big|_{\partial\mathbb{R}^n_+}^{\kappa-\mathrm{n.t.}} \text{ in } \mathcal{D}'(\mathbb{R}^{n-1}). \tag{2.1.5}$$

Proof From [129, (8.9.8)], [129, (8.9.44)], and (2.1.4) we see that, given $\alpha \in \mathbb{N}_0^n$ with $|\alpha| \le k$, the trace $(\partial^\alpha\omega)\big|_{\partial\mathbb{R}^n_+}^{\kappa-\mathrm{n.t.}}$ belongs to $L^p(\mathbb{R}^{n-1}, \mathcal{L}^{n-1})$. The membership of $\omega\big|_{\partial\mathbb{R}^n_+}^{\kappa-\mathrm{n.t.}}$ to $L^p_k(\mathbb{R}^{n-1}, \mathcal{L}^{n-1})$ follows by induction over k, since Lemma 2.1.1 yields

$$\partial_j\big[(\partial^\alpha\omega)\big|_{\partial\mathbb{R}^n_+}^{\kappa-\mathrm{n.t.}}\big] = (\partial^{\alpha+\mathbf{e}_j}\omega)\big|_{\partial\mathbb{R}^n_+}^{\kappa-\mathrm{n.t.}} \text{ in } \mathcal{D}'(\mathbb{R}^{n-1}). \tag{2.1.6}$$

for all $\alpha \in \mathbb{N}_0^n$ with $|\alpha| \le k-1$ and every $j \in \{1,\dots,n-1\}$. □

Let us now recall Calderón's Theorem regarding the nontangential traces of harmonic function in the upper half-space:

> Let u be a harmonic function in $\mathbb{R}^n_+$ which is nontangentially bounded at every point in a set $E \subseteq \mathbb{R}^{n-1} \equiv \partial\mathbb{R}^n_+$ (in the sense that for each given point $x^o \in E$ one can find some $\kappa_o \in (0,\infty)$ and $h_o \in (0,\infty)$ such that $\sup\{|u(x)| : x = (x', x_n) \in \Gamma_{\kappa_o}(x^o) \text{ and } x_n \in (0, h_o)\} < +\infty$). Then u has a nontangential limit at $\mathcal{L}^{n-1}$-a.e. point in E (in the sense that for each given aperture parameter $\kappa \in (0,\infty)$ the nontangential boundary trace $\big(u\big|_{\partial\mathbb{R}^n_+}^{\kappa-\mathrm{n.t.}}\big)(x)$ exists in $\mathbb{R}$ at $\mathcal{L}^{n-1}$-a.e. point $x \in E$). (2.1.7)

See [26], [198, Theorem 3, p. 201]. We conclude with a version of Corollary 2.1.3 for harmonic functions.

Corollary 2.1.4 *Fix $p \in (1,\infty)$, $\kappa > 0$, and $k \in \mathbb{N}_0$. Suppose $\omega \in \mathscr{C}^\infty(\mathbb{R}^n_+)$ is such that $\Delta\omega = 0$ in $\mathbb{R}^n_+$ and (2.1.4) holds for every multi-index $\alpha \in \mathbb{N}_0^n$ with $|\alpha| \le k$. Then the trace $\omega\big|^{\kappa-\mathrm{n.t.}}_{\partial\mathbb{R}^n_+}$ exists at $\mathcal{L}^{n-1}$-a.e. point in $\mathbb{R}^{n-1}$, belongs to $L^p_k(\mathbb{R}^{n-1},\mathcal{L}^{n-1})$, and (2.1.5) holds for every $\beta \in \mathbb{N}_0^{n-1}$ with $|\beta| \le k$.*

Proof This is a consequence of Corollary 2.1.3 and Calderón's Theorem (cf. (2.1.7)), which presently guarantees that if ω satisfies the current hypotheses then for each $\alpha \in \mathbb{N}_0^n$ with $|\alpha| \le k$ the trace $(\partial^\alpha\omega)\big|^{\kappa-\mathrm{n.t.}}_{\partial\mathbb{R}^n_+}$ exists $\mathcal{L}^{n-1}$-a.e. in $\mathbb{R}^{n-1}$. □

2.2 Conjugate Poisson Kernels

Recall the Poisson kernel for the Laplacian, i.e.,

$$P^\Delta(x') := \frac{2}{\omega_{n-1}}\frac{1}{(1+|x'|^2)^{n/2}} \quad \text{for all } x' \in \mathbb{R}^{n-1}, \tag{2.2.1}$$

and set

$$P_t^\Delta := t^{1-n}P^\Delta(\cdot/t) \text{ for each } t > 0. \tag{2.2.2}$$

Let E_Δ denote the standard fundamental solution for the Laplacian in $\mathbb{R}^n$. For each $r \in \{1,\dots,n\}$ define the conjugate kernel function $Q^{(r)} : \mathbb{R}^{n-1} \to \mathbb{R}$ by setting, at each $x' = (x_1,\dots,x_{n-1}) \in \mathbb{R}^{n-1}$,

$$Q^{(r)}(x') := (\partial_r E_\Delta)(x',1) = \begin{cases} \dfrac{1}{\omega_{n-1}}\cdot\dfrac{x_r}{\big(|x'|^2+1\big)^{n/2}} & \text{if } r \in \{1,\dots,n-1\}, \\[2ex] \dfrac{1}{\omega_{n-1}}\cdot\dfrac{1}{\big(|x'|^2+1\big)^{n/2}} & \text{if } r = n. \end{cases} \tag{2.2.3}$$

Then it follows that for each $r \in \{1,\dots,n\}$ and each $t > 0$, we have

$$Q_t^{(r)}(x') := t^{1-n}Q^{(r)}(x'/t) = (\partial_r E_\Delta)(x',t), \quad \forall\, x' \in \mathbb{R}^{n-1}. \tag{2.2.4}$$

Note that for any $g \in L^p_1(\mathbb{R}^{n-1},\mathcal{L}^{n-1})$ with $p \in (1,\infty)$ we have

$$\partial_t\big[P_t^\Delta * g\big] = -2\sum_{s=1}^{n-1} Q_t^{(s)} * (\partial_s g) \text{ in } \mathbb{R}^{n-1}. \tag{2.2.5}$$

Also, given $f \in L^p(\mathbb{R}^{n-1},\mathcal{L}^{n-1})$ with $p \in (1,\infty)$, if for $s \in \{1,\dots,n\}$ we set

$$u^{(s)} : \mathbb{R}^n_+ \to \mathbb{C}, \quad u^{(s)}(x',t) := \big(Q_t^{(s)} * f\big)(x'), \quad \forall\,(x',t) \in \mathbb{R}^n_+ \tag{2.2.6}$$

then, for each $\kappa > 0$, there exists a constant $C \in (0,\infty)$ independent of f such that

$$\|\mathcal{N}_\kappa u^{(s)}\|_{L^p(\mathbb{R}^{n-1},\mathcal{L}^{n-1})} \leq C\|f\|_{L^p(\mathbb{R}^{n-1},\mathcal{L}^{n-1})}. \tag{2.2.7}$$

Concerning the conjugate kernels from (2.2.4), henceforth we agree to employ the generic symbol Q_t^{I} for any of $Q_t^{(s)}$ with $s \in \{1, \dots, n-1\}$, and the notation Q_t^{II} for $Q_t^{(n)}$. In this language, work from [113] gives the following identities.

Proposition 2.2.1 *Let $p \in (1, \infty)$, $k \in \mathbb{N}_0$, and for some $f \in L_k^p(\mathbb{R}^{n-1}, \mathcal{L}^{n-1})$ define*

$$u(x', t) := (P_t^\Delta * f)(x'), \qquad \forall (x', t) \in \mathbb{R}_+^n. \tag{2.2.8}$$

Then, for every $(x', t) \in \mathbb{R}_+^n$ the following identity holds:

$$\nabla^k u(x', t) \equiv \big(P_t^\Delta * (\partial_\tau^k f)\big)(x') + \big(Q_t^{\mathrm{I}} * (\partial_\tau^k f)\big)(x'), \tag{2.2.9}$$

where the sign $\equiv$ in (2.2.9) is used to indicate that each derivative of order k of $u(x', t)$ may be written as a finite linear combination of terms of the form $\big(P_t^\Delta * (\partial_\tau^k f)\big)(x')$ and $\big(Q_t^{\mathrm{I}} * (\partial_\tau^k f)\big)(x')$, with ∂_τ denoting any of the derivatives $\partial_1, \dots, \partial_{n-1}$, and ∂_τ^k denoting its k-fold iteration.

2.3 Dirichlet-to-Neumann Operators in the Euclidean Space

We begin by considering the Dirichlet-to-Neumann operator, acting from a Sobolev space of order one into the corresponding Lebesgue space.

Lemma 2.3.1 *Let P^Δ be the Poisson kernel for the Laplacian in $\mathbb{R}^{n-1}$ (cf. (2.2.1)) and set $P_t^\Delta := t^{1-n} P^\Delta(\cdot/t)$ for each $t > 0$. Also, fix $p \in (1, \infty)$ and consider the Dirichlet-to-Neumann operator*

$$\Lambda_{\mathrm{DN}} : L_1^p(\mathbb{R}^{n-1}, \mathcal{L}^{n-1}) \longrightarrow L^p(\mathbb{R}^{n-1}, \mathcal{L}^{n-1}) \tag{2.3.1}$$

acting on each $f \in L_1^p(\mathbb{R}^{n-1}, \mathcal{L}^{n-1})$ by

$$\big(\Lambda_{\mathrm{DN}} f\big)(x') := \lim_{t \to 0^+} \Big[\partial_t \big(P_t^\Delta * f\big)(x')\Big] \quad \textit{for } \mathcal{L}^{n-1}\textit{-a.e. } x' \in \mathbb{R}^{n-1}. \tag{2.3.2}$$

Then this is a well-defined, linear, bounded, and injective operator.

Proof Fix an arbitrary function $f \in L_1^p(\mathbb{R}^{n-1}, \mathcal{L}^{n-1})$ and pick some aperture parameter $\kappa > 0$. Recall from [114] that the function

$$u(x', t) := \big(P_t^\Delta * f\big)(x') \text{ for all } x' \in \mathbb{R}^{n-1},\ t > 0, \tag{2.3.3}$$

is the unique solution of the Regularity Problem for the Laplacian in $\mathbb{R}_+^n$:

$$\begin{cases} u \in \mathscr{C}^\infty(\mathbb{R}^n_+), \quad \Delta u = 0 \text{ in } \mathbb{R}^n_+, \\ \mathcal{N}_\kappa u,\ \mathcal{N}_\kappa(\nabla u) \in L^p(\mathbb{R}^{n-1}, \mathcal{L}^{n-1}), \\ u\big|^{\kappa-\text{n.t.}}_{\partial\mathbb{R}^n_+} = f \text{ at } \mathcal{L}^{n-1}\text{-a.e. point on } \mathbb{R}^{n-1}. \end{cases} \tag{2.3.4}$$

Consequently, the expression in (2.3.2) may be written in terms of u as

$$\Lambda_{\rm DN} f = (\partial_n u)\big|^{\kappa-\text{n.t.}}_{\partial\mathbb{R}^n_+} \in L^p(\mathbb{R}^{n-1}, \mathcal{L}^{n-1}). \tag{2.3.5}$$

This proves that the Dirichlet-to-Neumann operator (2.3.1)-(2.3.2) is well defined, linear, and bounded. To show that it is also injective, suppose that f is such that $\Lambda_{\rm DN} f = 0$. In turn, this forces $(\partial_n u)\big|^{\kappa-\text{n.t.}}_{\partial\mathbb{R}^n_+} = 0$. As such, from the integral representation formula [132, Theorem 1.8.19, (1.8.200)] and [132, (1.8.8)] we conclude that u is of the form

$$u = \mathcal{D} f + c \text{ in } \mathbb{R}^n_+, \tag{2.3.6}$$

for some constant $c \in \mathbb{R}$. Taking the nontangential trace to $\partial\mathbb{R}^n_+$ in (2.3.6), and observing that in this setting $\big(\mathcal{D} f\big)\big|^{\kappa-\text{n.t.}}_{\partial\mathbb{R}^n_+} = \frac{1}{2} f$, we arrive at $f = \frac{1}{2} f + c$, thus

$$f \text{ is constant in } \mathbb{R}^{n-1}. \tag{2.3.7}$$

Upon recalling that $f \in L^p(\mathbb{R}^{n-1}, \mathcal{L}^{n-1})$, we conclude that $f = 0$. This proves that $\Lambda_{\rm DN}$ is indeed injective in the context of (2.3.1). □

As a consequence of the proof of this lemma we have the following result:

Corollary 2.3.2 *Let $p \in (1, \infty)$ and consider the Dirichlet-to-Neumann operator from* (2.3.1). *Then for each $f \in L^p_1(\mathbb{R}^{n-1}, \mathcal{L}^{n-1})$ one has*

$$\Lambda_{\rm DN} f = (\partial_n u)\big|^{\kappa-\text{n.t.}}_{\partial\mathbb{R}^n_+} \tag{2.3.8}$$

where u is the unique solution of the Regularity Problem for the Laplacian in $\mathbb{R}^n_+$ with boundary datum f (cf. (2.3.4)).

It is useful to know that each partial derivative factors as the composition between the Dirichlet-to-Neumann map and the corresponding Riesz transform. This is made precise in the next lemma.

Lemma 2.3.3 *Suppose $p \in (1, \infty)$ and let $f \in L^p_1(\mathbb{R}^{n-1}, \mathcal{L}^{n-1})$. Then for each $j \in \{1, \dots, n-1\}$ one has*

$$\Lambda_{\rm DN}(R_j f) = \partial_j f \ \text{ at } \mathcal{L}^{n-1}\text{-a.e. point in } \mathbb{R}^{n-1}, \tag{2.3.9}$$

where R_j is the j-th Riesz transform in $\mathbb{R}^{n-1}$ (defined as in (A.0.215) *with* $\Sigma := \mathbb{R}^{n-1}$).

Proof Fix $j \in \{1, \dots, n-1\}$ and define

$$u(x) := \frac{2}{\omega_{n-1}} \int_{\mathbb{R}^{n-1}} \frac{x_j - y_j}{|x - (y', 0)|^n} f(y')\,dy', \quad \forall x = (x_1, \dots, x_n) \in \mathbb{R}^n_+. \tag{2.3.10}$$

Then for each $k \in \{1, \dots, n\}$ we may integrate by parts and obtain, for each $x \in \mathbb{R}^n_+$,

$$\begin{aligned}(\partial_k u)(x) &= -2 \int_{\mathbb{R}^{n-1}} \partial_{y_j} \big[(\partial_k E_\Delta)(x - (y', 0))\big] f(y')\,\mathrm{d}y' \\ &= 2 \int_{\mathbb{R}^{n-1}} (\partial_k E_\Delta)(x - (y', 0))\,(\partial_j f)(y')\,\mathrm{d}y' = 2\partial_k \mathscr{S}_{\text{mod}}(\partial_j f)(x),\end{aligned} \tag{2.3.11}$$

with the last equality a particular case of [132, (1.5.51)]. Then, as seen from (2.3.10), (2.3.11), [131, (2.4.9), (2.5.4), (2.5.17)], and [132, (1.5.52), (1.5.57)],

$$\begin{aligned}&u \in \mathscr{C}^\infty(\mathbb{R}^n_+),\ \Delta u = 0 \text{ in } \mathbb{R}^n_+,\ \mathcal{N}_\kappa u,\ \mathcal{N}_\kappa(\nabla u) \in L^p(\mathbb{R}^{n-1}, \mathcal{L}^{n-1}), \\ &\qquad \text{and } u\big|^{\kappa-\text{n.t.}}_{\partial\mathbb{R}^n_+} = R_j f \text{ at } \mathcal{L}^{n-1}\text{-a.e. point in } \mathbb{R}^{n-1}.\end{aligned} \tag{2.3.12}$$

Hence, we may invoke Corollary 2.3.2 and (2.3.11) to write

$$\Lambda_{\text{DN}}(R_j f) = (\partial_n u)\big|^{\kappa-\text{n.t.}}_{\partial\mathbb{R}^n_+} = 2(\partial_n \mathscr{S}_{\text{mod}})(\partial_j f)\big|^{\kappa-\text{n.t.}}_{\partial\mathbb{R}^n_+} = -2\big(-\tfrac{1}{2}I + K^\#_\Delta\big)(\partial_j f) = \partial_j f \tag{2.3.13}$$

$\mathcal{L}^{n-1}$-a.e. in $\mathbb{R}^{n-1}$, since, as already noted, $K^\#_\Delta$ is identically zero in this case. □

Next we look at the Dirichlet-to-Neumann map mapping from homogeneous Sobolev spaces, as indicated below.

Lemma 2.3.4 *Let* $p \in (1, \infty)$. *Then the Dirichlet-to-Neumann map defined as in* (2.3.2) *when acting as an operator*

$$\Lambda_{\text{DN}} : \dot{L}^p_1(\mathbb{R}^{n-1}, \mathcal{L}^{n-1}) \longrightarrow L^p(\mathbb{R}^{n-1}, \mathcal{L}^{n-1}) \tag{2.3.14}$$

is well defined, linear, bounded, and surjective. The kernel of the operator (2.3.14) *is the collection of constant functions on* $\mathbb{R}^{n-1}$. *In particular, the mapping*

$$\dot{L}^p_1(\mathbb{R}^{n-1}, \mathcal{L}^{n-1})/\!\sim\, \ni [f] \longmapsto \Lambda_{\text{DN}} f \in L^p(\mathbb{R}^{n-1}, \mathcal{L}^{n-1}) \tag{2.3.15}$$

is an isomorphism.

Proof The fact that the operator in (2.3.14) is well defined and linear is a consequence of the solvability of the Homogeneous Regularity Problem

$$\begin{cases} u \in \mathscr{C}^\infty(\mathbb{R}^n_+), \quad \Delta u = 0 \text{ in } \mathbb{R}^n_+, \\ \mathcal{N}_\kappa(\nabla u) \in L^p(\mathbb{R}^{n-1}, \mathcal{L}^{n-1}), \\ u\big|^{\kappa-\text{n.t.}}_{\partial\Omega} = f \in \dot{L}^p_1(\mathbb{R}^{n-1}). \end{cases} \tag{2.3.16}$$

This problem has a solution, which is unique modulo constants, of the form

$$u(x', t) = \big(P^\Delta_t * f\big)(x') \text{ for each } x' \in \mathbb{R}^{n-1} \text{ and } t > 0. \tag{2.3.17}$$

In particular, this shows that

$$\Lambda_{\rm DN} f = (\partial_n u)\Big|_{\partial\mathbb{R}^n_+}^{\kappa-{\rm n.t.}} \tag{2.3.18}$$

for each $f \in \dot{L}^p_1(\mathbb{R}^{n-1}, \mathcal{L}^{n-1})$, where u is a solution of (2.3.16) with boundary datum f. Hence, $\Lambda_{\rm DN}$ is well defined, linear, and bounded in the context of (2.3.14). The fact that its kernel is the collection of constant functions on $\mathbb{R}^{n-1}$ follows from the argument in the proof of Lemma 2.3.1 that has led to (2.3.7). To show that this map is also surjective, pick $g \in L^p(\mathbb{R}^{n-1}, \mathcal{L}^{n-1})$ and set $u := 2\mathscr{S}_{\rm mod} g$ in $\mathbb{R}^n_+$. In view of [132, (1.5.52), (1.5.57), (1.5.80)], and [132, (1.8.123)], this is a solution of the Homogeneous Regularity Problem (formulated as in (2.3.16)) corresponding to the boundary datum $f = 2S_{\rm mod} g \in \dot{L}^p_1(\mathbb{R}^{n-1}, \mathcal{L}^{n-1})$. As such, based on (2.3.18) and [132, (1.5.58)] we may write

$$\Lambda_{\rm DN} f = \partial_n \big(2\mathscr{S}_{\rm mod} g\big)\Big|_{\partial\mathbb{R}^n_+}^{\kappa-{\rm n.t.}} = -2\big(-\tfrac{1}{2}I + K^{\#}_{\Delta}\big)g = g, \tag{2.3.19}$$

since $K^{\#}_{\Delta}$, the transpose double layer for the Laplacian in $\mathbb{R}^{n-1} \equiv \partial\mathbb{R}^n_+$, vanishes identically. This proves the surjectivity of (2.3.14). Finally, (2.3.15) is immediate from the properties of the mapping (2.3.14). □

A related perspective on the last two equalities in (2.3.19) is to bring to light the fact that, as seen from [132, (1.5.51)] and (2.2.1)-(2.2.2),

$$\begin{gathered} \text{if } \phi \in L^1\Big(\mathbb{R}^{n-1}, \tfrac{dx'}{1+|x'|^{n-1}}\Big) \text{ then} \\ \partial_n\big(2\mathscr{S}_{\rm mod}\phi\big)(x',t) = \big(P^{\Delta}_t * \phi\big)(x') \text{ at each point } (x',t) \in \mathbb{R}^n_+, \end{gathered} \tag{2.3.20}$$

and then recalling from Lemma 1.4.5 that the function $w(x',t) := \big(P^{\Delta}_t * \phi\big)(x')$ for each $(x',t) \in \mathbb{R}^n_+$ satisfies $w\Big|_{\partial\mathbb{R}^n_+}^{\kappa-{\rm n.t.}} = \phi$ at $\mathcal{L}^{n-1}$-a.e. point on $\mathbb{R}^{n-1}$.

Our last result in this section is a basic formula for the composition of the Dirichlet-to-Neumann map with a modified Riesz transform.

Lemma 2.3.5 *Suppose $p \in (1, \infty)$ and let $f \in \dot{L}^p_1(\mathbb{R}^{n-1}, \mathcal{L}^{n-1})$. Then for each $j \in \{1, \dots, n-1\}$ one has*

$$\Lambda_{\rm DN}(R^{\rm mod}_j f) = \partial_j f \ \text{ at } \mathcal{L}^{n-1}\text{-a.e. point in } \mathbb{R}^{n-1}, \tag{2.3.21}$$

where $R^{\rm mod}_j$ is the j-th modified Riesz transform in $\mathbb{R}^{n-1}$ (cf. (A.0.222)).

Proof Fix $j \in \{1, \dots, n-1\}$ and for each $x = (x_1, \dots, x_n) \in \mathbb{R}^n_+$ define

$$u(x) := \frac{2}{\omega_{n-1}} \int_{\mathbb{R}^{n-1}} \left\{ \frac{x_j - y_j}{|x-(y',0)|^n} - \frac{-y_j}{|(y',0)|^n} \mathbf{1}_{\mathbb{R}^{n-1}\setminus \overline{B_{n-1}(0',1)}}(y') \right\} f(y')\, dy'. \tag{2.3.22}$$

For each $k \in \{1, \dots, n\}$ we may integrate by parts to obtain

$$(\partial_k u)(x) = -2\int_{\mathbb{R}^{n-1}} \partial_{y_j}\big[(\partial_k E_\Delta)(x-(y',0))\big] f(y')\,dy' \tag{2.3.23}$$
$$= 2\int_{\mathbb{R}^{n-1}} (\partial_k E_\Delta)(x-(y',0))\,(\partial_j f)(y')\,dy' = 2\partial_k \mathscr{S}_{\rm mod}(\partial_j f)(x)$$

for all $x \in \mathbb{R}^n_+$. From this, (2.3.22), (2.3.23), [132, (1.5.52), (1.5.57)], and [131, (2.5.17), (2.5.31)] we then see that

$$u \in \mathscr{C}^\infty(\mathbb{R}^n_+),\ \Delta u = 0 \text{ in } \mathbb{R}^n_+,\ \mathcal{N}_\kappa(\nabla u) \in L^p(\mathbb{R}^{n-1}, \mathcal{L}^{n-1}),$$
$$\text{and } u\big|^{\kappa-\rm n.t.}_{\partial\mathbb{R}^n_+} = \overset{\rm mod}{R_j} f \text{ at } \mathcal{L}^{n-1}\text{-a.e. point in } \mathbb{R}^{n-1}. \tag{2.3.24}$$

As such, we may rely on (2.3.18) and (2.3.23) to write, $\mathcal{L}^{n-1}$-a.e. in $\mathbb{R}^{n-1}$,

$$\Lambda_{\rm DN}\big(\overset{\rm mod}{R_j} f\big) = (\partial_n u)\big|^{\kappa-\rm n.t.}_{\partial\mathbb{R}^n_+} = 2(\partial_n \mathscr{S}_{\rm mod})(\partial_j f) = -2\big(-\tfrac{1}{2}I + K^\#_\Delta\big)(\partial_j f) = \partial_j f \tag{2.3.25}$$

since, as already noted, $K^\#_\Delta$ is identically zero in this case. □

2.4 Spaces of Vector-Valued Diverge-Free Harmonic Functions

We begin by defining the space of admissible boundary data. Generally speaking, a boundary value problem consists of specifications regarding the size and smoothness of the solution (thus articulating the attributes of the class of functions from which a solution is sought), the actual partial differential equation and, finally, the boundary condition[1].

The boundary datum is selected from a space of functions defined on the boundary, agreed on a priori, hereby dubbed as the "(full) space of boundary data." Typically, the class of functions from which a solution is sought (characterized in terms of size/smoothness and the actual PDE) is such that all functions there have boundary traces belonging to the full space of boundary data. This brings into focus the collection of all boundary traces of functions in the class from which a solution is sought, which we shall refer to as the *space of admissible boundary data*.

Here we concretely identify the space of admissible boundary data for the Dirichlet Problem and the Regularity Problem for vector-valued diverge-free harmonic functions in the upper half-space. As opposed to just plain vector-valued harmonic functions, being diverge-free forces the boundary traces to satisfy a "compatibility condition" involving Riesz transforms, of the sort described in (2.4.1) below.

Proposition 2.4.1 *Fix an integer $n \in \mathbb{N}$ with $n \geq 2$, an integrability exponent $p \in (1,\infty)$, and an aperture parameter $\kappa > 0$. Then*

[1] the latter tacitly assumes the existence of nontangential boundary traces, if the class of functions from which a solution is sought does not imply that already

$$\Big\{\vec{v}\Big|^{\kappa-\mathrm{n.t.}}_{\partial\mathbb{R}^n_+} : \vec{v} \in \big[\mathscr{C}^\infty(\mathbb{R}^n_+)\big]^n,\ \Delta\vec{v} = 0 \ and \ \operatorname{div}\vec{v} = 0 \ in \ \mathbb{R}^n_+,\ \mathcal{N}_\kappa\vec{v} \in L^p(\mathbb{R}^{n-1}, \mathcal{L}^{n-1})\Big\}$$

$$= \Big\{(f_1, \dots, f_n) \in \big[L^p(\mathbb{R}^{n-1}, \mathcal{L}^{n-1})\big]^n : f_n = -\sum_{j=1}^{n-1} R_j f_j\Big\}, \tag{2.4.1}$$

where the R_j's with $1 \leq j \leq n-1$ are the Riesz transforms in $\mathbb{R}^{n-1}$ (defined as in (A.0.215) *with $\Sigma := \mathbb{R}^{n-1}$).*

There is a remarkable physical interpretation of Proposition 2.4.1. To elaborate, recall that the L^p-Dirichlet Problem for the Stokes system of hydrostatic in the upper-half space, modeling the flow of a fluid with velocity $\vec{v}$ and pressure π, reads

$$\begin{cases} \vec{v} \in \big[\mathscr{C}^\infty(\mathbb{R}^n_+)\big]^n, \quad \pi \in \mathscr{C}^\infty(\mathbb{R}^n_+), \\ \Delta\vec{v} - \nabla\pi = 0 \ \text{and} \ \operatorname{div}\vec{v} = 0 \ \text{in} \ \mathbb{R}^n_+, \\ \mathcal{N}_\kappa\vec{v} \in L^p(\mathbb{R}^{n-1}, \mathcal{L}^{n-1}) \\ \vec{v}\Big|^{\kappa-\mathrm{n.t.}}_{\partial\mathbb{R}^n_+} = \vec{f} = (f_1, \dots, f_n) \ \mathcal{L}^{n-1}\text{-a.e. on } \mathbb{R}^{n-1}. \end{cases} \tag{2.4.2}$$

Bearing this in mind, it follows that *formula* (2.4.1) *describes all admissible boundary values the velocity can take in the case when the pressure function is constant; in particular,* (2.4.1) *shows that for a fluid with constant pressure the velocity on the boundary cannot be arbitrary.* As an illustration, (2.4.1) confirms[2] the fact that the trace of the velocity of a fluid with constant pressure cannot be normal to the boundary, unless the fluid is stationary.

After this digression we present the proof of Proposition 2.4.1.

Proof of Proposition 2.4.1 Let $\vec{v} = (v_j)_{1\leq j\leq n} \in \big[\mathscr{C}^\infty(\mathbb{R}^n_+)\big]^n$ be such that $\Delta\vec{v} = 0$ and $\operatorname{div}\vec{v} = 0$ in $\mathbb{R}^n_+$, and $\mathcal{N}_\kappa\vec{v}$ belongs to $L^p(\mathbb{R}^{n-1}, \mathcal{L}^{n-1})$. Then Calderón's Theorem (recalled in (2.1.7)) guarantees the existence of the nontangential trace $\vec{v}\Big|^{\kappa-\mathrm{n.t.}}_{\partial\mathbb{R}^n_+}$ at $\mathcal{L}^{n-1}$-a.e. point in $\mathbb{R}^{n-1}$. Denote $\vec{f} := (f_j)_{1\leq j\leq n} := \vec{v}\Big|^{\kappa-\mathrm{n.t.}}_{\partial\mathbb{R}^n_+}$. Fix $\varepsilon > 0$ and define

$$\vec{v}_\varepsilon := \vec{v}\big(\cdot + \varepsilon\mathbf{e}_n\big) \in \big[\mathscr{C}^\infty\big(\overline{\mathbb{R}^n_+}\big)\big]^n, \quad f_j^{(\varepsilon)} := v_j(\cdot + \varepsilon\mathbf{e}_n)\big|_{\partial\mathbb{R}^n_+} \tag{2.4.3}$$

for each $j \in \{1, \dots, n\}$. Then $\Delta\vec{v}_\varepsilon = 0$ and $\operatorname{div}\vec{v}_\varepsilon = 0$ in $\mathbb{R}^n_+$. Upon observing that $\Gamma_\kappa(x) + \varepsilon\mathbf{e}_n \subseteq \Gamma_\kappa(x)$ for each $x \in \partial\mathbb{R}^n_+$, we also have $\mathcal{N}_\kappa\vec{v}_\varepsilon \leq \mathcal{N}_\kappa\vec{v}$, which implies $\mathcal{N}_\kappa\vec{v}_\varepsilon \in L^p(\mathbb{R}^{n-1}, \mathcal{L}^{n-1})$. Moreover, by Lebesgue's Dominated Convergence Theorem, with uniform domination provided by $\mathcal{N}_\kappa\vec{v} \in L^p(\mathbb{R}^{n-1}, \mathcal{L}^{n-1})$, we obtain

$$f_j^{(\varepsilon)} \longrightarrow f_j \ \text{in} \ L^p(\mathbb{R}^{n-1}, \mathcal{L}^{n-1}) \ \text{as} \ \varepsilon \to 0^+, \ \text{for each} \ j \in \{1, \dots, n\}. \tag{2.4.4}$$

Geometric considerations imply that

there exists $\widetilde{\kappa} > \kappa$ with the property that for each $z \in \partial\mathbb{R}^n_+$ and every $x = (x', x_n) \in \Gamma_\kappa(z)$ we have $B(x, x_n/2) \subseteq \Gamma_{\widetilde{\kappa}}(z)$. (2.4.5)

[2] also bearing in mind the uniqueness for the L^p-Dirichlet Problem for the Laplacian in the upper-half space

In particular,

$$\begin{aligned}&\text{if } z\in\partial\mathbb{R}^n_+ \text{ then whenever } x=(x',x_n)\in\Gamma_\kappa(z)\\ &\text{satisfies } x_n>\varepsilon \text{ we have } B(x,\varepsilon/2)\subseteq\Gamma_{\widetilde{\kappa}}(z).\end{aligned}\tag{2.4.6}$$

Hence, if $z\in\partial\mathbb{R}^n_+$ and $x\in\Gamma_\kappa(z)$ we may use interior estimates for the harmonic function $\vec{v}$ in $\mathbb{R}^n_+$ to write

$$\big|(\nabla\vec{v}_\varepsilon)(x)\big|=\big|(\nabla\vec{v})(x+\varepsilon\mathbf{e}_n)\big|\leq\frac{C}{\varepsilon}\fint_{B(x+\varepsilon\mathbf{e}_n,\varepsilon/2)}|\vec{v}\,|\,\mathrm{d}\mathcal{L}^n\leq C_\varepsilon\big(\mathcal{N}_{\widetilde{\kappa}}\vec{v}\big)(z).\tag{2.4.7}$$

Taking the supremum over all $x\in\Gamma_\kappa(z)$ in (2.4.7) we obtain

$$\mathcal{N}_\kappa\big(\nabla\vec{v}_\varepsilon\big)(z)\leq C_\varepsilon\big(\mathcal{N}_{\widetilde{\kappa}}\vec{v}\big)(z)\ \text{ for all } z\in\partial\mathbb{R}^n_+,\tag{2.4.8}$$

thus $\mathcal{N}_\kappa\big(\nabla\vec{v}_\varepsilon\big)\in L^p(\mathbb{R}^{n-1},\mathcal{L}^{n-1})$. This, Corollary 2.1.2, and (2.4.3) then yield

$$\big(f_j^{(\varepsilon)}\big)_{1\leq j\leq n}=\vec{v}_\varepsilon\big|_{\partial\mathbb{R}^n_+}\in\big[L^p_1(\mathbb{R}^{n-1},\mathcal{L}^{n-1})\big]^n.\tag{2.4.9}$$

Furthermore, at $\mathcal{L}^{n-1}$-a.e. point in $\mathbb{R}^{n-1}$ we have

$$\begin{aligned}\partial_1 f_1^{(\varepsilon)}+\cdots+\partial_{n-1}f_{n-1}^{(\varepsilon)}&=(\partial_1 v_1)(\cdot+\varepsilon\mathbf{e}_n)+\cdots+(\partial_{n-1}v_{n-1})(\cdot+\varepsilon\mathbf{e}_n)\\ &=(\operatorname{div}\vec{v})(\cdot+\varepsilon\mathbf{e}_n)-(\partial_n v_n)(\cdot+\varepsilon\mathbf{e}_n)=-\partial_n\big[v_n(\cdot+\varepsilon\mathbf{e}_n)\big]=-\Lambda_{\mathrm{DN}}\big(f_n^{(\varepsilon)}\big),\end{aligned}\tag{2.4.10}$$

where the last equality uses Corollary 2.3.2. In addition, from Lemma 2.3.3 we know that for each $j\in\{1,\dots,n-1\}$ we have

$$\partial_j f_j^{(\varepsilon)}=\Lambda_{\mathrm{DN}}(R_j f_j^{(\varepsilon)})\ \text{ at } \mathcal{L}^{n-1}\text{-a.e. point in } \mathbb{R}^{n-1}.\tag{2.4.11}$$

Combining (2.4.11) with (2.4.10) then implies

$$\Lambda_{\mathrm{DN}}\Big(\sum_{j=1}^{n-1}R_j f_j^{(\varepsilon)}\Big)=-\Lambda_{\mathrm{DN}}\big(f_n^{(\varepsilon)}\big)\ \text{ at } \mathcal{L}^{n-1}\text{-a.e. point in } \mathbb{R}^{n-1}.\tag{2.4.12}$$

The injectivity of Λ_{DN} (cf. Lemma 2.3.1) then forces

$$\sum_{j=1}^{n-1}R_j f_j^{(\varepsilon)}=-f_n^{(\varepsilon)}\ \text{ at } \mathcal{L}^{n-1}\text{-a.e. point in } \mathbb{R}^{n-1}.\tag{2.4.13}$$

That $\sum_{j=1}^{n-1}R_j f_j=-f_n$ holds at $\mathcal{L}^{n-1}$-a.e. point in $\mathbb{R}^{n-1}$ now follows from (2.4.13), (2.4.4), and the continuity of the Riesz transforms on $L^p(\mathbb{R}^{n-1},\mathcal{L}^{n-1})$. This proves the left-to-right inclusion in (2.4.1). Conversely, suppose

$$(f_1,\dots,f_n)\in\big[L^p(\mathbb{R}^{n-1},\mathcal{L}^{n-1})\big]^n\ \text{ are such that } f_n=-\sum_{j=1}^{n-1}R_j f_j.\tag{2.4.14}$$

Pick an arbitrary $\phi \in \mathscr{C}_c^\infty(\mathbb{R}^{n-1})$ with $\int_{\mathbb{R}^{n-1}} \phi \, d\mathcal{L}^{n-1} = 1$ and for each $\varepsilon > 0$ define $\phi_\varepsilon(x') := \varepsilon^{1-n}\phi(x'/\varepsilon)$, for all $x' \in \mathbb{R}^{n-1}$. Also, for $\varepsilon > 0$, set $f_j^{(\varepsilon)} := \phi_\varepsilon * f_j$ in $\mathbb{R}^{n-1}$, for each $j \in \{1, \dots, n\}$. Given the current assumptions, for each $j \in \{1, \dots, n\}$ we then have

$$f_j^{(\varepsilon)} \in L_1^p(\mathbb{R}^{n-1}, \mathcal{L}^{n-1}) \text{ and } f_j^{(\varepsilon)} \to f_j \text{ in } L^p(\mathbb{R}^{n-1}, \mathcal{L}^{n-1}) \text{ as } \varepsilon \to 0^+. \tag{2.4.15}$$

Moreover,

$$f_n^{(\varepsilon)} = \phi_\varepsilon * f_n = -\sum_{j=1}^{n-1} \phi_\varepsilon * R_j f_j. \tag{2.4.16}$$

We make the claim that

$$\phi_\varepsilon * (R_j f_j) = R_j\big(f_j^{(\varepsilon)}\big) \text{ for each } j \in \{1, \dots, n\}. \tag{2.4.17}$$

To see why this is true, denote by $\mathcal{F}'$ the Fourier transform in $\mathbb{R}^{n-1}$ (cf. (A.0.84)), and for each $\xi' \in \mathbb{R}^{n-1} \setminus \{0\}$, using standard properties of the Fourier transform (see [127, Theorem 4.35(c), p. 135] and [127, (4.9.15), p. 183]), compute

$$\begin{aligned} \mathcal{F}'\big(\phi_\varepsilon * (R_j f_j)\big)(\xi') &= \mathcal{F}'(\phi_\varepsilon)(\xi')\mathcal{F}'(R_j f_j)(\xi') = \mathcal{F}'(\phi_\varepsilon)(\xi')\frac{(-\mathrm{i})\xi_j}{|\xi'|}\mathcal{F}'(f_j)(\xi') \\ &= -\mathrm{i}\frac{\xi_j}{|\xi'|}\mathcal{F}'(\phi_\varepsilon * f_j)(\xi') = \mathcal{F}'\big(R_j(f_j^{(\varepsilon)})\big)(\xi'). \end{aligned} \tag{2.4.18}$$

Now (2.4.17) follows by applying the inverse Fourier transform to the resulting identity in (2.4.18). Combining (2.4.16) and (2.4.17) we arrive at

$$f_n^{(\varepsilon)} = -\sum_{j=1}^{n-1} R_j\big(f_j^{(\varepsilon)}\big), \quad \forall\, \varepsilon > 0. \tag{2.4.19}$$

Next, for each $\varepsilon > 0$ define

$$\vec{v}_\varepsilon(x', t) := \Big(\big(P_t^\Delta * f_j^{(\varepsilon)}\big)(x')\Big)_{1 \le j \le n}, \quad \forall x' \in \mathbb{R}^{n-1}, \ \forall t > 0, \tag{2.4.20}$$

$$\vec{v}(x', t) := \Big(\big(P_t^\Delta * f_j\big)(x')\Big)_{1 \le j \le n}, \quad \forall x' \in \mathbb{R}^{n-1}, \ \forall t > 0. \tag{2.4.21}$$

From the properties of the Poisson kernel for the Laplacian, these definitions, and (2.4.14), it follows that

$$\vec{v}, \vec{v}_\varepsilon \in \big[\mathscr{C}^\infty(\mathbb{R}^n_+)\big]^n, \ \Delta\vec{v} = 0 \text{ and } \Delta\vec{v}_\varepsilon = 0 \text{ in } \mathbb{R}^n_+, \tag{2.4.22}$$

$$\mathcal{N}_\kappa\vec{v}, \ \mathcal{N}_\kappa\vec{v}_\varepsilon, \ \mathcal{N}_\kappa\big(\nabla\vec{v}_\varepsilon\big) \in L^p(\mathbb{R}^{n-1}, \mathcal{L}^{n-1}), \tag{2.4.23}$$

$$\vec{v}\Big|_{\partial\mathbb{R}^n_+}^{\kappa-\mathrm{n.t.}} = \big(f_j\big)_{1 \le j \le n} \text{ at } \mathcal{L}^{n-1}\text{-a.e. point in } \mathbb{R}^{n-1}. \tag{2.4.24}$$

To conclude that $\vec{v}$ thus constructed is contained in the space in the left-hand side of (2.4.1), it remains to show that $\operatorname{div}\vec{v} = 0$ in $\mathbb{R}^n_+$. To this end, we claim:

$$\vec{v}_\varepsilon \longrightarrow \vec{v} \text{ uniformly on compact sets in } \mathbb{R}^n_+ \text{ as } \varepsilon \longrightarrow 0^+, \tag{2.4.25}$$

$$\text{and } \operatorname{div}\vec{v}_\varepsilon = 0 \text{ in } \mathbb{R}^n_+, \text{ for each } \varepsilon > 0. \tag{2.4.26}$$

Assume for now that these claims are true. Then for each $\varphi \in \mathscr{C}^\infty_c(\mathbb{R}^n_+)$ we may write

$$\begin{aligned} {}_{\mathcal{D}'(\mathbb{R}^n_+)}\langle \operatorname{div}\vec{v}, \varphi\rangle_{\mathcal{D}(\mathbb{R}^n_+)} &= -{}_{[\mathcal{D}'(\mathbb{R}^n_+)]^n}\langle \vec{v}, \nabla\varphi\rangle_{[\mathcal{D}(\mathbb{R}^n_+)]^n} = -\int_{\mathbb{R}^n_+} \vec{v}\cdot\nabla\varphi \,\mathrm{d}\mathcal{L}^n \\ &= -\lim_{\varepsilon\to 0^+}\int_{\mathbb{R}^n_+} \vec{v}_\varepsilon\cdot\nabla\varphi\,\mathrm{d}\mathcal{L}^n = \lim_{\varepsilon\to 0^+}\int_{\mathbb{R}^n_+}(\operatorname{div}\vec{v}_\varepsilon)\varphi\,\mathrm{d}\mathcal{L}^n = 0. \end{aligned} \tag{2.4.27}$$

In turn, (2.4.27) implies $\operatorname{div}\vec{v} = 0$ in $\mathcal{D}'(\mathbb{R}^n_+)$, which further yields $\operatorname{div}\vec{v} = 0$ pointwise in $\mathbb{R}^n_+$, as wanted. To complete the proof of the proposition we are left with justifying (2.4.25) and (2.4.26). We will first prove (2.4.25). Let K be a compact set in $\mathbb{R}^n_+$. Then there exists some $\varepsilon_0 > 0$ such that $K \subseteq \{(x', t) \in \mathbb{R}^n_+ : x' \in \mathbb{R}^{n-1}, t > \varepsilon_0\}$. If $g \in L^p(\mathbb{R}^{n-1}, \mathcal{L}^{n-1})$ is arbitrary, we may use Hölder's inequality to estimate

$$\begin{aligned} \sup_{(x',t)\in K}\big|(P^\Delta_t * g)(x')\big| &\le \sup_{(x',t)\in K}\Big[\|g\|_{L^p(\mathbb{R}^{n-1},\mathcal{L}^{n-1})}\big\|P^\Delta_t\big\|_{L^{p'}(\mathbb{R}^{n-1},\mathcal{L}^{n-1})}\Big] \\ &= \|g\|_{L^p(\mathbb{R}^{n-1},\mathcal{L}^{n-1})}\sup_{(x',t)\in K} t^{-(n-1)/p}\big\|P^\Delta\big\|_{L^{p'}(\mathbb{R}^{n-1},\mathcal{L}^{n-1})} \\ &\le C(n,p,\varepsilon_0)\|g\|_{L^p(\mathbb{R}^{n-1},\mathcal{L}^{n-1})}, \end{aligned} \tag{2.4.28}$$

where $p' := (1 - 1/p)^{-1}$. The estimate in (2.4.28) may then be used to write

$$\sup_K \big|\vec{v}_\varepsilon - \vec{v}\big| \le C(n,p,\varepsilon_0)\sum_{j=1}^n \|f^{(\varepsilon)}_j - f_j\|_{L^p(\mathbb{R}^{n-1},\mathcal{L}^{n-1})}, \quad \forall\,\varepsilon > \varepsilon_0. \tag{2.4.29}$$

Together, (2.4.29) and (2.4.15) yield (2.4.25). As for (2.4.26), start with (2.4.20), then invoke (2.3.2), Corollary 2.3.2, and Lemma 2.3.3 to write, for $\mathcal{L}^{n-1}$-a.e. $x' \in \mathbb{R}^{n-1}$,

$$\begin{aligned} (\operatorname{div}\vec{v}_\varepsilon)\Big|^{\kappa-\mathrm{n.t.}}_{\partial\mathbb{R}^n_+}(x',0) &= \sum_{j=1}^{n-1}\lim_{t\to 0^+}\Big[\big(P^\Delta_t * (\partial_j f^{(\varepsilon)}_j)\big)(x')\Big] + \lim_{t\to 0^+}\Big[\partial_t\big(P^\Delta_t * f^{(\varepsilon)}_n\big)(x')\Big] \\ &= \sum_{j=1}^{n-1}(\partial_j f^{(\varepsilon)}_j)(x') + \big(\Lambda_{\mathrm{DN}} f^{(\varepsilon)}_n\big)(x') = \sum_{j=1}^{n-1}\Lambda_{\mathrm{DN}}\big(R_j f^{(\varepsilon)}_j\big)(x') + \big(\Lambda_{\mathrm{DN}} f^{(\varepsilon)}_n\big)(x') = 0, \end{aligned} \tag{2.4.30}$$

with the last step using (2.4.19). Since $\operatorname{div}\vec{v}_\varepsilon \in \mathscr{C}^\infty(\mathbb{R}^n_+)$ is harmonic and

$$\mathcal{N}_\kappa\big(\operatorname{div}\vec{v}_\varepsilon\big) \le C\mathcal{N}_\kappa\big(\nabla\vec{v}_\varepsilon\big) \in L^p(\mathbb{R}^{n-1},\mathcal{L}^{n-1}), \tag{2.4.31}$$

now (2.4.26) follows by invoking the uniqueness of solution for the L^p Dirichlet Problem for the Laplacian in $\mathbb{R}^n_+$ (see, e.g., [114]). □

As a consequence of the Proposition 2.4.1 we have the following corollary, which squarely attributes the "compatibility condition" satisfied by the boundary data $(f_1, \dots, f_n)$ (appearing in the space in the right side of (2.4.1)) to the divergence-free property imposed on the harmonic vector field $\vec{v}$. Indeed, without said divergence-free property, the collection of boundary traces of harmonic vector fields in $\mathbb{R}^n_+$ possessing p-th power integrable nontangential maximal functions would simply be the entire space $\big[L^p(\mathbb{R}^{n-1}, \mathcal{L}^{n-1})\big]^n$ (in light of the well-posedness of the L^p Dirichlet Problem for the Laplacian in $\mathbb{R}^n_+$).

Corollary 2.4.2 *Fix an integer $n \in \mathbb{N}$ with $n \geq 2$, an exponent $p \in (1, \infty)$, and an aperture parameter $\kappa > 0$. Let $\vec{v} \in \big[\mathscr{C}^\infty(\mathbb{R}^n_+)\big]^n$ be such that $\Delta\vec{v} = 0$ in $\mathbb{R}^n_+$ and $\mathcal{N}_\kappa \vec{v} \in L^p(\mathbb{R}^{n-1}, \mathcal{L}^{n-1})$. Then $(f_1, \dots, f_n) := \vec{v}\big|^{\kappa-\mathrm{n.t.}}_{\partial\mathbb{R}^n_+}$ is a well-defined vector field with components in $L^p(\mathbb{R}^{n-1}, \mathcal{L}^{n-1})$ and*

$$\operatorname{div} \vec{v} = 0 \text{ in } \mathbb{R}^n_+ \iff f_n = -\sum_{j=1}^{n-1} R_j f_j. \tag{2.4.32}$$

In particular, for any $\vec{f} := (f_1, \dots, f_n) \in \big[L^p(\mathbb{R}^{n-1}, \mathcal{L}^{n-1})\big]^n$ one has

$$\operatorname{div}\big[(P^\Delta_{x_n} * \vec{f})(x')\big] = 0 \text{ for all } x = (x', x_n) \in \mathbb{R}^n_+ \iff f_n = -\sum_{j=1}^{n-1} R_j f_j. \tag{2.4.33}$$

Proof The existence of the nontangential trace $\vec{v}\big|^{\kappa-\mathrm{n.t.}}_{\partial\mathbb{R}^n_+}$ at $\mathcal{L}^{n-1}$-a.e. point in $\mathbb{R}^{n-1}$ is ensured by Calderón's Theorem (cf. (2.1.7)). As far as the equivalence (2.4.32) is concerned, the left-to-right implication is seen from (2.4.1). In the opposite direction, if $f_n = -\sum_{j=1}^{n-1} R_j f_j$ then Proposition 2.4.1 ensures that there exists a vector-valued function $\vec{v}_* \in \big[\mathscr{C}^\infty(\mathbb{R}^n_+)\big]^n$ satisfying $\Delta\vec{v}_* = 0$ and $\operatorname{div}\vec{v}_* = 0$ in $\mathbb{R}^n_+$, as well as $\mathcal{N}_\kappa \vec{v}_* \in L^p(\mathbb{R}^{n-1}, \mathcal{L}^{n-1})$ and $\vec{v}_*\big|^{\kappa-\mathrm{n.t.}}_{\partial\mathbb{R}^n_+} = (f_1, \dots, f_n)$. The latter property entails $\vec{v}_*\big|^{\kappa-\mathrm{n.t.}}_{\partial\mathbb{R}^n_+} = \vec{v}\big|^{\kappa-\mathrm{n.t.}}_{\partial\mathbb{R}^n_+}$ at $\mathcal{L}^{n-1}$-a.e. point in $\mathbb{R}^{n-1}$ which, by virtue of the uniqueness for the L^p Dirichlet Problem for the Laplacian in the upper half-space, forces $\vec{v} = \vec{v}_*$ in $\mathbb{R}^n_+$. Consequently, $\operatorname{div}\vec{v} = \operatorname{div}\vec{v}_* = 0$ in $\mathbb{R}^n_+$, as wanted.

Finally, (2.4.33) is a consequence of (2.4.32) applied to the vector-valued function $\vec{v}(x) := (P^\Delta_{x_n} * \vec{f})(x')$ for each $x = (x', x_n) \in \mathbb{R}^n_+$. □

Here is a version of Proposition 2.4.1 for diverge-free harmonic vector fields in $\mathbb{R}^n_+$ with the nontangential maximal function of their *gradients* p-th power integrable. As such, the "compatibility condition" satisfied by their boundary traces now involves the modified Riesz transforms (cf. (A.0.222)).

Proposition 2.4.3 *Pick an integer $n \in \mathbb{N}$ satisfying $n \geq 2$, along with an integrability exponent $p \in (1, \infty)$ and an aperture parameter $\kappa > 0$. Then*

$$\Big\{\vec{v}\big|^{\kappa-\mathrm{n.t.}}_{\partial\mathbb{R}^n_+} : \vec{v} \in \big[\mathscr{C}^\infty(\mathbb{R}^n_+)\big]^n,\ \Delta\vec{v} = 0 \textit{ and } \operatorname{div}\vec{v} = 0 \textit{ in } \mathbb{R}^n_+, \qquad\qquad \mathcal{N}_\kappa(\nabla\vec{v}) \in L^p(\mathbb{R}^{n-1}, \mathcal{L}^{n-1})\Big\}$$
$$= \Big\{(f_1, \dots, f_n) \in \big[\dot{L}^p_1(\mathbb{R}^{n-1}, \mathcal{L}^{n-1})\big]^n : f_n = -\sum_{j=1}^{n-1} R_j^{\text{mod}} f_j + c,\ c \in \mathbb{C}\Big\}. \tag{2.4.34}$$

Proof We start by recalling from Lemma 2.3.5 that

$$\Lambda_{\mathrm{DN}}(R_j^{\text{mod}} f) = \partial_j f \text{ at } \mathcal{L}^{n-1}\text{-a.e. point in } \mathbb{R}^{n-1} \text{ for each index } j \in \{1, \dots, n-1\} \text{ and each function } f \in \dot{L}^p_1(\mathbb{R}^{n-1}, \mathcal{L}^{n-1}). \tag{2.4.35}$$

Next let $\vec{v} \in \big[\mathscr{C}^\infty(\mathbb{R}^n_+)\big]^n$ be such that

$$\Delta\vec{v} = 0, \quad \operatorname{div}\vec{v} = 0 \text{ in } \mathbb{R}^n_+ \text{ and } \mathcal{N}_\kappa(\nabla\vec{v}) \in L^p(\mathbb{R}^{n-1}, \mathcal{L}^{n-1}). \tag{2.4.36}$$

Then [132, (1.8.199)] guarantees that

$$\vec{f} = (f_j)_{1\le j\le n} := \vec{v}\big|^{\kappa-\mathrm{n.t.}}_{\partial\mathbb{R}^n_+} \text{ exists and belongs to } \big[\dot{L}^p_1(\mathbb{R}, \mathcal{L}^{n-1})\big]^n. \tag{2.4.37}$$

Fix $j \in \{1, \dots, n-1\}$ and consider $\vec{\omega}^{(j)} := \partial_j\vec{v}$. Then $\vec{\omega}^{(j)} \in \big[\mathscr{C}^\infty(\mathbb{R}^n_+)\big]^n$ satisfies

$$\Delta\vec{\omega}^{(j)} = 0 \text{ and } \operatorname{div}\vec{\omega}^{(j)} = 0 \text{ in } \mathbb{R}^n_+, \quad \mathcal{N}_\kappa(\vec{\omega}^{(j)}) \in L^p(\mathbb{R}^{n-1}, \mathcal{L}^{n-1}), \tag{2.4.38}$$

and

$$\vec{\omega}^{(j)}\big|^{\kappa-\mathrm{n.t.}}_{\partial\mathbb{R}^n_+} = \begin{cases} \partial_j\vec{f} & \text{if } 1 \le j \le n-1, \\ \Lambda_{\mathrm{DN}}\vec{f} & \text{if } j = n, \end{cases} \tag{2.4.39}$$

where Λ_{DN} is the Dirichlet-to-Neumann map from (2.3.14). As such, Proposition 2.4.1 applies and gives that

$$\partial_j f_n = -\sum_{k=1}^{n-1} R_k(\partial_j f_k), \quad \forall j \in \{1, \dots, n-1\}. \tag{2.4.40}$$

In view of the fact that we have $R_k(\partial_j f_k) = \partial_j R_k^{\text{mod}} f_k$ for each $j, k \in \{1, \dots, n-1\}$ (cf. [132, (5.3.85)]), from (2.4.40) we ultimately obtain $f_n = -\sum_{j=1}^{n-1} R_j^{\text{mod}} f_j + c$ for some constant $c \in \mathbb{C}$.

Conversely, pick $(f_1, \dots, f_n) \in \big[\dot{L}^p_1(\mathbb{R}, \mathcal{L}^{n-1})\big]^n$ with $f_n = -\sum_{j=1}^{n-1} R_j^{\text{mod}} f_j + c$ for some $c \in \mathbb{C}$. Define $\vec{v}(x', t) := (P_t^\Delta * \vec{f})(x')$ for all $(x', t) \in \mathbb{R}^n_+$. Then $\vec{v} \in \big[\mathscr{C}^\infty(\mathbb{R}^n_+)\big]^n$, $\Delta\vec{v} = 0$ in $\mathbb{R}^n_+$, $\mathcal{N}_\kappa(\nabla\vec{v}) \in L^p(\mathbb{R}^{n-1}, \mathcal{L}^{n-1})$, $\vec{v}\big|^{\kappa-\mathrm{n.t.}}_{\partial\mathbb{R}^n_+} = \vec{f}$, and

$$
\begin{aligned}
(\operatorname{div}\vec{v})\Big|^{\kappa-\text{n.t.}}_{\partial\mathbb{R}^n_+} &= \sum_{j=1}^{n-1}(\partial_j v_j)\Big|^{\kappa-\text{n.t.}}_{\partial\mathbb{R}^n_+} + (\partial_n v_n)\Big|^{\kappa-\text{n.t.}}_{\partial\mathbb{R}^n_+} = \sum_{j-1}^{n-1}\partial_j\big(v_j\Big|^{\kappa-\text{n.t.}}_{\partial\mathbb{R}^n_+}\big) + \Lambda_{\text{DN}} f_n \\
&= \sum_{j-1}^{n-1}\partial_j f_j + \Lambda_{\text{DN}} f_n = \Lambda_{\text{DN}}\Big(\sum_{j-1}^{n-1} R_j^{\text{mod}} f_j + f_n\Big) = \Lambda_{\text{DN}}\, c = 0, \qquad (2.4.41)
\end{aligned}
$$

where for the fourth equality we used Lemma 2.3.5. Since also $\Delta(\operatorname{div}\vec{v}) = 0$ in $\mathbb{R}^n_+$ and $\mathcal{N}_\kappa(\operatorname{div}\vec{v})$ is contained in $L^p(\mathbb{R}^{n-1},\mathcal{L}^{n-1})$, uniqueness for the L^p Dirichlet Problem for the Laplacian in $\mathbb{R}^n_+$ implies $\operatorname{div}\vec{v} = 0$ in $\mathbb{R}^n_+$. Consequently, $\vec{f} = \vec{v}\big|^{\kappa-\text{n.t.}}_{\partial\mathbb{R}^n_+}$ for some function $\vec{v} \in \big[\mathscr{C}^\infty(\mathbb{R}^n_+)\big]^n$ satisfying $\Delta\vec{v} = 0$ and $\operatorname{div}\vec{u} = 0$ in $\mathbb{R}^n_+$, as well as $\mathcal{N}_\kappa(\nabla\vec{v}) \in L^p(\mathbb{R}^{n-1},\mathcal{L}^{n-1})$. □

Our next result describes the space of admissible boundary data for the Higher-Order Regularity Problem for divergence-free harmonic vector fields in $\mathbb{R}^n_+$.

Proposition 2.4.4 *Fix $n \in \mathbb{N}$ with $n \geq 2$, an integrability exponent $p \in (1,\infty)$, an aperture parameter $\kappa > 0$, and an integer $k \in \mathbb{N}_0$. Then the space*

$$
\Big\{\vec{v}\Big|^{\kappa-\text{n.t.}}_{\partial\mathbb{R}^n_+} : \vec{v} \in \big[\mathscr{C}^\infty(\mathbb{R}^n_+)\big]^n,\ \Delta\vec{v} = 0 \text{ and } \operatorname{div}\vec{v} = 0 \text{ in } \mathbb{R}^n_+, \qquad (2.4.42)
$$
$$
\mathcal{N}_\kappa(\partial^\alpha\vec{v}) \in L^p(\mathbb{R}^{n-1},\mathcal{L}^{n-1}) \text{ for all } \alpha \in \mathbb{N}_0^n \text{ with } |\alpha| \leq k\Big\}
$$

coincides with the space

$$
\Big\{(f_1,\dots,f_n) \in \big[L^p_k(\mathbb{R}^{n-1},\mathcal{L}^{n-1})\big]^n : f_n = -\sum_{j=1}^{n-1} R_j f_j\Big\}. \qquad (2.4.43)
$$

Proof That (2.4.42) is included in (2.4.43) follows from Corollary 2.1.4 and Proposition 2.4.1. To prove the opposite inclusion, pick $(f_1,\dots,f_n) \in \big[L^p_k(\mathbb{R}^{n-1},\mathcal{L}^{n-1})\big]^n$ satisfying $f_n = -\sum_{j=1}^{n-1} R_j f_j$, and define

$$
\vec{v}(x',t) := \Big((P^\Delta_t * f_j)(x')\Big)_{1\leq j\leq n} \quad \text{for all } x' \in \mathbb{R}^{n-1} \text{ and } t > 0. \qquad (2.4.44)
$$

Then

$$
\vec{v} \in \big[\mathscr{C}^\infty(\mathbb{R}^n_+)\big]^n,\ \Delta\vec{v} = 0 \text{ in } \mathbb{R}^n_+, \text{ and } \mathcal{N}_\kappa\vec{v} \in L^p(\mathbb{R}^{n-1},\mathcal{L}^{n-1}). \qquad (2.4.45)
$$

Next, let $\alpha \in \mathbb{N}_0^n$ be such that $|\alpha| \leq k$. By Proposition 2.2.1, we have

$$
(\partial^\alpha\vec{v})(x',t) \equiv \Big(\big(P^\Delta_t * (\partial^k_\tau f_j)\big)(x') + \big(Q^{\text{I}}_t * (\partial^k_\tau f_j)\big)(x')\Big)_{1\leq j\leq n} \qquad (2.4.46)
$$

for all $(x',t) \in \mathbb{R}^n_+$. Invoking (2.2.5), (2.2.7), and the fact that the n-tuple $(f_1,\dots,f_n)$ belongs to $\big[L^p_k(\mathbb{R}^{n-1},\mathcal{L}^{n-1})\big]^n$ we may conclude

$$\mathcal{N}_\kappa(\partial^\alpha \vec{v}) \in L^p(\mathbb{R}^{n-1}, \mathcal{L}^{n-1}), \text{ for all } \alpha \in \mathbb{N}_0^n, \ |\alpha| \leq k. \tag{2.4.47}$$

Also, for each $\ell \in \{1, \dots, n-1\}$, we have

$$(\partial_\ell \vec{v})(x', t) = \Big((P_t^\Delta * \partial_\ell f_j)(x') \Big)_{1 \leq j \leq n} \quad \text{for all } x' \in \mathbb{R}^{n-1}, \ t > 0. \tag{2.4.48}$$

In addition, recall that for each function $g \in L^p(\mathbb{R}^{n-1}, \mathcal{L}^{n-1})$, the Poisson convolution $u(x', t) := (P_t^\Delta * g)(x')$, for each $(x', t) \in \mathbb{R}^n_+$, satisfies $u\big|_{\partial \mathbb{R}^n_+}^{\kappa-\text{n.t.}} = g$. Based on this, (2.4.48), (2.3.2), and Lemma 2.3.3, we further see that

$$(\operatorname{div} \vec{v})\big|_{\partial \mathbb{R}^n_+}^{\kappa-\text{n.t.}} = \partial_1 f_1 + \cdots + \partial_{n-1} f_{n-1} + \Lambda_{\text{DN}} f_n = \sum_{j=1}^{n-1} \Lambda_{\text{DN}} \big(R_j f_j \big) + \Lambda_{\text{DN}} f_n = 0. \tag{2.4.49}$$

By the uniqueness of the L^p Dirichlet Problem for the Laplacian in $\mathbb{R}^n_+$ (cf., e.g., [114]) we conclude $\operatorname{div} \vec{v} = 0$ in $\mathbb{R}^n_+$. Since $\vec{v}\big|_{\partial \mathbb{R}^n_+}^{\kappa-\text{n.t.}} = (f_1, \dots, f_n)$, this completes the proof of Proposition 2.4.4. □

Our last theorem in this section shows that the boundary-to-domain vector Riesz transform is an isomorphism mapping functions defined on $\partial \mathbb{R}^n_+$ onto the space of gradients of harmonic functions in $\mathbb{R}^n_+$.

Theorem 2.4.5 *Recall the boundary-to-domain vector-Riesz transform in $\mathbb{R}^n_+$ defined for each $f \in L^p(\mathbb{R}^{n-1}, \mathcal{L}^{n-1})$ by*

$$\mathcal{R} f(x) := \frac{2}{\omega_{n-1}} \int_{\mathbb{R}^{n-1}} \frac{x - (y', 0)}{|x - (y', 0)|^n} f(y') \, dy', \quad \forall x \in \mathbb{R}^n_+. \tag{2.4.50}$$

Then $\mathcal{R}$ is an isomorphism from $L^p(\mathbb{R}^{n-1}, \mathcal{L}^{n-1})$ into

$$\Big\{ \nabla w : \ w \in \mathscr{C}^\infty(\mathbb{R}^n_+), \ \Delta w = 0 \ \textit{in} \ \mathbb{R}^n_+, \ \mathcal{N}_\kappa(\nabla w) \in L^p(\mathbb{R}^{n-1}, \mathcal{L}^{n-1}) \Big\}. \tag{2.4.51}$$

Proof Select an arbitrary function $f \in L^p(\mathbb{R}^{n-1}, \mathcal{L}^{n-1})$. Since thanks to (2.4.50) and [132, (1.5.51)] we have $\mathcal{R} f = \nabla(\mathscr{S}_{\text{mod}} f)$, the properties of the operator $\mathscr{S}_{\text{mod}}$ recorded in [132, (1.5.52), (1.5.57)] ensure that the image of $\mathcal{R}$ when acting on $L^p(\mathbb{R}^{n-1}, \mathcal{L}^{n-1})$ is contained in the set in (2.4.51). In the converse direction, given any function $w \in \mathscr{C}^\infty(\mathbb{R}^n_+)$ with $\Delta w = 0$ in $\mathbb{R}^n_+$ and $\mathcal{N}_\kappa(\nabla w) \in L^p(\mathbb{R}^{n-1}, \mathcal{L}^{n-1})$, we know from [132, (1.8.199)] that the trace $(\nabla w)\big|_{\partial \mathbb{R}^n_+}^{\kappa-\text{n.t.}}$ exists at $\mathcal{L}^{n-1}$-a.e. point in $\mathbb{R}^{n-1}$. If we now set $f := (\partial_n w)\big|_{\partial \mathbb{R}^n_+}^{\kappa-\text{n.t.}}$, then $f \in L^p(\mathbb{R}^{n-1}, \mathcal{L}^{n-1})$ and from the well-posedness of the Neumann Problem for the Laplacian in $\mathbb{R}^n_+$ we obtain $w = \mathscr{S}_{\text{mod}} f + c$ in $\mathbb{R}^n_+$. Hence, $\nabla w = \mathcal{R} f$ in $\mathbb{R}^n_+$ which proves that the image of $\mathcal{R}$ is the set in (2.4.51). It remain to prove that this operator is injective. To this end, let $f \in L^p(\mathbb{R}^{n-1}, \mathcal{L}^{n-1})$ be such that $\mathcal{R} f = 0$ in $\mathbb{R}^n_+$, thus $\nabla(\mathscr{S}_{\text{mod}} f) = 0$ in $\mathbb{R}^n_+$. Then $\mathscr{S}_{\text{mod}} f = c$ in $\mathbb{R}^n_+$, which after taking the nontangential trace to $\partial \mathbb{R}^n_+$ implies $S_{\text{mod}} f = c$ in $\mathbb{R}^{n-1}$. Then [132, Proposition 1.8.11] implies that $f = 0$, as wanted. □

2.5 A Special System $L_D = \Delta - 2\nabla\mathrm{div}$ and Structure Theorems

Recall from (1.6.109) the special system L_D, namely the homogeneous, constant real coefficient, symmetric, $n \times n$ second-order system acting on each vector-valued distribution $\vec{u} = (u_1, \dots, u_n)$ (defined in an open subset of $\mathbb{R}^n$) according to

$$L_D\vec{u} := \Delta\vec{u} - 2\nabla\mathrm{div}\,\vec{u}, \tag{2.5.1}$$

where the Laplacian is applied componentwise. Hence, a concrete representation of L_D using a particular coefficient tensor is

$$\begin{aligned} L_D &= \left(a_{jk}^{\alpha\beta}\partial_j\partial_k\right)_{1\leq\alpha,\beta\leq n} \text{ with} \\ a_{jk}^{\alpha\beta} &= \delta_{jk}\delta_{\alpha\beta} - 2\delta_{j\alpha}\delta_{k\beta} \text{ for all } \alpha, \beta, j, k \in \{1, \dots, n\}. \end{aligned} \tag{2.5.2}$$

As seen from (1.6.86), this is weakly elliptic. However, L_D fails to satisfy the Legendre-Hadamard strong ellipticity condition. Indeed, with $A = \left(a_{jk}^{\alpha\beta}\right)_{\substack{1\leq j,k\leq n\\ 1\leq\alpha,\beta\leq n}}$ as in (2.5.2), for each $\xi \in \mathbb{R}^n$ and $\zeta \in \mathbb{C}^n$ we have

$$a_{jk}^{\alpha\beta}\xi_j\xi_k\zeta_\beta\overline{\zeta_\alpha} = |\xi|^2|\zeta|^2 - 2\left|\langle\xi, \zeta\rangle\right|^2. \tag{2.5.3}$$

Since the latter expression fails to be strictly positive when, e.g., $\zeta = \xi \in \mathbb{R}^n \setminus \{0\}$, the Legendre-Hadamard strong ellipticity condition [131, (1.3.4) in Definition 1.3.2] presently fails to materialize.

One of the main goals in this section is to reveal the hidden structure of null-solutions of the system L_D by decomposing such vector fields into structurally simpler building blocks which, in turn, permit us to derive important information regarding the nature of the boundary value problems for the system L_D. We shall carry out the latter step in the next section, while here we focus on the task of identifying the aforementioned constitutive components and describing their basic properties. Our first such "structure theorem," which plays an important role in our subsequent work, reads as follows:

Theorem 2.5.1 *Fix $n \in \mathbb{N}$ with $n \geq 2$, along with an exponent $p \in (1, \infty)$, and an aperture parameter $\kappa \in (0, \infty)$. Then any vector-valued function $\vec{u}$ satisfying*

$$\begin{cases} \vec{u} \in \left[\mathscr{C}^\infty(\mathbb{R}^n_+)\right]^n, \quad \Delta\vec{u} - 2\nabla\mathrm{div}\,\vec{u} = 0 \text{ in } \mathbb{R}^n_+, \\ \mathcal{N}_\kappa(\nabla\vec{u}) \in L^p(\mathbb{R}^{n-1}, \mathcal{L}^{n-1}), \end{cases} \tag{2.5.4}$$

is of the form

$$\vec{u}(x) = \vec{v}(x) + x_n(\nabla w)(x), \quad \forall x = (x', x_n) \in \mathbb{R}^{n-1} \times (0, \infty), \tag{2.5.5}$$

for some scalar function w satisfying[3]

[3] Parenthetically, we note that if w is as in (2.5.6), then the scalar components $(\partial_j w)_{1\leq j\leq n}$ of ∇w satisfy the Moisil-Teodorescu system (or generalized Cauchy-Riemann equations) in $\mathbb{R}^n_+$.

$$\begin{cases} w \in \mathscr{C}^\infty(\mathbb{R}^n_+), \quad \Delta w = 0 \ \textit{in} \ \mathbb{R}^n_+, \\ \mathcal{N}_\kappa(\nabla w) \in L^p(\mathbb{R}^{n-1}, \mathcal{L}^{n-1}), \end{cases} \tag{2.5.6}$$

and some vector-valued function $\vec{v}$ *satisfying*

$$\begin{cases} \vec{v} \in \big[\mathscr{C}^\infty(\mathbb{R}^n_+)\big]^n, \quad \Delta\vec{v} = 0 \ \textit{and} \ \operatorname{div}\vec{v} = 0 \ \textit{in} \ \mathbb{R}^n_+, \\ \mathcal{N}_\kappa(\nabla\vec{v}) \in L^p(\mathbb{R}^{n-1}, \mathcal{L}^{n-1}). \end{cases} \tag{2.5.7}$$

Moreover, $\vec{u}$ *determines* ∇w *and* $\vec{v}$ *uniquely (in a quantitative fashion), the nontangential boundary trace* $\vec{u}\big|^{\kappa-\mathrm{n.t.}}_{\partial\mathbb{R}^n_+}$ *exists, and*

$$\vec{v}\big|^{\kappa-\mathrm{n.t.}}_{\partial\mathbb{R}^n_+} = \vec{u}\big|^{\kappa-\mathrm{n.t.}}_{\partial\mathbb{R}^n_+}. \tag{2.5.8}$$

Conversely, for each w *as in* (2.5.6) *and each* $\vec{v}$ *as in* (2.5.7)*, the vector-valued function* $\vec{u}$ *associated with* w *and* $\vec{v}$ *as in* (2.5.5) *has the properties listed in* (2.5.4) *and satisfies* (2.5.8)*. Furthermore,*

$$\begin{aligned} \mathcal{N}_\kappa\vec{u} \in L^p(\mathbb{R}^{n-1}, \mathcal{L}^{n-1}) \ \textit{if and only if} \ \mathcal{N}_\kappa\vec{v} \in L^p(\mathbb{R}^{n-1}, \mathcal{L}^{n-1}) \ \textit{and} \\ \mathcal{N}_\kappa\big(x \mapsto x_n(\nabla w)(x)\big) \in L^p(\mathbb{R}^{n-1}, \mathcal{L}^{n-1}). \end{aligned} \tag{2.5.9}$$

In particular,

$$\begin{aligned} \textit{if} \ \mathcal{N}_\kappa w \in L^p(\mathbb{R}^{n-1}, \mathcal{L}^{n-1}) \ \textit{and} \ \mathcal{N}_\kappa\vec{v} \in L^p(\mathbb{R}^{n-1}, \mathcal{L}^{n-1}) \\ \textit{then} \ \mathcal{N}_\kappa\vec{u} \in L^p(\mathbb{R}^{n-1}, \mathcal{L}^{n-1}). \end{aligned} \tag{2.5.10}$$

Proof Thanks to Calderón's Theorem (recalled in (2.1.7)) and [132, (1.8.199)],

$$\text{if } w \text{ satisfies (2.5.6) then } (\nabla w)\big|^{\kappa-\mathrm{n.t.}}_{\partial\mathbb{R}^n_+} \text{ exists and belongs to } \big[L^p(\mathbb{R}^{n-1}, \mathcal{L}^{n-1})\big]^n, \tag{2.5.11}$$

$$\text{if } \vec{v} \text{ satisfies (2.5.7) then } \vec{v}\big|^{\kappa-\mathrm{n.t.}}_{\partial\mathbb{R}^n_+} \text{ exists and belongs to } \big[\dot{L}^p_1(\mathbb{R}^{n-1}, \mathcal{L}^{n-1})\big]^n. \tag{2.5.12}$$

Now suppose $\vec{u}$ is as in (2.5.4). Then [132, (1.8.199)] tells us that

$$\begin{aligned} u\big|^{\kappa-\mathrm{n.t.}}_{\partial\mathbb{R}^n_+} \text{ exists } \mathcal{L}^{n-1}\text{-a.e. on } \mathbb{R}^{n-1} \text{ and is in } \big[\dot{L}^p_1(\mathbb{R}^{n-1}, \mathcal{L}^{n-1})\big]^n, \\ (\nabla u)\big|^{\kappa-\mathrm{n.t.}}_{\partial\mathbb{R}^n_+} \text{ exists } \mathcal{L}^{n-1}\text{-a.e. on } \mathbb{R}^{n-1} \text{ and is in } \big[L^p(\mathbb{R}^{n-1}, \mathcal{L}^{n-1})\big]^{n^2}. \end{aligned} \tag{2.5.13}$$

Also,

$$\begin{cases} \operatorname{div}\vec{u} \in \mathscr{C}^\infty(\mathbb{R}^n_+), \quad \Delta(\operatorname{div}\vec{u}) = 0 \text{ in } \mathbb{R}^n_+, \\ \mathcal{N}_\kappa(\operatorname{div}\vec{u}) \in L^p(\mathbb{R}^{n-1}, \mathcal{L}^{n-1}). \end{cases} \tag{2.5.14}$$

In addition, Calderón's Theorem (cf. (2.1.7)) implies that

$$\begin{aligned} &(\mathrm{div}\,\vec{u})\Big|^{\kappa-\mathrm{n.t.}}_{\partial\mathbb{R}^n_+} \text{ exists at } \mathcal{L}^{n-1}\text{-a.e. point in } \mathbb{R}^{n-1} \\ &\text{and, as a function, belongs to } L^p(\mathbb{R}^{n-1}, \mathcal{L}^{n-1}). \end{aligned} \tag{2.5.15}$$

Next, introduce

$$w := 2\mathscr{S}_{\mathrm{mod}}\Big((\mathrm{div}\,\vec{u})\Big|^{\kappa-\mathrm{n.t.}}_{\partial\mathbb{R}^n_+}\Big) \text{ in } \mathbb{R}^n_+. \tag{2.5.16}$$

The properties of the operator $\mathscr{S}_{\mathrm{mod}}$ recorded in [132, (1.5.52), (1.5.57), (1.5.58)] imply (bearing in mind that $K^\#_\Delta$ is identically zero in our setting)

$$\begin{cases} w \in \mathscr{C}^\infty(\mathbb{R}^n_+), \ \Delta w = 0 \text{ in } \mathbb{R}^n_+, \\ \mathcal{N}_\kappa(\nabla w) \in L^p(\mathbb{R}^{n-1}, \mathcal{L}^{n-1}), \\ (\partial_n w)\Big|^{\kappa-\mathrm{n.t.}}_{\partial\mathbb{R}^n_+} = (\mathrm{div}\,\vec{u})\Big|^{\kappa-\mathrm{n.t.}}_{\partial\mathbb{R}^n_+}. \end{cases} \tag{2.5.17}$$

Ergo, w satisfies all properties listed in (2.5.6), and $\partial_n w$ is a solution of the L^p Dirichlet Problem for the Laplacian in $\mathbb{R}^n_+$ with boundary datum $(\mathrm{div}\,\vec{u})\Big|^{\kappa-\mathrm{n.t.}}_{\partial\mathbb{R}^n_+}$. Uniqueness for this problem then implies (in view of (2.5.14))

$$\partial_n w = \mathrm{div}\,\vec{u} \text{ in } \mathbb{R}^n_+, \tag{2.5.18}$$

while from (2.5.16) and [132, (1.5.57)] we see that

$$\begin{aligned} \|\mathcal{N}_\kappa(\nabla w)\|_{L^p(\mathbb{R}^{n-1},\mathcal{L}^{n-1})} &\le C\Big\|(\mathrm{div}\,\vec{u})\Big|^{\kappa-\mathrm{n.t.}}_{\partial\mathbb{R}^n_+}\Big\|_{L^p(\mathbb{R}^{n-1},\mathcal{L}^{n-1})} \\ &\le C\|\mathcal{N}_\kappa(\nabla\vec{u})\|_{L^p(\mathbb{R}^{n-1},\mathcal{L}^{n-1})}. \end{aligned} \tag{2.5.19}$$

To proceed, define

$$\vec{v}(x) := \vec{u}(x) - x_n(\nabla w)(x), \quad \forall x \in \mathbb{R}^n_+. \tag{2.5.20}$$

Then (2.5.4), (2.5.17), (2.5.13), and (2.5.18) guarantee that

$$\vec{v} \in \big[\mathscr{C}^\infty(\mathbb{R}^n_+)\big]^n, \quad \vec{v}\Big|^{\kappa-\mathrm{n.t.}}_{\partial\mathbb{R}^n_+} = \vec{u}\Big|^{\kappa-\mathrm{n.t.}}_{\partial\mathbb{R}^n_+}, \text{ and } \mathrm{div}\,\vec{v} = \mathrm{div}\,\vec{u} - \partial_n w = 0 \text{ in } \mathbb{R}^n_+. \tag{2.5.21}$$

Also, using (2.5.20) and (2.5.18) we compute, for each $x \in \mathbb{R}^n_+$,

$$(\Delta\vec{v})(x) = (\Delta\vec{u})(x) - 2\nabla(\partial_n w)(x) - x_n\nabla(\Delta w)(x) = (\Delta\vec{u})(x) - 2\nabla(\mathrm{div}\,\vec{u})(x) = 0. \tag{2.5.22}$$

Let $\widetilde{\kappa}$ be as in (2.4.5). Interior estimates for the harmonic function ∇w give

$$\big|x_n(\nabla^2 w)(x)\big| \le c_n \fint_{B(x,x_n/2)} |\nabla w|\,d\mathcal{L}^n \le c_n\mathcal{N}_{\widetilde{\kappa}}(\nabla w)(z) \tag{2.5.23}$$

at each $x = (x', x_n) \in \Gamma_\kappa(z)$. Together, (2.5.20) and (2.5.23) yield

$$\begin{aligned} &\mathcal{N}_\kappa(\partial_j\vec{v})(z) \le \mathcal{N}_\kappa(\nabla\vec{u})(z) + c_n\mathcal{N}_{\widetilde{\kappa}}(\nabla w)(z), \\ &\text{for all } j \in \{1,\dots,n\} \text{ and all } z \in \partial\mathbb{R}^n_+. \end{aligned} \tag{2.5.24}$$

Collectively, [129, (8.2.28)], (2.5.24), (2.5.19), and [129, Proposition 8.4.1] imply

$$\begin{aligned}\|\mathcal{N}_\kappa(\nabla\vec{v})\|_{L^p(\mathbb{R}^{n-1},\mathcal{L}^{n-1})} &\leq C\|\mathcal{N}_\kappa(\nabla\vec{u})\|_{L^p(\mathbb{R}^{n-1},\mathcal{L}^{n-1})} + C\|\mathcal{N}_{\widetilde{\kappa}}(\nabla w)\|_{L^p(\mathbb{R}^{n-1},\mathcal{L}^{n-1})} \\ &\leq C\|\mathcal{N}_\kappa(\nabla\vec{u})\|_{L^p(\mathbb{R}^{n-1},\mathcal{L}^{n-1})}, \end{aligned} \tag{2.5.25}$$

for some $C = C(n,\kappa,\widetilde{\kappa},p) \in (0,\infty)$. Since $\mathcal{N}_\kappa(\nabla\vec{u}) \in L^p(\mathbb{R}^{n-1},\mathcal{L}^{n-1})$, we ultimately conclude that

$$\mathcal{N}_\kappa(\nabla\vec{v}) \in L^p(\mathbb{R}^{n-1},\mathcal{L}^{n-1}), \quad \|\mathcal{N}_\kappa(\nabla\vec{v})\|_{L^p(\mathbb{R}^{n-1},\mathcal{L}^{n-1})} \leq C\|\mathcal{N}_\kappa(\nabla\vec{u})\|_{L^p(\mathbb{R}^{n-1},\mathcal{L}^{n-1})}. \tag{2.5.26}$$

From (2.5.21), (2.5.22), and (2.5.26) it follows that $\vec{v}$ satisfies (2.5.7). Recalling (2.5.20), we see that $\vec{u}$ is of the form (2.5.5) and that (2.5.8) holds.

To prove that $\vec{u}$ determines ∇w and $\vec{v}$ uniquely, by linearity it suffices to show that if $\vec{u} = 0$ then $\vec{v}$ and ∇w are also zero. With this goal in mind, assume w satisfies (2.5.6), $\vec{v}$ satisfies (2.5.7), and $\vec{v}(x) + x_n(\nabla w)(x) = 0$ for each $x \in \mathbb{R}^n_+$. Applying div to this identity yields

$$0 = (\operatorname{div}\vec{v})(x) + x_n(\Delta w)(x) + (\partial_n w)(x) = (\partial_n w)(x) \text{ for each } x \in \mathbb{R}^n_+. \tag{2.5.27}$$

From this, (2.5.6), and [132, (1.8.200)] we then see that there exists $c \in \mathbb{C}$ such that

$$w = \mathcal{D}_{\text{mod}}\big(w\big|^{\kappa-\text{n.t.}}_{\partial\mathbb{R}^n_+}\big) + c \text{ in } \mathbb{R}^n_+. \tag{2.5.28}$$

After taking the nontangential trace to the boundary in (2.5.28), we have

$$w\big|^{\kappa-\text{n.t.}}_{\partial\mathbb{R}^n_+} = \big(\tfrac{1}{2}I + K_{\text{mod}}\big)\big(w\big|^{\kappa-\text{n.t.}}_{\partial\mathbb{R}^n_+}\big) + c \text{ in } \mathbb{R}^{n-1}. \tag{2.5.29}$$

Since as seen from [132, (1.8.26)] (bearing in mind that we now have $K = 0$), $K_{\text{mod}}\big(w\big|^{\kappa-\text{n.t.}}_{\partial\mathbb{R}^n_+}\big)$ is a constant, from (2.5.29) we obtain that $w\big|^{\kappa-\text{n.t.}}_{\partial\mathbb{R}^n_+}$ is constant in $\mathbb{R}^{n-1}$. When used in (2.5.28), this shows that w is constant in $\mathbb{R}^n_+$, thus $\nabla w = 0$, so $\vec{v} = 0$.

To prove the converse statement, assume w is as in (2.5.6), $\vec{v}$ is as in (2.5.7), and define $\vec{u}$ associated with w and $\vec{v}$ as described in (2.5.5). Then $\vec{u} \in \big[\mathscr{C}^\infty(\mathbb{R}^n_+)\big]^n$ and

$$\operatorname{div}\big[x_n(\nabla w)(x)\big] = (\partial_n w)(x) + x_n(\Delta w)(x) = (\partial_n w)(x) \text{ for all } x \in \mathbb{R}^n_+. \tag{2.5.30}$$

In view of (2.5.30) and the properties of $\vec{v}$ and w, we obtain

$$\begin{aligned}\Delta\vec{u} - 2\nabla\operatorname{div}\vec{u} &= \Delta\vec{v} + (\Delta x_n)\nabla w + 2\nabla(\partial_n w) + x_n\nabla(\Delta w) - 2\nabla\operatorname{div}\vec{v} - 2\nabla\operatorname{div}(x_n\nabla w) \\ &= 2\nabla(\partial_n w) - 2\nabla[\partial_n w] = 0 \text{ in } \mathbb{R}^n_+. \end{aligned} \tag{2.5.31}$$

In addition, for each $j \in \{1,\dots,n\}$, after applying ∂_j in (2.5.5) and using (2.4.5), (2.5.23), we may estimate

$$\mathcal{N}_\kappa(\partial_j\vec{u})(z) \leq \mathcal{N}_\kappa(\nabla\vec{v})(z) + c_n\mathcal{N}_{\widetilde{\kappa}}(\nabla w)(z), \quad \forall z \in \partial\mathbb{R}^n_+. \tag{2.5.32}$$

Together with [129, (8.2.28)], [129, Proposition 8.4.1], plus the memberships in the last lines in (2.5.6) and (2.5.7), this further implies $\mathcal{N}_\kappa(\nabla\vec{u}) \in L^p(\mathbb{R}^{n-1}, \mathcal{L}^{n-1})$. We have thus shown that $\vec{u}$ satisfies (2.5.4). That (2.5.8) also holds, follows from (2.5.5) and (2.5.11)-(2.5.12).

As far as the equivalence claimed in (2.5.9) is concerned, the right-to-left implication is immediate from (2.5.5). Consider next the left-to-right implication in (2.5.9). The fact that $\mathcal{N}_\kappa\vec{u} \in L^p(\mathbb{R}^{n-1}, \mathcal{L}^{n-1})$ together with the last line in (2.5.4), the existence of $\vec{u}\big|^{\kappa-\mathrm{n.t.}}_{\partial\mathbb{R}^n_+}$, the last part in Corollary 2.1.2, and (2.5.8) imply that

$$\vec{v}\big|^{\kappa-\mathrm{n.t.}}_{\partial\mathbb{R}^n_+} \text{ exists and belongs to } \big[L^p_1(\mathbb{R}^{n-1}, \mathcal{L}^{n-1})\big]^n. \tag{2.5.33}$$

Based on (2.5.33), (2.5.7), and the uniqueness of the Regularity Problem for the (vector) Laplacian in the upper half-space (see (8.4.10) with $\Omega := \mathbb{R}^n_+$ and L playing the role of the vector Laplacian) we then conclude that $\mathcal{N}_\kappa\vec{v} \in L^p(\mathbb{R}^{n-1}, \mathcal{L}^{n-1})$. From this and (2.5.5) we then see that the last membership in (2.5.9) is also true. The equivalence in (2.5.9) is therefore established.

Finally, we are left with justifying (2.5.10). To this end, make the additional assumptions that $\mathcal{N}_\kappa w$ and $\mathcal{N}_\kappa\vec{v}$ belong to $L^p(\mathbb{R}^{n-1}, \mathcal{L}^{n-1})$. Interior estimates for the harmonic function w in $\mathbb{R}^n_+$ imply that for each $z \in \partial\mathbb{R}^n_+$ we have

$$\big|x_n(\nabla w)(x)\big| \leq c_n \fint_{B(x,x_n/2)} |w|\,\mathrm{d}\mathcal{L}^n \leq c_n \mathcal{N}_{\widetilde{\kappa}} w(z) \tag{2.5.34}$$

at each point $x = (x', x_n) \in \Gamma_\kappa(z)$. In light of [129, (8.2.28)], [129, Proposition 8.4.1], and the current assumptions, this implies that the nontangential maximal function $\mathcal{N}_\kappa\big(x \mapsto x_n(\nabla w)(x)\big) \in L^p(\mathbb{R}^{n-1}, \mathcal{L}^{n-1})$. At this point we may invoke (2.5.9) to conclude that $\mathcal{N}_\kappa\vec{u}$ belong to $L^p(\mathbb{R}^{n-1}, \mathcal{L}^{n-1})$. □

We also wish to establish a version of Theorem 2.5.1 for null-solutions of the system L_D simply possessing a p-th power integrable nontangential maximal function (in place of the demand made in the last line of (2.5.4)). As a preamble, we first prove a couple of auxiliary lemmas, the first of which is an estimate in the spirit of lemma on page 213 of [198]. In view of the usefulness of this type of result in a variety of situations, below we state and prove a slightly more general version than the one we shall actually need here, in the proof of Lemma 2.5.3.

Lemma 2.5.2 *Suppose $\Gamma_\alpha \subset \Gamma_\beta \subset \mathbb{R}^n_+$ are two infinite, circular cones, with axis along the $\mathbf{e}_n$ direction, and vertex at $0 \in \mathbb{R}^n$, whose aperture angles $\alpha, \beta \in (0, \pi)$ satisfy $\alpha < \beta$. Also, fix $M, m \in \mathbb{N}$ and assume that $u \in \big[\mathscr{C}^\infty(\Gamma_\beta)\big]^M$ is a null-solution of a homogeneous, constant (complex) coefficient weakly elliptic $M \times M$ system L of order $2m$ in $\mathbb{R}^n$. Make the assumption that*

$$\lim_{t\to\infty} (\nabla u)(x', t) = 0 \text{ for each } x' \in \mathbb{R}^{n-1}. \tag{2.5.35}$$

Then for each $q \in [1, \infty)$ and $r \in (-\infty, n)$ there exists a constant $C \in (0, \infty)$ depending only on $\alpha, \beta, r, L, q, n$ such that

$$\int_{\Gamma_\alpha} \frac{|(\nabla u)(x)|^q}{x_n^r}\,\mathrm{d}x \leq C \int_{\Gamma_\beta} \frac{|(\partial_n u)(x)|^q}{x_n^r}\,\mathrm{d}x. \tag{2.5.36}$$

Proof Using the Fundamental Theorem of Calculus and the decay of u expressed in (2.5.35), for each $j \in \{1, \ldots, n\}$, $x = (x', x_n) \in \Gamma_\alpha$, and $a \in (x_n, \infty)$ we have

$$\begin{aligned} |(\partial_j u)(x)| &= \Big|(\partial_j u)(x', a) - \int_{x_n}^{a} (\partial_n \partial_j u)(x', t)\,\mathrm{d}t\Big| \\ &\leq \int_{x_n}^{\infty} |(\partial_j \partial_n u)(x', t)|\,\mathrm{d}t + |(\nabla u)(x', a)|. \end{aligned} \tag{2.5.37}$$

Passing to limit $a \to \infty$ then yields, on account of (2.5.35),

$$|(\partial_j u)(x)| \leq \int_{x_n}^{\infty} |(\partial_j \partial_n u)(x', t)|\,\mathrm{d}t. \tag{2.5.38}$$

On the other hand, using interior estimates, whenever $(x', t) \in \Gamma_\alpha$ we may write

$$|(\partial_j \partial_n u)(x', t)| \leq \frac{C}{t}\Big(\fint_{B((x',t),\lambda t)} |(\partial_n u)(y)|^q\,\mathrm{d}y\Big)^{\frac{1}{q}}, \tag{2.5.39}$$

where $\lambda \in (0, 1)$ is chosen to be small enough so that $B((x', t), \lambda t) \subset \Gamma_\beta$ whenever $(x', t) \in \Gamma_\alpha$. For each $t > 0$ define $D_t := \{y = (y', y_n) \in \Gamma_\beta : |y_n - t| < \lambda t\}$. Note that $B((x', t), \lambda t) \subset D_t$ for every $(x', t) \in \Gamma_\alpha$. As such, (2.5.39) gives

$$|(\partial_j \partial_n u)(x', t)| \leq C\, t^{-(n+q)/q} F(t)^{1/q} \ \text{ whenever } (x', t) \in \Gamma_\alpha, \tag{2.5.40}$$

where we have set $F(t) := \int_{D_t} |(\partial_n u)(y)|^q\,\mathrm{d}y$ for each $t \in (0, \infty)$. Consider the spherical cap $S_\alpha := S^{n-1} \cap \Gamma_\alpha$. Then, combining (2.5.38) and (2.5.40) and taking into account the fact that for each $\omega \in S_\alpha$ we have $(s\omega)_n = s\omega_n \geq cs$ for some $c = c(\alpha) \in (0, 1)$, we obtain $|(\partial_j u)(s\omega)| \leq C \int_{cs}^{\infty} t^{-(n+q)/q} F(t)^{1/q}\,\mathrm{d}t$ for all $\omega \in S_\alpha$ and $s \in (0, \infty)$. Hardy's inequality (cf. [130, Lemma 4.3.5]) gives

$$\int_0^{\infty} |(\partial_j u)(s\omega)|^q\, s^{n-1-r}\,\mathrm{d}s \leq C \int_0^{\infty} t^{-1-r} F(t)\,\mathrm{d}t \tag{2.5.41}$$

for all $\omega \in S_\alpha$. Let χ_t denote the characteristic function of D_t in Γ_β. Then, by Fubini-Tonelli's Theorem,

$$\begin{aligned} \int_0^{\infty} t^{-1-r} F(t)\,\mathrm{d}t &= \int_0^{\infty} t^{-1-r}\Big(\int_{D_t} |(\partial_n u)(y)|^q\,\mathrm{d}y\Big)\,\mathrm{d}t \\ &= \int_{\Gamma_\beta} |(\partial_n u)(y)|^q \Big(\int_0^{\infty} t^{-1-r} \chi_t(y)\,\mathrm{d}t\Big)\,\mathrm{d}y \\ &\leq \int_{\Gamma_\beta} |(\partial_n u)(y)|^q \Big(\int_{y_n/(1+\lambda)}^{y_n/(1-\lambda)} t^{-1-r}\,\mathrm{d}t\Big)\,\mathrm{d}y \leq C \int_{\Gamma_\beta} |(\partial_n u)(y)|^q\, y_n^{-r}\,\mathrm{d}y. \end{aligned} \tag{2.5.42}$$

Finally, using (2.5.41), (2.5.42), and polar coordinates, we have

$$\int_{\Gamma_\alpha} |(\partial_j u)(x)|^q\, x_n^{-r}\,\mathrm{d}x \le C\int_{S_\alpha}\int_0^\infty |(\partial_j u)(s\omega)|^q\, s^{n-1-r}\,\mathrm{d}s\,\mathrm{d}\omega$$
$$\le C\int_{\Gamma_\beta} |(\partial_n u)(x)|^q\, x_n^{-r}\,\mathrm{d}x. \tag{2.5.43}$$

Since $j \in \{1,\dots,n\}$ was arbitrary, this finishes the proof of the lemma. □

Lemma 2.5.2 is one of the ingredients in the proof of the following result, itself used later in the proof of Theorem 2.5.4.

Lemma 2.5.3 *Fix $n \in \mathbb{N}$ with $n \ge 2$ and assume w is a harmonic function in $\mathbb{R}^n_+$ with the property that*

$$\lim_{t\to\infty}(\nabla w)(x',t) = 0 \text{ and } \lim_{t\to\infty}(\nabla\nabla w)(x',t) = 0 \text{ for each } x' \in \mathbb{R}^{n-1}. \tag{2.5.44}$$

Then for each aperture parameter $\kappa \in (0,\infty)$ and each integrability exponent $p \in (0,\infty)$ there exists a finite constant $C = C(n,\kappa,p) > 0$ such that

$$\|\mathcal{N}_\kappa(\nabla w)\|_{L^p(\mathbb{R}^{n-1},\mathcal{L}^{n-1})} \le C\|\mathcal{N}_\kappa(\partial_n w)\|_{L^p(\mathbb{R}^{n-1},\mathcal{L}^{n-1})}. \tag{2.5.45}$$

Proof There are two basic ingredients used in the proof of this result. One is E. Stein's Lemma 2.5.2. The second one is the comparison between the nontangential maximal operator and the Lusin area operator in the class of harmonic functions in the upper half-space. The latter operator acts on any given function $u \in \mathscr{C}^1(\mathbb{R}^n_+)$ according to

$$(\mathcal{A}_\kappa u)(x') := \left(\int_{\Gamma_\kappa(x')} \frac{|(\nabla u)(y)|^2}{y_n^{n-2}}\,\mathrm{d}y\right)^{1/2}, \qquad \forall x' \in \mathbb{R}^{n-1}, \tag{2.5.46}$$

where $\Gamma_\kappa(x')$ is the nontangential approach region of aperture $\kappa > 0$ and with vertex at $x' \in \mathbb{R}^{n-1}$ (canonically identified with $(x',0) \in \partial\mathbb{R}^n_+$). Specifically, it has been proved by D. Burkholder and R. Gundy in [25] (see also the discussion in [66, 8.5, pp. 372–373]) that there exists some constant $C = C(n,p,\kappa) \in (0,\infty)$ with the property that for each function $u \in \mathscr{C}^\infty(\mathbb{R}^n_+)$ satisfying $\Delta u = 0$ in $\mathbb{R}^n_+$ one has

$$\|\mathcal{A}_\kappa u\|_{L^p(\mathbb{R}^{n-1},\mathcal{L}^{n-1})} \le C\|\mathcal{N}_\kappa u\|_{L^p(\mathbb{R}^{n-1},\mathcal{L}^{n-1})}. \tag{2.5.47}$$

Moreover,

if $u \in \mathscr{C}^\infty(\mathbb{R}^n)$ satisfies $\Delta u = 0$ in $\mathbb{R}^n_+$ and $\|\mathcal{A}_\kappa u\|_{L^p(\mathbb{R}^{n-1},\mathcal{L}^{n-1})} < +\infty$ then the limit $\lim_{t\to\infty} u(x',t)$ exists for each $x' \in \mathbb{R}^{n-1}$ and is actually a constant, independent of the point $x' \in \mathbb{R}^{n-1}$. (2.5.48)

Finally, if $u \in \mathscr{C}^\infty(\mathbb{R}^n)$ satisfies $\Delta u = 0$ in $\mathbb{R}^n_+$ as well as $\lim_{t\to\infty} u(x',t) = 0$ for each $x' \in \mathbb{R}^{n-1}$, then there exists a constant $C \in (0,\infty)$ independent of u such that

$$\|\mathcal{N}_\kappa u\|_{L^p(\mathbb{R}^{n-1},\mathcal{L}^{n-1})} \leq C\|\mathcal{A}_\kappa u\|_{L^p(\mathbb{R}^{n-1},\mathcal{L}^{n-1})}. \tag{2.5.49}$$

Suppose now that some harmonic function w in $\mathbb{R}^n_+$ satisfying (2.5.44) has been fixed. Pick a second aperture parameter $\widetilde{\kappa} > \kappa$. Granted the second property in (2.5.44), we may call upon Lemma 2.5.2 (presently used with $q := 2$ and $r := n-2$) applied to ∇w. In view of (2.5.46), for each $x' \in \mathbb{R}^{n-1}$ this permits us to write

$$\begin{aligned}\mathcal{A}_\kappa(\nabla w)(x')^2 &= \int_{\Gamma_\kappa(x')} \frac{|\nabla\nabla w(y)|^2}{y_n^{n-2}}\,dy \leq C\int_{\Gamma_{\widetilde{\kappa}}(x')} \frac{|\partial_n \nabla w(y)|^2}{y_n^{n-2}}\,dy \\ &= C\mathcal{A}_{\widetilde{\kappa}}(\partial_n w)(x')^2. \end{aligned}\tag{2.5.50}$$

We are now in a position to estimate

$$\begin{aligned}\|\mathcal{N}_\kappa(\nabla w)\|_{L^p(\mathbb{R}^{n-1},\mathcal{L}^{n-1})} &\leq C\|\mathcal{A}_\kappa(\nabla w)\|_{L^p(\mathbb{R}^{n-1},\mathcal{L}^{n-1})} \leq C\|\mathcal{A}_{\widetilde{\kappa}}(\partial_n w)\|_{L^p(\mathbb{R}^{n-1},\mathcal{L}^{n-1})} \\ &\leq C\|\mathcal{N}_{\widetilde{\kappa}}(\partial_n w)\|_{L^p(\mathbb{R}^{n-1},\mathcal{L}^{n-1})} \leq C\|\mathcal{N}_\kappa(\partial_n w)\|_{L^p(\mathbb{R}^{n-1},\mathcal{L}^{n-1})}. \end{aligned}\tag{2.5.51}$$

Above, the first inequality is a consequence of (2.5.47) (applied to $u := \nabla w$) and the first property in (2.5.44). The second inequality in (2.5.51) is implied by (2.5.50). The third inequality in (2.5.51) comes from (2.5.47) (used for $u := \partial_n w$), and the final inequality in (2.5.51) follows from [129, Proposition 8.4.1]. □

We are now prepared to present a companion result to Theorem 2.5.1, itself a structure theorem for vector fields which are null-solutions of the system L_D, now simply assumed to have a p-th power integrable nontangential maximal function (as opposed to the demand made in the last line of (2.5.4), which involves the gradient). Remarkably, we are still able to obtain the same type of decomposition for such vector fields, with the regularity properties of the constitutive components naturally adjusted. The lack of regularity prevents us from directly running the same argument as in the proof of Theorem 2.5.1, though we do employ the result established in Theorem 2.5.1 for a regularized version of the given vector field.

Theorem 2.5.4 *Select $n \in \mathbb{N}$ with $n \geq 2$, along with an exponent $p \in (1,\infty)$ and an aperture parameter $\kappa \in (0,\infty)$. Then any vector-valued function $\vec{u}$ satisfying*

$$\begin{cases} \vec{u} \in \left[\mathscr{C}^\infty(\mathbb{R}^n_+)\right]^n, \quad \Delta\vec{u} - 2\nabla \mathrm{div}\,\vec{u} = 0 \text{ in } \mathbb{R}^n_+, \\ \mathcal{N}_\kappa \vec{u} \in L^p(\mathbb{R}^{n-1},\mathcal{L}^{n-1}), \end{cases}\tag{2.5.52}$$

may be expressed as

$$\vec{u}(x) = \vec{v}(x) + x_n(\nabla w)(x), \quad \forall x = (x', x_n) \in \mathbb{R}^{n-1}\times(0,\infty), \tag{2.5.53}$$

for some scalar function w satisfying

$$\begin{cases} w \in \mathscr{C}^\infty(\mathbb{R}^n_+), \quad \Delta w = 0 \text{ in } \mathbb{R}^n_+, \\ \mathcal{N}_\kappa\big(x \mapsto x_n(\nabla w)(x)\big) \in L^p(\mathbb{R}^{n-1},\mathcal{L}^{n-1}), \end{cases}\tag{2.5.54}$$

and some vector-valued function $\vec{v}$ satisfying

$$\begin{cases} \vec{v} \in \big[\mathscr{C}^\infty(\mathbb{R}^n_+)\big]^n, & \Delta\vec{v} = 0 \textit{ and } \mathrm{div}\,\vec{v} = 0 \textit{ in } \mathbb{R}^n_+, \\ \mathcal{N}_\kappa\vec{v} \in L^p(\mathbb{R}^{n-1}, \mathcal{L}^{n-1}). \end{cases} \tag{2.5.55}$$

Furthermore, $\vec{u}$ determines ∇w and $\vec{v}$ uniquely. Also,

$$\vec{v}\Big|^{\kappa-\mathrm{n.t.}}_{\partial\mathbb{R}^n_+} \textit{ exists } \mathcal{L}^{n-1}\textit{-a.e. in } \mathbb{R}^{n-1} \tag{2.5.56}$$

and

$$\begin{gathered} \textit{if the nontangential boundary trace } \vec{u}\Big|^{\kappa-\mathrm{n.t.}}_{\partial\mathbb{R}^n_+} \textit{ exists then} \\ \vec{v}\Big|^{\kappa-\mathrm{n.t.}}_{\partial\mathbb{R}^n_+} = \vec{u}\Big|^{\kappa-\mathrm{n.t.}}_{\partial\mathbb{R}^n_+} \textit{ and } \big[x \mapsto x_n(\nabla w)(x)\big]\Big|^{\kappa-\mathrm{n.t.}}_{\partial\mathbb{R}^n_+} = 0. \end{gathered} \tag{2.5.57}$$

Moreover,

$$\begin{gathered} \textit{if the nontangential boundary trace } w\Big|^{\kappa-\mathrm{n.t.}}_{\partial\mathbb{R}^n_+} \textit{ exists then} \\ \textit{the nontangential boundary trace } \vec{u}\Big|^{\kappa-\mathrm{n.t.}}_{\partial\mathbb{R}^n_+} \textit{ exists.} \end{gathered} \tag{2.5.58}$$

Conversely, for each w as in (2.5.54) *and each $\vec{v}$ as in* (2.5.55)*, the vector-valued function $\vec{u}$ associated with w and $\vec{v}$ as in* (2.5.53) *has the properties listed in* (2.5.52).

In addition, in a quantitative fashion, one has the equivalence

$$\begin{gathered} \mathcal{N}_\kappa(\nabla\vec{u}) \textit{ belongs to } L^p(\mathbb{R}^{n-1}, \mathcal{L}^{n-1}) \textit{ if and only if} \\ \mathcal{N}_\kappa(\nabla\vec{v}) \in L^p(\mathbb{R}^{n-1}, \mathcal{L}^{n-1}) \textit{ and } \mathcal{N}_\kappa(\nabla w) \in L^p(\mathbb{R}^{n-1}, \mathcal{L}^{n-1}), \end{gathered} \tag{2.5.59}$$

and in each of the above two happenstances the decomposition given in (2.5.53) *agrees with the decomposition given in* (2.5.5).

Proof For each $\varepsilon > 0$ arbitrary define

$$\vec{u}_\varepsilon := \vec{u}(\cdot + \varepsilon\mathbf{e}_n) \in \big[\mathscr{C}^\infty(\overline{\mathbb{R}^n_+})\big]^n \tag{2.5.60}$$

and note that this satisfies

$$\Delta\vec{u}_\varepsilon - 2\nabla\mathrm{div}\,\vec{u}_\varepsilon = 0 \text{ in } \mathbb{R}^n_+, \ \mathcal{N}_\kappa\vec{u}_\varepsilon \in L^p(\mathbb{R}^{n-1}, \mathcal{L}^{n-1}), \tag{2.5.61}$$

as well as

$$\Big|\vec{u}_\varepsilon\big|_{\partial\mathbb{R}^n_+}\Big| \leq \mathcal{N}_\kappa\vec{u} \in L^p(\mathbb{R}^{n-1}, \mathcal{L}^{n-1}) \text{ for each } \varepsilon > 0. \tag{2.5.62}$$

In particular, $\big(\vec{u}_\varepsilon\big|_{\partial\mathbb{R}^n_+}\big)_{\varepsilon>0}$ is a bounded sequence in $\big[L^p(\mathbb{R}^{n-1}, \mathcal{L}^{n-1})\big]^n$, and so the Sequential Banach-Alaoglu Theorem (recalled in [129, (3.6.22)]) implies that there exist a sequence $\{\varepsilon_i\}_{i\in\mathbb{N}} \subseteq (0,\infty)$ with $\lim\limits_{i\to\infty}\varepsilon_i = 0$ and some vector-valued function $\vec{f} \in \big[L^p(\mathbb{R}^{n-1}, \mathcal{L}^{n-1})\big]^n$ with the property that

$$\vec{u}_{\varepsilon_i}\big|_{\partial\mathbb{R}^n_+} \longrightarrow \vec{f} \text{ weak-}* \text{ in } \big[L^p(\mathbb{R}^{n-1}, \mathcal{L}^{n-1})\big]^n \text{ as } i \to \infty. \tag{2.5.63}$$

On a different topic, we note that the present assumptions, interior estimates, and [129, Proposition 8.4.1] yield

$$\mathcal{N}_\kappa\big(\nabla\vec{u}_\varepsilon\big) \in L^p(\mathbb{R}^{n-1},\mathcal{L}^{n-1}) \text{ for each } \varepsilon > 0. \tag{2.5.64}$$

More specifically, the membership in (2.5.64) may be obtained via a reasoning similar to that used in (2.4.7)-(2.4.8), this time the first inequality in (2.4.7) being justified by invoking [129, Theorem 6.5.7]. Granted (2.5.60), (2.5.61), and (2.5.64), the "structure theorem" Theorem 2.5.1 applies to each $\vec{u}_\varepsilon$ and guarantees the existence of a scalar function w_ε satisfying

$$\begin{cases} w_\varepsilon \in \mathscr{C}^\infty(\mathbb{R}^n_+), \quad \Delta w_\varepsilon = 0 \text{ in } \mathbb{R}^n_+, \\ \mathcal{N}_\kappa(\nabla w_\varepsilon) \in L^p(\mathbb{R}^{n-1},\mathcal{L}^{n-1}), \end{cases} \tag{2.5.65}$$

together with some vector-valued function $\vec{v}_\varepsilon$ satisfying

$$\begin{cases} \vec{v}_\varepsilon \in \big[\mathscr{C}^\infty(\mathbb{R}^n_+)\big]^n, \\ \Delta\vec{v}_\varepsilon = 0 \text{ and } \operatorname{div}\vec{v}_\varepsilon = 0 \text{ in } \mathbb{R}^n_+, \\ \mathcal{N}_\kappa\vec{v}_\varepsilon,\ \mathcal{N}_\kappa(\nabla\vec{v}_\varepsilon) \in L^p(\mathbb{R}^{n-1},\mathcal{L}^{n-1}), \\ \vec{v}_\varepsilon\Big|^{\kappa-\text{n.t.}}_{\partial\mathbb{R}^n_+} = \vec{u}_\varepsilon\Big|_{\partial\mathbb{R}^n_+}, \end{cases} \tag{2.5.66}$$

and such that

$$\vec{u}_\varepsilon(x) = \vec{v}_\varepsilon(x) + x_n(\nabla w_\varepsilon)(x), \quad \forall\, x = (x', x_n) \in \mathbb{R}^{n-1}\times(0,\infty). \tag{2.5.67}$$

To proceed, bring in the Poisson kernel P^Δ for the Laplacian in $\mathbb{R}^n$ (cf. (2.2.1)) and for each $t > 0$ define $P^\Delta_t := t^{1-n}P^\Delta(\cdot/t)$. Also, recall (cf., e.g., [114]) that

$$\vec{v}(x',t) := \big(P^\Delta_t * \vec{f}\big)(x') \text{ for all } x' \in \mathbb{R}^{n-1},\ t > 0, \tag{2.5.68}$$

is the unique solution of the following Dirichlet Problem for the vector Laplacian:

$$\begin{cases} \vec{v} \in \big[\mathscr{C}^\infty(\mathbb{R}^n_+)\big]^n, \quad \Delta\vec{v} = 0 \text{ in } \mathbb{R}^n_+, \\ \mathcal{N}_\kappa\vec{v} \in L^p(\mathbb{R}^{n-1},\mathcal{L}^{n-1}), \\ \vec{v}\Big|^{\kappa-\text{n.t.}}_{\partial\mathbb{R}^n_+} = \vec{f} \text{ at } \mathcal{L}^{n-1}\text{-a.e. point on } \mathbb{R}^{n-1}. \end{cases} \tag{2.5.69}$$

By (2.5.66), the same well-posedness result implies that for each $\varepsilon > 0$ we have

$$\vec{v}_\varepsilon(x',t) = \Big(P^\Delta_t * \big(\vec{u}_\varepsilon\big|_{\partial\mathbb{R}^n_+}\big)\Big)(x') \text{ for all } x' \in \mathbb{R}^{n-1},\ t > 0. \tag{2.5.70}$$

We shall use this to show that

$$\vec{v}_{\varepsilon_i} \longrightarrow \vec{v} \text{ in } \big[\mathcal{D}'(\mathbb{R}^n_+)\big]^n \text{ as } i \to \infty. \tag{2.5.71}$$

Specifically, for each vector-valued test function $\vec{\varphi} \in \big[\mathscr{C}^\infty_c(\mathbb{R}^n_+)\big]^n$ we may write

$$\begin{aligned}
&\lim_{i\to\infty} {}_{[\mathcal{D}'(\mathbb{R}^n_+)]^n}\big\langle \vec{v}_{\varepsilon_i}, \vec{\varphi}\big\rangle_{[\mathcal{D}(\mathbb{R}^n_+)]^n} = \lim_{i\to\infty}\int_{\mathbb{R}^n_+}\big\langle \vec{v}_{\varepsilon_i}, \vec{\varphi}\big\rangle \,\mathrm{d}\mathcal{L}^n \qquad (2.5.72)\\
&= \lim_{i\to\infty}\int_{\mathbb{R}^n_+}\Big\langle\Big(P_t^{\Delta} * \big(\vec{u}_{\varepsilon_i}\big|_{\partial\mathbb{R}^n_+}\big)\Big)(x'), \vec{\varphi}(x',t)\Big\rangle\,\mathrm{d}x'\mathrm{d}t\\
&= \lim_{i\to\infty}\int_{\mathbb{R}^{n-1}}\Big\langle\int_{\mathbb{R}^n_+} P_t^{\Delta}(x'-y')\vec{\varphi}(x',t)\,\mathrm{d}x'\mathrm{d}t, \big(\vec{u}_{\varepsilon_i}\big|_{\partial\mathbb{R}^n_+}\big)(y')\Big\rangle\,\mathrm{d}y'\\
&= \int_{\mathbb{R}^{n-1}}\Big\langle\int_{\mathbb{R}^n_+} P_t^{\Delta}(x'-y')\vec{\varphi}(x',t)\,\mathrm{d}x'\mathrm{d}t, \vec{f}(y')\Big\rangle\,\mathrm{d}y'\\
&= \int_{\mathbb{R}^n_+}\Big\langle(P_t^{\Delta} * \vec{f})(x'), \vec{\varphi}(x',t)\Big\rangle\,\mathrm{d}x'\mathrm{d}t = \int_{\mathbb{R}^n_+}\big\langle\vec{v},\vec{\varphi}\big\rangle\,\mathrm{d}\mathcal{L}^n = {}_{[\mathcal{D}'(\mathbb{R}^n_+)]^n}\big\langle\vec{v},\vec{\varphi}\big\rangle_{[\mathcal{D}(\mathbb{R}^n_+)]^n}
\end{aligned}$$

where the fourth equality uses (2.5.63), and the sixth equality comes from (2.5.68). This establishes (2.5.71). Next, recall from (2.5.66) that

$$\mathrm{div}\,\vec{v}_\varepsilon = 0 \text{ in } \mathbb{R}^n_+ \text{ for each } \varepsilon > 0. \qquad (2.5.73)$$

For each $\varphi \in \mathscr{C}^\infty_c(\mathbb{R}^n_+)$ we may then combine (2.5.71) and (2.5.73) to write

$$\begin{aligned}
&{}_{\mathcal{D}'(\mathbb{R}^n_+)}\langle \mathrm{div}\,\vec{v}, \varphi\rangle_{\mathcal{D}(\mathbb{R}^n_+)} = -{}_{[\mathcal{D}'(\mathbb{R}^n_+)]^n}\langle\vec{v}, \nabla\varphi\rangle_{[\mathcal{D}(\mathbb{R}^n_+)]^n} \qquad (2.5.74)\\
&\quad = -\lim_{i\to\infty} {}_{[\mathcal{D}'(\mathbb{R}^n_+)]^n}\langle\vec{v}_{\varepsilon_i}, \nabla\varphi\rangle_{[\mathcal{D}(\mathbb{R}^n_+)]^n} = \lim_{i\to\infty} {}_{\mathcal{D}'(\mathbb{R}^n_+)}\langle \mathrm{div}\,\vec{v}_{\varepsilon_i}, \varphi\rangle_{\mathcal{D}(\mathbb{R}^n_+)} = 0,
\end{aligned}$$

which ultimately shows that, in a classical sense, $\mathrm{div}\,\vec{v} = 0$ in $\mathbb{R}^n_+$. It is also clear from (2.5.60) that $\vec{u}_\varepsilon \to \vec{u}$ uniformly on compact sets in $\mathbb{R}^n_+$ as $\varepsilon \to 0^+$. In particular,

$$\vec{u}_\varepsilon \longrightarrow \vec{u} \text{ in } \big[\mathcal{D}'(\mathbb{R}^n_+)\big]^n \text{ as } \varepsilon \to 0^+. \qquad (2.5.75)$$

Since $\mathbb{R}^n_+ \ni x \mapsto x_n^{-1} \in \mathbb{R}$ is a well-defined smooth function, we may define

$$\vec{F} := x_n^{-1}\vec{u} - x_n^{-1}\vec{v} \in \big[\mathscr{C}^\infty(\mathbb{R}^n_+)\big]^n \qquad (2.5.76)$$

and note that in the sense of (vector) distributions in $\mathbb{R}^n_+$ we have

$$\nabla w_{\varepsilon_i} = x_n^{-1}\vec{u}_{\varepsilon_i} - x_n^{-1}\vec{v}_{\varepsilon_i} \longrightarrow x_n^{-1}\vec{u} - x_n^{-1}\vec{v} = \vec{F} \text{ as } i \to \infty, \qquad (2.5.77)$$

thanks to (2.5.71) and (2.5.75)-(2.5.76). Hence, if $(F_j)_{1\le j\le n}$ are the scalar components of $\vec{F}$, for each $j \in \{1,\dots,n\}$ we obtain $\lim_{i\to\infty}\partial_j w_{\varepsilon_i} = F_j$ in $\mathcal{D}'(\mathbb{R}^n_+)$. In turn, for each $j,k \in \{1,\dots,n\}$ and each $\varphi \in \mathscr{C}^\infty_c(\mathbb{R}^n_+)$, this permits us to write

$$\begin{aligned}
&{}_{\mathcal{D}'(\mathbb{R}^n_+)}\langle\partial_k F_j, \varphi\rangle_{\mathcal{D}(\mathbb{R}^n_+)} = -{}_{\mathcal{D}'(\mathbb{R}^n_+)}\langle F_j, \partial_k\varphi\rangle_{\mathcal{D}(\mathbb{R}^n_+)} = -\lim_{i\to\infty} {}_{\mathcal{D}'(\mathbb{R}^n_+)}\langle\partial_j w_{\varepsilon_i}, \partial_k\varphi\rangle_{\mathcal{D}(\mathbb{R}^n_+)}\\
&\quad = \lim_{i\to\infty} {}_{\mathcal{D}'(\mathbb{R}^n_+)}\langle\partial_k\partial_j w_{\varepsilon_i}, \varphi\rangle_{\mathcal{D}(\mathbb{R}^n_+)} = \lim_{i\to\infty} {}_{\mathcal{D}'(\mathbb{R}^n_+)}\langle\partial_j\partial_k w_{\varepsilon_i}, \varphi\rangle_{\mathcal{D}(\mathbb{R}^n_+)}\\
&\quad = -\lim_{i\to\infty} {}_{\mathcal{D}'(\mathbb{R}^n_+)}\langle\partial_k w_{\varepsilon_i}, \partial_j\varphi\rangle_{\mathcal{D}(\mathbb{R}^n_+)} = -{}_{\mathcal{D}'(\mathbb{R}^n_+)}\langle F_k, \partial_j\varphi\rangle_{\mathcal{D}(\mathbb{R}^n_+)}\\
&\quad = {}_{\mathcal{D}'(\mathbb{R}^n_+)}\langle\partial_j F_k, \varphi\rangle_{\mathcal{D}(\mathbb{R}^n_+)} \qquad (2.5.78)
\end{aligned}$$

from which we conclude that, classically, $\partial_k F_j = \partial_j F_k$ in $\mathbb{R}^n$ for $j, k \in \{1, \ldots, n\}$. Thus, $\vec{F}$ is a smooth, curl-free vector field in $\mathbb{R}^n_+$. As is well known[4] this guarantees the existence of a scalar (potential) function

$$w \in \mathscr{C}^\infty(\mathbb{R}^n_+) \text{ with the property that } \nabla w = \vec{F} \text{ in } \mathbb{R}^n_+. \tag{2.5.79}$$

Together, (2.5.76) and (2.5.79) then prove that

$$\vec{u}(x) = x_n(\nabla w)(x) + \vec{v}(x) \text{ for each } x = (x', x_n) \in \mathbb{R}^n_+. \tag{2.5.80}$$

In addition, the convergence in (2.5.77) implies that in the sense of distributions in $\mathbb{R}^n_+$ we have $\Delta w = \operatorname{div}(\nabla w) = \operatorname{div} \vec{F} = \lim_{i\to\infty} \operatorname{div}(\nabla w_{\varepsilon_i}) = \lim_{i\to\infty} \Delta w_{\varepsilon_i} = 0$, with the last equality provided by (2.5.65). Finally, from (2.5.80), the last line in (2.5.52), and the third line in (2.5.69) we obtain $\mathcal{N}_\kappa\big(x \mapsto x_n(\nabla w)(x)\big) \in L^p(\mathbb{R}^{n-1}, \mathcal{L}^{n-1})$. At this stage, all properties claimed in (2.5.53)-(2.5.56) have been justified.

We now prove that $\vec{u}$ determines ∇w and $\vec{v}$ uniquely. By linearity, it suffices to show that if (2.5.53) holds for $\vec{u} = 0$ and with w and $\vec{v}$ as in (2.5.54)-(2.5.55) then $\vec{v}$ and ∇w are zero. To this end, assume w satisfies (2.5.54), $\vec{v}$ satisfies (2.5.55), and

$$\vec{v}(x) + x_n(\nabla w)(x) = 0 \text{ for each } x \in \mathbb{R}^n_+. \tag{2.5.81}$$

Applying the divergence operator to the latter identity yields

$$0 = (\operatorname{div} \vec{v})(x) + x_n(\Delta w)(x) + (\partial_n w)(x) = (\partial_n w)(x) \text{ for each } x \in \mathbb{R}^n_+. \tag{2.5.82}$$

In particular, the Mean Value Theorem implies that $w(x) = w(x', 1)$ for each point $x = (x', x_n) \in \mathbb{R}^{n-1} \times (0, \infty) = \mathbb{R}^n_+$. From this and (2.5.81) we then conclude that

$$\vec{v}(x) = \big(- x_n \nabla'[w(x', 1)], 0\big) \text{ for each } x = (x', x_n) \in \mathbb{R}^n_+, \tag{2.5.83}$$

where ∇' denotes the gradient operator in the variable $x' \in \mathbb{R}^{n-1}$. Consequently, $|\vec{v}(x)| = x_n \big|\nabla'[w(x', 1)]\big|$ for each $x = (x', x_n) \in \mathbb{R}^n_+$, which goes to show that for each $x' \in \mathbb{R}^{n-1}$ we have $\big(\mathcal{N}_\kappa \vec{v}\big)(x') = +\infty$ whenever $\big|\nabla'[w(x', 1)]\big| \neq 0$. Since $\mathcal{N}_\kappa \vec{v} \in L^p(\mathbb{R}^{n-1}, \mathcal{L}^{n-1})$, this forces $\nabla'[w(x', 1)] = 0$ for each $x' \in \mathbb{R}^{n-1}$, and in light of (2.5.82) we ultimately arrive at the conclusion that $\nabla w = 0$ in $\mathbb{R}^n_+$. From this and (2.5.81) we also see that $\vec{v} = 0$ in $\mathbb{R}^n_+$, as wanted.

Going further, in addition to (2.5.52) assume that the nontangential boundary trace $\vec{u}\big|^{\kappa-\mathrm{n.t.}}_{\partial\mathbb{R}^n_+}$ exists at $\mathcal{L}^{n-1}$-a.e. point in $\mathbb{R}^{n-1}$. Then Lebesgue's Dominated Convergence Theorem implies that the sequence

$$\big(\vec{u}_\varepsilon\big|_{\partial\mathbb{R}^n_+}\big)_{\varepsilon>0} \text{ converges to } \vec{u}\big|^{\kappa-\mathrm{n.t.}}_{\partial\mathbb{R}^n_+} \text{ in } \big[L^p(\mathbb{R}^{n-1}, \mathcal{L}^{n-1})\big]^n \text{ as } \varepsilon \to 0^+. \tag{2.5.84}$$

From this and (2.5.63) we then see that

[4] see, e.g., [140, Proposition 5.7] for general results of this flavor

$$\vec{f} = \vec{u}\Big|^{\kappa-\mathrm{n.t.}}_{\partial\mathbb{R}^n_+} \tag{2.5.85}$$

which, in concert with the last property in (2.5.69), shows that $\vec{v}\Big|^{\kappa-\mathrm{n.t.}}_{\partial\mathbb{R}^n_+} = \vec{u}\Big|^{\kappa-\mathrm{n.t.}}_{\partial\mathbb{R}^n_+}$ at $\mathcal{L}^{n-1}$-a.e. point on $\mathbb{R}^{n-1}$. Going nontangentially to the boundary in (2.5.53) also shows that $\big[x \mapsto x_n(\nabla w)(x)\big]\Big|^{\kappa-\mathrm{n.t.}}_{\partial\mathbb{R}^n_+} = 0$, finishing the proof of (2.5.57).

Next, assume w is as in (2.5.54), $\vec{v}$ is as in (2.5.55), and define $\vec{u}$ associated with w and $\vec{v}$ as in (2.5.53). Then it is immediate that $\vec{u} \in \big[\mathscr{C}^\infty(\mathbb{R}^n_+)\big]^n$ and, much as in (2.5.30)-(2.5.31), we see that $\Delta\vec{u} - 2\nabla\mathrm{div}\,\vec{u} = 0$ in $\mathbb{R}^n_+$. Moreover, from the last property in (2.5.54) and the last property in (2.5.55) we deduce that $\mathcal{N}_\kappa\vec{u}$ belongs to $L^p(\mathbb{R}^{n-1}, \mathcal{L}^{n-1})$. Thus, $\vec{u}$ has all the qualities specified in (2.5.52).

As far as the claim in (2.5.58) is concerned, if $w\Big|^{\kappa-\mathrm{n.t.}}_{\partial\mathbb{R}^n_+}$ exists at $\mathcal{L}^{n-1}$-a.e. point on $\mathbb{R}^{n-1}$, then the local Fatou theorem for the harmonic function w in the upper half-space (see, e.g., [198, Theorem 3, p.201]) guarantees that for any other aperture parameter $\widetilde{\kappa} \in (0,\infty)$ the nontangential boundary trace $w\Big|^{\widetilde{\kappa}-\mathrm{n.t.}}_{\partial\mathbb{R}^n_+}$ exists at $\mathcal{L}^{n-1}$-a.e. point on $\mathbb{R}^{n-1}$. Granted this, we may invoke the property in the last part of the statement of [129, Proposition 8.9.11] and conclude that

$$\big[x \mapsto x_n(\nabla w)(x)\big]\Big|^{\kappa-\mathrm{n.t.}}_{\partial\mathbb{R}^n_+} = 0 \text{ at } \mathcal{L}^{n-1}\text{-a.e. point on } \mathbb{R}^{n-1}. \tag{2.5.86}$$

Together, (2.5.86), (2.5.56), and (2.5.53) then show that the nontangential boundary trace $\vec{u}\Big|^{\kappa-\mathrm{n.t.}}_{\partial\mathbb{R}^n_+}$ exists, finishing the proof of (2.5.58).

To deal with the equivalence claimed in (2.5.59), assume first that, in addition to (2.5.52), we also have

$$\mathcal{N}_\kappa(\nabla\vec{u}) \text{ belongs to } L^p(\mathbb{R}^{n-1}, \mathcal{L}^{n-1}). \tag{2.5.87}$$

From what we have proved already, there exist w and $\vec{v}$ as in (2.5.54)-(2.5.55) such that the decomposition (2.5.53) holds. Applying the divergence operator to both sides of (2.5.53) leads to the conclusion that

$$\mathrm{div}\,\vec{u} = \partial_n w \text{ in } \mathbb{R}^n_+, \tag{2.5.88}$$

since both $\vec{v}$ and ∇w are divergence-free vector fields in $\mathbb{R}^n_+$ (cf. (2.5.54)-(2.5.55)). In concert with (2.5.87) this implies (also bearing in mind [129, (8.2.28)]) that

$$\begin{gathered}\mathcal{N}_\kappa(\partial_n w) \text{ belongs to the space } L^p(\mathbb{R}^{n-1}, \mathcal{L}^{n-1}), \text{ and} \\ \|\mathcal{N}_\kappa(\partial_n w)\|_{L^p(\mathbb{R}^{n-1},\mathcal{L}^{n-1})} = \|\mathcal{N}_\kappa(\mathrm{div}\,\vec{u})\|_{L^p(\mathbb{R}^{n-1},\mathcal{L}^{n-1})} \leq \|\mathcal{N}_\kappa(\nabla\vec{u})\|_{L^p(\mathbb{R}^{n-1},\mathcal{L}^{n-1})}.\end{gathered} \tag{2.5.89}$$

In addition,

$$\lim_{t\to\infty}(\nabla w)(z', t) = 0 \text{ for each fixed } z' \in \mathbb{R}^{n-1}. \tag{2.5.90}$$

To justify this, pick an arbitrary $z = (z', t) \in \mathbb{R}^{n-1} \times (0,\infty) = \mathbb{R}^n_+$ and notice that

$$\text{if } \varepsilon \in \big(0, \tfrac{\kappa}{4+2\kappa}\big) \text{ then } B(z, \varepsilon t) \subseteq \Gamma_\kappa(y) \text{ for each } y \in B_{n-1}(z', \kappa t/2). \tag{2.5.91}$$

As such,

$$|x_n(\nabla w)(x)| \leq \mathcal{N}_\kappa\big(x \mapsto x_n(\nabla w)(x)\big)(y), \;\; \forall x \in B(z, \varepsilon t), \;\; \forall y \in B_{n-1}(z', \kappa t/2). \tag{2.5.92}$$

Raising both sides to the p-th power and taking integral averages yields

$$\fint_{B(z,\varepsilon t)} |x_n(\nabla w)(x)|^p \, \mathrm{d}x \leq \fint_{B_{n-1}(z',\kappa t/2)} \mathcal{N}_\kappa\big(x \mapsto x_n(\nabla w)(x)\big)^p \, \mathrm{d}\mathcal{L}^{n-1}. \tag{2.5.93}$$

Based on interior estimates for harmonic functions and (2.5.93) we may now write

$$\begin{aligned} |(\nabla w)(z)| &\leq C\Big(\fint_{B(z,\varepsilon t)} |(\nabla w)(x)|^p \, \mathrm{d}x\Big)^{\frac{1}{p}} \leq \frac{C}{t}\Big(\fint_{B(z,\varepsilon t)} |x_n(\nabla w)(x)|^p \, \mathrm{d}x\Big)^{\frac{1}{p}} \\ &\leq \frac{C}{t}\Big(\fint_{B_{n-1}(z',\kappa t/2)} \mathcal{N}_\kappa\big(x \mapsto x_n(\nabla w)(x)\big)^p \, \mathrm{d}\mathcal{L}^{n-1}\Big)^{\frac{1}{p}} \\ &\leq \frac{C}{t^{1+(n-1)/p}} \big\|\mathcal{N}_\kappa\big(x \mapsto x_n(\nabla w)(x)\big)\big\|_{L^p(\mathbb{R}^{n-1},\mathcal{L}^{n-1})}. \end{aligned} \tag{2.5.94}$$

From (2.5.94) and the last property in (2.5.54) we then conclude that (2.5.90) holds. In a similar fashion, for each point $z = (z', t) \in \mathbb{R}^{n-1} \times (0, \infty) = \mathbb{R}^n_+$ we have

$$\begin{aligned} |(\nabla\nabla w)(z)| &\leq \frac{C}{t}\Big(\fint_{B(z,\varepsilon t)} |(\nabla w)(x)|^p \, \mathrm{d}x\Big)^{\frac{1}{p}} \leq \frac{C}{t^2}\Big(\fint_{B(z,\varepsilon t)} |x_n(\nabla w)(x)|^p \, \mathrm{d}x\Big)^{\frac{1}{p}} \\ &\leq \frac{C}{t^2}\Big(\fint_{B_{n-1}(z',\kappa t/2)} \mathcal{N}_\kappa\big(x \mapsto x_n(\nabla w)(x)\big)^p \, \mathrm{d}\mathcal{L}^{n-1}\Big)^{\frac{1}{p}} \\ &\leq \frac{C}{t^{2+(n-1)/p}} \big\|\mathcal{N}_\kappa\big(x \mapsto x_n(\nabla w)(x)\big)\big\|_{L^p(\mathbb{R}^{n-1},\mathcal{L}^{n-1})}. \end{aligned} \tag{2.5.95}$$

Hence, in addition to (2.5.90), we also have

$$\lim_{t\to\infty} (\nabla\nabla w)(z', t) = 0 \;\text{ for each }\; z' \in \mathbb{R}^{n-1}. \tag{2.5.96}$$

With (2.5.89), (2.5.90), (2.5.96) in hand, we next invoke Lemma 2.5.3 to conclude that there exists some $C \in (0, \infty)$ independent of u such that

$$\begin{gathered} \mathcal{N}_\kappa(\nabla w) \text{ belongs to the space } L^p(\mathbb{R}^{n-1}, \mathcal{L}^{n-1}) \text{ and} \\ \|\mathcal{N}_\kappa(\nabla w)\|_{L^p(\mathbb{R}^{n-1},\mathcal{L}^{n-1})} \leq C\|\mathcal{N}_\kappa(\partial_n w)\|_{L^p(\mathbb{R}^{n-1},\mathcal{L}^{n-1})} \leq C\|\mathcal{N}_\kappa(\nabla \vec{u})\|_{L^p(\mathbb{R}^{n-1},\mathcal{L}^{n-1})}, \end{gathered} \tag{2.5.97}$$

where the very last inequality comes from (2.5.89).

Pressing on, the working assumption in (2.5.87), the last property in (2.5.52), [129, (8.9.235) in Proposition 8.9.22], and [130, Proposition 11.3.4] imply

$$\vec{u}\Big|^{\kappa-\mathrm{n.t.}}_{\partial\mathbb{R}^n_+} \in \big[L^p_1(\mathbb{R}^{n-1}, \mathcal{L}^{n-1})\big]^n \tag{2.5.98}$$

and there exists a finite constant $C > 0$, independent of u, such that

$$\Big\|\vec{u}\Big|^{\kappa-\mathrm{n.t.}}_{\partial\mathbb{R}^n_+}\Big\|_{[L^p_1(\mathbb{R}^{n-1},\mathcal{L}^{n-1})]^n} \leq C\big\|\mathcal{N}_\kappa\vec{u}\big\|_{L^p(\mathbb{R}^{n-1},\mathcal{L}^{n-1})} + C\big\|\mathcal{N}_\kappa(\nabla\vec{u})\big\|_{L^p(\mathbb{R}^{n-1},\mathcal{L}^{n-1})}. \tag{2.5.99}$$

Since from (2.5.55) and (2.5.57) $\vec{v}$ is known to solve the Dirichlet Problem

$$\begin{cases} \vec{v} \in \big[\mathscr{C}^\infty(\mathbb{R}^n_+)\big]^n, \quad \Delta\vec{v} = 0 \text{ in } \mathbb{R}^n_+, \\ \mathcal{N}_\kappa\vec{v} \in L^p(\mathbb{R}^{n-1},\mathcal{L}^{n-1}), \\ \vec{v}\Big|^{\kappa-\mathrm{n.t.}}_{\partial\mathbb{R}^n_+} = \vec{u}\Big|^{\kappa-\mathrm{n.t.}}_{\partial\mathbb{R}^n_+} \text{ at } \mathcal{L}^{n-1}\text{-a.e. point on } \mathbb{R}^{n-1}, \end{cases} \tag{2.5.100}$$

and since the solution acquires more regularity if the boundary datum is more regular, from (2.5.100) and (2.5.98)-(2.5.99) we conclude (cf. [113, Theorem 4.1]) that

$$\begin{gathered} \mathcal{N}_\kappa(\nabla\vec{v}) \text{ belongs to the space } L^p(\mathbb{R}^{n-1},\mathcal{L}^{n-1}) \text{ and} \\ \|\mathcal{N}_\kappa(\nabla\vec{v})\|_{L^p(\mathbb{R}^{n-1},\mathcal{L}^{n-1})} \leq C\Big(\big\|\mathcal{N}_\kappa\vec{u}\big\|_{L^p(\mathbb{R}^{n-1},\mathcal{L}^{n-1})} + \big\|\mathcal{N}_\kappa(\nabla\vec{u})\big\|_{L^p(\mathbb{R}^{n-1},\mathcal{L}^{n-1})}\Big), \end{gathered} \tag{2.5.101}$$

with $C \in (0,\infty)$ independent of $\vec{u}$. In fact, a re-scaling argument (which entails working with $\vec{u}_\lambda(x) := \vec{u}(\lambda x)$ for each $\lambda \in (0,\infty)$ in place of $\vec{u}$, and with $\vec{v}_\lambda$ defined similarly in place of $\vec{v}$) shows that the above inequality self-improves (in view of the specific manner in which the terms involved behave under dilations) to just

$$\|\mathcal{N}_\kappa(\nabla\vec{v})\|_{L^p(\mathbb{R}^{n-1},\mathcal{L}^{n-1})} \leq C\big\|\mathcal{N}_\kappa(\nabla\vec{u})\big\|_{L^p(\mathbb{R}^{n-1},\mathcal{L}^{n-1})}. \tag{2.5.102}$$

This takes care of the left-to-right implication in (2.5.59).

Next, assume that w and $\vec{v}$ are as in (2.5.54)-(2.5.55) and define $\vec{u}$ as in (2.5.53). This time, work under the additional assumption that

$$\mathcal{N}_\kappa(\nabla\vec{v}) \in L^p(\mathbb{R}^{n-1},\mathcal{L}^{n-1}) \text{ and } \mathcal{N}_\kappa(\nabla w) \in L^p(\mathbb{R}^{n-1},\mathcal{L}^{n-1}). \tag{2.5.103}$$

Applying ∂_j with $j \in \{1,\dots,n\}$ to (2.5.53) shows that

$$(\partial_j\vec{u})(x) = (\partial_j\vec{v})(x) + \delta_{jn}(\nabla w)(x) + x_n(\nabla\partial_j w)(x), \quad \forall\, x = (x',x_n) \in \mathbb{R}^n_+. \tag{2.5.104}$$

From this, (2.5.103), and the last part in [129, Corollary 8.9.13] we then see that

$$\begin{gathered} \mathcal{N}_\kappa(\nabla\vec{u}) \text{ belongs to } L^p(\mathbb{R}^{n-1},\mathcal{L}^{n-1}) \text{ and} \\ \|\mathcal{N}_\kappa(\nabla\vec{u})\|_{L^p(\mathbb{R}^{n-1},\mathcal{L}^{n-1})} \leq C\Big(\|\mathcal{N}_\kappa(\nabla w)\|_{L^p(\mathbb{R}^{n-1},\mathcal{L}^{n-1})} + \|\mathcal{N}_\kappa(\nabla\vec{v})\|_{L^p(\mathbb{R}^{n-1},\mathcal{L}^{n-1})}\Big) \end{gathered} \tag{2.5.105}$$

for some constant $C \in (0,\infty)$ independent of $\vec{u}$. This finishes the proof of the equivalence claimed in (2.5.59) (including its quantitative aspect).

Finally, that in each of the above two happenstances described in (2.5.59) the decomposition in (2.5.53) agrees with the decomposition in (2.5.5) is clear from what we have just proved since, in each case, $\vec{u}$ determines ∇w and $\vec{v}$ uniquely. □

2.6 Spaces of Null-Solutions and Admissible Boundary Data for L_D in the Upper Half-Space

The principal aim in this section is to provide explicit descriptions for the spaces of null-solutions and the spaces of admissible boundary data for boundary value problems for the weakly elliptic $n\times n$ system $L_D = \Delta - 2\nabla\mathrm{div}$ in the upper half-space. Our main tools in this regard are the "structure theorems" established in the previous section (cf. Theorem 2.5.1 and Theorem 2.5.4). Tellingly, the two constitutive pieces, $x_n\nabla w$ and $\vec{v}$, appearing in the decompositions of a null-solution $\vec{u}$ of the system L_D in $\mathbb{R}^n_+$ given in (2.5.5) and (2.5.53), are solely and independently responsible for the format of said spaces of null-solutions and spaces of admissible boundary data.

We begin by providing a characterization of the space of null-solutions for the Homogeneous Regularity Problem for the system $L_D = \Delta - 2\nabla\mathrm{div}$ in the upper half-space, i.e., the space of vector-valued functions $\vec{u}$ satisfying (for some fixed exponent $p \in (1,\infty)$ and some background aperture parameter $\kappa > 0$)

$$\begin{gathered}\vec{u} \in \big[\mathscr{C}^\infty(\mathbb{R}^n_+)\big]^n, \quad \Delta\vec{u} - 2\nabla\mathrm{div}\,\vec{u} = 0 \text{ in } \mathbb{R}^n_+, \quad \mathcal{N}_\kappa(\nabla\vec{u}) \in L^p(\mathbb{R}^{n-1}, \mathcal{L}^{n-1}),\\ \text{and } \vec{u}\big|^{\kappa-\mathrm{n.t.}}_{\partial\mathbb{R}^n_+} = 0 \text{ at } \mathcal{L}^{n-1}\text{-a.e. point on } \mathbb{R}^{n-1}.\end{gathered} \tag{2.6.1}$$

Corollary 2.6.1 *Fix $n \in \mathbb{N}$ with $n \geq 2$, along with some $p \in (1,\infty)$ and $\kappa > 0$. Then the space of null-solutions for the Homogeneous Regularity Problem for the system $L_D = \Delta - 2\nabla\mathrm{div}$ in the upper half-space has the following description:*

$$\begin{aligned}&\Big\{\vec{u} \in \big[\mathscr{C}^\infty(\mathbb{R}^n_+)\big]^n : \Delta\vec{u} - 2\nabla\mathrm{div}\,\vec{u} = 0,\ \vec{u}\big|^{\kappa-\mathrm{n.t.}}_{\partial\mathbb{R}^n_+} = 0,\ \mathcal{N}_\kappa(\nabla\vec{u}) \in L^p(\mathbb{R}^{n-1}, \mathcal{L}^{n-1})\Big\}\\ &\quad = \Big\{x_n\nabla w : w \in \mathscr{C}^\infty(\mathbb{R}^n_+),\ \Delta w = 0,\ \mathcal{N}_\kappa(\nabla w) \in L^p(\mathbb{R}^{n-1}, \mathcal{L}^{n-1})\Big\}.\end{aligned} \tag{2.6.2}$$

As a consequence of this and Theorem 2.4.5,

the space of null-solutions for the Homogeneous Regularity Problem (2.6.2), for the system $L_D = \Delta - 2\nabla\mathrm{div}$ in the upper half-space, is isomorphic to the Lebesgue space $L^p(\mathbb{R}^{n-1}, \mathcal{L}^{n-1})$. (2.6.3)

In particular,

the space of null-solutions for the Homogeneous Regularity Problem (2.6.2), for the system L_D in the upper half-space (i.e., the space of functions as in (2.6.1)), is infinite dimensional. (2.6.4)

Proof If $\vec{u}$ belongs to the set in the left-hand side of (2.6.2), then by Theorem 2.5.1 we have that $\vec{u}(x) = \vec{v}(x) + x_n(\nabla w)(x)$ at each $x \in \mathbb{R}^n_+$, where w is an in (2.5.6) and $\vec{v}$ is as in (2.5.7)-(2.5.8). Consequently, $\vec{v}\big|^{\kappa-\mathrm{n.t.}}_{\partial\mathbb{R}^n_+} = 0$. Keeping this in mind, [132, Theorem 1.8.19] then gives that the nontangential trace $(\nabla\vec{v})\big|^{\kappa-\mathrm{n.t.}}_{\partial\mathbb{R}^n_+}$ exists at $\mathcal{L}^{n-1}$-a.e. point on $\mathbb{R}^{n-1}$ and belongs to $\big[L^p(\mathbb{R}^{n-1}, \mathcal{L}^{n-1})\big]^{n^2}$, and there exists $c \in \mathbb{C}^n$ so that

$$\vec{v} = \mathscr{S}_{\rm mod}\Big((\partial_n \vec{v})\Big|^{\kappa-{\rm n.t.}}_{\partial\mathbb{R}^n_+}\Big) + c \text{ in } \mathbb{R}^n_+. \tag{2.6.5}$$

Applying ∂_n, going nontangentially to the boundary then yields on account of [132, (1.5.58)] (bearing in mind that $K^{\#}_{\Delta}$ is presently identically zero)

$$(\partial_n \vec{v})\Big|^{\kappa-{\rm n.t.}}_{\partial\mathbb{R}^n_+} = \tfrac{1}{2}(\partial_n \vec{v})\Big|^{\kappa-{\rm n.t.}}_{\partial\mathbb{R}^n_+} \tag{2.6.6}$$

Hence, $(\partial_n \vec{v})\Big|^{\kappa-{\rm n.t.}}_{\partial\mathbb{R}^n_+} = 0$ which, when used back in (2.6.5), proves that $\vec{v} = c$. Finally, the fact that $\vec{v}\Big|^{\kappa-{\rm n.t.}}_{\partial\mathbb{R}^n_+} = 0$ forces $c = 0$, hence $\vec{v} \equiv 0$ in $\mathbb{R}^n_+$. As a consequence, $\vec{u} = x_n \nabla w$ in $\mathbb{R}^n_+$ proving the left-to-right inclusion in (2.6.2).

For the converse inclusion, we observe that if w is harmonic in $\mathbb{R}^n_+$ and $\mathcal{N}_\kappa(\nabla w)$ belongs to $L^p(\mathbb{R}^{n-1}, \mathcal{L}^{n-1})$, by Calderón's version of Fatou's theorem (cf. [26]) the trace $(\nabla w)\Big|^{\kappa-{\rm n.t.}}_{\partial\mathbb{R}^n_+}$ exists at $\mathcal{L}^{n-1}$-a.e. point in $\mathbb{R}^{n-1}$, hence $(x_n \nabla w)\Big|^{\kappa-{\rm n.t.}}_{\partial\mathbb{R}^n_+} = 0$. Now the desired conclusion follows from Theorem 2.5.1 used for this w and $\vec{v} := 0$. □

Next we include a characterization of the space of null-solutions for the Inhomogeneous Regularity Problem for the system $L_D = \Delta - 2\nabla{\rm div}$ in the upper half-space.

Corollary 2.6.2 *Pick $n \in \mathbb{N}$ with $n \geq 2$ along with an exponent $p \in (1, \infty)$ and an aperture parameter $\kappa > 0$. Then the space of null-solutions for the Inhomogeneous Regularity Problem for the system $L_D = \Delta - 2\nabla{\rm div}$ in the upper half-space, i.e.,*

$$\Big\{\vec{u} \in \big[\mathscr{C}^\infty(\mathbb{R}^n_+)\big]^n : \Delta\vec{u} - 2\nabla{\rm div}\,\vec{u} = 0 \text{ in } \mathbb{R}^n_+,\ \vec{u}\Big|^{\kappa-{\rm n.t.}}_{\partial\mathbb{R}^n_+} = 0, \\ \mathcal{N}_\kappa\vec{u} \in L^p(\mathbb{R}^{n-1}, \mathcal{L}^{n-1}) \text{ and } \mathcal{N}_\kappa(\nabla\vec{u}) \in L^p(\mathbb{R}^{n-1}, \mathcal{L}^{n-1})\Big\} \tag{2.6.7}$$

coincides with

$$\Big\{x_n \nabla w : w \in \mathscr{C}^\infty(\mathbb{R}^n_+),\ \Delta w = 0 \text{ in } \mathbb{R}^n_+,\ \mathcal{N}_\kappa(\nabla w) \in L^p(\mathbb{R}^{n-1}, \mathcal{L}^{n-1}) \\ \text{and } \mathcal{N}_\kappa\big(x \mapsto x_n(\nabla w)(x)\big) \in L^p(\mathbb{R}^{n-1}, \mathcal{L}^{n-1})\Big\}. \tag{2.6.8}$$

In particular, the space of null-solutions for the Inhomogeneous Regularity Problem for the system $L_D = \Delta - 2\nabla{\rm div}$ in the upper half-space, i.e., (2.6.7)*, contains*

$$\Big\{x_n \nabla w : w \in \mathscr{C}^\infty(\mathbb{R}^n_+),\ \Delta w = 0,\ \mathcal{N}_\kappa w,\ \mathcal{N}_\kappa(\nabla w) \in L^p(\mathbb{R}^{n-1}, \mathcal{L}^{n-1})\Big\}. \tag{2.6.9}$$

Moreover, the mapping

$$L^p_1(\mathbb{R}^{n-1}, \mathcal{L}^{n-1}) \ni f \longmapsto x_n \nabla\big[(P^{\Delta}_{x_n} * f)(x')\big], \qquad (x', x_n) \in \mathbb{R}^n_+, \tag{2.6.10}$$

establishes an isomorphism between the Sobolev space $L^p_1(\mathbb{R}^{n-1}, \mathcal{L}^{n-1})$ and the space described in (2.6.9)*. As a corollary,*

$$\text{the space of null-solutions for the Inhomogeneous Regularity Problem (2.6.7), for the system } L_D \text{ in the upper half-space (defined in (2.6.7)), is infinite dimensional.} \tag{2.6.11}$$

Proof The identification of (2.6.7) with (2.6.8) is seen from Corollary 2.6.1 and Theorem 2.5.1 (cf. (2.5.9)). The fact that (2.6.9) is contained in (2.6.8) follows from the last part in [129, Corollary 8.9.13]. There remains to show that the mapping described in (2.6.10) establishes an isomorphism between $L_1^p(\mathbb{R}^{n-1}, \mathcal{L}^{n-1})$ and the space in (2.6.9). This, however, readily follows from the fact that the Inhomogeneous Regularity Problem for the Laplacian in the upper half-space, i.e.,

$$\begin{aligned} w \in \mathscr{C}^\infty(\mathbb{R}^n_+), \quad \Delta w = 0 \text{ in } \mathbb{R}^n_+, \quad \mathcal{N}_\kappa w,\ \mathcal{N}_\kappa(\nabla w) \in L^p(\mathbb{R}^{n-1}, \mathcal{L}^{n-1}), \\ \text{and } w\Big|_{\partial\mathbb{R}^n_+}^{\kappa-\text{n.t.}} = f \in L_1^p(\mathbb{R}^{n-1}, \mathcal{L}^{n-1}), \end{aligned} \tag{2.6.12}$$

is well-posed, and for each boundary datum $f \in L_1^p(\mathbb{R}^{n-1}, \mathcal{L}^{n-1})$ its unique solution is given by the function $w(x', x_n) := (P^\Delta_{x_n} * f)(x')$ for all $(x', x_n) \in \mathbb{R}^n_+$. □

As a consequence of (2.6.11), we also see that

$$\text{the space of null-solutions for the } L^p \text{ Dirichlet Problem for the system } L_D \text{ in the upper half-space is infinite dimensional.} \tag{2.6.13}$$

An actual description of the aforementioned space is given in the corollary below.

Corollary 2.6.3 *Pick an integer $n \in \mathbb{N}$ with $n \geq 2$ and an exponent $p \in (1, \infty)$ along with an aperture parameter $\kappa > 0$. Then the space of null-solutions for the L^p Dirichlet Problem for $L_D = \Delta - 2\nabla\text{div}$ in $\mathbb{R}^n_+$, i.e.,*

$$\Big\{\vec{u} \in \big[\mathscr{C}^\infty(\mathbb{R}^n_+)\big]^n : \Delta\vec{u} - 2\nabla\text{div}\,\vec{u} = 0 \text{ in } \mathbb{R}^n_+, \quad \vec{u}\Big|_{\partial\mathbb{R}^n_+}^{\kappa-\text{n.t.}} = 0, \text{ and } \mathcal{N}_\kappa\vec{u} \in L^p(\mathbb{R}^{n-1}, \mathcal{L}^{n-1})\Big\} \tag{2.6.14}$$

coincides with

$$\Big\{x_n\nabla w : w \in \mathscr{C}^\infty(\mathbb{R}^n_+), \Delta w = 0 \text{ in } \mathbb{R}^n_+, \big[x \mapsto x_n(\nabla w)(x)\big]\Big|_{\partial\mathbb{R}^n_+}^{\kappa-\text{n.t.}} = 0, \text{ and } \mathcal{N}_\kappa\big(x \mapsto x_n(\nabla w)(x)\big) \in L^p(\mathbb{R}^{n-1}, \mathcal{L}^{n-1})\Big\}. \tag{2.6.15}$$

Proof This is a consequence of Theorem 2.5.4. Specifically, if $\vec{u}$ is as in (2.6.14) then from (2.5.53) and (2.5.57) we know that $\vec{u}$ may be written as

$$\vec{u} = \vec{v} + x_n\nabla w \text{ in } \mathbb{R}^n_+ \tag{2.6.16}$$

where

$$\begin{cases} w \in \mathscr{C}^\infty(\mathbb{R}^n_+), \quad \Delta w = 0 \text{ in } \mathbb{R}^n_+, \\ \mathcal{N}_\kappa\big(x \mapsto x_n(\nabla w)(x)\big) \in L^p(\mathbb{R}^{n-1}, \mathcal{L}^{n-1}), \\ \big[x \mapsto x_n(\nabla w)(x)\big]\Big|^{\kappa-\text{n.t.}}_{\partial\mathbb{R}^n_+} = 0, \end{cases} \tag{2.6.17}$$

and

$$\begin{cases} \vec{v} \in \big[\mathscr{C}^\infty(\mathbb{R}^n_+)\big]^n, \quad \Delta\vec{v} = 0 \text{ and } \operatorname{div}\vec{v} = 0 \text{ in } \mathbb{R}^n_+, \\ \mathcal{N}_\kappa\vec{v} \in L^p(\mathbb{R}^{n-1}, \mathcal{L}^{n-1}), \quad \vec{v}\Big|^{\kappa-\text{n.t.}}_{\partial\mathbb{R}^n_+} = 0. \end{cases} \tag{2.6.18}$$

In particular, the well-posedness of the L^p Dirichlet Problem for the Laplacian in the upper half-space forces $\vec{v} = 0$ in $\mathbb{R}^n_+$, so (2.6.16) simply reduces to $\vec{u} = x_n\nabla w$ in $\mathbb{R}^n_+$ with w as in (2.6.17). This proves that (2.6.14) is contained in (2.6.15). Since the opposite inclusion is clear, the desired conclusion follows. □

Changing topics, we now take up the task of describing the space of admissible boundary data for various natural boundary value problems in the upper half-space for the system $L_D = \Delta - 2\nabla\text{div}$. First we treat the L^p Dirichlet Problem.

Theorem 2.6.4 *Fix an integer $n \in \mathbb{N}$ with $n \geq 2$, along with an integrability exponent $p \in (1, \infty)$, and an aperture parameter $\kappa > 0$. Then the space of admissible boundary data for the L^p Dirichlet Problem for the system L_D in the upper half-space, namely*

$$\Big\{\vec{u}\Big|^{\kappa-\text{n.t.}}_{\partial\mathbb{R}^n_+} : \vec{u} \in \big[\mathscr{C}^\infty(\mathbb{R}^n_+)\big]^n,\ \Delta\vec{u} - 2\nabla\text{div}\,\vec{u} = 0 \text{ in } \mathbb{R}^n_+, \tag{2.6.19}$$
$$\mathcal{N}_\kappa\vec{u} \in L^p(\mathbb{R}^{n-1}, \mathcal{L}^{n-1}) \text{ and } \vec{u}\Big|^{\kappa-\text{n.t.}}_{\partial\mathbb{R}^n_+} \text{ exists } \mathcal{L}^{n-1}\text{-a.e. on } \mathbb{R}^{n-1}\Big\}$$

may be described as

$$\Big\{(f_1, \dots, f_n) \in \big[L^p(\mathbb{R}^{n-1}, \mathcal{L}^{n-1})\big]^n : f_n = -\sum_{j=1}^{n-1} R_j f_j\Big\}. \tag{2.6.20}$$

Proof This is a direct consequence of Theorem 2.5.4 and Proposition 2.4.1. □

A consequence of Theorem 2.6.4 is the description of the quotient space

$$\frac{\text{Full Space of Boundary Data}}{\text{Space of Admissible Boundary Data}} \tag{2.6.21}$$

in the case of the Dirichlet Problem for the system L_D in $\mathbb{R}^n_+$ (cf. (2.6.25)).

Corollary 2.6.5 *Fix an integer $n \in \mathbb{N}$ with $n \geq 2$, along with an integrability exponent $p \in (1, \infty)$, and an aperture parameter $\kappa > 0$. Then the space of admissible boundary data for the L^p Dirichlet Problem for the system L_D in $\mathbb{R}^n_+$,*

$$\mathscr{U}_p := \Big\{\vec{u}\Big|^{\kappa-\text{n.t.}}_{\partial\mathbb{R}^n_+} : \vec{u} \in \big[\mathscr{C}^\infty(\mathbb{R}^n_+)\big]^n,\ \Delta\vec{u} - 2\nabla\text{div}\,\vec{u} = 0 \text{ in } \mathbb{R}^n_+, \tag{2.6.22}$$
$$\vec{u}\Big|^{\kappa-\text{n.t.}}_{\partial\mathbb{R}^n_+} \text{ exists } \mathcal{L}^{n-1}\text{-a.e. in } \mathbb{R}^{n-1},\ \mathcal{N}_\kappa\vec{u} \in L^p(\mathbb{R}^{n-1}, \mathcal{L}^{n-1})\Big\},$$

is a closed subspace of $\big[L^p(\mathbb{R}^{n-1},\mathcal{L}^{n-1})\big]^n$, *and has*

$$\mathscr{V}_p := \Big\{(0,\dots,0,f) \in \big[L^p(\mathbb{R}^{n-1},\mathcal{L}^{n-1})\big]^n : f \in L^p(\mathbb{R}^{n-1},\mathcal{L}^{n-1})\Big\} \tag{2.6.23}$$

as a topological complement, i.e.,

$$\big[L^p(\mathbb{R}^{n-1},\mathcal{L}^{n-1})\big]^n = \mathscr{U}_p \oplus \mathscr{V}_p \text{ (direct topological sum)}. \tag{2.6.24}$$

In particular, one has the isomorphism

$$\big[L^p(\mathbb{R}^{n-1},\mathcal{L}^{n-1})\big]^n / \mathscr{U}_p \cong \mathscr{V}_p \cong L^p(\mathbb{R}^{n-1},\mathcal{L}^{n-1}) \tag{2.6.25}$$

and, as a consequence,

> *the codimension of the space of admissible boundary data for the* L^p *Dirichlet Problem for the system* L_D *in* $\mathbb{R}^n_+$ *(i.e.,* $\mathscr{U}_p$ *from* (2.6.22)) *into the full data space* $\big[L^p(\mathbb{R}^{n-1},\mathcal{L}^{n-1})\big]^n$ *is* $+\infty$. (2.6.26)

Proof The continuity of the Riesz transforms on $L^p(\mathbb{R}^{n-1},\mathcal{L}^{n-1})$ make (2.6.20), hence also $\mathscr{U}_p$, a closed subspace of $\big[L^p(\mathbb{R}^{n-1},\mathcal{L}^{n-1})\big]^n$. Obviously, so is $\mathscr{V}_p$. The identification made in Theorem 2.6.4 shows that $\mathscr{U}_p \cap \mathscr{V}_p$ is the singleton $\{(0,\dots,0)\}$. Given any n-tuple $f_1,\dots,f_n \in L^p(\mathbb{R}^{n-1},\mathcal{L}^{n-1})$, we may decompose

$$(f_1,\dots,f_n) = \Big(f_1,\dots,f_{n-1},-\sum_{j=1}^{n-1} R_j f_j\Big) + \Big(0,\dots,0,f_n + \sum_{j=1}^{n-1} R_j f_j\Big) \tag{2.6.27}$$

with

$$\Big(f_1,\dots,f_{n-1},-\sum_{j=1}^{n-1} R_j f_j\Big) \in \mathscr{U}_p \text{ and } \Big(0,\dots,0,f_n + \sum_{j=1}^{n-1} R_j f_j\Big) \in \mathscr{V}_p \tag{2.6.28}$$

depending continuously on $f_1,\dots,f_n$. This finishes the proof of (2.6.24). □

Here is a result of similar flavor to the one just presented above.

Corollary 2.6.6 *Pick* $n \in \mathbb{N}$ *with* $n \geq 2$, *an exponent* $p \in (1,\infty)$, *and* $\kappa \in (0,\infty)$. *Then* $\mathscr{U}_p$, *the space of admissible boundary data for the* L^p *Dirichlet Problem for the system* L_D *in the upper half-space, described in* (2.6.22), *has*

$$\mathscr{W}_p := \Big\{(R_1 f,\dots,R_{n-1} f, 0) \in \big[L^p(\mathbb{R}^{n-1},\mathcal{L}^{n-1})\big]^n : f \in L^p(\mathbb{R}^{n-1},\mathcal{L}^{n-1})\Big\} \tag{2.6.29}$$

as a topological complement, i.e.,

$$\big[L^p(\mathbb{R}^{n-1},\mathcal{L}^{n-1})\big]^n = \mathscr{U}_p \oplus \mathscr{W}_p \text{ (direct topological sum)}. \tag{2.6.30}$$

As a consequence, one has the isomorphism

$$\big[L^p(\mathbb{R}^{n-1},\mathcal{L}^{n-1})\big]^n/\mathscr{U}_p \cong \mathscr{W}_p \cong L^p(\mathbb{R}^{n-1},\mathcal{L}^{n-1}), \tag{2.6.31}$$

which offers another proof of (2.6.26).

Proof That $\mathscr{W}_p$ is a closed subspace of $\big[L^p(\mathbb{R}^{n-1},\mathcal{L}^{n-1})\big]^n$ is clear from (2.6.29) and the continuity of the Riesz transforms on $L^p(\mathbb{R}^{n-1},\mathcal{L}^{n-1})$. Since also

$$\sum_{j=1}^{n-1} R_j^2 = -I \text{ on } L^p(\mathbb{R}^{n-1},\mathcal{L}^{n-1}), \tag{2.6.32}$$

where I is the identity operator on $L^p(\mathbb{R}^{n-1},\mathcal{L}^{n-1})$, it readily follows from definitions and the identification made in Theorem 2.6.4 that $\mathscr{U}_p \cap \mathscr{V}_p$ is the singleton $\{(0,\dots,0)\}$. Next, given any n-tuple $f_1,\dots,f_n \in L^p(\mathbb{R}^{n-1},\mathcal{L}^{n-1})$, define

$$g := f_n + \sum_{j=1}^{n-1} R_j f_j \in L^p(\mathbb{R}^{n-1},\mathcal{L}^{n-1}), \tag{2.6.33}$$

and decompose

$$\begin{aligned}(f_1,\dots,f_n) =&\Big(f_1 + R_1 g, f_2 + R_2 g,\dots, f_{n-1} + R_{n-1} g, f_n\Big)\\ &+\Big(-R_1 g, -R_2 g,\dots,-R_{n-1} g, 0\Big).\end{aligned} \tag{2.6.34}$$

The identification made in Theorem 2.6.4, (2.6.32), (2.6.33), and (2.6.29) imply

$$\begin{aligned}&\Big(f_1 + R_1 g, f_2 + R_2 g,\dots, f_{n-1} + R_{n-1} g, f_n\Big) \in \mathscr{U}_p \text{ and}\\ &\Big(-R_1 g, -R_2 g,\dots,-R_{n-1} g, 0\Big) \in \mathscr{W}_p.\end{aligned} \tag{2.6.35}$$

Since the above vector-valued functions depend continuously in $\big[L^p(\mathbb{R}^{n-1},\mathcal{L}^{n-1})\big]^n$ on $f_1,\dots,f_n$, the claim made in (2.6.30) is established. □

We continue by providing a description of the space of admissible boundary data for the Higher-Order Inhomogeneous Regularity Problem in the upper half-space for the system $L_D = \Delta - 2\nabla\mathrm{div}$ (thus augmenting the result in Theorem 2.6.4).

Theorem 2.6.7 *Fix $n \in \mathbb{N}$ with $n \geq 2$, an integrability exponent $p \in (1,\infty)$, an aperture parameter $\kappa > 0$, and an integer $k \in \mathbb{N}$. Then the space*

$$\begin{aligned}\Big\{\vec{u}\big|^{\kappa-\mathrm{n.t.}}_{\partial\mathbb{R}^n_+} :\ &\vec{u} \in \big[\mathscr{C}^\infty(\mathbb{R}^n_+)\big]^n,\ \Delta\vec{u} - 2\nabla\mathrm{div}\,\vec{u} = 0 \text{ in } \mathbb{R}^n_+,\\ &\mathcal{N}_\kappa(\nabla^\ell \vec{u}) \in L^p(\mathbb{R}^{n-1},\mathcal{L}^{n-1}),\ \ell = 0,1,\dots,k\Big\}\end{aligned} \tag{2.6.36}$$

coincides with the space

$$\Big\{(f_1,\dots,f_n)\in\big[L^p_k(\mathbb{R}^{n-1},\mathcal{L}^{n-1})\big]^n:\ f_n=-\sum_{j=1}^{n-1}R_jf_j\Big\}. \tag{2.6.37}$$

In particular, corresponding to $k=1$, the space of admissible boundary data for the Inhomogeneous Regularity Problem for $L_D=\Delta-2\nabla\mathrm{div}$ in $\mathbb{R}^n_+$, i.e.,

$$\Big\{\vec{u}\big|^{\kappa-\mathrm{n.t.}}_{\partial\mathbb{R}^n_+}:\ \vec{u}\in\big[\mathscr{C}^\infty(\mathbb{R}^n_+)\big]^n,\ \Delta\vec{u}-2\nabla\mathrm{div}\,\vec{u}=0\ \text{in}\ \mathbb{R}^n_+,\qquad \mathcal{N}_\kappa\vec{u},\ \mathcal{N}_\kappa(\nabla\vec{u})\in L^p(\mathbb{R}^{n-1},\mathcal{L}^{n-1})\Big\} \tag{2.6.38}$$

coincides with the space

$$\Big\{(f_1,\dots,f_n)\in\big[L^p_1(\mathbb{R}^{n-1},\mathcal{L}^{n-1})\big]^n:\ f_n=-\sum_{j=1}^{n-1}R_jf_j\Big\}. \tag{2.6.39}$$

Proof Assume the function $\vec{u}\in\big[\mathscr{C}^\infty(\mathbb{R}^n_+)\big]^n$ solves $\Delta\vec{u}-2\nabla\mathrm{div}\,\vec{u}=0$ in $\mathbb{R}^n_+$ and has the property that $\mathcal{N}_\kappa(\nabla^\ell\vec{u})\in L^p(\mathbb{R}^{n-1},\mathcal{L}^{n-1})$ for each $\ell\in\{0,1,\dots,k\}$. Then from [132, (1.8.199)] we know that, for every multi-index $\alpha\in\mathbb{N}^n_0$ with $|\alpha|\le k$,

$$\text{the trace } (\partial^\alpha\vec{u})\big|^{\kappa-\mathrm{n.t.}}_{\partial\mathbb{R}^n_+} \text{ exists at } \mathcal{L}^{n-1}\text{-a.e. point in } \mathbb{R}^{n-1}. \tag{2.6.40}$$

Having established this, Corollary 2.1.3 applies and gives that the nontangential trace $\vec{u}\big|^{\kappa-\mathrm{n.t.}}_{\partial\mathbb{R}^n_+}$ belongs to $\big[L^p_k(\mathbb{R}^{n-1},\mathcal{L}^{n-1})\big]^n$. Theorem 2.6.4 then guarantees that the n-tuple $(f_1,\dots,f_n):=\vec{u}\big|^{\kappa-\mathrm{n.t.}}_{\partial\mathbb{R}^n_+}$ belongs to (2.6.37). This proves that the space in (2.6.36) is included in (2.6.37). The opposite inclusion is a consequence of Proposition 2.4.4.□

Here are the higher-order versions of Corollary 2.6.5 and Corollary 2.6.6.

Corollary 2.6.8 *Select some $n\in\mathbb{N}$ with $n\ge 2$, along with an integrability exponent $p\in(1,\infty)$, an aperture parameter $\kappa>0$, and an integer $k\in\mathbb{N}$. Then the space of admissible boundary data for the Higher-Order Inhomogeneous Regularity Problem for the system L_D in the upper half-space,*

$$\mathscr{U}_{p,k}:=\Big\{\vec{u}\big|^{\kappa-\mathrm{n.t.}}_{\partial\mathbb{R}^n_+}:\ \vec{u}\in\big[\mathscr{C}^\infty(\mathbb{R}^n_+)\big]^n,\ \Delta\vec{u}-2\nabla\mathrm{div}\,\vec{u}=0\ \text{in}\ \mathbb{R}^n_+,\qquad \mathcal{N}_\kappa(\nabla^\ell\vec{u})\in L^p(\mathbb{R}^{n-1},\mathcal{L}^{n-1}),\ \ell=0,1,\dots,k\Big\} \tag{2.6.41}$$

is a closed subspace of $\big[L^p_k(\mathbb{R}^{n-1},\mathcal{L}^{n-1})\big]^n$, and the following direct topological sum decompositions hold:

$$\big[L^p_k(\mathbb{R}^{n-1},\mathcal{L}^{n-1})\big]^n=\mathscr{U}_{p,k}\oplus\mathscr{V}_{p,k}=\mathscr{U}_{p,k}\oplus\mathscr{W}_{p,k} \tag{2.6.42}$$

where

$$\mathscr{V}_{p,k} := \Big\{(0,\dots,0,f) \in \big[L^p_k(\mathbb{R}^{n-1},\mathcal{L}^{n-1})\big]^n : f \in L^p_k(\mathbb{R}^{n-1},\mathcal{L}^{n-1})\Big\}, \tag{2.6.43}$$

$$\mathscr{W}_{p,k} := \Big\{(R_1 f,\dots,R_{n-1}f,0) \in \big[L^p_k(\mathbb{R}^{n-1},\mathcal{L}^{n-1})\big]^n : f \in L^p_k(\mathbb{R}^{n-1},\mathcal{L}^{n-1})\Big\}. \tag{2.6.44}$$

In particular, one has the isomorphisms

$$\big[L^p_k(\mathbb{R}^{n-1},\mathcal{L}^{n-1})\big]^n/\mathscr{U}_{p,k} \cong \mathscr{V}_{p,k} \cong L^p_k(\mathbb{R}^{n-1},\mathcal{L}^{n-1}), \tag{2.6.45}$$

$$\big[L^p_k(\mathbb{R}^{n-1},\mathcal{L}^{n-1})\big]^n/\mathscr{U}_{p,k} \cong \mathscr{W}_{p,k} \cong L^p_k(\mathbb{R}^{n-1},\mathcal{L}^{n-1}), \tag{2.6.46}$$

hence

the codimension of the space of admissible boundary data for the Higher-Order Inhomogeneous Regularity Problem in the upper half-space for the system $L_D = \Delta - 2\nabla\mathrm{div}$ *(i.e., the space* $\mathscr{U}_{p,k}$ *defined in* (2.6.41)) *into the full data space* $\big[L^p_k(\mathbb{R}^{n-1},\mathcal{L}^{n-1})\big]^n$ *is* $+\infty$. (2.6.47)

Proof Since the Riesz transforms are well-defined linear and bounded operators on the higher-order Sobolev space $L^p_k(\mathbb{R}^{n-1},\mathcal{L}^{n-1})$, and we continue to have (2.6.32) on $L^p_k(\mathbb{R}^{n-1},\mathcal{L}^{n-1})$, the same type of reasoning as in the proofs of Corollary 2.6.5 and Corollary 2.6.6 applies and yields all desired conclusions. □

A description of the space of admissible boundary data for the Homogeneous Regularity Problem for the system $L_D = \Delta - 2\nabla\mathrm{div}$ in the upper half-space is contained in our next theorem.

Theorem 2.6.9 *Fix an integer* $n \in \mathbb{N}$ *with* $n \geq 2$, *along with an integrability exponent* $p \in (1,\infty)$, *and an aperture parameter* $\kappa > 0$. *Then*

$$\Big\{\vec{u}\big|^{\kappa-\mathrm{n.t.}}_{\partial\mathbb{R}^n_+} : \vec{u} \in \big[\mathscr{C}^\infty(\mathbb{R}^n_+)\big]^n,\ \Delta\vec{u} - 2\nabla\mathrm{div}\,\vec{u} = 0,\ \mathcal{N}_\kappa(\nabla\vec{u}) \in L^p(\mathbb{R}^{n-1},\mathcal{L}^{n-1})\Big\}$$
$$= \Big\{(f_1,\dots,f_n) \in \big[\dot{L}^p_1(\mathbb{R}^{n-1},\mathcal{L}^{n-1})\big]^n : f_n = -\sum_{j=1}^{n-1} R_j^{\mathrm{mod}} f_j + c,\ c \in \mathbb{C}\Big\}. \tag{2.6.48}$$

Proof This is a direct consequence of Theorem 2.5.1 and Proposition 2.4.3. □

In the corollary below we identify the cokernel of the space of admissible boundary data for the Homogeneous Regularity Problem for $L_D = \Delta - 2\nabla\mathrm{div}$ in $\mathbb{R}^n_+$ (i.e., the quotient space between the full space of boundary data and the space of admissible boundary data for the L^p-Homogeneous Regularity Problem for L_D).

Corollary 2.6.10 *Fix some integer* $n \in \mathbb{N}$ *satisfying* $n \geq 2$, *along with some integrability exponent* $p \in (1,\infty)$, *and some aperture parameter* $\kappa \in (0,\infty)$. *Bring in the space of admissible boundary data for the* $\dot{L}^p_1$*-Homogeneous Regularity Problem for the system* $L_D = \Delta - 2\nabla\mathrm{div}$ *in the upper half-space, namely*

$$\dot{\mathscr{U}}_1^p := \Big\{ \vec{u}\big|_{\partial\mathbb{R}^n_+}^{\kappa-\mathrm{n.t.}} : \vec{u} \in \big[\mathscr{C}^\infty(\mathbb{R}^n_+)\big]^n,\ \Delta\vec{u} - 2\nabla\mathrm{div}\,\vec{u} = 0 \text{ in } \mathbb{R}^n_+,\ \text{and } \mathcal{N}_\kappa(\nabla\vec{u}) \in L^p(\mathbb{R}^{n-1}, \mathcal{L}^{n-1}) \Big\}. \tag{2.6.49}$$

Then the quotient space $\dot{\mathscr{U}}_1^p / \sim$, *i.e., the version of the above space of admissible boundary data modulo constants, is a closed subspace of* $\big[\dot{L}_1^p(\mathbb{R}^{n-1}, \mathcal{L}^{n-1})/\sim\big]^n$, *and the following direct topological sum decompositions hold:*

$$\big[\dot{L}_1^p(\mathbb{R}^{n-1}, \mathcal{L}^{n-1})/\sim\big]^n = \big(\dot{\mathscr{U}}_1^p/\sim\big) \oplus \big(\dot{\mathscr{V}}_{p,1}/\sim\big) = \big(\dot{\mathscr{U}}_1^p/\sim\big) \oplus \big(\dot{\mathscr{W}}_{p,1}/\sim\big) \tag{2.6.50}$$

where

$$\dot{\mathscr{V}}_{p,1} := \Big\{(0,\dots,0,f) \in \big[\dot{L}_1^p(\mathbb{R}^{n-1}, \mathcal{L}^{n-1})\big]^n : f \in \dot{L}_1^p(\mathbb{R}^{n-1}, \mathcal{L}^{n-1})\Big\} \tag{2.6.51}$$

$$\dot{\mathscr{W}}_{p,1} := \Big\{(\overset{\mathrm{mod}}{R_1} f,\dots,\overset{\mathrm{mod}}{R_{n-1}} f, 0) \in \big[\dot{L}_1^p(\mathbb{R}^{n-1}, \mathcal{L}^{n-1})\big]^n : f \in \dot{L}_1^p(\mathbb{R}^{n-1}, \mathcal{L}^{n-1})\Big\}. \tag{2.6.52}$$

Furthermore, one has the isomorphisms

$$\frac{\big[\dot{L}_1^p(\mathbb{R}^{n-1}, \mathcal{L}^{n-1})/\sim\big]^n}{\dot{\mathscr{U}}_1^p/\sim} \cong \dot{\mathscr{V}}_{p,1}/\sim\,\cong \dot{L}_1^p(\mathbb{R}^{n-1}, \mathcal{L}^{n-1})/\sim\, \cong L^p(\mathbb{R}^{n-1}, \mathcal{L}^{n-1}), \tag{2.6.53}$$

$$\frac{\big[\dot{L}_1^p(\mathbb{R}^{n-1}, \mathcal{L}^{n-1})/\sim\big]^n}{\dot{\mathscr{U}}_1^p/\sim} \cong \dot{\mathscr{W}}_{p,1}/\sim\,\cong \dot{L}_1^p(\mathbb{R}^{n-1}, \mathcal{L}^{n-1})/\sim\, \cong L^p(\mathbb{R}^{n-1}, \mathcal{L}^{n-1}). \tag{2.6.54}$$

In particular,

the codimension of the space of admissible boundary data, modulo constants, for the $\dot{L}_1^p$*-Homogeneous Regularity Problem for the system* L_D *in the upper half-space (i.e., the space* $\dot{\mathscr{U}}_1^p/\sim$*) into the full data space* $\big[\dot{L}_1^p(\mathbb{R}^{n-1}, \mathcal{L}^{n-1})/\sim\big]^n$ *is* $+\infty$. (2.6.55)

Proof As in the proof of Corollary 2.6.8, we need the fact that the modified Riesz transforms are well-defined linear and bounded operators on $\dot{L}_1^p(\mathbb{R}^{n-1}, \mathcal{L}^{n-1})/\sim$ (see [132, (5.3.57)]), and that

$$\sum_{j=1}^{n-1} \big[\overset{\mathrm{mod}}{R_j}\big]^2 = -I \text{ on } \dot{L}_1^p(\mathbb{R}^{n-1}, \mathcal{L}^{n-1})/\sim, \tag{2.6.56}$$

where now I is the identity operator on $\dot{L}_1^p(\mathbb{R}^{n-1}, \mathcal{L}^{n-1})/\sim$ (see [132, (5.3.90)]). This takes care of (2.6.50). The final isomorphisms in (2.6.53)-(2.6.54) are seen from (4.2.25) (presently written for $\Omega := \mathbb{R}^n_+$, $M := 1$, and $L := \Delta$). □

Remark 2.6.11 *Similar results to those described in this section are valid for Muckenhoupt weighted Lebesgue spaces, Morrey spaces, vanishing Morrey spaces, block spaces, and their (homogeneous and inhomogeneous) Sobolev space versions.*

We elaborate on the above remark, indicating the main changes in each case.

Muckenhoupt Weighted Lebesgue Spaces: The standard Riesz transform operators are bounded in the context of Muckenhoupt weighted Lebesgue spaces (cf. [132, (5.3.7)]), as well as Muckenhoupt weighted Sobolev spaces (cf. [132, (5.3.14)]). See [66, 7.2(b), p. 465], [77] for the version of (2.5.47)-(2.5.49) for Muckenhoupt weighted Lebesgue spaces. The equivalence of Muckenhoupt weighted Lebesgue norms for the nontangential maximal operator with various apertures has been established in [129, Corollary 8.4.4] (see also item *(3)* of [129, Corollary 8.4.8]). Given any weight $\omega \in A_p(\mathbb{R}^{n-1}, \mathcal{L}^{n-1})$, in place of (2.5.94) now use interior estimates for harmonic functions and [129, (7.7.10)] to write, for each point $z = (z', t) \in \mathbb{R}^{n-1} \times (0, \infty) = \mathbb{R}^n_+$,

$$
\begin{aligned}
|(\nabla w)(z)| &\leq C\Big(⨍_{B(z,\varepsilon t)} |(\nabla w)(x)|^p \, dx\Big)^{\frac{1}{p}} \leq \frac{C}{t}\Big(⨍_{B(z,\varepsilon t)} |x_n(\nabla w)(x)|^p \, dx\Big)^{\frac{1}{p}} \\
&\leq \frac{C}{t}\Big(⨍_{B_{n-1}(z',\kappa t/2)} \mathcal{N}_\kappa\big(x \mapsto x_n(\nabla w)(x)\big)^p \, d\mathcal{L}^{n-1}\Big)^{\frac{1}{p}} \\
&\leq \frac{C}{t}\left(⨍_{B_{n-1}(z',\kappa t/2)} \mathcal{N}_\kappa\big(x \mapsto x_n(\nabla w)(x)\big)^p \, d\omega\right)^{\frac{1}{p}} \\
&\leq \frac{C}{t \cdot \omega\big(B_{n-1}(z',\kappa t/2)\big)^{1/p}} \big\|\mathcal{N}_\kappa\big(x \mapsto x_n(\nabla w)(x)\big)\big\|_{L^p(\mathbb{R}^{n-1},\omega)},
\end{aligned}
\tag{2.6.57}
$$

then use this to conclude that

$$
\lim_{t\to\infty} (\nabla w)(z', t) = 0 \ \text{ for each } z' \in \mathbb{R}^{n-1}. \tag{2.6.58}
$$

The analogue of property (2.5.96) in the current case is justified analogously.

Morrey Spaces: The standard Riesz transforms are bounded in the context of Morrey spaces, Morrey-based Sobolev spaces, vanishing Morrey spaces, and vanishing Morrey-based Sobolev spaces (see [132, (5.3.16)-(5.3.19)]). See also [132, (5.3.58)-(5.3.61)] for similar results involving the modified Riesz transforms on Morrey-based homogeneous Sobolev spaces and vanishing Morrey-based homogeneous Sobolev spaces. The equivalence of Morrey norms for the nontangential maximal operator with various apertures has been proved in [129, Corollary 8.4.5] (see also item *(5)* of [129, Corollary 8.4.8]). The Morrey space versions of (2.5.47)-(2.5.49) may be established starting from the Muckenhoupt weighted Lebesgue space versions of these results (cf. [66, 7.2(b), p. 465], [77]). and then invoking [130, Proposition 6.2.12] (in the version recorded in Comment 2 following its statement).

The application of the Sequential Banach-Alaoglu Theorem (recalled in [129, (3.6.22)]) that has led to the weak-$*$ convergence result in (2.5.63) is still permissible in the context of Morrey spaces, thanks to [130, (6.2.71)] and [130, (6.2.80)]. Note that [130, Proposition 6.2.9] serves as a good substitute to the use of Lebesgue's Dominated Convergence Theorem in (2.5.84), so the same conclusion in (2.5.85) holds in the present setting. Finally, in place of (2.5.94) now use interior estimates for harmonic functions and (A.0.167) to write

$$\begin{aligned}|(\nabla w)(z)| &\leq C\Big(\fint_{B(z,\varepsilon t)}|(\nabla w)(x)|^p\,\mathrm{d}x\Big)^{\frac{1}{p}} \leq \frac{C}{t}\Big(\fint_{B(z,\varepsilon t)}|x_n(\nabla w)(x)|^p\,\mathrm{d}x\Big)^{\frac{1}{p}}\\ &\leq \frac{C}{t}\left(\fint_{B_{n-1}(z',\kappa t/2)}\mathcal{N}_\kappa\big(x\mapsto x_n(\nabla w)(x)\big)^p\,\mathrm{d}\mathcal{L}^{n-1}\right)^{\frac{1}{p}}\\ &\leq \frac{C}{t^{1+(n-1-\lambda)/p}}\big\|\mathcal{N}_\kappa\big(x\mapsto x_n(\nabla w)(x)\big)\big\|_{M^{p,\lambda}(\mathbb{R}^{n-1},\mathcal{L}^{n-1})},\end{aligned} \tag{2.6.59}$$

at each $z=(z',t)\in\mathbb{R}^{n-1}\times(0,\infty)=\mathbb{R}^n_+$. This allows us to conclude that

$$\lim_{t\to\infty}(\nabla w)(z',t)=0 \text{ for each } z'\in\mathbb{R}^{n-1}, \tag{2.6.60}$$

and the analogue of property (2.5.96) in the present setting is justified similarly.

Block Spaces: The standard Riesz transforms in the entire Euclidean ambient are bounded in the context of block spaces (cf. [132, (5.3.20)]) and block-based Sobolev spaces (cf. [132, (5.3.21)]). Similar results involving the modified Riesz transforms on block-based homogeneous Sobolev spaces are found in [132, (5.3.62), (5.3.63)]. The equivalence of block norms for the nontangential maximal operator with various apertures has been proved in item *(6)* of [129, Corollary 8.4.8]. Thanks to the fact that vanishing Morrey spaces are separable Banach spaces (cf. [130, (6.2.16)]), and [130, (6.2.155)], the application of the Sequential Banach-Alaoglu Theorem (cf. [129, (3.6.22)]) that has led to the weak-$*$ convergence result in (2.5.63) is still permissible in the context of block spaces. Finally, since [130, Proposition 6.2.19] is a satisfactory substitute to the use of Lebesgue's Dominated Convergence Theorem in (2.5.84), the same conclusion in (2.5.85) is valid the current setting.

Chapter 3
Quantifying Global and Infinitesimal Flatness

Quite often, geometry drives analysis. In the opposite direction, we would like to use analysis as a tool to study geometry. In particular, we would like to have methods for deriving geometrical information about an ambient from analytical conditions. A concrete issue is that of quantifying the flatness of a given set (heuristically thought of as some sort of "surface") using functional analytical machinery, such as John-Nirenberg space BMO, of functions with bounded mean oscillations.

To place this in a broader perspective, let us start from the realization that if $\Omega \subseteq \mathbb{R}^n$ is an open set which may be locally described as upper-graphs of real-valued functions defined on hyperplanes in $\mathbb{R}^n$, a natural way to quantify the flatness of $\partial\Omega$ is to look at the size of the gradients of said functions (should these exist in a reasonable sense). For example, this works great in the class of Lipschitz domains, the class of BMO_1 domains, and alike. However, this way of doing things no longer applies to more general categories of Euclidean sets, such as those who fail to be locally upper-graphs of real-valued functions defined on hyperplanes in $\mathbb{R}^n$. Assuming the sets in question are of locally finite perimeter, an alternative point of view is to look at the oscillations of the geometric measure theoretic outward unit normal. The size of these oscillations, suitably measured, then becomes a natural venue to quantify the flatness of the boundary.

More specifically, if $\Omega \subseteq \mathbb{R}^n$ is an Ahlfors regular domain, ν is the geometric measure theoretic outward unit normal to Ω, and $\sigma := \mathcal{H}^{n-1}\lfloor\partial\Omega$ is the "surface measure", then the BMO semi-norm of the Gauss map $\nu : \partial\Omega \to S^{n-1}$, i.e.,

$$\|\nu\|_{[\mathrm{BMO}(\partial\Omega,\sigma)]^n} := \sup_{\substack{x\in\partial\Omega \\ 0<r<2\,\mathrm{diam}\,\partial\Omega}} ⨍_{B(x,r)\cap\partial\Omega} \left|\nu - ⨍_{B(x,r)\cap\partial\Omega} \nu \, d\sigma\right| d\sigma, \tag{3.0.1}$$

serves as a good indicator of how flat $\partial\Omega$ is globally. For one thing, we have that $\|\nu\|_{[\mathrm{BMO}(\partial\Omega,\sigma)]^n} = 0$ if and only if ν is constant which, in turn, is equivalent with Ω being a half-space. Also, if Ω the upper-graph of a Lipschitz function $\phi : \mathbb{R}^{n-1} \to \mathbb{R}$, then $\|\nu\|_{[\mathrm{BMO}(\partial\Omega,\sigma)]^n}$ is controlled in terms of the Lipschitz constant $\|\nabla\phi\|_{[L^\infty(\mathbb{R}^{n-1},\mathcal{L}^{n-1})]^{n-1}}$, which is intuitively natural for something that has a chance at being considered a reasonable measure of flatness. Alternatively, in place of (3.0.1) one may wish to consider the (non-negative) quantity

D. Mitrea et al., *Geometric Harmonic Analysis V*, Developments in Mathematics 76,
https://doi.org/10.1007/978-3-031-31561-9_3

$$1 - \inf_{\substack{x\in\partial\Omega \\ 0<r<2\operatorname{diam}\partial\Omega}} \left| \fint_{B(x,r)\cap\partial\Omega} \nu \, d\sigma \right| \tag{3.0.2}$$

as an indicator of how flat $\partial\Omega$ is. See (4.1.3) in this regard. This is all good, but a peculiarity of this way of doing things, already noted in Lemma 4.1.1, is that having $\|\nu\|_{[\mathrm{BMO}(\partial\Omega,\sigma)]^n}$ small (specifically, strictly less than 1) forces $\partial\Omega$ to be unbounded. In fact, if $\partial\Omega$ is bounded then $\|\nu\|_{[\mathrm{BMO}(\partial\Omega,\sigma)]^n} = 1$ (cf. (4.1.4)), so (3.0.1) no longer serves the intended purpose. As a remedy, for Ahlfors regular domains with compact boundaries one may want to exclude in (3.0.1) balls with too large of a radius. More precisely, in this scenario we shall focus on the local BMO semi-norm of ν (defined as in (A.0.17)) and, at an infinitesimal level, we may want to consider

$$[\nu]_{\mathrm{Osc}} := \lim_{R\to 0^+} \left(\sup_{x\in\partial\Omega,\, 0<r<R} \fint_{B(x,r)\cap\partial\Omega} \left| \nu - \fint_{B(x,r)\cap\partial\Omega} \nu \, d\sigma \right| d\sigma \right). \tag{3.0.3}$$

In light of the last inequality in (4.1.1) we have

$$\begin{aligned} 1 - \lim_{R\to 0^+} \left(\inf_{x\in\partial\Omega,\, 0<r<R} \left| \fint_{B(x,r)\cap\partial\Omega} \nu \, d\sigma \right| \right) &\leq [\nu]_{\mathrm{Osc}} \\ &\leq \sqrt{2} \sqrt{1 - \lim_{R\to 0^+} \left(\inf_{x\in\partial\Omega,\, 0<r<R} \left| \fint_{B(x,r)\cap\partial\Omega} \nu \, d\sigma \right| \right)}. \end{aligned} \tag{3.0.4}$$

From [130, (3.1.53)] (cf. also [130, (3.1.55)]) we also know that the quantity $[\nu]_{\mathrm{Osc}}$ is equivalent (where the "proportionality" constant depend only on n and the Ahlfors regularity constants of $\partial\Omega$) with the distance from ν, measured in the John-Nirenberg space BMO, to the Sarason space VMO, i.e., with

$$\operatorname{dist}\big(\nu, [\mathrm{VMO}(\partial\Omega,\sigma)]^n\big) := \inf \Big\{ \|\nu - \phi\|_{[\mathrm{BMO}(\partial\Omega,\sigma)]^n} : \phi \in [\mathrm{VMO}(\partial\Omega,\sigma)]^n \Big\}. \tag{3.0.5}$$

This is the point of view we adopt in this work. In particular, our thesis is that Ahlfors regular domains with compact boundaries and whose geometric measure theoretic outward unit normals belong to[1] VMO is the "correct" generalization (from the point of view of Geometric Measure Theory) of the class of $\mathscr{C}^1$ domains.

3.1 Ahlfors Regular Domains and Flatness

To set the stage, we bring in some notation. The cylinder $C(x_0, r, w)$ with center at $x_0 \in \mathbb{R}^n$, radius $r \in (0, \infty)$, and axial direction $w \in S^{n-1}$ is defined as

$$C(x_0, r, w) := \Big\{ x \in \mathbb{R}^n : |\langle x - x_0, w\rangle| < r \text{ and } \big|x - x_0 - \langle x - x_0, w\rangle w\big| < r \Big\}. \tag{3.1.1}$$

[1] which, heuristically, should be thought of as an integral (as opposed to pointwise) version of uniform continuity

As in [110, p. 290], given a set of locally finite perimeter $\Omega \subseteq \mathbb{R}^n$, the `cylindrical excess` of Ω at the point $x_0 \in \partial\Omega$, for the scale $r \in (0, \infty)$, and with respect to the direction $w \in S^{n-1}$ is defined as

$$\mathbf{e}(\Omega, x_0, r, w) := \frac{1}{r^{n-1}} \int_{C(x_0,r,w)\cap\partial^*\Omega} \frac{|\nu(x) - w|^2}{2}\, d\mathcal{H}^{n-1}(x), \tag{3.1.2}$$

where ν is the geometric measure theoretic outward unit normal to Ω. Note that the cylindrical excess is a scale-invariant entity, in the sense that

$$\mathbf{e}(\lambda\Omega, \lambda x_0, \lambda r, w) = \mathbf{e}(\Omega, x_0, r, w) \text{ for each } \lambda \in (0, \infty). \tag{3.1.3}$$

This notion is studied at length in [110, Chapter 22], where a number of basic properties of the cylindrical excess (having to do with rescaling, change of direction, lower semi-continuity) are established. The proposition below contains an estimate of the cylindrical excess in terms of the local BMO semi-norm of the unit normal.

Proposition 3.1.1 *Suppose $\Omega \subseteq \mathbb{R}^n$ is an Ahlfors regular domain and abbreviate $\sigma := \mathcal{H}^{n-1}\lfloor\partial\Omega$. Denote by ν the geometric measure theoretic outward unit normal to Ω. Fix a point $x_0 \in \partial\Omega$ along with a radius $r \in \big(0, 2\operatorname{diam}\partial\Omega\big)$, and recall the piece of notation introduced in* (A.0.17). *Then*

$$\inf_{w\in S^{n-1}} \mathbf{e}(\Omega, x_0, r, w) \leq C\big[\|\nu\|_*(\Delta(x_0, \sqrt{2}r))\big]^2, \tag{3.1.4}$$

where $C \in [1, \infty)$ depends only on n and the Ahlfors regularity constants of $\partial\Omega$.

Proof At each point $y \in \partial^*\Omega$ we know from (A.0.197) that $\nu(y)$ is a well-defined vector belonging to S^{n-1}, and (3.1.1) implies that $C(x_0, r, \nu(y)) \subseteq B(x_0, \sqrt{2}r)$. Granted these properties, we conclude from (3.1.2) and [129, (7.4.56)] that

$$\mathbf{e}\big(\Omega, x_0, r, \nu(y)\big) \leq C \fint_{\Delta(x_0,\sqrt{2}r)} |\nu(x) - \nu(y)|^2\, d\sigma(x) \tag{3.1.5}$$

at each point $y \in \partial^*\Omega$, where $C \in [1, \infty)$ depends only on n and the Ahlfors regularity constants of $\partial\Omega$. Taking the integral average of both sides of (3.1.5) over $\Delta(x_0, \sqrt{2}r)$ and availing ourselves of [129, (7.4.59)] we obtain

$$\begin{aligned}\fint_{\Delta(x_0,\sqrt{2}r)} &\mathbf{e}\big(\Omega, x_0, r,\nu(y)\big)\, d\sigma(y) \\ &\leq C \fint_{\Delta(x_0,\sqrt{2}r)} \fint_{\Delta(x_0,\sqrt{2}r)} |\nu(x) - \nu(y)|^2\, d\sigma(x)\, d\sigma(y) \\ &\leq C\big[\|\nu\|_*(\Delta(x_0, \sqrt{2}r))\big]^2, \end{aligned} \tag{3.1.6}$$

where $C \in (0, \infty)$ depends only on n and the Ahlfors regularity constants of $\partial\Omega$. To proceed, abbreviate $\lambda := 2C\big[\|\nu\|_*(\Delta(x_0, \sqrt{2}r))\big]^2$. Then

$$\lambda \cdot \frac{\sigma\Big(\big\{y \in \Delta(x_0, \sqrt{2}r) : \mathbf{e}\big(\Omega, x_0, r, \nu(y)\big) > \lambda\big\}\Big)}{\sigma\big(\Delta(x_0, \sqrt{2}r)\big)}$$
$$\leq \fint_{\Delta(x_0,\sqrt{2}r)} \mathbf{e}\big(\Omega, x_0, r, \nu(y)\big)\, d\sigma(y) < \frac{\lambda}{2} \tag{3.1.7}$$

by (3.1.6) and Chebytcheff's inequality, which in turn implies

$$\sigma\Big(\big\{y \in \Delta(x_0, \sqrt{2}r) : \mathbf{e}\big(\Omega, x_0, r, \nu(y)\big) > \lambda\big\}\Big) \leq \frac{1}{2}\sigma\big(\Delta(x_0, \sqrt{2}r)\big). \tag{3.1.8}$$

Thus, there exists $y \in \partial^*\Omega \cap B(x_0, \sqrt{2}r)$ with $\mathbf{e}\big(\Omega, x_0, r, \nu(y)\big) \leq \lambda$, proving (3.1.4).□

Cylindrical excess is deemed relevant here because of the heuristic principle: *the smaller the cylindrical excess the flatter the set*. To make this precise, recall the class of (R, δ)-Reifenberg flat domains from [129, Definition 5.11.21].

Theorem 3.1.2 *Fix $n \in \mathbb{N}$ with $n \geq 2$. Given $C_A \in [1, \infty)$ there exist $\varepsilon_1 \in (0, 1)$ and $C_1 \in [1, \infty)$, depending only on n and C_A with the following significance: Whenever $\Omega \subseteq \mathbb{R}^n$ is an Ahlfors regular domain with (Ahlfors regularity) constants controlled by C_A and with the property that for some $R \in \big(0, 2\,\mathrm{diam}\,\partial\Omega\big)$ one has*

$$\delta_R := \sup_{x_0 \in \partial\Omega}\ \sup_{r \in (0,R)}\ \inf_{w \in S^{n-1}} \mathbf{e}(\Omega, x_0, r, w) < \varepsilon_1 \tag{3.1.9}$$

it follows that Ω is a $\big(R, C_1 \cdot \delta_R^{\frac{1}{2(n-1)}}\big)$-Reifenberg flat domain.

Proof This is established in [111], making use of work in [16] and [110, Section 22.2, pp. 294-302]. □

We may now combine Theorem 3.1.2 with Proposition 3.1.1 in order to be able to control how flat (the boundary of) an Ahlfors regular domain is directly in terms of the local BMO semi-norm of its geometric measure theoretic outward unit normal.

Corollary 3.1.3 *Fix $n \in \mathbb{N}$ with $n \geq 2$. Given $C_A \in [1, \infty)$ there exist $\varepsilon_2 \in (0, 1)$ and $C_2 \in [1, \infty)$, depending only on n and C_A with the following significance: Assume $\Omega \subseteq \mathbb{R}^n$ is an Ahlfors regular domain with (Ahlfors regularity) constants controlled by C_A and with the property that for some $R_0 \in \big(0, 2\,\mathrm{diam}\,\partial\Omega\big)$ one has (with ν denoting the geometric measure theoretic outward unit normal to Ω and with the piece of notation introduced in* (A.0.17) *relative to $\Sigma := \partial\Omega$ and $\mu := \mathcal{H}^{n-1}\lfloor\partial\Omega$)*

$$\sup_{x_0 \in \partial\Omega} \|\nu\|_*(\Delta(x_0, \sqrt{2}R_0)) < \varepsilon_2. \tag{3.1.10}$$

Then for $R \in (0, R_0]$, if one sets $\delta(R) := \sup_{x_0 \in \partial\Omega} \|\nu\|_*(\Delta(x_0, \sqrt{2}R))$, it follows that Ω is a $\big(R, C_2 \cdot \delta(R)^{\frac{1}{n-1}}\big)$-Reifenberg flat domain (cf. [129, Definition 5.11.21]).

Proof Having fixed $C_A \in [1, \infty)$, bring in the constants $\varepsilon_1 \in (0, 1)$ and $C_1 \in [1, \infty)$, depending only on n and C_A, from Theorem 3.1.2. Recall the constant $C \in [1, \infty)$ from Proposition 3.1.1 and set

$$\varepsilon_2 := \sqrt{\frac{\varepsilon_1}{C}} \in (0, 1). \tag{3.1.11}$$

If Ω is as in the statement, use (3.1.4), (3.1.10), (3.1.11), (A.0.17) to write

$$\begin{aligned} \delta_R &:= \sup_{x_0 \in \partial\Omega} \sup_{r \in (0,R)} \inf_{w \in S^{n-1}} \mathbf{e}(\Omega, x_0, r, w) \\ &\leq C \sup_{x_0 \in \partial\Omega} \sup_{r \in (0,R)} \big[\|\nu\|_*(\Delta(x_0, \sqrt{2}r))\big]^2 \leq C \sup_{x_0 \in \partial\Omega} \big[\|\nu\|_*(\Delta(x_0, \sqrt{2}R))\big]^2 \\ &\leq C \sup_{x_0 \in \partial\Omega} \big[\|\nu\|_*(\Delta(x_0, \sqrt{2}R_0))\big]^2 < \varepsilon_1, \quad \forall\, R \in (0, R_0]. \end{aligned} \tag{3.1.12}$$

Granted this, on the one hand we may invoke Theorem 3.1.2 to obtain that the set Ω is a $\big(R, C_1 \cdot \delta_R^{\frac{1}{2(n-1)}}\big)$-Reifenberg flat domain. On the other hand, with $\delta(R)$ as in the statement, the second inequality in (3.1.12) implies

$$\delta_R \leq C\delta(R)^2. \tag{3.1.13}$$

At this point, in view of [129, (5.11.98)], we may ultimately conclude that Ω is a $\big(R, C_2 \cdot \delta(R)^{\frac{1}{n-1}}\big)$-Reifenberg flat domain for the choice $C_2 := C_1 \cdot C^{\frac{1}{2(n-1)}} \in [1, \infty)$.□

Having linked the size of the local BMO semi-norm of the geometric measure theoretic outward unit normal to a given Ahlfors regular domain to Reifenberg flatness, other geometric attributes follow. We take a first look at this in the corollary below, then revisit the issue in Theorem 3.1.5 (for domains with unbounded boundaries) and Theorem 3.1.11 (for domains with compact boundaries); see also Theorem 3.4.2.

Corollary 3.1.4 *Fix $n \in \mathbb{N}$ with $n \geq 2$. Then for each given $C_A \in [1, \infty)$ there exists a constant $\varepsilon_3 \in (0, 1)$, which depends only on n and C_A, with the following significance. Let $\Omega \subseteq \mathbb{R}^n$ be an Ahlfors regular domain with (Ahlfors regularity) constants controlled by C_A and such that for some $r_0 \in \big(0, 2\,\mathrm{diam}\,\partial\Omega\big)$ one has (with ν denoting the geometric measure theoretic outward unit normal to Ω and with notation introduced in* (A.0.17) *relative to $\Sigma := \partial\Omega$ and $\mu := \mathcal{H}^{n-1}\lfloor\partial\Omega$)*

$$\sup_{x \in \partial\Omega} \|\nu\|_*(\Delta(x, r_0)) \leq \varepsilon_3. \tag{3.1.14}$$

Then Ω is a two-sided NTA *domain, in the sense of* [129, Definition 5.11.1], *with constants $R := r_0$ and $M = M(C_A, n) \in (0, \infty)$. As a corollary of this,* [129, (5.10.33)], *and* [129, Remark 5.11.2], *if $\partial\Omega$ is compact then the set Ω is a* UR *domain whose* UR *constants are controlled in terms of the dimension n, the constant C_A, and the ratio $r_0/\mathrm{diam}\,\partial\Omega$. Likewise, as a corollary of this,* [129, (5.11.28)], *and* [129, Definition 5.11.7] *(cf. also* (A.0.123)*), if $\partial\Omega$ is compact then the set Ω satisfies a two-sided local John condition, in the same quantitative fashion as before.*

Proof Fix $C_A \in [1, \infty)$ and consider the constants $\varepsilon_2 \in (0, 1)$ and $C_2 \in [1, \infty)$, associated with n and C_A as in Corollary 3.1.3. In addition, bring in the purely dimensional constant $\delta_n \in (0, 1)$ from [129, (5.11.99)], and define

$$\varepsilon_3 := \frac{1}{2} \cdot \min\left\{\varepsilon_2, \left(\frac{\delta_n}{C_2}\right)^{n-1}\right\} \in (0, 1). \tag{3.1.15}$$

Work under the assumption that $\Omega \subseteq \mathbb{R}^n$ is an Ahlfors regular domain with constants controlled by C_A and with the property that for some $r_0 \in \big(0, 2\,\mathrm{diam}\,\partial\Omega\big)$ we have

$$\delta_0 := \sup_{x\in\partial\Omega} \|\nu\|_*(\Delta(x, r_0)) \le \varepsilon_3. \tag{3.1.16}$$

Since by design $\varepsilon_3 < \varepsilon_2$, Corollary 3.1.3 (used with both R_0, R replaced by $r_0/\sqrt{2}$) guarantees that Ω is a $\big(r_0, C_2 \cdot \delta_0^{\frac{1}{n-1}}\big)$-Reifenberg flat domain. Given that, as seen from (3.1.16) and (3.1.15), we also have $C_2 \cdot \delta_0^{\frac{1}{n-1}} < \delta_n$, we conclude from [129, (5.11.99)] that there exists some $M(r_0, C_A, n) \in (0, \infty)$ such that Ω is an NTA domain, in the sense of [129, Definition 5.11.1], with constants $R := r_0$ and $M = M(r_0, C_A, n)$.

We claim that $M(r_0, C_A, n)$ may actually be taken to be independent of r_0, i.e., we may take $M = M(C_A, n)$. To see that this is the case, fix an arbitrary $\lambda \in (0, \infty)$ and observe that by running the same argument as above with Ω replaced by $\lambda\Omega$ and with r_0 replaced by λr_0 shows that

$$\begin{aligned}&\lambda\Omega \text{ is an NTA domain, in the sense of [129, Definition 5.11.1],}\\ &\text{with constants } R := \lambda r_0 \text{ and } M = M(\lambda r_0, C_A, n).\end{aligned} \tag{3.1.17}$$

Choose $\lambda := r_0^{-1}$ and use (3.1.17) in concert with [129, (5.11.3)] to conclude that $\Omega = \lambda^{-1}(\lambda\Omega)$ is an NTA domain with constants $R := r_0$ and $M = M(1, C_A, n)$. Re-denoting $M(1, C_A, n)$ simply as $M(C_A, n)$ then yields the desired conclusion.

Next, from item *(6)* in [129, Lemma 5.10.9] we know that $\Omega_- := \mathbb{R}^n \setminus \overline{\Omega}$ is also an Ahlfors regular domain (in the sense of (A.0.1)), whose topological boundary coincides with that of Ω, and whose geometric measure theoretic boundary agrees with that of Ω. Moreover, the geometric measure theoretic outward unit normal to Ω_- is $-\nu$ at σ-a.e. point on $\partial\Omega$. As a consequence, Ω_- satisfies the very same properties as Ω, so from what we have proved so far we conclude that Ω_- is also an NTA domain with constants $R := r_0$ and $M = M(C_A, n)$. Altogether, this shows that the original set Ω is a two-sided NTA domain with constants $R := r_0$ and $M = M(C_A, n)$. □

Let us recall from [105, Theorem 2 in 49.VI, 57.I.9(i), 57.III.1] (cf. also [97, Lemma 4(1) and Lemma 5, p. 1702]) that

$$\begin{aligned}&\text{if } \mathcal{O} \subseteq \mathbb{R}^n \text{ is some arbitrary connected open set, it follows that}\\ &\text{any connected component of } \mathbb{R}^n \setminus \overline{\mathcal{O}} \text{ has a connected boundary.}\end{aligned} \tag{3.1.18}$$

When used in concert, (3.1.18) and Corollary 3.1.4 imply the following result.

Theorem 3.1.5 *Let $\Omega \subseteq \mathbb{R}^n$ be an Ahlfors regular domain. Denote by ν the geometric measure theoretic outward unit normal to Ω and abbreviate $\sigma := \mathcal{H}^{n-1}\lfloor\partial\Omega$.*

Then there exists a threshold $\delta_* \in (0,1)$*, depending only on the Ahlfors regularity constants of* $\partial\Omega$ *and the dimension* n*, with the property that if*

$$\|\nu\|_{[\mathrm{BMO}(\partial\Omega,\sigma)]^n} < \delta_* \tag{3.1.19}$$

it follows that

$$\begin{array}{l}\Omega \text{ is an unbounded connected set, } \overline{\Omega} \text{ is an unbounded connected} \\ \text{set, } \partial\Omega \text{ is an unbounded connected set, } \mathbb{R}^n \setminus \overline{\Omega} \text{ is an unbounded} \\ \text{connected set, } \mathbb{R}^n \setminus \Omega \text{ is an unbounded connected set, and } \partial(\overline{\Omega}) = \partial\Omega, \\ \partial(\mathbb{R}^n \setminus \overline{\Omega}) = \partial\Omega,\ \partial(\mathbb{R}^n \setminus \Omega) = \partial\Omega,\end{array} \tag{3.1.20}$$

and

Ω *is a two-sided* NTA *domain (in the sense of* [129, Definition 5.11.1]) *with constants which depend only on the dimension* n *and the Ahlfors regularity constants of* $\partial\Omega$. (3.1.21)

In particular, (3.1.20)-(3.1.21) *and* [129, Proposition 5.11.14, *(4)*] *show that*

both Ω *and* $\mathbb{R}^n \setminus \overline{\Omega}$ *are uniform domains (in the sense of* [129, Definition 5.11.10]) *for a constant depending only on the dimension* n *and the Ahlfors regularity constants of* $\partial\Omega$*,* (3.1.22)

while as seen from (3.1.21) *and* [129, (5.11.28)],

Ω *satisfies a two-sided local John condition with constants which depend only on* n *and the Ahlfors regularity constants of* $\partial\Omega$*,* (3.1.23)

and, as seen from (3.1.23) *and* [129, (5.10.24)],

Ω *is a* UR *domain, with the* UR *constants of* $\partial\Omega$ *controlled solely in terms of the dimension* n *and the Ahlfors regularity constants of* $\partial\Omega$. (3.1.24)

In turn, Theorem 3.1.5 allows to further refine [130, Theorem 11.5.2] in the manner described below.

Corollary 3.1.6 *Suppose* $\Omega \subset \mathbb{R}^n$ *is an Ahlfors regular domain and abbreviate* $\sigma := \mathcal{H}^{n-1} \lfloor \partial\Omega$*. Denote by* ν *the geometric measure theoretic outward unit normal to* Ω*. Then there exists some constant* $C_* = C_*(\Omega) \in [1,\infty)$*, which depends only on* n *and the Ahlfors regularity constants of* $\partial\Omega$*, such that for each* $\gamma \in [1,\infty)$ *one has*

$$\sup_{z\in\partial\Omega}\sup_{R>0}\sup_{x,y\in\Delta(z,\gamma R)} R^{-1}\big|\langle x-y, \nu_{\Delta(z,R)}\rangle\big| \leq C_*\gamma\big(1+\log_2\gamma\big)\|\nu\|_{[\mathrm{BMO}(\partial\Omega,\sigma)]^n}. \tag{3.1.25}$$

In particular, with $C_* = C_*(\Omega) \in [1,\infty)$ *of the same nature as above, one has*

$$\sup_{x\in\partial\Omega}\sup_{R>0}\sup_{y\in\Delta(x,2R)} R^{-1}\big|\langle x-y, \nu_{\Delta(x,R)}\rangle\big| \leq C_*\|\nu\|_{[\mathrm{BMO}(\partial\Omega,\sigma)]^n}. \tag{3.1.26}$$

Proof Bring in the threshold $\delta_* \in (0,1)$ associated with the dimension n and the set Ω as in Theorem 3.1.5. In the case in which we have $\|\nu\|_{[\mathrm{BMO}(\partial\Omega,\sigma)]^n} \geq \delta_*$, use the Cauchy-Schwarz inequality to simply write

$$\begin{aligned}\sup_{z\in\partial\Omega}\sup_{R>0}\sup_{x,y\in\Delta(z,\gamma R)} R^{-1}|\langle x-y,\nu_{\Delta(z,R)}\rangle| &\leq 2\gamma \\ &\leq C_*\gamma\big(1+\log_2\gamma\big)\|\nu\|_{[\mathrm{BMO}(\partial\Omega,\sigma)]^n}\end{aligned} \tag{3.1.27}$$

provided $C_* \geq 2/\delta_*$. If $\|\nu\|_{[\mathrm{BMO}(\partial\Omega,\sigma)]^n} < \delta_*$ then Theorem 3.1.5 guarantees that $\partial\Omega$ is unbounded and that (3.1.23) holds. Granted these properties, [130, Theorem 11.5.2] applies and gives that [130, (11.5.34)] holds with $\widetilde{R} = \infty$, and with $\widetilde{C} \in [1,\infty)$ and $\lambda \in (32,\infty)$ now depending only on n and the Ahlfors regularity constants of $\partial\Omega$. In particular, [130, (11.5.34)] is presently valid for each $R \in (0,\infty)$. Estimating $\|\nu\|_*(\Delta(x,\gamma\lambda\varepsilon^{-1}R)) \leq \|\nu\|_{[\mathrm{BMO}(\partial\Omega,\sigma)]^n}$, then letting $\varepsilon \to 0^+$ in [130, (11.5.34)] yields (3.1.25). Finally, (3.1.26) is a particular case of (3.1.25). □

Remark 3.1.7 *An important geometric feature of a given Ahlfors regular domain $\Omega \subseteq \mathbb{R}^n$ is what we shall refer to as its "global tilt." We envision envision this as the maximal deviation of a chord $x-y$ with $x, y \in \Delta$ where Δ is an arbitrary surface ball on $\partial\Omega$ from being perpendicular to ν_Δ, the integral average in Δ of the geometric measure theoretic outward unit normal ν to Ω. It is natural to precisely quantify the global tilt of Ω with amplitude $\gamma \in [1,\infty)$ as*

$$\mathbf{t}_\gamma(\Omega) := \sup_{z\in\partial\Omega}\sup_{R>0}\sup_{x,y\in\Delta(z,\gamma R)}\Big|\Big\langle \nu_{\Delta(z,R)}, \frac{x-y}{\gamma R}\Big\rangle\Big| \tag{3.1.28}$$

where $\nu_{\Delta(z,R)} := \fint_{\Delta(z,R)} \nu\, d\sigma$ for each $z \in \partial\Omega$ and $R > 0$, with $\sigma := \mathcal{H}^{n-1}\lfloor\partial\Omega$.

To illustrate the scope of this notion, consider the case when Ω_θ is the cone in $\mathbb{R}^n$ of aperture $\theta \in (0,2\pi)$ with vertex at the origin and axis along $\mathbf{e}_n$, i.e.,

$$\Omega_\theta := \Big\{x \in \mathbb{R}^n \setminus \{0\} : \frac{x_n}{|x|} > \cos(\theta/2)\Big\}. \tag{3.1.29}$$

Then as seen directly from the definition given in (3.1.28), $\mathbf{t}_\gamma(\Omega_\theta) = |\cos(\theta/2)|$ for each $\gamma \in [1,\infty)$, while a direct computation shows that the outward unit normal vector ν to Ω_θ satisfies $\|\nu\|_{[\mathrm{BMO}(\partial\Omega_\theta,\sigma)]^n} = |\cos(\theta/2)|$. In particular, in this special case we simply have $\mathbf{t}_\gamma(\Omega_\theta) = \|\nu\|_{[\mathrm{BMO}(\partial\Omega_\theta,\sigma)]^n}$ for each $\gamma \in [1,\infty)$.

For a general Ahlfors regular domain $\Omega \subseteq \mathbb{R}^n$, the best we can hope for is to control the global tilt $\mathbf{t}(\Omega)$, for each fixed amplitude parameter $\lambda \in [1,\infty)$, in terms of the BMO*-seminorm of the geometric measure theoretic outward unit normal ν. Remarkably, this is possible, since* (3.1.25) *asserts that there exists some constant $C_* \in (0,\infty)$, depending only on the dimension and the Ahlfors regularity constants of $\partial\Omega$, such that for each amplitude $\gamma \in [1,\infty)$ we have*

$$\mathbf{t}_\gamma(\Omega) \leq C_*\big(1+\log_2\gamma\big)\|\nu\|_{[\mathrm{BMO}(\partial\Omega,\sigma)]^n}. \tag{3.1.30}$$

Theorem 3.1.5 suggests adopting, as in [111, Definition 2.15, p. 85], the following basic piece of terminology.

Definition 3.1.8 *Fix $n \in \mathbb{N}$ with $n \geq 2$ and consider a parameter $\delta > 0$. Call a nonempty, proper subset Ω of $\mathbb{R}^n$ a* δ-`flat Ahlfors regular domain` *(or* δ-`flat AR domain`*, or simply* δ-`AR domain`*) provided Ω is an Ahlfors regular domain (in the sense of Definition A.0.1) whose geometric measure theoretic outward unit normal ν satisfies (with $\sigma := \mathcal{H}^{n-1}\lfloor\partial\Omega$)*

$$\|\nu\|_{[\mathrm{BMO}(\partial\Omega,\sigma)]^n} < \delta. \tag{3.1.31}$$

Given that in the class of Ahlfors regular domains we always have the estimate $\|\nu\|_{[\mathrm{BMO}(\partial\Omega,\sigma)]^n} \leq 1$ (cf. (4.1.2)), condition (3.1.31) is redundant when $\delta > 1$. We will primarily be interested in the case when δ is small. In particular, when $\delta \in (0, 1)$, Lemma 4.1.1 guarantees that $\partial\Omega$ is unbounded.

If $\Omega \subseteq \mathbb{R}^n$ is a δ-AR domain, then our earlier results show that $\mathbb{R}^n \setminus \overline{\Omega}$ is also a δ-AR domain (having the same topological and measure theoretic boundaries as Ω, and whose geometric measure theoretic outward unit normal is the opposite of the one for Ω). Also, the class of δ-AR domain is preserved by rigid transformations.

Several examples and counterexamples of δ-AR domains in $\mathbb{R}^n$ are as follows. Of course, the set $\Omega := \mathbb{R}^n_+$ is a δ-AR domain for each $\delta > 0$. More generally, any half-space in $\mathbb{R}^n$ is a δ-AR domain for each $\delta > 0$. Next, denote the cone of (full) aperture angle $\theta \in (0, 2\pi)$ in $\mathbb{R}^n$ with vertex at the origin and axis along $\mathbf{e}_n$ by Ω_θ (cf. (3.1.29)). If we abbreviate $\sigma_\theta := \mathcal{H}^{n-1}\lfloor\partial\Omega_\theta$, then a direct computation shows that the outward unit normal vector ν to Ω_θ satisfies

$$\begin{gathered}\|\nu\|_{[\mathrm{BMO}(\partial\Omega_\theta,\sigma_\theta)]^n} = |\cos(\theta/2)| \text{ so, as a consequence,}\\ \Omega_\theta \text{ is a } \delta\text{-AR domain if and only if } \delta > |\cos(\theta/2)|.\end{gathered} \tag{3.1.32}$$

Given $\delta > 0$, the region $\Omega := \big\{(x', t) \in \mathbb{R}^{n-1} \times \mathbb{R} : t > \phi(x')\big\}$ above the graph of a Lipschitz function $\phi : \mathbb{R}^{n-1} \to \mathbb{R}$ whose Lipschitz constant is $< 2^{-3/2}\delta$ is a δ-AR domain (see [111, Example 2.3, p. 87]). Also, there exists a purely dimensional constant $C_n \in (1, \infty)$ with the property that if Ω is the region above the graph of some BMO_1 function $\phi : \mathbb{R}^{n-1} \to \mathbb{R}$, (i.e., a function $\phi \in L^1_{\mathrm{loc}}(\mathbb{R}^{n-1}, \mathcal{L}^{n-1})$ with $\nabla\phi$ belonging to $\big[\mathrm{BMO}(\mathbb{R}^{n-1}, \mathcal{L}^{n-1})\big]^{n-1}$) satisfying

$$\|\nabla\phi\|_{[\mathrm{BMO}(\mathbb{R}^{n-1},\mathcal{L}^{n-1})]^{n-1}} < \min\{1, \delta/C_n\} \tag{3.1.33}$$

is a δ-AR domain (see [111, Example 2.5, p. 89]). Lastly, we note that [111, Theorem 2.7, p. 122] establishes, in the two-dimensional Euclidean setting, the coincidence of the class of chord-arc domains with unbounded boundaries (in the sense of [129, Definition 5.9.13]) possessing small constants (i.e., C_o defined in [129, (5.9.92)] is sufficiently close to 1) with that of δ-AR domains with $\delta > 0$ small. We wrap up this discussion by including a couple of pictures of δ-AR domains.

The figure below highlights the fact that our notion of "flatness" for an Ahlfors regular domain Ω, interpreted as the quality of being a δ-AR domain with $\delta \in (0, 1)$ small, is more inclusive than the usual intuitive sense of the word "flat" would

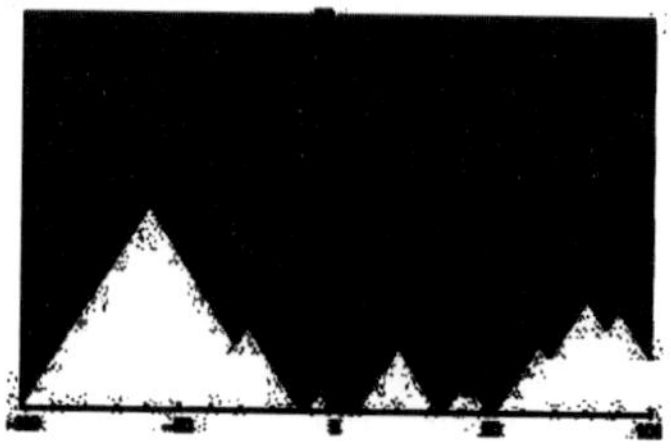

Fig. 3.1 Given any $\delta > 0$, the region above a zigzagged path, consisting of arbitrarily many line segments (of arbitrary, possibly infinite, lengths) with slopes $\pm\delta/2^{3/2}$, is a planar δ-AR domain

suggest[2]. Indeed, the δ-AR domain in the figure above is allowed to have as tall a peak, and as deep a valley, as wanted without affecting the value of δ. Heuristically speaking, the main attribute is the absence of "turns which are too sudden" of the boundary. In particular, as shown in [111, Example 2.7, pp. 92-95], δ-AR domains can develop spirals:

Fig. 3.2 An example of a δ-AR domains exhibiting a spiral point (in particular, failing to be of upper-graph type)

Hence, one can make the case that

heuristically, the category of δ-AR *domains in* $\mathbb{R}^n$ *can be thought of as the sharp version (from the point of view of Geometric Measure Theory) of the class of domains above the graphs of real-valued Lipschitz functions defined in* $\mathbb{R}^{n-1}$ *with a small Lipschitz constant.* (3.1.34)

Let us move on. In the definition below, which plays a major role in this work, we shall indicate how we may associate with any given Ahlfors regular domain with compact boundary $\Omega \subseteq \mathbb{R}^n$ a quantity $\wp(\Omega)$ which provides a reasonable answer to the question: *up to what scale is* $\partial\Omega$ *relatively*[3] *flat*?

[2] traditionally, flatness of a set is most commonly associated with the ability of containing the boundary of said set sandwiched inside a relatively narrow infinite strip, lying in between two infinite parallel planes sufficiently close to one another

[3] in relation to its Ahlfors regularity constants

Definition 3.1.9 *Let $\Omega \subseteq \mathbb{R}^n$, where $n \in \mathbb{N}$ with $n \geq 2$, be an Ahlfors regular domain with compact boundary. Abbreviate $\sigma := \mathcal{H}^{n-1}\lfloor\partial\Omega$ and denote by ν the geometric measure theoretic outward unit normal to Ω. Also, let $C_A \in [1, \infty)$ control the Ahlfors regularity constants of $\partial\Omega$, in the sense that*

$$r^{n-1}/C_A \leq \sigma\big(\Delta(x, r)\big) \leq C_A r^{n-1} \quad \text{for each } x \in \partial\Omega \text{ and } r \in \big(0, 2\,\mathrm{diam}\,\partial\Omega\big). \tag{3.1.35}$$

Finally, associate the threshold $\varepsilon_3 \in (0, 1)$ with C_A and n as in Corollary 3.1.4. In this context, define the `relative flatness radius of` Ω *to be the quantity (cf. also* (3.1.48) *below)*

$$\wp(\Omega) := \inf\Big\{r \in \big(0, 2\,\mathrm{diam}\,\partial\Omega\big) : \text{there exists } x \in \partial\Omega \text{ such that } ⨍_{\Delta(x,r)} \Big|\nu - ⨍_{\Delta(x,r)} \nu\,d\sigma\Big|\,d\sigma > \varepsilon_3\Big\}. \tag{3.1.36}$$

Roughly speaking, we define $\wp(\Omega)$ as the smallest scale at which the mean oscillation of the unit normal ν is large somewhere (at some location on the boundary). Here are several comments intended to provide a deeper insight.

`Comment 1.` If $r \in \big[\mathrm{diam}\,\partial\Omega, 2\,\mathrm{diam}\,\partial\Omega\big)$ then $\Delta(x, r) = \partial\Omega$ for each $x \in \partial\Omega$, hence $⨍_{\Delta(x,r)} \nu\,d\sigma = 0$ by the Divergence Theorem (keeping in mind that $\partial\Omega$ is compact; cf. [129, (1.1.8)]). Since σ-a.e. on $\partial\Omega$ we have $|\nu| = 1$, this further forces $⨍_{\Delta(x,r)} \big|\nu - ⨍_{\Delta(x,r)} \nu\,d\sigma\big| = 1$ for each $x \in \partial\Omega$. In view of the fact that $\varepsilon_3 < 1$, we ultimately conclude that

$$\wp(\Omega) \text{ is a well-defined number (as the infimum of a bounded nonempty subset of the real line), satisfying } 0 \leq \wp(\Omega) \leq \mathrm{diam}\,\partial\Omega. \tag{3.1.37}$$

In particular,

$$\text{the relative flatness ratio of } \Omega, \text{ defined as } \wp(\Omega)/\mathrm{diam}\,\partial\Omega, \text{ is a well-defined number belonging to the interval } [0, 1]. \tag{3.1.38}$$

`Comment 2.` Item *(6)* in [129, Lemma 5.10.9] tells us that $\Omega_- := \mathbb{R}^n \setminus \overline{\Omega}$ is also an Ahlfors regular domain (in the sense of (A.0.1)), whose topological boundary coincides with that of Ω (in particular, is compact), and whose geometric measure theoretic boundary agrees with that of Ω. In addition, the geometric measure theoretic outward unit normal to Ω_- is $-\nu$ at σ-a.e. point on $\partial\Omega$. As a result,

$$\wp\big(\mathbb{R}^n \setminus \overline{\Omega}\big) = \wp(\Omega). \tag{3.1.39}$$

`Comment 3.` In the context of Definition 3.1.9, one has

$$\sup_{x\in\partial\Omega} \|\nu\|_*(\Delta(x, r)) \leq \varepsilon_3 \text{ provided } 0 < r < C_A^{-2/(n-1)} \cdot \wp(\Omega). \tag{3.1.40}$$

To justify (3.1.40), fix a point $x \in \partial\Omega$ along with a scale $r \in \big(0, C_A^{-2/(n-1)}\wp(\Omega)\big)$. Observe that for each surface ball $\Delta' = B(x', r') \cap \partial\Omega$, with $x' \in \partial\Omega$ and radius $r' \in \big(0, 2\,\mathrm{diam}\,\partial\Omega\big)$, having $\Delta' \subseteq \Delta(x, r)$ forces $r' \leq C_A^{2/(n-1)} r < \wp(\Omega)$ (in view of (3.1.35) and assumptions). In concert with (3.1.36) this ultimately implies that $⨍_{\Delta'} \big|\nu - ⨍_{\Delta'} \nu\,\mathrm{d}\sigma\big|\,\mathrm{d}\sigma \leq \varepsilon_3$ for each surface ball $\Delta' \subseteq \Delta(x, r)$. In terms of the piece of notation introduced in (A.0.17), this translates into $\|\nu\|_*(\Delta(x, r)) \leq \varepsilon_3$. Given the arbitrariness of $x \in \partial\Omega$, we therefore conclude that (3.1.40) holds.

Comment 4. In the setting of Definition 3.1.9,

$$\begin{aligned} &\text{there exists } C \in (0, \infty) \text{ depending only on } n \text{ and } C_A \text{ such that} \\ &\wp(\Omega) > 0 \text{ if and only if } \mathrm{dist}\big(\nu, [\mathrm{VMO}(\partial\Omega, \sigma)]^n\big) < \varepsilon_3/C, \end{aligned} \tag{3.1.41}$$

where the distance is measured in the space $\big[\mathrm{BMO}(\partial\Omega, \sigma)\big]^n$. To prove (3.1.41), recall from [130, (3.1.55)] that there exists $C \in (0, \infty)$ which depends only on the dimension n and the constant C_A from (3.1.35) with the property that

$$\begin{aligned} &\sup_{x \in \partial\Omega} \|\nu\|_*(\Delta(x, R)) \searrow L \text{ as } R \searrow 0 \text{ for some number } L \geq 0 \\ &\qquad \text{satisfying } L \leq C \cdot \mathrm{dist}\big(\nu, [\mathrm{VMO}(\partial\Omega, \sigma)]^n\big). \end{aligned} \tag{3.1.42}$$

As such, whenever $\mathrm{dist}\big(\nu, [\mathrm{VMO}(\partial\Omega, \sigma)]^n\big) < \varepsilon_3/C$ it follows that there exists $R \in \big(0, 2\,\mathrm{diam}\,\partial\Omega\big)$ such that $\sup_{x \in \partial\Omega} \|\nu\|_*(\Delta(x, R)) < \varepsilon_3$. In turn, this readily implies that $⨍_{\Delta(x,r)} \big|\nu - ⨍_{\Delta(x,r)} \nu\,\mathrm{d}\sigma\big|\,\mathrm{d}\sigma < \varepsilon_3$ for each $x \in \partial\Omega$ and $r \in (0, R]$. Thanks to (3.1.36), this ultimately forces $\wp(\Omega) \geq R > 0$, proving the right-to-left implication in (3.1.41). The left-to-right implication in (3.1.41) is a consequence of (3.1.40) and (3.1.42). In the setting of Definition 3.1.9 we see from (3.1.41) that

$$\wp(\Omega) > 0 \text{ whenever } \nu \in \big[\mathrm{VMO}(\partial\Omega, \sigma)\big]^n. \tag{3.1.43}$$

Comment 5. If $\Omega \subseteq \mathbb{R}^n$ is a bounded open set with a smooth boundary, except for a cusp point at $x_0 \in \partial\Omega$, then $\wp(\Omega) = 0$. Indeed, in such a setting we have $\lim_{r \to 0^+} ⨍_{\Delta(x_0,r)} \nu\,\mathrm{d}\sigma = 0$ which, in turn, forces $\lim_{r \to 0^+} ⨍_{\Delta(x_0,r)} \big|\nu - ⨍_{\Delta(x_0,r)} \nu\,\mathrm{d}\sigma\big|\,\mathrm{d}\sigma = 1$, so the desired conclusion follows from (3.1.36) bearing in mind that $\varepsilon_3 \in (0, 1)$.

For similar reasons we have $\wp(\Omega) = 0$ whenever $\Omega \subseteq \mathbb{R}^n$ is a bounded open set with a smooth boundary, except for conical point at $x_0 \in \partial\Omega$ with a sufficiently small aperture (relative to n and the Ahlfors regularity constants of $\partial\Omega$).

Comment 6. Retain the context of Definition 3.1.9. Then, as seen from definitions,

$$\begin{aligned} &\wp\big(T(\Omega)\big) = \wp(\Omega) \text{ for each rigid transformation } T \text{ (composition of} \\ &\text{rotations and translations) of the Euclidean space } \mathbb{R}^n. \end{aligned} \tag{3.1.44}$$

Also, based on (3.1.36) the observation that both the mean oscillations of ν and the constant C_A from (3.1.35) are dilation invariant it follows that

$$\wp(\lambda\Omega) = \lambda\wp(\Omega) \text{ for each } \lambda \in (0, \infty). \tag{3.1.45}$$

As a consequence of this, we see that the relative flatness ratio $\wp(\Omega)/\operatorname{diam}\partial\Omega$ is a dilation invariant entity. Moreover, (3.1.45) implies

$$\wp\big(B(0,R)\big) = c\cdot R \text{ for each } R\in(0,\infty) \text{ where } c := \wp\big(B(0,1)\big)\in(0,2]. \tag{3.1.46}$$

Likewise, $\wp\big((-R,R)^n\big) = c\cdot R$ for each $R\in(0,\infty)$ where, this time, $c := \wp\big((-1,1)^n\big)$.

Comment 7. In the setting of Definition 3.1.9, for each point $x\in\partial\Omega$ define the relative flatness radius of x as

$$R_{\text{flat}}(x) := \inf\Big\{r\in\big(0,2\operatorname{diam}\partial\Omega\big) : \fint_{\Delta(x,r)}\Big|\nu-\fint_{\Delta(x,r)}\nu\,d\sigma\Big|\,d\sigma>\varepsilon_3\Big\}. \tag{3.1.47}$$

Hence, the relative flatness radius of a given point $x\in\partial\Omega$ measures in how large a neighborhood of x the boundary of Ω remains relatively flat. In relation to this, we claim that each $R_{\text{flat}}(x)$ is a well-defined number in $\big[0,2\operatorname{diam}\partial\Omega\big]$ and

$$\wp(\Omega) = \inf_{x\in\partial\Omega} R_{\text{flat}}(x). \tag{3.1.48}$$

That $R_{\text{flat}}(x)\in\big[0,2\operatorname{diam}\partial\Omega\big]$ for each $x\in\partial\Omega$ may be seen by reasoning as in (3.1.37). To justify (3.1.48), fix an arbitrary $\varepsilon\in\big(0,\operatorname{diam}\partial\Omega\big)$. Pick $x\in\partial\Omega$. Then $R_{\text{flat}}(x)+\varepsilon$ is strictly bigger than the infimum in the right-hand side of (3.1.47), hence there exists $r\in\big(0,R_{\text{flat}}(x)+\varepsilon\big)\subseteq\big(0,2\operatorname{diam}\partial\Omega\big)$ with $\Big|\nu-\fint_{\Delta(x,r)}\nu\,d\sigma\Big|\,d\sigma>\varepsilon_3$. This makes r a participant in the infimum game defining $\wp(\Omega)$ in (3.1.36). Consequently, $\wp(\Omega)\le r<R_{\text{flat}}(x)+\varepsilon$. In view of the arbitrariness of ε, the latter implies that $\wp(\Omega)\le\inf_{x\in\partial\Omega}R_{\text{flat}}(x)$. This proves the left-pointing inequality in (3.1.48). In the opposite direction, $\wp(\Omega)+\varepsilon$ is strictly bigger than the infimum in the right-hand side of (3.1.36). As such, there exist $x\in\partial\Omega$ and $r\in\big(0,\wp(\Omega)+\varepsilon\big)\subseteq\big(0,2\operatorname{diam}\partial\Omega\big)$ with $\Big|\nu-\fint_{\Delta(x,r)}\nu\,d\sigma\Big|\,d\sigma>\varepsilon_3$. Hence, r is a participant in the infimum game defining $R_{\text{flat}}(x)$ in (3.1.47), so $R_{\text{flat}}(x)\le r<\wp(\Omega)+\varepsilon$. Given the arbitrariness of ε, this forces $\wp(\Omega)\ge R_{\text{flat}}(x)\ge\inf_{x\in\partial\Omega}R_{\text{flat}}(x)$, finishing the proof of (3.1.48).

Convention 3.1.10 *Let $\Omega\subseteq\mathbb{R}^n$, where $n\in\mathbb{N}$ with $n\ge2$, be an Ahlfors regular domain with compact boundary. In view of* (3.1.47), *it makes sense to refer to R_{flat} as the* pointwise flatness radius *function for the set Ω. To better contrast this piece of terminology with the one introduced in Definition 3.1.9, we shall occasionally refer to $\wp(\Omega)$ as the* global flatness radius of Ω.

Comment 8. Retain the context of Definition 3.1.9. Observe that for each radius $r\in\big(0,2\operatorname{diam}\partial\Omega\big)$ and point $x\in\partial\Omega$ we have

$$\begin{aligned}1-\Big|\fint_{\Delta(x,r)}\nu\,d\sigma\Big| &= \fint_{\Delta(x,r)}\Big(|\nu|-\Big|\fint_{\Delta(x,r)}\nu\,d\sigma\Big|\Big)\,d\sigma\\ &\le \fint_{\Delta(x,r)}\Big|\nu-\fint_{\Delta(x,r)}\nu\,d\sigma\Big|\,d\sigma,\end{aligned} \tag{3.1.49}$$

thanks to the fact that $|\nu| = 1$ at σ-a.e. point on $\partial\Omega$ and the reverse triangle inequality. There are two consequences of (3.1.49) and (3.1.36) we wish to single out. First,

$$\begin{gathered}\wp(\Omega) \le r \ \text{ whenever } \ r \in \big(0, 2\operatorname{diam}\partial\Omega\big) \ \text{ has the property that} \\ \text{there exists a point } \ x \in \partial\Omega \ \text{ with } \ \Big|\fint_{\Delta(x,r)} \nu \, d\sigma\Big| < 1 - \varepsilon_3. \end{gathered} \tag{3.1.50}$$

Second,

$$\begin{gathered}\text{if } \ \wp(\Omega) > 0 \ \text{ then } \ \Big|\fint_{\Delta(x,r)} \nu \, d\sigma\Big| \ge 1 - \varepsilon_3 \\ \text{for all } \ r \in \big(0, \wp(\Omega)\big) \ \text{ and all } \ x \in \partial\Omega. \end{gathered} \tag{3.1.51}$$

Comment 9. Assume $\Omega \subseteq \mathbb{R}^n$ is a domain of class $\mathscr{C}^1$ with compact boundary. This implies that the outward unit normal ν to Ω is a uniformly continuous function, hence its modulus of continuity

$$\omega(t) := \sup\big\{|\nu(x) - \nu(y)| : \ x, y \in \partial\Omega \ \text{with} \ |x-y| \le t\big\} \ \text{ for } t \in [0,\infty) \tag{3.1.52}$$

is a non-decreasing function which vanishes continuously at zero. In relation to this, we note that since for each $r \in \big(0, 2\operatorname{diam}\partial\Omega\big)$ and $x \in \partial\Omega$ we have

$$\fint_{\Delta(x,r)} \Big|\nu - \fint_{\Delta(x,r)} \nu \, d\sigma\Big| \, d\sigma \le \fint_{\Delta(x,r)} \fint_{\Delta(x,r)} |\nu(y) - \nu(z)| \, d\sigma(y) \, d\sigma(z) \le \omega(2r),$$

where $\sigma := \mathcal{H}^{n-1} \lfloor \partial\Omega$, it follows that

$$\wp(\Omega) \ge \sup\Big\{r \in \big(0, 2\operatorname{diam}\partial\Omega\big) : \ \omega(2r) < \varepsilon_3\Big\}. \tag{3.1.53}$$

In particular,

$$\begin{gathered}\text{if } \Omega \subseteq \mathbb{R}^n \text{ is a domain of class } \mathscr{C}^1 \text{ with compact boundary whose} \\ \text{outward unit normal } \nu \text{ is Hölder continuous of order } \alpha \in (0,1] \\ \text{(which actually makes } \Omega \text{ a domain of class } \mathscr{C}^{1,\alpha}\text{; cf., e.g., [8])} \\ \text{then } \wp(\Omega) \ge 2^{-1}\big(\varepsilon_3 / \|\nu\|_{\dot{\mathscr{C}}^{\alpha}(\partial\Omega)}\big)^{1/\alpha}. \end{gathered} \tag{3.1.54}$$

To offer a concrete example, for each fixed $\varepsilon \in (0,1)$ consider $\Omega_\varepsilon \subseteq \mathbb{R}^2$ to be

$$\begin{gathered}\text{the rectangle } [-2,2] \times (-\varepsilon, \varepsilon) \text{ to which we glue, to the left and to the} \\ \text{right, an open half-disk of radius } \varepsilon \text{, to make it a domain of class } \mathscr{C}^{1,1}. \end{gathered} \tag{3.1.55}$$

If ν denotes the outward unit normal ν to Ω_ε, then the simple geometry of Ω_ε permits us to compute its Lipschitz constant as $\sup\limits_{\substack{x,y\in\partial\Omega \\ x\neq y}} \frac{|\nu(x)-\nu(y)|}{|x-y|} = \frac{1}{\varepsilon}$. With this in hand, (3.1.54) used with $\alpha := 1$ gives (bearing in mind that the Ahlfors regularity constants of $\partial\Omega_\varepsilon$ stay bounded from zero and infinity, uniformly in ε)

$$\wp(\Omega_\varepsilon) \geq c \cdot \varepsilon \text{ for some universal constant } c \in (0, \infty). \tag{3.1.56}$$

We claim that we also have

$$\wp(\Omega_\varepsilon) \leq C \cdot \varepsilon \text{ for some universal constant } C \in (0, \infty). \tag{3.1.57}$$

If ε stays away from zero, this is directly implied by the fact that $\wp(\Omega_\varepsilon) \leq 6$ (cf. (3.1.37)). In the case when the parameter $\varepsilon \in (0, 1)$ is sufficiently small, observe that if $x_o := (2 + \varepsilon, 0) \in \partial\Omega_\varepsilon$ and $r \in \big(\sqrt{2}\varepsilon, \sqrt{16 + 8\varepsilon + 2\varepsilon^2}\big)$ then

$$\left| \fint_{\Delta(x_o,r)} \nu \, d\sigma \right| = \frac{2\varepsilon}{\pi\varepsilon + 2\big(\sqrt{r^2 - \varepsilon^2} - \varepsilon\big)}. \tag{3.1.58}$$

Specializing this to the case when $r = \lambda\varepsilon$ for a large universal constant $\lambda \in (\sqrt{2}, \infty)$ and invoking (3.1.50) ultimately shows that (3.1.57) is valid in this case as well.

Moving on, we make the case that, for Ahlfors regular domains with compact boundary, relatively small flatness (measured via (3.0.5)) says a lot about the geometry of the set in question.

Theorem 3.1.11 *Fix $n \in \mathbb{N}$ with $n \geq 2$. Then for each given $C_A \in [1, \infty)$ there exists a constant $\varepsilon_4 \in (0, 1)$, which depends only on n and C_A, with the following significance. Suppose $\Omega \subseteq \mathbb{R}^n$ is an Ahlfors regular domain with compact boundary, whose Ahlfors regularity constants are controlled by C_A (in the sense of* (3.1.35)) *and with the property that the geometric measure theoretic outward unit normal ν to Ω satisfies*

$$\operatorname{dist}\Big(\nu, \big[\mathrm{VMO}(\partial\Omega, \sigma)\big]^n\Big) < \varepsilon_4 \tag{3.1.59}$$

where $\sigma := \mathcal{H}^{n-1}\lfloor\partial\Omega$ and the distance is measured in the space $\big[\mathrm{BMO}(\partial\Omega, \sigma)\big]^n$.

Then the relative flatness radius $\wp(\Omega)$ (Definition 3.1.9) is strictly positive, and

$$\begin{gathered}\textit{the set } \Omega \textit{ is a two-sided } \text{NTA} \textit{ domain, in the sense of } [129, \text{Definition 5.11.1}], \textit{ with constants } R := 2^{-1} C_A^{-2/(n-1)} \cdot \wp(\Omega), \textit{ and} \\ M = M(C_A, n) \in (0, \infty) \textit{ depending only on the dimension } n \textit{ and} \\ \textit{the Ahlfors regularity constant } C_A \textit{ of } \partial\Omega.\end{gathered} \tag{3.1.60}$$

As a corollary of this, [129, Remark 5.11.2], *and* [129, (5.10.33)], *the set Ω is a* UR *domain with constants controlled in terms of the dimension n, the constant C_A, and the relative flatness ratio $\wp(\Omega)/\operatorname{diam}\partial\Omega$. Moreover, as a consequence of* (3.1.60) *and* [129, (5.11.28)], *the set Ω satisfies a two-sided local John condition (again, in the same quantitative fashion). In concert,* (3.1.60) *and* [129, Lemma 5.11.3] *also imply that Ω has finitely many connected components, which are separated (i.e., have mutually disjoint closures), and the same is true for $\mathbb{R}^n \setminus \overline{\Omega}$. In particular, any connected component of Ω and $\mathbb{R}^n \setminus \overline{\Omega}$ is itself a two-sided* NTA *domain.*

As a consequence of (3.1.60) *and* [129, (5.11.66), (5.11.35)], *the set Ω is also an (ε, δ)-domain for some $\varepsilon, \delta \in (0, \infty)$ (cf.* (A.0.76)) *with* $\operatorname{rad}(\Omega) > 0$ *(cf.* (A.0.214)). *Finally, both Ω and $\mathbb{R}^n \setminus \overline{\Omega}$ are locally uniform domains in the sense of* [129,

Definition 5.11.12], *and each of their connected components is a uniform domain in the sense of* [129, Definition 5.11.10] (*thanks to what has been concluded already and items (1), (4) in* [129, Proposition 5.11.14]).

We wish to remark that, thanks to (3.1.45), the result described in (3.1.60) scales naturally. In addition, it is worth pointing out that the demand in (3.1.59) is satisfied if (with the piece of notation introduced in (A.0.17) relative to the set $\Sigma := \partial\Omega$ and measure $\mu := \mathcal{H}^{n-1}\lfloor\partial\Omega$) one has

$$\begin{array}{l}\sup_{x\in\partial\Omega}\|\nu\|_*(\Delta(x,r)) \leq \varepsilon_4/C \ \text{ for some } r \in \big(0, 2\operatorname{diam}\partial\Omega\big)\\ \text{where the constant } C \in (1,\infty) \text{ depends only on } n \text{ and } C_A.\end{array} \tag{3.1.61}$$

Indeed, [130, (3.1.55)] implies

$$\operatorname{dist}\Big(\nu, \big[\mathrm{VMO}(\partial\Omega,\sigma)\big]^n\Big) \leq C \cdot \lim_{r\to 0^+}\Big[\sup_{x\in\partial\Omega}\|\nu\|_*(\Delta(x,r))\Big], \tag{3.1.62}$$

where $C \in (1,\infty)$ depends only on n and the Ahlfors regularity constants of $\partial\Omega$. In view of the monotonicity of $\sup_{x\in\partial\Omega}\|\nu\|_*(\Delta(x,r))$ with respect to r, this proves that (3.1.59) is satisfied whenever (3.1.61) holds.

We also wish to remark that the involvement of the relative flatness radius $\wp(\Omega)$ in the quantitative aspect of the result formulated in (3.1.60) is indispensable. To illustrate this point, for each $\varepsilon \in (0,1)$ bring back the set $\Omega_\varepsilon \subseteq \mathbb{R}^2$ considered in (3.1.55). This is a domain of class $\mathscr{C}^{1,1}$ with compact boundary and whose relative flatness radius satisfies (see (3.1.56)-(3.1.57))

$$c\cdot\varepsilon \leq \wp(\Omega_\varepsilon) \leq C\cdot\varepsilon \ \text{ for two universal constants } c, C \in (0,\infty). \tag{3.1.63}$$

Moreover, as is apparent from (3.1.55),

$$\begin{array}{l}\text{the Ahlfors regularity constants of } \partial\Omega_\varepsilon \text{ stay bounded away from zero}\\ \text{and infinity, uniformly for } \varepsilon \in (0,1), \text{ while the outward unit normal } \nu\\ \text{to } \Omega_\varepsilon \text{ satisfies } \operatorname{dist}\big(\nu, \big[\mathrm{VMO}(\partial\Omega_\varepsilon,\sigma)\big]^2\big) = 0,\end{array} \tag{3.1.64}$$

where $\sigma := \mathcal{H}^1\lfloor\partial\Omega_\varepsilon$ and the distance is measured in the space $\big[\mathrm{BMO}(\partial\Omega_\varepsilon,\sigma)\big]^2$. However, while each Ω_ε is (a bounded $\mathscr{C}^{1,1}$ domain, hence) an NTA domain, the scale up to which Ω_ε satisfies an interior corkscrew condition (as in [129, (5.1.5)] with a constant $\theta \in (0,1)$ independent of ε) is of the order of ε, something impossible to predict based on the quantitative aspects noted in (3.1.64). In a nutshell, it is impossible to control the NTA character of an Ahlfors regular domain $\Omega \subseteq \mathbb{R}^n$ solely in terms of the Ahlfors regularity constants of $\partial\Omega$ and the distance (measured in the John-Nirenberg space) from its geometric measure theoretic outward unit normal ν to the Sarason space of functions of vanishing mean oscillations on $\partial\Omega$. This being said, (3.1.63) is in full agreement with the quantitative aspect of (3.1.60).

We now present the proof of Theorem 3.1.11.

Proof of Theorem 3.1.11 Let $\varepsilon_3 \in (0, 1)$ be associated with Ω as in Corollary 3.1.4, and let the constant $C \in (1, \infty)$ (depending only on n and C_A) be as in (3.1.41). Define $\varepsilon_4 := \varepsilon_3/C \in (0, 1)$. Then (3.1.59) and (3.1.41) guarantee that

$$\wp(\Omega) > 0. \tag{3.1.65}$$

Having established this, we may rely on (3.1.40) to conclude that

$$\sup_{x\in\partial\Omega} \|\nu\|_*(\Delta(x,\rho)) \leq \varepsilon_3 \text{ if } \rho := 2^{-1}C_A^{-2/(n-1)} \cdot \wp(\Omega). \tag{3.1.66}$$

Then (3.1.60) follows since (3.1.66) and Corollary 3.1.4 imply

$$\begin{aligned}&\Omega \text{ is a two-sided NTA domain, in the sense of [129, Definition 5.11.1],}\\ &\text{with constants } R := 2^{-1}C_A^{-2/(n-1)} \cdot \wp(\Omega) \text{ and } M = M(C_A, n) \in (0, \infty)\\ &\text{depending only on } n \text{ and the Ahlfors regularity constants of } \partial\Omega.\end{aligned} \tag{3.1.67}$$

All other claims are consequences of (3.1.60), as indicated in the statement. □

We are now in a position to prove the following companion result to [130, Theorem 11.5.2], for arbitrary Ahlfors regular domains with compact boundary.

Theorem 3.1.12 *Let $\Omega \subset \mathbb{R}^n$ be an Ahlfors regular domain with compact boundary. Set $\sigma := \mathcal{H}^{n-1}\lfloor\partial\Omega$, and let ν be the geometric measure theoretic outward unit normal to Ω.*

Then there exist a threshold $\widetilde{R} \in (0, \infty)$ and two numbers, $\widetilde{C} \in [1, \infty)$ and $\lambda \in (32, \infty)$, all of which depend only on the dimension n, the Ahlfors regularity constants of $\partial\Omega$, and the relative flatness ratio $\wp(\Omega)/\operatorname{diam}\partial\Omega$ (cf. Definition 3.1.9), with the property that for each scale $R \in (0, \widetilde{R})$, each threshold $\varepsilon \in (0, 1)$, and each dilation factor $\gamma \in [1, \infty)$ one has

$$\begin{aligned}\sup_{z\in\partial\Omega}\ \sup_{x,y\in\Delta(z,\gamma R)} & R^{-1}|\langle x-y, \nu_{\Delta(z,R)}\rangle| \\ &\leq \widetilde{C}\gamma(1+\log_2\gamma)\sup_{z\in\partial\Omega}\|\nu\|_*(\Delta(z,\gamma\lambda\varepsilon^{-1}R)) + \widetilde{C}\gamma\varepsilon,\end{aligned} \tag{3.1.68}$$

where $\nu_{\Delta(z,R)} := ⨍_{\Delta(z,R)} \nu\, d\sigma$. Also, for each dilation factor $\gamma \in [1, \infty)$ one has

$$\begin{aligned}\limsup_{R\to 0^+} & \Big[\sup_{z\in\partial\Omega}\ \sup_{x,y\in\Delta(z,\gamma R)} R^{-1}|\langle x-y, \nu_{\Delta(z,R)}\rangle|\Big] \\ &\leq \widetilde{C}\gamma(1+\log_2\gamma)\cdot \operatorname{dist}\big(\nu, \big[\mathrm{VMO}(\partial\Omega,\sigma)\big]^n\big),\end{aligned} \tag{3.1.69}$$

where the distance in the right-hand side is considered in the space $\big[\mathrm{BMO}(\partial\Omega,\sigma)\big]^n$.

Proof Associate the threshold $\varepsilon_3 \in (0, 1)$ with the given Ahlfors regular domain Ω as in Corollary 3.1.4, and bring in the constant $C \in (0, \infty)$ associated with Ω as in (3.1.41). We distinguish two cases.

`Case I:` *Assume* $\operatorname{dist}\big(\nu, \big[\mathrm{VMO}(\partial\Omega, \sigma)\big]^n\big) \geq \varepsilon_3/C$. Observe that [130, (3.1.55)] ensures the existence of some $c \in (0, \infty)$ which depends only on the Ahlfors regularity constants of $\partial\Omega$ with the property that

$$\sup_{x\in\partial\Omega} \|\nu\|_*(\Delta(x, R)) \geq c \cdot \operatorname{dist}\big(\nu, \big[\mathrm{VMO}(\partial\Omega, \sigma)\big]^n\big) \text{ for each } R > 0. \tag{3.1.70}$$

Observe that for each $\gamma \in [1, \infty)$ and $R > 0$, the Cauchy-Schwarz inequality implies

$$\sup_{z\in\partial\Omega}\ \sup_{x,y\in\Delta(z,\gamma R)} R^{-1}|\langle x - y, \nu_{\Delta(z,R)}\rangle| \leq 2\gamma. \tag{3.1.71}$$

Based on these two observations and the present working hypothesis, we conclude that (3.1.68) holds in the current case provided $\widetilde{C} \geq 2C/(c\varepsilon_3)$.

`Case II:` *Assume* $\operatorname{dist}\big(\nu, \big[\mathrm{VMO}(\partial\Omega, \sigma)\big]^n\big) < \varepsilon_3/C$. In concert with (3.1.41) this ensures that we presently have $\wp(\Omega) > 0$. Granted this, it follows from (3.1.40) that (with $C_A \in [1, \infty)$ as in (3.1.35)) we have

$$\sup_{x\in\partial\Omega} \|\nu\|_*(\Delta(x, r_0)) \leq \varepsilon_3 \text{ where } r_0 := 2^{-1}C_A^{-2/(n-1)} \cdot \wp(\Omega) \in \big(0, 2\operatorname{diam}\partial\Omega\big). \tag{3.1.72}$$

With this in hand, Corollary 3.1.4 then guarantees that

$$\begin{array}{l}\Omega \text{ satisfies a two-sided local John condition with constants controlled}\\ \text{in terms of the dimension } n\text{, the Ahlfors regularity constants of } \partial\Omega,\\ \text{and the ratio } \wp(\Omega)/\operatorname{diam}\partial\Omega.\end{array} \tag{3.1.73}$$

Once (3.1.73) has been proved we may invoke [130, Theorem 11.5.2] and conclude from [130, (11.5.34)] that (3.1.68) holds. The proof of (3.1.68) is therefore complete.

Having established (3.1.68), send $R \to 0^+$ and invoke [130, (3.1.55)], then finally send $\varepsilon \to 0^+$. In doing so, we arrive at (3.1.69). □

It is of interest to further refine the geometric estimates from [130, Theorem 11.5.2] and Theorem 3.1.12 by eliminating the parameter $\varepsilon \in (0, 1)$; see below.

Theorem 3.1.13 *Assume $\Omega \subset \mathbb{R}^n$ is an Ahlfors regular domain. Set $\sigma := \mathcal{H}^{n-1}\lfloor\partial\Omega$ and denote by ν the geometric measure theoretic outward unit normal to Ω. For each $z \in \partial\Omega$ and $R > 0$, abbreviate $\nu_{\Delta(z,R)} := ⨍_{\Delta(z,R)} \nu\, d\sigma$. Then the following two statements are true:*

(1) Under the additional assumption that Ω satisfies a two-sided local John condition, there exist two numbers $C, \lambda \in (1, \infty)$ which depend only on the local John constants of Ω and the Ahlfors regularity constants of $\partial\Omega$, with the property that for each point $z \in \partial\Omega$, each scale $R > 0$, and each dilation factor $\gamma \in [1, \infty)$ one has

$$\sup_{x,y\in\Delta(z,\gamma R)} R^{-1}|\langle x - y, \nu_{\Delta(z,R)}\rangle| \leq C\gamma\big(1 + \log_2\gamma\big)\|\nu\|_*\big(\Delta(z, \gamma\lambda R)\big). \tag{3.1.74}$$

(2) In the case when $\partial\Omega$ is compact, there exist two numbers $C, \lambda \in (1, \infty)$, which now depend only on the dimension n, the Ahlfors regularity constants of $\partial\Omega$, and the relative flatness ratio $\wp(\Omega)/\mathrm{diam}\,\partial\Omega$ (cf. Definition 3.1.9), with the property that for each scale $R > 0$ and each dilation factor $\gamma \in [1, \infty)$ one has

$$\sup_{z\in\partial\Omega}\Bigg(\sup_{x,y\in\Delta(z,\gamma R)} R^{-1}\big|\langle x-y, \nu_{\Delta(z,R)}\rangle\big|\Bigg) \leq C\gamma\big(1+\log_2\gamma\big)\sup_{z\in\partial\Omega}\|\nu\|_*\big(\Delta(z,\gamma\lambda R)\big). \tag{3.1.75}$$

Proof To prove the estimate claimed in item *(1)*, work under the assumption that $\Omega \subset \mathbb{R}^n$ is an Ahlfors regular domain satisfying a two-sided local John condition. Thanks to [132, (1.8.157)], this implies that Ω is a a two-sided NTA domain. Such a property permits us to invoke the Poincaré inequality proved in [201], according to which there exist two constants $C, \Lambda \in [1, \infty)$ (ultimately depending only on the local John constants of Ω and the Ahlfors regularity constants of $\partial\Omega$) such that for every Lipschitz function $f : \partial\Omega \to \mathbb{R}$ and every surface ball $\Delta := \Delta(z, R) = B(z, R) \cap \partial\Omega$ (with $z \in \partial\Omega$ and $R > 0$) we have

$$⨍_{\Delta}|f - f_\Delta|\,d\sigma \leq CR ⨍_{\Lambda\Delta}|\nabla_t f|\,d\sigma. \tag{3.1.76}$$

Above, $\nabla_t f$ is a tangential gradient of the following sort. Let $F : \mathbb{R}^n \to \mathbb{R}$ be any Lipschitz extension of f to the entire Euclidean ambient. Call f tangentially differentiable at a point $x \in \partial\Omega$ where an approximate tangent plane $T_x\partial\Omega$ exists if $F\big|_{x+T_x\partial\Omega}$ is differentiable (in an ordinary sense) at x, a scenario in which we define $(\nabla_t f)(x)$ to be the gradient of $F\big|_{x+T_x\partial\Omega}$ at x (this type of derivative is discussed in [110, §10-§11]). From [129, (5.11.27)] we know that $\partial\Omega$ is uniformly rectifiable, hence countably rectifiable of dimension $n-1$ (by [129, Proposition 5.10.5]). Granted this, [110, Theorem 11.4, p. 108] guarantees that $\nabla_t f$ exists at σ-a.e. point on $\partial\Omega$, while [110, Theorem 10.1 on p. 98, Proposition 10.5 on p. 101, Lemma 11.5 on p. 108] ensure that the actual choice of the Lipschitz extension F is immaterial as far as the definition of $\nabla_t f$ is concerned. Furthermore, it has been noted in [162, Lemma 6.4] that for any Lipschitz function $f : \partial\Omega \to \mathbb{R}$ there holds $\nabla_t f$ agrees σ-a.e. with (the opposite of) $\nabla_{\tan} f$, our tangential gradient from [130, §11.4] (cf. also [130, Corollary 11.1.18]). Strictly speaking, [162, Lemma 6.4] is stated for bounded domains, but the same proof works in the unbounded case if the Lipschitz function in question is compactly supported. In fact, it is possible to further eliminate the latter compact support requirement via a pedestrian localization argument (involving a smooth cutoff function), since differentiation is a local property. All things considered, this discussion shows that in place of (3.1.76) we have (with $C, \Lambda \in [1, \infty)$ as before)

$$⨍_{\Delta}|f - f_\Delta|\,d\sigma \leq CR ⨍_{\Lambda\Delta}|\nabla_{\tan} f|\,d\sigma, \tag{3.1.77}$$

for any Lipschitz function $f : \partial\Omega \to \mathbb{R}$. From this point we largely proceed as in the proof of [130, Theorem 11.5.2].

Concretely, fix an arbitrary point $x \in \partial\Omega$ along with a radius $R > 0$, and abbreviate $\Delta := \Delta(x, R) = B(x, R) \cap \partial\Omega$. The idea is to use (3.1.77) for the Lipschitz function $f : \partial\Omega \to \mathbb{R}$ defined by

$$f(y) := \langle x - y, \nu_\Delta \rangle \text{ for all } y \in \partial\Omega. \tag{3.1.78}$$

In this regard, we claim that for each exponent $\alpha \in (0, 1)$ there exist some constant $C = C(\Omega, n, \alpha) \in (0, \infty)$ and some geometric constant $\lambda \in (2, \infty)$ such that

$$|f(y) - f(y')| \leq CR^{1-\alpha}|y - y'|^\alpha \Big\{\|\nu\|_*(\Delta(x, \lambda\Lambda R))\Big\} \text{ for all } y, y' \in 2\Delta. \tag{3.1.79}$$

Accepting this for the time being and choosing $y' := x$ yields

$$|f(y)| \leq CR\|\nu\|_*(\Delta(x, \lambda\Lambda R)), \text{ for all } y \in 2\Delta, \tag{3.1.80}$$

which, after adjusting notation, gives (now with $\lambda \in (4\Lambda, \infty)$)

$$\sup_{y\in\Delta(x,2R)} R^{-1}|\langle x - y, \nu_{\Delta(x,R)}\rangle| \leq C\|\nu\|_*(\Delta(x, \lambda R)). \tag{3.1.81}$$

We may now use this to prove (3.1.74) as follows. Fix $\gamma \in [1, \infty)$ along with $z \in \partial\Omega$, $R \in (0, \infty)$, and $x, y \in \Delta(z, \gamma R)$. Then

$$\begin{aligned} y \in \Delta(x, 2\gamma R), \quad \Delta(x, 2\gamma R) \subseteq \Delta(z, 3\gamma R), \\ \Delta(x, \gamma\lambda R) \subseteq \Delta\big(z, \gamma(\lambda + 1)R\big), \end{aligned} \tag{3.1.82}$$

so for some $C \in (0, \infty)$, which depends only on the local John constants of Ω and the Ahlfors regularity constants of $\partial\Omega$, we may write

$$\begin{aligned} |\langle x - y, \nu_{\Delta(z,R)}\rangle| &\leq |\langle x - y, \nu_{\Delta(x,\gamma R)}\rangle| + |x - y||\nu_{\Delta(x,\gamma R)} - \nu_{\Delta(z,R)}| \\ &\leq C\gamma R\|\nu\|_*(\Delta(x, \gamma\lambda R)) + 2\gamma R|\nu_{\Delta(x,2\gamma R)} - \nu_{\Delta(z,3\gamma R)}| + 2\gamma R|\nu_{\Delta(z,3\gamma R)} - \nu_{\Delta(z,R)}| \\ &\leq C\gamma R\|\nu\|_*\big(\Delta(z, \gamma(\lambda + 1)R)\big) + C\gamma R\big(1 + \log_2 \gamma)\|\nu\|_*\big((\Delta(z, 3\gamma R)\big) \\ &\leq C\gamma R\big(1 + \log_2 \gamma)\|\nu\|_*\big(\Delta(z, \gamma(\lambda + 1)R)\big), \end{aligned} \tag{3.1.83}$$

thanks to (3.1.81) (written with γR in place of R), (3.1.82), [129, (7.4.56), (7.4.63)], and the fact that σ is a doubling measure. Work with the most extreme sides of (3.1.83). Dividing by R, and taking the supremum with respect to $x, y \in \Delta(z, \gamma R)$ and $z \in \partial\Omega$ establishes (3.1.74) (after estimating $\lambda + 1 \leq 2\lambda$ and relabeling 2λ as λ).

At this stage, there remains to prove (3.1.79). To this end, first note that much as in [129, (11.5.42)-(11.5.44)] we have, for σ-a.e. $y \in \partial\Omega$,

$$|\nabla_{\tan} f(y)| = |\nu_\Delta - \nu(y) - \langle \nu_\Delta - \nu(y), \nu(y)\rangle\nu(y)| \leq 2|\nu_\Delta - \nu(y)|. \tag{3.1.84}$$

In preparation for applying Poincaré's inequality recorded in (3.1.77) to the Lipschitz function f, fix an arbitrary surface ball Δ_r of radius $r \in \big(0, 2\,\mathrm{diam}\,(\partial\Omega)\big)$ such that $\Delta_r \subseteq \Delta$. In view of the Ahlfors regularity of $\partial\Omega$ this entails $r \leq CR$, for some geometric constant $C \in (0, \infty)$. Then (3.1.77) and Hölder's inequality imply

$$\frac{1}{r} \fint_{\Delta_r} |f - f_{\Delta_r}|\, d\sigma \leq C \Big(\fint_{\Lambda\Delta_r} |\nabla_{\tan} f|^p \, d\sigma \Big)^{1/p} \tag{3.1.85}$$

for each $p \in (1, \infty)$. Then (3.1.85), (3.1.84), the Ahlfors regularity of $\partial\Omega$, the fact that we have used the abbreviation $\Delta := \Delta(x, R)$, and John-Nirenberg's inequality [129, (7.4.67)] yield

$$\begin{aligned} \frac{1}{r} \fint_{\Delta_r} |f - f_{\Delta_r}|\, d\sigma &\leq C \Big(\fint_{\Lambda\Delta_r} |\nu - \nu_\Delta|^p \, d\sigma \Big)^{1/p} \\ &\leq C \Big(\frac{R}{r}\Big)^{(n-1)/p} \Big(\fint_{\Lambda\Delta} |\nu - \nu_\Delta|^p \, d\sigma \Big)^{1/p} \leq C \Big(\frac{R}{r}\Big)^{(n-1)/p} \|\nu\|_*\big(\Delta(x, \Lambda R)\big). \end{aligned} \tag{3.1.86}$$

Given $\alpha \in (0, 1)$, if $p \in (n-1, \infty)$ is chosen such that $\alpha = 1 - (n-1)/p$, then the above estimate shows that

$$r^{-\alpha} \fint_{\Delta_r} |f - f_{\Delta_r}|\, d\sigma \leq C R^{1-\alpha} \|\nu\|_*\big(\Delta(x, \Lambda R)\big) \tag{3.1.87}$$

for some $C = C(\Omega, n, \alpha) \in (0, \infty)$. Bearing in mind [129, Lemma 3.6.4], the criterion for (local) Hölder continuity from [129, Proposition 7.4.9] then gives that

$$\begin{aligned} \sup_{\substack{y, y' \in 2\Delta \\ y \neq y'}} \frac{|f(y) - f(y')|}{|y - y'|^\alpha} &\leq C \sup_{\Delta_r \subseteq \lambda\Delta} r^{-\alpha} \fint_{\Delta_r} |f - f_{\Delta_r}|\, d\sigma \\ &\leq C R^{1-\alpha} \|\nu\|_*\big(\Delta(x, \lambda\Lambda R)\big), \end{aligned} \tag{3.1.88}$$

for some geometric constant $\lambda \in (4, \infty)$. At this stage, estimate (3.1.88) justifies (3.1.79), and finishes the proof of (3.1.74). This concludes the treatment in item *(1)*.

To deal with item *(2)*, suppose the topological boundary $\partial\Omega$ is a compact set. Throughout, it is assumed that $C_A \in [1, \infty)$ controls the Ahlfors regularity constants of $\partial\Omega$ as in (3.1.35). We distinguish two cases. First, consider the situation when $\wp(\Omega) = 0$. From (3.1.41) we know that this implies the existence of some $C \in (0, \infty)$, depending only on n and C_A, such that

$$\mathrm{dist}\,\big(\nu,\, \big[\mathrm{VMO}(\partial\Omega, \sigma)\big]^n\big) \geq \varepsilon_3/C, \tag{3.1.89}$$

where the distance is measured in $\big[\mathrm{BMO}(\partial\Omega, \sigma)\big]^n$, and the threshold $\varepsilon_3 \in (0, 1)$ is associated with the constant C_A and the dimension n as in Corollary 3.1.4. In view of the monotonicity of the function $R \mapsto \sup_{z\in\partial\Omega} \|\nu\|_*\big(\Delta(x, R)\big)$, from [130, (3.1.55)] we also see that there exists a constant $c \in (0, \infty)$ such that

$$\sup_{z\in\partial\Omega} \|\nu\|_*\big(\Delta(x,R)\big) \geq c\cdot \mathrm{dist}\big(\nu,\ \big[\mathrm{VMO}(\partial\Omega,\sigma)\big]^n\big) \text{ for all } R>0. \tag{3.1.90}$$

Combining (3.1.89) and (3.1.90) then gives $\sup_{z\in\partial\Omega}\|\nu\|_*\big(\Delta(x,R)\big) \geq c\big(\varepsilon_3/C\big)$ for all $R>0$. Granted this, the inequality claimed in (3.1.75) follows by simply making a judicious choice of the multiplicative constant (appearing in the right-hand side).

The second case we need to consider corresponds to having $\wp(\Omega)>0$. If that is the case, then (3.1.40) implies that

$$\sup_{x\in\partial\Omega} \|\nu\|_*(\Delta(x,r_0)) \leq \varepsilon_3 \text{ provided } r_0 := 2^{-1}\cdot C_A^{-2/(n-1)}\cdot\wp(\Omega). \tag{3.1.91}$$

Thanks to (3.1.91) we may invoke Corollary 3.1.4 which guarantees that Ω satisfies a two-sided local John condition (in a quantitative fashion). Having established this, item *(1)* applies. Taking the supremum over $z\in\partial\Omega$ in (3.1.74) yields (3.1.75). □

We conclude with a comment pertaining to Theorem 3.1.12 which highlights a geometric characteristic of an Ahlfors regular domain we call "infinitesimal tilt."

Remark 3.1.14 *There is a natural variant of the notion of global tilt, introduced in Remark 3.1.7, which only takes into account the behavior of the boundary at small scales. Specifically, for a given Ahlfors regular domain $\Omega\subseteq\mathbb{R}^n$ we shall now consider the infimum of the maximal deviation of a chord $x-y$ with $x,y\in\Delta$, where Δ is a surface ball of arbitrary center on $\partial\Omega$, from being perpendicular to ν_Δ, the integral average in Δ of the geometric measure theoretic outward unit normal ν to Ω, as the radius of Δ shrinks to zero. In precise terms, we define the "infinitesimal tilt" of Ω with amplitude $\gamma\in[1,\infty)$ as the quantity*

$$\mathfrak{t}^*_\gamma(\Omega) := \limsup_{R\to 0^+}\left[\sup_{z\in\partial\Omega}\sup_{R>0}\sup_{x,y\in\Delta(z,\gamma R)}\left|\left\langle \nu_{\Delta(z,R)}, \frac{x-y}{\gamma R}\right\rangle\right|\right], \tag{3.1.92}$$

where for each point $z\in\partial\Omega$ and $R>0$ we have set $\nu_{\Delta(z,R)} := \fint_{\Delta(z,R)}\nu\, d\sigma$, with $\sigma := \mathcal{H}^{n-1}\lfloor\partial\Omega$ playing the role of surface measure on $\partial\Omega$.

With this piece of notation, (3.1.69) asserts that there exists a constant $\widetilde{C}\in(0,\infty)$, depending only on the dimension and the Ahlfors regularity constants of $\partial\Omega$, such that for each amplitude parameter $\gamma\in[1,\infty)$ we have

$$\mathfrak{t}^*_\gamma(\Omega) \leq \widetilde{C}\big(1+\log_2\gamma\big)\cdot \mathrm{dist}\big(\nu,\ \big[\mathrm{VMO}(\partial\Omega,\sigma)\big]^n\big), \tag{3.1.93}$$

where the distance in the right-hand side is considered in the space $\big[\mathrm{BMO}(\partial\Omega,\sigma)\big]^n$.

3.2 The Decomposition Theorem

In this section we discuss a basic decomposition theorem, of geometric flavor. The general idea is as follows: For a given "surface", working locally (inside a cylinder) one seeks a decomposition into an ample "good" piece, contained in a Lipschitz graph, and an "error" piece which is small (relative to the scale) and does not stray too much from the graph. This technique originated in [184, Proposition 5.1, p. 212] where such a decomposition result has been obtained for surfaces of class $\mathscr{C}^2$, via a proof which makes essential use of smoothness, though the main quantitative aspects only depend on the rough character of said surface. A formulation in which the $\mathscr{C}^2$ smoothness assumption is replaced by Reifenberg flatness appears in [95, Theorem 4.1, p. 398][4] without proof.

A significant step forward has been taken by S. Hofmann, M. Mitrea, and M. Taylor who succeeded in establishing a version of such a decomposition result (cf. [81, Theorem 4.16, p. 2701]) starting with a different set of hypotheses which, a priori, do not specifically require the domain in question to be Reifenberg flat. Below we present an even more inclusive version of this result, in the class Ahlfors regular domain with compact boundary. This degree of generality is quite desirable, and suits well the applications we have in mind.

Theorem 3.2.1 *Let $\Omega \subseteq \mathbb{R}^n$ be an Ahlfors regular domain with compact boundary. Set $\sigma := \mathcal{H}^{n-1}\lfloor\partial\Omega$ and denote by ν the geometric measure theoretic outward unit normal to Ω. Also, recall $\wp(\Omega)$ from Definition 3.1.9.*

Then there exist $C_0, C_1, C_2, C_3, C_4 \in (0, \infty)$, depending only on the dimension n, the Ahlfors regularity constants of $\partial\Omega$, and the relative flatness ratio $\wp(\Omega)/\operatorname{diam}\partial\Omega$, with the following significance. For each choice of a function

$$\phi : (0, 1) \longrightarrow (0, \infty) \quad \textit{with} \tag{3.2.1}$$

$$\lim_{t\to 0^+} \phi(t) = 0 \ \textit{and} \ \lim_{t\to 0^+} \frac{\phi(t)}{t} \in (1, \infty], \tag{3.2.2}$$

there exists a threshold $\delta_* \in \big(0, \min\{1, 1/C_0\}\big)$, depending only on the dimension n, the Ahlfors regularity constants of $\partial\Omega$, the ratio $\wp(\Omega)/\operatorname{diam}\partial\Omega$, and the function ϕ, such that if $\delta, r_0 \in (0, \infty)$ satisfy (with the piece of notation introduced in (A.0.17))

$$\sup_{x\in\partial\Omega} \|\nu\|_*(\Delta(x, r_0)) < \delta < \delta_* \tag{3.2.3}$$

then there exists some $R_\delta \in (0, \infty)$, depending only the dimension n, the Ahlfors regularity constants of $\partial\Omega$, $\wp(\Omega)/\operatorname{diam}\partial\Omega$, r_0, and δ, with the following property:

For every location $x_0 \in \partial\Omega$ and every scale $r \in (0, R_\delta)$ there exist a unit vector $\vec{n}_{x_0,r} \in S^{n-1}$ along with a Lipschitz function

$$h : H(x_0, r) := \langle\vec{n}_{x_0,r}\rangle^\perp \to \mathbb{R} \ \textit{with} \ \sup_{\substack{y_1,y_2\in H(x_0,r)\\ y_1\neq y_2}} \frac{|h(y_1) - h(y_2)|}{|y_1 - y_2|} \leq C_0\phi(\delta), \tag{3.2.4}$$

whose graph

$$\mathcal{G} := \big\{x = x_0 + x' + t\vec{n}_{x_0,r} : x' \in H(x_0, r),\ t = h(x')\big\} \tag{3.2.5}$$

[4] see also the comments in [30, p. 66]

(in the coordinate system $x = (x', t) \Leftrightarrow x = x_0 + x' + t\vec{n}_{x_0,r}$, $x' \in H(x_0, r)$, $t = h(x')$*) is a good approximation of* $\partial\Omega$ *in the cylinder*

$$C(x_0, r) := \{x_0 + x' + t\vec{n}_{x_0,r} : x' \in H(x_0, r),\ |x'| \le r,\ |t| \le r\} \tag{3.2.6}$$

in the following sense:

First, with υ_{n-1} *denoting the volume of the unit ball in* $\mathbb{R}^{n-1}$ *and with* $\vartriangle$ *denoting the symmetric set-theoretic difference, one has*

$$\mathcal{H}^{n-1}\big(C(x_0, r) \cap (\partial\Omega \vartriangle \mathcal{G})\big) \le C_1 \upsilon_{n-1} r^{n-1} \mathrm{e}^{-C_2\phi(\delta)/\delta}. \tag{3.2.7}$$

Second, there exist two disjoint σ*-measurable subsets* $G(x_0, r)$ *and* $E(x_0, r)$ *of* $\partial\Omega$ *such that*

$$C(x_0, r) \cap \partial\Omega = G(x_0, r) \cup E(x_0, r), \tag{3.2.8}$$

$$G(x_0, r) \subseteq \mathcal{G}, \quad \sigma\big(E(x_0, r)\big) \le C_1 \upsilon_{n-1} r^{n-1} \mathrm{e}^{-C_2\phi(\delta)/\delta}. \tag{3.2.9}$$

Third, if the function $\Pi : \mathbb{R}^n \to H(x_0, r)$ *is defined by* $\Pi(x) := x'$ *for each point* $x = x_0 + x' + t\vec{n}_{x_0,r} \in \mathbb{R}^n$ *with* $x' \in H(x_0, r)$ *and* $t \in \mathbb{R}$*, then*

$$\begin{gathered}\big|x - \big(x_0 + \Pi(x) + h(\Pi(x))\vec{n}_{x_0,r}\big)\big| \le 2C_0\phi(\delta) \cdot \mathrm{dist}\big(\Pi(x), \Pi(G(x_0, r))\big) \\ \textit{for each point } x \in E(x_0, r),\end{gathered} \tag{3.2.10}$$

and

$$C(x_0, r) \cap \partial\Omega \subseteq \{x_0 + x' + t\vec{n}_{x_0,r} : |t| \le C_0\delta r,\ x' \in H(x_0, r)\}, \tag{3.2.11}$$

$$\Pi\big(C(x_0, r) \cap \partial\Omega\big) = \{x' \in H(x_0, r) : |x'| < r\}. \tag{3.2.12}$$

Fourth, if

$$\begin{aligned}C^+(x_0, r) &:= \{x_0 + x' + t\,\vec{n}_{x_0,r} : x' \in H(x_0, r),\ |x'| \le r,\ -r < t < -C_0\delta r\},\\ C^-(x_0, r) &:= \{x_0 + x' + t\,\vec{n}_{x_0,r} : x' \in H(x_0, r),\ |x'| \le r,\ C_0\delta r < t < r\},\end{aligned} \tag{3.2.13}$$

(having $0 < \delta < \delta_* < 1/C_0$ *ensures that* $C^\pm \ne \varnothing$*) then*

$$C^+(x_0, r) \subseteq \Omega \textit{ and } C^-(x_0, r) \subseteq \mathbb{R}^n \setminus \overline{\Omega}. \tag{3.2.14}$$

Fifth,

$$\begin{gathered}\textit{any line in the direction of } \vec{n}_{x_0,r} \textit{ passing through a point} \\ \textit{on } G(x_0, r) \textit{ intersects } \partial\Omega \cap C(x_0, r) \textit{ only at said point.}\end{gathered} \tag{3.2.15}$$

Sixth, with $\Delta(x_0, r) := B(x_0, r) \cap \partial\Omega$ *one has*

$$\begin{aligned}&\Big(1 - C_3\delta - C_1 \exp\big(-C_2\phi(\delta)/\delta\big)\Big)\upsilon_{n-1} r^{n-1} \\ &\quad \le \sigma\big(\Delta(x_0, r)\big) \le \Big(1 + C_3\phi(\delta) + C_1 \exp\big(-C_2\phi(\delta)/\delta\big)\Big)\upsilon_{n-1} r^{n-1}.\end{aligned} \tag{3.2.16}$$

Finally, if $\widetilde{\nu}$ is the unit normal vector to the Lipschitz graph $\mathcal{G}$, pointing towards the upper-graph of the function h then

$$\begin{aligned}&\text{at } \mathcal{H}^{n-1}\text{-a.e. point } x \in \partial\Omega \cap \mathcal{G} \text{ one has}\\ &\text{either } \nu(x) = \widetilde{\nu}(x), \text{ or } \nu(x) = -\widetilde{\nu}(x),\end{aligned} \tag{3.2.17}$$

$$\nu(x) = \widetilde{\nu}(x) \ \text{ at } \mathcal{H}^{n-1}\text{-a.e. point } \ x \in G(x_0, r), \tag{3.2.18}$$

$$\sigma\Big(\big\{x \in \mathcal{G} \cap \Delta(x_0, 4r) : \ \nu(x) = -\widetilde{\nu}(x)\big\}\Big) \leq C_4 \cdot \phi(\delta) r^{n-1}, \tag{3.2.19}$$

and

$$⨍_{\Delta(x_0,4r)} \Big(\sup_{y\in\mathcal{G}} |\nu(x) - \widetilde{\nu}(y)|\Big)\, d\sigma(x) \leq C_4 \cdot \phi(\delta). \tag{3.2.20}$$

Proof Assume (3.2.3) holds for some $r_0 \in (0, \infty)$ and $\delta \in (0, \delta_*)$ with $\delta_* \in (0, 1)$ to be specified later. Throughout, for each given point $x \in \partial\Omega$ and each given radius $R > 0$ we agree to abbreviate $\Delta(x, R) := B(x, R) \cap \partial\Omega$ and $\nu_{\Delta(x,R)} := ⨍_{\Delta(x,R)} \nu\, d\sigma$. We shall also make frequent use of the piece of notation introduced in (A.0.17). Also, bring in the constant $C_A \in [1, \infty)$ as in (3.1.35) (which is entirely determined by the Ahlfors regularity constants of $\partial\Omega$). For starters, assume

$$0 < \delta_* < \varepsilon_4/N \tag{3.2.21}$$

where $\varepsilon_4 \in (0, 1)$ is associated with the Ahlfors regular domain Ω as in Theorem 3.1.11 and where $N \in (1, \infty)$ is a large constant depending only on n and C_A. Then Theorem 3.1.11 implies (cf. (3.1.65) and (3.1.67)) that

$$\begin{aligned}&\wp(\Omega) > 0 \text{ and } \Omega \text{ is a two-sided NTA domain, in the sense of [129,}\\ &\text{Definition 5.11.1], with constants } R := 2^{-1} C_A^{-2/(n-1)} \cdot \wp(\Omega) \text{ and}\\ &M \in (0, \infty) \text{ depending only on the dimension } n \text{ and the Ahlfors}\\ &\text{regularity constants of } \partial\Omega.\end{aligned} \tag{3.2.22}$$

Going further, Theorem 3.1.12 asserts that there exist $\widetilde{R} \in (0, \infty]$, $\widetilde{C} \in [1, \infty)$, and $\lambda \in (32, \infty)$, all of which depend only on the dimension n, the Ahlfors regularity constants of $\partial\Omega$, and the relative flatness ratio $\wp(\Omega)/\text{diam}\, \partial\Omega$, with the property that (3.1.68) holds for each scale $R \in (0, \widetilde{R})$, and each choice of parameters $\gamma \in [1, \infty)$ and $\varepsilon \in (0, 1)$. In relation to these parameters, define

$$R_\delta := \tfrac{1}{4} \cdot \min\big\{\widetilde{R}, (\delta r_0)/(40\lambda)\big\}. \tag{3.2.23}$$

(Note that, as required, R_δ defined above depends only on the dimension n, the Ahlfors regularity constants of $\partial\Omega$, $\wp(\Omega)/\text{diam}\, \partial\Omega$, r_0, and δ.) In turn, from (3.1.68) (used with $\varepsilon := \delta$, $\gamma := 4$, $x := z$) and (3.2.3) we then deduce that

$$\begin{aligned}&\sup\nolimits_{x\in\partial\Omega} \sup\nolimits_{y\in\Delta(x,4R)} R^{-1} |\langle x - y, \nu_{\Delta(x,R)}\rangle| \leq 16\widetilde{C}\delta\\ &\qquad\text{whenever } 0 < R < 4R_\delta.\end{aligned} \tag{3.2.24}$$

With $\widetilde{C}$ as in (3.2.24), choose

$$C_0 := \max\{20\widetilde{C} + 4, 72\widetilde{C}\} \tag{3.2.25}$$

and, for the remainder of the proof, make the assumption that

$$\delta_* \in (0, \min\{1/10, 1/C_0\}) \tag{3.2.26}$$

and that δ_* is also small enough, depending on ϕ, so that

$$\delta \leq \phi(\delta) \leq (20\widetilde{C} + 4)^{-1} \text{ for all } \delta \in (0, \delta_*). \tag{3.2.27}$$

That (3.2.27) may be accommodated is ensured by (3.2.2). (The choice made in (3.2.25) as well as the nature of the right-most expression in (3.2.27) are dictated by future considerations; see (3.2.56).)

Moving on, for each $x \in \partial\Omega$ and $R > 0$ set

$$\nu^*_{x,R}(y) := \sup_{0<\rho<R} ⨍_{\Delta(y,\rho)} |\nu(z) - \nu_{\Delta(x,2R)}| \, d\sigma(z), \qquad \forall y \in \partial\Omega. \tag{3.2.28}$$

Lebesgue's Differentiation Theorem guarantees that for each fixed $x \in \partial\Omega$ and $R > 0$ we have

$$|\nu(y) - \nu_{\Delta(x,2R)}| \leq \nu^*_{x,R}(y) \text{ at } \sigma\text{-a.e. } y \in \partial\Omega. \tag{3.2.29}$$

From (3.2.28) we also see that

$$\nu^*_{x,R}(y) \leq \mathcal{M}\Big(|\nu - \nu_{\Delta(x,2R)}| \cdot \mathbf{1}_{\Delta(x,2R)}\Big)(y), \qquad \forall y \in \Delta(x, R), \tag{3.2.30}$$

where $\mathcal{M}$ is the Hardy-Littlewood maximal operator on $\partial\Omega$. From (3.2.30), the boundedness of $\mathcal{M}$ on $L^2(\partial\Omega, \sigma)$, the local John-Nirenberg inequality recorded in [129, Lemma 7.4.10], and (3.2.3) we conclude that there exists $C \in [1, \infty)$ which depends only on n and the Ahlfors regularity constants of $\partial\Omega$ with the property that

$$\begin{aligned}\Big(⨍_{\Delta(x,R)} |\nu^*_{x,R}(y)|^2 \, d\sigma(y)\Big)^{1/2} &\leq C\Big(⨍_{\Delta(x,2R)} |\nu(y) - \nu_{\Delta(x,2R)}|^2 \, d\sigma(y)\Big)^{1/2} \\ &\leq C\|\nu\|_*(\Delta(x, 10R)) < C\delta \end{aligned} \tag{3.2.31}$$

for each $x \in \partial\Omega$ and $R \in (0, r_0/10)$. With the constant C as in (3.2.31), make the assumption that

$$0 < \delta_* < (20C)^{-1}. \tag{3.2.32}$$

Henceforth, fix a location $x_0 \in \partial\Omega$ along with a scale $r \in (0, R_\delta)$, where R_δ is as in (3.2.23). The fact that we are assuming $0 < \delta < \delta_* < (20C)^{-1}$ implies, with the help of (3.2.31) (used for $R := r \in (0, r_0/10)$) plus Chebytcheff's and Hölder's inequalities,

$$\frac{\sigma\big(\{y \in \Delta(x_0, r) : \nu^*_{x_0,r}(y) \geq 10^{-1}\}\big)}{10 \cdot \sigma\big(\Delta(x_0, r)\big)} \leq \fint_{\Delta(x_0,r)} \nu^*_{x,r}(y)\, d\sigma(y) \tag{3.2.33}$$
$$\leq \Big(\fint_{\Delta(x_0,r)} |\nu^*_{x,r}(y)|^2\, d\sigma(y)\Big)^{1/2} < \frac{1}{20}.$$

Hence, $\sigma\big(\{y \in \Delta(x_0, r) : \nu^*_{x_0,r}(y) \geq 10^{-1}\}\big) \leq \frac{1}{2}\sigma\big(\Delta(x_0, r)\big)$ which implies the inequality $\sigma\Big(\big\{y \in \Delta(x_0, r) : \nu^*_{x_0,r}(y) < 10^{-1}\big\}\Big) \geq \frac{1}{2}\sigma\big(\Delta(x_0, r)\big)$. This inequality, (3.2.29), and the fact that currently $\mathcal{H}^{n-1}(\partial\Omega \setminus \partial^*\Omega) = 0$, imply the existence of $y_r \in \Delta(x_0, r) \cap \partial^*\Omega$ such that $\nu^*_{x_0,r}(y_r) \leq 10^{-1}$ and $|\nu(y_r) - \nu_{\Delta(x_0,2r)}| \leq \nu^*_{x_0,r}(y_r)$. Given that $|\nu(y_r)| = 1$, this forces

$$\tfrac{9}{10} \leq |\nu_{\Delta(x_0,2r)}| \leq 1. \tag{3.2.34}$$

In a similar fashion we conclude that

$$\vec{n}_{x_0,r} := \frac{\nu_{\Delta(x_0,4r)}}{|\nu_{\Delta(x_0,4r)}|} \tag{3.2.35}$$

is a well-defined unit vector in $\mathbb{R}^n$. Set $H(x_0, r) := \big\{x \in \mathbb{R}^n : \langle x, \vec{n}_{x_0,r}\rangle = 0\big\}$ and introduce a new system of coordinates in $\mathbb{R}^n$ by setting

$$x = (\zeta, t) \Longleftrightarrow x = x_0 + t\,\vec{n}_{x_0,r} + \zeta, \quad t \in \mathbb{R},\ \zeta \in H(x_0, r). \tag{3.2.36}$$

We agree to write $\zeta(x), t(x)$ in place of ζ, t whenever necessary to stress the dependence of the new coordinates on the point $x \in \mathbb{R}^n$. Define the projection

$$\Pi : \mathbb{R}^n \to H(x_0, r) \text{ with } \Pi(x) := \zeta \text{ for each } x = (\zeta, t) \in \mathbb{R}^n. \tag{3.2.37}$$

Finally, consider the cylinder $C(x_0, r)$ defined as in (3.2.6) and introduce the "good" piece and the "error" piece, respectively as

$$\begin{aligned} G(x_0, r) &:= \big\{x \in C(x_0, r) \cap \partial\Omega : \nu^*_{x_0,2r}(x) \leq \phi(\delta)\big\}, \\ E(x_0, r) &:= \big(C(x_0, r) \cap \partial\Omega\big) \setminus G(x_0, r). \end{aligned} \tag{3.2.38}$$

Since the cylinder $C(x_0, r) \subseteq B(x_0, \sqrt{2}r)$ it follows from (3.2.38) that $G(x_0, r)$, $E(x_0, r)$ are disjoint σ-measurable subsets of $\Delta(x_0, \sqrt{2}r)$, and their union satisfies $G(x_0, r) \cup E(x_0, r) = C(x_0, r) \cap \partial\Omega$. In particular, (3.2.8) holds.

Next, we claim that there exist $c, C \in (0, \infty)$, which depend only on n and the Ahlfors regularity constants of $\partial\Omega$, with the property that

$$\fint_{\Delta(x_0,2r)} \exp\big(c\delta^{-1}\,\nu^*_{x_0,2r}\big)\, d\sigma \leq C. \tag{3.2.39}$$

Granted this, we may then conclude that

$$\exp\big(c\,\phi(\delta)/\delta\big)\,\frac{\sigma\big(E(x_0,r)\big)}{\sigma\big(\Delta(x_0,2r)\big)}\le\frac{1}{\sigma\big(\Delta(x_0,2r)\big)}\int_{E(x_0,r)}\exp\big(c\delta^{-1}\,\nu^{*}_{x_0,2r}\big)\,\mathrm{d}\sigma\le C.$$

This implies

$$\sigma\big(E(x_0,r)\big)\le C\exp\big(-c\,\phi(\delta)/\delta\big)\sigma\big(\Delta(x_0,2r)\big)\le 2^{n-1}C_A C\,r^{n-1}\exp\big(-c\,\phi(\delta)/\delta\big) \tag{3.2.40}$$

where C_A is as in (3.1.35). In particular, the estimate in (3.2.9) follows as long as

$$C_2:=c\ \text{ and }\ C_1\ge 2^{n-1}C_A C/\upsilon_{n-1}. \tag{3.2.41}$$

To justify (3.2.39), let us abbreviate

$$f:=\mathcal{M}\big(|\nu-\nu_{\Delta(x_0,4r)}|\cdot\mathbf{1}_{\Delta(x_0,4r)}\big) \tag{3.2.42}$$

and note that, thanks to (3.2.30), this entails

$$\nu^{*}_{x_0,2r}(x)\le f(x)\ \text{ whenever }\ x\in\Delta(x_0,2r). \tag{3.2.43}$$

From the John-Nirenberg inequality recorded in [129, Lemma 7.4.10] we know that there exist $A_1,A_2\in(0,\infty)$, depending only on the Ahlfors regularity constants of $\partial\Omega$, such that for each $p\in[1,\infty)$ we have

$$⨍_{\Delta(x_0,4r)}|\nu(x)-\nu_{\Delta(x_0,4r)}|^p\,\mathrm{d}\sigma(x)\le A_1\Gamma(p+1)\Big(A_2\|\nu\|_{*}(\Delta(x_0,20r))\Big)^p, \tag{3.2.44}$$

where Γ is the classical Gamma function (cf. [129, (7.4.68)]). We proceed to show that (3.2.39) holds for the choice

$$c:=2^{-1}A_2^{-1}\in(0,\infty) \tag{3.2.45}$$

and with $C\in(0,\infty)$ to be determined momentarily (see (3.2.49)). Using (3.2.43) and expanding the exponential function into an infinite power series we may then write, for some small $c>0$ to be specified momentarily,

$$
\begin{aligned}
\fint_{\Delta(x_0,2r)} \exp\big(c\delta^{-1} v^*_{x_0,r}\big)\,\mathrm{d}\sigma &\leq \fint_{\Delta(x_0,2r)} \exp\big(c\delta^{-1} f\big)\,\mathrm{d}\sigma \qquad (3.2.46)\\
&= \frac{1}{\sigma(\Delta(x_0,2r))}\int_0^\infty \sigma\Big(\big\{x\in\Delta(x_0,2r) : \exp\big(c\delta^{-1} f(x)\big) > \lambda\big\}\Big)\,\mathrm{d}\lambda\\
&\leq 1 + \frac{1}{\sigma(\Delta(x_0,2r))}\int_1^\infty \sigma\Big(\big\{x\in\Delta(x_0,2r) : \exp\big(c\delta^{-1} f(x)\big) > \lambda\big\}\Big)\,\mathrm{d}\lambda\\
&= 1 + \frac{1}{\sigma(\Delta(x_0,2r))}\int_0^\infty \sigma\Big(\big\{x\in\Delta(x_0,2r) : c\delta^{-1} f(x) > s\big\}\Big)\mathrm{e}^s\,\mathrm{d}s\\
&\leq e + \frac{1}{\sigma(\Delta(x_0,2r))}\sum_{k=0}^\infty \frac{1}{k!}\int_1^\infty \sigma\Big(\big\{x\in\Delta(x_0,2r) : f(x) > s\delta/c\big\}\Big)s^k\,\mathrm{d}s,
\end{aligned}
$$

where the equality is based on the change of variables $\lambda = \mathrm{e}^s$. To continue, fix an arbitrary exponent $p \in [2,\infty)$ along with an arbitrary number $s \in (0,\infty)$. Chebytcheff's inequality, (3.2.42), and the L^p-boundedness of the Hardy-Littlewood maximal operator (with bounds independent of p, as seen by interpolation) then allow us to estimate

$$
\begin{aligned}
\frac{\sigma\Big(\big\{x\in\Delta(x_0,2r) : f(x) > s\delta/c\big\}\Big)}{\sigma(\Delta(x_0,2r))} &\leq \Big(\frac{c}{s\delta}\Big)^p \fint_{\Delta(x_0,2r)} f(x)^p\,\mathrm{d}\sigma(x)\\
&\leq \Big(\frac{c}{s\delta}\Big)^p \frac{1}{\sigma(\Delta(x_0,2r))}\int_{\partial\Omega} \mathcal{M}\big(|v - v_{\Delta(x_0,4r)}|\cdot \mathbf{1}_{\Delta(x_0,4r)}\big)(x)^p\,\mathrm{d}\sigma(x)\\
&\leq \Big(\frac{c}{s\delta}\Big)^p \frac{C'}{\sigma(\Delta(x_0,2r))}\int_{\partial\Omega} \big(|v(x) - v_{\Delta(x_0,4r)}|\cdot \mathbf{1}_{\Delta(x_0,4r)}(x)\big)^p\,\mathrm{d}\sigma(x)\\
&\leq C''\Big(\frac{c}{s\delta}\Big)^p \fint_{\Delta(x_0,4r)} |v(x) - v_{\Delta(x_0,4r)}|^p\,\mathrm{d}\sigma(x), \qquad (3.2.47)
\end{aligned}
$$

where the constants $C', C'' \in (0,\infty)$ depend only on n and the Ahlfors regularity constants of $\partial\Omega$. We may now combine (3.2.44), (3.2.47), (3.2.45), and (3.2.3) (bearing in mind that $20r < r_0$) to obtain

$$
\frac{\sigma\Big(\big\{x\in\Delta(x_0,4r) : f(x) > s\delta/c\big\}\Big)}{\sigma(\Delta(x_0,4r))} \leq A_1\Gamma(p+1)\Big(\frac{c\,A_2}{s}\Big)^p = A_1\Gamma(p+1)\Big(\frac{1}{2s}\Big)^p, \tag{3.2.48}
$$

for each exponent $p \in [2,\infty)$ and each number $s \in (0,\infty)$. Utilizing (3.2.48), in which we take $p := k+2$ with $k = 0, 1, \dots$, back into (3.2.46) then yields (upon noting that $\Gamma(k+3) = (k+2)!$)

$$\fint_{\Delta(x_0,2r)} \exp\big(c\delta^{-1}\, \nu^*_{x_0,4r}\big)\, d\sigma \tag{3.2.49}$$
$$\leq e + C''\, A_1 \sum_{k=0}^{\infty} \frac{(k+1)(k+2)}{2^{k+2}} \Big(\int_1^{\infty} \frac{ds}{s^2}\Big) =: C < \infty.$$

with the finiteness condition guaranteed by the fact that $0 < c < 1/A_2$ (cf. (3.2.45)). Since both $c, C \in (0,\infty)$ depend only on n and the Ahlfors regularity constants of $\partial\Omega$, the proof of (3.2.39) is finished.

We now turn to the task of constructing the Lipschitz function h. As a preliminary matter, we remark that

$$\begin{gathered} |\langle x-y, \nu_{\Delta(x_0,4r)}\rangle| \leq \big(8\widetilde{C}\delta + \nu^*_{x_0,2r}(x)\big)|x-y| \\ \text{for each } x \in \partial\Omega \text{ and } y \in \Delta(x,4r). \end{gathered} \tag{3.2.50}$$

To justify this, observe that (3.2.50) is trivially true when $x = y$, so it suffices to consider the case when $x \in \partial\Omega$ and $y \in \Delta(x,4r)$ satisfy $x \neq y$. Assuming this is the case, based on (3.2.24) (used with $R := |x-y|/2$ which guarantees that $0 < R < 2r < 4R_\delta$ and that $y \in \Delta(x,4R)$) and (3.2.28) we may write

$$\begin{aligned} |\langle x-y, \nu_{\Delta(x_0,4r)}\rangle| &\leq |\langle x-y, \nu_{\Delta(x,|x-y|/2)}\rangle| + |x-y||\nu_{\Delta(x,|x-y|/2)} - \nu_{\Delta(x_0,4r)}| \\ &\leq 8\widetilde{C}\delta|x-y| + |x-y| \fint_{\Delta(x,|x-y|/2)} |\nu - \nu_{\Delta(x_0,4r)}|\, d\sigma \\ &\leq \big(8\widetilde{C}\delta + \nu^*_{x_0,2r}(x)\big)|x-y|, \end{aligned} \tag{3.2.51}$$

as desired. Moving on, observe from (3.2.36) that

$$t(x) = \langle x - x_0, \vec{n}_{x_0,r}\rangle \text{ for each } x \in \mathbb{R}^n. \tag{3.2.52}$$

In concert, (3.2.52), (3.2.34), (3.2.35), (3.2.50), (3.2.38), and (3.2.27) then allow us to control

$$\begin{aligned} |t(x)-t(y)| &= |\langle x-y, \vec{n}_{x_0,r}\rangle| \leq \tfrac{10}{9}|\langle x-y, \nu_{\Delta(x_0,4r)}\rangle| \leq \tfrac{10}{9}\big(8\widetilde{C}\delta + \phi(\delta)\big)|x-y| \\ &\leq \tfrac{10}{9}(8\widetilde{C}+1)\phi(\delta)|x-y| \leq (10C_* + 2)\phi(\delta)|x-y| \\ &\qquad \text{whenever } x \in G(x_0,r) \text{ and } y \in \Delta(x,4r). \end{aligned} \tag{3.2.53}$$

In turn, since for each $x, y \in \mathbb{R}^n$ we have

$$\zeta(x) - \zeta(y) = x - y - \big(t(x)-t(y)\big)\vec{n}_{x_0,r} \tag{3.2.54}$$

this permits us to estimate

$$\begin{gathered} |\zeta(x)-\zeta(y)| \geq |x-y| - |t(x)-t(y)| \geq \big(1 - (10\widetilde{C}+2)\phi(\delta)\big)|x-y|, \\ \text{for each } x \in G(x_0,r) \text{ and each } y \in \Delta(x,4r). \end{gathered} \tag{3.2.55}$$

Combining (3.2.53) and (3.2.55) (while keeping (3.2.27) and (3.2.25) in mind) then proves that

$$\begin{aligned}|t(x)-t(y)| &\le \frac{(10\widetilde{C}+2)\phi(\delta)}{1-(10\widetilde{C}+2)\phi(\delta)}|\zeta(x)-\zeta(y)| \le (20C_*+4)\phi(\delta)|\zeta(x)-\zeta(y)| \\ &\le C_0\phi(\delta)|\zeta(x)-\zeta(y)| \text{ for all } x\in G(x_0,r) \text{ and } y\in\Delta(x,4r).\end{aligned} \tag{3.2.56}$$

We now claim that

$$\begin{aligned}&\text{if } x\in C(x_0,r)\cap\partial\Omega \text{ and } \Pi(x)\in\Pi\big(G(x_0,r)\big)\\ &\text{then } x\in G(x_0,r); \text{ in particular, } x\in\mathcal{G} \text{ (cf. (3.2.9)).}\end{aligned} \tag{3.2.57}$$

Indeed, assume $x\in C(x_0,r)\cap\partial\Omega$ and $y\in G(x_0,r)$ are such that $\Pi(x)=\Pi(y)$. In view of (3.2.37), the latter condition means $\zeta(x)=\zeta(y)$. Since $x,y\in C(x_0,r)$, it follows that

$$|y-x|\le \operatorname{diam}\big(C(x_0,r)\big)=2\sqrt{2}r<4r, \tag{3.2.58}$$

hence $x\in\Delta(y,4r)$. We may invoke (3.2.56) (with the roles of x and y reversed) to conclude that $t(x)=t(y)$. Thus, $x=\big(\zeta(x),t(x)\big)=\big(\zeta(y),t(y)\big)=y\in G(x_0,r)$, ultimately proving (3.2.57).

The property established in (3.2.57) has several consequences. First, this implies (3.2.15). Also, as a consequence of the proof of (3.2.57) we see that

$$\text{the projection } \Pi \text{ is one-to-one on } G(x_0,r). \tag{3.2.59}$$

In turn, (3.2.59) guarantees that the mapping

$$\begin{aligned}&h:\Pi\big(G(x_0,r)\big)\longrightarrow\mathbb{R} \text{ given by}\\ &h\big(\zeta(x)\big):=t(x) \text{ for each } x\in G(x_0,r)\end{aligned} \tag{3.2.60}$$

is well defined. By (3.2.56), this mapping satisfies a Lipschitz condition with constant $\le C_0\phi(\delta)$ on the set $\Pi\big(G(x_0,r)\big)$. Indeed, given any $x,y\in G(x_0,r)$, the fact that $G(x_0,r)\subseteq\Delta(x_0,\sqrt{2}r)$ implies $|x-y|<2\sqrt{2}r<4r$, hence $y\in\Delta(x,4r)$. As such, (3.2.56) applies and, in view of (3.2.60), proves that

$$|h(x')-h(y')|\le C_0\phi(\delta)|x'-y'| \text{ for each } x',y'\in\Pi\big(G(x_0,r)\big).$$

It can be therefore extended (using Kirszbraun's theorem; see, e.g., the discussion in [135]) as a Lipschitz function, which we continue to denote by h, to the entire hyperplane $H(x_0,r)$, with Lipschitz constant $\le C\sqrt{\delta}$. Note that the graph of this extension (considered in the (ζ,t)-system of coordinates; cf. (3.2.36)), which we denote by $\mathcal{G}$, contains the set

$$\{(\zeta(x),t(x)) : x\in G(x_0,r)\}=G(x_0,r). \tag{3.2.61}$$

This proves the inclusion in (3.2.9). Together, (3.2.9) and (3.2.57) also prove that

$$\text{if } x \in C(x_0, r) \cap \partial\Omega \text{ and } \Pi(x) \in \Pi\big(G(x_0, r)\big) \text{ then } x \in \mathcal{G}. \tag{3.2.62}$$

In turn, the above property implies the claim made in (3.2.15). Specifically, assume $x \in G(x_0, r)$ and $y \in C(x_0, r) \cap \partial\Omega$ are such that $\Pi(y) = \Pi(x)$. Then $\Pi(y)$ belongs to $\Pi\big(G(x_0, r)\big)$ which, by virtue of (3.2.57), places y in $G(x_0, r)$. In particular, the points $x, y \in \mathcal{G}$ (cf. (3.2.9)) have the same projection. Thus, necessarily, $x = y$ since otherwise the Vertical Line Test would be violated for the graph $\mathcal{G}$.

To prove the inclusion claimed in (3.2.11), consider $x \in C(x_0, r) \cap \partial\Omega$ arbitrary. In particular, $x \in B(x_0, \sqrt{2}r) \cap \partial\Omega = \Delta(x_0, \sqrt{2}r)$. Also, employing the convention made in (3.2.36) allows us to express $x = x_0 + t(x)\vec{n}_{x_0,r} + \zeta(x)$, with $\zeta(x) \in H(x_0, r)$ satisfying $|\zeta(x)| < r$ (given that x belongs to $C(x_0, r)$) and with

$$t(x) = \langle x - x_0, \vec{n}_{x_0,r}\rangle = \frac{\langle x - x_0, \nu_{\Delta(x_0,4r)}\rangle}{|\nu_{\Delta(x_0,4r)}|}, \tag{3.2.63}$$

thanks to (3.2.52) and (3.2.35). In turn, (3.2.63), (3.2.34), and (3.2.24) (presently used with $R := 4r$, $x := x_0$, $y := x$) permit us to estimate

$$|t(x)| \leq \frac{\big|\langle x - x_0, \nu_{\Delta(x_0,4r)}\rangle\big|}{|\nu_{\Delta(x_0,4r)}|} \leq \tfrac{10}{9}(4r)16\widetilde{C}\delta \leq C_0\delta r, \tag{3.2.64}$$

since (3.2.25) guarantees that $C_0 \geq 72\widetilde{C}$. The proof of (3.2.11) is therefore complete.

From (3.2.11) it follows that $C^{\pm}(x_0, r)$ do not intersect $\partial\Omega$. As such, $\Omega_+ := \Omega$ and $\Omega_- := \mathbb{R}^n \setminus \overline{\Omega}$ constitute a disjoint open cover of the connected sets $C^{\pm}(x_0, r)$. Hence,

$$\begin{aligned} &C^+(x_0, r) \text{ is fully contained in either } \Omega_+ \text{ or } \Omega_-, \text{ and} \\ &C^-(x_0, r) \text{ is fully contained in either } \Omega_+ \text{ or } \Omega_-. \end{aligned} \tag{3.2.65}$$

From (3.2.22), [129, Definition 5.11.1], and [129, Remark 5.11.2] we see that Ω satisfies a two-sided corkscrew condition (in the sense of [129, Definition 5.1.3]). Specifically, there exists some parameter $\theta \in (0, 1)$ depending only on the dimension n, the Ahlfors regularity constants of $\partial\Omega$, and the relative flatness ratio $\wp(\Omega)/\text{diam}\,\partial\Omega$ such that for each $x \in \partial\Omega$ and each $r \in (0, 2\,\text{diam}\,\partial\Omega)$ one can find $x_r^{\pm} \in \Omega_{\pm}$ with the property that

$$|x_r^{\pm} - x_0| < r \text{ and } B\big(x_r^{\pm}, \theta r\big) \subseteq \Omega_{\pm}. \tag{3.2.66}$$

Assume, to begin with, that

$$0 < \delta_* < \theta/C_0 \tag{3.2.67}$$

(a condition which can be accommodated while preserving the nature of δ_* specified in the statement of the theorem). This makes it impossible to contain either of the balls $B\big(x_r^+, \theta r\big)$, $B\big(x_r^-, \theta r\big)$ in the strip $\big\{x_0 + x' + t\vec{n}_{x_0,r} : |t| \leq C_0\delta r,\ x' \in H(x_0, r)\big\}$. Since, as seen from (3.2.66), their centers $x_r^{\pm}$ belong to $B(x_0, r) \subset C(x_0, r)$, this forces one of the following alternatives to be true:

$$B(x_r^+, \theta r) \cap C^+(x_0, r) \neq \varnothing \text{ and } B(x_r^-, \theta r) \cap C^+(x_0, r) \neq \varnothing, \tag{3.2.68}$$

$$B(x_r^+, \theta r) \cap C^-(x_0, r) \neq \varnothing \text{ and } B(x_r^-, \theta r) \cap C^-(x_0, r) \neq \varnothing, \tag{3.2.69}$$

$$B(x_r^+, \theta r) \cap C^+(x_0, r) \neq \varnothing \text{ and } B(x_r^-, \theta r) \cap C^-(x_0, r) \neq \varnothing, \tag{3.2.70}$$

$$B(x_r^+, \theta r) \cap C^-(x_0, r) \neq \varnothing \text{ and } B(x_r^-, \theta r) \cap C^+(x_0, r) \neq \varnothing. \tag{3.2.71}$$

Note that the alternative described in (3.2.68) cannot possibly hold. Indeed, the existence of two points $z_1 \in B(x_r^+, \theta r) \cap C^+(x_0, r)$ and $z_2 \in B(x_r^-, \theta r) \cap C^+(x_0, r)$ would imply that the line segment $[z_1, z_2]$ lies in the convex set $C^+(x_0, r)$, hence also either in Ω_+ or in Ω_- by (3.2.65). However, the fact that $z_1 \in B(x_r^+, \theta r) \subseteq \Omega_+$ and $z_2 \in B(x_r^-, \theta r) \subseteq \Omega_-$ prevents either one of these eventualities form materializing. This contradiction therefore excludes (3.2.68). Reasoning in a similar fashion we may rule out (3.2.69). When (3.2.70) holds, from (3.2.65) and the fact that $B(x_r^\pm, \theta r) \subseteq \Omega_\pm$ (cf. (3.2.66)) we conclude that the inclusions in (3.2.14) hold as stated.

Finally, if (3.2.71) holds, from (3.2.65) and (3.2.66) we deduce $C^+(x_0, r) \subseteq \Omega_-$ and $C^-(x_0, r) \subseteq \Omega_+$. In such a scenario, we may ensure that the inclusions in (3.2.14) are valid simply by re-denoting $\vec{n}_{x_0,r}$ as $-\vec{n}_{x_0,r}$ which amounts to reversing the roles of $C^+(x_0, r)$ and $C^-(x_0, r)$. This concludes the proof of (3.2.14).

Next, observe that

$$\Pi\big(C(x_0, r) \cap \partial\Omega\big) \subseteq \Pi\big(C(x_0, r)\big) = \{\zeta \in H(x_0, r) : |\zeta| < r\}. \tag{3.2.72}$$

The opposite inclusion fails only when there exists a line segment parallel to $\vec{n}_{x_0,r}$ whose two end-points belong to $C^+(x_0, r)$ and to $C^-(x_0, r)$, respectively, and which does not intersect $\partial\Omega$ (here we implicitly use the fact that $C^\pm(x_0, r) \neq \varnothing$, itself a result of having imposed the condition that $0 < \delta < \delta_* < 1/C_0$; cf. (3.2.26)). However, (3.2.14) and simple connectivity arguments rule out this scenario, hence (3.2.12) is proved.

Going further, we note that (3.2.12) implies

$$\{\zeta \in H(x_0, r) : |\zeta| < r\} \setminus \Pi\big(G(x_0, r)\big) \subseteq \Pi\big(E(x_0, r)\big). \tag{3.2.73}$$

The fact that $\Pi : \mathbb{R}^n \to H(x_0, r)$ is a Lipschitz function, with Lipschitz constant 1, implies (cf., e.g., [49, Theorem 1, p. 75]) that

$$\mathcal{H}^{n-1}\big(\Pi(S)\big) \leq \mathcal{H}^{n-1}(S) \text{ for each Borel set } S \subseteq \mathbb{R}^n. \tag{3.2.74}$$

Based on (3.2.73), (3.2.74), the fact that $\sigma = \mathcal{H}^{n-1} \lfloor \partial\Omega$, (3.2.8), and (3.2.40) we then conclude that

$$\begin{aligned} &\mathcal{H}^{n-1}\Big(\{\zeta \in H(x_0, r) : |\zeta| < r\} \setminus \Pi\big(G(x_0, r)\big)\Big) \\ &\quad \leq \mathcal{H}^{n-1}\Big(\Pi\big(E(x_0, r)\big)\Big) \leq \mathcal{H}^{n-1}\big(E(x_0, r)\big) \leq 2^{n-1} C_A C r^{n-1} \cdot \exp\big(-C_2 \phi(\delta)/\delta\big). \end{aligned} \tag{3.2.75}$$

In addition, (3.2.57) gives

$$C(x_0,r)\cap\big(\mathcal{G}\setminus\partial\Omega\big)\subseteq\mathcal{G}\cap\Pi^{-1}\Big(\{\zeta\in H(x_0,r):\,|\zeta|<r\}\setminus\Pi\big(G(x_0,r)\big)\Big). \tag{3.2.76}$$

Keeping also in mind that

$$\mathcal{H}^{n-1}(S)\leq\sqrt{1+(C_0\phi(\delta))^2}\,\mathcal{H}^{n-1}(\Pi(S)),\ \text{ for each Borel set } S\subseteq\mathcal{G}, \tag{3.2.77}$$

we deduce that

$$\begin{aligned}
&\mathcal{H}^{n-1}(C(x_0,r)\cap(\mathcal{G}\setminus\partial\Omega))\\
&\leq\mathcal{H}^{n-1}\Big(\mathcal{G}\cap\Pi^{-1}(\{\zeta\in H(x_0,r):\,|\zeta|<r\}\setminus\Pi\big(G(x_0,r)\big))\Big)\\
&\leq\sqrt{1+(C_0\phi(\delta))^2}\times\\
&\qquad\times\mathcal{H}^{n-1}\Big(\Pi\Big(\mathcal{G}\cap\Pi^{-1}(\{\zeta\in H(x_0,r):\,|\zeta|<r\}\setminus\Pi\big(G(x_0,r)\big))\Big)\Big)\\
&\leq\sqrt{1+(C_0\phi(\delta))^2}\,\mathcal{H}^{n-1}\Big(\{\zeta\in H(x_0,r):\,|\zeta|<r\}\setminus\Pi\big(G(x_0,r)\big)\Big)\\
&\leq\sqrt{1+(C_0\phi(\delta))^2}\,2^{n-1}C_ACr^{n-1}\cdot\exp\big(-C_2\phi(\delta)/\delta\big)\\
&\leq\sqrt{1+C_0^2(20\widetilde{C}+4)^{-2}}\,2^{n-1}C_ACr^{n-1}\cdot\exp\big(-C_2\phi(\delta)/\delta\big),
\end{aligned}\tag{3.2.78}$$

by (3.2.75) and (3.2.27). Upon observing that $C(x_0,r)\cap\big(\partial\Omega\setminus\mathcal{G}\big)$ is contained in $E(x_0,r)$, the estimate claimed in (3.2.7) now follows from (3.2.78) and (3.2.40) if we choose (recall that $C_2:=c$; cf. (3.2.41))

$$C_1:=\sqrt{1+C_0^2(20\widetilde{C}+4)^{-2}}\,2^{n-1}C_AC/\upsilon_{n-1} \tag{3.2.79}$$

(a choice in line with the demand formulated in (3.2.41)), where C is as in (3.2.49), and where C_A is as in (3.1.35).

Let us now justify the proximity condition formulated in (3.2.10). To this end, fix a point $x\in E(x_0,r)=\big(C(x_0,r)\cap\partial\Omega\big)\setminus G(x_0,r)$ and pick some arbitrary $x^*\in G(x_0,r)$. In particular, $x,x^*\in C(x_0,r)$ hence $|x-x^*|<\operatorname{diam}C(x_0,r)=2\sqrt{2}r$. Given that we have $x^*\in G(x_0,r)$ and $x\in\Delta(x^*,4r)$, estimate (3.2.56) applies and presently gives

$$\big|t(x)-h\big(\Pi(x^*)\big)\big|=|t(x)-t(x^*)|\leq C_0\phi(\delta)|\Pi(x)-\Pi(x^*)|. \tag{3.2.80}$$

Consequently, since $x=\big(\Pi(x),t(x)\big)$, we may write

$$\begin{aligned}
\big|x-\big(\Pi(x),h(\Pi(x))\big)\big|&=\big|t(x)-h\big(\Pi(x)\big)\big|\\
&\leq\big|t(x)-h\big(\Pi(x^*)\big)\big|+\big|h\big(\Pi(x^*)\big)-h\big(\Pi(x)\big)\big|\\
&\leq 2C_0\phi(\delta)|\Pi(x)-\Pi(x^*)|,
\end{aligned}\tag{3.2.81}$$

by (3.2.80) and the Lipschitz condition on h (cf. (3.2.4)). Taking the infimum over $x^* \in G(x_0, r)$ now yields (3.2.10).

Let us now deal with (3.2.16). Recall that υ_{n-1} denotes the volume of the unit ball in $\mathbb{R}^{n-1}$. Using (3.2.7) and (3.2.77) we may estimate

$$\begin{aligned}\sigma\big(\Delta(x_0, r)\big) &= \mathcal{H}^{n-1}\big(B(x_0, r) \cap \partial\Omega\big) \leq \mathcal{H}^{n-1}\big(C(x_0, r) \cap \partial\Omega\big)\\ &\leq \mathcal{H}^{n-1}\big(C(x_0, r) \cap \mathcal{G}\big) + \mathcal{H}^{n-1}\Big(C(x_0, r) \cap (\partial\Omega \setminus \mathcal{G})\Big)\\ &\leq \sqrt{1 + (C_0\phi(\delta))^2}\,\mathcal{H}^{n-1}\big(\Pi(C(x_0, r) \cap \mathcal{G})\big)\\ &\quad + \mathcal{H}^{n-1}\Big(C(x_0, r) \cap (\partial\Omega \vartriangle \mathcal{G})\Big)\\ &\leq \big(1 + C_0\phi(\delta)\big)\mathcal{H}^{n-1}\big(\Pi(C(x_0, r))\big) + C_1\upsilon_{n-1}r^{n-1}\exp\big(-C_2\phi(\delta)/\delta\big)\\ &= \Big(1 + C_0\phi(\delta) + C_1\exp\big(-C_2\phi(\delta)/\delta\big)\Big)\upsilon_{n-1}\,r^{n-1}. \end{aligned} \tag{3.2.82}$$

Also, by employing (3.2.12), (3.2.74), (3.2.7), (3.2.14), and (3.2.77) we may write

$$\begin{aligned}\upsilon_{n-1}\,r^{n-1} &= \mathcal{H}^{n-1}\big(\{\zeta \in H(x_0, r) : |\zeta| < r\}\big) \leq \mathcal{H}^{n-1}\big(C(x_0, r) \cap \partial\Omega\big) \qquad (3.2.83)\\ &= \mathcal{H}^{n-1}\big(B(x_0, r) \cap \partial\Omega\big) + \mathcal{H}^{n-1}\Big(\big(C(x_0, r) \cap \partial\Omega\big) \setminus B(x_0, r)\Big)\\ &\leq \sigma\big(\Delta(x_0, r)\big) + \mathcal{H}^{n-1}\Big(C(x_0, r) \cap \big(\partial\Omega \setminus \mathcal{G}\big)\Big)\\ &\quad + \mathcal{H}^{n-1}\Big(\big(C(x_0, r) \cap \mathcal{G}\big) \setminus \big(B(x_0, r) \cup C^+(x_0, r) \cup C^-(x_0, r)\big)\Big)\\ &\leq \sigma\big(\Delta(x_0, r)\big) + \mathcal{H}^{n-1}\Big(C(x_0, r) \cap \big(\partial\Omega \vartriangle \mathcal{G}\big)\Big)\\ &\quad + \sqrt{1 + (C_0\phi(\delta))^2}\,\upsilon_{n-1}\,r^{n-1}\Big(1 - \big(\sqrt{1 - C_0^2\delta^2}\big)^{n-1}\Big)\\ &\leq \sigma\big(\Delta(x_0, r)\big) + C_1\upsilon_{n-1}\,r^{n-1}\exp\big(-C_2\phi(\delta)/\delta\big) + C_3\delta\upsilon_{n-1}\,r^{n-1}\end{aligned}$$

where $C_3 := C_n\,C_0\sqrt{1 + C_0^2(20\widetilde{C} + 4)^{-2}}$ with $C_n \in [1, \infty)$ depending only on the dimension n. This further implies

$$\Big(1 - C_3\delta - C_1\exp\big(-C_2\phi(\delta)/\delta\big)\Big)\,\upsilon_{n-1}\,r^{n-1} \leq \sigma\big(\Delta(x_0, r)\big). \tag{3.2.84}$$

Now, (3.2.16) follows from (3.2.82), (3.2.84), and (3.2.27).

Next, (3.2.17) is a direct consequence of [129, Proposition 5.6.6] applied to Ω and the upper-graph of the function h (both of which are Ahlfors regular domains). There remains to prove the claim made in (3.2.19). To get started, we make two observations. First, (3.2.3) implies

$$⨍_{\Delta(x_0,4r)} |\nu - \nu_{\Delta(x_0,4r)}|\,d\sigma \leq \delta. \tag{3.2.85}$$

Second, at σ-a.e. point on $\partial\Omega$ we may estimate

$$|\nu - \vec{n}_{x_0,r}| \le |\nu - \nu_{\Delta(x_0,4r)}| + |\nu_{\Delta(x_0,4r)} - \vec{n}_{x_0,r}| \tag{3.2.86}$$

and, thanks to (3.2.35), the fact that $|\nu| = 1$ at σ-a.e. point on $\partial\Omega$, and the reverse triangle inequality, we have

$$\begin{aligned}
|\nu_{\Delta(x_0,4r)} - \vec{n}_{x_0,r}| &= \left|\nu_{\Delta(x_0,4r)} - \frac{\nu_{\Delta(x_0,4r)}}{|\nu_{\Delta(x_0,4r)}|}\right| = \left|\left(1 - \frac{1}{|\nu_{\Delta(x_0,4r)}|}\right)\nu_{\Delta(x_0,4r)}\right| \\
&= \left|1 - \frac{1}{|\nu_{\Delta(x_0,4r)}|}\right| |\nu_{\Delta(x_0,4r)}| = \big|1 - |\nu_{\Delta(x_0,4r)}|\big| \\
&= \big||\nu| - |\nu_{\Delta(x_0,4r)}|\big| \le \big|\nu - \nu_{\Delta(x_0,4r)}\big|.
\end{aligned} \tag{3.2.87}$$

By combining (3.2.86) with (3.2.87) we arrive at the conclusion that

$$|\nu - \vec{n}_{x_0,r}| \le 2|\nu - \nu_{\Delta(x_0,4r)}| \text{ at } \sigma\text{-a.e. point on } \partial\Omega. \tag{3.2.88}$$

Recall that $\widetilde{\nu}$ denotes the unit normal vector to the Lipschitz graph $\mathcal{G}$, pointing towards the upper-graph of the function h. This is a well-defined vector at $\mathcal{H}^{n-1}$-a.e. point on $\mathcal{G}$, and we claim that

$$|\widetilde{\nu} - \vec{n}_{x_0,r}| \le C_0\phi(\delta) \text{ at } \mathcal{H}^{n-1}\text{-a.e. point on } \mathcal{G}. \tag{3.2.89}$$

To justify this, after performing a rotation, there is no loss of generality in assuming

$$\vec{n}_{x_0,r} = \mathbf{e}_n = (0,\dots,0,1) \in \mathbb{R}^n. \tag{3.2.90}$$

Then the hyperplane

$$H(x_0,r) = \langle \vec{n}_{x_0,r}\rangle^{\perp} = \langle \mathbf{e}_n\rangle^{\perp} = \mathbb{R}^{n-1} \times \{0\} \tag{3.2.91}$$

may be canonically identified with $\mathbb{R}^{n-1}$, a scenario in which

$$\widetilde{\nu}\big(x', h(x')\big) = \frac{\big((\nabla' h)(x'), 1\big)}{\sqrt{1 + |(\nabla' h)(x')|^2}} \text{ for } \mathcal{L}^{n-1}\text{-a.e. } x' \in \mathbb{R}^{n-1}, \tag{3.2.92}$$

where ∇' denotes the gradient operator in $\mathbb{R}^{n-1}$. From (3.2.90) and (3.2.92) we then see that at $\mathcal{H}^{n-1}$-a.e. point $x \in \mathcal{G}$, say $x = \big(x', h(x')\big)$ with $x' \in \mathbb{R}^{n-1}$, we have

$$\begin{aligned}
\big|\widetilde{\nu}(x) - \vec{n}_{x_0,r}\big|^2 &= 2 - 2\langle\widetilde{\nu}(x), \vec{n}_{x_0,r}\rangle = 2\left(1 - \frac{1}{\sqrt{1 + |(\nabla' h)(x')|^2}}\right) \\
&= \frac{2|(\nabla' h)(x')|^2}{1 + |(\nabla' h)(x')|^2 + \sqrt{1 + |(\nabla' h)(x')|^2}} \\
&\le |(\nabla' h)(x')|^2 \le \big(C_0\phi(\delta)\big)^2,
\end{aligned} \tag{3.2.93}$$

where the last inequality comes from (3.2.4). Ultimately, this establishes (3.2.89).

Collectively, (3.2.88) and (3.2.89) prove that

$$|\nu - \widetilde{\nu}| \leq 2|\nu - \nu_{\Delta(x_0,4r)}| + C_0\phi(\delta) \text{ at } \sigma\text{-a.e. point on } \mathcal{G} \cap \partial\Omega. \tag{3.2.94}$$

From (3.2.29) and (3.2.38) we also see that

$$|\nu(x) - \nu_{\Delta(x_0,4r)}| \leq \nu^*_{x_0,2r}(x) \leq \phi(\delta) \text{ for } \sigma\text{-a.e. } x \in G(x_0, r). \tag{3.2.95}$$

Combining (3.2.94) with (3.2.95) and keeping in mind that $G(x_0, r) \subseteq \mathcal{G} \cap \partial\Omega$ leads to the conclusion that

$$|\nu - \widetilde{\nu}| \leq (2 + C_0)\phi(\delta) \text{ at } \sigma\text{-a.e. point on } G(x_0, r). \tag{3.2.96}$$

If $\delta_* > 0$ is taken small enough so that $\phi(t) < 2(2 + C_0)^{-1}$ for all $t \in (0, \delta_*)$ (something that may always be arranged, thanks to (3.2.2)), we conclude from (3.2.96) and (3.2.17) (again, mindful of the fact that $G(x_0, r) \subseteq \mathcal{G} \cap \partial\Omega$) that

$$\nu(x) = \widetilde{\nu}(x) \text{ at } \sigma\text{-a.e. point } x \in G(x_0, r). \tag{3.2.97}$$

This proves (3.2.18). Let us now deal with (3.2.19). Together, (3.2.17), (3.2.94), (3.2.85), and the first inequality in (3.2.27) yield

$$\begin{aligned} \sigma\Big(\big\{x \in \mathcal{G} \cap \Delta(x_0, 4r) : \nu(x) = -\widetilde{\nu}(x)\big\}\Big) &= \frac{1}{2}\int_{\mathcal{G}\cap\Delta(x_0,4r)} |\nu - \widetilde{\nu}|\, d\sigma \\ &\leq \big(\delta + 2^{-1}C_0 \cdot \phi(\delta)\big) \cdot \sigma\big(\Delta(x_0, 4r)\big) \leq C_4 \cdot \phi(\delta)r^{n-1} \end{aligned} \tag{3.2.98}$$

provided $C_4 := 4^{n-1}(1 + 2^{-1}C_0)C_A$, where C_A is as in (3.1.35). Hence, (3.2.19) is established. There remains to prove (3.2.20). To this end, combine (3.2.88) and (3.2.89) to obtain

$$\begin{gathered} \sup\nolimits_{y\in\mathcal{G}} |\nu(x) - \widetilde{\nu}(y)| \leq 2|\nu(x) - \nu_{\Delta(x_0,4r)}| + C_0\phi(\delta) \\ \text{at } \sigma\text{-a.e. point } x \in \partial\Omega. \end{gathered} \tag{3.2.99}$$

Based on (3.2.99), (3.2.85), and (3.2.27) we then conclude that

$$⨍_{\Delta(x_0,4r)} \Big(\sup_{y\in\mathcal{G}} |\nu(x) - \widetilde{\nu}(y)|\Big)\, d\sigma(x) \leq 2\delta + C_0\phi(\delta) \leq C_4 \cdot \phi(\delta), \tag{3.2.100}$$

since our earlier choice of C_4 ensures that $C_4 \geq 2 + C_0$. This justifies (3.2.20), so the proof of Theorem 3.2.1 is now complete. □

3.3 Planar Chord-Arc Domains and Flatness

Throughout this section we shall work in the two-dimensional setting. The starting point is the observation that if $\Omega \subseteq \mathbb{R}^2$ is a chord-arc domain with compact boundary then the size of the chord-arc constant is a good measure of how flat $\partial\Omega$ is at small scales. We wish to reconcile this view with our earlier philosophy which has, so far, emphasized (3.0.5). This is accomplished in Theorem 3.3.1 below.

To set the stage, the reader is invited to recall the notion of chord-arc domain from [129, Definition 5.9.8]. By design, the boundary of any chord-arc domain is a simple curve, and this brings into focus the question: *when is the boundary of an open, connected, simply connected planar set a Jordan curve*? According to the classical Carathéodory theorem, this is the case if and only if some (or any) conformal mapping $\varphi : \mathbb{D} \to \Omega$ (where $\mathbb{D}$ is the unit disk in $\mathbb{C}$) extends to a homeomorphism $\varphi : \overline{\mathbb{D}} \to \overline{\Omega}$ (see, e.g., [68, Theorem 3.1, p. 13]). A characterization of bounded planar Jordan regions in terms of properties having no reference to their boundaries has been given by R.L. Moore in 1918. According to [161],

given an open, bounded, connected, simply connected set $\Omega \subseteq \mathbb{R}^2$, in order for $\partial\Omega$ to be a simple closed curve it is necessary and sufficient that Ω is *uniformly connected im kleinen* (i.e., if for every $\varepsilon_o > 0$ there exists $\delta_o > 0$ so that any two points $P, \widetilde{P} \in \Omega$ with $|P - \widetilde{P}| < \delta_o$ lie in a connected subset Γ of Ω satisfying $|P - Q| < \varepsilon_o$ for each $Q \in \Gamma$). (3.3.1)

A moment's reflection shows that the uniform connectivity condition (im kleinen) formulated above is equivalent to the demand that

for every $\varepsilon_o > 0$ there exists $\delta_o > 0$ such that any two points $P, \widetilde{P} \in \Omega$ with $|P - \widetilde{P}| < \delta_o$ lie in a connected subset Γ of Ω with $\operatorname{diam}(\Gamma) < \varepsilon_o$. (3.3.2)

This condition is meant to prevent the boundary of Ω to "branch out" (like in the case of a partially slit disk).

Our next theorem shows that, for sets with compact boundaries the plane, the class of chord-arc domain with small constant coincides with the class of Ahlfors regular domains with a relatively flat boundary (in the sense of (3.0.5) being small).

Theorem 3.3.1 *Assume $\Omega \subseteq \mathbb{R}^2$ is a chord-arc domain with compact boundary. Then Ω is an Ahlfors regular domain, which is simply connected when Ω is bounded and connected when Ω is unbounded. Moreover, if $\sigma := \mathcal{H}^1 \lfloor \partial\Omega$, the geometric measure theoretic outward unit normal to Ω is denoted by ν, and $\ell(\cdot,\cdot)$ stands for the shortest arc-length between points on $\partial\Omega$, then having*

$$\lim_{R \to 0^+} \left\{ \sup_{\substack{z_1, z_2 \in \partial\Omega \\ |z_1 - z_2| < R}} \left(\frac{\ell(z_1, z_2)}{|z_1 - z_2|} - 1 \right) \right\} < \varkappa \tag{3.3.3}$$

implies that (with the distance measured in the space $\left[\mathrm{BMO}(\partial\Omega, \sigma)\right]^2$)

$$\operatorname{dist}\Big(\nu, \big[\mathrm{VMO}(\partial\Omega, \sigma)\big]^2\Big) = O(\varkappa^{1/2}) \ \ as \ \varkappa \to 0^+, \tag{3.3.4}$$

where the implicit constant depends exclusively on the upper Ahlfors regularity constant of $\partial\Omega$.

Conversely, suppose $\Omega \subseteq \mathbb{R}^2$ is an Ahlfors regular domain with compact boundary, assumed to be simply connected when Ω is bounded and, respectively, connected when Ω is unbounded. Set $\sigma := \mathcal{H}^1 \lfloor \partial\Omega$ and denote by ν the geometric measure theoretic outward unit normal to Ω. Then there exists $\delta > 0$ with the property that if (with the distance measured in the John-Nirenberg space $\big[\mathrm{BMO}(\partial\Omega, \sigma)\big]^2$)

$$\operatorname{dist}\Big(\nu, \big[\mathrm{VMO}(\partial\Omega, \sigma)\big]^2\Big) < \delta \tag{3.3.5}$$

it follows that Ω is a chord-arc domain for which

$$\lim_{R \to 0^+} \left\{ \sup_{\substack{z_1, z_2 \in \partial\Omega \\ |z_1 - z_2| < R}} \left(\frac{\ell(z_1, z_2)}{|z_1 - z_2|} - 1 \right) \right\} = O\big(\delta \ln(1/\delta)\big) \ \ as \ \delta \to 0^+. \tag{3.3.6}$$

The case of chord-arc domains with unbounded boundaries is treated in [111]. This work refines earlier results in [184]-[185].

Proof of Theorem 3.3.1 Assume $\Omega \subseteq \mathbb{R}^2$ is a chord-arc domain with compact boundary. Then [129, Proposition 5.8.6] together with [129, (5.9.54)] imply that Ω is an Ahlfors regular domain. Moreover, the Jordan Curve Theorem (cf. [129, (5.8.8)]) implies that Ω is simply connected if it is bounded, and Ω is connected if it is unbounded. The next goal is to prove (3.3.4).

To this end, let L be the length of the curve $\Sigma := \partial\Omega$ and let $[0, L) \ni s \mapsto z(s) \in \Sigma$ be its arc-length parametrization. In particular, z is a bijection and

$$\begin{gathered} z(\cdot) \text{ is differentiable at } \mathcal{L}^1\text{-a.e. point in } (0, L), \\ \text{with } |z'(s)| = 1 \text{ for } \mathcal{L}^1\text{-a.e. } s \in (0, L). \end{gathered} \tag{3.3.7}$$

On the interval $[0, L)$ consider the distance

$$\begin{aligned} \rho(s_1, s_2) &:= \inf\big\{|t_1 - t_2| : t_j \in s_j + L\mathbb{Z},\ j = 1, 2\big\} \\ &= \min\big\{|s_1 - s_2|, L - |s_1 - s_2|\big\} \ \text{ for each } s_1, s_2 \in [0, L). \end{aligned} \tag{3.3.8}$$

Suppose (3.3.3) holds. We then conclude that there exists some $R_o > 0$ with the property that

$$\begin{gathered} |z_1 - z_2| \leq \ell(z_1, z_2) \text{ for all } z_1, z_2 \in \Sigma, \text{ and} \\ \ell(z_1, z_2) \leq (1 + \varkappa)|z_1 - z_2| \text{ for all } z_1, z_2 \in \Sigma \text{ with } |z_1 - z_2| < R_o, \end{gathered} \tag{3.3.9}$$

where $\ell(z_1, z_2)$ denotes the length of the shortest sub-arc of Σ joining z_1 and z_2. In the language of (3.3.8), this implies

$$|z(s_1) - z(s_2)| \leq \rho(s_1, s_2) \text{ for all } s_1, s_2 \in [0, L), \text{ and}$$
$$\rho(s_1, s_2) \leq (1 + \varkappa)|z(s_1) - z(s_2)| \text{ for all } s_1, s_2 \in [0, L) \text{ with } |z(s_1) - z(s_2)| < R_o. \tag{3.3.10}$$

Given any $z_o \in \Sigma$ and $r \in (0, \infty)$, abbreviate $\Delta_\Sigma(z_o, r) := B(z_o, r) \cap \Sigma$. Also, given any $s_o \in [0, L)$ and any $r \in (0, \infty)$, define $\Delta_\rho(s_o, r) := \{s \in [0, L) : \rho(s, s_o) < r\}$. As an elementary consequence of (3.3.10) we then have

$$\Delta_\rho(s_o, r) \subseteq z^{-1}\big(\Delta_\Sigma(z(s_o), r)\big) \subseteq \Delta_\rho\big(s_o, (1 + \varkappa)r\big). \\ \text{for each } s_o \in [0, L) \text{ and } r \in (0, R_o). \tag{3.3.11}$$

For further use let us also observe that for each $s_o \in [0, L)$ and each $r \in (0, \infty)$,

$$\mathcal{L}^1\big(\Delta_\rho(s_o, r)\big) = \begin{cases} 2r & \text{if } 0 < r < L/2, \\ L & \text{if } r \geq L/2. \end{cases} \tag{3.3.12}$$

In fact, given any $r \in (0, L/2)$ and any $s_o \in [0, L)$ we have

$$\Delta_\rho(s_o, r) = \begin{cases} [0, s_o + r) \cup (s_o + L - r, L) & \text{if } 0 \leq s_o < r, \\ (s_o - r, s_o + r) & \text{if } r \leq s_o \leq L - r, \\ [0, s_o - L + r) \cup (s_o - r, L) & \text{if } L - r < s_o < L. \end{cases} \tag{3.3.13}$$

Next, in view of (3.3.7), the area formula (cf. [49, Theorem 1, p. 96]) gives that for every Lebesgue measurable set $A \subseteq [0, L)$ we have

$$\mathcal{H}^1\big(z(A)\big) = \mathcal{L}^1(A). \tag{3.3.14}$$

This may be equivalently recast as the property that for each $\mathcal{H}^1$-measurable set $E \subseteq \Sigma$ we have

$$\mathcal{H}^1(E) = \mathcal{L}^1\big(z^{-1}(E)\big). \tag{3.3.15}$$

Also, the change of variable formula (cf. [49, Theorem 2, p. 99]) gives that for each σ-measurable set $E \subseteq \Sigma$, and each non-negative σ-measurable function g on E,

$$\int_E g \, d\sigma = \int_{z^{-1}(E)} g(z(s)) \, ds. \tag{3.3.16}$$

Finally, from [129, (5.8.44)] we know that the geometric measure theoretic outward unit normal ν to Ω satisfies

$$\nu(z(s)) = -iz'(s) \text{ for } \mathcal{L}^1\text{-a.e. } s \in (0, L). \tag{3.3.17}$$

To proceed, fix an arbitrary point $z_o \in \Sigma$ along with a radius $r \in (0, R_o)$ and, as in the past, define $\Delta_\Sigma(z_o, r) := B(z_o, r) \cap \Sigma$. Then there exists a unique number $s_o \in [0, L)$ such that $z_o = z(s_o) \in \Sigma$, and we abbreviate

$$\Delta_{\rho,\varkappa} := \Delta_\rho\big(s_o, (1 + \varkappa)r\big) = \big\{s \in [0, L) : \rho(s, s_o) < (1 + \varkappa)r\big\}. \tag{3.3.18}$$

In particular, (3.3.11), (3.3.15), and (3.3.12) imply

$$\sigma\big(\Delta_\Sigma(z_o,r)\big) = \mathcal{H}^1\big(\Delta_\Sigma(z(s_o),r)\big) = \mathcal{L}^1\Big(z^{-1}\big(\Delta_\Sigma(z(s_o),r)\big)\Big) \geq \mathcal{L}^1\big(\Delta_\rho(s_o,r)\big) \geq (1+\varkappa)^{-1}\mathcal{L}^1(\Delta_{\rho,\varkappa}). \tag{3.3.19}$$

With $c := -\mathrm{i} ⨍_{\Delta_{\rho,\varkappa}} z'(s)\,\mathrm{d}s \in \mathbb{C}$ we may then write

$$\begin{aligned} ⨍_{\Delta_\Sigma(z_o,r)} |\nu - c|\,\mathrm{d}\sigma &= \frac{1}{\sigma(\Delta_\Sigma(z_o,r))}\int_{\Delta_\Sigma(z_o,r)} |\nu - c|\,\mathrm{d}\sigma \\ &= \frac{1}{\sigma(\Delta_\Sigma(z_o,r))}\int_{z^{-1}(\Delta_\Sigma(z_o,r))} |\nu(z(s)) - c|\,\mathrm{d}s \\ &\leq \frac{1}{\sigma(\Delta_\Sigma(z_o,r))}\int_{\Delta_{\rho,\varkappa}} |\nu(z(s)) - c|\,\mathrm{d}s \\ &= \frac{\mathcal{L}^1(\Delta_{\rho,\varkappa})}{\sigma(\Delta_\Sigma(z_o,r))} ⨍_{\Delta_{\rho,\varkappa}} |\nu(z(s)) - c|\,\mathrm{d}s \\ &= \frac{\mathcal{L}^1(\Delta_{\rho,\varkappa})}{\sigma(\Delta_\Sigma(z_o,r))} ⨍_{\Delta_{\rho,\varkappa}} |z'(s) - \mathrm{i}c|\,\mathrm{d}s \\ &\leq (1+\varkappa) ⨍_{\Delta_{\rho,\varkappa}} |z'(s) - \mathrm{i}c|\,\mathrm{d}s \\ &\leq (1+\varkappa)\Big(⨍_{\Delta_{\rho,\varkappa}} |z'(s) - \mathrm{i}c|^2\,\mathrm{d}s\Big)^{1/2}, \end{aligned} \tag{3.3.20}$$

making use of (3.3.16), (3.3.11), (3.3.17), (3.3.19), and Cauchy-Schwarz inequality.

Keeping in mind that $|Z - W|^2 = |Z|^2 - 2\mathrm{Re}(Z \cdot \overline{W}) + |W|^2$ for each $Z, W \in \mathbb{C}$, the choice of c allows us to compute

$$\begin{aligned} ⨍_{\Delta_{\rho,\varkappa}} |z' - \mathrm{i}c|^2\,\mathrm{d}s &= ⨍_{\Delta_{\rho,\varkappa}} \Big|z' - ⨍_{\Delta_{\rho,\varkappa}} z'\,\mathrm{d}s\Big|^2\,\mathrm{d}s \\ &= ⨍_{\Delta_{\rho,\varkappa}} \Big(|z'|^2 - 2\mathrm{Re}\Big[z' \cdot \Big(⨍_{\Delta_{\rho,\varkappa}} \overline{z'}\,\mathrm{d}s\Big)\Big] + \Big|⨍_{\Delta_{\rho,\varkappa}} z'\,\mathrm{d}s\Big|^2\Big)\,\mathrm{d}s \\ &= ⨍_{\Delta_{\rho,\varkappa}} |z'|^2\,\mathrm{d}s - 2\mathrm{Re} ⨍_{\Delta_{\rho,\varkappa}} z' \cdot \Big(⨍_{\Delta_{\rho,\varkappa}} \overline{z'}\,\mathrm{d}s\Big)\,\mathrm{d}s + \Big|⨍_{\Delta_{\rho,\varkappa}} z'\,\mathrm{d}s\Big|^2 \\ &= 1 - \Big|⨍_{\Delta_{\rho,\varkappa}} z'\,\mathrm{d}s\Big|^2, \end{aligned} \tag{3.3.21}$$

where the last equality uses (3.3.7). Also, whenever $r \in (0, L/2)$, the Fundamental Theorem of Calculus and (3.3.13) give

$$\fint_{\Delta_{\rho,\varkappa}} z'\,\mathrm{d}s = \frac{1}{\sigma(\Delta_{\rho,\varkappa})}\int_{\Delta_{\rho,\varkappa}} z'\,\mathrm{d}s$$

$$= \begin{cases} \frac{1}{2r}\big(z(s_o+r)-z(s_o+L-r)\big) & \text{if } 0 \le s_o < r, \\ \frac{1}{2r}\big(z(s_o+r)-z(s_o-r)\big) & \text{if } r \le s_o \le L-r, \\ \frac{1}{2r}\big(z(s_o-L+r)-z(s_o-r)\big) & \text{if } L-r < s_o < L. \end{cases} \tag{3.3.22}$$

In concert with the observation that for each $r \in (0, L/4)$ we have (cf. (3.3.8))

$$2r = \begin{cases} \rho\big(s_o+r, s_o+L-r\big) & \text{if } 0 \le s_o < r, \\ \rho\big(s_o+r, s_o-r\big) & \text{if } r \le s_o \le L-r, \\ \rho\big(s_o-L+r, s_o-r\big) & \text{if } L-r < s_o < L, \end{cases} \tag{3.3.23}$$

we conclude that if $r \in (0, L/4)$ then

$$\fint_{\Delta_{\rho,\varkappa}} z'\,\mathrm{d}s = \begin{cases} \frac{z(s_o+r)-z(s_o+L-r)}{\rho\left(s_o+r,s_o+L-r\right)} & \text{if } 0 \le s_o < r, \\ \frac{z(s_o+r)-z(s_o-r)}{\rho\left(s_o+r,s_o-r\right)} & \text{if } r \le s_o \le L-r, \\ \frac{z(s_o-L+r)-z(s_o-r)}{\rho\left(s_o-L+r,s_o-r\right)} & \text{if } L-r < s_o < L. \end{cases} \tag{3.3.24}$$

Collectively, (3.3.24), (3.3.10), and (3.3.23) imply that whenever r is such that $0 < r < \min\{L/4, R_o/2\}$ we have

$$\fint_{\Delta_{\rho,\varkappa}} |z'-\mathrm{i}c|^2\,\mathrm{d}s \le 1 - \Big(\frac{1}{1+\varkappa}\Big)^2 = \frac{\varkappa(2+\varkappa)}{(1+\varkappa)^2}. \tag{3.3.25}$$

Combining (3.3.25) and (3.3.20) we arrive at the conclusion that

$$\fint_{\Delta_\Sigma(z_o,r)} \Big|\nu - \fint_{\Delta_\Sigma(z_o,r)} \nu\,\mathrm{d}\sigma\Big|\,\mathrm{d}\sigma \le 2\fint_{\Delta_\Sigma(z_o,r)} |\nu - c|\,\mathrm{d}\sigma \tag{3.3.26}$$

$$\le 2\sqrt{\varkappa(2+\varkappa)} \text{ if } 0 < r < \min\{L/4, R_o/2\}.$$

In turn, this implies that

$$\lim_{R\to 0^+}\Bigg\{\sup_{\substack{z_o\in\partial\Omega \\ r\in(0,R)}} \fint_{\Delta_\Sigma(z_o,r)} \Big|\nu - \fint_{\Delta_\Sigma(z_o,r)} \nu\,\mathrm{d}\sigma\Big|\,\mathrm{d}\sigma\Bigg\} \le 2\sqrt{\varkappa(2+\varkappa)}. \tag{3.3.27}$$

Based on this and [130, (3.1.53)] we finally conclude that there exists some constant $C \in (0, \infty)$ which depends exclusively on the upper Ahlfors regularity constant of $\partial\Omega$ such that

$$\operatorname{dist}\big(\nu, \big[\mathrm{VMO}(\partial\Omega,\sigma)\big]^2\big) \le C\sqrt{\varkappa(2+\varkappa)}, \tag{3.3.28}$$

where the distance in the left-hand side is measured in the John-Nirenberg space $\big[\mathrm{BMO}(\partial\Omega,\sigma)\big]^2$. This finishes the proof of (3.3.4).

To deal with the claim in the second part of the statement, suppose next that the set $\Omega \subseteq \mathbb{R}^2$ is an Ahlfors regular domain with compact boundary. Let $\sigma := \mathcal{H}^1 \lfloor \partial\Omega$ and denote by ν the geometric measure theoretic outward unit normal to Ω. To fix ideas, suppose first that Ω is bounded and simply connected (the case when Ω is unbounded and connected is treated later). Impose the condition that (3.3.5) holds for some $\delta \in (0, 1)$ small enough so that the conclusions in Theorem 3.1.11 hold. This ensures that Ω is a uniform domain (in the sense of [129, Definition 5.11.10]). In particular, there exists $c \in (0, \infty)$ with the property that for any given $x, y \in \Omega$ there exists a rectifiable simple curve γ joining x with y in Ω with $\text{length}(\gamma) \leq c|x - y|$. As such, the demand in (3.3.2) is satisfied. In view of (3.3.1), this proves that

$$\partial\Omega \text{ is a Jordan curve in the plane.} \tag{3.3.29}$$

Then Ω becomes the inner domain associated with this Jordan curve. Moreover, the fact that $\partial\Omega$ is compact and upper Ahlfors regular implies (cf. [129, (5.3.72)]) that $\partial\Omega$ is rectifiable. Let L stand for the length of $\partial\Omega$. Also, for each $z_1, z_2 \in \partial\Omega$ denote by $\ell(z_1, z_2)$ the length of the shortest sub-arc of $\partial\Omega$ joining z_1 and z_2. In particular, [129, (5.8.28)] guarantees that

$$\begin{gathered}\ell(\cdot,\cdot) : \partial\Omega \times \partial\Omega \longrightarrow [0, L/2] \text{ has the property that}\\ \text{for each } \varepsilon > 0 \text{ there exists } \eta > 0 \text{ such that } \ell(z_1, z_2) < \varepsilon\\ \text{whenever the points } z_1, z_2 \in \partial\Omega \text{ are such that } |z_1 - z_2| < \eta.\end{gathered} \tag{3.3.30}$$

Next, we shall prove that there exist $R \in (0, \infty)$ together with $\varkappa = \varkappa(\delta) \in (0, \infty)$ such that

$$\varkappa = O\big(\delta \ln(1/\delta)\big) \text{ as } \delta \to 0^+ \tag{3.3.31}$$

with the property that

$$\ell(z_1, z_2) \leq (1 + \varkappa)|z_1 - z_2| \text{ for all } z_1, z_2 \in \partial\Omega \text{ with } |z_1 - z_2| < R. \tag{3.3.32}$$

To this end, we shall employ Theorem 3.2.1 for the choice $\phi(t) := t \ln(1/t)$ in (3.2.1). By further decreasing the value of $\delta \in (0, 1)$ if necessary, we may invoke [130, (3.1.53)] to ensure that (3.2.3) is satisfied for some small radius $r_0 \in (0, \infty)$. Granted this, Theorem 3.2.1 then guarantees that there exists some $R_\delta \in \big(0, \text{diam}\,\partial\Omega\big)$, depending only the dimension n, the Ahlfors regularity constants of $\partial\Omega$, the relative flatness ratio $\wp(\Omega)/\text{diam}\,\partial\Omega$, r_0, and δ, with the property that we have (cf. (3.2.16) with $n = 2$)

$$\left| \frac{\mathcal{H}^1\big(B(z, r) \cap \partial\Omega\big)}{2r} - 1 \right| \leq C_o \delta \ln(1/\delta), \qquad \forall z \in \partial\Omega, \quad \forall r \in (0, R_\delta), \tag{3.3.33}$$

for a finite geometrical constant $C_o > 1$ which is independent of δ. Without loss of generality, for the remainder of the proof further assume that $\delta \in (0, 1)$ is small enough so that $0 < \delta \ln(1/\delta) < 1/(4C_o)$. From (3.3.30) we see that there exists $R \in (0, \infty)$ with the property that

$$\ell(z_1, z_2) < R_\delta \text{ whenever } z_1, z_2 \in \partial\Omega \text{ are such that } |z_1 - z_2| < R. \tag{3.3.34}$$

Pressing on, consider now two distinct points $z_1, z_2 \in \partial\Omega$ with $|z_1 - z_2| < R$. In particular, if we abbreviate $r := \ell(z_1, z_2)$ then (3.3.34) implies that $r \in (0, R_\delta)$. Hence, $r < \operatorname{diam} \partial\Omega$ which means that the curve $\partial\Omega$ goes out of $B(z_1, r)$. Let $z_3 \in \partial\Omega$ be the first such exit point such that the ordering z_1, z_2, z_3 conforms with the positive orientation of $\partial\Omega$. Then $|z_1 - z_3| = r$ and, moreover,

$$\text{the portion of } \partial\Omega \text{ between } z_1 \text{ and } z_3 \text{ is contained inside } B(z_1, r). \tag{3.3.35}$$

To proceed, introduce $\Delta := B(z_1, r) \cap \partial\Omega$ and decompose $\Delta \setminus \{z_1\} = \Delta^- \cup \Delta^+$ (disjoint union), where $\Delta^\pm$ denote the sets of points in Δ lying, respectively, to the left and to the right of z_1. Also, denote by $\ell(\Delta^\pm)$ the arc-lengths of $\Delta^\pm$. Then

$$\mathcal{H}^1\big(B(z_1, r) \cap \partial\Omega\big) = \ell(\Delta^-) + \ell(\Delta^+) \text{ and } \ell(\Delta^\pm) \geq r. \tag{3.3.36}$$

Making use of (3.3.33) and (3.3.36) we may therefore estimate

$$\begin{aligned} C_o \delta \ln(1/\delta) &\geq \left| \frac{\mathcal{H}^1\big(B(z, r) \cap \partial\Omega\big)}{2r} - 1 \right| = \left| \frac{\ell(\Delta^-) - r}{2r} + \frac{\ell(\Delta^+) - r}{2r} \right| \\ &= \frac{\ell(\Delta^-) - r}{2r} + \frac{\ell(\Delta^+) - r}{2r} \geq \frac{\ell(\Delta^+) - r}{2r}. \end{aligned} \tag{3.3.37}$$

Hence, in view of the significance of z_3 and (3.3.37),

$$|z_2 - z_3| \leq \ell(\Delta^+) - r \leq 2r\, C_o \delta \ln(1/\delta) \tag{3.3.38}$$

which further implies

$$\begin{aligned} |z_1 - z_2| &\geq |z_1 - z_3| - |z_2 - z_3| \geq r - 2r\, C_o \delta \ln(1/\delta) \\ &= \big(1 - 2C_o \delta \ln(1/\delta)\big) \ell(z_1, z_2). \end{aligned} \tag{3.3.39}$$

This proves that

$$\ell(z_1, z_2) \leq (1 + \varkappa)|z_1 - z_2| \text{ with } \varkappa := \frac{2C_o \delta \ln(1/\delta)}{1 - 2C_o \delta \ln(1/\delta)}, \tag{3.3.40}$$

which establishes (3.3.32) with $\varkappa$ as in (3.3.31). In turn, this has two consequences of relevance for us here. First, (3.3.32) and a simple argument involving reasoning by contradiction show that

$$\sup_{\substack{z_1 \neq z_2 \\ z_1, z_2 \in \Sigma}} \frac{\ell(z_1, z_2)}{|z_1 - z_2|} < +\infty. \tag{3.3.41}$$

With this in hand, we may now conclude that $\partial\Omega$ is a chord-arc curve (cf. [129, Definition 5.9.5]), hence Ω is a chord-arc domain (cf. [129, Definition 5.9.8]).

Second, (3.3.32) implies

$$\lim_{R\to 0^+}\left\{\sup_{\substack{z_1,z_2\in\partial\Omega\\ |z_1-z_2|<R}}\left(\frac{\ell(z_1,z_2)}{|z_1-z_2|}-1\right)\right\}\leq\varkappa, \tag{3.3.42}$$

from which (3.3.6) follows on account of (3.3.31).

To finish the proof of Theorem 3.3.1 there remains to consider the case when the Ahlfors regular domain with compact boundary Ω is unbounded and connected. To set the stage, recall from [14, Corollary 3, p. 352] that

$$\text{if } E\subseteq\mathbb{R}^2 \text{ is a closed set such that each connected component of } E \text{ is unbounded, then } \mathbb{R}^2\setminus E \text{ is a simply connected set.} \tag{3.3.43}$$

Since the closure of a connected set is always connected, it follows from (3.3.43) and [129, Lemma 5.10.10] that $\Omega_- := \mathbb{R}^2\setminus\overline{\Omega}$ is a bounded simply connected open set. Moreover, according to item *(6)* in [129, Lemma 5.10.9], Ω_- is also an Ahlfors regular domain, whose topological boundary coincides with that of Ω, whose geometric measure theoretic boundary agrees with that of Ω, and whose geometric measure theoretic outward unit normal is $-\nu$ at σ-a.e. point on $\partial\Omega$. Granted these properties, all desired conclusions follow from what we have proved already for bounded sets. □

In chord-arc domains with small constant, the relative flatness radius may be estimated from below in the manner described in our next corollary.

Corollary 3.3.2 *Assume $\Omega\subseteq\mathbb{R}^2$ is a chord-arc domain with compact boundary. In particular, Ω is an Ahlfors regular domain so it makes sense to consider the threshold $\varepsilon_3\in(0,1)$ associated with Ω as in Corollary 3.1.4. Make the assumption that there exists $R_o\in\big(0,2\,\mathrm{diam}\,\partial\Omega\big)$ such that*

$$\sup_{\substack{z_1,z_2\in\partial\Omega\\ |z_1-z_2|<R_o}}\left(\frac{\ell(z_1,z_2)}{|z_1-z_2|}-1\right)<\varkappa \tag{3.3.44}$$

for some $\varkappa\in(0,\infty)$ satisfying

$$2\sqrt{\varkappa(2+\varkappa)}<\varepsilon_3. \tag{3.3.45}$$

Then, if $L\in(0,\infty)$ denotes the length of $\partial\Omega$, it follows that the relative flatness radius of Ω satisfies

$$\wp(\Omega)\geq\min\{L/4,R_o/2\}. \tag{3.3.46}$$

Proof This is a consequence of (3.3.26) and (3.1.36). □

3.4 δ-Oscillating AR Domains and Infinitesimally Flat AR Domains

We start by making a definition which identifies a general class of sets providing a natural environment where the size of singular integral operators (of "chord-dot-normal" type) may be directly linked to geometry (as we shall eventually see in §7.1). Ultimately, this is going to play a major role in our work pertaining to boundary value problems (of the sort discussed in §8.1).

Definition 3.4.1 *Fix $n \in \mathbb{N}$ with $n \geq 2$ and consider an Ahlfors regular domain $\Omega \subseteq \mathbb{R}^n$ with compact boundary. Abbreviate $\sigma := \mathcal{H}^{n-1}\lfloor\partial\Omega$ and denote by ν the geometric measure theoretic outward unit normal to Ω.*

(1) Given $\delta \in (0,1)$, call Ω a δ-`oscillating` AR `domain` *(or* δ-`osc` AR*, for short) provided*

$$\operatorname{dist}\big(\nu, \big[\mathrm{VMO}(\partial\Omega, \sigma)\big]^n\big) < \delta, \tag{3.4.1}$$

where the distance is measured in the John-Nirenberg space $\big[\mathrm{BMO}(\partial\Omega, \sigma)\big]^n$.

(2) Call Ω an `infinitesimally flat` AR `domain` *(or* `inf-flat` AR*, for short) provided*[5]

$$\nu \in \big[\mathrm{VMO}(\partial\Omega, \sigma)\big]^n. \tag{3.4.2}$$

We emphasize that, by definition, any δ-oscillating AR domain is an open set with compact boundary. Clearly, an infinitesimally flat AR domain is a δ-oscillating AR domain for any $\delta \in (0,1)$. The fact that in item *(1)* of Definition 3.4.1 the threshold δ is taken to be a number in the interval $(0,1)$ has to do with the fact that, generally speaking,

$$0 \leq \operatorname{dist}\big(\nu, \big[\mathrm{VMO}(\partial\Omega, \sigma)\big]^n\big) \leq \|\nu\|_{[\mathrm{BMO}(\partial\Omega,\sigma)]^n} \leq 1, \tag{3.4.3}$$

where the last inequality comes from (4.1.2). As such, in the context of (3.4.1) one necessarily has $\delta > 0$, and (3.4.1) becomes a genuine demand only if $\delta < 1$.

This being said, we shall occasionally ignore the convention that δ must belong to the interval $(0,1)$ and, for any given $\delta \in (0,\infty)$, continue to call an Ahlfors regular domain with compact boundary $\Omega \subseteq \mathbb{R}^n$ a δ-oscillating AR domain provided (3.4.1) holds.

Heuristically,

> *the class of infinitesimally flat* AR *domains should be thought of as a sharp (i.e., significantly more inclusive, yet exhibiting similar properties) geometric measure theoretic version of the class of $\mathscr{C}^1$ domains with compact boundary.* (3.4.4)

In this regard, see the discussion in Example 3.4.6 leading up to (3.4.42), and also Proposition 3.4.3. For now we wish to point out that work in [80] shows that the class of infinitesimally flat AR domains is stable under $\mathscr{C}^1$ diffeomorphisms.

---

[5] i.e., the Gauss map has vanishing mean oscillations

Given an Ahlfors regular domain $\Omega \subseteq \mathbb{R}^n$, abbreviate $\sigma := \mathcal{H}^{n-1}\lfloor\partial\Omega$ and denote by ν the geometric measure theoretic outward unit normal to Ω. Then the distance, measured in the John-Nirenberg space $\big[\mathrm{BMO}(\partial\Omega,\sigma)\big]^n$, from ν to the Sarason space $\big[\mathrm{VMO}(\partial\Omega,\sigma)\big]^n$ is comparable with

$$[\nu]_{\mathrm{Osc}(\partial\Omega)} := \lim_{R\to 0^+}\left(\sup_{\substack{z\in\partial\Omega\\ 0<r<R}} \fint_{B(z,r)\cap\partial\Omega}\fint_{B(z,r)\cap\partial\Omega} \big|\nu(x)-\nu(y)\big|\,d\sigma(x)\,d\sigma(y)\right). \tag{3.4.5}$$

Indeed, as seen from [130, (3.1.53)] (used with $p := 1$), there exists some $C \in (1,\infty)$ which depends only on the Ahlfors regularity constants of $\partial\Omega$ such that

$$C^{-1}\cdot[\nu]_{\mathrm{Osc}(\partial\Omega)} \leq \mathrm{dist}\big(\nu, \big[\mathrm{VMO}(\partial\Omega,\sigma)\big]^n\big) \leq C\cdot[\nu]_{\mathrm{Osc}(\partial\Omega)}. \tag{3.4.6}$$

In turn, the quantity $[\nu]_{\mathrm{Osc}(\partial\Omega)}$ is controlled by

$$\lim_{r\to 0^+}\left(\sup_{\substack{|x-y|<r\\ x,y\in\partial^*\Omega}} |\nu(x)-\nu(y)|\right), \tag{3.4.7}$$

the pointwise local oscillation of ν as defined by A.P. Calderón (in the context of Lipschitz domains) in [28], [29]. Thus, it is of interest to remark that

$$\begin{gathered}\text{there exist Lipschitz domains for which } [\nu]_{\mathrm{Osc}(\partial\Omega)} \text{ is}\\ \text{as small compared to (3.4.7) as one wishes.}\end{gathered} \tag{3.4.8}$$

Specifically, for each fixed $\varepsilon \in (0, 1/2)$ bring in the function

$$\begin{gathered}\varphi_\varepsilon : \mathbb{R} \longrightarrow \mathbb{R} \text{ defined as } \varphi_\varepsilon(0) := 0 \text{ and}\\ \varphi_\varepsilon(x) := x\cdot\sin\big(\ln(|x|^{-\varepsilon})\big) \text{ for each } x \in \mathbb{R}\setminus\{0\},\end{gathered} \tag{3.4.9}$$

then consider the two-dimensional domain

$$\Omega_\varepsilon := \big\{(x,t)\in\mathbb{R}^2 : t > \varphi_\varepsilon(x)\big\}. \tag{3.4.10}$$

Note that

$$\varphi_\varepsilon'(x) = \sin\big(\ln(|x|^{-\varepsilon})\big) - \varepsilon\cdot\cos\big(\ln(|x|^{-\varepsilon})\big) \text{ for each } x \in \mathbb{R}\setminus\{0\}, \tag{3.4.11}$$

hence $|\varphi_\varepsilon'(x)| \leq 3/2$ for each $x \in \mathbb{R}\setminus\{0\}$ and $\|\varphi_\varepsilon'\|_{L^\infty(I,\mathcal{L}^1)} \approx 1$ uniformly in $\varepsilon \in (0, 1/2)$ and I neighborhood of 0. As such,

$$\begin{gathered}\text{each } \Omega_\varepsilon \text{ is a Lipschitz domain in the plane, and}\\ \text{the Lipschitz constant of } \Omega_\varepsilon \text{ stays} \approx 1 \text{ as } \varepsilon \to 0^+.\end{gathered} \tag{3.4.12}$$

Moreover, the outward unit normal to Ω_ε is the vector

$$\nu\big(x,\varphi_\varepsilon(x)\big)=\frac{\big(\varphi_\varepsilon'(x),-1\big)}{\sqrt{1+|\varphi_\varepsilon'(x)|^2}}\ \text{ for each } x\in\mathbb{R}\setminus\{0\}. \tag{3.4.13}$$

Since $\mathbb{R}\ni t\mapsto\frac{(t,-1)}{\sqrt{1+t^2}}\in\mathbb{R}^2$ is Lipschitz with constant $\leq\sqrt{31/27}$, this implies

$$\Big|\nu\big(x,\varphi_\varepsilon(x)\big)-\nu\big(y,\varphi_\varepsilon(y)\big)\Big|\leq\sqrt{\tfrac{31}{27}}\big|\varphi_\varepsilon'(x)-\varphi_\varepsilon'(y)\big|\ \text{ for each } x,y\in\mathbb{R}\setminus\{0\}. \tag{3.4.14}$$

In turn, from (3.4.14), (3.4.12), and (3.4.5) we see that there exists a universal constant $C\in(0,\infty)$ with the property that

$$[\nu]_{\mathrm{Osc}(\partial\Omega_\varepsilon)}\leq C\|\varphi_\varepsilon'\|_{\mathrm{BMO}(\mathbb{R},\mathcal{L}^1)}. \tag{3.4.15}$$

To proceed, note that (3.4.11) entails

$$\big|\varphi_\varepsilon'(x)-\varphi_\varepsilon'(y)\big|\leq\varepsilon\big|\ln|x|-\ln|y|\big|+2\varepsilon\ \text{ for each } x,y\in\mathbb{R}\setminus\{0\}, \tag{3.4.16}$$

hence

$$\|\varphi_\varepsilon'\|_{\mathrm{BMO}(\mathbb{R},\mathcal{L}^1)}\leq\varepsilon\Big(2+\big\|\ln|\cdot|\big\|_{\mathrm{BMO}(\mathbb{R},\mathcal{L}^1)}\Big). \tag{3.4.17}$$

Together, (3.4.15) and (3.4.17) prove $[\nu]_{\mathrm{Osc}(\partial\Omega_\varepsilon)}\leq\varepsilon\Big(2+\big\|\ln|\cdot|\big\|_{\mathrm{BMO}(\mathbb{R},\mathcal{L}^1)}\Big)$ hence

$$[\nu]_{\mathrm{Osc}(\partial\Omega_\varepsilon)}=O(\varepsilon)\ \text{ as }\ \varepsilon\to 0^+. \tag{3.4.18}$$

By way of contrast, we claim that

$$\lim_{r\to 0^+}\Bigg(\sup_{\substack{|x-y|<r\\ x,y\in\partial\Omega_\varepsilon\setminus\{0\}}}|\nu(x)-\nu(y)|\Bigg)\ \text{ stays}\approx 1\ \text{ as }\ \varepsilon\to 0^+. \tag{3.4.19}$$

Indeed, this follows from (3.4.13) upon observing that there exist two sequences $\{x_j^{(i)}\}_{j\in\mathbb{N}}$, with $i=0,1$, of nonzero real numbers converging to 0 and such that

$$\varphi_\varepsilon'\big(x_j^{(i)}\big)=\begin{cases}O(\varepsilon) & \text{for each } j\in\mathbb{N} \text{ if } i=0,\\ 1+O(\varepsilon) & \text{for each } j\in\mathbb{N} \text{ if } i=1.\end{cases} \tag{3.4.20}$$

This finishes the justification of (3.4.8). Ultimately, (3.4.8) points to the fact that the pointwise oscillations of the geometric measure theoretic outward unit normal ν to an Ahlfors regular domain Ω is a rougher measure of flatness for $\partial\Omega$ than the quantity $\mathrm{dist}\big(\nu,\big[\mathrm{VMO}(\partial\Omega,\sigma)\big]^n\big)$.

One of the relevant aspects of Definition 3.4.1 is that, for an open set with compact boundary $\Omega\subseteq\mathbb{R}^n$, being a δ-oscillating AR domain with δ small relative to the dimension n and the Ahlfors regular constants of $\partial\Omega$ says much about the geometry of Ω. This is made precise below.

Theorem 3.4.2 *Fix $n\in\mathbb{N}$ with $n\geq 2$. Then for each given $C_A\in[1,\infty)$ there exists a constant $\varepsilon_4\in(0,1)$, which depends only on n and C_A, with the following* *significance. Suppose $\Omega \subseteq \mathbb{R}^n$ is an open set with compact boundary, which is a δ-oscillating* AR *domain whose Ahlfors regularity constants are controlled by C_A (in the sense of* (3.1.35)) *and with $\delta \in (0, \varepsilon_4)$.*

Then the relative flatness radius $\wp(\Omega)$ (Definition 3.1.9) is strictly positive, and

$$\text{the set } \Omega \text{ is a two-sided NTA domain, in the sense of [129, Definition 5.11.1], with constants } R := 2^{-1}C_A^{-2/(n-1)} \cdot \wp(\Omega), \text{ and } M = M(C_A, n) \in (0, \infty) \text{ depending only on the dimension } n \text{ and the Ahlfors regularity constant } C_A \text{ of } \partial\Omega. \tag{3.4.21}$$

Moreover, the set Ω is a UR *domain with constants controlled in terms of the dimension n, the constant C_A, and the relative flatness ratio $\wp(\Omega)/\mathrm{diam}\,\partial\Omega$, and Ω satisfies a two-sided local John condition (again, in the same quantitative fashion). In addition, Ω has finitely many connected components, which are separated (i.e., have mutually disjoint closures), and the same is true for $\mathbb{R}^n \setminus \overline{\Omega}$. In particular, any connected component of Ω and $\mathbb{R}^n \setminus \overline{\Omega}$ is itself a two-sided* NTA *domain.*

The set Ω is an $(\varepsilon_o, \delta_o)$-domain for some $\varepsilon_o, \delta_o \in (0, \infty)$ (cf. (A.0.76)) *with* $\mathrm{rad}\,(\Omega) > 0$. *Finally, both Ω and $\mathbb{R}^n \setminus \overline{\Omega}$ are locally uniform domains in the sense of* [129, Definition 5.11.12], *and each of their connected components is a uniform domain in the sense of* [129, Definition 5.11.10].

Proof This is a direct consequence of Theorem 3.1.11 and Definition 3.4.1. □

Domains of class $\mathscr{C}^1$ fit nicely inside the category of infinitesimally flat AR domains, particularly in view of the description given in the proposition below.

Proposition 3.4.3 *Let $\Omega \subseteq \mathbb{R}^n$, where $n \in \mathbb{N}$ with $n \geq 2$, be an Ahlfors regular domain with compact boundary, and denote by ν its geometric measure theoretic outward unit normal. Then Ω is a domain of class $\mathscr{C}^1$ if and only if ν may be altered on a $\mathcal{H}^{n-1}$-nullset as to become a continuous vector field on $\partial\Omega$ (or, equivalently, a uniformly continuous vector field on $\partial\Omega$).*

Proof Assume $\nu \in \left[\mathscr{C}^0(\partial\Omega)\right]^n$. Then ν is a bounded, uniformly continuous, vector field on $\partial\Omega$. Hence, $\nu \in \left[\mathrm{VMO}(\partial\Omega, \sigma)\right]^n$ where $\sigma := \mathcal{H}^{n-1}\lfloor\partial\Omega$, so Ω is an infinitesimally flat AR domain. As such, Theorem 3.4.2 applies and gives that Ω satisfies an exterior corkscrew condition. In turn, this implies (cf. [129, (5.1.10)]) $\partial\Omega = \partial(\overline{\Omega})$. Having established this, we may then invoke [80, Theorem 2.12] to conclude that Ω is a domain of class $\mathscr{C}^1$. This takes care of one implication in the equivalence claimed in the statement. The remaining implication, i.e., that for a domain of class $\mathscr{C}^1$ its geometric measure theoretic outward unit normal is a continuous vector field is clear from definitions. □

We now proceed to catalog in a systematical fashion examples of δ-oscillating AR domains and infinitesimally flat AR domains.

Example 3.4.4 *Suppose $\Omega \subseteq \mathbb{R}^n$ is an open set with compact boundary. Thanks to item (6) in* [129, Lemma 5.10.9], *if $\Omega \subseteq \mathbb{R}^n$ is a δ-oscillating* AR *domain for some $\delta \in (0, 1)$, then $\mathbb{R}^n \setminus \overline{\Omega}$ is also a δ-oscillating* AR *domain (having the same* *topological and measure theoretic boundaries as Ω, and whose geometric measure theoretic outward unit normal is the opposite of the one for Ω). Also, any rigid transformation in $\mathbb{R}^n$ preserves the class of δ-oscillating* AR *domains. Finally, similar results hold in the class of infinitesimally flat* AR *domains.*

Lipschitz domains in $\mathbb{R}^n$ with compact boundary and small Lipschitz constant offer some basic examples of δ-oscillating AR domains, with the threshold δ linearly linked to the size of said Lipschitz constant. Specifically, fix $R, \delta \in (0, \infty)$ and assume $\Omega \subseteq \mathbb{R}^n$ is an open set with compact boundary having the property that Ω may be locally identified, up to some scale $R \in (0, \infty)$ and via a rigid transformation of the Euclidean space, with the upper-graph of a Lipschitz function $\phi : \mathbb{R}^{n-1} \to \mathbb{R}$ whose Lipschitz constant is $\leq \delta$. Then there exists some $C \in (0, \infty)$ which depends only on the Ahlfors regularity constants of $\partial\Omega$ with the property that Ω is a $(C\delta)$-oscillating AR domain whenever $0 < \delta < C^{-1}$. We elaborate on this issue below.

Example 3.4.5 *Suppose $\Omega \subseteq \mathbb{R}^n$ is a set with nonempty compact boundary, for which there exist a threshold $\delta \in (0, \infty)$ and a scale $R \in (0, \operatorname{diam} \partial\Omega)$ with the property that for each point $x_o \in \partial\Omega$ one may find a Lipschitz function $\phi : \mathbb{R}^{n-1} \to \mathbb{R}$ with $\phi(0') = 0$ and whose Lipschitz constant is $\leq \delta$ along with a rigid transformation*[6] *$T : \mathbb{R}^n \to \mathbb{R}^n$ such that $T(x_o) = 0$ and*

$$T\big(\Omega \cap B(x_o, R)\big) = \big\{(z', z_n) \in \mathbb{R}^{n-1} \times \mathbb{R} : z_n > \phi(z')\big\} \cap B(0, R). \tag{3.4.22}$$

Then Ω is a Lipschitz domain with compact boundary (cf. [129, Definition 2.8.12]*). In particular, Ω is an Ahlfors regular domain with compact boundary (see* (3.4.32)-(3.4.34) *below). Denote by ν the geometric measure theoretic outward unit normal to Ω and abbreviate $\sigma := \mathcal{H}^{n-1}\lfloor\partial\Omega$. Having fixed some location $x_o \in \partial\Omega$ and some scale $r \in (0, R)$, we claim that*

$$\fint_{B(x_o,r)\cap\partial\Omega} \Big|\nu - \fint_{B(x_o,r)\cap\partial\Omega} \nu \, d\sigma\Big| \, d\sigma \leq 2^{3/2}\delta. \tag{3.4.23}$$

To justify this, we find it convenient to identify points in the Euclidean space with their images under the rigid transformation $T : \mathbb{R}^n \to \mathbb{R}^n$ associated with x_o. Hence, for the goal we have in mind, there is no loss of generality in assuming that x_o is the origin in $\mathbb{R}^n$ and that (3.4.22) *holds with $T = I$, the identity in $\mathbb{R}^n$. That is, we shall assume that $x_o = 0 \in \mathbb{R}^n$ and that there exists a Lipschitz function $\phi : \mathbb{R}^{n-1} \to \mathbb{R}$ with $\phi(0') = 0$, whose Lipschitz constant is $\leq \delta$, and such that*

$$\Omega \cap B(0, R) = \big\{(z', z_n) \in \mathbb{R}^{n-1} \times \mathbb{R} : z_n > \phi(z')\big\} \cap B(0, R). \tag{3.4.24}$$

To proceed, note that any point $x \in \partial\Omega \cap B(0, R)$ may be expressed as $x = \big(x', \phi(x')\big)$ with x' belonging to $B_{n-1}(0, R)$, the $(n-1)$-dimensional ball centered at the origin in $\mathbb{R}^{n-1}$ of radius R. Moreover, if x' is a differentiability point for the function ϕ we have

[6] the composition between a translation and rotation about the origin

$$\nu(x) = \frac{(\nabla'\phi(x'), -1)}{\sqrt{1 + |\nabla'\phi(x')|^2}} \tag{3.4.25}$$

(with ∇' denoting the gradient operator in $\mathbb{R}^{n-1}$), a context in which we may compute

$$\begin{aligned}\nu(x) + \mathbf{e}_n &= \left(\frac{\nabla'\phi(x')}{\sqrt{1 + |\nabla'\phi(x')|^2}}, 1 - \frac{1}{\sqrt{1 + |\nabla'\phi(x')|^2}} \right) \\ &= \left(\frac{\nabla'\phi(x')}{\sqrt{1 + |\nabla'\phi(x')|^2}}, \frac{|\nabla'\phi(x')|^2}{\sqrt{1 + |\nabla'\phi(x')|^2}(1 + \sqrt{1 + |\nabla'\phi(x')|^2})} \right).\end{aligned} \tag{3.4.26}$$

As such, for each $r \in (0, R)$ we may estimate

$$\begin{aligned}&⨍_{B(0,r)\cap\partial\Omega} \left| \nu - ⨍_{B(0,r)\cap\partial\Omega} \nu \, d\sigma \right| d\sigma \\ &\quad = ⨍_{B(0,r)\cap\partial\Omega} \left| (\nu + \mathbf{e}_n) - ⨍_{B(0,r)\cap\partial\Omega} (\nu + \mathbf{e}_n) \, d\sigma \right| d\sigma \\ &\quad \leq 2 \|\nu + \mathbf{e}_n\|_{[L^\infty(B(0,r)\cap\partial\Omega,\sigma)]^n} \\ &\quad \leq 2^{3/2} \left\| \frac{|\nabla'\phi|}{(1 + |\nabla'\phi|^2)^{1/4}(1 + \sqrt{1 + |\nabla'\phi|^2})^{1/2}} \right\|_{L^\infty(\mathbb{R}^{n-1}, \mathcal{L}^{n-1})} \\ &\quad \leq 2^{3/2} \|\nabla'\phi\|_{[L^\infty(\mathbb{R}^{n-1}, \mathcal{L}^{n-1})]^{n-1}} \leq 2^{3/2}\delta.\end{aligned} \tag{3.4.27}$$

Ultimately, this establishes (3.4.23).

From (3.4.23) *and* [130, (3.1.53)] *we then conclude that there exists some constant $C \in (0, \infty)$ which depends only on n and the Ahlfors regularity constants of $\partial\Omega$ with the property that*

$$\operatorname{dist}\big(\nu, \big[\mathrm{VMO}(\partial\Omega, \sigma)\big]^n\big) < C\delta. \tag{3.4.28}$$

In view of item (1) in Definition 3.4.1, we then conclude that if $0 < \delta < C^{-1}$ then

$$\Omega \textit{ is a } (C\delta)\textit{-oscillating } \mathrm{AR} \textit{ domain.} \tag{3.4.29}$$

In relation to (3.4.29), *we wish to make two additional comments. First, from* (3.4.23) *and* (3.1.47) *we see that if δ is sufficiently small relative to n and the Ahlfors regularity constants of $\partial\Omega$ then the pointwise flatness radius satisfies*

$$R_{\text{flat}}(x_o) \geq R \textit{ for each } x_o \in \partial\Omega. \tag{3.4.30}$$

Together with (3.1.48), *this proves that the flatness radius of Ω satisfies*

$$\wp(\Omega) = \inf_{x_o \in \partial\Omega} R_{\text{flat}}(x_o) \geq R. \tag{3.4.31}$$

As such, the global flatness radius $\wp(\Omega)$ does not degenerate (to zero) as $\delta \to 0^+$.

The second comment we wish to make in relation to (3.4.29) *is that the present assumptions imply*

$$\upsilon_{n-1}(1+\delta^2)^{(1-n)/2} r^{n-1} \leq \sigma\big(B(x_o,r)\cap\partial\Omega\big) \leq \upsilon_{n-1}(1+\delta^2)^{1/2} r^{n-1}$$
$$\text{for any } x_o \in \partial\Omega \text{ and any } r \in (0,R), \tag{3.4.32}$$

where υ_{n-1} *denotes the volume of the unit ball in* $\mathbb{R}^{n-1}$. *In turn, this yields the Ahlfors regularity bounds*

$$c r^{n-1} \leq \sigma\big(B(x_o,r)\cap\partial\Omega\big) \leq C r^{n-1}$$
$$\text{for any } x_o \in \partial\Omega \text{ and any } r \in (0, \operatorname{diam}\partial\Omega) \tag{3.4.33}$$

with constants

$$c := \upsilon_{n-1}(1+\delta^2)^{(1-n)/2}\Big(\frac{R}{\operatorname{diam}\partial\Omega}\Big)^{n-1} \text{ and}$$
$$C := \max\Big\{\mathcal{H}^{n-1}(\partial\Omega)/R^{n-1},\, \upsilon_{n-1}(1+\delta^2)^{1/2}\Big\}. \tag{3.4.34}$$

In particular this shows that the Ahlfors regularity constants of $\partial\Omega$ *stay away from zero and infinity as* $\delta \to 0^+$, *as long as* R *does not go to zero and* $\operatorname{diam}\partial\Omega$ *stays uniformly bounded.*

We next refine the discussion in Example 3.4.5 by considering in place of Lipschitz domains the class of BMO_1-domains.

Example 3.4.6 *Recall the function space* $\mathrm{BMO}_1(\mathbb{R}^{n-1})$, *consisting of functions* ϕ *in* $L^1_{\mathrm{loc}}(\mathbb{R}^{n-1},\mathcal{L}^{n-1})$ *with* $\nabla'\phi$ *belonging to the space* $\big[\mathrm{BMO}(\mathbb{R}^{n-1},\mathcal{L}^{n-1})\big]^{n-1}$. *Also, let* $\mathrm{VMO}_1(\mathbb{R}^{n-1})$ *denote the subspace of* $\mathrm{BMO}_1(\mathbb{R}^{n-1})$ *consisting of functions* ϕ *in* $L^1_{\mathrm{loc}}(\mathbb{R}^{n-1},\mathcal{L}^{n-1})$ *with* $\nabla'\phi \in \big[\mathrm{VMO}(\mathbb{R}^{n-1},\mathcal{L}^{n-1})\big]^{n-1}$.

Suppose $\Omega \subseteq \mathbb{R}^n$ *is a set with nonempty compact boundary, for which there exist a threshold* $\delta \in (0,\infty)$ *and a scale* $R \in (0, \operatorname{diam}\partial\Omega)$ *such that for each point* $x_o \in \partial\Omega$ *one may find a real-valued function* $\phi \in \mathrm{BMO}_1(\mathbb{R}^{n-1})$ *with* $\phi(0')=0$ *and*

$$\operatorname{dist}\Big(\nabla'\phi,\, \big[\mathrm{VMO}(\mathbb{R}^{n-1},\mathcal{L}^{n-1})\big]^{n-1}\Big) < \delta \tag{3.4.35}$$

(*where the distance is measured in the space* $\big[\mathrm{BMO}(\mathbb{R}^n,\mathcal{L}^{n-1})\big]^{n-1}$), *along with a rigid transformation* $T:\mathbb{R}^n \to \mathbb{R}^n$ *with the property that* $T(x_o)=0$ *and*

$$T\big(\Omega\cap B(x_o,R)\big) = \big\{(z',z_n)\in\mathbb{R}^{n-1}\times\mathbb{R} :\, z_n > \phi(z')\big\} \cap B(0,R). \tag{3.4.36}$$

Then work in [81] *shows that* Ω *is an Ahlfors regular domain (in fact, a* UR *domain; cf.* [129, (5.10.31)]) *with compact boundary, and the Ahlfors regularity constants of* $\partial\Omega$ *stay bounded away from zero and infinity as* $\delta \to 0^+$, *as long as* R *does not go to zero and* $\operatorname{diam}\partial\Omega$ *stays uniformly bounded. In addition, the global flatness radius* $\wp(\Omega)$ *does not degenerate (to zero) as* $\delta \to 0^+$.

Moreover, if $\sigma := \mathcal{H}^{n-1}\lfloor\partial\Omega$ *and* ν *denotes the geometric measure theoretic outward unit normal to* Ω, *from the proof of* [81, Proposition 2.27, pp. 2622–2623] *we conclude that that there exists $C \in (0, \infty)$ which depends only on the Ahlfors regularity constants of $\partial\Omega$ with the property that*

$$\operatorname{dist}\Big(\nu, \big[\mathrm{VMO}(\partial\Omega, \sigma)\big]^n\Big) \leq C\delta, \tag{3.4.37}$$

where the distance is measured in $\big[\mathrm{BMO}(\partial\Omega, \sigma)\big]^n$. As a consequence of this we conclude (cf. item (1) in Definition 3.4.1) that if $0 < \delta < C^{-1}$ then

$$\Omega \textit{ is a } (C\delta)\textit{-oscillating } \mathrm{AR} \textit{ domain.} \tag{3.4.38}$$

Call $\Omega \subseteq \mathbb{R}^n$ a VMO_1**-domain** *provided in* (3.4.22) *we may actually take ϕ in* $\mathrm{VMO}_1(\mathbb{R}^{n-1})$. *The proof of* [81, Proposition 2.27, pp. 2622–2623] *shows that*

$$\begin{array}{c}\nu \in \big[\mathrm{VMO}(\partial\Omega, \sigma)\big]^n \textit{ whenever } \Omega \subseteq \mathbb{R}^n \textit{ is} \\ \textit{a } \mathrm{VMO}_1\textit{-domain with compact boundary.}\end{array} \tag{3.4.39}$$

In particular, (cf. item (2) in Definition 3.4.1)

$$\begin{array}{c}\textit{any } \mathrm{VMO}_1\textit{-domain with compact boundary} \\ \textit{is an infinitesimally flat } \mathrm{AR} \textit{ domain.}\end{array} \tag{3.4.40}$$

As a corollary,

any Lipschitz domain $\Omega \subseteq \mathbb{R}^n$ with compact boundary which may be locally described as the upper-graph of Lipschitz functions $\phi : \mathbb{R}^{n-1} \to \mathbb{R}$ with $\nabla'\phi \in \big[\mathrm{VMO}(\mathbb{R}^{n-1}, \mathcal{L}^{n-1})\big]^{n-1}$ is an infinitesimally flat AR *domain.* (3.4.41)

This, in particular, implies that

$$\begin{array}{c}\textit{any } \mathscr{C}^1 \textit{ domain in } \mathbb{R}^n \textit{ with compact boundary} \\ \textit{is an infinitesimally flat } \mathrm{AR} \textit{ domain.}\end{array} \tag{3.4.42}$$

The next example illustrates the heuristic principle that, in the class of Reifenberg flat domains with compact boundaries, the "flatness" of the surface measure implies the flatness of the boundary itself.

Example 3.4.7 *Combining* [94, Theorem 2.1, p. 515] *and* [94, Remark 2.2, pp. 514-515] *with* [130, (3.1.53)] *shows that there exists a purely dimensional constant $\delta_n \in (0, 1)$ with the following property: Assume $0 < \delta_o \leq \delta_n$ and $R \in (0, \operatorname{diam} \partial\Omega)$. Let $\Omega \subseteq \mathbb{R}^n$ be a $(\delta_o, 2R)$-Reifenberg flat domain, in the sense of* [94, Definition 1.2, pp. 509-510], *with compact boundary. Suppose $\sigma := \mathcal{H}^{n-1}\lfloor\partial\Omega$ satisfies*

$$\begin{array}{c}\sigma\big(B(x, r) \cap \partial\Omega\big) \leq (1 + \delta_o)\upsilon_{n-1} r^{n-1} \\ \textit{for each } x \in \partial\Omega \textit{ and } r > 0,\end{array} \tag{3.4.43}$$

with υ_{n-1} denoting the volume of the unit ball in $\mathbb{R}^{n-1}$. Then Ω is an Ahlfors regular domain whose geometric measure theoretic outward unit normal ν satisfies

$$\operatorname{dist}\Big(\nu,\, \big[\mathrm{VMO}(\partial\Omega,\sigma)\big]^n\Big) < C\sqrt{\delta_o} \tag{3.4.44}$$

where the constant $C \in (0,\infty)$ depends only on the dimension n and the Ahlfors regularity constants of $\partial\Omega$. See also [30, p. 11] *and* [184] *in this regard. Consequently, given any number $\delta > 0$ it follows that any $(\delta_o, 2R)$-Reifenberg flat domain with $0 < \delta_o < \min\{\delta_n, (\delta/C)^2\}$ which satisfies* (3.4.43) *is a δ-oscillating* AR *domain.*

Next, we wish to note that, in the two-dimensional setting, the class of chord-arc domains with vanishing constant and compact boundary agrees with that of infinitesimally flat AR domains. A perturbation of this result is also discussed below.

Example 3.4.8 *Let $\Omega \subseteq \mathbb{R}^2$ be a chord-arc domain with compact boundary. Denote by $\ell(\cdot,\cdot)$ the shortest arc-length between points on $\partial\Omega$ and abbreviate $\sigma := \mathcal{H}^1\lfloor\partial\Omega$. Since Ω is known to be an Ahlfors regular domain (cf.* [129, (5.9.76)]*), it also makes sense to consider the geometric measure theoretic outward unit normal ν to Ω. Finally, introduce*

$$\varkappa := \lim_{R\to 0^+}\left\{\sup_{\substack{z_1,z_2\in\partial\Omega\\ |z_1-z_2|<R}}\left(\frac{\ell(z_1,z_2)}{|z_1-z_2|}-1\right)\right\} \in [0,+\infty). \tag{3.4.45}$$

In this context, Theorem 3.3.1 implies that, with the distance measured in the John-Nirenberg space $\big[\mathrm{BMO}(\partial\Omega,\sigma)\big]^2$, we have

$$\operatorname{dist}\Big(\nu,\, \big[\mathrm{VMO}(\partial\Omega,\sigma)\big]^2\Big) = O(\varkappa^{1/2}) \ \text{ as } \varkappa\to 0^+, \tag{3.4.46}$$

where the implicit constant depends exclusively on the upper Ahlfors regularity constant of $\partial\Omega$. In particular,

$$\Omega \text{ is a } \delta\text{-oscillating AR domain with } \delta = O(\varkappa^{1/2}) \text{ as } \varkappa\to 0^+. \tag{3.4.47}$$

Recall that a chord-arc domain $\Omega \subseteq \mathbb{C}$ is said to have `vanishing constant` *provided $\varkappa$ in* (3.4.45) *is zero. Then, as a corollary of the above results,*

any chord-arc domain with vanishing constant and compact boundary is an infinitesimally flat AR *domain in the plane.* (3.4.48)

Conversely, suppose $\Omega \subseteq \mathbb{R}^2$ is a δ-oscillating AR *domain, assumed to be simply connected when Ω is bounded and, respectively, connected when Ω is unbounded. Set $\sigma := \mathcal{H}^1\lfloor\partial\Omega$ and denote by ν the geometric measure theoretic outward unit normal to Ω. Then Theorem 3.3.1 guarantees that Ω is a chord-arc domain with*

$$\lim_{R\to 0^+}\left\{\sup_{\substack{z_1,z_2\in\partial\Omega\\ |z_1-z_2|<R}}\left(\frac{\ell(z_1,z_2)}{|z_1-z_2|}-1\right)\right\}=O\big(\delta\ln(1/\delta)\big)\ \textit{as}\ \delta\to 0^+. \tag{3.4.49}$$

In particular,

any bounded, simply connected, infinitesimally flat AR *domain in the plane is a chord-arc domain with vanishing constant.* (3.4.50)

Finally, from (3.4.48) *and* (3.4.50) *we conclude that*

the class of bounded, simply connected, infinitesimally flat AR *domains in the plane coincides with the class of bounded chord-arc domains with vanishing constant.* (3.4.51)

Let us also note that, as seen from (3.4.40) and (3.4.50),

any bounded, simply connected, VMO_1-domain domain in the plane is a chord-arc domain with vanishing constant. (3.4.52)

The class of regular Semmes-Kenig-Toro (SKT) domains, in the parlance of [81], is also compatible with the category of infinitesimally flat AR domains.

Example 3.4.9 *Recall the class of regular* SKT *domains from* [129, (5.11.93)]. *Based on this,* [129, (5.2.4), (5.11.26)], (A.0.1), *Definition 3.4.1, and Theorem 3.4.2 we see that the class of regular* SKT *domains agrees with the class of infinitesimally flat* AR *domains. Describing the class of regular* SKT *domains in terms of infinitesimally flat* AR *domains is more economical, as it avoids redundancies in the geometric assumptions stipulated in the definition of said class. Indeed, as seen from Theorem 3.4.2, the two-sided local John condition is an automatic attribute of infinitesimally flat* AR *domains.*

In [81], *the authors also consider the class of δ-regular* SKT *domains, with $\delta>0$, defined as the collection of open sets $\Omega\subset\mathbb{R}^n$ with a compact Ahlfors regular boundary, satisfying a two-sided local John condition, and whose geometric measure theoretic unit normal ν satisfies* $\mathrm{dist}\Big(\nu,\big[\mathrm{VMO}(\partial\Omega,\sigma)\big]^n\Big)<\delta$, *where* $\sigma:=\mathcal{H}^{n-1}\lfloor\partial\Omega$. *From* [129, (5.2.4), (5.11.26)], (A.0.1), *Definition 3.4.1, and Theorem 3.4.2 we then conclude that any δ-regular* SKT *domain is also a δ-oscillating* AR *domain and, in the converse direction, that a δ-oscillating* AR *domain $\Omega\subset\mathbb{R}^n$ with $\delta\in(0,1)$ sufficiently small relative to the dimension n and the Ahlfors regularity constants of $\partial\Omega$ is a δ-regular* SKT *domain.*

From (3.4.48), (3.4.50), the first property discussed in Example 3.4.9, and [80, Theorem 4.5] we see that

the class of bounded chord-arc domains with vanishing constant is invariant under $\mathscr{C}^1$-diffeomorphisms of the plane. (3.4.53)

In Proposition 3.4.11 below, we provide an example of a family of δ-oscillating AR domains which are not of upper-graph type locally, and whose diameters, Ahlfors regularity constants, relative flatness radii, and relative flatness ratios are all bounded away from zero and infinity, in a uniform fashion as $\delta \to 0^+$. We first recall a result from [111].

Lemma 3.4.10 *Let* $b \in \mathrm{BMO}(\mathbb{R}, \mathcal{L}^1)$ *real-valued with* $\|b\|_{\mathrm{BMO}(\mathbb{R},\mathcal{L}^1)} < 1$ *and define*

$$z : \mathbb{R} \to \mathbb{C}, \quad z(s) := \int_0^s \mathrm{e}^{ib(t)}\,\mathrm{d}t \text{ for each } s \in \mathbb{R}. \tag{3.4.54}$$

Then $\Sigma := z(\mathbb{R})$, *the image of* $\mathbb{R}$ *under the mapping* $z(\cdot)$, *is a locally rectifiable Jordan curve passing through infinity, which contains the origin* $0 \in \mathbb{C}$, *and the mapping* $\mathbb{R} \ni s \mapsto z(s) \in \Sigma$ *is its arc-length parametrization. In addition, the (positively oriented) unit tangent* τ *to* Σ *satisfies*

$$\|\tau\|_{\mathrm{BMO}(\Sigma,\mathcal{H}^1\lfloor\Sigma)} \leq \frac{4\|b\|_{\mathrm{BMO}(\mathbb{R},\mathcal{L}^1)}}{1 - \|b\|_{\mathrm{BMO}(\mathbb{R},\mathcal{L}^1)}}. \tag{3.4.55}$$

Fig. 3.3 A prototype of a bounded δ-oscillating AR domain $\Omega \subseteq \mathbb{R}^2$ for which $\delta > 0$ may be made as small as desired, relative to the Ahlfors regularity constants of $\partial\Omega$, the diameter of Ω, the relative flatness radius of Ω, and relative flatness ratio of Ω. The spiral point precludes describing $\partial\Omega$ as the graph of a function, even locally. For more details, see Proposition 3.4.11.

Proposition 3.4.11 *For each* $\varepsilon > 0$ *there is a bounded, simply connected, Ahlfors regular domain* $\Omega_\varepsilon \subseteq \mathbb{R}^2$ *whose boundary is a rectifiable Jordan curve, and so that:*

(1) The diameter and the Ahlfors regularity constants of each Ω_ε *stay bounded away from zero and infinity, in a uniform fashion with respect to* $\varepsilon > 0$. *Also, the relative flatness radius* $\wp(\Omega_\varepsilon)$ *(cf. Definition 3.1.9) and the relative flatness ratio* $\wp(\Omega_\varepsilon)/\operatorname{diam}\partial\Omega_\varepsilon$ *(cf.* (3.1.38)) *of each* Ω_ε *stay bounded away from zero and infinity, in a uniform fashion with respect to* $\varepsilon > 0$. *Furthermore, each* Ω_ε *is a two-sided* NTA *domain (hence, in particular,* Ω_ε *is a* UR *domain satisfying a two-sided local John condition) with all constants involved bounded away from zero and infinity, uniformly with respect to* $\varepsilon > 0$.

(2) Each set Ω_ε *is a* $\delta(\varepsilon)$*-oscillating* AR *domain (cf. Definition 3.4.1) with*

$$\lim_{\varepsilon \to 0^+} \delta(\varepsilon) = 0. \tag{3.4.56}$$

(3) Each Ω_ε *has a point on the boundary near which* Ω_ε *cannot be described as the upper-graph of any function and, away from said point,* $\partial\Omega_\varepsilon$ *is a* $\mathscr{C}^\infty$ *curve.*

Proof For each $\varepsilon \in (0,1)$ fixed, consider the map $F_\varepsilon : \mathbb{C} \to \mathbb{C}$ defined as

$$F_\varepsilon(z) := \begin{cases} (\mathrm{i}\varepsilon + 1)^{-1} z\, \mathrm{e}^{\mathrm{i}\varepsilon \ln|z|} & \text{if } z \in \mathbb{C}\setminus\{0\}, \\ 0 \in \mathbb{C} & \text{if } z = 0. \end{cases} \quad \text{for each } z \in \mathbb{C}. \tag{3.4.57}$$

Note that F_ε is a bijective, odd function, with inverse $F_\varepsilon^{-1} : \mathbb{C} \to \mathbb{C}$ given by

$$F_\varepsilon^{-1}(\zeta) = \begin{cases} (\mathrm{i}\varepsilon + 1)\zeta\, \mathrm{e}^{-\mathrm{i}\varepsilon \ln(|\zeta|\sqrt{\varepsilon^2+1})} & \text{if } \zeta \in \mathbb{C}\setminus\{0\}, \\ 0 \in \mathbb{C} & \text{if } \zeta = 0. \end{cases} \quad \text{for each } \zeta \in \mathbb{C}. \tag{3.4.58}$$

Also, whenever $z_1, z_2 \in \mathbb{C}$ are such that $|z_1| \geq |z_2| > 0$ we may estimate

$$|F_\varepsilon(z_1) - F_\varepsilon(z_2)| \leq \frac{1}{\sqrt{\varepsilon^2+1}}\Big\{|z_1 - z_2| + |z_2|\big|\mathrm{e}^{\mathrm{i}\varepsilon\ln|z_1|} - \mathrm{e}^{\mathrm{i}\varepsilon\ln|z_2|}\big|\Big\} \tag{3.4.59}$$

and

$$\begin{aligned} \big|\mathrm{e}^{\mathrm{i}\varepsilon\ln|z_1|} - \mathrm{e}^{\mathrm{i}\varepsilon\ln|z_2|}\big| &= \big|\mathrm{e}^{\mathrm{i}\varepsilon(\ln|z_1| - \ln|z_2|)} - 1\big| \leq \varepsilon\big|\ln|z_1| - \ln|z_2|\big| \\ &= \varepsilon\ln\Big(\tfrac{|z_1|}{|z_2|}\Big) \leq \varepsilon\Big(\tfrac{|z_1|}{|z_2|} - 1\Big) = \varepsilon\Big(\tfrac{|z_1| - |z_2|}{|z_2|}\Big) \leq \varepsilon\tfrac{|z_1 - z_2|}{|z_2|}, \end{aligned} \tag{3.4.60}$$

using the fact that $|\mathrm{e}^{\mathrm{i}\theta} - 1| \leq |\theta|$ for each $\theta \in \mathbb{R}$ and $0 \leq \ln x \leq x - 1$ for each $x \in [1, \infty)$. From this we then eventually deduce that

$$|F_\varepsilon(z_1) - F_\varepsilon(z_2)| \leq \frac{\varepsilon + 1}{\sqrt{\varepsilon^2+1}}|z_1 - z_2| \text{ for all } z_1, z_2 \in \mathbb{C}, \tag{3.4.61}$$

hence F_ε is Lipschitz, with Lipschitz constant bounded with respect to $\varepsilon \in (0,1)$. The same type of argument also shows that F_ε^{-1} is also Lipschitz with Lipschitz constant bounded with respect to $\varepsilon \in (0,1)$, namely

$$|F_\varepsilon^{-1}(\zeta_1) - F_\varepsilon^{-1}(\zeta_2)| \leq (\varepsilon + 1)\sqrt{\varepsilon^2+1}\,|\zeta_1 - \zeta_2| \text{ for all } \zeta_1, \zeta_2 \in \mathbb{C}, \tag{3.4.62}$$

so we ultimately conclude that $F_\varepsilon : \mathbb{C} \to \mathbb{C}$ is an odd bi-Lipschitz homeomorphism of the complex plane $\mathbb{C}$, with Lipschitz constants for itself and its inverse bounded with respect to $\varepsilon \in (0,1)$. In addition, it is also visible from (3.4.57) that $F_\varepsilon(0) = 0$, F_ε is of class $\mathscr{C}^\infty$ in $\mathbb{C}\setminus\{0\}$, and a direct computation shows that DF_ε, the Jacobian matrix of F_ε, satisfies $\det\big((DF_\varepsilon)(z)\big) = 1$ for each $z \in \mathbb{C}\setminus\{0\}$. In summary:

For each $\varepsilon \in (0,1)$, the map $F_\varepsilon : \mathbb{C} \to \mathbb{C}$ defined as in (3.4.57) is an odd bi-Lipschitz homeomorphism of the complex plane $\mathbb{C}$ (its inverse $F_\varepsilon^{-1} : \mathbb{C} \to \mathbb{C}$ is described in (3.4.58)), with Lipschitz constants for F_ε, F_ε^{-1} bounded with respect to $\varepsilon \in (0,1)$. In addition, the function F_ε is of class $\mathscr{C}^\infty$ in $\mathbb{C}\setminus\{0\}$, satisfies $F_\varepsilon(0) = 0$ as well as $\det(DF_\varepsilon) = 1$ in $\mathbb{C}\setminus\{0\}$ (hence F_ε^{-1} is also of class $\mathscr{C}^\infty$ in $\mathbb{C}\setminus\{0\}$), and all partial derivatives (of any order) of F_ε and F_ε^{-1} stay bounded on compact subsets of $\mathbb{C}\setminus\{0\}$ uniformly in $\varepsilon \in (0,1)$. (3.4.63)

Next, for each fixed $\varepsilon \in (0,1)$ consider the curve

$$\Sigma_\varepsilon := F_\varepsilon(\partial\mathbb{C}_+), \tag{3.4.64}$$

where $\partial\mathbb{C}_+ \equiv \mathbb{R}\times\{0\} \equiv \mathbb{R}$. Hence, Σ_ε is a locally rectifiable Jordan curve passing through infinity in the plane. The fact that F_ε is odd ensures that

$$0 \in \Sigma_\varepsilon \text{ and } \Sigma_\varepsilon \text{ is symmetric with respect to the origin.} \tag{3.4.65}$$

Moreover, a global parametrization

$$\mathbb{R} \ni s \longmapsto z(s) \in \Sigma_\varepsilon \tag{3.4.66}$$

of the curve Σ_ε is given by

$$z(s) := F_\varepsilon(s) = \begin{cases} (\mathrm{i}\varepsilon+1)^{-1} s\, \mathrm{e}^{\mathrm{i}\varepsilon \ln|s|} & \text{for } s \in \mathbb{R}\setminus\{0\}, \\ 0 \in \mathbb{C} & \text{for } s = 0. \end{cases} \tag{3.4.67}$$

Since $\frac{d}{dt}\Big[(\mathrm{i}\varepsilon+1)^{-1} t\, \mathrm{e}^{\mathrm{i}\varepsilon\ln|t|}\Big] = \mathrm{e}^{\mathrm{i}\varepsilon\ln|t|}$ for each $t \in \mathbb{R}\setminus\{0\}$, the Fundamental Theorem of Calculus permits us to equivalently express this function as

$$z(s) := \int_0^s \mathrm{e}^{\mathrm{i}\varepsilon\ln|t|}\, dt \text{ for each } s \in \mathbb{R}. \tag{3.4.68}$$

In particular,

$$z'(s) = \mathrm{e}^{\mathrm{i}\varepsilon\ln|s|} \text{ for each } s \in \mathbb{R}\setminus\{0\}, \tag{3.4.69}$$

$$\text{and } |z'(s)| = 1 \text{ for each } s \in \mathbb{R}\setminus\{0\}. \tag{3.4.70}$$

As such, (3.4.66) is actually the arc-length parametrization of the curve Σ_ε.

Denote by $\Sigma_\varepsilon^+ := z\big((0,+\infty)\big)$ and $\Sigma_\varepsilon^- := z\big((-\infty,0)\big)$ the two connected components of $\Sigma_\varepsilon\setminus\{0\}$. Then Σ_ε^+, Σ_ε^- are symmetric to one another with respect to the origin, and we claim that these are two spirals spinning about the point $z(0) = 0 \in \mathbb{C}$. To see this is the case, for each $s \in (0,\infty)$ represent $z(s) = r\mathrm{e}^{\mathrm{i}\theta}$ with $(r,\theta) \in [0,\infty)\times\mathbb{R}$. Specifically, with $\omega := 2\pi - \arccos\big(\frac{1}{\sqrt{\varepsilon^2+1}}\big)$, we may take

$$\theta = \omega + \varepsilon\ln|s| \text{ and } r = |z(s)| = (\varepsilon^2+1)^{-1/2}\,|s| = (\varepsilon^2+1)^{-1/2}\mathrm{e}^{(\theta-\omega)/\varepsilon}. \tag{3.4.71}$$

Thus,

$$\begin{aligned} &\text{the curve } \Sigma_\varepsilon^+ \text{ has the parametrization } \mathbb{R} \ni \theta \mapsto \alpha \mathrm{e}^{\beta\theta}\mathrm{e}^{\mathrm{i}\theta} \text{ with} \\ &\alpha := (\varepsilon^2+1)^{-1/2}\mathrm{e}^{-\omega/\varepsilon} \in (0,\infty) \text{ and } \beta := \varepsilon^{-1} \in (0,\infty), \end{aligned} \tag{3.4.72}$$

which identifies Σ_ε^+ precisely as a logarithmic spiral. In a similar fashion,

$$\text{the curve } \Sigma_\varepsilon^- \text{ has the parametrization } \mathbb{R} \ni \theta \mapsto \alpha e^{\beta\theta} e^{i\theta} \text{ with} \\ \alpha := (\varepsilon^2+1)^{-1/2} e^{-(\omega+\pi)/\varepsilon} \in (0,\infty) \text{ and } \beta := \varepsilon^{-1} \in (0,\infty), \tag{3.4.73}$$

which once again identifies it as a logarithmic spiral.

To summarize, in (3.4.67) the factor $(i\varepsilon+1)^{-1}$ is merely a complex constant, s is the scaling factor that determines how far $z(s)$ is from the origin (specifically, $|z(s)| = |s|/\sqrt{\varepsilon^2+1}$), and $e^{i\varepsilon \ln|s|}$ is the factor that determines how the two spirals $\Sigma_\varepsilon^\pm$ (making up $\Sigma_\varepsilon \setminus \{0\}$) spin about the point $0 \in \mathbb{C}$. Note that $|z(s)|$ growths linearly (with respect to s) which is very fast compared to the spinning rate (which is logarithmic). Henceforth, make the assumption that

$$0 < \varepsilon < \tfrac{1}{2}\big\| \ln|\cdot| \big\|^{-1}_{\mathrm{BMO}(\mathbb{R},\mathcal{L}^1)}. \tag{3.4.74}$$

For example, the computation on [74, p. 520] shows that

$$\big\| \ln|\cdot| \big\|_{\mathrm{BMO}(\mathbb{R},\mathcal{L}^1)} \leq 3\ln(3/2), \tag{3.4.75}$$

so taking

$$0 < \varepsilon < 2^{-1}[3\ln(3/2)]^{-1} \approx 0.411, \tag{3.4.76}$$

will do, as far as (3.4.74) is concerned. Then $b : \mathbb{R} \to \mathbb{R}$ defined $\mathcal{L}^1$-a.e. as

$$b(s) := \varepsilon \ln|s| \text{ for each } s \in \mathbb{R} \setminus \{0\}, \tag{3.4.77}$$

is real-valued and, thanks to (3.4.74), satisfies

$$\|b\|_{\mathrm{BMO}(\mathbb{R},\mathcal{L}^1)} < \tfrac{1}{2}. \tag{3.4.78}$$

From (3.4.69) we also see that the positively oriented unit tangent vector τ to Σ_ε is

$$\tau(z(s)) = z'(s) = e^{ib(s)} \text{ for } \mathcal{L}^1\text{-a.e. } s \in \mathbb{R}. \tag{3.4.79}$$

With this in hand, Lemma 3.4.10 gives

$$\|\tau\|_{\mathrm{BMO}(\Sigma_\varepsilon,\mathcal{H}^1\lfloor\Sigma_\varepsilon)} \leq \frac{4\|b\|_{\mathrm{BMO}(\mathbb{R},\mathcal{L}^1)}}{1-\|b\|_{\mathrm{BMO}(\mathbb{R},\mathcal{L}^1)}} \tag{3.4.80}$$

which, in view of (3.4.74) and (3.4.75), entails

$$\|\tau\|_{\mathrm{BMO}(\Sigma_\varepsilon,\mathcal{H}^1\lfloor\Sigma_\varepsilon)} \leq 8\varepsilon\big\| \ln|\cdot| \big\|_{\mathrm{BMO}(\mathbb{R},\mathcal{L}^1)} < 24\varepsilon\ln(3/2). \tag{3.4.81}$$

Next, consider a bounded simply connected $\mathscr{C}^\infty$ domain $D \subseteq \mathbb{R}^2_+ \cap B(0,10)$ with

$$0 \in \partial D \text{ and } B(0,1) \cap D = B(0,1) \cap \mathbb{R}^2_+. \tag{3.4.82}$$

For each ε as in (3.4.74) define

$$\Omega_\varepsilon := F_\varepsilon(D). \tag{3.4.83}$$

Then Ω_ε is an open nonempty bounded simply connected subset of $\mathbb{C} \equiv \mathbb{R}^2$, with $0 \in \partial\Omega_\varepsilon = F_\varepsilon(\partial D)$, and there exists a universal constant $C \in (1, \infty)$ such that

$$C^{-1} \leq \operatorname{diam} \partial\Omega_\varepsilon \leq C \ \text{ for each } \varepsilon \in (0,1). \tag{3.4.84}$$

We also claim that

$$B(0, 2^{-3/2}) \cap \partial\Omega_\varepsilon = B(0, 2^{-3/2}) \cap \Sigma_\varepsilon. \tag{3.4.85}$$

Indeed, (3.4.62) implies that $F_\varepsilon^{-1}\big(B(0, 2^{-3/2})\big) \subseteq B(0,1)$ (recall that $\varepsilon \in (0,1)$), hence $B(0, 2^{-3/2}) \subseteq F_\varepsilon\big(B(0,1)\big)$ and (3.4.85) follows from this and (3.4.82).

In addition, from (3.4.63), (3.4.83), and the transformational properties of classes of domains under bi-Lipschitz maps established in [80] we conclude that

$$\begin{array}{l}\text{the set } \Omega_\varepsilon \text{ is a UR domain (in particular, an Ahlfors regular} \\ \text{domain in } \mathbb{R}^2) \text{ with constants bounded with respect to } \varepsilon.\end{array} \tag{3.4.86}$$

For similar reasons, each Ω_ε is a two-sided NTA domain (in particular, each Ω_ε is a UR domain satisfying a two-sided local John condition) with all constants involved bounded away from zero and infinity, uniformly with respect to $\varepsilon > 0$. Also, having F_ε is a local diffeomorphism in $\mathbb{C} \setminus \{0\}$ (as we do; cf. (3.4.63)) guarantees that Ω_ε is a $\mathscr{C}^\infty$ domain near each boundary point in $\partial\Omega_\varepsilon \setminus \{0\}$. If ν denotes the outward unit normal to Ω_ε and if $\sigma := \mathcal{H}^1 \lfloor \partial\Omega_\varepsilon$, then (3.4.85) implies $\nu = -i\tau$ at each point on $B(0, 2^{-3/2}) \cap \big(\partial\Omega_\varepsilon \setminus \{0\}\big) = B(0, 2^{-3/2}) \cap \big(\Sigma_\varepsilon \setminus \{0\}\big)$. As a consequence of this and (3.4.81), for each

$$z_o \in B(0, 2^{-5/2}) \cap \partial\Omega_\varepsilon = B(0, 2^{-5/2}) \cap \Sigma_\varepsilon \ \text{ and } \ r \in (0, 2^{-5/2}) \tag{3.4.87}$$

we have $B(z_o, r) \subseteq B(0, 2^{-3/2})$ hence

$$\begin{aligned} &\fint_{B(z_o,r)\cap\partial\Omega_\varepsilon} \Big| \nu - \fint_{B(z_o,r)\cap\partial\Omega_\varepsilon} \nu \, d\sigma \Big| d\sigma = \fint_{B(z_o,r)\cap\Sigma_\varepsilon} \Big| \tau - \fint_{B(z_o,r)\cap\Sigma_\varepsilon} \tau \, d\mathcal{H}^1 \Big| d\mathcal{H}^1 \\ &\qquad \leq \|\tau\|_{\mathrm{BMO}(\Sigma_\varepsilon, \mathcal{H}^1 \lfloor \Sigma_\varepsilon)} < 24\varepsilon \ln(3/2). \end{aligned} \tag{3.4.88}$$

Next, the properties listed in (3.4.63) make ν a Lipschitz function on the set $\partial\Omega_\varepsilon \setminus B(0, 2^{-7/2})$, in a uniform fashion with respect to ε. Specifically, there exists some universal constant $\varkappa \in (0, \infty)$ (in particular, independent of ε) such that

$$|\nu(z_1) - \nu(z_2)| \leq \varkappa |z_1 - z_2| \ \text{ for all } z_1, z_2 \in \partial\Omega_\varepsilon \setminus B(0, 2^{-7/2}). \tag{3.4.89}$$

Consequently, whenever

$$z_o \in \partial\Omega_\varepsilon \setminus B(0, 2^{-5/2}) \ \text{ and } \ r \in (0, 2^{-7/2}) \tag{3.4.90}$$

it follows that $B(z_o, r) \cap \partial\Omega_\varepsilon \subseteq \partial\Omega_\varepsilon \setminus B(0, 2^{-7/2})$ so we may estimate

$$\begin{aligned}&\fint_{B(z_o,r)\cap\partial\Omega_\varepsilon}\Big|\nu-\fint_{B(z_o,r)\cap\partial\Omega_\varepsilon}\nu\,d\sigma\Big|\,d\sigma \\ &\quad\le\fint_{B(z_o,r)\cap\partial\Omega_\varepsilon}\fint_{B(z_o,r)\cap\partial\Omega_\varepsilon}\big|\nu(z_1)-\nu(z_2)\big|\,d\sigma(z_1)\,d\sigma(z_2)\le 2\varkappa\cdot r.\end{aligned}\tag{3.4.91}$$

Next, Corollary 3.1.4 shows that the threshold ε_3 appearing in (3.1.47) depends exclusively on the Ahlfors regularity constants of the underlying domain. By (3.4.86), the Ahlfors regularity constants of Ω_ε are presently controlled in a uniform fashion with respect to ε. Hence ε_3 associated with the set Ω_ε may be taken to be independent of ε, say $\varepsilon_3\in[a,b]$ with $0<a<b<1$ universal constants. Henceforth restrict to

$$24\varepsilon\ln(3/2)<a.\tag{3.4.92}$$

Then (3.1.47) and (3.4.87)-(3.4.88) imply

$$R_{\rm flat}(z_o)\ge 2^{-5/2}\ \text{ for each }\ z_o\in B(0,2^{-5/2})\cap\partial\Omega_\varepsilon.\tag{3.4.93}$$

Let us also observe from (3.1.47) and (3.4.90)-(3.4.91) that the pointwise flatness radius may be estimated as

$$R_{\rm flat}(z_o)\ge\min\big\{2^{-7/2},a/(2\varkappa)\big\}\ \text{ for each }\ z_o\in\partial\Omega_\varepsilon\setminus B(0,2^{-5/2}).\tag{3.4.94}$$

Together, (3.4.93), (3.4.94), (3.1.48), (3.1.37), and (3.4.84) prove that

$$\begin{aligned}&\text{whenever }\varepsilon\in\big(0,a/(24\ln(3/2))\big)\text{ we have}\\ &C\ge\wp(\Omega_\varepsilon)=\inf_{z_o\in\partial\Omega_\varepsilon}R_{\rm flat}(z_o)\ge\min\big\{2^{-7/2},a/(2\varkappa)\big\},\end{aligned}\tag{3.4.95}$$

with constants independent of ε. Recall from [130, (3.1.53)] and (3.4.86) that

$$\begin{aligned}&{\rm dist}\Big(\nu,\ \big[{\rm VMO}\,(\partial\Omega_\varepsilon,\sigma)\big]^2\Big)\\ &\quad\approx\lim_{R\to0^+}\Big(\sup_{z_o\in\partial\Omega_\varepsilon,\,r\in(0,R)}\fint_{B(z_o,r)\cap\partial\Omega_\varepsilon}\Big|\nu-\fint_{B(z_o,r)\cap\partial\Omega_\varepsilon}\nu\,d\sigma\Big|\,d\sigma\Big),\end{aligned}\tag{3.4.96}$$

where the implicit proportionality constants are universal (in particular, independent of ε). In turn, from (3.4.96), (3.4.87)-(3.4.88), and (3.4.90)-(3.4.91) we conclude that for each ε as in (3.4.74) we have, for some universal constant $C\in(0,\infty)$,

$${\rm dist}\Big(\nu,\ \big[{\rm VMO}(\partial\Omega_\varepsilon,\sigma)\big]^2\Big)\le C\varepsilon.\tag{3.4.97}$$

Hence, each set Ω_ε is a $\delta(\varepsilon)$-oscillating AR domain (in the sense of item *(1)* in Definition 3.4.1) with $\delta(\varepsilon)=O(\varepsilon)$ as $\varepsilon\to0^+$ (so, in particular, (3.4.56) holds).

Finally, we remark that near $0\in\partial\Omega_\varepsilon$ one cannot describe Ω_ε as the upper-graph of any function (as the Vertical Line Test is violated at this spiral point). □

Chapter 4
Norm Estimates and Invertibility Results for SIO's on Unbounded Boundaries

One of the main goals in this chapter is to estimate singular integral operators of "chord-dot-normal" type, associated with Ahlfors regular domains $\Omega \subseteq \mathbb{R}^n$ with unbounded boundaries, in a manner which makes it clear how the flatness of said domains affects the size of these operators on various function spaces. A fitting way to quantify flatness in this setting is via the BMO semi-norm of the geometric measure theoretic outward unit normal ν to Ω. In broader terms, this is part of a central theme in Geometric Measure Theory concerned with how geometric properties translate into analytical ones. The challenge here is understanding the implications of demanding that Gauss' map

$$\partial_*\Omega \ni x \longmapsto \nu(x) \in S^{n-1} \tag{4.0.1}$$

has small BMO semi-norm, in the realm of singular integral operators. In turn, this paves the way for solving boundary value problems in such general sets Ω via boundary layer potentials, something we shall do later on.

The reason we shall focus on singular integral operators of "chord-dot-normal" type is that they are *sensitive to flatness*, in the sense that they all become identically zero when the underlying domain Ω is a half-space in $\mathbb{R}^n$. By way of contrast, SIO's like the Cauchy singular integral operator [132, (5.1.135)] do not. The difference maker is the algebraic structure of the integral kernel, manifested through the presence, or absence, of the inner product between the geometric measure theoretic outward unit normal ν and the "chord" $x - y$. In fact, this is not an accident, since here we shall show (see Theorem 4.1.10) that this is the only algebraic template (for SIO's whose integral kernel depends "linearly" on ν) guaranteeing "sensitivity to flatness" as interpreted above.

In order to be more specific, let us bring in some notation. Having fixed a UR domain $\Omega \subseteq \mathbb{R}^n$, abbreviate $\sigma := \mathcal{H}^{n-1}\lfloor\partial\Omega$ and denote by ν the geometric measure theoretic outward unit normal to Ω. For a sufficiently large integer $N = N(n) \in \mathbb{N}$, consider a complex-valued function $k \in \mathscr{C}^N(\mathbb{R}^n \setminus \{0\})$ which is even and positive homogeneous of degree $-n$. In such a setting, we shall define the principal-value singular integral operator of "chord-dot-normal" type acting on each given function $f \in L^1\big(\partial\Omega, \frac{\sigma(x)}{1+|x|^{n-1}}\big)$ at σ-a.e. point $x \in \partial\Omega$ according to

D. Mitrea et al., *Geometric Harmonic Analysis V*, Developments in Mathematics 76,
https://doi.org/10.1007/978-3-031-31561-9_4

$$Tf(x) := \lim_{\varepsilon\to 0^+} \int_{y\in\partial\Omega,\ |x-y|>\varepsilon} \langle x-y, \nu(y)\rangle k(x-y) f(y)\, d\sigma(y). \tag{4.0.2}$$

Then for each $p \in (1,\infty)$ and each $m \in \mathbb{N}$ there exists some $C_m \in (0,\infty)$, which depends only on m, n, p, and the UR constants of $\partial\Omega$ such that, with the piece of notation introduced in (4.1.12), we have

$$\|T\|_{L^p(\partial\Omega,\sigma)\to L^p(\partial\Omega,\sigma)} \le C_m\Big(\sum_{|\alpha|\le N} \sup_{S^{n-1}} |\partial^\alpha k|\Big)\|\nu\|^{\langle m\rangle}_{[\mathrm{BMO}(\partial\Omega,\sigma)]^n}. \tag{4.0.3}$$

See Theorem 4.1.3 for this and other results of this flavor. Here we only wish to note that estimate (4.0.3) is decidedly finer than the operator norm estimates from [131, Theorem 2.3.2]. Specifically, the boundedness result in [131, (2.3.18)] (presently invoked with $\Sigma := \partial\Omega$) gives that for T as in (4.0.2) we have

$$\|T\|_{L^p(\partial\Omega,\sigma)\to L^p(\partial\Omega,\sigma)} \le C(\partial\Omega, p)\big\|k\big|_{S^{n-1}}\big\|_{\mathscr{C}^N(S^{n-1})} \tag{4.0.4}$$

where $C(\partial\Omega, p) \in (0,\infty)$ depends on $\partial\Omega$ solely through its UR constants. In stark contrast to this, the estimate in (4.0.3) features in the right-hand side $\|\nu\|^{\langle m\rangle}_{[\mathrm{BMO}(\partial\Omega,\sigma)]^n}$ as a multiplicative factor, something which the UR constants of $\partial\Omega$ cannot control.

Indeed, [131, Theorem 2.3.2] has no provisions allowing one to take advantage of the specific algebraic format of the fact that we are dealing with a singular integral operator whose kernel is of "chord-dot-normal" type, i.e., $\langle x-y, \nu(y)\rangle k(x-y)$. For [131, Theorem 2.3.2] to apply, this integral kernel needs to be first disassembled into its most rudimentary building blocks, i.e., as $\sum_{j=1}^n k_j(x-y)\nu_j(y)$ where we have set $k_j(z) := z_j k(z)$ for each $z \in \mathbb{R}^n \setminus \{0\}$ and $j \in \{1,\dots,n\}$. Since multiplication by ν_j may be absorbed with the function f (without changing its membership, or increasing its size, in the Lebesgue space $L^p(\partial\Omega,\sigma)$), [131, Theorem 2.3.2] may then finally be invoked in relation to each SIO associated with the kernel k_j.

For example, while estimate (4.0.4) applied to the classical boundary-to-boundary harmonic double layer potential operator K_Δ from (A.0.114) in the case when the domain $\Omega := \mathbb{R}^n_+$, the upper half-space, only gives

$$\|K_\Delta\|_{L^p(\mathbb{R}^{n-1},\mathcal{L}^{n-1})\to L^p(\mathbb{R}^{n-1},\mathcal{L}^{n-1})} < \infty, \tag{4.0.5}$$

the estimate in (4.0.3) yields

$$\|K_\Delta\|_{L^p(\mathbb{R}^{n-1},\mathcal{L}^{n-1})\to L^p(\mathbb{R}^{n-1},\mathcal{L}^{n-1})} = 0 \tag{4.0.6}$$

given that the BMO-seminorm of the outward unit normal to $\mathbb{R}^n_+$ vanishes. This is precisely as it should be since, as is apparent from the algebraic format of its integral kernel, the operator K_Δ is identically zero when considered on $\partial\mathbb{R}^n_+$.

In summary, *there are much better estimates for singular integral operators of "chord-dot-normal" type on the boundaries of* UR *domains than for generic convolution type singular integral operators*. Establishing such estimates is a vastly intricate affair but, as noted above, the rewards are very palpable. The details, in the context of Muckenhoupt weighted Lebesgue spaces may be found in [111].

Here we use (4.0.3) as a stepping stone for proving similar estimates in other important function spaces, such as the "three-BMO" estimate from Theorem 4.1.8, to the effect that, for each $m \in \mathbb{N}$,

$$\big\|[T_{\rm mod}]\big\|_{\mathrm{BMO}(\partial\Omega,\sigma)/\sim\to\mathrm{BMO}(\partial\Omega,\sigma)/\sim} \leq C_m\Big(\sum_{|\alpha|\leq N}\sup_{S^{n-1}}|\partial^\alpha k|\Big)\|\nu\|^{\langle m\rangle}_{[\mathrm{BMO}(\partial\Omega,\sigma)]^n}, \tag{4.0.7}$$

where $T_{\rm mod}$ is the "modified" version of T from (4.0.2), and in the context of Hardy spaces, as done in Theorem 4.1.7, where we show that for each $p \in \big(\frac{n-1}{n}, 1\big]$ we have

$$\|T^\#\|_{H^p(\partial\Omega,\sigma)\to H^p(\partial\Omega,\sigma)} \leq C_m\Big(\sum_{|\alpha|\leq N}\sup_{S^{n-1}}|\partial^\alpha k|\Big)\|\nu\|^{\langle m\rangle}_{[\mathrm{BMO}(\partial\Omega,\sigma)]^n} \tag{4.0.8}$$

where $T^\#$ is the formal transpose of T from (4.0.2). We also establish analogous estimates on CMO, Hölder spaces, Muckenhoupt weighted Sobolev spaces, as well as homogeneous Sobolev spaces that are Lebesgue-based, Morrey-based, or block-based. Subsequently, we consider closely related issues for singular integral operators of double layer type, associated with UR domains and a given weakly elliptic, homogeneous, constant (complex) coefficient, second-order system L in $\mathbb{R}^n$. In this regard, we prove that estimates similar in spirit to those described above hold if and only if the double layer potential operator in question is associated with a distinguished coefficient tensor for the system L. This describes the contents of §4.1.

Moving on, in §4.2 we make use of the estimates in §4.1 in order to obtain invertibility results for singular integral operators of "chord-dot-normal" type in the aforementioned scales of spaces considered on the boundary of a UR domain $\Omega \subseteq \mathbb{R}^n$, under the assumption that $\|\nu\|_{[\mathrm{BMO}(\partial\Omega,\sigma)]^n}$ is sufficiently small (where ν is the geometric measure theoretic outward unit normal to Ω, and σ is the "surface measure" on $\partial\Omega$). In this vein, it is worth observing that having $\|\nu\|_{[\mathrm{BMO}(\partial\Omega,\sigma)]^n} < 1$ forces $\partial\Omega$ to be unbounded. The idea is that having $\|\nu\|_{[\mathrm{BMO}(\partial\Omega,\sigma)]^n}$ small implies, in light of (4.0.3), (4.0.7), (4.0.8) and similar estimates on other scales of spaces, that the singular integral operator of "chord-dot-normal" type in question has a correspondingly small norm when acting on said spaces. For a weakly elliptic, homogeneous, constant (complex) coefficient, second-order system L in $\mathbb{R}^n$ which has a distinguished coefficient tensor A, we may then conclude that the double layer potentials K_A, $K^\#_A$, associated with A in UR domains, have small operator norms. This permits us to invert $\pm\frac{1}{2}I + K_A$ and $\pm\frac{1}{2}I + K^\#_A$ (which are relevant in the context of boundary value problems for the system L) using a Neumann series argument.

Let us shed further light on the latter issue, by contemplating the task of solving the Dirichlet Problem in a UR domain $\Omega \subseteq \mathbb{R}^n$ for some given (homogeneous, constant complex coefficient) second-order weakly elliptic $M \times M$ system L in $\mathbb{R}^n$. To set the stage, bring in the surface measure $\sigma := \mathcal{H}^{n-1}\lfloor\partial\Omega$, and fix an integrability exponent $p \in (1,\infty)$ along with some aperture parameter $\kappa \in (0,\infty)$. The task at hand is finding a function u satisfying the following conditions:

$$\begin{aligned} u \in \big[\mathscr{C}^\infty(\Omega)\big]^M, \quad Lu = 0 \text{ in } \Omega, \quad \mathcal{N}_\kappa u \in L^p(\partial\Omega, \sigma), \\ u\Big|_{\partial\Omega}^{\kappa-\text{n.t.}} = f \text{ at } \sigma\text{-a.e. point on } \partial\Omega, \end{aligned} \tag{4.0.9}$$

where the boundary datum f is arbitrarily prescribed in $\big[L^p(\partial\Omega,\sigma)\big]^M$. Since any boundary-to-domain double layer for L is a mechanism for creating a multitude on null-solutions for the system L in Ω, a natural candidate is to consider the function $u := \mathcal{D}_A g$ for some coefficient tensor $A \in \mathfrak{A}_L$ and a density $g \in \big[L^p(\partial\Omega,\sigma)\big]^M$ yet to be determined. This has the distinct advantage of ensuring the membership of $\mathcal{N}_\kappa u$ to $L^p(\partial\Omega,\sigma)$, regardless of the choice of g. Moreover, the nontangential boundary trace of u may be explicitly computed as

$$u\Big|_{\partial\Omega}^{\kappa-\text{n.t.}} = \big(\tfrac{1}{2}I + K_A\big)g \text{ at } \sigma\text{-a.e. point on } \partial\Omega, \tag{4.0.10}$$

where I is the identity and K_A is the boundary-to-boundary double layer potential operator associated with A and Ω. As such, when faced with the task of finding a function as in (4.0.9) one is lead to solving the boundary integral equation

$$\big(\tfrac{1}{2}I + K_A\big)g = f. \tag{4.0.11}$$

A natural approach for accomplishing this task is to attempt making sense of

$$g = \big(\tfrac{1}{2}I + K_A\big)^{-1} f = 2\sum_{j=0}^{\infty} (-2K_A)^j f. \tag{4.0.12}$$

The first person to succeed in showing that the above series converges in the case when $L = \Delta$, the Laplacian in $\mathbb{R}^n$ and when the coefficient tensor A is simply the $n \times n$ identity matrix (in which scenario we shall write K_Δ in place of K_A), for each given $f \in L^2(\partial\Omega,\sigma)$, was Carl Gottfried Neumann in 1861[1] under the assumption that the domain Ω is convex, and two-dimensional. What is special about K_Δ associated with a convex domain is the fact that its integral kernel is non-negative, a feature fully exploited in Neumann's approach (presented in [168]). Subsequently, it became a priority to clarify whether the convexity assumption is really necessary in order to arrive at the same conclusion. In a genuine tour de force, Henri Poincaré succeeded (in a long and technical paper published in Acta Mathematica in 1897; cf. [173]) to replace the convexity hypothesis by suitable smoothness assumptions. More specifically, Poincaré accomplished the task of finding an alternative proof of the convergence of the Neumann series for nonconvex domains, via an approach requiring the domain Ω to be of class $\mathscr{C}^2$ and the density f to be fairly regular.

A sufficient condition guaranteeing the convergence of the Neumann series (4.0.12) in the space $L^2(\partial\Omega,\sigma)$ (when A is the identity) is that

[1] the series expressing $\big(\tfrac{1}{2}I + K_\Delta\big)^{-1}$ has eventually become known as a Neumann series

$$\|K_\Delta\|_{L^2(\partial\Omega,\sigma)\to L^2(\partial\Omega,\sigma)} < \tfrac{1}{2}. \tag{4.0.13}$$

Note that this necessarily forces $\partial\Omega$ to be unbounded. Indeed, if $\partial\Omega$ is bounded then $K_\Delta 1 = \frac{1}{2}$ forcing $\|K_\Delta\|_{L^2(\partial\Omega,\sigma)\to L^2(\partial\Omega,\sigma)} \geq \frac{1}{2}$ in this case. The demand in (4.0.13) is not unreasonable since, as noted earlier, $K_\Delta = 0$ if Ω is a half-space. These considerations point to the fact that the underlying set Ω needs to have an unbounded and sufficiently flat boundary for (4.0.13) to have a chance of materializing.

This brings us to one of the central results in this chapter. Whenever $A \in \mathfrak{A}_L^{\rm dis}$ it follows that K_A is a singular integral operator of "chord-dot-normal" type for which (4.0.3) applies. Hence, for each $m \in \mathbb{N}$ there exists a constant $C_m \in (0, \infty)$, which depends only on m, n, p, and the UR constants of $\partial\Omega$, such that

$$\|K_A\|_{[L^p(\partial\Omega,\sigma)]^M \to [L^p(\partial\Omega,\sigma)]^M} \leq C_m \|\nu\|_{[\mathrm{BMO}(\partial\Omega,\sigma)]^n}^{\langle m\rangle}. \tag{4.0.14}$$

The number $\|\nu\|_{[\mathrm{BMO}(\partial\Omega,\sigma)]^n}$ should be regarded as a way of quantifying the flatness of $\partial\Omega$, a point of view which confers (4.0.14) the following interpretation:

the flatter the boundary, the smaller the (operator) norm of any boundary-to-boundary double layer associated with a distinguished coefficient tensor for the given elliptic system. (4.0.15)

In particular, if $\|\nu\|_{[\mathrm{BMO}(\partial\Omega,\sigma)]^n}$ is sufficiently small then (4.0.13) holds, thus opening the door for solving the Dirichlet Problem for the Laplacian along the lines originally intended by C. Neumann and H. Poincaré more than 150 years ago.

Ultimately, our work here solidifies the connections between the regularity of a domain (measured in terms of the mean oscillations of its geometric measure theoretic outward unit normal), and singular integrals and boundary value problems for elliptic partial differential equations considered in said domain.

We close with some brief comments on the origins of integral equations. They go at least as far back as the work of Niels Henrick Abel who, in 1826, first reduced a problem in mechanics[2] to an integral equation. Toward the end of the 19th century, Vito Volterra considered one-dimensional integral equations with a fixed lower limit and a variable upper limit of integration. Such Volterra integral equations have been subsequently studied by Traian Lalescu[3] in his 1908 thesis[4], written under the direction of Èmile Picard. Lalescu then went on to write the very first book on integral equations in 1912, titled *Introduction à la théorie des équations intégrales* (see [107]). Other pioneers in the development of the theory of integral equations include Ivar Fredholm (1866-1943), David Hilbert (1862-1943), and Erhard Schmidt (1876-1959). Through the work of these and many other people, the theory of integral equations has grown into a broad, multifaceted area of mathematics, which has come to subsume boundary integral equations in potential theory, such as (4.0.11).

---

[2] specifically, the problem of finding the path of descent of a particle along a smooth curve in a vertical plane under the action of gravity in an interval of time

[3] Trajan Lalesco, in the French spelling of the epoch

[4] titled "*Sur les équations de Volterra*"

4.1 Norm Estimates for Chord-Dot-Normal SIO's on Unbounded Boundaries

We set the stage by reviewing a couple of results from [111]. First, the following lemma is a consequence of the De Giorgi-Federer version of the Divergence Formula from [129, Theorem 1.1.1].

Lemma 4.1.1 *Let $\Omega \subseteq \mathbb{R}^n$ be an Ahlfors regular domain. Abbreviate $\sigma := \mathcal{H}^{n-1}\lfloor\partial\Omega$ and denote by ν the geometric measure theoretic outward unit normal to Ω. Then for each surface ball $\Delta \subseteq \partial\Omega$ one has*

$$
\begin{aligned}
&\fint_\Delta \Big|\nu - \fint_\Delta \nu\, d\sigma\Big|^2 d\sigma = 1 - \Big|\fint_\Delta \nu\, d\sigma\Big|^2 \quad \textit{and} \\
&\Big(1 - \Big|\fint_\Delta \nu\, d\sigma\Big|\Big)^2 \leq \fint_\Delta \Big|\nu - \fint_\Delta \nu\, d\sigma\Big|^2 d\sigma \leq 2\Big(1 - \Big|\fint_\Delta \nu\, d\sigma\Big|\Big), \\
&0 \leq 1 - \Big|\fint_\Delta \nu\, d\sigma\Big| \leq \fint_\Delta \Big|\nu - \fint_\Delta \nu\, d\sigma\Big| d\sigma \leq \sqrt{2}\sqrt{1 - \Big|\fint_\Delta \nu\, d\sigma\Big|}.
\end{aligned}
\tag{4.1.1}
$$

Moreover,

$$
\|\nu\|_{[\mathrm{BMO}(\partial\Omega,\sigma)]^n} \leq 1, \tag{4.1.2}
$$

and

$$
1 - \inf_{\Delta\subseteq\partial\Omega} \Big|\fint_\Delta \nu\, d\sigma\Big| \leq \|\nu\|_{[\mathrm{BMO}(\partial\Omega,\sigma)]^n} \leq \sqrt{2}\sqrt{1 - \inf_{\Delta\subseteq\partial\Omega} \Big|\fint_\Delta \nu\, d\sigma\Big|}, \tag{4.1.3}
$$

where the two infima are taken over all surface balls $\Delta \subseteq \partial\Omega$. Also,

$$
\textit{if } \partial\Omega \textit{ is bounded then } \|\nu\|_{[\mathrm{BMO}(\partial\Omega,\sigma)]^n} = 1, \ \textit{so} \tag{4.1.4}
$$

$$
\partial\Omega \textit{ is unbounded whenever } \|\nu\|_{[\mathrm{BMO}(\partial\Omega,\sigma)]^n} < 1. \tag{4.1.5}
$$

In relation to (4.1.4) note that, in the class of Ahlfors regular domains, having the BMO semi-norm of its geometric measure theoretic outward unit normal precisely 1 is not an exclusive attribute of bounded domains. For example, a straightforward computation shows this is the case for an infinite strip in $\mathbb{R}^n$ (i.e., the region in between two parallel hyperplanes in $\mathbb{R}^n$), which is an unbounded Ahlfors regular domain. The same is true for an infinite half-strip. Another situation when this phenomenon occurs is described below.

Remark 4.1.2 *We note that*

the BMO *semi-norm of the geometric measure theoretic outward unit normal vector of an Ahlfors regular domain with a cusp is equal to* 1. (4.1.6)

To illustrate this, suppose $\Omega \subseteq \mathbb{R}^2$ is an Ahlfors regular domain with $0 \in \partial\Omega$ and such that near the origin Ω looks like the region above the graph of a continuous *function* $f : \mathbb{R} \to [0, \infty)$ *satisfying* $f(0) = 0$*,* f *is continuously differentiable on* $\mathbb{R} \setminus \{0\}$*,* $f' > 0$ *on* $(0, \infty)$*,* $f' < 0$ *on* $(-\infty, 0)$*, and* $\lim_{x \to 0^{\pm}} f'(x) = \pm\infty$*.*

Then for each $\theta \in (0, \pi)$ *there exists* $R > 0$ *with the property that, if* C_θ *denotes the upright sector with vertex at the origin and (full) aperture angle* θ*, then there holds* $\Omega \cap B(0, r) \subseteq C_\theta \cap B(0, r)$ *for each* $r \in (0, R)$*. As such, if we abbreviate* $\sigma := \mathcal{H}^1 \lfloor \partial\Omega$ *and for each* $r > 0$ *set* $\Delta_r := B(0, r) \cap \partial\Omega$ *and denote by* π *the projection onto the horizontal axis, then* $\mathcal{L}^1\big(\pi(\Delta_r)\big)/r \to 0$ *as* $r \to 0^+$*. Next, for each* $r > 0$ *small let* $a_r \in (0, \infty)$ *and* $b_r \in (-\infty, 0)$ *be such that* $\big|(a_r, f(a_r))\big| = \big|(b_r, f(b_r))\big| = r$*. Also, denote by* $\alpha_r, \beta_r \in (0, \pi/2)$ *the angles made by the vectors* $\big(a_r, f(a_r)\big)$ *and* $\big(b_r, f(b_r)\big)$ *with the horizontal axis. In light of these choices, we may compute* $\int_{\pi(\Delta_r)} f'(x)\, dx = f(a_r) - f(b_r) = r \cdot \sin\alpha_r - r \cdot \sin\beta_r$*. Since* $\alpha_r, \beta_r \to \pi/2$ *as* $r \to 0^+$*, it follows that* $\frac{1}{r}\int_{\pi(\Delta_r)} f'(x)\, dx \to 0$ *as* $r \to 0^+$*. All together, the above analysis shows that the geometric measure theoretic outward unit normal* ν *to* Ω *satisfies*

$$\fint_{\Delta_r} \nu\, d\sigma \approx \frac{1}{r}\int_{\pi(\Delta_r)} (f', -1)\, d\mathcal{L}^1 \to 0 \ \text{ as } \ r \to 0^+, \tag{4.1.7}$$

and since $1 - \Big|\fint_{\Delta_r} \nu\, d\sigma\Big|\, d\sigma \le \Big|\nu - \fint_{\Delta_r} \nu\, d\sigma\Big|\, d\sigma \le 1 + \Big|\fint_{\Delta_r} \nu\, d\sigma\Big|\, d\sigma$ *at* σ*-a.e. point on* $\partial\Omega$ *we obtain*

$$\fint_{\Delta_r} \Big|\nu - \fint_{\Delta_r} \nu\, d\sigma\Big|\, d\sigma \to 1 \ \text{ as } \ r \to 0^+. \tag{4.1.8}$$

Together with (4.1.2) *this shows that, in line with the claim made in* (4.1.6)*, we presently have*

$$\|\nu\|_{[\mathrm{BMO}(\partial\Omega,\sigma)]^2} = 1. \tag{4.1.9}$$

The same type of argument works in the higher-dimensional setting, namely for an Ahlfors regular domain $\Omega \subseteq \mathbb{R}^n$*, where* $n \ge 3$*, such that near the point* $0 \in \partial\Omega$ *the domain* Ω *coincides with the region above the graph of a continuous function* $f : \mathbb{R}^{n-1} \to [0, \infty)$ *satisfying* $f(0) = 0$*,* f *is continuously differentiable on* $\mathbb{R}^{n-1} \setminus \{0\}$*, and* $\lim_{x' \to 0} |\nabla f(x')| = \infty$*. For each* $r > 0$ *set* $\Delta_r := B(0, r) \cap \partial\Omega$*. Assume the cusp at the origin is "circular," so for each* $r > 0$ *the projection* $\pi(\Delta_r)$ *of* Δ_r *onto the horizontal plane* $\mathbb{R}^{n-1}$ *is an* $(n-2)$*-dimensional disk, with* $\mathcal{H}^{n-2}\big(\partial\pi(\Delta_r)\big) \approx r^{n-2}$*. Then* $\int_{\pi(\Delta_r)} \nabla f\, d\mathcal{L}^{n-1} = \int_{\partial\pi(\Delta_r)} N f\, d\mathcal{H}^{n-2}$ *by the Divergence Theorem, where* N *is the outward unit normal to* $\pi(\Delta_r)$*. Also,* $f(x') = r \cdot \sin\alpha(x')$ *for each* $x' \in \partial\pi(\Delta_r)$*, where* $\alpha(x') \in (0, \pi/2)$ *is the angle made by the vector* $\big(x', f(x')\big)$ *with the horizontal plane* $\mathbb{R}^{n-1}$*. Note that* $\alpha(x') \to \pi/2$ *as* $r \to 0^+$*, and* $\int_{\partial\pi(\Delta_r)} N\, d\mathcal{H}^{n-2} = 0$ *by the Divergence Theorem. Thus, in a uniform fashion, for* $x' \in \partial\pi(\Delta_r)$ *we have* $f(x')/r = \sin\alpha(x') = 1 + o(1)$ *as* $r \to 0^+$*. Much as before, if* C_θ *denotes the upright circular cone with vertex at the origin and (full) aperture angle* $\theta \in (0, \pi)$*, then for each* $\theta \in (0, \pi)$ *there exists some small threshold* $R \in (0, \infty)$ *with the property that* $\Omega \cap B(0, r) \subseteq C_\theta \cap B(0, r)$ *whenever* $r \in (0, R)$*. As a consequence, if* $\sigma := \mathcal{H}^{n-1} \lfloor \partial\Omega$ *then* $\mathcal{L}^{n-1}\big(\pi(\Delta_r)\big)/r^{n-1} \to 0$ *as* $r \to 0^+$*. We are now prepared to compute*

$$
\begin{aligned}
\fint_{\Delta_r} \nu \, \mathrm{d}\sigma &\approx \frac{1}{r^{n-1}} \int_{\pi(\Delta_r)} (\nabla f, -1) \, \mathrm{d}\mathcal{L}^{n-1} \\
&= \Big(\frac{1}{r^{n-1}} \int_{\partial\pi(\Delta_r)} N f \, \mathrm{d}\mathcal{H}^{n-2}, \, -\frac{\mathcal{L}^{n-1}\big(\pi(\Delta_r)\big)}{r^{n-1}} \Big) \\
&= \Big(\frac{1}{r^{n-2}} \int_{\partial\pi(\Delta_r)} N \big(1 + o(1)\big) \, \mathrm{d}\mathcal{H}^{n-2}, \, -\frac{\mathcal{L}^{n-1}\big(\pi(\Delta_r)\big)}{r^{n-1}} \Big) \\
&= \Big(\frac{1}{r^{n-2}} \int_{\partial\pi(\Delta_r)} o(1) \, \mathrm{d}\mathcal{H}^{n-2}, \, -\frac{\mathcal{L}^{n-1}\big(\pi(\Delta_r)\big)}{r^{n-1}} \Big) = 0 \in \mathbb{R}^n \text{ as } r \to 0^+.
\end{aligned}
\tag{4.1.10}
$$

With this in hand, by reasoning as before we may once again conclude that (4.1.8) *holds, from which we ultimately deduce that* $\|\nu\|_{[\mathrm{BMO}(\partial\Omega,\sigma)]^n} = 1$.

To facilitate stating the next result in this section we first introduce some notation and make some remarks. Specifically, with e denoting the base of natural logarithms, for each $m \in \mathbb{N}_0$ and $t \in [0, \infty)$ let us define

$$
t^{\langle 0 \rangle} := 1 \tag{4.1.11}
$$

and, if $m \geq 1$,

$$
t^{\langle m \rangle} := \begin{cases} 0 & \text{if } t = 0, \\ t \cdot \underbrace{\ln\Big(\cdots \ln\big(\ln(1/t)\big) \cdots \Big)}_{m \text{ natural logarithms}} & \text{if } 0 < t \leq ({}^m\mathrm{e})^{-1}, \\ ({}^m e)^{-1} & \text{if } t > ({}^m\mathrm{e})^{-1}, \end{cases} \tag{4.1.12}
$$

where ${}^m\mathrm{e}$ is the m-th tetration of e (involving m copies of e, combined by exponentiation), i.e.,

$$
{}^m\mathrm{e} := \underbrace{\mathrm{e}^{\mathrm{e}^{\cdot^{\cdot^{\cdot^{\mathrm{e}}}}}}}_{m \text{ copies of e}}, \quad \text{the } m\text{-th fold exponentiation of e.} \tag{4.1.13}
$$

We also agree to set ${}^0\mathrm{e} := 1$. Ergo, inductively, for each $m \in \mathbb{N}_0$ and each $t \in [0, \infty)$,

$$
t^{\langle m+1 \rangle} = \begin{cases} 0 & \text{if } t = 0, \\ t \cdot \ln\big(t^{\langle m \rangle}/t\big) & \text{if } 0 < t \leq ({}^{m+1}\mathrm{e})^{-1}, \\ ({}^{m+1}\mathrm{e})^{-1} & \text{if } t > ({}^{m+1}\mathrm{e})^{-1}. \end{cases} \tag{4.1.14}
$$

For further reference, it is useful to note that elementary calculus gives that this function enjoys the following properties:

$$
\begin{gathered}
[0, \infty) \ni t \longmapsto t^{\langle m \rangle} \in [0, \infty) \text{ is continuous, non-decreasing,} \\
\text{and vanishes continuously at the origin, for each } m \in \mathbb{N},
\end{gathered} \tag{4.1.15}
$$

$$t^{\langle m\rangle} \le t^{\langle m-1\rangle} \le \cdots \le t^{\langle 1\rangle} \le \left(e^{\varepsilon-1}/\varepsilon\right)\cdot t^{1-\varepsilon} \quad \text{for each } t\in[0,\infty),\ m\in\mathbb{N},\ \varepsilon\in(0,1), \tag{4.1.16}$$

$$t \le \max\{1, ({}^m e)t\}\cdot t^{\langle m\rangle} \quad \text{for all } t\in[0,\infty) \text{ and } m\in\mathbb{N}_0, \tag{4.1.17}$$

$$(\lambda t)^{\langle m\rangle} \le \lambda t^{\langle m\rangle} \quad \text{for all } t\in[0,\infty),\ m\in\mathbb{N}_0, \text{ and } \lambda\in[1,\infty), \tag{4.1.18}$$

$$(t^{\alpha})^{\langle m\rangle} \le t^{\alpha}\cdot \underbrace{\ln\Big(\cdots\ln\big(\ln(1/\min\{t, ({}^m e)^{-1}\})\big)\cdots\Big)}_{m \text{ natural logarithms}} \quad \text{for all numbers } t\in[0,\infty),\ \alpha\in(0,1], \text{ and } m\in\mathbb{N}, \tag{4.1.19}$$

(with the convention that the value at $t = 0$ for the function in the right-hand side of the inequality in (4.1.19) is its limit as $t \to 0^+$). In particular,

$$t^{\langle m\rangle} \le t\cdot \underbrace{\ln\Big(\cdots\ln\big(\ln({}^m e/t)\big)\cdots\Big)}_{m \text{ natural logarithms}} \quad \text{for all } t\in[0,1],\ m\in\mathbb{N}. \tag{4.1.20}$$

In fact, up to a multiplicative constant, the opposite inequality in (4.1.20) is true as well. Specifically,

$$({}^m e)^{-1}\cdot t\cdot \underbrace{\ln\Big(\cdots\ln\big(\ln({}^m e/t)\big)\cdots\Big)}_{m \text{ natural logarithms}} \le t^{\langle m\rangle} \quad \text{for all } t\in[0,1],\ m\in\mathbb{N}, \tag{4.1.21}$$

hence for each fixed $m \in \mathbb{N}$ we have

$$t^{\langle m\rangle} \approx t\cdot \underbrace{\ln\Big(\cdots\ln\big(\ln({}^m e/t)\big)\cdots\Big)}_{m \text{ natural logarithms}}, \quad \text{uniformly for } t\in[0,1]. \tag{4.1.22}$$

In general, say that a function $f : (0,\infty) \to (0,\infty)$ vanishes in an *asymptotically linear fashion* as $t \to 0^+$ provided

$$\lim_{t\to 0^+} \frac{\ln[f(t)]}{\ln t} = 1. \tag{4.1.23}$$

Also, say that $f(t)$ *vanishes in an at most asymptotically linear fashion* as $t \to 0^+$, if $f(t) = O(g(t))$ as $t \to 0^+$ for some function $g : (0,\infty) \to (0,\infty)$ which vanishes in an asymptotically linear fashion as $t \to 0^+$. It is then clear from (4.1.12) that for each given $m \in \mathbb{N}$ the function $(0,\infty) \ni t \mapsto t^{\langle m\rangle}$ vanishes in an asymptotically linear fashion as $t \to 0^+$. Hence, any non-negative function in the variable t majorized by a multiple of $t^{\langle m\rangle}$ vanishes in an at most asymptotically linear fashion as $t \to 0^+$.

The following fundamental result shows that *the operator norm of a chord-dot-normal singular integral operator acting on Muckenhoupt weighted Lebesgue spaces, as well as other Generalized Banach Function Spaces, depends in an at most asymptotically linear fashion on the* BMO *semi-norm of the geometric measure theoretic outward unit normal, as the latter quantity goes to zero.*

Theorem 4.1.3 *Let* $\Omega \subseteq \mathbb{R}^n$ *be a* UR *domain. Abbreviate* $\sigma := \mathcal{H}^{n-1}\lfloor\partial\Omega$ *and denote by* ν *the geometric measure theoretic outward unit normal to* Ω*. Also, consider a sufficiently large integer* $N = N(n) \in \mathbb{N}$ *and suppose* $k \in \mathscr{C}^N(\mathbb{R}^n \setminus \{0\})$ *is a complex-valued function which is even and positive homogeneous of degree* $-n$*. In this setting, for each* $\varepsilon > 0$ *and each* $f \in L^1\big(\partial\Omega, \frac{\sigma(x)}{1+|x|^{n-1}}\big)$ *define*

$$T_\varepsilon f(x) := \int_{y\in\partial\Omega,\ |x-y|>\varepsilon} \langle x-y, \nu(y)\rangle k(x-y) f(y)\, d\sigma(y) \text{ for all } x \in \partial\Omega, \tag{4.1.24}$$

$$T_\varepsilon^{\#} f(x) := \int_{y\in\partial\Omega,\ |x-y|>\varepsilon} \langle y-x, \nu(x)\rangle k(x-y) f(y)\, d\sigma(y) \text{ for } \sigma\text{-a.e. } x \in \partial\Omega. \tag{4.1.25}$$

Also, consider the maximal operators T_*, $T_*^{\#}$ *associated with these families of truncated SIO's, whose actions on each function* $f \in L^1\big(\partial\Omega, \frac{\sigma(x)}{1+|x|^{n-1}}\big)$ *are defined as*

$$\begin{aligned} T_* f(x) &:= \sup\nolimits_{\varepsilon>0} \big|T_\varepsilon f(x)\big| \text{ for each } x \in \partial\Omega, \\ T_*^{\#} f(x) &:= \sup\nolimits_{\varepsilon>0} \big|T_\varepsilon^{\#} f(x)\big| \text{ for } \sigma\text{-a.e. } x \in \partial\Omega, \end{aligned} \tag{4.1.26}$$

as well as the principal-value singular integral operators $T, T^{\#}$ *on* $\partial\Omega$ *acting on each* $f \in L^1\big(\partial\Omega, \frac{\sigma(x)}{1+|x|^{n-1}}\big)$ *at* σ*-a.e.* $x \in \partial\Omega$ *according to*

$$Tf(x) := \lim_{\varepsilon\to 0^+} \int_{y\in\partial\Omega,\ |x-y|>\varepsilon} \langle x-y, \nu(y)\rangle k(x-y) f(y)\, d\sigma(y), \tag{4.1.27}$$

$$T^{\#} f(x) := \lim_{\varepsilon\to 0^+} \int_{y\in\partial\Omega,\ |x-y|>\varepsilon} \langle y-x, \nu(x)\rangle k(x-y) f(y)\, d\sigma(y). \tag{4.1.28}$$

Finally, fix an integrability exponent $p \in (1,\infty)$ *along with a Muckenhoupt weight* $w \in A_p(\partial\Omega,\sigma)$*, and recall the earlier convention of using the same symbol* w *for the measure associated with the given weight* w *as in* [129, (7.7.1)].

Then for each $m \in \mathbb{N}$ *there exists some* $C_m \in (0,\infty)$*, which depends only on* m*,* n*,* p*,* $[w]_{A_p}$*, and the* UR *constants of* $\partial\Omega$ *such that, with the piece of notation introduced in* (4.1.12)*, one has*[5]

[5] recall the earlier convention of using the same symbol w for the measure associated with the given weight w as in [129, (7.7.1)]

$$\|T_*\|_{L^p(\partial\Omega,w)\to L^p(\partial\Omega,w)} \leq C_m\Big(\sum_{|\alpha|\leq N}\sup_{S^{n-1}}|\partial^\alpha k|\Big)\|\nu\|^{\langle m\rangle}_{[\mathrm{BMO}(\partial\Omega,\sigma)]^n}, \tag{4.1.29}$$

$$\|T_*^{\#}\|_{L^p(\partial\Omega,w)\to L^p(\partial\Omega,w)} \leq C_m\Big(\sum_{|\alpha|\leq N}\sup_{S^{n-1}}|\partial^\alpha k|\Big)\|\nu\|^{\langle m\rangle}_{[\mathrm{BMO}(\partial\Omega,\sigma)]^n}, \tag{4.1.30}$$

$$\|T\|_{L^p(\partial\Omega,w)\to L^p(\partial\Omega,w)} \leq C_m\Big(\sum_{|\alpha|\leq N}\sup_{S^{n-1}}|\partial^\alpha k|\Big)\|\nu\|^{\langle m\rangle}_{[\mathrm{BMO}(\partial\Omega,\sigma)]^n}, \tag{4.1.31}$$

$$\|T^{\#}\|_{L^p(\partial\Omega,w)\to L^p(\partial\Omega,w)} \leq C_m\Big(\sum_{|\alpha|\leq N}\sup_{S^{n-1}}|\partial^\alpha k|\Big)\|\nu\|^{\langle m\rangle}_{[\mathrm{BMO}(\partial\Omega,\sigma)]^n}. \tag{4.1.32}$$

In addition, when $\|\nu\|_{[\mathrm{BMO}(\partial\Omega,\sigma)]^n}$ is sufficiently small relative to n, p, $[w]_{A_p}$, and the Ahlfors regularity constants of $\partial\Omega$ one may take $C_m \in (0,\infty)$ appearing in (4.1.29)-(4.1.32) *to depend itself only on said entities (i.e., n, p, $[w]_{A_p}$, the Ahlfors regularity constant of $\partial\Omega$) and m.*

Similar estimates are valid on Morrey, vanishing Morrey, and block spaces. Specifically, given an integrability exponent $p \in (1,\infty)$ along with a parameter $\lambda \in (0, n-1)$, the following are well-defined, bounded operators:

$$T_*, T_*^{\#}, T, T^{\#} : M^{p,\lambda}(\partial\Omega,\sigma) \longrightarrow M^{p,\lambda}(\partial\Omega,\sigma), \tag{4.1.33}$$

$$T_*, T_*^{\#}, T, T^{\#} : \mathring{M}^{p,\lambda}(\partial\Omega,\sigma) \longrightarrow \mathring{M}^{p,\lambda}(\partial\Omega,\sigma). \tag{4.1.34}$$

Furthermore, for each $m \in \mathbb{N}$ there exists some $C_m \in (0,\infty)$, which depends only on m, n, p, λ, and the UR *constants of $\partial\Omega$ such that*

$$\begin{aligned}\max\Big\{&\|T_*\|_{M^{p,\lambda}(\partial\Omega,\sigma)\to M^{p,\lambda}(\partial\Omega,\sigma)},\ \|T_*\|_{\mathring{M}^{p,\lambda}(\partial\Omega,\sigma)\to \mathring{M}^{p,\lambda}(\partial\Omega,\sigma)},\\ &\|T_*^{\#}\|_{M^{p,\lambda}(\partial\Omega,\sigma)\to M^{p,\lambda}(\partial\Omega,\sigma)},\ \|T_*^{\#}\|_{\mathring{M}^{p,\lambda}(\partial\Omega,\sigma)\to \mathring{M}^{p,\lambda}(\partial\Omega,\sigma)}\Big\}\\ &\qquad\leq C_m\Big(\sum_{|\alpha|\leq N}\sup_{S^{n-1}}|\partial^\alpha k|\Big)\|\nu\|^{\langle m\rangle}_{[\mathrm{BMO}(\partial\Omega,\sigma)]^n},\end{aligned} \tag{4.1.35}$$

$$\begin{aligned}\max\Big\{&\|T\|_{M^{p,\lambda}(\partial\Omega,\sigma)\to M^{p,\lambda}(\partial\Omega,\sigma)},\ \|T\|_{\mathring{M}^{p,\lambda}(\partial\Omega,\sigma)\to \mathring{M}^{p,\lambda}(\partial\Omega,\sigma)}\Big\}\\ &\qquad\leq C_m\Big(\sum_{|\alpha|\leq N}\sup_{S^{n-1}}|\partial^\alpha k|\Big)\|\nu\|^{\langle m\rangle}_{[\mathrm{BMO}(\partial\Omega,\sigma)]^n},\end{aligned} \tag{4.1.36}$$

$$\begin{aligned}\max\Big\{&\|T^{\#}\|_{M^{p,\lambda}(\partial\Omega,\sigma)\to M^{p,\lambda}(\partial\Omega,\sigma)},\ \|T^{\#}\|_{\mathring{M}^{p,\lambda}(\partial\Omega,\sigma)\to \mathring{M}^{p,\lambda}(\partial\Omega,\sigma)}\Big\}\\ &\qquad\leq C_m\Big(\sum_{|\alpha|\leq N}\sup_{S^{n-1}}|\partial^\alpha k|\Big)\|\nu\|^{\langle m\rangle}_{[\mathrm{BMO}(\partial\Omega,\sigma)]^n}.\end{aligned} \tag{4.1.37}$$

Once again, when $\|\nu\|_{[\mathrm{BMO}(\partial\Omega,\sigma)]^n}$ is sufficiently small relative to n, p, λ, and the Ahlfors regularity constants of $\partial\Omega$ one may take $C_m \in (0,\infty)$ appearing in (4.1.35)-

(4.1.37) *to depend itself only on said entities (i.e., n, p, λ, the Ahlfors regularity constants of $\partial\Omega$) and m. In addition, for each $q \in (1,\infty)$ and $\lambda \in (0, n-1)$ the operators*

$$T, T^{\#} : \mathcal{B}^{q,\lambda}(\partial\Omega,\sigma) \longrightarrow \mathcal{B}^{q,\lambda}(\partial\Omega,\sigma), \tag{4.1.38}$$

$$T_*, T_*^{\#} : \mathcal{B}^{q,\lambda}(\partial\Omega,\sigma) \longrightarrow \mathcal{B}^{q,\lambda}(\partial\Omega,\sigma), \tag{4.1.39}$$

are well defined and bounded, and for each $m \in \mathbb{N}$ there exists some $C_m \in (0,\infty)$, which depends only on m, n, q, λ, and the UR *constants of $\partial\Omega$ such that*

$$\max\Big\{\|T\|_{\mathcal{B}^{q,\lambda}(\partial\Omega,\sigma)\to\mathcal{B}^{q,\lambda}(\partial\Omega,\sigma)}, \|T^{\#}\|_{\mathcal{B}^{q,\lambda}(\partial\Omega,\sigma)\to\mathcal{B}^{q,\lambda}(\partial\Omega,\sigma)}\Big\} \tag{4.1.40}$$
$$\leq C_m\Big(\sum_{|\alpha|\leq N}\sup_{S^{n-1}}|\partial^\alpha k|\Big)\|\nu\|^{\langle m\rangle}_{[\mathrm{BMO}(\partial\Omega,\sigma)]^n},$$

$$\max\Big\{\|T_*\|_{\mathcal{B}^{q,\lambda}(\partial\Omega,\sigma)\to\mathcal{B}^{q,\lambda}(\partial\Omega,\sigma)}, \|T_*^{\#}\|_{\mathcal{B}^{q,\lambda}(\partial\Omega,\sigma)\to\mathcal{B}^{q,\lambda}(\partial\Omega,\sigma)}\Big\} \tag{4.1.41}$$
$$\leq C_m\Big(\sum_{|\alpha|\leq N}\sup_{S^{n-1}}|\partial^\alpha k|\Big)\|\nu\|^{\langle m\rangle}_{[\mathrm{BMO}(\partial\Omega,\sigma)]^n}.$$

Much as before, when $\|\nu\|_{[\mathrm{BMO}(\partial\Omega,\sigma)]^n}$ is sufficiently small relative to n, q, and the Ahlfors regularity constants of $\partial\Omega$ one may take $C_m \in (0,\infty)$ appearing in (4.1.40)-(4.1.41) *to depend itself only on said entities (i.e., n, q, λ, the Ahlfors regularity constants of $\partial\Omega$) and m.*

In fact, more general results than the ones stated above are true. To formulate them, assume $\mathbb{X}$ is a Generalized Banach Function Space on $(\partial\Omega,\sigma)$ with the property that

$$\mathcal{M}_{\partial\Omega} : \mathbb{X} \to \mathbb{X} \text{ and } \mathcal{M}_{\partial\Omega} : \mathbb{X}' \to \mathbb{X}' \text{ are well-defined and bounded,} \tag{4.1.42}$$

where $\mathcal{M}_{\partial\Omega}$ is the Hardy-Littlewood maximal operator on $\partial\Omega$, and $\mathbb{X}'$ is the associated space of $\mathbb{X}$ (cf. [130, Definitions 5.1.4, 5.1.11]). *Then the following are well-defined, bounded operators:*

$$T_*, T_*^{\#}, T, T^{\#} : \mathbb{X} \longrightarrow \mathbb{X}, \tag{4.1.43}$$

$$T_*, T_*^{\#}, T, T^{\#} : \mathbb{X}' \longrightarrow \mathbb{X}', \tag{4.1.44}$$

$$T_*, T_*^{\#}, T, T^{\#} : \mathring{\mathbb{X}} \longrightarrow \mathring{\mathbb{X}}, \tag{4.1.45}$$

$$T_*, T_*^{\#}, T, T^{\#} : (\mathbb{X}')^\circ \longrightarrow (\mathbb{X}')^\circ. \tag{4.1.46}$$

Moreover, for each $m \in \mathbb{N}$ there exists some $C_m \in (0,\infty)$, which depends only on m, n, the UR *constants of $\partial\Omega$, and the operator norms*

$$\|\mathcal{M}_{\partial\Omega}\|_{\mathbb{X}\to\mathbb{X}} \text{ together with } \|\mathcal{M}_{\partial\Omega}\|_{\mathbb{X}'\to\mathbb{X}'} \tag{4.1.47}$$

such that

$$\max\Big\{\|T_*\|_{\mathbb{X}\to\mathbb{X}},\ \|T_*\|_{\mathbb{X}'\to\mathbb{X}'},\ \|T_*^{\#}\|_{\mathbb{X}\to\mathbb{X}},\ \|T_*^{\#}\|_{\mathbb{X}'\to\mathbb{X}'}, \tag{4.1.48}$$
$$\|T_*\|_{\mathring{\mathbb{X}}\to\mathring{\mathbb{X}}},\ \|T_*\|_{(\mathbb{X}')^\circ\to(\mathbb{X}')^\circ},\ \|T_*^{\#}\|_{\mathring{\mathbb{X}}\to\mathring{\mathbb{X}}},\ \|T_*^{\#}\|_{(\mathbb{X}')^\circ\to(\mathbb{X}')^\circ}\Big\}$$
$$\leq C_m\Big(\sum_{|\alpha|\leq N}\sup_{S^{n-1}}|\partial^\alpha k|\Big)\|\nu\|^{\langle m\rangle}_{[\mathrm{BMO}(\partial\Omega,\sigma)]^n}$$

and

$$\max\Big\{\|T\|_{\mathbb{X}\to\mathbb{X}},\ \|T\|_{\mathbb{X}'\to\mathbb{X}'},\ \|T^{\#}\|_{\mathbb{X}\to\mathbb{X}},\ \|T^{\#}\|_{\mathbb{X}'\to\mathbb{X}'}, \tag{4.1.49}$$
$$\|T\|_{\mathring{\mathbb{X}}\to\mathring{\mathbb{X}}},\ \|T\|_{(\mathbb{X}')^\circ\to(\mathbb{X}')^\circ},\ \|T^{\#}\|_{\mathring{\mathbb{X}}\to\mathring{\mathbb{X}}},\ \|T^{\#}\|_{(\mathbb{X}')^\circ\to(\mathbb{X}')^\circ}\Big\}$$
$$\leq C_m\Big(\sum_{|\alpha|\leq N}\sup_{S^{n-1}}|\partial^\alpha k|\Big)\|\nu\|^{\langle m\rangle}_{[\mathrm{BMO}(\partial\Omega,\sigma)]^n}.$$

Once more, when $\|\nu\|_{[\mathrm{BMO}(\partial\Omega,\sigma)]^n}$ *is sufficiently small relative to the Ahlfors regularity constants of* $\partial\Omega$ *and the operator norms* (4.1.47) *one may take* $C_m\in(0,\infty)$ *appearing in* (4.1.48)-(4.1.49) *to depend itself only on said entities and* m.

Before presenting the proof of Theorem 4.1.3, a few comments are in order.

`Comment 1.` It is of interest to compare the estimates in the above theorem with those from [131, Theorem 2.3.2]. Specifically, [131, (2.3.58), (2.3.59)] applied with $\Sigma:=\partial\Omega$ give that for T_* as in (4.1.26) we have

$$\|T_*\|_{L^p(\partial\Omega,w)\to L^p(\partial\Omega,w)}\leq C(\partial\Omega,p,[w]_{A_p})\big\|k\big|_{S^{n-1}}\big\|_{\mathscr{C}^N(S^{n-1})} \tag{4.1.50}$$

where $C(\partial\Omega,p,[w]_{A_p})\in(0,\infty)$ depends on $\partial\Omega$ solely through its UR constants. In sharp contrast to this estimate, (4.1.29) features $\|\nu\|^{\langle m\rangle}_{[\mathrm{BMO}(\partial\Omega,\sigma)]^n}$ as a multiplicative factor in the right-hand side, something which the UR constants of $\partial\Omega$ cannot control.

Indeed, for [131, (2.3.58), (2.3.59)] no provisions are in place to take advantage of the specific algebraic format of the present integral kernel $\langle x-y,\nu(y)\rangle k(x-y)$. For [131, Theorem 2.3.2] to apply, this integral kernel needs to be dismantled into its most primordial building blocks, i.e., as $\sum_{j=1}^n k_j(x-y)\nu_j(y)$ with $k_j(z):=z_jk(z)$ for each $z\in\mathbb{R}^n\setminus\{0\}$ and $j\in\{1,\dots,n\}$. Since multiplication by ν_j may be absorbed with the function f (without changing its membership, or increasing its size, in the Muckenhoupt weighted Lebesgue space $L^p(\partial\Omega,w)$), [131, Theorem 2.3.2] may then be invoked in relation to each maximal operator associated with the kernel k_j.

Estimate (4.1.50), the end-product of such an approach, is then rendered insensitive to the flatness of $\partial\Omega$. As an example, consider the scenario in which Ω is a half-space in $\mathbb{R}^n$. While is apparent from definitions (4.1.24), (4.1.26) that in this case we have $\|T_*\|_{L^p(\partial\Omega,w)\to L^p(\partial\Omega,w)} = 0$, estimate (4.1.50) only gives that $\|T_*\|_{L^p(\partial\Omega,w)\to L^p(\partial\Omega,w)} < +\infty$. By way of contrast, in the current scenario $\|\nu\|_{[\mathrm{BMO}(\partial\Omega,\sigma)]^n} = 0$, given that ν is a constant vector, and as such (4.1.29) accurately predicts $\|T_*\|_{L^p(\partial\Omega,w)\to L^p(\partial\Omega,w)} = 0$. Similar considerations apply to $T_*^\#$.

Comment 2. Of course, in the special case when $w \equiv 1$, Theorem 4.1.3 yields estimates on ordinary Lebesgue spaces, $L^p(\partial\Omega,\sigma)$ with $p \in (1,\infty)$. For further reference let us record here that (4.1.29)-(4.1.32) show that for each $m \in \mathbb{N}$ and $p \in (1,\infty)$ there exists some $C_m \in (0,\infty)$, which depends only on m, n, p, and the UR constants of $\partial\Omega$ such that, with the piece of notation introduced in (4.1.12),

$$\|T_*\|_{L^p(\partial\Omega,\sigma)\to L^p(\partial\Omega,\sigma)} \leq C_m\Big(\sum_{|\alpha|\leq N}\sup_{S^{n-1}}|\partial^\alpha k|\Big)\|\nu\|^{\langle m\rangle}_{[\mathrm{BMO}(\partial\Omega,\sigma)]^n}, \tag{4.1.51}$$

$$\|T_*^\#\|_{L^p(\partial\Omega,\sigma)\to L^p(\partial\Omega,\sigma)} \leq C_m\Big(\sum_{|\alpha|\leq N}\sup_{S^{n-1}}|\partial^\alpha k|\Big)\|\nu\|^{\langle m\rangle}_{[\mathrm{BMO}(\partial\Omega,\sigma)]^n}, \tag{4.1.52}$$

$$\|T\|_{L^p(\partial\Omega,\sigma)\to L^p(\partial\Omega,\sigma)} \leq C_m\Big(\sum_{|\alpha|\leq N}\sup_{S^{n-1}}|\partial^\alpha k|\Big)\|\nu\|^{\langle m\rangle}_{[\mathrm{BMO}(\partial\Omega,\sigma)]^n}, \tag{4.1.53}$$

$$\|T^\#\|_{L^p(\partial\Omega,\sigma)\to L^p(\partial\Omega,\sigma)} \leq C_m\Big(\sum_{|\alpha|\leq N}\sup_{S^{n-1}}|\partial^\alpha k|\Big)\|\nu\|^{\langle m\rangle}_{[\mathrm{BMO}(\partial\Omega,\sigma)]^n}. \tag{4.1.54}$$

Via real interpolation, these further imply similar estimates on the scale of Lorentz spaces on $\partial\Omega$. Specifically, from (4.1.51)-(4.1.54) and real interpolation (for sublinear operators) we conclude that for each $m \in \mathbb{N}$, $p \in (1,\infty)$, and $q \in (0,\infty]$ there exists some $C_m \in (0,\infty)$, which depends only on m, n, p, q, and the UR constants of $\partial\Omega$ with the property that, with the piece of notation introduced in (4.1.12),

$$\|T_*\|_{L^{p,q}(\partial\Omega,\sigma)\to L^{p,q}(\partial\Omega,\sigma)} \leq C_m\Big(\sum_{|\alpha|\leq N}\sup_{S^{n-1}}|\partial^\alpha k|\Big)\|\nu\|^{\langle m\rangle}_{[\mathrm{BMO}(\partial\Omega,\sigma)]^n}, \tag{4.1.55}$$

$$\|T_*^\#\|_{L^{p,q}(\partial\Omega,\sigma)\to L^{p,q}(\partial\Omega,\sigma)} \leq C_m\Big(\sum_{|\alpha|\leq N}\sup_{S^{n-1}}|\partial^\alpha k|\Big)\|\nu\|^{\langle m\rangle}_{[\mathrm{BMO}(\partial\Omega,\sigma)]^n}, \tag{4.1.56}$$

$$\|T\|_{L^{p,q}(\partial\Omega,\sigma)\to L^{p,q}(\partial\Omega,\sigma)} \leq C_m\Big(\sum_{|\alpha|\leq N}\sup_{S^{n-1}}|\partial^\alpha k|\Big)\|\nu\|^{\langle m\rangle}_{[\mathrm{BMO}(\partial\Omega,\sigma)]^n}, \tag{4.1.57}$$

$$\|T^\#\|_{L^{p,q}(\partial\Omega,\sigma)\to L^{p,q}(\partial\Omega,\sigma)} \leq C_m\Big(\sum_{|\alpha|\leq N}\sup_{S^{n-1}}|\partial^\alpha k|\Big)\|\nu\|^{\langle m\rangle}_{[\mathrm{BMO}(\partial\Omega,\sigma)]^n}. \tag{4.1.58}$$

Comment 3. In view of (4.1.2) and (4.1.22), in lieu of $\|\nu\|^{\langle m\rangle}_{[\mathrm{BMO}(\partial\Omega,\sigma)]^n}$ we may use

$$\|\nu\|_{[\mathrm{BMO}(\partial\Omega,\sigma)]^n}\cdot\underbrace{\ln\Big(\cdots\ln\big(\ln(^m e/\|\nu\|_{[\mathrm{BMO}(\partial\Omega,\sigma)]^n})\big)\cdots\Big)}_{m\text{ natural logarithms}} \tag{4.1.59}$$

in all estimates recorded in (4.1.29)-(4.1.32), (4.1.35)-(4.1.37), (4.1.40)-(4.1.41), (4.1.47)-(4.1.49), (4.1.51)-(4.1.54), and (4.1.55)-(4.1.58). Hence, if we abbreviate

$$\|\nu\|_* := \|\nu\|_{[\mathrm{BMO}(\partial\Omega,\sigma)]^n}, \tag{4.1.60}$$

then all these operator norms are

$$O\Big(\|\nu\|_* \ln\big(e/\|\nu\|_*\big)\Big) \quad \text{(corresponding to } m = 1\text{)}, \tag{4.1.61}$$

$$O\left(\|\nu\|_* \ln\Big(\ln\big(\mathrm{e}^e/\|\nu\|_*\big)\Big)\right) \quad \text{(corresponding to } m = 2\text{)}, \tag{4.1.62}$$

et cetera. Thus, all the aforementioned operator norms have at most linear growth in $\|\nu\|_*$, up to arbitrarily many iterated logarithms.

In the same vein, we may invoke (4.1.16) to conclude that all operator norms in (4.1.29)-(4.1.32), (4.1.35)-(4.1.37), (4.1.40)-(4.1.41), (4.1.47)-(4.1.49), (4.1.51)-(4.1.54), as well as (4.1.55)-(4.1.58) are

$$O\big(\|\nu\|_*^{1-\varepsilon}\big) \ \text{ for each fixed } \varepsilon \in (0,1). \tag{4.1.63}$$

Comment 4. The reason we have focused exclusively on singular integral operators of "chord-dot-normal" type in Theorem 4.1.3 becomes apparent from Theorem 4.1.10, discussed further below.

Comment 5. It is also possible to produce a version of Theorem 4.1.3 corresponding to chord-dot-normal singular integral operators with variable coefficient kernels. Specifically, assume

> $b(x,z)$ is a function which is even and positive homogeneous of degree $-n$ in the variable $z \in \mathbb{R}^n \setminus \{0\}$, and such that $\partial_z^\alpha b(x,z)$ is continuous and bounded on $\mathbb{R}^n \times S^{n-1}$ for each multi-index $\alpha \in \mathbb{N}_0^n$ satisfying $|\alpha| \le M$, where $M = M(n)$ is a large positive integer. (4.1.64)

Let $\Omega \subseteq \mathbb{R}^n$ be a UR domain. Abbreviate $\sigma := \mathcal{H}^{n-1}\lfloor\partial\Omega$ and denote by ν the geometric measure theoretic outward unit normal to the domain Ω. If for each function $f \in L^1\big(\partial\Omega, \frac{\sigma(x)}{1+|x|^{n-1}}\big)$ and σ-a.e. $x \in \partial\Omega$ we define

$$Tf(x) := \lim_{\varepsilon\to 0^+} \int_{y\in\partial\Omega,\ |x-y|>\varepsilon} \langle x-y, \nu(y)\rangle b(x, x-y) f(y)\, d\sigma(y), \tag{4.1.65}$$

then for each $m \in \mathbb{N}$ and $p \in (1,\infty)$ there exists some $C_m \in (0,\infty)$, which depends only on m, n, p, and the UR constants of $\partial\Omega$ with the property that, with the piece of notation introduced in (4.1.12),

$$\|T\|_{L^p(\partial\Omega,\sigma)\to L^p(\partial\Omega,\sigma)} \le C_m \cdot C_b \cdot \|\nu\|^{\langle m\rangle}_{[\mathrm{BMO}(\partial\Omega,\sigma)]^n}, \tag{4.1.66}$$

where

$$C_b := \sup_{\substack{(x,z)\in\mathbb{R}^n\times S^{n-1}\\ |\alpha|\leq M}} \big|(\partial_z^\alpha b)(x,z)\big| \in (0,\infty). \tag{4.1.67}$$

Indeed, this follows from the spherical harmonic decomposition technique employed in the proof of [132, Theorem 5.2.3] and the estimates for chord-dot-normal singular integral operators with constant coefficient kernels from Theorem 4.1.3. Concretely, [132, (5.2.312)] together with (4.1.53) (applied to each $T^{i\ell}$ in said decomposition), [132, (5.2.299), (5.2.304), (5.2.306)] readily imply (4.1.66).

Moreover, estimates similar to (4.1.66) are also valid for the class of chord-dot-normal singular integral operators with variable coefficient kernels, acting on functions $f \in L^1\big(\partial\Omega, \frac{\sigma(x)}{1+|x|^{n-1}}\big)$ at σ-a.e. $x \in \partial\Omega$ according to

$$T^{\#} f(x) := \lim_{\varepsilon\to 0^+} \int_{y\in\partial\Omega,\ |x-y|>\varepsilon} \langle \nu(x), y-x\rangle b(y, y-x) f(y)\, d\sigma(y), \tag{4.1.68}$$

$$\widetilde{T} f(x) := \lim_{\varepsilon\to 0^+} \int_{y\in\partial\Omega,\ |x-y|>\varepsilon} \langle \nu(y), x-y\rangle b(y, x-y) f(y)\, d\sigma(y), \tag{4.1.69}$$

$$\widetilde{T}^{\#} f(x) := \lim_{\varepsilon\to 0^+} \int_{y\in\partial\Omega,\ |x-y|>\varepsilon} \langle \nu(x), y-x\rangle b(x, y-x) f(y)\, d\sigma(y), \tag{4.1.70}$$

as well as for the class of their associated maximal operators. Finally, the aforementioned operator norm estimates continue to hold on a variety of other function spaces, as indicated earlier.

Comment 6. In the context of Theorem 4.1.3, estimates (4.1.29)-(4.1.32) continue to hold with a fixed constant $C_m \in (0,\infty)$ when the integrability exponent and the corresponding Muckenhoupt weight are allowed to vary with control. Specifically, an inspection of the proof of Theorem 4.1.3 given below shows that for each compact interval $I \subset (1,\infty)$ and each number $W \in (0,\infty)$ there exists a constant $C_m \in (0,\infty)$, which depends only on m, n, I, W, and the UR constants of $\partial\Omega$, with the property that (4.1.29)-(4.1.32) hold for each $p \in I$ and each $w \in A_p(\partial\Omega,\sigma)$ with $[w]_{A_p} \leq W$. In fact, similar considerations apply to all estimates in the conclusion of Theorem 4.1.3.

Here is the proof of Theorem 4.1.3.

Proof of Theorem 4.1.3 The key estimates recorded in (4.1.29)-(4.1.32) as well as the subsequent comment are proved in [111]. Granted these, whenever $\mathbb{X}$ is a Generalized Banach Function Space on $(\partial\Omega,\sigma)$ for which (4.1.42) holds, we may invoke the extrapolation result from [130, Corollary 5.2.3] to obtain that the operators (4.1.43)-(4.1.44) are well defined and bounded, and that for each $m \in \mathbb{N}$ we have

$$\max\Big\{\|T_*\|_{\mathbb{X}\to\mathbb{X}}, \|T_*\|_{\mathbb{X}'\to\mathbb{X}'}, \|T_*^\#\|_{\mathbb{X}\to\mathbb{X}}, \|T_*^\#\|_{\mathbb{X}'\to\mathbb{X}'}\Big\}$$
$$\leq C_m\Big(\sum_{|\alpha|\leq N}\sup_{S^{n-1}}|\partial^\alpha k|\Big)\|\nu\|^{\langle m\rangle}_{[\mathrm{BMO}(\partial\Omega,\sigma)]^n} \tag{4.1.71}$$

as well as

$$\max\Big\{\|T\|_{\mathbb{X}\to\mathbb{X}}, \|T\|_{\mathbb{X}'\to\mathbb{X}'}, \|T^\#\|_{\mathbb{X}\to\mathbb{X}}, \|T^\#\|_{\mathbb{X}'\to\mathbb{X}'}\Big\}$$
$$\leq C_m\Big(\sum_{|\alpha|\leq N}\sup_{S^{n-1}}|\partial^\alpha k|\Big)\|\nu\|^{\langle m\rangle}_{[\mathrm{BMO}(\partial\Omega,\sigma)]^n} \tag{4.1.72}$$

for a constant $C_m \in (0,\infty)$ having the nature specified in the statement of the theorem. That the operators in (4.1.45)-(4.1.46) are also well defined and bounded follows from [131, (2.3.68)] (applied both for $\mathbb{X}$ and $\mathbb{X}'$), keeping in mind the Lorentz-Luxemburg bi-dual theorem recorded in [130, Proposition 5.1.14] and the fact that $\mathring{\mathbb{X}}$ is stable under multiplication by essentially bounded functions (cf. [130, (5.2.85)]). In addition, all estimates in (4.1.48)-(4.1.49) not covered by (4.1.71)-(4.1.72) are implied by what we have proved already in (4.1.71)-(4.1.72) and abstract functional analysis (cf. [130, (1.2.20)]).

To proceed, recall that the Morrey and block spaces space have been identified in [130, Proposition 6.2.17] as Generalized Banach Function Spaces, which are actually mutually associated with one another. Also, thanks to [130, Corollaries 6.2.11, 6.2.13], the hypotheses assumed in (4.1.42) are valid for the aforementioned spaces. As such, the abstract results in the last portion of the statement of Theorem 4.1.3 imply all properties listed in (4.1.33)-(4.1.41) plus the final comment. □

We next record a two-dimensional consequence of Theorem 4.1.3.

Corollary 4.1.4 *Let Ω be a* UR *domain in $\mathbb{R}^2$. Abbreviate $\sigma := \mathcal{H}^1\lfloor\partial\Omega$ and denote by τ the geometric measure theoretic unit tangent vector along $\partial\Omega$ (cf.* [129, (5.6.29)-(5.6.31)]). *Also, assume $F : \mathbb{R}^2\setminus\{0\}\to\mathbb{R}$ is a function of class $\mathscr{C}^N$, for some sufficiently large $N\in\mathbb{N}$, which is even and positive homogeneous of degree zero. In this setting, for each $\varepsilon>0$ and each $f\in L^1\big(\partial\Omega,\frac{\sigma(x)}{1+|x|}\big)$ define*

$$T_\varepsilon f(x) := \int_{y\in\partial\Omega,\ |x-y|>\varepsilon}\partial_{\tau(y)}[F(x-y)]f(y)\,d\sigma(y)\ \textit{for all}\ x\in\partial\Omega, \tag{4.1.73}$$

$$T_\varepsilon^\# f(x) := \int_{y\in\partial\Omega,\ |x-y|>\varepsilon}\partial_{\tau(x)}[F(x-y)]f(y)\,d\sigma(y)\ \textit{for}\ \sigma\textit{-a.e.}\ x\in\partial\Omega, \tag{4.1.74}$$

Also, consider the maximal operators T_*, $T_*^\#$ associated with these families of truncated SIO's, *whose actions on each given function $f\in L^1\big(\partial\Omega,\frac{\sigma(x)}{1+|x|}\big)$ are defined as*

$$\begin{aligned} T_*f(x) &:= \sup\nolimits_{\varepsilon>0}\big|T_\varepsilon f(x)\big|\ \textit{for each}\ x\in\partial\Omega,\\ T_*^\# f(x) &:= \sup\nolimits_{\varepsilon>0}\big|T_\varepsilon^\# f(x)\big|\ \textit{for}\ \sigma\textit{-a.e.}\ x\in\partial\Omega, \end{aligned} \tag{4.1.75}$$

as well as the principal-value singular integral operators $T, T^{\#}$ on $\partial\Omega$ acting on each $f \in L^1\big(\partial\Omega, \frac{\sigma(x)}{1+|x|}\big)$ at σ-a.e. $x \in \partial\Omega$ according to

$$Tf(x) := \lim_{\varepsilon\to 0^+} \int_{y\in\partial\Omega,\ |x-y|>\varepsilon} \partial_{\tau(y)}[F(x-y)]f(y)\,d\sigma(y), \tag{4.1.76}$$

and, respectively,

$$T^{\#}f(x) := \lim_{\varepsilon\to 0^+} \int_{y\in\partial\Omega,\ |x-y|>\varepsilon} \partial_{\tau(x)}[F(x-y)]\,d\sigma(y). \tag{4.1.77}$$

Then similar results as in Theorem 4.1.3 are valid for these integral operators.

Proof This is a direct consequence of Theorem 4.1.3 and [132, Lemma 5.2.4]. □

From [131, Theorem 2.3.2] we know that principal-value singular integral operators of convolution-type on UR sets are of weak type $(1, 1)$, with operator norm bounded by the UR character of the set in question. Our next theorem shows that this may be further refined in the case of transpose chord-dot-normal SIO's. Specifically, the following version of the end-point case $p = 1$ of (4.1.54) holds:

Theorem 4.1.5 *Let $\Omega \subseteq \mathbb{R}^n$ be a* UR *domain. Abbreviate $\sigma := \mathcal{H}^{n-1}\lfloor\partial\Omega$ and denote by ν the geometric measure theoretic outward unit normal to Ω. Also, consider a sufficiently large integer $N = N(n) \in \mathbb{N}$ and suppose $k \in \mathscr{C}^N(\mathbb{R}^n \setminus \{0\})$ is a complex-valued function which is even and positive homogeneous of degree $-n$. In this setting, for each $f \in L^1\big(\partial\Omega, \frac{\sigma(x)}{1+|x|^{n-1}}\big)$ and σ-a.e. $x \in \partial\Omega$ define*

$$T^{\#}f(x) := \lim_{\varepsilon\to 0^+} \int_{y\in\partial\Omega,\ |x-y|>\varepsilon} \langle y-x, \nu(x)\rangle k(x-y)f(y)\,d\sigma(y). \tag{4.1.78}$$

Then for each $m \in \mathbb{N}$ there exists $C_m \in (0,\infty)$, depending only on m, n, and the UR *constants of $\partial\Omega$ so that, with the notation introduced in* (4.1.12),

$$\|T^{\#}\|_{L^1(\partial\Omega,\sigma)\to L^{1,\infty}(\partial\Omega,\sigma)} \leq C_m\Big(\sum_{|\alpha|\leq N} \sup_{S^{n-1}} |\partial^\alpha k|\Big)\|\nu\|^{\langle m\rangle}_{[\mathrm{BMO}(\partial\Omega,\sigma)]^n} \tag{4.1.79}$$

Also, when $\|\nu\|_{[\mathrm{BMO}(\partial\Omega,\sigma)]^n}$ is sufficiently small relative to n, and the Ahlfors regularity constants of $\partial\Omega$ one may take $C_m \in (0,\infty)$ appearing in (4.1.79) *to depend itself only on said entities (i.e., n, the Ahlfors regularity constants of $\partial\Omega$) and m.*

The remarks made in Comment 3 above, following the statement of Theorem 4.1.3, remain pertinent in the context of Theorem 4.1.5. For example, with $\|\nu\|_*$ abbreviating $\|\nu\|_{[\mathrm{BMO}(\partial\Omega,\sigma)]^n}$, they imply

$$\|T^{\#}\|_{L^1(\partial\Omega,\sigma)\to L^{1,\infty}(\partial\Omega,\sigma)} \leq C\Big(\sum_{|\alpha|\leq N} \sup_{S^{n-1}} |\partial^\alpha k|\Big)\|\nu\|_* \ln\big(e/\|\nu\|_*\big). \tag{4.1.80}$$

Also, Theorem 4.1.5 extends to chord-dot-normal singular integral operators with variable coefficient kernels. Concretely, assume that (4.1.64) holds and for each function $f \in L^1\big(\partial\Omega, \frac{\sigma(x)}{1+|x|^{n-1}}\big)$ define at σ-a.e. $x \in \partial\Omega$

$$T^{\#} f(x) := \lim_{\varepsilon\to 0^+} \int_{y\in\partial\Omega,\ |x-y|>\varepsilon} \langle x-y, \nu(x)\rangle b(y, x-y) f(y)\, d\sigma(y). \tag{4.1.81}$$

Then for each $m \in \mathbb{N}$ there exists a constant $C_m \in (0,\infty)$, depending only on m, n, and the UR constants of $\partial\Omega$ with the property that, with the piece of notation introduced in (4.1.12),

$$\|T^{\#}\|_{L^1(\partial\Omega,\sigma)\to L^{1,\infty}(\partial\Omega,\sigma)} \leq C_m \cdot C_b \cdot \|\nu\|^{\langle m\rangle}_{[\mathrm{BMO}(\partial\Omega,\sigma)]^n}, \tag{4.1.82}$$

where C_b is as in (4.1.67). This follows from (4.1.79) and the spherical harmonic decomposition technique employed in the proof of [132, Theorem 5.2.3]. Moreover, a similar result is valid for the singular integral operator acting on each function $f \in L^1\big(\partial\Omega, \frac{\sigma(x)}{1+|x|^{n-1}}\big)$ at σ-a.e. $x \in \partial\Omega$ according to

$$\widetilde{T}^{\#} f(x) := \lim_{\varepsilon\to 0^+} \int_{y\in\partial\Omega,\ |x-y|>\varepsilon} \langle x-y, \nu(x)\rangle b(x, x-y) f(y)\, d\sigma(y). \tag{4.1.83}$$

The proof of Theorem 4.1.5 is given next.

Proof of Theorem 4.1.5 Fix a number $m \in \mathbb{N}$. If the boundary $\partial\Omega$ is compact, then $\|\nu\|_{[\mathrm{BMO}(\partial\Omega,\sigma)]^n} = 1$ (see (4.1.4)) so (4.1.79) is directly implied by [131, (2.3.19)] in this scenario. Henceforth, assume that $\partial\Omega$ is unbounded. When dealing with (4.1.79), there is no loss of generality in assuming that

$$\sum_{|\alpha|\leq N} \sup_{S^{n-1}} |\partial^\alpha k| = 1, \tag{4.1.84}$$

a property which may be always arranged via linearly rescaling the kernel k. Pick an arbitrary function $f \in L^1(\partial\Omega,\sigma) \cap L^2(\partial\Omega,\sigma)$ with compact support. Throughout the proof, abbreviate

$$\delta := \|\nu\|_{[\mathrm{BMO}(\partial\Omega,\sigma)]^n}. \tag{4.1.85}$$

If $\delta = 0$, then ν is constant and Ω is a half-space. In such a scenario, $T^{\#} = 0$ so (4.1.79) is trivially true. Henceforth, assume $\delta > 0$. Then $\delta \in (0,1]$ by (4.1.2).

We need to show the existence of a constant $C_m \in (0,\infty)$ as specified in the statement with the property that for each $\lambda \in (0,\infty)$ we have

$$\sigma\Big(\big\{x \in \partial\Omega : |(T^{\#} f)(x)| > \lambda\big\}\Big) \leq C_m \delta^{\langle m\rangle} \frac{\|f\|_{L^1(\partial\Omega,\sigma)}}{\lambda}. \tag{4.1.86}$$

To this end, fix an arbitrary $\lambda \in (0,\infty)$. We shall perform a Calderón-Zygmund decomposition of f at level $\lambda/\delta^{\langle m\rangle}$. As in [131, (2.2.190)-(2.2.197)], this shows that there exist two finite constants $C > 0$, $N \in \mathbb{N}$ (depending only on n and the Ahlfors regularity constants of $\partial\Omega$), along with a family of surface balls $\{\Delta_j\}_{j\in\mathbb{N}}$, say $\Delta_j := B(x_j, r_j) \cap \partial\Omega$ with $x_j \in \partial\Omega$ and $r_j \in (0,\infty)$ for each $j \in \mathbb{N}$, and two functions $g, b : \partial\Omega \to \mathbb{R}$ satisfying the following properties:

$$f = g + b \text{ at } \sigma\text{-a.e. point in } \partial\Omega, \tag{4.1.87}$$

$$g \in L^1(\partial\Omega,\sigma) \cap L^\infty(\partial\Omega,\sigma) \text{ and } b \in L^1(\partial\Omega,\sigma) \cap L^2(\partial\Omega,\sigma), \tag{4.1.88}$$

$$\|g\|_{L^1(\partial\Omega,\sigma)} \le C\|f\|_{L^1(\partial\Omega,\sigma)}, \ |g(x)| \le C\lambda/\delta^{\langle m\rangle} \text{ for } \sigma\text{-a.e. } x \in \partial\Omega, \tag{4.1.89}$$

$$\text{if } b_j := \Big(f - \fint_{\Delta_j} f\,d\sigma\Big)\mathbf{1}_{\Delta_j} \text{ for each } j \in \mathbb{N}, \text{ then} \tag{4.1.90}$$

$$b = \sum_{j\in\mathbb{N}} b_j \text{ with convergence both in } L^1(\partial\Omega,\sigma) \text{ and } L^2(\partial\Omega,\sigma), \tag{4.1.91}$$

$$\sum_{j\in\mathbb{N}} \|b_j\|_{L^1(\partial\Omega,\sigma)} \le C\|f\|_{L^1(\partial\Omega,\sigma)}, \tag{4.1.92}$$

$$\operatorname{supp} b_j \subseteq \Delta_j, \ \int_{\partial\Omega} b_j\,d\sigma = 0, \text{ and } \fint_{\Delta_j} |b_j|\,d\sigma \le C\lambda/\delta^{\langle m\rangle}, \ \forall j \in \mathbb{N}, \tag{4.1.93}$$

$$\begin{aligned} &\text{if } \mathcal{O} := \textstyle\bigcup_{j\in\mathbb{N}} \Delta_j \text{ and } \mathcal{F} := \partial\Omega \setminus \mathcal{O}, \text{ then } \mathbf{1}_{\mathcal{O}} \le \textstyle\sum_{j\in\mathbb{N}} \mathbf{1}_{\Delta_j} \le N\mathbf{1}_{\mathcal{O}}, \\ &\sigma(\mathcal{O}) \le C\big(\lambda/\delta^{\langle m\rangle}\big)^{-1}\|f\|_{L^1(\partial\Omega,\sigma)}, \text{ and } \operatorname{dist}(\Delta_j,\mathcal{F}) \approx r_j \text{ for } j \in \mathbb{N}. \end{aligned} \tag{4.1.94}$$

Since $b = \sum_j b_j$ vanishes identically on $\partial\Omega \setminus \big(\bigcup_j \Delta_j\big) = \mathcal{F}$, it follows from (4.1.87) that $g = f$ on $\mathcal{F}$. Bearing this in mind, along with the pointwise estimate in (4.1.89), we may then write

$$\begin{aligned} \|g\|^2_{L^2(\partial\Omega,\sigma)} &= \int_{\partial\Omega} |g|^2\,d\sigma = \int_{\mathcal{F}} |g|^2\,d\sigma + \int_{\mathcal{O}} |g|^2\,d\sigma \\ &\le \int_{\mathcal{F}} \big(\lambda/\delta^{\langle m\rangle}\big)|f|\,d\sigma + C^2\big(\lambda/\delta^{\langle m\rangle}\big)^2\sigma(\mathcal{O}) \le C\big(\lambda/\delta^{\langle m\rangle}\big)\|f\|_{L^1(\partial\Omega,\sigma)}, \end{aligned} \tag{4.1.95}$$

where $C \in (0,\infty)$ depends only on n and the Ahlfors regularity constants of $\partial\Omega$. Using the L^2-theory (see (4.1.54) with $p := 2$) we therefore obtain

$$\begin{aligned} \sigma\Big(\big\{x \in \partial\Omega : |(T^\# g)(x)| > \lambda/2\big\}\Big) &\le (2/\lambda)^2 \int_{\partial\Omega} |T^\# g|^2\,d\sigma \\ &\le (2/\lambda)^2\big(C_m\delta^{\langle m\rangle}\big)^2\|g\|^2_{L^2(\partial\Omega,\sigma)} \le C_m\delta^{\langle m\rangle}\,\frac{\|f\|_{L^1(\partial\Omega,\sigma)}}{\lambda}, \end{aligned} \tag{4.1.96}$$

where $C_m \in (0,\infty)$ depends only on m, n, and the UR character of $\partial\Omega$.

To proceed, define

$$M_j := T^\# b_j \text{ for each } j \in \mathbb{N}. \tag{4.1.97}$$

Fix $j \in \mathbb{N}$ arbitrary and observe that, since b_j is supported in Δ_j, itself a subset of $\mathcal{O} = \partial\Omega \setminus \mathcal{F}$, we have

$$M_j(x) = \int_{\Delta_j} \langle y - x, \nu(x)\rangle k(x - y) b_j(y)\, d\sigma(y) \text{ for } \sigma\text{-a.e. } x \in \mathcal{F}. \tag{4.1.98}$$

Taking into account the vanishing moment condition satisfied by b_j (see (4.1.93)), we may further express

$$M_j(x) = -\int_{\Delta_j} \Big\{\langle \nu(x), x - y\rangle\, k(x - y) - \langle \nu(x), x - x_j\rangle\, k(x - x_j)\Big\} b_j(y)\, d\sigma(y) \tag{4.1.99}$$

at σ-a.e. point $x \in \mathcal{F}$. At each $x \in \mathcal{F}$ let us also define

$$M_{\Delta_j}(x) := -\int_{\Delta_j} \Big\{\langle \nu_{\Delta_j}, x - y\rangle\, k(x - y) - \langle \nu_{\Delta_j}, x - x_j\rangle\, k(x - x_j)\Big\} b_j(y)\, d\sigma(y), \tag{4.1.100}$$

where we have abbreviated $\nu_{\Delta_j} := \fint_{\Delta_j} \nu\, d\sigma$. For σ-a.e. point $x \in \mathcal{F}$, which we shall fix for now, write

$$M_{\Delta_j}(x) - M_j(x) = \int_{\Delta_j} \Big\{F_x(y) - F_x(x_j)\Big\} b_j(y)\, d\sigma(y), \tag{4.1.101}$$

where $F_x(y) := \langle \nu(x) - \nu_{\Delta_j}, x - y\rangle\, k(x - y)$ for each $y \in \Delta_j$. In view of this and the normalization in (4.1.84), it follows that

$$|\nabla_y F_x(y)| \leq C \frac{|\nu(x) - \nu_{\Delta_j}|}{|x - y|^n} \quad \text{for each } y \in \Delta_j, \tag{4.1.102}$$

for some purely dimensional constant $C = C(n) \in (0, \infty)$. Consequently, employing the Mean Value Theorem and using the fact that since $x \in \mathcal{F}$,

$$|z - x| \approx |x_j - x|, \qquad \text{uniformly for all } z \in \Delta_j \tag{4.1.103}$$

(cf. (4.1.94)), we obtain that there exists some constant $C \in (0, \infty)$, which depends only on n and the Ahlfors regularity constants of $\partial\Omega$, such that

$$|F_x(y) - F_x(x_j)| \leq C r_j \frac{|\nu(x) - \nu_{\Delta_j}|}{|x - x_j|^n} \quad \text{for each } y \in \Delta_j. \tag{4.1.104}$$

Together, (4.1.101), (4.1.104), the last property in (4.1.93), the definition on Δ_j, and the Ahlfors regularity of $\partial\Omega$ yield

$$\begin{aligned} \big|M_j(x) - M_{\Delta_j}(x)\big| &\leq C r_j \frac{|\nu(x) - \nu_{\Delta_j}|}{|x - x_j|^n} \int_{\Delta_j} |b_j(y)|\, d\sigma(y) \\ &\leq C\big(\lambda/\delta^{\langle m\rangle}\big)\, r_j\, \sigma(\Delta_j) \frac{|\nu(x) - \nu_{\Delta_j}|}{|x - x_j|^n} \quad \text{for } \sigma\text{-a.e. } x \in \mathcal{F}. \end{aligned} \tag{4.1.105}$$

Based on (4.1.105), we may then estimate

$$\int_{\mathcal{F}} \big|M_j(x) - M_{\Delta_j}(x)\big|\,\mathrm{d}\sigma(x) \le C(\lambda/\delta^{\langle m\rangle})r_j\,\sigma(\Delta_j)\int_{\mathcal{F}} \frac{|\nu(x) - \nu_{\Delta_j}|}{|x - x_j|^n}\,\mathrm{d}\sigma(x). \tag{4.1.106}$$

Also, the Ahlfors regularity of $\partial\Omega$ plus [129, (7.4.63)] and (4.1.85) imply

$$\begin{aligned}\int_{\mathcal{F}} \frac{|\nu(x) - \nu_{\Delta_j}|}{|x - x_j|^n}\,\mathrm{d}\sigma(x) &\le \int_{\partial\Omega\setminus\Delta_j} \frac{|\nu(x) - \nu_{\Delta_j}|}{|x - x_j|^n}\,\mathrm{d}\sigma(x)\\ &\le \sum_{\ell=1}^{\infty}\int_{[B(x_j,2^\ell r_j)\setminus B(x_j,2^{\ell-1}r_j)]\cap\partial\Omega} \frac{|\nu(x) - \nu_{\Delta_j}|}{|x - x_j|^n}\,\mathrm{d}\sigma(x)\\ &\le \sum_{\ell=1}^{\infty}(2^\ell r_j)^{-1}\fint_{B(x_j,2^\ell r_j)\cap\partial\Omega}|\nu(x) - \nu_{\Delta_j}|\,\mathrm{d}\sigma(x)\\ &\le C\delta\sum_{\ell=1}^{\infty}(2^\ell r_j)^{-1}\ell \le C\delta r_j^{-1}. \end{aligned} \tag{4.1.107}$$

Combining (4.1.106) with (4.1.107) then yields

$$\int_{\mathcal{F}} \big|M_j(x) - M_{\Delta_j}(x)\big|\,\mathrm{d}\sigma(x) \le C\lambda(\delta/\delta^{\langle m\rangle})\sigma(\Delta_j). \tag{4.1.108}$$

Next, for each $x \in \mathcal{F}$ we decompose

$$M_{\Delta_j}(x) = M^{(1)}_{\Delta_j}(x) + M^{(2)}_{\Delta_j}(x) \tag{4.1.109}$$

where

$$M^{(1)}_{\Delta_j}(x) := -\int_{\Delta_j}\langle\nu_{\Delta_j}, x - x_j\rangle\big[k(x-y) - k(x-x_j)\big]b_j(y)\,\mathrm{d}\sigma(y), \tag{4.1.110}$$

$$M^{(2)}_{\Delta_j}(x) := \int_{\Delta_j}\langle\nu_{\Delta_j}, y - x_j\rangle\,k(x-y)b_j(y)\,\mathrm{d}\sigma(y). \tag{4.1.111}$$

Pick $x \in \mathcal{F}$ and proceed with estimating $M^{(1)}_{\Delta_j}(x)$ and $M^{(2)}_{\Delta_j}(x)$. The Mean Value Theorem (plus (4.1.103) and (4.1.231)), and the last property in (4.1.93) imply

$$\begin{aligned}\big|M^{(1)}_{\Delta_j}(x)\big| &\le \big|\langle\nu_{\Delta_j}, x - x_j\rangle\big|\frac{Cr_j}{|x - x_j|^{n+1}}\int_{\Delta_j}|b_j(y)|\,\mathrm{d}\sigma(y)\\ &\le C\big|\langle\nu_{\Delta_j}, x - x_j\rangle\big|\frac{r_j}{|x - x_j|^{n+1}}(\lambda/\delta^{\langle m\rangle})\sigma(\Delta_j). \end{aligned} \tag{4.1.112}$$

Consequently,

$$\begin{aligned}\int_{\mathcal{F}}\big|M^{(1)}_{\Delta_j}(x)\big|\,d\sigma(x)&\leq C\,r_j\big(\lambda/\delta^{\langle m\rangle}\big)\sigma(\Delta_j)\int_{\partial\Omega\setminus\Delta_j}\frac{\big|\langle\nu_{\Delta_j},x-x_j\rangle\big|}{|x-x_j|^{n+1}}\,d\sigma(x)\\
&=C\,r_j\big(\lambda/\delta^{\langle m\rangle}\big)\sigma(\Delta_j)\sum_{\ell=1}^{\infty}\int_{[B(x_j,2^\ell r_j)\setminus B(x_j,2^{\ell-1}r_j)]\cap\partial\Omega}\frac{\big|\langle\nu_{\Delta_j},x-x_j\rangle\big|}{|x-x_j|^{n+1}}\,d\sigma(x)\\
&\leq C\,r_j\big(\lambda/\delta^{\langle m\rangle}\big)\sigma(\Delta_j)\sum_{\ell=1}^{\infty}\big(2^\ell r_j\big)^{-1-n}\sup_{x\in B(x_j,2^\ell r_j)\cap\partial\Omega}\big|\langle\nu_{\Delta_j},x-x_j\rangle\big|\times\\
&\qquad\times\sigma\big(B(x_j,2^\ell r_j)\cap\partial\Omega\big)\\
&\leq C\,r_j\big(\lambda/\delta^{\langle m\rangle}\big)\sigma(\Delta_j)\sum_{\ell=1}^{\infty}\big(2^\ell r_j\big)^{-2}\big(C\,r_j2^\ell(1+\ell)\delta\big)\\
&\leq C\lambda\big(\delta/\delta^{\langle m\rangle}\big)\sigma(\Delta_j)\sum_{\ell=1}^{\infty}2^{-\ell}(1+\ell)\leq C\lambda\big(\delta/\delta^{\langle m\rangle}\big)\sigma(\Delta_j),\end{aligned}\tag{4.1.113}$$

where the supremum has been estimated based on (3.1.25) (presently used with $z:=x_j$, $R:=r_j$, $y:=x_j$, and $\gamma:=2^\ell$) and the Ahlfors regularity of $\partial\Omega$.

Continue to assume that $x\in\mathcal{F}$. Focusing our attention on $M^{(2)}_{\Delta_j}(x)$, from (4.1.110), (3.1.25), and the last property in (4.1.93) we obtain

$$\begin{aligned}\big|M^{(2)}_{\Delta_j}(x)\big|&\leq\sup_{y\in\Delta_j}\big|\langle\nu_{\Delta_j},y-x_j\rangle\big|\frac{C}{|x-x_j|^n}\int_{\Delta_j}|b_j(y)|\,d\sigma(y)\\
&\leq\big(Cr_j\delta\big)\frac{1}{|x-x_j|^n}\,C\big(\lambda/\delta^{\langle m\rangle}\big)\sigma(\Delta_j).\end{aligned}\tag{4.1.114}$$

From (4.1.114) and [129, (7.2.5)] we then deduce that

$$\begin{aligned}\int_{\mathcal{F}}\big|M^{(2)}_{\Delta_j}(x)\big|\,d\sigma(x)&\leq C\lambda\big(\delta/\delta^{\langle m\rangle}\big)r_j\sigma(\Delta_j)\int_{\mathcal{F}}\frac{1}{|x-x_j|^n}\,d\sigma(x)\\
&\leq C\lambda\big(\delta/\delta^{\langle m\rangle}\big)r_j\sigma(\Delta_j)\int_{\partial\Omega\setminus\Delta_j}\frac{1}{|x-x_j|^n}\,d\sigma(x)\leq C\lambda\big(\delta/\delta^{\langle m\rangle}\big)\sigma(\Delta_j).\end{aligned}\tag{4.1.115}$$

Gathering (4.1.97), (4.1.108), (4.1.109), (4.1.113), and (4.1.115) we conclude

$$\int_{\mathcal{F}}\big|(T^\# b_j)(x)\big|\,d\sigma(x)\leq C\lambda\big(\delta/\delta^{\langle m\rangle}\big)\sigma(\Delta_j)\ \text{ for each } j\in\mathbb{N}.\tag{4.1.116}$$

From (4.1.91) and the continuity of $T^\#$ on $L^2(\partial\Omega,\sigma)$ we see that $T^\# b=\sum_{j\in\mathbb{N}}T^\# b_j$ in $L^2(\partial\Omega,\sigma)$. Then the sequence of partial sums for the latter series has a subsequence which converges σ-a.e. to $T^\# b$. In turn, this gives $|T^\# b|\leq\sum_{j\in\mathbb{N}}|T^\# b_j|$ at σ-a.e. point on $\partial\Omega$. In concert with (4.1.116) and (4.1.94), this implies

$$\int_{\mathcal{F}}\big|(T^\# b)(x)\big|\,d\sigma(x)\leq C\lambda\big(\delta/\delta^{\langle m\rangle}\big)\sigma(O)\leq C\delta\|f\|_{L^1(\partial\Omega,\sigma)}.\tag{4.1.117}$$

From this inequality it follows immediately that

$$\sigma\Big(\big\{x \in \mathcal{F} : \big|(T^{\#}b)(x)\big| > \lambda/2\big\}\Big) \leq C\delta\frac{\|f\|_{L^1(\partial\Omega,\sigma)}}{\lambda}. \tag{4.1.118}$$

We also know from (4.1.94) that

$$\sigma\big(\partial\Omega \setminus \mathcal{F}\big) = \sigma(O) \leq C\delta^{\langle m\rangle}\,\frac{\|f\|_{L^1(\partial\Omega,\sigma)}}{\lambda}. \tag{4.1.119}$$

As noted earlier, $\delta \in (0, 1]$ so from (4.1.17) we see that $\delta \leq ({}^{m}e) \cdot \delta^{\langle m\rangle}$. On account of this and (4.1.118)-(4.1.119) we arrive at the conclusion that

$$\sigma\Big(\big\{x \in (\partial\Omega : \big|(T^{\#}b)(x)\big| > \lambda/2\big\}\Big) \leq C_m\delta^{\langle m\rangle}\,\frac{\|f\|_{L^1(\partial\Omega,\sigma)}}{\lambda}. \tag{4.1.120}$$

At this stage, (4.1.87), (4.1.96), and (4.1.120) justify (4.1.86), from which (4.1.79) follows in view of the density of simple functions in $L^1(\partial\Omega, \sigma)$ and the fact that $T^{\#}$ is continuous from the latter space into $L^{1,\infty}(\partial\Omega, \sigma)$ (cf. [131, (2.3.19)]). The final claim in the statement is a consequence of Theorem 3.1.5 (cf. (3.1.24)). The proof of Theorem 4.1.5 is therefore complete. □

We next proceed to establish norm estimates similar to those in Theorem 4.1.3 for "chord-dot-normal" singular integral operators on boundary Sobolev spaces.

Theorem 4.1.6 *Let $\Omega \subseteq \mathbb{R}^n$ be a* UR *domain. Abbreviate $\sigma := \mathcal{H}^{n-1}\lfloor\partial\Omega$ and denote by ν the geometric measure theoretic outward unit normal to Ω. Also, consider a sufficiently large integer $N = N(n) \in \mathbb{N}$ and suppose $k \in \mathscr{C}^N(\mathbb{R}^n \setminus \{0\})$ is a complex-valued function which is even and positive homogeneous of degree $-n$. Bring in the principal-value singular integral operators $T, T^{\#}$ on $\partial\Omega$ acting from $L^1\big(\partial\Omega, \frac{\sigma(x)}{1+|x|^{n-1}}\big)$ as in* (4.1.27)-(4.1.28).

Then for each $m \in \mathbb{N}$, each $p \in (1,\infty)$, and each Muckenhoupt weight w in $A_p(\partial\Omega, \sigma)$ there is a constant $C_m \in (0,\infty)$, depending only on m, n, p, $[w]_{A_p}$, and the UR *constants of $\partial\Omega$, so that (with notation introduced in* (4.1.12))

$$\|T\|_{L^p_1(\partial\Omega,w)\to L^p_1(\partial\Omega,w)} \leq C_m\Big(\sum_{|\alpha|\leq N}\sup_{S^{n-1}}|\partial^\alpha k|\Big)\|\nu\|^{\langle m\rangle}_{[\mathrm{BMO}(\partial\Omega,\sigma)]^n}, \tag{4.1.121}$$

$$\|T^{\#}\|_{L^p_{-1}(\partial\Omega,w)\to L^p_{-1}(\partial\Omega,w)} \leq C_m\Big(\sum_{|\alpha|\leq N}\sup_{S^{n-1}}|\partial^\alpha k|\Big)\|\nu\|^{\langle m\rangle}_{[\mathrm{BMO}(\partial\Omega,\sigma)]^n}. \tag{4.1.122}$$

Under the stronger assumption that Ω is an NTA *domain with an Ahlfors regular boundary, for each $p \in (1,\infty)$ and each $m \in \mathbb{N}$ there exists $C_m \in (0,\infty)$, depending only on m, n, p, and the* NTA *constants of Ω, such that the operator $[T_{\rm mod}]$ associated with Ω and k as in* [132, (5.2.188)] *satisfies*

$$\big\|[T_{\rm mod}]\big\|_{\dot{L}^p_1(\partial\Omega,\sigma)/\sim\to\dot{L}^p_1(\partial\Omega,\sigma)/\sim} \leq C_m\Big(\sum_{|\alpha|\leq N}\sup_{S^{n-1}}|\partial^\alpha k|\Big)\|\nu\|^{\langle m\rangle}_{[\mathrm{BMO}(\partial\Omega,\sigma)]^n}. \tag{4.1.123}$$

Finally, similar estimates are valid on Morrey-based, block-based Sobolev spaces, and their homogeneous versions.

Proof Given any $f\in L^p_1(\partial\Omega,w)$ and $r,s\in\{1,\dots,n\}$, based on formula [132, (5.2.167)], Theorem 4.1.3, [131, Theorem 2.7.2], [132, (5.2.149)], [130, (11.4.8)], and (4.1.17) allow us to estimate

$$\begin{aligned}
\big\|\partial_{\tau_{rs}}(Tf)\big\|_{L^p(\partial\Omega,w)} &\leq \big\|T(\partial_{\tau_{rs}}f)\big\|_{L^p(\partial\Omega,w)} + \big\|\big[M_{\nu_r},T\big](\nabla_{\tan}f)_s\big\|_{L^p(\partial\Omega,w)}\\
&\quad+\big\|\big[M_{\nu_s},T\big](\nabla_{\tan}f)_r\big\|_{L^p(\partial\Omega,w)}+\big\|\big[M_{\nu_r},V\big](\nu_s\nabla_{\tan}f)\big\|_{L^p(\partial\Omega,w)}\\
&\quad+\big\|\big[M_{\nu_s},V\big](\nu_r\nabla_{\tan}f)\big\|_{L^p(\partial\Omega,w)}\\
&\leq C_m\Big(\sum_{|\alpha|\leq N}\sup_{S^{n-1}}|\partial^\alpha k|\Big)\|\nu\|^{\langle m\rangle}_{[\mathrm{BMO}(\partial\Omega,\sigma)]^n}\big\|\nabla_{\tan}f\big\|_{L^p(\partial\Omega,w)}
\end{aligned} \tag{4.1.124}$$

for some constant $C_m\in(0,\infty)$ which depends only on m, n, p, $[w]_{A_p}$, and the UR constants of $\partial\Omega$. Based on (A.0.140), (4.1.31), (4.1.124), and (A.0.88) for each $f\in L^p_1(\partial\Omega,w)$ we may write

$$\begin{aligned}
\|Tf\|_{L^p_1(\partial\Omega,w)} &= \|Tf\|_{L^p(\partial\Omega,w)}+\sum_{r,s=1}^n\big\|\partial_{\tau_{rs}}(Tf)\big\|_{L^p(\partial\Omega,w)}\\
&\leq C_m\Big(\sum_{|\alpha|\leq N}\sup_{S^{n-1}}|\partial^\alpha k|\Big)\|\nu\|^{\langle m\rangle}_{[\mathrm{BMO}(\partial\Omega,\sigma)]^n}\|f\|_{L^p_1(\partial\Omega,w)},
\end{aligned} \tag{4.1.125}$$

with $C_m\in(0,\infty)$ having the same nature as above. This establishes (4.1.121). Next, (4.1.122) is implied by (4.1.121) and duality (cf. [132, Theorem 1.5.1, item *(vi)*]). Also, the estimate in (4.1.123) may be justified based on [132, (5.2.193)-(5.2.195)] and [131, (2.7.24)]. Finally, the last claim in the statement of the theorem is dealt with similarly (reasoning as in the proof of [132, Theorem 5.2.2, item *(20)*]). □

We next take up the task of proving operator norm estimates of the same nature as before for "chord-dot-normal" singular integral operators acting on Hardy spaces.

Theorem 4.1.7 *Let $\Omega\subseteq\mathbb{R}^n$ (where $n\in\mathbb{N}$, $n\geq 2$) be a* UR *domain. Abbreviate $\sigma:=\mathcal{H}^{n-1}\lfloor\partial\Omega$ and denote by ν the geometric measure theoretic outward unit normal to Ω. Also, fix a large integer $N=N(n)\in\mathbb{N}$ and suppose $k\in\mathscr{C}^N(\mathbb{R}^n\setminus\{0\})$ is even and positive homogeneous of degree $-n$.*

Then for each $m\in\mathbb{N}$ and $p\in\big(\frac{n-1}{n},\infty\big)$ there exists $C_m\in(0,\infty)$ which depends only on m, n, p, the kernel k, and the UR *constants of $\partial\Omega$ with the property that the operator $T^\#$, originally acting on each function $f\in L^1\Big(\partial\Omega,\frac{\sigma(y)}{1+|y|^{n-1}}\Big)$ according to*

$$T^{\#}f(x) := \lim_{\varepsilon\to 0^+} \int_{y\in\partial\Omega,\ |x-y|>\varepsilon} \langle \nu(x), y-x\rangle k(y-x) f(y)\, d\sigma(y) \tag{4.1.126}$$

for σ-a.e. $x \in \partial\Omega$, extends to a linear and bounded mapping from the Hardy space $H^p(\partial\Omega, \sigma)$ into itself as in [132, (5.2.152)] *satisfying*

$$\|T^{\#}\|_{H^p(\partial\Omega,\sigma)\to H^p(\partial\Omega,\sigma)} \leq C_m\Big(\sum_{|\alpha|\leq N} \sup_{S^{n-1}} |\partial^\alpha k|\Big)\|\nu\|^{\langle m\rangle}_{[\mathrm{BMO}(\partial\Omega,\sigma)]^n}. \tag{4.1.127}$$

More generally, for each $m \in \mathbb{N}$, $p \in \big(\frac{n-1}{n}, \infty\big)$, and $q \in (0, \infty]$ there exists $C_m \in (0, \infty)$ depending only on m, n, p, q, the kernel k, and the UR *constants of $\partial\Omega$ so that $T^{\#}$, originally associated with k and Ω as in* (4.1.126), *extends to a linear and bounded mapping from the Lorentz-based Hardy space $H^{p,q}(\partial\Omega, \sigma)$ into itself as in* [132, (5.2.153)] *satisfying*

$$\|T^{\#}\|_{H^{p,q}(\partial\Omega,\sigma)\to H^{p,q}(\partial\Omega,\sigma)} \leq C_m\Big(\sum_{|\alpha|\leq N} \sup_{S^{n-1}} |\partial^\alpha k|\Big)\|\nu\|^{\langle m\rangle}_{[\mathrm{BMO}(\partial\Omega,\sigma)]^n}. \tag{4.1.128}$$

Thanks to (4.1.2) and (4.1.22), in the estimates recorded in (4.1.127)-(4.1.128) in place of $\|\nu\|^{\langle m\rangle}_{[\mathrm{BMO}(\partial\Omega,\sigma)]^n}$ we could employ

$$\|\nu\|_{[\mathrm{BMO}(\partial\Omega,\sigma)]^n}\cdot \underbrace{\ln\Big(\cdots \ln\big(\ln({}^{m}e/\|\nu\|_{[\mathrm{BMO}(\partial\Omega,\sigma)]^n})\big)\cdots\Big)}_{m \text{ natural logarithms}}. \tag{4.1.129}$$

In particular, abbreviating $\|\nu\|_* := \|\nu\|_{[\mathrm{BMO}(\partial\Omega,\sigma)]^n}$, these operator norms are

$$O\Big(\|\nu\|_* \ln\big(e/\|\nu\|_*\big)\Big) \quad \text{(corresponding to } m = 1\text{)}, \tag{4.1.130}$$

$$O\left(\|\nu\|_* \ln\Big(\ln\big(\mathrm{e}^e/\|\nu\|_*\big)\Big)\right) \quad \text{(corresponding to } m = 2\text{)}, \tag{4.1.131}$$

and so on. As such, all the aforementioned operator norms have at most linear growth in $\|\nu\|_*$, up to arbitrarily many iterated logarithms. In the same spirit, we may rely on (4.1.16) to conclude that the operator norms in (4.1.127)-(4.1.128) are

$$O\big(\|\nu\|_*^{1-\varepsilon}\big) \text{ for each fixed } \varepsilon \in (0,1). \tag{4.1.132}$$

We now turn to the task of providing the proof of Theorem 4.1.7.

Proof of Theorem 4.1.7 If $p \in (1, \infty)$ the estimate claimed in (4.1.127) is a direct consequence of (4.1.32) (used with $w \equiv 1$) and [129, (3.6.27)]. Henceforth, fix an integrability exponent $p \in \big(\frac{n-1}{n}, 1\big]$. If the set $\partial\Omega$ is compact, (4.1.4) gives that $\|\nu\|_{[\mathrm{BMO}(\partial\Omega,\sigma)]^n} = 1$, a scenario in which (4.1.127) is implied by [132, Theorem 5.1.1, item *(4)*]. There remains to deal with the case when $\partial\Omega$ is unbounded, which we will assume for the remainder of the proof. To simplify the presentation, observe that, via rescaling, there is no loss of generality in assuming that

$$\sum_{|\alpha|\le N}\sup_{S^{n-1}}|\partial^\alpha k| = 1. \tag{4.1.133}$$

The strategy is to show that there exists a constant $C_m \in (0,\infty)$ as in the statement of the theorem with the property that $T^\#$ maps atoms into a fixed multiple of molecules for the Hardy space in question, and such that the corresponding Hardy space quasi-norm of this molecule is bounded by $C_m \cdot \|\nu\|^{\langle m\rangle}_{[\mathrm{BMO}(\partial\Omega,\sigma)]^n}$. With this goal in mind, fix $q \in (1,\infty)$ and pick an arbitrary (p,q)-atom a on $\partial\Omega$. Recall from [130, (4.4.167)-(4.4.168)] that this means that $a : \partial\Omega \to \mathbb{C}$ is some σ-measurable function with the property that there exist $x_o \in \partial\Omega$ and $r \in \big(0, 2\,\mathrm{diam}(\partial\Omega)\big)$ such that

$$\begin{aligned}
&\operatorname{supp} a \subseteq B(x_o,r)\cap\partial\Omega,\\
&\|a\|_{L^q(\partial\Omega,\sigma)} \le \sigma\big(B(x_o,r)\cap\partial\Omega\big)^{1/q-1/p},\\
&\int_{\partial\Omega} a\,d\sigma = 0.
\end{aligned} \tag{4.1.134}$$

The key claim that we make in this regard is that for each given $\varepsilon \in \big(0,\frac{1}{n-1}\big)$ there exists some constant $C_m \in (0,\infty)$ depending only on ε, m, n, p, q, and the UR constants of $\partial\Omega$ such that

$$\begin{gathered}
\text{the function } M := T^\# a \text{ is a } C_m\cdot\|\nu\|^{\langle m\rangle}_{[\mathrm{BMO}(\partial\Omega,\sigma)]^n}\text{-multiple of a}\\
(p,q,\varepsilon)\text{-molecule centered near the ball } B(x_o,r) \text{ on } \partial\Omega\\
\text{(cf. [130, Definition 4.5.1]).}
\end{gathered} \tag{4.1.135}$$

To prove this, first note that according to [132, Theorem 5.2.2, item *(3)*] the function m is meaningfully defined and belongs to the space $L^q(\partial\Omega,\sigma)$. In fact, thanks to Theorem 4.1.3 we have

$$\begin{aligned}
\|M\|_{L^q(\partial\Omega,\sigma)} &= \|T^\# a\|_{L^q(\partial\Omega,\sigma)} \le C_m\cdot\|\nu\|^{\langle m\rangle}_{[\mathrm{BMO}(\partial\Omega,\sigma)]^n}\|a\|_{L^q(\partial\Omega,\sigma)}\\
&\le C_m\cdot\|\nu\|^{\langle m\rangle}_{[\mathrm{BMO}(\partial\Omega,\sigma)]^n}\,\sigma\big(B(x_o,r)\cap\partial\Omega\big)^{1/q-1/p},
\end{aligned} \tag{4.1.136}$$

for some $C_m \in (0,\infty)$ depending only on ε, m, n, p, q, and the UR constants of $\partial\Omega$, hence independent of the atom.

To proceed, introduce the notation $\Delta_\rho(x_o) := B(x_o,\rho)\cap\partial\Omega$ for each $\rho \in (0,\infty)$. Then (4.1.126) and (4.1.134) allow us to write, at σ-a.e. point $x \in \partial\Omega\setminus B(x_o,2r)$,

$$M(x) = -\int_{\Delta_r(x_0)}\Big\{\langle\nu(x),x-y\rangle\,k(x-y) - \langle\nu(x),x-x_o\rangle\,k(x-x_o)\Big\}a(y)\,d\sigma(y).$$

Going further, for each $\rho > 0$ define $\nu_{\Delta_\rho(x_o)} := \sigma(\Delta_\rho(x_o))^{-1}\int_{\Delta_\rho(x_o)}\nu\,d\sigma$ and, at each $x \in \partial\Omega\setminus B(x_o,2r)$, consider

$$M_\Delta(x) := -\int_{\Delta_r(x_0)} \Big\{\langle \nu_{\Delta_r(x_o)}, x-y\rangle\, k(x-y) \\ - \langle \nu_{\Delta_r(x_o)}, x-x_o\rangle\, k(x-x_o)\Big\} a(y)\,d\sigma(y). \tag{4.1.137}$$

For σ-a.e. point $x \in \partial\Omega \setminus B(x_o, 2r)$, which we shall fix for now, write

$$M_\Delta(x) - M(x) = \int_{\Delta_r(x_0)} \Big\{F_x(y) - F_x(x_o)\Big\} a(y)\,d\sigma(y), \tag{4.1.138}$$

where $F_x(y) := \langle \nu(x) - \nu_{\Delta_r(x_o)}, x-y\rangle\, k(x-y)$ for each $y \in \Delta_r(x_o)$. Based on this and the properties of the kernel k, it follows that

$$|\nabla_y F_x(y)| \le C\frac{|\nu(x) - \nu_{\Delta_r(x_o)}|}{|x-y|^n} \quad \text{for each } y \in \Delta_r(x_o), \tag{4.1.139}$$

where $C = C(n,k) \in (0,\infty)$. Consequently, employing the Mean Value Theorem and using the fact that since $x \in \partial\Omega \setminus B(x_o, 2r)$,

$$|z-x| \approx |x_o - x|, \quad \text{uniformly for all } z \in \Delta_r(x_o), \tag{4.1.140}$$

we obtain that there exists some $C = C(n,k) \in (0,\infty)$ such that

$$|F_x(y) - F_x(x_o)| \le Cr\frac{|\nu(x) - \nu_{\Delta_r(x_o)}|}{|x-x_o|^n} \quad \text{for each } y \in \Delta_r(x_o). \tag{4.1.141}$$

Together, (4.1.138), (4.1.141), Hölder's inequality, and (4.1.134) yield

$$\begin{aligned} \big|M(x) - M_\Delta(x)\big| &\le Cr\frac{|\nu(x) - \nu_{\Delta_r(x_o)}|}{|x-x_o|^n}\int_{\Delta_r(x_0)} |a(y)|\,d\sigma(y) \\ &\le Cr\frac{|\nu(x) - \nu_{\Delta_r(x_o)}|}{|x-x_o|^n}\sigma\big(\Delta_r(x_0)\big)^{1/q-1/p}\sigma\big(\Delta_r(x_0)\big)^{1-1/q} \\ &= Cr\frac{|\nu(x) - \nu_{\Delta_r(x_o)}|}{|x-x_o|^n}\sigma\big(\Delta_r(x_0)\big)^{1-1/p} \quad \text{for } \sigma\text{-a.e. } x \in \partial\Omega \setminus B(x_o, 2r). \end{aligned} \tag{4.1.142}$$

Next, for each $\ell \in \mathbb{N}$ define the boundary annulus

$$\widetilde{A}_\ell(x_o, r) := \big[B(x_o, 2^{\ell+1}r) \setminus B(x_o, 2^\ell r)\big] \cap \partial\Omega. \tag{4.1.143}$$

Fix $\ell \in \mathbb{N}$. Then (4.1.142), (4.1.143), and the Ahlfors regularity of $\partial\Omega$ we obtain

$$\begin{aligned} &\Big(\int_{\widetilde{A}_\ell(x_o,r)} |M - M_\Delta|^q\,d\sigma\Big)^{1/q} \\ &\quad \le C\frac{r}{(2^\ell r)^n}\sigma\big(\Delta_r(x_0)\big)^{1-1/p}\Big(\int_{\widetilde{A}_\ell(x_o,r)} |\nu - \nu_{\Delta_r(x_o)}|^q\,d\sigma\Big)^{1/q} \end{aligned} \tag{4.1.144}$$

for $C = C(\partial\Omega, k) \in (0,\infty)$ independent of a. In addition, a familiar estimate gives

$$
\begin{aligned}
\Big(\int_{\widetilde{A}_\ell(x_o,r)} |\nu - \nu_{\Delta_r(x_o)}|^q \, d\sigma\Big)^{1/q} &\leq C(2^\ell r)^{(n-1)/q}\Big(\fint_{\Delta_{2^{\ell+1}r}(x_o)} |\nu - \nu_{\Delta_r(x_o)}|^q \, d\sigma\Big)^{1/q} \\
&\leq C(2^\ell r)^{(n-1)/q}\Big(\fint_{\Delta_{2^{\ell+1}r}(x_o)} |\nu - \nu_{\Delta_{2^{\ell+1}r}(x_o)}|^q \, d\sigma\Big)^{1/q} \\
&\qquad + C(2^\ell r)^{(n-1)/q}\sum_{j=0}^{\ell}\big|\nu_{\Delta_{2^j r}(x_o)} - \nu_{\Delta_{2^{j+1}r}(x_o)}\big| \\
&\leq C(2^\ell r)^{(n-1)/q}\Big(\fint_{\Delta_{2^{\ell+1}r}(x_o)} |\nu - \nu_{\Delta_{2^{\ell+1}r}(x_o)}|^q \, d\sigma\Big)^{1/q} \\
&\qquad + C(2^\ell r)^{(n-1)/q}\sum_{j=0}^{\ell}\fint_{\Delta_{2^{j+1}r}(x_o)}\big|\nu - \nu_{\Delta_{2^{j+1}r}(x_o)}\big| \, d\sigma \\
&\leq C(2^\ell r)^{(n-1)/q}\sum_{j=0}^{\ell}\Big(\fint_{\Delta_{2^{j+1}r}(x_o)} |\nu - \nu_{\Delta_{2^{j+1}r}(x_o)}|^q \, d\sigma\Big)^{1/q} \\
&\leq C\ell(2^\ell r)^{(n-1)/q}\|\nu\|_{[\mathrm{BMO}(\partial\Omega,\sigma)]^n}.
\end{aligned}
\tag{4.1.145}
$$

In concert, (4.1.144) and (4.1.145) show that

$$
\begin{aligned}
\Big(\int_{\widetilde{A}_\ell(x_o,r)} |M - M_\Delta|^q \, d\sigma\Big)^{1/q} & \\
\leq C\ell 2^{\ell(n-1)[1/q-1-1/(n-1)]}\sigma\big(\Delta_r(x_o)\big)^{1/q-1/p}\|\nu\|_{[\mathrm{BMO}(\partial\Omega,\sigma)]^n}.
\end{aligned}
\tag{4.1.146}
$$

Fix $\varepsilon \in \big(0, \frac{1}{n-1}\big)$. Then the function $f(t) := t \cdot 2^{-t[1-\varepsilon(n-1)]}$, for all $t \geq 0$, is bounded. An elementary computation gives $0 \leq f(t) \leq C_1$ for all $t \geq 0$, where

$$
C_1 := \frac{2^{-1/\ln 2}}{[1 - \varepsilon(n-1)]\ln 2} > 0. \tag{4.1.147}
$$

Hence, $f(\ell) \leq C_1$ for all $\ell \in \mathbb{N}$, which further implies

$$
\ell 2^{\ell(n-1)[1/q-1-1/(n-1)]} \leq C_1 2^{\ell(n-1)[1/q-1-\varepsilon]} \quad \text{for all } \ell \in \mathbb{N}. \tag{4.1.148}
$$

Combined, estimates (4.1.146) and (4.1.148) imply that there exists some finite constant $C = C(n, \partial\Omega, k, \varepsilon) \in (0, \infty)$ such that

$$
\begin{aligned}
\Big(\int_{\widetilde{A}_\ell(x_o,r)} |M - M_\Delta|^q \, d\sigma\Big)^{1/q} & \\
\leq C\, 2^{\ell(n-1)[1/q-1-\varepsilon]}\sigma\big(\Delta_r(x_o)\big)^{1/q-1/p}\|\nu\|_{[\mathrm{BMO}(\partial\Omega,\sigma)]^n} \quad \text{for all } \ell \in \mathbb{N}.
\end{aligned}
\tag{4.1.149}
$$

Turning our attention to M_Δ, for each $x \in \partial\Omega \setminus B(x_o, 2r)$ we write

$$
M_\Delta(x) = M_\Delta^{(1)}(x) + M_\Delta^{(2)}(x) \text{ where} \tag{4.1.150}
$$

$$M_{\Delta}^{(1)}(x) := -\int_{\Delta_r(x_0)} \langle \nu_{\Delta_r(x_o)}, x - x_o \rangle \big[k(x-y) - k(x - x_o)\big] a(y)\, d\sigma(y),$$

$$M_{\Delta}^{(2)}(x) := \int_{\Delta_r(x_0)} \langle \nu_{\Delta_r(x_o)}, y - x_o \rangle\, k(x - x_o) a(y)\, d\sigma(y). \tag{4.1.151}$$

To estimate $M_{\Delta}^{(1)}(x)$, $M_{\Delta}^{(2)}(x)$ fix $x \in \partial\Omega \setminus B(x_o, 2r)$. The Mean Value Theorem (plus (4.1.140) and the homogeneity of the kernel), Hölder's inequality, and (4.1.134) imply

$$\begin{aligned} \big|M_{\Delta}^{(1)}(x)\big| &\leq \big|\big\langle \nu_{\Delta_r(x_o)}, x - x_o \big\rangle\big| \frac{Cr}{|x - x_o|^{n+1}} \int_{\Delta_r(x_0)} |a(y)|\, d\sigma(y) \\ &\leq \big|\big\langle \nu_{\Delta_r(x_o)}, x - x_o \big\rangle\big| \frac{Cr}{|x - x_o|^{n+1}} \|a\|_{L^q(\partial\Omega,\sigma)} \sigma(\Delta_r(x_0))^{1-1/q} \\ &\leq \big|\big\langle \nu_{\Delta_r(x_o)}, x - x_o \big\rangle\big| \frac{Cr}{|x - x_o|^{n+1}} \sigma\big(\Delta_r(x_0)\big)^{1-1/p}. \end{aligned} \tag{4.1.152}$$

Consequently, there exists some finite constant $C = C(\partial\Omega, k) > 0$ independent of the atom so that for each $\ell \in \mathbb{N}$ we may further compute

$$\begin{aligned} \Big(\int_{\widetilde{A}_\ell(x_o,r)} |M_{\Delta}^{(1)}|^q\, d\sigma\Big)^{1/q} &\leq \frac{Cr}{(2^\ell r)^{n+1}} \sigma\big(\Delta_r(x_0)\big)^{1-1/p} \cdot \sigma\big(\Delta_{2^{\ell+1}r}(x_o)\big)^{1/q} \times \\ &\qquad \times \sup_{x \in \Delta_{2^{\ell+1}r}(x_o)} \big|\big\langle \nu_{\Delta_r(x_o)}, x - x_o \big\rangle\big| \\ &\leq C(2+\ell) 2^{\ell(n-1)[1/q - 1 - 1/(n-1)]} \sigma\big(\Delta_r(x_o)\big)^{1/q - 1/p} \|\nu\|_{[\mathrm{BMO}(\partial\Omega,\sigma)]^n}, \end{aligned} \tag{4.1.153}$$

where the second inequality in (4.1.153) is based on (3.1.25) (presently used with $z := x_o$, $R := r$, $y := x_o$, and $\gamma := 2^{\ell+1}$) and the Ahlfors regularity of $\partial\Omega$. Now the reasoning that allowed us to obtain (4.1.149) from (4.1.146) applies to the resulting estimate in (4.1.153) and gives the existence of $C = C(\partial\Omega, k, \varepsilon) \in (0, \infty)$ so that

$$\begin{aligned} &\Big(\int_{\widetilde{A}_\ell(x_o,r)} |M_{\Delta}^{(1)}|^q\, d\sigma\Big)^{1/q} \\ &\quad \leq C\, 2^{\ell(n-1)[1/q - 1 - \varepsilon]} \sigma\big(\Delta_r(x_o)\big)^{1/q - 1/p} \|\nu\|_{[\mathrm{BMO}(\partial\Omega,\sigma)]^n} \text{ for all } \ell \in \mathbb{N}. \end{aligned} \tag{4.1.154}$$

Continue to assume that $x \in \partial\Omega \setminus B(x_o, 2r)$. Focusing our attention on $M_{\Delta}^{(2)}(x)$, from the homogeneity of the kernel, (3.1.25), Hölder's inequality, and the norm estimate for the atom (cf. (4.1.134)) we obtain

$$\begin{aligned} \big|M_{\Delta}^{(2)}(x)\big| &\leq \sup_{y \in \Delta_r(x_o)} \big|\big\langle \nu_{\Delta_r(x_o)}, y - x_o \big\rangle\big| \frac{C}{|x - x_o|^n} \int_{\Delta_r(x_0)} |a(y)|\, d\sigma(y) \\ &\leq Cr \|\nu\|_{[\mathrm{BMO}(\partial\Omega,\sigma)]^n} \frac{\sigma\big(\Delta_r(x_0)\big)^{1-1/p}}{|x - x_o|^n}. \end{aligned} \tag{4.1.155}$$

This, the Ahlfors regularity of $\partial\Omega$, and the condition $0 < \varepsilon < 1/(n-1)$ imply the existence of some finite constant $C = C(\partial\Omega, k) > 0$ independent of a such that

$$\begin{aligned}\Big(\int_{\widetilde{A}_\ell(x_o,r)}|M^{(2)}_\Delta|^q\,d\sigma\Big)^{1/q} &\le C\|\nu\|_{[\mathrm{BMO}(\partial\Omega,\sigma)]^n}\frac{r}{(2^\ell r)^n}\sigma\big(\Delta_r(x_0)\big)^{1-1/p}\sigma\big(\Delta_{2^{\ell+1}r}(x_0)\big)^{1/q}\\ &\le C\,2^{\ell(n-1)[1/q-1-1/(n-1)]}\sigma\big(\Delta_r(x_o)\big)^{1/q-1/p}\|\nu\|_{[\mathrm{BMO}(\partial\Omega,\sigma)]^n}\\ &\le C\,2^{\ell(n-1)[1/q-1-\varepsilon]}\sigma\big(\Delta_r(x_o)\big)^{1/q-1/p}\|\nu\|_{[\mathrm{BMO}(\partial\Omega,\sigma)]^n}\ \text{ for each } \ell\in\mathbb{N}.\end{aligned}\tag{4.1.156}$$

Together, the estimates in (4.1.149), (4.1.150), (4.1.154), and (4.1.156) imply that there exists $C = C(\partial\Omega, k, n, \varepsilon) \in (0,\infty)$ such that for each $\ell \in \mathbb{N}$ we have

$$\Big(\int_{\widetilde{A}_\ell(x_o,r)}|M|^q\,d\sigma\Big)^{1/q} \le C\,2^{\ell(n-1)[1/q-1-\varepsilon]}\sigma\big(\Delta_r(x_o)\big)^{1/q-1/p}\|\nu\|_{[\mathrm{BMO}(\partial\Omega,\sigma)]^n}.\tag{4.1.157}$$

Since from [132, (5.1.197)] (and [132, Example 5.1.7]) we know that $\int_{\partial\Omega} M\,d\sigma = 0$, the claim in (4.1.135) follows from (4.1.136), (4.1.157), and [130, Definition 4.5.1].

Let us record our progress: from (4.1.136), (4.1.157), [132, (2.1.13)] and (4.1.17) we deduce that, for each $\varepsilon \in \big(0, \frac{1}{n-1}\big)$ there exists a constant $C_m \in (0,\infty)$, depending only on ε, m, n, p, q, the kernel k, and the UR constants of $\partial\Omega$, such that whenever a is as in (4.1.134), the function $M := T^{\#}a$ is a $C_m \cdot \|\nu\|^{\langle m\rangle}_{[\mathrm{BMO}(\partial\Omega,\sigma)]^n}$-multiple of a (p, q, ε)-molecule centered near the ball $B(x_o, r)$ on $\partial\Omega$ (in the sense of [130, Definition 4.5.1]). Granted this, it follows from [130, (4.5.6)] that

$$M \in H^p(\partial\Omega,\sigma) \text{ and } \|M\|_{H^p(\partial\Omega,\sigma)} \le C_m\cdot\|\nu\|^{\langle m\rangle}_{[\mathrm{BMO}(\partial\Omega,\sigma)]^n}.\tag{4.1.158}$$

With (4.1.158) in hand, we may invoke [130, Theorem 4.4.7] (whose applicability in the present setting makes use of [132, (5.2.147)]) to conclude that, indeed, the mapping $T^{\#}$, originally considered as in (4.1.126), extends uniquely to a linear and bounded operator from the Hardy space $H^p(\partial\Omega,\sigma)$ into itself, with operator norm satisfying (4.1.127).

Finally, (4.1.128) follows from (4.1.127) and real interpolation (cf. [130, (4.3.3)]), bearing in mind the interpolation estimate from [130, Proposition 1.3.7, (1.3.64)].□

Moving on, the goal is to establish norm estimates of the same flavor as before for "chord-dot-normal" singular integral operators on BMO, VMO, CMO, "ordinary" homogeneous Hölder spaces, and "vanishing" homogeneous Hölder spaces (of the sort we introduced in [130, §3.2]).

Theorem 4.1.8 *Fix $n \in \mathbb{N}$ with the property that $n \ge 2$ and assume $\Omega \subseteq \mathbb{R}^n$ is a* UR *domain. Abbreviate $\sigma := \mathcal{H}^{n-1}\lfloor\partial\Omega$ and denote by ν the geometric measure theoretic outward unit normal to the set Ω. For a sufficiently large integer $N = N(n) \in \mathbb{N}$ consider a complex-valued function $k \in \mathscr{C}^N(\mathbb{R}^n \setminus \{0\})$ which is even and positive homogeneous of degree $-n$. Finally, consider the modified singular integral operator T_{mod} associated with Ω and k as in* (A.0.240) *(with $\Sigma := \partial\Omega$).*

Then for each $m \in \mathbb{N}$ there exists a constant $C_m \in (0, \infty)$, which depends only on m, n, the kernel k, and the UR *constants of $\partial\Omega$, with the property that operator $\big[T_{\text{mod}}\big]$ considered in the context of* [132, (5.2.208)] *satisfies*

$$\big\|\big[T_{\text{mod}}\big]\big\|_{\mathrm{BMO}(\partial\Omega,\sigma)/\sim\to\mathrm{BMO}(\partial\Omega,\sigma)/\sim} \leq C_m\|\nu\|^{\langle m\rangle}_{[\mathrm{BMO}(\partial\Omega,\sigma)]^n}, \tag{4.1.159}$$

the operator $\big[T_{\text{mod}}\big]$ considered in the context of [132, (5.2.209)] *satisfies*

$$\big\|\big[T_{\text{mod}}\big]\big\|_{\mathrm{VMO}(\partial\Omega,\sigma)/\sim\to\mathrm{VMO}(\partial\Omega,\sigma)/\sim} \leq C_m\|\nu\|^{\langle m\rangle}_{[\mathrm{BMO}(\partial\Omega,\sigma)]^n}. \tag{4.1.160}$$

while the operator $\big[T_{\text{mod}}\big]$ considered in the context of [132, (5.2.212)] *satisfies*

$$\big\|\big[T_{\text{mod}}\big]\big\|_{\mathrm{CMO}(\partial\Omega,\sigma)/\sim\to\mathrm{CMO}(\partial\Omega,\sigma)/\sim} \leq C_m\|\nu\|^{\langle m\rangle}_{[\mathrm{BMO}(\partial\Omega,\sigma)]^n}. \tag{4.1.161}$$

Also, for each exponent $\alpha \in (0, 1)$ and each $m \in \mathbb{N}$ there exists $C_m \in (0, \infty)$ depending only on m, n, α, the kernel k, and the UR *constants of $\partial\Omega$ such that the operator $\big[T_{\text{mod}}\big]$ considered in the context of* [132, (5.2.219)] *satisfies*

$$\big\|\big[T_{\text{mod}}\big]\big\|_{\dot{\mathscr{C}}^\alpha(\partial\Omega)/\sim\to\dot{\mathscr{C}}^\alpha(\partial\Omega)/\sim} \leq C_m\|\nu\|^{\langle m\rangle}_{[\mathrm{BMO}(\partial\Omega,\sigma)]^n}. \tag{4.1.162}$$

Lastly, for each exponent $\alpha \in (0, 1)$ and each $m \in \mathbb{N}$ there $C_m \in (0, \infty)$ depending only on m, n, α, the kernel k, and the UR *constants of $\partial\Omega$, such that the operator $\big[T_{\text{mod}}\big]$ considered in the context of* [132, (5.2.220)] *satisfies*

$$\big\|\big[T_{\text{mod}}\big]\big\|_{\dot{\mathscr{C}}^\alpha_{\text{van}}(\partial\Omega)/\sim\to\dot{\mathscr{C}}^\alpha_{\text{van}}(\partial\Omega)/\sim} \leq C_m\|\nu\|^{\langle m\rangle}_{[\mathrm{BMO}(\partial\Omega,\sigma)]^n}. \tag{4.1.163}$$

Proof Assume first that $\partial\Omega$ is unbounded. Then for each $f \in \mathrm{BMO}(\partial\Omega, \sigma)$ we have

$$\begin{aligned}
\big\|\big[T_{\text{mod}}\big][f]\big\|_{\mathrm{BMO}(\partial\Omega,\sigma)/\sim} &= \big\|\big[T_{\text{mod}}f\,\big]\big\|_{\mathrm{BMO}(\partial\Omega,\sigma)/\sim} = \big\|T_{\text{mod}}f\,\big\|_{\mathrm{BMO}(\partial\Omega,\sigma)} \\
&\leq C\cdot\sup\Big\{\big|\big\langle[T_{\text{mod}}f],g\big\rangle\big| : g\in H^1(\partial\Omega,\sigma) \text{ with } \|g\|_{H^1(\partial\Omega,\sigma)}=1\Big\} \\
&= C\cdot\sup\Big\{\big|\big\langle[f],T^{\#}g\big\rangle\big| : g\in H^1(\partial\Omega,\sigma) \text{ with } \|g\|_{H^1(\partial\Omega,\sigma)}=1\Big\} \\
&\leq C\cdot\sup\Big\{\|f\|_{\mathrm{BMO}(\partial\Omega,\sigma)}\cdot\|T^{\#}g\|_{H^1(\partial\Omega,\sigma)} : g\in H^1(\partial\Omega,\sigma), \\
&\qquad\qquad \|g\|_{H^1(\partial\Omega,\sigma)}=1\Big\} \\
&\leq C\|f\|_{\mathrm{BMO}(\partial\Omega,\sigma)}\cdot\|T^{\#}\|_{H^1(\partial\Omega,\sigma)\to H^1(\partial\Omega,\sigma)} \\
&\leq C_m\big\|[f]\big\|_{\mathrm{BMO}(\partial\Omega,\sigma)/\sim}\cdot\|\nu\|^{\langle m\rangle}_{[\mathrm{BMO}(\partial\Omega,\sigma)]^n},
\end{aligned} \tag{4.1.164}$$

where the first equality is provided by [132, (5.2.208)], the second equality comes from the format of the norm on $\widetilde{\mathrm{BMO}}(\partial\Omega, \sigma)$ (cf. [129, (7.4.95)]), the first inequality is a consequence of [130, (4.6.6)], the subsequent equality is implied by [132,

(5.2.210)], the second inequality is seen from [130, (4.6.9)], the penultimate inequality is guaranteed by [132, (5.2.152)], and the last inequality is a consequence of the format of the norm on $\widetilde{\mathrm{BMO}}(\partial\Omega, \sigma)$ (cf. [129, (7.4.95)]) and (4.1.127) (with $p = 1$). This establishes (4.1.159) in the case when $\partial\Omega$ is unbounded.

In turn, (4.1.160) is a consequence of [132, (5.2.209)], the properties of [130, (3.1.3)], (4.1.159), and [130, (1.2.20)], while (4.1.161) follows from (4.1.159), [132, (5.2.212)], and [130, (4.6.15)]. Next, the estimate in (4.1.162) may be established in a similar fashion, now relying on [132, (5.2.152), (5.2.219)], [129, (7.3.5)], [130, (4.6.7)], and (4.1.127) used with $p := \big(\frac{\alpha}{n-1} + 1\big)^{-1} \in \big(\frac{n-1}{n}, 1\big)$. Finally, (4.1.163) is a consequence of [132, (5.2.220)], (4.1.162), [130, Lemma 3.2.1], and [130, (1.2.20)].

Lastly, in the case when the set $\partial\Omega$ is bounded, all claims are seen from the fact that all operators considered in the statement are bounded and (4.1.4). □

We shall next establish a boundedness result for the "transpose chord-dot-normal" operator $T^{\#}$, originally defined in (4.1.28), similar in spirit to the earlier results in this section, now considering the action of $T^{\#}$ in the context of [132, (5.2.159)], on the pre-dual space of the Morrey-Campanato space introduced in [130, (6.1.16)].

Theorem 4.1.9 *Let $\Omega \subseteq \mathbb{R}^n$ (where $n \in \mathbb{N}$, $n \geq 2$) be a* UR *domain. Abbreviate $\sigma := \mathcal{H}^{n-1}\lfloor\partial\Omega$ and denote by ν the geometric measure theoretic outward unit normal to Ω. Also, pick large integer $N = N(n) \in \mathbb{N}$ and suppose $k \in \mathscr{C}^N(\mathbb{R}^n \setminus \{0\})$ is even and positive homogeneous of degree $-n$. Finally, select $\lambda \in (0, n-1)$ and $p, q \in (1, \infty)$ satisfying $1/p + 1/q = 1$.*

Then for each $m \in \mathbb{N}$ there exists $C_m \in (0, \infty)$ depending only on m, n, q, λ, and the UR *constants of $\partial\Omega$ such that the operator $T^{\#}$, originally associated with k and Ω as in* (4.1.28), *induces a linear and bounded mapping from the pre-dual space $\mathscr{H}^{q,\lambda}(\partial\Omega, \sigma)$ of the Morrey-Campanato space defined as in* (A.0.94) *(with $\Sigma := \partial\Omega$) into itself, as in* [132, (5.2.159)], *satisfying*

$$\|T^{\#}\|_{\mathscr{H}^{q,\lambda}(\partial\Omega,\sigma)\to\mathscr{H}^{q,\lambda}(\partial\Omega,\sigma)} \leq C_m \Big(\sum_{|\alpha|\leq N} \sup_{S^{n-1}} |\partial^\alpha k| \Big) \|\nu\|^{\langle m\rangle}_{[\mathrm{BMO}(\partial\Omega,\sigma)]^n}. \tag{4.1.165}$$

Also, $\big[T_{\mathrm{mod}}\big]$ acting on classes of equivalence of functions in the homogeneous Morrey-Campanato space modulo constants as in [132, (5.2.226)] *satisfies*

$$\big\|\big[T_{\mathrm{mod}}\big]\big\|_{\dot{\mathfrak{L}}^{p,\lambda}(\partial\Omega,\sigma)/\sim\to\dot{\mathfrak{L}}^{p,\lambda}(\partial\Omega,\sigma)/\sim} \leq C_m \Big(\sum_{|\alpha|\leq N} \sup_{S^{n-1}} |\partial^\alpha k| \Big) \|\nu\|^{\langle m\rangle}_{[\mathrm{BMO}(\partial\Omega,\sigma)]^n}. \tag{4.1.166}$$

Finally, the operator T, originally associated with k and Ω as in (4.1.27), *now acting on the inhomogeneous Morrey-Campanato space as in* [132, (5.2.225)] *has*

$$\|T\|_{\mathfrak{L}^{p,\lambda}(\partial\Omega,\sigma)\to\mathfrak{L}^{p,\lambda}(\partial\Omega,\sigma)} \leq C_m \Big(\sum_{|\alpha|\leq N} \sup_{S^{n-1}} |\partial^\alpha k| \Big) \|\nu\|^{\langle m\rangle}_{[\mathrm{BMO}(\partial\Omega,\sigma)]^n}. \tag{4.1.167}$$

Proof When the set $\partial\Omega$ is compact, (4.1.4) gives that $\|\nu\|_{[\mathrm{BMO}(\partial\Omega,\sigma)]^n} = 1$, a scenario in which (4.1.165), (4.1.166), and (4.1.167) become direct consequences of [132, (5.2.159), (5.1.105), (5.2.225)], respectively. As such, there remains to deal with the case when $\partial\Omega$ is unbounded, which we will assume for the remainder of the proof. In broad outline, we shall follow the proof of [132, Theorem 3.2.1], with some key input from Theorem 4.1.3 and the proof of Theorem 4.1.7. To simplify the presentation, observe that, via rescaling, there is no loss of generality in assuming

$$\sum_{|\alpha|\leq N} \sup_{S^{n-1}} |\partial^\alpha k| = 1. \tag{4.1.168}$$

To get started, fix an arbitrary number $m \in \mathbb{N}$, and pick an arbitrary $\mathscr{H}^{q,\lambda}$-atom $a \in L^q(\partial\Omega,\sigma)$ on $\partial\Omega$. From [130, (6.1.15)] we know that there exists a point $x_o \in \partial\Omega$ and a radius $R \in (0,\infty)$ such that

$$\operatorname{supp} a \subseteq B(x_o,R)\cap\partial\Omega, \quad \|a\|_{L^q(\partial\Omega,\sigma)} \leq R^{\lambda(\frac{1}{q}-1)}, \quad \int_{\partial\Omega} a\,\mathrm{d}\sigma = 0. \tag{4.1.169}$$

Also, much as in [132, (3.2.4)],

the function $M := R^{(\lambda-n+1)(1-1/q)}T^{\#}a$ is a fixed multiple of some $(1,q,\varepsilon)$-molecule on $\partial\Omega$, in the sense considered in [130, Definition 4.5.1] for the choice of parameter $\varepsilon := 1/(n-1)$. (4.1.170)

Using the boundedness of $T^{\#}$ on Lebesgue spaces and the specific format of the norm-estimate deduced in this setting (cf. (4.1.32) in Theorem 4.1.3), plus the normalization of the atom in (4.1.169), we may estimate

$$\begin{aligned}\|T^{\#}a\|_{L^q(\partial\Omega,\sigma)} &\leq \|T^{\#}\|_{L^q(\partial\Omega,\sigma)\to L^q(\partial\Omega,\sigma)}\cdot\|a\|_{L^q(\partial\Omega,\sigma)}\\ &\leq C_m\cdot\|\nu\|^{\langle m\rangle}_{[\mathrm{BMO}(\partial\Omega,\sigma)]^n}\cdot R^{\lambda(1/q-1)}\end{aligned} \tag{4.1.171}$$

for some constant $C_m \in (0,\infty)$ independent of the $\mathscr{H}^{q,\lambda}$-atom a.

To proceed, with $\varepsilon := 1/(n-1)$ as above, choose some

$$\theta \in \Big((n-1)(q-1),\,(n-1)[(1+\varepsilon)q-1]\Big). \tag{4.1.172}$$

Then (4.1.171) permits us to estimate

$$\begin{aligned}\left(\int_{B(x_o,2R)\cap\partial\Omega}|(T^{\#}a)(x)|^q|x-x_o|^\theta\,\mathrm{d}\sigma(x)\right)^{1/q} &\leq (2R)^{\theta/q}\|T^{\#}a\|_{L^q(\partial\Omega,\sigma)}\\ &\leq C_m\|\nu\|^{\langle m\rangle}_{[\mathrm{BMO}(\partial\Omega,\sigma)]^n}\cdot R^{\lambda(1/q-1)+\theta/q}.\end{aligned} \tag{4.1.173}$$

Also, if for each $\ell \in \mathbb{N}$ define the boundary annulus

$$\widetilde{A}_\ell(x_o, R) := \big[B(x_o, 2^{\ell+1}R) \setminus B(x_o, 2^\ell R)\big] \cap \partial\Omega, \tag{4.1.174}$$

then by relying on (4.1.170) and [130, (4.5.1)] we may write

$$\begin{aligned}\Big(\int_{\partial\Omega} |(T^\# a)(x)|^q |x - x_o|^\theta \, d\sigma(x)\Big)^{1/q} \qquad (4.1.175)\\ \leq \Big(\int_{B(x_o,2R)\cap\partial\Omega} |(T^\# a)(x)|^q |x - x_o|^\theta \, d\sigma(x)\Big)^{1/q}\\ + \sum_{\ell=1}^{\infty}\Big(\int_{\widetilde{A}_\ell(x_o,R)} |(T^\# a)(x)|^q |x - x_o|^\theta \, d\sigma(x)\Big)^{1/q}.\end{aligned}$$

In addition, thanks to the definition in (4.1.170) and (4.1.157) (used with $p := 1$), for each $\ell \in \mathbb{N}$ we have

$$\begin{aligned}\Big(\int_{\widetilde{A}_\ell(x_o,R)} |(T^\# a)(x)|^q |x - x_o|^\theta \, d\sigma(x)\Big)^{1/q} \qquad (4.1.176)\\ \leq R^{(n-1-\lambda)(1-1/q)} \cdot (2^{\ell+1}R)^{\theta/q}\Big(\int_{\widetilde{A}_\ell(x_o,R)} |M|^q \, d\sigma\Big)^{1/q}\\ \leq CR^{(n-1-\lambda)(1-1/q)+\theta/q} \cdot 2^{\ell(n-1)[1/q-1-\varepsilon]} 2^{\ell\theta/q} \times\\ \times \sigma\big(B(x_o, R) \cap \partial\Omega\big)^{1/q-1} \|\nu\|_{[\mathrm{BMO}(\partial\Omega,\sigma)]^n}\\ \leq CR^{(n-1-\lambda)(1-1/q)+\theta/q} \cdot 2^{\ell\{\theta/q-(n-1)[(1+\varepsilon)-1/q]\}} \times\\ \times R^{-(n-1)(1-1/q)} \|\nu\|_{[\mathrm{BMO}(\partial\Omega,\sigma)]^n}\\ = CR^{\lambda(1/q-1)+\theta/q} \cdot 2^{\ell\{\theta/q-(n-1)[(1+\varepsilon)-1/q]\}} \|\nu\|_{[\mathrm{BMO}(\partial\Omega,\sigma)]^n}\end{aligned}$$

for some $C \in (0, \infty)$ independent of ℓ and the $\mathscr{H}^{q,\lambda}$-atom a. In view of (4.1.172),

$$\sum_{\ell=1}^{\infty} 2^{\ell\{\theta/q-(n-1)[(1+\varepsilon)-1/q]\}} < +\infty. \tag{4.1.177}$$

Collectively, (4.1.175), (4.1.173), (4.1.176), (4.1.177), and (4.1.17) prove that

$$\Big(\int_{\partial\Omega} |(T^\# a)(x)|^q |x - x_o|^\theta \, d\sigma(x)\Big)^{1/q} \leq C_m \|\nu\|^{\langle m\rangle}_{[\mathrm{BMO}(\partial\Omega,\sigma)]^n} \cdot R^{\lambda(1/q-1)+\theta/q}. \tag{4.1.178}$$

In addition, from (4.1.170) and [132, (2.1.13)] we see that $\int_{\partial\Omega} T^\# a \, d\sigma = 0$. In concert with (4.1.171), (4.1.178), and [130, (6.1.33), (6.1.37)], this shows that

the function $T^\# a$ is a $C_m \|\nu\|^{\langle m\rangle}_{[\mathrm{BMO}(\partial\Omega,\sigma)]^n}$-multiple of a $\mathscr{H}^{q,\lambda,\theta}$-molecule on $\partial\Omega$, for θ chosen as in (4.1.172). (4.1.179)

In turn, from (4.1.179) and [130, (6.1.38)] we conclude that for each integer $m \in \mathbb{N}$ there exists a constant $C_m \in (0, \infty)$ independent of the $\mathscr{H}^{q,\lambda}$-atom a so that

$$T^{\#}a \in \mathscr{H}^{q,\lambda}(\partial\Omega,\sigma) \text{ and } \|T^{\#}a\|_{\mathscr{H}^{q,\lambda}(\partial\Omega,\sigma)} \leq C_m \|\nu\|^{\langle m \rangle}_{[\mathrm{BMO}(\partial\Omega,\sigma)]^n}. \tag{4.1.180}$$

With (4.1.180) in hand, the end-game in the proof of (4.1.165) is as follows. Assume some function $f \in \mathscr{H}^{q,\lambda}(\partial\Omega,\sigma)$ has been given. By definition, this means that there exist some numerical sequence $\{\lambda_j\}_{j\in\mathbb{N}} \in \ell^1(\mathbb{N})$ along with a sequence $\{a_j\}_{j\in\mathbb{N}}$ of $\mathscr{H}^{q,\lambda}$-atoms on $\partial\Omega$ such that $f = \sum_{j=1}^{\infty} \lambda_j a_j$ in the sense of distributions on $\partial\Omega$. Thanks to (A.0.95)-[130, (6.1.22)], the series $\sum_{j=1}^{\infty} \lambda_j a_j$ actually converges to f in $L^r(\partial\Omega,\sigma)$ with $r := \frac{q(n-1)}{n-1+\lambda(q-1)}$. Also, (4.1.180) readily implies that the sequence $\big\{T^{\#}\big(\sum_{j=1}^{J} \lambda_j a_j\big)\big\}_{J\in\mathbb{N}}$ is Cauchy, hence convergent, in the Banach space $\Big(\mathscr{H}^{q,\lambda}(\partial\Omega,\sigma), \|\cdot\|_{\mathscr{H}^{q,\lambda}(\partial\Omega,\sigma)}\Big)$. In light of [130, (6.1.22)], this latter convergence takes place in $L^r(\partial\Omega,\sigma)$ as well. Since $T^{\#}$ is continuous on $L^r(\partial\Omega,\sigma)$, it follows that $\big\{T^{\#}\big(\sum_{j=1}^{J} \lambda_j a_j\big)\big\}_{J\in\mathbb{N}}$ converges to $T^{\#}f$ in $\mathscr{H}^{q,\lambda}(\partial\Omega,\sigma)$. Keeping (4.1.180) in mind, this argument ultimately proves that

$$\begin{gathered} T^{\#}f \text{ belongs to } \mathscr{H}^{q,\lambda}(\partial\Omega,\sigma) \text{ and} \\ \|T^{\#}f\|_{\mathscr{H}^{q,\lambda}(\partial\Omega,\sigma)} \leq C_m \|\nu\|^{\langle m \rangle}_{[\mathrm{BMO}(\partial\Omega,\sigma)]^n} \|f\|_{\mathscr{H}^{q,\lambda}(\partial\Omega,\sigma)}, \end{gathered} \tag{4.1.181}$$

from which (4.1.165) follows, bearing in mind (4.1.168).

Next, the claim made in (4.1.166) follows from (4.1.165) and the duality result recorded in [132, (5.1.107)]. Finally, consider the claim made in (4.1.167). Once again, we may assume that the equality (4.1.168) holds. Pick a function $f \in \mathfrak{L}^{p,\lambda}(\partial\Omega,\sigma) = L^p(\partial\Omega,\sigma) \cap \dot{\mathfrak{L}}^{p,\lambda}(\partial\Omega,\sigma)$ and observe from [132, (5.2.172), (5.2.190), (5.2.138)] that the difference $T_{\mathrm{mod}} f - Tf$ is a constant on $\partial\Omega$. Thanks to this, [130, (6.1.8), (6.1.9)], and (4.1.166), for each $m \in \mathbb{N}$ we may then write

$$\begin{aligned} \|Tf\|_{\dot{\mathfrak{L}}^{p,\lambda}(\partial\Omega,\sigma)} &= \|T_{\mathrm{mod}} f\|_{\dot{\mathfrak{L}}^{p,\lambda}(\partial\Omega,\sigma)} = \big\| [T_{\mathrm{mod}} f] \big\|_{\dot{\mathfrak{L}}^{p,\lambda}(\partial\Omega,\sigma)} \\ &\leq C_m \|\nu\|^{\langle m \rangle}_{[\mathrm{BMO}(\partial\Omega,\sigma)]^n} \big\| [f] \big\|_{\dot{\mathfrak{L}}^{p,\lambda}(\partial\Omega,\sigma)} = C_m \|\nu\|^{\langle m \rangle}_{[\mathrm{BMO}(\partial\Omega,\sigma)]^n} \|f\|_{\dot{\mathfrak{L}}^{p,\lambda}(\partial\Omega,\sigma)}. \end{aligned} \tag{4.1.182}$$

In concert, (A.0.138), (4.1.182), and (4.1.53) then justify (4.1.167). □

In the class of singular integral operators whose integral kernel depends linearly on the geometric measure theoretic outward unit normal ν to the underlying domain, the validity of norm estimates controlling the operator norm in terms of the BMO semi-norm of ν turns out to be equivalent to the operator in question being of "chord-dot-normal" type. This is made precise in the theorem below.

Theorem 4.1.10 *Fix $n \in \mathbb{N}$ with $n \geq 2$, and pick a sufficiently large $N = N(n) \in \mathbb{N}$. Also, consider a vector-valued function*

$$\begin{gathered} \vec{k} \in \big[\mathscr{C}^N(\mathbb{R}^n \setminus \{0\})\big]^n \textit{ which is odd and} \\ \textit{positive homogeneous of degree } 1-n. \end{gathered} \tag{4.1.183}$$

For any

$$\Omega \subseteq \mathbb{R}^n \text{ UR } \textit{domain (and with } \sigma := \mathcal{H}^{n-1}\lfloor\partial\Omega \textit{ and } \nu \textit{ denoting the "surface measure" on } \partial\Omega \textit{ and, respectively, the geometric measure theoretic outward unit normal vector to } \Omega), \tag{4.1.184}$$

consider the principal-value singular integral operators $T, T^{\#}$ *on* $\partial\Omega$*, acting on each* $f \in L^1\Big(\partial\Omega, \frac{\sigma(y)}{1+|y|^{n-1}}\Big)$ *at* σ*-a.e.* $x \in \partial\Omega$ *according to*

$$Tf(x) := \lim_{\varepsilon\to 0^+} \int_{y\in\partial\Omega,\ |x-y|>\varepsilon} \langle \nu(y), \vec{k}(x-y)\rangle f(y)\,d\sigma(y), \tag{4.1.185}$$

$$T^{\#}f(x) := \lim_{\varepsilon\to 0^+} \int_{y\in\partial\Omega,\ |x-y|>\varepsilon} \langle \nu(x), \vec{k}(y-x)\rangle f(y)\,d\sigma(y). \tag{4.1.186}$$

Then the following statements are equivalent.

(1) For any $\Omega \subseteq \mathbb{R}^n$ *as in* (4.1.184)*, any (or some) exponent* $p \in (1,\infty)$*, any (or some) Muckenhoupt weight* $w \in A_p(\partial\Omega,\sigma)$*, and any (or some) integer* $m \in \mathbb{N}$ *there exists* $C_m \in (0,\infty)$*, which depends only on* m*,* n*,* p*,* $[w]_{A_p}$*,* $\vec{k}$*, and the* UR *constants of* $\partial\Omega$*, ensuring the validity of all (or one) of the following estimates:*

$$\|T\|_{L^p(\partial\Omega,w)\to L^p(\partial\Omega,w)} \leq C_m \|\nu\|^{\langle m\rangle}_{[\mathrm{BMO}(\partial\Omega,\sigma)]^n}, \tag{4.1.187}$$

$$\|T^{\#}\|_{L^p(\partial\Omega,w)\to L^p(\partial\Omega,w)} \leq C_m \|\nu\|^{\langle m\rangle}_{[\mathrm{BMO}(\partial\Omega,\sigma)]^n}. \tag{4.1.188}$$

(2) There exists a scalar-valued function $k \in \mathscr{C}^N(\mathbb{R}^n \setminus \{0\})$ *which is even and positive homogeneous of degree* $-n$ *with the property that*

$$\vec{k}(x) = x\,k(x) \textit{ for each } x \in \mathbb{R}^n \setminus \{0\}. \tag{4.1.189}$$

Moreover, either of the above items implies that $\vec{k}$ *is divergence-free in* $\mathbb{R}^n \setminus \{0\}$ *and, with this condition imposed (in addition to* (4.1.183)*), the following are also equivalent with (1)-(2):*

(3) For any $\Omega \subseteq \mathbb{R}^n$ *as in* (4.1.184)*, any (or some) exponent* $p \in (1,\infty)$*, any (or some) Muckenhoupt weight* $w \in A_p(\partial\Omega,\sigma)$*, and any (or some) integer* $m \in \mathbb{N}$ *there exists* $C_m \in (0,\infty)$*, which depends only on* m*,* n*,* p*,* $[w]_{A_p}$*,* $\vec{k}$*, and the* UR *constants of* $\partial\Omega$*, ensuring the validity of all (or one) of the following estimates:*

$$\|T\|_{L^p_1(\partial\Omega,w)\to L^p_1(\partial\Omega,w)} \leq C_m \|\nu\|^{\langle m\rangle}_{[\mathrm{BMO}(\partial\Omega,\sigma)]^n}, \tag{4.1.190}$$

$$\|T^{\#}\|_{L^p_{-1}(\partial\Omega,w)\to L^p_{-1}(\partial\Omega,w)} \leq C_m \|\nu\|^{\langle m\rangle}_{[\mathrm{BMO}(\partial\Omega,\sigma)]^n}, \tag{4.1.191}$$

(4) *For any $\Omega \subseteq \mathbb{R}^n$ as in* (4.1.184), *any* (*or some*) *exponent $p \in \left(\frac{n-1}{n}, \infty\right)$, and any* (*or some*) *integer $m \in \mathbb{N}$ there exists $C_m \in (0, \infty)$, which depends only on m, n, p, $\vec{k}$, and the* UR *constants of $\partial\Omega$, such that*

$$\|T^{\#}\|_{H^p(\partial\Omega,\sigma)\to H^p(\partial\Omega,\sigma)} \leq C_m \|\nu\|^{\langle m\rangle}_{[\mathrm{BMO}(\partial\Omega,\sigma)]^n}. \tag{4.1.192}$$

(5) *For any $\Omega \subseteq \mathbb{R}^n$ as in* (4.1.184), *any* (*or some*) *exponent $q \in (1, \infty)$ together with $\lambda \in (0, n-1)$, and any* (*or some*) *integer $m \in \mathbb{N}$ there exists $C_m \in (0, \infty)$, depending only on m, n, q, λ, $\vec{k}$, and the* UR *constants of $\partial\Omega$, so that*

$$\|T^{\#}\|_{\mathscr{H}^{q,\lambda}(\partial\Omega,\sigma)\to \mathscr{H}^{q,\lambda}(\partial\Omega,\sigma)} \leq C_m \|\nu\|^{\langle m\rangle}_{[\mathrm{BMO}(\partial\Omega,\sigma)]^n}. \tag{4.1.193}$$

(6) *For any $\Omega \subseteq \mathbb{R}^n$ as in* (4.1.184) *and any* (*or some*) *integer $m \in \mathbb{N}$ there exists $C_m \in (0, \infty)$, which depends only on m, n, $\vec{k}$, and the* UR *constants of $\partial\Omega$ such that the operator $[T_{\text{mod}}]$ associated with Ω and $\vec{k}$ as in* [132, (5.1.84)] *satisfies*

$$\big\|[T_{\text{mod}}]\big\|_{\mathrm{BMO}(\partial\Omega,\sigma)/\sim\to \mathrm{BMO}(\partial\Omega,\sigma)/\sim} \leq C_m \|\nu\|^{\langle m\rangle}_{[\mathrm{BMO}(\partial\Omega,\sigma)]^n}. \tag{4.1.194}$$

(7) *For any $\Omega \subseteq \mathbb{R}^n$ as in* (4.1.184) *and any* (*or some*) *integer $m \in \mathbb{N}$ there exists $C_m \in (0, \infty)$, which depends only on m, n, $\vec{k}$, and the* UR *constants of $\partial\Omega$ such that the operator $[T_{\text{mod}}]$ associated with Ω and $\vec{k}$ as in* [132, (5.1.88)] *satisfies*

$$\big\|[T_{\text{mod}}]\big\|_{\mathrm{CMO}(\partial\Omega,\sigma)/\sim\to \mathrm{CMO}(\partial\Omega,\sigma)/\sim} \leq C_m \|\nu\|^{\langle m\rangle}_{[\mathrm{BMO}(\partial\Omega,\sigma)]^n}. \tag{4.1.195}$$

(8) *For any $\Omega \subseteq \mathbb{R}^n$ as in* (4.1.184) *and any* (*or some*) *integer $m \in \mathbb{N}$ there exists $C_m \in (0, \infty)$, which depends only on m, n, $\vec{k}$, and the* UR *constants of $\partial\Omega$ such that the operator $[T_{\text{mod}}]$ associated with Ω and $\vec{k}$ as in* [132, (5.1.85)] *satisfies*

$$\big\|[T_{\text{mod}}]\big\|_{\mathrm{VMO}(\partial\Omega,\sigma)/\sim\to \mathrm{VMO}(\partial\Omega,\sigma)/\sim} \leq C_m \|\nu\|^{\langle m\rangle}_{[\mathrm{BMO}(\partial\Omega,\sigma)]^n}. \tag{4.1.196}$$

(9) *For any set $\Omega \subseteq \mathbb{R}^n$ as in* (4.1.184), *any* (*or some*) *exponent $\alpha \in (0, 1)$, and any* (*or some*) *integer $m \in \mathbb{N}$ there exists $C_m \in (0, \infty)$, which depends only on m, n, $\vec{k}$, α, and the* UR *constants of $\partial\Omega$ such that the operator $[T_{\text{mod}}]$ associated with Ω and $\vec{k}$ as in* [132, (5.1.97)] *satisfies*

$$\big\|[T_{\text{mod}}]\big\|_{\dot{\mathscr{C}}^{\alpha}(\partial\Omega)/\sim\to \dot{\mathscr{C}}^{\alpha}(\partial\Omega)/\sim} \leq C_m \|\nu\|^{\langle m\rangle}_{[\mathrm{BMO}(\partial\Omega,\sigma)]^n}, \tag{4.1.197}$$

and (*or*) *the operator $[T_{\text{mod}}]$ associated with Ω and $\vec{k}$ as in* [132, (5.1.98)] *satisfies*

$$\big\|[T_{\text{mod}}]\big\|_{\dot{\mathscr{C}}^{\alpha}_{\text{van}}(\partial\Omega)/\sim\to \dot{\mathscr{C}}^{\alpha}_{\text{van}}(\partial\Omega)/\sim} \leq C_m \|\nu\|^{\langle m\rangle}_{[\mathrm{BMO}(\partial\Omega,\sigma)]^n}. \tag{4.1.198}$$

(10) *For any $\Omega \subseteq \mathbb{R}^n$ as in* (4.1.184), *any* (*or some*) *exponent $p \in (1, \infty)$ together with $\lambda \in (0, n-1)$, and any* (*or some*) *$m \in \mathbb{N}$ there exists $C_m \in (0, \infty)$, depending only on m, n, p, λ, $\vec{k}$, and the* UR *constants of $\partial\Omega$, such that $[T_{\text{mod}}]$ associated with Ω and $\vec{k}$ as in* [132, (5.1.105)] *satisfies*

$$\big\|[T_{\text{mod}}]\big\|_{\dot{\mathfrak{L}}^{p,\lambda}(\partial\Omega,\sigma)/\sim\to\dot{\mathfrak{L}}^{p,\lambda}(\partial\Omega,\sigma)/\sim} \leq C_m\|\nu\|^{\langle m\rangle}_{[\mathrm{BMO}(\partial\Omega,\sigma)]^n}. \tag{4.1.199}$$

(11) For any $\Omega \subseteq \mathbb{R}^n$ *as in* (4.1.184)*, any (or some) exponent* $p \in (1,\infty)$ *together with* $\lambda \in (0, n-1)$*, and any (or some)* $m \in \mathbb{N}$ *there exists* $C_m \in (0,\infty)$ *depending only on* m*,* n*,* p*,* λ*,* $\vec{k}$*, and the* UR *constants of* $\partial\Omega$ *so that*

$$\|T\|_{\mathfrak{L}^{p,\lambda}(\partial\Omega,\sigma)\to\mathfrak{L}^{p,\lambda}(\partial\Omega,\sigma)} \leq C_m\Big(\sum_{|\alpha|\leq N}\sup_{S^{n-1}}|\partial^\alpha k|\Big)\|\nu\|^{\langle m\rangle}_{[\mathrm{BMO}(\partial\Omega,\sigma)]^n}. \tag{4.1.200}$$

(12) For any open set $\Omega \subseteq \mathbb{R}^n$ *satisfying a two-sided local John condition and having an Ahlfors regular boundary (with* ν *and* $\sigma := \mathcal{H}^{n-1}\lfloor\partial\Omega$ *denoting its geometric measure theoretic outward unit normal, and its surface measure, respectively), any (or some) exponents* $p, q \in (1,\infty)$*,* $\lambda \in (0, n-1)$*, and any (or some) integer* $m \in \mathbb{N}$ *there exists* $C_m \in (0,\infty)$*, which depends only on* m*,* n*,* $\vec{k}$*,* p*,* q*,* λ*, the two-sided local John constants of* Ω*, and the Ahlfors regularity constants of* $\partial\Omega$ *such that the operator* T_{mod} *associated with* Ω *and* $\vec{k}$ *as in* [132, (5.2.188)] *satisfies either (or all) of the following estimates:*

$$\big\|[T_{\text{mod}}]\big\|_{\dot{L}^p_1(\partial\Omega,\sigma)/\sim\to\dot{L}^p_1(\partial\Omega,\sigma)/\sim} \leq C_m\|\nu\|^{\langle m\rangle}_{[\mathrm{BMO}(\partial\Omega,\sigma)]^n}, \tag{4.1.201}$$

$$\big\|[T_{\text{mod}}]\big\|_{\dot{M}^{p,\lambda}_1(\partial\Omega,\sigma)/\sim\to\dot{M}^{p,\lambda}_1(\partial\Omega,\sigma)/\sim} \leq C_m\|\nu\|^{\langle m\rangle}_{[\mathrm{BMO}(\partial\Omega,\sigma)]^n}, \tag{4.1.202}$$

$$\big\|[T_{\text{mod}}]\big\|_{\dot{\mathbb{M}}^{p,\lambda}_1(\partial\Omega,\sigma)/\sim\to\dot{\mathbb{M}}^{p,\lambda}_1(\partial\Omega,\sigma)/\sim} \leq C_m\|\nu\|^{\langle m\rangle}_{[\mathrm{BMO}(\partial\Omega,\sigma)]^n}, \tag{4.1.203}$$

$$\big\|[T_{\text{mod}}]\big\|_{\dot{\mathcal{B}}^{q,\lambda}_1(\partial\Omega,\sigma)/\sim\to\dot{\mathcal{B}}^{q,\lambda}_1(\partial\Omega,\sigma)/\sim} \leq C_m\|\nu\|^{\langle m\rangle}_{[\mathrm{BMO}(\partial\Omega,\sigma)]^n}. \tag{4.1.204}$$

As a preamble to the proof of this theorem, we first establish the following companion result to Lemma 1.3.1.

Lemma 4.1.11 *Pick some* $n, N \in \mathbb{N}$ *and consider a vector-valued function*

$$\vec{k} \in \big[\mathscr{C}^N(\mathbb{R}^n \setminus \{0\})\big]^n, \text{ odd, positive homogeneous of degree } 1-n. \tag{4.1.205}$$

Then there exists a scalar-valued function $k \in \mathscr{C}^N(\mathbb{R}^n \setminus \{0\})$ *which is even and positive homogeneous of degree* $-n$ *with the property that*

$$\vec{k}(x) = x\,k(x) \text{ for each } x \in \mathbb{R}^n \setminus \{0\} \tag{4.1.206}$$

if and only if for each hyper-plane $H = \langle\omega\rangle^\perp \subseteq \mathbb{R}^n$*, with* $\omega \in S^{n-1}$*, each function* $\phi \in \mathscr{C}^\infty_c(\mathbb{R}^n)$*, and* $\mathcal{H}^{n-1}$*-a.e. point* $x \in H$ *one has*

$$\lim_{\varepsilon\to 0^+}\int_{y\in H,\ |x-y|>\varepsilon}\big\langle\omega, \vec{k}_\varepsilon(x-y) - \vec{k}_1(-y)\big\rangle\phi(y)\,d\mathcal{H}^{n-1}(y) = 0, \tag{4.1.207}$$

where $\vec{k}_\varepsilon := \vec{k}\cdot\mathbf{1}_{\mathbb{R}^n\setminus\overline{B(0,\varepsilon)}}$ *for each* $\varepsilon > 0$.

Proof In one direction, it is clear that (4.1.206) implies (4.1.207) (since the inner product is identically zero), so we shall focus on the opposite implication. In this regard, start by observing that (4.1.207) forces

$$\big\langle \omega, \vec{k}_\varepsilon(x-y) - \vec{k}_1(-y)\big\rangle = 0 \text{ for all } x, y \in \langle\omega\rangle^\perp \text{ with } x \neq y. \tag{4.1.208}$$

If for each fixed $y \in \langle\omega\rangle^\perp$ we send x to infinity from within $\langle\omega\rangle^\perp$, on account of the decay of $\vec{k}$ at infinity we arrive at the conclusion that $\big\langle \omega, -\vec{k}_1(-y)\big\rangle = 0$ for all $y \in \langle\omega\rangle^\perp$. Using this back into (4.1.208) we obtain $\big\langle \omega, \vec{k}(x-y)\big\rangle = 0$ for all $x, y \in \langle\omega\rangle^\perp$ with $x \neq y$. Hence, (1.3.3) holds, and this is known (cf. Lemma 1.3.1) to imply (4.1.206). □

We are now prepared to present the proof of Theorem 4.1.10.

Proof of Theorem 4.1.10 Let us prove that *(2)* is implied by the weakest version of *(1)*. To this end, note that any of the inequalities in (4.1.187), (4.1.188) forces

$$T = 0 \text{ whenever } \Omega \text{ is a half-space in } \mathbb{R}^n. \tag{4.1.209}$$

Invoking Lemma 1.3.1 we conclude that the conditions in item *(2)* are satisfied.

Conversely, if *(2)* holds, then the estimates in item *(1)* are guaranteed to hold by Theorem 4.1.3. Let us also observe that whenever (4.1.189) holds, the homogeneity property of k entails Euler's identity

$$\langle x, (\nabla k)(x)\rangle = -n\, k(x) \text{ for each } x \in \mathbb{R}^n \setminus \{0\}. \tag{4.1.210}$$

Consequently, for each $x \in \mathbb{R}^n \setminus \{0\}$,

$$(\mathrm{div}\vec{k})(x) = \partial_j\big(x_j\, k(z)\big) = n\, k(x) + x_j(\partial_j k)(x) = n\, k(x) + \langle x, (\nabla k)(x)\rangle = 0,$$

which ultimately shows that $\vec{k}$ is divergence-free in $\mathbb{R}^n \setminus \{0\}$. Bearing this property in mind, we then conclude from Theorem 4.1.6, Theorem 4.1.7, Theorem 4.1.8, and Theorem 4.1.9 that all estimates in items *(3)-(12)* hold. Finally, any of the estimates in items *(3)-(5)* implies (4.1.209) which, as before, implies *(2)*, whereas any of the estimates in items *(6)-(12)* implies

$$T_{\rm mod} = 0 \text{ whenever } \Omega \text{ is a half-space in } \mathbb{R}^n \tag{4.1.211}$$

which, in turn, implies (4.1.189), thanks to Lemma 4.1.11. □

Below we propose to prove a result similar in spirit to Theorem 4.1.10, for double layer potential operators associated with weakly elliptic systems. For the brand of norm estimates we seek in this section, this theorem brings to prominence the existence of a distinguished coefficient tensor for the given system.

Theorem 4.1.12 *Fix $n \in \mathbb{N}$ with $n \geq 2$, and let L be a homogeneous, weakly elliptic, constant (complex) coefficient, second-order $M \times M$ system in $\mathbb{R}^n$ (for some $M \in \mathbb{N}$). Also, pick $A \in \mathfrak{A}_L$ and, for each*

$$\Omega \subseteq \mathbb{R}^n \text{ UR } \textit{domain (and with } \sigma := \mathcal{H}^{n-1}\lfloor\partial\Omega \textit{ and } \nu \textit{ denoting, respectively, the "surface measure" on } \partial\Omega\textit{, and the geometric measure theoretic outward unit normal to } \Omega), \tag{4.1.212}$$

let $K, K^{\#}$ be the boundary-to-boundary double layer potential operators associated with Ω and the coefficient tensor A as in (A.0.116) *and* (A.0.118), *respectively. Then the following statements are equivalent.*

(i) For any set Ω as in (4.1.212), *any (or some) exponent $p \in (1,\infty)$, any (or some) Muckenhoupt weight $w \in A_p(\partial\Omega,\sigma)$, and any (or some) integer $m \in \mathbb{N}$ there exists $C_m \in (0,\infty)$, which depends only on m, n, p, $[w]_{A_p}$, A, and the* UR *constants of $\partial\Omega$, ensuring the validity of any (or all) of the following estimates:*

$$\|K\|_{[L^p(\partial\Omega,w)]^M \to [L^p(\partial\Omega,w)]^M} \leq C_m \|\nu\|^{\langle m\rangle}_{[\mathrm{BMO}(\partial\Omega,\sigma)]^n}, \tag{4.1.213}$$

$$\|K\|_{[L^p_1(\partial\Omega,w)]^M \to [L^p_1(\partial\Omega,w)]^M} \leq C_m \|\nu\|^{\langle m\rangle}_{[\mathrm{BMO}(\partial\Omega,\sigma)]^n}, \tag{4.1.214}$$

$$\|K^{\#}\|_{[L^p(\partial\Omega,w)]^M \to [L^p(\partial\Omega,w)]^M} \leq C_m \|\nu\|^{\langle m\rangle}_{[\mathrm{BMO}(\partial\Omega,\sigma)]^n}, \tag{4.1.215}$$

$$\|K^{\#}\|_{[L^p_{-1}(\partial\Omega,w)]^M \to [L^p_{-1}(\partial\Omega,w)]^M} \leq C_m \|\nu\|^{\langle m\rangle}_{[\mathrm{BMO}(\partial\Omega,\sigma)]^n}. \tag{4.1.216}$$

(ii) The coefficient tensor A is distinguished, i.e., $A \in \mathfrak{A}^{\mathrm{dis}}_L$.

Proof Recall the matrix-valued fundamental solution $E = (E_{\alpha\beta})_{1\leq\alpha,\beta\leq M}$ associated with the weakly elliptic system L as in [131, Theorem 1.4.2]. Also, denote by $\big(a^{\alpha\beta}_{rs}\big)_{\substack{1\leq r,s\leq n\\ 1\leq\alpha,\beta\leq M}}$ the entries of the given coefficient tensor $A \in \mathfrak{A}_L$.

To deal with the implication *(i)*$\Rightarrow$*(ii)*, assume one of the estimates in (4.1.213)-(4.1.216) holds for any set Ω as in (4.1.212), some weight $w \in A_p(\partial\Omega,\sigma)$ with $p \in (1,\infty)$, and some $m \in \mathbb{N}$. From the discussion in [132, Example 5.1.6] we know that $K = \big(T_{\alpha\gamma}\big)_{1\leq\gamma,\alpha\leq M}$ with each scalar operator associated as in (4.1.185) with the vector-valued kernel

$$\vec{k}_{\alpha\gamma} := -\big(a^{\beta\alpha}_{rs}\,\partial_r E_{\gamma\beta}\big)_{1\leq s\leq n}. \tag{4.1.217}$$

See [132, (5.1.159), (5.1.154)]. From [131, Theorem 1.4.2] we know that each such vector-valued kernel is odd and positive homogeneous of degree $1-n$. It has also been noted in [132, (5.1.155)] that each such kernel is divergence-free in $\mathbb{R}^n\setminus\{0\}$. In addition, we have $K^{\#} = \big(T^{\#}_{\alpha\gamma}\big)_{1\leq\alpha,\gamma\leq M}$ where each scalar operator is associated as in (4.1.186) with the vector-valued kernel $\vec{k}_{\alpha\gamma}$ from (4.1.217) (see [132, (5.1.161)]).

Then Theorem 4.1.10 applies and guarantees the existence of a family of scalar-valued functions $k_{\alpha\gamma} \in \mathscr{C}^N(\mathbb{R}^n\setminus\{0\})$ with $1 \leq \alpha,\gamma \leq M$, which are even and positive homogeneous of degree $-n$ and satisfy $\vec{k}_{\alpha\gamma}(x) = x\,k_{\alpha\gamma}(x)$ for each point $x \in \mathbb{R}^n\setminus\{0\}$ and $1 \leq \alpha,\gamma \leq M$. From this and (4.1.217) we then see that the property in item *(g)* in Proposition 1.2.1 holds. In view of Definition 1.2.2, we then conclude that $A \in \mathfrak{A}^{\mathrm{dis}}_L$.

To prove that *(ii)*$\Rightarrow$*(i)*, work under the assumption that $A \in \mathfrak{A}^{\mathrm{dis}}_L$. Then Proposition 1.2.1 and Definition 1.2.2 ensure the existence of a matrix-valued function

$$k = \{k_{\gamma\alpha}\}_{1\leq\gamma,\alpha\leq M} : \mathbb{R}^n \setminus \{0\} \longrightarrow \mathbb{C}^{M\times M} \tag{4.1.218}$$

with the property that for each $\gamma, \alpha \in \{1, \dots, M\}$ and $s \in \{1, \dots, n\}$ one has

$$a_{rs}^{\beta\alpha}(\partial_r E_{\gamma\beta})(x) = x_s k_{\gamma\alpha}(x) \text{ for all } x \in \mathbb{R}^n \setminus \{0\}. \tag{4.1.219}$$

Moreover, from (1.2.14) we know that

the entries of the matrix-valued function k from (4.1.218) belong to $\mathscr{C}^\infty(\mathbb{R}^n \setminus \{0\})$, are even, and positive homogeneous of degree $-n$. (4.1.220)

As such, from (A.0.116) and (4.1.218)-(4.1.219) we see that

$$Kf(x) = -\lim_{\varepsilon\to 0^+} \int_{y\in\partial\Omega,\ |x-y|>\varepsilon} \langle\nu(y), x-y\rangle k(x-y) f(y)\, d\sigma(y) \tag{4.1.221}$$

at σ-a.e. $x \in \partial\Omega$, for each function $f \in \Big[L^1\Big(\partial\Omega, \frac{\sigma(x)}{1+|x|^{n-1}}\Big)\Big]^M$. Likewise, with $k^\top$ denoting the transpose of k, the expression in (A.0.118) becomes

$$K^\# f(x) = \lim_{\varepsilon\to 0^+} \int_{y\in\partial\Omega,\ |x-y|>\varepsilon} \langle\nu(x), x-y\rangle\, k^\top(x-y) f(y)\, d\sigma(y) \tag{4.1.222}$$

at σ-a.e. $x \in \partial\Omega$, for each function $f \in \Big[L^1\Big(\partial\Omega, \frac{\sigma(x)}{1+|x|^{n-1}}\Big)\Big]^M$.

Granted these identifications, (4.1.121) in Theorem 4.1.6 gives (4.1.214). Also, (4.1.31) in Theorem 4.1.3 yields (4.1.213) while (4.1.32) in Theorem 4.1.3 gives (4.1.215). Finally, (4.1.122) in Theorem 4.1.6 justifies (4.1.216). □

Our next theorem contains operator norm estimates for the modified boundary-to-boundary double layer on a variety of homogeneous Sobolev spaces (modulo constants), in a manner which brings into play the BMO semi-norm of the geometric measure theoretic outward unit normal to the underlying domain. The main point is that the veracity of said estimates is equivalent to the coefficient tensor (with respect to which the double layer is defined) being distinguished.

Theorem 4.1.13 *Fix $n \in \mathbb{N}$ with $n \geq 2$, and let L be a homogeneous, weakly elliptic, constant (complex) coefficient, second-order $M \times M$ system in $\mathbb{R}^n$ (for some $M \in \mathbb{N}$). Also, pick $A \in \mathfrak{A}_L$ and, for each*

$\Omega \subseteq \mathbb{R}^n$ open set satisfying a two-sided local John condition and having an Ahlfors regular boundary (for which ν and $\sigma := \mathcal{H}^{n-1}\lfloor\partial\Omega$ stand, respectively, for its geometric measure theoretic outward unit normal, and its surface measure) (4.1.223)

let $K_{\rm mod}$ be the boundary-to-boundary modified double layer potential operators associated with Ω and the coefficient tensor A as in (A.0.117). *Then the following statements are equivalent.*

(i) For any set Ω as in (4.1.223), *any* (*or some*) *exponents $p, q \in (1,\infty)$, any* (*or some*) *parameter $\lambda \in (0, n-1)$, and any* (*or some*) *integer $m \in \mathbb{N}$ there exists $C_m \in (0,\infty)$, which depends only on m, n, p, λ, A, the two-sided local John constants of Ω, and the Ahlfors regularity constants of $\partial\Omega$ ensuring the validity of any* (*or all*) *of the following estimates:*

$$\big\|\big[K_{\rm mod}\big]\big\|_{[\dot{L}^p_1(\partial\Omega,\sigma)/\sim]^M\to[\dot{L}^p_1(\partial\Omega,\sigma)/\sim]^M} \leq C_m\|\nu\|^{\langle m\rangle}_{[\mathrm{BMO}(\partial\Omega,\sigma)]^n}, \tag{4.1.224}$$

$$\big\|\big[K_{\rm mod}\big]\big\|_{[\dot{M}^{p,\lambda}_1(\partial\Omega,\sigma)/\sim]^M\to[\dot{M}^{p,\lambda}_1(\partial\Omega,\sigma)/\sim]^M} \leq C_m\|\nu\|^{\langle m\rangle}_{[\mathrm{BMO}(\partial\Omega,\sigma)]^n}, \tag{4.1.225}$$

$$\big\|\big[K_{\rm mod}\big]\big\|_{[\dot{\mathbb{M}}^{p,\lambda}_1(\partial\Omega,\sigma)/\sim]^M\to[\dot{\mathbb{M}}^{p,\lambda}_1(\partial\Omega,\sigma)/\sim]^M} \leq C_m\|\nu\|^{\langle m\rangle}_{[\mathrm{BMO}(\partial\Omega,\sigma)]^n}, \tag{4.1.226}$$

$$\big\|\big[K_{\rm mod}\big]\big\|_{[\dot{\mathcal{B}}^{q,\lambda}_1(\partial\Omega,\sigma)/\sim]^M\to[\dot{\mathcal{B}}^{q,\lambda}_1(\partial\Omega,\sigma)/\sim]^M} \leq C_m\|\nu\|^{\langle m\rangle}_{[\mathrm{BMO}(\partial\Omega,\sigma)]^n}. \tag{4.1.227}$$

(ii) The coefficient tensor A is distinguished, i.e., $A \in \mathfrak{A}^{\rm dis}_L$.

Proof Bring back the matrix-valued fundamental solution $E = (E_{\alpha\beta})_{1\leq\alpha,\beta\leq M}$ associated with the given system L as in [131, Theorem 1.4.2], and denote by $\big(a^{\alpha\beta}_{rs}\big)_{\substack{1\leq r,s\leq n\\ 1\leq\alpha,\beta\leq M}}$ the entries of the given coefficient tensor $A \in \mathfrak{A}_L$. The key observation is that, as visible from definitions, $K_{\rm mod} = \big(T^{\alpha\gamma}_{\rm mod}\big)_{1\leq\gamma,\alpha\leq M}$ with each scalar operator $T^{\alpha\gamma}_{\rm mod}$ associated as in [132, (5.1.65)-(5.1.66)] with the vector-valued kernel

$$\vec{k}_{\alpha\gamma} := -\big(a^{\beta\alpha}_{rs}\,\partial_r E_{\gamma\beta}\big)_{1\leq s\leq n}. \tag{4.1.228}$$

From [131, Theorem 1.4.2] we know that each such vector-valued kernel is odd and positive homogeneous of degree $1-n$. It has also been noted in [132, (5.1.155)] that each such kernel is divergence-free in $\mathbb{R}^n \setminus \{0\}$.

Consider the implication *(i)*$\Rightarrow$*(ii)*. Specifically, suppose one of the estimates in (4.1.224)-(4.1.227) holds for any set Ω as in (4.1.223), for some $p, q \in (1,\infty)$, $\lambda \in (0, n-1)$, and $m \in \mathbb{N}$.

In view of the observation made earlier, Theorem 4.1.10 applies and guarantees the existence of a family of scalar-valued functions $k_{\alpha\gamma} \in \mathscr{C}^N(\mathbb{R}^n \setminus \{0\})$ with $1 \leq \alpha,\gamma \leq M$, which are even and positive homogeneous of degree $-n$ and satisfy $\vec{k}_{\alpha\gamma}(x) = x\,k_{\alpha\gamma}(x)$ for each point $x \in \mathbb{R}^n \setminus \{0\}$ and $1 \leq \alpha,\gamma \leq M$. Together with (4.1.228), this proves that the property in item *(g)* of Proposition 1.2.1 is true. According to Definition 1.2.2, we then have $A \in \mathfrak{A}^{\rm dis}_L$.

To prove that *(ii)*$\Rightarrow$*(i)*, suppose $A \in \mathfrak{A}^{\rm dis}_L$ to begin with. As such, Proposition 1.2.1 and Definition 1.2.2 guarantee the existence of a matrix-valued function

$$k = \big\{k_{\gamma\alpha}\big\}_{1\leq\gamma,\alpha\leq M} : \mathbb{R}^n \setminus \{0\} \longrightarrow \mathbb{C}^{M\times M} \tag{4.1.229}$$

with the property that for each $\gamma, \alpha \in \{1,\dots,M\}$ and $s \in \{1,\dots,n\}$ we have

$$a^{\beta\alpha}_{rs}(\partial_r E_{\gamma\beta})(x) = x_s k_{\gamma\alpha}(x) \text{ for all } x \in \mathbb{R}^n \setminus \{0\}. \tag{4.1.230}$$

In addition, from (1.2.14) we know that

the entries of the matrix-valued function k from (4.1.218) belong to $\mathscr{C}^\infty(\mathbb{R}^n \setminus \{0\})$, are even, and positive homogeneous of degree $-n$. (4.1.231)

Then from (A.0.117) and (4.1.229)-(4.1.230) we see that for each function

$$f = (f_\alpha)_{1\leq\alpha\leq M} \in \Big[L^1\big(\partial\Omega, \tfrac{\sigma(y)}{1+|y|^n}\big)\Big]^M \tag{4.1.232}$$

we have

$$K_{\text{mod}} f(x) = -\Bigg(\lim_{\varepsilon\to 0^+} \int\limits_{\substack{y\in\partial\Omega \\ |x-y|>\varepsilon}} \big\langle \nu(y), \vec{k}^{\gamma\alpha}_{\varepsilon}(x-y) - \vec{k}^{\gamma\alpha}_{1}(-y)\big\rangle f_\alpha(y)\, d\sigma(y)\Bigg)_{1\leq\gamma\leq M}$$

at σ-a.e. $x \in \partial\Omega$, where for each $\gamma, \alpha \in \{1, \dots, M\}$ we have set

$$\vec{k}^{\gamma\alpha}_{\varepsilon}(z) := z\, k_{\gamma\alpha}(z) \cdot \mathbf{1}_{\mathbb{R}^n\setminus\overline{B(0,\varepsilon)}}(z) \text{ for each } \varepsilon > 0 \text{ and } z \in \mathbb{R}^n. \tag{4.1.233}$$

Having made this identification, Theorem 4.1.6 gives all estimates in (4.1.224)-(4.1.227). □

Going further, we augment Theorem 4.1.12 with similar norm estimates on Hardy spaces, the John-Nirenberg space, Hölder spaces, as well as Morrey-Campanato spaces and their pre-duals, for double layer potential operators associated with weakly elliptic possessing distinguished coefficient tensors. As before, the latter property is actually *necessary* for the validity of said estimates.

Theorem 4.1.14 *Let $\Omega \subseteq \mathbb{R}^n$ (where $n \in \mathbb{N}$, $n \geq 2$) be a* UR *domain. Denote $\sigma := \mathcal{H}^{n-1}\lfloor\partial\Omega$ and let ν be the geometric measure theoretic outward unit normal to Ω. Also, let L be a homogeneous, weakly elliptic, constant (complex) coefficient, second-order $M \times M$ system in $\mathbb{R}^n$ (for some number $M \in \mathbb{N}$) with the property that $\mathfrak{A}^{\text{dis}}_L \neq \emptyset$. Pick $A \in \mathfrak{A}^{\text{dis}}_L$ and consider the boundary-to-boundary double layer potential operators $K, K^\#$ associated with Ω and the coefficient tensor A as in* (A.0.116) *and* (A.0.118)*, respectively. Finally, bring in the modified boundary-to-boundary double layer operator K_{mod} associated with the coefficient tensor A and the set Ω as in* (A.0.117).

Then for each given integer $m \in \mathbb{N}$ and each pair of exponents, $p \in \big(\frac{n-1}{n}, \infty\big)$ and $q \in (0, \infty]$, there exists a constant $C_m \in (0, \infty)$ which depends only on m, n, p, q, A, and the UR *constants of $\partial\Omega$ such that*

$$\|K^\#\|_{[H^{p,q}(\partial\Omega,\sigma)]^M \to [H^{p,q}(\partial\Omega,\sigma)]^M} \leq C_m \|\nu\|^{\langle m\rangle}_{[\text{BMO}(\partial\Omega,\sigma)]^n}. \tag{4.1.234}$$

In particular, for each $m \in \mathbb{N}$ and each $p \in \big(\frac{n-1}{n}, \infty\big)$ there exists a constant $C_m \in (0, \infty)$ of the same nature as above such that

$$\|K^{\#}\|_{[H^p(\partial\Omega,\sigma)]^M \to [H^p(\partial\Omega,\sigma)]^M} \le C_m \|\nu\|^{\langle m \rangle}_{[\mathrm{BMO}(\partial\Omega,\sigma)]^n}. \tag{4.1.235}$$

Also, for each $m \in \mathbb{N}$ there exists some constant $C_m \in (0,\infty)$, which depends only on the given integer m, the dimension n, the coefficient tensor A, and the UR *constants of $\partial\Omega$, such that the operator K_{mod} considered in the context of* [132, (2.1.93)] *satisfies*

$$\big\|\big[K_{\mathrm{mod}}\big]\big\|_{[\mathrm{BMO}(\partial\Omega,\sigma)/\sim]^M \to [\mathrm{BMO}(\partial\Omega,\sigma)/\sim]^M} \le C_m \|\nu\|^{\langle m \rangle}_{[\mathrm{BMO}(\partial\Omega,\sigma)]^n}, \tag{4.1.236}$$

while when considered in the context of [132, (2.1.156)] *satisfies*

$$\big\|\big[K_{\mathrm{mod}}\big]\big\|_{[\mathrm{VMO}(\partial\Omega,\sigma)/\sim]^M \to [\mathrm{VMO}(\partial\Omega,\sigma)/\sim]^M} \le C_m \|\nu\|^{\langle m \rangle}_{[\mathrm{BMO}(\partial\Omega,\sigma)]^n}, \tag{4.1.237}$$

and, finally, when considered in the context of [132, (2.1.172)] *satisfies*

$$\big\|\big[K_{\mathrm{mod}}\big]\big\|_{[\mathrm{CMO}(\partial\Omega,\sigma)/\sim]^M \to [\mathrm{CMO}(\partial\Omega,\sigma)/\sim]^M} \le C_m \|\nu\|^{\langle m \rangle}_{[\mathrm{BMO}(\partial\Omega,\sigma)]^n}. \tag{4.1.238}$$

Next, for each $m \in \mathbb{N}$ and each $\alpha \in (0,1)$ there exists some $C_m \in (0,\infty)$, which depends only on m, n, α, A, and the UR *constants of $\partial\Omega$, with the property that the operator K_{mod} considered in the context of* [132, (2.1.123)] *satisfies*

$$\big\|\big[K_{\mathrm{mod}}\big]\big\|_{[\dot{\mathscr{C}}^{\alpha}(\partial\Omega)/\sim]^M \to [\dot{\mathscr{C}}^{\alpha}(\partial\Omega)/\sim]^M} \le C_m \|\nu\|^{\langle m \rangle}_{[\mathrm{BMO}(\partial\Omega,\sigma)]^n}, \tag{4.1.239}$$

while the operator K_{mod} considered in the context of [132, (2.1.144)] *satisfies*

$$\big\|\big[K_{\mathrm{mod}}\big]\big\|_{[\dot{\mathscr{C}}^{\alpha}_{\mathrm{van}}(\partial\Omega)/\sim]^M \to [\dot{\mathscr{C}}^{\alpha}_{\mathrm{van}}(\partial\Omega)/\sim]^M} \le C_m \|\nu\|^{\langle m \rangle}_{[\mathrm{BMO}(\partial\Omega,\sigma)]^n}. \tag{4.1.240}$$

In addition, for each $m \in \mathbb{N}$, each exponent $p \in (1,\infty)$, and each parameter $\lambda \in (0, n-1)$, there exists some $C_m \in (0,\infty)$, which depends only on m, n, p, λ, A, and the UR *constants of $\partial\Omega$, with the property that the operator K_{mod} considered in the context of* [132, (3.2.14)] *satisfies*

$$\big\|\big[K_{\mathrm{mod}}\big]\big\|_{[\mathring{\mathfrak{L}}^{p,\lambda}(\partial\Omega,\sigma)/\sim]^M \to [\mathring{\mathfrak{L}}^{p,\lambda}(\partial\Omega,\sigma)/\sim]^M} \le C_m \|\nu\|^{\langle m \rangle}_{[\mathrm{BMO}(\partial\Omega,\sigma)]^n}, \tag{4.1.241}$$

while the operator K considered in the context of [132, (3.2.13)] *satisfies*

$$\|K\|_{[\mathfrak{L}^{p,\lambda}(\partial\Omega,\sigma)]^M \to [\mathfrak{L}^{p,\lambda}(\partial\Omega,\sigma)]^M} \le C_m \|\nu\|^{\langle m \rangle}_{[\mathrm{BMO}(\partial\Omega,\sigma)]^n}. \tag{4.1.242}$$

Finally, for each number $m \in \mathbb{N}$, each exponent $q \in (1,\infty)$, and each parameter $\lambda \in (0, n-1)$, there exists some $C_m \in (0,\infty)$, which depends only on m, n, q, λ, A, and the UR *constants of $\partial\Omega$, with the property that the operator $K^{\#}$ when considered in the context of* [132, (3.2.2)] *satisfies*

$$\|K^{\#}\|_{[\mathscr{H}^{q,\lambda}(\partial\Omega,\sigma)]^M \to [\mathscr{H}^{q,\lambda}(\partial\Omega,\sigma)]^M} \le C_m \|\nu\|^{\langle m \rangle}_{[\mathrm{BMO}(\partial\Omega,\sigma)]^n}. \tag{4.1.243}$$

Proof The present hypotheses guarantee that K may be expressed as in (4.1.221), and $K^{\#}$ may be expressed as in (4.1.222). As such, Theorem 4.1.7 applies and yields (4.1.234) (from which (4.1.235) follows; cf. [130, (4.2.25)]), while Theorem 4.1.9 applies and yields (4.1.243). Also, (4.1.242) become a consequence of (4.1.167).

Next, from (A.0.117) and [132, (5.2.188)-(5.2.189)] we see that K_{mod} may be thought of as a matrix operator in which each entry is of the form T_{mod}. As such, we may invoke Theorem 4.1.8 to conclude that (4.1.236)-(4.1.240) hold, and rely on Theorem 4.1.9 to obtain (4.1.241). □

We continue by revisiting Theorem 4.1.8 and [132, Theorem 5.1.15] for the purpose of proving a pointwise mean oscillation estimate for modified chord-dot-normal singular integral operators which is sensitive to the flatness of the boundary of the underlying domain.

Theorem 4.1.15 *Fix $n \in \mathbb{N}$ with the property that $n \geq 2$ and assume $\Omega \subseteq \mathbb{R}^n$ is a* UR *domain. Abbreviate $\sigma := \mathcal{H}^{n-1} \lfloor \partial\Omega$ and denote by ν the geometric measure theoretic outward unit normal to the set Ω. For a sufficiently large integer $N = N(n) \in \mathbb{N}$ consider $k \in \mathscr{C}^N(\mathbb{R}^n \setminus \{0\})$ which is even and positive homogeneous of degree $-n$.*

Let T_{mod} be the modified version of the chord-dot-normal singular integral operator associated with the kernel k, acting on each $f \in L^1\big(\partial\Omega, \frac{\sigma(x)}{1+|x|^n}\big)$ according to

$$T_{\text{mod}} f(x) := \lim_{\varepsilon \to 0^+} \int_{\partial\Omega} \big\langle \nu(y), \vec{k}_\varepsilon(x-y) - \vec{k}_1(-y) \big\rangle f(y)\, d\sigma(y) \tag{4.1.244}$$

at σ-a.e. $x \in \partial\Omega$, where

$$\begin{aligned} &\vec{k}(z) := z\, k(z) \text{ for each } z \in \mathbb{R}^n \text{ and} \\ &\vec{k}_\varepsilon := \vec{k} \cdot \mathbf{1}_{\mathbb{R}^n \setminus \overline{B(0,\varepsilon)}} \text{ for each } \varepsilon > 0. \end{aligned} \tag{4.1.245}$$

Then for each exponent $p \in (1, \infty)$, Muckenhoupt weight $w \in A_p(\partial\Omega, \sigma)$, and integer $m \in \mathbb{N}$ there exists a constant $C_m \in (0, \infty)$, which depends only on p, $[w]_{A_p}$, m, n, and the UR *constants of $\partial\Omega$, with the property that for each point $x_o \in \partial\Omega$, each radius $r \in \big(0, 2 \operatorname{diam} \partial\Omega\big)$, and each each function*

$$f \in L^1\Big(\partial\Omega, \frac{\sigma(x)}{1+|x|^n}\Big) \cap L^p_{\text{loc}}(\partial\Omega, w) \tag{4.1.246}$$

one has (with the piece of notation introduced in (4.1.12)*)*

$$\Big(\fint_{\Delta(x_o,r)}\Big|T_{\rm mod}f-\fint_{\Delta(x_o,r)}T_{\rm mod}f\,dw\Big|^p\,dw\Big)^{1/p} \tag{4.1.247}$$
$$\leq C_m\Big(\sum_{|\alpha|\leq N}\sup_{S^{n-1}}|\partial^\alpha k|\Big)\|\nu\|^{\langle m\rangle}_{[\mathrm{BMO}(\partial\Omega,\sigma)]^n}\times$$
$$\times\int_1^\infty\Big(\fint_{\Delta(x_o,\lambda r)}\Big|f-\fint_{\Delta(x_o,\lambda r)}f\,dw\Big|^p\,dw\Big)^{1/p}\frac{\ln\lambda}{\lambda^2}\,d\lambda.$$

As a consequence of (4.1.247) *and the definition of the Fefferman-Stein sharp maximal operator (cf.* (A.0.226)*), it follows that for each integrability exponent* $p\in(1,\infty)$ *and each integer* $m\in\mathbb{N}$ *there exists a constant* $C_m\in(0,\infty)$*, depending only on* p*,* m*,* n*, and the* UR *constants of* $\partial\Omega$*, such that the following pointwise inequality holds:*

$$\big(T_{\rm mod}f\big)^{\#}_p\leq C_m\Big(\sum_{|\alpha|\leq N}\sup_{S^{n-1}}|\partial^\alpha k|\Big)\|\nu\|^{\langle m\rangle}_{[\mathrm{BMO}(\partial\Omega,\sigma)]^n}\cdot f^{\#}_p\ \text{ on }\partial\Omega,$$
$$\text{for every given function } f\in L^1\Big(\partial\Omega,\tfrac{\sigma(x)}{1+|x|^n}\Big)\cap L^p_{\rm loc}(\partial\Omega,\sigma). \tag{4.1.248}$$

The same type of argument as in the proof of [131, (2.3.35)] shows that

$$T_{\rm mod}f\in L^p_{\rm loc}(\partial\Omega,w)\subseteq L^1_{\rm loc}(\partial\Omega,w). \tag{4.1.249}$$

As such, the expression in the left side of (4.1.247) is meaningful.

The estimate recorded in (4.1.247) has several key attributes. First, the estimate in question is 'space-less," in the sense that it does not involve any recognizable norm (employed in standard function spaces). It simply allows us to control an individual L^p-styled mean oscillation of the action of $T_{\rm mod}$ on any function f as in (4.1.246) on a given surface ball in terms of the L^p-styled mean oscillations of f over dilates of said surface ball, in a uniform fashion with respect to its center and radius.

Second, compared with [132, (5.1.312)], the current estimate involves the BMO semi-norm of the geometric measure theoretic outward unit normal ν. Thus, in contrast to [132, Theorem 5.1.15], the estimate in Theorem 4.1.15 is capable of taking into account the relative flatness of the boundary of the underlying domain.

Third, while our earlier estimate in [132, (5.1.312)] is valid for the larger class of modified generalized double layers, it only involves a typically large multiplicative geometric constant in the right side, which is insensitive to flatness. Restricting here to chord-dot-normal operators is justified, since this is the largest category of singular integral operators (with kernels depending linearly on the outward unit normal) for which an estimate like (4.1.247) is possible.

The aforementioned features ultimately lead to invertibility results for chord-dot-normal singular integral operators on a large variety of spaces whose norms are defined in terms of mean oscillations (as we shall see in future discussions).

Let us now present the proof of Theorem 4.1.15.

Proof of Theorem 4.1.15 Pick an integrability exponent $p \in (1, \infty)$ together with a Muckenhoupt weight $w \in A_p(\partial\Omega, \sigma)$, and fix a point $x_o \in \partial\Omega$. To set the stage, we shall show that, if $p' := p/(p-1)$ is the conjugate exponent of p and $w' := w^{1-p'}$ is the conjugate weight of w, then given any two σ-measurable functions $g, h : \partial\Omega \to \mathbb{R}$ the following Hölder inequality holds for each $R > 0$:

$$\fint_{\Delta(x_o,R)} |g||h|\, d\sigma \leq [w]_{A_p}^{1/p} \Big(\fint_{\Delta(x_o,R)} |g|^p\, dw \Big)^{1/p} \Big(\fint_{\Delta(x_o,R)} |h|^{p'}\, dw' \Big)^{1/p'}. \tag{4.1.250}$$

Indeed, for each given $R > 0$ we may estimate

$$\begin{aligned}
\fint_{\Delta(x_o,R)} |g||h|\, d\sigma &= \frac{1}{\sigma(\Delta(x_o,R))} \int_{\Delta(x_o,R)} \big(|g| w^{1/p}\big)\big(|h| w^{-1/p}\big)\, d\sigma \\
&\leq \frac{1}{\sigma(\Delta(x_o,R))} \Big(\int_{\Delta(x_o,R)} |g|^p w\, d\sigma \Big)^{1/p} \Big(\int_{\Delta(x_o,R)} |h|^{p'} w^{-p'/p}\, d\sigma \Big)^{1/p'} \\
&= \frac{w\big(\Delta(x_o,R)\big)^{1/p} w'\big(\Delta(x_o,R)\big)^{1/p'}}{\sigma\big(\Delta(x_o,R)\big)} \times \\
&\qquad \times \Big(\fint_{\Delta(x_o,R)} |g|^p\, dw \Big)^{1/p} \Big(\int_{\Delta(x_o,R)} |h|^{p'}\, dw' \Big)^{1/p'} \\
&= \Big[\Big(\fint_{\Delta(x_o,R)} w\, d\sigma \Big) \Big(\fint_{\Delta(x_o,R)} w'\, d\sigma \Big)^{p/p'} \Big]^{1/p} \times \\
&\qquad \times \Big(\fint_{\Delta(x_o,R)} |g|^p\, dw \Big)^{1/p} \Big(\int_{\Delta(x_o,R)} |h|^{p'}\, dw' \Big)^{1/p'} \\
&\leq [w]_{A_p}^{1/p} \Big(\fint_{\Delta(x_o,R)} |g|^p\, dw \Big)^{1/p} \Big(\fint_{\Delta(x_o,R)} |h|^{p'}\, dw' \Big)^{1/p'},
\end{aligned} \tag{4.1.251}$$

by Hölder's inequality and the definition of the characteristic of the Muckenhoupt weight w (cf. (A.0.3)). This establishes (4.1.250), which will play a role shortly.

To proceed in earnest, choose $r \in \big(0, 2\,\mathrm{diam}\,\partial\Omega\big)$ and abbreviate $\Delta := \Delta(x_o, r)$. Also, pick a function f as in (4.1.246) so, in particular, $f \in L^1_{\mathrm{loc}}(\partial\Omega, w)$. For each $\lambda \in (0, \infty)$ set $\lambda\Delta := \Delta(x_o, \lambda r)$ and define

$$f_{\lambda\Delta,w} := \fint_{\Delta(x_o,\lambda r)} f\, dw := \frac{1}{w\big(\Delta(x_o,\lambda r)\big)} \int_{\Delta(x_o,\lambda r)} f\, dw. \tag{4.1.252}$$

Next, decompose $f = (f - f_{2\Delta,w}) \cdot \mathbf{1}_{2\Delta} + (f - f_{2\Delta,w}) \cdot \mathbf{1}_{\partial\Omega \setminus 2\Delta} + f_{2\Delta,w}$. Note that $f_{2\Delta,w}$ is a constant, and

$$(f - f_{2\Delta,w}) \cdot \mathbf{1}_{2\Delta} \in L^p(\partial\Omega, w) \subseteq L^1\Big(\partial\Omega,\ \frac{\sigma(x)}{1+|x|^{n-1}}\Big) \tag{4.1.253}$$

thanks to the embedding of the Muckenhoupt weighted Lebesgue spaces from [129, (7.7.104)]. Also,

$$(f - f_{2\Delta,w}) \cdot \mathbf{1}_{\partial\Omega\setminus 2\Delta} \in L^1\Big(\partial\Omega,\, \frac{\sigma(x)}{1+|x|^n}\Big). \tag{4.1.254}$$

Consequently, [132, (5.1.69)], [131, (2.3.34)], and [129, Lemma 7.21] guarantee that

$$C^{(1)}_{f,\Delta,w} := T_{\rm mod} f_{2\Delta,w} \tag{4.1.255}$$

and

$$C^{(2)}_{f,\Delta,w} := T_{\rm mod}\big[(f - f_{2\Delta,w}) \cdot \mathbf{1}_{2\Delta}\big] - T\big[(f - f_{2\Delta,w}) \cdot \mathbf{1}_{2\Delta}\big], \tag{4.1.256}$$

as well as

$$C^{(3)}_{f,\Delta,w} := \int_{\partial\Omega\setminus 2\Delta} \big\langle \nu(y), (x_o - y)k(x_o - y) - \vec{k}_1(-y)\big\rangle \big(f(y) - f_{2\Delta,w}\big)\, d\sigma(y), \tag{4.1.257}$$

are all well-defined constants. Next, set $\nu_\Delta := ⨍_\Delta \nu\, d\sigma$, and for each $x \in \Delta$ define

$$\mathrm{I}_w(x) := \int_{\partial\Omega\setminus 2\Delta} \big\langle \nu(y) - \nu_\Delta,\, \vec{k}(x-y) - \vec{k}(x_o - y)\big\rangle \big(f(y) - f_{2\Delta,w}\big)\, d\sigma(y), \tag{4.1.258}$$

$$\mathrm{II}_w(x) := \int_{\partial\Omega\setminus 2\Delta} \big\langle \nu_\Delta,\, \vec{k}(x-y) - \vec{k}(x_o - y)\big\rangle \big(f(y) - f_{2\Delta,w}\big)\, d\sigma(y). \tag{4.1.259}$$

If we now introduce

$$C_{f,\Delta,w} := C^{(1)}_{f,\Delta,w} + C^{(2)}_{f,\Delta,w} + C^{(3)}_{f,\Delta,w} \tag{4.1.260}$$

then $C_{f,\Delta,w}$ is a well-defined constant and for each $x \in \Delta$ we have

$$\begin{aligned}(T_{\rm mod} f)(x) - C_{f,\Delta,w} &= \Big(T\big[(f - f_{2\Delta,w}) \cdot \mathbf{1}_{2\Delta}\big]\Big)(x) \\ &\quad + \int_{\partial\Omega\setminus 2\Delta} \big\langle \nu(y),\, \vec{k}(x-y) - \vec{k}(x_o - y)\big\rangle \big(f(y) - f_{2\Delta,w}\big)\, d\sigma(y) \\ &= \Big(T\big[(f - f_{2\Delta,w}) \cdot \mathbf{1}_{2\Delta}\big]\Big)(x) + \mathrm{I}_w(x) + \mathrm{II}_w(x). \end{aligned} \tag{4.1.261}$$

For each $m \in \mathbb{N}$, Theorem 4.1.3 (cf. (4.1.31)) gives, also bearing in mind that the measure w is doubling,

$$
\begin{aligned}
&\Big(⨍_{\Delta}\big|T\big[(f-f_{2\Delta,w})\cdot\mathbf{1}_{2\Delta}\big]\big|^{p}\,dw\Big)^{1/p} \qquad (4.1.262)\\
&\le C\|T\|_{L^p(\partial\Omega,w)\to L^p(\partial\Omega,w)}\Big(⨍_{2\Delta}|f-f_{2\Delta,w}|^{p}\,dw\Big)^{1/p}\\
&\le C_m\Big(\sum_{|\alpha|\le N}\sup_{S^{n-1}}|\partial^{\alpha}k|\Big)\|\nu\|^{(m)}_{[\mathrm{BMO}(\partial\Omega,\sigma)]^n}\sum_{j=1}^{\infty}\Big(\frac{j}{2^j}\Big)\Big(⨍_{2^j\Delta}|f-f_{2\Delta,w}|^{p}\,dw\Big)^{1/p}.
\end{aligned}
$$

Since $\vec{k}$ is smooth and positive homogeneous of degree $1-n$ off the origin, the Mean Value Theorem gives

$$
\big|\vec{k}(x-y)-\vec{k}(x_o-y)\big|\le C_n\Big(\sum_{|\alpha|\le 1}\sup_{S^{n-1}}|\partial^{\alpha}k|\Big)\frac{r}{|x_o-y|^n} \qquad (4.1.263)
$$

for each $x\in\Delta$ and $y\in\partial\Omega\setminus 2\Delta$, hence

$$
\begin{aligned}
|\mathrm{I}_w(x)| &\le C_n\Big(\sum_{|\alpha|\le 1}\sup_{S^{n-1}}|\partial^{\alpha}k|\Big)\int_{\partial\Omega\setminus 2\Delta}\frac{r}{|x_o-y|^n}|\nu(y)-\nu_{\Delta}||f(y)-f_{2\Delta,w}|\,d\sigma(y)\\
&\le C_n\Big(\sum_{|\alpha|\le 1}\sup_{S^{n-1}}|\partial^{\alpha}k|\Big)\sum_{j=1}^{\infty}\frac{1}{2^j}⨍_{2^{j+1}\Delta}|\nu(y)-\nu_{\Delta}||f(y)-f_{2\Delta,w}|\,d\sigma(y)\\
&\le C_n[w]^{1/p}_{A_p}\Big(\sum_{|\alpha|\le 1}\sup_{S^{n-1}}|\partial^{\alpha}k|\Big)\times \qquad (4.1.264)\\
&\quad\times\sum_{j=1}^{\infty}\frac{1}{2^j}\Big(⨍_{2^{j+1}\Delta}|\nu-\nu_{\Delta}|^{p'}\,dw'\Big)^{1/p'}\Big(⨍_{2^{j+1}\Delta}|f-f_{2\Delta,w}|^{p}\,dw\Big)^{1/p},
\end{aligned}
$$

where the last step uses the Hölder inequality deduced in (4.1.250). To proceed, for each fixed $j\in\mathbb{N}$ write

$$
\begin{aligned}
&\Big(⨍_{2^{j+1}\Delta}|\nu-\nu_{\Delta}|^{p'}\,dw'\Big)^{1/p'} \qquad (4.1.265)\\
&\le\Big(⨍_{2^{j+1}\Delta}\Big|\nu-⨍_{2^{j+1}\Delta}\nu\,dw'\Big|^{p'}\,dw'\Big)^{1/p'}+\Big|⨍_{2^{j+1}\Delta}\nu\,dw'-⨍_{\Delta}\nu\,d\sigma\Big|\\
&\le C\|\nu\|_{[\mathrm{BMO}(\partial\Omega,w')]^n}+⨍_{\Delta}\Big|\nu-⨍_{2^{j+1}\Delta}\nu\,dw'\Big|\,d\sigma\\
&\le C\|\nu\|_{[\mathrm{BMO}(\partial\Omega,\sigma)]^n}+[w']^{1/p'}_{A_{p'}}\Big(⨍_{\Delta}\Big|\nu-⨍_{2^{j+1}\Delta}\nu\,dw'\Big|^{p'}\,dw'\Big)^{1/p'}
\end{aligned}
$$

by the triangle inequality, [129, (7.4.71)], [129, Lemma 7.7.5], and [129, (7.7.12)] (see also (4.1.251)). In addition,

$$\Big(\fint_{\Delta}\Big|\nu-\fint_{2^{j+1}\Delta}\nu\,\mathrm{d}w'\Big|^{p'}\,\mathrm{d}w'\Big)^{1/p'} \leq \Big(\fint_{\Delta}\Big|\nu-\fint_{\Delta}\nu\,\mathrm{d}w'\Big|^{p'}\,\mathrm{d}w'\Big)^{1/p'}$$
$$+\sum_{\ell=0}^{j}\Big|\fint_{2^{\ell}\Delta}\nu\,\mathrm{d}w'-\fint_{2^{\ell+1}\Delta}\nu\,\mathrm{d}w'\Big|$$
$$\leq C\|\nu\|_{[\mathrm{BMO}(\partial\Omega,w')]^n}+\sum_{\ell=0}^{j}\fint_{2^{\ell}\Delta}\Big|\nu-\fint_{2^{\ell+1}\Delta}\nu\,\mathrm{d}w'\Big|\,\mathrm{d}w'$$
$$\leq C\|\nu\|_{[\mathrm{BMO}(\partial\Omega,w')]^n}+C\sum_{\ell=0}^{j}\fint_{2^{\ell+1}\Delta}\Big|\nu-\fint_{2^{\ell+1}\Delta}\nu\,\mathrm{d}w'\Big|\,\mathrm{d}w'$$
$$\leq Cj\|\nu\|_{[\mathrm{BMO}(\partial\Omega,w')]^n}\leq Cj\|\nu\|_{[\mathrm{BMO}(\partial\Omega,\sigma)]^n} \tag{4.1.266}$$

once again by the triangle inequality, [129, (7.4.71)], and [129, Lemma 7.7.5], while also keeping in mind that the measure w' is doubling (since there holds $w'=w^{1-p'}\in A_{p'}(\partial\Omega,\sigma)$). After gathering (4.1.264)-(4.1.266) we arrive at the conclusion that for each $x\in\Delta$ we have

$$|\mathrm{I}_w(x)|\leq C\Big(\sum_{|\alpha|\leq 1}\sup_{S^{n-1}}|\partial^{\alpha}k|\Big)\|\nu\|_{[\mathrm{BMO}(\partial\Omega,\sigma)]^n}\sum_{j=1}^{\infty}\frac{j}{2^j}\Big(\fint_{2^{j+1}\Delta}|f-f_{2\Delta,w}|^p\,\mathrm{d}w\Big)^{1/p} \tag{4.1.267}$$

with $C\in(0,\infty)$ depending only on n, p, $[w]_{A_p}$, and Ahlfors regularity.

As for (4.1.259), use the first definition in (4.1.245) and the Mean Value Theorem (together with the smoothness and homogeneity of the kernel k) to estimate

$$\Big|\big\langle\nu_{\Delta},\vec{k}(x-y)-\vec{k}(x_o-y)\big\rangle\Big| \tag{4.1.268}$$
$$\leq\big|\langle\nu_{\Delta},x-x_o\rangle\big||k(x-y)|+\big|\langle\nu_{\Delta},x_o-y\rangle\big||k(x-y)-k(x_o-y)|$$
$$\leq C_n\big|\langle\nu_{\Delta},x-x_o\rangle\big|\frac{\sup_{S^{n-1}}|k|}{|x_o-y|^n}+C_n\big|\langle\nu_{\Delta},x_o-y\rangle\big|\Big(\sup_{S^{n-1}}|\nabla k|\Big)\frac{r}{|x_o-y|^{n+1}}$$

for each $x\in\Delta$ and $y\in\partial\Omega\setminus 2\Delta$. In particular, from (4.1.268) and Corollary 3.1.6 (cf. also Theorem 3.1.13) we see that

$$\Big|\big\langle\nu_{\Delta},\vec{k}(x-y)-\vec{k}(x_o-y)\big\rangle\Big|\leq C\Big(\sum_{|\alpha|\leq 1}\sup_{S^{n-1}}|\partial^{\alpha}k|\Big)\frac{j\,r\|\nu\|_{[\mathrm{BMO}(\partial\Omega,\sigma)]^n}}{(2^j r)^n} \tag{4.1.269}$$

whenever $x\in\Delta$ and $y\in 2^{j+1}\Delta\setminus 2^j\Delta$ for some $j\in\mathbb{N}$. Thus, for each $x\in\Delta$,

$$
\begin{aligned}
|\mathrm{II}_w(x)| &\le \sum_{j=1}^{\infty} \int_{2^{j+1}\Delta\setminus 2^j\Delta} \Big|\big\langle \nu_\Delta, \vec{k}(x-y)-\vec{k}(x_o-y)\big\rangle\Big| \big|f(y)-f_{2\Delta,w}\big|\,\mathrm{d}\sigma(y) \\
&\le C\Big(\sum_{|\alpha|\le 1}\sup_{S^{n-1}}|\partial^\alpha k|\Big)\|\nu\|_{[\mathrm{BMO}(\partial\Omega,\sigma)]^n}\sum_{j=1}^{\infty}\Big(\frac{j}{2^j}\Big)⨍_{2^{j+1}\Delta}\big|f-f_{2\Delta,w}\big|\,\mathrm{d}\sigma \\
&\le C\Big(\sum_{|\alpha|\le 1}\sup_{S^{n-1}}|\partial^\alpha k|\Big)\|\nu\|_{[\mathrm{BMO}(\partial\Omega,\sigma)]^n}\sum_{j=1}^{\infty}\Big(\frac{j}{2^j}\Big)\Big(⨍_{2^{j+1}\Delta}\big|f-f_{2\Delta,w}\big|^p\,\mathrm{d}w\Big)^{1/p}
\end{aligned}
\tag{4.1.270}
$$

where the last step relies on [129, (7.7.12)] (see also (4.1.251)).

All together, (4.1.261), (4.1.262), and (4.1.17) permit us to write, for each $m\in\mathbb{N}$,

$$
\begin{aligned}
&\Big(⨍_{\Delta(x_o,r)}\Big|T_{\mathrm{mod}}f-⨍_{\Delta(x_o,r)}T_{\mathrm{mod}}f\,\mathrm{d}w\Big|^p\,\mathrm{d}w\Big)^{1/p} \\
&\quad\le 2\Big(⨍_{\Delta(x_o,r)}\big|T_{\mathrm{mod}}f-C_{f,\Delta,w}\big|^p\,\mathrm{d}w\Big)^{1/p} \\
&\quad\le C_m\Big(\sum_{|\alpha|\le N}\sup_{S^{n-1}}|\partial^\alpha k|\Big)\|\nu\|^{\langle m\rangle}_{[\mathrm{BMO}(\partial\Omega,\sigma)]^n}\sum_{j=1}^{\infty}\Big(\frac{j}{2^j}\Big)\Big(⨍_{2^j\Delta}\big|f-f_{2\Delta,w}\big|^p\,\mathrm{d}w\Big)^{1/p}.
\end{aligned}
\tag{4.1.271}
$$

Next, observe that for each $j\in\mathbb{N}$ the triangle inequality gives

$$
\Big(⨍_{2^j\Delta}\big|f-f_{2\Delta,w}\big|^p\,\mathrm{d}w\Big)^{1/p}\le\Big(⨍_{2^j\Delta}\big|f-f_{2^j\Delta,w}\big|^p\,\mathrm{d}w\Big)^{1/p}+\big|f_{2^j\Delta,w}-f_{2\Delta,w}\big|. \tag{4.1.272}
$$

In addition, [129, (7.4.125)] (corresponding to $\varepsilon=0$) presently guarantees that

$$
\Big(⨍_{2^j\Delta}\big|f-f_{2^j\Delta,w}\big|^p\,\mathrm{d}w\Big)^{1/p}\le C\int_{2^j}^{2^{j+1}}\Big(⨍_{\lambda\Delta}\big|f-f_{\lambda\Delta,w}\big|^p\,\mathrm{d}w\Big)^{1/p}\frac{\mathrm{d}\lambda}{\lambda}, \tag{4.1.273}
$$

while the first estimate in [131, Lemma 7.4.15] ensures that

$$
\begin{aligned}
\big|f_{2^j\Delta,w}-f_{2\Delta,w}\big| &\le C\int_{2r}^{2^{j+1}r}\Big(⨍_{\Delta(x_o,t)}\big|f-f_{\Delta(x_o,t),w}\big|^p\,\mathrm{d}w\Big)^{1/p}\frac{\mathrm{d}t}{t} \\
&= C\int_{2}^{2^{j+1}}\Big(⨍_{\lambda\Delta}\big|f-f_{\lambda\Delta,w}\big|^p\,\mathrm{d}w\Big)^{1/p}\frac{\mathrm{d}\lambda}{\lambda},
\end{aligned}
\tag{4.1.274}
$$

after the change of variables $\lambda:=t/r$. Collectively, (4.1.272)-(4.1.274) show that

$$
\Big(⨍_{2^j\Delta}\big|f-f_{2\Delta,w}\big|^p\,\mathrm{d}w\Big)^{1/p}\le C\int_{2}^{2^{j+1}}\Big(⨍_{\lambda\Delta}\big|f-f_{\lambda\Delta,w}\big|^p\,\mathrm{d}w\Big)^{1/p}\frac{\mathrm{d}\lambda}{\lambda}, \tag{4.1.275}
$$

hence

$$\sum_{j=1}^{\infty}\Big(\frac{j}{2^j}\Big)\Big(⨍_{2^j\Delta}|f-f_{2\Delta,w}|^p\,\mathrm{d}w\Big)^{1/p}$$
$$\leq C\sum_{j=1}^{\infty}\Big(\frac{j}{2^j}\Big)\int_2^{2^{j+1}}\Big(⨍_{\lambda\Delta}|f-f_{\lambda\Delta,w}|^p\,\mathrm{d}w\Big)^{1/p}\frac{\mathrm{d}\lambda}{\lambda}$$
$$=\int_2^{\infty}\Big(⨍_{\lambda\Delta}|f-f_{\lambda\Delta,w}|^p\,\mathrm{d}w\Big)^{1/p}\Big(\sum_{j=1}^{\infty}\Big(\frac{j}{2^j}\Big)\mathbf{1}_{2^{j+1}\geq\lambda}\Big)\frac{\mathrm{d}\lambda}{\lambda}. \tag{4.1.276}$$

Recall that for any real numbers the following summation-by-parts formula holds:

$$\sum_{j=N}^{M}a_j(b_{j+1}-b_j)=\big(a_M b_{M+1}-a_{N-1}b_N\big)-\sum_{j=N}^{M}(a_j-a_{j-1})b_j. \tag{4.1.277}$$

When used with $a_j := j$ and $b_j := 1/2^j$, this gives

$$\sum_{j=N}^{M}\frac{j}{2^{j+1}}=\Big(\frac{N-1}{2^N}-\frac{M}{2^{M+1}}\Big)+\sum_{j=N}^{M}\frac{1}{2^j} \tag{4.1.278}$$

for all $N, M \in \mathbb{N}$ with $M > N$. Passing to limit $M \to \infty$ then gives

$$\sum_{j=N}^{\infty}\frac{j}{2^j}=\frac{N+1}{2^{N-1}}\quad\text{for each } N\in\mathbb{N}_0. \tag{4.1.279}$$

In particular, there exists $C \in (0, \infty)$ such that

$$\sum_{j=1}^{\infty}\Big(\frac{j}{2^j}\Big)\mathbf{1}_{2^{j+1}\geq\lambda}\leq C\frac{\ln\lambda}{\lambda}\quad\text{for each } \lambda\in(2,\infty). \tag{4.1.280}$$

Combining (4.1.276) and (4.1.280) leads to the conclusion that

$$\sum_{j=1}^{\infty}\Big(\frac{j}{2^j}\Big)\Big(⨍_{2^j\Delta}|f-f_{2\Delta,w}|^p\,\mathrm{d}w\Big)^{1/p}\leq C\int_2^{\infty}\Big(⨍_{\lambda\Delta}|f-f_{\lambda\Delta,w}|^p\,\mathrm{d}w\Big)^{1/p}\frac{\ln\lambda}{\lambda^2}\,\mathrm{d}\lambda.$$

At this stage, the claim in (4.1.247) follows from the above estimate and (4.1.271). This finishes the proof of Theorem 4.1.15. □

Moving on, let $\Sigma \subseteq \mathbb{R}^n$ be an arbitrary closed Ahlfors regular set, and abbreviate $\sigma := \mathcal{H}^{n-1}\lfloor\Sigma$. Given an arbitrary function $\phi : (0,\infty) \to (0,\infty)$, an integrability exponent $p \in [1,\infty)$, and a weight function $w \in L^1_{\mathrm{loc}}(\Sigma,\sigma)$, define the ϕ-modulated weighted BMO space

$$\mathrm{BMO}_{\phi,p}(\Sigma,w):=\Big\{f\in L^1_{\mathrm{loc}}(\Sigma,w):\ \|f\|_{\mathrm{BMO}_{\phi,p}(\Sigma,w)}<+\infty\Big\}, \tag{4.1.281}$$

where for each $f \in L^1_{\rm loc}(\Sigma, w)$ we have set

$$\|f\|_{{\rm BMO}_{\phi,p}(\Sigma,w)} := \sup_{\substack{x\in\Sigma \text{ and} \\ r\in(0,\infty)}} \left\{ \frac{1}{\phi(r)} \Big(⨍_{\Delta(x,r)} \Big| f - ⨍_{\Delta(x,r)} f \, {\rm d}w \Big|^p \, {\rm d}w \Big)^{1/p} \right\}. \tag{4.1.282}$$

Compare with the un-weighted spaces introduced in [132, (5.1.343)-(5.1.344)].

Other, related and progressively larger, spaces of interest are as follows. First, we introduce the ϕ-modulated weighted BCMO (bounded central mean oscillations) space by employing in the above definition only surface balls with a fixed common center, at a given reference point $x_0 \in \Sigma$, i.e.,

$${\rm BCMO}_{\phi,p}(\Sigma, w) := \Big\{ f \in L^1_{\rm loc}(\Sigma, w) : \|f\|_{{\rm BCMO}_{\phi,p}(\Sigma,w)} < +\infty \Big\}, \tag{4.1.283}$$

where, for each $f \in L^1_{\rm loc}(\Sigma, w)$,

$$\|f\|_{{\rm BCMO}_{\phi,p}(\Sigma,w)} := \sup_{r\in(0,\infty)} \left\{ \frac{1}{\phi(r)} \Big(⨍_{\Delta(x_0,r)} \Big| f - ⨍_{\Delta(x_0,r)} f \, {\rm d}w \Big|^p \, {\rm d}w \Big)^{1/p} \right\} \tag{4.1.284}$$

Second, we may further restrict to large radii, namely consider

$$\dot{\rm B}{\rm CMO}_{\phi,p}(\Sigma, w) := \Big\{ f \in L^1_{\rm loc}(\Sigma, w) : \|f\|_{\dot{\rm B}{\rm CMO}_{\phi,p}(\Sigma,w)} < +\infty \Big\}, \tag{4.1.285}$$

where, for each $f \in L^1_{\rm loc}(\Sigma, w)$, we define

$$\|f\|_{\dot{\rm B}{\rm CMO}_{\phi,p}(\Sigma,{\rm w})} := \sup_{r\in(1,\infty)} \left\{ \frac{1}{\phi(r)} \Big(⨍_{\Delta(x_0,r)} \Big| f - ⨍_{\Delta(x_0,r)} f \, {\rm d}w \Big|^p \, {\rm d}w \Big)^{1/p} \right\}. \tag{4.1.286}$$

Then these are all Banach spaces modulo constants and, by design,

$${\rm BMO}_{\phi,p}(\Sigma, w) \subseteq {\rm BCMO}_{\phi,p}(\Sigma, w) \subseteq \dot{\rm B}{\rm CMO}_{\phi,p}(\Sigma, w). \tag{4.1.287}$$

We are particularly interested in the case when w is a Muckenhoupt weight, namely $w \in A_p(\Sigma, \sigma)$ with $p \in (1, \infty)$, and when the function $\phi : (0, \infty) \to (0, \infty)$ is $\mathcal{L}^1$-measurable and satisfies the following Dini condition at infinity:

$$\int_1^\infty \phi(t) \frac{{\rm d}t}{t^2} < +\infty. \tag{4.1.288}$$

In such a scenario, we claim that

$$\dot{\rm B}{\rm CMO}_{\phi,p}(\Sigma, w) \subseteq L^1\Big(\Sigma, \frac{\sigma(x)}{1 + |x|^n}\Big) \cap L^p_{\rm loc}(\Sigma, w). \tag{4.1.289}$$

To justify this claim, first observe from (4.1.285)-(4.1.286) and [129, (7.7.10)] that

$$\dot{\mathrm{B}}\mathrm{CMO}_{\phi,p}(\Sigma, w) \subseteq L^p_{\mathrm{loc}}(\Sigma, w) \subseteq L^1_{\mathrm{loc}}(\Sigma, \sigma). \tag{4.1.290}$$

Consequently, given $f \in \dot{\mathrm{B}}\mathrm{CMO}_{\phi,p}(\Sigma, w)$ it follows that for each $r > 0$ the number $⨍_{\Delta(x_0,r)} f\,d\sigma$ is a well defined, and we may estimate

$$\begin{aligned} ⨍_{\Delta(x_0,r)} \Big| f - ⨍_{\Delta(x_0,r)} f\,d\sigma \Big|\,d\sigma &\le 2 \cdot \inf_{c\in\mathbb{C}} ⨍_{\Delta(x_0,r)} |f - c|\,d\sigma \\ &\le 2 ⨍_{\Delta(x_0,r)} \Big| f - ⨍_{\Delta(x_0,r)} f\,dw \Big|\,d\sigma \\ &\le 2[w]_{A_p}^{1/p} \Big(⨍_{\Delta(x_0,r)} \Big| f - ⨍_{\Delta(x_0,r)} f\,dw \Big|^p\,dw \Big)^{1/p}, \end{aligned} \tag{4.1.291}$$

with the final inequality provided by [129, (7.7.12)]. Hence, if $r \ge 1$ we obtain

$$⨍_{\Delta(x_0,r)} \Big| f - ⨍_{\Delta(x_0,r)} f\,d\sigma \Big|\,d\sigma \le 2[w]_{A_p}^{1/p} \phi(r), \tag{4.1.292}$$

thanks to the membership of f to $\dot{\mathrm{B}}\mathrm{CMO}_{\phi,p}(\Sigma, w)$. In turn, (4.1.292) together with [129, (7.4.115)] (currently used with $X := \Sigma$, $\mu := \sigma$, $\rho := |\cdot - \cdot|$, $d := n-1$, $p := 1$, $q := 1$, and $\varepsilon := 1$) permit us to estimate for each $r \ge 1$

$$\int_\Sigma \frac{|f(x) - ⨍_{\Delta(x_0,r)} f\,d\sigma|}{r + |x - x_o|^n}\,d\sigma(x) \le \frac{C}{r} \int_1^\infty \phi(\lambda r)\,\frac{d\lambda}{\lambda^2} = C \int_r^\infty \phi(t)\,\frac{dt}{t^2} < +\infty, \tag{4.1.293}$$

with the last inequality coming from (4.1.288) (recall that $r \ge 1$). Now, the inclusion claimed in (4.1.289) is seen from (4.1.290) and (4.1.293).

It is also implicit in the above reasoning (cf. (4.1.291) in particular) that

$$\mathrm{BMO}_{\phi,p}(\Sigma, w) \subseteq \mathrm{BMO}_{\phi,1}(\Sigma, \sigma). \tag{4.1.294}$$

Let us explore the possibility of having an inclusion in the opposite direction. Since $w \in A_p(\Sigma, \sigma)$ it follows that w belongs to some Reverse Hölder class. Specifically, w satisfies the reverse Hölder inequality [129, (7.7.19)] for some $q \in (1, \infty)$, and we denote by $q' \in (1, \infty)$ the conjugate exponent of q. From [129, (7.7.39)] we see that

$$L^{q'}_{\mathrm{loc}}(\Sigma, \sigma) \subseteq L^1_{\mathrm{loc}}(\Sigma, w). \tag{4.1.295}$$

Hence, given any $f \in L^{q'}_{\mathrm{loc}}(\Sigma, \sigma)$, any $x \in \Sigma$, and any $r > 0$, the inclusion in (4.1.295) guarantees that $⨍_{\Delta(x_0,r)} f\,d\sigma$ is a well-defined number, and we may write

$$\begin{aligned}\Big(⨍_{\Delta(x,r)}\Big|f-⨍_{\Delta(x,r)}f\,dw\Big|^p\,dw\Big)^{1/p} &\leq 2\cdot\inf_{c\in\mathbb{C}}\Big(⨍_{\Delta(x,r)}|f-c|^p\,dw\Big)^{1/p}\\ &\leq 2\Big(⨍_{\Delta(x,r)}\Big|f-⨍_{\Delta(x,r)}f\,d\sigma\Big|^p\,dw\Big)^{1/p}\\ &\leq C\Big(⨍_{\Delta(x,r)}\Big|f-⨍_{\Delta(x,r)}f\,d\sigma\Big|^{pq'}\,d\sigma\Big)^{1/(pq')},\end{aligned}\tag{4.1.296}$$

with the last inequality from [129, (7.7.39)]. In view of definitions, this proves

$$\mathrm{BMO}_{\phi,pq'}(\Sigma,\sigma)\subseteq\mathrm{BMO}_{\phi,p}(\Sigma,w).\tag{4.1.297}$$

Ultimately, from (4.1.294), (4.1.297), and [132, (5.1.350)] we conclude that:

if the function $\phi:(0,\infty)\to(0,\infty)$ is $\mathcal{L}^1$-measurable, satisfies the Dini condition at infinity formulated in (4.1.288), and is quasi-increasing as well as doubling (cf. [132, (5.1.349)]), then the space $\mathrm{BMO}_{\phi,p}(\Sigma,w)$ is independent of the index $p\in(1,\infty)$ and weight $w\in A_p(\Sigma,\sigma)$. (4.1.298)

This extends the scope of the result proved in [129, Lemma 7.7.5].

We shall now employ the above brands of BMO spaces in the theorem below, which contains norm estimates for modified chord-dot-normal singular integral operators in these functional analytic settings.

Theorem 4.1.16 *Let $n\in\mathbb{N}$ be such that $n\geq 2$, and assume $\Omega\subseteq\mathbb{R}^n$ is an arbitrary* UR *domain. Set $\sigma:=\mathcal{H}^{n-1}\lfloor\partial\Omega$ and denote by ν the geometric measure theoretic outward unit normal to Ω. For a sufficiently large $N=N(n)\in\mathbb{N}$ consider a function $k\in\mathscr{C}^N(\mathbb{R}^n\setminus\{0\})$, even and positive homogeneous of degree $-n$.*

Let $T_{\rm mod}$ be the modified version of the chord-dot-normal singular integral operator associated with the kernel k, acting on each function $f\in L^1\big(\partial\Omega,\frac{\sigma(x)}{1+|x|^n}\big)$ as

$$T_{\rm mod}f(x):=\lim_{\varepsilon\to 0^+}\int_{\partial\Omega}\big\langle\nu(y),\vec{k}_\varepsilon(x-y)-\vec{k}_1(-y)\big\rangle f(y)\,d\sigma(y)\tag{4.1.299}$$

at σ-a.e. $x\in\partial\Omega$, where

$$\vec{k}(z):=z\,k(z)\text{ for each }z\in\mathbb{R}^n\text{ and}\qquad \vec{k}_\varepsilon:=\vec{k}\cdot\mathbf{1}_{\mathbb{R}^n\setminus\overline{B(0,\varepsilon)}}\text{ for each }\varepsilon>0.\tag{4.1.300}$$

Next, assume

$$\phi:(0,\infty)\to(0,\infty)\text{ is }\mathcal{L}^1\text{-measurable and there exists }C_*\in(0,\infty)\text{ such that } r\int_r^\infty\phi(t)\frac{\ln(t/r)}{t^2}\,dt\leq C_*\phi(r)\text{ for each }r\in(0,\infty).\tag{4.1.301}$$

Finally, fix $p\in(1,\infty)$ and pick a Muckenhoupt weight $w\in A_p(\partial\Omega,\sigma)$.

Then for each $m \in \mathbb{N}$ there exists $C_m \in (0, \infty)$, depending only on m, n, p, $[w]_{A_p}$, C_* from (4.1.301), *and the* UR *constants of $\partial\Omega$, with such that the operator*

$$T_{\text{mod}} : \text{BMO}_{\phi,p}(\partial\Omega, w) \longrightarrow \text{BMO}_{\phi,p}(\partial\Omega, w) \tag{4.1.302}$$

is well defined, linear, bounded, and satisfies (with notation introduced in (4.1.12))

$$\|T_{\text{mod}}\|_{\text{BMO}_{\phi,p}(\partial\Omega,w)\to\text{BMO}_{\phi,p}(\partial\Omega,w)} \leq C_m\Big(\sum_{|\alpha|\leq N}\sup_{S^{n-1}}|\partial^\alpha k|\Big)\|\nu\|^{\langle m\rangle}_{[\text{BMO}(\partial\Omega,\sigma)]^n}. \tag{4.1.303}$$

In particular, corresponding to $w \equiv 1$, the operator

$$T_{\text{mod}} : \text{BMO}_{\phi,p}(\partial\Omega, \sigma) \longrightarrow \text{BMO}_{\phi,p}(\partial\Omega, \sigma) \tag{4.1.304}$$

is well defined, linear, bounded, and for each $m \in \mathbb{N}$ one has

$$\|T_{\text{mod}}\|_{\text{BMO}_{\phi,p}(\partial\Omega,\sigma)\to\text{BMO}_{\phi,p}(\partial\Omega,\sigma)} \leq C_m\Big(\sum_{|\alpha|\leq N}\sup_{S^{n-1}}|\partial^\alpha k|\Big)\|\nu\|^{\langle m\rangle}_{[\text{BMO}(\partial\Omega,\sigma)]^n} \tag{4.1.305}$$

with $C_m \in (0, \infty)$ depending only on m, n, p, C_* from (4.1.301), *and the* UR *constants of $\partial\Omega$. Also, having fixed an arbitrary reference point $x_0 \in \partial\Omega$, the operators*[6]

$$T_{\text{mod}} : \text{BCMO}_{\phi,p}(\partial\Omega, w) \longrightarrow \text{BCMO}_{\phi,p}(\partial\Omega, w), \tag{4.1.306}$$

$$T_{\text{mod}} : \dot{\text{B}}\text{CMO}_{\phi,p}(\partial\Omega, w) \longrightarrow \dot{\text{B}}\text{CMO}_{\phi,p}(\partial\Omega, w), \tag{4.1.307}$$

are well defined, linear, bounded, and for each $m \in \mathbb{N}$ satisfy

$$\|T_{\text{mod}}\|_{\text{BCMO}_{\phi,p}(\partial\Omega,w)\to\text{BCMO}_{\phi,p}(\partial\Omega,w)} \leq C_m\Big(\sum_{|\alpha|\leq N}\sup_{S^{n-1}}|\partial^\alpha k|\Big)\|\nu\|^{\langle m\rangle}_{[\text{BMO}(\partial\Omega,\sigma)]^n}, \tag{4.1.308}$$

$$\|T_{\text{mod}}\|_{\dot{\text{B}}\text{MO}_{\phi,p}(\partial\Omega,w)\to\dot{\text{B}}\text{MO}_{\phi,p}(\partial\Omega,w)} \leq C_m\Big(\sum_{|\alpha|\leq N}\sup_{S^{n-1}}|\partial^\alpha k|\Big)\|\nu\|^{\langle m\rangle}_{[\text{BMO}(\partial\Omega,\sigma)]^n} \tag{4.1.309}$$

with $C_m \in (0, \infty)$ depending only on m, n, p, $[w]_{A_p}$, C_* from (4.1.301), *and the* UR *constants of $\partial\Omega$.*

As a corollary, the following three operators (acting on equivalence classes, modulo constants)

$$\begin{gathered}\big[T_{\text{mod}}\big] : \text{BMO}_{\phi,p}(\partial\Omega, w)/\sim \longrightarrow \text{BMO}_{\phi,p}(\partial\Omega, w)/\sim \ \text{ defined as}\\ \big[T_{\text{mod}}\big][f] := \big[T_{\text{mod}}f\big] \ \text{ for each function } f \in \text{BMO}_{\phi,p}(\partial\Omega, w),\end{gathered} \tag{4.1.310}$$

[6] for the operator in (4.1.307) it suffices to assume that the inequality in (4.1.301) holds only for $r \in (1, \infty)$

$$\begin{aligned}&\big[T_{\rm mod}\big] : \mathrm{BCMO}_{\phi,p}(\partial\Omega, w)/\sim \longrightarrow \mathrm{BCMO}_{\phi,p}(\partial\Omega, w)/\sim \ \textit{defined as}\\ &\quad\big[T_{\rm mod}\big][f] := \big[T_{\rm mod} f\big] \ \textit{for each function} \ f \in \mathrm{BCMO}_{\phi,p}(\partial\Omega, w),\end{aligned} \tag{4.1.311}$$

$$\begin{aligned}&\big[T_{\rm mod}\big] : \dot{\mathrm{B}}\mathrm{CMO}_{\phi,p}(\partial\Omega, w)/\sim \longrightarrow \dot{\mathrm{B}}\mathrm{CMO}_{\phi,p}(\partial\Omega, w)/\sim \ \textit{defined as}\\ &\quad\big[T_{\rm mod}\big][f] := \big[T_{\rm mod} f\big] \ \textit{for each function} \ f \in \dot{\mathrm{B}}\mathrm{CMO}_{\phi,p}(\partial\Omega, w),\end{aligned} \tag{4.1.312}$$

are well defined, linear, bounded, and for each given $m \in \mathbb{N}$ *their operator norms are* $\leq C_m\Big(\sum_{|\alpha|\leq N}\sup_{S^{n-1}}|\partial^\alpha k|\Big)\|\nu\|^{\langle m\rangle}_{[\mathrm{BMO}(\partial\Omega,\sigma)]^n}$, *for some* $C_m \in (0,\infty)$ *as before.*

This theorem encompasses many special cases of interest. For example,

$$\begin{aligned}&\phi : (0,\infty) \to (0,\infty) \text{ given by } \phi(t) := 1 \text{ for each } t \in (0,\infty)\\ &\quad\text{is quasi-increasing, doubling, satisfies (4.1.301), and}\\ &\quad\mathrm{BMO}_{\phi,p}(\partial\Omega, w) = \Big(\mathrm{BMO}(\partial\Omega,\sigma),\ \|\cdot\|_{\dot{\mathrm{B}}\mathrm{MO}(\partial\Omega,\sigma)}\Big).\end{aligned} \tag{4.1.313}$$

Also, for each $\alpha \in (0,1)$,

$$\begin{aligned}&\phi : (0,\infty) \to (0,\infty) \text{ given by } \phi(t) := t^\alpha \text{ for each } t \in (0,\infty)\\ &\quad\text{is quasi-increasing, doubling, satisfies (4.1.301), and}\\ &\quad\text{we have } \mathrm{BMO}_{\phi,p}(\partial\Omega, w) = \Big(\dot{\mathscr{C}}^\alpha(\partial\Omega),\ \|\cdot\|_{\dot{\mathscr{C}}^\alpha(\partial\Omega)}\Big).\end{aligned} \tag{4.1.314}$$

Next, for each $\lambda \in (0, n-1)$,

$$\begin{aligned}&\phi : (0,\infty) \to (0,\infty) \text{ with } \phi(t) := t^{(\lambda-n+1)/p} \text{ for each } t > 0\\ &\quad\text{is as in (4.1.301) and } \mathrm{BMO}_{\phi,p}(\partial\Omega,\sigma) = \dot{\mathfrak{L}}^{p,\lambda}(\partial\Omega,\sigma).\end{aligned} \tag{4.1.315}$$

More generally, $\mathrm{BMO}_{\phi,p}(\partial\Omega, w)$ corresponding to the choice of ϕ above may be regarded as a Muckenhoupt weighted version of the homogeneous Morrey-Campanato space from (A.0.135)-(A.0.136). Finally, "central" versions (i.e., when all mean oscillations are considered with respect to surface balls of a fixed common center) of the aforementioned spaces fall under the scope of Theorem 4.1.16 as well.

The proof of Theorem 4.1.16 is presented next.

Proof of Theorem 4.1.16 The properties in (4.1.301) imply that ϕ satisfies the Dini condition at infinity formulated in (4.1.288). Pick an arbitrary function $f \in \mathrm{BMO}_{\phi,p}(\partial\Omega, w)$. Then from (4.1.287) and (4.1.289) we conclude that

$$f \in L^1\Big(\partial\Omega,\ \frac{\sigma(x)}{1+|x|^n}\Big) \cap L^p_{\rm loc}(\partial\Omega, w). \tag{4.1.316}$$

Granted this, we may invoke (4.1.247) which guarantees that for each $m \in \mathbb{N}$ there exists some $C_m \in (0,\infty)$, depending only on m, n, p, $[w]_{A_p}$, and the UR constants of $\partial\Omega$, with the property that for each $x \in \partial\Omega$ and $r > 0$ we have

$$\left(\fint_{\Delta(x,r)}\left|T_{\rm mod}f-\fint_{\Delta(x,r)}T_{\rm mod}f\,\mathrm{d}w\right|^p\mathrm{d}w\right)^{1/p} \tag{4.1.317}$$
$$\leq C_m\Big(\sum_{|\alpha|\leq N}\sup_{S^{n-1}}|\partial^\alpha k|\Big)\|\nu\|^{\langle m\rangle}_{[\mathrm{BMO}(\partial\Omega,\sigma)]^n}\times$$
$$\times\int_1^\infty\left(\fint_{\Delta(x,\lambda r)}\left|f-\fint_{\Delta(x,\lambda r)}f\,\mathrm{d}w\right|^p\mathrm{d}w\right)^{1/p}\frac{\ln\lambda}{\lambda^2}\,\mathrm{d}\lambda$$
$$\leq C_m\Big(\sum_{|\alpha|\leq N}\sup_{S^{n-1}}|\partial^\alpha k|\Big)\|\nu\|^{\langle m\rangle}_{[\mathrm{BMO}(\partial\Omega,\sigma)]^n}\cdot\|f\|_{\mathrm{BMO}_{\phi,p}(\partial\Omega,w)}\times$$
$$\times\Big(\int_1^\infty\phi(\lambda r)\frac{\ln\lambda}{\lambda^2}\,\mathrm{d}\lambda\Big)$$
$$=C_m\Big(\sum_{|\alpha|\leq N}\sup_{S^{n-1}}|\partial^\alpha k|\Big)\|\nu\|^{\langle m\rangle}_{[\mathrm{BMO}(\partial\Omega,\sigma)]^n}\cdot\|f\|_{\mathrm{BMO}_{\phi,p}(\partial\Omega,w)}\times$$
$$\times\Big(r\int_r^\infty\phi(t)\frac{\ln(t/r)}{t^2}\,\mathrm{d}t\Big)$$
$$\leq C_*\cdot C_m\Big(\sum_{|\alpha|\leq N}\sup_{S^{n-1}}|\partial^\alpha k|\Big)\|\nu\|^{\langle m\rangle}_{[\mathrm{BMO}(\partial\Omega,\sigma)]^n}\cdot\|f\|_{\mathrm{BMO}_{\phi,p}(\partial\Omega,w)}\cdot\phi(r),$$

where we have made the change of variables $t=\lambda r$ and used (4.1.301). From (4.1.317) and (4.1.281)-(4.1.282) we then deduce that $T_{\rm mod}f$ belongs to the space $\mathrm{BMO}_{\phi,p}(\partial\Omega,w)$ and satisfies

$$\|T_{\rm mod}f\|_{\mathrm{BMO}_{\phi,p}(\partial\Omega,w)} \tag{4.1.318}$$
$$\leq C_*\cdot C_m\Big(\sum_{|\alpha|\leq N}\sup_{S^{n-1}}|\partial^\alpha k|\Big)\|\nu\|^{\langle m\rangle}_{[\mathrm{BMO}(\partial\Omega,\sigma)]^n}\cdot\|f\|_{\mathrm{BMO}_{\phi,p}(\partial\Omega,w)}.$$

In view of the arbitrariness of $f\in\mathrm{BMO}_{\phi,p}(\partial\Omega,w)$, the estimate claimed in (4.1.303) follows, and we also conclude that $T_{\rm mod}$ is a well-defined, linear, bounded operator in the context of (4.1.302). The claims pertaining to (4.1.304)-(4.1.309) are dealt with similarly. Finally, all claims regarding the operators in (4.1.310)-(4.1.312) are consequences of what we have proved so far and the fact that $T_{\rm mod}$ maps constants to constants (cf. [132, (5.2.191)]). □

Moving on, the goal is to introduce homogeneous Sobolev spaces styled after Generalized Banach Function Spaces, on boundaries of Ahlfors regular domains. To set the stage for this, consider an Ahlfors regular domain $\Omega\subseteq\mathbb{R}^n$ and abbreviate $\sigma:=\mathcal{H}^{n-1}\lfloor\partial\Omega$. Given a Generalized Banach Function Space $\mathbb{X}$ over $(\partial\Omega,\sigma)$ with the property that $\mathbb{X}\subseteq L^1_{\rm loc}(\partial\Omega,\sigma)$, we define the following GBFS-based homogeneous Sobolev space (see also item *(iii)* in Lemma 4.1.17 in this regard):

$$\dot{\mathbb{X}}_1:=\Big\{f\in L^1\big(\partial\Omega,\tfrac{\sigma(x)}{1+|x|^n}\big)\cap\mathbb{X}_{\rm loc}:\partial_{\tau_{jk}}f\in\mathbb{X},\ 1\leq j,k\leq n\Big\} \tag{4.1.319}$$

where $f \in \mathbb{X}_{\rm loc}$ means $f \cdot \mathbf{1}_\Delta \in \mathbb{X}$ for each surface ball $\Delta \subseteq \partial\Omega$. We thus have $\dot{\mathbb{X}}_1 \subseteq L^1_{1,{\rm loc}}(\partial\Omega, \sigma)$. We equip the space in (4.1.319) with the semi-norm

$$\dot{\mathbb{X}}_1 \ni f \longmapsto \|f\|_{\dot{\mathbb{X}}_1} := \sum_{j,k=1}^{n} \|\partial_{\tau_{jk}} f\|_{\mathbb{X}}. \tag{4.1.320}$$

Note that constant functions on $\partial\Omega$ belong to $\dot{\mathbb{X}}_1$ and have vanishing semi-norms. As such, we find it useful to also consider the quotient space $\dot{\mathbb{X}}_1/\sim$ of classes $[\cdot]$ of equivalence modulo constants of functions in $\dot{\mathbb{X}}_1$, equipped with the semi-norm

$$\dot{\mathbb{X}}_1/\sim \ni [f] \mapsto \big\|[f]\big\|_{\dot{\mathbb{X}}_1/\sim} := \sum_{j,k=1}^{n} \|\partial_{\tau_{jk}} f\|_{\dot{\mathbb{X}}_1}. \tag{4.1.321}$$

The following lemma is a useful tool in the study of this brand of Sobolev spaces.

Lemma 4.1.17 *Suppose $\Omega \subseteq \mathbb{R}^n$ is a two-sided* NTA *domain such that $\partial\Omega$ is an unbounded Ahlfors regular set, and abbreviate $\sigma := \mathcal{H}^{n-1}\lfloor\partial\Omega$. Also, denote by $\mathcal{M}_{\partial\Omega}$ the Hardy-Littlewood maximal operator on $(\partial\Omega, \sigma)$. Let $\mathbb{X}$ be a Generalized Banach Function Space over $(\partial\Omega, \sigma)$ with the property that*

$$\mathcal{M}_{\partial\Omega} : \mathbb{X} \to \mathbb{X} \text{ and } \mathcal{M}_{\partial\Omega} : \mathbb{X}' \to \mathbb{X}' \text{ are well-defined and bounded,} \tag{4.1.322}$$

where $\mathbb{X}'$ is the Köthe dual of $\mathbb{X}$. Finally, fix some reference point $x_0 \in \partial\Omega$ and, for each $r \in (0, \infty)$, define $\Delta_r := B(x_0, r) \cap \partial\Omega$. Then the following statements are true.

(i) There exists a constant $C \in (0, \infty)$, which depends only on Ω, x_0 and the operator norms $\|\mathcal{M}_{\partial\Omega}\|_{\mathbb{X}\to\mathbb{X}}$, $\|\mathcal{M}_{\partial\Omega}\|_{\mathbb{X}'\to\mathbb{X}'}$, such that for each function f belonging to $L^1\big(\partial\Omega, \frac{\sigma(x)}{1+|x|^n}\big) \cap L^1_{1,{\rm loc}}(\partial\Omega, \sigma)$ one has, with $f_{\Delta_r} := ⨍_{\Delta_r} f \, d\sigma$,

$$\sup_{r>0} \frac{1}{r} \big\| |f - f_{\Delta_r}| \cdot \mathbf{1}_{\Delta_r} \big\|_{\mathbb{X}} \leq C \sum_{j,k=1}^{n} \big\|\partial_{\tau_{jk}} f\big\|_{\mathbb{X}}. \tag{4.1.323}$$

(ii) For each $r \in (0, \infty)$ there exists a constant $C_r \in (0, \infty)$, which depends only on Ω, x_0, r, and the operator norms $\|\mathcal{M}_{\partial\Omega}\|_{\mathbb{X}\to\mathbb{X}}$, $\|\mathcal{M}_{\partial\Omega}\|_{\mathbb{X}'\to\mathbb{X}'}$, with the property that for each function f belonging to $L^1\big(\partial\Omega, \frac{\sigma(x)}{1+|x|^n}\big) \cap L^1_{1,{\rm loc}}(\partial\Omega, \sigma)$ one has

$$\int_{\partial\Omega} \frac{|f(x) - f_{\Delta_r}|}{1 + |x|^n} \, d\sigma(x) \leq \frac{C_r}{\|\mathbf{1}_{\Delta_r}\|_{\mathbb{X}}} \sum_{j,k=1}^{n} \big\|\partial_{\tau_{jk}} f\big\|_{\mathbb{X}}. \tag{4.1.324}$$

(iii) In the current setting, one has the following (equivalent) description of the space originally introduced in (4.1.319)*:*

$$\dot{\mathbb{X}}_1 = \Big\{ f \in L^1\big(\partial\Omega, \tfrac{\sigma(x)}{1+|x|^n}\big) : \partial_{\tau_{jk}} f \in \mathbb{X} \text{ for } 1 \leq j, k \leq n \Big\}. \tag{4.1.325}$$

Proof The claims in items *(i)-(ii)* may be established using the extrapolation theory for Generalized Banach Function Spaces developed in [130, §5.2] (cf. [130, Theorem 5.2.1] in particular), and the Muckenhoupt weighted Poincaré inequalities from [111, Proposition 2.25, pp. 154-155] (see the proof of [111, Lemma 8.1, pp. 515-516] for closely related results).

As regards item *(iii)*, it is apparent from definitions that (4.1.325) comes down to showing that for any function f belonging to the space in the right-hand side of (4.1.325) we have $f \in \mathbb{X}_{\rm loc}$. To justify this, recall that $f \in L^1_{1,\rm loc}(\partial\Omega, \sigma)$. Consequently, (4.1.323) holds, so for each $r > 0$ we have

$$\begin{aligned}\|f\mathbf{1}_{\Delta_r}\|_{\mathbb{X}} &\le \big\||f - f_{\Delta_r}|\cdot\mathbf{1}_{\Delta_r}\|_{\mathbb{X}} + |f_{\Delta_r}|\,\|\mathbf{1}_{\Delta_r}\big\|_{\mathbb{X}}\\ &\le Cr\sum_{j,k=1}^{n}\big\|\partial_{\tau_{jk}}f\big\|_{\mathbb{X}} + |f_{\Delta_r}|\,\|\mathbf{1}_{\Delta_r}\|_{\mathbb{X}} < \infty, \end{aligned}\tag{4.1.326}$$

thanks to [130, Lemma 5.2.5]. This ultimately shows that $f \in \mathbb{X}_{\rm loc}$, as desired. □

When considered on the boundaries of two-sided NTA domains, the quotient space $\dot{\mathbb{X}}_1/\sim$ turns out to actually be a genuine Banach space.

Proposition 4.1.18 *Suppose $\Omega \subseteq \mathbb{R}^n$ is a two-sided* NTA *domain with an unbounded Ahlfors regular boundary and abbreviate $\sigma := \mathcal{H}^{n-1}\lfloor\partial\Omega$. Denote by $\mathcal{M}_{\partial\Omega}$ the Hardy-Littlewood maximal operator on $(\partial\Omega, \sigma)$, and consider a Generalized Banach Function Space $\mathbb{X}$ over $(\partial\Omega, \sigma)$ with the property that*

$$\mathcal{M}_{\partial\Omega} : \mathbb{X} \to \mathbb{X} \text{ and } \mathcal{M}_{\partial\Omega} : \mathbb{X}' \to \mathbb{X}' \text{ are well-defined and bounded,}\tag{4.1.327}$$

where $\mathbb{X}'$ is the Köthe dual of $\mathbb{X}$. Finally, recall that $\dot{\mathbb{X}}_1/\sim$ denotes the quotient space of classes $[\,\cdot\,]$ of equivalence modulo constants of functions in $\dot{\mathbb{X}}_1$, equipped with the semi-norm (4.1.321).

Then (4.1.321) *is a genuine norm on $\dot{\mathbb{X}}_1/\sim$, and $\dot{\mathbb{X}}_1/\sim$ is a Banach space when equipped with the norm* (4.1.321).

Proof This is proved with the help of Lemma 4.1.17, reasoning much as in the proof of [130, Proposition 11.5.14] (compare with the proof of [111, Proposition 8.4, pp. 517-518]). □

We next state a trace result in the setting of Generalized Banach Function Spaces:

Proposition 4.1.19 *Let $\Omega \subseteq \mathbb{R}^n$ be an* NTA *domain with the property that $\partial\Omega$ is an unbounded Ahlfors regular set. Abbreviate $\sigma := \mathcal{H}^{n-1}\lfloor\partial\Omega$, and fix an aperture parameter $\kappa > 0$. Denote by $\mathcal{M}_{\partial\Omega}$ the Hardy-Littlewood maximal operator on $(\partial\Omega, \sigma)$, and consider a Generalized Banach Function Space $\mathbb{X}$ over $(\partial\Omega, \sigma)$ with the property that*

$$\mathcal{M}_{\partial\Omega} : \mathbb{X} \to \mathbb{X} \text{ and } \mathcal{M}_{\partial\Omega} : \mathbb{X}' \to \mathbb{X}' \text{ are well-defined and bounded,}\tag{4.1.328}$$

where $\mathbb{X}'$ is the Köthe dual of $\mathbb{X}$. Finally consider a function $u : \Omega \to \mathbb{C}$ satisfying $u \in \mathscr{C}^1(\Omega)$ and $\mathcal{N}_\kappa(\nabla u) \in \mathbb{X}$.

Then the nontangential boundary trace $u\big|_{\partial\Omega}^{\kappa-\mathrm{n.t.}}$ *exists* σ*-a.e. on* $\partial\Omega$*, belongs to the homogeneous Sobolev space space* $\dot{\mathbb{X}}_1$*, and there exists some constant* $C \in (0,\infty)$*, depending only on* Ω *and the operator norms* $\|\mathcal{M}_{\partial\Omega}\|_{\mathbb{X}\to\mathbb{X}}$, $\|\mathcal{M}_{\partial\Omega}\|_{\mathbb{X}'\to\mathbb{X}'}$*, such that*

$$\Big\|u\big|_{\partial\Omega}^{\kappa-\mathrm{n.t.}}\Big\|_{\dot{\mathbb{X}}_1} \leq C\big\|\mathcal{N}_\kappa(\nabla u)\big\|_{\mathbb{X}_1}. \tag{4.1.329}$$

Proof This may be justified based on the extrapolation result involving Generalized Banach Function Spaces from [130, Theorem 5.2.1], reasoning as in the proof of [111, Proposition 8.5, p. 520]. □

We continue by presenting a basic Fatou-type result and integral representation formula of the following sort:

Theorem 4.1.20 *Let* $\Omega \subseteq \mathbb{R}^n$ *(where* $n \in \mathbb{N}$, $n \geq 2$*) be an* NTA *domain such that* $\partial\Omega$ *is an unbounded Ahlfors regular set. Abbreviate* $\sigma := \mathcal{H}^{n-1}\lfloor\partial\Omega$ *and denote by* ν *the geometric measure theoretic outward unit normal to* Ω*. Let* $A = \big(a_{rs}^{\alpha\beta}\big)_{\substack{1\leq r,s\leq n\\ 1\leq\alpha,\beta\leq M}}$ *(where* $M \in \mathbb{N}$*) be a complex coefficient tensor with the property that* $L := L_A$ *is a weakly elliptic* $M \times M$ *system in* $\mathbb{R}^n$*. In this setting, recall the modified version of the double layer operator* $\mathcal{D}_{A,\mathrm{mod}}$ *from* (A.0.73)*, and the modified version of the single layer operator* $\mathcal{S}_{\mathrm{mod}}$ *from* (A.0.229)*.*

With $\mathcal{M}_{\partial\Omega}$ *denoting the Hardy-Littlewood maximal operator on* $(\partial\Omega,\sigma)$*, consider a Generalized Banach Function Space* $\mathbb{X}$ *over* $(\partial\Omega,\sigma)$ *with the property that*

$$\mathcal{M}_{\partial\Omega} : \mathbb{X} \to \mathbb{X} \text{ and } \mathcal{M}_{\partial\Omega} : \mathbb{X}' \to \mathbb{X}' \text{ are well-defined and bounded,} \tag{4.1.330}$$

where $\mathbb{X}'$ *is the Köthe dual of* $\mathbb{X}$*. Finally, fix an aperture parameter* $\kappa \in (0,\infty)$ *and consider a function* $u : \Omega \to \mathbb{C}^M$ *satisfying*

$$u \in \big[\mathscr{C}^\infty(\Omega)\big]^M, \quad Lu = 0 \text{ in } \Omega, \quad \mathcal{N}_\kappa(\nabla u) \in \mathbb{X}. \tag{4.1.331}$$

Then

$$\begin{aligned} &u\big|_{\partial\Omega}^{\kappa-\mathrm{n.t.}} \text{ exists } \sigma\text{-a.e. on } \partial\Omega \text{ and belongs to } \big[\dot{\mathbb{X}}_1\big]^M, \\ &(\nabla u)\big|_{\partial\Omega}^{\kappa-\mathrm{n.t.}} \text{ exists } \sigma\text{-a.e. on } \partial\Omega \text{ and } \partial_\nu^A u \in \big[\mathbb{X}\big]^M, \end{aligned} \tag{4.1.332}$$

and there exists some $\mathbb{C}^M$*-valued locally constant function* c_u *in* Ω *for which*

$$u = \mathcal{D}_{A,\mathrm{mod}}\Big(u\big|_{\partial\Omega}^{\kappa-\mathrm{n.t.}}\Big) - \mathcal{S}_{\mathrm{mod}}\big(\partial_\nu^A u\big) + c_u \text{ in } \Omega. \tag{4.1.333}$$

Also, if $\mathcal{D}_{\mathrm{mod}}$, $\mathcal{S}_{\mathrm{mod}}$ *are regarded as operators mapping into functions defined in* $\mathbb{R}^n \setminus \overline{\Omega}$*, then there exists a* $\mathbb{C}^M$*-valued locally constant function* $\widetilde{c}_u$ *in* $\mathbb{R}^n \setminus \overline{\Omega}$ *so that*

$$0 = \mathcal{D}_{A,\mathrm{mod}}\Big(u\big|_{\partial\Omega}^{\kappa-\mathrm{n.t.}}\Big) - \mathcal{S}_{\mathrm{mod}}\big(\partial_\nu^A u\big) + \widetilde{c}_u \text{ in } \mathbb{R}^n \setminus \overline{\Omega}. \tag{4.1.334}$$

Proof From Proposition 4.1.19 we know that $u\big|_{\partial\Omega}^{\kappa-\text{n.t.}}$ exists at σ-a.e. point on $\partial\Omega$ and belongs to $\big[\dot{\mathbb{X}}_1\big]^M$. The Fatou-type result established in [131, Theorem 3.3.4, item *(g)*] tells us that the nontangential boundary trace $(\nabla u)\big|_{\partial\Omega}^{\kappa-\text{n.t.}}$ exists (in $\mathbb{C}^{M\cdot n}$) at σ-a.e. point on $\partial\Omega$, hence the conormal derivative $\partial_\nu^A u$ is well defined and belongs $\big[\mathbb{X}\big]^M$. The integral representation formula claimed in (4.1.333) then follows with the help of [132, Theorem 1.8.17] and [130, Proposition 5.2.7]. □

Our next goal is to extend [132, Theorem 1.8.9] to modified single layer potential operators acting on Generalized Banach Function Spaces.

Theorem 4.1.21 *Assume $\Omega \subseteq \mathbb{R}^n$ (where $n \in \mathbb{N}$, $n \geq 2$) is a* UR *domain. Denote by $\nu = (\nu_1, \dots, \nu_n)$ the geometric measure theoretic outward unit normal to Ω and abbreviate $\sigma := \mathcal{H}^{n-1}\lfloor\partial\Omega$. Also, for some $M \in \mathbb{N}$, let $A = \big(a_{rs}^{\alpha\beta}\big)_{\substack{1\leq r,s\leq n\\ 1\leq\alpha,\beta\leq M}}$ be a complex coefficient tensor with the property that $L := L_A$ is a weakly elliptic $M \times M$ system in $\mathbb{R}^n$. Recall the modified boundary-to-boundary single layer operator S_{mod} associated with L and Ω as in* (A.0.232).

Bring in the Hardy-Littlewood maximal operator $\mathcal{M}_{\partial\Omega}$ on $(\partial\Omega, \sigma)$, and consider a Generalized Banach Function Space $\mathbb{X}$ over $(\partial\Omega, \sigma)$ with the property that

$$\mathcal{M}_{\partial\Omega} : \mathbb{X} \to \mathbb{X} \text{ and } \mathcal{M}_{\partial\Omega} : \mathbb{X}' \to \mathbb{X}' \text{ are well-defined and bounded,} \tag{4.1.335}$$

where $\mathbb{X}'$ is the Köthe dual of $\mathbb{X}$. Then the following properties are true.

(1) The modified boundary-to-boundary single layer operator induces a mapping

$$S_{\text{mod}} : \big[\mathbb{X}\big]^M \longrightarrow \big[\dot{\mathbb{X}}_1\big]^M \tag{4.1.336}$$

which is well defined, linear, and bounded, when the target space is endowed with the semi-norm (4.1.320). *In particular,*

$$\begin{aligned}&\text{for each function } f \in \big[\mathbb{X}\big]^M \text{ and each pair of indices}\\ &j,k \in \{1,\dots,n\}, \text{ one has } \partial_{\tau_{jk}}\big(S_{\text{mod}} f\big) \in \big[\mathbb{X}\big]^M.\end{aligned} \tag{4.1.337}$$

Also, for each function $f \in \big[\mathbb{X}\big]^M$, at σ-a.e. point $x \in \partial\Omega$ one has

$$\begin{aligned}&\big(\tfrac{1}{2}I + K^\#_{A^\top}\big)\Big(\big(-\tfrac{1}{2}I + K^\#_{A^\top}\big)f\Big)(x) \\ &= \Bigg(\lim_{\varepsilon\to 0^+}\int_{\substack{y\in\partial\Omega\\ |x-y|>\varepsilon}} \nu_i(x)a_{ij}^{\mu\gamma}a_{rs}^{\beta\alpha}(\partial_r E_{\gamma\beta})(x-y)\partial_{\tau_{js}}\big(S_{\text{mod}} f\big)_\alpha(y)\,d\sigma(y)\Bigg)_{1\leq\mu\leq M}\end{aligned} \tag{4.1.338}$$

where $K^\#_{A^\top}$ is the singular integral operator associated as in (A.0.118) *with the set Ω and the transpose coefficient tensor $A^\top$.*

(2) As a consequence of (4.1.336), *the following is a well-defined linear operator:*

$$\begin{aligned}&\big[S_{\text{mod}}\big] : \big[\mathbb{X}\big]^M \longrightarrow \big[\dot{\mathbb{X}}_1/\sim\big]^M \text{ defined as}\\&\big[S_{\text{mod}}\big]f := \big[S_{\text{mod}}f\big] \in \big[\dot{\mathbb{X}}_1/\sim\big]^M, \quad \forall\, f \in \big[\mathbb{X}\big]^M.\end{aligned} \tag{4.1.339}$$

Moreover, if actually $\Omega \subseteq \mathbb{R}^n$ *is an open set satisfying a two-sided local John condition and whose boundary is an unbounded Ahlfors regular set, then the operator* (4.1.339) *is also bounded when the quotient space is endowed with the norm introduced in* (4.1.321).

(3) *With* $\mathscr{S}_{\text{mod}}$ *denoting the modified version of the single layer operator acting on functions from* $\big[L^1\big(\partial\Omega, \frac{\sigma(x)}{1+|x|^{n-1}}\big)\big]^M$ *as in* (A.0.229), *for each given aperture parameter* $\kappa > 0$ *there exists some constant* $C \in (0,\infty)$ *with the property that for each given function* $f \in \big[\mathbb{X}\big]^M$ *one has*

$$\begin{aligned}&\mathscr{S}_{\text{mod}}f \in \big[\mathscr{C}^\infty(\Omega)\big]^M, \quad L\big(\mathscr{S}_{\text{mod}}f\big) = 0 \text{ in } \Omega,\\&\mathcal{N}_\kappa\big(\nabla\mathscr{S}_{\text{mod}}f\big) \in \mathbb{X} \text{ and } \big\|\mathcal{N}_\kappa\big(\nabla\mathscr{S}_{\text{mod}}f\big)\big\|_{\mathbb{X}} \le C\|f\|_{[\mathbb{X}]^M},\\&\Big(\big(\mathscr{S}_{\text{mod}}f\big)\Big|_{\partial\Omega}^{\kappa-\text{n.t.}}\Big)(x) = (S_{\text{mod}}f)(x) \text{ at } \sigma\text{-a.e. point } x \in \partial\Omega.\end{aligned} \tag{4.1.340}$$

Moreover, for each vector-valued function $f \in \big[\mathbb{X}\big]^M$ *the following jump formula holds (with* I *denoting the identity operator)*

$$\partial_\nu^A \mathscr{S}_{\text{mod}}f = \big(-\tfrac{1}{2}I + K^\#_{A^\top}\big)f \text{ at } \sigma\text{-a.e. point in } \partial\Omega, \tag{4.1.341}$$

where $K^\#_{A^\top}$ *is the singular integral operator associated as in* (A.0.118) *with the set* Ω *and the transpose coefficient tensor* $A^\top$.

Proof All claims may be justified in an analogous manner to the proof of [111, Theorem 8.5, pp. 522-524], making use of the extrapolation result involving Generalized Banach Function Spaces from [130, Theorem 5.2.1]. □

Our next theorem elaborates on some fundamental properties of the modified boundary-to-domain double layer potential operators and their conormal derivatives acting on homogeneous GBFS-based Sobolev spaces.

Theorem 4.1.22 *Assume* $\Omega \subseteq \mathbb{R}^n$ *(where* $n \in \mathbb{N}$, $n \ge 2$*) is a* UR *domain. Denote by* $\nu = (\nu_1, \dots, \nu_n)$ *the geometric measure theoretic outward unit normal to* Ω *and abbreviate* $\sigma := \mathcal{H}^{n-1}\lfloor\partial\Omega$. *In addition, for some* $M \in \mathbb{N}$, *let* $A = \big(a_{rs}^{\alpha\beta}\big)_{\substack{1\le r,s\le n\\1\le\alpha,\beta\le M}}$ *be a complex coefficient tensor with the property that* $L := L_A$ *is a weakly elliptic* $M \times M$ *system in* $\mathbb{R}^n$. *Also, let* $E = (E_{\gamma\beta})_{1\le\gamma,\beta\le M}$ *be the matrix-valued fundamental solution associated with* L *as in* [131, Theorem 1.4.2]. *In this setting, recall the modified version of the double layer operator* $\mathcal{D}_{A,\text{mod}}$ *acting on functions from* $\big[L^1\big(\partial\Omega, \frac{\sigma(x)}{1+|x|^n}\big)\big]^M$ *as in* (A.0.73).

Also, with $\mathcal{M}_{\partial\Omega}$ *denoting the Hardy-Littlewood maximal operator on* $(\partial\Omega, \sigma)$, *consider a Generalized Banach Function Space* $\mathbb{X}$ *over* $(\partial\Omega, \sigma)$ *such that*

$$\mathcal{M}_{\partial\Omega} : \mathbb{X} \to \mathbb{X} \text{ and } \mathcal{M}_{\partial\Omega} : \mathbb{X}' \to \mathbb{X}' \text{ are well-defined and bounded,} \tag{4.1.342}$$

where $\mathbb{X}'$ *is the Köthe dual of* $\mathbb{X}$*. Then the following statements are true.*

(1) There exists some constant $C \in (0, \infty)$ *with the property that for each function* $f \in [\dot{\mathbb{X}}_1]^M$ *one has*

$$\begin{gathered} \mathcal{D}_{A,\text{mod}} f \in [\mathscr{C}^\infty(\Omega)]^M, \quad L(\mathcal{D}_{A,\text{mod}} f) = 0 \text{ in } \Omega, \\ (\mathcal{D}_{A,\text{mod}} f)\big|_{\partial\Omega}^{\kappa-\text{n.t.}}, \ (\nabla \mathcal{D}_{A,\text{mod}} f)\big|_{\partial\Omega}^{\kappa-\text{n.t.}} \text{ exist } \sigma\text{-a.e. on } \partial\Omega, \\ \mathcal{N}_\kappa(\nabla \mathcal{D}_{A,\text{mod}} f) \in \mathbb{X} \text{ and } \|\mathcal{N}_\kappa(\nabla \mathcal{D}_{A,\text{mod}} f)\|_{\mathbb{X}} \leq C\|f\|_{[\dot{\mathbb{X}}_1]^M}. \end{gathered} \tag{4.1.343}$$

In fact, for each function $f \in [\dot{\mathbb{X}}_1]^M$ *one has*

$$(\mathcal{D}_{A,\text{mod}} f)\big|_{\partial\Omega}^{\kappa-\text{n.t.}} = \big(\tfrac{1}{2}I + K_{A,\text{mod}}\big)f \text{ at } \sigma\text{-a.e. point on } \partial\Omega, \tag{4.1.344}$$

where I *is the identity operator on* $[\dot{\mathbb{X}}_1]^M$*, and* $K_{A,\text{mod}}$ *is the modified boundary-to-boundary double layer potential operator from* (A.0.117).

(2) Given any function $f = (f_\alpha)_{1 \leq \alpha \leq M}$ *belonging to the homogeneous* GBFS*-styled Sobolev space* $[\dot{\mathbb{X}}_1]^M$*, for* $1 \leq \mu \leq M$ *and with the conormal derivative considered as in* (A.0.203)*, at* σ*-a.e. point* $x \in \partial\Omega$ *one has*

$$\big(\partial_\nu^A(\mathcal{D}_{A,\text{mod}} f)\big)_\mu(x) = \lim_{\varepsilon \to 0^+} \int_{\substack{y \in \partial\Omega \\ |x-y|>\varepsilon}} \nu_i(x) a_{ij}^{\mu\gamma} a_{rs}^{\beta\alpha} (\partial_r E_{\gamma\beta})(x-y)\big(\partial_{\tau_{js}} f_\alpha\big)(y)\, d\sigma(y). \tag{4.1.345}$$

(3) The operator

$$\begin{gathered} \partial_\nu^A \mathcal{D}_{A,\text{mod}} : [\dot{\mathbb{X}}_1]^M \longrightarrow [\mathbb{X}]^M \text{ defined as} \\ (\partial_\nu^A \mathcal{D}_{A,\text{mod}})f := \partial_\nu^A(\mathcal{D}_{A,\text{mod}} f) \text{ for each } f \in [\dot{\mathbb{X}}_1]^M \end{gathered} \tag{4.1.346}$$

is well defined, linear, and bounded, when the domain space is equipped with the semi-norm (4.1.320)*. As a consequence of* (4.1.346)*, the following is a well-defined linear operator:*

$$\begin{gathered} [\partial_\nu^A \mathcal{D}_{A,\text{mod}}] : [\dot{\mathbb{X}}_1/\sim]^M \longrightarrow [\mathbb{X}]^M \text{ given by} \\ [\partial_\nu^A \mathcal{D}_{A,\text{mod}}][f] := \partial_\nu^A(\mathcal{D}_{A,\text{mod}} f) \text{ for each } f \in [\dot{\mathbb{X}}_1]^M. \end{gathered} \tag{4.1.347}$$

If, in fact, $\Omega \subseteq \mathbb{R}^n$ *is an open set satisfying a two-sided local John condition with an unbounded Ahlfors regular boundary, then the operator* (4.1.347) *is also bounded when the quotient space is equipped with the norm* (4.1.321).

(4) With $K^\#_{A^\top}$ *denoting the singular integral operator associated as in* (A.0.118) *with the set* Ω *and the transpose coefficient tensor* $A^\top$*, one has*

$$\begin{gathered}\big(\tfrac{1}{2}I + K^{\#}_{A^{\top}}\big)\big(-\tfrac{1}{2}I + K^{\#}_{A^{\top}}\big) = \big[\partial^{A}_{\nu}\mathcal{D}_{A,\mathrm{mod}}\big]\big[S_{\mathrm{mod}}\big]\\ \text{as mappings acting from } \big[\mathbb{X}\big]^{M},\end{gathered} \tag{4.1.348}$$

and

$$\big[\partial^{A}_{\nu}\mathcal{D}_{A,\mathrm{mod}}\big]\big[K_{A,\mathrm{mod}}\big] = K^{\#}_{A^{\top}}\big[\partial^{A}_{\nu}\mathcal{D}_{A,\mathrm{mod}}\big] \text{ on } \big[\dot{\mathbb{X}}_1/\!\sim\big]^{M}. \tag{4.1.349}$$

Moreover, if $\partial\Omega$ *is connected then also*

$$\begin{gathered}\big(\tfrac{1}{2}I + \big[K_{A,\mathrm{mod}}\big]\big)\big(-\tfrac{1}{2}I + \big[K_{A,\mathrm{mod}}\big]\big) = \big[S_{\mathrm{mod}}\big]\big[\partial^{A}_{\nu}\mathcal{D}_{A,\mathrm{mod}}\big]\\ \text{as mappings acting from } \big[\dot{\mathbb{X}}_1/\!\sim\big]^{M},\end{gathered} \tag{4.1.350}$$

and

$$\big[S_{\mathrm{mod}}\big]K^{\#}_{A^{\top}} = \big[K_{A,\mathrm{mod}}\big]\big[S_{\mathrm{mod}}\big] \text{ when acting from } \big[\mathbb{X}\big]^{M}. \tag{4.1.351}$$

Proof In view of the extrapolation result involving Generalized Banach Function Spaces from [130, Theorem 5.2.1], the same type of argument as in the proof of [111, Theorem 8.6, pp. 525-527] applies and yields all desired conclusions. □

Next, we study mapping properties for modified boundary-to-boundary double layer potential operators acting on homogeneous GBFS-based Sobolev spaces.

Theorem 4.1.23 *Let* $\Omega \subseteq \mathbb{R}^n$ *(where* $n \in \mathbb{N}$, $n \geq 2$*) be an* NTA *domain such that* $\partial\Omega$ *is an Ahlfors regular set, and abbreviate* $\sigma := \mathcal{H}^{n-1}\lfloor\partial\Omega$. *Also, let* $L = \big(a^{\alpha\beta}_{rs}\partial_r\partial_s\big)_{1\leq\alpha,\beta\leq M}$ *be a homogeneous, weakly elliptic, constant (complex) coefficient, second-order* $M \times M$ *system in* $\mathbb{R}^n$ *(for some integer* $M \in \mathbb{N}$*). In this context, consider the modified boundary-to-boundary double layer potential operator* $K_{A,\mathrm{mod}}$ *from* (A.0.117).

Let $\mathcal{M}_{\partial\Omega}$ *be the Hardy-Littlewood maximal operator on* $(\partial\Omega, \sigma)$*, and consider a Generalized Banach Function Space* $\mathbb{X}$ *over* $(\partial\Omega, \sigma)$ *with the property that*

$$\mathcal{M}_{\partial\Omega} : \mathbb{X} \to \mathbb{X} \text{ and } \mathcal{M}_{\partial\Omega} : \mathbb{X}' \to \mathbb{X}' \text{ are well-defined and bounded,} \tag{4.1.352}$$

where $\mathbb{X}'$ *is the Köthe dual of* $\mathbb{X}$. *Then the following statements are valid.*

(1) The modified boundary-to-boundary double layer potential operator

$$K_{A,\mathrm{mod}} : \big[\dot{\mathbb{X}}_1\big]^{M} \longrightarrow \big[\dot{\mathbb{X}}_1\big]^{M} \tag{4.1.353}$$

is well defined, linear, and bounded, when the spaces involved are endowed with the semi-norm (4.1.320). *As a corollary of* (4.1.353)*, the following operator is well defined and linear:*

$$\begin{gathered}\big[K_{A,\mathrm{mod}}\big] : \big[\dot{\mathbb{X}}_1/\sim\big]^{M} \longrightarrow \big[\dot{\mathbb{X}}_1/\sim\big]^{M} \text{ given by}\\ \big[K_{A,\mathrm{mod}}\big][f] := \big[K_{A,\mathrm{mod}}f\big] \in \big[\dot{\mathbb{X}}_1/\sim\big]^{M} \text{ for each } f \in \big[\dot{\mathbb{X}}_1\big]^{M}.\end{gathered} \tag{4.1.354}$$

Moreover, if actually $\Omega \subseteq \mathbb{R}^n$ *is a two-sided* NTA *domain whose boundary is an unbounded Ahlfors regular set then the operator* (4.1.354) *is also bounded when all quotient spaces are endowed with the norm introduced in* (4.1.321).

(2) *If* U_{jk} *with* $j, k \in \{1, \dots, n\}$ *is the family of singular integral operators defined in* [132, (1.5.251)], *then*

$$\partial_{\tau_{jk}}\big(K_{A,\text{mod}} f\big) = K_A(\partial_{\tau_{jk}} f) + U_{jk}(\nabla_{\tan} f) \text{ at } \sigma\text{-a.e. point on } \partial\Omega \\ \text{for each } f \in \big[\dot{\mathbb{X}}_1\big]^M \text{ and each } j, k \in \{1, \dots, n\}. \tag{4.1.355}$$

Proof This is proved reasoning much as in the proof of [111, Theorem 8.7, pp. 527-528], now making of the extrapolation result involving Generalized Banach Function Spaces from [130, Theorem 5.2.1]. □

Our next goal is to present a version of Theorem 4.1.13 valid in the context of homogeneous GBFS-based Sobolev spaces, where again the key feature is the explicit dependence on the BMO semi-norm of the geometric measure theoretic outward unit normal to the underlying domain.

Theorem 4.1.24 *Let* $\Omega \subseteq \mathbb{R}^n$ *be a two-sided* NTA *domain whose boundary is an unbounded Ahlfors regular set. Abbreviate* $\sigma := \mathcal{H}^{n-1} \lfloor \partial\Omega$ *and denote by* ν *the geometric measure theoretic outward unit normal to* Ω*. Also, if* $\mathcal{M}_{\partial\Omega}$ *denotes the Hardy-Littlewood maximal operator on* $(\partial\Omega, \sigma)$*, consider a Generalized Banach Function Space* $\mathbb{X}$ *over* $(\partial\Omega, \sigma)$ *with the property that*

$$\mathcal{M}_{\partial\Omega} : \mathbb{X} \to \mathbb{X} \text{ and } \mathcal{M}_{\partial\Omega} : \mathbb{X}' \to \mathbb{X}' \text{ are well-defined and bounded,} \tag{4.1.356}$$

where $\mathbb{X}'$ *is the Köthe dual of* $\mathbb{X}$*. Next, let* L *be a homogeneous, second-order, constant complex coefficient, weakly elliptic* $M \times M$ *system in* $\mathbb{R}^n$ *for which* $\mathfrak{A}_L^{\text{dis}} \neq \emptyset$*. Finally, pick* $A \in \mathfrak{A}_L^{\text{dis}}$ *and consider the modified boundary-to-boundary double layer potential operator* $\big[K_{A,\text{mod}}\big]$ *associated with* Ω *and the coefficient tensor* A *as in Theorem 4.1.23.*

Then for each $m \in \mathbb{N}$ *there exists some* $C_m \in (0, \infty)$ *which depends only on* m, n, A*, the two-sided* NTA *constants of* Ω*, the Ahlfors regularity constant of* $\partial\Omega$*, the operator norms* $\|\mathcal{M}_{\partial\Omega}\|_{\mathbb{X}\to\mathbb{X}}$, $\|\mathcal{M}_{\partial\Omega}\|_{\mathbb{X}'\to\mathbb{X}'}$ *such that, with the piece of notation introduced in* (4.1.12)*, one has*

$$\big\|\big[K_{A,\text{mod}}\big]\big\|_{[\dot{\mathbb{X}}_1/\sim]^M \to [\dot{\mathbb{X}}_1/\sim]^M} \leq C_m \|\nu\|^{\langle m\rangle}_{[\text{BMO}(\partial\Omega,\sigma)]^n}. \tag{4.1.357}$$

In fact, a more general result is valid, to the effect that a similar estimate to (4.1.357) holds for the "modified" generalized double layer operator T_{mod} associated as in [132, (5.1.65)-(5.1.66)] with a kernel satisfying the properties listed in [132, (5.1.1)-(5.1.2)].

Proof of Theorem 4.1.24 We proceed much as in the proof of Theorem 4.1.13, making use of (4.1.354), (4.1.355), Proposition 1.3.2, the estimate in Theorem 4.1.3 recorded in (4.1.49), the intertwining identity [132, (5.2.167)], the embeddings from

[130, (5.2.90)] in [130, Proposition 5.2.7], the commutator estimates from [131, Corollary 2.7.9], and the fact that $\mathbb{X}$ is stable under multiplication by essentially bounded functions (cf. [130, (5.1.13)]). □

4.2 Invertibility Results for Chord-Dot-Normal SIO's on Unbounded Boundaries

In the introduction of his influential 1911 monograph [172] J. Plemelj writes that "*one of the most important tasks of modern Potential Theory is to explore the invertibility of boundary singular integral equations arising from single and double layer potentials.*"

With Theorem 4.1.12 and Theorem 4.1.14 in hand, in this section we establish invertibility results for certain types of boundary layer potentials on a variety of function spaces. The geometric context is that of an Ahlfors regular domain whose geometric measure theoretic outward unit normal has a suitably small BMO semi-norm relative to the dimension and the ADR constants of its boundary. On the analytic side, the key assumption is the existence of distinguished coefficient tensors for the given weakly elliptic system. This is made precise in the theorem below.

Theorem 4.2.1 *Assume that $\Omega \subseteq \mathbb{R}^n$, for $n \in \mathbb{N}$, is an Ahlfors regular domain. Abbreviate $\sigma := \mathcal{H}^{n-1} \lfloor \partial\Omega$ and denote by ν the geometric measure theoretic outward unit normal to Ω. Also, let L be a homogeneous, second-order, constant complex coefficient, weakly elliptic $M \times M$ system in $\mathbb{R}^n$. Pick $A \in \mathfrak{A}_L$ and consider the layer potential operators $\mathcal{D}_{\text{mod}}$, K_{mod}, K_A, $K_A^{\#}$, $\mathcal{S}$, $\mathcal{S}_{\text{mod}}$, associated with Ω and the coefficient tensor A as in* (A.0.73), (A.0.117), (A.0.116), (A.0.118), (A.0.231), *and* (A.0.232), *respectively. Then the following statements are true.*

(1) Suppose $A \in \mathfrak{A}_L^{\text{dis}}$. Then, having fixed an integrability exponent $p \in (1, \infty)$, a Muckenhoupt weight $w \in A_p(\partial\Omega, \sigma)$, and some number $\varepsilon \in (0, \infty)$, there exists a threshold $\delta \in (0, 1)$, which depends only on n, p, $[w]_{A_p}$, A, ε, and the ADR *constants of $\partial\Omega$, with the property that if $\|\nu\|_{[\text{BMO}(\partial\Omega,\sigma)]^n} < \delta$ (i.e., if Ω is a δ-*AR *domain in the sense of Definition 3.1.8) it follows that Ω is a* UR *domain and for each spectral parameter $z \in \mathbb{C}$ with $|z| \geq \varepsilon$ the following operators are invertible:*

$$zI + K_A : \big[L^p(\partial\Omega, w)\big]^M \longrightarrow \big[L^p(\partial\Omega, w)\big]^M, \tag{4.2.1}$$

$$zI + K_A : \big[L_1^p(\partial\Omega, w)\big]^M \longrightarrow \big[L_1^p(\partial\Omega, w)\big]^M, \tag{4.2.2}$$

$$zI + K_A^{\#} : \big[L^p(\partial\Omega, w)\big]^M \longrightarrow \big[L^p(\partial\Omega, w)\big]^M, \tag{4.2.3}$$

$$zI + K_A^{\#} : \big[L_{-1}^p(\partial\Omega, w)\big]^M \longrightarrow \big[L_{-1}^p(\partial\Omega, w)\big]^M. \tag{4.2.4}$$

As a corollary, corresponding to the particular case when $w \equiv 1$, the following operators are invertible:

$$zI + K_A : \big[L^p(\partial\Omega, \sigma)\big]^M \longrightarrow \big[L^p(\partial\Omega, \sigma)\big]^M, \tag{4.2.5}$$

$$zI + K_A : \big[L^p_1(\partial\Omega, \sigma)\big]^M \longrightarrow \big[L^p_1(\partial\Omega, \sigma)\big]^M, \tag{4.2.6}$$

$$zI + K^{\#}_A : \big[L^p(\partial\Omega, \sigma)\big]^M \longrightarrow \big[L^p(\partial\Omega, \sigma)\big]^M, \tag{4.2.7}$$

$$zI + K^{\#}_A : \big[L^p_{-1}(\partial\Omega, \sigma)\big]^M \longrightarrow \big[L^p_{-1}(\partial\Omega, \sigma)\big]^M. \tag{4.2.8}$$

Furthermore, similar invertibility results are valid for the operators $zI + K_A$, $zI + K^{\#}_A$ *acting on Morrey spaces, vanishing Morrey spaces, and block spaces on* $\partial\Omega$.

(2) *Assume* $A \in \mathfrak{A}^{\mathrm{dis}}_L$. *Then, given any integrability exponent* $p \in \big(\frac{n-1}{n}, 1\big]$ *and some number* $\varepsilon \in (0, \infty)$, *there exists some small threshold* $\delta \in (0, 1)$, *which depends only on* n, p, A, ε, *and the* ADR *constants of* $\partial\Omega$, *with the property that if* $\|\nu\|_{[\mathrm{BMO}(\partial\Omega,\sigma)]^n} < \delta$ (*i.e., if* Ω *is a* δ-AR *domain in the sense of Definition 3.1.8*) *it follows that* Ω *is a* UR *domain with unbounded boundary and for each spectral parameter* $z \in \mathbb{C}$ *with* $|z| \geq \varepsilon$ *the following operator is invertible:*

$$zI + K^{\#}_A : \big[H^p(\partial\Omega, \sigma)\big]^M \longrightarrow \big[H^p(\partial\Omega, \sigma)\big]^M. \tag{4.2.9}$$

Moreover, a similar invertibility result holds for the operator $zI + K^{\#}_A$ *acting on the Lorentz-based Hardy spaces introduced in* [130, Definition 4.2.3] (*cf.* (A.0.91)).

(3) *Assume* $A \in \mathfrak{A}^{\mathrm{dis}}_L$. *Then for any* $\varepsilon \in (0, \infty)$ *there exists some small threshold* $\delta \in (0, 1)$, *which depends only on* n, ε, A, *and the* ADR *constants of* $\partial\Omega$, *with the property that if* $\|\nu\|_{[\mathrm{BMO}(\partial\Omega,\sigma)]^n} < \delta$ (*i.e., if* Ω *is a* δ-AR *domain in the sense of Definition 3.1.8*) *it follows that* Ω *is a* UR *domain with unbounded boundary and for each spectral parameter* $z \in \mathbb{C}$ *with* $|z| \geq \varepsilon$ *the following operators* (*cf.* [132, (2.1.93), (2.1.156), (2.1.172)]) *are invertible:*

$$zI + \big[K_{\mathrm{mod}}\big] : \big[\mathrm{BMO}(\partial\Omega, \sigma)/\sim\big]^M \longrightarrow \big[\mathrm{BMO}(\partial\Omega, \sigma)/\sim\big]^M, \tag{4.2.10}$$

$$zI + \big[K_{\mathrm{mod}}\big] : \big[\mathrm{VMO}(\partial\Omega, \sigma)/\sim\big]^M \longrightarrow \big[\mathrm{VMO}(\partial\Omega, \sigma)/\sim\big]^M, \tag{4.2.11}$$

$$zI + \big[K_{\mathrm{mod}}\big] : \big[\mathrm{CMO}(\partial\Omega, \sigma)/\sim\big]^M \longrightarrow \big[\mathrm{CMO}(\partial\Omega, \sigma)/\sim\big]^M. \tag{4.2.12}$$

(4) *Assume* $A \in \mathfrak{A}^{\mathrm{dis}}_L$. *Then, given any* $\alpha \in (0, 1)$ *and some number* $\varepsilon \in (0, \infty)$, *there exists some small threshold* $\delta \in (0, 1)$, *which depends only on* n, α, ε, A, *and the* ADR *constants of* $\partial\Omega$, *with the property that if* $\|\nu\|_{[\mathrm{BMO}(\partial\Omega,\sigma)]^n} < \delta$ (*i.e., if* Ω *is a* δ-AR *domain in the sense of Definition 3.1.8*) *it follows that* Ω *is a* UR *domain with unbounded boundary and for each spectral parameter* $z \in \mathbb{C}$ *with* $|z| \geq \varepsilon$ *the following operators* (*cf.* [132, (2.1.123), (2.1.144)]) *are invertible:*

$$zI + \big[K_{\mathrm{mod}}\big] : \big[\dot{\mathscr{C}}^{\alpha}(\partial\Omega)/\sim\big]^M \longrightarrow \big[\dot{\mathscr{C}}^{\alpha}(\partial\Omega)/\sim\big]^M, \tag{4.2.13}$$

$$zI + \big[K_{\mathrm{mod}}\big] : \big[\dot{\mathscr{C}}^{\alpha}_{\mathrm{van}}(\partial\Omega)/\sim\big]^M \longrightarrow \big[\dot{\mathscr{C}}^{\alpha}_{\mathrm{van}}(\partial\Omega)/\sim\big]^M. \tag{4.2.14}$$

(5) Assume $A \in \mathfrak{A}_L^{\rm dis}$. Then, given any $p \in (1,\infty)$, $\lambda \in (0, n-1)$, and some number $\varepsilon \in (0,\infty)$, there exists some small threshold $\delta \in (0,1)$, which depends only on n, p, λ, ε, A, and the ADR *constants of $\partial\Omega$, with the property that if $\|\nu\|_{[\mathrm{BMO}(\partial\Omega,\sigma)]^n} < \delta$ (i.e., if Ω is a δ-*AR *domain in the sense of Definition 3.1.8) it follows that Ω is a* UR *domain with unbounded boundary and for each spectral parameter $z \in \mathbb{C}$ with $|z| \geq \varepsilon$ the following operators (cf.* [132, (3.2.14)] *and* [132, (3.2.13)], *respectively) are invertible:*

$$zI + \big[K_{\rm mod}\big] : \big[\dot{\mathfrak{L}}^{p,\lambda}(\partial\Omega,\sigma)/\sim\big]^M \longrightarrow \big[\dot{\mathfrak{L}}^{p,\lambda}(\partial\Omega,\sigma)/\sim\big]^M, \tag{4.2.15}$$

$$zI + K : \big[\mathfrak{L}^{p,\lambda}(\partial\Omega,\sigma)\big]^M \longrightarrow \big[\mathfrak{L}^{p,\lambda}(\partial\Omega,\sigma)\big]^M. \tag{4.2.16}$$

(6) Assume $A \in \mathfrak{A}_L^{\rm dis}$. Then, given any $q \in (1,\infty)$, $\lambda \in (0, n-1)$, and some number $\varepsilon \in (0,\infty)$, there exists some small threshold $\delta \in (0,1)$, which depends only on n, q, λ, ε, A, and the ADR *constants of $\partial\Omega$, with the property that if $\|\nu\|_{[\mathrm{BMO}(\partial\Omega,\sigma)]^n} < \delta$ (i.e., if Ω is a δ-*AR *domain in the sense of Definition 3.1.8) it follows that Ω is a* UR *domain with unbounded boundary and for each spectral parameter $z \in \mathbb{C}$ with $|z| \geq \varepsilon$ the following operator (cf.* [132, (3.2.2)]) *is invertible:*

$$zI + K^{\#} : \big[\mathscr{H}^{q,\lambda}(\partial\Omega,\sigma)\big]^M \longrightarrow \big[\mathscr{H}^{q,\lambda}(\partial\Omega,\sigma)\big]^M. \tag{4.2.17}$$

(7) Assume $\mathfrak{A}_L^{\rm dis} \neq \varnothing$ and $\mathfrak{A}_{L^\top}^{\rm dis} \neq \varnothing$. Then for any exponent $p \in \big(\frac{n-1}{n}, 1\big]$ there exists some small threshold $\delta \in (0,1)$, which depends only on n, p, A, and the ADR *constants of $\partial\Omega$, with the property that whenever $\|\nu\|_{[\mathrm{BMO}(\partial\Omega,\sigma)]^n} < \delta$ it follows that Ω is a* UR *domain with unbounded boundary and the following operator (cf.* [132, (2.3.93)]) *is invertible:*

$$[S] : \big[H^p(\partial\Omega,\sigma)\big]^M \longrightarrow \big[\dot{H}^p_1(\partial\Omega,\sigma)/\sim\big]^M. \tag{4.2.18}$$

(8) Assume $A \in \mathfrak{A}_L^{\rm dis}$ and $\mathfrak{A}_{L^\top}^{\rm dis} \neq \varnothing$. Then, given any exponent $p \in \big(\frac{n-1}{n}, 1\big]$ and some number $\varepsilon \in (0,\infty)$, there exists some small threshold $\delta \in (0,1)$, which depends only on n, p, A, ε, and the ADR *constants of $\partial\Omega$, with the property that if $\|\nu\|_{[\mathrm{BMO}(\partial\Omega,\sigma)]^n} < \delta$ it follows that Ω has an unbounded boundary and satisfies a two-sided local John condition, and for each spectral parameter $z \in \mathbb{C}$ with $|z| \geq \varepsilon$ the following operator (cf.* [132, (2.3.185)]) *is invertible:*

$$zI + \big[K_{\rm mod}\big] : \big[\dot{H}^p_1(\partial\Omega,\sigma)/\sim\big]^M \longrightarrow \big[\dot{H}^p_1(\partial\Omega,\sigma)/\sim\big]^M. \tag{4.2.19}$$

(9) Assume $A \in \mathfrak{A}_L^{\rm dis}$ and $\mathfrak{A}_{L^\top}^{\rm dis} \neq \varnothing$. Then for any exponent $p \in \big(\frac{n-1}{n}, 1\big]$ there exists some small threshold $\delta \in (0,1)$, which depends only on n, p, A, and the ADR *constants of $\partial\Omega$, with the property that whenever $\|\nu\|_{[\mathrm{BMO}(\partial\Omega,\sigma)]^n} < \delta$ it follows that Ω is a* UR *domain with unbounded boundary and the following operator (cf.* [132, (2.3.92)]) *is invertible:*

$$\big[\dot{\partial}^A_\nu \mathcal{D}_{\rm mod}\big] : \big[\dot{H}^p_1(\partial\Omega,\sigma)/\sim\big]^M \longrightarrow \big[H^p(\partial\Omega,\sigma)\big]^M. \tag{4.2.20}$$

(10) Assume $A \in \mathfrak{A}_L^{\rm dis}$. Then, given any $p, q \in (1, \infty)$, $\lambda \in (0, n-1)$, and some number $\varepsilon \in (0, \infty)$, there exists some small threshold $\delta \in (0, 1)$, which depends only on n, p, q, λ, ε, A, and the ADR *constants of $\partial\Omega$, with the property that if $\|\nu\|_{[\mathrm{BMO}(\partial\Omega,\sigma)]^n} < \delta$ (i.e., if Ω is a δ-*AR *domain in the sense of Definition 3.1.8) it follows that Ω has an unbounded boundary and satisfies a two-sided local John condition, and for each spectral parameter $z \in \mathbb{C}$ with $|z| \geq \varepsilon$ the following operators are invertible:*

$$zI + [K_{\rm mod}] : [\dot{L}_1^p(\partial\Omega, \sigma)/\sim]^M \longrightarrow [\dot{L}_1^p(\partial\Omega, \sigma)/\sim]^M, \tag{4.2.21}$$

$$zI + [K_{\rm mod}] : [\dot{M}_1^{p,\lambda}(\partial\Omega, \sigma)/\sim]^M \longrightarrow [\dot{M}_1^{p,\lambda}(\partial\Omega, \sigma)/\sim]^M, \tag{4.2.22}$$

$$zI + [K_{\rm mod}] : [\dot{\mathbb{M}}_1^{p,\lambda}(\partial\Omega, \sigma)/\sim]^M \longrightarrow [\dot{\mathbb{M}}_1^{p,\lambda}(\partial\Omega, \sigma)/\sim]^M, \tag{4.2.23}$$

$$zI + [K_{\rm mod}] : [\dot{\mathcal{B}}_1^{q,\lambda}(\partial\Omega, \sigma)/\sim]^M \longrightarrow [\dot{\mathcal{B}}_1^{q,\lambda}(\partial\Omega, \sigma)/\sim]^M. \tag{4.2.24}$$

(11) Assume that $\mathfrak{A}_L^{\rm dis} \neq \varnothing$ and that $\mathfrak{A}_{L^\top}^{\rm dis} \neq \varnothing$. Then for each $p, q \in (1, \infty)$ and each $\lambda \in (0, n-1)$ there exists some small threshold $\delta \in (0, 1)$, which depends only on n, p, q, λ, L, and the ADR *constants of $\partial\Omega$, with the property that whenever $\|\nu\|_{[\mathrm{BMO}(\partial\Omega,\sigma)]^n} < \delta$ it follows that Ω is a* UR *domain with unbounded boundary and the following operators (cf.* [132, (1.8.124)] *and* [132, Theorem 3.3.5, items *(2), (4), (5)*]) *are invertible:*

$$[S_{\rm mod}] : [L^p(\partial\Omega, \sigma)]^M \longrightarrow [\dot{L}_1^p(\partial\Omega, \sigma)/\sim]^M, \tag{4.2.25}$$

$$[S_{\rm mod}] : [M^{p,\lambda}(\partial\Omega, \sigma)]^M \longrightarrow [\dot{M}_1^{p,\lambda}(\partial\Omega, \sigma)/\sim]^M, \tag{4.2.26}$$

$$[S_{\rm mod}] : [\mathring{M}^{p,\lambda}(\partial\Omega, \sigma)]^M \longrightarrow [\dot{\mathbb{M}}_1^{p,\lambda}(\partial\Omega, \sigma)/\sim]^M, \tag{4.2.27}$$

$$[S_{\rm mod}] : [\mathcal{B}^{q,\lambda}(\partial\Omega, \sigma)]^M \longrightarrow [\dot{\mathcal{B}}_1^{q,\lambda}(\partial\Omega, \sigma)/\sim]^M. \tag{4.2.28}$$

(12) Assume that $A \in \mathfrak{A}_L^{\rm dis}$ and that $\mathfrak{A}_{L^\top}^{\rm dis} \neq \varnothing$. Then for any $p, q \in (1, \infty)$ and any $\lambda \in (0, n-1)$ there exists some small threshold $\delta \in (0, 1)$, which depends only on n, p, q, λ, A, and the ADR *constants of $\partial\Omega$, with the property that whenever $\|\nu\|_{[\mathrm{BMO}(\partial\Omega,\sigma)]^n} < \delta$ it follows that Ω is a* UR *domain with unbounded boundary and the following operators (cf.* [132, (1.8.148)] *and* [132, Theorem 3.3.6, items *(3)-(5)*]) *are invertible:*

$$[\partial_\nu^A \mathcal{D}_{\rm mod}] : [\dot{L}_1^p(\partial\Omega, \sigma)/\sim]^M \longrightarrow [L^p(\partial\Omega, \sigma)]^M, \tag{4.2.29}$$

$$[\partial_\nu^A \mathcal{D}_{\rm mod}] : [\dot{M}_1^{p,\lambda}(\partial\Omega, \sigma)/\sim]^M \longrightarrow [M^{p,\lambda}(\partial\Omega, \sigma)]^M, \tag{4.2.30}$$

$$[\partial_\nu^A \mathcal{D}_{\rm mod}] : [\dot{\mathbb{M}}_1^{p,\lambda}(\partial\Omega, \sigma)/\sim]^M \longrightarrow [\mathring{M}^{p,\lambda}(\partial\Omega, \sigma)]^M, \tag{4.2.31}$$

$$[\partial_\nu^A \mathcal{D}_{\rm mod}] : [\dot{\mathcal{B}}_1^{q,\lambda}(\partial\Omega, \sigma)/\sim]^M \longrightarrow [\mathcal{B}^{q,\lambda}(\partial\Omega, \sigma)]^M. \tag{4.2.32}$$

(13) Assume $A \in \mathfrak{A}_L^{\rm dis}$ and suppose

$$\begin{aligned} &\phi : (0,\infty) \to (0,\infty) \text{ is } \mathcal{L}^1\text{-measurable and there is } C_* \in (0,\infty) \\ &\text{so that } r\int_r^\infty \phi(t)\,\frac{\ln(t/r)}{t^2}\,dt \leq C_*\phi(r) \text{ for all } r \in (0,\infty). \end{aligned} \tag{4.2.33}$$

Also, fix an integrability exponent $p \in (1,\infty)$, select a Muckenhoupt weight $w \in A_p(\partial\Omega,\sigma)$, and pick some number $\varepsilon \in (0,\infty)$. Then there exists some small threshold $\delta \in (0,1)$, which depends only on n, p, $[w]_{A_p}$, ε, A, C_* from (4.2.33), *and the* ADR *constants of $\partial\Omega$, with the property that if $\|\nu\|_{[\mathrm{BMO}(\partial\Omega,\sigma)]^n} < \delta$ (i.e., if Ω is a δ-*AR *domain in the sense of Definition 3.1.8) it follows that Ω is a* UR *domain with an unbounded boundary, and for each spectral parameter $z \in \mathbb{C}$ with $|z| \geq \varepsilon$ the following operators are invertible:*

$$zI + [K_{\rm mod}] \text{ from } \big[\mathrm{BMO}_{\phi,p}(\partial\Omega, w)/\sim\big]^M \text{ to itself,} \tag{4.2.34}$$

$$zI + [K_{\rm mod}] \text{ from } \big[\mathrm{BCMO}_{\phi,p}(\partial\Omega, w)/\sim\big]^M \text{ to itself,} \tag{4.2.35}$$

$$zI + [K_{\rm mod}] \text{ from } \big[\dot{\mathrm{B}}\mathrm{CMO}_{\phi,p}(\partial\Omega, w)/\sim\big]^M \text{ to itself,} \tag{4.2.36}$$

where the above ϕ-modulated weighted BMO *and* BCMO *spaces are defined as in* (4.1.281)-(4.1.286) *with $\Sigma := \partial\Omega$. In particular, corresponding to $w \equiv 1$, under the aforementioned assumptions the following operators are invertible:*

$$zI + [K_{\rm mod}] \text{ from } \big[\mathrm{BMO}_{\phi,p}(\partial\Omega, \sigma)/\sim\big]^M \text{ to itself,} \tag{4.2.37}$$

$$zI + [K_{\rm mod}] \text{ from } \big[\mathrm{BCMO}_{\phi,p}(\partial\Omega, \sigma)/\sim\big]^M \text{ to itself,} \tag{4.2.38}$$

$$zI + [K_{\rm mod}] \text{ from } \big[\dot{\mathrm{B}}\mathrm{CMO}_{\phi,p}(\partial\Omega, \sigma)/\sim\big]^M \text{ to itself.} \tag{4.2.39}$$

(14) Suppose $\mathbb{X}$ is a Generalized Banach Function Space on $(\partial\Omega,\sigma)$ such that

$$\mathcal{M}_{\partial\Omega} : \mathbb{X} \to \mathbb{X} \text{ and } \mathcal{M}_{\partial\Omega} : \mathbb{X}' \to \mathbb{X}' \text{ are well-defined and bounded,} \tag{4.2.40}$$

where $\mathcal{M}_{\partial\Omega}$ is the Hardy-Littlewood maximal operator on $\partial\Omega$, and $\mathbb{X}'$ is the associated space of $\mathbb{X}$ (cf. [130, Definition 5.1.4] *and* [130, Definition 5.1.11]). *Recall that*

$$\mathring{\mathbb{X}} := \overline{L^\infty(\partial\Omega,\sigma)}^{\|\cdot\|_{\mathbb{X}}} \text{ and } (\mathbb{X}')^\circ := \overline{L^\infty(\partial\Omega,\sigma)}^{\|\cdot\|_{\mathbb{X}'}}. \tag{4.2.41}$$

In addition, consider the following $\mathbb{X}$-styled, $\mathbb{X}'$-styled, $\mathring{\mathbb{X}}$-styled and, respectively, $(\mathbb{X}')^\circ$-styled Sobolev spaces of order one on $\partial\Omega$:

$$\begin{aligned} \mathbb{X}_1 &:= \Big\{f \in \mathbb{X} : \partial_{\tau_{jk}} f \in \mathbb{X} \text{ for } 1 \leq j,k \leq n\Big\}, \\ (\mathbb{X}')_1 &:= \Big\{g \in \mathbb{X}' : \partial_{\tau_{jk}} g \in \mathbb{X}' \text{ for } 1 \leq j,k \leq n\Big\}, \\ \mathring{\mathbb{X}}_1 &:= \Big\{f \in \mathring{\mathbb{X}} : \partial_{\tau_{jk}} f \in \mathring{\mathbb{X}} \text{ for } 1 \leq j,k \leq n\Big\}, \\ (\mathbb{X}')_1^\circ &:= \Big\{g \in (\mathbb{X}')^\circ : \partial_{\tau_{jk}} g \in (\mathbb{X}')^\circ \text{ for } 1 \leq j,k \leq n\Big\}. \end{aligned} \tag{4.2.42}$$

Then $\mathbb{X}_1$ *and* $(\mathbb{X}')_1$ *are Banach spaces*[7] *when equipped with the natural norms*

$$\begin{aligned} \|f\|_{\mathbb{X}_1} &:= \|f\|_{\mathbb{X}} + \sum_{j,k=1}^{n} \|\partial_{\tau_{jk}} f\|_{\mathbb{X}} \text{ for each } f \in \mathbb{X}_1, \\ \|g\|_{(\mathbb{X}')_1} &:= \|g\|_{\mathbb{X}} + \sum_{j,k=1}^{n} \|\partial_{\tau_{jk}} g\|_{\mathbb{X}'}, \text{ for each } g \in (\mathbb{X}')_1, \end{aligned} \tag{4.2.43}$$

while $\mathring{\mathbb{X}}_1$ *and* $(\mathbb{X}')_1^{\circ}$ *are closed subspaces of* $\mathbb{X}_1$ *and* $(\mathbb{X}')_1$*, respectively (hence, Banach spaces in their own right, when equipped with the norms defined in* (4.2.43)*). Moreover, if* $A \in \mathfrak{A}_L^{\rm dis}$ *then for any given* $\varepsilon \in (0, \infty)$ *there exists some small threshold* $\delta \in (0, 1)$*, which depends only on* n*,* ε*,* A*, the* ADR *constants of* $\partial\Omega$*, and the operator norms* $\|\mathcal{M}_{\partial\Omega}\|_{\mathbb{X}\to\mathbb{X}}$*,* $\|\mathcal{M}_{\partial\Omega}\|_{\mathbb{X}'\to\mathbb{X}'}$*, with the property that whenever* $\|\nu\|_{[\mathrm{BMO}(\partial\Omega,\sigma)]^n} < \delta$ *it follows that* Ω *is a* UR *domain and for each spectral parameter* $z \in \mathbb{C}$ *with* $|z| \geq \varepsilon$ *the following operators are invertible:*

$$zI + K_A,\ zI + K_A^{\#} : \big[\mathbb{X}\big]^M \longrightarrow \big[\mathbb{X}\big]^M, \tag{4.2.44}$$

$$zI + K_A,\ zI + K_A^{\#} : \big[\mathbb{X}'\big]^M \longrightarrow \big[\mathbb{X}'\big]^M, \tag{4.2.45}$$

$$zI + K_A,\ zI + K_A^{\#} : \big[\mathring{\mathbb{X}}\big]^M \longrightarrow \big[\mathring{\mathbb{X}}\big]^n, \tag{4.2.46}$$

$$zI + K_A,\ zI + K_A^{\#} : \big[(\mathbb{X}')^{\circ}\big]^M \longrightarrow \big[(\mathbb{X}')^{\circ}\big]^M, \tag{4.2.47}$$

and

$$zI + K_A : \big[\mathbb{X}_1\big]^M \longrightarrow \big[\mathbb{X}_1\big]^M, \tag{4.2.48}$$

$$zI + K_A : \big[(\mathbb{X}')_1\big]^M \longrightarrow \big[(\mathbb{X}')_1\big]^M, \tag{4.2.49}$$

$$zI + K_A : \big[\mathring{\mathbb{X}}_1\big]^M \longrightarrow \big[\mathring{\mathbb{X}}_1\big]^M, \tag{4.2.50}$$

$$zI + K_A : \big[(\mathbb{X}')_1^{\circ}\big]^M \longrightarrow \big[(\mathbb{X}')_1^{\circ}\big]^M. \tag{4.2.51}$$

(15) Continue to assume that $\mathbb{X}$ *is a Generalized Banach Function Space on* $(\partial\Omega, \sigma)$ *satisfying* (4.2.40)*, and define the following* $\mathbb{X}$*-based homogeneous Sobolev space of order one on* $\partial\Omega$*:*

$$\dot{\mathbb{X}}_1 := \Big\{f \in L^1\Big(\partial\Omega, \tfrac{\sigma(x)}{1+|x|^n}\Big) : \partial_{\tau_{jk}} f \in \mathbb{X},\ 1 \leq j, k \leq n\Big\}, \tag{4.2.52}$$

equipped with the semi-norm

$$\dot{\mathbb{X}}_1 \ni f \longmapsto \|f\|_{\dot{\mathbb{X}}_1} := \sum_{j,k=1}^{n} \|\partial_{\tau_{jk}} f\|_{\mathbb{X}}. \tag{4.2.53}$$

(compare with (4.1.319)-(4.1.320)*). Then, under the assumption that* Ω *is a two-sided* NTA *domain with an unbounded Ahlfors regular boundary it follows that*

---

[7] this result only uses the fact that Ω is an Ahlfors regular domain

$$\dot{\mathbb{X}}_1 \subseteq L^p_{\rm loc}(\partial\Omega, \sigma) \;\; for\ some \;\; p \in (1, \infty), \tag{4.2.54}$$

and $\dot{\mathbb{X}}_1 / \sim$*, the space of equivalence classes of functions in* $\dot{\mathbb{X}}_1$ *modulo constants, is a Banach space when equipped with the norm*

$$[\dot{\mathbb{X}}_1/\sim] \ni [f] \longmapsto \|[f]\|_{\dot{\mathbb{X}}_1/\sim} := \|f\|_{\dot{\mathbb{X}}_1} = \sum_{j,k=1}^{n} \|\partial_{\tau_{jk}} f\|_{\mathbb{X}}. \tag{4.2.55}$$

Furthermore, under the assumption that $A \in \mathfrak{A}_L^{\rm dis}$*, for any* $\varepsilon \in (0, \infty)$ *there exists some small threshold* $\delta \in (0, 1)$*, which depends only on* n*,* A*,* ε*, the* ADR *constants of* $\partial\Omega$*, and the operator norms* $\|\mathcal{M}_{\partial\Omega}\|_{\mathbb{X}\to\mathbb{X}}$*,* $\|\mathcal{M}_{\partial\Omega}\|_{\mathbb{X}'\to\mathbb{X}'}$*, with the property that whenever* $\|\nu\|_{[{\rm BMO}(\partial\Omega,\sigma)]^n} < \delta$ *it follows that* Ω *is a two-sided* NTA *domain with an unbounded Ahlfors regular boundary and for each spectral parameter* $z \in \mathbb{C}$ *with* $|z| \geq \varepsilon$ *the following operator is invertible:*

$$zI + \big[K_{\rm mod}\big] : \big[\dot{\mathbb{X}}_1 / \sim\big]^M \longrightarrow \big[\dot{\mathbb{X}}_1 / \sim\big]^M, \tag{4.2.56}$$

If actually $A \in \mathfrak{A}_L^{\rm dis}$ *and* $\mathfrak{A}_{L^\top}^{\rm dis} \neq \varnothing$*, then taking* $\delta \in (0, 1)$ *small enough matters may also be arranged so that the following operators are invertible:*

$$\big[\partial_\nu^A \mathcal{D}_{\rm mod}\big] : \big[\dot{\mathbb{X}}_1 / \sim\big]^M \longrightarrow \big[\mathbb{X}\big]^M, \tag{4.2.57}$$

$$\big[S_{\rm mod}\big] : \big[\mathbb{X}\big]^M \longrightarrow \big[\dot{\mathbb{X}}_1 / \sim\big]^M. \tag{4.2.58}$$

More specifically, assuming $\delta \in (0, 1)$ *is sufficiently small, if* $A \in \mathfrak{A}_L^{\rm dis}$ *then the operator* (4.2.57) *is injective, and if* $A^\top \in \mathfrak{A}_{L^\top}^{\rm dis}$ *then the operator* (4.2.57) *is surjective. Also, under the assumption that* $\delta \in (0, 1)$ *is sufficiently small, if* $\mathfrak{A}_{L^\top}^{\rm dis} \neq \varnothing$ *it follows that the operator* (4.2.58) *is injective, while if* $\mathfrak{A}_L^{\rm dis} \neq \varnothing$ *it follows that the operator* (4.2.58) *is surjective. Finally, similar invertibility results are valid for the* $\mathring{\mathbb{X}}$*-based,* $\mathbb{X}'$*-based, and* $(\mathbb{X}')^\circ$*-based homogeneous Sobolev spaces of order one on* $\partial\Omega$*, defined analogously to* (4.2.52).

(16) The conclusions in items (14)-(15) are true whenever $\mathbb{X}$ *is a rearrangement invariant Banach function space whose Boyd indices satisfy* $1 < p_{\mathbb{X}} \leq q_{\mathbb{X}} < \infty$*. In particular, all conclusions in items (14)-(15) are valid if* $\mathbb{X} := L^\Phi(\partial\Omega, \sigma)$*, the Orlicz space associated as in* (A.0.133) *with the measure space* $(\partial\Omega, \sigma)$ *and any Young function* Φ *satisfying* $1 < i(\Phi) \leq I(\Phi) < \infty$ *(with the role of* $\mathbb{X}_1$ *from* (4.2.42) *now played by the Orlicz-based Sobolev space* $L_1^\Phi(\partial\Omega, \sigma)$*).*

We wish to comment on the significance of assuming that the system L has a distinguished coefficient tensor in the context of the aforementioned invertibility results. Specifically, recall from Proposition 1.6.8 that

$$L_D := \Delta - 2\nabla{\rm div} \tag{4.2.59}$$

is a homogeneous, second-order, real constant coefficient, $n \times n$ weakly elliptic system in $\mathbb{R}^n$ with the property that $\mathfrak{A}_{L_D}^{\rm dis} = \varnothing$. Also, consider $\Omega := \mathbb{R}^n_+$ and identify

$\partial\Omega \equiv \mathbb{R}^{n-1}$, and $\sigma \equiv \mathcal{L}^{n-1}$. Note that the unit normal to Ω is constant, hence its BMO semi-norm vanishes. In addition, fix an arbitrary integrability exponent $p \in (1, \infty)$. From the description of the space of admissible boundary data for the L^p Dirichlet Problem for the system L_D in the upper half-space, (2.6.19), given in (2.6.20) and the properties of generic double layers (cf. [132, (1.3.24)], and [132, Theorem 1.5.1, items *(i)*, *(iv)*]), we then see that the operator $\frac{1}{2}I + K_A$ cannot be Fredholm (hence, even less so an isomorphism) on the space $\big[L^p(\mathbb{R}^{n-1}, \mathcal{L}^{n-1})\big]^n$ for any coefficient tensor $A \in \mathfrak{A}_{L_D}$. Indeed, $\mathrm{Im}\Big(\frac{1}{2}I + K_A : \big[L^p(\mathbb{R}^{n-1}, \mathcal{L}^{n-1})\big]^n\Big)$ is a subspace of the space of admissible boundary data for the Dirichlet Problem from (2.6.19), and the latter has been shown in Corollary 2.6.5 to have an infinite codimension (in the full space of boundary data $\big[L^p(\mathbb{R}^{n-1}, \mathcal{L}^{n-1})\big]^n$).

This stands in contrast with the invertibility result from (4.2.5), and the only hypothesis in Theorem 4.2.1 which presently fails to materialize is the existence of a distinguished coefficient tensor for L_D.

For similar reasons, the fact that $A \in \mathfrak{A}_L^{\mathrm{dis}}$ is an essential assumption as far as the invertibility result from (4.2.6) is concerned (see Theorem 2.6.7 with $k := 1$ in this regard), and also for the invertibility results pertaining to the modified single layer recorded in item *(11)* of Theorem 4.2.1 (see Theorem 2.6.9 and Remark 2.6.11 in this respect). Here is an example which illustrates the rather dramatic failure of the latter operator to be invertible in the absence of distinguished coefficient tensors. Specifically, work in the two dimensional setting and identify $\mathbb{R}^2 \equiv \mathbb{C}$. Bring in Bitsadze's operator $L := \partial_{\bar{z}}^2$ and recall that this is a weakly elliptic operator whose (tempered distributions) fundamental solutions are of the form

$$E(z) = \frac{\bar{z}}{\pi z} + c \text{ for some } c \in \mathbb{C}. \tag{4.2.60}$$

Then the modified boundary-to-boundary single layer operator associated as in (A.0.232) with the L and the set $\Omega := \mathbb{C}_+$ presently becomes

$$S_{\mathrm{mod}} f(x) := \int_{\mathbb{R}} \{E(x-y) - E_*(-y)\} f(y)\,\mathrm{d}y \text{ at } \mathcal{L}^1\text{-a.e. } x \in \partial\mathbb{C}_+ \equiv \mathbb{R},$$
$$\text{for each } f \in L^1\big(\mathbb{R}, \tfrac{\mathrm{d}x}{1+|x|}\big), \text{ where } E_* := E \cdot \mathbf{1}_{\mathbb{C}\setminus B(0,1)}, \tag{4.2.61}$$

and since any E as in (4.2.60) reduces to a constant on $\mathbb{R} \setminus \{0\}$, we conclude from (4.2.61) that $\big[S_{\mathrm{mod}}\big]$ is actually the zero operator in the present scenario. Obviously, this is the antithesis of the invertibility results claimed in item *(11)* of Theorem 4.2.1, and the only hypothesis in Theorem 4.2.1 that is currently violated is the fact that Bitsadze's operator $L = \partial_{\bar{z}}^2$ is lacking a distinguished coefficient tensor (cf. the discussion in Example 1.6.14).

Finally, we make the following remark.

Remark 4.2.2 *A more precise formulation of the invertibility result corresponding to* (4.2.1) *in Theorem 4.2.1 goes as follows: Fix $n, M \in \mathbb{N}$ with $n \geq 2$ and consider a weakly elliptic homogeneous constant complex coefficient second-order $M \times M$*

system L in $\mathbb{R}^n$ with $\mathfrak{A}_L^{\rm dis} \neq \varnothing$. Then for each constants $C_A, C_W \in (0, \infty)$, each compact interval $I \subset (1, \infty)$, each spectral parameter $\varepsilon \in (0, \infty)$, and each coefficient tensor $A \in \mathfrak{A}_L^{\rm dis}$ there exists a threshold $\delta \in (0, 1)$ which depends only on n, C_A, C_W, I, ε, and A with the following significance. Assume $\Omega \subseteq \mathbb{R}^n$ is an Ahlfors regular domain such that the Ahlfors regularity constants of $\partial\Omega$ are controlled by C_A. Abbreviate $\sigma := \mathcal{H}^{n-1}\lfloor\partial\Omega$ and denote by ν the geometric measure theoretic outward unit normal to Ω. Also, fix an integrability exponent $p \in I$ and a Muckenhoupt weight $w \in A_p(\partial\Omega, \sigma)$ with $[w]_{A_p} \leq C_W$. Finally, consider the boundary-to-boundary double layer potential operator K_A, associated with the set Ω and the coefficient tensor A as in (A.0.116). *Then whenever $\|\nu\|_{[\mathrm{BMO}(\partial\Omega,\sigma)]^n} < \delta$ it follows that Ω is a* UR *domain and for each $z \in \mathbb{C}$ with $|z| \geq \varepsilon$ the operator $zI + K_A$ is invertible on $\big[L^p(\partial\Omega, w)\big]^M$.*

In fact, similar considerations apply to all invertibility results formulated in Theorem 4.2.1.

This is seen from Comment 6 following the statement of Theorem 4.1.3, and the proof of Theorem 4.2.1 presented below.

Proof of Theorem 4.2.1 Work under the assumption that $\|\nu\|_{[\mathrm{BMO}(\partial\Omega,\sigma)]^n} < \delta$ for some $\delta \in (0, \delta_*)$, where $\delta_* \in (0, 1)$ is as in Theorem 3.1.5. This guarantees that Ω is a two-sided NTA domain in the sense of [129, Definition 5.11.1] (hence, in particular, Ω satisfies a two-sided local John condition), $\partial\Omega$ is unbounded, the sets

$$\Omega_+ := \Omega \text{ and } \Omega_- := \mathbb{R}^n \setminus \overline{\Omega} \text{ are connected,} \tag{4.2.62}$$

and

$$\begin{array}{l}\text{the UR constants of } \partial\Omega \text{ may be controlled in terms of the} \\ \text{dimension } n \text{ and the Ahlfors regularity constants of } \partial\Omega.\end{array} \tag{4.2.63}$$

Consider next the claims made in item *(1)*. From (4.2.63), [131, (2.3.56)], (4.1.213), and (4.1.15) we see that there exists $\delta_0 \in (0, 1)$ which depends only on the n, p, $[w]_{A_p}$, and the Ahlfors regularity constants of $\partial\Omega$ such that if $\|\nu\|_{[\mathrm{BMO}(\partial\Omega,\sigma)]^n} < \delta$ with $\delta \in (0, \delta_0)$ then

$$\|K_A\|_{[L^p(\partial\Omega,w)]^M \to [L^p(\partial\Omega,w)]^M} < \varepsilon. \tag{4.2.64}$$

With this in hand, the claim pertaining to (4.2.1) becomes a consequence of [130, Lemma 1.2.12]. The operators (4.2.2)-(4.2.4) are handled in a similar manner, making use of (4.1.214)-(4.1.216).

The very last claim in item *(1)*, pertaining to invertibility results for the operators $zI + K_A$, $zI + K_A^\#$ acting on Morrey spaces, vanishing Morrey spaces, and block spaces on $\partial\Omega$, is justified in a similar fashion, making use of [131, Theorem 2.6.1], and (4.1.35)-(4.1.40).

As regards item *(2)*, observe that (4.2.63) together with [132, (2.1.4)], (4.1.235), and (4.1.15) imply that there exists $\delta_0 \in (0, 1)$ which depends only on the dimension n and the Ahlfors regularity constants of $\partial\Omega$ such that if $\|\nu\|_{[\mathrm{BMO}(\partial\Omega,\sigma)]^n} < \delta$ with $\delta \in (0, \delta_0)$ then

$$\|K^{\#}\|_{[H^p(\partial\Omega,\sigma)]^M \to [H^p(\partial\Omega,\sigma)]^M} < \varepsilon. \tag{4.2.65}$$

Granted this, the claim pertaining to (4.2.9) becomes a consequence of [130, Lemma 1.2.12]. Items *(3)* and *(4)* are handled similarly using (4.1.236)-(4.1.238) and (4.1.239)-(4.1.240), respectively. Likewise, the claims in items *(5)* and *(6)* may be justified based on (4.1.241)-(4.1.243).

Consider next the claim made in item *(7)*. To get started, recall from [132, (2.3.10)] that the operator S maps $\big[H^p(\partial\Omega,\sigma)\big]^M$ boundedly into $\big[\dot{H}^p_1(\partial\Omega,\sigma)\big]^M$. To show that the operator (4.2.18) is injective, suppose $f \in \big[H^p(\partial\Omega,\sigma)\big]^M$ is such that Sf is a constant on $\partial\Omega$. Pick a coefficient tensor $A \in \mathfrak{A}^{\rm dis}_L$ such that $A^\top \in \mathfrak{A}^{\rm dis}_{L^\top}$. That this is possible is seen from assumptions and Theorem 1.5.5. Since, as seen from [132, (1.8.10), (2.3.92)], and [130, Definition 10.2.18], the operator $\partial^A_\nu \mathcal{D}_{\rm mod}$ annihilates constants, we conclude from [132, (2.3.193)] that

$$\big(\tfrac{1}{2}I + K^{\#}_{A^\top}\big)\Big(\big(-\tfrac{1}{2}I + K^{\#}_{A^\top}\big)f\Big) = \big(\partial^A_\nu \mathcal{D}_{\rm mod}\big)(Sf) = 0. \tag{4.2.66}$$

From (4.2.66) and the invertibility of (4.2.9) (when $z = \pm 1/2$ and with $A^\top$ in place of A) we then conclude that $f = 0$. This proves the injectivity of the operator (4.2.18).

To show that the operator (4.2.18) is surjective, let us pick an arbitrary function $f \in \big[\dot{H}^p_1(\partial\Omega,\sigma)\big]^M$ and define

$$u_\pm := \mathcal{D}_{\rm mod} f \text{ in } \Omega_\pm, \tag{4.2.67}$$

where $\mathcal{D}_{\rm mod}$ are the modified boundary-to-domain double layer potential operators associated with L and $\Omega_\pm$ (cf. (4.2.62)) as in (A.0.73). Then, having fixed an aperture parameter $\kappa > 0$, from items [132, Theorem 2.3.1, items *(i)-(ii)*] we see that

$$\begin{gathered} u_\pm \in \big[\mathscr{C}^\infty(\Omega_\pm)\big]^M, \quad Lu_\pm = 0 \text{ in } \Omega_\pm, \\ \mathcal{N}_\kappa(\nabla u_\pm) \in L^p(\partial\Omega,\sigma), \quad \partial^A_\nu u_\pm \in \big[H^p(\partial\Omega,\sigma)\big]^M. \end{gathered} \tag{4.2.68}$$

Granted these properties, [132, Theorem 2.2.7, items *(a)-(c)*] guarantee that there exist two constants $b_\pm \in \mathbb{C}^M$ such that

$$\begin{gathered} u_\pm\Big|^{\kappa-\text{n.t.}}_{\partial\Omega} - b_\pm \text{ belongs to } \big[L^{p^*}(\partial\Omega,\sigma)\big]^M \text{ where } p^* := \big(\tfrac{1}{p} - \tfrac{1}{n-1}\big)^{-1} \in (1,\infty), \\ \text{if either } n \geq 3, \text{ or } n = 2 \text{ and } p \in \big(\tfrac{1}{2}, 1\big), \end{gathered} \tag{4.2.69}$$

while

$$u_\pm\Big|^{\kappa-\text{n.t.}}_{\partial\Omega} \in \big[\text{BMO}(\partial\Omega,\sigma)\big]^M \text{ if } n = 2 \text{ and } p = 1. \tag{4.2.70}$$

To proceed, define

$$w_\pm := u_\pm - \mathscr{S}\Big(\big(\mp\tfrac{1}{2}I + K^{\#}_{A^\top}\big)^{-1}\big(\partial^A_\nu u_\pm\big)\Big) \text{ in } \Omega_\pm, \tag{4.2.71}$$

where the existence of the inverse operators is guaranteed by (4.2.9) (used with $z = \pm 1/2$ and $A^\top$ in place of A). Then (4.2.68) and [132, Theorem 2.2.3] imply

$$\begin{aligned} &w_\pm \in \big[\mathscr{C}^\infty(\Omega_\pm)\big]^M, \quad Lw_\pm = 0 \text{ in } \Omega_\pm, \\ &\mathcal{N}_\kappa(\nabla w_\pm) \in L^p(\partial\Omega, \sigma), \quad \overset{\bullet}{\partial}{}^A_\nu w_\pm = 0. \end{aligned} \tag{4.2.72}$$

Also, from (4.2.69)-(4.2.70), and [132, (2.2.124), (2.2.127), (2.2.130), (2.2.132), (2.2.133)], we see that there exist two constants $b_\pm \in \mathbb{C}^M$ with the property that

$$\begin{gathered} w_\pm\big|_{\partial\Omega}^{\kappa-\text{n.t.}} - b_\pm \text{ belongs to } \big[L^{p^*}(\partial\Omega, \sigma)\big]^M \text{ where } p^* := \big(\tfrac{1}{p} - \tfrac{1}{n-1}\big)^{-1} \in (1, \infty), \\ \text{if either } n \geq 3, \text{ or } n = 2 \text{ and } p \in \big(\tfrac{1}{2}, 1\big), \end{gathered} \tag{4.2.73}$$

while

$$w_\pm\big|_{\partial\Omega}^{\kappa-\text{n.t.}} \in \big[\text{BMO}(\partial\Omega, \sigma)\big]^M \text{ if } n = 2 \text{ and } p = 1. \tag{4.2.74}$$

From (4.2.72)-(4.2.73) and the integral representation formula from [132, Theorem 2.2.7] we then conclude that there exist two constants, say $c_\pm \in \mathbb{C}^M$, with the property that

$$w_\pm = \pm\mathcal{D}_{\text{mod}}\Big(w_\pm\big|_{\partial\Omega}^{\kappa-\text{n.t.}}\Big) + c_\pm \text{ in } \Omega_\pm. \tag{4.2.75}$$

Recall from definitions and [129, (7.4.118)] that

$$\big[L^{p^*}(\partial\Omega, \sigma)\big]^M + \mathbb{C}^M \hookrightarrow \Big[L^1\Big(\partial\Omega, \frac{\sigma(x)}{1 + |x|^n}\Big)\Big]^M, \tag{4.2.76}$$

$$\big[\text{BMO}(\partial\Omega, \sigma)\big]^M \hookrightarrow \Big[L^1\Big(\partial\Omega, \frac{\sigma(x)}{1 + |x|^n}\Big)\Big]^M. \tag{4.2.77}$$

After taking nontangential boundary traces, on account of [132, (1.8.27)] while bearing in mind (4.2.76)-(4.2.77), we arrive at

$$\big(\mp\tfrac{1}{2}I + K_{\text{mod}}\big)\Big(w_\pm\big|_{\partial\Omega}^{\kappa-\text{n.t.}}\Big) = \mp c_\pm \text{ at } \sigma\text{-a.e. point on } \partial\Omega. \tag{4.2.78}$$

Assume that either $n \geq 3$, or $n = 2$ and $p \in \big(\tfrac{1}{2}, 1\big)$. Then from (4.2.73) and [132, (1.8.28)] we conclude that

$$\big(\mp\tfrac{1}{2}I + K_{\text{mod}}\big)\Big(w_\pm\big|_{\partial\Omega}^{\kappa-\text{n.t.}} - b_\pm\Big) \text{ are constants on } \partial\Omega, \tag{4.2.79}$$

which, in concert with (4.2.73) and [132, (1.8.26)], implies

$$\big(\mp\tfrac{1}{2}I + K\big)\Big(w_\pm\big|_{\partial\Omega}^{\kappa-\text{n.t.}} - b_\pm\Big) \text{ are constants on } \partial\Omega. \tag{4.2.80}$$

Given that $\partial\Omega$ is unbounded, and that $\big(\mp\tfrac{1}{2}I + K\big)\Big(w_\pm\big|_{\partial\Omega}^{\kappa-\text{n.t.}} - b_\pm\Big)$ belong to $\big[L^{p^*}(\partial\Omega, \sigma)\big]^M$ thanks to (4.2.73) and [132, Theorem 1.5.1, item *(iii)*], this forces

$$\big(\mp\tfrac{1}{2}I + K\big)\Big(w_\pm\big|_{\partial\Omega}^{\kappa-\text{n.t.}} - b_\pm\Big) = 0 \text{ on } \partial\Omega \tag{4.2.81}$$

hence, ultimately,

$$w_\pm\Big|_{\partial\Omega}^{\kappa-\mathrm{n.t.}} = b_\pm \tag{4.2.82}$$

by (4.2.73) and (4.2.5) (with $z = \pm 1/2$ and p replaced by p^*).

In the remaining case, when $n = 2$ and $p = 1$, from (4.2.78) we see that

$$\big(\mp\tfrac{1}{2}I + \big[K_{\rm mod}\big]\big)\Big[w_\pm\Big|_{\partial\Omega}^{\kappa-\mathrm{n.t.}}\Big] = 0 \ \text{ in } \big[\mathrm{BMO}(\partial\Omega,\sigma)/\!\sim\big]^M. \tag{4.2.83}$$

Together with (4.2.10) this proves that $\Big[w_\pm\Big|_{\partial\Omega}^{\kappa-\mathrm{n.t.}}\Big] = 0$ in $\big[\mathrm{BMO}(\partial\Omega,\sigma)/\!\sim\big]^M$, hence once again (4.2.82) holds for some constants $b_\pm \in \mathbb{C}^M$.

Thus, (4.2.82) is valid in all cases. When used back in (4.2.75), this and [132, (1.8.10)] lead to the conclusion that there exist two constants, say $a_\pm \in \mathbb{C}^M$, with the property that $w_\pm \equiv a_\pm$ in $\Omega_\pm$. From this, (4.2.71), and (4.2.67) we then see that

$$\mathcal{D}_{\rm mod} f = \mathscr{S}\Big(\big(\mp\tfrac{1}{2}I + K^{\#}_{A^\top}\big)^{-1}(\partial^{A}_{\nu}u_\pm)\Big) + a_\pm \ \text{ in } \Omega_\pm. \tag{4.2.84}$$

Taking nontangential boundary traces then yields, in view of (A.0.99), [132, (1.8.27)], and [132, (2.2.127), (2.2.132)],

$$\big(\pm\tfrac{1}{2}I + K_{\rm mod}\big)f = S\Big(\big(\pm\tfrac{1}{2}I + K^{\#}_{A^\top}\big)^{-1}(\partial^{A}_{\nu}u_\pm)\Big) + a_\pm \tag{4.2.85}$$

at σ-a.e. point on $\partial\Omega$. Hence,

$$\begin{aligned} f &= \big(\tfrac{1}{2}I + K_{\rm mod}\big)f - \big(-\tfrac{1}{2}I + K_{\rm mod}\big)f \\ &= S\Big(\big(-\tfrac{1}{2}I + K^{\#}_{A^\top}\big)^{-1}(\partial^{A}_{\nu}u_+)\Big) + a_+ - S\Big(\big(\tfrac{1}{2}I + K^{\#}_{A^\top}\big)^{-1}(\partial^{A}_{\nu}u_-)\Big) - a_- \end{aligned} \tag{4.2.86}$$

which proves that for $a := a_- - a_+ \in \mathbb{C}^M$ and

$$g := \big(-\tfrac{1}{2}I + K^{\#}_{A^\top}\big)^{-1}(\partial^{A}_{\nu}u_+) - \big(\tfrac{1}{2}I + K^{\#}_{A^\top}\big)^{-1}(\partial^{A}_{\nu}u_-) \in \big[H^p(\partial\Omega,\sigma)\big]^M \tag{4.2.87}$$

we have $Sg = f + a$. Ultimately, this establishes the fact that the operator (4.2.18) is surjective. This concludes the justification of the claim in item *(7)*.

To deal with the claim in item *(8)* start by observing that if $A \in \mathfrak{A}^{\rm dis}_{L}$ and $\mathfrak{A}^{\rm dis}_{L^\top} \neq \varnothing$ then $A^\top \in \mathfrak{A}^{\rm dis}_{L^\top}$ by Theorem 1.5.5. Next, [132, (2.3.192)] and the connectivity property in (4.2.62) imply

$$[S]\big(K^{\#}_{A^\top}f\big) = \big[K_{\rm mod}\big]\big([S]f\big) \ \text{ for each } f \in \big[H^p(\partial\Omega,\sigma)\big]^M. \tag{4.2.88}$$

Hence, we may express

$$zI + \big[K_{\rm mod}\big] = [S]\big(zI + K^{\#}_{A^\top}\big)[S]^{-1} \ \text{ on } \big[\dot{H}^p_1(\partial\Omega,\sigma)/\sim\big]^M, \tag{4.2.89}$$

where $[S]^{-1}$ is the inverse of (4.2.18). From (4.2.89), the invertibility of (4.2.18) (written with $A^\top$ in place of A), and the invertibility of (4.2.9) we then conclude that the operator $zI + \big[K_{\rm mod}\big]$ is invertible in the context of (4.2.19).

Pressing on, to deal with the claim in item *(9)* we first note that if $A \in \mathfrak{A}_L^{\rm dis}$ and $\mathfrak{A}_{L^\top}^{\rm dis} \neq \varnothing$ then $A^\top \in \mathfrak{A}_{L^\top}^{\rm dis}$ by Theorem 1.5.5. Next, observe that from [132, (2.3.190)] (and the connectivity property in (4.2.62)) we have

$$\begin{aligned} &\big(\tfrac{1}{2}I + \big[K_{\rm mod}\big]\big) \circ \big(-\tfrac{1}{2}I + \big[K_{\rm mod}\big]\big) = [S] \circ \big[\dot{\partial}_\nu^A \mathcal{D}_{\rm mod}\big] \\ &\qquad \text{as operators on } \big[\dot{H}_1^p(\partial\Omega, \sigma)/\sim\big]^M, \end{aligned} \tag{4.2.90}$$

while [132, (2.3.193)] implies

$$\begin{aligned} &\big(\tfrac{1}{2}I + K_{A^\top}^\#\big) \circ \big(-\tfrac{1}{2}I + K_{A^\top}^\#\big) = \big[\dot{\partial}_\nu^A \mathcal{D}_{\rm mod}\big] \circ [S] \\ &\qquad \text{as operators on } \big[H^p(\partial\Omega, \sigma)\big]^M. \end{aligned} \tag{4.2.91}$$

In concert, (4.2.90), the invertibility result in (4.2.19) (for $z = \pm 1/2$), and the invertibility of (4.2.18) guarantee that the operator $\big[\dot{\partial}_\nu^A \mathcal{D}_{\rm mod}\big]$ is injective in the context of (4.2.20). In addition, from (4.2.91), the invertibility result in (4.2.18) (written with $z = \pm 1/2$ and with $A^\top$ in place of A), and the invertibility of (4.2.18) we see that the operator $\big[\dot{\partial}_\nu^A \mathcal{D}_{\rm mod}\big]$ is surjective in the context of (4.2.20). Hence, (4.2.20) is an invertible operator.

Moving on, the first claim in item *(10)* to the effect that having the inequality $\|\nu\|_{[\mathrm{BMO}(\partial\Omega,\sigma)]^n} < \delta$ with δ sufficiently small guarantees that Ω has an unbounded boundary and satisfies a two-sided local John condition has been clarified at the beginning of the proof (where it has been noted that such a choice makes Ω a two-sided NTA domain with an unbounded boundary). The invertibility properties of the operators (4.2.21)-(4.2.24) are consequences of the estimates in (4.1.224)-(4.1.227) and the completeness results established in [130, Proposition 11.5.14], [130, Propositions 11.13.10 and 11.13.12], and the discussion around [130, (11.13.89)] (cf. also the last part in [130, Lemma 1.2.12]).

Consider next the claims made in item *(11)*. For starters, Theorem 1.5.5 guarantees that there exists some $A \in \mathfrak{A}_L$ such that $\mathfrak{A}_L^{\rm dis} = \{A\}$ and $\mathfrak{A}_{L^\top}^{\rm dis} = \{A^\top\}$. To deal with (4.2.25), fix some $p \in (1, \infty)$ and assume $\|\nu\|_{[\mathrm{BMO}(\partial\Omega,\sigma)]^n} < \delta$ for some $\delta \in (0, 1)$, depending only on n, p, A, and the ADR constants of $\partial\Omega$, small enough so that

$$\begin{aligned} &\Omega_\pm \text{ are connected NTA domains,} \\ &\text{with an unbounded Ahlfors regular boundary,} \end{aligned} \tag{4.2.92}$$

(see the discussion pertaining to (4.2.62)), and the following operators are invertible:

$$\pm\tfrac{1}{2}I + K_{A^\top}^\# : \big[L^p(\partial\Omega, \sigma)\big]^M \longrightarrow \big[L^p(\partial\Omega, \sigma)\big]^M, \tag{4.2.93}$$

$$\pm\tfrac{1}{2}I + \big[K_{\rm mod}\big] : \big[\dot{L}_1^p(\partial\Omega, \sigma)/\sim\big]^M \longrightarrow \big[\dot{L}_1^p(\partial\Omega, \sigma)/\sim\big]^M, \tag{4.2.94}$$

(see (4.2.7), (4.2.21)). To show that the operator (4.2.25) is injective, suppose the function $f \in \big[L^p(\partial\Omega, \sigma)\big]^M$ is such that $S_{\rm mod} f = c \in \mathbb{C}^M$. Then [132, Theorem 1.8.26, item *(1)*] implies (cf. [132, (1.8.251)])

$$\big(\tfrac{1}{2}I + K^{\#}_{A^{\top}}\big)\Big(\big(-\tfrac{1}{2}I + K^{\#}_{A^{\top}}\big)f\Big) = \big(\partial^A_\nu \mathcal{D}_{\rm mod}\big)\big(\mathcal{S}_{\rm mod} f\big) = \big(\partial^A_\nu \mathcal{D}_{\rm mod}\big)c = 0, \tag{4.2.95}$$

with the last equality a consequence of [132, (1.8.145)-(1.8.146)]. From this and (4.2.93) we then conclude that $f = 0$, as wanted.

To show that the operator (4.2.25) is surjective, let us pick an arbitrary function $g \in \big[\dot{L}^p_1(\partial\Omega,\sigma)\big]^M$. From (4.2.94) we know that there exist $h_\pm \in \big[\dot{L}^p_1(\partial\Omega,\sigma)\big]^M$ and $c_\pm \in \mathbb{C}^M$ with the property that $\big(\pm\tfrac{1}{2}I + K_{\rm mod}\big)h_\pm = g + c_\pm$. Use these to define $u_\pm := \pm\mathcal{D}_{\rm mod} h_\pm$ in $\Omega_\pm$, where $\mathcal{D}_{\rm mod}$ are the modified boundary-to-domain double layer potential operators associated with A and $\Omega_\pm$ (cf. (4.2.62)) as in (A.0.73). Then, having fixed an aperture parameter $\kappa > 0$, from [132, Theorem 1.8.2] we conclude that

$$\begin{gathered} u_\pm \in \big[\mathscr{C}^\infty(\Omega_\pm)\big]^M, \quad Lu_\pm = 0 \text{ in } \Omega_\pm, \\ u_\pm\big|^{\kappa-{\rm n.t.}}_{\partial\Omega} = \big(\pm\tfrac{1}{2}I + K_{\rm mod}\big)h_\pm = g + c_\pm, \\ \mathcal{N}_\kappa(\nabla u_\pm) \in L^p(\partial\Omega,\sigma), \quad \partial^A_{\nu_\pm} u_\pm \in \big[L^p(\partial\Omega,\sigma)\big]^M, \end{gathered} \tag{4.2.96}$$

where $\nu_\pm$ are the geometric measure theoretic outward unit normals to $\Omega_\pm$. Granted these properties, [132, Theorem 1.8.19] applies and the integral representation formula [132, (1.8.200)] presently gives

$$u_\pm = \mathcal{D}_{\rm mod}\big(u_\pm\big|^{\kappa-{\rm n.t.}}_{\partial\Omega}\big) - \mathscr{S}_{\rm mod}\big(\partial^A_{\nu_\pm} u_\pm\big) + \widetilde{c}_\pm \text{ in } \Omega_\pm \tag{4.2.97}$$

for some constants $\widetilde{c}_\pm \in \mathbb{C}^M$. Going nontangentially to the boundary in (4.2.97) then yields, on account of (4.2.96) and [132, (1.5.80)],

$$g + c_\pm = \big(\tfrac{1}{2}I \pm K_{\rm mod}\big)(g + c_\pm) - S_{\rm mod}\big(\partial^A_{\nu_\pm} u_\pm\big) + \widetilde{c}_\pm. \tag{4.2.98}$$

Adding the two versions, corresponding to the signs + and −, of (4.2.98) produces

$$\begin{aligned} 2g + c_+ + c_- = g &+ \big(\tfrac{1}{2}I + K_{\rm mod}\big)c_+ + \big(\tfrac{1}{2}I - K_{\rm mod}\big)c_- \\ &- S_{\rm mod}\big(\partial^A_{\nu_+} u_+ + \partial^A_{\nu_-} u_-\big) + \widetilde{c}_+ + \widetilde{c}_-. \end{aligned} \tag{4.2.99}$$

Abbreviate

$$f := -\big(\partial^A_{\nu_+} u_+ + \partial^A_{\nu_-} u_-\big) \in \big[L^p(\partial\Omega,\sigma)\big]^M. \tag{4.2.100}$$

Since $K_{\rm mod}$ maps constants to constants (cf. [132, (1.8.28)]), from (4.2.99)-(4.2.100) we conclude that there exists some $c \in \mathbb{C}^M$ such that $S_{\rm mod} f = g + c$. Hence, the operator (4.2.25) is indeed surjective, so ultimately an isomorphism.

The claims about (4.2.26)-(4.2.28) being isomorphisms are dealt with in a similar fashion, now relying on the very last part of the current item *(1)*, (4.2.22)-(4.2.24), and [132, Theorem 3.3.10].

Going further, the claims made in item *(12)* are consequences of the operator identities established in [132, Theorem 1.8.26] and [132, Theorem 3.3.12], together with the invertibility results from the current items *(1)* and *(11)*. The claims in item *(13)* are direct consequences of Theorem 3.1.5, (4.2.63), Proposition 1.3.2, and Theorem 4.1.16. As far as item *(14)* is concerned, that $\mathbb{X}_1$ and $(\mathbb{X}')_1$ are Banach spaces when equipped with the norms described in (4.2.43) follows the same pattern as the proof of item *(i)* in [130, Proposition 11.1.9], keeping in mind the embeddings recorded in [130, (2.5.90)] and the very last property in [130, Proposition 5.1.8]. Having established this, we may then conclude that $\mathring{\mathbb{X}}_1$ and $(\mathbb{X}')^\circ_1$ are indeed closed subspaces of $\mathbb{X}_1$ and $(\mathbb{X}')_1$, respectively (since, by design, $\mathring{\mathbb{X}}$ and $(\mathbb{X}')^\circ$ are closed subspaces of $\mathbb{X}$ and $\mathbb{X}'$, respectively). Next, all desired invertibility results in item *(14)* are seen from a Neumann series argument, relying on estimate (4.1.357) from Theorem 4.1.24, estimate (4.1.49) from Theorem 4.1.3, the intertwining identity [132, (5.2.167)], the embeddings from [130, (2.5.90) in Proposition 5.2.7], the commutator estimates from [130, Corollary 2.7.9], and the fact that $\mathbb{X}$, $\mathring{\mathbb{X}}$, $\mathbb{X}'$, $(\mathbb{X}')^\circ$ are all stable under multiplication by essentially bounded functions (cf. [130, (5.1.13)], [130, (5.2.85)]). Here, Proposition 1.3.2 is also relevant (given the current working assumptions).

Turning our attention to item *(15)*, the fact that whenever Ω is a two-sided NTA domain with an unbounded boundary it follows that the embedding in (4.2.54) holds and $\dot{\mathbb{X}}_1/\sim$ is a Banach space when equipped with the norm defined in (4.2.55) is implied by [111, Remark 8.3 and Proposition 8.4] together with [130, Proposition 5.2.7]. Consequently, under the assumption that $A \in \mathfrak{A}^{\rm dis}_L$, the invertibility result claimed in (4.2.56) is seen via a Neumann series argument, by once again relying on Proposition 1.3.2, the estimate in Theorem 4.1.3 recorded in (4.1.49), the intertwining identity [132, (5.2.167)], the embeddings from [130, (2.5.90)] in [130, Proposition 5.2.7], the commutator estimates from [130, Corollary 2.7.9], and the fact that $\mathbb{X}$ is stable under multiplication by essentially bounded functions (cf. [130, (5.1.13)]). The operators in (4.2.57)-(4.2.58) are dealt with in a manner similar to the treatment of (4.2.25) and (4.2.29), relying on Theorem 4.1.21, Theorem 4.1.22, and what we have proved so far The final claim in item *(15)* is handled by reasoning analogously.

Finally, the first claim in item *(16)* is clear from [130, Proposition 5.3.14] and [130, (5.3.165)]. This completes the proof of Theorem 4.2.1. □

Recall the family K_λ of principal-value double layer potential operators for the Stokes system, indexed by the parameter $\lambda \in \mathbb{C}$, recalled in (A.0.120). Odqvist and Lichtenstein have used the choice $\lambda = 1$ to produce an operator which is only weakly singular on smooth domains (see Chapter 3 of Ladyzhenskaya's book [106]). Here the goal is to establish invertibility results involving this specific double layer operator associated with the Stokes system, on a variety of function spaces, in the geometric setting described below.

Theorem 4.2.3 *Assume that $\Omega \subseteq \mathbb{R}^n$, for $n \in \mathbb{N}$, is an Ahlfors regular domain. Abbreviate $\sigma := \mathcal{H}^{n-1}\lfloor\partial\Omega$ and denote by $\nu = (\nu_j)_{1\leq j\leq n}$ the geometric measure theoretic outward unit normal to Ω. Denote by $K_{\rm slip}$ the singular integral operator corresponding to K_λ in* (A.0.120) *with $\lambda := 1$, i.e., $K_{\rm slip}$ acts on each vector-valued function*

$$\vec{f} \in \Big[L^1\Big(\partial\Omega, \frac{\sigma(x)}{1+|x|^{n-1}}\Big)\Big]^n \tag{4.2.101}$$

at σ-a.e. $x \in \partial\Omega$ according to

$$K_{\text{slip}}\vec{f}(x) = \lim_{\varepsilon\to 0^+} \int_{\substack{y\in\partial\Omega \\ |x-y|>\varepsilon}} \frac{n\langle y-x, \nu(y)\rangle\langle x-y, \vec{f}(y)\rangle(x-y)}{\omega_{n-1}|x-y|^{n+2}}\, d\sigma(y) \tag{4.2.102}$$

and also consider the version of $K_\lambda^\#$ from (A.0.121) *for $\lambda := 1$, i.e., define for each vector-valued function $\vec{f}$ as in* (4.2.101) *and σ-a.e. point $x \in \partial\Omega$*

$$K^\#_{\text{slip}}\vec{f}(x) = \lim_{\varepsilon\to 0^+} \int_{\substack{y\in\partial\Omega \\ |x-y|>\varepsilon}} \frac{n\langle x-y, \nu(x)\rangle\langle x-y, \vec{f}(y)\rangle(x-y)}{\omega_{n-1}|x-y|^{n+2}}\, d\sigma(y). \tag{4.2.103}$$

Next, consider the following modified version of K_{slip} from (4.2.101)-(4.2.102),

$$\vec{f} \in \Big[L^1\Big(\partial\Omega, \frac{\sigma(x)}{1+|x|^n}\Big)\Big]^n \tag{4.2.104}$$

at σ-a.e. $x \in \partial\Omega$ according to (compare with (A.0.240)*)*

$$\begin{gathered}(K^{\text{mod}}_{\text{slip}}\vec{f})(x) := \sum_{j=1}^n \lim_{\varepsilon\to 0^+} \int_{\partial\Omega} \Big\{k_j^{(\varepsilon)}(x-y) - k_j^{(1)}(-y)\Big\}\nu_j(y)\vec{f}(y)\, d\sigma(y), \\ \text{where } k_j(z) := -\frac{nz_j(z\otimes z)}{\omega_{n-1}|z|^{n+2}} \text{ for all } z \in \mathbb{R}^n \setminus \{0\} \text{ and} \\ k_j^{(\varepsilon)} := k_j \cdot \mathbf{1}_{\mathbb{R}^n\setminus\overline{B(0,\varepsilon)}} \text{ for each } \varepsilon > 0, \text{ for any } j \in \{1,\dots,n\}.\end{gathered} \tag{4.2.105}$$

Going further, denote by $\mathcal{D}_{\text{slip}}$ the boundary-to-domain double layer operator corresponding to $\mathcal{D}_\lambda$ from (A.0.75) *with $\lambda := 1$, i.e., $\mathcal{D}_{\text{slip}}$ acts on each vector-valued function $\vec{f}$ as in* (4.2.104) *according to (see* (A.0.120)*)*

$$\mathcal{D}_{\text{slip}}\vec{f}(x) = \int_{\partial\Omega} \frac{n\langle y-x, \nu(y)\rangle\langle x-y, \vec{f}(y)\rangle(x-y)}{\omega_{n-1}|x-y|^{n+2}}\, d\sigma(y) \text{ for all } x \in \Omega, \tag{4.2.106}$$

and define $\mathcal{D}^{\text{mod}}_{\text{slip}}$ to be its modified version (cf. (A.0.241)*), acting on each vector-valued function $\vec{f}$ as in* (4.2.104) *at each point $x \in \Omega$ according to*

$$\begin{gathered}\mathcal{D}^{\text{mod}}_{\text{slip}}\vec{f}(x) = \sum_{j=1}^n \int_{\partial\Omega} \Big\{k_j(x-y) - k_j^{(1)}(-y)\Big\}\nu_j(y)\vec{f}(y)\, d\sigma(y), \\ \text{where } k_j(z) := -\frac{nz_j(z\otimes z)}{\omega_{n-1}|z|^{n+2}} \text{ for all } z \in \mathbb{R}^n \setminus \{0\} \text{ and} \\ k_j^{(1)} := k_j \cdot \mathbf{1}_{\mathbb{R}^n\setminus\overline{B(0,1)}} \text{ for each } \varepsilon > 0, \text{ for any } j \in \{1,\dots,n\}.\end{gathered} \tag{4.2.107}$$

Furthermore, define $\mathcal{P}_{\text{slip}}$ *to be the (double layer) pressure potential* $\mathcal{P}_\lambda$ *from* (A.0.211) *corresponding to* $\lambda := 1$, *i.e., define the action of* $\mathcal{P}_{\text{slip}}$ *on functions* $\vec{f}$ *as in* (4.2.104) *to be*

$$\begin{aligned}\mathcal{P}_{\text{slip}}\vec{f}(x) &= (-2)\sum_{j=1}^{n}\int_{\partial\Omega} \nu_j(y)\langle(\partial_j\vec{q})(x-y),\vec{f}(y)\rangle\, d\sigma(y) \\ &= (-2)\int_{\partial\Omega}\big\langle \nu(y),(\nabla\vec{q})(x-y)\vec{f}(y)\big\rangle\, d\sigma(y) \ \textit{for}\ x\in\Omega,\end{aligned} \tag{4.2.108}$$

where the pressure vector $\vec{q}$ *is as in* (A.0.212). *Recall the boundary-to-boundary single layer operator for the Stokes system* $S = S_{\text{Stokes}}$ *defined as*

$$S\vec{f}(x) := \int_{\partial\Omega} E(x-y)\vec{f}(y)\, d\sigma(y) \ \textit{ for }\ x\in\partial\Omega, \tag{4.2.109}$$

(where $E = \big(E_{jk}\big)_{1\leq j,k\leq n}$ *is the Kelvin matrix-valued fundamental solution for the Stokes system in* $\mathbb{R}^n$ *from* [132, (6.2.1)]*), and denote its modified version by* S_{mod}, *so*

$$\begin{aligned}&S_{\text{mod}}\vec{f}(x) := \int_{\partial\Omega}\big\{E(x-y)-E_*(-y)\big\}f(y)\, d\sigma(y) \ \textit{ at } \sigma\textit{-a.e. }\ x\in\partial\Omega, \\ &\textit{for each }\ \vec{f}\in\Big[L^1\big(\partial\Omega,\tfrac{\sigma(x)}{1+|x|^{n-1}}\big)\Big]^n, \ \textit{ where }\ E_* := E\cdot \mathbf{1}_{\mathbb{R}^n\setminus B(0,1)}.\end{aligned} \tag{4.2.110}$$

It is also agreed that $\partial_\nu^{\text{slip}}$ *stands for the slip conormal derivative, i.e., the conormal derivative operator* ∂_ν^λ *defined as in* (A.0.208) *with* $\lambda := 1$. *More specifically, for any two functions*

$$\begin{aligned}&\vec{u}\in\big[W^{1,1}_{\text{loc}}(\Omega)\big]^n \ \textit{ such that }\ \big(\nabla\vec{u}\big)\Big|^{\kappa-\text{n.t.}}_{\partial\Omega} \ \textit{ exists } \sigma\textit{-a.e. on } \partial\Omega, \\ &\textit{and }\ \pi\in L^1_{\text{loc}}(\Omega,\mathcal{L}^n) \ \textit{ such that }\ \pi\Big|^{\kappa-\text{n.t.}}_{\partial\Omega} \ \textit{ exists } \sigma\textit{-a.e. on } \partial\Omega,\end{aligned} \tag{4.2.111}$$

(for some aperture parameter $\kappa > 0$*) one sets*

$$\partial_\nu^{\text{slip}}(\vec{u},\pi) = \Big[\nabla\vec{u}+(\nabla\vec{u})^\top\Big]\Big|^{\kappa-\text{n.t.}}_{\partial\Omega}\nu - \Big(\pi\Big|^{\kappa-\text{n.t.}}_{\partial\Omega}\Big)\nu, \ \textit{ at } \sigma\textit{-a.e. point on } \partial\Omega. \tag{4.2.112}$$

Finally, $\overset{\bullet}{\partial}{}_\nu^{\text{slip}}$ *denotes the weak slip conormal derivative as in* [132, Definition 6.2.9] *for the choice* $\lambda := 1$. *Then, in relation to these, the following results are true.*

(1) Given an exponent $p\in(1,\infty)$, *a Muckenhoupt weight* $w\in A_p(\partial\Omega,\sigma)$, *and some number* $\varepsilon\in(0,\infty)$, *there exists a threshold* $\delta\in(0,1)$, *which depends only on n, p,* $[w]_{A_p}$, ε, *and the* ADR *constants of* $\partial\Omega$, *with the property that if* $\|\nu\|_{[\mathrm{BMO}(\partial\Omega,\sigma)]^n} < \delta$ *it follows that* Ω *is a* UR *domain and for each spectral parameter* $z\in\mathbb{C}$ *with* $|z|\geq\varepsilon$ *the following operators are invertible on Muckenhoupt weighted Lebesgue and Muckenhoupt weighted Sobolev spaces:*

$$zI + K_{\text{slip}} : \big[L^p(\partial\Omega, w)\big]^n \longrightarrow \big[L^p(\partial\Omega, w)\big]^n, \tag{4.2.113}$$

$$zI + K_{\text{slip}} : \big[L^p_1(\partial\Omega, w)\big]^n \longrightarrow \big[L^p_1(\partial\Omega, w)\big]^n, \tag{4.2.114}$$

$$zI + K^{\#}_{\text{slip}} : \big[L^p(\partial\Omega, w)\big]^n \longrightarrow \big[L^p(\partial\Omega, w)\big]^n, \tag{4.2.115}$$

$$zI + K^{\#}_{\text{slip}} : \big[L^p_{-1}(\partial\Omega, w)\big]^n \longrightarrow \big[L^p_{-1}(\partial\Omega, w)\big]^n. \tag{4.2.116}$$

As a corollary, corresponding to the particular case when $w \equiv 1$, the following operators are invertible on "plain" Lebesgue and Sobolev spaces:

$$zI + K_{\text{slip}} : \big[L^p(\partial\Omega, \sigma)\big]^n \longrightarrow \big[L^p(\partial\Omega, \sigma)\big]^n, \tag{4.2.117}$$

$$zI + K_{\text{slip}} : \big[L^p_1(\partial\Omega, \sigma)\big]^n \longrightarrow \big[L^p_1(\partial\Omega, \sigma)\big]^n, \tag{4.2.118}$$

$$zI + K^{\#}_{\text{slip}} : \big[L^p(\partial\Omega, \sigma)\big]^n \longrightarrow \big[L^p(\partial\Omega, \sigma)\big]^n, \tag{4.2.119}$$

$$zI + K^{\#}_{\text{slip}} : \big[L^p_{-1}(\partial\Omega, \sigma)\big]^n \longrightarrow \big[L^p_{-1}(\partial\Omega, \sigma)\big]^n. \tag{4.2.120}$$

Furthermore, similar invertibility results are valid for the operators $zI + K_{\text{slip}}$, $zI + K^{\#}_{\text{slip}}$ acting on Morrey spaces, vanishing Morrey spaces, and block spaces on $\partial\Omega$ (cf. [130, §6.2]).

(2) *Given any integrability exponent $p \in \big(\frac{n-1}{n}, 1\big]$ and some number $\varepsilon \in (0, \infty)$, there exists some small threshold $\delta \in (0, 1)$, which depends only on n, p, ε, and the* ADR *constants of $\partial\Omega$, with the property that if $\|\nu\|_{[\text{BMO}(\partial\Omega,\sigma)]^n} < \delta$ (i.e., if Ω is a δ-*AR *domain in the sense of Definition 3.1.8) it follows that Ω is a* UR *domain with unbounded boundary and for each spectral parameter $z \in \mathbb{C}$ with $|z| \geq \varepsilon$ the following operator (cf.* [132, (6.2.250)]) *is invertible:*

$$zI + K^{\#}_{\text{slip}} : \big[H^p(\partial\Omega, \sigma)\big]^n \longrightarrow \big[H^p(\partial\Omega, \sigma)\big]^n, \tag{4.2.121}$$

where $H^p(\partial\Omega, \sigma)$ is the (Lebesgue-based) Hardy space from (A.0.89)-(A.0.90) *used with $\Sigma := \partial\Omega$. Moreover, a similar invertibility result holds for the operator $zI + K^{\#}_{\text{slip}}$ acting on the Lorentz-based Hardy spaces introduced in* [130, Definition 4.2.3] (*cf.* [132, (6.2.251)]).

(3) *For any $\varepsilon \in (0, \infty)$ there exists some small threshold $\delta \in (0, 1)$, which depends only on n, ε, and the* ADR *constants of $\partial\Omega$, with the property that if $\|\nu\|_{[\text{BMO}(\partial\Omega,\sigma)]^n} < \delta$ (i.e., if Ω is a δ-*AR *domain in the sense of Definition 3.1.8) it follows that Ω is a* UR *domain with unbounded boundary and for each spectral parameter $z \in \mathbb{C}$ with $|z| \geq \varepsilon$ the following operators are invertible:*

$$zI + \big[K^{\text{mod}}_{\text{slip}}\big] : \big[\text{BMO}(\partial\Omega, \sigma)/\sim\big]^n \longrightarrow \big[\text{BMO}(\partial\Omega, \sigma)/\sim\big]^n, \tag{4.2.122}$$

$$zI + \big[K^{\text{mod}}_{\text{slip}}\big] : \big[\text{VMO}(\partial\Omega, \sigma)/\sim\big]^n \longrightarrow \big[\text{VMO}(\partial\Omega, \sigma)/\sim\big]^n, \tag{4.2.123}$$

$$zI + \big[K^{\text{mod}}_{\text{slip}}\big] : \big[\text{CMO}(\partial\Omega, \sigma)/\sim\big]^n \longrightarrow \big[\text{CMO}(\partial\Omega, \sigma)/\sim\big]^n. \tag{4.2.124}$$

Above, $\mathrm{BMO}(\partial\Omega,\sigma)$ *is the John-Nirenberg space on* $\partial\Omega$, *while* $\mathrm{VMO}(\partial\Omega,\sigma)$ *is the Sarason space on* $\partial\Omega$, *and* $\mathrm{CMO}(\partial\Omega,\sigma)$ *is the closure in* BMO *of the space of continuous functions which vanish at infinity on* $\partial\Omega$ *(cf.* (A.0.53)).

(4) *Given any smoothness exponent* $\alpha \in (0,1)$ *and some number* $\varepsilon \in (0,\infty)$, *there exists some small threshold* $\delta \in (0,1)$, *which depends only on* n, α, ε, *and the* ADR *constants of* $\partial\Omega$, *with the property that if* $\|\nu\|_{[\mathrm{BMO}(\partial\Omega,\sigma)]^n} < \delta$ *(i.e., if* Ω *is a* δ-AR *domain in the sense of Definition 3.1.8) it follows that* Ω *is a* UR *domain with unbounded boundary and for each spectral parameter* $z \in \mathbb{C}$ *with* $|z| \geq \varepsilon$ *the following operators are invertible on homogeneous Hölder and vanishing Hölder spaces (cf.* [129, §7.3] *and* [130, §3.2]):

$$zI + \big[K^{\mathrm{mod}}_{\mathrm{slip}}\big] : \big[\dot{\mathscr{C}}^{\alpha}(\partial\Omega)/\sim\big]^n \longrightarrow \big[\dot{\mathscr{C}}^{\alpha}(\partial\Omega)/\sim\big]^n, \tag{4.2.125}$$

$$zI + \big[K^{\mathrm{mod}}_{\mathrm{slip}}\big] : \big[\dot{\mathscr{C}}^{\alpha}_{\mathrm{van}}(\partial\Omega)/\sim\big]^n \longrightarrow \big[\dot{\mathscr{C}}^{\alpha}_{\mathrm{van}}(\partial\Omega)/\sim\big]^n. \tag{4.2.126}$$

(5) *Given any* $p \in (1,\infty)$, $\lambda \in (0,n-1)$, *and some number* $\varepsilon \in (0,\infty)$, *there exists some small threshold* $\delta \in (0,1)$, *which depends only on* n, p, λ, ε, *and the* ADR *constants of* $\partial\Omega$, *with the property that if* $\|\nu\|_{[\mathrm{BMO}(\partial\Omega,\sigma)]^n} < \delta$ *(i.e., if* Ω *is a* δ-AR *domain in the sense of Definition 3.1.8) it follows that* Ω *is a* UR *domain with unbounded boundary and for each spectral parameter* $z \in \mathbb{C}$ *with* $|z| \geq \varepsilon$ *the following operators are invertible:*

$$zI + \big[K^{\mathrm{mod}}_{\mathrm{slip}}\big] : \big[\dot{\mathfrak{L}}^{p,\lambda}(\partial\Omega,\sigma)/\sim\big]^n \longrightarrow \big[\dot{\mathfrak{L}}^{p,\lambda}(\partial\Omega,\sigma)/\sim\big]^n, \tag{4.2.127}$$

$$zI + K_{\mathrm{slip}} : \big[\mathfrak{L}^{p,\lambda}(\partial\Omega,\sigma)\big]^n \longrightarrow \big[\mathfrak{L}^{p,\lambda}(\partial\Omega,\sigma)\big]^n, \tag{4.2.128}$$

where $\dot{\mathfrak{L}}^{p,\lambda}(\partial\Omega,\sigma)$ *is the homogeneous Morrey-Campanato space (defined as in* (A.0.135) *with* $\Sigma := \partial\Omega$), *and where* $\mathfrak{L}^{p,\lambda}(\partial\Omega,\sigma)$ *is the inhomogeneous Morrey-Campanato space (defined as in* (A.0.137) *with* $\Sigma := \partial\Omega$).

(6) *Given any exponents* $q \in (1,\infty)$, $\lambda \in (0,n-1)$, *and a number* $\varepsilon \in (0,\infty)$, *there exists some small threshold* $\delta \in (0,1)$, *which depends only on* n, q, λ, ε, *and the* ADR *constants of* $\partial\Omega$, *with the property that if* $\|\nu\|_{[\mathrm{BMO}(\partial\Omega,\sigma)]^n} < \delta$ *(i.e., if* Ω *is a* δ-AR *domain in the sense of Definition 3.1.8) it follows that* Ω *is a* UR *domain with unbounded boundary and for each spectral parameter* $z \in \mathbb{C}$ *with* $|z| \geq \varepsilon$ *the following operator is invertible:*

$$zI + K^{\#}_{\mathrm{slip}} : \big[\mathscr{H}^{q,\lambda}(\partial\Omega,\sigma)\big]^n \longrightarrow \big[\mathscr{H}^{q,\lambda}(\partial\Omega,\sigma)\big]^n, \tag{4.2.129}$$

where $\mathscr{H}^{q,\lambda}(\partial\Omega,\sigma)$ *is the pre-dual to the Morrey-Campanato space, defined as in* (A.0.94) *(with* $\Sigma := \partial\Omega$).

(7) *For any exponent* $p \in \big(\frac{n-1}{n}, 1\big]$ *there exists some small threshold* $\delta \in (0,1)$, *which depends only on* n, p, *and the* ADR *constants of* $\partial\Omega$, *such that whenever* $\|\nu\|_{[\mathrm{BMO}(\partial\Omega,\sigma)]^n} < \delta$ *it follows that* Ω *is a* UR *domain with unbounded boundary and the boundary-to-boundary single layer operator* S *for the Stokes system (mapping a Hardy space into a homogeneous Hardy-based Sobolev space on*

$\partial\Omega$; *cf.* [130, Definition 4.2.1], [130, Definition 11.10.5], *and* [132, (6.2.265)]) *is invertible:*

$$[S] : \big[H^p(\partial\Omega, \sigma)\big]^n \longrightarrow \big[\dot{H}^p_1(\partial\Omega, \sigma)/\sim\big]^n. \tag{4.2.130}$$

(8) Given any exponent $p \in \big(\frac{n-1}{n}, 1\big]$ *and some number* $\varepsilon \in (0, \infty)$, *there exists some small threshold* $\delta \in (0, 1)$, *which depends only on* n, p, ε, *and the* ADR *constants of* $\partial\Omega$, *with the property that if* $\|\nu\|_{[\mathrm{BMO}(\partial\Omega,\sigma)]^n} < \delta$ *(i.e., if* Ω *is a* δ-AR *domain in the sense of Definition 3.1.8) it follows that* Ω *has an unbounded boundary and satisfies a two-sided local John condition, and for each spectral parameter* $z \in \mathbb{C}$ *with* $|z| \geq \varepsilon$ *the following operator (cf.* (A.0.99)-(A.0.100) *and* [132, (6.2.261)]) *is invertible:*

$$zI + \big[K^{\rm mod}_{\rm slip}\big] : \big[\dot{H}^p_1(\partial\Omega, \sigma)/\sim\big]^n \longrightarrow \big[\dot{H}^p_1(\partial\Omega, \sigma)/\sim\big]^n. \tag{4.2.131}$$

(9) For any exponent $p \in \big(\frac{n-1}{n}, 1\big]$ *there exists some small threshold* $\delta \in (0, 1)$, *which depends only on* n, p, *and the* ADR *constants of* $\partial\Omega$, *such that whenever* $\|\nu\|_{[\mathrm{BMO}(\partial\Omega,\sigma)]^n} < \delta$ *it follows that* Ω *is a* UR *domain with unbounded boundary and the following operator (cf.* (A.0.99)-(A.0.100), (A.0.89)-(A.0.90), *and* [132, (6.2.258)]) *is invertible:*

$$\big[\dot{\partial}^{\rm slip}_\nu(\mathcal{D}^{\rm mod}_{\rm slip}, \mathcal{P}_{\rm slip})\big] : \big[\dot{H}^p_1(\partial\Omega, \sigma)/\sim\big]^n \longrightarrow \big[H^p(\partial\Omega, \sigma)\big]^n. \tag{4.2.132}$$

(10) Given any $p, q \in (1, \infty)$, $\lambda \in (0, n-1)$, *and some number* $\varepsilon \in (0, \infty)$, *there exists some small threshold* $\delta \in (0, 1)$, *which depends only on* n, p, q, λ, ε, *and the* ADR *constants of* $\partial\Omega$, *with the property that if* $\|\nu\|_{[\mathrm{BMO}(\partial\Omega,\sigma)]^n} < \delta$ *(i.e., if* Ω *is a* δ-AR *domain in the sense of Definition 3.1.8) it follows that* Ω *has an unbounded boundary and satisfies a two-sided local John condition, and for each spectral parameter* $z \in \mathbb{C}$ *with* $|z| \geq \varepsilon$ *the following operators are invertible on homogeneous Sobolev spaces, modulo constants, which are Lebesgue-based, Morrey-based, vanishing Morrey-based, and block-based (cf.* (A.0.145)-(A.0.146), (A.0.176)-(A.0.177), (A.0.178)-(A.0.179) *and, respectively,* (A.0.37)-(A.0.39)):

$$zI + \big[K^{\rm mod}_{\rm slip}\big] : \big[\dot{L}^p_1(\partial\Omega, \sigma)/\sim\big]^n \longrightarrow \big[\dot{L}^p_1(\partial\Omega, \sigma)/\sim\big]^n, \tag{4.2.133}$$

$$zI + \big[K^{\rm mod}_{\rm slip}\big] : \big[\dot{M}^{p,\lambda}_1(\partial\Omega, \sigma)/\sim\big]^n \longrightarrow \big[\dot{M}^{p,\lambda}_1(\partial\Omega, \sigma)/\sim\big]^n, \tag{4.2.134}$$

$$zI + \big[K^{\rm mod}_{\rm slip}\big] : \big[\dot{\mathbb{M}}^{p,\lambda}_1(\partial\Omega, \sigma)/\sim\big]^n \longrightarrow \big[\dot{\mathbb{M}}^{p,\lambda}_1(\partial\Omega, \sigma)/\sim\big]^n, \tag{4.2.135}$$

$$zI + \big[K^{\rm mod}_{\rm slip}\big] : \big[\dot{\mathcal{B}}^{q,\lambda}_1(\partial\Omega, \sigma)/\sim\big]^n \longrightarrow \big[\dot{\mathcal{B}}^{q,\lambda}_1(\partial\Omega, \sigma)/\sim\big]^n. \tag{4.2.136}$$

(11) Recall that $S_{\rm mod}$ *denotes the modified boundary-to-boundary single layer operator for the Stokes system associated with the set* Ω *as in* (4.2.110). *Then for each* $p, q \in (1, \infty)$ *and* $\lambda \in (0, n-1)$ *there exists some small threshold* $\delta \in (0, 1)$, *which depends only on* n, p, q, λ, *and the* ADR *constants of* $\partial\Omega$, *with the property that whenever* $\|\nu\|_{[\mathrm{BMO}(\partial\Omega,\sigma)]^n} < \delta$ *it follows that* Ω *is a* UR *domain with*

unbounded boundary and the following operators are invertible:

$$\big[S_{\rm mod}\big] : \big[L^p(\partial\Omega,\sigma)\big]^n \longrightarrow \big[\dot{L}^p_1(\partial\Omega,\sigma)/\sim\big]^n, \tag{4.2.137}$$

$$\big[S_{\rm mod}\big] : \big[M^{p,\lambda}(\partial\Omega,\sigma)\big]^n \longrightarrow \big[\dot{M}^{p,\lambda}_1(\partial\Omega,\sigma)/\sim\big]^n, \tag{4.2.138}$$

$$\big[S_{\rm mod}\big] : \big[\mathring{M}^{p,\lambda}(\partial\Omega,\sigma)\big]^n \longrightarrow \big[\dot{\mathbb{M}}^{p,\lambda}_1(\partial\Omega,\sigma)/\sim\big]^n, \tag{4.2.139}$$

$$\big[S_{\rm mod}\big] : \big[\mathcal{B}^{q,\lambda}(\partial\Omega,\sigma)\big]^n \longrightarrow \big[\dot{\mathcal{B}}^{q,\lambda}_1(\partial\Omega,\sigma)/\sim\big]^n. \tag{4.2.140}$$

(12) For any $p, q \in (1,\infty)$ and $\lambda \in (0, n-1)$ there exists some small threshold $\delta \in (0,1)$, which depends only on n, p, q, λ, and the ADR *constants of $\partial\Omega$, with the property that whenever $\|\nu\|_{[\mathrm{BMO}(\partial\Omega,\sigma)]^n} < \delta$ it follows that Ω is a* UR *domain with unbounded boundary and the following operators are invertible:*

$$\big[\partial_\nu^{\rm slip}(\mathcal{D}^{\rm mod}_{\rm slip}, \mathcal{P}_{\rm slip})\big] : \big[\dot{L}^p_1(\partial\Omega,\sigma)/\sim\big]^n \longrightarrow \big[L^p(\partial\Omega,\sigma)\big]^n. \tag{4.2.141}$$

$$\big[\partial_\nu^{\rm slip}(\mathcal{D}^{\rm mod}_{\rm slip}, \mathcal{P}_{\rm slip})\big] : \big[\dot{M}^{p,\lambda}_1(\partial\Omega,\sigma)/\sim\big]^n \longrightarrow \big[M^{p,\lambda}(\partial\Omega,\sigma)\big]^n, \tag{4.2.142}$$

$$\big[\partial_\nu^{\rm slip}(\mathcal{D}^{\rm mod}_{\rm slip}, \mathcal{P}_{\rm slip})\big] : \big[\dot{\mathbb{M}}^{p,\lambda}_1(\partial\Omega,\sigma)/\sim\big]^n \longrightarrow \big[\mathring{M}^{p,\lambda}(\partial\Omega,\sigma)\big]^n, \tag{4.2.143}$$

$$\big[\partial_\nu^{\rm slip}(\mathcal{D}^{\rm mod}_{\rm slip}, \mathcal{P}_{\rm slip})\big] : \big[\dot{\mathcal{B}}^{q,\lambda}_1(\partial\Omega,\sigma)/\sim\big]^n \longrightarrow \big[\mathcal{B}^{q,\lambda}(\partial\Omega,\sigma)\big]^n. \tag{4.2.144}$$

(13) Assume $\mathbb{X}$ is a Generalized Banach Function Space on $(\partial\Omega,\sigma)$ for which

$$\mathcal{M}_{\partial\Omega} : \mathbb{X} \to \mathbb{X} \text{ and } \mathcal{M}_{\partial\Omega} : \mathbb{X}' \to \mathbb{X}' \text{ are well-defined and bounded,} \tag{4.2.145}$$

where $\mathcal{M}_{\partial\Omega}$ is the Hardy-Littlewood maximal operator on $\partial\Omega$, and $\mathbb{X}'$ is the associated space of $\mathbb{X}$ (cf. [130, Definition 5.1.4] *and* [130, Definition 5.1.11]). *Define $\mathring{\mathbb{X}}$ and $(\mathbb{X}')^\circ$ as in* (4.2.41) *as well as the $\mathbb{X}$-styled, $\mathbb{X}'$-styled, $\mathring{\mathbb{X}}$-styled, and $(\mathbb{X}')^\circ$-styled Sobolev spaces of order one, $\mathbb{X}_1$, $(\mathbb{X}')_1$, $\mathring{\mathbb{X}}_1$, and $(\mathbb{X}')^\circ_1$ from* (4.2.42) *(these are known to be Banach spaces when equipped with natural norms; cf.* (5.4.84)). *Then, for any given parameter $\varepsilon \in (0,\infty)$ there exists some small threshold $\delta \in (0,1)$, which depends only on ε, the* ADR *constants of $\partial\Omega$, and the operator norms $\|\mathcal{M}_{\partial\Omega}\|_{\mathbb{X}\to\mathbb{X}}$, $\|\mathcal{M}_{\partial\Omega}\|_{\mathbb{X}'\to\mathbb{X}'}$, with the property that whenever $\|\nu\|_{[\mathrm{BMO}(\partial\Omega,\sigma)]^n} < \delta$ it follows that Ω is a* UR *domain and for each spectral parameter $z \in \mathbb{C}$ with $|z| \geq \varepsilon$ the following operators are invertible:*

$$zI + K_{\rm slip} : \big[\mathbb{X}\big]^n \longrightarrow \big[\mathbb{X}\big]^n, \quad zI + K_{\rm slip} : \big[\mathbb{X}'\big]^n \longrightarrow \big[\mathbb{X}'\big]^n, \tag{4.2.146}$$

$$zI + K_{\rm slip} : \big[\mathring{\mathbb{X}}\big]^n \longrightarrow \big[\mathring{\mathbb{X}}\big]^n, \quad zI + K_{\rm slip} : \big[(\mathbb{X}')^\circ\big]^n \longrightarrow \big[(\mathbb{X}')^\circ\big]^n, \tag{4.2.147}$$

and

$$zI + K_{\rm slip} : \big[\mathbb{X}_1\big]^n \longrightarrow \big[\mathbb{X}_1\big]^n, \quad zI + K_{\rm slip} : \big[(\mathbb{X}')_1\big]^n \longrightarrow \big[(\mathbb{X}')_1\big]^n, \tag{4.2.148}$$

$$zI + K_{\rm slip} : \big[\mathring{\mathbb{X}}_1\big]^n \longrightarrow \big[\mathring{\mathbb{X}}_1\big]^n, \quad zI + K_{\rm slip} : \big[(\mathbb{X}')^\circ_1\big]^n \longrightarrow \big[(\mathbb{X}')^\circ_1\big]^n, \tag{4.2.149}$$

as well as

$$zI + K^{\#}_{\text{slip}} : [\mathbb{X}]^n \longrightarrow [\mathbb{X}]^n, \quad zI + K^{\#}_{\text{slip}} : [\mathbb{X}']^n \longrightarrow [\mathbb{X}']^n, \tag{4.2.150}$$

$$zI + K^{\#}_{\text{slip}} : [\mathring{\mathbb{X}}]^n \longrightarrow [\mathring{\mathbb{X}}]^n, \quad zI + K^{\#}_{\text{slip}} : [(\mathbb{X}')^{\circ}]^n \longrightarrow [(\mathbb{X}')^{\circ}]^n. \tag{4.2.151}$$

(14) Continue to assume that $\mathbb{X}$ *is a Generalized Banach Function Space on* $(\partial\Omega, \sigma)$ *satisfying* (4.2.145)*, and bring back* $\dot{\mathbb{X}}_1$*, the* $\mathbb{X}$*-based homogeneous Sobolev space of order one on* $\partial\Omega$ *defined in* (4.2.52) (*recalled that this is equipped with the semi-norm introduced in* (4.2.53)). *Then for any* $\varepsilon \in (0, \infty)$ *there exists some small threshold* $\delta \in (0, 1)$*, which depends only on* ε*, the* ADR *constants of* $\partial\Omega$*, and the operator norms* $\|\mathcal{M}_{\partial\Omega}\|_{\mathbb{X}\to\mathbb{X}}$*,* $\|\mathcal{M}_{\partial\Omega}\|_{\mathbb{X}'\to\mathbb{X}'}$*, with the property that whenever* $\|\nu\|_{[\mathrm{BMO}(\partial\Omega,\sigma)]^n} < \delta$ *it follows that* Ω *is a two-sided* NTA *domain with an unbounded Ahlfors regular boundary and for each spectral parameter* $z \in \mathbb{C}$ *with* $|z| \geq \varepsilon$ *the following operators are invertible:*

$$zI + \big[K^{\text{mod}}_{\text{slip}}\big] : \big[\dot{\mathbb{X}}_1/\sim\big]^n \longrightarrow \big[\dot{\mathbb{X}}_1/\sim\big]^n, \tag{4.2.152}$$

$$\big[\partial^{\text{slip}}_{\nu}(\mathcal{D}^{\text{mod}}_{\text{slip}}, \mathcal{P}_{\text{slip}})\big] : \big[\dot{\mathbb{X}}_1/\sim\big]^n \longrightarrow [\mathbb{X}]^n, \tag{4.2.153}$$

$$\big[\mathcal{S}_{\text{mod}}\big] : [\mathbb{X}]^n \longrightarrow \big[\dot{\mathbb{X}}_1/\sim\big]^n, \tag{4.2.154}$$

(*where* $\mathcal{S}_{\text{mod}}$ *is the modified boundary-to-boundary single layer operator for the Stokes system associated with the set* Ω *as in* (4.2.110)). *Finally, analogous invertibility results are valid for the* $\mathring{\mathbb{X}}$*-based homogeneous Sobolev space of order one on* $\partial\Omega$*, defined analogously to* (4.2.52).

(15) The results in items (13)-(14) are valid whenever $\mathbb{X}$ *is a rearrangement invariant Banach function space whose Boyd indices satisfy* $1 < p_{\mathbb{X}} \leq q_{\mathbb{X}} < \infty$. *As a corollary, all conclusions in items (13)-(14) are true if* $\mathbb{X} := L^{\Phi}(\partial\Omega, \sigma)$*, the Orlicz space associated as in* (A.0.133) *with the measure space* $(\partial\Omega, \sigma)$ *and any Young function* Φ *satisfying* $1 < i(\Phi) \leq I(\Phi) < \infty$ (*when the role of* $\mathbb{X}_1$ *from* (4.2.42) *is now played by the Orlicz-based Sobolev space* $L^{\Phi}_1(\partial\Omega, \sigma)$).

(16) Suppose

$$\begin{aligned} &\phi : (0, \infty) \to (0, \infty) \text{ is } \mathcal{L}^1\text{-measurable and there is } C_* \in (0, \infty) \\ &\text{so that } r \int_r^{\infty} \phi(t) \frac{\ln(t/r)}{t^2}\, dt \leq C_*\phi(r) \text{ for all } r \in (0, \infty). \end{aligned} \tag{4.2.155}$$

In addition, select an integrability exponent $p \in (1, \infty)$*, choose a Muckenhoupt weight* $w \in A_p(\partial\Omega, \sigma)$*, and pick a number* $\varepsilon \in (0, \infty)$. *Then there exists some small threshold* $\delta \in (0, 1)$*, depending only on* n*,* p*,* $[w]_{A_p}$*,* ε*,* C_* *from* (4.2.155)*, and the* ADR *constants of* $\partial\Omega$*, with the property that if* $\|\nu\|_{[\mathrm{BMO}(\partial\Omega,\sigma)]^n} < \delta$ (*i.e., if* Ω *is a* δ-AR *domain in the sense of Definition 3.1.8*) *it follows that the set* Ω *is a* UR *domain with an unbounded boundary, and for each spectral parameter* $z \in \mathbb{C}$ *with* $|z| \geq \varepsilon$ *the following operators are invertible:*

$$zI + \big[K^{\mathrm{mod}}_{\mathrm{slip}}\big] : \big[\mathrm{BMO}_{\phi,p}(\partial\Omega, w)/\sim\big]^n \longrightarrow \big[\mathrm{BMO}_{\phi,p}(\partial\Omega, w)/\sim\big]^n, \quad (4.2.156)$$

$$zI + \big[K^{\mathrm{mod}}_{\mathrm{slip}}\big] : \big[\mathrm{BCMO}_{\phi,p}(\partial\Omega, w)/\sim\big]^n \longrightarrow \big[\mathrm{BCMO}_{\phi,p}(\partial\Omega, w)/\sim\big]^n, \quad (4.2.157)$$

$$zI + \big[K^{\mathrm{mod}}_{\mathrm{slip}}\big] : \big[\dot{\mathrm{B}}\mathrm{CMO}_{\phi,p}(\partial\Omega, w)/\sim\big]^n \longrightarrow \big[\dot{\mathrm{B}}\mathrm{CMO}_{\phi,p}(\partial\Omega, w)/\sim\big]^n, \quad (4.2.158)$$

where the above ϕ-modulated weighted BMO *and* BCMO *spaces are defined as in* (4.1.281)-(4.1.286) *with* $\Sigma := \partial\Omega$.

Proof The key aspect is that K_{slip} from (4.2.102) and $K^{\#}_{\mathrm{slip}}$ from (4.2.103) are singular integral operators of chord-dot-normal type (cf. the discussion in [132, §5.2]). Granted this, all claims may be justified by relying on the same type of arguments as in the proof of Theorem 4.2.1 and making use of Theorem 4.1.3, Theorem 4.1.6, Theorem 4.1.7, Theorem 4.1.9, [132, Theorem 6.2.6], [132, Theorem 6.2.8], [132, Theorem 6.2.11], [132, Theorem 6.3.8], and Theorem 4.1.16. □

The last portion of this section is reserved for comparing Theorem 4.2.1 with some precise invertibility results available in the special case when $L := \Delta$, the two-dimensional Laplacian, and for infinite sectors in the plane. Specifically, we have the following result:

Proposition 4.2.4 *Consider a sector of aperture $\theta \in (0, 2\pi)$ in the two-dimensional space, i.e., a planar set of the form*

$$\Omega_\theta = \Omega_{\theta,\xi} := \Big\{x \in \mathbb{R}^2 \setminus \{x_o\} : \tfrac{x-x_o}{|x-x_o|} \cdot \xi > \cos(\theta/2)\Big\} \quad \text{with } x_o \in \mathbb{R}^2,\ \theta \in (0, 2\pi),\ \text{and } \xi \in S^1, \quad (4.2.159)$$

and abbreviate $\sigma_\theta := \mathcal{H}^1\lfloor\partial\Omega_\theta$. In such a setting, the corresponding principal-value double layer K_Δ (associated with Ω_θ as in (A.0.116) *with $M = 1$ and $A = I_{2\times 2}$) has the following property:*

> *given any index $p \in (1, \infty)$, the operator $\frac{1}{2}I + K_\Delta$ is invertible on $L^p(\partial\Omega_\theta, \sigma_\theta)$ if and only if $p \neq p_\theta := 1 + |\pi - \theta|/\pi$ (which amounts to saying that $p \neq \frac{2\pi-\theta}{\pi}$ for $\theta \in (0, \pi)$ and $p \neq \frac{\theta}{\pi}$ for $\theta \in (\pi, 2\pi)$), and the operator $-\frac{1}{2}I + K_\Delta$ is invertible on $L^p(\partial\Omega_\theta, \sigma_\theta)$ if and only if $p \neq p_\theta := 1 + |\pi - \theta|/\pi$.* (4.2.160)

Moreover, corresponding to the critical exponent p_θ, the operators $\pm\frac{1}{2}I + K_\Delta$ acting on $L^{p_\theta}(\partial\Omega_\theta, \sigma_\theta)$ have dense images, but are not onto.

Corresponding to $\theta = \pi$, the result recorded in (4.2.160) implies that $\pm\frac{1}{2}I + K_\Delta$ are invertible on $L^p(\partial\Omega_\pi, \sigma_\pi)$ for each $p \in (1, \infty)$ which is, of course, the case since $K_\Delta = 0$ given that Ω_π is a half-plane. Let us also remark that the critical exponent $p_\theta := 1 + |\pi - \theta|/\pi$ from (4.2.160) belongs to the interval $[1, 2)$ and may be directly related to the BMO semi-norm of the outward unit normal vector ν_θ to Ω_θ. Indeed, regarding ν_θ as a complex-valued function, a straightforward computation (based on definitions) shows that

$$\begin{aligned}\|\nu_\theta\|_{\mathrm{BMO}(\partial\Omega_\theta,\sigma_\theta)} &= |\cos(\theta/2)| = \big|\sin\big(\pi/2-\theta/2\big)\big| \approx |\theta-\pi| \\ &\approx p_\theta - 1, \ \text{ uniformly for } \theta\in(0,2\pi). \end{aligned} \tag{4.2.161}$$

As a consequence, we have the following remark highlighting the relevance of the demand that the geometric measure theoretic outward unit normal has small BMO seminorm in the context of Theorem 4.2.1 (see more refined formulation given in Remark 4.2.2):

$$\begin{array}{l}\text{in the class of domains } \Omega_\theta \text{ with } \theta\in(0,2\pi), \text{ the ability of inverting} \\ \text{either } \tfrac12 I + K_\Delta, \text{ or } -\tfrac12 I + K_\Delta, \text{ on } L^p(\partial\Omega_\theta,\sigma_\theta) \text{ for each } p\in I := [a,2] \\ \text{with } a\in(1,2) \text{ close to 1 is equivalent to } \|\nu_\theta\|_{\mathrm{BMO}(\partial\Omega_\theta,\sigma_\theta)} \text{ being small.}\end{array} \tag{4.2.162}$$

We now turn to the proof of Proposition 4.2.4.

Proof of Proposition 4.2.4 Consider first the case when $\theta\in(0,\pi)$. Making a rotation and a translation, there is no loss of generality in assuming that Ω_θ is as in (4.2.159) with x_o the origin in $\mathbb{R}^2$ and $\xi := (0,1)\in S^1$, i.e.,

$$\Omega_\theta = \Big\{(x,y)\in\mathbb{R}^2\setminus\{0\} : \frac{y}{\sqrt{x^2+y^2}} > \cos(\theta/2)\Big\}. \tag{4.2.163}$$

In such a scenario, we may decompose $\partial\Omega_\theta\setminus\{0\}$ as $L_+\cup L_-$ where

$$L_\pm := \Big\{\big(\pm t\sin(\theta/2), t\cos(\theta/2)\big) : t\in(0,\infty)\Big\}. \tag{4.2.164}$$

Under the canonical identification of $L_\pm$ with $(0,\infty)$, we have

$$L^p(\partial\Omega_\theta,\sigma_\theta) = L^p\big((0,\infty),\mathcal{L}^1\big)\oplus L^p\big((0,\infty),\mathcal{L}^1\big) \ \text{ for each } p\in(1,\infty), \tag{4.2.165}$$

while the double layer potential operator K_Δ acquires the format

$$K_\Delta \equiv \begin{pmatrix} 0 & T_\theta \\ T_\theta & 0\end{pmatrix}. \tag{4.2.166}$$

Above, T_θ is the operator acting on each $f\in L^1\big((0,\infty),\frac{dx}{1+x^2}\big)$ according to

$$(T_\theta f)(x) := \int_0^\infty k_\theta(x,y)f(y)\,dy \ \text{ at each point } x\in(0,\infty), \tag{4.2.167}$$

with kernel k_θ given by

$$k_\theta(x,y) := \frac{\sin\theta}{2\pi}\cdot\frac{x}{x^2+y^2-2xy\cos\theta} \ \text{ for each } x,y\in(0,\infty). \tag{4.2.168}$$

Note that k_θ is positive homogeneous of degree -1, and for each $p \in (1, \infty)$ satisfies

$$\int_0^\infty x^{(1/p)-1} |k_\theta(x, 1)| \, \mathrm{d}x = \int_0^\infty y^{-1/p} |k_\theta(1, y)| \, \mathrm{d}y < \infty. \tag{4.2.169}$$

Hence, this is a Hardy kernel. As such, according to [50, Theorem, p. 189], [17, Corollary 2, p. 34], for each $p \in (1, \infty)$ it follows that the operator

$$T_\theta : L^p\big((0, \infty), \mathcal{L}^1\big) \longrightarrow L^p\big((0, \infty), \mathcal{L}^1\big) \tag{4.2.170}$$

is well defined, linear, bounded, with spectrum

$$\Big\{\widetilde{k_\theta}\big(\tfrac{1}{p} + \mathrm{i}\xi\big) : \xi \in \mathbb{R}\Big\} \cup \{0\}, \tag{4.2.171}$$

where $\widetilde{k_\theta}$ is the Mellin transform of k_θ given by (see [51, p. 98])

$$\widetilde{k_\theta}(z) := \frac{\sin[(\pi - \theta)z]}{2\sin(\pi z)} \quad \text{for each } z \in \mathbb{C} \text{ with } 0 < \operatorname{Re} z < 1. \tag{4.2.172}$$

As a consequence, the lack of invertibility for $\frac{1}{2}I - T_\theta$ acting on $L^p\big((0, \infty), \mathcal{L}^1\big)$ for some given $p \in (1, \infty)$ comes down to the issue of existence of some $\xi \in \mathbb{R}$ with the property that

$$\frac{\sin[(\pi - \theta)(\frac{1}{p} + \mathrm{i}\xi)]}{2\sin[\pi(\frac{1}{p} + \mathrm{i}\xi)]} = \frac{1}{2} \tag{4.2.173}$$

which, in view of the fact that $\sin A - \sin B = 2\cos\big(\frac{A+B}{2}\big)\sin\big(\frac{A-B}{2}\big)$, becomes

$$\cos\big[(\pi - \theta/2)(\tfrac{1}{p} + \mathrm{i}\xi)\big]\sin\big[\theta(\tfrac{1}{p} + \mathrm{i}\xi)\big] = 0. \tag{4.2.174}$$

Since the (complex) zeros of cos, sin are, respectively, $\big\{\frac{\pi}{2} + m\pi\big\}_{m\in\mathbb{Z}}$ and $\{m\pi\}_{m\in\mathbb{Z}}$, this further implies that there exists some $m \in \mathbb{Z}$ such as

$$\text{either } (\pi - \theta/2)(\tfrac{1}{p} + \mathrm{i}\xi) = (\pi/2) + m\pi, \text{ or } \theta(\tfrac{1}{p} + \mathrm{i}\xi) = m\pi. \tag{4.2.175}$$

This forces $\xi = 0$ and, further, $p = 2 - \theta/\pi$. Therefore, the conclusion is that

$$\begin{gathered} \tfrac{1}{2}I - T_\theta : L^p\big((0, \infty), \mathcal{L}^1\big) \longrightarrow L^p\big((0, \infty), \mathcal{L}^1\big) \\ \text{is invertible if and only if } p \in (1, \infty) \text{ satisfies } p \neq 2 - \theta/\pi. \end{gathered} \tag{4.2.176}$$

Consider next the issue as to whether $\frac{1}{2}I + T_\theta$ acting on $L^p\big((0, \infty), \mathcal{L}^1\big)$ for some given $p \in (1, \infty)$ fails to be invertible. This hinges on the existence of some $\xi \in \mathbb{R}$ with the property that

$$\frac{\sin[(\pi - \theta)(\frac{1}{p} + \mathrm{i}\xi)]}{2\sin[\pi(\frac{1}{p} + \mathrm{i}\xi)]} = -\frac{1}{2} \tag{4.2.177}$$

which, bearing in mind that $\sin A + \sin B = 2\cos\big(\frac{A-B}{2}\big)\sin\big(\frac{A+B}{2}\big)$, is equivalent to

$$\cos\left[(\theta/2)(\tfrac{1}{p}+\mathrm{i}\xi)\right]\sin\left[(\pi-\theta/2)(\tfrac{1}{p}+\mathrm{i}\xi)\right]=0. \tag{4.2.178}$$

Hence, there exists some $m \in \mathbb{Z}$ such as

$$\text{either } (\theta/2)(\tfrac{1}{p}+\mathrm{i}\xi)=(\pi/2)+m\pi, \text{ or } (\pi-\theta/2)(\tfrac{1}{p}+\mathrm{i}\xi)=m\pi. \tag{4.2.179}$$

This forces $\xi = 0$, and ultimately yields no solutions (since we are currently assuming $0 < \theta < \pi$). The conclusion is that

$$\begin{array}{l}\tfrac{1}{2}I + T_\theta : L^p\big((0,\infty),\mathcal{L}^1\big) \longrightarrow L^p\big((0,\infty),\mathcal{L}^1\big)\\ \text{is an invertible operator for each } p \in (1,\infty).\end{array} \tag{4.2.180}$$

To proceed, use (4.2.166) to write

$$\begin{aligned}\big(\tfrac{1}{2}I+K_\Delta\big)\big(-\tfrac{1}{2}I+K_\Delta\big) &= \begin{pmatrix}\tfrac{1}{2}I & T_\theta\\ T_\theta & \tfrac{1}{2}I\end{pmatrix}\cdot\begin{pmatrix}-\tfrac{1}{2}I & T_\theta\\ T_\theta & -\tfrac{1}{2}I\end{pmatrix} \\ &= \begin{pmatrix}-\tfrac{1}{4}I+T_\theta^2 & 0\\ 0 & -\tfrac{1}{4}I+T_\theta^2\end{pmatrix} = I_{2\times2}\big(\tfrac{1}{2}I+T_\theta\big)\big(-\tfrac{1}{2}I+T_\theta\big),\end{aligned} \tag{4.2.181}$$

where $I_{2\times2}$ is the 2×2 identity matrix. From (4.2.176), (4.2.180), and (4.2.181) we then conclude that

$$\begin{array}{l}\text{if } \theta\in(0,\pi), \text{ then given any } p\in(1,\infty) \text{ it follows that the operator}\\ \tfrac{1}{2}I+K_\Delta \text{ is invertible on } L^p(\partial\Omega_\theta,\sigma_\theta) \text{ if and only if } p\neq\tfrac{2\pi-\theta}{\pi}, \text{ and}\\ -\tfrac{1}{2}I+K_\Delta \text{ is invertible on } L^p(\partial\Omega_\theta,\sigma_\theta) \text{ if and only if } p\neq\tfrac{2\pi-\theta}{\pi}.\end{array} \tag{4.2.182}$$

In the case when $\theta\in(\pi,2\pi)$, using the fact that $\partial\Omega_{\theta,\xi}=\partial\Omega_{2\pi-\theta,-\xi}$ and that K_Δ associated with $\Omega_{\theta,\xi}$ is the opposite of K_Δ associated with $\Omega_{2\pi-\theta,-\xi}$, we may employ (4.2.182) to conclude that

$$\begin{array}{l}\text{if } \theta\in(\pi,2\pi), \text{ then given any } p\in(1,\infty) \text{ it follows that the operator}\\ \tfrac{1}{2}I+K_\Delta \text{ is invertible on } L^p(\partial\Omega_\theta,\sigma_\theta) \text{ if and only if } p\neq\tfrac{\theta}{\pi}, \text{ while also}\\ -\tfrac{1}{2}I+K_\Delta \text{ is invertible on } L^p(\partial\Omega_\theta,\sigma_\theta) \text{ if and only if } p\neq\tfrac{\theta}{\pi}.\end{array} \tag{4.2.183}$$

At this stage, (4.2.160) follows by combining (4.2.182)-(4.2.183).

To justify the very last claim in the statement, by reasoning as above we may reduce matters to the case when $\theta\in(0,\pi)$ (in this regard, it is also relevant to observe that changing θ into $2\pi-\theta$ leaves the critical exponent $p_\theta := 1+|\pi-\theta|/\pi$ invariant). In this scenario, in the proof of [51, Theorem 1.3, p. 98] it is mentioned that the operator $\tfrac{1}{2}I-T_\theta$ has a dense range on $L^{p_\theta}\big((0,\infty),\mathcal{L}^1\big)$ but is not onto. From this, (4.2.181), and (4.2.180) we then see that the operators $\pm\tfrac{1}{2}I+K_\Delta$ acting on $L^{p_\theta}(\partial\Omega_\theta,\sigma_\theta)$ have dense images, but are not onto. □

We continue by mentioning that, if Ω_θ is as in (4.2.159), then from the last claim in Proposition 4.2.4, (4.2.160), and [130, Lemma 1.2.7] (cf. also [88, Theorem 2.10]) we see that the operators

$$\pm\tfrac{1}{2}I + K_\Delta \text{ have dense images, of infinite co-dimension (hence not closed), when acting on } L^{p_\theta}(\partial\Omega_\theta, \sigma_\theta) \text{ with } \theta \in (0, 2\pi) \setminus \{\pi\} \text{ and } p_\theta := 1 + |\pi - \theta|/\pi. \tag{4.2.184}$$

From (4.2.160) and [132, Proposition 1.8.10] we next deduce that, if $K^{\#}_{\Delta}$ is the transpose harmonic double layer associated with Ω_θ as in (A.0.115), then

$$\text{given any } p \in (1,\infty), \text{ the operator } \tfrac{1}{2}I + K^{\#}_{\Delta} \text{ is invertible on the space } L^p(\partial\Omega_\theta, \sigma_\theta) \text{ if and only if } p \neq q_\theta := 1 + \pi/|\pi - \theta|, \text{ and the operator } -\tfrac{1}{2}I + K^{\#}_{\Delta} \text{ is invertible on } L^p(\partial\Omega_\theta, \sigma_\theta) \text{ if and only if there holds } p \neq q_\theta := 1 + \pi/|\pi - \theta|. \tag{4.2.185}$$

while from (4.2.184), [130, Lemma 1.2.7], [130, (2.1.45)-(2.1.46)] (or [181, Corollaries, p. 99]), and Banach's Closed Range Theorem (cf. [130, (2.1.47)]) we see that

$$\pm\tfrac{1}{2}I + K^{\#}_{\Delta} \text{ acting on } L^{q_\theta}(\partial\Omega_\theta, \sigma_\theta) \text{ with } \theta \in (0, 2\pi) \setminus \{\pi\} \text{ and } q_\theta := 1 + \pi/|\pi - \theta| \text{ are injective, but their ranges have infinite co-dimension and are not closed.} \tag{4.2.186}$$

In addition, with S_{mod} denoting the two-dimensional modified boundary-to-boundary harmonic single layer operator associated with Ω_θ as in (A.0.232), from (4.2.160) and [132, Proposition 1.8.10] we obtain:

$$\text{given any integrability exponent } p \in (1,\infty), \text{ it follows that the operator } \big[S_{\text{mod}}\big] : L^p(\partial\Omega_\theta, \sigma_\theta) \to \dot{L}^p_1(\partial\Omega_\theta, \sigma_\theta)/\sim \text{ (cf. [132, (1.8.124)]) is invertible if and only if } p \neq q_\theta := 1 + \pi/|\pi - \theta|. \tag{4.2.187}$$

Also, corresponding to $\theta \neq \pi$ and for the critical exponent $p = q_\theta := 1 + \pi/|\pi - \theta|$, from the operator identity [132, Lemma 1.6.10, (1.6.90)], the isomorphism [130, Proposition 11.5.15, (11.5.148)], the results recorded in (4.2.186), and basic functional analysis we see that

$$\text{for } \theta \in (0, 2\pi)\setminus\{\pi\} \text{ and exponent } q_\theta := 1+\pi/|\pi-\theta|, \text{ the operator } \big[S_{\text{mod}}\big] : L^{q_\theta}(\partial\Omega_\theta, \sigma_\theta) \to \dot{L}^{q_\theta}_1(\partial\Omega_\theta, \sigma_\theta)/\sim \text{ is injective, but its range has infinite co-dimension (hence is not closed).} \tag{4.2.188}$$

Moreover, from (4.2.185) and [132, Corollary 2.3.15] is apparent that, if $\big[K_{\Delta,\text{mod}}\big]$ denotes the modified boundary-to-boundary harmonic double layer operator acting on classes of functions modulo constants (cf. [132, (1.8.154)]), then

$$\text{given any index } p \in (1,\infty), \text{ the operator } \tfrac{1}{2}I + \big[K_{\Delta,\text{mod}}\big] \text{ is invertible on the space } \dot{L}^p_1(\partial\Omega_\theta, \sigma_\theta)/\sim \text{ if and only if } p \neq q_\theta := 1 + \pi/|\pi - \theta|, \text{ whereas the operator } -\tfrac{1}{2}I + \big[K_{\Delta,\text{mod}}\big] \text{ is invertible on the space } \dot{L}^p_1(\partial\Omega_\theta, \sigma_\theta)/\sim \text{ if and only if } p \neq q_\theta := 1 + \pi/|\pi - \theta|. \tag{4.2.189}$$

In this vein, corresponding to the case when the angle $\theta \neq \pi$, and for the critical exponent $p = q_\theta := 1+\pi/|\pi-\theta|$, from formula [132, Proposition 2.3.14, (2.3.224)], the isomorphism [130, Proposition 11.5.15, (11.5.148)], and the results in (4.2.186) we deduce that

$$\pm\frac{1}{2}I + \big[K_{\Delta,\mathrm{mod}}\big] \text{ acting on } \dot{L}_1^{q_\theta}(\partial\Omega_\theta, \sigma_\theta)/\!\sim \text{ corresponding to the angle } \theta \in (0, 2\pi) \setminus \{\pi\} \text{ and } q_\theta := 1 + \pi/|\pi - \theta| \text{ are injective, but their ranges have infinite co-dimension (hence are not closed).} \tag{4.2.190}$$

We may also consider the singular integral operator R associated with Ω_θ as in [131, Corollary 2.5.34]. In view of (4.2.160) and the last formula in [131, (2.5.334)] we then conclude that

$$\text{if } \theta \in (0, 2\pi), \text{ then given any } p \in (1, \infty) \text{ it follows that the operator } R \text{ associated with } \Omega_\theta \text{ as in [131, (2.5.333)] is invertible on } L^p(\partial\Omega_\theta, \sigma_\theta) \text{ if and only if } p \neq p_\theta := 1 + |\pi - \theta|/\pi. \tag{4.2.191}$$

In addition, from (4.2.184), the last formula in [131, (2.5.334)], and basic functional analysis we see that, corresponding to the critical exponent p_θ,

$$\text{if the angle } \theta \in (0, 2\pi), \text{ then the operator } R \text{ has a dense image, of infinite co-dimension (hence not closed), when acting on } L^{p_\theta}(\partial\Omega_\theta, \sigma_\theta) \text{ with } p_\theta := 1 + |\pi - \theta|/\pi. \tag{4.2.192}$$

As with (4.2.162), we may use the above invertibility results to highlight the significance of the demand that the geometric measure theoretic outward unit normal has small BMO seminorm in the context of Theorem 4.2.1 (see the more detailed formulation given in Remark 4.2.2). Concretely, (4.2.161) and (4.2.185) show that

$$\text{in the class of domains } \Omega_\theta \text{ with } \theta \in (0, 2\pi), \text{ the ability of inverting } \pm\frac{1}{2}I + K_\Delta^\# \text{ on } L^p(\partial\Omega_\theta, \sigma_\theta) \text{ for each } p \in I := [2, a] \text{ with } a \in (2, \infty) \text{ large forces } \|\nu_\theta\|_{\mathrm{BMO}(\partial\Omega_\theta,\sigma_\theta)} \text{ to be small,} \tag{4.2.193}$$

while from (4.2.161) and (4.2.189) we conclude that

$$\text{in the class of domains } \Omega_\theta \text{ with } \theta \in (0, 2\pi), \text{ the ability of inverting either } \frac{1}{2}I + \big[K_{\Delta,\mathrm{mod}}\big], \text{ or } -\frac{1}{2}I + \big[K_{\Delta,\mathrm{mod}}\big], \text{ on the space } \dot{L}_1^p(\partial\Omega_\theta, \sigma_\theta)/\sim \text{ for each } p \in I := [2, a] \text{ with } a \in (2, \infty) \text{ large forces } \|\nu_\theta\|_{\mathrm{BMO}(\partial\Omega_\theta,\sigma_\theta)} \text{ to be small.} \tag{4.2.194}$$

Similarly, (4.2.161) and (4.2.187) show that

$$\text{in the class of sets } \Omega_\theta \text{ with angle } \theta \in (0, 2\pi), \text{ the ability of inverting the operator } \big[S_{\mathrm{mod}}\big] : L^p(\partial\Omega_\theta, \sigma_\theta) \to \dot{L}_1^p(\partial\Omega_\theta, \sigma_\theta)/\sim \text{ for each index } p \in I := [2, a] \text{ with } a \in (2, \infty) \text{ large forces } \|\nu_\theta\|_{\mathrm{BMO}(\partial\Omega_\theta,\sigma_\theta)} \text{ to be small.} \tag{4.2.195}$$

Finally, we note that, in fact, a full description of description of the spectrum of the classical harmonic double layer potential operator on planar sectors is available. To state such a result, for each $\theta \in (0, 2\pi)$ and $p \in (1, \infty)$ define the following ("bow" shaped) closed contour given by the parametric representation

$$\Sigma_\theta(p) := \{0\} \cup \left\{ \pm \frac{\sin\big((\pi - \theta)z\big)}{\sin(\pi z)} : z \in \tfrac{1}{p} + \mathrm{i}\mathbb{R} \right\}. \tag{4.2.196}$$

The following proposition is proved in [188].

Proposition 4.2.5 *Let $\Omega_\theta \subseteq \mathbb{R}^2$ be an infinite sector with (full) aperture $\theta \in (0, 2\pi)$. Abbreviate $\sigma_\theta := \mathcal{H}^1 \lfloor \partial\Omega_\theta$ and denote by K_Δ the principal-value harmonic double layer potential operator associated with Ω_θ as in* (A.0.114). *Finally, fix an integrability exponent $p \in (1, \infty)$. Then* $\mathrm{Spec}\big(K_\Delta; L^p(\partial\Omega_\theta, \sigma_\theta)\big)$, *the spectrum of K_Δ acting from the space $L^p(\partial\Omega_\theta, \sigma_\theta)$ into itself is*

$$\mathrm{Spec}\Big(K_\Delta; L^p(\partial\Omega_\theta, \sigma_\theta)\Big) = \Sigma_\theta(p), \tag{4.2.197}$$

the bow-shaped curve associated with the given exponent p and angle θ as in (4.2.196). *Moreover, if $z \in \Sigma_\theta(p)$ then the operator $zI - K_\Delta$ is not Fredholm on $L^p(\partial\Omega_\theta, \sigma_\theta)$.*

Chapter 5
Estimating Chord-Dot-Normal SIO's on Domains with Compact Boundaries

In this chapter we shall embark on a study aimed at clarifying of how the flatness of a "surface" is related to the functional analytic properties of singular integral operators defined on it. We succeed in building a two-way street between geometry and analysis by addressing the following issues:

• *Failure of Compactness for Plain Convolution Type* SIO*'s:* Specifically, in Proposition 5.1.4 we prove that the only Calderón-Zygmund singular integral operator of convolution type which is compact, when acting on Lebesgue spaces considered on smooth bounded surfaces, is the zero operator. This points to the fact that the integral kernel should contain geometric information in order for the SIO in question to have a chance of becoming compact when considered on surfaces which are smooth and bounded. A natural way to allow for this to happen is to work with SIO's whose integral kernels depend linearly on the unit normal to the "surface" on which they are defined (more on this below).

• *Necessity of Chord-Dot-Normal Structure:* In the class of SIO's whose integral kernel depends linearly on the unit normal to the "surface" on which this is defined, only SIO's of chord-dot-normal type have a chance on inducing compact operators (on, say, Lebesgue spaces) when the surface in question is smooth and bounded. See Proposition 5.1.1 and Proposition 5.1.2.

• *Sufficiency of Chord-Dot-Normal Structure:* The essential norm (on, say, Lebesgue spaces) of SIO's of chord-dot-normal type may be estimated in, an almost linear fashion[1], in terms of the distance from the unit normal to the surface on which they are defined, measured in the John-Nirenberg space BMO, to the Sarason space VMO. See Theorem 5.2.3, Corollary 5.2.4, Theorem 5.3.1.

• *Compactness of Boundary-to-Boundary Double Layer Equivalent to Coefficient Tensor Being Distinguished:* A boundary-to-boundary double layer potential operator, associated with a coefficient tensor used to represent a given weakly elliptic homogeneous constant coefficient second-order system, becomes compact (on, say, Lebesgue spaces) on the boundary of any infinitesimally flat AR domain if and only if the coefficient tensor in question is distinguished. See Theorem 5.3.2.

[1] i.e., linear up to iterated logarithms

D. Mitrea et al., *Geometric Harmonic Analysis V*, Developments in Mathematics 76,
https://doi.org/10.1007/978-3-031-31561-9_5

• *The Cauchy-Clifford Operator and the Riesz Transforms Characterize Flatness:* Infinitesimal flatness, understood as the distance from the unit normal to the surface on which said operators are defined (measured in the John-Nirenberg space BMO) to the Sarason space VMO, may be controlled in terms of qualities such as how far is the Cauchy-Clifford operator from being symmetric, or by how much the "usual" Riesz transform identities fail to materialize. This goes in the opposite direction of the results in previous bullets, i.e., we now succeed in controlling flatness in terms of functional analytic features. See Theorems 5.5.8-5.5.9 and Theorems 5.5.11-5.5.12.

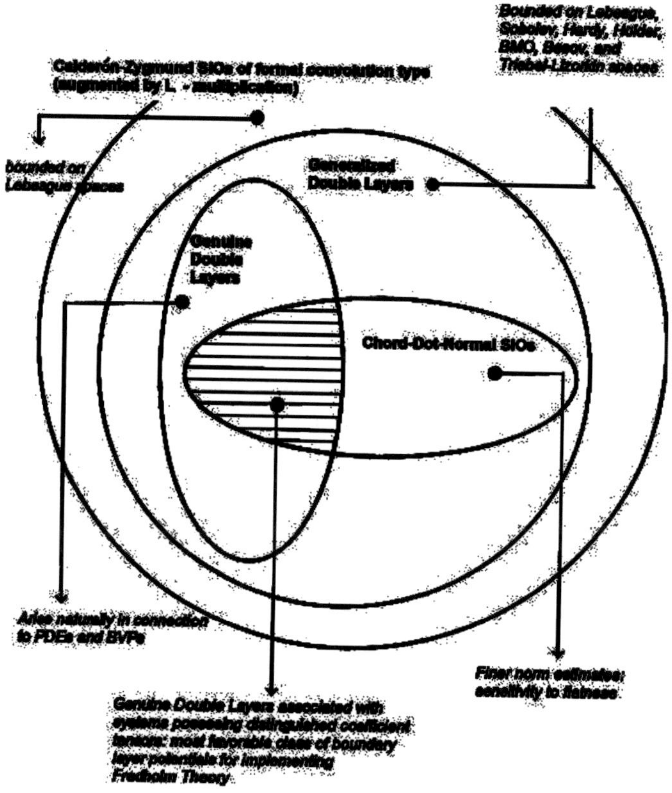

Fig. 5.1 The landscape of SIO's on UR sets

5.1 Only Chord-Dot-Normal SIO's May Induce Compact Operators on Smooth Bounded Surfaces

Here we shall consider singular integral operators on boundaries of UR domains in $\mathbb{R}^n$ whose kernels depend linearly on the geometric measure theoretic outward unit normal of the domain in question (see (5.1.2) below). In the class of all such singular integral operators, we are interested in identifying those who actually become *compact* when considered on bounded smooth domains.

Proposition 5.1.1 *Pick some $n \in \mathbb{N}$ with $n \geq 2$ along with $N = N(n) \in \mathbb{N}$ sufficiently large, and consider a vector-valued function*

$$\vec{k} \in \big[\mathscr{C}^N(\mathbb{R}^n \setminus \{0\})\big]^n, \text{ odd, positive homogeneous of degree } 1-n. \tag{5.1.1}$$

Also, for each UR *domain $\Omega \subseteq \mathbb{R}^n$ consider the principal-value singular integral operator*[2]

$$\begin{gathered} T_\Omega : L^2(\partial\Omega, \mathcal{H}^{n-1}) \longrightarrow L^2(\partial\Omega, \mathcal{H}^{n-1}) \text{ defined} \\ \text{for each } f \in L^2(\partial\Omega, \mathcal{H}^{n-1}) \text{ and } \mathcal{H}^{n-1}\text{-a.e. } x \in \partial\Omega \text{ as} \\ (T_\Omega f)(x) := \lim_{\varepsilon \to 0^+} \int_{y \in \partial\Omega,\ |x-y|>\varepsilon} \langle \nu_\Omega(y), \vec{k}(x-y)\rangle f(y)\, d\mathcal{H}^{n-1}(y) \end{gathered} \tag{5.1.2}$$

where ν_Ω is the geometric measure theoretic outward unit normal vector to Ω.

Then

$$\begin{gathered} \text{the operator } T_\Omega \text{ is compact whenever} \\ \Omega \subseteq \mathbb{R}^n \text{ is a bounded domain of class } \mathscr{C}^\infty \end{gathered} \tag{5.1.3}$$

if and only if there exists

$$\begin{gathered} k \in \mathscr{C}^N(\mathbb{R}^n \setminus \{0\}) \text{ scalar-valued function, even,} \\ \text{positive homogeneous of degree } -n, \text{ and satisfying} \\ \vec{k}(x) = x\, k(x) \text{ at each point } x \in \mathbb{R}^n \setminus \{0\}. \end{gathered} \tag{5.1.4}$$

In relation to this result we wish to make several comments.

Comment 1. It is well understood that the operator in (5.1.2) is *bounded* whenever $\Omega \subseteq \mathbb{R}^n$ is a bounded domain of class $\mathscr{C}^\infty$ (irrespective of whether there exists a function as in (5.1.4)). Here, the gist of the matter is that the *compactness* of said operator in the class of all smooth bounded domains implies the algebraic property of its kernel stipulated in (5.1.4).

Comment 2. Whenever one can find a function $k : \mathbb{R}^n \setminus \{0\} \to \mathbb{R}$ satisfying $\vec{k}(x) = x\, k(x)$ for each $x \in \mathbb{R}^n \setminus \{0\}$, the initial assumptions on $\vec{k}$ automatically imply that k is even, positive homogeneous of degree $-n$, and of class $\mathscr{C}^N$ in

---

[2] from item *(3)* of [131, Theorem 2.3.2] we know that this is well defined, linear, and bounded

$\mathbb{R}^n \setminus \{0\}$. Thus, as far as (5.1.3) implying the existence of a function as in (5.1.4) is concerned, the crux of the matter is the structural property in the last line of (5.1.4).

Comment 3. Whenever (5.1.4) holds it follows that $\vec{k}$ is divergence-free, hence T_Ω becomes what we have called a generalized double layer operator (whose main characteristic is that its integral kernel is the inner product of the outward unit normal with a divergence-free vector-valued kernel; see [132, (5.1.1)-(5.1.2), (5.1.5)-(5.1.6)]). In fact, if (5.1.4) holds then the operator T_Ω from (5.1.2) becomes

$$(T_\Omega f)(x) = \lim_{\varepsilon \to 0^+} \int_{y \in \partial\Omega,\ |x-y|>\varepsilon} \langle \nu_\Omega(y), x-y \rangle k(x-y) f(y)\, d\mathcal{H}^{n-1}(y) \tag{5.1.5}$$

for each $f \in L^2(\partial\Omega, \mathcal{H}^{n-1})$ and $\mathcal{H}^{n-1}$-a.e. $x \in \partial\Omega$, i.e., T_Ω is actually a chord-dot-normal singular integral operator.

Comment 4. Thanks to [129, Proposition 7.7.12] (or Krasnoselskiĭ's theorem; see, e.g., [15, Theorem 2.9, p. 203]), we may replace the integrability exponent 2 in (5.1.2) with any other $p \in (1, \infty)$.

Comment 5. A similar result is valid in the case when the kernel $\vec{k}$ is *matrix-valued* function[3] i.e., when for some $M \in \mathbb{N}$, we have

$$\vec{k} \in \big[\mathscr{C}^N(\mathbb{R}^n \setminus \{0\})\big]^{M \times n} \text{ is odd, positive homogeneous of degree } 1-n. \tag{5.1.6}$$

In such a scenario, the scalar-valued function k in (5.1.4) is actually a *vector-valued* function $k = (k_i)_{1\leq i\leq M} \in \big[\mathscr{C}^N(\mathbb{R}^n \setminus \{0\})\big]^M$ which is even and positive homogeneous of degree $-n$ and has the property that

$$\vec{k}(x) = \big(x_j k_i(x)\big)_{\substack{1\leq i\leq M \\ 1\leq j\leq n}} \text{ for each } x = (x_j)_{1\leq j\leq n} \in \mathbb{R}^n \setminus \{0\}. \tag{5.1.7}$$

This version is implied by Proposition 5.1.1 applied to the individual rows of (5.1.6).

We now present the proof of Proposition 5.1.1.

Proof of Proposition 5.1.1 In one direction, suppose the principal-value integral operator associated with each smooth bounded domain as in (5.1.2) is compact.

Pick an function arbitrary $\phi \in \mathscr{C}_c^\infty(\mathbb{R}^n)$. Also, fix a vector $\omega \in S^{n-1}$ and consider the hyperplane $H := \langle\omega\rangle^\perp$. Set $H^+ := \{x \in \mathbb{R}^n : x \cdot \omega < 0\}$. Next, pick $\psi \in \mathscr{C}_c^\infty(\mathbb{R}^n)$ which is identically one near the support of ϕ and choose a bounded domain $\Omega \subseteq \mathbb{R}^n$ of class $\mathscr{C}^\infty$ with the property that $H \cap \operatorname{supp}\psi \subseteq \partial\Omega$ and $\Omega \subseteq H^+$. Then the current working hypothesis entails that the operator

$$L^2(H, \mathcal{H}^{n-1}) \ni f \mapsto \psi T_\Omega(\phi f) \in L^2(H, \mathcal{H}^{n-1}) \text{ is compact} \tag{5.1.8}$$

---

[3] with the agreement that $\langle \nu_\Omega(y), \vec{k}(x-y) \rangle$ is now interpreted as the action of the matrix $\vec{k}(x-y)$ on the vector $\nu_\Omega(y)$

where we canonically identify functions from $L^2(H, \mathcal{H}^{n-1})$ supported in $H \cap \operatorname{supp} \phi$ with functions from $L^2(\partial\Omega, \mathcal{H}^{n-1})$ supported on $\partial\Omega \cap \operatorname{supp} \phi$ (via extension by zero outside the support), and we also canonically identify functions from $L^2(\partial\Omega, \mathcal{H}^{n-1})$ supported in $\partial\Omega \cap \operatorname{supp} \psi$ with functions from $L^2(H, \mathcal{H}^{n-1})$ supported on $H \cap \operatorname{supp} \psi$ (again, via extension by zero outside the support). Next we claim that the operator

$$\begin{gathered} R : L^2(H, \mathcal{H}^{n-1}) \longrightarrow L^2(H, \mathcal{H}^{n-1}) \text{ defined} \\ \text{for each } f \in L^2(H, \mathcal{H}^{n-1}) \text{ and } \mathcal{H}^{n-1}\text{-a.e. } x \in H \text{ as} \\ (Rf)(x) := \int_H (1 - \psi(x))\langle \omega, \vec{k}(x-y)\rangle (\phi f)(y)\, d\mathcal{H}^{n-1}(y) \end{gathered} \tag{5.1.9}$$

is a compact mapping. Indeed, this follows by observing that R is a Hilbert-Schmidt operator since its integral kernel, $K(x, y) := (1 - \psi(x))\langle \omega, \vec{k}(x-y)\rangle \phi(y)$ for each $(x, y) \in (H \times H) \setminus \text{diag}$ (where diag is the diagonal of the Cartesian product $H \times H$), satisfies

$$\begin{aligned} &\int_H \int_H |K(x,y)|^2 \, d\mathcal{H}^{n-1}(x)\, d\mathcal{H}^{n-1}(y) \\ &\qquad \le \int_H \int_H \frac{\mathbf{1}_{\operatorname{supp}\phi}(y)}{(1+|x|)^{2(n-1)}}\, d\mathcal{H}^{n-1}(x)\, d\mathcal{H}^{n-1}(y) < \infty. \end{aligned} \tag{5.1.10}$$

From (5.1.8) and the compactness of (5.1.9) we then conclude that the principal-value singular integral operator

$$\begin{gathered} Q : L^2(H, \mathcal{H}^{n-1}) \longrightarrow L^2(H, \mathcal{H}^{n-1}) \text{ defined} \\ \text{for each } f \in L^2(H, \mathcal{H}^{n-1}) \text{ and } \mathcal{H}^{n-1}\text{-a.e. } x \in H \text{ as} \\ (Qf)(x) := \lim_{\varepsilon \to 0^+} \int_{y \in H,\, |x-y|>\varepsilon} \langle \omega, \vec{k}(x-y)\rangle f(y)\, d\mathcal{H}^{n-1}(y) \end{gathered} \tag{5.1.11}$$

has the property that

$$\begin{gathered} \text{for each fixed function } \phi \in \mathscr{C}_c^\infty(\mathbb{R}^n) \text{ the assignment} \\ L^2(H, \mathcal{H}^{n-1}) \ni f \mapsto Q(\phi f) \in L^2(H, \mathcal{H}^{n-1}) \text{ is compact.} \end{gathered} \tag{5.1.12}$$

We next study in detail a special case, when $\omega := \mathbf{e}_n \in S^{n-1}$, a scenario in which $H = \langle \mathbf{e}_n \rangle^\perp = \mathbb{R}^{n-1} \times \{0\}$. In this setting, we agree to identify H with $\mathbb{R}^{n-1}$. Also, abbreviate $k := \langle \mathbf{e}_n, \vec{k} \rangle\big|_{\mathbb{R}^{n-1}\setminus\{0\}}$. With these conventions in mind, we have

$$k \in \mathscr{C}^N(\mathbb{R}^{n-1} \setminus \{0\}), \text{ odd, positive homogeneous of degree } 1-n, \tag{5.1.13}$$

and we may rephrase the compactness of Q in (5.1.11) simply as the statement that

$$\begin{gathered}\text{if for each } f \in L^2(\mathbb{R}^{n-1},\mathcal{L}^{n-1}) \text{ and } \mathcal{L}^{n-1}\text{-a.e. } x \in \mathbb{R}^{n-1} \text{ we set}\\ Tf(x) := \lim_{\varepsilon\to 0^+}\int_{y\in\mathbb{R}^{n-1},\,|x-y|>\varepsilon} k(x-y)f(y)\,dy\\ \text{then for each fixed } \phi \in \mathscr{C}_c^\infty(\mathbb{R}^n) \text{ the operator}\\ L^2(\mathbb{R}^{n-1},\mathcal{L}^{n-1}) \ni f \mapsto T(\phi f) \in L^2(\mathbb{R}^{n-1},\mathcal{L}^{n-1}) \text{ is compact.}\end{gathered} \tag{5.1.14}$$

Let $\mathscr{F}$ denote the Fourier transform in $\mathbb{R}^{n-1}$. Up to normalization, this is an isometry of the space $L^2(\mathbb{R}^{n-1},\mathcal{L}^{n-1})$ and we have

$$\mathscr{F}(Tf) = m\mathscr{F}f \text{ for each } f \in L^2(\mathbb{R}^{n-1},\mathcal{L}^{n-1}), \tag{5.1.15}$$

for some multiplier $m \in L^\infty(\mathbb{R}^{n-1},\mathcal{L}^{n-1})$ (see, e.g., [127, Theorem 4.100(c), p. 187], [74], [198]). In view of these properties and (5.1.14), we then conclude that

$$\begin{gathered}\text{for each fixed } R \in (0,\infty) \text{ the following is a compact mapping:}\\ \big\{f \in L^2(\mathbb{R}^{n-1},\mathcal{L}^{n-1}) : \operatorname{supp} f \subseteq B(0,R)\big\} \ni f \mapsto m\mathscr{F}f \in L^2(\mathbb{R}^{n-1},\mathcal{L}^{n-1}).\end{gathered} \tag{5.1.16}$$

To proceed, we agree to set

$$\begin{gathered}f_{(\varepsilon)}(x) := \varepsilon^{(1-n)/2} f(x/\varepsilon) \text{ for each}\\ f \in L^2(\mathbb{R}^{n-1},\mathcal{L}^{n-1}), \text{ each } x \in \mathbb{R}^{n-1}, \text{ and each } \varepsilon \in (0,\infty).\end{gathered} \tag{5.1.17}$$

Note that the non-standard normalization employed above is designed to ensure that

$$\begin{gathered}\big\|f_{(\varepsilon)}\big\|_{L^2(\mathbb{R}^{n-1},\mathcal{L}^{n-1})} = \|f\|_{L^2(\mathbb{R}^{n-1},\mathcal{L}^{n-1})}\\ \text{for each } f \in L^2(\mathbb{R}^{n-1},\mathcal{L}^{n-1}) \text{ and } \varepsilon \in (0,\infty).\end{gathered} \tag{5.1.18}$$

Fix now a function $\phi \in \mathscr{C}_c^\infty(\mathbb{R}^n)$ and set $\varphi := \phi\big|_{\mathbb{R}^{n-1}\times\{0\}}$, canonically identified with a function in $\mathscr{C}_c^\infty(\mathbb{R}^{n-1})$. Then for each $\varepsilon \in (0,\infty)$ the function $\mathscr{F}(\varphi_{(\varepsilon)})$ is continuous and satisfies

$$\mathscr{F}(\varphi_{(\varepsilon)})(\xi) = \varepsilon^{(n-1)/2}(\mathscr{F}\varphi)(\varepsilon\xi) \text{ for each } \xi \in \mathbb{R}^{n-1}. \tag{5.1.19}$$

Consequently,

$$\lim_{\varepsilon\to 0^+} \mathscr{F}(\varphi_{(\varepsilon)})(\xi) = 0 \text{ for each } \xi \in \mathbb{R}^{n-1}. \tag{5.1.20}$$

Next, pick $R \in (0,\infty)$ so that $\operatorname{supp}\varphi \subseteq B(0,R)$. From (5.1.17)-(5.1.18) we have

$$\begin{gathered}\big\|\varphi_{(\varepsilon)}\big\|_{L^2(\mathbb{R}^{n-1},\mathcal{L}^{n-1})} = \|\varphi\|_{L^2(\mathbb{R}^{n-1},\mathcal{L}^{n-1})} \text{ and}\\ \operatorname{supp}\varphi_{(\varepsilon)} \subseteq B(0,R) \text{ for each } \varepsilon \in (0,1).\end{gathered} \tag{5.1.21}$$

From this and (5.1.16) we then conclude that $\big\{m\mathscr{F}(\varphi_{(\varepsilon)})\big\}_{0<\varepsilon<1}$ is a relatively compact subset of the space $L^2(\mathbb{R}^{n-1},\mathcal{L}^{n-1})$. Hence, there exists a sequence

$\{\varepsilon_j\}_{j\in\mathbb{N}} \subseteq (0,1)$, which may also be assumed to satisfy $\lim_{j\to\infty} \varepsilon_j = 0$, with the property that $\{m\mathscr{F}(\varphi_{(\varepsilon_j)})\}_{j\in\mathbb{N}}$ converges in $L^2(\mathbb{R}^{n-1}, \mathcal{L}^{n-1})$. If we denote by $g \in L^2(\mathbb{R}^{n-1}, \mathcal{L}^{n-1})$ the limit function, by further restricting to a subsequence (something we omit to let the notation explicitly indicate), we may assume that we also have $\lim_{j\to\infty} m\mathscr{F}(\varphi_{(\varepsilon_j)})(\xi) = g(\xi)$ for $\mathcal{L}^{n-1}$-a.e. $\xi \in \mathbb{R}^{n-1}$. From this, (5.1.20), and the fact that m is essentially bounded, we then conclude that, in fact, $g(\xi) = 0$ for a.e. ξ in $\mathbb{R}^{n-1}$. Ultimately, this implies that $\lim_{j\to\infty} m\mathscr{F}(\varphi_{(\varepsilon_j)}) = 0$ in $L^2(\mathbb{R}^{n-1}, \mathcal{L}^{n-1})$ which, in view of (5.1.15) and the fact that $\mathscr{F}$ is an isomorphism of $L^2(\mathbb{R}^{n-1}, \mathcal{L}^{n-1})$, comes down to

$$\lim_{j\to\infty} T\varphi_{(\varepsilon_j)} = 0 \text{ in } L^2(\mathbb{R}^{n-1}, \mathcal{L}^{n-1}). \tag{5.1.22}$$

Since, as may be seen from (5.1.17) and the definition of T in (5.1.14) (bearing (5.1.13) in mind), we have

$$(Tf)_{(\varepsilon)} = Tf_{(\varepsilon)} \text{ for each } f \in L^2(\mathbb{R}^{n-1}, \mathcal{L}^{n-1}) \text{ and } \varepsilon \in (0,\infty), \tag{5.1.23}$$

we deduce from (5.1.22), (5.1.23), and (5.1.18) that

$$T\varphi = 0 \text{ for each function } \varphi \in \mathscr{C}_c^\infty(\mathbb{R}^{n-1}). \tag{5.1.24}$$

Recall that T is a particular embodiment of the operator Q from (5.1.11), corresponding to $\omega := \mathbf{e}_n \in S^{n-1}$. Via a rotation (which leave the Hausdorff measure invariant, preserves the Euclidean norm, and transforms $\vec{k}$ into a vector-field satisfying the same type of properties as in (5.1.1)), we then deduce from (5.1.24) that the operator Q from (5.1.11) also satisfies

$$Qf = 0 \text{ for each function } f \text{ of the form } \phi\big|_H \text{ with } \phi \in \mathscr{C}_c^\infty(\mathbb{R}^n). \tag{5.1.25}$$

In view of the arbitrariness of $\phi \in \mathscr{C}_c^\infty(\mathbb{R}^n)$, this further implies that, for each hyperplane $H := \langle\omega\rangle^\perp$ with $\omega \in S^{n-1}$, we have

$$\begin{gathered}\lim_{\varepsilon\to 0^+} \int_{y\in H,\ |x-y|>\varepsilon} \langle \omega, \vec{k}(x-y)\rangle \phi(y)\, d\mathcal{H}^{n-1}(y) = 0 \\ \text{for each } \phi \in \mathscr{C}_c^\infty(\mathbb{R}^n) \text{ and } \mathcal{H}^{n-1}\text{-a.e. } x \in H.\end{gathered} \tag{5.1.26}$$

With (5.1.26) in hand, Lemma 1.3.1 applies and proves the existence of a function k as in (5.1.4).

Parenthetically, we wish to note that we could have avoided the explicit use of the Fourier transform in the above argument and, instead, work with the operator Q from (5.1.11) throughout. Specifically, if in relation to (5.1.11) we now define

$$\begin{gathered} f_{(\varepsilon)}(x) := \varepsilon^{(1-n)/2} f(x/\varepsilon) \text{ for each} \\ f \in L^2(H, \mathcal{H}^{n-1}), \text{ each } x \in H, \text{ and each } \varepsilon \in (0,\infty), \end{gathered} \tag{5.1.27}$$

then for each $f \in L^2(H, \mathcal{H}^{n-1})$ and each $\varepsilon \in (0, \infty)$ we have

$$\big\|f_{(\varepsilon)}\big\|_{L^2(\mathbb{R}^{n-1},\mathcal{L}^{n-1})} = \|f\|_{L^2(\mathbb{R}^{n-1},\mathcal{L}^{n-1})}, \quad \operatorname{supp} f_{(\varepsilon)} \subseteq \varepsilon \cdot \operatorname{supp} f, \\ \text{and } Qf_{(\varepsilon)} = (Qf)_{(\varepsilon)}. \tag{5.1.28}$$

To proceed, observe that (5.1.12) is equivalent with the property that

$$\text{for each fixed } R \in (0, \infty) \text{ the following is a compact mapping:} \\ \big\{f \in L^2(H, \mathcal{H}^{n-1}) : \operatorname{supp} f \subseteq B(0, R)\big\} \ni f \mapsto Qf \in L^2(H, \mathcal{H}^{n-1}). \tag{5.1.29}$$

Let us now fix a function $\phi \in \mathscr{C}_c^\infty(\mathbb{R}^n)$ and set $\varphi := \phi\big|_H$. Also, pick $R \in (0, \infty)$ with the property that $\operatorname{supp}\phi \subseteq B(0, R)$. Then $\big\{\varphi_{(\varepsilon)}\big\}_{0<\varepsilon<1}$ is a bounded set in $\big\{f \in L^2(H, \mathcal{H}^{n-1}) : \operatorname{supp} f \subseteq B(0, R)\big\}$ and (5.1.29) implies that $\big\{Q(\varphi_{(\varepsilon)})\big\}_{0<\varepsilon<1}$ is a relatively compact subset of $L^2(H, \mathcal{H}^{n-1})$. As such, there exists a sequence $\{\varepsilon_j\}_{j\in\mathbb{N}} \subseteq (0, 1)$, for which we may also assume that $\lim\limits_{j\to\infty} \varepsilon_j = 0$, with the property that $\big\{Q(\varphi_{(\varepsilon_j)})\big\}_{j\in\mathbb{N}}$ converges to some function g in $L^2(H, \mathcal{H}^{n-1})$. Then

$$\begin{aligned} \|g\|_{L^2(H,\mathcal{H}^{n-1})} &= \lim_{j\to\infty} \big\|Q(\varphi_{(\varepsilon_j)})\big\|_{L^2(H,\mathcal{H}^{n-1})} = \lim_{j\to\infty} \big\|(Q\varphi)_{(\varepsilon_j)}\big\|_{L^2(H,\mathcal{H}^{n-1})} \\ &= \lim_{j\to\infty} \|Q\varphi\|_{L^2(H,\mathcal{H}^{n-1})} = \|Q\varphi\|_{L^2(H,\mathcal{H}^{n-1})}. \end{aligned} \tag{5.1.30}$$

Pick now an arbitrary $\Theta \in \mathscr{C}_c^\infty(\mathbb{R}^n)$ and define $\theta := \Theta\big|_H \in \mathscr{C}_c^\infty(H)$. Also, recall that $Q^\top$, the (real) transpose of Q in (5.1.11), is given by $Q^\top = -Q$ (as a very special case of [131, (2.3.25)]). Since Q is a classical convolution type singular integral operator, it preserves L^2-based Sobolev spaces on H (indeed, as a Fourier multiplier operator, Q commutes with differentiation). Hence, since θ belongs to any L^2-based Sobolev space, classical embedding theorems give $Q\theta \in \mathscr{C}^\infty(H) \cap L^2(H, \mathcal{H}^{n-1})$. Keeping this in mind we may then write

$$\begin{aligned} \int_H g\theta \,\mathrm{d}\mathcal{H}^{n-1} &= \lim_{j\to\infty} \int_H Q(\varphi_{(\varepsilon_j)})\theta \,\mathrm{d}\mathcal{H}^{n-1} = \lim_{j\to\infty} \int_H \varphi_{(\varepsilon_j)} Q^\top\theta \,\mathrm{d}\mathcal{H}^{n-1} \\ &= -\lim_{j\to\infty} \int_H \varphi_{(\varepsilon_j)}(x)(Q\theta)(x) \,\mathrm{d}\mathcal{H}^{n-1}(x) \\ &= -\lim_{j\to\infty} \varepsilon_j^{(1-n)/2} \int_H \varphi(x/\varepsilon_j)(Q\theta)(x) \,\mathrm{d}\mathcal{H}^{n-1}(x) \\ &= -\lim_{j\to\infty} \varepsilon_j^{(n-1)/2} \int_H \varphi(x)(Q\theta)(\varepsilon_j x) \,\mathrm{d}\mathcal{H}^{n-1}(x) = 0, \end{aligned} \tag{5.1.31}$$

where the last step may be justified, for example, based on Lebesgue's Dominated Convergence Theorem (whose applicability is ensured by the regularity of $Q\theta$ and the fact that φ is a test function). In turn, (5.1.31) forces $g = 0$ which, in concert with (5.1.30), proves (5.1.25). The rest of the argument leading up to the existence of a function k as in (5.1.4) is as before.

In the opposite direction, suppose $\Omega \subseteq \mathbb{R}^n$ is a bounded domain of class $\mathscr{C}^\infty$, and assume the existence of a function k as in (5.1.4). Then

$$\langle \nu_\Omega(y), \vec{k}(x-y)\rangle = \langle \nu_\Omega(y), x-y\rangle\, k(x-y) = O(|x-y|^{2-n}) \text{ for } x, y \in \partial\Omega, \quad (5.1.32)$$

so the principal-value singular integral operator T_Ω in (5.1.2) is actually weakly singular. In turn, this implies (cf., e.g., [81, Lemma 2.20, p. 2608]) that T_Ω is compact on $L^2(\partial\Omega, \mathcal{H}^{n-1})$. This finishes the proof of Proposition 5.1.1. □

Proposition 5.1.1 self-improves in the manner described in the result below.

Proposition 5.1.2 *Fix a number $n \in \mathbb{N}$ satisfying $n \geq 2$ along with some $N \in \mathbb{N}$, and consider two vector-valued functions*

$$\begin{gathered}\vec{k}_1, \vec{k}_2 \in \big[\mathscr{C}^N(\mathbb{R}^n \setminus \{0\})\big]^n, \text{ each of which is} \\ \text{odd and positive homogeneous of degree } 1-n.\end{gathered} \quad (5.1.33)$$

Next, for each UR *domain $\Omega \subseteq \mathbb{R}^n$ consider the principal-value singular integral operator*[4]

$$\begin{gathered}\mathcal{Q}_\Omega : L^2(\partial\Omega, \mathcal{H}^{n-1}) \longrightarrow L^2(\partial\Omega, \mathcal{H}^{n-1}) \text{ defined} \\ \text{for each } f \in L^2(\partial\Omega, \mathcal{H}^{n-1}) \text{ and } \mathcal{H}^{n-1}\text{-a.e. } x \in \partial\Omega \text{ as}\end{gathered}$$

$$(\mathcal{Q}_\Omega f)(x) := \lim_{\varepsilon\to 0^+} \int_{\substack{y\in\partial\Omega \\ |x-y|>\varepsilon}} \Big\{\langle \nu_\Omega(y), \vec{k}_1(x-y)\rangle + \langle \nu_\Omega(x), \vec{k}_2(x-y)\rangle\Big\} f(y)\, d\mathcal{H}^{n-1}(y) \quad (5.1.34)$$

where ν_Ω is the geometric measure theoretic outward unit normal vector to Ω. Then

$$\begin{gathered}\text{the operator } \mathcal{Q}_\Omega \text{ is compact whenever} \\ \Omega \subseteq \mathbb{R}^n \text{ is a bounded domain of class } \mathscr{C}^\infty\end{gathered} \quad (5.1.35)$$

if and only if there exists a scalar-valued function $k \in \mathscr{C}^N(\mathbb{R}^n \setminus \{0\})$ which is even and positive homogeneous of degree $-n$ with the property that

$$\vec{k}_1(x) + \vec{k}_2(x) = x\, k(x) \text{ for each } x \in \mathbb{R}^n \setminus \{0\}. \quad (5.1.36)$$

Comments 1-4 made in relation to Proposition 5.1.1 continue to be relevant here. In addition, we wish to note that Proposition 5.1.2 contains Proposition 5.1.1 (as the latter reduces precisely to the former in the case when $\vec{k}_2 = 0$).

The class of operators described in (5.1.34) is the linear span of all singular integral operators from (5.1.2) together with their transpositions.

Whenever (5.1.36) holds the operator $\mathcal{Q}_\Omega$ from (5.1.34) acquires the form

[4] from item *(3)* of [131, Theorem 2.3.2] we know that this is well defined, linear, and bounded

$$(Q_\Omega f)(x) = \lim_{\varepsilon\to 0^+} \int\limits_{y\in\partial\Omega,\ |x-y|>\varepsilon} \big\langle \nu_\Omega(x) - \nu_\Omega(y), \vec{k}_2(x-y)\big\rangle f(y)\, d\mathcal{H}^{n-1}(y)$$
$$+ \lim_{\varepsilon\to 0^+} \int\limits_{y\in\partial\Omega,\ |x-y|>\varepsilon} \langle \nu_\Omega(x), x-y\rangle k(x-y) f(y)\, d\mathcal{H}^{n-1}(y) \quad (5.1.37)$$

for each $f \in L^2(\partial\Omega, \mathcal{H}^{n-1})$ and $\mathcal{H}^{n-1}$-a.e. $x \in \partial\Omega$. Hence, Q_Ω becomes the sum of a chord-dot normal singular integral operator (cf. [132, Theorem 5.2.2]) and commutators between operators of pointwise multiplication with (the scalar components of) ν_Ω and standard convolution-like singular integral operators. In a nutshell, *these are the only types of singular integral operators within the class of operators considered in* (5.1.34) *which become compact whenever the underlying domain is smooth and bounded.*

Next, we present the proof of Proposition 5.1.2.

Proof of Proposition 5.1.2 In one direction, assume that the principal-value singular integral operator associated with each smooth bounded domain as in (5.1.34) is compact. We start by running the same argument as in the proof of Proposition 5.1.1 and conclude that the operator

$$L^2(H, \mathcal{H}^{n-1}) \ni f \mapsto \psi Q_\Omega(\phi f) \in L^2(H, \mathcal{H}^{n-1}) \text{ is compact.} \quad (5.1.38)$$

Since $\nu_\Omega(x) = \nu_\Omega(y) = \omega$ at all points $x, y \in H = \langle\omega\rangle^\perp$, the operator (5.1.38) corresponds precisely to the operator (5.1.8) for the kernel $\vec{k} := \vec{k}_1 + \vec{k}_2$. As such, the portion of the proof of Proposition 5.1.2 which, starting with (5.1.8), has led to the conclusion that there exists a function as in (5.1.4) presently gives the existence of a scalar-valued function $k \in \mathscr{C}^N(\mathbb{R}^n \setminus \{0\})$ which is even and positive homogeneous of degree $-n$ and such that (5.1.36) holds.

In the opposite direction, whenever (5.1.36) holds the operator Q_Ω from (5.1.34) takes the form (5.1.37). If we also assume that $\Omega \subseteq \mathbb{R}^n$ is a bounded domain of class $\mathscr{C}^\infty$, then [131, Theorem 2.7.3] guarantees that the (commutator) operator in the first line of (5.1.37) is compact on $L^2(\partial\Omega, \mathcal{H}^{n-1})$, while Proposition 5.1.1 ensures that the (chord-dot-normal) operator in the second line of (5.1.37) is compact on $L^2(\partial\Omega, \mathcal{H}^{n-1})$. Hence (5.1.35) is valid. □

We conclude by including the companion result to Proposition 5.1.1 formulated in Proposition 5.1.4 below, to the effect that Calderón-Zygmund singular integral operators of convolution type are never[5] compact on Lebesgue spaces. As a preamble, we record a useful observation in the next lemma.

Lemma 5.1.3 *Let $\Omega \subseteq \mathbb{R}^n$ be an Ahlfors regular domain. Abbreviate $\sigma := \mathcal{H}^{n-1}\lfloor\partial\Omega$ and denote by $\nu = (\nu_j)_{1\le j\le n}$ the geometric measure theoretic outward unit normal to Ω. Also, fix some exponent $p \in (1, \infty)$ and consider a linear and bounded operator $T : L^p(\partial\Omega, \sigma) \to L^p(\partial\Omega, \sigma)$. Then the operator T is actually compact if and*

[5] save for the zero operator

only if each $T_j : L^p(\partial\Omega, \sigma) \longrightarrow L^p(\partial\Omega, \sigma)$ *defined for every* $j \in \{1, \dots, n\}$ *as* $T_j f := T(\nu_j f)$ *for each* $f \in L^p(\partial\Omega, \sigma)$ *is a compact operator.*

Proof This follows since pointwise multiplication by each ν_j is a bounded operator on $L^p(\partial\Omega, \sigma)$, and $\sum_{j=1}^n T_j(\nu_j f) = Tf$ for each $f \in L^p(\partial\Omega, \sigma)$. □

Here is the result alluded to earlier, showing that the only Calderón-Zygmund singular integral operator of convolution type which is compact when acting on Lebesgue spaces considered on smooth compact surfaces is the zero operator.

Proposition 5.1.4 *Pick some* $n \in \mathbb{N}$ *with* $n \geq 2$ *along with* $N \in \mathbb{N}$, *and consider a scalar function* $k \in \mathscr{C}^N(\mathbb{R}^n \setminus \{0\})$ *which is odd and positive homogeneous of degree* $1 - n$. *Also, fix* $p \in (1, \infty)$ *and for each* UR *domain* $\Omega \subseteq \mathbb{R}^n$ *consider the principal-value singular integral operator*[6]

$$\begin{gathered} T_\Omega : L^p(\partial\Omega, \mathcal{H}^{n-1}) \longrightarrow L^p(\partial\Omega, \mathcal{H}^{n-1}) \textit{ defined} \\ \textit{for each } f \in L^p(\partial\Omega, \mathcal{H}^{n-1}) \textit{ and } \mathcal{H}^{n-1}\textit{-a.e. } x \in \partial\Omega \textit{ as} \\ (T_\Omega f)(x) := \lim_{\varepsilon \to 0^+} \int\limits_{y \in \partial\Omega,\ |x-y|>\varepsilon} k(x-y) f(y)\, d\mathcal{H}^{n-1}(y). \end{gathered} \tag{5.1.39}$$

Then

$$\begin{gathered} \textit{the operator } T_\Omega \textit{ is compact whenever} \\ \Omega \subseteq \mathbb{R}^n \textit{ is a bounded domain of class } \mathscr{C}^\infty \end{gathered} \tag{5.1.40}$$

if and only if $k = 0$.

Proof Assume (5.1.40) holds. Fix $\Omega \subseteq \mathbb{R}^n$, an arbitrary bounded domain of class $\mathscr{C}^\infty$ and denote by $\nu_\Omega = (\nu_{\Omega,j})_{1 \leq j \leq n}$ the outward unit normal to Ω. According to the result proved in Lemma 5.1.3, each $T_{\Omega,j} : L^p(\partial\Omega, \mathcal{H}^{n-1}) \to L^p(\partial\Omega, \mathcal{H}^{n-1})$ defined for every $j \in \{1, \dots, n\}$ as $T_{\Omega,j} f := T_\Omega(\nu_{\Omega,j} f)$ for each $f \in L^p(\partial\Omega, \mathcal{H}^{n-1})$ is then a compact operator. Since each $T_{\Omega,j}$ is as in (5.1.2) corresponding to the odd, positive homogeneous of degree $1 - n$, vector-valued kernel

$$\vec{k}_j := k\mathbf{e}_j \in \left[\mathscr{C}^N(\mathbb{R}^n \setminus \{0\})\right]^n, \tag{5.1.41}$$

from Proposition 5.1.1 (and Comment 4 following its statement) we conclude that for each index $j \in \{1, \dots, n\}$ there exists

$$\begin{gathered} \theta_j \in \mathscr{C}^N(\mathbb{R}^n \setminus \{0\}) \text{ scalar-valued function, even,} \\ \text{positive homogeneous of degree } -n, \text{ and satisfying} \\ k(x)\mathbf{e}_j = x\,\theta_j(x) \text{ at each point } x \in \mathbb{R}^n \setminus \{0\}. \end{gathered} \tag{5.1.42}$$

Taking the dot product with $\mathbf{e}_i$ of both sides in the above formula, for $i \in \{1, \dots, n\}$ such that $i \neq j$, then leads to the conclusion that $x_i\theta_j(x) = 0$ for each $x \in \mathbb{R}^n \setminus \{0\}$, whenever $1 \leq i \neq j \leq n$. Given that $n \geq 2$, this forces $\theta_j = 0$ in $\mathbb{R}^n \setminus \{0\}$ for each $j \in \{1, \dots, n\}$. Hence, ultimately $k = 0$ by (5.1.42). □

[6] from item *(3)* of [131, Theorem 2.3.2] we know that this is well defined, linear, and bounded

5.2 Chord-Dot-Normal SIO's on Lebesgue Spaces: the Flatter the Boundary, the Smaller the Essential Norm

The main goal in this section is to quantify the distance from a chord-dot-normal singular integral operator to the space of compact operators on Muckenhoupt weighted Lebesgue spaces in terms of the distance of the outward unit normal to the underlying domain to the Sarason space of functions with vanishing mean oscillations (on the boundary of said domain). The following lemma, which appears in [111], is going to be useful in this endeavor.

Lemma 5.2.1 *Pick $n \in \mathbb{N}$ with $n \geq 2$. Given a unit vector $\vec{n} \in S^{n-1}$, consider the hyperplane $H := \langle \vec{n} \rangle^{\perp} \subseteq \mathbb{R}^{n-1}$ and suppose $h : H \to \mathbb{R}$ is a function satisfying*

$$M := \sup_{x,y \in H,\, x \neq y} \frac{|h(x) - h(y)|}{|x - y|} < +\infty. \tag{5.2.1}$$

Let

$$\mathcal{G} := \big\{x + h(x)\vec{n} : x \in H\big\} \subseteq \mathbb{R}^n \tag{5.2.2}$$

denote the graph of h. Abbreviate $\sigma := \mathcal{H}^{n-1}\lfloor\mathcal{G}$ and denote by ν the unique unit normal to $\mathcal{G}$ satisfying $\nu \cdot \vec{n} < 0$ at σ-a.e. point on $\mathcal{G}$. Also, fix some integrability exponent $p \in (1, \infty)$. Given $k \in \mathscr{C}^{N+2}(\mathbb{R}^n \setminus \{0\})$, for some sufficiently large integer $N = N(n) \in \mathbb{N}$, which is even and positive homogeneous of degree $-n$, consider the maximal singular integral operator T acting on each $f \in L^p(\mathcal{G}, \sigma)$ as

$$T_* f(x) := \sup_{\varepsilon > 0} \Bigg| \int_{\substack{y \in \mathcal{G} \\ |x-y| > \varepsilon}} \langle x - y, \nu(y) \rangle k(x - y) f(y)\, d\sigma(y) \Bigg| \quad \textit{for all } x \in \mathcal{G}. \tag{5.2.3}$$

Then T_* is a well-defined continuous sub-linear mapping from $L^p(\mathcal{G}, \sigma)$ into itself and there exists $C(n, p) \in (0, \infty)$, depending only on n, p, such that

$$\|T_*\|_{L^p(\mathcal{G},\sigma) \to L^p(\mathcal{G},\sigma)} \leq C(n, p) M (1 + M)^{4n+N} \Bigg(\sum_{|\alpha| \leq N+2} \sup_{S^{n-1}} \big|\partial^{\alpha} k\big| \Bigg). \tag{5.2.4}$$

We also find it useful to state a particular case of [129, Proposition 7.5.4] for compact Ahlfors regular subsets of $\mathbb{R}^n$ which is going to be useful shortly.

Proposition 5.2.2 *Suppose $\Sigma \subseteq \mathbb{R}^n$ is a compact Ahlfors regular set, and abbreviate $\sigma := \mathcal{H}^{n-1}\lfloor\Sigma$. Also, fix an integer $\varkappa \in \mathbb{Z}$ with the property that*

$$2^{-\varkappa-1} < \operatorname{diam}(\Sigma) \leq 2^{-\varkappa}. \tag{5.2.5}$$

Then there are finite constants $a_1 \geq a_0 > 0$ such that, for each $i \in \mathbb{Z}$ with $i \geq \varkappa$, there exists a collection

$$\mathbb{D}_i(\Sigma) := \{Q^i_{\alpha}\}_{\alpha \in I_i} \tag{5.2.6}$$

of subsets of Σ *indexed by a nonempty, at most countable set* I_i*, as well as a family* $\{x^i_\alpha\}_{\alpha\in I_i}$ *of points in* Σ*, for which the collection of all dyadic cubes in* Σ*, i.e.,*

$$\mathbb{D}(\Sigma) := \bigcup_{i\in\mathbb{Z}} \mathbb{D}_i(\Sigma), \tag{5.2.7}$$

has the following properties:

(1) [All dyadic cubes are open] *For each* $i \in \mathbb{Z}$ *with* $i \geq \varkappa$ *and each* $\alpha \in I_i$ *the set* Q^i_α *is relatively open in* Σ*.*
(2) [Dyadic cubes are mutually disjoint within the same generation] *For each* $i \in \mathbb{Z}$ *with* $i \geq \varkappa$ *and each* $\alpha, \beta \in I_i$ *with* $\alpha \neq \beta$ *there holds* $Q^i_\alpha \cap Q^i_\beta = \varnothing$*;*
(3) [No partial overlap across generations] *For each* $i, \ell \in \mathbb{Z}$ *with* $\ell > i \geq \varkappa$ *and each* $\alpha \in I_i$*,* $\beta \in I_\ell$*, either* $Q^\ell_\beta \subseteq Q^i_\alpha$ *or* $Q^i_\alpha \cap Q^\ell_\beta = \varnothing$*.*
(4) [Any dyadic cube has a unique ancestor in any earlier generation] *For each integers* $i, \ell \in \mathbb{Z}$ *with* $i > \ell \geq \varkappa$ *and each* $\alpha \in I_i$ *there is a unique* $\beta \in I_\ell$ *such that* $Q^i_\alpha \subseteq Q^\ell_\beta$*. In particular, for each* $i \in \mathbb{Z}$ *with* $i \geq \varkappa$ *and each* $\alpha \in I_i$ *there exists a unique* $\beta \in I_{i-1}$ *such that* $Q^i_\alpha \subseteq Q^{i-1}_\beta$ *(a scenario in which* Q^{i-1}_β *is referred to as the parent of* Q^i_α*).*
(5) [The size is dyadically related to the generation] *For each* $i \in \mathbb{Z}$ *with* $i \geq \varkappa$ *and each* $\alpha \in I_i$ *one has*

$$\Delta(x^i_\alpha, a_0 2^{-i}) \subseteq Q^i_\alpha \subseteq \Delta_{Q^i_\alpha} := \Delta(x^i_\alpha, a_1 2^{-i}). \tag{5.2.8}$$

(6) [Control of the number of children] *There exists an integer* $M \in \mathbb{N}$ *with the property that for each* $i \in \mathbb{Z}$ *with* $i \geq \varkappa$ *and each* $\alpha \in I_i$ *one has*

$$\#\big\{\beta \in I_{i+1} : Q^{i+1}_\beta \subseteq Q^i_\alpha\big\} \leq M. \tag{5.2.9}$$

Moreover, for each $i \in \mathbb{Z}$ *with* $i \geq \varkappa$*, each* $x \in \Sigma$*, and each* $r \in (0, 2^{-i})$ *the number of* Q*'s in* $\mathbb{D}_i(\Sigma)$ *that intersect* $\Delta(x, r)$ *is at most* M*.*
(7) [Each generation covers the space σ-a.e.] *For each* $i \in \mathbb{Z}$ *with* $i \geq \varkappa$ *one has*

$$\sigma\Big(\Sigma \setminus \bigcup_{\alpha\in I_i} Q^i_\alpha\Big) = 0. \tag{5.2.10}$$

In particular,

$$N := \bigcup_{i\in\mathbb{Z},\, i\geq\varkappa} \Big(\Sigma \setminus \bigcup_{\alpha\in I_i} Q^i_\alpha\Big) \Longrightarrow \sigma(N) = 0, \tag{5.2.11}$$

and for each $i \in \mathbb{Z}$ *with* $i \geq \varkappa$ *and each* $\alpha \in I_i$ *one has*

$$\sigma\Big(Q^i_\alpha \setminus \bigcup_{\beta\in I_{i+1},\, Q^{i+1}_\beta \subseteq Q^i_\alpha} Q^{i+1}_\beta\Big) = 0. \tag{5.2.12}$$

(8) [Dyadic cubes have thin boundaries] *There exist constants, some small* $\vartheta \in (0, 1)$ *along with some large* $C \in (0, \infty)$, *such that for each* $i \in \mathbb{Z}$ *with* $i \geq \varkappa$, *each* $\alpha \in I_i$, *and each* $t > 0$ *one has*

$$\sigma\left(\left\{x \in Q_\alpha^i : \operatorname{dist}(x, \Sigma \setminus Q_\alpha^i) \leq t \cdot 2^{-i}\right\}\right) \leq Ct^\vartheta \cdot \sigma(Q_\alpha^i). \tag{5.2.13}$$

We now take up the task of estimating the essential norm (cf. (A.0.191)) of chord-dot-normal SIO's acting on Muckenhoupt weighted Lebesgue spaces in terms of the distance (measured in the John-Nirenberg space of functions with bounded mean oscillations) from the geometric measure theoretic outward unit normal to the Sarason space of functions with vanishing mean oscillations. In essence, this offers a quantified version of the heuristic principle stating that: *the flatter (at an infinitesimal level) the boundary, the closer the chord-dot-normal singular integral operator to being compact.* In such a qualitative fashion and for a more restrictive class of domains, this first appeared in [81, Theorem 4.36, pp. 2728-2729]. The following theorem is central for the present work. Its statement, which makes use of the notation introduced in (4.1.12), amounts to saying that

the essential operator norm of a chord-dot-normal singular integral operator acting on Muckenhoupt weighted Lebesgue spaces considered on the boundary of bounded Ahlfors regular domains, regarded as a function of the distance (measured in BMO*) of the geometric measure theoretic outward unit normal to the space* VMO, *vanishes in an at most asymptotically linear fashion (see* (4.1.23) *and the subsequent discussion) as the latter quantity goes to zero.* (5.2.14)

From a historical perspective, the fact that the classical (boundary-to-boundary) harmonic double layer operator K_Δ is compact on Lebesgue spaces on smooth bounded surfaces was at the core of the original development of Fredholm theory, and there has always been interest in understanding the nature of operators like K_Δ if the "surface" in question lacks regularity. For example, David Hilbert has posed this as a research question to Oliver Dimon Kellogg, while the latter was pursuing his doctorate in Göttigen in the early 1900's. More than a century later, Hilbert's pioneering vision retains its relevance, and our theorem below contributes to this line of research.

Theorem 5.2.3 *Let* $\Omega \subseteq \mathbb{R}^n$ *be a* UR *domain with compact boundary, denote by* ν *its geometric measure theoretic outward unit normal, and set* $\sigma := \mathcal{H}^{n-1}\lfloor\partial\Omega$. *Fix* $p \in (1, \infty)$ *along with a Muckenhoupt weight*[7] $w \in A_p(\partial\Omega, \sigma)$. *Also, consider a sufficiently large* $N = N(n) \in \mathbb{N}$. *Given* $k \in \mathscr{C}^N(\mathbb{R}^n \setminus \{0\})$ *which is even and positive homogeneous of degree* $-n$, *consider the chord-dot-normal singular integral operators* T, $T^\#$ *acting on each* $f \in L^p(\partial\Omega, w)$ *at* σ*-a.e.* $x \in \partial\Omega$, *respectively, as*

[7] recall the earlier convention of using the same symbol w for the measure associated with the given weight w as in [129, (7.7.1)]

$$Tf(x) := \lim_{\varepsilon\to 0^+} \int_{y\in\partial\Omega,\ |x-y|>\varepsilon} \langle x-y, \nu(y)\rangle k(x-y)f(y)\,d\sigma(y), \tag{5.2.15}$$

$$T^{\#}f(x) := \lim_{\varepsilon\to 0^+} \int_{y\in\partial\Omega,\ |x-y|>\varepsilon} \langle y-x, \nu(x)\rangle k(y-x)f(y)\,d\sigma(y). \tag{5.2.16}$$

Then for each $m \in \mathbb{N}$ there exists some $C_m \in (0,\infty)$ which depends only on m, n, p, $[w]_{A_p}$, the UR *constants of $\partial\Omega$, and the relative flatness ratio $\wp(\Omega)/\operatorname{diam}\partial\Omega$ (cf. Definition 3.1.9) so that, with the piece of notation introduced in* (4.1.12)*, one has*

$$\operatorname{dist}\big(T, \operatorname{Cp}(L^p(\partial\Omega, w))\big) \le C_m\Big(\sum_{|\alpha|\le N}\sup_{S^{n-1}}|\partial^\alpha k|\Big)\Big[\operatorname{dist}\big(\nu, \big[\mathrm{VMO}(\partial\Omega,\sigma)\big]^n\big)\Big]^{\langle m\rangle} \tag{5.2.17}$$

$$\operatorname{dist}\big(T^{\#}, \operatorname{Cp}(L^p(\partial\Omega, w))\big) \le C_m\Big(\sum_{|\alpha|\le N}\sup_{S^{n-1}}|\partial^\alpha k|\Big)\Big[\operatorname{dist}\big(\nu, \big[\mathrm{VMO}(\partial\Omega,\sigma)\big]^n\big)\Big]^{\langle m\rangle}. \tag{5.2.18}$$

Above, the distances in the left-hand side are measured in the operator norm in $\operatorname{Bd}(L^p(\partial\Omega, w))$*, while the distances in the right-hand side are measured in* $\big[\mathrm{BMO}(\partial\Omega,\sigma)\big]^n$*. In particular, corresponding to $w \equiv 1$,*

$$\operatorname{dist}\big(T, \operatorname{Cp}(L^p(\partial\Omega, \sigma))\big) \le C_m\Big(\sum_{|\alpha|\le N}\sup_{S^{n-1}}|\partial^\alpha k|\Big)\Big[\operatorname{dist}\big(\nu, \big[\mathrm{VMO}(\partial\Omega,\sigma)\big]^n\big)\Big]^{\langle m\rangle}, \tag{5.2.19}$$

$$\operatorname{dist}\big(T^{\#}, \operatorname{Cp}(L^p(\partial\Omega, \sigma))\big) \le C_m\Big(\sum_{|\alpha|\le N}\sup_{S^{n-1}}|\partial^\alpha k|\Big)\Big[\operatorname{dist}\big(\nu, \big[\mathrm{VMO}(\partial\Omega,\sigma)\big]^n\big)\Big]^{\langle m\rangle}. \tag{5.2.20}$$

Finally, when $\operatorname{dist}\big(\nu, \big[\mathrm{VMO}(\partial\Omega,\sigma)\big]^n\big)$ *is sufficiently small*[8] *relative to n and the Ahlfors regularity constants of $\partial\Omega$ one may take the constant $C_m \in (0,\infty)$ appearing in* (5.2.17)*-*(5.2.20) *to depend itself only on m, n, p, $[w]_{A_p}$, the Ahlfors regularity constants of $\partial\Omega$, and the relative flatness ratio $\wp(\Omega)/\operatorname{diam}\partial\Omega$.*

Before presenting the proof of this theorem, we make a number of remarks.

Remark 1. With the piece of notation introduced in (A.0.191), we may recast (5.2.17), (5.2.18) as

$$\|T\|^{\mathrm{ess}}_{L^p(\partial\Omega,w)\to L^p(\partial\Omega,w)} \le C_m\Big(\sum_{|\alpha|\le N}\sup_{S^{n-1}}|\partial^\alpha k|\Big)\Big[\operatorname{dist}\big(\nu, \big[\mathrm{VMO}(\partial\Omega,\sigma)\big]^n\big)\Big]^{\langle m\rangle}, \tag{5.2.21}$$

---

[8] that is, if Ω is a δ-oscillating AR domain with $\delta \in (0,1)$ sufficiently small relative to n and the Ahlfors regularity constants of $\partial\Omega$

and, respectively,

$$\|T^{\#}\|^{\text{ess}}_{L^p(\partial\Omega,w)\to L^p(\partial\Omega,w)} \leq C_m\Big(\sum_{|\alpha|\leq N}\sup_{S^{n-1}}|\partial^\alpha k|\Big)\Big[\operatorname{dist}\big(\nu,\big[\mathrm{VMO}(\partial\Omega,\sigma)\big]^n\big)\Big]^{\langle m\rangle}. \tag{5.2.22}$$

In turn, from (5.2.21)-(5.2.22) with $w\equiv 1$, real interpolation (cf. [129, (6.2.48)]), and [130, (1.4.71) in Proposition 1.4.24] (whose applicability for Lebesgue spaces is ensured by item *(3)* in [130, Proposition 7.3.6] together with [130, (7.1.55)]), we also conclude that whenever $p\in(1,\infty)$ and $q\in(0,\infty]$ we have

$$\begin{aligned}&\|T\|^{\text{ess}}_{L^{p,q}(\partial\Omega,\sigma)\to L^{p,q}(\partial\Omega,\sigma)}\\&\quad\leq C_m\Big(\sum_{|\alpha|\leq N}\sup_{S^{n-1}}|\partial^\alpha k|\Big)\Big[\operatorname{dist}\big(\nu,\big[\mathrm{VMO}(\partial\Omega,\sigma)\big]^n\big)\Big]^{\langle m\rangle},\end{aligned} \tag{5.2.23}$$

$$\begin{aligned}&\|T^{\#}\|^{\text{ess}}_{L^{p,q}(\partial\Omega,\sigma)\to L^{p,q}(\partial\Omega,\sigma)}\\&\quad\leq C_m\Big(\sum_{|\alpha|\leq N}\sup_{S^{n-1}}|\partial^\alpha k|\Big)\Big[\operatorname{dist}\big(\nu,\big[\mathrm{VMO}(\partial\Omega,\sigma)\big]^n\big)\Big]^{\langle m\rangle}.\end{aligned} \tag{5.2.24}$$

Remark 2. All these estimates are dilation invariant. Take, for example, the case of (5.2.17). Having fixed some $\lambda\in(0,\infty)$, define $\Omega_\lambda:=\lambda\Omega$, $\sigma_\lambda:=\mathcal{H}^{n-1}\lfloor\partial\Omega_\lambda$, and for each function $f:\partial\Omega\to\mathbb{C}$ set $(D_\lambda f)(x):=f(x/\lambda)$ for every point $x\in\partial\Omega_\lambda$. Finally, with the singular integral operator T as in (5.2.15), define $T_\lambda:=D_\lambda\circ T\circ D_{\lambda^{-1}}$. Then one may check directly from definitions that w_λ belongs to the class $A_p(\partial\Omega_\lambda,\sigma_\lambda)$ with $[w_\lambda]_{A_p}=[w]_{A_p}$ and $\operatorname{dist}\big(T_\lambda,\mathrm{Cp}(L^p(\partial\Omega_\lambda,w_\lambda))\big)$, measured in $\mathrm{Bd}(L^p(\partial\Omega_\lambda,w_\lambda))$, matches $\operatorname{dist}\big(T,\mathrm{Cp}(L^p(\partial\Omega,w))\big)$, measured in $\mathrm{Bd}(L^p(\partial\Omega,w))$. Also, if ν_λ is the geometric measure theoretic outward unit normal to Ω_λ then $\operatorname{dist}\big(\nu_\lambda,\big[\mathrm{VMO}(\partial\Omega_\lambda,\sigma_\lambda)\big]^n\big)$, considered in $\big[\mathrm{BMO}(\partial\Omega_\lambda,\sigma_\lambda)\big]^n$, matches $\operatorname{dist}\big(\nu,\big[\mathrm{VMO}(\partial\Omega,\sigma)\big]^n\big)$, considered in $\big[\mathrm{BMO}(\partial\Omega,\sigma)\big]^n$.

Remark 3. From (4.1.2) we see that

$$0\leq\operatorname{dist}\big(\nu,\big[\mathrm{VMO}(\partial\Omega,\sigma)\big]^n\big)\leq\|\nu\|_{[\mathrm{BMO}(\partial\Omega,\sigma)]^n}\leq 1. \tag{5.2.25}$$

Hence, if we abbreviate

$$\delta:=\operatorname{dist}\big(\nu,\big[\mathrm{VMO}(\partial\Omega,\sigma)\big]^n\big), \tag{5.2.26}$$

it follows that δ is a number belonging to the interval $[0,1]$. In view of this observation and (4.1.22), in place of $\Big[\operatorname{dist}\big(\nu,\big[\mathrm{VMO}(\partial\Omega,\sigma)\big]^n\big)\Big]^{\langle m\rangle}$ we could use

$$\delta\cdot\underbrace{\ln\Big(\cdots\ln\big(\ln({}^{m}e/\delta)\big)\cdots\Big)}_{m\text{ natural logarithms}} \tag{5.2.27}$$

in all estimates recorded in (5.2.17)-(5.2.18), (5.2.19)-(5.2.20), (5.2.21)-(5.2.24). In particular, all these essential operator norms are

$$O\Big(\delta\cdot\ln\big(e/\delta\big)\Big)\quad\text{(corresponding to } m=1\text{)},\tag{5.2.28}$$

$$O\Big(\delta\cdot\ln\big(\ln(e^e/\delta)\big)\Big)\quad\text{(corresponding to } m=2\text{)},\tag{5.2.29}$$

et cetera. Thus, all the aforementioned essential operator norms have at most linear growth in δ, up to arbitrarily many iterated logarithms.

In the same vein, we may invoke (4.1.16) to conclude that all essential operator norms in (5.2.17)-(5.2.18), (5.2.19)-(5.2.20), (5.2.21)-(5.2.24) are

$$O\big(\delta^{1-\varepsilon}\big)\ \text{ for each fixed } \varepsilon\in(0,1).\tag{5.2.30}$$

Remark 4. We can also prove a version of Theorem 5.2.3 in the class of chord-dot-normal singular integral operators with variable coefficient kernels. To elaborate, as in (5.2.31) assume that

> $b(x,z)$ is a function which is even and positive homogeneous of degree $-n$ in the variable $z\in\mathbb{R}^n\setminus\{0\}$, and such that $\partial_z^\alpha b(x,z)$ is continuous and bounded on $\mathbb{R}^n\times S^{n-1}$ for each multi-index $\alpha\in\mathbb{N}_0^n$ satisfying $|\alpha|\leq M$, where $M=M(n)$ is a sufficiently large integer. (5.2.31)

Given a UR domain $\Omega\subseteq\mathbb{R}^n$ with compact boundary, set $\sigma:=\mathcal{H}^{n-1}\lfloor\partial\Omega$ and denote by ν the geometric measure theoretic outward unit normal to Ω. Finally, for each $f\in L^1\big(\partial\Omega,\frac{\sigma(x)}{1+|x|^{n-1}}\big)$ define

$$Tf(x):=\lim_{\varepsilon\to0^+}\int_{\substack{y\in\partial\Omega,\ |x-y|>\varepsilon}}\langle x-y,\nu(y)\rangle b(x,x-y)f(y)\,d\sigma(y)\tag{5.2.32}$$

for σ-a.e. $x\in\partial\Omega$. Then for each integer $m\in\mathbb{N}$ and exponent $p\in(1,\infty)$ there exists some $C_m\in(0,\infty)$, depending only on m, n, p, and the UR constants of $\partial\Omega$ with the property that, with the piece of notation introduced in (4.1.12), the following essential norm estimate holds:

$$\begin{aligned}\|T\|^{\rm ess}_{L^p(\partial\Omega,\sigma)\to L^p(\partial\Omega,\sigma)}&=\operatorname{dist}\big(T,\operatorname{Cp}(L^p(\partial\Omega,\sigma))\big)\\&\leq C_m\cdot C_b\cdot\Big[\operatorname{dist}\big(\nu,\ [\mathrm{VMO}(\partial\Omega,\sigma)]^n\big)\Big]^{\langle m\rangle},\end{aligned}\tag{5.2.33}$$

where

$$C_b:=\sup_{\substack{(x,z)\in\mathbb{R}^n\times S^{n-1}\\|\alpha|\leq M}}\big|(\partial_z^\alpha b)(x,z)\big|\in(0,\infty).\tag{5.2.34}$$

To justify (5.2.33), recall the decomposition [132, (5.2.312)] and, for $\mu\in\mathbb{N}$ arbitrary, split

$$T=\sum_{\substack{\ell\in 2\mathbb{N}_0\\ \ell\leq\mu}}\sum_{1\leq i\leq H_\ell}a_{i\ell}T^{i\ell}+\sum_{\substack{\ell\in 2\mathbb{N}_0\\ \ell>\mu}}\sum_{1\leq i\leq H_\ell}a_{i\ell}T^{i\ell}. \tag{5.2.35}$$

Much as in [131, (2.5.505)], we may estimate

$$\Big\|\sum_{\substack{\ell\in 2\mathbb{N}_0\\ \ell>\mu}}\sum_{1\leq i\leq H_\ell}a_{i\ell}T^{i\ell}\Big\|_{L^p(\partial\Omega,\sigma)\to L^p(\partial\Omega,\sigma)}\leq\frac{C(\Omega,b,p,n)}{\mu}. \tag{5.2.36}$$

For each $\ell\in 2\mathbb{N}_0$ and $1\leq i\leq H_\ell$, let $C^{i\ell}\in\mathrm{Cp}\big(L^p(\partial\Omega,\sigma)\big)$ be such that

$$\|T^{i\ell}-C^{i\ell}\|_{L^p(\partial\Omega,\sigma)\to L^p(\partial\Omega,\sigma)}\leq 2\cdot\mathrm{dist}\,\big(T^{i\ell},\,\mathrm{Cp}(L^p(\partial\Omega,\sigma))\big). \tag{5.2.37}$$

Pick some $m\in\mathbb{N}$. By (5.2.17), for each $\ell\in 2\mathbb{N}_0$ and $1\leq i\leq H_\ell$ we have

$$\begin{aligned}&\mathrm{dist}\,\big(T^{i\ell},\,\mathrm{Cp}(L^p(\partial\Omega,\sigma))\big)\\&\quad\leq C_m\Big(\sum_{|\alpha|\leq N}\sup_{S^{n-1}}|\partial^\alpha k_{i\ell}|\Big)\Big[\mathrm{dist}\,\big(\nu,\,\big[\mathrm{VMO}(\partial\Omega,\sigma)\big]^n\big)\Big]^{\langle m\rangle}.\end{aligned} \tag{5.2.38}$$

Also, [132, (5.2.299), (5.2.304), (5.2.306)] ensure that

$$\sum_{\ell\in\mathbb{N}_0}\sum_{1\leq i\leq H_\ell}\Big(\sup_{\mathbb{R}^n}|a_{i\ell}|\Big)\Big(\sum_{|\alpha|\leq N}\sup_{S^{n-1}}|\partial^\alpha k_{i\ell}|\Big)\leq C_n\cdot C_b. \tag{5.2.39}$$

If we now define

$$C_\mu:=\sum_{\substack{\ell\in 2\mathbb{N}_0\\ \ell\leq\mu}}\sum_{1\leq i\leq H_\ell}a_{i\ell}C^{i\ell}\in\mathrm{Cp}\big(L^p(\partial\Omega,\sigma)\big), \tag{5.2.40}$$

then (5.2.35)-(5.2.40) imply

$$\begin{aligned}&\mathrm{dist}\,\big(T,\mathrm{Cp}(L^p(\partial\Omega,\sigma))\big)\leq\|T-C_\mu\|_{L^p(\partial\Omega,\sigma)\to L^p(\partial\Omega,\sigma)}\\&\quad\leq C_m\cdot C_b\cdot\Big[\mathrm{dist}\,\big(\nu,\,\big[\mathrm{VMO}(\partial\Omega,\sigma)\big]^n\big)\Big]^{\langle m\rangle}+\frac{C(\Omega,b,p,n)}{\mu}.\end{aligned} \tag{5.2.41}$$

Passing to limit $\mu\to\infty$ now yields (5.2.33). In addition, estimates similar to (5.2.33) are also valid for the chord-dot-normal singular integral operators with variable coefficient kernels acting on $f\in L^1\big(\partial\Omega,\frac{\sigma(x)}{1+|x|^{n-1}}\big)$ at σ-a.e. $x\in\partial\Omega$ according to

$$T^{\#}f(x):=\lim_{\varepsilon\to 0^+}\int_{y\in\partial\Omega,\ |x-y|>\varepsilon}\langle\nu(x),y-x\rangle b(y,y-x)f(y)\,d\sigma(y), \tag{5.2.42}$$

$$\widetilde{T}f(x):=\lim_{\varepsilon\to 0^+}\int_{y\in\partial\Omega,\ |x-y|>\varepsilon}\langle\nu(y),x-y\rangle b(y,x-y)f(y)\,d\sigma(y), \tag{5.2.43}$$

$$\widetilde{T}^{\#} f(x) := \lim_{\varepsilon \to 0^+} \int_{y \in \partial\Omega,\ |x-y|>\varepsilon} \langle \nu(x), y - x \rangle b(x, y - x) f(y)\, d\sigma(y). \tag{5.2.44}$$

Lastly, the aforementioned essential norm estimates remain valid on a variety of other function spaces, as indicated above.

In view of Definition 3.4.1 and [130, (1.2.53)], Theorem 5.2.3 implies the following compactness result on Muckenhoupt weighted Lebesgue spaces for chord-dot-normal singular integral operators on the boundaries of infinitesimally flat AR domains.

Corollary 5.2.4 *Let $\Omega \subseteq \mathbb{R}^n$ be an infinitesimally flat* AR *domain. That is, Ω is an Ahlfors regular domain with compact boundary and such that the geometric measure theoretic outward unit normal ν to Ω belongs to $\big[\mathrm{VMO}(\partial\Omega, \sigma)\big]^n$, where $\sigma := \mathcal{H}^{n-1} \lfloor \partial\Omega$ (cf. Definition 3.4.1). Fix a sufficiently large integer $N = N(n) \in \mathbb{N}$ and consider a complex-valued function $k \in \mathscr{C}^N(\mathbb{R}^n \setminus \{0\})$ which is even and positive homogeneous of degree $-n$. Bring in the chord-dot-normal singular integral operators T, $T^{\#}$ associated with k and Ω as in* (5.2.15), (5.2.16). *Then for each exponent $p \in (1, \infty)$ and each Muckenhoupt weight $w \in A_p(\partial\Omega, \sigma)$ one has*

$$T, T^{\#} \in \mathrm{Cp}\big(L^p(\partial\Omega, w)\big). \tag{5.2.45}$$

In particular, corresponding to $w \equiv 1$, (5.2.45) *implies*

$$T, T^{\#} \in \mathrm{Cp}\big(L^p(\partial\Omega, \sigma)\big) \text{ for each } p \in (1, \infty). \tag{5.2.46}$$

More generally, (5.2.23)-(5.2.24) *imply*

$$T, T^{\#} \in \mathrm{Cp}\big(L^{p,q}(\partial\Omega, \sigma)\big) \text{ for each } p \in (1, \infty) \text{ and } q \in (1, \infty]. \tag{5.2.47}$$

In the two-dimensional setting, we see from (5.2.46) and (3.4.48) that

if $\Omega \subseteq \mathbb{R}^2$ is a chord-arc domain with vanishing constant (i.e., $\varkappa$ in (3.4.45) is zero) and compact boundary then $T, T^{\#} \in \mathrm{Cp}\big(L^p(\partial\Omega, \sigma)\big)$ for each integrability exponent $p \in (1, \infty)$. (5.2.48)

A similar result is valid for Muckenhoupt weighted Lebesgue spaces (cf. (5.2.45)) and for Lorentz spaces (cf. (5.2.47)).

We also record the following two-dimensional consequence of Theorem 5.2.3.

Corollary 5.2.5 *Let Ω be a* UR *domain in $\mathbb{R}^2$. Abbreviate $\sigma := \mathcal{H}^1 \lfloor \partial\Omega$ and denote by τ the geometric measure theoretic unit tangent vector along $\partial\Omega$ (cf.* [129, (5.6.29)-(5.6.31)]). *Also, assume $F : \mathbb{R}^2 \setminus \{0\} \to \mathbb{R}$ is a function of class $\mathscr{C}^N$, for some sufficiently large $N \in \mathbb{N}$, which is even and positive homogeneous of degree zero. In this setting, consider the principal-value singular integral operators $T, T^{\#}$ on $\partial\Omega$ acting on each $f \in L^1\big(\partial\Omega, \frac{\sigma(x)}{1+|x|}\big)$ at σ-a.e. $x \in \partial\Omega$ according to*

$$Tf(x) := \lim_{\varepsilon\to 0^+} \int_{y\in\partial\Omega,\ |x-y|>\varepsilon} \partial_{\tau(y)}[F(x-y)]f(y)\,d\sigma(y) \tag{5.2.49}$$

and, respectively,

$$T^{\#}f(x) := \lim_{\varepsilon\to 0^+} \int_{y\in\partial\Omega,\ |x-y|>\varepsilon} \partial_{\tau(x)}[F(x-y)]\,d\sigma(y). \tag{5.2.50}$$

Then similar results as in Theorem 5.2.3 and Corollary 5.2.4 are valid for these singular integral operators.

Proof This is a direct consequence of Theorem 5.2.3 and Corollary 5.2.4, bearing in mind [132, Lemma 5.2.4]. □

Here is the proof of Theorem 5.2.3.

Proof of Theorem 5.2.3 We shall focus on establishing (5.2.17) since (5.2.18) follows from it, with the help of [130, Lemma 1.2.14] and the last part in [132, Theorem 5.2.2, item *(3)*]. We shall write the proof of (5.2.17) using an approach designed to shed light on the specific manner in which the right-hand side of (5.2.17) depends on the distance from ν to $\big[\mathrm{VMO}(\partial\Omega,\sigma)\big]^n$.

To begin, consider the maximal operator T_* acting on each $f \in L^p(\partial\Omega, w)$ as

$$T_*f(x) := \sup_{\varepsilon>0} \big|T_\varepsilon f(x)\big| \text{ for each } x \in \partial\Omega, \tag{5.2.51}$$

where, for each $\varepsilon > 0$,

$$T_\varepsilon f(x) := \int_{y\in\partial\Omega,\ |x-y|>\varepsilon} \langle x-y, \nu(y)\rangle k(x-y)f(y)\,d\sigma(y) \text{ for all } x \in \partial\Omega. \tag{5.2.52}$$

Then from [131, Theorem 2.3.2, *(10)*] we know that there exists $C \in (0,\infty)$, which depends only on n, p, $[w]_{A_p}$, and the UR constants of $\partial\Omega$ such that

$$\|T_*\|_{L^p(\partial\Omega,w)\to L^p(\partial\Omega,w)} \leq C\Big(\sum_{|\alpha|\leq N} \sup_{S^{n-1}} |\partial^\alpha k|\Big). \tag{5.2.53}$$

To proceed, recall the threshold $\varepsilon_4 \in (0,1)$ associated with the Ahlfors regular domain Ω as in Theorem 3.1.11 and bring in the parameter $\delta_* \in (0,1)$ appearing in the statement of Theorem 3.2.1. In addition, fix some sufficiently large number $C_o = C_o(\Omega) \in (1,\infty)$ depending only on n and the Ahlfors regularity constants of $\partial\Omega$. In the case when $\mathrm{dist}\big(\nu, \big[\mathrm{VMO}(\partial\Omega,\sigma)\big]^n\big) \geq \min\{\varepsilon_4, \delta_*\}/C_o$, the estimate claimed in (5.2.17) may be justified by making use of (5.2.53) to simply write

$$
\begin{aligned}
\operatorname{dist}\big(T, \mathrm{Cp}(L^p(\partial\Omega, w))\big) &\le \|T\|_{L^p(\partial\Omega,w)\to L^p(\partial\Omega,w)} \\
&\le \|T_*\|_{L^p(\partial\Omega,w)\to L^p(\partial\Omega,w)} \le C\Big(\sum_{|\alpha|\le N}\sup_{S^{n-1}}|\partial^\alpha k|\Big) \\
&\le \frac{C\cdot C_o}{\min\{\varepsilon_4,\delta_*\}}\Big(\sum_{|\alpha|\le N}\sup_{S^{n-1}}|\partial^\alpha k|\Big)\operatorname{dist}\big(\nu, \big[\mathrm{VMO}(\partial\Omega,\sigma)\big]^n\big). \qquad (5.2.54)
\end{aligned}
$$

In view of (4.1.17) and the fact that $\operatorname{dist}\big(\nu, \big[\mathrm{VMO}(\partial\Omega,\sigma)\big]^n\big) \le 1$, this readily implies (5.2.17). Thus, as far as the estimate claimed in (5.2.17) is concerned, there remains to consider the situation when

$$
C_o\cdot\operatorname{dist}\big(\nu, \big[\mathrm{VMO}(\partial\Omega,\sigma)\big]^n\big) < \min\{\varepsilon_4,\delta_*\}. \qquad (5.2.55)
$$

Assume this is the case, and observe that Theorem 3.4.2 implies

$\wp(\Omega) > 0$ and Ω is a two-sided NTA domain, in the sense of [129, Definition 5.11.1] with constants depending only on n, the Ahlfors regularity constants of $\partial\Omega$, and the relative flatness ratio $\wp(\Omega)/\operatorname{diam}\partial\Omega$; in particular, the UR constants of $\partial\Omega$ also depend solely on said entities. (5.2.56)

To proceed, pick some δ such that

$$
C_o\cdot\operatorname{dist}\big(\nu, \big[\mathrm{VMO}(\partial\Omega,\sigma)\big]^n\big) < \delta < \delta_*. \qquad (5.2.57)
$$

As visible from (5.2.15), the operator T depends in a homogeneous fashion on the kernel function k. As such, by working with the kernel k/K (in the case when k is not identically zero) where $K := \sum_{|\alpha|\le N}\sup_{S^{n-1}}|\partial^\alpha k|$, matters are reduced to proving that, whenever

$$
\sum_{|\alpha|\le N}\sup_{S^{n-1}}|\partial^\alpha k| \le 1, \qquad (5.2.58)
$$

for each $m\in\mathbb{N}$ there exists $C_m\in(0,\infty)$ which depends only on m, n, p, $[w]_{A_p}$, the ADR constants of $\partial\Omega$, and the relative flatness ratio $\wp(\Omega)/\operatorname{diam}\partial\Omega$ such that

$$
\operatorname{dist}\big(T, \mathrm{Cp}(L^p(\partial\Omega, w))\big) \le C_m\delta^{\langle m\rangle}. \qquad (5.2.59)
$$

Upon letting $\delta \searrow C_o\cdot\operatorname{dist}\big(\nu, \big[\mathrm{VMO}(\partial\Omega,\sigma)\big]^n\big)$ recovers (5.2.17) from (5.2.59) (cf. (5.2.57) and (4.1.15)). Henceforth, assume (5.2.55), (5.2.57), and (5.2.58).

To set the stage, let $\widetilde{R}\in(0,\infty)$ and $\lambda\in(32,\infty)$ be associated with the set Ω as in the statement of Theorem 3.1.12. In particular, both $\widetilde{R}$ and λ depend only on the dimension n, the Ahlfors regularity constants of $\partial\Omega$, and the relative flatness ratio $\wp(\Omega)/\operatorname{diam}\partial\Omega$. From [130, (3.1.55)] and (5.2.57) it follows that it is possible to select some $r^*_\delta\in\big(0,\operatorname{diam}\partial\Omega\big)$ such that

$$
\|\nu\|_*(\Delta(x,R)) < \delta, \qquad \forall x\in\partial\Omega, \quad \forall R\in(0, 20\lambda\, C r^*_\delta), \qquad (5.2.60)
$$

where $C = C(\Omega) \in (1, \infty)$ is a fixed, sufficiently large constant, depending only on the dimension n and the Ahlfors regularity constants of $\partial\Omega$. Choose r_δ such that

$$0 < r_\delta < \min\big\{\delta \cdot r_\delta^*, \widetilde{R}/(10C)\big\}. \tag{5.2.61}$$

We may then conclude from (3.1.68) (used with $\varepsilon := \delta \in (0, 1)$, $\gamma := 2$, and $x := z$) and (5.2.60) that there exists $C = C(\Omega) \in (1, \infty)$, depending only on n, the Ahlfors regularity constants of $\partial\Omega$, and the relative flatness ratio $\wp(\Omega)/\operatorname{diam}\partial\Omega$ such that

$$\sup_{x\in\partial\Omega}\ \sup_{y\in\Delta(x,2R)} R^{-1}|\langle x-y, \nu_{\Delta(x,R)}\rangle| \leq C\delta \text{ provided } 0 < R < 10\,Cr_\delta. \tag{5.2.62}$$

Furthermore, we shall assume that r_δ is small enough so the conclusions in Theorem 3.2.1 are valid, as stated, for the choice $R_\delta := 10Cr_\delta$. Going further, cover

$$\partial\Omega \subset \bigcup_{j=1}^{N} B(x_j, r_\delta), \qquad x_j \in \partial\Omega,\ 1 \leq j \leq N, \tag{5.2.63}$$

and assume that this has been refined (using Besicovitch's covering theorem; cf., e.g., [49, Theorem 2, p. 30]), so that the intervening balls have bounded overlap, independent of δ and N. That is,

there exists a purely dimensional constant $c(n) \in \mathbb{N}$ such that each point from $\mathbb{R}^n$ lies in at most $c(n)$ of the balls $\{B(x_j, r_\delta)\}_{1\leq j\leq N}$. (5.2.64)

Pick a family of functions $\{\varphi_j\}_{1\leq j\leq N}$ such that $\varphi_j \in \mathscr{C}_c^\infty(B(x_j, r_\delta))$, $0 \leq \varphi_j \leq 1$, for each j, and $\sum_{j=1}^N \varphi_j^2 = 1$ in a neighborhood of $\partial\Omega$. Also, for each $j \in \{1, \dots, N\}$, select $\psi_j \in \mathscr{C}_c^\infty(B(x_j, r_\delta))$ satisfying $0 \leq \psi_j \leq 1$ and which is identically one in a neighborhood of the support of φ_j. Finally, generally speaking, denote by M_b the operator of multiplication by the function b. We may then write

$$T = \sum_{j=1}^{N} M_{1-\psi_j} T M_{\varphi_j^2} + \sum_{j=1}^{N} M_{\psi_j} T M_{\varphi_j^2} \tag{5.2.65}$$

and note that the first sum in the right-hand side of (5.2.65) is a compact operator on $L^p(\partial\Omega, w)$. To see this, for every $j \in \{1, \dots, N\}$ decompose

$$M_{1-\psi_j} T M_{\varphi_j^2} = \sum_{i=1}^{n} (M_{1-\psi_j} T_i M_{\varphi_j^2}) M_{\nu_i} \tag{5.2.66}$$

where each T_i above is the (principal-value) singular integral operator on $\partial\Omega$ with integral kernel $(x_i - y_i)k(x-y)$. Since $(1-\psi_j(x))(x_i - y_i)k(x-y)\varphi_j(y)^2$ is continuous on the compact set $\partial\Omega \times \partial\Omega$, the operators $M_{1-\psi_j} T_i M_{\varphi_j^2}$ are compact as mappings from $L^1(\partial\Omega, \sigma)$ to $\mathscr{C}^0(\partial\Omega)$ by the Arzelà-Ascoli Theorem. Together with (5.2.66), this implies that

$$\sum_{j=1}^{N} M_{1-\psi_j} T M_{\varphi_j^2} \text{ is compact from } L^1(\partial\Omega, \sigma) \text{ into } L^\infty(\partial\Omega, \sigma). \tag{5.2.67}$$

Bearing in mind that $\partial\Omega$ is bounded, this further entails

$$L^\infty(\partial\Omega, \sigma) \hookrightarrow L^p(\partial\Omega, w) \hookrightarrow L^1(\partial\Omega, \sigma) \tag{5.2.68}$$

from which we ultimately conclude that

$$\sum_{j=1}^{N} M_{1-\psi_j} T M_{\varphi_j^2} \text{ is compact on } L^p(\partial\Omega, w). \tag{5.2.69}$$

As such, for the purpose we have in mind, it suffices to show that for each $m \in \mathbb{N}$ it is possible to find a constant $C_m \in (0, \infty)$ which depends only on m, n, p, $[w]_{A_p}$, the ADR constants of $\partial\Omega$, and the relative flatness ratio $\wp(\Omega)/\operatorname{diam}\partial\Omega$such that

$$\Big\| \sum_{j=1}^{N} M_{\psi_j} T M_{\varphi_j^2} \Big\|_{L^p(\partial\Omega, w) \to L^p(\partial\Omega, w)} \leq C_m \delta^{\langle m \rangle}. \tag{5.2.70}$$

Indeed, in light of (5.2.69) and (5.2.65) this implies (5.2.59), as wanted.

With the goal of proving (5.2.70), for an arbitrary $f \in L^p(\partial\Omega, w)$ we may write, making repeated use of the bounded overlap property (5.2.64),

$$\begin{aligned}
\Big(\int_{\partial\Omega} \Big| \sum_{j=1}^{N} \psi_j T(\varphi_j^2 f) \Big|^p \, dw \Big)^{1/p} &\leq C_{p,n} \Big(\int_{\partial\Omega} \sum_{j=1}^{N} |\psi_j T(\varphi_j^2 f)|^p \, dw \Big)^{1/p} \\
&\leq C_{p,n} \Big(\sum_{j=1}^{N} \| M_{\psi_j} T M_{\varphi_j} \|^p_{L^p(\partial\Omega, w) \to L^p(\partial\Omega, w)} \int_{\partial\Omega} |\varphi_j f|^p \, dw \Big)^{1/p} \\
&\leq C_{p,n} \max_{1 \leq j \leq N} \| M_{\psi_j} T M_{\varphi_j} \|_{L^p(\partial\Omega, w) \to L^p(\partial\Omega, w)} \Big(\int_{\partial\Omega} \sum_{j=1}^{N} |\varphi_j|^p |f|^p \, dw \Big)^{1/p} \\
&\leq C_{p,n} \max_{1 \leq j \leq N} \| M_{\psi_j} T M_{\varphi_j} \|_{L^p(\partial\Omega, w) \to L^p(\partial\Omega, w)} \Big(\int_{\partial\Omega} |f|^p \, dw \Big)^{1/p}.
\end{aligned} \tag{5.2.71}$$

Hence,

$$\begin{aligned}
&\Big\| \sum_{j=1}^{N} M_{\psi_j} T M_{\varphi_j^2} \Big\|_{L^p(\partial\Omega, w) \to L^p(\partial\Omega, w)} \\
&\qquad \leq C_{p,n} \max_{1 \leq j \leq N} \| M_{\psi_j} T M_{\varphi_j} \|_{L^p(\partial\Omega, w) \to L^p(\partial\Omega, w)}.
\end{aligned} \tag{5.2.72}$$

As such, (5.2.70) follows as soon as we show that for each $m \in \mathbb{N}$ there exists some $C_m \in (0, \infty)$ depending only on m, n, p, $[w]_{A_p}$, the ADR constants of $\partial\Omega$, and the relative flatness ratio $\wp(\Omega)/\operatorname{diam}\partial\Omega$ such that

$$\|M_{\psi_j} T M_{\varphi_j}\|_{L^p(\partial\Omega,w)\to L^p(\partial\Omega,w)} \leq C_m \delta^{\langle m \rangle}, \quad \text{for each } j \in \{1,\dots,N\}. \tag{5.2.73}$$

For the remaining of the proof we shall focus on establishing (5.2.73) for a fixed, arbitrary, index $j \in \{1,\dots,N\}$. We therefore find it convenient to re-denote $x_0 := x_j$ and, with r_δ as in (5.2.61), introduce

$$\Delta_0 := \Delta(x_0, r_\delta). \tag{5.2.74}$$

Since estimates on the gradients of φ_j and ψ_j are not used in subsequent calculations, we shall actually focus on proving a slightly more general result, involving two arbitrary functions satisfying

$$\begin{gathered} \varphi, \psi \in L^\infty(\partial\Omega,\sigma) \text{ with } \|\varphi\|_{L^\infty(\partial\Omega,\sigma)} \leq 1,\ \|\psi\|_{L^\infty(\partial\Omega,\sigma)} \leq 1 \\ \text{as well as } \operatorname{supp}\varphi \subseteq \Delta_0 \text{ and } \operatorname{supp}\psi \subseteq \Delta_0. \end{gathered} \tag{5.2.75}$$

Specifically, we seek to show that for each $m \in \mathbb{N}$ there exists $C_m \in (0,\infty)$ which depends only on m, n, p, $[w]_{A_p}$, the ADR constants of $\partial\Omega$, and the relative flatness ratio $\wp(\Omega)/\operatorname{diam}\partial\Omega$ such that

$$\|M_\psi T M_\varphi\|_{L^p(\partial\Omega,w)\to L^p(\partial\Omega,w)} \leq C_m \delta^{\langle m \rangle}. \tag{5.2.76}$$

The bulk of the proof of (5.2.76) is occupied by the justification of the following result (strongly reminiscent of an induction step, that allows us to boot-strap a weaker bound on the operator norm $\|M_\psi T_* M_\varphi\|_{L^p(\partial\Omega,w)\to L^p(\partial\Omega,w)}$ to a stronger one): knowing that there exists a function

$$\eta : [0,\infty) \longrightarrow [0,\infty) \tag{5.2.77}$$

which is quasi-increasing near the origin, i.e.,

$$\begin{gathered} \text{there exist } t_* > 0 \text{ and } C \in [1,\infty) \text{ such that} \\ \eta(t_0) \leq C\eta(t_1) \text{ whenever } 0 \leq t_0 < t_1 < t_*, \end{gathered} \tag{5.2.78}$$

such that for each integrability exponent $p \in (1,\infty)$ and each weight $w \in A_p(\partial\Omega,\sigma)$ there exists a constant $C \in (0,\infty)$, depending only on n, p, $[w]_{A_p}$, the ADR constants of $\partial\Omega$, the relative flatness ratio $\wp(\Omega)/\operatorname{diam}\partial\Omega$, and η, with the property that

$$\|M_\psi T_* M_\varphi\|_{L^p(\partial\Omega,w)\to L^p(\partial\Omega,w)} \leq C\eta(\delta), \tag{5.2.79}$$

implies that for each $p \in (1,\infty)$, each $w \in A_p(\partial\Omega,\sigma)$, and each function

$$\phi : [0,\infty) \longrightarrow [0,\infty) \tag{5.2.80}$$

satisfying

$$\begin{aligned}&\phi(t) > 0 \ \text{ for each } t > 0,\\ \phi(0) = \lim_{t\to 0^+}&\phi(t) = 0, \quad \phi'(0) = \lim_{t\to 0^+}\phi(t)/t = \infty,\\ \text{and } \eta(t)&\cdot\phi(t)^{-1}\cdot \mathrm{e}^{-\phi(t)/t} = O(1) \ \text{ as } t \to 0^+,\end{aligned} \tag{5.2.81}$$

there exists a constant $C \in (0,\infty)$ depending only on n, p, $[w]_{A_p}$, the ADR constants of $\partial\Omega$, the relative flatness ratio $\wp(\Omega)/\mathrm{diam}\,\partial\Omega$, η, and ϕ, such that we also have

$$\|M_\psi T_* M_\varphi\|_{L^p(\partial\Omega,w)\to L^p(\partial\Omega,w)} \leq C\phi(\delta). \tag{5.2.82}$$

Henceforth we shall summarize the above claim by simply saying that "(5.2.79) implies (5.2.82)."

In connection with (5.2.81) we wish to make two remarks. Our first remark pertains to the case when in addition to (5.2.77)-(5.2.78) we assume

$$\lim_{t\to 0^+} \eta(t)/t = \infty. \tag{5.2.83}$$

This additional hypothesis ensures that

$$t_e := \sup\big\{t_o \in (0,\infty) : \ \eta(t)/t > e \ \text{ for all } t \in (0,t_o)\big\} \in (0,\infty] \tag{5.2.84}$$

is well defined and we have

$$\eta(t)/t > e \ \text{ for all } t \in (0,t_e). \tag{5.2.85}$$

Then among all functions $\phi : [0,\infty) \to [0,\infty)$ satisfying the last property in (5.2.81) the smallest (up to multiplicative constants) in terms of behavior near the origin is actually the function

$$\begin{aligned}&\widehat{\eta} : [0,\infty) \longrightarrow [0,\infty) \text{ given for each } t \geq 0 \text{ by}\\ &\widehat{\eta}(0) := 0, \quad \widehat{\eta}(t) := t\ln(\eta(t)/t) \ \text{ if } t \in (0,t_e),\\ &\text{and } \widehat{\eta}(t) := t_e\ln(\eta(t_e)/t_e) \ \text{ for all } t \in [t_e,\infty).\end{aligned} \tag{5.2.86}$$

To justify the minimality of (5.2.86), observe that the property in the last line of (5.2.81) implies that there exist $t_*, M \in (0,\infty)$ such that

$$\eta(t) \leq M\phi(t)\cdot \mathrm{e}^{\phi(t)/t} \ \text{ for each } t \in (0,t_*). \tag{5.2.87}$$

Elementary calculus gives $x\mathrm{e}^x \leq \mathrm{e}^{2x-1}$ for each $x \in [0,\infty)$. From this used with $x := \phi(t)/t$ and (5.2.87) we then obtain $\eta(t)/t \leq M\mathrm{e}^{2\phi(t)/t-1}$ for each $t \in (0,t_*)$. In turn, this forces $\frac{1}{2}t\ln\big(e\eta(t)/(Mt)\big) \leq \phi(t)$ for each $t \in (0,t_*)$, and since thanks to (5.2.83) we have

$$\begin{aligned}\lim_{t\to 0^+}\frac{\frac{1}{2}t\ln\big(e\eta(t)/Mt\big)}{t\ln(\eta(t)/t)} &= \lim_{t\to 0^+}\frac{\frac{1}{2}\ln(e/M) + \frac{1}{2}\ln(\eta(t)/t)}{\ln(\eta(t)/t)}\\ &= \frac{1}{2} + \frac{1}{2}\ln(e/M)\lim_{t\to 0^+}\frac{1}{\ln(\eta(t)/t)} = \frac{1}{2},\end{aligned} \tag{5.2.88}$$

we ultimately conclude that

$$\text{given any } \phi : [0,\infty) \longrightarrow [0,\infty) \text{ satisfying the last property in (5.2.81) it follows that } \phi(t) \text{ dominates, up to a multiplicative constant, } \widehat{\eta}(t) \text{ for all } t \geq 0 \text{ sufficiently close to 0.} \tag{5.2.89}$$

This justifies the claim about the minimality of $\widehat{\eta}$ made in the previous paragraph.

The second remark we wish to make in connection with (5.2.81) is that if in addition to (5.2.77)-(5.2.78) and (5.2.83) we also assume that

$$\eta \text{ is continuous and } \lim_{t\to 0^+} t\ln(\eta(t)/t) = 0, \tag{5.2.90}$$

then

$$\text{the function } \widehat{\eta} \text{ defined in (5.2.86) is continuous, quasi-increasing near the origin (in the sense of (5.2.78)), } \lim_{t\to 0^+} t\ln(\widehat{\eta}(t)/t) = 0, \text{ and the function } \phi := \widehat{\eta} \text{ satisfies all properties listed in (5.2.81).} \tag{5.2.91}$$

That $\widehat{\eta}$ is continuous is clear from (5.2.86) and (5.2.90). To check the second claim made in (5.2.91), observe that

$$\text{for each fixed } x \in (0,\infty) \text{ the function } (0,\infty) \ni y \longmapsto x\ln(y/x) \text{ is strictly increasing and, for each fixed } y \in (0,\infty) \text{ the function } (0, y/e) \ni x \longmapsto x\ln(y/x) \text{ is also strictly increasing.} \tag{5.2.92}$$

If $t_* > 0$ and $C \in (0,\infty)$ are as in (5.2.78), if $t_e \in (0,\infty)$ is as in (5.2.85), and if $t^* > 0$ is small enough such that

$$\max\{C, e/C\} \leq \eta(t)/t \text{ for each } t \in (0, t^*), \tag{5.2.93}$$

(something we may always arrange, thanks to the assumption made in (5.2.83)) then whenever we have $0 \leq t_0 < t_1 < \min\{t_*, t^*, t_e\}$ we may write (using (5.2.86), (5.2.78), (5.2.85), (5.2.92), and (5.2.93))

$$\begin{aligned}\widehat{\eta}(t_0) &= t_0\ln(\eta(t_0)/t_0) \leq t_0\ln(C\eta(t_1)/t_0) < t_1\ln(C\eta(t_1)/t_1)\\ &= t_1\ln(C) + t_1\ln(\eta(t_1)/t_1) \leq 2t_1\ln(\eta(t_1)/t_1) = 2\widehat{\eta}(t_1),\end{aligned} \tag{5.2.94}$$

ultimately proving that $\widehat{\eta}$ is, as claimed, quasi-increasing near the origin. In fact, the same type of argument as in (5.2.94) (with $C := 1$) shows that

$$\text{if the original function } \eta \text{ is genuinely non-decreasing, then the function } \widehat{\eta} \text{ associated with } \eta \text{ as in (5.2.86) is strictly increasing on } (0, t_e) \text{ and constant thereafter.} \tag{5.2.95}$$

Next, (5.2.86) readily implies (bearing in mind that the function η is continuous; cf. (5.2.90)) that $\widehat{\eta}(t) > 0$ for each $t > 0$. The fact that $\widehat{\eta}$ is continuous at the origin is seen from (5.2.86) and (5.2.90). Furthermore, (5.2.83) implies

$$\lim_{t\to 0^+} \widehat{\eta}(t)/t = \lim_{t\to 0^+} \ln(\eta(t)/t) = \infty. \tag{5.2.96}$$

Let us also note here that (5.2.96), (5.2.86), the fact that $\ln(\ln x) \leq \ln x$ for each $x > 1$, and (5.2.90) allow us to write

$$\begin{aligned} 0 \leq \liminf_{t\to 0^+} t \ln(\widehat{\eta}(t)/t) &\leq \limsup_{t\to 0^+} t \ln(\widehat{\eta}(t)/t) = \limsup_{t\to 0^+} t \ln\big(\ln(\eta(t)/t)\big) \\ &\leq \limsup_{t\to 0^+} t \ln(\eta(t)/t) = 0, \end{aligned} \tag{5.2.97}$$

ultimately proving that, as claimed, $\lim_{t\to 0^+} t \ln(\widehat{\eta}(t)/t) = 0$. Finally, (5.2.86) and (5.2.83) give

$$\begin{aligned} \eta(t) \cdot \widehat{\eta}(t)^{-1} \cdot e^{-\widehat{\eta}(t)/t} &= \eta(t) \cdot \frac{1}{t \ln(\eta(t)/t)} \cdot e^{-\ln(\eta(t)/t)} \\ &= \frac{1}{\ln(\eta(t)/t)} = o(1) \text{ as } t \to 0^+. \end{aligned} \tag{5.2.98}$$

This completes the proof of (5.2.91).

Assuming for the time being that (5.2.79) implies (5.2.82), let us explain how this inductive step may be used to establish (5.2.76). From item *(10)* in [131, Theorem 2.3.2] (which guarantees that the maximal operator T_* is bounded in $L^p(\partial\Omega, w)$ for each $p \in (1, \infty)$ and each $w \in A_p(\partial\Omega, \sigma)$ with norm controlled solely in terms of n, p, $[w]_{A_p}$, and the ADR constants of $\partial\Omega$) we conclude that (5.2.79) holds for $\eta := \eta_0$ where the latter is the constant function

$$\eta_0(t) := 1 \text{ for each } t \in [0, \infty). \tag{5.2.99}$$

Incidentally, we may recast this as $\eta_0(t) = t^{\langle 0 \rangle}$ for each $t \in [0, \infty)$ (cf. (4.1.11)). This choice of function satisfies (5.2.78) (in fact, η_0 is non-decreasing), as well as (5.2.83) and (5.2.90). Granted these, we conclude from (5.2.91) and the working hypothesis, according to which (5.2.79) implies (5.2.82), that (5.2.82) holds with

$$\eta_1 := \widehat{\eta_0} \tag{5.2.100}$$

playing the role of the function ϕ. This selection of the function ϕ is actually optimal, since $\widehat{\eta_0}$ enjoys the minimality property described in (5.2.89). Specifically, given any $\phi : [0, \infty) \to [0, \infty)$ satisfying the last property in (5.2.81) with $\eta := \eta_0$ it follows that $\phi(t)$ dominates, up to a multiplicative constant, the quantity $\eta_1(t) = \widehat{\eta_0}(t)$ for all $t \geq 0$ sufficiently close to 0. In addition, from (5.2.91) and (5.2.95) we see that

η_1 is continuous, globally non-decreasing, strictly increasing near the origin, and satisfies $\lim_{t\to 0^+} \eta_1(t)/t = \infty$ as well as $\lim_{t\to 0^+} t \ln(\eta_1(t)/t) = 0$. (5.2.101)

In fact, according to (5.2.84)-(5.2.86), we have

$$\begin{aligned} &\eta_1 : [0,\infty) \longrightarrow [0,\infty) \text{ is given for each } t \geq 0 \text{ by} \\ &\eta_1(0) := 0, \quad \eta_1(t) := t\ln(1/t) \text{ if } t \in (0,1/e), \\ &\quad\text{and } \eta_1(t) := 1/e \text{ for all } t \in [1/e,\infty), \end{aligned} \tag{5.2.102}$$

hence (cf. (4.1.12))

$$\eta_1(t) = t^{\langle 1\rangle} \text{ for each } t \in [0,\infty). \tag{5.2.103}$$

In view of the aforementioned properties of η_1 (cf. (5.2.101)) and the fact that (5.2.82) holds with η_1 playing the role of the function ϕ, the present working hypothesis (according to which (5.2.79) implies (5.2.82)) shows that (5.2.82) also holds with $\eta_2 := \widehat{\eta_1}$ playing the role of the function ϕ, and that η_2 satisfies similar properties to those listed in (5.2.101). Actually, (5.2.102) and (5.2.84)-(5.2.86) yield a concrete description of η_2, namely:

$$\begin{aligned} &\eta_2 : [0,\infty) \longrightarrow [0,\infty) \text{ is given for each } t \geq 0 \text{ by} \\ &\eta_2(0) := 0, \quad \eta_2(t) := t\ln\big(\ln(1/t)\big) \text{ if } t \in (0,1/\mathrm{e}^{\mathrm{e}}), \\ &\quad\text{and } \eta_2(t) := 1/\mathrm{e}^{\mathrm{e}} \text{ for all } t \in [1/\mathrm{e}^{\mathrm{e}},\infty). \end{aligned} \tag{5.2.104}$$

Equivalently (cf. (4.1.12)), $\eta_2(t) = t^{\langle 2\rangle}$ for each $t \in [0,\infty)$. Iterating this scheme m times then proves (see (4.1.14)) that (5.2.82) holds with ϕ replaced by the function described (using notation introduced in (4.1.12)-(4.1.13)) as

$$\eta_m : [0,\infty) \longrightarrow [0,\infty), \quad \eta_m(t) = t^{\langle m\rangle} \text{ for each } t \in [0,\infty). \tag{5.2.105}$$

This induction establishes (5.2.76), modulo the proof of the fact that (5.2.79) implies (5.2.82) (which we shall deal with momentarily). The above line of reasoning explains the format of the conclusion in (5.2.76), while it also makes it clear that (5.2.76) is the best outcome one can produce working under the assumption that (5.2.79) implies (5.2.82).

On to the proof of the fact that (5.2.79) implies (5.2.82). Our working hypothesis is that there exists some function $\eta : [0,\infty) \to [0,\infty)$ which is quasi-increasing near the origin (in the sense of (5.2.78)) such that for each $p \in (1,\infty)$ and each $w \in A_p(\partial\Omega,\sigma)$ the estimate recorded in (5.2.79) holds for some constant $C \in (0,\infty)$ depending only on n, p, $[w]_{A_p}$, the ADR constants of $\partial\Omega$, the relative flatness ratio $\wp(\Omega)/\operatorname{diam}\partial\Omega$, and η. Having fixed a function ϕ as in (5.2.80)-(5.2.81), the goal is to prove (5.2.82). To get started, fix an integer $\varkappa \in \mathbb{Z}$ with the property that

$$2^{-\varkappa-1} < \operatorname{diam}(\partial\Omega) \leq 2^{-\varkappa}. \tag{5.2.106}$$

Since $r_\delta < \operatorname{diam}(\partial\Omega)$ it is possible to choose an integer

$$i \in \mathbb{Z} \text{ with } i \geq \varkappa \tag{5.2.107}$$

such that

$$2^{-i-1} < r_\delta \leq 2^{-i}. \tag{5.2.108}$$

Assuming this is the case we then have

$$\Delta(x_0, 2^{-i-1}) \subseteq \Delta_0 \subseteq \Delta(x_0, 2^{-i}). \tag{5.2.109}$$

Next, for each fixed $\gamma \geq 0$ bring in a brand of Hardy-Littlewood maximal operator which associates to each σ-measurable function f on $\partial\Omega$ the function $\mathcal{M}_\gamma f$ defined as

$$\mathcal{M}_\gamma f(x) := \sup_{\Delta \ni x} \left(⨍_\Delta |f|^{1+\gamma} \, d\sigma \right)^{1/(1+\gamma)} \quad \text{for each } x \in \partial\Omega, \tag{5.2.110}$$

with the supremum is taken over all surface balls $\Delta \subseteq \partial\Omega$ containing x. Also, consider a dyadic grid $\mathbb{D}(\partial\Omega)$ on the Ahlfors regular set $\partial\Omega$ (as in Proposition 5.2.2, presently used with $\Sigma := \partial\Omega$). Upon recalling (5.2.106)-(5.2.107) and (5.2.6), define

$$\mathcal{Q}_0 := \{ Q \in \mathbb{D}_i(\partial\Omega) : Q \cap 2\Delta_0 \neq \varnothing \} \tag{5.2.111}$$

then introduce

$$I_0 := \bigcup_{Q \in \mathcal{Q}_0} Q. \tag{5.2.112}$$

By design, I_0 is a relatively open subset of $\partial\Omega$. Recall the parameter $a_1 > 0$ appearing in (5.2.8) of Proposition 5.2.2. We claim that

$$I_0 \subseteq a\Delta_0 \text{ where } a := 4(1 + a_1). \tag{5.2.113}$$

Indeed, if $x \in I_0$ then $x \in Q$ for some $Q \in \mathcal{Q}_0$. In particular, $Q \cap 2\Delta_0 \neq \varnothing$ so we may pick some $y \in Q \cap 2\Delta_0$. Then $x, y \in Q \subseteq \Delta(x_Q, a_1 2^{-i})$ by (5.2.8), where x_Q denotes the "center" of the dyadic cube Q. Consequently, $|x - y| < a_1 2^{-i+1}$ which, in turn, permits us to estimate

$$|x - x_0| \leq |x - y| + |y - x_0| < a_1 2^{-i+1} + 2^{-i+1} = a \cdot 2^{-i-1}. \tag{5.2.114}$$

Thus $x \in B(x_0, a \cdot 2^{-i-1}) \cap \partial\Omega \subseteq a\Delta_0$ by (5.2.109), proving the inclusion in (5.2.113). We also claim that

$$\exists \text{ a } \sigma\text{-measurable set } N \subseteq \partial\Omega \text{ such that } \sigma(N) = 0 \text{ and } 2\Delta_0 \setminus N \subseteq I_0. \tag{5.2.115}$$

To justify this, recall from (5.2.10) that

$$\begin{array}{l} N := \partial\Omega \setminus \left(\bigcup_{Q \in \mathbb{D}_i(\partial\Omega)} Q \right) \text{ is a } \sigma\text{-measurable set} \\ \text{satisfying } \sigma(N) = 0 \text{ and } \partial\Omega \setminus N = \bigcup_{Q \in \mathbb{D}_i(\partial\Omega)} Q. \end{array} \tag{5.2.116}$$

Intersecting both sides of the last equality in (5.2.116) with $2\Delta_0$ while bearing in mind (5.2.111)-(5.2.112) then yields

$$2\Delta_0 \setminus N = \bigcup_{Q \in \mathbb{D}_i(\partial\Omega)} (Q \cap 2\Delta_0) = \bigcup_{Q \in \mathcal{Q}_0} (Q \cap 2\Delta_0) \subseteq \bigcup_{Q \in \mathcal{Q}_0} Q = I_0, \tag{5.2.117}$$

ultimately proving (5.2.115).

For reasons which are going to be clear momentarily, in addition to the truncated operators T_ε from (5.2.52) we shall need a version in which the truncation is performed using a smooth cutoff function (rather than a characteristic function). Specifically, fix $\zeta \in \mathscr{C}^\infty(\mathbb{R})$ satisfying $0 \leq \zeta \leq 1$ on $\mathbb{R}$ and with the property that $\zeta \equiv 0$ in $(-\infty, 1]$ and $\zeta \equiv 1$ in $[2, \infty)$. For each $\varepsilon > 0$ then define the action of the *smoothly* truncated operator $T_{(\varepsilon)}$ on $f \in L^p(\partial\Omega, w)$ by setting, for each $x \in \partial\Omega$,

$$T_{(\varepsilon)}f(x) := \int_{\partial\Omega} \zeta\Big(\frac{|x-y|}{\varepsilon}\Big)\langle x-y, \nu(y)\rangle\, k(x-y)f(y)\,d\sigma(y). \tag{5.2.118}$$

Let us also define a version of the maximal operator (5.2.51) in which the supremum is taken over smoothly truncated integrals by setting, i.e.,

$$T_{(*)}f(x) := \sup_{\varepsilon>0}\big|T_{(\varepsilon)}f(x)\big| \text{ at every point } x \in \partial\Omega. \tag{5.2.119}$$

For the time being, the goal is compare roughly truncated singular integral operators with their smoothly truncated counterparts. In a first stage, having fixed some function $f \in L^p(\partial\Omega, w)$, we make the observation that

$$T_\varepsilon(\varphi f)(x) = 0 \text{ and } T_{(\varepsilon)}(\varphi f)(x) = 0 \text{ if } x \in I_0 \text{ and } \varepsilon \geq (1+a)r_\delta. \tag{5.2.120}$$

Indeed, in the case when $x \in I_0$ and $(1+a)r_\delta \leq \varepsilon$ for each $y \in \Delta_0$ we may estimate (using (5.2.74) and (5.2.113)) $|y-x| \leq |y-x_0| + |x_0-x| < r_\delta + ar_\delta \leq \varepsilon$, thus $y \in \Delta(x,\varepsilon)$. This proves that $\Delta_0 \subseteq \Delta(x,\varepsilon)$, hence $\operatorname{supp}(\varphi f) \subseteq \Delta_0 \subseteq \Delta(x,\varepsilon)$, and (5.2.120) follows from this based on simple support considerations. On the other hand, if $\varepsilon < Cr_\delta$ then for each $f \in L^p(\partial\Omega, w)$ and each $x \in \partial\Omega$ we may estimate

$$\begin{aligned}
\big|T_\varepsilon(\varphi f)(x) - T_{(\varepsilon)}(\varphi f)(x)\big| &\leq \int_{\Delta(x,2\varepsilon)\setminus\overline{\Delta(x,\varepsilon)}} \big|\langle x-y, \nu(y)\rangle\big|\,|k(x-y)||(\varphi f)(y)|\,d\sigma(y) \\
&\leq C\varepsilon^{-1} \fint_{\Delta(x,2\varepsilon)} \big|\langle x-y, \nu(y)\rangle\big|\,|(\varphi f)(y)|\,d\sigma(y) \\
&\leq C\varepsilon^{-1} \fint_{\Delta(x,2\varepsilon)} \big|\big\langle x-y, \nu(y)-\nu_{\Delta(x,2\varepsilon)}\big\rangle\big||(\varphi f)(y)|\,d\sigma(y) \\
&\quad + C\varepsilon^{-1} \fint_{\Delta(x,2\varepsilon)} \big|\big\langle x-y, \nu_{\Delta(x,2\varepsilon)}\big\rangle\big||(\varphi f)(y)|\,d\sigma(y) \\
&\leq C\Big(\fint_{\Delta(x,2\varepsilon)} \big|\nu(y)-\nu_{\Delta(x,2\varepsilon)}\big|^{\frac{\gamma+1}{\gamma}}\,d\sigma(y)\Big)^{\frac{\gamma}{1+\gamma}}\Big(\fint_{\Delta(x,2\varepsilon)} |(\varphi f)(y)|^{1+\gamma}\,d\sigma(y)\Big)^{\frac{1}{1+\gamma}} \\
&\quad + C\Big(\sup_{y\in\Delta(x,2\varepsilon)} \varepsilon^{-1}\big|\big\langle x-y, \nu_{\Delta(x,2\varepsilon)}\big\rangle\big|\Big)\Big(\fint_{\Delta(x,2\varepsilon)} |f(y)|^{1+\gamma}\,d\sigma(y)\Big)^{\frac{1}{1+\gamma}} \\
&\leq C\delta\cdot \inf_{\Delta(x,2\varepsilon)} \mathcal{M}_\gamma f,
\end{aligned} \tag{5.2.121}$$

using Hölder's inequality, the consequence of the John-Nirenberg inequality recorded in [129, (7.4.71)], (5.2.57), (5.2.62), and (5.2.110). Ultimately, from (5.2.120) and (5.2.121) we deduce that there exists $C \in (0, \infty)$, which depends only on γ, the dimension n, the ADR constants of $\partial\Omega$, and the relative flatness ratio $\wp(\Omega)/\operatorname{diam}\partial\Omega$, with the property that for each function $f \in L^p(\partial\Omega, w)$ we have

$$\big|T_*(\varphi f)(x) - T_*(\varphi f)(x)\big| \leq C\delta \cdot \mathcal{M}_\gamma f(x) \ \text{ for each } x \in I_0. \tag{5.2.122}$$

Henceforth we agree to fix $\gamma \in (0, p-1)$, which depends only on n, p, $[w]_{A_p}$, and the ADR constants of $\partial\Omega$, such that $w \in A_{p/(1+\gamma)}(\partial\Omega, \sigma)$, with $[w]_{A_{p/(1+\gamma)}}$ controlled in terms of n, p, $[w]_{A_p}$, and the ADR constants of $\partial\Omega$. From item *(9)* in [129, Lemma 7.7.1] we know that such a choice is possible.

Going further, fix an arbitrary function $f \in L^p(\partial\Omega, w)$. Note that for each $\varepsilon > 0$ the function $\psi T_{(\varepsilon)}(\varphi f)$ is continuous on $\partial\Omega$, by Lebesgue's Dominated Convergence Theorem (whose applicability in the present setting is ensured by [129, Lemma 7.7.13]). Since the pointwise supremum of any collection of continuous functions is lower-semicontinuous, we conclude that for each $\lambda > 0$ the set

$$\big\{x \in \partial\Omega : |\psi(x) T_{(*)}(\varphi f)(x)| > \lambda\big\} \ \text{ is relatively open in } \partial\Omega. \tag{5.2.123}$$

Let us now define

$$A := \theta \cdot \phi(\delta)^{-1} \in (0, \infty) \ \text{ for some fixed small } \theta \in (0,1). \tag{5.2.124}$$

At various stages in the proof we shall make specific demands on the size of θ, though always in relation to the background geometric parameters, the weight, and the functions η, ϕ, namely n, p, $[w]_{A_p}$, η, ϕ, the Ahlfors regularity constants of $\partial\Omega$ and the relative flatness ratio $\wp(\Omega)/\operatorname{diam}\partial\Omega$ (the final demand of this nature is made in connection with (5.2.213)). We find it convenient to abbreviate

$$\mathfrak{Z}(\theta, \delta) := C\left\{\theta^{1+\gamma} + \theta^{1+\gamma/2}\Big(\frac{\eta(\delta)}{\phi(\delta)} \cdot \mathrm{e}^{-\phi(\delta)/\delta}\Big)^{1+\gamma/2} + \mathrm{e}^{-(3+\gamma+2/\gamma)\phi(\delta)/\delta}\right\}, \tag{5.2.125}$$

where $C \in (0, \infty)$ is a constant which depends only on n, p, $[w]_{A_p}$, the ADR constants of $\partial\Omega$, and the relative flatness ratio $\wp(\Omega)/\operatorname{diam}\partial\Omega$. We agree to retain the notation $\mathfrak{Z}(\theta, \delta)$ even when $C \in (0, \infty)$ may occasionally change in size (while retaining the same nature, however).

Since $w \in A_p(\partial\Omega, \sigma) \subseteq A_\infty(\partial\Omega, \sigma)$, there exists some small number $\tau > 0$ such that [129, (7.7.21)] holds. Our long-term goal is to obtain the following type of good-λ inequality: there exists a constant $C \in (0, \infty)$ as above (entering the makeup of the entity $\mathfrak{Z}(\theta, \delta)$ defined in (5.2.125)) such that for each $\lambda > 0$ we have

$$\begin{aligned} w\Big(\big\{x \in I_0 : |\psi(x) T_*(\varphi f)(x)| > 4\lambda \text{ and } \mathcal{M}_\gamma f(x) \leq A\lambda\big\}\Big) \\ \leq \mathfrak{Z}(\theta, \delta)^\tau \cdot w\Big(\big\{x \in I_0 : |\psi(x) T_{(*)}(\varphi f)(x)| > \lambda\big\}\Big). \end{aligned} \tag{5.2.126}$$

Here and elsewhere, we employ our earlier convention of using the same symbol w for the measure associated with the given weight w as in [129, (7.7.1)]. The reader is also alerted to the fact that the maximal operator appearing in the right-hand side of (5.2.126) employs smooth truncations (as in (5.2.119)).

To prove (5.2.126), fix an arbitrary $\lambda > 0$ and abbreviate

$$\mathcal{F}_\lambda := \{x \in I_0 : |\psi(x)T_*(\varphi f)(x)| > 4\lambda \text{ and } \mathcal{M}_\gamma f(x) \leq A\lambda\}. \tag{5.2.127}$$

From [131, (2.3.58)] we know that $\psi T_*(\varphi f)$ is a σ-measurable function. Since so is $\mathcal{M}_\gamma f$ (cf. [9]), it follows that $\mathcal{F}_\lambda$ is a σ-measurable set. From (5.2.123) and the fact that I_0 is a relatively open subset of $\partial\Omega$ we also conclude that $\{x \in I_0 : |\psi(x)T_{(*)}(\varphi f)(x)| > \lambda\}$ is a relatively open subset of $\partial\Omega$ (in particular, σ-measurable). As such, the good-λ inequality in (5.2.126) is meaningful.

Clearly, it is enough to consider the case $\mathcal{F}_\lambda \neq \varnothing$ since otherwise (5.2.126) is trivially satisfied by any choice of $C \in (0, \infty)$. For the remainder of the proof, assume this is the case. Since $\mathcal{F}_\lambda \subseteq I_0$ and $I_0 \subseteq a\Delta_0$, we conclude that

$$\mathcal{F}_\lambda \subseteq I_0 \subseteq a\Delta_0 \text{ and } \sup_{\mathcal{F}_\lambda} \mathcal{M}_\gamma f \leq A\lambda. \tag{5.2.128}$$

To proceed, decompose $I_0 = \mathcal{P}_\lambda \cup \mathcal{S}_\lambda$ (disjoint union) where, with the smoothly truncated maximal operator $T_{(*)}$ as in (5.2.119),

$$\begin{aligned} \mathcal{P}_\lambda &:= \{x \in I_0 : |\psi(x)T_{(*)}(\varphi f)(x)| \leq \lambda\} \text{ and} \\ \mathcal{S}_\lambda &:= \{x \in I_0 : |\psi(x)T_{(*)}(\varphi f)(x)| > \lambda\}. \end{aligned} \tag{5.2.129}$$

As a consequence of (5.2.123) and the fact that I_0 is a relatively open subset of $\partial\Omega$, the set $\mathcal{S}_\lambda$ is itself a relatively open subset of $\partial\Omega$. Moreover, using (5.2.122) and (5.2.128), for each point $x \in \mathcal{F}_\lambda$ we may estimate

$$\begin{aligned} 4\lambda &< |\psi(x)T_*(\varphi f)(x)| \leq |\psi(x)T_{(*)}(\varphi f)(x)| + C\delta \cdot \mathcal{M}_\gamma f(x) \\ &\leq |\psi(x)T_{(*)}(\varphi f)(x)| + C\delta A\lambda = |\psi(x)T_{(*)}(\varphi f)(x)| + C\theta\Big(\frac{\delta}{\phi(\delta)}\Big)\lambda \\ &< |\psi(x)T_{(*)}(\varphi f)(x)| + 3\lambda, \end{aligned} \tag{5.2.130}$$

by our choice of A in (5.2.124), the fact that $\theta \in (0, 1)$, and taking δ_* small enough to begin with (while keeping in mind that $\lim_{t\to 0^+} t/\phi(t) = 0$; cf. (5.2.81)). From (5.2.130) we see that $|\psi(x)T_{(*)}(\varphi f)(x)| > \lambda$, hence $x \in \mathcal{S}_\lambda$ which ultimately goes to show that $\mathcal{F}_\lambda \subseteq \mathcal{S}_\lambda$. Thus,

$$\begin{aligned} &\mathcal{S}_\lambda \text{ is a nonempty relatively open subset of } \partial\Omega, \\ &\text{with the property that we have } \mathcal{F}_\lambda \subseteq \mathcal{S}_\lambda \subseteq I_0. \end{aligned} \tag{5.2.131}$$

We first treat the case in which there exists $Q_0 \in \mathcal{Q}_0$ such that $\mathcal{P}_\lambda \cap Q_0 = \varnothing$ or, equivalently,

$$Q_0 \subseteq \mathcal{S}_\lambda. \tag{5.2.132}$$

The idea is now to apply Theorem 3.2.1 for the point x_0 and the radius $r := a2^{-i}$. Note that taking, say, $r_0 := r_\delta$, ensures that (3.2.3) is satisfied, thanks to (5.2.60)-(5.2.61). This guarantees the existence of constants $C_0, C_1, C_2 \in (0, \infty)$ of a purely geometric nature, depending only on n, the Ahlfors regularity constants of $\partial\Omega$, and the relative flatness ratio $\wp(\Omega)/\operatorname{diam}\partial\Omega$, with the following significance. Take

$$\widetilde{\phi} := \frac{(1+\gamma)(1+\gamma/2)}{C_2(\gamma/2)}\phi = \frac{3+\gamma+2/\gamma}{C_2}\phi \tag{5.2.133}$$

to play the role of the function in (3.2.1)-(3.2.2)). Assuming $\delta_* \in (0,1)$ to be sufficiently small to begin with (in the manner described in Theorem 3.2.1), we then have the decomposition

$$a\Delta_0 \subseteq G \cup E, \tag{5.2.134}$$

where G and E are disjoint σ-measurable subsets of $\partial\Omega$ satisfying properties implied by (3.2.4)-(3.2.9) (relative to the location x_0 and the scale $r := a2^{-i}$) in the present setting. Specifically, G is contained in the graph $\mathcal{G} = \{x_0 + x + h(x)\vec{n} : x \in H\}$ of a Lipschitz function $h : H \to \mathbb{R}$ (where $\vec{n} \in S^{n-1}$ is a unit vector and $H = \langle \vec{n} \rangle^{\perp}$ is the hyperplane in $\mathbb{R}^n$ orthogonal to $\vec{n}$) such that

$$\sup_{x,y\in H,\, x\neq y} \frac{|h(x)-h(y)|}{|x-y|} \leq C_0\widetilde{\phi}(\delta), \tag{5.2.135}$$

whereas E satisfies

$$\sigma(E) \leq C_1\,\mathrm{e}^{-C_2\widetilde{\phi}(\delta)/\delta}\sigma(a\Delta_0). \tag{5.2.136}$$

Since ϕ has a limit at 0, it follows that after possibly decreasing the value of δ_* we can assume that ϕ, hence also $\widetilde{\phi}$, is bounded on the interval $[0, \delta_*]$.

Next, from $\operatorname{supp}(\varphi f) \subseteq \Delta_0$ and $a > 2$ we conclude that

$$\varphi f = (\varphi f)\mathbf{1}_{a\Delta_0}. \tag{5.2.137}$$

Based on this observation and the fact that $I_0 \subseteq a\Delta_0$ (cf. (5.2.128)), we estimate

$$\sigma(\mathcal{F}_\lambda) \leq \sigma\Big(\big\{x \in a\Delta_0 : \big|\psi(x)T_*\big((\varphi f)\mathbf{1}_{a\Delta_0}\big)(x)\big| > 4\lambda\big\}\Big). \tag{5.2.138}$$

By further decomposing $(\varphi f)\mathbf{1}_{a\Delta_0} = (\varphi f)\mathbf{1}_G + (\varphi f)\mathbf{1}_E$ (cf. (5.2.134) and the fact that $f = f\mathbf{1}_{a\Delta_0}$), then using the sub-linearity of T_*, as well as (5.2.134), (5.2.136), and (5.2.133) we obtain

$$\begin{aligned} &\sigma\Big(\big\{x \in a\Delta_0 : \big|\psi(x)T_*\big((\varphi f)\mathbf{1}_{a\Delta_0}\big)(x)\big| > 4\lambda\big\}\Big) \\ &\leq \sigma\Big(\big\{x \in G : \big|\psi(x)T_*\big((\varphi f)\mathbf{1}_G\big)(x)\big| > 2\lambda\big\}\Big) \\ &\quad + \sigma\Big(\big\{x \in G : \big|\psi(x)T_*\big((\varphi f)\mathbf{1}_E\big)(x)\big| > 2\lambda\big\}\Big) + C_1\,\mathrm{e}^{-(3+\gamma+2/\gamma)\phi(\delta)/\delta}\sigma(a\Delta_0). \end{aligned} \tag{5.2.139}$$

To bound the first term in the right-hand side of (5.2.139), the idea is to use the fact that G is contained in the graph $\mathcal{G}$ of the function h, then employ Lemma 5.2.1 while taking advantage of (5.2.135). Turning to specifics, denote by $\widetilde{\sigma}$ the surface measure on $\mathcal{G}$, and by $\widetilde{T}_*$ the maximal operator associated with $\mathcal{G}$ as in (5.2.3) (much as T_* in (5.2.51)-(5.2.52) is associated with $\partial\Omega$). That is, for each $\widetilde{f} \in L^p(\mathcal{G},\widetilde{\sigma})$ set

$$\widetilde{T}_*\widetilde{f}(x) := \sup_{\varepsilon>0}\big|\widetilde{T}_\varepsilon\widetilde{f}(x)\big|, \quad \forall x \in \mathcal{G}, \tag{5.2.140}$$

where for each $\varepsilon > 0$ we have set

$$\widetilde{T}_\varepsilon\widetilde{f}(x) := \int_{y\in\mathcal{G},\ |x-y|>\varepsilon} \langle x-y, \widetilde{\nu}(y)\rangle k(x-y)\widetilde{f}(y)\,d\widetilde{\sigma}(y), \quad \forall x \in \mathcal{G}, \tag{5.2.141}$$

with $\widetilde{\nu}$ denoting the unit normal vector to the Lipschitz graph $\mathcal{G}$, pointing towards the upper-graph of the function h. From (3.2.18) we know that

$$\nu(x) = \widetilde{\nu}(x) \ \text{ at } \sigma\text{-a.e. point } x \in G. \tag{5.2.142}$$

We continue by fixing a point $\widetilde{x} \in \mathcal{F}_\lambda$ (which, according to (5.2.128), also places $\widetilde{x}$ into $a\Delta_0$). As regards the first term in the right-hand side of (5.2.139), we may rely on (5.2.142), the fact that the measures σ and $\widetilde{\sigma}$ agree on $\partial\Omega \cap \mathcal{G}$ (as they are both manifestations of $\mathcal{H}^{n-1}$), (5.2.140)-(5.2.141), (5.2.51)-(5.2.52), Chebytcheff's inequality, Lemma 5.2.1 (keeping in mind that $\widetilde{\phi}$ is bounded on the interval $[0,\delta_*]$), (5.2.133), (5.2.135), (5.2.137), (5.2.110), (5.2.128), and (5.2.124) to estimate

$$\begin{aligned}
&\sigma\Big(\big\{x \in G : \big|\psi(x)(T_*((\varphi f)\mathbf{1}_G))(x)\big| > 2\lambda\big\}\Big)\\
&\quad = \widetilde{\sigma}\Big(\big\{x \in G : \big|\psi(x)(\widetilde{T}_*((\varphi f)\mathbf{1}_G))(x)\big| > 2\lambda\big\}\Big)\\
&\quad \le \widetilde{\sigma}\Big(\big\{x \in \mathcal{G} : \widetilde{T}_*((\varphi f)\mathbf{1}_G)(x) > 2\lambda\big\}\Big)\\
&\quad \le \frac{1}{(2\lambda)^{1+\gamma}}\int_{\mathcal{G}} |\widetilde{T}_*((\varphi f)\mathbf{1}_G)|^{1+\gamma}\,d\widetilde{\sigma} \le C\frac{\widetilde{\phi}(\delta)^{1+\gamma}}{\lambda^{1+\gamma}}\int_{\mathcal{G}} |\varphi f\mathbf{1}_G|^{1+\gamma}\,d\widetilde{\sigma}\\
&\quad = C\frac{\phi(\delta)^{1+\gamma}}{\lambda^{1+\gamma}}\int_G |\varphi f|^{1+\gamma}\,d\sigma \le C\frac{\phi(\delta)^{1+\gamma}}{\lambda^{1+\gamma}}\int_{\partial\Omega} |\varphi f|^{1+\gamma}\,d\sigma\\
&\quad \le C\phi(\delta)^{1+\gamma}\frac{\sigma(a\Delta_0)}{\lambda^{1+\gamma}}\fint_{a\Delta_0} |f|^{1+\gamma}\,d\sigma \le C\phi(\delta)^{1+\gamma}\frac{\sigma(a\Delta_0)}{\lambda^{1+\gamma}}\Big[\mathcal{M}_\gamma f(\widetilde{x})\Big]^{1+\gamma}\\
&\quad \le C\,(A\phi(\delta))^{1+\gamma}\,\sigma(a\Delta_0) = C\theta^{1+\gamma}\,\sigma(a\Delta_0),
\end{aligned} \tag{5.2.143}$$

for some constant $C \in (0,\infty)$ which depends only on n, p, $[w]_{A_p}$, ϕ, the ADR constants of $\partial\Omega$, and the relative flatness ratio $\wp(\Omega)/\operatorname{diam}\partial\Omega$.

As regards the second term in the right-hand side of (5.2.139), once again fix a point $\widetilde{x} \in \mathcal{F}_\lambda$ (which then also belongs to $a\Delta_0$). We may then use Chebytcheff's inequality, the hypothesis made in (5.2.79) (used with $p := 1 + \gamma/2$ and $w \equiv 1$), the assumption (5.2.58), (5.2.57), Hölder's inequality, (5.2.136), (5.2.110), (5.2.133),

(5.2.128), and (5.2.124) to obtain

$$\begin{aligned}
&\sigma\Big(\big\{x \in G : \big|\psi(x)\big(T_*((\varphi f)\mathbf{1}_E)\big)(x)\big| > 2\lambda\big\}\Big) \\
&\quad \leq \sigma\Big(\big\{x \in \partial\Omega : \big|\psi(x)\big(T_*((\varphi f)\mathbf{1}_E)\big)(x)\big| > 2\lambda\big\}\Big) \\
&\quad \leq \frac{1}{(2\lambda)^{1+\gamma/2}} \int_{\partial\Omega} \big|\psi T_*((\varphi f)\mathbf{1}_E)\big|^{1+\gamma/2}\, d\sigma \\
&\quad \leq \frac{\big(\|M_\psi T_* M_\varphi\|_{L^{1+\gamma/2}(\partial\Omega,\sigma)\to L^{1+\gamma/2}(\partial\Omega,\sigma)}\big)^{1+\gamma/2}}{(2\lambda)^{1+\gamma/2}} \int_{\partial\Omega} \big(|f|\,\mathbf{1}_E\big)^{1+\gamma/2}\, d\sigma
\end{aligned} \tag{5.2.144}$$

and further[9]

$$\begin{aligned}
&\sigma\Big(\big\{x \in G : \big|\psi(x)\big(T_*((\varphi f)\mathbf{1}_E)\big)(x)\big| > 2\lambda\big\}\Big) \leq \frac{(C\eta(\delta))^{1+\gamma/2}}{\lambda^{1+\gamma/2}} \int_{a\Delta_0} |f|^{1+\gamma/2}\, \mathbf{1}_E \, d\sigma \\
&\quad \leq \frac{(C\eta(\delta))^{1+\gamma/2}}{\lambda^{1+\gamma/2}} \, \sigma(E)^{\frac{\gamma/2}{1+\gamma}} \Big(\int_{a\Delta_0} |f|^{1+\gamma}\, d\sigma\Big)^{\frac{1+\gamma/2}{1+\gamma}} \\
&\quad = \frac{(C\eta(\delta))^{1+\gamma/2}}{\lambda^{1+\gamma/2}} \Big(\frac{\sigma(E)}{\sigma(a\Delta_0)}\Big)^{\frac{\gamma/2}{1+\gamma}} \Big(\fint_{a\Delta_0} |f|^{1+\gamma}\, d\sigma\Big)^{\frac{1+\gamma/2}{1+\gamma}} \sigma(a\Delta_0) \\
&\quad \leq C\frac{\eta(\delta)^{1+\gamma/2}}{\lambda^{1+\gamma/2}} \exp\Big\{-\frac{C_2(\gamma/2)\widetilde{\phi}(\delta)}{(1+\gamma)\delta}\Big\} \Big[\mathcal{M}_\gamma f(\overline{x})\Big]^{1+\gamma/2} \sigma(a\Delta_0) \\
&\quad \leq C\big(A\eta(\delta)\big)^{1+\gamma/2} \cdot \exp\Big\{-\frac{(1+\gamma/2)\phi(\delta)}{\delta}\Big\} \sigma(a\Delta_0) \\
&\quad = C\theta^{1+\gamma/2}\Big[\eta(\delta)\phi(\delta)^{-1} \cdot \exp\Big\{-\frac{\phi(\delta)}{\delta}\Big\}\Big]^{1+\gamma/2} \sigma(a\Delta_0),
\end{aligned} \tag{5.2.145}$$

where $C \in (0, \infty)$ depends only on n, p, $[w]_{A_p}$, η, ϕ, the ADR constants of $\partial\Omega$, and the relative flatness ratio $\wp(\Omega)/\mathrm{diam}\,\partial\Omega$. Gathering (5.2.139), (5.2.143), and (5.2.145) then yields

$$\begin{aligned}
&\sigma\Big(\big\{x \in a\Delta_0 : \big|\psi(x)T_*\big((\varphi f)\mathbf{1}_{a\Delta_0}\big)(x)\big| > 4\lambda\big\}\Big) \\
&\quad \leq C\Big\{\theta^{1+\gamma} + \theta^{1+\gamma/2}\Big(\frac{\eta(\delta)}{\phi(\delta)} \cdot e^{-\phi(\delta)/\delta}\Big)^{1+\gamma/2} + e^{-(3+\gamma+2/\gamma)\phi(\delta)/\delta}\Big\} \sigma(a\Delta_0) \\
&\quad = \mathfrak{Z}(\theta, \delta) \cdot \sigma(a\Delta_0),
\end{aligned} \tag{5.2.146}$$

where $\mathfrak{Z}(\theta, \delta) \in (0, \infty)$ is as in (5.2.125). Finally, (5.2.146) and (5.2.138) imply

---

[9] It is from the format of (5.2.145) that the value of having the last property in (5.2.81) is most apparent. Indeed, since the left-most side of (5.2.145) is obviously dominated by $\sigma(G) \leq \sigma(a\Delta_0)$ (cf. (5.2.134)), the estimate derived in (5.2.145) is only useful if $\psi(\delta)\phi(\delta)^{-1} \cdot \exp\big\{-\frac{\phi(\delta)}{\delta}\big\}$ stays bounded for δ close to 0.

$$\sigma(\mathcal{F}_\lambda) \leq \mathfrak{Z}(\theta, \delta) \cdot \sigma(a\Delta_0), \tag{5.2.147}$$

where $\mathfrak{Z}(\theta, \delta) \in (0, \infty)$ is as in (5.2.125).

Moving on, observe that (5.2.8) implies that there exists $x_{Q_0} \in \partial\Omega$ such that

$$\Delta(x_{Q_0}, a_0 2^{-i}) \subseteq Q_0 \subseteq \Delta(x_{Q_0}, a_1 2^{-i}). \tag{5.2.148}$$

From this and (5.2.111) we then conclude that there exists some $c > 0$, which only depends on the ADR constants of $\partial\Omega$, with the property that $a\Delta_0 \subseteq c\Delta(x_{Q_0}, a_1 2^{-i})$. As a consequence of this inclusion we may write (for some $C \in (0, \infty)$ which depends only on n, p, $[w]_{A_p}$, and the ADR constants of $\partial\Omega$)

$$w(a\Delta_0) \leq w\big(c\Delta(x_{Q_0}, a_1 2^{-i})\big) \leq C\, w\big(\Delta(x_{Q_0}, a_0 2^{-i})\big) \leq C w(Q_0) \tag{5.2.149}$$

where we have also used the fact that w is a doubling measure (cf. [129, (7.7.16)]) and (5.2.148). With this in hand, we may now estimate

$$w(\mathcal{F}_\lambda) \leq \mathfrak{Z}(\theta, \delta)^\tau \cdot w(a\Delta_0) \leq \mathfrak{Z}(\theta, \delta)^\tau \cdot w(Q_0) \leq \mathfrak{Z}(\theta, \delta)^\tau \cdot w(S_\lambda), \tag{5.2.150}$$

where the first inequality uses [129, (7.7.21)], the fact that $\mathcal{F}_\lambda \subseteq a\Delta_0$ (cf. (5.2.128)), and (5.2.147), the second is based on (5.2.149), while the last is implied by (5.2.132). Thus, (5.2.126) holds whenever there exists $Q_0 \in \mathcal{Q}_0$ such that $\mathcal{P}_\lambda \cap Q_0 = \varnothing$.

To complete the proof of (5.2.126), it remains to consider the case when

$$\mathcal{P}_\lambda \cap Q \neq \varnothing \text{ for each } Q \in \mathcal{Q}_0. \tag{5.2.151}$$

In this scenario, take an arbitrary dyadic cube $Q \in \mathcal{Q}_0$. From (5.2.112) we know that $Q \subseteq I_0$. Subdivide Q dyadically, producing offsprings Q', and stop dividing Q' whenever $\mathcal{P}_\lambda \cap Q' = \varnothing$. This produces a family of pairwise disjoint (stopping time) dyadic cubes $\{Q^j\}_{j\in J_Q} \subset \mathbb{D}(\partial\Omega)$ such that $Q^j \cap \mathcal{P}_\lambda = \varnothing$, $Q^j \subseteq Q$ but $Q^j \neq Q$ (since $Q^j \cap \mathcal{P}_\lambda = \varnothing$ but $Q \cap \mathcal{P}_\lambda \neq \varnothing$), and $Q' \cap \mathcal{P}_\lambda \neq \varnothing$ for all $Q' \in \mathbb{D}(\partial\Omega)$ such that $Q^j \subsetneq Q' \subseteq Q$. In particular $Q^j \subsetneq Q$ for every $j \in J_Q$ and $\widetilde{Q^j}$, the dyadic parent of Q^j, satisfies $\widetilde{Q^j} \subseteq Q$. With the σ-nullset N as in (5.2.11), we now claim that

$$\bigcup_{j\in J_Q} Q^j \subseteq S_\lambda \cap Q \subseteq \Big(\bigcup_{j\in J_Q} Q^j\Big) \cup N. \tag{5.2.152}$$

To justify the first inclusion above, observe that if $j \in J_Q$ then $Q^j \subseteq S_\lambda \cap Q$, since $Q^j \subseteq Q \subseteq I_0$ and $Q^j \cap \mathcal{P}_\lambda = \varnothing$ imply that $Q_j \subseteq Q \setminus \mathcal{P}_\lambda = Q \cap S_\lambda$. This establishes the first inclusion in (5.2.152). As regards the second inclusion claimed in (5.2.152), consider an arbitrary point $x \in \big(S_\lambda \cap Q\big) \setminus N$. Then $|\psi(x)(T_{(*)}(\varphi f))(x)| > \lambda$ which, in view of (5.2.123), ensures that we may find a surface ball $\Delta_x := \Delta(x, r_x)$ such that $|\psi(y)(T_{(*)}(\varphi f))(y)| > \lambda$ for every $y \in \Delta_x$. Thanks to (5.2.8) and (5.2.10) we may then choose a dyadic cube $Q_x \in \mathbb{D}(\partial\Omega)$ such that $x \in Q_x$ and $Q_x \subseteq \Delta_x \cap Q \subseteq I_0$. This forces $Q_x \subseteq S_\lambda \cap Q$, hence $Q_x \cap \mathcal{P}_\lambda = \varnothing$. By the maximality of the family chosen above, $Q_x \subseteq Q^j$ for some $j \in J_Q$ which goes to show that $x \in Q^j$. Ultimately, this proves the second inclusion in (5.2.152).

Going further, the idea is to carry out the stopping-time argument just described for each dyadic cube Q in the family $\mathcal{Q}_0$. For ease of reference, organize the resulting (at most countable) collection of dyadic cubes $\{Q^j : Q \in \mathcal{Q}_0 \text{ and } j \in J_Q\}$ as a single-index family $\{Q_\ell\}_{\ell \in I}$ of mutually disjoint dyadic cubes; in particular,

$$\bigcup_{Q \in \mathcal{Q}_0} \bigcup_{j \in J_Q} Q^j = \bigcup_{\ell \in I} Q_\ell, \tag{5.2.153}$$

with the latter union comprised of pairwise disjoint dyadic cubes in $\partial\Omega$. Note that $\mathcal{S}_\lambda \cap Q$ might be empty for some $Q \in \mathcal{Q}_0$ and in this case $J_Q = \varnothing$ (i.e., the family of cubes $\{Q^j\}_{j \in J_Q}$ is empty, since there are no stopping time dyadic cubes produced in this case). However, (5.2.112) and (5.2.131) imply that $\mathcal{S}_\lambda \cap Q$ cannot be empty for every $Q \in \mathcal{Q}_0$ and, as a consequence, $I \neq \varnothing$. Going further, on account of (5.2.112) and the fact that $\mathcal{S}_\lambda \subseteq I_0$ (cf. (5.2.129)) we may write

$$\bigcup_{Q \in \mathcal{Q}_0} (\mathcal{S}_\lambda \cap Q) = \mathcal{S}_\lambda \tag{5.2.154}$$

which further entails, on account of (5.2.153) and (5.2.152), that

$$\bigcup_{\ell \in I} Q_\ell \subseteq \mathcal{S}_\lambda \subseteq \Big(\bigcup_{\ell \in I} Q_\ell\Big) \cup N. \tag{5.2.155}$$

By construction, for each index $\ell \in I$ there exists a point x_ℓ^* such that

$$x_\ell^* \in \widetilde{Q}_\ell \cap \mathcal{P}_\lambda = \widetilde{Q}_\ell \cap \big(I_0 \setminus \mathcal{S}_\lambda\big), \tag{5.2.156}$$

where $\widetilde{Q}_\ell$ denotes the dyadic parent of Q_ℓ (cf. item *(4)* in Proposition 5.2.2). For each $\ell \in I$ we let $\Delta_\ell := \Delta_{Q_\ell}$ and $\widetilde{\Delta}_\ell := \Delta_{\widetilde{Q}_\ell}$ be as in (5.2.8). Pressing on, split the collection $\{\Delta_\ell\}_{\ell \in I}$ into two sub-classes. Specifically, bring in

$$\begin{gathered} I_1 := \big\{\ell \in I : \text{ there exists } x_\ell^{**} \in \Delta_\ell \text{ such that } \mathcal{M}_\gamma f(x_\ell^{**}) \leq A\lambda\big\}, \\ \text{and } I_2 := I \setminus I_1. \end{gathered} \tag{5.2.157}$$

By design, $\mathcal{F}_\lambda \cap \Delta_\ell = \varnothing$ for each $\ell \in I_2$. Recall from (5.2.131) that $\mathcal{F}_\lambda \subseteq \mathcal{S}_\lambda$. From this, (5.2.155), and (5.2.8) we then obtain (since $\sigma(N) = 0$; cf. (5.2.11))

$$w(\mathcal{F}_\lambda) = \sum_{\ell \in I} w(\mathcal{F}_\lambda \cap Q_\ell) \leq \sum_{\ell \in I_1} w(\mathcal{F}_\lambda \cap \Delta_\ell). \tag{5.2.158}$$

Let us also consider

$$F_\ell := \big\{x \in \Delta_\ell : |\psi(x) T_*(\varphi f)(x)| > 4\lambda\big\} \text{ for each } \ell \in I_1, \tag{5.2.159}$$

and observe that this entails

$$\mathcal{F}_\lambda \cap \Delta_\ell \subseteq F_\ell \text{ for each } \ell \in \mathcal{I}_1, \tag{5.2.160}$$

Our next goal is to prove that

$$\sigma(F_\ell) \leq \mathfrak{Z}(\theta, \delta) \cdot \sigma(\Delta_\ell) \text{ for each } \ell \in \mathcal{I}_1. \tag{5.2.161}$$

Granted this, using [129, (7.7.21)] it would follow that

$$w(F_\ell) \leq \mathfrak{Z}(\theta, \delta)^\tau \cdot w(\Delta_\ell) \text{ for each } \ell \in \mathcal{I}_1 \tag{5.2.162}$$

which, in concert with (5.2.158), (5.2.160), (5.2.8) plus the fact that w is a doubling measure, and (5.2.155), would then finish the justification of (5.2.126) by writting

$$\begin{aligned} w(\mathcal{F}_\lambda) &\leq \sum_{\ell \in \mathcal{I}_1} w(\mathcal{F}_\lambda \cap \Delta_\ell) \leq \sum_{\ell \in \mathcal{I}_1} w(F_\ell) \leq \mathfrak{Z}(\theta, \delta)^\tau \cdot \sum_{\ell \in \mathcal{I}_1} w(\Delta_\ell) \\ &\leq \mathfrak{Z}(\theta, \delta)^\tau \cdot \sum_{\ell \in \mathcal{I}_1} w(Q_\ell) \leq \mathfrak{Z}(\theta, \delta)^\tau \cdot \sum_{\ell \in \mathcal{I}} w(Q_\ell) = \mathfrak{Z}(\theta, \delta)^\tau \cdot w(\mathcal{S}_\lambda). \end{aligned} \tag{5.2.163}$$

We now turn to the proof of (5.2.161). Fix $\ell \in \mathcal{I}_1$ and, in order to lighten notation, we suppress the dependence of Δ_ℓ, $\widetilde{\Delta}_\ell$, F_ℓ, x_ℓ^*, and x_ℓ^{**} on ℓ, and simply write Δ, $\widetilde{\Delta}$, F, x^*, and x^{**}, respectively. With this convention in mind, observe first that

$$\Delta \subseteq 2\widetilde{\Delta}. \tag{5.2.164}$$

To justify (5.2.164), recall from (5.2.8) that we may write $\Delta = B(x_Q, r_Q) \cap \partial\Omega$ and $\widetilde{\Delta} = B(x_{\widetilde{Q}}, r_{\widetilde{Q}}) \cap \partial\Omega$. Moreover, since $\widetilde{Q}$ is the parent of Q, we have $r_{\widetilde{Q}} = 2r_Q$. Then for each $x \in \Delta$ we have $|x - x_{\widetilde{Q}}| \leq |x - x_Q| + |x_Q - x_{\widetilde{Q}}| < r_Q + r_{\widetilde{Q}} = (3/2)r_{\widetilde{Q}} < 2r_{\widetilde{Q}}$ which ultimately proves (5.2.164). Going forward, let us consider the surface ball

$$\Delta^* := B(x^*, R) \cap \partial\Omega \text{ with } R := \Lambda \cdot r_Q \tag{5.2.165}$$

for a sufficiently large constant Λ satisfying

$$2 < \Lambda < (5C - a)/a_1 \tag{5.2.166}$$

(depending only on the implicit constants in the dyadic grid construction, which in turn depend only on the ADR constants of $\partial\Omega$) chosen so that

$$2\widetilde{\Delta} \subseteq \Delta^*. \tag{5.2.167}$$

For further reference let us remark here that since $\Delta = B(x_Q, r_Q) \cap \partial\Omega$ where $\Delta := \Delta_\ell := \Delta_{Q_\ell}$ and $Q_\ell \in \mathbb{D}_{i'}(\partial\Omega)$ with $i' \geq i$, we ultimately conclude from (5.2.8) that $r_Q \leq a_1 2^{-i}$ which, in concert with (5.2.108), gives

$$R \leq a_1 \Lambda 2^{-i} \leq 2a_1 \Lambda r_\delta < 10\, C r_\delta. \tag{5.2.168}$$

We then decompose

$$\varphi f = f_1 + f_2 \text{ where } f_1 := (\varphi f)\mathbf{1}_{\overline{2\Delta^*}} \text{ and } f_2 := (\varphi f)\mathbf{1}_{\partial\Omega\setminus\overline{2\Delta^*}}. \tag{5.2.169}$$

By virtue of the sub-linearity of T_*, (5.2.159), and the fact that $\Delta \subseteq \Delta^* \subseteq 4\Delta^*$ (cf. (5.2.164)-(5.2.167)) this implies

$$\begin{aligned}\sigma(F) &\leq \sigma\Big(\big\{x \in \Delta : |\psi(x)(T_* f_1)(x)| > 2\lambda\big\}\Big) + \sigma\Big(\big\{x \in \Delta : |\psi(x)(T_* f_2)(x)| > 2\lambda\big\}\Big)\\ &\leq \sigma\Big(\big\{x \in 4\Delta^* : |\psi(x)(T_* f_1)(x)| > 2\lambda\big\}\Big) + \sigma\Big(\big\{x \in \Delta : |\psi(x)(T_* f_2)(x)| > 2\lambda\big\}\Big).\end{aligned} \tag{5.2.170}$$

The contribution from f_1 in the last line above is handled as in (5.2.134)-(5.2.136), (5.2.139)-(5.2.146) by performing a decomposition of $4\Delta^*$ as in Theorem 3.2.1. Indeed, $a\Delta_0$, $\overline{x}$, f, and λ are replaced by $4\Delta^*$, x^{**}, f_1, and $\frac{1}{2}\lambda$, respectively, and we use that $\mathcal{M}_\gamma f(x^{**}) \leq A\lambda$ (cf. (5.2.157)), supp $f_1 \subseteq \overline{2\Delta^*} \subseteq 4\Delta^*$ (cf. (5.2.169)), and $\sigma(4\Delta^*) \leq c \cdot \sigma(\Delta)$ for some $c \in (0, \infty)$ depending only the ADR constants of $\partial\Omega$ (since $\partial\Omega$ is Ahlfors regular and the surface balls $4\Delta^*$, Δ have comparable radii) to run the same proof as before. The conclusion is that

$$\sigma\Big(\big\{x \in 4\Delta^* : |\psi(x)(T_* f_1)(x)| > 2\lambda\big\}\Big) \leq \mathfrak{Z}(\theta, \delta) \cdot \sigma(\Delta). \tag{5.2.171}$$

In view of the conclusion we seek (cf. (5.2.161)), this suits our purposes.

As regards f_2, recall that R is the radius of the surface ball Δ^*, and for each given $\varepsilon > 0$ define $\varepsilon' := \max\{\varepsilon, 2R\}$. Based on this choice of ε', the definition of the truncated singular integral operators in (5.2.52), the truncation in the definition of the function f_2, the estimate in (5.2.121) (presently used with x^* in place of x and ε' in place of ε), the fact that $x^{**} \in \Delta \subseteq \Delta^* \subseteq \Delta(x^*, 2\varepsilon')$ (cf. (5.2.157) and (5.2.164)-(5.2.167)), the fact that $\mathcal{M}_\gamma f(x^{**}) \leq A\lambda$ (cf. (5.2.157)), the definition of $T_{(*)}(\varphi f)(x^*)$ (cf. (5.2.119)), the membership of x^* to $\mathcal{P}_\lambda$ (cf. (5.2.156)), and the first formula in (5.2.129) we may write

$$\begin{aligned}&\big|\psi(x^*)(T_\varepsilon f_2)(x^*)\big| = \big|\psi(x^*)(T_{\varepsilon'}(\varphi f))(x^*)\big|\\ &\leq \big|\psi(x^*)(T_{\varepsilon'}(\varphi f))(x^*) - \psi(x^*)(T_{(\varepsilon')}(\varphi f))(x^*)\big| + \big|\psi(x^*)(T_{(\varepsilon')}(\varphi f))(x^*)\big|\\ &\leq C\delta\mathcal{M}_\gamma f(x^{**}) + |\psi(x^*)(T_{(*)}(\varphi f))(x^*)| \leq C\delta A\lambda + \lambda = C\theta\Big(\frac{\delta}{\phi(\delta)}\Big)\lambda + \lambda \leq \tfrac{3}{2}\lambda,\end{aligned} \tag{5.2.172}$$

with the last line a consequence of our choice of A in (5.2.124), the fact that $\theta \in (0, 1)$, and the ability of taking $\delta_* \in (0, 1)$ small enough to begin with (while bearing in mind that $\lim_{t\to 0^+} t/\phi(t) = 0$; cf. (5.2.81)). With $\varepsilon > 0$ momentarily fixed,

$$\text{pick an arbitrary } x \in \Delta \text{ and bound } \big|T_\varepsilon f_2(x) - T_\varepsilon f_2(x^*)\big| \leq \mathrm{I} + \mathrm{II} + \mathrm{III}, \tag{5.2.173}$$

where

$$\mathrm{I} := \int\limits_{\substack{y\in\partial\Omega\setminus\overline{2\Delta^*}\\ |x-y|>\varepsilon,\ |x^*-y|>\varepsilon}} \Big|\langle x-y,\nu(y)\rangle k(x-y) - \langle x^*-y,\nu(y)\rangle k(x^*-y)\Big| |(\varphi f)(y)|\,d\sigma(y),$$

$$\mathrm{II} := \int\limits_{\substack{y\in\partial\Omega\setminus\overline{2\Delta^*}\\ |x-y|>\varepsilon,\ |x^*-y|\leq\varepsilon}} |\langle x-y,\nu(y)\rangle||k(x-y)||(\varphi f)(y)|\,d\sigma(y), \tag{5.2.174}$$

$$\mathrm{III} := \int\limits_{\substack{y\in\partial\Omega\setminus\overline{2\Delta^*}\\ |x^*-y|>\varepsilon,\ |x-y|\leq\varepsilon}} |\langle x^*-y,\nu(y)\rangle||k(x^*-y)||(\varphi f)(y)|\,d\sigma(y). \tag{5.2.175}$$

In preparation for estimating the term I, we will first analyze the difference between I and a similar expression in which $\nu(y)$ has been replaced by the integral average $\nu_{\Delta^*} := ⨍_{\Delta^*} \nu\,d\sigma$. To set the stage, for each fixed $y \in \partial\Omega \setminus \overline{2\Delta^*}$ consider

$$F_y(z) := \langle z-y, \nu(y)-\nu_{\Delta^*}\rangle k(z-y) \ \text{ for each } z \in B(x^*, R). \tag{5.2.176}$$

Then

$$|(\nabla F_y)(z)| \leq \Big(\sum_{|\alpha|\leq 1}\sup_{S^{n-1}}|\partial^\alpha k|\Big)\frac{|\nu(y)-\nu_{\Delta^*}|}{|z-y|^n} \ \text{ for each } z \in B(x^*, R). \tag{5.2.177}$$

Since $x \in \Delta \subseteq \Delta^* = B(x^*, R) \cap \partial\Omega$ (cf. (5.2.164)-(5.2.167)), we have

$$|x - x^*| < R. \tag{5.2.178}$$

Also,

$$|x^* - y| \leq 2|\xi - y| \ \text{ for each } y \in \partial\Omega \setminus \overline{2\Delta^*} \text{ and each } \xi \in [x, x^*]. \tag{5.2.179}$$

We next introduce

$$\mathcal{J} := \big\{j \in \mathbb{N} : \Delta_0 \subseteq 2^j\Delta^*\big\}, \tag{5.2.180}$$

in relation to which we claim that

$$2^{j+1}R < 4ar_\delta \text{ and } 2^{j+1}\Delta^* \subseteq 6a\Delta_0 \text{ for all } j \in \mathbb{N}\setminus\mathcal{J}. \tag{5.2.181}$$

To justify this, observe first that $\mathcal{J}$ is a nonempty set of integers, bounded from below. As such, $j_\# := \min\mathcal{J}$ exists, and $\mathcal{J} = \big\{j \in \mathbb{N} : j \geq j_\#\big\}$. If $\mathbb{N}\setminus\mathcal{J} = \varnothing$ there is nothing to prove. Assume $\mathbb{N}\setminus\mathcal{J} \neq \varnothing$. Then $j_\# > 1$ and $j_\# - 1 \notin \mathcal{J}$. Hence,

$$\Delta_0 \not\subseteq 2^{j_\#-1}\Delta^* = B(x^*, 2^{j_\#-1}R) \cap \partial\Omega. \tag{5.2.182}$$

Since $x^* \in I_0 \subseteq a\Delta_0 = B(x_0, ar_\delta) \cap \partial\Omega$, we cannot have $2^{j_\#-1}R \geq 2ar_\delta$. Otherwise, this would imply $B(x_0, ar_\delta) \subseteq B(x^*, 2^{j_\#-1}R)$ which, after taking the intersection with $\partial\Omega$, would force $a\Delta_0 \subseteq 2^{j_\#-1}\Delta^*$, but $a > 1$ so this contradicts (5.2.182). Thus,

$$2^{j_\#-1}R < 2ar_\delta. \tag{5.2.183}$$

In turn, for each $j \in \mathbb{N} \setminus \mathcal{J}$ we may write $2^j R \leq 2^{j_\#-1}R < 2ar_\delta$, proving the inequality in (5.2.181). The estimate in (5.2.183) also permits us to conclude that, for each integer $j \in \mathbb{N} \setminus \mathcal{J}$ and each point $y \in 2^{j+1}\Delta^* = B(x^*, 2^{j+1}R) \cap \partial\Omega$,

$$|y - x_0| \leq |y - x^*| + |x^* - x_0| < 2^{j+1}R + ar_\delta \leq 2^{j_\#}R + ar_\delta < 5ar_\delta. \tag{5.2.184}$$

The inclusion in (5.2.181) now follows from (5.2.184).

At this stage, by (5.2.176), (5.2.177), the Mean Value Theorem (bearing in mind (5.2.58)), (5.2.178)-(5.2.179), (5.2.181), Hölder's inequality, and (5.2.60) (also taking into account (5.2.74)) it follows that

$$\begin{aligned}
&\int_{\partial\Omega\setminus\overline{2\Delta^*}} \Big|\langle x - y, \nu(y) - \nu_{\Delta^*}\rangle k(x-y) - \langle x^* - y, \nu(y) - \nu_{\Delta^*}\rangle k(x^* - y)\Big| \times \\
&\qquad\qquad \times |(\varphi f)(y)|\, d\sigma(y) = \int_{\partial\Omega\setminus\overline{2\Delta^*}} \big|F_y(x) - F_y(x^*)\big| |(\varphi f)(y)|\, d\sigma(y) \\
&\leq \int_{\partial\Omega\setminus\overline{2\Delta^*}} |x - x^*| \cdot \sup_{\xi\in[x,x^*]} \big|(\nabla F_y)(\xi)\big| |(\varphi f)(y)|\, d\sigma(y) \\
&\leq C \int_{\partial\Omega\setminus 2\Delta^*} \frac{R}{|x^* - y|^n} \big|\nu(y) - \nu_{\Delta^*}\big| |(\varphi f)(y)|\, d\sigma(y) \\
&\leq C \sum_{j\in\mathbb{N}} 2^{-j} ⨍_{2^{j+1}\Delta^*\setminus 2^j\Delta^*} \big|\nu(y) - \nu_{\Delta^*}\big| |(\varphi f)(y)|\, d\sigma(y) \\
&\leq C \sum_{\substack{j\in\mathbb{N} \\ 2^{j+1}\Delta^*\subseteq 6a\Delta_0}} 2^{-j} ⨍_{2^{j+1}\Delta^*\setminus 2^j\Delta^*} \big|\nu(y) - \nu_{\Delta^*}\big| |(\varphi f)(y)|\, d\sigma(y) \\
&\leq C \sum_{\substack{j\in\mathbb{N} \\ 2^{j+1}\Delta^*\subseteq 6a\Delta_0}} 2^{-j} \Big(⨍_{2^{j+1}\Delta^*} \Big(\big|\nu(y) - \nu_{2^{j+1}\Delta^*}\big| + \big|\nu_{2^{j+1}\Delta^*} - \nu_{\Delta^*}\big|\Big)^{\frac{1+\gamma}{\gamma}} d\sigma(y)\Big)^{\frac{\gamma}{1+\gamma}} \times \\
&\qquad\qquad \times \Big(⨍_{2^{j+1}\Delta^*} |f(y)|^{1+\gamma}\, d\sigma(y)\Big)^{\frac{1}{1+\gamma}} \\
&\leq C\Big(\sum_{j=1}^{\infty} (j+2)\, 2^{-j}\Big) \|\nu\|_*(6a\Delta_0) \mathcal{M}_\gamma f(x^{**}) \leq CA\delta\lambda,
\end{aligned} \tag{5.2.185}$$

for some $C \in (0, \infty)$ which depends only on n, p, $[w]_{A_p}$, and the ADR constants of $\partial\Omega$. Above, the sixth inequality relies on the consequence of the John-Nirenberg inequality recorded in [129, (7.4.71)] and the fact that

$$\big|\nu_{2^{j+1}\Delta^*} - \nu_{\Delta^*}\big| \leq C(j+1)\|\nu\|_*(2^{j+1}\Delta^*) \text{ for each } j \in \mathbb{N} \tag{5.2.186}$$

for $C \in (0, \infty)$ depending only on n and the Ahlfors regularity, which is a consequence of [129, (7.4.63)]. The sixth inequality in (5.2.185) also uses

$$x^{**} \in \Delta \subseteq \Delta^* \subseteq 2^{j+1}\Delta^* \text{ for each integer } j \in \mathbb{N}. \tag{5.2.187}$$

The last inequality in (5.2.185) uses $\mathcal{M}_\gamma f(x^{**}) \leq A\lambda$ (cf. (5.2.157)). Next,

$$\begin{aligned}
&\int_{\partial\Omega\setminus\overline{2\Delta^*}} \Big| \langle x-y, \nu_{\Delta^*}\rangle k(x-y) - \langle x^*-y, \nu_{\Delta^*}\rangle k(x^*-y) \Big| |(\varphi f)(y)| \, d\sigma(y) \\
&= \int_{\partial\Omega\setminus\overline{2\Delta^*}} \Big| \big(\langle x-y, \nu_{\Delta^*}\rangle - \langle x^*-y, \nu_{\Delta^*}\rangle\big) k(x^*-y) \\
&\qquad + \langle x-y, \nu_{\Delta^*}\rangle \big(k(x-y) - k(x^*-y)\big) \Big| |(\varphi f)(y)| \, d\sigma(y)
\end{aligned} \tag{5.2.188}$$

and thanks the properties of the kernel k, the Mean Value Theorem, (5.2.180)-(5.2.181), we may further bound the last term above by

$$\begin{aligned}
&\leq C_n \sum_{j\in\mathbb{N}\setminus\mathcal{J}} \int_{2^{j+1}\Delta^*\setminus 2^j\Delta^*} \Big(\frac{|\langle x-x^*, \nu_{\Delta^*}\rangle|}{|x^*-y|^n} + R\frac{|\langle x-y, \nu_{\Delta^*}\rangle|}{|x^*-y|^{n+1}} \Big) |(\varphi f)(y)| \, d\sigma(y) \\
&\leq C_n \sum_{j\in\mathbb{N}} \int_{2^{j+1}\Delta^*\setminus 2^j\Delta^*} \frac{|\langle x-x^*, \nu_{\Delta^*}\rangle|}{|x^*-y|^n} |(\varphi f)(y)| \, d\sigma(y) \\
&\quad + C_n R \sum_{\substack{j\in\mathbb{N} \\ 2^{j+1}\Delta^*\subseteq 6a\Delta_0}} \int_{2^{j+1}\Delta^*\setminus 2^j\Delta^*} \frac{|\langle x-y, \nu_{\Delta^*} - \nu_{2^{j+1}\Delta^*}\rangle|}{|x^*-y|^{n+1}} |f(y)| \, d\sigma(y) \\
&\quad + C_n R \sum_{\substack{j\in\mathbb{N} \\ 2^{j+1}R<4ar_\delta}} \int_{2^{j+1}\Delta^*\setminus 2^j\Delta^*} \frac{|\langle x-y, \nu_{2^{j+1}\Delta^*}\rangle|}{|x^*-y|^{n+1}} |(\varphi f)(y)| \, d\sigma(y) =: \mathrm{I}_1 + \mathrm{I}_2 + \mathrm{I}_3.
\end{aligned} \tag{5.2.189}$$

To estimate I_1, write

$$\begin{aligned}
\mathrm{I}_1 &\leq C_n R^{-1} |\langle x-x^*, \nu_{\Delta^*}\rangle| \sum_{j\in\mathbb{N}} ⨍_{2^{j+1}\Delta^*} |f(y)| \, d\sigma(y) \\
&\leq C\delta \sum_{j\in\mathbb{N}} 2^{-j} \mathcal{M}_\gamma f(x^{**}) \leq C\delta \mathcal{M}_\gamma f(x^{**}) \leq CA\delta\lambda,
\end{aligned} \tag{5.2.190}$$

where $C \in (0,\infty)$ depends only on n, the ADR constants of $\partial\Omega$, and the relative flatness ratio $\wp(\Omega)/\operatorname{diam}\partial\Omega$. The second inequality above is a consequence of (5.2.168) and (5.2.62) used here with $x := x^*$, $y := x$ (which constitutes a valid choice given that the point $x \in \Delta(x^*, 2R)$ since, as seen from (5.2.164)-(5.2.167), we have $x \in \Delta \subseteq \Delta^* = \Delta(x^*, R)$), as well as (5.2.187). The last inequality in (5.2.190) uses $\mathcal{M}_\gamma f(x^{**}) \leq A\lambda$ (cf. (5.2.157)).

To treat I_2, we write (for some $C \in (0,\infty)$ which depends only on n, and the ADR constants of $\partial\Omega$),

$$
\begin{aligned}
\mathrm{I}_2 &\leq CR \sum_{\substack{j\in\mathbb{N}\\ 2^{j+1}\Delta^*\subseteq 6a\Delta_0}} \int_{2^{j+1}\Delta^*\setminus 2^j\Delta^*} \frac{|\nu_{\Delta^*}-\nu_{2^{j+1}\Delta^*}|}{|x^*-y|^n}|f(y)|\,d\sigma(y)\\
&\leq C \sum_{\substack{j\in\mathbb{N}\\ 2^{j+1}\Delta^*\subseteq 6a\Delta_0}} (j+1)\|\nu\|_*(2^{j+1}\Delta^*)2^{-j} \fint_{2^{j+1}\Delta^*}|f(y)|\,d\sigma(y)\\
&\leq C\|\nu\|_*(6a\Delta_0)\,\mathcal{M}_\gamma f(x^{**}) \leq C\delta\mathcal{M}_\gamma f(x^{**}) \leq CA\delta\lambda, \qquad (5.2.191)
\end{aligned}
$$

where the first inequality uses the definition of I_2 and the estimate

$$
|x-y| \leq (3/2)|x^*-y| \quad \text{for each } y \in \partial\Omega\setminus 2\Delta^*, \qquad (5.2.192)
$$

the second inequality takes into account (5.2.186) and the Ahlfors regularity of $\partial\Omega$, the third inequality uses (5.2.187), the fourth inequality relies on (5.2.60), while the last inequality is justified as in (5.2.190).

As regards I_3, write (with $C \in (0,\infty)$ depending only on n, the ADR constants of $\partial\Omega$, and the relative flatness ratio $\wp(\Omega)/\mathrm{diam}\,\partial\Omega$)

$$
\begin{aligned}
\mathrm{I}_3 &\leq C \sum_{\substack{j\in\mathbb{N}\\ 2^{j+1}R<4ar_\delta}} 2^{-j} \fint_{2^{j+1}\Delta^*} \frac{|\langle x-y,\nu_{2^{j+1}\Delta^*}\rangle|}{2^{j+1}R}|f(y)|\,d\sigma(y)\\
&\leq C \sum_{\substack{j\in\mathbb{N}\\ 2^{j+1}R<4ar_\delta}} 2^{-j} \fint_{2^{j+1}\Delta^*} \frac{|\langle x^*-y,\nu_{2^{j+1}\Delta^*}\rangle|}{2^{j+1}R}|f(y)|\,d\sigma(y)\\
&\quad + C \sum_{\substack{j\in\mathbb{N}\\ 2^{j+1}R<4ar_\delta}} 2^{-j} \fint_{2^{j+1}\Delta^*} \frac{|\langle x-x^*,\nu_{2^{j+1}\Delta^*}\rangle|}{2^{j+1}R}|f(y)|\,d\sigma(y)\\
&\leq C\delta \sum_{j\in\mathbb{N}} 2^{-j} \fint_{2^{j+1}\Delta^*} |f(y)|\,d\sigma(y) \leq C\delta\mathcal{M}_\gamma f(x^{**}) \leq CA\delta\lambda. \qquad (5.2.193)
\end{aligned}
$$

The third inequality in (5.2.193) is based on (5.2.62) used twice, first with x replaced by x^* and R replaced by $2^{j+1}R$ and, second, with x replaced by x^*, y replaced by x, and R replaced by $2^{j+1}R$ (the fact that the latter is a valid choice in the context of (5.2.62) is ensured by the first inequality in (5.2.181)). The remaining inequalities in (5.2.193) are then justified much as in (5.2.190).

At this stage, by combining (5.2.185) and (5.2.188)-(5.2.193) we conclude that there exists some $C \in (0,\infty)$ which depends only on n, the ADR constants of $\partial\Omega$, and the relative flatness ratio $\wp(\Omega)/\mathrm{diam}\,\partial\Omega$ such that

$$
\mathrm{I} \leq CA\delta\lambda. \qquad (5.2.194)
$$

To bound II in (5.2.174), observe that if $\varepsilon > \mathrm{diam}\, I_0$ then since $x \in I_0$ if follows that for each $y \in \partial\Omega$ with $|x-y| > \varepsilon$ we have $y \notin I_0$ hence y cannot possibly belong to Δ_0, where the support of φf is contained. Consequently, $\mathrm{II} = 0$ in this case. There remains to consider the situation when $\varepsilon \leq \operatorname{diam} I_0$. In light of (5.2.113), this entails

$$\varepsilon \leq 2ar_\delta. \tag{5.2.195}$$

To proceed, recall that $x, x^{**} \in \Delta$ and assume $y \in \partial\Omega \setminus \overline{2\Delta^*}$ is such that $|x^* - y| \leq \varepsilon$ and $|x - y| > \varepsilon$. Then, $2R < |x^* - y| \leq \varepsilon$ and since $x, x^{**} \in \Delta \subseteq B(x_Q, r_Q)$ (where x_Q and r_Q are, respectively, the center and radius of the surface ball Δ) and $R = \Lambda \cdot r_Q$ with $\Lambda > 2$, we have $|x - x^{**}| < 2r_Q < R < \varepsilon/2$. Hence, the point x^{**} belongs to the surface ball $\Delta(x, \varepsilon/2)$. Moreover, on account of (5.2.178) we may write $|x - y| \leq |x - x^*| + |x^* - y| < R + \varepsilon < (3/2)\varepsilon$ which, in particular, guarantees that $y \in \Delta(x, 2\varepsilon)$. Consequently, $\varepsilon < |x - y| < 2\varepsilon$ hence $|k(x - y)| \leq \varepsilon^{-n}$ and (with $C \in (0, \infty)$ depending only on n and the ADR constants of $\partial\Omega$),

$$\begin{aligned}
\mathrm{II} &\leq C\varepsilon^{-1} \fint_{\Delta(x,2\varepsilon)} |\langle x - y, \nu(y)\rangle|\,|(\varphi f)(y)|\,d\sigma(y) \\
&\leq C\varepsilon^{-1} \fint_{\Delta(x,2\varepsilon)} |\langle x - y, \nu(y) - \nu_{\Delta(x,2\varepsilon)}\rangle|\,|(\varphi f)(y)|\,d\sigma(y) \\
&\quad + C\varepsilon^{-1} \fint_{\Delta(x,2\varepsilon)} |\langle x - y, \nu_{\Delta(x,2\varepsilon)}\rangle|\,|(\varphi f)(y)|\,d\sigma(y) =: \mathrm{II}_1 + \mathrm{II}_2.
\end{aligned} \tag{5.2.196}$$

Using Hölder's inequality, the consequence of the John-Nirenberg inequality recorded in [129, (7.4.71)], (5.2.157), and (5.2.57) we see that there exists $C \in (0, \infty)$ depending only on n, p, $[w]_{A_p}$, and the ADR constants of $\partial\Omega$ so that

$$\begin{aligned}
\mathrm{II}_1 &\leq C\Big(\fint_{\Delta(x,2\varepsilon)} |\nu(y) - \nu_{\Delta(x,2\varepsilon)}|^{\frac{1+\gamma}{\gamma}}\,d\sigma(y)\Big)^{\frac{\gamma}{1+\gamma}} \Big(\fint_{\Delta(x,2\varepsilon)} |f(y)|^{1+\gamma}\,d\sigma(y)\Big)^{\frac{1}{1+\gamma}} \\
&\leq C\|\nu\|_*(\Delta(x, 2\varepsilon))\mathcal{M}_\gamma f(x^{**}) \leq CA\delta\lambda,
\end{aligned} \tag{5.2.197}$$

thanks to (5.2.195), (5.2.60), and also taking into account that since the point x^{**} is contained in $\Delta(x, \varepsilon/2) \subseteq \Delta(x, 2\varepsilon)$ we have $\mathcal{M}_\gamma f(x^{**}) \leq CA\lambda$, as already noted earlier. As for II_2, invoking (5.2.62), Hölder's inequality, (5.2.195), and (5.2.157), it follows that (with $C \in (0, \infty)$ depending only on n, the ADR constants of $\partial\Omega$, and the relative flatness ratio $\wp(\Omega)/\operatorname{diam}\partial\Omega$)

$$\begin{aligned}
\mathrm{II}_2 &\leq C\Big(\sup_{y\in\Delta(x,2\varepsilon)} \varepsilon^{-1}|\langle x - y, \nu_{\Delta(x,2\varepsilon)}\rangle|\Big) \fint_{\Delta(x,2\varepsilon)} |f(y)|\,d\sigma(y) \\
&\leq C\delta\Big(\fint_{\Delta(x,2\varepsilon)} |f(y)|^{1+\gamma}\,d\sigma(y)\Big)^{\frac{1}{1+\gamma}} \leq C\delta \cdot \mathcal{M}_\gamma f(x^{**}) \leq CA\delta\lambda.
\end{aligned} \tag{5.2.198}$$

By (5.2.196)-(5.2.198), there exists $C \in (0, \infty)$ depending only on n, p, $[w]_{A_p}$, the ADR constants of $\partial\Omega$, and the relative flatness ratio $\wp(\Omega)/\operatorname{diam}\partial\Omega$ so that

$$\mathrm{II} \leq CA\delta\lambda. \tag{5.2.199}$$

Turning our attention to III from (5.2.175), we first remark that if $\varepsilon > \operatorname{diam} I_0$ then since $x^* \in I_0$ if follows that for each $y \in \partial\Omega$ with $|x^* - y| > \varepsilon$ we have $y \notin I_0$. Thus, y does not belong to Δ_0, where the support of φf is contained, so III $= 0$ in this case. There remains to treat the scenario in which $\varepsilon \leq \operatorname{diam} I_0$. By (5.2.113), this guarantees that (5.2.195) holds. In concert with (5.2.168), this further gives

$$\widetilde{R} := R + \varepsilon \leq 2a_1\Lambda r_\delta + 2ar_\delta < 10\,Cr_\delta, \tag{5.2.200}$$

where the last inequality comes from (5.2.166). Recall next that $x, x^{**} \in \Delta$ and suppose $y \in \partial\Omega \setminus \overline{2\Delta^*}$ is such that $|x^* - y| > \varepsilon$ and $|x - y| \leq \varepsilon$. Then (5.2.178) implies $|x^* - y| > 2R > R + |x - x^*|$, which further entails $\varepsilon \geq |x - y| \geq |x^* - y| - |x - x^*| > R$. In particular, $R < \varepsilon$. Then, on the one hand,

$$|x^* - y| \leq |x^* - x| + |x - y| < R + \varepsilon = \widetilde{R}, \tag{5.2.201}$$

while on the other hand, having $|x^* - y| > \varepsilon$ as well as $|x^* - y| > 2R$, permits us to estimate $|x^* - y| > R + (\varepsilon/2) > \frac{1}{2}\widetilde{R}$. As such, $|k(x^* - y)| \leq \widetilde{R}^{-n}$ and

$$\text{III} \leq C_n \widetilde{R}^{-1} \fint_{\Delta(x^*, \widetilde{R})} |\langle x^* - y, \nu(y)\rangle|\, |(\varphi f)(y)|\, d\sigma(y). \tag{5.2.202}$$

Granted this and (5.2.200), the same argument which starting with the first line in (5.2.196) has produced (5.2.199) (reasoning with $\widetilde{R}/2$ replacing ε and with x^* replacing x) will now yield (for some $C \in (0, \infty)$ which depends only on n, p, $[w]_{A_p}$, the ADR constants of $\partial\Omega$, and the relative flatness ratio $\wp(\Omega)/\operatorname{diam}\partial\Omega$)

$$\text{III} \leq CA\delta\lambda, \tag{5.2.203}$$

as soon as we show that we have $x^{**} \in \Delta(x^*, \widetilde{R})$. To justify this membership, start by recalling that $|x - x^{**}| < 2r_Q < R$ and then use (5.2.178), the triangle inequality, and the fact that $R < \varepsilon$ to estimate $|x^* - x^{**}| \leq |x - x^*| + |x - x^{**}| < 2R < \widetilde{R}$. The proof of (5.2.203) is therefore complete.

Let us summarize our progress. From (5.2.173), (5.2.194), (5.2.199), and (5.2.203) we conclude that there exists $C \in (0, \infty)$, depending only on n, p, $[w]_{A_p}$, the ADR constants of $\partial\Omega$, and the relative flatness ratio $\wp(\Omega)/\operatorname{diam}\partial\Omega$, such that

$$\Big|T_\varepsilon f_2(x) - T_\varepsilon f_2(x^*)\Big| \leq CA\delta\lambda = C\theta\Big(\frac{\delta}{\phi(\delta)}\Big)\lambda, \qquad \forall\, x \in \Delta, \quad \forall\, \varepsilon > 0. \tag{5.2.204}$$

Given that $\theta \in (0, 1)$, and taking $\delta_* \in (0, 1)$ small enough to begin with (keeping in mind that $\lim_{t \to 0^+} t/\phi(t) = 0$; cf. (5.2.81)), from (5.2.204) we see that

$$\Big|T_\varepsilon f_2(x) - T_\varepsilon f_2(x^*)\Big| \leq \tfrac{1}{2}\lambda, \qquad \forall\, x \in \Delta, \quad \forall\, \varepsilon > 0. \tag{5.2.205}$$

From (5.2.172), (5.2.205), and (5.2.51) we then obtain

$$|\psi(x)(T_* f_2)(x)| \leq T_* f_2(x) \leq 2\lambda \text{ for all } x \in \Delta, \tag{5.2.206}$$

whenever $\theta > 0$ is small enough. Therefore, for this choice of θ, we conclude that

$$\sigma\Big(\big\{x \in \Delta : |\psi(x)(T_* f_2)(x)| > 2\lambda\big\}\Big) = 0 \tag{5.2.207}$$

which, in concert with (5.2.170) and (5.2.171), establishes (5.2.161). This finishes the proof of the good-λ inequality (5.2.126).

Once (5.2.126) has been established, we proceed to prove the claim in (5.2.59). First, using (5.2.122), by our definition of A, and by possibly choosing a smaller $\delta_* \in (0,1)$ (again, bearing in mind that $\lim_{t\to 0^+} t/\phi(t) = 0$; cf. (5.2.81)), for each point $x \in I_0$ with $|\psi(x)(T_{(*)}(\varphi f))(x)| > \lambda$ and $\mathcal{M}_\gamma f(x) \leq A\lambda$ we may write

$$\begin{aligned}
\lambda &< |\psi(x)(T_{(*)}(\varphi f))(x)| \leq |\psi(x)(T_*(\varphi f))(x)| + C\delta \cdot \mathcal{M}_\gamma f(x) \\
&\leq |\psi(x)(T_*(\varphi f))(x)| + C\delta A\lambda = |\psi(x)(T_*(\varphi f))(x)| + C\theta\Big(\frac{\delta}{\phi(\delta)}\Big)\lambda \\
&< |\psi(x)(T_*(\varphi f))(x)| + \tfrac{1}{2}\lambda.
\end{aligned} \tag{5.2.208}$$

Hence, for such a choice of θ we have

$$\begin{aligned}
&\tfrac{1}{2}\lambda < |\psi(x)(T_*(\varphi f))(x)| \text{ whenever the point } x \in I_0 \text{ is} \\
&\text{such that } |\psi(x)(T_{(*)}(\varphi f))(x)| > \lambda \text{ and } \mathcal{M}_\gamma f(x) \leq A\lambda.
\end{aligned} \tag{5.2.209}$$

Consequently,

$$\begin{aligned}
&\big\{x \in I_0 : |\psi(x)(T_{(*)}(\varphi f))(x)| > \lambda \text{ and } \mathcal{M}_\gamma f(x) \leq A\lambda\big\} \\
&\qquad\qquad \subseteq \big\{x \in I_0 : |\psi(x)T_*(\varphi f))(x)| > \tfrac{\lambda}{2}\big\}
\end{aligned} \tag{5.2.210}$$

which, in turn, permits us to estimate

$$\begin{aligned}
&w\Big(\big\{x \in I_0 : |\psi(x)(T_{(*)}(\varphi f))(x)| > \lambda\big\}\Big) \\
&\quad \leq w\Big(\big\{x \in I_0 : |\psi(x)(T_{(*)}(\varphi f))(x)| > \lambda \text{ and } \mathcal{M}_\gamma f(x) \leq A\lambda\big\}\Big) \\
&\qquad + w\Big(\big\{x \in I_0 : \mathcal{M}_\gamma f(x) > A\lambda\big\}\Big) \\
&\quad \leq w\Big(\big\{x \in I_0 : |\psi(x)(T_*(\varphi f))(x)| > \tfrac{\lambda}{2}\big\}\Big) + w\Big(\big\{x \in I_0 : \mathcal{M}_\gamma f(x) > A\lambda\big\}\Big).
\end{aligned} \tag{5.2.211}$$

From (5.2.125) it is clear that for each fixed θ we have

$$\mathfrak{Z}(\theta,\delta) = C\big(\theta^{1+\gamma} + \theta^{1+\gamma/2} \cdot O(1) + o(1)\big) \text{ as } \delta \to 0^+. \tag{5.2.212}$$

This makes it is possible to first choose the threshold $\delta_* \in (0,1)$ and then pick the coefficient $\theta \in (0,1)$ small enough depending only on n, p, $[w]_{A_p}$, η, ϕ, the ADR constants of $\partial\Omega$, and the relative flatness ratio $\wp(\Omega)/\operatorname{diam}\partial\Omega$ so that

$$\mathfrak{Z}(\theta,\delta)^\tau < (2 \cdot 8^p)^{-1}. \tag{5.2.213}$$

This is the last demand imposed on δ_*, θ, and the totality of all these size specifications imply that the final choice of these parameters ultimately depends only on n, p, $[w]_{A_p}$, η, ϕ, the ADR constants of $\partial\Omega$, and the relative flatness ratio $\wp(\Omega)/\mathrm{diam}\,\partial\Omega$. Combining (5.2.211) with (5.2.126) and keeping (5.2.213) in mind we then get

$$\begin{aligned}
w\Big(\big\{x \in I_0 : |\psi(x)(T_*(\varphi f))(x)| > 4\lambda\big\}\Big)&\\
&\leq w\Big(\big\{x \in I_0 : |\psi(x)(T_*(\varphi f))(x)| > 4\lambda \text{ and } \mathcal{M}_\gamma f(x) \leq A\lambda\big\}\Big)\\
&\quad + w\Big(\big\{x \in I_0 : \mathcal{M}_\gamma f(x) > A\lambda\big\}\Big)\\
&\leq \mathfrak{Z}(\theta,\delta)^\tau \cdot w\Big(\big\{x \in I_0 : |\psi(x)(T_{(*)}(\varphi f))(x)| > \lambda\big\}\Big)\\
&\quad + w\Big(\big\{x \in I_0 : \mathcal{M}_\gamma f(x) > A\lambda\big\}\Big)\\
&< (2\cdot 8^p)^{-1}\, w\Big(\big\{x \in I_0 : |\psi(x)(T_*(\varphi f))(x)| > \tfrac{\lambda}{2}\big\}\Big)\\
&\quad + \big(1 + (2\cdot 8^p)^{-1}\big)\, w\Big(\big\{x \in I_0 : \mathcal{M}_\gamma f(x) > A\lambda\big\}\Big).
\end{aligned}\tag{5.2.214}$$

Recall that $\gamma \in (0, p-1)$ has been chosen so that $w \in A_{p/(1+\gamma)}(\partial\Omega, \sigma)$, hence $\mathcal{M}_\gamma$ is bounded on $L^p(\partial\Omega, w)$. Multiply the most extreme sides of (5.2.214) by $p\lambda^{p-1}$ and integrate over $\lambda \in (0,\infty)$. Bearing in mind that $A = \theta \cdot \phi(\delta)^{-1}$, after three natural changes of variables (namely, $\widetilde{\lambda} := 4\lambda$ in the first integral, $\widetilde{\lambda} := \frac{1}{2}\lambda$ in the second integral, and $\widetilde{\lambda} := \theta\,\phi(\delta)^{-1}\lambda$ in the third integral) we therefore obtain

$$\begin{aligned}
\int_{I_0} |\psi T_*(\varphi f)|^p \,\mathrm{d}w &\leq \frac{1}{2}\int_{I_0} |\psi T_*(\varphi f)|^p \,\mathrm{d}w\\
&\quad + \phi(\delta)^p \theta^{-p}\big(2^{2p} + 2^{-p-1}\big)\int_{I_0} (\mathcal{M}_\gamma f)^p \,\mathrm{d}w\\
&\leq \frac{1}{2}\int_{I_0} |\psi T_*(\varphi f)|^p \,\mathrm{d}w + C\phi(\delta)^p \int_{\partial\Omega} |\varphi f|^p \,\mathrm{d}w,
\end{aligned}\tag{5.2.215}$$

for some constant $C \in (0,\infty)$ which depends only on n, p, $[w]_{A_p}$, η, ϕ, the ADR constants of $\partial\Omega$, and the relative flatness ratio $\wp(\Omega)/\mathrm{diam}\,\partial\Omega$. Since $f \in L^p(\partial\Omega, w)$ and the operator T_* maps the space $L^p(\partial\Omega, w)$ into itself (cf. [131, (2.3.58)]), it follows that $\int_{I_0} |\psi T_*(\varphi f)|^p \,\mathrm{d}w \leq \|T_*(\varphi f)\|^p_{L^p(\partial\Omega,w)} < \infty$. Hence, the first integral in the right-most side of (5.2.215) may be absorbed in the left-most side. By also taking into account (5.2.115), we therefore obtain

$$\begin{aligned}
\int_{\partial\Omega} |\psi T_*(\varphi f)|^p \,\mathrm{d}w &\leq \int_{2\Delta_0} |\psi T_*(\varphi f)|^p \,\mathrm{d}w\\
&\leq \int_{I_0} |\psi T_*(\varphi f)|^p \,\mathrm{d}w \leq C\phi(\delta)^p \int_{\partial\Omega} |f|^p \,\mathrm{d}w.
\end{aligned}\tag{5.2.216}$$

From this, (5.2.82) follows on account of the arbitrariness of $f \in L^p(\partial\Omega, w)$. This finishes the proof of the fact that "(5.2.79) implies (5.2.82)." Hence, at long last, (5.2.17) is established.

Finally, the very last claim in the statement of Theorem 5.2.3 is a consequence of what we have proved so far and the fact that (5.2.55) implies (5.2.56). The proof of Theorem 5.2.3 is therefore complete. □

5.3 Essential Norm of Double Layer Estimable in Terms of Flatness if and only if Coefficient Tensor Distinguished

In Theorem 5.2.3 we have established essential norm estimates, on Muckenhoupt weighted Lebesgue spaces, for principal-value singular integral operators whose integral kernel has a special algebraic format, in that it involves the inner product between the outward unit normal and the chord, as a factor (aka, "chord-dot-normal" SIO's). It turns out that the latter algebraic property is actually *necessary* for the operator norm estimates to hold in the format specified (5.2.21)-(5.2.22) which, in particular, makes it clear that the infinitesimal flatness of the boundary has a direct effect on the size of the essential norm. Here is the statement of the theorem which bridges between those geometric and functional analytic aspects.

Theorem 5.3.1 *Fix a number $n \in \mathbb{N}$ satisfying $n \geq 2$, and pick a sufficiently large integer $N = N(n) \in \mathbb{N}$. Also, consider a vector-valued function*

$$\vec{k} \in \big[\mathscr{C}^N(\mathbb{R}^n \setminus \{0\})\big]^n \text{ which is odd and positive homogeneous of degree } 1-n. \tag{5.3.1}$$

For any given set

$$\Omega \subseteq \mathbb{R}^n \text{ which is a UR domain with compact boundary (and with } \sigma := \mathcal{H}^{n-1}\lfloor\partial\Omega \text{ and } \nu \text{ denoting the surface measure and the geometric measure theoretic outward unit normal to } \Omega\text{, respectively),} \tag{5.3.2}$$

consider the principal-value singular integral operators $T, T^{\#}$ on $\partial\Omega$, acting on each $f \in L^1(\partial\Omega, \sigma)$ at σ-a.e. $x \in \partial\Omega$ according to

$$Tf(x) := \lim_{\varepsilon\to 0^+} \int_{y\in\partial\Omega,\ |x-y|>\varepsilon} \langle \nu(y), \vec{k}(x-y)\rangle f(y)\, d\sigma(y) \tag{5.3.3}$$

and

$$T^{\#}f(x) := \lim_{\varepsilon\to 0^+} \int_{y\in\partial\Omega,\ |x-y|>\varepsilon} \langle \nu(x), \vec{k}(y-x)\rangle f(y)\, d\sigma(y). \tag{5.3.4}$$

Then the following statements are equivalent.

(1) For any set Ω as in (5.3.2), *any (or some) exponent $p \in (1, \infty)$, any (or some) Muckenhoupt weight $w \in A_p(\partial\Omega, \sigma)$, and any (or some) integer $m \in \mathbb{N}$ there exists $C_m \in (0, \infty)$, which depends only on m, n, p, $[w]_{A_p}$, $\vec{k}$, the* UR *constants of $\partial\Omega$, and the relative flatness ratio $\wp(\Omega)/\text{diam}\,\partial\Omega$, ensuring the validity of both (or one) of the following estimates:*

$$\text{dist}\big(T, \text{Cp}(L^p(\partial\Omega, w))\big) \leq C_m\Big[\text{dist}\big(\nu, \big[\text{VMO}(\partial\Omega, \sigma)\big]^n\big)\Big]^{\langle m\rangle}, \tag{5.3.5}$$

$$\text{dist}\big(T^{\#}, \text{Cp}(L^p(\partial\Omega, w))\big) \leq C_m\Big[\text{dist}\big(\nu, \big[\text{VMO}(\partial\Omega, \sigma)\big]^n\big)\Big]^{\langle m\rangle}, \tag{5.3.6}$$

with the distances in the left-hand sides measured in the operator norm in $\text{Bd}(L^p(\partial\Omega, w))$, *and the distances in the right-hand sides measured in* $\big[\text{BMO}(\partial\Omega, \sigma)\big]^n$.

(2) For any set Ω as in (5.3.2) *with the additional property that $\nu \in$* VMO$(\partial\Omega, \sigma)$, *that is, whenever Ω is an infinitesimally flat* AR *domain, both (or one of the) memberships*

$$T \in \text{Cp}\big(L^p(\partial\Omega, w)\big), \quad T^{\#} \in \text{Cp}\big(L^p(\partial\Omega, w)\big) \tag{5.3.7}$$

hold(s) for some (or all) $p \in (1, \infty)$ and some (or all) $w \in A_p(\partial\Omega, \sigma)$.

(3) There exists a scalar-valued function $k \in \mathscr{C}^N(\mathbb{R}^n \setminus \{0\})$ which is even and positive homogeneous of degree $-n$ with the property that

$$\vec{k}(x) = x\,k(x) \text{ for each } x \in \mathbb{R}^n \setminus \{0\}. \tag{5.3.8}$$

We wish to remark that a similar result is valid in the case when the kernel $\vec{k}$ is *matrix-valued* function. In such a scenario, we interpret the inner products appearing in (5.3.3), (5.3.4) as the action of the matrix $\vec{k}$ on the vector ν. Also, $L^p(\partial\Omega, w)$ is now replaced by $\big[L^p(\partial\Omega, w)\big]^M$. More specifically, if for some $M \in \mathbb{N}$ one assumes

$$\vec{k} \in \big[\mathscr{C}^N(\mathbb{R}^n \setminus \{0\})\big]^{M\times n} \text{ is odd, positive homogeneous of degree } 1-n, \tag{5.3.9}$$

then item *(1)* in the statement of Theorem 5.3.1 (with the natural adjustments in notation indicated above) is presently equivalent to the existence of some *vector-valued* function $k = (k_i)_{1\leq i\leq M}$ which belongs to $\big[\mathscr{C}^N(\mathbb{R}^n \setminus \{0\})\big]^M$, is even and positive homogeneous of degree $-n$, and has the property that

$$\vec{k}(x) = \big(x_j k_i(x)\big)_{\substack{1\leq i\leq M\\ 1\leq j\leq n}} \text{ for each } x = (x_j)_{1\leq j\leq n} \in \mathbb{R}^n \setminus \{0\}. \tag{5.3.10}$$

Such a version follows from Theorem 5.3.1 applied to the individual rows of the matrix-valued function (5.3.9).

The proof of Theorem 5.3.1 is presented below.

Proof of Theorem 5.3.1 It is clear from [130, (1.2.53)] and [129, (5.11.93)] that the weakest version of *(1)* implies the weakest version of *(2)*. Also, thanks to Proposition 5.1.1, [132, Theorem 5.2.2, item *(3)*], Schauder's theorem (cf. [130, (1.2.57)]), the extrapolation result established in [129, Proposition 7.7.12], and [129, (5.11.93)], the weakest version of *(2)* implies *(3)*. Finally, on account of Theorem 5.2.3, *(3)* implies the strongest version of *(1)* which, together with [129, (5.11.93)], implies the strongest version of *(2)*. □

For a given second-order, homogeneous, constant complex coefficient system L, and a given Ahlfors regular domain $\Omega \subseteq \mathbb{R}^n$, Theorem 5.2.3 applies to the boundary-to-boundary double layer potential operator K_A associated with a coefficient tensor $A \in \mathfrak{A}_L$, acting on Muckenhoupt weighted Lebesgue spaces on $\partial\Omega$, if and only if A is distinguished (in the sense of Definition 1.2.2), i.e., $A \in \mathfrak{A}_L^{\text{dis}}$.

Theorem 5.3.2 *Fix $n \in \mathbb{N}$ with $n \geq 2$, and let L be a homogeneous, weakly elliptic, constant (complex) coefficient, second-order $M \times M$ system in $\mathbb{R}^n$ (for some $M \in \mathbb{N}$). Also, pick $A \in \mathfrak{A}_L$ and, for each*

$$\text{UR } \textit{domain } \Omega \subseteq \mathbb{R}^n \textit{ with compact boundary (and with } \sigma := \mathcal{H}^{n-1}\lfloor\partial\Omega \textit{ and } \nu \textit{ denoting the surface measure and the geometric measure theoretic outward unit normal to } \Omega\textit{, respectively),} \tag{5.3.11}$$

let $K, K^\#$ be the boundary-to-boundary double layer potential operators associated with Ω and the coefficient tensor A as in (A.0.116) *and* (A.0.118), *respectively. Then the following statements are equivalent.*

(i) For any set Ω as in (5.3.11), *any (or some) exponent $p \in (1, \infty)$, any (or some) Muckenhoupt weight $w \in A_p(\partial\Omega, \sigma)$, and any (or some) integer $m \in \mathbb{N}$ there exists $C_m \in (0, \infty)$, which depends only on m, n, p, $[w]_{A_p}$, A, the* UR *constants of $\partial\Omega$, and the relative flatness ratio $\wp(\Omega)/\text{diam}\,\partial\Omega$ ensuring the validity of any (or both) of the following estimates:*

$$\text{dist}\Big(K, \text{Cp}\big([L^p(\partial\Omega, w)]^M\big)\Big) \leq C_m\Big[\text{dist}\big(\nu, \big[\text{VMO}(\partial\Omega, \sigma)\big]^n\big)\Big]^{\langle m\rangle}, \tag{5.3.12}$$

$$\text{dist}\Big(K^\#, \text{Cp}\big([L^p(\partial\Omega, w)]^M\big)\Big) \leq C_m\Big[\text{dist}\big(\nu, \big[\text{VMO}(\partial\Omega, \sigma)\big]^n\big)\Big]^{\langle m\rangle}, \tag{5.3.13}$$

where the distances in the left-hand sides are measured in the operator norm in the space $\text{Bd}\big([L^p(\partial\Omega, w)]^M\big)$, *while the distances in the right-hand sides are measured in the space* $\big[\text{BMO}(\partial\Omega, \sigma)\big]^n$.

(ii) For any set Ω as in (5.3.11) *with the additional property that $\nu \in \text{VMO}(\partial\Omega, \sigma)$, that is, whenever Ω is an infinitesimally flat* AR *domain, both (or one of the) memberships*

$$K \in \text{Cp}\big([L^p(\partial\Omega, w)]^M\big), \quad K^\# \in \text{Cp}\big([L^p(\partial\Omega, w)]^M\big) \tag{5.3.14}$$

hold(s) for some (or all) $p \in (1, \infty)$ and some (or all) $w \in A_p(\partial\Omega, \sigma)$.

(iii) The coefficient tensor A is distinguished, i.e., $A \in \mathfrak{A}_L^{\rm dis}$.

Proof This is a consequence of Proposition 1.3.2 and the version of Theorem 5.3.1 for matrix-valued kernels. □

5.4 Essential Norms and Fredholm Radii of Chord-Dot-Normal SIO's on Other Spaces

While it remains unclear how the essential norm of a chord-dot-normal SIO acting on Sobolev spaces may be estimated in terms of the flatness of the "surface" on which this is defined, we can do the next best thing (cf. [130, (2.2.16)]) and estimate the Fredholm radius of such an operator, in the same favorable fashion as before.

Theorem 5.4.1 *Suppose* $\Omega \subseteq \mathbb{R}^n$ *is a* UR *domain with compact boundary. Denote by* ν *the geometric measure theoretic outward unit normal to* Ω *and abbreviate* $\sigma := \mathcal{H}^{n-1}\lfloor\partial\Omega$. *Also, fix a sufficiently large integer* $N = N(n) \in \mathbb{N}$ *and consider* $k \in \mathscr{C}^N(\mathbb{R}^n \setminus \{0\})$ *which is even and positive homogeneous of degree* $-n$. *Finally, fix an exponent* $p \in (1, \infty)$ *along with a Muckenhoupt weight* $w \in A_p(\partial\Omega, \sigma)$, *and recall from* [132, Theorem 5.2.2, item *(7)*] *that the singular integral operator defined originally on* $f \in L^1(\partial\Omega, \sigma)$ *at* σ*-a.e.* $x \in \partial\Omega$ *as*

$$Tf(x) := \lim_{\varepsilon\to 0^+} \int_{y\in\partial\Omega,\ |x-y|>\varepsilon} \langle \nu(y), x-y\rangle k(x-y) f(y)\, d\sigma(y), \tag{5.4.1}$$

induces a linear and bounded mapping in the context of weighted Sobolev spaces

$$T : L_1^p(\partial\Omega, w) \longrightarrow L_1^p(\partial\Omega, w). \tag{5.4.2}$$

Then for each $m \in \mathbb{N}$ *there exists a constant* $C_m \in (0, \infty)$, *which depends only on* m, n, p, $[w]_{A_p}$, *the* UR *constants of* $\partial\Omega$, *and the relative flatness ratio* $\wp(\Omega)/\operatorname{diam}\partial\Omega$ *such that (with pieces of notation introduced in* (A.0.80) *and* (4.1.12))

$$\rho_{\rm Fred}\big(T; L_1^p(\partial\Omega, w)\big) \leq C_m\Big(\sum_{|\alpha|\leq N} \sup_{S^{n-1}} |\partial^\alpha k|\Big)\Big[\operatorname{dist}\big(\nu, \big[\mathrm{VMO}(\partial\Omega,\sigma)\big]^n\big)\Big]^{\langle m\rangle} \tag{5.4.3}$$

where the distance in the right-hand side is measured in $\big[\mathrm{BMO}(\partial\Omega,\sigma)\big]^n$. *In particular, corresponding to* $w \equiv 1$,

$$\rho_{\rm Fred}\big(T; L_1^p(\partial\Omega, \sigma)\big) \leq C_m\Big(\sum_{|\alpha|\leq N} \sup_{S^{n-1}} |\partial^\alpha k|\Big)\Big[\operatorname{dist}\big(\nu, \big[\mathrm{VMO}(\partial\Omega,\sigma)\big]^n\big)\Big]^{\langle m\rangle}. \tag{5.4.4}$$

Via duality, from (5.4.3), [130, (2.2.15)], (A.0.156), and [132, Theorem 5.2.2, item *(8)*] we also conclude that

$$\rho_{\text{Fred}}\big(T^{\#}; L^{p'}_{-1}(\partial\Omega, w')\big) \leq C_m\Big(\sum_{|\alpha|\leq N}\sup_{S^{n-1}}|\partial^\alpha k|\Big)\Big[\text{dist}\big(\nu, \big[\text{VMO}(\partial\Omega,\sigma)\big]^n\big)\Big]^{\langle m\rangle} \tag{5.4.5}$$

where $p' := (1 - 1/p)^{-1} \in (1, \infty)$ is the conjugate exponent of p, and w' is the conjugate weight of w defined by $w' := w^{1-p'} \in A_{p'}(\partial\Omega, \sigma)$ (cf. item *(2)* in [129, Lemma 7.7.1]).

Much as in Remark 3, following the statement of Theorem 5.2.3, from (5.4.3) we deduce that if

$$\delta := \text{dist}\big(\nu, \big[\text{VMO}(\partial\Omega,\sigma)\big]^n\big) \tag{5.4.6}$$

then

$$\rho_{\text{Fred}}\big(T; L^p_1(\partial\Omega, w)\big) \leq C\Big(\sum_{|\alpha|\leq N}\sup_{S^{n-1}}|\partial^\alpha k|\Big)\cdot\delta\cdot\ln\big(\text{e}/\delta\big). \tag{5.4.7}$$

Thus, the Fredholm radius of T on $L^p_1(\partial\Omega, w)$ has at most linear growth in δ, up to a logarithm; in fact, arbitrarily many iterated logarithms are allowed. In particular,

$$\rho_{\text{Fred}}\big(T; L^p_1(\partial\Omega, w)\big) \leq C_\varepsilon\Big(\sum_{|\alpha|\leq N}\sup_{S^{n-1}}|\partial^\alpha k|\Big)\cdot\delta^{1-\varepsilon}\ \text{ for each } \varepsilon \in (0,1). \tag{5.4.8}$$

Analogous results are valid for the operator $T^{\#}$ on the dual space.

Finally, we wish to point out that similar results to those described above are valid for the singular integral operators $T, T^{\#}$ defined in Corollary 5.2.5 in the two-dimensional setting.

We now turn to the task of providing the proof of Theorem 5.4.1.

Proof of Theorem 5.4.1 If $k \equiv 0$, there is nothing to prove. If k is not identically zero, working with k/A in place of k, where $A := \sum_{|\alpha|\leq N}\sup_{S^{n-1}}|\partial^\alpha k| \in (0,\infty)$ reduces matters to the case when

$$\sum_{|\alpha|\leq N}\sup_{S^{n-1}}|\partial^\alpha k| = 1. \tag{5.4.9}$$

Assume this is the case for the remainder of the proof. Our goal is to show that for each $m \in \mathbb{N}$ there exists $C_m \in (0,\infty)$ as in the statement of the theorem so that

$$\rho_{\text{Fred}}\big(T; L^p_1(\partial\Omega, w)\big) \leq C_m\Big[\text{dist}\big(\nu, \big[\text{VMO}(\partial\Omega,\sigma)\big]^n\big)\Big]^{\langle m\rangle}. \tag{5.4.10}$$

To this end, recall from [130, (2.2.13)] and [132, (5.2.164)] that

$$\rho_{\text{Fred}}\big(T; L^p_1(\partial\Omega, w)\big) \leq \|T\|_{L^p_1(\partial\Omega,w)\to L^p_1(\partial\Omega,w)} \leq C, \tag{5.4.11}$$

where, thanks to (5.4.9), the constant $C \in (0,\infty)$ depends only on n, p, $[w]_{A_p}$, and the UR constants of $\partial\Omega$. Fix an integer $m \in \mathbb{N}$ along with a large constant $C_o \in (1,\infty)$ depending only on m, n, p, $[w]_{A_p}$, the UR constants of $\partial\Omega$, and the relative flatness ratio $\wp(\Omega)/\operatorname{diam}\partial\Omega$, to be specified later. If $\operatorname{dist}\big(\nu, \big[\mathrm{VMO}(\partial\Omega,\sigma)\big]^n\big) \geq C_o^{-1}$ then (5.4.10) becomes a simple consequence of (5.4.11) and (4.1.17). Henceforth, we shall focus on the remaining case, when

$$\operatorname{dist}\big(\nu, \big[\mathrm{VMO}(\partial\Omega,\sigma)\big]^n\big) < C_o^{-1}. \tag{5.4.12}$$

To proceed, pick a real number $r > \rho_{\text{Fred}}\big(T; L^p(\partial\Omega, w)\big)$. Then for each complex number z with $|z| > r$ the operator $zI - T$ is Fredholm on $L^p(\partial\Omega, w)$. Fix such a number, $z \in \mathbb{C} \setminus \overline{B(0,r)}$. We may then find an operator $S \in \mathrm{Bd}\big(L^p(\partial\Omega, w)\big)$ together with $K \in \mathrm{Cp}\big(L^p(\partial\Omega, w)\big)$ such that $S \circ (zI - T) = I + K$ on $L^p(\partial\Omega, w)$. For each given function $f \in L^p(\partial\Omega, w)$ we then have

$$\begin{aligned}\|f\|_{L^p(\partial\Omega,w)} &\leq \|(I+K)f\|_{L^p(\partial\Omega,w)} + \|Kf\|_{L^p(\partial\Omega,w)}\\ &\leq \big\|S\big((zI-T)f\big)\big\|_{L^p(\partial\Omega,w)} + \|Kf\|_{L^p(\partial\Omega,w)}\\ &\leq C\|(zI-T)f\|_{L^p(\partial\Omega,w)} + \|Kf\|_{L^p(\partial\Omega,w)},\end{aligned} \tag{5.4.13}$$

where we have set $C := \|S\|_{L^p(\partial\Omega,w)\to L^p(\partial\Omega,w)} \in (0,\infty)$. Next, recall from [132, (5.2.167)] that for each $f \in L^p_1(\partial\Omega, w)$ and each $r, s \in \{1,\dots,n\}$ we have

$$\begin{aligned}(zI-T)\big(\partial_{\tau_{rs}} f\big) &= \partial_{\tau_{rs}}((zI-T)f) + \big[M_{\nu_r}, T\big](\nabla_{\tan} f)_s - \big[M_{\nu_s}, T\big](\nabla_{\tan} f)_r\\ &\quad - \big[M_{\nu_r}, V\big](\nu_s \nabla_{\tan} f) + \big[M_{\nu_s}, V\big](\nu_r \nabla_{\tan} f)\end{aligned} \tag{5.4.14}$$

at σ-a.e. point on $\partial\Omega$, where the operator V is as in [132, (5.2.149)]. Also, from the commutator estimate established in [131, Theorem 2.7.3] we know that for each index $j \in \{1,\dots,n\}$ we have (bearing in mind (5.4.9))

$$\operatorname{dist}\Big(\big[M_{\nu_j}, V\big], \mathrm{Cp}\big([L^p(\partial\Omega,w)]^n \to L^p(\partial\Omega,w)\big)\Big) \leq C \cdot \operatorname{dist}\big(\nu, [\mathrm{VMO}(\partial\Omega,\sigma)]^n\big) \tag{5.4.15}$$

and

$$\operatorname{dist}\Big(\big[M_{\nu_j}, T\big], \mathrm{Cp}\big(L^p(\partial\Omega,w)\big)\Big) \leq C \cdot \operatorname{dist}\big(\nu, [\mathrm{VMO}(\partial\Omega,\sigma)]^n\big) \tag{5.4.16}$$

with the constant $C \in (0,\infty)$ depending only on n, p, $[w]_{A_p}$, and the UR constants of $\partial\Omega$. In particular, for each index $j \in \{1,\dots,n\}$ there exist two operators

$$K_j \in \mathrm{Cp}\big([L^p(\partial\Omega,w)]^n \to L^p(\partial\Omega,w)\big) \text{ and } \widetilde{K}_j \in \mathrm{Cp}\big(L^p(\partial\Omega,w)\big) \tag{5.4.17}$$

such that

$$\big\|[M_{\nu_j}, V] - K_j\big\|_{[L^p(\partial\Omega,w)]^n \to L^p(\partial\Omega,w)} \leq C \cdot \operatorname{dist}\big(\nu, [\mathrm{VMO}(\partial\Omega,\sigma)]^n\big), \tag{5.4.18}$$

$$\big\|[M_{\nu_j}, T] - \widetilde{K}_j\big\|_{L^p(\partial\Omega,w) \to L^p(\partial\Omega,w)} \leq C \cdot \operatorname{dist}\big(\nu, [\mathrm{VMO}(\partial\Omega,\sigma)]^n\big), \tag{5.4.19}$$

where the constant $C \in (0,\infty)$ depends only on n, p, $[w]_{A_p}$, and the UR constants of $\partial\Omega$. In turn, from (5.4.14), (5.4.18), and (5.4.19) we conclude that for each function

$f \in L^p_1(\partial\Omega, w)$ and each pair of indices $r, s \in \{1, \dots, n\}$ we have

$$\begin{aligned}\big\|(zI - T)(\partial_{\tau_{rs}} f)\big\|_{L^p(\partial\Omega,w)} &\leq \big\|\partial_{\tau_{rs}}((zI - T)f)\big\|_{L^p(\partial\Omega,w)} \\ &\quad + \big\|\widetilde{K}_r(\nabla_{\tan} f)_s\big\|_{L^p(\partial\Omega,w)} + \big\|\widetilde{K}_s(\nabla_{\tan} f)_r\big\|_{L^p(\partial\Omega,w)} \\ &\quad + \big\|K_r(\nu_s \nabla_{\tan} f)\big\|_{L^p(\partial\Omega,w)} + \big\|K_s(\nu_r \nabla_{\tan} f)\big\|_{L^p(\partial\Omega,w)} \\ &\quad + C \cdot \operatorname{dist}\big(\nu, [\mathrm{VMO}(\partial\Omega, \sigma)]^n\big) \|\nabla_{\tan} f\|_{[L^p(\partial\Omega,w)]^n}. \end{aligned} \tag{5.4.20}$$

Combining (5.4.20) and (5.4.13) (written with f replaced by $\partial_{\tau_{rs}} f$) therefore gives

$$\begin{aligned}\big\|\partial_{\tau_{rs}} f\big\|_{L^p(\partial\Omega,w)} &\leq C\big\|(zI - T)(\partial_{\tau_{rs}} f)\big\|_{L^p(\partial\Omega,w)} + \big\|K(\partial_{\tau_{rs}} f)\big\|_{L^p(\partial\Omega,w)} \\ &\leq C\big\|\partial_{\tau_{rs}}((zI - T)f)\big\|_{L^p(\partial\Omega,w)} + \big\|K(\partial_{\tau_{rs}} f)\big\|_{L^p(\partial\Omega,w)} \\ &\quad + C\big\|\widetilde{K}_r(\nabla_{\tan} f)_s\big\|_{L^p(\partial\Omega,w)} + C\big\|\widetilde{K}_s(\nabla_{\tan} f)_r\big\|_{L^p(\partial\Omega,w)} \\ &\quad + C\big\|K_r(\nu_s \nabla_{\tan} f)\big\|_{L^p(\partial\Omega,w)} + C\big\|K_s(\nu_r \nabla_{\tan} f)\big\|_{L^p(\partial\Omega,w)} \\ &\quad + C \cdot \operatorname{dist}\big(\nu, [\mathrm{VMO}(\partial\Omega, \sigma)]^n\big) \|\nabla_{\tan} f\|_{[L^p(\partial\Omega,w)]^n} \end{aligned} \tag{5.4.21}$$

for each $f \in L^p_1(\partial\Omega, w)$ and each $r, s \in \{1, \dots, n\}$. Adding the most extreme sides in (5.4.13) and (5.4.21), and summing up over $r, s \in \{1, \dots, n\}$, we arrive at the conclusion that there exists an operator $\mathcal{K} \in \mathrm{Cp}\big(L^p_1(\partial\Omega, w) \to L^p(\partial\Omega, w)\big)$ with the property that for each function $f \in L^p_1(\partial\Omega, w)$ we have

$$\begin{aligned}\|f\|_{L^p_1(\partial\Omega,w)} &\leq C\|(zI - T)f\|_{L^p_1(\partial\Omega,w)} + \|\mathcal{K}f\|_{L^p(\partial\Omega,w)} \\ &\quad + C \cdot \operatorname{dist}\big(\nu, [\mathrm{VMO}(\partial\Omega, \sigma)]^n\big) \|f\|_{L^p_1(\partial\Omega,w)} \\ &\leq C\|(zI - T)f\|_{L^p_1(\partial\Omega,w)} + \|\mathcal{K}f\|_{L^p(\partial\Omega,w)} + (C/C_o)\|f\|_{L^p_1(\partial\Omega,w)}, \end{aligned} \tag{5.4.22}$$

where the last inequality makes use of (5.4.12). If we choose $C_o > 2C$ to begin with, from (5.4.22) we obtain (after absorbing the very last term in the left-most side)

$$\begin{aligned}\|f\|_{L^p_1(\partial\Omega,w)} \leq C\|(zI - T)f\|_{L^p_1(\partial\Omega,w)} + C\|\mathcal{K}f\|_{L^p(\partial\Omega,w)} \\ \text{for each function } f \in L^p_1(\partial\Omega, w), \end{aligned} \tag{5.4.23}$$

where the constant $C \in (0, \infty)$ is as in the statement of the theorem. In view of item *(4)* of [130, Theorem 2.1.2], this shows that $zI - T \in \Phi_+\big(L^p_1(\partial\Omega, w)\big)$. As a result,

$$\big\{z \in \mathbb{C} : |z| > r\big\} \ni z \longmapsto \operatorname{index}\big(zI - T : L^p_1(\partial\Omega, w) \to L^p_1(\partial\Omega, w)\big) \in \mathbb{Z} \tag{5.4.24}$$

is a continuous mapping, hence constant (cf. item *(8)* in [130, Theorem 2.1.2]), and takes the value zero whenever $|z|$ is large enough (cf. [130, Lemma 1.2.12]). As a consequence, the mapping (5.4.24) is identically zero. This implies that for each $z \in \mathbb{C}$ with $|z| > r$ we have

$$\dim\Big(L_1^p(\partial\Omega, w)/\mathrm{Im}\big(zI - T; L_1^p(\partial\Omega, w)\big)\Big) = \dim\Big(\mathrm{Ker}\big(zI - T; L_1^p(\partial\Omega, w)\big)\Big) < \infty, \tag{5.4.25}$$

thus $zI - T \in \Phi\big(L_1^p(\partial\Omega, w)\big)$ for each $z \in \mathbb{C}$ with $|z| > r$. Ultimately, this proves that $\rho_{\text{Fred}}\big(T; L_1^p(\partial\Omega, w)\big) \leq r$. Passing to the limit $r \searrow \rho_{\text{Fred}}\big(T; L^p(\partial\Omega, w)\big)$ we obtain

$$\rho_{\text{Fred}}\big(T; L_1^p(\partial\Omega, w)\big) \leq \rho_{\text{Fred}}\big(T; L^p(\partial\Omega, w)\big), \tag{5.4.26}$$

hence $\rho_{\text{Fred}}\big(T; L_1^p(\partial\Omega, w)\big) \leq \|T\|^{\text{ess}}_{L^p(\partial\Omega,w)\to L^p(\partial\Omega,w)} = \mathrm{dist}\big(T, \mathrm{Cp}\,(L^p(\partial\Omega, w))\big)$, thanks to [130, (2.2.16)] and (A.0.191). With this in hand, the claim in (5.4.3) follows on account of (5.2.17) (again, bearing in mind (5.4.9)). □

The following observation enhances the scope of Theorem 5.4.1.

Remark 5.4.2 *In the context of Theorem 5.4.1, an estimate like* (5.4.4) *also holds for Lorentz-based Sobolev spaces. Indeed, if for $p \in (1, \infty)$ and $q \in (0, \infty]$ we set*

$$L_1^{(p,q)}(\partial\Omega, \sigma) := \big\{f \in L^{p,q}(\partial\Omega, \sigma) : \partial_{\tau_{jk}} f \in L^{p,q}(\partial\Omega, \sigma),\ 1 \leq j, k \leq n\big\}, \tag{5.4.27}$$

equipped with the natural quasi-norm

$$L_1^{(p,q)}(\partial\Omega, \sigma) \ni f \longmapsto \|f\|_{L^{p,q}(\partial\Omega,\sigma)} + \sum_{j,k=1}^{n} \big\|\partial_{\tau_{jk}} f\big\|_{L^{p,q}(\partial\Omega,\sigma)}, \tag{5.4.28}$$

then for each $m \in \mathbb{N}$ there exists $C_m \in (0, \infty)$, which depends only on m, n, p, q, the UR *constants of $\partial\Omega$, and the relative flatness ratio $\wp(\Omega)/\mathrm{diam}\,\partial\Omega$ such that*

$$\rho_{\text{Fred}}\big(T; L_1^{(p,q)}(\partial\Omega, \sigma)\big) \leq C_m\Big(\sum_{|\alpha|\leq N} \sup_{S^{n-1}} |\partial^\alpha k|\Big)\Big[\mathrm{dist}\big(\nu, \big[\mathrm{VMO}(\partial\Omega, \sigma)\big]^n\big)\Big]^{\langle m\rangle}. \tag{5.4.29}$$

This is proved in a similar manner to (5.4.3), *now making use of* [131, (2.7.32)].

It is also worth to point out that, in the class of infinitesimally flat AR domains, we have the following remarkable compactness result for chord-dot-normal SIO's on Muckenhoupt weighted Sobolev spaces.

Corollary 5.4.3 *Assume $\Omega \subseteq \mathbb{R}^n$ is an infinitesimally flat* AR *domain, that is, Ω is an Ahlfors regular domain with compact boundary and such that the geometric measure theoretic outward unit normal ν to Ω belongs to $\big[\mathrm{VMO}(\partial\Omega, \sigma)\big]^n$, where $\sigma := \mathcal{H}^{n-1}\lfloor\partial\Omega$. Having fixed a sufficiently large integer $N = N(n) \in \mathbb{N}$, consider a complex-valued function $k \in \mathscr{C}^N(\mathbb{R}^n \setminus \{0\})$ which is even and positive homogeneous of degree $-n$. Finally, pick an integrability exponent $p \in (1, \infty)$ along with a Muckenhoupt weight $w \in A_p(\partial\Omega, \sigma)$, and recall from* [132, (5.2.164)] *that the operator T, associated with k and Ω as in* (5.2.15), *induces a well-defined, linear, and bounded mapping of the boundary Sobolev space $L_1^p(\partial\Omega, w)$. Then*

$$T \in \mathrm{Cp}\big(L_1^p(\partial\Omega, w)\big). \tag{5.4.30}$$

As a consequence,

$$T \in \mathrm{Cp}\big(L^p_1(\partial\Omega, \sigma)\big) \text{ for each } p \in (1, \infty). \tag{5.4.31}$$

Proof This follows from [132, (5.2.167)], Corollary 5.2.4, and [131, Theorem 2.7.3] (cf. [131, (2.7.33)]). □

In the context of Corollary 5.4.3, via duality, from (5.4.30), (A.0.156), and [132, Theorem 5.2.2, item *(8)*] we also conclude that

$$T^{\#} \in \mathrm{Cp}\big(L^{p'}_{-1}(\partial\Omega, w')\big) \tag{5.4.32}$$

where $p' := (1 - 1/p)^{-1} \in (1, \infty)$ is the conjugate exponent of p, and w is the conjugate weight of w defined as $w' := w^{1-p'} \in A_{p'}(\partial\Omega, \sigma)$ (cf. item *(2)* in [129, Lemma 7.7.1]). In particular,

$$T^{\#} \in \mathrm{Cp}\big(L^p_{-1}(\partial\Omega, \sigma)\big) \text{ for each } p \in (1, \infty). \tag{5.4.33}$$

In this vein, we also wish to note that similar results are valid for the singular integral operators $T, T^{\#}$ defined in Corollary 5.2.5 in the two-dimensional setting.

Next we shall prove essential norm estimates for chord-dot-normal SIO's, akin to those established earlier in Theorem 5.2.3, with Lebesgue spaces now replaced by Morrey spaces, vanishing Morrey spaces, and block spaces. This is done in two installments: First, in Theorem 5.4.4, we treat operators T as in (5.4.34). Second, in Corollary 5.4.5 we consider operators $T^{\#}$ as in (5.4.56).

Theorem 5.4.4 *Let $\Omega \subseteq \mathbb{R}^n$ be a* UR *domain with compact boundary. Denote by ν the geometric measure theoretic outward unit normal to Ω and abbreviate $\sigma := \mathcal{H}^{n-1}\lfloor\partial\Omega$. Also, fix a sufficiently large integer $N = N(n) \in \mathbb{N}$ and consider a complex-valued function $k \in \mathscr{C}^N(\mathbb{R}^n \setminus \{0\})$ which is even and positive homogeneous of degree $-n$. Finally, fix two integrability exponents $p, q \in (1, \infty)$ with $1/p + 1/q = 1$ along with a parameter $\lambda \in (0, n-1)$, and recall from* [131, Theorem 2.6.1] *(cf. also the final part in* [132, Theorem 5.2.2, item *(3)*]*) that the singular integral operator defined originally on functions $f \in L^1(\partial\Omega, \sigma)$ at σ-a.e. $x \in \partial\Omega$ as*

$$Tf(x) := \lim_{\varepsilon \to 0^+} \int_{\substack{y \in \partial\Omega,\ |x-y|>\varepsilon}} \langle \nu(y), x - y\rangle k(x - y) f(y)\, d\sigma(y) \tag{5.4.34}$$

induces well-defined, linear, and bounded mappings

$$T : M^{p,\lambda}(\partial\Omega, \sigma) \longrightarrow M^{p,\lambda}(\partial\Omega, \sigma), \tag{5.4.35}$$

$$T : \mathring{M}^{p,\lambda}(\partial\Omega, \sigma) \longrightarrow \mathring{M}^{p,\lambda}(\partial\Omega, \sigma), \tag{5.4.36}$$

$$T : \mathcal{B}^{q,\lambda}(\partial\Omega, \sigma) \longrightarrow \mathcal{B}^{q,\lambda}(\partial\Omega, \sigma). \tag{5.4.37}$$

Then for each $m \in \mathbb{N}$ there exists a constant $C_m \in (0, \infty)$, which depends only on m, n, p, λ, the UR *constants of $\partial\Omega$, and the relative flatness ratio $\wp(\Omega)/\mathrm{diam}\,\partial\Omega$*

such that (with the piece of notation introduced in (4.1.12))

$$\operatorname{dist}\Big(T, \operatorname{Cp}\big(M^{p,\lambda}(\partial\Omega,\sigma)\big)\Big) \\ \leq C_m\Big(\sum_{|\alpha|\leq N}\sup_{S^{n-1}}|\partial^\alpha k|\Big)\Big[\operatorname{dist}\big(\nu,\big[\mathrm{VMO}(\partial\Omega,\sigma)\big]^n\big)\Big]^{\langle m\rangle}, \tag{5.4.38}$$

$$\operatorname{dist}\Big(T, \operatorname{Cp}\big(\mathring{M}^{p,\lambda}(\partial\Omega,\sigma)\big)\Big) \\ \leq C_m\Big(\sum_{|\alpha|\leq N}\sup_{S^{n-1}}|\partial^\alpha k|\Big)\Big[\operatorname{dist}\big(\nu,\big[\mathrm{VMO}(\partial\Omega,\sigma)\big]^n\big)\Big]^{\langle m\rangle}, \tag{5.4.39}$$

$$\operatorname{dist}\Big(T, \operatorname{Cp}\big(\mathcal{B}^{q,\lambda}(\partial\Omega,\sigma)\big)\Big) \\ \leq C_m\Big(\sum_{|\alpha|\leq N}\sup_{S^{n-1}}|\partial^\alpha k|\Big)\Big[\operatorname{dist}\big(\nu,\big[\mathrm{VMO}(\partial\Omega,\sigma)\big]^n\big)\Big]^{\langle m\rangle}, \tag{5.4.40}$$

where distances in the left-hand side are measured in the spaces $\operatorname{Bd}\big(M^{p,\lambda}(\partial\Omega,\sigma)\big)$, $\operatorname{Bd}\big(\mathring{M}^{p,\lambda}(\partial\Omega,\sigma)\big)$, *and* $\operatorname{Bd}\big(\mathcal{B}^{q,\lambda}(\partial\Omega,\sigma)\big)$, *respectively, while the distances in the right-hand side are measured in the space* $\big[\mathrm{BMO}(\partial\Omega,\sigma)\big]^n$. *In particular, from* (5.4.38)-(5.4.40) *and* [130, (1.2.53)] *one concludes that*

$$\begin{array}{l} T \text{ is compact on } M^{p,\lambda}(\partial\Omega,\sigma),\ \mathring{M}^{p,\lambda}(\partial\Omega,\sigma), \text{ and } \mathcal{B}^{q,\lambda}(\partial\Omega,\sigma) \\ \text{if } \Omega \text{ is an infinitesimally flat AR domain (cf. Definition 3.4.1).} \end{array} \tag{5.4.41}$$

Proof We shall rework a portion of the proof of Theorem 5.2.3, with some natural alterations. First, recall the parameter $\delta_* > 0$ appearing in the statement of Theorem 3.2.1. In the case when $\operatorname{dist}\big(\nu,\big[\mathrm{VMO}(\partial\Omega,\sigma)\big]^n\big) \geq \delta_*$, the estimate claimed in (5.4.38) may be justified by using the fact that T is a bounded mapping on $M^{p,\lambda}(\partial\Omega,\sigma)$ (cf. the final part in [132, Theorem 5.2.2, item *(3)*]) by writing

$$\begin{aligned} \operatorname{dist}\big(T, \operatorname{Cp}\big(M^{p,\lambda}(\partial\Omega,\sigma)\big)\big) &\leq \|T\|_{M^{p,\lambda}(\partial\Omega,\sigma)\to M^{p,\lambda}(\partial\Omega,\sigma)} \\ &\leq \frac{C}{\delta_*}\Big(\sum_{|\alpha|\leq N}\sup_{S^{n-1}}|\partial^\alpha k|\Big)\operatorname{dist}\big(\nu,\big[\mathrm{VMO}(\partial\Omega,\sigma)\big]^n\big) \end{aligned} \tag{5.4.42}$$

and then invoking (4.1.17). Hence, as far as the estimate claimed in (5.4.38) is concerned, there remains to consider the case when $\operatorname{dist}\big(\nu,\big[\mathrm{VMO}(\partial\Omega,\sigma)\big]^n\big) < \delta_*$. Also, by homogeneity, there is no loss of generality in assuming that

$$\sum_{|\alpha|\leq N}\sup_{S^{n-1}}|\partial^\alpha k| = 1. \tag{5.4.43}$$

In this scenario, pick some δ such that

$$\operatorname{dist}\big(\nu,\big[\mathrm{VMO}(\partial\Omega,\sigma)\big]^n\big) < \delta < \min\{1,\delta_*\} \tag{5.4.44}$$

and recall from (5.2.70) that for each $m \in \mathbb{N}$ we have

$$\Big\| \sum_{j=1}^{N} M_{\psi_j} T M_{\varphi_j^2} \Big\|_{L^p(\partial\Omega,w) \to L^p(\partial\Omega,w)} \leq C_m \delta^{\langle m \rangle}. \tag{5.4.45}$$

Thanks to the extrapolation result established in [130, Proposition 6.2.12] (cf. Comment 2 following its statement), we deduce from (5.4.45) that

$$\Big\| \sum_{j=1}^{N} M_{\psi_j} T M_{\varphi_j^2} \Big\|_{M^{p,\lambda}(\partial\Omega,\sigma) \to M^{p,\lambda}(\partial\Omega,\sigma)} \leq C_m \delta^{\langle m \rangle}. \tag{5.4.46}$$

Next, from (5.2.67) and [130, (6.2.7)] we see that

$$\sum_{j=1}^{N} M_{1-\psi_j} T M_{\varphi_j^2} \text{ is compact on } M^{p,\lambda}(\partial\Omega,\sigma), \tag{5.4.47}$$

which, in light of (5.4.46) and (5.2.65), implies

$$\operatorname{dist}\big(T, \operatorname{Cp}(M^{p,\lambda}(\partial\Omega,\sigma))\big) \leq C_m \delta^{\langle m \rangle}. \tag{5.4.48}$$

After sending $\delta \searrow \operatorname{dist}\big(\nu, \big[\mathrm{VMO}(\partial\Omega,\sigma)\big]^n\big)$ (cf. (5.4.44)) and recalling (5.4.43), (4.1.15), this ultimately justifies (5.4.38).

The proof of the estimate claimed in (5.4.39) follows a similar blueprint. In the case when $\operatorname{dist}\big(\nu, \big[\mathrm{VMO}(\partial\Omega,\sigma)\big]^n\big) \geq \delta_*$ the desired conclusion is a direct consequence of the boundedness of T on $\mathring{M}^{p,\lambda}(\partial\Omega,\sigma)$ (cf. [131, Theorem 2.6.1] and the final part in [132, Theorem 5.2.2, item *(3)*]), much as in (5.4.42). When $\operatorname{dist}\big(\nu, \big[\mathrm{VMO}(\partial\Omega,\sigma)\big]^n\big) < \delta_*$ we find it convenient to assume that (5.4.43) holds (which may always be arranged by homogeneity). Pick δ as in (5.4.44). From (5.4.45), (5.4.46), and (A.0.168) we then see that for each $m \in \mathbb{N}$ we have

$$\Big\| \sum_{j=1}^{N} M_{\psi_j} T M_{\varphi_j^2} \Big\|_{\mathring{M}^{p,\lambda}(\partial\Omega,\sigma) \to \mathring{M}^{p,\lambda}(\partial\Omega,\sigma)} \leq C_m \delta^{\langle m \rangle}. \tag{5.4.49}$$

Also, from (5.2.67), [130, (6.2.15)], and [130, (6.2.7)] we see that

$$\sum_{j=1}^{N} M_{1-\psi_j} T M_{\varphi_j^2} \text{ is compact on } \mathring{M}^{p,\lambda}(\partial\Omega,\sigma). \tag{5.4.50}$$

Together with (5.4.49) and (5.2.65), this implies that for each $m \in \mathbb{N}$ we have

$$\operatorname{dist}\big(T, \operatorname{Cp}(\mathring{M}^{p,\lambda}(\partial\Omega,\sigma))\big) \leq C_m \delta^{\langle m \rangle}. \tag{5.4.51}$$

Sending $\delta \searrow \operatorname{dist}\big(\nu, \big[\mathrm{VMO}(\partial\Omega,\sigma)\big]^n\big)$ (cf. (5.4.44)) and recalling (5.4.43), (4.1.15) then establishes (5.4.39).

As far as (5.4.40) is concerned, from (5.4.39), [130, Lemma 1.2.14], [130, (6.2.155)], and duality we conclude that, for each $m \in \mathbb{N}$,

$$\begin{aligned}&\operatorname{dist}\Big(T^{\#}, \operatorname{Cp}\big(\mathcal{B}^{q,\lambda}(\partial\Omega, \sigma)\big)\Big)\\&\qquad\leq C_m\Big(\sum_{|\alpha|\leq N}\sup_{S^{n-1}}|\partial^{\alpha}k|\Big)\Big[\operatorname{dist}\big(\nu, \big[\operatorname{VMO}(\partial\Omega, \sigma)\big]^n\big)\Big]^{\langle m\rangle}, \qquad (5.4.52)\end{aligned}$$

where $T^{\#}$, the transpose of T, acts on each function $f \in L^1(\partial\Omega, \sigma)$ as

$$T^{\#}f(x) = -\lim_{\varepsilon\to 0^+}\int_{y\in\partial\Omega,\ |x-y|>\varepsilon}\langle\nu(x), x-y\rangle k(x-y)f(y)\,d\sigma(y) \qquad (5.4.53)$$

for σ-a.e. $x \in \partial\Omega$. As is apparent from (5.4.34) and (5.4.53),

$T + T^{\#}$ may be expressed as a finite sum of commutators between operators of pointwise multiplication by the scalar components of the unit normal ν and convolution-type singular integral operators with kernels of the form $x_j k(x)$, $1 \leq j \leq n$. (5.4.54)

To these commutators, [131, Theorem 2.7.6] applies and we therefore obtain

$$\operatorname{dist}\Big(T + T^{\#}, \operatorname{Cp}\big(\mathcal{B}^{q,\lambda}(\partial\Omega, \sigma)\big)\Big) \leq C\Big(\sum_{|\alpha|\leq N}\sup_{S^{n-1}}|\partial^{\alpha}k|\Big)\operatorname{dist}\big(\nu, \operatorname{VMO}(\partial\Omega, \sigma)\big). \qquad (5.4.55)$$

Finally, (5.4.40) follows from (5.4.52), (5.4.55), and (4.1.17). □

Here is the companion result to Theorem 5.4.4, alluded to earlier.

Corollary 5.4.5 *Let $\Omega \subseteq \mathbb{R}^n$ be a* UR *domain with a compact boundary. Denote by ν the geometric measure theoretic outward unit normal to Ω and abbreviate $\sigma := \mathcal{H}^{n-1}\lfloor\partial\Omega$. Next, pick a sufficiently large integer $N = N(n) \in \mathbb{N}$ and consider a complex-valued function $k \in \mathscr{C}^N(\mathbb{R}^n \setminus \{0\})$ which is even and positive homogeneous of degree $-n$. Finally, fix two arbitrary integrability exponents $p, q \in (1, \infty)$ along with a parameter $\lambda \in (0, n-1)$, and recall from* [131, Theorem 2.6.1] (*cf. also the final part in* [132, Theorem 5.2.2, item *(3)*]) *that the singular integral operator defined originally on functions $f \in L^1(\partial\Omega, \sigma)$ as*

$$T^{\#}f(x) := \lim_{\varepsilon\to 0^+}\int_{y\in\partial\Omega,\ |x-y|>\varepsilon}\langle\nu(x), y-x\rangle k(y-x)f(y)\,d\sigma(y) \qquad (5.4.56)$$

for σ-a.e. $x \in \partial\Omega$, induces well-defined, linear, and bounded mappings

$$T^{\#} : M^{p,\lambda}(\partial\Omega, \sigma) \longrightarrow M^{p,\lambda}(\partial\Omega, \sigma), \tag{5.4.57}$$

$$T^{\#} : \mathring{M}^{p,\lambda}(\partial\Omega, \sigma) \longrightarrow \mathring{M}^{p,\lambda}(\partial\Omega, \sigma), \tag{5.4.58}$$

$$T^{\#} : \mathcal{B}^{q,\lambda}(\partial\Omega, \sigma) \longrightarrow \mathcal{B}^{q,\lambda}(\partial\Omega, \sigma). \tag{5.4.59}$$

Then for each $m \in \mathbb{N}$ there exists a constant $C_m \in (0, \infty)$, which depends only on m, n, p, q, λ, the ADR *constants of $\partial\Omega$, and the relative flatness ratio $\wp(\Omega)/\operatorname{diam}\partial\Omega$ such that (with the piece of notation introduced in* (4.1.12))

$$\max\Big\{\operatorname{dist}\Big(T^{\#}, \operatorname{Cp}\big(M^{p,\lambda}(\partial\Omega, \sigma)\big)\Big), \operatorname{dist}\Big(T^{\#}, \operatorname{Cp}\big(\mathring{M}^{p,\lambda}(\partial\Omega, \sigma)\big)\Big),$$

$$\operatorname{dist}\Big(T^{\#}, \operatorname{Cp}\big(\mathcal{B}^{q,\lambda}(\partial\Omega, \sigma)\big)\Big)\Big\}$$

$$\leq C_m\Big(\sum_{|\alpha|\leq N} \sup_{S^{n-1}} |\partial^\alpha k|\Big) \Big[\operatorname{dist}\big(\nu, \big[\operatorname{VMO}(\partial\Omega, \sigma)\big]^n\big)\Big]^{\langle m\rangle}, \tag{5.4.60}$$

where distances in the left-hand side are measured in the spaces $\operatorname{Bd}\big(M^{p,\lambda}(\partial\Omega, \sigma)\big)$, $\operatorname{Bd}\big(\mathring{M}^{p,\lambda}(\partial\Omega, \sigma)\big)$, *and* $\operatorname{Bd}\big(\mathcal{B}^{q,\lambda}(\partial\Omega, \sigma)\big)$, *respectively, while the distance in the right-hand side is measured in the space* $\big[\operatorname{BMO}(\partial\Omega, \sigma)\big]^n$. *As a consequence, from* (5.4.60) *and* [130, (1.2.53)] *we conclude that*

$$\begin{array}{l} T^{\#} \text{ is compact on } M^{p,\lambda}(\partial\Omega, \sigma),\ \mathring{M}^{p,\lambda}(\partial\Omega, \sigma), \text{ and } \mathcal{B}^{q,\lambda}(\partial\Omega, \sigma) \\ \text{if } \Omega \text{ is an infinitesimally flat AR domain (cf. Definition 3.4.1).} \end{array} \tag{5.4.61}$$

Proof This is a consequence of Theorem 5.4.4, the commutator estimates established in [131, Theorem 2.7.6], and the observation made in (5.4.54). □

We next prove Fredholm radius estimates for chord-dot-normal SIO's on Morrey-based and block-based Sobolev spaces, of the sort established in Theorem 5.4.1.

Theorem 5.4.6 *Let $\Omega \subseteq \mathbb{R}^n$ be a* UR *domain with compact boundary. Denote by ν the geometric measure theoretic outward unit normal to Ω and abbreviate $\sigma := \mathcal{H}^{n-1}\lfloor\partial\Omega$. Also, fix a sufficiently large integer $N = N(n) \in \mathbb{N}$ and consider a complex-valued function $k \in \mathscr{C}^N(\mathbb{R}^n \setminus \{0\})$ which is even and positive homogeneous of degree $-n$. Finally, fix two arbitrary integrability exponents $p, q \in (1, \infty)$ along with a parameter $\lambda \in (0, n-1)$, and recall from* [132, (5.2.168)] *that the singular integral operator defined originally on $f \in L^1(\partial\Omega, \sigma)$ at σ-a.e. $x \in \partial\Omega$ as*

$$Tf(x) := \lim_{\varepsilon\to 0^+} \int_{y\in\partial\Omega,\ |x-y|>\varepsilon} \langle \nu(y), x-y\rangle k(x-y) f(y)\, d\sigma(y) \tag{5.4.62}$$

induces well-defined, linear, and bounded mappings

$$T : M_1^{p,\lambda}(\partial\Omega, \sigma) \longrightarrow M_1^{p,\lambda}(\partial\Omega, \sigma), \tag{5.4.63}$$

$$T : \mathring{M}_1^{p,\lambda}(\partial\Omega, \sigma) \longrightarrow \mathring{M}_1^{p,\lambda}(\partial\Omega, \sigma), \tag{5.4.64}$$

$$T : \mathcal{B}_1^{q,\lambda}(\partial\Omega, \sigma) \longrightarrow \mathcal{B}_1^{q,\lambda}(\partial\Omega, \sigma). \tag{5.4.65}$$

Then for each $m \in \mathbb{N}$ there exists a constant $C_m \in (0, \infty)$, which depends only on m, n, p, q, λ, the ADR *constants of $\partial\Omega$, and the relative flatness ratio $\wp(\Omega)/\mathrm{diam}\,\partial\Omega$ such that (with pieces of notation introduced in* (A.0.80) *and* (4.1.12))

$$\rho_{\mathrm{Fred}}\big(T; M_1^{p,\lambda}(\partial\Omega, \sigma)\big) \leq C_m\Big(\sum_{|\alpha|\leq N} \sup_{S^{n-1}} |\partial^\alpha k|\Big) \Big[\mathrm{dist}\big(\nu, \big[\mathrm{VMO}(\partial\Omega, \sigma)\big]^n\big)\Big]^{\langle m \rangle}, \tag{5.4.66}$$

$$\rho_{\mathrm{Fred}}\big(T; \mathring{M}_1^{p,\lambda}(\partial\Omega, \sigma)\big) \leq C_m\Big(\sum_{|\alpha|\leq N} \sup_{S^{n-1}} |\partial^\alpha k|\Big) \Big[\mathrm{dist}\big(\nu, \big[\mathrm{VMO}(\partial\Omega, \sigma)\big]^n\big)\Big]^{\langle m \rangle}, \tag{5.4.67}$$

$$\rho_{\mathrm{Fred}}\big(T; \mathcal{B}_1^{q,\lambda}(\partial\Omega, \sigma)\big) \leq C_m\Big(\sum_{|\alpha|\leq N} \sup_{S^{n-1}} |\partial^\alpha k|\Big) \Big[\mathrm{dist}\big(\nu, \big[\mathrm{VMO}(\partial\Omega, \sigma)\big]^n\big)\Big]^{\langle m \rangle}, \tag{5.4.68}$$

where the distances in the right-hand side are measured in $\big[\mathrm{BMO}(\partial\Omega, \sigma)\big]^n$.

Proof Thanks to Theorem 5.4.4, the availability of the formula in [132, (5.2.167)] for functions in the Sobolev spaces $M_1^{p,\lambda}(\partial\Omega, \sigma)$, $\mathring{M}_1^{p,\lambda}(\partial\Omega, \sigma)$, $\mathcal{B}_1^{q,\lambda}(\partial\Omega, \sigma)$, and the commutator estimates established in [131, Theorem 2.7.6], the same type of argument as in the proof of Theorem 5.4.1 applies and yields all desired conclusions. □

In turn, Theorem 5.4.6 implies the following remarkable compactness result for chord-dot-normal SIO's acting on Morrey-based and block-based Sobolev spaces defined on boundaries of infinitesimally flat AR domains.

Corollary 5.4.7 *Let $\Omega \subseteq \mathbb{R}^n$ be an infinitesimally flat* AR *domain (cf. Definition 3.4.1) and abbreviate $\sigma := \mathcal{H}^{n-1}\lfloor\partial\Omega$. Fix a sufficiently large integer $N = N(n) \in \mathbb{N}$ and consider a complex-valued function $k \in \mathscr{C}^N(\mathbb{R}^n \setminus \{0\})$ which is even and positive homogeneous of degree $-n$. Finally, pick two arbitrary integrability exponents $p, q \in (1, \infty)$ along with a parameter $\lambda \in (0, n-1)$ and recall that the operator T defined in* (5.4.62) *induces bounded linear mappings in the contexts described in* (5.4.63)-(5.4.65). *Then*

$$T \in \mathrm{Cp}\big(M_1^{p,\lambda}(\partial\Omega, \sigma)\big) \cap \mathrm{Cp}\big(\mathring{M}_1^{p,\lambda}(\partial\Omega, \sigma)\big) \cap \mathrm{Cp}\big(\mathcal{B}_1^{q,\lambda}(\partial\Omega, \sigma)\big). \tag{5.4.69}$$

Proof This is a consequence of [132, (5.2.167)], (5.4.61), and [131, (2.7.93)]. □

In fact, it is possible to extend a large portion of the results established so far by allowing certain Generalized Banach Function Spaces in place of Muckenhoupt weighted Lebesgue spaces, Lorentz spaces, Morrey spaces, or block spaces. This point of view is made precise in the theorem below.

Theorem 5.4.8 *Let $\Omega \subseteq \mathbb{R}^n$ be a* UR *domain with compact boundary. Denote by ν the geometric measure theoretic outward unit normal to Ω and abbreviate $\sigma := \mathcal{H}^{n-1}\lfloor\partial\Omega$. With $\mathcal{M}_{\partial\Omega}$ denoting the Hardy-Littlewood maximal operator on $\partial\Omega$, assume $\mathbb{X}$ is a Generalized Banach Function Space on $(\partial\Omega, \sigma)$ such that $\mathcal{M}_{\partial\Omega} : \mathbb{X} \to \mathbb{X}$ and $\mathcal{M}_{\partial\Omega} : \mathbb{X}' \to \mathbb{X}'$ are well-defined and bounded, where $\mathbb{X}'$ is the associated space of $\mathbb{X}$ (cf.* [130, Definitions 5.1.4, 5.1.11]). *Next, fix a sufficiently large integer $N = N(n) \in \mathbb{N}$ and consider a complex-valued function $k \in \mathscr{C}^N(\mathbb{R}^n \setminus \{0\})$ which is even and positive homogeneous of degree $-n$, and recall from* [130, Corollary 2.7.9] *(as well as* [130, (5.1.13)]*) that the singular integral operators defined originally on $f \in L^1(\partial\Omega, \sigma)$ at σ-a.e. $x \in \partial\Omega$ as*

$$Tf(x) := \lim_{\varepsilon\to 0^+} \int_{y\in\partial\Omega,\ |x-y|>\varepsilon} \langle\nu(y), x-y\rangle k(x-y) f(y)\, d\sigma(y) \tag{5.4.70}$$

and, respectively,

$$T^{\#}f(x) := \lim_{\varepsilon\to 0^+} \int_{y\in\partial\Omega,\ |x-y|>\varepsilon} \langle\nu(x), x-y\rangle k(x-y) f(y)\, d\sigma(y) \tag{5.4.71}$$

have $\mathbb{X}$, $\mathbb{X}'$, as well as $\mathring{\mathbb{X}} := \overline{L^\infty(\partial\Omega, \sigma)}^{\|\cdot\|_{\mathbb{X}}}$ and $(\mathbb{X}')^\circ := \overline{L^\infty(\partial\Omega, \sigma)}^{\|\cdot\|_{\mathbb{X}'}}$ as invariant subspaces (cf. [130, (5.2.76)]*), hence the following are well-defined, linear, bounded operators:*

$$T : \mathbb{X} \to \mathbb{X},\ \ T : \mathbb{X}' \to \mathbb{X}',\ \ T : \mathring{\mathbb{X}} \to \mathring{\mathbb{X}},\ \ T : (\mathbb{X}')^\circ \to (\mathbb{X}')^\circ, \tag{5.4.72}$$

$$T^{\#} : \mathbb{X} \to \mathbb{X},\ \ T^{\#} : \mathbb{X}' \to \mathbb{X}',\ \ T^{\#} : \mathring{\mathbb{X}} \to \mathring{\mathbb{X}},\ \ T^{\#} : (\mathbb{X}')^\circ \to (\mathbb{X}')^\circ. \tag{5.4.73}$$

Then for each $m \in \mathbb{N}$ there exists a constant $C_m \in (0, \infty)$, which depends only on m, the operator norms of $\mathcal{M}_{\partial\Omega}$ on $\mathbb{X}$ and $\mathbb{X}'$ the UR *constants of $\partial\Omega$, and the relative flatness ratio $\wp(\Omega)/\text{diam}\,\partial\Omega$ such that (with notation introduced in* (4.1.12))

$$\operatorname{dist}\Big(T, \mathrm{Cp}(\mathbb{X})\Big) \leq C_m\Big(\sum_{|\alpha|\leq N} \sup_{S^{n-1}} |\partial^\alpha k|\Big)\Big[\operatorname{dist}\big(\nu, [\mathrm{VMO}(\partial\Omega,\sigma)]^n\big)\Big]^{\langle m\rangle} \tag{5.4.74}$$

$$\operatorname{dist}\Big(T, \mathrm{Cp}(\mathbb{X}')\Big) \leq C_m\Big(\sum_{|\alpha|\leq N} \sup_{S^{n-1}} |\partial^\alpha k|\Big)\Big[\operatorname{dist}\big(\nu, [\mathrm{VMO}(\partial\Omega,\sigma)]^n\big)\Big]^{\langle m\rangle} \tag{5.4.75}$$

$$\operatorname{dist}\Big(T^\#, \mathrm{Cp}(\mathbb{X})\Big) \leq C_m\Big(\sum_{|\alpha|\leq N} \sup_{S^{n-1}} |\partial^\alpha k|\Big)\Big[\operatorname{dist}\big(\nu, [\mathrm{VMO}(\partial\Omega,\sigma)]^n\big)\Big]^{\langle m\rangle} \tag{5.4.76}$$

$$\operatorname{dist}\Big(T^\#, \mathrm{Cp}(\mathbb{X}')\Big) \leq C_m\Big(\sum_{|\alpha|\leq N} \sup_{S^{n-1}} |\partial^\alpha k|\Big)\Big[\operatorname{dist}\big(\nu, [\mathrm{VMO}(\partial\Omega,\sigma)]^n\big)\Big]^{\langle m\rangle} \tag{5.4.77}$$

$$\operatorname{dist}\Big(T, \mathrm{Cp}(\mathring{\mathbb{X}})\Big) \leq C_m\Big(\sum_{|\alpha|\leq N} \sup_{S^{n-1}} |\partial^\alpha k|\Big)\Big[\operatorname{dist}\big(\nu, [\mathrm{VMO}(\partial\Omega,\sigma)]^n\big)\Big]^{\langle m\rangle} \tag{5.4.78}$$

$$\operatorname{dist}\Big(T, \mathrm{Cp}\big((\mathbb{X}')^\circ\big)\Big) \leq C_m\Big(\sum_{|\alpha|\leq N} \sup_{S^{n-1}} |\partial^\alpha k|\Big)\Big[\operatorname{dist}\big(\nu, [\mathrm{VMO}(\partial\Omega,\sigma)]^n\big)\Big]^{\langle m\rangle} \tag{5.4.79}$$

$$\operatorname{dist}\Big(T^\#, \mathrm{Cp}(\mathring{\mathbb{X}})\Big) \leq C_m\Big(\sum_{|\alpha|\leq N} \sup_{S^{n-1}} |\partial^\alpha k|\Big)\Big[\operatorname{dist}\big(\nu, [\mathrm{VMO}(\partial\Omega,\sigma)]^n\big)\Big]^{\langle m\rangle} \tag{5.4.80}$$

$$\operatorname{dist}\Big(T^\#, \mathrm{Cp}\big((\mathbb{X}')^\circ\big)\Big) \leq C_m\Big(\sum_{|\alpha|\leq N} \sup_{S^{n-1}} |\partial^\alpha k|\Big)\Big[\operatorname{dist}\big(\nu, [\mathrm{VMO}(\partial\Omega,\sigma)]^n\big)\Big]^{\langle m\rangle} \tag{5.4.81}$$

where distances in the left-hand side are measured in either $\mathrm{Bd}(\mathbb{X})$ *or* $\mathrm{Bd}(\mathbb{X}')$ *(as appropriate), while distances in the right-hand side are measured in the space* $\big[\mathrm{BMO}(\partial\Omega,\sigma)\big]^n$. *In particular, from* (5.4.74)-(5.4.81) *and* [130, (1.2.53)] *one concludes (bearing in mind the very last conclusion in* [130, Proposition 5.1.8]*):*

$$\begin{array}{l}\textit{the operators } T \textit{ and } T^\# \textit{ are compact on } \mathbb{X},\ \mathbb{X}',\ \mathring{\mathbb{X}}, \textit{ and on } (\mathbb{X}')^\circ \\ \textit{if } \Omega \textit{ is an infinitesimally flat } \mathrm{AR} \textit{ domain (cf. Definition 3.4.1).}\end{array} \tag{5.4.82}$$

In addition, if one defines $\mathbb{X}$*-styled,* $\mathbb{X}'$*-styled,* $\mathring{\mathbb{X}}$*-styled, and* $(\mathbb{X}')^\circ$*-styled Sobolev spaces of order one on* $\partial\Omega$ *by setting*

$$\begin{aligned}
\mathbb{X}_1 &:= \Big\{f \in \mathbb{X} : \partial_{\tau_{jk}} f \in \mathbb{X} \textit{ for } 1 \leq j,k \leq n\Big\}, \\
(\mathbb{X}')_1 &:= \Big\{g \in \mathbb{X}' : \partial_{\tau_{jk}} g \in \mathbb{X}' \textit{ for } 1 \leq j,k \leq n\Big\}, \\
\mathring{\mathbb{X}}_1 &:= \Big\{f \in \mathring{\mathbb{X}} : \partial_{\tau_{jk}} f \in \mathring{\mathbb{X}} \textit{ for } 1 \leq j,k \leq n\Big\}, \\
(\mathbb{X}')^\circ_1 &:= \Big\{g \in (\mathbb{X}')^\circ : \partial_{\tau_{jk}} g \in (\mathbb{X}')^\circ \textit{ for } 1 \leq j,k \leq n\Big\},
\end{aligned} \tag{5.4.83}$$

then $\mathbb{X}_1$, $(\mathbb{X}')_1$ *are Banach spaces when equipped with the natural norms*

$$\begin{aligned}
\|f\|_{\mathbb{X}_1} &:= \|f\|_{\mathbb{X}} + \sum_{j,k=1}^n \|\partial_{\tau_{jk}} f\|_{\mathbb{X}} \textit{ for each } f \in \mathbb{X}_1, \\
\|g\|_{(\mathbb{X}')_1} &:= \|g\|_{\mathbb{X}} + \sum_{j,k=1}^n \|\partial_{\tau_{jk}} g\|_{\mathbb{X}'}, \textit{ for each } g \in (\mathbb{X}')_1,
\end{aligned} \tag{5.4.84}$$

while $\mathring{\mathbb{X}}_1$ *and* $(\mathbb{X}')_1^\circ$ *are closed subspaces of* $\mathbb{X}_1$ *and* $(\mathbb{X}')_1$*, respectively (hence, Banach spaces in their own right, when equipped with the norms defined in* (5.4.84)). *Moreover, the operators*

$$T : \mathbb{X}_1 \longrightarrow \mathbb{X}_1, \quad T : (\mathbb{X}')_1 \longrightarrow (\mathbb{X}')_1, \tag{5.4.85}$$

$$T : \mathring{\mathbb{X}}_1 \longrightarrow \mathring{\mathbb{X}}_1, \quad T : (\mathbb{X}')_1^\circ \longrightarrow (\mathbb{X}')_1^\circ, \tag{5.4.86}$$

are well defined, linear, bounded, and for each $m \in \mathbb{N}$ *there exists a constant* $C_m \in (0, \infty)$*, of the same nature as before, with the property that (with pieces of notation introduced in* (A.0.80) *and* (4.1.12))

$$\rho_{\text{Fred}}\big(T; \mathbb{X}_1\big) \leq C_m \Big(\sum_{|\alpha| \leq N} \sup_{S^{n-1}} |\partial^\alpha k| \Big) \Big[\text{dist}\big(\nu, \big[\text{VMO}(\partial\Omega, \sigma)\big]^n\big) \Big]^{\langle m \rangle}, \tag{5.4.87}$$

$$\rho_{\text{Fred}}\big(T; (\mathbb{X}')_1\big) \leq C_m \Big(\sum_{|\alpha| \leq N} \sup_{S^{n-1}} |\partial^\alpha k| \Big) \Big[\text{dist}\big(\nu, \big[\text{VMO}(\partial\Omega, \sigma)\big]^n\big) \Big]^{\langle m \rangle}, \tag{5.4.88}$$

$$\rho_{\text{Fred}}\big(T; \mathring{\mathbb{X}}_1\big) \leq C_m \Big(\sum_{|\alpha| \leq N} \sup_{S^{n-1}} |\partial^\alpha k| \Big) \Big[\text{dist}\big(\nu, \big[\text{VMO}(\partial\Omega, \sigma)\big]^n\big) \Big]^{\langle m \rangle}, \tag{5.4.89}$$

$$\rho_{\text{Fred}}\big(T; (\mathbb{X}')_1^\circ\big) \leq C_m \Big(\sum_{|\alpha| \leq N} \sup_{S^{n-1}} |\partial^\alpha k| \Big) \Big[\text{dist}\big(\nu, \big[\text{VMO}(\partial\Omega, \sigma)\big]^n\big) \Big]^{\langle m \rangle}, \tag{5.4.90}$$

where the distances in the right-hand side are measured in $\big[\text{BMO}(\partial\Omega, \sigma)\big]^n$. *Finally,*

$$\begin{array}{c}\textit{the operator } T \textit{ is compact on } \mathbb{X}_1,\ (\mathbb{X}')_1,\ \mathring{\mathbb{X}}_1, \textit{ and on } (\mathbb{X}')_1^\circ \\ \textit{if } \Omega \textit{ is an infinitesimally flat AR domain (cf. Definition 3.4.1).}\end{array} \tag{5.4.91}$$

Significantly, *all conclusions in Theorem 5.4.8 are valid for rearrangement invariant Banach function spaces* $\mathbb{X}$ *whose Boyd indices satisfy* $1 < p_{\mathbb{X}} \leq q_{\mathbb{X}} < \infty$. Indeed, this is seen from [130, Proposition 5.3.14]. As a particular case, the aforementioned conclusions are true if $\mathbb{X} := L^\Phi(\partial\Omega, \sigma)$, the Orlicz space associated as in (A.0.133) with the (non-atomic, sigma-finite) measure space $(\partial\Omega, \sigma)$ and any Young function Φ satisfying $1 < i(\Phi) \leq I(\Phi) < \infty$ (cf. [130, Proposition 5.3.15] and [130, (5.3.62)]). In the latter scenario, the role of $\mathbb{X}_1$ from (5.4.83) is now played by the `Orlicz-based Sobolev space` $L_1^\Phi(\partial\Omega, \sigma)$.

To define the latter variety of Sobolev space, assume $\Omega \subseteq \mathbb{R}^n$ is an Ahlfors regular domain and set $\sigma := \mathcal{H}^{n-1} \lfloor \partial\Omega$. For any Young function Φ, we then introduce

$$L_1^\Phi(\partial\Omega, \sigma) := \Big\{ f \in L^\Phi(\partial\Omega, \sigma) : \partial_{\tau_{jk}} f \in L^\Phi(\partial\Omega, \sigma),\ 1 \leq j, k \leq n \Big\}, \tag{5.4.92}$$

(here [130, (5.3.30)] is relevant, since this shows that $L^\Phi(\partial\Omega, \sigma) \subseteq L^1_{\text{loc}}(\partial\Omega, \sigma)$ for any Young function Φ), and endow this space with the natural norm, defined for each function $f \in L_1^\Phi(\partial\Omega, \sigma)$ as

$$\|f\|_{L_1^\Phi(\partial\Omega,\sigma)} := \|f\|_{L^\Phi(\partial\Omega,\sigma)} + \sum_{j,k=1}^{n} \|\partial_{\tau_{jk}} f\|_{L^\Phi(\partial\Omega,\sigma)}. \tag{5.4.93}$$

For example, having fixed $p \in (1,\infty)$ and $\alpha \in \mathbb{R}$, if $\Phi(t) \approx t^p[\ln(e+t)]^\alpha$ then the Orlicz space L^Φ actually becomes (as explained in [130, Example 5.3.10]) Zygmund's space $L^p(\log L)^\alpha$, i.e.,

$$L^p(\log L)^\alpha(\partial\Omega,\sigma) = \Big\{f \in \mathscr{M}(\partial\Omega,\sigma) : \int_{\partial\Omega} |f(x)|^p \big[\ln(e+|f(x)|)\big]^\alpha \, d\sigma(x) < \infty\Big\}$$

where $\mathscr{M}(\partial\Omega,\sigma)$ is the space of σ-measurable (complex-valued) functions on $\partial\Omega$. In this scenario, the corresponding Orlicz-based Sobolev space L_1^Φ becomes the `Zygmund-based Sobolev space` of order one on $\partial\Omega$, namely

$$L_1^p(\log L)^\alpha(\partial\Omega,\sigma) := \Big\{f \in L^p(\log L)^\alpha(\partial\Omega,\sigma) : \text{ for } 1 \le j,k \le n \\ \text{one has } \partial_{\tau_{jk}} f \in L^p(\log L)^\alpha(\partial\Omega,\sigma)\Big\}, \tag{5.4.94}$$

equipped with the natural norm, expressed for each $f \in L_1^p(\log L)^\alpha(\partial\Omega,\sigma)$ as

$$\|f\|_{L_1^p(\log L)^\alpha(\partial\Omega,\sigma)} := \|f\|_{L^p(\log L)^\alpha(\partial\Omega,\sigma)} + \sum_{j,k=1}^{n} \|\partial_{\tau_{jk}} f\|_{L^p(\log L)^\alpha(\partial\Omega,\sigma)}.$$

We now turn to the task of presenting the proof of Theorem 5.4.8.

Proof of Theorem 5.4.8 We already know that $\mathbb{X}_1$, $(\mathbb{X}')_1$ are Banach spaces when equipped with the norms described in (5.4.84), and that $\mathring{\mathbb{X}}_1$ and $(\mathbb{X}')_1^\circ$ are closed subspaces of $\mathbb{X}_1$ and $(\mathbb{X}')_1$, respectively. See item *(14)* in Theorem 4.2.1.

As regards the remaining claims in the statement, much as in the proof of Theorem 5.4.4, we shall rework a portion of the proof of Theorem 5.2.3, with some natural adjustments. To get started, recall the parameter $\delta_* > 0$ appearing in the statement of Theorem 3.2.1. In the case when $\operatorname{dist}(\nu, [\mathrm{VMO}(\partial\Omega,\sigma)]^n) \ge \delta_*$, the estimate claimed in (5.4.74) may be justified by using the fact that T is a bounded mapping on $\mathbb{X}$ and $\mathring{\mathbb{X}}$ (cf. (5.4.72)) by writing

$$\begin{aligned} \max\Big\{\operatorname{dist}(T, \mathrm{Cp}(\mathbb{X})), \ \operatorname{dist}(T, \mathrm{Cp}(\mathring{\mathbb{X}}))\Big\} &\le \max\big\{\|T\|_{\mathbb{X}\to\mathbb{X}}, \|T\|_{\mathring{\mathbb{X}}\to\mathring{\mathbb{X}}}\big\} \\ &\le \frac{C}{\delta_*}\Big(\sum_{|\alpha|\le N} \sup_{S^{n-1}} |\partial^\alpha k|\Big) \operatorname{dist}\big(\nu, [\mathrm{VMO}(\partial\Omega,\sigma)]^n\big) \end{aligned} \tag{5.4.95}$$

and then invoking (4.1.17). Thus, as far as (5.4.74) and (5.4.78) are concerned, there remains to consider the case when $\operatorname{dist}(\nu, [\mathrm{VMO}(\partial\Omega,\sigma)]^n) < \delta_*$. Also, by homogeneity, there is no loss of generality in assuming that

$$\sum_{|\alpha|\le N} \sup_{S^{n-1}} |\partial^\alpha k| = 1. \tag{5.4.96}$$

Suppose this is the case and pick some δ such that

$$\operatorname{dist}\big(\nu, \big[\mathrm{VMO}(\partial\Omega, \sigma)\big]^n\big) < \delta < \min\{1, \delta_*\}. \tag{5.4.97}$$

Recall from (5.2.70) that for each $m \in \mathbb{N}$ we have

$$\Big\| \sum_{j=1}^{N} M_{\psi_j} T M_{\varphi_j^2} \Big\|_{L^p(\partial\Omega, w) \to L^p(\partial\Omega, w)} \leq C_m \delta^{\langle m \rangle}. \tag{5.4.98}$$

Thanks to [130, (1.2.20)] and the extrapolation result established in [130, Corollary 5.2.3], we deduce from (5.4.98) (also keeping in mind [130, (5.2.85)]) that

$$\Big\| \sum_{j=1}^{N} M_{\psi_j} T M_{\varphi_j^2} \Big\|_{\mathring{\mathbb{X}} \to \mathring{\mathbb{X}}} \leq \Big\| \sum_{j=1}^{N} M_{\psi_j} T M_{\varphi_j^2} \Big\|_{\mathbb{X} \to \mathbb{X}} \leq C_m \delta^{\langle m \rangle}. \tag{5.4.99}$$

Next, from (5.2.67) and [130, (5.2.76), (5.2.84)] we see that

$$\sum_{j=1}^{N} M_{1-\psi_j} T M_{\varphi_j^2} \text{ is compact both on } \mathbb{X} \text{ and on } \mathring{\mathbb{X}}, \tag{5.4.100}$$

which, in light of (5.4.46) and (5.2.65), implies

$$\max\Big\{\operatorname{dist}\big(T, \mathrm{Cp}\,(\mathbb{X})\big), \operatorname{dist}\big(T, \mathrm{Cp}\,(\mathring{\mathbb{X}})\big)\Big\} \leq C_m \delta^{\langle m \rangle}. \tag{5.4.101}$$

After sending $\delta \searrow \operatorname{dist}\big(\nu, \big[\mathrm{VMO}(\partial\Omega, \sigma)\big]^n\big)$ (cf. (5.4.97)) and recalling (5.4.96), (4.1.15), this ultimately justifies (5.4.74) and (5.4.78). In turn, from these estimates written for the Generalized Banach Function Space $\mathbb{X}'$ (in place of $\mathbb{X}$, which satisfies the same properties as the latter; cf. [130, Proposition 5.1.14]) we conclude that (5.4.75) and (5.4.79) also hold. The claims in (5.4.75)-(5.4.77) and (5.4.79)-(5.4.81) are justified in a similar manner.

Pressing forward, from the definition of the Sobolev spaces $\mathbb{X}_1$, $(\mathbb{X}')_1$, $\mathring{\mathbb{X}}_1$, $(\mathbb{X}')_1^{\circ}$ given in (5.4.83), the embeddings in [130, (5.2.92), (5.2.101)], the identity described in [132, (5.2.167)], what we have proved already in (5.4.74) and (5.4.75), and the commutator estimates from [131, Theorem 2.7.8], the same type of argument as in the proof of Theorem 5.4.1 gives that the operator T induces well-defined, linear, bounded mappings in the contexts described in (5.4.85)-(5.4.86), and the estimates claimed in (5.4.87)-(5.4.90) hold. Finally, the claims made in (5.4.91) are consequences of what we have just proved, (5.4.82), the identity in [132, (5.2.167)] written for functions in the Sobolev spaces (5.4.83), and [131, (2.7.148)]. □

Moving on, we now obtain essential norm estimates for chord-dot-normal SIO's, in the spirit of those seen earlier in Theorem 5.2.3, now working on Hardy spaces.

Theorem 5.4.9 *Let $\Omega \subseteq \mathbb{R}^n$ be a* UR *domain with a compact boundary and denote by ν its geometric measure theoretic outward unit normal. Set $\sigma := \mathcal{H}^{n-1}\lfloor\partial\Omega$. Pick* *a sufficiently large* $N = N(n) \in \mathbb{N}$ *and consider* $k \in \mathscr{C}^N(\mathbb{R}^n \setminus \{0\})$ *which is even and positive homogeneous of degree* $-n$. *Having fixed an exponent*

$$p \in \left(\tfrac{n-1}{n}, 1\right], \tag{5.4.102}$$

recall from [132, Theorem 5.2.2, item *(4)*] *that the singular integral operator* $T^{\#}$, *defined originally on functions* $f \in L^1(\partial\Omega, \sigma)$ *at* σ-*a.e.* $x \in \partial\Omega$ *as*

$$T^{\#} f(x) := \lim_{\varepsilon \to 0^+} \int_{y \in \partial\Omega,\ |x-y|>\varepsilon} \langle \nu(x), y - x \rangle k(y - x) f(y)\, d\sigma(y), \tag{5.4.103}$$

induces a linear and bounded mapping

$$T^{\#} : H^p(\partial\Omega, \sigma) \longrightarrow H^p(\partial\Omega, \sigma). \tag{5.4.104}$$

Then this operator has the property that for each choice of a fractional power

$$\theta \in \left(0, n - \tfrac{n-1}{p}\right) \tag{5.4.105}$$

one can find a constant $C_\theta \in (0, \infty)$, *which depends only on* θ, n, p, *the* UR *constants of* $\partial\Omega$, *and the relative flatness ratio* $\wp(\Omega)/\mathrm{diam}\,\partial\Omega$ *such that*

$$\mathrm{dist}\big(T^{\#}, \mathrm{Cp}\,(H^p(\partial\Omega, \sigma))\big) \leq C_\theta \Big(\sum_{|\alpha| \leq N} \sup_{S^{n-1}} |\partial^\alpha k| \Big)\, \mathrm{dist}\big(\nu, \big[\mathrm{VMO}(\partial\Omega, \sigma)\big]^n\big)^\theta \tag{5.4.106}$$

where the distance in the left-hand side is measured in the operator norm in the space $\mathrm{Bd}(H^p(\partial\Omega, \sigma))$ *and the distance in the right-hand side is measured in the space* $\big[\mathrm{BMO}(\partial\Omega, \sigma)\big]^n$. *As a consequence,*

$$\begin{gathered} T^{\#} \text{ is a compact operator on } H^p(\partial\Omega, \sigma) \text{ whenever} \\ \Omega \text{ is an infinitesimally flat AR domain (cf. Definition 3.4.1).} \end{gathered} \tag{5.4.107}$$

Proof Fix θ as in (5.4.105). In particular, $\theta \in (0, 1)$ and there exist

$$p_0 \in \left(\tfrac{n-1}{n}, p\right) \text{ and } p_1 \in (1, \infty) \text{ such that } \tfrac{1}{p} = \tfrac{1-\theta}{p_0} + \tfrac{\theta}{p_1}. \tag{5.4.108}$$

The idea is to invoke [130, (1.4.71) in Proposition 1.4.24] for the operator $T^{\#}$ from (5.4.104) and

$$X_0 := H^{p_0}(\partial\Omega, \sigma), \qquad X_1 := L^{p_1}(\partial\Omega, \sigma). \tag{5.4.109}$$

Thanks to [130, (7.1.55), (7.1.60)] and item *(3)* in [130, Proposition 7.3.6], all hypotheses of [130, Proposition 1.4.24] are satisfied. Keeping in mind [129, (3.6.27)] and relying on the real interpolation result from [130, (4.3.4)] which presently gives

$$X_{\theta,p} := (X_0, X_1)_{\theta,p} = \Big(H^{p_0}(\partial\Omega, \sigma), H^{p_1}(\partial\Omega, \sigma)\Big)_{\theta,p} = H^p(\partial\Omega, \sigma), \tag{5.4.110}$$

we may then conclude from [130, (1.4.71) in Proposition 1.4.24] that

$$\operatorname{dist}\big(T^{\#}, \mathrm{Cp}(H^{p}(\partial\Omega, \sigma)\big) \leq \widetilde{C} \cdot \operatorname{dist}\big(T^{\#}, \mathrm{Cp}(H^{p_0}(\partial\Omega, \sigma))\big)^{1-\theta} \cdot \operatorname{dist}\big(T^{\#}, \mathrm{Cp}(L^{p_1}(\partial\Omega, \sigma))\big)^{\theta}, \tag{5.4.111}$$

where $\widetilde{C} \in (0, \infty)$ depends only on n, p_0, p_1, and the ADR constants of $\partial\Omega$ (cf. [130, (1.4.72) in Proposition 1.4.24]). We may also utilize [132, (5.2.154)] to estimate

$$\operatorname{dist}\big(T^{\#}, \mathrm{Cp}(H^{p_0}(\partial\Omega, \sigma))\big) \leq \|T^{\#}\|_{H^{p_0}(\partial\Omega,\sigma)\to H^{p_0}(\partial\Omega,\sigma)} \leq C\Big(\sum_{|\alpha|\leq N} \sup_{S^{n-1}} |\partial^{\alpha} k|\Big), \tag{5.4.112}$$

where the constant $C \in (0, \infty)$ depends only on n, p_0, and the UR character of $\partial\Omega$. On account of (5.4.111), (5.4.112), (5.2.18) (presently used with $p := p_1$ and $w \equiv 1$), and (4.1.16) the estimate claimed in (5.4.106) now follows. □

In our next corollary we estimate the essential norm of chord-dot-normal SIO's on the John-Nirenberg space, of functions of bounded mean oscillations, in terms of the flatness of the "surface" on which said operators are defined, in a manner strongly reminiscent of the estimates seen earlier in Theorem 5.2.3 on Lebesgue spaces. We also estimate the Fredholm radii of chord-dot-normal SIO's acting on the Sarason space of functions with vanishing mean oscillations, of a similar flavor.

Corollary 5.4.10 *Let $\Omega \subseteq \mathbb{R}^n$ be a* UR *domain with a compact boundary. Denote by ν the geometric measure theoretic outward unit normal to Ω and abbreviate $\sigma := \mathcal{H}^{n-1}\lfloor\partial\Omega$. Also, fix a sufficiently large integer $N = N(n) \in \mathbb{N}$ and consider a complex-valued function $k \in \mathscr{C}^N(\mathbb{R}^n \setminus \{0\})$ which is even and positive homogeneous of degree $-n$. Recall from* [132, Theorem 5.2.2, item *(14)*] *that the singular integral operator, defined originally on $f \in L^1(\partial\Omega, \sigma)$ at σ-a.e. $x \in \partial\Omega$ as*

$$Tf(x) := \lim_{\varepsilon\to 0^+} \int_{y\in\partial\Omega,\ |x-y|>\varepsilon} \langle \nu(y), x-y\rangle k(x-y) f(y)\, d\sigma(y), \tag{5.4.113}$$

induces bounded mappings both in the context of the John-Nirenberg spaces

$$T : \mathrm{BMO}(\partial\Omega, \sigma) \longrightarrow \mathrm{BMO}(\partial\Omega, \sigma), \tag{5.4.114}$$

and also in the context of the Sarason spaces

$$T : \mathrm{VMO}(\partial\Omega, \sigma) \longrightarrow \mathrm{VMO}(\partial\Omega, \sigma), \tag{5.4.115}$$

Then for each choice of a fractional power $\theta \in (0, 1)$ there exists a constant $C_\theta \in (0, \infty)$, which depends only on θ, n, the UR *constants of $\partial\Omega$, and the relative flatness ratio $\wp(\Omega)/\operatorname{diam}\partial\Omega$ such that*

$$\operatorname{dist}\big(T, \operatorname{Cp}(\mathrm{BMO}(\partial\Omega, \sigma))\big) \leq C_\theta \Big(\sum_{|\alpha| \leq N} \sup_{S^{n-1}} |\partial^\alpha k| \Big) \operatorname{dist}\big(\nu, \big[\mathrm{VMO}(\partial\Omega, \sigma)\big]^n\big)^\theta \tag{5.4.116}$$

$$\rho_{\text{Fred}}\big(T; \mathrm{VMO}(\partial\Omega, \sigma)\big) \leq C_\theta \Big(\sum_{|\alpha| \leq N} \sup_{S^{n-1}} |\partial^\alpha k| \Big) \operatorname{dist}\big(\nu, \big[\mathrm{VMO}(\partial\Omega, \sigma)\big]^n\big)^\theta, \tag{5.4.117}$$

where the distance in the left-hand side of (5.4.116) *is measured in the operator norm in the space* $\mathrm{Bd}(\mathrm{BMO}(\partial\Omega, \sigma))$ *and the distances in the right-hand sides of* (5.4.116)-(5.4.117) *are measured in* $\big[\mathrm{BMO}(\partial\Omega, \sigma)\big]^n$. *As a corollary,*

$$\begin{gathered} T \text{ is compact on } \mathrm{BMO}(\partial\Omega, \sigma) \text{ and on } \mathrm{VMO}(\partial\Omega, \sigma) \\ \text{if } \Omega \text{ is an infinitesimally flat AR domain (cf. Definition 3.4.1).} \end{gathered} \tag{5.4.118}$$

Proof The estimate recorded in (5.4.116) follows from Theorem 5.4.9, [130, Lemma 1.2.14], and [130, (4.6.2)] (with $p = 1$). In view of (A.0.191), this further implies that for each $\theta \in (0, 1)$ there exists a constant $C_\theta \in (0, \infty)$ as in the statement of the corollary with the property that

$$\|T\|^{\text{ess}}_{\mathrm{BMO}(\partial\Omega,\sigma)\to\mathrm{BMO}(\partial\Omega,\sigma)} \leq C_\theta \Big(\sum_{|\alpha| \leq N} \sup_{S^{n-1}} |\partial^\alpha k| \Big) \operatorname{dist}\big(\nu, \big[\mathrm{VMO}(\partial\Omega, \sigma)\big]^n\big)^\theta. \tag{5.4.119}$$

In concert with [130, (2.2.16)], this shows that

$$\rho_{\text{Fred}}\big(T; \mathrm{BMO}(\partial\Omega, \sigma)\big) \leq C_\theta \Big(\sum_{|\alpha| \leq N} \sup_{S^{n-1}} |\partial^\alpha k| \Big) \operatorname{dist}\big(\nu, \big[\mathrm{VMO}(\partial\Omega, \sigma)\big]^n\big)^\theta. \tag{5.4.120}$$

The estimate claimed in (5.4.117) then becomes a consequence of (5.4.120), [130, (2.2.18)], and the fact that $\mathrm{VMO}(\partial\Omega, \sigma)$ is a closed subspace of $\mathrm{BMO}(\partial\Omega, \sigma)$ which is invariant under T (cf. (A.0.244) and [132, (5.2.215)]). □

Here are essential norm estimates for chord-dot-normal SIO's, in the spirit of those proved earlier in Theorem 5.2.3, now working on Hölder spaces.

Theorem 5.4.11 *Let* $\Omega \subseteq \mathbb{R}^n$ *be a* UR *domain with a compact boundary. Denote by* ν *the geometric measure theoretic outward unit normal to* Ω *and abbreviate* $\sigma := \mathcal{H}^{n-1}\lfloor\partial\Omega$. *Next, fix a sufficiently large integer* $N = N(n) \in \mathbb{N}$ *and consider a complex-valued function* $k \in \mathscr{C}^N(\mathbb{R}^n \setminus \{0\})$ *which is even and positive homogeneous of degree* $-n$. *Having fixed an exponent* $\alpha \in (0, 1)$, *recall from* [132, Theorem 5.2.2, item *(15)*] *that the singular integral operator* T, *defined originally as*

$$Tf(x) := \lim_{\varepsilon\to0^+} \int_{\substack{y\in\partial\Omega,\ |x-y|>\varepsilon}} \langle\nu(y), x-y\rangle k(x-y) f(y)\, d\sigma(y) \tag{5.4.121}$$

for each $f \in L^1(\partial\Omega, \sigma)$ *and* σ*-a.e.* $x \in \partial\Omega$, *induces a linear and bounded mapping*

$$T : \mathscr{C}^\alpha(\partial\Omega) \longrightarrow \mathscr{C}^\alpha(\partial\Omega). \tag{5.4.122}$$

Then this operator has the property that for each choice of a fractional power

$$\theta \in (0, 1 - \alpha) \tag{5.4.123}$$

one can find a constant $C_\theta \in (0, \infty)$*, which depends only on* θ*,* n*,* α*, the* UR *constants of* $\partial\Omega$*, and the relative flatness ratio* $\wp(\Omega)/\operatorname{diam}\partial\Omega$ *such that*

$$\operatorname{dist}\big(T, \operatorname{Cp}(\mathscr{C}^\alpha(\partial\Omega))\big) \leq C_\theta\Big(\sum_{|\alpha|\leq N}\sup_{S^{n-1}}|\partial^\alpha k|\Big)\operatorname{dist}\big(\nu, \big[\mathrm{VMO}(\partial\Omega, \sigma)\big]^n\big)^\theta \tag{5.4.124}$$

where the distance in the left-hand side is measured in the operator norm in the space $\operatorname{Bd}\big(\mathscr{C}^\alpha(\partial\Omega)\big)$ *and the distance in the right-hand side is measured in the space* $\big[\mathrm{BMO}(\partial\Omega, \sigma)\big]^n$*. In particular,*

$$\begin{array}{c}T \text{ is a compact operator on } \mathscr{C}^\alpha(\partial\Omega) \text{ whenever}\\ \Omega \text{ is an infinitesimally flat AR domain (cf. Definition 3.4.1).}\end{array} \tag{5.4.125}$$

Proof Define

$$p := \big(\tfrac{\alpha}{n-1} + 1\big)^{-1} \in \big(\tfrac{n-1}{n}, 1\big). \tag{5.4.126}$$

As such, the estimate in (5.4.106) is valid. Its proof goes through (5.4.111) which, in turn, comes from [130, (1.4.83)]. When written for the operator $T^\#$ and $q := p$, the latter takes the form

$$\|T^\# - P_N T^\#\|_{\operatorname{Bd}(X_{\theta,p})} \leq \big((C_0 + C_0^2 C)\delta_0 + C_0^2\varepsilon\big)^{1-\theta}\big((C_1 + C_1^2 C)\delta_1 + C_1^2\varepsilon\big)^\theta, \tag{5.4.127}$$

where, with X_0, X_1 as in (5.4.109),

$$X_{\theta,p} := (X_0, X_1)_{\theta,p} = H^p(\partial\Omega, \sigma) \tag{5.4.128}$$

(cf. (5.4.110)), δ_0, δ_1 are arbitrary numbers satisfying $\delta_i > \operatorname{dist}\big(T^\#, \operatorname{Cp}(X_i)\big)$ for $i \in \{0, 1\}$, the threshold $\varepsilon > 0$ is arbitrary, $\theta \in \big(0, n - \frac{n-1}{p}\big) = (0, 1 - \alpha)$ (see (5.4.105) and (5.4.126)), the integer $N \in \mathbb{N}$ is a sufficiently large number depending on ε and δ_0, δ_1, the family of operators $\{P_N\}_{N\in\mathbb{N}}$ satisfying [130, (1.4.67)-(1.4.69)] is as in [130, Proposition 7.3.6], the constant C is as in [130, (1.4.68)], and $C_0, C_1 \in [1, \infty)$ are the moduli of concavity in X_0, X_1. Based on (5.4.127)-(5.4.128), duality (cf. [130, (4.6.2)]), [132, (5.2.224)], [130, (1.2.67)], we then conclude that

$$\|T - (P_N T^\#)^*\|_{\operatorname{Bd}(\mathscr{C}^\alpha(\partial\Omega))} \leq \big((C_0 + C_0^2 C)\delta_0 + C_0^2\varepsilon\big)^{1-\theta}\big((C_1 + C_1^2 C)\delta_1 + C_1^2\varepsilon\big)^\theta. \tag{5.4.129}$$

Upon observing that $P_N T^\#$ is a finite-rank operator on $H^p(\partial\Omega, \sigma)$ (since so is P_N; see [130, Proposition 7.3.6]), we conclude from [130, Lemma 1.1.4] and [130, (4.6.2)] that $(P_N T^\#)^*$ is a linear, continuous, finite-rank operator on $\mathscr{C}^\alpha(\partial\Omega)$ (here we have also used the continuity of $T^\#$ and P_N on $H^p(\partial\Omega, \sigma)$; see [132, (5.2.152)] and item *(3)* in [130, Proposition 7.3.6], keeping in mind [130, (7.1.60)]). As such,

$(P_N T^{\#})^*$ is a compact operator on $\mathscr{C}^\alpha(\partial\Omega)$ (cf. [130, (1.1.15)]). Bearing this in mind, we then deduce from (5.4.129) that

$$\operatorname{dist}\big(T, \operatorname{Cp}(\mathscr{C}^\alpha(\partial\Omega))\big) \le \big((C_0+C_0^2 C)\delta_0+C_0^2\varepsilon\big)^{1-\theta}\big((C_1+C_1^2 C)\delta_1+C_1^2\varepsilon\big)^\theta. \quad (5.4.130)$$

Sending $\varepsilon \searrow 0$ and $\delta_i \searrow \operatorname{dist}\big(T^{\#}, \operatorname{Cp}(X_i)\big)$ for $i \in \{0, 1\}$, then yields (5.4.124) on account of (5.4.112) and (5.2.18) (used here with $p := p_1$ and $w \equiv 1$). □

We next turn to the task of proving essential norm and Fredholm radii estimates for chord-dot-normal SIO's, in the same spirit as before, now working on Besov and Triebel-Lizorkin spaces. We do this in two installments: First, in Theorem 5.4.12, we deal with operators T as in (5.4.131). Second, in Theorem 5.4.13 we treat operators $T^{\#}$ as in (5.4.146).

Theorem 5.4.12 *Let $\Omega \subseteq \mathbb{R}^n$ be a* UR *domain with a compact boundary. Denote by ν the geometric measure theoretic outward unit normal to Ω and abbreviate $\sigma := \mathcal{H}^{n-1}\lfloor\partial\Omega$. Also, fix a sufficiently large integer $N = N(n) \in \mathbb{N}$ and consider a complex-valued function $k \in \mathscr{C}^N(\mathbb{R}^n \setminus \{0\})$ which is even and positive homogeneous of degree $-n$. Recall from* [132, Theorem 5.2.2, item *(19)*] *that the singular integral operator defined originally for $f \in L^1(\partial\Omega, \sigma)$ at σ-a.e. $x \in \partial\Omega$ as*

$$Tf(x) := \lim_{\varepsilon\to 0^+} \int_{\substack{y\in\partial\Omega,\ |x-y|>\varepsilon}} \langle x-y, \nu(y)\rangle k(x-y) f(y)\, d\sigma(y), \qquad (5.4.131)$$

induces linear and bounded mappings both in the context of Besov spaces

$$\begin{gathered} T : B^{p,q}_s(\partial\Omega,\sigma) \longrightarrow B^{p,q}_s(\partial\Omega,\sigma), \\ p \in \big(\tfrac{n-1}{n}, \infty\big), \quad q \in (0,\infty), \quad (n-1)\big(\tfrac{1}{p}-1\big)_+ < s < 1, \end{gathered} \qquad (5.4.132)$$

as well as in the context of Triebel-Lizorkin spaces

$$\begin{gathered} T : F^{p,q}_s(\partial\Omega,\sigma) \longrightarrow F^{p,q}_s(\partial\Omega,\sigma), \\ p \in \big(\tfrac{n-1}{n}, \infty\big), \quad q \in \big(\tfrac{n-1}{n}, \infty\big), \quad (n-1)\big(\tfrac{1}{\min\{p,q\}}-1\big)_+ < s < 1. \end{gathered} \qquad (5.4.133)$$

Then for each choice of exponents p, q, s as in (5.4.132) *there exists a power $\theta \in \big(0,1)$ which depends only on p, q, s along with a constant $C \in (0,\infty)$, which depends only on n, p, q, s, the* UR *constants of $\partial\Omega$, and the relative flatness ratio $\wp(\Omega)/\operatorname{diam}\partial\Omega$ such that*

$$\operatorname{dist}\big(T, \operatorname{Cp}(B^{p,q}_s(\partial\Omega,\sigma))\big) \le C\Big(\sum_{|\alpha|\le N}\sup_{S^{n-1}}|\partial^\alpha k|\Big)\operatorname{dist}\big(\nu, \big[\mathrm{VMO}(\partial\Omega,\sigma)\big]^n\big)^\theta, \qquad (5.4.134)$$

where the distance in the left-hand side of (5.4.134) *is measured in the operator norm in the space* $\operatorname{Bd}(B^{p,q}_s(\partial\Omega,\sigma))$ *and the distance in the right-hand side of* (5.4.134) *is measured in $\big[\mathrm{BMO}(\partial\Omega,\sigma)\big]^n$. In particular,*

$$\begin{aligned}&T \text{ is compact on } B^{p,q}_s(\partial\Omega,\sigma) \text{ whenever } p,q,s \text{ are as in (5.4.132),}\\&\text{provided } \Omega \text{ is an infinitesimally flat AR domain (cf. Definition 3.4.1).}\end{aligned} \tag{5.4.135}$$

Furthermore, if actually

$$p \in (1,\infty), \quad q \in [1,\infty], \quad s \in (0,1), \tag{5.4.136}$$

then for each $m \in \mathbb{N}$ there exists a constant $C_m \in (0,\infty)$, which depends only on m, n, p, q, s, the ADR *constants of $\partial\Omega$, and the relative flatness ratio $\wp(\Omega)/\mathrm{diam}\,\partial\Omega$ such that* (*with the pieces of notation introduced in* (A.0.80) *and* (4.1.12))

$$\rho_{\text{Fred}}\big(T; B^{p,q}_s(\partial\Omega,\sigma)\big) \leq C_m\Big(\sum_{|\alpha|\leq N}\sup_{S^{n-1}}|\partial^\alpha k|\Big)\Big[\mathrm{dist}\big(\nu,\big[\mathrm{VMO}(\partial\Omega,\sigma)\big]^n\big)\Big]^{\langle m\rangle}. \tag{5.4.137}$$

Likewise, for each choice of exponents p,q,s as in (5.4.133) *there exists a power $\theta \in (0,1)$ which depends only on p,q,s together with a constant $C \in (0,\infty)$, which depends only on n, p, q, s, the* UR *constants of $\partial\Omega$, and the relative flatness ratio $\wp(\Omega)/\mathrm{diam}\,\partial\Omega$ such that*

$$\mathrm{dist}\big(T,\mathrm{Cp}(F^{p,q}_s(\partial\Omega,\sigma))\big) \leq C\Big(\sum_{|\alpha|\leq N}\sup_{S^{n-1}}|\partial^\alpha k|\Big)\,\mathrm{dist}\big(\nu,\big[\mathrm{VMO}(\partial\Omega,\sigma)\big]^n\big)^\theta, \tag{5.4.138}$$

where the distance in the left-hand side of (5.4.138) *is measured in the operator norm in the space* $\mathrm{Bd}(F^{p,q}_s(\partial\Omega,\sigma))$ *and the distance in the right-hand side of* (5.4.138) *is measured in* $\big[\mathrm{BMO}(\partial\Omega,\sigma)\big]^n$. *As a consequence,*

$$\begin{aligned}&T \text{ is compact on } F^{p,q}_s(\partial\Omega,\sigma) \text{ whenever } p,q,s \text{ are as in (5.4.133),}\\&\text{provided } \Omega \text{ is an infinitesimally flat AR domain (cf. Definition 3.4.1).}\end{aligned} \tag{5.4.139}$$

Proof From Theorem 5.2.3 (cf. (5.2.17) with $p := 2$ and $w \equiv 1$), [130, (7.1.38), (7.1.55)], and (4.1.16) we know that for each $\varepsilon \in (0,1)$ we have

$$\begin{aligned}&\mathrm{dist}\big(T,\mathrm{Cp}\,(B^{2,2}_0(\partial\Omega))\big)\\&\quad\leq C\big(\mathrm{e}^{\varepsilon^{-1}}/\varepsilon\big)\Big(\sum_{|\alpha|\leq N}\sup_{S^{n-1}}|\partial^\alpha k|\Big)\,\mathrm{dist}\big(\nu,\big[\mathrm{VMO}(\partial\Omega,\sigma)\big]^n\big)^{1-\varepsilon}\end{aligned} \tag{5.4.140}$$

for some $C \in (0,\infty)$ which depends only on n, the UR constants of $\partial\Omega$, and the relative flatness ratio $\wp(\Omega)/\mathrm{diam}\,\partial\Omega$. Assume next that p, q, and s satisfy (5.4.132). Then we may choose $\theta_1 \in (0,1)$ sufficiently close to 1 such that if we set

$$s_1 := s/\theta_1,\ q_1 := \theta_1\big(1/q-(1-\theta_1)/2\big)^{-1},\ p_1 := \theta_1\big(1/p-(1-\theta)/2\big)^{-1} \tag{5.4.141}$$

then $\frac{n-1}{n} < p_1 < \infty$, $0 < q_1 < \infty$, and $(n-1)\big(\frac{1}{p}-1\big)_+ < s_1 < 1$. Note that choosing

$$\begin{aligned}&X_0 := B^{2,2}_0(\partial\Omega,\sigma) \text{ and } X_1 := B^{p_1,q_1}_{s_1}(\partial\Omega,\sigma) \text{ implies}\\&X_{\theta_1} := [X_0,X_1]_{\theta_1} = B^{p,q}_s(\partial\Omega,\sigma) \text{ (see [130, (7.5.8)]).}\end{aligned} \tag{5.4.142}$$

Invoke [130, (1.4.73) in Proposition 1.4.24] for this selection of X_0, X_1 and $\theta := \theta_1$, since the existence of a family of finite-rank operators $\{P_N\}_{N\in\mathbb{N}}$ from $X_0 + X_1$ into itself satisfying [130, (1.4.67)-(1.4.69)] is guaranteed by [130, Proposition 7.3.6]. Specifically, from [130, (1.4.71) in Proposition 1.4.24] we conclude that

$$\begin{aligned}\operatorname{dist}\big(T, \operatorname{Cp}(B^{p,q}_s(\partial\Omega,\sigma)\big) &\leq \widetilde{C}\cdot \operatorname{dist}\big(T, \operatorname{Cp}(B^{2,2}_0(\partial\Omega,\sigma))\big)^{1-\theta_1} \\ &\quad \cdot \operatorname{dist}\big(T, \operatorname{Cp}(B^{p_1,q_1}_{s_1}(\partial\Omega,\sigma))\big)^{\theta_1}, \end{aligned} \tag{5.4.143}$$

where $\widetilde{C} \in (0,\infty)$ depends only on n, p, q, s, and the ADR constants of $\partial\Omega$ (cf. [130, (1.4.72) in Proposition 1.4.24]). From [132, (5.2.241)] we also know that

$$\begin{aligned}\operatorname{dist}\big(T, \operatorname{Cp}(B^{p_1,q_1}_{s_1}(\partial\Omega,\sigma))\big) &\leq \|T\|_{B^{p_1,q_1}_{s_1}(\partial\Omega,\sigma)\to B^{p_1,q_1}_{s_1}(\partial\Omega,\sigma)} \\ &\leq C\Big(\sum_{|\alpha|\leq N}\sup_{S^{n-1}}|\partial^\alpha k|\Big), \end{aligned} \tag{5.4.144}$$

where $C \in (0,\infty)$ depends only on n, p, q, s, and the UR character of $\partial\Omega$. Combining (5.4.143), (5.4.140), and (5.4.144) then proves (5.4.134) for the choice

$$\theta := (1-\theta_1)(1-\varepsilon) \in (0,1). \tag{5.4.145}$$

Next, the fact that the Fredholm radius estimate claimed in (5.4.137) holds whenever p, q, s are as in (5.4.136) may be justified making use of (5.2.19), [130, (2.2.16)], (5.4.3), [130, Theorem 11.12.2], Theorem 3.4.2, [130, Proposition 2.2.7], and [132, Theorem 5.2.2, item *(19)*]. Finally, the proof of (5.4.138) is very similar (now making use of the complex interpolation result from [130, (7.5.11)]). □

Here is the companion result to Theorem 5.4.12, mentioned earlier.

Theorem 5.4.13 *Let $\Omega \subseteq \mathbb{R}^n$ be a* UR *domain with a compact boundary. Denote by ν the geometric measure theoretic outward unit normal to Ω and abbreviate $\sigma := \mathcal{H}^{n-1}\lfloor\partial\Omega$. Having fixed a large enough integer $N = N(n) \in \mathbb{N}$, consider a complex-valued function $k \in \mathscr{C}^N(\mathbb{R}^n \setminus \{0\})$ which is even and positive homogeneous of degree $-n$. Recall from* [132, Theorem 5.2.2, item *(19)*] *that the singular integral operator defined originally on functions $f \in L^1(\partial\Omega,\sigma)$ at σ-a.e. $x \in \partial\Omega$ as*

$$T^\# f(x) := \lim_{\varepsilon\to 0^+}\int_{\substack{y\in\partial\Omega,\ |x-y|>\varepsilon}} \langle y-x, \nu(x)\rangle k(y-x) f(y)\,d\sigma(y) \tag{5.4.146}$$

induces linear and bounded mappings both in the context of Besov spaces

$$\begin{aligned}&T^\# : B^{p,q}_{-s}(\partial\Omega,\sigma) \longrightarrow B^{p,q}_{-s}(\partial\Omega,\sigma) \\ &\textit{with } s \in (0,1), \quad p \in \big(\tfrac{n-1}{n-s},\infty\big), \quad q \in (0,\infty),\end{aligned} \tag{5.4.147}$$

as well as in the context of Triebel-Lizorkin spaces

$$T^{\#} : F^{p,q}_{-s}(\partial\Omega, \sigma) \longrightarrow F^{p,q}_{-s}(\partial\Omega, \sigma),$$
$$\text{with } s \in (0,1), \quad p \in \big(\tfrac{n-1}{n-s}, \infty\big), \quad q \in \big(\tfrac{n-1}{n-s}, \infty\big). \tag{5.4.148}$$

Then for each choice of exponents p, q, s as in (5.4.147) *there exists a power* $\theta \in (0,1)$ *which depends only on p, q, s along with a constant* $C \in (0,\infty)$, *which depends only on n, p, q, s, the* UR *constants of* $\partial\Omega$, *and the relative flatness ratio* $\wp(\Omega)/\operatorname{diam}\partial\Omega$ *such that*

$$\operatorname{dist}\big(T^{\#}, \operatorname{Cp}(B^{p,q}_{-s}(\partial\Omega,\sigma))\big) \leq C\Big(\sum_{|\alpha|\leq N} \sup_{S^{n-1}} |\partial^\alpha k|\Big) \operatorname{dist}\big(\nu, \big[\operatorname{VMO}(\partial\Omega,\sigma)\big]^n\big)^\theta, \tag{5.4.149}$$

where the distance in the left-hand side of (5.4.149) *is measured in the operator norm in the space* $\operatorname{Bd}(B^{p,q}_{-s}(\partial\Omega,\sigma))$ *and the distance in the right-hand side of* (5.4.149) *is measured in* $\big[\operatorname{BMO}(\partial\Omega,\sigma)\big]^n$. *As a corollary,*

$$T^{\#} \text{ is compact on } B^{p,q}_{-s}(\partial\Omega,\sigma) \text{ whenever } p,q,s \text{ are as in (5.4.147),}$$
$$\text{provided } \Omega \text{ is an infinitesimally flat AR domain (cf. Definition 3.4.1).} \tag{5.4.150}$$

Likewise, for each choice of exponents p, q, s as in (5.4.148) *there exists a power* $\theta \in (0,1)$ *which depends only on p, q, s together with a constant* $C \in (0,\infty)$, *which depends only on n, p, q, s, the* UR *constants of* $\partial\Omega$, *and the relative flatness ratio* $\wp(\Omega)/\operatorname{diam}\partial\Omega$ *such that*

$$\operatorname{dist}\big(T^{\#}, \operatorname{Cp}(F^{p,q}_{-s}(\partial\Omega,\sigma))\big) \leq C\Big(\sum_{|\alpha|\leq N} \sup_{S^{n-1}} |\partial^\alpha k|\Big) \operatorname{dist}\big(\nu, \big[\operatorname{VMO}(\partial\Omega,\sigma)\big]^n\big)^\theta, \tag{5.4.151}$$

where the distance in the left-hand side of (5.4.138) *is measured in the operator norm in the space* $\operatorname{Bd}(F^{p,q}_{-s}(\partial\Omega,\sigma))$ *and the distance in the right-hand side of* (5.4.151) *is measured in* $\big[\operatorname{BMO}(\partial\Omega,\sigma)\big]^n$. *In particular,*

$$T^{\#} \text{ is compact on } F^{p,q}_{-s}(\partial\Omega,\sigma) \text{ whenever } p,q,s \text{ are as in (5.4.148),}$$
$$\text{provided } \Omega \text{ is an infinitesimally flat AR domain (cf. Definition 3.4.1).} \tag{5.4.152}$$

Proof The argument is similar to the proof of Theorem 5.4.12. For starters, from Theorem 5.2.3 (cf. (5.2.18) with $p := 2$ and $w \equiv 1$), [130, (7.1.38), (7.1.55)], and (4.1.16) we deduce that for each $\varepsilon \in (0,1)$ we have

$$\begin{aligned} &\operatorname{dist}\big(T^{\#}, \operatorname{Cp}(B^{2,2}_0(\partial\Omega))\big) \\ &\quad \leq C\big(\mathrm{e}^{\varepsilon-1}/\varepsilon\big)\Big(\sum_{|\alpha|\leq N} \sup_{S^{n-1}} |\partial^\alpha k|\Big) \operatorname{dist}\big(\nu, \big[\operatorname{VMO}(\partial\Omega,\sigma)\big]^n\big)^{1-\varepsilon} \end{aligned} \tag{5.4.153}$$

with $C \in (0,\infty)$ depending only on n, the UR constants of $\partial\Omega$, and the relative flatness ratio $\wp(\Omega)/\operatorname{diam}\partial\Omega$. To proceed, pick p, q, s as in (5.4.147). Then there exists $\theta_1 \in (0,1)$ sufficiently close to 1 such that if we define

$$s_1 := s/\theta_1,\ q_1 := \theta_1\big(1/q - (1-\theta_1)/2\big)^{-1},\ p_1 := \theta_1\big(1/p - (1-\theta)/2\big)^{-1} \quad (5.4.154)$$

then $s_1 \in (0,1)$, $p_1 \in \big(\frac{n-1}{n-s_1}, \infty\big)$, and $q_1 \in (0,\infty)$. Observe that choosing

$$\begin{aligned} &X_0 := B_0^{2,2}(\partial\Omega, \sigma) \text{ and } X_1 := B_{-s_1}^{p_1,q_1}(\partial\Omega, \sigma) \text{ implies} \\ &X_{\theta_1} := [X_0, X_1]_{\theta_1} = B_{-s}^{p,q}(\partial\Omega, \sigma) \text{ (see [130, (7.5.8)]).} \end{aligned} \quad (5.4.155)$$

As such, we may use [130, (1.4.73) in Proposition 1.4.24] for this selection of X_0, X_1, and with $\theta := \theta_1$, since the existence of a family of finite-rank operators $\{P_N\}_{N\in\mathbb{N}}$ from $X_0 + X_1$ into itself satisfying [130, (1.4.67)-(1.4.69)] is ensured by [130, Proposition 7.3.6]. Concretely, first [130, (1.4.71) in Proposition 1.4.24] gives

$$\begin{aligned} \operatorname{dist}\big(T^{\#}, \mathrm{Cp}\,(B_{-s}^{p,q}(\partial\Omega,\sigma)\big) &\le \widetilde{C}\cdot \operatorname{dist}\big(T^{\#}, \mathrm{Cp}\,(B_0^{2,2}(\partial\Omega,\sigma))\big)^{1-\theta_1} \\ &\quad \cdot \operatorname{dist}\big(T^{\#}, \mathrm{Cp}\,(B_{-s_1}^{p_1,q_1}(\partial\Omega,\sigma))\big)^{\theta_1}, \end{aligned} \quad (5.4.156)$$

where $\widetilde{C} \in (0,\infty)$ depends only on n, p, q, s, and the ADR constants of $\partial\Omega$ (cf. [130, (1.4.72) in Proposition 1.4.24]). Second, [132, (5.2.241)] implies

$$\begin{aligned} \operatorname{dist}\big(T^{\#}, \mathrm{Cp}\,(B_{-s_1}^{p_1,q_1}(\partial\Omega,\sigma))\big) &\le \|T^{\#}\|_{B_{-s_1}^{p_1,q_1}(\partial\Omega,\sigma)\to B_{-s_1}^{p_1,q_1}(\partial\Omega,\sigma)} \\ &\le C\Big(\sum_{|\alpha|\le N}\sup_{S^{n-1}}|\partial^\alpha k|\Big), \end{aligned} \quad (5.4.157)$$

where the constant $C \in (0,\infty)$ depends only on n, p, q, s, and the UR character of $\partial\Omega$. Together, (5.4.156), (5.4.153), and (5.4.157) then prove (5.4.149) for the choice $\theta := (1-\theta_1)(1-\varepsilon) \in (0,1)$. Lastly, (5.4.151) is justified in a very similar fashion, now employing the complex interpolation result from [130, (7.5.11)]. □

5.5 Converse Inequalities: Quantifying Flatness in Terms of Essential Norms of SIO's

In [131, Theorem 2.7.3] we have established estimates which imply that, in any infinitesimally flat AR domain $\Omega \subseteq \mathbb{R}^n$, the commutator between the Riesz transforms on $\partial\Omega$ and the operators of pointwise multiplication by the scalar components of ν, the geometric measure theoretic outward unit normal to Ω, are compact on Muckenhoupt weighted Lebesgue spaces (see [131, (2.7.33)]). Also, Corollary 5.2.4 implies that the harmonic double layer potential operator is compact on Muckenhoupt weighted Lebesgue spaces on boundaries of infinitesimally flat AR domains. Here we shall show that, collectively, these results are sharp, in the sense that the compactness of the aforementioned operators, together with the background hypothesis that the domain in question is uniformly rectifiable, implies that its geometric measure theoretic outward unit normal ν belongs to the Sarason space of functions of vanishing mean oscillations, hence the underlying set is an infinitesimally flat AR domain (cf. Definition 3.4.1).

To state this result in a formal manner, we recall some notation. Given a UR domain $\Omega \subset \mathbb{R}^n$, abbreviate $\sigma := \mathcal{H}^{n-1}\lfloor\partial\Omega$ and denote by $\nu = (\nu_1, \dots, \nu_n)$ the geometric measure theoretic outward unit normal to Ω. In this setting, recall that the harmonic double layer is defined for each function $f \in L^1\big(\partial\Omega, \frac{\sigma(y)}{1+|y|^{n-1}}\big)$ as

$$K_\Delta f(x) := \lim_{\varepsilon\to0^+} \frac{1}{\omega_{n-1}} \int\limits_{y\in\partial\Omega,\ |x-y|>\varepsilon} \frac{\langle\nu(y), y-x\rangle}{|x-y|^n} f(y)\,\mathrm{d}\sigma(y) \tag{5.5.1}$$

for σ-a.e. $x \in \partial\Omega$. Also, consider the family of (boundary-to-boundary) Riesz transforms $(R_j)_{1\leq j\leq n}$, where each R_j, $j \in \{1, \dots, n\}$, acts on $f \in L^1\big(\partial\Omega, \frac{\sigma(y)}{1+|y|^{n-1}}\big)$ at σ-a.e. $x \in \partial\Omega$ according to

$$R_j f(x) := \lim_{\varepsilon\to0^+} \frac{2}{\omega_{n-1}} \int\limits_{y\in\partial\Omega,\ |x-y|>\varepsilon} \frac{x_j - y_j}{|x-y|^n} f(y)\,\mathrm{d}\sigma(y). \tag{5.5.2}$$

We are now ready to state the result advertised in the previous paragraph.

Theorem 5.5.1 *Given a* UR *domain with compact boundary* $\Omega \subseteq \mathbb{R}^n$, *denote by* $\nu = (\nu_1, \dots, \nu_n)$ *the geometric measure theoretic outward unit normal to* Ω *and set* $\sigma := \mathcal{H}^{n-1}\lfloor\partial\Omega$. *Consider the harmonic double layer* K_Δ *from* (5.5.1) *and the family of (boundary-to-boundary) Riesz transforms* $(R_j)_{1\leq j\leq n}$ *as in* (5.5.2). *Finally, for each* $j \in \{1, \dots, n\}$, *denote by* M_{ν_j} *the operator of pointwise multiplication by* ν_j.

Then Ω *is an infinitesimally flat* AR *domain (cf. Definition 3.4.1) if and only if the harmonic double layer* K_Δ *along with the commutators* $[M_{\nu_k}, R_j]$, *indexed by* $1 \leq j, k \leq n$, *are compact when acting from* $L^p(\partial\Omega, w)$ *into itself for some (or all)* $p \in (1, \infty)$ *and some (or all)* $w \in A_p(\partial\Omega, \sigma)$.

Bearing in mind the nature of the integral kernels for the principal-value harmonic double layer operator K_Δ (cf. (5.5.1)) and the commutators $[M_{\nu_k}, R_j]$ with $1 \leq j, k \leq n$, from Theorem 5.5.1, Corollary 5.2.4, and [131, (2.7.33)] we readily obtain the following result, bridging between geometry and (functional) analysis.

Theorem 5.5.2 *Let* $\Omega \subseteq \mathbb{R}^n$ *be a* UR *domain with compact boundary. Denote by* $\nu = (\nu_1, \dots, \nu_n)$ *the geometric measure theoretic outward unit normal to* Ω *and abbreviate* $\sigma := \mathcal{H}^{n-1}\lfloor\partial\Omega$. *If* $\nu \in \big[\mathrm{VMO}(\partial\Omega, \sigma)\big]^n$ *(i.e., if the Gauss map has vanishing mean oscillations) then all principal-value singular integral operators on* $\partial\Omega$ *with integral kernels of the form*

$$\begin{gathered}\langle\nu(y), x-y\rangle k_0(x-y) \ \text{with}\ x, y \in \partial\Omega, \\ \text{for } k_0 \in \mathscr{C}^\infty(\mathbb{R}^n \setminus \{0\}) \text{ even and positive homogeneous of degree } -n,\end{gathered} \tag{5.5.3}$$

and

$$\begin{aligned}&\big(\nu_j(y)-\nu_j(y)\big)k_1(x-y) \text{ with } x,y\in\partial\Omega, \text{ for } j\in\{1,\dots,n\}\\&\text{and } k_1\in\mathscr{C}^\infty(\mathbb{R}^n\setminus\{0\}) \text{ odd, and positive homogeneous of degree } 1-n,\end{aligned}\tag{5.5.4}$$

are compact operators on each Muckenhoupt weighted Lebesgue space $L^p(\partial\Omega, w)$ *with* $p\in(1,\infty)$ *and* $w\in A_p(\partial\Omega,\sigma)$.

Conversely, if there exists some integrability exponent $p\in(1,\infty)$ *and some weight* $w\in A_p(\partial\Omega,\sigma)$ *such that the principal-value singular integral operator on* $\partial\Omega$ *with integral kernel as in the first line of* (5.5.3) *corresponding to the choice*

$$k_0(z) := \frac{1}{|z|^n} \text{ for each } z\in\mathbb{R}^n\setminus\{0\},\tag{5.5.5}$$

together with the principal-value singular integral operators on $\partial\Omega$ *with integral kernels as in the first line of* (5.5.4) *corresponding to the choices*

$$k_1^{(i)}(z) := \frac{z_i}{|z|^n} \text{ for each } z\in\mathbb{R}^n\setminus\{0\},\quad i\in\{1,\dots,n\},\tag{5.5.6}$$

are compact on $L^p(\partial\Omega, w)$, *then necessarily* $\nu\in\big[\mathrm{VMO}(\partial\Omega,\sigma)\big]^n$.

There is also a quantitative version of Theorem 5.5.1. To state it, recall the piece of notation introduced in (A.0.191).

Theorem 5.5.3 *Let* $\Omega\subseteq\mathbb{R}^n$ *be a* UR *domain with compact boundary. Denote by* $\nu=(\nu_1,\dots,\nu_n)$ *be the geometric measure theoretic outward unit normal to* Ω *and set* $\sigma:=\mathcal{H}^{n-1}\lfloor\partial\Omega$. *For each* $j\in\{1,\dots,n\}$, *denote by* M_{ν_j} *the operator of pointwise multiplication by* ν_j. *Fix* $p\in(1,\infty)$ *along with some weight* $w\in A_p(\partial\Omega,\sigma)$. *Then there exists a constant* $C\in(0,\infty)$, *depending only on* n, p, $[w]_{A_p}$, *the* UR *constants of* $\partial\Omega$, *and the relative flatness ratio* $\wp(\Omega)/\mathrm{diam}\,\partial\Omega$ *such that*

$$\begin{aligned}\mathrm{dist}\Big(\nu, \big[\mathrm{VMO}(\partial\Omega,\sigma)\big]^n\Big) &\leq C\cdot \mathrm{dist}\Big(K_\Delta, \mathrm{Cp}\big(L^p(\partial\Omega,w)\big)\Big)\\&\quad + C\cdot\sum_{1\leq j,k\leq n}\mathrm{dist}\Big(\big[M_{\nu_k},R_j\big],\mathrm{Cp}\big(L^p(\partial\Omega,w)\big)\Big).\end{aligned}\tag{5.5.7}$$

As a corollary, the following holds. Assume $\Omega\subseteq\mathbb{R}^n$ *is a* UR *domain with compact boundary. Abbreviate* $\sigma:=\mathcal{H}^{n-1}\lfloor\partial\Omega$ *and denote by* $\nu=(\nu_1,\dots,\nu_n)$ *the geometric measure theoretic outward unit normal to* Ω. *For each* $j\in\{1,\dots,n\}$, *denote by* M_{ν_j} *the operator of pointwise multiplication by* ν_j. *Then for each exponent* $p\in(1,\infty)$ *and weight* $w\in A_p(\partial\Omega,\sigma)$ *there exists some* $C\in(0,\infty)$, *which depends only on* n, p, $[w]_{A_p}$, *the* UR *constants of* $\partial\Omega$, *and the relative flatness ratio* $\wp(\Omega)/\mathrm{diam}\,\partial\Omega$ *such that* Ω *is a* $(C\delta)$*-oscillating* AR *domain (cf. Definition 3.4.1) for*

$$\delta := \mathrm{dist}\Big(K_\Delta,\mathrm{Cp}\big(L^p(\partial\Omega,\sigma)\big)\Big)+\sum_{1\leq j,k\leq n}\mathrm{dist}\Big(\big[M_{\nu_k},R_j\big],\mathrm{Cp}\big(L^p(\partial\Omega,\sigma)\big)\Big).\tag{5.5.8}$$

It is significant that similar results can be phrased purely in terms of the Riesz transforms $(R_j)_{1\leq j\leq n}$, from (5.5.2). In contrast to the harmonic double layer K_Δ, the latter are operators whose kernels are universal (i.e., independent of the underlying domain). Specifically, we have the following theorem:

Theorem 5.5.4 *Assume $\Omega \subset \mathbb{R}^n$ is a* UR *domain with compact boundary. Denote by ν the geometric measure theoretic outward unit normal to Ω and let $\sigma := \mathcal{H}^{n-1}\lfloor\partial\Omega$. Fix $p \in (1,\infty)$ along with a weight $w \in A_p(\partial\Omega,\sigma)$. Then there exists $C \in (0,\infty)$, depending only on n, p, $[w]_{A_p}$, the* UR *constants of $\partial\Omega$, and the relative flatness ratio $\wp(\Omega)/\mathrm{diam}\,\partial\Omega$ such that (with I denoting the identity operator)*

$$\mathrm{dist}\Big(\nu, \big[\mathrm{VMO}(\partial\Omega,\sigma)\big]^n\Big) \leq C\cdot \mathrm{dist}\Big(I + \sum_{j=1}^{n} R_j^2, \mathrm{Cp}\big(L^p(\partial\Omega, w)\big)\Big) \qquad (5.5.9)$$
$$+ C\cdot \sum_{1\leq j,k\leq n} \mathrm{dist}\Big([R_j, R_k], \mathrm{Cp}\big(L^p(\partial\Omega, w)\big)\Big).$$

As a consequence, the following is true: Suppose $\Omega \subseteq \mathbb{R}^n$ is a UR *domain with compact boundary and abbreviate $\sigma := \mathcal{H}^{n-1}\lfloor\partial\Omega$. Then for each $p \in (1,\infty)$ and $w \in A_p(\partial\Omega,\sigma)$ there exists a constant $C \in (0,\infty)$, which depends only on n, p, $[w]_{A_p}$, the* UR *constants of $\partial\Omega$, and the relative flatness ratio $\wp(\Omega)/\mathrm{diam}\,\partial\Omega$ such that Ω is a $(C\delta)$-oscillating* AR *domain (cf. Definition 3.4.1) for*

$$\delta := \mathrm{dist}\Big(I + \sum_{j=1}^{n} R_j^2, \mathrm{Cp}\big(L^p(\partial\Omega, w)\big)\Big) + \sum_{1\leq j,k\leq n} \mathrm{dist}\Big([R_j, R_k], \mathrm{Cp}\big(L^p(\partial\Omega, w)\big)\Big). \qquad (5.5.10)$$

Theorem 5.5.5 *Let $\Omega \subseteq \mathbb{R}^n$ be an* UR *domain with compact boundary and abbreviate $\sigma := \mathcal{H}^{n-1}\lfloor\partial\Omega$. Then Ω is an infinitesimally flat* AR *domain if and only if*

$$I + \sum_{j=1}^{n} R_j^2 \in \mathrm{Cp}\big(L^p(\partial\Omega, w)\big) \text{ and } [R_j, R_k] \in \mathrm{Cp}\big(L^p(\partial\Omega, w)\big),\ 1 \leq j,k \leq n, \qquad (5.5.11)$$

for some (or all) $p \in (1,\infty)$ and some (or all) $w \in A_p(\partial\Omega,\sigma)$.

As the above results illustrate, there is significant geometric information encoded in the Riesz transforms. In this vein, it is interesting to compare Theorem 5.5.5 with a result from [79] to the effect that, if $\Omega \subset \mathbb{R}^n$ is a UR domain then

$$\partial\Omega \text{ is a sphere, or a } (n-1)\text{-plane} \iff$$
$$I + \sum_{j=1}^{n} R_j^2 = 0 \text{ and } [R_j, R_k] = 0 \text{ for } j,k \in \{1,\dots,n\}. \qquad (5.5.12)$$

To draw a parallel with the main result in [166] which, as recalled in [129, (5.10.7)], shows that the Riesz transforms may be used to characterize uniform rectifiability, our Theorem 5.5.4 proves that the Riesz transforms are also effective in characterizing infinitesimal flatness.

The proofs of Theorems 5.5.1-5.5.5, presented at the end of this section, make use of the Clifford algebra formalism discussed in [129, §6.4]. We begin by recalling the Cauchy-Clifford singular integral operator $\mathfrak{C}$ from (A.0.55) and its formal transpose $\mathfrak{C}^{\#}$ from (A.0.56). Here we are interested in the difference $\mathfrak{C} - \mathfrak{C}^{\#}$ which, up to multiplication by 2^{-1}, may be thought of as the antisymmetric part of the Cauchy-Clifford operator $\mathfrak{C}$. The following lemma, which refines [81, Lemma 4.45], elaborates on the relationship between the antisymmetric part of the Cauchy-Clifford operator, i.e., $\mathfrak{C} - \mathfrak{C}^{\#}$, and the harmonic boundary double layer potential (cf. (A.0.114)) together with commutators between Riesz transforms (cf. (A.0.215)) and operators of pointwise multiplication by scalar components of the unit vector.

Lemma 5.5.6 *Suppose $\Omega \subseteq \mathbb{R}^n$ is a* UR *domain, set $\sigma := \mathcal{H}^{n-1}\lfloor\partial\Omega$, and denote by $\nu = (\nu_1, \dots, \nu_n)$ the geometric measure theoretic outward unit normal to Ω. For each $j \in \{1, \dots, n\}$, denote by M_{ν_j} the operator of pointwise multiplication by ν_j. Also, recall the boundary-to-boundary harmonic double layer potential operator K_Δ from* (A.0.114) *and the family of Riesz transforms $\{R_j\}_{1\leq j\leq n}$ from* (A.0.215). *Finally, recall that $[A, B] := AB - BA$ is the usual commutator bracket. Then*

$$(\mathfrak{C}-\mathfrak{C}^{\#})f = 2\sum_{\ell=0}^{n}\sum_{|I|=\ell}{}'(K_\Delta f_I)\mathbf{e}_I + \frac{1}{2}\sum_{\ell=0}^{n}\sum_{|I|=\ell}{}'\sum_{j,k=1}^{n}\Big([M_{\nu_j}, R_k]f_I\Big)\,\mathbf{e}_j \odot \mathbf{e}_k \odot \mathbf{e}_I \tag{5.5.13}$$

for each Clifford algebra-valued function $f = \sum_{\ell=0}^{n}\sum_{|I|=\ell}' f_I \odot \mathbf{e}_I$ belonging to the weighted Lebesgue space $L^1\big(\partial\Omega, \frac{\sigma(x)}{1+|x|^{n-1}}\big) \otimes \mathcal{C}\ell_n$.

Proof From (A.0.55), (A.0.56), and the fact that $\mathcal{H}^{n-1}(\partial\Omega \setminus \partial_*\Omega) = 0$ we see that for each $f \in L^1\big(\partial\Omega, \frac{\sigma(x)}{1+|x|^{n-1}}\big) \otimes \mathcal{C}\ell_n$ we presently have, at σ-a.e. point $x \in \partial\Omega$,

$$\mathfrak{C}f(x) = \lim_{\varepsilon\to 0^+}\frac{1}{\omega_{n-1}}\int_{y\in\partial\Omega,\ |x-y|>\varepsilon}\frac{x-y}{|x-y|^n}\odot\nu(y)\odot f(y)\,d\sigma(y), \tag{5.5.14}$$

$$\mathfrak{C}^{\#}f(x) = -\lim_{\varepsilon\to 0^+}\frac{1}{\omega_{n-1}}\int_{y\in\partial\Omega,\ |x-y|>\varepsilon}\nu(x)\odot\frac{x-y}{|x-y|^n}\odot f(y)\,d\sigma(y). \tag{5.5.15}$$

Hence, for $f \in L^1\big(\partial\Omega, \frac{\sigma(x)}{1+|x|^{n-1}}\big) \otimes \mathcal{C}\ell_n$, at σ-a.e. point $x \in \partial\Omega$ we may write

$$\begin{aligned}(\mathfrak{C}-\mathfrak{C}^{\#})&f(x) \\ &= \lim_{\varepsilon\to 0^+}\frac{1}{\omega_{n-1}}\int_{\substack{y\in\partial\Omega\\ |x-y|>\varepsilon}}\Big[(x-y)\odot\nu(y)+\nu(x)\odot(x-y)\Big]\odot\frac{f(y)}{|x-y|^n}\,d\sigma(y).\end{aligned} \tag{5.5.16}$$

Based on [129, (6.4.29), (6.4.31), (6.4.32), (6.4.36)] we may re-write the expression in the brackets as

$$
\begin{aligned}
&(x-y)\odot\nu(y)+\nu(x)\odot(x-y)\\
&=\Big[(x-y)\odot\nu(y)+\overline{(x-y)\odot\nu(y)}\Big]+(\nu(x)-\nu(y))\odot(x-y)\\
&=2\Big((x-y)\odot\nu(y)\Big)_0+\sum_{j,k=1}^{n}(\nu_j(x)-\nu_j(y))(x_k-y_k)\mathbf{e}_j\odot\mathbf{e}_k\\
&=-2\langle x-y,\nu(y)\rangle+\sum_{j,k=1}^{n}(\nu_j(x)-\nu_j(y))(x_k-y_k)\mathbf{e}_j\odot\mathbf{e}_k. \qquad (5.5.17)
\end{aligned}
$$

Then the identity claimed in (A.0.55) readily follows from (5.5.17) and definitions.□

In turn, Lemma 5.5.6 is a basic ingredient in the proof of the following corollary.

Corollary 5.5.7 *Suppose $\Omega \subseteq \mathbb{R}^n$ is a* UR *domain with a compact boundary. Denote by ν the geometric measure theoretic outward unit normal to Ω and abbreviate $\sigma := \mathcal{H}^{n-1}\lfloor\partial\Omega$. Also, fix $p \in (1,\infty)$ and a Muckenhoupt weight $w \in A_p(\partial\Omega,\sigma)$.*

Then for each $m \in \mathbb{N}$ there exists some constant $C_m \in (0,\infty)$ which depends only on m, n, p, $[w]_{A_p}$, the UR *constants of $\partial\Omega$, and the relative flatness ratio $\wp(\Omega)/\operatorname{diam}\partial\Omega$ (cf. Definition 3.1.9) such that, with notation introduced in* (4.1.12),

$$
\operatorname{dist}\Big(\mathfrak{C}-\mathfrak{C}^{\#},\operatorname{Cp}\big(L^p(\partial\Omega,w)\otimes C\ell_n\big)\Big)\le C_m\Big[\operatorname{dist}\big(\nu,\big[\mathrm{VMO}(\partial\Omega,\sigma)\big]^n\big)\Big]^{\langle m\rangle} \qquad (5.5.18)
$$

where the distance in the left-hand side is measured in $\operatorname{Bd}\big(L^p(\partial\Omega,w)\otimes C\ell_n\big)$ *and the distances in the right-hand side is measured in* $\big[\mathrm{BMO}(\partial\Omega,\sigma)\big]^n$.

In particular, from (5.5.18), [129, (5.11.93)], *and* [130, (1.2.53)] *it follows that*

$$
\begin{array}{c}
\textit{the operator } \mathfrak{C}-\mathfrak{C}^{\#} \textit{ is compact on } L^p(\partial\Omega,w)\otimes C\ell_n \textit{ whenever}\\
\Omega \textit{ is an infinitesimally flat } \mathrm{AR} \textit{ domain (cf. Definition 3.4.1).}
\end{array} \qquad (5.5.19)
$$

Proof This is a consequence of Lemma 5.5.6, [129, Lemma 7.7.13], (A.0.114), Theorem 5.2.3, (A.0.215), and [131, Theorem 2.7.3]. □

Essentially, Corollary 5.5.7 says that the closer ν is to being in $\mathrm{VMO}(\partial\Omega,\sigma)$ the closer $\mathfrak{C}$ becomes to being symmetric, modulo compact operators. Remarkably, the opposite implication is also true, and this is made precise in the theorem below.

Theorem 5.5.8 *Let $\Omega \subset \mathbb{R}^n$ be a* UR *domain with compact boundary. Abbreviate $\sigma := \mathcal{H}^{n-1}\lfloor\partial\Omega$ and denote by ν the geometric measure theoretic outward unit normal to Ω. Fix $p \in (1,\infty)$ along with a Muckenhoupt weight $w \in A_p(\partial\Omega,\sigma)$. Then there exists $C \in (0,\infty)$, depending only on n, p, $[w]_{A_p}$, the Ahlfors regularity constants of $\partial\Omega$, and the relative flatness ratio $\wp(\Omega)/\operatorname{diam}\partial\Omega$ such that*

$$
\begin{aligned}
&\operatorname{dist}\Big(\nu,[\mathrm{VMO}(\partial\Omega,\sigma)]^n\Big) \qquad (5.5.20)\\
&\quad\le C\cdot\lim_{R\to0^+}\Big(\sup_{x\in\partial\Omega}\big\|\mathbf{1}_{\Delta(x,R)}\cdot(\mathfrak{C}-\mathfrak{C}^{\#})\big\|_{L^p(\partial\Omega,w)\otimes C\ell_n\to L^p(\partial\Omega,w)\otimes C\ell_n}\Big)
\end{aligned}
$$

where $\Delta(x, R) := B(x, R) \cap \partial\Omega$ *for each* $x \in \partial\Omega$ *and* $R \in (0, \infty)$.

As a corollary of this and [130, Lemma 1.2.19],

$$\operatorname{dist}\Big(\nu, [\mathrm{VMO}(\partial\Omega, \sigma)]^n\Big) \leq C \cdot \operatorname{dist}\Big(\mathfrak{C} - \mathfrak{C}^{\#}, \mathrm{Cp}\big(L^p(\partial\Omega, w) \otimes C\ell_n\big)\Big)$$
$$= C \|\mathfrak{C} - \mathfrak{C}^{\#}\|^{\mathrm{ess}}_{L^p(\partial\Omega, w)\otimes C\ell_n \to L^p(\partial\Omega, w)\otimes C\ell_n}. \tag{5.5.21}$$

In particular, if $\mathfrak{C} - \mathfrak{C}^{\#}$ *is a compact operator on* $L^p(\partial\Omega, w) \otimes C\ell_n$ *then* ν *belongs to* $\big[\mathrm{VMO}(\partial\Omega, \sigma)\big]^n$, *i.e.,* Ω *is an infinitesimally flat* AR *domain* (*cf. Definition 3.4.1*).

Proof Bring in the threshold $\widetilde{R} \in (0, \infty)$ as well as the two constants $\widetilde{C} \in [1, \infty)$ and $\lambda \in (32, \infty)$ from Theorem 3.1.12. All of these depend only on the dimension n, the Ahlfors regularity constants of $\partial\Omega$, and the relative flatness ratio $\wp(\Omega)/\operatorname{diam}\partial\Omega$. Next, let c be the lower ADR constant of $\partial\Omega$, and let C be the upper ADR constant of $\partial\Omega$ (cf. [129, Definition 5.9.1]). In addition, pick an arbitrary $\varepsilon \in (0, 1)$ and set

$$\eta := 2(C/c)^{1/(n-1)} \geq 2. \tag{5.5.22}$$

Pick a large $C_o \in [1, \infty)$ depending only on n, the Ahlfors regularity constants of $\partial\Omega$, and the relative flatness ratio $\wp(\Omega)/\operatorname{diam}\partial\Omega$, to be specified later. Also, fix a large dimensional constant $C_n \in [1, \infty)$ and choose $\Lambda \in (0, \infty)$ satisfying

$$\Lambda > \max\big\{4,\ 2C_n\eta^{n-1},\ 2C_o/\eta\big\} \ \text{ and } \ \Lambda^{-1}\ln\Lambda < (2C_o)^{-1}. \tag{5.5.23}$$

To proceed, fix a location $x_0 \in \partial\Omega$ along with a scale

$$0 < R < \min\Big\{\widetilde{R}, \frac{\operatorname{diam}\partial\Omega}{\eta\cdot\Lambda}\Big\}. \tag{5.5.24}$$

Since $\eta(\Lambda R) < \operatorname{diam}\partial\Omega$, the Ahlfors regularity of $\partial\Omega$ and (5.5.22) gives

$$\sigma\big(\Delta(x_0, \eta(\Lambda R)) \setminus \Delta(x_0, \Lambda R)\big) = \sigma\big(\Delta(x_0, \eta(\Lambda R))\big) - \sigma\big(\Delta(x_0, \Lambda R)\big)$$
$$\geq \big(c\,\eta^{n-1} - C\big)(\Lambda R)^{n-1} > 0. \tag{5.5.25}$$

This guarantees that $\Delta(x_0, \eta(\Lambda R)) \setminus \Delta(x_0, \Lambda R) \neq \varnothing$, hence we may choose a point y_0 in $\Delta(x_0, \eta(\Lambda R)) \setminus \Delta(x_0, \Lambda R)$. As a consequence, $\Lambda R \leq |x_0 - y_0| < \eta(\Lambda R)$.

Next, fix a point $x \in \Delta(x_0, R)$ and decompose

$$\int_{\Delta(y_0, R)} \Big\{\frac{x_0 - y}{|x_0 - y|^n} \odot \nu(y) + \nu(x) \odot \frac{x_0 - y}{|x_0 - y|^n}\Big\}\, d\sigma(y) = \mathrm{I} + \mathrm{II} + \mathrm{III}, \tag{5.5.26}$$

where

$$\mathrm{I} := \int_{\Delta(y_0,R)} \left\{ \frac{x_0 - y}{|x_0 - y|^n} \odot \nu(y) - \frac{x - y}{|x - y|^n} \odot \nu(y) \right\} d\sigma(y), \tag{5.5.27}$$

$$\mathrm{II} := \int_{\Delta(y_0,R)} \left\{ \frac{x - y}{|x - y|^n} \odot \nu(y) + \nu(x) \odot \frac{x - y}{|x - y|^n} \right\} d\sigma(y), \tag{5.5.28}$$

$$\mathrm{IIII} := \int_{\Delta(y_0,R)} \left\{ \nu(x) \odot \frac{x_0 - y}{|x_0 - y|^n} - \nu(x) \odot \frac{x - y}{|x - y|^n} \right\} d\sigma(y). \tag{5.5.29}$$

Based on definitions (cf. (A.0.55) and (A.0.56)) and the fact that $|x - y| > (\Lambda - 2)R$ for each point $y \in \Delta(y_0, R)$, the second term above is identified as

$$\mathrm{II} = \omega_{n-1}\,(\mathfrak{C} - \mathfrak{C}^{\#})\mathbf{1}_{\Delta(y_0,R)}(x). \tag{5.5.30}$$

If for each $u, w, z \in \mathbb{R}^n$ with $z \notin \{u, w\}$ we now abbreviate

$$E(u, w; z) := \frac{u - z}{|u - z|^n} - \frac{w - z}{|w - z|^n}, \tag{5.5.31}$$

and if $\nu_{\Delta(z,r)} := ⨍_{\partial\Omega \cap B(z,r)} \nu\, d\sigma$ for $z \in \partial\Omega$ and $r > 0$ then, thanks to [129, (6.4.6)],

$$\begin{aligned} \mathrm{I} + \mathrm{III} &= \int_{\Delta(y_0,R)} \big\{ E(x_0, x; y) \odot \nu(y) + \nu(x) \odot E(x_0, x; y) \big\} d\sigma(y) \\ &= -2 \int_{\Delta(y_0,R)} \big\langle E(x_0, x; y), \nu_{\Delta(x_0,R)} \big\rangle d\sigma(y) \\ &\quad + \int_{\Delta(y_0,R)} E(x_0, x; y) \odot \big(\nu(y) - \nu_{\Delta(x_0,R)}\big) d\sigma(y) \\ &\quad + \int_{\Delta(y_0,R)} \big(\nu(x) - \nu_{\Delta(x_0,R)}\big) \odot E(x_0, x; y)\, d\sigma(y) =: \mathrm{IV} + \mathrm{V} + \mathrm{VI}. \end{aligned} \tag{5.5.32}$$

Since $E(x_0, x; y) = \frac{x_0 - y}{|x_0 - y|^n} - \frac{(x - x_0) - (y - x_0)}{|x - y|^n}$ for each $y \in \Delta(y_0, R)$, it follows that

$$E(x_0, x; y) = -\frac{x - x_0}{|x - y|^n} + (x_0 - y)\Big(\frac{1}{|x_0 - y|^n} - \frac{1}{|x - y|^n} \Big) \tag{5.5.33}$$

for each $y \in \Delta(y_0, R)$, which permits us write

$$\begin{aligned} \mathrm{IV} &= 2 \int_{\Delta(y_0,R)} \frac{\big\langle x - x_0, \nu_{\Delta(x_0,R)} \big\rangle}{|x - y|^n} d\sigma(y) \\ &\quad - 2 \int_{\Delta(y_0,R)} \big\langle y - x_0, \nu_{\Delta(x_0,R)} \big\rangle \Big(\frac{1}{|x_0 - y|^n} - \frac{1}{|x - y|^n} \Big) d\sigma(y) =: \mathrm{IV}_a + \mathrm{IV}_b. \end{aligned} \tag{5.5.34}$$

Note that for each $y \in \Delta(y_0, R)$ we have

$$\Lambda R \le |x_0 - y_0| \le |x_0 - x| + |x - y| + |y - y_0| < |x - y| + 2R, \tag{5.5.35}$$

so $(\Lambda/2)R < (\Lambda-2)R < |x-y|$. Together with Theorem 3.1.12 (cf. (3.1.68) for $\gamma := 1$, $z := x_0$, $y := x_0$; here (5.5.24) is also relevant), this allows us to estimate

$$|\mathrm{IV}_a| = 2\big|\langle x - x_0, \nu_{\Delta(x_0,R)}\rangle\big| \Big|\int_{\Delta(y_0,R)} \frac{d\sigma(y)}{|x-y|^n}\Big| \leq \widetilde{C}\Lambda^{-n}\Big\{\sup_{z\in\partial\Omega}\|\nu\|_*(\Delta(z,\lambda\varepsilon^{-1}R)) + \varepsilon\Big\} \tag{5.5.36}$$

where $\widetilde{C} \in (0,\infty)$ depends only on n, the Ahlfors regularity constants of $\partial\Omega$, and the relative flatness ratio $\wp(\Omega)/\operatorname{diam}\partial\Omega$. Next, the Mean Value Theorem gives that

$$\Big|\frac{1}{|x_0-y|^n} - \frac{1}{|x-y|^n}\Big| \leq \frac{C_n R}{(\Lambda R)^{n+1}} = C_n\Lambda^{-n-1}R^{-n} \tag{5.5.37}$$

at each $y \in \Delta(y_0,R)$, for some purely dimensional constant $C_n \in (0,\infty)$. Since

$$|y-x_0| \leq |y-y_0| + |y_0-x_0| < R + \eta\Lambda R < 2\eta\Lambda R, \tag{5.5.38}$$

we may again invoke (3.1.68) in Theorem 3.1.12 (this time, with $\gamma := 2\eta\Lambda$, $z := x_0$, $x := x_0$; both (5.5.24) and (5.5.38) are relevant here) to conclude that

$$\begin{aligned}|\mathrm{IV}_b| &\leq \widetilde{C}R\Lambda\Big\{\big(1+\log_2(2\eta\Lambda)\big)\sup_{z\in\partial\Omega}\|\nu\|_*(\Delta(z,2\eta\Lambda\lambda\varepsilon^{-1}R)) + \varepsilon\Big\}\frac{\sigma\big(\Delta(x_0,R)\big)}{\Lambda^{n+1}R^n}\\ &\leq \widetilde{C}\Lambda^{-n}\Big\{(\ln\Lambda)\cdot\sup_{z\in\partial\Omega}\|\nu\|_*(\Delta(z,2\eta\Lambda\lambda\varepsilon^{-1}R)) + \varepsilon\Big\},\end{aligned} \tag{5.5.39}$$

where $\widetilde{C} \in (0,\infty)$ depends only on n, the Ahlfors regularity constants of $\partial\Omega$, and the relative flatness ratio $\wp(\Omega)/\operatorname{diam}\partial\Omega$. Next, the Mean Value Theorem shows that

$$|E(x_0,x;y)| = \Big|\frac{x_0-y}{|x_0-y|^n} - \frac{x-y}{|x-y|^n}\Big| \leq \frac{C_n R}{(\Lambda R)^n} = C_n\Lambda^{-n}R^{1-n} \tag{5.5.40}$$

at each $y \in \Delta(y_0,R)$, for some $C_n \in (0,\infty)$. Also, [129, (7.4.62)-(7.4.64)] imply

$$\begin{aligned}\big|\nu_{\Delta(x_0,R)} - \nu_{\Delta(y_0,R)}\big| &\leq \big|\nu_{\Delta(x_0,R)} - \nu_{\Delta(x_0,\eta\Lambda R)}\big| + \big|\nu_{\Delta(x_0,\eta\Lambda R)} - \nu_{\Delta(y_0,\eta\Lambda R)}\big|\\ &\quad + \big|\nu_{\Delta(y_0,\eta\Lambda R)} - \nu_{\Delta(y_0,R)}\big| \leq C(\ln\Lambda)\cdot\sup_{z\in\partial\Omega}\|\nu\|_*(\Delta(z,2\eta\Lambda R)).\end{aligned} \tag{5.5.41}$$

Based on (5.5.40) and (5.5.41) we may then estimate

$$\begin{aligned}|\mathrm{V}| &\leq \int_{\Delta(y_0,R)} |E(x_0,x;y)|\big|\nu(y) - \nu_{\Delta(x_0,R)}\big|\,d\sigma(y) \leq C\Lambda^{-n}\fint_{\Delta(y_0,R)}\big|\nu - \nu_{\Delta(x_0,R)}\big|\,d\sigma\\ &\leq C\Lambda^{-n}\fint_{\Delta(y_0,R)}\big|\nu - \nu_{\Delta(y_0,R)}\big|\,d\sigma + C\Lambda^{-n}\big|\nu_{\Delta(x_0,R)} - \nu_{\Delta(y_0,R)}\big|\\ &\leq C(\Lambda^{-n}\ln\Lambda)\cdot\sup_{z\in\partial\Omega}\|\nu\|_*(\Delta(z,2\eta\Lambda R)),\end{aligned} \tag{5.5.42}$$

where $C \in (0,\infty)$ depends only on n and the Ahlfors regularity constants of $\partial\Omega$. Finally, (5.5.40) implies that for a purely dimensional constant $C_n \in (0,\infty)$ we have

$$|\mathrm{VI}| \leq \int_{\Delta(y_0,R)} |E(x_0,x;y)|\big|\nu(x)-\nu_{\Delta(x_0,R)}\big|\,d\sigma(y) \leq C_n\Lambda^{-n}\big|\nu(x)-\nu_{\Delta(x_0,R)}\big|. \tag{5.5.43}$$

For further use, let us note here that [129, (7.7.20)] plus the consequence of the John-Nirenberg inequality recorded in [129, (7.4.71)] imply, reasoning much as in [129, (7.4.69)], that for some exponent $q' \in (1,\infty)$ which depends only on p, $[w]_{A_p}$, n, and the Ahlfors regularity constants of $\partial\Omega$ we have

$$\begin{aligned} ⨍_{\Delta(x_0,R)} \big|\nu-\nu_{\Delta(x_0,R)}\big|^p\,dw &= ⨍_{\Delta(x_0,R)} \Big|\nu - ⨍_{\Delta(x_0,R)} \nu\,d\sigma\Big|^p\,dw \\ &\leq C\bigg(⨍_{\Delta(x_0,R)} \Big|\nu - ⨍_{\Delta(x_0,R)} \nu\,d\sigma\Big|^{pq'}\,d\sigma\bigg)^{1/q'} \leq C\Big(\sup_{z\in\partial\Omega}\|\nu\|_*(\Delta(z,5R))\Big)^p \end{aligned} \tag{5.5.44}$$

with $C \in (0,\infty)$ of the same nature as before. We may use [129, (7.7.16)] to estimate

$$\begin{aligned} &⨍_{\Delta(x_0,R)} |(\mathfrak{C}-\mathfrak{C}^{\#})\mathbf{1}_{\Delta(y_0,R)}(x)|^p\,dw(x) \\ &\leq \frac{\|\mathbf{1}_{\Delta(y_0,R)}\|^p_{L^p(\partial\Omega,w)\otimes C\ell_n}}{w(\Delta(x_0,R))}\big\|\mathbf{1}_{\Delta(x_0,R)}\cdot(\mathfrak{C}-\mathfrak{C}^{\#})\big\|^p_{L^p(\partial\Omega,w)\otimes C\ell_n\to L^p(\partial\Omega,w)\otimes C\ell_n} \\ &= \frac{w(\Delta(y_0,R))}{w(\Delta(x_0,R))}\big\|\mathbf{1}_{\Delta(x_0,R)}\cdot(\mathfrak{C}-\mathfrak{C}^{\#})\big\|^p_{L^p(\partial\Omega,w)\otimes C\ell_n\to L^p(\partial\Omega,w)\otimes C\ell_n} \\ &\leq \frac{w(\Delta(x_0,2\lambda\Lambda R))}{w(\Delta(x_0,R))}\big\|\mathbf{1}_{\Delta(x_0,R)}\cdot(\mathfrak{C}-\mathfrak{C}^{\#})\big\|^p_{L^p(\partial\Omega,w)\otimes C\ell_n\to L^p(\partial\Omega,w)\otimes C\ell_n} \\ &\leq [w]_{A_p}\Big(\frac{\sigma(\Delta(x_0,2\lambda\Lambda R))}{\sigma(\Delta(x_0,R))}\Big)^p\big\|\mathbf{1}_{\Delta(x_0,R)}\cdot(\mathfrak{C}-\mathfrak{C}^{\#})\big\|^p_{L^p(\partial\Omega,w)\otimes C\ell_n\to L^p(\partial\Omega,w)\otimes C\ell_n} \\ &\leq C\Lambda^{(n-1)p}\big\|\mathbf{1}_{\Delta(x_0,R)}\cdot(\mathfrak{C}-\mathfrak{C}^{\#})\big\|^p_{L^p(\partial\Omega,w)\otimes C\ell_n\to L^p(\partial\Omega,w)\otimes C\ell_n}, \end{aligned} \tag{5.5.45}$$

where $C \in (0,\infty)$ depends only on n, p, $[w]_{A_p}$, and Ahlfors regularity. Observe that

$$⨍_{\Delta(x_0,R)}\bigg|\int_{\Delta(y_0,R)}\Big\{\frac{x_0-y}{|x_0-y|^n}\odot\nu(y)+\nu(x)\odot\frac{x_0-y}{|x_0-y|^n}\Big\}\,d\sigma(y)\bigg|^p\,dw(x)$$

may, thanks to (5.5.26), (5.5.30), (5.5.32), (5.5.34), (5.5.36), (5.5.39), (5.5.42), (5.5.43), (5.5.44), and (5.5.45), be estimated from above by

$$
\begin{aligned}
&C\Lambda^{-np}\Big\{(\ln\Lambda)\cdot\sup_{z\in\partial\Omega}\|\nu\|_*(\Delta(z,2\eta\Lambda\lambda\varepsilon^{-1}R))+\varepsilon\Big\}^p\\
&\quad+C\fint_{\Delta(x_0,R)}|(\mathfrak{C}-\mathfrak{C}^\#)\mathbf{1}_{\Delta(y_0,R)}|^p\,dw+C\Lambda^{-np}\fint_{\Delta(x_0,R)}\big|\nu-\nu_{\Delta(x_0,R)}\big|^p\,dw\\
&\leq C\Lambda^{-np}\Big\{(\ln\Lambda)\cdot\sup_{z\in\partial\Omega}\|\nu\|_*(\Delta(z,2\eta\Lambda\lambda\varepsilon^{-1}R))+\varepsilon\Big\}^p\\
&\quad+C\Lambda^{(n-1)p}\big\|\mathbf{1}_{\Delta(x_0,R)}\cdot(\mathfrak{C}-\mathfrak{C}^\#)\big\|^p_{L^p(\partial\Omega,w)\otimes C\ell_n\to L^p(\partial\Omega,w)\otimes C\ell_n}
\end{aligned}
\tag{5.5.46}
$$

where $C\in(0,\infty)$ depends only on n, p, $[w]_{A_p}$, the Ahlfors regularity constants of $\partial\Omega$, and the relative flatness ratio $\wp(\Omega)/\operatorname{diam}\partial\Omega$. Going further, define

$$a:=\fint_{\Delta(y_0,R)}\frac{x_0-y}{|x_0-y|^n}\,d\sigma(y)\in\mathbb{R}^n\hookrightarrow C\ell_n, \tag{5.5.47}$$

$$b:=\fint_{\Delta(y_0,R)}\frac{x_0-y}{|x_0-y|^n}\odot\nu(y)\,d\sigma(y)\in C\ell_n. \tag{5.5.48}$$

Note that

$$a=\frac{x_0-y_0}{|x_0-y_0|^n}+\fint_{\Delta(y_0,R)}\Big(\frac{x_0-y}{|x_0-y|^n}-\frac{x_0-y_0}{|x_0-y_0|^n}\Big)\,d\sigma(y) \tag{5.5.49}$$

and observe that the Mean Value Theorem gives, for some $C_n\in(0,\infty)$,

$$\Big|\frac{x_0-y}{|x_0-y|^n}-\frac{x_0-y_0}{|x_0-y_0|^n}\Big|\leq\frac{C_nR}{(\Lambda R)^n}=C_n\Lambda^{-n}R^{1-n}, \tag{5.5.50}$$

for each point $y\in\Delta(y_0,R)$. From formula (5.5.49) we then deduce the lower bound $|a|\geq\Big|\frac{x_0-y_0}{|x_0-y_0|^n}\Big|-\fint_{\Delta(y_0,R)}\Big|\frac{x_0-y}{|x_0-y|^n}-\frac{x_0-y_0}{|x_0-y_0|^n}\Big|\,d\sigma(y)$ which, by (5.5.50), shows

$$|a|\geq(\eta\Lambda R)^{1-n}-C_n\Lambda^{-n}R^{1-n}\geq(\eta^{1-n}/2)(\Lambda R)^{1-n}, \tag{5.5.51}$$

since $\Lambda>2C_n\eta^{n-1}$ (cf. (5.5.23)). Hence, if we also introduce

$$A:=b\odot\Big(\frac{a}{|a|^2}\Big)\in C\ell_n, \tag{5.5.52}$$

then, with A_{proj} denoting the vector part of $A\in C\ell_n$, we have

$$
\begin{aligned}
\fint_{\Delta(x_0,R)}|\nu(x)-A_{\text{proj}}|^p\,dw(x)&\leq\fint_{\Delta(x_0,R)}|\nu(x)-A|^p\,dw(x)\\
&=\fint_{\Delta(x_0,R)}\big|\nu(x)-b\odot(a/|a|^2)\big|^p\,dw(x)
\end{aligned}
\tag{5.5.53}
$$

and we may now bound, using (5.5.52), (5.5.51), [129, (6.4.1), (6.4.25), (6.4.35)], (5.5.47), (5.5.48), and (5.5.46), the last term above by

$$\leq C(\Lambda R)^{(n-1)p} \fint_{\Delta(x_0,R)} \big|\nu(x) - b \odot (a/|a|^2)\big|^p \, |a|^p \, dw(x)$$
$$= C(\Lambda R)^{(n-1)p} \fint_{\Delta(x_0,R)} \big|(\nu(x) - b \odot (a/|a|^2)) \odot a\big|^p \, dw(x)$$
$$= C(\Lambda R)^{(n-1)p} \fint_{\Delta(x_0,R)} |\nu(x) \odot a + b|^p \, dw(x)$$
$$= C(\Lambda R)^{(n-1)p} \fint_{\Delta(x_0,R)} \Bigg|\nu(x) \odot \Big(\fint_{\Delta(y_0,R)} \frac{x_0 - y}{|x_0 - y|^n} \, d\sigma(y)\Big) + \Big(\fint_{\Delta(y_0,R)} \frac{x_0 - y}{|x_0 - y|^n} \odot \nu(y) \, d\sigma(y)\Big)\Bigg|^p dw(x)$$
$$= C(\Lambda R)^{(n-1)p} \fint_{\Delta(x_0,R)} \Bigg|\fint_{\Delta(y_0,R)} \Big\{\nu(x) \odot \frac{x_0 - y}{|x_0 - y|^n} + \frac{x_0 - y}{|x_0 - y|^n} \odot \nu(y)\Big\} \, d\sigma(y)\Bigg|^p dw(x),$$
$$\leq C\Lambda^{-p}\Big\{(\ln \Lambda) \cdot \sup_{z \in \partial\Omega} \|\nu\|_*(\Delta(z, 2\eta\Lambda\lambda\varepsilon^{-1}R)) + \varepsilon\Big\}^p$$
$$+ C\Lambda^{2(n-1)p} \big\|\mathbf{1}_{\Delta(x_0,R)} \cdot (\mathfrak{C} - \mathfrak{C}^{\#})\big\|^p_{L^p(\partial\Omega,w)\otimes C\ell_n \to L^p(\partial\Omega,w)\otimes C\ell_n}, \tag{5.5.54}$$

with C depending only on n, p, $[w]_{A_p}$, the Ahlfors regularity constants of $\partial\Omega$, and the relative flatness ratio $\wp(\Omega)/\operatorname{diam} \partial\Omega$. Since [129, Lemmas 7.7.5, 3.1.1] imply

$$\operatorname{dist}\big(\nu, [\mathrm{VMO}(\partial\Omega, \sigma)]^n\big) \approx \operatorname{dist}\big(\nu, [\mathrm{VMO}(\partial\Omega, w)]^n\big), \tag{5.5.55}$$

with distances measured in $\big[\mathrm{BMO}(\partial\Omega, \sigma)\big]^n = \big[\mathrm{BMO}(\partial\Omega, w)\big]^n$, we have

$$\operatorname{dist}\Big(\nu, [\mathrm{VMO}(\partial\Omega, \sigma)]^n\Big) \approx \operatorname{dist}\Big(\nu, [\mathrm{VMO}(\partial\Omega, w)]^n\Big) \tag{5.5.56}$$
$$\approx \lim_{R\to 0^+} \Big(\sup_{x_0 \in \partial\Omega} \Big(\fint_{\Delta(x_0,R)} \Big|\nu(y) - \fint_{\Delta(x_0,R)} \nu \, dw\Big|^p \, dw(y)\Big)^{1/p}\Big)$$
$$\leq C \lim_{R\to 0^+} \Big(\sup_{x_0 \in \partial\Omega} \Big(\fint_{\Delta(x_0,R)} |\nu(y) - A_{\mathrm{proj}}|^p \, dw(y)\Big)^{1/p}\Big)$$
$$\leq C \lim_{R\to 0^+} \Lambda^{-1}\Big\{(\ln \Lambda) \cdot \sup_{z \in \partial\Omega} \|\nu\|_*(\Delta(z, 2\eta\Lambda\lambda\varepsilon^{-1}R)) + \varepsilon\Big\}$$
$$+ C\Lambda^{2(n-1)} \lim_{R\to 0^+} \sup_{x_0 \in \partial\Omega} \big\|\mathbf{1}_{\Delta(x_0,R)} \cdot (\mathfrak{C} - \mathfrak{C}^{\#})\big\|_{L^p(\partial\Omega,w)\otimes C\ell_n \to L^p(\partial\Omega,w)\otimes C\ell_n}$$
$$\leq C_o(\Lambda^{-1} \ln \Lambda) \cdot \operatorname{dist}\Big(\nu, [\mathrm{VMO}(\partial\Omega, \sigma)]^n\Big) + C\Lambda^{-1}\varepsilon$$
$$+ C\Lambda^{2(n-1)} \lim_{R\to 0^+} \Big(\sup_{x_0 \in \partial\Omega} \big\|\mathbf{1}_{\Delta(x_0,R)} \cdot (\mathfrak{C} - \mathfrak{C}^{\#})\big\|_{L^p(\partial\Omega,w)\otimes C\ell_n \to L^p(\partial\Omega,w)\otimes C\ell_n}\Big),$$

by (5.5.55), [130, (3.1.53)], the first inequality in [129, (7.4.57)], (5.5.54), and [130, (3.1.55)]. Here $C_o, C \in (0,\infty)$ depend only on n, p, $[w]_{A_p}$, the Ahlfors regularity constants of $\partial\Omega$, and the relative flatness ratio $\wp(\Omega)/\operatorname{diam}\partial\Omega$. Passing to limit as $\varepsilon \searrow 0$ then yields (5.5.20), on account of the last property in (5.5.23). □

To state our next result, recall the Clifford-Cauchy operator $\mathfrak{C}$ from (A.0.54), along with its "transpose" $\mathfrak{C}^{\#}$ from (A.0.56). Also, recall the harmonic double layer K_Δ defined in (5.5.1), and the Riesz transforms $(R_j)_{1\leq j\leq n}$ introduced in (5.5.2). Finally, recall that M_b denotes the operator of pointwise multiplication by the function b.

Theorem 5.5.9 *Let $\Omega \subseteq \mathbb{R}^n$ be a* UR *domain with compact boundary. Abbreviate $\sigma := \mathcal{H}^{n-1}\lfloor\partial\Omega$ and denote by $\nu = (\nu_1,\dots,\nu_n)$ the geometric measure theoretic outward unit normal to Ω. Then the following statements are equivalent:*

(i) The operator $\mathfrak{C} - \mathfrak{C}^{\#}$ is compact on $L^p(\partial\Omega, w)\otimes \mathcal{C}\ell_n$ for some (or all) $p \in (1,\infty)$ and some (or all) $w \in A_p(\partial\Omega,\sigma)$.
(ii) The commutator $[\mathfrak{C}, \mathfrak{C}^{\#}]$ is a compact operator on $L^p(\partial\Omega, w)\otimes \mathcal{C}\ell_n$ for some (or all) $p \in (1,\infty)$ and some (or all) $w \in A_p(\partial\Omega,\sigma)$.
(iii) The boundary-to-boundary harmonic double layer K_Δ (cf. (5.5.1)) and the commutators $[R_j, M_{\nu_k}]$, with $1 \leq j,k \leq n$, between the j-th Riesz transform R_j (cf. (5.5.2)) and the operator M_{ν_k} of pointwise multiplication by ν_k, are all compact on $L^p(\partial\Omega, w)$ for some (or any) $p \in (1,\infty)$ and some (or any) $w \in A_p(\partial\Omega,\sigma)$.
(iv) The operator $I + \sum_{j=1}^{n} R_j^2$ and the commutators $[R_j, R_k]$, with $1 \leq j,k \leq n$, are compact on $L^p(\partial\Omega, w)$ for some (or all) $p \in (1,\infty)$ and some (or all) $w \in A_p(\partial\Omega,\sigma)$.
(v) The commutators $[M_{\nu_k}, R_j]$ and $[R_j, R_k]$, with $1 \leq j,k \leq n$, are compact operators on $L^p(\partial\Omega, w)$ for some (or all) $p \in (1,\infty)$ and some (or all) weights $w \in A_p(\partial\Omega,\sigma)$.
(vi) The set Ω is an infinitesimally flat AR *domain (cf. Definition 3.4.1).*

Proof Recall from [131, (2.5.324)] and [132, (1.6.13)] that, in the current context, the boundary-to-boundary Cauchy-Clifford operator and its formal transpose satisfy

$$\mathfrak{C}^2 = \big(\mathfrak{C}^{\#}\big)^2 = \tfrac{1}{4}I \text{ on } L^p(\partial\Omega, w)\otimes \mathcal{C}\ell_n \quad \text{for all } p \in (1,\infty) \text{ and } w \in A_p(\partial\Omega,\sigma). \tag{5.5.57}$$

Hence, for each $p \in (1,\infty)$ and each $w \in A_p(\partial\Omega,\sigma)$ we may write

$$[\mathfrak{C}, \mathfrak{C}^{\#}] = (\mathfrak{C} - \mathfrak{C}^{\#})(\mathfrak{C} + \mathfrak{C}^{\#}) \text{ on } L^p(\partial\Omega, w)\otimes \mathcal{C}\ell_n. \tag{5.5.58}$$

We now make the observation that

$$\mathfrak{C} + \mathfrak{C}^{\#} \text{ is an invertible operator on } L^2(\partial\Omega,\sigma)\otimes \mathcal{C}\ell_n. \tag{5.5.59}$$

Indeed, on the one hand (5.5.57) implies

$$\mathfrak{C} + \mathfrak{C}^{\#} = \mathfrak{C}(I + 4\mathfrak{C}\mathfrak{C}^{\#}) \text{ on } L^2(\partial\Omega,\sigma)\otimes \mathcal{C}\ell_n. \tag{5.5.60}$$

On the other hand, $\mathfrak{C}\mathfrak{C}^{\#}$ is a non-negative self-adjoint operator on $L^2(\partial\Omega,\sigma)\otimes C\ell_n$ (cf. [132, Proposition 1.6.1, item *(ii)*]) so

$$I + 4\mathfrak{C}\mathfrak{C}^{\#} \text{ is invertible on } L^2(\partial\Omega,\sigma)\otimes C\ell_n. \tag{5.5.61}$$

Now, (5.5.59) readily follows from (5.5.60)-(5.5.61) and (5.5.57). In concert, (5.5.58) and (5.5.59) prove that

$$\begin{aligned}&[\mathfrak{C},\mathfrak{C}^{\#}] \text{ compact on } L^2(\partial\Omega,\sigma)\otimes C\ell_n\\ &\qquad\Longleftrightarrow\ \mathfrak{C}-\mathfrak{C}^{\#} \text{ compact on } L^2(\partial\Omega,\sigma)\otimes C\ell_n.\end{aligned} \tag{5.5.62}$$

In concert with [129, Proposition 7.7.12], this justifies the equivalence *(i)* $\Leftrightarrow$ *(ii)*.

Moving on, the implication *(iii)* $\Rightarrow$ *(i)* is a consequence of Lemma 5.5.6 and [129, Proposition 7.7.12]. If *(i)* holds, then Theorem 5.5.8 gives that the Gauss map has vanishing mean oscillations, i.e., $\nu\in\big[\mathrm{VMO}(\partial\Omega,\sigma)\big]^n$. Consequently, Ω is an infinitesimally flat AR domain by virtue of Definition 3.4.1. This proves the implication *(i)* $\Rightarrow$ *(vi)*. Assume next that *(vi)* holds and fix some $p\in(1,\infty)$ along with some $w\in A_p(\partial\Omega,\sigma)$. From [129, (5.11.93)], (5.5.1), and Theorem 5.2.3 we know that the harmonic double layer K_Δ is a compact operator on $L^p(\partial\Omega,w)$. Also, since ν belongs to $\big[\mathrm{VMO}(\partial\Omega,\sigma)\big]^n$, [131, Theorem 2.7.3] gives that $[M_{\nu_k},R_j]$, the commutator between the operator of multiplication by the k-th component of ν and the j-th Riesz transform on $\partial\Omega$, is also compact on $L^p(\partial\Omega,w)$, for each $j,k\in\{1,\dots,n\}$. In view of [129, Proposition 7.7.12], this establishes *(vi)* $\Rightarrow$ *(iii)*.

In summary, *(i)* $\Leftrightarrow$ *(ii)* $\Leftrightarrow$ *(iii)* $\Leftrightarrow$ *(vi)*. To show that these are also equivalent to *(iv)*, fix some $p\in(1,\infty)$ along with some $w\in A_p(\partial\Omega,\sigma)$. Also, given any $a\in C\ell_n$, denote by M_a the operator of pointwise Clifford algebra multiplication from the left by a. In particular, M_ν stands for the operator of pointwise left-multiplication in $C\ell_n$ by $\nu\equiv\nu_1\mathbf{e}_1+\cdots+\nu_n\mathbf{e}_n$. From [132, (1.6.21)] and (5.5.57) (applied twice) we see that, as operators on $L^p(\partial\Omega,w)\otimes C\ell_n$,

$$\mathfrak{C}(\mathfrak{C}^{\#}-\mathfrak{C}) = -\frac{1}{4}\Big(I+\sum_{k=1}^{n}R_k^2\Big)+\frac{1}{4}\sum_{1\le j<k\le n}[R_j,R_k]M_{\mathbf{e}_j\odot\mathbf{e}_k}, \tag{5.5.63}$$

$$\mathfrak{C}-\mathfrak{C}^{\#} = \mathfrak{C}\Big(I+\sum_{k=1}^{n}R_k^2\Big)-\sum_{1\le j<k\le n}\mathfrak{C}[R_j,R_k]M_{\mathbf{e}_j\odot\mathbf{e}_k}. \tag{5.5.64}$$

It is then clear from (5.5.63) and [129, Proposition 7.7.12] that *(i)* $\Rightarrow$ *(iv)*. Also, from (5.5.64) and [129, Proposition 7.7.12] we see that *(iv)* $\Rightarrow$ *(i)*. Hence, *(i)* $\Leftrightarrow$ *(iv)*.

Finally, let us prove that *(v)* is equivalent with *(i)*, *(ii)*, *(iii)*, *(iv)*, *(vi)*. To this end, assume the weakest version of *(v)* holds, i.e., the compactness properties specified there are valid for some $p\in(1,\infty)$ and some $w\in A_p(\partial\Omega,\sigma)$. We use [132, (1.6.20)] twice, [132, (1.6.21)], and $M_\nu M_\nu=-I$ (cf. [129, (6.4.1)]) to write

$$\begin{aligned}\mathfrak{C}^{\#}\mathfrak{C} &= -M_\nu \mathfrak{C} M_\nu \mathfrak{C} M_\nu M_\nu = -M_\nu \mathfrak{C}\mathfrak{C}^{\#} M_\nu \\ &= \frac{1}{4} M_\nu \Big(\sum_{k=1}^{n} R_k^2 \Big) M_\nu - \frac{1}{4} \sum_{1\leq j<k\leq n} M_\nu [R_j, R_k] M_{\mathbf{e}_j \odot \mathbf{e}_k} M_\nu \end{aligned} \tag{5.5.65}$$

on $L^p(\partial\Omega, w) \otimes C\ell_n$. We then deduce from (5.5.65) and [132, (1.6.21)] that

$$\begin{aligned}[\mathfrak{C}, \mathfrak{C}^{\#}] = \mathfrak{C}\mathfrak{C}^{\#} - \mathfrak{C}^{\#}\mathfrak{C} &= -\frac{1}{4} \sum_{k=1}^{n} R_k^2 - \frac{1}{4} M_\nu \Big(\sum_{k=1}^{n} R_k^2 \Big) M_\nu \\ &\quad + \frac{1}{4} \sum_{1\leq j<k\leq n} [R_j, R_k] M_{\mathbf{e}_j \odot \mathbf{e}_k} + \frac{1}{4} \sum_{1\leq j<k\leq n} M_\nu [R_j, R_k] M_{\mathbf{e}_j \odot \mathbf{e}_k} M_\nu \end{aligned} \tag{5.5.66}$$

on $L^p(\partial\Omega, w) \otimes C\ell_n$. Since $M_\nu M_\nu = -I$ (cf. [129, (6.4.1)]), we also have

$$\begin{aligned}&-\frac{1}{4} \sum_{k=1}^{n} R_k^2 - \frac{1}{4} M_\nu \Big(\sum_{k=1}^{n} R_k^2 \Big) M_\nu \\ &\qquad = -\frac{1}{4} \sum_{j,k=1}^{n} \Big\{ M_\nu R_k [R_k, M_{\nu_j}] M_{\mathbf{e}_j} + M_\nu [R_k, M_{\nu_j}] R_k M_{\mathbf{e}_j} \Big\} \end{aligned} \tag{5.5.67}$$

on $L^p(\partial\Omega, w) \otimes C\ell_n$. Given that we are presently assuming that the commutators $[M_{\nu_k}, R_j]$ and $[R_j, R_k]$, with $1 \leq j, k \leq n$, are compact on $L^p(\partial\Omega, w)$, we deduce from (5.5.66)-(5.5.67) that $[\mathfrak{C}, \mathfrak{C}^{\#}]$ is compact on $L^p(\partial\Omega, w) \otimes C\ell_n$. Together with [129, Proposition 7.7.12] this proves that *(v)*$\Rightarrow$*(ii)*. Lastly, it is clear that *(iii)*-*(iv)*$\Rightarrow$*(v)*, and this completes the proof of Theorem 5.5.9. □

We are now in a position to complete the proofs of Theorems 5.5.1-5.5.5:

Proof of Theorem 5.5.1 In view of [129, (5.11.93)], this is a direct consequence of the equivalence *(iii)*$\Leftrightarrow$*(v)* in Theorem 5.5.9. □

Proof of Theorem 5.5.2 The claim in the first half of the statement is seen from Corollary 5.2.4, and [131, (2.7.33)]. The claim in the second half of the statement is implied by Theorem 5.5.1. □

Proof of Theorem 5.5.3 This is an immediate consequence of Theorem 5.5.8 (cf. (5.5.21)) and Lemma 5.5.6. □

Proof of Theorem 5.5.4 This follows from (5.5.64) and (5.5.21), keeping in mind that the Cauchy-Clifford operator is bounded on Muckenhoupt weighted Lebesgue spaces of Clifford algebra-valued functions (cf. [131, (2.5.330)]). □

Proof of Theorem 5.5.5 This is implied by Theorem 5.5.9. □

Pressing on, the goal is to establish results similar to those in Theorem 5.5.8 and Theorem 5.5.9 on Morrey and block spaces. In anticipation to this, we first prove the following lemma. which should be seen as a version of [130, Lemma 1.2.19] with Lebesgue spaces replaced by vanishing Morrey spaces.

Lemma 5.5.10 *Let $\Sigma \subseteq \mathbb{R}^n$ (where $n \in \mathbb{N}$ with $n \geq 2$) be a closed Ahlfors regular set. Abbreviate $\sigma := \mathcal{H}^{n-1}\lfloor\Sigma$ and denote by $\mathfrak{M}$ the sigma-algebra of subsets of Σ which are σ-measurable. Finally, fix some $p \in (0,\infty)$ along with some $\lambda \in (0, n-1)$, and define $C_p := 2^{(1/p-1)_+}$. Then for each operator $T \in \mathrm{Bd}\big(\mathring{M}^{p,\lambda}(\Sigma,\sigma)\big)$ one has*

$$\lim_{\varepsilon\to 0^+}\Big(\sup_{\substack{A\in\mathfrak{M}\\ \mu(A)<\varepsilon}}\|\mathbf{1}_A\cdot T\|_{\mathring{M}^{p,\lambda}(\Sigma,\sigma)\to\mathring{M}^{p,\lambda}(\Sigma,\sigma)}\Big) \leq C_p\cdot\mathrm{dist}\big(T,\mathrm{Cp}\big(\mathring{M}^{p,\lambda}(\Sigma,\sigma)\big)\big), \tag{5.5.68}$$

where the distance in the right-hand side is measured in $\mathrm{Bd}\big(\mathring{M}^{p,\lambda}(\Sigma,\sigma)\big)$.

Proof We first claim that if $T \in \mathrm{Cp}\big(\mathring{M}^{p,\lambda}(\Sigma,\sigma)\big)$ then the left-hand side of (5.5.68) vanishes. The strategy for proving this is to reason by contradiction. Specifically, assume $T \in \mathrm{Cp}\big(\mathring{M}^{p,\lambda}(\Sigma,\sigma)\big)$ yet there exists some $\delta > 0$ together with a sequence

$$\{A_j\}_{j\in\mathbb{N}} \subseteq \mathfrak{M} \text{ with } \lim_{j\to\infty}\sigma(A_j) = 0 \tag{5.5.69}$$

and such that

$$\big\|\mathbf{1}_{A_j}\cdot T\big\|_{\mathring{M}^{p,\lambda}(\Sigma,\sigma)\to\mathring{M}^{p,\lambda}(\Sigma,\sigma)} > \delta \text{ for each } j \in \mathbb{N}. \tag{5.5.70}$$

As noted in [130, (1.2.95)-(1.2.98)], from (5.5.69) we conclude that

$$\lim_{j\to\infty}\mathbf{1}_{A_j}(x) = 0 \text{ for } \sigma\text{-a.e. } x \in \Sigma. \tag{5.5.71}$$

Also, (5.5.70) ensures that there exist functions

$$\begin{aligned} &f_j \in \mathring{M}^{p,\lambda}(\Sigma,\sigma) \text{ with } \|f_j\|_{M^{p,\lambda}(\Sigma,\sigma)} = 1 \text{ and}\\ &\big\|\mathbf{1}_{A_j}\cdot Tf_j\big\|_{M^{p,\lambda}(\Sigma,\sigma)} > \delta \text{ for each } j \in \mathbb{N}. \end{aligned} \tag{5.5.72}$$

Given we are presently working under the hypothesis that T is compact on $\mathring{M}^{p,\lambda}(\Sigma,\sigma)$, there is no loss of generality in assuming that

$$\{Tf_j\}_{j\in\mathbb{N}} \text{ converges in } M^{p,\lambda}(\Sigma,\sigma) \text{ to some function } g \in \mathring{M}^{p,\lambda}(\Sigma,\sigma). \tag{5.5.73}$$

We next claim that

$$\lim_{j\to\infty}\big\|\mathbf{1}_{A_j}\cdot g\big\|_{M^{p,\lambda}(\Sigma,\sigma)} = 0. \tag{5.5.74}$$

To justify this, fix an arbitrary $\varepsilon > 0$. From (A.0.168) we know that there exists $h \in L^s(\Sigma,\sigma)$ with $s := \frac{p(n-1)}{n-1-\lambda}$ such that

$$\|g - h\|_{M^{p,\lambda}(\Sigma,\sigma)} < \varepsilon. \tag{5.5.75}$$

In view of (5.5.71), Lebesgue's Dominated Convergence Theorem ensures that

$$\lim_{j\to\infty} \mathbf{1}_{A_j} \cdot h = 0 \ \text{ in } \ L^s(\Sigma, \sigma). \tag{5.5.76}$$

From (5.5.76) and the fact that $L^s(\Sigma, \sigma) \hookrightarrow \mathring{M}^{p,\lambda}(\Sigma, \sigma)$ continuously (cf. [130, (6.2.15)]), we then conclude that

$$\lim_{j\to\infty} \mathbf{1}_{A_j} \cdot h = 0 \ \text{ in } \ M^{p,\lambda}(\Sigma, \sigma). \tag{5.5.77}$$

Then we may write

$$\begin{aligned}
&\limsup_{j\to\infty} \big\|\mathbf{1}_{A_j} \cdot g\big\|_{M^{p,\lambda}(\Sigma,\sigma)} \\
&\quad \le C_p \Big\{ \limsup_{j\to\infty} \big\|\mathbf{1}_{A_j} \cdot (g-h)\big\|_{M^{p,\lambda}(\Sigma,\sigma)} + \limsup_{j\to\infty} \big\|\mathbf{1}_{A_j} \cdot h\big\|_{M^{p,\lambda}(\Sigma,\sigma)} \Big\} \\
&\quad \le C_p \|g-h\|_{M^{p,\lambda}(\Sigma,\sigma)} < C_p \cdot \varepsilon,
\end{aligned} \tag{5.5.78}$$

thanks to [130, (6.2.3), (6.2.8)], (5.5.75), and (5.5.77). Since $\varepsilon > 0$ has been arbitrarily chosen, this establishes (5.5.74). Moving on, for each $j \in \mathbb{N}$ we may now write, based on (5.5.72) and [130, (6.2.8)],

$$\begin{aligned}
\delta &< \big\|\mathbf{1}_{A_j} \cdot Tf_j\big\|_{M^{p,\lambda}(\Sigma,\sigma)} \le C_p \big\|\mathbf{1}_{A_j} \cdot (Tf_j - g)\big\|_{M^{p,\lambda}(\Sigma,\sigma)} + C_p \big\|\mathbf{1}_{A_j} \cdot g\big\|_{M^{p,\lambda}(\Sigma,\sigma)} \\
&\le C_p \|Tf_j - g\|_{M^{p,\lambda}(\Sigma,\sigma)} + C_p \big\|\mathbf{1}_{A_j} \cdot g\big\|_{M^{p,\lambda}(\Sigma,\sigma)}
\end{aligned} \tag{5.5.79}$$

which, after passing to limit $j \to \infty$, leads to a contradiction in view of (5.5.73) and (5.5.74). This establishes the fact that

$$\begin{aligned}
&\lim_{\varepsilon\to 0^+} \Big(\sup_{\substack{A\in\mathfrak{M} \\ \mu(A)<\varepsilon}} \big\|\mathbf{1}_A \cdot T\big\|_{\mathring{M}^{p,\lambda}(\Sigma,\sigma)\to \mathring{M}^{p,\lambda}(\Sigma,\sigma)} \Big) = 0 \\
&\qquad \text{for each } T \in \mathrm{Cp}\big(\mathring{M}^{p,\lambda}(\Sigma, \sigma)\big).
\end{aligned} \tag{5.5.80}$$

The end-game in the proof of the lemma is as follows. Fix an arbitrary operator $T \in \mathrm{Bd}\big(\mathring{M}^{p,\lambda}(\Sigma, \sigma)\big)$. Then for each $K \in \mathrm{Cp}\big(\mathring{M}^{p,\lambda}(\Sigma, \sigma)\big)$ we may estimate

$$\begin{aligned}
&\lim_{\varepsilon\to 0^+} \Big(\sup_{\substack{A\in\mathfrak{M} \\ \mu(A)<\varepsilon}} \big\|\mathbf{1}_A \cdot T\big\|_{\mathring{M}^{p,\lambda}(\Sigma,\sigma)\to \mathring{M}^{p,\lambda}(\Sigma,\sigma)} \Big) \\
&\qquad \le C_p \cdot \lim_{\varepsilon\to 0^+} \Big(\sup_{\substack{A\in\mathfrak{M} \\ \mu(A)<\varepsilon}} \big\|\mathbf{1}_A \cdot (T-K)\big\|_{\mathring{M}^{p,\lambda}(\Sigma,\sigma)\to \mathring{M}^{p,\lambda}(\Sigma,\sigma)} \Big) \\
&\qquad\quad + C_p \cdot \lim_{\varepsilon\to 0^+} \Big(\sup_{\substack{A\in\mathfrak{M} \\ \mu(A)<\varepsilon}} \big\|\mathbf{1}_A \cdot K\big\|_{\mathring{M}^{p,\lambda}(\Sigma,\sigma)\to \mathring{M}^{p,\lambda}(\Sigma,\sigma)} \Big) \\
&\qquad \le C_p \cdot \|T-K\|_{\mathring{M}^{p,\lambda}(\Sigma,\sigma)\to \mathring{M}^{p,\lambda}(\Sigma,\sigma)},
\end{aligned} \tag{5.5.81}$$

where the last inequality is based on (5.5.80). After taking the infimum over all $K \in \mathrm{Cp}\big(\mathring{M}^{p,\lambda}(\Sigma, \sigma)\big)$ we arrive at (5.5.68). □

Here is the analogue of Theorem 5.5.8 on vanishing Morrey spaces.

Theorem 5.5.11 *Let $\Omega \subset \mathbb{R}^n$ be a* UR *domain with compact boundary. Abbreviate $\sigma := \mathcal{H}^{n-1}\lfloor\partial\Omega$ and denote by ν the geometric measure theoretic outward unit normal to Ω. Fix some exponent $p \in (1,\infty)$ along with a parameter $\lambda \in (0, n-1)$. Then there exists $C \in (0,\infty)$, depending only on n, p, λ, the Ahlfors regularity constants of $\partial\Omega$, and the relative flatness ratio $\wp(\Omega)/\mathrm{diam}\,\partial\Omega$ such that*[10]

$$\begin{aligned}&\mathrm{dist}\Big(\nu, [\mathrm{VMO}(\partial\Omega,\sigma)]^n\Big) \\ &\quad\le C \lim_{R\to 0^+}\Big(\sup_{x\in\partial\Omega}\big\|\mathbf{1}_{\Delta(x,R)}\cdot(\mathfrak{C}-\mathfrak{C}^{\#})\big\|_{\mathring{M}^{p,\lambda}(\partial\Omega,\sigma)\otimes\mathcal{C}\ell_n\to\mathring{M}^{p,\lambda}(\partial\Omega,\sigma)\otimes\mathcal{C}\ell_n}\Big)\end{aligned} \tag{5.5.82}$$

where $\Delta(x,R) := B(x,R)\cap\partial\Omega$ for each $x \in \partial\Omega$ and $R \in (0,\infty)$.

As a consequence of this and Lemma 5.5.10,

$$\begin{aligned}\mathrm{dist}\Big(\nu, [\mathrm{VMO}(\partial\Omega,\sigma)]^n\Big) &\le C\cdot\mathrm{dist}\Big(\mathfrak{C}-\mathfrak{C}^{\#}, \mathrm{Cp}\big(\mathring{M}^{p,\lambda}(\partial\Omega,\sigma)\otimes\mathcal{C}\ell_n\big)\Big) \\ &= C\big\|\mathfrak{C}-\mathfrak{C}^{\#}\big\|^{\mathrm{ess}}_{\mathring{M}^{p,\lambda}(\partial\Omega,\sigma)\otimes\mathcal{C}\ell_n\to\mathring{M}^{p,\lambda}(\partial\Omega,\sigma)\otimes\mathcal{C}\ell_n}.\end{aligned} \tag{5.5.83}$$

In particular, if $\mathfrak{C}-\mathfrak{C}^{\#}$ is a compact operator on $\mathring{M}^{p,\lambda}(\partial\Omega,\sigma)\otimes\mathcal{C}\ell_n$ then the Gauss map has vanishing mean oscillations, i.e., $\nu \in \big[\mathrm{VMO}(\partial\Omega,\sigma)\big]^n$, so Ω is an infinitesimally flat AR *domain (cf. Definition 3.4.1).*

Proof The argument largely follows the proof of the unweighted version of Theorem 5.5.8, so we will only indicate the main changes. In place of (5.5.45) write

$$\begin{aligned}&⨍_{\Delta(x_0,R)}\big|(\mathfrak{C}-\mathfrak{C}^{\#})\mathbf{1}_{\Delta(y_0,R)}(x)\big|^p\,d\sigma(x) \\ &\quad\le R^{-(n-1-\lambda)}\big\|\mathbf{1}_{\Delta(x_0,R)}\cdot(\mathfrak{C}-\mathfrak{C}^{\#})\mathbf{1}_{\Delta(y_0,R)}\big\|^p_{M^{p,\lambda}(\partial\Omega,\sigma)\otimes\mathcal{C}\ell_n} \\ &\quad\le R^{-(n-1-\lambda)}\big\|\mathbf{1}_{\Delta(y_0,R)}\big\|^p_{M^{p,\lambda}(\partial\Omega,\sigma)}\times \\ &\qquad\qquad\times\big\|\mathbf{1}_{\Delta(x_0,R)}\cdot(\mathfrak{C}-\mathfrak{C}^{\#})\big\|^p_{\mathring{M}^{p,\lambda}(\partial\Omega,\sigma)\otimes\mathcal{C}\ell_n\to\mathring{M}^{p,\lambda}(\partial\Omega,\sigma)\otimes\mathcal{C}\ell_n} \\ &\quad\le C\cdot\frac{\sigma\big(\Delta(y_0,R)\big)^{(n-1-\lambda)/(n-1)}}{R^{n-1-\lambda}}\times \\ &\qquad\qquad\times\big\|\mathbf{1}_{\Delta(x_0,R)}\cdot(\mathfrak{C}-\mathfrak{C}^{\#})\big\|^p_{\mathring{M}^{p,\lambda}(\partial\Omega,\sigma)\otimes\mathcal{C}\ell_n\to\mathring{M}^{p,\lambda}(\partial\Omega,\sigma)\otimes\mathcal{C}\ell_n} \\ &\quad\le C\cdot\big\|\mathbf{1}_{\Delta(x_0,R)}\cdot(\mathfrak{C}-\mathfrak{C}^{\#})\big\|^p_{\mathring{M}^{p,\lambda}(\partial\Omega,\sigma)\otimes\mathcal{C}\ell_n\to\mathring{M}^{p,\lambda}(\partial\Omega,\sigma)\otimes\mathcal{C}\ell_n},\end{aligned} \tag{5.5.84}$$

thanks to [130, (6.2.9)], where $C \in (0,\infty)$ depends only on n, p, λ, and the Ahlfors regularity constants of $\partial\Omega$. Then in lieu of (5.5.56) we obtain

[10] where the distance is considered in the space $\big[\mathrm{BMO}(\partial\Omega,\sigma)\big]^n$

$$
\begin{aligned}
&\operatorname{dist}\Big(\nu, [\mathrm{VMO}(\partial\Omega,\sigma)]^n\Big) \\
&\leq C_o(\Lambda^{-1}\ln\Lambda)\cdot \operatorname{dist}\Big(\nu, [\mathrm{VMO}(\partial\Omega,\sigma)]^n\Big) + C\Lambda^{-1}\varepsilon \\
&\quad + C\Lambda^{n-1}\cdot \lim_{R\to 0^+}\Big(\sup_{x_0\in\partial\Omega}\big\|\mathbf{1}_{\Delta(x_0,R)}\cdot(\mathfrak{C}-\mathfrak{C}^\#)\big\|_{\mathring{M}^{p,\lambda}(\partial\Omega,\sigma)\otimes\mathcal{C}\ell_n\to\mathring{M}^{p,\lambda}(\partial\Omega,\sigma)\otimes\mathcal{C}\ell_n}\Big),
\end{aligned}
\tag{5.5.85}
$$

where $C_o, C \in (0,\infty)$ now depend only on n, p, λ, the Ahlfors regularity constants of $\partial\Omega$, and the relative flatness ratio $\wp(\Omega)/\operatorname{diam}\partial\Omega$. Keeping in mind the last property in (5.5.23), we then obtain (5.5.82) after passing to limit as $\varepsilon \searrow 0$ in (5.5.85). □

Finally, we prove an analogue of Theorem 5.5.9 for Morrey and block spaces.

Theorem 5.5.12 *Let $\Omega \subseteq \mathbb{R}^n$ be a* UR *domain with compact boundary. Abbreviate $\sigma := \mathcal{H}^{n-1}\lfloor\partial\Omega$ and denote by $\nu = (\nu_1,\dots,\nu_n)$ the geometric measure theoretic outward unit normal to Ω. Then the following statements are equivalent:*

(i) The operator $\mathfrak{C}-\mathfrak{C}^\#$ is compact on $\mathring{M}^{p,\lambda}(\partial\Omega,\sigma)\otimes\mathcal{C}\ell_n$ for some (or any) choice of $p\in(1,\infty)$ and $\lambda\in(0,n-1)$.
(ii) The operator $\mathfrak{C}-\mathfrak{C}^\#$ is compact on $M^{p,\lambda}(\partial\Omega,\sigma)\otimes\mathcal{C}\ell_n$ for some (or any) choice of $p\in(1,\infty)$ and $\lambda\in(0,n-1)$.
(iii) The operator $\mathfrak{C}-\mathfrak{C}^\#$ is compact on $\mathcal{B}^{q,\lambda}(\partial\Omega,\sigma)\otimes\mathcal{C}\ell_n$ for some (or any) choice of $q\in(1,\infty)$ and $\lambda\in(0,n-1)$.
(iv) The boundary-to-boundary harmonic double layer K_Δ (cf. (5.5.1)) and the commutators $[R_j, M_{\nu_k}]$, with $1\leq j,k\leq n$, between the j-th Riesz transform R_j (cf. (5.5.2)) and the operator M_{ν_k} of pointwise multiplication by ν_k, are all compact on $\mathring{M}^{p,\lambda}(\partial\Omega,\sigma)$ for some (or any) choice of $p\in(1,\infty)$ and $\lambda\in(0,n-1)$.
(v) The boundary-to-boundary harmonic double layer K_Δ and the commutators $[M_{\nu_k}, R_j]$, with $1\leq j,k\leq n$, are compact operators on $M^{p,\lambda}(\partial\Omega,\sigma)$ for some (or any) choice of the exponent $p\in(1,\infty)$ and parameter $\lambda\in(0,n-1)$.
(vi) The boundary-to-boundary harmonic double layer K_Δ and the commutators $[M_{\nu_k}, R_j]$, with $1\leq j,k\leq n$, are compact operators on $\mathcal{B}^{q,\lambda}(\partial\Omega,\sigma)$ for some (or any) choice of exponent $q\in(1,\infty)$ and parameter $\lambda\in(0,n-1)$.
(vii) The operator $I+\sum_{j=1}^n R_j^2$ and the commutators $[R_j,R_k]$, with $1\leq j,k\leq n$, are compact on $\mathring{M}^{p,\lambda}(\partial\Omega,\sigma)$ for some (or any) choice of $p\in(1,\infty)$ and $\lambda\in(0,n-1)$.
(viii) The operator $I+\sum_{j=1}^n R_j^2$ and the commutators $[R_j,R_k]$, with $1\leq j,k\leq n$, are compact on $M^{p,\lambda}(\partial\Omega,\sigma)$ for some (or any) choice of $p\in(1,\infty)$ and $\lambda\in(0,n-1)$.
(ix) The operator $I+\sum_{j=1}^n R_j^2$ and the commutators $[R_j,R_k]$, with $1\leq j,k\leq n$, are compact on $\mathcal{B}^{q,\lambda}(\partial\Omega,\sigma)$ for some (or any) choice of $q\in(1,\infty)$ and $\lambda\in(0,n-1)$.
(x) The set Ω is an infinitesimally flat AR *domain (cf. Definition 3.4.1).*

Proof For starters, observe that Schauder's theorem (cf. [130, (1.2.57)]) together with [130, Proposition 6.2.8] and [130, Proposition 6.2.16] imply (keeping in mind that $\mathfrak{C}-\mathfrak{C}^\#$ is skew-adjoint, as seen from [132, Proposition 3.3.4]) that the items *(i)*-*(iii)* are equivalent. Granted this, Lemma 5.5.6 implies that either of the items *(iv)*-*(vi)* is equivalent with *(i)*-*(iii)*. Thus, all items *(i)*-*(vi)* are equivalent.

Next, from [130, (6.2.25)], (A.0.114), Corollary 5.4.7, (A.0.215), and [131, Theorem 2.7.6] we see that item *(x)* implies the strongest versions of items *(iv)*-*(vi)*, hence also the strongest versions of items *(i)*-*(iii)*. In light of Theorem 5.5.11 and [129, (5.11.93)], the weakest version of item *(i)* implies item *(x)*. Hence, so far, items *(i)*-*(vi)* and *(x)* are all equivalent.

To proceed, recall from [132, (3.3.78)] and [130, (6.2.15)] that the boundary-to-boundary Cauchy-Clifford operator and its formal transpose satisfy

$$\begin{gathered}\mathfrak{C}^2 = \left(\mathfrak{C}^{\#}\right)^2 = \tfrac{1}{4}I \text{ as operators on either of the space} \\ \mathring{M}^{p,\lambda}(\partial\Omega,\sigma)\otimes \mathcal{C}\ell_n,\ M^{p,\lambda}(\partial\Omega,\sigma)\otimes \mathcal{C}\ell_n,\ \mathcal{B}^{q,\lambda}(\partial\Omega,\sigma)\otimes \mathcal{C}\ell_n \\ \text{for all } p,q\in(1,\infty) \text{ and all } \lambda\in(0,n-1).\end{gathered} \tag{5.5.86}$$

Also, much as in [132, (1.6.21)], we have

$$\begin{gathered}\mathfrak{C}\mathfrak{C}^{\#} = -\frac{1}{4}\sum_{k=1}^{n} R_k^2 + \frac{1}{4}\sum_{1\leq j<k\leq n}[R_j,R_k]M_{\mathbf{e}_j\odot\mathbf{e}_k} \text{ on either of the spaces} \\ \mathring{M}^{p,\lambda}(\partial\Omega,\sigma)\otimes \mathcal{C}\ell_n,\ M^{p,\lambda}(\partial\Omega,\sigma)\otimes \mathcal{C}\ell_n,\ \mathcal{B}^{q,\lambda}(\partial\Omega,\sigma)\otimes \mathcal{C}\ell_n \\ \text{for all } p,q\in(1,\infty) \text{ and all } \lambda\in(0,n-1).\end{gathered} \tag{5.5.87}$$

In turn, this and (5.5.86) (applied twice) imply that

$$\mathfrak{C}(\mathfrak{C}^{\#}-\mathfrak{C}) = -\frac{1}{4}\Big(I+\sum_{k=1}^{n}R_k^2\Big) + \frac{1}{4}\sum_{1\leq j<k\leq n}[R_j,R_k]M_{\mathbf{e}_j\odot\mathbf{e}_k}, \tag{5.5.88}$$

$$\mathfrak{C}-\mathfrak{C}^{\#} = \mathfrak{C}\Big(I+\sum_{k=1}^{n}R_k^2\Big) - \sum_{1\leq j<k\leq n}\mathfrak{C}[R_j,R_k]M_{\mathbf{e}_j\odot\mathbf{e}_k}, \tag{5.5.89}$$

as operators on either of the spaces $\mathring{M}^{p,\lambda}(\partial\Omega,\sigma)\otimes \mathcal{C}\ell_n$, $M^{p,\lambda}(\partial\Omega,\sigma)\otimes \mathcal{C}\ell_n$, $\mathcal{B}^{q,\lambda}(\partial\Omega,\sigma)\otimes \mathcal{C}\ell_n$ with $p,q\in(1,\infty)$ and $\lambda\in(0,n-1)$ arbitrary. Together, (5.5.88) and [132, Proposition 3.3.4] prove that *(i)*-*(iii)* ⟹ *(vii)*-*(ix)*. Finally, from (5.5.89) and [132, Proposition 3.3.4] we see that *(vii)*-*(ix)* ⟹ *(i)*-*(iii)*. □

Chapter 6
The Radon-Carleman Problem

In view of the pioneering work and fundamental contributions of Johann Karl August Radon and Torsten Carleman, we find it appropriate to label the following question as the Radon-Carleman Problem:

> **Radon-Carleman Problem:** Compute and/or estimate the essential norm and/or Fredholm radius of singular integral operators of double layer type, associated with elliptic PDE, on function spaces naturally intervening in the formulation of boundary problems for said PDE. (6.0.1)

Here we review the origins of this topic and establish new results, culminating with Theorem 6.3.3 (in effect, our main contribution to the Radon-Carleman Problem).

6.1 Radon Curves and Radon Domains

Let us start by recalling the notion of curve of bounded rotation ("Kurven beschränkte Drehung," in original), introduced by J. Radon in 1919.

Definition 6.1.1 *Assume $\Sigma \subset \mathbb{C}$ is a rectifiable Jordan curve. Let $L \in (0, \infty)$ be the length of the curve Σ and denote by $[0, L] \ni s \mapsto z(s) \in \Sigma$ its arc-length parametrization. Call Σ a* `curve of bounded rotation` *(or, simply, a* `Radon curve`*) provided there exists a function*

$$\theta : [0, L] \longrightarrow \mathbb{R} \tag{6.1.1}$$

which is of bounded variation (in particular, since θ is the difference of two monotonic functions, the one-sided limits $\theta(s^{\pm})$ exist for each $s \in [0, L]$ and differ for at most countably many values of $s \in [0, L]$), satisfies

$$\Theta_\Sigma := \sup_{0 \leq s \leq L} |\theta(s^+) - \theta(s^-)| < \pi, \tag{6.1.2}$$

(with the agreement that $0^- := L^-$ and $L^+ := 0^+$) as well as

$$\int_0^L e^{i\theta(t)}\, dt = 0 \tag{6.1.3}$$

D. Mitrea et al., *Geometric Harmonic Analysis V*, Developments in Mathematics 76,
https://doi.org/10.1007/978-3-031-31561-9_6

and

$$z(s) = z(0) + \int_0^s e^{i\theta(t)}\,dt \text{ for each } s \in [0, L]. \tag{6.1.4}$$

Retain the context of Definition 6.1.1. In view of (6.1.4), the demand in (6.1.3) simply ensures that the curve Σ is closed (i.e., $z(L) = z(0)$). The inequality in (6.1.2) prevents Σ for developing cusps. If Σ is actually a $\mathscr{C}^1$ curve then $\Theta_\Sigma = 0$.

Next, if Ω denotes the bounded connected component of $\mathbb{C} \setminus \Sigma$, then

$$\Omega \text{ is simply connected} \tag{6.1.5}$$

and item *(i)* in [129, Proposition 5.8.6] tells us that

$$\begin{aligned} &\Omega \text{ is an open set of finite perimeter which satisfies} \\ &\partial\Omega = \partial\big(\overline{\Omega}\big) = \Sigma \text{ and } \mathcal{H}^1(\partial\Omega \setminus \partial_*\Omega) = 0. \end{aligned} \tag{6.1.6}$$

In addition, from item *(ii)* in [129, Proposition 5.8.6] and (6.1.4) we see that if ν denotes the geometric measure theoretic outward unit normal to Ω then

$$\nu(z(s)) = (-i)e^{i\theta(s)} \text{ at } \mathcal{L}^1\text{-a.e. } s \in [0, L]. \tag{6.1.7}$$

Hence, the geometric measure theoretic positively oriented unit tangent to Ω is

$$\tau(z(s)) = i\nu(z(s)) = e^{i\theta(s)} \text{ at } \mathcal{L}^1\text{-a.e. } s \in [0, L]. \tag{6.1.8}$$

In particular, $\theta(s)$ is the angle which the geometric measure theoretic positively oriented unit tangent to Ω makes with the real axis at the point $z(s)$, for $\mathcal{L}^1$-a.e. $s \in [0, L]$. Finally, (6.1.7) and the that θ is a function of bounded variation imply

$$\begin{aligned} &\text{the one-sided limits } \nu(z(s^\pm)) \text{ exist for each } s \in [0, L] \\ &\text{and differ for at most countably many values of } s \in [0, L]. \end{aligned} \tag{6.1.9}$$

In view of Definition 6.1.1, it is natural to introduce Radon domains as follows:

Definition 6.1.2 *Call* $\Omega \subseteq \mathbb{C}$ *a* `Radon domain` *if* Ω *is the bounded connected component of the complement of a curve of bounded rotation in the plane.*

This definition is meaningful, since if $\Sigma \subset \mathbb{C}$ is a curve of bounded rotation then Σ is a simple closed curve, so it makes sense to consider the bounded connected component of $\mathbb{C} \setminus \Sigma$. Hence, if Ω stands for said set, then Ω is a Radon domain and (6.1.6) shows that Ω is a set of finite perimeter which satisfies $\partial\Omega = \partial\big(\overline{\Omega}\big) = \Sigma$.

Example 6.1.3 *Any open bounded convex set in the plane is a Radon domain.*

Indeed, if $\Omega \subseteq \mathbb{C}$ is an open bounded convex set then the arc-length parametrization of $\partial\Omega$ is (6.1.4) where, for $\mathcal{L}^1$-a.e. s, the quantity $\theta(s)$ is the angle made by the positively oriented tangent line to $\partial\Omega$ at the point $z(s)$ with the x-axis. Then θ is a non-decreasing function (hence of bounded variation), and (6.1.2) is satisfied by virtue of the following simple geometric fact:

given two concentric balls $B_0 \subset B_1$ in the plane, the angle made by any two oriented lines intersecting B_1 and leaving B_0 to their left stays away from π (by an amount which may be estimated in terms of B_0 and B_1). (6.1.10)

Directly from definitions one can also see the following:

Example 6.1.4 *Any open bounded set in the plane with a piecewise smooth boundary without cusps is a Radon domain.*

Since, so far, all examples of Radon domains turned out to be Lipschitz, it would be worthwhile to explore the causes of such happenstance.

Proposition 6.1.5 *Any Radon domain is a bounded simply connected Lipschitz domain in the plane (hence, a* UR *domain).*

Proof Let $\Sigma := \partial\Omega$. Then $\Sigma \subset \mathbb{C}$ be a curve of bounded rotation, and Ω is the bounded connected component of $\mathbb{C} \setminus \Sigma$. Hence, Ω is a bounded nonempty open subset of $\mathbb{R}^2$ which, as noted earlier, satisfies (6.1.5)-(6.1.6). To show that Ω is a Lipschitz domain we may therefore employ [80, Theorem 2.3], according to which matters come down to the local existence of a continuous transverse vector field. Concretely, having fixed an arbitrary $s_0 \in [0, L]$, define the constant vector field

$$h(z(s)) := \frac{e^{i\theta(s_0^+)} + e^{i\theta(s_0^-)}}{2i} \quad \text{for } s \text{ near } s_0. \tag{6.1.11}$$

Then, using (6.1.7) and (6.1.11), for $\mathcal{L}^1$-a.e. s near s_0 we may compute

$$\begin{aligned} \big\langle \nu(z(s)), h(z(s)) \big\rangle &= \frac{e^{i\theta(s_0^+)} + e^{i\theta(s_0^-)}}{2i} = \operatorname{Re}\big(\nu(z(s))\overline{h(z(s))}\big) \\ &= \operatorname{Re}\Big[e^{i\theta(s)}\Big(\frac{e^{-i\theta(s_0^+)} + e^{-i\theta(s_0^-)}}{2}\Big)\Big] = \tfrac{1}{2}\Big\{\cos\big(\theta(s) - \theta(s_0^+)\big) + \cos\big(\theta(s) - \theta(s_0^-)\big)\Big\}. \end{aligned} \tag{6.1.12}$$

In particular, this allows us to compute the one-sided limits

$$\lim_{s \to s_0^\pm} \big\langle \nu(z(s)), h(z(s)) \big\rangle = \tfrac{1}{2}\Big\{1 + \cos\big(\theta(s_0^+) - \theta(s_0^-)\big)\Big\}. \tag{6.1.13}$$

This goes to show that the two-sided limit

$$\lim_{s \to s_0} \big\langle \nu(z(s)), h(z(s)) \big\rangle \text{ exists and equals } \tfrac{1}{2}\Big\{1 + \cos\big(\theta(s_0^+) - \theta(s_0^-)\big)\Big\}. \tag{6.1.14}$$

In turn, from (6.1.14), (6.1.2), and [129, (5.8.17)-(5.8.18)] we conclude that

$$\langle \nu, h \rangle \geq \frac{1 + \cos\Theta}{2} > 0 \text{ for } \mathcal{H}^1\text{-a.e. point on } \Sigma \text{ near } z(s_0). \tag{6.1.15}$$

Hence, h is transverse to $\partial\Omega$ near $z(s_0)$, as wanted. □

According to Proposition 6.1.5, Radon domains make up a subclass of the family of bounded simply connected Lipschitz domains in the plane. As our next result shows, this inclusion is strict.

Proposition 6.1.6 *There exists a bounded simply connected Lipschitz domain in the plane which is not a Radon domain.*

Proof Let $\varphi : \mathbb{R} \to \mathbb{R}$ be the Lipschitz function defined at each $t \in \mathbb{R}$ by

$$\varphi(t) := \begin{cases} 3|t| - \frac{1}{2^{2k-1}} & \text{if } k \in \mathbb{Z} \text{ is such that } \frac{1}{2^{2k+1}} \leq |t| \leq \frac{1}{2^{2k}}, \\ -3|t| + \frac{1}{2^{2k}} & \text{if } k \in \mathbb{Z} \text{ is such that } \frac{1}{2^{2k+2}} \leq |t| \leq \frac{1}{2^{2k+1}}. \end{cases} \tag{6.1.16}$$

As visible from this definition, the graph of φ is a zigzagged line of slopes ± 3, as in the picture below:

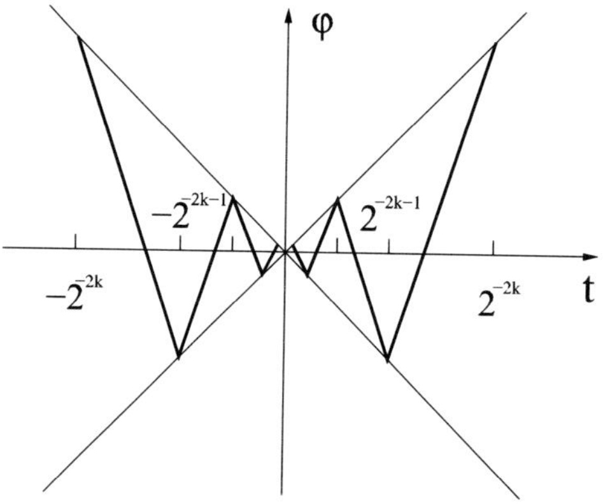

Suppose $\Omega \subseteq \mathbb{C}$ is a bounded simply connected Lipschitz domain with $0 \in \partial\Omega$ and such that Ω agrees with the upper-graph of the function φ near the origin. Let $s \mapsto z(s)$ be the arc-length parametrization of $\partial\Omega$ normalized so that $z(0) = 0$ and, for $\mathcal{L}^1$-a.e. s, denote by $\theta(s)$ the angle made by the positively oriented tangent line to $\partial\Omega$ at the point $z(s) \in \partial\Omega$ with the x-axis. Then $z'(s) = e^{i\theta(s)}$ for $\mathcal{L}^1$-a.e. s, so $z(s) = \int_0^s e^{i\theta(t)}\,dt$ for each s. As noted earlier, the "slope function" θ takes only the values $\arctan(1/3)$ and $\pi - \arctan(1/3)$, in an alternative fashion. Since the frequency at which θ changes between these two values increases indefinitely as s approaches the origin, we conclude that θ does not have bounded variation. As such, Ω fails to be a Radon domain. □

6.2 History of the Radon-Carleman Problem and Related Issues

To place matters in a proper, broader historical perspective, here we review the theory initiated by Radon and Carleman in its natural progression.

Consider the Dirichlet Problem for the Laplacian in a bounded UR domain Ω in the plane. In the classical boundary integral method one looks for solutions expressed as singular integral operators. The success of this enterprise hinges on the ability to solve a certain second kind integral equation involving the singular integral operator $\frac{1}{2}I + K_\Delta$ on the space of boundary data. Here, K_Δ is the harmonic double layer potential operator acting on each function $f \in L^1(\partial\Omega, \mathcal{H}^1)$ according to

$$K_\Delta f(x) := \lim_{\varepsilon\to 0^+} \frac{1}{2\pi} \int\limits_{y\in\partial\Omega,\ |x-y|>\varepsilon} \frac{\langle \nu(y), y-x\rangle}{|x-y|^2} f(y)\,\mathrm{d}\mathcal{H}^1(y) \ \text{ for } \mathcal{H}^1\text{-a.e. } x \in \partial\Omega. \tag{6.2.1}$$

Employing Fredholm theory then becomes a viable option when traveling down this road. The notions of Fredholm radius (recalled in (A.0.80)) and essential norm of an operator (recalled in (A.0.191)) have both been originally introduced in 1919 by J. Radon in [176] for this precise purpose. These have come to play a central role in the theory of boundary value problems via boundary integral methods.

To be specific, assume Ω to be a Radon domain in the plane (in the sense of Definition 6.1.2). That is, start with a curve $\Sigma \subset \mathbb{C}$ of bounded rotation (cf. Definition 6.1.1), and take Ω to be the bounded connected component of $\mathbb{C} \setminus \Sigma$. Also, recall the quantity $\Theta_{\partial\Omega} \in [0, \pi)$ associated with $\Sigma := \partial\Omega$ as in (6.1.2). In 1919 J. Radon proved in [177] that the harmonic double layer potential operator K_Δ is bounded on the space $\mathscr{C}^0(\partial\Omega)$ of continuous functions on $\partial\Omega$ (which becomes a Banach space when equipped with the supremum norm), with operator norm

$$\|K_\Delta\|_{\mathscr{C}^0(\partial\Omega)\to\mathscr{C}^0(\partial\Omega)} \leq \frac{1}{2\pi} \cdot \big(\text{total variation of } \theta\big) - \frac{1}{2}, \tag{6.2.2}$$

where θ is the function associated with $\Sigma := \partial\Omega$ as in Definition 6.1.1. In addition, Radon has succeeded in obtaining (cf. [177]) explicit formulas for the Fredholm radius and essential norm of K_Δ acting on $\mathscr{C}^0(\partial\Omega)$, namely

$$\rho_{\text{Fred}}\Big(K_\Delta : \mathscr{C}^0(\partial\Omega)\Big) = \|K_\Delta\|^{\text{ess}}_{\mathscr{C}^0(\partial\Omega)\to\mathscr{C}^0(\partial\Omega)} = \frac{1}{2\pi} \cdot \Theta_{\partial\Omega}. \tag{6.2.3}$$

These are remarkable in as much they link analytical objects to geometry. In particular, Radon's essential norm formula

$$\|K_\Delta\|^{\text{ess}}_{\mathscr{C}^0(\partial\Omega)\to\mathscr{C}^0(\partial\Omega)} = \frac{1}{2\pi} \cdot \Theta_{\partial\Omega} \tag{6.2.4}$$

shows that the singularities of the boundary have a concrete quantitative effect on how far K_Δ is from being a compact operator in this context.

Another profound aspect of Radon's results recorded in (6.2.3) is that these imply, upon recalling that $\Theta_{\partial\Omega} \in [0, \pi)$, that

$$\rho_{\text{Fred}}\Big(K_\Delta : \mathscr{C}^0(\partial\Omega)\Big) < \frac{1}{2}. \tag{6.2.5}$$

In particular, (6.2.5) shows that $\frac{1}{2}I + K_\Delta$ is a Fredholm operator with index zero on the space $\mathscr{C}^0(\partial\Omega)$. Radon then obtains the invertibility of this operator and goes on to solve classical Dirichlet Problem for the Laplacian in Ω with data in $\mathscr{C}^0(\partial\Omega)$:

$$\begin{cases} u \in \mathscr{C}^\infty(\Omega) \cap \mathscr{C}^0(\overline{\Omega}), & \Delta u = 0 \text{ in } \Omega, \\ u\big|_{\partial\Omega} = f \in \mathscr{C}^0(\partial\Omega). \end{cases} \tag{6.2.6}$$

Let us also note that, in concert, (6.2.3) and [130, (1.2.55)] imply that

$$K_\Delta \text{ is compact on } \mathscr{C}^0(\partial\Omega) \text{ if and only if } \Theta_{\partial\Omega} = 0. \tag{6.2.7}$$

In [101, p. 157], Král mentions an example (from [99]) of a bounded $\mathscr{C}^1$ domain Ω in the plane for which K_Δ is bounded on $\mathscr{C}^0(\partial\Omega)$ but it fails to be compact on this space (even though, $\Theta_{\partial\Omega} = 0$ in this case). Hence, Radon's essential norm formula (6.2.4) is no longer be expected to hold if the bounded rotation property for $\partial\Omega$ is simply dropped.

Radon has recognized the relevance of the notion of Fredholm radius as an abstract entity, and studied it as such. For example, given a linear and bounded operator T on a Banach space X, Radon has introduced the notion of continuity degree of T as follows. For each compact operator K on X, define $C_T(K)$ to be the radius of the largest disk centered at the origin in the z-plane in which the series $I + z(T-K) + z^2(T-K)^2 + \cdots$ converges uniformly. The continuity degree of T is defined as $C_T := \sup_K C_T(K)$, where the supremum runs over all compact operators K on X. It is then shown in [176] that

$$\rho_{\text{Fred}}(T; X) = \frac{1}{C_T}. \tag{6.2.8}$$

Radon's work has provided the foundation for an important branch of functional analysis to which many other people (such as S.M. Nikol'skii [169], F.V. Atkinson [12], I.Ts. Gohberg and M.G. Krein [72], to name just a few) have made remarkable contributions. For example, in [73] Gohberg and Marcus have shown that the Fredholm radius of a linear and bounded operator T on a Banach space $(X, \|\cdot\|)$ may be characterized as

$$\rho_{\text{Fred}}(T; X) = \inf\Big\{\|T\|^{\text{ess}}_{(X,\|\cdot\|')\to(X,\|\cdot\|')} : \|\cdot\|' \sim \|\cdot\|\Big\}, \tag{6.2.9}$$

where $\|\cdot\|' \sim \|\cdot\|$ indicates that said norms are topologically equivalent on X.

In 1916, hence three years before Radon's fundamental work in [176], [177], Torsten Carleman has proved in his Uppsala Ph.D. dissertation [31] a result which essentially amounts to a weighted version of (6.2.4). Concretely, assume Ω is a bounded domain in the plane whose boundary is piecewise smooth of class $\mathscr{C}^2$, i.e., $\partial\Omega$ is the union of a finite number N of closed $\mathscr{C}^2$ arcs L_j with end-points x_j and x_{j+1}, $1 \le j \le N$ (with $x_{N+1} = x_1$). For each $j \in \{1, \dots, N\}$, denote by α_j the internal angle (with respect to Ω) made by the curve $\partial\Omega$ at x_j (i.e., α_j is the angle made by the half-tangents to $\partial\Omega$ at the point x_j). Finally, given an N-tuple $m = (m_1, \dots, m_N) \in \mathbb{R}^N$, denote by $X_m(\partial\Omega)$ the space of continuous functions f defined on $\partial\Omega$ satisfying

$$\|f\|_{X_m(\partial\Omega)} := \max_{1\leq j\leq N} \sup_{x\in L_j} \left[|x-x_j|^{m_j}|f(x)|\right] < +\infty. \tag{6.2.10}$$

Carleman's result alluded to earlier, may be stated as the following estimate of the essential norm of K_Δ in the space $X_m(\partial\Omega)$:

$$\|K_\Delta\|^{\text{ess}}_{X_m(\partial\Omega)\to X_m(\partial\Omega)} \leq \max_{1\leq j\leq N} \left[\frac{\sin\left(m_j|\pi-\alpha_j|\right)}{2\sin(m_j\pi)}\right]. \tag{6.2.11}$$

It is worth mentioning that, in the unweighted case, i.e., when $m_1 = \cdots = m_N = 0$, Carleman's essential norm estimate (6.2.11) is in agreement with Radon's essential norm formula (6.2.4). Indeed,

$$\lim_{m\to 0}\max_{1\leq j\leq N} \left[\frac{\sin\left(m_j|\pi-\alpha_j|\right)}{2\sin(m_j\pi)}\right] = \max_{1\leq j\leq N}\left[\frac{|\pi-\alpha_j|}{2\pi}\right] = \frac{1}{2\pi}\cdot\Theta_{\partial\Omega}. \tag{6.2.12}$$

Carleman has also proved in [31] a version of the aforementioned result for a certain type of domains in the three-dimensional setting. Specifically, he has considered bounded open sets Ω with $\partial\Omega$ comprised of two surfaces S_1, S_2 of class $\mathscr{C}^2$ having a common curved edge Γ, itself assumed to be a curve of class $\mathscr{C}^2$ in $\mathbb{R}^3$. Carleman's result, dating back to 1916, in this setting is that the essential norm of the harmonic double layer potential operator K_Δ (defined as in (A.0.114) with $n = 3$) on the space $\mathscr{C}^0(\partial\Omega)$ of continuous functions on $\partial\Omega$ (equipped with the supremum norm) satisfies

$$\|K_\Delta\|^{\text{ess}}_{\mathscr{C}^0(\partial\Omega)\to\mathscr{C}^0(\partial\Omega)} \leq \sup_{x\in\Gamma}\left|1-\tfrac{\alpha(x)}{\pi}\right|, \tag{6.2.13}$$

where $\alpha(x)$ is the minimum of the two angles made by the tangent planes to S_1 and S_2 at the point $x\in\Gamma$.

At its core, the theory emerging from the work of Radon and Carleman is primarily motivated by the goal of solving boundary value problems in irregular domains. Early work on this subject (such as H. Poincaré's treaties [173], [174], O.D. Kellogg's monograph [90], F. Riesz and B. Sz.-Nagy's monograph [180], S.L. Sobolev's monograph [197]) only sufficiently smooth domains have been considered, but both the internal logic of the theory itself as well as the need for realistic modeling and applications require investigating non-smooth structures.

In their famous 1952 treatise (republished in [180]) on functional analysis, F. Riesz and B. Sz.-Nagy devote Section 81 to the Dirichlet and Neumann Problems in potential theory and their solutions by the Fredholm method. In relation this, they write (staring on p. 224): "*Despite the elegance of this method, it has the inconvenience of being applicable only under very restrictive hypotheses concerning the boundary of the domain considered. These restrictions are inherent in the method; in fact, since the double layers are made up of oriented dipoles, they can not be distributed over any but sufficiently regular curves (or surfaces).*

Nevertheless, we can somewhat weaken the hypotheses set down in the preceding section and thus extend the Fredholm method to domains of more general type. In the case of the plane, Radon has extended the generality of the method to what seems to us to be its natural limit. The domains considered by Radon are those bounded by curves of bounded rotation." At the very end of Section 91, F. Riesz and B. Sz.-Nagy note on page 226 of their book that "*in the case of the spatial problem an analogue of curves with bounded rotation has not yet been found.*"

While this has remained the prevalent opinion for almost 40 years after Radon's seminal work, the 1960's have seen a series of new remarkable developments in this area of mathematics. The novel point of view emerging from the work by Yu.D. Burago, V.G. Maz'ya, V.D. Sapozhnikova in [22], [23], [24], and J. Král in [100], [101], is that Radon's theory may be both refined (by allowing a larger class of curves) and extended to higher dimensions by replacing "bounded rotation" with suitable demands on the variation of the angle at which the subsets of the curve are seen from an arbitrary point on it.

In the aforementioned works, the basic results concerning the boundedness, Fredholmness, and invertibility of $\frac{1}{2}I + K_\Delta$ on $\mathscr{C}^0(\partial\Omega)$ have been obtained for domains $\Omega \subseteq \mathbb{R}^n$, with $n \geq 2$, subject to the following two conditions:

(A) $\sup_{x\in\partial\Omega} |\omega|\big(x; \partial\Omega \setminus \{x\}\big) < \infty$,

(B) $\lim\limits_{r\to 0^+} \sup_{x\in\partial\Omega} |\omega|\big(x; \partial\Omega \cap B(x,r)\big) < \omega_{n-1}/2$.

Here $\omega(x; E)$ is the solid angle at which a set $E \subseteq \partial\Omega$ is seen from the point $x \in \partial\Omega$, or rather its generalization to a certain set function (i.e., a charge; cf. [102, p. 154], [119, p. 186] for a precise definition), and $|\omega|$ denotes the total variation of said charge. For example, (A)-(B) hold for any bounded $\mathscr{C}^1$ domain $\Omega \subseteq \mathbb{R}^n$ with the property that the modulus of continuity θ of the outward unit normal ν to Ω, defined as

$$\theta(t) := \sup_{\substack{x,y\in\partial\Omega \\ |x-y|<t}} |\nu(x) - \nu(y)| \text{ for each } t \in \big(0, \operatorname{diam}\partial\Omega\big), \tag{6.2.14}$$

satisfies Dini's integrability condition

$$\int_0^{\operatorname{diam}\partial\Omega} \frac{\theta(t)}{t}\,\mathrm{d}t < +\infty. \tag{6.2.15}$$

This being said, there exist bounded $\mathscr{C}^1$ domains $\Omega \subseteq \mathbb{R}^n$ (with $n \geq 3$) for which condition (A) fails. See the discussion in [119, p. 185] in this regard.

Condition (A) ensures the fact that K_Δ is well-defined and bounded on $\mathscr{C}^0(\partial\Omega)$ (cf., e.g., [100, Theorem 2.15 on p. 528, and Lemma 3.4 on p. 529]; see also [119, Theorem 8, p. 187]). Significantly, J. Radon has already noted in [177] the validity of condition (A) for what we presently call Radon domains in the plane. In particular, K_Δ is well-defined and bounded on $\mathscr{C}^0(\partial\Omega)$ whenever Ω is a Radon domain. By way of contrast, condition (B) may fail to materialize in the class of Radon domains. Indeed, examples of $\mathscr{C}^1$ planar domains satisfying conditions (A)-(B) which are not Radon domains are given in [119, pp. 184–185]. The role of condition (B) is solely to guarantee Fredholmness of the operator $\frac{1}{2}I + K_\Delta$ on $\mathscr{C}^0(\partial\Omega)$. Specifically, Král

[100] and Burago and Maz'ya [22] have independently proved the following version of Radon's essential norm formula (6.2.4):

$$\|K_\Delta\|^{\mathrm{ess}}_{\mathscr{C}^0(\partial\Omega)\to\mathscr{C}^0(\partial\Omega)} = \frac{1}{\omega_{n-1}} \lim_{r\to 0^+} \sup_{x\in\partial\Omega} |\omega|\big(x;\partial\Omega\cap B(x,r)\big), \tag{6.2.16}$$

in the class of domains $\Omega \subseteq \mathbb{R}^n$, with $n \geq 2$, satisfying condition (A) (see also [119, Theorem 10, p. 188]). Hence, in such a setting, having condition (B) also fulfilled becomes equivalent to

$$\|K_\Delta\|^{\mathrm{ess}}_{\mathscr{C}^0(\partial\Omega)\to\mathscr{C}^0(\partial\Omega)} < \frac{1}{2}. \tag{6.2.17}$$

In concert with [130, (2.2.16)], this then entails

$$\rho_{\mathrm{Fred}}\Big(K_\Delta : \mathscr{C}^0(\partial\Omega)\Big) < \frac{1}{2} \tag{6.2.18}$$

which, as before, opens the door for using Fredholm theory in this setting. For more specialized geometries, formula (6.2.16) adjusts accordingly. For example, I. Netuka has proved in [167, Theorem, p. 376] that

$$\begin{gathered}\|K_\Delta\|^{\mathrm{ess}}_{\mathscr{C}^0(\partial\Omega)\to\mathscr{C}^0(\partial\Omega)} = \sup_{x\in\partial\Omega}\big(\tfrac{1}{2} - d(x)\big) < \tfrac{1}{2},\\ \text{whenever } \Omega \text{ is an open bounded convex subset of } \mathbb{R}^n,\end{gathered} \tag{6.2.19}$$

where $d(x)$ denotes the volume density of the set Ω at the point x. In relation to this body of results we also wish to remark that Problem 16 in [118] asks to "*Find a geometrical representation for the Fredholm radius of the harmonic double layer potential as an operator in* $\mathscr{C}^0(\partial\Omega)$."

References to other related results, including the case of the Lamé and Stokes systems, may be found in [102, Chapter 7]; see also [123]. As an example, consider the case of the Lamé system, that is, $L_{\lambda,\mu} = \mu\Delta + (\lambda+\mu)\nabla\mathrm{div}$ with moduli $\lambda, \mu \in \mathbb{R}$ satisfying $\mu > 0$ and $2\mu + \lambda > 0$ (which, according to [131, (1.3.11)], ensures that $L_{\lambda,\mu}$ satisfies the Legendre-Hadamard strongly ellipticity condition). Given a UR domain Ω, denote by K_ψ the (pseudo-traction) elastic double layer potential operator associated as in (A.0.116) with Ω and the coefficient tensor [132, (1.4.169)] (used to represent the Lamé system as in [132, (1.4.170)]) corresponding to the choice

$$t := \frac{\lambda+\mu}{\lambda+3\mu}. \tag{6.2.20}$$

For $n = 2$ and Ω a planar polygon with interior angles $\alpha_1, \dots, \alpha_N$, Kresin and Maz'ya have deduced the following formula:

$$\|K_\psi\|^{\mathrm{ess}}_{[\mathscr{C}^0(\partial\Omega)]^2\to[\mathscr{C}^0(\partial\Omega)]^2} = \frac{2(1+t)}{\pi} E\Big(\frac{\pi-\alpha_{\min}}{2}, \frac{2\sqrt{t}}{1+t}\Big) \tag{6.2.21}$$

where t is as in (6.2.20), E denotes the elliptic integral of the second kind, and

$$\alpha_{\min} := \min_{1\leq j\leq N} \min\{\alpha_j, 2\pi - \alpha_j\}. \tag{6.2.22}$$

Also, in the same setting, V. Shelepov [190] has identified the Fredholm radius of K_ψ on the space $\big[\mathscr{C}^0(\partial\Omega)\big]^2$ as being

$$\rho_{\text{Fred}}\Big(K_\psi;\, \big[\mathscr{C}^0(\partial\Omega)\big]^2\Big) = \frac{|\pi - \alpha_{\min}| + t\sin\big(|\pi - \alpha_{\min}|\big)}{\pi} \tag{6.2.23}$$

where t is as (6.2.20) and the convention made in (6.2.22) remains in effect. These two formulas imply the (somewhat unexpected) inequality (compare with [130, (2.2.16)] and (6.2.3))

$$\rho_{\text{Fred}}\Big(K_\psi;\, \big[\mathscr{C}^0(\partial\Omega)\big]^2\Big) < \big\|K_\psi\big\|^{\text{ess}}_{[\mathscr{C}^0(\partial\Omega)]^2\to[\mathscr{C}^0(\partial\Omega)]^2} \text{ whenever } t > 0. \tag{6.2.24}$$

The considerations so far have been limited to the space of continuous functions, though the very same issues, are meaningful on a variety of other function spaces. Prominently among these, the scale of Lebesgue spaces is worthy of special attention, particularly since, as noted earlier, all double layer potential operators associated with a UR domain and a weakly elliptic system are bounded on L^p with $p \in (1,\infty)$. In this regard, in 1969 V. Shelepov has proved in [188] (see also [189, Theorem 6, p. 192]) that, given any Radon domain $\Omega \subseteq \mathbb{R}^2$, if $\sigma := \mathcal{H}^1\lfloor\partial\Omega$ and $p \in (1,\infty)$, then the following spectral radius formula holds for the principal-value harmonic double layer potential operator K_Δ associated with Ω (as in (6.2.1)):

$$\rho_{\text{Fred}}\Big(K_\Delta;\, L^p(\partial\Omega,\sigma)\Big) = \max_{j\geq 1}\left[\frac{\sin\big(|\pi - \alpha_j|/p\big)}{2\sin(\pi/p)}\right]. \tag{6.2.25}$$

Above, if $\{x_j\}_{j\geq 1}$ are the discontinuity points for the outward unit normal ν to Ω (see Definition 6.1.1 and (6.1.7)), then for each $j \geq 1$ we have denoted by α_j the internal angle (with respect to Ω) made by the curve $\partial\Omega$ at x_j (i.e., α_j is the angle made by the half-tangents to $\partial\Omega$ at the point x_j). In particular, (6.2.25) gives that in the aforementioned setting

$$\rho_{\text{Fred}}\Big(K_\Delta;\, L^p(\partial\Omega,\sigma)\Big) < \frac{1}{2} \text{ whenever } 2 \leq p < \infty, \tag{6.2.26}$$

so $\frac{1}{2}I + K_\Delta$ is a Fredholm operator with index zero on $L^p(\partial\Omega,\sigma)$ with $p \in [2,\infty)$. More functional analytic information regarding the latter operator is provided by [189, Theorem 5, p. 192], which implies[1] the following result:

Proposition 6.2.1 *Assume $\Omega \subseteq \mathbb{R}^2$ is a bounded, simply connected, curvilinear polygon with internal angles $\theta_j \in (0,2\pi)$, labeled by $j \in \{1,\dots,N\}$. Abbreviate $\sigma := \mathcal{H}^1\lfloor\partial\Omega$ and denote by K_Δ the principal-value harmonic double layer potential operator associated with Ω_θ as in* (A.0.114)*. Then for each $p \in (1,\infty)$ the operator*

[1] [189, Theorem 5, p. 192] is stated for general Radon curves

$$\begin{gathered}\tfrac{1}{2}I + K_\Delta : L^p(\partial\Omega, \sigma) \longrightarrow L^p(\partial\Omega, \sigma) \text{ is Fredholm} \\ \text{if and only if } p \neq 1 + |\pi - \theta_j|/\pi \text{ for each } j \in \{1, \dots, N\}.\end{gathered} \tag{6.2.27}$$

If $p \in (1, \infty)$ *and* $p \neq 1 + |\pi - \theta_j|/\pi$ *for each* $j \in \{1, \dots, N\}$ *then*

$$\tfrac{1}{2}I + K_\Delta : L^p(\partial\Omega, \sigma) \to L^p(\partial\Omega, \sigma) \text{ is surjective} \tag{6.2.28}$$

and

$$\dim \operatorname{Ker}\Big(\tfrac{1}{2}I + K_\Delta; L^p(\partial\Omega, \sigma)\Big) = \sum_{j=1}^{N} \frac{1}{2}\Big\{1 - \operatorname{sgn}\big[p - (1 + |\pi - \theta_j|/\pi)\big]\Big\}. \tag{6.2.29}$$

Finally, if $p = 1 + |\pi - \theta_j|/\pi$ *for some* $j \in \{1, \dots, N\}$*, then* $\frac{1}{2}I + K_\Delta$ *has a finite dimensional kernel and a dense range on* $L^p(\partial\Omega, \sigma)$*, but is not onto.*

Also, formula (6.2.25) is part of a more explicit description of the spectrum of the classical harmonic double layer potential operator on curvilinear polygons, from [188]. To state it, recall the ("bow" shaped) closed contour described in (4.2.196).

Proposition 6.2.2 *Let* $\Omega \subseteq \mathbb{R}^2$ *be a bounded, simply connected, curvilinear polygon with internal angles* $\theta_j \in (0, 2\pi)$*,* $j \in \{1, \dots, N\}$*. Abbreviate* $\sigma := \mathcal{H}^1 \lfloor \partial\Omega$ *and fix an integrability exponent* $p \in (1, \infty)$*. For each* $j \in \{1, \dots, N\}$*, consider the bow-shaped curve* $\Sigma_{\theta_j}(p)$ *associated with the given* p *and the angle* θ_j *as in* (4.2.196)*. Also, denote by* $\widehat{\Sigma_{\theta_j}(p)}$ *the (closed) two-dimensional region encompassed by* $\Sigma_{\theta_j}(p)$*. Let* K_Δ *stand for the principal-value harmonic double layer potential operator associated with* Ω_θ *as in* (A.0.114)*. Then* $\operatorname{Spec}(K_\Delta; L^p(\partial\Omega, \sigma))$*, the spectrum of* K_Δ *acting on* $L^p(\partial\Omega, \sigma)$*, has the following structure:*

$$\operatorname{Spec}\Big(K_\Delta; L^p(\partial\Omega, \sigma)\Big) = \Bigg(\bigcup_{1 \leq j \leq N} \widehat{\Sigma_\theta(p)}\Bigg) \bigcup \{\lambda_k\}_k, \tag{6.2.30}$$

where $\{\lambda_k\}_k$ *is a finite subset of* $(-1/2, 1/2]$ *consisting of eigenvalues of* K_Δ *on* $L^p(\partial\Omega, \sigma)$*. In fact,* $1/2$ *is always an eigenvalue (corresponding to a constant eigenfunction; cf.* [132, (1.5.21)]*), and* [132, Corollary 1.6.9] *shows that the remaining (finitely many) eigenvalues are symmetrically displayed with respect to the origin.*

Moreover, if $z \in \bigcup_{j=1}^{N} \Sigma_{\theta_j}(p)$ *then the operator* $zI - K_\Delta$ *is not Fredholm on* $L^p(\partial\Omega, \sigma)$*, whereas if* $z \in \mathbb{C} \setminus \Big(\bigcup_{j=1}^{N} \Sigma_{\theta_j}(p)\Big)$ *then the operator* $zI - K_\Delta$ *is Fredholm on* $L^p(\partial\Omega, \sigma)$ *and its index is given by*

$$\operatorname{index}\Big(zI - K_\Delta; L^p(\partial\Omega, \sigma)\Big) = \sum_{j=1}^{N} W\big(z, \Sigma_{\theta_j}(p)\big), \tag{6.2.31}$$

where $W\big(z, \Sigma_{\theta_j}(p)\big)$ *is the winding number of* z *with respect to the curve* $\Sigma_{\theta_j}(p)$*.*

Subsequently, I. Mitrea has extended Shelepov's result to other partial differential operators. For example, work in [145] gives that, in the same geometric setting as above and with t as in (6.2.20), the Fredholm radius of K_ψ on $\big[L^p(\partial\Omega,\sigma)\big]^2$ with $2 \le p < \infty$ is given by

$$\rho_{\text{Fred}}\Big(K_\psi;\ \big[L^p(\partial\Omega,\sigma)\big]^2\Big) \tag{6.2.32}$$
$$= \max_{j\ge 1} \frac{\left|\frac{t}{p}\sin\alpha_j\cos\left(\frac{\pi-\alpha_j}{p}\right)+\sin\left(\frac{\pi-\alpha_j}{p}\right)\sqrt{1-t^2\,\frac{\sin^2\alpha_j}{p^2}}\,\right|}{2\sin\left(\frac{\pi}{p}\right)}$$

and satisfies

$$\rho_{\text{Fred}}\Big(K_\psi;\ \big[L^p(\partial\Omega,\sigma)\big]^2\Big) < \frac{1}{2}. \tag{6.2.33}$$

See also [144] and [146] for related results.

An open problem closely related to the Radon-Carleman Problem formulated in (6.0.1) is the Spectral Radius Conjecture. To state it, let Ω be a bounded connected Lipschitz domain in $\mathbb{R}^n$, whose complement is connected. Abbreviate $\sigma := \mathcal{H}^{n-1}\lfloor\partial\Omega$ and define

$$L^p_0(\partial\Omega,\sigma) := \Big\{f \in L^p(\partial\Omega,\sigma) : \int_{\partial\Omega} f\,d\sigma = 0\Big\} \text{ for each } p \in (1,\infty). \tag{6.2.34}$$

Also, recall the transpose harmonic double layer potential operator $K^\#_\Delta$ from (A.0.115), and the definition of the spectral radius from (A.0.227). Then the Spectral Radius Conjecture asks whether there exists $\varepsilon = \varepsilon(\Omega) > 0$ such that

$$\rho_{\text{inv}}\Big(K^\#_\Delta;\ L^p_0(\partial\Omega,\sigma)\Big) < \tfrac{1}{2} \text{ for } p \in (1, 2+\varepsilon). \tag{6.2.35}$$

This has been singled out by C. Kenig as an open problem in [93, p. 119], and also by G. Verchota in [41]. Its significance stems from the realization that whenever the estimate in (6.2.35) is true for some $p \in (1,\infty)$ then one may invert the operator $-\frac{1}{2}I + K^\#_\Delta$ on the space $L^p_0(\partial\Omega,\sigma)$ via a Neumann series, namely

$$\big(-\tfrac{1}{2}I + K^\#_\Delta\big)^{-1} = -2\sum_{j=0}^{\infty}(2K^\#_\Delta)^j, \tag{6.2.36}$$

with uniform convergence in $\text{Bd}\big(L^p_0(\partial\Omega,\sigma)\big)$. A number of special cases and partial results are known in this regard. It was Carl Gottfried Neumann who, in 1861, first succeeded in establishing that the above series converges[2] when applied to functions in $L^2(\partial\Omega,\sigma)$ under the assumption that the Ω is bounded open convex set in the plane. What is special about $K^\#_\Delta$ associated with a convex domain is the fact that its integral kernel is non-negative (as visible from (A.0.115)), a feature fully

[2] the series (6.2.36) has eventually become known as a Neumann series

exploited in C. Neumann's approach (presented in [168]). Subsequently, it became a priority to clarify whether the convexity assumption is really necessary in order to arrive at the same conclusion. In a genuine tour de force, H. Poincaré succeeded in 1897 (in his long and technical paper [173]) to replace the convexity hypothesis by suitable smoothness assumptions. More specifically, Poincaré accomplished the task of finding an alternative proof of the convergence of the Neumann series for nonconvex domains, via an approach which required the domain Ω to have a $\mathscr{C}^2$ boundary and the consideration of a function space involving a fair amount of regularity. Informative historical accounts may be found in [10] and [36].

The fact that the eigenvalues of $K^{\#}_{\Delta}$ on $L^2_0(\partial\Omega, \sigma)$ are real and belong to the interval $\big(-\frac{1}{2}, \frac{1}{2}\big)$ has been established by J. Plemelj in 1911 and O. Kellogg in 1929, initially in smooth domains but via an approach which works for more inclusive classes of sets. A general spectral result of this type appears in [136, Proposition 5.9, p. 194], where D. Mitrea, I. Mitrea, M. Mitrea, and M. Taylor have dealt with uniformly rectifiable subdomains of Riemannian manifolds. Earlier, this result has been noted in the class of Lipschitz domains by E. Fabes, L. Escauriaza, and G. Verchota in [47] (for special geometries, see [7]). Here we shall establish a similar result in the class of bounded two-sided NTA domains with Ahlfors regular boundaries in Proposition 7.2.13. E. Fabes, S. Sand, and J. Seo have proved in [58] the spectral radius estimate conjectured in (6.2.35) when $p = 2$ and Ω is actually convex. Subsequently, L. Escauriaza and M. Mitrea proved in [48] the spectral radius conjecture when Ω is convex, for the full range $p \in (1, 2 + \varepsilon)$. The range of p's in (6.2.35) comes from the work of B. Dahlberg and C. Kenig [38] who have shown that for each bounded Lipschitz domain $\Omega \subseteq \mathbb{R}^n$ there exists some small $\varepsilon = \varepsilon(\Omega) > 0$ with the property that $\frac{1}{2}I + K^{\#}_{\Delta}$ is invertible $L^p(\partial\Omega, \sigma)$ whenever $1 < p < 2 + \varepsilon$. This refines earlier work of G. Verchota who established in [203] this result in the more restrictive range $2 - \varepsilon < p < 2 + \varepsilon$. Let us also note that, when Ω is a Lipschitz polyhedron in $\mathbb{R}^3$, J. Elschner proved in [46] that the spectral radius estimate in (6.2.35) holds for $p = 2$, and A. Rathsfeld showed in [178] that the spectral radius of K_{Δ} on the space of continuous functions $\mathscr{C}^0(\partial\Omega)$ is $< \frac{1}{2}$. Lastly, we wish to point out that our Theorem 7.2.14 implies

$$\rho_{\text{inv}}\Big(K_{\Delta};\ B^{2,2}_{-1/2}(\partial\Omega, \sigma)/\mathbb{C}\Big) < \tfrac{1}{2} \tag{6.2.37}$$

assuming only that Ω is a bounded, connected, two-sided NTA domain with connected complement and Ahlfors regular boundary.

6.3 The Radon-Carleman Problem in UR Domains

Recall that V. Shelepov's Fredholm radius result in the case of the Laplacian recorded in (6.2.25) is formulated in the context of a Radon domain $\Omega \subseteq \mathbb{R}^2$, whose outward unit normal is denoted by ν and for which we have abbreviated $\sigma := \mathcal{H}^1 \lfloor \partial\Omega$. As in the past, we let L denote the length of $\partial\Omega$, let $\{x_j\}_{j\geq 1} \subseteq \partial\Omega$ stand for the discontinuity points of ν and, for each $j \geq 1$, denote by α_j the internal angle (with respect to Ω) made by the curve $\partial\Omega$ at x_j. That is, each α_j is the angle made by the half-tangents to $\partial\Omega$ at the point x_j, hence

$$|\pi - \alpha_j| = |\theta(s_j^+) - \theta(s_j^-)| \text{ where } s_j \in [0, L] \text{ is such that } \theta(s_j) = x_j. \tag{6.3.1}$$

Fix $p \in (1, \infty)$. Since $[0, \pi] \ni x \mapsto \phi(x) := \sin x/x$ is a decreasing function with $\phi(0) = 1$, we deduce from (6.3.1) and (6.1.2) that

$$\frac{\sin\left(\Theta_{\partial\Omega}/p\right)}{\Theta_{\partial\Omega}/p} \leq \frac{\sin\left(|\pi - \alpha_j|/p\right)}{|\pi - \alpha_j|/p} \leq 1 \text{ for each } j \geq 1. \tag{6.3.2}$$

In turn, from (6.3.2) and Shelepov's Fredholm radius formula (6.2.25) we conclude

$$\frac{\sin\left(\Theta_{\partial\Omega}/p\right)}{2\sin(\pi/p)} \leq \rho_{\text{Fred}}\Big(K_\Delta;\ L^p(\partial\Omega, \sigma)\Big) \leq \frac{\Theta_{\partial\Omega}/p}{2\sin(\pi/p)}. \tag{6.3.3}$$

Ultimately, from this we learn that the Fredholm radius of K_Δ on $L^p(\partial\Omega, \sigma)$ exhibits the behavior (with proportionality constants depending only on p)

$$\rho_{\text{Fred}}\Big(K_\Delta;\ L^p(\partial\Omega, \sigma)\Big) \approx \Theta_{\partial\Omega}. \tag{6.3.4}$$

In a very similar fashion, from I. Mitrea's Fredholm radius result in the case of the Lamé system recalled in (6.2.32) and (6.2.20) we see that for each $p \in [2, \infty)$

$$\rho_{\text{Fred}}\Big(K_\psi;\ \big[L^p(\partial\Omega, \sigma)\big]^2\Big) \approx \Theta_{\partial\Omega}, \tag{6.3.5}$$

where the proportionality constants now depend only on p, λ, and μ.

In search of appropriate analogues of (6.3.4) and (6.3.5) in the higher dimensional setting, we first first seek to describe the size of the entity $\Theta_{\partial\Omega}$ (originally defined as in (6.1.2) with $\Sigma := \partial\Omega$) in terms of other geometrical characteristics of Ω which have more immediate, natural counterparts, in higher dimensions. This is accomplished in the proposition below, where the distance from ν to the Sarason space VMO is our proposed alternative.

Proposition 6.3.1 *Let $\Omega \subseteq \mathbb{C}$ be a Radon domain (in particular, Ω is a Lipschitz domain, by Proposition 6.1.5). Denote by ν the outward unit normal vector to Ω and abbreviate $\sigma := \mathcal{H}^1\lfloor\partial\Omega$. Finally, let $\theta : [0, L] \to \mathbb{R}$ be the function associated with $\Sigma := \partial\Omega$ as in (6.1.1)-(6.1.4). Then*

$$\Theta_{\partial\Omega} = \sup_{0 \leq s \leq L} |\theta(s^+) - \theta(s^-)| \approx \text{dist}\Big(\nu, [\text{VMO}(\partial\Omega, \sigma)]^2\Big) \tag{6.3.6}$$

where the distance is measured in $\big[\text{BMO}(\partial\Omega, \sigma)\big]^2$ and the implicit proportionality constants depend only on the Ahlfors regularity constants of $\partial\Omega$.

Proof To set the stage, we note an elementary double inequality to the effect that

$$\frac{2}{\pi}\cdot \mathrm{dist}\big(\alpha, 2\pi\mathbb{Z}\big) \leq \big|e^{i\alpha}-1\big| \leq |\alpha| \text{ for each } \alpha\in\mathbb{R}. \tag{6.3.7}$$

Let us also recall from [130, (3.1.53)] that

$$\begin{aligned}\mathrm{dist}\Big(\nu,\ \big[\mathrm{VMO}(\partial\Omega,\sigma)\big]^2\Big) &\approx \limsup_{r\to 0^+}\left\{\sup_{x\in\partial\Omega} \fint_{\Delta(x,r)}\fint_{\Delta(x,r)} |\nu(y)-\nu(z)|\,d\sigma(y)\,d\sigma(z)\right\}\\ &\approx \limsup_{r\to 0^+}\left\{\sup_{x\in\partial\Omega}\fint_{\Delta}\Big|\nu-\fint_{\Delta}\nu\,d\sigma\Big|\,d\sigma\right\}\\ &\approx \lim_{R\to 0^+}\left(\sup_{x\in\partial\Omega,\ r\in(0,R)}\Big(\fint_{\Delta(x,r)}\Big|\nu-\fint_{\Delta(x,r)}\nu\,d\sigma\Big|\,d\sigma\Big)\right),\end{aligned} \tag{6.3.8}$$

where the implicit proportionality constants depend only on the Ahlfors regularity constants of $\partial\Omega$. Without loss of generality assume $0\in\partial\Omega$. Denote by L the length of $\partial\Omega$ and suppose the arc-length parametrization $z:[0,L]\to\partial\Omega$ satisfies $z(0)=0$.

We divide the proof into five steps (with the right-pointing inequality in (6.3.6) proved in Step III, and the left-pointing inequality in (6.3.6) proved in Step V).

`Step I:` *For each $s\in[0,L]$ we have (cf. (6.1.9))*

$$\tfrac{2}{\pi}\big|\theta(s^+)-\theta(s^-)\big| \leq \big|\nu\big(z(s^+)\big)-\nu\big(z(s^-)\big)\big| \leq \big|\theta(s^+)-\theta(s^-)\big|. \tag{6.3.9}$$

To justify this, for each given $s\in[0,L]$ use (6.1.7) to write

$$\big|\nu\big(z(s^+)\big)-\nu\big(z(s^-)\big)\big| = \big|e^{i\theta(s^+)}-e^{i\theta(s^-)}\big| = \big|e^{i(\theta(s^+)-\theta(s^-))}-1\big|. \tag{6.3.10}$$

In view of the second inequality in (6.3.7), the last inequality in (6.3.9) follows. From (6.3.10), the first inequality in (6.3.7), and the fact that $\theta(s^+)-\theta(s^-)$ belongs to the interval $(-\pi,\pi)$ (cf. (6.1.2)) we also see that

$$\big|\nu\big(z(s^+)\big)-\nu\big(z(s^-)\big)\big| \geq \tfrac{2}{\pi}\cdot\mathrm{dist}\big(\theta(s^+)-\theta(s^-), 2\pi\mathbb{Z}\big) = \tfrac{2}{\pi}\big|\theta(s^+)-\theta(s^-)\big|, \tag{6.3.11}$$

proving the first inequality in (6.3.9).

`Step II:` *If for each $x_0\in\partial\Omega\setminus\{0\}$ we set (cf. (6.1.9))*

$$\nu(x_0^\pm) := \lim_{s\to s_0^\pm}\nu\big(z(s)\big) \text{ if } s_0\in(0,L) \text{ is such that } z(s_0)=x_0, \tag{6.3.12}$$

$$\nu(L^+) := \nu(0^+) = \lim_{s\to 0^+}\nu\big(z(s)\big), \quad \nu(0^-) := \nu(L^-) = \lim_{s\to L^-}\nu\big(z(s)\big) \tag{6.3.13}$$

then

$$\lim_{R\to 0^+}\Big(\sup_{\substack{x\in\partial\Omega\\ y,z\in\Delta(x,R)}}\big|\nu(y)-\nu(z)\big|\Big) = \sup_{x\in\partial\Omega}\big|\nu(x^+)-\nu(x^-)\big|. \tag{6.3.14}$$

To prove the left-pointing inequality in (6.3.14) we reason by contradiction. Suppose said inequality fails. Then there exists $\varepsilon>0$ with the property that

$$\lim_{R\to 0^+}\Big(\sup_{\substack{x\in\partial\Omega\\ y,z\in\Delta(x,R)}}\big|\nu(y)-\nu(z)\big|\Big) > \sup_{x\in\partial\Omega}\big|\nu(x^+)-\nu(x^-)\big|+\varepsilon. \tag{6.3.15}$$

In turn, this implies that there exists some $R_* > 0$ with the property that for each $R\in(0,R_*)$ one may find $x_R\in\partial\Omega$ along with $y_R, z_R\in\Delta(x_R,R)$ such that

$$\big|\nu(y_R)-\nu(z_R)\big| > \sup_{x\in\partial\Omega}\big|\nu(x^+)-\nu(x^-)\big|+\varepsilon/2. \tag{6.3.16}$$

Specialize this to the case when $R=2^{-j}$ with $j\in\mathbb{N}$ sufficiently large and, for each such j, abbreviate $x_j := x_{2^{-j}}$, $y_j := y_{2^{-j}}$, $z_j := z_{2^{-j}}$. Since $\partial\Omega$ is compact, Bolzano-Weierstrass' theorem guarantees that after eventually passing to subsequences, there is no loss of generality in assuming that $\{x_j\}_j$ converges to some point $x_0\in\partial\Omega$. Given that $y_j, z_j\in\Delta(x_j,2^{-j})$ for each sufficiently large j, this also forces $\{y_j\}_j$ and $\{z_j\}_j$ to converge to x_0. For each large j, pick $s_j, t_j\in[0,L]$ such that $y_j=z(s_j)$ and $z_j=z(t_j)$. Also, consider $s_0\in[0,L]$ such that $x_0=z(s_0)$. Then (6.3.16) yields

$$\big|\nu(y_j)-\nu(z_j)\big| > \big|\nu(x_0^+)-\nu(x_0^-)\big|+\varepsilon/2 \text{ for each } j \text{ large}. \tag{6.3.17}$$

Suppose next that both $\{s_j\}_j$ and $\{t_j\}_j$ have subsequences either entirely located to the left of s_0, or entirely located to the right of s_0. By restricting attention to just these and after passing to limit in (6.3.17) we arrive at the conclusion that $0=\lim_{j\to\infty}\big|\nu(y_j)-\nu(z_j)\big|\geq\big|\nu(x_0^+)-\nu(x_0^-)\big|+\varepsilon/2$, a contradiction. In the remaining case, when one of the sequences $\{s_j\}_j$, $\{t_j\}_j$ has a subsequence entirely located to the left of s_0 while the other sequence has a subsequence entirely located to the right of s_0, passing to limit in (6.3.17) gives (keeping in mind (6.1.9))

$$\big|\nu(x_0^+)-\nu(x_0^-)\big| = \lim_{j\to\infty}\big|\nu(y_j)-\nu(z_j)\big| \geq \big|\nu(x_0^+)-\nu(x_0^-)\big|+\varepsilon/2, \tag{6.3.18}$$

again a contradiction. This establishes the left-pointing inequality in (6.3.14).

To justify the right-pointing inequality in (6.3.14), fix a point $x_0\in\partial\Omega\setminus\{0\}$ and consider an arbitrary sequence of positive numbers $\{R_j\}_j$ convergent to 0. Pick $s_0\in(0,L)$ such that $z(s_0)=x_0$ and, for each j, pick $s_j\in(s_0,L)$ sufficiently close to s_0 such that $y_j:=z(s_j)\in\Delta(x_0,R_j)$ along with $t_j\in(0,s_0)$ sufficiently close to s_0 such that $z_j:=z(t_j)\in\Delta(x_0,R_j)$. Then

$$\begin{aligned}
\lim_{R\to 0^+}\Big(\sup_{\substack{x\in\partial\Omega\\ y,z\in\Delta(x,R)}}\big|\nu(y)-\nu(z)\big|\Big) &= \lim_{j\to\infty}\Big(\sup_{\substack{x\in\partial\Omega\\ y,z\in\Delta(x,R_j)}}\big|\nu(y)-\nu(z)\big|\Big) \qquad (6.3.19)\\
&\geq \liminf_{j\to\infty}\Big(\sup_{y,z\in\Delta(x_0,R_j)}\big|\nu(y)-\nu(z)\big|\Big) \geq \lim_{j\to\infty}\big|\nu(y_j)-\nu(z_j)\big| = \big|\nu(x_0^+)-\nu(x_0^-)\big|.
\end{aligned}$$

The same type of argument works for $x_0=0$. Taking the supremum of the most extreme sides in (6.3.19) over $x_0\in\partial\Omega$ proves the left-pointing inequality in (6.3.14).

`Step III:` *There exists some $C \in (0, \infty)$ which depends only on the Ahlfors regularity constants of $\partial\Omega$ such that*

$$\operatorname{dist}\Big(\nu,\ \big[\mathrm{VMO}(\partial\Omega, \sigma)\big]^2\Big) \leq C \cdot \sup_{0 \leq s \leq L} |\theta(s^+) - \theta(s^-)|. \tag{6.3.20}$$

To prove this, based on (6.3.14) we estimate

$$\begin{aligned}
&\lim_{R\to 0^+}\left(\sup_{x\in\partial\Omega,\, r\in(0,R)}\left(\fint_{\Delta(x,r)}\left|\nu - \fint_{\Delta(x,r)} \nu\, d\sigma\right| d\sigma\right)\right) \qquad (6.3.21)\\
&\leq \lim_{R\to 0^+}\left(\sup_{x\in\partial\Omega,\, r\in(0,R)} \fint_{\Delta(x,r)}\fint_{\Delta(x,r)} \big|\nu(y) - \nu(z)\big|\, d\sigma(y)\, d\sigma(z)\right)\\
&\leq \lim_{R\to 0^+}\left(\sup_{x\in\partial\Omega,\, r\in(0,R)}\left(\sup_{y,z\in\Delta(x,r)} \big|\nu(y) - \nu(z)\big|\right)\right)\\
&\leq \lim_{R\to 0^+}\left(\sup_{x\in\partial\Omega}\left(\sup_{y,z\in\Delta(x,R)} \big|\nu(y) - \nu(z)\big|\right)\right) = \sup_{x\in\partial\Omega} \big|\nu(x^+) - \nu(x^-)\big|,
\end{aligned}$$

so (6.3.20) follows by combining (6.3.21) with (6.3.8) and (6.3.9).

`Step IV:` *There exists some $c \in (0, \infty)$ which depends only on the Ahlfors regularity constants of $\partial\Omega$ such that for each $x \in \partial\Omega$ we have*

$$\lim_{R\to 0^+} \fint_{\Delta(x,R)} \left|\nu - \fint_{\Delta(x,R)} \nu\, d\sigma\right| d\sigma \geq c\big|\nu(x^+) - \nu(x^-)\big|. \tag{6.3.22}$$

Fix some $x \in \partial\Omega$ and $R \in \big(0, \operatorname{diam}\partial\Omega\big)$. Define $\Delta_R^-(x)$ to be the portion of $\Delta(x, R)$ consisting of points lying to the left of x, and define $\Delta_R^+(x)$ to be the portion of $\Delta(x, R)$ consisting of points lying to the right of x (with respect to the counterclockwise orientation of the curve $\partial\Omega$). Then

$$\frac{\sigma(\Delta^+(x, R))}{\sigma(\Delta(x, R))} + \frac{\sigma(\Delta^-(x, R))}{\sigma(\Delta(x, R))} = 1 \ \text{ for each } R \in \big(0, \operatorname{diam}\partial\Omega\big), \tag{6.3.23}$$

$$\frac{\sigma(\Delta^\pm(x, R))}{\sigma(\Delta(x, R))} = O(1) \ \text{ as } R \to 0^+, \tag{6.3.24}$$

and there exists some $C \in [1, \infty)$ which depends only on the upper Ahlfors regularity constant of $\partial\Omega$ such that

$$R \leq \sigma(\Delta^\pm(x, R)) \leq \sigma(\Delta(x, R)) \leq CR \ \text{ for each } R \in \big(0, \operatorname{diam}\partial\Omega\big). \tag{6.3.25}$$

In addition, (6.1.9) ensures that

$$\nu(y) = \nu(x^\pm) + o(1) \ \text{ for } y \in \Delta^\pm(x, R) \ \text{ as } R \to 0^+, \tag{6.3.26}$$

hence

$$\fint_{\Delta(x,R)} \nu(y)\, d\sigma(y) = \frac{1}{\sigma(\Delta(x,R))} \int_{\Delta(x,R)} \nu(y)\, d\sigma(y) \tag{6.3.27}$$
$$= \frac{1}{\sigma(\Delta(x,R))} \left\{ \int_{\Delta^+(x,R)} \nu(y)\, d\sigma(y) + \int_{\Delta^-(x,R)} \nu(y)\, d\sigma(y) \right\}$$
$$= \frac{\sigma(\Delta^+(x,R))}{\sigma(\Delta(x,R))} \big[\nu(x^+) + o(1)\big] + \frac{\sigma(\Delta^-(x,R))}{\sigma(\Delta(x,R))} \big[\nu(x^-) + o(1)\big].$$

Next, decompose

$$\fint_{\Delta(x,R)} \Big| \nu - \fint_{\Delta(x,R)} \nu\, d\sigma \Big|\, d\sigma = \mathrm{I}_+ + \mathrm{I}_- \tag{6.3.28}$$

where

$$\mathrm{I}_{\pm} := \frac{1}{\sigma(\Delta(x,R))} \int_{\Delta^{\pm}(x,R)} \Big| \nu - \fint_{\Delta(x,R)} \nu\, d\sigma \Big|\, d\sigma. \tag{6.3.29}$$

From (6.3.29), (6.3.26), (6.3.27), (6.3.23), (6.3.24), and (6.3.25) we see that

$$\mathrm{I}_+ = \frac{1}{\sigma(\Delta(x,R))} \int_{\Delta^+(x,R)} \Big| \big[\nu(x^+) + o(1)\big] - \fint_{\Delta(x,R)} \nu\, d\sigma \Big|\, d\sigma$$
$$= \frac{\sigma(\Delta^-(x,R))}{\sigma(\Delta(x,R))^2} \int_{\Delta^+(x,R)} \Big| \big[\nu(x^+) - \nu(x^-)\big] + o(1) \Big|\, d\sigma$$
$$\geq \frac{\sigma(\Delta^+(x,R))\sigma(\Delta^-(x,R))}{\sigma(\Delta(x,R))^2} \big|\nu(x^+) - \nu(x^-)\big| + o(1)$$
$$\geq C^{-2} \big|\nu(x^+) - \nu(x^-)\big| + o(1) \text{ as } R \to 0^+, \tag{6.3.30}$$

plus a similar estimate for I_-. In concert with (6.3.28), these estimates imply (6.3.22)

`Step V:` *There exists some $c \in (0, \infty)$ which depends only on the upper Ahlfors regularity constant of $\partial\Omega$ such that*

$$\mathrm{dist}\Big(\nu, \big[\mathrm{VMO}(\partial\Omega, \sigma)\big]^2\Big) \geq c \cdot \sup_{0 \leq s \leq L} |\theta(s^+) - \theta(s^-)|. \tag{6.3.31}$$

Having fixed an arbitrary point $x_0 \in \partial\Omega$, we use (6.3.22) to write

$$\lim_{R \to 0^+} \left(\sup_{x \in \partial\Omega,\, r \in (0,R)} \Big(\fint_{\Delta(x,r)} \Big| \nu - \fint_{\Delta(x,r)} \nu\, d\sigma \Big|\, d\sigma \Big) \right) \tag{6.3.32}$$
$$\lim_{R \to 0^+} \Big(\fint_{\Delta(x_0,R/2)} \Big| \nu - \fint_{\Delta(x_0,R/2)} \nu\, d\sigma \Big|\, d\sigma \Big) \geq c \big|\nu(x^+) - \nu(x^-)\big|.$$

Then (6.3.31) follows from (6.3.32), (6.3.8), and (6.3.9). □

At this stage, by combining (6.3.4) and (6.3.5) with Proposition 6.3.1 we arrive at the conclusion that, given a Radon domain $\Omega \subseteq \mathbb{C}$, if $\sigma := \mathcal{H}^1 \lfloor \partial\Omega$ and ν denotes the outward unit normal vector to Ω, then for each $p \in (1, \infty)$ we have

$$\rho_{\text{Fred}}\Big(K_\Delta;\ L^p(\partial\Omega, \sigma)\Big) \approx \operatorname{dist}\Big(\nu, [\mathrm{VMO}(\partial\Omega, \sigma)]^2\Big) \tag{6.3.33}$$

where the proportionality constants depend only on p and the Ahlfors regularity constants of $\partial\Omega$, and, for each $p \in [2, \infty)$,

$$\rho_{\text{Fred}}\Big(K_\psi :\ \big[L^p(\partial\Omega, \sigma)\big]^2\Big) \approx \operatorname{dist}\Big(\nu, [\mathrm{VMO}(\partial\Omega, \sigma)]^2\Big) \tag{6.3.34}$$

where the proportionality constants now depend only on p, λ, μ, and the Ahlfors regularity constants of $\partial\Omega$. Here is another result of interest.

Proposition 6.3.2 *Let $\Omega \subseteq \mathbb{C}$ be a Radon domain. Denote by ν its outward unit normal and set $\sigma := \mathcal{H}^1 \lfloor \partial\Omega$. Also, consider the harmonic double layer potential operator K_Δ on $\partial\Omega$. Then for each $p \in (1, \infty)$ there exists some $c \in (0, \infty)$ depend only on p and the Ahlfors regularity constants of $\partial\Omega$ such that*

$$\operatorname{dist}\big(K_\Delta, \mathrm{Cp}(L^p(\partial\Omega, \sigma))\big) \geq c \cdot \operatorname{dist}\Big(\nu, [\mathrm{VMO}(\partial\Omega, \sigma)]^2\Big), \tag{6.3.35}$$

where the distance in the left-hand side is measured in $\mathrm{Bd}\big(L^p(\partial\Omega, \sigma)\big)$ and the distance in the right-hand side is measured in $\big[\mathrm{BMO}(\partial\Omega, \sigma)\big]^2$.

Proof This follows by writing

$$\begin{aligned} \operatorname{dist}\big(K_\Delta, \mathrm{Cp}(L^p(\partial\Omega, \sigma))\big) &= \big\|K_\Delta\big\|^{\text{ess}}_{L^p(\partial\Omega,\sigma)\to L^p(\partial\Omega,\sigma)} \\ &\geq \rho_{\text{Fred}}\Big(K_\Delta;\ L^p(\partial\Omega, \sigma)\Big) \approx \Theta_{\partial\Omega} \geq c \cdot \operatorname{dist}\Big(\nu, [\mathrm{VMO}(\partial\Omega, \sigma)]^2\Big), \end{aligned} \tag{6.3.36}$$

by the first equality in (A.0.191), the first inequality in [130, (2.2.16)], the equivalence in (6.3.4), and Proposition 6.3.1. □

If $\Omega \subseteq \mathbb{R}^2$ be a curvilinear polygon, then from [130, (3.1.53)] we know that

$$\begin{aligned} &\operatorname{dist}\big(\nu,\ \big[\mathrm{VMO}(\partial\Omega, \sigma)\big]^2\big) \\ &\quad\approx \limsup_{r\to 0^+} \bigg\{ \sup_{x\in\partial\Omega} ⨍_{\partial\Omega\cap B(x,r)} ⨍_{\partial\Omega\cap B(x,r)} |\nu(y) - \nu(z)|\, d\mathcal{H}^1(y)\, d\mathcal{H}^1(z) \bigg\} \end{aligned} \tag{6.3.37}$$

where the implicit proportionality constants depend only on the Ahlfors regularity constants of $\partial\Omega$. If $\theta_j \in (0, 2\pi)$ and $x_j \in \partial\Omega$ with $j \in \{1, \ldots, N\}$ are, respectively, the internal angles and vertices of Ω, then straightforward calculus yields

$$\lim_{r\to 0^+} ⨍_{\partial\Omega\cap B(x_j,r)} ⨍_{\partial\Omega\cap B(x_j,r)} |\nu(y) - \nu(z)|\, d\mathcal{H}^1(y)\, d\mathcal{H}^1(z) = \big|\cos(\theta_j/2)\big|,$$

for each $j \in \{1, \dots, N\}$. Combining this with (6.3.6) and (6.3.37) we arrive at the conclusion that for each $p \in (1, \infty)$ we have

$$\mathrm{dist}\big(K_\Delta, \mathrm{Cp}(L^p(\partial\Omega, \sigma))\big) \geq c \cdot \max_{1 \leq j \leq N} \big|\cos(\theta_j/2)\big|, \tag{6.3.38}$$

where the distance in the left-hand side is measured in the space $\mathrm{Bd}\big(L^p(\partial\Omega, \sigma)\big)$, and the constant $c \in (0, \infty)$ depends only on p and the Ahlfors regularity constants of $\partial\Omega$. In particular, (6.3.38) shows that if K_Δ is a compact operator on $L^p(\partial\Omega, \sigma)$ then necessarily the curvilinear polygon Ω is bounded domain of class $\mathscr{C}^1$ (the opposite implication has first been established in [52]).

Moving on, we note that in (6.3.33) and (6.3.34) we have Fredholm radius estimates, both for scalar operators and systems, written in a way which lends itself to higher dimensional generalizations. The philosophy emerging from the above considerations is that we should seek to estimate Fredholm spectral radii and essential norms of singular integral operators of double layer type associated with a given weakly elliptic system L and a UR domain $\Omega \subseteq \mathbb{R}^n$ with $n \geq 2$ arbitrary, in terms of the quantity $\mathrm{dist}\Big(\nu, [\mathrm{VMO}(\partial\Omega, \sigma)]^n\Big)$ where $\sigma := \mathcal{H}^{n-1} \lfloor \partial\Omega$ and ν is the geometric measure theoretic outward unit normal to Ω.

In our next theorem which, together with Theorem 6.3.4 (treating the important special case of the Stokes system), constitutes our solution to the higher dimensional Radon-Carleman Problem formulated in (6.0.1), we set to accomplish just that, working with a large variety of function spaces of interest defined on UR sets.

Theorem 6.3.3 *Let $\Omega \subseteq \mathbb{R}^n$ be a* UR *domain with compact boundary and denote by ν its geometric measure theoretic outward unit normal. Set $\sigma := \mathcal{H}^{n-1} \lfloor \partial\Omega$. Let L be a homogeneous, second-order, constant complex coefficient, weakly elliptic $M \times M$ system in $\mathbb{R}^n$ for which $\mathfrak{A}^{\mathrm{dis}}_L \neq \varnothing$. Finally, pick $A \in \mathfrak{A}^{\mathrm{dis}}_L$ and consider the boundary-to-boundary double layer potential operators $K_A, K^{\#}_A$ associated with Ω and the coefficient tensor A as in* (A.0.116) *and* (A.0.118), *respectively. Then, with notation introduced in* (A.0.191) *and* (4.1.12), *the following statements are true.*

(1) For each $m \in \mathbb{N}$, $p \in (1, \infty)$, and each Muckenhoupt weight $w \in A_p(\partial\Omega, \sigma)$ there exists $C_m \in (0, \infty)$, depending only on m, n, p, $[w]_{A_p}$, A, the UR *constants of $\partial\Omega$, and the relative flatness ratio $\wp(\Omega)/\mathrm{diam}\,\partial\Omega$ (cf. Definition 3.1.9) such that, with the distance measured in the John-Nirenberg space $\big[\mathrm{BMO}(\partial\Omega, \sigma)\big]^n$,*

$$\|K_A\|^{\mathrm{ess}}_{[L^p(\partial\Omega,w)]^M \to [L^p(\partial\Omega,w)]^M} \leq C_m \Big[\mathrm{dist}\big(\nu, \big[\mathrm{VMO}(\partial\Omega, \sigma)\big]^n\big)\Big]^{\langle m \rangle},$$

$$\|K^{\#}_A\|^{\mathrm{ess}}_{[L^p(\partial\Omega,w)]^M \to [L^p(\partial\Omega,w)]^M} \leq C_m \Big[\mathrm{dist}\big(\nu, \big[\mathrm{VMO}(\partial\Omega, \sigma)\big]^n\big)\Big]^{\langle m \rangle},$$

$$\rho_{\mathrm{Fred}}\Big(K_A;\ \big[L^p_1(\partial\Omega, w)\big]^M\Big) \leq C_m \Big[\mathrm{dist}\big(\nu, \big[\mathrm{VMO}(\partial\Omega, \sigma)\big]^n\big)\Big]^{\langle m \rangle},$$

$$\rho_{\mathrm{Fred}}\Big(K^{\#}_A;\ \big[L^p_{-1}(\partial\Omega, w)\big]^M\Big) \leq C_m \Big[\mathrm{dist}\big(\nu, \big[\mathrm{VMO}(\partial\Omega, \sigma)\big]^n\big)\Big]^{\langle m \rangle}$$

Also, whenever Ω is an infinitesimally flat AR *domain (cf. Definition 3.4.1) it follows that K_A is a compact operator on $\big[L^p(\partial\Omega, w)\big]^M$ and on $\big[L^p_1(\partial\Omega, w)\big]^M$, while $K^\#_A$ is a compact operator on $\big[L^p(\partial\Omega, w)\big]^M$ and on $\big[L^p_{-1}(\partial\Omega, w)\big]^M$, for any $p \in (1, \infty)$ and $w \in A_p(\partial\Omega, \sigma)$. In particular, corresponding to $w \equiv 1$, one obtains similar results involving "ordinary" (unweighted) Lebesgue and Sobolev spaces. Furthermore, similar Fredholm radius estimates are valid for the brand of Lorentz-based Sobolev spaces introduced in* (5.4.27)-(5.4.28), *namely for each $q \in (0, \infty]$ one has (with C_m now also depending on q)*

$$\rho_{\text{Fred}}\big(K_A; \big[L^{(p,q)}_1(\partial\Omega, \sigma)\big]^M\big) \leq C_m\Big[\operatorname{dist}\big(\nu, \big[\mathrm{VMO}(\partial\Omega, \sigma)\big]^n\big)\Big]^{\langle m\rangle}. \tag{6.3.39}$$

In addition, for each $q \in (0, \infty]$ one has (with C_m now also depending on q)

$$\big\|K_A\big\|^{\text{ess}}_{[L^{p,q}(\partial\Omega,\sigma)]^M \to [L^{p,q}(\partial\Omega,\sigma)]^M} \leq C_m\Big[\operatorname{dist}\big(\nu, \big[\mathrm{VMO}(\partial\Omega, \sigma)\big]^n\big)\Big]^{\langle m\rangle},$$
$$\big\|K^\#_A\big\|^{\text{ess}}_{[L^{p,q}(\partial\Omega,\sigma)]^M \to [L^{p,q}(\partial\Omega,\sigma)]^M} \leq C_m\Big[\operatorname{dist}\big(\nu, \big[\mathrm{VMO}(\partial\Omega, \sigma)\big]^n\big)\Big]^{\langle m\rangle}.$$

As a corollary, whenever Ω is an infinitesimally flat AR *domain it follows that K_A and $K^\#_A$ are compact on $\big[L^{p,q}(\partial\Omega, \sigma)\big]^M$ for any $p \in (1, \infty)$ and $q \in (0, \infty]$.*

(2) For each exponent $\alpha \in (0, 1)$ and each choice of a fractional power $\theta \in (0, 1-\alpha)$ there exists some $C_\theta \in (0, \infty)$, which depends only on n, α, θ, A, the UR *constants of $\partial\Omega$, and the relative flatness ratio $\wp(\Omega)/\operatorname{diam}\partial\Omega$ such that, with the distance measured in the John-Nirenberg space,*

$$\big\|K_A\big\|^{\text{ess}}_{[\mathscr{C}^\alpha(\partial\Omega)]^M \to [\mathscr{C}^\alpha(\partial\Omega)]^M} \leq C_\theta \cdot \operatorname{dist}\big(\nu, \big[\mathrm{VMO}(\partial\Omega, \sigma)\big]^n\big)^\theta, \tag{6.3.40}$$
$$\rho_{\text{Fred}}\Big(K_A; \big[\mathscr{C}^\alpha_{\text{van}}(\partial\Omega)\big]^M\Big) \leq C_\theta \cdot \operatorname{dist}\big(\nu, \big[\mathrm{VMO}(\partial\Omega, \sigma)\big]^n\big)^\theta. \tag{6.3.41}$$

In addition, whenever Ω is an infinitesimally flat AR *domain it follows that K_A is compact both on $\big[\mathscr{C}^\alpha(\partial\Omega)\big]^M$ and on $\big[\mathscr{C}^\alpha_{\text{van}}(\partial\Omega)\big]^M$, for each $\alpha \in (0, 1)$.*

(3) For each $p \in \big(\frac{n-1}{n}, 1\big]$ and each choice of a fractional power $\theta \in \big(0, n - \frac{n-1}{p}\big)$ there exists some $C_\theta \in (0, \infty)$, which depends only on p, n, θ, A, the UR *constants of $\partial\Omega$, and the relative flatness ratio $\wp(\Omega)/\operatorname{diam}\partial\Omega$ such that, with the distance measured in the John-Nirenberg space,*

$$\big\|K^\#_A\big\|^{\text{ess}}_{[H^p(\partial\Omega,\sigma)]^M \to [H^p(\partial\Omega,\sigma)]^M} \leq C_\theta \cdot \operatorname{dist}\big(\nu, \big[\mathrm{VMO}(\partial\Omega, \sigma)\big]^n\big)^\theta. \tag{6.3.42}$$

In particular, whenever Ω is an infinitesimally flat AR *domain it follows that $K^\#_A$ is a compact operator on $\big[H^p(\partial\Omega, \sigma)\big]^M$ for each $p \in \big(\frac{n-1}{n}, 1\big]$.*

(4) For each choice of a fractional power $\theta \in (0, 1)$ there exists a finite constant $C_\theta \in (0, \infty)$, which depends only on θ, n, A, the UR *constants of $\partial\Omega$, and the relative flatness ratio $\wp(\Omega)/\operatorname{diam}\partial\Omega$ such that, with the distances measured in*

the John-Nirenberg space,

$$\rho_{\text{Fred}}\Big(K_A;\ \big[\mathrm{VMO}(\partial\Omega,\sigma)\big]^M\Big) \leq C_\theta \cdot \operatorname{dist}\big(\nu, \big[\mathrm{VMO}(\partial\Omega,\sigma)\big]^n\big)^\theta,$$
$$\big\|K_A\big\|^{\mathrm{ess}}_{[\mathrm{BMO}(\partial\Omega,\sigma)]^M\to[\mathrm{BMO}(\partial\Omega,\sigma)]^M} \leq C_\theta \cdot \operatorname{dist}\big(\nu, \big[\mathrm{VMO}(\partial\Omega,\sigma)\big]^n\big)^\theta.$$

In addition, whenever Ω is an infinitesimally flat AR *domain it follows that K_A is a compact operator both on $\big[\mathrm{BMO}(\partial\Omega,\sigma)\big]^M$ and on $\big[\mathrm{VMO}(\partial\Omega,\sigma)\big]^M$.*

(5) For each choice of exponents

$$p \in \big(\tfrac{n-1}{n},\infty\big), \quad q \in (0,\infty), \quad (n-1)\big(\tfrac{1}{p}-1\big)_+ < s < 1, \tag{6.3.43}$$

there exists a power $\theta \in (0,1)$ which depends only on p, q, s along with a constant $C \in (0,\infty)$, which depends only on n, p, q, s, A, the UR *constants of $\partial\Omega$, and the relative flatness ratio $\wp(\Omega)/\operatorname{diam}\partial\Omega$ such that, with the distance measured in the John-Nirenberg space,*

$$\big\|K_A\big\|^{\mathrm{ess}}_{[B^{p,q}_s(\partial\Omega,\sigma)]^M\to[B^{p,q}_s(\partial\Omega,\sigma)]^M} \leq C \cdot \operatorname{dist}\big(\nu, \big[\mathrm{VMO}(\partial\Omega,\sigma)\big]^n\big)^\theta, \tag{6.3.44}$$

Moreover, whenever Ω is an infinitesimally flat AR *domain it follows that K_A is a compact operator on $\big[B^{p,q}_s(\partial\Omega,\sigma)\big]^M$ for any p, q, s as in* (6.3.43)*. Furthermore, a similar result is valid in the context of Triebel-Lizorkin spaces. Specifically, given any*

$$p \in \big(\tfrac{n-1}{n},\infty\big), \quad q \in \big(\tfrac{n-1}{n},\infty\big), \quad (n-1)\big(\tfrac{1}{\min\{p,q\}}-1\big)_+ < s < 1 \tag{6.3.45}$$

there exists a power $\theta \in (0,1)$ which depends only on p, q, s together with some constant $C \in (0,\infty)$, which depends only on n, p, q, s, A, the UR *constants of $\partial\Omega$, and the relative flatness ratio $\wp(\Omega)/\operatorname{diam}\partial\Omega$ such that, with the distance measured in the John-Nirenberg space,*

$$\big\|K_A\big\|^{\mathrm{ess}}_{[F^{p,q}_s(\partial\Omega,\sigma)]^M\to[F^{p,q}_s(\partial\Omega,\sigma)]^M} \leq C \cdot \operatorname{dist}\big(\nu, \big[\mathrm{VMO}(\partial\Omega,\sigma)\big]^n\big)^\theta. \tag{6.3.46}$$

Also, whenever Ω is an infinitesimally flat AR *domain it follows that K_A is a compact operator on $\big[F^{p,q}_s(\partial\Omega,\sigma)\big]^M$ for any p, q, s as in* (6.3.45)*.*

(6) For each choice of exponents

$$s \in (0,1), \quad p \in \big(\tfrac{n-1}{n-s},\infty\big), \quad q \in (0,\infty), \tag{6.3.47}$$

there exists a power $\theta \in (0,1)$ which depends only on p, q, s along with a constant $C \in (0,\infty)$, which depends only on n, p, q, s, A, the UR *constants of $\partial\Omega$, and the relative flatness ratio $\wp(\Omega)/\operatorname{diam}\partial\Omega$ such that, with the distance measured in the John-Nirenberg space,*

$$\big\|K^{\#}_A\big\|^{\mathrm{ess}}_{[B^{p,q}_{-s}(\partial\Omega,\sigma)]^M\to[B^{p,q}_{-s}(\partial\Omega,\sigma)]^M} \leq C \cdot \operatorname{dist}\big(\nu, \big[\mathrm{VMO}(\partial\Omega,\sigma)\big]^n\big)^\theta. \tag{6.3.48}$$

Moreover, whenever Ω *is an infinitesimally flat* AR *domain it follows that* $K_A^\#$ *is a compact operator on* $\big[B^{p,q}_{-s}(\partial\Omega,\sigma)\big]^M$ *for any* p,q,s *as in* (6.3.47).
Likewise, in the context of Triebel-Lizorkin spaces, for each choice of exponents

$$s\in(0,1),\quad p\in\big(\tfrac{n-1}{n-s},\infty\big),\quad q\in\big(\tfrac{n-1}{n-s},\infty\big), \tag{6.3.49}$$

there exists a power $\theta\in(0,1)$ *which depends only on* p,q,s *together with some constant* $C\in(0,\infty)$*, which depends only on* n*,* p*,* q*,* s*,* A*, the* UR *constants of* $\partial\Omega$*, and the relative flatness ratio* $\wp(\Omega)/\operatorname{diam}\partial\Omega$ *such that, with the distance measured in the John-Nirenberg space,*

$$\|K_A^\#\|^{\rm ess}_{[F^{p,q}_{-s}(\partial\Omega,\sigma)]^M\to[F^{p,q}_{-s}(\partial\Omega,\sigma)]^M}\le C\cdot\operatorname{dist}\big(\nu,\big[\mathrm{VMO}(\partial\Omega,\sigma)\big]^n\big)^\theta. \tag{6.3.50}$$

In particular, whenever Ω *is an infinitesimally flat* AR *domain it follows that* $K_A^\#$ *is a compact operator on* $\big[F^{p,q}_{-s}(\partial\Omega,\sigma)\big]^M$ *for any* p,q,s *as in* (6.3.49).

(7) *For each integer* $m\in\mathbb{N}$*, integrability exponents* $p,q\in(1,\infty)$*, and parameter* $\lambda\in(0,n-1)$ *there exists a constant* $C_m\in(0,\infty)$*, which depends only on* m*,* n*,* p*,* λ*,* A*, the* UR *constants of* $\partial\Omega$*, and the relative flatness ratio* $\wp(\Omega)/\operatorname{diam}\partial\Omega$ *such that, with the distance measured in the John-Nirenberg space,*

$$\|K_A\|^{\rm ess}_{[M^{p,\lambda}(\partial\Omega,\sigma)]^M\to[M^{p,\lambda}(\partial\Omega,\sigma)]^M}\le C_m\Big[\operatorname{dist}\big(\nu,\big[\mathrm{VMO}(\partial\Omega,\sigma)\big]^n\big)\Big]^{\langle m\rangle},$$

$$\|K_A\|^{\rm ess}_{[\mathring{M}^{p,\lambda}(\partial\Omega,\sigma)]^M\to[\mathring{M}^{p,\lambda}(\partial\Omega,\sigma)]^M}\le C_m\Big[\operatorname{dist}\big(\nu,\big[\mathrm{VMO}(\partial\Omega,\sigma)\big]^n\big)\Big]^{\langle m\rangle},$$

$$\|K_A\|^{\rm ess}_{[\mathcal{B}^{q,\lambda}(\partial\Omega,\sigma)]^M\to[\mathcal{B}^{q,\lambda}(\partial\Omega,\sigma)]^M}\le C_m\Big[\operatorname{dist}\big(\nu,\big[\mathrm{VMO}(\partial\Omega,\sigma)\big]^n\big)\Big]^{\langle m\rangle},$$

$$\|K_A^\#\|^{\rm ess}_{[M^{p,\lambda}(\partial\Omega,\sigma)]^M\to[M^{p,\lambda}(\partial\Omega,\sigma)]^M}\le C_m\Big[\operatorname{dist}\big(\nu,\big[\mathrm{VMO}(\partial\Omega,\sigma)\big]^n\big)\Big]^{\langle m\rangle},$$

$$\|K_A^\#\|^{\rm ess}_{[\mathring{M}^{p,\lambda}(\partial\Omega,\sigma)]^M\to[\mathring{M}^{p,\lambda}(\partial\Omega,\sigma)]^M}\le C_m\Big[\operatorname{dist}\big(\nu,\big[\mathrm{VMO}(\partial\Omega,\sigma)\big]^n\big)\Big]^{\langle m\rangle},$$

$$\|K_A^\#\|^{\rm ess}_{[\mathcal{B}^{q,\lambda}(\partial\Omega,\sigma)]^M\to[\mathcal{B}^{q,\lambda}(\partial\Omega,\sigma)]^M}\le C_m\Big[\operatorname{dist}\big(\nu,\big[\mathrm{VMO}(\partial\Omega,\sigma)\big]^n\big)\Big]^{\langle m\rangle},$$

as well as

$$\rho_{\rm Fred}\Big(K_A;\ \big[M^{p,\lambda}_1(\partial\Omega,\sigma)\big]^M\Big)\le C_m\Big[\operatorname{dist}\big(\nu,\big[\mathrm{VMO}(\partial\Omega,\sigma)\big]^n\big)\Big]^{\langle m\rangle},$$

$$\rho_{\rm Fred}\Big(K_A;\ \big[\mathring{M}^{p,\lambda}_1(\partial\Omega,\sigma)\big]^M\Big)\le C_m\Big[\operatorname{dist}\big(\nu,\big[\mathrm{VMO}(\partial\Omega,\sigma)\big]^n\big)\Big]^{\langle m\rangle},$$

$$\rho_{\rm Fred}\Big(K_A;\ \big[\mathcal{B}^{q,\lambda}_1(\partial\Omega,\sigma)\big]^M\Big)\le C_m\Big[\operatorname{dist}\big(\nu,\big[\mathrm{VMO}(\partial\Omega,\sigma)\big]^n\big)\Big]^{\langle m\rangle}.$$

Also, whenever Ω *is an infinitesimally flat* AR *domain (cf. Definition 3.4.1) it follows that* K_A *is a compact operator on* $\big[M^{p,\lambda}(\partial\Omega,\sigma)\big]^M$, $\big[\mathring{M}^{p,\lambda}(\partial\Omega,\sigma)\big]^M$, $\big[\mathcal{B}^{q,\lambda}(\partial\Omega,\sigma)\big]^M$, $\big[M_1^{p,\lambda}(\partial\Omega,\sigma)\big]^M$, $\big[\mathring{M}_1^{p,\lambda}(\partial\Omega,\sigma)\big]^M$, *and* $\big[\mathcal{B}_1^{q,\lambda}(\partial\Omega,\sigma)\big]^M$, *while* $K_A^{\#}$ *is a compact operator on* $\big[M^{p,\lambda}(\partial\Omega,\sigma)\big]^M$, $\big[\mathring{M}^{p,\lambda}(\partial\Omega,\sigma)\big]^M$, *and* $\big[\mathcal{B}^{q,\lambda}(\partial\Omega,\sigma)\big]^M$, *for any* $p,q\in(1,\infty)$ *and* $\lambda\in(0,n-1)$.

(8) *With* $\mathcal{M}_{\partial\Omega}$ *denoting the Hardy-Littlewood maximal operator on* $\partial\Omega$, *assume* $\mathbb{X}$ *is a Generalized Banach Function Space on* $(\partial\Omega,\sigma)$ *with the property that*

$$\mathcal{M}_{\partial\Omega}:\mathbb{X}\to\mathbb{X}\ \textit{and}\ \mathcal{M}_{\partial\Omega}:\mathbb{X}'\to\mathbb{X}'\ \textit{are well-defined and bounded,} \tag{6.3.51}$$

where $\mathbb{X}'$ *is the associated space of* $\mathbb{X}$ *(cf.* [130, Definitions 5.1.4, 5.1.11]). *Set*

$$\mathring{\mathbb{X}}:=\overline{L^\infty(\partial\Omega,\sigma)}^{\|\cdot\|_{\mathbb{X}}}\ \textit{and}\ (\mathbb{X}')^\circ:=\overline{L^\infty(\partial\Omega,\sigma)}^{\|\cdot\|_{\mathbb{X}'}}, \tag{6.3.52}$$

and consider the following $\mathbb{X}$*-styled,* $\mathbb{X}'$*-styled,* $\mathring{\mathbb{X}}$*-styled, and* $(\mathbb{X}')^\circ$*-styled Sobolev spaces of order one on* $\partial\Omega$*:*

$$\begin{aligned}\mathbb{X}_1&:=\Big\{f\in\mathbb{X}:\ \partial_{\tau_{jk}}f\in\mathbb{X}\ \textit{for}\ 1\leq j,k\leq n\Big\},\\(\mathbb{X}')_1&:=\Big\{g\in\mathbb{X}':\ \partial_{\tau_{jk}}g\in\mathbb{X}'\ \textit{for}\ 1\leq j,k\leq n\Big\},\\\mathring{\mathbb{X}}_1&:=\Big\{f\in\mathring{\mathbb{X}}:\ \partial_{\tau_{jk}}f\in\mathring{\mathbb{X}}\ \textit{for}\ 1\leq j,k\leq n\Big\},\\(\mathbb{X}')_1^\circ&:=\Big\{g\in(\mathbb{X}')^\circ:\ \partial_{\tau_{jk}}g\in(\mathbb{X}')^\circ\ \textit{for}\ 1\leq j,k\leq n\Big\},\end{aligned} \tag{6.3.53}$$

which are known to be Banach when equipped with natural norms (cf. (5.4.84)). *Then for each integer* $m\in\mathbb{N}$ *there exists a constant* $C_m\in(0,\infty)$*, which depends only on m, n, A, the operator norms of* $\mathcal{M}_{\partial\Omega}$ *on* $\mathbb{X}$ *and* $\mathbb{X}'$*, the* UR *constants of* $\partial\Omega$*, and the relative flatness ratio* $\wp(\Omega)/\mathrm{diam}\,\partial\Omega$ *such that, with the distance measured in the John-Nirenberg space,*

$$\|K_A\|^{\mathrm{ess}}_{[\mathbb{X}]^M\to[\mathbb{X}]^M}\leq C_m\Big[\mathrm{dist}\big(\nu,\big[\mathrm{VMO}(\partial\Omega,\sigma)\big]^n\big)\Big]^{\langle m\rangle}, \tag{6.3.54}$$

$$\|K_A\|^{\mathrm{ess}}_{[\mathbb{X}']^M\to[\mathbb{X}']^M}\leq C_m\Big[\mathrm{dist}\big(\nu,\big[\mathrm{VMO}(\partial\Omega,\sigma)\big]^n\big)\Big]^{\langle m\rangle}, \tag{6.3.55}$$

$$\|K_A\|^{\mathrm{ess}}_{[\mathring{\mathbb{X}}]^M\to[\mathring{\mathbb{X}}]^M}\leq C_m\Big[\mathrm{dist}\big(\nu,\big[\mathrm{VMO}(\partial\Omega,\sigma)\big]^n\big)\Big]^{\langle m\rangle}, \tag{6.3.56}$$

$$\|K_A\|^{\mathrm{ess}}_{[(\mathbb{X}')^\circ]^M\to[(\mathbb{X}')^\circ]^M}\leq C_m\Big[\mathrm{dist}\big(\nu,\big[\mathrm{VMO}(\partial\Omega,\sigma)\big]^n\big)\Big]^{\langle m\rangle}, \tag{6.3.57}$$

$$\|K_A^\#\|_{[\mathbb{X}]^M \to [\mathbb{X}]^M}^{\text{ess}} \le C_m \Big[\text{dist}\big(\nu, [\text{VMO}(\partial\Omega, \sigma)]^n\big)\Big]^{\langle m \rangle}, \tag{6.3.58}$$

$$\|K_A^\#\|_{[\mathbb{X}']^M \to [\mathbb{X}']^M}^{\text{ess}} \le C_m \Big[\text{dist}\big(\nu, [\text{VMO}(\partial\Omega, \sigma)]^n\big)\Big]^{\langle m \rangle}, \tag{6.3.59}$$

$$\|K_A^\#\|_{[\mathring{\mathbb{X}}]^M \to [\mathring{\mathbb{X}}]^M}^{\text{ess}} \le C_m \Big[\text{dist}\big(\nu, [\text{VMO}(\partial\Omega, \sigma)]^n\big)\Big]^{\langle m \rangle}, \tag{6.3.60}$$

$$\|K_A^\#\|_{[(\mathbb{X}')^\circ]^M \to [(\mathbb{X}')^\circ]^M}^{\text{ess}} \le C_m \Big[\text{dist}\big(\nu, [\text{VMO}(\partial\Omega, \sigma)]^n\big)\Big]^{\langle m \rangle}, \tag{6.3.61}$$

as well as

$$\rho_{\text{Fred}}\Big(K_A;\ [\mathbb{X}_1]^M\Big) \le C_m \Big[\text{dist}\big(\nu, [\text{VMO}(\partial\Omega, \sigma)]^n\big)\Big]^{\langle m \rangle}, \tag{6.3.62}$$

$$\rho_{\text{Fred}}\Big(K_A;\ [(\mathbb{X}')_1]^M\Big) \le C_m \Big[\text{dist}\big(\nu, [\text{VMO}(\partial\Omega, \sigma)]^n\big)\Big]^{\langle m \rangle}. \tag{6.3.63}$$

$$\rho_{\text{Fred}}\Big(K_A;\ [\mathring{\mathbb{X}}_1]^M\Big) \le C_m \Big[\text{dist}\big(\nu, [\text{VMO}(\partial\Omega, \sigma)]^n\big)\Big]^{\langle m \rangle}, \tag{6.3.64}$$

$$\rho_{\text{Fred}}\Big(K_A;\ [(\mathbb{X}')_1^\circ]^M\Big) \le C_m \Big[\text{dist}\big(\nu, [\text{VMO}(\partial\Omega, \sigma)]^n\big)\Big]^{\langle m \rangle}. \tag{6.3.65}$$

Finally, if Ω *is an infinitesimally flat* AR *domain (cf. Definition 3.4.1) then* K_A *is compact on* $[\mathbb{X}]^M$, $[\mathbb{X}']^M$, $[\mathbb{X}_1]^M$, $[(\mathbb{X}')_1]^M$, $[\mathring{\mathbb{X}}]^M$, $[(\mathbb{X}')^\circ]^M$, $[\mathring{\mathbb{X}}_1]^M$, $[(\mathbb{X}')_1^\circ]^M$, *while* $K_A^\#$ *is compact on* $[\mathbb{X}]^M$, $[\mathbb{X}']^M$, $[\mathring{\mathbb{X}}]^M$, $[(\mathbb{X}')^\circ]^M$.

(9) *All conclusions in item (8) are valid whenever* $\mathbb{X}$ *is a rearrangement invariant Banach function space whose Boyd indices satisfy* $1 < p_{\mathbb{X}} \le q_{\mathbb{X}} < \infty$. *In particular, all conclusions in item (8) hold if* $\mathbb{X} := L^\Phi(\partial\Omega, \sigma)$, *the Orlicz space associated as in* (A.0.133) *with the measure space* $(\partial\Omega, \sigma)$ *and any Young function* Φ *satisfying* $1 < i(\Phi) \le I(\Phi) < \infty$ *(when the role of* $\mathbb{X}_1$ *from* (5.4.83) *now played by the Orlicz-based Sobolev space* $L_1^\Phi(\partial\Omega, \sigma)$ *from* (5.4.92)).

Comment 1. In the two-dimensional setting, (3.3.4) allows us to obtain similar estimates in the class of chord-arc domains with compact boundaries, now formulated in terms of the entity $\varkappa$ appearing in (3.3.3). Furthermore, (3.4.48) implies that the various compactness results scattered throughout the statement of Theorem 6.3.3 remain valid if $\Omega \subseteq \mathbb{R}^2$ is a chord-arc domain with vanishing constant (i.e., $\varkappa$ in (3.4.45) is zero) and compact boundary. For example, as a very special case, the boundary-to-boundary harmonic double layer (cf. (5.5.1)) has the property that

$$\text{if } \Omega \subseteq \mathbb{R}^2 \text{ is a chord-arc domain with vanishing constant and compact boundary then } K_\Delta \in \text{Cp}\big(L^p(\partial\Omega, \sigma)\big) \text{ for each exponent } p \in (1, \infty). \tag{6.3.66}$$

Comment 2. The fact that $A \in \mathfrak{A}_L^{\text{dis}}$ is essential in ensuring that the claims in Theorem 6.3.3 are valid. Take, for example, the case when $L = \Delta$, the Laplacian in the two-dimensional setting, a scenario in which $\mathfrak{A}_L^{\text{dis}} = \{I_{2\times 2}\}$. Choosing

$$A := \begin{pmatrix} 1 & \mathrm{i} \\ -\mathrm{i} & 1 \end{pmatrix} \tag{6.3.67}$$

yields, as discussed in [132, (1.4.8)],

$$\begin{aligned} K_A f(z) = \lim_{\varepsilon\to 0^+} \frac{1}{2\pi\mathrm{i}} \int_{\partial\Omega\setminus\overline{B(z,\varepsilon)}} \frac{f(\zeta)}{\zeta - z}\,\mathrm{d}\zeta \ \text{ for } \sigma\text{-a.e. } z \in \partial\Omega, \\ \text{for each function } f \text{ belonging to the space } L^1(\partial\Omega,\sigma), \end{aligned} \tag{6.3.68}$$

i.e., the double layer potential operator K_A becomes the boundary-to-boundary Cauchy integral operator associated with the Ahlfors regular domain $\Omega \subseteq \mathbb{R}^2$ with compact boundary. Take, for example, Ω to be the open unit disk in the plane. If Theorem 6.3.3 were to hold for the choice (6.3.67), then (6.3.66) would imply that the boundary-to-boundary Cauchy integral operator on the unit circle $\mathbb{T} := \partial B(0,1)$ in the plane is a compact operator on $L^2(\mathbb{T},\mathcal{H}^1)$. This, however, is not the case since, as is well known, the Cauchy integral operator is invertible in this context.

This shows that the assumption $A \in \mathfrak{A}_L^{\mathrm{dis}}$ cannot be simply dropped. In particular, this explains the choice of coefficient tensor made in the case of the Lamé system, i.e., [132, (1.4.169)] used to represent the Lamé system as in [132, (1.4.170)] corresponding to t as in (6.2.20), which has produced the pseudo-traction elastic double layer potential K_ψ appearing in the formulation of the results recalled in (6.2.21), (6.2.23), (6.2.32). Indeed the coefficient tensor [132, (1.4.169)] corresponding to t as in (6.2.20) is the unique distinguished coefficient tensor of the Lamé system.

Comment 3. Assume Ω is a curvilinear polygon in $\mathbb{R}^2$ and abbreviate $\sigma := \mathcal{H}^1\lfloor\partial\Omega$. It is clear from Proposition 6.2.2 that the presence of any angle $\theta \neq \pi$ prevents the classical harmonic double layer potential operator K_Δ from being compact on $L^p(\partial\Omega,\sigma)$ for any $p \in (1,\infty)$. This failure of K_Δ to be compact can be quantified in a more precise fashion. Concretely, fix $\theta \in (0,\pi)$ and consider a curvilinear polygon Ω_θ with precisely one angular point located at the origin $0 \in \mathbb{R}^2$. Furthermore, assume that, in a neighborhood of $0 \in \partial\Omega_\theta$, the boundary of Ω_θ agrees with a sector of aperture θ with vertex at 0. In particular, the outward unit normal ν to Ω_θ is smooth on $\partial\Omega_\theta \setminus \{0\}$ and is piecewise constant near 0, where it assumes two values, say, ν_+ and ν_-. Bring in notation introduced in (3.4.5) and recall (3.4.6). Then

$$\begin{aligned} \operatorname{dist}\big(\nu, [\mathrm{VMO}(\partial\Omega_\theta,\sigma)]^2\big) &\approx [\nu]_{\mathrm{Osc}(\partial\Omega_\theta)} \\ &= \lim_{R\to 0^+}\Bigg(\sup_{\substack{z\in\partial\Omega_\theta \\ 0<r<R}} \fint_{B(z,r)\cap\partial\Omega_\theta}\fint_{B(z,r)\cap\partial\Omega_\theta} \big|\nu(x)-\nu(y)\big|\,\mathrm{d}\sigma(x)\,\mathrm{d}\sigma(y)\Bigg) \\ &\approx |\nu_+ - \nu_-| \approx \sqrt{1+\cos\theta}. \end{aligned} \tag{6.3.69}$$

In particular, $\operatorname{dist}\big(\nu, [\mathrm{VMO}(\partial\Omega_\theta,\sigma)]^2\big) \to 0$ if and only if $\theta \to \pi$. Based on this analysis, we may conclude the following result:

for each $\varepsilon > 0$ there exists a bounded Lipschitz domain $\Omega \subseteq \mathbb{R}^2$ (whose Lipschitz character is controlled by a universal constant) with the property that $\operatorname{dist}\big(\nu, \big[\mathrm{VMO}(\partial\Omega, \sigma)\big]^2\big) < \varepsilon$ and yet for each integrability index $p \in (1, \infty)$ the classical harmonic double layer potential operator K_Δ fails to be compact when acting on the space $L^p(\partial\Omega, \sigma)$. (6.3.70)

This shows that the hypotheses of Theorem 6.3.3 cannot be fundamentally relaxed.

Here is the proof of Theorem 6.3.3.

Proof of Theorem 6.3.3 All claims in item *(1)* are consequences of Theorem 3.4.2, Theorem 5.2.3, Theorem 5.4.1, (5.4.5), Corollary 5.2.4, Corollary 5.4.3, (5.4.32), Remark 5.4.2, and Proposition 1.3.2. The claims in item *(2)* are implied by Theorem 3.4.2, Theorem 5.4.11, Proposition 1.3.2, [132, (5.1.101)], and [130, Lemmas 2.2.6, 3.2.1]. Next, Theorem 3.4.2, Theorem 5.4.9, and Proposition 1.3.2 imply the claims in item *(3)*, while the claims in item *(4)* may be justified using Theorem 3.4.2 together with Corollary 5.4.10 and Proposition 1.3.2. Going further, item *(5)* is seen from Theorem 3.4.2, Theorem 5.4.12, and Proposition 1.3.2, while item *(6)* follows from Theorem 3.4.2, Theorem 5.4.13, and Proposition 1.3.2. The claims in item *(7)* are consequences of Theorem 3.4.2, Theorem 5.4.4, Corollary 5.4.5, Theorem 5.4.6, Corollary 5.4.7, and Proposition 1.3.2. Next, all conclusions in item *(8)* are justified based on Theorem 5.4.8, Theorem 3.4.2, and Proposition 1.3.2. Finally, the claims in item *(9)* follow from what we have proved in item *(8)* together with [130, Proposition 5.3.14], [130, Proposition 5.3.15], [130, (5.3.62)], and Proposition 1.3.2. □

For the Stokes system, Odqvist and Lichtenstein have shown that the boundary-to-boundary slip double layer K_{slip} from (4.2.102) is a compact operator when considered on boundaries of smooth bounded domains (cf. Chapter 3 of Ladyzhenskaya's book [106]). Our goal is to prove such compactness results in a much more inclusive geometric and functional analytic settings, as described in the theorem below.

Theorem 6.3.4 *Let $\Omega \subseteq \mathbb{R}^n$ be a* UR *domain with compact boundary. Denote by ν its geometric measure theoretic outward unit normal, and set $\sigma := \mathcal{H}^{n-1}\lfloor\partial\Omega$. In this context, consider the boundary-to-boundary slip double layer K_{slip} from* (4.2.102) *and the transpose slip double layer $K^{\#}_{\text{slip}}$ from* (4.2.103). *Then, with notation introduced in* (A.0.191) *and* (4.1.12), *the following statements are true.*

(1) For each $m \in \mathbb{N}$, $p \in (1, \infty)$, and Muckenhoupt weight $w \in A_p(\partial\Omega, \sigma)$ there exists $C_m \in (0, \infty)$, depending only on m, n, p, $[w]_{A_p}$, the UR *constants of $\partial\Omega$, and the relative flatness ratio $\wp(\Omega)/\operatorname{diam}\partial\Omega$ (cf. Definition 3.1.9) such that, with the distance measured in the John-Nirenberg space $\big[\mathrm{BMO}\,(\partial\Omega, \sigma)\big]^n$,*

$$\|K_{\rm slip}\|^{\rm ess}_{[L^p(\partial\Omega,w)]^n\to[L^p(\partial\Omega,w)]^n}\leq C_m\Big[{\rm dist}\,\big(\nu,\,\big[{\rm VMO}(\partial\Omega,\sigma)\big]^n\big)\Big]^{\langle m\rangle},$$

$$\|K^{\#}_{\rm slip}\|^{\rm ess}_{[L^p(\partial\Omega,w)]^n\to[L^p(\partial\Omega,w)]^n}\leq C_m\Big[{\rm dist}\,\big(\nu,\,\big[{\rm VMO}(\partial\Omega,\sigma)\big]^n\big)\Big]^{\langle m\rangle},$$

$$\rho_{\rm Fred}\Big(K_{\rm slip};\,\big[L^p_1(\partial\Omega,w)\big]^n\Big)\leq C_m\Big[{\rm dist}\,\big(\nu,\,\big[{\rm VMO}(\partial\Omega,\sigma)\big]^n\big)\Big]^{\langle m\rangle},$$

$$\rho_{\rm Fred}\Big(K^{\#}_{\rm slip};\,\big[L^p_{-1}(\partial\Omega,w)\big]^n\Big)\leq C_m\Big[{\rm dist}\,\big(\nu,\,\big[{\rm VMO}(\partial\Omega,\sigma)\big]^n\big)\Big]^{\langle m\rangle}$$

In addition, whenever Ω is an infinitesimally flat AR *domain (cf. Definition 3.4.1) it follows that $K_{\rm slip}$ is a compact operator on $\big[L^p(\partial\Omega,w)\big]^n$ and on $\big[L^p_1(\partial\Omega,w)\big]^n$, while $K^{\#}_{\rm slip}$ is a compact operator on $\big[L^p(\partial\Omega,w)\big]^n$ and on $\big[L^p_{-1}(\partial\Omega,w)\big]^n$, for any $p\in(1,\infty)$ and $w\in A_p(\partial\Omega,\sigma)$. Moreover, analogous Fredholm radius estimates are valid for the brand of Lorentz-based Sobolev spaces introduced in* (5.4.27)-(5.4.28), *namely for each $q\in(0,\infty]$ one has (where now C_m also depends on q)*

$$\rho_{\rm Fred}\big(K_{\rm slip};\,\big[L^{(p,q)}_1(\partial\Omega,\sigma)\big]^n\big)\leq C_m\Big[{\rm dist}\,\big(\nu,\,\big[{\rm VMO}(\partial\Omega,\sigma)\big]^n\big)\Big]^{\langle m\rangle}.$$

Furthermore, whenever the set Ω is an infinitesimally flat AR *domain it follows that for any $p\in(1,\infty)$ and $q\in(0,\infty]$ the operator $K_{\rm slip}$ is compact on $\big[L^p(\partial\Omega,\sigma)\big]^n$, $\big[L^p_1(\partial\Omega,\sigma)\big]^n$, and on $\big[L^{(p,q)}_1(\partial\Omega,\sigma)\big]^n$, whereas $K^{\#}_{\rm slip}$ is compact on $\big[L^p(\partial\Omega,\sigma)\big]^n$ and $\big[L^p_{-1}(\partial\Omega,\sigma)\big]^n$. Additionally, for each $q\in(0,\infty]$ one has (with C_m now also depending on q)*

$$\|K_{\rm slip}\|^{\rm ess}_{[L^{p,q}(\partial\Omega,\sigma)]^n\to[L^{p,q}(\partial\Omega,\sigma)]^n}\leq C_m\Big[{\rm dist}\,\big(\nu,\,\big[{\rm VMO}(\partial\Omega,\sigma)\big]^n\big)\Big]^{\langle m\rangle},$$

$$\|K^{\#}_{\rm slip}\|^{\rm ess}_{[L^{p,q}(\partial\Omega,\sigma)]^n\to[L^{p,q}(\partial\Omega,\sigma)]^n}\leq C_m\Big[{\rm dist}\,\big(\nu,\,\big[{\rm VMO}(\partial\Omega,\sigma)\big]^n\big)\Big]^{\langle m\rangle}.$$

In particular, if Ω is an infinitesimally flat AR *domain then $K_{\rm slip}$ and $K^{\#}_{\rm slip}$ are compact on $\big[L^{p,q}(\partial\Omega,\sigma)\big]^n$ for any $p\in(1,\infty)$ and $q\in(0,\infty]$.*

(2) *For each exponent $\alpha\in(0,1)$ and each choice of a fractional power $\theta\in(0,1-\alpha)$ there exists some $C_\theta\in(0,\infty)$, which depends only on n, α, θ, the* UR *constants of $\partial\Omega$, and the relative flatness ratio $\wp(\Omega)/{\rm diam}\,\partial\Omega$ such that, with the distance measured in the John-Nirenberg space,*

$$\rho_{\rm Fred}\Big(K_{\rm slip};\,\big[\mathscr{C}^\alpha_{\rm van}(\partial\Omega)\big]^n\Big)\leq C_\theta\cdot{\rm dist}\,\big(\nu,\,\big[{\rm VMO}(\partial\Omega,\sigma)\big]^n\big)^\theta,\qquad(6.3.71)$$

$$\|K_{\rm slip}\|^{\rm ess}_{[\mathscr{C}^\alpha(\partial\Omega)]^n\to[\mathscr{C}^\alpha(\partial\Omega)]^n}\leq C_\theta\cdot{\rm dist}\,\big(\nu,\,\big[{\rm VMO}(\partial\Omega,\sigma)\big]^n\big)^\theta.\qquad(6.3.72)$$

Moreover, whenever Ω is an infinitesimally flat AR *domain the operator $K_{\rm slip}$ is compact both on $\big[\mathscr{C}^\alpha(\partial\Omega)\big]^n$ and on $\big[\mathscr{C}^\alpha_{\rm van}(\partial\Omega)\big]^n$, for each $\alpha\in(0,1)$.*

(3) *For each* $p \in \left(\frac{n-1}{n}, 1\right]$ *and each choice of a fractional power* $\theta \in \left(0, n - \frac{n-1}{p}\right)$ *there exists some* $C_\theta \in (0, \infty)$*, which depends only on* p*,* n*,* θ*, the* UR *constants of* $\partial\Omega$*, and the relative flatness ratio* $\wp(\Omega)/\operatorname{diam}\partial\Omega$ *such that, with the distance measured in the John-Nirenberg space,*

$$\left\|K^{\#}_{\text{slip}}\right\|^{\text{ess}}_{[H^p(\partial\Omega,\sigma)]^n \to [H^p(\partial\Omega,\sigma)]^n} \leq C_\theta \cdot \operatorname{dist}\left(\nu, \left[\operatorname{VMO}(\partial\Omega, \sigma)\right]^n\right)^\theta. \quad (6.3.73)$$

As a corollary, whenever Ω *is an infinitesimally flat* AR *domain it follows that* $K^{\#}_{\text{slip}}$ *is a compact operator on* $\left[H^p(\partial\Omega, \sigma)\right]^n$ *for each* $p \in \left(\frac{n-1}{n}, 1\right]$.

(4) *For each fractional power* $\theta \in (0, 1)$ *there exists* $C_\theta \in (0, \infty)$*, depending only on* θ*,* n*, the* UR *constants of* $\partial\Omega$*, and the relative flatness ratio* $\wp(\Omega)/\operatorname{diam}\partial\Omega$ *such that, with the distances measured in the John-Nirenberg space,*

$$\rho_{\text{Fred}}\left(K_{\text{slip}}; \left[\operatorname{VMO}(\partial\Omega, \sigma)\right]^n\right) \leq C_\theta \cdot \operatorname{dist}\left(\nu, \left[\operatorname{VMO}(\partial\Omega, \sigma)\right]^n\right)^\theta,$$

$$\left\|K_{\text{slip}}\right\|^{\text{ess}}_{[\operatorname{BMO}(\partial\Omega,\sigma)]^n \to [\operatorname{BMO}(\partial\Omega,\sigma)]^n} \leq C_\theta \cdot \operatorname{dist}\left(\nu, \left[\operatorname{VMO}(\partial\Omega, \sigma)\right]^n\right)^\theta.$$

In addition, whenever Ω *is an infinitesimally flat* AR *domain it follows that* K_{slip} *is a compact operator both on* $\left[\operatorname{BMO}(\partial\Omega, \sigma)\right]^n$ *and on* $\left[\operatorname{VMO}(\partial\Omega, \sigma)\right]^n$.

(5) *For each choice of exponents*

$$p \in \left(\tfrac{n-1}{n}, \infty\right), \quad q \in (0, \infty), \quad (n-1)\left(\tfrac{1}{p} - 1\right)_+ < s < 1, \quad (6.3.74)$$

there exists a power $\theta \in (0, 1)$ *depending only on* p, q, s *along with some* $C \in (0, \infty)$*, depending only on* n*,* p*,* q*,* s*, the* UR *constants of* $\partial\Omega$*, and the relative flatness ratio* $\wp(\Omega)/\operatorname{diam}\partial\Omega$ *such that, with the distance measured in the John-Nirenberg space,*

$$\left\|K_{\text{slip}}\right\|^{\text{ess}}_{[B^{p,q}_s(\partial\Omega,\sigma)]^n \to [B^{p,q}_s(\partial\Omega,\sigma)]^n} \leq C \cdot \operatorname{dist}\left(\nu, \left[\operatorname{VMO}(\partial\Omega, \sigma)\right]^n\right)^\theta,$$

Furthermore, whenever Ω *is an infinitesimally flat* AR *domain it follows that* K_{slip} *is a compact operator on* $\left[B^{p,q}_s(\partial\Omega, \sigma)\right]^n$ *for any* p, q, s *as in* (6.3.74). *Moreover, a similar result is valid in the context of Triebel-Lizorkin spaces. Specifically, given any*

$$p \in \left(\tfrac{n-1}{n}, \infty\right), \quad q \in \left(\tfrac{n-1}{n}, \infty\right), \quad (n-1)\left(\tfrac{1}{\min\{p,q\}} - 1\right)_+ < s < 1 \quad (6.3.75)$$

there exists a power $\theta \in (0, 1)$ *which depends only on* p, q, s *together with some constant* $C \in (0, \infty)$*, which depends only on* n*,* p*,* q*,* s*, the* UR *constants of* $\partial\Omega$*, and the relative flatness ratio* $\wp(\Omega)/\operatorname{diam}\partial\Omega$ *such that, with the distance measured in the John-Nirenberg space,*

$$\left\|K_{\text{slip}}\right\|^{\text{ess}}_{[F^{p,q}_s(\partial\Omega,\sigma)]^n \to [F^{p,q}_s(\partial\Omega,\sigma)]^n} \leq C \cdot \operatorname{dist}\left(\nu, \left[\operatorname{VMO}(\partial\Omega, \sigma)\right]^n\right)^\theta.$$

In addition, whenever Ω *is an infinitesimally flat* AR *domain it follows that* K_{slip} *is a compact operator on* $\big[F^{p,q}_{s}(\partial\Omega,\sigma)\big]^n$ *for any* p, q, s *as in* (6.3.75).

(6) *For each choice of exponents*

$$s \in (0,1), \quad p \in \big(\tfrac{n-1}{n-s}, \infty\big), \quad q \in (0,\infty), \tag{6.3.76}$$

there exists a power $\theta \in (0,1)$ *which depends only on* p, q, s *along with a constant* $C \in (0,\infty)$*, which depends only on* n*,* p*,* q*,* s*, the* UR *constants of* $\partial\Omega$*, and the relative flatness ratio* $\wp(\Omega)/\operatorname{diam}\partial\Omega$ *so that, with the distance measured in the John-Nirenberg space,*

$$\big\|K^{\#}_{\text{slip}}\big\|^{\text{ess}}_{[B^{p,q}_{-s}(\partial\Omega,\sigma)]^n \to [B^{p,q}_{-s}(\partial\Omega,\sigma)]^n} \leq C \cdot \operatorname{dist}\big(\nu,\ \big[\mathrm{VMO}(\partial\Omega,\sigma)\big]^n\big)^{\theta}.$$

Also, whenever Ω *is an infinitesimally flat* AR *domain it follows that* $K^{\#}_{\text{slip}}$ *is a compact operator on* $\big[B^{p,q}_{-s}(\partial\Omega,\sigma)\big]^n$ *for any* p, q, s *as in* (6.3.76).
Similarly, in the context of Triebel-Lizorkin spaces, for each choice of exponents

$$s \in (0,1), \quad p \in \big(\tfrac{n-1}{n-s}, \infty\big), \quad q \in \big(\tfrac{n-1}{n-s}, \infty\big), \tag{6.3.77}$$

there exists a power $\theta \in (0,1)$ *which depends only on* p, q, s *together with some constant* $C \in (0,\infty)$*, which depends only on* n*,* p*,* q*,* s*, the* UR *constants of* $\partial\Omega$*, and the relative flatness ratio* $\wp(\Omega)/\operatorname{diam}\partial\Omega$ *such that, with the distance measured in the John-Nirenberg space,*

$$\big\|K^{\#}_{\text{slip}}\big\|^{\text{ess}}_{[F^{p,q}_{-s}(\partial\Omega,\sigma)]^b \to [F^{p,q}_{-s}(\partial\Omega,\sigma)]^n} \leq C \cdot \operatorname{dist}\big(\nu,\ \big[\mathrm{VMO}(\partial\Omega,\sigma)\big]^n\big)^{\theta}.$$

In particular, whenever Ω *is an infinitesimally flat* AR *domain it follows that* $K^{\#}_{\text{slip}}$ *is a compact operator on* $\big[F^{p,q}_{-s}(\partial\Omega,\sigma)\big]^n$ *for any* p, q, s *as in* (6.3.77).

(7) *For each* $m \in \mathbb{N}$*,* $p, q \in (1,\infty)$*, and* $\lambda \in (0, n-1)$ *there exists some constant* $C_m \in (0,\infty)$*, depending only on* m*,* n*,* p*,* λ*, the* UR *constants of* $\partial\Omega$*, and the relative flatness ratio* $\wp(\Omega)/\operatorname{diam}\partial\Omega$ *such that, with the distance measured in the John-Nirenberg space,*

$$\big\|K_{\text{slip}}\big\|^{\text{ess}}_{[M^{p,\lambda}(\partial\Omega,\sigma)]^n \to [M^{p,\lambda}(\partial\Omega,\sigma)]^n} \leq C_m \Big[\operatorname{dist}\big(\nu,\ \big[\mathrm{VMO}(\partial\Omega,\sigma)\big]^n\big)\Big]^{\langle m\rangle},$$

$$\big\|K_{\text{slip}}\big\|^{\text{ess}}_{[\mathring{M}^{p,\lambda}(\partial\Omega,\sigma)]^n \to [\mathring{M}^{p,\lambda}(\partial\Omega,\sigma)]^n} \leq C_m \Big[\operatorname{dist}\big(\nu,\ \big[\mathrm{VMO}(\partial\Omega,\sigma)\big]^n\big)\Big]^{\langle m\rangle},$$

$$\big\|K_{\text{slip}}\big\|^{\text{ess}}_{[\mathcal{B}^{q,\lambda}(\partial\Omega,\sigma)]^n \to [\mathcal{B}^{q,\lambda}(\partial\Omega,\sigma)]^n} \leq C_m \Big[\operatorname{dist}\big(\nu,\ \big[\mathrm{VMO}(\partial\Omega,\sigma)\big]^n\big)\Big]^{\langle m\rangle},$$

$$\|K^{\#}_{\text{slip}}\|^{\text{ess}}_{[M^{p,\lambda}(\partial\Omega,\sigma)]^n\to[M^{p,\lambda}(\partial\Omega,\sigma)]^n}\leq C_m\Big[\text{dist}\,\big(\nu,\,\big[\text{VMO}(\partial\Omega,\sigma)\big]^n\big)\Big]^{\langle m\rangle},$$

$$\|K^{\#}_{\text{slip}}\|^{\text{ess}}_{[\mathring{M}^{p,\lambda}(\partial\Omega,\sigma)]^n\to[\mathring{M}^{p,\lambda}(\partial\Omega,\sigma)]^n}\leq C_m\Big[\text{dist}\,\big(\nu,\,\big[\text{VMO}(\partial\Omega,\sigma)\big]^n\big)\Big]^{\langle m\rangle},$$

$$\|K^{\#}_{\text{slip}}\|^{\text{ess}}_{[\mathcal{B}^{q,\lambda}(\partial\Omega,\sigma)]^n\to[\mathcal{B}^{q,\lambda}(\partial\Omega,\sigma)]^n}\leq C_m\Big[\text{dist}\,\big(\nu,\,\big[\text{VMO}(\partial\Omega,\sigma)\big]^n\big)\Big]^{\langle m\rangle},$$

as well as

$$\rho_{\text{Fred}}\Big(K_{\text{slip}};\,\big[M^{p,\lambda}_1(\partial\Omega,\sigma)\big]^n\Big)\leq C_m\Big[\text{dist}\,\big(\nu,\,\big[\text{VMO}(\partial\Omega,\sigma)\big]^n\big)\Big]^{\langle m\rangle},$$

$$\rho_{\text{Fred}}\Big(K_{\text{slip}};\,\big[\mathring{M}^{p,\lambda}_1(\partial\Omega,\sigma)\big]^n\Big)\leq C_m\Big[\text{dist}\,\big(\nu,\,\big[\text{VMO}(\partial\Omega,\sigma)\big]^n\big)\Big]^{\langle m\rangle},$$

$$\rho_{\text{Fred}}\Big(K_{\text{slip}};\,\big[\mathcal{B}^{q,\lambda}_1(\partial\Omega,\sigma)\big]^n\Big)\leq C_m\Big[\text{dist}\,\big(\nu,\,\big[\text{VMO}(\partial\Omega,\sigma)\big]^n\big)\Big]^{\langle m\rangle}.$$

Also, whenever Ω is an infinitesimally flat AR *domain (cf. Definition 3.4.1) it follows that K_{slip} is a compact operator on $\big[M^{p,\lambda}(\partial\Omega,\sigma)\big]^n$, $\big[\mathring{M}^{p,\lambda}(\partial\Omega,\sigma)\big]^n$, $\big[\mathcal{B}^{q,\lambda}(\partial\Omega,\sigma)\big]^n$, $\big[M^{p,\lambda}_1(\partial\Omega,\sigma)\big]^n$, $\big[\mathring{M}^{p,\lambda}_1(\partial\Omega,\sigma)\big]^n$, and $\big[\mathcal{B}^{q,\lambda}_1(\partial\Omega,\sigma)\big]^n$, while $K^{\#}_{\text{slip}}$ is a compact operator on $\big[M^{p,\lambda}(\partial\Omega,\sigma)\big]^n$, $\big[\mathring{M}^{p,\lambda}(\partial\Omega,\sigma)\big]^n$, and $\big[\mathcal{B}^{q,\lambda}(\partial\Omega,\sigma)\big]^n$, for any $p,q\in(1,\infty)$ and $\lambda\in(0,n-1)$.*

(8) *Let $\mathcal{M}_{\partial\Omega}$ denote the Hardy-Littlewood maximal operator on $\partial\Omega$, and assume $\mathbb{X}$ is a Generalized Banach Function Space on $(\partial\Omega,\sigma)$ with the property that both $\mathcal{M}_{\partial\Omega}:\mathbb{X}\to\mathbb{X}$ and $\mathcal{M}_{\partial\Omega}:\mathbb{X}'\to\mathbb{X}'$ are well-defined and bounded, where $\mathbb{X}'$ is the associated space of $\mathbb{X}$ (cf.* [130, Definitions 5.1.4, 5.1.11]). *Define $\mathring{\mathbb{X}}$ and $(\mathbb{X}')^\circ:=\overline{L^\infty(\partial\Omega,\sigma)}^{\|\cdot\|_{\mathbb{X}'}}$ as in* (6.3.52), *then bring back the $\mathbb{X}$-styled, $\mathbb{X}'$-styled, $\mathring{\mathbb{X}}$-styled, and $(\mathbb{X}')^\circ$-styled Sobolev spaces of order one on $\partial\Omega$ from* (6.3.53). *Then for each $m\in\mathbb{N}$ there exists $C_m\in(0,\infty)$, depending only on m, n, the operator norms of $\mathcal{M}_{\partial\Omega}$ on $\mathbb{X}$ and $\mathbb{X}'$, the* UR *constants of $\partial\Omega$, and the relative flatness ratio $\wp(\Omega)/\text{diam}\,\partial\Omega$ such that, with the distance measured in the John-Nirenberg space,*

$$\|K_{\text{slip}}\|^{\text{ess}}_{[\mathbb{X}]^n\to[\mathbb{X}]^n}\leq C_m\Big[\text{dist}\,\big(\nu,\,\big[\text{VMO}(\partial\Omega,\sigma)\big]^n\big)\Big]^{\langle m\rangle},\tag{6.3.78}$$

$$\|K_{\text{slip}}\|^{\text{ess}}_{[\mathbb{X}']^n\to[\mathbb{X}']^n}\leq C_m\Big[\text{dist}\,\big(\nu,\,\big[\text{VMO}(\partial\Omega,\sigma)\big]^n\big)\Big]^{\langle m\rangle},\tag{6.3.79}$$

$$\|K_{\text{slip}}\|^{\text{ess}}_{[\mathring{\mathbb{X}}]^n\to[\mathring{\mathbb{X}}]^n}\leq C_m\Big[\text{dist}\,\big(\nu,\,\big[\text{VMO}(\partial\Omega,\sigma)\big]^n\big)\Big]^{\langle m\rangle},\tag{6.3.80}$$

$$\|K_{\text{slip}}\|^{\text{ess}}_{[(\mathbb{X}')^\circ]^n\to[(\mathbb{X}')^\circ]^n}\leq C_m\Big[\text{dist}\,\big(\nu,\,\big[\text{VMO}(\partial\Omega,\sigma)\big]^n\big)\Big]^{\langle m\rangle},\tag{6.3.81}$$

$$\|K^{\#}_{\text{slip}}\|^{\text{ess}}_{[\mathbb{X}]^n\to[\mathbb{X}]^n} \leq C_m\Big[\text{dist}\left(\nu,\ \left[\text{VMO}(\partial\Omega,\sigma)\right]^n\right)\Big]^{\langle m\rangle}, \tag{6.3.82}$$

$$\|K^{\#}_{\text{slip}}\|^{\text{ess}}_{[\mathbb{X}']^n\to[\mathbb{X}']^n} \leq C_m\Big[\text{dist}\left(\nu,\ \left[\text{VMO}(\partial\Omega,\sigma)\right]^n\right)\Big]^{\langle m\rangle}, \tag{6.3.83}$$

$$\|K^{\#}_{\text{slip}}\|^{\text{ess}}_{[\mathring{\mathbb{X}}]^n\to[\mathring{\mathbb{X}}]^n} \leq C_m\Big[\text{dist}\left(\nu,\ \left[\text{VMO}(\partial\Omega,\sigma)\right]^n\right)\Big]^{\langle m\rangle}, \tag{6.3.84}$$

$$\|K^{\#}_{\text{slip}}\|^{\text{ess}}_{[(\mathbb{X}')^{\circ}]^n\to[(\mathbb{X}')^{\circ}]^n} \leq C_m\Big[\text{dist}\left(\nu,\ \left[\text{VMO}(\partial\Omega,\sigma)\right]^n\right)\Big]^{\langle m\rangle}, \tag{6.3.85}$$

as well as

$$\rho_{\text{Fred}}\Big(K_{\text{slip}};\ \left[\mathbb{X}_1\right]^n\Big) \leq C_m\Big[\text{dist}\left(\nu,\ \left[\text{VMO}(\partial\Omega,\sigma)\right]^n\right)\Big]^{\langle m\rangle}, \tag{6.3.86}$$

$$\rho_{\text{Fred}}\Big(K_{\text{slip}};\ \left[(\mathbb{X}')_1\right]^n\Big) \leq C_m\Big[\text{dist}\left(\nu,\ \left[\text{VMO}(\partial\Omega,\sigma)\right]^n\right)\Big]^{\langle m\rangle}. \tag{6.3.87}$$

$$\rho_{\text{Fred}}\Big(K_{\text{slip}};\ \left[\mathring{\mathbb{X}}_1\right]^n\Big) \leq C_m\Big[\text{dist}\left(\nu,\ \left[\text{VMO}(\partial\Omega,\sigma)\right]^n\right)\Big]^{\langle m\rangle}, \tag{6.3.88}$$

$$\rho_{\text{Fred}}\Big(K_{\text{slip}};\ \left[(\mathbb{X}')^{\circ}_1\right]^n\Big) \leq C_m\Big[\text{dist}\left(\nu,\ \left[\text{VMO}(\partial\Omega,\sigma)\right]^n\right)\Big]^{\langle m\rangle}. \tag{6.3.89}$$

Finally, if Ω is an infinitesimally flat AR *domain (cf. Definition 3.4.1) then K_{slip} is compact on $[\mathbb{X}]^n$, $[\mathbb{X}']^n$, $[\mathbb{X}_1]^n$, $[(\mathbb{X}')_1]^n$, $[\mathring{\mathbb{X}}]^n$, $[(\mathbb{X}')^{\circ}]^n$, $[\mathring{\mathbb{X}}_1]^n$, $[(\mathbb{X}')^{\circ}_1]^n$, while $K^{\#}_{\text{slip}}$ is compact on $[\mathbb{X}]^n$, $[\mathbb{X}']^n$, $[\mathring{\mathbb{X}}]^n$, $[(\mathbb{X}')^{\circ}]^n$.*

(9) *All results in item (8) are true whenever $\mathbb{X}$ is a rearrangement invariant Banach function space whose Boyd indices satisfy $1 < p_{\mathbb{X}} \leq q_{\mathbb{X}} < \infty$. In particular, all conclusions in item (8)are valid for $\mathbb{X} := L^{\Phi}(\partial\Omega,\sigma)$, the Orlicz space associated as in* (A.0.133) *with the measure space $(\partial\Omega,\sigma)$ and any Young function Φ satisfying $1 < i(\Phi) \leq I(\Phi) < \infty$ (when the role of $\mathbb{X}_1$ from* (5.4.83) *is now played by the Orlicz-based Sobolev space $L^{\Phi}_1(\partial\Omega,\sigma)$ from* (5.4.92)).

Proof The claims in item *(1)* are consequences of Theorem 3.4.2, Theorem 5.2.3, Theorem 5.4.1, (5.4.5), Corollary 5.2.4, Corollary 5.4.3, (5.4.32), Remark 5.4.2, (4.2.102), and (4.2.103). The claims in item *(2)* are implied by Theorem 3.4.2, Theorem 5.4.11, (4.2.102), (4.2.103), [132, (5.1.101)], and [130, Lemmas 2.2.6, 3.2.1]. Also, Theorem 3.4.2 and Theorem 5.4.9 together with (4.2.102) and (4.2.103) imply the claims in item *(3)*, while the claims in item *(4)* may be justified using Theorem 3.4.2, Corollary 5.4.10, (4.2.102), and (4.2.103). Pressing on, item *(5)* is seen from Theorem 3.4.2, Theorem 5.4.12, (4.2.102), and (4.2.103), while item *(6)* follows from Theorem 3.4.2, Theorem 5.4.13, (4.2.102), and (4.2.103). The claims in item *(7)* are consequences of Theorem 3.4.2, Theorem 5.4.4, Corollary 5.4.5, Theorem 5.4.6, Corollary 5.4.7, (4.2.102), and (4.2.103). Next, all conclusions in item *(8)* are justified based on Theorem 5.4.8, Theorem 3.4.2, (4.2.102), and (4.2.103). Lastly, the claims in item *(9)* are consequences of what we have proved in item *(8)* together with [130, Proposition 5.3.14], [130, Proposition 5.3.15], [130, (5.3.62)], (4.2.102), and (4.2.103). □

Chapter 7
Fredholmness and Invertibility of Layer Potentials on Compact Boundaries

A central question in Geometric Measure Theory is whether (or understanding how) geometric qualities translate into analytic properties. Here we are concerned with the correlation between the infinitesimal flatness of a given set of locally finite perimeter, quantified in terms of the mean oscillations of its geometric measure theoretic outward unit normal, and the functional analytic nature of SIO's defined on the boundary of the set in question.

The emerging principle from the discussion in §5-§6 is that, for "chord-dot-normal" singular integral operators, the degree of proximity to compact operators on a variety of function spaces defined on boundaries of δ-oscillating AR domains is controlled by the size of δ. That is, in said class of SIO's, the flatter the domain the closer to being a compact operator. We have also learned (see Proposition 5.1.1) that "chord-dot-normal" singular integral operators make up the largest category of singular integral operators within which such a principle can take effect.

To illustrate the rather dramatic impact which geometry can have on the nature of a singular integral operator, consider the classical case of harmonic double layer potential operators in a ball. Specifically, fix $n \in \mathbb{N}$ with $n \geq 2$, and consider $\Omega := B(0,1)$, the unit ball in $\mathbb{R}^n$. Then $\nu(x) = x$ for each $x \in S^{n-1} = \partial\Omega$ so for any two distinct points $x, y \in S^{n-1} = \partial\Omega$ we may compute

$$\begin{aligned} \frac{1}{\omega_{n-1}} \frac{\langle \nu(y), y - x \rangle}{|x-y|^n} &= \frac{1}{\omega_{n-1}} \frac{\langle y, y-x \rangle}{\big(|x-y|^2\big)^{n/2}} = \frac{1}{\omega_{n-1}} \frac{1 - \langle x, y \rangle}{\big(2 - 2\langle x, y \rangle\big)^{n/2}} \\ &= \frac{1}{2^{n/2}\omega_{n-1}} \frac{1}{\big(1 - \langle x, y\rangle\big)^{(n-2)/2}} = \frac{1}{2\omega_{n-1}} \frac{1}{|x-y|^{n-2}}. \end{aligned} \tag{7.0.1}$$

This shows that the boundary layer potentials K_Δ, $K_\Delta^{\#}$, S_Δ associated with the Laplacian and the set $\Omega := B(0,1)$ satisfy, as operators acting on functions $f \in L^1(S^{n-1}, \mathcal{H}^{n-1})$,

$$K_\Delta f = K_\Delta^{\#} f = \big(\tfrac{2-n}{2}\big) S_\Delta f \quad \text{if } n \geq 3, \tag{7.0.2}$$

$$K_\Delta f = K_\Delta^{\#} f = \frac{1}{2} \fint_{S^1} f \, d\mathcal{H}^1 \quad \text{if } n = 2. \tag{7.0.3}$$

For example, from (7.0.2) and [132, (1.5.21)] we see that

D. Mitrea et al., *Geometric Harmonic Analysis V*, Developments in Mathematics 76,
https://doi.org/10.1007/978-3-031-31561-9_7

$$S_\Delta 1 = \big(\tfrac{2}{2-n}\big)K_\Delta 1 = \tfrac{1}{2-n} \text{ if } n \geq 3. \tag{7.0.4}$$

An alternative way of computing $S_\Delta 1$ in all dimensions $n \geq 2$ is to start from the observation that the boundary-to-domain harmonic single layer $\mathscr{S}_\Delta$ associated with the unit ball satisfies

$$(\mathscr{S}_\Delta 1)(x) = \int_{S^{n-1}} E_\Delta(x-y)\,\mathrm{d}\mathcal{H}^{n-1}(y) = \omega_{n-1}E_\Delta(x),\ \forall x \in \mathbb{R}^n \setminus \overline{B(0,1)}, \tag{7.0.5}$$

with the final equality provided by the Mean Value Formula for the harmonic function $E_\Delta(x-\cdot)$ in $B(0,1)$, and then conclude that

$$S_\Delta 1 = \mathscr{S}_\Delta 1\big|_{\partial B(0,1)} = \omega_{n-1}E_\Delta \text{ on } \partial B(0,1). \tag{7.0.6}$$

In view of the specific format of the fundamental solution E_Δ (cf. (A.0.77)), this ultimately shows that

$$S_\Delta 1 = \begin{cases} \frac{1}{2-n} & \text{if } n \geq 3, \\ 0 & \text{if } n \geq 2. \end{cases} \tag{7.0.7}$$

In turn, from (7.0.7) and the uniqueness in the Dirichlet Problem for the Laplacian we may conclude that, if $n \geq 3$,

$$\frac{1}{(2-n)\omega_{n-1}}\int_{\partial B(0,1)} \frac{\mathrm{d}\mathcal{H}^{n-1}(y)}{|x-y|^{n-2}} = \begin{cases} (2-n)^{-1} & \text{if } x \in B(0,1), \\ (2-n)^{-1}|x|^{2-n} & \text{if } x \in \mathbb{R}^n \setminus \overline{B(0,1)}. \end{cases}$$

In addition, corresponding to $n = 2$, from the uniqueness in the Dirichlet Problem for the Laplacian in the unit disk plus (7.0.7), and from (7.0.6), we see that

$$\frac{1}{2\pi}\int_{S^1} \ln|x-y|\,\mathrm{d}\mathcal{H}^1(y) = \begin{cases} 0 & \text{if } x \in B(0,1), \\ \ln|x| & \text{if } x \in \mathbb{R}^2 \setminus \overline{B(0,1)}. \end{cases} \tag{7.0.8}$$

Also, (7.0.3) implies that, given any integrability exponent $p \in (1,\infty)$ and aperture parameter $\kappa > 0$, for each datum $f \in L^p(S^1, \mathcal{H}^1)$ the unique solution of the Dirichlet Problem for the two-dimensional Laplacian in the unit disk, i.e.,

$$\begin{cases} u \in \mathscr{C}^\infty(B(0,1)),\ \Delta u = 0 \text{ in } B(0,1),\ \mathcal{N}_\kappa u \in L^p(S^1, \mathcal{H}^1), \\ u\big|_{S^1}^{\kappa-\mathrm{n.t.}} = f \text{ at } \mathcal{H}^1\text{-a.e. point on } S^1, \end{cases} \tag{7.0.9}$$

is given by

$$u := 2\mathcal{D}_\Delta f - \fint_{S^1} f\,\mathrm{d}\mathcal{H}^1 \text{ in } B(0,1). \tag{7.0.10}$$

In this vein, it is also worth noting that since at the level of integral kernels we have

$$\frac{2}{2\pi}\frac{\langle y, y-x\rangle}{|x-y|^2} - \frac{1}{2\pi} = \frac{1}{2\pi}\frac{1-|x|^2}{|x-y|^2} \text{ for every } x \in B(0,1) \text{ and } y \in S^1, \tag{7.0.11}$$

we may recast (7.0.10) as the more familiar Poisson integral representation formula

$$u(x) = \frac{1}{2\pi}\int_{S^1} \frac{1-|x|^2}{|x-y|^2} f(y)\,d\mathcal{H}^1(y) \text{ for each } x \in B(0,1). \tag{7.0.12}$$

Referring to the fact that "*The case of the Laplacian in the ball is relatively easy because of the existence of explicit formulas for the solution*", C. Kenig outlined in [92, p. 135] a program guided by the following key questions: "*What should we do in the case of a general domain, where explicit formulas are not available? What should we do to study systems of equations? What happens to our solutions as the domains become less smooth?*"

What we propose to show is that"good geometry" plus "chord-dot-normal" structure ensure favorable functional analytic properties for double layer potentials which, in turn, permit one to solve boundary value problems via layer potential operators associated with second-order weakly elliptic systems. We shall pursue this principle in its natural progression: in this chapter we establish Fredholm and invertibility properties of boundary layer potentials on compact surfaces, and in §8.1 we shall use these results to treat boundary value problems in domains with compact boundaries.

Collectively, these results point to the fact that: *the category of δ-oscillating* AR *domains, with $\delta \in (0,1)$ suitably small, is a most natural geometric environment where Fredholm theory, as originally envisioned, becomes applicable to boundary layer potential operators which, in turn, makes it possible to solve boundary value problems via layer potential methods for second-order weakly elliptic systems.*

As such, our work falls within the scope of the program outlined by A.P. Calderón in his 1978 ICM plenary address in which he advocates the use of layer potentials "*for much more general elliptic systems* [*than the Laplacian*]" – see [28, p. 90]. Our work in this chapter is also in line with the issue raised as an open problem by C. Kenig in [93, Problem 3.2.2, pp. 116-117], where he asked whether operators of layer potential type may be inverted on appropriate Lebesgue and Sobolev spaces in suitable subclasses on NTA domains with compact Ahlfors regular boundaries.

7.1 Fredholmness and Invertibility of Layer Potentials on δ-Oscillating AR Domains

The principal aim in this section is to prove the Fredholmness and invertibility of boundary layer potential operators in the class of δ-oscillating AR domains. Here is our first main result in this regard.

Theorem 7.1.1 *Let $\Omega \subseteq \mathbb{R}^n$ be an Ahlfors regular domain with compact boundary. Abbreviate $\sigma := \mathcal{H}^{n-1}\lfloor\partial\Omega$ and denote by ν the geometric measure theoretic outward unit normal to Ω. Next, let L be a homogeneous, second-order, constant complex coefficient, weakly elliptic $M \times M$ system in $\mathbb{R}^n$ for which $\mathfrak{A}_L^{\text{dis}} \neq \varnothing$. Pick $A \in \mathfrak{A}_L^{\text{dis}}$ and consider the boundary-to-boundary double layer potential operators K_A, $K_A^{\#}$ associ-*

ated with Ω *and the coefficient tensor* A *as in* (A.0.116) *and* (A.0.118), *respectively. Finally, choose some* $\varepsilon \in (0, \infty)$. *Then the following statements are true.*

(1) *For each exponent* $p \in (1, \infty)$ *and Muckenhoupt weight* $w \in A_p(\partial\Omega, \sigma)$ *there exists* $\delta_1 \in (0, 1)$ *depending only on* n, p, $[w]_{A_p}$, A, ε, *the* ADR *constants of* $\partial\Omega$, *and the relative flatness ratio* $\wp(\Omega)/\mathrm{diam}\,\partial\Omega$ *with the property that if*

$$\mathrm{dist}\Big(\nu, [\mathrm{VMO}(\partial\Omega, \sigma)]^n\Big) < \delta_1 \tag{7.1.1}$$

where the distance is considered in the space $\big[\mathrm{BMO}(\partial\Omega, \sigma)\big]^n$, *hence if* Ω *is a* δ_1*-oscillating* AR *domain* (*cf. Definition 3.4.1*), *then* Ω *becomes a* UR *domain and for each spectral parameter* $z \in \mathbb{C}$ *with* $|z| \geq \varepsilon$ *the following operators are Fredholm with index zero:*

$$zI + K_A : \big[L^p(\partial\Omega, w)\big]^M \longrightarrow \big[L^p(\partial\Omega, w)\big]^M, \tag{7.1.2}$$

$$zI + K_A : \big[L^p_1(\partial\Omega, w)\big]^M \longrightarrow \big[L^p_1(\partial\Omega, w)\big]^M, \tag{7.1.3}$$

$$zI + K^\#_A : \big[L^p(\partial\Omega, w)\big]^M \longrightarrow \big[L^p(\partial\Omega, w)\big]^M, \tag{7.1.4}$$

$$zI + K^\#_A : \big[L^p_{-1}(\partial\Omega, w)\big]^M \longrightarrow \big[L^p_{-1}(\partial\Omega, w)\big]^M. \tag{7.1.5}$$

In particular, corresponding to $w \equiv 1$, *the above operators are Fredholm with index zero when the spaces in question are considered with respect to the plain surface measure* σ. *In addition, for each* $q \in (0, \infty]$ *one may choose* $\delta_1 \in (0, 1)$ *small enough such that, whenever* (7.1.1) *holds, the following operators are also Fredholm with index zero:*

$$zI + K_A : \big[L^{p,q}(\partial\Omega, \sigma)\big]^M \longrightarrow \big[L^{p,q}(\partial\Omega, \sigma)\big]^M, \tag{7.1.6}$$

$$zI + K^\#_A : \big[L^{p,q}(\partial\Omega, \sigma)\big]^M \longrightarrow \big[L^{p,q}(\partial\Omega, \sigma)\big]^M, \tag{7.1.7}$$

$$zI + K_A : \big[L^{(p,q)}_1(\partial\Omega, \sigma)\big]^M \longrightarrow \big[L^{(p,q)}_1(\partial\Omega, \sigma)\big]^M, \tag{7.1.8}$$

where $L^{(p,q)}_1(\partial\Omega, \sigma)$ *is the brand of Lorentz-based Sobolev spaces introduced in* (5.4.27)-(5.4.28).

(2) *For each exponent* $\alpha \in (0, 1)$ *there exists some threshold* $\delta_2 \in (0, 1)$ *which depends only on* n, α, A, ε, *the* ADR *constants of* $\partial\Omega$, *and the relative flatness ratio* $\wp(\Omega)/\mathrm{diam}\,\partial\Omega$ *with the property that if*

$$\mathrm{dist}\Big(\nu, [\mathrm{VMO}(\partial\Omega, \sigma)]^n\Big) < \delta_2 \tag{7.1.9}$$

where the distance is considered in the space $\big[\mathrm{BMO}(\partial\Omega, \sigma)\big]^n$, *hence if* Ω *is a* δ_2*-oscillating* AR *domain* (*cf. Definition 3.4.1*), *then* Ω *becomes a* UR *domain and for each spectral parameter* $z \in \mathbb{C}$ *with* $|z| \geq \varepsilon$ *the following operators are Fredholm with index zero:*

$$zI + K_A : \big[\mathscr{C}^\alpha(\partial\Omega)\big]^M \longrightarrow \big[\mathscr{C}^\alpha(\partial\Omega)\big]^M, \tag{7.1.10}$$

$$zI + K_A : \big[\mathscr{C}^\alpha_{\text{van}}(\partial\Omega)\big]^M \longrightarrow \big[\mathscr{C}^\alpha_{\text{van}}(\partial\Omega)\big]^M. \tag{7.1.11}$$

(3) *For each* $p \in \big(\frac{n-1}{n}, 1\big]$ *there exists some threshold* $\delta_3 \in (0, 1)$ *which depends only on* n, p, A, ε, *the* ADR *constants of* $\partial\Omega$, *and the relative flatness ratio* $\wp(\Omega)/\text{diam}\,\partial\Omega$ *with the property that if*

$$\text{dist}\Big(\nu, [\text{VMO}(\partial\Omega, \sigma)]^n\Big) < \delta_3 \tag{7.1.12}$$

where the distance is considered in the space $\big[\text{BMO}(\partial\Omega, \sigma)\big]^n$, *hence if* Ω *is a* δ_3*-oscillating* AR *domain (cf. Definition 3.4.1), then* Ω *becomes a* UR *domain and for each spectral parameter* $z \in \mathbb{C}$ *with* $|z| \geq \varepsilon$ *the following operator is Fredholm with index zero:*

$$zI + K_A^\# : \big[H^p(\partial\Omega, \sigma)\big]^M \longrightarrow \big[H^p(\partial\Omega, \sigma)\big]^M. \tag{7.1.13}$$

(4) *There exists some* $\delta_4 \in (0, 1)$ *which depends only on* n, A, ε, *the* ADR *constants of* $\partial\Omega$, *and the relative flatness ratio* $\wp(\Omega)/\text{diam}\,\partial\Omega$ *with the property that if*

$$\text{dist}\Big(\nu, [\text{VMO}(\partial\Omega, \sigma)]^n\Big) < \delta_4 \tag{7.1.14}$$

where the distance is considered in the space $\big[\text{BMO}(\partial\Omega, \sigma)\big]^n$, *hence if* Ω *is a* δ_4*-oscillating* AR *domain (cf. Definition 3.4.1), then* Ω *becomes a* UR *domain and for each spectral parameter* $z \in \mathbb{C}$ *with* $|z| \geq \varepsilon$ *the following operators are Fredholm with index zero:*

$$zI + K_A : \big[\text{BMO}(\partial\Omega, \sigma)\big]^M \longrightarrow \big[\text{BMO}(\partial\Omega, \sigma)\big]^M, \tag{7.1.15}$$

$$zI + K_A : \big[\text{VMO}(\partial\Omega, \sigma)\big]^M \longrightarrow \big[\text{VMO}(\partial\Omega, \sigma)\big]^M. \tag{7.1.16}$$

(5) *Whenever*

$$p \in \big(\tfrac{n-1}{n}, \infty\big), \quad q \in (0, \infty), \quad (n-1)\big(\tfrac{1}{p} - 1\big)_+ < s < 1, \tag{7.1.17}$$

there exists some threshold $\delta_5 \in (0, 1)$ *which depends only on* n, p, q, s, A, ε, *the* ADR *constants of* $\partial\Omega$, *and the relative flatness ratio* $\wp(\Omega)/\text{diam}\,\partial\Omega$ *with the property that if*

$$\text{dist}\Big(\nu, [\text{VMO}(\partial\Omega, \sigma)]^n\Big) < \delta_5 \tag{7.1.18}$$

where the distance is considered in the space $\big[\text{BMO}(\partial\Omega, \sigma)\big]^n$, *hence if* Ω *is a* δ_5*-oscillating* AR *domain (cf. Definition 3.4.1), then* Ω *becomes a* UR *domain and for each spectral parameter* $z \in \mathbb{C}$ *with* $|z| \geq \varepsilon$ *the following operator is Fredholm with index zero:*

$$zI + K_A : \big[B^{p,q}_s(\partial\Omega, \sigma)\big]^M \longrightarrow \big[B^{p,q}_s(\partial\Omega, \sigma)\big]^M. \tag{7.1.19}$$

Furthermore, a similar result is valid in the context of Triebel-Lizorkin spaces. Specifically, given any

$$p \in \big(\tfrac{n-1}{n}, \infty\big), \quad q \in \big(\tfrac{n-1}{n}, \infty\big), \quad (n-1)\big(\tfrac{1}{\min\{p,q\}} - 1\big)_+ < s < 1 \tag{7.1.20}$$

there exists some threshold $\delta_5 \in (0,1)$ which depends only on n, p, q, s, A, ε, the ADR *constants of $\partial\Omega$, and the relative flatness ratio $\wp(\Omega)/\operatorname{diam}\partial\Omega$ with the property that if* (7.1.18) *holds (hence, if Ω is a δ_5-oscillating* AR *domain; cf. Definition 3.4.1), then for each spectral parameter $z \in \mathbb{C}$ with $|z| \geq \varepsilon$ the following operator is Fredholm with index zero:*

$$zI + K_A : \big[F^{p,q}_s(\partial\Omega, \sigma)\big]^M \longrightarrow \big[F^{p,q}_s(\partial\Omega, \sigma)\big]^M. \tag{7.1.21}$$

(6) Whenever

$$s \in (0,1), \quad p \in \big(\tfrac{n-1}{n-s}, \infty\big), \quad q \in (0, \infty), \tag{7.1.22}$$

there exists some threshold $\delta_6 \in (0,1)$ which depends only on n, p, q, s, A, ε, the ADR *constants of $\partial\Omega$, and the relative flatness ratio $\wp(\Omega)/\operatorname{diam}\partial\Omega$ with the property that if*

$$\operatorname{dist}\Big(\nu, [\mathrm{VMO}(\partial\Omega, \sigma)]^n\Big) < \delta_6 \tag{7.1.23}$$

where the distance is considered in the space $\big[\mathrm{BMO}(\partial\Omega, \sigma)\big]^n$, hence if Ω is a δ_6-oscillating AR *domain (cf. Definition 3.4.1), then Ω becomes a* UR *domain and for each spectral parameter $z \in \mathbb{C}$ with $|z| \geq \varepsilon$ the following operator is Fredholm with index zero:*

$$zI + K^\#_A : \big[B^{p,q}_{-s}(\partial\Omega, \sigma)\big]^M \longrightarrow \big[B^{p,q}_{-s}(\partial\Omega, \sigma)\big]^M. \tag{7.1.24}$$

In addition, a similar result is valid in the context of Triebel-Lizorkin spaces. Concretely, given any

$$s \in (0,1), \quad p \in \big(\tfrac{n-1}{n-s}, \infty\big), \quad q \in \big(\tfrac{n-1}{n-s}, \infty\big), \tag{7.1.25}$$

there exists some threshold $\delta_6 \in (0,1)$ which depends only on n, p, q, s, A, ε, the ADR *constants of $\partial\Omega$, and the relative flatness ratio $\wp(\Omega)/\operatorname{diam}\partial\Omega$ with the property that if* (7.1.23) *holds (hence, if Ω is a δ_6-oscillating* AR *domain; cf. Definition 3.4.1), then for each spectral parameter $z \in \mathbb{C}$ with $|z| \geq \varepsilon$ the following operator is Fredholm with index zero:*

$$zI + K^\#_A : \big[F^{p,q}_{-s}(\partial\Omega, \sigma)\big]^M \longrightarrow \big[F^{p,q}_{-s}(\partial\Omega, \sigma)\big]^M. \tag{7.1.26}$$

(7) For each $p, q \in (1, \infty)$ and $\lambda \in (0, n-1)$ there exists some threshold $\delta_7 \in (0,1)$ which depends only on n, p, q, λ, A, ε, the ADR *constants of $\partial\Omega$, and the relative flatness ratio $\wp(\Omega)/\operatorname{diam}\partial\Omega$ with the property that if*

$$\operatorname{dist}\Big(\nu, [\mathrm{VMO}(\partial\Omega,\sigma)]^n\Big) < \delta_7 \tag{7.1.27}$$

where the distance is considered in the space $\big[\mathrm{BMO}(\partial\Omega,\sigma)\big]^n$, *hence if* Ω *is a* δ_7*-oscillating* AR *domain (cf. Definition 3.4.1), then* Ω *becomes a* UR *domain and for each spectral parameter* $z \in \mathbb{C}$ *with* $|z| \geq \varepsilon$ *the following operators are Fredholm with index zero:*

$$zI + K_A : \big[M^{p,\lambda}(\partial\Omega,\sigma)\big]^M \longrightarrow \big[M^{p,\lambda}(\partial\Omega,\sigma)\big]^M, \tag{7.1.28}$$

$$zI + K_A : \big[M_1^{p,\lambda}(\partial\Omega,\sigma)\big]^M \longrightarrow \big[M_1^{p,\lambda}(\partial\Omega,\sigma)\big]^M, \tag{7.1.29}$$

$$zI + K_A : \big[\mathring{M}^{p,\lambda}(\partial\Omega,\sigma)\big]^M \longrightarrow \big[\mathring{M}^{p,\lambda}(\partial\Omega,\sigma)\big]^M, \tag{7.1.30}$$

$$zI + K_A : \big[\mathring{M}_1^{p,\lambda}(\partial\Omega,\sigma)\big]^M \longrightarrow \big[\mathring{M}_1^{p,\lambda}(\partial\Omega,\sigma)\big]^M, \tag{7.1.31}$$

$$zI + K_A : \big[\mathcal{B}^{q,\lambda}(\partial\Omega,\sigma)\big]^M \longrightarrow \big[\mathcal{B}^{q,\lambda}(\partial\Omega,\sigma)\big]^M, \tag{7.1.32}$$

$$zI + K_A : \big[\mathcal{B}_1^{q,\lambda}(\partial\Omega,\sigma)\big]^M \longrightarrow \big[\mathcal{B}_1^{q,\lambda}(\partial\Omega,\sigma)\big]^M, \tag{7.1.33}$$

and

$$zI + K_A^{\#} : \big[M^{p,\lambda}(\partial\Omega,\sigma)\big]^M \longrightarrow \big[M^{p,\lambda}(\partial\Omega,\sigma)\big]^M, \tag{7.1.34}$$

$$zI + K_A^{\#} : \big[\mathring{M}^{p,\lambda}(\partial\Omega,\sigma)\big]^M \longrightarrow \big[\mathring{M}^{p,\lambda}(\partial\Omega,\sigma)\big]^M, \tag{7.1.35}$$

$$zI + K_A^{\#} : \big[\mathcal{B}^{q,\lambda}(\partial\Omega,\sigma)\big]^M \longrightarrow \big[\mathcal{B}^{q,\lambda}(\partial\Omega,\sigma)\big]^M. \tag{7.1.36}$$

(8) With $\mathcal{M}_{\partial\Omega}$ *denoting the Hardy-Littlewood maximal operator on* $\partial\Omega$, *assume* $\mathbb{X}$ *is a Generalized Banach Function Space on* $(\partial\Omega,\sigma)$ *with the property that*

$$\mathcal{M}_{\partial\Omega} : \mathbb{X} \to \mathbb{X} \text{ and } \mathcal{M}_{\partial\Omega} : \mathbb{X}' \to \mathbb{X}' \text{ are well-defined and bounded,} \tag{7.1.37}$$

where $\mathbb{X}'$ *is the associated space of* $\mathbb{X}$ *(cf.* [130, Definitions 5.1.4, 5.1.11]). *Set*

$$\mathring{\mathbb{X}} := \overline{L^\infty(\partial\Omega,\sigma)}^{\|\cdot\|_{\mathbb{X}}} \text{ and } (\mathbb{X}')^\circ := \overline{L^\infty(\partial\Omega,\sigma)}^{\|\cdot\|_{\mathbb{X}'}}, \tag{7.1.38}$$

and consider the following $\mathbb{X}$*-styled,* $\mathbb{X}'$*-styled,* $\mathring{\mathbb{X}}$*-styled, and* $(\mathbb{X}')^\circ$*-styled Sobolev spaces of order one on* $\partial\Omega$:

$$\begin{aligned}
\mathbb{X}_1 &:= \Big\{f \in \mathbb{X} : \partial_{\tau_{jk}} f \in \mathbb{X} \text{ for } 1 \leq j,k \leq n\Big\}, \\
(\mathbb{X}')_1 &:= \Big\{g \in \mathbb{X}' : \partial_{\tau_{jk}} g \in \mathbb{X}' \text{ for } 1 \leq j,k \leq n\Big\}, \\
\mathring{\mathbb{X}}_1 &:= \Big\{f \in \mathring{\mathbb{X}} : \partial_{\tau_{jk}} f \in \mathring{\mathbb{X}} \text{ for } 1 \leq j,k \leq n\Big\}, \\
(\mathbb{X}')_1^\circ &:= \Big\{g \in (\mathbb{X}')^\circ : \partial_{\tau_{jk}} g \in (\mathbb{X}')^\circ \text{ for } 1 \leq j,k \leq n\Big\},
\end{aligned} \tag{7.1.39}$$

(which are]known to be Banach when equipped with natural norms; cf. (5.4.84)).

Then there exists a threshold $\delta_{\mathbb{X}} \in (0,1)$, which depends only on n, A, ε, the operator norms of $\mathcal{M}_{\partial\Omega}$ on $\mathbb{X}$ and $\mathbb{X}'$, the ADR *constants of $\partial\Omega$, and the relative flatness ratio $\wp(\Omega)/\mathrm{diam}\,\partial\Omega$, with the property that if*

$$\mathrm{dist}\Big(\nu, [\mathrm{VMO}(\partial\Omega,\sigma)]^n\Big) < \delta_{\mathbb{X}} \tag{7.1.40}$$

with the distance considered in $\big[\mathrm{BMO}(\partial\Omega,\sigma)\big]^n$ (hence if Ω is a $\delta_{\mathbb{X}}$-oscillating AR *domain; cf. Definition 3.4.1), it follows that Ω becomes a* UR *domain and for each spectral parameter $z \in \mathbb{C}$ with $|z| \geq \varepsilon$ the following operators are Fredholm with index zero:*

$$zI + K_A : [\mathbb{X}]^M \to [\mathbb{X}]^M, \quad zI + K_A : [\mathbb{X}']^M \to [\mathbb{X}']^M, \tag{7.1.41}$$

$$zI + K_A : [\mathring{\mathbb{X}}]^M \to [\mathring{\mathbb{X}}]^M, \quad zI + K_A : [(\mathbb{X}')^\circ]^M \to [(\mathbb{X}')^\circ]^M, \tag{7.1.42}$$

and

$$zI + K_A^\# : [\mathbb{X}]^M \to [\mathbb{X}]^M, \quad zI + K_A^\# : [\mathbb{X}']^M \to [\mathbb{X}']^M, \tag{7.1.43}$$

$$zI + K_A^\# : [\mathring{\mathbb{X}}]^M \to [\mathring{\mathbb{X}}]^M, \quad zI + K_A^\# : [(\mathbb{X}')^\circ]^M \to [(\mathbb{X}')^\circ]^M, \tag{7.1.44}$$

as well as

$$zI + K_A : \big[\mathbb{X}_1\big]^M \to \big[\mathbb{X}_1\big]^M, \quad zI + K_A : \big[(\mathbb{X}')_1\big]^M \to \big[(\mathbb{X}')_1\big]^M, \tag{7.1.45}$$

$$zI + K_A : \big[\mathring{\mathbb{X}}_1\big]^M \to \big[\mathring{\mathbb{X}}_1\big]^M, \quad zI + K_A : \big[(\mathbb{X}')_1^\circ\big]^M \to \big[(\mathbb{X}')_1^\circ\big]^M. \tag{7.1.46}$$

(9) All conclusions in item (8) are valid whenever $\mathbb{X}$ is a rearrangement invariant Banach function space whose Boyd indices satisfy $1 < p_{\mathbb{X}} \leq q_{\mathbb{X}} < \infty$. In particular, all conclusions in item (8) hold if $\mathbb{X} := L^\Phi(\partial\Omega,\sigma)$, the Orlicz space associated as in (A.0.133) *with the measure space $(\partial\Omega,\sigma)$ and any Young function Φ satisfying $1 < i(\Phi) \leq I(\Phi) < \infty$ (the role of $\mathbb{X}_1$ from* (5.4.83) *now played by the Orlicz-based Sobolev space $L_1^\Phi(\partial\Omega,\sigma)$ from* (5.4.92)*).*

(10) Completely analogous Fredholmness results[1] *to those recorded in items (1)-(9) above are valid for the slip double layer potential operators $K_{\rm slip}$, $K_{\rm slip}^\#$ associated with the Stokes system in Ω as in* (4.2.102) *and* (4.2.103)*, respectively.*

Proof All claims are consequences of Theorem 3.4.2, Theorem 6.3.3, [130, Theorem 2.1.2], and Theorem 6.3.4. □

We continue by making several remarks aimed to further clarify the nature and scope of Theorem 7.1.1. In particular, Remarks 2-3 below highlight the optimality of the hypotheses adopted in Theorem 7.1.1.

Remark 1. Observe that we may use (A.0.80) to recast the Fredholmness result in (7.1.2) (with $w \equiv 1$) as the Fredholm radius estimate

[1] formulated for $M := n$ and now disregarding the coefficient tensor A

$$\rho_{\text{Fred}}\Big(K_A;\ \big[L^p(\partial\Omega,\sigma)\big]^M\Big) < \varepsilon. \tag{7.1.47}$$

In fact, all Fredholmness results in Theorem 7.1.1 may be rephrased according to this template. Since (7.1.47) holds under the assumption (7.1.1), we ultimately have a the following result (valid under the background assumptions made in Theorem 7.1.1):

for each given number $\varepsilon \in (0,\infty)$ there exists a threshold $\delta \in (0,1)$ (which is dependent only on ε, n, p, A, the ADR constants of $\partial\Omega$, and the relative flatness ratio $\wp(\Omega)/\text{diam}\,\partial\Omega)$ with the property that $\rho_{\text{Fred}}\Big(K_A;\ \big[L^p(\partial\Omega,\sigma)\big]^M\Big) < \varepsilon$ if $\text{dist}\Big(\nu, [\text{VMO}(\partial\Omega,\sigma)]^n\Big) < \delta$. (7.1.48)

This amounts to an ε-δ criterion in which the geometry controls the functional analysis. In fact, this may be quantified more precisely since, according to item *(1)* in Theorem 6.3.3 and [130, (2.2.16)], for each integer $m \in \mathbb{N}$ there exists some $C_m \in (0,\infty)$, which depends only on m, n, p, A, the UR constants of $\partial\Omega$, and the relative flatness ratio $\wp(\Omega)/\text{diam}\,\partial\Omega$ such that, with the piece of notation introduced in (4.1.12), we have

$$\rho_{\text{Fred}}\Big(K_A;\ \big[L^p(\partial\Omega,\sigma)\big]^M\Big) \leq C_m\Big[\text{dist}\big(\nu, \big[\text{VMO}(\partial\Omega,\sigma)\big]^n\big)\Big]^{\langle m\rangle}. \tag{7.1.49}$$

At least for certain operators and certain domains there is an accompanying estimate going in the opposite direction. Specifically, work in the two-dimensional setting and suppose $\Omega \subseteq \mathbb{R}^2$ is a Radon domain. Denote by ν the outward unit normal vector to Ω and abbreviate $\sigma := \mathcal{H}^1\lfloor\partial\Omega$. Also, take L to be the Laplacian in the plane and consider the harmonic double layer potential operator K_Δ associated with Ω as in (6.2.1). Then from (6.3.3) and (6.3.6) we conclude that

$$c\cdot\text{dist}\Big(\nu, [\text{VMO}(\partial\Omega,\sigma)]^2\Big) \leq \rho_{\text{Fred}}\Big(K_\Delta;\ L^p(\partial\Omega,\sigma)\Big) \tag{7.1.50}$$

where $c \in (0,\infty)$ depends only on the p and the Ahlfors regularity constants of $\partial\Omega$. In particular, this shows that the smallness result formulated in (7.1.48) is optimal, in the sense that, generally speaking, the Fredholm radius can only be small provided $\text{dist}\Big(\nu, [\text{VMO}(\partial\Omega,\sigma)]^n\Big)$ is small. Ultimately, this points to the fact that Theorem 7.1.1 is optimal in this regard.

Remark 2. We wish to comment on the role of the hypothesis $A \in \mathfrak{A}_L^{\text{dis}}$ in the context of Theorem 7.1.1. Specifically, we claim that

if $\mathfrak{A}_L^{\text{dis}} = \varnothing$ then the conclusions in Theorem 7.1.1 may actually fail, even when Ω is a bounded domain of class $\mathscr{C}^\infty$ in $\mathbb{R}^n$, for each double layer potential operator K_A associated with said domain Ω and each coefficient tensor $A \in \mathfrak{A}_L$. (7.1.51)

To substantiate this claim, work in the case when $n = 2$ and bring in the second-order, homogeneous, real constant coefficient, weakly elliptic 2×2 system $L := \mathbb{L}_B$ defined in (1.6.110). From the discussion in (0.0.2)-(0.0.5) we know that if

$$\Omega := \{z \in \mathbb{C} : |z| < 1\} \tag{7.1.52}$$

then

$$\mathscr{X} := \Big\{u \in \big[\mathscr{C}^\infty(\overline{\Omega})\big]^2 : Lu = 0 \text{ in } \Omega \text{ and } u\big|_{\partial\Omega} = 0\Big\} \text{ is an infinite dimensional vector space.} \tag{7.1.53}$$

Next, with $L := \mathbb{L}_B$, where $\mathbb{L}_B$ is as in [132, (8.1.136)], and with Ω as in (7.1.52), associate the boundary-to-domain single layer potential operator $\mathscr{S}$ as in (A.0.228). Also, having fixed an arbitrary coefficient tensor $A \in \mathfrak{A}_L$, associate the boundary layer potential operator $K^{\#}_{A^\top}$ with Ω and $A^\top$ as in (A.0.118). Then the integral representation formula [132, (1.3.35)] used for each $u \in \mathscr{X}$ presently gives

$$u = -\mathscr{S}(\partial^A_\nu u) \text{ in } \Omega. \tag{7.1.54}$$

On account of [132, (1.5.66)], applying the conormal derivative ∂^A_ν to both sides yields $\partial^A_\nu u = -(-\frac{1}{2}I + K^{\#}_{A^\top})(\partial^A_\nu u)$, so $\partial^A_\nu u \in \operatorname{Ker}\big(\frac{1}{2}I + K^{\#}_{A^\top}; \big[L^p(\partial\Omega, \sigma)\big]^2\big)$ for each $p \in (1, \infty)$, where $\sigma := \mathcal{H}^1 \lfloor \partial\Omega$. From this and (7.1.54) we then see that

$$\mathscr{X} \ni u \longmapsto \partial^A_\nu u \in \operatorname{Ker}\Big(\tfrac{1}{2}I + K^{\#}_{A^\top}; \big[L^p(\partial\Omega, \sigma)\big]^2\Big) \tag{7.1.55}$$

is a well-defined, linear, injective mapping. Together with (7.1.53), this ultimately shows that $\operatorname{Ker}\big(\frac{1}{2}I + K^{\#}_{A^\top}; \big[L^p(\partial\Omega, \sigma)\big]^2\big)$ is an infinite dimensional vector space. In particular, this prevents the operator $\frac{1}{2}I + K^{\#}_{A^\top}$ from being Fredholm on $\big[L^p(\partial\Omega, \sigma)\big]^2$ for any $p \in (1, \infty)$. Since this contradicts (7.1.4) (written for $L^\top$ in place of L, for $A^\top$ in place of A, and with $z := 1/2$, $w \equiv 1$), the claim in (7.1.51) is therefore proved.

Remark 3. We wish to note that even in the class of bounded Lipschitz domains the conclusions of Theorem 7.1.1 may fail if the demand on the proximity of the outward unit normal to the Sarason space VMO space is violated. To illustrate this, work in the two-dimensional Euclidean space $\mathbb{R}^2 \equiv \mathbb{C}$. For each fixed angle $\beta \in (0, 2\pi)$, bring in the sector $D_\beta := \{z \in \mathbb{C} : |\arg z| < \beta/2\}$, then consider a bounded, connected, simply connected Lipschitz domain $\Omega_\beta \subseteq D_\beta$ with the property that $\Omega_\beta \cap B(0,1) = D_\beta \cap B(0,1)$ and $\partial\Omega_\beta \setminus \{0\}$ is a smooth curve[2].

In this setting, we claim that the operator (7.1.2) (with $n := 2$, $M := 1$, $L := \Delta$, $A := I_{2\times 2} \in \mathfrak{A}^{\text{dis}}_\Delta$, $z := 1/2$, $w \equiv 1$, and $\Omega := \Omega_\beta$) may fail to be Fredholm with index zero for certain integrability exponents $p \in (1, \infty)$ and certain Lipschitz domains with compact boundary Ω if the proximity condition (7.1.1) does not materialize. More precisely, Proposition 6.2.2 shows that for any exponent $p \in (1, \infty)$ and any angle $\beta \in (0, 2\pi)$ we have (with Ω_β the drop-shaped domain described above):

[2] e.g., take Ω_β to be shaped like a drop

$$\begin{array}{l}\text{for each } z \text{ belonging to } \mathrm{Spec}\big(K_\Delta; L^p(\partial\Omega_\beta,\sigma)\big), \text{ the spectrum of } K_\Delta \\ \text{acting on the space } L^p(\partial\Omega_\beta,\sigma), \text{ which is not an isolated point in} \\ \mathrm{Spec}\big(K_\Delta; L^p(\partial\Omega_\beta,\sigma)\big), \text{ the operator } zI + K_\Delta \text{ is not Fredholm with} \\ \text{index zero on } L^p(\partial\Omega_\beta,\sigma).\end{array} \tag{7.1.56}$$

In fact, using notation introduced in relation to Proposition 6.2.2 (cf. (4.2.196)),

$$\begin{array}{l}\text{given any angle } \beta \in (0,2\pi) \text{ and any exponent } p \in (1,\infty), \text{ it follows} \\ \text{that for each spectral parameter } z \in \Sigma_\beta(p) \text{ the operator } zI + K_\Delta \text{ fails} \\ \text{to be Fredholm on } L^p(\partial\Omega_\beta,\sigma) \text{ altogether.}\end{array} \tag{7.1.57}$$

Via duality (cf. [130, Theorem 2.1.2, *(3)*] and [132, Theorem 1.5.1, *(iii)*]), from this we also obtain that

$$\begin{array}{l}\text{given any angle } \beta \in (0,2\pi) \text{ and any two Hölder conjugate exponents} \\ p, p' \in (1,\infty), \text{ it follows that for each spectral parameter } z \in \Sigma_\beta(p') \\ \text{the operator } zI + K^\#_\Delta \text{ fails to be Fredholm when acting on } L^p(\partial\Omega_\beta,\sigma) \\ \text{(where } K^\#_\Delta \text{ denotes the transpose harmonic double layer potential} \\ \text{operator associated with } \Omega_\beta).\end{array} \tag{7.1.58}$$

From (7.1.57) and (7.1.58) we then see that the operators (7.1.2) and (7.1.4) (with $n := 2$, $M := 1$, $L := \Delta$, $A := I_{2\times 2} \in \mathfrak{A}^{\mathrm{dis}}_\Delta$, $w \equiv 1$, and $\Omega := \Omega_\beta$) may fail to be Fredholm for any given $p \in (1,\infty)$ and z as in (7.1.57) and, respectively, as in (7.1.58), for certain Lipschitz domains Ω with compact boundary if the proximity condition (7.1.1) is violated.

Finally, we wish to remark that similar counterexamples to the veracity of the conclusions in Theorem 7.1.1, in the case when the demand on the proximity of the outward unit normal to the Sarason space VMO fails to materialize, are valid for the principal-value (pseudo-traction) elastic double layer potential operator K_ψ (associated as in (A.0.116) with the coefficient tensor [132, (1.4.169)], used to represent the Lamé system as in [132, (1.4.170)], corresponding to the choice $t := \frac{\lambda+\mu}{\lambda+3\mu}$ which makes said coefficient tensor distinguished; cf. (1.6.116)). See the discussion in §6.

Remark 4. The functional analytic nature of the double layer operator on Orlicz spaces, which we address in item *(9)*, has attracted attention in the literature; see, e.g., [208, Question 3, p. 603].

Remark 5. A more detailed formulation of the Fredholmness result corresponding to (7.1.2) in Theorem 7.1.1 reads as follows: Fix $n, M \in \mathbb{N}$ with $n \geq 2$ and consider a weakly elliptic homogeneous constant complex coefficient second-order $M \times M$ system L in $\mathbb{R}^n$ with $\mathfrak{A}^{\mathrm{dis}}_L \neq \varnothing$. Then for each constants $C_A, C_R, C_W \in (0,\infty)$, each compact interval $I \subset (1,\infty)$, each spectral parameter $\varepsilon \in (0,\infty)$, and each coefficient tensor $A \in \mathfrak{A}^{\mathrm{dis}}_L$ there exists a threshold $\delta \in (0,1)$ which depends only on n, C_A, C_R, C_W, I, ε, and A with the following significance. Assume $\Omega \subseteq \mathbb{R}^n$ is an Ahlfors regular domain with compact boundary such that the Ahlfors regularity constants of $\partial\Omega$ are controlled by C_A, and the relative flatness ratio $\wp(\Omega)/\mathrm{diam}\,\partial\Omega$ is $\leq C_R$. Abbreviate $\sigma := \mathcal{H}^{n-1}\lfloor\partial\Omega$ and denote by ν the geometric measure theoretic outward unit normal to Ω. Also, fix an integrability exponent $p \in I$ and a Muckenhoupt weight

$w \in A_p(\partial\Omega, \sigma)$ with $[w]_{A_p} \leq C_W$. Finally, consider the boundary-to-boundary double layer potential operator K_A, associated with the set Ω and the coefficient tensor A as in (A.0.116). Then whenever $\operatorname{dist}\Big(\nu, [\mathrm{VMO}(\partial\Omega, \sigma)]^n\Big) < \delta$ it follows that Ω is a UR domain and for each $z \in \mathbb{C}$ with $|z| \geq \varepsilon$ the operator $zI + K_A$ is Fredholm with index zero when acting on $\big[L^p(\partial\Omega, w)\big]^M$. In fact, analogous considerations apply to all Fredholmness results in Theorem 7.1.1.

Moving on, the point of our next result is that in the class of δ-oscillating AR domains, with $\delta \in (0, 1)$ suitably small, the null-spaces of boundary layer potential operators are actually stable, in the sense that they do not change as the function spaces (on which said operators are considered) vary. At the heart of the matter is the fact that having Ω a δ-oscillating AR domain with $\delta \in (0, 1)$ suitably small is a geometric feature which forces functions/distributions from the kernels of $zI + K_A$, $zI + K_A^{\#}$ with $A \in \mathfrak{A}_L^{\mathrm{dis}}$ to be more regular than under ordinary circumstances.

Proposition 7.1.2 *Fix $n \in \mathbb{N}$ with $n \geq 2$ and suppose $\Omega \subseteq \mathbb{R}^n$ is an Ahlfors regular domain with compact boundary. Abbreviate $\sigma := \mathcal{H}^{n-1}\lfloor\partial\Omega$ and denote by ν the geometric measure theoretic outward unit normal to Ω. Next, let L be a homogeneous, second-order, constant complex coefficient, weakly elliptic $M \times M$ system in $\mathbb{R}^n$ for which $\mathfrak{A}_L^{\mathrm{dis}} \neq \emptyset$. Pick a coefficient tensor $A \in \mathfrak{A}_L^{\mathrm{dis}}$ and consider the layer potential operators K_A, $K_A^{\#}$ associated with A and Ω as in* (A.0.116) *and* (A.0.118), *respectively. Also, fix four compact intervals, $I_1 \subseteq \big(\frac{n-1}{n}, 1\big]$, $I_2 \subseteq (1, \infty)$, $I_3 \subseteq (0, 1)$, and $I_4 \subseteq (0, n-1)$. Finally, pick $\varepsilon \in (0, \infty)$.*

Then there exists a threshold $\delta \in (0, 1)$, depending on n, I_1, I_2, I_3, I_4, ε, A, the ADR *constants of $\partial\Omega$, and the relative flatness ratio $\wp(\Omega)/\operatorname{diam}\partial\Omega$ with the property that whenever*

$$\operatorname{dist}\Big(\nu, [\mathrm{VMO}(\partial\Omega, \sigma)]^n\Big) < \delta \tag{7.1.59}$$

with the distance considered in $\big[\mathrm{BMO}(\partial\Omega, \sigma)\big]^n$ (hence if Ω is a δ-oscillating AR *domain; cf. Definition 3.4.1), it follows that Ω is a* UR *domain and for each spectral parameter $z \in \mathbb{C}$ with $|z| \geq \varepsilon$ and each $r \in I_1$, $p, q \in I_2$, $\alpha \in I_3$, $\lambda \in I_4$, one has*

$$\begin{aligned}
\operatorname{Ker}\Big(zI + K_A;\ \big[L^p(\partial\Omega, \sigma)\big]^M\Big) &= \operatorname{Ker}\Big(zI + K_A;\ \big[L_1^q(\partial\Omega, \sigma)\big]^M\Big) \\
&= \operatorname{Ker}\Big(zI + K_A;\ \big[\mathscr{C}^\alpha(\partial\Omega)\big]^M\Big) = \operatorname{Ker}\Big(zI + K_A;\ \big[M^{p,\lambda}(\partial\Omega, \sigma)\big]^M\Big) \\
&= \operatorname{Ker}\Big(zI + K_A;\ \big[\mathring{M}^{p,\lambda}(\partial\Omega, \sigma)\big]^M\Big) = \operatorname{Ker}\Big(zI + K_A;\ \big[\mathcal{B}^{q,\lambda}(\partial\Omega, \sigma)\big]^M\Big) \\
&= \operatorname{Ker}\Big(zI + K_A;\ \big[M_1^{p,\lambda}(\partial\Omega, \sigma)\big]^M\Big) = \operatorname{Ker}\Big(zI + K_A;\ \big[\mathring{M}_1^{p,\lambda}(\partial\Omega, \sigma)\big]^M\Big) \\
&= \operatorname{Ker}\Big(zI + K_A;\ \big[\mathcal{B}_1^{q,\lambda}(\partial\Omega, \sigma)\big]^M\Big) = \operatorname{Ker}\Big(zI + K_A;\ \big[\mathrm{BMO}(\partial\Omega, \sigma)\big]^M\Big) \\
&= \operatorname{Ker}\Big(zI + K_A;\ \big[\mathrm{VMO}(\partial\Omega, \sigma)\big]^M\Big),
\end{aligned} \tag{7.1.60}$$

as well as

$$\begin{aligned}\operatorname{Ker}\Big(zI + K^{\#}_{A};\ \big[H^{r}(\partial\Omega,\sigma)\big]^{M}\Big) &= \operatorname{Ker}\Big(zI + K^{\#}_{A};\ \big[L^{p}(\partial\Omega,\sigma)\big]^{M}\Big) \\ = \operatorname{Ker}\Big(zI + K^{\#}_{A};\ \big[L^{q}_{-1}(\partial\Omega,\sigma)\big]^{M}\Big) &= \operatorname{Ker}\Big(zI + K^{\#}_{A};\ \big[M^{p,\lambda}(\partial\Omega,\sigma)\big]^{M}\Big) \\ = \operatorname{Ker}\Big(zI + K^{\#}_{A};\ \big[\mathring{M}^{p,\lambda}(\partial\Omega,\sigma)\big]^{M}\Big) &= \operatorname{Ker}\Big(zI + K^{\#}_{A};\ \big[\mathcal{B}^{q,\lambda}(\partial\Omega,\sigma)\big]^{M}\Big).\end{aligned} \tag{7.1.61}$$

Moreover,

$$\text{(7.1.60)-(7.1.61) } \textit{are finite-dimensional spaces, and they all have the same dimension.} \tag{7.1.62}$$

Finally, similar results[3] *to those stated above are valid for the slip double layer potential operators* K_{slip}, $K^{\#}_{\text{slip}}$ *associated with the Stokes system in* Ω *as in* (4.2.102) *and* (4.2.103), *respectively.*

Proof That choosing $\delta \in (0,1)$ small makes Ω a UR domain follows from Theorem 3.4.2. In fact, Theorem 3.4.2 also guarantees that if $\delta \in (0,1)$ is small enough then Ω also satisfies a two-sided local John condition. The equalities in (7.1.60)-(7.1.61) may be then justified based on Theorem 7.1.1, item *(f)* in [130, Theorem 2.2.4], [130, Lemma 2.1.3], [130, (11.7.25), (11.7.27)], and the fact that the following embeddings are continuous with dense range:

$$\mathscr{C}^{\alpha}(\partial\Omega) \hookrightarrow L^{p}(\partial\Omega,\sigma) \text{ for each } \alpha \in (0,1) \text{ and } p \in (1,\infty), \tag{7.1.63}$$

$$L^{p}_{1}(\partial\Omega,\sigma) \hookrightarrow L^{p}(\partial\Omega,\sigma) \hookrightarrow L^{p}_{-1}(\partial\Omega,\sigma) \text{ for all } p \in (1,\infty), \tag{7.1.64}$$

$$\begin{gathered} H^{r}(\partial\Omega,\sigma) \hookrightarrow L^{p}(\partial\Omega,\sigma) \hookrightarrow L^{q}(\partial\Omega,\sigma) \\ \text{if } \tfrac{n-1}{n} < r \leq 1 \text{ and } 1 < q \leq p < \infty, \end{gathered} \tag{7.1.65}$$

$$M^{p,\lambda}(\partial\Omega,\sigma) \hookrightarrow L^{p}(\partial\Omega,\sigma) \text{ for each } p \in (1,\infty) \text{ and } \lambda \in (0,n-1), \tag{7.1.66}$$

$$M^{p,\lambda}_{1}(\partial\Omega,\sigma) \hookrightarrow L^{p}_{1}(\partial\Omega,\sigma) \text{ for each } p \in (1,\infty) \text{ and } \lambda \in (0,n-1), \tag{7.1.67}$$

$$L^{s}(\partial\Omega,\sigma) \hookrightarrow \mathring{M}^{p,\lambda}(\partial\Omega,\sigma) \text{ if } p \in (1,\infty),\ \lambda \in (0,n-1),\ s = \tfrac{p(n-1)}{n-1-\lambda}, \tag{7.1.68}$$

$$\mathring{M}^{p,\lambda}_{1}(\partial\Omega,\sigma) \hookrightarrow L^{p}_{1}(\partial\Omega,\sigma) \text{ for each } p \in (1,\infty) \text{ and } \lambda \in (0,n-1), \tag{7.1.69}$$

$$\mathcal{B}^{q,\lambda}(\partial\Omega,\sigma) \hookrightarrow L^{r}(\partial\Omega,\sigma)\ q \in (1,\infty),\ \lambda \in (0,n-1),\ r = \tfrac{q(n-1)}{n-1+\lambda(q-1)}, \tag{7.1.70}$$

$$\mathrm{BMO}(\partial\Omega,\sigma) \hookrightarrow L^{p}(\partial\Omega,\sigma) \text{ for each } p \in (1,\infty), \tag{7.1.71}$$

$$\mathrm{VMO}(\partial\Omega,\sigma) \hookrightarrow L^{p}(\partial\Omega,\sigma) \text{ for each } p \in (1,\infty). \tag{7.1.72}$$

See [130, (3.1.50), (4.4.11), (6.2.71), (11.1.66), (11.8.4)], [129, (7.4.105)].

As far as (7.1.62) is concerned, the first claim made there is a direct consequence of the Fredholmness results from Theorem 7.1.1. Concerning the second claim in (7.1.62), from Theorem 7.1.1, [132, Theorem 1.5.1, item *(iii)*], and standard functional analysis we see that

[3] formulated for $M := n$ and now disregarding the coefficient tensor A

$$\operatorname{Im}\Big(zI + K_A^\#;\ \big[L^2(\partial\Omega,\sigma)\big]^M\Big) = \Big[\operatorname{Ker}\Big(zI + K_A;\ \big[L^2(\partial\Omega,\sigma)\big]^M\Big)\Big]^\top \tag{7.1.73}$$

where the orthogonal complement in the right-hand side is considered in the setting of $\big[L^2(\partial\Omega,\sigma)\big]^M$. As a result, on the one hand we have

$$\begin{aligned}\dim\left(\frac{\big[L^2(\partial\Omega,\sigma)\big]^M}{\operatorname{Im}\big(zI+K_A^\#;[L^2(\partial\Omega,\sigma)]^M\big)}\right) &= \dim\left(\frac{\big[L^2(\partial\Omega,\sigma)\big]^M}{\big[\operatorname{Ker}(zI+K_A;[L^2(\partial\Omega,\sigma)]^M)\big]^\top}\right)\\ &= \dim\operatorname{Ker}\Big(zI + K_A;\ \big[L^2(\partial\Omega,\sigma)\big]^M\Big).\end{aligned} \tag{7.1.74}$$

On the other hand, using the fact that the operator $zI + K_A^\#$ is Fredholm with index zero on $\big[L^2(\partial\Omega,\sigma)\big]^M$ (cf. Theorem 7.1.1), we may write

$$\begin{aligned}\dim\left(\frac{\big[L^2(\partial\Omega,\sigma)\big]^M}{\operatorname{Im}\big(zI+K_A^\#;[L^2(\partial\Omega,\sigma)]^M\big)}\right) &= \dim\operatorname{CoKer}\Big(zI + K_A^\#;\ \big[L^2(\partial\Omega,\sigma)\big]^M\Big)\\ &= \dim\operatorname{Ker}\Big(zI + K_A^\#;\ \big[L^2(\partial\Omega,\sigma)\big]^M\Big).\end{aligned} \tag{7.1.75}$$

At this stage, the fact that the dimension of any space from (7.1.61) is equal to the dimension of any space from (7.1.60) follows from (7.1.74), (7.1.75), (7.1.60), and (7.1.61). The final claim in the statement is dealt with in an analogous fashion. □

We may further augment Proposition 7.1.2 with similar results on Generalized Banach Function Spaces, described below.

Proposition 7.1.3 *Fix $n \in \mathbb{N}$ with $n \geq 2$ and suppose $\Omega \subseteq \mathbb{R}^n$ is an Ahlfors regular domain with compact boundary. Set $\sigma := \mathcal{H}^{n-1}\lfloor\partial\Omega$ and denote by ν the geometric measure theoretic outward unit normal to Ω. Let L be a homogeneous, second-order, constant complex coefficient, weakly elliptic $M \times M$ system in $\mathbb{R}^n$ for which $\mathfrak{A}_L^{\rm dis} \neq \varnothing$. Pick a coefficient tensor $A \in \mathfrak{A}_L^{\rm dis}$ and consider the layer potential operators K_A, $K_A^\#$ associated with A and Ω as in* (A.0.116) *and* (A.0.118)*, respectively. Next, assume $\mathbb{X}$ is a Generalized Banach Function Space on $(\partial\Omega,\sigma)$ such that*

$$\mathcal{M}_{\partial\Omega} : \mathbb{X} \to \mathbb{X} \text{ and } \mathcal{M}_{\partial\Omega} : \mathbb{X}' \to \mathbb{X}' \text{ are well-defined and bounded,} \tag{7.1.76}$$

where $\mathcal{M}_{\partial\Omega}$ is the Hardy-Littlewood maximal operator on $\partial\Omega$, and $\mathbb{X}'$ is the associated space of $\mathbb{X}$ (cf. [130, Definitions 5.1.4, 5.1.11]*). As before, define*

$$\mathring{\mathbb{X}} := \overline{L^\infty(\partial\Omega,\sigma)}^{\|\cdot\|_{\mathbb{X}}} \text{ and } (\mathbb{X}')^\circ := \overline{L^\infty(\partial\Omega,\sigma)}^{\|\cdot\|_{\mathbb{X}'}}, \tag{7.1.77}$$

and bring back $\mathbb{X}_1$, $(\mathbb{X}')_1$, $\mathring{\mathbb{X}}_1$, $(\mathbb{X}')_1^\circ$, the $\mathbb{X}$-styled, $\mathbb{X}'$-styled, $\mathring{\mathbb{X}}$-styled, and $(\mathbb{X}')^\circ$-styled Sobolev spaces of order one on $\partial\Omega$ from (7.1.39) *(these are known to be Banach spaces when equipped with natural norms; cf.* (5.4.84)*). Finally, pick some $\varepsilon \in (0,\infty)$ and fix a compact interval $I_o \subseteq (1,\infty)$.*

Then there exists a threshold $\delta_{\mathbb{X}} \in (0,1)$, which depends only on n, A, ε, I_o, the operator norms of $\mathcal{M}_{\partial\Omega}$ on $\mathbb{X}$ and $\mathbb{X}'$, the ADR *constants of $\partial\Omega$, and the relative* *flatness ratio* $\wp(\Omega)/\operatorname{diam}\partial\Omega$, *with the property that whenever*

$$\operatorname{dist}\Big(\nu, [\mathrm{VMO}(\partial\Omega,\sigma)]^n\Big) < \delta_{\mathbb{X}} \tag{7.1.78}$$

where the distance is considered in the space $\big[\mathrm{BMO}(\partial\Omega,\sigma)\big]^n$ *(hence if* Ω *is a* $\delta_{\mathbb{X}}$*-oscillating* AR *domain; cf. Definition 3.4.1), it follows that* Ω *is a* UR *domain and for each spectral parameter* $z \in \mathbb{C}$ *with* $|z| \geq \varepsilon$ *and each* $p \in I_o$ *one has*

$$\begin{aligned}\operatorname{Ker}\Big(zI + K^{\#}_A;\ [\mathbb{X}]^M\Big) &= \operatorname{Ker}\Big(zI + K^{\#}_A;\ [\mathbb{X}']^M\Big) = \operatorname{Ker}\Big(zI + K^{\#}_A;\ [\mathring{\mathbb{X}}]^M\Big)\\ &= \operatorname{Ker}\Big(zI + K^{\#}_A;\ [(\mathbb{X}')^\circ]^M\Big) = \operatorname{Ker}\Big(zI + K^{\#}_A;\ [L^p(\partial\Omega,\sigma)]^M\Big),\end{aligned} \tag{7.1.79}$$

as well as

$$\begin{aligned}\operatorname{Ker}\Big(zI + K_A;\ [\mathbb{X}]^M\Big) &= \operatorname{Ker}\Big(zI + K_A;\ [\mathbb{X}']^M\Big) = \operatorname{Ker}\Big(zI + K_A;\ [\mathring{\mathbb{X}}]^M\Big)\\ &= \operatorname{Ker}\Big(zI + K_A;\ [(\mathbb{X}')^\circ]^M\Big) = \operatorname{Ker}\Big(zI + K_A;\ [\mathbb{X}_1]^M\Big) = \operatorname{Ker}\Big(zI + K_A;\ [\mathring{\mathbb{X}}_1]^M\Big)\\ &= \operatorname{Ker}\Big(zI + K_A;\ [(\mathbb{X}')^\circ_1]^M\Big) = \operatorname{Ker}\Big(zI + K_A;\ [L^p(\partial\Omega,\sigma)]^M\Big).\end{aligned} \tag{7.1.80}$$

In addition,

$$\text{(7.1.79)-(7.1.80) are finite-dimensional spaces, and they all have the same dimension.} \tag{7.1.81}$$

Lastly, analogous results[4] *to those recorded above are valid for the slip double layer potential operators* K_{slip}, $K^{\#}_{\text{slip}}$ *associated with the Stokes system in* Ω *as in* (4.2.102) *and* (4.2.103), *respectively.*

Proof Theorem 3.4.2 ensures that choosing $\delta_{\mathbb{X}} \in (0,1)$ sufficiently small makes Ω a UR domain satisfying a two-sided local John condition. From [130, (5.2.92), (5.2.101), (11.5.200)] and (7.1.39), there exist $q \in (1,2)$ and $q' \in (2,\infty)$ satisfying $1/q + 1/q' = 1$, such that the following embeddings are continuous:

$$L^{q'}(\partial\Omega,\sigma) \hookrightarrow \mathring{\mathbb{X}} \hookrightarrow \mathbb{X}, \quad L^{q'}(\partial\Omega,\sigma) \hookrightarrow (\mathbb{X}')^\circ \hookrightarrow \mathbb{X}', \tag{7.1.82}$$

$$L^{q'}_1(\partial\Omega,\sigma) \hookrightarrow \mathring{\mathbb{X}}_1 \hookrightarrow \mathbb{X}_1, \quad L^{q'}_1(\partial\Omega,\sigma) \hookrightarrow (\mathbb{X}')^\circ_1 \hookrightarrow (\mathbb{X}')_1, \tag{7.1.83}$$

while the following embeddings are continuous with dense range:

$$\mathbb{X} \hookrightarrow L^q(\partial\Omega,\sigma), \quad \mathbb{X}' \hookrightarrow L^q(\partial\Omega,\sigma), \tag{7.1.84}$$

$$\mathbb{X}_1 \hookrightarrow L^q_1(\partial\Omega,\sigma), \quad (\mathbb{X}')_1 \hookrightarrow L^q_1(\partial\Omega,\sigma). \tag{7.1.85}$$

Then (7.1.79)-(7.1.80) follow from Theorem 7.1.1, [130, Lemma 2.1.3], Proposition 7.1.2, (7.1.82)-(7.1.83), and (7.1.84)-(7.1.85). Also, (7.1.81) is a consequence of (7.1.79)-(7.1.80) and (7.1.62). The final claim is dealt with similarly. □

---

[4] formulated for $M := n$ and now disregarding the coefficient tensor A

Combining Propositions 7.1.2-7.1.3 with the results about boundary layer potentials being Fredholm with index zero obtained in Theorem 7.1.1 and the injectivity results from [132, Theorem 1.7.2] yields the invertibility properties recorded in the theorem below. This should be compared with Theorem 7.1.11 highlighting the interplay between invertibility of boundary layer potentials and the topology of the underlying set.

Theorem 7.1.4 *Let $\Omega \subseteq \mathbb{R}^n$ (where $n \in \mathbb{N}$ with $n \geq 2$) be an Ahlfors regular domain with compact boundary. Abbreviate $\sigma := \mathcal{H}^{n-1}\lfloor\partial\Omega$ and denote by ν the geometric measure theoretic outward unit normal to Ω. Next, let L be a homogeneous, second-order, constant complex coefficient, weakly elliptic $M \times M$ system in $\mathbb{R}^n$ for which $\mathfrak{A}_L^{\rm dis} \neq \varnothing$. Pick $A \in \mathfrak{A}_L^{\rm dis}$ and consider the boundary-to-boundary double layer potential operators K_A, $K_A^{\#}$ associated with the domain Ω and the coefficient tensor A as in* (A.0.116) *and* (A.0.118)*, respectively. Finally, fix a spectral parameter $z \in \mathbb{C} \setminus [-\frac{1}{2}, \frac{1}{2}]$. Then the following properties hold.*

(1) For each given integrability exponent $p \in (1,\infty)$ there exists some threshold $\delta \in (0,1)$ which depends only on n, z, p, A, the ADR *constants of $\partial\Omega$, as well as the relative flatness ratio $\wp(\Omega)/\mathrm{diam}\,\partial\Omega$ with the property that if*

$$\mathrm{dist}\Big(\nu, [\mathrm{VMO}(\partial\Omega,\sigma)]^n\Big) < \delta \tag{7.1.86}$$

with the distance considered in $\big[\mathrm{BMO}(\partial\Omega,\sigma)\big]^n$, hence if Ω is a δ-oscillating AR *domain (cf. Definition 3.4.1), then Ω becomes a* UR *domain. Furthermore,*

$$\begin{array}{c}\text{if } A \text{ is positive semi-definite, Legendre-Hadamard elliptic,}\\ \text{and either } A \text{ is complex symmetric, or } z \text{ is real, it follows that}\\ zI + K_A^{\#} : \big[L^p(\partial\Omega,\sigma)\big]^M \longrightarrow \big[L^p(\partial\Omega,\sigma)\big]^M \text{ isomorphically,}\\ zI + K_A : \big[L^p(\partial\Omega,\sigma)\big]^M \longrightarrow \big[L^p(\partial\Omega,\sigma)\big]^M \text{ isomorphically,}\\ zI + K_A : \big[L^p_1(\partial\Omega,\sigma)\big]^M \longrightarrow \big[L^p_1(\partial\Omega,\sigma)\big]^M \text{ isomorphically,}\\ zI + K_A^{\#} : \big[L^p_{-1}(\partial\Omega,\sigma)\big]^M \longrightarrow \big[L^p_{-1}(\partial\Omega,\sigma)\big]^M \text{ isomorphically,}\end{array} \tag{7.1.87}$$

and

$$\begin{array}{c}\text{if } \Omega \text{ is bounded, } \mathbb{R}^n \setminus \overline{\Omega} \text{ is connected, and}\\ \text{the coefficient tensor } A \text{ is positive definite, then}\\ \tfrac{1}{2}I + K_A^{\#} : \big[L^p(\partial\Omega,\sigma)\big]^M \longrightarrow \big[L^p(\partial\Omega,\sigma)\big]^M \text{ isomorphically,}\\ \tfrac{1}{2}I + K_A : \big[L^p(\partial\Omega,\sigma)\big]^M \longrightarrow \big[L^p(\partial\Omega,\sigma)\big]^M \text{ isomorphically,}\\ \tfrac{1}{2}I + K_A : \big[L^p_1(\partial\Omega,\sigma)\big]^M \longrightarrow \big[L^p_1(\partial\Omega,\sigma)\big]^M \text{ isomorphically,}\\ \tfrac{1}{2}I + K_A^{\#} : \big[L^p_{-1}(\partial\Omega,\sigma)\big]^M \longrightarrow \big[L^p_{-1}(\partial\Omega,\sigma)\big]^M \text{ isomorphically.}\end{array} \tag{7.1.88}$$

In addition, with the piece of notation introduced in (A.0.159),

$$
\begin{aligned}
&\text{if } n \geq 3 \text{ and } A \text{ is compliant as well as complex symmetric, then}\\
&\pm\tfrac{1}{2}I + K_A^{\#} : L^p_{A,\pm 1/2}(\partial\Omega, \sigma) \to L^p_{A,\pm 1/2}(\partial\Omega, \sigma) \text{ isomorphically,}\\
&\text{and } \pm\tfrac{1}{2}I + \big[K_A\big] \text{ are isomorphisms of } \frac{[L^p(\partial\Omega,\sigma)]^M}{\mathrm{Ker}\big(\pm\frac{1}{2}I + K_A; [L^p(\partial\Omega,\sigma)]^M\big)},
\end{aligned}
\tag{7.1.89}
$$

where $\big[K_A\big]$ *is the quotient map induced by* K_A *(as in* [130, 2.1.56]*). In particular, the invertibility results in* (7.1.87)-(7.1.89) *are valid for any* $p \in (1, \infty)$ *provided* Ω *is an infinitesimally flat* AR *domain (cf. Definition 3.4.1).*

(2) *Similar invertibility results to those listed in* (7.1.87)-(7.1.88) *are valid for Morrey spaces, vanishing Morrey spaces, block spaces, Sobolev spaces adapted to these scales, Hölder and* BMO *spaces (for the boundary-to-boundary double layer), as well as Hardy spaces (in the case of the transpose double layer). Finally, invertibility results of a similar nature to those in* (7.1.89) *are also valid for Sobolev spaces, Morrey spaces, vanishing Morrey spaces, block spaces (and Sobolev spaces adapted to these scales), Hardy, Hölder, and* BMO *spaces (with natural definitions for the counterparts of* $L^p_{A,\pm 1/2}(\partial\Omega, \sigma)$ *in each case).*

(3) *Suppose* $\mathbb{X}$ *is a Generalized Banach Function Space on* $(\partial\Omega, \sigma)$ *such that*

$$
\mathcal{M}_{\partial\Omega} : \mathbb{X} \to \mathbb{X} \text{ and } \mathcal{M}_{\partial\Omega} : \mathbb{X}' \to \mathbb{X}' \text{ are well-defined and bounded,} \tag{7.1.90}
$$

where $\mathcal{M}_{\partial\Omega}$ *is the Hardy-Littlewood maximal operator on* $\partial\Omega$*, and* $\mathbb{X}'$ *is the associated space of* $\mathbb{X}$ *(cf.* [130, Definitions 5.1.4, 5.1.11]*). As in the past, define*

$$
\mathring{\mathbb{X}} := \overline{L^\infty(\partial\Omega, \sigma)}^{\|\cdot\|_{\mathbb{X}}} \text{ and } (\mathbb{X}')^\circ := \overline{L^\infty(\partial\Omega, \sigma)}^{\|\cdot\|_{\mathbb{X}'}}, \tag{7.1.91}
$$

and bring back $\mathbb{X}_1$, $(\mathbb{X}')_1$, $\mathring{\mathbb{X}}_1$, $(\mathbb{X}')^\circ_1$*, the* $\mathbb{X}$*-styled,* $\mathbb{X}'$*-styled,* $\mathring{\mathbb{X}}$*-styled, and* $(\mathbb{X}')^\circ$*-styled Sobolev spaces of order one on* $\partial\Omega$ *from* (7.1.39) *(these are known to be Banach spaces when equipped with natural norms; cf.* (5.4.84)*).*
Then there exists $\delta_{\mathbb{X}} \in (0, 1)$*, depending only on n, z, A, the operator norms of* $\mathcal{M}_{\partial\Omega}$ *on* $\mathbb{X}$ *and* $\mathbb{X}'$*, the* ADR *constants of* $\partial\Omega$*, and the relative flatness ratio* $\wp(\Omega)/\mathrm{diam}\,\partial\Omega$*, such that if*

$$
\mathrm{dist}\Big(\nu, [\mathrm{VMO}(\partial\Omega, \sigma)]^n\Big) < \delta_{\mathbb{X}} \tag{7.1.92}
$$

with the distance considered in $\big[\mathrm{BMO}(\partial\Omega, \sigma)\big]^n$ *(hence if* Ω *is a* $\delta_{\mathbb{X}}$*-oscillating* AR *domain; cf. Definition 3.4.1), it follows that* Ω *becomes a* UR *domain. Moreover,*

$$
\begin{gathered}
\text{if } A \text{ is positive semi-definite, Legendre-Hadamard elliptic,}\\
\text{and either } A \text{ is complex symmetric, or } z \text{ is real, it follows that}\\
zI + K_A^{\#} \text{ is an isomorphism when acting on either of the spaces}\\
[\mathbb{X}]^M,\ [\mathbb{X}']^M,\ [\mathring{\mathbb{X}}]^M,\ [(\mathbb{X}')^\circ]^M,\\
\text{while } zI + K_A \text{ is an isomorphism on either of the spaces}\\
[\mathbb{X}]^M, [\mathbb{X}']^M, [\mathring{\mathbb{X}}]^M, [(\mathbb{X}')^\circ]^M, [\mathbb{X}_1]^M, [\mathbb{X}'_1]^M, [\mathring{\mathbb{X}}_1]^M, [(\mathbb{X}')^\circ_1]^M,
\end{gathered}
\tag{7.1.93}
$$

and

$$\begin{array}{c}\text{if } \Omega \text{ is bounded, } \mathbb{R}^n\setminus\overline{\Omega} \text{ is connected, and } A \text{ is positive definite,}\\ \text{then } \tfrac{1}{2}I+K_A^{\#} \text{ is an isomorphism on either of the spaces}\\ \big[\mathbb{X}\big]^M,\ \big[\mathbb{X}'\big]^M,\ \big[\mathring{\mathbb{X}}\big]^M,\ \big[(\mathbb{X}')^{\circ}\big]^M,\\ \text{while } \tfrac{1}{2}I+K_A \text{ is an isomorphism on either of the spaces}\\ \big[\mathbb{X}\big]^M,\big[\mathbb{X}'\big]^M,\big[\mathring{\mathbb{X}}\big]^M,\big[(\mathbb{X}')^{\circ}\big]^M,\big[\mathbb{X}_1\big]^M,\big[\mathbb{X}_1'\big]^M,\big[\mathring{\mathbb{X}}_1\big]^M,\big[(\mathbb{X}')^{\circ}_1\big]^M.\end{array} \tag{7.1.94}$$

In particular, the above conclusions are true whenever $\mathbb{X}$ *is a rearrangement invariant Banach function space whose Boyd indices satisfy* $1 < p_{\mathbb{X}} \leq q_{\mathbb{X}} < \infty$. *As a corollary, the above conclusions hold if* $\mathbb{X} := L^{\Phi}(\partial\Omega, \sigma)$, *the Orlicz space associated as in* (A.0.133) *with the measure space* $(\partial\Omega, \sigma)$ *and any Young function* Φ *satisfying* $1 < i(\Phi) \leq I(\Phi) < \infty$.

Proof Theorem 3.4.2 tells us that it is possible to choose $\delta \in (0, 1)$ small enough as to guarantee that Ω is a UR domain satisfying a two-sided local John condition. By eventually decreasing the value of $\delta \in (0, 1)$ matters may be arranged so that the conclusions in Theorem 7.1.1 and Proposition 7.1.2 are valid (for suitable choices of the parameters involved in their statements).

Then (7.1.87) becomes a consequence of item *(1)* in Theorem 7.1.1 and [132, Theorem 1.7.2, item *(2)*]. More specifically, suppose A, z are as in (7.1.87). The fact that $zI + K_A^{\#}$ is a Fredholm operator with index zero in the context of (7.1.4) with $p := 2(n-1)/n$ and $w \equiv 1$ implies that $zI+K_A^{\#}$ is actually an isomorphism when acting on $\big[L^{2(n-1)/n}(\partial\Omega, \sigma)\big]^M$. Next, given a generic $p \in (1, \infty)$, if $1 < p < 2(n-1)/n$ then, as $L^{2(n-1)/n}(\partial\Omega, \sigma)$ embeds densely into $L^p(\partial\Omega, \sigma)$, we conclude from what we have just proved and (7.1.4) (with $w \equiv 1$) that $zI + K_A^{\#}$ has dense range and is Fredholm with index zero on $\big[L^p(\partial\Omega, \sigma)\big]^M$. Thus, $zI + K_A^{\#}$ is an isomorphism on $\big[L^p(\partial\Omega, \sigma)\big]^M$ in this case. Finally, if $2(n-1)/n \leq p < \infty$ then $L^p(\partial\Omega, \sigma)$ is contained in $L^{2(n-1)/n}(\partial\Omega, \sigma)$ so $zI + K_A^{\#}$ is injective and Fredholm with index zero, thus an isomorphism from the space $\big[L^p(\partial\Omega, \sigma)\big]^M$ onto itself. This establishes the first isomorphism in (7.1.87). The second isomorphism in (7.1.87) follows from the first (used with p replaced by its conjugate exponent) and duality (cf. [132, Theorem 1.5.1, item *(iii)*]). The third isomorphism in (7.1.87) is a consequence of what we have just proved and (7.1.3) (with $w \equiv 1$). Finally, the fourth isomorphism in (7.1.87) follows from the third isomorphism in (7.1.87) and duality (cf. [132, Theorem 1.5.1, item *(iv)*]).

Next, the claims in (7.1.88) are seen, much as above, using either item *(4)* or item *(6)* in [132, Theorem 1.7.2]. Moving on, the first isomorphism in (7.1.89) may be justified based on [132, Theorem 1.7.2, item *(3)*, (1.7.17)], the fact that $\pm\frac{1}{2}I + K_A^{\#}$ are Fredholm operators with index zero in the context of (7.1.4) (with $w \equiv 1$), and item *(c)* in [130, Theorem 2.2.4]. Specifically, in the case when $2(n-1)/n \leq p < \infty$, the same type of argument as before leads to the conclusion that $\pm\frac{1}{2}I + K_A^{\#}$ in (7.1.89) are both injective and Fredholm with index zero, hence ultimately invertible. The case when $1 < p < 2(n-1)/n$ the desired conclusion once again follows as before once we have established that

$$L^{2(n-1)/n}_{A,\pm 1/2}(\partial\Omega,\sigma) \hookrightarrow L^p_{A,\pm 1/2}(\partial\Omega,\sigma) \text{ densely.} \tag{7.1.95}$$

To justify this, work under the assumption that the sign is "plus" (the case when the sign is "minus" is handled similarly). If p' denotes the conjugate exponent of p, then

$$1 < p < \tfrac{2(n-1)}{n} < \tfrac{2(n-1)}{n-2} < p' < \infty. \tag{7.1.96}$$

Proposition 7.1.2 ensures that $\mathrm{Ker}\Big(\frac{1}{2}I+K_A : \big[L^{p'}(\partial\Omega,\sigma)\big]^M\Big)$ is a finite dimensional space. Let $\{\xi_1,\dots,\xi_N\} \subseteq \big[L^{p'}(\partial\Omega,\sigma)\big]^M$ be a basis for this space. For further use let us note that Proposition 7.1.2 additionally gives that

$$\begin{aligned}&\{\xi_1,\dots,\xi_N\} \text{ is also a basis for the space}\\ &\mathrm{Ker}\Big(\tfrac{1}{2}I + K_A : \big[L^{2(n-1)/(n-2)}(\partial\Omega,\sigma)\big]^M\Big).\end{aligned} \tag{7.1.97}$$

Since, as seen from (7.1.96) and the fact that $\partial\Omega$ has finite measure,

$$\xi_1,\dots,\xi_N \in \big[L^p(\partial\Omega,\sigma)\big]^M = \Big(\big[L^{p'}(\partial\Omega,\sigma)\big]^M\Big)^*, \tag{7.1.98}$$

From [130, Lemma 1.2.16] there exist $\eta_1,\dots,\eta_N \in \big[L^{p'}(\partial\Omega,\sigma)\big]^M$ satisfying $\int_{\partial\Omega}\langle\xi_j,\eta_k\rangle\,d\sigma = \delta_{jk}$ for all $j,k \in \{1,\dots,N\}$. If we now set

$$\begin{aligned}&\pi : \big[L^p(\partial\Omega,\sigma)\big]^M \longrightarrow \big[L^p(\partial\Omega,\sigma)\big]^M \text{ acting according to}\\ &\pi f := f - \sum_{j=1}^N \Big(\int_{\partial\Omega}\langle\xi_j, f\rangle\,d\sigma\Big)\cdot\eta_j \text{ for each } f \in \big[L^p(\partial\Omega,\sigma)\big]^M,\end{aligned} \tag{7.1.99}$$

then π is a well-defined, linear, and continuous mapping satisfying

$$\begin{aligned}&\pi f = f \text{ for each function } f \in L^p_{A,1/2}(\partial\Omega,\sigma), \text{ and}\\ &\pi g \in L^{2(n-1)/n}_{A,1/2}(\partial\Omega,\sigma) \text{ for each } g \in \big[L^{2(n-1)/n}(\partial\Omega,\sigma)\big]^M,\end{aligned} \tag{7.1.100}$$

where the justification of the last property above makes use of (7.1.97). Suppose now some arbitrary $f \in L^p_{A,1/2}(\partial\Omega,\sigma)$ has been given. Since $L^{2(n-1)/n}(\partial\Omega,\sigma)$ is a dense subset of $L^p(\partial\Omega,\sigma)$, there exists a sequence $\{f_j\}_{j\in\mathbb{N}} \subseteq \big[L^{2(n-1)/n}(\partial\Omega,\sigma)\big]^M$ with the property that $f_j \to f$ in $\big[L^p(\partial\Omega,\sigma)\big]^M$ as $j \to \infty$. Consequently, we have $\pi f_j \to \pi f = f$ in $\big[L^p(\partial\Omega,\sigma)\big]^M$ as $j \to \infty$, and since each πf_j belongs to the space $L^{2(n-1)/n}_{A,1/2}(\partial\Omega,\sigma)$, this finishes the proof of (7.1.95). Finally, the second isomorphism in (7.1.89) is a consequence of the first isomorphism in (7.1.89) (written with p' in place of p) and [130, Lemma 2.1.7].

The claims in the first part of item *(2)* are implied by (7.1.87)-(7.1.88), Proposition 7.1.2, and Theorem 7.1.1. The claims in the second part of item *(2)* follow from

(7.1.89), Proposition 7.1.2, Theorem 7.1.1, items *(c)-(d)* in [130, Theorem 2.2.4], [130, (2.1.57) in Lemma 2.1.7], plus the duality results in [130, Theorem 4.6.1], [130, Proposition 6.2.8], [130, Proposition 6.2.16], (A.0.154), [132, Theorem 1.5.1, item *(vi)*], and [132, Theorem 3.3.1, item *(iii)*]. Lastly, the claims in item *(3)* are implied by (7.1.87)-(7.1.88), Proposition 7.1.3, and Theorem 7.1.1. □

The proofs of the Fredholmness results stated in Theorem 7.1.1 take full advantage of the fact that the double layer operators considered in that context are of "chord-dot-normal" type (cf. (4.1.27)-(4.1.28)), which makes them sensitive to the flatness of the boundary of the underlying domain. Remarkably, the Fredholmness properties of the single layer potential operator, which has a "universal" integral kernel (i.e., entirely independent of the domain on which this is considered), continue to be strongly influenced by flatness of the boundary of the domain in relation to which this operator is considered.

Theorem 7.1.5 *Fix $n \in \mathbb{N}$ with $n \geq 2$ and let $\Omega \subseteq \mathbb{R}^n$ be an Ahlfors regular domain with compact boundary. Abbreviate $\sigma := \mathcal{H}^{n-1}\lfloor\partial\Omega$ and denote by ν the geometric measure theoretic outward unit normal to Ω. Also, let L be a homogeneous, second-order, constant complex coefficient, weakly elliptic $M \times M$ system in $\mathbb{R}^n$ for which*

$$\mathfrak{A}_L^{\mathrm{dis}} \neq \varnothing \text{ and } \mathfrak{A}_{L^\top}^{\mathrm{dis}} \neq \varnothing. \tag{7.1.101}$$

Consider the boundary-to-boundary single layer potential operator S associated with the set Ω and the system L as in (A.0.231). *Finally, fix three compact intervals, namely $I_1 \subset (1, \infty)$, $I_2 \subset (0, n-1)$, and $I_3 \subset (0, 1)$.*

Then there exists $\delta \in (0, 1)$ depending only on n, I_1, I_2, I_3, L, the ADR *constants of $\partial\Omega$, and the relative flatness ratio $\wp(\Omega)/\mathrm{diam}\,\partial\Omega$ with the property that if*

$$\mathrm{dist}\Big(\nu, [\mathrm{VMO}(\partial\Omega, \sigma)]^n\Big) < \delta \tag{7.1.102}$$

with the distance considered in $\big[\mathrm{BMO}(\partial\Omega, \sigma)\big]^n$ (hence if Ω is a δ-oscillating AR *domain; cf. Definition 3.4.1), it follows that Ω is a* UR *domain and for each $p, q \in I_1$, $\lambda \in I_2$, and $s \in I_3$ these operators are Fredholm:*

$$S : \big[L^p(\partial\Omega, \sigma)\big]^M \longrightarrow \big[L_1^p(\partial\Omega, \sigma)\big]^M, \tag{7.1.103}$$

$$S : \big[L_{-1}^p(\partial\Omega, \sigma)\big]^M \longrightarrow \big[L^p(\partial\Omega, \sigma)\big]^M, \tag{7.1.104}$$

$$S : \big[M^{p,\lambda}(\partial\Omega, \sigma)\big]^M \longrightarrow \big[M_1^{p,\lambda}(\partial\Omega, \sigma)\big]^M, \tag{7.1.105}$$

$$S : \big[\mathring{M}^{p,\lambda}(\partial\Omega, \sigma)\big]^M \longrightarrow \big[\mathring{M}_1^{p,\lambda}(\partial\Omega, \sigma)\big]^M, \tag{7.1.106}$$

$$S : \big[\mathcal{B}^{q,\lambda}(\partial\Omega, \sigma)\big]^M \longrightarrow \big[\mathcal{B}_1^{q,\lambda}(\partial\Omega, \sigma)\big]^M, \tag{7.1.107}$$

$$S : \big[B_{s-1}^{p,q}(\partial\Omega, \sigma)\big]^M \longrightarrow \big[B_s^{p,q}(\partial\Omega, \sigma)\big]^M, \tag{7.1.108}$$

$$S : \big[B_{s-1}^{\infty,\infty}(\partial\Omega, \sigma)\big]^M \longrightarrow \big[B_s^{\infty,\infty}(\partial\Omega, \sigma)\big]^M, \tag{7.1.109}$$

$$S : \big[B_{-s}^{1,1}(\partial\Omega, \sigma)\big]^M \longrightarrow \big[B_{1-s}^{1,1}(\partial\Omega, \sigma)\big]^M. \tag{7.1.110}$$

Moreover,

the indexes of S in (7.1.103)-(7.1.110) *are equal to one another and are actually independent of the parameters $p, q \in I_1$, $\lambda \in I_2$, and $s \in I_3$.* (7.1.111)

In fact, taking $\delta \in (0, 1)$ small enough also ensures that the operators

$$(S_{L^\top})^\top : \big[M^{p,\lambda}_{-1}(\partial\Omega, \sigma)\big]^M \longrightarrow \big[M^{p,\lambda}(\partial\Omega, \sigma)\big]^M, \tag{7.1.112}$$

$$(S_{L^\top})^\top : \big[\mathcal{B}^{q,\lambda}_{-1}(\partial\Omega, \sigma)\big]^M \longrightarrow \big[\mathcal{B}^{q,\lambda}(\partial\Omega, \sigma)\big]^M, \tag{7.1.113}$$

are Fredholm operators, and

the indexes of the operator $(S_{L^\top})^\top$ in any of the settings recorded in (7.1.112)-(7.1.113) *are equal to one another and are independent of the parameters $p, q \in I_1$ and $\lambda \in I_2$.* (7.1.114)

Finally, analogous results[5] *are valid for the boundary-to-boundary single layer potential operator associated with the Stokes system in Ω as in* (4.2.109).

Before presenting the proof of this theorem we make two comments.

Comment 1. In the case when $L = L^\top$ it follows that (7.1.101) reduces to just $\mathfrak{A}^{\rm dis}_L \neq \varnothing$ and the index of S in any of the settings recorded in (7.1.103)-(7.1.108) is now zero. Indeed, after eventually enlarging I_1, there is no loss of generality in assuming that I_1 is a compact sub-interval of $(1, \infty)$ which is stable under passing to Hölder conjugates. Granted this, the aforementioned index formula becomes a consequence of (7.1.111), the duality result in [132, Theorem 1.5.1, item *(ix)*], and item *(3)* of [130, Theorem 2.1.2].

Comment 2. As far as the role of the assumption made in (7.1.101) is concerned, we wish to note that there exists a weakly elliptic scalar operator L violating this assumption and for which the single layer potential operator S associated with L in a smooth domain with compact boundary $\Omega \subseteq \mathbb{R}^n$ fails to be Fredholm in the context of (7.1.103). Specifically, work in the two-dimensional setting (under the canonical identification $\mathbb{R}^2 \equiv \mathbb{C}$), where we consider $L := \partial_{\bar z}^2$, the square of the Cauchy-Riemann operator in the plane, and take $\Omega := \mathbb{D}$ where $\mathbb{D} := \big\{z \in \mathbb{C} : |z| < 1\big\}$ is the unit disk in the plane. Also, abbreviate $\sigma := \mathcal{H}^1 \lfloor \partial\mathbb{D}$. Since the image of

$$S : L^p(\partial\Omega, \sigma) \longrightarrow L^p_1(\partial\Omega, \sigma) \tag{7.1.115}$$

is contained in the space of admissible boundary data for the L^p_1 Regularity Problem for the operator $\partial_{\bar z}^2$ in the unit disk described in [132, Theorem 8.1.3], we see from [132, (8.1.88)] that the operator (7.1.115) cannot be Fredholm for any $p \in (1, \infty)$. This is in contrast with the positive result in Theorem 7.1.5. From Example 1.6.14 we know that the weakly elliptic operator $\partial_{\bar z}^2$ lacks a distinguished coefficient tensor, and this is the only assumption in Theorem 7.1.5 which is presently violated.

---

[5] formulated for $M := n$ and now disregarding the coefficient tensor A

As seen from [132, Remark 8.1.8], this lack of Fredholmness may also occur in relation to the single layer acting on other function spaces, if (7.1.101) is violated. In fact, we can be a bit more specific in regard to the nature of the failure of Fredholmness for the single layer operator. Recall that any tempered distribution which is a fundamental solution of Bitsadze's operator $L_B := \partial_{\bar z}^2$ is of the form $\bar z/(\pi z) + c$ for some $c \in \mathbb{C}$. Thus, for $c \in \mathbb{C}$ fixed, the single layer associated with Bitsadze's operator L_B in $\mathbb{D}$ acts on each $f \in L^p(\partial\mathbb{D}, \sigma)$ at σ-a.e. $z \in \partial\mathbb{D}$ as

$$\begin{aligned}(S_{L_B} f)(z) &= \frac{1}{\pi}\int_{\partial\mathbb{D}} \frac{\bar z - \bar\zeta}{z - \zeta} f(\zeta)\,d\sigma(\zeta) + c\int_{\partial\mathbb{D}} f(\zeta)\,d\sigma(\zeta) \\ &= \frac{1}{\pi i}\int_{\partial\mathbb{D}} \frac{\bar\zeta(\bar z - \bar\zeta)}{z - \zeta} f(\zeta)\,d\zeta - ic\int_{\partial\mathbb{D}} \bar\zeta f(\zeta)\,d\zeta,\end{aligned} \tag{7.1.116}$$

where we have used $d\zeta = i\zeta\,d\sigma(\zeta)$ on $\partial\mathbb{D}$. In particular, corresponding to the functions $f_k(z) := z^{k+2}$ for each $z \in \partial\mathbb{D}$, with $k \in \mathbb{N}_0$ arbitrary, we may compute

$$\begin{aligned}(S_{L_B} f_k)(z) &= \frac{1}{\pi i}\int_{\partial\mathbb{D}} \frac{\bar\zeta(\bar z - \bar\zeta)}{z - \zeta}\zeta^{k+2}\,d\zeta - ic\int_{\partial\mathbb{D}} \bar\zeta\zeta^{k+2}\,d\zeta = \frac{1}{\pi i}\int_{\partial\mathbb{D}} \frac{\bar z - \bar\zeta}{z - \zeta}\zeta^{k+1}\,d\zeta \\ &= \frac{1}{\pi i}\int_{\partial\mathbb{D}} \frac{\bar z - \bar\zeta}{z - \zeta}\big(\zeta^{k+1} - z^{k+1}\big)\,d\zeta + \frac{z^{k+1}}{\pi i}\int_{\partial\mathbb{D}} \frac{\bar z - \bar\zeta}{z - \zeta}\,d\zeta \\ &= \frac{-1}{\pi i}\int_{\partial\mathbb{D}} (\bar z - \bar\zeta)\big(\zeta^k + \zeta^{k-1}z + \cdots + z^k\big)\,d\zeta + \frac{z^{k+1}}{\pi}\int_{\partial\mathbb{D}} \frac{\zeta(\bar z - \bar\zeta)}{z(1 - \bar z\zeta)}\,d\sigma(\zeta) \\ &= \frac{1}{\pi i}\int_{\partial\mathbb{D}} \bar\zeta z^k\,d\zeta - \frac{z^k}{\pi}\int_{\partial\mathbb{D}} 1\,d\sigma(\zeta) = \frac{z^k}{\pi}\int_{\partial\mathbb{D}} 1\,d\sigma(\zeta) - \frac{z^k}{\pi}\int_{\partial\mathbb{D}} 1\,d\sigma(\zeta) = 0,\end{aligned} \tag{7.1.117}$$

where we have made repeated use of Cauchy's vanishing formula (aka Cauchy's integral theorem) and the fact that $d\zeta = i\zeta\,d\sigma(\zeta)$ on $\partial\mathbb{D}$. This proves that

$$\begin{aligned}&f_k(z) := z^{k+2} \text{ for each } z \in \partial\mathbb{D}, \text{ with } k \in \mathbb{N}_0 \text{ arbitrary,} \\ &\text{is an infinite family of linearly independent functions} \\ &\text{belonging to } \operatorname{Ker}\Big(S_{L_B} : L^p(\partial\mathbb{D}, \sigma) \to L^p_1(\partial\mathbb{D}, \sigma)\Big).\end{aligned} \tag{7.1.118}$$

As a consequence, the kernel of the single layer operator in this context is infinite dimensional, so this operator fails to be Fredholm.

We can also use (7.1.118) to derive information, in a more direct fashion (than in the first part of Comment 2), about the $\operatorname{Im}\Big(S_{L_B} : L^p(\partial\mathbb{D}, \sigma) \to L^p_1(\partial\mathbb{D}, \sigma)\Big)$. Specifically, from [130, (2.1.53)], the duality result in [132, Theorem 1.5.1, item *(ix)*], and (7.1.118), we first see that the image of S_{L_B} acting from $L^p_{-1}(\partial\mathbb{D}, \sigma)$ into $L^p(\partial\mathbb{D}, \sigma)$ is contained in the annihilator of the family of functions from (7.1.118) (regarded in $L^{p'}(\partial\mathbb{D}, \sigma)$ with $1/p + 1/p' = 1$), i.e., said image is contained in

$$\Big\{f \in L^p(\partial\mathbb{D}, \sigma) : \int_{\partial\mathbb{D}} f(\zeta)\zeta^{k+2}\,d\sigma(\zeta) = 0 \text{ for all } k \in \mathbb{N}_0\Big\}. \tag{7.1.119}$$

By virtue of [132, (8.1.14), (8.1.19)], the above space coincides with

$$\bar{z} \cdot \mathcal{H}^p(\partial\mathbb{D}_+, \sigma) = \mathcal{H}^p(\partial\mathbb{D}_+, \sigma) \oplus \langle \bar{z} \rangle, \tag{7.1.120}$$

where $\mathcal{H}^p(\partial\mathbb{D}_+, \sigma)$ is the "positive" boundary Hardy space defined in [132, (8.1.7)], and where $\langle \bar{z} \rangle$ stands for the (one-dimensional) linear span over $\mathbb{C}$ of the function $\partial\mathbb{D} \ni z \mapsto \bar{z} \in \mathbb{C}$. Hence,

$$\mathrm{Im}\Big(S_{L_B} : L^p_{-1}(\partial\mathbb{D}, \sigma) \to L^p(\partial\mathbb{D}, \sigma)\Big) \subseteq \mathcal{H}^p(\partial\mathbb{D}_+, \sigma) \oplus \langle \bar{z} \rangle. \tag{7.1.121}$$

Given that $\mathrm{Im}\Big(S_{L_B} : L^p(\partial\mathbb{D}, \sigma) \to L^p_1(\partial\mathbb{D}, \sigma)\Big)$ is contained in the intersection of the space appearing in the left side of (7.1.121) with $L^p_1(\partial\mathbb{D}, \sigma)$, from (7.1.121) and [132, (8.1.23)] we therefore conclude that

$$\mathrm{Im}\Big(S_{L_B} : L^p(\partial\mathbb{D}, \sigma) \to L^p_1(\partial\mathbb{D}, \sigma)\Big) \subseteq \mathcal{H}^p_1(\partial\mathbb{D}_+, \sigma) \oplus \langle \bar{z} \rangle, \tag{7.1.122}$$

where $\mathcal{H}^p_1(\partial\mathbb{D}_+, \sigma)$ is the "positive" boundary Hardy space with regularity, defined in [132, (8.1.22)]. Formula (7.1.122) is in agreement with the observation that $\mathrm{Im}\Big(S_{L_B} : L^p(\partial\mathbb{D}, \sigma) \to L^p_1(\partial\mathbb{D}, \sigma)\Big)$ is contained in the space of admissible boundary data for the L^p_1 Regularity Problem for the operator $\partial^2_{\bar{z}}$ in the unit disk, given the description of the latter space provided in [132, Theorem 8.1.3]. By virtue of [132, (8.1.27)], we may now conclude directly from formula (7.1.122) that the image of $S_{L_B} : L^p(\partial\mathbb{D}, \sigma) \to L^p_1(\partial\mathbb{D}, \sigma)$ has infinite codimension in $L^p_1(\partial\mathbb{D}, \sigma)$.

We now turn to the task of presenting the proof of Theorem 7.1.5.

Proof of Theorem 7.1.5 For starters, we note that Theorem 3.4.2 guarantees that choosing the threshold $\delta \in (0, 1)$ small enough makes Ω a UR domain which satisfies a two-sided local John condition (we shall tacitly employ these properties throughout the proof). In a first stage, the goal is to show that

$$S : \big[L^p(\partial\Omega, \sigma)\big]^M \to \big[L^p_1(\partial\Omega, \sigma)\big]^M, \quad S : \big[L^q_{-1}(\partial\Omega, \sigma)\big]^M \to \big[L^q(\partial\Omega, \sigma)\big]^M \text{ are Fredholm operators for each } p, q \in I_1, \tag{7.1.123}$$

and for each $p, q \in I_1$ we have

$$\begin{aligned} &\mathrm{index}\Big(S : \big[L^p(\partial\Omega, \sigma)\big]^M \to \big[L^p_1(\partial\Omega, \sigma)\big]^M\Big) \\ &\qquad = \mathrm{index}\Big(S : \big[L^q_{-1}(\partial\Omega, \sigma)\big]^M \to \big[L^q(\partial\Omega, \sigma)\big]^M\Big). \end{aligned} \tag{7.1.124}$$

To this end, pick $A_0 \in \mathfrak{A}^{\mathrm{dis}}_L$ along with $A_1 \in \mathfrak{A}_L$ such that $A_1^\top \in \mathfrak{A}^{\mathrm{dis}}_{L^\top}$. Theorem 7.1.1 then ensures that it is possible to pick some $\delta \in (0, 1)$ whose nature is as specified in the current statement with the property that if (7.1.102) holds then the following operators are Fredholm with index zero for each $p \in I_1$:

$$\pm\tfrac{1}{2}I + K_{A_0} : \big[L^p(\partial\Omega,\sigma)\big]^M \longrightarrow \big[L^p(\partial\Omega,\sigma)\big]^M, \tag{7.1.125}$$

$$\pm\tfrac{1}{2} + K_{A_0} : \big[L^p_1(\partial\Omega,\sigma)\big]^M \longrightarrow \big[L^p_1(\partial\Omega,\sigma)\big]^M, \tag{7.1.126}$$

$$\pm\tfrac{1}{2}I + K^{\#}_{A_1^\top} : \big[L^p(\partial\Omega,\sigma)\big]^M \longrightarrow \big[L^p(\partial\Omega,\sigma)\big]^M, \tag{7.1.127}$$

$$\pm\tfrac{1}{2}I + K^{\#}_{A_1^\top} : \big[L^p_{-1}(\partial\Omega,\sigma)\big]^M \longrightarrow \big[L^p_{-1}(\partial\Omega,\sigma)\big]^M. \tag{7.1.128}$$

We divide the remainder of the proof into a number of steps.

Step I: *For each $p,q \in I_1$ we have*

$$\begin{aligned}\mathrm{Ker}\Big(S : \big[L^p(\partial\Omega,\sigma)\big]^M \to \big[L^p_1(\partial\Omega,\sigma)\big]^M\Big)\\ = \mathrm{Ker}\Big(S : \big[L^q_{-1}(\partial\Omega,\sigma)\big]^M \to \big[L^q(\partial\Omega,\sigma)\big]^M\Big)\end{aligned} \tag{7.1.129}$$

and

$$\begin{aligned}\dim\mathrm{Ker}\Big(S : \big[L^p(\partial\Omega,\sigma)\big]^M \to \big[L^p_1(\partial\Omega,\sigma)\big]^M\Big)\\ = \dim\mathrm{Ker}\Big(S : \big[L^q_{-1}(\partial\Omega,\sigma)\big]^M \to \big[L^q(\partial\Omega,\sigma)\big]^M\Big) < +\infty.\end{aligned} \tag{7.1.130}$$

To justify (7.1.129), first observe that from [132, Theorem 1.5.1, item *(xiii)*] (cf. [132, (1.5.86)] and the very last comment in that item) it follows that

$$\begin{aligned}\mathrm{Ker}\Big(S : \big[L^q_{-1}(\partial\Omega,\sigma)\big]^M \to \big[L^q(\partial\Omega,\sigma)\big]^M\Big)\\ \subseteq \mathrm{Ker}\Big(\big(\tfrac{1}{2}I + K^{\#}_{A_1^\top}\big)\circ\big(-\tfrac{1}{2}I + K^{\#}_{A_1^\top}\big);\ \big[L^q_{-1}(\partial\Omega,\sigma)\big]^M\Big).\end{aligned} \tag{7.1.131}$$

When considered in concert, item *(2)* in [130, Theorem 2.1.2], (7.1.128), and (7.1.127) tell us that the composition $\big(\tfrac{1}{2}I + K^{\#}_{A_1^\top}\big)\circ\big(-\tfrac{1}{2}I + K^{\#}_{A_1^\top}\big)$ is a Fredholm operator with index zero both on $\big[L^q_{-1}(\partial\Omega,\sigma)\big]^M$ and on $\big[L^q(\partial\Omega,\sigma)\big]^M$. Since these are Banach spaces such that the latter is continuously and densely embedded into the former (cf. [130, (11.8.4)]), we then conclude from [130, Lemma 2.1.3] (cf. also [130, (2.1.16)] and item *(f)* in [130, Theorem 2.2.4]) that

$$\begin{aligned}\mathrm{Ker}\Big(\big(\tfrac{1}{2}I + K^{\#}_{A_1^\top}\big)\circ\big(-\tfrac{1}{2}I + K^{\#}_{A_1^\top}\big);\ \big[L^q_{-1}(\partial\Omega,\sigma)\big]^M\Big)\\ = \mathrm{Ker}\Big(\big(\tfrac{1}{2}I + K^{\#}_{A_1^\top}\big)\circ\big(-\tfrac{1}{2}I + K^{\#}_{A_1^\top}\big);\ \big[L^q(\partial\Omega,\sigma)\big]^M\Big)\end{aligned} \tag{7.1.132}$$

for each $q \in I_1$. In a similar fashion (keeping in mind that the Lebesgue scale on $\partial\Omega$ is nested, given that $\sigma(\partial\Omega) < \infty$) we also have

$$\begin{aligned}\mathrm{Ker}\Big(\big(\tfrac{1}{2}I + K^{\#}_{A_1^\top}\big)\circ\big(-\tfrac{1}{2}I + K^{\#}_{A_1^\top}\big);\ \big[L^p(\partial\Omega,\sigma)\big]^M\Big)\\ = \mathrm{Ker}\Big(\big(\tfrac{1}{2}I + K^{\#}_{A_1^\top}\big)\circ\big(-\tfrac{1}{2}I + K^{\#}_{A_1^\top}\big);\ \big[L^q(\partial\Omega,\sigma)\big]^M\Big)\end{aligned} \tag{7.1.133}$$

for each $p, q \in I_1$. Collectively, (7.1.131)-(7.1.133) prove that, for all $p, q \in I_1$,

$$\mathrm{Ker}\Big(S : \big[L^q_{-1}(\partial\Omega, \sigma)\big]^M \to \big[L^q(\partial\Omega, \sigma)\big]^M\Big) \subseteq \big[L^p(\partial\Omega, \sigma)\big]^M. \tag{7.1.134}$$

Together with [130, (11.8.4)] this readily implies (7.1.129). Finally, (7.1.130) is a consequence of (7.1.129), (7.1.131), and the fact that $\big(\frac{1}{2}I + K^{\#}_{A_1^\top}\big) \circ \big(-\frac{1}{2}I + K^{\#}_{A_1^\top}\big)$ is a Fredholm operator on $\big[L^q_{-1}(\partial\Omega, \sigma)\big]^M$.

Step II: *For each $p \in I_1$ we have*

$$\dim\left(\frac{\big[L^p_1(\partial\Omega,\sigma)\big]^M}{\mathrm{Im}\big(S:[L^p(\partial\Omega,\sigma)]^M \to [L^p_1(\partial\Omega,\sigma)]^M\big)}\right) < +\infty, \tag{7.1.135}$$

$$\dim\left(\frac{\big[L^p(\partial\Omega,\sigma)\big]^M}{\mathrm{Im}\big(S:[L^p_{-1}(\partial\Omega,\sigma)]^M \to [L^p(\partial\Omega,\sigma)]^M\big)}\right) < +\infty. \tag{7.1.136}$$

To prove (7.1.135) fix $p \in I_1$ and observe from [132, Theorem 1.5.1, item *(xiii)*] (cf. [132, (1.5.85)] and the very last comment in that item) and [132, (1.5.31)] that

$$\begin{aligned}&\mathrm{Im}\Big(S : \big[L^p(\partial\Omega, \sigma)\big]^M \to \big[L^p_1(\partial\Omega, \sigma)\big]^M\Big)\\ &\qquad \supseteq \mathrm{Im}\Big(\big(\tfrac{1}{2}I + K_{A_0}\big) \circ \big(-\tfrac{1}{2}I + K_{A_0}\big);\ \big[L^p_1(\partial\Omega, \sigma)\big]^M\Big).\end{aligned} \tag{7.1.137}$$

On the other hand, item *(2)* in [130, Theorem 2.1.2] together with [130, Definition 2.1.1] and (7.1.126) imply that $\big(\frac{1}{2}I+K_{A_0}\big)\circ\big(-\frac{1}{2}I+K_{A_0}\big)$ is a Fredholm operator (with index zero), hence an operator with (closed) range of finite codimension, when acting on $\big[L^p_1(\partial\Omega, \sigma)\big]^M$. Then (7.1.135) follows from this and (7.1.137).

The proof of (7.1.136) is similar, now making use of (7.1.125) and the observation

$$\begin{aligned}&\mathrm{Im}\Big(S : \big[L^p_{-1}(\partial\Omega, \sigma)\big]^M \to \big[L^p(\partial\Omega, \sigma)\big]^M\Big)\\ &\qquad \supseteq \mathrm{Im}\Big(\big(\tfrac{1}{2}I + K_{A_0}\big) \circ \big(-\tfrac{1}{2}I + K_{A_0}\big);\ \big[L^p(\partial\Omega, \sigma)\big]^M\Big),\end{aligned} \tag{7.1.138}$$

thanks to [132, (1.5.85)] and the last comment in [132, Theorem 1.5.1, item *(xiii)*].

Step III: *For each $p, q \in I_1$ we have*

$$f \in \big[L^q_{-1}(\partial\Omega, \sigma)\big]^M \text{ and } Sf \in \big[L^p_1(\partial\Omega, \sigma)\big]^M \Rightarrow f \in \big[L^p(\partial\Omega, \sigma)\big]^M. \tag{7.1.139}$$

In the proof of (7.1.139) we distinguish three cases.

Case 1: *Assume $p = q$.* From assumptions, [132, Theorem 1.5.1, item *(xiii)*] (cf. [132, (1.5.86)] and the last comment in that item), and [132, (1.5.31)] we see that

$$\Big(\big(\tfrac{1}{2}I + K^{\#}_{A_1^\top}\big) \circ \big(-\tfrac{1}{2}I + K^{\#}_{A_1^\top}\big)\Big)f = \big(\partial^{A_1}_\nu \mathcal{D}_{A_1}\big)(Sf) \in \big[L^p(\partial\Omega, \sigma)\big]^M. \tag{7.1.140}$$

As noted earlier, $\big(\frac{1}{2}I + K^{\#}_{A_1^\top}\big) \circ \big(-\frac{1}{2}I + K^{\#}_{A_1^\top}\big)$ is a Fredholm operator with index zero both on $\big[L^p_{-1}(\partial\Omega,\sigma)\big]^M$ and on $\big[L^p(\partial\Omega,\sigma)\big]^M$. Given that these are Banach spaces such that the latter is continuously and densely embedded into the former (cf. [130, (11.8.4)]), we then conclude from the abstract regularity result proved in [130, (2.1.14)] that $f \in \big[L^p(\partial\Omega,\sigma)\big]^M$. This justifies (7.1.139) in the current case.

`Case 2:` *Assume* $p \geq q$. Since we presently have $\big[L^p_1(\partial\Omega,\sigma)\big]^M \subseteq \big[L^q_1(\partial\Omega,\sigma)\big]^M$, Case 1 guarantees that $f \in \big[L^q(\partial\Omega,\sigma)\big]^M$. In place of (7.1.140) we now have

$$\Big(\big(\tfrac{1}{2}I + K^{\#}_{A_1^\top}\big) \circ \big(-\tfrac{1}{2}I + K^{\#}_{A_1^\top}\big)\Big)f = (\partial_\nu^{A_1}\mathcal{D}_{A_1})(Sf) \in \big[L^p(\partial\Omega,\sigma)\big]^M. \tag{7.1.141}$$

Then the abstract regularity result [130, (2.1.14)] gives that f is in $\big[L^p(\partial\Omega,\sigma)\big]^M$.

`Case 3:` *Assume* $q \geq p$. Since $\big[L^q_{-1}(\partial\Omega,\sigma)\big]^M \subseteq \big[L^p_{-1}(\partial\Omega,\sigma)\big]^M$, what we have proved in Case 1 implies that $f \in \big[L^p(\partial\Omega,\sigma)\big]^M$. This establishes (7.1.139).

`Step IV:` *Fix an arbitrary* $p \in I_1$, *and consider an algebraic complement* V *of* $\mathrm{Im}\Big(S : \big[L^p(\partial\Omega,\sigma)\big]^M \to \big[L^p_1(\partial\Omega,\sigma)\big]^M\Big)$ *into* $\big[L^p_1(\partial\Omega,\sigma)\big]^M$*; i.e.,* V *is a finite dimensional* (*thanks to* (7.1.135)) *linear subspace of* $\big[L^p_1(\partial\Omega,\sigma)\big]^M$ *such that*

$$V + S\Big(\big[L^p(\partial\Omega,\sigma)\big]^M\Big) = \big[L^p_1(\partial\Omega,\sigma)\big]^M \text{ and } V \cap S\Big(\big[L^p(\partial\Omega,\sigma)\big]^M\Big) = \{0\}. \tag{7.1.142}$$

Then

$$V \cap S\Big(\big[L^p_{-1}(\partial\Omega,\sigma)\big]^M\Big) = \{0\}, \tag{7.1.143}$$

$$V + S\Big(\big[L^p_{-1}(\partial\Omega,\sigma)\big]^M\Big) = \big[L^p(\partial\Omega,\sigma)\big]^M. \tag{7.1.144}$$

To prove (7.1.143), consider an arbitrary $g \in V \cap S\Big(\big[L^p_{-1}(\partial\Omega,\sigma)\big]^M\Big)$. Then there exists $f \in \big[L^p_{-1}(\partial\Omega,\sigma)\big]^M$ with the property that $Sf = g \in V \subseteq \big[L^p_1(\partial\Omega,\sigma)\big]^M$. Granted this, (7.1.139) forces f to belong to $\big[L^p(\partial\Omega,\sigma)\big]^M$ and, consequently, there holds $g = Sf \in S\Big(\big[L^p(\partial\Omega,\sigma)\big]^M\Big)$. At this point, (7.1.143) follows from

$$g \in V \cap S\Big(\big[L^p(\partial\Omega,\sigma)\big]^M\Big) = \{0\}. \tag{7.1.145}$$

Next, (7.1.136) and [130, Lemma 1.2.8] imply that, on the one hand,

$$V + S\Big(\big[L^p_{-1}(\partial\Omega,\sigma)\big]^M\Big) \text{ is a closed subspace of } \big[L^p(\partial\Omega,\sigma)\big]^M. \tag{7.1.146}$$

On the other hand, (7.1.142) gives

$$\big[L^p_1(\partial\Omega,\sigma)\big]^M = V + S\Big(\big[L^p(\partial\Omega,\sigma)\big]^M\Big) \subseteq V + S\Big(\big[L^p_{-1}(\partial\Omega,\sigma)\big]^M\Big) \tag{7.1.147}$$

which, in view of [130, (11.1.66)], implies that

$$V + S\Big(\big[L^p_{-1}(\partial\Omega,\sigma)\big]^M\Big) \text{ is a dense subspace of } \big[L^p(\partial\Omega,\sigma)\big]^M. \tag{7.1.148}$$

Together, (7.1.146) and (7.1.148) then prove (7.1.144).

`Step V:` *For each $p \in I_1$ we have*

$$\begin{aligned} &\dim \operatorname{CoKer}\Big(S : \big[L^p_{-1}(\partial\Omega,\sigma)\big]^M \to \big[L^p(\partial\Omega,\sigma)\big]^M\Big) \\ &\quad = \dim \operatorname{CoKer}\Big(S : \big[L^p(\partial\Omega,\sigma)\big]^M \to \big[L^p_1(\partial\Omega,\sigma)\big]^M\Big) < +\infty. \end{aligned} \tag{7.1.149}$$

To justify this, use (7.1.142) and (7.1.143)-(7.1.144) to write

$$\begin{aligned} &\dim \operatorname{CoKer}\Big(S : \big[L^p_{-1}(\partial\Omega,\sigma)\big]^M \to \big[L^p(\partial\Omega,\sigma)\big]^M\Big) \\ &\quad = \dim\left(\frac{\big[L^p(\partial\Omega,\sigma)\big]^M}{S\big([L^p_{-1}(\partial\Omega,\sigma)]^M\big)}\right) = \dim V = \dim\left(\frac{\big[L^p_1(\partial\Omega,\sigma)\big]^M}{S\big([L^p(\partial\Omega,\sigma)]^M\big)}\right) \\ &\quad = \dim \operatorname{CoKer}\Big(S : \big[L^p(\partial\Omega,\sigma)\big]^M \to \big[L^p_1(\partial\Omega,\sigma)\big]^M\Big). \end{aligned} \tag{7.1.150}$$

Now the desired conclusions follow bearing in mind (7.1.135)-(7.1.136).

`Step VI:` *For each given $p \in I_1$, both $S : \big[L^p(\partial\Omega,\sigma)\big]^M \to \big[L^p_1(\partial\Omega,\sigma)\big]^M$ and $S : \big[L^p_{-1}(\partial\Omega,\sigma)\big]^M \to \big[L^p(\partial\Omega,\sigma)\big]^M$ are Fredholm operators, and*

$$\begin{aligned} &\operatorname{index}\Big(S : \big[L^p_{-1}(\partial\Omega,\sigma)\big]^M \to \big[L^p(\partial\Omega,\sigma)\big]^M\Big) \\ &\quad = \operatorname{index}\Big(S : \big[L^p(\partial\Omega,\sigma)\big]^M \to \big[L^p_1(\partial\Omega,\sigma)\big]^M\Big). \end{aligned} \tag{7.1.151}$$

All claims are direct consequences of what we have proved in Step I and Step V (bearing in mind [130, Lemma 1.2.7]).

`Step VII:` *For each $p, q \in I_1$ we have* (7.1.124). Thanks to (7.1.151) it suffices to show that for each $p, q \in I_1$ we have

$$\begin{aligned} &\operatorname{index}\Big(S : \big[L^p(\partial\Omega,\sigma)\big]^M \to \big[L^p_1(\partial\Omega,\sigma)\big]^M\Big) \\ &\quad = \operatorname{index}\Big(S : \big[L^q(\partial\Omega,\sigma)\big]^M \to \big[L^q_1(\partial\Omega,\sigma)\big]^M\Big). \end{aligned} \tag{7.1.152}$$

In fact, since (7.1.129) readily implies that for each $p, q \in I_1$ we have

$$\mathrm{Ker}\Big(S : \big[L^p(\partial\Omega,\sigma)\big]^M \to \big[L^p_1(\partial\Omega,\sigma)\big]^M\Big)$$
$$= \mathrm{Ker}\Big(S : \big[L^q(\partial\Omega,\sigma)\big]^M \to \big[L^q_1(\partial\Omega,\sigma)\big]^M\Big), \tag{7.1.153}$$

it suffices to show that, for each $p, q \in I_1$,

$$\dim\left(\frac{\big[L^p_1(\partial\Omega,\sigma)\big]^M}{S\big([L^p(\partial\Omega,\sigma)]^M\big)}\right) = \dim\left(\frac{\big[L^q_1(\partial\Omega,\sigma)\big]^M}{S\big([L^q(\partial\Omega,\sigma)]^M\big)}\right). \tag{7.1.154}$$

By symmetry, there is no loss of generality in assuming that $p, q \in I_1$ are such that $p \geq q$. Next, choose a finite dimensional linear subspace V of $\big[L^p_1(\partial\Omega,\sigma)\big]^M$ satisfying (7.1.142). In particular, since $p \geq q$ and $\partial\Omega$ has finite measure,

$$V \text{ is a finite dimensional subspace of } \big[L^q_1(\partial\Omega,\sigma)\big]^M. \tag{7.1.155}$$

In relation to this, we claim that

$$V \cap S\Big(\big[L^q(\partial\Omega,\sigma)\big]^M\Big) = \{0\} \text{ and } V + S\Big(\big[L^q(\partial\Omega,\sigma)\big]^M\Big) = \big[L^q_1(\partial\Omega,\sigma)\big]^M. \tag{7.1.156}$$

To justify the first claim made in (7.1.156), pick $g \in V \cap S\big(\big[L^q(\partial\Omega,\sigma)\big]^M\big)$ arbitrary. Then there exists some $f \in \big[L^q(\partial\Omega,\sigma)\big]^M$ such that $Sf = g \in V \subseteq \big[L^p_1(\partial\Omega,\sigma)\big]^M$. With this in hand, (7.1.139) implies that, actually, the function f belongs to the space $\big[L^p(\partial\Omega,\sigma)\big]^M$. As a result, $g = Sf \in S\Big(\big[L^p(\partial\Omega,\sigma)\big]^M\Big)$, hence

$$g \in V \cap S\Big(\big[L^p(\partial\Omega,\sigma)\big]^M\Big) = \{0\} \tag{7.1.157}$$

which proves the first equality in (7.1.156). As regards the second equality in (7.1.156), observe that (7.1.135), (7.1.155), and [130, Lemma 1.2.8] imply

$$V + S\Big(\big[L^q(\partial\Omega,\sigma)\big]^M\Big) \text{ is a closed subspace of } \big[L^q_1(\partial\Omega,\sigma)\big]^M. \tag{7.1.158}$$

In addition, (7.1.142) and the fact that $p \geq q$ give

$$\big[\mathrm{Lip}\,(\partial\Omega)\big]^M \subseteq \big[L^p_1(\partial\Omega,\sigma)\big]^M = V + S\Big(\big[L^p(\partial\Omega,\sigma)\big]^M\Big)$$
$$\subseteq V + S\Big(\big[L^q(\partial\Omega,\sigma)\big]^M\Big). \tag{7.1.159}$$

Choosing $\delta \in (0,1)$ sufficiently small to begin with allows us to invoke Theorem 3.4.2 which, in turn, guarantees that the set Ω satisfies a two-sided local John condition. Granted this, the hypotheses of [130, Theorem 11.5.18] are satisfied. In view of [130, (11.5.200)], we then conclude from (7.1.159) that

$$V + S\Big(\big[L^q(\partial\Omega,\sigma)\big]^M\Big) \text{ is a dense subspace of } \big[L^q_1(\partial\Omega,\sigma)\big]^M. \tag{7.1.160}$$

Combining (7.1.159) and (7.1.160) then yields the second equality in (7.1.156). Finally, (7.1.154) follows from (7.1.142), (7.1.156), and

$$\dim\left(\frac{\left[L_1^p(\partial\Omega,\sigma)\right]^M}{S\big([L^p(\partial\Omega,\sigma)]^M\big)}\right)=\dim V=\dim\left(\frac{\left[L_1^q(\partial\Omega,\sigma)\right]^M}{S\big([L^q(\partial\Omega,\sigma)]^M\big)}\right). \tag{7.1.161}$$

Collectively, Steps I-V prove (7.1.123)-(7.1.124). Ergo, in any of the settings (7.1.103)-(7.1.104) the single layer S induces Fredholm operators of equal index, which is actually independent of $p \in I_1$. In a very similar fashion one can also show that in any of the settings (7.1.105), (7.1.106), (7.1.107), (7.1.108) the single layer S induces Fredholm operators having the same index, which is actually independent of $p, q \in I_1$, $\lambda \in I_2$, and $s \in I_3$.

To prove the Fredholmness of the operators (7.1.109)-(7.1.110), fix $s \in I_3$ and $p \in I_4$. By working with $\mathbb{R}^n \setminus \overline{\Omega}$ in place of Ω if necessary, there is no loss of generality in assuming that Ω is bounded (see item *(6)* in [129, Lemma 5.10.9] and [129, Lemma 5.10.10]). Take $\delta \in (0,1)$ sufficiently small so that we may invoke Theorem 3.4.2. This guarantees that the set Ω is both a locally uniform domain and an (ε,δ)-domain satisfying a two-sided local John condition. In turn, this permits us to use [132, (4.3.42)] to conclude that the operator (7.1.110) is well defined, linear, and bounded. In fact, since $L^\top$ (the real transpose of L) satisfies similar properties to those of L, we also have that $S_{L^\top}$ (the single layer potential operator associated with the system $L^\top$ in the same manner S has been associated with the original system L) is well defined, linear, and bounded in the context of (7.1.110). Hence,

$$\begin{gathered}\text{the operators } S, S_{L^\top} : \big[B^{1,1}_{-s}(\partial\Omega,\sigma)\big]^M \longrightarrow \big[B^{1,1}_{1-s}(\partial\Omega,\sigma)\big]^M \\ \text{are well defined, linear, and bounded.}\end{gathered} \tag{7.1.162}$$

We may also rely on [132, (4.3.50)] to conclude that

$$\begin{gathered}\text{the transpose of } S_{L^\top} : \big[B^{1,1}_{-s}(\partial\Omega,\sigma)\big]^M \longrightarrow \big[B^{1,1}_{1-s}(\partial\Omega,\sigma)\big]^M \\ \text{is the operator } S : \big[B^{\infty,\infty}_{s-1}(\partial\Omega,\sigma)\big]^M \longrightarrow \big[B^{\infty,\infty}_{s}(\partial\Omega,\sigma)\big]^M.\end{gathered} \tag{7.1.163}$$

To proceed, pick a coefficient tensor $A \in \mathfrak{A}^{\rm dis}_L$ and observe that [132, (4.4.55)] implies

$$\begin{aligned}&\mathrm{Im}\Big(S : \big[B^{\infty,\infty}_{s-1}(\partial\Omega,\sigma)\big]^M \to \big[B^{\infty,\infty}_{s}(\partial\Omega,\sigma)\big]^M\Big)\\ &\qquad \supseteq \mathrm{Im}\Big(\big(\tfrac{1}{2}I + K_A\big)\circ\big(-\tfrac{1}{2}I + K_A\big);\ \big[B^{\infty,\infty}_{s}(\partial\Omega,\sigma)\big]^M\Big).\end{aligned} \tag{7.1.164}$$

If the threshold $\delta \in (0,1)$ is sufficiently small to begin with, from item *(2)* in Theorem 7.1.1, item *(2)* in [130, Theorem 2.1.2], and [130, (7.1.59)] we deduce that $\big(\tfrac{1}{2}I + K_A\big)\circ\big(-\tfrac{1}{2}I + K_A\big)$ is a Fredholm operator on $\big[B^{\infty,\infty}_{s}(\partial\Omega,\sigma)\big]^M$. In particular, it has a range of finite codimension. From this and [130, Lemma 1.2.7] we see that

$$S \text{ has closed range of finite codimension in the context of (7.1.109).} \tag{7.1.165}$$

Next, pick $p \in (1, \infty)$ with $p > (n-1)/(1-s)$ along with $\widetilde{s} \in (0, s)$ and $\widetilde{p} \in (1, \infty)$. From [130, (7.7.8)], [130, Proposition 7.7.2, *(i)*] (used with $p_1 = q_1 = \infty$, $s_1 = s-1$, $p_2 = q_2 = \widetilde{p}$, $s_2 = \widetilde{s} - 1$), and [130, Proposition 11.11.3] we see that

$$L^p(\partial\Omega, \sigma) \hookrightarrow B^{\infty,\infty}_{s-1}(\partial\Omega, \sigma) \hookrightarrow B^{\widetilde{p},\widetilde{p}}_{\widetilde{s}-1}(\partial\Omega, \sigma) \hookrightarrow L^{\widetilde{p}}_{-1}(\partial\Omega, \sigma). \tag{7.1.166}$$

There is no loss of generality in assuming that I_1 is large enough so that $p, \widetilde{p} \in I_1$. Then (7.1.104) guarantees that

$$S : \left[L^{\widetilde{p}}_{-1}(\partial\Omega, \sigma)\right]^M \longrightarrow \left[L^{\widetilde{p}}(\partial\Omega, \sigma)\right]^M \text{ is Fredholm.} \tag{7.1.167}$$

Consequently, from (7.1.167) and (7.1.166) we conclude that

$$\dim \operatorname{Ker}\Big(S : \left[B^{\infty,\infty}_{s-1}(\partial\Omega, \sigma)\right]^M \to \left[B^{\infty,\infty}_{s}(\partial\Omega, \sigma)\right]^M\Big) < +\infty. \tag{7.1.168}$$

Ultimately, from (7.1.165) and (7.1.168) we see that S is a Fredholm operator in the context of (7.1.109). For further reference let us also note here that (7.1.166) and the version of (7.1.129) (written for the system $L^\top$ in place of L) give

$$\begin{aligned}
&\operatorname{Ker}\Big(S_{L^\top} : \left[B^{\infty,\infty}_{s-1}(\partial\Omega, \sigma)\right]^M \to \left[B^{\infty,\infty}_{s}(\partial\Omega, \sigma)\right]^M\Big) \\
&\quad = \operatorname{Ker}\Big(S_{L^\top} : \left[L^{p}(\partial\Omega, \sigma)\right]^M \to \left[L^{p}_{1}(\partial\Omega, \sigma)\right]^M\Big) \\
&\quad = \operatorname{Ker}\Big(S_{L^\top} : \left[L^{p}_{-1}(\partial\Omega, \sigma)\right]^M \to \left[L^{p}(\partial\Omega, \sigma)\right]^M\Big).
\end{aligned} \tag{7.1.169}$$

Moving on, pick $s_0 \in (s, 1)$ and $p_0 \in (1, \infty)$ such that $s_0 - s > (n-1)(1 - 1/p_0)$. This guarantees that item *(i)* in [130, Proposition 7.7.2] applies (with $p_1 = q_1 = 1$, $s_1 = -s$, $p_2 = q_2 = p_0$, $s_2 = -s_0$). If we also pick $p \in (1, \infty)$ then, together with [130, (7.7.8)] and [130, Proposition 11.11.3], this gives

$$L^p(\partial\Omega, \sigma) \hookrightarrow B^{1,1}_{-s}(\partial\Omega, \sigma) \hookrightarrow B^{p_0,p_0}_{-s_0}(\partial\Omega, \sigma) \hookrightarrow L^{p_0}_{-1}(\partial\Omega, \sigma). \tag{7.1.170}$$

There is no loss of generality in assuming that the interval I_1 is large enough so that $p, p_0 \in I_1$. Then (7.1.104) ensures that

$$S : \left[L^{p_0}_{-1}(\partial\Omega, \sigma)\right]^M \longrightarrow \left[L^{p_0}(\partial\Omega, \sigma)\right]^M \text{ is Fredholm.} \tag{7.1.171}$$

In particular, from (7.1.171) and (7.1.170) we see that

$$\dim \operatorname{Ker}\Big(S : \left[B^{1,1}_{-s}(\partial\Omega, \sigma)\right]^M \to \left[B^{1,1}_{1-s}(\partial\Omega, \sigma)\right]^M\Big) < +\infty. \tag{7.1.172}$$

Moreover, (7.1.170) and (7.1.129) imply

$$\operatorname{Ker}\Big(S : \big[B^{1,1}_{-s}(\partial\Omega,\sigma)\big]^M \to \big[B^{1,1}_{1-s}(\partial\Omega,\sigma)\big]^M\Big)$$
$$= \operatorname{Ker}\Big(S : \big[L^p(\partial\Omega,\sigma)\big]^M \to \big[L^p_1(\partial\Omega,\sigma)\big]^M\Big). \tag{7.1.173}$$

Next, from (7.1.165), (7.1.163), and Banach's Closed Range Theorem (recalled in [130, (2.1.47)]) we see that

$$S \text{ has closed range in the context of (7.1.110).} \tag{7.1.174}$$

Also, [130, (2.1.49)], (7.1.174), and (7.1.163) (with L replaced by $L^\top$) imply

$$\operatorname{Im}\Big(S : \big[B^{1,1}_{-s}(\partial\Omega,\sigma)\big]^M \to \big[B^{1,1}_{1-s}(\partial\Omega,\sigma)\big]^M\Big)$$
$$= {}^{\perp}\Big[\operatorname{Ker}\Big(S_{L^\top} : \big[B^{\infty,\infty}_{s-1}(\partial\Omega,\sigma)\big]^M \to \big[B^{\infty,\infty}_{s}(\partial\Omega,\sigma)\big]^M\Big)\Big]. \tag{7.1.175}$$

In concert, (7.1.175), [130, (2.1.26)], and (7.1.168) (used with $L^\top$ in place of L) give

$$\dim \operatorname{CoKer}\Big(S : \big[B^{1,1}_{-s}(\partial\Omega,\sigma)\big]^M \to \big[B^{1,1}_{1-s}(\partial\Omega,\sigma)\big]^M\Big) \tag{7.1.176}$$
$$= \dim\left(\frac{\big[B^{1,1}_{1-s}(\partial\Omega,\sigma)\big]^M}{S\big([B^{1,1}_{-s}(\partial\Omega,\sigma)]^M\big)}\right) = \dim\left(\frac{\big[B^{1,1}_{-s}(\partial\Omega,\sigma)\big]^M}{{}^{\perp}\Big[\operatorname{Ker}(S_{L^\top}:[B^{\infty,\infty}_{s-1}(\partial\Omega,\sigma)]^M \to [B^{\infty,\infty}_{s}(\partial\Omega,\sigma)]^M)\Big]}\right)$$
$$= \dim \operatorname{Ker}\Big(S_{L^\top} : \big[B^{\infty,\infty}_{s-1}(\partial\Omega,\sigma)\big]^M \to \big[B^{\infty,\infty}_{s}(\partial\Omega,\sigma)\big]^M\Big) < +\infty.$$

At this stage, from (7.1.172), (7.1.174), and (7.1.176) we conclude that S is a Fredholm operator in the context of (7.1.110). In addition, assuming (as we may) that $p' := (1-1/p)^{-1} \in I_1$, we may write

$$\dim \operatorname{CoKer}\Big(S : \big[B^{1,1}_{-s}(\partial\Omega,\sigma)\big]^M \to \big[B^{1,1}_{1-s}(\partial\Omega,\sigma)\big]^M\Big)$$
$$= \dim \operatorname{Ker}\Big(S_{L^\top} : \big[L^p_{-1}(\partial\Omega,\sigma)\big]^M \to \big[L^p(\partial\Omega,\sigma)\big]^M\Big)$$
$$= \dim\left(\frac{\big[L^{p'}_1(\partial\Omega,\sigma)\big]^M}{{}^{\perp}\Big[\operatorname{Ker}(S_{L^\top}:[L^p_{-1}(\partial\Omega,\sigma)]^M \to [L^p(\partial\Omega,\sigma)]^M)\Big]}\right)$$
$$= \dim\left(\frac{\big[L^{p'}_1(\partial\Omega,\sigma)\big]^M}{S\big([L^{p'}(\partial\Omega,\sigma)]^M\big)}\right) = \dim\left(\frac{\big[L^{p}_1(\partial\Omega,\sigma)\big]^M}{S\big([L^{p}(\partial\Omega,\sigma)]^M\big)}\right)$$
$$= \dim \operatorname{CoKer}\Big(S : \big[L^p(\partial\Omega,\sigma)\big]^M \to \big[L^p_1(\partial\Omega,\sigma)\big]^M\Big), \tag{7.1.177}$$

Above, the first equality is implied by (7.1.176) and (7.1.169). The second equality in (7.1.177) comes from [130, (2.1.26)] (bearing in mind (A.0.154) and (7.1.130)) The third equality in (7.1.177) is a consequence of the fact that

$$\operatorname{Im}\Big(S : \big[L^{p'}(\partial\Omega, \sigma)\big]^M \to \big[L^{p'}_1(\partial\Omega, \sigma)\big]^M\Big)$$
$$= {}^{\perp}\Big[\operatorname{Ker}\Big(S_{L^\top} : \big[L^p_{-1}(\partial\Omega, \sigma)\big]^M \to \big[L^p(\partial\Omega, \sigma)\big]^M\Big)\Big] \tag{7.1.178}$$

which, in turn, follows from [130, (2.1.49)], [132, Theorem 1.5.1, item *(ix)*], and the fact that S has closed range in the context of (7.1.103) (with p replaced by p'). The fourth equality in (7.1.177) is seen from (7.1.154), and the last equality in (7.1.177) is by definition. By combining (7.1.173) and (7.1.177) we arrive at

$$\operatorname{index}\Big(S : \big[B^{1,1}_{-s}(\partial\Omega, \sigma)\big]^M \to \big[B^{1,1}_{1-s}(\partial\Omega, \sigma)\big]^M\Big)$$
$$= \operatorname{index}\Big(S : \big[L^p(\partial\Omega, \sigma)\big]^M \to \big[L^p_1(\partial\Omega, \sigma)\big]^M\Big). \tag{7.1.179}$$

Running the same argument that produced (7.1.179) for $L^\top$ in place of L then invoking item *(3)* in [130, Theorem 2.1.2], [132, Theorem 1.5.1, item *(ix)*], and (7.1.163) also yields

$$\operatorname{index}\Big(S : \big[B^{\infty,\infty}_{s-1}(\partial\Omega, \sigma)\big]^M \to \big[B^{\infty,\infty}_{s}(\partial\Omega, \sigma)\big]^M\Big)$$
$$= \operatorname{index}\Big(S : \big[L^{p'}_{-1}(\partial\Omega, \sigma)\big]^M \to \big[L^{p'}(\partial\Omega, \sigma)\big]^M\Big). \tag{7.1.180}$$

The results in (7.1.179), (7.1.180) take care of the claims in (7.1.111) pertaining to the operators (7.1.109) and (7.1.110).

Next, the claims pertaining to the Fredholmness of the operator $(S_{L^\top})^\top$ in (7.1.112) and (7.1.113), as well as the properties of the indexes claimed in (7.1.114) are consequences of the fact that by taking $\delta \in (0, 1)$ sufficiently small we may ensure that $S_{L^\top}$ is Fredholm in the context of (7.1.105) and (7.1.107) (written for the conjugate exponents) and duality (cf. [130, Propositions 6.2.8 and 6.2.16], [130, Definition 11.8.9] (cf. (A.0.175), (A.0.36)), and item *(3)* in [130, Theorem 2.1.2]).

Finally, that similar results are valid for the boundary-to-boundary single layer potential operator S associated with the Stokes system is seen by reasoning in a completely analogous fashion. □

We have the following companion result to Theorem 7.1.5, identifying conditions ensuring that all Fredholm operators (7.1.103)-(7.1.110) have index zero.

Corollary 7.1.6 *Let $\Omega \subseteq \mathbb{R}^n$, where $n \in \mathbb{N}$ with $n \geq 2$, be an Ahlfors regular domain with compact boundary. Abbreviate $\sigma := \mathcal{H}^{n-1}\lfloor\partial\Omega$ and denote by ν the geometric measure theoretic outward unit normal to Ω. Also, let L be a homogeneous, second-order, constant complex coefficient, $M \times M$ system in $\mathbb{R}^n$ such that*

> *L satisfies the Legendre-Hadamard strong ellipticity condition (cf.* [131, Definition 1.3.2]) *and one has $\mathfrak{A}^{\rm dis}_L \neq \varnothing$.* (7.1.181)

In this context, consider the boundary-to-boundary single layer potential operator S associated with the set Ω and the system L as in (A.0.231)*. Finally, fix three compact intervals, namely $I_1 \subset (1, \infty)$, $I_2 \subset (0, n-1)$, and $I_3 \subset (0, 1)$.*

Then there exists $\delta \in (0,1)$ depending only on n, I_1, I_2, I_3, L, the ADR *constants of $\partial\Omega$, and the relative flatness ratio $\wp(\Omega)/\text{diam}\,\partial\Omega$ such that if*

$$\text{dist}\Big(\nu, [\text{VMO}(\partial\Omega,\sigma)]^n\Big) < \delta \tag{7.1.182}$$

where the distance is considered in the space $\big[\text{BMO}(\partial\Omega,\sigma)\big]^n$ (hence if Ω is a δ-oscillating AR *domain; cf. Definition 3.4.1), it follows that Ω is a* UR *domain and*

$$\begin{array}{l}\textit{for each } p,q \in I_1,\ \lambda \in I_2, \textit{ and } s \in I_3 \textit{ the operators}\\ \text{(7.1.103)-(7.1.110)} \textit{ are Fredholm with index zero.}\end{array} \tag{7.1.183}$$

Proof From Theorem 1.5.2 we know that (7.1.181) implies (7.1.101). Take $\delta \in (0,1)$ sufficiently small so that all conclusions in Theorem 7.1.5 hold. In particular, all operators (7.1.103)-(7.1.110) are Fredholm. Further decreasing the value of δ if necessary ensures that Ω is a two-sided NTA domain (cf. Theorem 3.4.2). Since $\partial\Omega$ is compact, after eventually replacing Ω by $\mathbb{R}^n \setminus \overline{\Omega}$, there is no loss of generality in assuming that Ω is bounded (cf. [129, Lemma 5.10.10]). Granted these properties, the fact that the indexes of the Fredholm operators (7.1.103)-(7.1.110) are all zero becomes a consequence of (7.1.111) and (7.2.28). □

Building on earlier work, we may now bring together a number of Fredholmness results established so far for the single layer on specific scales, under the more inclusive umbrella of Generalized Banach Function Spaces, as shown below.

Theorem 7.1.7 *Pick $n \in \mathbb{N}$ with $n \geq 2$ and let $\Omega \subseteq \mathbb{R}^n$ be an Ahlfors regular domain with compact boundary. Abbreviate $\sigma := \mathcal{H}^{n-1}\lfloor\partial\Omega$ and denote by ν the geometric measure theoretic outward unit normal to Ω. Also, let L be a homogeneous, second-order, constant complex coefficient, weakly elliptic $M \times M$ system in $\mathbb{R}^n$ for which*

$$\mathfrak{A}_L^{\text{dis}} \neq \varnothing \ \textit{and}\ \mathfrak{A}_{L^\top}^{\text{dis}} \neq \varnothing. \tag{7.1.184}$$

Next, let $\mathbb{X}$ be a Generalized Banach Function Space on $(\partial\Omega,\sigma)$ such that

$$\mathcal{M}_{\partial\Omega} : \mathbb{X} \to \mathbb{X} \textit{ and } \mathcal{M}_{\partial\Omega} : \mathbb{X}' \to \mathbb{X}' \textit{ are well-defined and bounded,} \tag{7.1.185}$$

where $\mathcal{M}_{\partial\Omega}$ is the Hardy-Littlewood maximal operator on $\partial\Omega$, and $\mathbb{X}'$ is the associated space of $\mathbb{X}$ (cf. [130, Definitions 5.1.4, 5.1.11]). *Define*

$$\mathring{\mathbb{X}} := \overline{L^\infty(\partial\Omega,\sigma)}^{\|\cdot\|_{\mathbb{X}}} \ \textit{and}\ (\mathbb{X}')^\circ := \overline{L^\infty(\partial\Omega,\sigma)}^{\|\cdot\|_{\mathbb{X}'}}, \tag{7.1.186}$$

and bring back $\mathbb{X}_1$, $(\mathbb{X}')_1$, $\mathring{\mathbb{X}}_1$, $(\mathbb{X}')^\circ_1$, the $\mathbb{X}$-styled, $\mathbb{X}'$-styled, $\mathring{\mathbb{X}}$-styled, and $(\mathbb{X}')^\circ$-styled Sobolev spaces of order one on $\partial\Omega$ from (7.1.39) *(these are known to be Banach spaces when equipped with natural norms; cf.* (5.4.84)*). Finally, consider the boundary-to-boundary single layer potential operator S associated with the set Ω and the system L as in* (A.0.231).

Then there exists a threshold $\delta_{\mathbb{X}} \in (0,1)$, which depends only on n, L, the operator norms of $\mathcal{M}_{\partial\Omega}$ on $\mathbb{X}$ and $\mathbb{X}'$, the ADR *constants of $\partial\Omega$, and the relative flatness ratio*

$\wp(\Omega)/\operatorname{diam}\partial\Omega$, *with the property that if*

$$\operatorname{dist}\Big(\nu,\ [\mathrm{VMO}\,(\partial\Omega,\sigma)]^n\Big) < \delta_{\mathbb{X}} \tag{7.1.187}$$

where the distance is considered in the space $\big[\mathrm{BMO}\,(\partial\Omega,\sigma)\big]^n$ *(hence if* Ω *is a* $\delta_{\mathbb{X}}$*-oscillating* AR *domain; cf. Definition 3.4.1), it follows that* Ω *is a* UR *domain and these operators are Fredholm:*

$$S : \big[\mathbb{X}\big]^M \to \big[\mathbb{X}_1\big]^M, \quad S : \big[\mathbb{X}'\big]^M \to \big[(\mathbb{X}')_1\big]^M, \tag{7.1.188}$$

$$S : \big[\mathring{\mathbb{X}}\big]^M \to \big[\mathring{\mathbb{X}}_1\big]^M, \quad S : \big[(\mathbb{X}')^\circ\big]^M \to \big[(\mathbb{X}')_1^\circ\big]^M. \tag{7.1.189}$$

Moreover, granted (7.1.187),

the operators (7.1.188)-(7.1.189) *are Fredholm if* $\mathbb{X}$ *is a rearrangement invariant Banach function space with Boyd indices* $1 < p_{\mathbb{X}} \leq q_{\mathbb{X}} < \infty$. (7.1.190)

As a corollary, under the assumption made in (7.1.187),

the operators in (7.1.188)-(7.1.189) *are Fredholm if* $\mathbb{X}$ *is* $L^\Phi(\partial\Omega,\sigma)$, *the Orlicz space associated as in* (A.0.133) *with the measure space* $(\partial\Omega,\sigma)$ *and any Young function* Φ *satisfying* $1 < i(\Phi) \leq I(\Phi) < \infty$. (7.1.191)

Finally, if (7.1.184) *is strengthened (cf. Theorem 1.5.2) to*

L *satisfies the Legendre-Hadamard strong ellipticity condition (cf.* [131, Definition 1.3.2]) *and one has* $\mathfrak{A}_L^{\mathrm{dis}} \neq \varnothing$, (7.1.192)

then all Fredholm operators mentioned above have index zero whenever (7.1.187) *holds. In particular, under the assumptions made in* (7.1.192), *all aforementioned operators are Fredholm with index zero whenever* Ω *is an infinitesimally flat* AR *domain (cf. Definition 3.4.1).*

Proof From Theorem 3.4.2 we know that choosing the threshold $\delta_{\mathbb{X}} \in (0,1)$ small enough makes Ω a UR domain satisfying a two-sided local John condition. Let us deal with the claims made in relation to the first operator in (7.1.188). To this end, first observe from [130, (5.2.92)], (7.1.39), and [130, (11.5.200)] that there exist $q \in (1,2)$ and $q' \in (2,\infty)$ satisfying $1/q + 1/q' = 1$, with the property that the following embeddings are continuous

$$L^{q'}(\partial\Omega,\sigma) \hookrightarrow \mathbb{X}, \quad L_1^{q'}(\partial\Omega,\sigma) \hookrightarrow \mathbb{X}_1, \tag{7.1.193}$$

and the following embeddings are continuous with dense range:

$$\mathbb{X} \hookrightarrow L^q(\partial\Omega,\sigma), \quad \mathbb{X}_1 \hookrightarrow L_1^q(\partial\Omega,\sigma). \tag{7.1.194}$$

Thanks to (7.1.194), (7.1.39), item *(11)* of [131, Theorem 2.3.2], and [130, (5.1.13)] it follows that the operator

$$S : [\mathbb{X}]^M \longrightarrow [\mathbb{X}_1]^M \text{is well defined, linear, and bounded.} \tag{7.1.195}$$

Next, (7.1.193) and (7.1.194) imply

$$\begin{aligned}&\mathrm{Ker}\Big(S : \big[L^{q'}(\partial\Omega,\sigma)\big]^M \to \big[L^{q'}_1(\partial\Omega,\sigma)\big]^M\Big) \\ &\quad\subseteq \mathrm{Ker}\Big(S : [\mathbb{X}]^M \to [\mathbb{X}_1]^M\Big) \subseteq \mathrm{Ker}\Big(S : \big[L^q(\partial\Omega,\sigma)\big]^M \to \big[L^q_1(\partial\Omega,\sigma)\big]^M\Big),\end{aligned} \tag{7.1.196}$$

and from (7.1.129)-(7.1.130) we know that

$$\begin{array}{c}\mathrm{Ker}\Big(S : \big[L^{q'}(\partial\Omega,\sigma)\big]^M \to \big[L^{q'}_1(\partial\Omega,\sigma)\big]^M\Big) \\ \text{and } \mathrm{Ker}\Big(S : \big[L^{q}(\partial\Omega,\sigma)\big]^M \to \big[L^{q}_1(\partial\Omega,\sigma)\big]^M\Big) \\ \text{are equal, finite dimensional, linear spaces.}\end{array} \tag{7.1.197}$$

Consequently,

$$\begin{array}{c}\mathrm{Ker}\Big(S : [\mathbb{X}]^M \to [\mathbb{X}_1]^M\Big) \text{ has finite dimension and} \\ \text{coincides with } \mathrm{Ker}\Big(S : \big[L^{q}(\partial\Omega,\sigma)\big]^M \to \big[L^{q}_1(\partial\Omega,\sigma)\big]^M\Big).\end{array} \tag{7.1.198}$$

Going further, in view of (7.1.184), we may select $A_0 \in \mathfrak{A}^{\mathrm{dis}}_L$ along with $A_1 \in \mathfrak{A}_L$ such that $A_1^\top \in \mathfrak{A}^{\mathrm{dis}}_{L^\top}$. After possibly further decreasing the value of the threshold $\delta_{\mathbb{X}} \in (0,1)$, item *(8)* in Theorem 7.1.1 ensures that the operators

$$\begin{array}{c}\pm\tfrac{1}{2} + K_{A_0} : [\mathbb{X}_1]^M \to [\mathbb{X}_1]^M, \ \pm\tfrac{1}{2}I + K^{\#}_{A_1^\top} : [\mathbb{X}]^M \to [\mathbb{X}]^M \\ \text{are Fredholm with index zero,}\end{array} \tag{7.1.199}$$

while item *(1)* in Theorem 7.1.1 guarantees that

$$\begin{array}{c}\pm\tfrac{1}{2}I + K^{\#}_{A_1^\top} : \big[L^{q}(\partial\Omega,\sigma)\big]^M \longrightarrow \big[L^{q}(\partial\Omega,\sigma)\big]^M. \\ \text{are Fredholm operators with index zero.}\end{array} \tag{7.1.200}$$

Thanks to (7.1.103), matters may also be arranged so that

$$S : \big[L^{q}(\partial\Omega,\sigma)\big]^M \to \big[L^{q}_1(\partial\Omega,\sigma)\big]^M \text{ is a Fredholm operator.} \tag{7.1.201}$$

We claim that

$$\dim\left(\frac{[\mathbb{X}_1]^M}{\mathrm{Im}\big(S:[\mathbb{X}]^M \to [\mathbb{X}_1]^M\big)}\right) < +\infty. \tag{7.1.202}$$

To prove (7.1.202) observe from [132, Theorem 1.5.1, item *(xiii)*] (specifically, [132, (1.5.85)] and the very last comment in that item) together with (7.1.194) that

$$\mathrm{Im}\Big(S : [\mathbb{X}]^M \to [\mathbb{X}_1]^M\Big) \supseteq \mathrm{Im}\Big(\big(\tfrac{1}{2}I + K_{A_0}\big) \circ \big(-\tfrac{1}{2}I + K_{A_0}\big);\ [\mathbb{X}_1]^M\Big). \tag{7.1.203}$$

At the same time, [130, Theorem 2.1.2, *(2)*] plus [130, Definition 2.1.1] and (7.1.199) imply that the composition $\big(\frac{1}{2}I + K_{A_0}\big) \circ \big(-\frac{1}{2}I + K_{A_0}\big)$ is a Fredholm operator (with index zero), hence an operator with (closed) range of finite codimension, on $\big[\mathbb{X}_1\big]^M$. Then (7.1.202) follows from this and (7.1.203). Next we claim that

$$f \in \big[L^q(\partial\Omega, \sigma)\big]^M \text{ and } Sf \in \big[\mathbb{X}_1\big]^M \Longrightarrow f \in \big[\mathbb{X}\big]^M. \tag{7.1.204}$$

To see this, note that the conormal derivative of the double layer,

$$\partial_\nu^{A_1}\mathcal{D}_{A_1} : \big[\mathbb{X}_1\big]^M \longrightarrow \big[\mathbb{X}\big]^M, \quad (\partial_\nu^{A_1}\mathcal{D}_{A_1})f := \partial_\nu^{A_1}(\mathcal{D}_{A_1}f), \tag{7.1.205}$$

is well defined, linear, and bounded. Indeed, this is clear from (7.1.194), [132, (1.5.29)], and [131, Theorem 2.3.2, *(11)*]. Pressing on, given any function f in $\big[L^q(\partial\Omega, \sigma)\big]^M$ with $Sf \in \big[\mathbb{X}_1\big]^M$, [132, Theorem 1.5.1, item *(xiii)*] (cf. [132, (1.5.86)] and the very last comment in that item), and (7.1.205) imply that

$$\Big(\big(\tfrac{1}{2}I + K^{\#}_{A_1^\top}\big) \circ \big(-\tfrac{1}{2}I + K^{\#}_{A_1^\top}\big)\Big)f = (\partial_\nu^{A_1}\mathcal{D}_{A_1})(Sf) \in \big[\mathbb{X}\big]^M. \tag{7.1.206}$$

By once again appealing to item *(2)* in [130, Theorem 2.1.2], [130, Definition 2.1.1], (7.1.199), and (7.1.200), we conclude that the composition $\big(\frac{1}{2}I+K^{\#}_{A_1^\top}\big)\circ\big(-\frac{1}{2}I+K^{\#}_{A_1^\top}\big)$ is a Fredholm operator with index zero both on $\big[L^q(\partial\Omega, \sigma)\big]^M$ and on $\big[\mathbb{X}\big]^M$. Given that these are Banach spaces such that the latter is continuously and densely embedded into the former (cf. (7.1.194)), we may invoke the abstract regularity result from [130, (2.1.14)] and obtain that, in fact, $f \in \big[\mathbb{X}\big]^M$. This proves (7.1.204).

Thanks to the result formulated in (7.1.202), there exists a finite dimensional algebraical complement of $\mathrm{Im}\Big(S : \big[\mathbb{X}\big]^M \to \big[\mathbb{X}_1\big]^M\Big)$ in $\mathbb{X}_1$, i.e., there exists a finite dimensional linear subspace V of $\big[\mathbb{X}_1\big]^M$ with the property that

$$V + S\Big(\big[\mathbb{X}\big]^M\Big) = \big[\mathbb{X}_1\big]^M \text{ and } V \cap S\Big(\big[\mathbb{X}\big]^M\Big) = \{0\}. \tag{7.1.207}$$

For any such choice of V we claim that

$$V \cap S\Big(\big[L^q(\partial\Omega, \sigma)\big]^M\Big) = \{0\} \tag{7.1.208}$$

and

$$V + S\Big(\big[L^q(\partial\Omega, \sigma)\big]^M\Big) = \big[L^q_1(\partial\Omega, \sigma)\big]^M. \tag{7.1.209}$$

To justify (7.1.208), pick an arbitrary $g \in V \cap S\Big(\big[L^q(\partial\Omega, \sigma)\big]^M\Big)$. Then there exists a function $f \in \big[L^q(\partial\Omega, \sigma)\big]^M$ with the property that $Sf = g \in V \subseteq \big[\mathbb{X}_1\big]^M$. In view of this, (7.1.204) places f in $\big[\mathbb{X}\big]^M$, hence $g = Sf \in S\Big(\big[\mathbb{X}\big]^M\Big)$. Altogether,

$$g \in V \cap S\Big(\big[\mathbb{X}\big]^M\Big) = \{0\} \tag{7.1.210}$$

which ultimately establishes (7.1.208).

Next, from (7.1.201) and [130, Lemma 1.2.8] we conclude that

$$V + S\Big(\big[L^q(\partial\Omega,\sigma)\big]^M\Big) \text{ is a closed subspace of } \big[L^q_1(\partial\Omega,\sigma)\big]^M. \tag{7.1.211}$$

Also, (7.1.39), (7.1.207), and (7.1.194) imply

$$\big[\mathrm{Lip}(\partial\Omega)\big]^M \subseteq \big[\mathbb{X}_1\big]^M = V + S\Big(\big[\mathbb{X}\big]^M\Big) \subseteq V + S\Big(\big[L^q(\partial\Omega,\sigma)\big]^M\Big) \tag{7.1.212}$$

which, in light of [130, (11.5.200)], permits us to conclude that

$$V + S\Big(\big[L^q(\partial\Omega,\sigma)\big]^M\Big) \text{ is a dense subspace of } \big[L^q_1(\partial\Omega,\sigma)\big]^M. \tag{7.1.213}$$

Collectively, (7.1.211) and (7.1.213) prove (7.1.209). We next propose to show that

$$\begin{aligned} &\dim \mathrm{CoKer}\Big(S : \big[\mathbb{X}\big]^M \to \big[\mathbb{X}_1\big]^M\Big) \\ &\quad = \dim \mathrm{CoKer}\Big(S : \big[L^q(\partial\Omega,\sigma)\big]^M \to \big[L^q_1(\partial\Omega,\sigma)\big]^M\Big) < +\infty. \end{aligned} \tag{7.1.214}$$

To prove this, use (7.1.207) and (7.1.208)-(7.1.209) to write

$$\begin{aligned} \dim \mathrm{CoKer}\Big(S : \big[\mathbb{X}\big]^M \to \big[\mathbb{X}_1\big]^M\Big) &= \dim\left(\frac{\big[\mathbb{X}_1\big]^M}{S\big([\mathbb{X}]^M\big)}\right) = \dim V \\ = \dim\left(\frac{\big[L^q_1(\partial\Omega,\sigma)\big]^M}{S\big([L^q(\partial\Omega,\sigma)]^M\big)}\right) &= \dim \mathrm{CoKer}\Big(S : \big[L^q(\partial\Omega,\sigma)\big]^M \to \big[L^q_1(\partial\Omega,\sigma)\big]^M\Big), \end{aligned} \tag{7.1.215}$$

from which (7.1.214) follows, bearing in mind (7.1.202) (cf. also (7.1.103)).

At this stage, if we combine (7.1.195), (7.1.202), the very last property in the statement of [130, Proposition 5.1.8], (5.4.84), [130, Lemma 1.2.7], (7.1.198), and (7.1.214), we see that the operator $S : \big[\mathbb{X}\big]^M \to \big[\mathbb{X}_1\big]^M$ is indeed Fredholm and

$$\begin{aligned} &\mathrm{index}\Big(S : \big[\mathbb{X}\big]^M \to \big[\mathbb{X}_1\big]^M\Big) \\ &\quad = \mathrm{index}\Big(S : \big[L^q(\partial\Omega,\sigma)\big]^M \to \big[L^q_1(\partial\Omega,\sigma)\big]^M\Big). \end{aligned} \tag{7.1.216}$$

This establishes that $S : \big[\mathbb{X}\big]^M \to \big[\mathbb{X}_1\big]^M$ is a Fredholm operator. The proof of the fact that $S : \big[\mathring{\mathbb{X}}\big]^M \to \big[\mathring{\mathbb{X}}_1\big]^M$ is also Fredholm follows the same pattern, mutatis mutandis (now using the embeddings in [130, (5.2.101)]).

Next, the Fredholmness of S in the last settings in (7.1.188) and (7.1.189) is a consequence of what we have proved, written for $\mathbb{X}'$ in place of $\mathbb{X}$ (which enjoys the same properties as the latter; cf. [130, Proposition 5.1.14]). The claims in (7.1.190) and (7.1.191) are consequences of [130, Proposition 5.3.14], [130, Proposition 5.3.15], and [130, (5.3.62)]. Finally, that all Fredholm operators in the statement have index zero under the stronger assumptions made in (7.1.192) follows from what we have proved so far and Corollary 7.1.6. This completes the proof of Theorem 7.1.7. □

Theorem 7.1.7 may be tweaked as to accommodate the boundary-to-boundary single layer for the Stokes system (which does not fall directly under its scope).

Theorem 7.1.8 *Pick $n \in \mathbb{N}$ with $n \geq 2$ and let $\Omega \subseteq \mathbb{R}^n$ be an Ahlfors regular domain with compact boundary. Abbreviate $\sigma := \mathcal{H}^{n-1}\lfloor\partial\Omega$ and denote by ν the geometric measure theoretic outward unit normal to Ω. Adopt the convention that*

$$S_{\text{Stokes}} \text{ stands for the boundary-to-boundary single layer potential operator for the Stokes system associated with the set } \Omega \text{ as in (4.2.109).} \tag{7.1.217}$$

Next, let $\mathbb{X}$ be a Generalized Banach Function Space on $(\partial\Omega, \sigma)$ such that

$$\mathcal{M}_{\partial\Omega} : \mathbb{X} \to \mathbb{X} \text{ and } \mathcal{M}_{\partial\Omega} : \mathbb{X}' \to \mathbb{X}' \text{ are well-defined and bounded,} \tag{7.1.218}$$

where $\mathcal{M}_{\partial\Omega}$ is the Hardy-Littlewood maximal operator on $\partial\Omega$, and $\mathbb{X}'$ is the associated space of $\mathbb{X}$ (cf. [130, Definitions 5.1.4, 5.1.11]). *Define $\mathring{\mathbb{X}}$, $(\mathbb{X}')^\circ$ as in* (7.1.186) *and bring back the Generalized Banach Function Space styled Sobolev spaces of order one $\mathbb{X}_1$, $(\mathbb{X}')_1$, $\mathring{\mathbb{X}}_1$, $(\mathbb{X}')_1^\circ$ from* (7.1.39) (*which are known to be Banach spaces when equipped with natural norms; cf.* (5.4.84)).

Then there exists $\delta_{\mathbb{X}} \in (0, 1)$, depending only on n, the operator norms of $\mathcal{M}_{\partial\Omega}$ on $\mathbb{X}$ and $\mathbb{X}'$, the ADR *constants of $\partial\Omega$, and the relative flatness ratio $\wp(\Omega)/\text{diam}\,\partial\Omega$, so that if*

$$\text{dist}\Big(\nu, [\text{VMO}(\partial\Omega, \sigma)]^n\Big) < \delta_{\mathbb{X}} \tag{7.1.219}$$

where the distance is considered in the space $\big[\text{BMO}(\partial\Omega, \sigma)\big]^n$ (hence if Ω is a $\delta_{\mathbb{X}}$-oscillating AR *domain; cf. Definition 3.4.1), it follows that Ω is a* UR *domain and the following operators are Fredholm with index zero:*

$$S_{\text{Stokes}} : \big[\mathbb{X}\big]^n \to \big[\mathbb{X}_1\big]^n, \quad S_{\text{Stokes}} : \big[\mathbb{X}'\big]^n \to \big[(\mathbb{X}')_1\big]^n, \tag{7.1.220}$$

$$S_{\text{Stokes}} : \big[\mathring{\mathbb{X}}\big]^n \to \big[\mathring{\mathbb{X}}_1\big]^n, \quad S_{\text{Stokes}} : \big[(\mathbb{X}')^\circ\big]^n \to \big[(\mathbb{X}')_1^\circ\big]^n. \tag{7.1.221}$$

Thus, all aforementioned operators are Fredholm with index zero whenever Ω is an infinitesimally flat AR *domain (cf. Definition 3.4.1). As a corollary,*

> *granted* (7.1.219), *the boundary-to-boundary single layer potential for the Stokes system induces operators which are Fredholm with index zero in each of the contexts specified in* (7.1.220)-(7.1.221) *whenever $\mathbb{X}$ is a rearrangement invariant Banach function space whose Boyd indices satisfy $1 < p_{\mathbb{X}} \leq q_{\mathbb{X}} < \infty$.* (7.1.222)

In particular, under the assumption made in (7.1.219),

the boundary-to-boundary single layer potential for the Stokes system induces operators which are Fredholm with index zero in each of the contexts specified in (7.1.220)-(7.1.221) *whenever* $\mathbb{X} := L^{\Phi}(\partial\Omega, \sigma)$, *the Orlicz space associated as in* (A.0.133) *with the measure space* $(\partial\Omega, \sigma)$ *and any Young function* Φ *satisfying* $1 < i(\Phi) \leq I(\Phi) < \infty$. (7.1.223)

Proof In a first stage, running the same argument as in the proof of Theorem 7.1.7 gives that the operators (7.1.220)-(7.1.221) are Fredholm. Let us show that the index of S_{Stokes} in the first context specified in (7.1.220) is zero. To this end, bring in the boundary-to-boundary single layer potential operator for the Lamé system with moduli $\mu := 1$ and $\lambda \in (0, \infty)$ arbitrary, which we agree to denote by S_λ. Hence, S_λ is the formal convolution integral operator on $\partial\Omega$ with the matrix-valued kernel $E_\lambda = (E^\lambda_{jk})_{1\leq j,k\leq n}$, whose (j, k) entry is defined at each point $x = (x_1, \dots, x_n)$ in $\mathbb{R}^n \setminus \{0\}$ according to (compare with [131, (1.4.72)] specialized to $\mu = 1$)

$$E^\lambda_{jk}(x) := \begin{cases} \dfrac{-1}{2\omega_{n-1}}\left[\left(\dfrac{3+\lambda}{2+\lambda}\right)\dfrac{\delta_{jk}}{(n-2)|x|^{n-2}} + \left(\dfrac{1+\lambda}{2+\lambda}\right)\dfrac{x_j x_k}{|x|^n}\right] & \text{if } n \geq 3, \\ \dfrac{1}{4\pi}\left[\left(\dfrac{3+\lambda}{2+\lambda}\right)\delta_{jk}\ln|x| - \left(\dfrac{1+\lambda}{2+\lambda}\right)\dfrac{x_j x_k}{|x|^2}\right] & \text{if } n = 2, \end{cases} \tag{7.1.224}$$

for every $j, k \in \{1, \dots, n\}$. By way of contrast, (7.1.217) and [132, (6.2.1)] imply that S_{Stokes} is the formal convolution integral operator on $\partial\Omega$ with the matrix-valued kernel $E = (E_{jk})_{1\leq j,k\leq n}$, whose (j, k) entry is defined by

$$E_{jk}(x) := \begin{cases} \dfrac{-1}{2\omega_{n-1}}\left(\dfrac{\delta_{jk}}{(n-2)|x|^{n-2}} + \dfrac{x_j x_k}{|x|^n}\right) & \text{if } n \geq 3, \\ \dfrac{1}{4\pi}\left(\delta_{jk}\ln|x| - \dfrac{x_j x_k}{|x|^2}\right) & \text{if } n = 2, \end{cases} \tag{7.1.225}$$

at each point $x = (x_1, \dots, x_n) \in \mathbb{R}^n \setminus \{0\}$, for every $j, k \in \{1, \dots, n\}$. It is apparent from (7.1.224) and (7.1.225) that for each fixed $N \in \mathbb{N}$ we have

$$\lim_{\lambda\to\infty}\left\{\sum_{|\alpha|\leq N}\left(\sup_{S^{n-1}}\left|(\partial^\alpha E) - (\partial^\alpha E_\lambda)\right|\right)\right\} = 0. \tag{7.1.226}$$

Recall the close relationship between fractional integral operators and the Hardy-Littlewood maximal operator, on a set of finite measure (cf., e.g., [129, (7.8.20)] in the proof of [129, Proposition 7.8.1]). This implies that there exists $C \in (0, \infty)$, which depends only on n and $\partial\Omega$, such that for each $\lambda \in (0, \infty)$ we have

$$\left|(S_{\text{Stokes}} - S_\lambda)\vec{f}\right| \leq C\Big(\sup_{S^{n-1}}|E - E_\lambda|\Big)\mathcal{M}_{\partial\Omega}|\vec{f}| \text{ on } \partial\Omega, \tag{7.1.227}$$

for any function $\vec{f} \in [\mathbb{X}]^n$. On account of (7.1.218) and [130, (5.1.12)], this implies

$$\left\|S_{\text{Stokes}} - S_\lambda\right\|_{[\mathbb{X}]^n \to [\mathbb{X}]^n} \leq C\Big(\sup_{S^{n-1}} |E - E_\lambda|\Big), \tag{7.1.228}$$

for each $\lambda \in (0, \infty)$. From (5.4.84), (7.1.228), [132, (1.2.30)], [131, (2.3.65)], plus the definitions of S_{Stokes} and S_λ we see that there exist a large integer $N = N(n) \in \mathbb{N}$ together with a constant $C \in (0, \infty)$ such that for each $\lambda \in (0, \infty)$ we have

$$\left\|S_{\text{Stokes}} - S_\lambda\right\|_{[\mathbb{X}]^n \to [\mathbb{X}_1]^n} \leq C \sum_{|\alpha| \leq N} \Big(\sup_{S^{n-1}} \big|(\partial^\alpha E) - (\partial^\alpha E_\lambda)\big|\Big). \tag{7.1.229}$$

Together, (7.1.229) and (7.1.226) lead to the conclusion that

$$\lim_{\lambda \to \infty} S_\lambda = S_{\text{Stokes}} \text{ in } \mathcal{B}\Big(\big[\mathbb{X}\big]^n \to \big[\mathbb{X}_1\big]^n\Big). \tag{7.1.230}$$

On the other hand, [131, (1.3.11)] and Remark 1.6.10 tell us that the Lamé system $L_{1,\lambda}$, defined as in (1.6.85) with $\mu := 1$ and $\lambda \in (0, \infty)$ arbitrary, is Legendre-Hadamard strongly elliptic and has $\mathfrak{A}^{\text{dis}}_{L_{1,\lambda}} \neq \varnothing$. Granted these properties, we conclude from the last part of Theorem 7.1.7 that the operator $S_\lambda : \big[\mathbb{X}\big]^n \to \big[\mathbb{X}_1\big]^n$ is Fredholm with index zero for each $\lambda \in (0, \infty)$. Based on this, (7.1.230), the fact that we already know that $S_{\text{Stokes}} : \big[\mathbb{X}\big]^n \to \big[\mathbb{X}_1\big]^n$ is a Fredholm operator (as pointed out at the beginning of the current proof), and the homotopic invariance of the index we finally conclude that S_{Stokes} is a Fredholm operator with index zero in the first context specified in (7.1.220). The remaining operators in (7.1.220)-(7.1.221) are dealt with using similar ideas. All other claims follow from what we have proved so far. □

We shall now systematically prove invertibility results for the boundary-to-boundary single layer potential operator on a variety of function spaces. This is done in several installments, starting with the case of Lebesgue and Sobolev spaces.

Theorem 7.1.9 *Fix $n \in \mathbb{N}$ with $n \geq 3$ and suppose $\Omega \subseteq \mathbb{R}^n$ is an Ahlfors regular domain with compact boundary. Abbreviate $\sigma := \mathcal{H}^{n-1} \lfloor \partial\Omega$ and denote by ν the geometric measure theoretic outward unit normal to Ω. Also, let L be a homogeneous, second-order, constant complex coefficient, $M \times M$ system in $\mathbb{R}^n$ such that*

L satisfies the Legendre-Hadamard strong ellipticity condition (cf. [131, Definition 1.3.2]) *and one has $\mathfrak{A}^{\text{dis}}_L \neq \varnothing$.* (7.1.231)

Consider the boundary-to-boundary single layer potential operator S associated with Ω and the system L as in (A.0.231). *Finally, fix a compact interval $I \subset (1, \infty)$.*

Then there exists $\delta \in (0, 1)$ depending only on n, I, L, the ADR *constants of $\partial\Omega$, and the relative flatness ratio $\wp(\Omega)/\text{diam}\,\partial\Omega$ such that if*

$$\text{dist}\Big(\nu, [\text{VMO}(\partial\Omega, \sigma)]^n\Big) < \delta \tag{7.1.232}$$

with the distance considered in $\big[\text{BMO}(\partial\Omega, \sigma)\big]^n$, (hence if Ω is a δ-oscillating AR *domain; cf. Definition 3.4.1), it follows that Ω is a* UR *domain and for each exponent $p \in I$ the operators*

$$S : \big[L^p(\partial\Omega,\sigma)\big]^M \longrightarrow \big[L^p_1(\partial\Omega,\sigma)\big]^M, \tag{7.1.233}$$
$$S : \big[L^p_{-1}(\partial\Omega,\sigma)\big]^M \longrightarrow \big[L^p(\partial\Omega,\sigma)\big]^M, \tag{7.1.234}$$

are well defined, linear, bounded, invertible, with bounded inverses.

As a corollary, (7.1.233)-(7.1.234) *are invertible for any integrability exponent* $p \in (1,\infty)$ *provided* Ω *is an infinitesimally flat* AR *domain (cf. Definition 3.4.1).*

Before presenting the proof of the above theorem, a few remarks are in order.

Remark 1. There is a natural version of the aforementioned invertibility results involving Muckenhoupt weighted Sobolev spaces. More specifically, given any constant $C_W \in (0,\infty)$, choosing the threshold $\delta \in (0,1)$ to also be sufficiently small relative to C_W, matters may be arranged so that the operators

$$S : \big[L^p(\partial\Omega,w)\big]^M \longrightarrow \big[L^p_1(\partial\Omega,w)\big]^M, \tag{7.1.235}$$
$$S : \big[L^p_{-1}(\partial\Omega,w)\big]^M \longrightarrow \big[L^p(\partial\Omega,w)\big]^M, \tag{7.1.236}$$

are invertible for each $p \in I$ and each weight $w \in A_p(\partial\Omega,\sigma)$ with $[w]_{A_p} \leq C_W$. This may be proved by showing that said operators are Fredholm (based on item *(1)* in Theorem 7.1.1 and [132, Theorem 1.5.1, item *(xiii)*]; here [129, Lemma 7.7.16] is also relevant), injective (thanks to the injectivity of (7.1.233), (7.1.234) and [129, (7.7.105)]), and duality (cf. (A.0.156) and [132, Theorem 1.5.1, item *(ix)*]).

Remark 2. From (7.1.183) we know that the operators (7.1.233)-(7.1.234) continue to be Fredholm with index zero in the context of Theorem 7.1.9 when $n = 2$. This being said, the two-dimensional case in Theorem 7.1.9 has been deliberately omitted. Indeed, we claim that there are smooth bounded domains Ω in the plane such that S_Δ, the classical logarithmic single layer (associated with Ω and the Laplacian) fails to yield invertible operators in the context of (7.1.233)-(7.1.234).

Indeed, if (7.1.233) for $L = \Delta$ fails to be invertible when $\Omega := B(0,1)$, the unit ball in $\mathbb{R}^2$, then there is nothing to prove. Hence, suppose (7.1.233) (for $L = \Delta$) is actually invertible when $\Omega := B(0,1) \subseteq \mathbb{R}^2$. Since $1 \in L^p_1(S^1,\mathcal{H}^1)$, this implies that there exists a function $f \in L^p(S^1,\mathcal{H}^1)$ such that $S_\Delta f = 1$. Then the function $u := \mathcal{S}_\Delta f \in \mathscr{C}^\infty\big(B(0,1)\big)$ satisfies, for any fixed aperture parameter $\kappa > 0$,

$$\begin{aligned} &\Delta u = 0 \text{ in } B(0,1), \quad \mathcal{N}_\kappa(\nabla u) \in L^p(S^1,\mathcal{H}^1), \\ &\text{and } u\big|_{S^1}^{\kappa-\text{n.t.}} = 1 \text{ at } \mathcal{H}^1\text{-a.e. point on } S^1. \end{aligned} \tag{7.1.237}$$

By the uniqueness for the Regularity Problem for the Laplacian in the disk, this forces $u \equiv 1$ in $B(0,1)$, hence

$$0 = \partial_\nu u = \partial_\nu \mathcal{S}_\Delta f = \big(-\tfrac{1}{2}I + K^\#_\Delta\big) f = -\frac{1}{2} f + \frac{1}{2} \fint_{S^1} f \, d\mathcal{H}^1, \tag{7.1.238}$$

by [132, (1.5.66)] and (7.0.2). In turn, from (7.1.238) and the fact that $S_\Delta f = 1$ we conclude that f is a non-zero constant on S^1, say $f \equiv c$ on S^1 for some $c \in \mathbb{R} \setminus \{0\}$. In particular, $S_\Delta c = 1$.

To proceed, for each $r > 0$ denote by S_r the boundary-to-boundary single layer potential operator associated with the Laplacian and the set $B(0,r) \subseteq \mathbb{R}^2$. Then for each $x \in \partial B(0,r)$ we may write

$$\begin{aligned}(S_r 1)(x) &= \frac{1}{2\pi}\int_{\partial B(0,r)} \ln|x-y|\,\mathcal{H}^1(y) = \frac{r}{2\pi}\int_{\partial B(0,1)} \ln|x-r\omega|\,\mathcal{H}^1(\omega)\\ &= \frac{r\ln r}{2\pi}\int_{\partial B(0,1)} 1\,\mathcal{H}^1(\omega) + \frac{r}{2\pi}\int_{\partial B(0,1)} \ln|(x/r)-\omega|\,\mathcal{H}^1(\omega)\\ &= r\ln r + r(S_\Delta 1)(x/r) = r\ln r + \frac{r}{c}(S_\Delta c)(x/r) = r\ln r + \frac{r}{c},\end{aligned} \tag{7.1.239}$$

where we have made the change of variables $\partial B(0,r) \ni y = r\omega$ with $\omega \in S^1$ in the second equality, and have used the fact that $S_\Delta c = 1$ in the last equality. Choosing $r := e^{-(1/c)} \in (0,\infty)$ then forces $(S_r 1)(x) = 0$ for each $x \in \partial B(0,r)$, hence S_r fails to be injective (hence invertible) on the space $L^p\big(\partial B(0,r), \mathcal{H}^1\big)$.

Remark 3. In the absence of any flatness demands imposed on the boundary, the invertibility results in Theorem 7.1.9 may fail. This is seen from Proposition 7.1.16, the real interpolation formulas [130, (11.12.11) in Theorem 11.12.2] [130, (11.12.60) in Corollary 11.12.3], and [56, Lemma 8.4].

Remark 4. It is worth emphasizing that in Theorem 7.1.9 no topological assumptions are made on the set Ω, no symmetry is required for the system L, and no positive definiteness (or positive semi-definiteness) is assumed for any of the coefficient tensors in $\mathfrak{A}_L$.

We now turn to the task of providing the proof Theorem 7.1.9.

Proof of Theorem 7.1.9 Recall from Theorem 3.4.2 that taking $\delta \in (0,1)$ small enough makes Ω a UR domain. Also, by eventually enlarging I, there is no loss of generality in assuming that I is a compact sub-interval of $(1,\infty)$ which is stable under passing to Hölder conjugates, and has the property that $2(n-1)/n \in I$. We shall tacitly use these properties in the remaining portion of the proof.

To proceed, observe that in light of the fact that $n \geq 3$, the system L satisfies the Legendre-Hadamard strong ellipticity condition, and we have $\mathfrak{A}_L^{\text{dis}} \neq \varnothing$, it follows that (7.1.101) holds; see Proposition 1.4.1. As a result, Theorem 7.1.5 applies and guarantees the existence of some threshold $\delta \in (0,1)$ which depends only on n, I, L, the ADR constants of $\partial\Omega$, and the relative flatness ratio $\wp(\Omega)/\text{diam}\,\partial\Omega$ with the property that if (7.1.232) holds then

$$\text{for each } p \in I \text{ the operators (7.1.233), (7.1.234) are both Fredholm and have the same index.} \tag{7.1.240}$$

Also, from [132, Theorem 1.7.2, item *(1)*] and [132, Theorem 1.5.1, item *(ix)*] we see that the operator

$$S : \left[L^{2(n-1)/n}(\partial\Omega,\sigma)\right]^M \longrightarrow \left[L_1^{2(n-1)/n}(\partial\Omega,\sigma)\right]^M \text{ is injective.} \tag{7.1.241}$$

From this and (7.1.129) we then conclude that the operator

$$S : \left[L^p(\partial\Omega,\sigma)\right]^M \longrightarrow \left[L_1^p(\partial\Omega,\sigma)\right]^M \text{ is injective for each } p \in I. \tag{7.1.242}$$

Next, from [132, Corollary 1.7.3, item *(1)*], the fact that we know that (7.1.234) is a Fredholm operator for $p := 2(n-1)/(n-2) \in I$, and (7.1.149) we deduce that

$$\begin{gathered} S : \left[L^{p_n}(\partial\Omega,\sigma)\right]^M \longrightarrow \left[L_1^{p_n}(\partial\Omega,\sigma)\right]^M \\ \text{is a surjective operator for } p_n := \tfrac{2(n-1)}{n-2} \in I. \end{gathered} \tag{7.1.243}$$

In concert with (7.1.242) this proves that $S : \left[L^{p_n}(\partial\Omega,\sigma)\right]^M \to \left[L_1^{p_n}(\partial\Omega,\sigma)\right]^M$ is an invertible operator. Together with (7.1.240) this shows that for each $p \in I$ the operators (7.1.233), (7.1.234) are Fredholm with index zero. A final appeal to (7.1.242) then shows that, in fact, for each $p \in I$ the operators (7.1.233), (7.1.234) are invertible. (Parenthetically, we wish to note that invoking (7.1.183) in conjunction with (7.1.242) yields an alternative proof of the fact that the operators (7.1.233)-(7.1.234) are invertible.) □

In Remark 2 following the statement of Theorem 7.1.9 we have seen that, in the two dimensional setting, it is not always possible to invert the single layer potential operator even for smooth bounded domains in the plane. To overcome this difficulty we shall introduce an "augmented" version of the single layer, in the manner described below.

Theorem 7.1.10 *Suppose $\Omega \subseteq \mathbb{R}^2$ is an Ahlfors regular domain with compact boundary. Abbreviate $\sigma := \mathcal{H}^1\lfloor\partial\Omega$ and denote by ν the geometric measure theoretic outward unit normal to Ω. Also, let L be a homogeneous, second-order, constant complex coefficient, $M \times M$ system in $\mathbb{R}^2$ such that*

$$\begin{gathered} L \textit{ satisfies the Legendre-Hadamard strong ellipticity con-} \\ \textit{dition (cf. [131, Definition 1.3.2]) and one has } \mathfrak{A}_L^{\rm dis} \neq \varnothing. \end{gathered} \tag{7.1.244}$$

Bring in the boundary-to-boundary single layer potential operator S associated with Ω and the system L as in (A.0.231). *Fix $p \in (1,\infty)$ and define the augmented versions of S to be*

$$\begin{gathered} S_{\rm aug} : \left[L^p(\partial\Omega,\sigma)\right]^M \oplus \mathbb{C}^M \longrightarrow \left[L_1^p(\partial\Omega,\sigma)\right]^M \oplus \mathbb{C}^M \\ \textit{given by } S_{\rm aug}(f,c) := \Big(Sf + c, \textstyle\int_{\partial\Omega} f\, d\sigma\Big) \textit{ for each} \\ \textit{function } f \in \left[L^p(\partial\Omega,\sigma)\right]^M \textit{ and constant } c \in \mathbb{C}^M, \end{gathered} \tag{7.1.245}$$

and

$$\begin{gathered}\widetilde{S}_{\text{aug}} : \big[L^p_{-1}(\partial\Omega,\sigma)\big]^M \oplus \mathbb{C}^M \longrightarrow \big[L^p(\partial\Omega,\sigma)\big]^M \oplus \mathbb{C}^M \\ \textit{given by } \widetilde{S}_{\text{aug}}(f,c) := \Big(Sf + c, (\langle f, \mathbf{e}_j\rangle)_{1\leq j\leq M}\Big) \textit{ for each} \\ \textit{functional } f \in \big[L^p_{-1}(\partial\Omega,\sigma)\big]^M \textit{ and each constant } c \in \mathbb{C}^M.\end{gathered} \tag{7.1.246}$$

Then there exists some threshold $\delta \in (0,1)$ *which depends only on* p, L, *the* ADR *constants of* $\partial\Omega$, *and the relative flatness ratio* $\wp(\Omega)/\text{diam}\,\partial\Omega$ *such that if*

$$\text{dist}\Big(\nu, [\text{VMO}(\partial\Omega,\sigma)]^2\Big) < \delta \tag{7.1.247}$$

where the distance is considered in the space $\big[\text{BMO}(\partial\Omega,\sigma)\big]^2$, *hence if* Ω *is a* δ*-oscillating* AR *domain (cf. Definition 3.4.1), then* Ω *is a* UR *domain and the operators* (7.1.245)-(7.1.246) *are well defined, linear, bounded, and invertible.*

As a corollary, the operators (7.1.245)-(7.1.246) *are invertible for any exponent* $p \in (1,\infty)$ *provided* Ω *is an infinitesimally flat* AR *domain (cf. Definition 3.4.1).*

From Example 3.4.8 we know that if Ω is a chord-arc domain with vanishing constant and compact boundary then Ω is an infinitesimally flat AR domain. As such, the operators (7.1.245)-(7.1.246) are invertible for any $p \in (1,\infty)$ in this case.

There are natural versions of the invertibility results pertaining to (7.1.245)-(7.1.246) in the context of Muckenhoupt weighted Sobolev spaces. Concretely, given any $C_W \in (0,\infty)$, choosing $\delta \in (0,1)$ to also be sufficiently small relative to C_W, matters may be arranged so that the augmented single layer potential operators

$$\begin{gathered}S_{\text{aug}} : \big[L^p(\partial\Omega,w)\big]^M \oplus \mathbb{C}^M \longrightarrow \big[L^p_1(\partial\Omega,w)\big]^M \oplus \mathbb{C}^M \\ \text{given by } S_{\text{aug}}(f,c) := \Big(Sf + c, \textstyle\int_{\partial\Omega} f\,d\sigma\Big) \text{ for each} \\ \text{function } f \in \big[L^p(\partial\Omega,w)\big]^M \text{ and constant } c \in \mathbb{C}^M,\end{gathered} \tag{7.1.248}$$

and

$$\begin{gathered}\widetilde{S}_{\text{aug}} : \big[L^p_{-1}(\partial\Omega,w)\big]^M \oplus \mathbb{C}^M \longrightarrow \big[L^p(\partial\Omega,w)\big]^M \oplus \mathbb{C}^M \\ \text{given by } \widetilde{S}_{\text{aug}}(f,c) := \Big(Sf + c, (\langle f, \mathbf{e}_j\rangle)_{1\leq j\leq M}\Big) \text{ for each} \\ \text{functional } f \in \big[L^p_{-1}(\partial\Omega,w)\big]^M \text{ and each constant } c \in \mathbb{C}^M,\end{gathered} \tag{7.1.249}$$

are invertible for each weight $w \in A_p(\partial\Omega,\sigma)$ with $[w]_{A_p} \leq C_W$. This may be proved in an analogous fashion to the argument for (7.1.245)-(7.1.246) presented below (see also Remark 1 following the statement of Theorem 7.1.9).

We now present the proof of Theorem 7.1.10.

Proof of Theorem 7.1.10 Pick $\delta \in (0,1)$ small enough so that having (7.1.247) valid ensures that Ω is a UR domain and the operators

$$S : \big[L^p(\partial\Omega,\sigma)\big]^M \longrightarrow \big[L^p_1(\partial\Omega,\sigma)\big]^M, \tag{7.1.250}$$

$$S : \big[L^p_{-1}(\partial\Omega,\sigma)\big]^M \longrightarrow \big[L^p(\partial\Omega,\sigma)\big]^M, \tag{7.1.251}$$

are Fredholm with index zero. That this is possible is ensured by Corollary 7.1.6. Observe that the augmented single layer operator (7.1.245) may be decomposed as

$$S_{\rm aug} = T + Q \tag{7.1.252}$$

where

$$\begin{gathered} T : \big[L^p(\partial\Omega,\sigma)\big]^M \oplus \mathbb{C}^M \longrightarrow \big[L^p_1(\partial\Omega,\sigma)\big]^M \oplus \mathbb{C}^M \text{ is given by} \\ T(f,c) := \big(Sf, 0\big) \text{ for each } f \in \big[L^p(\partial\Omega,\sigma)\big]^M \text{ and } c \in \mathbb{C}^M, \end{gathered} \tag{7.1.253}$$

and where

$$\begin{gathered} Q : \big[L^p(\partial\Omega,\sigma)\big]^M \oplus \mathbb{C}^M \longrightarrow \big[L^p_1(\partial\Omega,\sigma)\big]^M \oplus \mathbb{C}^M \\ \text{is defined as } Q(f,c) := \Big(c, \int_{\partial\Omega} f \, d\sigma\Big) \text{ for each} \\ \text{function } f \in \big[L^p(\partial\Omega,\sigma)\big]^M \text{ and constant } c \in \mathbb{C}^M. \end{gathered} \tag{7.1.254}$$

Both T, Q are well defined, linear, and bounded. Moreover, Q is also a finite-rank operator, hence actually compact, whereas

$$\begin{aligned} &\mathrm{Im}\Big(T : \big[L^p(\partial\Omega,\sigma)\big]^M \oplus \mathbb{C}^M \to \big[L^p_1(\partial\Omega,\sigma)\big]^M \oplus \mathbb{C}^M\Big) \\ &\qquad = \mathrm{Im}\Big(S : \big[L^p(\partial\Omega,\sigma) \to \big[L^p_1(\partial\Omega,\sigma)\big]^M\Big) \oplus \{0\}, \end{aligned} \tag{7.1.255}$$

and

$$\begin{aligned} &\mathrm{Ker}\Big(T : \big[L^p(\partial\Omega,\sigma)\big]^M \oplus \mathbb{C}^M \to \big[L^p_1(\partial\Omega,\sigma)\big]^M \oplus \mathbb{C}^M\Big) \\ &\qquad = \mathrm{Ker}\Big(S : \big[L^p(\partial\Omega,\sigma)\big]^M \to \big[L^p_1(\partial\Omega,\sigma)\big]^M\Big) \oplus \mathbb{C}^M. \end{aligned} \tag{7.1.256}$$

From (7.1.255)-(7.1.256) and the fact that S in (7.1.250) is Fredholm with index zero we conclude that T in (7.1.253) is also Fredholm with index zero. Since Q is compact, (7.1.252) and item *(6)* in [130, Theorem 2.1.2] imply that $S_{\rm aug}$ is a Fredholm operator with index zero. Moreover, from [132, (1.7.11)] we see that $S_{\rm aug}$ is also injective. Hence, ultimately, $S_{\rm aug}$ is an invertible operator.

The proof of the fact that the augmented single layer operator $\widetilde{S}_{\rm aug}$ defined in (7.1.246) is invertible proceeds, up to a point, along similar lines. Specifically, we now decompose

$$\widetilde{S}_{\rm aug} = \widetilde{T} + \widetilde{Q} \tag{7.1.257}$$

with

$$\begin{gathered} \widetilde{T} : \big[L^p_{-1}(\partial\Omega,\sigma)\big]^M \oplus \mathbb{C}^M \longrightarrow \big[L^p(\partial\Omega,\sigma)\big]^M \oplus \mathbb{C}^M \text{ given by} \\ \widetilde{T}(f,c) := \big(Sf, 0\big) \text{ for each } f \in \big[L^p_{-1}(\partial\Omega,\sigma)\big]^M \text{ and } c \in \mathbb{C}^M, \end{gathered} \tag{7.1.258}$$

and

$$\begin{aligned}&\widetilde{Q} : \big[L^p_{-1}(\partial\Omega,\sigma)\big]^M \oplus \mathbb{C}^M \longrightarrow \big[L^p(\partial\Omega,\sigma)\big]^M \oplus \mathbb{C}^M \\ &\quad\text{defined as } \widetilde{Q}(f,c) := \Big(c, (\langle f, \mathbf{e}_j\rangle)_{1\le j\le M}\Big) \text{ for each} \\ &\quad\text{functional } f \in \big[L^p_{-1}(\partial\Omega,\sigma)\big]^M \text{ and constant } c \in \mathbb{C}^M.\end{aligned} \tag{7.1.259}$$

Once again, $\widetilde{T}, \widetilde{Q}$ are well-defined, linear, and bounded, with $\widetilde{Q}$ a finite-rank operator, hence actually compact. Furthermore,

$$\begin{aligned}\mathrm{Im}\Big(\widetilde{T} : \big[L^p_{-1}(\partial\Omega,\sigma)\big]^M \oplus \mathbb{C}^M \to \big[L^p(\partial\Omega,\sigma)\big]^M \oplus \mathbb{C}^M\Big)& \\ = \mathrm{Im}\Big(S : \big[L^p_{-1}(\partial\Omega,\sigma) \to \big[L^p(\partial\Omega,\sigma)\big]^M\Big) \oplus \{0\},&\end{aligned} \tag{7.1.260}$$

and

$$\begin{aligned}\mathrm{Ker}\Big(\widetilde{T} : \big[L^p_{-1}(\partial\Omega,\sigma)\big]^M \oplus \mathbb{C}^M \to \big[L^p(\partial\Omega,\sigma)\big]^M \oplus \mathbb{C}^M\Big)& \\ = \mathrm{Ker}\Big(S : \big[L^p_{-1}(\partial\Omega,\sigma)\big]^M \to \big[L^p(\partial\Omega,\sigma)\big]^M\Big) \oplus \mathbb{C}^M,&\end{aligned} \tag{7.1.261}$$

which make $\widetilde{T}$ Fredholm with index zero, given that S in (7.1.251) is Fredholm with index zero. Thus, $\widetilde{S}_{\rm aug}$ is a Fredholm operator with index zero. Since the Banach space inclusions

$$L^p(\partial\Omega,\sigma) \hookrightarrow L^p_{-1}(\partial\Omega,\sigma) \text{ and } L^p_1(\partial\Omega,\sigma) \hookrightarrow L^p(\partial\Omega,\sigma) \tag{7.1.262}$$

have dense ranges (cf. [130, (11.1.66), (11.8.4)]), and since

$$\widetilde{S}_{\rm aug}(f,c) = S_{\rm aug}(f,c) \text{ whenever } f \in \big[L^p(\partial\Omega,\sigma)\big]^M \text{ and } c \in \mathbb{C}^M, \tag{7.1.263}$$

we then conclude from [130, Lemma 2.1.4] that $\widetilde{S}_{\rm aug}$ is also an invertible operator. □

Given an arbitrary open set $\Omega \subseteq \mathbb{R}^n$, define

$$\Omega_+ := \Omega \text{ and } \Omega_- := \mathbb{R}^n \setminus \overline{\Omega}. \tag{7.1.264}$$

In relation to these sets, the following function spaces arise naturally:

$$\begin{aligned}\mathbb{C}_{\Omega_\pm} &:= \mathrm{span}_{\mathbb{C}}\big\{\mathbf{1}_O : O \text{ bounded connected component of } \Omega_\pm\big\}, \\ \mathbb{C}_{\partial\Omega_\pm} &:= \mathrm{span}_{\mathbb{C}}\big\{\mathbf{1}_{\partial O} : O \text{ bounded connected component of } \Omega_\pm\big\}, \\ \mathbb{C}_{\partial\Omega} &:= \mathrm{span}_{\mathbb{C}}\big\{\mathbf{1}_\Sigma : \Sigma \text{ connected component of } \partial\Omega\big\}.\end{aligned} \tag{7.1.265}$$

Also, for any space X of distributions on $\partial\Omega$ we agree to abbreviate

$$X \cap \big(\mathbb{C}_{\partial\Omega_\pm}\big)^\perp := \Big\{f \in X : \langle f, \chi\rangle = 0 \text{ for each } \chi \in \mathbb{C}_{\partial\Omega_\pm}\Big\}, \tag{7.1.266}$$

$$X \cap \big(\mathbb{C}_{\partial\Omega}\big)^\perp := \Big\{f \in X : \langle f, \chi\rangle = 0 \text{ for each } \chi \in \mathbb{C}_{\partial\Omega}\Big\}. \tag{7.1.267}$$

In addition, we shall consider

$$\begin{aligned} \mathbb{C}^{\infty}_{\Omega_\pm} &:= \operatorname{span}_{\mathbb{C}} \{\mathbf{1}_O : O \text{ connected component of } \Omega_\pm\}, \\ \mathbb{C}^{\infty}_{\partial\Omega_\pm} &:= \operatorname{span}_{\mathbb{C}} \{\mathbf{1}_{\partial O} : O \text{ connected component of } \Omega_\pm\}, \end{aligned} \tag{7.1.268}$$

and, for any given space X of distributions on $\partial\Omega$, set

$$X \cap (\mathbb{C}^{\infty}_{\partial\Omega_\pm})^{\perp} := \Big\{ f \in X : \langle f, \chi \rangle = 0 \text{ for each } \chi \in \mathbb{C}^{\infty}_{\partial\Omega_\pm} \Big\}, \tag{7.1.269}$$

as well as

$$X \cap \mathbb{C}^{\perp} := \Big\{ f \in X : \langle f, c \rangle = 0 \text{ for each } c \in \mathbb{C} \Big\}. \tag{7.1.270}$$

For future use we want to observe that whenever

$\Omega \subseteq \mathbb{R}^n$, with ≥ 2, is a nonempty open set with a compact boundary satisfying $\partial\Omega = \partial(\overline{\Omega})$ and with the property that both the connected components of Ω as well as those of $\mathbb{R} \setminus \overline{\Omega}$ are separated (in the sense that they have mutually disjoint closures) (7.1.271)

it follows that Alexander's relation holds, i.e., we have

$$\mathbb{C}_{\partial\Omega} = \mathbb{C}_{\partial\Omega_+} \oplus \mathbb{C}_{\partial\Omega_-}, \tag{7.1.272}$$

where the sum is direct (but not necessarily orthogonal). Finally, for any given open set $\Omega \subseteq \mathbb{R}^n$ we agree to denote by $\mathrm{BCC}(\Omega_\pm)$ the collection of bounded connected components of $\Omega_\pm$, and note that if Ω is as in (7.1.271) then

$$\mathbf{1}_{\partial\Omega} = \sum_{O \in \mathrm{BCC}(\Omega_\pm)} \mathbf{1}_{\partial O} \text{ if } \Omega_\pm \text{ is bounded.} \tag{7.1.273}$$

Recall that the property of being positive definite for a coefficient tensor has been introduced in Definition 1.1.1. The following theorem should be compared with Theorem 7.1.4, and also with [126, Theorem 4.1, p. 326] where similar results have been obtained for the Laplacian in arbitrary Lipschitz domains, using Rellich identities.

Theorem 7.1.11 *Fix two integers $n, M \in \mathbb{N}$ with $n \geq 2$, and consider an integrability exponent $p \in (1, \infty)$. Let L be a homogeneous, second-order, constant complex coefficient, $M \times M$ system in $\mathbb{R}^n$, which has a positive definite coefficient tensor $A \in \mathfrak{A}^{\mathrm{dis}}_L$.*

Also, suppose $\Omega \subseteq \mathbb{R}^n$ is an Ahlfors regular domain with compact boundary. Denote by ν the geometric measure theoretic outward unit normal to Ω and abbreviate $\sigma := \mathcal{H}^{n-1} \lfloor \partial\Omega$. Define $\Omega_\pm$ as in (7.1.264) *and recall the spaces $\mathbb{C}_{\partial\Omega_\pm}$, $\mathbb{C}_{\partial\Omega}$ from* (7.1.265). *Bring in the boundary layer potential operators K_A, $K^{\#}_{A}$, S associated with the set Ω and the coefficient tensor A as in* (A.0.116), (A.0.118), *and* (A.0.231), *respectively. Finally, given $k \in \{0, 1\}$, the following convention is used* *throughout:* $L^p_k(\partial\Omega,\sigma)$ *corresponds to the Sobolev space* $L^p_1(\partial\Omega,\sigma)$ *when* $k=1$, *and to the Lebesgue space* $L^p(\partial\Omega,\sigma)$ *when* $k=0$, *while* $L^p_{k-1}(\partial\Omega,\sigma)$ *corresponds to the Lebesgue space* $L^p(\partial\Omega,\sigma)$ *when* $k=1$, *and to the negative Sobolev space* $L^p_{-1}(\partial\Omega,\sigma)$ *when* $k=0$.

Then there exists some threshold $\delta\in(0,1)$ *which depends only on* n, p, ε, L, *the* ADR *constants of* $\partial\Omega$, *and the relative flatness ratio* $\wp(\Omega)/\operatorname{diam}\partial\Omega$ *with the property that if*

$$\operatorname{dist}\Big(\nu,[\mathrm{VMO}(\partial\Omega,\sigma)]^n\Big)<\delta \tag{7.1.274}$$

where the distance is considered in the space $\big[\mathrm{BMO}(\partial\Omega,\sigma)\big]^n$ *(hence if* Ω *is a* δ*-oscillating* AR *domain; cf. Definition 3.4.1), it follows that* Ω *is a two-sided* NTA *domain and the results below are true for each* $k\in\{0,1\}$*:*

(1) One has

$$\operatorname{Ker}\Big(\pm\tfrac12 I+K_A;\ \big[L^p_k(\partial\Omega,\sigma)\big]^M\Big)=\big(\mathbb{C}_{\partial\Omega_\mp}\big)^M, \tag{7.1.275}$$

and

$$\dim\operatorname{Ker}\Big(\pm\tfrac12 I+K^\#_A;\ \big[L^p_{k-1}(\partial\Omega,\sigma)\big]^M\Big)=M\cdot\dim\mathbb{C}_{\partial\Omega_\mp}. \tag{7.1.276}$$

(2) One has

$$\operatorname{Im}\Big(\pm\tfrac12 I+K^\#_A;\ \big[L^p_{k-1}(\partial\Omega,\sigma)\big]^M\Big)=\big[L^p_{k-1}(\partial\Omega,\sigma)\cap\big(\mathbb{C}_{\partial\Omega_\mp}\big)^\perp\big]^M, \tag{7.1.277}$$

and

$$\dim\operatorname{CoKer}\Big(\pm\tfrac12 I+K_A;\ \big[L^p_k(\partial\Omega,\sigma)\big]^M\Big)=M\cdot\dim\mathbb{C}_{\partial\Omega_\mp}. \tag{7.1.278}$$

(3) The operators listed below are isomorphisms on the indicated spaces:

$$\pm\tfrac12 I+K^\#_A:\big[L^p_{k-1}(\partial\Omega,\sigma)\cap\big(\mathbb{C}_{\partial\Omega_\mp}\big)^\perp\big]^M\xrightarrow{\sim}\big[L^p_{k-1}(\partial\Omega,\sigma)\big]^M\cap\big(\mathbb{C}_{\partial\Omega_\mp}\big)^\perp\big]^M, \tag{7.1.279}$$

$$\pm\tfrac12 I+K_A:\ \frac{\big[L^p_k(\partial\Omega,\sigma)\big]^M}{\big(\mathbb{C}_{\partial\Omega_\mp}\big)^M}\xrightarrow{\sim}\frac{\big[L^p_k(\partial\Omega,\sigma)\big]^M}{\big(\mathbb{C}_{\partial\Omega_\mp}\big)^M}, \tag{7.1.280}$$

$$\pm\tfrac12 I+K_A:\ \frac{\big[L^p_k(\partial\Omega,\sigma)\big]^M}{\big(\mathbb{C}_{\partial\Omega}\big)^M}\xrightarrow{\sim}\frac{\big[L^p_k(\partial\Omega,\sigma)\big]^M}{\big(\mathbb{C}_{\partial\Omega}\big)^M}, \tag{7.1.281}$$

$$\pm\tfrac12 I+K^\#_A:\ \big[L^p_{k-1}(\partial\Omega,\sigma)\cap\big(\mathbb{C}_{\partial\Omega}\big)^\perp\big]^M\xrightarrow{\sim}\big[L^p_{k-1}(\partial\Omega,\sigma)\cap\big(\mathbb{C}_{\partial\Omega}\big)^\perp\big]^M. \tag{7.1.282}$$

(4) One has

$$\operatorname{Im}\Big(\partial_\nu^A \mathcal{D}_A : \big[L_k^p(\partial\Omega,\sigma)\big]^M \to \big[L_{k-1}^p(\partial\Omega,\sigma)\big]^M\Big) = \big[L_{k-1}^p(\partial\Omega,\sigma) \cap (\mathbb{C}_{\partial\Omega})^\perp\big]^M, \tag{7.1.283}$$

$$\operatorname{Ker}\Big(\partial_\nu^A \mathcal{D}_A : \big[L_k^p(\partial\Omega,\sigma)\big]^M \to \big[L_{k-1}^p(\partial\Omega,\sigma)\big]^M\Big) = (\mathbb{C}_{\partial\Omega})^M, \tag{7.1.284}$$

and the operator

$$\partial_\nu^A \mathcal{D}_A : \frac{\big[L_k^p(\partial\Omega,\sigma)\big]^M}{(\mathbb{C}_{\partial\Omega})^M} \xrightarrow{\sim} \big[L_{k-1}^p(\partial\Omega,\sigma) \cap (\mathbb{C}_{\partial\Omega})^\perp\big]^M \text{ is invertible.} \tag{7.1.285}$$

(5) The operators listed below are isomorphisms on the indicated spaces:

$$S_A : \operatorname{Ker}\Big(\pm\tfrac{1}{2}I + K_{A^\top}^\#;\ \big[L_{k-1}^p(\partial\Omega,\sigma)\big]^M\Big) \xrightarrow{\sim} (\mathbb{C}_{\partial\Omega_\mp})^M \text{ if } n \geq 3, \tag{7.1.286}$$

$$S_A : \operatorname{Ker}\Big(\tfrac{1}{2}I + K_{A^\top}^\#;\ \big[L_{k-1}^p(\partial\Omega,\sigma)\big]^M\Big) \xrightarrow{\sim} (\mathbb{C}_{\partial\Omega_-})^M \quad \text{if } n = 2 \text{ and } \Omega \text{ is bounded,} \tag{7.1.287}$$

$$S_A : \big[L_{k-1}^p(\partial\Omega,\sigma) \cap (\mathbb{C}_{\partial\Omega_\pm})^\perp\big]^M \xrightarrow{\sim} \frac{\big[L_k^p(\partial\Omega,\sigma)\big]^M}{(\mathbb{C}_{\partial\Omega_\pm})^M} \text{ if } n \geq 3, \tag{7.1.288}$$

the boundary-to-boundary single layer operator S_A acting from

$$\operatorname{Ker}\Big(\tfrac{1}{2}I + K_{A^\top}^\#;\ \big[L_{k-1}^p(\partial\Omega,\sigma)\big]^M\Big) \oplus \operatorname{Ker}\Big(-\tfrac{1}{2}I + K_{A^\top}^\#;\ \big[L_{k-1}^p(\partial\Omega,\sigma)\big]^M\Big) \text{ onto the space } (\mathbb{C}_{\partial\Omega})^M, \text{ provided } n \geq 3, \tag{7.1.289}$$

and

$$S_A : \big[L_{k-1}^p(\partial\Omega,\sigma) \cap (\mathbb{C}_{\partial\Omega})^\perp\big]^M \xrightarrow{\sim} \frac{\big[L_k^p(\partial\Omega,\sigma)\big]^M}{(\mathbb{C}_{\partial\Omega})^M} \text{ if } n \geq 3. \tag{7.1.290}$$

(6) Make the additional assumption that Ω is an exterior domain, and denote by Ω_∞ the (unique) unbounded connected component of Ω. Then

$$\big(-\tfrac{1}{2}I + K_A\big)\big(c\mathbf{1}_{\partial\Omega_\infty}\big) = -c\mathbf{1}_{\partial\Omega} \text{ for each } c \in \mathbb{C}, \tag{7.1.291}$$

and[6]

---

[6] with $\langle \cdots \rangle$ denoting the span (over $\mathbb{C}$) of $\{\dots\}$

$$\Big\{\big(-\tfrac{1}{2}I+K_A^{\#}\big)f : f\in\big[L^p_{k-1}(\partial\Omega,\sigma)\cap\mathbb{C}^{\perp}\big]^M\Big\}=\big[L^p_{k-1}(\partial\Omega,\sigma)\cap(\mathbb{C}^{\infty}_{\partial\Omega_+})^{\perp}\big]^M, \tag{7.1.292}$$

$$\Big\{f\in\big[L^p_k(\partial\Omega,\sigma)\big]^M : \big(-\tfrac{1}{2}I+K_A\big)f\in\big[\langle\mathbf{1}_{\partial\Omega}\rangle\big]^M\Big\}=\big(\mathbb{C}^{\infty}_{\partial\Omega_+}\big)^M. \tag{7.1.293}$$

Before presenting the proof of Theorem 7.1.11, a few comments are in order. First, a word of clarification regarding the action of operators involving quotient spaces. In all instances, this is to be understood in a natural fashion. Specifically, if for every function $f\in\big[L^p_k(\partial\Omega,\sigma)\big]^M$ with $k\in\{0,1\}$ we let $\big[f+(\mathbb{C}_{\partial\Omega_{\mp}})^M\big]$ stand for the class of f in the quotient space $\big[L^p_k(\partial\Omega,\sigma)\big]^M\big/(\mathbb{C}_{\partial\Omega_{\mp}})^M$ we define the action of the operators (7.1.280) according to

$$\big(\pm\tfrac{1}{2}I+K_A\big)\Big[f+(\mathbb{C}_{\partial\Omega_{\mp}})^M\Big]:=\Big[\big(\pm\tfrac{1}{2}I+K_A\big)f+(\mathbb{C}_{\partial\Omega_{\mp}})^M\Big] \tag{7.1.294}$$

for each $f\in\big[L^p_k(\partial\Omega,\sigma)\big]^M$. Thanks to (7.1.275), this definition is unambiguous. Likewise, the action of the operators (7.1.281) is to be naturally understood as

$$\big(\pm\tfrac{1}{2}I+K_A\big)\Big[f+(\mathbb{C}_{\partial\Omega})^M\Big]:=\Big[\big(\pm\tfrac{1}{2}I+K_A\big)f+(\mathbb{C}_{\partial\Omega})^M\Big] \tag{7.1.295}$$

for each function $f\in\big[L^p_k(\partial\Omega,\sigma)\big]^M$, where $\big[f+(\mathbb{C}_{\partial\Omega})^M\big]$ stands for the class of f in the quotient space $\big[L^p_k(\partial\Omega,\sigma)\big]^M\big/(\mathbb{C}_{\partial\Omega})^M$. That this is unambiguous follows from (7.1.275) and (7.1.272) (see (7.1.324)). Next, the operator in (7.1.285) is understood to act according to

$$\partial^A_\nu\mathcal{D}_A\Big[f+(\mathbb{C}_{\partial\Omega})^M\Big]:=\partial^A_\nu\mathcal{D}_Af,\qquad\forall\,f\in\big[L^p_k(\partial\Omega,\sigma)\big]^M, \tag{7.1.296}$$

which is unambiguous thanks to (7.1.284). Finally, the actions of the operators in (7.1.288) and (7.1.290) are interpreted as

$$\big[L^p_{k-1}(\partial\Omega,\sigma)\cap(\mathbb{C}_{\partial\Omega_\pm})^{\perp}\big]^M\ni f\longmapsto\Big[S_Af+(\mathbb{C}_{\partial\Omega_\pm})^M\Big]\in\frac{\big[L^p_k(\partial\Omega,\sigma)\big]^M}{(\mathbb{C}_{\partial\Omega_\pm})^M}, \tag{7.1.297}$$

and, respectively,

$$\big[L^p_{k-1}(\partial\Omega,\sigma)\cap(\mathbb{C}_{\partial\Omega})^{\perp}\big]^M\ni f\longmapsto\Big[S_Af+(\mathbb{C}_{\partial\Omega})^M\Big]\in\frac{\big[L^p_k(\partial\Omega,\sigma)\big]^M}{(\mathbb{C}_{\partial\Omega})^M}. \tag{7.1.298}$$

Second, Theorem 7.1.11 highlights the fundamental fact that there are natural obstructions, of a purely topological nature, to the invertibility of the operators

$\pm\frac{1}{2}I + K_A$, $\pm\frac{1}{2}I + K^{\#}_{A}$, and $\partial^A_\nu \mathcal{D}_A$ on Sobolev spaces (cf. (7.1.275)-(7.1.278) and (7.1.283)-(7.1.284)). It is remarkable that one can bridge between analysis and topology in a two-way fashion, without loss of information. For example, if the coefficient tensor A is positive definite, then under the initial background assumptions in Theorem 7.1.11 we have (cf. (7.1.275) and (7.1.280))

$$\begin{aligned} &\tfrac{1}{2}I + K_A : \big[L^p_k(\partial\Omega,\sigma)\big]^M \to \big[L^p_k(\partial\Omega,\sigma)\big]^M \text{ is an invertible operator} \\ &\quad\text{for some, or all, } k \in \{0,1\} \text{ if and only if } \mathbb{R}^n \setminus \overline{\Omega} \text{ is connected.} \end{aligned} \tag{7.1.299}$$

Such phenomena should be contrasted with the single layer S_A which, at least when $n \geq 3$, is always invertible on Sobolev spaces irrespective of the topological nature of the underlying domain (cf. Theorem 7.1.9).

We now turn to the proof of Theorem 7.1.11.

Proof of Theorem 7.1.11 Since the system L has a positive definite coefficient tensor A it follows that L satisfies satisfies the Legendre-Hadamard strong ellipticity condition (cf. (1.1.28)). Note that $A^\top$ continues to be positive definite (cf. (1.1.23)). In addition, given that $A \in \mathfrak{A}^{\rm dis}_L$ to being with, we also have $A^\top \in \mathfrak{A}^{\rm dis}_{L^\top}$, thanks to Theorem 1.5.2.

Taking $\delta \in (0,1)$ small ensures (cf. Theorem 3.4.2) that Ω is a two-sided NTA domain with a compact Ahlfors regular boundary, and each set $\Omega_\pm$ has finitely many connected components which are separated (i.e., have mutually disjoint closures; cf. [129, Lemma 5.11.3]). In this setting, any connected component of $\Omega_\pm$ is itself a two-sided NTA domain, and Alexander's relation (7.1.272) holds.

Let us now begin the proof in earnest. Throughout, all nontangential boundary traces are considered with respect to some fixed background aperture parameter $\kappa > 0$ (which is not explicitly used in the notation).

Proof of (1): Let us first prove (7.1.275) in the case when $n \geq 3$, $k = 1$, and $p = 2(n-1)/n$. To this end, denote by $\mathcal{D}^\pm_A$ the boundary-to-domain double layer potential operator associated with the set $\Omega_\pm$ and the coefficient tensor A as in (A.0.72). Also, recall the space $\mathbb{C}_{\Omega_\mp}$ from (7.1.265) and introduce the mappings

$$\begin{aligned} &\Phi_\pm : \mathrm{Ker}\Big(\pm\tfrac{1}{2}I + K_A;\ \big[L^{2(n-1)/n}_1(\partial\Omega,\sigma)\big]^M\Big) \longrightarrow \big(\mathbb{C}_{\Omega_\mp}\big)^M \\ &\quad\text{given by } \Phi_\pm(f) := \mathcal{D}^\mp_A f \text{ in } \Omega_\mp \text{ for each function} \\ &\quad f \in \big[L^{2(n-1)/n}_1(\partial\Omega,\sigma)\big]^M \text{ with } \Big(\pm\tfrac{1}{2}I + K_A\Big)f = 0, \end{aligned} \tag{7.1.300}$$

as well as

$$\begin{aligned} &\Psi_\pm : \big(\mathbb{C}_{\Omega_\mp}\big)^M \longrightarrow \mathrm{Ker}\Big(\pm\tfrac{1}{2}I + K_A;\ \big[L^{2(n-1)/n}_1(\partial\Omega,\sigma)\big]^M\Big) \\ &\quad\text{given by } \Psi_\pm(w) := w\big|^{\rm n.t.}_{\partial\Omega_\mp} \text{ for each } w \in \big(\mathbb{C}_{\Omega_\mp}\big)^M. \end{aligned} \tag{7.1.301}$$

To see that these mappings are indeed well defined, first let f be an arbitrary function in the domain of $\Phi_\pm$. Then $f \in \big[L^{2(n-1)/(n-2)}(\partial\Omega,\sigma)\big]^M$ by [130, (11.5.201)]. As such, we may rely on [131, Corollary 1.7.14], presently used with

$$u := \pm\mathcal{D}_A^{\pm} f \text{ in } \Omega_{\pm}, \text{ and} \\ p := 2(n-1)/n, \quad q := 2(n-1)/(n-2), \quad a := n, \quad b := n-1, \tag{7.1.302}$$

to write (bearing in mind that the outward unit normal for $\Omega_{\pm}$ is $\pm\nu$)

$$\begin{aligned} 0 &= \operatorname{Re}\int_{\partial\Omega} \langle \partial_\nu^A \mathcal{D}_A^{\pm} f, \overline{(\pm\tfrac{1}{2}I + K_A)f} \rangle \, d\sigma \\ &= \operatorname{Re}\int_{\Omega_{\pm}} \langle A\nabla\mathcal{D}_A^{\pm} f, \overline{\nabla\mathcal{D}_A^{\pm} f} \rangle \, d\mathcal{L}^n \geq c\int_{\Omega_{\pm}} |\nabla\mathcal{D}_A^{\pm} f|^2 \, d\mathcal{L}^n \end{aligned} \tag{7.1.303}$$

for some $c > 0$, in view of the fact that A is presently assumed to be positive definite. This forces $\nabla\mathcal{D}_A^{\pm} f = 0$ in $\Omega_{\pm}$. Hence $\partial_\nu^A \mathcal{D}_A^{\pm} f = 0$ on $\partial\Omega$, which further implies $\partial_\nu^A \mathcal{D}_A^{\mp} f = 0$ on $\partial\Omega$ (recall that the conormal of the double layer does not jump across $\partial\Omega$; cf. [132, (1.5.29)]). Having established this, one more application of [131, Corollary 1.7.14], this time over $\Omega_{\mp}$, produces

$$\begin{aligned} 0 &= \operatorname{Re}\int_{\partial\Omega} \langle \partial_\nu^A \mathcal{D}_A^{\mp} f, \overline{(\mp\tfrac{1}{2}I + K_A)f} \rangle \, d\sigma \\ &= \operatorname{Re}\int_{\Omega_{\mp}} \langle A\nabla\mathcal{D}_A^{\mp} f, \overline{\nabla\mathcal{D}_A^{\mp} f} \rangle \, d\mathcal{L}^n \geq c\int_{\Omega_{\mp}} |\nabla\mathcal{D}_A^{\mp} f|^2 \, d\mathcal{L}^n, \end{aligned} \tag{7.1.304}$$

which then implies $\nabla\mathcal{D}_A^{\mp} f = 0$ in $\Omega_{\mp}$. Thus, ultimately, $\mathcal{D}_A^{\mp} f \in (\mathbb{C}_{\Omega_{\mp}})^M$ (since $\mathcal{D}_A^{\mp} f$ vanishes at infinity if $\Omega_{\mp}$ is an exterior domain), which goes to show that the mappings $\Phi_{\pm}$ are well defined.

As regards the mappings $\Psi_{\pm}$, we first observe that

$$\text{for each function } w \in (\mathbb{C}_{\Omega_{\pm}})^M \text{ one has } w = \mathcal{D}_A^{\pm}\big(w\big|_{\partial\Omega_{\pm}}^{\text{n.t.}}\big) \text{ in } \Omega_{\pm}. \tag{7.1.305}$$

Indeed, viewing any given $w \in (\mathbb{C}_{\Omega_{\pm}})^M$ as zero in the unbounded connected component of $\Omega_{\mp}$ (if this is an exterior domain), the integral representation formula [132, (1.3.35)] simply reduces to the formula in the second line of (7.1.305), since $\partial_\nu^A w = 0$ on $\partial\Omega$. Taking the nontangential limit to the boundary in the formula in the last line of (7.1.305) eventually yields $(\mp\frac{1}{2}I + K_A)\big(w\big|_{\partial\Omega_{\pm}}^{\text{n.t.}}\big) = 0$ for each $w \in (\mathbb{C}_{\Omega_{\pm}})^M$. Together with the fact that for each such function w we have $\partial_{\tau_{ij}}\big(w\big|_{\partial\Omega_{\pm}}^{\text{n.t.}}\big) = 0$ for every pair of indices $i, j \in \{1, \dots, n\}$ (as seen from [130, (11.1.39)]), ultimately proves that the mappings $\Psi_{\pm}$ are also well defined.

We next claim that $\Phi_{\pm}$ and $\Psi_{\pm}$ are inverse to each other. In one direction, formula (7.1.305) may be re-interpreted as saying that $\Phi_{\pm}(\Psi_{\pm}(w)) = w$ for each function $w \in (\mathbb{C}_{\Omega_{\mp}})^M$. In the other direction, for each function $f \in \big[L_1^{2(n-1)/n}(\partial\Omega, \sigma)\big]^M$ with $(\pm\frac{1}{2}I + K_A)f = 0$ we have

$$\Psi_{\pm}(\Phi_{\pm}(f)) = \mathcal{D}_A^{\mp} f\big|_{\partial\Omega_{\mp}}^{\text{n.t.}} = \big(\tfrac{1}{2}I \mp K_A\big)f = f, \tag{7.1.306}$$

establishing that $\Phi_\pm$ and $\Psi_\pm$ are inverses to one other. In particular, $\Phi_\pm$ and $\Psi_\pm$ are bijections. Since we also have (see the discussion preceding (7.1.272))

$$\left\{w\Big|^{\text{n.t.}}_{\partial\Omega_\mp} : w \in \left(\mathbb{C}_{\Omega_\mp}\right)^M\right\} = \left(\mathbb{C}_{\partial\Omega_\mp}\right)^M, \tag{7.1.307}$$

we may now write

$$\begin{aligned}&\operatorname{Ker}\Big(\pm\tfrac{1}{2}I + K_A;\ \big[L_1^{2(n-1)/n}(\partial\Omega,\sigma)\big]^M\Big) \qquad (7.1.308)\\ &= \Psi_\pm\Big(\Phi_\pm\Big(\operatorname{Ker}\Big(\pm\tfrac{1}{2}I + K_A;\ \big[L_1^{2(n-1)/n}(\partial\Omega,\sigma)\big]^M\Big)\Big)\Big) = \Psi_\pm\Big(\big(\mathbb{C}_{\Omega_\mp}\big)^M\Big) = \big(\mathbb{C}_{\partial\Omega_\mp}\big)^M\end{aligned}$$

finishing the proof of (7.1.275) in the case when $n \geq 3$, $k = 1$, and $p = 2(n-1)/n$. The version of (7.1.275) in the case when $n \geq 3$, $k \in \{0, 1\}$, and $p \in (1, \infty)$ then follows from this and (7.1.60).

As far as formula (7.1.275) is concerned, there remains to treat the case when $n = 2$. However, in this scenario the same argument as before goes through provided the space $L_1^{2(n-1)/n}(\partial\Omega,\sigma)$ is replaced by $L_1^p(\partial\Omega,\sigma)$ with $p \in (1,\infty)$ arbitrary, and $L^{2(n-1)/(n-2)}(\partial\Omega,\sigma)$ is interpreted as $L^\infty(\partial\Omega,\sigma)$ (the embedding in [130, (11.5.201)] continues to be relevant, and we rely on [130, (10.2.253)-(10.2.254)] in this case as well). Finally, (7.1.276) is a consequence of (7.1.275) and (7.1.62). This takes care of item *(1)*.

Proof of (2): We shall first show that for each bounded connected component O of $\Omega_\pm$ we have

$$\int_{\partial O} \big(\mp\tfrac{1}{2}I + K_A^\#\big) f \, d\sigma = 0, \qquad \forall f \in \big[L^p(\partial\Omega,\sigma)\big]^M. \tag{7.1.309}$$

To prove this, pick an arbitrary function $f \in \big[L^p(\partial\Omega,\sigma)\big]^M$ and set $u^\pm := \mathscr{S}_{A^\top} f$ in $\Omega_\pm$. Then $u^\pm \in \big[\mathscr{C}^\infty(\Omega_\pm)\big]^M$ and $L^\top u^\pm = 0$ in $\Omega_\pm$. If for some constant $c \in \mathbb{C}^M$ we also consider the function $w := c \cdot \mathbf{1}_O$, then $w \in \big[\mathscr{C}^\infty(\Omega_\pm)\big]^M$ and vanishes at infinity. As such, the Green type formula [131, (1.7.78)] applies and yields

$$\Big\langle \int_{\partial O} \big(\mp\tfrac{1}{2}I + K_A^\#\big) f \, d\sigma, c\Big\rangle = \int_{\partial\Omega} \big\langle \partial_\nu^{A^\top} u^\pm, w\big|^{\text{n.t.}}_{\partial\Omega}\big\rangle \, d\sigma = \int_{\partial\Omega} \big\langle u^\pm\big|^{\text{n.t.}}_{\partial\Omega}, \partial_\nu^A w\big\rangle \, d\sigma = 0. \tag{7.1.310}$$

At this stage, (7.1.309) readily follows from this and the arbitrariness of $c \in \mathbb{C}^M$.

We next claim that for $k \in \{0, 1\}$ we have

$$\operatorname{Im}\Big(\pm\tfrac{1}{2}I + K_A^\#;\ \big[L_{k-1}^p(\partial\Omega,\sigma)\big]^M\Big) \subseteq \big[L_{k-1}^p(\partial\Omega,\sigma) \cap \big(\mathbb{C}_{\partial\Omega_\mp}\big)^\perp\big]^M. \tag{7.1.311}$$

Indeed, the version of (7.1.311) corresponding to $k = 1$ is implied by the fact that [132, (1.5.12)] is a well defined operator, plus (7.1.309) written for every bounded connected component O of $\Omega_\pm$. Also, the version of (7.1.311) corresponding to $k = 0$

follows from what we have just proved, the fact that [132, (1.5.27)] (written with p in place of p') is a well-defined continuous operator, the density result recorded in [130, (11.8.4)], and the compatibility result from [130, (11.8.16)].

To go further, for each $k \in \{0, 1\}$ write

$$\begin{aligned} M \cdot \dim \mathbb{C}_{\partial\Omega_{\mp}} &= \dim \mathrm{Ker}\Big(\pm \tfrac{1}{2}I + K^{\#}_{A};\ \big[L^{p}_{k-1}(\partial\Omega, \sigma)\big]^{M}\Big) \\ &= \dim \left(\frac{\big[L^{p}_{k-1}(\partial\Omega,\sigma)\big]^{M}}{\mathrm{Im}\big(\pm\frac{1}{2}I+K^{\#}_{A};[L^{p}_{k-1}(\partial\Omega,\sigma)]^{M}\big)} \right) \\ &\geq \dim \left(\frac{\big[L^{p}_{k-1}(\partial\Omega,\sigma)\big]^{M}}{\big[L^{p}_{k-1}(\partial\Omega,\sigma)\cap\big(\mathbb{C}_{\partial\Omega_{\mp}}\big)^{\perp}\big]^{M}} \right) = M \cdot \dim \mathbb{C}_{\partial\Omega_{\mp}}. \end{aligned} \tag{7.1.312}$$

Above, the first equality is (7.1.276), the second equality is implied by item *(1)* in Theorem 7.1.1, the subsequent inequality follows from (7.1.311), while the last equality is implied by simple functional analysis. In turn, since $\dim \mathbb{C}_{\partial\Omega_{\mp}} < \infty$, from (7.1.312) we deduce that

$$\dim \left(\frac{\big[L^{p}_{k-1}(\partial\Omega,\sigma)\big]^{M}}{\mathrm{Im}\big(\pm\frac{1}{2}I+K^{\#}_{A};[L^{p}_{k-1}(\partial\Omega,\sigma)]^{M}\big)} \right) = \dim \left(\frac{\big[L^{p}_{k-1}(\partial\Omega,\sigma)\big]^{M}}{\big[L^{p}_{k-1}(\partial\Omega,\sigma)\cap\big(\mathbb{C}_{\partial\Omega_{\mp}}\big)^{\perp}\big]^{M}} \right) < \infty. \tag{7.1.313}$$

With this in hand, [130, Lemma 1.1.1] applies and gives (7.1.277).

To conclude the treatment of item *(2)* there remains to observe that, thanks to item *(1)* in Theorem 7.1.1 and (7.1.275), for each $k \in \{0, 1\}$ we may write

$$\begin{aligned} &\dim \mathrm{CoKer}\Big(\pm \tfrac{1}{2}I + K_{A};\ \big[L^{p}_{k}(\partial\Omega, \sigma)\big]^{M}\Big) \\ &\quad = \dim \mathrm{Ker}\Big(\pm \tfrac{1}{2}I + K_{A};\ \big[L^{p}_{k}(\partial\Omega, \sigma)\big]^{M}\Big) = \dim\big(\mathbb{C}_{\partial\Omega_{\mp}}\big)^{M} = M \cdot \dim \mathbb{C}_{\partial\Omega_{\mp}}, \end{aligned} \tag{7.1.314}$$

proving (7.1.278).

<u>*Proof of (3)*</u>: We begin by considering the task of proving that the operators in (7.1.279) are isomorphisms. First, said operators are well defined, linear, and bounded, thanks to [132, Theorem 1.5.1, items *(iii)*, *(iv)*] and (7.1.277). To proceed, fix $k \in \{0, 1\}$ and consider the following commutative diagram:

$$\begin{array}{ccc} \big[L^{p}_{k-1}(\partial\Omega, \sigma)\big]^{M} & \xrightarrow{\pm\frac{1}{2}I+K^{\#}_{A}} & \big[L^{p}_{k-1}(\partial\Omega, \sigma)\big]^{M} \\ {\scriptstyle \iota_{\pm}}\big\uparrow & & {\scriptstyle \iota_{\pm}}\big\uparrow \\ \big[L^{p}_{k-1}(\partial\Omega, \sigma)\cap(\mathbb{C}_{\partial\Omega_{\mp}})^{\perp}\big]^{M} & \xrightarrow{\pm\frac{1}{2}I+K^{\#}_{A}} & \big[L^{p}_{k-1}(\partial\Omega, \sigma)\cap(\mathbb{C}_{\partial\Omega_{\mp}})^{\perp}\big]^{M} \end{array} \tag{7.1.315}$$

where $\iota_{\pm}$ are the canonical inclusions. In particular, $\iota_{\pm}$ are Fredholm operators. Since, by item *(1)* in Theorem 7.1.1, $\pm\frac{1}{2}I + K^{\#}_{A}$ are Fredholm with index zero on

$\big[L^p_{k-1}(\partial\Omega,\sigma)\big]^M$, from [130, Theorem 2.1.2] and (7.1.315) we see that $\pm\frac{1}{2}I + K^\#_A$ are Fredholm operators with index zero when considered acting from the space $\big[L^p_{k-1}(\partial\Omega,\sigma)\cap(\mathbb{C}_{\partial\Omega_\mp})^\perp\big]^M$ into itself. As such, matters have been reduced to proving injectivity for the operators in question. Thanks to Proposition 7.1.2 (cf. (7.1.61)), it suffices to do so when $p = 2$ and $k = 1$.

With this goal in mind, let the function $f \in \big[L^2(\partial\Omega,\sigma)\cap(\mathbb{C}_{\partial\Omega_\mp})^\perp\big]^M$ be such that $\big(\pm\frac{1}{2}I + K^\#_A\big)f = 0$. The goal is to show that $f = 0$. We first claim that

$$\int_{\partial\Omega} f\,\mathrm{d}\sigma = 0. \tag{7.1.316}$$

If $\Omega_\pm$ is unbounded, then $\Omega_\mp$ is bounded, in which case $\mathbf{1}_{\partial\Omega} \in \mathbb{C}_{\partial\Omega_\mp}$ (see (7.1.273)). As such, the orthogonality condition satisfied by f directly gives (7.1.316). If $\Omega_\pm$ is bounded, then (7.1.316) follows from [132, (1.7.59)]. Going further, (7.1.316) shows, in light of [132, Lemma 1.3.1], that if we consider the functions

$$u_\pm := \big(\mathscr{S}_{A^\top} f\big)\Big|_{\Omega_\pm} \tag{7.1.317}$$

then for each $\alpha \in \mathbb{N}_0^n$ we have

$$(\partial^\alpha u_\pm)(x) = O\big(|x|^{1-n-|\alpha|}\big) \text{ as } |x|\to\infty, \text{ if } \Omega_\pm \text{ is unbounded.} \tag{7.1.318}$$

Thanks to (7.1.318) and [132, Theorem 1.5.1], we may invoke [131, Corollary 1.7.14] (for the operator $L^\top$ and the set $\Omega_\mp$) which presently gives

$$\int_{\Omega_\mp}\big\langle A^\top\nabla u_\mp, \overline{\nabla u_\mp}\big\rangle\,\mathrm{d}\mathcal{L}^n = \int_{\partial\Omega}\Big\langle\big(\pm\tfrac{1}{2}I + K^\#_A\big)f, \overline{u_\mp}\Big|^{\text{n.t.}}_{\partial\Omega}\Big\rangle\,\mathrm{d}\sigma = 0, \tag{7.1.319}$$

since $\partial^{A^\top}_\nu u_\mp = \big(\pm\frac{1}{2}I + K^\#_A\big)f = 0$, thanks to [132, (1.5.59)] (written for the sets $\Omega_\mp$ and the coefficient tensor $A^\top$) and assumptions. Since, as already noted, the transposed coefficient tensor $A^\top$ is also positive definite, there exists some positive constant c such that $\operatorname{Re}\big\langle A^\top\nabla u_\mp, \overline{\nabla u_\mp}\big\rangle \geq c|\nabla u_\mp|^2$ in $\Omega_\mp$. In concert with (7.1.319), this proves that $\nabla u_\mp$ vanishes identically in $\Omega_\mp$. Bearing in mind (7.1.318), we ultimately conclude that $u_\mp \in \big(\mathbb{C}_{\Omega_\mp}\big)^M$. In view of [132, (1.5.44)], (7.1.317), and (7.1.265), the latter membership implies $u_\pm\big|^{\text{n.t.}}_{\partial\Omega} = u_\mp\big|^{\text{n.t.}}_{\partial\Omega} \in \big(\mathbb{C}_{\partial\Omega_\mp}\big)^M$. By once again relying on [131, Corollary 1.7.14] (used now in $\Omega_\pm$ with $A^\top$ in place of A) we may write

$$\begin{aligned}\int_{\Omega_\pm}\big\langle A^\top\nabla u_\pm, \overline{\nabla u_\pm}\big\rangle\,\mathrm{d}\mathcal{L}^n &= \int_{\partial\Omega}\Big\langle u_\pm\Big|^{\text{n.t.}}_{\partial\Omega}, \overline{\big(\mp\tfrac{1}{2}I + K^\#_A\big)f}\Big\rangle\,\mathrm{d}\sigma\\ &= \mp\int_{\partial\Omega}\Big\langle u_\pm\Big|^{\text{n.t.}}_{\partial\Omega}, \overline{f}\Big\rangle\,\mathrm{d}\sigma = 0,\end{aligned} \tag{7.1.320}$$

where the last equality follows from the fact that the components of f are orthogonal to $(\mathbb{C}_{\partial\Omega_\mp})^\perp$ and that $\overline{u_\pm}\big|^{\text{n.t.}}_{\partial\Omega}$ belongs to the space $\big(\mathbb{C}_{\partial\Omega_\mp}\big)^M$. Since $A^\top$ is positive definite, this forces $\nabla u_\pm = 0$ in $\Omega_\pm$. Based on this and the trace formulas for the normal derivative of the single layer (cf. [132, (1.5.59)]) we therefore obtain

$$\pm f = \big(\pm\tfrac{1}{2}I + K_A^\#\big)f - \big(\mp\tfrac{1}{2}I + K_A^\#\big)f = \partial_\nu^{A^\top} u_\mp - \partial_\nu^{A^\top} u_\pm = 0. \tag{7.1.321}$$

This proves that $\pm\frac{1}{2}I + K_A^\#$ is indeed injective on $\big[L^2(\partial\Omega,\sigma) \cap \big(\mathbb{C}_{\partial\Omega_\mp}\big)^\perp\big]^M$, hence ultimately an isomorphism on this space.

Next, the version $k = 0$ of (7.1.280) is obtained from (7.1.279) (written with $k := 1$ and with $p' := (1-1/p)^{-1}$ in place of p), and the abstract result from [130, (2.1.59) in Lemma 2.1.7] used with

$$X := \big[L^{p'}(\partial\Omega,\sigma)\big]^M, \quad V := (\mathbb{C}_{\partial\Omega_\mp})^M, \quad T := \pm\tfrac{1}{2}I + K_A. \tag{7.1.322}$$

In this vein we remark that the invariance of V under T follows from (7.1.275), while the identification of T^* with $\pm\frac{1}{2}I + K_A^\#$ is provided by [132, Theorem 1.5.1, *(iii)*].

Let us now deal with the operators in (7.1.280) when $k = 1$. That they are well defined follows from [132, Theorem 1.5.1, item *(v)*] and the fact that the operators in (7.1.280) with $k = 0$ are well defined. To prove that they are actually isomorphisms, consider the commutative diagram

$$\begin{array}{ccc} \big[L_1^p(\partial\Omega,\sigma)\big]^M & \xrightarrow{\pm\frac{1}{2}I+K_A} & \big[L_1^p(\partial\Omega,\sigma)\big]^M \\ {\scriptstyle\pi_\pm}\big\downarrow & & \big\downarrow{\scriptstyle\pi_\pm} \\ \big[L_1^p(\partial\Omega,\sigma)\big]^M/\big(\mathbb{C}_{\partial\Omega_\mp}\big)^M & \xrightarrow{\pm\frac{1}{2}I+K_A} & \big[L_1^p(\partial\Omega,\sigma)\big]^M/\big(\mathbb{C}_{\partial\Omega_\mp}\big)^M \end{array} \tag{7.1.323}$$

where $\pi_\pm$ are the canonical projections onto the corresponding quotient spaces. In particular, $\pi_\pm$ are Fredholm operators. From this and item *(1)* of Theorem 7.1.1, we deduce that the operators $\pm\frac{1}{2}I + K_A$ are Fredholm with index zero in the context of (7.1.280) with $k = 1$. Since we have already seen that they are also injective, we ultimately conclude that said operators are actually isomorphisms.

Next, we look at the operators in (7.1.281). They are well defined since, from (7.1.275) and (7.1.272), we see that

$$\begin{aligned} \big(\pm\tfrac{1}{2}I + K_A\big)\big(\mathbb{C}_{\partial\Omega}\big)^M &= \big(\pm\tfrac{1}{2}I + K_A\big)\big(\mathbb{C}_{\partial\Omega_\pm}\big)^M = \Big(\pm I + \big(\mp\tfrac{1}{2}I + K_A\big)\Big)\big(\mathbb{C}_{\partial\Omega_\pm}\big)^M \\ &= \big(\mathbb{C}_{\partial\Omega_\pm}\big)^M \subseteq \big(\mathbb{C}_{\partial\Omega}\big)^M. \end{aligned} \tag{7.1.324}$$

The fact that for each $k \in \{0,1\}$ we have the commutative diagram

$$\begin{array}{ccc} \big[L_k^p(\partial\Omega,\sigma)\big]^M & \xrightarrow{\pm\frac{1}{2}I+K_A} & \big[L_k^p(\partial\Omega,\sigma)\big]^M \\ {\scriptstyle\pi}\big\downarrow & & \big\downarrow{\scriptstyle\pi} \\ \big[L_k^p(\partial\Omega,\sigma)\big]^M/\big(\mathbb{C}_{\partial\Omega}\big)^M & \xrightarrow{\pm\frac{1}{2}I+K_A} & \big[L_k^p(\partial\Omega,\sigma)\big]^M/\big(\mathbb{C}_{\partial\Omega}\big)^M \end{array} \tag{7.1.325}$$

(in which the vertical arrows are canonical projections onto the quotient spaces) implies, much as before, that

$$\text{all operators in (7.1.281) are Fredholm with index zero.} \tag{7.1.326}$$

Using the isomorphisms in (7.1.280) and (7.1.272) it is immediate that the operators in (7.1.281) are onto, and, hence, isomorphisms (thanks to (7.1.326)).

Once we have proved that the operators in (7.1.281) are isomorphisms, we may conclude that the operators in (7.1.282) are isomorphisms by once again invoking the abstract result from [130, (2.1.59) in Lemma 2.1.7], now used with

$$X := \big[L^{p'}_k(\partial\Omega,\sigma)\big]^M, \quad V := (\mathbb{C}_{\partial\Omega})^M, \quad T := \pm\tfrac{1}{2}I + K_A. \tag{7.1.327}$$

The invariance of V under T follows from (7.1.275) and (7.1.272), whereas the identification of T^* with $\pm\frac{1}{2}I + K^{\#}_{A}$ in this context has been established in [132, Theorem 1.5.1, items *(iii)*, *(vi)*]. All claims in item *(3)* have now been justified.

<u>*Proof of (4)*</u>: To get started, recall from [132, Theorem 1.5.1, items *(vii)-(viii)*] that the operator

$$\partial^A_\nu \mathcal{D}_A : \big[L^p_k(\partial\Omega,\sigma)\big]^M \longrightarrow \big[L^p_{k-1}(\partial\Omega,\sigma)\big]^M \tag{7.1.328}$$

is well defined, linear, and bounded. Then the same type of argument as in the proof of (7.1.311) also gives that for $k \in \{0,1\}$ we have

$$\mathrm{Im}\Big(\partial^A_\nu \mathcal{D}_A : \big[L^p_k(\partial\Omega,\sigma)\big]^M \to \big[L^p_{k-1}(\partial\Omega,\sigma)\big]^M\Big) \subseteq \big[L^p_{k-1}(\partial\Omega,\sigma) \cap (\mathbb{C}_{\partial\Omega})^{\perp}\big]^M. \tag{7.1.329}$$

Let us also note here that, as seen from [132, Theorem 1.5.1, item *(xiii)*] (cf. [132, (1.5.85)]), (7.1.275), and (7.1.272), for each $k \in \{0,1\}$ we have

$$(\mathbb{C}_{\partial\Omega})^M \subseteq \mathrm{Ker}\Big(\partial^A_\nu \mathcal{D}_A : \big[L^p_k(\partial\Omega,\sigma)\big]^M \to \big[L^p_{k-1}(\partial\Omega,\sigma)\big]^M\Big). \tag{7.1.330}$$

Next, from (7.1.282) used with $A^\top$ in place of A, [132, (1.5.86)], and [132, Theorem 1.5.1, item *(ix)*] we deduce that for every $k \in \{0,1\}$ we have

$$\big[L^p_{k-1}(\partial\Omega,\sigma) \cap (\mathbb{C}_{\partial\Omega})^{\perp}\big]^M \subseteq \mathrm{Im}\Big(\partial^A_\nu \mathcal{D}_A : \big[L^p_k(\partial\Omega,\sigma)\big]^M \to \big[L^p_{k-1}(\partial\Omega,\sigma)\big]^M\Big). \tag{7.1.331}$$

At this stage, (7.1.283) follows from (7.1.329) and (7.1.331).

To proceed, denote by $p' \in (1,\infty)$ the Hölder conjugate exponent of p. By further decreasing the parameter $\delta > 0$ in the statement of the current theorem if necessary, we may assume that (7.1.283) also holds with p replaced by p'. Keeping this in mind we then write

$$
\begin{aligned}
\operatorname{Ker}\Big(\partial_\nu^A \mathcal{D}_A : \big[L^p(\partial\Omega,\sigma)\big]^M \to \big[L^p_{-1}(\partial\Omega,\sigma)\big]^M\Big)
&= \Big[\operatorname{Im}\Big(\partial_\nu^{A^\top} \mathcal{D}_{A^\top} : \big[L^{p'}_1(\partial\Omega,\sigma)\big]^M \to \big[L^{p'}(\partial\Omega,\sigma)\big]^M\Big)\Big]^\perp \\
&= \Big[\big[L^{p'}(\partial\Omega,\sigma)\cap(\mathbb{C}_{\partial\Omega})^\perp\big]^M\Big]^\perp = (\mathbb{C}_{\partial\Omega})^M. \qquad (7.1.332)
\end{aligned}
$$

Above, the first equality is a consequence of [132, Theorem 1.5.1, item *(viii)*] and [130, (2.1.45)], while the second equality is simply the version of (7.1.283) corresponding to $k = 0$, written with p' in place of p and $A^\top$ in place of A. The last equality in (7.1.332) is simple functional analysis.

Having proved (7.1.332), for every $k \in \{0, 1\}$ we obtain

$$
\begin{aligned}
&\operatorname{Ker}\Big(\partial_\nu^A \mathcal{D}_A : \big[L^p_k(\partial\Omega,\sigma)\big]^M \to \big[L^p_k(\partial\Omega,\sigma)\big]^M\Big) \qquad (7.1.333)\\
&\quad\subseteq \operatorname{Ker}\Big(\partial_\nu^A \mathcal{D}_A : \big[L^p(\partial\Omega,\sigma)\big]^M \to \big[L^p_{-1}(\partial\Omega,\sigma)\big]^M\Big) = (\mathbb{C}_{\partial\Omega})^M.
\end{aligned}
$$

When used in concert with (7.1.330), this finishes the proof of (7.1.284).

With (7.1.283)-(7.1.284) in hand, the First Isomorphism Theorem (or the Fundamental Theorem of Homomorphisms) in abstract algebra applies and gives that the operator in (7.1.285) is an isomorphism.

Proof of (5): The fact that the operator (7.1.286) is an isomorphism follows directly from (7.1.621), (7.1.275), and Proposition 7.1.2. Also, from [132, (1.7.11)], [132, (1.7.59)], (7.1.276), and Proposition 7.1.2 we see that the operator (7.1.287) is an isomorphism. Moving on to treat the operators in (7.1.288), fix $k \in \{0, 1\}$ and consider the commutative diagram

$$
\begin{array}{ccc}
\big[L^p_{k-1}(\partial\Omega,\sigma)\big]^M & \xrightarrow{\;S_A\;} & \big[L^p_k(\partial\Omega,\sigma)\big]^M \\
{\scriptstyle\iota}\big\uparrow & & \big\downarrow{\scriptstyle\pi} \\
\big[L^p_{k-1}(\partial\Omega,\sigma)\cap(\mathbb{C}_{\partial\Omega_\mp})^\perp\big]^M & \xrightarrow{\;S_A\;} & \big[L^p_k(\partial\Omega,\sigma)\big]^M\big/(\mathbb{C}_{\partial\Omega_\mp})^M
\end{array}
\qquad (7.1.334)
$$

where ι and π are the canonical inclusion and projection, respectively. Given that the top horizontal arrow is an invertible operator, it follows that the operators in (7.1.288) are Fredholm with index zero. Thus, as far as the task at hand is concerned, we are left with proving injectivity.

To this end, let $f \in \big[L^p_{k-1}(\partial\Omega,\sigma)\cap(\mathbb{C}_{\partial\Omega_\mp})^\perp\big]^M$ be such that $S_A f \in \big(\mathbb{C}_{\partial\Omega_\mp}\big)^M$. The fact that the operators in (7.1.286)-(7.1.287) are invertible plus Theorem 7.1.9 then places the function f in the space $\operatorname{Ker}\Big(\mp\frac{1}{2}I + K^\#_{A^\top};\ \big[L^p_{k-1}(\partial\Omega,\sigma)\big]^M\Big)$. This, however, forces $f = 0$ given that $\mp\frac{1}{2}I + K^\#_{A^\top}$ are isomorphisms when acting from $\big[L^p_{k-1}(\partial\Omega,\sigma)\cap(\mathbb{C}_{\partial\Omega_\mp})^\perp\big]^M$ into itself (see (7.1.279)). This takes care of (7.1.288).

Consider next the task of proving that the operator in (7.1.289) is an isomorphism. In this regard, thanks to Proposition 7.1.2 it suffices to only treat the case when $k = 1$, an assumption we enforce throughout. We begin by noting that the sum of null-spaces appearing in (7.1.289) is direct. To justify this claim, observe that if a function f happens to belong to $\mathrm{Ker}\Big(\frac{1}{2}I + K^{\#}_{A^\top}; \big[L^p(\partial\Omega,\sigma)\big]^M\Big)$ as well as to $\mathrm{Ker}\Big(-\frac{1}{2}I + K^{\#}_{A^\top}; \big[L^p(\partial\Omega,\sigma)\big]^M\Big)$, then $f = \big(\frac{1}{2}I + K^{\#}_{A^\top}\big)f - \big(-\frac{1}{2}I + K^{\#}_{A^\top}\big)f = 0$. In particular, from this, (7.1.276), and (7.1.272) we conclude that

$$\begin{aligned}
&\dim\Big[\mathrm{Ker}\Big(\tfrac{1}{2}I + K^{\#}_{A^\top}; \big[L^p(\partial\Omega,\sigma)\big]^M\Big) \oplus \mathrm{Ker}\Big(-\tfrac{1}{2}I + K^{\#}_{A^\top}; \big[L^p(\partial\Omega,\sigma)\big]^M\Big)\Big] \\
&\quad = \dim\Big[\mathrm{Ker}\Big(\tfrac{1}{2}I + K^{\#}_{A^\top}; \big[L^p(\partial\Omega,\sigma)\big]^M\Big)\Big] \\
&\qquad + \dim\Big[\mathrm{Ker}\Big(-\tfrac{1}{2}I + K^{\#}_{A^\top}; \big[L^p(\partial\Omega,\sigma)\big]^M\Big)\Big] = M\cdot\dim\mathbb{C}_{\Omega_-} + M\cdot\dim\mathbb{C}_{\Omega_+} \\
&\quad = \dim\Big[\big(\mathbb{C}_{\partial\Omega_-}\big)^M\Big] + \dim\Big[\big(\mathbb{C}_{\partial\Omega_+}\big)^M\Big] = \dim\Big[\big(\mathbb{C}_{\partial\Omega}\big)^M\Big] < \infty.
\end{aligned} \tag{7.1.335}$$

Having proved (7.1.335), the fact that the operator in (7.1.289) is an isomorphism follows immediately from the injectivity of S_A in the context of Theorem 7.1.9.

As regards the invertibility of the operator in (7.1.290), the fact that this is Fredholm with index zero is seen as before, relying on Theorem 7.1.9 and making use of a commutative diagram similar to (7.1.334). To establish its injectivity, consider $f \in \big[L^p_{k-1}(\partial\Omega,\sigma) \cap (\mathbb{C}_{\partial\Omega})^{\perp}\big]^M$ with the property that $S_A f \in \big(\mathbb{C}_{\partial\Omega}\big)^M$. Thanks to the fact that the operator in (7.1.289) is actually invertible, it is then possible to find $g_\pm \in \mathrm{Ker}\Big(\pm\frac{1}{2}I + K^{\#}_{A^\top}; \big[L^p(\partial\Omega,\sigma)\big]^M\Big)$ such that $f = g_+ + g_-$. Consequently,

$$\begin{aligned}
\big(\tfrac{1}{2}I + K^{\#}_{A^\top}\big)\big(-\tfrac{1}{2}I + K^{\#}_{A^\top}\big)f &= \big(-\tfrac{1}{2}I + K^{\#}_{A^\top}\big)\big(\tfrac{1}{2}I + K^{\#}_{A^\top}\big)g_+ \\
&\quad + \big(\tfrac{1}{2}I + K^{\#}_{A^\top}\big)\big(-\tfrac{1}{2}I + K^{\#}_{A^\top}\big)g_- = 0 + 0 = 0.
\end{aligned} \tag{7.1.336}$$

In turn, this forces $f = 0$ since the composition $\big(\frac{1}{2}I + K^{\#}_{A^\top}\big)\big(-\frac{1}{2}I + K^{\#}_{A^\top}\big)$ is an isomorphism on $\big[L^p_{k-1}(\partial\Omega,\sigma) \cap (\mathbb{C}_{\partial\Omega})^{\perp}\big]^M$ (see (7.1.282)). Hence, the operator in (7.1.290) is an isomorphism.

Proof of (6): Work under the assumption that Ω is an exterior domain, and denote by Ω_∞ the unbounded connected component of Ω. Let $\mathcal{D}_A^{(\Omega_\infty)^c}$ be the double layer potential operator associated with the coefficient tensor A and the set $\mathbb{R}^n \setminus \overline{\Omega_\infty}$. Then $\partial\Omega \setminus \partial\Omega_\infty \subseteq \mathbb{R}^n \setminus \overline{\Omega_\infty}$ and for each point $x \in \partial\Omega \setminus \partial\Omega_\infty$ we have

$$K_A\big(c\mathbf{1}_{\partial\Omega_\infty}\big)(x) = -\big(\mathcal{D}_A^{(\Omega_\infty)^c} c\big)(x) = -c \ \text{ for each } c \in \mathbb{C}^M, \tag{7.1.337}$$

where the first equality is a direct consequence of definitions (cf. (A.0.116), (A.0.72)), and [132, (1.3.45)] (also bearing in mind that the geometric measure theoretic outward unit normal to $\mathbb{R}^n \setminus \overline{\Omega_\infty}$ is $-\nu\big|_{\partial\Omega_\infty}$). If we also denote by $\mathcal{D}_A^{\Omega_\infty}$ be the double layer potential operator associated with the coefficient tensor A and the set Ω_∞, then for each $c \in \mathbb{C}^M$ we have $\mathcal{D}_A^{\Omega_\infty} c = 0$ in Ω_∞ (cf. [132, (1.3.45)]). Going nontangentially to the boundary then gives $\big(\frac{1}{2}I + K_A^{\Omega_\infty}\big)c = 0$ on $\partial\Omega_\infty$, where $K_A^{\Omega_\infty}$ is the boundary-to-boundary double layer potential operator associated as in (A.0.116) with the coefficient tensor A and the set Ω_∞. Hence, $K_A^{\Omega_\infty} = -\frac{1}{2}c$ on $\partial\Omega_\infty$ which, in view of definitions (cf. (A.0.116)), further translates into $K_A\big(c\mathbf{1}_{\partial\Omega_\infty}\big) = -\frac{1}{2}c$ on $\partial\Omega_\infty$. From this and (7.1.337) we therefore conclude that for each $c \in \mathbb{C}^M$ we have

$$K_A\big(c\mathbf{1}_{\partial\Omega_\infty}\big) = \begin{cases} -c & \text{on } \partial\Omega \setminus \partial\Omega_\infty, \\ -\frac{1}{2}c & \text{on } \partial\Omega_\infty. \end{cases} \tag{7.1.338}$$

which may then be re-interpreted as in (7.1.291).

Finally, we will establish the equality claimed in (7.1.292) by double inclusion. In one direction, given any $g \in \big[L^p_{k-1}(\partial\Omega,\sigma) \cap (\mathbb{C}^\infty_{\partial\Omega_+})^\perp\big]^M$, it follows that

$$g \in \big[L^p_{k-1}(\partial\Omega,\sigma) \cap (\mathbb{C}_{\partial\Omega_+})^\perp\big]^M = \mathrm{Im}\Big(-\tfrac{1}{2}I + K_A^{\#};\ \big[L^p_{k-1}(\partial\Omega,\sigma)\big]^M\Big), \tag{7.1.339}$$

thanks to (7.1.277) and the fact that we have $\mathbb{C}^\infty_{\partial\Omega_+} \supseteq \mathbb{C}_{\partial\Omega_+}$. As such, there exists some $f \in \big[L^p_{k-1}(\partial\Omega,\sigma)\big]^M$ with the property that $g = \big(-\frac{1}{2}I + K_A^{\#}\big)f$. Having proved this, for each $c \in \mathbb{C}^M$ we then compute

$$\langle f, c\rangle = -\Big\langle f, \big(-\tfrac{1}{2}I + K_A\big)\big(c\mathbf{1}_{\partial\Omega_\infty}\big)\Big\rangle = -\Big\langle \big(-\tfrac{1}{2}I + K_A^{\#}\big)f, c\mathbf{1}_{\partial\Omega_\infty}\Big\rangle = \big\langle g, c\mathbf{1}_{\partial\Omega_\infty}\big\rangle = 0. \tag{7.1.340}$$

Above, the first equality is a consequence of (7.1.291), the second equality is implied by the duality results from [132, Theorem 1.5.1, items *(iii)*, *(vi)*], and the third equality follows from the fact that $g \in \big[L^p_{k-1}(\partial\Omega,\sigma) \cap (\mathbb{C}^\infty_{\partial\Omega_+})^\perp\big]^M$ plus the observation that $\mathbf{1}_{\partial\Omega_\infty} \in \mathbb{C}^\infty_{\partial\Omega_+}$ (cf. (7.1.268)). Hence, $f \in \big[L^p_{k-1}(\partial\Omega,\sigma) \cap \mathbb{C}^\perp\big]^M$ thus establishing the right-to-left inclusion in (7.1.292).

In the opposite direction, for any $f \in \big[L^p_{k-1}(\partial\Omega,\sigma) \cap \mathbb{C}^\perp\big]^M$ and $c \in \mathbb{C}^M$,

$$\Big\langle \big(-\tfrac{1}{2}I + K_A^{\#}\big)f, c\mathbf{1}_{\partial\Omega_\infty}\Big\rangle = \Big\langle f, \big(-\tfrac{1}{2}I + K_A\big)\big(c\mathbf{1}_{\partial\Omega_\infty}\big)\Big\rangle = -\big\langle f, c\mathbf{1}_{\partial\Omega}\big\rangle = -\langle f, c\rangle = 0, \tag{7.1.341}$$

with similar justifications as in (7.1.340). Also, (7.1.277) gives

$$\big(-\tfrac{1}{2}I + K_A^{\#}\big)f \in \big[L^p_{k-1}(\partial\Omega,\sigma) \cap (\mathbb{C}_{\partial\Omega_+})^\perp\big]^M. \tag{7.1.342}$$

Since we presently have

$$\mathbb{C}^\infty_{\partial\Omega_+} = \mathbb{C}_{\partial\Omega_+} \oplus \big\langle \mathbf{1}_{\partial\Omega_\infty}\big\rangle, \tag{7.1.343}$$

we ultimately conclude from (7.1.341) and (7.1.342) that $\big(-\frac{1}{2}I + K_A^{\#}\big)f$ actually belongs to the smaller space $\big[L^p_{k-1}(\partial\Omega,\sigma) \cap (\mathbb{C}^\infty_{\partial\Omega_+})^\perp\big]^M$. This gives the left-to-right inclusion in (7.1.292), and finishes the proof of (7.1.292).

Finally, consider the claim in (7.1.293). By (7.1.291), any function in $\big[\langle \mathbf{1}_{\partial\Omega}\rangle\big]^M$ is of the form $\big(-\frac{1}{2}I + K_A\big)\big(c\mathbf{1}_{\partial\Omega_\infty}\big)$ for some $c \in \mathbb{C}$. As such, given any f in

$\big[L^p_k(\partial\Omega,\sigma)\big]^M$, it follows that $\big(-\frac{1}{2}I+K_A\big)f$ belongs to $\big[\langle \mathbf{1}_{\partial\Omega}\rangle\big]^M$ if and only if

$$\big(-\tfrac{1}{2}I+K_A\big)f=\big(-\tfrac{1}{2}I+K_A\big)\big(c\mathbf{1}_{\partial\Omega_\infty}\big)\ \text{ for some } c\in\mathbb{C}, \tag{7.1.344}$$

which, in light of (7.1.275), is further equivalent to the existence of some $c\in\mathbb{C}$ such that $f-c\mathbf{1}_{\partial\Omega_\infty}\in\operatorname{Ker}\Big(-\frac{1}{2}I+K_A;\ \big[L^p_k(\partial\Omega,\sigma)\big]^M\Big)=\big(\mathbb{C}_{\partial\Omega_+}\big)^M$. In turn, this is equivalent to having $f\in\Big[\operatorname{span}_{\mathbb{C}}\big\{\mathbb{C}_{\partial\Omega_+},\mathbf{1}_{\partial\Omega_\infty}\big\}\Big]^M=\big(\mathbb{C}^\infty_{\partial\Omega_+}\big)^M$, as wanted. This finishes the proof of Theorem 7.1.11. □

Before going any further, we make the following observation, additionally enhancing the scope of Theorem 7.1.11.

Remark 7.1.12 *Theorem 7.1.11 implies similar results for a variety of other function spaces in place of $L^p_k(\partial\Omega,\sigma)$ and $L^p_{k-1}(\partial\Omega,\sigma)$ with $k\in\{0,1\}$, namely Muckenhoupt weighted Lebesgue and Sobolev spaces, Lorentz spaces, Morrey and block spaces, Hölder spaces, Hardy spaces, the John-Nirenberg space, the Sarason space, Besov spaces, and Triebel-Lizorkin spaces. This is seen from Theorem 7.1.11 and its proof, Theorem 7.1.1,* [130, Theorem 2.1.2], *Proposition 7.1.2,* [130, Lemma 2.1.4], *embeddings (cf.* (7.1.63)-(7.1.72)), *plus mapping properties and duality results on these scales of spaces.*

Moving on, we establish invertibility properties for the boundary-to-boundary single layer on end-point Besov spaces. In Theorem 7.1.13 below we treat dimensions ≥ 3, while in Theorem 7.1.14 stated a little further down the road we consider the two-dimensional case.

Theorem 7.1.13 *Fix $n\in\mathbb{N}$ with $n\geq 3$ and suppose $\Omega\subseteq\mathbb{R}^n$ is an Ahlfors regular domain with compact boundary. Abbreviate $\sigma:=\mathcal{H}^{n-1}\lfloor\partial\Omega$ and denote by ν the geometric measure theoretic outward unit normal to Ω. Also, let L be a homogeneous, second-order, constant complex coefficient, $M\times M$ system in $\mathbb{R}^n$ such that*

$$\text{L satisfies the Legendre-Hadamard strong ellipticity condition (cf. [131, Definition 1.3.2]) and one has } \mathfrak{A}^{\rm dis}_L\neq\varnothing. \tag{7.1.345}$$

Consider the boundary-to-boundary single layer potential operator S associated with Ω and the system L as in (A.0.231). *Finally, fix some $s\in(0,1)$.*

Then there exists some threshold $\delta_0\in(0,1)$ which depends only on n, s, L, the ADR *constants of $\partial\Omega$, and the relative flatness ratio $\wp(\Omega)/\operatorname{diam}\partial\Omega$ such that if*

$$\operatorname{dist}\Big(\nu,[\mathrm{VMO}(\partial\Omega,\sigma)]^n\Big)<\delta_0 \tag{7.1.346}$$

with the distance considered in $\big[\mathrm{BMO}(\partial\Omega,\sigma)\big]^n$ (hence if Ω is a δ_0-oscillating AR *domain; cf. Definition 3.4.1), it follows that Ω is a* UR *domain and the operators*

$$S:\big[B^{1,1}_{-s}(\partial\Omega,\sigma)\big]^M\longrightarrow\big[B^{1,1}_{1-s}(\partial\Omega,\sigma)\big]^M \tag{7.1.347}$$

$$S:\big[B^{\infty,\infty}_{s-1}(\partial\Omega,\sigma)\big]^M\longrightarrow\big[B^{\infty,\infty}_{s}(\partial\Omega,\sigma)\big]^M \tag{7.1.348}$$

are well defined, linear, bounded, invertible, with bounded inverses.

As a corollary, the operators (7.1.347)-(7.1.348) *are invertible for any* $s \in (0,1)$ *provided* Ω *is an infinitesimally flat* AR *domain* (*cf. Definition 3.4.1*).

Via interpolation and known counterexamples to the solvability of the Poisson problem for the Laplacian in Lipschitz domains (cf. [86, Theorem A(1), p. 163]) one can show that the invertibility results from Theorem 7.1.13 may fail to materialize for an arbitrary given smoothness exponent $s \in (0,1)$ if one no longer insists on imposing any flatness assumptions on the underlying domain.

We also wish to stress that in Theorem 7.1.13 no topological assumptions are made on the set Ω, no symmetry is required for the system L, and no positive definiteness (or positive semi-definiteness) is assumed for any of the coefficient tensors in $\mathfrak{A}_L$.

Proof of Theorem 7.1.13 Choose $\delta_0 \in (0,1)$ small so that the conclusions in Theorem 3.4.2 hold (in particular, Ω is a UR domain), Theorem 7.1.5 is applicable (both for L and $L^\top$), and the invertibility results from Theorem 7.1.9 hold (all formulated in some sufficiently large intervals of indices). Next, recall from Proposition 1.4.1 that we presently have $\mathfrak{A}_L^{\rm dis} \neq \varnothing$ and $\mathfrak{A}_{L^\top}^{\rm dis} \neq \varnothing$. In concert, these ensure that the operators (7.1.347)-(7.1.348) are Fredholm with index zero. Since from (7.1.173) and (7.1.169) (written for L in place of $L^\top$) we know that these operators are also injective, it follows that (7.1.347)-(7.1.348) are, in fact, invertible. □

Here is the two-dimensional counterpart of Theorem 7.1.13, dealing with the issue of invertibility for the boundary-to-boundary single layer on end-point Besov space on boundaries of δ-oscillating AR domains in the plane.

Theorem 7.1.14 *Suppose* $\Omega \subseteq \mathbb{R}^2$ *is an Ahlfors regular domain with compact boundary. Abbreviate* $\sigma := \mathcal{H}^1 \lfloor \partial\Omega$ *and denote by* ν *the geometric measure theoretic outward unit normal to* Ω. *Also, let* L *be a homogeneous, second-order, constant complex coefficient,* $M \times M$ *system in* $\mathbb{R}^2$ *such that*

$$L \text{ satisfies the Legendre-Hadamard strong ellipticity condition (cf. [131, Definition 1.3.2]) and one has } \mathfrak{A}_L^{\rm dis} \neq \varnothing. \tag{7.1.349}$$

Consider the boundary-to-boundary single layer S *associated with* Ω *and* L *as in* (A.0.231). *Fix* $s \in (0,1)$ *and define the following augmented versions of* S:

$$\begin{gathered} S_{\rm aug} : \big[B^{1,1}_{-s}(\partial\Omega,\sigma)\big]^M \oplus \mathbb{C}^M \longrightarrow \big[B^{1,1}_{1-s}(\partial\Omega,\sigma)\big]^M \oplus \mathbb{C}^M \\ \text{given by } S_{\rm aug}(f,c) := \Big(Sf + c, (\langle f, \mathbf{e}_j\rangle)_{1\leq j\leq M}\Big) \text{ for each} \\ f \in \big[B^{1,1}_{-s}(\partial\Omega,\sigma)\big]^M \text{ and } c \in \mathbb{C}^M, \end{gathered} \tag{7.1.350}$$

$$\begin{gathered} S_{\rm aug} : \big[B^{\infty,\infty}_{s-1}(\partial\Omega,\sigma)\big]^M \oplus \mathbb{C}^M \longrightarrow \big[B^{\infty,\infty}_{s}(\partial\Omega,\sigma)\big]^M \oplus \mathbb{C}^M \\ \text{given by } S_{\rm aug}(f,c) := \Big(Sf + c, (\langle f, \mathbf{e}_j\rangle)_{1\leq j\leq M}\Big) \text{ for each} \\ f \in \big[B^{\infty,\infty}_{s-1}(\partial\Omega,\sigma)\big]^M \text{ and } c \in \mathbb{C}^M. \end{gathered} \tag{7.1.351}$$

Then there exists some $\delta_0 \in (0,1)$ *depending only on* s, L, *the* ADR *constants of* $\partial\Omega$, *and the relative flatness ratio* $\wp(\Omega)/\text{diam}\,\partial\Omega$ *with the property that if*

$$\mathrm{dist}\Big(\nu, [\mathrm{VMO}(\partial\Omega,\sigma)]^2\Big) < \delta_0 \tag{7.1.352}$$

with the distance considered in $\big[\mathrm{BMO}(\partial\Omega,\sigma)\big]^2$, *hence if* Ω *is a* δ_0*-oscillating* AR *domain (cf. Definition 3.4.1), then* Ω *is a* UR *domain and the operators* (7.1.350)-(7.1.351) *are well defined, linear, bounded, invertible, with bounded inverses.*

As a corollary, the operators (7.1.350)-(7.1.351) *are invertible for any* $s \in (0,1)$ *provided* Ω *is an infinitesimally flat* AR *domain (cf. Definition 3.4.1).*

As noted in Example 3.4.8, if Ω is a chord-arc domain with vanishing constant and compact boundary then Ω is an infinitesimally flat AR domain. Hence, the operators (7.1.350)-(7.1.351) are invertible for any $s \in (0,1)$ in this case.

Proof of Theorem 7.1.14 Select $\delta_0 \in (0,1)$ small so that the conclusions in Theorem 3.4.2 are valid (in particular, Ω is a UR domain, satisfying a two-sided local John condition), Theorem 7.1.5 is applicable (both for L and $L^\top$), and the invertibility results from Theorem 7.1.9 and Theorem 7.1.10 hold. Also, recall from Proposition 1.4.1 that we presently have $\mathfrak{A}^{\rm dis}_L \neq \varnothing$ and $\mathfrak{A}^{\rm dis}_{L^\top} \neq \varnothing$. Granted these results and reasoning as in the proof of Theorem 7.1.10 then shows that the operators (7.1.350)-(7.1.351) are Fredholm with index zero.

Next, from [130, Proposition 11.11.3] and item *(i)* in [130, Proposition 7.7.2] we see that it is possible to choose $s^* \in (s,1)$ and $p \in (1,\infty)$ such that the embeddings $B^{1,1}_{-s}(\partial\Omega,\sigma) \hookrightarrow B^{p,p}_{-s^*}(\partial\Omega,\sigma) \hookrightarrow L^p_{-1}(\partial\Omega,\sigma)$ are well-defined and continuous. In turn, from this and the fact that the operator (7.1.246) is injective, we conclude that $S_{\rm aug}$ is injective in the context of (7.1.350) as well. Since this is also Fredholm with index zero, $S_{\rm aug}$ is ultimately invertible in the context of (7.1.350).

From [130, Proposition 11.11.3] and item *(i)* in [130, Proposition 7.7.2] we also see that $B^{\infty,\infty}_{s-1}(\partial\Omega,\sigma) \hookrightarrow B^{p,p}_{s_*-1}(\partial\Omega,\sigma) \hookrightarrow L^p_{-1}(\partial\Omega,\sigma)$ are are well-defined continuous embeddings for each $s_* \in (0,s)$ and $p \in (1,\infty)$. Granted this, the fact that the operator (7.1.246) is injective, we conclude that $S_{\rm aug}$ is injective in the context of (7.1.351). Given that, as already noted, the latter operator is Fredholm with index zero, it must be invertible. □

We next prove the invertibility of the single layer potential operator on the scale of Besov spaces on boundaries of δ-oscillating AR domains, with δ sufficiently small. Later on, in Theorem 7.2.4, we shall augment this result by establishing the invertibility of the single layer on energy spaces (and small perturbations thereof) in the class of bounded two-sided NTA domains with Ahlfors regular boundaries, without imposing any flatness assumptions on the underlying domain.

Theorem 7.1.15 *Fix* $n \in \mathbb{N}$ *with* $n \geq 3$ *and suppose* $\Omega \subseteq \mathbb{R}^n$ *is an Ahlfors regular domain with compact boundary. Abbreviate* $\sigma := \mathcal{H}^{n-1}\lfloor\partial\Omega$ *and denote by* ν *the geometric measure theoretic outward unit normal to* Ω*. Also, let* L *be a homogeneous, second-order, constant complex coefficient,* $M \times M$ *system in* $\mathbb{R}^n$ *such that*

L satisfies the Legendre-Hadamard strong ellipticity condition (cf. [131, Definition 1.3.2]*) and one has* $\mathfrak{A}^{\rm dis}_L \neq \varnothing$. (7.1.353)

Let S be the boundary-to-boundary single layer potential operator associated with Ω and L as in (A.0.231). *Finally, fix $p \in (1, \infty)$, $q \in (0, \infty)$, and $s \in (0, 1)$.*

Then there exists a threshold $\delta_0 \in (0, 1)$ depending only on n, p, q, s, L, the ADR *constants of $\partial\Omega$, and the relative flatness ratio $\wp(\Omega)/\mathrm{diam}\,\partial\Omega$ such that if*

$$\mathrm{dist}\Big(\nu, [\mathrm{VMO}(\partial\Omega, \sigma)]^n\Big) < \delta_0 \tag{7.1.354}$$

with the distance considered in $\big[\mathrm{BMO}(\partial\Omega, \sigma)\big]^n$, (hence if Ω is a δ_0-oscillating AR *domain; cf. Definition 3.4.1), it follows that Ω is a* UR *domain and*

$$S : \big[B^{p,q}_{s-1}(\partial\Omega, \sigma)\big]^M \longrightarrow \big[B^{p,q}_{s}(\partial\Omega, \sigma)\big]^M \tag{7.1.355}$$

is a well-defined, linear, bounded, invertible operator, with a bounded inverse.

As a corollary, the operator (7.1.355) *is invertible for any $p \in (1, \infty)$, $q \in (0, \infty)$, and $s \in (0, 1)$, provided Ω is an infinitesimally flat* AR *domain (cf. Definition 3.4.1).*

A version of Theorem 7.1.15 for bounded Lipschitz domains in $\mathbb{R}^n$ satisfying (7.1.354) appears in [148, Theorem 6.14, pp. 316–317], where the case of Hermitian self-adjoint, homogeneous, constant complex coefficient $M \times M$ systems of order $2m$, with $n > 2m$, satisfying the Legendre-Hadamard strong ellipticity condition is considered (compare with (7.1.353)); see also [148, Theorem 6.12, p. 313].

We also note that in Theorem 7.1.15 no topological assumptions are made on the set Ω, no symmetry is required for the system L, and no positive definiteness (or positive semi-definiteness) is assumed for any of the coefficient tensors in $\mathfrak{A}_L$.

Proof of Theorem 7.1.15 Note that (7.1.101) holds, thanks to Proposition 1.4.1. Take $\delta_0 \in (0, 1)$ sufficiently small as to ensure that the conclusions in Theorem 3.4.2 hold (in particular, Ω is a UR domain), the conclusions in Theorem 7.1.5 are valid, and that the invertibility results from Theorem 7.1.9 hold (all formulated in some sufficiently large intervals of indices). Together, these prove that the operator (7.1.355) is Fredholm with index zero. Fix some $s_0 \in (0, s)$. By eventually decreasing the value of δ we may also guarantee that S is injective on $\big[B^{1,1}_{s_0-1}(\partial\Omega, \sigma)\big]^M$ (cf. (7.1.347)). Via embeddings (cf. item *(i)* in [130, Proposition 7.7.2]), this further implies that the operator S is injective in the context of (7.1.355). Ultimately, this shows that (7.1.355) is actually invertible. □

Our next proposition shows that, in the absence of any flatness demands imposed on the boundary, the invertibility result in Theorem 7.1.15 may fail.

Proposition 7.1.16 *Fix $n \in \mathbb{N}$ with $n \geq 3$. Pick $p \in (1, \infty)$ and $s \in (0, 1)$ so that*

$$\textit{either}\ \max\big\{\tfrac{2}{p}, \tfrac{1}{2}\big\} < s, \ \textit{or}\ s < \min\big\{\tfrac{2}{p} - 1, \tfrac{1}{2}\big\}. \tag{7.1.356}$$

Then there exists a bounded Lipschitz domain $\Omega \subseteq \mathbb{R}^n$ such that, if $\sigma := \mathcal{H}^{n-1}\lfloor\partial\Omega$ and S_Δ is the boundary-to-boundary single layer potential operator associated with Ω and the Laplacian, then the operator

$$S_\Delta : B^{p,p}_{s-1}(\partial\Omega, \sigma) \longrightarrow B^{p,p}_{s}(\partial\Omega, \sigma) \tag{7.1.357}$$

(which is well defined, linear, and bounded by [132, (4.3.40)]*) fails to be invertible.*

Proof To fix ideas, suppose first that $\max\left\{\frac{2}{p}, \frac{1}{2}\right\} < s$. Then [86, Theorem 1.2(a), p. 167] guarantees the existence of a bounded Lipschitz domain $\Omega \subseteq \mathbb{R}^n$ and some function $f \in \mathscr{C}^\infty(\overline{\Omega})$ with the property that the unique solution of

$$u \in H^1(\Omega) \text{ with } \Delta u = f \text{ in } \Omega \text{ and } \operatorname{Tr}_{\Omega\to\partial\Omega} u = 0 \text{ on } \partial\Omega \tag{7.1.358}$$

does not belong to $F^{p,2}_{s+1/p}(\Omega)$. Seeking a contradiction, assume the operator (7.1.357) is invertible. From [56] (cf. also (7.2.29)) we know that

$$S_\Delta : B^{2,2}_{-1/2}(\partial\Omega, \sigma) \longrightarrow B^{2,2}_{1/2}(\partial\Omega, \sigma) \tag{7.1.359}$$

is invertible. Since the $B^{p,p}_{s-1}(\partial\Omega, \sigma) \hookrightarrow B^{2,2}_{-1/2}(\partial\Omega, \sigma)$ by the present assumptions, simple functional analysis ensures that

$$\begin{aligned}&\text{the inverse of (7.1.359) agrees with the inverse of (7.1.357)}\\&\text{on the common domain (i.e., on the space } B^{p,p}_{s-1}(\partial\Omega, \sigma)).\end{aligned} \tag{7.1.360}$$

To proceed, let $F \in \mathscr{C}^\infty_c(\mathbb{R}^n)$ be such that $f = F\big|_\Omega$. Let $\mathscr{S}_\Delta$ be the boundary-to-domain single layer potential operator associated with Ω and the Laplacian. Finally, let E_Δ be the standard fundamental solution for the Laplacian (cf. [132, (1.1.1)]). Then from (7.1.359), [132, (4.3.26)] (with $s := 1/2$, $p := 2$, $q := 2$), [130, (9.2.22)] (with $s := 1$), and [132, (4.3.44)] it follows that the function

$$u := (F * E_\Delta)\big|_\Omega - \mathscr{S}_\Delta\Big(S_\Delta^{-1}\Big(\operatorname{Tr}_{\Omega\to\partial\Omega}\big((F * E_\Delta)\big|_\Omega\big)\Big)\Big) \text{ in } \Omega, \tag{7.1.361}$$

with S_Δ^{-1} the inverse of (7.1.359), is the unique solution of (7.1.358). Hence, on the one hand, u from (7.1.361) does not belong to $F^{p,2}_{s+1/p}(\Omega)$. On the other hand, in view of the fact that the function $(F * E_\Delta)\big|_\Omega$ belongs to $\mathscr{C}^\infty(\overline{\Omega})$ (which also entails $\operatorname{Tr}_{\Omega\to\partial\Omega}\big((F * E_\Delta)\big|_\Omega\big) \in \operatorname{Lip}(\partial\Omega)$), the compatibility result in (7.1.360), and the mapping properties of $\mathscr{S}_\Delta$ (cf. [132, (4.3.26)]) we conclude that u from (7.1.361) belongs to $F^{p,2}_{s+1/p}(\Omega)$. This contradiction shows that S_Δ fails to be invertible in the context of (7.1.357) when $\max\left\{\frac{2}{p}, \frac{1}{2}\right\} < s$. Finally, the failure of S_Δ to be invertible in the context of (7.1.357) when $s < \min\left\{\frac{2}{p} - 1, \frac{1}{2}\right\}$ is seen from what we have just proved and duality. □

Our next result deals with the issue of invertibility for the augmented single layer potential operator on the scale of Besov spaces on boundaries of two-dimensional δ-oscillating AR domains, with δ sufficiently small.

Theorem 7.1.17 *Suppose $\Omega \subseteq \mathbb{R}^2$ is an Ahlfors regular domain with compact boundary. Abbreviate $\sigma := \mathcal{H}^1\lfloor\partial\Omega$ and denote by ν the geometric measure theoretic* *outward unit normal to* Ω. *Also, let* L *be a homogeneous, second-order, constant complex coefficient,* $M \times M$ *system in* $\mathbb{R}^2$ *such that*

$$L \text{ satisfies the Legendre-Hadamard strong ellipticity condition (cf. [131, Definition 1.3.2]) and one has } \mathfrak{A}_L^{\rm dis} \neq \emptyset. \tag{7.1.362}$$

Bring in the boundary-to-boundary single layer potential operator S *associated with* Ω *and the system* L *as in* (A.0.231). *Fix* $p \in (1, \infty)$, $q \in (0, \infty)$, *and* $s \in (0, 1)$. *Finally, consider the following augmented version of* S:

$$\begin{gathered} S_{\rm aug} : \big[B^{p,q}_{s-1}(\partial\Omega, \sigma)\big]^M \oplus \mathbb{C}^M \longrightarrow \big[B^{p,q}_{s}(\partial\Omega, \sigma)\big]^M \oplus \mathbb{C}^M \\ \text{given by } S_{\rm aug}(f, c) := \Big(Sf + c, (\langle f, \mathbf{e}_j\rangle)_{1\leq j\leq M}\Big) \text{ for each} \\ f \in \big[B^{p,q}_{s-1}(\partial\Omega, \sigma)\big]^M \text{ and } c \in \mathbb{C}^M, \end{gathered} \tag{7.1.363}$$

Then there exists $\delta_0 \in (0, 1)$ *depending only on* p, q, s, L, *the* ADR *constants of* $\partial\Omega$, *and the relative flatness ratio* $\wp(\Omega)/\mathrm{diam}\,\partial\Omega$ *with the property that if*

$$\mathrm{dist}\Big(\nu, [\mathrm{VMO}(\partial\Omega, \sigma)]^2\Big) < \delta_0 \tag{7.1.364}$$

with the above distance considered in $\big[\mathrm{BMO}(\partial\Omega, \sigma)\big]^2$, *hence if* Ω *is a* δ_0*-oscillating* AR *domain (cf. Definition 3.4.1), then* Ω *is a* UR *domain and the operator* (7.1.363) *is well defined, linear, bounded, invertible, with a bounded inverse.*

As a corollary, the operator (7.1.363) *is invertible for any* $p \in (1, \infty)$, $q \in (0, \infty)$, $s \in (0, 1)$ *provided* Ω *is an infinitesimally flat* AR *domain (cf. Definition 3.4.1).*

It was pointed out in Example 3.4.8 that if Ω is a chord-arc domain with vanishing constant and compact boundary then Ω is an infinitesimally flat AR domain. Consequently, the operator (7.1.363) is invertible for any $p \in (1, \infty)$, $q \in (0, \infty)$, $s \in (0, 1)$ in such a case.

Proof of Theorem 7.1.17 That the augmented single layer (7.1.363) is a Fredholm operator with index zero is seen from an argument very similar to the one employed in the proof of Theorem 7.1.14. To establish injectivity, we now make use of the embedding $B^{p,q}_{s-1}(\partial\Omega, \sigma) \hookrightarrow L^p_{-1}(\partial\Omega, \sigma)$, from [130, Proposition 11.11.3]. □

We now present invertibility results for the boundary-to-boundary single layer potential operator in the context of Morrey and block type spaces. We treat dimensions ≥ 3 in Theorem 7.1.18 below, then consider the two-dimensional case (for the augmented version of the single layer) in Theorem 7.1.19.

Theorem 7.1.18 *Fix* $n \in \mathbb{N}$ *with* $n \geq 3$ *and suppose* $\Omega \subseteq \mathbb{R}^n$ *is an Ahlfors regular domain with compact boundary. Abbreviate* $\sigma := \mathcal{H}^{n-1}\lfloor\partial\Omega$ *and denote by* ν *the geometric measure theoretic outward unit normal to* Ω. *Also, let* L *be a homogeneous, second-order, constant complex coefficient,* $M \times M$ *system in* $\mathbb{R}^n$ *such that*

$$L \text{ satisfies the Legendre-Hadamard strong ellipticity condition (cf. [131, Definition 1.3.2]) and one has } \mathfrak{A}_L^{\rm dis} \neq \emptyset. \tag{7.1.365}$$

Let S be the boundary-to-boundary single layer associated with Ω and L as in (A.0.231). *Finally, fix $p, q \in (1, \infty)$ with $1/p + 1/q = 1$, and $\lambda \in (0, n-1)$.*

Then there exists $\delta_0 \in (0, 1)$ depending only on n, p, λ, L, the ADR *constants of $\partial\Omega$, and the relative flatness ratio $\wp(\Omega)/\mathrm{diam}\,\partial\Omega$ such that if*

$$\mathrm{dist}\Big(\nu, [\mathrm{VMO}(\partial\Omega, \sigma)]^n\Big) < \delta_0 \tag{7.1.366}$$

with the distance considered in $\big[\mathrm{BMO}(\partial\Omega, \sigma)\big]^n$ (hence if Ω is a δ_0-oscillating AR *domain; cf. Definition 3.4.1), it follows that Ω is a* UR *domain and*

$$S : \big[M^{p,\lambda}(\partial\Omega, \sigma)\big]^M \longrightarrow \big[M_1^{p,\lambda}(\partial\Omega, \sigma)\big]^M, \tag{7.1.367}$$

$$S : \big[\mathring{M}^{p,\lambda}(\partial\Omega, \sigma)\big]^M \longrightarrow \big[\mathring{M}_1^{p,\lambda}(\partial\Omega, \sigma)\big]^M, \tag{7.1.368}$$

$$S : \big[\mathcal{B}^{q,\lambda}(\partial\Omega, \sigma)\big]^M \longrightarrow \big[\mathcal{B}_1^{q,\lambda}(\partial\Omega, \sigma)\big]^M, \tag{7.1.369}$$

are well-defined, linear, bounded, invertible operators, with bounded inverses. Also,

$$(S_{L^\top})^\top : \big[M_{-1}^{p,\lambda}(\partial\Omega, \sigma)\big]^M \longrightarrow \big[M^{p,\lambda}(\partial\Omega, \sigma)\big]^M, \tag{7.1.370}$$

$$(S_{L^\top})^\top : \big[\mathring{M}_{-1}^{p,\lambda}(\partial\Omega, \sigma)\big]^M \longrightarrow \big[\mathring{M}^{p,\lambda}(\partial\Omega, \sigma)\big]^M, \tag{7.1.371}$$

$$(S_{L^\top})^\top : \big[\mathcal{B}_{-1}^{q,\lambda}(\partial\Omega, \sigma)\big]^M \longrightarrow \big[\mathcal{B}^{q,\lambda}(\partial\Omega, \sigma)\big]^M, \tag{7.1.372}$$

are well-defined, linear, bounded, invertible operators, with bounded inverses.

As a corollary, the operators (7.1.367)-(7.1.372) *are invertible for any exponents $p \in (1, \infty)$, $q \in (1, \infty)$, $\lambda \in (0, n-1)$ provided Ω is an infinitesimally flat* AR *domain (cf. Definition 3.4.1).*

Proof As seen from Proposition 1.4.1, (7.1.101) holds. Take $\delta_0 \in (0, 1)$ sufficiently small so that the conclusions in Theorem 3.4.2 are valid (in particular, Ω is a UR domain), Theorem 7.1.5 is applicable, and that the invertibility results from Theorem 7.1.9 hold (all formulated in some sufficiently large intervals of indices). These ensure that the operators (7.1.367)-(7.1.369) are all Fredholm with index zero. Since from (7.1.233), (7.1.234) [130, (6.2.25)], and [130, (6.2.71)] we see that these operators are also injective, it follows that (7.1.367)-(7.1.369) are in fact invertible.

Next, the invertibility results claimed in (7.1.370) and (7.1.372) are corollaries of the invertibility results in (7.1.367) and (7.1.369) (considered for $L^\top$ in place of L, which satisfies the same properties as the original system; cf. Proposition 1.4.1) and duality (cf. [130, Propositions 6.2.8 and 6.2.16], and [130, Definition 11.8.9]; see (A.0.175), (A.0.36)). Finally, the invertibility result claimed in (7.1.371) follows from (7.1.370), (7.1.234) (with p replaced by $s := \frac{p(n-1)}{n-1-\lambda}$), [130, Lemma 1.2.20], (A.0.168), and [130, Definition 11.8.12] (cf. (A.0.168)). □

Here is the two-dimensional counterpart of Theorem 7.1.18, dealing with the issue of invertibility for the augmented boundary-to-boundary single layer potential operator in the context of Morrey and block type spaces on the boundaries of δ-oscillating AR domains in the plane.

Theorem 7.1.19 *Suppose $\Omega \subseteq \mathbb{R}^2$ is an Ahlfors regular domain with compact boundary. Abbreviate $\sigma := \mathcal{H}^1 \lfloor \partial\Omega$ and denote by ν the geometric measure theoretic outward unit normal to Ω. Also, let L be a homogeneous, second-order, constant complex coefficient, $M \times M$ system in $\mathbb{R}^2$ such that*

$$L \text{ satisfies the Legendre-Hadamard strong ellipticity condition (cf. [131, Definition 1.3.2]) and one has } \mathfrak{A}_L^{\text{dis}} \neq \emptyset. \tag{7.1.373}$$

Recall the boundary-to-boundary single layer potential operator S associated with Ω and the system L as in (A.0.231). *Fix $p, q \in (1, \infty)$ along with $\lambda \in (0, 1)$. Consider the following augmented versions of S:*

$$S_{\text{aug}} : \big[M^{p,\lambda}(\partial\Omega, \sigma)\big]^M \oplus \mathbb{C}^M \longrightarrow \big[M_1^{p,\lambda}(\partial\Omega, \sigma)\big]^M \oplus \mathbb{C}^M, \tag{7.1.374}$$

$$S_{\text{aug}} : \big[\mathring{M}^{p,\lambda}(\partial\Omega, \sigma)\big]^M \oplus \mathbb{C}^M \longrightarrow \big[\mathring{M}_1^{p,\lambda}(\partial\Omega, \sigma)\big]^M \oplus \mathbb{C}^M, \tag{7.1.375}$$

$$S_{\text{aug}} : \big[\mathcal{B}^{q,\lambda}(\partial\Omega, \sigma)\big]^M \oplus \mathbb{C}^M \longrightarrow \big[\mathcal{B}_1^{q,\lambda}(\partial\Omega, \sigma)\big]^M \oplus \mathbb{C}^M, \tag{7.1.376}$$

acting, in all cases, according to

$$S_{\text{aug}}(f, c) := \Big(Sf + c, \int_{\partial\Omega} f \, d\sigma\Big) \tag{7.1.377}$$

where S is the corresponding operator from (7.1.367)-(7.1.369). *In addition, consider the following augmented versions of S:*

$$\widetilde{S}_{\text{aug}} : \big[M_{-1}^{p,\lambda}(\partial\Omega, \sigma)\big]^M \oplus \mathbb{C}^M \longrightarrow \big[M^{p,\lambda}(\partial\Omega, \sigma)\big]^M \oplus \mathbb{C}^M, \tag{7.1.378}$$

$$\widetilde{S}_{\text{aug}} : \big[\mathring{M}_{-1}^{p,\lambda}(\partial\Omega, \sigma)\big]^M \oplus \mathbb{C}^M \longrightarrow \big[\mathring{M}^{p,\lambda}(\partial\Omega, \sigma)\big]^M \oplus \mathbb{C}^M, \tag{7.1.379}$$

$$\widetilde{S}_{\text{aug}} : \big[\mathcal{B}_{-1}^{q,\lambda}(\partial\Omega, \sigma)\big]^M \oplus \mathbb{C}^M \longrightarrow \big[\mathcal{B}^{q,\lambda}(\partial\Omega, \sigma)\big]^M \oplus \mathbb{C}^M, \tag{7.1.380}$$

acting, in all cases, according to

$$\widetilde{S}_{\text{aug}}(f, c) := \Big((S_{L^\top})^\top f + c, (\langle f, \mathbf{e}_j \rangle)_{1 \leq j \leq M}\Big) \tag{7.1.381}$$

where $(S_{L^\top})^\top$ is the corresponding operator from (7.1.370)-(7.1.372).

Then there exists a threshold $\delta_0 \in (0, 1)$ which depends only on p, q, λ, L, the ADR *constants of $\partial\Omega$, and the relative flatness ratio $\wp(\Omega)/\text{diam}\, \partial\Omega$ so that if*

$$\text{dist}\Big(\nu, [\text{VMO}(\partial\Omega, \sigma)]^2\Big) < \delta_0 \tag{7.1.382}$$

where the distance is considered in the space $\big[\text{BMO}(\partial\Omega, \sigma)\big]^2$, hence if Ω is a δ_0-oscillating AR *domain (cf. Definition 3.4.1), then Ω is a* UR *domain and the operators* (7.1.374)-(7.1.376), (7.1.378)-(7.1.380) *are all well defined, linear, bounded, invertible, with bounded inverses.*

As a corollary, the operators (7.1.374)-(7.1.376), (7.1.378)-(7.1.380) *are all invertible for any choice of* $p, q \in (1, \infty)$ *and* $\lambda \in (0, 1)$ *provided* Ω *is an infinitesimally flat* AR *domain* (*cf. Definition 3.4.1*).

Given that any chord-arc domain with vanishing constant and compact boundary is an infinitesimally flat AR domain (see Example 3.4.8), in the former class of domains the operators (7.1.374)-(7.1.376), (7.1.378)-(7.1.380) are invertible for any $p, q \in (1, \infty)$ and $\lambda \in (0, 1)$.

Proof of Theorem 7.1.19 That the augmented single layers (7.1.374)-(7.1.376), (7.1.378), and (7.1.380) are all Fredholm operators with index zero may be justified reasoning in a very similar fashion to the proof of Theorem 7.1.10. These operators are also injective, thanks to the embeddings in [130, (6.2.7)], [130, (6.2.15)], [130, (6.2.71)], and the injectivity of the augmented single layer in the context of (7.1.245). As such, said operators are all invertible. Finally, that the augmented single layer is also invertible in the context of (7.1.379) follows from (7.1.378), (7.1.246) (with p replaced by $s := \frac{p(n-1)}{n-1-\lambda}$), [130, Lemma 1.2.20], (A.0.168), and [130, Definition 11.8.12] (cf. (A.0.168)). □

Remarkably, there are invertibility results for the single layer operator acting on Generalized Banach Function Spaces, in the context made precise below.

Theorem 7.1.20 *Fix* $n \in \mathbb{N}$ *with*[7] $n \geq 3$ *and let* $\Omega \subseteq \mathbb{R}^n$ *be an Ahlfors regular domain with compact boundary. Abbreviate* $\sigma := \mathcal{H}^{n-1}\lfloor\partial\Omega$ *and denote by* ν *the geometric measure theoretic outward unit normal to* Ω. *Also, let* L *be a homogeneous, second-order, constant complex coefficient,* $M \times M$ *system in* $\mathbb{R}^n$ *such that*

$$L \text{ satisfies the Legendre-Hadamard strong ellipticity condition (cf. [131, Definition 1.3.2]) and one has } \mathfrak{A}_L^{\mathrm{dis}} \neq \varnothing. \tag{7.1.383}$$

Let $\mathbb{X}$ *be a Generalized Banach Function Space on* $(\partial\Omega, \sigma)$ *such that*

$$\mathcal{M}_{\partial\Omega} : \mathbb{X} \to \mathbb{X} \text{ and } \mathcal{M}_{\partial\Omega} : \mathbb{X}' \to \mathbb{X}' \text{ are well-defined and bounded,} \tag{7.1.384}$$

where $\mathcal{M}_{\partial\Omega}$ *is the Hardy-Littlewood maximal operator on* $\partial\Omega$, *and* $\mathbb{X}'$ *is the associated space of* $\mathbb{X}$ (*cf.* [130, Definitions 5.1.4, 5.1.11]). *Define*

$$\mathring{\mathbb{X}} := \overline{L^\infty(\partial\Omega, \sigma)}^{\|\cdot\|_{\mathbb{X}}} \text{ and } (\mathbb{X}')^\circ := \overline{L^\infty(\partial\Omega, \sigma)}^{\|\cdot\|_{\mathbb{X}'}}, \tag{7.1.385}$$

and bring back $\mathbb{X}_1$, $(\mathbb{X}')_1$, $\mathring{\mathbb{X}}_1$, $(\mathbb{X}')^\circ_1$, *the* $\mathbb{X}$*-styled,* $\mathbb{X}'$*-styled,* $\mathring{\mathbb{X}}$*-styled, and* $(\mathbb{X}')^\circ$*-styled Sobolev spaces of order one on* $\partial\Omega$ *from* (7.1.39) (*these are known to be Banach spaces when equipped with natural norms; cf.* (5.4.84)). *Finally, consider the boundary-to-boundary single layer potential operator* S *associated with the set* Ω *and the system* L *as in* (A.0.231).

Then there exists a threshold $\delta_{\mathbb{X}} \in (0, 1)$, *which depends only on* n, L, *the operator norms of* $\mathcal{M}_{\partial\Omega}$ *on* $\mathbb{X}$ *and* $\mathbb{X}'$, *the* ADR *constants of* $\partial\Omega$, *and the relative flatness ratio* $\wp(\Omega)/\mathrm{diam}\,\partial\Omega$, *with the property that if*

[7] see the last part in the statement for the two-dimensional setting

$$\operatorname{dist}\Big(\nu, [\mathrm{VMO}(\partial\Omega, \sigma)]^n\Big) < \delta_{\mathbb{X}} \tag{7.1.386}$$

where the distance is considered in the space $\big[\mathrm{BMO}(\partial\Omega, \sigma)\big]^n$ *(hence if* Ω *is a* $\delta_{\mathbb{X}}$*-oscillating* AR *domain; cf. Definition 3.4.1), it follows that* Ω *is a* UR *domain and the operators below are invertible:*

$$S : \big[\mathbb{X}\big]^M \to \big[\mathbb{X}_1\big]^M, \quad S : \big[\mathbb{X}'\big]^M \to \big[(\mathbb{X}')_1\big]^M, \tag{7.1.387}$$

$$S : \big[\mathring{\mathbb{X}}\big]^M \to \big[\mathring{\mathbb{X}}_1\big]^M, \quad S : \big[(\mathbb{X}')^{\circ}\big]^M \to \big[(\mathbb{X}')^{\circ}_1\big]^M. \tag{7.1.388}$$

As a corollary, the operators (7.1.387)-(7.1.388) *are invertible whenever* Ω *is an infinitesimally flat* AR *domain (cf. Definition 3.4.1).*

In the case when $n = 2$*, under the assumption made in* (7.1.386) *the following augmented versions of* S *are invertible operators:*

$$S_{\text{aug}} : \big[\mathbb{X}\big]^M \oplus \mathbb{C}^M \longrightarrow \big[\mathbb{X}_1\big]^M \oplus \mathbb{C}^M, \tag{7.1.389}$$

$$S_{\text{aug}} : \big[\mathbb{X}'\big]^M \oplus \mathbb{C}^M \longrightarrow \big[(\mathbb{X}')_1\big]^M \oplus \mathbb{C}^M, \tag{7.1.390}$$

$$S_{\text{aug}} : \big[\mathring{\mathbb{X}}\big]^M \oplus \mathbb{C}^M \longrightarrow \big[\mathring{\mathbb{X}}_1\big]^M \oplus \mathbb{C}^M, \tag{7.1.391}$$

$$S_{\text{aug}} : \big[(\mathbb{X}')^{\circ}\big]^M \oplus \mathbb{C}^M \longrightarrow \big[(\mathbb{X}')^{\circ}_1\big]^M \oplus \mathbb{C}^M, \tag{7.1.392}$$

acting, in all cases, according to

$$S_{\text{aug}}(f, c) := \Big(Sf + c, \int_{\partial\Omega} f \, d\sigma\Big) \tag{7.1.393}$$

where S *is the corresponding operator from* (7.1.387)-(7.1.388).

Finally, all conclusions above hold whenever $\mathbb{X}$ *is a rearrangement invariant Banach function space whose Boyd indices satisfy* $1 < p_{\mathbb{X}} \leq q_{\mathbb{X}} < \infty$*. As a corollary, said conclusions are true whenever* $\mathbb{X} := L^{\Phi}(\partial\Omega, \sigma)$*, the Orlicz space associated as in* (A.0.133) *with the measure space* $(\partial\Omega, \sigma)$ *and any Young function* Φ *satisfying* $1 < i(\Phi) \leq I(\Phi) < \infty$*.*

Proof It has been noted in Theorem 3.4.2 that taking $\delta_{\mathbb{X}} \in (0, 1)$ small enough makes Ω a UR domain. First, assume that $n \geq 3$. In view of this and (7.1.383), we may invoke Proposition 1.4.1 to conclude that (7.1.184) holds. As such, Theorem 7.1.7 applies and gives that it is possible to take $\delta_{\mathbb{X}} \in (0, 1)$ small enough as to guarantee that the operators (7.1.387)-(7.1.388) are Fredholm with index zero. With this in hand, and also relying on (7.1.198), as well as the invertibility results from Theorem 7.1.9, we finally conclude that we may take $\delta_{\mathbb{X}} \in (0, 1)$ small enough as to ensure that the operators (7.1.387)-(7.1.388) are actually invertible.

There remains to treat the case $n = 2$. Take $\delta_{\mathbb{X}} \in (0, 1)$ small enough so that the conclusions in Theorem 7.1.7 are valid. To show that (7.1.389) is an isomorphism, the idea is to decompose the augmented single layer operator (7.1.389) as

$$S_{\text{aug}} = T + Q \tag{7.1.394}$$

with

$$T : [\mathbb{X}]^M \oplus \mathbb{C}^M \longrightarrow [\mathbb{X}_1]^M \oplus \mathbb{C}^M \text{ given by}$$
$$T(f, c) := (Sf, 0) \text{ for each } f \in [\mathbb{X}]^M \text{ and } c \in \mathbb{C}^M, \tag{7.1.395}$$

$$Q : [\mathbb{X}]^M \oplus \mathbb{C}^M \longrightarrow [\mathbb{X}_1]^M \oplus \mathbb{C}^M \text{ defined as}$$
$$Q(f, c) := \Big(c, \int_{\partial\Omega} f \, d\sigma\Big) \text{ for each } f \in [\mathbb{X}]^M \text{ and } c \in \mathbb{C}^M. \tag{7.1.396}$$

As is apparent from the above definitions, T, Q are well-defined, linear, and bounded operators. In addition, Q is a finite-rank operator, hence actually compact. Also,

$$\mathrm{Im}\Big(T : [\mathbb{X}]^M \oplus \mathbb{C}^M \to [\mathbb{X}_1]^M \oplus \mathbb{C}^M\Big) = \mathrm{Im}\Big(S : [\mathbb{X} \to [\mathbb{X}_1]^M\Big) \oplus \{0\},$$
$$\mathrm{Ker}\Big(T : [\mathbb{X}]^M \oplus \mathbb{C}^M \to [\mathbb{X}_1]^M \oplus \mathbb{C}^M\Big) = \mathrm{Ker}\Big(S : [\mathbb{X}]^M \to [\mathbb{X}_1]^M\Big) \oplus \mathbb{C}^M.$$

Based on these and the fact that S in (7.1.188) is Fredholm with index zero we conclude that T in (7.1.395) is Fredholm with index zero. Keeping in mind that Q is compact, from this, (7.1.394), and item *(6)* in [130, Theorem 2.1.2] we then see that S_{aug} is a Fredholm operator with index zero. In addition, [132, (1.7.11)] and [130, (5.2.92)] imply that S_{aug} is injective. Ultimately, this shows that S_{aug} is an invertible operator in the context of (7.1.389). That S_{aug} is also invertible in each of the contexts described in (7.1.390)-(7.1.392) is established in a similar manner.

Finally, the claims in the very last paragraph in the statement are consequences of [130, Proposition 5.3.14], [130, Proposition 5.3.15], and [130, (5.3.62)]. □

Let $\Omega \subseteq \mathbb{R}^n$ be an Ahlfors regular domain with compact boundary. Denote by ν the geometric measure theoretic outward unit normal to Ω and abbreviate $\sigma := \mathcal{H}^{n-1}\lfloor\partial\Omega$. Recall from (7.1.265) the space

$$\mathbb{C}_{\partial\Omega} := \mathrm{span}_{\mathbb{C}} \{\mathbf{1}_\Sigma : \Sigma \text{ connected component of } \partial\Omega\} \tag{7.1.397}$$

and, for any given space $X \subseteq [L^1(\partial\Omega, \sigma)]^n$, set

$$X \cap \langle \nu\mathbb{C}_{\partial\Omega}\rangle^{\perp} := \Big\{\vec{f} \in X : \int_{\partial\Omega} \langle \vec{f}, \eta\rangle \, d\sigma = 0, \ \forall \eta \in \nu\mathbb{C}_{\partial\Omega}\Big\}. \tag{7.1.398}$$

Likewise, with $\mathbb{C}_{\partial\Omega_\pm}$ as in (7.1.265), for any $X \subseteq [L^1(\partial\Omega, \sigma)]^n$ we define

$$X \cap \langle \nu\mathbb{C}_{\partial\Omega_\pm}\rangle^{\perp} := \Big\{\vec{f} \in X : \int_{\partial\Omega} \langle \vec{f}, \eta\rangle \, d\sigma = 0, \ \forall \eta \in \nu\mathbb{C}_{\partial\Omega_\pm}\Big\}. \tag{7.1.399}$$

In particular, the piece of notation introduced in (7.1.398) features in the statement of the following invertibility result for the boundary-to-boundary single layer potential operator for the Stokes system acting from a suitable class of Generalized Banach Function Spaces.

Theorem 7.1.21 *Pick $n \in \mathbb{N}$ with $n \geq 2$ and let $\Omega \subseteq \mathbb{R}^n$ be an Ahlfors regular domain with compact boundary. Abbreviate $\sigma := \mathcal{H}^{n-1}\lfloor\partial\Omega$ and denote by ν the* *geometric measure theoretic outward unit normal to* Ω. *As in the past, adopt the convention that*

$$S_{\text{Stokes}} \text{ stands for the boundary-to-boundary single layer potential operator for the Stokes system associated with } \Omega \text{ as in (4.2.109).} \tag{7.1.400}$$

Let $\mathbb{X}$ *be a Generalized Banach Function Space on* $(\partial\Omega, \sigma)$ *so that* $\mathcal{M}_{\partial\Omega} : \mathbb{X} \to \mathbb{X}$ *and* $\mathcal{M}_{\partial\Omega} : \mathbb{X}' \to \mathbb{X}'$ *are well-defined and bounded, where* $\mathcal{M}_{\partial\Omega}$ *is the Hardy-Littlewood maximal operator on* $\partial\Omega$, *and* $\mathbb{X}'$ *is the associated space of* $\mathbb{X}$ *(cf.* [130, Definitions 5.1.4, 5.1.11]*). Define* $\mathring{\mathbb{X}}$, $(\mathbb{X}')^{\circ}$ *as in* (7.1.186) *and bring back the Generalized Banach Function Space styled Sobolev spaces of order one* $\mathbb{X}_1$, $(\mathbb{X}')_1$, $\mathring{\mathbb{X}}_1$, $(\mathbb{X}')_1^{\circ}$ *from* (7.1.39) *(which are known to be Banach spaces when equipped with natural norms; cf.* (5.4.84)*).*

Then there exists a threshold $\delta_{\mathbb{X}} \in (0, 1)$, *which depends only on n, the operator norms of* $\mathcal{M}_{\partial\Omega}$ *on* $\mathbb{X}$ *and* $\mathbb{X}'$, *the* ADR *constants of* $\partial\Omega$, *and the relative flatness ratio* $\wp(\Omega)/\operatorname{diam}\partial\Omega$, *with the property that if*

$$\operatorname{dist}\Big(\nu,\ [\mathrm{VMO}(\partial\Omega,\sigma)]^n\Big) < \delta_{\mathbb{X}} \tag{7.1.401}$$

with the distance considered in $\big[\mathrm{BMO}(\partial\Omega,\sigma)\big]^n$ *(hence if* Ω *is a* $\delta_{\mathbb{X}}$*-oscillating* AR *domain; cf. Definition 3.4.1), it follows that* Ω *is a two-sided* NTA *domain,*

$$\operatorname{Ker}\Big(S_{\text{Stokes}} : [\mathbb{X}]^n \to \big[\mathbb{X}_1\big]^n\Big) = \nu\mathcal{C}_{\partial\Omega} \ \text{ if } n \geq 3, \tag{7.1.402}$$

and

$$\begin{gathered} \big[S_{\text{Stokes}}\big] : [\mathbb{X}]^n \big/ \nu\mathcal{C}_{\partial\Omega} \longrightarrow \big[\mathbb{X}_1\big]^n \cap \big\langle \nu\mathcal{C}_{\partial\Omega}\big\rangle^{\perp} \\ \text{defined as } \big[S_{\text{Stokes}}\big][\vec{f}\,] := S_{\text{Stokes}}\vec{f} \text{ for each } \vec{f} \in [\mathbb{X}]^n \\ \text{is a well-defined linear isomorphism provided } n \geq 3. \end{gathered} \tag{7.1.403}$$

Furthermore, under the assumption made in (7.1.401), *the following augmented version of* S_{Stokes} *is an invertible operator in the case when* $n = 2$*:*

$$\begin{gathered} S^{\text{aug}}_{\text{Stokes}} : \Big([\mathbb{X}]^2 \big/ \nu\mathcal{C}_{\partial\Omega}\Big) \oplus \mathbb{C}^2 \longrightarrow \Big(\big[\mathbb{X}_1\big]^2 \cap \big\langle \nu\mathcal{C}_{\partial\Omega}\big\rangle^{\perp}\Big) \oplus \mathbb{C}^2 \\ \text{given by } S^{\text{aug}}_{\text{Stokes}}([\vec{f}\,], c) := \Big(S_{\text{Stokes}}\vec{f} + c,\ \int_{\partial\Omega} \vec{f}\, d\sigma\Big) \\ \text{for each function } \vec{f} \in [\mathbb{X}]^2 \text{ and constant } c \in \mathbb{C}^2. \end{gathered} \tag{7.1.404}$$

In particular, (7.1.402)-(7.1.403) *are true and* (7.1.404) *is invertible for* $n = 2$ *if* Ω *is an infinitesimally flat* AR *domain (cf. Definition 3.4.1). As a corollary,*

$$\text{granted (7.1.401), it follows that (7.1.402)-(7.1.403) are true and (7.1.404) is invertible for } n = 2 \text{ if } \mathbb{X} \text{ is a rearrangement invariant Banach function space with Boyd indices satisfying } 1 < p_{\mathbb{X}} \leq q_{\mathbb{X}} < \infty. \tag{7.1.405}$$

In particular,

under the assumption (7.1.401)*, it follows that* (7.1.402)-(7.1.403) *are true and* (7.1.404) *is invertible for* $n = 2$ *if* $\mathbb{X} := L^{\Phi}(\partial\Omega, \sigma)$*, the Orlicz space associated as in* (A.0.133) *with the measure space* $(\partial\Omega, \sigma)$ *and any Young function* Φ *with* $1 < i(\Phi) \leq I(\Phi) < \infty$. (7.1.406)

In addition, similar results as above are true with $\mathring{\mathbb{X}} := \overline{L^{\infty}(\partial\Omega, \sigma)}^{\|\cdot\|_{\mathbb{X}}}$ *replacing* $\mathbb{X}$*. The conclusions in* (7.1.402)-(7.1.404) *are also true if* $\mathbb{X}$ *is any of these spaces:*

$$
\begin{aligned}
&L^{p,q}(\partial\Omega, \sigma) \text{ with } p \in (1, \infty) \text{ and } q \in (1, \infty),\\
&L^{p}(\partial\Omega, w) \text{ with } p \in (1, \infty) \text{ and } w \in A_p(\partial\Omega, \sigma),\\
&\mathring{M}^{p,\lambda}(\partial\Omega, \sigma) \text{ with } p \in (1, \infty) \text{ and } \lambda \in (0, n-1),\\
&M^{p,\lambda}(\partial\Omega, \sigma) \text{ with } p \in (1, \infty) \text{ and } \lambda \in (0, n-1),\\
&\mathcal{B}^{q,\lambda}(\partial\Omega, \sigma) \text{ with } q \in (1, \infty) \text{ and } \lambda \in (0, n-1).
\end{aligned}
\tag{7.1.407}
$$

Finally, via duality one obtains invertibility results for the boundary-to-boundary single layer potential operator for the Stokes system S_{Stokes} *when* $n \geq 3$*, and its augmented version* $S^{\text{aug}}_{\text{Stokes}}$ *when* $n = 2$*, now acting from negative Sobolev spaces on* $\partial\Omega$ *(styled after the spaces in* (7.1.407)*).*

Proof Theorem 3.4.2 ensures that taking $\delta_{\mathbb{X}} \in (0, 1)$ small enough makes Ω a two-sided NTA domain. Assuming this is the case, the identification of the kernel in (7.1.402) is proved much as in [160, (5.127), p. 97] (where the case of Lipschitz domains is treated), making use of [130, (5.2.101)] and the last claim in Theorem 7.1.5 (see also the proof of Theorem 7.1.20). In turn, (7.1.402) implies that the operator in (7.1.403) is injective, while Theorem 7.1.8 implies that said operator is Fredholm with index zero. Hence, the claim in (7.1.403) follows. The fact that (7.1.404) is an invertible operator for $n = 2$ may be shown based on Theorem 7.1.8, proceeding as in the proof of Theorem 7.1.10 (see also [160, Theorem 5.4.4, pp. 106-107]). Here we only wish to mention that the last three choices in (7.1.407) are allowable thanks to (A.0.168) and [130, (6.2.7)], as well as [130, Proposition 6.2.17], and [130, Corollaries 6.2.11, 6.2.13]. □

It is also of interest to prove invertibility results in the spirit of Theorem 7.1.21 for the boundary-to-boundary single layer potential operator for the Stokes system acting on Besov spaces.

Theorem 7.1.22 *Choose* $n \in \mathbb{N}$ *with* $n \geq 2$ *and assume* $\Omega \subseteq \mathbb{R}^n$ *is an Ahlfors regular domain with compact boundary. Abbreviate* $\sigma := \mathcal{H}^{n-1}\lfloor\partial\Omega$ *and denote by* ν *the geometric measure theoretic outward unit normal to* Ω*. Fix* $p, q \in (1, \infty)$ *and* $s \in (0, 1)$ *and, once more, adopt the convention made in* (7.1.400).

Then there exists a threshold $\delta \in (0, 1)$*, which depends only on* n*,* p*,* q*,* s*, the* ADR *constants of* $\partial\Omega$*, and the relative flatness ratio* $\wp(\Omega)/\operatorname{diam}\partial\Omega$*, such that if*

$$
\operatorname{dist}\Big(\nu, [\mathrm{VMO}(\partial\Omega, \sigma)]^n\Big) < \delta \tag{7.1.408}
$$

with the distance considered in $\big[\mathrm{BMO}(\partial\Omega,\sigma)\big]^n$ *(hence if* Ω *is a* δ*-oscillating* AR *domain; cf. Definition 3.4.1), it follows that* Ω *is a two-sided* NTA *domain,*

$$\operatorname{Ker}\Big(S_{\text{Stokes}}:\big[B^{p,q}_{s-1}(\partial\Omega,\sigma)\big]^n\to\big[B^{p,q}_{s}(\partial\Omega,\sigma)\big]^n\Big)=\nu\mathbb{C}_{\partial\Omega}\ \textit{if } n\geq 3, \tag{7.1.409}$$

and

$$\begin{gathered}S_{\text{Stokes}}:\big[B^{p,q}_{s-1}(\partial\Omega,\sigma)\big]^n\big/\nu\mathbb{C}_{\partial\Omega}\longrightarrow\big[B^{p,q}_{s}(\partial\Omega,\sigma)\big]^n\cap\big\langle\nu\mathbb{C}_{\partial\Omega}\big\rangle^{\perp}\\ \textit{defined as } S_{\text{Stokes}}[\vec{f}\,]:=S_{\text{Stokes}}\vec{f}\ \textit{for each } \vec{f}\in\big[B^{p,q}_{s-1}(\partial\Omega,\sigma)\big]^n\\ \textit{is a well-defined linear isomorphism provided } n\geq 3.\end{gathered} \tag{7.1.410}$$

In addition, granted the assumption made in (7.1.408)*, the following augmented version of* S_{Stokes} *is an invertible operator in the case when* $n=2$*:*

$$\begin{gathered}S^{\text{aug}}_{\text{Stokes}}:\Big(\big[B^{p,q}_{s-1}(\partial\Omega,\sigma)\big]^2\big/\nu\mathbb{C}_{\partial\Omega}\Big)\oplus\mathbb{C}^2\to\Big(\big[B^{p,q}_{s}(\partial\Omega,\sigma)\big]^2\cap\big\langle\nu\mathbb{C}_{\partial\Omega}\big\rangle^{\perp}\Big)\oplus\mathbb{C}^2\\ \textit{given by } S^{\text{aug}}_{\text{Stokes}}([\vec{f}\,],c):=\Big(S_{\text{Stokes}}\vec{f}+c\,,\,(\langle\vec{f},\mathbf{e}_j\rangle)_{1\leq j\leq 2}\Big)\\ \textit{for each given } \vec{f}\in\big[B^{p,q}_{s-1}(\partial\Omega,\sigma)\big]^2 \textit{ and } c\in\mathbb{C}^2.\end{gathered} \tag{7.1.411}$$

Finally, similar results are true for the end-point cases $p=q=1$ *and* $p=q=\infty$*.*

Proof According to Theorem 3.4.2, choosing $\delta\in(0,1)$ small enough guarantees that Ω a two-sided NTA domain. By eventually decreasing the value of $\delta\in(0,1)$, we may additionally ensure the validity of the very last claim in Theorem 7.1.5 (as well as its proof), and also that the Fredholmness plus index zero results from Theorem 7.1.8 hold. In concert, these permit us to deduce that

$$\begin{gathered}S_{\text{Stokes}}:\big[B^{p,q}_{s-1}(\partial\Omega,\sigma)\big]^n\longrightarrow\big[B^{p,q}_{s}(\partial\Omega,\sigma)\big]^n\\ \text{is a Fredholm operator with index zero.}\end{gathered} \tag{7.1.412}$$

From (7.1.412), (7.1.402), and the abstract regularity result from [136, Lemma 3.10, pp. 136-137] we then conclude that (7.1.409) holds. In turn, this may be used in concert with (7.1.412) to show that the operator in (7.1.410) is, first, Fredholm with index zero and, second, injective, hence ultimately an isomorphism. That (7.1.411) is an isomorphism when $n=2$ may be justified by reasoning as in the proof of Theorem 7.1.10. The end-point cases $p=q=1$ and $p=q=\infty$ are similar. □

We continue our discussion by introducing notation which is going to be useful in the formulation of Theorem 7.1.23, a little further below. Let Ψ be the complex $n(n+1)/2$-dimensional linear space consisting of all smooth $\mathbb{C}^n$-valued functions $\psi=(\psi_j)_{1\leq j\leq n}$ defined in $\mathbb{R}^n$ and satisfying

$$\partial_j\psi_k+\partial_k\psi_j=0\ \text{ in } \mathbb{R}^n \text{ for each } j,k\in\{1,\dots,n\}. \tag{7.1.413}$$

This may be concretely identified as the space of functions

$$\Psi = \Big\{\psi(x) = Ax + \vec{a} : A \in \mathbb{C}^{n\times n} \text{ antisymmetric matrix, and } \vec{a} \in \mathbb{C}^n\Big\}. \quad (7.1.414)$$

Note that all functions in Ψ are $\mathbb{C}^n$-valued divergence-free harmonic functions.

Given an arbitrary open set $\Omega \subseteq \mathbb{R}^n$, define

$$\Omega_+ := \Omega \text{ and } \Omega_- := \mathbb{R}^n \setminus \overline{\Omega}. \quad (7.1.415)$$

In the treatment of the slip double layer potential operator, the following function spaces arise naturally:

$$\Psi(\Omega_\pm) := \operatorname{span}_{\mathbb{C}} \Big\{\psi \cdot \mathbf{1}_O : \psi \in \Psi,\ O \text{ bounded connected component of } \Omega_\pm\Big\},$$
$$\Psi(\partial\Omega_\pm) := \operatorname{span}_{\mathbb{C}} \Big\{\psi \cdot \mathbf{1}_{\partial O} : \psi \in \Psi,\ O \text{ bounded connected component of } \Omega_\pm\Big\}. \quad (7.1.416)$$

Also, for any space X of distributions on $\partial\Omega$ we agree to abbreviate

$$[X]^n \cap \Psi(\partial\Omega_\pm)^\perp := \Big\{\vec{f} \in [X]^n : \langle \vec{f}, \chi\rangle = 0 \text{ for each } \chi \in \Psi(\partial\Omega_\pm)\Big\}. \quad (7.1.417)$$

Finally, we set

$$\Psi(\partial\Omega) := \operatorname{span}_{\mathbb{C}} \Big\{\psi \cdot \mathbf{1}_\Sigma : \psi \in \Psi,\ \Sigma \text{ connected component of } \partial\Omega\Big\}. \quad (7.1.418)$$

Assume next that

$\Omega \subseteq \mathbb{R}^n$, with ≥ 2, is an Ahlfors regular domain with compact boundary and with the property that both the connected components of Ω as well as those of $\mathbb{R} \setminus \overline{\Omega}$ are separated (in the sense that they have mutually disjoint closures). (7.1.419)

Then Ω is as in (7.1.271), thanks to item *(4)* in [129, Lemma 5.10.9], so Alexander's relation holds (cf. (7.1.272)). We claim that

$$\Psi(\partial\Omega_+) \oplus \Psi(\partial\Omega_-) = \Psi(\partial\Omega) \quad (7.1.420)$$

where the sum is direct (but not necessarily orthogonal). Indeed, expect for the fact that the sum is direct, this is a consequence of (7.1.272) and the above definitions. There remains to show that the sum in (7.1.420) is direct. To this end, assume $\psi \in \Psi(\partial\Omega_+) \cap \Psi(\partial\Omega_-)$ is arbitrary, and denote by $\psi_\pm \in \Psi(\Omega_\pm)$ the natural extensions of ψ in $\Omega_\pm$. If we now define $\widetilde{\psi} := \psi_\pm$ in $\Omega_\pm$, the fact that $\psi_+|_{\partial\Omega} = \psi_-|_{\partial\Omega}$ ensures that (7.1.413) is satisfied by this function in $\mathbb{R}^n$, in the sense of distributions. Hence, $\widetilde{\psi} \in \Psi$, and since it has compact support, ψ must vanish in $\mathbb{R}^n$. This forces $\psi = 0$ on $\partial\Omega$, finishing the proof of (7.1.420).

For further use, let us also note here that, given Ω as in (7.1.419), if ν denotes its geometric measure theoretic outward unit normal and $\sigma := \mathcal{H}^{n-1}\lfloor\partial\Omega$, then

$$\int_{\partial\Omega} \langle \chi, \eta\rangle \, d\sigma = 0, \ \forall\, \chi \in \Psi(\partial\Omega), \quad \forall\, \eta \in \nu\mathbb{C}_{\partial\Omega}. \quad (7.1.421)$$

This follows from (7.1.420), the De Giorgi-Federer version of the Divergence Formula recalled in [129, Theorem 1.1.1], and the fact that all vector fields in Ψ are divergence-free. It is also useful to remark that

$$\text{if } \Omega \text{ is as in (7.1.419) then for any } \psi \in \Psi(\Omega_\pm) \text{ the pair } (\psi, 0) \text{ is a nullsolution of the Stokes system in } \Omega_\pm \text{ with the property that } \partial_\nu^{\text{slip}}(\psi, 0) = 0 \text{ on } \partial\Omega. \tag{7.1.422}$$

As a consequence of this and the integral representation formulas from [132, Theorem 6.2.1] we therefore obtain

$$\mathcal{D}_{\text{slip}}\big(\psi|_{\partial\Omega}\big) = \pm\psi \quad \text{for each} \quad \psi \in \Psi(\Omega_\pm) \tag{7.1.423}$$

(with the convention that $\mathcal{D}_{\text{slip}}$ is the slip double layer associated with $\Omega = \Omega_+$). Finally, we observe that

$$\text{if } \Omega \text{ is as in (7.1.419) and } (\vec{u}, \pi) \text{ is a nullsolution of the Stokes system in } \Omega_\pm \text{ with the property that } \vec{u} \in \Psi(\Omega_\pm) \text{ then } \pi \in \mathbb{C}_{\Omega_\pm} \text{ (cf. (7.1.265)) and the slip conormal } \partial_\nu^{\text{slip}}(\vec{u}, \pi) \text{ belongs to } \nu\mathbb{C}_{\partial\Omega_\pm}. \tag{7.1.424}$$

We are now prepared to state and prove a result which may be regarded as the counterpart of Theorem 7.1.11 for the Stokes system.

Theorem 7.1.23 *Pick $n \in \mathbb{N}$ with $n \geq 2$ and assume $\Omega \subseteq \mathbb{R}^n$ is an Ahlfors regular domain with compact boundary. Abbreviate $\sigma := \mathcal{H}^{n-1}\lfloor\partial\Omega$ and denote by ν the geometric measure theoretic outward unit normal to Ω. In this setting, consider the slip double layer potential operators K_{slip}, $K^{\#}_{\text{slip}}$ associated with the Stokes system in Ω as in* (4.2.102) *and* (4.2.103)*, respectively. Finally, recall notation introduced in* (7.1.416), (7.1.417).

Then for each integrability exponent $p \in (1, \infty)$ there exists a threshold $\delta \in (0, 1)$, which depends only on n, p, the ADR *constants of $\partial\Omega$, and the relative flatness ratio $\wp(\Omega)/\text{diam}\,\partial\Omega$, with the property that whenever*

$$\text{dist}\Big(\nu\,,\,[\text{VMO}\,(\partial\Omega, \sigma)]^n\Big) < \delta \tag{7.1.425}$$

with the distance considered in $\big[\text{BMO}\,(\partial\Omega, \sigma)\big]^n$ (hence if Ω is a δ-oscillating AR *domain; cf. Definition 3.4.1), it follows that Ω is a two-sided* NTA *domain and*

$$\text{Ker}\Big(\pm\tfrac{1}{2}I + K_{\text{slip}};\ \big[L^p_1(\partial\Omega, \sigma)\big]^n \cap \big\langle \nu\mathbb{C}_{\partial\Omega_\pm}\big\rangle^{\perp}\Big) = \Psi(\partial\Omega_\mp), \tag{7.1.426}$$

$$\text{Ker}\Big(\pm\tfrac{1}{2}I + K_{\text{slip}};\ \big[L^p(\partial\Omega, \sigma)\big]^n \cap \big\langle \nu\mathbb{C}_{\partial\Omega_\pm}\big\rangle^{\perp}\Big) = \Psi(\partial\Omega_\mp), \tag{7.1.427}$$

$$\text{Ker}\Big(\pm\tfrac{1}{2}I + K^{\#}_{\text{slip}} : \big[L^p(\partial\Omega, \sigma)\big]^n \cap \Psi(\partial\Omega_\mp)^{\perp}\Big) = \nu\mathbb{C}_{\partial\Omega_\pm}, \tag{7.1.428}$$

$$\text{Ker}\Big(\pm\tfrac{1}{2}I + K^{\#}_{\text{slip}} : \big[L^p_{-1}(\partial\Omega, \sigma)\big]^n \cap \Psi(\partial\Omega_\mp)^{\perp}\Big) = \nu\mathbb{C}_{\partial\Omega_\pm}. \tag{7.1.429}$$

Also,

$$\mathrm{Im}\Big(\pm\tfrac{1}{2}I+K_{\mathrm{slip}};\,\big[L^p_1(\partial\Omega,\sigma)\big]^n\Big)\subseteq\big[L^p_1(\partial\Omega,\sigma)\big]^n\cap\big\langle\nu\mathbb{C}_{\partial\Omega_\pm}\big\rangle^{\perp},\tag{7.1.430}$$

$$\mathrm{Im}\Big(\pm\tfrac{1}{2}I+K_{\mathrm{slip}};\,\big[L^p(\partial\Omega,\sigma)\big]^n\Big)\subseteq\big[L^p(\partial\Omega,\sigma)\big]^n\cap\big\langle\nu\mathbb{C}_{\partial\Omega_\pm}\big\rangle^{\perp},\tag{7.1.431}$$

$$\mathrm{Im}\Big(\pm\tfrac{1}{2}I+K^{\#}_{\mathrm{slip}};\,\big[L^p(\partial\Omega,\sigma)\big]^n\Big)\subseteq\big[L^p(\partial\Omega,\sigma)\big]^n\cap\Psi(\partial\Omega_\mp)^{\perp},\tag{7.1.432}$$

$$\mathrm{Im}\Big(\pm\tfrac{1}{2}I+K^{\#}_{\mathrm{slip}}:\,\big[L^p_{-1}(\partial\Omega,\sigma)\big]^n\Big)\subseteq\big[L^p_{-1}(\partial\Omega,\sigma)\big]^n\cap\Psi(\partial\Omega_\mp)^{\perp}.\tag{7.1.433}$$

Moreover, under the assumption made in (7.1.425), *the operators*

$$\begin{aligned}\pm\tfrac{1}{2}I+\big[K_{\mathrm{slip}}\big]:\Big(\big[L^p_1(\partial\Omega,\sigma)\big]^n\cap\big\langle\nu\mathbb{C}_{\partial\Omega_\pm}\big\rangle^{\perp}\Big)\Big/\Psi(\partial\Omega_\mp)&\\ \longrightarrow\Big(\big[L^p_1(\partial\Omega,\sigma)\big]^n\cap\big\langle\nu\mathbb{C}_{\partial\Omega_\pm}\big\rangle^{\perp}\Big)\Big/\Psi(\partial\Omega_\mp),&\end{aligned}\tag{7.1.434}$$

and

$$\begin{aligned}\pm\tfrac{1}{2}I+\big[K_{\mathrm{slip}}\big]:\Big(\big[L^p(\partial\Omega,\sigma)\big]^n\cap\big\langle\nu\mathbb{C}_{\partial\Omega_\pm}\big\rangle^{\perp}\Big)\Big/\Psi(\partial\Omega_\mp)&\\ \longrightarrow\Big(\big[L^p(\partial\Omega,\sigma)\big]^n\cap\big\langle\nu\mathbb{C}_{\partial\Omega_\pm}\big\rangle^{\perp}\Big)\Big/\Psi(\partial\Omega_\mp),&\end{aligned}\tag{7.1.435}$$

as well as

$$\begin{aligned}\pm\tfrac{1}{2}I+\big[K^{\#}_{\mathrm{slip}}\big]:\Big(\big[L^p(\partial\Omega,\sigma)\big]^n\cap\Psi(\partial\Omega_\mp)^{\perp}\Big)\Big/\nu\mathbb{C}_{\partial\Omega_\pm}&\\ \longrightarrow\Big(\big[L^p(\partial\Omega,\sigma)\big]^n\cap\Psi(\partial\Omega_\mp)^{\perp}\Big)\Big/\nu\mathbb{C}_{\partial\Omega_\pm},&\end{aligned}\tag{7.1.436}$$

and

$$\begin{aligned}\pm\tfrac{1}{2}I+\big[K^{\#}_{\mathrm{slip}}\big]:\Big(\big[L^p_{-1}(\partial\Omega,\sigma)\big]^n\cap\Psi(\partial\Omega_\mp)^{\perp}\Big)\Big/\nu\mathbb{C}_{\partial\Omega_\pm}&\\ \longrightarrow\Big(\big[L^p_{-1}(\partial\Omega,\sigma)\big]^n\cap\Psi(\partial\Omega_\mp)^{\perp}\Big)\Big/\nu\mathbb{C}_{\partial\Omega_\pm},&\end{aligned}\tag{7.1.437}$$

(defined by sending the equivalence class of a function into the equivalence class of the operator in question acting on said function) are all well-defined isomorphisms.

Similar results are valid on other scales of spaces, including Morrey spaces, vanishing Morrey spaces, block spaces, together with their Sobolev space versions, as well as Hölder spaces, Besov spaces, the John-Nirenberg space BMO, *the Sarason space* VMO *on* $\partial\Omega$, *and also for Generalized Banach Function Spaces* $\mathbb{X}$ *on* $(\partial\Omega,\sigma)$ *such that the Hardy-Littlewood maximal operator is bounded on* $\mathbb{X}$ *and* $\mathbb{X}'$.

Finally, analogous results are valid for the operators K_{slip} *and* $K^{\#}_{\mathrm{slip}}$ *considered on Besov and Triebel-Lizorkin spaces on* $\partial\Omega$ (*cf.* [132, Theorem 6.4.1]).

Proof Theorem 3.4.2 guarantees that if (7.1.425) holds for a sufficiently small $\delta\in(0,1)$ then Ω actually a two-sided NTA domain. By eventually further decreasing the value of δ we may also ensure that the very last claim in the statement of Proposition 7.1.2 is true. Granted these properties, the identifications made in (7.1.426)-(7.1.429) can be verified by reasoning as in the proof of [163, Proposition 5.3.5, p. 97] (where the case of Lipschitz domains has been treated), now availing ourselves of the Green-type formulas for the Stokes system from [132, §6.1], and the Calderón-Zygmund theory for the boundary layer potentials associated with the Stokes system in UR domains from [132, Theorem 6.2.4]. Once (7.1.426)-(7.1.429) have been established we may rely on the Fredholmness results from item *(10)* of Theorem 7.1.1 to conclude that the operators in (7.1.434)-(7.1.437) are isomorphisms (see also [163, Theorem 5.3.6, pp. 102-103] in this regard).

All other claims in the theorem are dealt with similarly, also using the properties boundary layer potentials associated with the Stokes system on Besov and Triebel-Lizorkin spaces established in [132, §6.4] to deal with the very last statement. □

Below we identify the kernels and images of the conormal derivative of the double layer operator on a variety of function spaces considered on boundaries of δ-oscillating AR domains.

Proposition 7.1.24 *Fix $n \in \mathbb{N}$ with $n \geq 2$ and suppose $\Omega \subseteq \mathbb{R}^n$ is an Ahlfors regular domain with compact boundary. Abbreviate $\sigma := \mathcal{H}^{n-1}\lfloor\partial\Omega$ and denote by ν the geometric measure theoretic outward unit normal to Ω. Also, let L be a homogeneous, second-order, constant complex coefficient, $M \times M$ system in $\mathbb{R}^n$ such that*

$$\textit{L satisfies the Legendre-Hadamard strong ellipticity condition (cf. [131, Definition 1.3.2]) and one has } \mathfrak{A}_L^{\rm dis} \neq \varnothing. \tag{7.1.438}$$

Pick a coefficient tensor $A \in \mathfrak{A}_L^{\rm dis}$ and consider the boundary-to-boundary double layer potential operator K associated with A and Ω as in (A.0.116), *the transpose double layer potential operator $K^{\#}_{A^\top}$ associated with $A^\top$ and Ω as in* (A.0.118), *as well as the conormal derivative of the double layer potential operator $\partial_\nu^A \mathcal{D}_A$ associated with A and Ω as in* [132, (1.5.29)]. *Finally, for a threshold $\delta_0 \in (0,1)$ to be specified shortly, assume that*

$$\operatorname{dist}\Big(\nu, [\mathrm{VMO}(\partial\Omega,\sigma)]^n\Big) < \delta_0, \tag{7.1.439}$$

with the distance considered in $\big[\mathrm{BMO}(\partial\Omega,\sigma)\big]^n$ (hence, assume Ω is a δ_0-oscillating AR *domain; cf. Definition 3.4.1). Then the following statements are true.*

(1) Given any $p \in (1,\infty)$, choosing $\delta_0 \in (0,1)$ sufficiently small, depending on n, p, A, the ADR *constants of $\partial\Omega$, and the relative flatness ratio $\wp(\Omega)/\operatorname{diam}\partial\Omega$ ensures that Ω is a* UR *domain,*

$$\begin{aligned} &\operatorname{Im}\Big(\partial_\nu^A\mathcal{D}_A : \big[L_1^p(\partial\Omega,\sigma)\big]^M \to \big[L^p(\partial\Omega,\sigma)\big]^M\Big) \\ &= \operatorname{Im}\Big(\big(\tfrac12 I + K^{\#}_{A^\top}\big)\circ\big(-\tfrac12 I + K^{\#}_{A^\top}\big);\ \big[L^p(\partial\Omega,\sigma)\big]^M\Big) \\ &= \operatorname{Im}\Big(\tfrac12 I + K^{\#}_{A^\top};\ \big[L^p(\partial\Omega,\sigma)\big]^M\Big) \cap \operatorname{Im}\Big(-\tfrac12 I + K^{\#}_{A^\top};\ \big[L^p(\partial\Omega,\sigma)\big]^M\Big) \end{aligned} \tag{7.1.440}$$

and, with $\boxplus$ standing for algebraic direct sum,

$$\operatorname{Ker}\Big(\partial_\nu^A \mathcal{D}_A : \big[L_1^p(\partial\Omega,\sigma)\big]^M \to \big[L^p(\partial\Omega,\sigma)\big]^M\Big) \\ = \operatorname{Ker}\Big(\tfrac{1}{2}I + K^{\#}_{A^\top};\ \big[L^p(\partial\Omega,\sigma)\big]^M\Big) \boxplus \operatorname{Ker}\Big(-\tfrac{1}{2}I + K^{\#}_{A^\top};\ \big[L^p(\partial\Omega,\sigma)\big]^M\Big). \tag{7.1.441}$$

More generally, given any integrability exponent $p \in (1,\infty)$ along with any Muckenhoupt weight $w \in A_p(\partial\Omega,\sigma)$, choosing $\delta_0 \in (0,1)$ sufficiently small, depending on n, p, $[w]_{A_p}$, A, the ADR *constants of $\partial\Omega$, and the relative flatness ratio $\wp(\Omega)/\operatorname{diam}\partial\Omega$ ensures that Ω is a* UR *domain,*

$$\operatorname{Im}\Big(\partial_\nu^A \mathcal{D}_A : \big[L_1^p(\partial\Omega,w)\big]^M \to \big[L^p(\partial\Omega,w)\big]^M\Big) \\ = \operatorname{Im}\Big(\big(\tfrac{1}{2}I + K^{\#}_{A^\top}\big)\circ\big(-\tfrac{1}{2}I + K^{\#}_{A^\top}\big);\ \big[L^p(\partial\Omega,w)\big]^M\Big) \\ = \operatorname{Im}\Big(\tfrac{1}{2}I + K^{\#}_{A^\top};\ \big[L^p(\partial\Omega,w)\big]^M\Big) \cap \operatorname{Im}\Big(-\tfrac{1}{2}I + K^{\#}_{A^\top};\ \big[L^p(\partial\Omega,w)\big]^M\Big), \tag{7.1.442}$$

and

$$\operatorname{Ker}\Big(\partial_\nu^A \mathcal{D}_A : \big[L_1^p(\partial\Omega,w)\big]^M \to \big[L^p(\partial\Omega,w)\big]^M\Big) \\ = \operatorname{Ker}\Big(\tfrac{1}{2}I + K;\ \big[L_1^p(\partial\Omega,w)\big]^M\Big) \boxplus \operatorname{Ker}\Big(-\tfrac{1}{2}I + K;\ \big[L_1^p(\partial\Omega,w)\big]^M\Big). \tag{7.1.443}$$

(2) Given any $p,q \in (1,\infty)$ and $\lambda \in (0,n-1)$, choosing $\delta_0 \in (0,1)$ sufficiently small, depending on n, p, q, λ, A, the ADR *constants of $\partial\Omega$, and the relative flatness ratio $\wp(\Omega)/\operatorname{diam}\partial\Omega$ ensures that Ω is a* UR *domain,*

$$\operatorname{Im}\Big(\partial_\nu^A \mathcal{D}_A : \big[M_1^{p,\lambda}(\partial\Omega,\sigma)\big]^M \to \big[M^{p,\lambda}(\partial\Omega,\sigma)\big]^M\Big) \\ = \operatorname{Im}\Big(\big(\tfrac{1}{2}I + K^{\#}_{A^\top}\big)\circ\big(-\tfrac{1}{2}I + K^{\#}_{A^\top}\big);\ \big[M^{p,\lambda}(\partial\Omega,\sigma)\big]^M\Big) \\ = \operatorname{Im}\Big(\tfrac{1}{2}I + K^{\#}_{A^\top};\ \big[M^{p,\lambda}(\partial\Omega,\sigma)\big]^M\Big) \cap \operatorname{Im}\Big(-\tfrac{1}{2}I + K^{\#}_{A^\top};\ \big[M^{p,\lambda}(\partial\Omega,\sigma)\big]^M\Big) \tag{7.1.444}$$

$$\operatorname{Im}\Big(\partial_\nu^A \mathcal{D}_A : \big[\mathring{M}_1^{p,\lambda}(\partial\Omega,\sigma)\big]^M \to \big[\mathring{M}^{p,\lambda}(\partial\Omega,\sigma)\big]^M\Big) \\ = \operatorname{Im}\Big(\big(\tfrac{1}{2}I + K^{\#}_{A^\top}\big)\circ\big(-\tfrac{1}{2}I + K^{\#}_{A^\top}\big);\ \big[\mathring{M}^{p,\lambda}(\partial\Omega,\sigma)\big]^M\Big) \\ = \operatorname{Im}\Big(\tfrac{1}{2}I + K^{\#}_{A^\top};\ \big[\mathring{M}^{p,\lambda}(\partial\Omega,\sigma)\big]^M\Big) \cap \operatorname{Im}\Big(-\tfrac{1}{2}I + K^{\#}_{A^\top};\ \big[\mathring{M}^{p,\lambda}(\partial\Omega,\sigma)\big]^M\Big), \tag{7.1.445}$$

$$\operatorname{Im}\Big(\partial_\nu^A \mathcal{D}_A : \big[\mathcal{B}_1^{q,\lambda}(\partial\Omega,\sigma)\big]^M \to \big[\mathcal{B}^{q,\lambda}(\partial\Omega,\sigma)\big]^M\Big) \\ = \operatorname{Im}\Big(\big(\tfrac{1}{2}I + K^{\#}_{A^\top}\big)\circ\big(-\tfrac{1}{2}I + K^{\#}_{A^\top}\big);\ \big[\mathcal{B}^{q,\lambda}(\partial\Omega,\sigma)\big]^M\Big) \\ = \operatorname{Im}\Big(\tfrac{1}{2}I + K^{\#}_{A^\top};\ \big[\mathcal{B}^{q,\lambda}(\partial\Omega,\sigma)\big]^M\Big) \cap \operatorname{Im}\Big(-\tfrac{1}{2}I + K^{\#}_{A^\top};\ \big[\mathcal{B}^{q,\lambda}(\partial\Omega,\sigma)\big]^M\Big), \tag{7.1.446}$$

and that

$$\begin{aligned}&\mathrm{Ker}\Big(\partial_\nu^A\mathcal{D}_A : \big[M_1^{p,\lambda}(\partial\Omega,\sigma)\big]^M \to \big[M^{p,\lambda}(\partial\Omega,\sigma)\big]^M\Big) \\ &\quad = \mathrm{Ker}\Big(\tfrac12 I + K;\ \big[M_1^{p,\lambda}(\partial\Omega,\sigma)\big]^M\Big) \boxplus \mathrm{Ker}\Big(-\tfrac12 I + K;\ \big[M_1^{p,\lambda}(\partial\Omega,\sigma)\big]^M\Big),\end{aligned} \tag{7.1.447}$$

$$\begin{aligned}&\mathrm{Ker}\Big(\partial_\nu^A\mathcal{D}_A : \big[\mathring{M}_1^{p,\lambda}(\partial\Omega,\sigma)\big]^M \to \big[\mathring{M}^{p,\lambda}(\partial\Omega,\sigma)\big]^M\Big) \\ &\quad = \mathrm{Ker}\Big(\tfrac12 I + K;\ \big[\mathring{M}_1^{p,\lambda}(\partial\Omega,\sigma)\big]^M\Big) \boxplus \mathrm{Ker}\Big(-\tfrac12 I + K;\ \big[\mathring{M}_1^{p,\lambda}(\partial\Omega,\sigma)\big]^M\Big),\end{aligned} \tag{7.1.448}$$

$$\begin{aligned}&\mathrm{Ker}\Big(\partial_\nu^A\mathcal{D}_A : \big[\mathcal{B}_1^{q,\lambda}(\partial\Omega,\sigma)\big]^M \to \big[\mathcal{B}^{q,\lambda}(\partial\Omega,\sigma)\big]^M\Big) \\ &\quad = \mathrm{Ker}\Big(\tfrac12 I + K;\ \big[\mathcal{B}_1^{q,\lambda}(\partial\Omega,\sigma)\big]^M\Big) \boxplus \mathrm{Ker}\Big(-\tfrac12 I + K;\ \big[\mathcal{B}_1^{q,\lambda}(\partial\Omega,\sigma)\big]^M\Big).\end{aligned} \tag{7.1.449}$$

(3) Given any $p, q \in (1,\infty)$ and $s \in (0,1)$, choosing $\delta_0 \in (0,1)$ sufficiently small, depending on n, p, q, s, A, the ADR *constants of $\partial\Omega$, and the relative flatness ratio $\wp(\Omega)/\mathrm{diam}\,\partial\Omega$ ensures that Ω is a* UR *domain,*

$$\begin{aligned}&\mathrm{Im}\Big(\partial_\nu^A\mathcal{D}_A : \big[B_{1-s}^{p,q}(\partial\Omega,\sigma)\big]^M \to \big[B_{-s}^{p,q}(\partial\Omega,\sigma)\big]^M\Big) \\ &\quad = \mathrm{Im}\Big(\big(\tfrac12 I + K^\#_{A^\top}\big)\circ\big(-\tfrac12 I + K^\#_{A^\top}\big);\ \big[B_{1-s}^{p,q}(\partial\Omega,\sigma)\big]^M\Big) \\ &\quad = \mathrm{Im}\Big(\tfrac12 I + K^\#_{A^\top};\ \big[B_{1-s}^{p,q}(\partial\Omega,\sigma)\big]^M\Big) \cap \mathrm{Im}\Big(-\tfrac12 I + K^\#_{A^\top};\ \big[B_{1-s}^{p,q}(\partial\Omega,\sigma)\big]^M\Big),\end{aligned} \tag{7.1.450}$$

and

$$\begin{aligned}&\mathrm{Ker}\Big(\partial_\nu^A\mathcal{D}_A : \big[B_{1-s}^{p,q}(\partial\Omega,\sigma)\big]^M \to \big[B_{-s}^{p,q}(\partial\Omega,\sigma)\big]^M\Big) \\ &\quad = \mathrm{Ker}\Big(\tfrac12 I + K;\ \big[B_{1-s}^{p,q}(\partial\Omega,\sigma)\big]^M\Big) \boxplus \mathrm{Ker}\Big(-\tfrac12 I + K;\ \big[B_{1-s}^{p,q}(\partial\Omega,\sigma)\big]^M\Big).\end{aligned} \tag{7.1.451}$$

Proof Assume $\delta_0 \in (0,1)$ is small enough as to ensure the validity of the conclusions in Theorem 3.4.2; in particular, Ω is a UR domain. Let S be the boundary-to-boundary single layer potential operator associated with Ω and the system L as in (A.0.231). To deal with item *(1)*, fix $p \in (1,\infty)$, $w \in A_p(\partial\Omega,\sigma)$ and, if necessary, further decrease the value of $\delta_0 \in (0,1)$. From [129, (7.7.105)] and [132, (1.5.86)] we conclude that

$$\big(\tfrac12 I + K^\#_{A^\top}\big)\circ\big(-\tfrac12 I + K^\#_{A^\top}\big) = \big(\partial_\nu^A\mathcal{D}_A\big)\circ S \ \text{ on } \ \big[L^p(\partial\Omega,w)\big]^M. \tag{7.1.452}$$

Since S maps $\big[L^p(\partial\Omega,w)\big]^M$ into $\big[L_1^p(\partial\Omega,w)\big]^M$, this implies

$$\begin{aligned}&\mathrm{Im}\Big(\partial_\nu^A\mathcal{D}_A : \big[L_1^p(\partial\Omega,w)\big]^M \to \big[L^p(\partial\Omega,w)\big]^M\Big) \\ &\qquad\qquad \supseteq \mathrm{Im}\Big(\big(\tfrac12 I + K^\#_{A^\top}\big)\circ\big(-\tfrac12 I + K^\#_{A^\top}\big);\ \big[L^p(\partial\Omega,w)\big]^M\Big).\end{aligned} \tag{7.1.453}$$

To prove the opposite inclusion, we distinguish two cases. First, if $n \geq 3$, we know that (7.1.235) is an invertible operator. If S^{-1} denotes its inverse we may write

$$\partial_\nu^A \mathcal{D}_A = \big(\tfrac{1}{2}I + K^{\#}_{A^\top}\big) \circ \big(-\tfrac{1}{2}I + K^{\#}_{A^\top}\big) \circ S^{-1} \text{ on } \big[L^p_1(\partial\Omega, w)\big]^M, \qquad (7.1.454)$$

which gives the opposite inclusion in (7.1.453). Second, if $n = 2$, the augmented single layer potential operator S_{aug} is invertible in the context of (7.1.248). As such, given any function f in $\big[L^p_1(\partial\Omega, w)\big]^M$ we may find $g \in \big[L^p(\partial\Omega, w)\big]^M$ and $c \in \mathbb{C}^M$ with $Sg + c = f$ and $\int_{\partial\Omega} g\, d\sigma = 0$. We may therefore use this and the fact that $\partial_\nu^A \mathcal{D}_A$ annihilates constants (which is seen from [132, (1.5.29)]) to write

$$\partial_\nu^A \mathcal{D}_A f = \partial_\nu^A \mathcal{D}_A (f - c) = \partial_\nu^A \mathcal{D}_A (Sg) = \big(\tfrac{1}{2}I + K^{\#}_{A^\top}\big)\Big(\big(-\tfrac{1}{2}I + K^{\#}_{A^\top}\big)g\Big), \quad (7.1.455)$$

from which we conclude that

$$\partial_\nu^A \mathcal{D}_A f \in \operatorname{Im}\Big(\big(\tfrac{1}{2}I + K^{\#}_{A^\top}\big) \circ \big(-\tfrac{1}{2}I + K^{\#}_{A^\top}\big);\ \big[L^p(\partial\Omega, w)\big]^M\Big). \qquad (7.1.456)$$

This establishes the opposite inclusion in (7.1.453). Hence, the opposite inclusion in (7.1.453) holds in all dimensions $n \geq 2$, and the first equality in (7.1.442) follows.

To prove the second equality in (7.1.442), let $p' \in (1, \infty)$ be the conjugate exponent of p, and set $w' := w^{1-p'} \in A_{p'}(\partial\Omega, \sigma)$. Observe that any $f \in \big[L^{p'}(\partial\Omega, w')\big]^M$ may be decomposed as $\big(\tfrac{1}{2}I + K_{A^\top}\big)f + \big(\tfrac{1}{2}I - K_{A^\top}\big)f$. Bearing this in mind, as well as the fact that $\tfrac{1}{2}I + K_{A^\top}$ and $-\tfrac{1}{2}I + K_{A^\top}$ commute, we conclude that

$$\begin{aligned} &\operatorname{Ker}\Big(\big(\tfrac{1}{2}I + K_{A^\top}\big) \circ \big(-\tfrac{1}{2}I + K_{A^\top}\big);\ \big[L^{p'}(\partial\Omega, w')\big]^M\Big) \qquad (7.1.457)\\ &\quad = \operatorname{Ker}\Big(\tfrac{1}{2}I + K_{A^\top};\ \big[L^{p'}(\partial\Omega, w')\big]^M\Big) \boxplus \operatorname{Ker}\Big(-\tfrac{1}{2}I + K_{A^\top};\ \big[L^{p'}(\partial\Omega, w')\big]^M\Big). \end{aligned}$$

In addition, from [132, Theorem 1.5.1, item *(iii)*] we know that the (real) transpose of $\pm\tfrac{1}{2}I + K^{\#}_{A^\top}$ acting on $\big[L^p(\partial\Omega, w)\big]^M$ is $\pm\tfrac{1}{2}I + K_{A^\top}$ acting on $\big[L^{p'}(\partial\Omega, w')\big]^M$; in particular, the aforementioned identification implies that the (real) transpose of the operator $\big(\tfrac{1}{2}I + K^{\#}_{A^\top}\big) \circ \big(-\tfrac{1}{2}I + K^{\#}_{A^\top}\big)$ acting on the space $\big[L^p(\partial\Omega, w)\big]^M$ is the operator $\big(\tfrac{1}{2}I + K_{A^\top}\big) \circ \big(-\tfrac{1}{2}I + K_{A^\top}\big)$ acting on the space $\big[L^{p'}(\partial\Omega, w')\big]^M$. Recalling (A.0.7) we then have (using [130, (2.1.49)] and Theorem 7.1.1 multiple times)

$$\begin{aligned} &\operatorname{Im}\Big(\big(\tfrac{1}{2}I + K^{\#}_{A^\top}\big) \circ \big(-\tfrac{1}{2}I + K^{\#}_{A^\top}\big);\ \big[L^p(\partial\Omega, w)\big]^M\Big) \qquad (7.1.458)\\ &= {}^{\perp}\Big[\operatorname{Ker}\Big(\big(\tfrac{1}{2}I + K_{A^\top}\big) \circ \big(-\tfrac{1}{2}I + K_{A^\top}\big);\ \big[L^{p'}(\partial\Omega, w')\big]^M\Big)\Big]\\ &= {}^{\perp}\Big[\operatorname{Ker}\Big(\tfrac{1}{2}I + K_{A^\top};\big[L^{p'}(\partial\Omega, w')\big]^M\Big)\Big] \cap {}^{\perp}\Big[\operatorname{Ker}\Big(-\tfrac{1}{2}I + K_{A^\top};\big[L^{p'}(\partial\Omega, w')\big]^M\Big)\Big]\\ &= \operatorname{Im}\Big(\tfrac{1}{2}I + K^{\#}_{A^\top};\ \big[L^p(\partial\Omega, w)\big]^M\Big) \cap \operatorname{Im}\Big(-\tfrac{1}{2}I + K^{\#}_{A^\top};\ \big[L^p(\partial\Omega, w)\big]^M\Big), \end{aligned}$$

proving the second equality in (7.1.442). By [129, (7.7.105)] and [132, (1.5.85)],

$$\big(\tfrac{1}{2}I + K\big) \circ \big(-\tfrac{1}{2}I + K\big) = S \circ \big(\partial_\nu^A \mathcal{D}_A\big) \text{ on } \big[L_1^p(\partial\Omega, w)\big]^M. \tag{7.1.459}$$

Given that $\partial_\nu^A \mathcal{D}_A$ maps $\big[L_1^p(\partial\Omega, w)\big]^M$ into $\big[L^p(\partial\Omega, w)\big]^M$, this implies

$$\begin{aligned}\operatorname{Ker}\Big(\partial_\nu^A \mathcal{D}_A : \big[L_1^p(\partial\Omega, w)\big]^M \to \big[L^p(\partial\Omega, w)\big]^M\Big)& \\ \subseteq \operatorname{Ker}\Big(\big(\tfrac{1}{2}I + K\big) \circ \big(-\tfrac{1}{2}I + K\big);\ \big[L_1^p(\partial\Omega, w)\big]^M\Big).&\end{aligned} \tag{7.1.460}$$

As for the opposite inclusion, consider two cases. Assume first $n \geq 3$. Then the operator (7.1.235) is invertible and, with S^{-1} denotes its inverse, (7.1.459) gives

$$\partial_\nu^A \mathcal{D}_A = S^{-1} \circ \big(\tfrac{1}{2}I + K\big) \circ \big(-\tfrac{1}{2}I + K\big) \text{ on } \big[L_1^p(\partial\Omega, w)\big]^M, \tag{7.1.461}$$

from which the opposite inclusion in (7.1.460) follows. Assume next that $n = 2$ and consider $f \in \big[L_1^p(\partial\Omega, w)\big]^M$ such that $\big(\tfrac{1}{2}I + K\big)\Big(\big(-\tfrac{1}{2}I + K\big)f\Big) = 0$. Then the function $g := \partial_\nu^A \mathcal{D}_A f$ belongs to $\big[L^p(\partial\Omega, w)\big]^M$ and the present assumptions together with (7.1.459) guarantee that $Sg = 0$. In addition, from [131, (1.7.67)] (whose applicability in the present setting is ensured by [132, (1.3.24), (1.5.9)], [129, (7.7.104)], (A.0.1), and the fact that $\partial\Omega$ is bounded) we see that

$$\int_{\partial\Omega} g \, d\sigma = 0. \tag{7.1.462}$$

From these properties and (7.1.248) we conclude that $S_{\text{aug}}(g, 0) = (0, 0)$. Given that the augmented single layer operator is invertible in the functional analytic context specified in (7.1.248), this forces $g = 0$ which, in turn, translated into saying that $f \in \operatorname{Ker}\Big(\partial_\nu^A \mathcal{D}_A : \big[L_1^p(\partial\Omega, w)\big]^M \to \big[L^p(\partial\Omega, w)\big]^M\Big)$. In view of the arbitrariness of f, this ultimately shows that the opposite inclusion in (7.1.460) holds when $n = 2$ as well. This finishes the proof of the fact that, in all cases,

$$\begin{aligned}\operatorname{Ker}\Big(\partial_\nu^A \mathcal{D}_A : \big[L_1^p(\partial\Omega, w)\big]^M \to \big[L^p(\partial\Omega, w)\big]^M\Big)& \\ = \operatorname{Ker}\Big(\big(\tfrac{1}{2}I + K\big) \circ \big(-\tfrac{1}{2}I + K\big);\ \big[L_1^p(\partial\Omega, w)\big]^M\Big).&\end{aligned} \tag{7.1.463}$$

Since, much as in (7.1.457), we have

$$\begin{aligned}\operatorname{Ker}\Big(\big(\tfrac{1}{2}I + K\big) \circ \big(-\tfrac{1}{2}I + K\big);\ \big[L_1^p(\partial\Omega, w)\big]^M\Big)& \\ = \operatorname{Ker}\Big(\tfrac{1}{2}I + K;\ \big[L_1^p(\partial\Omega, w)\big]^M\Big) \boxplus \operatorname{Ker}\Big(-\tfrac{1}{2}I + K;\ \big[L_1^p(\partial\Omega, w)\big]^M\Big),&\end{aligned} \tag{7.1.464}$$

we conclude from (7.1.463) and (7.1.464) that (7.1.443) holds.

The proof of (7.1.444) is similar to that of (7.1.442), though now we use [130, (2.1.48)] in place of [130, (2.1.49)], rely on the duality results from [130, (6.2.80)] and [132, Theorem 3.3.1, item *(iii)*], employ Theorem 7.1.18, and make systematic use of [132, Theorems 3.3.1-3.3.2]. The proofs of (7.1.445) and (7.1.446) fully mirror the justification of (7.1.442), now keeping in mind [130, (6.2.80), (6.2.155)], [132, Theorem 3.3.1, item *(iii)*], Theorem 7.1.18, Theorem 7.1.19, as well as [132, Theorems 3.3.1-3.3.2]. The proofs of (7.1.447)-(7.1.449) follow the template of (7.1.443) (now using Theorem 7.1.18, Theorem 7.1.19 and also keeping in mind [132, Theorems 3.3.1-3.3.2]). Finally, (7.1.450) and (7.1.451) are dealt with in a similar fashion now using Theorem 7.1.15, Theorem 7.1.17, [132, (4.1.29), (4.3.71)], and [132, Theorems 4.1.1, 4.1.5, 4.3.5, 4.4.2]. □

Moving on, we make the following definition, which is going to be relevant in the treatment of the Neumann Problem with data in Hardy spaces from Theorem 8.1.12 (as well as in the buildup to that result).

Definition 7.1.25 *Let $\Omega \subseteq \mathbb{R}^n$ (where $n \in \mathbb{N}$, $n \geq 2$) be a* UR *domain with compact boundary. Set $\sigma := \mathcal{H}^{n-1}\lfloor\partial\Omega$ and consider*

$$p \in \left(\tfrac{n-1}{n}, 1\right] \text{ and } z \in \mathbb{C}. \tag{7.1.465}$$

Also, let L be a homogeneous, weakly elliptic, constant (complex) coefficient, second-order $M \times M$ system in $\mathbb{R}^n$ (for some $M \in \mathbb{N}$). Finally, pick a coefficient tensor $A \in \mathfrak{A}_L$ and recall the double layer potential operator K_A associated with A and Ω as in (A.0.116). *This is a well-defined linear and bounded mapping on Hölder spaces (cf.* [132, (2.1.125)]), *and on the John-Nirenberg space (cf.* [132, (2.1.95)]). *In the above setting, denote by*

$$H^p_{A,z}(\partial\Omega, \sigma) \tag{7.1.466}$$

the collection of distributions $f \in \left[H^p(\partial\Omega, \sigma)\right]^M$ having the property that (with the brackets indicating duality in the sense of [130, Theorem 4.6.1])

$$\begin{gathered} {}_{[H^p(\partial\Omega,\sigma)]^M}\big\langle f, g\big\rangle_{[\mathscr{C}^\eta(\partial\Omega)]^M} = 0 \text{ for each function} \\ g \in \left[\mathscr{C}^\eta(\partial\Omega)\right]^M \text{ satisfying } (zI + K_A)g = 0 \\ \text{if } p \in \left(\tfrac{n-1}{n}, 1\right) \text{ and } \eta := (n-1)\left(\tfrac{1}{p} - 1\right) \in (0, 1), \end{gathered} \tag{7.1.467}$$

and, corresponding to the case when $p = 1$,

$$\begin{gathered} {}_{[H^1(\partial\Omega,\sigma)]^M}\big\langle f, g\big\rangle_{[\mathrm{BMO}(\partial\Omega,\sigma)]^M} = 0 \text{ for each function} \\ g \in \left[\mathrm{BMO}(\partial\Omega, \sigma)\right]^M \text{ satisfying } (zI + K_A)g = 0. \end{gathered} \tag{7.1.468}$$

The space (7.1.466) is linked to the action of the transpose double layer operator on Hardy spaces, via the inclusion described next.

Proposition 7.1.26 *Let $\Omega \subseteq \mathbb{R}^n$ be a* UR *domain with compact boundary. Abbreviate $\sigma := \mathcal{H}^{n-1}\lfloor\partial\Omega$ and fix*

$$p \in \left(\tfrac{n-1}{n}, 1\right] \text{ and } z \in \mathbb{C}. \tag{7.1.469}$$

Next, let L be a homogeneous, weakly elliptic, constant (complex) coefficient, second-order $M \times M$ system in $\mathbb{R}^n$ (for some $M \in \mathbb{N}$). Finally, pick a coefficient tensor $A \in \mathfrak{A}_L$ and recall the transpose double layer $K^\#_A$ associated with A and Ω as in (A.0.118), *acting on the Hardy space* $\big[H^p(\partial\Omega,\sigma)\big]^M$ *as in* [132, Theorem 2.1.1]. *Then*

$$\mathrm{Im}\Big(zI+K_A^{\#};\ \big[H^p(\partial\Omega,\sigma)\big]^M\Big)\subseteq H^p_{A,z}(\partial\Omega,\sigma). \tag{7.1.470}$$

Proof The inclusion claimed in (7.1.470) is a direct consequence of [132, Theorem 2.1.1], Definition 7.1.25, and [132, (2.1.98), (2.1.128)]. □

That the inclusion from Proposition 7.1.26 becomes equality in the context of δ-oscillating AR domains makes the object of our next proposition.

Proposition 7.1.27 *Fix* $n\in\mathbb{N}$ *with* $n\geq 2$ *and suppose* $\Omega\subseteq\mathbb{R}^n$ *is an Ahlfors regular domain with compact boundary. Abbreviate* $\sigma:=\mathcal{H}^{n-1}\lfloor\partial\Omega$ *and denote by* ν *the geometric measure theoretic outward unit normal to* Ω. *Also, let* L *be a homogeneous, second-order, constant complex coefficient, weakly elliptic* $M\times M$ *system in* $\mathbb{R}^n$, *with* $\mathfrak{A}_L^{\mathrm{dis}}\neq\varnothing$, *and fix some* $p\in\big(\frac{n-1}{n},1\big]$ *and* $z\in\mathbb{C}\setminus\{0\}$. *Finally, pick a coefficient tensor* $A\in\mathfrak{A}_L^{\mathrm{dis}}$, *and let* $K_A^{\#}$ *be the transpose double layer associated with* A *and* Ω, *acting on the Hardy space* $\big[H^p(\partial\Omega,\sigma)\big]^M$ *as in* [132, Theorem 2.1.1].

Then there exists $\delta\in(0,1)$ *depending only on* n, p, z, A, *the* ADR *constants of* $\partial\Omega$, *and the relative flatness ratio* $\wp(\Omega)/\mathrm{diam}\,\partial\Omega$ *with the property that when*

$$\mathrm{dist}\Big(\nu,[\mathrm{VMO}(\partial\Omega,\sigma)]^n\Big)<\delta \tag{7.1.471}$$

with the distance considered in the space $\big[\mathrm{BMO}(\partial\Omega,\sigma)\big]^n$, *(hence if* Ω *is a* δ*-oscillating* AR *domain; cf. Definition 3.4.1), it follows that the set* Ω *is a* UR *domain and one has equality in* (7.1.470), *i.e.,*

$$\mathrm{Im}\Big(zI+K_A^{\#};\ \big[H^p(\partial\Omega,\sigma)\big]^M\Big)=H^p_{A,z}(\partial\Omega,\sigma). \tag{7.1.472}$$

Proof Suppose (7.1.471) holds for a sufficiently small $\delta\in(0,1)$. Then Theorem 3.4.2 implies that Ω is a UR domain. Also, Theorem 7.1.1 guarantees that $zI+K_A^{\#}$ is a Fredholm operator with index zero on $\big[H^p(\partial\Omega,\sigma)\big]^M$. As a result,

$$\begin{aligned}&\dim\left(\frac{\big[H^p(\partial\Omega,\sigma)\big]^M}{\mathrm{Im}\big(zI+K_A^{\#};[H^p(\partial\Omega,\sigma)]^M\big)}\right) \qquad (7.1.473)\\ &=\dim\mathrm{CoKer}\Big(zI+K_A^{\#};\ \big[H^p(\partial\Omega,\sigma)\big]^M\Big)=\dim\mathrm{Ker}\Big(zI+K_A^{\#};\ \big[H^p(\partial\Omega,\sigma)\big]^M\Big).\end{aligned}$$

On the other hand, Definition 7.1.25 implies that

$$\dim\left(\frac{\big[H^p(\partial\Omega,\sigma)\big]^M}{H^p_{A,z}(\partial\Omega,\sigma)}\right)=\begin{cases}\dim\mathrm{Ker}\Big(zI+K_A;\ \big[\mathscr{C}^\eta(\partial\Omega)\big]^M\Big)\\ \text{if } p<1 \text{ and } \eta:=(n-1)(\tfrac{1}{p}-1)\in(0,1),\\ \dim\mathrm{Ker}\Big(zI+K_A;\ \big[\mathrm{BMO}(\partial\Omega,\sigma)\big]^M\Big)\\ \text{in the case when } p=1.\end{cases} \tag{7.1.474}$$

Having established (7.1.473), (7.1.474), we may invoke (7.1.62) to conclude that $\mathrm{Im}\Big(zI + K_A^{\#};\, \big[H^p(\partial\Omega,\sigma)\big]^M\Big)$ and $H^p_{A,z}(\partial\Omega,\sigma)$ have the same (finite) codimension in the ambient $\big[H^p(\partial\Omega,\sigma)\big]^M$. Since, according to Proposition 7.1.26 (whose present applicability is ensured by Theorem 3.4.2 and (7.1.471)), the former space is contained into the latter, [130, Remark 1.1.2] ultimately yields (7.1.472). □

Pressing on, we prove the following regularity result for the transpose double layer operator on Hardy spaces on boundaries of δ-oscillating AR domains.

Proposition 7.1.28 *Fix $n \in \mathbb{N}$ with $n \geq 2$ and suppose $\Omega \subseteq \mathbb{R}^n$ is an Ahlfors regular domain with compact boundary. Abbreviate $\sigma := \mathcal{H}^{n-1}\lfloor\partial\Omega$ and denote by ν the geometric measure theoretic outward unit normal to Ω. Also, let L be a homogeneous, second-order, constant complex coefficient, weakly elliptic $M \times M$ system in $\mathbb{R}^n$, with $\mathfrak{A}_L^{\mathrm{dis}} \neq \varnothing$, and pick a coefficient tensor $A \in \mathfrak{A}_L^{\mathrm{dis}}$. Finally, fix some*

$$p \in \begin{cases} \big(\frac{n-1}{n}, 1\big] & \text{if } n \geq 3, \\ \big(\frac{1}{2}, 1\big) & \text{if } n = 2, \end{cases} \tag{7.1.475}$$

and define

$$p^* := \big(\tfrac{1}{p} - \tfrac{1}{n-1}\big)^{-1} \in \big(1, \tfrac{n-1}{n-2}\big]. \tag{7.1.476}$$

Then for each $z \in \mathbb{C} \setminus \{0\}$ there exists a threshold $\delta \in (0,1)$ which depends only on n, p, z, A, the ADR *constants of $\partial\Omega$, and the relative flatness ratio $\wp(\Omega)/\mathrm{diam}\,\partial\Omega$ with the property that when*

$$\mathrm{dist}\Big(\nu, [\mathrm{VMO}(\partial\Omega,\sigma)]^n\Big) < \delta \tag{7.1.477}$$

where the distance is considered in the space $\big[\mathrm{BMO}(\partial\Omega,\sigma)\big]^n$, (hence if Ω is a δ-oscillating AR *domain; cf. Definition 3.4.1), the set Ω is a* UR *domain and the following regularity holds*[8]*:*

$$\left.\begin{array}{r} f \in \big[L^{p^*}_{-1}(\partial\Omega,\sigma)\big]^M \\ (zI + K_A^{\#})f \in \big[H^p(\partial\Omega,\sigma)\big]^M \end{array}\right\} \Longrightarrow f \in \big[H^p(\partial\Omega,\sigma)\big]^M, \tag{7.1.478}$$

where $K_A^{\#}$ is the transpose double layer associated with A and Ω, acting on the negative Sobolev space $\big[L^{p^*}_{-1}(\partial\Omega,\sigma)\big]^M$ as in [132, Theorem 1.5.1, item *(vi)*].

Proof Choose a sufficiently small $\delta \in (0,1)$ as to guarantee that the set Ω is a UR domain satisfying a two-sided local John condition (cf. Theorem 3.4.2). Next, suppose $f \in \big[L^{p^*}_{-1}(\partial\Omega,\sigma)\big]^M$ has $(zI + K_A^{\#})f \in \big[H^p(\partial\Omega,\sigma)\big]^M$. We claim that

$$(zI + K_A^{\#})f \in \mathrm{Im}\Big(zI + K_A^{\#};\, \big[H^p(\partial\Omega,\sigma)\big]^M\Big). \tag{7.1.479}$$

[8] in this regard, it is instructive to recall [130, (11.11.30)]

To prove this, thanks to (7.1.472) it suffices to show that

$$\big(zI + K_A^{\#}\big)f \text{ belongs to } H^p_{A,z}(\partial\Omega,\sigma), \tag{7.1.480}$$

the space introduced in Definition 7.1.25. Since, by hypothesis, $\big(zI + K_A^{\#}\big)f$ belongs to the Hardy space $\big[H^p(\partial\Omega,\sigma)\big]^M$, there remains to check (7.1.467)-(7.1.468). To fix ideas, first treat the case when $p \in \big(\frac{n-1}{n},1\big)$. Set $\eta := (n-1)\big(\frac{1}{p}-1\big) \in (0,1)$ and pick a function $g \in \big[\mathscr{C}^\eta(\partial\Omega)\big]^M$ satisfying $\big(zI + K_A\big)g = 0$.

By eventually further decreasing the value of $\delta \in (0,1)$ we may ensure that (7.1.60)-(7.1.61) hold. As a result, we have $g \in \big[L_1^{(p^*)'}(\partial\Omega,\sigma)\big]^M$, where

$$(p^*)' \in \begin{cases} [n-1,\infty) & \text{if } n \geq 3, \\ (1,\infty) & \text{if } n = 2, \end{cases} \tag{7.1.481}$$

denotes the Hölder conjugate exponent of p^*. We may then write

$$\begin{aligned} &{}_{[H^p(\partial\Omega,\sigma)]^M}\Big\langle \big(zI + K_A^{\#}\big)f, g\Big\rangle_{[\mathscr{C}^\eta(\partial\Omega)]^M} \\ &\quad = {}_{[L^{p^*}_{-1}(\partial\Omega,\sigma)]^M}\Big\langle \big(zI + K_A^{\#}\big)f, g\Big\rangle_{[L_1^{(p^*)'}(\partial\Omega,\sigma)]^M} \\ &\quad = {}_{[L^{p^*}_{-1}(\partial\Omega,\sigma)]^M}\Big\langle f, \big(zI + K_A\big)g\Big\rangle_{[L_1^{(p^*)'}(\partial\Omega,\sigma)]^M} = 0. \end{aligned} \tag{7.1.482}$$

Above, the first equality comes from [130, (11.11.35)], while the second equality is implied by the duality result from [132, Theorem 1.5.1, item *(vi)*]. Finally, the last equality in (7.1.482) is a consequence of our assumption that $\big(zI + K_A\big)g = 0$.

This proves (7.1.480) in the case when $p \in \big(\frac{n-1}{n},1\big)$. The case when $p = 1$ is dealt with similarly, now making use of [130, (11.11.36)]. Hence, (7.1.480) holds for all $p \in \big(\frac{n-1}{n},1\big]$. In turn, as already pointed out, (7.1.480) implies (7.1.479). As a consequence of (7.1.479) and [130, (11.11.30)] we see that there exists some h in $\big[H^p(\partial\Omega,\sigma)\big]^M \subseteq \big[L^{p^*}_{-1}(\partial\Omega,\sigma)\big]^M$ such that $\big(zI+K_A^{\#}\big)f = \big(zI+K_A^{\#}\big)h$. In particular, $f - h \in \mathrm{Ker}\Big(zI + K_A^{\#};\ \big[L^{p^*}_{-1}(\partial\Omega,\sigma)\big]^M\Big)$. From this and (7.1.61) we see that $f-h$ belongs to $\mathrm{Ker}\Big(zI + K_A^{\#};\ \big[H^p(\partial\Omega,\sigma)\big]^M\Big)$. Thus, we have $f - h \in \big[H^p(\partial\Omega,\sigma)\big]^M$, and (7.1.478) follows. □

We next introduce a space of Hölder functions via a scheme which is similar in spirit to the one employed earlier in Definition 7.1.25.

Definition 7.1.29 *Let $\Omega \subseteq \mathbb{R}^n$ (where $n \in \mathbb{N}$, $n \geq 2$) be a* UR *domain with compact boundary. Set $\sigma := \mathcal{H}^{n-1}\lfloor\partial\Omega$ and consider*

$$\alpha \in (0,1), \quad p := \big(1 + \tfrac{\alpha}{n-1}\big)^{-1} \in \big(\tfrac{n-1}{n},1\big), \ \textit{and } z \in \mathbb{C}. \tag{7.1.483}$$

Also, let L be a homogeneous, weakly elliptic, constant (complex) coefficient, second-order $M\times M$ system in $\mathbb{R}^n$ (for some $M \in \mathbb{N}$). Finally, pick a coefficient tensor $A \in \mathfrak{A}_L$

and recall the transpose double layer $K_A^{\#}$, associated with A and Ω, acting on Hardy spaces as in [132, Theorem 2.1.1]. *In the above setting, denote by*

$$\mathscr{C}^{\alpha}_{A,z}(\partial\Omega) \tag{7.1.484}$$

the collection of functions $f \in \big[\mathscr{C}^{\alpha}(\partial\Omega)\big]^M$ having the property that (*with the brackets indicating duality in the sense of* [130, Theorem 4.6.1])

$$\begin{aligned} {}_{[\mathscr{C}^{\alpha}(\partial\Omega)]^M}\big\langle f, g\big\rangle_{[H^p(\partial\Omega,\sigma)]^M} = 0 \textit{ for each distribution} \\ g \in \big[H^p(\partial\Omega,\sigma)\big]^M \textit{ satisfying } \big(zI + K_A^{\#}\big)g = 0. \end{aligned} \tag{7.1.485}$$

The space (7.1.484) is linked to the action of the double layer operator on Hölder spaces, via the following inclusion valid in UR domains with compact boundaries.

Proposition 7.1.30 *Suppose $\Omega \subseteq \mathbb{R}^n$ (where $n \in \mathbb{N}$, $n \geq 2$) is a* UR *domain with compact boundary. Abbreviate $\sigma := \mathcal{H}^{n-1}\lfloor\partial\Omega$ and fix $\alpha \in (0,1)$ and $z \in \mathbb{C}$. Next, let L be a homogeneous, weakly elliptic, constant (complex) coefficient, second-order $M \times M$ system in $\mathbb{R}^n$ (for some $M \in \mathbb{N}$). Finally, pick a coefficient tensor $A \in \mathfrak{A}_L$ and recall the boundary-to-boundary double layer potential operator K_A associated with A and Ω as in* (A.0.118), *acting on Hölder spaces as in* [132, (2.1.125)]. *Then*

$$\mathrm{Im}\Big(zI + K_A;\ \big[\mathscr{C}^{\alpha}(\partial\Omega)\big]^M\Big) \subseteq \mathscr{C}^{\alpha}_{A,z}(\partial\Omega). \tag{7.1.486}$$

Proof This is a consequence of Definition 7.1.29 and [132, (2.1.128)]. □

The inclusion from Proposition 7.1.30 further improves, to an actual equality, in the context of δ-oscillating AR domains.

Proposition 7.1.31 *Fix $n \in \mathbb{N}$ with $n \geq 2$ and suppose $\Omega \subseteq \mathbb{R}^n$ is an Ahlfors regular domain with compact boundary. Abbreviate $\sigma := \mathcal{H}^{n-1}\lfloor\partial\Omega$ and denote by ν the geometric measure theoretic outward unit normal to Ω. Also, let L be a homogeneous, second-order, constant complex coefficient, weakly elliptic $M \times M$ system in $\mathbb{R}^n$, with $\mathfrak{A}_L^{\rm dis} \neq \varnothing$, and fix $\alpha \in (0,1)$ and $z \in \mathbb{C}\setminus\{0\}$. Pick a coefficient tensor $A \in \mathfrak{A}_L^{\rm dis}$, and consider the boundary-to-domain double layer potential operator K_A associated with A and Ω, acting on Hölder space as in* [132, (2.1.125)].

Then there exists a threshold $\delta \in (0,1)$ which depends only on n, α, z, A, the ADR *constants of $\partial\Omega$, and the relative flatness ratio $\wp(\Omega)/\mathrm{diam}\,\partial\Omega$ such that when*

$$\mathrm{dist}\Big(\nu, [\mathrm{VMO}(\partial\Omega,\sigma)]^n\Big) < \delta \tag{7.1.487}$$

where the distance is considered in the space $\big[\mathrm{BMO}(\partial\Omega,\sigma)\big]^n$ (hence if the set Ω is a δ-oscillating AR *domain; cf. Definition 3.4.1), it follows that Ω is a* UR *domain and one has equality in* (7.1.486), *i.e.,*

$$\mathrm{Im}\Big(zI + K_A;\ \big[\mathscr{C}^{\alpha}(\partial\Omega)\big]^M\Big) = \mathscr{C}^{\alpha}_{A,z}(\partial\Omega). \tag{7.1.488}$$

Proof Assume (7.1.487) holds for some $\delta \in (0,1)$ small enough. In particular, Theorem 3.4.2 guarantees that Ω is a UR domain. Also, Theorem 7.1.1 ensures that $zI + K_A$ is a Fredholm operator with index zero on $\big[\mathscr{C}^\alpha(\partial\Omega)\big]^M$. Hence,

$$\begin{aligned}\dim&\left(\frac{\big[\mathscr{C}^\alpha(\partial\Omega)\big]^M}{\mathrm{Im}\big(zI+K_A;[\mathscr{C}^\alpha(\partial\Omega)]^M\big)}\right)\qquad(7.1.489)\\&=\dim\mathrm{CoKer}\Big(zI+K_A;\ \big[\mathscr{C}^\alpha(\partial\Omega)\big]^M\Big)=\dim\mathrm{Ker}\Big(zI+K_A;\ \big[\mathscr{C}^\alpha(\partial\Omega)\big]^M\Big).\end{aligned}$$

Next, Definition 7.1.25 implies

$$\dim\left(\frac{\big[\mathscr{C}^\alpha(\partial\Omega)\big]^M}{\mathscr{C}^\alpha_{A,z}(\partial\Omega)}\right)=\dim\mathrm{Ker}\Big(zI+K_A^\#;\ \big[H^p(\partial\Omega,\sigma)\big]^M\Big)\qquad(7.1.490)$$

where $p := \big(1+\frac{\alpha}{n-1}\big)^{-1} \in \big(\frac{n-1}{n},1\big)$. With (7.1.489) and (7.1.490) in hand, we may invoke (7.1.62) to conclude that the linear spaces $\mathrm{Im}\big(zI+K_A;\ [\mathscr{C}^\alpha(\partial\Omega)]^M\big)$ and $\mathscr{C}^\alpha_{A,z}(\partial\Omega)$ have the same (finite) codimension in $[\mathscr{C}^\alpha(\partial\Omega)]^M$. Given that, according to Proposition 7.1.30 (whose applicability in the present setting is ensured by Theorem 3.4.2 and (7.1.487)), the former space is contained into the latter, we may rely on [130, Remark 1.1.2] to conclude that (7.1.488) holds. □

The stage is set to prove that the single layer operator is Fredholm (with "stable" index) when acting from Hardy spaces on boundaries of δ-oscillating AR domains.

Theorem 7.1.32 *Fix $n \in \mathbb{N}$ with $n \geq 2$ and let $\Omega \subseteq \mathbb{R}^n$ be an Ahlfors regular domain with compact boundary. Abbreviate $\sigma := \mathcal{H}^{n-1}\lfloor\partial\Omega$ and denote by ν the geometric measure theoretic outward unit normal to Ω. Also, let L be a homogeneous, second-order, constant complex coefficient, weakly elliptic $M \times M$ system in $\mathbb{R}^n$ for which*

$$\mathfrak{A}_L^{\mathrm{dis}} \neq \varnothing \ \text{ and } \ \mathfrak{A}_{L^\top}^{\mathrm{dis}} \neq \varnothing. \qquad (7.1.491)$$

Consider the boundary-to-boundary single layer potential operator S associated with the set Ω and the system L as in (A.0.231)*. Finally, fix an exponent*

$$p \in \big(\tfrac{n-1}{n},1\big]. \qquad (7.1.492)$$

Then there exists $\delta \in (0,1)$ depending only on n, p, L, the ADR *constants of $\partial\Omega$, and the relative flatness ratio $\wp(\Omega)/\mathrm{diam}\,\partial\Omega$ such that if*

$$\mathrm{dist}\Big(\nu,\ [\mathrm{VMO}(\partial\Omega,\sigma)]^n\Big) < \delta \qquad (7.1.493)$$

with the distance considered in the space $\big[\mathrm{BMO}(\partial\Omega,\sigma)\big]^n$ (hence if Ω is a δ-oscillating AR *domain; cf. Definition 3.4.1), it follows that Ω is a* UR *domain and*

$$S : \big[H^p(\partial\Omega,\sigma)\big]^M \to \big[H_1^{p^*,p}(\partial\Omega,\sigma)\big]^M,\ \ p^* := \big(\tfrac{1}{p}-\tfrac{1}{n-1}\big)^{-1} \in (1,\infty] \qquad (7.1.494)$$

is a Fredholm operator. Moreover,

$$\text{if } \delta \in (0,1) \text{ is sufficiently small, the index of } S \text{ in (7.1.494) is equal to the index of } S \text{ in any of the settings described in (7.1.103)-(7.1.110).} \tag{7.1.495}$$

Proof Pick $\delta \in (0,1)$ small enough so that Ω is a UR domain (cf. Theorem 3.4.2). Also, select a coefficient tensor $A \in \mathfrak{A}_L^{\text{dis}}$ such that $A^\top \in \mathfrak{A}_{L^\top}^{\text{dis}}$. That the latter choice is possible is ensured by Theorem 1.5.5. First consider the case when

$$p \in \left(\tfrac{n-1}{n}, 1\right] \text{ if } n \geq 3, \text{ and } p \in \left(\tfrac{1}{2}, 1\right) \text{ if } n = 2. \tag{7.1.496}$$

Note that in such a setting we actually have $p^* := \left(\frac{1}{p} - \frac{1}{n-1}\right)^{-1} \in (1, \infty)$. From [132, (2.3.12)] and [132, (2.3.6)] we know that the following operators are bounded:

$$S : \left[H^p(\partial\Omega, \sigma)\right]^M \longrightarrow \left[H_1^{p^*,p}(\partial\Omega, \sigma)\right]^M, \tag{7.1.497}$$

$$\dot{\partial}_\nu^A \mathcal{D} : \left[H_1^{p^*,p}(\partial\Omega, \sigma)\right]^M \longrightarrow \left[H^p(\partial\Omega, \sigma)\right]^M. \tag{7.1.498}$$

Also, in view of the fact that $\partial\Omega$ is compact, [132, (2.3.283)] guarantees that

$$\left(\tfrac{1}{2}I + K_{A^\top}^\#\right) \circ \left(-\tfrac{1}{2}I + K_{A^\top}^\#\right) = \left(\dot{\partial}_\nu^A \mathcal{D}\right) \circ S \text{ on } \left[H^p(\partial\Omega, \sigma)\right]^M. \tag{7.1.499}$$

Collectively, (7.1.497)-(7.1.499) imply

$$\begin{aligned} &\operatorname{Im}\Big(\dot{\partial}_\nu^A \mathcal{D} : \left[H_1^{p^*,p}(\partial\Omega, \sigma)\right]^M \longrightarrow \left[H^p(\partial\Omega, \sigma)\right]^M\Big) \\ &\qquad \supseteq \operatorname{Im}\Big(\left(\tfrac{1}{2}I + K_{A^\top}^\#\right) \circ \left(-\tfrac{1}{2}I + K_{A^\top}^\#\right); \left[H^p(\partial\Omega, \sigma)\right]^M\Big). \end{aligned} \tag{7.1.500}$$

By eventually further decreasing the value of the threshold $\delta \in (0,1)$, we are able to invoke item *(3)* in Theorem 7.1.1 together with item *(b)* in [130, Theorem 2.2.4] in order to conclude that $\left(\frac{1}{2}I + K_{A^\top}^\#\right) \circ \left(-\frac{1}{2}I + K_{A^\top}^\#\right)$ is a Fredholm operator when acting on the space $\left[H^p(\partial\Omega, \sigma)\right]^M$. As such, it has a range of finite codimension in this setting. Having established this, [130, Lemma 1.2.7] gives that

$$\dot{\partial}_\nu^A \mathcal{D} \text{ has closed range of finite codimension in (7.1.498).} \tag{7.1.501}$$

Given that the kernel of $\dot{\partial}_\nu^A \mathcal{D}$ from (7.1.498) is a subspace of $\left[H_1^{p^*,p}(\partial\Omega, \sigma)\right]^M$ which, in turn, is contained in $\left[L^{p^*}(\partial\Omega, \sigma)\right]^M$, it follows that

$$\begin{aligned} &\operatorname{Ker}\Big(\dot{\partial}_\nu^A \mathcal{D} : \left[H_1^{p^*,p}(\partial\Omega, \sigma)\right]^M \longrightarrow \left[H^p(\partial\Omega, \sigma)\right]^M\Big) \\ &\qquad \subseteq \operatorname{Ker}\Big(\partial_\nu^A \mathcal{D} : \left[L^{p^*}(\partial\Omega, \sigma)\right]^M \longrightarrow \left[L_{-1}^{p^*}(\partial\Omega, \sigma)\right]^M\Big) \\ &\qquad \subseteq \operatorname{Ker}\Big(\left(\tfrac{1}{2}I + K_A\right) \circ \left(-\tfrac{1}{2}I + K_A\right); \left[L^{p^*}(\partial\Omega, \sigma)\right]^M\Big), \end{aligned} \tag{7.1.502}$$

where we have used the compatibility property formulated in [132, (2.3.8)] (presently used with the choice $q := p^* \in (1, \infty)$) as well as [132, Theorem 1.5.1, item *(xiii)*, (1.5.85)]. By eventually decreasing the value of $\delta \in (0, 1)$ so that the operators obtained by specializing (7.1.2) to the case when $w \equiv 1$, $z := \pm 1/2$, and $p := p^*$ are Fredholm, we conclude from (7.1.502) and item *(2)* in [130, Theorem 2.1.2] that $\mathring{\partial}_\nu^A \mathcal{D}$ has a finite dimensional kernel in the context of (7.1.498). This together with (7.1.501) prove (cf. [130, Definition 2.2.1] and [130, Remark 2.2.2, *(iii)*]) that

$$\mathring{\partial}_\nu^A \mathcal{D} \text{ is a Fredholm operator in the context of (7.1.498).} \tag{7.1.503}$$

From (7.1.499), (7.1.13) (with A replaced by $A^\top$ and $z = \pm 1/2$), item *(b)* in [130, Theorem 2.2.4], (7.1.503), and [130, (2.2.10)] we see that S is Fredholm in (7.1.494).

Choose $\delta \in (0, 1)$ sufficiently small so that Ω also satisfies a two-sided local John condition (cf. Theorem 3.4.2). Since by [130, (4.2.13), (4.4.114)], and [130, (11.11.30)] we have the following continuous embeddings with dense ranges

$$L^{p^*}(\partial\Omega, \sigma) \hookrightarrow H^p(\partial\Omega, \sigma) \hookrightarrow L^{p^*}_{-1}(\partial\Omega, \sigma), \tag{7.1.504}$$

we conclude, with the help of (7.1.129), that for any $p \in I$ we have

$$\begin{aligned} &\mathrm{Ker}\Big(S : \big[H^p(\partial\Omega, \sigma)\big]^M \to \big[H_1^{p^*,p}(\partial\Omega, \sigma)\big]^M\Big) \\ &\quad = \mathrm{Ker}\Big(S : \big[L^{p^*}_{-1}(\partial\Omega, \sigma)\big]^M \to \big[L^{p^*}(\partial\Omega, \sigma)\big]^M\Big). \end{aligned} \tag{7.1.505}$$

Next, given that, as proved earlier, the single layer potential operator S is Fredholm in the context of (7.1.494), it follows that $S\Big(\big[H^p(\partial\Omega, \sigma)\big]^M\Big)$ is a closed subspace of finite codimension in $\big[H_1^{p^*,p}(\partial\Omega, \sigma)\big]^M$. Let $V \subseteq \big[H_1^{p^*,p}(\partial\Omega, \sigma)\big]^M$ be an algebraic complement of the former space into the latter. Hence,

$$\dim V < +\infty, \tag{7.1.506}$$

$$V \cap S\Big(\big[H^p(\partial\Omega, \sigma)\big]^M\Big) = \{0\}, \tag{7.1.507}$$

$$V + S\Big(\big[H^p(\partial\Omega, \sigma)\big]^M\Big) = \big[H_1^{p^*,p}(\partial\Omega, \sigma)\big]^M. \tag{7.1.508}$$

From (7.1.506) and the fact that $H_1^{p^*,p}(\partial\Omega, \sigma) \hookrightarrow L^{p^*}(\partial\Omega, \sigma)$ we see that V is a finite dimensional subspace of $\big[L^{p^*}(\partial\Omega, \sigma)\big]^M$. Bearing this in mind, from (7.1.136) (used with p^* in place of p) and [130, Lemma 1.2.8] we see that, on the one hand,

$$V + S\Big(\big[L^{p^*}_{-1}(\partial\Omega, \sigma)\big]^M\Big) \text{ is a closed subspace of } \big[L^{p^*}(\partial\Omega, \sigma)\big]^M. \tag{7.1.509}$$

On the other hand, (7.1.508) and (7.1.504) permit us to conclude that

$$\big[H_1^{p^*,p}(\partial\Omega, \sigma)\big]^M = V + S\Big(\big[H^p(\partial\Omega, \sigma)\big]^M\Big) \subseteq V + S\Big(\big[L^{p^*}_{-1}(\partial\Omega, \sigma)\big]^M\Big) \tag{7.1.510}$$

which, in light of [130, (11.10.39)], implies that

$$V + S\Big(\big[L^{p^*}_{-1}(\partial\Omega,\sigma)\big]^M\Big) \text{ is a dense subspace of } \big[L^{p^*}(\partial\Omega,\sigma)\big]^M. \tag{7.1.511}$$

Together, (7.1.509) and (7.1.511) then prove

$$V + S\Big(\big[L^{p^*}_{-1}(\partial\Omega,\sigma)\big]^M\Big) = \big[L^{p^*}(\partial\Omega,\sigma)\big]^M. \tag{7.1.512}$$

Also, if $g \in V \cap S\Big(\big[L^{p^*}_{-1}(\partial\Omega,\sigma)\big]^M\Big)$ then there exists $f \in \big[L^{p^*}_{-1}(\partial\Omega,\sigma)\big]^M$ such that

$$g = Sf \in V \subseteq \big[L^{p^*}(\partial\Omega,\sigma)\big]^M. \tag{7.1.513}$$

Combining the version of [132, (1.5.86)] for compact boundaries with (7.1.513) and keeping in mind (7.1.498) then yields

$$\Big(\big(\tfrac{1}{2}I + K^{\#}_{A^\top}\big) \circ \big(-\tfrac{1}{2}I + K^{\#}_{A^\top}\big)\Big)f = (\dot{\partial}^A_\nu \mathcal{D})(Sf) = (\dot{\partial}^A_\nu \mathcal{D})g \in \big[H^p(\partial\Omega,\sigma)\big]^M. \tag{7.1.514}$$

We may rely on (7.1.478) (used twice) to conclude from (7.1.514) that actually $f \in \big[H^p(\partial\Omega,\sigma)\big]^M$. Hence,

$$g = Sf \in V \cap S\Big(\big[H^p(\partial\Omega,\sigma)\big]^M\Big) = \{0\}, \tag{7.1.515}$$

by (7.1.507), which proves that $V \cap S\Big(\big[L^{p^*}_{-1}(\partial\Omega,\sigma)\big]^M\Big) = \{0\}$. From this and (7.1.512) we then conclude that

$$\dim\left(\frac{\big[H^{p^*,p}_1(\partial\Omega,\sigma)\big]^M}{S\big([H^p(\partial\Omega,\sigma)]^M\big)}\right) = \dim V = \dim\left(\frac{\big[L^{p^*}(\partial\Omega,\sigma)\big]^M}{S\big([L^{p^*}_{-1}(\partial\Omega,\sigma)]^M\big)}\right). \tag{7.1.516}$$

Together with (7.1.505) this shows that

$$\begin{aligned} &\operatorname{index}\Big(S : \big[H^p(\partial\Omega,\sigma)\big]^M \to \big[H^{p^*,p}_1(\partial\Omega,\sigma)\big]^M\Big) \\ &\qquad = \operatorname{index}\Big(S : \big[L^{p^*}_{-1}(\partial\Omega,\sigma)\big]^M \to \big[L^{p^*}(\partial\Omega,\sigma)\big]^M\Big). \end{aligned} \tag{7.1.517}$$

With this in hand, Theorem 7.1.5 applies and gives (7.1.495). This finishes the treatment of the case when p is as in (7.1.496).

There remains to consider the case when $n = 2$ and $p = 1$, a scenario in which p^* becomes ∞. We shall indicate the changes so that the earlier argument still applies. For starters, fix $q \in (1,\infty)$. In place of (7.1.497) we now consider the operators

$$S : \big[H^1(\partial\Omega,\sigma)\big]^M \longrightarrow \big[H^{\infty,1}_1(\partial\Omega,\sigma)\big]^M, \tag{7.1.518}$$

$$\dot{\partial}^A_\nu \mathcal{D} : \big[H^{q,1}_1(\partial\Omega,\sigma)\big]^M \longrightarrow \big[H^1(\partial\Omega,\sigma)\big]^M, \tag{7.1.519}$$

which are well defined, linear, and bounded, thanks to [132, (2.3.12)] (with $n = 2$ and $p = 1$) and [132, (2.3.6)]. Much as in (7.1.499)-(7.1.501), we now have

$$\overset{\bullet}{\partial}{}^A_\nu \mathcal{D} \text{ as in (7.1.519) has closed range of finite codimension.} \tag{7.1.520}$$

Since the kernel of $\overset{\bullet}{\partial}{}^A_\nu \mathcal{D}$ from (7.1.519) is a subspace of $\big[H_1^{q,1}(\partial\Omega,\sigma)\big]^M$ which, in turn, is contained in $\big[L^q(\partial\Omega,\sigma)\big]^M$, it follows that

$$\begin{aligned}\operatorname{Ker}&\Big(\overset{\bullet}{\partial}{}^A_\nu \mathcal{D} : \big[H_1^{q,1}(\partial\Omega,\sigma)\big]^M \longrightarrow \big[H^1(\partial\Omega,\sigma)\big]^M\Big)\\ &\subseteq \operatorname{Ker}\Big(\partial^A_\nu \mathcal{D} : \big[L^q(\partial\Omega,\sigma)\big]^M \longrightarrow \big[L^q_{-1}(\partial\Omega,\sigma)\big]^M\Big)\\ &\subseteq \operatorname{Ker}\Big(\big(\tfrac{1}{2}I + K_A\big)\circ\big(-\tfrac{1}{2}I + K_A\big);\ \big[L^q(\partial\Omega,\sigma)\big]^M\Big),\end{aligned} \tag{7.1.521}$$

thanks to the compatibility property formulated in [132, (2.3.8)] (with $p = 1$) and [132, Theorem 1.5.1, item *(xiii)*] (cf. [132, (1.5.85)]). By eventually decreasing the value of $\delta \in (0,1)$ so that the operators obtained by specializing (7.1.2) to the case when $w \equiv 1$, $z := \pm 1/2$, and $p := q$ are Fredholm, we conclude from (7.1.521) and item *(2)* in [130, Theorem 2.1.2] that $\overset{\bullet}{\partial}{}^A_\nu \mathcal{D}$ has a finite dimensional kernel in the context of (7.1.519). This together with (7.1.520) then prove that

$$\overset{\bullet}{\partial}{}^A_\nu \mathcal{D} \text{ is a Fredholm operator in the context of (7.1.519).} \tag{7.1.522}$$

From (7.1.499), (7.1.13) (with A replaced by $A^\top$ and $z = \pm 1/2$), item *(b)* in [130, Theorem 2.2.4], (7.1.522), and [130, (2.2.10)] with $p = 1$ we conclude that S is Fredholm in the context of (7.1.518).

We are left with justifying (7.1.495) in the case when $n = 2$ and $p = 1$. Again, we reason along previous lines, with some adjustments. As before, choose some $\delta \in (0,1)$ sufficiently small so that Ω also satisfies a two-sided local John condition (cf. Theorem 3.4.2). Fix some $p \in \big(\tfrac{1}{2},1\big)$ and define $p^* := \big(\tfrac{1}{p} - \tfrac{1}{n-1}\big)^{-1} \in (1,\infty)$. Since $\partial\Omega$ is compact, from [129, (3.6.27)], [130, (4.4.11)], and [130, (11.11.30)] we have the following continuous embeddings with dense ranges:

$$L^{p^*}(\partial\Omega,\sigma) \hookrightarrow H^1(\partial\Omega,\sigma) \hookrightarrow H^p(\partial\Omega,\sigma) \hookrightarrow L^{p^*}_{-1}(\partial\Omega,\sigma). \tag{7.1.523}$$

In concert with (7.1.129), this proves

$$\begin{aligned}\operatorname{Ker}&\Big(S : \big[H^1(\partial\Omega,\sigma)\big]^M \to \big[H_1^{\infty,1}(\partial\Omega,\sigma)\big]^M\Big)\\ &= \operatorname{Ker}\Big(S : \big[L^{p^*}_{-1}(\partial\Omega,\sigma)\big]^M \to \big[L^{p^*}(\partial\Omega,\sigma)\big]^M\Big).\end{aligned} \tag{7.1.524}$$

Since the single layer potential operator S is Fredholm in the context of (7.1.518), it follows that $S\Big(\big[H^1(\partial\Omega,\sigma)\big]^M\Big)$ is a closed subspace of finite codimension in $\big[H_1^{\infty,1}(\partial\Omega,\sigma)\big]^M$. Let us pick $V \subseteq \big[H_1^{\infty,1}(\partial\Omega,\sigma)\big]^M$, an algebraic complement of the former space into the latter. This satisfies:

$$\dim V < +\infty, \tag{7.1.525}$$

$$V \cap S\Big(\big[H^1(\partial\Omega,\sigma)\big]^M\Big) = \{0\}, \tag{7.1.526}$$

$$V + S\Big(\big[H^1(\partial\Omega,\sigma)\big]^M\Big) = \big[H_1^{\infty,1}(\partial\Omega,\sigma)\big]^M. \tag{7.1.527}$$

Based on (7.1.525) and the fact that $H_1^{\infty,1}(\partial\Omega,\sigma) \hookrightarrow L^{p^*}(\partial\Omega,\sigma)$ we see that V is a finite dimensional subspace of $\big[L^{p^*}(\partial\Omega,\sigma)\big]^M$. Keeping this in mind, from (7.1.136) (used with p^* in place of p) and [130, Lemma 1.2.8] we conclude that

$$V + S\Big(\big[L^{p^*}_{-1}(\partial\Omega,\sigma)\big]^M\Big) \text{ is a closed subspace of } \big[L^{p^*}(\partial\Omega,\sigma)\big]^M. \tag{7.1.528}$$

Also, (7.1.527) and (7.1.523) give

$$\big[H_1^{\infty,1}(\partial\Omega,\sigma)\big]^M = V + S\Big(\big[H^1(\partial\Omega,\sigma)\big]^M\Big) \subseteq V + S\Big(\big[L^{p^*}_{-1}(\partial\Omega,\sigma)\big]^M\Big) \tag{7.1.529}$$

which, together with [130, (11.10.39)], implies that

$$V + S\Big(\big[L^{p^*}_{-1}(\partial\Omega,\sigma)\big]^M\Big) \text{ is a dense subspace of } \big[L^{p^*}(\partial\Omega,\sigma)\big]^M. \tag{7.1.530}$$

Collectively, (7.1.528) and (7.1.530) then show that

$$V + S\Big(\big[L^{p^*}_{-1}(\partial\Omega,\sigma)\big]^M\Big) = \big[L^{p^*}(\partial\Omega,\sigma)\big]^M. \tag{7.1.531}$$

In addition, if $g \in V \cap S\Big(\big[L^{p^*}_{-1}(\partial\Omega,\sigma)\big]^M\Big)$ then there exists $f \in \big[L^{p^*}_{-1}(\partial\Omega,\sigma)\big]^M$ such that

$$g = Sf \in V \subseteq \big[L^{p^*}(\partial\Omega,\sigma)\big]^M. \tag{7.1.532}$$

Combining the version of [132, (1.5.86)] for compact boundaries, with (7.1.532), and keeping in mind (7.1.519) and that $g \in \big[H_1^{\infty,1}(\partial\Omega,\sigma)\big]^M \subseteq \big[H_1^{q,1}(\partial\Omega,\sigma)\big]^M$, then yields

$$\begin{aligned}\Big(\big(\tfrac{1}{2}I + K^{\#}_{A^\top}\big) \circ \big(-\tfrac{1}{2}I + K^{\#}_{A^\top}\big)\Big) f &= (\dot{\partial}^A_\nu \mathcal{D})(Sf) \\ &= (\dot{\partial}^A_\nu \mathcal{D}) g \in \big[H^1(\partial\Omega,\sigma)\big]^M \subseteq \big[H^p(\partial\Omega,\sigma)\big]^M.\end{aligned} \tag{7.1.533}$$

Invoking (7.1.478) (twice) shows that, in fact, $f \in \big[H^p(\partial\Omega,\sigma)\big]^M$. We actually desire to place f in $\big[H^1(\partial\Omega,\sigma)\big]^M$. This follows from the abstract regularity result [130, (2.1.14)] from [130, Lemma 2.1.4], used here for the Banach spaces

$$X_0 := \big[H^1(\partial\Omega,\sigma)\big]^M, \qquad X_1 := \big[L^{p^*}_{-1}(\partial\Omega,\sigma)\big]^M, \tag{7.1.534}$$

and for $T := \big(\tfrac{1}{2}I + K^{\#}_{A^{\top}}\big) \circ \big(-\tfrac{1}{2}I + K^{\#}_{A^{\top}}\big)$ which is a Fredholm operator with index zero both on X_0 and on X_1 (cf. Theorem 7.1.1 and [130, Theorem 2.1.2]). Once we have established that $f \in \big[H^1(\partial\Omega,\sigma)\big]^M$, we inkove (7.1.526) to obtain

$$g = Sf \in V \cap S\Big(\big[H^1(\partial\Omega,\sigma)\big]^M\Big) = \{0\}. \tag{7.1.535}$$

Hence, $V \cap S\Big(\big[L^{p^*}_{-1}(\partial\Omega,\sigma)\big]^M\Big) = \{0\}$. This and (7.1.531) then imply

$$\dim\left(\frac{\big[H^{\infty,1}_1(\partial\Omega,\sigma)\big]^M}{S\big([H^1(\partial\Omega,\sigma)]^M\big)}\right) = \dim V = \dim\left(\frac{\big[L^{p^*}(\partial\Omega,\sigma)\big]^M}{S\big([L^{p^*}_{-1}(\partial\Omega,\sigma)]^M\big)}\right). \tag{7.1.536}$$

Together with (7.1.524) this shows that

$$\begin{aligned}&\text{index}\Big(S : \big[H^1(\partial\Omega,\sigma)\big]^M \to \big[H^{\infty,1}_1(\partial\Omega,\sigma)\big]^M\Big)\\ &\quad = \text{index}\Big(S : \big[L^{p^*}_{-1}(\partial\Omega,\sigma)\big]^M \to \big[L^{p^*}(\partial\Omega,\sigma)\big]^M\Big).\end{aligned} \tag{7.1.537}$$

With this in hand, Theorem 7.1.5 applies and gives (7.1.495) in the case when $n = 2$ and $p = 1$. The proof of Theorem 7.1.32 is therefore complete. □

Moving on, we make the following observation, further enhancing the scope of Theorem 7.1.32.

Remark 7.1.33 *Virtually the same argument used to prove (the case $p = 1$ of) Theorem 7.1.32 shows that, in the same setting and under the same assumptions, the boundary-to-boundary single layer potential operator has the property that*

$$S : \big[H^1(\partial\Omega,\sigma)\big]^M \longrightarrow \big[H^{q,1}_1(\partial\Omega,\sigma)\big]^M \text{ is Fredholm for } q \in (1,\infty], \tag{7.1.538}$$

and the index of S in (7.1.538) *is actually equal to the index of S in any of the settings described in* (7.1.103)-(7.1.110).

Having just dealt with the issue of Fredholmness, we are now able to actually invert the boundary-to-boundary single layer potential operator associated with Legendre-Hadamard elliptic systems, acting from Hardy spaces defined on boundaries of δ-oscillating AR domains. We first treat dimensions ≥ 3 in Theorem 7.1.34 below, then deal with the two-dimensional case in Theorem 7.1.35, a little later.

Theorem 7.1.34 *Fix $n \in \mathbb{N}$ with $n \geq 3$ and suppose $\Omega \subseteq \mathbb{R}^n$ is an Ahlfors regular domain with compact boundary. Abbreviate $\sigma := \mathcal{H}^{n-1}\lfloor\partial\Omega$ and denote by ν the geometric measure theoretic outward unit normal to Ω. Also, let L be a homogeneous, second-order, constant complex coefficient, $M \times M$ system in $\mathbb{R}^n$ such that*

$$\text{L satisfies the Legendre-Hadamard strong ellipticity condition (cf. [131, Definition 1.3.2]) and one has } \mathfrak{A}^{\text{dis}}_L \neq \varnothing. \tag{7.1.539}$$

Consider the boundary-to-boundary single layer potential operator S associated with Ω and the system L as in (A.0.231). *Finally, let*

$$p \in \left(\tfrac{n-1}{n}, 1\right] \text{ and } p^* := \left(\tfrac{1}{p} - \tfrac{1}{n-1}\right)^{-1}. \tag{7.1.540}$$

Then there exists $\delta_0 \in (0,1)$ depending only on n, p, L, the ADR *constants of $\partial\Omega$, and the relative flatness ratio $\wp(\Omega)/\operatorname{diam}\partial\Omega$ such that if*

$$\operatorname{dist}\Big(\nu, [\mathrm{VMO}(\partial\Omega,\sigma)]^n\Big) < \delta_0 \tag{7.1.541}$$

with the distance considered in $\big[\mathrm{BMO}(\partial\Omega,\sigma)\big]^n$ (hence if Ω is a δ_0-oscillating AR *domain; cf. Definition 3.4.1), it follows that Ω is a* UR *domain and*

$$S : \big[H^p(\partial\Omega,\sigma)\big]^M \longrightarrow \big[H_1^{p^*,p}(\partial\Omega,\sigma)\big]^M \tag{7.1.542}$$

is a well-defined, linear, bounded, invertible operator, with a bounded inverse.

In particular, the operator (7.1.542) *is invertible for any exponent $p \in \left(\tfrac{n-1}{n}, 1\right]$ provided Ω is an infinitesimally flat* AR *domain (cf. Definition 3.4.1).*

As a corollary of duality (cf. [130, Theorem 4.6.1]) *and the fact that* (7.1.542) *is an isomorphism when*[9] $p \in \left(\tfrac{n-1}{n}, 1\right)$ *it follows that the transpose single layer is an isomorphism in the context of*

$$S^\top : \Big(\big[H_1^{p^*,p}(\partial\Omega,\sigma)\big]^M\Big)^* \longrightarrow \big[\mathscr{C}^\eta(\partial\Omega)\big]^M \quad \text{where } \eta := (n-1)\left(\tfrac{1}{p} - 1\right) \in (0,1). \tag{7.1.543}$$

Proof In view of Proposition 1.4.1, all demands in (7.1.101) are presently satisfied. Take $\delta_0 \in (0,1)$ sufficiently small so that all conclusions in Theorem 3.4.2, Theorem 7.1.32, Theorem 7.1.5, and Theorem 7.1.9 hold. Together, these prove that the operator (7.1.542) is Fredholm with index zero. Also, we may assume that Ω satisfies a two-sided local John condition. In turn, this guarantees the validity of the embedding in [130, (11.11.30)]. In concert with the invertibility of (7.1.234) (used with p^* in place of p), this shows that the operator (7.1.542) is also injective. Ultimately, we conclude that (7.1.542) is actually invertible. □

Here is the two-dimensional version of Theorem 7.1.34, dealing with the issue of invertibility for the boundary-to-boundary single layer potential operator, acting from Hardy spaces, associated with Legendre-Hadamard elliptic systems and δ-oscillating AR domains in the plane.

Theorem 7.1.35 *Suppose $\Omega \subseteq \mathbb{R}^2$ is an Ahlfors regular domain with compact boundary. Abbreviate $\sigma := \mathcal{H}^1 \lfloor \partial\Omega$ and denote by ν the geometric measure theoretic outward unit normal to Ω. Also, let L be a homogeneous, second-order, constant complex coefficient, $M \times M$ system in $\mathbb{R}^2$ such that*

---

[9] the case $p = 1$ is discussed separately in Theorem 7.1.39

$$L \textit{ satisfies the Legendre-Hadamard strong ellipticity condition (cf. [131, Definition 1.3.2]) and one has } \mathfrak{A}_L^{\rm dis} \neq \varnothing. \tag{7.1.544}$$

Consider the boundary-to-boundary single layer potential operator S associated with Ω and the system L as in (A.0.231). *Finally, fix*

$$p \in \left(\tfrac{1}{2}, 1\right] \tag{7.1.545}$$

and define

$$p^* := \begin{cases} \left(\frac{1}{p} - 1\right)^{-1} \in (1, \infty) & \textit{if } p < 1, \\ q \in (1, \infty] \textit{ arbitrary} & \textit{if } p = 1. \end{cases} \tag{7.1.546}$$

Then there exists some threshold $\delta_0 \in (0, 1)$ which depends only on p, L, the ADR *constants of $\partial\Omega$, and the relative flatness ratio $\wp(\Omega)/\mathrm{diam}\,\partial\Omega$ such that if*

$$\mathrm{dist}\Big(\nu, [\mathrm{VMO}(\partial\Omega, \sigma)]^2\Big) < \delta_0 \tag{7.1.547}$$

where the distance is considered in the space $\big[\mathrm{BMO}(\partial\Omega, \sigma)\big]^2$ (hence if Ω is a δ_0-oscillating AR *domain; cf. Definition 3.4.1), it follows that Ω is a* UR *domain, and the augmented version of the single layer potential operator,*

$$\begin{gathered} S_{\rm aug} : \big[H^p(\partial\Omega, \sigma)\big]^M \oplus \mathbb{C}^M \longrightarrow \big[H_1^{p^*,p}(\partial\Omega, \sigma)\big]^M \oplus \mathbb{C}^M \\ \textit{given by } S_{\rm aug}(f, c) := \Big(Sf + c, (\langle f, \mathbf{e}_j\rangle)_{1\leq j\leq M}\Big) \textit{ for each} \\ \textit{distribution } f \in \big[H^p(\partial\Omega, \sigma)\big]^M \textit{ and each constant } c \in \mathbb{C}^M, \end{gathered} \tag{7.1.548}$$

is well defined, linear, bounded, invertible, with a bounded inverse.

In particular, the operator (7.1.548) *is invertible for any $p \in \left(\frac{1}{2}, 1\right]$ and p^* as in* (7.1.546) *provided Ω is an infinitesimally flat* AR *domain (cf. Definition 3.4.1).*

A byproduct of the invertibility result (7.1.548) (used with, say, $M := 1$ and $L := \Delta$) and [132, (2.2.130)] is that

$$\begin{gathered} \text{if } \Omega \subseteq \mathbb{R}^2 \text{ is a } \delta\text{-oscillating AR domain for a sufficiently small } \delta > 0, \\ \text{then the space } H_1^{q,1}(\partial\Omega, \sigma) \text{ is independent of the exponent } q \in (1, \infty] \\ \text{and, in fact, } H_1^{q,1}(\partial\Omega, \sigma) = \big\{f \in \mathscr{C}^0(\partial\Omega) : \partial_{\tau_{12}} f \in H^1(\partial\Omega, \sigma)\big\}. \end{gathered} \tag{7.1.549}$$

As a corollary of the fact that (7.1.548) is an isomorphism when $p \in \left(\frac{1}{2}, 1\right)$ and duality (cf. [130, Theorem 4.6.1]) it follows that the transpose augmented single layer is also an isomorphism in the context of

$$\begin{gathered} S^\top_{\rm aug} : \Big(\big[H_1^{p^*,p}(\partial\Omega, \sigma)\big]^M\Big)^* \oplus \mathbb{C}^M \longrightarrow \big[\mathscr{C}^\eta(\partial\Omega)\big]^M \oplus \mathbb{C}^M \\ \text{given by } S^\top_{\rm aug}(\Lambda, c) := \Big(S^\top\Lambda + c, (\langle \Lambda, \mathbf{e}_j\rangle)_{1\leq j\leq M}\Big) \text{ for each} \\ \text{functional } \Lambda \in \Big(\big[H_1^{p^*,p}(\partial\Omega, \sigma)\big]^M\Big)^* \text{ and each constant } c \in \mathbb{C}^M, \end{gathered} \tag{7.1.550}$$

where $\eta := \frac{1}{p} - 1 \in (0,1)$ and $S^\top : \big([H_1^{p^*,p}(\partial\Omega,\sigma)]^M\big)^* \to [\mathscr{C}^\eta(\partial\Omega)]^M$ is the transpose of S in [132, (2.3.12)] with $n := 2$.

Here is the proof of Theorem 7.1.35.

Proof of Theorem 7.1.35 Since the system L satisfies the Legendre-Hadamard strong ellipticity condition and we are presently assuming $\mathfrak{A}_L^{\rm dis} \neq \varnothing$, it follows from Proposition 1.4.1 that

$$\mathfrak{A}_L^{\rm dis} \neq \varnothing \text{ and } \mathfrak{A}_{L^\top}^{\rm dis} \neq \varnothing. \tag{7.1.551}$$

To proceed, choose $\delta_0 \in (0,1)$ sufficiently enough so that having (7.1.547) ensures that Ω is a UR domain and the operator

$$S : \big[H^p(\partial\Omega,\sigma)\big]^M \longrightarrow \big[H_1^{p^*,p}(\partial\Omega,\sigma)\big]^M \tag{7.1.552}$$

is Fredholm with index zero. That this is possible is ensured by Theorem 7.1.32, Remark 7.1.33 (whose applicability require (7.1.551)), and Corollary 7.1.6. In what follows, we reserve the right to further decrease the value $\delta_0 \in (0,1)$ as needed (to ensure that the results invoked apply). Observe that the augmented single layer operator (7.1.548) may be decomposed as

$$S_{\rm aug} = T + Q \tag{7.1.553}$$

where

$$\begin{gathered} T : \big[H^p(\partial\Omega,\sigma)\big]^M \oplus \mathbb{C}^M \longrightarrow \big[H_1^{p^*,p}(\partial\Omega,\sigma)\big]^M \oplus \mathbb{C}^M \text{ is given} \\ \text{by } T(f,c) := \big(Sf, 0\big) \text{ for each } f \in \big[H^p(\partial\Omega,\sigma)\big]^M \text{ and } c \in \mathbb{C}^M, \end{gathered} \tag{7.1.554}$$

and where

$$\begin{gathered} Q : \big[H^p(\partial\Omega,\sigma)\big]^M \oplus \mathbb{C}^M \longrightarrow \big[H_1^{p^*,p}(\partial\Omega,\sigma)\big]^M \oplus \mathbb{C}^M \\ \text{is defined as } Q(f,c) := \Big(c, (\langle f, \mathbf{e}_j\rangle)_{1\leq j\leq M}\Big) \text{ for each} \\ \text{distribution } f \in \big[H^p(\partial\Omega,\sigma)\big]^M \text{ and constant } c \in \mathbb{C}^M. \end{gathered} \tag{7.1.555}$$

Then both T, Q are well-defined, linear, bounded. In fact, since Q is also a finite-rank operator, Q is actually compact. In addition, it is clear from definitions that

$$\begin{aligned} &{\rm Im}\Big(T : \big[H^p(\partial\Omega,\sigma)\big]^M \oplus \mathbb{C}^M \to \big[H_1^{p^*,p}(\partial\Omega,\sigma)\big]^M \oplus \mathbb{C}^M\Big) \\ &\quad = {\rm Im}\Big(S : \big[H^p(\partial\Omega,\sigma) \to \big[H_1^{p^*,p}(\partial\Omega,\sigma)\big]^M\Big) \oplus \{0\}, \end{aligned} \tag{7.1.556}$$

and

$$\begin{aligned} &{\rm Ker}\Big(T : \big[H^p(\partial\Omega,\sigma)\big]^M \oplus \mathbb{C}^M \to \big[H_1^{p^*,p}(\partial\Omega,\sigma)\big]^M \oplus \mathbb{C}^M\Big) \\ &\quad = {\rm Ker}\Big(S : \big[H^p(\partial\Omega,\sigma)\big]^M \to \big[H_1^{p^*,p}(\partial\Omega,\sigma)\big]^M\Big) \oplus \mathbb{C}^M. \end{aligned} \tag{7.1.557}$$

From (7.1.556)-(7.1.557) and since S in (7.1.552) is Fredholm with index zero we conclude that T in (7.1.554) is Fredholm with index zero. Since Q is compact, (7.1.553) implies that S_{aug} is a Fredholm operator with index zero in (7.1.548).

To show that S_{aug} is actually an invertible operator, there remains to prove that S_{aug} is injective. To this end, assume $f \in \big[H^p(\partial\Omega, \sigma)\big]^M$ and $c \in \mathbb{C}^M$ are such that $Sf + c = 0$ and $\langle f, \mathbf{e}_j\rangle = 0$ for $1 \le j \le M$. Then (7.1.499) presently gives

$$\big(\tfrac{1}{2}I + K^{\#}_{A^\top}\big)\big(-\tfrac{1}{2}I + K^{\#}_{A^\top}\big)f = -(\dot{\partial}^A_\nu \mathcal{D})c = 0, \tag{7.1.558}$$

where the last equality is based on the compatibility property formulated at the end of [132, Theorem 2.3.1, item *(iii)*], [132, Proposition 1.3.6], and (A.0.203). As such,

$$\big(-\tfrac{1}{2}I + K^{\#}_{A^\top}\big)f \in \operatorname{Ker}\Big(\tfrac{1}{2}I + K^{\#}_{A^\top};\ \big[H^p(\partial\Omega, \sigma)\big]^M\Big) \tag{7.1.559}$$

which, in concert with the first equality in (7.1.61) (used with $z := 1/2$, $r := p$, $p := 2$, and $A^\top$ in place of A) gives that

$$\big(-\tfrac{1}{2}I + K^{\#}_{A^\top}\big)f \in \big[L^2(\partial\Omega, \sigma)\big]^M. \tag{7.1.560}$$

Returning to (7.1.558) and interchanging the operators in the left-most side gives

$$\big(\tfrac{1}{2}I + K^{\#}_{A^\top}\big)f \in \operatorname{Ker}\Big(-\tfrac{1}{2}I + K^{\#}_{A^\top};\ \big[H^p(\partial\Omega, \sigma)\big]^M\Big). \tag{7.1.561}$$

Together with the first equality in (7.1.61) (now used with $z := -1/2$, $r := p$, $p := 2$, and $A^\top$ in place of A) this yields

$$\big(\tfrac{1}{2}I + K^{\#}_{A^\top}\big)f \in \big[L^2(\partial\Omega, \sigma)\big]^M. \tag{7.1.562}$$

By combining (7.1.560) and (7.1.562) we arrive at

$$f = \big(\tfrac{1}{2}I + K^{\#}_{A^\top}\big)f - \big(-\tfrac{1}{2}I + K^{\#}_{A^\top}\big)f \in \big[L^2(\partial\Omega, \sigma)\big]^M. \tag{7.1.563}$$

As such, the pair (f, c) belongs to the null-space of the augmented single layer operator S_{aug} from (7.1.245) with $p := 2$. However, Theorem 7.1.10 guarantees that said operator is invertible, forcing $f = 0$. In turn, the fact that $Sf + c = 0$ also gives $c = 0$. Hence, $(f, c) = (0, 0)$ proving that S_{aug} is injective in the context of (7.1.548). As already mentioned, all desired conclusions now follow. □

In turn, Theorem 7.1.34 is a key ingredient in the proof of the essential norm estimates for the boundary-to-boundary double layer associated with a Legendre-Hadamard elliptic system and a δ-oscillating AR domain, as an operator on Hardy-based Sobolev spaces. We accomplish this in dimensions ≥ 3 in Theorem 7.1.36 below, and in the two-dimensional setting in Theorem 7.1.37 stated a little later.

Theorem 7.1.36 *Fix $n \in \mathbb{N}$ with $n \ge 3$ and suppose $\Omega \subseteq \mathbb{R}^n$ is a* UR *domain with compact boundary. Abbreviate $\sigma := \mathcal{H}^{n-1}\lfloor\partial\Omega$ and denote by ν the geometric* *measure theoretic outward unit normal to Ω. Also, let L be a homogeneous, second-order, constant complex coefficient, $M \times M$ system in $\mathbb{R}^n$ such that*

$$\text{L satisfies the Legendre-Hadamard strong ellipticity condition (cf. [131, Definition 1.3.2]) and one has } \mathfrak{A}_L^{\rm dis} \neq \varnothing. \tag{7.1.564}$$

Pick a coefficient tensor $A \in \mathfrak{A}_L^{\rm dis}$ and consider

$$p \in \left(\tfrac{n-1}{n}, 1\right], \ \theta \in \left(0, n - \tfrac{n-1}{p}\right), \ \text{ and } p^* := \left(\tfrac{1}{p} - \tfrac{1}{n-1}\right)^{-1}. \tag{7.1.565}$$

Finally, recall from [132, (2.3.9)] *that the boundary-to-boundary double layer potential operator K_A associated with the set Ω and the coefficient tensor A induces a well-defined, linear, and bounded mapping*

$$K_A : \left[H_1^{p^*,p}(\partial\Omega, \sigma)\right]^M \longrightarrow \left[H_1^{p^*,p}(\partial\Omega, \sigma)\right]^M. \tag{7.1.566}$$

Then there exists a constant $C_\theta \in (0, \infty)$, which depends only on θ, n, p, A, the UR *constants of $\partial\Omega$, and the relative flatness ratio $\wp(\Omega)/\mathrm{diam}\,\partial\Omega$ such that*

$$\mathrm{dist}\Big(K_A, \mathrm{Cp}\big(\left[H_1^{p^*,p}(\partial\Omega, \sigma)\right]^M\big)\Big) \leq C_\theta \Big[\mathrm{dist}\big(\nu, \left[\mathrm{VMO}(\partial\Omega, \sigma)\right]^n\big)\Big]^\theta \tag{7.1.567}$$

where distance in the left-hand side is measured in $\mathrm{Bd}\big(\left[H_1^{p^*,p}(\partial\Omega, \sigma)\right]^M\big)$, *while the distance in the right-hand side is measured in the space* $\left[\mathrm{BMO}(\partial\Omega, \sigma)\right]^n$. *In particular, from* (7.1.567) *and* [130, (1.2.53)] *one concludes that*

$$\text{the operator K_A is compact on } \left[H_1^{p^*,p}(\partial\Omega, \sigma)\right]^M \text{ whenever Ω is an infinitesimally flat AR domain (cf. Definition 3.4.1).} \tag{7.1.568}$$

Proof From [132, Theorem 2.3.1, item *(iii)*] we know that there exists some constant $C_0 \in (0, \infty)$, which depends only on n, p, A, and the UR constants of $\partial\Omega$ such that

$$\|K_A\|_{[H_1^{p^*,p}(\partial\Omega,\sigma)]^M \to [H_1^{p^*,p}(\partial\Omega,\sigma)]^M} \leq C_0. \tag{7.1.569}$$

Recall the parameter $\delta_0 \in (0, 1)$ from the statement of Theorem 7.1.34. When we actually have $\mathrm{dist}\big(\nu, \left[\mathrm{VMO}(\partial\Omega, \sigma)\right]^n\big) \geq \delta_0$, the estimate claimed in (7.1.567) is justified by using [130, (1.2.56)] and (7.1.569) to simply write

$$\begin{aligned}\mathrm{dist}\Big(K_A, \mathrm{Cp}\big(\left[H_1^{p^*,p}(\partial\Omega, \sigma)\right]^M\big)\Big) &\leq \|K_A\|_{[H_1^{p^*,p}(\partial\Omega,\sigma)]^M \to [H_1^{p^*,p}(\partial\Omega,\sigma)]^M} \\ &\leq C_0 \leq (C_0/\delta_0) \cdot \mathrm{dist}\big(\nu, \left[\mathrm{VMO}(\partial\Omega, \sigma)\right]^n\big).\end{aligned} \tag{7.1.570}$$

In view of (4.1.17) and the fact that $\mathrm{dist}\big(\nu, \left[\mathrm{VMO}(\partial\Omega, \sigma)\right]^n\big) \leq 2$, this readily implies (7.1.567). Thus, as far as the estimate claimed in (7.1.567) is concerned, there remains to consider the situation when

$$\mathrm{dist}\big(\nu, \left[\mathrm{VMO}(\partial\Omega, \sigma)\right]^n\big) < \delta_0. \tag{7.1.571}$$

In such a scenario, Theorem 7.1.34 implies that (7.1.542) is a well-defined, linear, bounded operator with a bounded inverse

$$S^{-1} : \big[H_1^{p^*,p}(\partial\Omega,\sigma)\big]^M \longrightarrow \big[H^p(\partial\Omega,\sigma)\big]^M. \tag{7.1.572}$$

Since from [132, (2.3.284)] we also know that

$$S \circ K^{\#}_{A^\top} = K_A \circ S \text{ on } \big[H^p(\partial\Omega,\sigma)\big]^M, \tag{7.1.573}$$

it follows that we may represent the operator K_A from (7.1.566) as

$$K_A = S \circ K^{\#}_{A^\top} \circ S^{-1} \text{ on } \big[H_1^{p^*,p}(\partial\Omega,\sigma)\big]^M. \tag{7.1.574}$$

Granted this, [130, Lemma 1.2.10] then yields

$$\|K_A\|^{\text{ess}}_{[H_1^{p^*,p}(\partial\Omega,\sigma)]^M \to [H_1^{p^*,p}(\partial\Omega,\sigma)]^M} \leq C\|K^{\#}_{A^\top}\|^{\text{ess}}_{[H^p(\partial\Omega,\sigma)]^M \to [H^p(\partial\Omega,\sigma)]^M} \tag{7.1.575}$$

where $C \in (0,\infty)$ is given by

$$C := \|S\|_{[H^p(\partial\Omega,\sigma)]^M \to [H_1^{p^*,p}(\partial\Omega,\sigma)]^M} \cdot \|S^{-1}\|_{[H_1^{p^*,p}(\partial\Omega,\sigma)]^M \to [H^p(\partial\Omega,\sigma)]^M}.$$

In particular, C depends only on n, p, L, the UR constants of $\partial\Omega$, and the relative flatness ratio $\wp(\Omega)/\text{diam}\,\partial\Omega$. The current assumptions and Proposition 1.4.1 guarantee that $A^\top \in \mathfrak{A}^{\text{dis}}_{L^\top}$. As such, from item *(vi)* in Proposition 1.3.2 and Theorem 5.4.9 we conclude that there exists a constant $C_\theta \in (0,\infty)$, which depends only on θ, n, p, A, the UR constants of $\partial\Omega$, and the relative flatness ratio $\wp(\Omega)/\text{diam}\,\partial\Omega$ such that

$$\|K^{\#}_{A^\top}\|^{\text{ess}}_{[H^p(\partial\Omega,\sigma)]^M \to [H^p(\partial\Omega,\sigma)]^M} \leq C_\theta \cdot \Big[\text{dist}\big(\nu, \big[\text{VMO}(\partial\Omega,\sigma)\big]^n\big)\Big]^\theta \tag{7.1.576}$$

where the distance in the right-hand side is measured in $\big[\text{BMO}(\partial\Omega,\sigma)\big]^n$. The estimate claimed in (7.1.567) now follows from (7.1.575) and (7.1.576). □

Here is the two-dimensional counterpart of Theorem 7.1.36, dealing with essential norm estimates on Hardy-based Sobolev spaces for the boundary-to-boundary double layer associated with a Legendre-Hadamard elliptic system and a δ-oscillating AR domain in the plane.

Theorem 7.1.37 *Suppose $\Omega \subseteq \mathbb{R}^2$ is a* UR *domain with compact boundary. Abbreviate $\sigma := \mathcal{H}^1\lfloor\partial\Omega$ and denote by ν the geometric measure theoretic outward unit normal to Ω. Also, let L be a homogeneous, second-order, constant complex coefficient, $M \times M$ system in $\mathbb{R}^2$ such that*

L satisfies the Legendre-Hadamard strong ellipticity condition (cf. [131, Definition 1.3.2]) *and one has $\mathfrak{A}^{\text{dis}}_L \neq \varnothing$.* (7.1.577)

Consider

$$p \in \left(\tfrac{1}{2}, 1\right], \qquad \theta \in \left(0, 2 - \tfrac{1}{p}\right), \tag{7.1.578}$$

and define

$$p^* := \begin{cases} \left(\frac{1}{p} - 1\right)^{-1} \in (1, \infty) & \text{if } p < 1, \\ q \in (1, \infty) \text{ arbitrary} & \text{if } p = 1. \end{cases} \tag{7.1.579}$$

Finally, pick a coefficient tensor $A \in \mathfrak{A}_L^{\text{dis}}$ *and recall from* [132, (2.3.9)] *that the boundary-to-boundary double layer potential operator* K_A *associated with the set* Ω *and the coefficient tensor A induces a well-defined, linear, and bounded mapping*

$$K_A : \left[H_1^{p^*,p}(\partial\Omega, \sigma)\right]^M \longrightarrow \left[H_1^{p^*,p}(\partial\Omega, \sigma)\right]^M. \tag{7.1.580}$$

Then there exists a constant $C_\theta \in (0, \infty)$*, which depends only on* θ*,* p*,* A*, the* UR *constants of* $\partial\Omega$*, and the relative flatness ratio* $\wp(\Omega)/\operatorname{diam}\partial\Omega$ *such that*

$$\operatorname{dist}\Big(K_A, \operatorname{Cp}\big(\left[H_1^{p^*,p}(\partial\Omega, \sigma)\right]^M\big)\Big) \leq C_\theta \Big[\operatorname{dist}\big(\nu, \left[\mathrm{VMO}(\partial\Omega, \sigma)\right]^2\big)\Big]^\theta \tag{7.1.581}$$

where distance in the left-hand side is measured in $\operatorname{Bd}\big(\left[H_1^{p^*,p}(\partial\Omega, \sigma)\right]^M\big)$*, while the distance in the right-hand side is measured in the space* $\left[\mathrm{BMO}(\partial\Omega, \sigma)\right]^2$*. In particular, from* (7.1.581) *and* [130, (1.2.53)] *one concludes that*

$$\begin{aligned} &\textit{the operator } K_A \textit{ is compact on } \left[H_1^{p^*,p}(\partial\Omega, \sigma)\right]^M \textit{ whenever} \\ &\Omega \textit{ is an infinitesimally flat } \text{AR} \textit{ domain (cf. Definition 3.4.1).} \end{aligned} \tag{7.1.582}$$

Proof As in the proof of Theorem 7.1.36 the crux of the matter is establishing that there exists some constant $C \in (0, \infty)$ is as in the current statement

$$\|K_A\|^{\text{ess}}_{[H_1^{p^*,p}(\partial\Omega,\sigma)]^M \to [H_1^{p^*,p}(\partial\Omega,\sigma)]^M} \leq C \|K^\#_{A^\top}\|^{\text{ess}}_{[H^p(\partial\Omega,\sigma)]^M \to [H^p(\partial\Omega,\sigma)]^M} \tag{7.1.583}$$

whenever

$$\operatorname{dist}\big(\nu, \left[\mathrm{VMO}(\partial\Omega, \sigma)\right]^2\big) < \delta_0 \text{ with } \delta_0 \in (0, 1) \text{ sufficiently small.} \tag{7.1.584}$$

To this end, assume (7.1.584). Also, by eventually replacing Ω by $\mathbb{R}^2 \setminus \overline{\Omega}$, there is no loss of generality in assuming that Ω is bounded. See [129, Lemma 5.10.9] and [129, Lemma 5.10.10] in this regard. To proceed, define

$$\begin{aligned} &\mathbb{K}_A : \left[H_1^{p^*,p}(\partial\Omega, \sigma)\right]^M \oplus \mathbb{C}^M \longrightarrow \left[H_1^{p^*,p}(\partial\Omega, \sigma)\right]^M \oplus \mathbb{C}^M \\ &\mathbb{K}_A(f, c) := \big(K_A f, c/2\big) \text{ for } f \in \left[H_1^{p^*,p}(\partial\Omega, \sigma)\right]^M \text{ and } c \in \mathbb{C}^M, \end{aligned} \tag{7.1.585}$$

as well as

$$\begin{aligned}&\mathbb{K}^{\#}_{A^\top}:\big[H^p(\partial\Omega,\sigma)\big]^M\oplus\mathbb{C}^M\longrightarrow\big[H^p(\partial\Omega,\sigma)\big]^M\oplus\mathbb{C}^M\ \text{ given by}\\&\mathbb{K}^{\#}_{A^\top}(f,c):=\big(K^{\#}_{A^\top}f,c/2\big)\ \text{ for each } f\in\big[H^p(\partial\Omega,\sigma)\big]^M \text{ and } c\in\mathbb{C}^M.\end{aligned}\tag{7.1.586}$$

From assumptions, Theorem 3.4.2, [132, Theorem 2.3.1, item *(iii)*], and [132, Theorem 2.1.1] we know that these are well-defined, linear, and bounded operators. Also, assuming δ_0 small enough, matters may be arranged so that the augmented version of the single layer potential operator $S_{\rm aug}$ is invertible in the context of (7.1.548). For each $f\in\big[H^p(\partial\Omega,\sigma)\big]^M$ and $c\in\mathbb{C}^M$ we may then make use of the transposition results from [132, Theorem 1.5.1, item *(xiii)*] together with formula [132, (1.5.21)] (both for $A^\top$ and A) to compute

$$\begin{aligned}\big(S_{\rm aug}\circ\mathbb{K}^{\#}_{A^\top}\big)(f,c)&=S_{\rm aug}\big(\mathbb{K}^{\#}_{A^\top}(f,c)\big)=S_{\rm aug}\big(K^{\#}_{A^\top}f,c/2\big)\\&=\Big(S(K^{\#}_{A^\top}f)+c/2,(\langle K^{\#}_{A^\top}f,\mathbf{e}_j\rangle)_{1\le j\le M}\Big)\\&=\Big(K_A(Sf)+c/2,(\langle f,K_{A^\top}\mathbf{e}_j\rangle)_{1\le j\le M}\Big)\\&=\Big(K_A(Sf)+c/2,(\langle f,\mathbf{e}_j/2\rangle)_{1\le j\le M}\Big),\end{aligned}\tag{7.1.587}$$

and

$$\begin{aligned}\big(\mathbb{K}_A\circ S_{\rm aug}\big)(f,c)&=\mathbb{K}_A\big(S_{\rm aug}(f,c)\big)=\mathbb{K}_A\Big(Sf+c,(\langle f,\mathbf{e}_j\rangle)_{1\le j\le M}\Big)\\&=\Big(K_A(Sf+c),(\langle f,\mathbf{e}_j/2\rangle)_{1\le j\le M}\Big)=\Big(K_A(Sf)+c/2,(\langle f,\mathbf{e}_j/2\rangle)_{1\le j\le M}\Big).\end{aligned}\tag{7.1.588}$$

From (7.1.587) and (7.1.588) we see that $S_{\rm aug}\circ\mathbb{K}^{\#}_{A^\top}=\mathbb{K}_A\circ S_{\rm aug}$ as operators from the space $\big[H^p(\partial\Omega,\sigma)\big]^M\oplus\mathbb{C}^M$ into itself. Upon recalling that $S_{\rm aug}$ is invertible in the context of (7.1.548), from this we further deduce that

$$\mathbb{K}_A=S_{\rm aug}\circ\mathbb{K}^{\#}_{A^\top}\circ S^{-1}_{\rm aug}\tag{7.1.589}$$

as operators from the space $\big[H^{p^*,p}_1(\partial\Omega,\sigma)\big]^M\oplus\mathbb{C}^M$ into itself. Let us also introduce the inclusion and projection operators defined as

$$\begin{aligned}&\iota:\big[H^{p^*,p}_1(\partial\Omega,\sigma)\big]^M\longrightarrow\big[H^{p^*,p}_1(\partial\Omega,\sigma)\big]^M\oplus\mathbb{C}^M\\&\iota f:=(f,0)\ \text{ for each } f\in\big[H^{p^*,p}_1(\partial\Omega,\sigma)\big]^M,\end{aligned}\tag{7.1.590}$$

and, respectively,

$$\begin{aligned}&\pi:\big[H^{p^*,p}_1(\partial\Omega,\sigma)\big]^M\oplus\mathbb{C}^M\longrightarrow\big[H^{p^*,p}_1(\partial\Omega,\sigma)\big]^M\\&\pi(f,c):=f\ \text{ for each } f\in\big[H^{p^*,p}_1(\partial\Omega,\sigma)\big]^M \text{ and } c\in\mathbb{C}^M.\end{aligned}\tag{7.1.591}$$

Again, these are both linear and bounded operators. In particular, (7.1.589) together with (7.1.585) imply

$$K_A=\pi\circ S_{\rm aug}\circ\mathbb{K}^{\#}_{A^\top}\circ S^{-1}_{\rm aug}\circ\iota\tag{7.1.592}$$

as operators from the space $\big[H_1^{p^*,p}(\partial\Omega,\sigma)\big]^M$ into itself. Granted this, [130, Lemma 1.2.10] then yields

$$\begin{aligned}\|K_A\|^{\mathrm{ess}}_{[H_1^{p^*,p}(\partial\Omega,\sigma)]^M\to[H_1^{p^*,p}(\partial\Omega,\sigma)]^M}\\ &= \big\|\pi\circ S_{\mathrm{aug}}\circ\mathbb{K}^{\#}_{A^\top}\circ S^{-1}_{\mathrm{aug}}\circ\iota\big\|^{\mathrm{ess}}_{[H_1^{p^*,p}(\partial\Omega,\sigma)]^M\to[H_1^{p^*,p}(\partial\Omega,\sigma)]^M}\\ &\le C\|\mathbb{K}^{\#}_{A^\top}\|^{\mathrm{ess}}_{[H^p(\partial\Omega,\sigma)]^M\oplus\mathbb{C}^M\to[H^p(\partial\Omega,\sigma)]^M\oplus\mathbb{C}^M}.\end{aligned}\tag{7.1.593}$$

Next, fix an arbitrary compact operator $Q\in\mathrm{Cp}\Big(\big[H^p(\partial\Omega,\sigma)\big]^M\Big)$ and define

$$\begin{aligned}&\widetilde{Q}:\big[H^p(\partial\Omega,\sigma)\big]^M\oplus\mathbb{C}^M\longrightarrow\big[H^p(\partial\Omega,\sigma)\big]^M\oplus\mathbb{C}^M\ \text{ given by}\\ &\widetilde{Q}(f,c):=\big(Qf,c/2\big)\ \text{ for } f\in\big[H^p(\partial\Omega,\sigma)\big]^M \text{ and } c\in\mathbb{C}^M.\end{aligned}\tag{7.1.594}$$

Then $\widetilde{Q}\in\mathrm{Cp}\Big(\big[H^p(\partial\Omega,\sigma)\big]^M\oplus\mathbb{C}^M\Big)$ and

$$(\mathbb{K}^{\#}_{A^\top}-\widetilde{Q})(f,c)=\big((K^{\#}_{A^\top}-Q)f,0\big)\ \text{ for }\ f\in\big[H^p(\partial\Omega,\sigma)\big]^M \text{ and } c\in\mathbb{C}^M.$$

Consequently, for each $f\in\big[H^p(\partial\Omega,\sigma)\big]^M$ and $c\in\mathbb{C}^M$ we may estimate

$$\begin{aligned}&\big\|(\mathbb{K}^{\#}_{A^\top}-\widetilde{Q})(f,c)\big\|_{[H^p(\partial\Omega,\sigma)]^M\oplus\mathbb{C}^M}\\ &\quad=\big\|\big((K^{\#}_{A^\top}-Q)f,0\big)\big\|_{[H^p(\partial\Omega,\sigma)]^M\oplus\mathbb{C}^M}=\big\|(K^{\#}_{A^\top}-Q)f\big\|_{[H^p(\partial\Omega,\sigma)]^M}\\ &\quad\le\|K^{\#}_{A^\top}-Q\|_{[H^p(\partial\Omega,\sigma)]^M\to[H^p(\partial\Omega,\sigma)]^M}\cdot\|f\|_{[H^p(\partial\Omega,\sigma)]^M}\\ &\quad\le\|K^{\#}_{A^\top}-Q\|_{[H^p(\partial\Omega,\sigma)]^M\to[H^p(\partial\Omega,\sigma)]^M}\cdot\|(f,c)\|_{[H^p(\partial\Omega,\sigma)]^M\oplus\mathbb{C}^M}.\end{aligned}\tag{7.1.595}$$

Bearing in mind that $\widetilde{Q}\in\mathrm{Cp}\Big(\big[H^p(\partial\Omega,\sigma)\big]^M\oplus\mathbb{C}^M\Big)$, this further implies

$$\begin{aligned}&\|\mathbb{K}^{\#}_{A^\top}\|^{\mathrm{ess}}_{[H^p(\partial\Omega,\sigma)]^M\oplus\mathbb{C}^M\to[H^p(\partial\Omega,\sigma)]^M\oplus\mathbb{C}^M}\\ &\qquad\le\big\|\mathbb{K}^{\#}_{A^\top}-\widetilde{Q}\big\|_{[H^p(\partial\Omega,\sigma)]^M\oplus\mathbb{C}^M\to[H^p(\partial\Omega,\sigma)]^M\oplus\mathbb{C}^M}\\ &\qquad\le\|K^{\#}_{A^\top}-Q\|_{[H^p(\partial\Omega,\sigma)]^M\to[H^p(\partial\Omega,\sigma)]^M}\end{aligned}\tag{7.1.596}$$

which, after taking the infimum over all $Q\in\mathrm{Cp}\Big(\big[H^p(\partial\Omega,\sigma)\big]^M\Big)$ yields

$$\begin{aligned}&\|\mathbb{K}^{\#}_{A^\top}\|^{\mathrm{ess}}_{[H^p(\partial\Omega,\sigma)]^M\oplus\mathbb{C}^M\to[H^p(\partial\Omega,\sigma)]^M\oplus\mathbb{C}^M}\\ &\qquad\le C\|K^{\#}_{A^\top}\|^{\mathrm{ess}}_{[H^p(\partial\Omega,\sigma)]^M\to[H^p(\partial\Omega,\sigma)]^M}.\end{aligned}\tag{7.1.597}$$

Collectively, (7.1.593) and (7.1.597) establish (7.1.583). With this in hand, the same argument as in the end-game of the proof of Theorem 7.1.36 yields all desired conclusions. □

The essential norm estimate (7.1.567) yields the following Fredholmness result, involving the boundary-to-boundary double layer associated with a Legendre-Hadamard elliptic system and a δ-oscillating AR domain, as an operator on Hardy-based Sobolev spaces.

Corollary 7.1.38 *Fix $n \in \mathbb{N}$ with $n \geq 2$ and suppose $\Omega \subseteq \mathbb{R}^n$ is an Ahlfors regular domain with compact boundary. Abbreviate $\sigma := \mathcal{H}^{n-1}\lfloor\partial\Omega$ and denote by ν the geometric measure theoretic outward unit normal to Ω. Also, let L be a homogeneous, second-order, constant complex coefficient, $M \times M$ system in $\mathbb{R}^n$ such that*

$$\text{L satisfies the Legendre-Hadamard strong ellipticity condition (cf. [131, Definition 1.3.2]) and one has } \mathfrak{A}_L^{\text{dis}} \neq \emptyset. \tag{7.1.598}$$

Pick a coefficient tensor $A \in \mathfrak{A}_L^{\text{dis}}$. Finally, fix

$$p \in \left(\tfrac{n-1}{n}, 1\right] \tag{7.1.599}$$

and define

$$p^* := \left(\tfrac{1}{p} - \tfrac{1}{n-1}\right)^{-1} \in \left(1, \tfrac{n-1}{n-2}\right]. \tag{7.1.600}$$

Then for each $\varepsilon \in (0, \infty)$ there exists some threshold $\delta \in (0, 1)$ which depends only on ε, n, p, A, the ADR constants of $\partial\Omega$, and the relative flatness ratio $\wp(\Omega)/\text{diam}\,\partial\Omega$ with the property that if

$$\text{dist}\Big(\nu, [\text{VMO}(\partial\Omega, \sigma)]^n\Big) < \delta \tag{7.1.601}$$

(where the distance is considered in the space $\left[\text{BMO}(\partial\Omega, \sigma)\right]^n$) it follows that Ω is a UR domain, the conclusion in (7.1.549) holds, and for each spectral parameter $z \in \mathbb{C}$ with $|z| \geq \varepsilon$ the operator

$$zI + K_A : \left[H_1^{p^*,p}(\partial\Omega, \sigma)\right]^M \longrightarrow \left[H_1^{p^*,p}(\partial\Omega, \sigma)\right]^M \tag{7.1.602}$$

is Fredholm with index zero.

Proof This is a consequence of Theorem 3.4.2, Theorem 7.1.36, Theorem 7.1.37, (7.1.549), [130, Lemma 1.2.12], and item *(a)* in [130, Theorem 2.2.4]. □

We continue by deducing invertibility results for the (transpose) single layer acting as a smoothing operator of order one from the negative Sobolev spaces BMO_{-1} and VMO_{-1}, respectively, defined on boundaries of δ-oscillating AR domains (cf. [130, Definition 11.10.9]; see also (A.0.31) and (A.0.245)).

Theorem 7.1.39 *Fix $n \in \mathbb{N}$ with $n \geq 3$ and suppose $\Omega \subseteq \mathbb{R}^n$ is an Ahlfors regular domain with compact boundary. Abbreviate $\sigma := \mathcal{H}^{n-1}\lfloor\partial\Omega$ and denote by ν the geometric measure theoretic outward unit normal to Ω. Also, let L be a homogeneous, second-order, constant complex coefficient, $M \times M$ system in $\mathbb{R}^n$ such that*

$$\text{L satisfies the Legendre-Hadamard strong ellipticity condition (cf. [131, Definition 1.3.2]) and one has } \mathfrak{A}_L^{\text{dis}} \neq \emptyset. \tag{7.1.603}$$

Then there exists some threshold $\delta_0 \in (0,1)$ which depends only on n, p, L, the ADR *constants of $\partial\Omega$, and the relative flatness ratio $\wp(\Omega)/\mathrm{diam}\,\partial\Omega$ with the property that when*

$$\mathrm{dist}\Big(\nu, [\mathrm{VMO}(\partial\Omega,\sigma)]^n\Big) < \delta_0 \tag{7.1.604}$$

where the distance is considered in the space $\big[\mathrm{BMO}(\partial\Omega,\sigma)\big]^n$ (hence if Ω is a δ_0-oscillating AR *domain; cf. Definition 3.4.1), it follows that Ω is a* UR *domain and $S^\top$, the (real) transpose of the operator* [132, (2.3.12)] *corresponding to $p := 1$, induces isomorphisms in the settings below:*

$$S^\top : \big[\mathrm{BMO}_{-1}(\partial\Omega,\sigma)\big]^M \longrightarrow \big[\mathrm{BMO}(\partial\Omega,\sigma)\big]^M, \tag{7.1.605}$$

$$S^\top : \big[\mathrm{VMO}_{-1}(\partial\Omega,\sigma)\big]^M \longrightarrow \big[\mathrm{VMO}(\partial\Omega,\sigma)\big]^M. \tag{7.1.606}$$

As a corollary, the operators (7.1.605)-(7.1.606) *are isomorphisms whenever Ω is an infinitesimally flat* AR *domain (cf. Definition 3.4.1).*

Proof Take $\delta_0 \in (0,1)$ sufficiently small so that the invertibility results in Theorem 7.1.34 and (7.1.233) with $p := n-1$ are true. By eventually further decreasing δ_0, we may also ensure that Ω is a UR domain satisfying a two-sided local John condition (cf. Theorem 3.4.2); this guarantees that [130, (11.5.203)] holds. Now, the fact that $S^\top$ is an isomorphism in the context of (7.1.605) follows from (7.1.542) with $p := 1$ and (A.0.31). Next, that $S^\top$ is an isomorphism in the context of (7.1.606) is implied by (7.1.606), (7.1.233) with $p := n-1$, (A.0.245), [130, (11.5.203)], and [130, Lemma 1.2.20]. □

In our next theorem, the transpose of the augmented single layer is shown to act as as a smoothing operator of order one mapping BMO_{-1} isomorphically onto BMO, in the two-dimensional setting.

Theorem 7.1.40 *Suppose $\Omega \subseteq \mathbb{R}^2$ is an Ahlfors regular domain with compact boundary. Abbreviate $\sigma := \mathcal{H}^1\lfloor\partial\Omega$ and denote by ν the geometric measure theoretic outward unit normal to Ω. Also, let L be a homogeneous, second-order, constant complex coefficient, $M \times M$ system in $\mathbb{R}^2$ such that*

$$\text{L satisfies the Legendre-Hadamard strong ellipticity condition (cf. [131, Definition 1.3.2]) and one has } \mathfrak{A}_L^{\mathrm{dis}} \neq \varnothing. \tag{7.1.607}$$

Then there exists $\delta_0 \in (0,1)$ which depends only on L, the ADR *constants of $\partial\Omega$, and the relative flatness ratio $\wp(\Omega)/\mathrm{diam}\,\partial\Omega$ with the property that if*

$$\mathrm{dist}\Big(\nu, [\mathrm{VMO}(\partial\Omega,\sigma)]^2\Big) < \delta_0 \tag{7.1.608}$$

where the distance is considered in the space $\big[\mathrm{BMO}(\partial\Omega,\sigma)\big]^2$, hence if Ω is a δ_0-oscillating AR *domain (cf. Definition 3.4.1), then Ω is a two-sided* NTA *domain and the (real) transpose of the augmented single layer potential operator* (7.1.548) *considered for $p := 1$ is the isomorphism*

$$\begin{aligned}&S^{\top}_{\rm aug} : \big[\mathrm{BMO}_{-1}(\partial\Omega,\sigma)\big]^M \oplus \mathbb{C}^M \longrightarrow \big[\mathrm{BMO}(\partial\Omega,\sigma)\big]^M \oplus \mathbb{C}^M\\ &\textit{given by } S^{\top}_{\rm aug}(f,c) := \Big(S^{\top} f + c, (\langle f, \mathbf{e}_j\rangle)_{1\le j\le M}\Big) \textit{ for each}\\ &\textit{functional } f \in \big[\mathrm{BMO}_{-1}(\partial\Omega,\sigma)\big]^M \textit{ and each constant } c \in \mathbb{C}^M,\end{aligned} \tag{7.1.609}$$

where

$$\begin{aligned}&S^{\top} : \big[\mathrm{BMO}_{-1}(\partial\Omega,\sigma)\big]^M \longrightarrow \big[\mathrm{BMO}(\partial\Omega,\sigma)\big]^M\\ &\textit{is the (real) transpose of the operator (cf. [132, (2.3.12)])}\\ &S : \big[H^1(\partial\Omega,\sigma)\big]^M \longrightarrow \big[H^{q,1}_1(\partial\Omega,\sigma)\big]^M \textit{ with } q \in (1,\infty).\end{aligned} \tag{7.1.610}$$

Furthermore, the operator $S^{\top}_{\rm aug}$ *from* (7.1.609) *maps* $\big[\mathrm{VMO}_{-1}(\partial\Omega,\sigma)\big]^M \oplus \mathbb{C}^M$ *into* $\big[\mathrm{VMO}(\partial\Omega,\sigma)\big]^M \oplus \mathbb{C}^M$*, and the operator thus induced, i.e.,*

$$\begin{aligned}&S^{\top}_{\rm aug} : \big[\mathrm{VMO}_{-1}(\partial\Omega,\sigma)\big]^M \oplus \mathbb{C}^M \longrightarrow \big[\mathrm{VMO}(\partial\Omega,\sigma)\big]^M \oplus \mathbb{C}^M\\ &\textit{given by } S^{\top}_{\rm aug}(f,c) := \Big(S^{\top} f + c, (\langle f, \mathbf{e}_j\rangle)_{1\le j\le M}\Big) \textit{ for each}\\ &\textit{functional } f \in \big[\mathrm{VMO}_{-1}(\partial\Omega,\sigma)\big]^M \textit{ and each constant } c \in \mathbb{C}^M,\end{aligned} \tag{7.1.611}$$

is actually an isomorphism.

As a corollary, the operators (7.1.609)-(7.1.611) *are isomorphisms whenever* Ω *is an infinitesimally flat* AR *domain (cf. Definition 3.4.1).*

Proof That Ω is a two-sided NTA domain whenever if (7.1.608) holds for $\delta_0 \in (0,1)$ sufficiently small is implied by Theorem 3.4.2. Henceforth, assume this is the case. That the operator (7.1.609) is an isomorphism when (7.1.608) holds for $\delta_0 \in (0,1)$ sufficiently small is a consequence of the fact that (7.1.548) is an isomorphism when $p = 1$ and duality (cf. [130, Theorem 4.6.1] and [132, Remark 2.3.5]).

To deal with (7.1.611), fix $q \in (1,\infty)$ and bring in the augmented version of the single layer operator associated with $L^{\top}$ as in (7.1.245) with p replaced by q, i.e.,

$$\begin{aligned}&\big(S_{L^{\top}}\big)_{\rm aug} : \big[L^q(\partial\Omega,\sigma)\big]^M \oplus \mathbb{C}^M \longrightarrow \big[L^q_1(\partial\Omega,\sigma)\big]^M \oplus \mathbb{C}^M\\ &\text{given by } \big(S_{L^{\top}}\big)_{\rm aug}(f,c) := \Big(S_{L^{\top}} f + c, \textstyle\int_{\partial\Omega} f\,d\sigma\Big) \text{ for each}\\ &\text{function } f \in \big[L^q(\partial\Omega,\sigma)\big]^M \text{ and constant } c \in \mathbb{C}^M.\end{aligned} \tag{7.1.612}$$

From Theorem 7.1.10 (used for $L^{\top}$ in place of L; here Theorem 1.5.2 is relevant), we know that this is an isomorphism if (7.1.608) holds with $\delta_0 \in (0,1)$ small enough. We claim that $\big(S_{L^{\top}}\big)_{\rm aug}$ from (7.1.612) agrees with $S^{\top}_{\rm aug}$ from (7.1.611). As is apparent from definitions, this comes down to checking that for each function $f \in \big[L^q(\partial\Omega,\sigma)\big]^M \hookrightarrow \big[\mathrm{BMO}_{-1}(\partial\Omega,\sigma)\big]^M$ (cf. [132, (2.3.135)]) we have

$$\begin{aligned}&(S_L)^{\top} f, \text{ with } (S_L)^{\top} \text{ as defined in (7.1.610) and with } f \text{ viewed in}\\ &\big[\mathrm{BMO}_{-1}(\partial\Omega,\sigma)\big]^M, \text{ agrees with } S_{L^{\top}} f, \text{ with } S_{L^{\top}} \text{ as in (A.0.231)}\\ &\text{with } L \text{ replaced by } L^{\top} \text{ and } f \text{ viewed in } \big[L^q(\partial\Omega,\sigma)\big]^M,\end{aligned} \tag{7.1.613}$$

and

$$\big(\langle f, \mathbf{e}_j \rangle\big)_{1 \le j \le M} = \int_{\partial\Omega} f \, d\sigma. \tag{7.1.614}$$

To justify (7.1.613), denote by $q' \in (1, \infty)$ the Hölder conjugate exponent of q, and pick a vector-valued $(q', 1)$-atom $a \in \big[H^1(\partial\Omega, \sigma)\big]^M$. We may then write

$$\begin{aligned}
\int_{\partial\Omega} \big\langle (S_L)^\top f, a \big\rangle \, d\sigma &= {}_{[\mathrm{BMO}(\partial\Omega,\sigma)]^M}\big\langle (S_L)^\top f, a \big\rangle_{[H^1(\partial\Omega,\sigma)]^M} \\
&= {}_{[\mathrm{BMO}_{-1}(\partial\Omega,\sigma)]^M}\big\langle f, S_L a \big\rangle_{[H_1^{q',1}(\partial\Omega,\sigma)]^M} \\
&= \int_{\partial\Omega} \big\langle f, S_L a \big\rangle \, d\sigma = \int_{\partial\Omega} \big\langle S_{L^\top} f, a \big\rangle \, d\sigma,
\end{aligned} \tag{7.1.615}$$

using the duality result from [130, Theorem 4.6.1, (4.6.8)], (7.1.610) with q' in place of q, [132, (2.3.133)], [132, (2.3.138)] with the roles of q, q' interchanged, as well as [132, (1.3.65)] and the subsequent comments. In view of [130, Lemma 4.6.9], this establishes (7.1.613). Also, (7.1.614) is a consequence of [132, (2.3.138)].

Granted the compatibility of $\big(S_{L^\top}\big)_{\mathrm{aug}}$ from (7.1.612) with $S^\top_{\mathrm{aug}}$ from (7.1.611), and the fact that said operators are invertible, we may now invoke [130, Lemma 1.2.20] to conclude that the operator $S^\top_{\mathrm{aug}}$ induces an isomorphism when considered from

$$\text{the closure of } \big[L^q(\partial\Omega, \sigma)\big]^M \oplus \mathbb{C}^M \text{ in } \big[\mathrm{BMO}_{-1}(\partial\Omega, \sigma)\big]^M \oplus \mathbb{C}^M \tag{7.1.616}$$

onto

$$\text{the closure of } \big[L_1^q(\partial\Omega, \sigma)\big]^M \oplus \mathbb{C}^M \text{ in } \big[\mathrm{BMO}(\partial\Omega, \sigma)\big]^M \oplus \mathbb{C}^M. \tag{7.1.617}$$

However, from [132, (2.3.134)] we know that (7.1.616) is $\big[\mathrm{VMO}_{-1}(\partial\Omega, \sigma)\big]^M \oplus \mathbb{C}^M$. Also, since $\mathrm{Lip}\,(\partial\Omega)$ is a dense subspace of $L_1^q(\partial\Omega, \sigma)$ (cf. [130, (11.5.199)]), we conclude from [130, (3.1.51)] that the closure of the space $L_1^q(\partial\Omega, \sigma)$ in $\mathrm{BMO}(\partial\Omega, \sigma)$ is $\mathrm{VMO}(\partial\Omega, \sigma)$. This argument shows that the operator $S^\top_{\mathrm{aug}}$ from (7.1.609) maps $\big[\mathrm{VMO}_{-1}(\partial\Omega, \sigma)\big]^M \oplus \mathbb{C}^M$ into $\big[\mathrm{VMO}(\partial\Omega, \sigma)\big]^M \oplus \mathbb{C}^M$, and the operator thus induced in (7.1.611) is an isomorphism. □

For later purposes, it is useful to know that the single layer operator induces an isomorphism at the level of kernels of double layers. This is made precise in the proposition below.

Proposition 7.1.41 *Fix $n \in \mathbb{N}$ with $n \ge 3$ and suppose $\Omega \subseteq \mathbb{R}^n$ is an Ahlfors regular domain with compact boundary. Abbreviate $\sigma := \mathcal{H}^{n-1}\lfloor\partial\Omega$ and denote by ν the geometric measure theoretic outward unit normal to Ω. Also, let L be a homogeneous, second-order, constant complex coefficient, $M \times M$ system in $\mathbb{R}^n$ such that*

L satisfies the Legendre-Hadamard strong ellipticity condition (cf. [131, Definition 1.3.2]) *and one has $\mathfrak{A}^{\mathrm{dis}}_L \neq \emptyset$.* (7.1.618)

Finally, pick an arbitrary coefficient tensor $A \in \mathfrak{A}_L$, *and fix some* $z \in \mathbb{C}$ *along with some exponent* $p \in (1, \infty)$.

Then there exists a threshold $\delta \in (0, 1)$ *depending on* n, p, L, *the* ADR *constants of* $\partial\Omega$, *and the relative flatness ratio* $\wp(\Omega)/\operatorname{diam}\partial\Omega$ *with the property that whenever*

$$\operatorname{dist}\Big(\nu, [\mathrm{VMO}(\partial\Omega, \sigma)]^n\Big) < \delta \tag{7.1.619}$$

where the distance is considered in the space $\big[\mathrm{BMO}(\partial\Omega, \sigma)\big]^n$ *(hence if* Ω *is a* δ*-oscillating* AR *domain; cf. Definition 3.4.1), it follows that* Ω *is a* UR *domain and*

$$S\Big[\operatorname{Ker}\Big(zI + K^{\#}_{A^{\top}}; \big[L^p(\partial\Omega, \sigma)\big]^M\Big)\Big] = \operatorname{Ker}\Big(zI + K_A; \big[L^p_1(\partial\Omega, \sigma)\big]^M\Big), \tag{7.1.620}$$

where S *is the boundary-to-boundary single layer potential operator associated with* Ω *and the system* L *as in* (A.0.231), K_A *is the boundary-to-boundary double layer potential operator associated with* A *and* Ω *as in* (A.0.116), *and* $K^{\#}_A$ *is the transpose double layer potential operator associated with* $A^{\top}$ *and* Ω *as in* (A.0.118). *In fact,*

$$S : \operatorname{Ker}\Big(zI + K^{\#}_{A^{\top}}; \big[L^p(\partial\Omega, \sigma)\big]^M\Big) \to \operatorname{Ker}\Big(zI + K_A; \big[L^p_1(\partial\Omega, \sigma)\big]^M\Big) \quad \text{is an isomorphism,} \tag{7.1.621}$$

hence

$$\dim \operatorname{Ker}\Big(zI + K^{\#}_{A^{\top}}; \big[L^p(\partial\Omega, \sigma)\big]^M\Big) = \dim \operatorname{Ker}\Big(zI + K_A; \big[L^p_1(\partial\Omega, \sigma)\big]^M\Big). \tag{7.1.622}$$

Proof Take $\delta \in (0, 1)$ sufficiently small so as to ensure that Ω is a UR domain (cf. Theorem 3.4.2) and that S is an isomorphism in the context of (7.1.233) (Theorem 7.1.9 guarantees that this is possible). To deal with the left-to-right inclusion in (7.1.620), pick some $f \in \big[L^p(\partial\Omega, \sigma)\big]^M$ satisfying $(zI + K^{\#}_{A^{\top}})f = 0$. Then $Sf \in \big[L^p_1(\partial\Omega, \sigma)\big]^M$ (cf. [132, Theorem 1.5.1, *(ix)*]), and [132, Theorem 1.5.1, *(xiii)*] allows us to write $(zI + K_A)(Sf) = S\big[(zI + K^{\#}_{A^{\top}})f\big] = 0$. Hence, Sf is in $\operatorname{Ker}(zI + K_A; [L^p_1(\partial\Omega, \sigma)]^M)$, proving the left-to-right inclusion in (7.1.620).

To prove the right-to-left inclusion in (7.1.620), pick some $f \in \big[L^p_1(\partial\Omega, \sigma)\big]^M$ with the property that $(zI + K_A)f = 0$. The fact that (7.1.233) is an isomorphism guarantees that there exists a function $g \in \big[L^p(\partial\Omega, \sigma)\big]^M$ such that $Sg = f$. Using [132, Theorem 1.5.1, item *(xiii)*] we may then compute

$$S\big[(zI + K^{\#}_{A^{\top}})g\big] = (zI + K_A)(Sg) = (zI + K_A)f = 0. \tag{7.1.623}$$

In light of the fact that the single layer operator is an isomorphism in the context of (7.1.233), this forces $(zI + K^{\#}_{A^{\top}})g = 0$, hence $g \in \operatorname{Ker}\Big(zI + K^{\#}_{A^{\top}}; \big[L^p(\partial\Omega, \sigma)\big]^M\Big)$. Consequently,

$$f = Sg \in S\Big[\operatorname{Ker}\Big(zI + K^{\#}_{A^{\top}}; \big[L^p(\partial\Omega, \sigma)\big]^M\Big)\Big], \tag{7.1.624}$$

proving the right-to-left inclusion in (7.1.620).

Finally, the claim made in (7.1.621) is a consequence of (7.1.620) (which gives surjectivity) and (7.1.233) (which gives injectivity). □

Next we take a look at the action of the augmented single layer on the null-space of the augmented double layer on Sobolev spaces on boundaries of two-dimensional Ahlfors regular domains.

Proposition 7.1.42 *Suppose $\Omega \subseteq \mathbb{R}^2$ is an Ahlfors regular domain with compact boundary. Abbreviate $\sigma := \mathcal{H}^1 \lfloor \partial\Omega$ and denote by ν the geometric measure theoretic outward unit normal to Ω. Also, let L be a homogeneous, second-order, constant complex coefficient, $M \times M$ system in $\mathbb{R}^2$ such that*

$$\text{L satisfies the Legendre-Hadamard strong ellipticity condition (cf. [131, Definition 1.3.2]) and one has } \mathfrak{A}_L^{\text{dis}} \neq \varnothing. \tag{7.1.625}$$

Pick an arbitrary coefficient tensor $A \in \mathfrak{A}_L$ and fix some exponent $p \in (1,\infty)$. Define the augmented double layer potential operators

$$\begin{aligned} &\mathbb{K}_A : \big[L_1^p(\partial\Omega,\sigma)\big]^M \oplus \mathbb{C}^M \longrightarrow \big[L_1^p(\partial\Omega,\sigma)\big]^M \oplus \mathbb{C}^M \text{ given by} \\ &\mathbb{K}_A(f,c) := \big(K_A f, \pm c/2\big) \text{ for each } f \in \big[L_1^p(\partial\Omega,\sigma)\big]^M \text{ and } c \in \mathbb{C}^M, \end{aligned} \tag{7.1.626}$$

with the sign "+" if Ω is bounded and the sign "−" if Ω is unbounded, as well as

$$\begin{aligned} &\mathbb{K}^{\#}_{A^\top} : \big[L^p(\partial\Omega,\sigma)\big]^M \oplus \mathbb{C}^M \longrightarrow \big[L^p(\partial\Omega,\sigma)\big]^M \oplus \mathbb{C}^M \text{ given by} \\ &\mathbb{K}^{\#}_{A^\top}(f,c) := \big(K^{\#}_{A^\top} f, \pm c/2\big) \text{ for each } f \in \big[L^p(\partial\Omega,\sigma)\big]^M \text{ and } c \in \mathbb{C}^M. \end{aligned} \tag{7.1.627}$$

Then there exists a threshold $\delta \in (0,1)$ depending on p, L, the ADR constants of $\partial\Omega$, and the relative flatness ratio $\wp(\Omega)/\operatorname{diam}\partial\Omega$ with the property that whenever

$$\operatorname{dist}\Big(\nu, [\mathrm{VMO}(\partial\Omega,\sigma)]^2\Big) < \delta \tag{7.1.628}$$

where the distance is considered in the space $\big[\mathrm{BMO}(\partial\Omega,\sigma)\big]^2$, hence if Ω is a δ-oscillating AR domain (cf. Definition 3.4.1), it follows that Ω is a UR domain and for each $z \in \mathbb{C}$ one has

$$S_{\text{aug}}\Big[\operatorname{Ker}\Big(z\mathbb{I} + \mathbb{K}^{\#}_{A^\top}; \big[L^p(\partial\Omega,\sigma)\big]^M \oplus \mathbb{C}^M\Big)\Big] = \operatorname{Ker}\Big(z\mathbb{I} + \mathbb{K}_A; \big[L_1^p(\partial\Omega,\sigma)\big]^M \oplus \mathbb{C}^M\Big), \tag{7.1.629}$$

where S_{aug} is the augmented version of the boundary-to-boundary single layer potential operator associated with Ω and the system L as in (7.1.245), and $\mathbb{I}$ is the identity operator (acting on pairs). In fact, for each $z \in \mathbb{C}$ one has

$$\begin{aligned} S_{\text{aug}} &: \operatorname{Ker}\Big(z\mathbb{I} + \mathbb{K}^{\#}_{A^\top}; \big[L^p(\partial\Omega,\sigma)\big]^M \oplus \mathbb{C}^M\Big) \\ &\to \operatorname{Ker}\Big(z\mathbb{I} + \mathbb{K}_A; \big[L_1^p(\partial\Omega,\sigma)\big]^M \oplus \mathbb{C}^M\Big) \text{ is an isomorphism,} \end{aligned} \tag{7.1.630}$$

and, as a corollary,

$$\dim \operatorname{Ker}\Big(zI + K^{\#}_{A^{\top}};\, \big[L^p(\partial\Omega,\sigma)\big]^M\Big) = \dim \operatorname{Ker}\Big(zI + K_A;\, \big[L^p_1(\partial\Omega,\sigma)\big]^M\Big). \tag{7.1.631}$$

Proof Theorem 3.4.2 ensures that taking $\delta \in (0,1)$ sufficiently small makes Ω a UR domain. From Theorem 7.1.10 we also know that, after eventually further decreasing the value of $\delta \in (0,1)$, the augmented single layer potential operator $S_{\rm aug}$ becomes an isomorphism in the context of (7.1.245). Fix an arbitrary $z \in \mathbb{C}$. From definitions it is clear that, with the convention that the "top" sign is adopted if Ω is bounded and the "bottom" sign is adopted if Ω is unbounded,

$$\begin{aligned}\operatorname{Ker}\Big(z\mathbb{I} + \mathbb{K}^{\#}_{A^{\top}};\big[L^p(\partial\Omega,\sigma)\big]^M \oplus \mathbb{C}^M\Big)&\\ = \operatorname{Ker}\Big(zI + K^{\#}_{A^{\top}};\, \big[L^p(\partial\Omega,\sigma)\big]^M\Big) &\oplus \begin{cases}\{0\} & \text{if } z \neq \mp\frac{1}{2},\\ \mathbb{C}^M & \text{if } z = \mp\frac{1}{2},\end{cases}\end{aligned} \tag{7.1.632}$$

and

$$\begin{aligned}\operatorname{Ker}\Big(z\mathbb{I} + \mathbb{K}_A;\big[L^p_1(\partial\Omega,\sigma)\big]^M \oplus \mathbb{C}^M\Big)&\\ = \operatorname{Ker}\Big(zI + K_A;\, \big[L^p_1(\partial\Omega,\sigma)\big]^M\Big) &\oplus \begin{cases}\{0\} & \text{if } z \neq \mp\frac{1}{2},\\ \mathbb{C}^M & \text{if } z = \mp\frac{1}{2}.\end{cases}\end{aligned} \tag{7.1.633}$$

Also, for each $f \in \big[L^p(\partial\Omega,\sigma)\big]^M$ the transposition results from [132, Theorem 1.5.1, *(iii)*] and [132, (1.5.21)] (with $A^{\top}$ in place of A) allow us to compute

$$\begin{aligned}\int_{\partial\Omega} K^{\#}_{A^{\top}} f\, d\sigma &= \Big(\Big\langle \int_{\partial\Omega} K^{\#}_{A^{\top}} f\, d\sigma, \mathbf{e}_j\Big\rangle\Big)_{1\le j\le M} = \Big(\int_{\partial\Omega} \langle K^{\#}_{A^{\top}} f, \mathbf{e}_j\rangle\, d\sigma\Big)_{1\le j\le M}\\ &= \Big(\int_{\partial\Omega} \langle f, K_{A^{\top}}\mathbf{e}_j\rangle\, d\sigma\Big)_{1\le j\le M} = \pm\frac{1}{2}\Big(\int_{\partial\Omega} \langle f, \mathbf{e}_j\rangle\, d\sigma\Big)_{1\le j\le M} = \pm\frac{1}{2}\int_{\partial\Omega} f\, d\sigma.\end{aligned}$$

Given any $f \in \big[L^p(\partial\Omega,\sigma)\big]^M$ and $c \in \mathbb{C}^M$, we may then use this, (7.1.627), and (7.1.245) to write

$$\begin{aligned}(S_{\rm aug} \circ \mathbb{K}^{\#}_{A^{\top}})(f,c) &= S_{\rm aug}\big(\mathbb{K}^{\#}_{A^{\top}}(f,c)\big) = S_{\rm aug}\big(K^{\#}_{A^{\top}} f, \pm c/2\big)\\ &= \Big(S(K^{\#}_{A^{\top}} f) \pm c/2, \textstyle\int_{\partial\Omega} K^{\#}_{A^{\top}} f\, d\sigma\Big) = \Big(K_A(Sf) \pm c/2, \pm\frac{1}{2}\textstyle\int_{\partial\Omega} f\, d\sigma\Big),\end{aligned} \tag{7.1.634}$$

and use (7.1.245), (7.1.626), together with formula [132, (1.5.21)] to write

$$\begin{aligned}(\mathbb{K}_A \circ S_{\rm aug})(f,c) &= \mathbb{K}_A\big(S_{\rm aug}(f,c)\big) = \mathbb{K}_A\Big(Sf + c, \textstyle\int_{\partial\Omega} f\, d\sigma\Big)\\ &= \Big(K_A(Sf + c), \pm\frac{1}{2}\textstyle\int_{\partial\Omega} f\, d\sigma\Big) = \Big(K_A(Sf) \pm c/2, \pm\frac{1}{2}\textstyle\int_{\partial\Omega} f\, d\sigma\Big).\end{aligned} \tag{7.1.635}$$

From (7.1.634) and (7.1.635) we see that

$$S_{\rm aug} \circ \mathbb{K}^{\#}_{A^\top} = \mathbb{K}_A \circ S_{\rm aug} \tag{7.1.636}$$

as operators from the space $\big[L^p(\partial\Omega, \sigma)\big]^M \oplus \mathbb{C}^M$ into itself.

To show the left-to-right inclusion in (7.1.629), pick some $f \in \big[L^p(\partial\Omega, \sigma)\big]^M$ and $c \in \mathbb{C}^M$ satisfying $(z\mathbb{I} + \mathbb{K}^{\#}_{A^\top})(f, c) = (0, 0)$. Then (7.1.636) gives

$$\big(z\mathbb{I} + \mathbb{K}_A\big)\big(S_{\rm aug}(f, c)\big) = S_{\rm aug}\big[(z\mathbb{I} + \mathbb{K}^{\#}_{A^\top})(f, c)\big] = (0, 0). \tag{7.1.637}$$

Consequently, $S_{\rm aug}(f, c) \in {\rm Ker}\Big(z\mathbb{I} + \mathbb{K}_A;\, \big[L^p_1(\partial\Omega, \sigma)\big]^M \oplus \mathbb{C}^M\Big)$, proving the left-to-right inclusion in (7.1.629).

To justify the right-to-left inclusion in (7.1.629), pick some pair (f, c) in the space $\big[L^p_1(\partial\Omega, \sigma)\big]^M \oplus \mathbb{C}^M$ with the property that $(z\mathbb{I} + \mathbb{K}_A)(f, c) = (0, 0)$. The fact that (7.1.245) is an isomorphism guarantees that there exist $g \in \big[L^p(\partial\Omega, \sigma)\big]^M$ and $d \in \mathbb{C}^M$ such that $S_{\rm aug}(g, d) = (f, c)$. Using (7.1.636) we may then compute

$$S_{\rm aug}\big[(z\mathbb{I}+\mathbb{K}^{\#}_{A^\top})(g, d)\big] = \big(z\mathbb{I}+\mathbb{K}_A\big)\big(S_{\rm aug}(g, d)\big) = \big(z\mathbb{I}+\mathbb{K}_A\big)(f, c) = (0, 0). \tag{7.1.638}$$

In light of the fact that the augmented single layer operator $S_{\rm aug}$ is an isomorphism in the context of (7.1.245), this forces $\big(z\mathbb{I} + \mathbb{K}^{\#}_{A^\top}\big)(g, d) = (0, 0)$, which may be interpreted as saying that $(g, d) \in {\rm Ker}\Big(z\mathbb{I} + \mathbb{K}^{\#}_{A^\top};\, \big[L^p(\partial\Omega, \sigma)\big]^M \oplus \mathbb{C}^M\Big)$. As such,

$$(f, c) = S_{\rm aug}(g, d) \in S_{\rm aug}\Big[{\rm Ker}\Big(z\mathbb{I} + \mathbb{K}^{\#}_{A^\top};\, \big[L^p(\partial\Omega, \sigma)\big]^M \oplus \mathbb{C}^M\Big)\Big], \tag{7.1.639}$$

proving the right-to-left inclusion in (7.1.629). This finishes the proof of (7.1.629). Going further, the claim made in (7.1.630) is a consequence of (7.1.629) (which gives surjectivity) and (7.1.245) (which gives injectivity). Finally, formula (7.1.631) is seen from (7.1.630), (7.1.632), and (7.1.633). □

In anticipation of dealing with the Neumann Problem with data in Hardy spaces later on, in Theorem 8.1.12, here we take a first look at the space of admissible boundary data for this problem. In Proposition 7.1.43 we treat the case when the dimension is ≥ 3, while in Proposition 7.1.44 we consider the two-dimensional case.

Proposition 7.1.43 *Fix $n \in \mathbb{N}$ with $n \geq 3$ and suppose $\Omega \subseteq \mathbb{R}^n$ is an Ahlfors regular domain with compact boundary. Abbreviate $\sigma := \mathcal{H}^{n-1}\lfloor\partial\Omega$ and denote by ν the geometric measure theoretic outward unit normal to Ω. Also, let L be a homogeneous, second-order, constant complex coefficient, $M \times M$ system in $\mathbb{R}^n$ such that*

L satisfies the Legendre-Hadamard strong ellipticity condition (cf. [131, Definition 1.3.2]) *and one has $\mathfrak{A}^{\rm dis}_L \neq \varnothing$.* (7.1.640)

Finally, pick a coefficient tensor $A \in \mathfrak{A}^{\rm dis}_L$, and fix some aperture parameter $\kappa > 0$ along with some exponent

$$p \in \big(\tfrac{n-1}{n}, 1\big]. \tag{7.1.641}$$

Then there exists a threshold $\delta \in (0,1)$ depending on n, p, A, the ADR *constants of $\partial\Omega$, and the relative flatness ratio $\wp(\Omega)/\mathrm{diam}\,\partial\Omega$ with the property that whenever*

$$\mathrm{dist}\Big(\nu, [\mathrm{VMO}(\partial\Omega,\sigma)]^n\Big) < \delta \tag{7.1.642}$$

with the distance considered in $\big[\mathrm{BMO}(\partial\Omega,\sigma)\big]^n$ (hence if Ω is a δ-oscillating AR *domain; cf. Definition 3.4.1), it follows that Ω is a* UR *domain and any*

$$u \in \big[\mathscr{C}^\infty(\Omega)\big]^M \text{ with } Lu = 0 \text{ in } \Omega \text{ and } \mathcal{N}_\kappa(\nabla u) \in L^p(\partial\Omega,\sigma), \tag{7.1.643}$$

satisfying, in the case when Ω is an exterior domain, the additional property that there exists some $\lambda \in (1,\infty)$ such that

$$\fint_{B(0,\lambda R)\setminus B(0,R)} |u|\,\mathrm{d}\mathcal{L}^n = o(1) \text{ as } R \to \infty \tag{7.1.644}$$

has (with the weak conormal derivative considered in the sense of [130, Definition 10.2.18] *and with the piece of notation introduced in Definition 7.1.25)*

$$\overset{\bullet}{\partial}{}^A_\nu u \in H^p_{A^\top,-1/2}(\partial\Omega,\sigma). \tag{7.1.645}$$

As a corollary of (7.1.645) *and* (7.1.472) *(used with $A^\top$ in place of A), any function u as in* (7.1.643)-(7.1.644) *has the property that*

$$\overset{\bullet}{\partial}{}^A_\nu u \in \mathrm{Im}\Big(-\tfrac{1}{2}I + K^\#_{A^\top};\ \big[H^p(\partial\Omega,\sigma)\big]^M\Big). \tag{7.1.646}$$

where $K^\#_{A^\top}$ is the transpose double layer associated with $A^\top$ and Ω as in (A.0.118).

From [132, (2.2.47)] (and the fact that we are presently assuming $n \geq 3$) we also see that anything in $\mathrm{Im}\Big(-\tfrac{1}{2}I + K^\#_{A^\top};\ \big[H^p(\partial\Omega,\sigma)\big]^M\Big)$ is the weak conormal derivative $\overset{\bullet}{\partial}{}^A_\nu u$ of a function u as in (7.1.643)-(7.1.644). Hence, the space of weak conormal derivatives $\overset{\bullet}{\partial}{}^A_\nu u$ of functions u as in (7.1.643)-(7.1.644) actually coincides with $\mathrm{Im}\Big(-\tfrac{1}{2}I + K^\#_{A^\top};\ \big[H^p(\partial\Omega,\sigma)\big]^M\Big)$. This points to the naturalness of the layer potential method in the treatment of boundary value problems of Neumann type.

Proof of Proposition 7.1.43 Suppose $\delta \in (0,1)$ is small enough so that Ω is a two-sided NTA domain, hence also a UR domain (cf. Theorem 3.4.2). We shall also tacitly assume that δ is sufficiently small so that Proposition 7.1.2 and Proposition 7.1.41 are suitably applicable. For starters, [130, Theorem 10.2.24] guarantees that the weak conormal derivative of u associated with A as in [130, Definition 10.2.18] satisfies

$$\overset{\bullet}{\partial}{}^A_\nu u \in \big[H^p(\partial\Omega,\sigma)\big]^M. \tag{7.1.647}$$

Introduce

$$p^* := \Big(\tfrac{1}{p} - \tfrac{1}{n-1}\Big)^{-1} \in \big(1, \tfrac{n-1}{n-2}\big], \quad (p^*)' := \big(1 - \tfrac{1}{p^*}\big)^{-1} \in [n-1,\infty). \tag{7.1.648}$$

Then [132, Theorem 2.2.7] guarantees (bearing in mind [131, Lemma 1.5.6]) that the nontangential boundary trace $u\big|_{\partial\Omega}^{\kappa-\mathrm{n.t.}}$ exists at σ-a.e. point on $\partial\Omega$ and

$$u\big|_{\partial\Omega}^{\kappa-\mathrm{n.t.}} \in \big[L^{p^*}(\partial\Omega,\sigma)\big]^M, \quad \mathcal{N}_\kappa u \in L^{p^*}(\partial\Omega,\sigma). \tag{7.1.649}$$

Granted these properties, [132, Theorem 2.3.16] applies and gives that

$$u = \mathcal{D}\big(u\big|_{\partial\Omega}^{\kappa-\mathrm{n.t.}}\big) - \mathscr{S}(\partial_\nu^A u) \text{ in } \Omega. \tag{7.1.650}$$

Going nontangentially to the boundary then gives (cf. [132, Theorem 1.5.1])

$$S_A(\partial_\nu^A u) = \big(-\tfrac{1}{2}I + K_A\big)\big(u\big|_{\partial\Omega}^{\kappa-\mathrm{n.t.}}\big) \text{ on } \partial\Omega. \tag{7.1.651}$$

To proceed, consider the case when we have $p < 1$. In this scenario, define the exponent $\eta := (n-1)(\frac{1}{p}-1) \in (0,1)$ and fix an arbitrary function

$$g \in \big[\mathscr{C}^\eta(\partial\Omega)\big]^M \text{ satisfying } \big(-\tfrac{1}{2}I + K_{A^\top}\big)g = 0. \tag{7.1.652}$$

Proposition 1.4.1 ensures that in the present setting we have $A^\top \in \mathfrak{A}^{\mathrm{dis}}_{L^\top}$. We may then invoke (7.1.60) (used with $A^\top$ in place of A) to conclude that $g \in \big[L_1^{(p^*)'}(\partial\Omega,\sigma)\big]^M$. Granted this, (7.1.621) applies (with p replaced by $(p^*)'$ and A replaced by $A^\top$) and gives that there exists

$$\phi \in \mathrm{Ker}\Big(-\tfrac{1}{2}I + K_A^\#;\ \big[L^{(p^*)'}(\partial\Omega,\sigma)\big]^M\Big) \text{ with } S_{A^\top}\phi = g. \tag{7.1.653}$$

Then we may write (with $S_{A^\top}$ denoting the boundary-to-boundary single layer potential operator associated with Ω and the system $L^\top$ as in (A.0.231))

$$\begin{aligned}
{}_{[H^p(\partial\Omega,\sigma)]^M}\big\langle \partial_\nu^A u, g\big\rangle_{[\mathscr{C}^\eta(\partial\Omega)]^M} &= {}_{[H^p(\partial\Omega,\sigma)]^M}\big\langle \partial_\nu^A u, S_{A^\top}\phi\big\rangle_{[\mathscr{C}^\eta(\partial\Omega)]^M} \\
&= {}_{[L^{p^*}(\partial\Omega,\sigma)]^M}\big\langle S_A(\partial_\nu^A u), \phi\big\rangle_{[L^{(p^*)'}(\partial\Omega,\sigma)]^M} \\
&= {}_{[L^{p^*}(\partial\Omega,\sigma)]^M}\Big\langle \big(-\tfrac{1}{2}I + K_A\big)\big(u\big|_{\partial\Omega}^{\kappa-\mathrm{n.t.}}\big), \phi\Big\rangle_{[L^{(p^*)'}(\partial\Omega,\sigma)]^M} \\
&= {}_{[L^{p^*}(\partial\Omega,\sigma)]^M}\Big\langle u\big|_{\partial\Omega}^{\kappa-\mathrm{n.t.}}, \big(-\tfrac{1}{2}I + K_A^\#\big)\phi\Big\rangle_{[L^{(p^*)'}(\partial\Omega,\sigma)]^M} = 0.
\end{aligned} \tag{7.1.654}$$

Above, the first equality (in which all brackets indicate duality in the sense of [130, Theorem 4.6.1]) is a consequence of the last formula in (7.1.653). The second equality in (7.1.654) comes from [132, (2.2.120)] (the last pair of brackets here stand for ordinary duality in Lebesgue spaces). The third equality in (7.1.654) is implied by (7.1.651). The fourth equality in (7.1.654) is a consequence of the duality result in [132, Theorem 1.5.1, item *(iii)*]. The final equality in (7.1.654) follows from the membership in (7.1.653). Having established (7.1.654), in view of the arbitrariness of g in (7.1.652) we conclude from Definition 7.1.25 that (7.1.645) holds if $p < 1$.

Finally, the case $p = 1$ is dealt with similarly, now selecting $g \in \big[\mathrm{BMO}(\partial\Omega, \sigma)\big]^M$ with $\big(-\frac{1}{2}I + K_{A^\top}\big)g = 0$ in place of (7.1.652), employing the case $p = 1$ of [130, (4.6.2)], and making use of [132, (2.2.123)] in place of [132, (2.2.120)]. □

Here is the two-dimensional counterpart of Proposition 7.1.43, pertaining to the nature of the space of admissible boundary data for the Neumann Problem with data in Hardy spaces.

Proposition 7.1.44 *Suppose $\Omega \subseteq \mathbb{R}^2$ is an Ahlfors regular domain with compact boundary. Abbreviate $\sigma := \mathcal{H}^1\lfloor\partial\Omega$ and denote by ν the geometric measure theoretic outward unit normal to Ω. Also, let L be a homogeneous, second-order, constant complex coefficient, $M \times M$ system in $\mathbb{R}^2$ such that*

$$L \text{ satisfies the Legendre-Hadamard strong ellipticity condition (cf. [131, Definition 1.3.2]) and one has } \mathfrak{A}_L^{\mathrm{dis}} \neq \varnothing. \tag{7.1.655}$$

Finally, pick a coefficient tensor $A \in \mathfrak{A}_L^{\mathrm{dis}}$, and fix some aperture parameter $\kappa > 0$ along with some exponent

$$p \in \big(\tfrac{1}{2}, 1\big]. \tag{7.1.656}$$

Then there exists a threshold $\delta \in (0,1)$ depending on p, A, the ADR constants of $\partial\Omega$, and the relative flatness ratio $\wp(\Omega)/\mathrm{diam}\,\partial\Omega$ with the property that whenever

$$\mathrm{dist}\Big(\nu, [\mathrm{VMO}(\partial\Omega, \sigma)]^2\Big) < \delta \tag{7.1.657}$$

where the distance is considered in the space $\big[\mathrm{BMO}(\partial\Omega, \sigma)\big]^2$, hence if Ω is a δ-oscillating AR domain (cf. Definition 3.4.1), the following properties are true. First, Ω is a UR domain and, as a result, the transpose double layer potential $K^\#_{A^\top}$ operator associated with $A^\top$ and Ω is well defined in the context of Hardy spaces (cf. (A.0.118)). Second, any function

$$\begin{aligned} &u \in \big[\mathscr{C}^\infty(\Omega)\big]^M \text{ with } Lu = 0 \text{ in } \Omega,\ \mathcal{N}_\kappa(\nabla u) \in L^p(\partial\Omega, \sigma), \\ &\text{and } u(x) = O(1) \text{ as } |x| \to \infty \text{ if } \Omega \text{ is an exterior domain,} \end{aligned} \tag{7.1.658}$$

has (with the weak conormal derivative as in [130, Definition 10.2.18])

$$\overset{\bullet}{\partial}{}^A_\nu u \in \mathrm{Im}\Big(-\tfrac{1}{2}I + K^\#_{A^\top};\ \big[H^p(\partial\Omega, \sigma)\big]^M\Big) \text{ if } \Omega \text{ is a bounded set in } \mathbb{R}^2, \tag{7.1.659}$$

and, respectively,

$$\begin{aligned} &\text{if } \Omega \text{ is an exterior domain in } \mathbb{R}^2 \text{ the distribution } \overset{\bullet}{\partial}{}^A_\nu u \text{ belongs to the space} \\ &\Big\{\big(-\tfrac{1}{2}I + K^\#_{A^\top}\big)g :\ g \in \big[H^p(\partial\Omega, \sigma)\big]^M \text{ with } \langle g, \mathbf{e}_j\rangle = 0 \text{ for } 1 \leq j \leq M\Big\}. \end{aligned} \tag{7.1.660}$$

Proof Make the assumption that $\delta \in (0,1)$ is small enough so that Ω is a two-sided NTA domain, hence also a UR domain satisfying a two-sided local John condition (cf. Theorem 3.4.2). We shall also tacitly assume that δ is sufficiently small so that Proposition 7.1.2 and Proposition 7.1.41 are suitably applicable.

For starters, [130, Theorem 10.2.24] guarantees that the weak conormal derivative of u associated with A in the sense of [130, Definition 10.2.18] satisfies

$$\dot{\partial}^A_\nu u \in \big[H^p(\partial\Omega,\sigma)\big]^M. \tag{7.1.661}$$

In the case when Ω is an exterior domain, from (7.1.658) and [131, (1.5.212)] we conclude that for each fixed $\lambda \in (1,\infty)$ we have

$$\fint_{B(0,\lambda R)\setminus B(0,R)} |\nabla u|\,d\mathcal{L}^2 = O(R^{-2}) \text{ as } R \to \infty. \tag{7.1.662}$$

Granted this, [130, Lemma 10.2.19] applies and gives that, in all cases,

$${}_{[\mathrm{Lip}(\partial\Omega)']^M}\big\langle \dot{\partial}^A_\nu u, c\big\rangle_{[\mathrm{Lip}(\partial\Omega)]^M} = 0 \text{ for each } c \in \mathbb{C}^M. \tag{7.1.663}$$

To proceed, work under the assumption that

$$p \in \big(\tfrac{1}{2}, 1\big) \text{ and introduce the exponents} \tag{7.1.664}$$

$$p^* := \big(\tfrac{1}{p} - 1\big)^{-1} \in (1,\infty) \text{ and } (p^*)' := \big(1 - \tfrac{1}{p^*}\big)^{-1} \in (1,\infty). \tag{7.1.665}$$

Then [132, Theorem 2.2.7] guarantees (bearing in mind [131, Lemma 1.5.6]) that the nontangential boundary trace $u\big|_{\partial\Omega}^{\kappa-\mathrm{n.t.}}$ exists at σ-a.e. point on $\partial\Omega$ and

$$u\big|_{\partial\Omega}^{\kappa-\mathrm{n.t.}} \in \big[L^{p^*}(\partial\Omega,\sigma)\big]^M, \quad \mathcal{N}_\kappa u \in L^{p^*}(\partial\Omega,\sigma). \tag{7.1.666}$$

From (7.1.658) and [131, (1.5.211)] we also know that

$$\begin{gathered}\text{if } \Omega \text{ is an exterior domain there exists } c_0 \in \mathbb{C}^M \\ \text{such that } u(x) = c_0 + O(|x|^{-1}) \text{ as } |x| \to \infty.\end{gathered} \tag{7.1.667}$$

Let us now consider the case when Ω is bounded. In view of the aforementioned properties, [132, Theorem 2.3.16] (cf. also [132, Corollary 2.3.17]) yields

$$u = \mathcal{D}\big(u\big|_{\partial\Omega}^{\kappa-\mathrm{n.t.}}\big) - \mathcal{S}(\dot{\partial}^A_\nu u) \text{ in } \Omega. \tag{7.1.668}$$

Going nontangentially to the boundary (cf. [132, (1.5.20)] and [132, (2.2.132)]) gives

$$S_A(\dot{\partial}^A_\nu u) = \big(-\tfrac{1}{2}I + K_A\big)\big(u\big|_{\partial\Omega}^{\kappa-\mathrm{n.t.}}\big) \text{ on } \partial\Omega. \tag{7.1.669}$$

To proceed, define $\eta := \frac{1}{p} - 1 \in (0,1)$ and fix an arbitrary function

$$g \in \big[\mathscr{C}^\eta(\partial\Omega)\big]^M \text{ satisfying } \big(-\tfrac{1}{2}I + K_{A^\top}\big)g = 0. \tag{7.1.670}$$

Proposition 1.4.1 ensures that in the present setting we have $A^\top \in \mathfrak{A}^{\rm dis}_{L^\top}$. We may then invoke (7.1.60) (used with $A^\top$ in place of A) to conclude that $g \in \big[L_1^{(p^*)'}(\partial\Omega,\sigma)\big]^M$. Hence, the function g belongs to $\mathrm{Ker}\Big(-\frac12 I + K_{A^\top};\ \big[L_1^{(p^*)'}(\partial\Omega,\sigma)\big]^M\Big)$ which in view of (7.1.633) further implies $(g,0) \in \mathrm{Ker}\Big(-\frac12 \mathbb{I} + \mathbb{K}_{A^\top};\ \big[L_1^{(p^*)'}(\partial\Omega,\sigma)\big]^M \oplus \mathbb{C}^M\Big)$. Granted this, we may invoke (7.1.630) (with p replaced by $(p^*)'$, the coefficient tensor A replaced by $A^\top$, and with $z := -1/2$) to conclude that there exists

$$
\begin{aligned}
&(\phi, c) \in \mathrm{Ker}\Big(-\tfrac12 \mathbb{I} + \mathbb{K}^\#_A;\ \big[L^{(p^*)'}(\partial\Omega,\sigma)\big]^M \oplus \mathbb{C}^M\Big)\\
&\quad\text{with the property that } S_{\rm aug}(\phi, c) = (g, 0).
\end{aligned}
\tag{7.1.671}
$$

Bearing in mind (7.1.632) we see, after unraveling definitions, that

$$
\phi \in \mathrm{Ker}\Big(-\tfrac12 I + K^\#_A;\ \big[L^{(p^*)'}(\partial\Omega,\sigma)\big]^M\Big) \text{ has } \int_{\partial\Omega} \phi\, d\sigma = 0 \text{ and } S_{A^\top}\phi + c = g.
\tag{7.1.672}
$$

Then we may write (with $S_{A^\top}$ denoting the boundary-to-boundary single layer potential operator associated with Ω and the system $L^\top$ as in (A.0.231))

$$
\begin{aligned}
{}_{[H^p(\partial\Omega,\sigma)]^M}\big\langle \partial^A_\nu u, g\big\rangle_{[\mathscr{C}^\eta(\partial\Omega)]^M} &= {}_{[H^p(\partial\Omega,\sigma)]^M}\big\langle \partial^A_\nu u, S_{A^\top}\phi + c\big\rangle_{[\mathscr{C}^\eta(\partial\Omega)]^M}\\
&= {}_{[H^p(\partial\Omega,\sigma)]^M}\big\langle \partial^A_\nu u, S_{A^\top}\phi\big\rangle_{[\mathscr{C}^\eta(\partial\Omega)]^M}\\
&= {}_{[L^{p^*}(\partial\Omega,\sigma)]^M}\big\langle S_A(\partial^A_\nu u), \phi\big\rangle_{[L^{(p^*)'}(\partial\Omega,\sigma)]^M}\\
&= {}_{[L^{p^*}(\partial\Omega,\sigma)]^M}\Big\langle \big(-\tfrac12 I + K_A\big)\big(u\big|^{\kappa-\mathrm{n.t.}}_{\partial\Omega}\big), \phi\Big\rangle_{[L^{(p^*)'}(\partial\Omega,\sigma)]^M}\\
&= {}_{[L^{p^*}(\partial\Omega,\sigma)]^M}\Big\langle u\big|^{\kappa-\mathrm{n.t.}}_{\partial\Omega}, \big(-\tfrac12 I + K^\#_A\big)\phi\Big\rangle_{[L^{(p^*)'}(\partial\Omega,\sigma)]^M} = 0.
\end{aligned}
\tag{7.1.673}
$$

Above, the first equality (in which all brackets indicate duality in the sense of [130, Theorem 4.6.1]) is a consequence of the last formula in (7.1.672). The second equality in (7.1.673) is implied by (7.1.663) and [130, (4.6.31)]. The third equality in (7.1.673) (with the last pair of brackets standing for ordinary duality in Lebesgue spaces) comes from [132, (2.2.136)-(2.2.138)]. The fourth equality in (7.1.673) is implied by (7.1.669). The fifth equality in (7.1.673) is a consequence of the duality result in [132, Theorem 1.5.1, item *(iii)*]. The final equality in (7.1.673) follows from the membership in (7.1.672).

Once (7.1.673) has been established, in view of the arbitrariness of g in (7.1.670) we conclude from Definition 7.1.25 that, if Ω is bounded,

$$
\partial^A_\nu u \in H^p_{A^\top,-1/2}(\partial\Omega,\sigma).
\tag{7.1.674}
$$

As a corollary of (7.1.674) and (7.1.472) (used with $A^\top$ in place of A), any function u as in (7.1.658) with $p \in \big(\frac12, 1\big)$ has the property that, if Ω is bounded,

$$
\partial^A_\nu u \in \mathrm{Im}\Big(-\tfrac12 I + K^\#_{A^\top};\ \big[H^p(\partial\Omega,\sigma)\big]^M\Big).
\tag{7.1.675}
$$

We claim that (7.1.675) also holds when $p = 1$ and Ω is bounded. Assume this is the case. Choose some $p_o \in \left(\frac{1}{2}, 1\right)$ and define $p_o^* := \left(\frac{1}{p_o} - 1\right)^{-1} \in (1, \infty)$. Since $\partial\Omega$ is a compact set, we have $L^1(\partial\Omega, \sigma) \subseteq L^{p_o}(\partial\Omega, \sigma)$ and $H^1(\partial\Omega, \sigma) \subseteq H^{p_o}(\partial\Omega, \sigma)$ (cf. [130, (4.2.21)]). As such, the hypotheses in Proposition 7.1.44 stated with $p := 1$ imply the hypotheses in Proposition 7.1.44 stated with $p := p_o$. Then from what we have already established in the first part of the proof we conclude that

$$\overset{\bullet}{\partial}{}^A_\nu u \in \left[H^1(\partial\Omega, \sigma)\right]^M \text{ and } \overset{\bullet}{\partial}{}^A_\nu u \in \operatorname{Im}\Big(-\tfrac{1}{2}I + K^\#_{A^\top};\ \left[H^{p_o}(\partial\Omega, \sigma)\right]^M\Big). \quad (7.1.676)$$

Hence, there exists $f \in \left[H^{p_o}(\partial\Omega, \sigma)\right]^M$ which, according to [130, (11.11.30)], also places f in the space $\left[L^{p_o^*}_{-1}(\partial\Omega, \sigma)\right]^M$, with the property that

$$\left(-\tfrac{1}{2}I + K^\#_{A^\top}\right)f = \overset{\bullet}{\partial}{}^A_\nu u \in \left[H^1(\partial\Omega, \sigma)\right]^M. \quad (7.1.677)$$

From this and the abstract regularity result from [130, Lemma 2.1.4, (2.1.14)], used here for the Banach spaces $X_0 := \left[H^1(\partial\Omega, \sigma)\right]^M$ and $\left[L^{p_o^*}_{-1}(\partial\Omega, \sigma)\right]^M$ and for the operator $T := -\frac{1}{2}I + K^\#_{A^\top}$ which is Fredholm with index zero both on X_0 and on X_1 (cf. Theorem 7.1.1 and [130, Theorem 2.1.2]), we then conclude that, in fact, $f \in \left[H^1(\partial\Omega, \sigma)\right]^M$. Thus, as wanted,

$$\overset{\bullet}{\partial}{}^A_\nu u = \left(-\tfrac{1}{2}I + K^\#_{A^\top}\right)f \in \operatorname{Im}\Big(-\tfrac{1}{2}I + K^\#_{A^\top};\ \left[H^1(\partial\Omega, \sigma)\right]^M\Big). \quad (7.1.678)$$

Hence, (7.1.675) is true for any $p \in \left(\frac{1}{2}, 1\right]$ if Ω is bounded, proving (7.1.659).

There remains to prove (7.1.660). Henceforth, assume Ω is an exterior domain. In view of (7.1.658), (7.1.666), (7.1.667), [132, Theorem 2.3.16] (cf. also [132, Corollary 2.3.17]) applies to the function $u - c_0$ and, by [132, (1.3.45)], yields

$$u - c_0 = \mathcal{D}\Big(u\Big|^{\kappa-\mathrm{n.t.}}_{\partial\Omega}\Big) - \mathcal{S}(\overset{\bullet}{\partial}{}^A_\nu u) \text{ in } \Omega. \quad (7.1.679)$$

On account of [132, (1.5.20)] and [132, (2.2.132)], going nontangentially to the boundary in (7.1.679) gives

$$S_A(\overset{\bullet}{\partial}{}^A_\nu u) - c_0 = \left(-\tfrac{1}{2}I + K_A\right)\Big(u\Big|^{\kappa-\mathrm{n.t.}}_{\partial\Omega}\Big) \text{ on } \partial\Omega. \quad (7.1.680)$$

Let us first treat the case when $p \in \left(\frac{1}{2}, 1\right)$. Retain notation introduced in (7.1.665) and let us once again define $\eta := \frac{1}{p} - 1 \in (0, 1)$. Fix an arbitrary function g as in (7.1.670). As before (cf. (7.1.672)), we may find

$$\begin{gathered}\phi \in \operatorname{Ker}\Big(-\tfrac{1}{2}I + K^\#_A;\ \left[L^{(p^*)'}(\partial\Omega, \sigma)\right]^M\Big) \text{ and } c \in \mathbb{C}^M \\ \text{with } \int_{\partial\Omega} \phi \, d\sigma = 0 \text{ and } S_{A^\top}\phi + c = g.\end{gathered} \quad (7.1.681)$$

If $S_{A^\top}$ denoting the boundary-to-boundary single layer potential operator associated with Ω and the system $L^\top$ as in (A.0.231), we may then write

$$
\begin{aligned}
{}_{[H^p(\partial\Omega,\sigma)]^M}\big\langle \overset{\bullet}{\partial}{}^A_\nu u, g\big\rangle_{[\mathscr{C}^\eta(\partial\Omega)]^M} &= {}_{[H^p(\partial\Omega,\sigma)]^M}\big\langle \overset{\bullet}{\partial}{}^A_\nu u, S_{A^\top}\phi + c\big\rangle_{[\mathscr{C}^\eta(\partial\Omega)]^M} \\
&= {}_{[H^p(\partial\Omega,\sigma)]^M}\big\langle \overset{\bullet}{\partial}{}^A_\nu u, S_{A^\top}\phi\big\rangle_{[\mathscr{C}^\eta(\partial\Omega)]^M} \qquad (7.1.682)\\
&= {}_{[L^{p^*}(\partial\Omega,\sigma)]^M}\big\langle S_A(\overset{\bullet}{\partial}{}^A_\nu u), \phi\big\rangle_{[L^{(p^*)'}(\partial\Omega,\sigma)]^M} \\
&= {}_{[L^{p^*}(\partial\Omega,\sigma)]^M}\Big\langle c_0 + \big(-\tfrac{1}{2}I + K_A\big)\big(u\big|^{\kappa-\text{n.t.}}_{\partial\Omega}\big), \phi\Big\rangle_{[L^{(p^*)'}(\partial\Omega,\sigma)]^M} \\
&= {}_{[L^{p^*}(\partial\Omega,\sigma)]^M}\Big\langle \big(-\tfrac{1}{2}I + K_A\big)\big(u\big|^{\kappa-\text{n.t.}}_{\partial\Omega}\big), \phi\Big\rangle_{[L^{(p^*)'}(\partial\Omega,\sigma)]^M} \\
&= {}_{[L^{p^*}(\partial\Omega,\sigma)]^M}\Big\langle u\big|^{\kappa-\text{n.t.}}_{\partial\Omega}, \big(-\tfrac{1}{2}I + K^\#_A\big)\phi\Big\rangle_{[L^{(p^*)'}(\partial\Omega,\sigma)]^M} = 0,
\end{aligned}
$$

with the same justifications as in (7.1.673), except that we now use (7.1.680) in place of (7.1.669), and employ the cancellation property in (7.1.681) in the fifth equality.

Having established (7.1.682), much as before we obtain (7.1.674) hence, ultimately, (7.1.675) holds in the case when Ω is an exterior domain and $p \in \big(\frac{1}{2}, 1\big)$. That (7.1.675) also holds in the case when Ω is an exterior domain and $p = 1$. is then seen by reasoning as in the proof of (7.1.678).

The end-game in the proof of (7.1.660) is as follows. From (7.1.675) we have

$$
\overset{\bullet}{\partial}{}^A_\nu u = \big(-\tfrac{1}{2}I + K^\#_{A^\top}\big)f \ \text{ for some } f \in \big[H^p(\partial\Omega,\sigma)\big]^M. \qquad (7.1.683)
$$

Then for each $j \in \{1, \dots, M\}$ we may write, based on the duality results from [132, (2.1.98)] and [132, (2.1.128)] together with the cancellation property (7.1.663) and formula [132, (1.5.21)] (written for $A^\top$ in place of A, and bearing in mind that Ω is presently assumed to be unbounded),

$$
\begin{aligned}
\langle f, \mathbf{e}_j\rangle &= \big\langle \big(\tfrac{1}{2}I + K^\#_{A^\top}\big)f, \mathbf{e}_j\big\rangle - \big\langle \big(-\tfrac{1}{2}I + K^\#_{A^\top}\big)f, \mathbf{e}_j\big\rangle \\
&= \big\langle \big(\tfrac{1}{2}I + K^\#_{A^\top}\big)f, \mathbf{e}_j\big\rangle - \big\langle \overset{\bullet}{\partial}{}^A_\nu u, \mathbf{e}_j\big\rangle = \big\langle \big(\tfrac{1}{2}I + K^\#_{A^\top}\big)f, \mathbf{e}_j\big\rangle \\
&= \big\langle f, \big(\tfrac{1}{2}I + K_{A^\top}\big)\mathbf{e}_j\big\rangle = \langle f, 0\rangle = 0. \qquad (7.1.684)
\end{aligned}
$$

Finally, from (7.1.683) and (7.1.684) we then conclude that (7.1.660) holds. □

Regarding the space appearing in (7.1.660) we make the following observation.

Proposition 7.1.45 *Suppose $\Omega \subseteq \mathbb{R}^2$ is an unbounded Ahlfors regular domain with compact boundary. Abbreviate $\sigma := \mathcal{H}^1\lfloor\partial\Omega$ and denote by ν the geometric measure theoretic outward unit normal to Ω. Also, let L be a homogeneous, second-order, constant complex coefficient, $M\times M$ system in $\mathbb{R}^2$, satisfying the Legendre-Hadamard strong ellipticity condition (cf.* [131, Definition 1.3.2]) *and such that $\mathfrak{A}^{\text{dis}}_L \neq \emptyset$. Pick a coefficient tensor $A \in \mathfrak{A}^{\text{dis}}_L$, and fix some exponent*

$$
p \in \big(\tfrac{1}{2}, 1\big]. \qquad (7.1.685)
$$

Then there exists a threshold $\delta \in (0, 1)$ *depending on* p, A, *the* ADR *constants of* $\partial\Omega$, *and the relative flatness ratio* $\wp(\Omega)/\mathrm{diam}\,\partial\Omega$ *with the property that whenever*

$$\mathrm{dist}\Big(\nu, [\mathrm{VMO}(\partial\Omega, \sigma)]^2\Big) < \delta \tag{7.1.686}$$

it follows that any given distribution $f \in \big[H^p(\partial\Omega, \sigma)\big]^M$ *belongs to the space appearing in* (7.1.660) *if and only if* (*with the brackets indicating duality in the sense of* [130, Theorem 4.6.1])

$$\begin{gathered} {}_{[H^p(\partial\Omega,\sigma)]^M}\big\langle f, h\big\rangle_{[\mathscr{C}^\eta(\partial\Omega)]^M} = 0 \textit{ for each function} \\ h \in \big[\mathscr{C}^\eta(\partial\Omega)\big]^M \textit{ satisfying } \big(-\tfrac{1}{2}I + K_{A^\top}\big)h \in \mathbb{C}^M \\ \textit{provided } p \in \big(\tfrac{1}{2}, 1\big) \textit{ and } \eta := \tfrac{1}{p} - 1 \in (0, 1), \end{gathered} \tag{7.1.687}$$

and, corresponding to the case when $p = 1$,

$$\begin{gathered} {}_{[H^1(\partial\Omega,\sigma)]^M}\big\langle f, h\big\rangle_{[\mathrm{BMO}(\partial\Omega,\sigma)]^M} = 0 \textit{ for each function} \\ h \in \big[\mathrm{BMO}(\partial\Omega, \sigma)\big]^M \textit{ satisfying } \big(-\tfrac{1}{2}I + K_{A^\top}\big)h \in \mathbb{C}^M. \end{gathered} \tag{7.1.688}$$

Proof Assume first that $p \in \big(\tfrac{1}{2}, 1\big)$ and set $\eta := \tfrac{1}{p} - 1 \in (0, 1)$. In one direction, if a given distribution $f \in \big[H^p(\partial\Omega, \sigma)\big]^M$ belongs to the space appearing in (7.1.660), then $f = \big(-\tfrac{1}{2}I + K^{\#}_{A^\top}\big)g$ for some $g \in \big[H^p(\partial\Omega, \sigma)\big]^M$ satisfying $\langle g, \mathbf{e}_j\rangle = 0$ for each $j \in \{1, \dots, M\}$. In particular, $\langle g, c\rangle = 0$ for each $c \in \mathbb{C}^M$. As such, for any function $h \in \big[\mathscr{C}^\eta(\partial\Omega)\big]^M$ with $\big(-\tfrac{1}{2}I + K_{A^\top}\big)h \in \mathbb{C}^M$ we may use the duality formula [132, (2.1.128)] (with $A^\top$ in place of A) to write

$$\begin{aligned} {}_{[H^p(\partial\Omega,\sigma)]^M}\big\langle f, h\big\rangle_{[\mathscr{C}^\eta(\partial\Omega)]^M} &= {}_{[H^p(\partial\Omega,\sigma)]^M}\big\langle \big(-\tfrac{1}{2}I + K^{\#}_{A^\top}\big)g, h\big\rangle_{[\mathscr{C}^\eta(\partial\Omega)]^M} \\ &= {}_{[H^p(\partial\Omega,\sigma)]^M}\big\langle g, \big(-\tfrac{1}{2}I + K_{A^\top}\big)h\big\rangle_{[\mathscr{C}^\eta(\partial\Omega)]^M} = 0, \end{aligned} \tag{7.1.689}$$

as wanted. In the opposite direction, if $f \in \big[H^p(\partial\Omega, \sigma)\big]^M$ satisfies (7.1.687), then Definition 7.1.25 and Proposition 7.1.27 imply

$$f \in H^p_{A^\top,-1/2}(\partial\Omega, \sigma) = \mathrm{Im}\Big(-\tfrac{1}{2}I + K^{\#}_{A^\top};\ \big[H^p(\partial\Omega, \sigma)\big]^M\Big). \tag{7.1.690}$$

Hence, $f = \big(-\tfrac{1}{2}I + K^{\#}_{A^\top}\big)g$ for some $g \in \big[H^p(\partial\Omega, \sigma)\big]^M$. Let us fix an arbitrary $j \in \{1, \dots, M\}$. When regarded as a constant function on $\partial\Omega$, it follows that $\mathbf{e}_j$ belongs to $\big[\mathscr{C}^\eta(\partial\Omega)\big]^M$ and satisfies $\big(-\tfrac{1}{2}I + K_{A^\top}\big)\mathbf{e}_j = -\mathbf{e}_j \in \mathbb{C}^M$, thanks to [132, (1.5.21)] (used with $A^\top$ in place of A) and the fact that Ω is an unbounded set. Granted this, (7.1.687) guarantees that

$${}_{[H^p(\partial\Omega,\sigma)]^M}\big\langle f, \mathbf{e}_j\big\rangle_{[\mathscr{C}^\eta(\partial\Omega)]^M} = 0. \tag{7.1.691}$$

In turn, this allows us to write

$$\begin{aligned}\langle g, \mathbf{e}_j\rangle &= \big\langle\big(\tfrac{1}{2}I + K^{\#}_{A^\top}\big)g, \mathbf{e}_j\big\rangle - \big\langle\big(-\tfrac{1}{2}I + K^{\#}_{A^\top}\big)g, \mathbf{e}_j\big\rangle \\ &= \big\langle\big(\tfrac{1}{2}I + K^{\#}_{A^\top}\big)g, \mathbf{e}_j\big\rangle - \big\langle f, \mathbf{e}_j\big\rangle = \big\langle\big(\tfrac{1}{2}I + K^{\#}_{A^\top}\big)g, \mathbf{e}_j\big\rangle \\ &= \big\langle g, \big(\tfrac{1}{2}I + K_{A^\top}\big)\mathbf{e}_j\big\rangle = \langle g, 0\rangle = 0,\end{aligned} \tag{7.1.692}$$

thanks again to [132, (1.5.21)] (used with $A^\top$ in place of A) and the fact that Ω is an unbounded set. This shows that

$${}_{[H^p(\partial\Omega,\sigma)]^M}\big\langle g, \mathbf{e}_j\big\rangle_{[\mathscr{C}^\eta(\partial\Omega)]^M} = 0 \text{ for each } j \in \{1, \dots, M\}, \tag{7.1.693}$$

which ultimately places f in the space appearing in (7.1.660).

Finally, the case when $p = 1$ (in which scenario (7.1.688) is involved) is dealt with similarly (now using the duality result in [132, (2.1.98)]). □

We continue by inspecting the action of the double layer potential operator on BMO and VMO spaces. This is going to be relevant later, particularly in the solution of the Dirichlet Problem with data in BMO and VMO discussed in Theorem 8.1.11.

Proposition 7.1.46 *Fix two integers $n, M \in \mathbb{N}$ with $n \geq 2$ and suppose $\Omega \subseteq \mathbb{R}^n$ is an Ahlfors regular domain with compact boundary. Abbreviate $\sigma := \mathcal{H}^{n-1}\lfloor\partial\Omega$ and denote by ν the geometric measure theoretic outward unit normal to Ω. Let L be a homogeneous, second-order, constant complex coefficient, weakly elliptic $M \times M$ system in $\mathbb{R}^n$, with $\mathfrak{A}^{\rm dis}_L \neq \varnothing$. Finally, pick a coefficient tensor $A \in \mathfrak{A}^{\rm dis}_L$, and fix $z \in \mathbb{C} \setminus \{0\}$.*

Then there exists a threshold $\delta \in (0, 1)$ depending only on n, z, A, the ADR *constants of $\partial\Omega$, and the relative flatness ratio $\wp(\Omega)/\mathrm{diam}\,\partial\Omega$ such that when*

$$\mathrm{dist}\Big(\nu, [\mathrm{VMO}(\partial\Omega, \sigma)]^n\Big) < \delta \tag{7.1.694}$$

where the distance in the left-hand side is considered in the space $\big[\mathrm{BMO}(\partial\Omega, \sigma)\big]^n$, (hence if Ω is a δ-oscillating AR *domain; cf. Definition 3.4.1), it follows that Ω is a* UR *domain and the property below holds:*

Consider the boundary-to-boundary double layer potential operator K_A associated with A and Ω, acting on the John-Nirenberg space (cf. [132, (2.1.95)]) *as well as on the Sarason space (cf.* [132, (2.1.157)]). *Also, recall the transpose double layer potential operator $K^{\#}_A$ associated with A and Ω, acting on the Hardy space $\big[H^1(\partial\Omega, \sigma)\big]^M$ as in* [132, Theorem 2.1.1].

Then the space $\mathrm{Im}\Big(zI+K_A;\ \big[\mathrm{BMO}(\partial\Omega, \sigma)\big]^M\Big)$ may be described as the collection of all functions $f \in \big[\mathrm{BMO}(\partial\Omega, \sigma)\big]^M$ having the property that (with the brackets indicating duality in the sense of [130, Theorem 4.6.1])

$$\begin{aligned}&{}_{[\mathrm{BMO}(\partial\Omega,\sigma)]^M}\big\langle f, g\big\rangle_{[H^1(\partial\Omega,\sigma)]^M} = 0 \textit{ for each function} \\ &g \in \big[H^1(\partial\Omega, \sigma)\big]^M \textit{ satisfying } \big(zI + K^{\#}_A\big)g = 0,\end{aligned} \tag{7.1.695}$$

while the space $\mathrm{Im}\Big(zI + K_A;\ \big[\mathrm{VMO}(\partial\Omega,\sigma)\big]^M\Big)$ *may be described as the collection of all functions* $f \in \big[\mathrm{VMO}(\partial\Omega,\sigma)\big]^M$ *such that* (7.1.695) *holds. As a corollary,*

$$\begin{aligned}\mathrm{Im}\Big(zI + K_A;\big[\mathrm{VMO}(\partial\Omega,\sigma)\big]^M\Big)\\ = \mathrm{Im}\Big(zI + K_A;\ \big[\mathrm{BMO}(\partial\Omega,\sigma)\big]^M\Big) \cap \big[\mathrm{VMO}(\partial\Omega,\sigma)\big]^M. \quad (7.1.696)\end{aligned}$$

Proof Recall from Theorem 3.4.2 that choosing $\delta \in (0,1)$ small enough ensures that Ω is a UR domain. From [132, Theorem 2.1.7, (2.1.96)] and [130, (2.1.48)] we conclude that the collection of all functions $f \in \big[\mathrm{BMO}(\partial\Omega,\sigma)\big]^M$ enjoying the property (7.1.695) is the weak*-closure of the image of $zI + K_A$ acting on $\big[\mathrm{BMO}(\partial\Omega,\sigma)\big]^M$. Further decreasing the value of δ guarantees that $zI + K_A^\#$ is a Fredholm operator on $\big[H^1(\partial\Omega,\sigma)\big]^M$ (cf. Theorem 7.1.1). As such, [132, (2.1.96)] and the Closed Range Theorem (cf. [130, (2.1.47)]) imply that the image of the operator $zI + K_A$ acting on $\big[\mathrm{BMO}(\partial\Omega,\sigma)\big]^M$ is weak*-closed. From this, the description of the space $\mathrm{Im}\Big(zI + K_A;\ \big[\mathrm{BMO}(\partial\Omega,\sigma)\big]^M\Big)$ given in the statement follows. Going further, staring from [132, (2.1.158)] and [130, (2.1.49)], we see that the collection of all functions $f \in \big[\mathrm{VMO}(\partial\Omega,\sigma)\big]^M$ with the property (7.1.695) is the norm-closure of the image of $zI + K_A$ acting on $\big[\mathrm{VMO}(\partial\Omega,\sigma)\big]^M$. Assuming δ is sufficiently small, Theorem 7.1.1 guarantees that $zI + K_A$ is a Fredholm operator when acting on the space $\big[\mathrm{VMO}(\partial\Omega,\sigma)\big]^M$. In particular, this operator has a closed range to begin with. This gives the description of the space $\mathrm{Im}\Big(zI + K_A;\ \big[\mathrm{VMO}(\partial\Omega,\sigma)\big]^M\Big)$ claimed in the statement. □

We conclude this section by re-visiting the boundary layer potentials associated with the Stokes system. First consider the issues of Fredholmness and invertibility for S_{Stokes} (cf. (7.1.400)) acting on Hardy spaces.

Theorem 7.1.47 *Pick* $n \in \mathbb{N}$ *with* $n \geq 2$ *and assume* $\Omega \subseteq \mathbb{R}^n$ *is an Ahlfors regular domain with compact boundary. Abbreviate* $\sigma := \mathcal{H}^{n-1}\lfloor\partial\Omega$ *and denote by* ν *the geometric measure theoretic outward unit normal to* Ω. *Also, fix an exponent*

$$p \in \big(\tfrac{n-1}{n}, 1\big]. \quad (7.1.697)$$

Finally, recall the convention made in (7.1.400).

Then there exists a threshold $\delta_0 \in (0,1)$, *which depends only on* n, p, *the* ADR *constants of* $\partial\Omega$, *and the relative flatness ratio* $\wp(\Omega)/\mathrm{diam}\,\partial\Omega$, *such that if*

$$\mathrm{dist}\Big(\nu,\ [\mathrm{VMO}\,(\partial\Omega,\sigma)]^n\Big) < \delta_0 \quad (7.1.698)$$

where the distance is considered in the space $\big[\mathrm{BMO}\,(\partial\Omega,\sigma)\big]^n$ *(hence if* Ω *is a* δ*-oscillating* AR *domain; cf. Definition 3.4.1), it follows that* Ω *is a two-sided* NTA *domain and the following operators are Fredholm with index zero:*

$$S_{\text{Stokes}} : \big[H^p(\partial\Omega,\sigma)\big]^n \longrightarrow \big[H_1^{p^*,p}(\partial\Omega,\sigma)\big]^n \quad \text{where } p^* := \big(\tfrac{1}{p} - \tfrac{1}{n-1}\big)^{-1} \in (1,\infty], \tag{7.1.699}$$

$$S_{\text{Stokes}} : \big[H^1(\partial\Omega,\sigma)\big]^2 \longrightarrow \big[H_1^{q,1}(\partial\Omega,\sigma)\big]^2 \quad \text{for each } q \in (1,\infty] \text{ if } n=2. \tag{7.1.700}$$

Also, granted the assumption made in (7.1.408) *and with* $p^* := \big(\frac{1}{p} - \frac{1}{n-1}\big)^{-1}$, *one has*

$$\operatorname{Ker}\Big(S_{\text{Stokes}} : \big[H^p(\partial\Omega,\sigma)\big]^n \to \big[H_1^{p^*,p}(\partial\Omega,\sigma)\big]^n\Big) = \nu\mathbb{C}_{\partial\Omega} \text{ if } n \geq 3, \tag{7.1.701}$$

and the operator

$$\begin{gathered} S_{\text{Stokes}} : \big[H^p(\partial\Omega,\sigma)\big]^n \big/ \nu\mathbb{C}_{\partial\Omega} \longrightarrow \big[H_1^{p^*,p}(\partial\Omega,\sigma)\big]^n \cap \big\langle \nu\mathbb{C}_{\partial\Omega}\big\rangle^{\perp} \\ \text{defined as } S_{\text{Stokes}}[\vec{f}\,] := S_{\text{Stokes}}\vec{f} \text{ for each } \vec{f} \in \big[H^p(\partial\Omega,\sigma)\big]^n \\ \text{is a well-defined linear isomorphism provided } n \geq 3, \end{gathered} \tag{7.1.702}$$

while in the case when $n = 2$ *the following augmented version of* S_{Stokes} *is an invertible operator for each* $q \in (1,\infty]$:

$$\begin{gathered} S^{\text{aug}}_{\text{Stokes}} : \Big(\big[H^p(\partial\Omega,\sigma)\big]^2 \big/ \nu\mathbb{C}_{\partial\Omega}\Big) \oplus \mathbb{C}^2 \to \Big(\big[H_1^{q,p}(\partial\Omega,\sigma)\big]^2 \cap \big\langle \nu\mathbb{C}_{\partial\Omega}\big\rangle^{\perp}\Big) \oplus \mathbb{C}^2 \\ \text{given by } S^{\text{aug}}_{\text{Stokes}}([\vec{f}\,], c) := \Big(S_{\text{Stokes}}\vec{f} + c\,,\, (\langle \vec{f}, \mathbf{e}_j\rangle)_{1\leq j\leq 2}\Big) \\ \text{for each given } \vec{f} \in \big[H^p(\partial\Omega,\sigma)\big]^2 \text{ and } c \in \mathbb{C}^2. \end{gathered} \tag{7.1.703}$$

Proof The justification of the fact that S_{Stokes} is a Fredholm operator in the context of (7.1.699) proceeds exactly as the proof of the Fredholmness result stated in Theorem 7.1.32, as long as all ingredients are available. Let us go through the list of said ingredients one by one. First, from [132, Theorem 6.2.11, item *(7)*] we know that the operator (7.1.699) is well defined, linear, and bounded. Second, [132, Theorem 6.2.11, item *(5)*] with $\lambda := 1$ guarantees that

$$\overset{\bullet\text{slip}}{\partial_\nu}\big(\mathcal{D}_{\text{slip}}, \mathcal{P}_{\text{slip}}\big) : \big[H_1^{p^*,p}(\partial\Omega,\sigma)\big]^n \longrightarrow \big[H^p(\partial\Omega,\sigma)\big]^n \tag{7.1.704}$$

is well defined, linear, and bounded. Third, from item *(10)* in Theorem 7.1.1 we know that the operators $\pm\frac{1}{2}I + K^{\#}_{\text{slip}}$ are Fredholm (with index zero) when acting from $\big[H^p(\partial\Omega,\sigma)\big]^n$ into itself. Fourth, [132, Theorem 6.3.7] tells us that

$$\big(\tfrac{1}{2}I + K^{\#}_{\text{slip}}\big) \circ \big(-\tfrac{1}{2}I + K^{\#}_{\text{slip}}\big) = \big(\overset{\bullet\text{slip}}{\partial_\nu}(\mathcal{D}_{\text{slip}}, \mathcal{P}_{\text{slip}})\big) \circ S_{\text{Stokes}} \text{ on } \big[H^p(\partial\Omega,\sigma)\big]^n. \tag{7.1.705}$$

Much as in the proof of Theorem 7.1.32, these ingredients may be used to show that

$$S_{\text{Stokes}} : \big[H^p(\partial\Omega,\sigma)\big]^n \longrightarrow \big[H_1^{p^*,p}(\partial\Omega,\sigma)\big]^n \text{ is a Fredholm operator.} \tag{7.1.706}$$

Going further, let S_λ denote the boundary-to-boundary single layer potential operator on $\partial\Omega$ associated with the Lamé system with moduli $\mu := 1$ and $\lambda \in (0, \infty)$ arbitrary. From [131, (1.3.11)] and Remark 1.6.10 it is known that the Lamé system $L_{1,\lambda}$, defined as in (1.6.85) with $\mu := 1$ and $\lambda \in (0, \infty)$ arbitrary, is Legendre-Hadamard strongly elliptic and has $\mathfrak{A}^{\rm dis}_{L_{1,\lambda}} \neq \varnothing$. Granted these properties, we conclude from Theorem 7.1.34 (when $n \geq 3$) and Theorem 7.1.35 (when $n = 2$) that, in all space dimensions $n \geq 2$, the operator

$$S_\lambda : \big[H^p(\partial\Omega, \sigma)\big]^n \longrightarrow \big[H_1^{p^*,p}(\partial\Omega, \sigma)\big]^n \text{ is Fredholm with index zero.} \tag{7.1.707}$$

Finally, from the analogue of the first property in [132, Theorem 2.3.1, item *(iv)*] for $S_{\rm Stokes}$, [132, Theorem 2.1.4] (or, alternatively, the estimate in [132, Theorem 5.1.1, item *(4)*], plus [132, Example 5.1.4]), and (7.1.227) we deduce that

$$\lim_{\lambda\to\infty} S_\lambda = S_{\rm Stokes} \text{ in } \mathscr{B}\Big(\big[H^p(\partial\Omega, \sigma)\big]^n \to \big[H_1^{p^*,p}(\partial\Omega, \sigma)\big]^n\Big). \tag{7.1.708}$$

Since the index is homotopic invariant within the class of Fredholm operators on quasi-Banach spaces (cf. the discussion in [130, §2.2]), we finally conclude from (7.1.706), (7.1.707), and (7.1.708) that $S_{\rm Stokes}$ is, as claimed, a Fredholm operator with index zero in the context of (7.1.699).

Virtually the same argument used to prove the Fredholmness with index zero of $S_{\rm Stokes}$ in (7.1.699) shows that $S_{\rm Stokes}$ is also a Fredholm operator with index zero in the context of (7.1.700). Once these Fredholmness results have been established, the rest of the claims in the statement of the theorem follow based on Theorem 7.1.21, the abstract regularity result from [130, Lemma 2.1.4], and Fredholm theory in quasi-Banach spaces (cf. [130, §2.2]). □

Theorem 7.1.47 turns out to play a key role in the proof of essential norm estimates for the boundary-to-boundary slip double layer operator for the Stokes system acting on Hardy-based Sobolev spaces on δ-oscillating AR domains.

Theorem 7.1.48 *Fix $n \in \mathbb{N}$ with $n \geq 2$ and suppose $\Omega \subseteq \mathbb{R}^n$ is a* UR *domain with compact boundary. Abbreviate $\sigma := \mathcal{H}^{n-1}\lfloor\partial\Omega$ and denote by ν the geometric measure theoretic outward unit normal to Ω. Consider*

$$p \in \big(\tfrac{n-1}{n}, 1\big], \quad \theta \in \big(0, n - \tfrac{n-1}{p}\big), \tag{7.1.709}$$

and define

$$p^* := \begin{cases} \big(\frac{1}{p} - \frac{1}{n-1}\big)^{-1} \in (1, \infty) \text{ if either } n \geq 3, \text{ or } n = 2 \text{ and } p < 1, \\ q \in (1, \infty) \text{ arbitrary if } n = 2 \text{ and } p = 1. \end{cases} \tag{7.1.710}$$

Finally, recall from [132, Theorem 6.2.11, item *(6)*] *that the boundary-to-boundary slip double layer potential operator $K_{\rm slip}$ (originally defined as in* (4.2.102)) *induces a well-defined, linear, and bounded mapping*

$$K_{\text{slip}} : \big[H_1^{p^*,p}(\partial\Omega,\sigma)\big]^n \longrightarrow \big[H_1^{p^*,p}(\partial\Omega,\sigma)\big]^n. \tag{7.1.711}$$

Then there exists some $C_\theta \in (0,\infty)$, which depends only on θ, n, p, the UR *constants of $\partial\Omega$, and the relative flatness ratio $\wp(\Omega)/\operatorname{diam}\partial\Omega$ such that*

$$\operatorname{dist}\Big(K_{\text{slip}}, \operatorname{Cp}\big(\big[H_1^{p^*,p}(\partial\Omega,\sigma)\big]^n\big)\Big) \leq C_\theta\Big[\operatorname{dist}\big(\nu, \big[\operatorname{VMO}(\partial\Omega,\sigma)\big]^n\big)\Big]^\theta \tag{7.1.712}$$

where the distance in the left-hand side is measured in Bd $\big(\big[H_1^{p^*,p}(\partial\Omega,\sigma)\big]^n\big)$, *while the distance in the right-hand side is measured in* $\big[\operatorname{BMO}(\partial\Omega,\sigma)\big]^n$. *In particular, from* (7.1.712) *and* [130, (1.2.53)] *one concludes that*

$$\begin{array}{c}\textit{the operator } K_{\text{slip}} \textit{ is compact on } \big[H_1^{p^*,p}(\partial\Omega,\sigma)\big]^n \textit{ whenever}\\ \Omega \textit{ is an infinitesimally flat AR domain (cf. Definition 3.4.1).}\end{array} \tag{7.1.713}$$

Proof To set the stage, recall from [132, Theorem 6.2.11, *(6)*] that there exists $C_0 \in (0,\infty)$ depending only on n, p, and the UR constants of $\partial\Omega$ so that

$$\|K_{\text{slip}}\|_{[H_1^{p^*,p}(\partial\Omega,\sigma)]^n \to [H_1^{p^*,p}(\partial\Omega,\sigma)]^n} \leq C_0. \tag{7.1.714}$$

Bring in the parameter $\delta_0 \in (0,1)$ featuring in the statement of Theorem 7.1.47. When actually dist $\big(\nu, \big[\operatorname{VMO}(\partial\Omega,\sigma)\big]^n\big) \geq \delta_0$, the estimate claimed in (7.1.712) may be seen by making use of [130, (1.2.56)] and (7.1.714) to simply write

$$\begin{aligned}\operatorname{dist}\Big(K_{\text{slip}}, \operatorname{Cp}\big(\big[H_1^{p^*,p}(\partial\Omega,\sigma)\big]^n\big)\Big) &\leq \|K_{\text{slip}}\|_{[H_1^{p^*,p}(\partial\Omega,\sigma)]^n \to [H_1^{p^*,p}(\partial\Omega,\sigma)]^n}\\ &\leq C_0 \leq (C_0/\delta_0)\cdot \operatorname{dist}\big(\nu, \big[\operatorname{VMO}(\partial\Omega,\sigma)\big]^n\big).\end{aligned} \tag{7.1.715}$$

On account of (4.1.17) and the fact that dist $\big(\nu, \big[\operatorname{VMO}(\partial\Omega,\sigma)\big]^n\big) \leq 1$, this readily implies (7.1.712). Hence, as regards the estimate claimed in (7.1.712), there remains to treat the situation when

$$\operatorname{dist}\big(\nu, \big[\operatorname{VMO}(\partial\Omega,\sigma)\big]^n\big) < \delta_0. \tag{7.1.716}$$

In such a scenario, Theorem 7.1.47 and [130, Theorem 2.2.3] imply that there exists a linear bounded operator

$$\begin{array}{c}R : \big[H_1^{p^*,p}(\partial\Omega,\sigma)\big]^n \longrightarrow \big[H^p(\partial\Omega,\sigma)\big]^n \text{ such that}\\ Q := S_{\text{Stokes}} \circ R - I \text{ belongs to } \operatorname{Cp}\big(\big[H_1^{p^*,p}(\partial\Omega,\sigma)\big]^n\big).\end{array} \tag{7.1.717}$$

Upon recalling from [132, Theorem 6.3.7] that

$$S_{\text{Stokes}} \circ K^\#_{\text{slip}} = K_{\text{slip}} \circ S_{\text{Stokes}} \text{ on } \big[H^p(\partial\Omega,\sigma)\big]^n, \tag{7.1.718}$$

it follows that we may represent the operator K_{slip} from (7.1.711) as

$$K_{\text{slip}} = S_{\text{Stokes}} \circ K^{\#}_{\text{slip}} \circ R - K^{\#}_{\text{slip}} \circ Q \text{ on } \big[H^{p^*,p}_1(\partial\Omega,\sigma)\big]^n. \tag{7.1.719}$$

Since $K^{\#}_{\text{slip}} \circ Q \in \mathrm{Cp}\big(\big[H^{p^*,p}_1(\partial\Omega,\sigma)\big]^n\big)$, we may invoke [130, Lemma 1.2.10] to conclude

$$\|K_{\text{slip}}\|^{\text{ess}}_{[H^{p^*,p}_1(\partial\Omega,\sigma)]^n \to [H^{p^*,p}_1(\partial\Omega,\sigma)]^n} \leq C\|K^{\#}_{\text{slip}}\|^{\text{ess}}_{[H^p(\partial\Omega,\sigma)]^n \to [H^p(\partial\Omega,\sigma)]^n} \tag{7.1.720}$$

where the constant $C \in (0,\infty)$ is given by

$$C := \|S_{\text{Stokes}}\|_{[H^p(\partial\Omega,\sigma)]^n \to [H^{p^*,p}_1(\partial\Omega,\sigma)]^n} \cdot \|R\|_{[H^{p^*,p}_1(\partial\Omega,\sigma)]^n \to [H^p(\partial\Omega,\sigma)]^n}.$$

The nature of C suits our purposes. In addition, from (4.2.103) and Theorem 5.4.9 we see that there exists a constant $C_\theta \in (0,\infty)$, depending only on θ, n, p, the UR constants of $\partial\Omega$, and the relative flatness ratio $\wp(\Omega)/\operatorname{diam}\partial\Omega$, such that

$$\|K^{\#}_{\text{slip}}\|^{\text{ess}}_{[H^p(\partial\Omega,\sigma)]^n \to [H^p(\partial\Omega,\sigma)]^n} \leq C_\theta \cdot \Big[\operatorname{dist}\big(\nu, \big[\mathrm{VMO}(\partial\Omega,\sigma)\big]^n\big)\Big]^\theta \tag{7.1.721}$$

with the distance in the right-hand side in $\big[\mathrm{BMO}(\partial\Omega,\sigma)\big]^n$. At this stage, the estimate in (7.1.712) is obtained by combining (7.1.720) with (7.1.721). □

We conclude this section by augmenting Theorem 7.1.23 with an invertibility result involving the boundary-to-boundary slip double layer operator for the Stokes system acting on Hardy-based Sobolev spaces on Ahlfors regular domains which are sufficiently flat at an infinitesimal level.

Theorem 7.1.49 *Fix $n \in \mathbb{N}$ with $n \geq 2$ and assume $\Omega \subseteq \mathbb{R}^n$ is an Ahlfors regular domain with compact boundary. Abbreviate $\sigma := \mathcal{H}^{n-1}\lfloor\partial\Omega$ and denote by ν the geometric measure theoretic outward unit normal to Ω. In this setting, consider the boundary-to-boundary slip double layer potential operator K_{slip} associated with the Stokes system in Ω as in* (4.2.102). *Also, select some*

$$p \in \big(\tfrac{n-1}{n}, 1\big] \tag{7.1.722}$$

and define

$$p^* := \begin{cases} \big(\frac{1}{p} - \frac{1}{n-1}\big)^{-1} \in (1,\infty) \text{ if either } n \geq 3, \text{ or } n = 2 \text{ and } p < 1, \\ q \in (1,\infty) \text{ arbitrary } \text{ if } n = 2 \text{ and } p = 1. \end{cases} \tag{7.1.723}$$

Finally, recall notation introduced in (7.1.399) *and* (7.1.416).

Then there exists $\delta \in (0,1)$, depending only on n, p, the ADR *constants of $\partial\Omega$, and the relative flatness ratio $\wp(\Omega)/\operatorname{diam}\partial\Omega$, with the property that whenever*

$$\operatorname{dist}\Big(\nu, [\mathrm{VMO}(\partial\Omega,\sigma)]^n\Big) < \delta \tag{7.1.724}$$

where the distance is considered in the space $\big[\mathrm{BMO}(\partial\Omega,\sigma)\big]^n$ (hence if Ω is a δ-oscillating AR *domain; cf. Definition 3.4.1), it follows that Ω is a two-sided* NTA *domain,*

$$\operatorname{Ker}\Big(\pm\tfrac{1}{2}I+K_{\text{slip}};\,\big[H_1^{p^*,p}(\partial\Omega,\sigma)\big]^n\cap\big\langle\nu\mathbb{C}_{\partial\Omega_\pm}\big\rangle^{\perp}\Big)=\Psi(\partial\Omega_\mp), \tag{7.1.725}$$

$$\operatorname{Im}\Big(\pm\tfrac{1}{2}I+K_{\text{slip}};\,\big[H_1^{p^*,p}(\partial\Omega,\sigma)\big]^n\Big)\subseteq\big[H_1^{p^*,p}(\partial\Omega,\sigma)\big]^n\cap\big\langle\nu\mathbb{C}_{\partial\Omega_\pm}\big\rangle^{\perp}, \tag{7.1.726}$$

and the operators

$$\begin{aligned}\pm\tfrac{1}{2}I+\big[K_{\text{slip}}\big]:&\Big(\big[H_1^{p^*,p}(\partial\Omega,\sigma)\big]^n\cap\big\langle\nu\mathbb{C}_{\partial\Omega_\pm}\big\rangle^{\perp}\Big)\Big/\Psi(\partial\Omega_\mp)\\&\longrightarrow\Big(\big[H_1^{p^*,p}(\partial\Omega,\sigma)\big]^n\cap\big\langle\nu\mathbb{C}_{\partial\Omega_\pm}\big\rangle^{\perp}\Big)\Big/\Psi(\partial\Omega_\mp)\end{aligned} \tag{7.1.727}$$

(naturally defined, by sending the equivalence class of a function into the equivalence class of the operator in question acting on said function) are all well-defined linear isomorphisms.

Finally,

$$\operatorname{Ker}\Big(\pm\tfrac{1}{2}I+K^{\#}_{\text{slip}}:\,\big[H^{p}(\partial\Omega,\sigma)\big]^n\cap\Psi(\partial\Omega_\mp)^{\perp}\Big)=\nu\mathbb{C}_{\partial\Omega_\pm}, \tag{7.1.728}$$

$$\operatorname{Im}\Big(\pm\tfrac{1}{2}I+K^{\#}_{\text{slip}};\,\big[H^{p}(\partial\Omega,\sigma)\big]^n\Big)\subseteq\big[H^{p}(\partial\Omega,\sigma)\big]^n\cap\Psi(\partial\Omega_\mp)^{\perp}, \tag{7.1.729}$$

and the operators

$$\begin{aligned}\pm\tfrac{1}{2}I+\big[K^{\#}_{\text{slip}}\big]:&\Big(\big[H^{p}(\partial\Omega,\sigma)\big]^n\cap\Psi(\partial\Omega_\mp)^{\perp}\Big)\Big/\nu\mathbb{C}_{\partial\Omega_\pm}\\&\longrightarrow\Big(\big[H^{p}(\partial\Omega,\sigma)\big]^n\cap\Psi(\partial\Omega_\mp)^{\perp}\Big)\Big/\nu\mathbb{C}_{\partial\Omega_\pm}\end{aligned} \tag{7.1.730}$$

(once more naturally defined, by sending the equivalence class of a function into the equivalence class of the operator in question acting on said function) are all well-defined linear isomorphisms.

Proof All claims up to (and including) (7.1.727) are consequences of Theorem 3.4.2, Theorem 7.1.48, Theorem 7.1.23, and Fredholm theory in quasi-Banach spaces (cf. [130, §2.2]). The claims pertaining to (7.1.728)-(7.1.730) may be justified based on item *(10)* in Theorem 7.1.1, Theorem 7.1.23, and the Fredholm theory in quasi-Banach spaces reviewed in [130, §2.2]. □

7.2 Fredholm and Invertibility Properties of Boundary Layer Potentials in Energy Spaces

The work in this section complements the Fredholmness and invertibility results for boundary layer potential operators from §7.1. The main difference is that while we now restrict ourselves to "energy spaces" (see (7.2.1)-(7.2.3) below), no flatness assumptions on the boundary of the underlying domain are made.

Suppose $\Omega \subseteq \mathbb{R}^n$, where $n \in \mathbb{N}$ satisfies $n \geq 2$, is an Ahlfors regular domain and abbreviate $\sigma := \mathcal{H}^{n-1}\lfloor\partial\Omega$. The energy spaces[10] $H^{\pm 1/2}(\partial\Omega, \sigma)$ are then defined, demanding the finiteness of a Slobodeckij semi-norm, as

$$H^{1/2}(\partial\Omega,\sigma) := \left\{ f \in L^2(\partial\Omega,\sigma) : \int_{\partial\Omega}\int_{\partial\Omega} \frac{|f(x)-f(y)|^2}{|x-y|^n}\, \mathrm{d}\sigma(x)\,\mathrm{d}\sigma(y) < \infty \right\}, \tag{7.2.1}$$

equipped with the norm

$$\|f\|_{H^{1/2}(\partial\Omega,\sigma)} := \|f\|_{L^2(\partial\Omega,\sigma)} + \left(\int_{\partial\Omega}\int_{\partial\Omega} \frac{|f(x)-f(y)|^2}{|x-y|^n}\, \mathrm{d}\sigma(x)\,\mathrm{d}\sigma(y) \right)^{\frac{1}{2}} \tag{7.2.2}$$

for each $f \in H^{1/2}(\partial\Omega,\sigma)$, and, respectively, via duality, as

$$H^{-1/2}(\partial\Omega,\sigma) := \big(H^{1/2}(\partial\Omega,\sigma)\big)^*. \tag{7.2.3}$$

From (7.2.1)-(7.2.2) and [130, (7.9.10)-(7.9.11)] we then see that

$$\begin{array}{c} B^{2,2}_{1/2}(\partial\Omega,\sigma) \text{ and } H^{1/2}(\partial\Omega,\sigma) \text{ coincides as sets, and} \\ \|f\|_{B^{2,2}_{1/2}(\partial\Omega,\sigma)} \approx \|f\|_{H^{1/2}(\partial\Omega,\sigma)} \text{ uniformly for} \\ f \in B^{2,2}_{1/2}(\partial\Omega,\sigma) = H^{1/2}(\partial\Omega,\sigma). \end{array} \tag{7.2.4}$$

In concert with (7.2.3) and item *(i)* in [130, Proposition 7.6.1], this also implies that

$$\begin{array}{c} B^{2,2}_{-1/2}(\partial\Omega,\sigma) \text{ and } H^{-1/2}(\partial\Omega,\sigma) \text{ coincides as sets, and} \\ \|f\|_{B^{2,2}_{-1/2}(\partial\Omega,\sigma)} \approx \|f\|_{H^{-1/2}(\partial\Omega,\sigma)} \text{ uniformly for} \\ f \in B^{2,2}_{-1/2}(\partial\Omega,\sigma) = H^{-1/2}(\partial\Omega,\sigma). \end{array} \tag{7.2.5}$$

The manner in which the energy spaces (7.2.1), (7.2.3) related to ordinary Lebesgue spaces is elucidated in the lemma below.

Lemma 7.2.1 *Let $\Omega \subseteq \mathbb{R}^n$, where $n \in \mathbb{N}$ satisfies $n \geq 2$, be an Ahlfors regular domain and abbreviate $\sigma := \mathcal{H}^{n-1}\lfloor\partial\Omega$. Then the following are well-defined continuous inclusions, with dense ranges if $n \geq 3$:*

$$H^{1/2}(\partial\Omega,\sigma) \hookrightarrow L^{2(n-1)/(n-2)}(\partial\Omega,\sigma), \tag{7.2.6}$$

$$L^{2(n-1)/n}(\partial\Omega,\sigma) \hookrightarrow H^{-1/2}(\partial\Omega,\sigma). \tag{7.2.7}$$

If $n = 2$ and $\partial\Omega$ is compact then the following are also well-defined continuous inclusions, with dense ranges for each $p \in (1,\infty)$:

---

[10] not to be confused with Hardy spaces $H^p(\partial\Omega,\sigma)$ with $\frac{n-1}{n} < p < \infty$, for which the values $p = \pm 1/2$ are prohibited

$$H^{1/2}(\partial\Omega,\sigma) \hookrightarrow L^p(\partial\Omega,\sigma) \hookrightarrow H^{-1/2}(\partial\Omega,\sigma). \tag{7.2.8}$$

The spaces $L^{2(n-1)/(n-2)}(\partial\Omega,\sigma)$ and $L^{2(n-1)/n}(\partial\Omega,\sigma)$ in (7.2.6)-(7.2.7) have been encountered earlier, in [132, Theorem 1.7.2], dealing with injectivity properties for the boundary-to-boundary versions of the single and double layer potential operators associated with a Legendre-Hadamard elliptic system L and a UR domain Ω. This is one way in which the material in this section links up with [132, §1.7] (for more on this see the comment right after the statement of Theorem 7.2.4).

Proof of Lemma 7.2.1 First assume that $n \geq 3$. Observe that

$$B^{2,2}_{1/2}(\partial\Omega,\sigma) \hookrightarrow L^{2(n-1)/(n-2)}(\partial\Omega,\sigma) \tag{7.2.9}$$

is a well-defined, continuous embedding, with dense range. This is seen from [130, Theorem 7.7.4] (used with $p_0 = q_0 := 2$, $s_0 := 1/2$, $p_1 := 2(n-1)/(n-2)$, $q_1 := 2$, and $s_1 := 0$), [130, (7.7.5)] (used with $p := 2(n-1)/(n-2)$, $q := 2 = \min\{p, 2\}$, $s := 0$), and [130, (7.1.55)] (used with $p := 2(n-1)/(n-2)$) which, in concert, give

$$\begin{aligned} B^{2,2}_{1/2}(\partial\Omega,\sigma) &\hookrightarrow B^{2(n-1)/(n-2),2}_{0}(\partial\Omega,\sigma) \\ &\hookrightarrow F^{2(n-1)/(n-2),2}_{0}(\partial\Omega,\sigma) = L^{2(n-1)/(n-2)}(\partial\Omega,\sigma). \end{aligned} \tag{7.2.10}$$

Granted (7.2.9), we may then use [130, Lemma 1.2.1] and duality (cf. [130, Proposition 7.6.1]) to conclude that

$$L^{2(n-1)/n}(\partial\Omega,\sigma) \hookrightarrow B^{2,2}_{-1/2}(\partial\Omega,\sigma) \tag{7.2.11}$$

is a well-defined, continuous embedding. Moreover, since compactly supported Lipschitz functions belong to $L^{2(n-1)/n}(\partial\Omega,\sigma)$, from [130, Lemma 7.1.10] we also see that the embedding in (7.2.11) has dense range. In view of (7.2.4), (7.2.5), all desired conclusions pertaining to (7.2.6)-(7.2.7) follow.

Finally, the fact that the inclusions in (7.2.8) are also well-defined, continuous, with dense ranges for each $p \in (1,\infty)$, if $n = 2$ and $\partial\Omega$ is compact may be justified in a similar fashion, now employing [130, (7.7.6)] in place of [130, Theorem 7.7.4]. □

We begin in earnest by proving that, in an appropriate context, the single layer potential operator is bounded from below, modulo compacts.

Lemma 7.2.2 *Let $\Omega \subseteq \mathbb{R}^n$, where $n \in \mathbb{N}$, $n \geq 2$, be a two-sided* NTA *domain with a compact Ahlfors regular boundary, abbreviate $\sigma := \mathcal{H}^{n-1}\lfloor\partial\Omega$, and let L be a homogeneous, second-order, constant complex coefficient, $M \times M$ system in $\mathbb{R}^n$ satisfying the Legendre-Hadamard strong ellipticity condition* (*cf.* [131, Definition 1.3.2]). *Recall the boundary-to-boundary single layer potential operator S associated with L and Ω, acting on Besov spaces as in* [132, Theorem 4.3.4, item *(2)*].

Then there exists some constant $C \in (0,\infty)$ together with a linear compact operator Comp *mapping $\big[H^{-1/2}(\partial\Omega,\sigma)\big]^M$ into a Banach space, such that*

$$-\mathrm{Re}\,\big\langle f, \overline{Sf}\big\rangle + \|\mathrm{Comp}\, f\|^2 \geq C\|f\|^2_{[H^{-1/2}(\partial\Omega,\sigma)]^M} \tag{7.2.12}$$

for every functional $f \in \big[H^{-1/2}(\partial\Omega,\sigma)\big]^M$, *where* $\langle\cdot,\cdot\rangle$ *above is the duality pairing between the spaces* $\big[H^{-1/2}(\partial\Omega,\sigma)\big]^M$ *and* $\big[H^{1/2}(\partial\Omega,\sigma)\big]^M$.

Proof Without loss of generality (cf. [129, Lemma 5.10.10]) assume Ω is bounded. Recall the boundary-to-domain single layer potential operator $\mathscr{S}$ associated with L and Ω, acting from the scale of Besov spaces as in [132, Proposition 4.2.8]. Fix some $f \in \big[H^{-1/2}(\partial\Omega,\sigma)\big]^M$. Define $\Omega_+ := \Omega$ and $\Omega_- := \mathbb{R}^n \setminus \overline{\Omega}$ and consider

$$u := u^\pm \text{ in } \Omega_\pm, \text{ where } u^\pm := \mathscr{S} f \text{ in } \Omega_\pm. \tag{7.2.13}$$

Thus, $u^\pm = u|_{\Omega_\pm}$. Fix $\psi \in \mathscr{C}_c^\infty(\mathbb{R}^n)$ with $\psi \equiv 1$ near $\overline{\Omega}$. Then

$$\psi u^\pm \in \big[W^{1,2}(\Omega_\pm)\big]^M \tag{7.2.14}$$

(cf. [132, Theorem 4.2.10]) while the trace formula [132, (4.3.44)] in the context described in [132, (4.3.45)] gives

$$\mathrm{Tr}_{\Omega_+\to\partial\Omega}(\psi u^+) = \mathrm{Tr}_{\Omega_-\to\partial\Omega}(\psi u^-) \text{ in } \big[H^{1/2}(\partial\Omega,\sigma)\big]^M. \tag{7.2.15}$$

With this in hand, we may invoke [18, Theorem 5.13, p. 4381] which permits us to glue together Sobolev functions with matching traces and conclude that

$$\psi u \in \big[W^{1,2}(\mathbb{R}^n)\big]^M. \tag{7.2.16}$$

Next, pick a coefficient tensor which yields the system L, i.e.,

$$A = \big(a_{rs}^{\alpha\beta}\big)_{\substack{1\le r,s\le n\\ 1\le\alpha,\beta\le M}} \text{ such that } L = L_A. \tag{7.2.17}$$

Observe that the function $\phi := L(\psi u^-)$ belongs to $\big[\mathscr{C}_c^\infty(\Omega_-,\mathcal{L}^n)\big]^M$, and the latter space embeds into $\big(\big[W^{1,2}(\Omega_-)\big]^M\big)^*$. As such, [130, Proposition 8.5.3] gives

$$\begin{aligned} &\|\partial_\nu^A(u^+,0)\|_{[H^{-1/2}(\partial\Omega,\sigma)]^M} \le C\|u^+\|_{[W^{1,2}(\Omega_+)]^M}, \text{ and} \\ &\|\partial_\nu^A(\psi u^-,\phi)\|_{[H^{-1/2}(\partial\Omega,\sigma)]^M} \le C\|\psi u^-\|_{[W^{1,2}(\Omega_-)]^M} + C\|\phi\|_{([W^{1,2}(\Omega_-)]^M)^*}. \end{aligned} \tag{7.2.18}$$

In addition, the jump-formula [132, (4.3.83)] in the context described in [132, Theorem 4.3.6, item *(a)*] gives

$$\partial_\nu^A(u^+,0) = \big(-\tfrac{1}{2}I + K^\#_{A^\top}\big)f \quad\text{and}\quad \partial_\nu^A(\psi u^-,\phi) = \big(\tfrac{1}{2}I + K^\#_{A^\top}\big)f. \tag{7.2.19}$$

Based on these and the triangle inequality we may thus estimate

$$\begin{aligned} \|f\|_{[H^{-1/2}(\partial\Omega,\sigma)]^M} &\le \|\partial_\nu^A(u^+,0)\|_{[H^{-1/2}(\partial\Omega,\sigma)]^M} + \|\partial_\nu^A(\psi u^-,\phi)\|_{[H^{-1/2}(\partial\Omega,\sigma)]^M} \\ &\le C\Big(\|\psi u\|_{[W^{1,2}(\mathbb{R}^n)]^M} + \|\phi\|_{([W^{1,2}(\Omega_-)]^M)^*}\Big). \end{aligned} \tag{7.2.20}$$

As already noted, the function ϕ is supported in a compact subset of Ω_- (since the cutoff function ψ is identically 1 near $\partial\Omega$). In turn, this observation readily leads to the conclusion that the assignment

$$\big[H^{-1/2}(\partial\Omega,\sigma)\big]^M \ni f \mapsto \phi \in \big(\big[W^{1,2}(\Omega_-)\big]^M\big)^* \text{ is compact.} \tag{7.2.21}$$

Next, we employ again the jump-formulas in (7.2.19), the trace formula [132, (4.3.44)], and the generalized "half" Green's formula [130, (8.5.18)] to write

$$\begin{aligned} -\mathrm{Re}\,\langle f, \overline{Sf}\rangle &= -\mathrm{Re}\Big\langle (\tfrac{1}{2}I + K^{\#}_{A^\top})f, \overline{Sf}\Big\rangle + \mathrm{Re}\Big\langle (-\tfrac{1}{2}I + K^{\#}_{A^\top})f, \overline{Sf}\Big\rangle \\ &= -\mathrm{Re}\Big\langle \partial^A_\nu(\psi u^-,\phi), \overline{\mathrm{Tr}_{\Omega_-\to\partial\Omega}(\psi u^-)}\Big\rangle + \mathrm{Re}\Big\langle \partial^A_\nu(u^+,0), \overline{\mathrm{Tr}_{\Omega_-\to\partial\Omega}(u^+)}\Big\rangle \\ &= \mathrm{Re}\int_{\Omega_-} a^{\alpha\beta}_{rs}\,\partial_s(\psi u^-_\beta)\overline{\partial_r(\psi u^-_\alpha)}\,d\mathcal{L}^n + \mathrm{Re}\int_{\Omega_-} \langle \phi, \overline{\psi u^-}\rangle\,d\mathcal{L}^n \\ &\quad + \mathrm{Re}\int_{\Omega_+} a^{\alpha\beta}_{rs}\,(\partial_s u^+_\beta)\overline{\partial_r u^+_\alpha}\,d\mathcal{L}^n \\ &= \mathrm{Re}\int_{\mathbb{R}^n} a^{\alpha\beta}_{rs}\,\partial_s(\psi u_\beta)\overline{\partial_r(\psi u_\alpha)}\,d\mathcal{L}^n + \mathrm{Re}\int_{\Omega_-} \langle \phi, \overline{\psi u^-}\rangle\,d\mathcal{L}^n. \end{aligned} \tag{7.2.22}$$

On the other hand, the coercivity estimate (1.1.29) gives that there exists some constant $c \in (0,\infty)$ with the property that

$$\mathrm{Re}\int_{\mathbb{R}^n} a^{\alpha\beta}_{rs}(\partial_s\eta_\beta)\overline{\partial_r\eta_\alpha}\,d\mathcal{L}^n \geq c\int_{\mathbb{R}^n}|\nabla\eta|^2\,d\mathcal{L}^n, \quad \forall\,\eta\in\big[W^{1,2}(\mathbb{R}^n)\big]^M. \tag{7.2.23}$$

In particular, there exists $c = c(\Omega, L) \in (0,\infty)$ such that, for each $\eta \in \big[W^{1,2}(\mathbb{R}^n)\big]^M$,

$$\mathrm{Re}\int_{\mathbb{R}^n} a^{\alpha\beta}_{rs}(\partial_s\eta_\beta)\overline{\partial_r\eta_\alpha}\,d\mathcal{L}^n + c\int_{\mathbb{R}^n}|\eta|^2\,d\mathcal{L}^n \geq c\|\eta\|^2_{[W^{1,2}(\mathbb{R}^n)]^M}. \tag{7.2.24}$$

We shall use (7.2.24) for $\eta := \psi u$. In this regard, it is also useful to note that [132, Theorem 4.2.10] implies that $\big[H^{-1/2}(\partial\Omega,\sigma)\big]^M \ni f \mapsto \psi u \in \big[L^2(\mathbb{R}^n,\mathcal{L}^n)\big]^M$ is a compact assignment. Based on this, (7.2.22), (7.2.20) and the various comments made about compact operators, the claim made in (7.2.12) follows. □

It is useful to further isolate the estimate contained in the lemma below.

Lemma 7.2.3 *Let $\Omega \subseteq \mathbb{R}^n$, where $n \in \mathbb{N}$, $n \geq 2$, be a two-sided* NTA *domain with a compact Ahlfors regular boundary, and abbreviate $\sigma := \mathcal{H}^{n-1}\lfloor\partial\Omega$. Let L be a homogeneous, second-order, constant complex coefficient, $M \times M$ system in $\mathbb{R}^n$ satisfying the Legendre-Hadamard strong ellipticity condition (cf.* [131, Definition 1.3.2]*). Recall the boundary-to-boundary single layer potential operator S associated with L and Ω, acting on Besov spaces as in* [132, Theorem 4.3.4, *(2)*].

Then there exist $C \in (0,\infty)$ and some linear compact operator Comp *mapping $\big[H^{-1/2}(\partial\Omega,\sigma)\big]^M$ into a Banach space, such that for every $f \in \big[H^{-1/2}(\partial\Omega,\sigma)\big]^M$,*

$$\|Sf\|_{[H^{1/2}(\partial\Omega,\sigma)]^M} + \|\mathrm{Comp}\, f\| \geq C\|f\|_{[H^{-1/2}(\partial\Omega,\sigma)]^M}. \tag{7.2.25}$$

Hence, by (7.2.25) *and* [130, Theorem 2.1.2, *(4)*], *the operator*

$$S : \left[H^{-1/2}(\partial\Omega,\sigma)\right]^M \longrightarrow \left[H^{1/2}(\partial\Omega,\sigma)\right]^M \tag{7.2.26}$$

has closed range and finite dimensional kernel.

Proof This is an immediate consequence of estimate (7.2.12) and the fact that, given any $\varepsilon > 0$, there exists $C_\varepsilon \in (0,\infty)$ such that

$$\begin{aligned} \left|\langle f, \overline{Sf}\rangle\right| &\leq \|f\|_{[H^{-1/2}(\partial\Omega,\sigma)]^M} \cdot \|Sf\|_{[H^{1/2}(\partial\Omega,\sigma)]^M} \\ &\leq \varepsilon\|f\|^2_{[H^{-1/2}(\partial\Omega,\sigma)]^M} + C_\varepsilon\|Sf\|^2_{[H^{1/2}(\partial\Omega,\sigma)]^M}, \end{aligned} \tag{7.2.27}$$

for every $f \in \left[H^{-1/2}(\partial\Omega,\sigma)\right]^M$. □

The next theorem contains the main result on to the Fredholm and invertibility properties of the single layer operator on energy spaces. What is remarkable is that no flatness assumptions are imposed on the underlying domain. This complements the invertibility results on more general Besov spaces obtained in Theorem 7.1.15 but for the more restrictive class of δ-oscillating AR domains (cf. Definition 3.4.1).

Theorem 7.2.4 *Let* $\Omega \subseteq \mathbb{R}^n$, *where* $n \in \mathbb{N}$ *with* $n \geq 2$, *be a two-sided* NTA *domain with a compact Ahlfors regular boundary. Set* $\sigma := \mathcal{H}^{n-1}\lfloor\partial\Omega$. *Also, let L be a homogeneous, second-order, constant complex coefficient,* $M \times M$ *system in* $\mathbb{R}^n$ *satisfying the Legendre-Hadamard strong ellipticity condition (cf.* [131, Definition 1.3.2]). *Finally, recall the boundary-to-boundary single layer potential operator S associated with L and* Ω, *acting on Besov spaces as in* [132, Theorem 4.3.4, item *(2)*].

Then there exists some small $\varepsilon > 0$ *with the property that*

$$\begin{gathered} \text{whenever } |p-2| + |q-2| + \left|s - \tfrac{1}{2}\right| < \varepsilon, \text{ the operator} \\ S : \left[B^{p,q}_{s-1}(\partial\Omega,\sigma)\right]^M \to \left[B^{p,q}_{s}(\partial\Omega,\sigma)\right]^M \text{ is Fredholm with index zero,} \\ \text{and is actually invertible if } n \geq 3. \end{gathered} \tag{7.2.28}$$

As a corollary of (7.2.28) *and* (7.2.4)-(7.2.5),

$$\begin{gathered} \text{the operator } S : \left[H^{-1/2}(\partial\Omega,\sigma)\right]^M \longrightarrow \left[H^{1/2}(\partial\Omega,\sigma)\right]^M \\ \text{is Fredholm with index zero, and is actually invertible if } n \geq 3. \end{gathered} \tag{7.2.29}$$

In particular, (7.2.29) implies that in the above geometric setting, if $n \geq 3$ then the single layer operator $S : \left[H^{-1/2}(\partial\Omega,\sigma)\right]^M \to \left[H^{1/2}(\partial\Omega,\sigma)\right]^M$ is injective. This result should be compared with [132, (1.7.10)], according to which $S : \left[L^{2(n-1)/n}(\partial\Omega,\sigma)\right]^M \to \left[L^{2(n-1)/(n-2)}(\partial\Omega,\sigma)\right]^M$ is an injective operator. While from Lemma 7.2.1 we know that $\left[H^{-1/2}(\partial\Omega,\sigma)\right]^M$ is a larger space, [132, (1.7.10)] only requires Ω to be a UR domain.

Proof of Theorem 7.2.4 It suffices to deal only with (7.2.29), since the more general result in (7.2.28) follows from this and abstract perturbation results (cf. [88]) on complex interpolation scales (here, [130, Theorem 7.5.2] is relevant). Thus, assume $p = q = 2$, $s = 1/2$ and, in order to stress the dependence of S on L, temporarily use the notation S_L. The first order of business is to show that the operator

$$S_L : \big[H^{-1/2}(\partial\Omega,\sigma)\big]^M \to \big[H^{1/2}(\partial\Omega,\sigma)\big]^M \text{ is Fredholm.} \tag{7.2.30}$$

With this in mind, we note that if the system L is as in the statement of the theorem, then so is its Hermitian adjoint $L^* := \overline{L}^{\top}$. Lemma 7.2.3 used for L^* in place of L then shows that the operator $S_{L^*} : \big[H^{-1/2}(\partial\Omega,\sigma)\big]^M \to \big[H^{1/2}(\partial\Omega,\sigma)\big]^M$ has closed range and a finite dimensional kernel. This, duality, and [132, (4.3.41)] then give that $S_L : \big[H^{-1/2}(\partial\Omega,\sigma)\big]^M \to \big[H^{1/2}(\partial\Omega,\sigma)\big]^M$ has closed range, of finite codimension (cf. [89, Theorem 5.13, p. 234]). In concert with (7.2.26), this establishes the claim in (7.2.30). Next, we claim that

$$\begin{gathered}\text{the operator } S_L : \big[H^{-1/2}(\partial\Omega,\sigma)\big]^M \longrightarrow \big[H^{1/2}(\partial\Omega,\sigma)\big]^M \\ \text{is Fredholm with index zero provided the system } L \text{ is self-adjoint.}\end{gathered} \tag{7.2.31}$$

Indeed, this is a consequence of (7.2.38), [132, Theorem 4.3.4, item *(2)*] (which ensures that, in our context, S_L is self-adjoint), and [89, Corollary 5.14, p. 234].

Moving on, let us pick a coefficient tensor which yields the original system L, say $A = \big(a^{\alpha\beta}_{rs}\big)_{\substack{1\le r,s\le n\\ 1\le\alpha,\beta\le M}}$ such that $L = L_A$, and define

$$\begin{gathered}A_t := \big(a^{\alpha\beta}_{rs}(t)\big)_{\substack{1\le r,s\le n\\ 1\le\alpha,\beta\le M}} \text{ where} \\ a^{\alpha\beta}_{rs}(t) := t a^{\alpha\beta}_{rs} + (1-t)\delta_{\alpha\beta}\delta_{rs} \text{ for each } t \in [0,1].\end{gathered} \tag{7.2.32}$$

Then each A_t yields the system $L_{A_t} = tL + (1-t)\Delta I_{M\times M}$ where Δ is the Laplacian in $\mathbb{R}^n$ and $I_{M\times M}$ is the $M\times M$ identity matrix. Each L_t is a homogeneous, second-order, constant complex coefficient, $M\times M$ system in $\mathbb{R}^n$, which satisfies the Legendre-Hadamard strong ellipticity condition. Indeed, for each $t \in [0,1]$, $\xi \in \mathbb{R}^n$, and $\zeta \in \mathbb{C}^M$ we may write (with $c \in (0,\infty)$ as in [131, (1.3.4) in Definition 1.3.2])

$$\begin{aligned}\mathrm{Re}\Big[a^{\alpha\beta}_{rs}(t)\xi_r\xi_s\zeta_\beta\overline{\zeta_\alpha}\Big] &= \mathrm{Re}\Big[t a^{\alpha\beta}_{rs}\xi_r\xi_s\zeta_\beta\overline{\zeta_\alpha}\Big] + (1-t)|\xi|^2|\zeta|^2 \\ &\ge ct|\xi|^2|\zeta|^2 + (1-t)|\xi|^2|\zeta|^2 \ge \min\{1,c\}|\xi|^2|\zeta|^2,\end{aligned} \tag{7.2.33}$$

which proves that [131, (1.3.4) in Definition 1.3.2] is presently satisfied by each coefficient tensor A_t. Granted this, from (7.2.30) we conclude that the operator

$$S_{L_t} : \big[H^{-1/2}(\partial\Omega,\sigma)\big]^M \to \big[H^{1/2}(\partial\Omega,\sigma)\big]^M \text{ is Fredholm, } \forall t \in [0,1]. \tag{7.2.34}$$

Next, we claim that

$$[0,1] \ni t \mapsto S_{L_t} \in \mathrm{Bd}\Big(\big[H^{-1/2}(\partial\Omega,\sigma)\big]^M \to \big[H^{1/2}(\partial\Omega,\sigma)\big]^M\Big) \quad \text{is a continuous operator-valued function.} \tag{7.2.35}$$

To see that this the case, for each $t_0, t_1 \in [0,1]$ use [130, (11.12.11) in Theorem 11.12.2], [130, (11.12.60) in Corollary 11.12.3], and the interpolation estimate from [130, Proposition 1.3.7, (1.3.64)] to write

$$\begin{aligned} &\|S_{L_{t_0}} - S_{L_{t_1}}\|_{[H^{-1/2}(\partial\Omega,\sigma)]^M \to [H^{1/2}(\partial\Omega,\sigma)]^M} \\ &\leq C\|S_{L_{t_0}} - S_{L_{t_1}}\|^{1/2}_{[L^2_{-1}(\partial\Omega,\sigma)]^M \to [L^2(\partial\Omega,\sigma)]^M} \|S_{L_{t_0}} - S_{L_{t_1}}\|^{1/2}_{[L^2(\partial\Omega,\sigma)]^M \to [L^2_1(\partial\Omega,\sigma)]^M}. \end{aligned} \tag{7.2.36}$$

From this and [132, (1.5.102), (1.5.103)] we conclude that (7.2.35) holds. Next, a combination of (7.2.34), (7.2.35), and item *(8)* in [130, Theorem 2.1.2] proves that $\mathrm{index}\Big(S_{L_t} : \big[H^{-1/2}(\partial\Omega,\sigma)\big]^M \to \big[H^{1/2}(\partial\Omega,\sigma)\big]^M\Big)$ is independent of $\in [0,1]$. In particular,

$$\begin{aligned} 0 &= \mathrm{index}\Big(S_{L_0} : \big[H^{-1/2}(\partial\Omega,\sigma)\big]^M \to \big[H^{1/2}(\partial\Omega,\sigma)\big]^M\Big) \\ &= \mathrm{index}\Big(S_{L_1} : \big[H^{-1/2}(\partial\Omega,\sigma)\big]^M \to \big[H^{1/2}(\partial\Omega,\sigma)\big]^M\Big), \end{aligned} \tag{7.2.37}$$

where the first equality comes from (7.2.31) upon realizing that $L_0 = \Delta I_{M\times M}$ is a self-adjoint system. Given that $L_1 = L$, then (7.2.37) shows that the operator

$$S : \big[H^{-1/2}(\partial\Omega,\sigma)\big]^M \to \big[H^{1/2}(\partial\Omega,\sigma)\big]^M \text{ is Fredholm with index zero.} \tag{7.2.38}$$

To finish the proof of (7.2.28), there remains to show that the operator

$$S : \big[H^{-1/2}(\partial\Omega,\sigma)\big]^M \longrightarrow \big[H^{1/2}(\partial\Omega,\sigma)\big]^M \text{ is injective if } n \geq 3. \tag{7.2.39}$$

To this end, consider $f \in \big[H^{-1/2}(\partial\Omega,\sigma)\big]^M$ such that $Sf = 0$ and introduce u as in (7.2.13). Since we are assuming that $n \geq 3$, from [132, (1.5.41), (4.2.210)] we see that for each $\gamma \in \mathbb{N}_0^n$ there exists some $C_\gamma \in (0,\infty)$ such that

$$|(\partial^\gamma u)(x)| \leq \frac{C_\gamma}{|x|^{n-2+|\gamma|}} \text{ if } |x| \text{ is large.} \tag{7.2.40}$$

Next, recall from (7.2.14) that for each $\psi \in \mathscr{C}_c^\infty(\mathbb{R}^n)$ with $\psi \equiv 1$ near $\overline{\Omega}$ we have $\psi u \in \big[W^{1,2}(\mathbb{R}^n)\big]^M$. In particular, this implies

$$u \in \big[W^{1,2}_{\mathrm{loc}}(\mathbb{R}^n)\big]^M. \tag{7.2.41}$$

For further use, let us also observe that (7.2.40) and (7.2.41) guarantee that

$$\nabla u \in \big[L^2(\mathbb{R}^n, \mathcal{L}^n)\big]^{n\cdot M}. \tag{7.2.42}$$

Also, there exists some $C = C(A) \in (0, \infty)$ with the property

$$\phi := L(\psi u^-) \in \big[\mathscr{C}^\infty_c(\Omega_-, \mathcal{L}^n)\big]^M \text{ has } |\phi| \leq C|\nabla^2\psi||u| + C|\nabla\psi||\nabla u|, \tag{7.2.43}$$

and (7.2.22) presently reduces to

$$0 = \mathrm{Re}\int_{\mathbb{R}^n} a^{\alpha\beta}_{rs}\,\partial_s(\psi u_\beta)\overline{\partial_r(\psi u_\alpha)}\,\mathrm{d}\mathcal{L}^n + \mathrm{Re}\int_{\Omega_-}\langle\phi, \overline{\psi u}\rangle\,\mathrm{d}\mathcal{L}^n. \tag{7.2.44}$$

To proceed, fix a real-valued function $\varphi \in \mathscr{C}^\infty_c(\mathbb{R}^n)$ which is identically one in a neighborhood of the origin. For each $j \in \mathbb{N}$ set $\psi_j(x) := \varphi(x/j)$ for every $x \in \mathbb{R}^n$. Note that since $|\nabla\psi_j| \leq C/j$, $|\nabla^2\psi_j| \leq C/j^2$, and $|x| \approx j$ on the support of $\nabla\psi_j$, uniformly in $j \in \mathbb{N}$, based on (7.2.40) we may estimate, as $j \to \infty$,

$$\int_{\mathbb{R}^n} |\psi_j||\nabla\psi_j||u||\nabla u|\,\mathrm{d}\mathcal{L}^n \leq Cj^{-1}\int_{x\in\mathbb{R}^n,\,|x|\approx j} |u||\nabla u|\,\mathrm{d}\mathcal{L}^n = O(j^{2-n}), \tag{7.2.45}$$

$$\int_{\mathbb{R}^n} |\nabla\psi_j|^2|u|^2\,\mathrm{d}\mathcal{L}^n \leq Cj^{-2}\int_{x\in\mathbb{R}^n,\,|x|\approx j} |u|^2\,\mathrm{d}\mathcal{L}^n = O(j^{2-n}), \tag{7.2.46}$$

$$\int_{\mathbb{R}^n} |\nabla^2\psi_j||\psi_j||u|^2\,\mathrm{d}\mathcal{L}^n \leq Cj^{-2}\int_{x\in\mathbb{R}^n,\,|x|\approx j} |u|^2\,\mathrm{d}\mathcal{L}^n = O(j^{2-n}). \tag{7.2.47}$$

For each large $j \in \mathbb{N}$ set $\phi_j := L(\psi_j u^-)$ and write (7.2.44) with ψ replaced by ψ_j:

$$0 = \mathrm{Re}\int_{\mathbb{R}^n} a^{\alpha\beta}_{rs}\,\partial_s(\psi_j u_\beta)\overline{\partial_r(\psi_j u_\alpha)}\,\mathrm{d}\mathcal{L}^n + \mathrm{Re}\int_{\Omega_-}\langle\phi_j, \overline{\psi_j u}\rangle\,\mathrm{d}\mathcal{L}^n. \tag{7.2.48}$$

Since (7.2.41) implies that for each $j \in \mathbb{N}$ we have $\psi_j u \in \big[W^{1,2}(\mathbb{R}^n)\big]^M$, in concert with the coercivity estimate (1.1.29) this gives, for some $c \in (0, \infty)$,

$$\begin{aligned}
0 &= \lim_{j\to\infty}\Big\{\mathrm{Re}\int_{\mathbb{R}^n} a^{\alpha\beta}_{rs}\,\partial_s(\psi_j u_\beta)\overline{\partial_r(\psi_j u_\alpha)}\,\mathrm{d}\mathcal{L}^n + \mathrm{Re}\int_{\Omega_-}\langle\phi_j, \overline{\psi_j u}\rangle\,\mathrm{d}\mathcal{L}^n\Big\}\\
&\geq c\lim_{j\to\infty}\int_{\mathbb{R}^n}|\nabla(\psi_j u)|^2\mathrm{d}\mathcal{L}^n + \lim_{j\to\infty}\int_{\mathbb{R}^n} O\big(|\nabla^2\psi_j||\psi_j||u|^2 + |\nabla\psi_j||\psi_j||u||\nabla u|\big)\mathrm{d}\mathcal{L}^n\\
&= \lim_{j\to\infty}\int_{\mathbb{R}^n} O\big(|\nabla\psi_j|^2|u|^2 + |\psi_j||\nabla\psi_j||u||\nabla u| + |\nabla^2\psi_j||\psi_j||u|^2\big)\,\mathrm{d}\mathcal{L}^n\\
&\quad + c\lim_{j\to\infty}\int_{\mathbb{R}^n}\psi_j^2|\nabla u|^2\,\mathrm{d}\mathcal{L}^n = c\int_{\mathbb{R}^n}|\nabla u|^2\,\mathrm{d}\mathcal{L}^n,
\end{aligned} \tag{7.2.49}$$

where the last equality makes use of (7.2.45), (7.2.46), (7.2.42), (7.2.47), the fact that $n \geq 3$, and Lebesgue's Dominated Convergence Theorem. This ultimately proves that $\nabla u = 0$ in $\mathbb{R}^n$, hence u is a constant in $\mathbb{R}^n$. In concert with (7.2.40) this finally implies (bearing in mind that $n \geq 3$) $u = 0$ in $\mathbb{R}^n$. In turn, the latter and (7.2.19) give

$$0 = \partial_\nu^A(u^+, 0) = (-\tfrac{1}{2}I + K^\#_{A^\top})f \text{ and } 0 = \partial_\nu^A(\psi u^-, \phi) = (\tfrac{1}{2}I + K^\#_{A^\top})f. \tag{7.2.50}$$

This allows us to conclude that

$$f = (\tfrac{1}{2}I + K^\#_{A^\top})f - \big(-\tfrac{1}{2}I + K^\#_{A^\top}\big)f = 0, \tag{7.2.51}$$

which finishes the proof of (7.2.39). □

As may be seen from an inspection of Remark 2, following the statement of Theorem 7.1.9, the single layer operator $S : \big[H^{-1/2}(\partial\Omega, \sigma)\big]^M \to \big[H^{1/2}(\partial\Omega, \sigma)\big]^M$ may fail to be invertible in the two-dimensional setting, even for such basic systems and domains as the Laplacian in the unit disk in the plane. As a remedy, we shall modify the single layer in the manner described below.

Theorem 7.2.5 *Let $\Omega \subseteq \mathbb{R}^2$ be a two-sided* NTA *domain with a compact Ahlfors regular boundary, and abbreviate $\sigma := \mathcal{H}^1\lfloor\partial\Omega$. Also, let L be a homogeneous, second-order, constant complex coefficient, $M \times M$ system in $\mathbb{R}^n$ satisfying the Legendre-Hadamard strong ellipticity condition (cf.* [131, Definition 1.3.2]*). Finally, recall the boundary-to-boundary single layer potential operator S associated with L and Ω, acting on Besov spaces as in* [132, Theorem 4.3.4, item *(2)*].

Then there exists a small $\varepsilon > 0$ such that the following augmented version of S:

$$\begin{aligned} &S_{\text{aug}} : \big[B^{p,q}_{s-1}(\partial\Omega, \sigma)\big]^M \oplus \mathbb{C}^M \longrightarrow \big[B^{p,q}_{s}(\partial\Omega, \sigma)\big]^M \oplus \mathbb{C}^M \\ &\quad \textit{given by } S_{\text{aug}}(f, c) := \Big(Sf + c, (\langle f, \mathbf{e}_j\rangle)_{1\leq j\leq M}\Big) \textit{ for each} \\ &\qquad f \in \big[B^{p,q}_{s-1}(\partial\Omega, \sigma)\big]^M \textit{ and } c \in \mathbb{C}^M, \end{aligned} \tag{7.2.52}$$

is an isomorphism whenever

$$|p-2| + |q-2| + \big|s - \tfrac{1}{2}\big| < \varepsilon, \tag{7.2.53}$$

As a corollary of this and (7.2.4)-(7.2.5)*, the following is an isomorphism:*

$$\begin{aligned} &S_{\text{aug}} : \big[H^{-1/2}(\partial\Omega, \sigma)\big]^M \oplus \mathbb{C}^M \longrightarrow \big[H^{1/2}(\partial\Omega, \sigma)\big]^M \oplus \mathbb{C}^M \\ &\quad \textit{given by } S_{\text{aug}}(f, c) := \Big(Sf + c, (\langle f, \mathbf{e}_j\rangle)_{1\leq j\leq M}\Big) \textit{ for each} \\ &\qquad f \in \big[H^{-1/2}(\partial\Omega, \sigma)\big]^M \textit{ and } c \in \mathbb{C}^M. \end{aligned} \tag{7.2.54}$$

Proof It suffices to show that the augmented single layer S_{aug} is an isomorphism in the context of (7.2.54), since this further implies the claim made in relation to (7.2.52) via stability results on complex interpolation scales (cf. [88]; here, [130, Theorem 7.5.2] is relevant). With this goal in mind, we first note that, thanks to (7.2.29), the same type of argument as in the proof of Theorem 7.1.14 shows that the augmented single layer (7.2.54) is a Fredholm operator with index zero. As such, there remains to show that S_{aug} is an injective operator in the context of (7.2.54). After unraveling definition, this comes down to showing that if

$$\begin{aligned} & f \in \left[H^{-1/2}(\partial\Omega, \sigma)\right]^M \text{ and } c \in \mathbb{C}^M \text{ are such that} \\ & Sf = c \text{ and } \langle f, \mathbf{e}_j \rangle = 0 \text{ for } 1 \leq j \leq M, \end{aligned} \tag{7.2.55}$$

then necessarily $f = 0$. To this end, introduce u as in (7.2.13). The additional cancellation property satisfied by f ensures that u continues to exhibit the decay as in (7.2.40). Granted this, the same argument which has produced (7.2.22) presently gives that for each cutoff function $\psi \in \mathscr{C}_c^\infty(\mathbb{R}^n)$, identically 1 near $\partial\Omega$, we have

$$-\mathrm{Re}\langle f, \overline{Sf}\rangle = \mathrm{Re}\int_{\mathbb{R}^n} a_{rs}^{\alpha\beta}\, \partial_s(\psi u_\beta)\overline{\partial_r(\psi u_\alpha)}\, d\mathcal{L}^n + \mathrm{Re}\int_{\mathbb{R}^n} \langle L(\psi u), \overline{\psi u}\rangle\, d\mathcal{L}^n.$$

In relation to this, note that $\langle f, \overline{Sf}\rangle = \langle f, \overline{c}\rangle = 0$ by (7.2.55). Consequently,

$$0 = \mathrm{Re}\int_{\mathbb{R}^n} a_{rs}^{\alpha\beta}\, \partial_s(\psi u_\beta)\overline{\partial_r(\psi u_\alpha)}\, d\mathcal{L}^n + \mathrm{Re}\int_{\mathbb{R}^n} \langle L(\psi u), \overline{\psi u}\rangle\, d\mathcal{L}^n. \tag{7.2.56}$$

With this in hand, the same argument which, starting with (7.2.44), has led to (7.2.51) then proves that $f = 0$, as wanted. □

Basic spectral theory results for the transpose double layer are given below.

Theorem 7.2.6 *Let $\Omega \subseteq \mathbb{R}^n$, where $n \in \mathbb{N}$ with $n \geq 2$, be a two-sided* NTA *domain with a compact Ahlfors regular boundary, and abbreviate $\sigma := \mathcal{H}^{n-1}\lfloor\partial\Omega$. Also, let L be a homogeneous, second-order, constant complex coefficient, $M \times M$ system in $\mathbb{R}^n$ satisfying the Legendre-Hadamard strong ellipticity condition (cf.* [131, Definition 1.3.2]*). Fix a coefficient tensor $A \in \mathfrak{A}_L$ and bring in the double layer potential operator K_A, associated with A and Ω, acting on Besov spaces as in* [132, Theorem 4.1.1]. *In addition, associate with $A^\top$ and Ω the transpose double layer potential operator $K^\#_{A^\top}$ acting on Besov spaces as in* [132, Theorem 4.1.5].

For $z \in \mathbb{C}$ and Comp *a linear, compact operator mapping $\left[H^{-1/2}(\partial\Omega, \sigma)\right]^M$ into a Banach space, consider the estimate*

$$\begin{aligned} & \left\|(zI - K^\#_{A^\top})f\right\|_{[H^{-1/2}(\partial\Omega,\sigma)]^M} + \|\mathrm{Comp}\, f\| \geq C\|f\|_{[H^{-1/2}(\partial\Omega,\sigma)]^M}, \\ & \textit{for all functionals } f \in \left[H^{-1/2}(\partial\Omega, \sigma)\right]^M. \end{aligned} \tag{7.2.57}$$

Then, in relation to this, the following claims are true:

(i) If A is Hermitian self-adjoint (i.e., $A^* = A$) then there exists Comp *as above and a finite constant $C > 0$ such that* (7.2.57) *holds for every $z \in \mathbb{C}$ with* $\mathrm{Im}\, z \neq 0$.
(ii) If the tensor coefficient A is both Hermitian self-adjoint and positive semi-definite, then there exists Comp *as above and finite constant $C > 0$ such that* (7.2.57) *holds for every complex number $z \in \mathbb{C} \setminus [-1/2, 1/2]$.*
(iii) If A is both Hermitian self-adjoint and positive definite then there exists Comp *as above and finite constant $C > 0$ such that* (7.2.57) *holds for every complex number $z \in \mathbb{C} \setminus (-1/2, 1/2)$.*

Furthermore, in each of the situations (i)-(iii) described above, there exists some small $\varepsilon > 0$ with the property that

$$\begin{aligned}&zI - K_A : \left[B^{p,q}_s(\partial\Omega,\sigma)\right]^M \to \left[B^{p,q}_s(\partial\Omega,\sigma)\right]^M \text{ and}\\&zI - K^{\#}_{A^\top} : \left[B^{p,q}_{s-1}(\partial\Omega,\sigma)\right]^M \to \left[B^{p,q}_{s-1}(\partial\Omega,\sigma)\right]^M\\&\text{are both Fredholm operators, with index zero,}\\&\text{whenever } |p-2|+|q-2|+|s-1/2|<\varepsilon.\end{aligned} \tag{7.2.58}$$

Proof Without loss of generality (cf. [129, Lemma 5.10.10]), we may assume that Ω is bounded. Consider the scenario described in *(i)*. Fix an arbitrary complex number $z \in \mathbb{C}$ with $\operatorname{Im} z \neq 0$ and pick an arbitrary $f \in \left[H^{-1/2}(\partial\Omega,\sigma)\right]^M$. Thanks to [131, (1.4.32)] and [132, (4.1.29), (4.3.41), (4.4.53)], the operators

$$S_L \circ K^{\#}_{A^\top} : \left[H^{-1/2}(\partial\Omega,\sigma)\right]^M \longrightarrow \left[H^{1/2}(\partial\Omega,\sigma)\right]^M \tag{7.2.59}$$

$$S_L : \left[H^{-1/2}(\partial\Omega,\sigma)\right]^M \longrightarrow \left[H^{1/2}(\partial\Omega,\sigma)\right]^M \tag{7.2.60}$$

are self-adjoint. In particular,

$$\left\langle K^{\#}_{A^\top} f, \overline{Sf}\right\rangle \in \mathbb{R} \text{ and } \left\langle f, \overline{Sf}\right\rangle \in \mathbb{R}, \tag{7.2.61}$$

hence, on the one hand,

$$\operatorname{Im}\left\langle (zI - K^{\#}_{A^\top})f, \overline{Sf}\right\rangle = (\operatorname{Im} z)\left\langle f, \overline{Sf}\right\rangle. \tag{7.2.62}$$

On the other hand,

$$\left|\left\langle (zI - K^{\#}_{A^\top})f, \overline{Sf}\right\rangle\right| \le \left\|(zI - K^{\#}_{A^\top})f\right\|_{[H^{-1/2}(\partial\Omega,\sigma)]^M} \|Sf\|_{[H^{1/2}(\partial\Omega,\sigma)]^M}. \tag{7.2.63}$$

Thus, for each $\varepsilon > 0$ there exists a finite constant $C_{\varepsilon,z} > 0$ such that

$$\begin{aligned}C_\varepsilon&\left\|(zI - K^{\#}_{A^\top})f\right\|^2_{[H^{-1/2}(\partial\Omega,\sigma)]^M} + \varepsilon C\|f\|^2_{[H^{-1/2}(\partial\Omega,\sigma)]^M}\\&\ge C_\varepsilon\left\|(zI - K^{\#}_{A^\top})f\right\|^2_{[H^{-1/2}(\partial\Omega,\sigma)]^M} + \varepsilon\|Sf\|^2_{[H^{1/2}(\partial\Omega,\sigma)]^M}\\&\ge \left|\operatorname{Im}\left\langle (zI - K^{\#}_{A^\top})f, \overline{Sf}\right\rangle\right| \ge |\operatorname{Im} z|\left(-\left\langle f, \overline{Sf}\right\rangle\right)\\&\ge C\|f\|^2_{[H^{-1/2}(\partial\Omega,\sigma)]^M} - \|\operatorname{Comp} f\|^2,\end{aligned} \tag{7.2.64}$$

by [132, (4.3.40)], (7.2.63), Cauchy-Schwarz with ε, (7.2.62), and (7.2.12) (also bearing in mind that $|\operatorname{Im} z| > 0$). Then (7.2.57) follows from (7.2.64) (for some $C = C(z) > 0$, finite constant depending only on z, Ω, and L) by choosing $\varepsilon > 0$ small enough. This finishes the treatment of case (i).

Consider next the scenario described in *(ii)*. Given what we have proved already, it suffices to treat the case when $z \in \mathbb{R} \setminus [-1/2, 1/2]$. In this situation, write

$$\mp\left\langle (zI - K^{\#}_{A^\top})f, \overline{Sf}\right\rangle = -\left\langle (\tfrac{1}{2}I \mp K^{\#}_{A^\top})f, \overline{Sf}\right\rangle - \left(-\tfrac{1}{2} \pm z\right)\left\langle f, \overline{Sf}\right\rangle. \tag{7.2.65}$$

We claim that in the current scenario there exists Comp mapping $\left[H^{-1/2}(\partial\Omega,\sigma)\right]^M$ linearly and compactly into some Banach space, such that for each choice of the sign,

$$\|\mathrm{Comp}\, f\|^2 - \big\langle (\tfrac{1}{2}I \pm K^{\#}_{A^\top})f, \overline{Sf}\,\big\rangle \geq 0. \tag{7.2.66}$$

To justify this, set $u := \mathscr{S}f$ in Ω_-, and pick some real-valued function $\psi \in \mathscr{C}^\infty_c(\mathbb{R}^n)$ satisfying $\psi \equiv 1$ near $\partial\Omega$. Also, suppose the coefficient tensor is explicitly written as $A = \big(a^{\alpha\beta}_{rs}\big)_{\substack{1\leq r,s\leq n\\ 1\leq\alpha,\beta\leq M}}$. Based on (7.2.61), the second jump-formula in (7.2.19), the trace formula [132, (4.3.44)] (in the context of [132, (4.3.47)]), and the definition of the conormal derivative from [130, (8.5.16)], we may then write

$$\begin{aligned}
&-\big\langle (\tfrac{1}{2}I + K^{\#}_{A^\top})f, \overline{Sf}\,\big\rangle\\
&\quad= -\mathrm{Re}\,\big\langle (\tfrac{1}{2}I + K^{\#}_{A^\top})f, \overline{Sf}\,\big\rangle = -\mathrm{Re}\Big\langle \partial^A_\nu(\psi u, L(\psi u)), \overline{\mathrm{Tr}_{\Omega_-\to\partial\Omega}(\psi u)}\Big\rangle\\
&\quad= \mathrm{Re}\int_{\Omega_-} a^{\alpha\beta}_{rs}\partial_s(\psi u_\beta)\overline{\partial_r(\psi u_\alpha)}\,\mathrm{d}\mathscr{L}^n - \int_{\Omega_-}\big\langle L(\psi u), \overline{\psi u}\big\rangle\,\mathrm{d}\mathscr{L}^n.
\end{aligned} \tag{7.2.67}$$

Since, as before, the last integral in (7.2.67) contributes terms of the form $\|\mathrm{Comp}\, f\|^2$, estimate (7.2.66) corresponding to the "top" choice of signs readily follows from the above identity bearing in mind that A is positive semi-definite. Its other version, corresponding to the "bottom" choice of signs, is proved analogously, by working in Ω_+ rather than Ω_-. Thus, from (7.2.65) and (7.2.66) we obtain

$$\mp\big\langle (zI - K^{\#}_{A^\top})f, \overline{Sf}\,\big\rangle + \|\mathrm{Comp}\, f\|^2 \geq -\big(-\tfrac{1}{2} \pm z\big)\big\langle f, \overline{Sf}\,\big\rangle. \tag{7.2.68}$$

Since the membership of z to $\mathbb{R}\setminus[-1/2, 1/2]$ is equivalent with the existence of a choice of the sign for which $\big(-\tfrac{1}{2} \pm z\big) > 0$, it follows from the Cauchy-Schwarz inequality with ε, [132, (4.3.40)], (7.2.68), and (7.2.12) that, for every $\varepsilon > 0$,

$$\begin{aligned}
C_\varepsilon\big\|(zI - K^{\#}_{A^\top})f\big\|^2_{[H^{-1/2}(\partial\Omega,\sigma)]^M} &+ \varepsilon C\|f\|^2_{[H^{-1/2}(\partial\Omega,\sigma)]^M} + \|\mathrm{Comp}\, f\|^2\\
&\geq \big\|(zI - K^{\#}_{A^\top})f\big\|_{[H^{-1/2}(\partial\Omega,\sigma)]^M}\cdot\|Sf\|_{[H^{1/2}(\partial\Omega,\sigma)]^M} + \|\mathrm{Comp}\, f\|^2\\
&\geq \Big|\big\langle (zI - K^{\#}_{A^\top})f, \overline{Sf}\,\big\rangle\Big| + \|\mathrm{Comp}\, f\|^2 \geq -\big(-\tfrac{1}{2} \pm z\big)\big\langle f, \overline{Sf}\,\big\rangle\\
&\geq c\|f\|^2_{[H^{-1/2}(\partial\Omega,\sigma)]^M} - \|\mathrm{Comp}\, f\|^2,
\end{aligned} \tag{7.2.69}$$

for some finite $C_\varepsilon > 0$, and $C, c > 0$ independent of ε. Choosing $\varepsilon > 0$ sufficiently small, we may conclude that (7.2.57) holds in the situation described in *(ii)*.

In the final part of the proof, we consider the situation described in case *(iii)* in the statement of the theorem. Granted what we have proved so far, it suffices to only treat the values $z = \pm 1/2$. In this scenario, (7.2.25) is useful. To be specific, fix $f \in \big[H^{-1/2}(\partial\Omega,\sigma)\big]^M$ and set $u := \mathscr{S}f$ in Ω_-. Then, based on (7.2.67) and the comments right after we may write

$$
\begin{aligned}
-\big\langle (\tfrac{1}{2}I + K^{\#}_{A^{\top}})f, \overline{Sf}\big\rangle &= \mathrm{Re}\int_{\Omega_-} a^{\alpha\beta}_{rs}\partial_s(\psi u_\beta)\overline{\partial_r(\psi u_\alpha)}\,d\mathcal{L}^n - \mathrm{Re}\int_{\Omega_-}\big\langle L(\psi u), \overline{\psi u}\big\rangle\,d\mathcal{L}^n \\
&\geq C\|\psi u\|^2_{[W^{1,2}(\Omega_-)]^M} - \|\mathrm{Comp}\, f\|^2 \\
&\geq C\big\|\mathrm{Tr}_{\Omega_-\to\partial\Omega}(\psi u)\big\|^2_{[H^{1/2}(\partial\Omega,\sigma)]^M} - \|\mathrm{Comp}\, f\|^2 \qquad (7.2.70)\\
&= C\|Sf\|^2_{[H^{1/2}(\partial\Omega,\sigma)]^M} - \|\mathrm{Comp}\, f\|^2 \geq C\|f\|^2_{[H^{-1/2}(\partial\Omega,\sigma)]^M} - \|\mathrm{Comp}\, f\|^2.
\end{aligned}
$$

Above, we have also used the fact that the coefficient tensor A is positive definite, the boundedness of the trace operator (cf. [130, (8.3.38)]), the trace formula [132, (4.3.44)] (in the context of [132, (4.3.47)]), as well as (7.2.25). Proceeding as before and estimating the first term in (7.2.70) from above using the Cauchy-Schwarz inequality, the using the Cauchy inequality with ε, and finally employing the boundedness of the operator S we conclude that (7.2.57) holds for $z = 1/2$. The case when $z = -1/2$ is analogous (and simpler), finishing the proof of case *(iii)*. All in all, the above reasoning shows that (7.2.57) holds in each of the cases *(i)*-*(iii)*.

As for (7.2.58), once again it suffices to consider only the case $p = q = 2$, $s = 1/2$ since, as before, the more general case discussed in (7.2.58) follows from this, abstract perturbation results (cf. [88]), and our interpolation results from [130, Theorem 7.5.2]. In this scenario, (7.2.57) yields that $zI - K^{\#}_{A^{\top}}$ is semi-Fredholm when acting from $\big[H^{-1/2}(\partial\Omega,\sigma)\big]^M$ into itself for each z belonging to an unbounded, connected subset of $\mathbb{C}$ (described concretely in each of the cases *(i)*-*(iii)*). Since, for $|z|$ large, $zI - K^{\#}_{A^{\top}}$ can be inverted using a Neumann series, it follows that the index of $zI - K^{\#}_{A^{\top}}$ is zero whenever $|z|$ is sufficiently large. Given that the index is homotopic invariant and that the operator-valued assignment $z \mapsto zI - K^{\#}_{A^{\top}}$ is continuous, we finally see that $zI - K^{\#}_{A^{\top}} : \big[H^{-1/2}(\partial\Omega,\sigma)\big]^M \to \big[H^{-1/2}(\partial\Omega,\sigma)\big]^M$ is Fredholm with index zero for each z belonging to each of the subsets of $\mathbb{C}$ described in the cases *(i)*-*(iii)*. The corresponding claim for $zI - K_A$ follows from this and duality. □

In relation to the notion of Fredholm radius from (A.0.80), we have the following:

Corollary 7.2.7 *Let $\Omega \subseteq \mathbb{R}^n$, where $n \in \mathbb{N}$ with $n \geq 2$, be a two-sided* NTA *domain with a compact Ahlfors regular boundary, and set $\sigma := \mathcal{H}^{n-1}\lfloor\partial\Omega$. Also, let L be a homogeneous, second-order, constant complex coefficient, $M \times M$ system in $\mathbb{R}^n$. In addition, assume L has a Hermitian self-adjoint, positive definite, coefficient tensor $A \in \mathfrak{A}_L$. Recall the double layer potential operator K_A, associated with A and Ω, acting on Besov spaces as in* [132, Theorem 4.1.1]. *Then there exists $\varepsilon > 0$ such that*

$$
\begin{aligned}
&\textit{the Fredholm radius of } K_A \textit{ on } \big[B^{p,q}_s(\partial\Omega,\sigma)\big]^M \textit{ is} < \tfrac{1}{2} \\
&\textit{whenever } |p-2| + |q-2| + |s-1/2| < \varepsilon.
\end{aligned}
\qquad (7.2.71)
$$

Proof This is a direct consequence of Theorem 7.2.6 (cf. (7.2.58)). □

In Theorems 7.2.9-7.2.10 we determine the kernels and ranges of the operators $\pm\frac{1}{2}I + K_A$, $\pm\frac{1}{2}I + K^{\#}_A$ acting on energy spaces. As a preliminary matter, we establish the cancellation result below.

Lemma 7.2.8 *Let $\Omega \subseteq \mathbb{R}^n$, where $n \in \mathbb{N}$ with $n \geq 2$, be a* UR *domain with a compact boundary, and abbreviate $\sigma := \mathcal{H}^{n-1} \lfloor \partial\Omega$. Also, let L be a homogeneous, second-order, constant complex coefficient, weakly elliptic $M \times M$ system in $\mathbb{R}^n$, and pick an arbitrary coefficient tensor $A \in \mathfrak{A}_L$. In this context, let $f \in \big[H^{-1/2}(\partial\Omega, \sigma)\big]^M$ be such that $(zI + K_A^{\#})f = 0$ for some $z \in \mathbb{C}$ and assume that*

$$z \neq -\tfrac{1}{2} \text{ if } \Omega \text{ is bounded, and } z \neq \tfrac{1}{2} \text{ if } \Omega \text{ is unbounded.} \tag{7.2.72}$$

Then $\langle f, c\rangle = 0$ for every $c \in \mathbb{C}^M$.

Proof Assume Ω is bounded. Then, by [132, (1.5.21), (4.1.29)], we may write

$$\langle f, c\rangle = \big\langle f, \big(\tfrac{1}{2}I + K_A\big)c\big\rangle = \big\langle \big(\tfrac{1}{2}I + K_A^{\#}\big)f, c\big\rangle = \big(\tfrac{1}{2} - z\big)\langle f, c\rangle, \tag{7.2.73}$$

for every $c \in \mathbb{C}^M$. Hence, $\big(\tfrac{1}{2} + z\big)\langle f, c\rangle = 0$ and, ultimately, $\langle f, c\rangle = 0$ if $z \neq -\tfrac{1}{2}$. The case when Ω is unbounded is very similar. □

Here is a result pertaining to the kernels of $\pm\tfrac{1}{2}I + K_A$, $\pm\tfrac{1}{2}I + K_A^{\#}$ on energy spaces.

Theorem 7.2.9 *Let $\Omega \subseteq \mathbb{R}^n$, where $n \in \mathbb{N}$ with $n \geq 2$, be a two-sided* NTA *domain with a compact Ahlfors regular boundary. Abbreviate $\sigma := \mathcal{H}^{n-1} \lfloor \partial\Omega$ and recall the spaces $\mathbb{C}_{\partial\Omega_\pm}$ from* (7.1.265). *Also, let L be a homogeneous, second-order, constant complex coefficient, $M \times M$ system in $\mathbb{R}^n$, which has a positive definite coefficient tensor $A \in \mathfrak{A}_L$. Finally, associate with A and Ω the double layer potential operators K_A and $K_A^{\#}$ acting on Besov spaces as in* [132, Theorem 4.1.1] *and* [132, Theorem 4.1.5], *respectively. Then*

$$\mathrm{Ker}\Big(\pm\tfrac{1}{2}I + K_A;\ \big[H^{1/2}(\partial\Omega, \sigma)\big]^M\Big) = \big(\mathbb{C}_{\partial\Omega_\mp}\big)^M \text{ and} \tag{7.2.74}$$

$$\dim \mathrm{Ker}\Big(\pm\tfrac{1}{2}I + K_A^{\#};\ \big[H^{-1/2}(\partial\Omega, \sigma)\big]^M\Big) = M \cdot \dim \mathbb{C}_{\partial\Omega_\mp}. \tag{7.2.75}$$

For example, in the context of Theorem 7.2.9, the operator

$$\begin{gathered}\tfrac{1}{2}I + K_A^{\#} : \big[H^{-1/2}(\partial\Omega, \sigma)\big]^M \longrightarrow \big[H^{-1/2}(\partial\Omega, \sigma)\big]^M \text{ is one-to-one}\\ \text{if } \Omega \text{ is bounded and } \mathbb{R}^n \setminus \overline{\Omega} \text{ is connected.}\end{gathered} \tag{7.2.76}$$

This follows from (7.2.74) since $\mathbb{C}_{\partial\Omega_\mp} = \{0\}$ if Ω is bounded and $\mathbb{R}^n \setminus \overline{\Omega}$ is connected.

Proof Assume the given system is expressed as $L = L_A$ for some positive definite coefficient tensor $A = \Big(a_{rs}^{\alpha\beta}\Big)_{\substack{1\leq r,s\leq n\\ 1\leq\alpha,\beta\leq M}}$. Also, define $\Omega_+ := \Omega$ and $\Omega_- := \mathbb{R}^n \setminus \overline{\Omega}$, and denote by $\mathcal{D}_A^{\pm}$ the boundary-to-domain double layer potential operator associated with the set $\Omega_\pm$ and the coefficient tensor A. In addition, we shall denote by $\mathrm{Tr}_{\Omega_\pm\to\partial\Omega}$ the boundary trace operators associated with $\Omega_\pm$. To proceed, introduce the mappings

$$\begin{aligned}\Phi_\pm :\ &\mathrm{Ker}\Big(\pm\tfrac{1}{2}I + K_A;\ \big[H^{1/2}(\partial\Omega,\sigma)\big]^M\Big) \longrightarrow \big(\mathbb{C}_{\Omega_\mp}\big)^M\\ &\text{given by } \Phi_\pm(f) := \mathcal{D}_A^\mp f \text{ in } \Omega_\mp \text{ for each function}\\ &f \in \big[H^{1/2}(\partial\Omega,\sigma)\big]^M \text{ with } \Big(\pm\tfrac{1}{2}I + K_A\Big) f = 0,\end{aligned} \tag{7.2.77}$$

and

$$\begin{aligned}\Psi_\pm :\ &\big(\mathbb{C}_{\Omega_\mp}\big)^M \longrightarrow \mathrm{Ker}\Big(\pm\tfrac{1}{2}I + K_A;\ \big[H^{1/2}(\partial\Omega,\sigma)\big]^M\Big)\\ &\text{given by } \Psi_\pm(w) := \mathrm{Tr}_{\Omega_\mp\to\partial\Omega}\, w \text{ for each } w \in \big(\mathbb{C}_{\Omega_\mp}\big)^M.\end{aligned} \tag{7.2.78}$$

To see that these mappings are indeed well defined, first let f be an arbitrary function in the domain of $\Phi_\pm$ and set

$$u^\pm := \pm\mathcal{D}_A^\pm f \in \big[W^{1,2}_{\mathrm{bdd}}(\Omega_\pm)\big]^M, \tag{7.2.79}$$

where the membership is provided by [132, Theorem 4.2.3]. Then for each given $\gamma \in \mathbb{N}_0^n$ there exists $C_\gamma \in (0,\infty)$ such that

$$\big|(\partial^\gamma u^\pm)(x)\big| \leq C_\gamma |x|^{1-n-|\gamma|} \text{ if } |x| \text{ is large,} \tag{7.2.80}$$

while the jump-formula [132, (4.3.22)] and the choice of f give

$$\mathrm{Tr}_{\Omega_\mp\to\partial\Omega}\, u^\pm = \big(\pm\tfrac{1}{2}I + K_A\big) f = 0. \tag{7.2.81}$$

Furthermore, (7.2.80) and (7.2.79) ensure that

$$\nabla u^\pm \in \big[L^2(\Omega_\pm, \mathcal{L}^n)\big]^{n\cdot M}. \tag{7.2.82}$$

Pick an arbitrary cutoff function $\psi \in \mathscr{C}_c^\infty(\mathbb{R}^n)$ which is identically 1 near $\partial\Omega$ and observe that $L(\psi u^\pm) \in \big[\mathscr{C}_c^\infty(\Omega_\pm)\big]^M$. Bearing in mind that the outward unit normal for $\Omega_\pm$ is $\pm\nu$ we may use (7.2.81) together with [130, Proposition 8.5.3] to write

$$\begin{aligned}0 &= \Big\langle \partial_\nu^A(\psi u^\pm, L(\psi u^\pm)), \overline{\big(\pm\tfrac{1}{2}I + K_A\big) f}\Big\rangle = \Big\langle \partial_\nu^A(\psi u^\pm, L(\psi u^\pm)), \overline{\mathrm{Tr}_{\Omega_\pm\to\partial\Omega}(u^\pm)}\Big\rangle\\ &= \int_{\Omega_\pm} a_{rs}^{\alpha\beta}\,\partial_s(\psi u_\beta^\pm)\overline{\partial_r(\psi u_\alpha^\pm)}\,\mathrm{d}\mathcal{L}^n + \int_{\Omega_\pm} \big\langle L(\psi u^\pm), \overline{\psi u^\pm}\big\rangle\,\mathrm{d}\mathcal{L}^n.\end{aligned} \tag{7.2.83}$$

Let us now fix a real-valued function $\varphi \in \mathscr{C}_c^\infty(\mathbb{R}^n)$ which is identically one in a neighborhood of the origin. For each $j \in \mathbb{N}$ set $\psi_j(x) := \varphi(x/j)$ for every $x \in \mathbb{R}^n$. Since $|\nabla\psi_j| \leq C/j$, $|\nabla^2\psi_j| \leq C/j^2$, and $|x| \approx j$ on the support of $\nabla\psi_j$, uniformly in $j \in \mathbb{N}$, based on (7.2.40) we may estimate, as $j \to \infty$,

$$\int_{\Omega_\pm} |\psi_j||\nabla\psi_j||u^\pm||\nabla u^\pm|\,\mathrm{d}\mathcal{L}^n \le Cj^{-1}\int_{x\in\Omega_\pm,\,|x|\approx j} |u^\pm||\nabla u^\pm|\,\mathrm{d}\mathcal{L}^n = O(j^{-n}), \tag{7.2.84}$$

$$\int_{\Omega_\pm} |\nabla\psi_j|^2|u^\pm|^2\,\mathrm{d}\mathcal{L}^n \le Cj^{-2}\int_{x\in\Omega_\pm,\,|x|\approx j} |u^\pm|^2\,\mathrm{d}\mathcal{L}^n = O(j^{-n}), \tag{7.2.85}$$

$$\int_{\Omega_\pm} |\nabla^2\psi_j||\psi_j||u^\pm|^2\,\mathrm{d}\mathcal{L}^n \le Cj^{-2}\int_{x\in\Omega_\pm,\,|x|\approx j} |u^\pm|^2\,\mathrm{d}\mathcal{L}^n = O(j^{-n}). \tag{7.2.86}$$

For each large $j \in \mathbb{N}$ write (7.2.83) with ψ replaced by ψ_j, and obtain

$$0 = \int_{\Omega_\pm} a_{rs}^{\alpha\beta}\,\partial_s(\psi_j u_\beta^\pm)\overline{\partial_r(\psi_j u_\alpha^\pm)}\,\mathrm{d}\mathcal{L}^n + \int_{\Omega_\pm} \big\langle L(\psi_j u^\pm), \overline{\psi_j u^\pm}\big\rangle\,\mathrm{d}\mathcal{L}^n. \tag{7.2.87}$$

Based on this, (7.2.82), Lebesgue's Dominated Convergence Theorem, and (7.2.84)-(7.2.86) we may then compute

$$\begin{aligned}
0 &= \lim_{j\to\infty}\Big\{\int_{\Omega_\pm} a_{rs}^{\alpha\beta}\,\partial_s(\psi_j u_\beta^\pm)\overline{\partial_r(\psi_j u_\alpha^\pm)}\,\mathrm{d}\mathcal{L}^n + \int_{\Omega_-} \big\langle L(\psi_j u^\pm), \overline{\psi_j u^\pm}\big\rangle\,\mathrm{d}\mathcal{L}^n\Big\}\\
&= \lim_{j\to\infty}\int_{\Omega_\pm} O\big(|\nabla\psi_j|^2|u^\pm|^2 + |\psi_j||\nabla\psi_j||u^\pm||\nabla u^-| + |\nabla^2\psi_j||\psi_j||u^\pm|^2\big)\,\mathrm{d}\mathcal{L}^n\\
&\quad + \lim_{j\to\infty}\int_{\Omega_\pm} \psi_j^2 a_{rs}^{\alpha\beta}\,(\partial_s u_\beta^\pm)\overline{\partial_r u_\alpha^\pm}\,\mathrm{d}\mathcal{L}^n = \int_{\Omega_\pm} a_{rs}^{\alpha\beta}\,(\partial_s u_\beta^\pm)\overline{\partial_r u_\alpha^\pm}\,\mathrm{d}\mathcal{L}^n.
\end{aligned} \tag{7.2.88}$$

Since the fact that A is positive definite implies the pointwise inequality

$$\mathrm{Re}\Big(a_{rs}^{\alpha\beta}(\partial_s u_\beta^\pm)\overline{\partial_r u_\alpha^\pm}\Big) \ge c|\nabla u^\pm|^2 \ \text{ in } \Omega_\pm \tag{7.2.89}$$

for some $c \in (0,\infty)$, we deduce from (7.2.88) that

$$\nabla u^\pm \equiv 0 \ \text{ in } \Omega_\pm. \tag{7.2.90}$$

In concert with the compatibility result proved in [132, (4.3.72)] and the lack of jump of the conormal derivative of the double layer across the boundary (cf. [132, (1.5.29)]), this further implies

$$0 = \partial_\nu^A\big(\psi\mathcal{D}_A^\pm f, L(\psi\mathcal{D}_A^\pm f)\big) = \partial_\nu^A\mathcal{D}_A^\pm f = \partial_\nu^A\mathcal{D}_A^\mp f = \mp\partial_\nu^A\big(\psi u^\mp, L(\psi u^\mp)\big). \tag{7.2.91}$$

Since $\partial_\nu^A\big(\psi u^\mp, L(\psi u^\mp)\big) = 0$, much as in the third equality in (7.2.83) we have

$$\begin{aligned}
0 &= \Big\langle \partial_\nu^A\big(\psi u^\mp, L(\psi u^\mp)\big), \overline{\mathrm{Tr}_{\Omega_\mp\to\partial\Omega}(u^\mp)}\Big\rangle\\
&= \int_{\Omega_\mp} a_{rs}^{\alpha\beta}\,\partial_s(\psi u_\beta^\mp)\overline{\partial_r(\psi u_\alpha^\mp)}\,\mathrm{d}\mathcal{L}^n + \int_{\Omega_\mp} \big\langle L(\psi u^\mp), \overline{\psi u^\mp}\big\rangle\,\mathrm{d}\mathcal{L}^n.
\end{aligned} \tag{7.2.92}$$

With this in hand, we may reason as in (7.2.84)-(7.2.89) and conclude (again, based on the fact that A is positive definite), that $\nabla u^{\mp} \equiv 0$ in $\Omega_{\mp}$. Thus, ultimately, $\mathcal{D}_A^{\mp} f \in (\mathbb{C}_{\Omega_{\mp}})^M$ (since $\mathcal{D}_A^{\mp} f$ vanishes at infinity if $\Omega_{\mp}$ is an exterior domain), which goes to show that the mappings $\Phi_{\pm}$ are well defined.

Concerning the mappings $\Psi_{\pm}$, for each function $w \in (\mathbb{C}_{\Omega_{\mp}})^M$ (viewed as zero on the unbounded connected component of $\Omega_{\mp}$ if this is an exterior domain) the integral representation formula [132, (4.4.2)] presently reduces to

$$w = \mathcal{D}_A^{\mp}\big(\mathrm{Tr}_{\Omega_{\mp}\to\partial\Omega}\, w\big) \text{ in } \Omega_{\mp} \tag{7.2.93}$$

given that $\partial_\nu^A(w, 0) = 0$ on $\partial\Omega$. Taking boundary traces in (7.2.93) and invoking the jump-formula [132, (4.3.22)] then yields $\big(\pm\frac{1}{2}I + K_A\big)\big(\mathrm{Tr}_{\Omega_{\mp}\to\partial\Omega}\, w\big) = 0$ which ultimately proves that the mappings $\Psi_{\pm}$ are also well defined.

Let us now show that $\Phi_{\pm}$ and $\Psi_{\pm}$ are inverse to each other. In one direction, formula (7.2.93) may be re-interpreted as $\Phi_{\pm}\big(\Psi_{\pm}(w)\big) = w$ for each $w \in (\mathbb{C}_{\Omega_{\mp}})^M$. In the opposite direction, for each $f \in \big[H^{1/2}(\partial\Omega, \sigma)\big]^M$ with $\big(\pm\frac{1}{2}I + K_A\big)f = 0$,

$$\Psi_{\pm}\big(\Phi_{\pm}(f)\big) = \mathrm{Tr}_{\Omega_{\mp}\to\partial\Omega}\,(\mathcal{D}_A^{\mp} f) = \big(\tfrac{1}{2}I \mp K_A\big)f = f, \tag{7.2.94}$$

establishing that $\Phi_{\pm}$ and $\Psi_{\pm}$ are inverses to one other. In particular, $\Phi_{\pm}$ and $\Psi_{\pm}$ are bijections. Since we also have $\big\{\mathrm{Tr}_{\Omega_{\mp}\to\partial\Omega}\, w : w \in (\mathbb{C}_{\Omega_{\mp}})^M\big\} = (\mathbb{C}_{\partial\Omega_{\mp}})^M$, we may now finish the proof of (7.2.74) by writing

$$\begin{aligned} &\mathrm{Ker}\Big(\pm\tfrac{1}{2}I + K_A;\, \big[H^{1/2}(\partial\Omega,\sigma)\big]^M\Big) \\ &\quad = \Psi_{\pm}\Big(\Phi_{\pm}\Big(\mathrm{Ker}\Big(\pm\tfrac{1}{2}I + K_A;\, \big[H^{1/2}(\partial\Omega,\sigma)\big]^M\Big)\Big)\Big) = \Psi_{\pm}\Big((\mathbb{C}_{\Omega_{\mp}})^M\Big) = (\mathbb{C}_{\partial\Omega_{\mp}})^M. \end{aligned} \tag{7.2.95}$$

Next, from [132, (4.4.53)] we know that $S \circ K^{\#}_{A^{\top}} = K_A \circ S$ on $\big[H^{-1/2}(\partial\Omega,\sigma)\big]^M$. Granted this, reasoning as in the proof of (7.1.621) (now relying on the invertibility result from (7.2.29)) we see that

$$S : \mathrm{Ker}\Big(\pm\tfrac{1}{2}I + K^{\#}_{A^{\top}};\, \big[H^{-1/2}(\partial\Omega,\sigma)\big]^M\Big) \to \mathrm{Ker}\Big(\pm\tfrac{1}{2}I + K_A;\, \big[H^{1/2}(\partial\Omega,\sigma)\big]^M\Big)$$

is an isomorphism if $n \geq 3$. Hence, if $n \geq 3$,

$$\dim\mathrm{Ker}\Big(\pm\tfrac{1}{2}I + K^{\#}_{A^{\top}};\, \big[H^{-1/2}(\partial\Omega,\sigma)\big]^M\Big) = \dim\mathrm{Ker}\Big(\pm\tfrac{1}{2}I + K_A;\, \big[H^{1/2}(\partial\Omega,\sigma)\big]^M\Big).$$

In fact, the above formula is also true when $n = 2$, as may be seen by reasoning as in the proof of (7.1.631), now relying on the invertibility result for the augmented single layer in the context of (7.2.54). Having proved above formula in all dimensions $n \geq 2$, we conclude from this and (7.2.74) that

$$\dim\mathrm{Ker}\Big(\pm\tfrac{1}{2}I + K^{\#}_{A^{\top}};\, \big[H^{-1/2}(\partial\Omega,\sigma)\big]^M\Big) = M \cdot \dim\mathbb{C}_{\partial\Omega_{\mp}}. \tag{7.2.96}$$

Writing (7.2.96) with A replaced by $A^\top$ (which has the same properties as A), we finally arrive at (7.2.74). □

We next look at the images of $\pm\frac{1}{2}I + K_A$, $\pm\frac{1}{2}I + K_A^\#$ on energy spaces.

Theorem 7.2.10 *Let $\Omega \subseteq \mathbb{R}^n$, where $n \in \mathbb{N}$ with $n \geq 2$, be a two-sided* NTA *domain with a compact Ahlfors regular boundary. Set $\sigma := \mathcal{H}^{n-1}\lfloor\partial\Omega$. Recall the spaces $\mathbb{C}_{\partial\Omega_\pm}$ from* (7.1.265) *and the convention made in* (7.1.266)*. Next, let L be a homogeneous, second-order, constant complex coefficient, $M \times M$ system in $\mathbb{R}^n$, which has a positive definite, Hermitian self-adjoint, coefficient tensor $A \in \mathfrak{A}_L$. Finally, associate with A and Ω the double layers K_A and $K_A^\#$ acting on Besov spaces as in* [132, Theorem 4.1.1] *and* [132, Theorem 4.1.5], *respectively. Then*

$$\mathrm{Im}\Big(\pm\tfrac{1}{2}I + K_A^\#;\, \big[H^{-1/2}(\partial\Omega,\sigma)\big]^M\Big) = \big[H^{-1/2}(\partial\Omega,\sigma) \cap \big(\mathbb{C}_{\partial\Omega_\mp}\big)^\perp\big]^M, \tag{7.2.97}$$

$$\dim \mathrm{CoKer}\Big(\pm\tfrac{1}{2}I + K_A;\, \big[H^{1/2}(\partial\Omega,\sigma)\big]^M\Big) = M \cdot \dim \mathbb{C}_{\partial\Omega_\mp}. \tag{7.2.98}$$

Proof To get started, observe that for each $f \in \big[H^{-1/2}(\partial\Omega,\sigma)\big]^M$ and each function $g \in \big[\mathbb{C}_{\partial\Omega_\mp}\big]^M$ we have $g \in \big[H^{1/2}(\partial\Omega,\sigma)\big]^M$ and

$$\big\langle (\mp\tfrac{1}{2}I + K_A^\#)f, g\big\rangle = \big\langle f, (\mp\tfrac{1}{2}I + K_A)g\big\rangle = 0, \tag{7.2.99}$$

by (7.2.74) and (7.1.265). In concert with [132, Theorem 4.1.5], this implies

$$\mathrm{Im}\Big(\pm\tfrac{1}{2}I + K_A^\#;\, \big[H^{-1/2}(\partial\Omega,\sigma)\big]^M\Big) \subseteq \big[H^{-1/2}(\partial\Omega,\sigma) \cap \big(\mathbb{C}_{\partial\Omega_\mp}\big)^\perp\big]^M. \tag{7.2.100}$$

Next, write

$$\begin{aligned} M\cdot\dim \mathbb{C}_{\partial\Omega_\mp} &= \dim \mathrm{Ker}\Big(\pm\tfrac{1}{2}I + K_A^\#;\, \big[H^{-1/2}(\partial\Omega,\sigma)\big]^M\Big) \\ &= \dim\left(\frac{\big[H^{-1/2}(\partial\Omega,\sigma)\big]^M}{\mathrm{Im}\big(\pm\frac{1}{2}I+K_A^\#;[H^{-1/2}(\partial\Omega,\sigma)]^M\big)}\right) \geq \dim\left(\frac{\big[H^{-1/2}(\partial\Omega,\sigma)\big]^M}{\big[H^{-1/2}(\partial\Omega,\sigma)\cap\big(\mathbb{C}_{\partial\Omega_\mp}\big)^\perp\big]^M}\right) \\ &= M\cdot\dim\mathbb{C}_{\partial\Omega_\mp}. \end{aligned} \tag{7.2.101}$$

The first equality above is (7.2.75), the second equality is implied by (7.2.58), the subsequent inequality follows from (7.2.100), while the last equality is rudimentary functional analysis. In turn, since $\dim \mathbb{C}_{\partial\Omega_\mp} < \infty$, from (7.2.101) we conclude

$$\dim\left(\frac{\big[H^{-1/2}(\partial\Omega,\sigma)\big]^M}{\mathrm{Im}\big(\pm\frac{1}{2}I+K_A^\#;[H^{-1/2}(\partial\Omega,\sigma)]^M\big)}\right) = \dim\left(\frac{\big[H^{-1/2}(\partial\Omega,\sigma)\big]^M}{\big[H^{-1/2}(\partial\Omega,\sigma)\cap\big(\mathbb{C}_{\partial\Omega_\mp}\big)^\perp\big]^M}\right) < \infty. \tag{7.2.102}$$

Having established this, [130, Lemma 1.1.1] applies and proves(7.2.97).

Finally, thanks to (7.2.58) and (7.2.74), we may write

$$
\begin{aligned}
&\dim \operatorname{CoKer}\Big(\pm \tfrac{1}{2}I + K_A; \big[H^{1/2}(\partial\Omega,\sigma)\big]^M\Big) \qquad (7.2.103)\\
&\quad = \dim \operatorname{Ker}\Big(\pm \tfrac{1}{2}I + K_A; \big[H^{1/2}(\partial\Omega,\sigma)\big]^M\Big) = \dim\big(\mathbb{C}_{\partial\Omega_\mp}\big)^M = M \cdot \dim \mathbb{C}_{\partial\Omega_\mp},
\end{aligned}
$$

proving (7.2.98). □

We next discuss invertibility results for $\pm\frac{1}{2}I + K_A$, $\pm\frac{1}{2}I + K_A^{\#}$ on energy spaces.

Theorem 7.2.11 *Let $\Omega \subseteq \mathbb{R}^n$, where $n \in \mathbb{N}$ with $n \geq 2$, be a two-sided* NTA *domain with a compact Ahlfors regular boundary, and abbreviate $\sigma := \mathcal{H}^{n-1}\lfloor\partial\Omega$. Recall the spaces $\mathbb{C}_{\partial\Omega_\pm}$, $\mathbb{C}_{\partial\Omega}$ from* (7.1.265), *and recall the conventions made in* (7.1.266)-(7.1.267). *Next, let L be a homogeneous, second-order, constant complex coefficient, $M\times M$ system in $\mathbb{R}^n$, which has a positive definite, Hermitian self-adjoint, coefficient tensor $A \in \mathfrak{A}_L$. Finally, associate with A and Ω the double layer potential operators K_A and $K_A^{\#}$ acting on Besov spaces as in* [132, Theorem 4.1.1] *and* [132, Theorem 4.1.5], *respectively. Then the following operators are isomorphisms:*

$$\pm\tfrac{1}{2}I + K_A^{\#} : \big[H^{-1/2}(\partial\Omega,\sigma) \cap (\mathbb{C}_{\partial\Omega_\mp})^{\perp}\big]^M \xrightarrow{\sim} \big[H^{-1/2}(\partial\Omega,\sigma) \cap (\mathbb{C}_{\partial\Omega_\mp})^{\perp}\big]^M, \tag{7.2.104}$$

$$\pm\tfrac{1}{2}I + K_A : \frac{\big[H^{1/2}(\partial\Omega,\sigma)\big]^M}{\big(\mathbb{C}_{\partial\Omega_\mp}\big)^M} \xrightarrow{\sim} \frac{\big[H^{1/2}(\partial\Omega,\sigma)\big]^M}{\big(\mathbb{C}_{\partial\Omega_\mp}\big)^M}, \tag{7.2.105}$$

$$\pm\tfrac{1}{2}I + K_A : \frac{\big[H^{1/2}(\partial\Omega,\sigma)\big]^M}{\big(\mathbb{C}_{\partial\Omega}\big)^M} \xrightarrow{\sim} \frac{\big[H^{1/2}(\partial\Omega,\sigma)\big]^M}{\big(\mathbb{C}_{\partial\Omega}\big)^M}, \tag{7.2.106}$$

$$\pm\tfrac{1}{2}I + K_A^{\#} : \big[H^{-1/2}(\partial\Omega,\sigma) \cap (\mathbb{C}_{\partial\Omega})^{\perp}\big]^M \xrightarrow{\sim} \big[H^{-1/2}(\partial\Omega,\sigma) \cap (\mathbb{C}_{\partial\Omega})^{\perp}\big]^M, \tag{7.2.107}$$

all operators acting on quotient spaces in a natural manner (cf. (7.1.294)-(7.1.295)).

Proof Let us show that the operators in (7.2.104) are isomorphisms. First, the operators in question are well defined, linear, and bounded, thanks to [132, Theorem 4.1.5] and (7.2.97). To proceed, consider the following commutative diagram:

$$
\begin{array}{ccc}
\big[H^{-1/2}(\partial\Omega,\sigma)\big]^M & \xrightarrow{\pm\frac{1}{2}I+K_A^{\#}} & \big[H^{-1/2}(\partial\Omega,\sigma)\big]^M \\
\iota_\pm \Big\uparrow & & \iota_\pm \Big\uparrow \\
\big[H^{-1/2}(\partial\Omega,\sigma) \cap (\mathbb{C}_{\partial\Omega_\mp})^{\perp}\big]^M & \xrightarrow{\pm\frac{1}{2}I+K_A^{\#}} & \big[H^{-1/2}(\partial\Omega,\sigma) \cap (\mathbb{C}_{\partial\Omega_\mp})^{\perp}\big]^M
\end{array}
$$

where $\iota_\pm$ are the canonical inclusions. In particular, $\iota_\pm$ are Fredholm operators. Since, by (7.2.58) and assumptions, $\pm\frac{1}{2}I + K_A^{\#}$ are Fredholm with index zero on $\big[H^{-1/2}(\partial\Omega,\sigma)\big]^M$, from [130, Theorem 2.1.2] and the commutativity of the above diagram we see that $\pm\frac{1}{2}I + K_A^{\#}$ are Fredholm operators with index zero when considered acting from the space $\big[H^{-1/2}(\partial\Omega,\sigma) \cap (\mathbb{C}_{\partial\Omega_\mp})^{\perp}\big]^M$ into itself.

As such, matters have been reduced to proving injectivity for the operators in question. With this goal in mind, let $f \in \big[H^{-1/2}(\partial\Omega,\sigma) \cap (\mathbb{C}_{\partial\Omega_{\mp}})^{\perp}\big]^M$ be such that $\big(\pm\frac{1}{2}I + K_A^{\#}\big)f = 0$. The goal is to show that $f = 0$. We first observe that

$$\langle f, c\rangle = 0 \text{ for each } c \in \mathbb{C}^M. \tag{7.2.108}$$

Indeed, if $\Omega_{\pm}$ is unbounded, then $\Omega_{\mp}$ is bounded, in which case $\mathbf{1}_{\partial\Omega} \in \mathbb{C}_{\partial\Omega_{\mp}}$ (see (7.1.273)). As such, the orthogonality condition satisfied by f directly gives (7.2.108). If $\Omega_{\pm}$ is bounded, then (7.2.108) follows from Lemma 7.2.8.

Going further, assume that the given system is expressed as $L = L_A$ for some Hermitian self-adjoint, positive definite, coefficient tensor $A = \Big(a_{rs}^{\alpha\beta}\Big)_{\substack{1\leq r,s\leq n\\ 1\leq\alpha,\beta\leq M}}$. As in the past, define $\Omega_+ := \Omega$ and $\Omega_- := \mathbb{R}^n \setminus \overline{\Omega}$, and denote by $\mathrm{Tr}_{\Omega_{\pm}\to\partial\Omega}$ the boundary trace operators associated with $\Omega_{\pm}$. Next, consider the functions

$$u^{\pm} := \big(\mathscr{S}_{A^{\top}} f\big)\Big|_{\Omega_{\pm}} \in \big[\mathscr{C}^{\infty}(\Omega_{\pm})\big]^M, \tag{7.2.109}$$

with the membership provided by [132, (4.2.207)]. Also, [132, (4.2.209)] gives that $L^{\top}u^{\pm} = 0$ in $\Omega_{\pm}$. In light of (7.2.108), for each $\alpha \in \mathbb{N}_0^n$ we have

$$(\partial^{\alpha}u^{\pm})(x) = O\big(|x|^{1-n-|\alpha|}\big) \text{ as } |x| \to \infty, \text{ if } \Omega_{\pm} \text{ is unbounded.} \tag{7.2.110}$$

Also, as seen from [132, (4.2.218)], we have $u^{\pm} \in \big[W^{1,2}_{\mathrm{bdd}}(\Omega_{\pm})\big]^M$. Together with (7.2.110) this also proves that

$$\nabla u^{\pm} \in \big[L^2(\Omega_{\pm}, \mathcal{L}^n)\big]^{n\cdot M}. \tag{7.2.111}$$

Pick an arbitrary function $\psi \in \mathscr{C}_c^{\infty}(\mathbb{R}^n)$ which is identically 1 near $\partial\Omega$ and observe that $L^{\top}(\psi u^{\pm}) \in \big[\mathscr{C}_c^{\infty}(\Omega_{\pm})\big]^M$. Since the outward unit normal for $\Omega_{\pm}$ is $\pm\nu$, we may use the assumption that $\big(\pm\frac{1}{2}I + K_A^{\#}\big)f = 0$ together with [130, Proposition 8.5.3] (applied with $A^{\top}$ in place of A) to write

$$\begin{aligned} 0 &= \Big\langle \big(\pm\tfrac{1}{2}I + K_A^{\#}\big)f, \overline{\mathrm{Tr}_{\Omega_{\mp}\to\partial\Omega}(u^{\mp})}\Big\rangle = \Big\langle \partial_{\nu}^{A^{\top}}(\psi u^{\mp}, L^{\top}(\psi u^{\mp})), \overline{\mathrm{Tr}_{\Omega_{\mp}\to\partial\Omega}(u^{\mp})}\Big\rangle \\ &= \int_{\Omega_{\mp}} a_{sr}^{\beta\alpha}\,\partial_s(\psi u_{\beta}^{\mp})\overline{\partial_r(\psi u_{\alpha}^{\mp})}\,\mathrm{d}\mathcal{L}^n + \int_{\Omega_{\mp}} \big\langle L^{\top}(\psi u^{\mp}), \overline{\psi u^{\mp}}\big\rangle\,\mathrm{d}\mathcal{L}^n. \end{aligned} \tag{7.2.112}$$

With this in hand, the same reasoning which, starting with (7.2.83), has produced (7.2.90), currently gives $\nabla u^{\mp} \equiv 0$ in $\Omega_{\mp}$ (note that $A^{\top}$ is positive definite given that A is). Since $u^{\pm}$ also decays at infinity if $\Omega_{\pm}$ is unbounded (as seen from (7.2.110) with $|\alpha| = 0$), from this we ultimately conclude that $u^{\mp} \in (\mathbb{C}_{\Omega_{\mp}})^M$. The latter implies $\mathrm{Tr}_{\Omega_{\pm}\to\partial\Omega}(u^{\pm}) = \mathrm{Tr}_{\Omega_{\mp}\to\partial\Omega}(u^{\mp}) \in (\mathbb{C}_{\partial\Omega_{\mp}})^M$, in view of (7.2.109), [132, (4.3.44)], and (7.1.265). By once again relying on [130, Proposition 8.5.3] (once more applied with $A^{\top}$ in place of A), we may then compute

$$\int_{\Omega_\pm} d_{sr}^{\beta\alpha}\partial_s(\psi u_\beta^\pm)\overline{\partial_r(\psi u_\alpha^\pm)}\,\mathrm{d}\mathcal{L}^n + \int_{\Omega_\pm}\langle L^\top(\psi u^\pm),\overline{\psi u^\pm}\rangle\,\mathrm{d}\mathcal{L}^n$$
$$= \Big\langle \partial_\nu^{A^\top}(\psi u^\pm, L^\top(\psi u^\pm)), \overline{\mathrm{Tr}_{\Omega_\pm\to\partial\Omega}(u^\pm)}\Big\rangle \tag{7.2.113}$$
$$= \Big\langle \big(\mp\tfrac{1}{2}I + K_A^\#\big)f, \overline{\mathrm{Tr}_{\Omega_\pm\to\partial\Omega}(u^\pm)}\Big\rangle = \mp\Big\langle f, \overline{\mathrm{Tr}_{\Omega_\pm\to\partial\Omega}(u^\pm)}\Big\rangle = 0,$$

where the last equality follows from the fact that the components of f are orthogonal to $(\mathbb{C}_{\partial\Omega_\mp})^\perp$ and that $\overline{\mathrm{Tr}_{\Omega_\pm\to\partial\Omega}(u^\pm)}$ belongs to the space $(\mathbb{C}_{\partial\Omega_\mp})^M$. Since $A^\top$ is positive definite, we may once again run the same type of argument as in (7.2.83)-(7.2.90) and conclude that $\nabla u^\pm = 0$ in $\Omega_\pm$. Based on this and the trace formulas for the normal derivative of the single layer (cf. [132, (4.3.83)]) we therefore obtain

$$\pm f = \big(\pm\tfrac{1}{2}I + K_A^\#\big)f - \big(\mp\tfrac{1}{2}I + K_A^\#\big)f$$
$$= \partial_\nu^{A^\top}(\psi u^\mp, L^\top(\psi u^\pm)) - \partial_\nu^{A^\top}(\psi u^\pm, L^\top(\psi u^\pm)). \tag{7.2.114}$$

We claim that the above (two) conormal derivatives vanish as functionals in $\big[H^{-1/2}(\partial\Omega,\sigma)\big]^M$. For example, take the case of $\partial_\nu^{A^\top}(\psi u^\pm, L^\top(\psi u^\pm))$. According to [130, (9.5.8)], its action on an arbitrary test function $\varphi \in \big[\mathrm{Lip}\,(\partial\Omega)\big]^M$ is

$${}_{[H^{-1/2}(\partial\Omega,\sigma)]^M}\Big\langle \partial_\nu^{A^\top}(\psi u^\pm, L^\top(\psi u^\pm)), \varphi\Big\rangle_{[H^{1/2}(\partial\Omega,\sigma)]^M}$$
$$= \big\langle d_{rs}^{\beta\alpha}\partial_s u_\beta, \partial_r\Phi_\alpha\big\rangle + \big\langle L^\top(\psi u^\pm), \Phi\big\rangle \tag{7.2.115}$$

for any $\Phi = (\Phi_\alpha)_\alpha \in \big[\mathrm{Lip}_c(\mathbb{R}^n)\big]^M$ with $\Phi\big|_{\partial\Omega} = \varphi$. Since matters may be arranged so that the support of Φ is contained in the set where $\psi \equiv 1$, using the fact that $\nabla u^\pm = 0$ in $\Omega_\pm$ it follows that both pairings above vanish, thanks to support considerations. Hence, $\partial_\nu^{A^\top}(\psi u^\pm, L^\top(\psi u^\pm)) = 0$ as a (vector-valued) distribution on $\partial\Omega$. Similarly, $\partial_\nu^{A^\top}(\psi u^\mp, L^\top(\psi u^\mp)) = 0$ as a (vector-valued) distribution on $\partial\Omega$. Based on this and (7.2.114) we finally conclude that $f = 0$. This proves that $\pm\frac{1}{2}I + K_A^\#$ is indeed injective on $\big[H^{-1/2}(\partial\Omega,\sigma) \cap (\mathbb{C}_{\partial\Omega_\mp})^\perp\big]^M$, hence ultimately an isomorphism on this space. This takes care of (7.2.104).

Going further, (7.2.105) is obtained from (7.2.104) and the abstract result from [130, (2.1.59) in Lemma 2.1.7] used with

$$X := \big[H^{1/2}(\partial\Omega,\sigma)\big]^M, \quad V := (\mathbb{C}_{\partial\Omega_\mp})^M, \quad T := \pm\tfrac{1}{2}I + K_A. \tag{7.2.116}$$

In this vein we wish to remark that the invariance of V under T follows from (7.2.74), while the identification of T^* with $\pm\frac{1}{2}I + K_A^\#$ follows from [132, (4.1.29)].

Next, consider the operators in (7.2.106). They are well defined since from (7.2.74) and (7.1.272) we have

$$\big(\pm\tfrac{1}{2}I + K_A\big)(\mathbb{C}_{\partial\Omega})^M = \big(\pm\tfrac{1}{2}I + K_A\big)(\mathbb{C}_{\partial\Omega_\pm})^M = \Big(\pm I + \big(\mp\tfrac{1}{2}I + K_A\big)\Big)(\mathbb{C}_{\partial\Omega_\pm})^M$$
$$= (\mathbb{C}_{\partial\Omega_\pm})^M \subseteq (\mathbb{C}_{\partial\Omega})^M. \tag{7.2.117}$$

The fact that the following diagram

$$\begin{array}{ccc} \big[H^{1/2}(\partial\Omega,\sigma)\big]^M & \xrightarrow{\pm\frac{1}{2}I+K_A} & \big[H^{1/2}(\partial\Omega,\sigma)\big]^M \\ \pi\big\downarrow & & \big\downarrow\pi \\ \big[H^{1/2}(\partial\Omega,\sigma)\big]^M/\big(\mathbb{C}_{\partial\Omega}\big)^M & \xrightarrow{\pm\frac{1}{2}I+K_A} & \big[H^{1/2}(\partial\Omega,\sigma)\big]^M/\big(\mathbb{C}_{\partial\Omega}\big)^M \end{array} \tag{7.2.118}$$

in which the vertical arrows are canonical projections onto the quotient spaces, is commutative implies, much as before, that the operators in (7.2.106) are Fredholm with index zero. Using the isomorphisms in (7.2.105) and (7.1.272) it is immediate that the operators in (7.2.106) are surjective, hence, isomorphisms.

Having established that the operators in (7.2.106) are isomorphisms, we may conclude that the operators in (7.2.107) are isomorphisms by once again invoking the abstract result from [130, (2.1.59) in Lemma 2.1.7], now used with

$$X := \big[H^{1/2}(\partial\Omega,\sigma)\big]^M, \quad V := (\mathbb{C}_{\partial\Omega})^M, \quad T := \pm\tfrac{1}{2}I + K_A. \tag{7.2.119}$$

The invariance of V under T follows from (7.2.74) and (7.1.272), whereas the identification of T^* with $\pm\frac{1}{2}I + K_A^\#$ in this context follows from [132, (4.1.29)]. □

We are now prove invertibility results similar to Theorem 7.2.11 on Besov scales.

Theorem 7.2.12 *Let $\Omega \subseteq \mathbb{R}^n$, where $n \in \mathbb{N}$ with $n \geq 2$, be a two-sided* NTA *domain with a compact Ahlfors regular boundary, and abbreviate $\sigma := \mathcal{H}^{n-1}\lfloor\partial\Omega$. Recall the spaces $\mathbb{C}_{\partial\Omega_\pm}$, $\mathbb{C}_{\partial\Omega}$ from* (7.1.265), *and recall the conventions made in* (7.1.266)-(7.1.267). *Next, let L be a homogeneous, second-order, constant complex coefficient, $M\times M$ system in $\mathbb{R}^n$, which has a positive definite, Hermitian self-adjoint, coefficient tensor $A \in \mathfrak{A}_L$. Finally, associate with A and Ω the double layer potential operators K_A and $K_A^\#$ acting on Besov spaces as in* [132, Theorem 4.1.1] *and* [132, Theorem 4.1.5], *respectively. Then there exists $\varepsilon > 0$ with the property that whenever*

$$|p-2| + |q-2| + |s-1/2| < \varepsilon \tag{7.2.120}$$

the following operators are isomorphisms:

$$\pm\tfrac{1}{2}I + K_A^\# : \big[B^{p,q}_{-s}(\partial\Omega,\sigma)\cap(\mathbb{C}_{\partial\Omega_\mp})^\perp\big]^M \xrightarrow{\sim} \big[B^{p,q}_{-s}(\partial\Omega,\sigma)\cap(\mathbb{C}_{\partial\Omega_\mp})^\perp\big]^M, \tag{7.2.121}$$

$$\pm\tfrac{1}{2}I + K_A : \frac{\big[B^{p,q}_{s}(\partial\Omega,\sigma)\big]^M}{\big(\mathbb{C}_{\partial\Omega_\mp}\big)^M} \xrightarrow{\sim} \frac{\big[B^{p,q}_{s}(\partial\Omega,\sigma)\big]^M}{\big(\mathbb{C}_{\partial\Omega_\mp}\big)^M}, \tag{7.2.122}$$

$$\pm\tfrac{1}{2}I + K_A : \frac{\big[B^{p,q}_{s}(\partial\Omega,\sigma)\big]^M}{\big(\mathbb{C}_{\partial\Omega}\big)^M} \xrightarrow{\sim} \frac{\big[B^{p,q}_{s}(\partial\Omega,\sigma)\big]^M}{\big(\mathbb{C}_{\partial\Omega}\big)^M}, \tag{7.2.123}$$

$$\pm\tfrac{1}{2}I + K_A^\# : \big[B^{p,q}_{-s}(\partial\Omega,\sigma)\cap(\mathbb{C}_{\partial\Omega})^\perp\big]^M \xrightarrow{\sim} \big[B^{p,q}_{-s}(\partial\Omega,\sigma)\cap(\mathbb{C}_{\partial\Omega})^\perp\big]^M, \tag{7.2.124}$$

all operators acting on quotient spaces in a natural manner (*cf.* (7.1.294)-(7.1.295)).

As a particular case, in the context of the above theorem we see that the operators

$$\tfrac{1}{2}I + K_A : \big[B^{p,q}_s(\partial\Omega,\sigma)\big]^M \longrightarrow \big[B^{p,q}_s(\partial\Omega,\sigma)\big]^M, \tag{7.2.125}$$

$$-\tfrac{1}{2}I + K_A : \big[B^{p,q}_s(\partial\Omega,\sigma)\big]^M/\mathbb{C}^M \longrightarrow \big[B^{p,q}_s(\partial\Omega,\sigma)\big]^M/\mathbb{C}^M, \tag{7.2.126}$$

$$\tfrac{1}{2}I + K_A^{\#} : \big[B^{p,q}_{-s}(\partial\Omega,\sigma)\big]^M \longrightarrow \big[B^{p,q}_{-s}(\partial\Omega,\sigma)\big]^M, \tag{7.2.127}$$

$$-\tfrac{1}{2}I + K_A^{\#} : \big[B^{p,q}_{-s}(\partial\Omega,\sigma)\cap\mathbb{C}^{\perp}\big]^M \longrightarrow \big[B^{p,q}_{-s}(\partial\Omega,\sigma)\cap\mathbb{C}^{\perp}\big]^M, \tag{7.2.128}$$

are invertible provided Ω is also bounded, connected, with a connected complement. Indeed, in such a scenario, the set Ω has a connected boundary (cf. (3.1.18)), so $\mathbb{C}_{\partial\Omega_+} = \mathbb{C}_{\partial\Omega} = \mathbb{C}$ and $\mathbb{C}_{\partial\Omega_-} = \{0\}$. Here is the proof of Theorem 7.2.12.

Proof of Theorem 7.2.12 The invertibility of the operators in (7.2.121)-(7.2.124) when $p = q = 2$ and $s = 1/2$ is a consequence of Theorem 7.2.11 and (7.2.4)-(7.2.5). With this in hand, the fact that these operators continue to be invertible for the larger range of indices specified in (7.2.120) follows from our complex interpolation results from [130, Theorem 7.5.2] and [88, Theorem 2.7]. □

The next goal is to estimate the spectral radius of the double layer operator on energy spaces. We begin with the following result, refining classical work of Josip Plemelj in [172] and Oliver Kellogg in [90, § 11, pp. 309–311], pertaining to the nature of eigenvalues.

Proposition 7.2.13 *Let $\Omega \subseteq \mathbb{R}^n$, where $n \in \mathbb{N}$ with $n \geq 2$, be a two-sided* NTA *domain with a compact Ahlfors regular boundary, and set $\sigma := \mathcal{H}^{n-1}\lfloor\partial\Omega$. Also, let L be a homogeneous, second-order, constant complex coefficient, $M \times M$ system in $\mathbb{R}^n$. Assume L is Hermitian self-adjoint, satisfies the Legendre-Hadamard strong ellipticity condition* (*cf.* [131, Definition 1.3.2]), *and that there exists a Hermitian self-adjoint, positive semi-definite, coefficient tensor $A \in \mathfrak{A}_L$. Associate with A and Ω the transpose double layer operator $K_A^{\#}$ acting on Besov spaces as in* [132, Theorem 4.1.5]. *Finally, suppose that $z \in \mathbb{C}$ is such that there exists*

$$f \in \big[H^{-1/2}(\partial\Omega,\sigma)\big]^M, \quad f \neq 0, \ \text{with} \ \big(zI - K_A^{\#}\big)f = 0. \tag{7.2.129}$$

Then necessarily $z \in [-\frac{1}{2}, \frac{1}{2}]$.

Proof Without loss of generality (cf. [129, Lemma 5.10.10]), we may assume that Ω is bounded. If $z = \frac{1}{2}$ there is nothing to prove. Suppose in what follows that $z \in \mathbb{C} \setminus \{\frac{1}{2}\}$. Granted this, for f as in (7.2.129), we deduce from Lemma 7.2.8 that

$$\langle f, c\rangle = 0 \ \text{ for every } \ c \in \mathbb{C}^M. \tag{7.2.130}$$

To proceed, express the given system as $L = L_A$ for some positive semi-definite Hermitian self-adjoint coefficient tensor $A = \Big(a_{rs}^{\alpha\beta}\Big)_{\substack{1\leq r,s\leq n\\ 1\leq\alpha,\beta\leq M}}$. Bring in the boundary-to-domain single layer potential operator $\mathscr{S}_{A^\top}$, associated with $A^\top$ and Ω, then define

$$
\begin{aligned}
& u := u^{\pm} \text{ in } \Omega_{\pm}, \text{ where } u^{\pm} := \mathscr{S}_{A^{\top}} f \text{ in } \Omega_{\pm}, \\
& \quad \text{with } \Omega_{+} := \Omega \text{ and } \Omega_{-} := \mathbb{R}^n \setminus \overline{\Omega}.
\end{aligned} \tag{7.2.131}
$$

Hence, $u^{\pm} = u|_{\Omega_{\pm}}$. Then (7.2.130) ensures that for each multi-index $\gamma \in \mathbb{N}_0^n$ there exists some constant $C_{\gamma} \in (0, \infty)$ such that

$$
|(\partial^{\gamma} u^{-})(x)| \leq \frac{C_{\gamma}}{|x|^{n-1+|\gamma|}} \text{ if } |x| \text{ is large.} \tag{7.2.132}
$$

In particular, this ensures that

$$
\nabla u^{-} \in \big[L^2(\Omega_{-}, \mathcal{L}^n)\big]^{n \cdot M}. \tag{7.2.133}
$$

Also, (7.2.16) implies

$$
u \in \big[W^{1,2}_{\text{loc}}(\mathbb{R}^n)\big]^M. \tag{7.2.134}
$$

Let $\text{Tr}_{\Omega_{\pm} \to \partial\Omega}$ stand, respectively, for the boundary trace operators associated with $\Omega_{\pm}$. Then [132, (4.3.44)] implies $\text{Tr}_{\Omega_{+} \to \partial\Omega}(u^{+}) = \text{Tr}_{\Omega_{-} \to \partial\Omega}(u^{-}) = S_{A^{\top}} f$. Reasoning analogously to (7.2.88) presently yields

$$
-\int_{\Omega_{-}} a_{sr}^{\beta\alpha} (\partial_s u_{\beta}^{-}) \overline{\partial_r u_{\alpha}^{-}} \, d\mathcal{L}^n = \Big\langle \big(\tfrac{1}{2} I + K_A^{\#}\big) f, \overline{S_{A^{\top}} f} \Big\rangle, \tag{7.2.135}
$$

$$
\int_{\Omega_{+}} a_{sr}^{\beta\alpha} (\partial_s u_{\beta}^{+}) \overline{\partial_r u_{\alpha}^{+}} \, d\mathcal{L}^n = \Big\langle \big(-\tfrac{1}{2} I + K_A^{\#}\big) f, \overline{S_{A^{\top}} f} \Big\rangle. \tag{7.2.136}
$$

Using the last property in (7.2.129) we may then write

$$
\begin{aligned}
0 &= \Big\langle \big(zI - K_A^{\#}\big) f, \overline{S_{A^{\top}} f} \Big\rangle \\
&= -\big(z + \tfrac{1}{2}\big) \Big\langle \big(-\tfrac{1}{2} I + K_A^{\#}\big) f, \overline{S_{A^{\top}} f} \Big\rangle + \big(z - \tfrac{1}{2}\big) \Big\langle \big(\tfrac{1}{2} I + K_A^{\#}\big) f, \overline{S_{A^{\top}} f} \Big\rangle \\
&= \big(z + \tfrac{1}{2}\big) \int_{\Omega_{-}} a_{sr}^{\beta\alpha} (\partial_s u_{\beta}^{-}) \overline{\partial_r u_{\alpha}^{-}} \, d\mathcal{L}^n + \big(z - \tfrac{1}{2}\big) \int_{\Omega_{+}} a_{sr}^{\beta\alpha} (\partial_s u_{\beta}^{+}) \overline{\partial_r u_{\alpha}^{+}} \, d\mathcal{L}^n.
\end{aligned} \tag{7.2.137}
$$

Since A is Hermitian self-adjoint, the integrands $a_{sr}^{\beta\alpha} (\partial_s u_{\beta}^{\pm}) \overline{\partial_r u_{\alpha}^{\pm}}$ are (pointwise) real in $\Omega_{\pm}$. Taking the imaginary parts of the most extreme sides of (7.2.137) yields

$$
(\text{Im}\, z) \int_{\mathbb{R}^n} a_{sr}^{\beta\alpha} (\partial_s u_{\beta}) \overline{\partial_r u_{\alpha}} \, d\mathcal{L}^n = 0. \tag{7.2.138}
$$

To proceed observe that, thanks to (7.2.133)-(7.2.134),

$$
a_{sr}^{\beta\alpha} (\partial_s u_{\beta}) \overline{\partial_r u_{\alpha}} \text{ is absolutely integrable in } \mathbb{R}^n. \tag{7.2.139}
$$

As in the past, fix a real-valued function $\varphi \in \mathscr{C}_c^{\infty}(\mathbb{R}^n)$ which is identically one in a neighborhood of the origin, and for each $j \in \mathbb{N}$ define $\psi_j(x) := \varphi(x/j)$ for every $x \in \mathbb{R}^n$. For each $j \in \mathbb{N}$ decompose

$$(\operatorname{Im} z) \int_{\mathbb{R}^n} a_{sr}^{\beta\alpha} \partial_s(\psi_j u_\beta) \overline{\partial_r(\psi_j u_\alpha)} \, d\mathcal{L}^n = \mathrm{I}_j + \mathrm{II}_j, \tag{7.2.140}$$

where

$$\mathrm{I}_j := (\operatorname{Im} z) \int_{\mathbb{R}^n} \psi_j^2 a_{sr}^{\beta\alpha} (\partial_s u_\beta) \overline{\partial_r u_\alpha} \, d\mathcal{L}^n, \tag{7.2.141}$$

$$\mathrm{II}_j := \int_{\mathbb{R}^n} O\big(|\nabla\psi_j||\psi_j||u||\nabla u| + |\nabla\psi_j|^2|u|^2\big) \, d\mathcal{L}^n. \tag{7.2.142}$$

Making use of (7.2.139), Lebesgue's Dominated Convergence Theorem, and (7.2.138) we may then write

$$\lim_{j\to\infty} \mathrm{I}_j = (\operatorname{Im} z) \int_{\mathbb{R}^n} a_{sr}^{\beta\alpha} (\partial_s u_\beta) \overline{\partial_r u_\alpha} \, d\mathcal{L}^n. \tag{7.2.143}$$

Also, since $|\nabla\psi_j| \leq C/j$ and $|x| \approx j$ on the support of $\nabla\psi_j$, uniformly in $j \in \mathbb{N}$, based on (7.2.132) we may estimate, as $j \to \infty$,

$$\int_{\mathbb{R}^n} |\psi_j||\nabla\psi_j||u||\nabla u| \, d\mathcal{L}^n \leq C j^{-1} \int_{x\in\mathbb{R}^n,\ |x|\approx j} |u||\nabla u| \, d\mathcal{L}^n = O(j^{-n}), \tag{7.2.144}$$

$$\int_{\mathbb{R}^n} |\nabla\psi_j|^2|u|^2 \, d\mathcal{L}^n \leq C j^{-2} \int_{x\in\mathbb{R}^n,\ |x|\approx j} |u|^2 \, d\mathcal{L}^n = O(j^{-n}). \tag{7.2.145}$$

Hence,

$$\lim_{j\to\infty} \mathrm{II}_j = 0. \tag{7.2.146}$$

Together, (7.2.143) and (7.2.146) imply that

$$\lim_{j\to\infty} \left\{ (\operatorname{Im} z) \int_{\mathbb{R}^n} a_{sr}^{\beta\alpha} \partial_s(\psi_j u_\beta) \overline{\partial_r(\psi_j u_\alpha)} \, d\mathcal{L}^n \right\} = 0. \tag{7.2.147}$$

Since from (7.2.134) we know that $\psi_j u \in W^{1,2}(\mathbb{R}^n)$ for every $j \in \mathbb{N}$, it follows from (7.2.147) and (7.2.23) that

$$\lim_{j\to\infty} \left\{ |\operatorname{Im} z| \int_{\mathbb{R}^n} |\nabla(\psi_j u)|^2 \, d\mathcal{L}^n \right\} = 0. \tag{7.2.148}$$

Writing $\nabla(\psi_j u) = (\nabla\psi_j)u + \psi_j\nabla u$, a reasoning very similar to the one that has led to (7.2.146) shows that $\lim\limits_{j\to\infty} \int_{\mathbb{R}^n} |\nabla(\psi_j u)|^2 \, d\mathcal{L}^n = \int_{\mathbb{R}^n} |\nabla u|^2 \, d\mathcal{L}^n$. In concert with (7.2.148), this proves that $|\operatorname{Im} z| \int_{\mathbb{R}^n} |\nabla u|^2 \, d\mathcal{L}^n = 0$. If $\operatorname{Im} z \neq 0$, the latter forces $\nabla u \equiv 0$ in $\mathbb{R}^n$, hence $u \equiv 0$ since u vanishes at infinity. Granted this, we may finally conclude by reasoning as in (7.2.50)-(7.2.51) that

$$f = 0, \tag{7.2.149}$$

in contradiction with the second condition in (7.2.129). This argument shows that we necessarily have $\operatorname{Im} z = 0$, i.e.,

$$z \in \mathbb{R}. \tag{7.2.150}$$

If we now define $\Phi_\pm := \int_{\Omega_\pm} a_{sr}^{\beta\alpha}(\partial_s u_\beta)\overline{\partial_r u_\alpha}\, d\mathcal{L}^n$, it follows that

$$\Phi_+ \text{ and } \Phi_- \text{ cannot be simultaneously zero} \tag{7.2.151}$$

since otherwise the same type of argument, starting with (7.2.138) and leading to (7.2.149), would yield a contradiction as before. Moreover, the fact that A is positive semi-definite and Hermitian self-adjoint imply that, on the one hand,

$$\Phi_\pm \in [0, \infty). \tag{7.2.152}$$

On the other hand, (7.2.137) entails

$$\big(z + \tfrac{1}{2}\big)\Phi_- + \big(z - \tfrac{1}{2}\big)\Phi_+ = 0. \tag{7.2.153}$$

Note that if $z \in \mathbb{R} \setminus [-\frac{1}{2}, \frac{1}{2}]$ then $z + \frac{1}{2}$ and $z - \frac{1}{2}$ are nonzero real numbers of the same sign. However, in light of (7.2.153), (7.2.152), and (7.2.151), this is an impossibility. Together with (7.2.150), this contraction proves that $z \in [-\frac{1}{2}, \frac{1}{2}]$, as desired. □

Our main spectral result for the double layer potential operator is given next.

Theorem 7.2.14 *Let $\Omega \subseteq \mathbb{R}^n$, where $n \in \mathbb{N}$ with $n \geq 2$, be a bounded, connected, two-sided* NTA *domain with an Ahlfors regular boundary, whose complement is connected. Abbreviate $\sigma := \mathcal{H}^{n-1}\lfloor\partial\Omega$. Also, let L be a homogeneous, second-order, constant complex coefficient, $M \times M$ system in $\mathbb{R}^n$, which has a Hermitian self-adjoint, positive definite, coefficient tensor $A \in \mathfrak{A}_L$. Recall the double layer potential operators K_A, $K_A^\#$ associated with A and Ω, acting on Besov spaces as in* [132, Theorem 4.1.1] *and* [132, Theorem 4.1.5], *respectively. Then*

$$\begin{array}{l}\textit{the spectrum of } K_A \textit{ acting on } \big[H^{1/2}(\partial\Omega,\sigma)\big]^M \big/ \big(\mathbb{C}_{\partial\Omega}\big)^M \textit{ and the} \\ \textit{spectrum of } K_A^\# \textit{ acting on the space } \big[H^{-1/2}(\partial\Omega,\sigma) \cap (\mathbb{C}_{\partial\Omega})^\perp\big]^M \\ \textit{are both included in the interval } (-1/2, 1/2).\end{array} \tag{7.2.154}$$

Moreover, there exists some small $\varepsilon > 0$ with the property that whenever

$$|p - 2| + |q - 2| + |s - 1/2| < \varepsilon \tag{7.2.155}$$

one has (with the piece of notation introduced in (A.0.227)*),*

$$\rho_{\text{inv}}\Big(K_A; \big[B_s^{p,q}(\partial\Omega,\sigma)\big]^M \Big/ \big(\mathbb{C}_{\partial\Omega}\big)^M\Big) < \tfrac{1}{2}, \tag{7.2.156}$$

$$\rho_{\text{inv}}\Big(K_A^\#; \big[B_{-s}^{p,q}(\partial\Omega,\sigma) \cap \big(\mathbb{C}_{\partial\Omega}\big)^\perp\big]^M\Big) < \tfrac{1}{2}. \tag{7.2.157}$$

Proof For any $z \in \mathbb{C} \setminus (-\frac{1}{2}, \frac{1}{2})$, item *(iii)* in Theorem 7.2.6 together with items *(4)* and *(8)* in [130, Theorem 2.1.2] imply that the operator

$$zI - K_A^\# \text{ is Fredholm with index zero on } \big[H^{-1/2}(\partial\Omega, \sigma)\big]^M. \tag{7.2.158}$$

If we denote by

$$\iota : \big[H^{-1/2}(\partial\Omega, \sigma) \cap (\mathbb{C}_{\partial\Omega})^\perp\big]^M \longrightarrow \big[H^{-1/2}(\partial\Omega, \sigma)\big]^M \tag{7.2.159}$$

the canonical inclusion operator, we have the commutative diagram:

$$\begin{array}{ccc} \big[H^{-1/2}(\partial\Omega, \sigma)\big]^M & \xrightarrow{zI-K_A^\#} & \big[H^{-1/2}(\partial\Omega, \sigma)\big]^M \\ {\scriptstyle\iota}\big\uparrow & & {\scriptstyle\iota}\big\uparrow \\ \big[H^{-1/2}(\partial\Omega, \sigma) \cap (\mathbb{C}_{\partial\Omega})^\perp\big]^M & \xrightarrow{zI-K_A^\#} & \big[H^{-1/2}(\partial\Omega, \sigma) \cap (\mathbb{C}_{\partial\Omega})^\perp\big]^M \end{array} \tag{7.2.160}$$

Based on this, (7.2.158), as well as items *(2)* and *(7)* in [130, Theorem 2.1.2] we conclude that for any $z \in \mathbb{C} \setminus (-\frac{1}{2}, \frac{1}{2})$ the operator

$$zI - K_A^\# \text{ is Fredholm with index zero on } \big[H^{-1/2}(\partial\Omega, \sigma) \cap (\mathbb{C}_{\partial\Omega})^\perp\big]^M. \tag{7.2.161}$$

Assume next that $z \in \mathbb{C} \setminus (-\frac{1}{2}, \frac{1}{2})$ belongs to the spectrum of the operator

$$K_A^\# : \big[H^{-1/2}(\partial\Omega, \sigma) \cap (\mathbb{C}_{\partial\Omega})^\perp\big]^M \longrightarrow \big[H^{-1/2}(\partial\Omega, \sigma) \cap (\mathbb{C}_{\partial\Omega})^\perp\big]^M. \tag{7.2.162}$$

In concert with (7.2.161) this implies that

$$zI - K_A^\# \text{ is not injective on } \big[H^{-1/2}(\partial\Omega, \sigma) \cap (\mathbb{C}_{\partial\Omega})^\perp\big]^M. \tag{7.2.163}$$

Consequently, there exists $f \in \big[H^{-1/2}(\partial\Omega, \sigma) \cap (\mathbb{C}_{\partial\Omega})^\perp\big]^M$ with $f \neq 0$ and satisfying $\big(zI - K_A^\#\big)f = 0$. As such, Proposition 7.2.13 applies and gives that, necessarily, $z \in [-\frac{1}{2}, \frac{1}{2}]$. Given that we are assuming $z \in \mathbb{C} \setminus (-\frac{1}{2}, \frac{1}{2})$ to begin with, this forces $z \in \{\pm\frac{1}{2}\}$. This, however, is in contradiction with the fact that (7.2.124) are invertible.

In summary, the proof so far shows that the spectrum of the operator $K_A^\#$ acting on the space $\big[H^{-1/2}(\partial\Omega, \sigma) \cap (\mathbb{C}_{\partial\Omega})^\perp\big]^M$ is contained in the interval $(-\frac{1}{2}, \frac{1}{2})$. Via duality (cf. [130, (2.1.58)], [132, (4.1.29)], (7.2.4)-(7.2.5), and item *(i)* in [130, Proposition 7.6.1]), this yields the claim made in (7.2.154). With this in hand, the claims made in (7.2.156)-(7.2.157) follow when $p = q = 2$ and $s = 1/2$. Finally, the more general versions of the aforementioned claims, as stated in the theorem, are consequences of what we have proved so far and the stability theory in [88]. □

In the last part of this section we employ the invertibility results for boundary layer potentials deduced earlier to prove well-posedness results for Legendre-Hadamard elliptic systems in bounded two-sided NTA domains with Ahlfors regular boundaries. We start by treating the Inhomogeneous Dirichlet Problem, in the theorem below.

Theorem 7.2.15 (The Inhomogeneous Dirichlet Problem) *Consider two integers $n, M \in \mathbb{N}$ with $n \geq 2$. Suppose $\Omega \subseteq \mathbb{R}^n$ is a bounded two-sided* NTA *domain with an Ahlfors regular boundary, and abbreviate $\sigma := \mathcal{H}^{n-1}\lfloor\partial\Omega$. Also, let L be a homogeneous, second-order, constant complex coefficient, $M \times M$ system in $\mathbb{R}^n$ satisfying the Legendre-Hadamard strong ellipticity condition (cf.* [131, Definition 1.3.2]). *Then there exists some small $\varepsilon > 0$ such that the following statements are true.*

(1) If $p, q \in (1, \infty)$ and $s \in (0, 1)$ are such that

$$|p-2| + |q-2| + \left|s - \tfrac{1}{2}\right| < \varepsilon \tag{7.2.164}$$

then the following Inhomogeneous Dirichlet Problem (with the boundary trace considered as in [130, Theorem 9.4.5, *(ii)*, (9.4.91)]) *is well posed:*

$$\begin{cases} u \in \big[B^{p,q}_{s+\frac{1}{p}}(\Omega)\big]^M, \quad Lu = f \in \big[B^{p,q}_{s+\frac{1}{p}-2}(\Omega)\big]^M, \\ \mathrm{Tr}_{\Omega\to\partial\Omega}\, u = g \in \big[B^{p,q}_s(\partial\Omega, \sigma)\big]^M. \end{cases} \tag{7.2.165}$$

That is, a solution always exists, this solution is unique, and the unique solution of (7.2.165) *satisfies, for some $C = C(\Omega, p, q, s, L) \in (0, \infty)$,*

$$\|u\|_{[B^{p,q}_{s+\frac{1}{p}}(\Omega)]^M} \leq C\Big\{\|f\|_{[B^{p,q}_{s+\frac{1}{p}-2}(\Omega)]^M} + \|g\|_{[B^{p,q}_s(\partial\Omega,\sigma)]^M}\Big\}. \tag{7.2.166}$$

(2) A similar well-posedness result as in item (1) above is true for the version of (7.2.165) *in which a solution is sought in a Triebel-Lizorkin space, i.e., for the boundary value problem*

$$\begin{cases} u \in \big[F^{p,q}_{s+\frac{1}{p}}(\Omega)\big]^M, \quad Lu = f \in \big[F^{p,q}_{s+\frac{1}{p}-2}(\Omega)\big]^M, \\ \mathrm{Tr}_{\Omega\to\partial\Omega}\, u = g \in \big[B^{p,p}_s(\partial\Omega, \sigma)\big]^M, \end{cases} \tag{7.2.167}$$

with the trace operator now considered in the sense of [130, Theorem 9.4.5, *(ii)*, (9.4.93)].

(3) An analogous well-posedness result to those in items (1)-(2) above is also true for the version of (7.2.165) *in which a solution is sought in a weighted Sobolev space, i.e., for the boundary value problem*

$$\begin{cases} u \in \big[W^{1,p}_a(\Omega)\big]^M, \quad Lu = f \in \big[W^{-1,p}_a(\Omega)\big]^M, \\ \mathrm{Tr}_{\Omega\to\partial\Omega}\, u = g \in \big[B^{p,p}_s(\partial\Omega, \sigma)\big]^M, \end{cases} \tag{7.2.168}$$

this time demanding

$$p \in (1, \infty), \quad s \in (0, 1), \quad a := 1 - s - \tfrac{1}{p}, \quad |p-2| + \left|s - \tfrac{1}{2}\right| < \varepsilon, \tag{7.2.169}$$

and now considering the trace operator as in [130, Theorem 8.3.6, (8.3.38)].

Proof Choose $\varepsilon > 0$ small enough so that the conclusion in Theorem 7.2.4 is true. Work under the assumption specified in (7.2.164), and fix $f \in \big[B^{p,q}_{s+\frac{1}{p}-2}(\Omega)\big]^M$ arbitrary. We claim that there exists

$$\begin{aligned} & u_0 \in \big[B^{p,q}_{s+\frac{1}{p}}(\Omega)\big]^M \text{ satisfying} \\ & Lu_0 = f \text{ in } \Omega \text{ and } \|u_0\|_{[B^{p,q}_{s+\frac{1}{p}}(\Omega)]^M} \leq C\|f\|_{[B^{p,q}_{s+\frac{1}{p}-2}(\Omega)]^M} \end{aligned} \tag{7.2.170}$$

for some $C = C(\Omega, L, n, p, q, s) \in (0, \infty)$. To construct such a function u_0, consider the Fourier multiplier operator $L^{-1}\varphi := \mathscr{F}^{-1}\big(L(\xi)^{-1}(\mathscr{F}\varphi)\big)$, for all $\varphi \in \big[\mathscr{S}'(\mathbb{R}^n)\big]^M$. Also, recall from [130, (9.2.1)] that there exists some $f_0 \in \big[B^{p,q}_{s+\frac{1}{p}-2}(\mathbb{R}^n)\big]^M$ with $f_0\big|_\Omega = f$ and such that $\|f_0\|_{[B^{p,q}_{s+\frac{1}{p}-2}(\mathbb{R}^n)]^M} \leq 2\|f\|_{[B^{p,q}_{s+\frac{1}{p}-2}(\Omega)]^M}$. Having picked two cutoff functions $\psi_1, \psi_2 \in \mathscr{C}^\infty_c(\mathbb{R}^n)$ which are identically one near the compact $\overline{\Omega}$, set

$$u_0 := \big[(\psi_1 L^{-1}\psi_2)f_0\big]\Big|_\Omega \text{ in } \Omega. \tag{7.2.171}$$

Then [87, Corollary 4.2] implies that (7.2.170) holds.

To proceed, with $g \in \big[B^{p,q}_s(\partial\Omega, \sigma)\big]^M$ as in (7.2.165) define

$$g_0 := g - \mathrm{Tr}_{\Omega\to\partial\Omega}\, u_0 \in \big[B^{p,q}_s(\partial\Omega, \sigma)\big]^M, \tag{7.2.172}$$

where the trace is considered in the sense of [130, (9.4.91) from item *(ii)* of Theorem 9.4.5]. If $n \geq 3$, the function $u := u_0 + \mathscr{S}(S^{-1}g_0)$ in Ω, solves (7.2.165) and satisfies (7.2.166), thanks to (7.2.170), [132, (4.2.209), (4.3.25)] and the trace formula [132, (4.3.44)-(4.3.45)]. If $n = 2$, then the fact that the augmented single layer S_{aug} is an isomorphism in the context of (7.2.52) guarantees the existence of

$$\begin{aligned} & g \in \big[B^{p,q}_{s-1}(\partial\Omega, \sigma)\big]^M \text{ and } c \in \mathbb{C}^M \text{ such that} \\ & Sg + c = f, \quad \langle g, \mathbf{e}_j\rangle = 0 \text{ for } 1 \leq j \leq M, \text{ and} \\ & |c| + \|g\|_{[B^{p,q}_{s-1}(\partial\Omega,\sigma)]^M} \leq C\|f\|_{[B^{p,q}_{s}(\partial\Omega,\sigma)]^M}, \end{aligned} \tag{7.2.173}$$

for some constant $C \in (0, \infty)$ independent of f. Then, for similar reasons as before, the function $u := u_0 + \mathscr{S}g + c$ in Ω, solves (7.2.165) and satisfies (7.2.166).

To prove uniqueness for (7.2.165), assume the function $u \in \big[B^{p,q}_{s+\frac{1}{p}}(\Omega)\big]^M$ is such that $Lu = 0$ in Ω and $\mathrm{Tr}_{\Omega\to\partial\Omega}\, u = 0$ on $\partial\Omega$. Pick an arbitrary coefficient tensor $A \in \mathfrak{A}_L$. Then the integral representation formula [132, (4.4.2)] in the context described in [132, Theorem 4.4.1, item *(1)*] presently reduces to

$$u = -\mathscr{S}\big(\partial^A_\nu(u, 0)\big) \text{ in } \Omega. \tag{7.2.174}$$

Consequently,

$$0 = \mathrm{Tr}_{\Omega\to\partial\Omega}\, u = -\mathrm{Tr}_{\Omega\to\partial\Omega}\Big(\mathscr{S}\big(\partial_\nu^A(u,0)\big)\Big) = -S\big(\partial_\nu^A(u,0)\big), \tag{7.2.175}$$

by the trace formula [132, (4.3.44)] in the version specified in [132, (4.3.45)] and assumptions. In particular, $\partial_\nu^A(u,0)$ belongs to $\big[B^{p,q}_{s-1}(\partial\Omega,\sigma)\big]^M$ (cf. [130, Proposition 9.5.2]). If $n \geq 3$, we conclude from (7.2.28) that $\partial_\nu^A(u,0) = 0$. When used back in (7.2.174), this finally proves that $u = 0$, as wanted. If $n = 2$, from the generalized "half" Green's formula [130, (9.5.10)] presently gives

$${}_{[B^{p,q}_{s-1}(\partial\Omega,\sigma)]^M}\big\langle \partial_\nu^A(u,0), c\big\rangle_{[B^{p',q'}_{1-s}(\partial\Omega,\sigma)]^M} = 0 \text{ for each } c \in \mathbb{C}^M, \tag{7.2.176}$$

where p', q' are the Hölder conjugate exponents of p, q. Granted this, we conclude from (7.2.52) and (7.2.175) that $S_{\rm aug}\big(\partial_\nu^A(u,0), 0\big) = (0,0)$. Recalling that the augmented single layer $S_{\rm aug}$ is an isomorphism in the context of (7.2.52), this forces $\partial_\nu^A(u,0) = 0$. Using this back in (7.2.174) we arrive at the conclusion that $u = 0$, hence we have uniqueness in the two-dimensional setting as well. The above reasoning proves that the problem (7.2.165) is indeed well posed.

Finally, the well-posedness results claimed in items *(2)-(3)* are justified in a similar manner, now making use of the trace formula [132, (4.3.44)] in the versions specified in [132, (4.3.46), (4.3.47)], the mapping properties of the boundary-to-domain single layer from [132, (4.2.218), (4.3.26)], and the integral representation formula [132, (4.4.2)] in the context described in [132, Theorem 4.4.1, items *(2)-(3)*]. □

Lastly, we discuss the Inhomogeneous Neumann Problem in bounded two-sided NTA domains with Ahlfors regular boundaries.

Theorem 7.2.16 (The Inhomogeneous Neumann Problem) *Consider two integers $n, M \in \mathbb{N}$ with $n \geq 2$. Let $\Omega \subseteq \mathbb{R}^n$ be a bounded two-sided* NTA *domain with an Ahlfors regular boundary. Abbreviate $\sigma := \mathcal{H}^{n-1}\lfloor\partial\Omega$ and denote by ν the geometric measure theoretic outward unit normal to Ω. Recall the spaces $\mathbb{C}_{\Omega_\pm}$, $\mathbb{C}_{\partial\Omega_\pm}$ from* (7.1.265). *Also, let L be a homogeneous, second-order, constant complex coefficient, $M\times M$ system in $\mathbb{R}^n$, which has a Hermitian self-adjoint, positive definite, coefficient tensor $A \in \mathfrak{A}_L$. Finally, select $p, q \in (1,\infty)$ and $s \in (0,1)$. Then there exists a small number $\varepsilon > 0$ with the property that the following statements are true:*

(1) Consider the Inhomogeneous Neumann Problem which, given any data

$$f \in \big[B^{p,q}_{s+\frac{1}{p}-2,0}(\Omega)\big]^M \text{ and } g \in \big[B^{p,q}_{s-1}(\partial\Omega,\sigma)\big]^M, \tag{7.2.177}$$

is concerned with finding a function u such that[11]

$$\begin{cases} u \in \big[B^{p,q}_{s+\frac{1}{p}}(\Omega)\big]^M, \quad Lu = f\big|_\Omega \text{ in } \Omega, \\ \partial_\nu^A(u,f) = g \text{ on } \partial\Omega, \end{cases} \tag{7.2.178}$$

[11] with both equalities understood in the sense of distributions

where the conormal derivative is meaningfully defined as in [130, Proposition 9.5.2] *with* $\mathbb{A} := B$*, now regarding* f *as a functional in* $\Big(\big[B^{p',q'}_{1-s-\frac{1}{p'}}(\Omega)\big]^M\Big)^*$ *(where* $1/p + 1/p' = 1/q + 1/q' = 1$*; cf.* [130, (9.2.123)]) *and keeping in mind the identification from* [130, Lemma 9.5.1, (9.5.4)]. *Then whenever*

$$|p-2| + |q-2| + \big|s - \tfrac{1}{2}\big| < \varepsilon \tag{7.2.179}$$

the inhomogeneous boundary value problem (7.2.178) *is solvable if and only if the data* f, g *satisfy the compatibility condition*

$$\begin{gathered} {}_{[(B^{p',q'}_{1-s+1/p'}(\Omega))^*]^M}\big\langle f, \chi\big\rangle_{[B^{p',q'}_{1-s+1/p'}(\Omega)]^M} = {}_{[B^{p,q}_{1-s}(\partial\Omega,\sigma)]^M}\big\langle g, \chi\big|_{\partial\Omega}\big\rangle_{[B^{p',q'}_{s-1}(\partial\Omega,\sigma)]^M} \\ \text{for each function } \chi \in \big(\mathbb{C}_{\Omega_+}\big)^M, \end{gathered} \tag{7.2.180}$$

and if (7.2.180) *holds then the solution of* (7.2.178) *is unique modulo additive functions from* $\big(\mathbb{C}_{\Omega_+}\big)^M$ *and satisfies, for some* $C = C(\Omega, n, p, q, s, A) \in (0, \infty)$,

$$\inf_{\chi \in (\mathbb{C}_{\Omega_+})^M} \|u - \chi\|_{[B^{p,q}_{s+\frac{1}{p}}(\Omega)]^M} \leq C\|f\|_{[B^{p,q}_{s+\frac{1}{p}-2}(\mathbb{R}^n)]^M} + C\|g\|_{[B^{p,q}_{s-1}(\partial\Omega,\sigma)]^M}. \tag{7.2.181}$$

(2) *Analogous results as in item (1) are valid for the version of the Inhomogeneous Neumann Problem* (7.2.178) *formulated for functions in Triebel-Lizorkin spaces, a scenario in which given any*

$$f \in \big[F^{p,q}_{s+\frac{1}{p}-2,0}(\Omega)\big]^M \ \text{ and } \ g \in \big[B^{p,p}_{s-1}(\partial\Omega, \sigma)\big]^M, \tag{7.2.182}$$

one is concerned with finding a function u such that[12]

$$\begin{cases} u \in \big[F^{p,q}_{s+\frac{1}{p}}(\Omega)\big]^M, \quad Lu = f\big|_{\Omega} \text{ in } \Omega, \\ \partial^A_\nu(u, f) = g \text{ on } \partial\Omega, \end{cases} \tag{7.2.183}$$

(where the conormal derivative is taken as in [130, Proposition 9.5.2] *with* $\mathbb{A} := F$*), and for the version of the Inhomogeneous Neumann Problem* (7.2.178) *formulated for functions in weighted Sobolev spaces in which case, given any*

$$\begin{gathered} f \in \Big(\big[W^{1,p'}_{-a}(\Omega)\big]^M\Big)^* \text{ with } a := 1 - s - \tfrac{1}{p} \text{ and } p' := \big(1 - \tfrac{1}{p}\big)^{-1} \\ \text{and } g \in \big[B^{p,p}_{s-1}(\partial\Omega, \sigma)\big]^M, \end{gathered} \tag{7.2.184}$$

one is seeking a function u satisfying[13]

[12] again, with both equalities understood in the sense of distributions

[13] with both equalities understood in the sense of distributions

$$\begin{cases} u \in \big[W^{1,p}_a(\Omega)\big]^M, \quad Lu = f\lfloor_\Omega \text{ in } \Omega, \\ \partial^A_\nu(u, f) = g \text{ on } \partial\Omega, \end{cases} \tag{7.2.185}$$

where the conormal derivative is now considered as in [130, Proposition 8.5.3].

Proof Pick $\varepsilon > 0$ small enough so that the conclusions in Theorem 7.2.12 are valid. In particular, (7.2.121) implies that the operator

$$-\tfrac{1}{2}I + K^\#_A : \big[B^{p,q}_{-s}(\partial\Omega,\sigma) \cap (\mathbb{C}_{\partial\Omega_+})^\perp\big]^M \xrightarrow{\sim} \big[B^{p,q}_{-s}(\partial\Omega,\sigma)\big]^M \cap (\mathbb{C}_{\partial\Omega_+})^\perp\big]^M \tag{7.2.186}$$

is invertible. To prove the necessity of the compatibility condition formulated in (7.2.180), observe that, whenever f, g, u are as in (7.2.177)-(7.2.178), for each given function $\chi \in (\mathbb{C}_{\Omega_+})^M$ we may write

$$\begin{aligned} &{}_{[B^{p,q}_{1-s}(\partial\Omega,\sigma)]^M}\big\langle g, \chi\big|_{\partial\Omega}\big\rangle_{[B^{p',q'}_{s-1}(\partial\Omega,\sigma)]^M} \\ &\quad = {}_{[B^{p,q}_{1-s}(\partial\Omega,\sigma)]^M}\big\langle \partial^A_\nu(u, f), \chi\big\rangle_{[B^{p',q'}_{s-1}(\partial\Omega,\sigma)]^M} \\ &\quad = {}_{[(B^{p',q'}_{1-s+1/p'}(\Omega))^*]^M}\big\langle f, \chi\big\rangle_{[B^{p',q'}_{1-s+1/p'}(\Omega)]^M} \end{aligned} \tag{7.2.187}$$

thanks to the generalized "half" Green's formula [130, (9.5.10)] (presently used with $w := \chi$). Thus, (7.2.180) holds. Suppose next that f, g are as in (7.2.177) and satisfy the compatibility condition formulated in (7.2.180). In particular, f belongs to $\big[B^{p,q}_{s+\frac{1}{p}-2}(\mathbb{R}^n)\big]^M$ and is supported in $\overline{\Omega}$. Convolving f with E, the matrix-valued fundamental solution of the system L from [131, Theorem 1.4.2], and then restricting to Ω yields a function

$$\begin{aligned} u_0 \in \big[B^{p,q}_{s+\frac{1}{p}}(\Omega)\big]^M \text{ satisfying} \\ Lu_0 = f\big|_\Omega \text{ in } \Omega \text{ and } \|u_0\|_{[B^{p,q}_{s+\frac{1}{p}}(\Omega)]^M} \leq C\|f\|_{[B^{p,q}_{s+\frac{1}{p}-2}(\mathbb{R}^n)]^M} \end{aligned} \tag{7.2.188}$$

for some $C = C(\Omega, n, p, q, s, A) \in (0, \infty)$. Then [130, Proposition 9.5.2] guarantees

$$\begin{aligned} \partial^A_\nu(u_0, f) \in \big[B^{p,q}_{s-1}(\partial\Omega,\sigma)\big]^M \text{ and} \\ \big\|\partial^A_\nu(u_0, f)\big\|_{[B^{p,q}_{s-1}(\partial\Omega,\sigma)]^M} \leq C\|f\|_{[B^{p,q}_{s+\frac{1}{p}-2}(\mathbb{R}^n)]^M} \end{aligned} \tag{7.2.189}$$

for some $C = C(\Omega, n, p, q, s, A) \in (0, \infty)$. Hence,

$$\begin{aligned} g_0 := g - \partial^A_\nu(u_0, f) \in \big[B^{p,q}_{s-1}(\partial\Omega,\sigma)\big]^M \text{ and} \\ \|g_0\|_{[B^{p,q}_{s-1}(\partial\Omega,\sigma)]^M} \leq \|g\|_{[B^{p,q}_{s-1}(\partial\Omega,\sigma)]^M} + C\|f\|_{[B^{p,q}_{s+\frac{1}{p}-2}(\mathbb{R}^n)]^M} \end{aligned} \tag{7.2.190}$$

for some $C = C(\Omega, n, p, q, s, A) \in (0, \infty)$. Also, for each $\chi \in (\mathbb{C}_{\Omega_+})^M$ we may write

$$\begin{aligned}
&{}_{[B^{p,q}_{1-s}(\partial\Omega,\sigma)]^M}\big\langle g_0, \chi|_{\partial\Omega}\big\rangle_{[B^{p',q'}_{s-1}(\partial\Omega,\sigma)]^M} \\
&= {}_{[B^{p,q}_{1-s}(\partial\Omega,\sigma)]^M}\big\langle g, \chi|_{\partial\Omega}\big\rangle_{[B^{p',q'}_{s-1}(\partial\Omega,\sigma)]^M} \\
&\qquad - {}_{[B^{p,q}_{1-s}(\partial\Omega,\sigma)]^M}\big\langle \partial^A_\nu(u_0, f), \chi\big\rangle_{[B^{p',q'}_{s-1}(\partial\Omega,\sigma)]^M} \\
&= {}_{[(B^{p',q'}_{1-s+1/p'}(\Omega))^*]^M}\big\langle f, \chi\big\rangle_{[B^{p',q'}_{1-s+1/p'}(\Omega)]^M} - {}_{[(B^{p',q'}_{1-s+1/p'}(\Omega))^*]^M}\big\langle f, \chi\big\rangle_{[B^{p',q'}_{1-s+1/p'}(\Omega)]^M} = 0
\end{aligned} \tag{7.2.191}$$

by (7.2.190), the generalized "half" Green's formula [130, (9.5.10)] (presently used with $w := \chi$), and the compatibility condition (7.2.180). The fact that the operator (7.2.186) is invertible then permits us to define

$$u_1 := u_0 + \mathscr{S}\Big(\big(-\tfrac{1}{2}I + K^{\#}_{A^\top}\big)^{-1} g_0\Big) \text{ in } \Omega. \tag{7.2.192}$$

Then $u_1 \in \big[B^{p,q}_{s+\frac{1}{p}}(\Omega)\big]^M$ and, for some $C = C(\Omega, n, p, q, s, A) \in (0, \infty)$, satisfies

$$\|u_1\|_{[B^{p,q}_{s+\frac{1}{p}}(\Omega)]^M} \leq C\|f\|_{[B^{p,q}_{s+\frac{1}{p}-2}(\mathbb{R}^n)]^M} + C\|g\|_{[B^{p,q}_{s-1}(\partial\Omega,\sigma)]^M}, \tag{7.2.193}$$

thanks to (7.2.188), (7.2.186), and the mapping properties of the single layer operator $\mathscr{S}$ in the context of [132, (4.3.25)]. In addition,

$$\begin{aligned}
\partial^A_\nu(u_1, f) &= \partial^A_\nu(u_0, f) + \partial^A_\nu\Big(\mathscr{S}\Big(\big(-\tfrac{1}{2}I + K^{\#}_{A^\top}\big)^{-1} g_0\Big), 0\Big) \\
&= \partial^A_\nu(u_0, f) + \big(-\tfrac{1}{2}I + K^{\#}_{A^\top}\big)\big(-\tfrac{1}{2}I + K^{\#}_{A^\top}\big)^{-1} g_0 = \partial^A_\nu(u_0, f) + g_0 = g,
\end{aligned} \tag{7.2.194}$$

by (7.2.192), the linearity of the conormal derivative operator in the context of [130, Proposition 9.5.2], the jump-formula [132, (4.3.83)] considered as in [132, Theorem 4.3.6, item *(b)*], and (7.2.190). Hence, u_1 is a solution of (7.2.178) which satisfies (7.2.193).

Suppose next u is a null-solution of (7.2.178), i.e., $u \in \big[B^{p,q}_{s+\frac{1}{p}}(\Omega)\big]^M$ is such that $Lu = 0$ in Ω and $\partial^A_\nu(u, 0) = 0$ on $\partial\Omega$. Then the integral representation formula [132, (4.4.2)], in the context of [132, Theorem 4.4.1, item *(1)*], presently reduces to

$$u = \mathcal{D}_A\big(\mathrm{Tr}_{\Omega\to\partial\Omega}\, u\big) \text{ in } \Omega. \tag{7.2.195}$$

In view of the jump-formula [132, (4.3.22)] (considered in the context of [132, Theorem 4.3.2, item *(1)*]) we may then write

$$\mathrm{Tr}_{\Omega\to\partial\Omega}\, u = \mathrm{Tr}_{\Omega\to\partial\Omega}\Big(\mathcal{D}_A\big(\mathrm{Tr}_{\Omega\to\partial\Omega}\, u\big)\Big) = \big(\tfrac{1}{2}I + K_A\big)\big(\mathrm{Tr}_{\Omega\to\partial\Omega}\, u\big). \tag{7.2.196}$$

hence, $\mathrm{Tr}_{\Omega\to\partial\Omega}\, u \in \big[B^{p,q}_s(\partial\Omega, \sigma)\big]^M$ satisfies $\big(-\frac{1}{2}I + K_A\big)\big(\mathrm{Tr}_{\Omega\to\partial\Omega}\, u\big) = 0$. Upon recalling from (7.2.122) that the operator $-\frac{1}{2}I + K_A$ is invertible on when acting on the space $\big[B^{p,q}_s(\partial\Omega, \sigma)\big]^M / \big(\mathscr{C}_{\partial\Omega_+}\big)^M$, this forces$\mathrm{Tr}_{\Omega\to\partial\Omega}\, u$ to belong to $\big(\mathscr{C}_{\partial\Omega_+}\big)^M$. When used back in (7.2.195), this ultimately shows (in view of the fact that (7.2.93) holds for each function $w \in (\mathbb{C}_{\Omega_\mp})^M)$ that u belongs to the space $(\mathbb{C}_{\Omega_+})^M$. In the converse direction, it is clear from [129, Lemma 5.11.3] (which implies that Ω has finitely many connected components, which are separated) that any function in $(\mathbb{C}_{\Omega_+})^M$ is a null-solution of (7.2.178).

Hence, null-solution of (7.2.178) is precisely $(\mathbb{C}_{\Omega_+})^M$. Finally, from this and the fact that we have constructed a solution u_1 of (7.2.178) satisfying (7.2.193), we conclude that any solution u of (7.2.178) satisfies (7.2.181). This takes care of all claims in item *(1)*.

Lastly, the claims in item *(2)* are dealt with similarly, now making use of the mapping properties for the single layer operator $\mathscr{S}$ in the context of [132, (4.3.26)], [132, (4.2.218)], and the jump-formula [132, (4.3.22)] considered in the context of [132, Theorem 4.3.2, items *(2)-(3)*]. □

Chapter 8
Boundary Value Problems for Elliptic Systems in Rough Domains

One may argue that the topic of boundary value problems is as old and venerable as mathematics itself. After all, the Greeks' realization that there is a unique line passing through any two given distinct points in the plane is the modern-day equivalent of the well-posedness of the Dirichlet Problem for the 1-dimensional Laplacian $\Delta = \frac{d^2}{dx^2}$ in an interval on the real axis[1].

Here we study boundary value problems for elliptic systems in Ahlfors regular domains of a very general geometric nature, best described in terms of the nature of the mean oscillations of their (geometric measure theoretic) outward unit normal vectors. Nowadays, PDE's in rough settings is a well-established field of research, though this brand of mathematics did not get instantaneous recognition. Indeed, a certain degree of skepticism has long accompanied efforts to understand non-smooth calculus. One well-known quotation attributed to H. Poincaré in 1908, which typifies the aforementioned distrust, goes as follows[2]: "*La logique parfois engendre des monstres. On vit surgir toute une foule de fonctions bizarres qui semblaient s'efforcer de ressembler aussi peu que possible aux honnêtes fonctions qui servent à quelque chose. Plus de continuité, ou bien de la continuité mais pas de dérivées* [...] *Autrefois quand on invantait une fonction nouvelle, c'était en vue de quelque but practique; aujourd'hui on les invente tout exprès pour mettre en défaut les raisonnements de nos péres et on n'en tirera jamais que celà.*" Cf. [175]. Such a point of view was by no means isolated. For example, in a letter to Thomas Stieltjes dated 1893, Charles Hermite expressed himself in even stronger terms[3]: "*Je me détourne avec effroi et horreur de cette plaie lementable des fonctions qui n'on pas de dérivées.*" Cf. [78, Vol. II, p. 318].

As S. Saks describes it in the preface of his book [182], "*Researches dealing with non-analytic functions and with functions violating laws which one hoped were*

[1] Indeed, given $-\infty < x_0 < x_1 < +\infty$ along with $y_0, y_1 \in \mathbb{R}$, the boundary value problem $y'' = 0$ in (x_0, x_1) and $y(x_i) = y_i$ for $i = 0, 1$, has a unique solution, of the form $y = mx + n$.

[2] Logic sometimes engenders monsters. Masses of strange functions emerged and seemed to endeavor to look as little as possible to the honest functions that are of some use. No more continuity, or else continuity but no derivatives [...] In the past, when a new function was invented, it was with practical perspectives; nowadays, they are invented on purpose to show our ancestors' reasoning is at fault, and we shall never get anything more out of them.

[3] I turn away with fear and horror from this lamentable plague of continuous functions that do not have a derivative.

D. Mitrea et al., *Geometric Harmonic Analysis V*, Developments in Mathematics 76,
https://doi.org/10.1007/978-3-031-31561-9_8

universal, were regarded almost as the propagation of anarchy and chaos where past generations had sought order and harmony. Even the first attempts to establish a positive theory were rather sceptically received: it was feared that an excessive pedantic exactitude in formulating hypotheses would spoil the elegance of classical methods, and that discussions of details would end by obscuring the main ideas of analysis." In one form or another, these attitudes have persisted for quite a while: years later, S.G. Mikhlin referring to the perspective of studying PDE's under low regularity assumptions on the underling geometry, opinionated to the effect that "*no mother would ever let her child play in such ravines.*" Cf. [71], [117].

On the flip side, one can make the case that results in this degree of generality represent outposts on the frontier separating what's possible from what's impossible. Indeed, the true nature of phenomenon, the very fabric of a universe, only reveals itself when things are pushed to the limit, by stripping away all inessential structural assumptions and working under minimal hypotheses[4].

The topic of boundary value problems in irregular domains interfaces tightly with many aspects of non-smooth calculus and non-smooth geometry. For example, by constructing what has come to be known as Lebesgue's spine[5] (or Lebesgue's thorn), Henri Léon Lebesgue has settled in 1913 (cf. [108]) a long-running debate about whether the nature of the boundary of a region affects the solvability of the classical Dirichlet Problem[6]. See Dieudonné's history of functional analysis [43, p. 46]. In fact, much of Carl Neumann's 1877 memoir on logarithmic and Newtonian potentials[7] is devoted to an in-depth study of the geometric conditions which must be satisfied by a so-called general region in space and in the plane as to guarantee the veracity of the basic theorems in potential theory. It was therefore clear from the very outset that the regularity conditions on the bounding surfaces and the functions involved are rather complicated, and that the influence geometry has on such proceedings is both subtle and decisive[8].

Another example comes from F. Riesz and B. Sz.-Nagy's influential 1952 treatise on functional analysis, in which they pay a great deal of attention to the Dirichlet and Neumann Problems in potential theory and their solutions by the Fredholm method. In regard to this subject, they write (starting on p. 224 of the edition republished in [180]): "*Despite the elegance of this method, it has the inconvenience of being applicable only under very restrictive hypotheses concerning the boundary of the*

[4] as an abundance of (ultimately redundant) structures makes it difficult to identify the active ingredients in a proof

[5] defined as $S := \{(x_1, \dots, x_n) \in \mathbb{R}^n : x_n > 0,\ x_1^2 + x_2^2 + \cdots + x_{n-1}^2 \leq \exp(-1/x_n^2)\}$ for $n \geq 3$; cf., e.g., the discussion in [42, § 7.2, pp. 59-60]

[6] spoiler alert: it does, particularly when the boundary has a sufficiently sharp edge protruding into the interior of the region, like for $\Omega := B(0, 1) \setminus S$, in $\mathbb{R}^n$ with $n \geq 3$, where S is Lebesgue's spine described earlier

[7] where he succeeded in showing that, provided the region in question is convex, the Dirichlet problem does indeed possess a solution

[8] O. Hölder, a student of Weierstrass, sought to place potential theory of firmer ground in terms of analytic rigor, while H.A. Schwarz (another student of Weierstrass) together with E.B. Christoffel continued the work of Neumann by further extending the class of regions for which constructive solutions to the Dirichlet problem were possible

domain considered. These restrictions are inherent in the method; in fact, since the double layers are made up of oriented dipoles, they can not be distributed over any but sufficiently regular curves (or surfaces)."

With the wisdom of hindsight, one may attempt to paraphrase Mark Twain's famous quip about his own death, and claim that the reports of the ineffectiveness of the method of boundary layer potentials have been greatly exaggerated. Our aim here is to substantiate this claim.

Broadly speaking, there are two types of questions pertaining to the well-posedness of boundary value problems in, say, an Ahlfors regular domain $\Omega \subseteq \mathbb{R}^n$, for a given homogeneous, second-order, constant complex coefficient, elliptic $M \times M$ system L in $\mathbb{R}^n$.

`Question I.` *Identify spaces of boundary data for which said boundary value problems are Fredholm solvable and/or well posed for a given domain* Ω.

`Question II.` *Alternatively, having fixed a certain space of boundary data, characterize the geometric properties of* Ω *guaranteeing that the aforementioned problems are Fredholm solvable and/or well posed.*

These, as well as other related issues, have been a driving force behind many major developments in partial differential equations and allied fields. An authoritative account of their impact is given by C. Kenig in [93] where he describes the state of the art in this field of research up to mid 1990's.

For example, one problem which falls under the scope of `Question I` is that of determining optimal scales of spaces on which the Dirichlet Problem for a given elliptic system of second order is solvable in an arbitrary Lipschitz domain Ω in $\mathbb{R}^n$. The most basic case, involving the Laplacian in $\mathbb{R}^n$, is now well understood, thanks to the work of B. Dahlberg and C. Kenig [38], in the case of L^p-data, and D. Jerison and C. Kenig [86] in the case of Besov data. A program aimed at extending the scope of this body of results to Lipschitz sub-domains of Riemannian manifolds (now involving variable coefficient elliptic operators, such as the Laplace-Beltrami operator) has been initiated by M. Mitrea and M. Taylor in the late 1990's and has produced [153], [154], [156], [157], [158]. This body of work contributes to the understanding of an intricate picture in which, at one end, the Lax-Milgram Lemma says that the Poisson Problem for strongly elliptic systems in energy spaces is always solvable in rather very general domains, while at the other end of the spectrum the Agmon-Douglis-Nirenberg theory asserts that one can accommodate Sobolev-Besov-Triebel-Lizorkin spaces of arbitrary nature if the domain is smooth.

Even in standard geometric settings, such as the upper half-space, `Question I` has motivated much work. For example, in [200, Chapter II, p. 37] E.M. Stein and G. Weiss write "*A basic tool in the study of a function, f, on E_n is obtained by passage to a harmonic function, defined on the $(n+1)$-dimensional upper half-space, which has f as its boundary value.*" A moment's reflection reveals that, in fact, this principle calls for clarifying the nature of function spaces which yield well posed boundary value problems for the Laplacian in the upper half-space. Such a doctrine underpins a significant portion of Stein's vastly influential 1970 book [198] where, among other things, he provides characterizations of Besov spaces of functions in $\mathbb{R}^n$ in terms of their harmonic extension to $\mathbb{R}^n_+$ (cf., e.g., [198, Proposition 7', p. 151]).

Concerning `Question II` formulated above, it has long been observed that the nature of the domain strongly influences the solvability of the classical Dirichlet Problem for the Laplacian (in which the boundary data are arbitrarily prescribed in the space of continuous functions). Early work supporting the thesis that a solution always exists involved informal arguments[9]. In his influential treatise on potential theory, Kellogg summarized this state of affairs by writing in [90, p. 278] that *In fact, the Dirichlet problem is not always solvable, and no "proof" can be valid unless it places some restriction on the region.*

As regards modern work on `Question II`, E. Fabes, M. Jodeit, and N. Rivière, in their fundamental paper [52], have proved that given any $p \in (1, \infty)$ the Dirichlet Problem for the Laplacian with boundary data from L^p is well posed provided the underlying set Ω is a bounded domain of class $\mathscr{C}^1$. The case when the boundary data is selected from Besov spaces exhibiting higher regularity has been dealt with by V. Maz'ya and T. Shaposhnikova in [122] where geometric conditions guaranteeing solvability are found in terms of the properties of the outward unit normal ν to Ω as a Sobolev space multiplier.

A classical strategy for dealing with `Question II` consists of "flattening", i.e., formulating and solving the analogue of the original problem in a standard case, typically when $\Omega = \mathbb{R}^n_+$, and then deviating from this most standard setting by allowing perturbations of a certain magnitude. A paradigm result in this regard, going back to the work of S. Agmon, A. Douglis, and L. Nirenberg, as well as V.A. Solonnikov in the 1950's and 1960's, is that the Dirichlet Problem is solvable in the context of Sobolev-Besov spaces if Ω is a sufficiently smooth domain. This is a topic which has been investigated by many authors, including L. Gårding [67], M.I. Višik [204], F.E. Browder [19], S. Agmon, A. Douglis and L. Nirenberg [2], [3], [4], [5], V. Solonnikov [195], [196], L. Hörmander [83], G. Grubb and N.J. Kokholm [76].

In all these works, the underlying domain Ω is locally of upper-graph type, i.e., near each boundary point the set Ω may be identified (up to a rigid transformation of the ambient Euclidean space) with $\{(x', x_n) : x' \in \mathbb{R}^{n-1} \text{ and } x_n > \varphi(x')\}$, where $\varphi : \mathbb{R}^{n-1} \to \mathbb{R}$ is a (sufficiently smooth) function.

In the first five sections of this chapter, §§8.1-8.5, the goal is to address `Question II` by allowing more general classes of sets Ω, which are not necessarily of upper-graph type. First, in §8.1 and §8.2, we consider boundary value problems for "generic" elliptic homogeneous constant (complex) coefficient second-order systems L and, respectively, the Stokes system of liniarized hydrostatic, in domains $\Omega \subseteq \mathbb{R}^n$ with *compact boundaries*. In such a setting, we shall impose the condition

[9] For example, George Green attempted in [75] to construct a solution for the classical Dirichlet Problem for the Laplacian relying on the concept of Green function, but the existence of the latter was only supported by heuristic physical intuition. Carl Friedrich Gauss employed in [69, p. 222] a variational argument, but the functional being minimized was assumed to attain its infimum without a proper justification (in fact, the existence of a minimizer may fail without additional assumptions). Difficulties with early versions of the Dirichlet principle were famously pointed out by Weierstrass who, in 1870, constructed an example showing its unreliability in the format used at that time.

that Ω is an Ahlfors regular domain, with "surface measure" $\sigma := \mathcal{H}^{n-1}\lfloor\partial\Omega$, whose geometric measure theoretic outward unit normal ν has the property that its distance to the Sarason space VMO, i.e.,

$$\operatorname{dist}\Big(\nu, [\mathrm{VMO}(\partial\Omega, \sigma)]^n\Big) := \inf_{\psi\in[\mathrm{VMO}(\partial\Omega,\sigma)]^n} \|\nu - \psi\|_{[\mathrm{BMO}(\partial\Omega,\sigma)]^n}, \tag{8.0.1}$$

is sufficiently small relative to the dimension n, the Ahlfors regularity constants of $\partial\Omega$, the relative flatness ratio $\wp(\Omega)/\operatorname{diam}\partial\Omega$ (defined in §3.4), the system L, and the smoothness/integrability parameters characterizing the space of boundary data. In a nutshell, we are asking that Ω is a δ-oscillating AR domain with δ suitably small, in the sense of Definition 3.4.1. This is an infinitesimal flatness requirement which, in particular, is satisfied by all bounded $\mathscr{C}^1$ domains, Lipschitz domains with small Lipschitz constant, BMO_1-domains with small constant, and chord-arc domains with small constant in the plane (cf. Example 3.4.5, Example 3.4.6, Example 3.4.8). Having said that, the boundary of a δ-oscillating AR domain Ω is allowed to develop certain spirals even when δ is as small as one pleases (relative to the other geometric characteristics of Ω), which prevent Ω from being locally of upper-graph type. See the example in Proposition 3.4.11 in this regard.

Our treatment here makes essential use of the Fredholm theory for boundary layer potentials in the class of infinitesimally flat domains which has been developed earlier in Chapter 7, the uniqueness results from [131, Chapter 4], and the Calderón-Zygmund theory for hydrostatic boundary layer potentials from [132, Chapter 6].

The Fredholmness/well-posedness results in §§8.1-8.2 are sharp, in the sense that they may fail to materialize for any arbitrary given integrability/smoothness exponents used to describe the space of boundary data if one no longer insists on imposing flatness assumptions on the underlying domain, as we do by working within the class of δ-oscillating AR domains with δ sufficiently small (relative to the environment). In the case of §8.1 this already follows from known counterexamples to the solvability of boundary value problems for the Laplacian in Lipschitz domains.

Section §8.3 is reserved for discussing basic boundary problems in scattering theory by rough obstacles: Exterior Dirichlet Problem and the Exterior Regularity Problem for the Helmholtz operator in Theorem 8.3.6, and the Exterior Neumann Problem for the Helmholtz operator in Theorem 8.3.7.

Complementing the work in §§8.1-8.2, we then consider boundary value problems for elliptic homogeneous constant (complex) coefficient second-order systems in domains with *unbounded boundaries* in §8.4 for "generic" elliptic systems, and in §8.5 for the Stokes system. Throughout, we shall work with Ahlfors regular domains $\Omega \subseteq \mathbb{R}^n$ whose geometric measure theoretic outward unit normals have small BMO semi-norms (a demand which necessarily implies that $\partial\Omega$ is unbounded). This time, we employ the invertibility results from Chapter 4 to prove solvability results in the aforementioned setting.

In all cases, our boundary layer potential based approach yields a large variety of Fredholmness and well-posedness results. Moreover, the same approach can handle a multitude of other function spaces (quantifying size and/or smoothness) as boundary data such as Lorentz spaces, Lorentz-based Hardy spaces, Lorentz-based Sobolev spaces, and off-diagonal Sobolev spaces. We also wish to mention that all the tools are in place for treating a variety of other basic boundary value problems in similar classes of domains, such as transmission boundary value problems, Robin boundary value problems, oblique derivative boundary value problems, boundary value problems for the Stokes system, and boundary value for Maxwell's equations.

In the final section in this chapter, §8.6, we discuss non-Fredholm boundary value problems for weakly elliptic systems, a topic in sharp contrast to the nature of the results established earlier in §§8.1-8.5. This lack of Fredholmness manifests itself at all levels, namely both the space of null-solutions is infinite dimensional and the space of admissible boundary data has infinite codimension in the full space of boundary data. The aforementioned failure of Fredholm solvability is not attributable to the underlying set having a wild geometry: indeed, in the class of domains with unbounded boundaries we restrict ourselves to the upper half-space, while in the class of domains with compact boundaries we consider the unit disk in the plane (and the complement of its closure). In view of the Fredholm solvability theory developed in §§8.1-8.5, this means that we necessarily have to restrict attention to systems lacking distinguished coefficient tensors. Specifically, we shall treat boundary value problems in the upper half-space for the system $L_D = \Delta - 2\nabla \mathrm{div}$ which we have introduced and studied in §2.5-§2.6, and interior/exterior boundary value problems for the Bitsadze operator $\partial_{\bar{z}}^2$ in the unit disk. Remarkably, in spite of the lack of Fredholm solvability, we are able to precisely describe both the space of null-solutions and the space of admissible boundary data, by providing intrinsic characterizations in each case.

8.1 Boundary Problems in Domains with Compact Boundaries for Elliptic Systems

In this section the goal is to treat boundary value problems for elliptic homogeneous constant (complex) coefficient second-order systems in domains with compact boundaries, of a general geometric nature. More specifically, we shall employ the method of boundary layer potentials to prove Fredholmness and well-posedness results for the following (interior/exterior) boundary value problems:

Dirichlet Problem with data in Lebesgue spaces (Theorem 8.1.6); Dirichlet Problem with data in Morrey and block spaces (Theorem 8.1.8); Dirichlet Problem with data in Generalized Banach Function Spaces (Theorem 8.1.9); Dirichlet Problem with data in BMO and VMO (Theorem 8.1.11); Dirichlet Problem with data in ordinary Hölder spaces (Theorem 8.1.18); Dirichlet Problem with data in vanishing Hölder spaces (Theorem 8.1.19); Dirichlet Problem with data in Besov spaces (Theorem 8.1.20); Poisson Problem in Besov/Triebel-Lizorkin/weighted Sobolev spaces (Theorem 8.1.21); Regularity Problem with data in Sobolev spaces (Theorem 8.1.1); Regularity Problem with data in Morrey and block styled Sobolev spaces (Theorem 8.1.4); Regularity Problem with data in GBFS[10] styled Sobolev

[10] Generalized Banach Function Spaces

spaces (Theorem 8.1.5); Regularity Problem with data in Hardy-Sobolev spaces (Theorem 8.1.10); Neumann Problem with data in Hardy spaces (Theorem 8.1.12); Neumann Problem with data in Lebesgue spaces (Theorem 8.1.15); Neumann Problem with data in Morrey and block spaces (Theorem 8.1.16); Neumann Problem with data in Generalized Banach Function Spaces (Theorem 8.1.17); Neumann Problem with data in Besov spaces (Theorem 8.1.22); Inhomogeneous Neumann Problem with boundary data in Besov spaces (Theorem 8.1.23).

We wish to stress that, in all cases, our principal results are valid in settings where no specific information about the topology of the underlying domain Ω is available, no symmetry conditions are imposed for the complex coefficient system L, and where no positive definiteness (or positive semi-definiteness) is assumed for any of the coefficient tensors in $\mathfrak{A}_L$. The new philosophy that emerges from the current work may be succinctly (yet remarkably accurately) stated as follows:

given a second-order, weakly elliptic, constant complex coefficient, homogeneous system L in $\mathbb{R}^n$ such that $\mathfrak{A}_L^{\rm dis} \neq \varnothing \neq \mathfrak{A}_{L^\top}^{\rm dis}$, along with an Ahlfors regular domain $\Omega \subseteq \mathbb{R}^n$ whose boundary is compact and sufficiently flat at small scales, all basic boundary value problems for L in Ω (such as Dirichlet, Regularity, and Neumann) are Fredholm solvable when formulated on a wide variety of functions spaces (including large classes of Generalized Banach Function Spaces on $\partial\Omega$), and actually become well posed if L satisfies a strong Legendre-Hadamard ellipticity condition; moreover, the aforementioned Fredholm solvability and well-posedness results may fail in the absence of distinguished coefficient tensors for L, even when Ω has a $\mathscr{C}^\infty$ boundary. (8.1.1)

Finally, in Theorem 8.1.24 we use techniques developed in the process of dealing with the Poisson Problem formulated in Besov/Sobolev spaces, as well as the machinery of boundary layer potentials developed earlier, to prove the following result:

given any $n, M \in \mathbb{N}$ with $n \geq 2$, it follows that any second-order, homogeneous, constant complex coefficient, weakly elliptic $M \times M$ system L in $\mathbb{R}^n$ with $\mathfrak{A}_L^{\rm dis} \neq \varnothing$ and $\mathfrak{A}_{L^\top}^{\rm dis} \neq \varnothing$ satisfies the Shapiro-Lopatinskiĭ condition (for the Dirichlet boundary condition, relative to any given smooth bounded domain in $\mathbb{R}^n$); cf. (8.1.920). (8.1.2)

The Shapiro-Lopatinskiĭ condition has been originally introduced in [109], [187], and further extended in [4], [5] (see, for example, [6, (6.5), p. 53], [98, Definition 4.2.2, p. 113], [207, Definition 9.28, p. 389]). This classical condition plays a key role in the study of boundary value problems for elliptic systems in smooth domains, formulated in suitable Sobolev spaces.

For logistical reasons, we prefer to start by proving the well-posedness of the Regularity Problem for weakly elliptic systems in δ-oscillating AR domains in $\mathbb{R}^n$ with boundary data in Sobolev spaces L_1^p with $1 < p < \infty$. There are size conditions imposed on the solution, formulated in terms of the L^p-integrability of the nontangential maximal operator, both for the function in question and its gradient. Such size conditions may be construed as having mixed character, as that they involve taking an L^∞-norm "vertically" (inside nontangential approach regions) and an L^p-norm "horizontally" (with respect to the vertex of said nontangential approach regions, along the boundary of the domain). Suitably nuanced, such a feature remains pertinent to a multitude of other boundary problems treated here.

Theorem 8.1.1, which is our main result regarding the aforementioned Regularity Problem, refines and extends work of S. Hofmann, M. Mitrea, and M. Taylor in [81] where the case of Laplacian and the Lamé system in regular SKT domains with compact boundaries has been considered. Another way to think of Theorem 8.1.1 is as the Geometric Measure Theoretic sharp version of [52, Theorem 2.4, p. 184] where E. Fabes, M. Jodeit, and N. Rivière have solved the Regularity Problem for the Laplacian in bounded $\mathscr{C}^1$ domains with boundary data in Sobolev spaces L^p_1 with $p \in (1, \infty)$ arbitrary. Theorem 8.1.1 sheds light on the issue singled out as Question 2.5 in [202]. A similar well-posedness result for the Regularity Problem for the Laplacian in bounded Lipschitz domains with boundary data in Sobolev spaces L^p_1 has been subsequently established by G. Verchota in [203, Theorem 5.1, p. 598] when $p \in (1, 2]$ and by B. Dahlberg and C. Kenig in [38, Theorem 4.14, p. 461] when $p \in (1, 2 + \varepsilon)$ for some small $\varepsilon > 0$ which depends on the Lipschitz character of the domain in question. The Regularity Problem for the Laplace-Beltrami operator on Lipschitz subdomains of Riemannian manifolds with boundary data in Sobolev spaces has been treated by M. Mitrea and M. Taylor in [153], [154], [156]. At the level of genuine systems, in [40, Theorem 3.7, p. 814] the Regularity Problem for the Lamé system in bounded Lipschitz domains has been solved with boundary in data in L^2_1. Analogous well-posedness results for the Regularity Problem for generic Legendre-Hadamard elliptic systems with real coefficients in bounded Lipschitz domains may be found in [65, Theorem 4.2, p. 398]. See also [81], and [82] in this regard. For additional references the reader may wish to consult [93].

Lastly, we wish to remark that the two-dimensional Regularity Problem for the Laplacian in a bounded chord-arc domain Ω in the plane with boundary data in $L^p_1(\partial\Omega, \sigma)$ with $p > 1$ sufficiently small so that the arc-length measure $\sigma := \mathcal{H}^1 \lfloor \partial\Omega$ belongs to the Muckenhoupt weight class $A_{p'}$ with respect to the harmonic measure on $\partial\Omega$ (where $1/p + 1/p' = 1$) is mentioned by D. Jerison and C. Kenig in [85, p. 245]. In contrast to their result, our Theorem 8.1.1 particularized to the case when $L = \Delta$ is the two-dimensional Laplacian and when $\Omega \subseteq \mathbb{R}^2 \equiv \mathbb{C}$ is a chord-arc domain shows (in view of Theorem 3.3.1 and [130, Proposition 11.5.19]) that any exponent $p \in (1, \infty)$ may be allowed provided the local chord-arc constant of Ω becomes sufficiently small at small scales (i.e., if $\varkappa$ from (3.3.3) is small enough).

Theorem 8.1.1 (The Regularity Problem with Data in Sobolev Spaces) *Fix two integers $n, M \in \mathbb{N}$ with $n \geq 2$ and suppose $\Omega \subseteq \mathbb{R}^n$ is an Ahlfors regular domain with compact boundary. Abbreviate $\sigma := \mathcal{H}^{n-1} \lfloor \partial\Omega$ and denote by ν the geometric measure theoretic outward unit normal to Ω. Also, let L be a homogeneous, second-order, constant complex coefficient, $M \times M$ system in $\mathbb{R}^n$. Finally, fix some integrability exponent $p \in (1, \infty)$ along with arbitrary aperture parameter $\kappa > 0$,* *and consider the Regularity Problem*[11]

$$\begin{cases} u \in \big[\mathscr{C}^\infty(\Omega)\big]^M, \quad Lu = 0 \text{ in } \Omega, \\ \mathcal{N}_\kappa u \in L^p(\partial\Omega, \sigma), \ \mathcal{N}_\kappa(\nabla u) \in L^p(\partial\Omega, \sigma), \\ u\big|_{\partial\Omega}^{\kappa-\text{n.t.}} = f \in \big[L^p_1(\partial\Omega, \sigma)\big]^M, \\ u(x) = O(|x|^{2-n}) \text{ as } |x| \to \infty \text{ if } \Omega \text{ is unbounded.} \end{cases} \tag{8.1.3}$$

Then there exists $\delta_0 \in (0,1)$ *depending only on n, p, L, the* ADR *constants of* $\partial\Omega$, *and the relative flatness ratio* $\wp(\Omega)/\text{diam}\,\partial\Omega$ *such that if*

$$\text{dist}\Big(\nu, [\text{VMO}(\partial\Omega, \sigma)]^n\Big) < \delta_0 \tag{8.1.4}$$

where the distance is considered in the space $\big[\text{BMO}(\partial\Omega, \sigma)\big]^n$ *(hence if* Ω *is a* δ_0*-oscillating* AR *domain*[12]*; cf. Definition 3.4.1), the following results are true:*

(1) If one assumes that

$$L \text{ is weakly elliptic with } \mathfrak{A}_L^{\text{dis}} \neq \varnothing \text{ and } \mathfrak{A}_{L^\top}^{\text{dis}} \neq \varnothing, \tag{8.1.5}$$

then the Regularity Problem (8.1.3) *is Fredholm solvable, in the following precise sense: First, the space of admissible boundary data for the problem* (8.1.3)*, i.e.,*

$$\Big\{u\big|_{\partial\Omega}^{\kappa-\text{n.t.}} : u \in \big[\mathscr{C}^\infty(\Omega)\big]^M, \ Lu = 0 \text{ in } \Omega, \ \mathcal{N}_\kappa u, \ \mathcal{N}_\kappa(\nabla u) \in L^p(\partial\Omega, \sigma), \\ \text{and } u(x) = O(|x|^{2-n}) \text{ as } |x| \to \infty \text{ if } \Omega \text{ is unbounded}\Big\}, \tag{8.1.6}$$

is a closed linear subspace of finite codimension in $\big[L^p_1(\partial\Omega, \sigma)\big]^M$*. Second, a solution u of* (8.1.3) *exists if and only if the boundary datum f belongs to* (8.1.6)*, in which scenario a solution u may be found such that*

$$\big\|\mathcal{N}_\kappa u\big\|_{L^p(\partial\Omega,\sigma)} + \big\|\mathcal{N}_\kappa(\nabla u)\big\|_{L^p(\partial\Omega,\sigma)} \approx \|f\|_{[L^p_1(\partial\Omega,\sigma)]^M}. \tag{8.1.7}$$

Third, the space of null-solutions of the boundary problem (8.1.3)*, i.e.,*

$$\Big\{u \in \big[\mathscr{C}^\infty(\Omega)\big]^M : Lu = 0, \ \mathcal{N}_\kappa u, \ \mathcal{N}_\kappa(\nabla u) \in L^p(\partial\Omega, \sigma), \ u\big|_{\partial\Omega}^{\kappa-\text{n.t.}} = 0, \\ \text{and } u(x) = O(|x|^{2-n}) \text{ as } |x| \to \infty \text{ if } \Omega \text{ is unbounded}\Big\}, \tag{8.1.8}$$

is finite dimensional. Moreover, if with $p' := (1 - 1/p)^{-1}$ *denoting the Hölder conjugate exponent of p and having fixed a coefficient tensor* $A \in \mathfrak{A}_L^{\text{dis}}$*, for any given* $f \in \big[L^p_1(\partial\Omega, \sigma)\big]^M$ *the Regularity Problem* (8.1.3) *with boundary datum f is solvable in the case when*

---

[11] in view of work in [129, §8.5], one may replace the nontangential maximal operator by the tangential maximal operator, associated as in [129, Definition 8.5.1] with a sufficiently large power

[12] in particular, if Ω is an infinitesimally flat AR domain

$$\text{either } n \geq 3, \text{ or } n = 2 \text{ and } \Omega \text{ is bounded,} \tag{8.1.9}$$

if and only if

$$\begin{aligned} {}_{L_1^p(\partial\Omega,\sigma)}\big\langle f, \phi \big\rangle_{L_{-1}^{p'}(\partial\Omega,\sigma)} = 0 \text{ for each } \phi \in \big[L_{-1}^{p'}(\partial\Omega, \sigma)\big]^M \\ \text{with } \big(\tfrac{1}{2}I + K_A^{\#}\big)\phi = 0 \text{ and } S_{L^\top}\phi = 0, \end{aligned} \tag{8.1.10}$$

where $K_A^{\#}$ *is the transpose double layer potential operator associated with* Ω *and* A*, acting on negative Sobolev spaces as in* [132, Theorem 1.5.1, item *(vi)*], *and where* $S_{L^\top}$ *is the boundary-to-boundary single layer potential operator associated with* Ω *and* $L^\top$*, acting from negative Sobolev spaces as in* [132, Theorem 1.5.1, item *(ix)*]. *In fact, with the aforementioned operators now considered on Lebesgue spaces,* (8.1.10) *is further equivalent to*

$$\begin{aligned} \int_{\partial\Omega} \langle f, \phi \rangle \, d\sigma = 0 \text{ for each } \phi \in \big[L^{p'}(\partial\Omega, \sigma)\big]^M \\ \text{with } \big(\tfrac{1}{2}I + K_A^{\#}\big)\phi = 0 \text{ and } S_{L^\top}\phi = 0. \end{aligned} \tag{8.1.11}$$

Finally, in the case complementing (8.1.9)*, i.e., when* $n = 2$ *and* Ω *is unbounded (hence,* Ω *is a two-dimensional exterior domain), for any given function* $f \in \big[L_1^p(\partial\Omega, \sigma)\big]^M$ *the Regularity Problem* (8.1.3) *with boundary datum* f *is solvable if and only if*

$$\begin{aligned} {}_{L_1^p(\partial\Omega,\sigma)}\big\langle f, \phi \big\rangle_{L_{-1}^{p'}(\partial\Omega,\sigma)} = 0 \text{ for each } \phi \in \big[L_{-1}^{p'}(\partial\Omega, \sigma)\big]^M \\ \text{with } \big(\tfrac{1}{2}I + K_A^{\#}\big)\phi = 0 \text{ and } \big(\widetilde{S}_{L^\top}\big)_{\rm aug}(\phi, c) = (0, 0) \\ \text{for some } c \in \Big(\int_{S^1} \big[L^\top(\xi)\big]^{-1} \, d\mathcal{H}^1(\xi)\Big)\mathbb{C}^M, \end{aligned} \tag{8.1.12}$$

where $\big(\widetilde{S}_{L^\top}\big)_{\rm aug}$ *is the augmented single layer potential operator associated with* $L^\top$ *and* Ω *as in* (7.1.246)*, a condition further equivalent to*

$$\begin{aligned} \int_{\partial\Omega} \langle f, \phi \rangle \, d\sigma = 0 \text{ for each } \phi \in \big[L^{p'}(\partial\Omega, \sigma)\big]^M \\ \text{with } \big(\tfrac{1}{2}I + K_A^{\#}\big)\phi = 0 \text{ and } \big(S_{L^\top}\big)_{\rm aug}(\phi, c) = (0, 0) \\ \text{for some } c \in \Big(\int_{S^1} \big[L^\top(\xi)\big]^{-1} \, d\mathcal{H}^1(\xi)\Big)\mathbb{C}^M, \end{aligned} \tag{8.1.13}$$

where $\big(S_{L^\top}\big)_{\rm aug}$ *is the augmented single layer associated with* $L^\top$ *as in* (7.1.245).

(2) Retain the hypotheses imposed in (8.1.5) *on the system* L*. Then if* Ω *is bounded it follows that any solution* u *of the boundary value problem* (8.1.3) *satisfies*

$$\begin{aligned} u \in \big[W^{2,p}_{1-\frac{1}{p},\star}(\Omega)\big]^M \cap \Big[B^{p,q_0}_{1+\frac{1}{p}}(\Omega)\Big]^M \cap \Big[F^{p,q_1}_{1+\frac{1}{p}}(\Omega)\Big]^M \\ \text{whenever } p \leq q_0 \leq \infty \text{ and } \tfrac{n}{n+1+1/p} < q_1 \leq \infty, \end{aligned} \tag{8.1.14}$$

as well as

$$u \in \left[H^{3/2}(\Omega)\right]^M \quad \text{if } p = 2. \tag{8.1.15}$$

(3) *If* (8.1.5) *is strengthened to*[13]

$$\text{L satisfies the Legendre-Hadamard strong ellipticity condition (cf. [131, Definition 1.3.2]) and one has } \mathfrak{A}_L^{\mathrm{dis}} \neq \varnothing, \tag{8.1.16}$$

the Regularity Problem (8.1.3) *is actually well posed. More specifically, the boundary value problem* (8.1.3) *has a unique solution, and this solution enjoys the additional property that* (8.1.7) *holds.*

(4) *Similar results to those described in items (1) and (3) above also hold for the version of the Regularity Problem formulated for boundary data in Muckenhoupt weighted Sobolev spaces, namely*[14]

$$\begin{cases} u \in \left[\mathscr{C}^\infty(\Omega)\right]^M, \quad Lu = 0 \text{ in } \Omega, \\ \mathcal{N}_\kappa u \in L^p(\partial\Omega, w), \ \mathcal{N}_\kappa(\nabla u) \in L^p(\partial\Omega, w), \\ u\big|_{\partial\Omega}^{\kappa-\mathrm{n.t.}} = f \in \left[L_1^p(\partial\Omega, w)\right]^M, \\ u(x) = O(|x|^{2-n}) \text{ as } |x| \to \infty \text{ if } \Omega \text{ is unbounded,} \end{cases} \tag{8.1.17}$$

for each $w \in A_p(\partial\Omega, \sigma)$ *(now allowing* δ_0 *in* (8.1.4) *to also depend on* $[w]_{A_p}$*).*

(5) *There exist symmetric weakly elliptic systems* L *with* $\mathfrak{A}_L^{\mathrm{dis}} = \varnothing$ *(compare to* (8.1.5)*) for which the Regularity Problems* (8.1.3), (8.1.17) *fail to be Fredholm solvable even when* Ω *is the unit disk in the plane (or the complement of its closure). In fact, the aforementioned failure of Fredholm solvability may be total, in the sense that for each of these problems, both the space of admissible boundary data may have infinite codimension in the full space of boundary data, and the space of null-solutions may be infinite dimensional.*

It is of significance to remark that, as seen from Proposition 1.6.12 and [131, Proposition 1.4.39],

$$\text{in the scalar case, i.e., if } M = 1, \text{ in dimensions } n \geq 3, \text{ both (8.1.5) and (8.1.16) may be replaced by the mere demand that } L \text{ is weakly elliptic.} \tag{8.1.18}$$

Moreover, in the scalar case in the two-dimensional setting (i.e., for $M = 1$ and $n = 2$), if L satisfies the Legendre-Hadamard strong ellipticity condition or, more generally (cf. [131, Lemma 1.4.24]), if $L = \mathrm{div} A \nabla$ for some 2×2 complex matrix A with $\mathrm{sym}\, A := \frac{1}{2}(A + A^\top) \in M_2^0$ (cf. [131, (1.4.187)]), then Proposition 1.6.11 and Proposition 1.6.13 guarantee that $\mathfrak{A}_L^{\mathrm{dis}} \neq \varnothing$ and $\mathfrak{A}_{L^\top}^{\mathrm{dis}} \neq \varnothing$. Also, according to

---

[13] Proposition 1.4.1 ensures that if L is as in (8.1.16) then $\mathfrak{A}_L^{\mathrm{dis}} \neq \varnothing$ and $\mathfrak{A}_{L^\top}^{\mathrm{dis}} \neq \varnothing$

[14] once again, thanks to work in [129, §8.5] one may replace the nontangential maximal operator by the tangential maximal operator, associated as in [129, Definition 8.5.1] with a sufficiently large power (cf. (A.0.165))

(1.6.141) in Proposition 1.6.13, in the scalar case in the two-dimensional setting (i.e., for $M = 1$ and $n = 2$) we may omit (8.1.16) altogether if L has real coefficients.

The situation for genuine systems is more nuanced. To elaborate on this, suppose

$$\lambda, \mu \in \mathbb{C} \text{ are such that } \mu \neq 0 \text{ and } \frac{\lambda}{\mu} \in (-\infty, -3) \cup (-3, -2). \tag{8.1.19}$$

As observed in (1.6.86) and (1.6.116), for this choice of the Lamé moduli λ, μ, the corresponding Lamé system $L_{\lambda,\mu} = \mu\Delta + (\lambda + \mu)\nabla\text{div}$ in $\mathbb{R}^n$, with $n \geq 2$, is weakly elliptic and $\mathfrak{A}^{\text{dis}}_{L_{\lambda,\mu}} \neq \varnothing$. Next, we claim that we can choose

$$a_1, \ldots, a_n \in (0, \infty) \text{ satisfying } \mu \sum_{j=1}^{n} a_j + (\lambda + \mu)a_1 = 0. \tag{8.1.20}$$

Indeed, since $\mu \neq 0$, the identity in (8.1.20) is equivalent with

$$\frac{\lambda}{\mu} = -\frac{1}{a_1} \sum_{j=1}^{n} a_j - 1 = -2 - \sum_{j=2}^{n} \frac{a_j}{a_1}. \tag{8.1.21}$$

Since the last condition in (8.1.19) guarantees that $-2 - \frac{\lambda}{\mu} \in (0, \infty)$, the demand in (8.1.21) can always be accommodated by choosing $a_1, \ldots, a_n \in (0, \infty)$ such that

$$\sum_{j=2}^{n} \frac{a_j}{a_1} = -2 - \frac{\lambda}{\mu}. \tag{8.1.22}$$

Having selected a family of numbers $a_1, \ldots, a_n$ as in (8.1.20), pick an arbitrary $b \in (0, \infty)$ and define the vector field $\vec{u} = (u_1, 0, \ldots, 0)$ where

$$u_1(x) := b - \sum_{j=1}^{n} a_j x_j^2 \text{ for all } x = (x_1, \ldots, x_n) \in \mathbb{R}^n. \tag{8.1.23}$$

Hence, $\vec{u} \in \big[\mathscr{C}^\infty(\mathbb{R}^n)\big]^n$ and $L_{\lambda,\mu}\vec{u} = \mu\Delta\vec{u} + (\lambda + \mu)\nabla\text{div}\vec{u} = 0$ in $\mathbb{R}^n$, thanks to (8.1.20). In addition, define $\Omega := \big\{x = (x_1, \ldots, x_n) \in \mathbb{R}^n : \sum_{j=1}^{n} a_j x_j^2 < b\big\}$, which is a smooth open set in $\mathbb{R}^n$, whose boundary $\partial\Omega = \big\{x \in \mathbb{R}^n : \sum_{j=1}^{n} a_j x_j^2 = b\big\}$ is an ellipsoid. By design, the vector field $\vec{u}$ vanishes on $\partial\Omega$.

To summarize, if λ, μ are as in (8.1.19), we have constructed a bounded $\mathscr{C}^\infty$ smooth domain Ω in $\mathbb{R}^n$ and a vector field $\vec{u}$ satisfying

$$\vec{u} \in \big[\mathscr{C}^\infty(\overline{\Omega})\big]^n, \quad L_{\lambda,\mu}\vec{u} = 0 \text{ in } \Omega, \quad \vec{u}\big|_{\partial\Omega} = 0 \text{ on } \partial\Omega, \tag{8.1.24}$$

and such that $\vec{u}$ is never zero in Ω. This shows that

$$\text{if } \lambda, \mu \text{ are as in (8.1.19), then the Lamé system } L_{\lambda,\mu} = \mu\Delta + (\lambda+\mu)\nabla\text{div} \text{ in } \mathbb{R}^n, \text{ with } n \geq 2, \text{ is weakly elliptic, has } \mathfrak{A}^{\text{dis}}_{L_{\lambda,\mu}} \neq \varnothing, \text{ and there exists a nontrivial null-solution of the Regularity Problem (8.1.3) formulated for } L := L_{\lambda,\mu} \text{ in some bounded } \mathscr{C}^\infty \text{ smooth domain in } \mathbb{R}^n. \tag{8.1.25}$$

In particular, even though item *(1)* of Theorem 8.1.1 guarantees that the Regularity Problem (8.1.3) is Fredholm solvable, said problem fails to be well posed in this case. As seen from item *(3)* of Theorem 8.1.1, the source of the aforementioned failure is the lack of strong ellipticity, in the sense of Legendre-Hadamard, for $L_{\lambda,\mu}$.

In this vein, it is interesting to check directly that there are no Lamé moduli λ, μ as in (8.1.19) which make the Lamé system $L_{\lambda,\mu}$ strongly elliptic, in the sense of Legendre-Hadamard. Recall from [131, (1.3.11)] that the latter property is equivalent to having $\text{Re}\,\mu > 0$ and $\text{Re}\,(2\mu + \lambda) > 0$. Then, since $\mu \neq 0$ and $2 + \lambda/\mu$ is real and negative, $\text{Re}(2\mu+\lambda) = \text{Re}\big(\mu(2+\lambda/\mu)\big) = \big(2+\lambda/\mu\big)\cdot\text{Re}\,\mu$ which shows that $\text{Re}\,\mu$ and $\text{Re}\,(2\mu + \lambda)$ have opposite signs, so they cannot be simultaneously strictly positive.

In a broader perspective, under the geometric assumptions stipulated in Theorem 8.1.1, the following statements hold: First,

$$\text{in the class of weakly elliptic systems } L \text{ with } \mathfrak{A}^{\text{dis}}_{L} \neq \varnothing \text{ and } \mathfrak{A}^{\text{dis}}_{L^\top} \neq \varnothing \text{ the Regularity Problem (8.1.3) is always Fredholm solvable (cf. item } (1) \text{ in Theorem 8.1.1), though said problem may not necessarily be well posed (cf. (8.1.25)).} \tag{8.1.26}$$

Second,

$$\text{within the class of weakly elliptic systems } L \text{ with either } \mathfrak{A}^{\text{dis}}_{L} = \varnothing \text{ or } \mathfrak{A}^{\text{dis}}_{L^\top} = \varnothing \text{ the Regularity Problem (8.1.3) may fail to be Fredholm solvable (cf. [132, §§8.1-8.2]).} \tag{8.1.27}$$

Last,

$$\text{in the class of systems } L \text{ satisfying the Legendre-Hadamard strong ellipticity condition and with } \mathfrak{A}^{\text{dis}}_{L} \neq \varnothing, \text{ the Regularity Problem (8.1.3) is always well posed (cf. item } (3) \text{ in Theorem 8.1.1).} \tag{8.1.28}$$

Before presenting its proof, we also wish to make a few comments on the very nature of Theorem 8.1.1. First, by choosing $\delta_0 \in (0, 1)$ sufficiently small, Theorem 3.4.2 ensures that Ω is actually a two-sided NTA domain, hence also a UR domain (cf. [129, (5.10.33)]). Granted this, the Fatou-type result established in [131, (3.3.62) in Theorem 3.3.4] ensures (while also bearing [130, (11.5.60)] in mind) that any function u as in the first two lines of (8.1.3) has the property that

$$\begin{gathered} u\Big|_{\partial\Omega}^{\kappa-\text{n.t.}} \text{ exists (in } \mathbb{C}^M) \text{ at } \sigma\text{-a.e. point on } \partial\Omega \\ \text{and } u\Big|_{\partial\Omega}^{\kappa-\text{n.t.}} \text{ belongs to the space } \big[L^p_1(\partial\Omega, \sigma)\big]^M, \end{gathered} \tag{8.1.29}$$

and also

$$\begin{aligned}&(\nabla u)\Big|_{\partial\Omega}^{\kappa-\mathrm{n.t.}} \text{ exists (in } \mathbb{C}^{n\cdot M}) \text{ at } \sigma\text{-a.e. point on } \partial\Omega \text{ and}\\ &(\nabla u)\Big|_{\partial\Omega}^{\kappa-\mathrm{n.t.}} \text{ belongs to the space } \big[L^p(\partial\Omega,\sigma)\big]^{n\cdot M}, \text{ hence}\\ &\quad \partial_\nu^A u \text{ is well defined and belongs to } \big[L^p(\partial\Omega,\sigma)\big]^M.\end{aligned} \tag{8.1.30}$$

See also [130, Corollary 11.3.5] in this regard[15]. The membership in (8.1.29) points to the fact that prescribing boundary data in the Sobolev space $\big[L_1^p(\partial\Omega,\sigma)\big]^M$ is natural in the context of the Regularity Problem (8.1.3).

Second, in the case in which Ω is an exterior domain, [131, Theorem 1.5.9] guarantees that the decay condition imposed in (8.1.3) is further equivalent with any of the following conditions if $n \geq 3$:

$$⨍_{B(0,2R)\setminus B(0,R)} |u|\,\mathrm{d}\mathcal{L}^n = o(1) \text{ as } R \to \infty, \tag{8.1.31}$$

$$u(x) = o(1) \text{ as } |x| \to \infty, \tag{8.1.32}$$

$$⨍_{\partial B(0,R)} |u|\,\mathrm{d}\mathcal{H}^{n-1} = o(1) \text{ as } R \to \infty. \tag{8.1.33}$$

Moreover, if $n = 2$ and

$$\text{the } M \times M \text{ matrix } \int_{S^1} \big[L(\xi)\big]^{-1}\,\mathrm{d}\mathcal{H}^1(\xi) \text{ is invertible} \tag{8.1.34}$$

(which is always the case if the system L is assumed to actually satisfy the Legendre-Hadamard strong ellipticity condition, and if $M = 1$ this is true if and only if $L = \nabla \cdot A\nabla$ for some $A \in M_2^0$; cf. [131, Lemma 1.4.19] and [131, (1.4.186)]) then [131, Theorem 1.5.9] also tells us that the decay condition imposed in (8.1.3) is equivalent with any of the following conditions:

$$u(x) = o(\ln|x|) \text{ as } |x| \to \infty; \tag{8.1.35}$$

$$\text{the limit } \lim_{x\to\infty} u(x) \text{ exists (in } \mathbb{C}^M); \tag{8.1.36}$$

$$u(x) = c + O(|x|^{-1}) \text{ as } |x| \to \infty \text{ for some } c \in \mathbb{C}^M. \tag{8.1.37}$$

Third, as is apparent from [129, Proposition 8.9.8] and [129, Corollary 8.9.9], the actual value of the aperture parameter κ employed in the formulation of the boundary value problems considered in Theorem 8.1.1 is insignificant (in the sense that said problems enjoy the same solvability properties discussed in Theorem 8.1.1 irrespective of the choice of κ). In fact, similar considerations apply to all boundary value problems treated in this work whose formulations involve a size condition formulated in terms of a nontangential maximal operator, and for which the boundary condition employs a nontangential trace.

We also wish to note that the optimal range of integrability exponents p for which the Regularity Problem for the Laplacian in a bounded Lipschitz domain $\Omega \subseteq \mathbb{R}^n$

[15] this also implies that the first membership in the second line of (8.1.3) is implied by the second

is $(1, 2 + \varepsilon_\Omega)$; see [38]. Here, $\varepsilon_\Omega > 0$ is a parameter which depends only on the set Ω, and $\varepsilon_\Omega \to 0^+$ as the Lipschitz character of Ω deteriorates. This points to the fact that well-posedness results in Theorem 8.1.1 may fail to materialize for a generic exponent $p \in (1, \infty)$ if one no longer insists on imposing flatness assumptions on the underlying domain, as we do (8.1.4). Finally, we make two remarks.

Remark 8.1.2 *A more detailed formulation of the well-posedness result in item (4) of Theorem 8.1.1 reads as follows: Fix $n, M \in \mathbb{N}$ with $n \geq 2$ and consider a homogeneous constant complex coefficient second-order $M \times M$ system L in $\mathbb{R}^n$ satisfying the Legendre-Hadamard strong ellipticity condition and with the property that $\mathfrak{A}_L^{\rm dis} \neq \varnothing$. Then for each constants $C_A, C_R, C_W \in (0, \infty)$ and each compact interval $I \subset (1, \infty)$, there exists a threshold $\delta \in (0, 1)$ which depends only on n, C_A, C_R, C_W, L, and I with the following significance. Assume $\Omega \subseteq \mathbb{R}^n$ is an Ahlfors regular domain with compact boundary such that the Ahlfors regularity constants of $\partial\Omega$ are controlled by C_A, and the relative flatness ratio $\wp(\Omega)/{\rm diam}\,\partial\Omega$ is $\leq C_R$. Abbreviate $\sigma := \mathcal{H}^{n-1}\lfloor\partial\Omega$ and denote by ν the geometric measure theoretic outward unit normal to Ω. Also, fix an integrability exponent $p \in I$ and a Muckenhoupt weight $w \in A_p(\partial\Omega, \sigma)$ with $[w]_{A_p} \leq C_W$. Then whenever* ${\rm dist}\Big(\nu, [{\rm VMO}(\partial\Omega, \sigma)]^n\Big) < \delta$ *it follows that the Regularity Problem with data in Muckenhoupt weighted Sobolev spaces stated in* (8.1.17) *is well posed (in the sense that it possesses a unique solution, which also satisfies a naturally accompanying estimate). In fact, similar considerations apply to all results formulated in Theorem 8.1.1.*

To formulate the second remark promised earlier, we first introduce an equivalence relation on the set of all homogeneous, second-order, constant complex coefficient, $M \times M$ systems in $\mathbb{R}^n$. Specifically, we shall say that two such systems L_1, L_2 are equivalent, and write $L_1 \approx L_2$, if L_2 may be obtained from L_1 after finitely many transformations of the following three types:

(i) multiply a given system L to the left by a non-singular matrix $P \in \mathbb{C}^{M\times M}$, i.e., change L into PL;
(ii) multiply a given system L to the right by a non-singular matrix $Q \in \mathbb{C}^{M\times M}$, i.e., change L into LQ;
(iii) perform a linear change of variables in the Euclidean space via $W : \mathbb{R}^n \to \mathbb{R}^n$ for some non-singular matrix $W \in \mathbb{R}^{n\times n}$, i.e., change $L = \big(a_{rs}^{\alpha\beta}\partial_r\partial_s\big)_{1\leq\alpha,\beta\leq M}$ into $L \circ W$, formally defined as $\Big(a_{jk}^{\alpha\beta}(W\nabla)_j(W\nabla)_k\Big)_{1\leq\alpha,\beta\leq M}$.

The second remark expands the scope of the well-posedness result in item *(4)* of Theorem 8.1.1 beyond the class of Legendre-Hadamard strongly elliptic systems:

Remark 8.1.3 *Retaining the same background assumptions as in Theorem 8.1.1 (up to, and including,* (8.1.4)*), we actually have well-posedness for the Regularity Problem* (8.1.3) *whenever the system L has the property that $L \approx \widetilde{L}$, where $\widetilde{L}$ is a homogeneous, second-order, constant complex coefficient, $M \times M$ systems in $\mathbb{R}^n$ satisfying the Legendre-Hadamard strong ellipticity condition and with $\mathfrak{A}_{\widetilde{L}}^{\rm dis} \neq \varnothing$.*

Indeed, this is a consequence of the well-posedness result in item (4) of Theorem 8.1.1, [131, Proposition 1.4.3], *Proposition 1.2.3, Proposition 1.2.4,* [130, Propositions 11.5.20, 11.6.1] (*see also the proof of* [130, Proposition 11.6.3]), *and the transformational properties of Euclidean domains established in* [80].

Here is the proof of Theorem 8.1.1:

Proof of Theorem 8.1.1 As explained in the preamble to the current proof, by taking the threshold $\delta_0 \in (0,1)$ sufficiently small, depending on n, p, L, the ADR constants of $\partial\Omega$, and the relative flatness ratio $\wp(\Omega)/\operatorname{diam}\partial\Omega$, we may ensure that Ω is an (ε,δ)-domain as well as a two-sided NTA domain (hence also a UR domain; cf. [129, (5.10.33)]). By eventually further decreasing the value of $\delta_0 \in (0,1)$ we may also guarantee that the conclusions in Theorem 7.1.1 hold (for suitable choices of the parameters involved in its statement). Work under the assumptions made in (8.1.5) and pick $A \in \mathfrak{A}_L^{\rm dis}$. Then Theorem 1.5.5 guarantees that we also have $A^\top \in \mathfrak{A}_{L^\top}^{\rm dis}$. Bearing this in mind, item *(1)* in Theorem 7.1.1 used with $z := 1/2$ implies that

$$\begin{aligned} &\tfrac{1}{2}I + K_A : \big[L_1^p(\partial\Omega,\sigma)\big]^M \longrightarrow \big[L_1^p(\partial\Omega,\sigma)\big]^M \\ &\text{is a Fredholm operator with index zero,} \end{aligned} \tag{8.1.38}$$

$$\begin{aligned} &\tfrac{1}{2}I + K^{\#}_{A^\top} : \big[L^p(\partial\Omega,\sigma)\big]^M \longrightarrow \big[L^p(\partial\Omega,\sigma)\big]^M \\ &\text{is a Fredholm operator with index zero.} \end{aligned} \tag{8.1.39}$$

As such,

$$\begin{aligned} &\operatorname{Im}\Big(\tfrac{1}{2}I + K_A; \big[L_1^p(\partial\Omega,\sigma)\big]^M\Big) \text{ is a closed space} \\ &\text{finite codimension in } \big[L_1^p(\partial\Omega,\sigma)\big]^M, \end{aligned} \tag{8.1.40}$$

$$\dim\operatorname{Ker}\Big(\tfrac{1}{2}I + K^{\#}_{A^\top}; \big[L^p(\partial\Omega,\sigma)\big]^M\Big) < +\infty. \tag{8.1.41}$$

At this point we find it useful to focus on the case when either $n \geq 3$, or $n = 2$ and Ω is bounded (cf. (8.1.9)). Bring in the boundary-to-domain single layer $\mathscr{S}$ associated with L and Ω as in (A.0.228). Then for any function satisfying

$$\begin{aligned} &u \in \big[\mathscr{C}^\infty(\Omega)\big]^M, \ Lu = 0 \text{ in } \Omega, \ \mathcal{N}_\kappa u, \ \mathcal{N}_\kappa(\nabla u) \in L^p(\partial\Omega,\sigma), \\ &\text{and also } u(x) = O(|x|^{2-n}) \text{ as } |x| \to \infty \text{ if } \Omega \text{ is unbounded,} \end{aligned} \tag{8.1.42}$$

the trace results in (8.1.29)-(8.1.30) hold, and [132, Theorem 1.3.3] gives

$$u = \mathcal{D}_A\Big(u\big|_{\partial\Omega}^{\kappa-\text{n.t.}}\Big) - \mathscr{S}(\partial_\nu^A u) \text{ in } \Omega. \tag{8.1.43}$$

Going nontangentially to the boundary (and using [132, (1.5.20), (1.5.44)]) gives

$$u\big|_{\partial\Omega}^{\kappa-\text{n.t.}} = \big(\tfrac{1}{2}I + K_A\big)\Big(u\big|_{\partial\Omega}^{\kappa-\text{n.t.}}\Big) - S(\partial_\nu^A u) \text{ on } \partial\Omega. \tag{8.1.44}$$

This shows that the space of admissible boundary data for the boundary problem (8.1.3) described in (8.1.6) is contained in

$$\mathrm{Im}\Big(\tfrac{1}{2}I + K_A; \big[L^p_1(\partial\Omega,\sigma)\big]^M\Big) + \mathrm{Im}\Big(S : \big[L^p(\partial\Omega,\sigma)\big]^M \to \big[L^p_1(\partial\Omega,\sigma)\big]^M\Big).$$

In the converse direction, given any function f belonging to the above space, say $f = \big(\tfrac{1}{2}I + K_A\big) f_0 + S f_1$ where $f_0 \in \big[L^p_1(\partial\Omega,\sigma)\big]^M$ and $f_1 \in \big[L^p(\partial\Omega,\sigma)\big]^M$, it follows that the function

$$u := \mathcal{D}_A f_0 + \mathcal{S} f_1 \text{ in } \Omega \tag{8.1.45}$$

satisfies all properties in (8.1.42) as well as $u\big|^{\kappa-\mathrm{n.t.}}_{\partial\Omega} = \big(\tfrac{1}{2}I + K_A\big) f_0 + S f_1 = f$. Ultimately, this goes to show that

$$\begin{gathered}\text{the space of admissible boundary data described in (8.1.6) is equal to}\\ \mathrm{Im}\Big(\tfrac{1}{2}I + K_A; \big[L^p_1(\partial\Omega,\sigma)\big]^M\Big) + \mathrm{Im}\Big(S : \big[L^p(\partial\Omega,\sigma)\big]^M \to \big[L^p_1(\partial\Omega,\sigma)\big]^M\Big).\end{gathered} \tag{8.1.46}$$

From this, (8.1.40), and item *(3)* in [130, Theorem 1.1.3] we then conclude that

$$\begin{gathered}\text{the space of admissible boundary data for the boundary problem}\\ \text{(8.1.3), described in (8.1.6), is a closed linear subspace of finite}\\ \text{codimension in the boundary Sobolev space } \big[L^p_1(\partial\Omega,\sigma)\big]^M.\end{gathered} \tag{8.1.47}$$

Let us formulate (8.1.46) more precisely. Define

$$\begin{aligned} T : \big[L^p_1(\partial\Omega,\sigma)\big]^M \times \big[L^p(\partial\Omega,\sigma)\big]^M &\longrightarrow \big[L^p_1(\partial\Omega,\sigma)\big]^M,\\ T(f_0, f_1) := \big(\tfrac{1}{2}I + K_A\big) f_0 + S f_1 &\text{ for all}\\ (f_0, f_1) \in \big[L^p_1(\partial\Omega,\sigma)\big]^M \times \big[L^p(\partial\Omega,\sigma)\big]^M&. \end{aligned} \tag{8.1.48}$$

Then T is a well-defined, linear, and bounded operator, with a closed range of finite codimension and (8.1.46) shows that

$$\begin{gathered}\text{the image of } T \text{ is precisely the space of admissible boundary}\\ \text{data for the boundary problem formulated in (8.1.3).}\end{gathered} \tag{8.1.49}$$

Granted these properties, [130, Proposition 1.2.13] applies and guarantees the existence of some constant $C \in (0,\infty)$ with the property that, given any f in the space of admissible boundary data for the boundary problem (8.1.3), we may find

$$\begin{gathered} f_0 \in \big[L^p_1(\partial\Omega,\sigma)\big]^M, \quad f_1 \in \big[L^p(\partial\Omega,\sigma)\big]^M \text{ with } T(f_0, f_1) = f\\ \text{and } \|f_0\|_{[L^p_1(\partial\Omega,\sigma)]^M} + \|f_1\|_{[L^p(\partial\Omega,\sigma)]^M} \le C\|f\|_{[L^p_1(\partial\Omega,\sigma)]^M}. \end{gathered} \tag{8.1.50}$$

Then u defined in (8.1.45) solves the Regularity Problem (8.1.3) for the given datum f and, thanks to (8.1.50), [132, (1.5.1)], [132, (1.5.47)], [132, (1.5.48)] (plus the very last statement in [132, Theorem 1.5.1, item *(x)*]), satisfies (8.1.7).

As may be seen from (8.1.48), [132, Theorem 1.5.1, items *(vi), (ix)*], and definitions, the adjoint of T in (8.1.48) is the operator

$$\begin{aligned} &T^* : \big[L^{p'}_{-1}(\partial\Omega,\sigma)\big]^M \longrightarrow \big[L^{p'}_{-1}(\partial\Omega,\sigma)\big]^M \times \big[L^{p'}(\partial\Omega,\sigma)\big]^M, \\ &T^*\phi = \Big(\big(\tfrac{1}{2}I + K^{\#}_{A}\big)\phi, S_{L^\top}\phi\Big) \text{ for each } \phi \in \big[L^{p'}_{-1}(\partial\Omega,\sigma)\big]^M. \end{aligned} \tag{8.1.51}$$

From (8.1.49), (8.1.47), (8.1.51), and [130, (2.1.49)] we then see that (8.1.10) holds. The fact that (8.1.10) is further equivalent to (8.1.11) follows from Proposition 7.1.2.

Next, let u be an arbitrary null-solution of the boundary problem (8.1.3) (i.e., u belongs to the space defined in (8.1.8)). Then the integral representation formula (8.1.43) reduces to

$$u = -\mathscr{S}(\partial^A_\nu u) \text{ in } \Omega. \tag{8.1.52}$$

Taking the conormal derivative of both sides then yields (on account of the jump-formula [132, (1.5.59)]) $\big(\tfrac{1}{2}I + K^{\#}_{A^\top}\big)(\partial^A_\nu u) = 0$, hence

$$\partial^A_\nu u \in \mathrm{Ker}\Big(\tfrac{1}{2}I + K^{\#}_{A^\top}; \big[L^p(\partial\Omega,\sigma)\big]^M\Big). \tag{8.1.53}$$

Note that, as seen from (8.1.52), the assignment taking any null-solution u of (8.1.3) into $\partial^A_\nu u$ is linear and injective. From this and (8.1.41) we then conclude that

$$\begin{aligned} &\text{the dimension of the space described in (8.1.8) is} \\ &\leq \dim \mathrm{Ker}\Big(\tfrac{1}{2}I + K^{\#}_{A^\top}; \big[L^p(\partial\Omega,\sigma)\big]^M\Big) < +\infty, \end{aligned} \tag{8.1.54}$$

hence the space of null-solutions of the boundary problem (8.1.3), described in (8.1.8), is indeed finite dimensional.

To complete the treatment of item *(1)*, we are left with the case when $n = 2$ and $\Omega \subseteq \mathbb{R}^2$ is an unbounded Ahlfors regular domain with compact boundary (in particular, Ω is an exterior domain). To get started, fix a function satisfying

$$\begin{aligned} u \in \big[\mathscr{C}^\infty(\Omega)\big]^M,\ Lu = 0 \text{ in } \Omega,\ \mathcal{N}_\kappa u,\ \mathcal{N}_\kappa(\nabla u) \in L^p(\partial\Omega,\sigma), \\ \text{and } u(x) = O(1) \text{ as } |x| \to \infty. \end{aligned} \tag{8.1.55}$$

Recall that the properties recorded in (8.1.29)-(8.1.30) are presently valid. Granted all these attributes, [131, Corollary 1.5.4] applies and guarantees the existence of a constant $c \in \mathbb{C}^M$ with the property that (cf. [131, (1.5.81)])

$$u = \mathcal{D}\Big(u\big|^{\kappa-\mathrm{n.t.}}_{\partial\Omega}\Big) - \mathscr{S}\big(\partial^A_\nu u\big) + c \text{ in } \Omega. \tag{8.1.56}$$

Together with [132, (1.5.20), (1.5.44)], (8.1.29), this implies that

$$u\big|^{\kappa-\mathrm{n.t.}}_{\partial\Omega} = \big(\tfrac{1}{2}I + K_A\big)\Big(u\big|^{\kappa-\mathrm{n.t.}}_{\partial\Omega}\Big) - S(\partial^A_\nu u) + c \text{ on } \partial\Omega. \tag{8.1.57}$$

Keeping in mind the last property in (8.1.55) and the fact that (as seen from [132, (1.3.26)] and (8.1.29)), $\mathcal{D}\Big(u\big|^{\kappa-\mathrm{n.t.}}_{\partial\Omega}\Big)(x) = O\big(|x|^{-1}\big)$ as $|x| \to \infty$, from (8.1.56) we also deduce that $\mathscr{S}(\partial_\nu^A u) = O(1)$ as $|x| \to \infty$. Combining this with (8.1.30) and [132, (1.3.11)] then leads to the conclusion that

$$\Big(\int_{S^1} \big[L(\xi)\big]^{-1}\, d\mathcal{H}^1(\xi)\Big)\Big(\int_{\partial\Omega} \partial_\nu^A u\, d\sigma\Big) = 0 \in \mathbb{C}^M. \tag{8.1.58}$$

Introduce the linear subspace of $\mathbb{C}^M$

$$W := \Big(\int_{S^1} \big[L(\xi)\big]^{-1}\, d\mathcal{H}^1(\xi)\Big)^{\top}\mathbb{C}^M = \Big(\int_{S^1} \big[L^{\top}(\xi)\big]^{-1}\, d\mathcal{H}^1(\xi)\Big)\mathbb{C}^M, \tag{8.1.59}$$

and canonically regard W as a space of constant functions in $\big[L^{p'}(\partial\Omega,\sigma)\big]^M$, where p' stands for the Hölder conjugate exponent of p. In turn, upon regarding the $\big[L^{p'}(\partial\Omega,\sigma)\big]^M$ as the dual space $\Big(\big[L^{p}(\partial\Omega,\sigma)\big]^M\Big)^*$ and recalling (A.0.7), this identification permits us to define

$$L^p_W(\partial\Omega,\sigma) := {}^{\perp}W = \Big\{\phi \in \big[L^p(\partial\Omega,\sigma)\big]^M : \int_{\partial\Omega} \langle \phi, w\rangle\, d\sigma = 0 \text{ for each } w \in W\Big\}.$$

Then linear algebra allows us to recast (8.1.58) simply as

$$\partial_\nu^A u \in L^p_W(\partial\Omega,\sigma). \tag{8.1.60}$$

For further use let us also note here that, since W is finite dimensional, [130, (2.1.21), (2.1.50), (2.1.51)] and (A.0.6) guarantee that

$$W = ({}^{\perp}W)^{\perp} = \big(L^p_W(\partial\Omega,\sigma)\big)^{\perp} \tag{8.1.61}$$

which, together with [130, (2.1.22)], yields the identification

$$\big(L^p_W(\partial\Omega,\sigma)\big)^* = \frac{\Big(\big[L^p(\partial\Omega,\sigma)\big]^M\Big)^*}{\big(L^p_W(\partial\Omega,\sigma)\big)^{\perp}} = \frac{\big[L^{p'}(\partial\Omega,\sigma)\big]^M}{W}. \tag{8.1.62}$$

From (8.1.57), (8.1.29), and (8.1.60) we see that the space of admissible boundary data for the boundary problem (8.1.3) (described in (8.1.6)) is contained in

$$\mathrm{Im}\Big(\tfrac12 I + K_A;\, \big[L^p_1(\partial\Omega,\sigma)\big]^M\Big) + \mathrm{Im}\Big(S : L^p_W(\partial\Omega,\sigma) \to \big[L^p_1(\partial\Omega,\sigma)\big]^M\Big) + \mathbb{C}^M. \tag{8.1.63}$$

Conversely, given an arbitrary function f belonging to the space described in (8.1.63), say $f = \big(\tfrac12 I + K_A\big)f_0 + Sf_1 + c$ where $f_0 \in \big[L^p_1(\partial\Omega,\sigma)\big]^M$, $f_1 \in L^p_W(\partial\Omega,\sigma)$, and $c \in \mathbb{C}^M$, consider the function

$$u := \mathcal{D}_A f_0 + \mathscr{S} f_1 + c \text{ in } \Omega. \tag{8.1.64}$$

Much as in (8.1.58)-(8.1.60), the membership of f_1 to $L^p_W(\partial\Omega,\sigma)$ entails

$$\Big(\int_{S^1}\big[L(\xi)\big]^{-1}\,\mathrm{d}\mathcal{H}^1(\xi)\Big)\Big(\int_{\partial\Omega} f_1\,\mathrm{d}\sigma\Big) = 0 \in \mathbb{C}^M \tag{8.1.65}$$

which, in view of [132, (1.3.11)], guarantees that $\mathscr{S} f_1(x) = O(1)$ as $|x| \to \infty$. In turn, from this, (8.1.64), [132, (1.5.8)], the last statement in [132, Theorem 1.5.1, item *(x)*], and [132, (1.3.8), (1.3.24), (1.3.26)] we see that u satisfies all properties listed in (8.1.55) as well as

$$u\Big|_{\partial\Omega}^{\kappa-\mathrm{n.t.}} = \big(\tfrac{1}{2}I + K_A\big) f_0 + S f_1 + c = f. \tag{8.1.66}$$

Ultimately, this proves that the space of admissible boundary data for the boundary problem (8.1.3) (described in (8.1.6)) coincides with (8.1.63). From this, (8.1.40), and item *(3)* in [130, Theorem 1.1.3] we then conclude that (8.1.47) continues to hold in the present setting. Next, if we let $\iota : L^p_W(\partial\Omega,\sigma) \hookrightarrow \big[L^p(\partial\Omega,\sigma)\big]^M$ be the canonical inclusion, (8.1.62) implies that its adjoint is given by

$$\big[L^{p'}(\partial\Omega,\sigma)\big]^M \ni f \mapsto [f]_W \in \frac{\big[L^{p'}(\partial\Omega,\sigma)\big]^M}{W}, \tag{8.1.67}$$

where $[f]_W$ denotes the equivalence class of f in $\big[L^{p'}(\partial\Omega,\sigma)\big]^M$ modulo W. Define

$$\begin{gathered} T : \big[L^p_1(\partial\Omega,\sigma)\big]^M \times L^p_W(\partial\Omega,\sigma)\times\mathbb{C}^M \longrightarrow \big[L^p_1(\partial\Omega,\sigma)\big]^M \\ T(f_0,f_1,c) := \big(\tfrac{1}{2}I + K_A\big) f_0 + (S\circ\iota) f_1 + c \text{ for all} \\ (f_0,f_1,c) \in \big[L^p_1(\partial\Omega,\sigma)\big]^M \times L^p_W(\partial\Omega,\sigma)\times\mathbb{C}^M. \end{gathered} \tag{8.1.68}$$

Then T is a well-defined, linear, and bounded operator, with a closed range of finite codimension and, in fact,

the image of T is precisely the space of admissible boundary data for the boundary problem formulated in (8.1.3). (8.1.69)

Granted these properties, [130, Proposition 1.2.13] applies and guarantees the existence of some constant $C \in (0,\infty)$ with the property that, given any f in the space of admissible boundary data for the boundary problem (8.1.3), we may find

$$\begin{gathered} f_0 \in \big[L^p_1(\partial\Omega,\sigma)\big]^M,\ \ f_1 \in L^p_W(\partial\Omega,\sigma),\ \ c \in \mathbb{C} \text{ with } T(f_0,f_1,c) = f \\ \text{and } \|f_0\|_{[L^p_1(\partial\Omega,\sigma)]^M} + \|f_1\|_{[L^p(\partial\Omega,\sigma)]^M} + |c| \le C\|f\|_{[L^p_1(\partial\Omega,\sigma)]^M} \end{gathered} \tag{8.1.70}$$

Then the function u defined as in (8.1.64) solves the Regularity Problem (8.1.3) for the given datum f and, thanks to (8.1.70), [132, (1.5.1), (1.5.47), (1.5.48)] (plus the very last statement in [132, Theorem 1.5.1, item *(x)*]), satisfies (8.1.7).

The next goal is to establish the characterization claimed in (8.1.12) of the space of admissible boundary data in the present setting. To this end, using (8.1.68), [132, Theorem 1.5.1, item *(vi)*], [132, Theorem 1.5.1, item *(ix)*], (8.1.67), and definitions, we see that the adjoint of T in (8.1.68) is the operator

$$T^*: \big[L^{p'}_{-1}(\partial\Omega,\sigma)\big]^M \to \big[L^{p'}_{-1}(\partial\Omega,\sigma)\big]^M \times \Big(\big[L^{p'}(\partial\Omega,\sigma)\big]^M/W\Big)\times\mathbb{C}^M$$
$$T^*\phi = \Big(\big(\tfrac12 I + K^\#_A\big)\phi,\ \big[S_{L^\top}\phi\big]_W,\ (\langle\phi,\mathbf{e}_j\rangle)_{1\le j\le M}\Big) \text{ for each } \phi\in\big[L^{p'}_{-1}(\partial\Omega,\sigma)\big]^M.$$

From this, (8.1.69), (8.1.47), and [130, (2.1.49)] we then see that the Regularity Problem (8.1.3) with a given boundary datum $f\in\big[L^p_1(\partial\Omega,\sigma)\big]^M$ is solvable if and only if

$$\begin{aligned}&{}_{L^p_1(\partial\Omega,\sigma)}\big\langle f,\phi\big\rangle_{L^{p'}_{-1}(\partial\Omega,\sigma)} = 0 \text{ for each } \phi\in\big[L^{p'}_{-1}(\partial\Omega,\sigma)\big]^M\\ &\qquad\text{satisfying } \big(\tfrac12 I+K^\#_A\big)\phi=0,\quad S_{L^\top}\phi\in W, \text{ and}\\ &\qquad\langle\phi,\mathbf{e}_j\rangle = 0 \text{ for each } j\in\{1,\dots,M\}.\end{aligned}\tag{8.1.71}$$

In view of (7.1.246), this condition may be equivalently rephrased as in (8.1.12).

Let us now consider the space of null-solution of the boundary problem (8.1.3) in the case when Ω is a two-dimensional exterior domain. The integral representation formula (8.1.56) tells that for each u belonging to the space defined as in (8.1.8) in the present setting there exists $c\in\mathbb{C}^M$ such that

$$u = -\mathscr{S}(\partial^A_\nu u) + c \text{ in } \Omega.\tag{8.1.72}$$

After taking the conormal derivative of both sides we arrive (on account of the jump-formula [132, (1.5.59)]) at $\big(\tfrac12 I + K^\#_{A^\top}\big)(\partial^A_\nu u) = 0$, which shows that

$$\partial^A_\nu u \in \operatorname{Ker}\Big(\tfrac12 I + K^\#_{A^\top};\ \big[L^p(\partial\Omega,\sigma)\big]^M\Big).\tag{8.1.73}$$

Note that the assignment

$$\Phi: \operatorname{Ker}\Big(\tfrac12 I + K^\#_{A^\top};\ \big[L^p(\partial\Omega,\sigma)\big]^M\Big)\oplus\mathbb{C}^M \to \big[\mathscr{C}^\infty(\Omega)\big]^M,\quad \Phi(f,c) := -\mathscr{S}f + c$$
$$\text{for each } f\in\operatorname{Ker}\Big(\tfrac12 I + K^\#_{A^\top};\ \big[L^p(\partial\Omega,\sigma)\big]^M\Big) \text{ and } c\in\mathbb{C}^M$$

is linear, and maps from a finite dimensional linear space, hence has a finite dimensional image. Since, as seen from (8.1.72), the space of null-solutions of (8.1.3) (defined as in (8.1.8) in the present setting) is contained in said image, we conclude that (8.1.8) is finite dimensional in this case as well. This takes care of item *(1)*.

As regards item *(2)*, under the additional assumption that Ω is bounded, the regularity properties claimed in (8.1.14)-(8.1.15) are seen from (8.1.43), the last lines in (8.1.29) and (8.1.30), [132, Theorem 4.3.1, item *(4)*], [132, Theorem 4.3.3, item *(4)*], [132, Theorem 4.2.3], and [132, Theorem 4.2.10].

Consider next the well-posedness result claimed in item *(3)*, working under the stronger hypotheses imposed in (8.1.16). First assume that $n\ge 3$. After further decreasing the value of $\delta_0\in(0,1)$, if necessary, we may assume that the conclusions in Theorem 7.1.9 hold. In particular, the invertibility result in (7.1.233) implies that there exists a constant $C\in(0,\infty)$ such that, given any $f\in\big[L^p_1(\partial\Omega,\sigma)\big]^M$, we may find a function $g \in \left[L^p(\partial\Omega, \sigma)\right]^M$ with

$$Sg = f \text{ and } \|g\|_{[L^p(\partial\Omega,\sigma)]^M} \leq C\|f\|_{[L_1^p(\partial\Omega,\sigma)]^M}. \tag{8.1.74}$$

With the boundary-to-domain single layer potential operator $\mathscr{S}$ associated with L and Ω as in (A.0.228), consider $u := \mathscr{S}g$ in Ω. Then from this, [132, (1.3.8), (1.5.44), (1.5.47), (1.5.48)], (8.1.74), the very last property in [132, Theorem 1.5.1, item *(x)*], and [132, Lemma 1.3.1] we conclude that u satisfies all properties demanded in (8.1.3) and that (8.1.7) holds.

Moving on, let us prove uniqueness in the case when $n \geq 3$. To this end, in the present geometric setting the Fatou-type result established in [131, Theorem 3.3.4] ensures that (8.1.30) holds. Granted this, [132, Theorem 1.3.3] applies and, for any $A \in \mathfrak{A}_L$, gives that any null-solution u of (8.1.3) may be represented as $u = -\mathscr{S}(\partial_\nu^A u)$ in Ω. Going nontangentially to the boundary and, once again, using that $u\big|_{\partial\Omega}^{\kappa-\text{n.t.}} = 0$ at σ-a.e. point on $\partial\Omega$ yields (cf. [132, (1.5.44)])

$$S(\partial_\nu^A u) = 0 \text{ at } \sigma\text{-a.e. point on } \partial\Omega. \tag{8.1.75}$$

Since, as noted in (8.1.30), we have $\partial_\nu^A u \in \left[L^p(\partial\Omega, \sigma)\right]^M$ we then conclude from (8.1.75) and the invertibility result from (7.1.233) that $\partial_\nu^A u = 0$. Using this back in $u = -\mathscr{S}(\partial_\nu^A u)$ ultimately gives that $u = 0$ in Ω.

As far as item *(3)* is concerned, there remains to treat the case when $n = 2$. By further decreasing the value of $\delta_0 \in (0, 1)$, if needed, we may assume that the conclusions in Theorem 7.1.10 are true. In particular, the invertibility result recorded in (7.1.245) guarantees the existence of a constant $C \in (0, \infty)$ such that, given any $f \in \left[L_1^p(\partial\Omega, \sigma)\right]^M$, we may find $g \in \left[L^p(\partial\Omega, \sigma)\right]^M$ and $c \in \mathbb{C}^M$ so that

$$Sg + c = f, \quad \int_{\partial\Omega} g \, d\sigma = 0, \quad |c| + \|g\|_{[L^p(\partial\Omega,\sigma)]^M} \leq C\|f\|_{[L_1^p(\partial\Omega,\sigma)]^M}. \tag{8.1.76}$$

Consider

$$u := \mathscr{S}g + c \text{ in } \Omega, \tag{8.1.77}$$

where $\mathscr{S}$ is the boundary-to-domain single layer potential operator associated with L and Ω as in (A.0.228). In light of [132, (1.3.11)], the vanishing moment property in (8.1.76) ensures that $u(x) = O(1)$ as $|x| \to \infty$. Based on this, (8.1.77), the very last statement in [132, Theorem 1.5.1, item *(x)*], [132, (1.3.8), (1.5.44), (1.5.48)], and (8.1.76), we conclude that u satisfies (8.1.3) when $n = 2$, and that (8.1.7) holds.

To prove uniqueness when $n = 2$, assume u is a null-solution of (8.1.3) in the two-dimensional setting. First consider the case when Ω is an exterior domain. The Fatou-type result established in [131, Theorem 3.3.4] ensures that (8.1.30) holds. Granted this, [131, Corollary 1.5.4] applies and shows that there exists a constant $c \in \mathbb{C}^M$ with the property that (cf. [131, (1.5.81)])

$$u = -\mathscr{S}\big(\partial_\nu^A u\big) + c \text{ in } \Omega. \tag{8.1.78}$$

This has two remarkable consequences. First, (8.1.78) proves that $\mathscr{S}(\partial_\nu^A u) = O(1)$ as $|x| \to \infty$ which, in concert with [132, Lemma 1.3.1, *(iii)*] forces

$$\int_{\partial\Omega} \partial_\nu^A u \, d\sigma = 0 \in \mathbb{C}^M. \tag{8.1.79}$$

Second, going nontangentially to the boundary in (8.1.78) leads to the conclusion that $0 = u\big|_{\partial\Omega}^{\kappa-\text{n.t.}} = -S(\partial_\nu^A u) + c$ at σ-a.e. point on $\partial\Omega$, hence

$$S(\partial_\nu^A u) - c = 0 \text{ on } \partial\Omega. \tag{8.1.80}$$

From (8.1.79), (8.1.80), and (7.1.245) we then conclude that $S_{\text{aug}}(\partial_\nu^A u, -c) = (0,0)$. As such, the invertibility result recorded in (7.1.245) implies that $\partial_\nu^A u = 0$ and $c = 0$. When used back in (8.1.78), this leads to the conclusion that $u = 0$ in Ω.

This establishes uniqueness in the case when Ω is an exterior domain in the two-dimensional setting. Lastly, the case when $\Omega \subseteq \mathbb{R}^n$ is bounded, is dealt with very similarly (the reasoning is simpler, as the behavior at infinity is no longer an issue). This completes the treatment of item *(3)*.

Next, that the Regularity Problem with boundary data in Muckenhoupt weighted Sobolev spaces formulated in (8.1.17) also enjoys Fredholm solvability and well-posedness results as in items *(1)* and *(3)* may be proved is a similar manner (making use of (7.1.236) and (7.1.248)). Finally, the claim in item *(5)* is seen from Theorem 8.6.6, Theorem 8.6.8, Example 1.6.14, and Remark 8.6.9. □

Next we discuss the Regularity Problem for weakly elliptic systems in δ-oscillating AR domains, with boundary data in Morrey-based Sobolev spaces, vanishing Morrey-based Sobolev spaces, and block-based Sobolev spaces. Choosing δ appropriately small allows us to target any space in the aforementioned scales. The Regularity Problem with data in Morrey-based Sobolev spaces with $p = 2$, for a bounded Lipschitz domain Ω with connected boundary, and a homogeneous, second-order, constant real coefficient, symmetric system L in $\mathbb{R}^n$, satisfying the Legendre-Hadamard strong ellipticity condition, has been dealt with by Z. Shen in [192, Theorem 3.7] using different techniques (which make essential use of the setting and hypotheses just highlighted). Our main result on this topic, Theorem 8.1.4 stated below, is new even in the case of the Laplacian on a bounded $\mathscr{C}^1$ domain.

Theorem 8.1.4 (The Regularity Problem with Data in Morrey and Block Styled Sobolev Spaces) *Fix two integers $n, M \in \mathbb{N}$ with $n \geq 2$ and suppose $\Omega \subseteq \mathbb{R}^n$ is an Ahlfors regular domain with compact boundary. Abbreviate $\sigma := \mathcal{H}^{n-1}\lfloor\partial\Omega$ and denote by ν the geometric measure theoretic outward unit normal to Ω. Also, let L be a homogeneous, second-order, constant complex coefficient, $M \times M$ system in $\mathbb{R}^n$. Finally, fix some integrability exponents $p, q \in (1, \infty)$ along with some $\lambda \in (0, n-1)$ and an arbitrary aperture parameter $\kappa > 0$.*

Then there exists $\delta_0 \in (0,1)$ depending only on n, p, q, λ, L, the ADR *constants of $\partial\Omega$, and the relative flatness ratio $\wp(\Omega)/\text{diam}\,\partial\Omega$ such that if*

$$\operatorname{dist}\Big(\nu, [\mathrm{VMO}(\partial\Omega, \sigma)]^n\Big) < \delta_0 \tag{8.1.81}$$

where the distance is considered in the space $\big[\mathrm{BMO}(\partial\Omega, \sigma)\big]^n$ *(hence if* Ω *is a* δ_0*-oscillating* AR *domain*[16]*; cf. Definition 3.4.1), the following results are true:*

(1) If one assumes that

$$L \text{ is weakly elliptic with } \mathfrak{A}_L^{\mathrm{dis}} \neq \varnothing \text{ and } \mathfrak{A}_{L^\top}^{\mathrm{dis}} \neq \varnothing, \tag{8.1.82}$$

then the Regularity Problem[17]

$$\begin{cases} u \in \big[\mathscr{C}^\infty(\Omega)\big]^M, \quad Lu = 0 \text{ in } \Omega, \\ \mathcal{N}_\kappa u \in M^{p,\lambda}(\partial\Omega, \sigma), \ \mathcal{N}_\kappa(\nabla u) \in M^{p,\lambda}(\partial\Omega, \sigma), \\ u\big|_{\partial\Omega}^{\kappa-\mathrm{n.t.}} = f \in \big[M_1^{p,\lambda}(\partial\Omega, \sigma)\big]^M, \\ u(x) = O(|x|^{2-n}) \text{ as } |x| \to \infty \text{ if } \Omega \text{ is unbounded,} \end{cases} \tag{8.1.83}$$

is Fredholm solvable, in the following precise sense: First, the space of admissible boundary data for the boundary problem (8.1.83)*, i.e.,*

$$\Big\{u\big|_{\partial\Omega}^{\kappa-\mathrm{n.t.}} : u \in \big[\mathscr{C}^\infty(\Omega)\big]^M, \ Lu = 0 \text{ in } \Omega, \ \mathcal{N}_\kappa u, \ \mathcal{N}_\kappa(\nabla u) \in M^{p,\lambda}(\partial\Omega, \sigma), \\ \text{and } u(x) = O(|x|^{2-n}) \text{ as } |x| \to \infty \text{ if } \Omega \text{ is unbounded}\Big\}, \tag{8.1.84}$$

is a closed linear subspace of finite codimension in $\big[M_1^{p,\lambda}(\partial\Omega, \sigma)\big]^M$*. Second, a solution u of* (8.1.83) *exists if and only if the boundary datum f belongs to* (8.1.84)*, in which scenario a solution u may be found such that*

$$\begin{aligned} \|f\|_{[M_1^{p,\lambda}(\partial\Omega,\sigma)]^M} &\leq \big\|\mathcal{N}_\kappa u\big\|_{M^{p,\lambda}(\partial\Omega,\sigma)} + \big\|\mathcal{N}_\kappa(\nabla u)\big\|_{M^{p,\lambda}(\partial\Omega,\sigma)} \\ &\leq C\|f\|_{[M_1^{p,\lambda}(\partial\Omega,\sigma)]^M} \end{aligned} \tag{8.1.85}$$

for some constant $C \in (0, \infty)$ *independent of f. Third, the space of null-solutions of the boundary problem* (8.1.83)*, i.e.,*

$$\Big\{u \in \big[\mathscr{C}^\infty(\Omega)\big]^M : Lu = 0, \ \mathcal{N}_\kappa u, \ \mathcal{N}_\kappa(\nabla u) \in M^{p,\lambda}(\partial\Omega, \sigma), \ u\big|_{\partial\Omega}^{\kappa-\mathrm{n.t.}} = 0, \\ \text{and } u(x) = O(|x|^{2-n}) \text{ as } |x| \to \infty \text{ if } \Omega \text{ is unbounded}\Big\}, \tag{8.1.86}$$

is finite dimensional.

---

[16] in particular, if Ω is an infinitesimally flat AR domain

[17] thanks to work in [129, §8.5], one may replace the nontangential maximal operator by the tangential maximal operator, associated as in [129, Definition 8.5.1] with a sufficiently large power (cf. (A.0.165))

(2) If (8.1.82) *is strengthened to*[18]

$$\text{L satisfies the Legendre-Hadamard strong ellipticity condition (cf. [131, Definition 1.3.2]) and one has } \mathfrak{A}_L^{\rm dis} \neq \varnothing, \tag{8.1.87}$$

the Regularity Problem (8.1.83) *is actually well posed. More specifically, the boundary value problem* (8.1.83) *has a unique solution, and this solution enjoys the additional property that there exists a constant* $C = C(\Omega, L, p, \lambda) \in (0, \infty)$ *such that* (8.1.85) *holds.*

(3) Similar results to those described in items (1) and (2) above also hold for the version of the Regularity Problem formulated for boundary data in vanishing Morrey-based Sobolev spaces, i.e.,

$$\begin{cases} u \in \big[\mathscr{C}^\infty(\Omega)\big]^M, \quad Lu = 0 \text{ in } \Omega, \\ \mathcal{N}_\kappa u \in \mathring{M}^{p,\lambda}(\partial\Omega, \sigma), \ \mathcal{N}_\kappa(\nabla u) \in \mathring{M}^{p,\lambda}(\partial\Omega, \sigma), \\ u\big|_{\partial\Omega}^{\kappa-\text{n.t.}} = f \in \big[\mathring{M}_1^{p,\lambda}(\partial\Omega, \sigma)\big]^M, \\ u(x) = O(|x|^{2-n}) \text{ as } |x| \to \infty \text{ if } \Omega \text{ is unbounded,} \end{cases} \tag{8.1.88}$$

and for the version of the Regularity Problem formulated for boundary data in block-based Sobolev spaces, i.e.,

$$\begin{cases} u \in \big[\mathscr{C}^\infty(\Omega)\big]^M, \quad Lu = 0 \text{ in } \Omega, \\ \mathcal{N}_\kappa u \in \mathcal{B}^{q,\lambda}(\partial\Omega, \sigma), \ \mathcal{N}_\kappa(\nabla u) \in \mathcal{B}^{q,\lambda}(\partial\Omega, \sigma), \\ u\big|_{\partial\Omega}^{\kappa-\text{n.t.}} = f \in \big[\mathcal{B}_1^{q,\lambda}(\partial\Omega, \sigma)\big]^M, \\ u(x) = O(|x|^{2-n}) \text{ as } |x| \to \infty \text{ if } \Omega \text{ is unbounded.} \end{cases} \tag{8.1.89}$$

Once again, in light of work in [129, §8.5], *one may replace the nontangential maximal operator employed in the formulation of* (8.1.88)-(8.1.89) *by the tangential maximal operator, associated as in* [129, Definition 8.5.1] *with a sufficiently large power* (*cf.* (A.0.165)).

(4) There exist symmetric weakly elliptic systems L *with* $\mathfrak{A}_L^{\rm dis} = \varnothing$ *(compare to* (8.1.82)) *for which the Regularity Problems* (8.1.83), (8.1.88), (8.1.89) *fail to be Fredholm solvable even when* Ω *is the unit disk in the plane (or the complement of its closure). In fact, the aforementioned failure of Fredholm solvability may be total, in the sense that for each of these problems, both the space of admissible boundary data may have infinite codimension in the full space of boundary data, and the space of null-solutions may be infinite dimensional.*

It is of relevance to point out that, as seen from Proposition 1.6.12 and [131, Proposition 1.4.39], in the scalar case, i.e., if $M = 1$, in dimensions $n \geq 3$, both (8.1.82) and (8.1.87) may be replaced by the mere demand that the system L is weakly elliptic. Furthermore, in the scalar case in the two-dimensional setting (i.e.,

[18] Proposition 1.4.1 ensures that if L is as in (8.1.87) then $\mathfrak{A}_L^{\rm dis} \neq \varnothing$ and $\mathfrak{A}_{L^\top}^{\rm dis} \neq \varnothing$

for $M = 1$ and $n = 2$), if L satisfies the Legendre-Hadamard strong ellipticity condition or, more generally (cf. [131, Lemma 1.4.24]), if $L = \mathrm{div}A\nabla$ for some 2×2 complex matrix A with $\mathrm{sym}\, A := \frac{1}{2}(A + A^{\top}) \in M_2^0$ (cf. [131, (1.4.187)]), then Proposition 1.6.11 and Proposition 1.6.13 ensure that $\mathfrak{A}_L^{\mathrm{dis}} \neq \varnothing$ and $\mathfrak{A}_{L^{\top}}^{\mathrm{dis}} \neq \varnothing$. Also, according to (1.6.141) in Proposition 1.6.13, in the scalar case in the two-dimensional setting (i.e., for $M = 1$ and $n = 2$) we may omit (8.1.87) altogether if L is a weakly elliptic operator with real coefficients. Finally, we wish to note that, as far as the formulation of Theorem 8.1.4 is concerned, similar considerations to those made in Remark 8.1.2 are presently applicable.

Proof of Theorem 8.1.4 The argument proceeds in a similar fashion to the proof of Theorem 8.1.1, now also making use of [132, Theorem 3.3.1], [132, Theorem 3.3.2], item *(7)* in Theorem 7.1.1, Theorem 7.1.18, Theorem 7.1.19, and [132, Remark 8.1.8]. □

Many of the Regularity Problems considered so far in this section involve function spaces (in relation to which the size conditions for the solution are formulated) that are particular cases of Generalized Banach Function Spaces in the sense of [130, Definition 5.1.4]. These include ordinary Lebesgue spaces and Muckenhoupt weighted Lebesgue spaces in Theorem 8.1.1, Lorentz spaces in (8.1.106), as well as Morrey and block spaces in Theorem 8.1.4 (in this regard, see [130, Proposition 6.2.17] where Morrey and block spaces space have been identified as being Generalized Banach Function Spaces). This invites the question whether an all-inclusive result of this nature holds for the Regularity Problem phrased in terms of Generalized Banach Function Spaces (GBFS) of a more generic nature. Our next theorem provides a positive answer to this question. Following its statement, we discuss other (new) examples, including the Regularity Problem with data in Sobolev spaces styled on rearrangement invariant Banach function spaces. These contain Orlicz-based Sobolev spaces so, as a corollary, we obtain the well-posedness of the Regularity Problem with boundary data in Zygmund-based Sobolev spaces.

Theorem 8.1.5 (The Regularity Problem with Boundary Data in GBFS-Styled Sobolev Spaces) *Fix two integers $n, M \in \mathbb{N}$ with $n \geq 2$ and suppose $\Omega \subseteq \mathbb{R}^n$ is an Ahlfors regular domain with compact boundary. Abbreviate $\sigma := \mathcal{H}^{n-1}\lfloor\partial\Omega$ and denote by ν the geometric measure theoretic outward unit normal to Ω. In this context, suppose $\mathbb{X}$ is a Generalized Banach Function Space on $(\partial\Omega, \sigma)$ so that*

$$\mathcal{M}_{\partial\Omega} : \mathbb{X} \to \mathbb{X} \text{ and } \mathcal{M}_{\partial\Omega} : \mathbb{X}' \to \mathbb{X}' \text{ are well-defined and bounded,} \tag{8.1.90}$$

where $\mathcal{M}_{\partial\Omega}$ is the Hardy-Littlewood maximal operator on $\partial\Omega$, and $\mathbb{X}'$ is the associated space of $\mathbb{X}$ (cf. [130, Definitions 5.1.4, 5.1.11]). *Introduce*

$$\mathring{\mathbb{X}} := \overline{L^\infty(\partial\Omega, \sigma)}^{\|\cdot\|_{\mathbb{X}}} \text{ and } (\mathbb{X}')^\circ := \overline{L^\infty(\partial\Omega, \sigma)}^{\|\cdot\|_{\mathbb{X}'}}, \tag{8.1.91}$$

and recall the $\mathbb{X}$-styled, $\mathbb{X}'$-styled, $\mathring{\mathbb{X}}$-styled, and $(\mathbb{X}')^\circ$-styled Sobolev spaces of order one on $\partial\Omega$, defined respectively as:

$$\begin{aligned}
\mathbb{X}_1 &:= \Big\{ f \in \mathbb{X} : \partial_{\tau_{jk}} f \in \mathbb{X} \textit{ for } 1 \le j,k \le n \Big\}, \\
(\mathbb{X}')_1 &:= \Big\{ g \in \mathbb{X}' : \partial_{\tau_{jk}} g \in \mathbb{X}' \textit{ for } 1 \le j,k \le n \Big\}, \\
\mathring{\mathbb{X}}_1 &:= \Big\{ f \in \mathring{\mathbb{X}} : \partial_{\tau_{jk}} f \in \mathring{\mathbb{X}} \textit{ for } 1 \le j,k \le n \Big\}, \\
(\mathbb{X}')_1^\circ &:= \Big\{ g \in (\mathbb{X}')^\circ : \partial_{\tau_{jk}} g \in (\mathbb{X}')^\circ \textit{ for } 1 \le j,k \le n \Big\}
\end{aligned} \tag{8.1.92}$$

(these are known to be Banach spaces when equipped with natural norms; cf. (5.4.84)*). Finally, consider a homogeneous, second-order, constant complex coefficient, $M \times M$ system L in $\mathbb{R}^n$, and fix an arbitrary aperture parameter $\kappa > 0$.*

Then there exists some threshold $\delta_{\mathbb{X}} \in (0,1)$, which depends only on n, the operator norms of $\mathcal{M}_{\partial\Omega}$ on $\mathbb{X}$ and $\mathbb{X}'$, the system L, the ADR *constants of $\partial\Omega$, and the relative flatness ratio $\wp(\Omega)/\mathrm{diam}\,\partial\Omega$, with the property that when*

$$\mathrm{dist}\Big(\nu, [\mathrm{VMO}(\partial\Omega,\sigma)]^n\Big) < \delta_{\mathbb{X}} \tag{8.1.93}$$

where the distance is considered in the space $\big[\mathrm{BMO}(\partial\Omega,\sigma)\big]^n$, (hence if Ω is a $\delta_{\mathbb{X}}$-oscillating AR *domain*[19]*; cf. Definition 3.4.1), the following results are true:*

(1) If one assumes that

$$L \textit{ is weakly elliptic with } \mathfrak{A}_L^{\mathrm{dis}} \neq \varnothing \textit{ and } \mathfrak{A}_{L^\top}^{\mathrm{dis}} \neq \varnothing, \tag{8.1.94}$$

then the Regularity Problem[20]

$$\begin{cases}
u \in \big[\mathscr{C}^\infty(\Omega)\big]^M, \quad Lu = 0 \textit{ in } \Omega, \\
\mathcal{N}_\kappa u \in \mathbb{X},\ \mathcal{N}_\kappa(\nabla u) \in \mathbb{X}, \quad u\big|_{\partial\Omega}^{\kappa-\mathrm{n.t.}} = f \in \big[\mathbb{X}_1\big]^M, \\
u(x) = O(|x|^{2-n}) \textit{ as } |x| \to \infty \textit{ if } \Omega \textit{ is unbounded},
\end{cases} \tag{8.1.95}$$

is Fredholm solvable, in the following precise sense: First, the space of admissible boundary data for the boundary problem (8.1.95)*, i.e.,*

$$\Big\{ u\big|_{\partial\Omega}^{\kappa-\mathrm{n.t.}} : u \in \big[\mathscr{C}^\infty(\Omega)\big]^M,\ Lu = 0 \textit{ in } \Omega,\ \mathcal{N}_\kappa u,\ \mathcal{N}_\kappa(\nabla u) \in \mathbb{X}, \\ \textit{and } u(x) = O(|x|^{2-n}) \textit{ as } |x| \to \infty \textit{ if } \Omega \textit{ is unbounded} \Big\}, \tag{8.1.96}$$

is a closed linear subspace of finite codimension in $\big[\mathbb{X}_1\big]^M$. Second, a solution u of (8.1.95) *exists if and only if the boundary datum f belongs to* (8.1.96)*, in which scenario a solution u may be found such that*

$$\|f\|_{[\mathbb{X}_1]^M} \le \big\|\mathcal{N}_\kappa u\big\|_{\mathbb{X}} + \big\|\mathcal{N}_\kappa(\nabla u)\big\|_{\mathbb{X}} \le C\|f\|_{[\mathbb{X}_1]^M} \tag{8.1.97}$$

---

[19] in particular, if Ω is an infinitesimally flat AR domain

[20] thanks to work in [129, §8.5], one may replace the nontangential maximal operator by the tangential maximal operator, associated as in [129, Definition 8.5.1] with a sufficiently large power (cf. (A.0.165))

for some constant $C \in (0, \infty)$ *independent of* f. *Third, the space of null-solutions of the boundary problem* (8.1.95), *i.e.,*

$$\Big\{u \in \big[\mathscr{C}^\infty(\Omega)\big]^M : Lu = 0 \text{ in } \Omega,\ \mathcal{N}_\kappa u,\ \mathcal{N}_\kappa(\nabla u) \in \mathbb{X},\ u\Big|_{\partial\Omega}^{\kappa-\text{n.t.}} = 0, \\ \text{and } u(x) = O(|x|^{2-n}) \text{ as } |x| \to \infty \text{ if } \Omega \text{ is unbounded}\Big\}, \tag{8.1.98}$$

is finite dimensional.

(2) *If* (8.1.94) *is strengthened to*

> *L satisfies the Legendre-Hadamard strong ellipticity condition* (*cf.* [131, Definition 1.3.2]) *and one has* $\mathfrak{A}_L^{\rm dis} \neq \varnothing$, (8.1.99)

the Regularity Problem (8.1.95) *is actually well posed. More specifically, the boundary value problem* (8.1.95) *has a unique solution, and this solution enjoys the additional property that there exists a constant* $C = C(\Omega, L, \mathbb{X}) \in (0, \infty)$ *such that* (8.1.97) *holds.*

(3) *Similar results to those described in items (1) and (2) above also hold for the Regularity Problem with boundary data in* $(\mathbb{X}')_1$, *i.e.,*

$$\begin{cases} u \in \big[\mathscr{C}^\infty(\Omega)\big]^M, \quad Lu = 0 \text{ in } \Omega, \\ \mathcal{N}_\kappa u \in \mathbb{X}',\ \mathcal{N}_\kappa(\nabla u) \in \mathbb{X}', \quad u\big|_{\partial\Omega}^{\kappa-\text{n.t.}} = f \in \big[(\mathbb{X}')_1\big]^M, \\ u(x) = O(|x|^{2-n}) \text{ as } |x| \to \infty \text{ if } \Omega \text{ is unbounded,} \end{cases} \tag{8.1.100}$$

as well as for the Regularity Problem with boundary data in $\mathring{\mathbb{X}}_1$, *i.e.,*

$$\begin{cases} u \in \big[\mathscr{C}^\infty(\Omega)\big]^M, \quad Lu = 0 \text{ in } \Omega, \\ \mathcal{N}_\kappa u \in \mathring{\mathbb{X}},\ \mathcal{N}_\kappa(\nabla u) \in \mathring{\mathbb{X}}, \quad u\big|_{\partial\Omega}^{\kappa-\text{n.t.}} = f \in \big[\mathring{\mathbb{X}}_1\big]^M, \\ u(x) = O(|x|^{2-n}) \text{ as } |x| \to \infty \text{ if } \Omega \text{ is unbounded,} \end{cases} \tag{8.1.101}$$

and the Regularity Problem with boundary data in $(\mathbb{X}')_1^\circ$, *i.e.,*

$$\begin{cases} u \in \big[\mathscr{C}^\infty(\Omega)\big]^M, \quad Lu = 0 \text{ in } \Omega, \\ \mathcal{N}_\kappa u \in (\mathbb{X}')^\circ,\ \mathcal{N}_\kappa(\nabla u) \in (\mathbb{X}')^\circ, \quad u\big|_{\partial\Omega}^{\kappa-\text{n.t.}} = f \in \big[(\mathbb{X}')_1^\circ\big]^M, \\ u(x) = O(|x|^{2-n}) \text{ as } |x| \to \infty \text{ if } \Omega \text{ is unbounded.} \end{cases} \tag{8.1.102}$$

Also, in light of work in [129, §8.5] (*cf.* [129, Theorem 8.5.5], *whose applicability in the present setting is ensured by* (8.1.90)), *one may replace the nontangential maximal operator employed in the formulation of* (8.1.100)-(8.1.102) *by the tangential maximal operator, associated as in* [129, Definition 8.5.1] *with a sufficiently large power* (*cf.* (A.0.165)).

(4) *There exist symmetric weakly elliptic systems L with* $\mathfrak{A}_L^{\rm dis} = \varnothing$ (*compare to* (8.1.82)) *and Generalized Banach Function Spaces* $\mathbb{X}$ *satisfying* (8.1.90) *for*

which the Regularity Problems (8.1.95), (8.1.100) *fail to be Fredholm solvable even when the underlying domain* Ω *is the unit disk in the plane (or the complement of its closure). In fact, the aforementioned failure of Fredholm solvability may be total, in the sense that for each of these problems, both the space of admissible boundary data may have infinite codimension in the full space of boundary data, and the space of null-solutions may be infinite dimensional.*

In relation to Theorem 8.1.5 we wish to remark that, as far as the statement of Theorem 8.1.5 is concerned, similar considerations to those made in Remark 8.1.2 are presently applicable. Also, as is apparent from Proposition 1.6.12 and [131, Proposition 1.4.39], in the scalar case, i.e., if $M = 1$, in dimensions $n \geq 3$, both (8.1.94) and (8.1.99) may be replaced by the mere demand that L is weakly elliptic.

Moreover, in the scalar case in the two-dimensional setting (i.e., for $M = 1$ and $n = 2$), if L satisfies the Legendre-Hadamard strong ellipticity condition or, more generally (cf. [131, Lemma 1.4.24]), if $L = \operatorname{div} A \nabla$ for some 2×2 complex matrix A with $\operatorname{sym} A := \frac{1}{2}(A + A^{\top}) \in M_2^0$ (cf. [131, (1.4.187)]), then Proposition 1.6.11 and Proposition 1.6.13 guarantee that $\mathfrak{A}_L^{\text{dis}} \neq \varnothing$ and $\mathfrak{A}_{L^{\top}}^{\text{dis}} \neq \varnothing$. Also, according to (1.6.141) in Proposition 1.6.13, in the scalar case in the two-dimensional setting (i.e., for $M = 1$ and $n = 2$) we may omit (8.1.99) altogether if L is a weakly elliptic operator with real coefficients.

Regarding Theorem 8.1.5 we also want to note that, as seen from [130, Proposition 5.3.14],

$$\text{the hypotheses formulated in (8.1.90) are automatically verified in the case when } \mathbb{X} \text{ is a rearrangement invariant Banach function space on the (non-atomic, sigma-finite measure space } (\partial\Omega, \sigma)) \text{ with Boyd indices satisfying } 1 < p_{\mathbb{X}} \leq q_{\mathbb{X}} < \infty. \tag{8.1.103}$$

Hence, *all conclusions of Theorem 8.1.5 are true in the class of rearrangement invariant Banach function spaces* $\mathbb{X}$ *with Boyd indices* $1 < p_{\mathbb{X}} \leq q_{\mathbb{X}} < \infty$. Significantly, this is the case when the Generalized Banach Function Space in question is the Orlicz space $L^{\Phi}(\partial\Omega, \sigma)$, associated with the (non-atomic, sigma-finite) measure space $(\partial\Omega, \sigma)$ and a Young function Φ satisfying $1 < i(\Phi) \leq I(\Phi) < \infty$; see [130, Proposition 5.3.15] and [130, (5.3.62)] in this regard.

To elaborate on the latter scenario, retain the initial background assumptions made in Theorem 8.1.5, and pick a Young function Φ whose dilation indices satisfy $1 < i(\Phi) \leq I(\Phi) < \infty$. In such a setting, bring back the Orlicz-based Sobolev space $L_1^{\Phi}(\partial\Omega, \sigma)$ introduced in (5.4.92). Then the Regularity Problem with boundary data in $\big[L_1^{\Phi}(\partial\Omega, \sigma)\big]^M$, formulated as

$$\begin{cases} u \in \big[\mathscr{C}^{\infty}(\Omega)\big]^M, \quad Lu = 0 \text{ in } \Omega, \\ \mathcal{N}_{\kappa} u,\ \mathcal{N}_{\kappa}(\nabla u) \in L^{\Phi}(\partial\Omega, \sigma), \quad u\big|_{\partial\Omega}^{\kappa-\text{n.t.}} = f \in \big[L_1^{\Phi}(\partial\Omega, \sigma)\big]^M, \\ u(x) = O(|x|^{2-n}) \text{ as } |x| \to \infty \text{ if } \Omega \text{ is unbounded,} \end{cases} \tag{8.1.104}$$

is well-posed whenever the system L satisfies the Legendre-Hadamard strong ellipticity condition and $\mathfrak{A}_L^{\rm dis} \neq \varnothing$, provided $\operatorname{dist}\big(\nu, [\mathrm{VMO}(\partial\Omega,\sigma)]^n\big) < \delta_\Phi$ for a sufficiently small threshold $\delta_\Phi \in (0,1)$. Indeed, this is a consequence of item *(2)* in Theorem 8.1.5, [130, (5.3.62), (5.3.63)], and [130, Proposition 5.3.15].

In particular, the result above concerning (8.1.104) contains the well-posedness result for the Regularity Problem with data in the Zygmund-based Sobolev space $\big[L^p_1(\log L)^\alpha(\partial\Omega,\sigma)\big]^M$ introduced in (5.4.94), namely

$$\begin{cases} u \in \big[\mathscr{C}^\infty(\Omega)\big]^M, \quad Lu = 0 \text{ in } \Omega, \\ \mathcal{N}_\kappa u,\ \mathcal{N}_\kappa(\nabla u) \in L^p(\log L)^\alpha(\partial\Omega,\sigma), \\ u\big|_{\partial\Omega}^{\kappa-\mathrm{n.t.}} = f \in \big[L^p_1(\log L)^\alpha(\partial\Omega,\sigma)\big]^M, \\ u(x) = O(|x|^{2-n}) \text{ as } |x| \to \infty \text{ if } \Omega \text{ is unbounded,} \end{cases} \tag{8.1.105}$$

once again under the assumption that the system L satisfies the Legendre-Hadamard strong ellipticity condition, that $\mathfrak{A}_L^{\rm dis} \neq \varnothing$, and that $\operatorname{dist}\big(\nu, [\mathrm{VMO}(\partial\Omega,\sigma)]^n\big) < \delta$ for some sufficiently small threshold $\delta = \delta(L,\Omega,p,\alpha) \in (0,1)$.

Of course, this is a very accommodating template. For example, in light of the discussion in [130, Example 5.3.11], whenever $p \in (1,\infty)$ and $\theta \in [0,1)$ a similar well-posedness result holds for the Regularity Problem with data in the Sobolev space $\big[L^p_1\exp(\log^\theta L)(\partial\Omega,\sigma)\big]^M$, which is naturally defined as the Orlicz-based Sobolev space $\big[L^\Phi_1(\partial\Omega,\sigma)\big]^M$ corresponding to a Young function $\Phi(t) \approx t^p\exp\big([\ln(e+t)]^\theta\big)$.

This being said, there are solvability results for the Regularity Problem involving boundary data not necessarily in GBFS-styled Sobolev spaces. For example, recall the brand of Lorentz-based Sobolev spaces $L^{(p,q)}_1(\partial\Omega,\sigma)$ introduced in (5.4.27)-(5.4.28). Then for each $p \in (1,\infty)$ and $q \in (0,\infty]$ the following Regularity Problem, with boundary data in Lorentz-based Sobolev spaces,

$$\begin{cases} u \in \big[\mathscr{C}^\infty(\Omega)\big]^M, \quad Lu = 0 \text{ in } \Omega, \\ \mathcal{N}_\kappa u,\ \mathcal{N}_\kappa(\nabla u) \in L^{p,q}(\partial\Omega,\sigma), \\ u\big|_{\partial\Omega}^{\kappa-\mathrm{n.t.}} = g \in \big[L^{(p,q)}_1(\partial\Omega,\sigma)\big]^M, \\ u(x) = O(|x|^{2-n}) \text{ as } |x| \to \infty \text{ if } \Omega \text{ is unbounded,} \end{cases} \tag{8.1.106}$$

possesses analogous solvability properties to those described in Theorem 8.1.5 (with similar proofs), even though $L^{(p,q)}_1(\partial\Omega,\sigma)$ with $p \in (1,\infty)$ and $q \in (0,\infty]$ is not a GBFS-styled Sobolev space[21].

We now turn to the task of presenting the proof of Theorem 8.1.5.

Proof of Theorem 8.1.5 First note that Theorem 3.4.2 guarantees that taking the threshold $\delta_{\mathbb{X}} \in (0,1)$ appearing in (8.1.93) sufficiently small renders Ω a two-sided NTA domain, hence also a UR domain (cf. [129, (5.10.33)]). Let us first work under the assumptions made in (8.1.94). Select $A \in \mathfrak{A}_L^{\rm dis}$, then use Theorem 1.5.5 to conclude that $A^\top \in \mathfrak{A}_{L^\top}^{\rm dis}$. Granted (8.1.90), by eventually further decreasing the value

[21] see the discussion in relation to (8.1.271)

of $\delta_{\mathbb{X}} \in (0,1)$ we may also enures that the conclusions in item *(8)* of Theorem 7.1.1 hold with $z := 1/2$. In particular,

$$\tfrac{1}{2}I + K_A : \big[\mathbb{X}_1\big]^M \longrightarrow \big[\mathbb{X}_1\big]^M \text{ is a Fredholm operator with index zero,} \tag{8.1.107}$$

$$\tfrac{1}{2}I + K^{\#}_{A^\top} : \big[\mathbb{X}\big]^M \longrightarrow \big[\mathbb{X}\big]^M \text{ is a Fredholm operator with index zero.} \tag{8.1.108}$$

Consequently,

$$\begin{gathered}\operatorname{Im}\Big(\tfrac{1}{2}I + K_A;\ \big[\mathbb{X}_1\big]^M\Big) \text{ is a closed space}\\ \text{of finite codimension in the Sobolev space } \big[\mathbb{X}_1\big]^M,\end{gathered} \tag{8.1.109}$$

$$\text{and } \dim \operatorname{Ker}\Big(\tfrac{1}{2}I + K^{\#}_{A^\top};\ \big[\mathbb{X}\big]^M\Big) < +\infty. \tag{8.1.110}$$

For ease of presentation, let us now focus on the case when either $n \geq 3$, or $n = 2$ and Ω is bounded. Then for any function satisfying

$$\begin{gathered} u \in \big[\mathscr{C}^\infty(\Omega)\big]^M,\ \ Lu = 0 \text{ in } \Omega,\ \ \mathcal{N}_\kappa u,\ \mathcal{N}_\kappa(\nabla u) \in \mathbb{X},\\ \text{and also } u(x) = O(|x|^{2-n}) \text{ as } |x| \to \infty \text{ if } \Omega \text{ is unbounded,}\end{gathered} \tag{8.1.111}$$

the Fatou-type result established in [131, Theorem 3.3.4] (cf. [131, (3.3.55)]) implies, while also bearing in mind [130, (5.2.92)] and [130, (5.1.2), (11.5.60)], that

$$\begin{gathered}(\nabla u)\Big|^{\kappa-\text{n.t.}}_{\partial\Omega} \text{ exists (in } \mathbb{C}^{n\cdot M}) \text{ at } \sigma\text{-a.e. point on } \partial\Omega \text{ and}\\ (\nabla u)\Big|^{\kappa-\text{n.t.}}_{\partial\Omega} \in \big[\mathbb{X}\big]^{n\cdot M}, \text{ hence } \partial^A_\nu u \in \big[\mathbb{X}\big]^M.\end{gathered} \tag{8.1.112}$$

Moreover, in view of [131, (3.3.62) in Theorem 3.3.4], [130, (5.2.92)], [130, (5.1.12), (11.5.60)], and [130, Proposition 11.3.2], we have

$$u\Big|^{\kappa-\text{n.t.}}_{\partial\Omega} \text{ exists (in } \mathbb{C}^M) \text{ at } \sigma\text{-a.e. point on } \partial\Omega \text{ and } u\Big|^{\kappa-\text{n.t.}}_{\partial\Omega} \in \big[\mathbb{X}_1\big]^M. \tag{8.1.113}$$

The membership in (8.1.113) shows that prescribing boundary data in the GBFS-styled Sobolev space $\big[\mathbb{X}_1\big]^M$ is natural for the Regularity Problem (8.1.95).

Bring in the boundary-to-domain single layer $\mathscr{S}$ associated with L and Ω as in (A.0.228). Also, denote by $\mathcal{D}_A$ the boundary-to-domain double layer associated with the coefficient tensor A and the set Ω as in (A.0.72). By (8.1.111)-(8.1.113), [130, (5.2.92)], [132, Theorem 1.3.3] we have $u = \mathcal{D}_A\big(u\big|^{\kappa-\text{n.t.}}_{\partial\Omega}\big) - \mathscr{S}(\partial^A_\nu u)$ in Ω. Going nontangentially to the boundary (and using [132, (1.5.20), (1.5.44)], [130, (5.2.92)], and (8.1.112)-(8.1.113)) yields $u\big|^{\kappa-\text{n.t.}}_{\partial\Omega} = \big(\tfrac{1}{2}I + K_A\big)\big(u\big|^{\kappa-\text{n.t.}}_{\partial\Omega}\big) - S(\partial^A_\nu u)$ on $\partial\Omega$. This shows that the space of admissible boundary data for the boundary problem (8.1.95) described in (8.1.96) is contained in

$$\operatorname{Im}\Big(\tfrac{1}{2}I + K_A;\ \big[\mathbb{X}_1\big]^M\Big) + \operatorname{Im}\Big(S : \big[\mathbb{X}\big]^M \to \big[\mathbb{X}_1\big]^M\Big). \tag{8.1.114}$$

We claim that in fact we have equality. To deal with the converse inclusion, we first observe that [129, (7.7.120) in Lemma 7.7.16] together with [131, (2.5.548)] in [131, Proposition 2.5.39] imply (bearing in mind that $\partial\Omega$ is compact) the existence of a constant $C \in (0, \infty)$ for which we have the pointwise inequality

$$\mathcal{N}_\kappa(\mathcal{S} f) \le C\, \mathcal{M}_{\partial\Omega} f \ \text{ on } \partial\Omega, \ \text{ for each } f \in \big[L^1(\partial\Omega, \sigma)\big]^M. \tag{8.1.115}$$

In concert with (8.1.90), [130, (5.2.92)], [130, (5.1.12)], and [129, (8.2.28)], this ultimately shows that, in a quantitative fashion,

$$\mathcal{N}_\kappa(\mathcal{S} f) \in \mathbb{X} \text{ for each function } f \in \big[\mathbb{X}\big]^M. \tag{8.1.116}$$

Now, given any f belonging to the space (8.1.114), say $f = \big(\frac{1}{2}I + K_A\big)f_0 + S f_1$ where $f_0 \in \big[\mathbb{X}_1\big]^M$ and $f_1 \in \big[\mathbb{X}\big]^M$, it follows from [130, (5.1.13), (5.2.92)], [131, (2.4.20)], [132, (1.3.8), (1.3.24), (1.3.26), (1.5.20), (1.5.44)], (8.1.116), and [132, Lemmas 1.3.1, 1.3.2] that the function $u := \mathcal{D}_A f_0 + \mathcal{S} f_1$ in Ω satisfies all properties in (8.1.111) as welll as $u\big|_{\partial\Omega}^{\kappa-\text{n.t.}} = \big(\frac{1}{2}I + K_A\big)f_0 + S f_1 = f$. Altogether, this proves that

$$\begin{gathered}\text{the space of admissible boundary data (8.1.96) is equal to}\\ \operatorname{Im}\Big(\tfrac{1}{2}I + K_A;\ \big[\mathbb{X}_1\big]^M\Big) + \operatorname{Im}\Big(S : \big[\mathbb{X}\big]^M \to \big[\mathbb{X}_1\big]^M\Big).\end{gathered} \tag{8.1.117}$$

Based on this, (8.1.109), and [130, Theorem 1.1.3, item *(3)*] we then conclude that

the space of admissible boundary data for the boundary problem (8.1.95), described in (8.1.96), is a closed linear subspace of finite codimension in $\big[\mathbb{X}_1\big]^M$. (8.1.118)

In fact, it is possible to quantify (8.1.117). To do so, define

$$\begin{gathered} T : \big[\mathbb{X}_1\big]^M \times \big[\mathbb{X}\big]^M \longrightarrow \big[\mathbb{X}_1\big]^M,\\ T(f_0, f_1) := \big(\tfrac{1}{2}I + K_A\big)f_0 + S f_1 \ \text{ for all } (f_0, f_1) \in \big[\mathbb{X}_1\big]^M \times \big[\mathbb{X}\big]^M. \end{gathered} \tag{8.1.119}$$

Then T is a well-defined, linear, and bounded operator, with a closed range of finite codimension, and (8.1.117) shows that

the image of T is precisely the space of admissible boundary data for the boundary problem formulated in (8.1.95). (8.1.120)

Granted these properties, [130, Proposition 1.2.13] applies and guarantees the existence of some constant $C \in (0, \infty)$ with the property that, given any f in the space of admissible boundary data for the boundary problem (8.1.95), we may find

$$\begin{gathered} f_0 \in \big[\mathbb{X}_1\big]^M \text{ and } f_1 \in \big[\mathbb{X}\big]^M \text{ with } T(f_0, f_1) = f\\ \text{and } \|f_0\|_{[\mathbb{X}_1]^M} + \|f_1\|_{[\mathbb{X}]^M} \le C\|f\|_{[\mathbb{X}_1]^M}. \end{gathered} \tag{8.1.121}$$

Then the function $u := \mathcal{D}_A f_0 + \mathcal{S} f_1$ for this choice of f_0, f_1 solves the Regularity Problem (8.1.95) for the given datum f and satisfies (8.1.97), thanks to (8.1.121), [130, (5.1.13), (5.2.92)], [131, (2.4.20)], [132, (1.3.8), (1.3.24), (1.3.26), (1.5.20), (1.5.44)], (8.1.116), and [132, Lemmas 1.3.1, 1.3.2].

Next, let u be a null-solution of the boundary problem (8.1.95) (i.e., u belongs to the space in (8.1.98)). Then the integral representation formula (8.1.43) becomes

$$u = -\mathcal{S}(\partial_\nu^A u) \text{ in } \Omega. \tag{8.1.122}$$

After taking the conormal derivative of both sides of (8.1.122) we arrive at $\big(\frac{1}{2}I + K^{\#}_{A^\top}\big)(\partial_\nu^A u) = 0$ (keeping in mind [130, (5.2.92)] and the jump-formula [132, (1.5.59)]), from which we conclude that

$$\partial_\nu^A u \in \operatorname{Ker}\Big(\tfrac{1}{2}I + K^{\#}_{A^\top};\ \big[\mathbb{X}\big]^M\Big). \tag{8.1.123}$$

In view of this and (8.1.122), the assignment taking any null-solution u of (8.1.95) into $\partial_\nu^A u$ is well defined, linear, and injective. In concert with (8.1.110), this forces

$$\text{the dimension of (8.1.98) is } \leq \dim \operatorname{Ker}\Big(\tfrac{1}{2}I + K^{\#}_{A^\top};\ \big[\mathbb{X}\big]^M\Big) < \infty. \tag{8.1.124}$$

Thus, the space of null-solutions of the boundary problem (8.1.95), described in (8.1.98), is indeed finite dimensional.

To complete the treatment of item *(1)*, there remains to consider the case when $n = 2$ and $\Omega \subseteq \mathbb{R}^2$ is an unbounded NTA domain with compact boundary (in particular, Ω is an exterior domain). To get started, suppose u is a function satisfying

$$\begin{gathered} u \in \big[\mathscr{C}^\infty(\Omega)\big]^M,\ Lu = 0 \text{ in } \Omega,\ \mathcal{N}_\kappa u,\ \mathcal{N}_\kappa(\nabla u) \in \mathbb{X}, \\ \text{and } u(x) = O(1) \text{ as } |x| \to \infty. \end{gathered} \tag{8.1.125}$$

Then the properties recorded in (8.1.112)-(8.1.113) continue to be valid. Granted all these attributes and keeping in mind [130, (5.2.92)], [131, Corollary 1.5.4] applies and yields a constant $c \in \mathbb{C}^M$ with the property that (cf. [131, (1.5.81)])

$$u = \mathcal{D}_A\big(u\big|_{\partial\Omega}^{\kappa-\text{n.t.}}\big) - \mathcal{S}\big(\partial_\nu^A u\big) + c \text{ in } \Omega. \tag{8.1.126}$$

Together with [130, (5.2.92)], [132, (1.5.20), (1.5.44)], and (8.1.113), this implies

$$u\big|_{\partial\Omega}^{\kappa-\text{n.t.}} = \big(\tfrac{1}{2}I + K_A\big)\big(u\big|_{\partial\Omega}^{\kappa-\text{n.t.}}\big) - S(\partial_\nu^A u) + c \text{ on } \partial\Omega. \tag{8.1.127}$$

From the last property in (8.1.125), the fact that (as seen from [132, (1.3.26)], (8.1.113), and [130, (5.2.92)]),

$$\mathcal{D}_A\big(u\big|_{\partial\Omega}^{\kappa-\text{n.t.}}\big)(x) = O\big(|x|^{-1}\big) \text{ as } |x| \to \infty, \tag{8.1.128}$$

and (8.1.126), we see that $\mathcal{S}\big(\partial_\nu^A u\big) = O(1)$ as $|x| \to \infty$. Together with (8.1.112) and [132, (1.3.11)], this prove that

$$\Big(\int_{S^1}\big[L(\xi)\big]^{-1}\,\mathrm{d}\mathcal{H}^1(\xi)\Big)\Big(\int_{\partial\Omega}\partial_\nu^A u\,\mathrm{d}\sigma\Big)=0\in\mathbb{C}^M. \tag{8.1.129}$$

Collectively, (8.1.127), (8.1.112), (8.1.113), and (8.1.129) show that the space of admissible boundary data described in (8.1.96) is, in the current case, contained in

$$\begin{aligned}&\mathbb{C}^M+\mathrm{Im}\Big(\tfrac12 I+K_A;\ \big[\mathbb{X}_1\big]^M\Big) \\ &\quad+\Big\{Sf:\ f\in\big[\mathbb{X}\big]^M \text{ with } \Big(\int_{S^1}\big[L(\xi)\big]^{-1}\,\mathrm{d}\mathcal{H}^1(\xi)\Big)\Big(\int_{\partial\Omega}f\,\mathrm{d}\sigma\Big)=0\Big\}.\end{aligned} \tag{8.1.130}$$

Conversely, given any function belonging to the space described in (8.1.130), that is, given any $f=c+(\frac12 I+K_A)f_0+Sf_1$ with $c\in\mathbb{C}^M$, $f_0\in\big[\mathbb{X}_1\big]^M$, and with

$$f_1\in\big[\mathbb{X}\big]^M \text{ satisfying } \Big(\int_{S^1}\big[L(\xi)\big]^{-1}\,\mathrm{d}\mathcal{H}^1(\xi)\Big)\Big(\int_{\partial\Omega}f\,\mathrm{d}\sigma\Big)=0, \tag{8.1.131}$$

it follows that $u:=c+\mathcal{D}_A f_0+\mathcal{S}f_1$ enjoys all properties specified in (8.1.125) and, in addition, has $u\big|_{\partial\Omega}^{\kappa-\mathrm{n.t.}}=c+\big(\frac12 I+K_A\big)f_0+Sf_1=f$. Altogether, this proves that

if $n=2$ and Ω is an exterior domain, then the space of admissible boundary data described in (8.1.96) is equal to the space in (8.1.130). (8.1.132)

Having established this, (8.1.109) and item *(3)* in [130, Theorem 1.1.3] imply that (8.1.118) continues to hold in the present case. Moreover, a similar argument as the one used in (8.1.119)-(8.1.121) shows that, for any admissible boundary datum f, the Regularity Problem (8.1.95) has a solution u which satisfies (8.1.97).

Going further, we now consider the space of null-solution of the boundary problem (8.1.95) in the case when Ω is a two-dimensional exterior domain. The integral representation formula (8.1.126) gives that for each u belonging to the space defined as in (8.1.98) in the present setting there exists $c\in\mathbb{C}^M$ such that

$$u=-\mathcal{S}(\partial_\nu^A u)+c \text{ in } \Omega. \tag{8.1.133}$$

Applying conormal derivatives to both sides of (8.1.133) we obtain (on account of [130, (5.2.92)] and the jump-formula [132, (1.5.59)]) $\big(\frac12 I+K^{\#}_{A^\top}\big)(\partial_\nu^A u)=0$, from which we conclude (also recalling (8.1.112)) that

$$\partial_\nu^A u\in\mathrm{Ker}\Big(\tfrac12 I+K^{\#}_{A^\top};\ \big[\mathbb{X}\big]^M\Big). \tag{8.1.134}$$

Note that the operator $\Phi:\mathrm{Ker}\Big(\frac12 I+K^{\#}_{A^\top};\ \big[\mathbb{X}\big]^M\Big)\oplus\mathbb{C}^M\to\big[\mathscr{C}^\infty(\Omega)\big]^M$ given by

$$\Phi(f,c):=-\mathcal{S}f+c \text{ for each } f\in\mathrm{Ker}\Big(\tfrac12 I+K^{\#}_{A^\top};\ \big[\mathbb{X}\big]^M\Big) \text{ and } c\in\mathbb{C}^M \tag{8.1.135}$$

is linear, and maps from a finite dimensional linear space, hence has a finite dimensional image. Given that, as seen from (8.1.133) and (8.1.134), the space of null-solutions of (8.1.95) (defined as in (8.1.98) in the present setting) is contained in said image, we conclude that (8.1.98) is a finite dimensional space in this case as well. This completes the treatment of item *(1)*.

Consider next the well-posedness result claimed in item *(2)*, working under the stronger hypotheses imposed in (8.1.99). First assume that $n \geq 3$. After further decreasing the value of $\delta_{\mathbb{X}} \in (0,1)$ if necessary, we may assume that the conclusions in Theorem 7.1.20 hold. In particular, the invertibility of the first operator in (7.1.387) implies that there exists a constant $C \in (0,\infty)$ such that, given any $f \in \big[\mathbb{X}_1\big]^M$, we may find a function $g \in \big[\mathbb{X}\big]^M$ with

$$Sg = f \text{ and } \|g\|_{[\mathbb{X}]^M} \leq C\|f\|_{[\mathbb{X}_1]^M}. \tag{8.1.136}$$

With the boundary-to-domain single layer potential operator $\mathscr{S}$ associated with L and Ω as in (A.0.228), consider $u := \mathscr{S}g$ in Ω. At this stage, from this definition, [130, (5.2.92)], [131, (2.4.20)], [132, (1.3.8), (1.5.44)], (8.1.136), (8.1.116), and [132, Lemma 1.3.1] we conclude that u satisfies all properties imposed in (8.1.95) and that (8.1.97) holds.

The next step is to prove uniqueness when $n \geq 3$. With this goal in mind, we first note that, in the present geometric setting, the Fatou-type result established in [131, Theorem 3.3.4] ensures that the properties listed in (8.1.112)-(8.1.113) are valid. In view of these and [130, (5.2.92)], [132, Theorem 1.3.3] applies and, for any $A \in \mathfrak{A}_L$, gives that any null-solution u of (8.1.95) may be represented as

$$u = -\mathscr{S}(\partial_\nu^A u) \text{ in } \Omega. \tag{8.1.137}$$

Taking nontangential boundary traces and, once again, using that $u\big|_{\partial\Omega}^{\kappa-\text{n.t.}} = 0$ at σ-a.e. point on $\partial\Omega$, yields (cf. [132, (1.5.44)] and [130, (5.2.92)])

$$S(\partial_\nu^A u) = 0 \text{ at } \sigma\text{-a.e. point on } \partial\Omega. \tag{8.1.138}$$

From (8.1.112) we know that $\partial_\nu^A u \in \big[\mathbb{X}\big]^M$, so (8.1.138) together with the invertibility result from (7.1.233) force $\partial_\nu^A u = 0$. Plugging this back in (8.1.137) ultimately shows that $u = 0$ in Ω, as wanted.

To finish the treatment of item *(2)*, there remains to consider the case when $n = 2$. We may assume that the conclusions in Theorem 7.1.20 are true in the two-dimensional setting, by further decreasing the value of $\delta_{\mathbb{X}} \in (0,1)$ if necessary. In particular, the fact that (7.1.389) is an isomorphism guarantees the existence of a constant $C \in (0,\infty)$ such that, given any $f \in \big[\mathbb{X}_1\big]^M$, we may find a function $g \in \big[\mathbb{X}\big]^M$ and a constant $c \in \mathbb{C}^M$ with the property that

$$Sg + c = f, \quad \int_{\partial\Omega} g\,d\sigma = 0, \text{ and } |c| + \|g\|_{[\mathbb{X}]^M} \leq C\|f\|_{[\mathbb{X}_1]^M}. \tag{8.1.139}$$

With $\mathscr{S}$ denoting the boundary-to-domain single layer potential operator associated with L and Ω as in (A.0.228), define

$$u := \mathscr{S}g + c \text{ in } \Omega. \tag{8.1.140}$$

Thanks to [132, (1.3.11)] and the cancellation condition in (8.1.76), we have

$$u(x) = O(1) \text{ as } |x| \to \infty. \tag{8.1.141}$$

Based on (8.1.140), (8.1.141), [130, (5.2.92)], [131, (2.4.20)], [132, (1.3.8), (1.5.44)], (8.1.139), (8.1.116) (which continues to hold even when Ω is an exterior domain, thanks to the cancellation property of g; see (8.1.139)), and [132, Lemma 1.3.1] we conclude that u satisfies all properties in (8.1.95) with $n = 2$, and that (8.1.97) holds.

Turning our attention to the issue of uniqueness when $n = 2$, suppose u is a null-solution of (8.1.95) in the two-dimensional setting. Let us first assume that Ω is an exterior domain. Once again, the Fatou-type result established in [131, Theorem 3.3.4] ensures that (8.1.112)-(8.1.113) hold. Thanks to these properties and [130, (5.2.92)], we may invoke [131, Corollary 1.5.4] which shows that there exists a constant $c \in \mathbb{C}^M$ with the property that (cf. [131, (1.5.81)])

$$u = -\mathscr{S}\big(\partial_\nu^A u\big) + c \text{ in } \Omega. \tag{8.1.142}$$

Together, (8.1.141) and (8.1.142) imply $\mathscr{S}\big(\partial_\nu^A u\big) = O(1)$ as $|x| \to \infty$ which, by the last part of [132, Lemma 1.3.1, *(iii)*] (whose validity is ensured by (8.1.99)), forces

$$\int_{\partial\Omega} \partial_\nu^A u \, d\sigma = 0 \in \mathbb{C}^M. \tag{8.1.143}$$

If we also take nontangential boundary traces in (8.1.142), at σ-a.e. point on $\partial\Omega$ we obtain $0 = u\big|_{\partial\Omega}^{\kappa-\text{n.t.}} = -S\big(\partial_\nu^A u\big) + c$, hence

$$S\big(\partial_\nu^A u\big) = c \text{ on } \partial\Omega. \tag{8.1.144}$$

From (8.1.143), (8.1.144), and (7.1.393) we then obtain $S_{\text{aug}}\big(\partial_\nu^A u, -c\big) = (0, 0)$. Granted this, the invertibility result recorded in (7.1.389) forces $\partial_\nu^A u = 0$ and $c = 0$. Using this back in (8.1.142) then shows that $u = 0$ in Ω. Hence, uniqueness in the case when Ω is an exterior domain in the two-dimensional setting has been established. The case when $\Omega \subseteq \mathbb{R}^2$ is bounded is handled in a very similar fashion (the reasoning is actually simpler, since the behavior at infinity is no longer an issue). This completes the treatment of item *(2)*.

Turning our attention to item *(3)*, the claims pertaining to the Regularity Problem (8.1.101) are dealt with in a very similar fashion to what we have proved already. Granted this, all desired conclusions for the Regularity Problems (8.1.100) and (8.1.102) become consequence of what we have proved so far, with $\mathbb{X}'$ in place of $\mathbb{X}$ (a permissible choice, thanks to [130, Proposition 5.1.14]). Finally, the claim in item *(4)* is seen from Theorem 8.6.6, Theorem 8.6.8, Example 1.6.14, and Remark 8.6.9.□

Moving on, the goal is to prove the well-posedness of the Dirichlet Problem[22] for weakly elliptic systems in δ-oscillating AR domains with boundary data in Lebesgue spaces L^p with $1 < p < \infty$. Our main result in this regard (i.e., Theorem 8.1.6 stated shortly) refines work of S. Hofmann, M. Mitrea, and M. Taylor in [81] where the case of Laplacian and the Lamé system in regular SKT domains with compact boundaries has been considered. The special case when Ω is an infinitesimally flat AR domain (cf. Definition 3.4.1) may also be thought of a sharp version of [52, Theorem 2.3, p. 183] where E. Fabes, M. Jodeit, and N. Rivière have solved the Dirichlet Problem for the Laplacian in bounded $\mathscr{C}^1$ domains with boundary data in L^p for $p \in (1, \infty)$ arbitrary. The well-posedness of the Dirichlet Problem for the Laplacian in bounded Lipschitz domains with data in L^2 has via the method of layer potentials has been first established by G. Verchota in [203, Corollary 3.2, p. 588] and subsequently extended by B. Dahlberg and C. Kenig in [38, Theorem 4.18, p. 463] to allow for data in L^p with p in the optimal range[23] $(2 - \varepsilon, \infty)$ for some small $\varepsilon > 0$ depending only the Lipschitz character of the domain in question. The Dirichlet Problem for the Laplace-Beltrami operator on Lipschitz subdomains of Riemannian manifolds with boundary data in Lebesgue spaces has been treated by M. Mitrea and M. Taylor in [153], [154], [156]. As for genuine systems, [40, Theorem 3.6, p. 813] contains the solution of the Dirichlet Problem for the Lamé system in bounded Lipschitz domains with boundary in data in L^2. A similar well-posedness result for the Dirichlet Problem for generic Legendre-Hadamard elliptic systems with real coefficients in bounded Lipschitz domains may be found in [65, Theorem 2.1, p. 384]. See also [193] for further results of this flavor. Additional references may be found in [93].

To close this preamble, we note that the interest in the correlation between the regularity of the underlying domain and the solvability of the Dirichlet Problem goes back a long way. Shortly after Erik Ivar Fredholm succeeded in proving the existence for the Dirichlet Problem for the Laplacian in smooth domains with compact boundary via integral methods[24] in 1899-1900, David Hilbert raised the issue[26] of studying the Dirichlet Problem in domains with corners, a scenario in which Fredholm's theory no longer applies directly.

---

[22] named after Johann Peter Gustav Lejeune Dirichlet (1805-1859) who, in the process of investigating Laplace's problem on the stability of the solar system, was led to consider (in lectures he gave in 1856-1857) the problem of finding a harmonic function in a closed region with a prescribed continuous boundary trace

[23] relative to the class of Lipschitz domains

[24] In a letter dated August 8, 1899, Fredholm informed Mittag-Leffler about his novel method for solving integral equations, regarded as functional equations. Fredholm immediately applied this method to prove existence for the Dirichlet Problem. This was communicated to Poincaré in December 1899, and subsequently published[25] in 1900; cf. [61]. Fredholm also gave a lecture on this topic at the 1900 International Congress of Mathematicians held in Paris, on August 9, titled *L'inversion des intégrales définies et son application aux problèmes de la Physique mathématique*. A summary of his talk was subsequently published in 1902 in Comptes Rendus; cf. [62]. The full version of Fredholm's work on the theory which now bears his name appeared in 1903 in an issue of Acta Mathematica dedicated to Abel (interestingly, Fredholm proposes calling generic integral equations *équation fonctionnelle abélienne*; cf. [63, p. 365]).

[26] Hilbert posed this as a research question to Oliver Dimon Kellogg, while the latter was pursuing his doctorate in Göttigen in the early 1900's.

Theorem 8.1.6 (The Dirichlet Problem with Data in Lebesgue Spaces) *Fix two integers $n, M \in \mathbb{N}$ with $n \geq 2$ and suppose $\Omega \subseteq \mathbb{R}^n$ is an Ahlfors regular domain with compact boundary. Abbreviate $\sigma := \mathcal{H}^{n-1}\lfloor\partial\Omega$ and denote by ν the geometric measure theoretic outward unit normal to Ω. Also, let L be a homogeneous, second-order, constant complex coefficient, $M \times M$ system in $\mathbb{R}^n$. Finally, fix some integrability exponent $p \in (1, \infty)$ along with arbitrary aperture parameter $\kappa > 0$, and consider the Dirichlet Problem*[27]

$$\begin{cases} u \in \left[\mathscr{C}^\infty(\Omega)\right]^M, \quad Lu = 0 \text{ in } \Omega, \\ \mathcal{N}_\kappa u \in L^p(\partial\Omega, \sigma), \quad u\big|_{\partial\Omega}^{\kappa-\text{n.t.}} = f \in \left[L^p(\partial\Omega, \sigma)\right]^M, \\ u(x) = O(|x|^{2-n}) \text{ as } |x| \to \infty \text{ if } \Omega \text{ is unbounded.} \end{cases} \tag{8.1.145}$$

Then there exists $\delta_0 \in (0, 1)$ depending only on n, p, L, the ADR *constants of $\partial\Omega$, and the relative flatness ratio $\wp(\Omega)/\text{diam}\,\partial\Omega$ with the property that when*

$$\text{dist}\left(\nu, \, [\text{VMO}\,(\partial\Omega, \sigma)]^n\right) < \delta_0 \tag{8.1.146}$$

where the distance is considered in the space $\left[\text{BMO}\,(\partial\Omega, \sigma)\right]^n$ (hence if Ω is a δ_0-oscillating AR *domain*[28] *cf. Definition 3.4.1), the following results are true:*

(1) If one assumes that

$$L \text{ is weakly elliptic with } \mathfrak{A}_L^{\text{dis}} \neq \varnothing \text{ and } \mathfrak{A}_{L^\top}^{\text{dis}} \neq \varnothing, \tag{8.1.147}$$

then the Dirichlet Problem (8.1.145) *is Fredholm solvable, in the following precise sense: First, if the system L is weakly elliptic with $\mathfrak{A}_L^{\text{dis}} \neq \varnothing$, the space of admissible boundary data for the Dirichlet Problem* (8.1.145)*, i.e.,*

$$\Big\{u\big|_{\partial\Omega}^{\kappa-\text{n.t.}} : u \in \left[\mathscr{C}^\infty(\Omega)\right]^M, \; Lu = 0 \text{ in } \Omega, \; \mathcal{N}_\kappa u \in L^p(\partial\Omega, \sigma), \; u\big|_{\partial\Omega}^{\kappa-\text{n.t.}} \text{ exists,}$$
$$\text{and } u(x) = O(|x|^{2-n}) \text{ as } |x| \to \infty \text{ if } \Omega \text{ is unbounded}\Big\}, \tag{8.1.148}$$

is a closed linear subspace of finite codimension in $\left[L^p(\partial\Omega, \sigma)\right]^M$, and a solution u of (8.1.145) *exists if and only if the boundary datum f belongs to* (8.1.148)*. Second, if L is weakly elliptic with $\mathfrak{A}_{L^\top}^{\text{dis}} \neq \varnothing$, the space of null-solutions of the boundary problem* (8.1.145)*, i.e.,*

$$\Big\{u \in \left[\mathscr{C}^\infty(\Omega)\right]^M : Lu = 0, \; \mathcal{N}_\kappa u \in L^p(\partial\Omega, \sigma), \; u\big|_{\partial\Omega}^{\kappa-\text{n.t.}} = 0, \tag{8.1.149}$$
$$\text{and } u(x) = O(|x|^{2-n}) \text{ as } |x| \to \infty \text{ if } \Omega \text{ is unbounded}\Big\},$$

is finite dimensional; in addition,

---

[27] in view of work in [129, §8.5], one may replace the nontangential maximal operator by the tangential maximal operator, associated as in [129, Definition 8.5.1] with a sufficiently large power

[28] in particular, if Ω is an infinitesimally flat AR domain

$$\text{for any null-solution } u \text{ of the boundary problem (8.1.145) (i.e., for any function } u \text{ belonging to the space in (8.1.149)), } \mathcal{N}_\kappa(\nabla u) \text{ belongs to } L^p(\partial\Omega,\sigma) \text{ and } (\nabla u)\Big|_{\partial\Omega}^{\kappa-\mathrm{n.t.}} \text{ exists } \sigma\text{-a.e. on } \partial\Omega, \tag{8.1.150}$$

so, in particular,

$$\text{the space of null-solution for the Dirichlet Problem (8.1.145) (cf. (8.1.149)) coincides with the space of null-solution for the Regularity Problem (8.1.3) (cf. (8.1.8)).} \tag{8.1.151}$$

Finally, the Fredholm solvability result just described is optimal, since in the class of operators L for which $\mathfrak{A}_L^{\rm dis} = \varnothing$ *or* $\mathfrak{A}_{L^\top}^{\rm dis} = \varnothing$ *it may happen that the Dirichlet Problem* (8.1.145) *fails to be Fredholm solvable even when the underlying domain* Ω *is the unit disk in the plane (or the complement of its closure). In fact, the aforementioned failure of Fredholm solvability may be total, in the sense that for the problem* (8.1.145), *both the space of admissible boundary data may have infinite codimension in the full space of boundary data, and the space of null-solutions may be infinite dimensional.*

(2) If (8.1.147) *is strengthened to*[29]

$$L \text{ satisfies the Legendre-Hadamard strong ellipticity condition (cf. [131, Definition 1.3.2]) and one has } \mathfrak{A}_L^{\rm dis} \neq \varnothing. \tag{8.1.152}$$

the Dirichlet Problem (8.1.145) *is actually well posed. More specifically, the boundary problem* (8.1.145) *has a unique solution, and this solution enjoys the additional property that, for a constant* $C = C(\Omega, L, p, \kappa) \in (0, \infty)$,

$$\|f\|_{[L^p(\partial\Omega,\sigma)]^M} \leq \big\|\mathcal{N}_\kappa u\big\|_{L^p(\partial\Omega,\sigma)} \leq C\|f\|_{[L^p(\partial\Omega,\sigma)]^M}. \tag{8.1.153}$$

Lastly, under the additional assumption that Ω *is bounded, the unique solution of the boundary value problem* (8.1.145) *satisfies (in a quantitative sense)*

$$u \in \big[W^{1,p}_{1-\frac{1}{p},\star}(\Omega)\big]^M \cap \Big[B^{p,q_0}_{\frac{1}{p}}(\Omega)\Big]^M \cap \Big[F^{p,q_1}_{\frac{1}{p}}(\Omega)\Big]^M \text{ whenever } p \leq q_0 \leq \infty \text{ and } \tfrac{n}{n+1+1/p} < q_1 \leq \infty, \tag{8.1.154}$$

as well as

$$u \in \big[H^{1/2}(\Omega)\big]^M \text{ if } p = 2. \tag{8.1.155}$$

(3) Similar solvability properties to the ones described in item (1) and the first part of item (2) are true for the version of the Dirichlet Problem formulated for boundary data in Muckenhoupt weighted Lebesgue spaces, namely[30]

---

29 Proposition 1.4.1 ensures that if L is as in (8.1.152) then $\mathfrak{A}_L^{\rm dis} \neq \varnothing$ and $\mathfrak{A}_{L^\top}^{\rm dis} \neq \varnothing$

30 once again, one may replace the nontangential maximal operator by the tangential maximal operator, associated as in [129, Definition 8.5.1] with a sufficiently large power; cf. [129, §8.5]

$$\begin{cases} u \in \big[\mathscr{C}^\infty(\Omega)\big]^M, \quad Lu = 0 \ \textit{in}\ \Omega, \\ \mathcal{N}_\kappa u \in L^p(\partial\Omega, w), \quad u\big|_{\partial\Omega}^{\kappa-\mathrm{n.t.}} = f \in \big[L^p(\partial\Omega, w)\big]^M, \\ u(x) = O(|x|^{2-n}) \ \textit{as}\ |x| \to \infty \ \textit{if}\ \Omega \ \textit{is unbounded}, \end{cases} \tag{8.1.156}$$

for each given weight $w \in A_p(\partial\Omega, \sigma)$ *(now allowing the threshold* δ_0 *in* (8.1.146) *to also depend on* $[w]_{A_p}$*).*

In relation to this theorem it is important to remark that, as seen from Proposition 1.6.12 and [131, Proposition 1.4.39],

for the scalar case, i.e., if $M = 1$, and for dimensions $n \geq 3$, in place of (8.1.152) one may simply ask that the operator L is weakly elliptic. (8.1.157)

Also, in the scalar case in the two-dimensional setting (i.e., for $M = 1$ and $n = 2$), if L satisfies the Legendre-Hadamard strong ellipticity condition then Proposition 1.6.11 implies $\mathfrak{A}_L^{\mathrm{dis}} \neq \varnothing$. Moreover, according to (1.6.141) in Proposition 1.6.13 in the scalar case in the two-dimensional setting (i.e., for $M = 1$ and $n = 2$), we may omit (8.1.152) altogether if the weakly elliptic operator L has real coefficients.

It is also relevant to note that the optimal range of integrability exponents p for which the Dirichlet Problem for the Laplacian in a bounded Lipschitz domain $\Omega \subseteq \mathbb{R}^n$ is $(2-\varepsilon_\Omega, \infty)$; see [38]. Here, $\varepsilon_\Omega \in (0, 1]$ is a parameter which depends only on the set Ω, and $\varepsilon_\Omega \to 0^+$ as the Lipschitz character of Ω deteriorates. This points to the fact that the well-posedness results in Theorem 8.1.6 may fail to materialize for a generic exponent $p \in (1, \infty)$ in the absence of suitable flatness assumptions on the underlying domain, of the sort we hypothesize in (8.1.146).

Concerning the very formulation of Theorem 8.1.6, similar considerations to those made in Remark 8.1.2 are presently applicable. Also, as in the case of the Regularity Problem (see Remark 8.1.3) we may further expand the scope of the well-posedness result in item *(2)* of Theorem 8.1.6 beyond the class of Legendre-Hadamard strongly elliptic systems, as indicated below:

Remark 8.1.7 *Under the same background assumptions as in Theorem 8.1.6 (up to, and including,* (8.1.146)*), we actually have well-posedness for the Dirichlet Problem* (8.1.145) *whenever the system* L *has the property that* $L \approx \widetilde{L}$ *(in the sense explained just above Remark 8.1.3), where* $\widetilde{L}$ *is a homogeneous, second-order, constant complex coefficient,* $M \times M$ *systems in* $\mathbb{R}^n$ *satisfying the Legendre-Hadamard strong ellipticity condition and with* $\mathfrak{A}_{\widetilde{L}}^{\mathrm{dis}} \neq \varnothing$.

This follows from the well-posedness result in item (2) of Theorem 8.1.6, [131, Proposition 1.4.3], *Proposition 1.2.3, Proposition 1.2.4,* [130, Proposition 11.6.1], [130, Proposition 11.5.20] *(cf. the "surface-to-surface" change of variable formula* [129, (11.6.4)]*), and the transformational properties of Euclidean domains in* [80].

A final comment has to do with harmonic functions in the unit disk $\mathbb{D}$ in the plane possessing a vanishing nontangential boundary trace at $\mathcal{H}^1$-a.e. point on $\mathbb{T} := \partial\mathbb{D}$. Of course, if the *unrestricted* limit of a harmonic function u in $\mathbb{D}$ at each point on $\mathbb{T}$ is

0, then the Maximum Principle[31] guarantees that $u = 0$ in $\mathbb{D}$. Examples of non-zero harmonic functions with vanishing nontangential boundary traces at *every* point on $\mathbb{T}$ may be found in [11] (albeit their nontangential maximal functions exhibit stronger singularities than [131, (4.4.27)]). The proof of Theorem 8.1.6 is presented next:

Proof of Theorem 8.1.6 Pick $\delta_0 \in (0,1)$ sufficiently small, depending on n, p, L, the ADR constants of $\partial\Omega$, and the relative flatness ratio $\wp(\Omega)/\mathrm{diam}\,\partial\Omega$, such that the conclusions in Theorem 7.1.9 are valid. By further decreasing $\delta_0 \in (0,1)$ if necessary, Theorem 3.4.2 ensures that Ω is both an (ε,δ)-domain, a two-sided NTA domain, and a UR domain satisfying a two-sided local John condition.

To deal with the claims in item *(1)*, first work under the assumptions that the system L is weakly elliptic and has the property that $\mathfrak{A}_L^{\mathrm{dis}} \neq \varnothing$. Pick $A \in \mathfrak{A}_L^{\mathrm{dis}}$. Taking $\delta_0 \in (0,1)$ small, Theorem 7.1.1 guarantees that the operator

$$\tfrac{1}{2}I + K_A : \big[L^p(\partial\Omega,\sigma)\big]^M \longrightarrow \big[L^p(\partial\Omega,\sigma)\big]^M \text{ is Fredholm with index zero.} \tag{8.1.158}$$

In particular,

$$\begin{aligned} &\mathrm{Im}\Big(\tfrac{1}{2}I + K_A; \big[L^p(\partial\Omega,\sigma)\big]^M\Big) \text{ is a closed subspace} \\ &\qquad \text{of finite codimension in } \big[L^p(\partial\Omega,\sigma)\big]^M. \end{aligned} \tag{8.1.159}$$

Thanks to [132, (1.3.24), (1.3.26)], [129, (7.7.106)], [132, Theorem 1.5.1, items *(iv)*], and definitions, it follows that $\mathrm{Im}\Big(\tfrac{1}{2}I + K_A; \big[L^p(\partial\Omega,\sigma)\big]^M\Big)$ is contained in the space of admissible boundary data for the Dirichlet Problem (8.1.145). Granted this, we conclude from (8.1.159) and [130, Theorem 1.1.3, item *(3)*] that, in the case when L is weakly elliptic with $\mathfrak{A}_L^{\mathrm{dis}} \neq \varnothing$, the space described in (8.1.148) has, indeed, closed and of finite codimension in $\big[L^p(\partial\Omega,\sigma)\big]^M$.

Next, work under the assumption that L is weakly elliptic with $\mathfrak{A}_{L^\top}^{\mathrm{dis}} \neq \varnothing$, and pick $A \in \mathfrak{A}_L$ such that $A^\top \in \mathfrak{A}_{L^\top}^{\mathrm{dis}}$. Also, set $p' := (1-1/p)^{-1} \in (1,\infty)$. Assuming $\delta_0 \in (0,1)$ to be sufficiently small, Theorem 7.1.1 ensures that

$$\tfrac{1}{2}I + K_{A^\top} : \big[L^{p'}(\partial\Omega,\sigma)\big]^M \longrightarrow \big[L^{p'}(\partial\Omega,\sigma)\big]^M \text{ is Fredholm with index zero,} \tag{8.1.160}$$

$$\tfrac{1}{2}I + K_{A^\top} : \big[L_1^{p'}(\partial\Omega,\sigma)\big]^M \longrightarrow \big[L_1^{p'}(\partial\Omega,\sigma)\big]^M \text{ is Fredholm with index zero.} \tag{8.1.161}$$

The claim that we make is that there exists a finite (possibly vacuous) family of functions $\phi_1,\dots,\phi_N \in \big[\mathscr{C}_c^\infty(\mathbb{R}^n)\big]^M$ satisfying the following three properties: First,

$$\begin{aligned} &\phi_1\big|_{\partial\Omega},\dots,\phi_N\big|_{\partial\Omega} \text{ are linearly independent,} \\ &\text{modulo the space } \mathrm{Im}\Big(\tfrac{1}{2}I + K_{A^\top}; \big[L^{p'}(\partial\Omega,\sigma)\big]^M\Big), \end{aligned} \tag{8.1.162}$$

[31] whose applicability is ensured by the observation that if $f(z) := \lim_{\mathbb{D}\ni\zeta\to z} u(\zeta)$ for each $z \in \mathbb{T}$ is a well-defined continuous function on $\mathbb{T}$ then u has a continuous extension to $\overline{\mathbb{D}}$

Second, for any $f \in \big[L^{p'}(\partial\Omega,\sigma)\big]^M$ there exists $\{\lambda_j\}_{1\le j\le N} \in \mathbb{C}^N$ uniquely determined by f, and $g \in \big[L^{p'}(\partial\Omega,\sigma)\big]^M$ not necessarily unique, such that

$$f = \big(\tfrac{1}{2}I + K_{A^\top}\big)g + \sum_{j=1}^{N} \lambda_j \phi_j\big|_{\partial\Omega}. \tag{8.1.163}$$

Third, the assignment

$$\big[L^{p'}(\partial\Omega,\sigma)\big]^M \ni f \longmapsto \{\lambda_j\}_{1\le j\le N} \in \mathbb{C}^N \tag{8.1.164}$$

(which, by the previous property, is unambiguously defined) is linear and bounded.

To justify the above claim, first recall from [129, Corollary 3.7.3] that

$$\Big\{\phi\big|_{\partial\Omega} : \phi \in \big[\mathscr{C}_c^\infty(\mathbb{R}^n)\big]^M\Big\} \text{ is a dense subspace of } \big[L^{p'}(\partial\Omega,\sigma)\big]^M. \tag{8.1.165}$$

Also, as a consequence of (8.1.160),

$$\begin{array}{c}\mathrm{Im}\Big(\tfrac{1}{2}I + K_{A^\top};\big[L^{p'}(\partial\Omega,\sigma)\big]^M\Big) \text{ is a closed subspace} \\ \text{of finite codimension in } \big[L^{p'}(\partial\Omega,\sigma)\big]^M.\end{array} \tag{8.1.166}$$

From (8.1.165), (8.1.166), and [130, Theorem 1.1.3, item *(3)*] we then conclude that

$$\big[L^{p'}(\partial\Omega,\sigma)\big]^M = \Big\{\phi\big|_{\partial\Omega} : \phi \in \big[\mathscr{C}_c^\infty(\mathbb{R}^n)\big]^M\Big\} + \mathrm{Im}\Big(\tfrac{1}{2}I + K_{A^\top};\big[L^{p'}(\partial\Omega,\sigma)\big]^M\Big).$$

Thus $\big\{\big[\phi\big|_{\partial\Omega}\big] : \phi \in \big[\mathscr{C}_c^\infty(\mathbb{R}^n)\big]^M\big\}$ coincides with the finite dimensional vector space

$$\big[L^{p'}(\partial\Omega,\sigma)\big]^M \Big/ \mathrm{Im}\Big(\tfrac{1}{2}I + K_{A^\top};\big[L^{p'}(\partial\Omega,\sigma)\big]^M\Big), \tag{8.1.167}$$

so we may choose $\phi_1, \dots \phi_N \in \big[\mathscr{C}_c^\infty(\mathbb{R}^n)\big]^M$ such that the equivalence classes $\big[\phi_1\big|_{\partial\Omega}\big], \dots \big[\phi_1\big|_{\partial\Omega}\big]$ constitute a basis for the quotient space in (8.1.167). With this in hand, the properties described in (8.1.162) and (8.1.163) follow. The fact that, in the context of (8.1.163), the function $f \in \big[L^{p'}(\partial\Omega,\sigma)\big]^M$ determines the coefficients $\{\lambda_j\}_{1\le j\le N} \in \mathbb{C}^N$ uniquely makes the assignment in (8.1.164) linear. To show that said assignment is also bounded, consider the operator

$$\begin{array}{l} T : \big[L^{p'}(\partial\Omega,\sigma)\big]^M \oplus \mathbb{C}^N \to \big[L^{p'}(\partial\Omega,\sigma)\big]^M \\ T\big(g, \{\lambda_j\}_{1\le j\le N}\big) := \big(\tfrac{1}{2}I + K_{A^\top}\big)g + \sum_{j=1}^N \lambda_j \phi_j\big|_{\partial\Omega}, \\ \text{for all } g \in \big[L^{p'}(\partial\Omega,\sigma)\big]^M \text{ and } \{\lambda_j\}_{1\le j\le N} \in \mathbb{C}^N. \end{array} \tag{8.1.168}$$

Then T is well defined, linear, bounded and, thanks to (8.1.163), also surjective. Since the operator T acts between two Banach spaces, the latter implies "ontoness with control" (cf. [130, Proposition 1.2.13]), so there exists some $C \in (0,\infty)$ with the property that any function $f \in \big[L^{p'}(\partial\Omega,\sigma)\big]^M$ may be written as in (8.1.163) for a choice of $g \in \big[L^{p'}(\partial\Omega,\sigma)\big]^M$ and $\{\lambda_j\}_{1\leq j\leq N} \in \mathbb{C}^N$ satisfying

$$\|g\|_{[L^{p'}(\partial\Omega,\sigma)]^M} + \sum_{j=1}^{N} |\lambda_j| \leq C\|f\|_{[L^{p'}(\partial\Omega,\sigma)]^M}. \tag{8.1.169}$$

Thus, $\sum_{j=1}^{N} |\lambda_j| \leq C\|f\|_{[L^{p'}(\partial\Omega,\sigma)]^M}$ hence the mapping (8.1.164) is bounded.

Next, we claim that there exist functions $\psi_1,\dots,\psi_N \in \big[L^p(\partial\Omega,\sigma)\big]^M$ such that

$$\int_{\partial\Omega} \big\langle \phi_k\big|_{\partial\Omega}, \psi_j\big\rangle\, d\sigma = \delta_{jk} \text{ for any } j,k \in \{1,\dots,N\}, \tag{8.1.170}$$

and for any $j \in \{1,\dots,N\}$ we have

$$\int_{\partial\Omega} \big\langle (\tfrac{1}{2}I + K_{A^\top})g, \psi_j\big\rangle\, d\sigma = 0 \text{ for all } g \in \big[L^{p'}(\partial\Omega,\sigma)\big]^M. \tag{8.1.171}$$

To produce such a family, fix $j_o \in \{1,\dots,N\}$ and consider the assignment

$$\big[L^{p'}(\partial\Omega,\sigma)\big]^M \ni f \longmapsto \lambda_{j_o} \in \mathbb{C}, \tag{8.1.172}$$

where the coefficients $\{\lambda_j\}_{1\leq j\leq N} \in \mathbb{C}^N$ are associated with $f \in \big[L^{p'}(\partial\Omega,\sigma)\big]^M$ as in (8.1.163). This assignment is well defined, linear, and bounded, since (8.1.164) is so. As such, it belongs to $\Big(\big[L^{p'}(\partial\Omega,\sigma)\big]^M\Big)^* = \big[L^p(\partial\Omega,\sigma)\big]^M$. Consequently, there exists a function $\psi_{j_o} \in \big[L^p(\partial\Omega,\sigma)\big]^M$ with the property that

$$\int_{\partial\Omega} \langle f, \psi_{j_o}\rangle\, d\sigma = \lambda_{j_o} \text{ for any } f \in \big[L^{p'}(\partial\Omega,\sigma)\big]^M \tag{8.1.173}$$

where $\{\lambda_j\}_{1\leq j\leq N} \in \mathbb{C}^N$ are associated with $f \in \big[L^{p'}(\partial\Omega,\sigma)\big]^M$ as in (8.1.163). Varying j_o yields a family of functions $\psi_1,\dots,\psi_N \in \big[L^p(\partial\Omega,\sigma)\big]^M$. In view of (8.1.173) and the fact that we obviously have

$$\phi_k\big|_{\partial\Omega} = (\tfrac{1}{2}I + K_{A^\top})0 + \sum_{j=1}^{N} \delta_{jk}\phi_j\big|_{\partial\Omega} \text{ for any } k \in \{1,\dots,N\}, \tag{8.1.174}$$

$$(\tfrac{1}{2}I + K_{A^\top})g = (\tfrac{1}{2}I + K_{A^\top})g + \sum_{j=1}^{N} 0\cdot\phi_j\big|_{\partial\Omega} \text{ for all } g \in \big[L^{p'}(\partial\Omega,\sigma)\big]^M \tag{8.1.175}$$

the properties demanded in (8.1.170)-(8.1.171) follow.

Moving on, denote by E_L and $E_{L^\top}$ the fundamental solutions associated with the systems L and $L^\top$ as in [131, Theorem 1.4.2]. Let us also fix a point $x_* \in \mathbb{R}^n \setminus \overline{\Omega}$. Finally, pick an arbitrary point $x_0 \in \Omega$ along with an index $\beta \in \{1,\dots,M\}$. Then

(8.1.163) guarantees that there exist a function $g^{(\beta)}_{x_0} \in \big[L^{p'}(\partial\Omega,\sigma)\big]^M$ and a family of numbers $\{\lambda^{(\beta)}_j(x_0)\}_{1\leq j\leq N} \in \mathbb{C}^N$ with the property that

$$(E_{L^\top})_{\bullet\beta}(\cdot - x_*)\big|_{\partial\Omega} - (E_{L^\top})_{\bullet\beta}(\cdot - x_0)\big|_{\partial\Omega} = \big(\tfrac{1}{2}I + K_{A^\top}\big)g^{(\beta)}_{x_0} + \sum_{j=1}^{N} \lambda^{(\beta)}_j(x_0)\phi_j\big|_{\partial\Omega}. \tag{8.1.176}$$

Since $\phi\big|_{\partial\Omega} \in \big[L^{p'}_1(\partial\Omega,\sigma)\big]^M$ for each $\phi \in \big[\mathscr{C}^\infty_c(\mathbb{R}^n)\big]^M$, this entails

$$\big(\tfrac{1}{2}I + K_{A^\top}\big)g^{(\beta)}_{x_0} \in \big[L^{p'}_1(\partial\Omega,\sigma)\big]^M. \tag{8.1.177}$$

Together, (8.1.177), (8.1.160), (8.1.161), the abstract regularity result from [130, Lemma 2.1.4] then show that we actually have

$$g^{(\beta)}_{x_0} \in \big[L^{p'}_1(\partial\Omega,\sigma)\big]^M. \tag{8.1.178}$$

To proceed, in $\Omega\setminus\{x_0\}$ we define

$$G^{x_0}_{\bullet\beta} := (E_{L^\top})_{\bullet\beta}(\cdot - x_*) - (E_{L^\top})_{\bullet\beta}(\cdot - x_0) - \mathcal{D}_{A^\top}g^{(\beta)}_{x_0} - \sum_{j=1}^{N}\lambda^{(\beta)}_j(x_0)\phi_j. \tag{8.1.179}$$

Finally, for each $j \in \{1,\dots,N\}$ set $\lambda_j(x_0) := \big(\lambda^{(\beta)}_j(x_0)\big)_{1\leq\beta\leq M} \in \mathbb{C}^M$, and introduce the matrix-valued function $G^{x_0} := \big(G^{x_0}_{\alpha\beta}\big)_{1\leq\alpha,\beta\leq M}$. Then G^{x_0} may be thought of as a "pseudo Green function" for $L^\top$ in Ω with pole at x_0 since, thanks to (8.1.176), [131, Theorem 1.4.2], [132, Theorem 1.5.1, item *(ii)*], and (8.1.178), it satisfies:

$$\begin{cases} G^{x_0} \in \big[L^1_{\rm loc}(\Omega,\mathcal{L}^n)\big]^{M\times M}, \\ L^\top G^{x_0} = -\delta_{x_0}\cdot I_{M\times M} - \sum_{j=1}^{N}(L^\top\phi_j)\big|_\Omega \otimes \lambda_j(x_0) \text{ in } \big[\mathcal{D}'(\Omega)\big]^{M\times M}, \\ \big(\nabla G^{x_0}\big)\Big|^{\kappa-{\rm n.t.}}_{\partial\Omega} \text{ exists (in } \mathbb{C}^{n\cdot M^2}) \text{ at } \sigma\text{-a.e. point on } \partial\Omega, \\ G^{x_0}\Big|^{\kappa-{\rm n.t.}}_{\partial\Omega} = 0 \in \mathbb{C}^{M\times M} \text{ at } \sigma\text{-a.e. point on } \partial\Omega, \\ \mathcal{N}^\rho_\kappa G^{x_0} \in L^{p'}(\partial\Omega,\sigma) \text{ and } \mathcal{N}^\rho_\kappa(\nabla G^{x_0}) \in L^{p'}(\partial\Omega,\sigma), \\ G^{x_0}(x) = O(|x|^{1-n}) \text{ and } \nabla G^{x_0}(x) = O(|x|^{-n}) \text{ as } |x| \longrightarrow \infty \\ \text{whenever the set } \Omega \text{ happens to be an exterior domain,} \end{cases} \tag{8.1.180}$$

for some truncation parameter $0 < \rho < \operatorname{dist}(x_0,\partial\Omega)$. Consider now a null-solution u of the Dirichlet Problem (8.1.145), i.e., a function satisfying

$$\begin{cases} u \in \big[\mathscr{C}^\infty(\Omega)\big]^M, \quad Lu = 0 \text{ in } \Omega, \\ \mathcal{N}_\kappa u \in L^p(\partial\Omega,\sigma), \quad u\Big|^{\kappa-{\rm n.t.}}_{\partial\Omega} = 0 \text{ at } \sigma\text{-a.e. point on } \partial\Omega, \\ u(x) = O(|x|^{2-n}) \text{ as } |x| \to \infty \text{ if } \Omega \text{ is unbounded.} \end{cases} \tag{8.1.181}$$

In particular, [129, Proposition 8.6.3] implies that

$$u \in \big[L^{np/(n-1)}_{\rm bdd}(\Omega, \mathcal{L}^n)\big]^M \subseteq \big[L^1_{\rm bdd}(\Omega, \mathcal{L}^n)\big]^M. \tag{8.1.182}$$

The idea is now to re-run the proof of [131, Theorem 4.4.1] for the current function $u = (u_\alpha)_{1\leq\alpha\leq M}$ and for $G := G^{x_0}$ with $x_0 \in \Omega$ arbitrary. For these choices, and having fixed $\beta \in \{1, \dots, M\}$, one defines the vector field $\vec{F}$ as in [131, (4.4.47)]. Observe that this enjoys similar properties as before expect the actual formula for its divergence. The only difference is that now, in place of [131, (4.4.52)], we have

$$\begin{aligned} I_2 &= u_\alpha (L_{A^\top} G_{\cdot\beta})_\alpha = u_\alpha (L^\top G_{\cdot\beta})_\alpha = u_\alpha\Big(-\delta_{\alpha\beta}\delta_{x_0} - \sum_{j=1}^N \lambda_j^{(\beta)}(x_0)\big(L^\top \phi_j\big|_\Omega\big)_\alpha\Big) \\ &= -u_\beta(x_0)\,\delta_{x_0} - \sum_{j=1}^N \lambda_j^{(\beta)}(x_0)\big(L^\top \phi_j\big|_\Omega\big)_\alpha u_\alpha. \end{aligned} \tag{8.1.183}$$

In view of (8.1.182), this implies that in place of [131, (4.4.53)] we now have

$$\operatorname{div}\vec{F} = -u_\beta(x_0)\,\delta_{x_0} - \sum_{j=1}^N \lambda_j^{(\beta)}(x_0)\big(L^\top \phi_j\big|_\Omega\big)_\alpha u_\alpha \in \mathscr{E}'(\Omega) + L^1(\Omega, \mathcal{L}^n).$$

In addition, when Ω is an exterior domain, from (8.1.180), (8.1.181), and [131, Theorem 1.5.9] it follows that as $|x| \to \infty$ we also have

$$\begin{aligned} G(x) &= O(|x|^{1-n}), \quad (\nabla G)(x) = O(|x|^{-n}), \\ u(x) &= O(|x|^{2-n}), \quad (\nabla u)(x) = O(|x|^{1-n}). \end{aligned} \tag{8.1.184}$$

Together with [131, (4.4.47)], this implies that $\vec{F}(x) = O(|x|^{2-2n})$ as $|x| \to \infty$. Thus, since $n \geq 2$, for each $\lambda \in (1, \infty)$ we have

$$\int_{B(0,\lambda R)\setminus B(0,R)} |\vec{F}(x)|\,\mathrm{d}\mathcal{L}^n(x) = o(R) \text{ as } R \to \infty. \tag{8.1.185}$$

In view of these properties and since we are currently assuming that $u\big|_{\partial\Omega}^{\kappa-\mathrm{n.t.}} = 0$, an application of the Divergence Theorem gives, in place of [131, (4.4.80)],

$$\begin{aligned} -u_\beta(x_0) - \sum_{j=1}^N \lambda_j^{(\beta)}(x_0) \int_\Omega \langle L^\top \phi_j, u\rangle\,\mathrm{d}\mathcal{L}^n &= {}_{(\mathscr{C}_b^\infty(\Omega))^*}\big(\operatorname{div}\vec{F}, 1\big)_{\mathscr{C}_b^\infty(\Omega)} \\ &= \int_{\partial\Omega} \nu\cdot\big(\vec{F}\big|_{\partial\Omega}^{\kappa-\mathrm{n.t.}}\big)\,\mathrm{d}\sigma = \int_{\partial\Omega} \big\langle u\big|_{\partial\Omega}^{\kappa-\mathrm{n.t.}}, \partial_\nu^{A^\top} G^{x_0}_{\cdot\beta}\big\rangle\,\mathrm{d}\sigma = 0, \end{aligned} \tag{8.1.186}$$

where ν is the geometric measure theoretic outward unit normal to the set Ω. Since $\beta \in \{1, \dots, M\}$ has been arbitrarily chosen, this leads to the conclusion that

$$u(x_0) = -\sum_{j=1}^N \lambda_j(x_0) \int_\Omega \langle L^\top \phi_j, u\rangle\,\mathrm{d}\mathcal{L}^n. \tag{8.1.187}$$

Given that $x_0 \in \Omega$ is arbitrary, this shows that

$$u = -\sum_{j=1}^{N} \lambda_j(\cdot) \int_\Omega \langle L^\top \phi_j, u\rangle \, d\mathcal{L}^n \text{ as functions in } \Omega, \tag{8.1.188}$$

so u necessarily belongs to the linear span of the $\mathbb{C}^M$-valued functions $\{\lambda_j(\cdot)\}_{1\leq j\leq N}$ which, of course, is finite dimensional.

To justify the additional properties claimed in (8.1.150), pick an arbitrary null-solution u of the boundary problem (8.1.145) and run the argument that has produced (8.1.188). For each fixed $j \in \{1, \dots, N\}$ we may (meaningfully) pair both sides of (8.1.176) with ψ_j and obtain

$$\begin{aligned}\int_{\partial\Omega} \big\langle (E_{L^\top})_{\cdot\beta}(\cdot - x_*)\big|_{\partial\Omega}, \psi_j\big\rangle \, d\sigma - \int_{\partial\Omega} \big\langle (E_{L^\top})_{\cdot\beta}(\cdot - x_0)\big|_{\partial\Omega}, \psi_j\big\rangle \, d\sigma \\ = \int_{\partial\Omega} \big\langle (\tfrac{1}{2}I + K_{A^\top}) g^{(\beta)}_{x_0}, \psi_j\big\rangle \, d\sigma + \lambda_j^{(\beta)}(x_0) = \lambda_j^{(\beta)}(x_0),\end{aligned} \tag{8.1.189}$$

thanks to (8.1.170)-(8.1.171). Based on [131, (1.4.32)] and (A.0.231), we may recast the second term in the left-most side as

$$\int_{\partial\Omega} \big\langle (E_{L^\top})_{\cdot\beta}(\cdot - x_0)\big|_{\partial\Omega}, \psi_j\big\rangle \, d\sigma = \int_{\partial\Omega} \big\langle (E_L)_{\beta\cdot}(x_0 - \cdot), \psi_j\big\rangle \, d\sigma = (\mathscr{S}_L \psi_j)_\beta(x_0),$$

where $\mathscr{S}_L$ is the boundary-to-domain single layer potential operator associated with the system L and the set Ω. The first term in the left-most side of (8.1.189) is a well-defined complex number, independent of x_0. In view of the arbitrariness of $x_0 \in \Omega$ and $\beta \in \{1, \dots, M\}$, we eventually conclude that for each $j \in \{1, \dots, N\}$ there exists $c_j \in \mathbb{C}^M$ such that, as functions in Ω, we have $\lambda_j(\cdot) = \mathscr{S}_L \psi_j + c_j$. Ultimately, this and (8.1.188) imply that there exist $\psi \in \big[L^p(\partial\Omega, \sigma)\big]^M$ and $c \in \mathbb{C}^M$ such that $u = \mathscr{S}_L \psi + c$ in Ω. At this stage, the additional properties claimed in (8.1.150) follow from this integral representation formula and Calderón-Zygmund theory (cf. [132, Theorem 1.5.1]). In turn, (8.1.150) readily implies (8.1.151).

Finally, the very first claim in item *(1)* is a consequence of what we have proved so far, while the very last claim in item *(1)* may be seen (working in the two-dimensional setting) from Theorem 8.6.5, Theorem 8.6.7, Example 1.6.14, and Remark 8.6.9. This completes the treatment of item *(1)* in the statement of the theorem.

To deal with item *(2)*, work under the assumptions made in (8.1.152). Assume first that $n \geq 3$. Then the invertibility result recorded in (7.1.234) guarantees that there exists $C \in (0, \infty)$ such that, given any $f \in \big[L^p(\partial\Omega, \sigma)\big]^M$,

$$\begin{aligned}&\text{there exists } g \in \big[L^p_{-1}(\partial\Omega, \sigma)\big]^M \text{ with} \\ &Sg = f \text{ and } \|g\|_{[L^p_{-1}(\partial\Omega,\sigma)]^M} \leq C\|f\|_{[L^p(\partial\Omega,\sigma)]^M}.\end{aligned} \tag{8.1.190}$$

With the boundary-to-domain single layer potential operator associated with L and Ω as in [132, (1.5.40)], consider

$$u := \mathscr{S}g \text{ in } \Omega. \tag{8.1.191}$$

Then from (8.1.191), [132, (1.5.41), (1.5.43), (1.5.44), (1.5.46)], (8.1.190), the very last property in [132, Theorem 1.5.1, item *(ix)*], [132, Lemma 1.3.1], [129, (8.9.8)], and [130, Proposition 11.8.2] we conclude that u satisfies all demands in (8.1.145) and that (8.1.153) holds. Also, the regularity properties formulated in (8.1.154)-(8.1.155) are implied by (8.1.191), (8.1.190), and [132, Theorem 4.3.3, *(4)*].

When $n = 2$, by further decreasing the value of $\delta \in (0, 1)$ if needed, we may assume that the conclusions in Theorem 7.1.10 are true. In particular, the invertibility result recorded in (7.1.246) guarantees the existence of a constant $C \in (0, \infty)$ such that, given any function $f \in \big[L^p(\partial\Omega, \sigma)\big]^M$, we may find $g \in \big[L^p_{-1}(\partial\Omega, \sigma)\big]^M$ and a constant $c \in \mathbb{C}^M$ with the property that

$$\begin{gathered} Sg + c = f, \quad |c| + \|g\|_{[L^p_{-1}(\partial\Omega,\sigma)]^M} \leq C\|f\|_{[L^p(\partial\Omega,\sigma)]^M} \\ \text{and } \langle g, \mathbf{e}_j\rangle = 0 \text{ for each } j \in \{1, \dots, M\}. \end{gathered} \tag{8.1.192}$$

If we define

$$u := \mathscr{S}g + c \text{ in } \Omega, \tag{8.1.193}$$

we may employ the vanishing moment property in (8.1.192) much as we have done in the proof of [132, (1.3.11)] to obtain

$$u(x) = O(1) \text{ as } |x| \to \infty. \tag{8.1.194}$$

Based on (8.1.193), (8.1.194), [132, (1.5.43), (1.5.44)], the very last statement in [132, Theorem 1.5.1, item *(x)*], and (8.1.192), we conclude that u satisfies all properties demanded in (8.1.145) and that (8.1.153) is true.

Moving on, uniqueness for the boundary value problem (8.1.145) follows from [131, Theorem 4.3.1] and the solvability aspect of Theorem 8.1.1 (used with $L^\top$ in place of L, and with $p' := (1 - 1/p)^{-1}$ in place of p). In this regard, is relevant to note that, thanks to (8.1.152), from Theorem 1.5.2 and [131, (1.3.5)] we see that

$$\begin{gathered} L^\top \text{ also satisfies the Legendre-Hadamard} \\ \text{strong ellipticity condition and } \mathfrak{A}^{\text{dis}}_{L^\top} \neq \varnothing. \end{gathered} \tag{8.1.195}$$

Finally, that the Dirichlet Problem with boundary data in Muckenhoupt weighted Lebesgue spaces from (8.1.156) also enjoys similar solvability properties to the ones described in item *(1)* and the first part of item *(2)* is proved analogously, now using (7.1.235) and (7.1.249). The proof of Theorem 8.1.6 is therefore complete. □

Next, we discuss the Dirichlet Problem for weakly elliptic systems in δ-oscillating AR domains, with boundary data in Morrey spaces, vanishing Morrey spaces, and block spaces. Our main result in this regard, Theorem 8.1.8 stated below, allows us to target any space in the aforementioned scales by choosing δ small enough. The Dirichlet Problem with data in Morrey spaces with $p = 2$, for a bounded Lipschitz domain Ω with connected boundary, and a homogeneous, second-order, constant real coefficient, symmetric system L in $\mathbb{R}^n$, satisfying the Legendre-Hadamard strong ellipticity condition, has been treated by Z. Shen in [192, Theorem 1.5] via an approach which make essential use of the setting and hypotheses just described. Our main result in this regard, Theorem 8.1.8 stated below, is new even in the case of the Laplacian on a $\mathscr{C}^1$ domain with compact boundary.

Theorem 8.1.8 (The Dirichlet Problem with Data in Morrey and Block Spaces) *Fix two integers $n, M \in \mathbb{N}$ with $n \geq 2$ and suppose $\Omega \subseteq \mathbb{R}^n$ is an Ahlfors regular domain with compact boundary. Abbreviate $\sigma := \mathcal{H}^{n-1}\lfloor\partial\Omega$ and denote by ν the geometric measure theoretic outward unit normal to Ω. Also, let L be a homogeneous, second-order, constant complex coefficient, $M \times M$ system in $\mathbb{R}^n$. Finally, fix some integrability exponents $p, q \in (1, \infty)$ satisfying $1/p + 1/q = 1$ along with some $\lambda \in (0, n-1)$ and an arbitrary aperture parameter $\kappa > 0$.*

Then there exists $\delta \in (0, 1)$ depending only on n, p, λ, L, the ADR *constants of $\partial\Omega$, and the relative flatness ratio $\wp(\Omega)/\mathrm{diam}\,\partial\Omega$ with the property that when*

$$\operatorname{dist}\Big(\nu\,,\,[\mathrm{VMO}\,(\partial\Omega,\sigma)]^n\Big) < \delta \tag{8.1.196}$$

with the distance considered in the space $\big[\mathrm{BMO}\,(\partial\Omega,\sigma)\big]^n$, (hence if Ω is a δ-oscillating AR *domain*[32]*; cf. Definition 3.4.1), the following statements are valid:*

(1) If one assumes that

$$L \text{ is weakly elliptic with } \mathfrak{A}^{\mathrm{dis}}_{L} \neq \varnothing \text{ and } \mathfrak{A}^{\mathrm{dis}}_{L^\top} \neq \varnothing, \tag{8.1.197}$$

then the Dirichlet Problem[33]

$$\begin{cases} u \in \big[\mathscr{C}^\infty(\Omega)\big]^M, \quad Lu = 0 \text{ in } \Omega, \\ \mathcal{N}_\kappa u \in M^{p,\lambda}(\partial\Omega,\sigma), \quad u\big|^{\kappa-\mathrm{n.t.}}_{\partial\Omega} = f \in \big[M^{p,\lambda}(\partial\Omega,\sigma)\big]^M, \\ u(x) = O(|x|^{2-n}) \text{ as } |x| \to \infty \text{ if } \Omega \text{ is unbounded,} \end{cases} \tag{8.1.198}$$

is Fredholm solvable, in the following precise sense: First, if the system L is weakly elliptic with $\mathfrak{A}^{\mathrm{dis}}_{L} \neq \varnothing$, the space of admissible boundary data for the Dirichlet Problem (8.1.198), *i.e.,*

$$\Big\{u\big|^{\kappa-\mathrm{n.t.}}_{\partial\Omega} : u \in \big[\mathscr{C}^\infty(\Omega)\big]^M,\ Lu = 0 \text{ in } \Omega,\ \mathcal{N}_\kappa u \in M^{p,\lambda}(\partial\Omega,\sigma),\ u\big|^{\kappa-\mathrm{n.t.}}_{\partial\Omega} \text{ exists,} \\ \text{and } u(x) = O(|x|^{2-n}) \text{ as } |x| \to \infty \text{ if } \Omega \text{ is unbounded}\Big\}, \tag{8.1.199}$$

is a closed linear subspace of finite codimension in $\big[M^{p,\lambda}(\partial\Omega,\sigma)\big]^M$, and a solution u of (8.1.198) *exists if and only if the boundary datum f belongs to* (8.1.199). *Second, if L is weakly elliptic with $\mathfrak{A}^{\mathrm{dis}}_{L^\top} \neq \varnothing$, the space of null-solutions of the boundary problem* (8.1.198), *i.e.,*

[32] in particular, if Ω is an infinitesimally flat AR domain

[33] in view of work in [129, §8.5], one may replace the nontangential maximal operator by the tangential maximal operator, associated as in [129, Definition 8.5.1] with a sufficiently large power

$$\Big\{u \in [\mathscr{C}^\infty(\Omega)]^M : Lu = 0,\ \mathcal{N}_\kappa u \in M^{p,\lambda}(\partial\Omega, \sigma),\ u\big|_{\partial\Omega}^{\kappa-\text{n.t.}} = 0, \quad \text{and } u(x) = O(|x|^{2-n}) \text{ as } |x| \to \infty \text{ if } \Omega \text{ is unbounded}\Big\}, \tag{8.1.200}$$

is finite dimensional; in addition,

for any null-solution u of the boundary problem (8.1.198) (*i.e., for any function u belonging to the space in* (8.1.200)), $\mathcal{N}_\kappa(\nabla u)$ *belongs to* $M^{p,\lambda}(\partial\Omega, \sigma)$ *and* $(\nabla u)\big|_{\partial\Omega}^{\kappa-\text{n.t.}}$ *exists* σ*-a.e. on* $\partial\Omega$, (8.1.201)

so, in particular,

the space of null-solution for the Dirichlet Problem (8.1.198) (*cf.* (8.1.200)) *coincides with the space of null-solution for the Regularity Problem* (8.1.83) (*cf.* (8.1.86)). (8.1.202)

Finally, the Fredholm solvability result just described is optimal, since in the class of operators L for which $\mathfrak{A}_L^{\text{dis}} = \varnothing$ *or* $\mathfrak{A}_{L^\top}^{\text{dis}} = \varnothing$ *it may happen that the Dirichlet Problem* (8.1.198) *fails to be Fredholm solvable even when the underlying domain* Ω *is the unit disk in the plane (or the complement of its closure). In fact, the aforementioned failure of Fredholm solvability may be total, in the sense that for the problem* (8.1.198), *both the space of admissible boundary data may have infinite codimension in the full space of boundary data, and the space of null-solutions may be infinite dimensional.*

(2) *If* (8.1.197) *is strengthened to*[34]

L satisfies the Legendre-Hadamard strong ellipticity condition (cf. [131, Definition 1.3.2]) *and one has* $\mathfrak{A}_L^{\text{dis}} \neq \varnothing$, (8.1.203)

then the Dirichlet Problem (8.1.198) *is actually well posed. More specifically,* (8.1.198) *has a unique solution, and this solution enjoys the additional property that there exists a constant* $C = C(\Omega, L, p, \lambda, \kappa) \in (0, \infty)$ *such that*

$$\|f\|_{[M^{p,\lambda}(\partial\Omega,\sigma)]^M} \leq \big\|\mathcal{N}_\kappa u\big\|_{M^{p,\lambda}(\partial\Omega,\sigma)} \leq C\|f\|_{[M^{p,\lambda}(\partial\Omega,\sigma)]^M}. \tag{8.1.204}$$

(3) *The version of the Dirichlet Problem formulated for boundary data in vanishing Morrey spaces, i.e.,*

$$\begin{cases} u \in [\mathscr{C}^\infty(\Omega)]^M, \quad Lu = 0 \text{ in } \Omega, \\ \mathcal{N}_\kappa u \in \mathring{M}^{p,\lambda}(\partial\Omega, \sigma), \quad u\big|_{\partial\Omega}^{\kappa-\text{n.t.}} = f \in \big[\mathring{M}^{p,\lambda}(\partial\Omega, \sigma)\big]^M, \\ u(x) = O(|x|^{2-n}) \text{ as } |x| \to \infty \text{ if } \Omega \text{ is unbounded}, \end{cases} \tag{8.1.205}$$

as well as the version of the Dirichlet Problem formulated for boundary data in block spaces, i.e.,

[34] Proposition 1.4.1 ensures that if L is as in (8.1.203) then $\mathfrak{A}_L^{\text{dis}} \neq \varnothing$ and $\mathfrak{A}_{L^\top}^{\text{dis}} \neq \varnothing$

$$\begin{cases} u \in \left[\mathscr{C}^\infty(\Omega)\right]^M, \quad Lu = 0 \ \textit{in} \ \Omega, \\ \mathcal{N}_\kappa u \in \mathcal{B}^{q,\lambda}(\partial\Omega, \sigma), \quad u\big|_{\partial\Omega}^{\kappa-\mathrm{n.t.}} = f \in \left[\mathcal{B}^{q,\lambda}(\partial\Omega, \sigma)\right]^M, \\ u(x) = O(|x|^{2-n}) \ \textit{as} \ |x| \to \infty \ \textit{if} \ \Omega \ \textit{is unbounded}, \end{cases} \tag{8.1.206}$$

enjoy similar solvability properties as those described in items (1)-(2) above. Once more, in the formulation of the boundary value problems (8.1.205)-(8.1.206) *one may replace the nontangential maximal operator by the tangential maximal operator* (*associated as in* [129, Definition 8.5.1] *with a sufficiently large power; cf.* [129, §8.5]).

(4) *Under the assumptions made in* (8.1.203), *if*

$$n \geq 3 \ \textit{and} \ \max\left\{\tfrac{n-1-\lambda}{n-2}, 1\right\} < p < \infty \tag{8.1.207}$$

then there exists a constant $C = C(\Omega, n, L, p, \lambda) \in (0, \infty)$ *with the property that the unique solution u of the boundary value problem* (8.1.198) *satisfies the fractional Carleson measure estimate*

$$\sup_{x\in\partial\Omega,\, r>0} \left(r^{-\lambda} \int_{B(x,r)\cap\Omega} |\nabla u|^p \operatorname{dist}(\cdot, \partial\Omega)^{p-1} \, d\mathcal{L}^n\right)^{\frac{1}{p}} \leq C\|f\|_{[M^{p,\lambda}(\partial\Omega,\sigma)]^M} \tag{8.1.208}$$

as well as

$$\lim_{R\to 0^+} \left\{ \sup_{x\in\partial\Omega,\, r\in(0,R)} r^{-\lambda} \int_{B(x,r)\cap\Omega} |\nabla u|^p \operatorname{dist}(\cdot, \partial\Omega)^{p-1} \, d\mathcal{L}^n \right\}^{1/p} \tag{8.1.209}$$
$$\leq C \cdot \operatorname{dist}\left(f, \left[\mathring{M}^{p,\lambda}(\partial\Omega, \sigma)\right]^M\right),$$

where the distance in the right-hand side is measured in $\left[M^{p,\lambda}(\partial\Omega, \sigma)\right]^M$. *In particular, if in fact* $f \in \left[\mathring{M}^{p,\lambda}(\partial\Omega, \sigma)\right]^M$ *with n and p as in* (8.1.207), *then*

$$\lim_{R\to 0^+} \left\{ \sup_{x\in\partial\Omega,\, r\in(0,R)} r^{-\lambda} \int_{B(x,r)\cap\Omega} |\nabla u|^p \operatorname{dist}(\cdot, \partial\Omega)^{p-1} \, d\mathcal{L}^n \right\}^{1/p} = 0. \tag{8.1.210}$$

Regarding this theorem, we wish to remark that, as seen from Proposition 1.6.12 and [131, Proposition 1.4.39], for the scalar case, i.e., if $M = 1$, and for dimensions $n \geq 3$, in place of (8.1.203) one may simply demand that the operator L is weakly elliptic. Moreover, in the scalar case in the two-dimensional setting (that is, for $M = 1$ and $n = 2$), if L satisfies the Legendre-Hadamard strong ellipticity condition then Proposition 1.6.11 tells us that $\mathfrak{A}_L^{\mathrm{dis}} \neq \varnothing$. Also, according to (1.6.141) in Proposition 1.6.13, in the scalar case in the two-dimensional setting (i.e., for $M = 1$ and $n = 2$) we may omit (8.1.203) altogether if the weakly elliptic operator L has real coefficients.

Finally, we note that, concerning the very formulation of Theorem 8.1.8, similar considerations to those made in Remark 8.1.2 are presently applicable.

Here is the proof of Theorem 8.1.8:

Proof of Theorem 8.1.8 The fact that the spaces of admissible boundary data (8.1.199) has finite codimension in $\big[M^{p,\lambda}(\partial\Omega,\sigma)\big]^M$ whenever L is weakly elliptic with $\mathfrak{A}_L^{\rm dis} \neq \varnothing$ is justified much as in the first part of the proof of Theorem 8.1.6, using of Theorem 3.4.2, item *(7)* in Theorem 7.1.1, [130, Theorem 1.1.3, item *(3)*].

In the case when L is weakly elliptic with $\mathfrak{A}_{L^\top}^{\rm dis} \neq \varnothing$, making use of the fact that $M^{p,\lambda}(\partial\Omega,\sigma)$ is contained in $L^p(\partial\Omega,\sigma)$ since $\partial\Omega$ is bounded, we conclude from the finite dimensionality property of the space of null-solutions of the Dirichlet Problem (8.1.145) (cf. item *(1)* in Theorem 8.1.6) that the space of null-solutions of the boundary problem (8.1.198), described in (8.1.200), is also finite dimensional. In addition, this also shows that any null-solution u of the boundary problem (8.1.198) enjoys the property that the nontangential trace $(\nabla u)\big|_{\partial\Omega}^{\kappa-{\rm n.t.}}$ exists at σ-a.e. point on $\partial\Omega$, and $\mathcal{N}_\kappa(\nabla u)$ belongs to $L^p(\partial\Omega,\sigma)$. In relation to the null-solution u of (8.1.198), we now claim that there exists a constant $c \in \mathbb{C}^M$ such that

$$u = -\mathscr{S}(\partial_\nu^A u) + c \ \text{ in } \ \Omega. \tag{8.1.211}$$

Keeping in mind that we currently have $u\big|_{\partial\Omega}^{\kappa-{\rm n.t.}} = 0$, in the case when either $n \geq 3$, or $n = 2$ and Ω is bounded, this follows (with $c = 0$) from [132, Theorem 1.3.3], while in the case when $n = 2$ and $\Omega \subseteq \mathbb{R}^2$ is an exterior domain this is guaranteed by [131, Corollary 1.5.4]. After taking the conormal derivative ∂_ν^A of both sides in (8.1.211) we arrive (on account of the jump-formula [132, (1.5.59)]) at the conclusion that

$$\partial_\nu^A u \in \operatorname{Ker}\Big(\tfrac{1}{2}I + K^{\#}_{A^\top};\ \big[L^p(\partial\Omega,\sigma)\big]^M\Big). \tag{8.1.212}$$

By further decreasing the value of $\delta \in (0,1)$ as to ensure that the conclusions in Proposition 7.1.2 are valid, we deduce from (8.1.212) that $\partial_\nu^A u \in \big[M^{p,\lambda}(\partial\Omega,\sigma)\big]^M$. In concert with (8.1.211) and [132, (3.3.19)], this membership proves that $\mathcal{N}_\kappa(\nabla u)$ lies in $M^{p,\lambda}(\partial\Omega,\sigma)$. As such, the additional properties claimed in (8.1.201) have been verified. In turn, (8.1.202) follows from (8.1.201) and definitions. Finally, the very first claim in item *(1)* is implied by what we have proved so far, while the very last claim in item *(1)* is a consequence of Theorem 8.6.5, Theorem 8.6.7, Example 1.6.14, and Remark 8.6.9. This completes the treatment of the present item *(1)*. In fact, analogous Fredholm solvability results are true (with similar proofs) for the Dirichlet Problem with boundary data in vanishing Morrey spaces, formulated in (8.1.205), and for the Dirichlet Problem with boundary data in block spaces, formulated in (8.1.206).

For the remaining of the proof, work under the assumptions made in (8.1.203). The (quantitative) existence for (8.1.198), (8.1.205), and (8.1.206) may be justified by reasoning along the lines of the proof of Theorem 8.1.6, now also making use of [132, Theorem 3.3.1], [132, Theorem 3.3.2], Theorem 7.1.18, and Theorem 7.1.19. Uniqueness for (8.1.198), (8.1.205), and (8.1.206) follows from [131,

Theorem 4.3.1], and Theorem 8.1.4 formulated for $L^\top$ (bearing in mind that (8.1.203) implies that $L^\top$ also satisfies the Legendre-Hadamard strong ellipticity condition and has $\mathfrak{A}^{\rm dis}_{L^\top} \neq \varnothing$, thanks to Theorem 1.5.2 and [131, (1.3.5)]). Alternatively, we may rely on the uniqueness part in Theorem 8.1.6 and embeddings of Morrey and block spaces into Lebesgue spaces (cf. [130, (6.2.25)] and [130, (6.2.71)]).

Let us now justify the claims pertaining to (8.1.207)-(8.1.210). To get started, suppose n and p are as in (8.1.207) and define

$$p_\lambda := \frac{p(n-1)}{n-1-\lambda+p} \in (1, n-1). \tag{8.1.213}$$

Then, for suitable choices of the parameters involved, [131, (2.4.46)] gives that there exists a constant $C = C(\Omega, n, p, \lambda, L) \in (0, \infty)$ such that

$$\sup_{x\in\partial\Omega,\, r>0} \Big(r^{-\lambda} \int_{B(x,r)\cap\Omega} |\nabla \mathscr{S} f|^p \operatorname{dist}(\cdot, \partial\Omega)^{p-1} \, d\mathcal{L}^n\Big)^{\frac{1}{p}} \leq C\|f\|_{[L^{p_\lambda}(\partial\Omega,\sigma)]^M} \tag{8.1.214}$$

for each function $f \in \big[L^{p_\lambda}(\partial\Omega, \sigma)\big]^M$. For further use, let us also note here that

$$\lim_{R\to 0^+} \Big\{ \sup_{x\in\partial\Omega,\, r\in(0,R)} r^{-\lambda} \int_{B(x,r)\cap\Omega} |\nabla \mathscr{S} f|^p \operatorname{dist}(\cdot, \partial\Omega)^{p-1} \, d\mathcal{L}^n \Big\}^{1/p} = 0 \tag{8.1.215}$$

for each function $f \in \big[L^{p_\lambda}(\partial\Omega, \sigma)\big]^M$. To justify (8.1.215), pick some arbitrary exponent $q \in (n-1, \infty)$ and define $\eta := 1-(n-1)/q \in (0, 1)$. In addition select two arbitrary functions, $f \in \big[L^{p_\lambda}(\partial\Omega, \sigma)\big]^M$ and $g \in \big[L^q(\partial\Omega, \sigma)\big]^M$. Then g belongs to the space $\big[L^{p_\lambda}(\partial\Omega, \sigma)\big]^M$ (since $\partial\Omega$ is bounded), and for each $r > 0$ and $x \in \partial\Omega$ we may use (8.1.214) to estimate

$$\begin{aligned} &\Big(r^{-\lambda} \int_{B(x,r)\cap\Omega} \big|\nabla\big(\mathscr{S}(f-g)\big)\big|^p \operatorname{dist}(\cdot,\partial\Omega)^{p-1} \, d\mathcal{L}^n\Big)^{\frac{1}{p}} \\ &\qquad\qquad \leq C\|f-g\|_{[L^{p_\lambda}(\partial\Omega,\sigma)]^M}. \end{aligned} \tag{8.1.216}$$

In addition, we have

$$\begin{aligned} &\Big(r^{-\lambda} \int_{B(x,r)\cap\Omega} \big|\nabla(\mathscr{S} g)\big|^p \operatorname{dist}(\cdot, \partial\Omega)^{p-1} \, d\mathcal{L}^n\Big)^{\frac{1}{p}} \\ &\quad \leq C\Big(r^{-\lambda} \int_{B(x,r)\cap\Omega} \operatorname{dist}(\cdot, \partial\Omega)^{p\eta-1} \, d\mathcal{L}^n\Big)^{\frac{1}{p}} \|g\|_{[L^q(\partial\Omega,\sigma)]^M} \\ &\quad \leq C r^{n-1-\lambda+\eta} \|g\|_{[L^q(\partial\Omega,\sigma)]^M}. \end{aligned} \tag{8.1.217}$$

The first inequality above uses the version of [132, (1.5.57)] written for $\mathscr{S}$ in place of $\mathscr{S}_{\rm mod}$ (and for g in place of f), while the second inequality is based on [129, (8.6.101)] used with $\lambda := 1-p\eta$, $\alpha := 1$, $\beta := n-1$, and $E := B(x, r)\cap\Omega$. Collectively, (8.1.216) and (8.1.217) give

$$\lim_{R\to 0^+}\sup_{\substack{x\in\partial\Omega \text{ and}\\ r\in(0,R)}}\Big(r^{-\lambda}\int_{B(x,r)\cap\Omega}\big|\nabla(\mathscr{S}f)\big|^p \operatorname{dist}(\cdot,\partial\Omega)^{p-1}\,d\mathcal{L}^n\Big)^{\frac{1}{p}} \leq C\|f-g\|_{[L^{p_\lambda}(\partial\Omega,\sigma)]^M}, \tag{8.1.218}$$

for some constant $C\in(0,\infty)$ independent of f and g. With this in hand, (8.1.215) now follows, on account of the arbitrariness of $g\in\big[L^q(\partial\Omega,\sigma)\big]^M$ and the density of $L^q(\partial\Omega,\sigma)$ in $L^{p_\lambda}(\partial\Omega,\sigma)$. Next, let us recall the Hölder conjugate exponents

$$s:=\frac{p(n-1)}{n-1-\lambda}\in(p,\infty)\ \text{ and }\ r:=\frac{q(n-1)}{n-1+\lambda(q-1)}\in(1,q) \tag{8.1.219}$$

appearing in [130, (6.2.7)] and [130, (6.2.71)], respectively. Choose $\delta\in(0,1)$ small enough so that the conclusions in Theorem 7.1.1, Proposition 7.1.2, Proposition 7.1.46 (all stated with $z:=1/2$), and Theorem 7.1.9 are valid. By further decreasing $\delta\in(0,1)$ if necessary, we may invoke Theorem 3.4.2 to also ensure that Ω is a UR domain satisfying a two-sided local John condition. As a consequence of this, [130, (11.5.200), (11.5.201)], (8.1.213), and (8.1.219) we have a embedding

$$L^{p_\lambda}_1(\partial\Omega,\sigma)\hookrightarrow L^s(\partial\Omega,\sigma) \text{ continuously and densely.} \tag{8.1.220}$$

From this and [130, (6.2.7)] we then see that

$$L^{p_\lambda}_1(\partial\Omega,\sigma)\hookrightarrow M^{p,\lambda}(\partial\Omega,\sigma) \text{ continuously.} \tag{8.1.221}$$

Together, (8.1.220), [130, Lemma 1.2.1], and (A.0.154) imply

$$L^r(\partial\Omega,\sigma)\hookrightarrow L^{(p_\lambda)'}_{-1}(\partial\Omega,\sigma) \text{ continuously,} \tag{8.1.222}$$

where $(p_\lambda)'$ denotes the Hölder conjugate exponent of p_λ. In addition, the abstract compatibility result from [130, (1.2.25)] presently gives

$$\int_{\partial\Omega} fg\,d\sigma = {}_{L^r(\partial\Omega,\sigma)}\langle f,g\rangle_{L^s(\partial\Omega,\sigma)} = {}_{L^{(p_\lambda)'}_{-1}(\partial\Omega,\sigma)}\langle f,g\rangle_{L^{p_\lambda}_1(\partial\Omega,\sigma)} \text{ for all}$$
$$f\in L^r(\partial\Omega,\sigma)\hookrightarrow L^{(p_\lambda)'}_{-1}(\partial\Omega,\sigma) \text{ and } g\in L^{p_\lambda}_1(\partial\Omega,\sigma)\hookrightarrow L^s(\partial\Omega,\sigma). \tag{8.1.223}$$

In concert, (8.1.222) and [130, (6.2.71)] prove that

$$\mathcal{B}^{q,\lambda}(\partial\Omega,\sigma)\hookrightarrow L^{(p_\lambda)'}_{-1}(\partial\Omega,\sigma) \text{ continuously.} \tag{8.1.224}$$

As is apparent from [130, (6.2.69)] and that $L^q(\partial\Omega,\sigma)\subseteq L^r(\partial\Omega,\sigma)$, we have

$$L^r(\partial\Omega,\sigma)\cap\mathcal{B}^{q,\lambda}(\partial\Omega,\sigma)\hookrightarrow\mathcal{B}^{q,\lambda}(\partial\Omega,\sigma) \text{ densely.} \tag{8.1.225}$$

From (8.1.223), (8.1.225), [130, Proposition 6.2.8], (8.1.224), (8.1.221) we get

$$\int_{\partial\Omega} fg\,\mathrm{d}\sigma = {}_{\mathcal{B}^{q,\lambda}(\partial\Omega,\sigma)}\langle f, g\rangle_{M^{p,\lambda}(\partial\Omega,\sigma)} = {}_{L_{-1}^{(p_\lambda)'}(\partial\Omega,\sigma)}\langle f, g\rangle_{L_1^{p_\lambda}(\partial\Omega,\sigma)} \text{ for all}$$
$$f \in \mathcal{B}^{q,\lambda}(\partial\Omega,\sigma) \hookrightarrow L_{-1}^{(p_\lambda)'}(\partial\Omega,\sigma) \text{ and } g \in L_1^{p_\lambda}(\partial\Omega,\sigma) \hookrightarrow M^{p,\lambda}(\partial\Omega,\sigma). \tag{8.1.226}$$

To proceed, fix a coefficient tensor $A \in \mathfrak{A}_L^{\mathrm{dis}}$. In view of the current assumptions, Proposition 7.1.2 applies, and tells us that $\mathrm{Ker}\Big(\frac{1}{2}I + K_A^{\#} : \big[\mathcal{B}^{q,\lambda}(\partial\Omega,\sigma)\big]^M\Big)$ is a finite dimensional space. Let $\{\xi_1,\dots,\xi_N\} \subseteq \big[\mathcal{B}^{q,\lambda}(\partial\Omega,\sigma)\big]^M$ be a basis for this space. In view of (8.1.224) and (A.0.154) we conclude that

$$\begin{gathered}\{\xi_1,\dots,\xi_N\} \text{ are linearly independent functionals in} \\ \Big(\big[L_1^{p_\lambda}(\partial\Omega,\sigma)\big]^M\Big)^* = \big[L_{-1}^{(p_\lambda)'}(\partial\Omega,\sigma)\big]^M.\end{gathered} \tag{8.1.227}$$

Granted this, we may rely on [130, Lemma 1.2.16] to guarantee the existence

$$\eta_1,\dots,\eta_N \in \big[L_1^{p_\lambda}(\partial\Omega,\sigma)\big]^M \text{ with the property that} \tag{8.1.228}$$
$${}_{[L_{-1}^{(p_\lambda)'}(\partial\Omega,\sigma)]^M}\big\langle \xi_j, \eta_k\big\rangle_{[L_1^{p_\lambda}(\partial\Omega,\sigma)]^M} = \delta_{jk} \text{ for all } j,k \in \{1,\dots,N\}. \tag{8.1.229}$$

Going further, introduce

$$\begin{gathered}\pi : \big[M^{p,\lambda}(\partial\Omega,\sigma)\big]^M \longrightarrow \big[M^{p,\lambda}(\partial\Omega,\sigma)\big]^M \text{ defined as} \\ \pi f := \sum_{j=1}^N {}_{[M^{p,\lambda}(\partial\Omega,\sigma)]^M}\big\langle f, \xi_j\big\rangle_{[\mathcal{B}^{q,\lambda}(\partial\Omega,\sigma)]^M} \cdot \eta_j \\ \text{for each function } f \in \big[M^{p,\lambda}(\partial\Omega,\sigma)\big]^M.\end{gathered} \tag{8.1.230}$$

In view of (8.1.228) and (8.1.221) we deduce that π is a well-defined, linear, and continuous mapping both in the context of (8.1.230) and also in the setting

$$\pi : \big[M^{p,\lambda}(\partial\Omega,\sigma)\big]^M \longrightarrow \big[L_1^{p_\lambda}(\partial\Omega,\sigma)\big]^M. \tag{8.1.231}$$

In addition, for each $f \in \big[M^{p,\lambda}(\partial\Omega,\sigma)\big]^M$ and $k \in \{1,\dots,N\}$ we compute

$$
\begin{aligned}
&{}_{[M^{p,\lambda}(\partial\Omega,\sigma)]^M}\big\langle f-\pi f,\xi_k\big\rangle_{[\mathcal{B}^{q,\lambda}(\partial\Omega,\sigma)]^M} = {}_{[M^{p,\lambda}(\partial\Omega,\sigma)]^M}\big\langle f,\xi_k\big\rangle_{[\mathcal{B}^{q,\lambda}(\partial\Omega,\sigma)]^M}\\
&\quad -\sum_{j=1}^{N} {}_{[M^{p,\lambda}(\partial\Omega,\sigma)]^M}\big\langle f,\xi_j\big\rangle_{[\mathcal{B}^{q,\lambda}(\partial\Omega,\sigma)]^M}\,{}_{[M^{p,\lambda}(\partial\Omega,\sigma)]^M}\big\langle \eta_j,\xi_k\big\rangle_{[\mathcal{B}^{q,\lambda}(\partial\Omega,\sigma)]^M}\\
&= {}_{[M^{p,\lambda}(\partial\Omega,\sigma)]^M}\big\langle f,\xi_k\big\rangle_{[\mathcal{B}^{q,\lambda}(\partial\Omega,\sigma)]^M}\\
&\quad -\sum_{j=1}^{N} {}_{[M^{p,\lambda}(\partial\Omega,\sigma)]^M}\big\langle f,\xi_j\big\rangle_{[\mathcal{B}^{q,\lambda}(\partial\Omega,\sigma)]^M}\,{}_{[L_1^{p_\lambda}(\partial\Omega,\sigma)]^M}\big\langle \eta_j,\xi_k\big\rangle_{[L_{-1}^{(p_\lambda)'}(\partial\Omega,\sigma)]^M}\\
&= {}_{[M^{p,\lambda}(\partial\Omega,\sigma)]^M}\big\langle f,\xi_k\big\rangle_{[\mathcal{B}^{q,\lambda}(\partial\Omega,\sigma)]^M}\\
&\quad -\sum_{j=1}^{N} {}_{[M^{p,\lambda}(\partial\Omega,\sigma)]^M}\big\langle f,\xi_j\big\rangle_{[\mathcal{B}^{q,\lambda}(\partial\Omega,\sigma)]^M}\cdot\delta_{jk}=0,
\end{aligned}
\tag{8.1.232}
$$

thanks to (8.1.230), (8.1.226), and (8.1.229). Since, by definition, $\{\xi_1,\dots,\xi_N\}$ is a basis in $\operatorname{Ker}\Big(\tfrac12 I+K_A^{\#}:\,\big[\mathcal{B}^{q,\lambda}(\partial\Omega,\sigma)\big]^M\Big)$, from (8.1.232), (A.0.6) we see that

$$
f-\pi f \text{ belongs to } \Big[\operatorname{Ker}\Big(\tfrac12 I+K_A^{\#}:\,\big[\mathcal{B}^{q,\lambda}(\partial\Omega,\sigma)\big]^M\Big)\Big]^{\perp}. \tag{8.1.233}
$$

From this, item *(7)* in Theorem 7.1.1 (presently used with $z:=1/2$), [130, (2.1.47), (2.1.48)], and [132, Theorem 3.3.1, item *(iii)*] we deduce that

$$
\begin{aligned}
&\text{for each function } f\in\big[M^{p,\lambda}(\partial\Omega,\sigma)\big]^M \text{ it follows that}\\
&f-\pi f \text{ belongs to } \operatorname{Im}\Big(\tfrac12 I+K_A;\,\big[M^{p,\lambda}(\partial\Omega,\sigma)\big]^M\Big).
\end{aligned}
\tag{8.1.234}
$$

Going further, suppose now that an arbitrary function $f\in\big[M^{p,\lambda}(\partial\Omega,\sigma)\big]^M$ has been fixed. Then, (8.1.234) together with [130, Proposition 1.2.13] (whose applicability is presently ensured by item *(7)* in Theorem 7.1.1) and the fact that π is bounded in the context of (8.1.230) guarantee that there exists

$$
\begin{aligned}
&g_0\in\big[M^{p,\lambda}(\partial\Omega,\sigma)\big]^M \text{ such that } \big(\tfrac12 I+K_A\big)g_0=f-\pi f\\
&\text{and } \|g_0\|_{[M^{p,\lambda}(\partial\Omega,\sigma)]^M}\le C\|f\|_{[M^{p,\lambda}(\partial\Omega,\sigma)]^M}
\end{aligned}
\tag{8.1.235}
$$

for some constant $C\in(0,\infty)$ independent of f. Also, as seen from (8.1.230),

$$
\pi f\in\operatorname{span}\{\eta_1,\dots,\eta_N\}\subseteq\big[L_1^{p_\lambda}(\partial\Omega,\sigma)\big]^M. \tag{8.1.236}
$$

The invertibility result from (7.1.233) (used with $p:=p_\lambda$) then guarantees (bearing in mind that π in (8.1.231) is bounded) that there exists

$$
\begin{aligned}
&g_1\in\big[L^{p_\lambda}(\partial\Omega,\sigma)\big]^M \text{ such that } Sg_1=\pi f\\
&\text{and } \|g_1\|_{[L^{p_\lambda}(\partial\Omega,\sigma)]^M}\le C\|f\|_{[M^{p,\lambda}(\partial\Omega,\sigma)]^M}
\end{aligned}
\tag{8.1.237}
$$

for some constant $C\in(0,\infty)$ independent of f. If we now set

$$u := \mathcal{D}_A g_0 + \mathscr{S} g_1 \text{ in } \Omega, \tag{8.1.238}$$

we deduce from (8.1.238), (8.1.235), (8.1.237), [132, (1.3.8), (1.3.24), (1.3.26), (1.5.20), (1.5.44)], and [132, Lemma 1.3.1] that

$$\begin{cases} u \in \left[\mathscr{C}^\infty(\Omega)\right]^M, \quad Lu = 0 \text{ in } \Omega, \\ u\Big|_{\partial\Omega}^{\kappa-\text{n.t.}} = f \in \left[M^{p,\lambda}(\partial\Omega,\sigma)\right]^M, \\ u(x) = O(|x|^{2-n}) \text{ as } |x| \to \infty \text{ if } \Omega \text{ is unbounded.} \end{cases} \tag{8.1.239}$$

Let us also note that [130, (6.2.7)], [132, (1.5.47)], and (8.1.237) presently give

$$\begin{aligned} \big\|\mathcal{N}_\kappa(\mathscr{S} g_1)\big\|_{M^{p,\lambda}(\partial\Omega,\sigma)} &\leq C\big\|\mathcal{N}_\kappa(\mathscr{S} g_1)\big\|_{L^s(\partial\Omega,\sigma)} \\ &\leq C\|g_1\|_{[L^{p,\lambda}(\partial\Omega,\sigma)]^M} \leq C\|f\|_{[M^{p,\lambda}(\partial\Omega,\sigma)]^M}, \end{aligned} \tag{8.1.240}$$

while from [132, (1.5.1)] and (8.1.235) we deduce

$$\big\|\mathcal{N}_\kappa(\mathcal{D}_A g_0)\big\|_{M^{p,\lambda}(\partial\Omega,\sigma)} \leq C\|g_0\|_{[M^{p,\lambda}(\partial\Omega,\sigma)]^M} \leq C\|f\|_{[M^{p,\lambda}(\partial\Omega,\sigma)]^M}.$$

Together, the above estimate, (8.1.238), and (8.1.240) prove

$$\mathcal{N}_\kappa u \in M^{p,\lambda}(\partial\Omega,\sigma) \text{ and } \big\|\mathcal{N}_\kappa u\big\|_{M^{p,\lambda}(\partial\Omega,\sigma)} \leq C\|f\|_{[M^{p,\lambda}(\partial\Omega,\sigma)]^M}. \tag{8.1.241}$$

It is now apparent from (8.1.239), (8.1.241), and the well posedness of (8.1.198) that

the function u from (8.1.239) is the unique solution of the boundary value problem (8.1.198) with datum f. (8.1.242)

Use (8.1.238), [132, (3.3.69)], (8.1.214), (8.1.235), and (8.1.237) to estimate

$$\begin{aligned} &\sup_{x\in\partial\Omega,\, r>0} \Big(r^{-\lambda}\int_{B(x,r)\cap\Omega} |\nabla u|^p \operatorname{dist}(\cdot,\partial\Omega)^{p-1}\, d\mathcal{L}^n\Big)^{\frac{1}{p}} \\ &\leq \sup_{x\in\partial\Omega,\, r>0} \Big(r^{-\lambda}\int_{B(x,r)\cap\Omega} |\nabla \mathcal{D}_A g_0|^p \operatorname{dist}(\cdot,\partial\Omega)^{p-1}\, d\mathcal{L}^n\Big)^{\frac{1}{p}} \\ &\quad + \sup_{x\in\partial\Omega,\, r>0} \Big(r^{-\lambda}\int_{B(x,r)\cap\Omega} |\nabla \mathscr{S} g_1|^p \operatorname{dist}(\cdot,\partial\Omega)^{p-1}\, d\mathcal{L}^n\Big)^{\frac{1}{p}} \\ &\leq C\|g_0\|_{[M^{p,\lambda}(\partial\Omega,\sigma)]^M} + C\|g_1\|_{[L^{p,\lambda}(\partial\Omega,\sigma)]^M} \leq C\|f\|_{[M^{p,\lambda}(\partial\Omega,\sigma)]^M}. \end{aligned} \tag{8.1.243}$$

At this point, from (8.1.242) and (8.1.243) we conclude that (8.1.208) holds.

To prove (8.1.209), fix some function $f \in \left[M^{p,\lambda}(\partial\Omega,\sigma)\right]^M$ and consider an arbitrary $\phi \in \left[L_1^{p,\lambda}(\partial\Omega,\sigma)\right]^M$. In particular, (8.1.221) implies that $f - \phi$ belongs to $\left[M^{p,\lambda}(\partial\Omega,\sigma)\right]^M$. As such, we may invoke (8.1.234) which shows that the function $(f-\phi) - \pi(f-\phi)$ belongs to $\operatorname{Im}\Big(\frac{1}{2}I + K_A;\ \left[M^{p,\lambda}(\partial\Omega,\sigma)\right]^M\Big)$. Reasoning as in (8.1.235)-(8.1.242) we see that the unique solution of the boundary value problem

(8.1.198) with datum f may be expressed as $u = \mathcal{D}_A h_0 + \mathcal{S} h_1$ in Ω, where

$$\begin{aligned} h_0 \in \big[M^{p,\lambda}(\partial\Omega,\sigma)\big]^M \text{ satisfies } \big(\tfrac{1}{2}I + K_A\big)h_0 = (f-\phi) - \pi(f-\phi) \\ \text{as well as } \|h_0\|_{[M^{p,\lambda}(\partial\Omega,\sigma)]^M} \leq C\|f-\phi\|_{[M^{p,\lambda}(\partial\Omega,\sigma)]^M}, \end{aligned} \tag{8.1.244}$$

and

$$\begin{aligned} h_1 \in \big[L^{p,\lambda}(\partial\Omega,\sigma)\big]^M \text{ is such that } Sh_1 = \phi + \pi(f-\phi) \\ \text{and } \|h_1\|_{[L^{p,\lambda}(\partial\Omega,\sigma)]^M} \leq C\|\phi + \pi(f-\phi)\|_{[M^{p,\lambda}(\partial\Omega,\sigma)]^M}. \end{aligned} \tag{8.1.245}$$

Based on these, much as in (8.1.243) we may now write, making use of (8.1.215),

$$\begin{aligned} &\lim_{R\to 0^+} \sup_{\substack{x\in\partial\Omega \text{ and} \\ r\in(0,R)}} \Big(r^{-\lambda}\int_{B(x,r)\cap\Omega} |\nabla u|^p \operatorname{dist}(\cdot,\partial\Omega)^{p-1}\, d\mathcal{L}^n\Big)^{\frac{1}{p}} \\ &\quad \leq \lim_{R\to 0^+} \sup_{\substack{x\in\partial\Omega \text{ and} \\ r\in(0,R)}} \Big(r^{-\lambda}\int_{B(x,r)\cap\Omega} |\nabla \mathcal{D}_A h_0|^p \operatorname{dist}(\cdot,\partial\Omega)^{p-1}\, d\mathcal{L}^n\Big)^{\frac{1}{p}} \\ &\quad\quad + \lim_{R\to 0^+} \sup_{\substack{x\in\partial\Omega \text{ and} \\ r\in(0,R)}} \Big(r^{-\lambda}\int_{B(x,r)\cap\Omega} |\nabla \mathcal{S} h_1|^p \operatorname{dist}(\cdot,\partial\Omega)^{p-1}\, d\mathcal{L}^n\Big)^{\frac{1}{p}} \\ &\quad \leq C\|h_0\|_{[M^{p,\lambda}(\partial\Omega,\sigma)]^M} \leq C\|f-\phi\|_{[M^{p,\lambda}(\partial\Omega,\sigma)]^M}. \end{aligned} \tag{8.1.246}$$

Let us also observe that, thanks to (8.1.220) and [130, (6.2.15)],

$$L_1^{p,\lambda}(\partial\Omega,\sigma) \hookrightarrow L^s(\partial\Omega,\sigma) \hookrightarrow \mathring{M}^{p,\lambda}(\partial\Omega,\sigma) \tag{8.1.247}$$

where both inclusions are continuous with dense range, hence

$$L_1^{p,\lambda}(\partial\Omega,\sigma) \hookrightarrow \mathring{M}^{p,\lambda}(\partial\Omega,\sigma) \text{ continuously and densely.} \tag{8.1.248}$$

Taking the infimum in (8.1.246) over all $\phi \in \big[L_1^{p,\lambda}(\partial\Omega,\sigma)\big]^M$ and keeping in mind the density result recorded in (8.1.248) ultimately yields (8.1.209). Finally, (8.1.210) is a direct consequence of (8.1.209). The proof of Theorem 8.1.8 is now complete.□

The fact that the theory of weights plays a basic role in relation to the L^p-Dirichlet Problem for the Laplacian is well documented (see, e.g., [93] and the references therein; cf. also [131, Theorem 5.7.6]). The remarkable aspect is that the theory of weights continues to play a significant role in relation to boundary value problems for weakly elliptic systems with constant complex coefficients. The linchpin is the Hardy-Littlewood maximal operator and the versatile extrapolation result established in [130, Theorem 5.2.1], acting as a portal between Muckenhoupt weighted Lebesgue spaces and Generalized Banach Function Spaces (on which the Hardy-Littlewood maximal operator happens to be bounded). Altogether, this machinery yields the well-posedness of entire (infinite) classes of boundary value problems.

We illustrate this philosophy by formulating and solving the Dirichlet Problem with data in Generalized Banach Function Spaces (GBFS) in the sense of [130, Definition 5.1.4], thus extending a number of earlier results fitting under this inclusive label. Other (new) special cases are discussed explicitly, right after the statement of the theorem.

Theorem 8.1.9 (The Dirichlet Problem with Boundary Data in GBFS) *Fix two integers $n, M \in \mathbb{N}$ with $n \geq 2$ and suppose $\Omega \subseteq \mathbb{R}^n$ is an Ahlfors regular domain with compact boundary. Abbreviate $\sigma := \mathcal{H}^{n-1}\lfloor\partial\Omega$ and denote by ν the geometric measure theoretic outward unit normal to Ω. In this context, suppose $\mathbb{X}$ is a Generalized Banach Function Space on $(\partial\Omega, \sigma)$ with the property that*

$$\mathcal{M}_{\partial\Omega} : \mathbb{X} \to \mathbb{X} \text{ and } \mathcal{M}_{\partial\Omega} : \mathbb{X}' \to \mathbb{X}' \text{ are well-defined and bounded,} \tag{8.1.249}$$

where $\mathcal{M}_{\partial\Omega}$ is the Hardy-Littlewood maximal operator on $\partial\Omega$, and $\mathbb{X}'$ is the associated space of $\mathbb{X}$ (cf. [130, Definitions 5.1.4, 5.1.11]). Finally, suppose L is a homogeneous, second-order, constant complex coefficient, $M \times M$ system in $\mathbb{R}^n$ and, for some fixed arbitrary aperture parameter $\kappa > 0$, consider the boundary value problem[35]

$$\begin{cases} u \in \big[\mathscr{C}^\infty(\Omega)\big]^M, \quad Lu = 0 \text{ in } \Omega, \\ \mathcal{N}_\kappa u \in \mathbb{X}, \quad u\big|_{\partial\Omega}^{\kappa-\mathrm{n.t.}} = f \in \big[\mathbb{X}\big]^M, \\ u(x) = O(|x|^{2-n}) \text{ as } |x| \to \infty \text{ if } \Omega \text{ is unbounded.} \end{cases} \tag{8.1.250}$$

Then there exists some threshold $\delta_{\mathbb{X}} \in (0, 1)$, which depends only on n, the operator norms of $\mathcal{M}_{\partial\Omega}$ on $\mathbb{X}$ and $\mathbb{X}'$, the system L, the ADR *constants of $\partial\Omega$, and the relative flatness ratio $\wp(\Omega)/\mathrm{diam}\,\partial\Omega$, with the property that when*

$$\mathrm{dist}\Big(\nu, [\mathrm{VMO}(\partial\Omega, \sigma)]^n\Big) < \delta_{\mathbb{X}} \tag{8.1.251}$$

where the distance is considered in the space $\big[\mathrm{BMO}(\partial\Omega, \sigma)\big]^n$, hence if Ω is a $\delta_{\mathbb{X}}$-oscillating AR *domain*[36] *(cf. Definition 3.4.1), the following statements are true:*

(1) If one assumes that

$$L \text{ is weakly elliptic with } \mathfrak{A}_L^{\mathrm{dis}} \neq \emptyset \text{ and } \mathfrak{A}_{L^\top}^{\mathrm{dis}} \neq \emptyset, \tag{8.1.252}$$

then the Dirichlet Problem (8.1.250) is Fredholm solvable, in the following precise sense: First, if the system L is weakly elliptic with $\mathfrak{A}_L^{\mathrm{dis}} \neq \emptyset$, the space of admissible boundary data for the Dirichlet Problem (8.1.250), i.e.,

---

[35] thanks to work in [129, §8.5], in the formulation of (8.1.250) one can replace the nontangential maximal operator by the tangential maximal operator, associated as in [129, Definition 8.5.1] with a sufficiently large power (cf. (A.0.165))

[36] in particular, if Ω is an infinitesimally flat AR domain

$$\Big\{u\Big|_{\partial\Omega}^{\kappa-\mathrm{n.t.}} : u \in \big[\mathscr{C}^\infty(\Omega)\big]^M,\ Lu = 0 \text{ in } \Omega,\ \mathcal{N}_\kappa u \in \mathbb{X},\ u\Big|_{\partial\Omega}^{\kappa-\mathrm{n.t.}} \text{ exists } \sigma\text{-a.e.},$$
$$\text{and } u(x) = O(|x|^{2-n}) \text{ as } |x| \to \infty \text{ if } \Omega \text{ is unbounded}\Big\}, \tag{8.1.253}$$

is a closed linear subspace of finite codimension in $\big[\mathbb{X}\big]^M$*, and a solution* u *of* (8.1.250) *exists if and only if the boundary datum* f *belongs to* (8.1.253). *Second, if* L *is weakly elliptic with* $\mathfrak{A}^{\mathrm{dis}}_{L^\top} \neq \varnothing$*, the space of null-solutions of the boundary problem* (8.1.250)*, i.e.,*

$$\Big\{u \in \big[\mathscr{C}^\infty(\Omega)\big]^M : Lu = 0 \text{ in } \Omega,\ \mathcal{N}_\kappa u \in \mathbb{X},\ u\Big|_{\partial\Omega}^{\kappa-\mathrm{n.t.}} = 0, \tag{8.1.254}$$
$$\text{and } u(x) = O(|x|^{2-n}) \text{ as } |x| \to \infty \text{ if } \Omega \text{ is unbounded}\Big\},$$

is finite dimensional; in addition,

for any null-solution u *of the boundary problem* (8.1.250) *(i.e., for any function* u *belonging to the space in* (8.1.254)), $\mathcal{N}_\kappa(\nabla u)$ *belongs to* $\mathbb{X}$ *and* $(\nabla u)\big|_{\partial\Omega}^{\kappa-\mathrm{n.t.}}$ *exists* σ*-a.e. on* $\partial\Omega$*,* (8.1.255)

so, in particular,

the space of null-solution for the Dirichlet Problem (8.1.250) *(cf.* (8.1.254)) *coincides with the space of null-solution for the Regularity Problem* (8.1.95) *(cf.* (8.1.98)). (8.1.256)

Lastly, the Fredholm solvability result stated above is optimal, since in the class of operators L *for which* $\mathfrak{A}^{\mathrm{dis}}_{L} = \varnothing$ *or* $\mathfrak{A}^{\mathrm{dis}}_{L^\top} = \varnothing$ *it may happen that the Dirichlet Problem* (8.1.250) *fails to be Fredholm solvable even when the underlying domain* Ω *is the unit disk in the plane (or the complement of its closure). In fact, the aforementioned failure of Fredholm solvability may be total, in the sense that for the problem* (8.1.250)*, both the space of admissible boundary data may have infinite codimension in the full space of boundary data, and the space of null-solutions may be infinite dimensional.*

(2) If (8.1.252) *is strengthened to*[37] *assuming that*

L *satisfies the Legendre-Hadamard strong ellipticity condition (cf.* [131, Definition 1.3.2]) *and one has* $\mathfrak{A}^{\mathrm{dis}}_{L} \neq \varnothing$*,* (8.1.257)

then the Dirichlet Problem (8.1.250) *is actually well posed. More specifically, the boundary problem* (8.1.250) *has a unique solution, and this solution enjoys the additional property that there exists* $C = C(\Omega, L, \mathbb{X}, \kappa) \in (0, \infty)$ *so that*

$$\|f\|_{[\mathbb{X}]^M} \leq \big\|\mathcal{N}_\kappa u\big\|_{\mathbb{X}} \leq C\|f\|_{[\mathbb{X}]^M}. \tag{8.1.258}$$

[37] according to Proposition 1.4.1, whenever L is as in (8.1.257) we have $\mathfrak{A}^{\mathrm{dis}}_{L} \neq \varnothing$ and $\mathfrak{A}^{\mathrm{dis}}_{L^\top} \neq \varnothing$

(3) Similar properties like the ones discussed in items (1)-(2) above are valid for the Dirichlet Problem with data in the associated space $\mathbb{X}'$, *i.e.,*

$$\begin{cases} u \in \big[\mathscr{C}^\infty(\Omega)\big]^M, \quad Lu = 0 \ \text{in} \ \Omega, \\ \mathcal{N}_\kappa u \in \mathbb{X}', \quad u\big|_{\partial\Omega}^{\kappa-\text{n.t.}} = f \in \big[\mathbb{X}'\big]^M, \\ u(x) = O(|x|^{2-n}) \ \text{as} \ |x| \to \infty \ \text{if} \ \Omega \ \text{is unbounded}, \end{cases} \tag{8.1.259}$$

as well as for the Dirichlet Problems with data in the spaces

$$\mathring{\mathbb{X}} := \overline{L^\infty(\partial\Omega, \sigma)}^{\|\cdot\|_{\mathbb{X}}} \ \text{and} \ (\mathbb{X}')^\circ := \overline{L^\infty(\partial\Omega, \sigma)}^{\|\cdot\|_{\mathbb{X}'}}, \tag{8.1.260}$$

i.e.,

$$\begin{cases} u \in \big[\mathscr{C}^\infty(\Omega)\big]^M, \quad Lu = 0 \ \text{in} \ \Omega, \\ \mathcal{N}_\kappa u \in \mathring{\mathbb{X}}, \quad u\big|_{\partial\Omega}^{\kappa-\text{n.t.}} = f \in \big[\mathring{\mathbb{X}}\big]^M, \\ u(x) = O(|x|^{2-n}) \ \text{as} \ |x| \to \infty \ \text{if} \ \Omega \ \text{is unbounded}, \end{cases} \tag{8.1.261}$$

and, respectively,

$$\begin{cases} u \in \big[\mathscr{C}^\infty(\Omega)\big]^M, \quad Lu = 0 \ \text{in} \ \Omega, \\ \mathcal{N}_\kappa u \in (\mathbb{X}')^\circ, \quad u\big|_{\partial\Omega}^{\kappa-\text{n.t.}} = f \in \big[(\mathbb{X}')^\circ\big]^M, \\ u(x) = O(|x|^{2-n}) \ \text{as} \ |x| \to \infty \ \text{if} \ \Omega \ \text{is unbounded}. \end{cases} \tag{8.1.262}$$

Finally, in all cases one may replace the nontangential maximal operator by the tangential maximal operator, associated as in [129, Definition 8.5.1] *with a sufficiently large power* (*cf.* (A.0.165)).

Before presenting the proof of this theorem, we wish make two comments. First, as far as the formulation of Theorem 8.1.9 is concerned, similar considerations to those made in Remark 8.1.2 are presently applicable. Second, as seen from Proposition 1.6.12 and [131, Proposition 1.4.39], for the scalar case, i.e., if $M = 1$, and for dimensions $n \geq 3$, in place of (8.1.257) one may simply demand that the operator L is weakly elliptic.

In addition, in the scalar two-dimensional setting (i.e., for $M = 1$ and $n = 2$), if L satisfies the Legendre-Hadamard strong ellipticity condition then Proposition 1.6.11 guarantees that $\mathfrak{A}_L^{\text{dis}} \neq \varnothing$. Also, according to (1.6.141) in Proposition 1.6.13, in the scalar case in the two-dimensional setting (i.e., for $M = 1$ and $n = 2$) we may omit (8.1.257) altogether if L is a weakly elliptic operator with real coefficients.

Our third comment regarding Theorem 8.1.9 is that, as noted in [130, Proposition 5.3.14],

the conditions demanded in (8.1.249) automatically hold in the case when $\mathbb{X}$ is a rearrangement invariant Banach function space on the (non-atomic, sigma-finite measure space $(\partial\Omega, \sigma)$) whose Boyd indices satisfy $1 < p_{\mathbb{X}} \leq q_{\mathbb{X}} < \infty$. (8.1.263)

As such, *all conclusions in Theorem 8.1.9 are valid in the class of rearrangement invariant Banach function spaces* $\mathbb{X}$ *whose Boyd indices satisfy* $1 < p_{\mathbb{X}} \leq q_{\mathbb{X}} < \infty$. In particular, this is the case when the Generalized Banach Function Space $\mathbb{X}$ is actually an Orlicz space, $\mathbb{X} := L^{\Phi}(\partial\Omega, \sigma)$, associated with the (non-atomic, sigma-finite) measure space $(\partial\Omega, \sigma)$ and a Young function Φ whose dilation indices satisfy $1 < i(\Phi) \leq I(\Phi) < \infty$; see [130, Proposition 5.3.15] and [130, (5.3.62)].

For example, under the initial background assumptions made in Theorem 8.1.9, given any Young function Φ whose dilation indices satisfy $1 < i(\Phi) \leq I(\Phi) < \infty$ it follows that the Dirichlet Problem with data in the Orlicz space L^{Φ} (cf. (A.0.133)), formulated as

$$\begin{cases} u \in \big[\mathscr{C}^\infty(\Omega)\big]^M, \quad Lu = 0 \text{ in } \Omega, \\ \mathcal{N}_\kappa u \in L^{\Phi}(\partial\Omega, \sigma), \quad u\big|_{\partial\Omega}^{\kappa-\text{n.t.}} = f \in \big[L^{\Phi}(\partial\Omega, \sigma)\big]^M, \\ u(x) = O(|x|^{2-n}) \text{ as } |x| \to \infty \text{ if } \Omega \text{ is unbounded,} \end{cases} \tag{8.1.264}$$

is well-posed whenever the system L satisfies the Legendre-Hadamard strong ellipticity condition and $\mathfrak{A}_L^{\text{dis}} \neq \varnothing$, provided $\text{dist}\big(\nu, [\text{VMO}(\partial\Omega, \sigma)]^n\big) < \delta_\Phi$ for a sufficiently small threshold $\delta_\Phi \in (0, 1)$. Also, the solution satisfies natural Orlicz space estimates, namely

$$\|f\|_{[L^{\Phi}(\partial\Omega,\sigma)]^M} \leq \|\mathcal{N}_\kappa u\|_{L^{\Phi}(\partial\Omega,\sigma)} \leq C\|f\|_{[L^{\Phi}(\partial\Omega,\sigma)]^M} \tag{8.1.265}$$

for any function $f \in \big[L^{\Phi}(\partial\Omega, \sigma)\big]^M$, with a constant $C \in (0, \infty)$ independent of f. Indeed, all these are consequences of [130, (5.3.31), (5.3.62)], (8.1.258), and [130, Proposition 5.3.15].

In fact, by taking the threshold $\delta_\Phi \in (0, 1)$ sufficiently small, matters may be arranged so that the unique solution of (8.1.264) also satisfies the following modular estimates

$$\begin{gathered} \int_{\partial\Omega} \Phi\big(|f(x)|\big)\, d\sigma(x) \leq \int_{\partial\Omega} \Phi\big((\mathcal{N}_\kappa u)(x)\big)\, d\sigma(x) \leq C \int_{\partial\Omega} \Phi\big(|f(x)|\big)\, d\sigma(x) \\ \text{for each function } f \text{ belonging to the Orlicz space } \big[L^{\Phi}(\partial\Omega, \sigma)\big]^M, \end{gathered} \tag{8.1.266}$$

with a constant $C \in (0, \infty)$ independent of f. To justify this, note that the first inequality in (8.1.266) comes directly from [130, (5.1.12)]. As far as the second inequality in (8.1.266) concerned, pick p, q such that $1 < p < i(\Phi) \leq I(\Phi) < q < \infty$ and assume $\delta_\Phi \in (0, 1)$ is small enough so that the Dirichlet Problems for L in Ω with data in $\big[L^p(\partial\Omega, \sigma)\big]^M$ and, respectively, $\big[L^q(\partial\Omega, \sigma)\big]^M$, are both well posed (see Theorem 8.1.6). The idea is to rely on the modular interpolation result from [130, Theorem 5.3.16], presently invoked with T the sub-linear operator assigning to each $f \in \big[L^p(\partial\Omega, \sigma)\big]^M$ the function $\mathcal{N}_\kappa u_f$, where u_f is the unique solution of the Dirichlet Problem with data in $\big[L^p(\partial\Omega, \sigma)\big]^M$. Thanks to (8.1.153), the operator T is bounded in this context. Since $\big[L^q(\partial\Omega, \sigma)\big]^M$ is a subspace of $\big[L^p(\partial\Omega, \sigma)\big]^M$ (given that $\partial\Omega$ is compact), the well-posedness of the aforementioned Dirichlet Problems ensures that $\big[L^q(\partial\Omega,\sigma)\big]^M$ is an invariant subspace of T and its restriction to the latter subspace is also bounded. Granted these boundedness properties, the second inequality in (8.1.266) follows from [130, (5.3.177)].

Significantly, a Muckenhoupt weighted version of the well-posedness result formulated in relation to (8.1.264) is actually valid. To elaborate, retain the initial background assumptions made in Theorem 8.1.9, and once again assume that Φ is a Young function whose dilation indices satisfy $1 < i(\Phi) \leq I(\Phi) < \infty$. In addition, we now pick a Muckenhoupt weight $w \in A_{i(\Phi)}(\partial\Omega,\sigma)$. Then the Dirichlet Problem with data in the Muckenhoupt weighted Orlicz space L^{Φ}_{w}, formulated as

$$\begin{cases} u \in \big[\mathscr{C}^\infty(\Omega)\big]^M, \quad Lu = 0 \text{ in } \Omega, \\ \mathcal{N}_\kappa u \in L^\Phi(\partial\Omega, w), \quad u\big|_{\partial\Omega}^{\kappa-\mathrm{n.t.}} = f \in \big[L^\Phi(\partial\Omega,w)\big]^M, \\ u(x) = O(|x|^{2-n}) \text{ as } |x| \to \infty \text{ if } \Omega \text{ is unbounded}, \end{cases} \tag{8.1.267}$$

is well-posed whenever the system L satisfies the Legendre-Hadamard strong ellipticity condition and $\mathfrak{A}^{\mathrm{dis}}_L \neq \varnothing$, provided $\mathrm{dist}\big(\nu, [\mathrm{VMO}(\partial\Omega,\sigma)]^n\big) < \delta_{\Phi,w}$ for some sufficiently small threshold $\delta_{\Phi,w} \in (0,1)$, which now also depends on $[w]_{A_{i(\Phi)}(\partial\Omega,\sigma)}$. In addition, there exists a constant $C \in (0,\infty)$ with the property that for each function f belonging to the weighted Orlicz space $\big[L^\Phi(\partial\Omega,w)\big]^M$ the unique solution u of (8.1.267) satisfies the following weighted Orlicz space estimates

$$\|f\|_{[L^\Phi(\partial\Omega,w)]^M} \leq \|\mathcal{N}_\kappa u\|_{L^\Phi(\partial\Omega,w)} \leq C\|f\|_{[L^\Phi(\partial\Omega,w)]^M}, \tag{8.1.268}$$

as well as weighted modular estimates

$$\begin{aligned} \int_{\partial\Omega} \Phi\big(|f(x)|\big)\, w(x)\, d\sigma(x) &\leq \int_{\partial\Omega} \Phi\big((\mathcal{N}_\kappa u)(x)\big)\, w(x)\, d\sigma(x) \\ &\leq C \int_{\partial\Omega} \Phi\big(|f(x)|\big) w(x)\, d\sigma(x). \end{aligned} \tag{8.1.269}$$

This may be justified based on the same strategy as before, using the modular interpolation result from [130, Theorem 5.3.16] for the solution operator now acting on Muckenhoupt weighted Lebesgue spaces (here the well-posedness of (8.1.156) is relevant), with the last estimate in (8.1.269) obtained by reasoning as in the proof of [130, Corollary 5.3.17]. Then (8.1.268) follows from (8.1.269) and [130, Lemma 5.3.3] applied to the aforementioned solution operator.

It is interesting to compare our well-posedness result for the Dirichlet Problem with data in the Orlicz space L^Φ, formulated in (8.1.264) (with the naturally accompanying modular estimates obtained in (8.1.266)) with the result from [210, Theorem 4.4, pp. 677-678], where the solvability of the Dirichlet problem with data in Orlicz spaces (with modular estimates) when $\Omega := B(0,1)$ is the unit ball in $\mathbb{R}^n$ and $L := \mathrm{div}A\nabla$ is a scalar divergence-form elliptic operator with bounded measurable coefficients, is related to a reverse Hölder condition for the Radon-Nikodym derivative $k := d\omega_L/d\sigma$ of the harmonic measure ω_L (associated with the operator L) and the surface measure σ on $S^{n-1} = \partial B(0,1)$.

To further exemplify the scope of our well-posedness result in relation to the Dirichlet Problem (8.1.264), take $\Phi(t) \approx t^p[\ln(\mathrm{e}+t)]^\alpha$ for some fixed $p \in (1,\infty)$ and $\alpha \in \mathbb{R}$. For such a choice, the Orlicz space L^Φ actually becomes (see the discussion in [130, Example 5.3.10]) Zygmund's space $L^p(\log L)^\alpha$, i.e.,

$$L^p(\log L)^\alpha(\partial\Omega,\sigma) = \Big\{f \in \mathscr{M}(\partial\Omega,\sigma) : \int_{\partial\Omega} |f(x)|^p\big[\ln(\mathrm{e}+|f(x)|)\big]^\alpha \,d\sigma(x) < \infty\Big\}.$$

Then the result in the previous paragraph becomes a well-posedness result for the Dirichlet Problem with data in Zygmund's space $L^p(\log L)^\alpha$, i.e.,

$$\begin{cases} u \in \big[\mathscr{C}^\infty(\Omega)\big]^M, \quad Lu = 0 \text{ in } \Omega, \\ \mathcal{N}_\kappa u \in L^p(\log L)^\alpha(\partial\Omega,\sigma), \\ u\big|_{\partial\Omega}^{\kappa-\mathrm{n.t.}} = f \in \big[L^p(\log L)^\alpha(\partial\Omega,\sigma)\big]^M, \\ u(x) = O(|x|^{2-n}) \text{ as } |x| \to \infty \text{ if } \Omega \text{ is unbounded}, \end{cases} \tag{8.1.270}$$

once again assuming that the system L satisfies the Legendre-Hadamard strong ellipticity condition, that $\mathfrak{A}_L^{\mathrm{dis}} \neq \emptyset$, and that $\mathrm{dist}\big(\nu, [\mathrm{VMO}(\partial\Omega,\sigma)]^n\big) < \delta$ for some sufficiently small threshold $\delta = \delta(L,\Omega,p,\alpha) \in (0,1)$. In regard to (8.1.270), we wish to note that the aforementioned well-posedness result may fail for the end-point case $p = 1$, even when $n := 2$, $M := 1$, $L := \Delta$, and $\Omega := B(0,1)$ (see [209]).

Many other interesting choices present themselves. For example, in view of the discussion in [130, Example 5.3.11], a similar well-posedness result holds in relation to the Dirichlet Problem with data in the Orlicz space $\big[L^p\exp(\log^\theta L)(\partial\Omega,\sigma)\big]^M$ (see [130, (5.3.149)]) whenever $p \in (1,\infty)$ and $\theta \in [0,1)$. Lastly, we note that, as noted above, we may also consider Muckenhoupt weighted versions of all these Orlicz spaces.

In addition, we obtain Fredholmness and well-posedness results as in Theorem 8.1.9 for the Dirichlet Problem with boundary data (and the size condition for the solution) in variable exponent Lebesgue spaces by choosing $\mathbb{X} := L^{p(\cdot)}(\partial\Omega,\sigma)$ where the exponent function $p(\cdot) : \partial\Omega \to (1,\infty)$ satisfies a log-Hölder continuity condition.

There are also well-posedness results for the Dirichlet Problem involving function spaces outside the realm of Generalized Banach Function Spaces. As an example, for each $p \in (1,\infty)$ and $q \in (0,\infty]$ the following Dirichlet Problem, with boundary data in Lorentz spaces,

$$\begin{cases} u \in \big[\mathscr{C}^\infty(\Omega)\big]^M, \quad Lu = 0 \text{ in } \Omega, \\ \mathcal{N}_\kappa u \in L^{p,q}(\partial\Omega,\sigma), \quad u\big|_{\partial\Omega}^{\kappa-\mathrm{n.t.}} = f \in \big[L^{p,q}(\partial\Omega,\sigma)\big]^M, \\ u(x) = O(|x|^{2-n}) \text{ as } |x| \to \infty \text{ if } \Omega \text{ is unbounded}, \end{cases} \tag{8.1.271}$$

enjoys similar solvability properties to those in Theorem 8.1.9 (with very similar proofs), even though Lorentz spaces are not Generalized Banach Function Spaces[38].

In relation to the Dirichlet Problem with boundary data in Lorentz spaces we wish to remark that

$$\text{well-posedness for the boundary value problem formulated in (8.1.271) fails at the end-point values } p = 1 \text{ and } q = \infty. \tag{8.1.272}$$

To prove this is the case, work in the complex plane $\mathbb{R}^2 \equiv \mathbb{C}$ and considering the function $u : \Omega \to \mathbb{R}$ defined for at each $z = (x, y) \equiv x + \mathrm{i}y \in \Omega := B(0, 1) \cap \mathbb{R}^2_+$ as

$$u(z) := \mathrm{Im}\big(z + z^{-1}\big) = y\Big(1 - \frac{1}{x^2 + y^2}\Big). \tag{8.1.273}$$

Then it may be checked without difficulty that, with $\kappa > 0$ arbitrary,

$$\begin{aligned} &u \in \mathscr{C}^\infty\big(\overline{\Omega} \setminus \{0\}\big), \quad \Delta u = 0 \text{ in } \Omega, \quad u\big|_{\partial\Omega\setminus\{0\}} = 0, \\ &\text{and } \mathcal{N}_\kappa u \in L^{1,\infty}\big(\partial\Omega, \mathcal{H}^1\big) \setminus L^1\big(\partial\Omega, \mathcal{H}^1\big), \end{aligned} \tag{8.1.274}$$

since $\big(\mathcal{N}_\kappa u\big)(x, 0) \approx |x|^{-1}$ uniformly for x near 0. From (8.1.274) it is then clear that uniqueness for the Dirichlet Problem (8.1.271) fails in the case when Ω is the open upper-half disk in the plane, i.e., $\Omega := B(0, 1) \cap \mathbb{R}^2_+$, $L := \Delta$, $p := 1$, and $q := \infty$.

To offer another interesting example proving (8.1.272) take $\Omega := \mathbb{D} := B(0, 1)$, the unit disk in the plane $\mathbb{R}^2 \equiv \mathbb{C}$, and consider the function

$$u : \mathbb{D} \to \mathbb{R} \text{ defined for each } z \in \mathbb{D} \text{ as } u(z) := \mathrm{Re}\Big(\frac{z + 1}{z - 1}\Big) = \frac{|z|^2 - 1}{|z - 1|^2}. \tag{8.1.275}$$

It has been noted in [131, (4.4.17), (4.4.18)] that

$$\begin{aligned} &u \in \mathscr{C}^\infty\big(\overline{\mathbb{D}} \setminus \{1\}\big), \quad \Delta u = 0 \text{ in } \mathbb{D}, \quad u\big|_{\mathbb{T}\setminus\{1\}} = 0, \text{ and} \\ &\mathcal{N}_\kappa u \in L^{1,\infty}\big(\mathbb{T}, \mathcal{H}^1\lfloor\mathbb{T}\big) \setminus L^1\big(\mathbb{T}, \mathcal{H}^1\lfloor\mathbb{T}\big), \end{aligned} \tag{8.1.276}$$

where $\mathbb{T} := \partial\mathbb{D}$ and $\kappa > 0$ is an arbitrary aperture parameter. In turn, from (8.1.276) we see that uniqueness for the Dirichlet Problem (8.1.271) fails in the case when $\Omega := \mathbb{D}$, $L := \Delta$, $p := 1$, and $q := \infty$.

Finally, we note that the assumption made in (8.1.249) has a "structural" character, as far as the integrity of Theorem 8.1.9 is concerned. Indeed, even for $L := \Delta$ in the unit disk $\mathbb{D}$ in the plane, there exists a Generalized Banach Function Space $\mathbb{X}$ over $(\partial\mathbb{D}, \mathcal{H}^1)$ such that the corresponding problem (8.1.250) is not solvable, which is solely attributed to the fact that a portion of (8.1.249) fails in this case. Specifically, taking $\mathbb{X} := L^1(\partial\mathbb{D}, \mathcal{H}^1)$ leads to a Dirichlet Problem whose space of

[38] from (A.0.129)-(A.0.128), [129, Proposition 6.2.1], Lebesgue's Monotone Convergence Theorem, and [129, Lemma 6.2.4] we see that all axioms *(P1)-(P5)* from [130, Definition 5.1.1] (which are subsequently used to define Generalized Banach Function Spaces) are satisfied except for the subadditivity of the function norm, which now involves a multiplicative constant (strictly bigger than 1) in the right side

admissible boundary data $\Big\{u\Big|_{\partial\mathbb{D}}^{\kappa-\mathrm{n.t.}} : u \in \mathscr{C}^\infty(\mathbb{D}),\ \Delta u = 0,\ \mathcal{N}_\kappa u \in L^1(\partial\mathbb{D}, \mathcal{H}^1)\Big\}$ is precisely the Hardy space $H^1(\partial\mathbb{D}, \mathcal{H}^1)$ (see the discussion in [54, p. 20]), which is strictly smaller than the full space of boundary data $L^1(\partial\mathbb{D}, \mathcal{H}^1)$. As regards the hypothesis (8.1.249), while the Hardy-Littlewood maximal operator is bounded on $\big(L^1(\partial\mathbb{D}, \mathcal{H}^1)\big)' = L^\infty(\partial\mathbb{D}, \mathcal{H}^1)$, it fails to map $L^1(\partial\mathbb{D}, \mathcal{H}^1)$ to itself, and this is the root of the aforementioned failure of solvability.

We next present the proof of Theorem 8.1.9.

Proof of Theorem 8.1.9 To justify the claim in item *(1)*, first assume that the system L is weakly elliptic and has the property that $\mathfrak{A}_L^{\mathrm{dis}} \neq \varnothing$. Pick some $A \in \mathfrak{A}_L^{\mathrm{dis}}$. From the current hypotheses and item *(8)* in Theorem 7.1.1 we know that if $\delta_{\mathbb{X}} \in (0, 1)$ is sufficiently small, then the operator

$$\tfrac{1}{2}I + K_A : \big[\mathbb{X}\big]^M \to \big[\mathbb{X}\big]^M \text{ is Fredholm (with index zero).} \tag{8.1.277}$$

As such,

$$\begin{gathered}\mathrm{Im}\Big(\tfrac{1}{2}I + K_A;\, \big[\mathbb{X}\big]^M\Big) \text{ is a closed subspace}\\ \text{of finite codimension in } \big[\mathbb{X}\big]^M.\end{gathered} \tag{8.1.278}$$

In view of [130, (5.1.13), (5.2.92)], [131, (2.4.20)], [132, (1.3.24), (1.3.26), (1.5.20)], and (A.0.72), it follows that $\mathrm{Im}\Big(\tfrac{1}{2}I + K_A;\, \big[\mathbb{X}\big]^M\Big)$ is contained in the space of admissible boundary data for the Dirichlet Problem (8.1.250) described in (8.1.253). Once this has been established, we conclude from (8.1.278) and [130, Theorem 1.1.3, item *(3)*] that, in the case when L is weakly elliptic with $\mathfrak{A}_L^{\mathrm{dis}} \neq \varnothing$, the latter space is, as claimed, closed and of finite codimension in $\big[\mathbb{X}\big]^M$.

Next, assume that L is weakly elliptic with $\mathfrak{A}_{L^\top}^{\mathrm{dis}} \neq \varnothing$, and pick a coefficient tensor $A \in \mathfrak{A}_L$ with $A^\top \in \mathfrak{A}_{L^\top}^{\mathrm{dis}}$. Recall from [130, (5.2.92)] that there exists an integrability exponent $q \in (1, 2)$ with the property that

$$L^{q'}(\Sigma, \sigma) \hookrightarrow \mathbb{X} \hookrightarrow L^q(\Sigma, \sigma), \tag{8.1.279}$$

where $q' := (1 - 1/q)^{-1} \in (1, \infty)$. After further decreasing the value of $\delta_{\mathbb{X}} \in (0, 1)$ if necessary, we may assume that the conclusions in Theorem 8.1.6 hold for the Dirichlet Problem (8.1.145) formulated with $p := q'$. Since (8.1.279) ensures that (8.1.254) embeds into the space of null-solutions for said problem (cf. (8.1.149)), we conclude from item *(1)* of Theorem 8.1.6 that the space of null-solutions of the boundary problem (8.1.250), introduced in (8.1.254), is indeed is finite dimensional. From (8.1.150) we also conclude that for any null-solution u of the boundary problem (8.1.250) the nontangential boundary trace $(\nabla u)\Big|_{\partial\Omega}^{\kappa-\mathrm{n.t.}}$ exists at σ-a.e. point on $\partial\Omega$. We claim that there exists a constant $c \in \mathbb{C}^M$ such that

$$u = -\mathscr{S}(\partial_\nu^A u) + c \text{ in } \Omega. \tag{8.1.280}$$

Bearing in mind that we currently have $u\big|_{\partial\Omega}^{\kappa-\mathrm{n.t.}} = 0$, in the case when either $n \geq 3$, or $n = 2$ and Ω is bounded, this follows (with $c = 0$) from [132, Theorem 1.3.3], while in the case when $n = 2$ and $\Omega \subseteq \mathbb{R}^2$ is an exterior domain this is guaranteed by [131, Corollary 1.5.4]. Applying the conormal the derivative ∂_ν^A to (8.1.280) yields (thanks to the jump-formula [132, (1.5.59)]) $\big(\frac{1}{2}I + K^{\#}_{A^\top}\big)(\partial_\nu^A u) = 0$, hence

$$\partial_\nu^A u \in \mathrm{Ker}\Big(\tfrac{1}{2}I + K^{\#}_{A^\top};\ \big[L^q(\partial\Omega,\sigma)\big]^M\Big). \tag{8.1.281}$$

By further decreasing $\delta_{\mathbb{X}}$ as to ensure that the conclusions in Proposition 7.1.2 are valid (for the interval $[q', q]$), we conclude from (8.1.281) and (8.1.279) that

$$\partial_\nu^A u \in \big[L^{q'}(\partial\Omega,\sigma)\big]^M. \tag{8.1.282}$$

Collectively, (8.1.280), (8.1.282), (8.1.279), Calderón-Zygmund theory (cf. [132, Theorem 1.5.1]) then prove that $\mathcal{N}_\kappa(\nabla u) \in \mathbb{X}$. Thus, the additional properties claimed in (8.1.255) are proved. Note that (8.1.256) follows from (8.1.255) and definitions. Finally, the very first claim in item *(1)* is implied by what we have proved so far, while the very last claim in item *(1)* is a consequence of Theorem 8.6.5, Theorem 8.6.7, Example 1.6.14, and Remark 8.6.9. This takes care of the current item *(1)*.

Turning our attention to item *(2)*, work under the assumptions specified in (8.1.257). Pick a coefficient tensor $A \in \mathfrak{A}_L^{\mathrm{dis}}$ and recall the exponent $q \in (1, 2)$ appearing in (8.1.279). The goal is to show that the Dirichlet Problem (8.1.250) is well posed. Take $\delta_{\mathbb{X}} \in (0, 1)$ small enough so that the conclusions in Theorem 8.1.6 are true for $p := q$. Then (8.1.279) plus the well-posedness result in item *(2)* of Theorem 8.1.6 imply that the Dirichlet Problem (8.1.250) has at most one solution. To show that this problem is always solvable, assume $\delta_{\mathbb{X}} \in (0, 1)$ is sufficiently small so that Theorem 7.1.1 applies. This ensures that

$$\begin{aligned}&\tfrac{1}{2}I + K_A \text{ is a Fredholm operator with index zero when acting}\\ &\text{both from } \big[\mathbb{X}\big]^M \text{ into itself, and from } \big[L^q(\partial\Omega,\sigma)\big]^M \text{ into itself.}\end{aligned} \tag{8.1.283}$$

In particular, the latter property implies that

$$\begin{aligned}&\mathrm{Im}\Big(\tfrac{1}{2}I + K_A;\ \big[L^q(\partial\Omega,\sigma)\big]^M\Big) \text{ is a closed subspace}\\ &\text{of finite codimension in } \big[L^q(\partial\Omega,\sigma)\big]^M.\end{aligned} \tag{8.1.284}$$

Also, if we set

$$N := \dim \mathrm{Ker}\Big(\tfrac{1}{2}I + K_A;\ \big[\mathbb{X}\big]^M\Big), \tag{8.1.285}$$

then $N \in \mathbb{N}_0$ and

$$\begin{aligned}&\mathrm{Im}\Big(\tfrac{1}{2}I + K_A;\ \big[\mathbb{X}\big]^M\Big) \text{ is a closed subspace}\\ &\text{of codimension } N \text{ in } \big[\mathbb{X}\big]^M.\end{aligned} \tag{8.1.286}$$

If $N = 0$, then $\frac{1}{2}I + K_A$ is invertible as an operator from $\big[\mathbb{X}\big]^M$ into itself, so a solution to the Dirichlet Problem (8.1.250) satisfying (8.1.258) is given by

$$u = \mathcal{D}_A\Big(\big(\tfrac{1}{2}I + K_A\big)^{-1} f\Big) \text{ in } \Omega. \tag{8.1.287}$$

Assume next that $N = 1$. We claim that

$$\text{the Sobolev space } \big[\mathbb{X}_1\big]^M \text{ is } \textit{not} \text{ contained in } \operatorname{Im}\Big(\tfrac{1}{2}I + K_A; \big[\mathbb{X}\big]^M\Big). \tag{8.1.288}$$

Seeking a contradiction, write

$$\begin{aligned} \Big\{\phi\big|_{\partial\Omega} : \phi \in \big[\mathscr{C}_c^\infty(\mathbb{R}^n)\big]^M\Big\} &\subseteq \big[L_1^{q'}(\partial\Omega, \sigma)\big]^M \subseteq \big[\mathbb{X}_1\big]^M \\ &\subseteq \operatorname{Im}\Big(\tfrac{1}{2}I + K_A; \big[\mathbb{X}\big]^M\Big) \subseteq \operatorname{Im}\Big(\tfrac{1}{2}I + K_A; \big[L^q(\partial\Omega, \sigma)\big]^M\Big), \end{aligned} \tag{8.1.289}$$

with the first inclusion provided by [130, Lemma 11.1.3, item *(3)*], the second inclusion implied by (8.1.279) and definitions (cf. (8.1.92)), the third inclusion a consequence of the assumed failure of (8.1.288), and with the last inclusion coming from (8.1.279). If we also recall from [129, Corollary 3.7.3] that

$$\Big\{\phi\big|_{\partial\Omega} : \phi \in \big[\mathscr{C}_c^\infty(\mathbb{R}^n)\big]^M\Big\} \text{ is a dense subspace of } \big[L^q(\partial\Omega, \sigma)\big]^M, \tag{8.1.290}$$

we deduce from (8.1.284), (8.1.289), and (8.1.290) that

$$\operatorname{Im}\Big(\tfrac{1}{2}I + K_A; \big[L^q(\partial\Omega, \sigma)\big]^M\Big) = \big[L^q(\partial\Omega, \sigma)\big]^M. \tag{8.1.291}$$

On account of (8.1.283), this forces $\operatorname{Ker}\Big(\tfrac{1}{2}I + K_A; \big[L^q(\partial\Omega, \sigma)\big]^M\Big)$ to be trivial, hence

$$\operatorname{Ker}\Big(\tfrac{1}{2}I + K_A; \big[\mathbb{X}\big]^M\Big) = \{0\}, \tag{8.1.292}$$

by the second inclusion in (8.1.279). However, (8.1.292) stands in conflict with the fact that we are currently assuming $N = 1$. This contradiction proves the claim made in (8.1.288). In turn, this shows that there exists a function

$$\phi_1 \in \big[\mathbb{X}_1\big]^M \text{ such that } \phi_1 \notin \operatorname{Im}\Big(\tfrac{1}{2}I + K_A; \big[\mathbb{X}\big]^M\Big). \tag{8.1.293}$$

As a consequence, $[\phi_1]$, the equivalence class of the function ϕ_1 in the quotient space $\big[\mathbb{X}\big]^M \Big/ \operatorname{Im}\Big(\tfrac{1}{2}I + K_A; \big[\mathbb{X}\big]^M\Big)$, satisfies

$$0 \neq [\phi_1] \in \big[\mathbb{X}\big]^M \Big/ \operatorname{Im}\Big(\tfrac{1}{2}I + K_A; \big[\mathbb{X}\big]^M\Big). \tag{8.1.294}$$

Since, thanks to (8.1.283), (8.1.285), and the current assumptions, the quotient space in (8.1.294) is one-dimensional, it follows that $[\phi_1]$ actually spans the space in question. In particular, the operator

$$T : \big[\mathbb{X}\big]^M \oplus \mathbb{C} \to \big[\mathbb{X}\big]^M \text{ defined as}$$
$$T(g,\lambda) := \big(\tfrac{1}{2}I + K_A\big)g + \lambda\phi_1, \text{ for all } g \in \big[\mathbb{X}\big]^M \text{ and } \lambda \in \mathbb{C}, \tag{8.1.295}$$

is well defined, linear, bounded, and surjective. Since T acts between Banach spaces, [130, Proposition 1.2.13] guarantees the existence of some $C \in (0,\infty)$ with the property that any $f \in \big[\mathbb{X}\big]^M$ may be written as

$$f = \big(\tfrac{1}{2}I + K_A\big)g + \lambda\phi_1 \text{ with } g \in \big[\mathbb{X}\big]^M \text{ and } \lambda \in \mathbb{C}$$
$$\text{satisfying } \|g\|_{[\mathbb{X}]^M} + |\lambda| \leq C\|f\|_{[\mathbb{X}]^M}. \tag{8.1.296}$$

To continue, we find it useful to first focus on the case when $n \geq 3$. In such a scenario, we know from Theorem 7.1.20 that

$$S : \big[\mathbb{X}\big]^M \longrightarrow \big[\mathbb{X}_1\big]^M \text{ is an invertible operator.} \tag{8.1.297}$$

In concert with (8.1.296), this shows that any $f \in \big[\mathbb{X}\big]^M$ may be written as

$$f = \big(\tfrac{1}{2}I + K_A\big)g + Sh \text{ for some } g, h \in \big[\mathbb{X}\big]^M$$
$$\text{satisfying } \|g\|_{[\mathbb{X}]^M} + \|h\|_{[\mathbb{X}]^M} \leq C\|f\|_{[\mathbb{X}]^M}. \tag{8.1.298}$$

Granted this, we conclude that for any boundary datum $f \in \big[\mathbb{X}\big]^M$ a solution to the Dirichlet Problem (8.1.250) obeying (8.1.258) is given by $u = \mathcal{D}_A g + \mathcal{S}h$ in Ω. In the case when $n = 2$, in place of (8.1.297) we employ the invertibility result from (7.1.389). This guarantees that any boundary datum $f \in \big[\mathbb{X}\big]^M$ may be written as

$$f = \big(\tfrac{1}{2}I + K_A\big)g + Sh + c \text{ for some } g, h \in \big[\mathbb{X}\big]^M \text{ and } c \in \mathbb{C}^M$$
$$\text{such that } \int_{\partial\Omega} h\,d\sigma = 0 \text{ and } \|g\|_{[\mathbb{X}]^M} + \|h\|_{[\mathbb{X}]^M} + |c| \leq C\|f\|_{[\mathbb{X}]^M}. \tag{8.1.299}$$

Also bearing in mind (8.1.116) and [132, Lemma 1.3.1, item *(ii)*], it follows that

$$u = \mathcal{D}_A g + \mathcal{S}h + c \text{ in } \Omega \tag{8.1.300}$$

is a solution of the Dirichlet Problem (8.1.250) satisfying (8.1.258). This finishes the treatment of the case when $N = 1$.

Assume next that $N = 2$. As before, there exists a function $\phi_1 \in \big[L^{q'}(\partial\Omega,\sigma)\big]^M$ as in (8.1.293). With $\langle\phi_1\rangle$ denoting its linear span, we claim that the Sobolev space

$$\big[\mathbb{X}_1\big]^M \text{ is not contained in } \langle\phi_1\rangle + \mathrm{Im}\Big(\tfrac{1}{2}I + K_A; \big[\mathbb{X}\big]^M\Big). \tag{8.1.301}$$

For otherwise much as in (8.1.289) we would have the sequence of inclusions

$$\Big\{\phi\big|_{\partial\Omega} : \phi \in \big[\mathscr{C}_c^\infty(\mathbb{R}^n)\big]^M\Big\} \subseteq \big[L_1^{q'}(\partial\Omega,\sigma)\big]^M \subseteq \big[\mathbb{X}_1\big]^M \tag{8.1.302}$$
$$\subseteq \langle\phi_1\rangle + \mathrm{Im}\Big(\tfrac{1}{2}I + K_A; \big[\mathbb{X}\big]^M\Big)\langle\phi_1\rangle + \subseteq \mathrm{Im}\Big(\tfrac{1}{2}I + K_A; \big[L^q(\partial\Omega,\sigma)\big]^M\Big)$$

which, thanks to (8.1.290), (8.1.284), and [130, Theorem 1.1.3, item *(3)*], implies

$$\langle\phi_1\rangle + \mathrm{Im}\Big(\tfrac{1}{2}I + K_A; \big[L^q(\partial\Omega,\sigma)\big]^M\Big) = \big[L^q(\partial\Omega,\sigma)\big]^M. \tag{8.1.303}$$

In turn, (8.1.303), (8.1.283), (8.1.279), (8.1.285) imply

$$\begin{aligned} 1 &\geq \dim\left[\frac{\big[L^q(\partial\Omega,\sigma)\big]^M}{\mathrm{Im}\big(\frac{1}{2}I+K_A;[L^q(\partial\Omega,\sigma)]^M\big)}\right] \\ &= \dim\mathrm{Ker}\Big(\tfrac{1}{2}I + K_A; \big[L^q(\partial\Omega,\sigma)\big]^M\Big) \geq \dim\mathrm{Ker}\Big(\tfrac{1}{2}I + K_A; \big[\mathbb{X}\big]^M\Big) = N, \end{aligned} \tag{8.1.304}$$

in contradiction with the fact that we are presently assuming $N = 2$. This contradiction proves (8.1.301). As a result, there exists a function

$$\phi_2 \in \big[\mathbb{X}_1\big]^M \text{ such that } \phi_2 \notin \langle\phi_1\rangle + \mathrm{Im}\Big(\tfrac{1}{2}I + K_A; \big[\mathbb{X}\big]^M\Big). \tag{8.1.305}$$

Since

$$\dim\left[\frac{\big[\mathbb{X}\big]^M}{\mathrm{Im}\big(\frac{1}{2}I+K_A;[\mathbb{X}]^M\big)}\right] = \dim\mathrm{Ker}\Big(\tfrac{1}{2}I + K_A; \big[\mathbb{X}\big]^M\Big) = N = 2, \tag{8.1.306}$$

this shows that $[\phi_1]$ and $[\phi_2]$, the equivalence classes of the functions ϕ_1 and ϕ_2 in the quotient space $\big[\mathbb{X}\big]^M \big/ \mathrm{Im}\Big(\tfrac{1}{2}I + K_A; \big[\mathbb{X}\big]^M\Big)$, span said space. Thus,

$$\begin{aligned} T : \big[\mathbb{X}\big]^M \oplus \mathbb{C}^2 \longrightarrow \big[\mathbb{X}\big]^M &\text{ given for any } g \in \big[\mathbb{X}\big]^M \text{ and } \lambda_1, \lambda_2 \in \mathbb{C} \\ &\text{by } T(g, \lambda_1, \lambda_2) := \big(\tfrac{1}{2}I + K_A\big)g + \lambda_1\phi_1 + \lambda_2\phi_2, \end{aligned} \tag{8.1.307}$$

is a well-defined, linear, bounded, and surjective operator. In view of the fact that T acts between Banach spaces, [130, Proposition 1.2.13] ensures the existence of a constant $C \in (0, \infty)$ such that any $f \in \big[\mathbb{X}\big]^M$ may be written as

$$\begin{aligned} f = \big(\tfrac{1}{2}I + K_A\big)g + \lambda_1\phi_1 + \lambda_2\phi_2 &\text{ with } g \in \big[\mathbb{X}\big]^M \text{ and } \lambda_1, \lambda_2 \in \mathbb{C} \\ &\text{satisfying } \|g\|_{[\mathbb{X}]^M} + |\lambda_1| + |\lambda_2| \leq C\|f\|_{[\mathbb{X}]^M}. \end{aligned} \tag{8.1.308}$$

In the case when $n \geq 3$, we may avail ourselves of (8.1.297) to conclude that if a given boundary datum $f \in \big[\mathbb{X}\big]^M$ is expressed as above, and if

$$h \in \big[\mathbb{X}\big]^M \text{ is such that } Sh = \lambda_1\phi_1 + \lambda_2\phi_2, \tag{8.1.309}$$

then the function $u = \mathcal{D}_A g + \mathscr{S} h$ in Ω is a solution of the Dirichlet Problem (8.1.250) satisfying (8.1.258). The case $n = 2$ is dealt with similarly, now relying on the invertibility result from (7.1.389) to write, in place of (8.1.309),

$$Sh + c = \lambda_1\phi_1 + \lambda_2\phi_2 \text{ for some } c \in \mathbb{C}^M,\ h \in \big[\mathbb{X}\big]^M, \text{ satisfying} \quad \int_{\partial\Omega} h\,d\sigma = 0 \text{ and } \|h\|_{[\mathbb{X}]^M} + |c| \leq C\|f\|_{[\mathbb{X}]^M}. \tag{8.1.310}$$

The proof in the case $N = 2$ is therefore complete. Finally, the above scheme may be iterated to cover any value of $N \in \mathbb{N}_0$. This finishes the treatment of item *(2)* in the statement of the theorem. Finally, all claims in item *(3)* are dealt with similarly.□

Going further, we study the well-posedness of the Regularity Problem for weakly elliptic systems in δ-oscillating AR domains in $\mathbb{R}^n$ with boundary data in Hardy-Sobolev spaces $\dot{H}^p_1$ with $\frac{n-1}{n} < p \leq 1$. In the case of the Laplacian in bounded Lipschitz domains, the well-posedness of the Regularity Problem with boundary data in Hardy-Sobolev spaces $\dot{H}^p_1$ has been proved by B. Dahlberg and C. Kenig in [38, Theorem 4.12, p. 461] when $p = 1$, and by R. Brown in [20, Theorem D, p. 393] when $1 - \varepsilon < p < 1$, where $\varepsilon > 0$ is a (typically small) threshold depending only on the Lipschitz character of the underlying domain. A different approach based on stability of the quality of being an isomorphism on complex interpolation scales of quasi-Banach spaces has been developed in N. Kalton and M. Mitrea in [88]. This covers the case when $1 - \varepsilon < p < 1$, with $\varepsilon > 0$ as before, for Laplacian in bounded Lipschitz domains, and also applies to the Regularity Problem with data in Hardy-Sobolev spaces for the three-dimensional Lamé system in bounded Lipschitz domains. The aforementioned results are in dimensions ≥ 3. For two dimensional bounded Lipschitz domains, D. Mitrea has proved in [125, Proposition 3.5, p. 90] the well-posedness of the Regularity Problem with boundary data in Hardy-Sobolev spaces $\dot{H}^p_1$ with $\frac{2}{3} - \varepsilon < p \leq 1$ for some small $\varepsilon > 0$ of the same nature as above. In [156, Theorem 6.3], M. Mitrea and M. Taylor have solved the Regularity Problem for the Laplace-Beltrami operator on Lipschitz subdomains of Riemannian manifolds with boundary data in the Hardy space $\dot{H}^1_1$. The reader is also referred to the work of A. Barton and S. Mayboroda in [13, Chapter 3] for related problems. All these results involve Lipschitz domains, so our theorem below is new even in the case when $L := \Delta$, the Laplace operator.

Theorem 8.1.10 (The Regularity Problem with Data in Hardy-Sobolev Spaces) *Assume $n, M \in \mathbb{N}$ with $n \geq 2$ and suppose $\Omega \subseteq \mathbb{R}^n$ is an Ahlfors regular domain with compact boundary. Denote by ν the geometric measure theoretic outward unit normal to Ω and abbreviate $\sigma := \mathcal{H}^{n-1}\lfloor\partial\Omega$. Also, let L be a homogeneous, second-order, constant complex coefficient, $M \times M$ system in $\mathbb{R}^n$. Finally, pick an arbitrary aperture parameter $\kappa > 0$ and fix some exponent*

$$p \in \big(\tfrac{n-1}{n}, 1\big]. \tag{8.1.311}$$

Then there exists $\delta \in (0, 1)$ depending only on n, p, L, the ADR *constants of $\partial\Omega$, and the relative flatness ratio $\wp(\Omega)/\mathrm{diam}\,\partial\Omega$ with the property that when*

$$\operatorname{dist}\Big(\nu, [\mathrm{VMO}(\partial\Omega, \sigma)]^n\Big) < \delta \tag{8.1.312}$$

where the distance is considered in the space $\big[\mathrm{BMO}(\partial\Omega,\sigma)\big]^n$ *(hence if* Ω *is a* δ*-oscillating* AR *domain*[39]*; cf. Definition 3.4.1), the following statements are true:*

(1) If one assumes that

$$L \text{ is weakly elliptic with } \mathfrak{A}^{\rm dis}_{L} \neq \varnothing \text{ and } \mathfrak{A}^{\rm dis}_{L^\top} \neq \varnothing, \tag{8.1.313}$$

then the Regularity Problem[40]

$$\begin{cases} u \in \big[\mathscr{C}^\infty(\Omega)\big]^M, \quad Lu = 0 \text{ in } \Omega, \\ \mathcal{N}_\kappa(\nabla u) \in L^p(\partial\Omega,\sigma), \quad u\big|_{\partial\Omega}^{\kappa-\mathrm{n.t.}} = f \in \big[\dot{H}^p_1(\partial\Omega,\sigma)\big]^M, \\ u(x) = O(|x|^{2-n}) \text{ as } |x| \to \infty \text{ if } \Omega \text{ is unbounded}, \end{cases} \tag{8.1.314}$$

is Fredholm solvable, in the following precise sense: First, the space of admissible boundary data for the Regularity Problem (8.1.314)*, i.e.,*

$$\Big\{ u\big|_{\partial\Omega}^{\kappa-\mathrm{n.t.}} : u \in \big[\mathscr{C}^\infty(\Omega)\big]^M,\ Lu = 0 \text{ in } \Omega,\ \mathcal{N}_\kappa(\nabla u) \in L^p(\partial\Omega,\sigma), \text{ and} \\ u(x) = O(|x|^{2-n}) \text{ as } |x| \to \infty \text{ if } \Omega \text{ is unbounded} \Big\}, \tag{8.1.315}$$

is a closed linear subspace of finite codimension in $\big[\dot{H}^p_1(\partial\Omega,\sigma)\big]^M$*, and a solution* u *of* (8.1.314) *exists if and only if the boundary datum* f *belongs to* (8.1.315)*. Second, the space of null-solutions of the boundary problem* (8.1.314)*,*

$$\Big\{ u \in \big[\mathscr{C}^\infty(\Omega)\big]^M : Lu = 0,\ \mathcal{N}_\kappa(\nabla u) \in L^p(\partial\Omega,\sigma),\ u\big|_{\partial\Omega}^{\kappa-\mathrm{n.t.}} = 0, \\ \text{and } u(x) = O(|x|^{2-n}) \text{ as } |x| \to \infty \text{ if } \Omega \text{ is unbounded} \Big\}, \tag{8.1.316}$$

is finite dimensional. Lastly, the Fredholm solvability result stated above is optimal, since in the class of operators L *for which* $\mathfrak{A}^{\rm dis}_{L} = \varnothing$ *or* $\mathfrak{A}^{\rm dis}_{L^\top} = \varnothing$ *it may happen that the Regularity Problem* (8.1.314) *fails to be Fredholm solvable even when the underlying domain* Ω *is the unit disk in the plane (or the complement of its closure). In fact, the aforementioned failure of Fredholm solvability may be total, in the sense that for the problem* (8.1.314)*, both the space of admissible boundary data may have infinite codimension in the full space of boundary data, and the space of null-solutions may be infinite dimensional.*

(2) If (8.1.252) *is strengthened to*[41] *assuming that*

$$L \text{ satisfies the Legendre-Hadamard strong ellipticity condition (cf. [131, Definition 1.3.2]) and one has } \mathfrak{A}^{\rm dis}_{L} \neq \varnothing \tag{8.1.317}$$

[39] in particular, if Ω is an infinitesimally flat AR domain

[40] in light of work in [129, §8.5], one may replace the nontangential maximal operator by the tangential maximal operator, associated as in [129, Definition 8.5.1] with a sufficiently large power

[41] according to Proposition 1.4.1, whenever L is as in (8.1.317) we have $\mathfrak{A}^{\rm dis}_{L} \neq \varnothing$ and $\mathfrak{A}^{\rm dis}_{L^\top} \neq \varnothing$

then the Regularity Problem (8.1.314) *is actually well posed. More specifically,* (8.1.314) *has a unique solution, and this unique solution has the additional property that if either* $n \geq 3$, *or* $n = 2$ *and* $p \in \left(\frac{1}{2}, 1\right)$, *then*

$$\mathcal{N}_\kappa u \in L^{p^*}(\partial\Omega, \sigma) \text{ where } p^* := \left(\tfrac{1}{p} - \tfrac{1}{n-1}\right)^{-1} \in (1, \infty) \text{ and} \\ \|\mathcal{N}_\kappa u\|_{L^{p^*}(\partial\Omega,\sigma)} + \|\mathcal{N}_\kappa(\nabla u)\|_{L^p(\partial\Omega,\sigma)} \leq C\|f\|_{[H_1^{p^*,p}(\partial\Omega,\sigma)]^M} < \infty, \tag{8.1.318}$$

for some constant $C = C(\Omega, n, L, p) \in (0, \infty)$, *while if* $n = 2$ *and* $p = 1$ *then*

$$\mathcal{N}_\kappa u \text{ belongs to } L^\infty(\partial\Omega, \sigma) \text{ and for each } q \in (1, \infty) \text{ one has} \\ \|\mathcal{N}_\kappa u\|_{L^\infty(\partial\Omega,\sigma)} + \|\mathcal{N}_\kappa(\nabla u)\|_{L^1(\partial\Omega,\sigma)} \leq C\|f\|_{[H_1^{q,1}(\partial\Omega,\sigma)]^M} < \infty, \tag{8.1.319}$$

for some constant $C = C(\Omega, L, q) \in (0, \infty)$.

It is of interest to observe that, as seen from Proposition 1.6.12 and [131, Proposition 1.4.39], for the scalar case, i.e., if $M = 1$, and for dimensions $n \geq 3$, in place of (8.1.317) one may simply now ask that the operator L is weakly elliptic. In addition, in the scalar case in the two-dimensional setting (i.e., for $M = 1$ and $n = 2$), if L satisfies the Legendre-Hadamard strong ellipticity condition then Proposition 1.6.11 gives $\mathfrak{A}_L^{\text{dis}} \neq \varnothing$. We also wish to observe that, according to (1.6.141) in Proposition 1.6.13, in the scalar case in the two-dimensional setting (i.e., for $M = 1$ and $n = 2$) we may omit (8.1.317) altogether if L is a weakly elliptic operator with real coefficients. Lastly, as far as the statement of Theorem 8.1.10 is concerned, similar considerations to those made in Remark 8.1.2 are presently applicable.

Proof of Theorem 8.1.10 Pick $\delta \in (0, 1)$ sufficiently small, depending on n, the ADR constants of $\partial\Omega$, and the relative flatness ratio $\wp(\Omega)/\operatorname{diam}\partial\Omega$. Then Theorem 3.4.2 ensures that Ω is actually a two-sided NTA domain (hence also a UR domain satisfying a two-sided local John condition). Then [132, Theorem 2.2.8] applies and gives that for any function u as in the first two lines of (8.1.314) the nontangential boundary trace $u\big|_{\partial\Omega}^{\kappa-\text{n.t.}}$ exists at σ-a.e. point on $\partial\Omega$ and, as a function, belongs to the homogeneous Hardy-based Sobolev space $\left[\dot{H}_1^p(\partial\Omega, \sigma)\right]^M$. As a consequence, the space in (8.1.315) is meaningfully defined, and a a solution u of (8.1.314) exists if and only if the boundary datum f belongs to (8.1.315).

Work under the assumptions made in (8.1.313). By further decreasing the threshold $\delta \in (0, 1)$ if necessary, we can also ensure that the conclusions in Theorem 7.1.32 and Remark 7.1.33 are true. Based on these, [132, (2.3.97)] together with [132, Proposition 2.3.4], [132, Theorem 2.2.3], and [130, Theorem 1.1.3, item *(3)*] we deduce that the space of admissible boundary data for the Regularity Problem (8.1.314) (defined in (8.1.315)) is a closed linear subspace of finite codimension in $\left[\dot{H}_1^p(\partial\Omega, \sigma)\right]^M$.

Next, let u be an arbitrary null-solution of the boundary problem (8.1.314), i.e., assume u belongs to the space defined in (8.1.316). Since [132, Theorem 2.2.7] implies (also bearing [131, Theorem 1.5.9] in mind) that

$$\mathcal{N}_\kappa u \in L^{p^*}(\partial\Omega, \sigma) \text{ if either } n \geq 3, \text{ or } n = 2 \text{ and } p \in \left(\tfrac{1}{2}, 1\right), \tag{8.1.320}$$

while (8.1.320) and [130, (4.2.21)] further yield

$$\mathcal{N}_\kappa u \in L^q(\partial\Omega, \sigma) \text{ for any } q \in (1, \infty) \text{ if } n = 2 \text{ and } p = 1, \tag{8.1.321}$$

we see that the space of null-solutions of the boundary problem (8.1.314) is contained in the space of null-solutions for the Dirichlet Problem with data in Lebesgue spaces. Since from item *(1)* in Theorem 8.1.6 we know that the latter is finite dimensional, it follows that (8.1.316) is finite dimensional as well. At this stage, the very first claim in the current item *(1)* follows from what we have proved so far. Finally, the very last claim in the current item *(1)* is seen from Theorem 8.6.5, Theorem 8.6.7, and Example 1.6.14. This finishes the treatment of item *(1)*.

To deal with item *(2)*, assume $\delta \in (0, 1)$ is small enough so that the conclusions in Theorem 7.1.34 are valid (for a suitable choice of the parameters involved in its statement). Fix an arbitrary function $f \in \big[\dot{H}^p_1(\partial\Omega, \sigma)\big]^M$. From [132, Theorem 2.3.3] we know that in this case f belongs to $\big[H^{p^*,p}_1(\partial\Omega, \sigma)\big]^M$ if either $n \geq 3$ or $p \in \big(\frac{1}{2}, 1\big)$, while [132, Proposition 2.3.4] together with [132, Remark 2.3.5] tell us that f belongs to $\big[H^{q,1}_1(\partial\Omega, \sigma)\big]^M$ for each $q \in [1, \infty)$ if $n = 2$ and $p = 1$. Granted these, if $n \geq 3$ then the invertibility result recorded in (7.1.542) guarantees that there exists some $g \in \big[H^p(\partial\Omega, \sigma)\big]^M$ with

$$Sg = f \text{ and } \|g\|_{[H^p(\partial\Omega,\sigma)]^M} \leq C\|f\|_{[H^{p^*,p}_1(\partial\Omega,\sigma)]^M}, \tag{8.1.322}$$

for some $C \in (0, \infty)$ independent of f. With the boundary-to-domain single layer potential operator $\mathscr{S}$ associated with L and Ω as in [132, (2.2.31)], consider

$$u := \mathscr{S}g \text{ in } \Omega. \tag{8.1.323}$$

Then from (8.1.323) and [132, (2.2.32), (2.2.33), (2.2.35), (2.2.39), (2.2.44), (2.2.127)] we conclude that u satisfies all properties demanded in (8.1.314) and that (8.1.318) holds. This takes care of existence and estimates when $n \geq 3$.

In the case when $n = 2$ and $p = 1$, the invertibility result pertaining to the augmented single layer operator in the context of (7.1.548) gives that for any given $f \in \big[H^{q,1}_1(\partial\Omega, \sigma)\big]^M$ with $q \in (1, \infty)$ we may find $g \in \big[H^1(\partial\Omega, \sigma)\big]^M$ and a constant $c \in \mathbb{C}^M$ with the property that

$$\begin{gathered} Sg + c = f, \quad |c| + \|g\|_{[H^1(\partial\Omega,\sigma)]^M} \leq C\|f\|_{[H^{q,1}_1(\partial\Omega,\sigma)]^M} \\ \text{and } \langle g, \mathbf{e}_j\rangle = 0 \text{ for each } j \in \{1, \dots, M\}. \end{gathered} \tag{8.1.324}$$

We then define $u := \mathscr{S}g + c$ in Ω, and note that if Ω is an exterior domain then the vanishing moment condition in (8.1.324) ensures that $u(x) = O(1)$ as $|x| \to \infty$. Granted this, much as before, it follows that u solves (8.1.314) and satisfies (8.1.319). This completes the proof of existence and estimates in all cases.

To prove uniqueness, suppose u is a null-solution of the boundary value problem (8.1.314). In particular, the nontangential boundary trace $u\big|^{\kappa-\mathrm{n.t.}}_{\partial\Omega}$ vanishes at σ-a.e. point on $\partial\Omega$. Upon recalling (8.1.320)-(8.1.321), the uniqueness part in Theorem 8.1.6 applies and gives that $u = 0$. □

We next treat the Dirichlet Problem for Legendre-Hadamard elliptic second-order systems with constant complex coefficients in δ-oscillating AR domains with boundary data from BMO and VMO. The Dirichlet Problem for the Laplacian in a bounded Lipschitz domain Ω, with boundary data in BMO, has been treated by E. Fabes and U. Neri in [57], where they proved that a harmonic function u in Ω has boundary trace in $\mathrm{BMO}(\partial\Omega,\sigma)$ if and only if the Littlewood-Paley measure associated with u, i.e., $|\nabla u|^2\,\mathrm{dist}(\cdot,\partial\Omega)\,d\mathcal{L}^n$, is a Carleson measure in Ω. Later, in [171], S. Piro refined the approach in [57], and proved that a harmonic function u in a bounded Lipschitz domain Ω has boundary trace in $\mathrm{VMO}(\partial\Omega,\sigma)$ if and only if said Littlewood-Paley measure associated with u is actually a vanishing Carleson measure in Ω. We note that, the proofs in [57] and [171] make use of positivity and maximum principles and, as such, do not apply to systems. In [54, Theorem 1.8, p. 11], E. Fabes and C. Kenig have used boundary layer potentials to solve the Dirichlet Problem for the Laplacian with boundary data in BMO in bounded $\mathscr{C}^1$ domains, an approach extended by M. Mitrea and M. Taylor in [156, Proposition 7.3] to the case of the Laplace-Beltrami operator in Lipschitz subdomains of Riemannian manifolds.

Theorem 8.1.11 (**The Dirichlet Problem with Data in** BMO **and** VMO) *Fix two integers $n, M \in \mathbb{N}$ with $n \geq 2$ and suppose $\Omega \subseteq \mathbb{R}^n$ is an Ahlfors regular domain with compact boundary. Abbreviate $\sigma := \mathcal{H}^{n-1}\lfloor\partial\Omega$ and denote by ν the geometric measure theoretic outward unit normal to Ω. Also, let L be a homogeneous, second-order, constant complex coefficient, $M\times M$ system in $\mathbb{R}^n$. Assume that*

$$\text{L satisfies the Legendre-Hadamard strong ellipticity condition (cf. [131, Definition 1.3.2]) and one has } \mathfrak{A}_L^{\mathrm{dis}} \neq \emptyset. \tag{8.1.325}$$

Finally, fix an arbitrary aperture parameter $\kappa > 0$.

Then there exists a threshold $\delta \in (0,1)$ which depends only on n, L, the ADR *constants of $\partial\Omega$, and the relative flatness ratio $\wp(\Omega)/\mathrm{diam}\,\partial\Omega$ such that when*

$$\mathrm{dist}\Big(\nu, [\mathrm{VMO}(\partial\Omega,\sigma)]^n\Big) < \delta \tag{8.1.326}$$

where the distance is considered in the space $\big[\mathrm{BMO}(\partial\Omega,\sigma)\big]^n$ (hence if Ω is a δ-oscillating AR *domain*[42]*; cf. Definition 3.4.1), it follows that the boundary problem*

$$\begin{cases} u \in \big[\mathscr{C}^\infty(\Omega)\big]^M, \quad Lu = 0 \text{ in } \Omega, \\ |\nabla u|^p\,\mathrm{dist}(\cdot,\partial\Omega)^{p-1}\,d\mathcal{L}^n \text{ is a Carleson} \\ \text{measure in } \Omega \text{ for each } p \in (1,\infty), \\ \mathcal{N}_\kappa u \in L^p(\partial\Omega,\sigma) \text{ for each } p \in (1,\infty), \\ u\big|_{\partial\Omega}^{\kappa-\mathrm{n.t.}} = f \in \big[\mathrm{BMO}(\partial\Omega,\sigma)\big]^M, \\ u(x) = O(|x|^{2-n}) \text{ as } |x| \to \infty \text{ if } \Omega \text{ is unbounded,} \end{cases} \tag{8.1.327}$$

is well posed. More specifically, the boundary value problem (8.1.327) *has a unique solution, and this solution enjoys the additional property that for each exponent*

[42] in particular, if Ω is an infinitesimally flat AR domain

$p \in (1,\infty)$ one may find a constant $C = C(\Omega, L, p) \in (0,\infty)$ such that

$$\int_{\partial\Omega} (\mathcal{N}_\kappa u)^p \, d\sigma + \sup_{x\in\partial\Omega,\, r>0} \frac{1}{\sigma\big(B(x,r)\cap\partial\Omega\big)} \int_{B(x,r)\cap\Omega} |\nabla u|^p \operatorname{dist}(\cdot,\partial\Omega)^{p-1} \, d\mathcal{L}^n \\ \leq C\|f\|^p_{[\mathrm{BMO}(\partial\Omega,\sigma)]^M}, \tag{8.1.328}$$

as well as

$$\lim_{R\to 0^+} \left\{ \sup_{x\in\partial\Omega,\, r\in(0,R)} \frac{1}{\sigma\big(B(x,r)\cap\partial\Omega\big)} \int_{B(x,r)\cap\Omega} |\nabla u|^p \operatorname{dist}(\cdot,\partial\Omega)^{p-1} \, d\mathcal{L}^n \right\}^{1/p} \\ \leq C\cdot \operatorname{dist}\Big(f, \big[\mathrm{VMO}(\partial\Omega,\sigma)\big]^M\Big), \tag{8.1.329}$$

where the distance in the right-hand side is measured in $\big[\mathrm{BMO}(\partial\Omega,\sigma)\big]^M$.

In particular, if in fact $f \in \big[\mathrm{VMO}(\partial\Omega,\sigma)\big]^M$ then actually

$$|\nabla u|^p \operatorname{dist}(\cdot,\partial\Omega)^{p-1}\, d\mathcal{L}^n \text{ is a vanishing Carleson} \\ \text{measure in } \Omega, \text{ for each exponent } p\in(1,\infty). \tag{8.1.330}$$

As a corollary, the following boundary value problem is well posed:

$$\begin{cases} u \in \big[\mathscr{C}^\infty(\Omega)\big]^M, \quad Lu = 0 \text{ in } \Omega, \\ |\nabla u|^p \operatorname{dist}(\cdot,\partial\Omega)^{p-1}\, d\mathcal{L}^n \text{ is a vanishing Carleson} \\ \text{measure in } \Omega \text{ for each } p\in(1,\infty), \\ \mathcal{N}_\kappa u \in L^p(\partial\Omega,\sigma) \text{ for each } p\in(1,\infty), \\ u\big|^{\kappa-\mathrm{n.t.}}_{\partial\Omega} = f \in \big[\mathrm{VMO}(\partial\Omega,\sigma)\big]^M, \\ u(x) = O(|x|^{2-n}) \text{ as } |x|\to\infty \text{ if } \Omega \text{ is unbounded.} \end{cases} \tag{8.1.331}$$

We wish to highlight the fact that, as seen from Proposition 1.6.12 and [131, Proposition 1.4.39], in the scalar case, i.e., for $M = 1$, and dimensions $n \geq 3$, in place of (8.1.325) one may just demand that the operator L is weakly elliptic. Furthermore, in the scalar case in the two-dimensional setting (i.e., for $M = 1$ and $n = 2$), if L satisfies the Legendre-Hadamard strong ellipticity condition then Proposition 1.6.11 guarantees that the condition $\mathfrak{A}^{\mathrm{dis}}_L \neq \varnothing$ is automatically satisfied. Finally, according to (1.6.141) in Proposition 1.6.13, in the scalar case in the two-dimensional setting (i.e., for $M = 1$ and $n = 2$) we may omit (8.1.325) altogether if L is a weakly elliptic operator with real coefficients.

Proof of Theorem 8.1.11 Choose $\delta \in (0,1)$ small enough, depending on n, L, the ADR constants of $\partial\Omega$, and the relative flatness ratio $\wp(\Omega)/\operatorname{diam}\partial\Omega$, such that the conclusions in Theorem 7.1.39 hold (for a suitable choice of the parameters involved in its statement). By further decreasing $\delta \in (0,1)$ if necessary, we may invoke Theorem 3.4.2 to also ensure that Ω is a UR domain satisfying a two-sided local John condition.

If $n \geq 3$, the invertibility result in (7.1.605) used for the system $L^{\top}$ in place of L (which satisfies the same properties as the original system; cf. Proposition 1.4.1) ensures the existence of a constant $C \in (0, \infty)$ with the property that, given any function $f \in \big[\mathrm{BMO}(\partial\Omega, \sigma)\big]^M$, one may find $g \in \big[\mathrm{BMO}_{-1}(\partial\Omega, \sigma)\big]^M$ satisfying

$$(S_{L^{\top}})^{\top} g = f \text{ and } \|g\|_{[\mathrm{BMO}_{-1}(\partial\Omega,\sigma)]^M} \leq C\|f\|_{[\mathrm{BMO}(\partial\Omega,\sigma)]^M}. \tag{8.1.332}$$

With the boundary-to-domain single layer potential operator associated with L and Ω as in [132, (2.3.20)], consider

$$u := \mathscr{S} g \text{ in } \Omega. \tag{8.1.333}$$

Then from (8.1.333), [132, (2.3.19), (2.3.21), (2.3.22), (2.3.24)], (8.1.332), [132, Lemma 1.3.1], [130, Proposition 11.10.7], and [130, Lemma 4.6.5] we conclude that u satisfies all properties demanded in (8.1.327).

Furthermore, that the function u defined in (8.1.333) satisfies (8.1.328) is clear from [132, (1.5.1), (2.3.22), (2.3.24)]. Also, the fact that u enjoys (8.1.329) follows from (8.1.333), (8.1.332), [132, (2.3.25)], and the fact that $(S_{L^{\top}})^{\top}$ maps $\big[\mathrm{VMO}_{-1}(\partial\Omega, \sigma)\big]^M$ isomorphically onto $\big[\mathrm{VMO}(\partial\Omega, \sigma)\big]^M$ (cf. (7.1.606)).

If $n = 2$, we shall employ the invertibility of (7.1.609) (written for the system $L^{\top}$ in place of L) to find $g \in \big[\mathrm{BMO}_{-1}(\partial\Omega, \sigma)\big]^M$ and $c \in \mathbb{C}^M$ so that

$$\begin{gathered}(S_{L^{\top}})^{\top} g + c = f, \quad |c| + \|g\|_{[\mathrm{BMO}_{-1}(\partial\Omega,\sigma)]^M} \leq C\|f\|_{[\mathrm{BMO}(\partial\Omega,\sigma)]^M} \\ \text{and } \langle g, \mathbf{e}_j\rangle = 0 \text{ for each } j \in \{1, \dots, M\}.\end{gathered} \tag{8.1.334}$$

Bring in the function $u := \mathscr{S} g + c$ in Ω and note that, if Ω is an exterior domain then, much as in the proof of [132, (1.3.11)], the vanishing moment property in (8.1.334) ensures that $u(x) = O(1)$ as $|x| \to \infty$. Thanks to this and [132, Theorem 2.3.7] (cf. also [132, (2.3.159)-(2.3.162)]), we then see that u solves (8.1.327) and satisfies (8.1.328)-(8.1.329). Finally, uniqueness for the boundary value problem (8.1.327) is implied by the uniqueness part in Theorem 8.1.6. □

We next discuss Neumann Problems in various geometric and functional analytic settings. In stark contrast to the Dirichlet Problems treated earlier, the very format of a Neumann Problem is affected, through the conormal derivative employed in the formulation of the boundary condition, by the choice of the coefficient tensor used to write the given weakly elliptic second-order system of differential operators. In fact, as noted in the comments leading up to the example in [132, (1.4.208)], even in the case of the (scalar) Laplacian there exist infinitely many conormal derivatives, hence infinitely many Neumann Problems. Remarkably, the Neumann Problems associated with distinguished coefficient tensors turned out to be always solvable, modulo finite dimensional spaces, if the underlying domain is flat enough at infinitesimal scales.

First we treat the Neumann Problem for weakly elliptic systems in δ-oscillating AR domains in $\mathbb{R}^n$, with boundary data in Hardy spaces H^p with $\frac{n-1}{n} < p \leq 1$. Our main result in this regard, Theorem 8.1.12 stated shortly, may be viewed as the sharp version of [54, Theorem 1.4, p. 8], where E. Fabes and C. Kenig have solved the Neumann Problem for the Laplacian with boundary data in the Hardy space H^1 in a bounded $\mathscr{C}^1$ domain. As regards the Neumann Problem for the Laplacian in bounded Lipschitz domains with boundary data in the Hardy space H^p have dealt in [38, Theorem 4.3, p. 458] with the case $p = 1$, and R. Brown has treated in [20, Theorem C, p. 393] the case when $1 - \varepsilon < p < 1$, where the (typically small) threshold $\varepsilon > 0$ depends only on the Lipschitz character of the underlying domain. The latter range is also covered by the functional analytic approach developed by N. Kalton and M. Mitrea in [88]. In addition, the methods in [88] can also treat the Neumann Problem with boundary data in Hardy spaces for the three-dimensional Lamé system in bounded Lipschitz domains. For two dimensional bounded Lipschitz domains, D. Mitrea has solved in [125, Proposition 3.5, p. 90] the Neumann Problem for the Laplacian with boundary data in the Hardy space H^p when $\frac{2}{3}-\varepsilon < p \leq 1$, with $\varepsilon > 0$ of the same nature as above. In [156, Proposition 5.3], [156, Theorem 7.9.], [154, Theorem 6.2], M. Mitrea and M. Taylor have solved the Neumann Problem for the Laplace-Beltrami operator on Lipschitz subdomains of Riemannian manifolds with boundary data in the Hardy space H^p with $1-\varepsilon < p \leq 1$, for some small $\varepsilon > 0$. The work in [125] was subsequently extended to systems on Lipschitz subdomains of two-dimensional Riemannian manifolds in [138, Proposition 3.2]. See also the work of A. Barton and S. Mayboroda in [13, Chapter 3] for related problems.

Theorem 8.1.12 (The Neumann Problem with Data in Hardy Spaces) *Consider two integers $n, M \in \mathbb{N}$ with $n \geq 2$ and suppose $\Omega \subseteq \mathbb{R}^n$ is an Ahlfors regular domain with compact boundary. Abbreviate $\sigma := \mathcal{H}^{n-1}\lfloor\partial\Omega$ and denote by ν the geometric measure theoretic outward unit normal to Ω. Also, let L be a homogeneous, second-order, constant complex coefficient, $M\times M$ system in $\mathbb{R}^n$, and pick a coefficient tensor A for which $L = L_A$ (i.e., $A \in \mathfrak{A}_L$). Finally, fix an arbitrary aperture parameter $\kappa > 0$, and select an exponent*

$$p \in \left(\tfrac{n-1}{n}, 1\right]. \tag{8.1.335}$$

In this setting, consider the Neumann Problem[43]

$$\begin{cases} u \in \left[\mathscr{C}^\infty(\Omega)\right]^M, \quad Lu = 0 \text{ in } \Omega, \\ \mathcal{N}_\kappa(\nabla u) \in L^p(\partial\Omega, \sigma), \quad \partial^A_\nu u = f \in \left[H^p(\partial\Omega, \sigma)\right]^M, \\ u(x) = O(|x|^{2-n}) \text{ as } |x| \to \infty \text{ if } \Omega \text{ is unbounded,} \end{cases} \tag{8.1.336}$$

where the weak conormal derivative is interpreted as in [130, Definition 10.2.18].

Then there exists a threshold $\delta \in (0, 1)$, which depends only on n, p, A, the ADR *constants of $\partial\Omega$, and the relative flatness ratio $\wp(\Omega)/\operatorname{diam}\partial\Omega$, so that if*

$$\operatorname{dist}\Big(\nu, [\mathrm{VMO}(\partial\Omega, \sigma)]^n\Big) < \delta \tag{8.1.337}$$

[43] thanks to work in [129, §8.5], one may replace the nontangential maximal operator by the tangential maximal operator, associated as in [129, Definition 8.5.1] with a sufficiently large power (cf. (A.0.165))

where the distance is considered in the space $\big[\mathrm{BMO}(\partial\Omega,\sigma)\big]^n$ *(hence whenever* Ω *is a* δ*-oscillating* AR *domain*[44]*; cf. Definition 3.4.1), the set* Ω *becomes a* UR *domain and the following statements are true:*

(1) If one assumes that

$$L \text{ is weakly elliptic with } \mathfrak{A}^{\rm dis}_L \neq \varnothing \text{ and } \mathfrak{A}^{\rm dis}_{L^\top} \neq \varnothing, \tag{8.1.338}$$

then whenever $A \in \mathfrak{A}^{\rm dis}_L$ *it follows that the boundary value problem* (8.1.336) *is Fredholm solvable, in the following precise sense: First, the space of admissible boundary data for the boundary problem* (8.1.336)*, i.e.,*

$$\begin{aligned}\Big\{\partial^A_\nu u : u \in \big[\mathscr{C}^\infty(\Omega)\big]^M,\ Lu = 0 \text{ in } \Omega,\ \mathcal{N}_\kappa(\nabla u) \in L^p(\partial\Omega,\sigma),\\ \text{and } u(x) = O(|x|^{2-n}) \text{ as } |x| \to \infty \text{ if } \Omega \text{ is unbounded}\Big\}\end{aligned} \tag{8.1.339}$$

is a closed linear subspace of finite codimension in $\big[H^p(\partial\Omega,\sigma)\big]^M$*. Second, a solution* u *of* (8.1.336) *exists if and only if the boundary datum* f *belongs to* (8.1.339)*, in which scenario a solution* u *may be found such that*

$$\big\|\mathcal{N}_\kappa(\nabla u)\big\|_{L^p(\partial\Omega,\sigma)} \leq C\|f\|_{[H^p(\partial\Omega,\sigma)]^M} \tag{8.1.340}$$

for some constant $C \in (0,\infty)$ *independent of* f*. Third, the space of null-solutions of the boundary problem* (8.1.336)*, i.e.,*

$$\begin{aligned}\Big\{u \in \big[\mathscr{C}^\infty(\Omega)\big]^M : Lu = 0,\ \mathcal{N}_\kappa(\nabla u) \in L^p(\partial\Omega,\sigma),\ \partial^A_\nu u = 0,\\ \text{and } u(x) = O(|x|^{2-n}) \text{ as } |x| \to \infty \text{ if } \Omega \text{ is unbounded}\Big\}\end{aligned} \tag{8.1.341}$$

is finite dimensional.

(2) If (8.1.338) *is strengthened to*[45]

$$\begin{aligned}&L \text{ satisfies the Legendre-Hadamard strong ellipticity condition (cf. [131, Definition 1.3.2]) and one has } \mathfrak{A}^{\rm dis}_L \neq \varnothing\end{aligned} \tag{8.1.342}$$

then for each $A \in \mathfrak{A}^{\rm dis}_L$ *the index of the boundary value problem* (8.1.336)*, i.e., the dimension of* (8.1.341) *minus the codimension of* (8.1.339) *in* $\big[H^p(\partial\Omega,\sigma)\big]^M$*, is equal to zero. Moreover,*

---

[44] in particular, if Ω is an infinitesimally flat AR domain

[45] according to Proposition 1.4.1, whenever L is as in (8.1.342) we have $\mathfrak{A}^{\rm dis}_L \neq \varnothing$ and $\mathfrak{A}^{\rm dis}_{L^\top} \neq \varnothing$

the space of admissible boundary data for the Neumann Problem (8.1.336), *originally defined as in* (8.1.339), *is equal to* $\mathrm{Im}\Big(-\tfrac{1}{2}I + K^{\#}_{A^\top};\ \big[H^p(\partial\Omega,\sigma)\big]^M\Big)$ *if either* $n \geq 3$, *or* $n = 2$ *and the set* Ω *is bounded, and may be described as* $\Big\{\big(-\tfrac{1}{2}I + K^{\#}_{A^\top}\big)g :\ g \in \big[H^p(\partial\Omega,\sigma)\big]^M$ *with* $\langle g, \mathbf{e}_j\rangle = 0$ *for* $1 \leq j \leq M\Big\}$ *if* $n = 2$ *and* Ω *is an exterior domain.* (8.1.343)

Put differently, in the case when either $n \geq 3$, *or* $n = 2$ *and the set* Ω *is bounded, the Neumann Problem* (8.1.336) *has a solution if and only if the boundary datum* $f \in \big[H^p(\partial\Omega,\sigma)\big]^M$ *satisfies (with the brackets indicating duality in the sense of* [130, Theorem 4.6.1])

$$\begin{gathered} {}_{[H^p(\partial\Omega,\sigma)]^M}\big\langle f, g\big\rangle_{[\mathscr{C}^\eta(\partial\Omega)]^M} = 0 \text{ for each function} \\ g \in \big[\mathscr{C}^\eta(\partial\Omega)\big]^M \text{ satisfying } \big(-\tfrac{1}{2}I + K_{A^\top}\big)g = 0 \\ \text{if } p \in \big(\tfrac{n-1}{n}, 1\big) \text{ and } \eta := (n-1)\big(\tfrac{1}{p} - 1\big) \in (0,1), \end{gathered} \tag{8.1.344}$$

and, corresponding to the case when $p = 1$,

$$\begin{gathered} {}_{[H^1(\partial\Omega,\sigma)]^M}\big\langle f, g\big\rangle_{[\mathrm{BMO}(\partial\Omega,\sigma)]^M} = 0 \text{ for each function} \\ g \in \big[\mathrm{BMO}(\partial\Omega,\sigma)\big]^M \text{ satisfying } \big(-\tfrac{1}{2}I + K_{A^\top}\big)g = 0. \end{gathered} \tag{8.1.345}$$

Also, if $n = 2$ *and* Ω *is an exterior domain, the Neumann Problem* (8.1.336) *has a solution if and only if the boundary datum* $f \in \big[H^p(\partial\Omega,\sigma)\big]^M$ *satisfies (again, with the brackets indicating duality in the sense of* [130, Theorem 4.6.1])

$$\begin{gathered} {}_{[H^p(\partial\Omega,\sigma)]^M}\big\langle f, g\big\rangle_{[\mathscr{C}^\eta(\partial\Omega)]^M} = 0 \text{ for each function} \\ g \in \big[\mathscr{C}^\eta(\partial\Omega)\big]^M \text{ satisfying } \big(-\tfrac{1}{2}I + K_{A^\top}\big)g \in \mathbb{C}^M \\ \text{provided } p \in \big(\tfrac{1}{2}, 1\big) \text{ and } \eta := \tfrac{1}{p} - 1 \in (0,1), \end{gathered} \tag{8.1.346}$$

and, corresponding to the case when $p = 1$,

$$\begin{gathered} {}_{[H^1(\partial\Omega,\sigma)]^M}\big\langle f, g\big\rangle_{[\mathrm{BMO}(\partial\Omega,\sigma)]^M} = 0 \text{ for each function} \\ g \in \big[\mathrm{BMO}(\partial\Omega,\sigma)\big]^M \text{ satisfying } \big(-\tfrac{1}{2}I + K_{A^\top}\big)g \in \mathbb{C}^M. \end{gathered} \tag{8.1.347}$$

Furthermore, there exists a constant $C \in (0,\infty)$ *with the property that whenever the boundary datum* f *belongs to the space of admissible boundary data, defined in* (8.1.339), *then a solution* u *of the boundary problem* (8.1.336) *may be found such that* (8.1.340) *holds.*

Moreover,

$$\begin{gathered} \text{if either } n \geq 3 \text{ and } p^* := \big(\tfrac{1}{p} - \tfrac{1}{n-1}\big)^{-1} \in (1,\infty), \\ \text{or if } n = 2,\ \Omega \text{ is bounded, } p \in \big(\tfrac{1}{2}, 1\big), \text{ and } p^* := \big(\tfrac{1}{p} - 1\big)^{-1} \in (1,\infty), \\ \text{or if } n = 2, \text{ the set } \Omega \text{ is bounded, } p = 1, \text{ and } p^* := q \in (1,\infty) \text{ is arbitrary,} \end{gathered} \tag{8.1.348}$$

with $\mathcal{D}_A$ *denoting the boundary-to-domain double layer potential operator associated with A and* Ω *as in* (A.0.72)*, the following properties are true:*

$$\begin{array}{l}\textit{the space of null-solutions of the Neumann Problem (8.1.336)}\\ \textit{(as originally described in (8.1.341)) is actually equal to the}\\ \textit{space } \Big\{\mathcal{D}_A\phi : \phi \in \mathrm{Ker}\Big(-\tfrac{1}{2}I + K_A;\ \big[H_1^{p^*,p}(\partial\Omega,\sigma)\big]^M\Big)\Big\},\end{array} \tag{8.1.349}$$

the mapping associating to each function u belonging to the space of null-solutions of the Neumann Problem (8.1.336) (*as originally described in* (8.1.341)) *its nontangential boundary trace* $u\big|_{\partial\Omega}^{\kappa-\mathrm{n.t.}}$ *is in fact a linear bounded isomorphism onto the finite dimensional space* $\mathrm{Ker}\Big(-\tfrac{1}{2}I + K_A;\ \big[H_1^{p^*,p}(\partial\Omega,\sigma)\big]^M\Big)$, (8.1.350)

the inverse of the isomorphism $u \mapsto u\big|_{\partial\Omega}^{\kappa-\mathrm{n.t.}}$ *described in* (8.1.350) *is the boundary-to-domain double layer potential operator* $\mathcal{D}_A$ *mapping* $\mathrm{Ker}\Big(-\tfrac{1}{2}I + K_A;\ \big[H_1^{p^*,p}(\partial\Omega,\sigma)\big]^M\Big)$ *isomorphically onto the space of null-solutions of the Neumann Problem* (8.1.336) (*described in* (8.1.341)). (8.1.351)

Finally, in the case when $n = 2$*, the set* Ω *is an exterior domain and*

$$\begin{array}{c}\textit{either } p \in \big(\tfrac{1}{2}, 1\big), \textit{ and } p^* := \big(\tfrac{1}{p} - 1\big)^{-1} \in (1, \infty),\\ \textit{or } p = 1 \textit{ and } p^* := q \in (1, \infty) \textit{ is arbitrary,}\end{array} \tag{8.1.352}$$

it follows that:

the space of null-solutions of the Neumann Problem (8.1.336) *equals* $\Big\{\mathcal{D}_A\phi + c : \phi \in \big[H_1^{p^*,p}(\partial\Omega,\sigma)\big]^M,\ c \in \mathbb{C}^M,\ \big(-\tfrac{1}{2}I + K_A\big)\phi = -c\Big\}$, (8.1.353)

the mapping associating to each function u belonging to the space of null-solutions of the Neumann Problem (8.1.336) (*from* (8.1.341)) *the pair* $\Big(u\big|_{\partial\Omega}^{\kappa-\mathrm{n.t.}},\ \lim_{|x|\to\infty} u(x)\Big)$ *is a well-defined isomorphism onto the space* $\Big\{(\phi, c) \in \big[H_1^{p^*,p}(\partial\Omega,\sigma)\big]^M \oplus \mathbb{C}^M : \big(-\tfrac{1}{2}I + K_A\big)\phi = -c\Big\}$, (8.1.354)

the inverse of the linear isomorphism $u \mapsto \Big(u\big|_{\partial\Omega}^{\kappa-\mathrm{n.t.}},\ \lim_{|x|\to\infty} u(x)\Big)$ *in the above context is the assignment* $(\phi, c) \longmapsto \mathcal{D}_A\phi + c$ *mapping the last space in* (8.1.354) *isomorphically onto the space of null-solutions of the Neumann Problem* (8.1.336). (8.1.355)

(3) If (8.1.342) *is further strengthened to*[46]

$$A \in \mathfrak{A}_L^{\mathrm{dis}} \text{ is positive definite,} \tag{8.1.356}$$

then the Neumann Problem (8.1.336) *has the following additional properties (with the spaces* $\mathbb{C}_{\Omega_\pm}$, $\mathbb{C}_{\partial\Omega_\pm}$, $\mathbb{C}^\infty_{\Omega_\pm}$, $\mathbb{C}^\infty_{\partial\Omega_\pm}$ *defined as in* (7.1.265), (7.1.268), *and with the conventions* (7.1.266), (7.1.269) *in effect). First, if either* $n \geq 3$, *or* $n = 2$ *and the set* Ω *is bounded, then*

> *the space of admissible boundary data for the Neumann Problem* (8.1.336) *(defined in* (8.1.339)) *is equal to* $\big[H^p(\partial\Omega,\sigma) \cap (\mathbb{C}_{\partial\Omega_+})^{\perp}\big]^M$, *and the space of null-solutions of the Neumann Problem* (8.1.336) *(described in* (8.1.341)) *is equal to* $(\mathbb{C}_{\Omega_+})^M$. (8.1.357)

Second, if $n = 2$ *and* Ω *is an exterior domain, then*

> *the space of admissible boundary data for the Neumann Problem* (8.1.336) *(defined in* (8.1.339)) *is equal to* $\big[H^p(\partial\Omega,\sigma) \cap (\mathbb{C}^\infty_{\partial\Omega_+})^{\perp}\big]^M$, *and the space of null-solutions of the Neumann Problem* (8.1.336) *(described in* (8.1.341)) *is equal to* $(\mathbb{C}^\infty_{\Omega_+})^M$. (8.1.358)

(4) In the case $p = 1$ *similar results to those in items (1)-(3) above are valid for the Neumann Problem with a strong conormal derivative (defined in a pointwise sense, as in* (A.0.203)), *when one seeks a function* $u = (u_\beta)_{1\leq\beta\leq M}$ *such that*[47]

$$\begin{cases} u \in \big[\mathscr{C}^\infty(\Omega)\big]^M,\ Lu = 0 \text{ in } \Omega,\ \mathcal{N}_\kappa(\nabla u) \in L^1(\partial\Omega,\sigma), \\ \partial^A_\nu u := \Big(\nu_r\big(a^{\alpha\beta}_{rs}\partial_s u_\beta\big)\Big|^{\kappa-\mathrm{n.t.}}_{\partial\Omega}\Big)_{1\leq\alpha\leq M} = f \in \big[H^1(\partial\Omega,\sigma)\big]^M \\ \text{at } \sigma\text{-a.e. point on } \partial\Omega, \\ u(x) = O(|x|^{2-n}) \text{ as } |x| \to \infty \text{ if } \Omega \text{ is unbounded,} \end{cases} \tag{8.1.359}$$

where $\big(a^{\alpha\beta}_{rs}\big)_{\substack{1\leq r,s\leq n \\ 1\leq\alpha,\beta\leq M}}$ *are the entries of the coefficient tensor A, and* $(\nu_1,\dots,\nu_n)$ *are the scalar components of the vector* ν.

Prior to giving the proof of the above theorem, we make a number of comments designed to shed further light on the scope and nature of this result. First, we wish to emphasize that, as seen from Proposition 1.6.12 and [131, Proposition 1.4.39],

> in the scalar case, i.e., if $M = 1$, in dimensions $n \geq 3$, both (8.1.338) and (8.1.342) may be replaced by the mere demand of weak ellipticity. (8.1.360)

---

[46] according to (1.1.28), whenever L has a positive definite coefficient tensor it follows that L satisfies the Legendre-Hadamard strong ellipticity condition

[47] once again, in view of work in [129, §8.5], one may replace the nontangential maximal operator by the tangential maximal operator, associated as in [129, Definition 8.5.1] with a sufficiently large power (cf. (A.0.165))

Moreover, in the scalar case in the two-dimensional setting (i.e., for $M = 1$ and $n = 2$), if L satisfies the Legendre-Hadamard strong ellipticity condition or, more generally (cf. [131, Lemma 1.4.24]), if $L = \mathrm{div} A\nabla$ for some 2×2 complex matrix A with $\mathrm{sym}\, A := \frac{1}{2}(A + A^{\top}) \in M_2^0$ (cf. [131, (1.4.187)]), then Proposition 1.6.11 and Proposition 1.6.13 ensure that $\mathfrak{A}_L^{\rm dis} \neq \varnothing$ and $\mathfrak{A}_{L^{\top}}^{\rm dis} \neq \varnothing$. Also, according to (1.6.141) in Proposition 1.6.13, in the scalar case in the two-dimensional setting (i.e., for $M = 1$ and $n = 2$) we may omit (8.1.342) altogether if L is a weakly elliptic operator with real coefficients.

Second, [130, Theorem 10.2.24] guarantees that, whenever

$$u \in \big[\mathscr{C}^\infty(\Omega)\big]^M,\ Lu = 0 \text{ in } \Omega,\ \mathcal{N}_\kappa(\nabla u) \in L^p(\partial\Omega, \sigma),\ p \in \big(\tfrac{n-1}{n}, 1\big], \tag{8.1.361}$$

it follows that the weak conormal derivative of u associated with any coefficient tensor $A \in \mathfrak{A}_L$ in the sense of [130, Definition 10.2.18] satisfies

$$\dot{\partial}_\nu^A u \in \big[H^p(\partial\Omega, \sigma)\big]^M. \tag{8.1.362}$$

This shows that the existence of the weak conormal derivative as a meaningfully defined distribution in $\big[H^p(\partial\Omega, \sigma)\big]^M$ is automatically guaranteed by the conditions imposed in the first two lines of (8.1.336). The membership in (8.1.362) also points to the fact that prescribing boundary data in the Hardy space $\big[H^p(\partial\Omega, \sigma)\big]^M$ is natural in the context of (8.1.336). Finally, (8.1.362) makes (8.1.339) a well-defined linear subspace of $\big[H^p(\partial\Omega, \sigma)\big]^M$.

The boundary condition in (8.1.359) is also naturally formulated. Indeed, choosing $\delta \in (0, 1)$ small permits us to invoke Theorem 3.4.2 which ensures that Ω is a UR domain. Granted this, the Fatou-type result proved in [131, Theorem 3.3.4] then guarantees that the nontangential boundary trace $(\nabla u)\big|_{\partial\Omega}^{\kappa-\text{n.t.}}$ exists (in $\mathbb{C}^{n\cdot M}$) at σ-a.e. point on $\partial\Omega$. As such, for any function satisfying the properties listed in the first two lines of (8.1.359) has a well-defined strong conormal derivative, considered in a pointwise sense, as in (A.0.203). Moreover, [130, Corollary 10.1.9] implies that said strong conormal derivative lies in the Hardy space $\big[H^1(\partial\Omega, \sigma)\big]^M$ (the space from which the boundary data for the problem (8.1.359) are selected).

We also wish to make the following remark, regarding the solvability of a related brand of Neumann Problem.

Remark 8.1.13 *In the context of Theorem 8.1.12, other (more general) types of behavior at infinity may be imposed in the formulation of the Neumann Problem in exterior domains. For example, assuming that $n = 2$ and $\Omega \subseteq \mathbb{R}^2$ is an exterior domain, for each $p \in \big(\frac{1}{2}, 1\big]$ consider the following the Neumann Problem:*

$$\begin{cases} u \in \big[\mathscr{C}^\infty(\Omega)\big]^M, \quad Lu = 0 \ \text{in}\ \Omega, \\ \mathcal{N}_\kappa(\nabla u) \in L^p(\partial\Omega, \sigma), \quad \dot{\partial}_\nu^A u = f \in \big[H^p(\partial\Omega, \sigma)\big]^M, \\ u(x) = E(x)c + O(1) \ \text{as}\ |x| \to \infty, \end{cases} \tag{8.1.363}$$

where $c \in \mathbb{C}^M$ is a given constant, and E is the matrix-valued fundamental solution associated with L as in [131, Theorem 1.4.2]. *Hence, the "old" version* (8.1.336) *corresponds to* (8.1.363) *with $c := 0 \in \mathbb{C}^M$ (and $n = 2$). In relation to this brand of Neumann Problem we then have:*

$$\begin{aligned}&\textit{if } A \in \mathfrak{A}_L^{\mathrm{dis}} \textit{ is positive definite then } (8.1.363) \textit{ is solvable if and only if}\\ &f \in \big[H^p(\partial\Omega,\sigma) \cap (\mathbb{C}_{\partial\Omega_+})^{\perp}\big]^M \textit{ and } c = -\Big(\big\langle f, \mathbf{e}_j \mathbf{1}_{\partial\Omega_\infty}\big\rangle\Big)_{1\leq j\leq M}.\end{aligned} \tag{8.1.364}$$

To justify (8.1.364), pick an arbitrary point $x_0 \in \Omega_- := \mathbb{R}^2 \setminus \overline{\Omega}$ and define

$$g := f - \partial_\nu^A\big[E(\cdot - x_0)c\big] \in \big[H^p(\partial\Omega,\sigma)\big]^M. \tag{8.1.365}$$

Given an arbitrary function $w \in \big(\mathbb{C}_{\Omega_+}^\infty\big)^M$, denote by $w\big|_{\partial\Omega}^{\text{n.t.}}$ its nontangential boundary trace (with respect to some background aperture parameter $\kappa > 0$), and denote by $w_\infty \in \mathbb{C}^M$ the value of w in Ω_∞. Then with $\mathcal{D}_{A^\top}$ denoting the boundary-to-domain double layer potential operator associated with Ω and $A^\top$ as in (A.0.72) we have (with the convention made in [132, (1.3.21)])

$$\int_{\partial\Omega}\Big\langle \partial_\nu^A\big[E(\cdot - x_0)c\big], w\big|_{\partial\Omega}^{\text{n.t.}}\Big\rangle\, d\sigma = \Big\langle \mathcal{D}_{A^\top}\big(w\big|_{\partial\Omega}^{\text{n.t.}}\big)(x_0), c\Big\rangle = -\langle w_\infty, c\rangle, \tag{8.1.366}$$

where the first equality is seen from (A.0.203) and (A.0.72), while the second equality is a consequence of [132, (1.3.46)]. Based on this, (7.1.268)-(7.1.269), and (8.1.365) we conclude that

$$\big\langle g, w\big|_{\partial\Omega}^{\text{n.t.}}\big\rangle = \big\langle f, w\big|_{\partial\Omega}^{\text{n.t.}}\big\rangle + \langle c, w_\infty\rangle \text{ for each } w \in \big(\mathbb{C}_{\Omega_+}^\infty\big)^M. \tag{8.1.367}$$

From this and (7.1.343) we then see (specializing (8.1.367) first to the case when $w \in \big(\mathbb{C}_{\Omega_+}\big)^M$, then to the case when $w \in \big[\langle \mathbf{1}_{\Omega_\infty}\rangle\big]^M$) that g belongs to the space $\big[H^p(\partial\Omega,\sigma) \cap (\mathbb{C}_{\partial\Omega_+}^\infty)^{\perp}\big]^M$ if and only if

$$f \in \big[H^p(\partial\Omega,\sigma) \cap (\mathbb{C}_{\partial\Omega_+})^{\perp}\big]^M \text{ and } c = -\Big(\big\langle f, \mathbf{e}_j \mathbf{1}_{\partial\Omega_\infty}\big\rangle\Big)_{1\leq j\leq M}. \tag{8.1.368}$$

Let us summarize our progress: Whenever (8.1.368) holds, it follows that g belongs to $\big[H^p(\partial\Omega,\sigma) \cap (\mathbb{C}_{\partial\Omega_+}^\infty)^{\perp}\big]^M$ which, according to (8.1.358), is the space of admissible boundary data for the Neumann Problem (8.1.336). Hence, we may find u_0 with

$$\begin{cases} u_0 \in \big[\mathscr{C}^\infty(\Omega)\big]^M, \quad Lu_0 = 0 \text{ in } \Omega,\\ \mathcal{N}_\kappa(\nabla u_0) \in L^p(\partial\Omega,\sigma), \quad \mathring{\partial}_\nu^A u_0 = g \text{ in } \big[H^p(\partial\Omega,\sigma)\big]^M,\\ u_0(x) = O(1) \text{ as } |x| \to \infty. \end{cases} \tag{8.1.369}$$

Then $u := u_0 + E(\cdot - x_0)c$ in Ω is a solution (8.1.363). This shows that the Neumann Problem (8.1.363) is solvable whenever the properties listed in (8.1.368) are satisfied.

Conversely, if the Neumann Problem (8.1.363) has a solution u, then the function $u_0 := u - E(\cdot - x_0)c$ solves (8.1.369) for the boundary datum g defined in (8.1.365). Then (8.1.358) tells us that g necessarily belongs to $\big[H^p(\partial\Omega,\sigma) \cap (\mathbb{C}^\infty_{\partial\Omega_+})^\perp\big]^M$ which, as proved earlier, is equivalent to having (8.1.368). This finishes the proof of the claim made in (8.1.364).

Before presenting the proof Theorem 8.1.12, it is instructive to consider the special case considered in following example.

Example 8.1.14 *Retain the context of item (2) in Theorem 8.1.12, with $n \geq 3$ and $p \in \big(\frac{n-1}{n}, 1\big)$. For each fixed $x_0, x_1 \in \partial\Omega$, we know from* [130, (4.2.17)] *that the distribution $\delta_{x_0} - \delta_{x_1} \in \big(\mathrm{Lip}(\partial\Omega)\big)'$ belongs to the Hardy space $H^p(\partial\Omega,\sigma)$. In particular, having picked some arbitrary index $\beta \in \{1,\dots,M\}$, it follows that $(\delta_{x_0} - \delta_{x_1})\mathbf{e}_\beta \in \big[H^p(\partial\Omega,\sigma)\big]^M$. As such, it makes sense to consider the issue as to whether the Neumann Problem*

$$\begin{cases} u \in \big[\mathscr{C}^\infty(\Omega)\big]^M, \ Lu = 0 \text{ in } \Omega, \ \mathcal{N}_\kappa(\nabla u) \in L^p(\partial\Omega,\sigma), \\ \partial^A_\nu u = (\delta_{x_0} - \delta_{x_1})\mathbf{e}_\beta \text{ in } \big[H^p(\partial\Omega,\sigma)\big]^M, \\ u(x) = O(|x|^{2-n}) \text{ as } |x| \to \infty \text{ if } \Omega \text{ is unbounded,} \end{cases} \tag{8.1.370}$$

is solvable. With $\eta := (n-1)\big(\frac{1}{p} - 1\big) \in (0,1)$, from (8.1.344) *we see that the Neumann Problem* (8.1.370) *has a solution (assuming* (8.1.337) *holds for some sufficiently small δ) if and only if*

$$\begin{aligned} &g_\beta(x_0) = g_\beta(x_1) \text{ for each function} \\ &g \in \mathrm{Ker}\Big(-\tfrac{1}{2}I + K_{A^\top} : \big[\mathscr{C}^\eta(\partial\Omega)\big]^M \Big). \end{aligned} \tag{8.1.371}$$

In particular, the boundary value problem (8.1.370) *is solvable for each choice of a pair of points $x_0, x_1 \in \partial\Omega$ and each index $\beta \in \{1,\dots,M\}$ if and only if the kernel of $-\frac{1}{2}I + K_{A^\top}$ on $\big[\mathscr{C}^\eta(\partial\Omega)\big]^M$ consists of only constant functions, that is,* $\mathrm{Ker}\Big(-\frac{1}{2}I + K_{A^\top} : \big[\mathscr{C}^\eta(\partial\Omega)\big]^M \Big) = \mathbb{C}^M$.

Going further, let us strengthen the original hypotheses by now assuming (8.1.356). *From* (8.1.371) *and the version of* (7.1.275) *for Hölder functions (cf. Remark 7.1.12) we then see that, in this case, the Neumann Problem* (8.1.370) *is solvable (assuming* (8.1.337) *holds for some sufficiently small δ) for each choice of $x_0, x_1 \in \partial\Omega$ and each index $\beta \in \{1,\dots,M\}$ if and only if Ω has precisely one bounded connected component.*

We next present the proof of Theorem 8.1.12.

Proof of Theorem 8.1.12 Choose $\delta \in (0,1)$ small enough, depending on n, p, A, the ADR constants of $\partial\Omega$, and the relative flatness ratio $\wp(\Omega)/\mathrm{diam}\,\partial\Omega$, such that the conclusions in Theorem 7.1.1, Proposition 7.1.2, Proposition 7.1.43, and Theorem 7.1.34 hold (for suitable choices of the parameters involved in their statements). By further decreasing the value of $\delta \in (0,1)$, if necessary, Theorem 3.4.2 applies and gives that Ω is actually a two-sided NTA domain, hence also a UR domain (cf. [129, (5.10.33)]).

Let us work under the assumptions made in (8.1.338). Then Theorem 1.5.5 guarantees that $A^{\top} \in \mathfrak{A}^{\rm dis}_{L^{\top}}$. Granted this, from item *(3)* in Theorem 7.1.1 used with $z := -1/2$ we conclude that

$$-\tfrac{1}{2}I + K^{\#}_{A^{\top}} : \big[H^p(\partial\Omega,\sigma)\big]^M \longrightarrow \big[H^p(\partial\Omega,\sigma)\big]^M \text{ is a Fredholm operator with index zero.} \tag{8.1.372}$$

In particular,

$$\operatorname{Im}\Big(-\tfrac{1}{2}I + K^{\#}_{A^{\top}};\ \big[H^p(\partial\Omega,\sigma)\big]^M\Big) \text{ is a closed space of finite codimension in the space } \big[H^p(\partial\Omega,\sigma)\big]^M. \tag{8.1.373}$$

Let $\mathscr{S}$ be the boundary-to-domain single layer potential operator acting on Hardy spaces as in [132, (2.2.31)]. Then for each $f \in \big[H^p(\partial\Omega,\sigma)\big]^M$, the function $u := \mathscr{S}f$ in Ω satisfies

$$u \in \big[\mathscr{C}^\infty(\Omega)\big]^M,\ Lu = 0 \text{ in } \Omega,\ \mathcal{N}_\kappa(\nabla u) \in L^p(\partial\Omega,\sigma) \text{ as well as } u(x) = O(|x|^{2-n}) \text{ as } |x| \to \infty \text{ if } n \geq 3 \text{ and } \Omega \text{ is unbounded,} \tag{8.1.374}$$

thanks to [132, (2.2.32), (2.2.33), (2.2.35), (2.2.44)]. Moreover, much as in the proof of [132, (1.3.11)],

$$u(x) = O(1) \text{ as } |x| \to \infty \text{ if } n = 2 \text{ and } \Omega \text{ is unbounded, provided } \langle f, \mathbf{e}_j\rangle = 0 \text{ for each } j \in \{1,\dots,M\}. \tag{8.1.375}$$

In view of (8.1.374) and [132, (2.2.47)] we then conclude that

$$\big(-\tfrac{1}{2}I + K^{\#}_{A^{\top}}\big)f = \partial^{A}_{\nu}u \text{ belongs to the space of admissible boundary data for the Neumann Problem (8.1.336), defined in (8.1.339), for each given } f \in \big[H^p(\partial\Omega,\sigma)\big]^M, \text{ with the additional property that } \langle f, \mathbf{e}_j\rangle = 0 \text{ for all indices } j \in \{1,\dots,M\} \text{ in the case when } n = 2 \text{ and } \Omega \text{ is unbounded.} \tag{8.1.376}$$

In turn, (8.1.376) implies that

$$\text{the space (8.1.339) of admissible boundary data for the Neumann Problem (8.1.336) contains } \operatorname{Im}\Big(-\tfrac{1}{2}I + K^{\#}_{A^{\top}};\ \big[H^p(\partial\Omega,\sigma)\big]^M\Big) \text{ if either } n \geq 3, \text{ or } n = 2 \text{ and the set } \Omega \text{ is bounded, whereas the space (8.1.339) of admissible boundary data for the Neumann Problem (8.1.336) contains } \Big\{\big(-\tfrac{1}{2}I + K^{\#}_{A^{\top}}\big)g :\ g \in \big[H^p(\partial\Omega,\sigma)\big]^M,\ \langle g, \mathbf{e}_j\rangle = 0 \text{ for } 1 \leq j \leq M\Big\} \text{ if } n = 2 \text{ and } \Omega \text{ is an exterior domain.} \tag{8.1.377}$$

Finally, from (8.1.377), (8.1.373), and item *(3)* in [130, Theorem 1.1.3] we see that

$$\text{the space of admissible boundary data for the boundary problem (8.1.336), described in (8.1.339), is a closed linear subspace of finite codimension in the space } \big[H^p(\partial\Omega,\sigma)\big]^M. \tag{8.1.378}$$

By design, a solution u of (8.1.336) exists if and only if the boundary datum f belongs to (8.1.339). Let us show that, whenever this is the case, a solution u of (8.1.336) may be found such that (8.1.340) holds. Consider first the case when $n \geq 3$. Define the operator

$$\begin{aligned} T : \big[H_1^{p^*,p}(\partial\Omega,\sigma)\big]^M \times \big[H^p(\partial\Omega,\sigma)\big]^M &\longrightarrow \big[H^p(\partial\Omega,\sigma)\big]^M, \\ T(f_0, f_1) := \partial_\nu^A \mathcal{D} f_0 - \big(-\tfrac{1}{2}I + K^\#_{A^\top}\big) f_1 \ \text{ for all} \\ (f_0, f_1) \in \big[H_1^{p^*,p}(\partial\Omega,\sigma)\big]^M \times \big[H^p(\partial\Omega,\sigma)\big]^M. \end{aligned} \tag{8.1.379}$$

In the above context, T is a well-defined, linear, and bounded operator. Since its image contains the space $\mathrm{Im}\Big(-\frac{1}{2}I + K^\#_{A^\top};\ \big[H^p(\partial\Omega,\sigma)\big]^M\Big)$, we conclude from (8.1.373) and item *(3)* in [130, Theorem 1.1.3] that

$$\text{the image of } T \text{ is a closed subspace of finite codimension in the space } \big[H^p(\partial\Omega,\sigma)\big]^M. \tag{8.1.380}$$

In relation to this, we claim that

$$\text{the image of } T \text{ is precisely the space of admissible boundary data for the Neumann Problem formulated in (8.1.336).} \tag{8.1.381}$$

In one direction, assume

$$\begin{aligned} &u \in \big[\mathscr{C}^\infty(\Omega)\big]^M, \ Lu = 0 \ \text{in } \Omega, \ \mathcal{N}_\kappa(\nabla u) \in L^p(\partial\Omega,\sigma), \\ &\text{and } u(x) = O(|x|^{2-n}) \ \text{as } |x| \to \infty \ \text{if } \Omega \text{ is unbounded.} \end{aligned} \tag{8.1.382}$$

Whenever Ω is unbounded, the last property above together with [131, (1.5.206)] imply $(\nabla u)(x) = O(|x|^{1-n})$ as $|x| \to \infty$. Granted this, [132, Theorem 2.2.7] applies and gives (bearing in mind that $\partial\Omega$ is compact) that the nontangential trace $u\big|_{\partial\Omega}^{\kappa-\text{n.t.}}$ exists and belongs to $\big[L^{p^*}(\partial\Omega,\sigma)\big]^M$, and (keeping in mind [132, (1.8.8)]) that there exists some $\mathbb{C}^M$-valued locally constant function c_u in Ω such that

$$u = \mathcal{D}\big(u\big|_{\partial\Omega}^{\kappa-\text{n.t.}}\big) - \mathcal{S}\big(\partial_\nu^A u\big) + c_u \ \text{ in } \Omega. \tag{8.1.383}$$

Moreover, from [130, Theorem 11.10.10] (which requires $\mathcal{N}_\kappa^\varepsilon u \in L^1_{\text{loc}}(\partial\Omega,\sigma)$ for some $\varepsilon > 0$, a property presently guaranteed by [132, (2.2.170)]) and [130, Definition 11.10.6] we see that, in fact,

$$u\big|_{\partial\Omega}^{\kappa-\text{n.t.}} \in \big[H_1^{p^*,p}(\partial\Omega,\sigma)\big]^M. \tag{8.1.384}$$

Taking the weak conormal derivative of both sides of (8.1.383) then gives, on account of [130, Theorem 10.2.24] and [132, (2.2.55), (2.3.6)],

$$\dot{\partial}^A_\nu u = \dot{\partial}^A_\nu \mathcal{D}\big(u\big|^{\kappa-\mathrm{n.t.}}_{\partial\Omega}\big) - \big(-\tfrac{1}{2}I + K^{\#}_{A^\top}\big)\big(\dot{\partial}^A_\nu u\big) \text{ in } \big[H^p(\partial\Omega,\sigma)\big]^M. \tag{8.1.385}$$

Hence,

$$\dot{\partial}^A_\nu u = T\big(u\big|^{\kappa-\mathrm{n.t.}}_{\partial\Omega}, \dot{\partial}^A_\nu u\big), \tag{8.1.386}$$

which ultimately shows that the space of admissible boundary data for the Neumann Problem (8.1.336) is contained in the image of the operator T in (8.1.379).

In the opposite direction, pick an arbitrary distribution in the image of the operator T, say, $T(f_0, f_1)$ for some $f_0 \in \big[H^{p^*,p}_1(\partial\Omega,\sigma)\big]^M$ and $f_1 \in \big[H^p(\partial\Omega,\sigma)\big]^M$. Then the function defined by $u := \mathcal{D} f_0 - \mathcal{S} f_1$ is as in (8.1.382) and satisfies $\dot{\partial}^A_\nu u = T(f_0, f_1)$. This shows that $T(f_0, f_1)$ is an admissible boundary datum for the Neumann Problem (8.1.336). The proof of (8.1.381) is therefore complete.

Having established (8.1.381), the fact that a solution u of (8.1.336) exists if and only if the boundary datum f belongs to (8.1.339), in which scenario a solution u may be found such that the estimate in (8.1.340) is true, becomes a consequence of (8.1.380) and [130, Proposition 1.2.13]. This finishes the analysis when $n \geq 3$.

To proceed, we find it convenient to abbreviate

$$\big[H^p_V(\partial\Omega,\sigma)\big]^M := \Big\{f_1 \in \big[H^p(\partial\Omega,\sigma)\big]^M : V\Big(\big(\langle f_1, \mathbf{e}_j\rangle\big)_{1\leq j\leq M}\Big) = 0\Big\}, \tag{8.1.387}$$

where

$$V := \int_{S^1} \big[L(\xi)\big]^{-1}\, d\mathcal{H}^1(\xi) \in \mathbb{C}^{M\times M}, \tag{8.1.388}$$

Let us now find the analogue of (8.1.381) in the case when

$$n = 2, \text{ the set } \Omega \text{ is an exterior domain, and } p \in \big(\tfrac{1}{2}, 1\big). \tag{8.1.389}$$

In such a case, we define the operator

$$\begin{gathered}\widetilde{T} : \big[H^{p^*,p}_1(\partial\Omega,\sigma)\big]^M \times \big[H^p_V(\partial\Omega,\sigma)\big]^M \to \big[H^p(\partial\Omega,\sigma)\big]^M \\ \text{given by } \widetilde{T}(f_0, f_1) := \dot{\partial}^A_\nu \mathcal{D} f_0 - \big(-\tfrac{1}{2}I + K^{\#}_{A^\top}\big) f_1 \\ \text{for each } (f_0, f_1) \in \big[H^{p^*,p}_1(\partial\Omega,\sigma)\big]^M \times \big[H^p_V(\partial\Omega,\sigma)\big]^M.\end{gathered} \tag{8.1.390}$$

We claim that

the image of $\widetilde{T}$ is precisely the space of admissible boundary data for the Neumann Problem formulated in (8.1.336). (8.1.391)

To prove one inclusion, assume

$$\begin{gathered} u \in \big[\mathscr{C}^\infty(\Omega)\big]^M,\ Lu = 0 \text{ in } \Omega,\ \mathcal{N}_\kappa(\nabla u) \in L^p(\partial\Omega,\sigma), \\ \text{and } u(x) = O(1) \text{ as } |x| \to \infty. \end{gathered} \tag{8.1.392}$$

Recall from [131, (1.5.205)] that the last property above entails $(\nabla u)(x) = O(|x|^{-1})$ as $|x| \to \infty$. As such, [132, Theorem 2.2.7] applies and gives (keeping in mind that $\partial\Omega$ is compact) that $u\big|_{\partial\Omega}^{\kappa-\mathrm{n.t.}}$ exists and belongs to $\big[L^{p^*}(\partial\Omega,\sigma)\big]^M$, and (also taking [132, (1.8.8)] into account) that there exists some $\mathbb{C}^M$-valued locally constant function c_u in Ω such that the integral representation formula (8.1.383) holds. Also, from [130, Theorem 11.10.10] and [130, Definition 11.10.6] we see that actually $u\big|_{\partial\Omega}^{\kappa-\mathrm{n.t.}} \in \big[H_1^{p^*,p}(\partial\Omega,\sigma)\big]^M$. After taking the conormal derivative of both sides of (8.1.383) we arrive, on account of [130, Theorem 10.2.24] and [132, (2.2.55), (2.3.6)], at

$$\overset{\bullet}{\partial}{}^A_\nu u = \overset{\bullet}{\partial}{}^A_\nu \mathcal{D}\big(u\big|_{\partial\Omega}^{\kappa-\mathrm{n.t.}}\big) - \big(-\tfrac{1}{2}I + K^{\#}_{A^\top}\big)\big(\overset{\bullet}{\partial}{}^A_\nu u\big) \ \text{ in } \big[H^p(\partial\Omega,\sigma)\big]^M. \tag{8.1.393}$$

We next claim that

$$\Big(\int_{S^1} \big[L(\xi)\big]^{-1}\, d\mathcal{H}^1(\xi)\Big)\Big(\big(\langle\overset{\bullet}{\partial}{}^A_\nu u, \mathbf{e}_j\rangle\big)_{1\le j\le M}\Big) = 0. \tag{8.1.394}$$

Indeed, from (8.1.383), the last property in (8.1.392), the fact that $c_u \equiv c \in \mathbb{C}^M$ in the unbounded connected component of the set Ω, and the observation that one has $\mathcal{D}\big(u\big|_{\partial\Omega}^{\kappa-\mathrm{n.t.}}\big)(x) = O(|x|^{-1})$ as $|x| \to \infty$ (as seen from [132, (1.3.26)] and (8.1.29)) we conclude that $\mathscr{S}\big(\overset{\bullet}{\partial}{}^A_\nu u\big) = O(1)$ as $|x| \to \infty$. With this in hand and reasoning much as in the proof of [132, (1.3.12)] we see that we necessarily have (8.1.394). Next, (8.1.393), (8.1.394), and the definition of the operator $\widetilde{T}$ in (8.1.390) give that

$$\overset{\bullet}{\partial}{}^A_\nu u = \widetilde{T}\big(u\big|_{\partial\Omega}^{\kappa-\mathrm{n.t.}}, \overset{\bullet}{\partial}{}^A_\nu u\big) \in \operatorname{Im}\widetilde{T}, \tag{8.1.395}$$

the image of the operator $\widetilde{T}$. This proves one inclusion in (8.1.391). To establish the opposite inclusion in (8.1.391), start with a distribution in the image of $\widetilde{T}$, say, $\widetilde{T}(f_0, f_1)$ with (f_0, f_1) belonging to the space $\big[H_1^{p^*,p}(\partial\Omega,\sigma)\big]^M \times \big[H^p_\nu(\partial\Omega,\sigma)\big]^M$. Then $u := \mathcal{D}f_0 - \mathscr{S}f_1$ in Ω is as in (8.1.392) and satisfies $\overset{\bullet}{\partial}{}^A_\nu u = \widetilde{T}(f_0, f_1)$. Hence $\widetilde{T}(f_0, f_1)$ is an admissible boundary datum for the Neumann Problem (8.1.336).

This finishes the proof of (8.1.391). Once this has been established, the same functional analytic argument (based on [130, Proposition 1.2.13]) then shows that a solution u of (8.1.336) exists if and only if the boundary datum f belongs to (8.1.339), in which scenario a solution u may be found such that the estimate in (8.1.340) is true in the case (8.1.389).

Next, in the case when

$$n = 2, \text{ the set } \Omega \text{ is an exterior domain, and } p = 1, \tag{8.1.396}$$

the same argument as above works provided we fix an arbitrary integrability exponent $q \in (1, \infty)$ and now define the operator $\widetilde{T}$ as the mapping

$$\begin{aligned}\widetilde{T}:\big[H^{q,1}_{1}(\partial\Omega,\sigma)\big]^M\times\big[H^{1}_{V}(\partial\Omega,\sigma)\big]^M&\to\big[H^{1}(\partial\Omega,\sigma)\big]^M\\ \text{given by } \widetilde{T}(f_0,f_1):=\mathring{\partial}^{A}_{\nu}\mathcal{D}f_0-\big(-\tfrac{1}{2}I+K^{\#}_{A^{\top}}\big)f_1&\\ \text{for each } (f_0,f_1)\in\big[H^{q,1}_{1}(\partial\Omega,\sigma)\big]^M\times\big[H^{1}_{V}(\partial\Omega,\sigma)\big]^M.&\end{aligned}\tag{8.1.397}$$

The only difference is that, in the case described in (8.1.396), [132, Theorem 2.2.7] gives that the nontangential boundary trace $u\big|^{\kappa-\text{n.t.}}_{\partial\Omega}$ belongs to the John-Nirenberg space $\big[\text{BMO}(\partial\Omega,\sigma)\big]^M$. Since $\partial\Omega$ is compact, this and [129, (7.4.105)] imply that the latter space further embeds into $\big[L^{q}(\partial\Omega,\sigma)\big]^M$ and this permits us to carry out the same argument as in the past and conclude that the image of $\widetilde{T}$ is precisely the space of admissible boundary data for the Neumann Problem (8.1.336) in this case as well. Once again, this implies the same desired conclusions as before.

Going further, the goal is to show that the space null-solutions of the Neumann Problem (8.1.336) is finite dimensional. First, we analyze the case when either $n\geq 3$, or $n=2$ and $p\in\big(\frac{1}{2},1\big)$ and Ω is bounded. Let u be an arbitrary null-solution of the boundary problem (8.1.336) (i.e., u belongs to the space defined in (8.1.341)) in this setting. Then [132, Theorem 2.2.7] guarantees (bearing in mind [131, (1.5.205)]) that the nontangential boundary trace $u\big|^{\kappa-\text{n.t.}}_{\partial\Omega}$ exists at σ-a.e. point on $\partial\Omega$ and

$$\begin{aligned}u\big|^{\kappa-\text{n.t.}}_{\partial\Omega}\in\big[L^{p^*}(\partial\Omega,\sigma)\big]^M,\quad \mathcal{N}_\kappa u\in L^{p^*}(\partial\Omega,\sigma)\\ \text{where } p^*:=\big(\tfrac{1}{p}-\tfrac{1}{n-1}\big)^{-1}\in(1,\infty).\end{aligned}\tag{8.1.398}$$

Having established these properties, [132, Theorem 2.3.16] applies and, in view of the fact that we presently have $\mathring{\partial}^{A}_{\nu}u=0$, gives

$$u=\mathcal{D}_A\big(u\big|^{\kappa-\text{n.t.}}_{\partial\Omega}\big)\text{ in }\Omega.\tag{8.1.399}$$

Going nontangentially to the boundary, this leads (on account of [132, (1.5.20)]) to the conclusion that

$$\big(-\tfrac{1}{2}I+K_A\big)\big(u\big|^{\kappa-\text{n.t.}}_{\partial\Omega}\big)=0,\tag{8.1.400}$$

hence

$$u\big|^{\kappa-\text{n.t.}}_{\partial\Omega}\in\text{Ker}\Big(-\tfrac{1}{2}I+K_A;\ \big[L^{p^*}(\partial\Omega,\sigma)\big]^M\Big).\tag{8.1.401}$$

Note that, as seen from (8.1.399), the assignment taking u into $u\big|^{\kappa-\text{n.t.}}_{\partial\Omega}$ is linear and injective. From this and (8.1.401) we then conclude that

$$\text{the dimension of (8.1.341) is }\leq\dim\text{Ker}\Big(-\tfrac{1}{2}I+K_A;\ \big[L^{p^*}(\partial\Omega,\sigma)\big]^M\Big).\tag{8.1.402}$$

Bearing in mind (7.1.62), we then deduce from (8.1.402) that the space of null-solutions of the boundary problem (8.1.336), described in (8.1.341), is indeed finite dimensional in this case.

Next, consider the situation when $n=2$ and $p=1$ and Ω is bounded. This is handled in a similar fashion as above, the only difference being that, in place of

(8.1.398), [132, Theorem 2.2.7] now gives that any null-solution u of the boundary problem (8.1.336) satisfies

$$u\big|_{\partial\Omega}^{\kappa-\mathrm{n.t.}} \in \big[L^q(\partial\Omega,\sigma)\big]^M, \quad \mathcal{N}_\kappa u \in L^q(\partial\Omega,\sigma) \text{ for any } q \in (1,\infty). \tag{8.1.403}$$

Going further, suppose $n = 2$ and $p \in \big(\frac{1}{2}, 1\big)$ and Ω is an exterior domain. Once again, assume u is a null-solution of the boundary problem (8.1.336). From [131, (1.5.205)] and [132, Theorem 2.2.7] we then see that

$$\begin{aligned} &(\nabla u)(x) = O(|x|^{-1}) \text{ as } |x| \to \infty, \\ &u\big|_{\partial\Omega}^{\kappa-\mathrm{n.t.}} \in \big[L^{p^*}(\partial\Omega,\sigma)\big]^M \text{ where } p^* := \big(\tfrac{1}{p} - 1\big)^{-1} \in (1,\infty), \end{aligned} \tag{8.1.404}$$

and (also bearing [132, (1.8.8)] in mind) that there exists some $\mathbb{C}^M$-valued locally constant function c_u in Ω with the property that

$$u = \mathcal{D}\big(u\big|_{\partial\Omega}^{\kappa-\mathrm{n.t.}}\big) + c_u \text{ in } \Omega. \tag{8.1.405}$$

If $N \in \mathbb{N}$ is the number of connected components of the set Ω, then $\big(\mathbb{C}^M\big)^N$ may be canonically identified with the space of locally constant $\mathbb{C}^M$-valued functions in Ω. After going nontangentially to the boundary, (8.1.405) yields, on account of [132, (1.5.20)] and the convention just made, $\big(-\frac{1}{2}I + K_A\big)\big(u\big|_{\partial\Omega}^{\kappa-\mathrm{n.t.}}\big) \in \big(\mathbb{C}^M\big)^N$, from which we further conclude that

$$u\big|_{\partial\Omega}^{\kappa-\mathrm{n.t.}} \text{ belongs to } \Big\{\phi \in \big[L^{p^*}(\partial\Omega,\sigma)\big]^M : \big(-\tfrac{1}{2}I + K_A\big)\phi \in \big(\mathbb{C}^M\big)^N\Big\}. \tag{8.1.406}$$

Since taking $\delta \in (0,1)$ small enough in (8.1.337) ensures that $-\frac{1}{2}I+K_A$ is a Fredholm operator on $\big[L^{p^*}(\partial\Omega,\sigma)\big]^M$, from item *(g)* in [130, Theorem 2.2.4] we see that the space appearing above is finite dimensional. Bearing this in mind, from (8.1.406), (8.1.405), and the fact that $c_u \in \big(\mathbb{C}^M\big)^N$ it follows that the space of null-solutions of the boundary problem (8.1.336) is finite dimensional in this case.

Finally, the case when $n = 2$, $p = 1$, and the set Ω is an exterior domain is dealt with very similarly, the only difference being that in (8.1.404) we presently have $u\big|_{\partial\Omega}^{\kappa-\mathrm{n.t.}} \in \big[L^q(\partial\Omega,\sigma)\big]^M$ for any $q \in (1,\infty)$. This takes care of item *(1)*.

Turning attention to item *(2)*, work under the stronger hypotheses imposed in (8.1.342). In particular, having chosen some $A \in \mathfrak{A}_L^{\mathrm{dis}}$, Proposition 1.4.1 ensures that $A^\top \in \mathfrak{A}_{L^\top}^{\mathrm{dis}}$. Then the claim in (8.1.343) follows from (8.1.377), Proposition 7.1.43, and Proposition 7.1.44. Next, in the case when either $n \geq 3$, or $n = 2$ and the set Ω is bounded, the fact that the Neumann Problem (8.1.336) has a solution if and only if the boundary datum $f \in \big[H^p(\partial\Omega,\sigma)\big]^M$ satisfies (8.1.344) if $p < 1$ and (8.1.345) if $p = 1$ is a consequence of (8.1.343), (7.1.472), and Definition 7.1.25. Also, in the case when $n = 2$ and Ω is unbounded, the fact that the Neumann Problem (8.1.336) has a solution if and only if the boundary datum $f \in \big[H^p(\partial\Omega,\sigma)\big]^M$ satisfies (8.1.346) if $p < 1$ and (8.1.347) if $p = 1$ is seen from (8.1.343) and Proposition 7.1.45.

Pressing on, work under the assumptions made in (8.1.348). Consider first the claim made in (8.1.350). From [132, (2.3.100), (2.3.132)] and [132, Theorem 2.2.8] we see that $u\big|_{\partial\Omega}^{\kappa-\mathrm{n.t.}} \in \big[H_1^{p^*,p}(\partial\Omega,\sigma)\big]^M$. Keeping this membership in mind and reasoning as in (8.1.399)-(8.1.400) then proves that the assignment taking u into $u\big|_{\partial\Omega}^{\kappa-\mathrm{n.t.}}$ is linear and injective from the space (8.1.341) of null-solutions for the boundary problem (8.1.336) into the space $\mathrm{Ker}\Big(-\tfrac{1}{2}I+K_A;\ \big[H_1^{p^*,p}(\partial\Omega,\sigma)\big]^M\Big)$. We claim that this assignment is also surjective. To justify this, pick an arbitrary

$$\phi \in \mathrm{Ker}\Big(-\tfrac{1}{2}I+K_A;\ \big[H_1^{p^*,p}(\partial\Omega,\sigma)\big]^M\Big) \tag{8.1.407}$$

and define $u := \mathcal{D}_A\phi$ in Ω. Then $\phi \in \big[L^{p^*}(\partial\Omega,\sigma)\big]^M$ and u satisfies

$$\begin{gathered} u \in \big[\mathscr{C}^\infty(\Omega)\big]^M,\ Lu=0,\ \mathcal{N}_\kappa(\nabla u)\in L^p(\partial\Omega,\sigma),\ \mathcal{N}_\kappa u \in L^{p^*}(\partial\Omega,\sigma), \\ \text{and } u(x)=O(|x|^{1-n}) \text{ as } |x|\to\infty \text{ if } \Omega \text{ is unbounded,} \end{gathered} \tag{8.1.408}$$

by virtue of [132, (1.3.24), (1.3.26), (1.5.1), (1.8.8), (2.3.4), (2.3.100)]. In addition, [132, (1.5.20)] and (8.1.407) give

$$u\big|_{\partial\Omega}^{\kappa-\mathrm{n.t.}} = \big(\tfrac{1}{2}I+K_A\big)\phi = \phi. \tag{8.1.409}$$

As such, the claim we seek to prove follows as soon as we show that

$$\dot{\partial}_\nu^A u = 0. \tag{8.1.410}$$

To this end, first observe that, thanks to the properties listed in (8.1.408), [130, Theorem 10.2.24] guarantees that

$$\dot{\partial}_\nu^A u \in \big[H^p(\partial\Omega,\sigma)\big]^M. \tag{8.1.411}$$

The properties recorded in (8.1.408) also ensure that [132, Theorem 2.3.16] is presently applicable. The integral representation formula [132, (2.3.236)] then gives $u = \mathcal{D}\big(u\big|_{\partial\Omega}^{\kappa-\mathrm{n.t.}}\big) - \mathcal{S}\big(\dot{\partial}_\nu^A u\big)$ in the set Ω. Upon noting that formula (8.1.409) implies $\mathcal{D}\big(u\big|_{\partial\Omega}^{\kappa-\mathrm{n.t.}}\big) = \mathcal{D}_A\phi = u$, this reduces to $\mathcal{S}\big(\dot{\partial}_\nu^A u\big)=0$ in Ω. Going nontangentially to the boundary, we arrive at

$$S\big(\dot{\partial}_\nu^A u\big) = 0, \tag{8.1.412}$$

on account of (8.1.411), [132, (2.2.127)], and (A.0.1). In concert with the invertibility result proved in Theorem 7.1.34, this implies (8.1.410) when $n \geq 3$. If $n=2$ and Ω is bounded, from (8.1.411), (8.1.412), [130, Lemma 10.2.19], and (7.1.548) we get

$$S_{\mathrm{aug}}\big(\dot{\partial}_\nu^A u, 0\big) = (0,0). \tag{8.1.413}$$

With this in hand, the invertibility result proved in Theorem 7.1.35 then once again proves (8.1.410).

Let us summarize our progress so far. The above reasoning shows that the assignment taking the function u into its nontangential trace $u\big|_{\partial\Omega}^{\kappa-\mathrm{n.t.}}$ is a linear isomorphism from the space (8.1.341) of null-solutions for the boundary problem (8.1.336) into the space $\mathrm{Ker}\Big(-\frac{1}{2}I+K_A;\ \big[H_1^{p^*,p}(\partial\Omega,\sigma)\big]^M\Big)$. This proves (8.1.350). In addition, from this we also conclude that

$$\begin{aligned}&\text{the dimension of the space (8.1.341) of null-solutions for the boundary}\\&\text{problem (8.1.336) is precisely } \dim\mathrm{Ker}\Big(-\tfrac{1}{2}I+K_A;\ \big[H_1^{p^*,p}(\partial\Omega,\sigma)\big]^M\Big)\\&\text{if either } n\geq 3, \text{ or if } n=2 \text{ and the set } \Omega \text{ is bounded.}\end{aligned}\tag{8.1.414}$$

Finally, (8.1.399) justifies the claim made in (8.1.351) which, in turn, gives (8.1.349).

Moving on, work in the case when $n=2$, the set Ω is an exterior domain, and (8.1.352) holds. Let us first justify (8.1.354). With this goal in mind, pick a function u belonging to the space of null-solutions of the Neumann Problem (8.1.336). Then [131, Theorem 1.5.9] and (8.1.342) ensure

$$\text{the limit } c_0:=\lim_{|x|\to\infty}u(x) \text{ exists in } \mathbb{C}^M,\tag{8.1.415}$$

$$u(x)=c_0+O(|x|^{-1}) \text{ and } (\nabla u)(x)=O(|x|^{-2}) \text{ as } |x|\to\infty.\tag{8.1.416}$$

In addition, [132, Theorem 2.2.7, items *(a)-(b)*] guarantee that the nontangential boundary trace $u\big|_{\partial\Omega}^{\kappa-\mathrm{n.t.}}$ exists at σ-a.e. point on $\partial\Omega$ and, as a function, belongs to $\big[\mathrm{BMO}(\partial\Omega,\sigma)\big]^M$. In concert with [129, (7.4.105)] and [132, Theorem 2.2.8] this further implies that

$$u\big|_{\partial\Omega}^{\kappa-\mathrm{n.t.}}\in\big[H_1^{p^*,p}(\partial\Omega,\sigma)\big]^M.\tag{8.1.417}$$

We also have $\mathcal{N}_\kappa u\in L^1(\partial\Omega,\sigma)$, thanks to (8.1.416) and [132, Theorem 2.2.7, *(d)*]. Granted these properties, [132, Theorem 2.3.16] applies to the function $\widetilde{u}:=u-c_0$ and gives (bearing in mind that $\partial_\nu^A\widetilde{u}=\partial_\nu^A u=0$ and [132, (1.3.45)])

$$u-c_0=\mathcal{D}\big(u\big|_{\partial\Omega}^{\kappa-\mathrm{n.t.}}\big) \text{ in } \Omega.\tag{8.1.418}$$

Going nontangentially to the boundary then yields $\big(-\frac{1}{2}I+K_A\big)\big(u\big|_{\partial\Omega}^{\kappa-\mathrm{n.t.}}\big)=-c_0$. In concert with (8.1.417) and (8.1.415), this shows that the pair $\Big(u\big|_{\partial\Omega}^{\kappa-\mathrm{n.t.}},\ \lim_{|x|\to\infty}u(x)\Big)$ belongs to the space

$$\Big\{(\phi,c)\in\big[H_1^{p^*,p}(\partial\Omega,\sigma)\big]^M\oplus\mathbb{C}^M:\ \big(-\tfrac{1}{2}I+K_A\big)\phi=-c\Big\}.\tag{8.1.419}$$

Also, it is clear from (8.1.415) and (8.1.418) that the mapping associating to each null-solution u of the Neumann Problem (8.1.336) the pair $\Big(u\big|_{\partial\Omega}^{\kappa-\mathrm{n.t.}},\ \lim_{|x|\to\infty}u(x)\Big)$ is injective. Let us show that said map is also surjective, i.e., its range is the space (8.1.419). Given $\phi\in\big[H_1^{p^*,p}(\partial\Omega,\sigma)\big]^M$ and $c\in\mathbb{C}^M$ with $\big(-\frac{1}{2}I+K_A\big)\phi=-c$, set

$$u := \mathcal{D}\phi + c \text{ in } \Omega. \tag{8.1.420}$$

Thanks to [132, (1.3.24), (1.3.26), (1.5.1), (1.8.8), (2.3.4), (2.3.100)], we have

$$\begin{aligned} &u \in \big[\mathscr{C}^\infty(\Omega)\big]^M, \ \ \mathcal{N}_\kappa(\nabla u) \in L^p(\partial\Omega,\sigma), \ \ \mathcal{N}_\kappa u \in L^{p^*}(\partial\Omega,\sigma), \\ &Lu = 0, \ \ u(x) = c + O(|x|^{-1}) \text{ and } (\nabla u)(x) = O(|x|^{-2}) \text{ as } |x| \to \infty. \end{aligned} \tag{8.1.421}$$

In addition, [132, (1.5.20)] and $K_A\phi = \frac{1}{2}\phi - c$ give $u\big|_{\partial\Omega}^{\kappa-\mathrm{n.t.}} = \big(\frac{1}{2}I + K_A\big)\phi + c = \phi$. Granted these properties, the surjectivity claim we presently seek follows as soon as we justify the formula $\dot{\partial}_\nu^A u = 0$. This, however, is obtained by running the same argument which has established (8.1.410) in the case $n = 2$ (based on the invertibility result proved in Theorem 7.1.35), now working with the function $u - c$ in place of u. This takes care of (8.1.354). In turn, from (8.1.354) we conclude that

$$\begin{aligned} &\text{the dimension of the space (8.1.341) of null-solutions for the Neumann} \\ &\text{Problem (8.1.336) when } n = 2 \text{ and } \Omega \text{ is unbounded is equal to the} \\ &\text{dimension of } \Big\{(\phi, c) \in \big[H_1^{p^*,p}(\partial\Omega,\sigma)\big]^M \oplus \mathbb{C}^M : \big(-\tfrac{1}{2}I + K_A\big)\phi = -c\Big\}. \end{aligned} \tag{8.1.422}$$

Lastly, (8.1.418) implies (8.1.355) which further justifies (8.1.353).

We are now ready to prove that the index of the boundary value problem (8.1.336), i.e., the dimension of (8.1.341) minus the codimension of (8.1.339) in $\big[H^p(\partial\Omega,\sigma)\big]^M$, is equal to zero. With this goal in mind, we shall analyze several cases. Suppose first that either $n \geq 3$, or $n = 2$ and $p \in \big(\frac{1}{2}, 1\big)$ and Ω is bounded. Then $p^* := \big(\frac{1}{p} - \frac{1}{n-1}\big)^{-1} \in (1, \infty)$ and

$$\begin{aligned} \dim\left(\frac{\big[H^p(\partial\Omega,\sigma)\big]^M}{\mathrm{Im}\big(-\frac{1}{2}I + K^\#_{A^\top}; [H^p(\partial\Omega,\sigma)]^M\big)}\right) &= \dim \mathrm{Ker}\Big(-\tfrac{1}{2}I + K^\#_{A^\top};\ \big[H^p(\partial\Omega,\sigma)\big]^M\Big) \\ &= \dim \mathrm{Ker}\Big(-\tfrac{1}{2}I + K^\#_{A^\top};\ \big[L^{p^*}(\partial\Omega,\sigma)\big]^M\Big) \\ &= \dim \mathrm{Ker}\Big(-\tfrac{1}{2}I + K_A;\ \big[L_1^{p^*}(\partial\Omega,\sigma)\big]^M\Big) \\ &= \dim \mathrm{Ker}\Big(-\tfrac{1}{2}I + K_A;\ \big[L^{p^*}(\partial\Omega,\sigma)\big]^M\Big) < +\infty. \end{aligned} \tag{8.1.423}$$

Indeed, the first equality is implied by (8.1.372), the second equality comes from (7.1.61), the third equality is a consequence of (7.1.622) if $n \geq 3$, and of (7.1.631) if $n = 2$, the fourth equality is seen from (7.1.60), and the final inequality follows from (8.1.372). Moreover, since [130, (11.10.38)] and (A.0.103) give

$$L_1^{p^*}(\partial\Omega,\sigma) \subseteq H_1^{p^*,p}(\partial\Omega,\sigma) \subseteq L^{p^*}(\partial\Omega,\sigma), \tag{8.1.424}$$

we conclude that

$$\mathrm{Ker}\Big(-\tfrac{1}{2}I+K_A;\ \big[L_1^{p^*}(\partial\Omega,\sigma)\big]^M\Big)\subseteq \mathrm{Ker}\Big(-\tfrac{1}{2}I+K_A;\ \big[H_1^{p^*,p}(\partial\Omega,\sigma)\big]^M\Big)$$
$$\subseteq \mathrm{Ker}\Big(-\tfrac{1}{2}I+K_A;\ \big[L^{p^*}(\partial\Omega,\sigma)\big]^M\Big) \tag{8.1.425}$$

hence,

$$\dim\mathrm{Ker}\Big(-\tfrac{1}{2}I+K_A;\ \big[L_1^{p^*}(\partial\Omega,\sigma)\big]^M\Big)$$
$$\leq \dim\mathrm{Ker}\Big(-\tfrac{1}{2}I+K_A;\ \big[H_1^{p^*,p}(\partial\Omega,\sigma)\big]^M\Big)$$
$$\leq \dim\mathrm{Ker}\Big(-\tfrac{1}{2}I+K_A;\ \big[L^{p^*}(\partial\Omega,\sigma)\big]^M\Big). \tag{8.1.426}$$

Combining (8.1.423) with (8.1.426) leads to the conclusion that

$$\dim\left(\frac{\big[H^p(\partial\Omega,\sigma)\big]^M}{\mathrm{Im}\big(-\frac{1}{2}I+K^{\#}_{A^\top};[H^p(\partial\Omega,\sigma)]^M\big)}\right)$$
$$=\dim\mathrm{Ker}\Big(-\tfrac{1}{2}I+K_A;\ \big[H_1^{p^*,p}(\partial\Omega,\sigma)\big]^M\Big)<+\infty. \tag{8.1.427}$$

Finally, from (8.1.427), (8.1.343), and (8.1.414) we see that the index of the boundary value problem (8.1.336) is equal to zero in this case.

Suppose next that $n=2$ and $p=1$ and Ω is bounded. Having fixed an arbitrary $q\in(1,\infty)$ we may run the same argument as above with p^* replaced by q and arrive at the same conclusion, namely that the index of the boundary value problem (8.1.336) is equal to zero in this case as well.

Next we treat the case when $n=2$ and $p\in\big(\frac{1}{2},1\big]$ and Ω is an exterior domain. In this scenario, define $p^*:=\big(\frac{1}{p}-1\big)^{-1}\in(1,\infty)$ if $p\in\big(\frac{1}{2},1\big)$ and take $p^*:=q\in(1,\infty)$ arbitrary if $p=1$. Consider the canonical inclusion

$$\iota:\Big\{g\in\big[H^p(\partial\Omega,\sigma)\big]^M:\ \langle g,\mathbf{e}_j\rangle=0\ \text{ for }\ 1\leq j\leq M\Big\}\to\big[H^p(\partial\Omega,\sigma)\big]^M \tag{8.1.428}$$

and abbreviate

$$T:=\big(-\tfrac{1}{2}I+K^{\#}_{A^\top}\big)\circ\iota,$$
$$T:\Big\{g\in\big[H^p(\partial\Omega,\sigma)\big]^M:\ \langle g,\mathbf{e}_j\rangle=0\ \text{ for }\ 1\leq j\leq M\Big\}\to\big[H^p(\partial\Omega,\sigma)\big]^M. \tag{8.1.429}$$

Since ι is a Fredholm operator with index $-M$ and $-\frac{1}{2}I+K^{\#}_{A^\top}$ is a Fredholm operator with index zero from $\big[H^p(\partial\Omega,\sigma)\big]^M$ into itself, we may invoke [130, Theorem 2.2.4] to conclude that the operator

$$T\text{ is Fredholm with index }-M\text{ in the context of (8.1.429).} \tag{8.1.430}$$

Then the codimension on the space of admissible boundary data for the Neumann Problem (8.1.336) in the "full" space of boundary data may be computed as follows:

$$\dim\left(\frac{\big[H^p(\partial\Omega,\sigma)\big]^M}{\big\{(-\frac{1}{2}I+K^{\#}_{A^\top})g:g\in[H^p(\partial\Omega,\sigma)]^M\text{ with }\langle g,\mathbf{e}_j\rangle=0\text{ for }1\leq j\leq M\big\}}\right)$$

$$= \dim\left(\frac{\big[H^p(\partial\Omega,\sigma)\big]^M}{\operatorname{Im}T}\right) = M + \dim\operatorname{Ker}T \tag{8.1.431}$$

$$= M + \dim\operatorname{Ker}\Big(-\tfrac{1}{2}I + K^{\#}_{A^\top};\ \big[L^{p^*}(\partial\Omega,\sigma)\big]^M\Big)$$

$$= \dim\Big\{(\phi,c)\in\big[H_1^{p^*,p}(\partial\Omega,\sigma)\big]^M\oplus\mathbb{C}^M:$$

$$\phi - c\in\operatorname{Ker}\Big(-\tfrac{1}{2}I+K_A;\ \big[H_1^{p^*,p}(\partial\Omega,\sigma)\big]^M\Big)\Big\}$$

$$= \dim\Big\{(\phi,c)\in\big[H_1^{p^*,p}(\partial\Omega,\sigma)\big]^M\oplus\mathbb{C}^M:\ \big(-\tfrac{1}{2}I+K_A\big)\phi=-c\Big\}.$$

Above, the first equality comes from (8.1.428)-(8.1.429), while the second equality is implied by (8.1.430). The third equality in (8.1.431) is a consequence of

$$\Big\{g\in\operatorname{Ker}\Big(-\tfrac{1}{2}I+K^{\#}_{A^\top};\ \big[H^p(\partial\Omega,\sigma)\big]^M\Big):\ \langle g,\mathbf{e}_j\rangle=0\text{ for }1\leq j\leq M\Big\}$$
$$=\operatorname{Ker}\Big(-\tfrac{1}{2}I+K^{\#}_{A^\top};\ \big[L^{p^*}(\partial\Omega,\sigma)\big]^M\Big). \tag{8.1.432}$$

Indeed, the left-to-right inclusion in (8.1.432) is seen from Proposition 7.1.2 (used with $A^\top$ in place of A), while the right-to-left inclusion in (8.1.432) follows from [129, (3.6.27)], [130, (4.2.21)], and [132, (1.7.59)] (keeping in mind the compatibility results from [130, Lemmas 4.6.5-4.6.6]).

The fourth equality in (8.1.431) is based on the fact that, much as before (see (8.1.423)-(8.1.426)) we have

$$\dim\operatorname{Ker}\Big(-\tfrac{1}{2}I+K^{\#}_{A^\top};\ \big[L^{p^*}(\partial\Omega,\sigma)\big]^M\Big)$$
$$=\dim\operatorname{Ker}\Big(-\tfrac{1}{2}I+K_A;\ \big[H_1^{p^*,p}(\partial\Omega,\sigma)\big]^M\Big), \tag{8.1.433}$$

plus an abstract result to the effect that if V is a vector space and F_0, F_1 are to finite dimensional linear subspaces of V then

$$\dim\big\{(x,y)\in V\times F_0:\ x-y\in F_1\big\}=\dim F_0+\dim F_1. \tag{8.1.434}$$

This is justified by observing that

$$\begin{aligned}&\Phi: F_0\times F_1\longrightarrow\big\{(x,y)\in X\times F_0:\ x-y\in F_1\big\}\\ &\Phi(y_0,y_1):=(y_0,y_0+y_1)\text{ for each }y_0\in F_0\text{ and }y_1\in F_1,\end{aligned} \tag{8.1.435}$$

is a linear isomorphism. The fourth equality in (8.1.431) is then obtained from (8.1.433) and (8.1.434), taking

$$V := \big[H^{p^*,p}_1(\partial\Omega,\sigma)\big]^M, \quad F_0 := \mathbb{C}^M, \text{ and}$$
$$F_1 := \operatorname{Ker}\Big(-\tfrac{1}{2}I + K_A;\ \big[H^{p^*,p}_1(\partial\Omega,\sigma)\big]^M\Big). \tag{8.1.436}$$

The final equality in (8.1.431) is a direct corollary of [132, (1.5.21)]. Having established (8.1.431), from (8.1.343) and (8.1.354) we conclude that the index of the Neumann Problem (8.1.336) is also equal to zero in this last case. This completes the treatment of item *(2)*.

Moving on to item *(3)* work under the assumption that $A \in \mathfrak{A}^{\rm dis}_L$ is positive definite. In the case when either $n \geq 3$, or $n = 2$ and the set Ω is bounded, we conclude from (8.1.343) and the version of (7.1.277) written for Hardy spaces (cf. Remark 7.1.12) that the space of admissible boundary data for the Neumann Problem (8.1.336) is

$$\operatorname{Im}\Big(-\tfrac{1}{2}I + K^{\#}_{A^\top};\ \big[H^p(\partial\Omega,\sigma)\big]^M\Big) = \big[H^p(\partial\Omega,\sigma)\cap(\mathbb{C}_{\partial\Omega_+})^{\perp}\big]^M. \tag{8.1.437}$$

This takes care of the first claim in (8.1.357). The second claim in (8.1.357) follows from (8.1.349), the version of (7.1.275) for Hardy-based Sobolev spaces (cf. Remark 7.1.12), and (7.1.305).

Suppose now that $n = 2$ and Ω is an exterior domain. Then the last property in (8.1.343) combines with the version of (7.1.292) written for Hardy spaces (cf. Remark 7.1.12) to give that the space of admissible boundary data for the Neumann Problem (8.1.336) is, in this case, equal to

$$\Big\{\big(-\tfrac{1}{2}I + K^{\#}_A\big)g\ :\ g \in \big[H^p(\partial\Omega,\sigma)\cap\mathbb{C}^{\perp}\big]^M\Big\} = \big[H^p(\partial\Omega,\sigma)\cap(\mathbb{C}^{\infty}_{\partial\Omega_+})^{\perp}\big]^M.$$

This justifies the first claim in (8.1.358). As regards the second claim in (8.1.358), from (8.1.353) and the version of (7.1.293) for Hardy-based Sobolev spaces (cf. Remark 7.1.12) we see that the space of null-solutions of the Neumann Problem (8.1.336) is, in this case, contained in

$$\Big\{\mathcal{D}_A\phi + c\ :\ \phi \in \big(\mathbb{C}^{\infty}_{\partial\Omega_+}\big)^M \text{ and } c \in \mathbb{C}^M\Big\}. \tag{8.1.438}$$

Note that, thanks to (7.1.343), each given function $\phi \in \big(\mathbb{C}^{\infty}_{\partial\Omega_+}\big)^M$ may be decomposed as $\phi = \phi_+ + \lambda\mathbf{1}_{\partial\Omega_\infty}$ for some $\lambda \in \mathbb{C}^M$. Also, (7.1.305) shows that

$$\mathcal{D}_A\phi_+ \in \big(\mathbb{C}_{\Omega_+}\big)^M, \tag{8.1.439}$$

while [132, (1.3.46)] used with $\eta := \lambda\mathbf{1}_{\partial\Omega_\infty}$ gives

$$\mathcal{D}_A\big(\lambda\mathbf{1}_{\partial\Omega_\infty}\big) = -\lambda\mathbf{1}_{\Omega\setminus\Omega_\infty} \in \big(\mathbb{C}_{\Omega_+}\big)^M. \tag{8.1.440}$$

Altogether, (8.1.438)-(8.1.440) show that the space of null-solutions of the Neumann Problem (8.1.336) is presently contained in

$$\big(\mathbb{C}_{\Omega_+}\big)^M + \mathbb{C}^M = \big(\mathbb{C}^{\infty}_{\Omega_+}\big)^M. \tag{8.1.441}$$

Conversely, it is clear from definitions that any function in $\left(\mathbb{C}^{\infty}_{\Omega_+}\right)^M$ is a null-solution of the Neumann Problem (8.1.336). Hence, the space of null-solutions of the Neumann Problem (8.1.336) is, in the current case, precisely $\left(\mathbb{C}^{\infty}_{\Omega_+}\right)^M$. This completes the treatment of item *(3)*.

Lastly, item *(4)* is a consequence pf what we have proved so far and item *(3)* in [130, Theorem 10.2.24]. The proof of Theorem 8.1.12 is therefore complete. □

Going further, we treat the Neumann Problem for weakly elliptic systems in δ-oscillating AR domains with data in Lebesgue spaces L^p with $1 < p < \infty$. Theorem 8.1.15, which is our main result in this regard, extends work of S. Hofmann, M. Mitrea, and M. Taylor in [81] where the case of Laplacian and the Lamé system in regular SKT domains with compact boundaries has been considered. Theorem 8.1.15 furthers the understanding of the open problem formulated in [202, Question 2.5]. The special case when Ω is an infinitesimally flat AR domain (cf. Definition 3.4.1) may be considered to be the sharp version of [52, Theorem 2.6, p. 185] where E. Fabes, M. Jodeit, and N. Rivière have solved the Neumann Problem for the Laplacian in bounded $\mathscr{C}^1$ domains with boundary data in L^p for $p \in (1, \infty)$ arbitrary. In the class of bounded Lipschitz domains, G. Verchota has dealt in [203, Corollary 3.4, p. 590] Neumann Problem for the Laplacian with boundary data in L^2, a result which has been subsequently extended by B. Dahlberg and C. Kenig in [38, Theorem 4.15, p. 462] to include boundary data from L^p with p in the optimal range[48] $(1, 2 + \varepsilon)$ where $\varepsilon > 0$ is a (typically small) threshold depending only on the Lipschitz character of the domain in question. The Neumann Problem for the Laplace-Beltrami operator on Lipschitz subdomains of Riemannian manifolds with boundary data in Lebesgue spaces has been solved by M. Mitrea and M. Taylor in [153], [154], [156]. The solution to the traction boundary value problem for the system of elastostatics is given in [40, Theorem 2.7, p. 811] for bounded Lipschitz domains with boundary in data in L^p with p near 2. In the three-dimensional setting, this has been extended by B. Dahlberg and C. Kenig in [39, Theorem 1.5, p. 625] to the range $1 < p < 2 + \varepsilon$ for some small $\varepsilon > 0$. This has been further augmented by work of Z. Shen who has dealt in [194, Theorem 1.2, p. 217] with the traction boundary value problem for elastostatics in Lipschitz domains in dimensions $n \geq 4$ for boundary data in L^p assuming $\frac{2(n-1)}{n+1} < p < 2$. In the very same setting, Z. Shen has solved in [194, Theorem 1.1, p. 214] the Neumann Problem for generic systems possessing positive definite symmetric (constant) real coefficient tensors. Further references pertaining to this and other related topics may be found in [93].

Finally, we wish to mention that the two-dimensional Neumann Problem for the Laplacian in a bounded chord-arc domain Ω in the plane has been solved for boundary data in $L^p(\partial\Omega, \sigma)$ with $p > 1$ sufficiently small so that the arc-length measure $\sigma := \mathcal{H}^1 \lfloor \partial\Omega$ belongs to the Muckenhoupt weight class $A_{p'}$ with respect to the harmonic measure on $\partial\Omega$ (where $1/p + 1/p' = 1$) by D. Jerison and C. Kenig in [85, p. 245]. By way of contrast, our Theorem 8.1.15 specialized to the case $L = \Delta$ in the setting of a chord-arc domain $\Omega \subseteq \mathbb{R}^2 \equiv \mathbb{C}$ shows (in light of Theorem 3.3.1) that any integrability exponent $p \in (1, \infty)$ may be allowed provided the local chord-arc

[48] relative to the class of Lipschitz domains

constant of Ω is sufficiently small at small scales (i.e., if the quantity $\varkappa$ described in (3.3.3) is small enough).

Theorem 8.1.15 (The Neumann Problem with Data in Lebesgue Spaces) *Select $n, M \in \mathbb{N}$ with $n \geq 2$ and suppose $\Omega \subseteq \mathbb{R}^n$ is an Ahlfors regular domain with compact boundary. Abbreviate $\sigma := \mathcal{H}^{n-1}\lfloor\partial\Omega$ and denote by ν the geometric measure theoretic outward unit normal to Ω. Also, let L be a homogeneous, second-order, constant complex coefficient, $M\times M$ system in $\mathbb{R}^n$, and pick a coefficient tensor A for which $L = L_A$ (i.e., $A \in \mathfrak{A}_L$). Finally, fix an arbitrary aperture parameter $\kappa > 0$, and select an exponent*

$$p \in (1, \infty). \tag{8.1.442}$$

In this setting, consider the Neumann Problem[49]

$$\begin{cases} u \in \big[\mathscr{C}^\infty(\Omega)\big]^M, \quad Lu = 0 \text{ in } \Omega, \\ \mathcal{N}_\kappa(\nabla u) \in L^p(\partial\Omega, \sigma), \quad \partial^A_\nu u = f \in \big[L^p(\partial\Omega, \sigma)\big]^M, \\ u(x) = O(|x|^{2-n}) \text{ as } |x| \to \infty \text{ if } \Omega \text{ is unbounded,} \end{cases} \tag{8.1.443}$$

where the strong conormal derivative is defined in a pointwise sense as in (A.0.203).
Then there exists a threshold $\delta_0 \in (0, 1)$, which depends only on n, p, A, the ADR *constants of $\partial\Omega$, and the relative flatness ratio $\wp(\Omega)/\mathrm{diam}\,\partial\Omega$, so that when*

$$\mathrm{dist}\Big(\nu, [\mathrm{VMO}(\partial\Omega, \sigma)]^n\Big) < \delta_0 \tag{8.1.444}$$

where the distance is considered in the space $\big[\mathrm{BMO}(\partial\Omega, \sigma)\big]^n$ (hence whenever Ω is a δ_0-oscillating AR *domain*[50]*; cf. Definition 3.4.1), the set Ω becomes a* UR *domain and the following statements are true:*

(1) If one assumes that

$$L \text{ is weakly elliptic with } \mathfrak{A}^{\mathrm{dis}}_L \neq \varnothing \text{ and } \mathfrak{A}^{\mathrm{dis}}_{L^\top} \neq \varnothing, \tag{8.1.445}$$

then whenever $A \in \mathfrak{A}^{\mathrm{dis}}_L$ it follows that the boundary value problem (8.1.443) *is Fredholm solvable, in the following precise sense: First, the space of admissible boundary data for the boundary problem* (8.1.443)*, i.e.,*

$$\Big\{\partial^A_\nu u : u \in \big[\mathscr{C}^\infty(\Omega)\big]^M, \; Lu = 0 \text{ in } \Omega, \; \mathcal{N}_\kappa(\nabla u) \in L^p(\partial\Omega, \sigma), \\ \text{and } u(x) = O(|x|^{2-n}) \text{ as } |x| \to \infty \text{ if } \Omega \text{ is unbounded}\Big\} \tag{8.1.446}$$

is a closed linear subspace of finite codimension in $\big[L^p(\partial\Omega, \sigma)\big]^M$. Second, a solution u of (8.1.443) *exists if and only if the boundary datum f belongs to*

---

[49] thanks to work in [129, §8.5], one may replace the nontangential maximal operator by the tangential maximal operator, associated as in [129, Definition 8.5.1] with a sufficiently large power (cf. (A.0.165))

[50] in particular, if Ω is an infinitesimally flat AR domain

(8.1.446), *in which scenario a solution u may be found such that*

$$\big\|\mathcal{N}_\kappa(\nabla u)\big\|_{L^p(\partial\Omega,\sigma)} \leq C\|f\|_{[L^p(\partial\Omega,\sigma)]^M}, \tag{8.1.447}$$

for some constant $C \in (0,\infty)$ *independent of* f. *Third, the space of null-solutions of the boundary problem* (8.1.443), *i.e.,*

$$\begin{aligned}\Big\{u \in \big[\mathscr{C}^\infty(\Omega)\big]^M &: Lu = 0,\ \mathcal{N}_\kappa(\nabla u) \in L^p(\partial\Omega,\sigma),\ \partial^A_\nu u = 0, \\ &\text{and } u(x) = O(|x|^{2-n}) \text{ as } |x| \to \infty \text{ if } \Omega \text{ is unbounded}\Big\}\end{aligned} \tag{8.1.448}$$

is finite dimensional.

(2) *Suppose* (8.1.445) *is strengthened (cf. Proposition 1.4.1) to the demand that*

> *L satisfies the Legendre-Hadamard strong ellipticity condition (cf.* [131, Definition 1.3.2]) *and one has* $\mathfrak{A}^{\rm dis}_L \neq \varnothing$. (8.1.449)

Then for each coefficient tensor $A \in \mathfrak{A}^{\rm dis}_L$ *the index of the Neumann Problem* (8.1.443), *i.e., the dimension of* (8.1.448) *minus the codimension of* (8.1.446) *in* $\big[L^p(\partial\Omega,\sigma)\big]^M$, *is equal to zero.*

Moreover, in the case when either $n \geq 3$, *or* $n = 2$ *and the set* Ω *is bounded, the Neumann Problem* (8.1.443) *has a solution if and only if the boundary datum* $f \in \big[L^p(\partial\Omega,\sigma)\big]^M$ *satisfies*

$$\begin{aligned}&\int_{\partial\Omega}\langle f,g\rangle\,d\sigma = 0 \text{ for each } g \in \big[L^{p'}(\partial\Omega,\sigma)\big]^M \\ &\text{with } p' := (1-1/p)^{-1}, \text{ such that } \big(-\tfrac12 I + K_{A^\top}\big)g = 0.\end{aligned} \tag{8.1.450}$$

Also, corresponding to the case when $n = 2$ *and* Ω *is an exterior domain, the Neumann Problem* (8.1.443) *has a solution if and only if the boundary datum* $f \in \big[L^p(\partial\Omega,\sigma)\big]^M$ *satisfies*

$$\begin{aligned}&\int_{\partial\Omega}\langle f,g\rangle\,d\sigma = 0 \text{ for each } g \in \big[L^{p'}(\partial\Omega,\sigma)\big]^M \\ &\text{with } p' := (1-1/p)^{-1}, \text{ such that } \big(-\tfrac12 I + K_{A^\top}\big)g \in \mathbb{C}^M.\end{aligned} \tag{8.1.451}$$

In fact,

> *the space of admissible boundary data for the Neumann Problem* (8.1.443), *originally defined as in* (8.1.446), *is equal to* $\mathrm{Im}\Big(-\tfrac12 I + K^\#_{A^\top};\ \big[L^p(\partial\Omega,\sigma)\big]^M\Big)$ *if either* $n \geq 3$, *or* $n = 2$ *and the set* Ω *is bounded, and may be described as the space* $\Big\{\big(-\tfrac12 I + K^\#_{A^\top}\big)g : g \in \big[L^p(\partial\Omega,\sigma)\big]^M \text{ with } \int_{\partial\Omega} g\,d\sigma = 0\Big\}$ *if* $n = 2$ *and* Ω *is an exterior domain.* (8.1.452)

Furthermore, there exists a constant $C \in (0, \infty)$ *with the property that whenever the boundary datum* f *belongs to the space of admissible boundary data, defined in* (8.1.446)*, then a solution* u *of the boundary problem* (8.1.443) *may be found such that* (8.1.447) *holds.*

Also, with $\mathcal{D}_A$ *denoting the boundary-to-domain double layer potential operator associated with* A *and* Ω *as in* (A.0.72)*, when either* $n \geq 3$*, or* $n = 2$ *and the set* Ω *is bounded, it follows that:*

$$\begin{array}{l}\textit{the space of null-solutions of the Neumann Problem (8.1.443)}\\ \textit{(originally described as in (8.1.448)) is actually equal to the}\\ \textit{space } \Big\{\mathcal{D}_A\phi : \phi \in \mathrm{Ker}\Big(-\tfrac{1}{2}I + K_A;\ \big[L^p_1(\partial\Omega, \sigma)\big]^M\Big)\Big\},\end{array} \tag{8.1.453}$$

$$\begin{array}{l}\textit{the mapping associating to each function } u \textit{ belonging to the space}\\ \textit{of null-solutions of the Neumann Problem (8.1.443) (described in}\\ \textit{(8.1.448)) its nontangential boundary trace } u\big|^{\kappa-\mathrm{n.t.}}_{\partial\Omega} \textit{ is an isomorphism}\\ \textit{onto the finite dimensional space } \mathrm{Ker}\Big(-\tfrac{1}{2}I + K_A;\ \big[L^p_1(\partial\Omega, \sigma)\big]^M\Big),\end{array} \tag{8.1.454}$$

$$\begin{array}{l}\textit{the inverse of the isomorphism } u \mapsto u\big|^{\kappa-\mathrm{n.t.}}_{\partial\Omega} \textit{ described in (8.1.454) is}\\ \textit{the boundary-to-domain double layer potential operator } \mathcal{D}_A \textit{ mapping}\\ \mathrm{Ker}\Big(-\tfrac{1}{2}I + K_A;\ \big[L^p_1(\partial\Omega, \sigma)\big]^M\Big) \textit{ isomorphically onto the space of null-}\\ \textit{solutions of the Neumann Problem (8.1.443) (described in (8.1.448)).}\end{array} \tag{8.1.455}$$

Finally, in the case when $n = 2$ *and* Ω *is an exterior domain it follows that:*

$$\begin{array}{l}\textit{the space of null-solutions of the Neumann Problem (8.1.443) equals}\\ \Big\{\mathcal{D}_A\phi + c : \phi \in \big[L^p_1(\partial\Omega, \sigma)\big]^M,\ c \in \mathbb{C}^M,\ \big(-\tfrac{1}{2}I + K_A\big)\phi = -c\Big\},\end{array} \tag{8.1.456}$$

$$\begin{array}{l}\textit{the mapping associating to each function } u \textit{ belonging to the space}\\ \textit{of null-solutions of the Neumann Problem (8.1.443) (described in}\\ \textit{(8.1.448)) the pair } \Big(u\big|^{\kappa-\mathrm{n.t.}}_{\partial\Omega},\ \lim\limits_{|x|\to\infty} u(x)\Big) \textit{ is a well-defined isomorphism}\\ \textit{onto the space } \Big\{(\phi, c) \in \big[L^p_1(\partial\Omega, \sigma)\big]^M \oplus \mathbb{C}^M : \big(-\tfrac{1}{2}I + K_A\big)\phi = -c\Big\},\end{array} \tag{8.1.457}$$

$$\begin{array}{l}\textit{the inverse of the linear isomorphism } u \mapsto \Big(u\big|^{\kappa-\mathrm{n.t.}}_{\partial\Omega},\ \lim\limits_{|x|\to\infty} u(x)\Big) \textit{ in the}\\ \textit{above context is the assignment } (\phi, c) \longmapsto \mathcal{D}_A\phi + c \textit{ mapping the last}\\ \textit{space in (8.1.457) isomorphically onto the space of null-solutions of}\\ \textit{the Neumann Problem (8.1.443) (described in (8.1.448)).}\end{array} \tag{8.1.458}$$

(3) If (8.1.449) *is further strengthened to*[51]

$$A \in \mathfrak{A}_L^{\mathrm{dis}} \text{ is positive definite,} \tag{8.1.459}$$

then the Neumann Problem (8.1.443) *has the following additional properties (with the spaces* $\mathbb{C}_{\Omega_\pm}$, $\mathbb{C}_{\partial\Omega_\pm}$, $\mathbb{C}^\infty_{\Omega_\pm}$, $\mathbb{C}^\infty_{\partial\Omega_\pm}$ *defined as in* (7.1.265), (7.1.268), *and with the conventions* (7.1.266), (7.1.269) *in effect). First, if either* $n \geq 3$, *or* $n = 2$ *and the set* Ω *is bounded, then*

> *the space of admissible boundary data for the Neumann Problem* (8.1.443) *(defined in* (8.1.446)*) is equal to* $\big[L^p(\partial\Omega,\sigma) \cap \big(\mathbb{C}_{\partial\Omega_+}\big)^\perp\big]^M$, *and the space of null-solutions of the Neumann Problem* (8.1.443) *(described in* (8.1.448)*) is equal to* $\big(\mathbb{C}_{\Omega_+}\big)^M$. (8.1.460)

Second, if $n = 2$ *and* Ω *is an exterior domain, then*

> *the space of admissible boundary data for the Neumann Problem* (8.1.443) *(defined in* (8.1.446)*) is equal to* $\big[L^p(\partial\Omega,\sigma) \cap \big(\mathbb{C}^\infty_{\partial\Omega_+}\big)^\perp\big]^M$, *and the space of null-solutions of the Neumann Problem* (8.1.443) *(described in* (8.1.448)*) is equal to* $\big(\mathbb{C}^\infty_{\Omega_+}\big)^M$. (8.1.461)

(4) Reinstate the hypotheses imposed in (8.1.449) *and work under the additional assumption that* Ω *is bounded. Then any solution of the Neumann Problem* (8.1.443) *satisfies*

$$\begin{gathered} u \in \big[W^{2,p}_{1-\frac{1}{p},\star}(\Omega)\big]^M \cap \Big[B^{p,q_0}_{1+\frac{1}{p}}(\Omega)\Big]^M \cap \Big[F^{p,q_1}_{1+\frac{1}{p}}(\Omega)\Big]^M \\ \text{whenever } p \leq q_0 \leq \infty \text{ and } \tfrac{n}{n+1+1/p} < q_1 \leq \infty, \end{gathered} \tag{8.1.462}$$

as well as

$$u \in \big[H^{3/2}(\Omega)\big]^M \text{ if } p = 2. \tag{8.1.463}$$

(5) Similar results as in items (1)-(3) above are true for the Neumann Problem with boundary data in Muckenhoupt weighted Lebesgue spaces, namely[52]

$$\begin{cases} u \in \big[\mathscr{C}^\infty(\Omega)\big]^M, \quad Lu = 0 \text{ in } \Omega, \\ \mathcal{N}_\kappa(\nabla u) \in L^p(\partial\Omega, w), \quad \partial^A_\nu u = f \in \big[L^p(\partial\Omega, w)\big]^M, \\ u(x) = O(|x|^{2-n}) \text{ as } |x| \to \infty \text{ if } \Omega \text{ is unbounded,} \end{cases} \tag{8.1.464}$$

---

[51] according to (1.1.28), whenever L has a positive definite coefficient tensor it follows that L satisfies the Legendre-Hadamard strong ellipticity condition

[52] once more, work in [129, §8.5] shows that one may replace the nontangential maximal operator by the tangential maximal operator, associated as in [129, Definition 8.5.1] with a sufficiently large power (cf. (A.0.165))

where $w \in A_p(\partial\Omega, \sigma)$ *(now allowing* δ_0 *in* (8.1.444) *to also depend on* $[w]_{A_p}$*).*

(6) Even when $L := \Delta$*, the Laplace operator (which obviously satisfies* (8.1.445)*), if* $A \notin \mathfrak{A}_L^{\rm dis}$ *then it may happen that the Neumann Problems* (8.1.443), (8.1.464) *fail to be Fredholm solvable even when the underlying domain* Ω *is the unit disk in the plane (or the complement of its closure). Indeed, the aforementioned failure of Fredholm solvability may be total, in the sense that for each of these problems, both the space of admissible boundary data may have infinite codimension in the full space of boundary data, and the space of null-solutions may be infinite dimensional. Finally, a similar lack of Fredholm solvability occurs in higher dimensions.*

We wish to make four remarks in relation to Theorem 8.1.15. First, as seen from Proposition 1.6.12 and [131, Proposition 1.4.39], in the scalar case, i.e., if $M = 1$, in dimensions $n \geq 3$, both (8.1.445) and (8.1.449) may be replaced by the mere demand that L is weakly elliptic. Also, in the scalar case in the two-dimensional setting (i.e., for $M = 1$ and $n = 2$), if L satisfies the Legendre-Hadamard strong ellipticity condition or, more generally (cf. [131, Lemma 1.4.24]), if $L = {\rm div} A \nabla$ for some 2×2 complex matrix A with ${\rm sym}\, A := \frac{1}{2}(A + A^\top) \in M_2^0$ (cf. [131, (1.4.187)]), then Proposition 1.6.11 and Proposition 1.6.13 guarantee that we automatically have $\mathfrak{A}_L^{\rm dis} \neq \varnothing$ and $\mathfrak{A}_{L^\top}^{\rm dis} \neq \varnothing$. Moreover, according to (1.6.141) in Proposition 1.6.13, in the scalar case in the two-dimensional setting (i.e., for $M = 1$ and $n = 2$) we may omit (8.1.449) altogether if L is a weakly elliptic operator with real coefficients.

Second, the boundary condition in (8.1.443) is naturally formulated. Indeed, choosing the threshold $\delta_0 \in (0, 1)$ small permits us to invoke Theorem 3.4.2 which ensures that Ω is a two-sided NTA domain. Granted this, the Fatou-type result proved in item *(a)* in [131, Theorem 3.3.4] gives that any function satisfying

$$u \in \big[\mathscr{C}^\infty(\Omega)\big]^M, \ Lu = 0 \text{ in } \Omega, \ \mathcal{N}_\kappa(\nabla u) \in L^p(\partial\Omega, \sigma), \ p \in (1, \infty), \tag{8.1.465}$$

has the property that the nontangential boundary trace

$$\begin{gathered} (\nabla u)\big|_{\partial\Omega}^{\kappa-{\rm n.t.}} \text{ exists (in } \mathbb{C}^{n\cdot M}) \text{ at } \sigma\text{-a.e. point on } \partial\Omega \text{ and} \\ \text{the function } (\nabla u)\big|_{\partial\Omega}^{\kappa-{\rm n.t.}} \text{ belongs to } \big[L^p(\partial\Omega, \sigma)\big]^{n\cdot M}. \end{gathered} \tag{8.1.466}$$

In particular, the strong conormal derivative considered in a pointwise sense as in (A.0.203) is meaningful and satisfies

$$\partial_\nu^A u \in \big[L^p(\partial\Omega, \sigma)\big]^M. \tag{8.1.467}$$

This points to the fact that any function satisfying the properties in the first two lines of (8.1.443) has a well-defined strong conormal derivative, considered in a pointwise sense, as in (A.0.203), and said conormal derivative belongs to $\big[L^p(\partial\Omega, \sigma)\big]^M$ (the space from which the boundary data for the problem (8.1.443) are selected). From [130, Proposition 11.5.12] we also see that, for any u as in (8.1.465),

$$u\Big|_{\partial\Omega}^{\kappa-\mathrm{n.t.}} \text{ exists } \sigma\text{-a.e. on } \partial\Omega, \text{ and belongs to } \big[L_1^p(\partial\Omega,\sigma)\big]^M. \tag{8.1.468}$$

The third remark we wish to make in relation to Theorem 8.1.15 is as follows: Under the assumption (8.1.449), the Neumann Problem (8.1.443) has a solution if and only if the boundary datum $f \in \big[L^p(\partial\Omega,\sigma)\big]^M$ satisfies

$$\int_{\partial\Omega} \Big\langle f, w\Big|_{\partial\Omega}^{\kappa-\mathrm{n.t.}} \Big\rangle \, d\sigma = 0 \tag{8.1.469}$$

for each w null-solution of the $L^{p'}$-Neumann Problem for the transposed system $L^\top$ in Ω, where $p' := (1 - 1/p)^{-1} \in (1,\infty)$, i.e., for each function w satisfying

$$\begin{cases} w \in \big[\mathscr{C}^\infty(\Omega)\big]^M, \quad L^\top w = 0 \text{ in } \Omega, \\ \mathcal{N}_\kappa(\nabla w) \in L^{p'}(\partial\Omega,\sigma), \ \ \partial_\nu^{A^\top} w = 0 \text{ at } \sigma\text{-a.e. point on } \partial\Omega, \\ w(x) = O(|x|^{2-n}) \text{ as } |x| \to \infty \text{ if } \Omega \text{ is unbounded.} \end{cases} \tag{8.1.470}$$

This is seen from (8.1.450) together with (8.1.454) and (8.1.456)-(8.1.457) (used with $A^\top$ in place of A and with p' in place of p), also bearing in mind (7.1.60). Note that the existence of the nontangential boundary trace $w\Big|_{\partial\Omega}^{\kappa-\mathrm{n.t.}}$ for any function w as in (8.1.470) is guaranteed by the fact that (8.1.465) implies (8.1.466) (presently used for $L^\top$ in place of L). The claim in (8.1.469) may also be justified (under less demanding assumptions on the set Ω) based on the version of Green's formula proved in [131, (1.7.81)]. Ultimately, (8.1.469) provides an intrinsic description of the space of admissible boundary data for the Neumann Problem (8.1.443), which avoids the use of boundary layer potentials.

The fourth remark we wish to make pertains to the Neumann Problem in exterior domains whose formulation involves other types of behavior at infinity. Specifically, assuming that $n = 2$ and that $\Omega \subseteq \mathbb{R}^2$ is an exterior domain, we are interested in the following brand of Neumann Problem:

$$\begin{cases} u \in \big[\mathscr{C}^\infty(\Omega)\big]^M, \quad Lu = 0 \text{ in } \Omega, \\ \mathcal{N}_\kappa(\nabla u) \in L^p(\partial\Omega,\sigma), \quad \partial_\nu^A u = f \in \big[L^p(\partial\Omega,\sigma)\big]^M, \\ u(x) = E(x)c + O(1) \text{ as } |x| \to \infty, \end{cases} \tag{8.1.471}$$

where $c \in \mathbb{C}^M$ is a given constant, and E is the matrix-valued fundamental solution associated with L as in [131, Theorem 1.4.2]. This contains the version of (8.1.336) (with $n = 2$ and Ω exterior domain) as a special case, corresponding to $c := 0 \in \mathbb{C}^M$. Then, with the same type of justification as in the case of Remark 8.1.13, we have:

$$\begin{aligned} &\text{if } A \in \mathfrak{A}_L^{\mathrm{dis}} \text{ is positive definite then (8.1.471) is solvable if and only if} \\ &f \in \big[L^p(\partial\Omega,\sigma) \cap (\mathbb{C}_{\partial\Omega_+})^\perp\big]^M \text{ and } c = -\Big(\big\langle f, \mathbf{e}_j \mathbf{1}_{\partial\Omega_\infty}\big\rangle\Big)_{1\le j\le M}. \end{aligned} \tag{8.1.472}$$

Finally, similar considerations are valid for the Neumann Problem (8.1.464).

Let us now present the proof of Theorem 8.1.15.

Proof of Theorem 8.1.15 Take $\delta_0 \in (0,1)$ sufficiently small, depending on n, p, A, the ADR constants of $\partial\Omega$, and the relative flatness ratio $\wp(\Omega)/\operatorname{diam}\partial\Omega$, such that the conclusions in Theorem 7.1.1, Proposition 7.1.2, Proposition 7.1.24, and Proposition 7.1.41 are all valid (for suitable choices of the parameters involved in their statements). By further decreasing the value of $\delta_0 \in (0,1)$, if necessary, Theorem 3.4.2 applies and gives that Ω is both an (ε,δ)-domain as well as a two-sided NTA domain, hence also a UR domain (cf. [129, (5.10.33)]).

For the time being, work under the assumptions made in (8.1.445). Theorem 1.5.5 ensures that $A^\top \in \mathfrak{A}^{\rm dis}_{L^\top}$. With this in hand, item *(1)* in Theorem 7.1.1 used with $z := -1/2$ guarantees that

$$
\begin{gathered}
-\tfrac{1}{2}I + K^{\#}_{A^\top} : \big[L^p(\partial\Omega,\sigma)\big]^M \longrightarrow \big[L^p(\partial\Omega,\sigma)\big]^M \\
\text{is a Fredholm operator with index zero.}
\end{gathered}
\tag{8.1.473}
$$

As a consequence,

$$
\begin{gathered}
\operatorname{Im}\Big(-\tfrac{1}{2}I + K^{\#}_{A^\top};\ \big[L^p(\partial\Omega,\sigma)\big]^M\Big) \text{ is a closed space} \\
\text{of finite codimension in the space } \big[L^p(\partial\Omega,\sigma)\big]^M.
\end{gathered}
\tag{8.1.474}
$$

Bring in the boundary-to-domain single layer potential operator $\mathscr{S}$ associated with L and Ω as in (A.0.228). Then, as seen from [132, (1.3.8), (1.5.63)] and [132, Lemma 1.3.1], if either $n \geq 3$ or $n = 2$ and Ω is bounded it follows that for each $g \in \big[L^p(\partial\Omega,\sigma)\big]^M$ the function $u := \mathscr{S}g$ in Ω satisfies

$$
\begin{gathered}
u \in \big[\mathscr{C}^\infty(\Omega)\big]^M,\ \ Lu = 0 \text{ in } \Omega,\ \ \mathcal{N}_\kappa(\nabla u) \in L^p(\partial\Omega,\sigma), \\
\text{and } u(x) = O(|x|^{2-n}) \text{ as } |x| \to \infty \text{ if } \Omega \text{ is unbounded.}
\end{gathered}
\tag{8.1.475}
$$

In fact, the same properties are true when $n = 2$ and Ω is an exterior domain provided $\int_{\partial\Omega} g\,d\sigma = 0$ (see [132, Lemma 1.3.1, item *(iii)*]). From (8.1.475) and [132, (1.5.59)] we then conclude that

$$
\begin{gathered}
\big(-\tfrac{1}{2}I + K^{\#}_{A^\top}\big)g = \partial^A_\nu u \text{ belongs to the space of admissible boundary} \\
\text{data for the boundary problem (8.1.443), defined in (8.1.446), for each} \\
g \in \big[L^p(\partial\Omega,\sigma)\big]^M \text{ if either } n \geq 3 \text{ or } n = 2 \text{ and } \Omega \text{ is bounded, plus} \\
\text{also } \textstyle\int_{\partial\Omega} g\,d\sigma = 0 \text{ if } n = 2 \text{ and } \Omega \text{ is an exterior domain.}
\end{gathered}
\tag{8.1.476}
$$

Based on (8.1.476) we then conclude that

$$
\begin{gathered}
\text{the space of admissible boundary data for the Neumann Problem} \\
\text{(8.1.443), defined as indicated in (8.1.446), contains the image} \\
\operatorname{Im}\Big(-\tfrac{1}{2}I + K^{\#}_{A^\top};\ \big[L^p(\partial\Omega,\sigma)\big]^M\Big) \text{ if either } n \geq 3 \text{ or } n = 2 \text{ and the} \\
\text{set } \Omega \text{ is bounded, while (8.1.446) contains the space } \Big\{\big(-\tfrac{1}{2}I + \\
K^{\#}_{A^\top}\big)g : g \in \big[L^p(\partial\Omega,\sigma)\big]^M \text{ with } \textstyle\int_{\partial\Omega} g\,d\sigma = 0\Big\} \text{ if } n = 2 \text{ and} \\
\Omega \text{ is an exterior domain.}
\end{gathered}
\tag{8.1.477}
$$

Collectively, (8.1.477), (8.1.474), and [130, Theorem 1.1.3] then imply that

the space of admissible boundary data for the boundary problem (8.1.443), described in (8.1.446), is a closed linear subspace of finite codimension in $\big[L^p(\partial\Omega,\sigma)\big]^M$. (8.1.478)

By design, a solution u of (8.1.443) exists if and only if the boundary datum f belongs to (8.1.446). Let us show that, whenever this is the case, a solution u of (8.1.443) may be found such that (8.1.447) holds. Consider first the case when either $n \geq 3$, or $n = 2$ and Ω is bounded. In this setting define the operator

$$\begin{aligned} T : \big[L^p_1(\partial\Omega,\sigma)\big]^M \times \big[L^p(\partial\Omega,\sigma)\big]^M &\longrightarrow \big[L^p(\partial\Omega,\sigma)\big]^M, \\ T(f_0, f_1) := \partial^A_\nu \mathcal{D}_A f_0 - \big(-\tfrac{1}{2}I + K^\#_{A^\top}\big) f_1 &\text{ for all} \\ (f_0, f_1) \in \big[L^p_1(\partial\Omega,\sigma)\big]^M \times \big[L^p(\partial\Omega,\sigma)\big]^M&. \end{aligned} \tag{8.1.479}$$

In the above context, T is a well-defined, linear, and bounded operator. Since its image contains the space $\mathrm{Im}\Big(-\tfrac{1}{2}I + K^\#_{A^\top};\ \big[L^p(\partial\Omega,\sigma)\big]^M\Big)$, we conclude from (8.1.474) and [130, Theorem 1.1.3, item *(3)*] that

the image of T is a closed subspace of finite codimension in the space $\big[L^p(\partial\Omega,\sigma)\big]^M$. (8.1.480)

In relation to this, we claim that

the image of T is precisely the space of admissible boundary data for the Neumann Problem (8.1.443) when either $n \geq 3$, or $n = 2$ and the set Ω happens to be bounded. (8.1.481)

In one direction, assume

$$\begin{aligned} &u \in \big[\mathscr{C}^\infty(\Omega)\big]^M,\ Lu = 0 \text{ in } \Omega,\ \mathcal{N}_\kappa(\nabla u) \in L^p(\partial\Omega,\sigma), \\ &\text{and } u(x) = O(|x|^{2-n}) \text{ as } |x| \to \infty \text{ if } \Omega \text{ is unbounded.} \end{aligned} \tag{8.1.482}$$

Then, as noted in (8.1.468), the nontangential boundary trace $u\big|^{\kappa-\mathrm{n.t.}}_{\partial\Omega}$ also exists at σ-a.e. point on $\partial\Omega$ and, as a function, belongs to $\big[L^p_1(\partial\Omega,\sigma)\big]^M$. Furthermore, [129, Proposition 8.4.9] presently implies that $\mathcal{N}_\kappa u \in L^p(\partial\Omega,\sigma)$. Granted these properties, [132, Theorem 1.3.3] gives $u = \mathcal{D}_A\big(u\big|^{\kappa-\mathrm{n.t.}}_{\partial\Omega}\big) - \mathscr{S}\big(\partial^A_\nu u\big)$ in Ω. Taking the conormal derivative of both sides of this integral representation formula then gives, on account of [132, (1.5.30)-(1.5.31), (1.5.66)],

$$\partial^A_\nu u = \partial^A_\nu \mathcal{D}_A\big(u\big|^{\kappa-\mathrm{n.t.}}_{\partial\Omega}\big) - \big(-\tfrac{1}{2}I + K^\#_{A^\top}\big)\big(\partial^A_\nu u\big) \text{ in } \big[L^p(\partial\Omega,\sigma)\big]^M. \tag{8.1.483}$$

Hence, $\partial^A_\nu u = T\big(u\big|^{\kappa-\mathrm{n.t.}}_{\partial\Omega}, \partial^A_\nu u\big)$, proving that the space of admissible boundary data for the Neumann Problem (8.1.443) is contained in the image of T from (8.1.479).

In the opposite direction, pick an arbitrary function in the image of the operator T, say, $T(f_0, f_1)$ for some $f_0 \in \left[L^p_1(\partial\Omega, \sigma)\right]^M$ and $f_1 \in \left[L^p(\partial\Omega, \sigma)\right]^M$. Then the function defined in Ω as $u := \mathcal{D} f_0 - \mathcal{S} f_1$ is as in (8.1.482) and satisfies $\partial^A_\nu u = T(f_0, f_1)$. This shows that $T(f_0, f_1)$ is an admissible boundary datum for the Neumann Problem (8.1.443). The proof of (8.1.481) is therefore complete.

Having established (8.1.481), the fact that in the case when either $n \geq 3$, or $n = 2$ and Ω is bounded, a solution u of (8.1.443) exists if and only if the boundary datum f belongs to (8.1.446), in which scenario a solution u may be found such that the estimate in (8.1.447) is true, becomes a consequence of (8.1.480) and [130, Proposition 1.2.13]. This finishes the analysis in the case when either $n \geq 3$, or $n = 2$ and Ω is bounded.

The next goal is to find the analogue of (8.1.481) when $n = 2$ and Ω is an exterior domain. In such a case, we define the operator $\widetilde{T}$ mapping the space

$$\left[L^p_1(\partial\Omega, \sigma)\right]^M \times \left\{f_1 \in \left[L^p(\partial\Omega, \sigma)\right]^M : \left(\int_{S^1} \left[L(\xi)\right]^{-1} d\mathcal{H}^1(\xi)\right)\left(\int_{\partial\Omega} f_1 \, d\sigma\right) = 0\right\} \tag{8.1.484}$$

into the space $\left[L^p(\partial\Omega, \sigma)\right]^M$ according to

$$\begin{aligned} &\widetilde{T}(f_0, f_1) := \partial^A_\nu \mathcal{D}_A f_0 - \left(-\tfrac{1}{2}I + K^\#_{A^\top}\right) f_1 \\ &\text{for each } (f_0, f_1) \text{ in the space (8.1.484).} \end{aligned} \tag{8.1.485}$$

The claim that we make in this regard is that

$$\begin{aligned} &\text{the image of } \widetilde{T} \text{ is precisely the space of admissible bound-} \\ &\text{ary data for the Neumann Problem (8.1.443) in the case} \\ &\text{when } n = 2 \text{ and the set } \Omega \text{ is an exterior domain.} \end{aligned} \tag{8.1.486}$$

To establish one inclusion, assume

$$\begin{aligned} &u \in \left[\mathscr{C}^\infty(\Omega)\right]^M, \ Lu = 0 \text{ in } \Omega, \ \mathcal{N}_\kappa(\nabla u) \in L^p(\partial\Omega, \sigma), \\ &\text{and } u(x) = O(1) \text{ as } |x| \to \infty. \end{aligned} \tag{8.1.487}$$

Then, as noted in (8.1.468), the nontangential boundary trace $u\big|^{\kappa-\text{n.t.}}_{\partial\Omega}$ exists at σ-a.e. point on $\partial\Omega$ and, as a function, it belongs to $\left[L^p_1(\partial\Omega, \sigma)\right]^M$. Also, from [129, Proposition 8.4.9], [130, (5.1.12)], [130, (5.2.92)], and the fact that u is bounded at infinity we also see (keeping in mind that $\partial\Omega$ is compact) that $\mathcal{N}_\kappa u \in L^p(\partial\Omega, \sigma)$. As such, [131, Corollary 1.5.4] applies and gives that there exists some constant $c \in \mathbb{C}^M$ such that the following integral representation formula holds:

$$u = \mathcal{D}_A\left(u\big|^{\kappa-\text{n.t.}}_{\partial\Omega}\right) - \mathcal{S}\left(\partial^A_\nu u\right) + c \text{ in } \Omega. \tag{8.1.488}$$

After taking the conormal derivative of both sides of (8.1.488) we arrive, on account of [132, (1.5.30)-(1.5.31), (1.5.66)], at

$$\partial_\nu^A u = \partial_\nu^A \mathcal{D}_A\big(u\big|_{\partial\Omega}^{\kappa-\mathrm{n.t.}}\big) - \big(-\tfrac{1}{2}I + K_{A^\top}^{\#}\big)(\partial_\nu^A u) \text{ in } \big[L^p(\partial\Omega,\sigma)\big]^M. \tag{8.1.489}$$

We next claim that

$$\Big(\int_{S^1} \big[L(\xi)\big]^{-1}\, d\mathcal{H}^1(\xi)\Big)\Big(\int_{\partial\Omega} \partial_\nu^A u \, d\sigma\Big) = 0. \tag{8.1.490}$$

Indeed, from (8.1.488), the last property in (8.1.487), and the fact that (as seen from [132, (1.3.26)] and (8.1.468)) $\mathcal{D}_A\big(u\big|_{\partial\Omega}^{\kappa-\mathrm{n.t.}}\big)(x) = O(|x|^{-1})$ as $|x| \to \infty$, we conclude that $\mathscr{S}(\partial_\nu^A u) = O(1)$ as $|x| \to \infty$. With this in hand, [132, (1.3.12)] shows that we necessarily have (8.1.490). Next, (8.1.489), (8.1.490), and the definition of the operator $\widetilde{T}$ in (8.1.484)-(8.1.485) give $\partial_\nu^A u = \widetilde{T}\big(u\big|_{\partial\Omega}^{\kappa-\mathrm{n.t.}}, \partial_\nu^A u\big) \in \operatorname{Im}\widetilde{T}$, the image of the operator $\widetilde{T}$. This proves one inclusion in (8.1.486). To establish the opposite inclusion in (8.1.486), start with a function in the image of $\widetilde{T}$, say, $\widetilde{T}(f_0, f_1)$ with (f_0, f_1) belonging to the space (8.1.484). Then the function $u := \mathcal{D}_A f_0 - \mathscr{S} f_1$ in Ω is as in (8.1.487) (here [132, (1.3.11)] is relevant) and satisfies $\partial_\nu^A u = \widetilde{T}(f_0, f_1)$. Hence, $\widetilde{T}(f_0, f_1)$ is an admissible boundary datum for the Neumann Problem (8.1.443).

This completes the proof of (8.1.486). Once this is available, the same functional analytic argument (based on [130, Proposition 1.2.13]) then shows that a solution u of (8.1.443) exists if and only if the boundary datum f belongs to (8.1.446), in which scenario a solution u may be found such that the estimate in (8.1.447) is true in this final case as well.

Moving on, consider an arbitrary null-solution u of the boundary problem (8.1.443) (i.e., a function u belonging to the space defined in (8.1.448)). As noted in (8.1.468), the nontangential boundary trace $u\big|_{\partial\Omega}^{\kappa-\mathrm{n.t.}}$ exists at σ-a.e. point on $\partial\Omega$ and, as a function, belongs to $\big[L_1^p(\partial\Omega,\sigma)\big]^M$. Furthermore, [129, Proposition 8.4.9] presently implies that $\mathcal{N}_\kappa u \in L^p(\partial\Omega,\sigma)$. Granted these properties, [132, Theorem 1.3.3] gives (bearing in mind that we currently have $\partial_\nu^A u = 0$)

$$u = \mathcal{D}_A\big(u\big|_{\partial\Omega}^{\kappa-\mathrm{n.t.}}\big) \text{ in } \Omega \tag{8.1.491}$$

provided either $n \geq 3$ or $n = 2$ and Ω is bounded. Assume this is the case. After going nontangentially to the boundary in (8.1.491) and using the jump-formula [132, (1.5.20)], we arrive at the conclusion that $\big(-\tfrac{1}{2}I + K_A\big)\big(u\big|_{\partial\Omega}^{\kappa-\mathrm{n.t.}}\big) = 0$. Thus,

$$u\big|_{\partial\Omega}^{\kappa-\mathrm{n.t.}} \in \operatorname{Ker}\Big(-\tfrac{1}{2}I + K_A;\ \big[L_1^p(\partial\Omega,\sigma)\big]^M\Big). \tag{8.1.492}$$

Next, from (8.1.491) we see that the assignment taking u into $u\big|_{\partial\Omega}^{\kappa-\mathrm{n.t.}}$ is linear and injective. Based on this and (8.1.492) we then deduce that if either $n \geq 3$ or $n = 2$ and Ω is bounded we have

$$\text{the dimension of (8.1.448) is } \leq \dim \operatorname{Ker}\Big(-\tfrac{1}{2}I + K_A;\ \big[L_1^p(\partial\Omega,\sigma)\big]^M\Big). \tag{8.1.493}$$

In view of (7.1.62), we conclude from (8.1.493) that the space of null-solutions of the boundary problem (8.1.443), described in (8.1.448), is finite dimensional, if either $n \geq 3$ or $n = 2$ and Ω.

Finally, in the case when $n = 2$ and Ω is an exterior domain we proceed analogously, starting from the observation that the integral representation formula (8.1.488) presently reduces to

$$u = \mathcal{D}_A\big(u\big|_{\partial\Omega}^{\kappa-\mathrm{n.t.}}\big) + c \text{ in } \Omega. \tag{8.1.494}$$

After going to the boundary, (8.1.494) yields, on account of [132, (1.5.20)],

$$\big(-\tfrac{1}{2}I + K_A\big)\big(u\big|_{\partial\Omega}^{\kappa-\mathrm{n.t.}}\big) \in \mathbb{C}^M, \tag{8.1.495}$$

from which we further conclude that

$$u\big|_{\partial\Omega}^{\kappa-\mathrm{n.t.}} \text{ belongs to } \Big\{\phi \in \big[L_1^p(\partial\Omega,\sigma)\big]^M : \big(-\tfrac{1}{2}I + K_A\big)\phi \in \mathbb{C}^M\Big\}. \tag{8.1.496}$$

Since taking the threshold $\delta \in (0, 1)$ small enough in (8.1.444) ensures that $-\frac{1}{2}I + K_A$ is a Fredholm operator on $\big[L_1^p(\partial\Omega,\sigma)\big]^M$, from item *(g)* in [130, Theorem 2.2.4] we see that the space appearing above is finite dimensional. Bearing this in mind, from (8.1.496) and (8.1.494) we then conclude that the space of null-solutions of the boundary problem (8.1.443) is finite dimensional in this case as well. This finishes the treatment of item *(1)*.

Let us now turn our attention to item *(2)*. Henceforth, work under the stronger hypotheses demanded in (8.1.449). In particular, Proposition 1.4.1 guarantees that, having chosen some coefficient tensor $A \in \mathfrak{A}_L^{\mathrm{dis}}$, we also have $A^\top \in \mathfrak{A}_{L^\top}^{\mathrm{dis}}$.

First, we treat the case when either $n \geq 3$, or $n = 2$ and Ω is bounded. Pick a function f in the space of admissible boundary data for the boundary problem (8.1.443) (defined in (8.1.446)). Then there exists u as in (8.1.475) such that $f = \partial_\nu^A u$. As noted earlier, the properties in (8.1.443) further imply that the nontangential boundary trace $u\big|_{\partial\Omega}^{\kappa-\mathrm{n.t.}}$ is well defined as a function in the space $\big[L_1^p(\partial\Omega,\sigma)\big]^M$, and $\mathcal{N}_\kappa u \in L^p(\partial\Omega,\sigma)$. With these in hand, the integral representation formula from [132, Theorem 1.3.3] gives

$$u = \mathcal{D}_A\big(u\big|_{\partial\Omega}^{\kappa-\mathrm{n.t.}}\big) - \mathcal{S}f \text{ in } \Omega. \tag{8.1.497}$$

Applying the conormal derivative of both sides of (8.1.497) then yields (cf. [132, (1.5.59)]) $f = \partial_\nu^A \mathcal{D}_A\big(u\big|_{\partial\Omega}^{\kappa-\mathrm{n.t.}}\big) - \big(-\frac{1}{2}I + K_{A^\top}^\#\big)f$. From this and (7.1.440) we then conclude that

$$f \in \mathrm{Im}\Big(-\tfrac{1}{2}I + K_{A^\top}^\#;\ \big[L^p(\partial\Omega,\sigma)\big]^M\Big). \tag{8.1.498}$$

Together, (8.1.477) and (8.1.498) prove the portion of (8.1.452) concerning the case when either $n \geq 3$, or $n = 2$ and Ω is bounded.

Next, whenever f belongs to the space of admissible boundary data defined in (8.1.446), it follows from (8.1.452) that (8.1.498) holds. Based on this, (8.1.474), and [130, Proposition 1.2.13] we see that there exists $g \in \big[L^p(\partial\Omega,\sigma)\big]^M$ such that

$$\big(-\tfrac{1}{2}I+K^{\#}_{A^{\top}}\big)g=f \text{ and } \|g\|_{[L^p(\partial\Omega,\sigma)]^M}\leq C\|f\|_{[L^p(\partial\Omega,\sigma)]^M} \tag{8.1.499}$$

for some $C\in(0,\infty)$ independent of f. Then, as before, $u:=\mathscr{S}g$ satisfies (8.1.475), as well as (8.1.447), thanks to the estimate in (8.1.499). Finally, the equality in (8.1.499) together with [132, (1.5.59)] give $\partial^A_\nu u=\big(-\frac{1}{2}I+K^{\#}_{A^{\top}}\big)g=f$, implying that the function u is a solution of the boundary problem (8.1.443).

Going further, from (8.1.474), [130, (2.1.49)], (A.0.7), and [132, Theorem 1.5.1, item *(iii)*] (with the coefficient tensor $A^{\top}$ used in place of A) we conclude that the space $\mathrm{Im}\Big(-\frac{1}{2}I+K^{\#}_{A^{\top}};\ \big[L^p(\partial\Omega,\sigma)\big]^M\Big)$ may be described as the collection of all functions $f\in\big[L^p(\partial\Omega,\sigma)\big]^M$ satisfying (8.1.450). From this, and the portion of (8.1.452) corresponding to the case when either $n\geq 3$, or when $n=2$ and the set Ω happens to be bounded, we then see that the Neumann Problem (8.1.443) has a solution (again, when either $n\geq 3$ or $n=2$ and Ω is bounded) if and only if the boundary datum $f\in\big[L^p(\partial\Omega,\sigma)\big]^M$ satisfies (8.1.450).

Let us now deal with the claim made in (8.1.454) when either $n\geq 3$, or when $n=2$ and the set Ω is actually bounded. As seen from (8.1.491)-(8.1.492), the assignment associating to each function u from the space defined in (8.1.448) its nontangential trace $u\big|^{\kappa-\mathrm{n.t.}}_{\partial\Omega}\in\mathrm{Ker}\Big(-\frac{1}{2}I+K_A;\ \big[L^p_1(\partial\Omega,\sigma)\big]^M\Big)$ is well defined, linear, and injective. To see that this assignment is also surjective, pick an arbitrary function

$$\phi\in\mathrm{Ker}\Big(-\tfrac{1}{2}I+K_A;\ \big[L^p_1(\partial\Omega,\sigma)\big]^M\Big) \tag{8.1.500}$$

and define $u:=\mathcal{D}_A\phi$ in Ω. Then [132, (1.3.24), (1.3.26), (1.5.8)] imply

$$\begin{aligned}&u\in\big[\mathscr{C}^\infty(\Omega)\big]^M,\ Lu=0 \text{ in } \Omega,\ \mathcal{N}_\kappa u,\ \mathcal{N}_\kappa(\nabla u)\in L^p(\partial\Omega,\sigma),\\ &\quad\text{and } u(x)=O(|x|^{1-n}) \text{ as } |x|\to\infty \text{ if } \Omega \text{ is unbounded.}\end{aligned} \tag{8.1.501}$$

Also, thanks to (8.1.500) and (7.1.441) we have $\partial^A_\nu u=\partial^A_\nu\mathcal{D}_A\phi=0$. Moreover, [132, (1.5.20)] gives $u\big|^{\kappa-\mathrm{n.t.}}_{\partial\Omega}=\big(\frac{1}{2}I+K_A\big)\phi=\phi$, so the claim we seek to prove follows.

In summary, the above reasoning shows that the assignment taking u into $u\big|^{\kappa-\mathrm{n.t.}}_{\partial\Omega}$ is a linear isomorphism from the space (8.1.448) of null-solutions for the boundary problem (8.1.443) into the space $\mathrm{Ker}\Big(-\frac{1}{2}I+K_A;\ \big[L^p_1(\partial\Omega,\sigma)\big]^M\Big)$. This proves (8.1.454). In addition, from this we also conclude that:

$$\begin{aligned}&\text{the dimension of the space (8.1.448) of null-solutions for the boundary}\\ &\text{problem (8.1.443) is precisely } \dim\mathrm{Ker}\Big(-\tfrac{1}{2}I+K_A;\ \big[L^p_1(\partial\Omega,\sigma)\big]^M\Big)\\ &\text{in the case when either } n\geq 3, \text{ or } n=2 \text{ and the set } \Omega \text{ is bounded.}\end{aligned} \tag{8.1.502}$$

Finally, (8.1.491) justifies the claim made in (8.1.455) which, in turn, gives (8.1.453).

Retaining the assumption that either $n\geq 3$, or $n=2$ and Ω is bounded, let us now prove that the index of the boundary value problem (8.1.443), i.e., the dimension of (8.1.448) minus the codimension of (8.1.446) in $\big[L^p(\partial\Omega,\sigma)\big]^M$, is equal to zero. With this goal in mind, observe that

$$\dim\left(\frac{\left[L^p(\partial\Omega,\sigma)\right]^M}{\operatorname{Im}(-\frac{1}{2}I+K^{\#}_{A^\top};[L^p(\partial\Omega,\sigma)]^M)}\right)=\dim\operatorname{Ker}\Big(-\tfrac{1}{2}I+K^{\#}_{A^\top};\ \left[L^p(\partial\Omega,\sigma)\right]^M\Big)$$
$$=\dim\operatorname{Ker}\Big(-\tfrac{1}{2}I+K_A;\ \left[L^p_1(\partial\Omega,\sigma)\right]^M\Big)<+\infty. \tag{8.1.503}$$

Above, the first equality is implied by (8.1.473), the second equality comes from (7.1.622), and the final inequality follows from (8.1.473). From (8.1.503), (8.1.454), and (8.1.452) we then see that the index of the boundary value problem (8.1.443) is indeed zero when either $n \geq 3$, or $n = 2$ and Ω is bounded.

To complete the treatment of item *(2)*, there remains to consider the case when $n = 2$ and Ω is an exterior domain. Suppose f belongs to the space of admissible boundary data for the Neumann Problem (8.1.443) (cf. (8.1.446)). Then there exists u as in (8.1.475) with

$$f=\partial^A_\nu u. \tag{8.1.504}$$

This time, the fact that $u(x)=O(1)$ as $|x|\to\infty$ entails that

$$c_0:=\lim_{|x|\to\infty}u(x)\text{ exists (in }\mathbb{C}^M\text{) and }u(x)=c_0+O(|x|^{-1})\text{ as }|x|\to\infty \tag{8.1.505}$$

(see [131, Theorem 1.5.9] and (8.1.449)). As remarked earlier, the properties in (8.1.443) further imply that the nontangential boundary trace $u\big|^{\kappa-\text{n.t.}}_{\partial\Omega}$ is well defined as a function in $\left[L^p_1(\partial\Omega,\sigma)\right]^M$, and we have $\mathcal{N}_\kappa u\in L^p(\partial\Omega,\sigma)$. With these in hand, the integral representation formula from [132, Theorem 1.3.3] applies to the function $\widetilde{u}:=u-c_0$ so, bearing in mind that $\partial^A_\nu\widetilde{u}=\partial^A_\nu u=f$ and [132, (1.3.45)],

$$u-c_0=\mathcal{D}_A\big(u\big|^{\kappa-\text{n.t.}}_{\partial\Omega}\big)-\mathcal{S}f\ \text{ in }\Omega. \tag{8.1.506}$$

Taking conormal derivatives in (8.1.506) and using [132, (1.5.59)] then leads to $f=(\partial^A_\nu\mathcal{D}_A)\big(u\big|^{\kappa-\text{n.t.}}_{\partial\Omega}\big)-\big(-\tfrac{1}{2}I+K^{\#}_{A^\top}\big)f$. In concert with (7.1.440), this proves

$$f\in\operatorname{Im}\Big(-\tfrac{1}{2}I+K^{\#}_{A^\top};\ \left[L^p(\partial\Omega,\sigma)\right]^M\Big) \tag{8.1.507}$$

hence $f=\big(-\tfrac{1}{2}I+K^{\#}_{A^\top}\big)g$ for some $g\in\left[L^p(\partial\Omega,\sigma)\right]^M$. We claim that

$$\int_{\partial\Omega}g\,d\sigma=0. \tag{8.1.508}$$

To justify (8.1.508), first observe that since we are assuming $u(x)=O(1)$ as $|x|\to\infty$, from [131, (1.5.212)] we conclude that

$$(\nabla u)(x)=O\big(|x|^{-2}\big)\ \text{ as }|x|\to\infty. \tag{8.1.509}$$

To proceed, let us explicitly write the given coefficient tensor A as $\big(a^{\alpha\beta}_{jk}\big)_{\substack{1\leq j,k\leq n\\ 1\leq\alpha,\beta\leq M}}$ and, for each index $\alpha\in\{1,\dots,M\}$, define the vector field

$$\vec{F}^{\alpha} := \left(a_{jk}^{\alpha\beta}\partial_k u_\beta\right)_{1\leq j\leq 2} \quad \text{in } \Omega. \tag{8.1.510}$$

Then each vector field $\vec{F}^{\alpha}$ is divergence-free in Ω, and from [129, (1.3.5)], (8.1.510), (8.1.509) we see that there exist $C \in (0,\infty)$ and $\lambda \in (1,\infty)$ such that

$$\big|[\vec{F}^{\alpha}]_\infty\big| \leq C \lim_{R\to\infty}\left\{R\cdot \fint_{B(0,\lambda R)\setminus B(0,R)} |\nabla u|\,\mathrm{d}\mathcal{L}^2\right\} = 0. \tag{8.1.511}$$

Consequently, the Divergence Formula [129, (1.3.8)] presently gives (bearing in mind (8.1.466))

$$\int_{\partial\Omega} \partial_\nu^A u\,\mathrm{d}\sigma = -\left([\vec{F}^{\alpha}]_\infty\right)_{1\leq\alpha\leq M} = 0. \tag{8.1.512}$$

Then for each $c \in \mathbb{C}^M$ we may write, based on the duality results from [132, Theorem 1.5.1, item *(iii)*] together with formula [132, (1.5.21)] (written for $A^\top$ in place of A, and bearing in mind that Ω is presently assumed to be unbounded), (8.1.504), and the cancellation property (8.1.512),

$$\begin{aligned}\int_{\partial\Omega}\langle g, c\rangle\,\mathrm{d}\sigma &= \int_{\partial\Omega}\big\langle\big(\tfrac12 I + K^{\#}_{A^\top}\big)g, c\big\rangle\,\mathrm{d}\sigma - \int_{\partial\Omega}\big\langle\big(-\tfrac12 I + K^{\#}_{A^\top}\big)g, c\big\rangle\,\mathrm{d}\sigma\\ &= \int_{\partial\Omega}\big\langle g, \big(\tfrac12 I + K_{A^\top}\big)c\big\rangle\,\mathrm{d}\sigma - \int_{\partial\Omega}\big\langle f, c\big\rangle\,\mathrm{d}\sigma\\ &= \int_{\partial\Omega}\langle g, 0\rangle\,\mathrm{d}\sigma - \int_{\partial\Omega}\big\langle \partial_\nu^A u, c\big\rangle\,\mathrm{d}\sigma = 0 - 0 = 0.\end{aligned} \tag{8.1.513}$$

In view of the arbitrariness of $c \in \mathbb{C}^M$, we then conclude that (8.1.508) holds. At this point, (8.1.477), (8.1.507), and (8.1.508) prove the portion of (8.1.452) concerning the case when $n = 2$ and Ω is an exterior domain.

Let us prove that, in the case when $n = 2$ and Ω is an exterior domain, the Neumann Problem (8.1.443) has a solution if and only if the boundary datum $f \in \big[L^p(\partial\Omega,\sigma)\big]^M$ satisfies (8.1.451). In one direction, if the Neumann Problem (8.1.443) has a solution for a given boundary datum $f \in \big[L^p(\partial\Omega,\sigma)\big]^M$ then (8.1.452) implies that $f = \big(-\tfrac12 I + K^{\#}_{A^\top}\big)h$ for some $h \in \big[L^p(\partial\Omega,\sigma)\big]^M$ satisfying $\int_{\partial\Omega} h\,\mathrm{d}\sigma = 0$. As such, for any function $g \in \big[L^{p'}(\partial\Omega,\sigma)\big]^M$ with the property that $\big(-\tfrac12 I + K_{A^\top}\big)g \in \mathbb{C}^M$ we may use the duality results from [132, Theorem 1.5.1, *(iii)*] (with $A^\top$ in place of A) to write

$$\int_{\partial\Omega}\langle f, g\rangle\,\mathrm{d}\sigma = \int_{\partial\Omega}\big\langle\big(-\tfrac12 I + K^{\#}_{A^\top}\big)h, g\big\rangle\,\mathrm{d}\sigma = \int_{\partial\Omega}\big\langle h, \big(-\tfrac12 I + K_{A^\top}\big)g\big\rangle\,\mathrm{d}\sigma = 0,$$

as wanted. In the opposite direction, if $f \in \big[L^p(\partial\Omega,\sigma)\big]^M$ satisfies (8.1.451), then (A.0.159), [132, (1.7.15)], (both used with $A^\top$ in place of A), and (8.1.474) imply $f \in L^p_{A^\top,-1/2}(\partial\Omega,\sigma) = \mathrm{Im}\Big(-\tfrac12 I + K^{\#}_{A^\top}; \big[L^p(\partial\Omega,\sigma)\big]^M\Big)$. Thus $f = \big(-\tfrac12 I + K^{\#}_{A^\top}\big)h$

for some $h \in \big[L^p(\partial\Omega,\sigma)\big]^M$. Let us fix an arbitrary $c \in \mathbb{C}^M$. When regarded as a constant function on $\partial\Omega$, it follows that this belongs to $\big[L^{p'}(\partial\Omega,\sigma)\big]^M$ and satisfies $\big(-\frac{1}{2}I + K_{A^\top}\big)c = -c \in \mathbb{C}^M$, thanks to [132, (1.5.21)] (used with $A^\top$ in place of A) and the fact that Ω is an unbounded set. Granted this, (8.1.451) (used with $g := c$) guarantees that $\int_{\partial\Omega}\langle f, c\rangle\,d\sigma = 0$. In turn, this allows us to write

$$\begin{aligned}\int_{\partial\Omega}\langle h, c\rangle\,d\sigma &= \int_{\partial\Omega}\big\langle\big(\tfrac{1}{2}I + K^{\#}_{A^\top}\big)h, c\big\rangle\,d\sigma - \int_{\partial\Omega}\big\langle\big(-\tfrac{1}{2}I + K^{\#}_{A^\top}\big)h, c\big\rangle\,d\sigma \\ &= \int_{\partial\Omega}\big\langle\big(\tfrac{1}{2}I + K^{\#}_{A^\top}\big)h, c\big\rangle\,d\sigma - \int_{\partial\Omega}\langle f, c\rangle\,d\sigma = \int_{\partial\Omega}\big\langle\big(\tfrac{1}{2}I + K^{\#}_{A^\top}\big)h, c\big\rangle\,d\sigma \\ &= \int_{\partial\Omega}\big\langle h, \big(\tfrac{1}{2}I + K_{A^\top}\big)c\big\rangle\,d\sigma = \int_{\partial\Omega}\langle h, 0\rangle\,d\sigma = 0, \end{aligned} \tag{8.1.514}$$

thanks again to [132, (1.5.21)] (used with $A^\top$ in place of A) and the fact that Ω is an unbounded set. This shows that $\int_{\partial\Omega}\langle h, c\rangle\,d\sigma = 0$ for each $c \in \mathbb{C}^M$ or, equivalently, $\int_{\partial\Omega} h\,d\sigma = 0$. The latter property ultimately places f in the space

$$\Big\{\big(-\tfrac{1}{2}I + K^{\#}_{A^\top}\big)h : h \in \big[L^p(\partial\Omega,\sigma)\big]^M \text{ with } \int_{\partial\Omega} h\,d\sigma = 0\Big\} \tag{8.1.515}$$

which, according to (8.1.452), makes f an admissible boundary data for the Neumann Problem (8.1.443). This finishes the proof of the claim pertaining to (8.1.451).

Consider next an arbitrary null-solution u of the boundary problem (8.1.443) (i.e., a function u belonging to the space defined in (8.1.448)) in the case when $n = 2$ and Ω is an exterior domain. In particular, [131, Theorem 1.5.9] and (8.1.449) ensure the limit $c_0 := \lim_{|x|\to\infty} u(x)$ exists in $\mathbb{C}^M$ and we have $u(x) = c_0 + O(|x|^{-1})$ as well as $(\nabla u)(x) = O(|x|^{-2})$ when $|x| \to \infty$. We also know that the properties in (8.1.443) further imply that the nontangential boundary trace $u\big|^{\kappa-\text{n.t.}}_{\partial\Omega}$ is well defined as a function in $\big[L^p_1(\partial\Omega,\sigma)\big]^M$, and we have $\mathcal{N}_\kappa u \in L^p(\partial\Omega,\sigma)$. Then [132, Theorem 2.3.16] applies to the function $\widetilde{u} := u - c_0$ and gives (bearing in mind [132, (1.3.45)] and that now $\partial^A_\nu \widetilde{u} = \partial^A_\nu u = 0$)

$$u - c_0 = \mathcal{D}\big(u\big|^{\kappa-\text{n.t.}}_{\partial\Omega}\big) \text{ in } \Omega. \tag{8.1.516}$$

Going nontangentially to the boundary then yields $\big(-\frac{1}{2}I + K_A\big)\big(u\big|^{\kappa-\text{n.t.}}_{\partial\Omega}\big) = -c_0$. From this, (8.1.468), and (8.1.505) we then conclude that $\Big(u\big|^{\kappa-\text{n.t.}}_{\partial\Omega}, \lim_{|x|\to\infty} u(x)\Big)$ is a pair belonging to the space

$$\Big\{(\phi, c) \in \big[L^p_1(\partial\Omega,\sigma)\big]^M \oplus \mathbb{C}^M : \big(-\tfrac{1}{2}I + K_A\big)\phi = -c\Big\}. \tag{8.1.517}$$

Also, it is clear from (8.1.516) that the mapping associating to each null-solution u of the Neumann Problem (8.1.443) the pair $\Big(u\big|_{\partial\Omega}^{\kappa-\mathrm{n.t.}}, \lim_{|x|\to\infty} u(x)\Big)$ is injective.

Let us show that said map is also surjective, i.e., its range is the space (8.1.517). Pick $\phi \in \big[L^p_1(\partial\Omega,\sigma)\big]^M$ and $c \in \mathbb{C}^M$ satisfying $\big(-\frac{1}{2}I + K_A\big)\phi = -c$ then define

$$u := \mathcal{D}_A\phi + c \text{ in } \Omega. \tag{8.1.518}$$

Thanks to [132, (1.3.24), (1.3.26), (1.5.1), (1.8.8), (2.3.4), (2.3.100)], this satisfies

$$\begin{gathered} u \in \big[\mathscr{C}^\infty(\Omega)\big]^M,\ \ \mathcal{N}_\kappa(\nabla u) \in L^p(\partial\Omega,\sigma),\ \ \mathcal{N}_\kappa u \in L^p(\partial\Omega,\sigma), \\ Lu = 0,\ \ u(x) = c + O(|x|^{-1}) \text{ and } (\nabla u)(x) = O(|x|^{-2}) \text{ as } |x| \to \infty. \end{gathered} \tag{8.1.519}$$

In addition, [132, (1.5.20)] and the fact that $K_A\phi = \frac{1}{2}\phi - c$ ensure that

$$u\big|_{\partial\Omega}^{\kappa-\mathrm{n.t.}} = \big(\tfrac{1}{2}I + K_A\big)\phi + c = \phi. \tag{8.1.520}$$

Granted these properties, the surjectivity claim we presently seek follows as soon as we show that

$$\partial^A_\nu u = 0. \tag{8.1.521}$$

With this goal in mind, first observe that the properties in (8.1.443) further imply that the nontangential boundary trace $u\big|_{\partial\Omega}^{\kappa-\mathrm{n.t.}}$ is well defined as a function in $\big[L^p_1(\partial\Omega,\sigma)\big]^M$, and $\mathcal{N}_\kappa u \in L^p(\partial\Omega,\sigma)$. Granted these, the integral representation formula from [132, Theorem 1.3.3] applies to the function $\widetilde{u} := u - c$ and, thanks to [132, (1.3.45)] and the fact $\partial^A_\nu\widetilde{u} = \partial^A_\nu u$, gives $u - c = \mathcal{D}\big(u\big|_{\partial\Omega}^{\kappa-\mathrm{n.t.}}\big) - \mathscr{S}\big(\partial^A_\nu u\big)$ in Ω. Since (8.1.518) and (8.1.520) imply $\mathcal{D}\big(u\big|_{\partial\Omega}^{\kappa-\mathrm{n.t.}}\big) = \mathcal{D}_A\phi = u - c$, this reduces to

$$\mathscr{S}\big(\partial^A_\nu u\big) = 0 \text{ in } \Omega. \tag{8.1.522}$$

Going nontangentially to the boundary, we therefore arrive at

$$S\big(\partial^A_\nu u\big) = 0, \tag{8.1.523}$$

on account of (8.1.467), [132, (1.5.44)], and (A.0.1). Also, much as in (8.1.490), we presently have $\int_{\partial\Omega}\partial^A_\nu u\, d\sigma = 0$. Based on this, (8.1.467), (8.1.523), and (7.1.245) we then see that $S_{\mathrm{aug}}\big(\partial^A_\nu u, 0\big) = (0,0)$. Having justified this, the invertibility result proved in Theorem 7.1.10 in relation to the operator (7.1.245) then establishes (8.1.521). This takes care of (8.1.457). In turn, from (8.1.457) we conclude that

the dimension of the space (8.1.448) of null-solutions for the Neumann Problem (8.1.443) when $n = 2$ and the set Ω is unbounded is equal to the dimension of $\Big\{(\phi, c) \in \big[L^p_1(\partial\Omega,\sigma)\big]^M \oplus \mathbb{C}^M : \big(-\frac{1}{2}I + K_A\big)\phi = -c\Big\}$. (8.1.524)

Lastly, (8.1.516) implies (8.1.458) which further justifies (8.1.456).

Next we shall show that the index of the boundary value problem (8.1.443) continues to be zero in the case when $n = 2$ and Ω is an exterior domain. Consider the canonical inclusion

$$\iota : \Big\{g \in \big[L^p(\partial\Omega,\sigma)\big]^M : \int_{\partial\Omega} g\,d\sigma = 0\Big\} \to \big[L^p(\partial\Omega,\sigma)\big]^M \tag{8.1.525}$$

and abbreviate

$$\begin{gathered} T := \big(-\tfrac{1}{2}I + K^{\#}_{A^\top}\big)\circ\iota, \\ T : \Big\{g \in \big[L^p(\partial\Omega,\sigma)\big]^M : \int_{\partial\Omega} g\,d\sigma = 0\Big\} \to \big[L^p(\partial\Omega,\sigma)\big]^M. \end{gathered} \tag{8.1.526}$$

Since ι is a Fredholm operator with index $-M$ and $-\frac{1}{2}I + K^{\#}_{A^\top}$ is a Fredholm operator with index zero from $\big[L^p(\partial\Omega,\sigma)\big]^M$ into itself, we may invoke [130, Theorem 2.2.4] to conclude that,

$$T \text{ is Fredholm with index } -M \text{ in the context of (8.1.526).} \tag{8.1.527}$$

Then the codimension on the space of admissible boundary data for the Neumann Problem (8.1.443) in the "full" space of boundary data may be computed as follows:

$$\begin{aligned} \dim&\left(\frac{\big[L^p(\partial\Omega,\sigma)\big]^M}{\big\{(-\frac{1}{2}I+K^{\#}_{A^\top})g : g\in[L^p(\partial\Omega,\sigma)]^M,\ \int_{\partial\Omega} g\,d\sigma=0\big\}}\right) = \dim\left(\frac{\big[L^p(\partial\Omega,\sigma)\big]^M}{\operatorname{Im} T}\right) \\ &= M + \dim \operatorname{Ker} T = M + \dim \operatorname{Ker}\Big(-\tfrac{1}{2}I + K^{\#}_{A^\top};\ \big[L^p(\partial\Omega,\sigma)\big]^M\Big) \\ &= \dim\Big\{(\phi, c) \in \big[L^p_1(\partial\Omega,\sigma)\big]^M \oplus \mathbb{C}^M : \\ &\qquad\qquad \phi - c \in \operatorname{Ker}\Big(-\tfrac{1}{2}I + K_A;\ \big[L^p_1(\partial\Omega,\sigma)\big]^M\Big)\Big\} \\ &= \dim\Big\{(\phi, c) \in \big[L^p_1(\partial\Omega,\sigma)\big]^M \oplus \mathbb{C}^M : \big(-\tfrac{1}{2}I + K_A\big)\phi = -c\Big\}. \end{aligned} \tag{8.1.528}$$

Above, the first equality comes from (8.1.525)-(8.1.526), while the second equality is implied by (8.1.527). The third equality in (8.1.528) is a consequence of

$$\begin{aligned} \Big\{g \in \operatorname{Ker}\Big(-\tfrac{1}{2}I + K^{\#}_{A^\top};\ \big[L^p(\partial\Omega,\sigma)\big]^M\Big) : \int_{\partial\Omega} g\,d\sigma = 0\Big\} \\ = \operatorname{Ker}\Big(-\tfrac{1}{2}I + K^{\#}_{A^\top};\ \big[L^p(\partial\Omega,\sigma)\big]^M\Big). \end{aligned} \tag{8.1.529}$$

Indeed, the left-to-right inclusion in (8.1.529) is trivial, while the right-to-left inclusion in (8.1.529) follows from [132, (1.7.59)] and the fact that Ω is presently assumed to be unbounded. The fourth equality in (8.1.528) is based on (7.1.631) used with $z := -1/2$, plus (8.1.434) used with

$$V := \big[L^p_1(\partial\Omega,\sigma)\big]^M, \quad F_0 := \mathbb{C}^M, \text{ and}$$
$$F_1 := \mathrm{Ker}\Big(-\tfrac{1}{2}I + K_A;\ \big[L^p_1(\partial\Omega,\sigma)\big]^M\Big). \tag{8.1.530}$$

The final equality in (8.1.528) is a direct corollary of [132, (1.5.21)]. Having established (8.1.528), from (8.1.452) and (8.1.457) we conclude that the index of the Neumann Problem (8.1.443) is also equal to zero in this last case. This completes the treatment of item *(2)*.

Turning our attention to item *(3)*, work under the assumption that $A \in \mathfrak{A}^{\mathrm{dis}}_L$ is positive definite. In the case when either $n \geq 3$, or $n = 2$ and the set Ω is bounded, we conclude from (8.1.343) and (7.1.277) that the space of admissible boundary data for the Neumann Problem (8.1.443) is equal to

$$\mathrm{Im}\Big(-\tfrac{1}{2}I + K^{\#}_{A^\top};\ \big[L^p(\partial\Omega,\sigma)\big]^M\Big) = \big[L^p(\partial\Omega,\sigma) \cap (\mathbb{C}_{\partial\Omega_+})^{\perp}\big]^M. \tag{8.1.531}$$

This takes care of the first claim in (8.1.460). The second claim in (8.1.460) follows from (8.1.453), (7.1.275), and (7.1.305).

Assume now that $n = 2$ and Ω is an exterior domain. Then the last property in (8.1.452) combines with (7.1.292) to give that the space of admissible boundary data for the Neumann Problem (8.1.443) is, in this case, equal to

$$\Big\{\big(-\tfrac{1}{2}I + K^{\#}_A\big)g :\ g \in \big[L^p(\partial\Omega,\sigma) \cap \mathbb{C}^{\perp}\big]^M\Big\} = \big[L^p(\partial\Omega,\sigma) \cap (\mathbb{C}^{\infty}_{\partial\Omega_+})^{\perp}\big]^M.$$

The first claim in (8.1.461) is now justified. As for the second claim in (8.1.461), from (8.1.456), (7.1.293), and (8.1.439)-(8.1.441) we see that the space of null-solutions of the Neumann Problem (8.1.443) is, in this case, contained in

$$\Big\{\mathcal{D}_A\phi + c :\ \phi \in \big(\mathbb{C}^{\infty}_{\partial\Omega_+}\big)^M \text{ and } c \in \mathbb{C}^M\Big\} \subseteq \big(\mathbb{C}^{\infty}_{\Omega_+}\big)^M. \tag{8.1.532}$$

Conversely, it is clear from definitions that any function in $\big(\mathbb{C}^{\infty}_{\Omega_+}\big)^M$ is a null-solution of the Neumann Problem (8.1.443). Consequently, the space of null-solutions of the Neumann Problem (8.1.443) is, in the current case, precisely $\big(\mathbb{C}^{\infty}_{\Omega_+}\big)^M$. This finishes the treatment of item *(3)*.

Moving on, the regularity claims formulated in item *(4)* are seen by observing from (8.1.452) and (8.1.453) that any solution of the Neumann Problem (8.1.443) when Ω is bounded may be expressed as

$$u = \mathscr{S}g + \mathcal{D}_A\phi \text{ for some } g \in \big[L^p(\partial\Omega,\sigma)\big]^M,$$
$$\text{and some } \phi \in \mathrm{Ker}\Big(-\tfrac{1}{2}I + K_A;\ \big[L^p_1(\partial\Omega,\sigma)\big]^M\Big). \tag{8.1.533}$$

Granted this, (8.1.462)-(8.1.463) become consequences of the mapping properties of the boundary-to-domain single and double layer potentials from [132, Theorem 4.3.1, item *(4)*] and [132, Theorem 4.3.3, item *(4)*], respectively.

Pressing on, the claims in the current item *(5)*, pertaining to the Neumann Problem with boundary data in Muckenhoupt weighted Lebesgue spaces formulated in (8.1.464), are dealt with in an analogous fashion.

There remains to justify the claims made in item *(6)*. Assume $n = 2$, $M = 1$, and $L := \Delta$, the two-dimensional Laplacian, a scenario in which $\mathfrak{A}_L^{\rm dis} = \{I_{2\times 2}\}$. In particular, all conditions in (8.1.445) are satisfied. Write $L = \nabla \cdot A\nabla$ for the 2×2 matrix $A = (a_{rs})_{1\leq r,s\leq 2}$ given by

$$A := \begin{pmatrix} 1 & -\mathrm{i} \\ \mathrm{i} & 1 \end{pmatrix} \in \mathbb{C}^{2\times 2} \setminus \mathfrak{A}_L^{\rm dis}. \tag{8.1.534}$$

To proceed, take $\Omega := \mathbb{D} := \big\{z \in \mathbb{C} : |z| < 1\big\}$, the unit disk in the plane, and abbreviate $\sigma := \mathcal{H}^1 \lfloor \partial\mathbb{D}$. Note that the outward unit normal to $\mathbb{D}$ is given by $\nu(z) := z$ at each point $z \in \partial\mathbb{D}$. Also, fix an aperture parameter $\kappa > 0$ and an arbitrary exponent $p \in (1, \infty)$.

We claim that the space of admissible boundary data for the L^p-Neumann Problem associated with the domain $\mathbb{D}$ and the writing of the Laplacian in the form ${\rm div}A\nabla$ (with A as in (8.1.534)) may be described as follows:

$$\begin{aligned} &\Big\{\partial_\nu^A u : u \in \mathscr{C}^\infty(\mathbb{D}),\ \Delta u = 0 \text{ in } \mathbb{D},\ \mathcal{N}_\kappa(\nabla u) \in L^p(\partial\mathbb{D}, \sigma)\Big\} = z \cdot \mathcal{H}^p(\partial\mathbb{D}, \sigma) \\ &\qquad = \Big\{f \in L^p(\partial\mathbb{D}, \sigma) : \int_{\partial\mathbb{D}} f(z) z^k \, d\sigma(z) = 0 \text{ for all } k \in \mathbb{N}_0\Big\}, \end{aligned} \tag{8.1.535}$$

where $\mathcal{H}^p(\partial\mathbb{D}, \sigma)$ is the classical L^p-based boundary Hardy space associated with the Cauchy-Riemann operator, on the boundary of the unit disk (see [132, (8.1.7)]). Since the last space above has infinite codimension in $L^p(\partial\mathbb{D}, \sigma)$, the full space of boundary data, this suits our current purposes.

To justify the left-to-right inclusion in the first equality in (8.1.535), fix a complex-valued function u satisfying

$$u \in \mathscr{C}^\infty(\mathbb{D}), \quad \Delta u = 0 \text{ in } \mathbb{D}, \quad \mathcal{N}_\kappa(\nabla u) \in L^p(\partial\mathbb{D}, \sigma). \tag{8.1.536}$$

In view of the Fatou-type result established in [131, Theorem 3.3.4], these properties ensure that $(\nabla u)\big|_{\partial\mathbb{D}}^{\kappa-{\rm n.t.}}$ exists at σ-a.e. point on $\partial\mathbb{D}$. As such, from (A.0.203) we get

$$\begin{aligned} \partial_\nu^A u &= \nu_r a_{rs} \big(\partial_s u\big)\big|_{\partial\mathbb{D}}^{\kappa-{\rm n.t.}} \\ &= \nu_1 \big(\partial_1 u\big)\big|_{\partial\mathbb{D}}^{\kappa-{\rm n.t.}} - \mathrm{i}\nu_1 \big(\partial_2 u\big)\big|_{\partial\mathbb{D}}^{\kappa-{\rm n.t.}} + \mathrm{i}\nu_2 \big(\partial_1 u\big)\big|_{\partial\mathbb{D}}^{\kappa-{\rm n.t.}} + \nu_2 \big(\partial_2 u\big)\big|_{\partial\mathbb{D}}^{\kappa-{\rm n.t.}} \\ &= (\nu_1 + \mathrm{i}\nu_2)\Big((\partial_1 u)\big|_{\partial\mathbb{D}}^{\kappa-{\rm n.t.}} - \mathrm{i}(\partial_2 u)\big|_{\partial\mathbb{D}}^{\kappa-{\rm n.t.}}\Big) = 2\nu(\partial_z u)\big|_{\partial\mathbb{D}}^{\kappa-{\rm n.t.}} \text{ at } \sigma\text{-a.e. point on } \partial\mathbb{D}, \end{aligned} \tag{8.1.537}$$

where $\partial_z := \frac{1}{2}(\partial_x - \mathrm{i}\partial_y)$ is the complex conjugate of the Cauchy-Riemann operator $\partial_{\bar z} := \frac{1}{2}(\partial_x + \mathrm{i}\partial_y)$. Hence, if we define

$$w := 2\partial_z u \text{ in } \mathbb{D}, \tag{8.1.538}$$

upon recalling that $\Delta = 4\partial_{\bar{z}}\partial_z$, the properties in (8.1.536) imply

$$w \in \mathscr{C}^\infty(\mathbb{D}), \quad \partial_{\bar{z}} w = 0 \text{ in } \mathbb{D}, \quad \mathcal{N}_\kappa w \in L^p(\partial\mathbb{D}, \sigma). \tag{8.1.539}$$

These properties simply amount to stating that w is a holomorphic function belonging to the classical Hardy space $\mathcal{H}^p(\mathbb{D})$ (associated with the Cauchy-Riemann operator in the unit disk; see [132, (8.1.2)]). In addition, (8.1.537) tells us that

$$(\partial_\nu^A u)(z) = z\big(w\big|_{\partial\mathbb{D}}^{\kappa-\text{n.t.}}\big)(z) \text{ at } \sigma\text{-a.e. point } z \in \partial\mathbb{D}. \tag{8.1.540}$$

Since $\mathcal{H}^p(\partial\mathbb{D}, \sigma)$ is the space of nontangential boundary traces of functions belonging to $\mathcal{H}^p(\mathbb{D})$, we ultimately conclude from (8.1.540) that $\partial_\nu^A u$ belongs to $z \cdot \mathcal{H}^p(\partial\mathbb{D}, \sigma)$, as wanted.

In the opposite direction, any function in $z \cdot \mathcal{H}^p(\partial\mathbb{D}, \sigma)$ is of the form $z \cdot \big(U\big|_{\partial\mathbb{D}}^{\kappa-\text{n.t.}}\big)$ for some U in the Hardy space $\mathcal{H}^p(\mathbb{D})$. In turn, any function $U \in \mathcal{H}^p(\mathbb{D})$ is of the form $U = \mathscr{C} f$ for some $f \in L^p(\partial\mathbb{D}, \sigma)$ where $\mathscr{C}$ is the boundary-to-domain Cauchy operator associated with $\mathbb{D}$ (see, e.g., [132, (8.1.11)]). Based on these observations and formula [132, (1.6.45)] (written for $\Omega := \mathbb{D}$) we then conclude that any function in $z \cdot \mathcal{H}^p(\partial\mathbb{D}, \sigma)$ is of the form $z \cdot \big(U\big|_{\partial\mathbb{D}}^{\kappa-\text{n.t.}}\big)$ where $U = \partial_z \mathscr{S} f$ for some $f \in L^p(\partial\mathbb{D}, \sigma)$, with $\mathscr{S}$ denoting the boundary-to-domain harmonic single layer potential operator in the unit disk. If we now set $u := \frac{1}{2}\mathscr{S} f$, then $u \in \mathscr{C}^\infty(\mathbb{D})$ is a harmonic function with $\mathcal{N}_\kappa(\nabla u) \in L^p(\partial\mathbb{D}, \sigma)$ and, as seen from (8.1.537), $\partial_\nu^A u = z \cdot \big(U\big|_{\partial\mathbb{D}}^{\kappa-\text{n.t.}}\big)$. This proves the right-to-left inclusion in the first equality claimed in (8.1.535). The first equality in (8.1.535) is therefore fully justified. Finally, the second equality in (8.1.535) readily follows from [132, (8.1.14)].

Next we claim that the space of null-solutions for the L^p-Neumann Problem associated with the domain $\mathbb{D}$ and the writing of the Laplacian in the form $\operatorname{div} A \nabla$ (with A as in (8.1.534)) is the infinite dimensional space

$$\begin{aligned} &\Big\{u \in \mathscr{C}^\infty(\mathbb{D}) : \Delta u = 0 \text{ in } \mathbb{D},\ \mathcal{N}_\kappa(\nabla u) \in L^p(\partial\mathbb{D}, \sigma),\ \partial_\nu^A u = 0\Big\} \\ &\qquad = \big\{\overline{w} : w \in \mathcal{H}_1^p(\mathbb{D})\big\}, \end{aligned} \tag{8.1.541}$$

where $\mathcal{H}_1^p(\mathbb{D})$ is the Hardy space with regularity in the unit disk (cf. [132, (8.1.20)]). Indeed, given any $w \in \mathcal{H}_1^p(\mathbb{D})$, the function $u := \overline{w}$ is harmonic in the disk $\mathbb{D}$, has $\mathcal{N}_\kappa(\nabla u) \in L^p(\partial\mathbb{D}, \sigma)$, and based on (8.1.537) we may compute

$$\partial_\nu^A u = 2\nu(\partial_z u)\big|_{\partial\mathbb{D}}^{\kappa-\text{n.t.}} = 2\nu\overline{(\partial_{\bar{z}} w)\big|_{\partial\mathbb{D}}^{\kappa-\text{n.t.}}} = 0 \text{ at } \sigma\text{-a.e. point on } \partial\mathbb{D}. \tag{8.1.542}$$

This places u in the space on the left side of (8.1.541), thus establishing the right-to-left inclusion in (8.1.541). To prove the opposite inclusion, start with a harmonic function u in $\mathbb{D}$ satisfying $\mathcal{N}_\kappa(\nabla u) \in L^p(\partial\mathbb{D}, \sigma)$ and such that $\partial_\nu^A u = 0$. In view of (8.1.537), the latter implies $(\partial_z u)\big|_{\partial\mathbb{D}}^{\kappa-\text{n.t.}} = 0$ at σ-a.e. point on $\partial\mathbb{D}$. Since the current working hypotheses also guarantee that $\partial_z u \in \mathcal{H}^p(\mathbb{D})$, Cauchy's reproducing formula (cf. [132, (8.1.11)]) permits us to write

$$\partial_z u = \mathscr{C}\Big((\partial_z u)\Big|_{\partial\mathbb{D}}^{\kappa-\text{n.t.}}\Big) = 0 \text{ in } \mathbb{D}. \tag{8.1.543}$$

Hence, the function $w := \overline{u}$ satisfies $\partial_{\bar{z}} w = \overline{\partial_z u} = 0$ in $\mathbb{D}$ and $\mathcal{N}_\kappa(\nabla w)$ is in $L^p(\partial\mathbb{D}, \sigma)$. Simply put, $w \in \mathcal{H}_1^p(\mathbb{D})$ (cf. [132, (8.1.20)]). Thus, $u = \overline{w}$ with w belonging to the space in the left side of (8.1.541). This establishes the left-to-right inclusion, and concludes the proof of (8.1.541).

The fact that the L^p-Neumann Problem in $\mathbb{C}\setminus\overline{\mathbb{D}}$ also fails to be Fredholm solvable for the writing of the Laplacian in the form $\operatorname{div} A\nabla$ with A as in (8.1.534) may be justified in a similar fashion. In fact, analogous results concerning the failure of Fredholm solvability may be established via similar methods for the interior/exterior Neumann Problems with boundary data in Muckenhoupt weighted Lebesgue spaces. Finally, that a similar lack of Fredholm solvability occurs in higher dimensions is apparent from Remark 8.6.14. The treatment of item *(6)* is therefore complete. This finishes the proof of Theorem 8.1.15. □

Next, we discuss the Neumann Problem in δ-oscillating AR domains, with boundary data in Morrey spaces, vanishing Morrey spaces, and block spaces. Our principal result in this regard, Theorem 8.1.16 stated a little below, allows us to pinpoint any space in the aforementioned scales, simply by making δ sufficiently small. The Neumann Problem with data in Morrey spaces with $p = 2$, for a bounded Lipschitz domain Ω with connected boundary, and a homogeneous, second-order, constant real coefficient, symmetric system L in $\mathbb{R}^n$, possessing a positive definite coefficient tensor A, has been dealt with by Z. Shen in [192, Theorem 1.12] employing techniques which are strongly dependent on the setting and hypotheses just mentioned.

Theorem 8.1.16 (The Neumann Problem with Boundary Data in Morrey and Block Spaces) *Fix two integers $n, M \in \mathbb{N}$ with $n \geq 2$ and suppose $\Omega \subseteq \mathbb{R}^n$ is an Ahlfors regular domain with compact boundary. Abbreviate $\sigma := \mathcal{H}^{n-1}\lfloor\partial\Omega$ and denote by ν the geometric measure theoretic outward unit normal to Ω. Also, let L be a homogeneous, second-order, constant complex coefficient, $M \times M$ system in $\mathbb{R}^n$, and pick a coefficient tensor A for which $L = L_A$ (i.e., $A \in \mathfrak{A}_L$). Finally, fix an arbitrary aperture parameter $\kappa > 0$, and select*

$$p, q \in (1, \infty) \text{ with } \tfrac{1}{p} + \tfrac{1}{q} = 1, \text{ and } \lambda \in (0, n-1). \tag{8.1.544}$$

Then there exists a threshold $\delta_0 \in (0, 1)$, which depends only on n, p, A, the ADR *constants of $\partial\Omega$, and the relative flatness ratio $\wp(\Omega)/\operatorname{diam}\partial\Omega$, so that if*

$$\operatorname{dist}\Big(\nu, [\mathrm{VMO}(\partial\Omega, \sigma)]^n\Big) < \delta_0 \tag{8.1.545}$$

where the distance is considered in the space $\big[\mathrm{BMO}(\partial\Omega, \sigma)\big]^n$ (hence whenever Ω is a δ_0-oscillating AR *domain*[53]*; cf. Definition 3.4.1), the set Ω becomes a* UR *domain and the following statements are true:*

[53] in particular, if Ω is an infinitesimally flat AR domain

(1) *If one assumes that*

$$L \text{ is weakly elliptic with } \mathfrak{A}_L^{\mathrm{dis}} \neq \varnothing \text{ and } \mathfrak{A}_{L^\top}^{\mathrm{dis}} \neq \varnothing, \tag{8.1.546}$$

then whenever $A \in \mathfrak{A}_L^{\mathrm{dis}}$ *it follows that the Neumann Problem with data in Morrey spaces, namely*[54]

$$\begin{cases} u \in \left[\mathscr{C}^\infty(\Omega)\right]^M, \quad Lu = 0 \text{ in } \Omega, \\ \mathcal{N}_\kappa(\nabla u) \in M^{p,\lambda}(\partial\Omega, \sigma), \quad \partial_\nu^A u = f \in \left[M^{p,\lambda}(\partial\Omega, \sigma)\right]^M, \\ u(x) = O(|x|^{2-n}) \text{ as } |x| \to \infty \text{ if } \Omega \text{ is unbounded,} \end{cases} \tag{8.1.547}$$

(where the strong conormal derivative is defined pointwise, as in (A.0.203)) *is Fredholm solvable, in the following precise sense: First, the space of admissible boundary data for the boundary problem* (8.1.547), *i.e.,*

$$\Big\{\partial_\nu^A u : u \in \left[\mathscr{C}^\infty(\Omega)\right]^M, \ Lu = 0 \text{ in } \Omega, \ \mathcal{N}_\kappa(\nabla u) \in M^{p,\lambda}(\partial\Omega, \sigma), \\ \text{and } u(x) = O(|x|^{2-n}) \text{ as } |x| \to \infty \text{ if } \Omega \text{ is unbounded}\Big\} \tag{8.1.548}$$

is a closed linear subspace of finite codimension in $\left[M^{p,\lambda}(\partial\Omega, \sigma)\right]^M$. *Second, a solution u of* (8.1.547) *exists if and only if the boundary datum f belongs to* (8.1.548), *in which scenario a solution u may be found such that*

$$\left\|\mathcal{N}_\kappa(\nabla u)\right\|_{M^{p,\lambda}(\partial\Omega,\sigma)} \leq C\|f\|_{[M^{p,\lambda}(\partial\Omega,\sigma)]^M}, \tag{8.1.549}$$

for some constant $C \in (0, \infty)$ *independent of f. Third, the space of null-solutions of the boundary problem* (8.1.547), *i.e.,*

$$\Big\{u \in \left[\mathscr{C}^\infty(\Omega)\right]^M : Lu = 0 \ \mathcal{N}_\kappa(\nabla u) \in M^{p,\lambda}(\partial\Omega, \sigma), \ \partial_\nu^A u = 0, \\ \text{and } u(x) = O(|x|^{2-n}) \text{ as } |x| \to \infty \text{ if } \Omega \text{ is unbounded}\Big\} \tag{8.1.550}$$

is finite dimensional.

(2) *Suppose* (8.1.546) *is strengthened (cf. Proposition 1.4.1) to the demand that*

$$L \text{ satisfies the Legendre-Hadamard strong ellipticity condition (cf. [131, Definition 1.3.2]) and one has } \mathfrak{A}_L^{\mathrm{dis}} \neq \varnothing. \tag{8.1.551}$$

Then for each coefficient tensor $A \in \mathfrak{A}_L^{\mathrm{dis}}$ *the index of the Neumann Problem* (8.1.547), *i.e., the dimension of* (8.1.550) *minus the codimension of* (8.1.548) *in* $\left[M^{p,\lambda}(\partial\Omega, \sigma)\right]^M$, *is zero.*

---

[54] from work in [129, §8.5] it is known that one may replace the nontangential maximal operator by the tangential maximal operator, associated as in [129, Definition 8.5.1] with a sufficiently large power (cf. (A.0.165))

Moreover, in the case when either $n \geq 3$, *or* $n = 2$ *and the set* Ω *is bounded, the Neumann Problem* (8.1.547) *has a solution if and only if the boundary datum* $f \in \big[M^{p,\lambda}(\partial\Omega, \sigma)\big]^M$ *satisfies*

$$\int_{\partial\Omega} \langle f, g\rangle \, \mathrm{d}\sigma = 0 \text{ for each } g \in \big[\mathcal{B}^{q,\lambda}(\partial\Omega, \sigma)\big]^M \\ \text{with the property that } \big(-\tfrac{1}{2}I + K_{A^\top}\big)g = 0, \tag{8.1.552}$$

while in the case when $n = 2$ *and* Ω *is an exterior domain, the Neumann Problem* (8.1.547) *has a solution if and only if the boundary datum* $f \in \big[M^{p,\lambda}(\partial\Omega, \sigma)\big]^M$ *satisfies*

$$\int_{\partial\Omega} \langle f, g\rangle \, \mathrm{d}\sigma = 0 \text{ for each } g \in \big[\mathcal{B}^{q,\lambda}(\partial\Omega, \sigma)\big]^M \\ \text{with the property that } \big(-\tfrac{1}{2}I + K_{A^\top}\big)g \in \mathbb{C}^M. \tag{8.1.553}$$

In fact,

> *the space of admissible boundary data for the Neumann Problem* (8.1.547), *defined as indicated in* (8.1.548), *is equal to the space* $\mathrm{Im}\Big(-\tfrac{1}{2}I + K^{\#}_{A^\top};\ \big[M^{p,\lambda}(\partial\Omega, \sigma)\big]^M\Big)$ *if either* $n \geq 3$, *or* $n = 2$ *and the set* Ω *is bounded, and may be described as the space* $\Big\{\big(-\tfrac{1}{2}I + K^{\#}_{A^\top}\big)g :\ g \in \big[M^{p,\lambda}(\partial\Omega, \sigma)\big]^M$ *with* $\int_{\partial\Omega} g \, \mathrm{d}\sigma = 0\Big\}$ *if* $n = 2$ *and* Ω *is an exterior domain.* (8.1.554)

Furthermore, there exists a constant $C \in (0, \infty)$ *with the property that whenever the boundary datum* f *belongs to the space of admissible boundary data, defined in* (8.1.548), *then a solution* u *of the boundary problem* (8.1.547) *may be found such that* (8.1.549) *holds.*

In addition, with $\mathcal{D}_A$ *denoting the boundary-to-domain double layer potential operator associated with* A *and* Ω *as in* (A.0.72), *when either* $n \geq 3$, *or* $n = 2$ *and the set* Ω *is bounded, it follows that:*

> *the space of null-solutions of the Neumann Problem* (8.1.547) (*originally described as in* (8.1.550)) *is actually equal to the space* $\Big\{\mathcal{D}_A\phi :\ \phi \in \mathrm{Ker}\Big(-\tfrac{1}{2}I + K_A;\ \big[M^{p,\lambda}_1(\partial\Omega, \sigma)\big]^M\Big)\Big\}$, (8.1.555)

> *the map associating to each function* u *belonging to the space of null-solutions of the Neumann Problem* (8.1.547) (*described in* (8.1.550)) *its nontangential boundary trace* $u\big|^{\kappa-\mathrm{n.t.}}_{\partial\Omega}$ *is an isomorphism onto the finite dimensional space* $\mathrm{Ker}\Big(-\tfrac{1}{2}I + K_A;\ \big[M^{p,\lambda}_1(\partial\Omega, \sigma)\big]^M\Big)$, (8.1.556)

the inverse of the isomorphism $u \mapsto u\big|_{\partial\Omega}^{\kappa-\mathrm{n.t.}}$ *described in* (8.1.556) *is the boundary-to-domain double layer potential operator* $\mathcal{D}_A$ *mapping* $\mathrm{Ker}\Big(-\frac{1}{2}I + K_A;\ \big[M_1^{p,\lambda}(\partial\Omega,\sigma)\big]^M\Big)$ *isomorphically onto the space of null-solutions of the Neumann Problem* (8.1.547) (*from* (8.1.550)). (8.1.557)

Finally, in the case when $n = 2$ *and* Ω *is an exterior domain it follows that:*

the space of null-solutions of the Neumann Problem (8.1.547) *equals* $\Big\{\mathcal{D}_A\phi + c :\ \phi \in \big[M_1^{p,\lambda}(\partial\Omega,\sigma)\big]^M,\ c \in \mathbb{C}^M,\ \big(-\frac{1}{2}I + K_A\big)\phi = -c\Big\}$, (8.1.558)

the mapping associating to each function u *belonging to the space of null-solutions of the Neumann Problem* (8.1.547) (*described in* (8.1.550)) *the pair* $\Big(u\big|_{\partial\Omega}^{\kappa-\mathrm{n.t.}},\ \lim\limits_{|x|\to\infty} u(x)\Big)$ *is a well-defined isomorphism onto* $\Big\{(\phi, c) \in \big[M_1^{p,\lambda}(\partial\Omega,\sigma)\big]^M \oplus \mathbb{C}^M :\ \big(-\frac{1}{2}I + K_A\big)\phi = -c\Big\}$, (8.1.559)

the inverse of the linear isomorphism $u \mapsto \Big(u\big|_{\partial\Omega}^{\kappa-\mathrm{n.t.}},\ \lim\limits_{|x|\to\infty} u(x)\Big)$ *in the above context is the assignment* $(\phi, c) \longmapsto \mathcal{D}_A\phi + c$ *mapping the last space in* (8.1.559) *isomorphically onto the space of null-solutions of the Neumann Problem* (8.1.547) (*described in* (8.1.550)). (8.1.560)

(3) If (8.1.551) *is further strengthened to*[55]

$$A \in \mathfrak{A}_L^{\mathrm{dis}} \text{ is positive definite,} \tag{8.1.561}$$

then the Neumann Problem (8.1.547) *has the following additional properties (with the spaces* $\mathbb{C}_{\Omega_\pm}$, $\mathbb{C}_{\partial\Omega_\pm}$, $\mathbb{C}^\infty_{\Omega_\pm}$, $\mathbb{C}^\infty_{\partial\Omega_\pm}$ *defined as in* (7.1.265), (7.1.268), *and with the conventions* (7.1.266), (7.1.269) *in effect). First, if either* $n \geq 3$, *or* $n = 2$ *and the set* Ω *is bounded, then*

the space of admissible boundary data for the Neumann Problem (8.1.547) (*originally defined as in* (8.1.548)) *is actually equal to the space* $\big[M^{p,\lambda}(\partial\Omega,\sigma) \cap (\mathbb{C}_{\partial\Omega_+})^{\perp}\big]^M$, *and the space of null-solutions of the Neumann Problem* (8.1.547) (*cf.* (8.1.550)) *is* $(\mathbb{C}_{\Omega_+})^M$. (8.1.562)

Second, if $n = 2$ *and* Ω *is an exterior domain, then*

[55] according to (1.1.28), whenever L has a positive definite coefficient tensor it follows that L satisfies the Legendre-Hadamard strong ellipticity condition

the space of admissible boundary data for the Neumann Problem (8.1.547) (*originally defined as in* (8.1.548)) *is actually equal to* $\big[M^{p,\lambda}(\partial\Omega,\sigma)\cap\big(\mathbb{C}^{\infty}_{\partial\Omega_+}\big)^{\perp}\big]^M$, *and the space of null-solutions of the Neumann Problem* (8.1.547) (*cf.* (8.1.550)) *is equal to* $\big(\mathbb{C}^{\infty}_{\Omega_+}\big)^M$. (8.1.563)

(4) *Similar results as in items (1)-(3) above are true for the version of the Neumann Problem* (8.1.547) *with boundary data in vanishing Morrey spaces, i.e.,*

$$\begin{cases} u\in\big[\mathscr{C}^\infty(\Omega)\big]^M, \quad Lu=0 \text{ in } \Omega,\\ \mathcal{N}_\kappa(\nabla u)\in \mathring{M}^{p,\lambda}(\partial\Omega,\sigma), \quad \partial^A_\nu u=f\in\big[\mathring{M}^{p,\lambda}(\partial\Omega,\sigma)\big]^M,\\ u(x)=O(|x|^{2-n}) \text{ as } |x|\to\infty \text{ if } \Omega \text{ is unbounded,}\end{cases} \tag{8.1.564}$$

as well as for the version of the Neumann Problem (8.1.547) *with boundary data in block spaces, i.e.,*

$$\begin{cases} u\in\big[\mathscr{C}^\infty(\Omega)\big]^M, \quad Lu=0 \text{ in } \Omega,\\ \mathcal{N}_\kappa(\nabla u)\in \mathcal{B}^{q,\lambda}(\partial\Omega,\sigma), \quad \partial^A_\nu u=f\in\big[\mathcal{B}^{q,\lambda}(\partial\Omega,\sigma)\big]^M,\\ u(x)=O(|x|^{2-n}) \text{ as } |x|\to\infty \text{ if } \Omega \text{ is unbounded.}\end{cases} \tag{8.1.565}$$

Once again, work in [129, §8.5] *shows that one may replace the nontangential maximal operator in the formulation of the boundary value problems* (8.1.564)-(8.1.565) *by the tangential maximal operator* (*associated as in* [129, Definition 8.5.1] *with a sufficiently large power; cf.* (A.0.165)).

(5) *Even for* $L:=\Delta$, *the Laplace operator* (*which clearly satisfies* (8.1.546)), *if* $A\notin\mathfrak{A}^{\rm dis}_L$ *then it may happen that the Neumann Problems* (8.1.547), (8.1.564), (8.1.565) *fail to be Fredholm solvable even when the underlying domain* Ω *is the unit disk in the plane* (*or the complement of its closure*). *In fact, the aforementioned failure of Fredholm solvability may be total, in the sense that for each of these problems, both the space of admissible boundary data may have infinite codimension in the full space of boundary data, and the space of null-solutions may be infinite dimensional.*

Before presenting its proof, we wish to elaborate on the nature of Theorem 8.1.16. The first comment in this regard is that, as seen from Proposition 1.6.12 and [131, Proposition 1.4.39], in the scalar case (i.e., if $M=1$) in dimensions $n\geq 3$, both (8.1.546) and (8.1.551) may be replaced by the condition that L is weakly elliptic. Moreover, in the scalar case in the two-dimensional setting (i.e., for $M=1$ and $n=2$), if L satisfies the Legendre-Hadamard strong ellipticity condition or, more generally (cf. [131, Lemma 1.4.24]), if $L={\rm div}A\nabla$ for some 2×2 complex matrix A with ${\rm sym}\,A:=\frac{1}{2}(A+A^\top)\in M^0_2$ (cf. [131, (1.4.187)]), then Proposition 1.6.11 and Proposition 1.6.13 ensure that $\mathfrak{A}^{\rm dis}_L\neq\varnothing$ and $\mathfrak{A}^{\rm dis}_{L^\top}\neq\varnothing$. Also, according to (1.6.141) in Proposition 1.6.13, in the scalar case in the two-dimensional setting (i.e., for $M=1$ and $n=2$) we may omit (8.1.551) altogether if L is a weakly elliptic operator with real coefficients.

The second comment is that the boundary conditions in (8.1.547), (8.1.564), and (8.1.565) are all naturally formulated. To see that this is the case, observe that choosing $\delta_0 \in (0,1)$ small permits us to invoke Theorem 3.4.2 which guarantees that Ω is a two-sided NTA domain. Granted this, the Fatou-type result proved in [131, Theorem 3.3.7] implies that, given any function

$$\begin{gathered} u \in \big[\mathscr{C}^\infty(\Omega)\big]^M \text{ with } Lu = 0 \text{ in } \Omega \text{ and} \\ \mathcal{N}_\kappa(\nabla u) \in M^{p,\lambda}(\partial\Omega,\sigma) \text{ for some } p \in (1,\infty),\ \lambda \in (0,n-1), \end{gathered} \tag{8.1.566}$$

the nontangential boundary trace

$$\begin{gathered} (\nabla u)\big|_{\partial\Omega}^{\kappa-\mathrm{n.t.}} \text{ exists (in } \mathbb{C}^{n\cdot M}) \text{ at } \sigma\text{-a.e. point on } \partial\Omega, \\ (\nabla u)\big|_{\partial\Omega}^{\kappa-\mathrm{n.t.}} \text{ belongs to the Morrey space } \big[M^{p,\lambda}(\partial\Omega,\sigma)\big]^{n\cdot M}, \\ \text{and } \Big\|(\nabla u)\big|_{\partial\Omega}^{\kappa-\mathrm{n.t.}}\Big\|_{[M^{p,\lambda}(\partial\Omega,\sigma)]^{n\cdot M}} \leq \|\mathcal{N}_\kappa(\nabla u)\|_{M^{p,\lambda}(\partial\Omega,\sigma)}. \end{gathered} \tag{8.1.567}$$

In particular, from (8.1.567) and [130, (6.2.5)] we conclude that for each $A \in \mathfrak{A}_L$ there exists some constant $C = C(A,n) \in (0,\infty)$ such that

$$\begin{gathered} \partial_\nu^A u \text{ belongs to the Morrey space } \big[M^{p,\lambda}(\partial\Omega,\sigma)\big]^M \\ \text{and } \big\|\partial_\nu^A u\big\|_{[M^{p,\lambda}(\partial\Omega,\sigma)]^M} \leq C\|\mathcal{N}_\kappa(\nabla u)\|_{M^{p,\lambda}(\partial\Omega,\sigma)}. \end{gathered} \tag{8.1.568}$$

Hence, any function satisfying the properties listed in the first two lines of (8.1.547) has a well-defined strong conormal derivative, considered in a pointwise sense, as in (A.0.203), and that said conormal derivative belongs to the Morrey space $\big[M^{p,\lambda}(\partial\Omega,\sigma)\big]^M$, from which the boundary data for the problem (8.1.547) are selected. In addition, [130, Proposition 11.13.7] and [130, (11.13.34)] imply that for any function as in (8.1.566) we have

$$u\big|_{\partial\Omega}^{\kappa-\mathrm{n.t.}} \text{ exists } \sigma\text{-a.e. on } \partial\Omega, \text{ and belongs to } \big[M_1^{p,\lambda}(\partial\Omega,\sigma)\big]^M. \tag{8.1.569}$$

Similar trace results are valid in relation to the problem (8.1.564), thanks to the last part in [131, Theorem 3.3.7] and [130, Proposition 11.13.16], as well as in relation to the problem (8.1.565), thanks to [131, Theorem 3.3.8] and [130, Proposition 11.13.13].

We are ready to present the proof of Theorem 8.1.16.

Proof of Theorem 8.1.16 Granted the trace results discussed in the preamble to the proof of Theorem 8.1.16, all claims may be justified using the same blueprint as in the proof of Theorem 8.1.15, now making use of item *(7)* in Theorem 7.1.1, item *(2)* in Proposition 7.1.24, [130, Proposition 6.2.8], [130, Proposition 6.2.16], [132, Theorem 3.3.1], [132, Theorem 3.3.2], [130, (6.2.7)], and [130, (6.2.15), (6.2.71)].□

A large number of earlier results for the Neumann Problem formulated with boundary data prescribed in a variety of concrete spaces (such as ordinary Lebesgue spaces, Muckenhoupt weighted Lebesgue spaces, Morrey spaces, vanishing Morrey spaces, and block spaces) may be actually subsumed under the considerably more inclusive umbrella of Neumann Problems with data in Generalized Banach Function Spaces (GBFS), as described in our next theorem.

Theorem 8.1.17 (The Neumann Problem with Boundary Data in GBFS) *Select $n, M \in \mathbb{N}$ with $n \geq 2$ and suppose $\Omega \subseteq \mathbb{R}^n$ is an Ahlfors regular domain with compact boundary. Abbreviate $\sigma := \mathcal{H}^{n-1}\lfloor\partial\Omega$ and denote by ν the geometric measure theoretic outward unit normal to Ω. Assume $\mathbb{X}$ is a Generalized Banach Function Space on $(\partial\Omega, \sigma)$ with the property that*

$$\mathcal{M}_{\partial\Omega} : \mathbb{X} \to \mathbb{X} \text{ and } \mathcal{M}_{\partial\Omega} : \mathbb{X}' \to \mathbb{X}' \text{ are well-defined and bounded,} \tag{8.1.570}$$

where $\mathcal{M}_{\partial\Omega}$ is the Hardy-Littlewood maximal operator on $\partial\Omega$, and $\mathbb{X}'$ is the associated space of $\mathbb{X}$ (cf. [130, Definitions 5.1.4, 5.1.11]*). Recall the $\mathbb{X}$-styled Sobolev space of order one on $\partial\Omega$, defined as*

$$\mathbb{X}_1 := \Big\{ f \in \mathbb{X} : \partial_{\tau_{jk}} f \in \mathbb{X} \text{ for } 1 \leq j, k \leq n \Big\}, \tag{8.1.571}$$

(which is known to be Banach space when equipped with the natural norm; cf. (5.4.84)*). Finally, let L be a homogeneous, second-order, constant complex coefficient, $M \times M$ system in $\mathbb{R}^n$, and pick a coefficient tensor A for which $L = L_A$ (i.e., $A \in \mathfrak{A}_L$). In this setting, having fixed an arbitrary aperture parameter $\kappa > 0$, consider the Neumann Problem*[56]

$$\begin{cases} u \in \big[\mathscr{C}^\infty(\Omega)\big]^M, \quad Lu = 0 \text{ in } \Omega, \\ \mathcal{N}_\kappa(\nabla u) \in \mathbb{X}, \quad \partial_\nu^A u = f \in \big[\mathbb{X}\big]^M, \\ u(x) = O(|x|^{2-n}) \text{ as } |x| \to \infty \text{ if } \Omega \text{ is unbounded,} \end{cases} \tag{8.1.572}$$

where the strong conormal derivative is defined in a pointwise sense as in (A.0.203)*.*

Then there exists some threshold $\delta_{\mathbb{X}} \in (0,1)$, which depends only on n, A, the operator norms of $\mathcal{M}_{\partial\Omega}$ on $\mathbb{X}$ and $\mathbb{X}'$, the ADR *constants of $\partial\Omega$, and the relative flatness ratio $\wp(\Omega)/\mathrm{diam}\,\partial\Omega$, with the property that whenever*

$$\mathrm{dist}\Big(\nu, [\mathrm{VMO}(\partial\Omega,\sigma)]^n\Big) < \delta_{\mathbb{X}} \tag{8.1.573}$$

where the distance is considered in the space $\big[\mathrm{BMO}(\partial\Omega,\sigma)\big]^n$, hence whenever Ω is a $\delta_{\mathbb{X}}$-oscillating AR *domain*[57] *(cf. Definition 3.4.1), the set Ω becomes a* UR *domain and the following statements are true:*

(1) If one assumes that

$$L \text{ is weakly elliptic with } \mathfrak{A}_L^{\mathrm{dis}} \neq \emptyset \text{ and } \mathfrak{A}_{L^\top}^{\mathrm{dis}} \neq \emptyset, \tag{8.1.574}$$

[56] thanks to work in [129, §8.5], one may replace the nontangential maximal operator by the tangential maximal operator, associated as in [129, Definition 8.5.1] with a sufficiently large power (cf. (A.0.165))

[57] in particular, if Ω is an infinitesimally flat AR domain

then whenever $A \in \mathfrak{A}_L^{\rm dis}$ *it follows that the boundary value problem* (8.1.572) *is Fredholm solvable, in the following precise sense: First, the space of admissible boundary data for the boundary problem* (8.1.572), *i.e.,*

$$\Big\{\partial_\nu^A u : u \in \big[\mathscr{C}^\infty(\Omega)\big]^M,\ Lu = 0 \text{ in } \Omega,\ \mathcal{N}_\kappa(\nabla u) \in \mathbb{X}, \\ \text{and } u(x) = O(|x|^{2-n}) \text{ as } |x| \to \infty \text{ if } \Omega \text{ is unbounded}\Big\} \tag{8.1.575}$$

is a closed linear subspace of finite codimension in $\big[\mathbb{X}\big]^M$. *Second, a solution u of* (8.1.572) *exists if and only if the boundary datum f belongs to* (8.1.575), *in which scenario a solution u may be found such that*

$$\big\|\mathcal{N}_\kappa(\nabla u)\big\|_{\mathbb{X}} \le C\|f\|_{[\mathbb{X}]^M}, \tag{8.1.576}$$

for some constant $C \in (0, \infty)$ *independent of f. Third, the space of null-solutions of the boundary problem* (8.1.572), *i.e.,*

$$\Big\{u \in \big[\mathscr{C}^\infty(\Omega)\big]^M : Lu = 0 \text{ in } \Omega,\ \mathcal{N}_\kappa(\nabla u) \in \mathbb{X},\ \partial_\nu^A u = 0, \\ \text{and } u(x) = O(|x|^{2-n}) \text{ as } |x| \to \infty \text{ if } \Omega \text{ is unbounded}\Big\} \tag{8.1.577}$$

is finite dimensional.

(2) *Suppose* (8.1.574) *is strengthened* (*cf. Proposition 1.4.1*) *to the demand that*

$$\text{L satisfies the Legendre-Hadamard strong ellipticity condition (cf. [131, Definition 1.3.2]) and one has } \mathfrak{A}_L^{\rm dis} \neq \varnothing. \tag{8.1.578}$$

Then for each coefficient tensor $A \in \mathfrak{A}_L^{\rm dis}$ *the index of the Neumann Problem* (8.1.572), *i.e., the dimension of* (8.1.577) *minus the codimension of* (8.1.575) *in* $\big[\mathbb{X}\big]^M$, *is equal to zero. Moreover,*

$$\text{the space of admissible boundary data for the Neumann Problem (8.1.572), defined in (8.1.575), is equal to } \mathrm{Im}\Big(-\tfrac{1}{2}I + K^\#_{A^\top};\ \big[\mathbb{X}\big]^M\Big) \text{ if either } n \ge 3, \text{ or } n = 2 \text{ and the set } \Omega \text{ is bounded, while if } n = 2 \text{ and } \Omega \text{ is an exterior domain it may be described as being } \Big\{\big(-\tfrac{1}{2}I + K^\#_{A^\top}\big)g : g \in \big[\mathbb{X}\big]^M \text{ with } \int_{\partial\Omega} g\,d\sigma = 0\Big\}. \tag{8.1.579}$$

Furthermore, there exists a constant $C \in (0, \infty)$ *with the property that whenever the boundary datum f belongs to the space of admissible boundary data, defined in* (8.1.575), *then a solution u of the boundary problem* (8.1.572) *may be found such that* (8.1.576) *holds.*

Also, with $\mathcal{D}_A$ *denoting the boundary-to-domain double layer potential operator associated with A and* Ω *as in* (A.0.72), *when either* $n \ge 3$, *or* $n = 2$ *and the set* Ω *is bounded, it follows that:*

the space of null-solutions of the Neumann Problem (8.1.572) (*originally described as in* (8.1.577)) *is actually equal to the space* $\Big\{\mathcal{D}_A\phi : \phi \in \mathrm{Ker}\Big(-\tfrac{1}{2}I + K_A;\ \big[\mathbb{X}_1\big]^M\Big)\Big\}$, (8.1.580)

the mapping associating to each function u belonging to the space of null-solutions of the Neumann Problem (8.1.572) (*described in* (8.1.577)) *its nontangential boundary trace* $u\big|_{\partial\Omega}^{\kappa-\mathrm{n.t.}}$ *is an isomorphism onto the finite dimensional space* $\mathrm{Ker}\Big(-\tfrac{1}{2}I + K_A;\ \big[\mathbb{X}_1\big]^M\Big)$, (8.1.581)

the inverse of the isomorphism $u \mapsto u\big|_{\partial\Omega}^{\kappa-\mathrm{n.t.}}$ *described in* (8.1.581) *is the boundary-to-domain double layer potential operator* $\mathcal{D}_A$ *mapping* $\mathrm{Ker}\Big(-\tfrac{1}{2}I + K_A;\ \big[\mathbb{X}_1\big]^M\Big)$ *isomorphically onto the space of null-solutions of the Neumann Problem* (8.1.572) (*described in* (8.1.577)). (8.1.582)

Finally, in the case when $n = 2$ *and* Ω *is an exterior domain it follows that:*

the space of null-solutions of the Neumann Problem (8.1.572) *is equal to the space* $\Big\{\mathcal{D}_A\phi + c : \phi \in \big[\mathbb{X}_1\big]^M,\ c \in \mathbb{C}^M,\ \big(-\tfrac{1}{2}I + K_A\big)\phi = -c\Big\}$, (8.1.583)

the mapping associating to each function u belonging to the space of null-solutions of the Neumann Problem (8.1.572) (*described in* (8.1.577)) *the pair* $\Big(u\big|_{\partial\Omega}^{\kappa-\mathrm{n.t.}},\ \lim_{|x|\to\infty} u(x)\Big)$ *is a well-defined isomorphism onto the space* $\Big\{(\phi, c) \in \big[\mathbb{X}_1\big]^M \oplus \mathbb{C}^M : \big(-\tfrac{1}{2}I + K_A\big)\phi = -c\Big\}$, (8.1.584)

the inverse of the linear isomorphism $u \mapsto \Big(u\big|_{\partial\Omega}^{\kappa-\mathrm{n.t.}},\ \lim_{|x|\to\infty} u(x)\Big)$ *in the above context is the assignment* $(\phi, c) \longmapsto \mathcal{D}_A\phi + c$ *mapping the last space in* (8.1.584) *isomorphically onto the space of null-solutions of the Neumann Problem* (8.1.572) (*described in* (8.1.577)). (8.1.585)

(3) If (8.1.578) *is further strengthened to*[58]

$$A \in \mathfrak{A}_L^{\mathrm{dis}} \text{ is positive definite,} \tag{8.1.586}$$

[58] according to (1.1.28), whenever L has a positive definite coefficient tensor it follows that L satisfies the Legendre-Hadamard strong ellipticity condition

then the Neumann Problem (8.1.572) *has the following additional properties (with the spaces* $\mathbb{C}_{\Omega_\pm}$, $\mathbb{C}_{\partial\Omega_\pm}$, $\mathbb{C}^\infty_{\Omega_\pm}$, $\mathbb{C}^\infty_{\partial\Omega_\pm}$ *defined as in* (7.1.265), (7.1.268), *and with the conventions* (7.1.266), (7.1.269) *in effect). First, if either* $n \geq 3$, *or* $n = 2$ *and the set* Ω *is bounded, then*

$$\text{the space of admissible boundary data for the Neumann Problem (8.1.572) (defined in (8.1.575)) is equal to } \big[\mathbb{X} \cap (\mathbb{C}_{\partial\Omega_+})^\perp\big]^M, \text{ and the space of null-solutions of the Neumann Problem (8.1.572) (described in (8.1.577)) is equal to } (\mathbb{C}_{\Omega_+})^M. \tag{8.1.587}$$

Second, if $n = 2$ *and* Ω *is an exterior domain, then*

$$\text{the space of admissible boundary data for the Neumann Problem (8.1.572) (defined in (8.1.575)) is equal to } \big[\mathbb{X} \cap (\mathbb{C}^\infty_{\partial\Omega_+})^\perp\big]^M, \text{ and the space of null-solutions of the Neumann Problem (8.1.572) (described in (8.1.577)) is equal to } (\mathbb{C}^\infty_{\Omega_+})^M. \tag{8.1.588}$$

(4) *Similar results as in items (1)-(3) above are true for the Neumann Problem with boundary data in* $\mathbb{X}'$, *the associated space of* $\mathbb{X}$ *(in the sense of* [130, Definition 5.1.11]), *namely*

$$\begin{cases} u \in \big[\mathscr{C}^\infty(\Omega)\big]^M, & Lu = 0 \text{ in } \Omega, \\ \mathcal{N}_\kappa(\nabla u) \in \mathbb{X}', & \partial^A_\nu u = f \in \big[\mathbb{X}'\big]^M, \\ u(x) = O(|x|^{2-n}) & \text{as } |x| \to \infty \text{ if } \Omega \text{ is unbounded,} \end{cases} \tag{8.1.589}$$

as well as for the Neumann Problems with boundary data in the spaces

$$\mathring{\mathbb{X}} := \overline{L^\infty(\partial\Omega, \sigma)}^{\|\cdot\|_{\mathbb{X}}} \text{ and } (\mathbb{X}')^\circ := \overline{L^\infty(\partial\Omega, \sigma)}^{\|\cdot\|_{\mathbb{X}'}}, \tag{8.1.590}$$

i.e., for

$$\begin{cases} u \in \big[\mathscr{C}^\infty(\Omega)\big]^M, & Lu = 0 \text{ in } \Omega, \\ \mathcal{N}_\kappa(\nabla u) \in \mathring{\mathbb{X}}, & \partial^A_\nu u = f \in \big[\mathring{\mathbb{X}}\big]^M, \\ u(x) = O(|x|^{2-n}) & \text{as } |x| \to \infty \text{ if } \Omega \text{ is unbounded,} \end{cases} \tag{8.1.591}$$

and, respectively,

$$\begin{cases} u \in \big[\mathscr{C}^\infty(\Omega)\big]^M, & Lu = 0 \text{ in } \Omega, \\ \mathcal{N}_\kappa(\nabla u) \in (\mathbb{X}')^\circ, & \partial^A_\nu u = f \in \big[(\mathbb{X}')^\circ\big]^M, \\ u(x) = O(|x|^{2-n}) & \text{as } |x| \to \infty \text{ if } \Omega \text{ is unbounded.} \end{cases} \tag{8.1.592}$$

In all cases, work in [129, §8.5] *shows that one may replace the nontangential maximal operator by the tangential maximal operator, associated as in* [129, Definition 8.5.1] *with a sufficiently large power* (*cf.* (A.0.165)).

(5) *Even when* $L := \Delta$, *the Laplace operator* (*which obviously satisfies* (8.1.574)), *if* $A \notin \mathfrak{A}_L^{\rm dis}$ *then it may happen that the Neumann Problems* (8.1.572), (8.1.589)-(8.1.592) *fail to be Fredholm solvable even when the underlying domain* Ω *is the unit disk in the plane* (*or the complement of its closure*). *Indeed, the aforementioned failure of Fredholm solvability may be total, in the sense that for each of these problems, both the space of admissible boundary data may have infinite codimension in the full space of boundary data, and the space of null-solutions may be infinite dimensional. Finally, a similar lack of Fredholm solvability occurs in higher dimensions.*

We wish to make three remarks in relation to Theorem 8.1.17. First, as seen from Proposition 1.6.12 and [131, Proposition 1.4.39], in the scalar case, i.e., if $M = 1$, in dimensions $n \geq 3$, both (8.1.574) and (8.1.578) may be replaced by the mere demand that L is weakly elliptic. Also, in the scalar case in the two-dimensional setting (i.e., for $M = 1$ and $n = 2$), if L satisfies the Legendre-Hadamard strong ellipticity condition or, more generally (cf. [131, Lemma 1.4.24]), if $L = {\rm div} A\nabla$ for some 2×2 complex matrix A with ${\rm sym}\, A := \frac{1}{2}(A + A^\top) \in M_2^0$ (cf. [131, (1.4.187)]), then Proposition 1.6.11 and Proposition 1.6.13 guarantee that we automatically have $\mathfrak{A}_L^{\rm dis} \neq \varnothing$ and $\mathfrak{A}_{L^\top}^{\rm dis} \neq \varnothing$. Furthermore, according to (1.6.141) in Proposition 1.6.13, in the scalar case in the two-dimensional setting (i.e., for $M = 1$ and $n = 2$) we may omit (8.1.578) altogether if L is a weakly elliptic operator with real coefficients.

Second, the boundary condition in (8.1.572) is naturally formulated. Indeed, choosing the threshold $\delta_{\mathbb{X}} \in (0, 1)$ small permits us to invoke Theorem 3.4.2 which ensures that Ω is a two-sided NTA domain. Granted this, the Fatou-type result proved in item *(a)* in [131, Theorem 3.3.4] together with the embeddings in [130, (5.2.92)] and the lattice property [130, (5.1.12)] give that any function satisfying

$$u \in \big[\mathscr{C}^\infty(\Omega)\big]^M \text{ with } Lu = 0 \text{ in } \Omega \text{ and } \mathcal{N}_\kappa(\nabla u) \in \mathbb{X} \tag{8.1.593}$$

has the property that the nontangential boundary trace

$$(\nabla u)\Big|_{\partial\Omega}^{\kappa-{\rm n.t.}} \text{ exists (in } \mathbb{C}^{n\cdot M}) \ \sigma\text{-a.e. on } \partial\Omega \text{ and } (\nabla u)\Big|_{\partial\Omega}^{\kappa-{\rm n.t.}} \in \big[\mathbb{X}\big]^{n\cdot M}. \tag{8.1.594}$$

In particular, the strong conormal derivative considered in a pointwise sense as in (A.0.203) is meaningful in this case, and satisfies

$$\partial_\nu^A u \in \big[\mathbb{X}\big]^M. \tag{8.1.595}$$

In summary, any function satisfying the properties listed in the first two lines of (8.1.572) has a well-defined strong conormal derivative, considered in a pointwise sense, as in (A.0.203), and that said conormal derivative belongs to $\big[\mathbb{X}\big]^M$ (the space from which the boundary data for the problem (8.1.572) are selected). From

[130, Proposition 11.5.12], [130, (5.2.92)], [130, (5.1.12)], [129, (8.9.8)], [130, Proposition 11.3.2], and [130, (5.1.13)] we also see that, for any function u as in (8.1.593),

$$u\Big|_{\partial\Omega}^{\kappa-\mathrm{n.t.}} \text{ exists } \sigma\text{-a.e. on } \partial\Omega, \text{ and belongs to } \big[\mathbb{X}_1\big]^M. \tag{8.1.596}$$

Thanks to [130, (5.2.101)] and [130, Proposition 5.1.14], similar considerations to those described in (8.1.593)-(8.1.596) are valid with $\mathbb{X}$ replaced by $\mathbb{X}'$, $\mathring{\mathbb{X}}$, or $(\mathbb{X}')^{\circ}$.

Our third (and final) remark regarding Theorem 8.1.17 is that [130, Proposition 5.3.14] guarantees that

the conditions demanded in (8.1.570) automatically hold in the case when $\mathbb{X}$ is a rearrangement invariant Banach function space on the (non-atomic, sigma-finite measure space $(\partial\Omega, \sigma)$) whose Boyd indices satisfy $1 < p_{\mathbb{X}} \leq q_{\mathbb{X}} < \infty$. (8.1.597)

Consequently, *all conclusions in Theorem 8.1.17 are true in the class of rearrangement invariant Banach function spaces $\mathbb{X}$ whose Boyd indices satisfy* $1 < p_{\mathbb{X}} \leq q_{\mathbb{X}} < \infty$. Significantly, this is the case when the Generalized Banach Function Space $\mathbb{X}$ is actually an Orlicz space, $\mathbb{X} := L^{\Phi}(\partial\Omega, \sigma)$, associated with the (non-atomic, sigma-finite) measure space $(\partial\Omega, \sigma)$ and a Young function Φ satisfying $1 < i(\Phi) \leq I(\Phi) < \infty$ (see [130, Proposition 5.3.15] and [130, (5.3.62)]).

To further elaborate on this aspect, work under the initial background assumptions made in Theorem 8.1.17 and pick a Young function Φ whose dilation indices satisfy $1 < i(\Phi) \leq I(\Phi) < \infty$. Then the Neumann Problem with data in the Orlicz space L^{Φ} (cf. (A.0.133)), formulated as

$$\begin{cases} u \in \big[\mathscr{C}^{\infty}(\Omega)\big]^M, \quad Lu = 0 \text{ in } \Omega, \\ \mathcal{N}_{\kappa}(\nabla u) \in L^{\Phi}(\partial\Omega, \sigma), \quad \partial_{\nu}^{A} u = f \in \big[L^{\Phi}(\partial\Omega, \sigma)\big]^M, \\ u(x) = O(|x|^{2-n}) \text{ as } |x| \to \infty \text{ if } \Omega \text{ is unbounded,} \end{cases} \tag{8.1.598}$$

is Fredholm solvable, with index zero, whenever the system L satisfies the Legendre-Hadamard strong ellipticity condition and the tensor coefficient A belongs to $\mathfrak{A}_L^{\mathrm{dis}}$, if $\mathrm{dist}\big(\nu, [\mathrm{VMO}(\partial\Omega, \sigma)]^n\big) < \delta_{\Phi}$ for a sufficiently small threshold $\delta_{\Phi} \in (0, 1)$. Indeed, this is a consequence of item *(2)* of Theorem 8.1.17, [130, (5.3.62), (5.3.63)], and [130, Proposition 5.3.15].

To offer a concrete example, take $\Phi(t) \approx t^p[\ln(e + t)]^{\alpha}$ for some fixed exponent $p \in (1, \infty)$ and $\alpha \in \mathbb{R}$, in which case the Orlicz space L^{Φ} actually is Zygmund's space $L^p(\log L)^{\alpha}$ discussed in [130, Example 5.3.10]. For this choice, the result in the previous paragraph becomes a Fredholm solvability result (with index zero) for the Neumann Problem with data in Zygmund's space $L^p(\log L)^{\alpha}$, formulated as

$$\begin{cases} u \in \big[\mathscr{C}^\infty(\Omega)\big]^M, \quad Lu = 0 \text{ in } \Omega, \\ \mathcal{N}_\kappa(\nabla u) \in L^p(\log L)^\alpha(\partial\Omega, \sigma), \\ \partial_\nu^A u = f \in \big[L^p(\log L)^\alpha(\partial\Omega, \sigma)\big]^M, \\ u(x) = O(|x|^{2-n}) \text{ as } |x| \to \infty \text{ if } \Omega \text{ is unbounded,} \end{cases} \tag{8.1.599}$$

once again assuming that the system L satisfies the Legendre-Hadamard strong ellipticity condition, that $A \in \mathfrak{A}_L^{\rm dis}$, and that we have $\operatorname{dist}\big(\nu, [\mathrm{VMO}(\partial\Omega, \sigma)]^n\big) < \delta$ for some sufficiently small threshold $\delta = \delta(L, \Omega, p, \alpha) \in (0, 1)$.

Many other choices are, of course, possible. For example, in light of the discussion in [130, Example 5.3.11], a similar Fredholm solvability result is valid for the Neumann Problem with data in the Orlicz space $\big[L^p \exp(\log^\theta L)(\partial\Omega, \sigma)\big]^M$ (see [130, (5.3.149)]) whenever $p \in (1, \infty)$ and $\theta \in [0, 1)$.

After this preamble, we now turn to the proof of Theorem 8.1.17.

Proof of Theorem 8.1.17 Take $\delta_{\mathbb{X}} \in (0, 1)$ sufficiently small, depending on n, A, the operator norms of $\mathcal{M}_{\partial\Omega}$ on $\mathbb{X}$ and $\mathbb{X}'$, the ADR constants of $\partial\Omega$, and the relative flatness ratio $\wp(\Omega)/\operatorname{diam}\partial\Omega$, such that the conclusions in Theorem 7.1.1, and Proposition 7.1.3. By further decreasing the value of $\delta_{\mathbb{X}} \in (0, 1)$, if necessary, Theorem 3.4.2 applies and gives that Ω is a two-sided NTA domain, hence also a UR domain (cf. [129, (5.10.33)]).

For now, we shall work under the assumptions made in (8.1.574). In this scenario, Theorem 1.5.5 ensures that $A^\top \in \mathfrak{A}_{L^\top}^{\rm dis}$. Granted this, item *(8)* in Theorem 7.1.1 used with $z := -1/2$ guarantees that the operator

$$-\tfrac{1}{2}I + K^\#_{A^\top} : [\mathbb{X}]^M \longrightarrow [\mathbb{X}]^M \text{ is Fredholm with index zero.} \tag{8.1.600}$$

Consequently,

$$\begin{gathered} \operatorname{Im}\Big(-\tfrac{1}{2}I + K^\#_{A^\top};\ [\mathbb{X}]^M\Big) \text{ is a closed space} \\ \text{of finite codimension in } [\mathbb{X}]^M. \end{gathered} \tag{8.1.601}$$

Bring in the boundary-to-domain single layer potential operator $\mathscr{S}$ associated with L and Ω as in (A.0.228). Then, as seen from [130, (5.2.92)], [132, (1.3.8)], [131, Theorem 2.4.1, item *(4)*], and [132, Lemma 1.3.1], if either $n \geq 3$ or $n = 2$ and Ω is bounded it follows that for each $g \in [\mathbb{X}]^M$ the function $u := \mathscr{S}g$ in Ω satisfies

$$\begin{gathered} u \in \big[\mathscr{C}^\infty(\Omega)\big]^M,\ Lu = 0 \text{ in } \Omega,\ \mathcal{N}_\kappa(\nabla u) \in \mathbb{X}, \\ \text{and } u(x) = O(|x|^{2-n}) \text{ as } |x| \to \infty \text{ if } \Omega \text{ is unbounded.} \end{gathered} \tag{8.1.602}$$

Moreover, the same properties remain true when $n = 2$ and Ω is an exterior domain under the additional assumption that $\int_{\partial\Omega} g\, d\sigma = 0$ (cf. [132, Lemma 1.3.1, item *(iii)*]). Combining (8.1.602), [130, (5.2.92)], and [132, (1.5.59)] then leads to the conclusion that

$$\begin{gathered}\big(-\tfrac{1}{2}I + K^{\#}_{A^{\top}}\big)g = \partial^{A}_{\nu}u \text{ belongs to the space of admissible boundary} \\ \text{data for the boundary problem (8.1.572), defined in (8.1.575), for each} \\ g \in \big[\mathbb{X}\big]^{M} \text{ if either } n \geq 3 \text{ or } n = 2 \text{ and } \Omega \text{ is bounded, plus the additional} \\ \text{assumption } \int_{\partial\Omega} g\,d\sigma = 0 \text{ if } n = 2 \text{ and } \Omega \text{ is an exterior domain.}\end{gathered} \tag{8.1.603}$$

Also, based on (8.1.603) we see that

$$\begin{gathered}\text{the space of admissible boundary data for the Neumann Problem} \\ \text{(8.1.572), defined as indicated in (8.1.575), contains the image} \\ \mathrm{Im}\Big(-\tfrac{1}{2}I + K^{\#}_{A^{\top}};\ \big[\mathbb{X}\big]^{M}\Big) \text{ in the case when either } n \geq 3 \text{ or } n = 2 \text{ and} \\ \text{the set } \Omega \text{ is bounded, and the space (8.1.575) contains the space} \\ \Big\{\big(-\tfrac{1}{2}I + K^{\#}_{A^{\top}}\big)g : g \in \big[\mathbb{X}\big]^{M} \text{ with } \int_{\partial\Omega} g\,d\sigma = 0\Big\} \text{ if } n = 2 \text{ and } \Omega \\ \text{is an exterior domain.}\end{gathered} \tag{8.1.604}$$

In concert, (8.1.604), (8.1.601), and [130, Theorem 1.1.3] then imply that

$$\begin{gathered}\text{the space of admissible boundary data for the boundary problem} \\ \text{(8.1.572), described in (8.1.575), is a closed linear subspace of finite} \\ \text{codimension in } \big[\mathbb{X}\big]^{M}.\end{gathered} \tag{8.1.605}$$

By design, a solution u of the boundary value problem (8.1.572) exists if and only if the boundary datum f belongs to the space (8.1.575). We shall show that, whenever this is the case, a solution u of (8.1.572) may be found with the additional property that (8.1.576) holds. Let us first look at the case when either $n \geq 3$, or $n = 2$ and Ω is bounded. In this setting, introduce the operator

$$\begin{gathered}T : \big[\mathbb{X}_1\big]^{M} \times \big[\mathbb{X}\big]^{M} \longrightarrow \big[\mathbb{X}\big]^{M} \text{ given for } (f_0, f_1) \in \big[\mathbb{X}_1\big]^{M} \times \big[\mathbb{X}\big]^{M} \\ \text{by } T(f_0, f_1) := \partial^{A}_{\nu}\mathcal{D}_{A} f_0 - \big(-\tfrac{1}{2}I + K^{\#}_{A^{\top}}\big) f_1.\end{gathered} \tag{8.1.606}$$

By virtue of [132, (1.5.29)], [130, (5.2.92)], and item *(11)* in [131, Theorem 2.3.2], it follows that T is a well-defined, linear, and bounded operator in the above context. Since its image contains the space $\mathrm{Im}\Big(-\tfrac{1}{2}I + K^{\#}_{A^{\top}};\ \big[\mathbb{X}\big]^{M}\Big)$, we conclude from (8.1.601) and item *(3)* in [130, Theorem 1.1.3] that

$$\text{the image of } T \text{ is a closed subspace of finite codimension in } \big[\mathbb{X}\big]^{M}. \tag{8.1.607}$$

In this regard, we claim that

$$\begin{gathered}\text{the image of } T \text{ is precisely the space of admissible bound-} \\ \text{ary data for the Neumann Problem (8.1.572) when either} \\ n \geq 3, \text{ or when } n = 2 \text{ and the set } \Omega \text{ is bounded.}\end{gathered} \tag{8.1.608}$$

To deal with one inclusion, suppose

$$\begin{gathered}u \in \big[\mathscr{C}^{\infty}(\Omega)\big]^{M},\ Lu = 0 \text{ in } \Omega,\ \mathcal{N}_{\kappa}(\nabla u) \in \mathbb{X}, \\ \text{and } u(x) = O(|x|^{2-n}) \text{ as } |x| \to \infty \text{ if } \Omega \text{ is unbounded.}\end{gathered} \tag{8.1.609}$$

As remarked in (8.1.596), these properties enure that the nontangential boundary trace $u\big|_{\partial\Omega}^{\kappa-\mathrm{n.t.}}$ exists at σ-a.e. point on $\partial\Omega$ and, as a function, belongs to the space $\big[\mathbb{X}_1\big]^M$. Moreover, [129, Proposition 8.4.9] together with [130, (5.1.12)] and [130, (5.2.92)] presently imply that $\mathcal{N}_\kappa u \in \mathbb{X}$. Granted these properties and once again keeping in mind [130, (5.2.92)], we may invoke [132, Theorem 1.3.3] which gives $u = \mathcal{D}_A\big(u\big|_{\partial\Omega}^{\kappa-\mathrm{n.t.}}\big) - \mathscr{S}\big(\partial_\nu^A u\big)$ in Ω. Taking conormal derivatives then yields, on account of [130, (5.2.92)] and [132, (1.5.30)-(1.5.31), (1.5.66)],

$$\partial_\nu^A u = \partial_\nu^A \mathcal{D}_A\big(u\big|_{\partial\Omega}^{\kappa-\mathrm{n.t.}}\big) - \big(-\tfrac{1}{2}I + K^{\#}_{A^\top}\big)\big(\partial_\nu^A u\big). \tag{8.1.610}$$

As such, $\partial_\nu^A u = T\big(u\big|_{\partial\Omega}^{\kappa-\mathrm{n.t.}}, \partial_\nu^A u\big)$, from which we ultimately conclude that the space of admissible boundary data for the Neumann Problem (8.1.572) is contained in the image of the operator T in (8.1.606).

To prove the opposite inclusion, select an arbitrary function in the image of the operator T, say, $T(f_0, f_1)$ for some $f_0 \in \big[\mathbb{X}_1\big]^M$ and $f_1 \in \big[\mathbb{X}\big]^M$. Then the function defined as $u := \mathcal{D}_A f_0 - \mathscr{S} f_1$ in Ω has all the properties listed in (8.1.609), and satisfies $\partial_\nu^A u = T(f_0, f_1)$. This implies that $T(f_0, f_1)$ is an admissible boundary datum for the Neumann Problem (8.1.572), so the proof of (8.1.608) is now finished.

Once (8.1.608) has been established, the fact that in the case when either $n \geq 3$, or $n = 2$ and Ω is bounded, a solution u of (8.1.572) exists if and only if the boundary datum f belongs to (8.1.575), in which scenario a solution u may be found such that the estimate in (8.1.576) is true, becomes a consequence of (8.1.607) and [130, Proposition 1.2.13]. This completes the analysis in the case when either $n \geq 3$, or $n = 2$ and Ω is bounded.

Next, the goal is to find the analogue of (8.1.608) in the case when $n = 2$ and Ω is an exterior domain. In such a scenario, define the operator $\widetilde{T}$ mapping the space

$$\big[\mathbb{X}_1\big]^M \times \Big\{f_1 \in \big[\mathbb{X}\big]^M : \Big(\int_{S^1} \big[L(\xi)\big]^{-1}\, d\mathcal{H}^1(\xi)\Big)\Big(\int_{\partial\Omega} f_1\, d\sigma\Big) = 0\Big\} \tag{8.1.611}$$

into the space $\big[\mathbb{X}\big]^M$ according to

$$\widetilde{T}(f_0, f_1) := \partial_\nu^A \mathcal{D}_A f_0 - \big(-\tfrac{1}{2}I + K^{\#}_{A^\top}\big) f_1 \text{ for each } (f_0, f_1) \text{ in (8.1.611).} \tag{8.1.612}$$

In relation to this, we claim that

> the image of $\widetilde{T}$ is precisely the space of admissible boundary data for the Neumann Problem (8.1.572) in the case when $n = 2$ and the set Ω is an exterior domain. (8.1.613)

To establish one inclusion, assume

$$u \in \big[\mathscr{C}^\infty(\Omega)\big]^M,\ Lu = 0,\ \mathcal{N}_\kappa(\nabla u) \in \mathbb{X},\ u(x) = O(1) \text{ as } |x| \to \infty. \tag{8.1.614}$$

Under these hypotheses, it has been noted in (8.1.596) that the nontangential boundary trace $u\big|_{\partial\Omega}^{\kappa-\mathrm{n.t.}}$ exists at σ-a.e. point on $\partial\Omega$ and, as a function, it belongs to $\big[\mathbb{X}_1\big]^M$. Also, from [129, Proposition 8.4.9], [130, (5.1.12)], [130, (5.2.92)], and the fact that u is bounded at infinity we also see (keeping in mind that $\partial\Omega$ is compact) that $\mathcal{N}_\kappa u \in \mathbb{X}$. These properties allow us to invoke [131, Corollary 1.5.4] which guarantees the existence of some $c \in \mathbb{C}^M$ validating the integral representation formula

$$u = \mathcal{D}_A\big(u\big|_{\partial\Omega}^{\kappa-\mathrm{n.t.}}\big) - \mathcal{S}\big(\partial_\nu^A u\big) + c \text{ in } \Omega. \tag{8.1.615}$$

Taking the conormal derivative of both sides of (8.1.615) and availing ourselves of [130, (5.2.92)], [132, (1.5.30)-(1.5.31), (1.5.66)], we obtain

$$\partial_\nu^A u = \partial_\nu^A \mathcal{D}_A\big(u\big|_{\partial\Omega}^{\kappa-\mathrm{n.t.}}\big) - \big(-\tfrac{1}{2}I + K_{A^\top}^\#\big)\big(\partial_\nu^A u\big). \tag{8.1.616}$$

The next step is to show that

$$\Big(\int_{S^1} \big[L(\xi)\big]^{-1}\,d\mathcal{H}^1(\xi)\Big)\Big(\int_{\partial\Omega} \partial_\nu^A u\,d\sigma\Big) = 0. \tag{8.1.617}$$

Concretely, from (8.1.615), the last property in (8.1.614), and the fact that (as seen from [132, (1.3.26)], (8.1.596), and [130, (5.2.92)]) $\mathcal{D}_A\big(u\big|_{\partial\Omega}^{\kappa-\mathrm{n.t.}}\big)(x) = O\big(|x|^{-1}\big)$ as $|x| \to \infty$, we see that $\mathcal{S}\big(\partial_\nu^A u\big) = O(1)$ as $|x| \to \infty$. Having proved this, [132, (1.3.12)] then shows that we necessarily have (8.1.617). Going further, from (8.1.616), (8.1.617), and the definition of the operator $\widetilde{T}$ in (8.1.611)-(8.1.612) we deduce that $\partial_\nu^A u = \widetilde{T}\big(u\big|_{\partial\Omega}^{\kappa-\mathrm{n.t.}}, \partial_\nu^A u\big) \in \operatorname{Im}\widetilde{T}$, the image of the operator $\widetilde{T}$. This proves one inclusion in (8.1.613). To establish the opposite inclusion in (8.1.613), select an arbitrary function in the image of $\widetilde{T}$, say, $\widetilde{T}(f_0, f_1)$ with (f_0, f_1) belonging to the space (8.1.611). Upon recalling [132, (1.3.11)], it follows that $u := \mathcal{D}_A f_0 - \mathcal{S} f_1$ is as in (8.1.614) and satisfies $\partial_\nu^A u = \widetilde{T}(f_0, f_1)$. Hence, $\widetilde{T}(f_0, f_1)$ is an admissible boundary datum for the Neumann Problem (8.1.572).

At this stage, the proof of (8.1.613) is complete. With this in hand, the same functional analytic argument (based on [130, Proposition 1.2.13]) then proves that a solution u of (8.1.572) exists if and only if the boundary datum f belongs to (8.1.575), in which scenario a solution u may be found such that the estimate in (8.1.576) holds as well.

Pressing on, consider an arbitrary null-solution u of the boundary problem (8.1.572) (that is, a function u belonging to the space defined in (8.1.577)). As observed in (8.1.596), the nontangential boundary trace $u\big|_{\partial\Omega}^{\kappa-\mathrm{n.t.}}$ exists at σ-a.e. point on $\partial\Omega$ and, as a function, belongs to $\big[\mathbb{X}_1\big]^M$. Moreover, [129, Proposition 8.4.9] together with [130, (5.1.12)] and [130, (5.2.92)] currently imply that $\mathcal{N}_\kappa u \in \mathbb{X}$. In light of these properties, [132, Theorem 1.3.3] gives (keeping in mind that we presently have $\partial_\nu^A u = 0$)

$$u = \mathcal{D}_A\big(u\big|_{\partial\Omega}^{\kappa-\mathrm{n.t.}}\big) \text{ in } \Omega, \tag{8.1.618}$$

provided either $n \geq 3$, or $n = 2$ and Ω is bounded. Suppose this is the case. Going nontangentially to the boundary in (8.1.618) and using the jump-formula [132, (1.5.20)] (while bearing [130, (5.2.92)] in mind), we obtain $\big(-\frac{1}{2}I + K_A\big)\big(u\big|_{\partial\Omega}^{\kappa-\mathrm{n.t.}}\big) = 0$. As a consequence,

$$u\Big|_{\partial\Omega}^{\kappa-\mathrm{n.t.}} \in \mathrm{Ker}\Big(-\tfrac{1}{2}I + K_A;\ \big[\mathbb{X}_1\big]^M\Big). \tag{8.1.619}$$

Also, from (8.1.618) it is clear that the assignment taking u into $u\big|_{\partial\Omega}^{\kappa-\mathrm{n.t.}}$ is linear and injective. In view of this and (8.1.619), we deduce that if either $n \geq 3$, or $n = 2$ and Ω is bounded, then

$$\text{the dimension of (8.1.577) is } \leq \dim \mathrm{Ker}\Big(-\tfrac{1}{2}I + K_A;\ \big[\mathbb{X}_1\big]^M\Big). \tag{8.1.620}$$

Based on (7.1.81) and (8.1.620), we deduce that the space of null-solutions of the boundary problem (8.1.572), described in (8.1.577), is finite dimensional, if either $n \geq 3$, or $n = 2$ and Ω.

When $n = 2$ and Ω is an exterior domain we proceed analogously, starting from the observation that the integral representation formula (8.1.615) now reduces to

$$u = \mathcal{D}_A\Big(u\Big|_{\partial\Omega}^{\kappa-\mathrm{n.t.}}\Big) + c \ \text{ in } \Omega. \tag{8.1.621}$$

Taking nontangential boundary traces and using (8.1.621) we obtain, thanks to [132, (1.5.20)] and [130, (5.2.92)], $\big(-\frac{1}{2}I + K_A\big)\big(u\big|_{\partial\Omega}^{\kappa-\mathrm{n.t.}}\big) \in \mathbb{C}^M$. In turn, from this we further conclude that

$$u\Big|_{\partial\Omega}^{\kappa-\mathrm{n.t.}} \text{ belongs to } \Big\{\phi \in \big[\mathbb{X}_1\big]^M : \big(-\tfrac{1}{2}I + K_A\big)\phi \in \mathbb{C}^M\Big\}. \tag{8.1.622}$$

From item *(8)* in Theorem 7.1.1 we know that taking the threshold $\delta_{\mathbb{X}} \in (0, 1)$ small enough in (8.1.573) guarantees that $-\frac{1}{2}I + K_A$ is a Fredholm operator on $\big[\mathbb{X}_1\big]^M$. Assuming this is the case, item *(g)* in [130, Theorem 2.2.4] implies that the space appearing above is finite dimensional. From this, (8.1.622), and (8.1.621) we then conclude that the space of null-solutions of the boundary problem (8.1.572) is finite dimensional in this case, as well. This completes the treatment of item *(1)*.

Turning our attention to item *(2)*, work under the stronger hypotheses stipulated in (8.1.578). In particular, Proposition 1.4.1 tells that, having chosen a coefficient tensor $A \in \mathfrak{A}_L^{\mathrm{dis}}$, we also have $A^\top \in \mathfrak{A}_{L^\top}^{\mathrm{dis}}$. We shall first treat the case when either $n \geq 3$, or $n = 2$ and Ω is bounded. Choose a function f in the space of admissible boundary data for the boundary problem (8.1.572) (defined in (8.1.575)). This means that there exists u as in (8.1.602) such that $f = \partial_\nu^A u$. As noted earlier, the properties in (8.1.572) further imply that the nontangential boundary trace $u\big|_{\partial\Omega}^{\kappa-\mathrm{n.t.}}$ is well defined as a function in the space $\big[\mathbb{X}_1\big]^M$, and $\mathcal{N}_\kappa u \in \mathbb{X}$. With these in hand, the integral representation formula from [132, Theorem 1.3.3] gives $u = \mathcal{D}_A\big(u\big|_{\partial\Omega}^{\kappa-\mathrm{n.t.}}\big) - \mathcal{S}f$ in Ω. After taking the conormal derivative of both sides (see [132, (1.5.59)] and [130,

(5.2.92)]) we arrive at

$$f = \partial_\nu^A \mathcal{D}_A\big(u\big|_{\partial\Omega}^{\kappa-\text{n.t.}}\big) - \big(-\tfrac{1}{2}I + K^\#_{A^\top}\big)f. \tag{8.1.623}$$

At this point, let us momentarily digress for the purpose of proving that, in the current context,

$$\begin{aligned}\operatorname{Im}\Big(\partial_\nu^A \mathcal{D}_A : \big[\mathbb{X}_1\big]^M \to \big[\mathbb{X}\big]^M\Big) &= \operatorname{Im}\Big(\big(\tfrac{1}{2}I + K^\#_{A^\top}\big)\circ\big(-\tfrac{1}{2}I + K^\#_{A^\top}\big);\ \big[\mathbb{X}\big]^M\Big)\\ &= \operatorname{Im}\Big(\tfrac{1}{2}I + K^\#_{A^\top};\ \big[\mathbb{X}\big]^M\Big) \cap \operatorname{Im}\Big(-\tfrac{1}{2}I + K^\#_{A^\top};\ \big[\mathbb{X}\big]^M\Big).\end{aligned} \tag{8.1.624}$$

To this end, bring in the boundary-to-boundary single layer potential operator S, associated with Ω and the system L as in (A.0.231). If necessary, further decrease the value of $\delta_{\mathbb{X}} \in (0,1)$ as to ensure that the conclusions in Theorem 7.1.20 are presently valid. From [130, (5.2.92)] and [132, (1.5.86)] we conclude that

$$\big(\tfrac{1}{2}I + K^\#_{A^\top}\big)\circ\big(-\tfrac{1}{2}I + K^\#_{A^\top}\big) = \big(\partial_\nu^A \mathcal{D}_A\big)\circ S \text{ on } \big[\mathbb{X}\big]^M. \tag{8.1.625}$$

Given that S maps $\big[\mathbb{X}\big]^M$ into $\big[\mathbb{X}_1\big]^M$ (cf. item *(11)* in [131, Theorem 2.3.2], (8.1.571), [130, (5.2.92)], and [132, Theorem 1.5.1, item *(ix)*]), this implies

$$\begin{aligned}&\operatorname{Im}\Big(\partial_\nu^A \mathcal{D}_A : \big[\mathbb{X}_1\big]^M \to \big[\mathbb{X}\big]^M\Big)\\ &\qquad\supseteq \operatorname{Im}\Big(\big(\tfrac{1}{2}I + K^\#_{A^\top}\big)\circ\big(-\tfrac{1}{2}I + K^\#_{A^\top}\big);\ \big[\mathbb{X}\big]^M\Big).\end{aligned} \tag{8.1.626}$$

To prove the opposite inclusion, we distinguish two cases. First, if $n \geq 3$, the first operator (7.1.387) is invertible. With S^{-1} denoting its inverse, we may then write

$$\partial_\nu^A \mathcal{D}_A = \big(\tfrac{1}{2}I + K^\#_{A^\top}\big)\circ\big(-\tfrac{1}{2}I + K^\#_{A^\top}\big)\circ S^{-1} \text{ on } \big[\mathbb{X}_1\big]^M, \tag{8.1.627}$$

proving the opposite inclusion in (8.1.626). Second, if $n = 2$, the augmented single layer potential operator S_{aug} is invertible in the context of (7.1.389). Hence, for any function $f \in \big[\mathbb{X}_1\big]^M$ there exist $g \in \big[\mathbb{X}\big]^M$ and $c \in \mathbb{C}^M$ with $Sg + c = f$ and $\int_{\partial\Omega} g\, d\sigma = 0$. Using this and the fact that $\partial_\nu^A \mathcal{D}_A$ annihilates constants (as may be seen from [132, (1.5.29)]), we then write

$$\partial_\nu^A \mathcal{D}_A f = \partial_\nu^A \mathcal{D}_A(f - c) = \partial_\nu^A \mathcal{D}_A(Sg) = \big(\tfrac{1}{2}I + K^\#_{A^\top}\big)\Big(\big(-\tfrac{1}{2}I + K^\#_{A^\top}\big)g\Big). \tag{8.1.628}$$

In turn, this implies $\partial_\nu^A \mathcal{D}_A f \in \operatorname{Im}\Big(\big(\tfrac{1}{2}I + K^\#_{A^\top}\big)\circ\big(-\tfrac{1}{2}I + K^\#_{A^\top}\big);\ \big[\mathbb{X}\big]^M\Big)$. Since the function $f \in \big[\mathbb{X}_1\big]^M$ was arbitrary, this establishes the opposite inclusion in (8.1.626). Thus, the opposite inclusion in (8.1.626) holds in all dimensions $n \geq 2$, so the first equality in (8.1.624) is justified.

To prove the second equality in (8.1.624), observe that the left-to-right inclusion is a simple consequence of the fact that $\frac{1}{2}I + K^\#_{A^\top}$ and $-\frac{1}{2}I + K^\#_{A^\top}$ commute (as operators on $[\mathbb{X}]^M$). There remains to justify the right-to-left inclusion in the second equality in (8.1.624). With this aim in mind, pick an arbitrary function

$$f \in \mathrm{Im}\Big(\tfrac{1}{2}I + K^{\#}_{A^\top};\ [\mathbb{X}]^M\Big) \cap \mathrm{Im}\Big(-\tfrac{1}{2}I + K^{\#}_{A^\top};\ [\mathbb{X}]^M\Big). \tag{8.1.629}$$

Then there exist $g_\pm \in [\mathbb{X}]^M$ such that $\big(\tfrac{1}{2}I + K^{\#}_{A^\top}\big)g_+ = f = \big(-\tfrac{1}{2}I + K^{\#}_{A^\top}\big)g_-$ and if we now define $g := -Sg_+ + Sg_- \in [\mathbb{X}_1]^M$, then based on [130, (5.2.92)] and [132, (1.5.85)-(1.5.87)] we may write

$$\begin{aligned}(S \circ \partial^A_\nu \mathcal{D}_A)g &= \big[\big(\tfrac{1}{2}I + K_A\big) \circ \big(-\tfrac{1}{2}I + K_A\big)\big]g \\ &= -\big[\big(-\tfrac{1}{2}I + K_A\big) \circ \big(\tfrac{1}{2}I + K_A\big)\big]Sg_+ + \big[\big(\tfrac{1}{2}I + K_A\big) \circ \big(-\tfrac{1}{2}I + K_A\big)\big]Sg_- \\ &= -\big[\big(-\tfrac{1}{2}I + K_A\big) \circ S\big]\big(\tfrac{1}{2}I + K^{\#}_{A^\top}\big)g_+ + \big[\big(\tfrac{1}{2}I + K_A\big) \circ S\big]\big(-\tfrac{1}{2}I + K^{\#}_{A^\top}\big)g_- \\ &= -\big[\big(-\tfrac{1}{2}I + K_A\big) \circ S\big]f + \big[\big(\tfrac{1}{2}I + K_A\big) \circ S\big]f = Sf.\end{aligned} \tag{8.1.630}$$

If $n \geq 3$, then from (8.1.630) and the fact that the first operator in (7.1.387) is invertible we conclude that

$$f = (\partial^A_\nu \mathcal{D}_A)g \in \mathrm{Im}\Big(\partial^A_\nu \mathcal{D}_A : [\mathbb{X}_1]^M \to [\mathbb{X}]^M\Big), \tag{8.1.631}$$

as wanted. This establishes the second equality in (8.1.624) when ≥ 3. To treat the case when $n = 2$, we first claim that

$$\int_{\partial\Omega} (\partial^A_\nu \mathcal{D}_A)g \,\mathrm{d}\sigma = 0 \tag{8.1.632}$$

and

$$\int_{\partial\Omega} f \,\mathrm{d}\sigma = 0. \tag{8.1.633}$$

To prove these formulas, there is no loss of generality in assuming that Ω is bounded, since otherwise we shall consider the bounded set $\mathbb{R}^2 \setminus \overline{\Omega}$ in place of Ω (here, the lack of jump for $\partial^A_\nu \mathcal{D}_A$ is relevant; see [132, Theorem 1.5.1, item *(vii)*]). Then (8.1.632) is a consequence of [130, (5.2.92)] and [131, (1.7.67)] (which is applicable in the present setting, thanks to [132, (1.3.24)], [130, (5.2.92)], [130, (5.1.13)], item *(4)* in [131, Theorem 2.4.1], and (A.0.1)). Also, (8.1.633) follows in a similar manner, using the fact that $f = \big(-\tfrac{1}{2}I + K^{\#}_{A^\top}\big)g_- = \partial^A_\nu \mathscr{S}g_+$. Having established (8.1.632)-(8.1.633), from (8.1.630) and (7.1.393) we then conclude that

$$S_{\text{aug}}\big((\partial^A_\nu \mathcal{D}_A)g, 0\big) = S_{\text{aug}}(f, 0). \tag{8.1.634}$$

Upon recalling that the augmented single layer potential operator S_{aug} is invertible in the context of (7.1.389), from (8.1.634) we ultimately see that we once again have (8.1.631). Hence, the second equality in (8.1.624) continues to be true in the two-dimensional setting, as well. This finishes the proof of (8.1.624) in all dimensions.

Returning to the mainstream discussion, from (8.1.623) and (8.1.624) we then conclude that

$$f \in \operatorname{Im}\Big(-\tfrac{1}{2}I + K^{\#}_{A^{\top}};\ [\mathbb{X}]^M \Big). \tag{8.1.635}$$

Collectively, (8.1.604) and (8.1.635) prove the portion of (8.1.579) concerning the case when either $n \geq 3$, or $n = 2$ and Ω is bounded.

Going further, whenever the boundary datum f belongs to the space of admissible boundary data defined in (8.1.575), it follows from (8.1.579) that (8.1.635) holds. From this, (8.1.601), and [130, Proposition 1.2.13] we conclude that there exists $g \in [\mathbb{X}]^M$ such that

$$\big(-\tfrac{1}{2}I + K^{\#}_{A^{\top}} \big) g = f \ \text{ and } \ \|g\|_{[\mathbb{X}]^M} \leq C\|f\|_{[\mathbb{X}]^M} \tag{8.1.636}$$

for some constant $C \in (0, \infty)$ independent of f. Then, as before, $u := \mathscr{S}g$ satisfies (8.1.602), as well as (8.1.576), thanks to the estimate recorded in (8.1.636). Lastly, the equality in (8.1.636) together with [130, (5.2.92)] and [132, (1.5.59)] give that $\partial^A_\nu u = \big(-\tfrac{1}{2}I + K^{\#}_{A^{\top}} \big) g = f$, from which we conclude that u is a solution of the boundary problem (8.1.572).

Let us once again temporarily digress for the purpose of establishing that, in the current setting,

$$\begin{aligned} &\operatorname{Ker}\Big(\partial^A_\nu \mathcal{D}_A : [\mathbb{X}_1]^M \to [\mathbb{X}]^M \Big) \\ &\quad = \operatorname{Ker}\Big(\tfrac{1}{2}I + K;\ [\mathbb{X}_1]^M \Big) \boxplus \operatorname{Ker}\Big(-\tfrac{1}{2}I + K;\ [\mathbb{X}_1]^M \Big), \end{aligned} \tag{8.1.637}$$

where $\boxplus$ stands for algebraic direct sum. To justify (8.1.637), use [130, (5.2.92)] and [132, (1.5.85)] to write

$$\big(\tfrac{1}{2}I + K \big) \circ \big(-\tfrac{1}{2}I + K \big) = S \circ \big(\partial^A_\nu \mathcal{D}_A \big) \ \text{ on } \ [\mathbb{X}_1]^M. \tag{8.1.638}$$

As noted earlier, $\partial^A_\nu \mathcal{D}_A$ maps $[\mathbb{X}_1]^M$ into $[\mathbb{X}]^M$, so this further implies

$$\operatorname{Ker}\Big(\partial^A_\nu \mathcal{D}_A : [\mathbb{X}_1]^M \to [\mathbb{X}]^M \Big) \subseteq \operatorname{Ker}\Big(\big(\tfrac{1}{2}I+K\big) \circ \big(-\tfrac{1}{2}I+K\big);\ [\mathbb{X}_1]^M \Big). \tag{8.1.639}$$

To prove the opposite inclusion, we shall consider two cases. Assume first that $n \geq 3$. Then we know that the first operator (7.1.387) is invertible. If S^{-1} denotes its inverse, we see from (8.1.638) that

$$\partial^A_\nu \mathcal{D}_A = S^{-1} \circ \big(\tfrac{1}{2}I + K \big) \circ \big(-\tfrac{1}{2}I + K \big) \ \text{ on } \ [\mathbb{X}_1]^M, \tag{8.1.640}$$

from which the opposite inclusion in (8.1.639) follows. Assume next that $n = 2$ and pick $f \in [\mathbb{X}_1]^M$ with the property that $\big(\tfrac{1}{2}I+K\big)\Big(\big(-\tfrac{1}{2}I+K\big)f\Big) = 0$. Then the function $g := \partial^A_\nu \mathcal{D}_A f$ belongs to $[\mathbb{X}]^M$ and the present assumptions together with (8.1.638) guarantee that $Sg = 0$. In addition, from [131, (1.7.67)] (whose applicability in the present setting is ensured by [132, (1.3.24)], [130, (5.2.92)], [130, (5.1.13)], item *(4)* in [131, Theorem 2.4.1], (A.0.1), and the fact that $\partial\Omega$ is bounded) we see that

$$\int_{\partial\Omega} g \, d\sigma = 0. \tag{8.1.641}$$

From these properties and (7.1.393) we conclude that $S_{\text{aug}}(g, 0) = (0, 0)$. Given that the augmented single layer operator is invertible in the context of (7.1.389), this forces $g = 0$ which, in turn, translated into saying that $f \in \operatorname{Ker}\Big(\partial_\nu^A \mathcal{D}_A : \big[\mathbb{X}_1\big]^M \to \big[\mathbb{X}\big]^M\Big)$. In view of the arbitrariness of f, this ultimately shows that the opposite inclusion in (8.1.639) also holds for $n = 2$. This finishes the proof of the fact that, always,

$$\begin{aligned}\operatorname{Ker}\Big(\partial_\nu^A \mathcal{D}_A : \big[\mathbb{X}_1\big]^M \to \big[\mathbb{X}\big]^M\Big) \\ = \operatorname{Ker}\Big(\big(\tfrac{1}{2}I + K\big) \circ \big(-\tfrac{1}{2}I + K\big);\, \big[\mathbb{X}_1\big]^M\Big).\end{aligned} \tag{8.1.642}$$

Since any $f \in \big[\mathbb{X}_1\big]^M$ may be decomposed as $\big(\tfrac{1}{2}I + K_{A^\top}\big)f + \big(\tfrac{1}{2}I - K_{A^\top}\big)f$, and since the operators $\tfrac{1}{2}I + K_{A^\top}$ and $-\tfrac{1}{2}I + K_{A^\top}$ commute, we conclude that

$$\begin{aligned}\operatorname{Ker}\Big(\big(\tfrac{1}{2}I + K\big) \circ \big(-\tfrac{1}{2}I + K\big);\, \big[\mathbb{X}_1\big]^M\Big) \\ = \operatorname{Ker}\Big(\tfrac{1}{2}I + K;\, \big[\mathbb{X}_1\big]^M\Big) \boxplus \operatorname{Ker}\Big(-\tfrac{1}{2}I + K;\, \big[\mathbb{X}_1\big]^M\Big).\end{aligned} \tag{8.1.643}$$

Finally, from (8.1.642) and (8.1.643) we see that (8.1.637) holds.

Returning to the mainstream argument, we shall now deal with the claim made in (8.1.581) when either $n \geq 3$, or $n = 2$ and Ω is bounded. As seen from (8.1.618)-(8.1.619), the assignment associating to each function u from the space defined in (8.1.577) its nontangential trace $u\big|_{\partial\Omega}^{\kappa-\text{n.t.}} \in \operatorname{Ker}\Big(-\tfrac{1}{2}I + K_A;\, \big[\mathbb{X}_1\big]^M\Big)$ is well defined, linear, and injective. To show that this assignment is also surjective, choose an arbitrary function

$$\phi \in \operatorname{Ker}\Big(-\tfrac{1}{2}I + K_A;\, \big[\mathbb{X}_1\big]^M\Big) \tag{8.1.644}$$

and set $u := \mathcal{D}_A\phi$ in Ω. Together, [130, (5.1.13), (5.2.92)], [132, (1.3.24), (1.3.26)], [132, Lemma 1.3.2], (8.1.571), and [131, Theorem 2.4.1, item *(4)*] imply

$$\begin{aligned}&u \in \big[\mathscr{C}^\infty(\Omega)\big]^M,\ Lu = 0 \text{ in } \Omega,\ \mathcal{N}_\kappa u,\ \mathcal{N}_\kappa(\nabla u) \in \mathbb{X}, \\ &\text{and } u(x) = O(|x|^{1-n}) \text{ as } |x| \to \infty \text{ if } \Omega \text{ is unbounded.}\end{aligned} \tag{8.1.645}$$

In addition, thanks to (8.1.644) and (7.1.441) we have $\partial_\nu^A u = \partial_\nu^A \mathcal{D}_A\phi = 0$. Finally, [130, (5.2.92)] and [132, (1.5.20)] give $u\big|_{\partial\Omega}^{\kappa-\text{n.t.}} = \big(\tfrac{1}{2}I + K_A\big)\phi = \phi$, so the desired claim follows.

In summary, the above reasoning shows that the assignment taking u into $u\big|_{\partial\Omega}^{\kappa-\text{n.t.}}$ is a linear isomorphism from the space (8.1.577) of null-solutions for the boundary problem (8.1.572) into the space $\operatorname{Ker}\Big(-\tfrac{1}{2}I + K_A;\, \big[\mathbb{X}_1\big]^M\Big)$. This proves (8.1.581).

Moreover, from this we also conclude that:

$$\begin{aligned}&\text{the dimension of the space (8.1.577) of null-solutions for the boundary}\\ &\text{problem (8.1.572) is precisely } \dim \mathrm{Ker}\Big(-\tfrac{1}{2}I+K_A;\ \big[\mathbb{X}_1\big]^M\Big) \text{ in the case}\\ &\text{when either } n \geq 3, \text{ or } n=2 \text{ and the set } \Omega \text{ is bounded.}\end{aligned} \tag{8.1.646}$$

Finally, (8.1.618) justifies the claim made in (8.1.582) which, in turn, gives (8.1.580).

The next goal is to derive a useful identity, valid in all dimensions $n \geq 2$, and without any additional constraints on the nature of the set Ω. To set the stage, pick an arbitrary integrability exponent $p \in (1, \infty)$. Then, after possibly further decreasing the value of the threshold $\delta_{\mathbb{X}} \in (0, 1)$, we may write

$$\begin{aligned}\dim\left(\frac{\big[\mathbb{X}\big]^M}{\mathrm{Im}\big(-\frac{1}{2}I+K^{\#}_{A^\top};[\mathbb{X}]^M\big)}\right) &= \dim \mathrm{Ker}\Big(-\tfrac{1}{2}I+K^{\#}_{A^\top};\ \big[\mathbb{X}\big]^M\Big)\\ &= \dim \mathrm{Ker}\Big(-\tfrac{1}{2}I+K^{\#}_{A^\top};\ \big[L^p(\partial\Omega,\sigma)\big]^M\Big)\\ &= \dim \mathrm{Ker}\Big(-\tfrac{1}{2}I+K_A;\ \big[L^p_1(\partial\Omega,\sigma)\big]^M\Big) = \dim \mathrm{Ker}\Big(-\tfrac{1}{2}I+K_A;\ \big[\mathbb{X}_1\big]^M\Big) < \infty.\end{aligned} \tag{8.1.647}$$

Above, the first equality is implied by (8.1.600), the second comes from (7.1.79), the third has been noted in (7.1.622), the fourth is seen from (7.1.80) and (7.1.60), while the final inequality follows from item *(8)* in Theorem 7.1.1.

The above considerations readily yield that the index of the boundary value problem (8.1.572) (i.e., the dimension of (8.1.577) minus the codimension of (8.1.575) in $\big[\mathbb{X}\big]^M$) is equal to zero, provided either $n \geq 3$, or $n = 2$ and Ω is bounded. Indeed, this is clear from (8.1.647), (8.1.581), and (8.1.579).

As far as item *(2)* is concerned, there remains to consider the case when $n = 2$ and Ω is an exterior domain. Assume f is a function belonging to the space of admissible boundary data for the Neumann Problem (8.1.572) (cf. (8.1.575)). Then there exists some u as in (8.1.602) for which $f = \partial^A_\nu u$. The fact that $u(x) = O(1)$ as $|x| \to \infty$ entails (see [131, Theorem 1.5.9] and (8.1.578)) that

$$c_0 := \lim_{x\to\infty} u(x) \text{ exists (in } \mathbb{C}^M) \text{ and } u(x) = c_0 + O(|x|^{-1}) \text{ as } |x| \to \infty. \tag{8.1.648}$$

As remarked earlier, the properties in (8.1.572) further imply that the nontangential boundary trace $u\big|^{\kappa-\mathrm{n.t.}}_{\partial\Omega}$ is well defined as a function in $\big[\mathbb{X}_1\big]^M$, and $\mathcal{N}_\kappa u \in \mathbb{X}$. In view of these and [130, (5.2.92)], the integral representation formula from [132, Theorem 1.3.3] applies to the function $\widetilde{u} := u - c_0$ and, bearing in mind that $\partial^A_\nu \widetilde{u} = \partial^A_\nu u = f$ and [132, (1.3.45)], gives $u - c_0 = \mathcal{D}_A\big(u\big|^{\kappa-\mathrm{n.t.}}_{\partial\Omega}\big) - \mathcal{S}f$ in Ω. After applying the conormal derivative to both sides and using [132, (1.5.59)], [130, (5.2.92)], we end up with $f = (\partial^A_\nu \mathcal{D}_A)\big(u\big|^{\kappa-\mathrm{n.t.}}_{\partial\Omega}\big) - \big(-\frac{1}{2}I + K^{\#}_{A^\top}\big)f$. Together with (8.1.624), this shows that

$$f \in \mathrm{Im}\Big(-\tfrac{1}{2}I + K^{\#}_{A^\top};\ \big[\mathbb{X}\big]^M\Big) \tag{8.1.649}$$

thus, $f = \left(-\frac{1}{2}I + K^{\#}_{A^{\top}}\right)g$ for some $g \in \left[\mathbb{X}\right]^M$. In addition, the same type of argument used to justify (8.1.508) presently gives $\int_{\partial\Omega} g\,d\sigma = 0$. The latter property together with (8.1.604) and (8.1.649) prove the portion of (8.1.579) concerning the case when $n = 2$ and Ω is an exterior domain.

Consider next an arbitrary null-solution u of the boundary problem (8.1.572) (i.e., a function u belonging to the space defined in (8.1.577)) in the case when $n = 2$ and Ω is an exterior domain. Then [131, Theorem 1.5.9] and (8.1.578) guarantee that

$$\text{the limit } c_0 := \lim_{|x|\to\infty} u(x) \text{ exists in } \mathbb{C}^M \text{ and} \tag{8.1.650}$$

$$u(x) = c_0 + O(|x|^{-1}),\ (\nabla u)(x) = O(|x|^{-2}) \text{ as } |x| \to \infty. \tag{8.1.651}$$

It is known that the properties listed in (8.1.572) imply that the nontangential boundary trace $u\big|^{\kappa-\text{n.t.}}_{\partial\Omega}$ is a well-defined function in $\left[\mathbb{X}_1\right]^M$, and $\mathcal{N}_\kappa u \in \mathbb{X}$. Then [132, Theorem 2.3.16] applied to the function $\widetilde{u} := u - c_0$ gives (keeping in mind [132, (1.3.45)] and that now $\partial^A_\nu \widetilde{u} = \partial^A_\nu u = 0$)

$$u - c_0 = \mathcal{D}\big(u\big|^{\kappa-\text{n.t.}}_{\partial\Omega}\big) \text{ in } \Omega. \tag{8.1.652}$$

Taking nontangential boundary traces then yields (in view of the jump-formula [132, (1.5.20)] and the embeddings in [130, (5.2.92)])

$$\left(-\tfrac{1}{2}I + K_A\right)\big(u\big|^{\kappa-\text{n.t.}}_{\partial\Omega}\big) = -c_0. \tag{8.1.653}$$

Together, (8.1.596), (8.1.653), (8.1.648) show that $\Big(u\big|^{\kappa-\text{n.t.}}_{\partial\Omega}, \lim_{|x|\to\infty} u(x)\Big)$ is a pair belonging to the space

$$\Big\{(\phi, c) \in \left[\mathbb{X}_1\right]^M \oplus \mathbb{C}^M : \left(-\tfrac{1}{2}I + K_A\right)\phi = -c\Big\}. \tag{8.1.654}$$

Additionally, it is apparent from (8.1.652) that the mapping associating to each null-solution u of the Neumann Problem (8.1.572) the pair $\Big(u\big|^{\kappa-\text{n.t.}}_{\partial\Omega}, \lim_{|x|\to\infty} u(x)\Big)$ is injective. We claim that said map is also surjective, i.e., its range is the space (8.1.654). To justify this, choose some $\phi \in \left[\mathbb{X}_1\right]^M$ along with some $c \in \mathbb{C}^M$ satisfying $\left(-\frac{1}{2}I + K_A\right)\phi = -c$, and introduce

$$u := \mathcal{D}_A\phi + c \text{ in } \Omega. \tag{8.1.655}$$

As seen from [130, (5.1.13), (5.2.92)], [132, (1.3.24), (1.3.26)], [132, Lemma 1.3.2], (8.1.571), and [131, Theorem 2.4.1, item *(4)*], this function satisfies

$$\begin{gathered} u \in \left[\mathscr{C}^\infty(\Omega)\right]^M,\ Lu = 0 \text{ in } \Omega,\ \mathcal{N}_\kappa(\nabla u) \in \mathbb{X},\ \mathcal{N}_\kappa u \in \mathbb{X}, \\ u(x) = c + O(|x|^{-1}) \text{ and } (\nabla u)(x) = O(|x|^{-2}) \text{ as } |x| \to \infty. \end{gathered} \tag{8.1.656}$$

Moreover, [132, (1.5.20)], [130, (5.2.92)], and $K_A\phi = \frac{1}{2}\phi - c$ ensure that

$$u\Big|_{\partial\Omega}^{\kappa-\text{n.t.}} = \big(\tfrac{1}{2}I + K_A\big)\phi + c = \phi. \tag{8.1.657}$$

Thanks to these properties, the surjectivity claim we presently seek follows as soon as we show that

$$\partial_\nu^A u = 0. \tag{8.1.658}$$

With this goal in mind, first observe that the properties in (8.1.572) further imply that the nontangential boundary trace $u\big|_{\partial\Omega}^{\kappa-\text{n.t.}}$ is a well-defined function in $\big[\mathbb{X}_1\big]^M$, and $\mathcal{N}_\kappa u \in \mathbb{X}$. Granted these properties, and keeping in mind [130, (5.2.92)], the integral representation formula from [132, Theorem 1.3.3] applies to $\widetilde{u} := u - c$ and, thanks to [132, (1.3.45)] and the fact $\partial_\nu^A \widetilde{u} = \partial_\nu^A u$, gives $u - c = \mathcal{D}_A\big(u\big|_{\partial\Omega}^{\kappa-\text{n.t.}}\big) - \mathcal{S}\big(\partial_\nu^A u\big)$ in Ω. Since (8.1.655) and (8.1.657) imply $\mathcal{D}_A\big(u\big|_{\partial\Omega}^{\kappa-\text{n.t.}}\big) = \mathcal{D}_A\phi = u - c$, this formula actually reduces to $\mathcal{S}\big(\partial_\nu^A u\big) = 0$ in Ω. After taking nontangential boundary traces, we therefore arrive at

$$S\big(\partial_\nu^A u\big) = 0, \tag{8.1.659}$$

on account of [130, (5.2.92)], (8.1.595), [132, (1.5.44)], and (A.0.1). Also, much as in (8.1.617), we presently have $\int_{\partial\Omega} \partial_\nu^A u \, d\sigma = 0$. Based on this, (8.1.595), (8.1.659), and (7.1.393), we then conclude that

$$S_{\text{aug}}\big(\partial_\nu^A u, 0\big) = (0, 0). \tag{8.1.660}$$

Having shown this, the invertibility result proved in Theorem 7.1.20 in relation to the operator (7.1.389) then establishes (8.1.658). This argument takes care of (8.1.584). In turn, from (8.1.584) we conclude that

the dimension of the space (8.1.577) of null-solutions for the Neumann Problem (8.1.572) when $n = 2$ and the set Ω is unbounded is equal to the dimension of $\Big\{(\phi, c) \in \big[\mathbb{X}_1\big]^M \oplus \mathbb{C}^M : \big(-\tfrac{1}{2}I + K_A\big)\phi = -c\Big\}$. (8.1.661)

Lastly, (8.1.652) implies (8.1.585) which further justifies (8.1.583).

Next, we propose to show that the index of the boundary value problem (8.1.572) continues to be zero in the case when $n = 2$ and Ω is an exterior domain. To this aim, consider the canonical inclusion

$$\iota : \Big\{g \in \big[\mathbb{X}\big]^M : \int_{\partial\Omega} g \, d\sigma = 0\Big\} \to \big[\mathbb{X}\big]^M \tag{8.1.662}$$

and abbreviate

$$T := \big(-\tfrac{1}{2}I + K^\#_{A^\top}\big) \circ \iota : \Big\{g \in \big[\mathbb{X}\big]^M : \int_{\partial\Omega} g \, d\sigma = 0\Big\} \to \big[\mathbb{X}\big]^M. \tag{8.1.663}$$

In view of the fact that ι is a Fredholm operator with index $-M$ and $-\frac{1}{2}I + K^\#_{A^\top}$ is a Fredholm operator with index zero from $\big[\mathbb{X}\big]^M$ into itself, we may invoke [130, Theorem 2.2.4] to conclude that the operator

$$T \text{ is Fredholm with index } -M \text{ in the context of (8.1.663).} \tag{8.1.664}$$

As such, the codimension on the space of admissible boundary data for the Neumann Problem (8.1.572) in the "full" space of boundary data may be computed as follows:

$$\begin{aligned}
&\dim\left(\frac{\big[\mathbb{X}\big]^M}{\big\{(-\frac{1}{2}I+K^{\#}_{A^\top})g : g\in\big[\mathbb{X}\big]^M,\ \int_{\partial\Omega} g\,d\sigma=0\big\}}\right) = \dim\left(\frac{\big[\mathbb{X}\big]^M}{\operatorname{Im}T}\right) = M + \dim\operatorname{Ker}T \\
&\quad = M + \dim\operatorname{Ker}\Big(-\tfrac{1}{2}I + K^{\#}_{A^\top};\ \big[\mathbb{X}\big]^M\Big) = M + \dim\operatorname{Ker}\Big(-\tfrac{1}{2}I + K_A;\ \big[\mathbb{X}_1\big]^M\Big) \\
&\quad = \dim\Big\{(\phi,c)\in\big[\mathbb{X}_1\big]^M\oplus\mathbb{C}^M :\ \phi - c\in\operatorname{Ker}\Big(-\tfrac{1}{2}I+K_A;\ \big[\mathbb{X}_1\big]^M\Big)\Big\} \\
&\quad = \dim\Big\{(\phi,c)\in\big[\mathbb{X}_1\big]^M\oplus\mathbb{C}^M :\ \big(-\tfrac{1}{2}I+K_A\big)\phi = -c\Big\}.
\end{aligned} \tag{8.1.665}$$

Above, the first equality is seen from (8.1.662)-(8.1.663), while the second equality is implied by (8.1.664). The third equality in (8.1.665) is a consequence of

$$\Big\{g\in\operatorname{Ker}\Big(-\tfrac{1}{2}I+K^{\#}_{A^\top};\ \big[\mathbb{X}\big]^M\Big) : \int_{\partial\Omega} g\,d\sigma = 0\Big\} = \operatorname{Ker}\Big(-\tfrac{1}{2}I+K^{\#}_{A^\top};\ \big[\mathbb{X}\big]^M\Big).$$

Indeed, the left-to-right inclusion is trivial, while the right-to-left inclusion follows from [132, (1.7.59)], [130, (5.2.92)], and the fact that Ω is presently assumed to be unbounded. The fourth equality in (8.1.665) is contained in (8.1.647). The fifth equality in (8.1.665) is a consequence of (8.1.434) used with

$$V := \big[\mathbb{X}_1\big]^M,\quad F_0 := \mathbb{C}^M,\ \text{ and } F_1 := \operatorname{Ker}\Big(-\tfrac{1}{2}I+K_A;\ \big[\mathbb{X}_1\big]^M\Big). \tag{8.1.666}$$

The sixth equality in (8.1.665) is directly implied by [132, (1.5.21)]. Having established (8.1.665), from (8.1.579) and (8.1.584) we conclude that the index of the Neumann Problem (8.1.572) is also equal to zero in this last case. This finishes the treatment of item *(2)*.

Turning our attention to item *(3)*, work under the assumption that $A\in\mathfrak{A}^{\rm dis}_L$ is positive definite. In the case when either $n\geq 3$, or $n=2$ and the set Ω is bounded, we conclude from (8.1.343) and (7.1.277) that the space of admissible boundary data for the Neumann Problem (8.1.572) is equal to

$$\operatorname{Im}\Big(-\tfrac{1}{2}I+K^{\#}_{A^\top};\ \big[\mathbb{X}\big]^M\Big) = \big[\mathbb{X}\cap(\mathbb{C}_{\partial\Omega_+})^{\perp}\big]^M. \tag{8.1.667}$$

This takes care of the first claim in (8.1.587). The second claim in (8.1.587) follows from (8.1.580), (7.1.60), (7.1.80), and (7.1.305).

Suppose now $n=2$ and Ω is an exterior domain. The last property in (8.1.579) then combines with (7.1.292) to give (bearing in mind [130, (5.2.92)]) that the space of admissible boundary data for the Neumann Problem (8.1.572) is equal to

$$\Big\{\big(-\tfrac{1}{2}I+K^{\#}_A\big)g :\ g\in\big[\mathbb{X}\cap\mathbb{C}^{\perp}\big]^M\Big\} = \big[\mathbb{X}\cap(\mathbb{C}^{\infty}_{\partial\Omega_+})^{\perp}\big]^M. \tag{8.1.668}$$

The first claim in (8.1.588) is therefore justified. As regards the second claim in (8.1.588), from (8.1.583), (7.1.293), [130, (5.2.92)], and (8.1.439)-(8.1.441) we see that the space of null-solutions of the Neumann Problem (8.1.572) is contained in

$$\Big\{\mathcal{D}_A\phi + c : \phi \in \big(\mathbb{C}^{\infty}_{\partial\Omega_+}\big)^M \text{ and } c \in \mathbb{C}^M\Big\} \subseteq \big(\mathbb{C}^{\infty}_{\Omega_+}\big)^M. \tag{8.1.669}$$

In the converse direction, it is clear from definitions that any function in $\big(\mathbb{C}^{\infty}_{\Omega_+}\big)^M$ is a null-solution of the Neumann Problem (8.1.572). Consequently, the space of null-solutions of the Neumann Problem (8.1.572) is, in the current case, precisely $\big(\mathbb{C}^{\infty}_{\Omega_+}\big)^M$. This completes the treatment of item *(3)*.

Moving on, the claims in the current item *(4)*, pertaining to the Neumann Problem (8.1.591), with boundary data $\mathring{\mathbb{X}}$ are established in a very analogous manner (now using [130, (5.2.101)], as well as [130, (5.2.85), (5.2.86)]). In turn, all claims made in relation to the Neumann Problems (8.1.589) and (8.1.592) follow from what we have proved so far with $\mathbb{X}'$ in place of $\mathbb{X}$ (which is a permissible choice, thanks to [130, Proposition 5.1.14]). Finally, the claims made in the current item *(5)* are seen from item *(6)* in Theorem 8.1.15. This finishes the proof of Theorem 8.1.17. □

We next discuss the Dirichlet Problem for weakly elliptic systems in δ-oscillating AR domains with boundary data in Hölder spaces $\mathscr{C}^\alpha$, with $0 < \alpha < 1$. Taking $\delta > 0$ sufficiently small allows us to target any desired exponent $\alpha \in (0, 1)$. In particular, in the class of infinitesimally flat AR domains, any $\alpha \in (0, 1)$ will do. The Dirichlet Problem for the system of elastostatics in three-dimensional bounded Lipschitz domains for boundary data in Hölder spaces (with some small exponent) has been solved by B. Dahlberg and C. Kenig in [39, Theorem 1.5, p. 625]. The well-posedness of the Dirichlet Problem for the Laplacian in arbitrary bounded Lipschitz domains with boundary data in Hölder spaces (again, for some small exponent) goes back to the work of O. Kellogg (see the comments in relation to [86, Theorem 5.2, p. 192]). Alternative proofs of the aforementioned results have been given by N. Kalton and M. Mitrea in [88, Theorem 4.7, p. 3920] and [88, Theorem 4.1, p. 3916], respectively. Let us also mention that M. Mitrea and M. Taylor have dealt in [156, Corollary 7.8] with the Dirichlet Problem for the Laplace-Beltrami operator with boundary data in Hölder spaces on Lipschitz subdomains of Riemannian manifolds.

Theorem 8.1.18 (The Dirichlet Problem with Data in Hölder Spaces) *Fix two integers $n, M \in \mathbb{N}$ with $n \geq 2$ and suppose $\Omega \subseteq \mathbb{R}^n$ is an Ahlfors regular domain with compact boundary. Abbreviate $\sigma := \mathcal{H}^{n-1}\lfloor\partial\Omega$ and denote by ν the geometric measure theoretic outward unit normal to Ω. Also, let L be a homogeneous, second-order, constant complex coefficient, $M\times M$ system in $\mathbb{R}^n$. Finally, fix an arbitrary $\alpha \in (0, 1)$.*

Then there exists a threshold $\delta \in (0, 1)$ which depends only on n, α, L, the ADR *constants of $\partial\Omega$, and the relative flatness ratio $\wp(\Omega)/\mathrm{diam}\,\partial\Omega$ so that when*

$$\mathrm{dist}\Big(\nu\,,\,[\mathrm{VMO}\,(\partial\Omega,\sigma)]^n\Big) < \delta \tag{8.1.670}$$

where the distance is considered in the space $\left[\mathrm{BMO}(\partial\Omega,\sigma)\right]^n$ *(hence if* Ω *is a* δ*-oscillating* AR *domain*[59]*; cf. Definition 3.4.1), the following statements are true:*

(1) If one assumes that

$$L \text{ is weakly elliptic with } \mathfrak{A}_L^{\mathrm{dis}} \neq \varnothing \text{ and } \mathfrak{A}_{L^\top}^{\mathrm{dis}} \neq \varnothing, \tag{8.1.671}$$

then the Dirichlet Problem with data in Hölder spaces

$$\begin{cases} u \in \left[\mathscr{C}^\alpha(\overline{\Omega}) \cap \mathscr{C}^\infty(\Omega)\right]^M, \quad Lu = 0 \text{ in } \Omega, \\ u\big|_{\partial\Omega} = f \in \left[\mathscr{C}^\alpha(\partial\Omega)\right]^M, \\ u(x) = O(|x|^{2-n}) \text{ as } |x| \to \infty \text{ if } \Omega \text{ is unbounded}, \end{cases} \tag{8.1.672}$$

is Fredholm solvable, in the following precise sense: First, if the system L is weakly elliptic with $\mathfrak{A}_L^{\mathrm{dis}} \neq \varnothing$*, then the space of admissible boundary data for the boundary problem* (8.1.672)*, i.e.,*

$$\Big\{u\big|_{\partial\Omega} : u \in \left[\mathscr{C}^\alpha(\overline{\Omega}) \cap \mathscr{C}^\infty(\Omega)\right]^M, \ Lu = 0 \text{ in } \Omega, \\ \text{and } u(x) = O(|x|^{2-n}) \text{ as } |x| \to \infty \text{ if } \Omega \text{ is unbounded}\Big\} \tag{8.1.673}$$

is a closed linear subspace of finite codimension in $\left[\mathscr{C}^\alpha(\partial\Omega)\right]^M$*, and a solution u of* (8.1.672) *exists if and only if the boundary datum f belongs to* (8.1.673)*. Second, if L is weakly elliptic with* $\mathfrak{A}_{L^\top}^{\mathrm{dis}} \neq \varnothing$ *then the space of null-solutions of the boundary problem* (8.1.673)*, i.e.,*

$$\Big\{u \in \left[\mathscr{C}^\alpha(\overline{\Omega}) \cap \mathscr{C}^\infty(\Omega)\right]^M : \ Lu = 0 \text{ in } \Omega, \ u\big|_{\partial\Omega} = 0, \\ \text{and } u(x) = O(|x|^{2-n}) \text{ as } |x| \to \infty \text{ if } \Omega \text{ is unbounded}\Big\} \tag{8.1.674}$$

is finite dimensional. Lastly, the Fredholm solvability result just described is optimal, since in the class of operators L for which $\mathfrak{A}_L^{\mathrm{dis}} = \varnothing$ *or* $\mathfrak{A}_{L^\top}^{\mathrm{dis}} = \varnothing$ *it may happen that the Dirichlet Problem* (8.1.672) *fails to be Fredholm solvable even when the underlying domain* Ω *is the unit disk in the plane (or the complement of its closure). In fact, the aforementioned failure of Fredholm solvability may be total, in the sense that for the problem* (8.1.672)*, both the space of admissible boundary data may have infinite codimension in the full space of boundary data, and the space of null-solutions may be infinite dimensional.*

(2) If (8.1.671) *is strengthened to*[60] *assuming that*

$$L \text{ satisfies the Legendre-Hadamard strong ellipticity condition (cf. [131, Definition 1.3.2]) and one has } \mathfrak{A}_L^{\mathrm{dis}} \neq \varnothing \tag{8.1.675}$$

[59] in particular, if Ω is an infinitesimally flat AR domain

[60] according to Proposition 1.4.1, whenever L is as in (8.1.675) we have $\mathfrak{A}_L^{\mathrm{dis}} \neq \varnothing$ and $\mathfrak{A}_{L^\top}^{\mathrm{dis}} \neq \varnothing$

then the boundary value problem (8.1.672) *is actually well posed. More specifically,* (8.1.672) *has a unique solution, and this solution enjoys the additional property that there exists a constant* $C = C(\Omega, \alpha, L, p) \in (0, \infty)$ *such that*

$$\sup_{x\in\overline{\Omega}} |u(x)| + \sup_{\substack{x,y\in\overline{\Omega}\\ x\neq y}} \frac{|u(x)-u(y)|}{|x-y|^\alpha} \leq C\|f\|_{[\mathscr{C}^\alpha(\partial\Omega)]^M}. \tag{8.1.676}$$

We wish to underscore the fact that, as seen from Proposition 1.6.12 and [131, Proposition 1.4.39], in the scalar case, i.e., for $M = 1$, and dimensions $n \geq 3$, in place of (8.1.675) one may just demand that the operator L is weakly elliptic. Furthermore, in the scalar case in the two-dimensional setting (i.e., for $M = 1$ and $n = 2$), if L satisfies the Legendre-Hadamard strong ellipticity condition then Proposition 1.6.11 implies that we always have $\mathfrak{A}_L^{\rm dis} \neq \varnothing$. In addition, according to (1.6.141) in Proposition 1.6.13, in the scalar case in the two-dimensional setting (i.e., for $M = 1$ and $n = 2$) we may omit (8.1.675) altogether if L is a weakly elliptic operator with real coefficients. Finally, regarding the formulation of Theorem 8.1.18, similar considerations to those made in Remark 8.1.2 are currently applicable.

Proof of Theorem 8.1.18 To deal with item *(1)*, first work under the assumption that L is weakly elliptic with $\mathfrak{A}_L^{\rm dis} \neq \varnothing$ Choose $\delta \in (0, 1)$ small enough, depending on n, α, L, the ADR constants of $\partial\Omega$, and the relative flatness ratio $\wp(\Omega)/\operatorname{diam}\partial\Omega$, such that the conclusions in Theorem 3.4.2 and item *(2)* of Theorem 7.1.1 are valid. Keeping in mind [132, Theorem 1.8.2] and [130, Theorem 1.1.3, item *(3)*], we then conclude that the space of admissible boundary data for the boundary problem (8.1.672), described in (8.1.673), is a closed linear subspace of finite codimension in $\big[\mathscr{C}^\alpha(\partial\Omega)\big]^M$. Obviously, (8.1.672) is solvable if and only if the boundary datum belongs to said space. Suppose next that L is weakly elliptic with $\mathfrak{A}_{L^\top}^{\rm dis} \neq \varnothing$. This ensures the finite dimensionality of the space of nullsolutions for the Dirichlet Problem with boundary data in Lebesgue spaces (cf. item *(1)* in Theorem 8.1.6). From this and simple embeddings, we also see that the space of nullsolutions for the Dirichlet Problem (8.1.672) is finite dimensional. Finally, the very first claim in the current item *(1)* is a simple consequence of what we have proved so far, while the very last claim in the current item *(1)* is seen from Theorem 8.6.5, Theorem 8.6.7, and Example 1.6.14. This takes care of item *(1)*.

Turning attention to item *(2)*, further decrease the value of $\delta \in (0, 1)$ so that the conclusions in Theorem 7.1.13 and Theorem 7.1.14 hold (for suitable choices of the smoothness parameter s, involved in their statements). Suppose first that $n \geq 3$. Bearing [130, (7.1.59)] in mind, the invertibility result in (7.1.348) then ensures that there exists $C \in (0, \infty)$ such that, given any $f \in \big[\mathscr{C}^\alpha(\partial\Omega)\big]^M$, there exists some function $g \in \big[B^{\infty,\infty}_{\alpha-1}(\partial\Omega, \sigma)\big]^M$ with

$$Sg = f \text{ and } \|g\|_{[B^{\infty,\infty}_{\alpha-1}(\partial\Omega,\sigma)]^M} \leq C\|f\|_{[\mathscr{C}^\alpha(\partial\Omega)]^M}. \tag{8.1.677}$$

With the boundary-to-domain single layer potential operator associated with L and Ω as in [132, (4.2.205)], consider

$$u := \mathscr{S}g \text{ in } \Omega. \tag{8.1.678}$$

By further decreasing the value of $\delta \in (0, 1)$, if necessary, Theorem 3.4.2 applies and gives that Ω is a locally uniform domain satisfying a two-sided local John condition. In particular, the latter property allows us to combine [130, (7.7.6)] with [130, (11.11.12)] to arrive at the conclusion that

$$\begin{array}{c} B^{\infty,\infty}_{\alpha-1}(\partial\Omega,\sigma) \hookrightarrow B^{p,p}_{s-1}(\partial\Omega,\sigma) \hookrightarrow L^p_{-1}(\partial\Omega,\sigma) \\ \text{whenever } 1 < p < \infty \text{ and } 0 < s < \alpha < 1. \end{array} \tag{8.1.679}$$

From (8.1.679), [132, (4.2.210)], [130, Proposition 11.8.2], and [132, Lemma 1.3.1] we see that if Ω is an exterior domain then $u(x) = O(|x|^{2-n})$ as $|x| \to \infty$. Based on this decay property, (8.1.678), [132, (4.2.207), (4.2.209), (4.2.219), (4.3.49)], and (8.1.677) we conclude that u satisfies all demands in (8.1.672).

Suppose next that $n = 2$ and Ω is an exterior domain. Then the invertibility of (7.1.351) guarantees the existence of some $C \in (0, \infty)$ such that for any given $f \in \big[\mathscr{C}^\alpha(\partial\Omega)\big]^M$ we may find $g \in \big[B^{\infty,\infty}_{\alpha-1}(\partial\Omega,\sigma)\big]^M$ and $c \in \mathbb{C}^M$ such that

$$\begin{array}{c} Sg + c = f, \quad \langle g, \mathbf{e}_j\rangle = 0 \text{ for } 1 \leq j \leq M, \text{ and} \\ |c| + \|g\|_{[B^{\infty,\infty}_{\alpha-1}(\partial\Omega,\sigma)]^M} \leq C\|f\|_{[\mathscr{C}^\alpha(\partial\Omega)]^M}. \end{array} \tag{8.1.680}$$

Then, much as before, it follows that the function $u := \mathscr{S}g + c$ in Ω solves (8.1.672) and satisfies (8.1.676). In fact, the same procedure as above gives a solution of (8.1.672) satisfying (8.1.676) in the case when $n = 2$ and Ω is bounded.

Finally, uniqueness for the boundary value problem (8.1.672) (in all dimensions $n \geq 2$) follows by observing that any function $u \in \big[\mathscr{C}^\alpha(\overline{\Omega})\big]^M$ with the additional property that $u(x) = O(1)$ as $|x| \to \infty$ if Ω is unbounded is, necessarily, a bounded function. Hence, for any aperture parameter $\kappa > 0$, we have $\mathcal{N}_\kappa u \in L^\infty(\partial\Omega, \sigma)$ (cf. [129, Lemma 8.3.2]). As such we may invoke the uniqueness part in Theorem 8.1.6 to conclude that (8.1.672) has at most one solution. □

The Dirichlet Problem with boundary data in Hölder spaces treated in Theorem 8.1.18 has a natural "sub-problem", in which the boundary data is now prescribed in *vanishing* Hölder spaces (of the sort introduced in [130, §3.2]) and a solution is sought satisfying a similar vanishing property. This is all made precise in the theorem below, which appears to be entirely new (in any type of analytic and geometric context).

Theorem 8.1.19 (The Dirichlet Problem with Data in Vanishing Hölder Spaces) *Fix two integers $n, M \in \mathbb{N}$ with $n \geq 2$ and suppose $\Omega \subseteq \mathbb{R}^n$ is an Ahlfors regular domain with compact boundary. Abbreviate $\sigma := \mathcal{H}^{n-1}\lfloor\partial\Omega$ and denote by ν the geometric measure theoretic outward unit normal to Ω. Also, let L be a homogeneous, second-order, constant complex coefficient, $M \times M$ system in $\mathbb{R}^n$. Finally, fix an arbitrary exponent*

$$\alpha \in (0, 1) \tag{8.1.681}$$

and consider the Dirichlet Problem with data in vanishing Hölder spaces

$$\begin{cases} u \in \left[\mathscr{C}^{\alpha}_{\text{van}}(\overline{\Omega}) \cap \mathscr{C}^{\infty}(\Omega)\right]^M, \quad Lu = 0 \text{ in } \Omega, \\ u\big|_{\partial\Omega} = f \in \left[\mathscr{C}^{\alpha}_{\text{van}}(\partial\Omega)\right]^M, \\ u(x) = O(|x|^{2-n}) \text{ as } |x| \to \infty \text{ if } \Omega \text{ is unbounded,} \end{cases} \tag{8.1.682}$$

where the homogeneous vanishing Hölder spaces intervening above are defined as in (A.0.49) *(with* $\Sigma := \partial\Omega$ *and* $\Sigma := \overline{\Omega}$*, respectively).*

Then there exists $\delta \in (0, 1)$ *depending only on n,* α*, L, the* ADR *constants of* $\partial\Omega$*, and the relative flatness ratio* $\wp(\Omega)/\mathrm{diam}\,\partial\Omega$ *with the property that when*

$$\mathrm{dist}\Big(\nu, \, [\mathrm{VMO}\,(\partial\Omega, \sigma)]^n\Big) < \delta \tag{8.1.683}$$

where the distance is considered in the space $\left[\mathrm{BMO}\,(\partial\Omega, \sigma)\right]^n$ *(hence if* Ω *is a* δ*-oscillating* AR *domain*[61]*; cf. Definition 3.4.1), the following statements are true:*

(1) If one assumes that

$$L \text{ is weakly elliptic with } \mathfrak{A}^{\text{dis}}_{L} \neq \varnothing \text{ and } \mathfrak{A}^{\text{dis}}_{L^{\top}} \neq \varnothing, \tag{8.1.684}$$

then the boundary value problem (8.1.682) *is Fredholm solvable, in the following precise sense: First, if L is weakly elliptic with* $\mathfrak{A}^{\text{dis}}_{L} \neq \varnothing$ *then the space of admissible boundary data for the boundary problem* (8.1.682)*, i.e.,*

$$\Big\{u\big|_{\partial\Omega} : u \in \left[\mathscr{C}^{\alpha}_{\text{van}}(\overline{\Omega}) \cap \mathscr{C}^{\infty}(\Omega)\right]^M, \; Lu = 0 \text{ in } \Omega, \\ \text{and } u(x) = O(|x|^{2-n}) \text{ as } |x| \to \infty \text{ if } \Omega \text{ is unbounded}\Big\} \tag{8.1.685}$$

is a closed linear subspace of finite codimension in $\left[\mathscr{C}^{\alpha}(\partial\Omega)\right]^M$*, and a solution u of* (8.1.682) *exists if and only if the boundary datum f belongs to* (8.1.685)*. Second, if L is weakly elliptic with* $\mathfrak{A}^{\text{dis}}_{L^{\top}} \neq \varnothing$ *then the space of null-solutions of the boundary problem* (8.1.682)*, i.e.,*

$$\Big\{u \in \left[\mathscr{C}^{\alpha}(\overline{\Omega})_{\text{van}} \cap \mathscr{C}^{\infty}(\Omega)\right]^M : \; Lu = 0 \text{ in } \Omega, \; u\big|_{\partial\Omega} = 0, \text{ and} \\ u(x) = O(|x|^{2-n}) \text{ as } |x| \to \infty \text{ if } \Omega \text{ is unbounded}\Big\} \tag{8.1.686}$$

is finite dimensional. Lastly, the Fredholm solvability result just described is optimal, since in the class of operators L for which $\mathfrak{A}^{\text{dis}}_{L} = \varnothing$ *or* $\mathfrak{A}^{\text{dis}}_{L^{\top}} = \varnothing$ *it may happen that the Dirichlet Problem* (8.1.682) *fails to be Fredholm solvable even when the underlying domain* Ω *is the unit disk in the plane (or the complement of its closure). In fact, the aforementioned failure of Fredholm solvability may be total, in the sense that for the problem* (8.1.682)*, both the space of admissible boundary data may have infinite codimension in the full space of boundary data, and the space of null-solutions may be infinite dimensional.*

[61] in particular, if Ω is an infinitesimally flat AR domain

(2) If (8.1.684) *is strengthened to*[62] *assuming that*

$$L \text{ satisfies the Legendre-Hadamard strong ellipticity condition (cf. [131, Definition 1.3.2]) and one has } \mathfrak{A}_L^{\rm dis} \neq \varnothing \tag{8.1.687}$$

then the boundary value problem (8.1.682) *is actually well posed. More specifically,* (8.1.682) *has a unique solution, this solution satisfies a naturally accompanying estimate (cf.* (8.1.676)*), and enjoys the additional property that for each* $p \in (1, \infty)$ *one has*

$$\lim_{R\to 0^+}\left\{\sup_{\substack{x\in\partial\Omega\\ r\in(0,R)}}\left(\frac{1}{r^{n-1+\alpha p}}\int_{B(x,r)\cap\Omega}|\nabla u|^p \operatorname{dist}(\cdot,\partial\Omega)^{p-1}\,d\mathcal{L}^n\right)^{1/p}\right\}=0. \tag{8.1.688}$$

Once again, Proposition 1.6.12 and [131, Proposition 1.4.39] guarantee that

$$\text{in the scalar case, i.e., for } M=1, \text{ and dimensions } n\geq 3, \text{ in place of (8.1.687) one may simply ask that the operator } L \text{ is weakly elliptic.} \tag{8.1.689}$$

Also, in the scalar case in the two-dimensional setting (i.e., for $M = 1$ and $n = 2$), if L satisfies the Legendre-Hadamard strong ellipticity condition then Proposition 1.6.11 guarantees that we automatically have $\mathfrak{A}_L^{\rm dis} \neq \varnothing$. Furthermore, according to (1.6.141) in Proposition 1.6.13, in the scalar case in the two-dimensional setting (i.e., for $M = 1$ and $n = 2$) we may omit (8.1.687) altogether if L is a weakly elliptic operator with real coefficients. Finally, as far as the formulation of Theorem 8.1.19 is concerned, similar considerations to those made in Remark 8.1.2 are currently applicable.

Proof of Theorem 8.1.19 All claims in item *(1)* either follow directly from, or may be dealt with analogously to, the corresponding claims in Theorem 8.1.18. To deal with item *(2)*, assume first that $n \geq 3$. In addition to the index α fixed in (8.1.681), pick some $\beta \in (\alpha, 1)$. Select $\delta \in (0, 1)$ small enough, depending on n, α, β, L, the ADR constants of $\partial\Omega$, and the relative flatness ratio $\wp(\Omega)/\operatorname{diam}\partial\Omega$, such that the conclusions in Theorem 7.1.13 and Theorem 7.1.14 hold for suitable choices of the smoothness parameter $s \in (0, 1)$ involved in their statements, including $s := \alpha$ and $s := \beta$. In particular, the operators

$$S : \big[B^{\infty,\infty}_{\alpha-1}(\partial\Omega,\sigma)\big]^M \to \big[\mathscr{C}^\alpha(\partial\Omega)\big]^M, \quad S : \big[B^{\infty,\infty}_{\beta-1}(\partial\Omega,\sigma)\big]^M \to \big[\mathscr{C}^\beta(\partial\Omega)\big]^M$$

are well defined, linear, bounded, and invertible. Based on these, item *(2)* in [130, Lemma 1.2.20], and [130, (3.2.15)] we conclude that S is also a well-defined, linear, bounded, invertible operator in the context

$$S : \overline{\big[B^{\infty,\infty}_{\beta-1}(\partial\Omega,\sigma)\big]^M}^{[B^{\infty,\infty}_{\alpha-1}(\partial\Omega,\sigma)]^M} \longrightarrow \big[\mathscr{C}^\alpha_{\rm van}(\partial\Omega)\big]^M, \tag{8.1.690}$$

[62] according to Proposition 1.4.1, whenever L is as in (8.1.687) we have $\mathfrak{A}_L^{\rm dis} \neq \varnothing$ and $\mathfrak{A}_{L^\top}^{\rm dis} \neq \varnothing$

where the space in the right is the closure of $\big[B^{\infty,\infty}_{\beta-1}(\partial\Omega,\sigma)\big]^M$ in $\big[B^{\infty,\infty}_{\alpha-1}(\partial\Omega,\sigma)\big]^M$. As such, given a boundary datum f as in (8.1.682), we may find a function

$$g \in \overline{\big[B^{\infty,\infty}_{\beta-1}(\partial\Omega,\sigma)\big]^M}^{[B^{\infty,\infty}_{\alpha-1}(\partial\Omega,\sigma)]^M} \quad \text{satisfying } Sg = f. \tag{8.1.691}$$

With the boundary-to-domain single layer $\mathscr{S}$ associated with L and Ω as in [132, (4.2.205)], define

$$u := \mathscr{S}g \text{ in } \Omega. \tag{8.1.692}$$

Much as in the proof of Theorem 8.1.18, this enjoys the following properties:

$$\begin{aligned} &u \in \big[\mathscr{C}^\alpha(\overline{\Omega}) \cap \mathscr{C}^\infty(\Omega)\big]^M, \quad Lu = 0 \text{ in } \Omega, \\ &u\big|_{\partial\Omega} = f, \quad u(x) = O(|x|^{2-n}) \text{ as } |x| \to \infty \text{ if } \Omega \text{ is unbounded.} \end{aligned} \tag{8.1.693}$$

There remains to show that u also belongs to the vanishing Hölder space $\big[\mathscr{C}^\alpha_{\text{van}}(\overline{\Omega})\big]^M$, something which follows as soon as we show that the boundary-to-domain single layer potential operator is well defined in the context

$$\mathscr{S} : \overline{\big[B^{\infty,\infty}_{\beta-1}(\partial\Omega,\sigma)\big]^M}^{[B^{\infty,\infty}_{\alpha-1}(\partial\Omega,\sigma)]^M} \longrightarrow \big[\mathscr{C}^\alpha_{\text{van}}(\overline{\Omega})\big]^M. \tag{8.1.694}$$

In turn, this may be justified much as in the proof of [132, (1.8.16)] provided we check the availability of two key tools employed on that occasion (see the argument in [132, (1.8.69)-(1.8.79)]). Specifically, we need to know that for each $s \in (0,1)$ matters may be arranged (taking $\delta > 0$ sufficiently small) so that the operator

$$\mathscr{S} : \big[B^{\infty,\infty}_{s-1}(\partial\Omega,\sigma)\big]^M \longrightarrow \big[\mathscr{C}^s(\overline{\Omega})\big]^M \tag{8.1.695}$$

is well defined, linear, and continuous, and that there exists a finite constant $C > 0$ (which depends only on Ω, L, and s) such that, for each $h \in \big[B^{\infty,\infty}_{s-1}(\partial\Omega,\sigma)\big]^M$,

$$\sup_{x\in\Omega} \Big\{\operatorname{dist}(x,\partial\Omega)^{1-s}\big|\nabla(\mathscr{S}h)(x)\big| \le C\|h\|_{[B^{\infty,\infty}_{s-1}(\partial\Omega,\sigma)]^M}. \tag{8.1.696}$$

However, (8.1.695) is implied by [132, (4.2.219)], while (8.1.696) is a consequence of [132, (4.2.216)]. These ingredients may also be used to show, via the same type of argument as in the proof of [132, (1.8.19)], that for each $p \in (1,\infty)$ we have

$$\begin{aligned} &\lim_{R\to 0^+} \left\{ \sup_{\substack{x\in\partial\Omega \\ r\in(0,R)}} \left(\frac{1}{r^{n-1+\eta p}} \int_{B(x,r)\cap\Omega} \big|\nabla\mathscr{S}g\big|^p \operatorname{dist}(\cdot,\partial\Omega)^{p-1}\, d\mathcal{L}^n \right)^{1/p} \right\} = 0 \\ &\qquad \text{whenever } g \in \overline{\big[B^{\infty,\infty}_{\beta-1}(\partial\Omega,\sigma)\big]^M}^{[B^{\infty,\infty}_{\alpha-1}(\partial\Omega,\sigma)]^M}. \end{aligned} \tag{8.1.697}$$

In view of (8.1.692), this justifies (8.1.688). Moreover, uniqueness for the boundary value problem (8.1.682) is implied by the corresponding uniqueness statement in Theorem 8.1.18. This completes the treatment of the case when $n \geq 3$.

Finally, the case when $n = 2$ is dealt with along similar lines, now employing an augmented single layer potential operator (as in the proof of Theorem 8.1.18). □

Next we shall discuss the Dirichlet Problem for weakly elliptic systems in δ-oscillating AR domains with boundary data in Besov spaces. The smoothness of the solution is measured in Besov, Triebel-Lizorkin, or weighted Sobolev spaces. In all cases, it becomes possible to prescribe arbitrary data from a desired function space by choosing the threshold $\delta > 0$ to be sufficiently small. In particular, for bounded infinitesimally flat AR domains all the aforementioned scenarios lead to well-posed problems if the given system satisfies the Legendre-Hadamard strong ellipticity condition and has a distinguished coefficient tensor. In the case of the Laplacian in a bounded $\mathscr{C}^1$ domain, this has been shown to be true by D. Jerison and C. Kenig, as part of [86, Theorem 1.1, p. 166], and Theorem 8.1.20 below may be regarded as a sharp version of their result. Other related work has been done by M. Mitrea and M. Taylor in [155], V. Maz'ya and T. Shaposhnikova in [122], V. Maz'ya, M. Mitrea, and T. Shaposhnikova in [121], D. Mitrea, M. Mitrea, and L. Yan in [143], X. Duong, S. Hofmann, D. Mitrea, M. Mitrea, and L. Yan in [45], S. Hofmann, M. Mitrea, and M. Taylor in [82], A. Barton and S. Mayboroda in [13].

All these works deal with domains which are locally described as upper-graphs of real-valued functions of various degree of regularity, no worse than Lipschitz. On the boundaries of such domains Besov and Triebel-Lizorkin spaces may be defined by lifting the corresponding scales from the Euclidean space. This being said, Besov and Triebel-Lizorkin spaces have been extensively studied in the general setting of spaces of homogeneous type, but so far it has been difficult to take proper advantage of this level of generality in the realm of boundary value problems for elliptic systems of PDE's since much of the necessary machinery associated with integral operators was not available. Here we register, for the first time, significant progress in this direction. Specifically, in Theorems 8.1.20, 8.1.21, 8.1.22, 8.1.23 we deal with boundary value problems for elliptic systems of PDE's in Euclidean domains which may *not* be locally of upper-graph type, and by allowing as boundary data *arbitrary* spaces on the Besov scale in such a context.

Theorem 8.1.20 (The Dirichlet Problem with Data in Besov Spaces) *Fix two integers $n, M \in \mathbb{N}$ with $n \geq 2$ and suppose*

$$p \in \big(\tfrac{n-1}{n}, \infty\big), \quad q \in (0, \infty), \quad (n-1)\big(\tfrac{1}{p} - 1\big)_+ < s < 1. \tag{8.1.698}$$

Also, let $\Omega \subseteq \mathbb{R}^n$ be an Ahlfors regular domain with compact boundary. Denote by ν the geometric measure theoretic outward unit normal to Ω and abbreviate $\sigma := \mathcal{H}^{n-1}\lfloor\partial\Omega$. Finally, let L be a homogeneous, second-order, constant complex coefficient, $M \times M$ system in $\mathbb{R}^n$.

Then there exists a threshold $\delta_0 \in (0, 1)$ depending only on n, p, q, s, L, the ADR *constants of $\partial\Omega$, and the relative flatness ratio $\wp(\Omega)/\mathrm{diam}\,\partial\Omega$ so that when*

$$\operatorname{dist}\Big(\nu, [\mathrm{VMO}(\partial\Omega, \sigma)]^n\Big) < \delta_0 \qquad (8.1.699)$$

with the distance considered in the space $\big[\mathrm{BMO}(\partial\Omega, \sigma)\big]^n$*, hence when* Ω *is a* δ_0*-oscillating* AR *domain*[63] *(cf. Definition 3.4.1), the set* Ω *is both n-thick*[64] *as well as an* (ε, δ)*-domain with* $\mathrm{rad}\,(\Omega) > 0$*, and the following statements are true:*

(1) If the system

$$L \text{ is weakly elliptic with } \mathfrak{A}_L^{\mathrm{dis}} \neq \emptyset \text{ and } \mathfrak{A}_{L^\top}^{\mathrm{dis}} \neq \emptyset, \qquad (8.1.700)$$

then the boundary value problem (with the trace considered in the sense of [130, (9.4.91) in item *(ii)* of Theorem 9.4.5])

$$\begin{cases} u \in \big[B^{p,q}_{s+\frac{1}{p}}(\Omega)_{\mathrm{bdd}}\big]^M, \quad Lu = 0 \text{ in } \Omega, \\ \mathrm{Tr}_{\Omega\to\partial\Omega}\, u = f \in \big[B^{p,q}_s(\partial\Omega, \sigma)\big]^M, \\ u(x) = O(|x|^{2-n}) \text{ as } |x| \to \infty \text{ if } \Omega \text{ is unbounded,} \end{cases} \qquad (8.1.701)$$

is Fredholm solvable in the following precise sense: First, the space of null-solutions of the boundary problem (8.1.701)*, i.e.,*

$$\Big\{u \in \big[B^{p,q}_{s+\frac{1}{p}}(\Omega)_{\mathrm{bdd}}\big]^M : Lu = 0 \text{ in } \Omega,\ \mathrm{Tr}_{\Omega\to\partial\Omega}\, u = 0 \text{ on } \partial\Omega, \qquad (8.1.702)$$
$$\text{and } u(x) = O(|x|^{2-n}) \text{ as } |x| \to \infty \text{ if } \Omega \text{ is unbounded}\Big\},$$

is finite dimensional. Second, the space of admissible boundary data for the boundary problem (8.1.701)*, i.e.,*

$$\Big\{\mathrm{Tr}_{\Omega\to\partial\Omega}\, u : u \in \big[B^{p,q}_{s+\frac{1}{p}}(\Omega)_{\mathrm{bdd}}\big]^M,\ Lu = 0 \text{ in } \Omega, \qquad (8.1.703)$$
$$\text{and } u(x) = O(|x|^{2-n}) \text{ as } |x| \to \infty \text{ if } \Omega \text{ is unbounded}\Big\},$$

is a closed linear subspace of finite codimension in $\big[B^{p,q}_s(\partial\Omega, \sigma)\big]^M$*. Third, a solution u of* (8.1.701) *exists if and only if the boundary datum f belongs to* (8.1.703).

Moreover, if $p, q \in (1, \infty)$ *and* $s \in (0, 1)$*, then for any f belonging to* (8.1.703) *a solution u of* (8.1.701) *may be found satisfying, for each cutoff function* $\psi \in \mathscr{C}^\infty_c(\mathbb{R}^n)$*,*

$$C^{-1}\|f\|_{[B^{p,q}_s(\partial\Omega,\sigma)]^M} \leq \|\psi u\|_{[B^{p,q}_{s+\frac{1}{p}}(\Omega)]^M} \leq C\|f\|_{[B^{p,q}_s(\partial\Omega,\sigma)]^M}, \qquad (8.1.704)$$

where $C = C(\Omega, p, q, s, L, \psi) \in (1, \infty)$ *is a constant independent of f.*

[63] in particular, if Ω is an infinitesimally flat AR domain

[64] hence the trace operator $\mathrm{Tr}_{\Omega\to\partial\Omega}$ considered in [130, (8.3.43)] as well as [130, (9.4.105), (9.4.106)] is meaningful

(2) Suppose (8.1.700) *is strengthened* (*cf.* *Proposition 1.4.1*) *to the demand that*

$$\begin{array}{l} L \textit{ satisfies the Legendre-Hadamard strong ellipticity con-} \\ \textit{dition (cf. [131, Definition 1.3.2]) and one has } \mathfrak{A}_L^{\text{dis}} \neq \varnothing \end{array} \tag{8.1.705}$$

and, in place of (8.1.698), *assume now*

$$p \in (1, \infty), \quad q \in (0, \infty), \quad s \in (0, 1). \tag{8.1.706}$$

Then the boundary value problem (8.1.701) *is actually well posed. More specifically,* (8.1.701) *has a unique solution, and this solution enjoys the additional property that* (8.1.704) *holds.*

(3) Similar results as in items (1)-(2) above are true for the version of (8.1.701) *in which a solution is sought in a Triebel-Lizorkin space, i.e., for the boundary value problem*

$$\begin{cases} u \in \left[F^{p,q}_{s+\frac{1}{p}}(\Omega)\right]^M, \quad Lu = 0 \textit{ in } \Omega, \\ \mathrm{Tr}_{\Omega\to\partial\Omega}\, u = f \in \left[B^{p,p}_s(\partial\Omega, \sigma)\right]^M, \\ u(x) = O(|x|^{2-n}) \textit{ as } |x| \to \infty \textit{ if } \Omega \textit{ is unbounded,} \end{cases} \tag{8.1.707}$$

(*with the trace operator now considered in the sense of* [130, (9.4.93) in item *(ii)* of Theorem 9.4.5]), *if in place of* (8.1.698) *one now starts by assuming*

$$\tfrac{n-1}{n} < p < \infty, \quad (n-1)\big(\tfrac{1}{p} - 1\big)_+ < s < 1, \quad \tfrac{n}{n+s+1/p} < q < \infty, \tag{8.1.708}$$

and if in place of (8.1.706) *one now assumes*

$$1 < p < \infty, \quad 0 < s < 1, \quad \tfrac{n}{n+s+1/p} < q < \infty. \tag{8.1.709}$$

(4) Analogous results to those in items (1)-(2) above are also true for the version of (8.1.701) *in which a solution is sought in a weighted Sobolev space, i.e., for the boundary value problem*

$$\begin{cases} u \in \left[W^{1,p}_a(\Omega)_{\mathrm{bdd}}\right]^M, \quad Lu = 0 \textit{ in } \Omega, \\ \mathrm{Tr}_{\Omega\to\partial\Omega}\, u = f \in \left[B^{p,p}_s(\partial\Omega, \sigma)\right]^M, \\ u(x) = O(|x|^{2-n}) \textit{ as } |x| \to \infty \textit{ if } \Omega \textit{ is unbounded,} \end{cases} \tag{8.1.710}$$

this time demanding

$$p \in (1, \infty), \quad s \in (0, 1), \quad a := 1 - s - \tfrac{1}{p}, \tag{8.1.711}$$

throughout, and now considering the trace operator in the sense of [130, Theorem 8.3.6] (*cf.* [130, (8.3.38)]).

As seen from Proposition 1.6.12 and [131, Proposition 1.4.39], in the scalar case (i.e., if $M = 1$) in dimensions $n \geq 3$, both (8.1.700) and (8.1.705) may be replaced by the plain demand that L is weakly elliptic. Moreover, in the scalar case in the two-dimensional setting (i.e., for $M = 1$ and $n = 2$), if L satisfies the Legendre-Hadamard strong ellipticity condition or, more generally (cf. [131, Lemma 1.4.24]), if $L = \operatorname{div} A\nabla$ for some $2{\times}2$ complex matrix A with $\operatorname{sym} A := \frac{1}{2}(A+A^{\top}) \in M_2^0$ (cf. [131, (1.4.187)]), then Proposition 1.6.11 and Proposition 1.6.13 ensure that we always have $\mathfrak{A}_L^{\rm dis} \neq \varnothing$ and $\mathfrak{A}_{L^{\top}}^{\rm dis} \neq \varnothing$. Also, according to (1.6.141) in Proposition 1.6.13, in the scalar case in the two-dimensional setting (i.e., for $M = 1$ and $n = 2$) we may omit (8.1.705) altogether if L is a weakly elliptic operator with real coefficients. As regards the formulation of Theorem 8.1.20, similar considerations to those made in Remark 8.1.2 are presently applicable.

By elliptic regularity (cf. [129, Theorem 6.5.7]), any u as in the first line of either (8.1.701), (8.1.707), or (8.1.710) is necessarily $\mathscr{C}^\infty$ smooth in Ω.

Choosing $\delta_0 \in (0, 1)$ small enough guarantees, thanks to Theorem 3.4.2, that Ω is a two-sided NTA domain. Granted this, from [131, Theorem 3.4.1] we see that, in the context of Theorem 8.1.20, the trace considered in the sense of Besov/Triebel-Lizorkin/Sobolev spaces actually agrees with the nontangential boundary trace, and the nontangential maximal operator belongs to a Lebesgue space. More specifically,

in any of the scenarios described in (8.1.701), (8.1.707), (8.1.710), for each aperture parameter $\kappa > 0$ it follows that the nontangential boundary trace $u\big|_{\partial\Omega}^{\kappa-{\rm n.t.}}$ exists at σ-a.e. point on $\partial\Omega$ and agrees with $\operatorname{Tr}_{\Omega\to\partial\Omega} u$, and there exists some $p_o \in (1, \infty)$ such that $\mathcal{N}_\kappa u \in L^{p_o}(\partial\Omega, \sigma)$. (8.1.712)

We now turn to the task of presenting the proof of Theorem 8.1.20.

Proof of Theorem 8.1.20 Choose $\delta_0 \in (0, 1)$ small enough, depending on n, p, q, s, L, the ADR constants of $\partial\Omega$, and the relative flatness ratio $\wp(\Omega)/\operatorname{diam}\partial\Omega$, such that the conclusions in Theorem 7.1.1, Theorem 7.1.15, and Theorem 7.1.17 hold (for suitable choices of the parameters involved in their statements). By eventually further decreasing the value of $\delta_0 \in (0, 1)$ we may also arrange, thanks to Theorem 3.4.2, that Ω is both an (ε, δ)-domain with $\operatorname{rad}(\Omega) > 0$ as well as a two-sided NTA domain, a scenario in which Ω is guaranteed to be n-thick (cf. [129, (5.1.6)]).

Work under the assumptions specified in (8.1.698) and (8.1.700). Pick $A \in \mathfrak{A}_L^{\rm dis}$ and consider the boundary-to-boundary double layer K_A associated with A and Ω as in (A.0.116). From (7.1.19) (used with $z := 1/2$) we know that

$$\begin{array}{c}\tfrac{1}{2}I + K_A : \big[B_s^{p,q}(\partial\Omega, \sigma)\big]^M \longrightarrow \big[B_s^{p,q}(\partial\Omega, \sigma)\big]^M \\ \text{is a Fredholm operator with index zero.}\end{array} \tag{8.1.713}$$

As such,

$$\begin{array}{l}\operatorname{Im}\Big(\tfrac{1}{2}I + K_A;\ \big[B_s^{p,q}(\partial\Omega, \sigma)\big]^M\Big) \text{ is a closed space of} \\ \text{finite codimension in the space } \big[B_s^{p,q}(\partial\Omega, \sigma)\big]^M.\end{array} \tag{8.1.714}$$

Let $\mathcal{D}_A$ be the boundary-to-domain double layer potential operator associated with A and Ω, acting on Besov spaces as in [132, Theorem 4.3.1]. Then for each choice of $\phi \in \big[B^{p,q}_s(\partial\Omega,\sigma)\big]^M$, the function $u := \mathcal{D}_A\phi$ in Ω satisfies

$$\begin{aligned} &u \in \big[B^{p,q}_{s+\frac{1}{p}}(\Omega)_{\rm bdd}\big]^M, \ \ Lu = 0 \ \text{in} \ \Omega, \ \ \mathrm{Tr}_{\Omega\to\partial\Omega}u = \big(\tfrac{1}{2}I + K_A\big)\phi, \\ &\text{and } u(x) = O(|x|^{1-n}) \ \text{as} \ |x| \to \infty \ \text{if } \Omega \text{ is unbounded,} \end{aligned} \tag{8.1.715}$$

thanks to [132, (1.3.24), (1.3.26), (4.3.1), (4.3.22)]. We then conclude that

$$\begin{aligned} &\big(\tfrac{1}{2}I + K_A\big)\phi \text{ belongs to the space of admissible boundary data} \\ &\text{for the boundary problem (8.1.701), defined in (8.1.703),} \end{aligned} \tag{8.1.716}$$

hence

$$\begin{aligned} &\text{the space (8.1.703) of admissible boundary data for the boundary value} \\ &\text{problem (8.1.701) contains the space } \mathrm{Im}\Big(\tfrac{1}{2}I + K_A;\ \big[B^{p,q}_s(\partial\Omega,\sigma)\big]^M\Big). \end{aligned} \tag{8.1.717}$$

In turn, combining (8.1.717), (8.1.714), and item *(3)* in [130, Theorem 1.1.3] we arrive at the conclusion that

$$\begin{aligned} &\text{the space of admissible boundary data for the boundary problem} \\ &\text{(8.1.701), described in (8.1.703), is a closed linear subspace of finite} \\ &\text{codimension in } \big[B^{p,q}_s(\partial\Omega,\sigma)\big]^M. \end{aligned} \tag{8.1.718}$$

It is apparent from (8.1.701) and (8.1.703) that a solution u of (8.1.701) exists if and only if the boundary datum f belongs to (8.1.703). To study the latter space in greater detail, let us define the operator

$$Q : \big[B^{p,q}_s(\partial\Omega,\sigma)\big]^M \times \big[B^{p,q}_{s-1}(\partial\Omega,\sigma)\big]^M \longrightarrow \big[B^{p,q}_s(\partial\Omega,\sigma)\big]^M, \tag{8.1.719}$$

$$Q(f_0, f_1) := \big(\tfrac{1}{2}I + K_A\big)f_0 - Sf_1, \ \ \forall (f_0, f_1) \in \big[B^{p,q}_s(\partial\Omega,\sigma)\big]^M \times \big[B^{p,q}_{s-1}(\partial\Omega,\sigma)\big]^M,$$

where K_A is as in [132, (4.1.1)], and S is as in [132, (4.3.42)]. In the above context, Q is a well-defined, linear, and bounded operator. Since its image contains the space $\mathrm{Im}\Big(\tfrac{1}{2}I + K_A;\ \big[B^{p,q}_s(\partial\Omega,\sigma)\big]^M\Big)$, by (8.1.714) and [130, Theorem 1.1.3, *(3)*] we get

$$\begin{aligned} &\text{the image of } Q \text{ is a closed subspace of} \\ &\text{finite codimension in } \big[B^{p,q}_s(\partial\Omega,\sigma)\big]^M. \end{aligned} \tag{8.1.720}$$

Going further, select an arbitrary function in the image of the operator Q, say, $Q(f_0, f_1)$ for some $f_0 \in \big[B^{p,q}_s(\partial\Omega,\sigma)\big]^M$ and $f_1 \in \big[B^{p,q}_{s-1}(\partial\Omega,\sigma)\big]^M$. Then the function defined in Ω as $u := \mathcal{D}f_0 - \mathcal{S}f_1$ belongs to the space $\big[B^{p,q}_{s+\frac{1}{p}}(\Omega)_{\rm bdd}\big]^M$, with naturally accompanying estimates (thanks to [132, (4.3.1), (4.3.25)]). In addition, this function satisfies $Lu = 0$ in Ω (thanks to [132, (1.3.24), (4.2.209)]), as well as $u(x) = O(|x|^{1-n})$ as $|x| \to \infty$ if Ω is unbounded (cf. [132, (1.3.26)]), and we also have $\mathrm{Tr}_{\Omega\to\partial\Omega}\, u = Q(f_0, f_1)$ on $\partial\Omega$ (thanks to [132, (4.3.22)] and [132, (4.3.44)]). This shows that $Q(f_0, f_1)$ is an admissible boundary datum for the boundary value problem (8.1.701). Hence, the image of Q is contained in the space of admissible boundary data for the boundary problem (8.1.701). In relation to this, we claim that

$$\begin{array}{l}\text{if either } n \geq 3, \text{ or } n = 2 \text{ and } \Omega \text{ is bounded, and if } p, q \in (1, \infty),\\ s \in (0, 1), \text{ then the image of } Q \text{ is precisely the space of admissible}\\ \text{boundary data for the boundary problem (8.1.701).}\end{array} \tag{8.1.721}$$

One inclusion has already been established, so we shall focus on the opposite one, assuming $n \geq 3$. To get started, suppose

$$\begin{array}{c} u \in \big[B^{p,q}_{s+\frac{1}{p}}(\Omega)_{\mathrm{bdd}}\big]^M \text{ with } p, q \in (1, \infty) \text{ and } s \in (0, 1),\\ Lu = 0 \text{ in } \Omega, \text{ and } u(x) = O(|x|^{2-n}) \text{ as } |x| \to \infty \text{ if } \Omega \text{ unbounded.}\end{array} \tag{8.1.722}$$

In particular,

$$\mathrm{Tr}_{\Omega\to\partial\Omega}\, u \in \big[B^{p,q}_s(\partial\Omega, \sigma)\big]^M \text{ and } \partial^A_\nu(u, 0) \in \big[B^{p,q}_{s-1}(\partial\Omega, \sigma)\big]^M, \tag{8.1.723}$$

by [130, (9.4.105)] and [130, Remark 9.5.3] (with $\mathbb{A} := B$). In addition, [132, Theorem 4.4.1] applies and gives the integral representation formula

$$u = \mathcal{D}\big(\mathrm{Tr}_{\Omega\to\partial\Omega}\, u\big) - \mathcal{S}\big(\partial^A_\nu(u, 0)\big) \text{ in } \Omega. \tag{8.1.724}$$

Taking the boundary trace of both sides of (8.1.724) then gives, on account of [132, (4.3.22)] and [132, (4.3.44)],

$$\mathrm{Tr}_{\Omega\to\partial\Omega}\, u = \big(\tfrac{1}{2}I + K_A\big)\big(\mathrm{Tr}_{\Omega\to\partial\Omega}\, u\big) - S\big(\partial^A_\nu(u, 0)\big) \text{ in } \big[B^{p,q}_s(\partial\Omega, \sigma)\big]^M. \tag{8.1.725}$$

In view of (8.1.723) and (8.1.719), we get $\mathrm{Tr}_{\Omega\to\partial\Omega}\, u = Q\big(\mathrm{Tr}_{\Omega\to\partial\Omega}\, u, \partial^A_\nu(u, 0)\big)$, which ultimately shows that the space of admissible boundary data for the boundary value problem (8.1.701) is contained in the image of the operator Q in (8.1.719). The proof of (8.1.721) is therefore complete.

Having established (8.1.721), the fact that if either $n \geq 3$, or $n = 2$ and Ω is bounded, then whenever f belongs to (8.1.703) a solution u of (8.1.701) may be found such that the second estimate in (8.1.704) is true, becomes a consequence of (8.1.720) and [130, Proposition 1.2.13]. Since, the first estimate in (8.1.704) is always valid (as a direct consequence of the boundedness of the trace operator), the proof of (8.1.704) is complete when $n \geq 3$.

The case $n = 2$ and Ω an exterior domain is similar, now replacing (8.1.719) by

$$\begin{array}{c}\widetilde{Q} : \big[B^{p,q}_s(\partial\Omega, \sigma)\big]^M \times \big[B^{p,q}_{s-1}(\partial\Omega, \sigma)\big]^M \times \mathbb{C}^M \to \big[B^{p,q}_s(\partial\Omega, \sigma)\big]^M,\\ \widetilde{Q}(f_0, f_1, c) := \big(\tfrac{1}{2}I + K_A\big) f_0 - S f_1 + c \text{ for every}\\ (f_0, f_1) \in \big[B^{p,q}_s(\partial\Omega, \sigma)\big]^M \times \big[B^{p,q}_{s-1}(\partial\Omega, \sigma)\big]^M \text{ and } c \in \mathbb{C}^M\end{array} \tag{8.1.726}$$

and (8.1.724) by the integral representation formula [132, (4.4.19)].

To show that the space (8.1.702) is finite dimensional, let $u \in \big[B^{p,q}_{s+\frac{1}{p}}(\Omega)\big]^M$ be such that $Lu = 0$ in Ω and $\mathrm{Tr}_{\Omega\to\partial\Omega}\, u = 0$ on $\partial\Omega$. The fact that $(n-1)(\frac{1}{p}-1) < s$, itself a consequence of the bounds for s in (8.1.698), implies that the inequality

$$\frac{1}{p} - \frac{s+\frac{1}{p}}{n} < \frac{1}{p_*} - \frac{s_* + \frac{1}{p_*}}{n} \tag{8.1.727}$$

is true in the limiting case $p_* = 1$ and $s_* = 0$. By continuity, this guarantees that (8.1.727) remains valid for some

$$p_* \in (1,\infty) \text{ and } s_* \in (0,1). \tag{8.1.728}$$

In particular, from (8.1.727) and [130, (9.2.13)] we conclude that

$$B^{p,q}_{s+\frac{1}{p}}(\Omega)_{\rm bdd} \hookrightarrow B^{p_*,p_*}_{s_*+\frac{1}{p_*}}(\Omega)_{\rm bdd}. \tag{8.1.729}$$

Hence $u \in \big[B^{p_*,p_*}_{s_*+\frac{1}{p_*}}(\Omega)_{\rm bdd}\big]^M$ with p_*, s_* as in (8.1.728). Granted this, in the case when $n \geq 3$ we may invoke [132, Theorem 4.4.1] to write the integral representation formula [132, (4.4.2)] in the setting [132, (4.4.3)-(4.4.7)] with $p = q := p_*$ and $s := s_*$. Since the trace of u is presently assumed to vanish, this reduces to

$$u = -\mathscr{S}\big(\partial^A_\nu(u,0)\big) \text{ in } \Omega \tag{8.1.730}$$

where the conormal derivative $\partial^A_\nu(u,0) \in \big[B^{p_*,p_*}_{s_*-1}(\partial\Omega,\sigma)\big]^M$ is considered as in [130, Proposition 9.5.2] (with $A := B$). After applying the conormal derivative to both sides in (8.1.730) and keeping in mind the jump-formula [132, (4.3.83)] (as described in [132, Theorem 4.3.6, item *(b)*]) we arrive at

$$\partial^A_\nu(u,0) = -\big(-\tfrac{1}{2}I + K^{\#}_{A^\top}\big)\big(\partial^A_\nu(u,0)\big) \text{ in } \big[B^{p_*,p_*}_{s_*-1}(\partial\Omega,\sigma)\big]^M. \tag{8.1.731}$$

In particular, this shows that

$$\partial^A_\nu(u,0) \in \mathrm{Ker}\Big(\tfrac{1}{2}I + K^{\#}_{A^\top};\ \big[B^{p_*,p_*}_{s_*-1}(\partial\Omega,\sigma)\big]^M\Big). \tag{8.1.732}$$

At this stage in the proof, from (8.1.732) and (8.1.730) we conclude that the assignment taking any function u belonging to the space of null-solutions of the boundary problem (8.1.701), defined as in (8.1.702), into the conormal derivative $\partial^A_\nu(u,0) \in \mathrm{Ker}\Big(\tfrac{1}{2}I + K^{\#}_{A^\top};\ \big[B^{p_*,p_*}_{s_*-1}(\partial\Omega,\sigma)\big]^M\Big)$ is a well-defined injection. Hence,

$$\text{the dimension of (8.1.702) is } \leq \dim \mathrm{Ker}\Big(\tfrac{1}{2}I + K^{\#}_{A^\top};\ \big[B^{p_*,p_*}_{s_*-1}(\partial\Omega,\sigma)\big]^M\Big) < \infty, \tag{8.1.733}$$

where the last inequality follows from item *(6)* in Theorem 7.1.1. Finally, the case when $n = 2$ is handled similarly, now using the integral representation formula [132, (4.4.19)]. At this point, all claims in item *(1)* have been justified.

To deal with item *(2)* strengthen (8.1.700) to (8.1.705) and assume (8.1.706) in place of (8.1.698). First, assume $n \geq 3$. Then (7.1.355) tells us that

$$\begin{gathered} \text{the operator } S : \big[B^{p,q}_{s-1}(\partial\Omega, \sigma)\big]^M \longrightarrow \big[B^{p,q}_{s}(\partial\Omega, \sigma)\big]^M \\ \text{is bounded, invertible, with a bounded inverse.} \end{gathered} \tag{8.1.734}$$

Hence, there exists $C \in (0, \infty)$ such that, given any $f \in \big[B^{p,q}_{s}(\partial\Omega, \sigma)\big]^M$, it is possible to find $g \in \big[B^{p,q}_{s-1}(\partial\Omega, \sigma)\big]^M$ with

$$Sg = f \text{ and } \|g\|_{[B^{p,q}_{s-1}(\partial\Omega,\sigma)]^M} \leq C\|f\|_{[B^{p,q}_{s}(\partial\Omega,\sigma)]^M}. \tag{8.1.735}$$

With the boundary-to-domain single layer potential operator associated with L and Ω acting on Besov spaces as in [132, Theorem 4.3.3], consider

$$u := \mathscr{S}g \text{ in } \Omega. \tag{8.1.736}$$

Then from (8.1.736), (8.1.735), [132, (4.2.209), (4.3.25), (4.3.44), (4.3.45)], and the boundedness of the trace operator from [130, (9.4.91) in item *(ii)* of Theorem 9.4.5] we see that u satisfies all demands in (8.1.701) and (8.1.704) in the case when $n \geq 3$.

Assume now that $n = 2$. Then Theorem 7.1.17 guarantees that the augmented single layer operator S_{aug} is invertible in the context of (7.1.363). Hence, there exists a constant $C \in (0, \infty)$ with the property that, given any $f \in \big[B^{p,q}_{s}(\partial\Omega, \sigma)\big]^M$, we can find $g \in \big[B^{p,q}_{s-1}(\partial\Omega, \sigma)\big]^M$ and $c \in \mathbb{C}^M$ with

$$\begin{gathered} Sg + c = f, \quad \langle g, \mathbf{e}_j\rangle = 0 \text{ for } 1 \leq j \leq M, \text{ and} \\ |c| + \|g\|_{[B^{p,q}_{s-1}(\partial\Omega,\sigma)]^M} \leq C\|f\|_{[B^{p,q}_{s}(\partial\Omega,\sigma)]^M}. \end{gathered} \tag{8.1.737}$$

If in place of (8.1.736) we now take $u := \mathscr{S}g + c$ in Ω, then once again u solves (8.1.701) and satisfies (8.1.704).

To prove uniqueness for the boundary value problem (8.1.701), let us suppose $u \in \big[B^{p,q}_{s+\frac{1}{p}}(\Omega)_{\text{bdd}}\big]^M$ is such that $Lu = 0$ in Ω, $\operatorname{Tr}_{\Omega\to\partial\Omega} u = 0$ on $\partial\Omega$, and also satisfies $u(x) = O(1)$ as $|x| \to \infty$ if Ω is unbounded. Assume first $n \geq 3$. Then, since we are now assuming (8.1.706), the same reasoning that has produced (8.1.730) now yields

$$\begin{gathered} u = -\mathscr{S}\big(\partial^A_\nu(u, 0)\big) \text{ in } \Omega \text{ with } \partial^A_\nu(u, 0) \in \big[B^{p,q}_{s-1}(\partial\Omega, \sigma)\big]^M \\ \text{considered as in [130, Remark 9.5.3] (with } \mathbb{A} := B). \end{gathered} \tag{8.1.738}$$

Applying the boundary trace operator, in the sense of [130, (9.4.91) in item *(ii)* of Theorem 9.4.5], to both sides in (8.1.738) yields (in view of [132, (4.3.44)-(4.3.45)])

$$S\big(\partial^A_\nu(u, 0)\big) = 0. \tag{8.1.739}$$

In concert with (8.1.734) and the membership in (8.1.738), this implies $\partial_\nu^A(u,0) = 0$ which then forces $u = 0$ in Ω, by the formula in (8.1.738).

Let us now consider the case when $n = 2$ and Ω is bounded. As before, we have (8.1.739), and the generalized "half" Green's formula [130, (9.5.10)] used with $f := 0$ and w a constant in $\mathbb{C}^M$ gives

$$
{}_{([B^{p',q'_*}_{1-s}(\partial\Omega,\sigma)]^{\widetilde{M}})^*}\big\langle \partial_\nu^A(u,0), c\big\rangle_{[B^{p',q'_*}_{1-s}(\partial\Omega,\sigma)]^{\widetilde{M}}} = 0 \text{ for each } c \in \mathbb{C}^M. \tag{8.1.740}
$$

Granted these properties, we then conclude from (7.1.363) that $S_{\text{aug}}\big(\partial_\nu^A(u,0)\big) = 0$. In view of the invertibility result established in Theorem 7.1.17, this once again forces $\partial_\nu^A(u,0) = 0$ which, as before, leads to the conclusion that $u = 0$ in Ω.

Finally, consider the case when $n = 2$ and Ω is an exterior domain. For starters, in place of (8.1.738) the integral representation formula [132, (4.4.19)] presently gives

$$
\begin{aligned}
u = -\mathscr{S}\big(\partial_\nu^A(u,0)\big) + c \text{ in } \Omega \text{ for some } c \in \mathbb{C}^M \text{ and with}\\
\partial_\nu^A(u,0) \in \big[B^{p,q}_{s-1}(\partial\Omega,\sigma)\big]^M \text{ as in [130, Remark 9.5.3] (with } \mathbb{A} := B).
\end{aligned}
\tag{8.1.741}
$$

Since $u = O(1)$ at infinity, this implies (much as in the proof of the last part in [132, Lemma 1.3.1, item *(iii)*]) that $\big\langle \partial_\nu^A(u,0), \mathbf{e}_j\big\rangle = 0$ for each $j \in \{1,\dots,M\}$. Also, taking traces in the first formula in (8.1.741) yields $0 = -S\big(\partial_\nu^A(u,0)\big) + c$. Based on these two properties and (7.1.363) we next conclude that

$$
S_{\text{aug}}\big(\partial_\nu^A(u,0), -c\big) = (0,0). \tag{8.1.742}
$$

In concert with the membership $\partial_\nu^A(u,0) \in \big[B^{p,q}_{s-1}(\partial\Omega,\sigma)\big]^M$ (cf. (8.1.741)) and the invertibility result in Theorem 7.1.17, this forces $\partial_\nu^A(u,0) = 0$ and $c = 0$. Hence, ultimately, $u = 0$ thanks to (8.1.741). This finishes the proof of uniqueness in all cases. The treatment of item *(2)* is therefore complete.

The claims in item *(3)* are dealt with similarly, based on the trace results from [130, (9.4.93) in item *(ii)* of Theorem 9.4.5], the mapping properties of the boundary-to-domain single layer from [132, (4.3.26)], the trace formula [132, (4.3.44)] in the context of [132, (4.3.46)], the integral representation formula [132, (4.4.2)] in the context of [132, (4.4.8)-(4.4.12)], the definition of the conormal derivative from [130, Remark 9.5.3] (with $\mathbb{A} := F$), the jump-formula [132, (4.3.83)] in the context of [132, Theorem 4.3.6, item *(c)*], and the embedding result in [130, (9.2.12)].

Finally, the claims in item *(4)* are established in an analogous fashion, now relying on the trace results from [130, Theorem 8.3.6], the mapping properties of the boundary-to-domain single layer from [132, (4.2.218)], the trace formula [132, (4.3.44)] in the context of [132, (4.3.47)], the integral representation formula [132, (4.4.2)] in the context of [132, Theorem 4.4.1, item *(3)*], the definition of the conormal derivative from [130, Remark 8.5.4], and the jump-formula [132, (4.3.83)] in the context of [132, Theorem 4.3.6, item *(a)*]; see also [132, Corollary 4.4.3]. □

We next consider the Poisson Problem[65] in Besov, Triebel-Lizorkin, and weighted Sobolev spaces, formulated for weakly elliptic systems and δ-oscillating AR domains with δ sufficiently small. This augments the well-posedness result for the Inhomogeneous Dirichlet Problem formulated in Theorem 7.2.15 for Legendre-Hadamard elliptic systems in bounded two-sided NTA domains with Ahlfors regular boundaries and for (small perturbations of) energy spaces. The Poisson Problem (8.1.755) in the case when $M := 1$, $L := \Delta$ (the Laplacian in $\mathbb{R}^n$), Ω a bounded domain of class $\mathscr{C}^1$ in $\mathbb{R}^n$, and $q := 2$ (corresponding to Bessel potential spaces; cf. [130, (9.1.54)]) has been dealt with by D. Jerison and C. Kenig in [86, Theorem 1.1, Theorem 5.1] using harmonic measure techniques. The Poisson Problem (8.1.755) with $q = 2$ for $L := \Delta$ and Ω a bounded domain of class $\mathscr{C}^1$ in $\mathbb{R}^n$ has been subsequently treated by E. Fabes, O. Mendez, and M. Mitrea in [56] using boundary layer methods. Also, when L is the Laplace-Beltrami operator on a Lipschitz subdomain of a compact Riemannian manifold equipped with a rough metric tensor, M. Mitrea and M. Taylor have treated (8.1.755) with $q = 2$ in [155], and (8.1.758) in [159]. The Poisson problem with mixed boundary conditions in Sobolev and Besov spaces in (creased) Lipschitz domains has been solved by I. Mitrea and M. Mitrea in [147]. Related work may be found in [155], [122], [121], [143], [45], [148], [82], [138], [13], and the references therein. Here we only wish to point out that known counterexamples to the solvability of the Poisson problem for the Laplacian in Lipschitz domains (cf. [86, Theorem A(1), p. 163], [86, Theorem 1.2(a), p. 167]) show that the well-posedness results from Theorem 8.1.21 may fail to materialize for an arbitrary given integrability/smoothness exponents if one no longer insists on imposing flatness assumptions on the underlying domain, as we do in (8.1.744).

Theorem 8.1.21 (The Poisson Problem Formulated in Besov, Triebel-Lizorkin, and Weighted Sobolev Spaces) *Fix two integers $n, M \in \mathbb{N}$ with $n \geq 2$ and suppose*

$$p \in \big(\tfrac{n-1}{n}, \infty\big), \quad q \in (0, \infty), \quad (n-1)\big(\tfrac{1}{p} - 1\big)_+ < s < 1. \tag{8.1.743}$$

Also, let $\Omega \subseteq \mathbb{R}^n$ be a bounded Ahlfors regular domain. Denote by ν its geometric measure theoretic outward unit normal, and set $\sigma := \mathcal{H}^{n-1}\lfloor\partial\Omega$. Finally, let L be a homogeneous, second-order, constant complex coefficient, $M \times M$ system in $\mathbb{R}^n$.

Then there exists a threshold $\delta_0 \in (0, 1)$ depending only on n, p, q, s, L, the ADR *constants of $\partial\Omega$, and the relative flatness ratio $\wp(\Omega)/\mathrm{diam}\,\partial\Omega$ so that when*

$$\mathrm{dist}\Big(\nu, [\mathrm{VMO}(\partial\Omega, \sigma)]^n\Big) < \delta_0 \tag{8.1.744}$$

[65] named after Simeon-Denis Poisson (1781-1840), who first derived the equation of force potential ϕ for points interior to a body with mass density ρ as $\Delta\phi = -4\pi\rho$, a result published in Bulletin de la Société Philomatique in 1813

with the distance considered in the space $\big[\mathrm{BMO}(\partial\Omega,\sigma)\big]^n$ *(hence when* Ω *is a bounded* δ_0*-oscillating* AR *domain*[66]*; cf. Definition 3.4.1), the set* Ω *is n-thick*[67] *as well as an* (ε,δ)*-domain with* $\mathrm{rad}\,(\Omega)>0$*, and the following statements are true.*

(1) If the system

$$L \text{ is weakly elliptic with } \mathfrak{A}^{\rm dis}_L \neq \emptyset \text{ and } \mathfrak{A}^{\rm dis}_{L^\top} \neq \emptyset, \tag{8.1.745}$$

then the Poisson Problem (with the boundary trace considered as in [130, (9.4.91) from item *(ii)* of Theorem 9.4.5]*)*

$$\begin{cases} u \in \big[B^{p,q}_{s+\frac{1}{p}}(\Omega)\big]^M, \quad Lu = f \in \big[B^{p,q}_{s+\frac{1}{p}-2}(\Omega)\big]^M, \\ \mathrm{Tr}_{\Omega\to\partial\Omega}\, u = g \in \big[B^{p,q}_s(\partial\Omega,\sigma)\big]^M, \end{cases} \tag{8.1.746}$$

is Fredholm solvable, in the following precise sense: First, the space of admissible data for the problem (8.1.746)*, i.e.,*

$$\Big\{\big(Lu, \mathrm{Tr}_{\Omega\to\partial\Omega}\, u\big) : u \in \big[B^{p,q}_{s+\frac{1}{p}}(\Omega)\big]^M\Big\} \tag{8.1.747}$$

happens to be a closed linear subspace of finite codimension in the Cartesian product space $\big[B^{p,q}_{s+\frac{1}{p}-2}(\Omega)\big]^M \times \big[B^{p,q}_s(\partial\Omega,\sigma)\big]^M$*. Second, a solution u of* (8.1.746) *exists if and only if the data* (f,g) *belongs to* (8.1.747)*. Third, the space of null-solutions of the problem* (8.1.746)*, i.e.,*

$$\Big\{u \in \big[B^{p,q}_{s+\frac{1}{p}}(\Omega)\big]^M : Lu = 0 \text{ in } \Omega \text{ and } \mathrm{Tr}_{\Omega\to\partial\Omega}\, u = 0 \text{ on } \partial\Omega\Big\} \tag{8.1.748}$$

is finite dimensional. In fact, the data map

$$\begin{gathered} \Phi : \big[B^{p,q}_{s+\frac{1}{p}}(\Omega)\big]^M \longrightarrow \big[B^{p,q}_{s+\frac{1}{p}-2}(\Omega)\big]^M \times \big[B^{p,q}_s(\partial\Omega,\sigma)\big]^M, \\ \Phi u := \big(Lu, \mathrm{Tr}_{\Omega\to\partial\Omega}\, u\big) \text{ for each } u \in \big[B^{p,q}_{s+\frac{1}{p}}(\Omega)\big]^M \end{gathered} \tag{8.1.749}$$

is a Fredholm operator.

(2) In place of (8.1.743)*, assume now*

$$p \in (1,\infty), \quad q \in (0,\infty), \quad s \in (0,1). \tag{8.1.750}$$

Then the dimension of the cokernel of the data map Φ *from* (8.1.749) *is*

$$\dim \mathrm{Coker}\,\Phi = \dim\left(\frac{\big[B^{p,q}_s(\partial\Omega,\sigma)\big]^M}{\Big\{\mathrm{Tr}_{\Omega\to\partial\Omega}\, w : w \in [B^{p,q}_{s+\frac{1}{p}}(\Omega)]^M : Lw=0\Big\}}\right) < \infty. \tag{8.1.751}$$

---

[66] in particular, if Ω is a bounded infinitesimally flat AR domain

[67] hence the trace operator $\mathrm{Tr}_{\Omega\to\partial\Omega}$ considered in [130, Theorem 9.4.5] is meaningful

Also, for each coefficient tensor $A \in \mathfrak{A}_L^{\rm dis}$ *and each* $q_o \in (1,\infty)$ *with* $q_o \geq q$*, the dimension of the kernel of the data map* (8.1.749) *may be identified as*

$$\begin{aligned} &\dim \operatorname{Ker} \Phi \\ &= \dim\Big\{\partial_\nu^A(w,0) : w \in \big[B^{p,q_o}_{s+\frac{1}{p}}(\Omega)\big]^M,\ Lw = 0,\ \operatorname{Tr}_{\Omega\to\partial\Omega} w = 0\Big\} < \infty, \end{aligned} \tag{8.1.752}$$

where the conormal derivative $\partial_\nu^A(w,0)$ *is as in* [130, Proposition 9.5.2].

(3) *Continue to assume that* p, q, s *are as in* (8.1.750)*, and suppose* (8.1.745) *is strengthened* (*cf. Proposition 1.4.1*) *to the demand that*

$$L \text{ satisfies the Legendre-Hadamard strong ellipticity condition (cf. [131, Definition 1.3.2]) and one has } \mathfrak{A}_L^{\rm dis} \neq \varnothing. \tag{8.1.753}$$

Then the Poisson Problem (8.1.746) *is actually well posed. Moreover, there exists a constant* $C = C(\Omega, p, q, s, L) \in (0,\infty)$ *such that the unique solution of* (8.1.746) *satisfies*

$$\|u\|_{[B^{p,q}_{s+\frac{1}{p}}(\Omega)]^M} \leq C\Big\{\|f\|_{[B^{p,q}_{s+\frac{1}{p}-2}(\Omega)]^M} + \|g\|_{[B^{p,q}_s(\partial\Omega,\sigma)]^M}\Big\}, \tag{8.1.754}$$

and the data map (8.1.749) *is an isomorphism.*

(4) *Similar results as in items (1)-(3) above are true for the version of* (8.1.746) *in which a solution is sought in a Triebel-Lizorkin space, i.e., for the boundary value problem*

$$\begin{cases} u \in \big[F^{p,q}_{s+\frac{1}{p}}(\Omega)\big]^M, \quad Lu = f \in \big[F^{p,q}_{s+\frac{1}{p}-2}(\Omega)\big]^M, \\ \operatorname{Tr}_{\Omega\to\partial\Omega} u = g \in \big[B^{p,p}_s(\partial\Omega,\sigma)\big]^M, \end{cases} \tag{8.1.755}$$

(*with the trace operator now considered in the sense of* [130, (9.4.93) in item (*ii*) of Theorem 9.4.5]), *if in place of* (8.1.743) *one now starts by assuming*

$$\tfrac{n-1}{n} < p < \infty, \quad (n-1)\big(\tfrac{1}{p} - 1\big)_+ < s < 1, \quad \tfrac{n}{n+s+1/p} < q < \infty, \tag{8.1.756}$$

and if in place of (8.1.750) *one now assumes*

$$1 < p < \infty, \quad 0 < s < 1, \quad \tfrac{n}{n+s+1/p} < q < \infty. \tag{8.1.757}$$

(5) *Analogous results to those in items (1)-(3) above are also true for the version of* (8.1.746) *in which a solution is sought in a weighted Sobolev space, i.e., for the boundary value problem*

$$\begin{cases} u \in \big[W^{1,p}_a(\Omega)\big]^M, \quad Lu = f \in \big[W^{-1,p}_a(\Omega)\big]^M, \\ \operatorname{Tr}_{\Omega\to\partial\Omega} u = g \in \big[B^{p,p}_s(\partial\Omega,\sigma)\big]^M, \end{cases} \tag{8.1.758}$$

this time demanding

$$p \in (1, \infty), \quad s \in (0, 1), \quad a := 1 - s - \tfrac{1}{p}, \tag{8.1.759}$$

throughout, and now considering the trace operator in the sense of [130, Theorem 8.3.6] (*cf.* [130, (8.3.38)]).

We want to highlight the fact that, as seen from Proposition 1.6.12 and [131, Proposition 1.4.39], in the scalar case (i.e., if $M = 1$) in dimensions $n \geq 3$, both (8.1.745) and (8.1.753) may be replaced by plainly demanding that L is weakly elliptic. In addition, in the scalar case in the two-dimensional setting (i.e., for $M = 1$ and $n = 2$), if L satisfies the Legendre-Hadamard strong ellipticity condition or, more generally (cf. [131, Lemma 1.4.24]), if $L = \operatorname{div} A\nabla$ for some 2×2 complex matrix A with $\operatorname{sym} A := \frac{1}{2}(A + A^\top) \in M_2^0$ (cf. [131, (1.4.187)]), then Proposition 1.6.11 and Proposition 1.6.13 guarantee that we automatically have $\mathfrak{A}_L^{\rm dis} \neq \varnothing$ and $\mathfrak{A}_{L^\top}^{\rm dis} \neq \varnothing$. Also, according to (1.6.141) in Proposition 1.6.13, in the scalar case in the two-dimensional setting (i.e., for $M = 1$ and $n = 2$) we may omit (8.1.753) altogether if L is a weakly elliptic operator with real coefficients. Finally, we wish to note that, as far as the formulation of Theorem 8.1.21 is concerned, similar considerations to those made in Remark 8.1.2 are presently applicable.

Proof of Theorem 8.1.21 Choose $\delta_0 \in (0, 1)$ small enough, depending on n, p, q, s, L, the ADR constants of $\partial\Omega$, and the relative flatness ratio $\wp(\Omega)/\operatorname{diam} \partial\Omega$, such that the conclusions in Theorem 7.1.1, Proposition 7.1.2, and Theorem 8.1.20 apply. Thanks to Theorem 3.4.2, by eventually further decreasing the value of $\delta_0 \in (0, 1)$ we may also arrange so that Ω becomes both an (ε, δ)-domain with $\operatorname{rad}(\Omega) > 0$ as well as a two-sided NTA domain, a scenario in which Ω is guaranteed to be n-thick (cf. [129, (5.1.6)]).

Work under the assumptions specified in (8.1.743) and (8.1.700). That the data map is a well-defined, linear, and bounded operator in the context of (8.1.749) follows from item *(ii)* in [130, Theorem 9.4.5] and [130, (9.2.8)]. Let us prove that said operator is in fact Fredholm. To this end, pick $A \in \mathfrak{A}_L^{\rm dis}$. Theorem 1.5.5 ensures that we also have $A^\top \in \mathfrak{A}_{L^\top}^{\rm dis}$. Consider the boundary-to-boundary double layer potential operator K_A associated with Ω and the coefficient tensor A as in (A.0.116). From (7.1.19) (used with $z := 1/2$) we know that

$$\begin{gathered} \tfrac{1}{2}I + K_A : \big[B_s^{p,q}(\partial\Omega, \sigma)\big]^M \longrightarrow \big[B_s^{p,q}(\partial\Omega, \sigma)\big]^M \\ \text{is a Fredholm operator with index zero.} \end{gathered} \tag{8.1.760}$$

Then [130, Theorem 2.2.3] guarantees the existence of two linear operators,

$$\begin{aligned} R &: \big[B_s^{p,q}(\partial\Omega, \sigma)\big]^M \to \big[B_s^{p,q}(\partial\Omega, \sigma)\big]^M \ \text{ bounded,} \\ C_0 &: \big[B_s^{p,q}(\partial\Omega, \sigma)\big]^M \to \big[B_s^{p,q}(\partial\Omega, \sigma)\big]^M \ \text{ compact,} \end{aligned} \tag{8.1.761}$$

such that, with I denoting the identity operator,

$$\big(\tfrac{1}{2}I + K_A\big) \circ R = I + C_0 \ \text{ on } \ \big[B_s^{p,q}(\partial\Omega, \sigma)\big]^M. \tag{8.1.762}$$

Consider the manifestation of the system L as a linear bounded map in the context

$$L : \big[B^{p,q}_{s+\frac{1}{p}}(\Omega)\big]^M \longrightarrow \big[B^{p,q}_{s+\frac{1}{p}-2}(\Omega)\big]^M. \tag{8.1.763}$$

We claim that this has an inverse-from-the-right, i.e., there exists an operator

$$\begin{aligned} &\Psi : \big[B^{p,q}_{s+\frac{1}{p}-2}(\Omega)\big]^M \longrightarrow \big[B^{p,q}_{s+\frac{1}{p}}(\Omega)\big]^M \text{ liner, bounded,} \\ &\text{and with } L(\Psi f) = f \text{ for each } f \in \big[B^{p,q}_{s+\frac{1}{p}-2}(\Omega)\big]^M. \end{aligned} \tag{8.1.764}$$

To construct it, consider the Fourier multiplier operator $L^{-1}\varphi := \mathscr{F}^{-1}\big(L(\xi)^{-1}(\mathscr{F}\varphi)\big)$ for all $\varphi \in \big[\mathscr{S}'(\mathbb{R}^n)\big]^M$. Also, for some sufficiently small $\varepsilon \in (0,1)$ (so that $\varepsilon < p$ and $-\frac{1}{\varepsilon} < s + \frac{1}{p} - 2$), bring in the linear and bounded extension operator

$$E^{\varepsilon}_{\Omega\to\mathbb{R}^n} : \big[B^{p,q}_{s+\frac{1}{p}-2}(\Omega)\big]^M \longrightarrow \big[B^{p,q}_{s+\frac{1}{p}-2}(\mathbb{R}^n)\big]^M \tag{8.1.765}$$

from [130, Theorem 9.2.19]. Finally, pick two cutoff functions $\psi_1, \psi_2 \in \mathscr{C}^\infty_c(\mathbb{R}^n)$ which are identically one near the compact $\overline{\Omega}$. If we now set

$$\Psi f := \big[(\psi_1 L^{-1}\psi_2)(E^{\varepsilon}_{\Omega\to\mathbb{R}^n} f)\big]\Big|_{\Omega} \text{ for each } f \in \big[B^{p,q}_{s+\frac{1}{p}-2}(\Omega)\big]^M, \tag{8.1.766}$$

then [87, Corollary 4.2] implies that (8.1.764) holds. To proceed, define

$$\begin{aligned} &Q : \big[B^{p,q}_{s+\frac{1}{p}-2}(\Omega)\big]^M \longrightarrow \big[B^{p,q}_s(\partial\Omega,\sigma)\big]^M \\ &Qf := \mathrm{Tr}_{\Omega\to\partial\Omega}(\Psi f) \text{ for each } f \in \big[B^{p,q}_{s+\frac{1}{p}-2}(\Omega)\big]^M, \end{aligned} \tag{8.1.767}$$

$$\begin{aligned} &T : \big[B^{p,q}_{s+\frac{1}{p}-2}(\Omega)\big]^M \times \big[B^{p,q}_s(\partial\Omega,\sigma)\big]^M \to \big[B^{p,q}_{s+\frac{1}{p}-2}(\Omega)\big]^M \times \big[B^{p,q}_s(\partial\Omega,\sigma)\big]^M \\ &T(f,\phi) := (f, Qf + \phi),\ \forall (f,\phi) \in \big[B^{p,q}_{s+\frac{1}{p}-2}(\Omega)\big]^M \times \big[B^{p,q}_s(\partial\Omega,\sigma)\big]^M, \end{aligned} \tag{8.1.768}$$

$$\begin{aligned} &C_1 : \big[B^{p,q}_{s+\frac{1}{p}-2}(\Omega)\big]^M \times \big[B^{p,q}_s(\partial\Omega,\sigma)\big]^M \to \big[B^{p,q}_{s+\frac{1}{p}-2}(\Omega)\big]^M \times \big[B^{p,q}_s(\partial\Omega,\sigma)\big]^M \\ &C_1(f,\phi) := (0, C_0\phi),\ \forall (f,\phi) \in \big[B^{p,q}_{s+\frac{1}{p}-2}(\Omega)\big]^M \times \big[B^{p,q}_s(\partial\Omega,\sigma)\big]^M. \end{aligned} \tag{8.1.769}$$

These are all linear and bounded operators. Moreover, it is clear that T is invertible and C_1 is compact. As such, item *(a)* in [130, Theorem 2.2.4] gives that

$$\begin{aligned} T + C_1 : \big[B^{p,q}_{s+\frac{1}{p}-2}(\Omega)\big]^M \times \big[B^{p,q}_s(\partial\Omega,\sigma)\big]^M \to \big[B^{p,q}_{s+\frac{1}{p}-2}(\Omega)\big]^M \times \big[B^{p,q}_s(\partial\Omega,\sigma)\big]^M \\ \text{is a Fredholm operator (with index zero).} \end{aligned} \tag{8.1.770}$$

In view of [130, Definition 2.2.1], this further implies that

$$\begin{aligned}&\mathrm{Im}\Big(T + C_1;\ \big[B^{p,q}_{s+\frac{1}{p}-2}(\Omega)\big]^M \times \big[B^{p,q}_s(\partial\Omega,\sigma)\big]^M\Big) \text{ is a closed subspace}\\&\text{of } \big[B^{p,q}_{s+\frac{1}{p}-2}(\Omega)\big]^M \times \big[B^{p,q}_s(\partial\Omega,\sigma)\big]^M, \text{ of finite codimension.}\end{aligned} \tag{8.1.771}$$

If $\mathcal{D}_A$ is the boundary-to-domain double layer operator acting on Besov spaces as in [132, Theorem 4.3.1] (cf. [132, (4.3.1)]) then (8.1.749), (8.1.764), (8.1.767), (8.1.768), (8.1.769), (8.1.762), and the jump-formula described in [132, Theorem 4.3.2, *(1)*] imply that, for each $(f,\phi) \in \big[B^{p,q}_{s+\frac{1}{p}-2}(\Omega)\big]^M \times \big[B^{p,q}_s(\partial\Omega,\sigma)\big]^M$,

$$\Phi(\Psi f + \mathcal{D}_A(R\phi)) = \Big(f, Qf + \big(\tfrac{1}{2} + K_A\big)(R\phi)\Big) = \Big(f, Qf + \phi + C_0\phi\Big) = (T + C_1)(f,\phi).$$

Based on this and the mapping properties of the operators involved we conclude that

$$\begin{aligned}&\text{the image of the data map } \Phi, \text{ defined in (8.1.749), contains}\\&\mathrm{Im}\Big(T + C_1;\ \big[B^{p,q}_{s+\frac{1}{p}-2}(\Omega)\big]^M \times \big[B^{p,q}_s(\partial\Omega,\sigma)\big]^M\Big).\end{aligned} \tag{8.1.772}$$

In turn, (8.1.772), (8.1.771), and [130, Theorem 1.1.3, *(3)*] finally imply

$$\begin{aligned}&\text{the image of the data map } \Phi, \text{ from (8.1.749), is a closed space}\\&\text{of finite codimension in } \big[B^{p,q}_{s+\frac{1}{p}-2}(\Omega)\big]^M \times \big[B^{p,q}_s(\partial\Omega,\sigma)\big]^M.\end{aligned} \tag{8.1.773}$$

As is apparent from (8.1.749) and the fact that (8.1.702) is finite dimensional,

$$\begin{aligned}&\text{the kernel of the data map } \Phi, \text{ in the context of (8.1.749)},\\&\text{is finite dimensional subspace of } \big[B^{p,q}_{s+\frac{1}{p}}(\Omega)\big]^M.\end{aligned} \tag{8.1.774}$$

Collectively, (8.1.773), (8.1.774), and item *(iii)* of [130, Remark 2.2.2] prove that the data map is actually a Fredholm operator in the context of (8.1.749). From this, all other claims in item *(1)* follow.

Next, let us deal with the claims in item *(2)*, now working under the assumption that p, q, s are as in (8.1.750). Keeping in mind that the data map is a Fredholm operator in the context of (8.1.749), we may invoke [130, (2.1.49)] to write

$$\dim \mathrm{Coker}\, \Phi = \dim \mathrm{Ker}\, \Phi^*, \tag{8.1.775}$$

where Φ^* is the adjoint of Φ in (8.1.749). Concretely, if $p', q' \in (1,\infty)$ are the Hölder conjugate exponents of p, q, then (cf. [130, Proposition 7.6.1])

$$\Phi^* : \Big(\big[B^{p,q}_{s+\frac{1}{p}-2}(\Omega)\big]^M\Big)^* \times \big[B^{p',q'}_{-s}(\partial\Omega,\sigma)\big]^M \longrightarrow \Big(\big[B^{p,q}_{s+\frac{1}{p}}(\Omega)\big]^M\Big)^*$$
$$\text{given by } \big\langle \Phi^*(\Lambda,\psi), u\big\rangle = \big\langle \Lambda, Lu\big\rangle + \big\langle \psi, \mathrm{Tr}_{\Omega\to\partial\Omega}\, u\big\rangle \text{ for every} \tag{8.1.776}$$
$$\Lambda \in \Big(\big[B^{p,q}_{s+\frac{1}{p}-2}(\Omega)\big]^M\Big)^*, \ \psi \in \big[B^{p',q'}_{-s}(\partial\Omega,\sigma)\big]^M, \ u \in \big[B^{p,q}_{s+\frac{1}{p}}(\Omega)\big]^M.$$

Let us now select an arbitrary pair (Λ, ψ) belonging to the space $\mathrm{Ker}\,\Phi^*$. Then we have $\Lambda \in \Big(\big[B^{p,q}_{s+\frac{1}{p}-2}(\Omega)\big]^M\Big)^*$, $\psi \in \big[B^{p',q'}_{-s}(\partial\Omega,\sigma)\big]^M$, and

$$\big\langle \Lambda, Lu\big\rangle + \big\langle \psi, \mathrm{Tr}_{\Omega\to\partial\Omega}\, u\big\rangle = 0 \text{ for every } u \in \big[B^{p,q}_{s+\frac{1}{p}}(\Omega)\big]^M. \tag{8.1.777}$$

Specializing this to the case when $u \in \big[B^{p,q}_{s+\frac{1}{p}}(\Omega)\big]^M$ has $Lu = 0$ in Ω proves that

$$\psi \in \Big\{\mathrm{Tr}_{\Omega\to\partial\Omega}\, u : u \in \big[B^{p,q}_{s+\frac{1}{p}}(\Omega)\big]^M \text{ and } Lu = 0\Big\}^{\perp} \tag{8.1.778}$$

with the annihilator understood as in (A.0.6) with $X := \big[B^{p,q}_{s}(\partial\Omega,\sigma)\big]^M$, a scenario in which [130, Proposition 7.6.1] gives $X^* = \big[B^{p',q'}_{-s}(\partial\Omega,\sigma)\big]^M$. Going further, consider an arbitrary $f \in \big[B^{p,q}_{s+\frac{1}{p}-2}(\Omega)\big]^M$ and specialize (8.1.777) to the case when $u := \Psi f$, with Ψ as in (8.1.764). Then $Lu = f$ in Ω and (8.1.777) yields

$$\big\langle \Lambda, f\big\rangle = -\big\langle \psi, \mathrm{Tr}_{\Omega\to\partial\Omega}(\Psi f)\big\rangle = -\big\langle \big(\mathrm{Tr}_{\Omega\to\partial\Omega}\circ\Psi\big)^*\psi, f\big\rangle, \tag{8.1.779}$$

where

$$\big(\mathrm{Tr}_{\Omega\to\partial\Omega}\circ\Psi\big)^* : \big[B^{p',q'}_{-s}(\partial\Omega,\sigma)\big]^M \longrightarrow \Big(\big[B^{p,q}_{s+\frac{1}{p}-2}(\Omega)\big]^M\Big)^* \tag{8.1.780}$$

is the adjoint of

$$\mathrm{Tr}_{\Omega\to\partial\Omega}\circ\Psi : \big[B^{p,q}_{s+\frac{1}{p}-2}(\Omega)\big]^M \longrightarrow \big[B^{p,q}_{s}(\partial\Omega,\sigma)\big]^M. \tag{8.1.781}$$

In turn, from (8.1.779) and the arbitrariness of $f \in \big[B^{p,q}_{s+\frac{1}{p}-2}(\Omega)\big]^M$ we see that

$$\Lambda = -\big(\mathrm{Tr}_{\Omega\to\partial\Omega}\circ\Psi\big)^*\psi. \tag{8.1.782}$$

Conversely, assume ψ is as in (8.1.778) and Λ is as in (8.1.782). Also, pick an arbitrary $u \in \big[B^{p,q}_{s+\frac{1}{p}}(\Omega)\big]^M$ and define $\omega := \Psi(Lu) \in \big[B^{p,q}_{s+\frac{1}{p}}(\Omega)\big]^M$, with the membership guaranteed by the first line in (8.1.764). The second line in (8.1.764) also gives $L\omega = Lu$, hence the function $w := u - \omega$ belongs to $\big[B^{p,q}_{s+\frac{1}{p}}(\Omega)\big]^M$ and satisfies $Lw = 0$ in Ω. As such, we may write

$$\begin{aligned}\langle \Lambda, Lu\rangle + \langle \psi, \mathrm{Tr}_{\Omega\to\partial\Omega}\, u\rangle &= -\big\langle (\mathrm{Tr}_{\Omega\to\partial\Omega}\circ\Psi)^*\psi, Lu\big\rangle + \langle \psi, \mathrm{Tr}_{\Omega\to\partial\Omega}\, u\rangle \\ &= -\big\langle \psi, \mathrm{Tr}_{\Omega\to\partial\Omega}\big(\Psi(Lu)\big)\big\rangle + \langle \psi, \mathrm{Tr}_{\Omega\to\partial\Omega}\, u\rangle \\ &= -\langle \psi, \mathrm{Tr}_{\Omega\to\partial\Omega}\, \omega\rangle + \langle \psi, \mathrm{Tr}_{\Omega\to\partial\Omega}\, u\rangle = \langle \psi, \mathrm{Tr}_{\Omega\to\partial\Omega}\, w\rangle = 0,\end{aligned} \tag{8.1.783}$$

thanks to the properties of w and (8.1.778). Together with what we have just proved, (8.1.778) and (8.1.782) ultimately show that

$$\begin{aligned}&\text{the space } \mathrm{Ker}\,\Phi^* \text{ consists of all pairs } \Big(-\big(\mathrm{Tr}_{\Omega\to\partial\Omega}\circ\Psi\big)^*\psi, \psi\Big) \\ &\text{with } \psi \in \Big\{\mathrm{Tr}_{\Omega\to\partial\Omega}\, u : \ u \in \big[B^{p,q}_{s+\frac{1}{p}}(\Omega)\big]^M \text{ and } Lu = 0\Big\}^{\perp} \text{ arbitrary.}\end{aligned} \tag{8.1.784}$$

From (8.1.775) and (8.1.784) we then conclude that (8.1.751) holds.

To justify (8.1.752), fix a coefficient tensor $A \in \mathfrak{A}^{\rm dis}_L$ along with an exponent $q_o \in (1,\infty)$ satisfying $q_o \geq q$. Theorem 1.5.5 ensures that $A^\top \in \mathfrak{A}^{\rm dis}_{L^\top}$. By definition, the kernel of the data map is

$$\mathrm{Ker}\,\Phi = \Big\{u \in \big[B^{p,q}_{s+\frac{1}{p}}(\Omega)\big]^M, \ Lu = 0, \ \mathrm{Tr}_{\Omega\to\partial\Omega}\, u = 0\Big\}, \tag{8.1.785}$$

so [130, (9.2.10)] gives

$$\mathrm{Ker}\,\Phi \subseteq \Big\{u \in \big[B^{p,q_o}_{s+\frac{1}{p}}(\Omega)\big]^M, \ Lu = 0, \ \mathrm{Tr}_{\Omega\to\partial\Omega}\, u = 0\Big\}. \tag{8.1.786}$$

We claim that the opposite inclusion in (8.1.786) is also true. To see that this is indeed the case, pick a function $u \in \big[B^{p,q_o}_{s+\frac{1}{p}}(\Omega)\big]^M$ satisfying $Lu = 0$ and $\mathrm{Tr}_{\Omega\to\partial\Omega}\, u = 0$. Then the integral representation formula [132, (4.4.2)] in the setting described in [132, Theorem 4.4.1, item *(1)*] (with $q := q_o$) presently reduces to

$$u = -\mathscr{S}\big(\partial^A_\nu(u,0)\big) \ \text{ in } \Omega \tag{8.1.787}$$

Applying conormal derivatives and relying on the jump-formula [132, (4.3.83)] in the setting described in [132, Theorem 4.3.6, item *(b)*] (with $q := q_o$) then gives

$$\partial^A_\nu(u,0) = -\partial^A_\nu\Big(\mathscr{S}\big(\partial^A_\nu(u,0)\big), 0\Big) = -\big(-\tfrac{1}{2}I + K^\#_{A^\top}\big)\big(\partial^A_\nu(u,0)\big), \tag{8.1.788}$$

hence

$$\partial^A_\nu(u,0) \in \mathrm{Ker}\Big(\tfrac{1}{2}I + K^\#_{A^\top}; \big[B^{p,q_o}_{s-1}(\partial\Omega,\sigma)\big]^M\Big). \tag{8.1.789}$$

To proceed, recall from [130, (11.11.12)] that

$$L^p(\partial\Omega,\sigma) \hookrightarrow B^{p,q_o}_{s-1}(\partial\Omega,\sigma) \hookrightarrow L^p_{-1}(\partial\Omega,\sigma). \tag{8.1.790}$$

From this and (7.1.62) we then conclude that

$$\mathrm{Ker}\Big(\tfrac{1}{2}I + K^\#_{A^\top}; \big[B^{p,q_o}_{s-1}(\partial\Omega,\sigma)\big]^M\Big) = \mathrm{Ker}\Big(\tfrac{1}{2}I + K^\#_{A^\top}; \big[B^{p,q}_{s-1}(\partial\Omega,\sigma)\big]^M\Big). \tag{8.1.791}$$

Together with (8.1.788) this proves that $\partial_\nu^A(u,0) \in \big[B^{p,q}_{s-1}(\partial\Omega,\sigma)\big]^M$. Finally, from this, (8.1.787), and [132, Theorem 4.3.3, item *(1)*] we conclude that u belongs to $\big[B^{p,q}_{s+\frac{1}{p}}(\Omega)\big]^M$. In view of (8.1.785), this finishes the proof of the fact that the opposite inclusion in (8.1.786) is also true. Consequently,

$$\operatorname{Ker}\Phi = \Big\{u \in \big[B^{p,q_o}_{s+\frac{1}{p}}(\Omega)\big]^M,\ Lu = 0,\ \operatorname{Tr}_{\Omega\to\partial\Omega} u = 0\Big\}. \tag{8.1.792}$$

To complete the proof of (8.1.752) there remains to observe that, thanks to (8.1.792) and (8.1.787), the map $\operatorname{Ker}\Phi \ni u \mapsto \partial_\nu^A(u,0)$ is well defined, linear, and injective.

Turning our attention to item *(3)*, work under the assumption specified in (8.1.745), and fix some arbitrary $f \in \big[B^{p,q}_{s+\frac{1}{p}-2}(\Omega)\big]^M$. From (8.1.764) we know that

$$\begin{array}{c} u_0 := \Psi f \text{ belongs to the space } \big[B^{p,q}_{s+\frac{1}{p}}(\Omega)\big]^M \text{ and satisfies} \\ Lu_0 = f \text{ in } \Omega \text{ as well as } \|u_0\|_{[B^{p,q}_{s+\frac{1}{p}}(\Omega)]^M} \leq C\|f\|_{[B^{p,q}_{s+\frac{1}{p}-2}(\Omega)]^M} \end{array} \tag{8.1.793}$$

for some $C = C(\Omega, L, n, p, q, s) \in (0,\infty)$. To proceed, let us also bring an arbitrary $g \in \big[B^{p,q}_s(\partial\Omega,\sigma)\big]^M$ and define

$$g_1 := g - \operatorname{Tr}_{\Omega\to\partial\Omega} u_0 \in \big[B^{p,q}_s(\partial\Omega,\sigma)\big]^M, \tag{8.1.794}$$

with the trace in the sense of [130, Theorem 9.4.5, *(ii)*]. Since Theorem 8.1.20 guarantees that (8.1.701) is a well-posed problem, we may find u_1 satisfying

$$\begin{cases} u_1 \in \big[B^{p,q}_{s+\frac{1}{p}}(\Omega)\big]^M, \quad Lu_1 = 0 \text{ in } \Omega, \\ \operatorname{Tr}_{\Omega\to\partial\Omega} u_1 = g_1 \in \big[B^{p,q}_s(\partial\Omega,\sigma)\big]^M, \end{cases} \tag{8.1.795}$$

and such that, for some constant $C = C(\Omega, p, q, s, L) \in (1,\infty)$,

$$\|u_1\|_{[B^{p,q}_{s+\frac{1}{p}}(\Omega)]^M} \leq C\|g_1\|_{[B^{p,q}_s(\partial\Omega,\sigma)]^M}. \tag{8.1.796}$$

Then $u := u_0 + u_1$ solves (8.1.746) and satisfies (8.1.754). Finally, uniqueness for (8.1.746) is implied by uniqueness for (8.1.701) in the context discussed in item *(2)* of Theorem 8.1.20.

Next, the claims in item *(4)* pertaining to (8.1.755) are handled similarly, once again relying on [130, (9.2.1)], [87, Corollary 4.2], and Theorem 8.1.20. Finally, the claims in item *(5)* regarding the Poisson Problem (8.1.758) may be established reasoning along similar lines, now using (A.0.251), [130, Remark 8.5.1], [129, Proposition 8.7.4], [124, Theorem 3.1, p. 103], [87, Theorem 4.3], and Theorem 8.1.20. □

The Neumann Problem for elliptic PDE's with boundary data in Besov spaces, when the global regularity of the solution is measured on Besov, Triebel-Lizorkin, and weighted Sobolev spaces, has received a considerable amount of attention. Work on this topic in rough domains includes [56], [125], [155], [159], [45], [147], [13], and the references cited there. Our goal here is to treat the brand of Neumann Problem mentioned earlier for weakly elliptic systems in δ-oscillating AR domains. Our main result in this regard, Theorem 8.1.22 stated below, which refines and/or tightly interfaces with the aforementioned results, is the first result of its kind in domains which are not necessarily of upper-graph type.

Theorem 8.1.22 (The Neumann Problem with Data in Besov Spaces) *Fix two integers $n, M \in \mathbb{N}$ with $n \geq 2$ and suppose $\Omega \subseteq \mathbb{R}^n$ is an Ahlfors regular domain with compact boundary. Denote by ν the geometric measure theoretic outward unit normal to Ω and abbreviate $\sigma := \mathcal{H}^{n-1}\lfloor\partial\Omega$. Also, let L be a homogeneous, second-order, constant complex coefficient, $M\times M$ system in $\mathbb{R}^n$, and pick a coefficient tensor A for which $L = L_A$ (i.e., $A \in \mathfrak{A}_L$). Finally, select $p \in (1,\infty)$, $q \in (1,\infty)$, and $s \in (0,1)$.*

Then there exists a threshold $\delta_0 \in (0,1)$, depending only on n, p, q, s, A, the ADR *constants of $\partial\Omega$, and the relative flatness ratio $\wp(\Omega)/\mathrm{diam}\,\partial\Omega$, so that when*

$$\mathrm{dist}\Big(\nu, [\mathrm{VMO}(\partial\Omega,\sigma)]^n\Big) < \delta_0 \tag{8.1.797}$$

where the distance in the left-hand side is considered in the John-Nirenberg space $\big[\mathrm{BMO}(\partial\Omega,\sigma)\big]^n$, hence whenever the set Ω is a δ_0-oscillating AR *domain*[68] *(cf. Definition 3.4.1), the set Ω becomes both an (ε,δ)-domain as well as a two-sided* NTA *domain (hence also a* UR *domain) and the following statements are true:*

(1) If the system

$$L \text{ is weakly elliptic with } \mathfrak{A}_L^{\mathrm{dis}} \neq \emptyset \text{ and } \mathfrak{A}_{L^\top}^{\mathrm{dis}} \neq \emptyset, \tag{8.1.798}$$

whenever $A \in \mathfrak{A}_L^{\mathrm{dis}}$ the Neumann Problem

$$\begin{cases} u \in \big[B^{p,q}_{s+\frac{1}{p}}(\Omega)_{\mathrm{bdd}}\big]^M, \quad Lu = 0 \text{ in } \Omega, \\ \partial^A_\nu(u,0) = f \in \big[B^{p,q}_{s-1}(\partial\Omega,\sigma)\big]^M, \\ u(x) = O(|x|^{2-n}) \text{ as } |x| \to \infty \text{ if } \Omega \text{ is unbounded}, \end{cases} \tag{8.1.799}$$

(where the conormal derivative is taken as in [130, Remark 9.5.3] *with $\mathbb{A} := B$) is Fredholm solvable, in the following precise sense: First, the space of admissible boundary data for the boundary value problem* (8.1.799)*, i.e.,*

$$\Big\{\partial^A_\nu(u,0) : u \in \big[B^{p,q}_{s+\frac{1}{p}}(\Omega)_{\mathrm{bdd}}\big]^M,\ Lu = 0 \text{ in } \Omega, \text{ and } u(x) = O(|x|^{2-n}) \text{ as } |x| \to \infty \text{ if } \Omega \text{ is unbounded}\Big\} \tag{8.1.800}$$

is a closed linear subspace of finite codimension in $\big[B^{p,q}_{s-1}(\partial\Omega,\sigma)\big]^M$. Second, a solution u of (8.1.799) *exists if and only if the boundary datum f belongs to* (8.1.800)*, in which scenario a solution u may be found satisfying, for each cutoff function $\psi \in \mathscr{C}^\infty_c(\mathbb{R}^n)$,*

[68] in particular, if Ω is an infinitesimally flat AR domain

$$C^{-1}\|f\|_{[B^{p,q}_s(\partial\Omega,\sigma)]^M} \leq \|\psi u\|_{[B^{p,q}_{s+\frac{1}{p}}(\Omega)]^M} \leq C\|f\|_{[B^{p,q}_s(\partial\Omega,\sigma)]^M}, \tag{8.1.801}$$

where $C = C(\Omega, p, q, s, L, \psi) \in (1, \infty)$ *is a constant independent of* f. *Third, the space of null-solutions of the boundary value problem* (8.1.799), *i.e.,*

$$\Big\{u \in \big[B^{p,q}_{s+\frac{1}{p}}(\Omega)_{\text{bdd}}\big]^M : Lu = 0 \text{ in } \Omega,\ \partial^A_\nu(u, 0) = 0, \text{ and} \\ u(x) = O(|x|^{2-n}) \text{ as } |x| \to \infty \text{ if } \Omega \text{ is unbounded}\Big\}, \tag{8.1.802}$$

is finite dimensional.

(2) *Suppose* (8.1.798) *is strengthened (cf. Proposition 1.4.1) to the demand that*

$$L \text{ satisfies the Legendre-Hadamard strong ellipticity condition (cf. [131, Definition 1.3.2]) and one has } \mathfrak{A}^{\text{dis}}_L \neq \varnothing. \tag{8.1.803}$$

Then for each coefficient tensor $A \in \mathfrak{A}^{\text{dis}}_L$ *the boundary value problem* (8.1.799) *is Fredholm solvable and its index, i.e., the dimension of the space of null-solutions* (8.1.802) *minus the codimension in* $\big[B^{p,q}_{s-1}(\partial\Omega, \sigma)\big]^M$ *of the space of admissible boundary data* (8.1.800) *is equal to zero.*

Moreover, in the case when either $n \geq 3$, *or* $n = 2$ *and the set* Ω *is bounded, the Neumann Problem* (8.1.799) *has a solution if and only if the boundary datum* $f \in \big[B^{p,q}_{s-1}(\partial\Omega, \sigma)\big]^M$ *is such that*

$$\begin{gathered} {}_{[B^{p,q}_{s-1}(\partial\Omega,\sigma)]^M}\big\langle f, g\big\rangle_{[B^{p',q'}_{1-s}(\partial\Omega,\sigma)]^M} = 0 \text{ for all } g \in \big[B^{p',q'}_{1-s}(\partial\Omega, \sigma)\big]^M, \\ \text{where } p' := (1 - 1/p)^{-1} \in (1, \infty) \text{ and } q' := (1 - 1/q)^{-1} \in (1, \infty), \\ \text{such that } \big(-\tfrac{1}{2}I + K_{A^\top}\big)g = 0 \text{ (with } K_{A^\top} \text{ as in [132, Theorem 4.1.1]).} \end{gathered} \tag{8.1.804}$$

Also, when $n = 2$ *and* Ω *is an exterior domain, the Neumann Problem* (8.1.799) *has a solution if and only if the boundary datum* $f \in \big[B^{p,q}_{s-1}(\partial\Omega, \sigma)\big]^M$ *satisfies*

$$\begin{gathered} {}_{[B^{p,q}_{s-1}(\partial\Omega,\sigma)]^M}\big\langle f, g\big\rangle_{[B^{p',q'}_{1-s}(\partial\Omega,\sigma)]^M} = 0 \text{ for each } g \in \big[B^{p',q'}_{1-s}(\partial\Omega, \sigma)\big]^M \\ \text{with } p' := (1 - 1/p)^{-1} \in (1, \infty) \text{ and } q' := (1 - 1/q)^{-1} \in (1, \infty), \\ \text{with the property that } \big(-\tfrac{1}{2}I + K_{A^\top}\big)g \in \mathbb{C}^M. \end{gathered} \tag{8.1.805}$$

Furthermore,

> *the space of admissible boundary data for the Neumann Problem* (8.1.443), *originally defined as in* (8.1.446), *is equal to the image* $\mathrm{Im}\Big(-\frac{1}{2}I + K^{\#}_{A^\top};\ \big[B^{p,q}_{s-1}(\partial\Omega,\sigma)\big]^M\Big)$ *if either* $n \geq 3$, *or* $n = 2$ *and the set* Ω *is bounded, and may be described as the space* $\Big\{\big(-\frac{1}{2}I + K^{\#}_{A^\top}\big)g :\ g \in \big[B^{p,q}_{s-1}(\partial\Omega,\sigma)\big]^M$ *with* $\langle g, \mathbf{e}_j\rangle = 0$ *for* $1 \leq j \leq M\Big\}$ *if* $n = 2$ *and* Ω *is an exterior domain.* (8.1.806)

In addition, there exists a constant $C \in (0,\infty)$ *with the property that whenever the boundary datum* f *belongs to the space of admissible boundary data, defined in* (8.1.800), *then a solution* u *of the boundary problem* (8.1.799) *may be found such that* (8.1.801) *holds.*

Also, with $\mathcal{D}_A$ *denoting the boundary-to-domain double layer potential operator associated with* A *and* Ω *acting on Besov spaces as in* [132, Theorem 4.3.1, *(1)*], *when either* $n \geq 3$, *or* $n = 2$ *and the set* Ω *is bounded, it follows that:*

> *the space of null-solutions of the Neumann Problem* (8.1.799) (*originally described as in* (8.1.802)) *is actually equal to the space* $\Big\{\mathcal{D}_A\phi :\ \phi \in \mathrm{Ker}\Big(-\frac{1}{2}I + K_A;\ \big[B^{p,q}_{s}(\partial\Omega,\sigma)\big]^M\Big)\Big\}$, (8.1.807)

> *the map associating to each function* u *belonging to the space of null-solutions of the Neumann Problem* (8.1.799) (*described in* (8.1.802)) *its trace* $\mathrm{Tr}_{\Omega\to\partial\Omega}\, u$ (*defined as in* [130, (9.4.105)]) *is a isomorphism onto the finite dimensional space* $\mathrm{Ker}\Big(-\frac{1}{2}I + K_A;\ \big[B^{p,q}_{s}(\partial\Omega,\sigma)\big]^M\Big)$, (8.1.808)

> *the inverse of the isomorphism* $u \mapsto \mathrm{Tr}_{\Omega\to\partial\Omega}\, u$ *described in* (8.1.808) *is the boundary-to-domain double layer potential operator* $\mathcal{D}_A$ *mapping* $\mathrm{Ker}\Big(-\frac{1}{2}I + K_A;\ \big[B^{p,q}_{s}(\partial\Omega,\sigma)\big]^M\Big)$ *isomorphically onto the space of null-solutions of the Neumann Problem* (8.1.799) (*from* (8.1.802)). (8.1.809)

Finally, in the case when $n = 2$ *and* Ω *is an exterior domain it follows that:*

> *the space of null-solutions of the Neumann Problem* (8.1.799) *equals* $\Big\{\mathcal{D}_A\phi + c :\ \phi \in \big[B^{p,q}_{s}(\partial\Omega,\sigma)\big]^M,\ c \in \mathbb{C}^M,\ \big(-\frac{1}{2}I + K_A\big)\phi = -c\Big\}$, (8.1.810)

the map associating to each function u belonging to the space of null-solutions of the Neumann Problem (8.1.799) *(described in* (8.1.802)*) the pair* $\left(\mathrm{Tr}_{\Omega\to\partial\Omega}\, u,\ \lim_{|x|\to\infty} u(x)\right)$ *is a well-defined isomorphism onto the space* $\left\{(\phi, c) \in \left[B^{p,q}_{s}(\partial\Omega,\sigma)\right]^M \oplus \mathbb{C}^M : \left(-\tfrac{1}{2}I + K_A\right)\phi = -c\right\}$, (8.1.811)

the inverse of the linear isomorphism $u \mapsto \left(\mathrm{Tr}_{\Omega\to\partial\Omega}\, u,\ \lim_{|x|\to\infty} u(x)\right)$ *in the above context is the assignment* $(\phi, c) \longmapsto \mathcal{D}_A\phi + c$ *mapping the last space in* (8.1.811) *isomorphically onto the space of null-solutions of the Neumann Problem* (8.1.799) *(described in* (8.1.802)*).* (8.1.812)

(3) *If* (8.1.803) *is further strengthened to*[69]

$$A \in \mathfrak{A}^{\mathrm{dis}}_{L} \text{ is positive definite,} \tag{8.1.813}$$

then the Neumann Problem (8.1.799) *has the following additional properties (with the spaces* $\mathbb{C}_{\Omega_\pm}$, $\mathbb{C}_{\partial\Omega_\pm}$, $\mathbb{C}^{\infty}_{\Omega_\pm}$, $\mathbb{C}^{\infty}_{\partial\Omega_\pm}$ *defined as in* (7.1.265), (7.1.268)*, and with the conventions* (7.1.266), (7.1.269) *in effect). First, if either* $n \geq 3$*, or* $n = 2$ *and the set* Ω *is bounded, then*

the space of admissible boundary data for the Neumann Problem (8.1.799) *(originally defined as in* (8.1.800)*) is equal to* $\left[B^{p,q}_{s}(\partial\Omega,\sigma) \cap \left(\mathbb{C}_{\partial\Omega_+}\right)^{\perp}\right]^M$*, and the space of null-solutions of the Neumann Problem* (8.1.799) *(from* (8.1.448)*) is equal to* $\left(\mathbb{C}_{\Omega_+}\right)^M$. (8.1.814)

Second, if $n = 2$ *and* Ω *is an exterior domain then*

the space of admissible boundary data for the Neumann Problem (8.1.799) *(originally defined as in* (8.1.800)*) is equal to* $\left[B^{p,q}_{s}(\partial\Omega,\sigma) \cap \left(\mathbb{C}^{\infty}_{\partial\Omega_+}\right)^{\perp}\right]^M$*, and the space of null-solutions of the Neumann Problem* (8.1.799) *(from* (8.1.802)*) is equal to* $\left(\mathbb{C}^{\infty}_{\Omega_+}\right)^M$. (8.1.815)

(4) *Similar results as in items (1)-(3) above are true for the version of the Neumann Problem* (8.1.799) *formulated for functions in Triebel-Lizorkin spaces, i.e.,*

$$\begin{cases} u \in \left[F^{p,q}_{s+\frac{1}{p}}(\Omega)_{\mathrm{bdd}}\right]^M, \quad Lu = 0 \text{ in } \Omega, \\ \partial^A_\nu(u,0) = f \in \left[B^{p,p}_{s-1}(\partial\Omega,\sigma)\right]^M, \\ u(x) = O(|x|^{2-n}) \text{ as } |x| \to \infty \text{ if } \Omega \text{ is unbounded,} \end{cases} \tag{8.1.816}$$

[69] according to (1.1.28), if L has a positive definite coefficient tensor then L satisfies the Legendre-Hadamard strong ellipticity condition

(where the conormal derivative is taken as in [130, Remark 9.5.3] *with* $\mathbb{A} := F$*), and for the version of the Neumann Problem* (8.1.799) *formulated for functions in weighted Sobolev spaces, i.e.,*

$$\begin{cases} u \in \big[W^{1,p}_a(\Omega)_{\text{bdd}}\big]^M, & Lu = 0 \ \ in \ \Omega, \\ \partial^A_\nu(u,0) = f \in \big[B^{p,p}_{s-1}(\partial\Omega,\sigma)\big]^M, \\ u(x) = O(|x|^{2-n}) \ \ as \ \ |x| \to \infty \ \ if \ \Omega \text{ is unbounded,} \end{cases} \tag{8.1.817}$$

where $a := 1-s-\frac{1}{p}$ *and the conormal derivative is now as in* [130, Remark 8.5.4].

We wish to note that, as seen from Proposition 1.6.12 and [131, Proposition 1.4.39], if $M = 1$ and $n \geq 3$, then (8.1.798) and (8.1.803) may be replaced by the simpler demand that L is weakly elliptic. Also, in the scalar two-dimensional case (i.e., for $M = 1$ and $n = 2$), if L satisfies the Legendre-Hadamard strong ellipticity condition or, more generally (cf. [131, Lemma 1.4.24]), if $L = \operatorname{div} A \nabla$ for some 2×2 complex matrix A with $\operatorname{sym} A := \frac{1}{2}(A + A^\top) \in M^0_2$ (cf. [131, (1.4.187)]), then Proposition 1.6.11 and Proposition 1.6.13 imply that we automatically have $\mathfrak{A}^{\text{dis}}_L \neq \varnothing$ and $\mathfrak{A}^{\text{dis}}_{L^\top} \neq \varnothing$. Finally, according to (1.6.141) in Proposition 1.6.13, in the scalar two-dimensional case (i.e., for $M = 1$ and $n = 2$) we may omit (8.1.803) altogether if L is a weakly elliptic operator with real coefficients.

Proof of Theorem 8.1.22 Take $\delta_0 \in (0,1)$ sufficiently small, depending on n, p, A, the ADR constants of $\partial\Omega$, and the relative flatness ratio $\wp(\Omega)/\operatorname{diam}\partial\Omega$, such that the conclusions in Theorem 7.1.1, Proposition 7.1.2, Theorem 7.1.15, Theorem 7.1.17, and Proposition 7.1.41 are valid (for suitable choices of the parameters involved in their statements). After further decreasing the value of the threshold $\delta_0 \in (0,1)$, if necessary, Theorem 3.4.2 applies and ensures that Ω is both an (ε,δ)-domain as well as a two-sided NTA domain, hence also a UR domain (cf. [129, (5.10.33)]) satisfying a two-sided local John condition (cf. [129, (5.11.28)]).

To proceed in earnest, work under the assumptions made in (8.1.798) and pick some $A \in \mathfrak{A}^{\text{dis}}_L$. Theorem 1.5.5 ensures that $A^\top \in \mathfrak{A}^{\text{dis}}_{L^\top}$. Bearing this in mind, items *(5)-(6)* in Theorem 7.1.1 used with $z := -1/2$ guarantee that

$$-\tfrac{1}{2}I + K_A : \big[B^{p,q}_s(\partial\Omega,\sigma)\big]^M \longrightarrow \big[B^{p,q}_s(\partial\Omega,\sigma)\big]^M \text{ is Fredholm with index zero,} \tag{8.1.818}$$

$$-\tfrac{1}{2}I + K^\#_{A^\top} : \big[B^{p,q}_{s-1}(\partial\Omega,\sigma)\big]^M \longrightarrow \big[B^{p,q}_{s-1}(\partial\Omega,\sigma)\big]^M \text{ is Fredholm with index zero.} \tag{8.1.819}$$

In particular,

$$\dim\operatorname{Ker}\Big(-\tfrac{1}{2}I + K_A;\ \big[B^{p,q}_s(\partial\Omega,\sigma)\big]^M\Big) < +\infty, \ \text{ and} \tag{8.1.820}$$

$$\begin{aligned} &\operatorname{Im}\Big(-\tfrac{1}{2}I + K^\#_{A^\top};\ \big[B^{p,q}_{s-1}(\partial\Omega,\sigma)\big]^M\Big) \text{ is a closed space} \\ &\text{of finite codimension in the space } \big[B^{p,q}_{s-1}(\partial\Omega,\sigma)\big]^M. \end{aligned} \tag{8.1.821}$$

Bring in the boundary-to-domain single layer potential operator $\mathscr{S}$ (associated with L and Ω) acting on Besov spaces as in [132, Theorem 4.3.3, item *(1)*]. Then, as seen from [132, (4.2.209), (4.3.25)], if either $n \geq 3$ or $n = 2$ and Ω is bounded it follows that for each $g \in \big[B^{p,q}_{s-1}(\partial\Omega, \sigma)\big]^M$ the function $u := \mathscr{S}g$ in Ω satisfies

$$\begin{aligned} & u \in \big[B^{p,q}_{s+\frac{1}{p}}(\Omega)_{\rm bdd} \cap \mathscr{C}^\infty(\Omega)\big]^M, \ Lu = 0 \text{ in } \Omega, \text{ and} \\ & u(x) = O(|x|^{2-n}) \text{ as } |x| \to \infty \text{ if } \Omega \text{ is unbounded.} \end{aligned} \tag{8.1.822}$$

In fact, the same properties are true when $n = 2$ and Ω is an exterior domain if we also demand that $\langle g, \mathbf{e}_j \rangle = 0$ for $1 \leq j \leq M$. From (8.1.822) and the jump-formula [132, (4.3.83)] in the setting of [132, Theorem 4.3.6, *(b)*] we then see that

$$\begin{aligned} & \big(-\tfrac{1}{2}I + K^{\#}_{A^\top}\big)g = \partial^A_\nu(u, 0) \text{ belongs to the space of admissible boundary} \\ & \text{data for the boundary problem (8.1.799), defined in (8.1.800), for each} \\ & g \in \big[B^{p,q}_{s-1}(\partial\Omega, \sigma)\big]^M \text{ if either } n \geq 3 \text{ or } n = 2 \text{ and } \Omega \text{ is bounded, and} \\ & \text{under the additional assumption that } \langle g, \mathbf{e}_j \rangle = 0 \text{ for } 1 \leq j \leq M \text{ if } n = 2 \\ & \text{and } \Omega \text{ is an exterior domain.} \end{aligned} \tag{8.1.823}$$

Based on (8.1.823) we then conclude that

$$\begin{aligned} & \text{the space (8.1.800) of admissible boundary data for the boundary} \\ & \text{problem (8.1.799) contains } \mathrm{Im}\Big(-\tfrac{1}{2}I + K^{\#}_{A^\top};\ \big[B^{p,q}_{s-1}(\partial\Omega, \sigma)\big]^M\Big) \text{ when} \\ & \text{either } n \geq 3 \text{ or } n = 2 \text{ and } \Omega \text{ is bounded, and the space (8.1.800) contains} \\ & \Big\{\big(-\tfrac{1}{2}I + K^{\#}_{A^\top}\big)g : g \in \big[L^p(\partial\Omega, \sigma)\big]^M,\ \langle g, \mathbf{e}_j \rangle = 0 \text{ for } 1 \leq j \leq M\Big\} \text{ if} \\ & n = 2 \text{ and } \Omega \text{ is an exterior domain.} \end{aligned} \tag{8.1.824}$$

Collectively, (8.1.824), (8.1.821), and item *(3)* in [130, Theorem 1.1.3] imply that

$$\begin{aligned} & \text{the space of admissible boundary data for the boundary problem} \\ & \text{(8.1.799), described in (8.1.800), is a closed linear subspace of} \\ & \text{finite codimension in the space } \big[B^{p,q}_{s-1}(\partial\Omega, \sigma)\big]^M. \end{aligned} \tag{8.1.825}$$

By design, (8.1.799) is solvable if and only if the boundary datum f belongs to (8.1.800), in which scenario the first estimate in (8.1.801) automatically holds (by virtue of the continuity of the conormal derivative operator; see [130, (9.5.9)]). Let us show that if the boundary datum belongs to (8.1.800) then a solution of (8.1.799) may be found which also satisfies the second estimate in (8.1.801). Consider first the case when either $n \geq 3$, or $n = 2$ and Ω is bounded. Define the operator

$$\begin{aligned} T : \big[B^{p,q}_{s}(\partial\Omega, \sigma)\big]^M \times \big[B^{p,q}_{s-1}(\partial\Omega, \sigma)\big]^M & \longrightarrow \big[B^{p,q}_{s-1}(\partial\Omega, \sigma)\big]^M, \\ T(f_0, f_1) := (\partial^A_\nu \mathcal{D})f_0 - \big(-\tfrac{1}{2}I + K^{\#}_{A^\top}\big)f_1 & \text{ for all} \\ (f_0, f_1) \in \big[B^{p,q}_{s}(\partial\Omega, \sigma)\big]^M \times \big[B^{p,q}_{s-1}(\partial\Omega, \sigma)\big]^M&, \end{aligned} \tag{8.1.826}$$

where $\partial^A_\nu \mathcal{D} : \big[B^{p,q}_{s}(\partial\Omega, \sigma)\big]^M \to \big[B^{p,q}_{s-1}(\partial\Omega, \sigma)\big]^M$ is as in [132, (4.3.70)]. As such, T is a well-defined, linear, and bounded operator in the above context. In addition, it is rather apparent from definitions that the image of the operator T contains the space $\mathrm{Im}\Big(-\tfrac{1}{2}I+K^{\#}_{A^{\top}};\ \big[B^{p,q}_{s-1}(\partial\Omega,\sigma)\big]^M\Big)$. As such, we conclude from (8.1.821) and item *(3)* in [130, Theorem 1.1.3] that

$$\text{the image of } T \text{ is a closed subspace of finite codimension in the space } \big[B^{p,q}_{s-1}(\partial\Omega,\sigma)\big]^M. \tag{8.1.827}$$

In relation to this, we claim that

the image of T is precisely the space of admissible boundary data for the Neumann Problem (8.1.799) when either $n\geq 3$, or when $n=2$ and the set Ω is bounded. (8.1.828)

To justify this, assume first that

$$u\in\big[B^{p,q}_{s+\frac{1}{p}}(\Omega)_{\mathrm{bdd}}\big]^M,\ Lu=0,\ u(x)=O(|x|^{2-n}) \text{ as } |x|\to\infty \text{ if } \Omega \text{ is unbounded.} \tag{8.1.829}$$

Granted this, [132, Theorem 4.4.1] applies, and the integral representation formula [132, (4.4.2)] reads $u=\mathcal{D}(\mathrm{Tr}_{\Omega\to\partial\Omega}u)-\mathcal{S}(\partial^A_\nu(u,0))$ in Ω. Taking the conormal derivative of both sides then gives, on account of [132, (4.3.83)] and [132, (4.3.72)],

$$\begin{aligned}\partial^A_\nu(u,0)&=\partial^A_\nu\Big(\mathcal{D}(\mathrm{Tr}_{\Omega\to\partial\Omega}u,0)\Big)-\big(-\tfrac{1}{2}I+K^{\#}_{A^{\top}}\big)\big(\partial^A_\nu(u,0)\big)\\&=(\partial^A_\nu\mathcal{D})(\mathrm{Tr}_{\Omega\to\partial\Omega}u)-\big(-\tfrac{1}{2}I+K^{\#}_{A^{\top}}\big)\big(\partial^A_\nu(u,0)\big)\ \text{ in } \big[B^{p,q}_{s-1}(\partial\Omega,\sigma)\big]^M.\end{aligned} \tag{8.1.830}$$

This may be recast as $\partial^A_\nu(u,0)=T\big(\mathrm{Tr}_{\Omega\to\partial\Omega}u,\partial^A_\nu(u,0)\big)$, which ultimately shows that the space of admissible boundary data for the Neumann Problem (8.1.799) is contained in the image of the operator T in (8.1.826).

In the opposite direction, pick an arbitrary distribution in the image of the operator T, say, $T(f_0,f_1)$ for some $f_0\in\big[B^{p,q}_s(\partial\Omega,\sigma)\big]^M$ and $f_1\in\big[B^{p,q}_{s-1}(\partial\Omega,\sigma)\big]^M$. Then the function defined by $u:=\mathcal{D}f_0-\mathcal{S}f_1$ enjoys all the properties demanded in (8.1.829) and satisfies $\partial^A_\nu(u,0)=T(f_0,f_1)$, thanks to [132, (1.3.24), (4.2.209), (4.3.1), (4.3.25), (4.3.72), (4.3.83)]. This shows that $T(f_0,f_1)$ is an admissible boundary datum for the Neumann Problem (8.1.799). The proof of (8.1.828) is complete.

With (8.1.828) in hand, the fact that when either $n\geq 3$, or $n=2$ and Ω is bounded, a solution u of (8.1.799) exists if and only if the boundary datum f belongs to (8.1.339), in which case a solution u may be found so that the second estimate in (8.1.801) is valid, becomes a consequence of (8.1.827) and [130, Proposition 1.2.13]. This finishes the case when either $n\geq 3$, or $n=2$ and Ω is bounded.

Our next goal is to find the analogue of (8.1.828) when $n=2$ and Ω is an exterior domain. In such a case, we abbreviate

$$V:=\int_{S^1}\big[L(\xi)\big]^{-1}\,d\mathcal{H}^1(\xi)\in\mathbb{C}^{M\times M}, \tag{8.1.831}$$

then define the operator $\widetilde{T}$ mapping the space

$$\big[B^{p,q}_{s}(\partial\Omega,\sigma)\big]^M \times \Big\{f_1 \in \big[B^{p,q}_{s-1}(\partial\Omega,\sigma)\big]^M : V\Big(\big(\langle f_1, \mathbf{e}_j\rangle\big)_{1\leq j\leq M}\Big) = 0\Big\} \tag{8.1.832}$$

into the space $\big[B^{p,q}_{s-1}(\partial\Omega,\sigma)\big]^M$ according to

$$\widetilde{T}(f_0, f_1) := \partial^A_\nu \mathcal{D} f_0 - \big(-\tfrac{1}{2}I + K^{\#}_{A^\top}\big) f_1 \text{ for each } (f_0, f_1) \text{ in the space (8.1.832).} \tag{8.1.833}$$

The claim that we make in this regard is that

when $n = 2$ and Ω is an exterior domain, the image of $\widetilde{T}$ is the space of admissible boundary data for the Neumann Problem (8.1.799). (8.1.834)

To establish one inclusion, assume

$$u \in \big[B^{p,q}_{s+\frac{1}{p}}(\Omega)_{\text{bdd}}\big]^M, \ Lu = 0 \text{ in } \Omega, \text{ and } u(x) = O(1) \text{ as } |x| \to \infty. \tag{8.1.835}$$

Then [132, Theorem 4.4.1] applies and from [132, (4.4.18)-(4.4.19)] we see that there exists some $c \in \mathbb{C}^M$ so that the integral representation formula

$$u = \mathcal{D}\big(\text{Tr}_{\Omega\to\partial\Omega} u\big) - \mathcal{S}\big(\partial^A_\nu(u,0)\big) + c \text{ in } \Omega \tag{8.1.836}$$

holds. On account of [132, (4.3.72), (4.3.83)], after taking the conormal derivative of both sides of (8.1.836) we arrive at

$$\begin{aligned}\partial^A_\nu(u,0) &= \partial^A_\nu\Big(\mathcal{D}\big(\text{Tr}_{\Omega\to\partial\Omega} u\big), 0\Big) - \big(-\tfrac{1}{2}I + K^{\#}_{A^\top}\big)\big(\partial^A_\nu(u,0)\big) \\ &= \big(\partial^A_\nu \mathcal{D}\big)\big(\text{Tr}_{\Omega\to\partial\Omega} u\big) - \big(-\tfrac{1}{2}I + K^{\#}_{A^\top}\big)\big(\partial^A_\nu(u,0)\big) \text{ in } \big[B^{p,q}_{s-1}(\partial\Omega,\sigma)\big]^M.\end{aligned} \tag{8.1.837}$$

We next claim that

$$V\Big(\big(\langle \partial^A_\nu(u,0), \mathbf{e}_j\rangle\big)_{1\leq j\leq M}\Big) = 0. \tag{8.1.838}$$

Indeed, from (8.1.836), the last property in (8.1.835), and the fact that (as seen from [132, (1.3.26)]) $\mathcal{D}\big(\text{Tr}_{\Omega\to\partial\Omega} u\big)(x) = O\big(|x|^{-1}\big)$ as $|x| \to \infty$, we conclude that $\mathcal{S}\big(\partial^A_\nu(u,0)\big) = O(1)$ as $|x| \to \infty$. With this in hand, reasoning as in the proof of [132, (1.3.12)] shows that we necessarily have (8.1.838). Next, (8.1.837), (8.1.838), and the definition of the operator $\widetilde{T}$ in (8.1.832)-(8.1.833) give that there holds $\partial^A_\nu(u,0) = \widetilde{T}\big(\text{Tr}_{\Omega\to\partial\Omega} u, \partial^A_\nu(u,0)\big) \in \text{Im}\,\widetilde{T}$, the image of the operator $\widetilde{T}$. This proves one inclusion in (8.1.834). To establish the opposite inclusion in (8.1.834), start with a function in the image of $\widetilde{T}$, say, $\widetilde{T}(f_0, f_1)$ with (f_0, f_1) belonging to the space described in (8.1.832). Then the function $u := \mathcal{D} f_0 - \mathcal{S} f_1$ in Ω is as in (8.1.835) and satisfies $\partial^A_\nu(u,0) = \widetilde{T}(f_0, f_1)$. This shows that $\widetilde{T}(f_0, f_1)$ is an admissible boundary datum for the Neumann Problem (8.1.799).

At this stage, the proof of (8.1.834) is complete. With this in hand, the same functional analytic argument (based on [130, Proposition 1.2.13]) then shows that a solution u of (8.1.799) exists if and only if the boundary datum f belongs to (8.1.800), in which scenario a solution u may be found such that the estimate in (8.1.801) is true in this final case as well.

Consider next an arbitrary null-solution u of the inhomogeneous boundary problem (8.1.799) (i.e., a function u belonging to the space defined in (8.1.802)) in the case when either $n \geq 3$, or $n = 2$ and Ω is bounded. From [130, (9.4.105)] we know that the boundary trace $\mathrm{Tr}_{\Omega\to\partial\Omega}\, u$ is a well-defined function in $\big[B^{p,q}_s(\partial\Omega,\sigma)\big]^M$, while from [132, Theorem 4.4.1] we know that the integral representation formula [132, (4.4.2)] holds in the context described in [132, (4.4.3)-(4.4.7)]. Given that we currently have $\partial^A_\nu(u,0) = 0$, this reduces to

$$u = \mathcal{D}_A\big(\mathrm{Tr}_{\Omega\to\partial\Omega}\, u\big) \text{ in } \Omega. \tag{8.1.839}$$

Taking the boundary traces of both sides and using the jump-formula described in [132, Theorem 4.3.2, item *(1)*] leads to the conclusion that

$$\big(-\tfrac{1}{2}I + K_A\big)\big(\mathrm{Tr}_{\Omega\to\partial\Omega}\, u\big) = 0. \tag{8.1.840}$$

Since having $u \in \big[B^{p,q}_{s+\frac{1}{p}}(\Omega)\big]^M$ ensures, in light of [130, (9.4.91) in item *(ii)* of Theorem 9.4.5], that $\mathrm{Tr}_{\Omega\to\partial\Omega}\, u \in \big[B^{p,q}_s(\partial\Omega,\sigma)\big]^M$, we may recast (8.1.840) as

$$\mathrm{Tr}_{\Omega\to\partial\Omega}\, u \in \mathrm{Ker}\Big(-\tfrac{1}{2}I + K_A;\ \big[B^{p,q}_s(\partial\Omega,\sigma)\big]^M\Big). \tag{8.1.841}$$

Next, from (8.1.839) we see that the assignment taking u into $\mathrm{Tr}_{\Omega\to\partial\Omega}\, u$ is linear and injective. Based on this, (8.1.841), and (8.1.820) we then deduce that

$$\begin{gathered}\text{the dimension of the space described in (8.1.802) is}\\ \leq \dim \mathrm{Ker}\Big(-\tfrac{1}{2}I + K_A;\ \big[B^{p,q}_s(\partial\Omega,\sigma)\big]^M\Big) < +\infty.\\ \text{if either } n \geq 3, \text{ or } n = 2 \text{ and } \Omega \text{ is an exterior domain.}\end{gathered} \tag{8.1.842}$$

Hence, the space of null-solutions of the boundary problem (8.1.799), described in (8.1.802), is finite dimensional, if either $n \geq 3$ or $n = 2$ and Ω.

When $n = 2$ and Ω is an exterior domain we proceed analogously, starting from the observation that the integral representation formula (8.1.836) now reduces to

$$u = \mathcal{D}\big(\mathrm{Tr}_{\Omega\to\partial\Omega}\, u\big) + c \text{ in } \Omega. \tag{8.1.843}$$

On account of the jump-formula [132, (4.3.22)], after taking boundary traces (8.1.843) yields $\big(-\tfrac{1}{2}I + K_A\big)\big(\mathrm{Tr}_{\Omega\to\partial\Omega}\, u\big) \in \mathbb{C}^M$, from which we further see that

$$\mathrm{Tr}_{\Omega\to\partial\Omega}\, u \text{ belongs to } \Big\{\phi \in \big[B^{p,q}_s(\partial\Omega,\sigma)\big]^M : \big(-\tfrac{1}{2}I + K_A\big)\phi \in \mathbb{C}^M\Big\}. \tag{8.1.844}$$

Since taking the threshold $\delta_0 \in (0,1)$ small enough in (8.1.797) guarantees that $-\frac{1}{2}I + K_A$ is a Fredholm operator on $\big[B^{p,q}_s(\partial\Omega,\sigma)\big]^M$, from item *(g)* in [130, Theorem 2.2.4] we see that the space appearing in (8.1.844) is finite dimensional. Keeping this in mind, from (8.1.844) and (8.1.843) we then conclude that the space of null-solutions of the Neumann Problem (8.1.799) is finite dimensional in this case as well. This finishes the treatment of item *(1)*.

Let us now deal with item *(2)*. Henceforth, work under the stronger hypotheses demanded in (8.1.803). In particular, Proposition 1.4.1 guarantees that, having chosen some coefficient tensor $A \in \mathfrak{A}_L^{\mathrm{dis}}$, we also have $A^\top \in \mathfrak{A}_{L^\top}^{\mathrm{dis}}$.

Select a function f in the space of admissible boundary data for the boundary problem (8.1.799) (defined in (8.1.800)). Thus, there exists u as in (8.1.822) such that $f = \partial_\nu^A(u, 0)$. If either $n \geq 3$, or $n = 2$ and Ω is bounded, the integral representation formula [132, (4.4.2)] gives $u = \mathcal{D}_A\big(\mathrm{Tr}_{\Omega\to\partial\Omega}\, u\big) - \mathscr{S} f$ in Ω. If $n = 2$ and Ω is an exterior domain, we employ the integral representation formula [132, (4.4.19)], according to which there exists some constant $c \in \mathbb{C}^M$ with the property that $u = \mathcal{D}_A\big(\mathrm{Tr}_{\Omega\to\partial\Omega}\, u\big) - \mathscr{S} f + c$ in Ω, In either scenario, take conormal derivatives in the sense of [130, Remark 9.5.3] (with $\mathbb{A} := B$ and $f := 0$). Based on the jump-formula [132, (4.3.83)] considered in the setting described in [132, Theorem 4.3.6, item *(b)*] and [132, Theorem 4.3.5], in all cases we may then write

$$f = \partial_\nu^A(u, 0) = (\partial_\nu^A \mathcal{D}_A)\big(\mathrm{Tr}_{\Omega\to\partial\Omega}\, u\big) - \big(-\tfrac{1}{2}I + K_{A^\top}^{\#}\big) f. \tag{8.1.845}$$

If $n \geq 3$, use the invertibility result in (7.1.355) together with [132, (4.4.52)] to write

$$\begin{aligned}(\partial_\nu^A \mathcal{D}_A)\big(\mathrm{Tr}_{\Omega\to\partial\Omega}\, u\big) &= \big(\partial_\nu^A \mathcal{D}_A\big) \circ S\big(S^{-1}\mathrm{Tr}_{\Omega\to\partial\Omega}\, u\big) \\ &= \big(-\tfrac{1}{2}I + K_{A^\top}^{\#}\big)\Big(\big(\tfrac{1}{2}I + K_{A^\top}^{\#}\big)\big(S^{-1}\mathrm{Tr}_{\Omega\to\partial\Omega}\, u\big)\Big). \end{aligned}\tag{8.1.846}$$

From (8.1.845) and (8.1.846) we then conclude that, in the case when $n \geq 3$,

$$f \in \mathrm{Im}\Big(-\tfrac{1}{2}I + K_{A^\top}^{\#};\ \big[B_{s-1}^{p,q}(\partial\Omega, \sigma)\big]^M\Big). \tag{8.1.847}$$

Together, (8.1.824) and (8.1.847) then prove (8.1.806) in the case when $n \geq 3$.

Let us show that the latter result remains true when $n = 2$ and Ω is bounded. In this situation, in place of the invertibility result in (7.1.355) we shall employ Theorem 7.1.17 which presently guarantees the existence of some

$$\begin{gathered} g \in \big[B_{s-1}^{p,q}(\partial\Omega, \sigma)\big]^M \ \text{ and } c \in \mathbb{C}^M \ \text{ such that} \\ Sg + c = \mathrm{Tr}_{\Omega\to\partial\Omega}\, u \ \text{ and } \langle g, \mathbf{e}_j\rangle = 0 \ \text{ for } \ 1 \leq j \leq M. \end{gathered}\tag{8.1.848}$$

As may be seen from the fact that $\partial_\nu^A \mathcal{D}$ on the Besov scale is compatible with [132, (1.5.31)] and [132, (1.5.29)], we have

$$(\partial_\nu^A \mathcal{D}_A) c = 0. \tag{8.1.849}$$

Based on (8.1.848), (8.1.849), and [132, (4.4.52)] we may now compute

$$\begin{aligned}(\partial_\nu^A \mathcal{D}_A)\big(\mathrm{Tr}_{\Omega\to\partial\Omega}\, u\big) &= (\partial_\nu^A \mathcal{D}_A)\big(Sg + c\big) = \big(\partial_\nu^A \mathcal{D}_A\big)(Sg) \\ &= \big(-\tfrac{1}{2}I + K_{A^\top}^{\#}\big)\Big(\big(\tfrac{1}{2}I + K_{A^\top}^{\#}\big)g\Big) \in \mathrm{Im}\Big(-\tfrac{1}{2}I + K_{A^\top}^{\#};\ \big[B_{s-1}^{p,q}(\partial\Omega, \sigma)\big]^M\Big). \end{aligned}\tag{8.1.850}$$

Collectively, (8.1.845) and (8.1.850) prove that (8.1.847) continues to be true when $n = 2$ and Ω is bounded. Together with (8.1.824) this establishes the portion of (8.1.806) pertaining to the case when $n = 2$ and Ω is bounded.

Let us now justify the portion of (8.1.806) referring to the case when $n = 2$ and Ω is an exterior domain. As before, select a function f in the space of admissible boundary data for the boundary problem (8.1.799) (defined in (8.1.800)). Then there exists u as in (8.1.822) such that

$$f = \partial_\nu^A(u, 0). \tag{8.1.851}$$

In the current scenario, we still have (8.1.845) and (8.1.848). In concert with (8.1.849), the latter once again permits us to write (8.1.850). Based on this and (8.1.845) we then conclude that (8.1.847) continues to hold. As a consequence, $f = \big(-\frac{1}{2}I + K^{\#}_{A^\top}\big)h$ for some $h \in \big[B^{p,q}_{s-1}(\partial\Omega, \sigma)\big]^M$, and we claim that also

$$\langle h, \mathbf{e}_j\rangle = 0 \quad \text{for each } j \in \{1, \dots, M\}. \tag{8.1.852}$$

To justify (8.1.852), first observe that since we are assuming $u(x) = O(1)$ as $|x| \to \infty$, from [131, (1.5.212)] we conclude that

$$(\nabla u)(x) = O\big(|x|^{-2}\big) \quad \text{as } |x| \to \infty. \tag{8.1.853}$$

Fix a cutoff function $\psi \in \mathscr{C}_c^\infty(\mathbb{R}^n)$ with $\psi \equiv 1$ near the origin in $\mathbb{R}^n$ and for each $R \in (0, \infty)$ define $\psi_R := \psi(\cdot/R)$. Also, write the coefficient tensor A explicitly as $\big(a_{rs}^{\alpha\beta}\big)_{\substack{1\leq r,s\leq n\\ 1\leq\alpha,\beta\leq M}}$. Then for each $j \in \{1, \dots, M\}$ the generalized "half" Green's formula [130, (9.5.18)] written for $w := \psi_R \mathbf{e}_j$ (and $f := 0$) presently gives

$${}_{[B^{p,q}_{s-1}(\partial\Omega,\sigma)]^M}\big\langle \partial_\nu^A(u, 0), \mathbf{e}_j \big\rangle_{[B^{p',q'}_{1-s}(\partial\Omega,\sigma)]^M} = \int_\Omega \big\langle a_{rs}^{\alpha\beta}\partial_s u_\beta, (\partial_r\psi_R)\mathbf{e}_j\big\rangle \, d\mathcal{L}^2 \tag{8.1.854}$$

for each $R > 0$ large enough. Since (8.1.853) entails

$$\lim_{R\to\infty} \int_\Omega \big\langle a_{rs}^{\alpha\beta}\partial_s u_\beta, (\partial_r\psi_R)\mathbf{e}_j\big\rangle \, d\mathcal{L}^2 = 0, \tag{8.1.855}$$

from (8.1.854)-(8.1.855) we conclude that for each $j \in \{1, \dots, M\}$ we have

$${}_{[B^{p,q}_{s-1}(\partial\Omega,\sigma)]^M}\big\langle \partial_\nu^A(u, 0), \mathbf{e}_j \big\rangle_{[B^{p',q'}_{1-s}(\partial\Omega,\sigma)]^M} = 0. \tag{8.1.856}$$

Then for each $j \in \{1, \dots, M\}$ we may write, based on the duality result from [132, (4.1.29)] together with (8.1.856), formula [132, (1.5.21)] (written for $A^\top$ in place of A, and bearing in mind that Ω is presently assumed to be unbounded), and (8.1.851),

$$\begin{aligned}\langle h, \mathbf{e}_j\rangle &= \big\langle\big(\tfrac{1}{2}I + K^{\#}_{A^\top}\big)h, \mathbf{e}_j\big\rangle - \big\langle\big(-\tfrac{1}{2}I + K^{\#}_{A^\top}\big)h, \mathbf{e}_j\big\rangle \\ &= \big\langle\big(\tfrac{1}{2}I + K^{\#}_{A^\top}\big)h, \mathbf{e}_j\big\rangle - \big\langle\partial^A_\nu(u,0), \mathbf{e}_j\big\rangle = \big\langle\big(\tfrac{1}{2}I + K^{\#}_{A^\top}\big)h, \mathbf{e}_j\big\rangle \\ &= \big\langle h, \big(\tfrac{1}{2}I + K_{A^\top}\big)\mathbf{e}_j\big\rangle = \langle h, 0\rangle = 0. \end{aligned} \tag{8.1.857}$$

This establishes (8.1.852). Finally, from (8.1.851)-(8.1.852) we then see that the portion of (8.1.806) referring to the case when $n = 2$ and Ω is an exterior domain is true. At this point, (8.1.806) is fully justified.

Going further, whenever the boundary datum f belongs to the space of admissible boundary data defined in (8.1.800), it follows from (8.1.806), (8.1.821), and [130, Proposition 1.2.13] that there exists some $g \in \big[B^{p,q}_{s-1}(\partial\Omega, \sigma)\big]^M$ such that

$$\big(-\tfrac{1}{2}I + K^{\#}_{A^\top}\big)g = f \text{ and } \|g\|_{[B^{p,q}_{s-1}(\partial\Omega,\sigma)]^M} \leq C\|f\|_{[B^{p,q}_{s-1}(\partial\Omega,\sigma)]^M} \tag{8.1.858}$$

for a constant $C \in (0, \infty)$ independent of f, and with the additional property that

$$\langle g, \mathbf{e}_j\rangle = 0 \text{ for } j \in \{1, \dots, M\}, \text{ if } n = 2 \text{ and } \Omega \text{ is an exterior domain.} \tag{8.1.859}$$

Then, much as before, the function $u := \mathscr{S}g$ (with the single layer considered as in [132, (4.3.25)]) satisfies (8.1.822), as well as (8.1.801), thanks to the estimate in (8.1.858). Finally, (8.1.858) and the jump-formula [132, (4.3.83)] (considered as in [132, Theorem 4.3.6, *(b)*]), give $\partial^A_\nu(u, 0) = \big(-\tfrac{1}{2}I + K^{\#}_{A^\top}\big)g = f$, implying that the function u is indeed a solution of the boundary problem (8.1.799).

Going further, by combining (8.1.821), [130, (2.1.49)], (A.0.7), item *(i)* in [130, Proposition 7.6.1], and [132, (4.1.29)] we conclude that when either $n \geq 3$, or $n = 2$ and the set Ω is bounded, the space $\mathrm{Im}\Big(-\tfrac{1}{2}I + K^{\#}_{A^\top};\ \big[B^{p,q}_{s-1}(\partial\Omega, \sigma)\big]^M\Big)$ may be described as the collection of all $f \in \big[B^{p,q}_{s-1}(\partial\Omega, \sigma)\big]^M$ satisfying (8.1.804). In turn, from this and (8.1.806) we then see that in the case when either $n \geq 3$, or $n = 2$ and the set Ω is bounded, the Neumann Problem (8.1.799) has a solution if and only if the boundary datum $f \in \big[B^{p,q}_{s-1}(\partial\Omega, \sigma)\big]^M$ satisfies (8.1.804).

Let us now prove that, in the case when $n = 2$ and the set Ω is an exterior domain, the Neumann Problem (8.1.799) has a solution if and only if the boundary datum $f \in \big[B^{p,q}_{s-1}(\partial\Omega, \sigma)\big]^M$ satisfies (8.1.805). In one direction, if the Neumann Problem (8.1.799) has a solution for a given boundary datum $f \in \big[B^{p,q}_{s-1}(\partial\Omega, \sigma)\big]^M$ then (8.1.806) implies that $f = \big(-\tfrac{1}{2}I + K^{\#}_{A^\top}\big)h$ for some $h \in \big[B^{p,q}_{s-1}(\partial\Omega, \sigma)\big]^M$ satisfying $\langle h, \mathbf{e}_j\rangle = 0$ for $1 \leq j \leq M$. As a consequence, for any $g \in \big[B^{p',q'}_{1-s}(\partial\Omega, \sigma)\big]^M$ with $\big(-\tfrac{1}{2}I + K_{A^\top}\big)g \in \mathbb{C}^M$, duality (cf. [132, (4.1.29)] with $A^\top$ in place of A) gives

$$\begin{aligned} &{}_{[B^{p,q}_{s-1}(\partial\Omega,\sigma)]^M}\big\langle f, g\big\rangle_{[B^{p',q'}_{1-s}(\partial\Omega,\sigma)]^M} \\ &\quad = {}_{[B^{p,q}_{s-1}(\partial\Omega,\sigma)]^M}\big\langle\big(-\tfrac{1}{2}I + K^{\#}_{A^\top}\big)h, g\big\rangle_{[B^{p',q'}_{1-s}(\partial\Omega,\sigma)]^M} \\ &\quad = {}_{[B^{p,q}_{s-1}(\partial\Omega,\sigma)]^M}\big\langle h, \big(-\tfrac{1}{2}I + K_{A^\top}\big)g\big\rangle_{[B^{p',q'}_{1-s}(\partial\Omega,\sigma)]^M} = 0, \end{aligned} \tag{8.1.860}$$

as wanted. In the opposite direction, if $f \in \big[B^{p,q}_{s-1}(\partial\Omega,\sigma)\big]^M$ satisfies (8.1.805), then from the duality result in [132, (4.1.29)] (with $A^\top$ in place of A), [130, (2.1.49)], and (8.1.821) we conclude that $f \in \mathrm{Im}\Big(-\frac{1}{2}I + K^\#_{A^\top};\ \big[B^{p,q}_{s-1}(\partial\Omega,\sigma)\big]^M\Big)$. Hence, $f = \big(-\frac{1}{2}I + K^\#_{A^\top}\big)h$ for some $h \in \big[B^{p,q}_{s-1}(\partial\Omega,\sigma)\big]^M$. Let us fix an arbitrary c in $\mathbb{C}^M$. When regarded as a constant function on $\partial\Omega$, it follows that this belongs to $\big[B^{p',q'}_{1-s}(\partial\Omega,\sigma)\big]^M$ and satisfies $\big(-\frac{1}{2}I + K_{A^\top}\big)c = -c \in \mathbb{C}^M$, thanks to [132, (1.5.21)] (used with $A^\top$ in place of A) and the fact that Ω is an unbounded set. Granted this, (8.1.805) (used with $g := c$) guarantees that

$$ {}_{[B^{p,q}_{s-1}(\partial\Omega,\sigma)]^M}\big\langle f, c\big\rangle_{[B^{p',q'}_{1-s}(\partial\Omega,\sigma)]^M} = 0. \tag{8.1.861}$$

In turn, this allows us to write

$$\begin{aligned}
&{}_{[B^{p,q}_{s-1}(\partial\Omega,\sigma)]^M}\big\langle h,c\big\rangle_{[B^{p',q'}_{1-s}(\partial\Omega,\sigma)]^M} \\
&\quad = {}_{[B^{p,q}_{s-1}(\partial\Omega,\sigma)]^M}\big\langle \big(\tfrac{1}{2}I + K^\#_{A^\top}\big)h, c\big\rangle_{[B^{p',q'}_{1-s}(\partial\Omega,\sigma)]^M} \\
&\qquad - {}_{[B^{p,q}_{s-1}(\partial\Omega,\sigma)]^M}\big\langle \big(-\tfrac{1}{2}I + K^\#_{A^\top}\big)h, c\big\rangle_{[B^{p',q'}_{1-s}(\partial\Omega,\sigma)]^M} \\
&\quad = {}_{[B^{p,q}_{s-1}(\partial\Omega,\sigma)]^M}\big\langle h, \big(\tfrac{1}{2}I + K_{A^\top}\big)c\big\rangle_{[B^{p',q'}_{1-s}(\partial\Omega,\sigma)]^M} \\
&\qquad - {}_{[B^{p,q}_{s-1}(\partial\Omega,\sigma)]^M}\big\langle f, c\big\rangle_{[B^{p',q'}_{1-s}(\partial\Omega,\sigma)]^M} = 0 - 0 = 0,
\end{aligned} \tag{8.1.862}$$

thanks again to [132, (1.5.21)] (used with $A^\top$ in place of A) and the fact that Ω is an unbounded set. This shows that ${}_{[B^{p,q}_{s-1}(\partial\Omega,\sigma)]^M}\big\langle h, c\big\rangle_{[B^{p',q'}_{1-s}(\partial\Omega,\sigma)]^M} = 0$ for each $c \in \mathbb{C}^M$, or, equivalently,

$$ {}_{[B^{p,q}_{s-1}(\partial\Omega,\sigma)]^M}\big\langle h, \mathbf{e}_j\big\rangle_{[B^{p',q'}_{1-s}(\partial\Omega,\sigma)]^M} = 0 \text{ for each } j \in \{1,\dots,M\}. \tag{8.1.863}$$

This ultimately places f in the space

$$\Big\{\big(-\tfrac{1}{2}I + K^\#_{A^\top}\big)h :\ h \in \big[B^{p,q}_{s-1}(\partial\Omega,\sigma)\big]^M \text{ with } \langle h, \mathbf{e}_j\rangle = 0 \text{ for } 1 \le j \le M\Big\}$$

which, according to (8.1.806), makes f an admissible boundary data for the Neumann Problem (8.1.799). This finishes the proof of the claim pertaining to (8.1.805).

Let us now deal with the claim made in (8.1.808), working under the assumption that either $n \ge 3$, or $n = 2$ and the set Ω is actually bounded. For starters we note that, as may be seen from (8.1.839)-(8.1.841), the assignment associating to each given function u belonging to the space defined in (8.1.802) its boundary trace $\mathrm{Tr}_{\Omega\to\partial\Omega}\, u \in \mathrm{Ker}\Big(-\frac{1}{2}I + K_A;\ \big[B^{p,q}_{s}(\partial\Omega,\sigma)\big]^M\Big)$ is well defined, linear, and injective. To see that this assignment is also surjective in this setting, pick an arbitrary function

$$\phi \in \mathrm{Ker}\Big(-\tfrac{1}{2}I + K_A;\ \big[B^{p,q}_{s}(\partial\Omega,\sigma)\big]^M\Big) \text{ and define } u := \mathcal{D}_A\phi \text{ in } \Omega, \tag{8.1.864}$$

with the boundary-to-domain double layer potential operator understood as in [132, Theorem 4.3.1, item *(1)*]. Then [130, (7.7.9)] and [132, (1.3.24), (4.3.1)] imply

$$u \in \big[B^{p,q}_{s+\frac{1}{p}}(\Omega)_{\rm bdd}\big]^M, \; Lu = 0 \text{ in } \Omega, \; u(x) = O(|x|^{1-n}) \text{ as } |x| \to \infty. \tag{8.1.865}$$

In addition, the jump-formula described in [132, Theorem 4.3.2, item *(1)*] gives

$$\mathrm{Tr}_{\Omega\to\partial\Omega}\, u = \big(\tfrac{1}{2}I + K_A\big)\phi = \phi. \tag{8.1.866}$$

In the case when $n \geq 3$, the claim we seek to prove follows upon observing that, thanks to [132, Theorem 4.3.5], the invertibility result in (7.1.355), formula [132, (4.4.52)], and the intertwining identity [132, (4.4.53)], we may write

$$\begin{aligned}\partial^A_\nu(u,0) &= (\partial^A_\nu \mathcal{D}_A)\phi = \big(\partial^A_\nu \mathcal{D}_A\big)\circ S\big(S^{-1}\phi\big) \qquad (8.1.867)\\ &= \big(\tfrac{1}{2}I + K^{\#}_{A^\top}\big)\big(-\tfrac{1}{2}I + K^{\#}_{A^\top}\big)\big(S^{-1}\phi\big) = \big(\tfrac{1}{2}I + K^{\#}_{A^\top}\big)S^{-1}\Big(\big(-\tfrac{1}{2}I + K_A\big)\phi\Big) = 0.\end{aligned}$$

To prove a similar result when $n = 2$ and Ω is bounded, we invoke Theorem 7.1.17 which presently guarantees the existence of some $g \in \big[B^{p,q}_{s-1}(\partial\Omega,\sigma)\big]^M$ and some $c \in \mathbb{C}^M$ satisfying $Sg + c = \phi$ and $\langle g, \mathbf{e}_j\rangle = 0$ for $1 \leq j \leq M$. Note that, thanks to this, (8.1.864), [132, (1.5.21)] (bearing in mind that Ω is a bounded set), and the intertwining property from [132, (4.4.53)] (with $A^\top$ in place of A), we have

$$\begin{aligned}&0 = \big(-\tfrac{1}{2}I + K_A\big)\phi = \big(-\tfrac{1}{2}I + K_A\big)(Sg + c) = \big(-\tfrac{1}{2}I + K_A\big)(Sg) = S\big(-\tfrac{1}{2}I + K^{\#}_{A^\top}\big)g\\ &\text{and } \Big\langle \big(-\tfrac{1}{2}I + K^{\#}_{A^\top}\big)g, \mathbf{e}_j\Big\rangle = \Big\langle g, \big(-\tfrac{1}{2}I + K_{A^\top}\big)\mathbf{e}_j\Big\rangle = \langle g, 0\rangle = 0 \qquad (8.1.868)\end{aligned}$$

for each $j \in \{1,\dots,M\}$, thanks to [132, (1.5.21), (4.1.29)], both written with $A^\top$ in place of A (and also keeping in mind that Ω is bounded). In turn, from (8.1.868), and (7.1.363) we conclude that $S_{\rm aug}\big(\big(-\tfrac{1}{2}I + K^{\#}_{A^\top}\big)g, 0\big) = (0,0)$. In view of the invertibility result established in Theorem 7.1.17, this forces $\big(-\tfrac{1}{2}I + K^{\#}_{A^\top}\big)g = 0$. Using this, (8.1.849), and [132, (4.4.52)] we may then compute

$$\begin{aligned}\partial^A_\nu(u,0) &= (\partial^A_\nu \mathcal{D}_A)\phi = \big(\partial^A_\nu \mathcal{D}_A\big)(Sg + c) = \big(\partial^A_\nu \mathcal{D}_A\big)(Sg)\\ &= \big(\tfrac{1}{2}I + K^{\#}_{A^\top}\big)\big(-\tfrac{1}{2}I + K^{\#}_{A^\top}\big)g = 0. \qquad (8.1.869)\end{aligned}$$

Hence, as long as either $n \geq 3$, or $n = 2$ and the set Ω is bounded, we have $\partial^A_\nu(u,0) = 0$, placing the function u, defined as in (8.1.864), in the space (8.1.802) of null-solutions for the boundary problem (8.1.799). In view of (8.1.866), this establishes the surjectivity property we sought to prove.

To summarize, in the case when either $n \geq 3$, or $n = 2$ and Ω is bounded, the above reasoning shows that the assignment taking the function u into its boundary trace $\mathrm{Tr}_{\Omega\to\partial\Omega}\, u$ is a linear isomorphism from the space (8.1.802) of null-solutions for the boundary problem (8.1.799) into the space $\mathrm{Ker}\Big(-\tfrac{1}{2}I + K_A;\, \big[B^{p,q}_s(\partial\Omega,\sigma)\big]^M\Big)$. This proves (8.1.808). In addition, from this we also conclude that the dimension of the space (8.1.802) of null-solutions for the boundary problem (8.1.799) is precisely $\dim \mathrm{Ker}\Big(-\frac{1}{2}I+K_A;\ \big[B^{p,q}_s(\partial\Omega,\sigma)\big]^M\Big)$ (8.1.870) whenever either $n\geq 3$, or $n=2$ and the set Ω is bounded.

Finally, (8.1.839) justifies the claim made in (8.1.809) which, in turn, gives (8.1.807).

Let us prove that the index of the boundary problem (8.1.799), i.e., the dimension of (8.1.802) minus the codimension of (8.1.800) in $\big[B^{p,q}_{s-1}(\partial\Omega,\sigma)\big]^M$, is equal to zero when either $n\geq 3$, or $n=2$ and Ω is bounded. With this in mind, observe that

$$\dim\left(\frac{\big[B^{p,q}_{s-1}(\partial\Omega,\sigma)\big]^M}{\mathrm{Im}\big(-\frac{1}{2}I+K^{\#}_{A^\top};[B^{p,q}_{s-1}(\partial\Omega,\sigma)]^M\big)}\right)=\dim \mathrm{Ker}\Big(-\frac{1}{2}I+K^{\#}_{A^\top};\ \big[B^{p,q}_{s-1}(\partial\Omega,\sigma)\big]^M\Big)$$

$$=\dim \mathrm{Ker}\Big(-\frac{1}{2}I+K_A;\ \big[B^{p,q}_{s}(\partial\Omega,\sigma)\big]^M\Big)<+\infty. \qquad (8.1.871)$$

Above, the first equality is implied by (8.1.819). The second equality is justified by observing that, thanks to [130, (11.11.12)], we have

$$L^p_1(\partial\Omega,\sigma)\hookrightarrow B^{p,q}_s(\partial\Omega,\sigma)\hookrightarrow L^p(\partial\Omega,\sigma)\hookrightarrow B^{p,q}_{s-1}(\partial\Omega,\sigma)\hookrightarrow L^p_{-1}(\partial\Omega,\sigma)$$

then invoking (7.1.62) and (7.1.622). The final inequality in (8.1.871) follows from (8.1.819). By (8.1.871), (8.1.808), and (8.1.806), the index of the boundary value problem (8.1.799) is indeed zero when either $n\geq 3$, or $n=2$ and Ω is bounded.

Let us justify (8.1.810)-(8.1.812) when $n=2$ and Ω is an exterior domain. First, consider an arbitrary null-solution u of the boundary problem (8.1.799) (i.e., a function u belonging to the space (8.1.802)) when $n=2$ and Ω is an exterior domain. In particular, [131, Theorem 1.5.9] and (8.1.803) ensure the limit

$$c_0:=\lim_{x\to\infty}u(x)\ \text{ exists in } \mathbb{C}^M \qquad (8.1.872)$$

and we have $u(x)=c_0+O(|x|^{-1})$ and $(\nabla u)(x)=O(|x|^{-2})$ as $|x|\to\infty$. From [130, (9.4.105)] we also know that (8.1.799) further implies that

$$\mathrm{Tr}_{\Omega\to\partial\Omega}\,u \text{ is a well-defined function in } \big[B^{p,q}_s(\partial\Omega,\sigma)\big]^M. \qquad (8.1.873)$$

Then [132, Theorem 4.4.1] applies to the function $\widetilde{u}:=u-c_0$ and (bearing in mind [132, (1.3.45)] and the fact that $\partial^A_\nu(\widetilde{u},0)=\partial^A_\nu(u,0)=0$) gives

$$u-c_0=\mathcal{D}\big(\mathrm{Tr}_{\Omega\to\partial\Omega}\,u\big)\ \text{ in } \Omega. \qquad (8.1.874)$$

Taking boundary traces then yields

$$\big(-\tfrac{1}{2}I+K_A\big)\big(\mathrm{Tr}_{\Omega\to\partial\Omega}\,u\big)=-c_0. \qquad (8.1.875)$$

Collectively, the properties established in (8.1.873), (8.1.875), and (8.1.872) then show that the pair $\Big(\mathrm{Tr}_{\Omega\to\partial\Omega}\,u,\ \lim\limits_{|x|\to\infty}u(x)\Big)$ belongs to the space

$$\Big\{(\phi, c) \in \big[B^{p,q}_{s}(\partial\Omega, \sigma)\big]^{M} \oplus \mathbb{C}^{M} : \big(-\tfrac{1}{2}I + K_A\big)\phi = -c\Big\}. \tag{8.1.876}$$

Thanks to (8.1.872) and (8.1.874), the mapping associating to each null-solution u of the Neumann Problem (8.1.799) the pair $\Big(\mathrm{Tr}_{\Omega\to\partial\Omega}\, u, \lim\limits_{|x|\to\infty} u(x)\Big)$ is injective.

Let us show that said map is also surjective, i.e., its range is the space (8.1.876). To this end, pick $\phi \in \big[B^{p,q}_{s}(\partial\Omega, \sigma)\big]^{M}$ and $c \in \mathbb{C}^{M}$ with $\big(-\tfrac{1}{2}I + K_A\big)\phi = -c$. If

$$u := \mathcal{D}\phi + c \text{ in } \Omega, \tag{8.1.877}$$

then, thanks to [132, (1.3.24), (1.3.26), (4.3.1)], this function satisfies

$$\begin{aligned} &u \in \big[B^{p,q}_{s+\frac{1}{p}}(\Omega)_{\mathrm{bdd}} \cap \mathscr{C}^{\infty}(\Omega)\big]^{M}, \ Lu = 0 \text{ in } \Omega, \\ &u(x) = c + O(|x|^{-1}) \text{ and } (\nabla u)(x) = O(|x|^{-2}) \text{ as } |x| \to \infty. \end{aligned} \tag{8.1.878}$$

In addition, [132, (4.3.22)] and the fact that $K_A\phi = \frac{1}{2}\phi - c$ ensure that

$$\mathrm{Tr}_{\Omega\to\partial\Omega}\, u = \big(\tfrac{1}{2}I + K_A\big)\phi + c = \phi. \tag{8.1.879}$$

Granted these properties, the desired surjectivity follows as soon as we show that

$$\partial^{A}_{\nu}(u, 0) = 0. \tag{8.1.880}$$

With this goal in mind, first observe that (8.1.799) implies that [132, Theorem 4.4.1] applies to the function $\widetilde{u} := u - c$. Thanks to [132, (1.3.45)] and the fact that $\partial^{A}_{\nu}(\widetilde{u}, 0) = \partial^{A}_{\nu}(u, 0)$, the integral representation formula [132, (4.4.2)] gives

$$u - c = \mathcal{D}\big(\mathrm{Tr}_{\Omega\to\partial\Omega}\, u\big) - \mathscr{S}\big(\partial^{A}_{\nu}(u, 0)\big) \text{ in } \Omega. \tag{8.1.881}$$

Since (8.1.877) and (8.1.879) imply $\mathcal{D}\big(\mathrm{Tr}_{\Omega\to\partial\Omega}\, u\big) = \mathcal{D}_A\phi = u - c$, this reduces to $\mathscr{S}\big(\partial^{A}_{\nu}(u, 0)\big) = 0$ in Ω. After taking boundary traces we arrive at

$$S\big(\partial^{A}_{\nu}(u, 0)\big) = 0, \tag{8.1.882}$$

on account of [132, (4.3.44)] and

$$\partial^{A}_{\nu}(u, 0) \in \big[B^{p,q}_{s-1}(\partial\Omega, \sigma)\big]^{M}. \tag{8.1.883}$$

Also, from (8.1.878), (8.1.881), and [132, (1.3.26)] we see that $\mathscr{S}\big(\partial^{A}_{\nu}(u, 0)\big)(x)$ is $O(|x|^{-1})$ as $|x| \to \infty$. Based on this and reasoning as in the proof of [132, (1.3.12)] we arrive at the conclusion that $\big\langle \partial^{A}_{\nu}(u, 0), \mathbf{e}_j \big\rangle = 0$ for each $j \in \{1, \dots, M\}$. From this, (8.1.883), (8.1.882), and (7.1.363) we obtain $S_{\rm aug}\big(\partial^{A}_{\nu}(u, 0), 0\big) = (0, 0)$. Having justified this, the invertibility result proved in Theorem 7.1.17 in relation to the operator (7.1.363) then establishes (8.1.880). This takes care of (8.1.811). In turn, from (8.1.811) we conclude that

$$\text{the dimension of the space (8.1.802) of null-solutions for the Neumann Problem (8.1.799) when } n=2 \text{ and the set } \Omega \text{ is unbounded is equal to the dimension of } \Big\{(\phi,c)\in\big[B^{p,q}_{s}(\partial\Omega,\sigma)\big]^M\oplus\mathbb{C}^M:\ \big(-\tfrac{1}{2}I+K_A\big)\phi=-c\Big\}. \tag{8.1.884}$$

Lastly, (8.1.874) implies (8.1.812) which further justifies (8.1.810).

Next we shall show that the index of the boundary value problem (8.1.799) continues to be zero in the case when $n=2$ and Ω is an exterior domain. Let

$$\iota:\Big\{g\in\big[B^{p,q}_{s-1}(\partial\Omega,\sigma)\big]^M:\ \langle g,\mathbf{e}_j\rangle=0\ \text{ for }\ 1\leq j\leq M\Big\}\to\big[B^{p,q}_{s-1}(\partial\Omega,\sigma)\big]^M \tag{8.1.885}$$

be the canonical inclusion, and introduce

$$T:\Big\{g\in\big[B^{p,q}_{s-1}(\partial\Omega,\sigma)\big]^M:\ \langle g,\mathbf{e}_j\rangle=0\ \text{ for }\ 1\leq j\leq M\Big\}\to\big[B^{p,q}_{s-1}(\partial\Omega,\sigma)\big]^M,$$
$$T:=\big(-\tfrac{1}{2}I+K^{\#}_{A^{\top}}\big)\circ\iota. \tag{8.1.886}$$

Since ι is a Fredholm operator with index $-M$ and $-\frac{1}{2}I+K^{\#}_{A^{\top}}$ is a Fredholm operator with index zero from $\big[B^{p,q}_{s-1}(\partial\Omega,\sigma)\big]^M$ into itself, we may invoke [130, Theorem 2.2.4] to conclude that the operator

$$T \text{ is Fredholm with index } -M \text{ in the context of (8.1.886).} \tag{8.1.887}$$

Then the codimension on the space of admissible boundary data for the Neumann Problem (8.1.799) in the "full" space of boundary data may be computed as follows:

$$\begin{aligned}
&\dim\left(\frac{\big[B^{p,q}_{s-1}(\partial\Omega,\sigma)\big]^M}{\big\{(-\frac{1}{2}I+K^{\#}_{A^{\top}})g:\ g\in[B^{p,q}_{s-1}(\partial\Omega,\sigma)]^M \text{ with } \langle g,\mathbf{e}_j\rangle=0 \text{ for } 1\leq j\leq M\big\}}\right)=\dim\left(\frac{\big[B^{p,q}_{s-1}(\partial\Omega,\sigma)\big]^M}{\operatorname{Im}T}\right)\\
&=M+\dim\operatorname{Ker}T=M+\dim\operatorname{Ker}\Big(-\tfrac{1}{2}I+K^{\#}_{A^{\top}};\ \big[B^{p,q}_{s-1}(\partial\Omega,\sigma)\big]^M\Big)\\
&=\dim\Big\{(\phi,c)\in\big[B^{p,q}_{s}(\partial\Omega,\sigma)\big]^M\oplus\mathbb{C}^M:\\
&\qquad\qquad \phi-c\in\operatorname{Ker}\Big(-\tfrac{1}{2}I+K_A;\ \big[B^{p,q}_{s}(\partial\Omega,\sigma)\big]^M\Big)\Big\}\\
&=\dim\Big\{(\phi,c)\in\big[B^{p,q}_{s}(\partial\Omega,\sigma)\big]^M\oplus\mathbb{C}^M:\ \big(-\tfrac{1}{2}I+K_A\big)\phi=-c\Big\}.
\end{aligned} \tag{8.1.888}$$

Above, the first equality comes from (8.1.885)-(8.1.886), while the second is implied by (8.1.887). The third equality in (8.1.888) is a consequence of the fact that

$$\begin{aligned}
&\Big\{g\in\operatorname{Ker}\Big(-\tfrac{1}{2}I+K^{\#}_{A^{\top}};\ \big[B^{p,q}_{s-1}(\partial\Omega,\sigma)\big]^M\Big):\ \langle g,\mathbf{e}_j\rangle=0\ \text{ for }\ 1\leq j\leq M\Big\}\\
&\qquad=\operatorname{Ker}\Big(-\tfrac{1}{2}I+K^{\#}_{A^{\top}};\ \big[B^{p,q}_{s-1}(\partial\Omega,\sigma)\big]^M\Big).
\end{aligned} \tag{8.1.889}$$

Indeed, the left-to-right inclusion in (8.1.889) is trivial, while the right-to-left inclusion in (8.1.889), is seen by writing

$$
\begin{aligned}
&\operatorname{Ker}\Big(-\tfrac{1}{2}I+K^{\#}_{A^{\top}};\ \big[L^{p}(\partial\Omega,\sigma)\big]^{M}\Big) \qquad (8.1.890)\\
&=\Big\{g\in\operatorname{Ker}\Big(-\tfrac{1}{2}I+K^{\#}_{A^{\top}};\ \big[L^{p}(\partial\Omega,\sigma)\big]^{M}\Big):\ \int_{\partial\Omega}g\,\mathrm{d}\sigma=0\Big\}\\
&\subseteq\Big\{g\in\operatorname{Ker}\Big(-\tfrac{1}{2}I+K^{\#}_{A^{\top}};\ \big[B^{p,q}_{s-1}(\partial\Omega,\sigma)\big]^{M}\Big):\ \langle g,\mathbf{e}_j\rangle=0\ \text{for}\ 1\leq j\leq M\Big\},
\end{aligned}
$$

thanks to [130, (11.11.12)], (7.1.61), and (8.1.529). The fourth equality in (8.1.528) is based on two ingredients. First, (7.1.631) used with $z := -1/2$, [130, (11.11.12)], and (7.1.62) imply

$$
\begin{aligned}
&\dim\operatorname{Ker}\Big(-\tfrac{1}{2}I+K^{\#}_{A^{\top}};\big[B^{p,q}_{s-1}(\partial\Omega,\sigma)\big]^{M}\Big)=\dim\operatorname{Ker}\Big(-\tfrac{1}{2}I+K^{\#}_{A^{\top}};\big[L^{p}(\partial\Omega,\sigma)\big]^{M}\Big)\\
&=\dim\operatorname{Ker}\Big(-\tfrac{1}{2}I+K_{A};\big[L^{p}_{1}(\partial\Omega,\sigma)\big]^{M}\Big)=\dim\operatorname{Ker}\Big(-\tfrac{1}{2}I+K_{A};\big[B^{p,q}_{s}(\partial\Omega,\sigma)\big]^{M}\Big).
\end{aligned}
$$

The second ingredient in the fourth equality in (8.1.528) is (8.1.434) used with

$$
\begin{aligned}
&V:=\big[B^{p,q}_{s}(\partial\Omega,\sigma)\big]^{M},\quad F_0:=\mathbb{C}^{M},\ \text{and}\\
&F_1:=\operatorname{Ker}\Big(-\tfrac{1}{2}I+K_{A};\ \big[B^{p,q}_{s}(\partial\Omega,\sigma)\big]^{M}\Big).
\end{aligned}
\qquad (8.1.891)
$$

The final equality in (8.1.888) is a direct corollary of [132, (1.5.21)]. Having established (8.1.888), from (8.1.452) and (8.1.457) we conclude that the index of the Neumann Problem (8.1.799) is also equal to zero in this last case. This completes the treatment of item *(2)*.

As far as item *(3)* is concerned, suppose $A \in \mathfrak{A}^{\mathrm{dis}}_{L}$ is positive definite. When either $n \geq 3$, or $n = 2$ and the set Ω is bounded, we conclude from (8.1.343) and (7.1.277) for Besov spaces (cf. Remark 7.1.12) that the space of admissible boundary data for the Neumann Problem (8.1.799) is equal to

$$
\operatorname{Im}\Big(-\tfrac{1}{2}I+K^{\#}_{A^{\top}};\ \big[B^{p,q}_{s-1}(\partial\Omega,\sigma)\big]^{M}\Big)=\big[B^{p,q}_{s-1}(\partial\Omega,\sigma)\cap(\mathbb{C}_{\partial\Omega_+})^{\perp}\big]^{M}. \qquad (8.1.892)
$$

This proves the first claim in (8.1.814), while the second claim follows from (8.1.807), the version of (7.1.275) for Besov spaces (cf. Remark 7.1.12), and (7.1.305).

Suppose now that $n = 2$ and Ω is an exterior domain. Then the last property in (8.1.806) combines with the version of (7.1.292) written for Besov spaces (cf. Remark 7.1.12) to give that the space of admissible boundary data for the Neumann Problem (8.1.799) is, in this case, equal to

$$
\Big\{\big(-\tfrac{1}{2}I+K^{\#}_{A}\big)g:\ g\in\big[B^{p,q}_{s}(\partial\Omega,\sigma)\cap\mathbb{C}^{\perp}\big]^{M}\Big\}=\big[B^{p,q}_{s-1}(\partial\Omega,\sigma)\cap(\mathbb{C}^{\infty}_{\partial\Omega_+})^{\perp}\big]^{M}.
$$

The first claim in (8.1.815) is therefore justified. As far as the second claim in (8.1.815) is concerned, from (8.1.810), the version of (7.1.293) for Besov spaces (cf. Remark 7.1.12), and (8.1.439)-(8.1.441) we see that the space of null-solutions of the Neumann Problem (8.1.799) is, in this case, contained in

$$\Big\{\mathcal{D}_A\phi + c : \phi \in \big(\mathbb{C}^{\infty}_{\partial\Omega_+}\big)^M \text{ and } c \in \mathbb{C}^M\Big\} \subseteq \big(\mathbb{C}^{\infty}_{\Omega_+}\big)^M. \tag{8.1.893}$$

In the converse direction, it is clear from definitions that any function in $\big(\mathbb{C}^{\infty}_{\Omega_+}\big)^M$ is a null-solution of the Neumann Problem (8.1.799). Consequently, the space of null-solutions of the Neumann Problem (8.1.799) is, in the current case, precisely $\big(\mathbb{C}^{\infty}_{\Omega_+}\big)^M$. This finishes the treatment of item *(3)*.

Lastly, the claims in the current item *(4)*, pertaining to the versions of the Neumann Problem for functions in Triebel-Lizorkin spaces and weighted Sobolev spaces formulated as in (8.1.816) and (8.1.817), respectively, are dealt with in an analogous fashion. This finishes the proof of Theorem 8.1.22. □

We next consider the Inhomogeneous Neumann Problem with boundary data in Besov spaces, formulated for weakly elliptic systems and δ-oscillating AR domains with δ sufficiently small. This augments the well-posedness result for the Inhomogeneous Neumann Problem formulated for Legendre-Hadamard elliptic systems in bounded two-sided NTA domains with Ahlfors regular boundaries and connected complements in Theorem 7.2.16. The Inhomogeneous Neumann Problem (8.1.898) with $q = 2$ for the Laplacian in bounded Lipschitz and $\mathscr{C}^1$ domains has been solved by E. Fabes, O. Mendez, and M. Mitrea in [56]. This answered the question raised by C. Kenig in [93, Problem 3.2.21, p. 121]. Theorem 8.1.23 may be viewed as the Geometric Measure Theoretic sharp version of the work in [56] pertaining to bounded $\mathscr{C}^1$ domains and bounded Lipschitz with small constant. It complements work done earlier in Theorem 7.2.16 in the context of bounded NTA domains with Ahlfors regular boundaries in (small perturbations of) energy spaces. Other related results are in [125], [155], [159], [45], [147], [13], and the references there.

Theorem 8.1.23 (The Inhomogeneous Neumann Problem with Boundary Data in Besov Spaces) *Fix two integers $n, M \in \mathbb{N}$ with $n \geq 2$ and suppose $\Omega \subseteq \mathbb{R}^n$ is a bounded Ahlfors regular domain. Abbreviate $\sigma := \mathcal{H}^{n-1}\lfloor\partial\Omega$ and denote by ν the geometric measure theoretic outward unit normal to Ω. Also, let L be a homogeneous, second-order, constant complex coefficient, weakly elliptic $M \times M$ system in $\mathbb{R}^n$, satisfying*

$$\mathfrak{A}^{\mathrm{dis}}_L \neq \varnothing \text{ and } \mathfrak{A}^{\mathrm{dis}}_{L^\top} \neq \varnothing. \tag{8.1.894}$$

Finally, pick a coefficient tensor A for which $L = L_A$ (i.e., $A \in \mathfrak{A}_L$) and select

$$p \in (1,\infty), \quad q \in (1,\infty), \quad s \in (0,1). \tag{8.1.895}$$

Then there exists a threshold $\delta_0 \in (0,1)$, depending only on n, p, q, s, A, the ADR *constants of $\partial\Omega$, and the relative flatness ratio $\wp(\Omega)/\mathrm{diam}\,\partial\Omega$, so that when*

$$\mathrm{dist}\Big(\nu, [\mathrm{VMO}(\partial\Omega,\sigma)]^n\Big) < \delta_0 \tag{8.1.896}$$

where the distance in the left-hand side is considered in the space $\big[\mathrm{BMO}(\partial\Omega,\sigma)\big]^n$ *(hence whenever* Ω *is a bounded* δ_0*-oscillating* AR *domain*[70]*; cf. Definition 3.4.1), the set* Ω *becomes both a two-sided* NTA *domain*[71] *as well as an* (ε,δ)*-domain, and the following statements are true:*

(1) Consider the Inhomogeneous Neumann Problem which, given any data

$$f\in\big[B^{p,q}_{s+\frac{1}{p}-2,0}(\Omega)\big]^M \text{ and } g\in\big[B^{p,q}_{s-1}(\partial\Omega,\sigma)\big]^M, \tag{8.1.897}$$

is concerned with finding a function u such that[72]

$$\begin{cases} u\in\big[B^{p,q}_{s+\frac{1}{p}}(\Omega)\big]^M, & Lu=f\big|_{\Omega} \text{ in } \Omega,\\ \partial^A_\nu(u,f)=g \text{ on } \partial\Omega, \end{cases} \tag{8.1.898}$$

where the conormal derivative is meaningfully defined as in [130, Proposition 9.5.2] *with* $\mathbb{A}:=B$*, now regarding* f *as a functional in* $\Big(\big[B^{p',q'}_{1-s-\frac{1}{p'}}(\Omega)\big]^M\Big)^*$ (*cf.* [130, (9.2.123)]) *and keeping in mind the identification from* [130, (9.5.4) in Lemma 9.5.1].

Then, whenever $A\in\mathfrak{A}^{\rm dis}_L$*, the inhomogeneous boundary value problem* (8.1.898) *is Fredholm solvable, in the following precise sense: First, the space of admissible boundary data for the inhomogeneous boundary problem* (8.1.898)*, i.e.,*

$$\Big\{(f,g)\in\big[B^{p,q}_{s+\frac{1}{p}-2,0}(\Omega)\big]^M\times\big[B^{p,q}_{s-1}(\partial\Omega,\sigma)\big]^M : \text{there is } u\in\big[B^{p,q}_{s+\frac{1}{p}}(\Omega)\big]^M$$
$$\text{such that } Lu=f\big|_{\Omega} \text{ and } \partial^A_\nu(u,f)=g\Big\} \tag{8.1.899}$$

happens to be a closed linear subspace of finite codimension in Cartesian product space $\big[B^{p,q}_{s+\frac{1}{p}-2,0}(\Omega)\big]^M\times\big[B^{p,q}_{s-1}(\partial\Omega,\sigma)\big]^M$*. Second, a solution u of* (8.1.898) *exists if and only if the data* (f,g) *belongs to* (8.1.899)*. Third, the space of null-solutions of the inhomogeneous boundary problem* (8.1.898)*, i.e.,*

$$\Big\{u\in\big[B^{p,q}_{s+\frac{1}{p}}(\Omega)\big]^M : Lu=0 \text{ in } \Omega \text{ and } \partial^A_\nu(u,0)=0\Big\} \tag{8.1.900}$$

is finite dimensional.

(2) Analogous results as in item (1) are valid for the version of the Inhomogeneous Neumann Problem (8.1.898) *formulated for functions in Triebel-Lizorkin spaces, a scenario in which given any data*

---

[70] in particular, if Ω is a bounded infinitesimally flat AR domain

[71] hence Ω is also a UR domain satisfying a two-sided corkscrew condition, which further implies that $\partial\Omega=\partial(\overline{\Omega})$

[72] with both equalities understood in the sense of distributions

$$f \in \left[F^{p,q}_{s+\frac{1}{p}-2,0}(\Omega)\right]^M \text{ and } g \in \left[B^{p,p}_{s-1}(\partial\Omega,\sigma)\right]^M, \tag{8.1.901}$$

one is concerned with finding a function u such that

$$\begin{cases} u \in \left[F^{p,q}_{s+\frac{1}{p}}(\Omega)\right]^M, \quad Lu = f\big|_\Omega \text{ in } \Omega, \\ \partial^A_\nu(u,f) = g \text{ on } \partial\Omega, \end{cases} \tag{8.1.902}$$

(where the conormal derivative is taken as in [130, Proposition 9.5.2] *with* $\mathbb{A} := F$*), and for the version of the Inhomogeneous Neumann Problem* (8.1.898) *formulated for functions in weighted Sobolev spaces when, given any data*

$$f \in \left(\left[W^{1,p'}_{-a}(\Omega)\right]^M\right)^* \text{ with } a := 1-s-\tfrac{1}{p} \text{ and } p' := \left(1-\tfrac{1}{p}\right)^{-1} \\ \text{and } g \in \left[B^{p,p}_{s-1}(\partial\Omega,\sigma)\right]^M, \tag{8.1.903}$$

one is seeking a function u satisfying

$$\begin{cases} u \in \left[W^{1,p}_{a}(\Omega)\right]^M, \quad Lu = f\lfloor_\Omega \text{ in } \Omega, \\ \partial^A_\nu(u,f) = g \text{ on } \partial\Omega, \end{cases} \tag{8.1.904}$$

where the conormal derivative is now considered as in [130, Proposition 8.5.3].

As seen from Proposition 1.6.12, if $M = 1$ (i.e., in the scalar case) and $n \geq 3$ then (8.1.894) is automatically satisfied. Furthermore, in the scalar case in the two-dimensional setting (i.e., for $M = 1$ and $n = 2$), if L satisfies the Legendre-Hadamard strong ellipticity condition or, more generally (cf. [131, Lemma 1.4.24]), if $L = \operatorname{div} A\nabla$ for some 2×2 complex matrix A with $\operatorname{sym} A := \frac{1}{2}(A + A^\top) \in M^0_2$ (cf. [131, (1.4.187)]), then Proposition 1.6.11 and Proposition 1.6.13 ensure that (8.1.894) holds. Finally, according to (1.6.141) in Proposition 1.6.13, in the scalar case in the two-dimensional setting (i.e., for $M = 1$ and $n = 2$) we may omit (8.1.894) altogether if L is a weakly elliptic operator with real coefficients.

Proof of Theorem 8.1.23 Take $\delta_0 \in (0,1)$ sufficiently small, depending on n, p, A, the ADR constants of $\partial\Omega$, and the relative flatness ratio $\wp(\Omega)/\operatorname{diam}\partial\Omega$, such that the conclusions in Theorem 7.1.1, Proposition 7.1.2, Theorem 7.1.15, Proposition 7.1.41, and Theorem 8.1.22 are valid (for suitable choices of the parameters involved in their statements). After further decreasing the value of $\delta_0 \in (0,1)$, if necessary, Theorem 3.4.2 applies and ensures that Ω is both an (ε,δ)-domain as well as a two-sided NTA domain, hence also a UR domain (cf. [129, (5.10.33)]) satisfying a two-sided local John condition (cf. [129, (5.11.28)]).

For now, work under the assumptions made in (8.1.798) and pick $A \in \mathfrak{A}^{\text{dis}}_L$. Theorem 1.5.5 ensures that $A^\top \in \mathfrak{A}^{\text{dis}}_{L^\top}$. Consider the transpose double layer potential operator $K^\#_{A^\top}$ associated with Ω and the coefficient tensor $A^\top$ acting on Besov spaces as in [132, (4.1.27)]. From (7.1.24) (used with $z := -1/2$) we know that

$$-\tfrac{1}{2}I + K^{\#}_{A^{\top}} : \big[B^{p,q}_{s-1}(\partial\Omega,\sigma)\big]^M \longrightarrow \big[B^{p,q}_{s-1}(\partial\Omega,\sigma)\big]^M \quad \text{is a Fredholm operator with index zero.} \tag{8.1.905}$$

Then [130, Theorem 2.2.3] guarantees the existence of two linear operators,

$$\begin{aligned} R &: \big[B^{p,q}_{s-1}(\partial\Omega,\sigma)\big]^M \to \big[B^{p,q}_{s-1}(\partial\Omega,\sigma)\big]^M \text{ bounded,} \\ C_0 &: \big[B^{p,q}_{s-1}(\partial\Omega,\sigma)\big]^M \to \big[B^{p,q}_{s-1}(\partial\Omega,\sigma)\big]^M \text{ compact,} \end{aligned} \tag{8.1.906}$$

such that, with I denoting the identity operator,

$$\big(-\tfrac{1}{2}I + K^{\#}_{A^{\top}}\big) \circ R = I + C_0 \text{ on } \big[B^{p,q}_{s-1}(\partial\Omega,\sigma)\big]^M. \tag{8.1.907}$$

We next claim that there exists a liner and bounded operator

$$\begin{aligned} &\Theta : \big[B^{p,q}_{s+\frac{1}{p}-2,0}(\Omega)\big]^M \longrightarrow \big[B^{p,q}_{s+\frac{1}{p}}(\Omega)\big]^M \\ &\text{with } L(\Theta f) = f\big|_{\Omega} \text{ for each } f \in \big[B^{p,q}_{s+\frac{1}{p}-2,0}(\Omega)\big]^M. \end{aligned} \tag{8.1.908}$$

To construct it, consider the Fourier multiplier operator $L^{-1}\varphi := \mathscr{F}^{-1}\big(L(\xi)^{-1}(\mathscr{F}\varphi)\big)$ for all $\varphi \in \big[\mathcal{S}'(\mathbb{R}^n)\big]^M$, and set

$$\Theta f := (L^{-1}f)\big|_{\Omega} \text{ for each } f \in \big[B^{p,q}_{s+\frac{1}{p}-2,0}(\Omega)\big]^M. \tag{8.1.909}$$

Then [87, Corollary 4.2] implies that (8.1.908) holds. Together with [130, Proposition 9.5.2], this permits us to define the following linear and bounded operator:

$$\begin{aligned} &Q : \big[B^{p,q}_{s+\frac{1}{p}-2,0}(\Omega)\big]^M \longrightarrow \big[B^{p,q}_{s-1}(\partial\Omega,\sigma)\big]^M \\ &Qf := \partial^A_\nu\big(\Theta f, f\big) \text{ for each } f \in \big[B^{p,q}_{s+\frac{1}{p}-2,0}(\Omega)\big]^M. \end{aligned} \tag{8.1.910}$$

Let us also introduce

$$\begin{aligned} &T : \big[B^{p,q}_{s+\frac{1}{p}-2,0}(\Omega)\big]^M \times \big[B^{p,q}_{s-1}(\partial\Omega,\sigma)\big]^M \to \big[B^{p,q}_{s+\frac{1}{p}-2,0}(\Omega)\big]^M \times \big[B^{p,q}_{s-1}(\partial\Omega,\sigma)\big]^M \\ &T(f,\phi) := (f, Qf + \phi),\ \forall (f,\phi) \in \big[B^{p,q}_{s+\frac{1}{p}-2,0}(\Omega)\big]^M \times \big[B^{p,q}_{s-1}(\partial\Omega,\sigma)\big]^M, \end{aligned} \tag{8.1.911}$$

$$\begin{aligned} &C_1 : \big[B^{p,q}_{s+\frac{1}{p}-2,0}(\Omega)\big]^M \times \big[B^{p,q}_{s-1}(\partial\Omega,\sigma)\big]^M \to \big[B^{p,q}_{s+\frac{1}{p}-2,0}(\Omega)\big]^M \times \big[B^{p,q}_{s-1}(\partial\Omega,\sigma)\big]^M \\ &C_1(f,\phi) := (0, C_0\phi),\ \forall (f,\phi) \in \big[B^{p,q}_{s+\frac{1}{p}-2,0}(\Omega)\big]^M \times \big[B^{p,q}_{s-1}(\partial\Omega,\sigma)\big]^M. \end{aligned} \tag{8.1.912}$$

By design, T and C_1 are linear and bounded operators. Moreover, it is clear that T is invertible and C_1 is compact. As such, item *(a)* in [130, Theorem 2.2.4] gives that

$$T + C_1 : \big[B^{p,q}_{s+\frac{1}{p}-2,0}(\Omega)\big]^M \times \big[B^{p,q}_{s-1}(\partial\Omega,\sigma)\big]^M \to \big[B^{p,q}_{s+\frac{1}{p}-2,0}(\Omega)\big]^M \times \big[B^{p,q}_{s-1}(\partial\Omega,\sigma)\big]^M$$
$$\text{is a Fredholm operator (with index zero).} \tag{8.1.913}$$

In view of [130, Definition 2.2.1], this further implies that

$$\operatorname{Im}\Big(T + C_1;\ \big[B^{p,q}_{s+\frac{1}{p}-2,0}(\Omega)\big]^M \times \big[B^{p,q}_{s-1}(\partial\Omega,\sigma)\big]^M\Big) \text{ is a closed subspace}$$
$$\text{of } \big[B^{p,q}_{s+\frac{1}{p}-2,0}(\Omega)\big]^M \times \big[B^{p,q}_{s-1}(\partial\Omega,\sigma)\big]^M, \text{ of finite codimension.} \tag{8.1.914}$$

To proceed, bring in the boundary-to-domain single layer operator $\mathscr{S}$ acting on Besov spaces as in [132, Theorem 4.3.3, item *(1)*]. Then given any pair (f,ϕ) belonging to $\big[B^{p,q}_{s+\frac{1}{p}-2,0}(\Omega)\big]^M \times \big[B^{p,q}_{s-1}(\partial\Omega,\sigma)\big]^M$ define

$$u := \Theta f + \mathscr{S}(R\phi) \text{ in } \Omega. \tag{8.1.915}$$

Then (8.1.908), (8.1.906), and [132, (4.2.209), (4.3.25)] imply that

$$u \in \big[B^{p,q}_{s+\frac{1}{p}}(\Omega)\big]^M \text{ and } Lu = f\big|_{\Omega} \text{ in } \Omega. \tag{8.1.916}$$

From (8.1.915), (8.1.916), (8.1.910), the jump-formula [132, (4.3.83)] in the context of [132, Theorem 4.3.6, item *(b)*], (8.1.907), (8.1.911), and (8.1.912) we see that

$$\begin{aligned} \partial^A_\nu(u,f) &= \partial^A_\nu(\Theta f, f) + \partial^A_\nu(\mathscr{S}(R\phi),0) = Qf + \big(-\tfrac{1}{2}I + K^{\#}_{A^\top}\big)(R\phi) \\ &= Qf + \phi + C_0\phi = (T + C_1)(f,\phi). \end{aligned} \tag{8.1.917}$$

Collectively, (8.1.916) and (8.1.917) prove that

$$\text{the space of admissible data, defined in (8.1.899), contains}$$
$$\operatorname{Im}\Big(T + C_1;\ \big[B^{p,q}_{s+\frac{1}{p}-2,0}(\Omega)\big]^M \times \big[B^{p,q}_{s-1}(\partial\Omega,\sigma)\big]^M\Big). \tag{8.1.918}$$

In turn, from (8.1.918), (8.1.914), [130, Theorem 1.1.3, *(3)*] we conclude that

$$\text{the space of admissible data, defined in (8.1.899), is a closed}$$
$$\text{space of finite codimension in } \big[B^{p,q}_{s+\frac{1}{p}-2,0}(\Omega)\big]^M \times \big[B^{p,q}_{s-1}(\partial\Omega,\sigma)\big]^M. \tag{8.1.919}$$

Since the fact that (8.1.900), the space of null-solutions of the inhomogeneous boundary problem (8.1.898), is finite dimensional has been already established in Theorem 8.1.22, at this stage all claims in item *(1)* have been justified.

Finally, the claims in item *(2)*, regarding the versions of the Inhomogeneous Neumann Problem for functions in Triebel-Lizorkin spaces and weighted Sobolev spaces as in (8.1.901)-(8.1.902) and (8.1.903)-(8.1.904) are established similarly. □

Our last result in this section elaborates on (8.1.2). Given a weakly elliptic, homogeneous, second-order, constant complex coefficient, $M \times M$ system L in $\mathbb{R}^n$, we shall say that L satisfies the `Shapiro-Lopatinskiĭ condition` (relative to the Dirichlet boundary condition) provided for any two vectors $\xi \in \mathbb{R}^n \setminus \{0\}$ and $\eta \in S^{n-1}$ such that $\langle \xi, \eta \rangle = 0$ the ODE formally written as

$$\begin{cases} L\big(\xi + \eta(1/\mathrm{i})(\mathrm{d}/\mathrm{d}t)\big)w(t) = 0, & t > 0, \\ w(0) = 0 \text{ and } \lim\limits_{t\to\infty} w(t) = 0, \end{cases} \tag{8.1.920}$$

has only the trivial solution $w \equiv 0$. See, for example, [6, (6.5), p. 53], [98, Definition 4.2.2, p. 113], [207, Definition 9.28, p. 389]. Originally introduced in [109], [187] (and further extended in [4], [5]), this condition plays a key role in the study of boundary value problems for elliptic systems in smooth domains, formulated in suitable Sobolev spaces. By adapting the proof of Theorem 8.1.21 and employing the machinery of boundary layer potentials developed earlier, here we succeed in linking it to the property of having distinguished coefficient tensors.

Theorem 8.1.24 *Pick two integers $n, M \in \mathbb{N}$ with $n \geq 2$ and suppose L is a weakly elliptic, homogeneous, second-order, constant complex coefficient, $M \times M$ system in $\mathbb{R}^n$, with $\mathfrak{A}^{\rm dis}_L \neq \varnothing$ and $\mathfrak{A}^{\rm dis}_{L^\top} \neq \varnothing$. Then L satisfies the Shapiro-Lopatinskiĭ condition.*

Proof Let $\Omega \subseteq \mathbb{R}^n$ be a bounded $\mathscr{C}^\infty$ domain. Denote by $\nu = (\nu_1, \dots, \nu_n)$ the geometric measure theoretic outward unit normal to Ω and abbreviate $\sigma := \mathcal{H}^{n-1} \lfloor \partial\Omega$. The strategy of the proof is to show that the (higher regularity) data map[73]

$$\begin{gathered} \Phi : \big[H^2(\Omega)\big]^M \longrightarrow \big[L^2(\Omega, \mathcal{L}^n)\big]^M \times \big[H^{3/2}(\partial\Omega, \sigma)\big]^M, \\ \Phi u := \big(Lu, \operatorname{Tr}_{\Omega\to\partial\Omega} u\big) \text{ for all } u \in \big[H^2(\Omega)\big]^M, \text{ is Fredholm.} \end{gathered} \tag{8.1.921}$$

Once this is proved, the fact that L satisfies the Shapiro-Lopatinskiĭ condition follows on account of well-known, classical results (the so-called Main Theorem for elliptic BVP's; cf., e.g., [207, Theorem 9.32, p. 394], or [6, Theorem 6.3.1, p. 58]).

In turn, (8.1.921) is established in a series of steps. First, recall the matrix-valued fundamental solution $E = (E_{\alpha\beta})_{1\leq\alpha,\beta\leq M}$ associated with L as in [131, Theorem 1.4.2]. Also, pick a coefficient tensor $A = \big(a^{\alpha\beta}_{jk}\big)_{\substack{1\leq\alpha,\beta\leq M \\ 1\leq j,k\leq n}} \in \mathfrak{A}^{\rm dis}_L$. Denote by $\mathcal{D}_A$ the double layer potential operator associated with A and Ω as in (A.0.72). Next, the boundary-to-domain single layer potential operator $\mathscr{S}$ associated with L and Ω as in (A.0.228) is $\mathscr{S} = \big(\mathscr{S}_{\alpha\beta}\big)_{1\leq\alpha,\beta\leq M}$ where, for each pair of indices $\alpha, \beta \in \{1, \dots, M\}$ and each function $f \in L^1(\partial\Omega, \sigma)$, we have set

$$\mathscr{S}_{\alpha\beta} f(x) := \int_{\partial\Omega} E_{\alpha\beta}(x - y) f(y)\, \mathrm{d}\sigma(y), \qquad \forall x \in \Omega. \tag{8.1.922}$$

We agree to further extend the action of $\mathscr{S}$ and of each $\mathscr{S}_{\alpha\beta}$ on distributions as in [132, Proposition 4.2.8, (4.2.202)].

[73] compare with (8.1.749)

Step I: *Given an arbitrary function* $f = (f_\alpha)_{1\le\alpha\le M} \in \big[L^2_1(\partial\Omega,\sigma)\big]^M$, *for each* $\gamma \in \{1,\dots.M\}$ *and each* $j \in \{1,\dots,n\}$ *there holds*

$$\partial_j\big(\mathcal{D}_A f\big)_\gamma = a^{\beta\alpha}_{rs}\,\partial_r \mathcal{S}_{\gamma\beta}\big(\nu_j(\nabla_{\tan} f_\alpha)_s\big) + \big(\mathcal{D}_A\big((\nabla_{\tan} f)^j\big)\big)_\gamma \text{ in } \Omega, \tag{8.1.923}$$

where $(\nabla_{\tan} f)^j$ *is the vector whose* α*-th component is* $(\nabla_{\tan} f_\alpha)_j$. (8.1.924)

Moreover, if $\mathrm{Sym}(L;\xi) := -\big(a^{\alpha\beta}_{jk}\xi_j\xi_k\big)_{1\le\alpha,\beta\le M}$, *for all* $\xi = (\xi_1,\dots,\xi_n) \in \mathbb{R}^n$, *is the principal symbol of the system L, given an arbitrary* $f \in L^2(\partial\Omega,\sigma)$ *then for each* $\gamma,\beta \in \{1,\dots,M\}$ *and* $r \in \{1,\dots,n\}$ *one has (with the single layer in the right-hand side understood in the sense of* [132, (4.2.202)]*):*

$$\begin{aligned}\partial_r\big(\mathcal{S}_{\gamma\beta} f\big) = &-\mathcal{S}_{\gamma\eta}\big(\partial_{\tau_{sr}}\big(\nu_i a^{\eta\mu}_{si}\big(\mathrm{Sym}(L;\nu)^{-1}\big)_{\mu\beta} f\big)\big)\\ &+\big(\mathcal{D}_A\big(\big(\big(\mathrm{Sym}(L;\nu)^{-1}\big)_{\alpha\beta}\nu_r f\big)_\alpha\big)\big)_\gamma \text{ in } \Omega.\end{aligned} \tag{8.1.925}$$

To see this, select an arbitrary $f = (f_\alpha)_{1\le\alpha\le M} \in \big[L^2_1(\partial\Omega,\sigma)\big]^M$. Also, fix an index $\gamma \in \{1,\dots,M\}$, a point $x \in \Omega$, and an index $j \in \{1,\dots,n\}$. Then [132, Lemma 1.3.2, (1.3.28)], [130, Proposition 11.4.2, (11.4.8)], (8.1.922), (8.1.924) give

$$\begin{aligned}\partial_j\big(\mathcal{D}_A f\big)_\gamma(x) &= \int_{\partial\Omega} a^{\beta\alpha}_{rs}(\partial_r E_{\gamma\beta})(x-y)(\partial_{\tau_{js}} f_\alpha)(y)\,d\sigma(y)\\ &= \int_{\partial\Omega} \nu_j(y) a^{\beta\alpha}_{rs}(\partial_r E_{\gamma\beta})(x-y)(\nabla_{\tan} f_\alpha)_s(y)\,d\sigma(y)\\ &\quad - \int_{\partial\Omega} \nu_s(y) a^{\beta\alpha}_{rs}(\partial_r E_{\gamma\beta})(x-y)(\nabla_{\tan} f_\alpha)_j(y)\,d\sigma(y)\\ &= a^{\beta\alpha}_{rs}\,\partial_r\mathcal{S}_{\gamma\beta}\big(\nu_j(\nabla_{\tan} f_\alpha)_s\big)(x) + \big(\mathcal{D}_A\big((\nabla_{\tan} f)^j\big)\big)_\gamma(x),\end{aligned} \tag{8.1.926}$$

proving (8.1.923). Moving on to (8.1.925), fix some $f \in L^2(\partial\Omega,\sigma)$. Then, given any indices $\gamma,\beta \in \{1,\dots,M\}$ and $r \in \{1,\dots,n\}$, at any point $x \in \Omega$ we have

$$\begin{aligned}\partial_r\big(\mathcal{S}_{\gamma\beta} f\big)(x) &= -\int_{\partial\Omega} \partial_{y_r}[E_{\gamma\beta}(x-y)] f(y)\,d\sigma(y)\\ &= -\int_{\partial\Omega} \langle \partial_{y_r}[E_{\gamma\bullet}(x-y)], f(y)\delta_{\beta\bullet}\rangle\,d\sigma(y)\\ &= -\int_{\partial\Omega} \langle D_y[E_{\gamma\bullet}(x-y)], f(y)\delta_{\beta\bullet}\delta_{r\bullet}\rangle\,d\sigma(y),\end{aligned} \tag{8.1.927}$$

where $D(u_\alpha)_{1\le\alpha\le M} := (\partial_s u_\beta)_{\substack{1\le\beta\le M\\ 1\le s\le n}}$. Hence, Du is the Jacobian matrix of the vector-valued function u, and

$$D^*(v_{\beta s})_{\beta,s} = D^\top(v_{\beta s})_{\beta,s} = -(\partial_s v_{\beta s})_\beta, \tag{8.1.928}$$

$$\mathrm{Sym}(D^*;\xi)(v_{\beta s})_{\beta,s} = -\sqrt{-1}\,(\xi_s v_{\beta s})_\beta \tag{8.1.929}$$

(recall that the principal symbol of a first-order system has been introduced in (A.0.223)). Going further, use D to create the tangential operator

$$\partial_{\tau_D} := D - \sqrt{-1}\,\mathrm{Sym}\,(D;\nu)\,\mathrm{Sym}\,(L^\top;\nu)^{-1}\partial_\nu^{A^\top}. \tag{8.1.930}$$

A more explicit formula for ∂_{τ_D} may be obtained by further expending the expression in (8.1.930). Specifically, given $u = (u_\alpha)_{1\le\alpha\le M}$ we write (with $\mathrm{i} := \sqrt{-1}$)

$$\begin{aligned}
\partial_{\tau_D} u &= Du - \mathrm{i}\,\mathrm{Sym}\,(D;\nu)\,\mathrm{Sym}\,(L^\top;\nu)^{-1}\partial_\nu^{A^\top} u \\
&= \big(\partial_k u + \mathrm{i}\,\nu_k\,\mathrm{Sym}\,(L^\top;\nu)^{-1}\mathrm{Sym}\,(D^*;\nu)A^\top Du\big)_{1\le k\le n} \\
&= \big(\mathrm{Sym}\,(L^\top;\nu)^{-1}\big(\mathrm{Sym}\,(L^\top;\nu)\partial_k u + \mathrm{i}\,\nu_k\,\mathrm{Sym}\,(D^*;\nu)A^\top Du\big)\big)_{1\le k\le n} \\
&= \Big(\mathrm{Sym}(L^\top;\nu)^{-1}\big(-\,\mathrm{Sym}(D^*;\nu)A^\top\mathrm{Sym}(D;\nu)\partial_k u + \mathrm{i}\nu_k\mathrm{Sym}(D^*;\nu)A^\top Du\big)\Big)_{1\le k\le n} \\
&= \big(\mathrm{Sym}\,(L^\top;\nu)^{-1}\mathrm{Sym}\,(D^*;\nu)A^\top\big(-\,\mathrm{Sym}\,(D;\nu)\partial_k u + \mathrm{i}\,\nu_k Du\big)\big)_{1\le k\le n} \\
&= -\mathrm{i}\,\big(\mathrm{Sym}\,(L^\top;\nu)^{-1}\,\mathrm{Sym}\,(D^*;\nu)\,A^\top\big(\nu_s\partial_k u_\beta - \nu_k\partial_s u_\beta\big)_{\beta,s}\big)_{1\le k\le n} \\
&= -\mathrm{i}\,\big(\mathrm{Sym}\,(L^\top;\nu)^{-1}\,\mathrm{Sym}\,(D^*;\nu)\,A^\top\big(\partial_{\tau_{sk}} u_\beta\big)_{\beta,s}\big)_{1\le k\le n},
\end{aligned} \tag{8.1.931}$$

where k above is used to index columns. Hence, (8.1.931) gives

$$\partial_{\tau_D} u = -\,\mathrm{Sym}\,(L^\top;\nu)^{-1}\,\big(\nu_r a_{sr}^{\beta\alpha}\partial_{\tau_{sk}} u_\beta\big)_{\substack{1\le\alpha\le M\\ 1\le k\le n}}. \tag{8.1.932}$$

Compare with the discussion in [131, (2.8.347)-(2.8.361)]. At this point, a combination of (8.1.927), (8.1.930), (8.1.932), (8.1.929), and [129, (1.7.17)] give that (bearing in mind that $D^* = D^\top$ since D has real coefficients)

$$\begin{aligned}
\partial_r(\mathcal{S}_{\gamma\beta} f)(x) &= -\int_{\partial\Omega}\big\langle\partial_{\tau_D(y)}\big[E_{\gamma\bullet}(x-y)\big],\ f(y)\delta_{\beta\bullet}\delta_{r\bullet}\big\rangle\,d\sigma(y) \\
&\quad -\mathrm{i}\int_{\partial\Omega}\big\langle\mathrm{Sym}(D;\nu(y))\mathrm{Sym}\,(L^\top;\nu(y))^{-1}\partial_{\nu(y)}^{A^\top}\big[E_{\gamma\bullet}(x-y)\big], f(y)\delta_{\beta\bullet}\delta_{r\bullet}\big\rangle\,d\sigma(y) \\
&= \int_{\partial\Omega}\big\langle\mathrm{Sym}\,(L^\top;\nu(y))^{-1}\big(\nu_i(y)a_{si}^{\eta\mu}\partial_{\tau_{sk}(y)}[E_{\gamma\eta}(x-y)]\big)_{\mu k}, f(y)\delta_{\beta\bullet}\delta_{r\bullet}\big\rangle\,d\sigma(y) \\
&\quad +\mathrm{i}\int_{\partial\Omega}\Big\langle\partial_{\nu(y)}^{A^\top}[E_{\gamma\bullet}(x-y)], \mathrm{Sym}\,(L;\nu(y))^{-1}\mathrm{Sym}\,(D^*;\nu(y))[f(y)\delta_{\beta\bullet}\delta_{r\bullet}]\Big\rangle\,d\sigma(y) \\
&= \int_{\partial\Omega}\partial_{\tau_{sr}(y)}[E_{\gamma\eta}(x-y)]\,\nu_i(y)a_{si}^{\eta\mu}\big(\mathrm{Sym}\,(L;\nu(y))^{-1}\big)_{\mu\beta}f(y)\,d\sigma(y) \\
&\quad +\int_{\partial\Omega}\Big\langle\partial_{\nu(y)}^{A^\top}\big[E_{\gamma\bullet}(x-y)\big], \mathrm{Sym}\,(L;\nu(y))^{-1}\big(\nu_r(y)f(y)\delta_{\beta\mu}\big)_\mu\Big\rangle\,d\sigma(y) \\
&= -\int_{\partial\Omega}E_{\gamma\eta}(x-y)\,\partial_{\tau_{sr}(y)}\big[\nu_i(y)a_{si}^{\eta\mu}\big(\mathrm{Sym}\,(L;\nu(y))^{-1}\big)_{\mu\beta}f(y)\big]\,d\sigma(y) \\
&\quad +\int_{\partial\Omega}\Big\langle\partial_{\nu(y)}^{A^\top}\big[E_{\gamma\bullet}(x-y)\big], \mathrm{Sym}\,(L;\nu(y))^{-1}\big(\nu_r(y)f(y)\delta_{\beta\mu}\big)_\mu\Big\rangle\,d\sigma(y),
\end{aligned} \tag{8.1.933}$$

where the last equality above employs an integration by parts on the boundary. Keeping in mind [132, (4.2.202)], formula (8.1.925) now follows by combining (8.1.933), (8.1.922), and [132, (1.3.22)]. This finishes the treatment of Step I.

To proceed, we adopt two conventions:

(i) We agree to let ∂_τ stand for a generic tangential derivative operator on $\partial\Omega$ of the form $\partial_{\tau_{jk}}$ with $j, k \in \{1, \dots, n\}$, and denote by M_ν (possibly repeated compositions of) multiplication operators with either components of ν, or entries in the matrix $\mathrm{Sym}\,(L; \nu)^{-1}$.

(ii) Given any two vector-valued functions f, g, we write $f \equiv g$ to indicate that each component of f may be written as a finite linear combination of the components of g, with coefficients depending only on the entries of the tensor A.

In terms of these conventions, we may succinctly summarize the identities in (8.1.925) and (a combination of (8.1.925) with) (8.1.923) as follows:

$$\nabla \mathscr{S} f \equiv \mathscr{S}(\partial_\tau(M_\nu f)) + \mathcal{D}_A(M_\nu f), \quad \forall\, f \in \big[L^2(\partial\Omega, \sigma)\big]^M, \tag{8.1.934}$$

$$\nabla \mathcal{D}_A f \equiv \mathscr{S}(\partial_\tau(M_\nu\, \partial_\tau f)) + \mathcal{D}_A(M_\nu\, \partial_\tau f), \quad \forall\, f \in \big[L^2_1(\partial\Omega, \sigma)\big]^M. \tag{8.1.935}$$

Step II: *In the current setting, the operator*

$$\mathcal{D}_A : \big[H^{3/2}(\partial\Omega, \sigma)\big]^M \to \big[H^2(\Omega)\big]^M \text{ is well defined and bounded.} \tag{8.1.936}$$

Indeed, (8.1.934)-(8.1.935) show that, for each $f \in \big[H^{3/2}(\partial\Omega, \sigma)\big]^M$,

$$\nabla^2 \mathcal{D}_A f \equiv \nabla \mathscr{S}(\partial_\tau(M_\nu\, \partial_\tau f)) + \nabla \mathcal{D}_A(M_\nu\, \partial_\tau f) \text{ in } \Omega. \tag{8.1.937}$$

Since $\partial_\tau(M_\nu\, \partial_\tau f) \in \big[H^{-1/2}(\partial\Omega, \sigma)\big]^M$ and $M_\nu\, \partial_\tau f \in \big[H^{1/2}(\partial\Omega, \sigma)\big]^M$, we conclude from (8.1.935), (8.1.937) plus the mapping properties of the single and double layer potential operators acting from Besov spaces into Sobolev spaces established in [132, Theorem 4.2.10] and [132, Theorem 4.2.3] that

$$\mathcal{D}_A f,\ \nabla \mathcal{D}_A f,\ \nabla^2 \mathcal{D}_A f \in \big[L^2(\Omega, \mathcal{L}^n)\big]^M \text{ and} \tag{8.1.938}$$

$$\sum_{j=0}^{2} \|\nabla^j \mathcal{D}_A f\|_{[L^2(\Omega, \mathcal{L}^n)]^M} \leq C \|f\|_{[H^{3/2}(\partial\Omega, \sigma)]^M}. \tag{8.1.939}$$

With this in hand, the claims in (8.1.936) follow. This takes care of Step II.

Moving on, bring in the boundary-to-boundary double layer potential operator K_A associated with Ω and the coefficient tensor A as in (A.0.116). Also, recall the boundary-to-boundary single layer potential operator S, associated with Ω and the system L as in (A.0.231). Hence, $S = \big(S_{\alpha\beta}\big)_{1 \leq \alpha, \beta \leq M}$ where, for each indices $\alpha, \beta \in \{1, \dots, M\}$ and each function $f \in L^1(\partial\Omega, \sigma)$, we have defined

$$(S_{\alpha\beta} f)(x) := \int_{\partial\Omega} E_{\alpha\beta}(x - y) f(y)\, d\sigma(y) \text{ for } \sigma\text{-a.e. } x \in \partial\Omega. \tag{8.1.940}$$

As in [132, Theorem 1.5.1, item *(ix)*], we may extend the action of each $S_{\alpha\beta}$ to the negative Sobolev space $L^2_{-1}(\partial\Omega,\sigma)$. Finally, for each $\alpha,\beta\in\{1,\dots,M\}$, each $r\in\{1,\dots,n\}$, each $f\in L^2(\partial\Omega,\sigma)$, and each σ-a.e. $x\in\partial\Omega$, define

$$(\partial_r S_{\alpha\beta})f(x) := \lim_{\varepsilon\to 0^+} \int_{y\in\partial\Omega,\ |x-y|>\varepsilon} (\partial_r E_{\alpha\beta})(x-y)f(y)\,d\sigma(y). \tag{8.1.941}$$

The claim we make in this regard is as follows:

Step III: *Given any $f\in L^2(\partial\Omega,\sigma)$, for each indices $\gamma,\beta\in\{1,\dots,M\}$ and $r\in\{1,\dots,n\}$ there holds*

$$\begin{aligned}(\partial_r S_{\gamma\beta})f &= -S_{\gamma\eta}\big(\partial_{\tau_{sr}}\big(\nu_i a^{\eta\mu}_{si}\big(\mathrm{Sym}\,(L;\nu)^{-1}\big)_{\mu\beta}f\big)\big)\\ &\quad+\big(K_A\big(\big(\big(\mathrm{Sym}\,(L;\nu)^{-1}\big)_{\alpha\beta}\nu_r f\big)_\alpha\big)\big)_\gamma.\end{aligned} \tag{8.1.942}$$

As a consequence, for each $\gamma,\beta\in\{1,\dots,M\}$ and $r\in\{1,\dots,n\}$ the operator

$$\partial_r S_{\gamma\beta}: L^2_1(\partial\Omega,\sigma)\longrightarrow L^p_1(\partial\Omega,\sigma) \text{ is well-defined and bounded.} \tag{8.1.943}$$

Using the jump-formula for generic derivatives of the single layer from [131, Corollary 2.9.5], (8.1.941) and [132, (1.2.51), (1.5.20), (1.5.242)] we obtain

$$\begin{aligned}\tfrac12\nu_r\big(\mathrm{Sym}\,(L;\nu)^{-1}\big)_{\gamma\beta}f+(\partial_r S_{\gamma\beta})f &= \partial_r\mathscr{S}_{\gamma\beta}f\Big|^{\kappa-\mathrm{n.t.}}_{\partial\Omega}\\ &= -\mathscr{S}_{\gamma\eta}\big(\partial_{\tau_{sr}}\big(\nu_i a^{\eta\mu}_{si}\big(\mathrm{Sym}\,(L;\nu)^{-1}\big)_{\mu\beta}f\big)\big)\Big|^{\kappa-\mathrm{n.t.}}_{\partial\Omega}\\ &\quad+\big(\mathcal{D}_A\big(\big(\big(\mathrm{Sym}\,(L;\nu)^{-1}\big)_{\alpha\beta}\nu_r f\big)_\alpha\big)\big)_\gamma\Big|^{\kappa-\mathrm{n.t.}}_{\partial\Omega}\\ &= -S_{\gamma\eta}\big(\partial_{\tau_{sr}}\big(\nu_i a^{\eta\mu}_{si}\big(\mathrm{Sym}\,(L;\nu)^{-1}\big)_{\mu\beta}f\big)\big)\\ &\quad+\tfrac12\nu_r\big(\mathrm{Sym}\,(L;\nu)^{-1}\big)_{\gamma\beta}f+\big(K_A\big(\big(\big(\mathrm{Sym}\,(L;\nu)^{-1}\big)_{\alpha\beta}\nu_r f\big)_\alpha\big)\big)_\gamma\end{aligned} \tag{8.1.944}$$

at σ-a.e. point on $\partial\Omega$. Canceling like-terms in the most extreme sides of (8.1.944) then yields formula (8.1.942). Finally, (8.1.943) is a consequence of (8.1.942), [132, Theorem 1.5.1, item *(ix)*], and [132, Theorem 4.1.1] (bearing in mind that ν is presently a smooth function).

Step IV: *In the current setting, for each $\gamma,\beta\in\{1,\dots,M\}$ and $r,s\in\{1,\dots,n\}$ the commutator*

$$\begin{gathered}\big[M_{\nu_s},\partial_r S_{\gamma\beta}\big]: H^{1/2}(\partial\Omega,\sigma)\longrightarrow H^{1/2}(\partial\Omega,\sigma)\\ \text{is a well-defined, linear, and compact operator.}\end{gathered} \tag{8.1.945}$$

Since ν is presently a smooth function, Step III enures that

$$\begin{gathered}\big[M_{\nu_s},\partial_r S_{\gamma\beta}\big]: L^2_1(\partial\Omega,\sigma)\longrightarrow L^2_1(\partial\Omega,\sigma)\\ \text{is a well-defined, linear, and bounded operator.}\end{gathered} \tag{8.1.946}$$

The smoothness of ν also implies that $\big[M_{\nu_s}, \partial_r S_{\gamma\beta}\big]$ has a weakly singular kernel and, as such, [81, Lemma 2.20, p. 2608] guarantees that

$$\begin{gathered}\big[M_{\nu_s}, \partial_r S_{\gamma\beta}\big] : L^2(\partial\Omega, \sigma) \longrightarrow L^2(\partial\Omega, \sigma)\\ \text{is a well-defined, linear, and compact operator.}\end{gathered} \tag{8.1.947}$$

From (8.1.946), (8.1.947) the one-sided compactness property for the real method of interpolation recorded in [129, Proposition 7.7.10], and [130, Theorem 11.12.2] then justify (8.1.945). This takes care of Step IV.

Step V: *In the current setting,*

$$\begin{gathered}K_A : \big[H^{3/2}(\partial\Omega, \sigma)\big]^M \longrightarrow \big[H^{3/2}(\partial\Omega, \sigma)\big]^M\\ \textit{is a well-defined, linear, compact operator.}\end{gathered} \tag{8.1.948}$$

As a corollary,

$$\begin{gathered}\tfrac{1}{2}I + K_A : \big[H^{3/2}(\partial\Omega, \sigma)\big]^M \longrightarrow \big[H^{3/2}(\partial\Omega, \sigma)\big]^M\\ \textit{is a well-defined Fredholm operator with index zero.}\end{gathered} \tag{8.1.949}$$

To justify these claims, first observe from (5.4.135) and the equivalence *(i)* ⇔ *(v)* in Proposition 1.3.2 that

$$\begin{gathered}K_A : \big[H^{1/2}(\partial\Omega, \sigma)\big]^M \longrightarrow \big[H^{1/2}(\partial\Omega, \sigma)\big]^M\\ \text{is a well-defined, linear, compact operator.}\end{gathered} \tag{8.1.950}$$

Then (8.1.948) follows from [132, Proposition 1.5.6, (1.5.253)], the fact that each operator in [132, (1.5.251)] is a linear combinations of commutators of the form $\big[M_{\nu_s}, \partial_r S_{\gamma\beta}\big]$, (8.1.945), and (8.1.950). Finally, (8.1.949) is a consequence of (8.1.949) and Fredholm theory.

Step VI: *The end-game in the proof of* (8.1.921). Granted (8.1.949), we may invoke [130, Theorem 2.2.3] to find two linear operators,

$$\begin{aligned}R &: \big[H^{3/2}(\partial\Omega, \sigma)\big]^M \to \big[H^{3/2}(\partial\Omega, \sigma)\big]^M \text{ bounded,}\\ C_0 &: \big[H^{3/2}(\partial\Omega, \sigma)\big]^M \to \big[H^{3/2}(\partial\Omega, \sigma)\big]^M \text{ compact,}\end{aligned} \tag{8.1.951}$$

with the property that, if I is the identity operator on $\big[H^{3/2}(\partial\Omega, \sigma)\big]^M$, then

$$\big(\tfrac{1}{2}I + K_A\big) \circ R = I + C_0 \text{ on } \big[H^{3/2}(\partial\Omega, \sigma)\big]^M. \tag{8.1.952}$$

We claim that the linear, bounded mapping, induced by the system L in the context

$$L : \big[H^2(\Omega)\big]^M \longrightarrow \big[L^2(\Omega, \mathcal{L}^n)\big]^M, \tag{8.1.953}$$

has an inverse-from-the-right, i.e., there is a smoothing liner bounded operator

$$\begin{aligned} \Psi : \big[L^2(\Omega, \mathcal{L}^n)\big]^M &\longrightarrow \big[H^2(\Omega)\big]^M \\ \text{with } L(\Psi f) = f \text{ for each } f &\in \big[L^2(\Omega, \mathcal{L}^n)\big]^M. \end{aligned} \tag{8.1.954}$$

To justify this, bring in the Fourier multiplier operator $L^{-1}\varphi := \mathscr{F}^{-1}\big(L(\xi)^{-1}(\mathscr{F}\varphi)\big)$ for all $\varphi \in \big[\mathcal{S}'(\mathbb{R}^n)\big]^M$. If we now set

$$\Psi f := \big(L^{-1}\widetilde{f}\big)\Big|_{\Omega} \quad \text{for each } f \in \big[L^2(\Omega, \mathcal{L}^n)\big]^M \tag{8.1.955}$$

(where $\widetilde{f}$ is the extension of f by zero outside Ω to the entire $\mathbb{R}^n$), then [87, Corollary 4.2] implies that (8.1.954) holds. Let us now introduce

$$\begin{aligned} Q : \big[L^2(\Omega, \mathcal{L}^n)\big]^M &\longrightarrow \big[H^{3/2}(\partial\Omega, \sigma)\big]^M \\ Qf := \mathrm{Tr}_{\Omega\to\partial\Omega}(\Psi f) \ \text{ for each } f &\in \big[L^2(\Omega, \mathcal{L}^n)\big]^M, \end{aligned} \tag{8.1.956}$$

$$\begin{aligned} &T : \big[L^2(\Omega, \mathcal{L}^n)\big]^M \times \big[H^{3/2}(\partial\Omega, \sigma)\big]^M \to \big[L^2(\Omega, \mathcal{L}^n)\big]^M \times \big[H^{3/2}(\partial\Omega, \sigma)\big]^M \\ &T(f, \phi) := (f, Qf + \phi), \ \forall (f, \phi) \in \big[L^2(\Omega, \mathcal{L}^n)\big]^M \times \big[H^{3/2}(\partial\Omega, \sigma)\big]^M, \end{aligned} \tag{8.1.957}$$

and, with C_0 as in (8.1.951),

$$\begin{aligned} &C_1 : \big[L^2(\Omega, \mathcal{L}^n)\big]^M \times \big[H^{3/2}(\partial\Omega, \sigma)\big]^M \to \big[L^2(\Omega, \mathcal{L}^n)\big]^M \times \big[H^{3/2}(\partial\Omega, \sigma)\big]^M \\ &C_1(f, \phi) := (0, C_0\phi), \ \forall (f, \phi) \in \big[L^2(\Omega, \mathcal{L}^n)\big]^M \times \big[H^{3/2}(\partial\Omega, \sigma)\big]^M. \end{aligned} \tag{8.1.958}$$

These are all well defined, linear, and bounded. In addition, it is clear that T is invertible, and C_1 is compact. Consequently, [130, Theorem 2.1.2, *(6)*] tells us that

$$\begin{aligned} T + C_1 : \big[L^2(\Omega, \mathcal{L}^n)\big]^M \times \big[H^{3/2}(\partial\Omega, \sigma)\big]^M &\to \big[L^2(\Omega, \mathcal{L}^n)\big]^M \times \big[H^{3/2}(\partial\Omega, \sigma)\big]^M \\ &\text{is a Fredholm operator (with index zero).} \end{aligned} \tag{8.1.959}$$

In light of [130, Definition 2.1.1], this then implies that

$$\begin{aligned} &\mathrm{Im}\Big(T + C_1;\ \big[L^2(\Omega, \mathcal{L}^n)\big]^M \times \big[H^{3/2}(\partial\Omega, \sigma)\big]^M\Big) \text{ is a closed subspace} \\ &\quad \text{of } \big[L^2(\Omega, \mathcal{L}^n)\big]^M \times \big[H^{3/2}(\partial\Omega, \sigma)\big]^M, \text{ of finite codimension.} \end{aligned} \tag{8.1.960}$$

Next, observe that if $\mathcal{D}_A$ is the boundary-to-domain double layer operator in the context of (8.1.936), then the definition of Φ in (8.1.921), (8.1.954), (8.1.956), (8.1.957), (8.1.958), (8.1.952), and the jump-formula from [132, Theorem 4.3.2, *(1)*] imply that for any pair $(f, \phi) \in \big[L^2(\Omega, \mathcal{L}^n)\big]^M \times \big[H^{3/2}(\partial\Omega, \sigma)\big]^M$ we have

$$\begin{aligned} \Phi(\Psi f + \mathcal{D}_A(R\phi)) &= \Big(f,\ Qf + \big(\tfrac{1}{2} + K_A\big)(R\phi)\Big) \\ &= \Big(f,\ Qf + \phi + C_0\phi\Big) = (T + C_1)(f, \phi). \end{aligned} \tag{8.1.961}$$

On account of (8.1.961) and the mapping properties of the operators involved (cf. (8.1.936), (8.1.954), and (8.1.951)) we conclude that

$$\begin{gathered}\text{the image of the data map } \Phi, \text{ defined in (8.1.921), contains}\\ \mathrm{Im}\Big(T + C_1;\ \big[L^2(\Omega, \mathcal{L}^n)\big]^M \times \big[H^{3/2}(\partial\Omega, \sigma)\big]^M\Big).\end{gathered} \tag{8.1.962}$$

Based on (8.1.962), (8.1.960), [130, Theorem 1.1.3, *(3)*] we then deduce that

$$\begin{gathered}\text{the image of the data map } \Phi, \text{ from (8.1.921), is a closed}\\ \text{space of finite codimension in } \big[L^2(\Omega, \mathcal{L}^n)\big]^M \times \big[H^{3/2}(\partial\Omega, \sigma)\big]^M.\end{gathered} \tag{8.1.963}$$

As may be seen from the definition of Φ in (8.1.921) and the fact that the space (8.1.702) is already known to be finite dimensional,

$$\begin{gathered}\text{the kernel of the data map } \Phi, \text{ in the context of (8.1.921)},\\ \text{is finite dimensional subspace of } \big[H^2(\Omega)\big]^M.\end{gathered} \tag{8.1.964}$$

Together, (8.1.963) and (8.1.964) prove that the data map is actually a Fredholm operator in the context of (8.1.921). As remarked earlier, this further implies that the system L satisfies the Shapiro-Lopatinskiĭ condition. □

8.2 Boundary Problems in Domains with Compact Boundaries for the Stokes System

There is a large amount of literature devoted to the topic of boundary value problems for the Stokes system of hydrostatic in domains with compact boundaries. The classical results of Odqvist and Lichtenstein, obtained for smooth domains, are described in Chapter 3 of Ladyzhenskaya's book [106]. In [55], E. Fabes, C. Kenig, and G. Verchota have initiated the study of boundary problems for the Stokes system in Lipschitz domains. Using Rellich identities, they succeeded in solving the Dirichlet Problem, Regularity Problem, and Neumann Problem with data in L^2 in Lipschitz domains with connected boundaries (in dimensions $n \geq 3$), measuring the size of the solution in terms of the nontangential maximal operator. The L^p Dirichlet Problem for the Stokes system in Lipschitz domains has been considered by Z. Shen in [191], and by J. Kilty in [96]. The Transmission Problem for the Stokes system in Lipschitz domains of arbitrary topology in $\mathbb{R}^n$ (where $n \geq 2$) with boundary data in Lebesgue/Hardy/Sobolev spaces has been solved by M. Mitrea and M. Wright in [160] (building on the earlier work in [48] for the Laplacian). Along the way, the L^p Dirichlet, Regularity, and Neumann Problems have also been treated in Lipschitz domains with arbitrary topology in [160]. Additionally, the monograph [160] contains Regularity and Neumann Problems for the Stokes system in Lipschitz domains whose formulations involve Hardy spaces, in dimensions $n = 2, 3$. Moreover, R. Brown, I. Mitrea, M. Mitrea, and M. Wright have treated in [21] the mixed boundary value problem for the Stokes system in a class of Lipschitz domains, while V. Maz'ya, M. Mitrea, and T. Shaposhnikova have solved in [120] the Poisson Problem for the Stokes system in Lipschitz domains with outward unit normals close to VMO. In yet another direction, M. Mitrea and M. Taylor have employed in [157] boundary layer methods to deal with the L^2 Dirichlet Problem and the L^2_1 Regularity Problem for the Stokes system in Lipschitz subdomains of Riemannian manifolds. This topic has been revisited by M. Dindoš and M. Mitrea in [44]. Another thing we wish to note in this brief review is that work by S. Hofmann, M. Mitrea, and M. Taylor in [81] is applicable to boundary value problems (Dirichlet, Regularity, Neumann, and Transmission) for the Stokes system of hydrostatics in regular SKT domains with boundary data in Lebesgue/Sobolev spaces. For more references on the subject of boundary value problems for the Stokes system the reader is referred to the literature cited in the aforementioned sources.

Here the goal is to solve boundary value problems for the Stokes system in Ahlfors regular domains with a compact boundary which is sufficiently flat at an infinitesimal level. The domains in question are of a general topological nature. More specifically, we shall employ the method of boundary layer potentials to prove solvability and well-posedness results for the following (interior/exterior) boundary value problems:

> the Dirichlet Problem with data in Lebesgue spaces and, in a more comprehensive fashion, for boundary data in Generalized Banach Function Spaces (including Morrey, block, and Orlicz, and rearrangement invariant Banach function spaces) in Theorem 8.2.3; the Dirichlet Problem with data in Besov spaces in Theorem 8.2.4; the Poisson Problem in Besov/Triebel-Lizorkin/weighted Sobolev spaces in Theorem 8.2.5; the Regularity Problem with data in Lebesgue, Hardy, and GBFS[74] styled Sobolev spaces (including Morrey, block, and Orlicz styled Sobolev spaces) in Theorem 8.2.1; the Neumann Problem with data in Lebesgue, Hardy, and GBFS spaces (including Morrey, block, and Orlicz spaces) in Theorem 8.2.2.

A multitude of other boundary value problems for the Stokes system in domains with compact infinitesimally flat boundaries, well within the scope of our present techniques, are reviewed in Remark 8.2.6. These include the Dirichlet Problem with boundary data in BMO and VMO, the Dirichlet Problem with data in ordinary Hölder spaces $\mathscr{C}^\alpha$ with $0 < \alpha < 1$, the Dirichlet Problem with data in vanishing Hölder spaces $\mathscr{C}^\alpha_{\text{van}}$ with $0 < \alpha < 1$, the Neumann Problem with data in Besov spaces, the Inhomogeneous Neumann Problem with boundary data in Besov spaces, and the Transmission Problem with data in Lebesgue, Hardy, and Generalized Banach Function Spaces for the Stokes system.

We begin in earnest by discussing the Regularity Problem for the Stokes system. This is currently formulated in the class of Ahlfors regular domain with compact boundary, which are subsequently assumed to be sufficiently flat at an infinitesimal level. In this setting, we succeed in proving well-posedness results for boundary data

[74] Generalized Banach Function Spaces

in a large variety of function spaces, namely ordinary (i.e., Lebesgue-based) Sobolev spaces and, more generally, GBFS-based Sobolev spaces, as well as Hardy-based Sobolev spaces.

Theorem 8.2.1 (The Regularity Problem for the Stokes System) *Fix $n \in \mathbb{N}$ with $n \geq 2$ and suppose $\Omega \subseteq \mathbb{R}^n$ is an Ahlfors regular domain with compact boundary. Abbreviate $\sigma := \mathcal{H}^{n-1}\lfloor\partial\Omega$ and denote by ν the geometric measure theoretic outward unit normal to Ω. Also, fix some arbitrary aperture parameter $\kappa > 0$.*

(1) Given any integrability exponent $p \in (1, \infty)$, there exists some small threshold $\delta \in (0, 1)$ which depends only on n, p, the ADR *constants of $\partial\Omega$, and the relative flatness ratio $\wp(\Omega)/\mathrm{diam}\,\partial\Omega$ with the property that when*

$$\mathrm{dist}\Big(\nu,\ [\mathrm{VMO}\,(\partial\Omega, \sigma)]^n\Big) < \delta \tag{8.2.1}$$

where the distance is considered in the space $\big[\mathrm{BMO}\,(\partial\Omega, \sigma)\big]^n$ (hence if Ω is a δ-oscillating AR *domain*[75]*; cf. Definition 3.4.1), the following results are true. First, the Regularity Problem*[76]

$$\begin{cases} \vec{u} \in \big[\mathscr{C}^\infty(\Omega)\big]^n, \quad \pi \in \mathscr{C}^\infty(\Omega), \\ \Delta\vec{u} - \nabla\pi = 0 \ \text{and}\ \mathrm{div}\,\vec{u} = 0 \ \text{in}\ \Omega, \\ \mathcal{N}_\kappa\vec{u},\ \mathcal{N}_\kappa(\nabla\vec{u}),\ \mathcal{N}_\kappa\pi \in L^p(\partial\Omega, \sigma), \\ \vec{u}\big|_{\partial\Omega}^{\kappa-\mathrm{n.t.}} = \vec{f} \ \text{at}\ \sigma\text{-a.e. point on}\ \partial\Omega, \\ \vec{u}(x) = O(|x|^{2-n}) \ \text{and}\ \pi(x) = O(|x|^{1-n}) \\ \text{as}\ |x| \to \infty \ \text{if}\ \Omega \ \text{is an exterior domain,} \end{cases} \tag{8.2.2}$$

has a solution if and only if

$$\vec{f} \in \big[L^p_1(\partial\Omega, \sigma)\big]^n \cap \big\langle \nu\mathbb{C}_{\partial\Omega_+}\big\rangle^\perp. \tag{8.2.3}$$

Second, this solution is unique (modulo adding locally constant functions to the pressure) and matters may be arranged so that, for some constant $C \in (0, \infty)$ independent of $\vec{f}$, one has

$$\|\mathcal{N}_\kappa\vec{u}\|_{L^p(\partial\Omega,\sigma)} + \|\mathcal{N}_\kappa(\nabla\vec{u})\|_{L^p(\partial\Omega,\sigma)} + \|\mathcal{N}_\kappa\pi\|_{L^p(\partial\Omega,\sigma)} \leq C\|\vec{f}\|_{[L^p_1(\partial\Omega,\sigma)]^n}. \tag{8.2.4}$$

Third, in the case when Ω is bounded it also follows that any solution $(\vec{u}, \pi)$ of the boundary value problem (8.2.2) *satisfies*

---

[75] in particular, if Ω is an infinitesimally flat AR domain

[76] in view of work in [129, §8.5], one may replace the nontangential maximal operator by the tangential maximal operator, associated as in [129, Definition 8.5.1] with a sufficiently large power (cf. (A.0.165))

$$\vec{u} \in \left[B^{p,q_0}_{1+\frac{1}{p}}(\Omega)\right]^n \cap \left[F^{p,q_1}_{1+\frac{1}{p}}(\Omega)\right]^n \text{ and } \pi \in B^{p,q_0}_{\frac{1}{p}}(\Omega) \cap F^{p,q_1}_{\frac{1}{p}}(\Omega) \quad \text{if } p \leq q_0 \leq \infty \text{ and } \tfrac{n}{n+1+1/p} < q_1 \leq \infty, \tag{8.2.5}$$

as well as

$$\vec{u} \in \left[H^{3/2}(\Omega)\right]^n \text{ and } \pi \in \left[H^{1/2}(\Omega)\right]^n \text{ if } p = 2. \tag{8.2.6}$$

(2) *A more general well-posedness result than the one stated in item (1) holds for the Regularity Problem formulated with boundary data in certain Generalized Banach Function Spaces. Specifically, assume* $\mathbb{X}$ *is a Generalized Banach Function Space on* $(\partial\Omega, \sigma)$ *with the property that* $\mathcal{M}_{\partial\Omega} : \mathbb{X} \to \mathbb{X}$ *and* $\mathcal{M}_{\partial\Omega} : \mathbb{X}' \to \mathbb{X}'$ *are well-defined and bounded, where* $\mathcal{M}_{\partial\Omega}$ *is the Hardy-Littlewood maximal operator on* $\partial\Omega$*, and* $\mathbb{X}'$ *is the associated space of* $\mathbb{X}$ *(cf.* [130, Definitions 5.1.4, 5.1.11]*). Recall the* $\mathbb{X}$*-styled Sobolev space of order one on* $\partial\Omega$*, defined as*

$$\mathbb{X}_1 := \Big\{f \in \mathbb{X} : \partial_{\tau_{jk}} f \in \mathbb{X} \text{ for } 1 \leq j, k \leq n\Big\}, \tag{8.2.7}$$

(which is known to be Banach space when equipped with the natural norm; cf. (5.4.84)*). Then there exists some threshold* $\delta_{\mathbb{X}} \in (0, 1)$ *which depends only on n, the operator norms of* $\mathcal{M}_{\partial\Omega}$ *on* $\mathbb{X}$ *and* $\mathbb{X}'$*, as well as on the* ADR *constants of* $\partial\Omega$ *and the relative flatness ratio* $\wp(\Omega)/\mathrm{diam}\,\partial\Omega$*, with the property that when*

$$\mathrm{dist}\Big(\nu,\ [\mathrm{VMO}(\partial\Omega, \sigma)]^n\Big) < \delta_{\mathbb{X}} \tag{8.2.8}$$

where the distance is considered in the space $\left[\mathrm{BMO}(\partial\Omega, \sigma)\right]^n$*, it follows that the Regularity Problem*[77]

$$\begin{cases} \vec{u} \in \left[\mathscr{C}^\infty(\Omega)\right]^n, \quad \pi \in \mathscr{C}^\infty(\Omega), \\ \Delta\vec{u} - \nabla\pi = 0 \text{ and } \operatorname{div}\vec{u} = 0 \text{ in } \Omega, \\ \mathcal{N}_\kappa\vec{u},\ \mathcal{N}_\kappa(\nabla\vec{u}),\ \mathcal{N}_\kappa\pi \in \mathbb{X}, \\ \vec{u}\big|^{\kappa-\mathrm{n.t.}}_{\partial\Omega} = \vec{f} \text{ at } \sigma\text{-a.e. point on } \partial\Omega, \\ \vec{u}(x) = O(|x|^{2-n}) \text{ and } \pi(x) = O(|x|^{1-n}) \\ \text{as } |x| \to \infty \text{ if } \Omega \text{ is an exterior domain,} \end{cases} \tag{8.2.9}$$

has a solution if and only if

$$\vec{f} \in \left[\mathbb{X}_1\right]^n \cap \left\langle \nu \mathbb{C}_{\partial\Omega_+}\right\rangle^\perp. \tag{8.2.10}$$

Furthermore, this solution is unique, modulo adding locally constant functions to the pressure, and matters may be arranged so that

---

[77] once again, thanks to work in [129, §8.5] one may replace the nontangential maximal operator by the tangential maximal operator, associated as in [129, Definition 8.5.1] with a sufficiently large power (cf. (A.0.165))

$$\|\mathcal{N}_\kappa \vec{u}\|_{\mathbb{X}} + \|\mathcal{N}_\kappa(\nabla\vec{u})\|_{\mathbb{X}} + \|\mathcal{N}_\kappa \pi\|_{\mathbb{X}} \leq C\|\vec{f}\|_{[\mathbb{X}_1]^n} \tag{8.2.11}$$

for some constant $C \in (0, \infty)$ *independent of* $\vec{f}$. *Finally, a similar well-posedness result is true with* $\mathbb{X}$ *replaced by* $\mathring{\mathbb{X}} := \overline{L^\infty(\partial\Omega, \sigma)}^{\|\cdot\|_{\mathbb{X}}}$.

(3) *A related well-posedness result to the one in item (1) holds for the Regularity Problem for the Stokes system with boundary data in Hardy-based Sobolev spaces. Specifically, for each exponent*

$$p \in \left(\tfrac{n-1}{n}, 1\right] \tag{8.2.12}$$

there exists a threshold $\delta \in (0, 1)$ *depending only on* n, p, *the* ADR *constants of* $\partial\Omega$, *and the relative flatness ratio* $\wp(\Omega)/\mathrm{diam}\,\partial\Omega$ *so that when*

$$\mathrm{dist}\left(\nu, [\mathrm{VMO}(\partial\Omega, \sigma)]^n\right) < \delta \tag{8.2.13}$$

with the distance considered in $\left[\mathrm{BMO}(\partial\Omega, \sigma)\right]^n$ *(hence if* Ω *is a* δ*-oscillating* AR *domain*[78]*; cf. Definition 3.4.1), it follows that the boundary problem*[79]

$$\begin{cases} \vec{u} \in \left[\mathscr{C}^\infty(\Omega)\right]^n, \quad \pi \in \mathscr{C}^\infty(\Omega), \\ \Delta\vec{u} - \nabla\pi = 0 \text{ and } \operatorname{div}\vec{u} = 0 \text{ in } \Omega, \\ \mathcal{N}_\kappa\vec{u},\ \mathcal{N}_\kappa(\nabla\vec{u}),\ \mathcal{N}_\kappa\pi \in L^p(\partial\Omega, \sigma), \\ \vec{u}\big|_{\partial\Omega}^{\kappa-\mathrm{n.t.}} = \vec{f} \text{ at } \sigma\text{-a.e. point on } \partial\Omega, \\ \vec{u}(x) = O(|x|^{2-n}) \text{ and } \pi(x) = O(|x|^{1-n}) \\ \text{as } |x| \to \infty \text{ if } \Omega \text{ is an exterior domain,} \end{cases} \tag{8.2.14}$$

has a solution if and only if

$$\vec{f} \in \left[\dot{H}^p_1(\partial\Omega, \sigma)\right]^n \cap \left\langle \nu\mathbb{C}_{\partial\Omega_+}\right\rangle^\perp. \tag{8.2.15}$$

Furthermore, this solution is unique, modulo adding locally constant functions to the pressure, and any solution $(\vec{u}, \pi)$ *of* (8.2.14) *has the property that*

$$\begin{gathered} \mathcal{N}_\kappa\vec{u} \in L^{p^*}(\partial\Omega, \sigma) \text{ where } p^* := \left(\tfrac{1}{p} - \tfrac{1}{n-1}\right)^{-1} \in (1, \infty) \\ \text{if either } n \geq 3 \text{ or } n = 2 \text{ and } p \in \left(\tfrac{1}{2}, 1\right), \end{gathered} \tag{8.2.16}$$

while

$$\mathcal{N}_\kappa\vec{u} \in L^q(\partial\Omega, \sigma) \text{ for any } q \in (1, \infty), \text{ if } n = 2 \text{ and } p = 1. \tag{8.2.17}$$

---

[78] in particular, if Ω is an infinitesimally flat AR domain

[79] in light of work in [129, §8.5], one may replace the nontangential maximal operator by the tangential maximal operator, associated as in [129, Definition 8.5.1] with a sufficiently large power (cf. (A.0.165))

As a corollary of item *(2)* in Theorem 8.2.1 and [130, Proposition 5.3.14],

granted (8.2.8), it follows that the Regularity Problem (8.2.9) is uniquely (modulo locally constant functions for the pressure) solvable whenever the boundary datum satisfies (8.2.10) provided $\mathbb{X}$ is a rearrangement invariant Banach function space whose Boyd indices satisfy $1 < p_{\mathbb{X}} \leq q_{\mathbb{X}} < \infty$. (8.2.18)

In particular (see [130, (5.3.165)] and [130, Proposition 5.3.15]),

under the assumption made in (8.2.8), it follows that the Regularity Problem (8.2.9) is uniquely (modulo locally constant functions for the pressure) solvable whenever the boundary datum satisfies (8.2.10) provided $\mathbb{X} := L^{\Phi}(\partial\Omega, \sigma)$, the Orlicz space associated as in (A.0.133) with the measure space $(\partial\Omega, \sigma)$ and any Young function Φ satisfying $1 < i(\Phi) \leq I(\Phi) < \infty$. (8.2.19)

In fact, a multitude of other Regularity Problems formulated with boundary data prescribed in a variety of concrete Sobolev spaces, including Muckenhoupt weighted Sobolev spaces, Morrey-based Sobolev spaces, vanishing Morrey-based Sobolev spaces, and block-based Sobolev spaces, are contained under the inclusive umbrella of Regularity Problems with data in Generalized Banach Function Spaces, as described above.

Finally, we wish to note that the usefulness of our brand of Divergence Theorem (developed in [129]) is apparent from the very formulation of the boundary value problems for the Stokes system in Theorem 8.2.1 since the "compatibility" (annihilator) conditions imposed in (8.2.3), (8.2.10), (8.2.15) may be traced back (bearing in mind the divergence-free property of the velocity field) directly to the aforementioned version of the Divergence Theorem.

We now turn to the proof of Theorem 8.2.1.

Proof of Theorem 8.2.1 Select $\delta \in (0, 1)$ small enough, depending on n, p, the ADR constants of $\partial\Omega$, and the relative flatness ratio $\wp(\Omega)/\operatorname{diam}\partial\Omega$, such that the conclusions in Theorem 7.1.21 and Theorem 7.1.23 hold. In view of Theorem 3.4.2, by eventually further decreasing the value of $\delta \in (0, 1)$ we may also ensure that Ω is a two-sided NTA domain. In particular, Ω is also a UR domain, and satisfies the properties in (7.1.419) (see [129, (5.11.6)] and [129, Lemma 5.11.3]). Granted this, the Fatou-type result for null-solutions of the Stokes system established in [132, Theorem 6.3.2] applies and, together with [129, Proposition 8.9.22], gives that

the nontangential pointwise boundary traces $\vec{u}\big|_{\partial\Omega}^{\kappa-\mathrm{n.t.}}$, $\nabla\vec{u}\big|_{\partial\Omega}^{\kappa-\mathrm{n.t.}}$, and $\pi\big|_{\partial\Omega}^{\kappa-\mathrm{n.t.}}$ exist σ-a.e. on $\partial\Omega$ (8.2.20)

and, in fact, these traces are actually independent of the aperture parameter $\kappa \in (0, \infty)$. Having established these properties, we may invoke [130, Proposition 11.3.2] and conclude that

$$\vec{u}\big|_{\partial\Omega}^{\kappa-\mathrm{n.t.}} \in \big[L^p_1(\partial\Omega, \sigma)\big]^n. \tag{8.2.21}$$

Let us also observe that for each bounded connected component O of Ω we have

$$\int_{\partial O} \nu \cdot \Big(\vec{u}\big|_{\partial\Omega}^{\kappa-\mathrm{n.t.}}\Big)\, d\sigma = 0, \tag{8.2.22}$$

thanks to the Divergence Formula from [129, Theorem 1.2.1] and the fact that $\vec{u}$ is divergence-free. From (8.2.21), (8.2.22), and (7.1.399) we see that the membership claimed in (8.2.3) is indeed true.

Conversely, suppose a boundary datum $\vec{f}$ as in (8.2.10) has been fixed. The goal is to show that the boundary value problem (8.2.2) has a solution. To this end, let us first assume that $n \geq 3$. Define the operator

$$T : \Big(\big[L_1^p(\partial\Omega, \sigma)\big]^n \cap \big\langle \nu \mathbb{C}_{\partial\Omega_+}\big\rangle^{\perp}\Big) \oplus \big[L^p(\partial\Omega, \sigma)\big]^n \to \big[L_1^p(\partial\Omega, \sigma)\big]^n \cap \big\langle \nu \mathbb{C}_{\partial\Omega_+}\big\rangle^{\perp}$$

by setting, for each $\vec{g}_1 \in \big[L_1^p(\partial\Omega, \sigma)\big]^n \cap \big\langle \nu \mathbb{C}_{\partial\Omega_+}\big\rangle^{\perp}$ and each $\vec{g}_2 \in \big[L^p(\partial\Omega, \sigma)\big]^n$,

$$T\big(\vec{g}_1, \vec{g}_2\big) := \big(\tfrac{1}{2}I + K_{\mathrm{slip}}\big)\vec{g}_1 + S\vec{g}_2 \tag{8.2.23}$$

where K_{slip} is the boundary-to-boundary double layer for the Stokes system (corresponding to the slip conormal derivative) defined in (4.2.102), and S is the boundary-to-boundary single layer for the Stokes system (cf. (4.2.109)). Based on the Calderón-Zygmund theory for the boundary layer operators associated with Stokes system developed in [132, Theorem 6.2.4] (including mapping properties, nontangential maximal function estimates, and trace formulas) as well as the divergence-free properties from [132, (6.2.15), (6.2.22)], the definition in (7.1.399), and our Divergence Formula from [129, Theorem 1.2.1], we conclude that T is a well-defined, linear, and bounded operator.

In relation to this, we claim that T is also surjective. To prove this claim, recall first from Theorem 7.1.23 (and (8.2.1)) that the operator

$$\begin{aligned} \tfrac{1}{2}I + \big[K_{\mathrm{slip}}\big] : \Big(\big[L_1^p(\partial\Omega, \sigma)\big]^n \cap \big\langle \nu \mathbb{C}_{\partial\Omega_+}\big\rangle^{\perp}\Big)\Big/\Psi(\partial\Omega_-) \\ \longrightarrow \Big(\big[L_1^p(\partial\Omega, \sigma)\big]^n \cap \big\langle \nu \mathbb{C}_{\partial\Omega_+}\big\rangle^{\perp}\Big)\Big/\Psi(\partial\Omega_-) \end{aligned} \tag{8.2.24}$$

is a well-defined linear isomorphism. Thus, given any $\vec{h} \in \big[L_1^p(\partial\Omega, \sigma)\big]^n \cap \big\langle \nu \mathbb{C}_{\partial\Omega_+}\big\rangle^{\perp}$ we may find some vector-valued function $\vec{h}_1 \in \big[L_1^p(\partial\Omega, \sigma)\big]^n \cap \big\langle \nu \mathbb{C}_{\partial\Omega_+}\big\rangle^{\perp}$ such that

$$\vec{h} - \big(\tfrac{1}{2}I + K_{\mathrm{slip}}\big)\vec{h}_1 \text{ belongs to } \Psi(\partial\Omega_-). \tag{8.2.25}$$

However, (7.1.420), (7.1.421), (7.1.418), and [130, Proposition 11.3.2] imply that

$$\Psi(\partial\Omega_-) \subseteq \Psi(\partial\Omega) \subseteq \big[L_1^p(\partial\Omega, \sigma)\big]^n \cap \big\langle \nu \mathbb{C}_{\partial\Omega}\big\rangle^{\perp}, \tag{8.2.26}$$

while Theorem 7.1.21 implies (bearing in mind (8.2.1) and that we are presently assuming $n \geq 3$) that the operator

$$[S] : \big[L^p(\partial\Omega,\sigma)\big]^n \big/ \nu\mathbb{C}_{\partial\Omega} \longrightarrow \big[L^p_1(\partial\Omega,\sigma)\big]^n \cap \big\langle \nu\mathbb{C}_{\partial\Omega}\big\rangle^{\perp} \tag{8.2.27}$$

is an isomorphism. As such, there exists $\vec{h}_2 \in \big[L^p(\partial\Omega,\sigma)\big]^n$ with the property that $S\vec{h}_2 = \vec{h} - \big(\frac{1}{2}I + K_{\text{slip}}\big)\vec{h}_1$. This shows that $T\big(\vec{h}_1,\vec{h}_2\big) = \vec{h}$, ultimately proving that the operator T is, as claimed, surjective. Once this has been established, [130, Proposition 1.2.13] also ensures that there exists a constant $C \in (0,\infty)$, independent of the given $\vec{h}$, such that

$$\|\vec{h}_1\|_{[L^p_1(\partial\Omega,\sigma)]^n} + \|\vec{h}_2\|_{[L^p(\partial\Omega,\sigma)]^n} \leq C\|\vec{h}\|_{[L^p_1(\partial\Omega,\sigma)]^n}. \tag{8.2.28}$$

When used in relation to the boundary datum $\vec{f}$ as in (8.2.3), these considerations prove that there exist $\vec{g}_1 \in \big[L^p_1(\partial\Omega,\sigma)\big]^n \cap \big\langle \nu\mathbb{C}_{\partial\Omega_+}\big\rangle^{\perp}$ and $\vec{g}_2 \in \big[L^p(\partial\Omega,\sigma)\big]^n$ with

$$\|\vec{g}_1\|_{[L^p_1(\partial\Omega,\sigma)]^n} + \|\vec{g}_2\|_{[L^p(\partial\Omega,\sigma)]^n} \leq C\|\vec{f}\|_{[L^p_1(\partial\Omega,\sigma)]^n}, \tag{8.2.29}$$

$$\text{and } \vec{f} = \big(\tfrac{1}{2}I + K_{\text{slip}}\big)\vec{g}_1 + S\vec{g}_2. \tag{8.2.30}$$

Consequently, the pair $(\vec{u},\pi)$ with

$$\vec{u} := \mathcal{D}_{\text{slip}}\vec{g}_1 + \mathcal{S}\vec{g}_2 \text{ and } \pi := \mathcal{P}_{\text{slip}}\vec{g}_1 + \mathcal{Q}\vec{g}_2 \text{ in } \Omega \tag{8.2.31}$$

is a solution of the boundary value problem (8.2.2) and satisfies (8.2.4) (thanks to [132, (6.2.15), (6.2.22)], [132, Theorem 6.2.4], (8.2.29), and [132, (6.2.1)-(6.2.2)]).

Proving existence in the case when $n = 2$ requires some adjustments. Specifically, in place of (8.2.23) we now consider the augmented operator

$$T_{\text{aug}}: \Big(\big[L^p_1(\partial\Omega,\sigma)\big]^2 \cap \big\langle \nu\mathbb{C}_{\partial\Omega_+}\big\rangle^{\perp}\Big) \oplus \big[L^p_0(\partial\Omega,\sigma)\big]^2 \oplus \mathbb{C}^2 \to \big[L^p_1(\partial\Omega,\sigma)\big]^2 \cap \big\langle \nu\mathbb{C}_{\partial\Omega_+}\big\rangle^{\perp}$$

where $L^p_0(\partial\Omega,\sigma) := \Big\{\phi \in L^p_0(\partial\Omega,\sigma) : \int_{\partial\Omega} \phi\, d\sigma = 0\Big\}$, defined as

$$T_{\text{aug}}\big(\vec{g}_1,\vec{g}_2,\vec{c}\big) := \big(\tfrac{1}{2}I + K_{\text{slip}}\big)\vec{g}_1 + S\vec{g}_2 + c \tag{8.2.32}$$

for each $\vec{g}_1 \in \big[L^p_1(\partial\Omega,\sigma)\big]^2 \cap \big\langle \nu\mathbb{C}_{\partial\Omega_+}\big\rangle^{\perp}$, each $\vec{g}_2 \in \big[L^p_0(\partial\Omega,\sigma)\big]^2$, and each $c \in \mathbb{C}^2$. As before, this is a well-defined, linear, and continuous mapping. We claim that T_{aug} is also surjective. To justify this, fix an arbitrary $\vec{h} \in \big[L^p_1(\partial\Omega,\sigma)\big]^2 \cap \big\langle \nu\mathbb{C}_{\partial\Omega_+}\big\rangle^{\perp}$. Since (8.2.24) continues to be a well-defined linear isomorphism, we may find some function $\vec{h}_1 \in \big[L^p_1(\partial\Omega,\sigma)\big]^2 \cap \big\langle \nu\mathbb{C}_{\partial\Omega_+}\big\rangle^{\perp}$ such that (8.2.25) holds. Granted this, we may now rely on (8.2.26) and the fact that (7.1.404) is an isomorphism when we take $\mathbb{X} := L^p(\partial\Omega,\sigma)$ in order to conclude that there exist some vector-valued function $\vec{h}_2 \in \big[L^p(\partial\Omega,\sigma)\big]^2$ and constant $c \in \mathbb{C}^2$ with the property that

$$S\vec{h}_2 + c = \vec{h} - \big(\tfrac{1}{2}I + K_{\text{slip}}\big)\vec{h}_1 \text{ and } \int_{\partial\Omega} \vec{h}_2\, d\sigma = 0. \tag{8.2.33}$$

This places $\vec{h}_2$ in $\big[L^p_0(\partial\Omega,\sigma)\big]^2$ and shows that $T_{\rm aug}\big(\vec{h}_1,\vec{h}_2,\vec{c}\big)=\vec{h}$. Thus, $T_{\rm aug}$ is indeed surjective. Having established this, [130, Proposition 1.2.13] once again ensures that this ontoness property is also quantitative. Thus, given any

$$\vec{f}\in\big[L^p_1(\partial\Omega,\sigma)\big]^2\cap\big\langle\nu\mathbb{C}_{\partial\Omega_+}\big\rangle^{\perp}, \tag{8.2.34}$$

there exist $\vec{g}_1\in\big[L^p_1(\partial\Omega,\sigma)\big]^2\cap\big\langle\nu\mathbb{C}_{\partial\Omega_+}\big\rangle^{\perp}$, $\vec{g}_2\in\big[L^p_0(\partial\Omega,\sigma)\big]^2$, $c\in\mathbb{C}^2$ such that

$$\|\vec{g}_1\|_{[L^p_1(\partial\Omega,\sigma)]^2}+\|\vec{g}_2\|_{[L^p(\partial\Omega,\sigma)]^2}+|c|\leq C\|\vec{f}\|_{[L^p_1(\partial\Omega,\sigma)]^2} \tag{8.2.35}$$

for a constant $C\in(0,\infty)$ independent of $\vec{f}$, and such that

$$\vec{f}=\big(\tfrac{1}{2}I+K_{\rm slip}\big)\vec{g}_1+S\vec{g}_2+c. \tag{8.2.36}$$

If we now define $\vec{u}:=\mathcal{D}_{\rm slip}\vec{g}_1+\mathcal{S}\vec{g}_2+c$ and $\pi:=\mathcal{P}_{\rm slip}\vec{g}_1+Q\vec{g}_2$ in Ω, then the pair $(\vec{u},\pi)$ is a solution of the boundary value problem (8.2.2) when $n=2$, which also satisfies (8.2.4) (by [132, (6.2.15), (6.2.22)], [132, Theorem 6.2.4], the quantitative aspect of the ontoness of $T_{\rm aug}$, and [132, (6.2.1)-(6.2.2)], keeping in mind that the vanishing moment property of $\vec{g}_2$). This establishes the solvability of the boundary value problem (8.2.2) in all space dimensions $n\geq 2$, with natural estimates.

Moving on, the goal is to prove uniqueness in the case when $n\geq 3$. With this in mind, assume the pair $(\vec{u},\pi)$ solves the boundary value problem (8.2.2) formulated with $\vec{f}=0$. Thanks to (8.2.20) the hypotheses in [132, Theorem 6.2.1, item *(a)*] are satisfied, so we may bring in the boundary layer integral representation formula [132, (6.2.25)]. When written for $\lambda:=1$, this currently reduces to

$$\vec{u}=-\mathcal{S}\Big(\partial^{\rm slip}_\nu(\vec{u},\pi)\Big)\ \text{ in }\Omega. \tag{8.2.37}$$

Going nontangentially to the boundary and recalling that $\vec{u}\big|^{\kappa-\rm n.t.}_{\partial\Omega}=0$ at σ-a.e. point on $\partial\Omega$, we therefore obtain $S\big(\partial^{\rm slip}_\nu(\vec{u},\pi)\big)=0$. Hence,

$$\partial^{\rm slip}_\nu(\vec{u},\pi)\in\operatorname{Ker}\Big(S:\big[L^p(\partial\Omega,\sigma)\big]^n\to\big[L^p_1(\partial\Omega,\sigma)\big]^n\Big)=\nu\mathbb{C}_{\partial\Omega}, \tag{8.2.38}$$

with the equality a consequence of (8.2.1) and (7.1.402) used with $\mathbb{X}:=L^p(\partial\Omega,\sigma)$ (recall that we are presently assuming $n\geq 3$). On the other hand, from the definition in (A.0.233), the divergence-free property recorded in [132, (6.2.5)], and our Divergence Formula from [129, Theorem 1.2.1] we deduce that

$$\mathcal{S}\big(\nu\mathbb{C}_{\partial\Omega}\big)=\{0\}. \tag{8.2.39}$$

In concert, (8.2.37), (8.2.38), and (8.2.39) prove that $\vec{u}$ vanishes identically in Ω. This further forces $\nabla\pi=\Delta\vec{u}=0$ in Ω, which shows that π is locally constant.

Let us now prove uniqueness in the case when $n=2$. First consider the case when Ω is an exterior domain. Thus, we shall work under the assumptions that

$$\begin{cases} \vec{u} \in \left[\mathscr{C}^\infty(\Omega)\right]^2, \quad \pi \in \mathscr{C}^\infty(\Omega), \\ \Delta\vec{u} - \nabla\pi = 0 \text{ and } \operatorname{div}\vec{u} = 0 \text{ in } \Omega, \\ \mathcal{N}_\kappa\vec{u},\ \mathcal{N}_\kappa(\nabla\vec{u}),\ \mathcal{N}_\kappa\pi \in L^p(\partial\Omega,\sigma), \\ \vec{u}\big|_{\partial\Omega}^{\kappa-\text{n.t.}} = 0 \text{ at } \sigma\text{-a.e. point on } \partial\Omega, \\ \vec{u}(x) = O(1) \text{ and } \pi(x) = O(|x|^{-1}) \text{ as } |x| \to \infty. \end{cases} \tag{8.2.40}$$

As such, all hypotheses in [132, Theorem 6.2.1, item *(c)*] are satisfies, and the integral representation formula [132, (6.2.32)] written with $\lambda := 1$ presently reduces to

$$\vec{u} = -\mathscr{S}\big(\partial_\nu^{\text{slip}}(\vec{u},\pi)\big) + c \text{ in } \Omega, \tag{8.2.41}$$

for some $c \in \mathbb{C}^n$. In turn, from (8.2.41) and the last line in (8.2.40) we see that $\Big(\mathscr{S}\big(\partial_\nu^{\text{slip}}(\vec{u},\pi)\big)\Big)(x) = O(1)$ as $|x| \to \infty$. In concert with (A.0.233) and [132, (6.2.1)], this forces

$$\int_{\partial\Omega} \partial_\nu^{\text{slip}}(\vec{u},\pi)\, d\sigma = 0. \tag{8.2.42}$$

Next, going nontangentially to the boundary in (8.2.41) gives $S\big(\partial_\nu^{\text{slip}}(\vec{u},\pi)\big) - c = 0$ which, in light of (8.2.42) and the invertibility result recorded in (7.1.404), ultimately shows that $\partial_\nu^{\text{slip}}(\vec{u},\pi)$ belongs to $\nu\mathbb{C}_{\partial\Omega}$. Together with (8.2.39) and (8.2.41), this proves that $\vec{u}$ is constant in Ω. Given that its nontangential boundary trace is currently assumed to vanish, this finally forces $\vec{u} = 0$ in Ω. Finally, this implies $\nabla\pi = \Delta\vec{u} = 0$ in Ω, hence the pressure function is locally constant, as wanted.

There remains to prove uniqueness in the case when $n = 2$ and Ω is bounded. However, this is handled in a similar and (simpler) fashion as above, now disregarding the issues pertaining to behavior at infinity.

At this stage, we conclude that under the necessary assumption made in (8.2.3) the boundary value problem (8.2.2) is solvable uniquely (modulo adding locally constant functions to the pressure), and a solution may be found which also satisfies the naturally accompanying estimate formulated in (8.2.4). Finally, the regularity properties stated in (8.2.5)-(8.2.6) follow from the integral representation formulas from [132, Theorem 6.2.1], the fact that we presently have

$$\vec{u}\big|_{\partial\Omega}^{\kappa-\text{n.t.}} \in \left[L_1^p(\partial\Omega,\sigma)\right]^n, \quad \partial_\nu^{\text{slip}}(\vec{u},\pi) \in \left[L^p(\partial\Omega,\sigma)\right]^n, \tag{8.2.43}$$

plus the mapping properties of the boundary layer potential operators involved from [132, Theorem 6.4.3, item *(4)*]. The treatment of item *(1)* in the statement of the theorem is now complete.

The treatment of item *(2)* proceeds along similar lines, now availing ourselves of the results described in the last portion of Theorem 7.1.23, the embeddings from [130, Proposition 5.2.7] (cf. also the Fatou-type result from [132, Theorem 6.3.3]), and now using the full force of Theorem 7.1.21.

As regards the item *(3)*, we may choose $\delta \in (0,1)$ small enough so that Ω becomes a two-sided NTA domain (cf. Theorem 3.4.2). Granted this, [132, Theorem 6.3.6] guarantees that for any solution $(\vec{u}, \pi)$ of the Regularity problem (8.2.14) one has

$$\begin{aligned} &\vec{u}\big|_{\partial\Omega}^{\kappa-\mathrm{n.t.}},\ \nabla\vec{u}\big|_{\partial\Omega}^{\kappa-\mathrm{n.t.}},\ \text{and } \pi\big|_{\partial\Omega}^{\kappa-\mathrm{n.t.}} \text{ exist at } \sigma\text{-a.e. point on } \partial\Omega, \\ &\vec{u}\big|_{\partial\Omega}^{\kappa-\mathrm{n.t.}} \in \big[\dot{H}_1^p(\partial\Omega,\sigma)\big]^n \text{ and } \partial_\nu^A(\vec{u},\pi) \in \big[H^p(\partial\Omega,\sigma)\big]^n, \end{aligned} \tag{8.2.44}$$

and (also bearing in mind the decay condition at infinity if Ω is an exterior domain)

$$\begin{aligned} &\mathcal{N}_\kappa\vec{u} \in L^{p^*}(\partial\Omega,\sigma) \text{ with } p^* := \big(\tfrac{1}{p} - \tfrac{1}{n-1}\big)^{-1} \in (1,\infty) \\ &\quad \text{if either } n \geq 3, \text{ or } n = 2 \text{ and } p \in \big(\tfrac{1}{2}, 1\big), \end{aligned} \tag{8.2.45}$$

as well as

$$\mathcal{N}_\kappa\vec{u} \in L^q(\partial\Omega,\sigma) \text{ for each } q \in (1,\infty) \text{ if } n = 2 \text{ and } p = 1. \tag{8.2.46}$$

This takes care of the properties stated in (8.2.16)-(8.2.17). With these in hand, the fact that $\vec{f} := \vec{u}\big|_{\partial\Omega}^{\kappa-\mathrm{n.t.}}$ is necessarily as in (8.2.15) follows from (8.2.44) and (8.2.22) (with (8.2.45)-(8.2.46) ensuring that the Divergence Theorem is still applicable in the present setting).

Conversely, suppose some function $\vec{f} \in \big[\dot{H}_1^p(\partial\Omega,\sigma)\big]^n \cap \big\langle \nu\mathbb{C}_{\partial\Omega_+}\big\rangle^\perp$ has been given. Then [132, Theorem 2.3.3] tells us that $\vec{f}$ belongs to $\big[H_1^{p^*,p}(\partial\Omega,\sigma)\big]^n$ if either $n \geq 3$ or $p \in \big(\tfrac{1}{2}, 1\big)$, while according to [132, Remark 2.3.5] $\vec{f}$ belongs to $\big[H_1^{q,1}(\partial\Omega,\sigma)\big]^n$ for each $q \in (1,\infty)$ if $n = 2$ and $p = 1$. Granted these properties, we may now employ the invertibility results from Theorem 7.1.47 and Theorem 7.1.49 to show, much as in the proof of the corresponding solvability result in current item *(1)*, that the Regularity Problem (8.2.14) has a solution $(\vec{u}, \pi)$.

Finally, uniqueness (modulo adding locally constant functions to the pressure) may be established by reasoning similarly to the proof of the uniqueness statement in the current item *(1)*, now availing ourselves of the integral representation formula [132, (6.3.94)], the properties of the single layer operator for the Stokes system from Theorem 7.1.47, and the Calderón-Zygmund theory for boundary layer operators associated with the Stokes system on Hardy spaces from [132, Theorem 6.2.11]. This concludes the proof of Theorem 8.2.1. □

We continue by proving the well-posedness of the Neumann Problem involving the slip conormal derivative for the Stokes system on any Ahlfors regular domain with a compact boundary which is sufficiently flat at an infinitesimal level. Our treatment allows for a large variety of function spaces, such as ordinary Lebesgue as well as Muckenhoupt weighted Lebesgue spaces, Hardy spaces, Morrey spaces, vanishing Morrey spaces, block spaces, Orlicz spaces, and also certain Generalized Banach Function Spaces, including rearrangement invariant Banach function space with Boyd indices in $(1,\infty)$.

Theorem 8.2.2 (The Neumann Problem for the Stokes System) *Pick $n \in \mathbb{N}$ with $n \geq 2$ and assume $\Omega \subseteq \mathbb{R}^n$ is an Ahlfors regular domain with compact boundary.*

Denote by ν *the geometric measure theoretic outward unit normal to* Ω *and abbreviate* $\sigma := \mathcal{H}^{n-1}\lfloor\partial\Omega$. *Also, fix some arbitrary aperture parameter* $\kappa > 0$.

(1) For any integrability exponent $p \in (1, \infty)$ *there exists a threshold* $\delta \in (0, 1)$ *which depends only on* n, p, *the* ADR *constants of* $\partial\Omega$, *and the relative flatness ratio* $\wp(\Omega)/\operatorname{diam}\partial\Omega$ *with the property that when*

$$\operatorname{dist}\Big(\nu, [\operatorname{VMO}(\partial\Omega, \sigma)]^n\Big) < \delta \tag{8.2.47}$$

where the distance is considered in the space $\big[\operatorname{BMO}(\partial\Omega, \sigma)\big]^n$ *(hence if* Ω *is a* δ*-oscillating* AR *domain*[80]*; cf. Definition 3.4.1), the following results hold. First, the Neumann Problem*[81]

$$\begin{cases} \vec{u} \in \big[\mathscr{C}^\infty(\Omega)\big]^n, \quad \pi \in \mathscr{C}^\infty(\Omega), \\ \Delta\vec{u} - \nabla\pi = 0 \text{ and } \operatorname{div}\vec{u} = 0 \text{ in } \Omega, \\ \mathcal{N}_\kappa(\nabla\vec{u}),\ \mathcal{N}_\kappa\pi \in L^p(\partial\Omega, \sigma), \\ \partial_\nu^{\text{slip}}\big(\vec{u}, \pi\big) = \vec{f} \text{ at } \sigma\text{-a.e. point on } \partial\Omega, \\ \vec{u}(x) = O(|x|^{2-n}) \text{ and } \pi(x) = O(|x|^{1-n}) \\ \text{as } |x| \to \infty \text{ if } \Omega \text{ is an exterior domain,} \end{cases} \tag{8.2.48}$$

has a solution if and only if the boundary datum satisfies finitely many linearly independent constraints. More specifically, the Neumann Problem (8.2.48) *has a solution if and only if*

$$\begin{aligned} &\vec{f} \in \operatorname{Im}\Big(-\tfrac{1}{2}I + K^{\#}_{\text{slip}}\,;\, \big[L^p(\partial\Omega, \sigma)\big]^n \cap \Psi(\partial\Omega_+)^\perp\Big) \\ &\text{if either } n \geq 3, \text{ or } n = 2 \text{ and } \Omega \text{ is bounded,} \end{aligned} \tag{8.2.49}$$

and

$$\begin{aligned} &\vec{f} \in \operatorname{Im}\Big(-\tfrac{1}{2}I + K^{\#}_{\text{slip}}\,;\, \big[L^p_0(\partial\Omega, \sigma)\big]^2\Big) \\ &\text{if } n = 2 \text{ and } \Omega \text{ is an exterior domain,} \end{aligned} \tag{8.2.50}$$

where $L^p_0(\partial\Omega, \sigma) := \Big\{\phi \in L^p_0(\partial\Omega, \sigma) : \int_{\partial\Omega}\phi\,d\sigma = 0\Big\}$. *Second, as far as uniqueness is concerned,*

$$\begin{aligned} &\text{if either } n \geq 3, \text{ or } n = 2 \text{ and } \Omega \text{ is bounded, the solution of the Neumann} \\ &\text{Problem (8.2.48) is unique modulo adding functions from } \Psi(\Omega_+) \text{ to} \\ &\text{the velocity field,} \end{aligned} \tag{8.2.51}$$

whereas

---

[80] in particular, if Ω is an infinitesimally flat AR domain

[81] in view of work in [129, §8.5], one may replace the nontangential maximal operator by the tangential maximal operator, associated as in [129, Definition 8.5.1] with a sufficiently large power (cf. (A.0.165))

if $n = 2$ and Ω is an exterior domain, the solution of the Neumann Problem (8.2.48) *is unique modulo adding functions from $\mathbb{C}^2 + \Psi(\Omega_+)$ to the velocity field.*

(8.2.52)

Third, there exists $C \in (0,\infty)$ independent of $\vec{f}$ (assumed to be as in (8.2.49)-(8.2.50)) *such that any solution $(\vec{u},\pi)$ of the Neumann Problem* (8.2.48) *has*

$$\big\|\mathcal{N}_\kappa(\nabla\vec{u})\big\|_{L^p(\partial\Omega,\sigma)} + \big\|\mathcal{N}_\kappa\pi\big\|_{L^p(\partial\Omega,\sigma)} \leq C\|\vec{f}\|_{[L^p(\partial\Omega,\sigma)]^n}. \tag{8.2.53}$$

Fourth, in the case when Ω is bounded, any solution $(\vec{u},\pi)$ of the boundary value problem (8.2.48) *satisfies*

$$\begin{gathered}\vec{u} \in \Big[B^{p,q_0}_{1+\frac{1}{p}}(\Omega)\Big]^n \cap \Big[F^{p,q_1}_{1+\frac{1}{p}}(\Omega)\Big]^n \ \text{ and } \pi \in B^{p,q_0}_{\frac{1}{p}}(\Omega)\cap F^{p,q_1}_{\frac{1}{p}}(\Omega) \\ \text{if } p \leq q_0 \leq \infty \text{ and } \tfrac{n}{n+1+1/p} < q_1 \leq \infty,\end{gathered} \tag{8.2.54}$$

as well as

$$\vec{u} \in \big[H^{3/2}(\Omega)\big]^n \ \text{ and } \pi \in \big[H^{1/2}(\Omega)\big]^n \ \text{ if } p = 2. \tag{8.2.55}$$

(2) *A more general well-posedness result than the one in item (1) above may be formulated in relation to the version of the Neumann Problem formulated for boundary data in certain Generalized Banach Function Spaces. To elaborate, suppose $\mathbb{X}$ is a Generalized Banach Function Space on $(\partial\Omega,\sigma)$ for which*

$$\mathcal{M}_{\partial\Omega} : \mathbb{X} \to \mathbb{X} \text{ and } \mathcal{M}_{\partial\Omega} : \mathbb{X}' \to \mathbb{X}' \text{ are well-defined and bounded,} \tag{8.2.56}$$

where $\mathcal{M}_{\partial\Omega}$ is the Hardy-Littlewood maximal operator on $\partial\Omega$, and $\mathbb{X}'$ is the associated space of $\mathbb{X}$ (cf. [130, Definitions 5.1.4, 5.1.11]). *Then there exists some threshold $\delta_{\mathbb{X}} \in (0,1)$ which depends only on n, the operator norms of $\mathcal{M}_{\partial\Omega}$ on $\mathbb{X}$ and $\mathbb{X}'$, as well as on the* ADR *constants of $\partial\Omega$ and the relative flatness ratio $\wp(\Omega)/\mathrm{diam}\,\partial\Omega$, with the property that when*

$$\mathrm{dist}\Big(\nu\,,\,[\mathrm{VMO}\,(\partial\Omega,\sigma)]^n\Big) < \delta_{\mathbb{X}} \tag{8.2.57}$$

with the distance in $\big[\mathrm{BMO}\,(\partial\Omega,\sigma)\big]^n$, it follows that the Neumann Problem[82]

[82] once again, thanks to work in [129, §8.5] one may replace the nontangential maximal operator by the tangential maximal operator, associated as in [129, Definition 8.5.1] with a sufficiently large power (cf. (A.0.165))

$$\begin{cases} \vec{u} \in \big[\mathscr{C}^\infty(\Omega)\big]^n, \quad \pi \in \mathscr{C}^\infty(\Omega), \\ \Delta\vec{u} - \nabla\pi = 0 \textit{ and } \operatorname{div}\vec{u} = 0 \textit{ in } \Omega, \\ \mathcal{N}_\kappa(\nabla\vec{u}),\ \mathcal{N}_\kappa\pi \in \mathbb{X}, \\ \partial_\nu^{\text{slip}}(\vec{u},\pi) = \vec{f} \textit{ at } \sigma\textit{-a.e. point on } \partial\Omega, \\ \vec{u}(x) = O(|x|^{2-n}) \textit{ and } \pi(x) = O(|x|^{1-n}) \\ \textit{as } |x| \to \infty \textit{ if } \Omega \textit{ is an exterior domain,} \end{cases} \tag{8.2.58}$$

has a solution if and only if

$$\vec{f} \in \operatorname{Im}\Big(-\tfrac{1}{2}I + K^{\#}_{\text{slip}};\ [\mathbb{X}]^n \cap \Psi(\partial\Omega_+)^{\perp}\Big) \quad \textit{if either } n \geq 3, \textit{ or } n = 2 \textit{ and } \Omega \textit{ is bounded,} \tag{8.2.59}$$

and

$$\vec{f} \in \operatorname{Im}\Big(-\tfrac{1}{2}I + K^{\#}_{\text{slip}};\ [\mathbb{X}_0]^2\Big) \quad \textit{if } n = 2 \textit{ and } \Omega \textit{ is an exterior domain,} \tag{8.2.60}$$

where $\mathbb{X}_0 := \Big\{\phi \in \mathbb{X} : \int_{\partial\Omega} \phi \, d\sigma = 0\Big\}$. *Regarding the issue of uniqueness,*

> *if either* $n \geq 3$, *or* $n = 2$ *and* Ω *is bounded, the solution of the Neumann Problem* (8.2.58) *is unique modulo adding functions from* $\Psi(\Omega_+)$ *to the velocity field,* (8.2.61)

whereas

> *if* $n = 2$ *and* Ω *is an exterior domain, the solution of the Neumann Problem* (8.2.58) *is unique modulo adding functions from* $\mathbb{C}^2 + \Psi(\Omega_+)$ *to the velocity field.* (8.2.62)

In addition, there exists a constant $C \in (0, \infty)$ *independent of* $\vec{f}$ *(assumed to be as in* (8.2.59)-(8.2.60)*) with the property that any solution* $(\vec{u}, \pi)$ *of the Neumann Problem* (8.2.58) *satisfies*

$$\big\|\mathcal{N}_\kappa(\nabla\vec{u})\big\|_{\mathbb{X}} + \big\|\mathcal{N}_\kappa\pi\big\|_{\mathbb{X}} \leq C\|\vec{f}\|_{[\mathbb{X}]^n}. \tag{8.2.63}$$

Finally, a similar well-posedness result to the one described above is true with $\mathbb{X}$ *replaced by* $\mathring{\mathbb{X}} := \overline{L^\infty(\partial\Omega, \sigma)}^{\|\cdot\|_{\mathbb{X}}}$.

(3) *The well-posedness result in item (1) for the Neumann Problem for the Stokes system may be further extended to the case when the boundary data now belong to Hardy spaces. Specifically, for each exponent*

$$p \in \big(\tfrac{n-1}{n}, 1\big] \tag{8.2.64}$$

there exists a threshold $\delta \in (0, 1)$ *which depends only on* n, p, *the* ADR *constants of* $\partial\Omega$, *and the relative flatness ratio* $\wp(\Omega)/\operatorname{diam}\partial\Omega$ *so that when*

$$\operatorname{dist}\Big(\nu,\,[\mathrm{VMO}\,(\partial\Omega,\sigma)]^n\Big) < \delta \tag{8.2.65}$$

with the distance considered in $\big[\mathrm{BMO}\,(\partial\Omega,\sigma)\big]^n$ *(hence if* Ω *is a* δ*-oscillating* AR *domain*[83]*; cf. Definition 3.4.1), it follows that the boundary problem*[84]

$$\begin{cases} \vec{u} \in \big[\mathscr{C}^\infty(\Omega)\big]^n, \quad \pi \in \mathscr{C}^\infty(\Omega), \\ \Delta\vec{u} - \nabla\pi = 0 \ \textit{and} \ \operatorname{div}\vec{u} = 0 \ \textit{in} \ \Omega, \\ \mathcal{N}_\kappa\vec{u},\ \mathcal{N}_\kappa(\nabla\vec{u}),\ \mathcal{N}_\kappa\pi \in L^p(\partial\Omega,\sigma), \\ \overset{\bullet}{\partial}{}_\nu^{\mathrm{slip}}(\vec{u},\pi) = \vec{f} \ \textit{as distributions on} \ \partial\Omega, \\ \vec{u}(x) = O(|x|^{2-n}) \ \textit{and} \ \pi(x) = O(|x|^{1-n}) \\ \textit{as} \ |x| \to \infty \ \textit{if} \ \Omega \ \textit{is an exterior domain}, \end{cases} \tag{8.2.66}$$

has a solution if and only if

$$\begin{array}{c} \vec{f} \in \operatorname{Im}\Big(-\tfrac{1}{2}I + K^{\#}_{\mathrm{slip}};\ \big[H^p(\partial\Omega,\sigma)\big]^n \cap \Psi(\partial\Omega_+)^{\perp}\Big) \\ \textit{if either } n \geq 3, \textit{ or } n = 2 \textit{ and } \Omega \textit{ is bounded}, \end{array} \tag{8.2.67}$$

and

$$\begin{array}{c} \vec{f} \in \operatorname{Im}\Big(-\tfrac{1}{2}I + K^{\#}_{\mathrm{slip}};\ \big[H^p_0(\partial\Omega,\sigma)\big]^2\Big) \\ \textit{if } n = 2 \textit{ and } \Omega \textit{ is an exterior domain}, \end{array} \tag{8.2.68}$$

where $H^p_0(\partial\Omega,\sigma) := \big\{\phi \in H^p_0(\partial\Omega,\sigma) : \langle\phi, 1\rangle = 0\big\}$. *Furthermore,*

> *if either* $n \geq 3$*, or* $n = 2$ *and* Ω *is bounded, the solution of the Neumann Problem* (8.2.66) *is unique modulo adding functions from* $\Psi(\Omega_+)$ *to the velocity field,* (8.2.69)

whereas

> *if* $n = 2$ *and* Ω *is an exterior domain, the solution of the Neumann Problem* (8.2.66) *is unique modulo adding functions from* $\mathbb{C}^2 + \Psi(\Omega_+)$ *to the velocity field.* (8.2.70)

Finally, any solution $(\vec{u},\pi)$ *of* (8.2.66) *has the property that*

$$\begin{array}{c} \mathcal{N}_\kappa\vec{u} \in L^{p^*}(\partial\Omega,\sigma) \ \textit{where} \ p^* := \big(\tfrac{1}{p} - \tfrac{1}{n-1}\big)^{-1} \in (1,\infty) \\ \textit{if either } n \geq 3 \textit{ or } n = 2 \textit{ and } p \in \big(\tfrac{1}{2}, 1\big), \end{array} \tag{8.2.71}$$

while

$$\mathcal{N}_\kappa\vec{u} \in L^q(\partial\Omega,\sigma) \ \textit{for any} \ q \in (1,\infty), \ \ \textit{if } n = 2 \textit{ and } p = 1. \tag{8.2.72}$$

---

[83] in particular, if Ω is an infinitesimally flat AR domain

[84] in light of work in [129, §8.5], one may replace the nontangential maximal operator by the tangential maximal operator, associated as in [129, Definition 8.5.1] with a sufficiently large power (cf. (A.0.165))

From Theorem 8.2.2 and [130, Proposition 5.3.14] we see that

> granted (8.2.57), it follows that the Neumann Problem (8.2.58) is uniquely (modulo functions from $\Psi(\Omega_+)$ for the velocity field) solvable whenever the boundary datum satisfies (8.2.59) provided $\mathbb{X}$ is a rearrangement invariant Banach function space whose Boyd indices satisfy $1 < p_{\mathbb{X}} \leq q_{\mathbb{X}} < \infty$. (8.2.73)

As a corollary (cf. [130, (5.3.165)] and [130, Proposition 5.3.15]),

> under the assumption made in (8.2.57), it follows that the Neumann Problem (8.2.58) is uniquely (modulo functions from $\Psi(\Omega_+)$ for the velocity field) solvable whenever the boundary datum satisfies (8.2.59) provided $\mathbb{X} := L^{\Phi}(\partial\Omega, \sigma)$, the Orlicz space associated as in (A.0.133) with the measure space $(\partial\Omega, \sigma)$ and any Young function Φ satisfying $1 < i(\Phi) \leq I(\Phi) < \infty$. (8.2.74)

Many other Neumann Problems fall under the scope of item *(2)* in Theorem 8.2.1, such as those formulated with boundary data prescribed in Muckenhoupt weighted Lebesgue spaces, Morrey spaces, vanishing Morrey spaces, and block spaces (all of which are Generalized Banach Function Spaces satisfying (8.2.56)).

Finally, we wish to note that in the formulation of (8.2.66) when $p = 1$ we may allow the slip conormal derivative to be considered in a strong, pointwise sense (see [132, Theorem 6.2.10, item *(2)*] in this regard).

We now turn to the proof of Theorem 8.2.2.

Proof of Theorem 8.2.2 Pick $\delta \in (0, 1)$ sufficiently small, depending on n, p, the ADR constants of $\partial\Omega$, and the relative flatness ratio $\wp(\Omega)/\operatorname{diam}\partial\Omega$, so that Ω becomes a two-sided NTA domain (cf. Theorem 3.4.2). In particular, Ω is also a UR domain, and satisfies the properties in (7.1.419) (see [129, (5.11.6)] and [129, Lemma 5.11.3]). By eventually further decreasing the value of $\delta \in (0, 1)$ we may also guarantee that the conclusions in Theorem 7.1.21 and Theorem 7.1.23 are valid.

Suppose that either $n \geq 3$, or $n = 2$ and Ω is bounded. Given a boundary datum $\vec{f}$ as in (8.2.49)-(8.2.50), from item *(10)* in Theorem 7.1.1 and [130, Proposition 1.2.13] we conclude that

$$\begin{aligned} &\vec{f} = \big(-\tfrac{1}{2}I + K^{\#}_{\text{slip}}\big)\vec{g} \text{ for some } \vec{g} \in \big[L^p(\partial\Omega, \sigma)\big]^n \cap \Psi(\partial\Omega_+)^{\perp} \\ &\qquad \text{with } \|\vec{g}\|_{[L^p(\partial\Omega,\sigma)]^n} \leq C\|\vec{f}\|_{[L^p(\partial\Omega,\sigma)]^n}, \end{aligned} \tag{8.2.75}$$

for some constant $C \in (0, \infty)$ independent of $\vec{f}$. Then the pair $(\vec{u}, \pi)$ with

$$\vec{u} := \mathscr{S}\vec{g} \text{ and } \pi := \mathcal{Q}\vec{g} \text{ in } \Omega \tag{8.2.76}$$

is a solution of the boundary value problem (8.2.48) and satisfies (8.2.53) (thanks to [132, (6.2.15)], [132, Theorem 6.2.4], (8.2.75), and [132, (6.2.1)-(6.2.2)]).

In the opposite direction, continue to assume that either $n \geq 3$, or $n = 2$ and Ω is bounded. Suppose $\vec{f} \in \big[L^p(\partial\Omega, \sigma)\big]^n$ is such that the Neumann Problem (8.2.48) has a solution $(\vec{u}, \pi)$. The goal is to prove the membership in (8.2.49). To set the stage, observe that the Fatou-type result for null-solutions of the Stokes system established in [132, Theorem 6.3.2] applies and, together with [129, Proposition 8.9.22], guarantees that

$$\text{the nontangential pointwise boundary traces } \vec{u}\Big|_{\partial\Omega}^{\kappa-\mathrm{n.t.}}, \nabla\vec{u}\Big|_{\partial\Omega}^{\kappa-\mathrm{n.t.}}, \text{ and } \pi\Big|_{\partial\Omega}^{\kappa-\mathrm{n.t.}} \text{ exist } \sigma\text{-a.e. on } \partial\Omega \tag{8.2.77}$$

and these traces are also independent of the aperture parameter $\kappa \in (0, \infty)$. Granted these properties, we may invoke [130, Proposition 11.3.2] and conclude that

$$\vec{u}\Big|_{\partial\Omega}^{\kappa-\mathrm{n.t.}} \in \big[L_1^p(\partial\Omega, \sigma)\big]^n. \tag{8.2.78}$$

In addition, for each bounded connected component O of Ω we may write

$$\int_{\partial O} \nu \cdot \Big(\vec{u}\Big|_{\partial\Omega}^{\kappa-\mathrm{n.t.}}\Big)\, d\sigma = 0, \tag{8.2.79}$$

thanks to the Divergence Formula from [129, Theorem 1.2.1] and the fact that $\vec{u}$ is divergence-free. From (8.2.78), (8.2.79), and (7.1.399) we see that actually

$$\vec{u}\Big|_{\partial\Omega}^{\kappa-\mathrm{n.t.}} \in \big[L_1^p(\partial\Omega, \sigma)\big]^n \cap \big\langle \nu \mathbb{C}_{\partial\Omega_+}\big\rangle^{\perp}. \tag{8.2.80}$$

Pushing ahead, pick an arbitrary $\psi \in \Psi(\partial\Omega_+)$, say $\psi = \widetilde{\psi}|_{\partial\Omega}$ for some $\widetilde{\psi} \in \Psi(\Omega_+)$, and write

$$\int_{\partial\Omega} \langle \psi, \vec{f}\rangle\, d\sigma = \int_{\partial\Omega} \Big\langle \widetilde{\psi}|_{\partial\Omega}, \partial_\nu^{\mathrm{slip}}(\vec{u}, \pi)\Big\rangle\, d\sigma = \int_{\partial\Omega} \Big\langle \partial_\nu^{\mathrm{slip}}(\widetilde{\psi}, 0), \vec{u}\Big|_{\partial\Omega}^{\kappa-\mathrm{n.t.}}\Big\rangle\, d\sigma = 0,$$

using the boundary condition in the Neumann Problem (8.2.48), the Green-type formula for the Stokes system proved in [132, Theorem 6.1.2], and (7.1.422). Thanks to the arbitrariness of ψ, this shows that $\vec{f} \in \big[L^p(\partial\Omega, \sigma)\big]^n \cap \Psi(\partial\Omega_+)^{\perp}$.

Having established this, we now use the fact that (7.1.436) is an isomorphism in order to find a vector-valued function $\vec{g} \in \big[L^p(\partial\Omega, \sigma)\big]^n \cap \Psi(\partial\Omega_+)^{\perp}$ such that $\big(-\frac{1}{2}I + K^{\#}_{\mathrm{slip}}\big)\vec{g} - \vec{f} = \nu\chi$, for some $\chi \in \mathbb{C}_{\partial\Omega_-}$. If we set $\vec{w} := \mathscr{S}\vec{g}$ and $\rho := Q\vec{g}$ in Ω, then the pair $(\vec{w} - \vec{u}, \rho - \pi)$ solves the Neumann Problem (8.2.48) with datum $\nu\chi$. We will now make a claim which implies that, necessarily, $\chi = 0$. This, of course, entails $\vec{f} = \big(-\frac{1}{2}I + K^{\#}_{\mathrm{slip}}\big)\vec{g}$, finishing the proof of (8.2.49). The claim just alluded to above is the following:

$$\text{assuming either } n \geq 3, \text{ or } n = 2 \text{ and } \Omega \text{ is bounded, if } (\vec{u}, \pi) \text{ solves the Neumann Problem (8.2.48) for the boundary datum } \vec{f} := \nu\chi \text{ with } \chi \in \mathbb{C}_{\partial\Omega_-}, \text{ then } \chi = 0. \tag{8.2.81}$$

To justify this claim, write [132, (6.2.25)] (cf. also [132, Theorem 6.2.1, item *(d)*]) and make use of (8.2.39) together with (7.1.272) to conclude that

$$\vec{u} = \mathcal{D}_{\text{slip}}\big(\vec{u}\big|^{\kappa-\text{n.t.}}_{\partial\Omega}\big) \text{ in } \Omega. \tag{8.2.82}$$

Going nontangentially to the boundary in (8.2.82) then yields

$$\big(\vec{u}\big|^{\kappa-\text{n.t.}}_{\partial\Omega}\big) \in \operatorname{Ker}\Big(-\tfrac{1}{2}I + K_{\text{slip}};\ \big[L^p_1(\partial\Omega,\sigma)\big]^n \cap \big\langle \nu\mathbb{C}_{\partial\Omega_+}\big\rangle^{\perp}\Big) = \Psi(\partial\Omega_+), \tag{8.2.83}$$

by [132, (6.2.109)] (written with $\lambda := 1$), (8.2.80), and (7.1.426). Utilizing this back into (8.2.82) and relying on (7.1.423) further gives $\vec{u} \in \Psi(\Omega_+)$. As a consequence, $\nu\chi = \partial^{\text{slip}}_\nu(\vec{u},\pi) \in \nu\mathbb{C}_{\partial\Omega_+}$ by (7.1.424) and, ultimately, $\chi = 0$ given that the sum in (7.1.272) is direct. This establishes the claim made in (8.2.81), and concludes the proof of the fact that (8.2.48) is equivalent to the membership in (8.2.49) if either $n \geq 3$, or $n = 2$ and Ω is bounded.

We next propose to show that (8.2.48) is equivalent to the membership in (8.2.50) when $n = 2$ and Ω is an exterior domain. Work in the latter geometric setting. In one direction, if $\vec{f}$ is as in (8.2.49) with $n = 2$ and the membership in (8.2.50) holds, item *(10)* in Theorem 7.1.1 together with [130, Proposition 1.2.13] imply

$$\begin{gathered}\vec{f} = \big(-\tfrac{1}{2}I + K^{\#}_{\text{slip}}\big)\vec{g} \text{ for some } \vec{g} \in \big[L^p_0(\partial\Omega,\sigma)\big]^2 \\ \text{with } \|\vec{g}\|_{[L^p(\partial\Omega,\sigma)]^2} \leq C\|\vec{f}\|_{[L^p(\partial\Omega,\sigma)]^2},\end{gathered} \tag{8.2.84}$$

where the constant $C \in (0,\infty)$ is independent of $\vec{f}$. As such, the pair $(\vec{u},\pi)$ with

$$\vec{u} := \mathcal{S}\vec{g} \text{ and } \pi := \mathcal{Q}\vec{g} \text{ in } \Omega \tag{8.2.85}$$

is a solution of the boundary value problem (8.2.48) and satisfies (8.2.53), thanks to [132, (6.2.15)], [132, Theorem 6.2.4], (8.2.75), [132, (6.2.1)-(6.2.2)] with $n = 2$, and the vanishing moment property satisfied by $\vec{g}$ (implicit in (8.2.84)).

To deal with the converse direction, continue to work in the case when $n = 2$ and $\Omega \subseteq \mathbb{R}^2$ is an exterior domain. The goal is to show that if

$$\begin{cases} \vec{u} \in \big[\mathscr{C}^\infty(\Omega)\big]^2, \quad \pi \in \mathscr{C}^\infty(\Omega), \\ \Delta\vec{u} - \nabla\pi = 0 \text{ and } \operatorname{div}\vec{u} = 0 \text{ in } \Omega, \\ \mathcal{N}_\kappa(\nabla\vec{u}),\ \mathcal{N}_\kappa\pi \in L^p(\partial\Omega,\sigma), \\ \vec{u}(x) = O(1) \text{ and } \pi(x) = O(|x|^{-1}) \text{ as } |x| \to \infty, \end{cases} \tag{8.2.86}$$

then

$$\begin{gathered}\vec{f} := \partial^{\text{slip}}_\nu(\vec{u},\pi) \text{ is a well-defined function} \\ \text{belonging to } \operatorname{Im}\Big(-\tfrac{1}{2}I + K^{\#}_{\text{slip}};\ \big[L^p_0(\partial\Omega,\sigma)\big]^2\Big).\end{gathered} \tag{8.2.87}$$

For one thing, the Fatou-type result for null-solutions of the Stokes system established in [132, Theorem 6.3.2] together with [129, Proposition 8.9.22] ensure that

$$\begin{gathered}\text{the nontangential pointwise boundary traces} \\ \vec{u}\big|^{\kappa-\text{n.t.}}_{\partial\Omega},\ \nabla\vec{u}\big|^{\kappa-\text{n.t.}}_{\partial\Omega}, \text{ and } \pi\big|^{\kappa-\text{n.t.}}_{\partial\Omega} \text{ exist } \sigma\text{-a.e. on } \partial\Omega.\end{gathered} \tag{8.2.88}$$

As such, $\vec{f} := \partial_\nu^{\text{slip}}(\vec{u}, \pi)$ is a well-defined function belonging to $\big[L^p(\partial\Omega, \sigma)\big]^2$, while $\vec{u}\big|_{\partial\Omega}^{\kappa-\text{n.t.}}$ is a well-defined function belonging to $\big[L_1^p(\partial\Omega, \sigma)\big]^2$.

Properties (8.2.86) and (8.2.88) guarantee that all hypotheses made in [132, Theorem 6.2.1, item *(c)*] are presently satisfied. As such, the integral representation formulas [132, (6.2.32)-(6.2.33)] are valid and, when written with $\lambda := 1$, they presently give

$$\vec{u} = \mathcal{D}_{\text{slip}}\big(\vec{u}\big|_{\partial\Omega}^{\kappa-\text{n.t.}}\big) - \mathcal{S}\vec{f} + c \text{ in } \Omega, \tag{8.2.89}$$

for some constant $c \in \mathbb{C}^2$, and

$$\pi = \mathcal{P}_{\text{slip}}\Big(\vec{u}\big|_{\partial\Omega}^{\kappa-\text{n.t.}}\Big) - \mathcal{Q}\vec{f} \text{ in } \Omega. \tag{8.2.90}$$

In turn, from (8.2.89) and the last line in (8.2.86) we see that

$$\big(\mathcal{S}\vec{f}\big)(x) = O(1) \text{ as } |x| \to \infty. \tag{8.2.91}$$

Together with (A.0.233) and [132, (6.2.1)], this forces

$$\int_{\partial\Omega} \vec{f}\, \mathrm{d}\sigma = 0. \tag{8.2.92}$$

Since, as noted earlier, $\vec{u}\big|_{\partial\Omega}^{\kappa-\text{n.t.}}$ belongs to $\big[L_1^p(\partial\Omega, \sigma)\big]^2$, the invertibility result stated in (7.1.404) presently used with $\mathbb{X} := L^p(\partial\Omega, \sigma)$ implies that there exist $\vec{h} \in \big[L_0^p(\partial\Omega, \sigma)\big]^2$ and $c \in \mathbb{C}^2$ such that

$$S\vec{h} + c = \vec{u}\big|_{\partial\Omega}^{\kappa-\text{n.t.}}. \tag{8.2.93}$$

Based on (8.2.93), the fact that $\partial_\nu^{\text{slip}}(\mathcal{D}_{\text{slip}}, \mathcal{P}_{\text{slip}})$ annihilates constants (cf. [132, (6.2.118)]), and the operator theoretic identity recorded in [132, (6.2.124)], we compute

$$\begin{aligned}\partial_\nu^{\text{slip}}(\mathcal{D}_{\text{slip}}, \mathcal{P}_{\text{slip}})\big(\vec{u}\big|_{\partial\Omega}^{\kappa-\text{n.t.}}\big) &= \partial_\nu^{\text{slip}}(\mathcal{D}_{\text{slip}}, \mathcal{P}_{\text{slip}})\big(S\vec{h} + c\big) = \partial_\nu^{\text{slip}}(\mathcal{D}_{\text{slip}}, \mathcal{P}_{\text{slip}})(S\vec{h}) \\ &= \big(-\tfrac{1}{2}I + K^{\#}_{\text{slip}}\big)\Big(\big(\tfrac{1}{2}I + K^{\#}_{\text{slip}}\big)\vec{h}\Big). \end{aligned} \tag{8.2.94}$$

Making use of (8.2.87), (8.2.89), (8.2.90), (8.2.94), and the jump-formula [132, (6.2.113)] written for $\lambda := 1$, we may now write

$$\begin{aligned}\vec{f} = \partial_\nu^{\text{slip}}(\vec{u}, \pi) &= \partial_\nu^{\text{slip}}(\mathcal{D}_{\text{slip}}, \mathcal{P}_{\text{slip}})\big(\vec{u}\big|_{\partial\Omega}^{\kappa-\text{n.t.}}\big) - \partial_\nu^{\text{slip}}\big(\mathcal{S}\vec{f} - c, \mathcal{Q}\vec{f}\big) \\ &= \big(-\tfrac{1}{2}I + K^{\#}_{\text{slip}}\big)\Big(\big(\tfrac{1}{2}I + K^{\#}_{\text{slip}}\big)\vec{h}\Big) - \big(-\tfrac{1}{2}I + K^{\#}_{\text{slip}}\big)\vec{f} = \big(-\tfrac{1}{2}I + K^{\#}_{\text{slip}}\big)\vec{g}, \end{aligned} \tag{8.2.95}$$

where

$$\vec{g} := -\vec{f} + \big(\tfrac{1}{2}I + K^{\#}_{\text{slip}}\big)\vec{h} \in \big[L^p(\partial\Omega, \sigma)\big]^2. \tag{8.2.96}$$

Note that $\left(\frac{1}{2}I + K^{\#}_{\text{slip}}\right)\vec{h}$ belongs to $\operatorname{Im}\left(\frac{1}{2}I + K^{\#}_{\text{slip}}\,;\,\left[L^p(\partial\Omega,\sigma)\right]^2\right)$ which, by (7.1.432), is contained in $\left[L^p(\partial\Omega,\sigma)\right]^2 \cap \Psi(\partial\Omega_-)^{\perp}$. Since $\Omega_- = \mathbb{R}^2 \setminus \overline{\Omega}$ is bounded, from (7.1.273) and the definition of $\Psi(\partial\Omega_-)$ in (7.1.416) we see that

$$\left[L^p(\partial\Omega,\sigma)\right]^2 \cap \Psi(\partial\Omega_-)^{\perp} \subseteq \left[L^p_0(\partial\Omega,\sigma)\right]^2, \tag{8.2.97}$$

hence

$$\left(\tfrac{1}{2}I + K^{\#}_{\text{slip}}\right)\vec{h} \in \left[L^p_0(\partial\Omega,\sigma)\right]^2. \tag{8.2.98}$$

Collectively, (8.2.95), (8.2.92), (8.2.96), and (8.2.98) prove that

$$\vec{f} = \left(-\tfrac{1}{2}I + K^{\#}_{\text{slip}}\right)\vec{g} \text{ for some } \vec{g} \in \left[L^p_0(\partial\Omega,\sigma)\right]^2. \tag{8.2.99}$$

This finishes the proof of (8.2.87), thus completing the proof of the fact that (8.2.48) is equivalent to the membership in (8.2.50) when $n = 2$ and Ω is an exterior domain.

At this stage, the solvability claims made in item *(1)* as well as the fact that a solution may be found which satisfies a natural estimate (cf. (8.2.53)) have all been established. As far as item *(1)* is concerned, there remains to deal with the issue of uniqueness.

We shall first establish uniqueness in the case when either $n \geq 3$, or $n = 2$ and Ω is bounded. Assume this is the case and suppose the functions $\vec{u}$ and π solve the homogeneous version of the Neumann Problem (8.2.48). Then [132, Theorem 6.2.1] implies $\vec{u} = \mathcal{D}_{\text{slip}}\big(\vec{u}\big|^{\kappa-\text{n.t.}}_{\partial\Omega}\big)$ in Ω. Going nontangentially to the boundary then once again yields (8.2.83). As before, this places $\vec{u}$ in $\Psi(\Omega_+)$. The converse direction is covered by (7.1.422). This proves the uniqueness statement made in (8.2.51).

Next, consider the issue of uniqueness in the case when $n = 2$ and Ω is an exterior domain. Assume the pair $(\vec{u},\pi)$ solves the homogeneous version of the Neumann Problem (8.2.48) in the aforementioned geometric setting. This time, [132, Theorem 6.2.1, item *(c)*] implies that there exists some $c \in \mathbb{C}^2$ such that

$$\vec{u} = \mathcal{D}_{\text{slip}}\big(\vec{u}\big|^{\kappa-\text{n.t.}}_{\partial\Omega}\big) + c \text{ in } \Omega. \tag{8.2.100}$$

After going nontangentially to the boundary we now arrive (cf. [132, (6.2.109)]) at

$$\left(-\tfrac{1}{2}I + K_{\text{slip}}\right)\big(\vec{u}\big|^{\kappa-\text{n.t.}}_{\partial\Omega}\big) + c = 0. \tag{8.2.101}$$

From (8.2.80), definitions, and the Divergence Formula from [129, Theorem 1.1.1] we also see that

$$\vec{u}\big|^{\kappa-\text{n.t.}}_{\partial\Omega} - c \text{ belongs to } \left[L^p_1(\partial\Omega,\sigma)\right]^2 \cap \left\langle \nu\mathbb{C}_{\partial\Omega_+}\right\rangle^{\perp}. \tag{8.2.102}$$

Since Ω is presently assumed to be unbounded, from (8.2.101) and [132, (6.2.110)] we obtain

$$\left(-\tfrac{1}{2}I + K_{\text{slip}}\right)\big(\vec{u}\big|^{\kappa-\text{n.t.}}_{\partial\Omega} - c\big) = 0 \tag{8.2.103}$$

which, in concert with (8.2.102) shows that

$$\Big(\vec{u}\Big|_{\partial\Omega}^{\kappa-\mathrm{n.t.}} - c\Big) \in \mathrm{Ker}\Big(-\tfrac{1}{2}I + K_{\mathrm{slip}};\ \big[L_1^p(\partial\Omega,\sigma)\big]^2 \cap \big\langle \nu\mathbb{C}_{\partial\Omega_+}\big\rangle^{\perp}\Big) = \Psi(\partial\Omega_+), \quad (8.2.104)$$

with the last equality provided by (7.1.426). (8.2.100), Combining (8.2.104), [132, (6.2.19)], and (7.1.423) ultimately leads to the conclusion that

$$\vec{u} \text{ belongs to the space } \mathbb{C}^2 + \Psi(\Omega_+). \qquad (8.2.105)$$

With this in hand, the uniqueness claim made in (8.2.52) readily follows.

Lastly, the regularity properties in (8.2.54)-(8.2.55) are implied by the integral representation formulas from [132, Theorem 6.2.1], the fact that we presently have

$$\vec{u}\Big|_{\partial\Omega}^{\kappa-\mathrm{n.t.}} \in \big[L_1^p(\partial\Omega,\sigma)\big]^n, \quad \partial_\nu^{\mathrm{slip}}(\vec{u},\pi) \in \big[L^p(\partial\Omega,\sigma)\big]^n, \qquad (8.2.106)$$

and the mapping properties of the boundary layer potential operators involved from [132, Theorem 6.4.3, item *(4)*]. This completes the treatment of item *(1)* in the statement of the theorem.

The treatment of item *(2)* is very similar, now making use of the results included in the last portion of Theorem 7.1.23, the embeddings from [130, Proposition 5.2.7] (see also the Fatou-type result from [132, Theorem 6.3.3]), and now employing the full force of Theorem 7.1.21.

Finally, all claims in item *(3)* may be justified reasoning along the lines of the treatment of item *(1)*, now relying on the Fatou-type results and integral representation formulas for null-solutions of the Stokes system from [132, Theorem 6.3.6], the Calderón-Zygmund theory for boundary layer operators associated with the Stokes system on Hardy spaces from [132, Theorem 6.2.11], the results contained in Theorem 7.1.48, and the operator identities from [132, Theorem 6.3.7], with two significant differences. First, if $n = 2$ and $\Omega \subseteq \mathbb{R}^2$ is an exterior domain, when proving (8.2.66) implies (8.2.68), in place of [132, Theorem 6.2.1, item *(c)*] (which was used to obtain (8.2.89)) we now employ [132, Theorem 6.3.6] which gives a slightly weaker result than (8.2.89), namely that

$$\vec{u} = \mathcal{D}_{\mathrm{slip}}\Big(\vec{u}\Big|_{\partial\Omega}^{\kappa-\mathrm{n.t.}}\Big) - \mathcal{S}\vec{f} + c_{\vec{u},\pi} \text{ in } \Omega \qquad (8.2.107)$$

for some $\mathbb{C}^n$-valued locally constant function $c_{\vec{u},\pi}$ (in lieu of a genuine constant). However, (8.2.107) is still good enough to conclude that $\langle \vec{f}, 1\rangle = 0$. This is the natural analogue of (8.2.92) and the proof proceeds from here much as before.

The second aspect which requires closer attention occurs when dealing with the issue of uniqueness for the Neumann Problem (8.2.66) when $n = 2$ and $\Omega \subseteq \mathbb{R}^2$ is an exterior domain. Specifically, in place of [132, Theorem 6.2.1, item *(c)*] (which was used to write (8.2.100)) we now rely on [132, Theorem 6.3.6] and this only gives

$$\vec{u} = \mathcal{D}_{\mathrm{slip}}\Big(\vec{u}\Big|_{\partial\Omega}^{\kappa-\mathrm{n.t.}}\Big) + c_{\vec{u},\pi} \text{ in } \Omega \qquad (8.2.108)$$

for some $\mathbb{C}^n$-valued locally constant function $c_{\vec{u},\pi}$ (in lieu of a genuine constant). Nonetheless, (8.2.107) implies that there exists some $c \in \mathbb{C}^2$ (namely the value of

$c_{\vec{u},\pi}$ in the unbounded connected component of Ω) such that $\vec{u}(x) = c + O(|x|^{-1})$ as $|x| \to \infty$. Thanks to this and (8.2.71)-(8.2.72) (which are proved independently of the present considerations) we may now apply [132, Theorem 6.2.1, item *(c)*] for the pair $(\vec{u} - c, \pi)$ which remains a nullsolution of the Stokes system in Ω. This ultimately shows that (8.2.100) continues to be valid in the current context, and the goes on from here much as before. The proof of Theorem 8.2.2 is therefore complete. □

We continue by discussing the solvability of the Dirichlet Problem for the Stokes system in in Ahlfors regular domains with a compact boundary which is sufficiently flat at an infinitesimal level. As in past, we allow for a large variety of function spaces, such as ordinary Lebesgue as well as Muckenhoupt weighted Lebesgue spaces, Hardy spaces, Morrey spaces, vanishing Morrey spaces, block spaces, Orlicz spaces, and also certain Generalized Banach Function Spaces, including rearrangement invariant Banach function space with Boyd indices in $(1, \infty)$.

Theorem 8.2.3 (The Dirichlet Problem for the Stokes System) *Fix $n \in \mathbb{N}$ with $n \geq 2$ and suppose $\Omega \subseteq \mathbb{R}^n$ is an Ahlfors regular domain with compact boundary. Abbreviate $\sigma := \mathcal{H}^{n-1} \lfloor \partial\Omega$ and denote by ν the geometric measure theoretic outward unit normal to Ω. Also, fix some arbitrary aperture parameter $\kappa > 0$.*

(1) Given any integrability exponent $p \in (1, \infty)$, there exists some small threshold $\delta \in (0, 1)$ which depends only on n, p, the ADR *constants of $\partial\Omega$, and the relative flatness ratio $\wp(\Omega)/\mathrm{diam}\, \partial\Omega$ with the property that when*

$$\mathrm{dist}\Big(\nu\,,\ [\mathrm{VMO}\,(\partial\Omega, \sigma)]^n\Big) < \delta \tag{8.2.109}$$

where the distance is considered in the space $\big[\mathrm{BMO}\,(\partial\Omega, \sigma)\big]^n$ (hence if Ω is a δ-oscillating AR *domain*[85]*; cf. Definition 3.4.1), the following results are true. First, the Dirichlet Problem*[86]

$$\begin{cases} \vec{u} \in \big[\mathscr{C}^\infty(\Omega)\big]^n, \quad \pi \in \mathscr{C}^\infty(\Omega), \\ \Delta\vec{u} - \nabla\pi = 0 \ \textit{and}\ \mathrm{div}\,\vec{u} = 0 \ \textit{in}\ \Omega, \\ \mathcal{N}_\kappa \vec{u} \in L^p(\partial\Omega, \sigma), \\ \vec{u}\Big|_{\partial\Omega}^{\kappa-\mathrm{n.t.}} = \vec{f} \ \textit{at}\ \sigma\textit{-a.e. point on}\ \partial\Omega, \\ \vec{u}(x) = O(|x|^{2-n}) \ \textit{and}\ \pi(x) = O(|x|^{1-n}) \\ \textit{as}\ |x| \to \infty \ \textit{if}\ \Omega \ \textit{is an exterior domain,} \end{cases} \tag{8.2.110}$$

has a solution if and only if

$$\vec{f} \in \big[L^p(\partial\Omega, \sigma)\big]^n \cap \big\langle \nu\mathbb{C}_{\partial\Omega_+}\big\rangle^{\perp}. \tag{8.2.111}$$

[85] in particular, if Ω is an infinitesimally flat AR domain

[86] in view of work in [129, §8.5], one may replace the nontangential maximal operator by the tangential maximal operator, associated as in [129, Definition 8.5.1] with a sufficiently large power (cf. (A.0.165))

Second, whenever (8.2.111) *holds a solution* $(\vec{u}, \pi)$ *of* (8.2.110) *may be found such that, for some constant* $C \in (0, \infty)$ *independent of* $\vec{f}$, *one has*

$$\|\mathcal{N}_\kappa \vec{u}\|_{L^p(\partial\Omega,\sigma)} \leq C\|\vec{f}\|_{[L_1^p(\partial\Omega,\sigma)]^n} \tag{8.2.112}$$

and, in the case when Ω *is bounded, also satisfies*

$$\begin{gathered}\vec{u} \in \Big[B^{p,q_0}_{\frac{1}{p}}(\Omega)\Big]^n \cap \Big[F^{p,q_1}_{\frac{1}{p}}(\Omega)\Big]^n \\ \text{if } p \leq q_0 \leq \infty \text{ and } \tfrac{n}{n+1+1/p} < q_1 \leq \infty,\end{gathered} \tag{8.2.113}$$

as well as

$$\vec{u} \in \big[H^{3/2}(\Omega)\big]^n \text{ if } p = 2. \tag{8.2.114}$$

(2) *A more general well-posedness result than the one stated in item (1) holds for the Dirichlet Problem formulated with boundary data in certain Generalized Banach Function Spaces. Specifically, assume* $\mathbb{X}$ *is a Generalized Banach Function Space on* $(\partial\Omega, \sigma)$ *with the property that* $\mathcal{M}_{\partial\Omega} : \mathbb{X} \to \mathbb{X}$ *and* $\mathcal{M}_{\partial\Omega} : \mathbb{X}' \to \mathbb{X}'$ *are well-defined and bounded, where* $\mathcal{M}_{\partial\Omega}$ *is the Hardy-Littlewood maximal operator on* $\partial\Omega$, *and* $\mathbb{X}'$ *is the associated space of* $\mathbb{X}$ (*cf.* [130, Definitions 5.1.4, 5.1.11]). *Then there exists some threshold* $\delta_{\mathbb{X}} \in (0, 1)$ *which depends only on n, the operator norms of* $\mathcal{M}_{\partial\Omega}$ *on* $\mathbb{X}$ *and* $\mathbb{X}'$, *as well as on the* ADR *constants of* $\partial\Omega$ *and the relative flatness ratio* $\wp(\Omega)/\text{diam}\,\partial\Omega$, *with the property that when*

$$\text{dist}\Big(\nu\,,\,[\text{VMO}\,(\partial\Omega,\sigma)]^n\Big) < \delta_{\mathbb{X}} \tag{8.2.115}$$

with the distance considered in $\big[\text{BMO}\,(\partial\Omega,\sigma)\big]^n$, *it follows that the Dirichlet Problem*[87]

$$\begin{cases}\vec{u} \in \big[\mathscr{C}^\infty(\Omega)\big]^n, \quad \pi \in \mathscr{C}^\infty(\Omega), \\ \Delta\vec{u} - \nabla\pi = 0 \text{ and } \text{div}\,\vec{u} = 0 \text{ in } \Omega, \\ \mathcal{N}_\kappa\vec{u} \in \mathbb{X}, \quad \vec{u}\Big|^{\kappa-\text{n.t.}}_{\partial\Omega} = \vec{f} \text{ at } \sigma\text{-a.e. point on } \partial\Omega, \\ \vec{u}(x) = O(|x|^{2-n}) \text{ and } \pi(x) = O(|x|^{1-n}) \\ \text{as } |x| \to \infty \text{ if } \Omega \text{ is an exterior domain,}\end{cases} \tag{8.2.116}$$

has a solution if and only if

$$\vec{f} \in \big[\mathbb{X}\big]^n \cap \big\langle \nu\mathbb{C}_{\partial\Omega_+}\big\rangle^\perp. \tag{8.2.117}$$

Furthermore, granted (8.2.117), *a solution may always be found so that*

$$\|\mathcal{N}_\kappa\vec{u}\|_{\mathbb{X}} \leq C\|\vec{f}\|_{[\mathbb{X}]^n} \tag{8.2.118}$$

[87] once again, thanks to work in [129, §8.5] one may replace the nontangential maximal operator by the tangential maximal operator, associated as in [129, Definition 8.5.1] with a sufficiently large power (cf. (A.0.165))

for some constant $C \in (0, \infty)$ independent of $\vec{f}$. Finally, a similar solvability result to the one described above is true with $\mathbb{X}$ replaced by $\mathring{\mathbb{X}} := \overline{L^\infty(\partial\Omega, \sigma)}^{\|\cdot\|_{\mathbb{X}}}$.

As in the case of the Regularity Problem (discussed in Theorem 8.2.1), the significance of the brand of Divergence Theorem produced in [129] is highlighted by the very formulation of the boundary value problems for the Stokes system in Theorem 8.2.3 since the "compatibility" (annihilator) conditions imposed in (8.2.111) and (8.2.117) may be traced back (bearing in mind the divergence-free property of the velocity field) directly to the aforementioned version of the Divergence Theorem.

Proof of Theorem 8.2.3 All claims are established by reasoning in a very similar fashion to the proof of the existence results from Theorem 8.2.1. □

Our next theorem contains well-posedness results for the Dirichlet Problem for in which a nullsolution of the Stokes system in an Ahlfors regular domain with compact boundary, sufficiently flat at an infinitesimal level, is sought in Besov/Triebel-Lizorkin/weighted Sobolev spaces, while the boundary data in prescribed in Besov spaces. In this regard, the reader is advised to recall the piece of notation introduced in [130, Convention 8.3.7] (cf. also (A.0.256)).

Theorem 8.2.4 (The Dirichlet Problem for the Stokes System with Boundary Data in Besov Spaces) *Fix an integer $n \in \mathbb{N}$ with $n \geq 2$ and suppose*

$$p \in \left(\tfrac{n-1}{n}, \infty\right), \quad q \in (0, \infty), \quad (n-1)\left(\tfrac{1}{p} - 1\right)_+ < s < 1. \tag{8.2.119}$$

Also, let $\Omega \subseteq \mathbb{R}^n$ be an Ahlfors regular domain with compact boundary. Denote its geometric measure theoretic outward unit normal by ν and set $\sigma := \mathcal{H}^{n-1}\lfloor\partial\Omega$.

Then there exists a threshold $\delta_0 \in (0, 1)$ depending only on n, p, q, s, the ADR *constants of $\partial\Omega$, and the relative flatness ratio $\wp(\Omega)/\operatorname{diam}\partial\Omega$ so that when*

$$\operatorname{dist}\Big(\nu, [\operatorname{VMO}(\partial\Omega, \sigma)]^n\Big) < \delta_0 \tag{8.2.120}$$

with the distance considered in $\big[\operatorname{BMO}(\partial\Omega, \sigma)\big]^n$ (hence when Ω is a δ_0-oscillating AR *domain*[88]*; cf. Definition 3.4.1), it follows that the set Ω is both n-thick*[89] *as well as an (ε, δ)-domain with* $\operatorname{rad}(\Omega) > 0$, *and all statements below hold:*

(1) The boundary value problem (with the trace considered in the sense of [130, Theorem 9.4.5, *(ii)*, (9.4.91)])

[88] in particular, if Ω is an infinitesimally flat AR domain

[89] hence the trace operator $\operatorname{Tr}_{\Omega\to\partial\Omega}$ considered in [130, (8.3.43)] as well as [130, (9.4.105), (9.4.106)] is meaningful

$$\begin{cases} \vec{u} \in \big[B^{p,q}_{s+\frac{1}{p}}(\Omega)_{\mathrm{bdd}}\big]^n, \quad \pi \in B^{p,q}_{s+\frac{1}{p}-1}(\Omega)_{\mathrm{bdd}} \\ \Delta\vec{u} - \nabla\pi = 0 \ \textit{and} \ \operatorname{div}\vec{u} = 0 \ \textit{in} \ \Omega, \\ \operatorname{Tr}_{\Omega\to\partial\Omega}\vec{u} = \vec{f} \in \big[B^{p,q}_s(\partial\Omega,\sigma)\big]^n, \\ \vec{u}(x) = O(|x|^{2-n}) \ \textit{and} \ \pi(x) = O(|x|^{1-n}) \\ \textit{as} \ |x| \to \infty \ \textit{if} \ \Omega \ \textit{is an exterior domain}, \end{cases} \tag{8.2.121}$$

is Fredholm solvable in the following precise sense: First, the space of null-solutions of the boundary problem (8.2.121), *i.e.*,

$$\Big\{(\vec{u},\pi) \in \big[B^{p,q}_{s+\frac{1}{p}}(\Omega)_{\mathrm{bdd}}\big]^n \times B^{p,q}_{s+\frac{1}{p}-1}(\Omega)_{\mathrm{bdd}} : \Delta\vec{u} - \nabla\pi = 0, \ \operatorname{div}\vec{u} = 0, \\ \operatorname{Tr}_{\Omega\to\partial\Omega}\vec{u} = 0 \ \textit{on} \ \partial\Omega, \quad \vec{u}(x) = O(|x|^{2-n}), \ \pi(x) = O(|x|^{1-n}) \\ \textit{as} \ |x| \to \infty \ \textit{if} \ \Omega \ \textit{is an exterior domain}\Big\}, \tag{8.2.122}$$

is finite dimensional. Second, the space of admissible boundary data for the boundary problem (8.2.121), *i.e.*,

$$\Big\{\operatorname{Tr}_{\Omega\to\partial\Omega}\vec{u} : (\vec{u},\pi) \in \big[B^{p,q}_{s+\frac{1}{p}}(\Omega)_{\mathrm{bdd}}\big]^n \times B^{p,q}_{s+\frac{1}{p}-1}(\Omega)_{\mathrm{bdd}} \ \textit{satisfying} \\ \Delta\vec{u} - \nabla\pi = 0 \ \textit{and} \ \operatorname{div}\vec{u} = 0 \ \textit{in} \ \Omega, \ \textit{as well as} \\ \vec{u}(x) = O(|x|^{2-n}) \ \textit{and} \ \pi(x) = O(|x|^{1-n}) \\ \textit{as} \ |x| \to \infty \ \textit{if} \ \Omega \ \textit{is an exterior domain}\Big\}, \tag{8.2.123}$$

is a closed linear subspace of finite codimension in $\big[B^{p,q}_s(\partial\Omega,\sigma)\big]^n$. *Third, a solution* $(\vec{u},\pi)$ *of* (8.2.121) *exists if and only if the datum* $\vec{f}$ *belongs to* (8.2.123). *Moreover, if* $p, q \in (1,\infty)$ *and* $s \in (0,1)$, *then for any* f *belonging to* (8.2.123) *a solution* $(\vec{u},\pi)$ *of* (8.2.121) *may be found satisfying, for each cutoff function* $\psi \in \mathscr{C}^\infty_c(\mathbb{R}^n)$,

$$\|\psi\vec{u}\|_{[B^{p,q}_{s+\frac{1}{p}}(\Omega)]^n} + \|\psi\pi\|_{B^{p,q}_{s+\frac{1}{p}-1}(\Omega)} \leq C\|\vec{f}\|_{[B^{p,q}_s(\partial\Omega,\sigma)]^n}, \tag{8.2.124}$$

where $C = C(\Omega, p, q, s, \psi) \in (1,\infty)$ *is a constant independent of* $\vec{f}$.

(2) *If in place of* (8.2.119) *one now assumes*

$$p \in (1,\infty), \quad q \in (0,\infty), \quad s \in (0,1), \tag{8.2.125}$$

then the boundary value problem (8.2.121) *is actually well posed. More specifically,* (8.2.121) *has a unique solution (modulo modulo locally constant functions in* Ω *for the pressure), and a solution may be found which enjoys the additional property stated in* (8.2.124).

(3) *Similar results as in items (1)-(2) above are true for the version of* (8.2.121) *in which a solution is sought on the Triebel-Lizorkin scale, i.e., for the boundary value problem*

$$\begin{cases} \vec{u} \in \big[F^{p,q}_{s+\frac{1}{p}}(\Omega)_{\text{bdd}}\big]^n, \quad \pi \in F^{p,q}_{s+\frac{1}{p}-1}(\Omega)_{\text{bdd}} \\ \Delta\vec{u} - \nabla\pi = 0 \ \textit{and} \ \operatorname{div}\vec{u} = 0 \ \textit{in} \ \Omega, \\ \operatorname{Tr}_{\Omega\to\partial\Omega}\vec{u} = \vec{f} \in \big[B^{p,p}_s(\partial\Omega,\sigma)\big]^n, \\ \vec{u}(x) = O(|x|^{2-n}) \ \textit{and} \ \pi(x) = O(|x|^{1-n}) \\ \textit{as} \ |x| \to \infty \ \textit{if} \ \Omega \ \textit{is an exterior domain}, \end{cases} \tag{8.2.126}$$

(*with the trace operator now considered in the sense of* [130, (9.4.93) in item *(ii)* of Theorem 9.4.5]), *if in place of* (8.2.119) *one now starts by assuming*

$$\tfrac{n-1}{n} < p < \infty, \quad (n-1)\big(\tfrac{1}{p} - 1\big)_+ < s < 1, \quad \tfrac{n}{n+s+1/p} < q < \infty, \tag{8.2.127}$$

and if in place of (8.2.125) *one now assumes*

$$1 < p < \infty, \quad 0 < s < 1, \quad \tfrac{n}{n+s+1/p} < q < \infty. \tag{8.2.128}$$

(4) *Analogous results to those in items (1)-(2) above are also true for the version of* (8.2.121) *in which a solution is sought in a weighted Sobolev space, i.e., for the boundary value problem*[90]

$$\begin{cases} \vec{u} \in \big[W^{1,p}_a(\Omega)_{\text{bdd}}\big]^n, \quad \pi \in L^p_{\text{bdd}}(\Omega, \delta^{ap}_{\partial\Omega}\mathcal{L}^n), \\ \Delta\vec{u} - \nabla\pi = 0 \ \textit{and} \ \operatorname{div}\vec{u} = 0 \ \textit{in} \ \Omega, \\ \operatorname{Tr}_{\Omega\to\partial\Omega}\vec{u} = \vec{f} \in \big[B^{p,p}_s(\partial\Omega,\sigma)\big]^n, \\ \vec{u}(x) = O(|x|^{2-n}) \ \textit{and} \ \pi(x) = O(|x|^{1-n}) \\ \textit{as} \ |x| \to \infty \ \textit{if} \ \Omega \ \textit{is an exterior domain}, \end{cases} \tag{8.2.129}$$

this time demanding throughout that

$$p \in (1,\infty), \quad s \in (0,1), \quad a := 1 - s - \tfrac{1}{p}, \tag{8.2.130}$$

and now considering the boundary trace operator in the sense of [130, Theorem 8.3.6] (*cf.* [130, (8.3.38)]).

Proof This may be proved employing the same strategy as in the proof of Theorem 8.1.20, now making use of the properties of the layer potentials (and conormal derivatives) associated with the Stokes system on Besov, Triebel-Lizorkin, and weighted Sobolev spaces from [132, §6.1], the Fredholmness results from item *(10)* in Theorem 7.1.1, as well as the invertibility results from Theorem 7.1.23 and Theorem 7.1.22. □

---

[90] recall that $\delta_{\partial\Omega}$ denotes the distance function to $\partial\Omega$

We next discuss the inhomogeneous versions of the problems treated in Theorem 8.2.4, i.e., present well-posedness results for the Poisson Problem for the Stokes system in the setting of Ahlfors regular domains with compact boundaries that are sufficiently flat at an infinitesimal level, formulated in the context of Besov/Triebel-Lizorkin/weighted Sobolev spaces.

Theorem 8.2.5 (The Poisson Problem Formulated in Besov, Triebel-Lizorkin, and Weighted Sobolev Spaces) *Fix an integer $n \in \mathbb{N}$ with $n \geq 2$ and suppose*

$$p \in \big(\tfrac{n-1}{n}, \infty\big), \quad q \in (0, \infty), \quad (n-1)\big(\tfrac{1}{p} - 1\big)_+ < s < 1. \tag{8.2.131}$$

Also, let $\Omega \subseteq \mathbb{R}^n$ be a bounded Ahlfors regular domain. Denote by ν the geometric measure theoretic outward unit normal to Ω and abbreviate $\sigma := \mathcal{H}^{n-1}\lfloor\partial\Omega$.

Then there exists a threshold $\delta_0 \in (0,1)$ depending only on n, p, q, s, the ADR *constants of $\partial\Omega$, and the relative flatness ratio $\wp(\Omega)/\operatorname{diam}\partial\Omega$ so that when*

$$\operatorname{dist}\Big(\nu\,,\,[\mathrm{VMO}\,(\partial\Omega,\sigma)]^n\Big) < \delta_0 \tag{8.2.132}$$

with the distance considered in the space $\big[\mathrm{BMO}\,(\partial\Omega,\sigma)\big]^n$ (hence when Ω is a δ_0-oscillating AR *domain*[91]*; cf. Definition 3.4.1), the set Ω is both n-thick*[92] *as well as an (ε,δ)-domain with* $\operatorname{rad}(\Omega) > 0$*, and the following statements are true:*

(1) The Poisson Problem for the Stokes system on Besov spaces

$$\begin{cases} \vec{u} \in \big[B^{p,q}_{s+\frac{1}{p}}(\Omega)\big]^n, \quad \pi \in B^{p,q}_{s+\frac{1}{p}-1}(\Omega), \\ \Delta\vec{u} - \nabla\pi = \vec{f} \in \big[B^{p,q}_{s+\frac{1}{p}-2}(\Omega)\big]^n, \\ \operatorname{div}\vec{u} = g \in B^{p,q}_{s+\frac{1}{p}-1}(\Omega), \\ \operatorname{Tr}_{\Omega\to\partial\Omega}\vec{u} = \vec{h} \in \big[B^{p,q}_{s}(\partial\Omega,\sigma)\big]^n, \end{cases} \tag{8.2.133}$$

becomes well posed when made subject to the (*necessary*) *compatibility condition*

$$\int_{\partial O} \langle \nu, \vec{h}\rangle \, d\sigma = \big\langle g, \mathbf{1}_O \big\rangle \text{ for every component } O \text{ of } \Omega, \tag{8.2.134}$$

with the brackets in the right-hand side denoting duality pairing (cf. [130, Proposition 9.2.29]*). Uniqueness is understood modulo locally constant functions in Ω for the pressure, and any solution satisfies, for some constant $C = C(\Omega, p, q, s, n) \in (0,\infty)$,*

[91] in particular, if Ω is an infinitesimally flat AR domain

[92] hence the trace operator $\operatorname{Tr}_{\Omega\to\partial\Omega}$ considered in [130, (8.3.43)] as well as [130, (9.4.105), (9.4.106)] is meaningful

$$
\begin{aligned}
&\|\vec{u}\|_{[B^{p,q}_{s+\frac{1}{p}}(\Omega)]^n} + \inf_{c} \|\pi - c\|_{B^{p,q}_{s+\frac{1}{p}-1}(\Omega)} \\
&\qquad \leq C\|\vec{f}\|_{[B^{p,q}_{s+\frac{1}{p}-2}(\Omega)]^n} + C\|g\|_{B^{p,q}_{s+\frac{1}{p}-1}(\Omega)} + C\|\vec{h}\|_{[B^{p,q}_{s}(\partial\Omega,\sigma)]^n}
\end{aligned}
\tag{8.2.135}
$$

where the infimum is taken over all locally constant functions c in Ω.

(2) If in place of (8.2.131) *one assumes*

$$
1 < p < \infty, \quad 0 < s < 1, \quad \tfrac{n}{n+s+1/p} < q < \infty, \tag{8.2.136}
$$

then an analogous well-posedness result holds for the Poisson Problem for the Stokes system on the Triebel-Lizorkin scale, i.e., for the problem

$$
\begin{cases}
\vec{u} \in \big[F^{p,q}_{s+\frac{1}{p}}(\Omega)\big]^n, \quad \pi \in F^{p,q}_{s+\frac{1}{p}-1}(\Omega), \\
\Delta\vec{u} - \nabla\pi = \vec{f} \in \big[F^{p,q}_{s+\frac{1}{p}-2}(\Omega)\big]^n, \\
\operatorname{div}\vec{u} = g \in F^{p,q}_{s+\frac{1}{p}-1}(\Omega), \\
\operatorname{Tr}_{\Omega\to\partial\Omega}\vec{u} = \vec{g} \in \big[B^{p,p}_{s}(\partial\Omega,\sigma)\big]^n,
\end{cases}
\tag{8.2.137}
$$

where the data is, once again, made subject to the (necessary) compatibility condition (8.2.134).

(3) If in place of (8.2.131) *one assumes*

$$
1 < p < \infty, \quad 0 < s < 1, \quad a := 1 - s - \tfrac{1}{p}, \tag{8.2.138}
$$

then an analogous well-posedness result holds for the Poisson Problem for the Stokes system on the weighted Sobolev spaces, i.e., for the problem

$$
\begin{cases}
\vec{u} \in \big[W^{1,p}_{a}(\Omega)\big]^n, \quad \pi \in L^p\big(\Omega, \delta^{ap}_{\partial\Omega}\mathcal{L}^n\big), \\
\Delta\vec{u} - \nabla\pi = \vec{f} \in \big[W^{-1,p}_{a}(\Omega)\big]^n, \\
\operatorname{div}\vec{u} = g \in L^p\big(\Omega, \delta^{ap}_{\partial\Omega}\mathcal{L}^n\big), \\
\operatorname{Tr}_{\Omega\to\partial\Omega}\vec{u} = \vec{g} \in \big[B^{p,p}_{s}(\partial\Omega,\sigma)\big]^n,
\end{cases}
\tag{8.2.139}
$$

where the data is, as before, made subject to the (necessary) compatibility condition (8.2.134).

Proof Recall from [132, (6.2.1)] the Kelvin matrix-valued fundamental solution $E = \big(E_{jk}\big)_{1\leq j,k\leq n}$ for the Stokes system in $\mathbb{R}^n$, and from (A.0.212) the pressure vector $\vec{q} = (q_j)_{1\leq j\leq n}$. Consider the integral operators

$$
\Pi\vec{U}(x) := \int_{\mathbb{R}^n} E(x-y)\vec{U}(y)\,dy, \quad \Theta\vec{U}(x) := \int_{\mathbb{R}^n} \langle\vec{q}(x-y), \vec{U}(y)\rangle\,dy, \quad x \in \mathbb{R}^n. \tag{8.2.140}
$$

Then these are smoothing operators of order two and one, respectively, both on the Besov and Triebel-Lizorkin scale. Furthermore,

$$\Delta\Pi - \nabla\Theta = I, \qquad \operatorname{div}\Pi = 0, \qquad \Delta\Pi_\Delta = I, \tag{8.2.141}$$

where I stands for the identity operator. To proceed, consider

$$\vec{\omega} \in \big[B^{p,q}_{s+\frac{1}{p}}(\Omega)\big]^n \text{ with } \operatorname{div}\vec{\omega} = g \text{ in } \Omega \text{ and } \|\vec{\omega}\|_{[B^{p,q}_{s+\frac{1}{p}}(\Omega)]^n} \leq C\|g\|_{B^{p,q}_{s+\frac{1}{p}-1}(\Omega)}, \tag{8.2.142}$$

for some constant $C \in (0,\infty)$ independent of g. For example, the very definition of the Besov scale in the open set Ω ensures that we may extend tg $\in B^{p,q}_{s+\frac{1}{p}-1}(\Omega)$ to a compactly supported distribution $G \in B^{p,q}_{s+\frac{1}{p}-1}(\mathbb{R}^n)$ with norm control, and then take $\vec{\omega} := \nabla\big(\Pi_\Delta G\big)|_\Omega$ where Π_Δ is the harmonic Newtonian potential in $\mathbb{R}^n$ (i.e., the operator of convolution with the standard fundamental solution E_Δ of the Laplacian; cf. from (A.0.77)). Since Π_Δ is smoothing of order two on the Besov scale, all desired properties in (8.2.142) follow. Next, consider

$$\begin{gathered}(\vec{w},\rho) \in \big[B^{p,q}_{s+\frac{1}{p}}(\Omega)\big]^n \oplus B^{p,q}_{s+\frac{1}{p}-1}(\Omega) \text{ with} \\ \Delta\vec{w} - \nabla\rho = \vec{f} - \Delta\vec{\omega} \text{ and } \operatorname{div}\vec{w} = 0 \text{ in } \Omega, \text{ and such that} \\ \|\vec{w}\|_{[B^{p,q}_{s+\frac{1}{p}}(\Omega)]^n} + \|\rho\|_{B^{p,q}_{s+\frac{1}{p}-1}(\Omega)} \leq C\|\vec{f}\|_{[B^{p,q}_{s+\frac{1}{p}-2}(\Omega)]^n} + C\|g\|_{B^{p,q}_{s+\frac{1}{p}-1}(\Omega)}\end{gathered} \tag{8.2.143}$$

where $C \in (0,\infty)$ is independent of $\vec{f}$ and g. For example, we may extend $\vec{f}$, $\vec{\omega}$ to compactly supported distributions in $\mathbb{R}^n$, in a quantitative fashion and with preservation of class on the Besov scale, and then take

$$\vec{w} := \Pi(\vec{f} - \Delta\vec{\omega})\big|_\Omega \text{ and } \rho := \Theta(\vec{f} - \Delta\vec{\omega})\big|_\Omega, \tag{8.2.144}$$

where Π, Θ are as in (8.2.140) with the integral interpreted as duality pairing. We now claim that

$$\operatorname{Tr}_{\Omega\to\partial\Omega}\vec{\omega} + \operatorname{Tr}_{\Omega\to\partial\Omega}\vec{w} - \vec{h} \in \big[B^{p,q}_s(\partial\Omega,\sigma)\big]^n \cap \big\langle \nu\mathbb{C}_{\partial\Omega_+}\big\rangle^\perp. \tag{8.2.145}$$

To see this, observe first that $\operatorname{Tr}_{\Omega\to\partial\Omega}\vec{\omega} + \operatorname{Tr}_{\Omega\to\partial\Omega}\vec{w} - \vec{h} \in \big[B^{p,q}_s(\partial\Omega,\sigma)\big]^n$, thanks to (8.2.142), (8.2.143), and [130, Theorem 9.4.5]. To verify the orthogonality condition in (8.2.145), it suffices to note that for every connected component O of Ω we have

$$\int_{\partial O}\Big\langle \operatorname{Tr}_{\Omega\to\partial\Omega}\vec{\omega} + \operatorname{Tr}_{\Omega\to\partial\Omega}\vec{w},\, \nu\Big\rangle\, d\sigma = \big\langle \operatorname{div}(\vec{\omega}+\vec{w}),\, \mathbf{1}_O\big\rangle = \big\langle g, \mathbf{1}_O\big\rangle = \int_{\partial O}\langle \nu, \vec{h}\rangle\, d\sigma,$$

by the Divergence Theorem (which may be justified here based on the density property [130, (9.2.7)] and a limiting argument), (8.2.142), (8.2.143), and the compatibility condition assumed in (8.2.134). This proves the claim made in (8.2.145).

Let us next assume that $n \geq 3$ and define the operator

$$T:\Big(\big[B^{p,q}_s(\partial\Omega,\sigma)\big]^n\cap\big\langle\nu\mathbb{C}_{\partial\Omega_+}\big\rangle^{\perp}\Big)\oplus\big[B^{p,q}_{s-1}(\partial\Omega,\sigma)\big]^n \qquad (8.2.146)$$
$$\longrightarrow\big[B^{p,q}_s(\partial\Omega,\sigma)\big]^n\cap\big\langle\nu\mathbb{C}_{\partial\Omega_+}\big\rangle^{\perp},$$

by setting, for each $\vec{\phi}_1\in\big[B^{p,q}_s(\partial\Omega,\sigma)\big]^n\cap\big\langle\nu\mathbb{C}_{\partial\Omega_+}\big\rangle^{\perp}$ and $\vec{\phi}_2\in\big[B^{p,q}_{s-1}(\partial\Omega,\sigma)\big]^n$,

$$T(\vec{\phi}_1,\vec{\phi}_2):=\big(\tfrac{1}{2}I+K_{\text{slip}}\big)\vec{\phi}_1+S\vec{\phi}_2 \qquad (8.2.147)$$

where K_{slip} is the boundary-to-boundary double layer for the Stokes system (corresponding to the slip conormal derivative) defined in (4.2.102), and S is the boundary-to-boundary single layer for the Stokes system (cf. (4.2.109)). Based on the properties boundary layer potentials associated with the Stokes system on Besov and Triebel-Lizorkin spaces established in [132, §6.1] (cf. [132, Theorem 6.4.1] and [132, Remark 6.4.4] in particular), and Theorem 7.1.23 we conclude that T is a well-defined, linear, and bounded operator.

We make the claim that T is onto in the context of (8.2.146)-(8.2.147). To justify this claim, we first recall from the very last statement in Theorem 7.1.23 (and (8.2.132)) that the operator

$$\tfrac{1}{2}I+\big[K_{\text{slip}}\big]:\Big(\big[B^{p,q}_s(\partial\Omega,\sigma)\big]^n\cap\big\langle\nu\mathbb{C}_{\partial\Omega_+}\big\rangle^{\perp}\Big)\Big/\Psi(\partial\Omega_-) \qquad (8.2.148)$$
$$\longrightarrow\Big(\big[B^{p,q}_s(\partial\Omega,\sigma)\big]^n\cap\big\langle\nu\mathbb{C}_{\partial\Omega_+}\big\rangle^{\perp}\Big)\Big/\Psi(\partial\Omega_-)$$

is a well-defined linear bounded isomorphism. Keeping this property in mind, it follows that given any $\vec{\phi}\in\big[B^{p,q}_s(\partial\Omega,\sigma)\big]^n\cap\big\langle\nu\mathbb{C}_{\partial\Omega_+}\big\rangle^{\perp}$ we may find some vector-valued function $\vec{\phi}_1\in\big[B^{p,q}_s(\partial\Omega,\sigma)\big]^n\cap\big\langle\nu\mathbb{C}_{\partial\Omega_+}\big\rangle^{\perp}$ such that

$$\vec{\phi}-\big(\tfrac{1}{2}I+K_{\text{slip}}\big)\vec{\phi}_1 \text{ belongs to } \Psi(\partial\Omega_-). \qquad (8.2.149)$$

However, (7.1.420), (7.1.421), (7.1.418), and [130, Theorem 9.4.5] imply that

$$\Psi(\partial\Omega_-)\subseteq\Psi(\partial\Omega)\subseteq\big[B^{p,p}_s(\partial\Omega,\sigma)\big]^n\cap\big\langle\nu\mathbb{C}_{\partial\Omega}\big\rangle^{\perp}, \qquad (8.2.150)$$

while Theorem 7.1.22 implies (bearing in mind (8.2.132) and that we are presently assuming $n\geq 3$) that the operator

$$[S]:\big[B^{p,q}_{s-1}(\partial\Omega,\sigma)\big]^n\Big/\nu\mathbb{C}_{\partial\Omega}\longrightarrow\big[B^{p,q}_s(\partial\Omega,\sigma)\big]^n\cap\big\langle\nu\mathbb{C}_{\partial\Omega}\big\rangle^{\perp} \qquad (8.2.151)$$

is an isomorphism. As such, there exists $\vec{\phi}_2\in\big[B^{p,q}_{s-1}(\partial\Omega,\sigma)\big]^n$ with the property that $S\vec{\phi}_2=\vec{\phi}-\big(\tfrac{1}{2}I+K_{\text{slip}}\big)\vec{\phi}_1$. This shows that $T(\vec{\phi}_1,\vec{\phi}_2)=\vec{\phi}$, ultimately proving that the operator T is, as claimed, surjective. Once this has been established, [130, Proposition 1.2.13] also ensures that there exists a constant $C\in(0,\infty)$, independent of the given $\vec{\phi}$, such that

$$\|\vec{\phi}_1\|_{[B^{p,q}_s(\partial\Omega,\sigma)]^n}+\|\vec{\phi}_2\|_{[B^{p,q}_{s-1}(\partial\Omega,\sigma)]^n}\leq C\|\vec{\phi}\|_{[B^{p,q}_s(\partial\Omega,\sigma)]^n}. \qquad (8.2.152)$$

When used in relation to the vector-valued function given in (8.2.145), these considerations prove that there exists $\vec{h}_1 \in \big[B^{p,q}_s(\partial\Omega,\sigma)\big]^n \cap \big\langle \nu\mathbb{C}_{\partial\Omega_+}\big\rangle^{\perp}$ along with some $\vec{h}_2 \in \big[B^{p,q}_{s-1}(\partial\Omega,\sigma)\big]^n$ such that

$$\begin{aligned}\|\vec{h}_1\|_{[B^{p,q}_s(\partial\Omega,\sigma)]^n} &+ \|\vec{h}_2\|_{[B^{p,q}_{s-1}(\partial\Omega,\sigma)]^n} \\ &\leq C\Big\|\mathrm{Tr}_{\Omega\to\partial\Omega}\,\vec{\omega} + \mathrm{Tr}_{\Omega\to\partial\Omega}\,\vec{w} - \vec{h}\Big\|_{[B^{p,q}_s(\partial\Omega,\sigma)]^n} \\ &\leq C\|\vec{h}\|_{[B^{p,q}_s(\partial\Omega,\sigma)]^n} + C\|\vec{f}\|_{[B^{p,q}_{s+\frac{1}{p}-2}(\Omega)]^n} + C\|g\|_{B^{p,q}_{s+\frac{1}{p}-1}(\Omega)},\end{aligned} \tag{8.2.153}$$

with the last inequality coming from (8.2.142) and (8.2.143), and also

$$\vec{h} - \mathrm{Tr}_{\Omega\to\partial\Omega}\,\vec{\omega} - \mathrm{Tr}_{\Omega\to\partial\Omega}\,\vec{w} = \big(\tfrac{1}{2}I + K_{\mathrm{slip}}\big)\vec{h}_1 + S\vec{h}_2. \tag{8.2.154}$$

Consequently, the pair $(\vec{u},\pi)$ with

$$\vec{u} := \vec{\omega} + \vec{w} + \mathcal{D}_{\mathrm{slip}}\vec{h}_1 + \mathcal{S}\vec{h}_2 \ \text{ and } \ \pi := \rho + \mathcal{P}_{\mathrm{slip}}\vec{h}_1 + Q\vec{h}_2 \ \text{ in } \Omega \tag{8.2.155}$$

is a solution of the Poisson Problem (8.2.133) and satisfies (8.2.135) (thanks to [132, (6.2.15), (6.2.22)], (8.2.154), (8.2.142), (8.2.143), (8.2.153), [132, Theorem 6.4.3], and [132, Remark 6.4.4]).

We may also use the same line of reasoning to prove existence in the case when $n = 2$, with some adjustments. Specifically, in place of the operator T in (8.2.146)-(8.2.147) we now define

$$\begin{aligned}T_{\mathrm{aug}} : \Big(\big[B^{p,p}_s(\partial\Omega,\sigma)\big]^2 \cap \big\langle \nu\mathbb{C}_{\partial\Omega_+}\big\rangle^{\perp}\Big) \oplus \big[B^{p,p}_{s-1}(\partial\Omega,\sigma)\big]^2 \oplus \mathbb{C}^2 \\ \longrightarrow \big[B^{p,p}_s(\partial\Omega,\sigma)\big]^2 \cap \big\langle \nu\mathbb{C}_{\partial\Omega_+}\big\rangle^{\perp},\end{aligned} \tag{8.2.156}$$

defined as

$$T_{\mathrm{aug}}\big(\vec{\phi}_1,\vec{\phi}_2,\vec{c}\big) := \big(\tfrac{1}{2}I + K_{\mathrm{slip}}\big)\vec{\phi}_1 + S\vec{\phi}_2 + c \tag{8.2.157}$$

for each $\vec{\phi}_1 \in \big[B^{p,p}_s(\partial\Omega,\sigma)\big]^2 \cap \big\langle \nu\mathbb{C}_{\partial\Omega_+}\big\rangle^{\perp}$, each $\vec{\phi}_2 \in \big[B^{p,p}_{s-1}(\partial\Omega,\sigma)\big]^2$, and each $c \in \mathbb{C}^2$. Much as before, this is a well-defined, linear, and continuous mapping, and using the invertibility of (7.1.411) we can show that T_{aug} is also surjective. With this in hand, the rest of the proof of existence proceeds as in the past.

Finally, in all space dimensions $n \geq 2$, uniqueness (modulo locally constant functions in Ω for the pressure) is seen from the corresponding uniqueness property in item *(2)* of Theorem 8.2.4 and embedding results (so as to ensure that the new indices satisfy the conditions stated in (8.2.125)). This completes the treatment of item *(1)*. For the Triebel-Lizorkin scale considered in item *(2)* a very similar approach works as well. The same applies to the scale of weighted Sobolev scale from item *(3)*, now also bearing in mind [129, Proposition 8.7.4]. □

We conclude with a remark highlighting other significant boundary value problems for the Stokes system in domains with compact, infinitesimally flat boundaries, which our present techniques can easily handle.

Remark 8.2.6 *In view of the Fredholm and invertibility results for hydrostatic layer potentials established in Theorem 7.1.21, Theorem 7.1.23, Theorem 7.1.47, Theorem 7.1.49, item (10) in Theorem 7.1.1, and Theorem 7.1.22, as well as the Calderón-Zygmund theory for boundary layer operators associated with the Stokes system developed in* [132, Theorem 6.2.4], [132, Theorem 6.2.6], [132, Theorem 6.2.11], *and also the properties of the layer potentials for the Stokes system on Besov, Triebel-Lizorkin, and weighted Sobolev spaces obtained in* [132, §6.1], *similar solvability results to those presented in Theorem 8.1.11 for the Dirichlet Problem with boundary data in* BMO *and* VMO, *in Theorem 8.1.18 for the Dirichlet Problem with data in ordinary Hölder spaces $\mathscr{C}^\alpha$ with $0 < \alpha < 1$, in Theorem 8.1.19 for the Dirichlet Problem with data in vanishing Hölder spaces $\mathscr{C}^\alpha_{\mathrm{van}}$ with $0 < \alpha < 1$, in Theorem 8.1.22 for the Neumann Problem with data in Besov spaces, in Theorem 8.1.23 for the Inhomogeneous Neumann Problem with boundary data in Besov spaces, and also for the Transmission Problem with data in Lebesgue, Hardy, and Generalized Banach Function Spaces, are all valid for the Stokes system, in the same geometric settings.*

8.3 Boundary Value Problems in Scattering Theory

Here we are concerned with boundary value problems in scattering theory by rough obstacles in $\mathbb{R}^2$ with $n \geq 2$. In this regard, it is worth mentioning that the issue of smoothness (or rather lack thereof) for the underlying domain has played an important role in scattering theory. The classical theory typically requires these domains to be at least of class $\mathscr{C}^2$. See [27], [164], [165], [183], [206], as well as the discussion in the monographs [34], [35]. Subsequently, boundary value problems for the Helmholtz operator and the Maxwell system in Lipschitz domains have been treated in [150], [84], [151], [137], [141], [152], [139], [112]. From this perspective, our goal is to initiate a program aimed at developing scattering theory in more general domains which, in particular, may not necessarily be locally described as upper-graphs of functions.

To set the stage, consider $n \in \mathbb{N}$ satisfying $n \geq 2$, and fix an arbitrary wave number $k \in (0, \infty)$. Recall that the (only) radiating fundamental solution of the Helmholtz operator $\Delta + k^2$ in $\mathbb{R}^n$ which is defined at each $x \in \mathbb{R}^n \setminus \{0\}$ as

$$\Phi_k(x) := c_n k^{(n-2)/2} \frac{H^{(1)}_{(n-2)/2}(k|x|)}{|x|^{(n-2)/2}}, \quad \text{with } c_n := \frac{1}{4\mathrm{i}(2\pi)^{(n-2)/2}}. \tag{8.3.1}$$

Here, $H^{(1)}_\lambda(\cdot)$ is the Hankel function of the first kind with index $\lambda \in \mathbb{R}$ (cf. [1, § 9.1]).

The first order of business is to discuss a higher dimensional version of Rellich's uniqueness lemma. Indeed, one of the most striking differences between the null-solutions of the Helmholtz operator $\Delta + k^2$ with wave number $k > 0$, on the one hand, and the null-solutions of the Laplace operator (viewed as a limiting case of the Helmholtz operator $\Delta + k^2$ corresponding to $k = 0$), on the other hand, is the nature of their behavior at infinity. Specifically, given any exterior proper domain in $\mathbb{R}^n$, there is a nonzero harmonic function which decays faster than any a priori prescribed polynomial rate at infinity (as seen by taking derivatives of sufficiently large order of the fundamental solution for the Laplacian, appropriately shifted so that its singularity lies outside the given domain). By way of contrast, a null-solution of the Helmholtz operator $\Delta + k^2$ with $k \in (0, \infty)$ in an exterior domain in $\mathbb{R}^n$ cannot exhibit a faster decay than the critical threshold $O(|x|^{-(n-1)/2})$ as $|x| \to \infty$ without vanishing identically near infinity. The latter result makes the object of the classical Rellich lemma. This was originally proved in [179], under stronger assumptions. See also [103], [104] in this regard. For the convenience of the reader, we provide a self-contained proof of the n-dimensional version stated below.

Lemma 8.3.1 (Rellich's lemma [179]) *Fix $k \in (0, \infty)$ along with $n \in \mathbb{N}$, $n \geq 2$, and suppose Ω is an exterior domain in $\mathbb{R}^n$. Assume $u \in \mathscr{C}^\infty(\Omega)$ satisfies $(\Delta + k^2)u = 0$ in Ω as well as*

$$\int_{|x|=R} |u(x)|^2 \, d\mathcal{H}^{n-1}(x) = o(1) \text{ as } R \to \infty. \tag{8.3.2}$$

Then u vanishes identically in the unbounded component of Ω.

Proof In a first stage, suppose Ω is the complement of the closure of a ball centered at the origin in $\mathbb{R}^n$. Let $\{Y_{j,l}\}_{j,l}$ be an orthonormal basis in $L^2(S^{n-1}, \mathcal{H}^{n-1})$ consisting of spherical harmonics. In particular, if $\Delta_{S^{n-1}}$ denotes the Laplace-Beltrami operator on the unit sphere $S^{n-1} \subseteq \mathbb{R}^n$, then for each j, l we have

$$\Delta_{S^{n-1}} Y_{j,l} = -j(j + n - 2) Y_{j,l} \text{ on } S^{n-1}. \tag{8.3.3}$$

Next, if $r > 0$ is large enough (so that the sphere $|x| = r$ is included in the set Ω), it follows that the function $S^{n-1} \ni \omega \mapsto u(r\omega) \in \mathbb{C}$ is well defined and continuous. Expanding this in spherical harmonics then yields the representation

$$u(x) = \sum \sum a_{j,l}(|x|) Y_{j,l}(\widehat{x}), \text{ for } |x| \text{ large}, \tag{8.3.4}$$

where $\widehat{x} := x/|x|$ and for each j, l we have set

$$a_{j,l}(r) := \int_{S^{n-1}} u(r\omega) \overline{Y_{j,l}(\omega)} \, d\mathcal{H}^{n-1}(\omega) \text{ for all } r > 0 \text{ sufficiently large}. \tag{8.3.5}$$

Recall that the Laplace operator in $\mathbb{R}^n$ is related to the Laplacian on the unit sphere $S^{n-1} \subseteq \mathbb{R}^n$ via

$$\Delta_{\mathbb{R}^n} = \frac{\partial^2}{\partial r^2} + \frac{n-1}{r} \frac{\partial}{\partial r} + \frac{1}{r^2} \Delta_{S^{n-1}}. \tag{8.3.6}$$

Fix j, l arbitrary. Based on the eigenfunction property (8.3.3), integration by parts on the unit sphere, (8.3.6), and the fact that $\Delta_{\mathbb{R}^n} u = -k^2 u$, for $r > 0$ sufficiently large we may compute

$$\begin{aligned}
-\frac{j(j+n-2)}{r^2} a_{j,l}(r) &= \frac{1}{r^2} \int_{S^{n-1}} u(r\omega) \overline{(\Delta_{S^{n-1}} Y_{j,l})(\omega)} \, \mathrm{d}\mathcal{H}^{n-1}(\omega) \\
&= \frac{1}{r^2} \int_{S^{n-1}} \Delta_{S^{n-1}} \big[u(r\omega)\big] \overline{Y_{j,l}(\omega)} \, \mathrm{d}\mathcal{H}^{n-1}(\omega) \\
&= \int_{S^{n-1}} \Big[-k^2 - \frac{\partial^2}{\partial r^2} - \frac{n-1}{r} \frac{\partial}{\partial r} \Big] \big[u(r\omega)\big] \overline{Y_{j,l}(\omega)} \, \mathrm{d}\mathcal{H}^{n-1}(\omega) \\
&= \Big[-k^2 - \frac{\mathrm{d}^2}{\mathrm{d}r^2} - \frac{n-1}{r} \frac{\mathrm{d}}{\mathrm{d}r} \Big] \int_{S^{n-1}} u(r\omega) \overline{Y_{j,l}(\omega)} \, \mathrm{d}\mathcal{H}^{n-1}(\omega) \\
&= \Big[-k^2 - \frac{\mathrm{d}^2}{\mathrm{d}r^2} - \frac{n-1}{r} \frac{\mathrm{d}}{\mathrm{d}r} \Big] a_{j,l}(r). \qquad (8.3.7)
\end{aligned}$$

This proves that each $a_{j,l}$ satisfies the following ODE in the variable r:

$$\frac{\mathrm{d}^2 a_{j,l}}{\mathrm{d}r^2} + \frac{n-1}{r} \frac{\mathrm{d}a_{j,l}}{\mathrm{d}r} + \left(k^2 - \frac{j(j+n-2)}{r^2} \right) a_{j,l} = 0. \qquad (8.3.8)$$

To proceed, we find it convenient to introduce, for each j, l,

$$b_{j,l}(r) := \left(\frac{r}{k} \right)^{\frac{n-2}{2}} \cdot a_{j,l}(r/k), \quad \text{for all } r > 0 \text{ sufficiently large.} \qquad (8.3.9)$$

A direct computation then shows that each $b_{j,l}$ solves Bessel's differential equation

$$r^2 \frac{\mathrm{d}^2 y}{\mathrm{d}r^2} + r \frac{\mathrm{d}y}{\mathrm{d}r} + \left(r^2 - \lambda^2 \right) y = 0 \text{ if } \lambda := \big[(n/2-1)^2 + j(j+n-2) \big]^{1/2}. \quad (8.3.10)$$

As is well known (cf., e.g., [32, (20)-(23), p. 96]) the solutions of (8.3.10) are linear combinations of regular and irregular Bessel functions J_λ and Y_λ, themselves related to the Hankel function of the first kind with index $\lambda \in \mathbb{R}$ (cf. [1, § 9.1]) via

$$H_\lambda^{(1)}(\cdot) = J_\lambda(\cdot) + \mathrm{i} Y_\lambda(\cdot). \qquad (8.3.11)$$

Thus, for each j, l there exist $c_{j,l}^{(1)}, c_{j,l}^{(2)} \in \mathbb{C}$ with the property that for all $r > 0$ sufficiently large we may express

$$b_{j,l}(r) = c_{j,l}^{(1)} J_\lambda(r) + c_{j,l}^{(2)} Y_\lambda(r). \qquad (8.3.12)$$

In turn, from (8.3.9) and (8.3.12) we deduce that for all $r > 0$ sufficiently large we have

$$a_{j,l}(r) = r^{-(n-2)/2} \big\{ c_{j,l}^{(1)} J_\lambda(kr) + c_{j,l}^{(2)} Y_\lambda(kr) \big\} \qquad (8.3.13)$$

for each j, l. As a consequence of this, the fact that (as is apparent from (8.3.11))

$$\operatorname{Re} H_\lambda^{(1)}(r) = J_\lambda(r) \text{ and } \operatorname{Im} H_\lambda^{(1)}(r) = Y_\lambda(r), \qquad \forall\, r > 0, \tag{8.3.14}$$

and the well-known asymptotic formula (cf., e.g., [170])

$$H_\lambda^{(1)}(r) = \Big(\frac{2}{\pi r}\Big)^{1/2} e^{\mathrm{i}(r-\lambda\pi/2-\pi/4)} + O(r^{-3/2}) \text{ as } r \to \infty, \tag{8.3.15}$$

we conclude that each coefficient function $a_{j,l}$ has the asymptotic behavior

$$\begin{aligned} a_{j,l}(r) = r^{-(n-1)/2}\Big(\frac{2}{\pi k}\Big)^{1/2}\big\{c_{j,l}^{(1)}\cos(kr-\lambda\pi/2-\pi/4) \\ + c_{j,l}^{(2)}\sin(kr-\lambda\pi/2-\pi/4)\big\} + O\big(r^{-(n+1)/2}\big) \text{ as } r \to \infty. \end{aligned} \tag{8.3.16}$$

On the other hand, Parseval's identity and a change of variables give

$$\sum\sum |a_{j,l}(r)|^2 = \int_{S^{n-1}} |u(r\omega)|^2 \,\mathrm{d}\mathcal{H}^{n-1}(\omega) = r^{1-n}\int_{|x|=r} |u(x)|^2 \,\mathrm{d}\mathcal{H}^{n-1}(x). \tag{8.3.17}$$

Thanks to (8.3.2), this implies that $r^{n-1}|a_{j,l}(r)|^2 = o(1)$ as $r \to \infty$, or

$$a_{j,l}(r) = o\big(r^{-(n-1)/2}\big) \text{ as } r \to \infty. \tag{8.3.18}$$

Comparing (8.3.18) with (8.3.16) then reveals that, as $r \to \infty$,

$$c_{j,l}^{(1)}\cos(kr-\lambda\pi/2-\pi/4) + c_{j,l}^{(2)}\sin(kr-\lambda\pi/2-\pi/4) = o(1). \tag{8.3.19}$$

Upon letting r approach infinity via two suitably chosen sequences then forces $c_{j,l}^{(1)} = c_{j,l}^{(2)} = 0$. Utilizing this back in (8.3.16) finally yields $a_{j,l} = 0$ for each j, l. This proves that u vanishes identically in the case when Ω is the complement of the closure of a ball centered at the origin in $\mathbb{R}^n$.

The case when Ω is an arbitrary exterior domain is dealt with based on what we have just proved and unique continuation. More specifically, it is well known that any null-solution u of the Helmholtz operator $\Delta + k^2$ in an open set $O \subseteq \mathbb{R}^n$ is real-analytic. In particular, this implies that whenever $u \equiv 0$ in a nonempty open subset O' of O then u vanishes identically in any connected component of O which intersects O'. From this, the desired conclusion readily follows. □

Moving on, suppose $\Omega \subseteq \mathbb{R}^n$ is an Ahlfors regular domain with compact boundary. Abbreviate $\sigma := \mathcal{H}^{n-1}\lfloor\partial\Omega$ and denote by $\nu = (\nu_1, \dots, \nu_n)$ the geometric measure theoretic outward unit normal to Ω. Then the action of the boundary-to-domain (scalar) acoustic single layer potential operator and of the acoustic double layer potential operator on any function $f \in L^1(\partial\Omega, \sigma)$ is given by

$$\mathcal{S}_k f(x) := \int_{\partial\Omega} \Phi_k(x-y) f(y)\,\mathrm{d}\sigma(y) \text{ for all } x \in \mathbb{R}^n \setminus \partial\Omega, \tag{8.3.20}$$

and, respectively,

$$\mathcal{D}_k f(x) := \int_{\partial\Omega} \partial_{\nu(y)}\big[\Phi_k(x-y)\big] f(y)\,d\sigma(y) \text{ for all } x \in \mathbb{R}^n \setminus \partial\Omega. \tag{8.3.21}$$

In the same setting, the action of the principal-value boundary-to-boundary acoustic double layer potential operator on any function $f \in L^1(\partial\Omega, \sigma)$ is defined as

$$K_k f(x) := \lim_{\varepsilon\to 0^+} \int_{y\in\partial\Omega,\,|x-y|>\varepsilon} \partial_{\nu(y)}\big[\Phi_k(x-y)\big] f(y)\,d\sigma(y) \tag{8.3.22}$$

at σ-a.e. $x \in \partial\Omega$. In addition, define the "transpose" acoustic double later by setting, for each $f \in L^1(\partial\Omega, \sigma)$,

$$K_k^{\#} f(x) := \lim_{\varepsilon\to 0^+} \int_{y\in\partial\Omega,\,|x-y|>\varepsilon} \partial_{\nu(x)}\big[\Phi_k(x-y)\big] f(y)\,d\sigma(y), \tag{8.3.23}$$

at σ-a.e. $x \in \partial\Omega$. Finally, consider the boundary-to-boundary single layer potential operator, acting on each given function $f \in L^1(\partial\Omega, \sigma)$ according to

$$S_k f(x) := \int_{\partial\Omega} \Phi_k(x-y) f(y)\,d\sigma(y) \text{ for } \sigma\text{-a.e. } x \in \partial\Omega. \tag{8.3.24}$$

Throughout, we let K_0, $K_0^{\#}$, S_0 be the versions of K_k, $K_k^{\#}$, S_k formally corresponding to $k = 0$, a scenario in which Φ_k is replaced in (8.3.20)-(8.3.24) by $\Phi_0 = E_\Delta$, the standard fundamental solution for the Laplacian in $\mathbb{R}^n$ (cf. [132, (1.1.1)]).

Lemma 8.3.2 *Let $\Omega \subseteq \mathbb{R}^n$, where $n \in \mathbb{N}$ with $n \geq 2$, be a* UR *domain with a compact boundary, and abbreviate $\sigma := \mathcal{H}^{n-1}\lfloor\partial\Omega$. Also, pick an arbitrary wave number $k \in (0, \infty)$ and an integrability exponent $p \in (1, \infty)$. Then the operator*

$$K_k - K_0 : L^p(\partial\Omega, \sigma) \longrightarrow L_1^p(\partial\Omega, \sigma) \text{ is compact.} \tag{8.3.25}$$

Proof This is a consequence of definitions and [131, Lemma 6.3.1]. □

Next, we deal with invertibility properties of acoustic boundary layer potential operators.

Theorem 8.3.3 *Let $n \in \mathbb{N}$ satisfying $n \geq 2$ and pick an arbitrary wave number $k \in (0, \infty)$. Also, suppose $\Omega \subset \mathbb{R}^n$ is a connected exterior Ahlfors regular domain. Abbreviate $\sigma := \mathcal{H}^{n-1}\lfloor\partial\Omega$ and denote by ν the geometric measure theoretic outward unit normal to Ω. Finally, select an integrability exponent $p \in (1, \infty)$.*

Then there exists a threshold $\delta \in (0, 1)$ depending only on n, p, k, the ADR *constants of $\partial\Omega$, and the relative flatness ratio $\wp(\Omega)/\operatorname{diam}\partial\Omega$ so that when*

$$\operatorname{dist}\Big(\nu,\ [\mathrm{VMO}(\partial\Omega, \sigma)]^n\Big) < \delta \tag{8.3.26}$$

with the distance considered in the space $\big[\mathrm{BMO}\,(\partial\Omega,\sigma)\big]^n$ *(hence when* Ω *is a* δ*-oscillating* AR *domain*[93]*; cf. Definition 3.4.1), it follows that the set* Ω *is a* UR *domain, and the operators below are invertible:*

$$\tfrac{1}{2}I + K_k - \mathrm{i}S_k : L^p(\partial\Omega,\sigma) \longrightarrow L^p(\partial\Omega,\sigma), \tag{8.3.27}$$

$$\tfrac{1}{2}I + K_k - \mathrm{i}S_k : L^p_1(\partial\Omega,\sigma) \longrightarrow L^p_1(\partial\Omega,\sigma). \tag{8.3.28}$$

Proof Take $\delta \in (0,1)$ small enough so that the conclusions in Theorem 3.4.2 and Theorem 7.1.1 (stated for $L = \Delta$ and with $z := 1/2$) are valid. Write

$$\tfrac{1}{2}I + K_k - \mathrm{i}S_k = \tfrac{1}{2}I + K_0 + \big\{(K_k - K_0) - \mathrm{i}S_k\big\} \tag{8.3.29}$$

and note that, by Lemma 8.3.2, [132, Theorem 1.5.5], and [130, Theorem 11.5.18, item *(iii)*], the expression in the curly brackets is a compact operator, both from $L^p(\partial\Omega,\sigma)$ into itself, and from $L^p_1(\partial\Omega,\sigma)$ into itself. Thanks to Theorem 7.1.1 and standard Fredholm theory, we then conclude that the operators (8.3.27)-(8.3.28) are Fredholm with index zero, both in the original context and for the choice of integrability index $p = 2$. Granted this, the abstract regularity result from [130, Lemma 2.1.4] (whose present applicability also relies on the density results from [129, (3.1.13)] and [130, Theorem 11.5.18, item *(i)*]) shows that in order to conclude that the operators (8.3.27)-(8.3.28) are invertible it suffices to prove that

$$\tfrac{1}{2}I + K_k - \mathrm{i}S_k : L^2_1(\partial\Omega,\sigma) \longrightarrow L^2_1(\partial\Omega,\sigma) \text{ is injective.} \tag{8.3.30}$$

To this end, suppose $f \in L^2_1(\partial\Omega,\sigma)$ is such that

$$(\tfrac{1}{2}I + K_k - \mathrm{i}S_k)f = 0. \tag{8.3.31}$$

Also, fix a background constant $\kappa \in (0,\infty)$, serving the role of the aperture parameter, and fix some (height) truncation parameter $\rho \in (0,\infty)$. If we define

$$u := \mathcal{D}_k f - \mathrm{i}\mathcal{S}_k f \text{ in } \Omega, \tag{8.3.32}$$

then based on the properties of the acoustic boundary layer potentials from [132, Theorem 1.5.5] and (8.3.31) we see that

$$\begin{cases} u \in \mathscr{C}^\infty(\Omega), \quad (\Delta + k^2)u = 0 \text{ in } \Omega, \\ \mathcal{N}^\rho_\kappa u \in L^2(\partial\Omega,\sigma), \quad \mathcal{N}^\rho_\kappa(\nabla u) \in L^2(\partial\Omega,\sigma), \\ u\big|^{\kappa-\mathrm{n.t.}}_{\partial\Omega} = 0 \text{ at } \sigma\text{-a.e. point belonging to } \partial\Omega, \\ u \text{ satisfies Sommerfeld's radiation condition.} \end{cases} \tag{8.3.33}$$

To proceed, pick $R \in (0,\infty)$ sufficiently large so that $\partial\Omega \subseteq B(0,R)$ and define $D_R := B(0,R) \cap \Omega$. Also, recall that $\widehat{x} := x/|x|$ for each $x \in \mathbb{R}^n \setminus \{0\}$. Then

[93] in particular, if Ω is an infinitesimally flat AR domain

$$\int_{|x|=R} \overline{u} \cdot \partial_{\vec{x}} u \, d\mathcal{H}^{n-1} = \int_{\partial D_R} \Big(\overline{u}\Big|_{\partial\Omega}^{\kappa-\mathrm{n.t.}}\Big) \cdot \partial_\nu u \, d\mathcal{H}^{n-1} = \int_{D_R} \Big\{ |\nabla u|^2 - k^2 |u|^2 \Big\} \, d\mathcal{L}^n, \tag{8.3.34}$$

thanks to (8.3.33) and the Divergence Theorem, in the version recorded in [129, Theorem 1.2.1], applied to the vector field $\vec{F} := \overline{u} \cdot \nabla u$ in the set D_R. In particular, (8.3.34) implies that

$$\int_{|x|=R} \mathrm{Im}\, \big(\overline{u} \cdot \partial_{\vec{x}} u\big) \, d\mathcal{H}^{n-1} = 0. \tag{8.3.35}$$

Keeping this in mind we then compute

$$\begin{aligned}
&\int_{|x|=R} \Big\{ |\partial_{\vec{x}} u|^2 + k^2 |u|^2 \Big\} \, d\mathcal{H}^{n-1} \\
&\quad = \int_{|x|=R} \Big\{ |\partial_{\vec{x}} u|^2 + k^2 |u|^2 + 2k \cdot \mathrm{Im}\, \big(\overline{u} \cdot \partial_{\vec{x}} u\big) \Big\} \, d\mathcal{H}^{n-1} \\
&\quad = \int_{|x|=R} \big|\partial_{\vec{x}} u - \mathrm{i} k u\big|^2 \, d\mathcal{H}^{n-1} = o(1) \ \text{ as } \ R \to \infty,
\end{aligned} \tag{8.3.36}$$

since u satisfies the Sommerfeld radiation condition. In turn, (8.3.36) forces

$$\int_{|x|=R} |u|^2 \, d\mathcal{H}^{n-1} = o(1) \ \text{ as } \ R \to \infty \tag{8.3.37}$$

which, in concert with Rellich's uniqueness lemma (recorded in Lemma 8.3.1), ultimately shows that $u \equiv 0$ in the connected set Ω.

At this stage, bring in the function

$$w := \mathcal{D}_k f - \mathrm{i} \mathcal{S}_k f \ \text{ in } \ \Omega_- := \mathbb{R}^n \setminus \overline{\Omega}, \tag{8.3.38}$$

and note that this satisfies

$$\begin{cases}
w \in \mathscr{C}^\infty(\Omega_-), \quad (\Delta + k^2) w = 0 \ \text{ in } \ \Omega_-, \\
\mathcal{N}_\kappa w \in L^2(\partial\Omega, \sigma), \quad \mathcal{N}_\kappa(\nabla w) \in L^2(\partial\Omega, \sigma), \\
w\Big|_{\partial\Omega_-}^{\kappa-\mathrm{n.t.}} \ \text{exists at } \sigma\text{-a.e. point on } \partial\Omega, \\
(\nabla w)\Big|_{\partial\Omega_-}^{\kappa-\mathrm{n.t.}} \ \text{exists at } \sigma\text{-a.e. point on } \partial\Omega.
\end{cases} \tag{8.3.39}$$

In fact, since u vanishes identically in Ω, from the trace formulas for acoustic boundary layer potentials proved in [132, Theorem 1.5.5] we obtain

$$w\Big|_{\partial\Omega_-}^{\kappa-\mathrm{n.t.}} = w\Big|_{\partial\Omega_-}^{\kappa-\mathrm{n.t.}} - u\Big|_{\partial\Omega}^{\kappa-\mathrm{n.t.}} = -f \ \text{ on } \ \partial\Omega, \tag{8.3.40}$$

$$\text{and } \ \partial_\nu w = \partial_\nu w - \partial_\nu u = \mathrm{i} f \ \text{ on } \ \partial\Omega. \tag{8.3.41}$$

Let us also note that (8.3.39) and [129, Proposition 8.6.3] imply

$$w, \ |\nabla w| \in L^{2n/(n-1)}(\Omega_-, \mathcal{L}^n) \hookrightarrow L^2(\Omega_-, \mathcal{L}^n), \tag{8.3.42}$$

since Ω_- is bounded. Based on (8.3.39), (8.3.40), (8.3.41), and the version of Green's first identity from [131, Theorem 1.7.12] written for $L := \Delta$ (whose applicability in the present setting uses (8.3.42)) we obtain

$$\begin{aligned}\int_{\Omega_-} |\nabla w|^2 \, d\mathcal{L}^n &= -\int_{\partial\Omega} \big(\overline{w}\big|_{\partial\Omega}^{\kappa-\mathrm{n.t.}}\big) \cdot \partial_\nu w \, d\sigma + k^2 \int_{\Omega_-} |w|^2 \, d\mathcal{L}^n \\ &= \mathrm{i} \int_{\partial\Omega} |f|^2 \, d\sigma + k^2 \int_{\Omega_-} |w|^2 \, d\mathcal{L}^n. \qquad (8.3.43)\end{aligned}$$

Taking imaginary parts then yields $\int_{\partial\Omega} |f|^2 \, d\sigma = 0$, hence $f = 0$. This establishes (8.3.30) and finishes the proof of Theorem 8.3.3. □

The invertibility results from Theorem 8.3.3 are a key ingredient in the construction of a Green function for the Helmholtz operator in an exterior domain.

Theorem 8.3.4 *Let $n \in \mathbb{N}$ satisfying $n \geq 2$ and pick an arbitrary wave number $k \in (0, \infty)$. Also, suppose $\Omega \subset \mathbb{R}^n$ is a connected exterior Ahlfors regular domain. Abbreviate $\sigma := \mathcal{H}^{n-1} \lfloor \partial\Omega$ and denote by ν the geometric measure theoretic outward unit normal to Ω. Finally, select an integrability exponent $q \in (1, \infty)$.*

Then there exists a threshold $\delta \in (0, 1)$ depending only on n, q, k, the ADR *constants of $\partial\Omega$, and the relative flatness ratio $\wp(\Omega)/\mathrm{diam}\, \partial\Omega$ so that when*

$$\mathrm{dist}\Big(\nu, [\mathrm{VMO}(\partial\Omega, \sigma)]^n\Big) < \delta \qquad (8.3.44)$$

with the distance considered in the space $\big[\mathrm{BMO}(\partial\Omega, \sigma)\big]^n$ (hence when Ω is a δ-oscillating AR *domain*[94]*; cf. Definition 3.4.1), it follows that for each point $x_0 \in \Omega$ there exists a function G^{x_0} such that, for each truncation $\rho \in \big(0, \frac{1}{2}\mathrm{dist}(x_0, \partial\Omega)\big)$ and each aperture parameter $\kappa \in (0, \infty)$, one has:*

$$\begin{cases} G^{x_0} \in \mathscr{C}^\infty(\Omega \setminus \{x_0\}) \cap W^{1,1}_{\mathrm{loc}}(\Omega, \mathcal{L}^n), \\ (\Delta + k^2) G^{x_0} = -\delta_{x_0} \ \text{ in } \mathcal{D}'(\Omega), \\ \mathcal{N}^\rho_\kappa G^{x_0} \in L^q(\partial\Omega, \sigma), \quad \mathcal{N}^\rho_\kappa(\nabla G^{x_0}) \in L^q(\partial\Omega, \sigma), \\ G^{x_0}\big|_{\partial\Omega}^{\kappa-\mathrm{n.t.}} = 0 \ \text{ at } \sigma\text{-a.e. point on } \partial\Omega, \\ (\nabla G^{x_0})\big|_{\partial\Omega}^{\kappa-\mathrm{n.t.}} \ \text{ exists at } \sigma\text{-a.e. point on } \partial\Omega, \\ G^{x_0}(x) = O\big(|x|^{-(n-1)/2}\big) \ \text{ as } |x| \to \infty, \\ G^{x_0} \ \text{satisfies Sommerfeld's radiation condition.} \end{cases} \qquad (8.3.45)$$

Proof Fix an arbitrary point $x_0 \in \Omega$ and take $\rho \in \big(0, \frac{1}{2}\mathrm{dist}(x_0, \partial\Omega)\big)$. Choose $\delta \in (0, 1)$ so that the conclusion in Theorem 8.3.3 holds for the given integrability exponent $q \in (1, \infty)$. Then the operator $\frac{1}{2}I + K_k - \mathrm{i}S_k$ is invertible on $L^q_1(\partial\Omega, \sigma)$, and since $\Phi_k(\cdot - x_0)\big|_{\partial\Omega}$ belongs to the latter space it makes sense to define

[94] in particular, if Ω is an infinitesimally flat AR domain

$$w := \big(\mathcal{D}_k - \mathrm{i}\mathcal{S}_k\big)\Big[\big(\tfrac{1}{2}I + K_k - \mathrm{i}S_k\big)^{-1}\Phi_k(\cdot - x_0)\big|_{\partial\Omega}\Big] \text{ in } \Omega. \tag{8.3.46}$$

Constructed as such, this function enjoys the following properties:

$$\begin{cases} w \in \mathscr{C}^\infty(\Omega), \quad (\Delta + k^2)w = 0 \text{ in } \Omega, \\ \mathcal{N}_\kappa^\rho w \in L^q(\partial\Omega, \sigma), \quad \mathcal{N}_\kappa^\rho(\nabla w) \in L^q(\partial\Omega, \sigma), \\ w\big|_{\partial\Omega}^{\kappa-\mathrm{n.t.}} = \Phi_k(\cdot - x_0)\big|_{\partial\Omega} \text{ at } \sigma\text{-a.e. point on } \partial\Omega, \\ (\nabla w)\big|_{\partial\Omega}^{\kappa-\mathrm{n.t.}} \text{ exists at } \sigma\text{-a.e. point on } \partial\Omega, \\ w \text{ satisfies Sommerfeld's radiation condition,} \\ w(x) = O\big(|x|^{-(n-1)/2}\big) \text{ as } |x| \to \infty, \end{cases} \tag{8.3.47}$$

with the last property coming from [131, Lemma 6.3.2]. Hence, if we now set

$$G^{x_0}(x) := w(x) - \Phi_k(x - x_0) \text{ for each } x \in \Omega, \tag{8.3.48}$$

then G^{x_0} satisfies all stipulations in (8.3.45) (here again [131, Lemma 6.3.2] is relevant). □

Theorem 8.3.4 is an essential tool in dealing with the issue of uniqueness for the Dirichlet Problem for the Helmholtz operator in an exterior domain. In preparation to stating and proving such a result, below we record a useful version of interior estimates for the Laplacian but involving functions which are not necessarily harmonic.

Lemma 8.3.5 *Let $n \in \mathbb{N}$ satisfying $n \geq 2$. Pick an arbitrary point $x_0 \in \mathbb{R}^n$ along with some radius $r \in (0, \infty)$, and suppose $u \in W^{2,2}(B(x_0, 2r))$. Then*

$$\sup_{B(x_0,r)} |\nabla u| \leq \frac{C}{r} \fint_{B(x_0,2r)} |u| \, d\mathcal{L}^n + C_n r \cdot \sup_{B(x_0,2r)} |\Delta u| \textit{ if } n \geq 3, \tag{8.3.49}$$

and

$$\sup_{B(x_0,r)} |\nabla u| \leq \frac{C}{r} \fint_{B(x_0,2r)} |u| \, d\mathcal{L}^2 + Cr(1 + |\ln r|) \cdot \sup_{B(x_0,2r)} |\Delta u| \textit{ if } n = 2. \tag{8.3.50}$$

Proof Denote by $\Pi_{x_0,2r}$ the standard Newtonian (volume) potential in the ball $B(x_0, 2r)$. Let us abbreviate $h := \Delta u$ in $B(x_0, 2r)$. Also, define $w := u - \Pi_{x_0,2r} h$, so

$$\Delta w = 0 \text{ in } B(x_0, 2r) \text{ and } u = w + \Pi_{x_0,2r} h \text{ in } B(x_0, 2r). \tag{8.3.51}$$

In particular, $\nabla u = \nabla w + \nabla \Pi_{x_0,2r} h$ in $B(x_0, 2r)$. Consequently, using interior estimates for harmonic functions we may estimate

$$\begin{aligned}\sup_{B(x_0,r)} |\nabla u| &\leq \sup_{B(x_0,r)} |\nabla w| + \sup_{B(x_0,r)} |\nabla \Pi_{x_0,2r} h| \\ &\leq \frac{C}{r} \fint_{B(x_0,2r)} |w| \, \mathrm{d}\mathcal{L}^n + \sup_{B(x_0,r)} |\nabla \Pi_{x_0,2r} h| \\ &\leq \frac{C}{r} \fint_{B(x_0,2r)} |u| \, \mathrm{d}\mathcal{L}^n + \frac{C}{r} \cdot \sup_{B(x_0,2r)} |\Pi_{x_0,2r} h| + \sup_{B(x_0,r)} |\nabla \Pi_{x_0,2r} h|.\end{aligned} \tag{8.3.52}$$

Also,

$$\begin{aligned}\sup_{B(x_0,r)} |\nabla \Pi_{x_0,2r} h| &\leq C_n \cdot \sup_{|x-x_0|<r} \int_{B(x_0,2r)} \frac{|h(y)|}{|x-y|^{n-1}} \, \mathrm{d}y \\ &\leq C_n \Big(\sup_{B(x_0,2r)} |h| \Big) \cdot \sup_{|x-x_0|<r} \int_{B(x_0,2r)} \frac{dy}{|x-y|^{n-1}} \\ &\leq C_n \Big(\sup_{B(x_0,2r)} |h| \Big) \cdot \sup_{|x-x_0|<r} \int_{B(x,3r)} \frac{dy}{|x-y|^{n-1}} \\ &= C_n \cdot r \cdot \Big(\sup_{B(x_0,2r)} |h| \Big).\end{aligned} \tag{8.3.53}$$

In a similar fashion,

$$\sup_{B(x_0,r)} |\Pi_{x_0,2r} h| \leq C_n \cdot r^2 \cdot \Big(\sup_{B(x_0,2r)} |h| \Big) \text{ if } n \geq 3, \tag{8.3.54}$$

and

$$\sup_{B(x_0,r)} |\Pi_{x_0,2r} h| \leq C \cdot r^2 (1 + |\ln r|) \cdot \Big(\sup_{B(x_0,2r)} |h| \Big) \text{ if } n = 2. \tag{8.3.55}$$

At this point, (8.3.49)-(8.3.50) follow by combining (8.3.52)-(8.3.55). □

We are now ready to state and prove our first main result in this section, asserting the well-posedness of the Dirichlet Problem and the Regularity Problem for the Helmholtz operator in (connected) exterior domains, wrapped into one, with the latter as a sub-problem of the former.

Theorem 8.3.6 *Let $n \in \mathbb{N}$ satisfying $n \geq 2$ and pick an arbitrary wave number $k \in (0, \infty)$. Also, suppose $\Omega \subset \mathbb{R}^n$ is a connected exterior Ahlfors regular domain. Abbreviate $\sigma := \mathcal{H}^{n-1} \lfloor \partial\Omega$ and denote by ν the geometric measure theoretic outward unit normal to Ω. Finally, select an integrability exponent $p \in (1, \infty)$, along with a truncation $\rho \in (0, \infty)$ and an aperture parameter $\kappa \in (0, \infty)$.*

Then there exists a threshold $\delta \in (0, 1)$ depending only on n, p, k, the ADR *constants of $\partial\Omega$, and the relative flatness ratio $\wp(\Omega)/\operatorname{diam} \partial\Omega$ so that when*

$$\operatorname{dist}\Big(\nu, [\mathrm{VMO}(\partial\Omega, \sigma)]^n \Big) < \delta \tag{8.3.56}$$

with the distance considered in $\big[\mathrm{BMO}(\partial\Omega,\sigma)\big]^n$ *(hence when* Ω *is a* δ*-oscillating* AR *domain*[95]*; cf. Definition 3.4.1), it follows that for each* $f\in L^p(\partial\Omega,\sigma)$ *the Exterior Dirichlet Problem for the Helmholtz operator in* Ω*,*

$$\begin{cases} u\in\mathscr{C}^\infty(\Omega), \quad (\Delta+k^2)u=0 \ \textit{in}\ \Omega,\\ \mathcal{N}_\kappa^\rho u\in L^p(\partial\Omega,\sigma),\\ u\big|_{\partial\Omega}^{\kappa-\mathrm{n.t.}}=f \ \textit{at}\ \sigma\textit{-a.e. point on}\ \partial\Omega,\\ u \ \textit{satisfies Sommerfeld's radiation condition,}\end{cases} \tag{8.3.57}$$

has a unique solution. Moreover, this solution satisfies the regularity property:

$$\mathcal{N}_\kappa^\rho(\nabla u)\in L^p(\partial\Omega,\sigma)\Longleftrightarrow f\in L^p_1(\partial\Omega,\sigma). \tag{8.3.58}$$

Proof Choose the parameter $\delta\in(0,1)$ sufficiently small so that the conclusions in Theorem 8.3.3 hold for the given integrability exponent $p\in(1,\infty)$, and the conclusion in Theorem 8.3.4 is valid for $q:=(1-1/p)^{-1}$, the Hölder conjugate exponent of p. Then, using the inverse of (8.3.27), the function

$$u:=\big(\mathcal{D}_k-\mathrm{i}\mathscr{S}_k\big)\Big[\big(\tfrac12 I+K_k-\mathrm{i}S_k\big)^{-1}f\Big] \ \text{ in } \Omega \tag{8.3.59}$$

solves (8.3.57). This takes care of existence.

To prove uniqueness, suppose u solves the version of (8.3.57) formulated for $f=0$. From [131, Lemma 6.3.2] and the final integral representation formula in [131, Theorem 6.4.1] applied to u, considered in the open set $\mathbb{R}^n\setminus\overline{B(0,R)}$ for some large $R>0$, we see that

$$u(x)=O\big(|x|^{-(n-1)/2}\big) \ \text{ as } |x|\to\infty. \tag{8.3.60}$$

Going further, fix an arbitrary point $x_0\in\Omega$ and, assuming $\rho\in\big(0,\frac12\,\mathrm{dist}\,(x_0,\partial\Omega)\big)$, bring in the Green function G^{x_0}, with pole at x_0, constructed in Theorem 8.3.4 for $q:=(1-1/p)^{-1}\in(1,\infty)$. The idea is to apply the Divergence Theorem, in the version recorded in [129, Corollary 1.5.2], to the vector field

$$\vec{F}:=G^{x_0}\cdot\nabla u-u\cdot\nabla G^{x_0} \ \text{ in } \Omega. \tag{8.3.61}$$

Note that

$$\vec{F}\in\big[\mathscr{C}^\infty(\Omega\setminus\{x_0\})\cap L^1_{\mathrm{loc}}(\Omega,\mathcal{L}^n)\big]^n \tag{8.3.62}$$

and that, in the sense of distributions in Ω, we have

$$\begin{aligned}\mathrm{div}\vec{F}&=\nabla G^{x_0}\cdot\nabla u+G^{x_0}\cdot\Delta u-\nabla u\cdot\nabla G^{x_0}-u\cdot\Delta G^{x_0}\\ &=G^{x_0}\cdot(\Delta+k^2)u-u\cdot(\Delta+k^2)G^{x_0}=u\delta_{x_0}=u(x_0)\delta_{x_0}\in\mathscr{E}'(\Omega).\end{aligned} \tag{8.3.63}$$

Also,

[95] in particular, if Ω is an infinitesimally flat AR domain

$$\begin{aligned}\widehat{x}\cdot\vec{F} &= G^{x_0}\,\widehat{x}\cdot\nabla u - u\,\widehat{x}\cdot\nabla G^{x_0} = G^{x_0}\,\partial_{\widehat{x}}u - u\,\partial_{\widehat{x}}G^{x_0} \\ &= G^{x_0}\big[\partial_{\widehat{x}}u - iku\big] - u\big[\partial_{\widehat{x}}G^{x_0} - ikG^{x_0}\big]. \end{aligned} \tag{8.3.64}$$

Based on this, [131, Theorem 6.6.1], (8.3.60), and the penultimate condition in (8.3.45) we conclude that

$$\widehat{x}\cdot\vec{F}(x) = O\big(|x|^{-(n-1)/2}\big)\cdot o\big(|x|^{-(n-1)/2}\big) = o\big(|x|^{-(n-1)}\big) \text{ as } |x|\to\infty. \tag{8.3.65}$$

As a result,

$$\begin{aligned}&\int_{B(0,2R)\setminus B(0,R)}|x\cdot\vec{F}(x)|\,\mathrm{d}\mathcal{L}^n(x) \\ &\qquad\approx R\cdot\int_{B(0,2R)\setminus B(0,R)}|\widehat{x}\cdot\vec{F}(x)|\,\mathrm{d}\mathcal{L}^n(x) = o(R^2) \text{ as } R\to\infty,\end{aligned} \tag{8.3.66}$$

so condition [129, (1.5.22)] is presently satisfied.

The fact that there exists an aperture parameter $\kappa'\in(0,\kappa)$ with the property that

$$\vec{F}\Big|_{\partial\Omega}^{\kappa'-\mathrm{n.t.}} \text{ exists at } \sigma\text{-a.e. point on } \partial\Omega \tag{8.3.67}$$

and, in fact,

$$\vec{F}\Big|_{\partial\Omega}^{\kappa'-\mathrm{n.t.}} = -\Big(u\Big|_{\partial\Omega}^{\kappa'-\mathrm{n.t.}}\Big)\Big(\nabla G^{x_0}\Big|_{\partial\Omega}^{\kappa'-\mathrm{n.t.}}\Big) = 0 \text{ at } \sigma\text{-a.e. point on } \partial\Omega, \tag{8.3.68}$$

(since we are currently assuming that $u\big|_{\partial\Omega}^{\kappa-\mathrm{n.t.}} = 0$ at σ-a.e. point on $\partial\Omega$), plus the fact that there exists some height $\varepsilon_o>0$ such that the truncated nontangential maximal function of $\vec{F}$ satisfies

$$\mathcal{N}_{\kappa'}^{\varepsilon_o}\vec{F}\in L^1(\partial\Omega,\sigma) \tag{8.3.69}$$

are established as in the proof of [131, Theorem 4.4.1] (see [131, (4.4.70)-(4.4.71), (4.4.76)]), but with one caveat. Specifically, the interior estimates (cf. [129, Theorem 6.5.7]) used in [131, (4.4.65), (4.4.73)] are for null-solutions of a *homogeneous* elliptic system, and the Helmholtz operator $\Delta+k^2$ lacks homogeneity. This being said, the fact that we currently have $\Delta u = -k^2u$ in Ω permits us to invoke Lemma 8.3.5 which shows that for any $x\in\Omega$ and any $r\in\big(0,\tfrac12\operatorname{dist}(x,\partial\Omega)\big)$ we have

$$\sup_{B(x,r)}|\nabla u| \le \frac{C}{r}\fint_{B(x,2r)}|u|\,\mathrm{d}\mathcal{L}^n + Cr\cdot\sup_{B(x,2r)}|u| \text{ if } n\ge 3, \tag{8.3.70}$$

and

$$\sup_{B(x,r)}|\nabla u| \le \frac{C}{r}\fint_{B(x,2r)}|u|\,\mathrm{d}\mathcal{L}^2 + Cr(1+|\ln r|)\cdot\sup_{B(x,2r)}|u| \text{ if } n=2. \tag{8.3.71}$$

For the goals we have in mind, these are effective substitutes (keeping in mind that we are currently assuming $u\big|_{\partial\Omega}^{\kappa-\mathrm{n.t.}} = 0$ at σ-a.e. point on $\partial\Omega$), permitting us justify (8.3.67)-(8.3.68) and (8.3.69) reasoning much as in the proof of [131, Theorem 4.4.1].

Granted (8.3.62), (8.3.63), (8.3.66), (8.3.67), (8.3.69), we may invoke the Divergence Formula [129, (1.5.20)] which, thanks to (8.3.68) and (8.3.63), simply reduces to $u(x_0) = 0$. Given the arbitrariness of $x_0 \in \Omega$, this ultimately shows that $u \equiv 0$ in Ω. This establishes uniqueness for the solution of Exterior Dirichlet Problem for the Helmholtz operator in Ω, formulated in (8.3.57).

Finally, the right-pointing implication in (8.3.58) is a consequence of [130, Proposition 11.3.2], while the left-pointing implication in (8.3.58) follows from the invertibility result recorded in (8.3.28), the integral representation formula (8.3.59), and the regularity properties of the intervening acoustic boundary layer potentials, proved in [132, Theorem 1.5.5]. □

It is of significance to point out that, thanks to the uniqueness part in (8.3.57), the Green function G^{x_0} constructed in Theorem 8.3.4 is uniquely determined by the properties specified in (8.3.45).

Our final principal result in this section concerns the well-posedness of the Neumann Problem for the Helmholtz operator in (connected) exterior domains.

Theorem 8.3.7 *Let $n \in \mathbb{N}$ satisfying $n \geq 2$ and pick an arbitrary wave number $k \in (0, \infty)$. Also, suppose $\Omega \subset \mathbb{R}^n$ is a connected exterior Ahlfors regular domain. Abbreviate $\sigma := \mathcal{H}^{n-1}\lfloor\partial\Omega$ and denote by ν the geometric measure theoretic outward unit normal to Ω. Finally, select an integrability exponent $p \in (1, \infty)$, along with a truncation $\rho \in (0, \infty)$ and an aperture parameter $\kappa \in (0, \infty)$.*

Then there exists a threshold $\delta \in (0, 1)$ depending only on n, p, k, the ADR *constants of $\partial\Omega$, and the relative flatness ratio $\wp(\Omega)/\mathrm{diam}\,\partial\Omega$ so that when*

$$\mathrm{dist}\Big(\nu, [\mathrm{VMO}(\partial\Omega, \sigma)]^n\Big) < \delta \tag{8.3.72}$$

with the distance considered $\big[\mathrm{BMO}(\partial\Omega, \sigma)\big]^n$ (hence when Ω is a δ-oscillating AR *domain*[96]*; cf. Definition 3.4.1), it follows that for $f \in L^p(\partial\Omega, \sigma)$ the Exterior Neumann Problem for the Helmholtz operator in Ω,*

$$\begin{cases} u \in \mathscr{C}^\infty(\Omega), \quad (\Delta + k^2)u = 0 \ \text{in} \ \Omega, \\ \mathcal{N}_\kappa^\rho u \in L^p(\partial\Omega, \sigma), \quad \mathcal{N}_\kappa^\rho(\nabla u) \in L^p(\partial\Omega, \sigma), \\ \partial_\nu u := \nu \cdot (\nabla u)\big|_{\partial\Omega}^{\kappa-\mathrm{n.t.}} = f \ \text{at } \sigma\text{-a.e. point on } \partial\Omega, \\ u \ \text{satisfies Sommerfeld's radiation condition}, \end{cases} \tag{8.3.73}$$

has a unique solution.

Proof Choose $\delta \in (0, 1)$ sufficiently small so that the conclusions in Theorem 3.4.2 are valid. To prove uniqueness, suppose u solves the version of (8.3.73) corresponding

[96] in particular, if Ω is an infinitesimally flat AR domain

to $f = 0$. From [129, Proposition 8.9.22] we know that $u\big|_{\partial\Omega}^{\kappa-\text{n.t.}}$ exists at σ-a.e. point on $\partial\Omega$. This permits us to run the same argument as in (8.3.34)-(8.3.36) and conclude

$$\int_{|x|=R} |u|^2 \, d\mathcal{H}^{n-1} = o(1) \text{ as } R \to \infty. \tag{8.3.74}$$

Thanks to Rellich's uniqueness lemma (recorded in Lemma 8.3.1), this proves that $u \equiv 0$ in the connected set Ω.

To prove existence for (8.3.73), we shall look for a candidate of the form

$$u := \mathscr{S}_k g + \mathrm{i}\mathcal{D}_k(S_0^2 g) \text{ in } \Omega, \tag{8.3.75}$$

for some function $g \in L^p(\partial\Omega, \sigma)$ yet to be determined. Thanks to the properties of the acoustic boundary layer potentials from [132, Theorem 1.5.5], any such candidate satisfies all properties demanded in (8.3.73) save, perhaps, for the boundary condition. By once again relying on [132, Theorem 1.5.5], we see that the latter amounts to solving

$$\big(-\tfrac{1}{2}I + K_k\big)g + \mathrm{i}(\partial_\nu \mathcal{D}_k)(S_0^2 g) = f. \tag{8.3.76}$$

In this regard, we make the claim that the operator

$$T := -\tfrac{1}{2}I + K_k^\# + \mathrm{i}(\partial_\nu \mathcal{D}_k) \circ S_0^2 : L^p(\partial\Omega, \sigma) \to L^p(\partial\Omega, \sigma) \text{ is invertible.} \tag{8.3.77}$$

From Lemma 8.3.2, [132, Theorem 1.5.5], and [130, Theorem 11.5.18, item *(iii)*] we see that the operator

$$(K_k^\# - K_0^\#) + \mathrm{i}(\partial_\nu \mathcal{D}_k) \circ S_0^2 : L^p(\partial\Omega, \sigma) \to L^p(\partial\Omega, \sigma) \text{ is compact,} \tag{8.3.78}$$

while Theorem 7.1.1 (stated for $L = \Delta$ and with $z := -1/2$) tells us that $-\frac{1}{2}I + K_0^\#$ is a Fredholm operator with index zero on $L^p(\partial\Omega, \sigma)$. This together with (8.3.78) then shows that T in (8.3.77) is a Fredholm operator with index zero on $L^p(\partial\Omega, \sigma)$. Moreover, after possibly further decreasing the value of $\delta \in (0, 1)$, we may rely on the abstract regularity result from [130, Lemma 2.1.4] (whose present applicability also relies on the density results from [129, (3.1.13)]) to arrange matters so that the kernel of T on $L^p(\partial\Omega, \sigma)$ coincides with the kernel of T on $L^2(\partial\Omega, \sigma)$. Granted this, (8.3.77) will follow the moment we show that

$$T = -\tfrac{1}{2}I + K_k^\# + \mathrm{i}(\partial_\nu \mathcal{D}_k) \circ S_0^2 : L^2(\partial\Omega, \sigma) \to L^2(\partial\Omega, \sigma) \text{ is injective.} \tag{8.3.79}$$

To this end, assume $\phi \in L^2(\partial\Omega, \sigma)$ is such that $T\phi = 0$, and define

$$w := \mathscr{S}_k \phi + \mathrm{i}\mathcal{D}_k(S_0^2 \phi) \text{ in } \Omega. \tag{8.3.80}$$

Then the current working assumptions and [132, Theorem 1.5.5] imply that

$$\begin{cases} w \in \mathscr{C}^\infty(\Omega), \quad (\Delta + k^2)u = 0 \text{ in } \Omega, \\ \mathcal{N}_\kappa^\rho w \in L^2(\partial\Omega, \sigma), \quad \mathcal{N}_\kappa^\rho(\nabla w) \in L^2(\partial\Omega, \sigma), \\ \partial_\nu w = 0 \text{ at } \sigma\text{-a.e. point on } \partial\Omega, \\ w \text{ satisfies Sommerfeld's radiation condition.} \end{cases} \tag{8.3.81}$$

Thanks to the uniqueness for the Exterior Neumann Problem already established in the first portion of the current proof, this forces $w \equiv 0$ in Ω. Next, define

$$\widetilde{w} := \mathscr{S}_k\phi + \mathrm{i}\mathcal{D}_k(S_0^2\phi) \text{ in } \Omega_- := \mathbb{R}^n \setminus \overline{\Omega}, \tag{8.3.82}$$

and observe that this function satisfies

$$\begin{cases} \widetilde{w} \in \mathscr{C}^\infty(\Omega_-), \quad (\Delta + k^2)\widetilde{w} = 0 \text{ in } \Omega_-, \\ \mathcal{N}_\kappa\widetilde{w} \in L^2(\partial\Omega, \sigma), \quad \mathcal{N}_\kappa(\nabla\widetilde{w}) \in L^2(\partial\Omega, \sigma), \\ \widetilde{w}\big|_{\partial\Omega_-}^{\kappa-\mathrm{n.t.}} \text{ exists at } \sigma\text{-a.e. point on } \partial\Omega, \\ (\nabla\widetilde{w})\big|_{\partial\Omega_-}^{\kappa-\mathrm{n.t.}} \text{ exists at } \sigma\text{-a.e. point on } \partial\Omega. \end{cases} \tag{8.3.83}$$

Bearing in mind that $w \equiv 0$ in Ω, the trace formulas for acoustic boundary layer potentials proved in [132, Theorem 1.5.5] then permit us to write

$$\widetilde{w}\big|_{\partial\Omega_-}^{\kappa-\mathrm{n.t.}} = \widetilde{w}\big|_{\partial\Omega_-}^{\kappa-\mathrm{n.t.}} - w\big|_{\partial\Omega}^{\kappa-\mathrm{n.t.}} = -\mathrm{i}S_0^2\phi \text{ on } \partial\Omega, \tag{8.3.84}$$

$$\text{and } \partial_\nu\widetilde{w} = \partial_\nu\widetilde{w} - \partial_\nu w = \phi \text{ on } \partial\Omega. \tag{8.3.85}$$

Thanks to (8.3.83), [129, Proposition 8.6.3], and the fact that Ω_- is bounded, we also have

$$\widetilde{w},\ |\nabla\widetilde{w}| \in L^{2n/(n-1)}(\Omega_-, \mathcal{L}^n) \hookrightarrow L^2(\Omega_-, \mathcal{L}^n). \tag{8.3.86}$$

At this stage, we may rely on (8.3.83), (8.3.84), (8.3.85), and the version of Green's first identity from [131, Theorem 1.7.12] written for $L := \Delta$ (whose applicability in the present setting uses (8.3.86)), and the fact that S_0 is symmetric to obtain

$$\begin{aligned} \int_{\Omega_-} |\nabla\widetilde{w}|^2 \, d\mathcal{L}^n &= -\int_{\partial\Omega} \overline{\Big(\widetilde{w}\big|_{\partial\Omega}^{\kappa-\mathrm{n.t.}}\Big)} \cdot \partial_\nu\widetilde{w} \, d\sigma + k^2 \int_{\Omega_-} |\widetilde{w}|^2 \, d\mathcal{L}^n \\ &= \mathrm{i}\int_{\partial\Omega} |S_0\phi|^2 \, d\sigma + k^2 \int_{\Omega_-} |\widetilde{w}|^2 \, d\mathcal{L}^n. \end{aligned} \tag{8.3.87}$$

By taking imaginary parts we arrive at the conclusion that

$$\int_{\partial\Omega} |S_0\phi|^2 \, d\sigma = 0. \tag{8.3.88}$$

In the case when $n \geq 3$, after adjusting yet again the size of the parameter $\delta \in (0, 1)$, matters may be arranged so that the conclusions in Theorem 7.1.9 are true (for $L := \Delta$). In concert with (8.3.88), this forces $\phi = 0$, proving (8.3.79) when $n \geq 3$.

To complete the proof of (8.3.77), there remains to treat the case when $n = 2$. In this scenario, in place of (8.3.77) we consider

$$\begin{aligned} &\widetilde{T} : L^p(\partial\Omega, \sigma) \to L^p(\partial\Omega, \sigma) \text{ given for each } \phi \in L^p(\partial\Omega, \sigma) \\ &\text{by } \widetilde{T}\phi := \big(-\tfrac{1}{2}I + K_k^{\#}\big)\phi + \mathrm{i}(\partial_\nu \mathcal{D}_k)\big(S_0^2\phi + \textstyle\int_{\partial\Omega} \phi \, \mathrm{d}\sigma\big). \end{aligned} \tag{8.3.89}$$

The same type of argument as before reduces the matter of inverting $\widetilde{T}$ in the context of (8.3.89) to the issue of showing that said operator is injective for $p = 2$. With this aim in mind, let $\phi \in L^2(\partial\Omega, \sigma)$ be such that $\widetilde{T}\phi = 0$. Reasoning much as in (8.3.80)-(8.3.88), the task at hand comes down to showing that

$$\int_{\partial\Omega} |S_0\phi|^2 \, \mathrm{d}\sigma + \Big| \int_{\partial\Omega} \phi \, \mathrm{d}\sigma \Big|^2 = 0 \tag{8.3.90}$$

forces $\phi = 0$. This, however, is implied by Theorem 7.1.10 (used with $L := \Delta$), granted $\delta \in (0, 1)$ is small enough. Ultimately, this establishes that $\widetilde{T}$ is invertible in the context of (8.3.89). As such, a solution to the Exterior Neumann Problem (8.3.73) when $n = 2$ may be obtained by taking

$$u := \mathscr{S}_k(\widetilde{T}^{-1} f) + \mathrm{i}\mathcal{D}_k\Big(S_0^2(\widetilde{T}^{-1} f) + \textstyle\int_{\partial\Omega} \widetilde{T}^{-1} f \, \mathrm{d}\sigma\Big) \text{ in } \Omega. \tag{8.3.91}$$

The proof of Theorem 8.3.7 is therefore complete. □

8.4 Boundary Problems in Domains with Unbounded Boundaries for Elliptic Systems

Complementing work carried out in §8.1 for Ahlfors regular domains with compact boundaries, here we deal with Ahlfors regular domains $\Omega \subseteq \mathbb{R}^n$ whose geometric measure theoretic outward unit normal ν has a small BMO semi-norm, and according to (4.1.5) this forces $\partial\Omega$ to be unbounded.

In the first part of this section we shall treat the Dirichlet, Regularity, and Neumann Problems in such domains with boundary data in Lebesgue spaces, Morrey spaces, vanishing Morrey spaces, block spaces, and their associated (homogeneous and inhomogeneous) Sobolev-styled spaces. For these boundary value problems we shall establish well-posedness results in Ahlfors regular domains whose geometric measure theoretic outward unit normals have a suitably small BMO semi-norm, as a consequence of: (1) the invertibility results from Theorem 4.2.1, (2) traces/jump-formulas and estimates for boundary-to-domain single and double layer potential operators as those in [132, Theorem 1.5.1] and [132, Theorem 1.8.2], (3) the uniqueness results for Dirichlet and Regularity Problems in [131, Chapter 4], and (4) integral representation formulas of the sort discussed in [132, Theorem 1.8.17] and [132, Theorem 1.8.19] (which are useful in dealing with the issue of uniqueness for Homogeneous Regularity Problem and also the Neumann Problem).

Subsequently, in the second part of this section, we treat the Neumann Problem with data from Hardy spaces H^p with $\frac{n-1}{n} < p \leq 1$ in Theorem 8.4.25, the Homogeneous Regularity Problem with boundary data from homogeneous Hardy-based Sobolev spaces $\dot{H}^p_1$ with $\frac{n-1}{n} < p \leq 1$ in Theorem 8.4.29, the Tangential Derivative Problem with data from Hardy spaces H^p with $\frac{n-1}{n} < p \leq 1$ in Theorem 8.4.31, the Transmission Problem with data from Hardy spaces H^p and homogeneous Hardy-based Sobolev spaces $\dot{H}^p_1$ with $\frac{n-1}{n} < p \leq 1$ in Theorem 8.4.32, the Dirichlet Problem with boundary data from the John-Nirenberg space BMO in Theorem 8.4.33, the Dirichlet Problem with boundary data from the Sarason space VMO in Theorem 8.4.34, the Dirichlet Problem with boundary data from the Coifman-Weiss space CMO in Theorem 8.4.35, the Dirichlet Problem with boundary data from "ordinary" Hölder spaces in Theorem 8.4.36, the Dirichlet Problem with boundary data from "vanishing" Hölder spaces in Theorem 8.4.37, the Dirichlet Problem with boundary data from the ϕ-modulated space of functions of bounded mean oscillations $\mathrm{BMO}_{\phi,p}$ in Theorem 8.4.38, and the Dirichlet Problem with boundary data from the homogeneous and inhomogeneous ϕ-modulated space of functions of bounded central mean oscillations, $\mathrm{BCMO}_{\phi,p}$ and $\dot{\mathrm{B}}\mathrm{CMO}_{\phi,p}$, in Theorem 8.4.39. Finally, in Theorems 8.4.40-8.4.44 we treat boundary value problems in δ-AR domains with data in Generalized Banach Function spaces.

We wish to emphasize that, in all cases, our main results are valid for weakly elliptic, complex coefficient, systems L which may not satisfy the strong Legendre-Hadamard ellipticity condition, nor possess any symmetry properties, and without assuming any positive definiteness (or positive semi-definiteness) for any of the coefficient tensors in $\mathfrak{A}_L$. The novel philosophy emerging from the present work may be succinctly (yet remarkably accurately) phrased as follows:

given a second-order, weakly elliptic, constant complex coefficient, homogeneous system L in $\mathbb{R}^n$ such that $\mathfrak{A}_L^{\mathrm{dis}} \neq \emptyset \neq \mathfrak{A}_{L^\top}^{\mathrm{dis}}$, together with an Ahlfors regular domain $\Omega \subseteq \mathbb{R}^n$ whose boundary is sufficiently flat at all scales (large and small, measured through the integral oscillations of the unit normal), the basic boundary value problems for L in Ω, such as Dirichlet, Regularity, Neumann, Transmission, are all well posed when formulated on a wide variety of functions spaces (including large classes of Generalized Banach Function Spaces on $\partial\Omega$), and these well-posedness results may no longer be valid (as even the Fredholm solvability may fail to materialize) in the absence of distinguished coefficient tensors for L, even if Ω is a half-space in $\mathbb{R}^n$. (8.4.1)

The notion of "flatness" referred to above is more inclusive than the usual intuitive sense of the word "flat" would suggest. Indeed, traditionally flatness of a set is most commonly associated with the ability of containing the boundary of said set sandwiched inside a relatively narrow infinite strip, lying in between two parallel planes sufficiently close to one another. By way of contrast, the main attribute of our class of "relatively flat domains" is the absence of "turns which are too sudden" of the boundary. In particular, arbitrarily high peaks as well as arbitrarily deep valleys are allowed as long as they develop along sufficiently small slopes. Remarkably, boundary spiral points are also permissible if the branches of the spiral in question are sufficiently gently chasing one another.

The internal logic of our approach dictates that we begin by considering the Homogeneous Regularity Problem with boundary data in homogeneous Sobolev spaces $\dot{L}^p_1$ with $1 < p < \infty$. This is formulated for weakly elliptic systems and sufficiently flat Ahlfors regular domains with unbounded boundaries. The class of functions in which a solution is sought involves a size condition formulated in terms of the L^p-integrability of the nontangential maximal operator of the gradient of the function in question. The aforementioned size condition may therefore be thought as having mixed character, in the sense that they entail taking an L^∞-norm "vertically" (inside nontangential approach regions) and an L^p-norm "horizontally" (with respect to the vertex of said nontangential approach regions, along the boundary of the domain). Naturally nuanced, such a feature is germane to a large number of boundary problems discussed in this section.

Theorem 8.4.1 *Let $\Omega \subseteq \mathbb{R}^n$ be an Ahlfors regular domain, where $n \in \mathbb{N}$ satisfies $n \geq 2$. Abbreviate $\sigma := \mathcal{H}^{n-1}\lfloor\partial\Omega$ and denote by ν the geometric measure theoretic outward unit normal to Ω. Also, fix an aperture parameter $\kappa > 0$ and pick an integrability exponent $p \in (1, \infty)$. Given a homogeneous, second-order, constant complex coefficient, weakly elliptic $M \times M$ system L in $\mathbb{R}^n$, consider the Homogeneous Regularity Problem*

$$\begin{cases} u \in \big[\mathscr{C}^\infty(\Omega)\big]^M, \quad Lu = 0 \text{ in } \Omega, \\ \mathcal{N}_\kappa(\nabla u) \in L^p(\partial\Omega, \sigma), \quad u\big|_{\partial\Omega}^{\kappa-\text{n.t.}} = f \in \big[\dot{L}^p_1(\partial\Omega, \sigma)\big]^M, \end{cases} \tag{8.4.2}$$

where $\dot{L}^p_1(\partial\Omega, \sigma)$ is the homogeneous boundary Sobolev space introduced in [130, Definition 11.5.3] (*cf.* (A.0.145)-(A.0.146)). *Then the following statements are true:*

(a) [Existence, Estimates, and Integral Representations] *If $\mathfrak{A}^{\text{dis}}_L \neq \varnothing$ then there exists $\delta \in (0, 1)$ which depends only on n, p, L, and the Ahlfors regularity constants of $\partial\Omega$ with the property that if $\|\nu\|_{[\mathrm{BMO}(\partial\Omega,\sigma)]^n} < \delta$ (i.e., if Ω is a δ*-AR *domain in the sense of Definition 3.1.8) then the following properties hold. First,*

$$\begin{aligned} &\big[S_{\text{mod}}\big] : \big[L^p(\partial\Omega, \sigma)\big]^M \longrightarrow \big[\dot{L}^p_1(\partial\Omega, \sigma)/\sim\big]^M, \\ &\big[S_{\text{mod}}\big]g := \big[S_{\text{mod}}g\big] \text{ for each } g \in \big[L^p(\partial\Omega, \sigma)\big]^M, \end{aligned} \tag{8.4.3}$$

is a surjective operator, and the Homogeneous Regularity Problem (8.4.2) *is solvable. More specifically, with $[f] \in \big[\dot{L}^p_1(\partial\Omega, \sigma)/\sim\big]^M$ denoting the equivalence class (modulo constants) of the boundary datum f, and with*

$$g \in \big[L^p(\partial\Omega, \sigma)\big]^M \text{ chosen so that } \big[S_{\text{mod}}\big]g = [f], \tag{8.4.4}$$

there exists a constant $c \in \mathbb{C}^M$ such that the function

$$u := \mathcal{S}_{\text{mod}}g + c \text{ in } \Omega \tag{8.4.5}$$

is a solution of the Homogeneous Regularity Problem (8.4.2). *In addition, this solution satisfies (with implicit constants independent of f)*

$$\big\|\mathcal{N}_\kappa(\nabla u)\big\|_{L^p(\partial\Omega,\sigma)} \approx \|\nabla_{\tan} f\|_{[L^p(\partial\Omega,\sigma)]^{n\cdot M}}. \tag{8.4.6}$$

Second, for each coefficient tensor $A \in \mathfrak{A}_L^{\rm dis}$ the operator

$$\tfrac{1}{2}I + \big[K_{\rm mod}\big] : \big[\dot{L}_1^p(\partial\Omega,\sigma)/\sim\big]^M \longrightarrow \big[\dot{L}_1^p(\partial\Omega,\sigma)/\sim\big]^M \tag{8.4.7}$$

is an isomorphism, and the Homogeneous Regularity Problem (8.4.2) *has*

$$u := \mathcal{D}_{\rm mod} h + c \text{ in } \Omega, \tag{8.4.8}$$

as a solution, for a suitable constant $c \in \mathbb{C}^M$ and with

$$h \in \big[\dot{L}_1^p(\partial\Omega,\sigma)\big]^M \text{ such that } [h] = \Big(\tfrac{1}{2}I + \big[K_{\rm mod}\big]\Big)^{-1}[f]. \tag{8.4.9}$$

Moreover, this solution continues to satisfy (8.4.6). *Finally,*

> *if the boundary datum f actually belongs to the inhomogeneous Sobolev space $\big[L_1^p(\partial\Omega,\sigma)\big]^M$ then a solution u of the Homogeneous Regularity Problem* (8.4.2) *may be found with the additional property that $\mathcal{N}_\kappa u$ belongs to $L^p(\partial\Omega,\sigma)$.* (8.4.10)

(b) [Uniqueness] *Whenever $\mathfrak{A}_{L^\top}^{\rm dis} \neq \varnothing$, there exists $\delta \in (0,1)$ which depends only on n, p, L, and the Ahlfors regularity constants of $\partial\Omega$ with the property that if $\|\nu\|_{[{\rm BMO}(\partial\Omega,\sigma)]^n} < \delta$ (i.e., if Ω is a δ-*AR *domain in the sense of Definition 3.1.8) then the Homogeneous Regularity Problem* (8.4.2) *has at most one solution.*

(c) [Well-Posedness and Additional Integrability/Regularity] *Whenever $\mathfrak{A}_L^{\rm dis} \neq \varnothing$ and $\mathfrak{A}_{L^\top}^{\rm dis} \neq \varnothing$ it follows that there exists $\delta \in (0,1)$ which depends only on n, p, L, and the Ahlfors regularity constants of $\partial\Omega$ with the property that if $\|\nu\|_{[{\rm BMO}(\partial\Omega,\sigma)]^n} < \delta$ (i.e., if Ω is a δ-*AR *domain in the sense of Definition 3.1.8) then the Homogeneous Regularity Problem* (8.4.2) *is uniquely solvable. Moreover, the unique solution u of* (8.4.2) *satisfies (in a quantitative fashion)*

$$\mathcal{N}_\kappa u \in L^p(\partial\Omega,\sigma) \Longleftrightarrow f \in \big[L_1^p(\partial\Omega,\sigma)\big]^M. \tag{8.4.11}$$

(d) [Sharpness] *If $\mathfrak{A}_L^{\rm dis} = \varnothing$ the Homogeneous Regularity Problem* (8.4.2) *may fail to be solvable (actually for boundary data belonging to an infinite dimensional subspace of the corresponding weighted homogeneous Sobolev space), and if $\mathfrak{A}_{L^\top}^{\rm dis} = \varnothing$ the Homogeneous Regularity Problem* (8.4.2) *may posses more than one solution (in fact, the linear space of null-solutions may actually be infinite dimensional), even in the case when $\Omega = \mathbb{R}^n_+$. To sum up, if either $\mathfrak{A}_L^{\rm dis} = \varnothing$, or $\mathfrak{A}_{L^\top}^{\rm dis} = \varnothing$, then the Homogeneous Regularity Problem* (8.4.2) *may fail to be well posed, again, even in the case when $\Omega = \mathbb{R}^n_+$.*

Moreover, the solvability result described in item (a) above (hence also the well-posedness from item (c)) may actually fail in the absence of any specifications regarding the flatness of the boundary. To illustrate this, consider the class of infinite planar sectors, i.e., sets of the form

$$\begin{aligned}&\Omega_\theta := \Big\{x \in \mathbb{R}^2 \setminus \{x_o\} : \tfrac{x-x_o}{|x-x_o|} \cdot \xi > \cos(\theta/2)\Big\}\\ &\textit{for some } x_o \in \mathbb{R}^2,\ \theta \in (0, 2\pi), \textit{ and } \xi \in S^1.\end{aligned} \tag{8.4.12}$$

Then, as shown a little later in Theorem 8.4.12,

the $\dot{L}^p_1$*-Homogeneous Regularity Problem formulated as in* (8.4.2) *for* $n = 2$, $L := \Delta$ *(the two-dimensional Laplacian), and* $\Omega := \Omega_\theta$ *fails to be Fredholm solvable in the situation when* $\theta \in (\pi, 2\pi)$ *and* $p = q_\theta := 1 + \pi/|\pi - \theta| = \frac{\theta}{\theta-\pi}$. (8.4.13)

(e) [Other Spaces of Boundary Data] *Similar results to those described in items (a)-(c) and the first part of item (d) are valid with* $\dot{L}^p_1(\partial\Omega, \sigma)$ *replaced by any of the spaces* $\dot{M}^{p,\lambda}_1(\partial\Omega, \sigma)$, $\dot{\mathbb{M}}^{p,\lambda}_1(\partial\Omega, \sigma)$, *and* $\dot{\mathcal{B}}^{q,\lambda}_1(\partial\Omega, \sigma)$ *with* $p, q \in (1, \infty)$ *and* $\lambda \in (0, n - 1)$.

It is significant to remark that, as seen from Proposition 1.6.12, in the scalar case (that is, when $M = 1$) in dimensions $n \geq 3$, the mere fact that the operator L is weakly elliptic implies that $\mathfrak{A}^{\rm dis}_L \neq \emptyset$ and $\mathfrak{A}^{\rm dis}_{L^\top} \neq \emptyset$. Moreover, in the scalar case in the two-dimensional setting (i.e., for $M = 1$ and $n = 2$), if L satisfies the Legendre-Hadamard strong ellipticity condition or, more generally (cf. [131, Lemma 1.4.24]), if $L = {\rm div}A\nabla$ for some 2×2 complex matrix A with ${\rm sym}\,A := \frac{1}{2}(A + A^\top) \in M^0_2$ (cf. [131, (1.4.187)]), then Proposition 1.6.11 and Proposition 1.6.13 guarantee that $\mathfrak{A}^{\rm dis}_L \neq \emptyset$ and $\mathfrak{A}^{\rm dis}_{L^\top} \neq \emptyset$. The statement of Theorem 8.4.1 may be significantly streamlined, since we have genuine well-posedness, in each of these scenarios.

We also wish to remark that Theorem 3.1.5 and [130, Proposition 11.5.12] imply that if $\delta > 0$ is small enough then for any function u satisfying the properties in the first three lines of (8.4.2) it follows that

$$u\big|^{\kappa-{\rm n.t.}}_{\partial\Omega} \text{ exists } \sigma\text{-a.e. on } \partial\Omega, \text{ and belongs to } \big[\dot{L}^p_1(\partial\Omega, \sigma)\big]^M, \tag{8.4.14}$$

a Fatou-type result ensuring that the space of boundary data in (8.4.2) is optimal.

We also note that the solution u of the Homogeneous Regularity Problem (8.4.2) with boundary datum f satisfies other size estimates of interest. For example, the equivalence in (8.4.6) together with [129, Proposition 8.6.10] imply

$$\sup_{\rho \in (0,\infty)} \Big(\frac{1}{\rho}\int_{O_\rho} |\nabla u|^p \,{\rm d}\mathcal{L}^n\Big)^{1/p} \leq C\|\nabla_{\tan} f\|_{[L^p(\partial\Omega,\sigma)]^{n\cdot M}} \tag{8.4.15}$$

where $O_\rho := \big\{x \in \Omega : {\rm dist}(x, \partial\Omega) < \rho\big\}$ for each $\rho > 0$, while the equivalence in (8.4.6) in concert with the last estimate in [129, Proposition 8.6.3] (presently used with $E := \Omega$) yield

$$\Big(\int_{\Omega}|\nabla u|^{np/(n-1)}\,\mathrm{d}\mathcal{L}^n\Big)^{(n-1)/(np)} \leq C\|\nabla_{\tan}f\|_{[L^p(\partial\Omega,\sigma)]^{n\cdot M}}. \tag{8.4.16}$$

In addition, work in [129, §8.5] shows that one may replace the nontangential maximal operator in the formulation of the Homogeneous Regularity Problem (8.4.2) by the tangential maximal operator, associated as in [129, Definition 8.5.1] with a sufficiently large power (cf. (A.0.165)); a similar comment is relevant for the Homogeneous Regularity Problems mentioned in item *(e)*.

Finally, we make the following remark.

Remark 8.4.2 *A more precise formulation of the well-posedness result in item (c) of Theorem 8.4.1 reads as follows: Fix $n, M \in \mathbb{N}$ with $n \geq 2$ and consider a weakly elliptic homogeneous constant complex coefficient second-order $M \times M$ system L in $\mathbb{R}^n$ with $\mathfrak{A}_L^{\rm dis} \neq \varnothing$ and $\mathfrak{A}_{L^\top}^{\rm dis} \neq \varnothing$. Then for each constant $C_A \in (0, \infty)$ and each compact interval $I \subset (1, \infty)$ there exists a threshold $\delta \in (0, 1)$ which depends only on n, C_A, I, and L with the following significance. Assume $\Omega \subseteq \mathbb{R}^n$ is an Ahlfors regular domain such that the Ahlfors regularity constants of $\partial\Omega$ are controlled by C_A. Abbreviate $\sigma := \mathcal{H}^{n-1}\lfloor\partial\Omega$ and denote by ν the geometric measure theoretic outward unit normal to Ω. Also, fix an integrability exponent $p \in I$. Then whenever $\|\nu\|_{[\mathrm{BMO}(\partial\Omega,\sigma)]^n} < \delta$ it follows that the Homogeneous Regularity Problem* (8.4.2) *is well posed (in the sense that it is uniquely solvable, and the solution also satisfies a naturally accompanying estimate). Moreover, similar considerations apply to all results formulated in Theorem 8.4.1.*

This follows from the invertibility results from Remark 4.2.2 and the proof of Theorem 8.4.1 presented next.

Proof of Theorem 8.4.1 Throughout, we shall work under the assumption that we have $\|\nu\|_{[\mathrm{BMO}(\partial\Omega,\sigma)]^n} < \delta$ with $\delta \in (0, 1)$ sufficiently small as to guarantee that Ω is a connected two-sided NTA domain with an unbounded boundary. This may be arranged, thanks to Theorem 3.1.5. By further decreasing the value of δ if necessary, matters may be arranges so that the conclusions in Theorem 4.2.1 hold (for suitable choices of the parameters involved in its statement).

To deal with the claims in item *(a)*, assume $\mathfrak{A}_L^{\rm dis} \neq \varnothing$ and pick $A \in \mathfrak{A}_L^{\rm dis}$. In particular, Theorem 4.2.1 ensures that the operators

$$\pm\tfrac{1}{2}I + \big[K_{\rm mod}\big] : \big[\dot{L}^p_1(\partial\Omega,\sigma)/\sim\big]^M \longrightarrow \big[\dot{L}^p_1(\partial\Omega,\sigma)/\sim\big]^M \tag{8.4.17}$$

are invertible. Based on this and [132, (1.8.146), (1.8.254)] we then conclude that the operator $\big[S_{\rm mod}\big]$ is surjective in the context of (8.4.3). Then the fact that u defined as in (8.4.5) is a solution of the Homogeneous Regularity Problem (8.4.2) is seen from [132, (1.5.52), (1.5.57), (1.5.80)], and (8.4.4). In addition, the left-pointing inequality in (8.4.6) is implied by the fact that g in (8.4.4) may be chosen so that

$$\|g\|_{[L^p(\partial\Omega,\sigma)]^M} \leq C\big\|[f]\big\|_{[\dot{L}^p_1(\partial\Omega,\sigma)/\sim]^M} \tag{8.4.18}$$

for some constant $C \in (0, \infty)$ independent of f (cf. [130, Proposition 1.2.13], whose applicability uses the completeness results established in [130, Proposition 11.5.14]), together with (8.4.5), [130, (11.4.9)], [132, (1.5.57)], and (A.0.146). Also, the right-pointing inequality in (8.4.6) follows from [130, Proposition 11.3.2] (whose current applicability makes use of [132, (1.5.69)]) and (A.0.88).

The fact that a solution to the Homogeneous Regularity Problem (8.4.2) which satisfies (8.4.6) may, alternatively, be constructed as indicated in (8.4.8)-(8.4.9) follows from [132, (1.8.142), (1.8.143)] and (8.4.17).

To completes the treatment of item *(a)*, there remains to deal with (8.4.10). To this end, since $A \in \mathfrak{A}_L^{\rm dis}$ we may rely on Theorem 4.2.1 to conclude that the operator $\frac{1}{2}I + K_A : \big[L_1^p(\partial\Omega,\sigma)\big]^M \to \big[L_1^p(\partial\Omega,\sigma)\big]^M$ is invertible. Then, with the inverse interpreted in this setting, we conclude from [132, (1.5.8)] and [132, (1.5.20)] that the function $u := \mathcal{D}_A\Big(\big(\frac{1}{2}I + K_A\big)^{-1} f\Big)$ in Ω, is a solution of the Homogeneous Regularity Problem (8.4.2) with the additional property that $\mathcal{N}_\kappa u$ belongs to $L^p(\partial\Omega,\sigma)$.

To deal with the uniqueness issue from item *(b)* retain the assumptions on δ and assume now that $\mathfrak{A}_{L^\top}^{\rm dis} \neq \varnothing$. The goal is to show that if u is a function satisfying

$$\begin{gathered} u \in \big[\mathscr{C}^\infty(\Omega)\big]^M, \quad Lu = 0 \text{ in } \Omega, \quad \mathcal{N}_\kappa(\nabla u) \in L^p(\partial\Omega,\sigma), \\ \text{and } u\Big|_{\partial\Omega}^{\kappa-\text{n.t.}} = 0 \text{ at } \sigma\text{-a.e. point on } \partial\Omega, \end{gathered} \tag{8.4.19}$$

then necessarily $u = 0$ in Ω. To this end, pick a coefficient tensor $A \in \mathfrak{A}_L$ with $A^\top \in \mathfrak{A}_{L^\top}^{\rm dis}$. Then Theorem 4.2.1 guarantees that the operator

$$\tfrac{1}{2}I + K_{A^\top}^\# : \big[L^p(\partial\Omega,\sigma)\big]^M \longrightarrow \big[L^p(\partial\Omega,\sigma)\big]^M \tag{8.4.20}$$

is invertible. Also, [132, Theorem 1.8.19] may be used to conclude (keeping in mind that Ω is connected and $\partial\Omega$ is presently unbounded) that

$$(\nabla u)\Big|_{\partial\Omega}^{\kappa-\text{n.t.}} \text{ exists } \sigma\text{-a.e. on } \partial\Omega, \text{ and } \partial_\nu^A u \text{ belongs to } \big[L^p(\partial\Omega,\sigma)\big]^M, \tag{8.4.21}$$

and that there exists some $c \in \mathbb{C}^M$ with the property that

$$u = -\mathscr{S}_{\rm mod}\big(\partial_\nu^A u\big) + c \text{ in } \Omega. \tag{8.4.22}$$

Taking the conormal derivative of both sides in (8.4.22) and using [132, (1.5.58)], (8.4.21) we arrive at $\big(\frac{1}{2}I + K_{A^\top}^\#\big)\big(\partial_\nu^A u\big) = 0$. In view of (8.4.20) and (8.4.21) this forces $\partial_\nu^A u$ which, when used back in (8.4.22), gives $u = c$ in Ω. Upon recalling the last property in (8.4.19), we finally see that $u = 0$ in Ω, as wanted. This finishes the proof of the claim in item *(b)*.

As regards the claims in item *(c)*, the fact that the Homogeneous Regularity Problem (8.4.2) is well-posed whenever $\mathfrak{A}_L^{\rm dis} \neq \varnothing$ and $\mathfrak{A}_{L^\top}^{\rm dis} \neq \varnothing$ is clear from items *(a)-(b)*. Also, that the unique solution u of (8.4.2) satisfies (in a quantitative fashion) the additional integrability property claimed in (8.4.11) is seen from (4.2.6) with $z = 1/2$, and [129, (8.9.8)].

Next, the claim that if $\mathfrak{A}_L^{\rm dis} = \varnothing$ then the Homogeneous Regularity Problem (8.4.2) may fail to be solvable (actually for boundary data belonging to an infinite dimensional subspace of $\big[\dot{L}^p_1(\partial\Omega,\sigma)/\sim\big]^M$), is seen from (2.6.55). Also, that the Homogeneous Regularity Problem (8.4.2) for L may have more than one solution if $\mathfrak{A}^{\rm dis}_{L^\top}=\varnothing$ is seen from (2.6.4).

Finally, the fact that similar results to those described in items *(a)-(c)* and the first part of item *(d)* are also valid for the spaces listed in item *(e)* may be justified by reasoning as before (now replacing [132, Theorem 1.8.19] by [132, Theorem 1.8.17], and making use of the Fatou-type results from [131, Theorem 3.3.7] and [131, Theorem 3.3.8]). The proof of Theorem 8.4.1 is therefore complete. □

Having dealt with the Homogeneous Regularity Problem in Theorem 8.4.1, we now take up the task of treating the Dirichlet Problem with data in Lebesgue spaces, for weakly elliptic systems and sufficiently flat Ahlfors regular domains with unbounded boundaries. Our principal result in this regard, Theorem 8.4.3 stated below, can be regarded as a Geometric Measure Theoretic sharp version of similar results obtained for the upper half-space in [114].

Theorem 8.4.3 *Pick $n\in\mathbb{N}$ with $n\geq 2$, and let $\Omega\subseteq\mathbb{R}^n$ be an Ahlfors regular domain. Denote by ν its geometric measure theoretic outward unit normal, and abbreviate $\sigma:=\mathcal{H}^{n-1}\lfloor\partial\Omega$. Also, fix an aperture parameter $\kappa>0$ and select an integrability exponent $p\in(1,\infty)$. Given a homogeneous, second-order, constant complex coefficient, weakly elliptic $M\times M$ system L in $\mathbb{R}^n$, consider the Dirichlet Problem*

$$\begin{cases} u\in\big[\mathscr{C}^\infty(\Omega)\big]^M, & Lu=0 \text{ in } \Omega,\\ \mathcal{N}_\kappa u\in L^p(\partial\Omega,\sigma), & u\big|^{\kappa-\mathrm{n.t.}}_{\partial\Omega}=f\in\big[L^p(\partial\Omega,\sigma)\big]^M. \end{cases} \tag{8.4.23}$$

Then, in relation to this, the following claims are true:

(a) [Existence, Estimates, and Integral Representation] *If $\mathfrak{A}^{\rm dis}_L\neq\varnothing$ and $A\in\mathfrak{A}^{\rm dis}_L$, then there exists $\delta\in(0,1)$ which depends only on n, p, A, and the Ahlfors regularity constants of $\partial\Omega$ with the property that if $\|\nu\|_{[\mathrm{BMO}(\partial\Omega,\sigma)]^n}<\delta$ (i.e., if Ω is a δ-*AR *domain in the sense of Definition 3.1.8) then $\frac{1}{2}I+K_A$ is an invertible operator on $\big[L^p(\partial\Omega,\sigma)\big]^M$ and the function $u:\Omega\to\mathbb{C}^M$ defined as*

$$u:=\mathcal{D}_A\Big(\Big(\tfrac{1}{2}I+K_A\Big)^{-1}f\Big) \text{ in } \Omega, \tag{8.4.24}$$

is a solution of the Dirichlet Problem (8.4.23). *Moreover, there exists $C\in(0,\infty)$ independent of f with the property that*

$$\|f\|_{[L^p(\partial\Omega,\sigma)]^M}\leq\big\|\mathcal{N}_\kappa u\big\|_{L^p(\partial\Omega,\sigma)}\leq C\|f\|_{[L^p(\partial\Omega,\sigma)]^M}. \tag{8.4.25}$$

(b) [Uniqueness] *Whenever $\mathfrak{A}^{\rm dis}_{L^\top}\neq\varnothing$, there exists $\delta\in(0,1)$ which depends only on n, p, L, and the Ahlfors regularity constants of $\partial\Omega$ with the property that if $\|\nu\|_{[\mathrm{BMO}(\partial\Omega,\sigma)]^n}<\delta$ (i.e., if Ω is a δ-*AR *domain in the sense of Definition 3.1.8) then the Dirichlet Problem* (8.4.23) *has at most one solution.*

(c) [Well-Posedness] *If* $\mathfrak{A}_L^{\mathrm{dis}} \neq \varnothing$ *and* $\mathfrak{A}_{L^\top}^{\mathrm{dis}} \neq \varnothing$ *then there exists* $\delta \in (0,1)$ *which depends only on n, p, L, and the Ahlfors regularity constants of* $\partial\Omega$ *such that if* $\|\nu\|_{[\mathrm{BMO}(\partial\Omega,\sigma)]^n} < \delta$ *(i.e., if* Ω *is a* δ-AR *domain in the sense of Definition 3.1.8) then the Dirichlet Problem* (8.4.23) *is well posed (i.e., it is uniquely solvable and the solution satisfies the estimate formulated in* (8.4.25)).

(d) [Sharpness] *If* $\mathfrak{A}_L^{\mathrm{dis}} = \varnothing$ *then the Dirichlet Problem* (8.4.23) *may not be solvable. Also, if* $\mathfrak{A}_{L^\top}^{\mathrm{dis}} = \varnothing$ *then the Dirichlet Problem* (8.4.23) *may have more than one solution. In fact, there exists a homogeneous, second-order, constant real coefficient, weakly elliptic* $n \times n$ *system L in* $\mathbb{R}^n$ *with* $\mathfrak{A}_L^{\mathrm{dis}} = \mathfrak{A}_{L^\top}^{\mathrm{dis}} = \varnothing$ *and which satisfies the following two properties: (i) the Dirichlet Problem formulated for this system as in* (8.4.23) *with* $\Omega := \mathbb{R}^n_+$ *fails to have a solution for each non-zero boundary datum belonging to an infinite-dimensional linear subspace of* $\big[L^p(\partial\Omega,\sigma)\big]^n$, *and (ii) the linear space of null-solutions for the Dirichlet Problem formulated for the system L as in* (8.4.23) *with* $\Omega := \mathbb{R}^n_+$ *is actually infinite dimensional.*

In addition, the solvability result described in item (a) above (hence also the well-posedness from item (c)) may actually fail in the absence of any specifications regarding the flatness of the boundary. Specifically, as indicated a bit later in Theorem 8.4.8,

$$\text{the } L^p\text{-Dirichlet Problem formulated as in (8.4.23) for } n=2,\ L:=\Delta \text{ (the two-dimensional Laplacian), and } \Omega:=\Omega_\theta \text{ the infinite planar sector of full aperture angle } \theta \text{ defined in (8.4.12), fails to be Fredholm solvable in the scenario when } \theta\in(\pi,2\pi) \text{ and } p=p_\theta:=1+|\pi-\theta|/\pi=\tfrac{\theta}{\pi}. \tag{8.4.26}$$

(e) [Other Spaces of Boundary Data] *Similar results to those described in items (a)-(d) are valid with* $L^p(\partial\Omega,\sigma)$ *replaced by any of the spaces* $M^{p,\lambda}(\partial\Omega,\sigma)$, $\mathring{M}^{p,\lambda}(\partial\Omega,\sigma)$, *and* $\mathcal{B}^{q,\lambda}(\partial\Omega,\sigma)$ *with* $p,q \in (1,\infty)$ *and* $\lambda \in (0, n-1)$.

Moreover, if $A \in \mathfrak{A}_L^{\mathrm{dis}}$ *then there exist* $\delta \in (0,1)$ *and* $C \in (0,\infty)$ *which depend only on n, p,* λ, *A, and the Ahlfors regularity constants of* $\partial\Omega$ *with the property that if* $\|\nu\|_{[\mathrm{BMO}(\partial\Omega,\sigma)]^n} < \delta$ *then the Dirichlet Problem with data in Morrey spaces*

$$\begin{cases} u \in \big[\mathscr{C}^\infty(\Omega)\big]^M, \quad Lu = 0 \ \text{in} \ \Omega, \\ \mathcal{N}_\kappa u \in M^{p,\lambda}(\partial\Omega,\sigma), \quad u\big|_{\partial\Omega}^{\kappa-\mathrm{n.t.}} = f \in \big[M^{p,\lambda}(\partial\Omega,\sigma)\big]^M, \end{cases} \tag{8.4.27}$$

has a solution satisfying the uniform estimate (for each multi-index $\alpha \in \mathbb{N}_0^n$)

$$\sup_{x\in\Omega}\Big\{\operatorname{dist}(x,\partial\Omega)^{|\alpha|+\frac{n-1-\lambda}{p}}\big|(\partial^\alpha u)(x)\big|\Big\} \leq C\|f\|_{[M^{p,\lambda}(\partial\Omega,\sigma)]^M}, \tag{8.4.28}$$

the fractional Carleson measure estimate

$$\sup_{\substack{x\in\partial\Omega \text{ and}\\ r\in(0,\infty)}} \left\{ r^{-\lambda} \int_{B(x,r)\cap\Omega} |(\nabla u)(y)|^p \operatorname{dist}(y,\partial\Omega)^{p-1}\, d\mathcal{L}^n(y) \right\}^{\frac{1}{p}} \tag{8.4.29}$$
$$\leq C\|f\|_{[M^{p,\lambda}(\partial\Omega,\sigma)]^M}$$

and with the property that

$$\lim_{R\to 0^+} \sup_{\substack{x\in\partial\Omega \text{ and}\\ r\in(0,R)}} \Big(r^{-\lambda} \int_{B(x,r)\cap\Omega} |\nabla u|^p \operatorname{dist}(\cdot,\partial\Omega)^{p-1}\, d\mathcal{L}^n \Big)^{\frac{1}{p}} \tag{8.4.30}$$
$$\leq C \operatorname{dist}\big(f, [\mathring{M}^{p,\lambda}(\partial\Omega,\sigma)]^M\big),$$

where the last distance above is measured in $\big[M^{p,\lambda}(\partial\Omega,\sigma)\big]^M$. *In particular, for each boundary datum* f *belonging to the space* $\big[\mathring{M}^{p,\lambda}(\partial\Omega,\sigma)\big]^M$ *a solution of the boundary problem* (8.4.27) *may be found with the additional property that*

$$\lim_{R\to 0^+} \sup_{\substack{x\in\partial\Omega \text{ and}\\ r\in(0,R)}} \Big(r^{-\lambda} \int_{B(x,r)\cap\Omega} |\nabla u|^p \operatorname{dist}(\cdot,\partial\Omega)^{p-1}\, d\mathcal{L}^n \Big)^{\frac{1}{p}} = 0. \tag{8.4.31}$$

In relation to this theorem it should be noted that, as seen from Proposition 1.6.12, in the scalar case (when $M = 1$) in dimensions $n \geq 3$, the operator L being weakly elliptic always implies $\mathfrak{A}^{\rm dis}_{L} \neq \varnothing$ and $\mathfrak{A}^{\rm dis}_{L^\top} \neq \varnothing$. In addition, in the scalar case in the two-dimensional setting (i.e., for $M = 1$ and $n = 2$), if L satisfies the Legendre-Hadamard strong ellipticity condition or, more generally (cf. [131, Lemma 1.4.24]), if $L = \operatorname{div} A\nabla$ for some 2×2 complex matrix A with $\operatorname{sym} A := \frac{1}{2}(A+A^\top) \in M^0_2$ (cf. [131, (1.4.187)]), then Proposition 1.6.11 and Proposition 1.6.13 ensure that $\mathfrak{A}^{\rm dis}_{L} \neq \varnothing$ and $\mathfrak{A}^{\rm dis}_{L^\top} \neq \varnothing$. As such, the statement of Theorem 8.4.3 may be significantly streamlined, since genuine well-posedness holds in each of the aforementioned situations.

It should be noted that the solution u of the Dirichlet Problem (8.4.23) satisfies other relevant size estimates. Here we wish to record three such examples. First, the estimate in (8.4.25) together with [129, Proposition 8.6.10] imply

$$\sup_{\rho\in(0,\infty)} \Big(\frac{1}{\rho} \int_{O_\rho} |u|^p\, d\mathcal{L}^n \Big)^{1/p} \leq C\|f\|_{[L^p(\partial\Omega,\sigma)]^M} \tag{8.4.32}$$

where $O_\rho := \{x \in \Omega : \operatorname{dist}(x,\partial\Omega) < \rho\}$ for each $\rho > 0$. Second, from the equivalence in (8.4.25) and the last estimate in [129, Proposition 8.6.3] (currently used with $E := \Omega$) we see that

$$\Big(\int_\Omega |u|^{np/(n-1)}\, d\mathcal{L}^n \Big)^{(n-1)/(np)} \leq C\|f\|_{[L^p(\partial\Omega,\sigma)]^M}. \tag{8.4.33}$$

Third, from work in [129, §8.5] it is known that we may replace the nontangential maximal operator (employed in the formulation of the Dirichlet Problem (8.4.23), as well as its versions described in item *(e)*) by the tangential maximal operator, associated as in [129, Definition 8.5.1] with a sufficiently large power; cf. (A.0.165).

Lastly, concerning the very formulation of Theorem 8.4.3, similar considerations to those made in Remark 8.4.2 are presently applicable. Specifically, we have:

Remark 8.4.4 *A more refined statement of the well-posedness result in item (c) of Theorem 8.4.3 goes as follows: Pick $n, M \in \mathbb{N}$ with $n \geq 2$ and consider a weakly elliptic homogeneous constant complex coefficient second-order $M \times M$ system L in $\mathbb{R}^n$ with $\mathfrak{A}_L^{\rm dis} \neq \emptyset$ and $\mathfrak{A}_{L^\top}^{\rm dis} \neq \emptyset$. Then for each constant $C_A \in (0, \infty)$ and each compact interval $I \subset (1, \infty)$ there exists a threshold $\delta \in (0, 1)$ which depends only on n, C_A, I, and L with the following significance. Suppose $\Omega \subseteq \mathbb{R}^n$ is an Ahlfors regular domain such that the Ahlfors regularity constants of $\partial\Omega$ are controlled by C_A. Abbreviate $\sigma := \mathcal{H}^{n-1}\lfloor\partial\Omega$ and denote by ν the geometric measure theoretic outward unit normal to Ω. Lastly, fix an integrability exponent $p \in I$. Then whenever $\|\nu\|_{[\mathrm{BMO}(\partial\Omega,\sigma)]^n} < \delta$ it follows that the Dirichlet Problem (8.4.23) is well posed (in the sense that it is uniquely solvable, and the solution also satisfies a naturally accompanying estimate). Furthermore, analogous considerations apply to all results in the statement of Theorem 8.4.3.*

This follows from the invertibility results from Remark 4.2.2 and the proof of Theorem 8.4.3 given below.

Proof of Theorem 8.4.3 Suppose that $\|\nu\|_{[\mathrm{BMO}(\partial\Omega,\sigma)]^n} < \delta$ for some $\delta \in (0, 1)$ small enough as to guarantee that Ω is a connected two-sided NTA domain with an unbounded boundary (cf. Theorem 3.1.5). After eventually further decreasing the value of δ, we may also assume that the conclusions in Theorem 4.2.1 are valid (for suitable choices of the parameters involved in its statement).

Consider the claims in item *(a)*, working under the assumption that $\mathfrak{A}_L^{\rm dis} \neq \emptyset$. Then for $A \in \mathfrak{A}_L^{\rm dis}$ Theorem 4.2.1 ensures that the operator $\frac{1}{2}I + K_A$ is invertible on $\big[L^p(\partial\Omega, \sigma)\big]^M$. With this in hand, the fact that the function u defined as in (8.4.24) is a solution of the Dirichlet Problem (8.4.23) which satisfies (8.4.25) follows from [132, (1.3.24), (1.5.1), (1.5.20)], and [129, (8.9.8)]. Next we shall deal with the uniqueness claim made in item *(b)*, assuming that $\mathfrak{A}_{L^\top}^{\rm dis} \neq \emptyset$. Let $p' \in (1, \infty)$ denote the Hölder conjugate exponent of p. Then from item *(a)* in Theorem 8.4.1 (for $L^\top$ in place of L and p' in place of p) we know that the Inhomogeneous Regularity Problem

$$\begin{cases} u \in \big[\mathscr{C}^\infty(\Omega)\big]^M, \quad L^\top u = 0 \text{ in } \Omega, \\ \mathcal{N}_\kappa u,\ \mathcal{N}_\kappa(\nabla u) \in L^p(\partial\Omega, \sigma), \quad u\big|_{\partial\Omega}^{\kappa-\mathrm{n.t.}} = f \in \big[L_1^{p'}(\partial\Omega, \sigma)\big]^M, \end{cases} \tag{8.4.34}$$

is always solvable. Granted this, [131, Theorem 4.3.1] applies and proves that the Dirichlet Problem (8.4.23) has at most one solution. Next, *(c)* is implied by *(a)*-*(b)*, while *(d)* follows from (2.6.13) and (2.6.26). Finally, the claims in item *(e)* may be justified by reasoning in a similar fashion as above, now also making use of [132, Theorem 3.3.1], and availing ourselves of [132, Theorem 3.3.3]. □

Remark 8.4.5 *It is of interest to compare Theorems 8.4.1-8.4.3 with explicit well-posedness results in planar infinite sectors, and for the two-dimensional Laplacian. Specifically, let*

$$\begin{gathered}\Omega_\theta := \Big\{x \in \mathbb{R}^2 \setminus \{x_o\} : \tfrac{x-x_o}{|x-x_o|} \cdot \xi > \cos(\theta/2)\Big\} \\ \textit{for some } x_o \in \mathbb{R}^2,\ \theta \in (0, 2\pi),\ \textit{and } \xi \in S^1,\end{gathered} \tag{8.4.35}$$

and abbreviate $\sigma_\theta := \mathcal{H}^1 \lfloor \partial\Omega_\theta$. *Also, fix some* $\kappa \in (0, \infty)$. *In this setting, for each given* $p \in (1, \infty)$ *consider the* L^p*-Dirichlet Problem in* Ω_θ *for* $L := \Delta$, *i.e.,*

$$\begin{cases} u \in \mathscr{C}^\infty(\Omega_\theta), \quad \Delta u = 0 \ \textit{in}\ \Omega_\theta, \\ \mathcal{N}_\kappa u \in L^p(\partial\Omega_\theta, \sigma_\theta), \quad u\big|_{\partial\Omega_\theta}^{\kappa-\mathrm{n.t.}} = f \in L^p(\partial\Omega_\theta, \sigma_\theta), \end{cases} \tag{8.4.36}$$

as well as the $\dot{L}^p_1$*-Regularity Problem in* Ω_θ *for* $L := \Delta$, *i.e.,*

$$\begin{cases} u \in \mathscr{C}^\infty(\Omega_\theta), \quad \Delta u = 0 \ \textit{in}\ \Omega_\theta, \\ \mathcal{N}_\kappa(\nabla u) \in L^p(\partial\Omega_\theta, \sigma_\theta), \quad u\big|_{\partial\Omega_\theta}^{\kappa-\mathrm{n.t.}} = f \in \dot{L}^p_1(\partial\Omega_\theta, \sigma_\theta). \end{cases} \tag{8.4.37}$$

Employing boundary layer potentials (much as in the proofs of Theorems 8.4.1-8.4.3), from (4.2.160) *we then see that*

$$\begin{gathered}\textit{the } L^p\textit{-Dirichlet Problem } (8.4.36) \textit{ has a solution} \\ \textit{whenever } p \neq p_\theta := 1 + |\pi - \theta|/\pi\end{gathered} \tag{8.4.38}$$

while from (4.2.187) (*or, alternatively,* (4.2.189)) *we conclude that*

$$\begin{gathered}\textit{the } \dot{L}^p_1\textit{-Regularity Problem } (8.4.37) \textit{ has a solution} \\ \textit{whenever } p \neq q_\theta := 1 + \pi/|\pi - \theta|.\end{gathered} \tag{8.4.39}$$

Granted these solvability results, and bearing in mind that $1/p_\theta + 1/q_\theta = 1$, *from* [131, Theorem 4.3.1] *and* [131, Theorem 4.3.2] *we then conclude that, in fact,*

$$\begin{gathered}\textit{the } L^p\textit{-Dirichlet Problem } (8.4.36) \textit{ is well-posed} \\ \textit{whenever } p \neq p_\theta := 1 + |\pi - \theta|/\pi,\end{gathered} \tag{8.4.40}$$

$$\begin{gathered}\textit{the } \dot{L}^p_1\textit{-Regularity Problem } (8.4.37) \textit{ is well-posed} \\ \textit{whenever } p \neq q_\theta := 1 + \pi/|\pi - \theta|.\end{gathered} \tag{8.4.41}$$

See also [51, Theorems 1.2-1.3, pp. 97-98] *in this regard*[97].

It is also instructive to compare the efficacy of the method of boundary layer potentials, which has produced the well-posedness result recorded in (8.4.40), with the alternative approach (based on conformal mappings and weighted spaces) described in the remark below.

---

[97] all these results should be contrasted with the last claim in [91, Theorem 4.4, p. 162]

Remark 8.4.6 *Given $\theta \in (0, 2\pi)$, consider the infinite planar sector*

$$\Omega_\theta := \Big\{(x, y) \in \mathbb{R}^2 \setminus \{0\} : \frac{y}{\sqrt{x^2 + y^2}} > \cos(\theta/2)\Big\} \tag{8.4.42}$$

of (full) aperture θ, and symmetry line along the vertical axis. Henceforth abbreviate $\sigma_\theta := \mathcal{H}^1\lfloor\partial\Omega_\theta$ and fix some $\kappa \in (0, \infty)$. In this setting, for a given $p \in (1, \infty)$, we are interested in the well-posedness of the L^p-Dirichlet Problem in Ω_θ for $L := \Delta$, i.e.,

$$\begin{cases} u \in \mathscr{C}^\infty(\Omega_\theta), \quad \Delta u = 0 \text{ in } \Omega_\theta, \\ \mathcal{N}_\kappa u \in L^p(\partial\Omega_\theta, \sigma_\theta), \quad u\big|^{\kappa-\mathrm{n.t.}}_{\partial\Omega_\theta} = f \in L^p(\partial\Omega_\theta, \sigma_\theta). \end{cases} \tag{8.4.43}$$

To this end, observe that the function[98]

$$\Phi : \mathbb{C}_+ \to \Omega_\theta, \quad \Phi(z) := \mathrm{i}\big[(-\mathrm{i})z\big]^{\theta/\pi} \text{ for each } z \in \mathbb{C}_+, \tag{8.4.44}$$

maps $\mathbb{C}_+$ conformally onto the sector Ω_θ. In addition, Φ extends to a homeomorphism $\Phi : \overline{\mathbb{R}^2_+} \to \overline{\Omega_\theta}$, with $\Phi(\partial\mathbb{R}^2_+) = \partial\Omega_\theta$. Also, under the canonical identification $\partial\mathbb{R}^2_+ \equiv \mathbb{R}$, the function $\Phi : \mathbb{R} \to \partial\Omega_\theta$ becomes a parametrization of the curve $\partial\Omega_\theta$, explicitly given by

$$\Phi(x) = \mathrm{e}^{-\mathrm{i}(\operatorname{sgn} x)(\theta/2)} \cdot |x|^{\theta/\pi} \text{ for each } x \in \mathbb{R}. \tag{8.4.45}$$

Furthermore, regarded as such, we have $\Phi \in \mathscr{C}^\infty\big(\mathbb{R} \setminus \{0\}\big)$ and

$$\Phi'(x) = (\theta/\pi)(\operatorname{sgn} x)\mathrm{e}^{-\mathrm{i}(\operatorname{sgn} x)(\theta/2)} \cdot |x|^{(\theta/\pi)-1} \text{ for each } x \in \mathbb{R} \setminus \{0\}, \tag{8.4.46}$$

hence,

$$|\Phi'(x)| = (\theta/\pi)|x|^{(\theta/\pi)-1} \text{ for each } x \in \mathbb{R} \setminus \{0\}. \tag{8.4.47}$$

Note that, for each $p \in (1, \infty)$, the assignment

$$\begin{array}{c} L^p(\partial\Omega_\theta, \sigma_\theta) \ni f \longmapsto \widetilde{f} := f \circ \Phi \in L^p\big(\mathbb{R}, |\Phi'(x)|\mathrm{d}x\big) \\ \text{is a well-defined linear isometric isomorphism.} \end{array} \tag{8.4.48}$$

In relation to this assignment, from (8.4.46) *and* [129, Example 8.7.6] *we see that $|\Phi'| \in A_p(\mathbb{R}, \mathcal{L}^1)$ if and only if $-1 < (\theta/\pi) - 1 < p - 1$, hence*

$$|\Phi'| \in A_p(\mathbb{R}, \mathcal{L}^1) \iff p > \theta/\pi. \tag{8.4.49}$$

Another way to put this is to write

$$\begin{array}{c} \text{if } \theta \in (0, \pi] \text{ then } |\Phi'| \in A_p(\mathbb{R}, \mathcal{L}^1) \text{ for each } p \in (1, \infty), \text{ while} \\ \text{if } \theta \in (\pi, 2\pi) \text{ then } |\Phi'| \in A_p(\mathbb{R}, \mathcal{L}^1) \text{ if and only if } p \in \big(\theta/\pi, \infty\big). \end{array} \tag{8.4.50}$$

---

[98] each complex number $\zeta \in \mathbb{C} \setminus [0, -\infty)$ may be uniquely written as $\zeta = \rho\mathrm{e}^{\mathrm{i}\alpha}$ for some $\rho \in (0, \infty)$ and $\alpha \in (-\pi, \pi)$, and we agree to define $\zeta^t := \rho^a \cdot \mathrm{e}^{\mathrm{i}\alpha t} \in \mathbb{C}$ for any given exponent $t \in (0, \infty)$

As a consequence,

$$\text{for every angle } \theta \in (0, 2\pi) \text{ we have } |\Phi'| \in A_2(\mathbb{R}, \mathcal{L}^1) \text{ so, in particular, } |\Phi'(x)|\,dx \text{ is a doubling Borel measure on } \mathbb{R}. \tag{8.4.51}$$

From (8.4.51), (8.4.48), [129, Lemma 8.1.6], [129, Proposition 8.4.1], *and Calderón's Theorem (recalled in* (2.1.7)) *we see that, under the transformation* $u \mapsto \widetilde{u} := u \circ \Phi$, *for each given* $\theta \in (0, 2\pi)$ *and* $p \in (1, \infty)$ *the* L^p*-Dirichlet Problem* (8.4.43) *becomes equivalent to the following weighted Dirichlet Problem for the Laplacian in the upper half-plane:*

$$\begin{cases} \widetilde{u} \in \mathscr{C}^\infty(\mathbb{R}^2_+), \quad \Delta\widetilde{u} = 0 \ \text{in } \mathbb{R}^2_+, \\ \mathcal{N}_\kappa \widetilde{u} \in L^p\big(\mathbb{R}, |\Phi'(x)|\,dx\big), \\ \widetilde{u}\big|^{\kappa-\text{n.t.}}_{\partial\mathbb{R}^2_+} = \widetilde{f} \in L^p\big(\mathbb{R}, |\Phi'(x)|\,dx\big). \end{cases} \tag{8.4.52}$$

Since in the regime where $|\Phi'|$ *is a Muckenhoupt* A_p *weight the latter is known to be well-posed (cf.* [114], [111]), *we ultimately conclude from* (8.4.50) *that*

$$\text{the } L^p\text{-Dirichlet Problem (8.4.43) is well-posed whenever } \theta \in (0, 2\pi) \text{ and } p \in \big(\max\{1, \theta/\pi\}, \infty\big). \tag{8.4.53}$$

Regarding the Dirichlet Problem for the Laplacian in two-dimensional sectors, it is natural to contrast the method of boundary layer potentials (from Remark 8.4.5) with the approach based on conformal mappings (from Remark 8.4.6); see below.

Remark 8.4.7 *From* (8.4.40) *and* (8.4.53) *we see that, as far as the Dirichlet Problem is concerned, the method of boundary layer potentials (cf. Remark 8.4.5) is more effective in the case when* $\theta \in (\pi, 2\pi)$, *while the method described in Remark 8.4.6 (based on conformal mappings and weighted spaces) is more effective in the situation when* $\theta \in (0, \pi)$. *Ultimately, a combination of* (8.4.40) *and* (8.4.53) *shows that, for each* $\theta \in (0, 2\pi)$ *and* $p \in (1, \infty)$,

$$\text{the } L^p\text{-Dirichlet Problem (8.4.36) is well-posed if either } p > \max\{1, \theta/\pi\}, \text{ or } p \neq p_\theta := 1 + |\pi - \theta|/\pi. \tag{8.4.54}$$

Note that the condition in the last line of (8.4.54) amounts to having any exponent $p \in (1, \infty)$ if $\theta \in (0, \pi]$, and any exponent $p \in (1, \infty) \setminus \{p_\theta\}$ if $\theta \in (\pi, 2\pi)$. In particular, the latter result shows that there exist convex domains with boundary as far from being flat as wanted, yet for which the corresponding L^p-Dirichlet Problem for the Laplacian is well posed for each $p \in (1, \infty)$.

In fact, we can further sharpen (8.4.54) as follows:

Theorem 8.4.8 *Fix* $\theta \in (0, 2\pi)$ *arbitrary, and consider the infinite planar sector* $\Omega_\theta \subseteq \mathbb{R}^2$ *of (full) aperture angle* θ *and symmetry line along the vertical axis, described in* (8.4.42). *Abbreviate* $\sigma_\theta := \mathcal{H}^1 \lfloor \partial\Omega_\theta$ *and fix a background parameter* $\kappa \in (0, \infty)$. *Then the following more refined version on* (8.4.54) *holds:*

$$\text{if } \theta \in (0,\pi] \text{ then the } L^p\text{-Dirichlet Problem (8.4.36) for the Laplacian in } \Omega_\theta \text{ is well-posed for any } p \in (1,\infty), \text{ while in the scenario in which } \theta \in (\pi,2\pi) \text{ it follows that the } L^p\text{-Dirichlet Problem (8.4.36) (8.4.43) for the Laplacian in } \Omega_\theta \text{ is well-posed if and only if the integrability exponent satisfies } p \in (1,\infty)\setminus\{p_\theta\} = (1,\infty)\setminus\{\tfrac{\theta}{\pi}\}. \tag{8.4.55}$$

Moreover, corresponding to the case when one has $\theta \in (\pi, 2\pi)$ *and consider the exponent* $p = p_\theta := 1 + |\pi - \theta|/\pi = \frac{\theta}{\pi}$*, the space of admissible boundary data for the Dirichlet Problem* (8.4.43) *(i.e., the collection of functions from* $L^{p_\theta}(\partial\Omega_\theta, \sigma_\theta)$ *for which said problem is solvable when the functions in question are used as boundary data) coincides with the image of* R *acting on* $L^{p_\theta}(\partial\Omega_\theta, \sigma_\theta)$*, namely*

$$\begin{aligned}\Big\{u\big|^{\kappa-\mathrm{n.t.}}_{\partial\Omega_\theta} : u \in \mathscr{C}^\infty(\Omega_\theta),\ \Delta u = 0 \text{ in } \Omega_\theta,\ \mathcal{N}_\kappa u \in L^{p_\theta}(\partial\Omega_\theta, \sigma_\theta)\Big\} \\ = \Big\{Rg : g \in L^{p_\theta}(\partial\Omega_\theta, \sigma_\theta)\Big\},\end{aligned} \tag{8.4.56}$$

where R *is the boundary-to-boundary singular integral operator associated as in* [131, (2.5.333)] *with the set* Ω_θ^+*, i.e.,*

$$\begin{aligned}Rg(x) := \lim_{\varepsilon\to 0^+} \int_{y\in\partial\Omega_\theta,\ |x-y|>\varepsilon} \Big[(\nu_\theta)_1(y)(\partial_2 E_\Delta)(y-x) \\ - (\nu_\theta)_2(y)(\partial_1 E_\Delta)(y-x)\Big] g(y)\, d\sigma_\theta(y),\end{aligned} \tag{8.4.57}$$

for each function $g \in L^1\big(\partial\Omega_\theta, \frac{\sigma_\theta(x)}{1+|x|}\big)$ *and* σ_θ*-a.e.* $x \in \partial\Omega_\theta$*, where* $(\nu_\theta)_1$, $(\nu_\theta)_2$ *are the components of the outward unit normal* ν_θ *to* Ω_θ*, and* E_Δ *is the standard fundamental solution for the two-dimensional Laplacian (cf.* (A.0.77)*).*

As a corollary of (8.4.56) *and* (4.2.192)*, it follows that*

$$\text{corresponding to } \theta \in (\pi,2\pi) \text{ and } p = p_\theta := 1 + |\pi-\theta|/\pi = \tfrac{\theta}{\pi}, \text{ the space of admissible boundary data for the Dirichlet Problem (8.4.43) is dense and of infinite co-dimension (in particular, not closed) in } L^{p_\theta}(\partial\Omega_\theta, \sigma_\theta). \tag{8.4.58}$$

In particular, this shows that

$$\text{the Dirichlet Problem (8.4.43) fails to be Fredholm solvable in the case when } \theta \in (\pi,2\pi) \text{ and } p = p_\theta := 1 + |\pi-\theta|/\pi = \tfrac{\theta}{\pi}. \tag{8.4.59}$$

Finally, regarding the space of nullsolutions of the Dirichlet Problem (8.4.43) *corresponding to* $\theta \in (\pi, 2\pi)$ *and* $p = p_\theta := 1 + |\pi - \theta|/\pi = \frac{\theta}{\pi}$*, the operator acting on each* $g \in L^1\big(\partial\Omega_\theta, \frac{\sigma_\theta(x)}{1+|x|}\big)$ *at each* $x \in \Omega_\theta$ *according to*

$$\mathcal{R}g(x) := \int_{\partial\Omega} \Big[(\nu_\theta)_1(y)(\partial_2 E_\Delta)(y-x) - (\nu_\theta)_2(y)(\partial_1 E_\Delta)(y-x)\Big] g(y)\, d\sigma(y), \tag{8.4.60}$$

maps Ker $\big(R; L^{p_\theta}(\partial\Omega_\theta, \sigma_\theta)\big)$ *in a linear and bijective fashion onto the space*

$$\Big\{u \in \mathscr{C}^\infty(\Omega_\theta) : \Delta u = 0 \text{ in } \Omega_\theta,\ \mathcal{N}_\kappa u \in L^{p_\theta}(\partial\Omega_\theta, \sigma_\theta),\ u\Big|_{\partial\Omega_\theta}^{\kappa-\text{n.t.}} = 0\Big\}. \tag{8.4.61}$$

Proof In view of (8.4.54), the claims made in (8.4.55) once we establish (8.4.56). To this end, fix $\theta \in (\pi, 2\pi)$ and set $\Omega_\theta^+ := \Omega_\theta$ and $\Omega_\theta^- := \mathbb{R}^2 \setminus \overline{\Omega_\theta}$. Then $\Omega_\theta^+, \Omega_\theta^-$ share a common boundary, $\partial\Omega_\theta$, and Ω_θ^- is an infinite sector of (full) aperture angle $2\pi - \theta \in (0, \pi)$. Next, pick an arbitrary function $f \in L^{p_\theta}(\partial\Omega_\theta, \sigma_\theta)$ belonging to the space in the left side of (8.4.56). Given the goals we presently have in mind, there is no loss of generality in assuming that f is real-valued. Then there exists a function $u_+ : \Omega_\theta^+ \to \mathbb{R}$ satisfying

$$\begin{cases} u_+ \in \mathscr{C}^\infty(\Omega_\theta^+), \quad \Delta u_+ = 0 \text{ in } \Omega_\theta^+, \\ \mathcal{N}_\kappa u_+ \in L^{p_\theta}(\partial\Omega_\theta, \sigma_\theta), \\ u_+\Big|_{\partial\Omega_\theta^+}^{\kappa-\text{n.t.}} = f \text{ at } \sigma_\theta\text{-a.e. point on } \partial\Omega_\theta. \end{cases} \tag{8.4.62}$$

Since $\theta \in (\pi, 2\pi)$ forces $p_\theta = \frac{\theta}{\pi} > \max\{1, (2\pi - \theta)/\pi\}$, we conclude from (8.4.54) (for Ω_θ^- in place of Ω_θ) that there exists a (unique) function $u_- : \Omega_\theta^- \to \mathbb{R}$ satisfying

$$\begin{cases} u_- \in \mathscr{C}^\infty(\Omega_\theta^-), \quad \Delta u_- = 0 \text{ in } \Omega_\theta^-, \\ \mathcal{N}_\kappa u_- \in L^{p_\theta}(\partial\Omega_\theta, \sigma_\theta), \\ u_-\Big|_{\partial\Omega_\theta^-}^{\kappa-\text{n.t.}} = f \text{ at } \sigma_\theta\text{-a.e. point on } \partial\Omega_\theta. \end{cases} \tag{8.4.63}$$

Let $w_\pm : \Omega_\theta^\pm \to \mathbb{R}$ be harmonic conjugate functions of $u_\pm$ with the property that $\mathcal{N}_\kappa w_\pm \in L^{p_\theta}(\partial\Omega_\theta, \sigma_\theta)$ and $g_\pm := w_\pm\Big|_{\partial\Omega_\theta^\pm}^{\kappa-\text{n.t.}}$ exist σ_θ-a.e. on $\partial\Omega_\theta$. The existence of such functions is guaranteed by [91, Theorems 4.1-4.2, p. 161] (and the subsequent remark). This implies that $F_\pm := u_\pm + \mathrm{i}w_\pm : \Omega_\theta^\pm \to \mathbb{C}$ are holomorphic functions with the property that $\mathcal{N}_\kappa F_\pm \in L^{p_\theta}(\partial\Omega_\theta, \sigma_\theta)$ and $F_\pm\Big|_{\partial\Omega_\theta^\pm}^{\kappa-\text{n.t.}} = f + \mathrm{i}g_\pm$ at σ_θ-a.e. point on $\partial\Omega_\theta$. As such, [131, Theorem 1.1.1] guarantees that the following Cauchy reproducing formula is valid:

$$F_+ = \mathscr{C}(f + \mathrm{i}g_+) \text{ in } \Omega_\theta^+, \tag{8.4.64}$$

where $\mathscr{C}$ is the boundary-to-domain Cauchy operator associated with Ω_θ^+ (cf. (A.0.57)). A similar Cauchy reproducing formula is valid for F_-. Bring in the boundary-to-domain harmonic double layer potential operator $\mathcal{D}$ associated with Ω_θ^+, and recall the boundary-to-domain integral operator $\mathcal{R}$ from (8.4.60). Then, as noted in [132, (1.6.83)], we have $\mathscr{C} = \mathcal{D} - \mathrm{i}\mathcal{R}$. Using this back in (8.4.64) and taking real parts yields

$$u_+ = \mathcal{D}f + \mathcal{R}g_+ \text{ in } \Omega_\theta^+. \tag{8.4.65}$$

After going nontangentially to the boundary in (8.4.65), on account of (8.4.62), the jump-formula for the harmonic double layer, and the trace formula in [132, (1.6.82)] we arrive at

$$f = \left(\tfrac{1}{2}I + K_\Delta\right)f + Rg_+ \text{ on } \partial\Omega_\theta. \tag{8.4.66}$$

Carrying out a similar analysis in relation to u_- also gives (bearing in mind that the outward unit normal for Ω_θ^- is the opposite of the outward unit normal for Ω_θ^+)

$$f = \left(\tfrac{1}{2}I - K_\Delta\right)f - Rg_- \text{ on } \partial\Omega_\theta. \tag{8.4.67}$$

Adding up (8.4.66) and (8.4.67) proves that $f = R(g_+ - g_-)$ on $\partial\Omega_\theta$. Since $g_\pm$ are in $L^{p_\theta}(\partial\Omega_\theta, \sigma_\theta)$, we ultimately conclude that the function f belongs to the image of R acting on $L^{p_\theta}(\partial\Omega_\theta, \sigma_\theta)$. This gives the left-to-right inclusion in (8.4.56).

As regards the opposite direction, pick an arbitrary $g \in L^{p_\theta}(\partial\Omega_\theta, \sigma_\theta)$. The right-to-left inclusion in (8.4.56) follows upon noting that if $u := \mathcal{R}g$ in Ω_θ then, by (8.4.60) and the trace formula in [132, (1.6.82)], we have $u \in \mathscr{C}^\infty(\Omega_\theta)$, $\Delta u = 0$ in Ω_θ, $\mathcal{N}_\kappa u \in L^{p_\theta}(\partial\Omega_\theta, \sigma_\theta)$, and $u\big|_{\partial\Omega_\theta}^{\kappa-\mathrm{n.t.}} = Rg$ at σ_θ-a.e. point on $\partial\Omega_\theta$. As far as the very final claim in the statement of the theorem is concerned, fix an arbitrary angle $\theta \in (\pi, 2\pi)$ and set $p_\theta := 1 + |\pi - \theta|/\pi = \frac{\theta}{\pi}$. From (8.4.60) and [132, (1.6.82)] we see that the linear operator $\mathcal{R}$ indeed maps $\operatorname{Ker}\big(R; L^{p_\theta}(\partial\Omega_\theta, \sigma_\theta)\big)$ into the space of nullsolutions of the Dirichlet Problem (8.4.43) corresponding to $p := p_\theta$, described in (8.4.61). The fact that this is surjective follows from (8.4.65) with $f = 0$. Finally, to show that the operator in question is also injective, suppose $g \in \operatorname{Ker}\big(R; L^{p_\theta}(\partial\Omega_\theta, \sigma_\theta)\big)$ is such that $\mathcal{R}g = 0$ in Ω_θ. Without loss of generality we may assume that g is real-valued. Define $F := \mathscr{C}g$ in Ω_θ. It follows that F is holomorphic and from [132, (1.6.83)] we see that $F = \mathcal{D}g - i\mathcal{R}g = \mathcal{D}g$ in Ω_θ. This makes F real-valued, hence a constant. Since we also have $\mathcal{N}_\kappa F \in L^{p_\theta}(\partial\Omega_\theta, \sigma_\theta)$, we ultimately conclude that $F = 0$ in Ω_θ. Thus, $\mathcal{D}g = 0$ in Ω_θ which, after taking nontangentially boundary traces yields $\left(\frac{1}{2}I + K_\Delta\right)g = 0$ on $\partial\Omega_\theta$. Next, for each $x \in \Omega_\theta^-$ define

$$w(x) := \int_{\partial\Omega} \Big[(\nu_\theta)_1(y)(\partial_2 E_\Delta)(y - x) - (\nu_\theta)_2(y)(\partial_1 E_\Delta)(y - x)\Big] g(y)\, d\sigma(y)$$

and note that [132, (1.6.81)] implies

$$\begin{aligned} & w \in \mathscr{C}^\infty(\Omega_\theta^-), \quad \Delta w = 0 \text{ in } \Omega_\theta^-, \quad \mathcal{N}_\kappa w \in L^{p_\theta}(\partial\Omega_\theta, \sigma_\theta), \\ & \text{and } w\big|_{\partial\Omega_\theta^-}^{\kappa-\mathrm{n.t.}} = Rg = 0 \text{ at } \sigma_\theta\text{-a.e. point on } \partial\Omega_\theta. \end{aligned} \tag{8.4.68}$$

Also, if $\mathcal{D}^-$ denotes the boundary-to-domain harmonic double layer potential operator associated with Ω_θ^-, it has been noted in the proof of [132, Corollary 1.6.8] that that w and $-\mathcal{D}^- g$ are conjugate harmonic functions in Ω_θ^-. However, from (8.4.68) and the uniqueness of the L^{p_θ}-Dirichlet Problem for the Laplacian in Ω_θ^- (cf. Remark 8.4.7) we deduce that $w = 0$ in Ω_θ^-. This makes $\mathcal{D}^- g$ a constant in Ω_θ^-, and since $\mathcal{N}_\kappa(\mathcal{D}^- g) \in L^{p_\theta}(\partial\Omega_\theta, \sigma_\theta)$ we see that actually $\mathcal{D}^- g = 0$ in Ω_θ^-. Going nontangentially to the boundary we obtain $\left(-\frac{1}{2}I + K_\Delta\right)g = 0$ on $\partial\Omega_\theta$ so, finally, $g = \left(\frac{1}{2}I + K_\Delta\right)g - \left(-\frac{1}{2}I + K_\Delta\right)g = 0 - 0 = 0$ on $\partial\Omega_\theta$. This shows that the operator

(8.4.60) acting from $\mathrm{Ker}\,\big(R; L^{p_\theta}(\partial\Omega_\theta, \sigma_\theta)\big)$ is invective, as wanted. The proof of Theorem 8.4.8 is now finished. □

We may also use Theorem 8.4.8 to emphasize the relevance of the smallness condition imposed on the BMO seminorm of the geometric measure theoretic outward unit normal in the statement of Theorem 8.4.3.

Remark 8.4.9 *Once again, associate the infinite sector $\Omega_\theta \subseteq \mathbb{R}^2$ with any given angle $\theta \in (\pi, 2\pi)$ as in* (8.4.35)*. Also, abbreviate $\sigma_\theta := \mathcal{H}^1 \lfloor \partial\Omega_\theta$ and denote by ν_θ the outward unit normal to Ω_θ. Note that $p_\theta := 1+|\pi-\theta|/\pi \in (1, 2)$ and $p_\theta \to 1$ if and only if $\theta \to \pi$. Also, recall from* (4.2.161) *that $\|\nu_\theta\|_{\mathrm{BMO}(\partial\Omega_\theta,\sigma_\theta)} \approx p_\theta - 1$ uniformly in $\theta \in (0, 2\pi)$. Then, as seen from this and the result described in Theorem 8.4.8, for the class of domains Ω_θ with $\theta \in (\pi, 2\pi)$, the well-posedness of the L^p-Dirichlet Problem for the Laplacian for each $p \in I := [a, 2]$ with $a \in (1, 2)$ close to 1 is equivalent to the demand that $\|\nu_\theta\|_{\mathrm{BMO}(\partial\Omega_\theta,\sigma_\theta)}$ is small. This highlights the significance of the latter property in the context of Theorem 8.4.3 (see the formulation given in Remark 8.4.4).*

Next we wish to carry out a similar analysis to the one above but in relation to the Homogeneous Regularity Problem for the Laplacian (in infinite planar sectors). To this end, retain the notation introduced in Remark 8.4.6. In particular, since $\Phi : \mathbb{R} \to \partial\Omega_\theta$ is a parametrization of the curve $\partial\Omega_\theta$, it follows that

$$\int_{\partial\Omega_\theta} f\,\mathrm{d}\sigma_\theta = \int_{\mathbb{R}} (f \circ \Phi) \cdot |\Phi'|\,\mathrm{d}\mathcal{L}^1 \tag{8.4.69}$$

for each σ_θ-measurable function $f : \partial\Omega_\theta \to \mathbb{R}$. Also, the outward unit normal to Ω_θ is given by

$$\nu_\theta\big(\Phi(x)\big) = (-\mathrm{i})\Phi'(x)/|\Phi'(x)| \ \text{ for } \mathcal{L}^1\text{-a.e. } x \in \mathbb{R}, \tag{8.4.70}$$

or, with (Φ_1, Φ_2) denoting the scalar components of Φ,

$$\begin{gathered}(\nu_\theta)_1\big(\Phi(x)\big) = \Phi_2'(x)/|\Phi'(x)| \ \text{ and } \ (\nu_\theta)_2\big(\Phi(x)\big) = -\Phi_1'(x)/|\Phi'(x)| \\ \text{at } \mathcal{L}^1\text{-a.e. point } x \in \mathbb{R}.\end{gathered} \tag{8.4.71}$$

Consequently, for each $\varphi \in \mathscr{C}_c^\infty(\mathbb{R}^2)$ and $\mathcal{L}^1$-a.e. point $x \in \mathbb{R}$ we have

$$\begin{aligned}(\partial_\tau\varphi)\big(\Phi(x)\big) &= (\nu_\theta)_1\big(\Phi(x)\big)(\partial_2\varphi)\big(\Phi(x)\big) - (\nu_\theta)_2\big(\Phi(x)\big)(\partial_1\varphi)\big(\Phi(x)\big) \\ &= \Big\{\Phi_2'(x)(\partial_2\varphi)\big(\Phi(x)\big) + \Phi_1'(x)\big(\Phi(x)\big)(\partial_1\varphi)\big(\Phi(x)\big)\Big\} \cdot |\Phi'(x)|^{-1} \\ &= \frac{\mathrm{d}}{\mathrm{d}x}\big[\varphi\big(\Phi(x)\big)\big] \cdot |\Phi'(x)|^{-1},\end{aligned} \tag{8.4.72}$$

by the definition of the tangential derivative operator ∂_τ and the Chain Rule, hence

$$\big[(\partial_\tau\varphi) \circ \Phi\big] \cdot |\Phi'| = \frac{\mathrm{d}}{\mathrm{d}x}\big[\varphi \circ \Phi\big] \ \text{ at } \mathcal{L}^1\text{-a.e. point on } \mathbb{R}. \tag{8.4.73}$$

Based on (8.4.69), (8.4.73), and definitions, we may then check (see also the proof of [130, Proposition 11.5.15] in this regard) that for each $f \in L^1_{1,\mathrm{loc}}(\partial\Omega_\theta, \sigma_\theta)$ one has $f \circ \Phi \in L^1_{1,\mathrm{loc}}(\mathbb{R}, \mathcal{L}^1)$ and

$$\big[(\partial_\tau f) \circ \Phi\big] \cdot |\Phi'| = \frac{\mathrm{d}}{\mathrm{d}x}\big[f \circ \Phi\big] \text{ at } \mathcal{L}^1\text{-a.e. point on } \mathbb{R}. \tag{8.4.74}$$

Collectively, (8.4.69) and (8.4.74) further show that

$$\begin{aligned} \int_{\partial\Omega_\theta} |\partial_\tau f|^p \, \mathrm{d}\sigma_\theta &= \int_{\mathbb{R}} \big[(\partial_\tau f) \circ \Phi\big]^p \cdot |\Phi'| \cdot |\Phi'| \, \mathrm{d}\mathcal{L}^1 \\ &= \int_{\mathbb{R}} \Big|\frac{\mathrm{d}}{\mathrm{d}x}\big[f \circ \Phi\big]\Big|^p \cdot |\Phi'|^{1-p} \, \mathrm{d}\mathcal{L}^1 \end{aligned} \tag{8.4.75}$$

for any $p \in (0, \infty)$. Ultimately, this analysis proves that for any σ_θ-measurable function $f : \partial\Omega_\theta \to \mathbb{R}$ and any $p \in (1, \infty)$ we have

$$\begin{aligned} & f \in \dot{L}^p_1(\partial\Omega_\theta, \sigma_\theta) \iff f \circ \Phi \in \dot{L}^p_1(\mathbb{R}, \mathcal{L}^1) \\ & \text{and } \|f\|_{\dot{L}^p_1(\partial\Omega_\theta, \sigma_\theta)} = \|f \circ \Phi\|_{\dot{L}^p_1(\mathbb{R}, \mathcal{L}^1)}. \end{aligned} \tag{8.4.76}$$

For the remaining portion of this discussion, fix an integrability exponent $p \in (1, \infty)$ together with some aperture parameter $\kappa \in (0, \infty)$. Consider now a harmonic function $u : \Omega_\theta \to \mathbb{R}$ with the property that $\mathcal{N}_\kappa(\nabla u) \in L^p(\partial\Omega_\theta, \sigma_\theta)$, and note that $u \circ \Phi$ is a real-valued harmonic function in $\mathbb{R}^2_+$. Denote by (X, Y) the variables in Ω_θ, and by (x, y) the variables in $\mathbb{R}^2_+$. Then

$$F := 2\partial u = \partial_X u + \mathrm{i}\partial_Y u \text{ and } \widetilde{F} := 2\partial(u \circ \Phi) = \partial_x(u \circ \Phi) + \mathrm{i}\partial_y(u \circ \Phi) \tag{8.4.77}$$

satisfy $\overline{\partial} F = 0$ in Ω_θ and $\overline{\partial}\widetilde{F} = 0$ in $\mathbb{R}^2_+$. Thus, F is holomorphic in Ω_θ, while $\widetilde{F}$ is holomorphic in $\mathbb{R}^2_+$. In addition, $\mathcal{N}_\kappa F \in L^p(\partial\Omega_\theta, \sigma_\theta)$, which places F in the Hardy space $\mathcal{H}^p(\Omega_\theta, \sigma_\theta)$, associated with the Cauchy-Riemann operator $\overline{\partial}$ in Ω_θ. Thanks to [129, Lemma 8.1.6] and (8.4.69), this further implies that

$$F \circ \Phi \text{ belongs to the Hardy space } \mathcal{H}^p(\mathbb{R}^2_+, |\Phi'(x)| \, \mathrm{d}x), \tag{8.4.78}$$

consisting of holomorphic functions in $\mathbb{R}^2_+$ with a nontangential maximal function in the weighted Lebesgue space $L^p(\mathbb{R}, |\Phi'(x)| \, \mathrm{d}x)$. We claim that

$$|F \circ \Phi| \cdot |\Phi'| \approx |\widetilde{F}|, \tag{8.4.79}$$

in a uniform pointwise fashion in $\mathbb{R}^2_+$. To prove this, first observe that, by Chain Rule,

$$\begin{aligned} \partial_x(u \circ \Phi) &= \big[(\partial_X u) \circ \Phi\big] \cdot \partial_x \Phi_1 + \big[(\partial_Y u) \circ \Phi\big] \cdot \partial_x \Phi_2, \\ \partial_y(u \circ \Phi) &= \big[(\partial_X u) \circ \Phi\big] \cdot \partial_y \Phi_1 + \big[(\partial_Y u) \circ \Phi\big] \cdot \partial_y \Phi_2, \end{aligned} \tag{8.4.80}$$

while the fact that Φ is holomorphic implies (via the Cauchy-Riemann equations)

$$\Phi' = \partial\Phi = \partial_x\Phi = (-\mathrm{i})\partial_y\Phi. \tag{8.4.81}$$

Together, (8.4.80) and (8.4.81) establish the right-pointing inequality in (8.4.79). To justify the left-pointing inequality in (8.4.79), bring in $\Psi := \Phi^{-1} : \Omega_\theta \to \mathbb{R}^2_+$, the inverse of Φ, which is a holomorphic function satisfying $|\Phi' \circ \Psi| = |\Psi'|^{-1}$. Writing $u = (u \circ \Phi) \circ \Psi$ and using the Chain Rule then permits us to estimate

$$|F| = |\nabla u| \leq 2|\widetilde{F} \circ \Psi| \cdot |\Psi'| \tag{8.4.82}$$

which, after composition with Φ, yields the left-pointing inequality in (8.4.79). We next claim that the following equivalences hold:

$$\begin{aligned} F \circ \Phi \in \mathcal{H}^p(\mathbb{R}^2_+, |\Phi'(x)|\,\mathrm{d}x) &\Longleftrightarrow (F \circ \Phi) \cdot \Phi' \in \mathcal{H}^p(\mathbb{R}^2_+, |\Phi'(x)|^{1-p}\,\mathrm{d}x) \\ &\Longleftrightarrow \widetilde{F} \cdot \Phi' \in \mathcal{H}^p(\mathbb{R}^2_+, |\Phi'(x)|^{1-p}\,\mathrm{d}x) \\ &\Longleftrightarrow \mathcal{N}_\kappa\big(\nabla(u \circ \Phi)\big) \in L^p(\mathbb{R}, |\Phi'(x)|^{1-p}\,\mathrm{d}x). \end{aligned} \tag{8.4.83}$$

Indeed, the second equivalence in (8.4.83) is implied by (8.4.79), while the third equivalence in (8.4.83) is clear from (8.4.77). The first equivalence in (8.4.83) follows from [91, Corollary 2.18, p. 148], presently invoked with the sets $\mathcal{O}_1 = \mathcal{O}_2 := \mathbb{R}^2_+$, $\sigma_{1,2}$ the identity map in $\mathbb{R}^2_+$, the measures $\nu := |\Phi'(x)|\,\mathrm{d}x$ and $\mu := |\Phi'(x)|^{1-p}\,\mathrm{d}x$, with F replaced by $F \circ \Phi$, and with $G := (\Phi')^p$. For these choices to be valid, several conditions need to be verified. First, we need $|\Phi'| \in A_\infty(\mathbb{R}, \mathcal{L}^1)$, which is seen from (8.4.51). Second, we need $|\Phi'|^{1-p} \in A_\infty(\mathbb{R}, \mathcal{L}^1)$. To justify this, we observe that the membership in (8.4.51) entails $|\Phi'|^{-1} \in A_2(\mathbb{R}, \mathcal{L}^1) \subseteq A_\infty(\mathbb{R}, \mathcal{L}^1)$. Granted this, the desired conclusion follows on account of the fact that having $w \in A_\infty(\mathbb{R}, \mathcal{L}^1)$ implies $w^a \in A_\infty(\mathbb{R}, \mathcal{L}^1)$ for each $a \in (0, \infty)$; cf. [129, Proposition 7.7.9, item *(3)*]. The third (and final) condition which we need is that $G \in \mathrm{AE}(\nu, \mu)$, the class of functions defined in [91, Definition 2.11, p. 144]. The key property in this regard is the existence of a function H which is holomorphic and never zero in $\mathbb{R}^2_+$, has nontangential pointwise traces $\mathcal{L}^1$-a.e. on $\mathbb{R} \equiv \partial\mathbb{R}^2_+$ which are equivalent to $|\Phi'|^{1-p}$ in a pointwise uniform fashion on $\mathbb{R}$, and for which there exist $m_1, m_2 \geq 0$ such that

$$\frac{(\Phi'(z))^p \cdot H(z)}{(\mathrm{i} + z)^{m_1}} \in \mathcal{H}^p(\mathbb{R}^2_+, \mathcal{L}^1) \text{ and } \frac{H(z)}{(\mathrm{i} + z)^{m_2}} \in \mathcal{H}^p(\mathbb{R}^2_+, \mathcal{L}^1). \tag{8.4.84}$$

Save for the second membership in (8.4.84), taking $H(z) := \big(\Phi'(z)\big)^{1-p}$ for each $z \in \mathbb{R}^2_+$ satisfies all desired requirements (with the first membership in (8.4.84) guaranteed by [91, Theorem 1.10(iii), p. 134] with $m_1 := 2$). In view of (8.4.47) and [129, Lemma 8.3.7], the choice we have made for H guarantees the validity of the second membership in (8.4.84) (for $m_2 > 0$ large enough) provided $\big|\Phi'(x)\big|^{1-p}$ does not have a singularity at $x = 0 \in \mathbb{R}$ which lacks local integrability with respect to $\mathcal{L}^1$. The latter condition is satisfied for any $p \in (1, \infty)$ if $\theta \in (0, \pi]$, and requires that we restrict p to the interval $\big(1, \theta/(\theta - \pi)\big)$ if $\theta \in (\pi, 2\pi)$. More economically stated, the entire analysis above goes through for any given $\theta \in (0, 2\pi)$ provided $1 < p < \big(\max\{0, 1 - \pi/\theta\}\big)^{-1}$.

In summary, we have proved the following result:

Remark 8.4.10 *Fix $\theta \in (0, 2\pi)$ arbitrary, and consider the infinite planar sector $\Omega_\theta \subseteq \mathbb{R}^2$ of (full) aperture angle θ and symmetry line along the vertical axis, described in* (8.4.42). *Abbreviate $\sigma_\theta := \mathcal{H}^1 \lfloor \partial\Omega_\theta$ and fix a background parameter $\kappa \in (0, \infty)$. Finally, select an integrability exponent*

$$1 < p < \big(\max\{0, 1 - \pi/\theta\}\big)^{-1} = \big(\max\{1, \theta/\pi\}\big)' \tag{8.4.85}$$

(where "prime" stands for Hölder conjugation). Then, under the transformation $u \mapsto \widetilde{u} := u \circ \Phi$, the $\dot{L}^p_1$-Homogeneous Regularity Problem for $L := \Delta$ in Ω_θ, i.e.,

$$\begin{cases} u \in \mathscr{C}^\infty(\Omega_\theta), \quad \Delta u = 0 \text{ in } \Omega_\theta, \\ \mathcal{N}_\kappa(\nabla u) \in L^p(\partial\Omega_\theta, \sigma_\theta), \quad u\big|_{\partial\Omega_\theta}^{\kappa-\mathrm{n.t.}} = f \in \dot{L}^p_1(\partial\Omega_\theta, \sigma_\theta), \end{cases} \tag{8.4.86}$$

becomes equivalent to the following weighted Homogeneous Regularity Problem for the Laplacian in the upper half-plane:

$$\begin{cases} \widetilde{u} \in \mathscr{C}^\infty(\mathbb{R}^2_+), \quad \Delta\widetilde{u} = 0 \text{ in } \mathbb{R}^2_+, \\ \mathcal{N}_\kappa(\nabla\widetilde{u}) \in L^p\big(\mathbb{R}, |\Phi'(x)|^{1-p}\,dx\big), \\ \widetilde{u}\big|_{\partial\mathbb{R}^2_+}^{\kappa-\mathrm{n.t.}} = \widetilde{f} := f \circ \Phi \in L^p\big(\mathbb{R}, |\Phi'(x)|^{1-p}\,dx\big). \end{cases} \tag{8.4.87}$$

As a special case of more general results from [111] *we know* (8.4.87) *to be well posed if $|\Phi'|^{1-p} \in A_p(\mathbb{R}, \mathcal{L}^1)$, while from* (8.4.46) *and* [129, Example 8.7.6] *we see that $|\Phi'|^{1-p} \in A_p(\mathbb{R}, \mathcal{L}^1)$ if and only if $-1 < (1-p)(\theta/\pi - 1) < p - 1$, hence*

$$|\Phi'|^{1-p} \in A_p(\mathbb{R}, \mathcal{L}^1) \Longleftrightarrow p \text{ is as in (8.4.85)}. \tag{8.4.88}$$

The bottom line is that

$$\text{the } \dot{L}^p_1\text{-Homogeneous Regularity Problem (8.4.86) is well-posed for} \\ \theta \in (0, 2\pi) \text{ and } 1 < p < \big(\max\{0, 1 - \pi/\theta\}\big)^{-1} = \big(\max\{1, \theta/\pi\}\big)'. \tag{8.4.89}$$

Lastly, we remark that the above range of p's is precisely the dual of the range of indices for which the L^p-Dirichlet Problem (8.4.43) *is well-posed (cf.* (8.4.53)).

As regards the Homogeneous Regularity Problem for the Laplacian in planar sectors, we also wish to compare the method of boundary layer potentials (from Remark 8.4.5) with the method of conformal mappings (from Remark 8.4.10).

Remark 8.4.11 *From* (8.4.40) *and* (8.4.53) *we see that, as far as the Homogeneous Regularity Problem is concerned, the method of boundary layer potentials (cf. Remark 8.4.5) is more effective in the case when $\theta \in (\pi, 2\pi)$, while the method described in Remark 8.4.10 (based on conformal mappings and weighted spaces) is* *more effective in the situation when $\theta \in (0,\pi)$. Together,* (8.4.41) *and* (8.4.89) *show that, for each $\theta \in (0,2\pi)$ and $p \in (1,\infty)$,*

$$\begin{aligned}&\textit{if either } p < \big(\max\{0, 1-\pi/\theta\}\big)^{-1} = \big(\max\{1,\theta/\pi\}\big)', \textit{ or } p \neq q_\theta := 1+\pi/|\pi-\theta|,\\ &\textit{then the } \dot{L}^p_1\textit{-Homogeneous Regularity Problem (8.4.86) is well-posed.}\end{aligned} \tag{8.4.90}$$

Incidentally, the above range of p's is precisely the dual of the range of exponents ensuring the well-posedness of the L^p-Dirichlet Problem (8.4.36) *from* (8.4.54).

Let us also observe that the condition in the last line of (8.4.90) allows any exponent $p \in (1,\infty)$ if $\theta \in (0,\pi]$, and any exponent $p \in (1,\infty) \setminus \{q_\theta\}$ if $\theta \in (\pi, 2\pi)$. In fact, we may establish a more refined variant of (8.4.90) as indicated below:

Theorem 8.4.12 *Pick $\theta \in (0,2\pi)$ arbitrary, and consider the infinite planar sector $\Omega_\theta \subseteq \mathbb{R}^2$ of (full) aperture angle θ and symmetry line along the vertical axis, considered in* (8.4.42). *Abbreviate $\sigma_\theta := \mathcal{H}^1 \lfloor \partial\Omega_\theta$ and fix a background parameter $\kappa \in (0,\infty)$. Finally, for any given $p \in (1,\infty)$ consider the $\dot{L}^p_1$-Homogeneous Regularity Problem for $L := \Delta$ in Ω_θ, i.e.,*

$$\begin{cases} u \in \mathscr{C}^\infty(\Omega_\theta), \quad \Delta u = 0 \ \textit{in} \ \Omega_\theta, \\ \mathcal{N}_\kappa(\nabla u) \in L^p(\partial\Omega_\theta, \sigma_\theta), \quad u\Big|^{\kappa-\mathrm{n.t.}}_{\partial\Omega_\theta} = f \in \dot{L}^p_1(\partial\Omega_\theta, \sigma_\theta). \end{cases} \tag{8.4.91}$$

Then the following stronger version of (8.4.90) *is valid:*

$$\begin{aligned}&\textit{if } \theta \in (0,\pi] \textit{ then the } \dot{L}^p_1\textit{-Homogeneous Regularity Problem (8.4.91) is}\\ &\textit{well-posed for any } p \in (1,\infty)\textit{, while in the case } \theta \in (\pi,2\pi) \textit{ it follows}\\ &\textit{that the } \dot{L}^p_1\textit{-Homogeneous Regularity Problem (8.4.91) is well-posed}\\ &\textit{if and only if } p \in (1,\infty)\setminus\{q_\theta\} = (1,\infty)\setminus\{\tfrac{\theta}{\theta-\pi}\}.\end{aligned} \tag{8.4.92}$$

Furthermore, additional information is available in the case when $\theta \in (\pi,2\pi)$ for the critical exponent $p = q_\theta := 1+\pi/|\pi-\theta| = \frac{\theta}{\theta-\pi}$. As for uniqueness,

$$\begin{aligned}&\textit{if } \theta \in (\pi,2\pi) \textit{ then the } \dot{L}^p_1\textit{-Homogeneous Regularity Problem (8.4.91)}\\ &\textit{corresponding to } p = q_\theta := 1+\pi/|\pi-\theta| = \tfrac{\theta}{\theta-\pi} \textit{ continues to have at}\\ &\textit{most one solution.}\end{aligned} \tag{8.4.93}$$

To elaborate on the issue of solvability, consider the two-dimensional modified boundary-to-boundary harmonic single layer operator $S_{\rm mod}$, associated with Ω_θ as in (A.0.232). *In addition, recall that the space of admissible boundary data for the $\dot{L}^p_1$-Homogeneous Regularity Problem* (8.4.91) *corresponding to $p := q_\theta$ is the collection of functions from $\dot{L}^{q_\theta}_1(\partial\Omega_\theta,\sigma_\theta)$ for which said problem is solvable when the functions in question are used as boundary data. Then said space coincides, modulo constants, with the image of $\big[S_{\rm mod}\big]$ acting from the space $L^{q_\theta}(\partial\Omega_\theta,\sigma_\theta)$ into the space $\dot{L}^{q_\theta}_1(\partial\Omega_\theta,\sigma_\theta)/\sim$, i.e., for $\theta \in (\pi,2\pi)$ and $q_\theta := 1+\pi/|\pi-\theta| = \frac{\theta}{\theta-\pi}$ one has*

$$\Big\{\Big[u\big|^{\kappa-\text{n.t.}}_{\partial\Omega_\theta}\Big] : u \in \mathscr{C}^\infty(\Omega_\theta),\ \Delta u = 0 \ \text{in}\ \Omega_\theta,\ \mathcal{N}_\kappa(\nabla u) \in L^{q_\theta}(\partial\Omega_\theta, \sigma_\theta)\Big\} \\ = \Big\{\big[\mathcal{S}_{\text{mod}} g\big] : g \in L^{q_\theta}(\partial\Omega_\theta, \sigma_\theta)\Big\}. \tag{8.4.94}$$

As a corollary of (8.4.94) *and* (4.2.188), *it follows that*

> *for* $\theta \in (\pi, 2\pi)$ *and* $p = q_\theta := 1 + \pi/|\pi - \theta| = \frac{\theta}{\theta-\pi}$, *the space of admissible boundary data for the Homogeneous Regularity Problem* (8.4.91) *has infinite co-dimension (in particular, is not closed) in* $\dot{L}^{q_\theta}_1(\partial\Omega_\theta, \sigma_\theta)$. (8.4.95)

In particular, this shows that

> *the* $\dot{L}^p_1$*-Homogeneous Regularity Problem formulated in* (8.4.91) *fails to be Fredholm solvable in the scenario when* $\theta \in (\pi, 2\pi)$ *and* $p = q_\theta := 1 + \pi/|\pi - \theta| = \frac{\theta}{\theta-\pi}$. (8.4.96)

As a preamble to the proof of the above theorem, we record the following lemma, which establishes a two-way street between invertibility of the single layer potential operator and the well-posedness of the Homogeneous Regularity Problem.

Lemma 8.4.13 *Let* $\Omega \subseteq \mathbb{R}^n$ *be a two-sided* NTA *domain with an unbounded Ahlfors regular boundary, and abbreviate* $\sigma := \mathcal{H}^{n-1}\lfloor\partial\Omega$. *Fix some aperture parameter* $\kappa > 0$ *and an integrability exponent* $p \in (1, \infty)$. *Let L be a homogeneous, second-order, constant complex coefficient, weakly elliptic* $M \times M$ *system in* $\mathbb{R}^n$ *and denote by* (HRP_+) *and* (HRP_-) *the Homogeneous Regularity Problems formulated as in* (8.4.2) *corresponding to* $\Omega_+ := \Omega$ *and* $\Omega_- := \mathbb{R}^n \setminus \overline{\Omega}$, *respectively. Then the operator*

$$\begin{gathered}\big[\mathcal{S}_{\text{mod}}\big] : \big[L^p(\partial\Omega, \sigma)\big]^M \longrightarrow \big[\dot{L}^p_1(\partial\Omega, \sigma)/\sim\big]^M \\ \big[\mathcal{S}_{\text{mod}}\big] f := \big[\mathcal{S}_{\text{mod}} f\big] \ \text{for each}\ f \in \big[L^p(\partial\Omega, \sigma)\big]^M,\end{gathered} \tag{8.4.97}$$

has the following properties:

(a) The operator $\big[\mathcal{S}_{\text{mod}}\big]$ *is injective in the context of* (8.4.97) *if and only if* (HRP_+) *and* (HRP_-) *have at most one solution.*
(b) The operator $\big[\mathcal{S}_{\text{mod}}\big]$ *is surjective in the context of* (8.4.97) *if and only if* (HRP_+) *and* (HRP_-) *have a solution. In fact, if*

$$\text{ABD}_\pm := \Big\{u\big|^{\kappa-\text{n.t.}}_{\partial\Omega_\pm} : u \in \big[\mathscr{C}^\infty(\Omega_\pm)\big]^M,\ Lu = 0 \ \text{in}\ \Omega_\pm,\ \mathcal{N}_\kappa(\nabla u) \in L^p(\partial\Omega, \sigma)\Big\}$$

are the space of admissible boundary data for the Homogeneous Regularity Problems $(\text{HRP}_\pm)$ *(recall from the Fatou-type result from* [130, Proposition 11.5.12] *that* $\text{ABD}_\pm$ *are well-defined subspaces of* $\big[\dot{L}^p_1(\partial\Omega, \sigma)\big]^M$*), and if* $\big[\text{ABD}_\pm\big]$ *denote the spaces of equivalence classes of functions in* $\text{ABD}_\pm$ *modulo constants (in* $\mathbb{C}^M$*), then*

> *the image of the operator* $\big[\mathcal{S}_{\text{mod}}\big]$ *in the context of* (8.4.97) *is precisely* $\big[\text{ABD}_+\big] \cap \big[\text{ABD}_-\big]$. (8.4.98)

(c) *The operator* $\left[S_{\text{mod}}\right]$ *is an isomorphism in the context of* (8.4.97) *if and only if* (HRP$_+$) *and* (HRP$_-$) *are well-posed.*

Proof All claims may be justified by revisiting the proof of [111, Proposition 4.3, pp. 319-320]. □

We now turn to the task of providing the proof of Theorem 8.4.12.

Proof of Theorem 8.4.12 To deal with the claim made in (8.4.93), fix $\theta \in (\pi, 2\pi)$ and consider some arbitrary nullsolution u of the Homogeneous Regularity Problem (8.4.91) corresponding to $p = q_\theta := 1 + \pi/|\pi - \theta| = \frac{\theta}{\theta-\pi}$. If ν_θ denotes the outward unit normal to Ω_θ, then integral representation formula [132, (1.8.200)] from [132, Theorem 1.8.19] presently implies (since $u\big|_{\partial\Omega_\theta}^{\kappa-\text{n.t.}} = 0$) that

$$u = -\mathscr{S}_{\text{mod}}\big(\partial_{\nu_\theta} u\big) + c \ \text{ in } \Omega_\theta \tag{8.4.99}$$

for some constant $c \in \mathbb{C}$, where $\mathscr{S}_{\text{mod}}$ is the modified boundary-to-domain harmonic single layer potential operator associated with Ω_θ. After taking the normal derivatives of both sides in (8.4.99) and utilizing the jump-formula [132, (1.5.58)] (for $L := \Delta$) we arrive at $\big(\frac{1}{2}I + K_\Delta^\#\big)\big(\partial_{\nu_\theta} u\big) = 0$ on $\partial\Omega_\theta$. In view of the fact that the function $\partial_{\nu_\theta} u$ belongs to $L^{q_\theta}(\partial\Omega_\theta, \sigma_\theta)$, we conclude from this and (4.2.186) that $\partial_{\nu_\theta} u = 0$. Plugging this back in (8.4.99) then yields $u \equiv c$ in Ω_θ. Upon recalling that $u\big|_{\partial\Omega_\theta}^{\kappa-\text{n.t.}} = 0$, this forces $c = 0$ hence, ultimately, $u \equiv 0$ in Ω_θ. The proof of (8.4.93) is therefore complete.

Next, the equality claimed in (8.4.94) is seen from (8.4.98) and the well-posedness result contained in (8.4.90). All other desired conclusions follow from (8.4.94) and (4.2.188) (as indicated in the statement). □

Theorem 8.4.12 may also be used to highlight the significance of the smallness condition imposed on the BMO seminorm of the geometric measure theoretic outward unit normal in the formulation of Theorem 8.4.1.

Remark 8.4.14 *Given an angle* $\theta \in (\pi, 2\pi)$, *define the infinite sector* $\Omega_\theta \subseteq \mathbb{R}^2$ *as in* (8.4.35). *Also, abbreviate* $\sigma_\theta := \mathcal{H}^1\lfloor\partial\Omega_\theta$ *and denote by* ν_θ *the outward unit normal to* Ω_θ. *Then* $q_\theta := 1 + \pi/|\pi - \theta| \in (2, \infty)$ *and, much as in* (4.2.161),

$$\begin{aligned}\|\nu_\theta\|_{\text{BMO}(\partial\Omega_\theta,\sigma_\theta)} &= |\cos(\theta/2)| = \big|\sin\big(\pi/2 - \theta/2\big)\big| \approx |\theta - \pi| \\ &\approx (q_\theta - 1)^{-1}, \ \textit{uniformly for } \theta \in (\pi, 2\pi).\end{aligned} \tag{8.4.100}$$

As a consequence of this and Theorem 8.4.12, for the class of domains Ω_θ *with* $\theta \in (\pi, 2\pi)$, *the well-posedness of the* $\dot{L}^p_1$*-Regularity Problem for the Laplacian for each* $p \in I := [2, a]$ *with* $a \in (2, \infty)$ *large is equivalent to the demand that* $\|\nu_\theta\|_{\text{BMO}(\partial\Omega_\theta,\sigma_\theta)}$ *is small. This underscores the relevance of the latter property in the context of Theorem 8.4.1 (see the formulation given in Remark 8.4.2).*

Moving on, the next goal is to establish the well-posedness of the Dirichlet Problem with data in inhomogeneous Morrey-Campanato spaces, for weakly elliptic systems and sufficiently flat Ahlfors regular domains with unbounded boundaries. The solution we produce satisfies both a nontangential maximal function estimate and a fractional Carleson measure estimate, as formulated in the theorem below.

Theorem 8.4.15 *Pick $n \in \mathbb{N}$ with $n \geq 2$ and let $\Omega \subseteq \mathbb{R}^n$ be an Ahlfors regular domain. Abbreviate $\sigma := \mathcal{H}^{n-1}\lfloor\partial\Omega$ and denote by ν the geometric measure theoretic outward unit normal to Ω. Also, fix an aperture parameter $\kappa > 0$ and pick an integrability exponent $p \in (1, \infty)$ along with a number $\lambda \in (0, n-1)$. Given a homogeneous, second-order, constant complex coefficient, weakly elliptic $M \times M$ system L in $\mathbb{R}^n$, consider the following Dirichlet Problem (with boundary data in inhomogeneous Morrey-Campanato spaces):*

$$\begin{cases} u \in \left[\mathscr{C}^\infty(\Omega)\right]^M, \quad Lu = 0 \text{ in } \Omega, \quad \mathcal{N}_\kappa u \in L^p(\partial\Omega, \sigma), \\ \sup\limits_{\substack{x\in\partial\Omega \text{ and} \\ r\in(0,\infty)}} \left\{ r^{-\lambda} \displaystyle\int_{B(x,r)\cap\Omega} |(\nabla u)(y)|^p \operatorname{dist}(y, \partial\Omega)^{p-1} \, d\mathcal{L}^n(y) \right\}^{\frac{1}{p}} < \infty, \\ u\big|_{\partial\Omega}^{\kappa-\text{n.t.}} = f \in \left[\mathfrak{L}^{p,\lambda}(\partial\Omega, \sigma)\right]^M. \end{cases} \tag{8.4.101}$$

Then, in relation to this, the following claims are true:

(a) [Existence, Estimates, and Integral Representation] *If $\mathfrak{A}_L^{\text{dis}} \neq \varnothing$ and $A \in \mathfrak{A}_L^{\text{dis}}$, then there exists some $\delta \in (0, 1)$ which depends only on n, p, A, and the Ahlfors regularity constants of $\partial\Omega$ with the property that if $\|\nu\|_{[\text{BMO}(\partial\Omega,\sigma)]^n} < \delta$ (i.e., if Ω is a δ-AR domain in the sense of Definition 3.1.8) then $\frac{1}{2}I + K_A$ is an invertible operator on the inhomogeneous Morrey-Campanato space $\left[\mathfrak{L}^{p,\lambda}(\partial\Omega, \sigma)\right]^M$ and the function $u : \Omega \to \mathbb{C}^M$ defined as*

$$u := \mathcal{D}_A\Big(\left(\tfrac{1}{2}I + K_A\right)^{-1} f\Big) \text{ in } \Omega, \tag{8.4.102}$$

is a solution of the Dirichlet Problem (8.4.101). Moreover, there exists some constant $C \in (0, \infty)$ independent of f with the property that

$$\left\|\mathcal{N}_\kappa u\right\|_{L^p(\partial\Omega,\sigma)} \leq C\|f\|_{[L^p(\partial\Omega,\sigma)]^M} \tag{8.4.103}$$

and

$$\sup_{\substack{x\in\partial\Omega \text{ and} \\ r\in(0,\infty)}} \left\{ r^{-\lambda} \int_{B(x,r)\cap\Omega} |(\nabla u)(y)|^p \operatorname{dist}(y, \partial\Omega)^{p-1} \, d\mathcal{L}^n(y) \right\}^{\frac{1}{p}} \\ \leq C\|f\|_{[\dot{\mathfrak{L}}^{p,\lambda}(\partial\Omega,\sigma)]^M}. \tag{8.4.104}$$

(b) [Uniqueness] *Whenever $\mathfrak{A}_{L^\top}^{\text{dis}} \neq \varnothing$, there exists $\delta \in (0, 1)$ which depends only on n, p, L, and the Ahlfors regularity constants of $\partial\Omega$ with the property that if $\|\nu\|_{[\text{BMO}(\partial\Omega,\sigma)]^n} < \delta$ (i.e., if Ω is a δ-AR domain in the sense of Definition 3.1.8) then the Dirichlet Problem (8.4.101) has at most one solution.*

(c) [Well-Posedness] *If* $\mathfrak{A}_L^{\rm dis} \neq \varnothing$ *and* $\mathfrak{A}_{L^\top}^{\rm dis} \neq \varnothing$ *then there exists* $\delta \in (0,1)$ *which depends only on n, p, L, and the Ahlfors regularity constants of* $\partial\Omega$ *such that if* $\|\nu\|_{[\mathrm{BMO}(\partial\Omega,\sigma)]^n} < \delta$ *(i.e., if* Ω *is a* δ-AR *domain in the sense of Definition 3.1.8) then the Dirichlet Problem* (8.4.101) *is well posed (i.e., it is uniquely solvable and the solution satisfies the estimate formulated in* (8.4.103)).

As may be seen from Proposition 1.6.12, in the scalar case (that is, when $M = 1$) in dimensions $n \geq 3$, just by virtue of having the operator L weakly elliptic it follows that $\mathfrak{A}_L^{\rm dis} \neq \varnothing$ and $\mathfrak{A}_{L^\top}^{\rm dis} \neq \varnothing$. Furthermore, in the scalar case in the two-dimensional setting (i.e., for $M = 1$ and $n = 2$), if L satisfies the Legendre-Hadamard strong ellipticity condition or, more generally (cf. [131, Lemma 1.4.24]), if $L = \mathrm{div}A\nabla$ for some 2×2 complex matrix A with $\mathrm{sym}\,A := \frac{1}{2}(A + A^\top) \in M_2^0$ (cf. [131, (1.4.187)]), then Proposition 1.6.11 and Proposition 1.6.13 guarantee that $\mathfrak{A}_L^{\rm dis} \neq \varnothing$ and $\mathfrak{A}_{L^\top}^{\rm dis} \neq \varnothing$. Consequently, the statement of Theorem 8.4.15 may be significantly streamlined, as we have genuine well-posedness in each of these scenarios.

We also wish to remark that Theorem 8.4.15 is sharp in the sense that if $\mathfrak{A}_L^{\rm dis} = \varnothing$ then the Dirichlet Problem (8.4.23) may not be solvable, and if $\mathfrak{A}_{L^\top}^{\rm dis} = \varnothing$ then the Dirichlet Problem (8.4.101) may have more than one solution. This may be justified much as in the proof of item *(d)* of Theorem 8.4.3.

Proof of Theorem 8.4.15 In the context of item *(a)*, assume $\delta \in (0,1)$ is sufficiently small. The fact that the operator $\frac{1}{2}I + K$ is then invertible when acting both on the Lebesgue space $\big[L^p(\partial\Omega,\sigma)\big]^M$ and on the inhomogeneous Morrey-Campanato space $\big[\mathfrak{L}^{p,\lambda}(\partial\Omega,\sigma)\big]^M$ is implied by the results proved in items *(1)* and *(5)* of Theorem 4.2.1 (cf. (4.2.5) and (4.2.16)). Furthermore the inverses of $\frac{1}{2}I + K$ these two contexts are compatible with one another. In particular,

$$g := \big(\tfrac{1}{2}I + K\big)^{-1} f \in \big[\mathfrak{L}^{p,\lambda}(\partial\Omega,\sigma)\big]^M \hookrightarrow \big[L^p(\partial\Omega,\sigma)\big]^M \tag{8.4.105}$$

is unambiguously defined and satisfies

$$\begin{aligned} \|g\|_{[\mathfrak{L}^{p,\lambda}(\partial\Omega,\sigma)]^M} &\leq C\|f\|_{[\mathfrak{L}^{p,\lambda}(\partial\Omega,\sigma)]^M}, \\ \|g\|_{[L^p(\partial\Omega,\sigma)]^M} &\leq C\|f\|_{[L^p(\partial\Omega,\sigma)]^M}, \end{aligned} \tag{8.4.106}$$

for some $C \in (0,\infty)$ independent of f. Granted these properties, the same type of argument as in the proof of Theorem 8.4.3 shows that (8.4.102) satisfying all conditions in the first and last lines of (8.4.101), as well as (8.4.103).

In addition, with $K_{\rm mod}$ denoting the modified version of the principal-value double layer operator associated with the coefficient tensor A and the set Ω as in (A.0.117), matters may be arranged (by eventually further decreasing δ) so that

$$\tfrac{1}{2}I + \big[K_{\rm mod}\big] : \big[\dot{\mathfrak{L}}^{p,\lambda}(\partial\Omega,\sigma)/\sim\big]^M \longrightarrow \big[\dot{\mathfrak{L}}^{p,\lambda}(\partial\Omega,\sigma)/\sim\big]^M \tag{8.4.107}$$

is also invertible. See (4.2.15) in item *(5)* of Theorem 4.2.1 in this regard. Since $(\frac{1}{2}I + K)g = f$ in $\big[L^p(\partial\Omega,\sigma)\big]^M$, from [132, (1.8.26)] we deduce that there exists some $c \in \mathbb{C}^M$ with the property that $\big(\frac{1}{2}I + K_{\text{mod}}\big)g = f + c$. In turn, based on this and [132, (3.2.14)] we may write

$$\big(\tfrac{1}{2}I + \big[K_{\text{mod}}\big]\big)[g] = [f] \text{ in } \big[\dot{\mathfrak{L}}^{p,\lambda}(\partial\Omega, \sigma)/\sim\big]^M. \tag{8.4.108}$$

Since $\frac{1}{2}I + \big[K_{\text{mod}}\big]$ is invertibile in the context of (8.4.107), this further entails

$$[g] = \big(\tfrac{1}{2}I + \big[K_{\text{mod}}\big]\big)^{-1}[f] = \text{ in } \big[\dot{\mathfrak{L}}^{p,\lambda}(\partial\Omega, \sigma)/\sim\big]^M. \tag{8.4.109}$$

As such, we may write

$$\begin{aligned}\|g\|_{[\dot{\mathfrak{L}}^{p,\lambda}(\partial\Omega,\sigma)]^M} &= \big\|[g]\big\|_{[\dot{\mathfrak{L}}^{p,\lambda}(\partial\Omega,\sigma)/\sim]^M} = \big\|\big(\tfrac{1}{2}I + \big[K_{\text{mod}}\big]\big)^{-1}[f]\big\|_{[\dot{\mathfrak{L}}^{p,\lambda}(\partial\Omega,\sigma)/\sim]^M} \\ &\leq C\big\|[f]\big\|_{[\dot{\mathfrak{L}}^{p,\lambda}(\partial\Omega,\sigma)/\sim]^M} = C\|f\|_{[\dot{\mathfrak{L}}^{p,\lambda}(\partial\Omega,\sigma)]^M},\end{aligned} \tag{8.4.110}$$

based on [130, (6.1.9)] (applied twice) and (8.4.109). We are now in a position to conclude, from (8.4.102), (8.4.105), [132, (1.8.9), (3.2.39)], and (8.4.110), that the function u given in (8.4.102) satisfies

$$\begin{aligned}&\sup_{\substack{x\in\partial\Omega \text{ and}\\ r\in(0,\infty)}} \left\{r^{-\lambda}\int_{B(x,r)\cap\Omega} |(\nabla u)(y)|^p \operatorname{dist}(y,\partial\Omega)^{p-1}\, d\mathcal{L}^n(y)\right\}^{\frac{1}{p}} \\ &\quad = \sup_{\substack{x\in\partial\Omega \text{ and}\\ r\in(0,\infty)}} \left\{r^{-\lambda}\int_{B(x,r)\cap\Omega} \big|(\nabla \mathcal{D}_{\text{mod}} g)(y)\big|^p \operatorname{dist}(y,\partial\Omega)^{p-1}\, d\mathcal{L}^n(y)\right\}^{\frac{1}{p}} \\ &\quad \leq C\|g\|_{[\dot{\mathfrak{L}}^{p,\lambda}(\partial\Omega,\sigma)]^M} \leq C\|f\|_{[\dot{\mathfrak{L}}^{p,\lambda}(\partial\Omega,\sigma)]^M},\end{aligned} \tag{8.4.111}$$

proving (8.4.104). This takes care of all claims made in item *(a)*. The uniqueness in item *(b)* is a direct consequence of the corresponding uniqueness statement in Theorem 8.4.3, since (8.4.101) is a sub-problem of (8.4.23). Finally, the well-posedness result claimed in item *(c)* follows from what we have proved in items *(a)*-*(b)*. □

We wish to remark that similar results to those described in Theorem 8.4.15 are also valid for the Dirichlet Problem with boundary data in *homogeneous* Morrey-Campanato spaces, formulated as

$$\begin{cases} u \in \big[\mathscr{C}^\infty(\Omega)\big]^M, \quad Lu = 0 \text{ in } \Omega, \\ \mathcal{N}_\kappa(u - c) \text{ belongs to } M^{p,\lambda}(\partial\Omega, \sigma) \text{ for some } c \in \mathbb{C}^M, \\ \sup\limits_{\substack{x\in\partial\Omega \text{ and}\\ r\in(0,\infty)}} \left\{r^{-\lambda}\displaystyle\int_{B(x,r)\cap\Omega} |(\nabla u)(y)|^p \operatorname{dist}(y,\partial\Omega)^{p-1}\, d\mathcal{L}^n(y)\right\}^{\frac{1}{p}} < \infty, \\ u\big|_{\partial\Omega}^{\kappa-\text{n.t.}} = f \in \big[\dot{\mathfrak{L}}^{p,\lambda}(\partial\Omega, \sigma)\big]^M, \end{cases} \tag{8.4.112}$$

or, equivalently, as

$$u \in \big[\mathscr{C}^\infty(\Omega)\big]^M, \quad Lu = 0 \text{ in } \Omega, \quad \mathcal{N}_\kappa u \in M^{p,\lambda}(\partial\Omega, \sigma),$$
$$\sup_{\substack{x\in\partial\Omega \text{ and} \\ r\in(0,\infty)}} \Big\{ r^{-\lambda} \int_{B(x,r)\cap\Omega} |(\nabla u)(y)|^p \operatorname{dist}(y, \partial\Omega)^{p-1}\, d\mathcal{L}^n(y) \Big\}^{\frac{1}{p}} < \infty, \tag{8.4.113}$$
$$\Big[u\big|_{\partial\Omega}^{\kappa-\text{n.t.}} \Big] = [f] \in \big[\dot{\mathfrak{L}}^{p,\lambda}(\partial\Omega, \sigma)/\sim\big]^M,$$

where $[\cdot]$ indicates the equivalence class modulo constants.

This may be seen from either using the invertibility result recorded in (4.2.15) (from item *(5)* of Theorem 4.2.1), the fractional Carleson measure estimate from [132, Theorem 3.2.4], and exploiting the relationship between the Morrey spaces and the Morrey-Campanato spaces from [130, Proposition 6.2.4]. Alternatively, we may directly rely on the Dirichlet Problem with boundary data in Morrey spaces, and [130, Proposition 6.2.4].

Moving on, we shall now consider the Neumann Problem with data from L^p with $1 < p < \infty$, for weakly elliptic systems in sufficiently flat Ahlfors regular domains with unbounded boundaries. The formulation of this boundary value problem involves the conormal derivative operator defined pointwise, as in (A.0.203). In contrast to the earlier Dirichlet Problems, the very statement of a Neumann Problem is strongly affected, through the conormal derivative employed in the formulation of the boundary condition, by the choice of the coefficient tensor used to write the given weakly elliptic second-order system of differential operators. As a matter of fact, it has been mentioned in the comments leading up to the example in [132, (1.4.208)] that even in the case of the (scalar) Laplacian there exist infinitely many conormal derivatives, hence infinitely many Neumann Problems. Remarkably, the Neumann Problems associated with distinguished coefficient tensors turned out to be always solvable, provided the underlying domain is sufficiently flat. The case of the Laplacian in upper-graph Lipschitz domains has been treated by B. Dahlberg and C. Kenig in [38, Theorem 2.13, p. 448] when $1 < p < 2+\varepsilon$ for some small threshold $\varepsilon > 0$ depending only on the Lipschitz constant.

Theorem 8.4.16 *Let $\Omega \subseteq \mathbb{R}^n$ be a* UR *domain. Denote by ν the geometric measure theoretic outward unit normal to Ω and abbreviate $\sigma := \mathcal{H}^{n-1}\lfloor\partial\Omega$. Also, fix an aperture parameter $\kappa > 0$ and pick an integrability exponent $p \in (1, \infty)$.*

Suppose L is a homogeneous, second-order, constant complex coefficient, weakly elliptic $M \times M$ system in $\mathbb{R}^n$. Select $A \in \mathfrak{A}_L$ and consider the Neumann Problem

$$\begin{cases} u \in \big[\mathscr{C}^\infty(\Omega)\big]^M, \quad Lu = 0 \text{ in } \Omega, \\ \mathcal{N}_\kappa(\nabla u) \in L^p(\partial\Omega, \sigma), \quad \partial_\nu^A u = f \in \big[L^p(\partial\Omega, \sigma)\big]^M. \end{cases} \tag{8.4.114}$$

Then, in relation to this, the following statements are valid:

(a) [Existence, Estimates, and Integral Representations] *Whenever $A^\top \in \mathfrak{A}^{\text{dis}}_{L^\top}$ there exists some threshold $\delta \in (0, 1)$ which depends only on n, p, A, and the Ahlfors regularity constants of $\partial\Omega$ such that if $\|\nu\|_{[\text{BMO}(\partial\Omega,\sigma)]^n} < \delta$ then $-\frac{1}{2}I + K^\#_{A^\top}$ is an invertible operator on $\big[L^p(\partial\Omega, \sigma)\big]^M$ and the function defined as*

$$u := \mathscr{S}_{\rm mod}\Big(\Big(-\tfrac{1}{2}I + K^{\#}_{A^{\top}}\Big)^{-1} f\Big) \text{ in } \Omega, \tag{8.4.115}$$

is a solution of the Neumann Problem (8.4.114) *which satisfies*

$$\big\|\mathcal{N}_\kappa(\nabla u)\big\|_{L^p(\partial\Omega,\sigma)} \leq C\|f\|_{[L^p(\partial\Omega,\sigma)]^M}, \tag{8.4.116}$$

for some constant $C \in (0, \infty)$ *independent of* f.

(b) [Uniqueness (modulo constants)] *Assume* $A \in \mathfrak{A}^{\rm dis}_{L}$. *Then there exists* $\delta \in (0, 1)$ *which depends only on* n, p, A, *and the Ahlfors regularity constants of* $\partial\Omega$ *such that if* $\|\nu\|_{[{\rm BMO}(\partial\Omega,\sigma)]^n} < \delta$ *(i.e., if* Ω *is a* δ-AR *domain in the sense of Definition 3.1.8) then any two solutions of the Neumann Problem* (8.4.114) *differ by a constant from* $\mathbb{C}^M$.

(c) [Well-Posedness] *Whenever* $A \in \mathfrak{A}^{\rm dis}_{L}$ *and* $A^{\top} \in \mathfrak{A}^{\rm dis}_{L^{\top}}$ *there exists* $\delta \in (0, 1)$ *which depends only on* n, p, A, *and the Ahlfors regularity constants of* $\partial\Omega$ *such that if* $\|\nu\|_{[{\rm BMO}(\partial\Omega,\sigma)]^n} < \delta$ *(i.e., if* Ω *is a* δ-AR *domain in the sense of Definition 3.1.8) then the Neumann Problem* (8.4.114) *is solvable, the solution is unique modulo constants from* $\mathbb{C}^M$, *and each solution satisfies* (8.4.116).

(d) [Sharpness] *If* $A^{\top} \notin \mathfrak{A}^{\rm dis}_{L^{\top}}$ *then the Neumann Problem* (8.4.114) *may not be solvable. Also, if* $A \notin \mathfrak{A}^{\rm dis}_{L}$ *then the Neumann Problem* (8.4.114) *may have more than one solution. In fact, even the two-dimensional Laplacian may be written as* $\Delta = {\rm div}A\nabla$ *for some matrix* $A \in \mathbb{C}^{2\times 2}$ *(not belonging to* $\mathfrak{A}^{\rm dis}_{\Delta} = \{I_{2\times 2}\}$*) such that the Neumann Problem formulated for this as in* (8.4.114) *for this choice of* A *and with* $\Omega := \mathbb{R}^2_+$ *fails to have a solution for each non-zero boundary datum belonging to an infinite-dimensional linear subspace of* $L^p(\partial\Omega, \sigma)$, *and the linear space of null-solutions for the Neumann Problem formulated as in* (8.4.114) *for this choice of* A *and with* $\Omega := \mathbb{R}^2_+$ *is actually infinite dimensional. Also, a similar lack of Fredholm solvability is present in higher dimensions.*

Furthermore, the solvability result described in item (a) above (hence also the well-posedness from item (c)) may actually fail in the absence of any specifications regarding the flatness of the boundary. To elaborate on this issue, consider the class of infinite planar sectors, i.e., sets of the form

$$\begin{gathered}\Omega_\theta := \Big\{x \in \mathbb{R}^2 \setminus \{x_o\} : \tfrac{x-x_o}{|x-x_o|} \cdot \xi > \cos(\theta/2)\Big\} \\ \text{for some } x_o \in \mathbb{R}^2,\ \theta \in (0, 2\pi), \text{ and } \xi \in S^1.\end{gathered} \tag{8.4.117}$$

Then, as shown a bit later in Theorem 8.4.22,

$$\begin{gathered}\text{the "standard" } L^p\text{-Neumann Problem (8.4.114) formulated for } n = 2, \\ L := \Delta,\ A := I_{2\times 2}, \text{ and } \Omega := \Omega_\theta \text{ fails to be Fredholm solvable when} \\ \theta \in (\pi, 2\pi) \text{ and } p = q_\theta := 1 + \pi/|\pi - \theta| = \tfrac{\theta}{\theta-\pi}.\end{gathered} \tag{8.4.118}$$

(e) [Other Spaces of Boundary Data] *Similar results to those described in items (a)-(c) and the first part of item (d) are valid with* $L^p(\partial\Omega, \sigma)$ *replaced by any*

of the spaces $M^{p,\lambda}(\partial\Omega,\sigma)$, $\mathring{M}^{p,\lambda}(\partial\Omega,\sigma)$, *and* $\mathcal{B}^{q,\lambda}(\partial\Omega,\sigma)$ *with* $p,q \in (1,\infty)$ *and* $\lambda \in (0,n-1)$.

In special cases, the statement of Theorem 8.4.16 may be further streamlined. For example, Proposition 1.4.1 implies that if the system L actually satisfies the Legendre-Hadamard strong ellipticity condition then for the well-posedness result formulated in item *(c)* it suffices to assume that $A \in \mathfrak{A}_L^{\rm dis}$, while if $M = 1$ (i.e., in the scalar case) and $n \geq 3$ then Proposition 1.6.12 implies that for the aforementioned well-posedness result it suffices to assume that the matrix $A \in \mathfrak{A}_L$ is symmetric. Furthermore, in the scalar case in the two-dimensional setting (i.e., for $M = 1$ and $n = 2$), we may rely on Proposition 1.6.13 to see that the well-posedness result formulated in item *(c)* of Theorem 8.4.16 holds if the matrix A belongs to the class M_2^0 (defined in [131, (1.4.187)]).

We also wish to note that, as seen from (A.0.203) and the Fatou-type result described in [131, Theorem 3.3.5], the conormal derivative $\partial_\nu^A u$ is well defined in the context of (8.4.114). As a consequence of this and Theorem 8.4.1, whenever $\delta \in (0,1)$ is small enough, $\mathfrak{A}_L^{\rm dis} \neq \varnothing$, and $\mathfrak{A}_{L^\top}^{\rm dis} \neq \varnothing$, it follows that for every $A \in \mathfrak{A}_L$ the Dirichlet-to-Neumann map

$$\Lambda_{\rm DN} : \big[\dot{L}_1^p(\partial\Omega,\sigma)\big]^M \longrightarrow \big[L^p(\partial\Omega,\sigma)\big]^M \tag{8.4.119}$$

defined for each $f \in \big[\dot{L}_1^p(\partial\Omega,\sigma)\big]^M$ as $\Lambda_{\rm DN} f := \partial_\nu^A u$ where u is the unique solution of the Homogeneous Regularity Problem (8.4.2) (written for the given f as boundary datum), is well defined, linear, and bounded.

Likewise, (8.4.14) and Theorem 1.5.5 imply that whenever $\delta \in (0,1)$ is small enough, $A \in \mathfrak{A}_L^{\rm dis}$, and $\mathfrak{A}_{L^\top}^{\rm dis} \neq \varnothing$, it follows that the Neumann-to-Dirichlet map

$$\Lambda_{\rm ND} : \big[L^p(\partial\Omega,\sigma)\big]^M \longrightarrow \big[\dot{L}_1^p(\partial\Omega,\sigma)/\sim\big]^M \tag{8.4.120}$$

defined for each $f \in \big[L^p(\partial\Omega,\sigma)\big]^M$ as $\Lambda_{\rm ND} f := \big[u\big|_{\partial\Omega}^{\kappa-{\rm n.t.}}\big]$ where u is a solution of the Neumann Problem (8.4.114) (written for the given f as boundary datum), is also well defined, linear, and bounded.

A more refined statement of Theorem 8.4.16 is recorded below.

Remark 8.4.17 *A more precise formulation of the well-posedness result in item (c) of Theorem 8.4.16 reads as follows: Fix* $n, M \in \mathbb{N}$ *with* $n \geq 2$ *and consider a weakly elliptic homogeneous constant complex coefficient second-order* $M \times M$ *system* L *in* $\mathbb{R}^n$. *Suppose there exists a coefficient tensor* $A \in \mathfrak{A}_L^{\rm dis}$ *with the property that* $A^\top \in \mathfrak{A}_{L^\top}^{\rm dis}$. *Then for each constant* $C_A \in (0,\infty)$ *and each compact interval* $I \subset (1,\infty)$ *there exists a threshold* $\delta \in (0,1)$ *depending only on* n, C_A, I, *and* A *with the following significance. Assume* $\Omega \subseteq \mathbb{R}^n$ *is an Ahlfors regular domain such that the Ahlfors regularity constants of* $\partial\Omega$ *are controlled by* C_A. *Set* $\sigma := \mathcal{H}^{n-1}\lfloor\partial\Omega$ *and denote by* ν *the geometric measure theoretic outward unit normal to* Ω. *Also, fix an integrability exponent* $p \in I$. *Then whenever* $\|\nu\|_{[{\rm BMO}(\partial\Omega,\sigma)]^n} < \delta$ *it follows that the Neumann Problem* (8.4.114) *is well posed (in the sense that it has a solution satisfying* *a naturally accompanying estimate, which is also unique modulo constants). In addition, similar considerations apply to all results formulated in Theorem 8.4.16.*

This follows from the invertibility results from Remark 4.2.2 and the proof of Theorem 8.4.16. However, before taking up the task of proving Theorem 8.4.16 we also want to comment on the fact that for scalar operators in a half-space the boundary between the Neumann and Dirichlet problems is blurred, in that the Neumann Problem may be used to solve a Dirichlet Problem, as indicated below:

Remark 8.4.18 *Let L be a scalar, second-order, homogeneous, constant (complex) coefficient, weakly elliptic operator in* $\mathbb{R}^n$*, say,* $L = \operatorname{div} A\nabla$ *for a weakly elliptic matrix* $A = (a_{jk})_{1\leq j,k\leq n} \in \mathbb{C}^{n\times n}$*. Fix* $\kappa \in (0,\infty)$*,* $p \in (1,\infty)$*, and consider a solution of the* L^p*-Neumann Problem for L in the upper half-space,*

$$\begin{cases} u \in \mathscr{C}^\infty(\mathbb{R}^n_+), \quad Lu = 0 \text{ in } \mathbb{R}^n_+, \\ \mathcal{N}_\kappa(\nabla u) \in L^p(\mathbb{R}^{n-1}, \mathcal{L}^{n-1}), \\ \partial^A_\nu u := -a_{nk}(\partial_k u)\Big|^{\kappa-\text{n.t.}}_{\partial\mathbb{R}^n_+} = f \in L^p(\mathbb{R}^{n-1}, \mathcal{L}^{n-1}). \end{cases} \tag{8.4.121}$$

Then the function

$$w := -a_{nk}\partial_k u \text{ in } \mathbb{R}^n_+ \tag{8.4.122}$$

is a solution of the L^p*-Dirichlet Problem for L in the upper half-space, specifically,*

$$\begin{cases} w \in \mathscr{C}^\infty(\mathbb{R}^n_+), \quad Lw = 0 \text{ in } \mathbb{R}^n_+, \\ \mathcal{N}_\kappa w \in L^p(\mathbb{R}^{n-1}, \mathcal{L}^{n-1}), \quad w\Big|^{\kappa-\text{n.t.}}_{\partial\mathbb{R}^n_+} = f \in L^p(\mathbb{R}^{n-1}, \mathcal{L}^{n-1}). \end{cases} \tag{8.4.123}$$

In particular, the L^p*-Dirichlet Problem* (8.4.123) *is solvable, as a consequence of the solvability of the* L^p*-Neumann Problem* (8.4.121)*. See*[99] *Theorem 8.4.16 where, in addition to ordinary Lebesgue spaces, the Neumann problem has been shown to be solvable in a variety of other function spaces, including Muckenhoupt weighted Lebesgue spaces, Lorentz spaces, Morrey spaces, vanishing Morrey spaces, block spaces, etc.*

We now turn to the task of presenting the proof of Theorem 8.4.16.

Proof of Theorem 8.4.16 Work under the assumption that $\|\nu\|_{[\mathrm{BMO}(\partial\Omega,\sigma)]^n} < \delta$ with $\delta \in (0,1)$ sufficiently small as to guarantee that Ω is a connected two-sided NTA domain with an unbounded boundary. That this may be arranged is guaranteed by Theorem 3.1.5. After further decreasing the value of δ if necessary, we may also assume that the conclusions in Theorem 4.2.1 are valid (for suitable choices of the parameters involved in its statement).

Then the operator $-\frac{1}{2}I + K^{\#}_{A^\top}$ is invertible on the space $\big[L^p(\partial\Omega,\sigma)\big]^M$ and the function u defined in (8.4.115) solves the Neumann Problem (8.4.114) and satisfies (8.4.116), thanks to [132, (1.5.52), (1.5.57), (1.5.58)]. Let us prove that if $A \in \mathfrak{A}^{\text{dis}}_L$

[99] concerning the hypotheses made in Theorem 8.4.16, it is helpful to recall Propositions 1.6.12-1.6.13

then any two solutions of the Neumann Problem (8.4.114) differ by a constant. This comes down to showing that any function satisfying

$$\begin{aligned} u \in \big[\mathscr{C}^\infty(\Omega)\big]^M, \quad Lu = 0 \text{ in } \Omega, \quad \mathcal{N}_\kappa(\nabla u) \in L^p(\partial\Omega, \sigma), \\ \text{and } \partial_\nu^A u = 0 \text{ at } \sigma\text{-a.e. point on } \partial\Omega, \end{aligned} \tag{8.4.124}$$

is necessarily constant in Ω. With this goal in mind, we first invoke Theorem 4.2.1 to conclude that the operator

$$-\tfrac{1}{2}I + \big[K_{\text{mod}}\big] : \big[\dot{L}_1^p(\partial\Omega, \sigma)/\sim\big]^M \longrightarrow \big[\dot{L}_1^p(\partial\Omega, \sigma)/\sim\big]^M \tag{8.4.125}$$

is invertible. In a second step, we observe that [132, Theorem 1.8.19] is applicable and presently gives (keeping in mind that Ω is connected and $\partial\Omega$ is unbounded) that

$$u\big|_{\partial\Omega}^{\kappa-\text{n.t.}} \text{ exists } \sigma\text{-a.e. on } \partial\Omega, \text{ belongs to } \big[\dot{L}_1^p(\partial\Omega, \sigma)\big]^M \tag{8.4.126}$$

and there exists some $c \in \mathbb{C}^M$ with the property that

$$u = \mathcal{D}_{\text{mod}}\big(u\big|_{\partial\Omega}^{\kappa-\text{n.t.}}\big) + c \text{ in } \Omega. \tag{8.4.127}$$

Taking nontangential boundary traces of both sides in (8.4.127) then yields, on account of [132, (1.8.27)], $\big(-\tfrac{1}{2}I + \big[K_{\text{mod}}\big]\big)\big[u\big|_{\partial\Omega}^{\kappa-\text{n.t.}}\big] = [0]$ in $\big[\dot{L}_1^p(\partial\Omega, \sigma)/\sim\big]^M$. In view of (8.4.125) this forces $u\big|_{\partial\Omega}^{\kappa-\text{n.t.}}$ to be a constant on $\partial\Omega$. When used back in (8.4.127) this ultimately shows (keeping in mind [132, (1.8.10)]) that u is constant.

Next, the well-posed claimed in item *(c)* is a consequence of what we have proved in items *(a)-(b)*. To justify the claims made in item *(d)*, consider the Laplacian Δ in $\mathbb{R}^2 \equiv \mathbb{C}$, in the writing $\Delta = a_{rs}\partial_r\partial_s$, where the coefficient tensor $A = \big(a_{rs}\big)_{1\le r,s\le 2}$ is the 2×2 complex matrix

$$A := \begin{pmatrix} 1 & -\mathrm{i} \\ \mathrm{i} & 1 \end{pmatrix}. \tag{8.4.128}$$

Let $\kappa > 0$ be fixed, and let $p \in (1, \infty)$. We are interested in describing the space of admissible boundary data for the L^p-Neumann Problem for the Laplacian in the upper half-plane where the prescribed conormal derivative is the one associated with the matrix A. In this regard, we make the claim

$$\begin{aligned} &\Big\{\partial_\nu^A u : u \in \mathscr{C}^\infty(\mathbb{R}^2_+), \ \Delta u = 0 \text{ in } \mathbb{R}^2_+, \ \mathcal{N}_\kappa(\nabla u) \in L^p(\mathbb{R}, \mathcal{L}^1)\Big\} \\ &\quad = \Big\{f \in L^p(\mathbb{R}, \mathcal{L}^1) : Hf = -\mathrm{i}f\Big\}, \end{aligned} \tag{8.4.129}$$

where H is the Hilbert transform on the real line (cf. (1.0.6)). Since the latter space has infinite codimension in $L^p(\mathbb{R}, \mathcal{L}^1)$ (given that $H^2 = -I$ on this space), the identification in (8.4.129) suits our current goals.

To prove the left-to-right inclusion in (8.4.129), pick a complex-valued function

$$u \in \mathscr{C}^\infty(\mathbb{R}^2_+) \text{ with } \Delta u = 0 \text{ in } \mathbb{R}^2_+ \text{ and } \mathcal{N}_\kappa(\nabla u) \in L^p(\mathbb{R}, \mathcal{L}^1). \tag{8.4.130}$$

Thanks to the Fatou-type result established in [131, Theorem 3.3.4], these properties guarantee that $(\nabla u)\big|^{\kappa-\mathrm{n.t.}}_{\partial\mathbb{R}^2_+}$ exists at $\mathcal{L}^1$-a.e. point on $\partial\mathbb{R}^2_+$. In particular, $f := \partial^A_\nu u$ is a well-defined function in $L^p(\mathbb{R},\mathcal{L}^1)$. More specifically, bearing in mind that the outward unit normal for the upper half-plane is $\nu = (\nu_1,\nu_2) = (0,-1) \equiv -\mathrm{i}$, from (A.0.203) we see that, at $\mathcal{L}^1$-a.e. point on $\partial\mathbb{R}^2_+ \equiv \mathbb{R}$

$$\begin{aligned} f = \partial^A_\nu u &= \nu_r a_{rs}\big(\partial_s u\big)\big|^{\kappa-\mathrm{n.t.}}_{\partial\mathbb{R}^2_+} \\ &= \nu_1\big(\partial_1 u\big)\big|^{\kappa-\mathrm{n.t.}}_{\partial\mathbb{R}^2_+} - \mathrm{i}\nu_1\big(\partial_2 u\big)\big|^{\kappa-\mathrm{n.t.}}_{\partial\mathbb{R}^2_+} + \mathrm{i}\nu_2\big(\partial_1 u\big)\big|^{\kappa-\mathrm{n.t.}}_{\partial\mathbb{R}^2_+} + \nu_2\big(\partial_2 u\big)\big|^{\kappa-\mathrm{n.t.}}_{\partial\mathbb{R}^2_+} \\ &= (\nu_1+\mathrm{i}\nu_2)\Big((\partial_1 u)\big|^{\kappa-\mathrm{n.t.}}_{\partial\mathbb{R}^2_+} - \mathrm{i}(\partial_2 u)\big|^{\kappa-\mathrm{n.t.}}_{\partial\mathbb{R}^2_+}\Big) = 2\nu(\partial_z u)\big|^{\kappa-\mathrm{n.t.}}_{\partial\mathbb{R}^2_+} = -2\mathrm{i}(\partial_z u)\big|^{\kappa-\mathrm{n.t.}}_{\partial\mathbb{R}^2_+} \end{aligned} \tag{8.4.131}$$

where $\partial_z := \frac{1}{2}(\partial_x - \mathrm{i}\partial_y)$ is the complex conjugate of the Cauchy-Riemann operator $\partial_{\bar z} := \frac{1}{2}(\partial_x + \mathrm{i}\partial_y)$. Hence, if we define $w := 2\partial_z u$ in $\mathbb{R}^2_+$, upon recalling that $\Delta = 4\partial_{\bar z}\partial_z$, the properties in (8.4.130) imply $w \in \mathscr{C}^\infty(\mathbb{R}^2_+)$, $\partial_{\bar z} w = 0$ in $\mathbb{R}^2_+$, $\mathcal{N}_\kappa w \in L^p(\mathbb{R},\mathcal{L}^1)$. These simply amount to saying that w is a holomorphic function belonging to the classical Hardy space $\mathcal{H}^p(\mathbb{R}^2_+)$ (associated with the Cauchy-Riemann operator in the upper half-plane; see the discussion in [131, §3.2]). In addition, (8.4.131) tells us that

$$w\big|^{\kappa-\mathrm{n.t.}}_{\partial\mathbb{R}^2_+} = \mathrm{i}f \ \text{ at } \mathcal{L}^1\text{-a.e. point on } \partial\mathbb{R}^2_+ \equiv \mathbb{R}. \tag{8.4.132}$$

Together with Cauchy's reproducing formula for holomorphic functions in the Hardy space associated with the upper half-plane, this gives $w = \mathscr{C}\Big(w\big|^{\kappa-\mathrm{n.t.}}_{\partial\mathbb{R}^2_+}\Big) = \mathrm{i}\mathscr{C}f$ in $\mathbb{R}^2_+$. After taking the nontangential trace to the boundary and using (8.4.132), we arrive at the conclusion that $\mathrm{i}\mathrm{i}f = \mathrm{i}\big(\frac{1}{2}I + C\big)f$ at $\mathcal{L}^1$-a.e. point in $\mathbb{R}$, where C is the principal-value Cauchy singular integral operator on the real line. Since $C = -\frac{1}{2\mathrm{i}}H$, we ultimately see that f must satisfy the compatibility condition $Hf = -\mathrm{i}f$ at $\mathcal{L}^1$-a.e. point in $\mathbb{R}$. This proves the left-to-right inclusion in (8.4.129).

For the converse inclusion, consider $f \in L^p(\mathbb{R},\mathcal{L}^1)$ satisfying $Hf = -\mathrm{i}f$ at $\mathcal{L}^1$-a.e. point in $\mathbb{R}$. Bring $\mathscr{S}_{\mathrm{mod}}$, the modified boundary-to-domain harmonic single layer potential operator associated with the Laplacian in the upper half-plane (cf. (A.0.229)), and define $u := -\mathscr{S}_{\mathrm{mod}} f$ in $\mathbb{R}^2_+$. Then this function belongs to $\mathscr{C}^\infty(\mathbb{R}^2_+)$, satisfies $\Delta u = 0$ in $\mathbb{R}^2_+$, has $\mathcal{N}_\kappa(\nabla u) \in L^p(\mathbb{R},\mathcal{L}^1)$, and (8.4.131) together with [132, (1.6.45)] permits us to compute

$$\begin{aligned} \partial^A_\nu u &= -2\mathrm{i}\big(\partial_z u\big)\big|^{\kappa-\mathrm{n.t.}}_{\partial\mathbb{R}^2_+} = 2\mathrm{i}(\partial_z \mathscr{S}_{\mathrm{mod}} f)\big|^{\kappa-\mathrm{n.t.}}_{\partial\mathbb{R}^2_+} \\ &= (\mathscr{C}f)\big|^{\kappa-\mathrm{n.t.}}_{\partial\mathbb{R}^2_+} = \big(\tfrac{1}{2}I + C\big)f = \tfrac{1}{2}f - \tfrac{1}{2\mathrm{i}}Hf = \tfrac{1}{2}f + \tfrac{1}{2}f = f, \end{aligned} \tag{8.4.133}$$

as wanted. As regards the space of null-solutions for the L^p-Neumann Problem (8.4.114) in the case when $n = 2$, $M = 1$, $L = \Delta$ (the two-dimensional Laplacian), $\Omega = \mathbb{R}^2_+$, and A as in (8.4.128), we claim that

$$\Big\{u \in \mathscr{C}^\infty(\mathbb{R}^2_+) : \Delta u = 0 \text{ in } \mathbb{R}^2_+,\ \mathcal{N}_\kappa(\nabla u) \in L^p(\mathbb{R}, \mathcal{L}^1),\ \partial^A_\nu u = 0\Big\} \tag{8.4.134}$$
$$= \Big\{\overline{w} : w \text{ holomorphic in } \mathbb{R}^2_+, \text{ with } \mathcal{N}_\kappa(\nabla w) \in L^p(\mathbb{R}, \mathcal{L}^1)\Big\}.$$

Indeed, if u is any function belonging to the space in the left side of (8.4.134), then $\partial_z u$ is holomorphic in $\mathbb{R}^2_+$ (since $\partial_{\bar z}\partial_z = \frac{1}{4}\Delta$), and satisfies $\mathcal{N}_\kappa(\partial_z u) \in L^p(\mathbb{R}, \mathcal{L}^1)$. These properties place $\partial_z u$ in $\mathcal{H}^p(\mathbb{R}^2_+)$, the classical Hardy space in the upper half-plane for the Cauchy-Riemann operator. Since from (8.4.131) we have

$$(\partial_z u)\Big|^{\kappa-\text{n.t.}}_{\partial\mathbb{R}^2_+} = 0 \text{ at } \mathcal{L}^1\text{-a.e. point on } \partial\mathbb{R}^2_+ \equiv \mathbb{R}, \tag{8.4.135}$$

Cauchy's reproducing formula then forces $\partial_z u$ to vanish identically in $\mathbb{R}^2_+$. The latter property may be interpreted simply as saying that $w := \overline{u}$ is holomorphic in $\mathbb{R}^2_+$. This puts $\overline{w}$ (and, ultimately, u) in the space in the right side of (8.4.134). Conversely, given any holomorphic function w in $\mathbb{R}^2_+$ satisfying $\mathcal{N}_\kappa(\nabla w) \in L^p(\mathbb{R}, \mathcal{L}^1)$, the function $u := \overline{w}$ is harmonic in $\mathbb{R}^2_+$, has $\mathcal{N}_\kappa(\nabla u) \in L^p(\mathbb{R}, \mathcal{L}^1)$ and, much as in (8.4.131), we see that at $\mathcal{L}^1$-a.e. point on $\partial\mathbb{R}^2_+ \equiv \mathbb{R}$ we have

$$\partial^A_\nu u = -2\mathrm{i}(\partial_z u)\Big|^{\kappa-\text{n.t.}}_{\partial\mathbb{R}^2_+} = -2\mathrm{i}\overline{(\partial_{\bar z}\overline{u})\Big|^{\kappa-\text{n.t.}}_{\partial\mathbb{R}^2_+}} = -2\mathrm{i}\overline{(\partial_{\bar z} w)\Big|^{\kappa-\text{n.t.}}_{\partial\mathbb{R}^2_+}} = 0, \tag{8.4.136}$$

since w is holomorphic in $\mathbb{R}^2_+ \equiv \mathbb{C}_+$. The proof of (8.4.134) is therefore complete. Note that the space in the right side of (8.4.134) is infinite dimensional since, for example, for each $m \in \mathbb{N}$ the function $\mathbb{C}_+ \ni z \mapsto (\bar z - \mathrm{i})^{-m} \in \mathbb{C}$ belongs to said space. In view of this, we conclude that the space of null-solutions for the L^p-Neumann Problem (8.4.114) is, as claimed, infinite dimensional.

That a similar lack of Fredholm solvability is present in higher dimensions may be seen from Remark 8.6.14. These involve the vector Laplacian, but there is also a way of identifying Neumann Problems for the ordinary Laplacian in $\mathbb{R}^n_+$ which fail to be Fredholm solvable. To elaborate on this aspect, fix $n \in \mathbb{N}$ with $n \geq 2$, and write $\Delta = a_{jk}\partial_j\partial_k = \operatorname{div} A\nabla$ for some matrix $A = (a_{jk})_{1\leq j,k\leq n} \in \mathbb{C}^{n\times n}$. Incidentally, this demand on the coefficient matrix A is equivalent to asking that $a_{jk} + a_{kj} = 2\delta_{jk}$ for each $j,k \in \{1,\dots,n\}$, and further equivalent to having $a_{jk} = \delta_{jk} + c_{kj}$ with $c_{kj} = -c_{jk}$ for each $j,k \in \{1,\dots,n\}$. Having also picked $\kappa > 0$ and $p \in (1,\infty)$, consider the Neumann Problem for the Laplacian in the upper half-space in which the conormal derivative is associated with the above coefficient matrix A, i.e.,

$$\begin{cases} u \in \mathscr{C}^\infty(\mathbb{R}^n_+), \quad \Delta u = 0 \text{ in } \mathbb{R}^n_+, \\ \mathcal{N}_\kappa(\nabla u) \in L^p(\mathbb{R}^{n-1}, \mathcal{L}^{n-1}), \quad \partial^A_\nu u = f \in L^p(\mathbb{R}^{n-1}, \mathcal{L}^{n-1}). \end{cases} \tag{8.4.137}$$

From Corollary 1.4.7 and [132, Theorem 1.5.1, item *(xi)*] we know that the first three conditions above are equivalent to the existence of a function $g \in L^p(\mathbb{R}^{n-1}, \mathcal{L}^{n-1})$ and a constant $c \in \mathbb{C}$ such that

$$u = \mathscr{S}_{\text{mod}}\, g + c \text{ in } \mathbb{R}^n_+, \tag{8.4.138}$$

where $\mathscr{S}_{\rm mod}$ is the modified boundary-to-domain harmonic single layer potential operator, associated with $L := \Delta$ and $\Omega := \mathbb{R}^n_+$ as in (A.0.229). Also, whenever u is as in (8.4.138), the boundary condition in (8.4.137) becomes, thanks to the jump-formula (1.4.56) and the specific format which the singular integral operator (A.0.118) now acquires,

$$\begin{aligned} f(x') = \big(\partial^A_\nu u\big)(x') = \big(\partial^A_\nu \mathscr{S}_{\rm mod} g\big)(x') = \big[\big(-\tfrac{1}{2}I + K^\#_{A^\top}\big)g\big](x') \\ = -\tfrac{1}{2}g(x') + \lim_{\varepsilon\to 0^+} \int\limits_{y'\in\mathbb{R}^{n-1},\ |x'-y'|>\varepsilon} a_{jn}(\partial_j E_\Delta)(x'-y',0)g(y')\,dy' \end{aligned} \tag{8.4.139}$$

at $\mathcal{L}^{n-1}$-a.e. $x' \in \mathbb{R}^{n-1}$, where E_Δ is the standard fundamental solution for the Laplacian from (A.0.77). In view of (A.0.77) and (A.0.215) (with $\Sigma := \mathbb{R}^{n-1}$), this is further equivalent to the demand that

$$\Big(-I + \sum_{j=1}^{n-1} a_{jn}R_j\Big)g = 2f \ \text{ at } \mathcal{L}^{n-1}\text{-a.e. point on } \mathbb{R}^{n-1}, \tag{8.4.140}$$

where R_j is the j-th Riesz transform in $\mathbb{R}^{n-1}$ (defined as in (A.0.215) with $\Sigma := \mathbb{R}^{n-1}$).

From the above considerations and [132, (5.3.25)] we then see that

if $n \geq 3$ and $p = 2$, then the space of admissible boundary data for the Neumann Problem (8.4.137) is dense in the full space of boundary data $L^2(\mathbb{R}^{n-1}, \mathcal{L}^{n-1})$, and any two solutions of the Neumann Problem (8.4.137) differ by an additive constant. (8.4.141)

In concert with (8.4.140) and the equivalence *(b)*$\Leftrightarrow$*(c)* in [132, Proposition 5.3.1], this also shows that

if $n \geq 3$ and $p = 2$, then the space of admissible boundary data for the Neumann Problem (8.4.137) is closed if and only if it coincides with the full space of boundary data $L^2(\mathbb{R}^{n-1}, \mathcal{L}^{n-1})$, which is further equivalent to having $\sum_{j=1}^{n-1} a_{jn}\xi_j \neq \mathrm{i}|\xi'|$ for each $\xi' \in \mathbb{R}^{n-1}\setminus\{0\}$. (8.4.142)

In particular, (8.4.142) tells us that it may well happen that the space of admissible boundary data for the Neumann Problem (8.4.137) is *not* closed (hence the Neumann Problem (8.4.137) is not always solvable). For example, this is the case whenever

$$\begin{aligned} A = I_{n\times n} + C \ \text{ with } C = (c_{jk})_{1\leq j,k\leq n} \in \mathbb{C}^{n\times n} \\ \text{satisfying } C^\top = -C \text{ as well as } (c_{jn})_{1\leq j\leq n-1} \in \mathrm{i}S^{n-2}. \end{aligned} \tag{8.4.143}$$

This being said, from (8.4.140) and the equivalence *(a)*$\Leftrightarrow$*(c)* in [132, Proposition 5.3.1] we see that for any $n \geq 2$ and $p \in (1, \infty)$, the Neumann Problem (8.4.137) is well posed (with uniqueness understood modulo constants) if and only if the coefficient matrix $A = (a_{jk})_{1 \leq j,k \leq n} \in \mathbb{C}^{n \times n}$ satisfies $\sum_{j=1}^{n-1} a_{jn}\xi_j \neq \mathrm{i}|\xi'|$ for each $\xi' = (\xi_1, \dots, \xi_{n-1}) \in \mathbb{R}^{n-1} \setminus \{0\}$. (8.4.144)

Lastly, we wish to note that in the case when $n = 2$, the algebraic condition in (8.4.144) simply reduces to $a_{12} \neq \pm\mathrm{i}$, and our earlier choice in (8.4.128) was deliberately made as to violate it. This completes the treatment of item *(d)*.

Finally, the claim in item *(e)* may be justified by reasoning in a similar fashion as above, now also making use of [132, Theorem 3.3.1]. □

Remark 8.4.19 *Let us compare Theorem 8.4.16 with the explicit well-posedness result for the two-dimensional Laplacian in a planar infinite sector* $\Omega_\theta \subseteq \mathbb{R}^2$ *with aperture* $\theta \in (0, 2\pi)$*, as in* (8.4.35)*. To be specific, abbreviate* $\sigma_\theta := \mathcal{H}^1 \lfloor \partial\Omega_\theta$ *and denote by* ν_θ *the outward unit normal to* Ω_θ*. Also, fix an aperture parameter* $\kappa \in (0, \infty)$*. In this setting, for each given integrability exponent* $p \in (1, \infty)$ *consider the* L^p*-Neumann Problem in* Ω_θ *for* $L := \Delta$*, i.e.,*

$$\begin{cases} u \in \mathscr{C}^\infty(\Omega_\theta), \quad \Delta u = 0 \ \text{in } \Omega_\theta, \\ \mathcal{N}_\kappa(\nabla u) \in L^p(\partial\Omega_\theta, \sigma_\theta), \\ \nu_\theta \cdot \Big((\nabla u)\Big|_{\partial\Omega_\theta}^{\kappa-\mathrm{n.t.}} \Big) = f \in L^p(\partial\Omega_\theta, \sigma_\theta). \end{cases} \tag{8.4.145}$$

Relying on boundary layer potentials (much as in the proof of Theorem 8.4.16), from (4.2.185)-(4.2.187) *we then see that*

whenever $p \neq q_\theta := 1 + \pi/|\pi - \theta|$*, the* L^p*-Neumann Problem* (8.4.145) *has a solution, which is unique modulo constants, and satisfies a naturally accompanying estimate.* (8.4.146)

In Theorem 8.4.22, stated a little later, we shall further refine the result recorded in Remark 8.4.19. En route, we need a couple of auxiliary facts of independent interest. First, complementing [130, Proposition 11.5.15], we have the following result, regarding the nullspace of the tangential derivative operator acting from local Sobolev spaces on boundaries of chord-arc domains.

Lemma 8.4.20 *Let* $\Omega \subseteq \mathbb{R}^2$ *be a chord-arc domain with unbounded boundary and set* $\sigma := \mathcal{H}^1 \lfloor \partial\Omega$*. Let* ∂_τ *stand for the tangential derivative operator* $\partial_{\tau_{12}}$ *on* $\partial\Omega$ *(cf.* (A.0.202)*), and fix some* $p \in [1, \infty]$*. Then for each* $f \in L^p_{1,\mathrm{loc}}(\partial\Omega, \sigma)$ *one has*

$$\partial_\tau f = 0 \ \text{at } \sigma\text{-a.e. point on } \partial\Omega \iff f \text{ is constant on } \partial\Omega. \tag{8.4.147}$$

Proof From [129, (5.9.93)] and [129, Definition 5.9.13] we know that Ω is a connected, simply connected, unbounded, two-sided NTA domain with a connected Ahlfors regular boundary (hence also an Ahlfors regular domain satisfying a two-sided local John condition and, in particular, a UR domain), such that $\partial\Omega$ is a locally rectifiable Jordan curve passing through infinity in $\mathbb{C} \equiv \mathbb{R}^2$.

The locality of the tangential differential operator ∂_τ (cf. [130, Lemma 11.1.6]) shows that the left-pointing implication in (8.4.147) is true. To justify the right-pointing implication in (8.4.147), assume $f \in L^p_{1,\mathrm{loc}}(\partial\Omega, \sigma)$ is such that $\partial_\tau f = 0$. Let $\mathbb{R} \ni s \longmapsto z(s) \in \partial\Omega$ be the arc-length parametrization of $\partial\Omega$, and denote by $z^{-1}(\cdot) : \partial\Omega \to \mathbb{R}$ its inverse. We claim that

$$\partial_\tau(\psi \circ z^{-1}) = \psi' \circ z^{-1} \text{ for each } \psi \in \mathscr{C}^\infty_c(\mathbb{R}). \tag{8.4.148}$$

To justify this, pick an arbitrary $\psi \in \mathscr{C}^\infty_c(\mathbb{R})$ and note that

$$\psi \circ z^{-1} \in \mathrm{Lip}_c(\partial\Omega) \subseteq L^\infty_1(\partial\Omega, \sigma), \tag{8.4.149}$$

thanks to the fact that $z^{-1}(\cdot)$ is bi-Lipschitz (cf. [129, (5.9.95)]) and [130, Proposition 11.1.11]. For each function $\varphi \in \mathscr{C}^\infty_c(\mathbb{R}^2)$ we may then compute

$$\begin{aligned}\int_{\partial\Omega} (\psi \circ z^{-1})\partial_\tau\varphi \, \mathrm{d}\sigma &= \int_{\mathbb{R}} \psi(s)(\partial_\tau\varphi)\big(z(s)\big) \, \mathrm{d}s = \int_{\mathbb{R}} \psi(s)\frac{\mathrm{d}}{\mathrm{d}s}\big[\varphi(z(s))\big] \, \mathrm{d}s \\ &= -\int_{\mathbb{R}} \psi'(s)\varphi\big(z(s)\big) \, \mathrm{d}s = -\int_{\partial\Omega} (\psi' \circ z^{-1})\varphi \, \mathrm{d}\sigma, \end{aligned} \tag{8.4.150}$$

based on the change of variable formula [129, (5.9.100)] (used twice), [130, (11.5.156)], and integration by parts on the real line. In light of the arbitrariness of the test function $\varphi \in \mathscr{C}^\infty_c(\mathbb{R}^2)$, we conclude from (8.4.150) and [130, Definition 11.1.2] that (8.4.148) holds.

Having established (8.4.148), and upon observing that $f\big(z(\cdot)\big) \in L^p_{\mathrm{loc}}(\partial\Omega, \sigma)$ (as may be seen from assumptions and the change of variable formula [129, (5.9.100)]), for each given $\psi \in \mathscr{C}^\infty_c(\mathbb{R})$ we may now write

$$\begin{aligned}\int_{\mathbb{R}} f\big(z(s)\big)\psi'(s) \, \mathrm{d}s &= \int_{\partial\Omega} f \cdot (\psi' \circ z^{-1}) \, \mathrm{d}\sigma = \int_{\partial\Omega} f \cdot \partial_\tau(\psi \circ z^{-1}) \, \mathrm{d}\sigma \\ &= -\int_{\partial\Omega} (\partial_\tau f) \cdot (\psi \circ z^{-1}) \, \mathrm{d}\sigma = 0, \end{aligned} \tag{8.4.151}$$

making use of the change of variable formula [129, (5.9.100)], the identity in (8.4.148), the membership in (8.4.149), the integration by parts formula on the boundary from [130, Proposition 11.1.14], and the fact that we are presently assuming $\partial_\tau f = 0$ at σ-a.e. point on $\partial\Omega$. In turn, (8.4.151) shows that, in the distributional sense, $\frac{\mathrm{d}}{\mathrm{d}s}\big[f(z(s))\big] = 0$ on $\mathbb{R}$. Hence, $f\big(z(\cdot)\big)$ is constant on $\mathbb{R}$ (cf., e.g., [127]), which further implies that f itself is constant on $\partial\Omega$. □

We shall next use Lemma 8.4.20 to prove that, in any chord-arc domain with unbounded boundary and for any integrability exponent $p \in (1, \infty)$, having the L^p-Neumann Problem (for the Laplacian) well posed is equivalent with having the $\dot{L}^p_1$-Homogeneous Regularity Problem (for the Laplacian) well posed.

Proposition 8.4.21 *Let $\Omega \subseteq \mathbb{R}^2$ be a chord-arc domain with unbounded boundary and abbreviate $\sigma := \mathcal{H}^1 \lfloor \partial\Omega$. Also, denote by ν the geometric measure theoretic* *outward unit normal to* Ω. *Finally, fix an integrability exponent* $p \in (1,\infty)$ *and an aperture parameter* $\kappa \in (0,\infty)$. *Then the ("standard")* L^p*-Neumann Problem for the Laplacian in* Ω, *i.e.,*

$$\begin{cases} u \in \mathscr{C}^\infty(\Omega), \quad \Delta u = 0 \ \text{in } \Omega, \\ \mathcal{N}_\kappa(\nabla u) \in L^p(\partial\Omega,\sigma), \\ \partial_\nu u := \nu \cdot \Big((\nabla u)\Big|_{\partial\Omega}^{\kappa-\text{n.t.}} \Big) = f \in L^p(\partial\Omega,\sigma), \end{cases} \tag{8.4.152}$$

is solvable (respectively, has a solution satisfying a naturally accompanying estimate, or any two solutions differ by a constant) if and only if the $\dot{L}^p_1$*-Homogeneous Regularity Problem for the Laplacian in* Ω, *i.e.,*

$$\begin{cases} w \in \mathscr{C}^\infty(\Omega), \quad \Delta w = 0 \ \text{in } \Omega, \\ \mathcal{N}_\kappa(\nabla w) \in L^p(\partial\Omega,\sigma), \quad w\Big|_{\partial\Omega}^{\kappa-\text{n.t.}} = g \in \dot{L}^p_1(\partial\Omega,\sigma), \end{cases} \tag{8.4.153}$$

is solvable (respectively, has a solution satisfying a naturally accompanying estimate, or any two solutions coincide).

As a corollary, the ("standard") L^p*-Neumann Problem* (8.4.152) *is well posed (in the sense that it always solvable, the solution is unique modulo constants, and there is a naturally accompanying estimate) if and only if the* $\dot{L}^p_1$*-Homogeneous Regularity Problem* (8.4.153) *is well posed (in the sense that this boundary problem is always uniquely solvable, and the solution satisfies a natural estimate).*

Finally, at the level of admissible boundary data for the two boundary value problems considered above,

$$\begin{aligned} &\Big\{\partial_\tau\Big(w\Big|_{\partial\Omega}^{\kappa-\text{n.t.}}\Big) : w \in \mathscr{C}^\infty(\Omega),\ \Delta w = 0 \ \text{in } \Omega,\ \mathcal{N}_\kappa(\nabla w) \in L^p(\partial\Omega,\sigma)\Big\} \\ &\qquad = \Big\{\partial_\nu u : u \in \mathscr{C}^\infty(\Omega),\ \Delta u = 0 \ \text{in } \Omega,\ \mathcal{N}_\kappa(\nabla u) \in L^p(\partial\Omega,\sigma)\Big\}. \end{aligned} \tag{8.4.154}$$

Proof To set the stage, we observe from [129, (5.9.93)] and [129, Definition 5.9.13] that Ω is a connected, simply connected, unbounded, two-sided NTA domain with a connected Ahlfors regular boundary (thus also an Ahlfors regular domain satisfying a two-sided local John condition and, in particular, a UR domain), with the property that $\partial\Omega$ is a locally rectifiable Jordan curve passing through infinity in $\mathbb{C} \equiv \mathbb{R}^2$.

To start in earnest, first assume the L^p-Neumann Problem (8.4.152) is (always) solvable. Given any $g \in \dot{L}^p_1(\partial\Omega,\sigma)$, define $f := \partial_\tau g \in L^p(\partial\Omega,\sigma)$ (with the membership provided by [130, Proposition 11.5.15]), and let u solve (8.4.152) for this boundary datum. In view of the topological properties of Ω discussed in the preamble (see the equivalence *(1)* $\Leftrightarrow$ *(10)* in [129, Proposition 5.8.1]), there exists a harmonic conjugate function $\widetilde{u}$ of u in Ω, i.e., $\widetilde{u} \in \mathscr{C}^\infty(\Omega)$ such that

$$\partial_x u = \partial_y \widetilde{u} \text{ and } \partial_y u = -\partial_x \widetilde{u} \text{ in } \Omega. \tag{8.4.155}$$

As a consequence, $|\nabla \widetilde{u}| = |\nabla u|$ in Ω, hence $\mathcal{N}_\kappa(\nabla \widetilde{u}) = \mathcal{N}_\kappa(\nabla u) \in L^p(\partial\Omega,\sigma)$. In concert with the Fatou-type result established in [131, Theorem 3.3.4], this guarantees that $(\nabla\widetilde{u})\big|_{\partial\Omega}^{\kappa-\mathrm{n.t.}}$ exists at σ-a.e. point on $\partial\Omega$. Furthermore, [129, Proposition 8.9.22] guarantees that $\widetilde{u}\big|_{\partial\Omega}^{\kappa-\mathrm{n.t.}}$ exists at σ-a.e. point on $\partial\Omega$, while [129, Proposition 8.4.9] ensures that $\mathcal{N}_\kappa^\varepsilon\widetilde{u}\in L^p(\partial\Omega,\sigma)$ for any truncation parameter $\varepsilon\in(0,\infty)$. Subsequently, [130, Proposition 11.3.2] shows that $\widetilde{u}\big|_{\partial\Omega}^{\kappa-\mathrm{n.t.}}\in L^p_{1,\mathrm{loc}}(\partial\Omega,\sigma)$. Based on these properties, (8.4.155), and [130, (11.3.26)] we may therefore write

$$\begin{aligned}\partial_\tau\Big(\widetilde{u}\big|_{\partial\Omega}^{\kappa-\mathrm{n.t.}}\Big)&=\nu_1\Big((\partial_y\widetilde{u})\big|_{\partial\Omega}^{\kappa-\mathrm{n.t.}}\Big)-\nu_2\Big((\partial_x\widetilde{u})\big|_{\partial\Omega}^{\kappa-\mathrm{n.t.}}\Big)\\&=\nu_1\Big((\partial_x u)\big|_{\partial\Omega}^{\kappa-\mathrm{n.t.}}\Big)+\nu_2\Big((\partial_y u)\big|_{\partial\Omega}^{\kappa-\mathrm{n.t.}}\Big)=\nu\cdot\Big((\nabla u)\big|_{\partial\Omega}^{\kappa-\mathrm{n.t.}}\Big)=f=\partial_\tau g.\end{aligned}\tag{8.4.156}$$

In view of Lemma 8.4.20, this implies that $\widetilde{u}\big|_{\partial\Omega}^{\kappa-\mathrm{n.t.}}$ and g differ by a constant. Thus, there exists $c\in\mathbb{C}$ such that $\widetilde{u}\big|_{\partial\Omega}^{\kappa-\mathrm{n.t.}}=g+c$ at σ-a.e. point on $\partial\Omega$. If we now define $w:=\widetilde{u}-c$, then w is a solution of the $\dot{L}^p_1$-Homogeneous Regularity Problem (8.4.153). Moreover, if u satisfies a naturally accompanying estimate we get

$$\begin{aligned}\|\mathcal{N}_\kappa(\nabla w)\|_{L^p(\partial\Omega,\sigma)}&=\|\mathcal{N}_\kappa(\nabla u)\|_{L^p(\partial\Omega,\sigma)}\leq C\|f\|_{L^p(\partial\Omega,\sigma)}\\&=C\|\partial_\tau g\|_{L^p(\partial\Omega,\sigma)}=\big\|[g]\big\|_{\dot{L}^p_1(\partial\Omega,\sigma)/\sim}\end{aligned}\tag{8.4.157}$$

for some constant $C\in(0,\infty)$ independent of g.

As far as the issue of uniqueness is concerned, assume that any two solutions of L^p-Neumann Problem (8.4.152) differ by a constant. If w solves the version of the $\dot{L}^p_1$-Homogeneous Regularity Problem (8.4.153) corresponding to the zero boundary datum and $\widetilde{w}$ is a harmonic conjugate of w, then much as before (see (8.4.155)-(8.4.156)) we have that $\widetilde{w}$ solves the version of (8.4.152) corresponding to the zero boundary datum. This forces $\widetilde{w}$ to be constant in Ω. Hence, w itself is a constant in Ω, whose nontangential trace vanishes, so ultimately $w\equiv 0$ in Ω. This finishes the proof of the fact that uniqueness modulo constants for the L^p-Neumann Problem implies uniqueness for the $\dot{L}^p_1$-Homogeneous Regularity Problem.

In the opposite direction, assume the $\dot{L}^p_1$-Homogeneous Regularity Problem (8.4.153) is (always) solvable, and fix an arbitrary $f\in L^p(\partial\Omega,\sigma)$. From [130, Proposition 11.5.15] we know that there exists $g\in\dot{L}^p_1(\partial\Omega,\sigma)$ such that $\partial_\tau g=f$. Let w solve the $\dot{L}^p_1$-Homogeneous Regularity Problem (8.4.153) for this boundary datum g, and let $\widetilde{w}$ be a harmonic conjugate of w in Ω. Then, much as above (cf. (8.4.155)-(8.4.157)), it follows that $\widetilde{w}$ solves the L^p-Neumann Problem (8.4.152) with boundary datum f. Moreover, $\widetilde{w}$satisfies a naturally accompanying estimate if w does so. Also, uniqueness (modulo constants) for the L^p-Neumann Problem (8.4.152) follows from uniqueness for the $\dot{L}^p_1$-Homogeneous Regularity Problem (8.4.153) reasoning as before. Finally, (8.4.154) is implicit in the above reasoning.□

We are now prepared to state and prove the following refinement of the result recorded in Remark 8.4.19, which has been advertised earlier (also compare with Theorem 8.4.16). For an arbitrary planar infinite sector, this explicitly links the values of the integrability exponent p for which the L^p-Neumann Problem for the two-dimensional Laplacian is well posed to the aperture of said sector.

Theorem 8.4.22 *With a given arbitrary angle $\theta \in (0, 2\pi)$, associate the infinite sector $\Omega_\theta \subseteq \mathbb{R}^2$ as in* (8.4.35)*. Abbreviate $\sigma_\theta := \mathcal{H}^1 \lfloor \partial\Omega_\theta$ and denote by ν_θ the outward unit normal to Ω_θ. Also, fix an aperture parameter $\kappa \in (0, \infty)$.*

Then the ("standard") L^p-Neumann Problem for the Laplacian in Ω_θ, i.e.,

$$\begin{cases} u \in \mathscr{C}^\infty(\Omega_\theta), \quad \Delta u = 0 \text{ in } \Omega_\theta, \\ \mathcal{N}_\kappa(\nabla u) \in L^p(\partial\Omega_\theta, \sigma_\theta), \\ \nu_\theta \cdot \Big((\nabla u)\big|^{\kappa-\text{n.t.}}_{\partial\Omega_\theta} \Big) = f \in L^p(\partial\Omega_\theta, \sigma_\theta), \end{cases} \tag{8.4.158}$$

is well posed (in the sense that this is always solvable, the solution is unique modulo constants, and there is a naturally accompanying estimate) for any integrability exponent $p \in (1, \infty)$ if $\theta \in (0, \pi]$, while if $\theta \in (\pi, 2\pi)$ then the L^p-Neumann Problem (8.4.158) *is well posed (in the same sense as above) if and only if*

$$p \in (1, \infty) \setminus \{q_\theta\} = (1, \infty) \setminus \{\tfrac{\theta}{\theta - \pi}\}. \tag{8.4.159}$$

Furthermore, supplementary information is available in the case when one has $\theta \in (\pi, 2\pi)$ for the critical exponent $p = q_\theta := 1 + \pi/|\pi - \theta| = \frac{\theta}{\theta - \pi}$. As far as uniqueness is concerned,

$$\text{if } \theta \in (\pi, 2\pi) \text{ then any two solutions of the } L^p\text{-Neumann Problem (8.4.158) for } p = q_\theta := 1 + \pi/|\pi - \theta| = \tfrac{\theta}{\theta - \pi} \text{ differ by a constant.} \tag{8.4.160}$$

To shed more light on the issue of solvability, bring in the two-dimensional modified boundary-to-boundary harmonic single layer operator S_{mod}, associated with Ω_θ as in (A.0.232)*. In addition, recall that the space of admissible boundary data for the L^p-Neumann Problem* (8.4.158) *corresponding to $p := q_\theta$ is the collection of functions from $L^{q_\theta}(\partial\Omega_\theta, \sigma_\theta)$ for which said problem is solvable when the functions in question are used as boundary data. Then said space coincides with the image of the operator $\partial_\tau S_{\text{mod}}$ acting from the space $L^{q_\theta}(\partial\Omega_\theta, \sigma_\theta)$ into itself, i.e., for $\theta \in (\pi, 2\pi)$ and $q_\theta := 1 + \pi/|\pi - \theta| = \frac{\theta}{\theta - \pi}$ one has*

$$\begin{aligned} \Big\{ \nu_\theta \cdot \Big((\nabla u)\big|^{\kappa-\text{n.t.}}_{\partial\Omega_\theta} \Big) : u \in \mathscr{C}^\infty(\Omega_\theta),\ \Delta u = 0 \text{ in } \Omega_\theta,\ \mathcal{N}_\kappa(\nabla u) \in L^{q_\theta}(\partial\Omega_\theta, \sigma_\theta) \Big\} \\ = \Big\{ \big(\partial_\tau S_{\text{mod}}\big) g : g \in L^{q_\theta}(\partial\Omega_\theta, \sigma_\theta) \Big\}. \end{aligned} \tag{8.4.161}$$

A consequence of (8.4.161), [130, Proposition 11.5.15], *and* (4.2.188) *is that*

$$\text{for } \theta \in (\pi, 2\pi) \text{ and } p = q_\theta := 1 + \pi/|\pi - \theta| = \tfrac{\theta}{\theta - \pi}, \text{ the space of admissible boundary data for the Neumann Problem (8.4.158) has infinite co-dimension (in particular, is not closed) in } L^{q_\theta}(\partial\Omega_\theta, \sigma_\theta). \tag{8.4.162}$$

As a corollary,

the L^p-Neumann Problem formulated in (8.4.158) *fails to be Fredholm solvable in the scenario when $\theta \in (\pi, 2\pi)$ and for $p = q_\theta := 1 + \pi/|\pi - \theta| = \frac{\theta}{\theta - \pi}$.* (8.4.163)

Proof All claims follow from Proposition 8.4.21, Theorem 8.4.12, and (8.4.92). □

We may further use Theorem 8.4.22 to underscore the significance of the smallness condition imposed on the BMO seminorm of the geometric measure theoretic outward unit normal in the formulation of Theorem 8.4.16.

Remark 8.4.23 *As before, associate the infinite sector $\Omega_\theta \subseteq \mathbb{R}^2$ with any given angle $\theta \in (\pi, 2\pi)$ as in* (8.4.35)*. Also, abbreviate $\sigma_\theta := \mathcal{H}^1\lfloor\partial\Omega_\theta$ and denote by ν_θ the outward unit normal to Ω_θ. Then $q_\theta := 1 + \pi/|\pi - \theta| \in (2, \infty)$ and*

$$\begin{aligned}\|\nu_\theta\|_{\mathrm{BMO}(\partial\Omega_\theta,\sigma_\theta)} &= |\cos(\theta/2)| = \big|\sin\big(\pi/2 - \theta/2\big)\big| \approx |\theta - \pi| \\ &\approx (q_\theta - 1)^{-1}, \quad \textit{uniformly for } \theta \in (\pi, 2\pi)\end{aligned} \tag{8.4.164}$$

(*cf.* (8.4.100)). *As a consequence of this and Theorem 8.4.22, for the class of domains Ω_θ with $\theta \in (\pi, 2\pi)$, the well-posedness of the* (*standard*) *L^p-Neumann Problem for the Laplacian for each exponent $p \in I := [2, a]$ with $a \in (2, \infty)$ large is equivalent to the demand that $\|\nu_\theta\|_{\mathrm{BMO}(\partial\Omega_\theta,\sigma_\theta)}$ is small. This points to the significance of the latter property in the context of Theorem 8.4.16* (*see the more refined statement given in Remark 8.4.17*).

We next consider the Neumann Problem with boundary data in pre-duals of Morrey-Campanato spaces, $\mathscr{H}^{q,\lambda}$ with $1 < q < \infty$ and $0 < \lambda < n - 1$, for weakly elliptic systems in sufficiently flat Ahlfors regular domains with unbounded boundaries. As in the case of Lebesgue spaces, the formulation of this boundary value problem involves the conormal derivative operator defined in a pointwise sense, as in (A.0.203). Our result, described in the theorem below, is entirely novel (e.g., is new even in the case of the Laplacian in the upper half-space). Before stating this, the reader is reminded that, for any closed Ahlfors regular set $\Sigma \subseteq \mathbb{R}^n$, the pre-dual $\mathscr{H}^{q,\lambda}(\Sigma, \sigma)$ of a Morrey-Campanato space has been introduced in (A.0.94)-(A.0.95). Also, recall that the (q, λ)-midway space $\mathscr{M}^{q,\lambda}(\Sigma, \sigma)$ has been defined in [131, (6.2.249), (6.2.250)].

Theorem 8.4.24 *Let $\Omega \subseteq \mathbb{R}^n$ be a* UR *domain. Denote by ν the geometric measure theoretic outward unit normal to Ω and abbreviate $\sigma := \mathcal{H}^{n-1}\lfloor\partial\Omega$. Also, fix an aperture parameter $\kappa > 0$ and pick $q \in (1, \infty)$ together with $\lambda \in (0, n - 1)$.*

Suppose L is a homogeneous, second-order, constant complex coefficient, weakly elliptic $M \times M$ system in $\mathbb{R}^n$. Select $A \in \mathfrak{A}_L$ and consider the Neumann Problem

$$\begin{cases} u \in \big[\mathscr{C}^\infty(\Omega)\big]^M, \quad Lu = 0 \textit{ in } \Omega, \\ \mathcal{N}_\kappa(\nabla u) \in \mathscr{M}^{q,\lambda}(\partial\Omega, \sigma), \quad \partial_\nu^A u = f \in \big[\mathscr{H}^{q,\lambda}(\partial\Omega, \sigma)\big]^M. \end{cases} \tag{8.4.165}$$

Then, in relation to this, the following statements are valid:

(a) [Existence, Estimates, and Integral Representations] *Whenever* $A^\top \in \mathfrak{A}^{\text{dis}}_{L^\top}$ *there exists some threshold* $\delta \in (0,1)$ *which depends only on* n, q, λ, A, *and the Ahlfors regularity constants of* $\partial\Omega$ *such that if* $\|\nu\|_{[\mathrm{BMO}(\partial\Omega,\sigma)]^n} < \delta$ *then* $-\frac{1}{2}I + K^{\#}_{A^\top}$ *is an invertible operator on the space* $\big[\mathscr{H}^{q,\lambda}(\partial\Omega,\sigma)\big]^M$ *and the function defined*

$$u := \mathscr{S}_{\text{mod}}\Big(\Big(-\tfrac{1}{2}I + K^{\#}_{A^\top}\Big)^{-1} f\Big) \text{ in } \Omega, \tag{8.4.166}$$

is a solution of the Neumann Problem (8.4.165) *which satisfies*

$$\big\|\mathcal{N}_\kappa(\nabla u)\big\|_{\mathscr{M}^{q,\lambda}(\partial\Omega,\sigma)} \leq C\|f\|_{[\mathscr{H}^{q,\lambda}(\partial\Omega,\sigma)]^M}, \tag{8.4.167}$$

for some constant $C \in (0,\infty)$ *independent of* f.

(b) [Uniqueness (modulo constants)] *Assume* $A \in \mathfrak{A}^{\text{dis}}_L$. *Then there exists* $\delta \in (0,1)$ *which depends only on* n, q, λ, A, *and the Ahlfors regularity constants of* $\partial\Omega$ *such that if* $\|\nu\|_{[\mathrm{BMO}(\partial\Omega,\sigma)]^n} < \delta$ *(i.e., if* Ω *is a* δ-AR *domain in the sense of Definition 3.1.8) then any two solutions of the Neumann Problem* (8.4.165) *differ by a constant from* $\mathbb{C}^M$.

(c) [Well-Posedness] *Whenever* $A \in \mathfrak{A}^{\text{dis}}_L$ *and* $A^\top \in \mathfrak{A}^{\text{dis}}_{L^\top}$ *there exists* $\delta \in (0,1)$ *which depends only on* n, q, λ, A, *and the Ahlfors regularity constants of* $\partial\Omega$ *such that if* $\|\nu\|_{[\mathrm{BMO}(\partial\Omega,\sigma)]^n} < \delta$ *(i.e., if* Ω *is a* δ-AR *domain in the sense of Definition 3.1.8) then the Neumann Problem* (8.4.165) *is solvable, the solution is unique modulo constants from* $\mathbb{C}^M$, *and each solution satisfies* (8.4.167).

The statement of Theorem 8.4.24 may be further streamlined in certain special situations of interest. For instance, Proposition 1.4.1 guarantees that if the system L actually satisfies the Legendre-Hadamard strong ellipticity condition then for the well-posedness result formulated in item *(c)* it suffices to assume that $A \in \mathfrak{A}^{\text{dis}}_L$, whereas if $M = 1$ (i.e., in the scalar case) and $n \geq 3$ then Proposition 1.6.12 shows that for the aforementioned well-posedness result it suffices to assume that the matrix $A \in \mathfrak{A}_L$ is symmetric. Moreover, in the scalar case in the two-dimensional setting (i.e., for $M = 1$ and $n = 2$), Proposition 1.6.13 proves that the well-posedness result from item *(c)* of Theorem 8.4.24 is true if $A \in M^0_2$ (cf. [131, (1.4.187)]).

Let us also point out that (A.0.203), the Fatou-type result described in [131, Theorem 3.3.5], and the embedding in [130, (6.2.254)] ensure that the conormal derivative $\partial^A_\nu u$ is well defined in a pointwise sense in the context of (8.4.165).

Proof of Theorem 8.4.24 That, in the context of item *(a)*, the operator $-\frac{1}{2}I + K^{\#}_{A^\top}$ becomes invertible when acting on the space $\big[\mathscr{H}^{q,\lambda}(\partial\Omega,\sigma)\big]^M$ is seen from item *(6)* in Theorem 4.2.1 (cf. (4.2.17)). As such, $u := \mathscr{S}_{\text{mod}}\Big(\Big(-\frac{1}{2}I + K^{\#}_{A^\top}\Big)^{-1} f\Big)$ is a well-defined function. More specifically, with

$$r := \frac{q(n-1)}{n-1+\lambda(q-1)} \in (1,q) \tag{8.4.168}$$

we have (cf. [130, Lemma 6.2.24])

$$g := \left(-\tfrac{1}{2}I + K^{\#}_{A^{\top}}\right)^{-1} f \in \left[\mathscr{H}^{q,\lambda}(\partial\Omega,\sigma)\right]^M \hookrightarrow \left[L^r(\partial\Omega,\sigma)\right]^M, \tag{8.4.169}$$

$$\text{and } \|g\|_{[\mathscr{H}^{q,\lambda}(\partial\Omega,\sigma)]^M} \leq C\|f\|_{[\mathscr{H}^{q,\lambda}(\partial\Omega,\sigma)]^M} \text{ for some constant } C \in (0,\infty). \tag{8.4.170}$$

Thanks to (8.4.169) and [132, (1.5.55)], for each $j \in \{1,\dots,n\}$ we have

$$(\partial_j u)(x) = \int_{\partial\Omega} (\partial_j E)(x-y)g(y)\,d\sigma(y) \text{ for all } x \in \Omega, \tag{8.4.171}$$

where E is the matrix-valued fundamental solution associated with the system L as in [131, Theorem 1.4.2]. Granted this, we may invoke [131, (2.6.21)] (with $\sigma := \partial\Omega$ and $k := \nabla E$) to conclude that there exists $C \in (0,\infty)$ independent of f such that

$$\left\|\mathcal{N}_\kappa(\nabla u)\right\|_{\mathscr{M}^{q,\lambda}(\partial\Omega,\sigma)} \leq C\|g\|_{[\mathscr{H}^{q,\lambda}(\partial\Omega,\sigma)]^M} \leq C\|f\|_{[\mathscr{H}^{q,\lambda}(\partial\Omega,\sigma)]^M}, \tag{8.4.172}$$

where the last inequality comes from (8.4.170). In addition, the fact that the function g belongs to the Lebesgue space $\left[L^r(\partial\Omega,\sigma)\right]^M$ for some $r \in (1,\infty)$ implies, much as in the proof of Theorem 8.4.16 that u also satisfies $u \in \left[\mathscr{C}^\infty(\Omega)\right]^M$, $Lu = 0$ in Ω, and $\partial^A_\nu u = f$ at σ-a.e. point on $\partial\Omega$. This finishes the treatment of item *(a)*.

Next, in view of the embeddings

$$\left[\mathscr{H}^{q,\lambda}(\partial\Omega,\sigma)\right]^M \hookrightarrow \left[\mathscr{M}^{q,\lambda}(\partial\Omega,\sigma)\right]^M \hookrightarrow \left[L^r(\partial\Omega,\sigma)\right]^M \tag{8.4.173}$$

with $r \in (1,\infty)$ as in (8.4.168), it follows that any null-solution of the Neumann Problem (8.4.165) is also a null-solution of the Neumann Problem (8.4.114). As such, the claim in the current item *(b)* is implied by the uniqueness result established in item *(b)* of Theorem 8.4.16. Finally, the claim in the current item *(c)* is a consequence of what we have proved so far. □

Moving on, the goal is to complement the body of results established so far in this section by treating the Neumann Problem with boundary data prescribed in Hardy spaces H^p with $\frac{n-1}{n} < p \leq 1$ in Theorem 8.4.25, the Homogeneous Regularity Problem with boundary data from homogeneous Hardy-based Sobolev spaces $\dot{H}^p_1$ with $\frac{n-1}{n} < p \leq 1$ in Theorem 8.4.29, the Tangential Derivative Problem with data from Hardy spaces H^p with $\frac{n-1}{n} < p \leq 1$ in Theorem 8.4.31, the Transmission Problem with data from Hardy spaces H^p and homogeneous Hardy-based Sobolev spaces $\dot{H}^p_1$ with $\frac{n-1}{n} < p \leq 1$ in Theorem 8.4.32, the Dirichlet Problem with boundary data from BMO in Theorem 8.4.33, the Dirichlet Problem with boundary data from VMO in Theorem 8.4.34, the Dirichlet Problem with boundary data from CMO in Theorem 8.4.35, the Dirichlet Problem with boundary data from homogeneous Hölder spaces $\dot{\mathscr{C}}^\alpha$ with $\alpha \in (0,1)$ in Theorem 8.4.36, and the Dirichlet Problem with boundary data from homogeneous "vanishing Hölder spaces" $\dot{\mathscr{C}}^\alpha_{\mathrm{van}}$ with $\alpha \in (0,1)$.

We begin by considering the Neumann Problem with data from the Hardy space H^p with $\frac{n-1}{n} < p \leq 1$, for weakly elliptic systems in sufficiently flat Ahlfors regular domains with unbounded boundaries in $\mathbb{R}^n$. The formulation of this boundary value problem involves the *weak* conormal derivative operator considered in a distributional sense, as in [130, Definition 10.2.18]. In the special case $p = 1$ we also allow the conormal derivative operator to be taken in a strong pointwise sense, as in (A.0.203). In this regard, we wish to note that the case of the Laplacian in upper-graph Lipschitz domains has been treated by B. Dahlberg and C. Kenig in [38, Theorem 2.12, p. 446] when $p = 1$.

Theorem 8.4.25 *Fix $n \in \mathbb{N}$ with $n \geq 2$, and let $\Omega \subseteq \mathbb{R}^n$ be an Ahlfors regular domain. Denote by ν the geometric measure theoretic outward unit normal to Ω, abbreviate $\sigma := \mathcal{H}^{n-1}\lfloor\partial\Omega$, and fix an aperture parameter $\kappa > 0$. Also, fix an exponent*

$$p \in \left(\tfrac{n-1}{n}, 1\right]. \tag{8.4.174}$$

Suppose L is a homogeneous, second-order, constant complex coefficient, weakly elliptic $M \times M$ system in $\mathbb{R}^n$. Select $A \in \mathfrak{A}_L$ and consider the Neumann Problem[100]

$$\begin{cases} u \in \left[\mathscr{C}^\infty(\Omega)\right]^M, \quad Lu = 0 \text{ in } \Omega, \\ \mathcal{N}_\kappa(\nabla u) \in L^p(\partial\Omega, \sigma), \quad \partial^A_\nu u = f \in \left[H^p(\partial\Omega, \sigma)\right]^M. \end{cases} \tag{8.4.175}$$

Then the following statements are valid:

(a) [Existence, Estimates, and Integral Representations] *Whenever $A^\top \in \mathfrak{A}^{\mathrm{dis}}_{L^\top}$ there exists some threshold $\delta \in (0,1)$ which depends only on n, p, A, and the Ahlfors regularity constants of $\partial\Omega$ such that if $\|\nu\|_{[\mathrm{BMO}(\partial\Omega,\sigma)]^n} < \delta$ then $-\frac{1}{2}I + K^\#_{A^\top}$ is an invertible operator on the Hardy space $\left[H^p(\partial\Omega, \sigma)\right]^M$ and*

$$u(x) := \left(\mathscr{S}\left(-\tfrac{1}{2}I + K^\#_{A^\top}\right)^{-1} f\right)(x) \text{ for all } x \in \Omega, \tag{8.4.176}$$

is a solution of the Neumann Problem (8.4.175) *which satisfies*

$$\left\|\mathcal{N}_\kappa(\nabla u)\right\|_{L^p(\partial\Omega,\sigma)} \leq C\|f\|_{[H^p(\partial\Omega,\sigma)]^M}, \tag{8.4.177}$$

for some constant $C \in (0, \infty)$ independent of f.

(b) [Uniqueness (modulo constants)] *Assume $A \in \mathfrak{A}^{\mathrm{dis}}_L$. Then there exists $\delta \in (0,1)$ which depends only on n, p, $[w]_{A_p}$, A, and the Ahlfors regularity constants of $\partial\Omega$ such that whenever $\|\nu\|_{[\mathrm{BMO}(\partial\Omega,\sigma)]^n} < \delta$ it follows that any two solutions of the Neumann Problem* (8.4.175) *differ by a constant from $\mathbb{C}^M$.*

(c) [Well-Posedness] *Whenever $A \in \mathfrak{A}^{\mathrm{dis}}_L$ and $\mathfrak{A}^{\mathrm{dis}}_{L^\top} \neq \varnothing$ there exists $\delta \in (0,1)$ which depends only on n, p, A, and the Ahlfors regularity constants of $\partial\Omega$ such that if*

[100] from [129, §8.5] it is known that one may replace the nontangential maximal operator by the tangential maximal operator, associated as in [129, Definition 8.5.1] with a sufficiently large power (cf. (A.0.165))

$\|\nu\|_{[\mathrm{BMO}(\partial\Omega,\sigma)]^n} < \delta$ (*i.e., if* Ω *is a* δ-AR *domain in the sense of Definition 3.1.8*) *then the Neumann Problem* (8.4.175) *is solvable, the solution is unique modulo constants from* $\mathbb{C}^M$, *and each solution satisfies* (8.4.177).

(d) [The Neumann Problem with a Strong Conormal Derivative in H^1] *In the case* $p = 1$ *similar results to those in items (a)-(c) above are valid for the Neumann Problem with a strong conormal derivative (defined in a pointwise sense, as in* (A.0.203)), *i.e., for the boundary problem*[101] *seeking* $u = (u_\beta)_{1\leq\beta\leq M}$ *so that*

$$\begin{cases} u \in \big[\mathscr{C}^\infty(\Omega)\big]^M, \ Lu = 0 \text{ in } \Omega, \ \mathcal{N}_\kappa(\nabla u) \in L^1(\partial\Omega,\sigma), \\ \partial_\nu^A u := \Big(\nu_r \big(a_{rs}^{\alpha\beta}\partial_s u_\beta\big)\Big|_{\partial\Omega}^{\kappa-\mathrm{n.t.}}\Big)_{1\leq\alpha\leq M} = f \in \big[H^1(\partial\Omega,\sigma)\big]^M \\ \text{at } \sigma\text{-a.e. point on } \partial\Omega, \end{cases} \tag{8.4.178}$$

where $\big(a_{rs}^{\alpha\beta}\big)_{\substack{1\leq r,s\leq n\\ 1\leq\alpha,\beta\leq M}}$ *are the entries of the coefficient tensor A, and* $(\nu_1,\dots,\nu_n)$ *are the scalar components of the vector* ν.

From item *(1)* of [130, Theorem 10.2.24] we know that the first three properties in (8.4.175) imply that the weak conormal derivative $\overset{\bullet}{\partial}{}_\nu^A u$ is well defined as a distribution in the Hardy space $\big[H^p(\partial\Omega,\sigma)\big]^M$. In particular, this explains the choice of the space of boundary data in the Neumann Problem (8.4.175).

As regards the well-posedness result in item *(c)* we wish to note that in the case when $A \in \mathfrak{A}_L^{\mathrm{dis}}$ and the system L satisfies the Legendre-Hadamard strong ellipticity condition (cf. [131, (1.3.4) in Definition 1.3.2]) we automatically have $\mathfrak{A}_{L^\top}^{\mathrm{dis}} \neq \varnothing$. See Theorem 1.5.2.

We also wish to note that the statement of Theorem 8.4.25 further simplifies in a number of special cases of interest. For example, Proposition 1.4.1 ensures that if the system L actually satisfies the Legendre-Hadamard strong ellipticity condition then for the well-posedness result formulated in item *(c)* it suffices to assume that $A \in \mathfrak{A}_L^{\mathrm{dis}}$, while if $M = 1$ (i.e., in the scalar case) and $n \geq 3$ then Proposition 1.6.12 guarantees that for the aforementioned well-posedness result it suffices to assume that the matrix $A \in \mathfrak{A}_L$ is symmetric. Furthermore, in the scalar case in the two-dimensional setting (i.e., for $M = 1$ and $n = 2$), Proposition 1.6.13 shows that the well-posedness result from item *(c)* of Theorem 8.4.25 is true if the matrix A belongs to the class M_2^0 (cf. [131, (1.4.187)]).

Before presenting the proof of Theorem 8.4.25 we want to elaborate on a peculiar feature, namely that for scalar operators in a half-space one may use the solvability of the Neumann Problem with a strong conormal derivative prescribed in the Hardy space H^1 to solve a Dirichlet Problem with boundary data in the Hardy space H^1.

Remark 8.4.26 *Assume L is a scalar, second-order, homogeneous, constant (complex) coefficient, weakly elliptic operator in* $\mathbb{R}^n$. *Concretely, consider* $L = \mathrm{div}A\nabla$,

[101] once again, in view of work in [129, §8.5], one may replace the nontangential maximal operator by the tangential maximal operator, associated as in [129, Definition 8.5.1] with a sufficiently large power (cf. (A.0.165))

for some weakly elliptic matrix $A = (a_{jk})_{1\leq j,k\leq n} \in \mathbb{C}^{n\times n}$. *Fix an aperture parameter* $\kappa \in (0,\infty)$, *and suppose* u *is a solution of the following Neumann Problem for* L *in the upper half-space:*

$$\begin{cases} u \in \mathscr{C}^\infty(\mathbb{R}^n_+),\ Lu = 0 \ in\ \mathbb{R}^n_+,\ \mathcal{N}_\kappa(\nabla u) \in L^1(\mathbb{R}^{n-1},\mathcal{L}^{n-1}),\\ \partial^A_\nu u := -a_{nk}(\partial_k u)\Big|^{\kappa-\text{n.t.}}_{\partial\mathbb{R}^n_+} = f \in H^1(\mathbb{R}^{n-1},\mathcal{L}^{n-1}). \end{cases} \tag{8.4.179}$$

Then the function

$$w := -a_{nk}\partial_k u \ \ in\ \mathbb{R}^n_+ \tag{8.4.180}$$

is a solution of the following Dirichlet Problem with data in a Hardy space:

$$\begin{cases} w \in \mathscr{C}^\infty(\mathbb{R}^n_+), \quad Lw = 0 \ \ in\ \mathbb{R}^n_+,\\ \mathcal{N}_\kappa w \in L^1(\mathbb{R}^{n-1},\mathcal{L}^{n-1}), \quad w\Big|^{\kappa-\text{n.t.}}_{\partial\mathbb{R}^n_+} = f \in H^1(\mathbb{R}^{n-1},\mathcal{L}^{n-1}). \end{cases} \tag{8.4.181}$$

In particular, the Dirichlet Problem (8.4.181) *is solvable, as a consequence of the solvability of the Neumann Problem* (8.4.179) *(cf. item (d) in Theorem 8.4.25, and Propositions 1.6.12-1.6.13 in this regard).*

We now turn to the proof of Theorem 8.4.25.

Proof of Theorem 8.4.25 Throughout, we shall assume $\|\nu\|_{[\mathrm{BMO}(\partial\Omega,\sigma)]^n} < \delta$ for some $\delta \in (0,\delta_*)$, where the threshold $\delta_* \in (0,1)$ is as in Theorem 3.1.5. In particular, this implies that the set Ω is a connected two-sided NTA domain with an unbounded boundary. Also, Ω is a UR domain and the UR constants of $\partial\Omega$ are controlled in terms of the dimension n and the Ahlfors regularity constants of $\partial\Omega$.

By eventually decreasing the value of δ we may also ensure that the conclusions in Theorem 4.2.1 are true (with $z = \pm 1/2$). Then $-\frac{1}{2}I + K^\#_{A^\top}$ is an invertible operator on $\big[H^p(\partial\Omega,\sigma)\big]^M$ (cf. (4.2.9)) and, thanks to [132, (2.2.32), (2.2.33), (2.2.44)], [132, (2.2.47)], the function defined in (8.4.176) is a solution of the Neumann Problem (8.4.175) which satisfies (8.4.177). This takes care of the claims in item *(a)*.

To deal with the claim that the solution of the Neumann Problem (8.4.175) is unique (modulo constants), suppose u satisfies

$$u \in \big[\mathscr{C}^\infty(\Omega)\big]^M,\ Lu = 0 \text{ in } \Omega,\ \mathcal{N}_\kappa(\nabla u) \in L^p(\partial\Omega,\sigma),\ \overset{\bullet}{\partial}{}^A_\nu u = 0. \tag{8.4.182}$$

Then [132, Theorem 2.2.7, items *(a)*-*(c)*] guarantee that the nontangential trace $u\big|^{\kappa-\text{n.t.}}_{\partial\Omega}$ exists at σ-a.e. point on $\partial\Omega$ and there exists a constant $c \in \mathbb{C}^M$ such that

$$u\Big|^{\kappa-\text{n.t.}}_{\partial\Omega} - c \text{ belongs to } \big[L^{p^*}(\partial\Omega,\sigma)\big]^M \text{ where } p^* := \big(\tfrac{1}{p} - \tfrac{1}{n-1}\big)^{-1} \in (1,\infty),$$
$$\text{if either } n \geq 3, \text{ or } n = 2 \text{ and } p \in \big(\tfrac{1}{2},1\big), \tag{8.4.183}$$

while

$$u\Big|^{\kappa-\text{n.t.}}_{\partial\Omega} \in \big[\mathrm{BMO}(\partial\Omega,\sigma)\big]^M \text{ if } n = 2 \text{ and } p = 1. \tag{8.4.184}$$

From (8.4.182), (8.4.183), (8.4.184), and the integral representation formula from [132, Theorem 2.2.7] we then conclude that there exists a constant $\widetilde{c} \in \mathbb{C}^M$ so that

$$u = \mathcal{D}_{\text{mod}}\Big(u\big|_{\partial\Omega}^{\kappa-\text{n.t.}}\Big) + \widetilde{c} \text{ in } \Omega. \tag{8.4.185}$$

Observe that

$$\big[L^{p^*}(\partial\Omega,\sigma)\big]^M + \mathbb{C}^M \hookrightarrow \Big[L^1\Big(\partial\Omega, \frac{\sigma(x)}{1+|x|^n}\Big)\Big]^M, \tag{8.4.186}$$

$$\big[\text{BMO}(\partial\Omega,\sigma)\big]^M \hookrightarrow \Big[L^1\Big(\partial\Omega, \frac{\sigma(x)}{1+|x|^n}\Big)\Big]^M \tag{8.4.187}$$

(cf. [129, (7.4.118)]). Take nontangential boundary traces in (8.4.185) and make use of [132, (1.8.27)] as well as (8.4.186)-(8.4.187) to arrive at

$$\big(-\tfrac{1}{2}I + K_{\text{mod}}\big)\Big(u\big|_{\partial\Omega}^{\kappa-\text{n.t.}}\Big) = -\widetilde{c} \text{ at } \sigma\text{-a.e. point on } \partial\Omega. \tag{8.4.188}$$

The discussion now branches out into two cases. First, assume that either $n \geq 3$, or $n = 2$ and $p \in \big(\tfrac{1}{2}, 1\big)$. Then from (8.4.183) and [132, (1.8.28)] we conclude that the function $\big(-\tfrac{1}{2}I + K_{\text{mod}}\big)\Big(u\big|_{\partial\Omega}^{\kappa-\text{n.t.}} - c\Big)$ is constant on $\partial\Omega$ which, in concert with (8.4.183) and [132, (1.8.26)], implies $\big(-\tfrac{1}{2}I + K\big)\Big(u\big|_{\partial\Omega}^{\kappa-\text{n.t.}} - c\Big)$ is constant on $\partial\Omega$. Upon recalling that $\partial\Omega$ is unbounded, and that $\big(-\tfrac{1}{2}I + K\big)\Big(u\big|_{\partial\Omega}^{\kappa-\text{n.t.}} - c\Big)$ belongs to $\big[L^{p^*}(\partial\Omega,\sigma)\big]^M$ (thanks to (8.4.183) and [132, Theorem 1.5.1, item *(iii)*]), this forces

$$\big(-\tfrac{1}{2}I + K\big)\Big(u\big|_{\partial\Omega}^{\kappa-\text{n.t.}} - c\Big) = 0 \text{ on } \partial\Omega. \tag{8.4.189}$$

In turn, from (8.4.189), (8.4.183), and (4.2.5), whose applicability with $z = -1/2$ and p replaced by p^* uses the assumption that $A \in \mathfrak{A}_L^{\text{dis}}$, we conclude that

$$u\big|_{\partial\Omega}^{\kappa-\text{n.t.}} = c. \tag{8.4.190}$$

In the remaining scenario, when $n = 2$ and $p = 1$, from (8.4.188) we see that $\big(-\tfrac{1}{2}I + [K_{\text{mod}}]\big)\Big[u\big|_{\partial\Omega}^{\kappa-\text{n.t.}}\Big] = 0$ in $\big[\text{BMO}(\partial\Omega,\sigma)/\sim\big]^M$. Together with (4.2.10) (whose validity in the present setting uses the assumption that $A \in \mathfrak{A}_L^{\text{dis}}$), this proves that $\Big[u\big|_{\partial\Omega}^{\kappa-\text{n.t.}}\Big] = 0$ in $\big[\text{BMO}(\partial\Omega,\sigma)/\sim\big]^M$, hence again (8.4.190) holds for some $c \in \mathbb{C}^M$.

All things considered, (8.4.190) is always valid. When used back in (8.4.185), this and [132, (1.8.10)] lead to the conclusion that there exists a constant $C \in \mathbb{C}^M$ with the property that $u \equiv C$ in Ω. This finishes the proof of the claim in item *(b)*. The well-posedness result formulated in item *(c)* is then a consequence of items *(a)*-*(b)* and the fact that whenever $A \in \mathfrak{A}_L^{\text{dis}}$ and $\mathfrak{A}_{L^\top}^{\text{dis}} \neq \varnothing$ we also have $A^\top \in \mathfrak{A}_{L^\top}^{\text{dis}}$ (see Theorem 1.5.5). Finally, when $p = 1$, the Fatou-type result established in [131, Theorem 3.3.4, *(a)*] ensures that for any function satisfying the first three conditions in (8.4.178) the nontangential trace $(\nabla u)\big|_{\partial\Omega}^{\kappa-\mathrm{n.t.}}$ exists (in $\mathbb{C}^{n\cdot M}$) at σ-a.e. point on $\partial\Omega$. Granted this, [130, Theorem 10.2.24, *(2)*] then guarantees that the weak conormal derivative in (8.4.175) agrees with the conormal derivative considered in a pointwise sense (cf. [130, (10.2.184)]). All desired conclusions now follow from this. □

The scope of the above theorem is illustrated by considering some special cases.

Corollary 8.4.27 *Fix $n \in \mathbb{N}$ with $n \geq 2$, and let $\Omega \subseteq \mathbb{R}^n$ be an Ahlfors regular domain. Denote by ν the geometric measure theoretic outward unit normal to Ω, set $\sigma := \mathcal{H}^{n-1}\lfloor\partial\Omega$, and pick some $\kappa > 0$. Also, fix an exponent $p \in \big(\frac{n-1}{n}, 1\big)$.*

Then there exists some threshold $\delta \in (0, 1)$ which depends only on n, p, and the Ahlfors regularity constants of $\partial\Omega$ such that if $\|\nu\|_{[\mathrm{BMO}(\partial\Omega,\sigma)]^n} < \delta$ it follows that for each $x_0, x_1 \in \partial\Omega$ the Neumann Problem

$$\begin{cases} u \in \mathscr{C}^\infty(\Omega), \ \Delta u = 0 \ \text{in} \ \Omega, \ \mathcal{N}_\kappa(\nabla u) \in L^p(\partial\Omega, \sigma), \\ \overset{\bullet}{\partial}_\nu u = \delta_{x_0} - \delta_{x_1} \ \text{in} \ \big(\mathrm{Lip}_c(\partial\Omega)\big)', \end{cases} \tag{8.4.191}$$

is unique solvable, where the weak conormal derivative is as in [130, Definition 10.2.18] *with $A := I_{n\times n}$, the $n \times n$ identity matrix, i.e., $\overset{\bullet}{\partial}_\nu u := \nu \bullet (\nabla u)$.*

Proof This is a consequence of Theorem 8.4.25 and [130, (4.2.17)]. □

Below we take another look at Theorem 8.4.25 and its proof in a special case.

Example 8.4.28 *Work in the context of item (a) in Theorem 8.4.25. Fix an index $\beta \in \{1, \dots, M\}$ and recall from* [132, (2.2.101)] *that, for σ-a.e. $x_0, x_1 \in \partial\Omega$,*

$$f := \overset{\bullet}{\partial}{}^A_\nu\big[E_{\cdot\beta}(\cdot - x_0)\big] - \overset{\bullet}{\partial}{}^A_\nu\big[E_{\cdot\beta}(\cdot - x_1)\big] \ \text{belongs to} \bigcap_{\frac{n-1}{n}<p<\infty} \big[H^p(\partial\Omega, \sigma)\big]^M.$$

In particular, it makes sense to attempt to solve the Neumann Problem (8.4.191) *with this boundary datum. One can see directly that a solution to said boundary problem is given by $u := E_{\cdot\beta}(\cdot - x_0) - E_{\cdot\beta}(\cdot - x_1)$ in Ω, where $E = (E_{\gamma\beta})_{1\leq\gamma,\beta\leq M}$ is the matrix-valued fundamental solution associated with L as in* [131, Theorem 1.4.2]. *Here, the goal is to obtain independent confirmation of this fact, using formula* (8.4.176). *Concretely,* [132, (2.2.105)] *gives*

$$\big(-\tfrac{1}{2}I + K^\#_{A^\top}\big)\big((\delta_{x_0} - \delta_{x_1})\mathbf{e}_\beta\big) = f \ \text{in} \ \big[H^p(\partial\Omega, \sigma)\big]^M, \tag{8.4.192}$$

hence $\big(-\frac{1}{2}I + K^\#_{A^\top}\big)^{-1} f = (\delta_{x_0} - \delta_{x_1})\mathbf{e}_\beta$ in $\big[H^p(\partial\Omega, \sigma)\big]^M$. This, (8.4.176), *and* [132, (2.2.103)] *then show that, as wanted,*

$$u = \mathscr{S}\Big(\big(-\tfrac{1}{2}I + K^\#_{A^\top}\big)^{-1} f\Big) = \mathscr{S}\big((\delta_{x_0} - \delta_{x_1})\mathbf{e}_\beta\big) = E_{\cdot\beta}(\cdot - x_0) - E_{\cdot\beta}(\cdot - x_1) \ \text{in} \ \Omega.$$

We next solve the `Homogeneous Regularity Problem` with data in homogeneous Hardy-based Sobolev spaces $\dot{H}^p_1$ with $\frac{n-1}{n} < p \leq 1$, for weakly elliptic systems in sufficiently flat Ahlfors regular domains with unbounded boundaries in $\mathbb{R}^n$.

Theorem 8.4.29 *Fix $n \in \mathbb{N}$ with $n \geq 2$ and let $\Omega \subseteq \mathbb{R}^n$ be an Ahlfors regular domain. Abbreviate $\sigma := \mathcal{H}^{n-1} \lfloor \partial\Omega$ and denote by ν the geometric measure theoretic outward unit normal to Ω. Also, fix an aperture parameter $\kappa > 0$ and pick an exponent*

$$p \in \left(\tfrac{n-1}{n}, 1\right]. \tag{8.4.193}$$

Given a homogeneous, second-order, constant complex coefficient, weakly elliptic $M \times M$ system L in $\mathbb{R}^n$ with the property that

$$\mathfrak{A}_L^{\text{dis}} \neq \varnothing \text{ and } \mathfrak{A}_{L^\top}^{\text{dis}} \neq \varnothing, \tag{8.4.194}$$

consider the Homogeneous Regularity Problem[102]

$$\begin{cases} u \in \left[\mathscr{C}^\infty(\Omega)\right]^M, \quad Lu = 0 \text{ in } \Omega, \\ \mathcal{N}_\kappa(\nabla u) \in L^p(\partial\Omega, \sigma), \quad u\Big|_{\partial\Omega}^{\kappa-\text{n.t.}} = f \in \left[\dot{H}_1^p(\partial\Omega, \sigma)\right]^M, \end{cases} \tag{8.4.195}$$

where $\dot{H}_1^p(\partial\Omega, \sigma)$ is the homogeneous Hardy-based Sobolev space introduced in [130, Definition 11.10.5] (*cf.* (A.0.99)-(A.0.100)).

Then there exists $\delta \in (0, 1)$ which depends only on n, p, L, and the Ahlfors regularity constants of $\partial\Omega$ with the property that if $\|\nu\|_{[\mathrm{BMO}(\partial\Omega,\sigma)]^n} < \delta$ (i.e., if Ω is a δ-AR domain in the sense of Definition 3.1.8) the Homogeneous Regularity Problem (8.4.195) *is solvable, uniquely modulo constants. More specifically,*

$$[S] : \left[H^p(\partial\Omega, \sigma)\right]^M \longrightarrow \left[\dot{H}_1^p(\partial\Omega, \sigma)/ \sim\right]^M \tag{8.4.196}$$

is an invertible operator and, if $[f] \in \left[\dot{H}_1^p(\partial\Omega, \sigma)/ \sim\right]^M$ is the equivalence class (modulo constants) of the boundary datum f, then there exists $c \in \mathbb{C}^M$ such that

$$u := \mathscr{S}\left([S]^{-1}[f]\right) + c \text{ in } \Omega \tag{8.4.197}$$

is the unique solution of the Homogeneous Regularity Problem (8.4.195). *In addition, this solution satisfies, for some constant $C \in (0, \infty)$ independent of f,*

$$\|\mathcal{N}_\kappa(\nabla u)\|_{L^p(\partial\Omega,\sigma)} \leq C \sum_{j,k=1}^n \left\|\partial_{\tau_{jk}} f\right\|_{[H^p(\partial\Omega,\sigma)]^M}. \tag{8.4.198}$$

Alternatively, if K_{mod} is associated with Ω and a coefficient tensor $A \in \mathfrak{A}_L^{\text{dis}}$ as in (A.0.117), *then the operator*

$$\tfrac{1}{2}I + \left[K_{\text{mod}}\right] : \left[\dot{H}_1^p(\partial\Omega, \sigma)/\sim\right]^M \longrightarrow \left[\dot{H}_1^p(\partial\Omega, \sigma)/\sim\right]^M \tag{8.4.199}$$

[102] as noted in [129, §8.5], one may replace the nontangential maximal operator by the tangential maximal operator, associated as in [129, Definition 8.5.1] with a sufficiently large power (cf. (A.0.165))

(*cf.* [132, (2.3.185)]) *is an isomorphism, and the Homogeneous Regularity Problem* (8.4.195) *may also be solved* (*uniquely*) *as*

$$u := \mathcal{D}_{\text{mod}} g + c \ \text{ in } \Omega, \tag{8.4.200}$$

where $g \in \big[\dot{H}^p_1(\partial\Omega, \sigma)\big]^M$ *is an arbitrarily chosen representative of the equivalence class* $\big(\frac{1}{2}I + \big[K_{\text{mod}}\big]\big)^{-1}[f] \in \big[\dot{H}^p_1(\partial\Omega, \sigma)/\sim\big]^M$ *and* $c \in \mathbb{C}^M$ *is a suitable constant.*

When Ω is an NTA domain, [132, Theorem 2.2.8] guarantees that for any function u satisfying the first three properties in (8.4.195) the nontangential boundary trace $u\big|^{\kappa-\text{n.t.}}_{\partial\Omega}$ exists (in $\mathbb{C}^M$) at σ-a.e. point on $\partial\Omega$ and, as a vector-valued function, belongs to the Hardy-based Sobolev space $\big[\dot{H}^p_1(\partial\Omega, \sigma)\big]^M$. This explains the choice of the space of boundary data for the Homogeneous Regularity Problem (8.4.195).

We may further streamline the statement of Theorem 8.4.29 in some particular cases of interest. For one thing, Proposition 1.4.1 guarantees that if the system L actually satisfies the Legendre-Hadamard strong ellipticity condition then in place of (8.4.194) it suffices to simply assume that $\mathfrak{A}^{\text{dis}}_L \neq \varnothing$, whereas if $M = 1$ (i.e., in the scalar case) and $n \geq 3$ then Proposition 1.6.12 proves that (8.4.194) is automatically satisfied. For another thing, in the scalar case in the two-dimensional setting (i.e., for $M = 1$ and $n = 2$), Proposition 1.6.13 shows that condition (8.4.194) is satisfied whenever $\mathfrak{A}_L \cap M^0_2 \neq \varnothing$ (recall that the class M^0_2 is defined in [131, (1.4.187)]).

Proof of Theorem 8.4.29 Work under the assumption that $\|\nu\|_{[\text{BMO}(\partial\Omega,\sigma)]^n} < \delta$ for some $\delta \in (0, \delta_*)$, where $\delta_* \in (0, 1)$ is as in Theorem 3.1.5. This guarantees that Ω is a connected two-sided NTA domain with an unbounded boundary, and that Ω is also a UR domain and the UR constants of $\partial\Omega$ are controlled in terms of the dimension n and the Ahlfors regularity constants of $\partial\Omega$. Decreasing further the value of δ if necessary, also ensures that the conclusions in Theorem 4.2.1 are true (with $z = \pm 1/2$). Finally, pick a coefficient tensor $A \in \mathfrak{A}^{\text{dis}}_L$. Theorem 1.5.5 implies that, in fact, $\mathfrak{A}^{\text{dis}}_L = \{A\}$ and $\mathfrak{A}^{\text{dis}}_{L^\top} = \{A^\top\}$.

Then from (4.2.18) we know that the operator $[S]$ maps the space $\big[H^p(\partial\Omega, \sigma)\big]^M$ isomorphically onto $\big[\dot{H}^p_1(\partial\Omega, \sigma)/\sim\big]^M$. Also, according to [132, (2.3.93)] we have $\big[S([S]^{-1}[f])\big] = [S]([S]^{-1}[f]) = [f]$, hence there exists $c \in \mathbb{C}^M$ such that $S([S]^{-1}[f]) + c = f$ at σ-a.e. point on $\partial\Omega$. For this choice of c consider u as in (8.4.197). Then u is well defined, and from [132, (2.2.32), (2.2.33), (2.2.44), (2.2.127)], and (A.0.1) we see that u solves the Homogeneous Regularity Problem (8.4.195) and satisfies (8.4.198) (here (A.0.100) is also relevant).

Alternatively, from (4.2.19) we know that the operator $\frac{1}{2}I + \big[K_{\text{mod}}\big]$ maps $\big[\dot{H}^p_1(\partial\Omega, \sigma)/\sim\big]^M$ isomorphically onto itself and, if $g \in \big[\dot{H}^p_1(\partial\Omega, \sigma)\big]^M$ is a representative of the equivalence class $\big(\frac{1}{2}I + \big[K_{\text{mod}}\big]\big)^{-1}[f] \in \big[\dot{H}^p_1(\partial\Omega, \sigma)/\sim\big]^M$, based on [132, (2.3.185)] we may write

$$\Big[\big(\tfrac{1}{2}I + K_{\text{mod}}\big)g\Big] = \big(\tfrac{1}{2}I + \big[K_{\text{mod}}\big]\big)[g] = \big(\tfrac{1}{2}I + \big[K_{\text{mod}}\big]\big)\big(\tfrac{1}{2}I + \big[K_{\text{mod}}\big]\big)^{-1}[f] = [f].$$

As such, there exists $c \in \mathbb{C}^M$ with the property that $\big(\frac{1}{2}I + K_{\text{mod}}\big)g + c = f$ at σ-a.e. point on $\partial\Omega$. Granted this, it follows from (A.0.99) and [132, (1.8.7), (1.8.27), (2.3.4)] that the function u, defined as in (8.4.200) for the choice of $c \in \mathbb{C}^M$ as above, solves the Homogeneous Regularity Problem (8.4.195) (and also satisfies (8.4.198)).

There remains to show that the solution of the Homogeneous Regularity Problem (8.4.195) is unique, modulo constants. To this end, suppose u satisfies

$$u \in \big[\mathscr{C}^\infty(\Omega)\big]^M, \ Lu = 0 \text{ in } \Omega, \ \mathcal{N}_\kappa(\nabla u) \in L^p(\partial\Omega, \sigma), \ u\big|_{\partial\Omega}^{\kappa-\text{n.t.}} = 0. \tag{8.4.201}$$

Then [132, Theorem 2.2.7] presently tells us that the weak conormal derivative $\dot{\partial}^A_\nu u$ belongs to the Hardy space $\big[H^p(\partial\Omega, \sigma)\big]^M$ and that there exists some constant $c \in \mathbb{C}^M$ with the property that

$$u = -\mathscr{S}\big(\dot{\partial}^A_\nu u\big) + c \text{ in } \Omega, \tag{8.4.202}$$

where the single layer potential operator acts on distributions in the Hardy space $\big[H^p(\partial\Omega, \sigma)\big]^M$ as in [132, (2.2.32)]. Taking the weak conormal derivative $\dot{\partial}^A_\nu$ of both sides then yields, on account of the jump-formula [132, (2.2.47)],

$$\big(\tfrac{1}{2}I + K^\#_{A^\top}\big)\big(\dot{\partial}^A_\nu u\big) = 0 \text{ in } \big[H^p(\partial\Omega, \sigma)\big]^M, \tag{8.4.203}$$

where I is the identity and $K^\#_{A^\top}$ is the operator associated as in [132, (2.1.4)] with the transpose coefficient tensor $A^\top$ (in place of A). Bearing in mind that we have $\dot{\partial}^A_\nu u \in \big[H^p(\partial\Omega, \sigma)\big]^M$ and that the operator $\frac{1}{2}I + K^\#_{A^\top}$ is invertible when acting on the space $\big[H^p(\partial\Omega, \sigma)\big]^M$ (cf. (4.2.9)), we conclude from (8.4.203) that $\dot{\partial}^A_\nu u = 0$ in the space $\big[H^p(\partial\Omega, \sigma)\big]^M$. When used back in (8.4.202) this ultimately proves that $u \equiv c$ in Ω, as wanted. The proof of Theorem 8.4.29 is therefore complete. □

The next example sheds further light on the inner-workings of Theorem 8.4.29.

Example 8.4.30 *Retain the context of Theorem 8.4.29. Pick an index $\beta \in \{1, \dots, M\}$ and recall from* [130, Proposition 11.10.13] *that for σ-a.e. $x_0, x_1 \in \partial\Omega$ we have*

$$f := E_{\cdot\beta}(\cdot - x_0)\big|_{\partial\Omega} - E_{\cdot\beta}(\cdot - x_1)\big|_{\partial\Omega} \textit{ belongs to } \bigcap_{\frac{n-1}{n}<p<\infty} \big[\dot{H}^p_1(\partial\Omega, \sigma)\big]^M.$$

As such, it makes sense to attempt to solve the Homogeneous Regularity Problem (8.4.195) *with this boundary datum. It is apparent that the solution to the latter boundary value problem is given by $u := E_{\cdot\beta}(\cdot - x_0) - E_{\cdot\beta}(\cdot - x_1) + c$ in Ω, where $E = (E_{\gamma\beta})_{1\leq\gamma,\beta\leq M}$ is the matrix-valued fundamental solution associated with the given system L as in* [131, Theorem 1.4.2], *and $c \in \mathbb{C}^M$. Here, the goal is to obtain independent confirmation of this fact, using formula* (8.4.197). *To this end, observe from* [132, (2.2.139)] *that*

$$S\big((\delta_{x_0} - \delta_{x_1})\mathbf{e}_\beta\big) = f \textit{ at } \sigma\textit{-a.e. point on } \partial\Omega, \tag{8.4.204}$$

hence $[S]^{-1}[f] = (\delta_{x_0} - \delta_{x_1})\mathbf{e}_\beta$ *in* $\big[H^p(\partial\Omega,\sigma)\big]^M$. *In concert with* (8.4.197) *and* [132, (2.2.103)] *this shows that, as anticipated above,*

$$u = \mathscr{S}\big([S]^{-1}[f]\big) + c = \mathscr{S}\big((\delta_{x_0} - \delta_{x_1})\mathbf{e}_\beta\big) + c = E_{\cdot\beta}(\cdot - x_0) - E_{\cdot\beta}(\cdot - x_1) + c \text{ in } \Omega.$$

Moving on, there is a closely related version on the Homogeneous Regularity Problem treated in Theorem 8.4.29, which we shall call the `Tangential Derivative Problem` for weakly elliptic systems in sufficiently flat Ahlfors regular domains with unbounded boundaries in $\mathbb{R}^n$. The trademark feature of this boundary value problem is that it asks for a null-solution of a given system whose (weak) tangential derivatives and prescribed in the Hardy space H^p with $\frac{n-1}{n} < p < \infty$. The case of the Laplacian in upper-graph Lipschitz domains B. Dahlberg and C. Kenig in [38, Theorem 3.6, p. 452] when $p = 1$, and in [38, Theorem 3.8, p. 454] when $1 < p < 2 + \varepsilon$ for some small threshold $\varepsilon > 0$ depending only on the Lipschitz constant. The Tangential Derivative Problem with $p \in (1,\infty)$ has been treated in [111], so here we shall focus on the regime $p \in \big(\frac{n-1}{n}, 1\big]$. The main result pertaining to the well-posedness of this brand of boundary value problem is stated as follows.

Theorem 8.4.31 *Fix* $n \in \mathbb{N}$ *with* $n \geq 2$ *and let* $\Omega \subseteq \mathbb{R}^n$ *be an Ahlfors regular domain. Abbreviate* $\sigma := \mathcal{H}^{n-1}\lfloor\partial\Omega$ *and denote by* ν *the geometric measure theoretic outward unit normal to* Ω. *Also, fix some* $\kappa > 0$ *and pick an exponent* $p \in \big(\frac{n-1}{n}, 1\big]$. *In addition, suppose* L *is a homogeneous, second-order, constant complex coefficient, weakly elliptic* $M \times M$ *system in* $\mathbb{R}^n$ *with the property that*

$$\mathfrak{A}_L^{\text{dis}} \neq \varnothing \text{ and } \mathfrak{A}_{L^\top}^{\text{dis}} \neq \varnothing. \tag{8.4.205}$$

In this setting, consider the Tangential Derivative Problem which, given any boundary datum $f \in \big[\dot{H}^p_1(\partial\Omega,\sigma)\big]^M$, *is concerned with finding a function* u *satisfying*[103]

$$\begin{cases} u \in \big[\mathscr{C}^\infty(\Omega)\big]^M, \ Lu = 0 \text{ in } \Omega, \ \mathcal{N}_\kappa(\nabla u) \in L^p(\partial\Omega,\sigma), \\ \overset{\bullet}{\partial}_{\tau_{jk}} u = \partial_{\tau_{jk}} f \text{ in } \big[H^p(\partial\Omega,\sigma)\big]^M \text{ for } j,k \in \{1,\dots,n\}. \end{cases} \tag{8.4.206}$$

Then there exists $\delta \in (0,1)$ *which depends only on* n, p, L, *and the Ahlfors regularity constants of* $\partial\Omega$ *with the property that if* $\|\nu\|_{[\mathrm{BMO}(\partial\Omega,\sigma)]^n} < \delta$ *(i.e., if* Ω *is a* δ-AR *domain in the sense of Definition 3.1.8) the Tangential Derivative Problem* (8.4.206) *is solvable, uniquely modulo constants. Moreover, any solution satisfies, for some constant* $C \in (0,\infty)$ *independent of* f,

$$\|\mathcal{N}_\kappa(\nabla u)\|_{L^p(\partial\Omega,\sigma)} \leq C \sum_{j,k=1}^{n} \big\|\partial_{\tau_{jk}} f\big\|_{[H^p(\partial\Omega,\sigma)]^M}. \tag{8.4.207}$$

Finally, in the special case $p = 1$, *there exists* $\delta \in (0,1)$ *which depends only on* n, L, *and the Ahlfors regularity constants of* $\partial\Omega$ *with the property that if*

---

[103] as noted in [129, §8.5], one may replace the nontangential maximal operator by the tangential maximal operator, associated as in [129, Definition 8.5.1] with a sufficiently large power (cf. (A.0.165))

$\|\nu\|_{[\mathrm{BMO}(\partial\Omega,\sigma)]^n} < \delta$ *(i.e., if* Ω *is a* δ-AR *domain in the sense of Definition 3.1.8) then the following version of the Tangential Derivative Problem*

$$\begin{cases} u \in \left[\mathscr{C}^\infty(\Omega)\right]^M, \ Lu = 0 \ \textit{in} \ \Omega, \ \mathcal{N}_\kappa(\nabla u) \in L^1(\partial\Omega, \sigma), \\ \nu_j(\partial_k u)\big|_{\partial\Omega}^{\kappa-\mathrm{n.t.}} - \nu_k(\partial_j u)\big|_{\partial\Omega}^{\kappa-\mathrm{n.t.}} = \partial_{\tau_{jk}} f \ \textit{in} \ \left[H^1(\partial\Omega,\sigma)\right]^M, \\ \textit{for each pair of indices } j,k \in \{1,\dots,n\}, \end{cases} \tag{8.4.208}$$

is solvable, uniquely modulo constants, for any given

$$f \in \left[\dot{H}^1_1(\partial\Omega,\sigma)\right]^M, \tag{8.4.209}$$

any solution satisfies, for some constant $C \in (0,\infty)$ *independent of* f,

$$\|\mathcal{N}_\kappa(\nabla u)\|_{L^1(\partial\Omega,\sigma)} \le C \sum_{j,k=1}^n \left\|\partial_{\tau_{jk}} f\right\|_{[H^1(\partial\Omega,\sigma)]^M}. \tag{8.4.210}$$

A word regarding the nature of the boundary conditions in (8.4.206) is in order. From [130, (10.2.14) in Example 10.2.2] we know that for any function u satisfying the first and third conditions in (8.4.206) we have $\dot{\partial}_{\tau_{jk}} u \in \left[H^p(\partial\Omega,\sigma)\right]^M$ for all $j,k \in \{1,\dots,n\}$. Given that for each boundary datum $f \in \left[\dot{H}^p_1(\partial\Omega,\sigma)\right]^M$ we also have $\partial_{\tau_{jk}} f \in \left[H^p(\partial\Omega,\sigma)\right]^M$ for all $j,k \in \{1,\dots,n\}$ (cf. (A.0.99)), it follows that the boundary conditions in (8.4.206) are meaningful.

As regards the boundary conditions in (8.4.208), taking $\delta \in (0,1)$ sufficiently small forces Ω to be an NTA domain (cf. Theorem 3.1.5), a scenario in which the Fatou-type result from item *(a)* of [131, Theorem 3.3.4] guarantees that for any function satisfying the first three conditions in (8.4.208) the nontangential boundary trace $(\nabla u)\big|_{\partial\Omega}^{\kappa-\mathrm{n.t.}}$ exists at σ-a.e. point on $\partial\Omega$. Granted this, we then conclude from [130, (10.1.55)-(10.1.56)] that $\nu_j(\partial_k u)\big|_{\partial\Omega}^{\kappa-\mathrm{n.t.}} - \nu_k(\partial_j u)\big|_{\partial\Omega}^{\kappa-\mathrm{n.t.}}$ belongs to $\left[H^1(\partial\Omega,\sigma)\right]^M$ for all pairs of indices $j,k \in \{1,\dots,n\}$. Since whenever the function f is as in (8.4.209) we also have $\partial_{\tau_{jk}} f \in \left[H^1(\partial\Omega,\sigma)\right]^M$ for all $j,k \in \{1,\dots,n\}$ (cf. (A.0.99)), we see that the boundary conditions in (8.4.206) are meaningful as well.

The statement of Theorem 8.4.31 takes a simpler form in a number of relevant special cases. First, if the system L actually satisfies the Legendre-Hadamard strong ellipticity condition then Proposition 1.4.1 permits us to replace (8.4.205) simply by $\mathfrak{A}^{\mathrm{dis}}_L \neq \varnothing$. Also, if $M = 1$ (i.e., in the scalar case) and $n \ge 3$ then Proposition 1.6.12 shows that (8.4.205) is automatically satisfied. Finally, corresponding to the scalar case in the two-dimensional setting (i.e., for $M = 1$ and $n = 2$), condition (8.4.205) may be replaced by $\mathfrak{A}_L \cap M^0_2 \neq \varnothing$ (with the class M^0_2 defined in [131, (1.4.187)]); this is seen from Proposition 1.6.13.

The proof of Theorem 8.4.31 is given next.

Proof of Theorem 8.4.31 Throughout, assume $\|\nu\|_{[\mathrm{BMO}(\partial\Omega,\sigma)]^n} < \delta$ for some $\delta \in (0,\delta_*)$, where the threshold $\delta_* \in (0,1)$ is as in Theorem 3.1.5. In particular, this implies that the set Ω is a connected two-sided NTA domain with an unbounded connected boundary. Granted this, [132, Theorem 2.2.8] guarantees that the nontangential boundary trace $u\big|_{\partial\Omega}^{\kappa-\mathrm{n.t.}}$ exists (in $\mathbb{C}^M$) at σ-a.e. point on $\partial\Omega$ and, as a vector-valued function, belongs to the Hardy-based Sobolev space $\big[\dot{H}^p_1(\partial\Omega,\sigma)\big]^M$. In addition, for each $j,k\in\{1,\dots,n\}$ we have (cf. [132, (2.2.217)])

$$\partial_{\tau_{jk}}\Big(u\big|_{\partial\Omega}^{\kappa-\mathrm{n.t.}}\Big)=\dot{\partial}_{\tau_{jk}}u\in\big[H^p(\partial\Omega,\sigma)\big]^M. \tag{8.4.211}$$

Let us also assume that $\delta\in(0,1)$ is sufficiently small so that the conclusions in Theorem 8.4.29 hold. Fix an arbitrary $f\in\big[\dot{H}^p_1(\partial\Omega,\sigma)\big]^M$ and let u solve the Homogeneous Regularity Problem (8.4.195) (formulated for this boundary datum). Then, as seen from (8.4.211), u solves the Tangential Derivative Problem (8.4.206).

To prove that the solution of the Tangential Derivative Problem (8.4.206) is unique modulo constants, consider a function u satisfying

$$\begin{aligned}&u\in\big[\mathscr{C}^\infty(\Omega)\big]^M,\quad Lu=0\ \text{in}\ \Omega,\quad \mathcal{N}_\kappa(\nabla u)\in L^p(\partial\Omega,\sigma),\\&\text{and}\ \dot{\partial}_{\tau_{jk}}u=0\ \text{in}\ \big[H^p(\partial\Omega,\sigma)\big]^M\ \text{for each}\ j,k\in\{1,\dots,n\}.\end{aligned} \tag{8.4.212}$$

From [132, Theorem 2.2.7, item *(a)*] it follows that the nontangential boundary trace $u\big|_{\partial\Omega}^{\kappa-\mathrm{n.t.}}$ exists at σ-a.e. point on $\partial\Omega$ and, as function, it belongs to the weighted Lebesgue space $\big[L^1\big(\partial\Omega,\frac{\sigma(x)}{1+|x|^n}\big)\big]^M$. To proceed, we consider two cases. First, assume that either $n\geq 3$, or $n=2$ and $p\in\big(\frac{1}{2},1\big)$. Then [132, Theorem 2.2.7, item *(b)*] implies that there exists a constant $c\in\mathbb{C}^M$ with the property that

$$u\big|_{\partial\Omega}^{\kappa-\mathrm{n.t.}}-c\ \text{belongs to}\ \big[L^{p^*}(\partial\Omega,\sigma)\big]^M\ \text{where}\ p^*:=\big(\tfrac{1}{p}-\tfrac{1}{n-1}\big)^{-1}. \tag{8.4.213}$$

Since (8.4.211) and (8.4.212) imply that for each $j,k\in\{1,\dots,n\}$ we have

$$\partial_{\tau_{jk}}\Big(u\big|_{\partial\Omega}^{\kappa-\mathrm{n.t.}}-c\Big)=\partial_{\tau_{jk}}\Big(u\big|_{\partial\Omega}^{\kappa-\mathrm{n.t.}}\Big)=\dot{\partial}_{\tau_{jk}}u=0, \tag{8.4.214}$$

we conclude from (8.4.213), (8.4.214), and (A.0.88) that

$$\begin{aligned}&\text{the function}\ u\big|_{\partial\Omega}^{\kappa-\mathrm{n.t.}}-c\ \text{belongs to the Sobolev space}\ \big[L^{p^*}_1(\partial\Omega,\sigma)\big]^M\\&\text{(with}\ p^*\ \text{as in (8.4.213)), and its tangential gradient is identically zero.}\end{aligned} \tag{8.4.215}$$

Granted this, [130, Proposition 11.4.3] applies and gives that the function $u\big|_{\partial\Omega}^{\kappa-\mathrm{n.t.}}-c$ is locally constant on $\partial\Omega$. Since, as noted earlier, $\partial\Omega$ is connected, this ultimately implies that $u\big|_{\partial\Omega}^{\kappa-\mathrm{n.t.}}$ is constant on $\partial\Omega$. Then the uniqueness, modulo constants, for the Homogeneous Regularity Problem (8.4.195) gives that u is constant in Ω.

Next, consider the case remaining case, when $n=2$ and $p=1$. Then from [132, Theorem 2.2.7, *(b)*] we know that the nontangential boundary trace $u\big|_{\partial\Omega}^{\kappa-\mathrm{n.t.}}$ belongs to the John-Nirenberg space $\big[\mathrm{BMO}(\partial\Omega,\sigma)\big]^M$. In particular,

$$u\Big|_{\partial\Omega}^{\kappa-\text{n.t.}} \in \Big[L^1\Big(\partial\Omega, \frac{\sigma(x)}{1+|x|^n}\Big) \cap L^q_{\text{loc}}(\partial\Omega, \sigma)\Big]^M \text{ for each } q \in (1, \infty), \tag{8.4.216}$$

by [129, (7.4.118)] and [129, (7.4.105)]. Moreover, (8.4.211) and (8.4.212) imply

$$\partial_{\tau_{jk}}\Big(u\Big|_{\partial\Omega}^{\kappa-\text{n.t.}}\Big) = \overset{\bullet}{\partial}_{\tau_{jk}} u = 0 \text{ for each } j, k \in \{1, \dots, n\}. \tag{8.4.217}$$

Based on (8.4.216), (8.4.217), [130, Proposition 11.4.3], and the fact that $\partial\Omega$ is connected, we then conclude that $u\big|_{\partial\Omega}^{\kappa-\text{n.t.}}$ is constant. Then, as before, the uniqueness modulo constants for the Homogeneous Regularity Problem (8.4.195) guarantees that actually u is constant in Ω. Finally, the claims made in relation to (8.4.208) (with f as in (8.4.209)) are consequences of what we have proved so far, [132, Theorem 2.2.8], and [130, Example 10.2.10]. □

We continue by stating and proving a result pertaining to the well-posedness (modulo constants) of the Transmission Problem for weakly elliptic systems in sufficiently flat Ahlfors regular domains with unbounded boundaries in $\mathbb{R}^n$. In its current formulation, this asks for two functions defined on either side of a common interface, on which one prescribes the jumps of the traces in the Hardy-based Sobolev space $\dot{H}^p_1$ together with the coupling of the (weak) conormal derivatives in the Hardy space H^p, with $\frac{n-1}{n} < p < \infty$. The Transmission Problem with $p \in (1, \infty)$ has been treated in [111] and is also covered, as a particular case, by Theorem 8.4.44 (stated later), so here we focus on the regime $p \in \big(\frac{n-1}{n}, 1\big]$.

Theorem 8.4.32 *Fix $n \in \mathbb{N}$ with $n \geq 2$, and let $\Omega \subseteq \mathbb{R}^n$ be a two-sided* NTA *domain with an Ahlfors regular boundary. Denote by ν the geometric measure theoretic outward unit normal to Ω, abbreviate $\sigma := \mathcal{H}^{n-1}\lfloor\partial\Omega$, and set*

$$\Omega_+ := \Omega, \quad \Omega_- := \mathbb{R}^n \setminus \overline{\Omega}. \tag{8.4.218}$$

Also, fix an aperture parameter $\kappa > 0$, select a transmission (or coupling) parameter $\eta \in \mathbb{C} \setminus \{-1\}$, and pick an exponent

$$p \in \big(\tfrac{n-1}{n}, 1\big]. \tag{8.4.219}$$

In addition, suppose L is a homogeneous, second-order, constant complex coefficient, weakly elliptic $M \times M$ system in $\mathbb{R}^n$ with the property that

$$\mathfrak{A}^{\text{dis}}_L \neq \varnothing \text{ and } \mathfrak{A}^{\text{dis}}_{L^\top} \neq \varnothing. \tag{8.4.220}$$

Lastly, pick a coefficient tensor $A \in \mathfrak{A}^{\text{dis}}_L$ and consider the Transmission Problem[104]*:*

[104] work in [129, §8.5] shows that one may replace the nontangential maximal operator by the tangential maximal operator, associated as in [129, Definition 8.5.1] with a sufficiently large power (cf. (A.0.165))

$$\begin{cases} u^{\pm} \in \big[\mathscr{C}^{\infty}(\Omega_{\pm})\big]^M, \ Lu^{\pm} = 0 \ in \ \Omega_{\pm}, \ \mathcal{N}_{\kappa}(\nabla u^{\pm}) \in L^p(\partial\Omega, \sigma), \\ u^{+}\big|_{\partial\Omega}^{\kappa-\mathrm{n.t.}} - u^{-}\big|_{\partial\Omega}^{\kappa-\mathrm{n.t.}} = f \in \big[\dot{H}^p_1(\partial\Omega, \sigma)\big]^M, \\ \dot{\partial}^A_\nu u^{+} - \eta \cdot \dot{\partial}^A_\nu u^{-} = g \in \big[H^p(\partial\Omega, \sigma)\big]^M. \end{cases} \tag{8.4.221}$$

Then there exists $\delta \in (0,1)$ *which depends only on* n, p, L, *and the Ahlfors regularity constants of* $\partial\Omega$ *with the property that if* $\|\nu\|_{[\mathrm{BMO}(\partial\Omega,\sigma)]^n} < \delta$ *(i.e., if* Ω *is a* δ-AR *domain in the sense of Definition 3.1.8) the* Transmission Problem (8.4.221) *is solvable, uniquely modulo constants. Moreover, any solution satisfies, for some constant* $C \in (0,\infty)$ *independent of* f *and* g,

$$\big\|\mathcal{N}_{\kappa}(\nabla u^{\pm})\big\|_{L^p(\partial\Omega,\sigma)} \leq C\Big(\|f\|_{[\dot{H}^p_1(\partial\Omega,\sigma)]^M} + \|g\|_{[H^p(\partial\Omega,\sigma)]^M}\Big). \tag{8.4.222}$$

A few clarifications pertaining to the nature of the above theorem are in order here. First, [132, Theorem 2.2.8] guarantees that the nontangential boundary traces $u^{\pm}\big|_{\partial\Omega}^{\kappa-\mathrm{n.t.}}$ exist (in $\mathbb{C}^M$) at σ-a.e. point on $\partial\Omega$ and, as a vector-valued functions, they belong to the Hardy-based Sobolev space $\big[\dot{H}^p_1(\partial\Omega,\sigma)\big]^M$. Also, from item *(1)* of [130, Theorem 10.2.24] we know that the first three properties in (8.4.221) imply that the weak conormal derivatives $\dot{\partial}^A_\nu u^{\pm}$ are well-defined distributions in the Hardy space $\big[H^p(\partial\Omega,\sigma)\big]^M$. In particular, this explains the choice of the spaces of boundary data in the Transmission Problem (8.4.221).

Second, the weak conormal derivative $\dot{\partial}^A_\nu u^{+}$ is defined as in [130, Definition 10.2.18] relative to the set Ω_{+}, while in light of the fact that the geometric measure theoretic outward unit normal to Ω_{-} is $-\nu$, at σ-a.e. point on $\partial\Omega$, we take $\dot{\partial}^A_\nu u^{-}$ to be the opposite of (i.e.,-1 times) the weak conormal derivative operator from [130, Definition 10.2.18] for the set Ω_{-}, acting on the function u^{-} which is defined in said set, i.e., we make the convention that

$$\dot{\partial}^A_\nu u^{-} := -\dot{\partial}^A_{(-\nu)} u^{-}. \tag{8.4.223}$$

Third, we observe that in certain particular cases of interest the statement of Theorem 8.4.32 may be further streamlined. For example, if the given system L actually satisfies the Legendre-Hadamard strong ellipticity condition then Proposition 1.4.1 allows us to replace (8.4.220) simply by $\mathfrak{A}^{\mathrm{dis}}_L \neq \varnothing$. In addition, if $M = 1$ (i.e., in the scalar case) and $n \geq 3$ then Proposition 1.6.12 guarantees that (8.4.220) is automatically satisfied. Lastly, in the scalar case in the two-dimensional setting (i.e., for $M = 1$ and $n = 2$), Proposition 1.6.13 shows that condition (8.4.220) may be replaced by $\mathfrak{A}_L \cap M^0_2 \neq \varnothing$ (with the class M^0_2 defined in [131, (1.4.187)]).

Finally, we wish to note that, corresponding to the case when $p = 1$, much as in the case of the Neumann Problem (8.4.178), there exists a natural version of the Transmission Problem (8.4.221) in which the weak conormal derivatives are replaced by conormal derivatives taken in a strong pointwise sense (cf. (A.0.203)).

We now turn to the task of proving Theorem 8.4.32.

Proof of Theorem 8.4.32 Work under the assumption that $\|\nu\|_{[\mathrm{BMO}(\partial\Omega,\sigma)]^n} < \delta$ for some $\delta \in (0, \delta_*)$, where $\delta_* \in (0, 1)$ is as in Theorem 3.1.5. In particular, this implies that $\Omega_\pm$ are connected sets, and that $\partial\Omega$ is connected and unbounded. After eventually further decreasing $\delta \in (0, 1)$, we may also assume that the conclusions in Theorem 4.2.1 are valid. To prove that the solution of the Transmission Problem (8.4.221) is unique modulo constants, suppose

$$\begin{aligned} &u^\pm \in \big[\mathscr{C}^\infty(\Omega_\pm)\big]^M, \quad Lu^\pm = 0 \text{ in } \Omega_\pm, \quad \mathcal{N}_\kappa(\nabla u^\pm) \in L^p(\partial\Omega, \sigma), \\ &u^+\big|_{\partial\Omega}^{\kappa-\mathrm{n.t.}} = u^-\big|_{\partial\Omega}^{\kappa-\mathrm{n.t.}} \text{ and } \mathring{\partial}_\nu^A u^+ = \eta \cdot \mathring{\partial}_\nu^A u^-. \end{aligned} \tag{8.4.224}$$

As noted earlier, the weak conormal derivatives $\mathring{\partial}_\nu^A u^\pm$ belong to the Hardy space $\big[H^p(\partial\Omega, \sigma)\big]^M$. Together with the first line of (8.4.224), this shows that $u^\pm$ solve a Neumann Problem (of the sort considered in (8.4.175)) in $\Omega_\pm$. As such, Theorem 8.4.25 implies that there exist $h_\pm \in \big[H^p(\partial\Omega, \sigma)\big]^M$ and $c_\pm \in \mathbb{C}^M$ such that

$$u^\pm = \mathscr{S} h_\pm + c_\pm \text{ in } \Omega_\pm. \tag{8.4.225}$$

In view of the trace formulas [132, (2.2.127), (2.2.132)] (plus (A.0.1)), the jump-formula [132, (2.2.47)], and the convention made in (8.4.223), the properties in the second line of (8.4.224) entail

$$Sh_+ + c_+ = Sh_- + c_- \tag{8.4.226}$$

and, respectively,

$$\big(-\tfrac{1}{2}I + K^\#_{A^\top}\big)h_+ = \eta\big(\tfrac{1}{2}I + K^\#_{A^\top}\big)h_-. \tag{8.4.227}$$

In turn, (8.4.226) allows us to write $S(h_+ - h_-) = c_- - c_+$, from which we further conclude that $[S](h_+ - h_-) = 0$ in $\big[\dot{H}_1^p(\partial\Omega, \sigma)/\sim\big]^M$. Granted this, the invertibility of (4.2.18) then forces $h_+ - h_- = 0$, hence $h_+ = h_-$, which in concert with (8.4.226) further implies $c_+ = c_-$. Henceforth, we agree to abbreviate $h := h_\pm$ and $c := c_\pm$. In this notation, from (8.4.227) we deduce that $\big(-\frac{\eta+1}{2}I + (1-\eta)K^\#_{A^\top}\big)h = 0$. Since $\eta \in \mathbb{C}\setminus\{-1\}$, from this and the invertibility result proved in (4.2.9) we conclude that $h = 0$. Ultimately, (8.4.225) simply reduces to $u^\pm \equiv c$ in $\Omega_\pm$, as desired. This proves that any two solutions of the Transmission Problem (8.4.221) differ by a constant.

To show that the Transmission Problem (8.4.221) is always solvable, consider the functions $u^\pm : \Omega_\pm \to \mathbb{C}^M$ defined as

$$u^+ := \mathscr{S}^+ h_0 + \mathscr{S}^+ h_1 - c_* \text{ in } \Omega_+, \quad u^- := \mathscr{S}^- h_0 \text{ in } \Omega_-, \tag{8.4.228}$$

where the superscripts $\pm$ indicate that the boundary-to-domain single layer potentials are associated with the sets $\Omega_\pm$, and where

$$\begin{aligned} &h_1 := [S]^{-1}[f] \in \big[H^p(\partial\Omega, \sigma)\big]^M, \quad c_* := Sh_1 - f \in \mathbb{C}^M, \\ &h_0 := \Big(-\tfrac{\eta+1}{2}I + (1-\eta)K^\#_{A^\top}\Big)^{-1}\Big(g - \big(-\tfrac{1}{2}I + K^\#_{A^\top}\big)h_1\Big). \end{aligned} \tag{8.4.229}$$

Then from [132, (2.2.32), (2.2.33), (2.2.44)], the trace formula [132, (2.2.127)] (as well as (A.0.1)), and the jump-formula [132, (2.2.47)] we see that $u^{\pm}$ thus defined solve the Transmission Problem (8.4.221) and satisfy (8.4.222). □

Originally, work on boundary problems with data in BMO has been motivated by basic questions in Harmonic Analysis. Indeed, in his ground breaking 1971 article [59], C. Fefferman writes "*The main idea in proving* [*that the dual of the Hardy space* H^1 *is the John-Nirenberg space* BMO] *is to study the Poisson integral of a function in* BMO." Subsequently, in 1976 E. Fabes, R. Johnson, and U. Neri have considered in [53] the BMO Dirichlet Problem for the Laplacian in the upper half-space in the class of harmonic functions satisfying a Carleson measure condition. In [115] the authors have developed machinery that permits replacing the Laplacian in the work of Fabes-Johnson-Neri with general second-order Legendre-Hadamard elliptic systems with complex coefficients. Also, the Dirichlet Problem with VMO data in upper-graph Lipschitz domains has been solved in [134]. Here we treat the Dirichlet Problem with boundary data in BMO for weakly elliptic systems and sufficiently flat Ahlfors regular domains with an unbounded boundary, and also discuss the implications of actually having the boundary data in the space CMO (cf. (A.0.53)).

Theorem 8.4.33 *Fix two integers $n, M \in \mathbb{N}$ with $n \geq 2$ and suppose $\Omega \subseteq \mathbb{R}^n$ is an Ahlfors regular domain. Abbreviate $\sigma := \mathcal{H}^{n-1}\lfloor\partial\Omega$ and denote by ν the geometric measure theoretic outward unit normal to Ω. Also, let L be a homogeneous, second-order, constant complex coefficient, weakly elliptic $M \times M$ system in $\mathbb{R}^n$, satisfying with the property that $\mathfrak{A}_L^{\mathrm{dis}} \neq \emptyset$. Finally, fix an arbitrary aperture parameter $\kappa > 0$.*

Then there exists a threshold $\delta \in (0,1)$ depending only on n, L, and the ADR *constants of $\partial\Omega$ with the property that whe $\|\nu\|_{[\mathrm{BMO}(\partial\Omega,\sigma)]^n} < \delta$ it follows that Ω is a* UR *domain with an unbounded boundary, and the boundary value problem*

$$\begin{cases} u \in \big[\mathscr{C}^\infty(\Omega)\big]^M, \quad Lu = 0 \text{ in } \Omega, \\ |\nabla u|^p \operatorname{dist}(\cdot,\partial\Omega)^{p-1}\, d\mathcal{L}^n \text{ is a Carleson} \\ \text{measure in } \Omega \text{ for each } p \in (1,\infty), \\ u\big|_{\partial\Omega}^{\kappa-\mathrm{n.t.}} = f \in \big[\mathrm{BMO}(\partial\Omega,\sigma)\big]^M, \end{cases} \tag{8.4.230}$$

is solvable. More specifically, if K_{mod} is associated with Ω and a coefficient tensor $A \in \mathfrak{A}_L^{\mathrm{dis}}$ as in (A.0.117)*, then the operator (cf.* [132, (2.1.93)]*)*

$$\tfrac{1}{2}I + \big[K_{\text{mod}}\big] : \big[\mathrm{BMO}(\partial\Omega,\sigma)/\sim\big]^M \longrightarrow \big[\mathrm{BMO}(\partial\Omega,\sigma)/\sim\big]^M \tag{8.4.231}$$

is an isomorphism, and the boundary value problem (8.4.230) *is solved by*

$$u := \mathcal{D}_{\text{mod}}\, g + c \text{ in } \Omega, \tag{8.4.232}$$

where $g \in \big[\mathrm{BMO}(\partial\Omega,\sigma)\big]^M$ is an arbitrarily chosen representative of the equivalence class $\big(\frac{1}{2}I + \big[K_{\text{mod}}\big]\big)^{-1}[f] \in \big[\mathrm{BMO}(\partial\Omega,\sigma)/\sim\big]^M$ and $c \in \mathbb{C}^M$ is a suitable constant. Furthermore, this solution has the property that for each $p \in (1,\infty)$ there *exists a constant* $C = C(\Omega, L, p) \in (0,\infty)$ *such that*

$$\sup_{x\in\partial\Omega,\, r>0} \frac{1}{\sigma\big(B(x,r)\cap\partial\Omega\big)} \int_{B(x,r)\cap\Omega} |\nabla u|^p \operatorname{dist}(\cdot,\partial\Omega)^{p-1}\, d\mathcal{L}^n \leq C\|f\|^p_{[\mathrm{BMO}(\partial\Omega,\sigma)]^M}. \tag{8.4.233}$$

Moreover, for each $p \in (1,\infty)$ *there exists a constant* $C \in (0,\infty)$ *with the property that the solution* u *given in* (8.4.232) *of the boundary value problem* (8.4.230) *satisfies*

$$\lim_{R\to 0^+} \sup_{\substack{x\in\partial\Omega \text{ and}\\ r\in(0,R)}} \left(\frac{1}{\sigma\big(B(x,r)\cap\partial\Omega\big)} \int_{B(x,r)\cap\Omega} |\nabla u|^p \operatorname{dist}(\cdot,\partial\Omega)^{p-1}\, d\mathcal{L}^n\right)^{\frac{1}{p}} \leq C \operatorname{dist}\big(f, [\mathrm{VMO}(\partial\Omega,\sigma)]^M\big), \tag{8.4.234}$$

with the distance in the right-hand side considered in $\big[\mathrm{BMO}(\partial\Omega,\sigma)\big]^M$.

As a corollary, for each $p \in (1,\infty)$ *the solution* u *given in* (8.4.232) *of the boundary value problem* (8.4.230) *satisfies*

$$\begin{aligned} &|\nabla u|^p \operatorname{dist}(\cdot,\partial\Omega)^{p-1}\, d\mathcal{L}^n \text{ is a vanishing Carleson measure in } \Omega\\ &\text{whenever the boundary datum } f \text{ belongs to } \big[\mathrm{VMO}(\partial\Omega,\sigma)\big]^M. \end{aligned} \tag{8.4.235}$$

Finally, for each $p \in (1,\infty)$ *there exists a constant* $C \in (0,\infty)$, *which depends only on* n, p, L, *and the* ADR *constants of* $\partial\Omega$, *with the property that the solution* u *from* (8.4.232) *of the boundary value problem* (8.4.230) *satisfies*

$$\begin{aligned} \max\Bigg\{ &\lim_{R\to\infty} \sup_{\substack{x\in\partial\Omega,\, r>0\\ B(x,r)\subseteq\mathbb{R}^n\setminus B(0,R)}} \left(\frac{1}{\sigma\big(B(x,r)\cap\partial\Omega\big)} \int_{B(x,r)\cap\Omega} |\nabla u|^p \operatorname{dist}(\cdot,\partial\Omega)^{p-1}\, d\mathcal{L}^n\right)^{1/p},\\ &\lim_{R\to 0^+} \sup_{x\in\partial\Omega,\, r\in(0,R)} \left(\frac{1}{\sigma\big(B(x,r)\cap\partial\Omega\big)} \int_{B(x,r)\cap\Omega} |\nabla u|^p \operatorname{dist}(\cdot,\partial\Omega)^{p-1}\, d\mathcal{L}^n\right)^{1/p},\\ &\lim_{R\to\infty} \sup_{x\in\partial\Omega,\, r>R} \left(\frac{1}{\sigma\big(B(x,r)\cap\partial\Omega\big)} \int_{B(x,r)\cap\Omega} |\nabla u|^p \operatorname{dist}(\cdot,\partial\Omega)^{p-1}\, d\mathcal{L}^n\right)^{1/p} \Bigg\}\\ &\leq C \operatorname{dist}\Big(f,\ \big[\mathrm{CMO}(\partial\Omega,\sigma)\big]^M\Big), \end{aligned} \tag{8.4.236}$$

where the distance in the right-hand side is considered in the space $[\mathrm{BMO}(\partial\Omega,\sigma)]^M$ *(recall that* $\mathrm{CMO}(\partial\Omega,\sigma)$ *has been introduced in* (A.0.53)*). As a consequence of* (8.4.236) *and* [132, Definition 1.8.1], *given any* $p \in (1,\infty)$ *it follows that the solution* u *from* (8.4.232) *of the boundary problem* (8.4.230) *has the property that*

$$\begin{aligned} &|\nabla u|^p \operatorname{dist}(\cdot,\partial\Omega)^{p-1}\, d\mathcal{L}^n \text{ is a super vanishing Carleson measure in } \Omega,\\ &\text{whenever the boundary datum } f \text{ belongs to } \big[\mathrm{CMO}(\partial\Omega,\sigma)\big]^M. \end{aligned} \tag{8.4.237}$$

Corresponding to the scalar case (i.e., if $M = 1$), the statement of Theorem 8.4.33 may be further simplified. Specifically, if $M = 1$ and $n \geq 3$ then Proposition 1.6.12 ensures that condition $\mathfrak{A}_L^{\mathrm{dis}} \neq \varnothing$ is automatically satisfied. Furthermore, in the scalar case in the two-dimensional setting (i.e., for $M = 1$ and $n = 2$), Proposition 1.6.13 guarantees that condition $\mathfrak{A}_L^{\mathrm{dis}} \neq \varnothing$ may be replaced by the demand that $\mathfrak{A}_L \cap M_2^0 \neq \varnothing$ (with M_2^0 defined in [131, (1.4.187)]).

We also want to mention that even when $\Omega = \mathbb{R}^n_+$ our solvability result in Theorem 8.4.33 differs from the one in [115] since, as opposed to the latter work in which the system in question is required to satisfy the strong Legendre-Hadamard ellipticity condition, we currently only ask that the system L is weakly elliptic and has the property that $\mathfrak{A}_L^{\mathrm{dis}} \neq \varnothing$. This is an important distinction. For example, in the case of the complex Lamé system $L_{\mu,\lambda}$ from (1.6.85) our Theorem 8.4.33 works (see Remark 1.6.10) whenever $\mu, \lambda \in \mathbb{C}$ satisfy $\mu \neq 0$, $2\mu + \lambda \neq 0$, $3\mu + \lambda \neq 0$, whereas the solvability result in [115] would require (cf. [131, (1.3.11)]) the more restrictive set of conditions $\mathrm{Re}\,\mu > 0$ and $\mathrm{Re}\,(2\mu + \lambda) > 0$. We also note that if $\Omega = \mathbb{R}^n_+$ and the system L satisfies the strong Legendre-Hadamard ellipticity condition it has been shown in [115] that one has uniqueness for the Dirichlet Problem with data in BMO.

Let us now turn to the proof of Theorem 8.4.33.

Proof of Theorem 8.4.33 Adopting the assumption that $\|\nu\|_{[\mathrm{BMO}(\partial\Omega,\sigma)]^n} < \delta$ for some sufficiently small $\delta \in (0, 1)$ guarantees (cf. Theorem 3.1.5) that Ω is a UR domain with an unbounded boundary. That we may further decrease the value of $\delta \in (0, 1)$ so that the operator (8.4.231) is an isomorphism follows from (4.2.10). In addition, from (4.2.11) we see that matters may also be arranged so that the operator

$$\tfrac{1}{2}I + \big[K_{\mathrm{mod}}\big] : \big[\mathrm{VMO}(\partial\Omega, \sigma)/\sim\big]^M \longrightarrow \big[\mathrm{VMO}(\partial\Omega, \sigma)/\sim\big]^M \tag{8.4.238}$$

is an isomorphism. Granted the invertibility of (8.4.231) and (8.4.238), the fact that u defined as in (8.4.232) solves the boundary value problem (8.4.230) and satisfies the properties claimed in (8.4.233)-(8.4.237) is seen from [129, (7.4.118)], [132, (1.8.7), (1.8.27), (1.8.29), (1.8.31), (1.8.34)], and the Open Mapping Theorem. □

It is also of interest to single out the Dirichlet Problem with data in the Sarason space, of functions of vanishing mean oscillations, in the context of δ-AR domains with $\delta \in (0, 1)$ sufficiently small. We do so in the theorem below, which generalizes the solvability result obtained in [115] for strongly elliptic systems L (in the sense of Legendre-Hadamard) and when the underlying domain Ω is the upper half-space. See also [134] for related work in the case when Ω upper-graph Lipschitz domains and L is the Laplacian, or the three-dimensional Lamé system.

Theorem 8.4.34 *Fix two integers $n, M \in \mathbb{N}$ with $n \geq 2$ and suppose $\Omega \subseteq \mathbb{R}^n$ is an Ahlfors regular domain. Abbreviate $\sigma := \mathcal{H}^{n-1}\lfloor\partial\Omega$ and denote by ν the geometric measure theoretic outward unit normal to Ω. Also, let L be a homogeneous, second-order, constant complex coefficient, weakly elliptic $M \times M$ system in $\mathbb{R}^n$, with the property that $\mathfrak{A}_L^{\mathrm{dis}} \neq \varnothing$. Finally, fix an arbitrary aperture parameter $\kappa > 0$.*

Then there exists some threshold $\delta \in (0,1)$ *which depends only on n, L, and the* ADR *constants of* $\partial\Omega$ *with the property that whenever* $\|\nu\|_{[\mathrm{BMO}(\partial\Omega,\sigma)]^n} < \delta$ *it follows that the boundary value problem*

$$\begin{cases} u \in \left[\mathscr{C}^\infty(\Omega)\right]^M, \quad Lu = 0 \text{ in } \Omega, \\ |\nabla u|^p \operatorname{dist}(\cdot,\partial\Omega)^{p-1}\, d\mathcal{L}^n \text{ is a vanishing Carleson} \\ \text{measure in } \Omega \text{ for each } p \in (1,\infty), \\ u\big|_{\partial\Omega}^{\kappa-\mathrm{n.t.}} = f \in \left[\mathrm{VMO}(\partial\Omega,\sigma)\right]^M, \end{cases} \tag{8.4.239}$$

is solvable. More specifically, if K_{mod} *is associated with* Ω *and a coefficient tensor* $A \in \mathfrak{A}_L^{\mathrm{dis}}$ *as in* (A.0.117)*, then the operator*

$$\tfrac{1}{2}I + \left[K_{\text{mod}}\right] : \left[\mathrm{VMO}(\partial\Omega,\sigma)/\sim\right]^M \longrightarrow \left[\mathrm{VMO}(\partial\Omega,\sigma)/\sim\right]^M \tag{8.4.240}$$

(*cf.* (4.2.11)) *is an isomorphism, and the boundary problem* (8.4.239) *is solved by*

$$u := \mathcal{D}_{\text{mod}} g + c \text{ in } \Omega, \tag{8.4.241}$$

where the function $g \in \left[\mathrm{VMO}(\partial\Omega,\sigma)\right]^M$ *is an arbitrarily chosen representative of the equivalence class* $\left(\tfrac{1}{2}I + \left[K_{\text{mod}}\right]\right)^{-1}[f] \in \left[\mathrm{VMO}(\partial\Omega,\sigma)/\sim\right]^M$ *and* $c \in \mathbb{C}^M$ *is a suitable constant. Furthermore, this solution has the property that for each* $p \in (1,\infty)$ *there exists a constant* $C = C(\Omega, L, p) \in (0,\infty)$ *such that*

$$\sup_{x\in\partial\Omega,\, r>0} \frac{1}{\sigma\big(B(x,r)\cap\partial\Omega\big)} \int_{B(x,r)\cap\Omega} |\nabla u|^p \operatorname{dist}(\cdot,\partial\Omega)^{p-1}\, d\mathcal{L}^n \leq C\|f\|^p_{[\mathrm{BMO}(\partial\Omega,\sigma)]^M}. \tag{8.4.242}$$

The statement of Theorem 8.4.34 may be further simplified in the scalar case (i.e., for $M = 1$). Indeed, if $M = 1$ and $n \geq 3$ then condition $\mathfrak{A}_L^{\mathrm{dis}} \neq \varnothing$ is automatically satisfied, thanks to Proposition 1.6.12. In addition, in the scalar case in the two-dimensional setting (i.e., for $M = 1$ and $n = 2$), Proposition 1.6.13 ensures that condition $\mathfrak{A}_L^{\mathrm{dis}} \neq \varnothing$ may be replaced by the demand that $\mathfrak{A}_L \cap M_2^0 \neq \varnothing$ (with M_2^0 defined in [131, (1.4.187)]). As in the case Theorem 8.4.33, we also have uniqueness for the Dirichlet Problem with data in VMO when $\Omega = \mathbb{R}^n_+$ and the system L satisfies the strong Legendre-Hadamard ellipticity condition.

Proof of Theorem 8.4.34 All claims made in the statement may be justified by reasoning as in the proof of Theorem 8.4.33. □

We next treat the Dirichlet Problem with data in the Coifman-Weiss space CMO, defined in (A.0.53), in the setting of sufficiently flat domains. The novel aspect is that now the solution is sought in the class of super vanishing Carleson measures, introduced in [132, Definition 1.8.1]. Here is a precise statement:

Theorem 8.4.35 *Select two integers n,* $M \in \mathbb{N}$ *with* $n \geq 2$ *and assume* $\Omega \subseteq \mathbb{R}^n$ *is an Ahlfors regular domain. Abbreviate* $\sigma := \mathcal{H}^{n-1}\lfloor\partial\Omega$ *and denote by* ν *the geometric* *measure theoretic outward unit normal to Ω. Next, let L be a homogeneous, second-order, constant complex coefficient, weakly elliptic $M \times M$ system in $\mathbb{R}^n$, with the property that $\mathfrak{A}_L^{\rm dis} \neq \varnothing$. Finally, pick an arbitrary aperture parameter $\kappa > 0$.*

Then there exists a threshold $\delta \in (0,1)$ depending only on n, L, and the ADR *constants of $\partial\Omega$ with the property that when $\|\nu\|_{[\mathrm{BMO}(\partial\Omega,\sigma)]^n} < \delta$ it follows that Ω is a* UR *domain with unbounded boundary and the boundary value problem*

$$\begin{cases} u \in \big[\mathscr{C}^\infty(\Omega)\big]^M, \quad Lu = 0 \ \text{ in } \ \Omega, \\ |\nabla u|^p \operatorname{dist}(\cdot, \partial\Omega)^{p-1} \, d\mathcal{L}^n \ \text{ is a super vanishing} \\ \text{Carleson measure in } \Omega \text{ for each } \ p \in (1, \infty), \\ u\big|_{\partial\Omega}^{\kappa-\mathrm{n.t.}} = f \in \big[\mathrm{CMO}(\partial\Omega, \sigma)\big]^M, \end{cases} \tag{8.4.243}$$

is solvable. More specifically, if $K_{\rm mod}$ is associated with Ω and a coefficient tensor $A \in \mathfrak{A}_L^{\rm dis}$ as in (A.0.117)*, then the operator*

$$\tfrac{1}{2}I + \big[K_{\rm mod}\big] : \big[\mathrm{CMO}(\partial\Omega, \sigma)/\sim\big]^M \longrightarrow \big[\mathrm{CMO}(\partial\Omega, \sigma)/\sim\big]^M \tag{8.4.244}$$

(cf. (4.2.12)*) is an isomorphism, and the boundary problem* (8.4.239) *is solved by*

$$u := \mathcal{D}_{\rm mod} g + c \ \text{ in } \ \Omega, \tag{8.4.245}$$

where the function $g \in \big[\mathrm{CMO}(\partial\Omega, \sigma)\big]^M$ is an arbitrarily chosen representative of the equivalence class $\big(\tfrac{1}{2}I + \big[K_{\rm mod}\big]\big)^{-1}[f] \in \big[\mathrm{CMO}(\partial\Omega, \sigma)/\sim\big]^M$ and $c \in \mathbb{C}^M$ is a suitable constant. In addition, this solution has the property that for each $p \in (1, \infty)$ there exists a constant $C = C(\Omega, L, p) \in (0, \infty)$ such that

$$\begin{aligned} \sup_{x \in \partial\Omega,\, r>0} \frac{1}{\sigma\big(B(x,r) \cap \partial\Omega\big)} \int_{B(x,r)\cap\Omega} |\nabla u|^p \operatorname{dist}(\cdot, \partial\Omega)^{p-1} \, d\mathcal{L}^n \\ \leq C \|f\|^p_{[\mathrm{BMO}(\partial\Omega,\sigma)]^M}. \end{aligned} \tag{8.4.246}$$

It is worth emphasizing that we are not requiring the system L to satisfy the Legendre-Hadamard strong ellipticity condition, but rather ask that L is weakly elliptic and possesses a distinguished coefficient tensor. For example, in the family of complex Lamé systems $L_{\mu,\lambda}$ indexed by $\mu, \lambda \in \mathbb{C}$ from (1.6.85) our Theorem 8.4.35 works in the range $\mu \neq 0$, $2\mu + \lambda \neq 0$, $3\mu + \lambda \neq 0$, whereas the Legendre-Hadamard strong ellipticity condition would require (cf. [131, (1.3.11)]) placing ourselves in the more restrictive range $\operatorname{Re}\mu > 0$ and $\operatorname{Re}(2\mu + \lambda) > 0$.

We also wish to note that it is possible to further simplify the statement of Theorem 8.4.35 in the scalar case (i.e., for $M = 1$). Specifically, if $M = 1$ and $n \geq 3$ then condition $\mathfrak{A}_L^{\rm dis} \neq \varnothing$ is automatically satisfied, thanks to Proposition 1.6.12. Moreover, in the scalar case in the two-dimensional setting (that is, when $M = 1$ and $n = 2$), Proposition 1.6.13 guarantees that the condition $\mathfrak{A}_L^{\rm dis} \neq \varnothing$ may be replaced by the demand that $\mathfrak{A}_L \cap M_2^0 \neq \varnothing$ (with M_2^0 defined in [131, (1.4.187)]).

Theorem 8.4.35 answers a question formulated in a more restrictive setting in [116, Open Question 8]. As noted in relation to Theorem 8.4.34, if $\Omega = \mathbb{R}^n_+$ and the system L satisfies the strong Legendre-Hadamard ellipticity condition work in [115] also guarantees that one has uniqueness for the Dirichlet Problem with data in CMO.

Here is the proof of Theorem 8.4.35.

Proof of Theorem 8.4.35 Since $g \in \big[\mathrm{CMO}(\partial\Omega,\sigma)\big]^M \subseteq \big[\mathrm{BMO}(\partial\Omega,\sigma)\big]^M$, we know from Theorem 8.4.33 that u defined in (8.4.245) satisfies

$$u \in \big[\mathscr{C}^\infty(\Omega)\big]^M,\ \ Lu = 0 \ \text{ in } \ \Omega,\ \ u\big|_{\partial\Omega}^{\kappa-\mathrm{n.t.}} = \big(\tfrac{1}{2}I + K_{\mathrm{mod}}\big)g + c = f, \tag{8.4.247}$$

and (8.4.246) holds. That $|\nabla u|^p \operatorname{dist}(\cdot,\partial\Omega)^{p-1}\,\mathrm{d}\mathcal{L}^n$ is a super vanishing Carleson measure in Ω for each $p \in (1,\infty)$ is seen from [132, Theorem 1.8.2, *(4)*]. □

We next consider the solvability of the Dirichlet Problem with boundary data in homogeneous Hölder spaces for weakly elliptic systems L in sufficiently flat Ahlfors regular domains Ω with unbounded boundaries in $\mathbb{R}^n$. The case when L satisfies the Legendre-Hadamard strong ellipticity condition and when $\Omega = \mathbb{R}^n_+$ has been dealt with in [115]. Related results in the case of the Lamé system and three-dimensional upper-graph Lipschitz domains may be found in [39] and [88].

Theorem 8.4.36 *Fix two integers $n, M \in \mathbb{N}$ with $n \geq 2$ and suppose $\Omega \subseteq \mathbb{R}^n$ is an Ahlfors regular domain. Abbreviate $\sigma := \mathcal{H}^{n-1}\lfloor\partial\Omega$ and denote by ν the geometric measure theoretic outward unit normal to Ω. Also, let L be a homogeneous, second-order, constant complex coefficient, weakly elliptic $M \times M$ system in $\mathbb{R}^n$, with the property that $\mathfrak{A}^{\mathrm{dis}}_L \neq \varnothing$. Finally, fix an arbitrary exponent $\alpha \in (0,1)$.*

Then there exists some threshold $\delta \in (0,1)$ which depends only on n, α, L, and the ADR *constants of $\partial\Omega$ with the property that whenever $\|\nu\|_{[\mathrm{BMO}(\partial\Omega,\sigma)]^n} < \delta$ it follows that the boundary value problem*

$$\begin{cases} u \in \big[\dot{\mathscr{C}}^\alpha(\overline{\Omega}) \cap \mathscr{C}^\infty(\Omega)\big]^M, & Lu = 0 \ \text{in} \ \Omega, \\ u\big|_{\partial\Omega} = f \in \big[\dot{\mathscr{C}}^\alpha(\partial\Omega)\big]^M, \end{cases} \tag{8.4.248}$$

is solvable. Concretely, if K_{mod} is associated with Ω and a coefficient tensor $A \in \mathfrak{A}^{\mathrm{dis}}_L$ as in (A.0.117)*, then the operator (cf.* [132, (2.3.185)]*)*

$$\tfrac{1}{2}I + \big[K_{\mathrm{mod}}\big] : \big[\dot{\mathscr{C}}^\alpha(\partial\Omega)/\!\sim\big]^M \longrightarrow \big[\dot{\mathscr{C}}^\alpha(\partial\Omega)/\!\sim\big]^M \tag{8.4.249}$$

is an isomorphism, and the boundary problem (8.4.248) *is solved by*

$$u := \mathcal{D}_{\mathrm{mod}}\, g + c \ \text{ in } \ \Omega, \tag{8.4.250}$$

where the function $g \in \big[\dot{\mathscr{C}}^\alpha(\partial\Omega)\big]^M$ is an arbitrarily chosen representative of the equivalence class $\big(\tfrac{1}{2}I + \big[K_{\mathrm{mod}}\big]\big)^{-1}[f] \in \big[\dot{\mathscr{C}}^\alpha(\partial\Omega)/\sim\big]^M$ and $c \in \mathbb{C}^M$ is a suitable constant. Furthermore, this solution has the property that there exist two constants $C_1 = C_1(\Omega,\alpha) \in (0,\infty)$ and $C_2 = C_2(\Omega, L, \alpha) \in (0,\infty)$ such that

$$\sup_{\substack{x,y\in\overline{\Omega}\\ x\neq y}} \frac{|u(x)-u(y)|}{|x-y|^{\alpha}} \leq C_1 \cdot \sup_{x\in\Omega} \Big\{ |(\nabla u)(x)| \cdot \operatorname{dist}(x,\partial\Omega)^{1-\alpha} \Big\} \leq C_2 \|f\|_{[\dot{\mathscr{C}}^{\alpha}(\partial\Omega)]^M}, \tag{8.4.251}$$

and also

$$\sup_{\substack{x\in\partial\Omega\\ r\in(0,\infty)}} \left(\frac{1}{r^{n-1+\alpha p}} \int_{B(x,r)\cap\Omega} |\nabla u|^p \operatorname{dist}(\cdot,\partial\Omega)^{p-1}\, d\mathcal{L}^n \right)^{1/p} \leq C_2 \|f\|_{[\dot{\mathscr{C}}^{\alpha}(\partial\Omega)]^M} \tag{8.4.252}$$

as well as

$$\lim_{R\to 0^+} \left\{ \sup_{\substack{x\in\partial\Omega\\ r\in(0,R)}} \left(\frac{1}{r^{n-1+\alpha p}} \int_{B(x,r)\cap\Omega} |\nabla u|^p \operatorname{dist}(\cdot,\partial\Omega)^{p-1}\, d\mathcal{L}^n \right)^{1/p} \right\} \tag{8.4.253}$$
$$\leq C_2 \operatorname{dist}\big(f, \big[\dot{\mathscr{C}}^{\alpha}_{\mathrm{van}}(\partial\Omega)\big]^M\big),$$

where the distance is measured in the space $\Big(\big[\dot{\mathscr{C}}^{\alpha}(\partial\Omega)\big]^M, \|\cdot\|_{[\dot{\mathscr{C}}^{\alpha}(\partial\Omega)]^M}\Big)$.

We may further simplify the statement of Theorem 8.4.36 when $M = 1$ (i.e., in the scalar case). Concretely, Proposition 1.6.12 tells us that if $M = 1$ and $n \geq 3$ then condition $\mathfrak{A}_L^{\mathrm{dis}} \neq \varnothing$ is automatically satisfied. Also, in the scalar case in the two-dimensional setting (i.e., for $M = 1$ and $n = 2$), Proposition 1.6.13 shows that we may replace the demand that $\mathfrak{A}_L^{\mathrm{dis}} \neq \varnothing$ by the condition $\mathfrak{A}_L \cap M_2^0 \neq \varnothing$ (where M_2^0 is defined in [131, (1.4.187)]).

Proof of Theorem 8.4.36 Take $\delta \in (0,1)$ small enough so that Ω is a uniform domain (in the sense of [129, Definition 5.11.10]), for a constant depending only on the dimension n and the Ahlfors regularity constants of $\partial\Omega$, and that the operator (8.4.249) is an isomorphism. That this is possible is ensured by (3.1.22) and (4.2.13). Having established this, that u defined in (8.4.250) solves the boundary value problem (8.4.248) and satisfies (8.4.251), (8.4.252), (8.4.253), follows from [129, (5.11.78), (7.4.119)] and [132, (1.8.7), (1.8.15), (1.8.17), (1.8.18), (1.8.27)]. □

It is also of interest to note that a solution to the Dirichlet Problem (8.4.248) may be found in the homogeneous *vanishing* Hölder space $\big[\dot{\mathscr{C}}^{\alpha}_{\mathrm{van}}(\overline{\Omega})\big]^M$ (which is a strict subspace of $\big[\dot{\mathscr{C}}^{\alpha}(\overline{\Omega})\big]^M$, used in the formulation of the boundary value problem (8.4.248)) if and only if the boundary datum f belongs to the homogeneous vanishing Hölder space $\big[\dot{\mathscr{C}}^{\alpha}_{\mathrm{van}}(\partial\Omega)\big]^M$. Indeed, one implication is clear from the fact that, as seen from the definition given in (A.0.49), membership to a homogeneous vanishing Hölder space is hereditary with respect to restriction, while the other implication is contained in the theorem below, dealing with the issue of solvability for the Dirichlet Problem with data in homogeneous *vanishing* Hölder spaces.

Theorem 8.4.37 *Fix two integers $n, M \in \mathbb{N}$ with $n \geq 2$ and suppose $\Omega \subseteq \mathbb{R}^n$ is an Ahlfors regular domain. Abbreviate $\sigma := \mathcal{H}^{n-1}\lfloor\partial\Omega$ and denote by ν the geometric* *measure theoretic outward unit normal to* Ω. *Also, let* L *be a homogeneous, second-order, constant complex coefficient, weakly elliptic* $M \times M$ *system in* $\mathbb{R}^n$, *with the property that* $\mathfrak{A}_L^{\rm dis} \neq \varnothing$. *Finally, fix an arbitrary exponent* $\alpha \in (0, 1)$.

Then there exists some threshold $\delta \in (0, 1)$ *which depends only on* n, α, L, *and the* ADR *constants of* $\partial\Omega$ *with the property that whenever* $\|\nu\|_{[\mathrm{BMO}(\partial\Omega,\sigma)]^n} < \delta$ *it follows that the boundary value problem*

$$\begin{cases} u \in \big[\dot{\mathscr{C}}^{\alpha}_{\rm van}(\overline{\Omega}) \cap \mathscr{C}^{\infty}(\Omega)\big]^M, & Lu = 0 \text{ in } \Omega, \\ u\big|_{\partial\Omega} = f \in \big[\dot{\mathscr{C}}^{\alpha}_{\rm van}(\partial\Omega)\big]^M, \end{cases} \tag{8.4.254}$$

is solvable. More precisely, regarding the function f *in the space* $\big[\dot{\mathscr{C}}^{\alpha}(\partial\Omega)\big]^M$, *the solution* u *of the boundary value problem* (8.4.248) *corresponding to this boundary datum actually belongs to homogeneous vanishing Hölder space* $\big[\dot{\mathscr{C}}^{\alpha}_{\rm van}(\overline{\Omega})\big]^M$ *and for each* $p \in (1, \infty)$ *satisfies*

$$\lim_{R\to 0^+} \left\{ \sup_{\substack{x\in\partial\Omega \\ r\in(0,R)}} \left(\frac{1}{r^{n-1+\alpha p}} \int_{B(x,r)\cap\Omega} |\nabla u|^p \operatorname{dist}(\cdot, \partial\Omega)^{p-1}\, d\mathcal{L}^n \right)^{1/p} \right\} = 0. \tag{8.4.255}$$

The statement of Theorem 8.4.37 takes a simpler form in the scalar case (i.e., when $M = 1$). Specifically, from Proposition 1.6.12 we know that if $M = 1$ and $n \geq 3$ then the demand $\mathfrak{A}_L^{\rm dis} \neq \varnothing$ is automatically satisfied. In addition, in the scalar case in the two-dimensional setting (i.e., for $M = 1$ and $n = 2$), Proposition 1.6.13 guarantees that in place of the demand $\mathfrak{A}_L^{\rm dis} \neq \varnothing$ we may now ask that $\mathfrak{A}_L \cap M_2^0 \neq \varnothing$ (where M_2^0 is defined in [131, (1.4.187)]).

Proof of Theorem 8.4.37 If the modified double layer $K_{\rm mod}$ is associated with Ω and a coefficient tensor $A \in \mathfrak{A}_L^{\rm dis}$ as in (A.0.117), then taking $\delta \in (0, 1)$ small enough ensures that the operator

$$\tfrac{1}{2}I + \big[K_{\rm mod}\big] : \big[\dot{\mathscr{C}}^{\alpha}_{\rm van}(\partial\Omega)/\sim\big]^M \longrightarrow \big[\dot{\mathscr{C}}^{\alpha}(\partial\Omega)_{\rm van}/\sim\big]^M \tag{8.4.256}$$

is an isomorphism (cf. (4.2.14)). Then, much as in the case of Theorem 8.4.36, the boundary value problem (8.4.254) is solved by $u := \mathcal{D}_{\rm mod} g + c$ in Ω, where the function $g \in \big[\dot{\mathscr{C}}^{\alpha}_{\rm van}(\partial\Omega)\big]^M$ is an arbitrarily chosen representative of the equivalence class $\big(\frac{1}{2}I + \big[K_{\rm mod}\big]\big)^{-1}[f] \in \big[\dot{\mathscr{C}}^{\alpha}_{\rm van}(\partial\Omega)/\sim\big]^M$ and $c \in \mathbb{C}^M$ is a suitable constant. Finally, that the solution constructed above satisfies (8.4.255) is clear from [132, (1.8.19)], and this completes the proof of the theorem. □

In the last portion of this section we present several generalizations of the solvability result presented in Theorem 8.4.33 involving ϕ-modulated BMO and BCMO spaces, defined as in (4.1.281)-(4.1.286).

Theorem 8.4.38 *Fix two integers* $n, M \in \mathbb{N}$ *with* $n \geq 2$ *and suppose* $\Omega \subseteq \mathbb{R}^n$ *is an Ahlfors regular domain. Abbreviate* $\sigma := \mathcal{H}^{n-1}\lfloor\partial\Omega$ *and denote by* ν *the* *geometric measure theoretic outward unit normal to* Ω. *Also, fix an arbitrary aperture parameter* $\kappa > 0$, *together with an integrability exponent* $p \in (1, \infty)$, *and suppose*

$$\begin{aligned} &\phi : (0,\infty) \to (0,\infty) \text{ is } \mathcal{L}^1\text{-measurable and there exists } C_* \in (0,\infty) \\ &\text{such that } r \int_r^\infty \phi(t)\,\frac{\ln(2t/r)}{t^2}\,dt \leq C_*\phi(r) \text{ for each } r \in (0,\infty). \end{aligned} \tag{8.4.257}$$

Use this to define (*cf.* (4.1.281)-(4.1.282)) *the* ϕ*-modulated space of functions of bounded* L^p*-styled mean oscillations*

$$\mathrm{BMO}_{\phi,p}(\partial\Omega,\sigma) := \Big\{\psi \in L^1_{\mathrm{loc}}(\partial\Omega,\sigma) : \|\psi\|_{\mathrm{BMO}_{\phi,p}(\partial\Omega,\sigma)} < +\infty\Big\}, \tag{8.4.258}$$

where, for each $\psi \in L^1_{\mathrm{loc}}(\partial\Omega,\sigma)$,

$$\|\psi\|_{\mathrm{BMO}_{\phi,p}(\partial\Omega,\sigma)} := \sup_{\substack{x\in\partial\Omega\\ r\in(0,\infty)}} \left\{\frac{1}{\phi(r)}\left(⨍_{\Delta(x,r)}\Big|\psi - ⨍_{\Delta(x,r)}\psi\,d\sigma\Big|^p d\sigma\right)^{1/p}\right\}.$$

Finally, let L be a homogeneous, second-order, constant complex coefficient, weakly elliptic $M \times M$ *system in* $\mathbb{R}^n$, *with the property that* $\mathfrak{A}_L^{\mathrm{dis}} \neq \varnothing$.

Then there exists a threshold $\delta \in (0,1)$, *depending only on n, p, L, the constant* C_* *from* (8.4.257), *and the* ADR *constants of* $\partial\Omega$, *with the following significance. If* $\|\nu\|_{[\mathrm{BMO}(\partial\Omega,\sigma)]^n} < \delta$ *then* Ω *is a* UR *domain with an unbounded boundary, and the boundary value problem asking to find a function satisfying*

$$\begin{cases} u \in \big[\mathscr{C}^\infty(\Omega)\big]^M,\ Lu = 0 \text{ in } \Omega,\ u\Big|_{\partial\Omega}^{\kappa-\mathrm{n.t.}} = f \in \big[\mathrm{BMO}_{\phi,p}(\partial\Omega,\sigma)\big]^M, \text{ and} \\ \sup\limits_{\substack{x\in\partial\Omega \text{ and}\\ r\in(0,\infty)}} \left\{\frac{1}{\phi(r)}\left(\frac{1}{\sigma\big(\Delta(x,r)\big)}\int_{B(x,r)\cap\Omega}|\nabla u|^p \operatorname{dist}(\cdot,\partial\Omega)^{p-1}\,d\mathcal{L}^n\right)^{1/p}\right\} < +\infty, \end{cases}$$

is solvable. More specifically, if the modified boundary-to-boundary double layer K_{mod} *is associated with the set* Ω *and a coefficient tensor* $A \in \mathfrak{A}_L^{\mathrm{dis}}$ *as in* (A.0.117), *then the operator* (*cf.* (4.1.310))

$$\tfrac{1}{2}I + \big[K_{\mathrm{mod}}\big] : \big[\mathrm{BMO}_{\phi,p}(\partial\Omega,\sigma)/\sim\big]^M \longrightarrow \big[\mathrm{BMO}_{\phi,p}(\partial\Omega,\sigma)/\sim\big]^M \tag{8.4.259}$$

is an isomorphism, and the above boundary value problem is solved by

$$u := \mathcal{D}_{\mathrm{mod}}\,g + c \text{ in } \Omega, \tag{8.4.260}$$

where the function $g \in \big[\mathrm{BMO}_{\phi,p}(\partial\Omega,\sigma)\big]^M$ *is an arbitrarily chosen representative of the equivalence class* $\big(\tfrac{1}{2}I + \big[K_{\mathrm{mod}}\big]\big)^{-1}[f] \in \big[\mathrm{BMO}_{\phi,p}(\partial\Omega,\sigma)/\sim\big]^M$ *and* $c \in \mathbb{C}^M$ *is a suitable constant. Furthermore, this solution has the property that*

$$\sup_{\substack{x\in\partial\Omega \text{ and}\\ r\in(0,\infty)}} \left\{ \frac{1}{\phi(r)} \left(\frac{1}{\sigma(\Delta(x,r))} \int_{B(x,r)\cap\Omega} |\nabla u|^p \operatorname{dist}(\cdot,\partial\Omega)^{p-1}\, d\mathcal{L}^n \right)^{1/p} \right\}$$
$$\leq C\|f\|_{[\mathrm{BMO}_{\phi,p}(\partial\Omega,\sigma)]^M} \tag{8.4.261}$$

for some constant $C = C(\Omega, L, p, C_*) \in (0,\infty)$ *independent of the datum* f.

In the scalar case (i.e., for $M = 1$), the statement of Theorem 8.4.38 may be further streamlined. Concretely, if $M = 1$ and $n \geq 3$ then Proposition 1.6.12 guarantees that condition $\mathfrak{A}_L^{\mathrm{dis}} \neq \varnothing$ is always satisfied. Moreover, in the scalar case in the two-dimensional setting (i.e., for $M = 1$ and $n = 2$), Proposition 1.6.13 ensures that condition $\mathfrak{A}_L^{\mathrm{dis}} \neq \varnothing$ may be replaced by the demand that $\mathfrak{A}_L \cap M_2^0 \neq \varnothing$ (with M_2^0 defined in [131, (1.4.187)]).

Recall from (4.1.313)-(4.1.315) that the above theorem encompasses many special cases of interest (some of which have been treated directed earlier), depending on the choice of the function $\phi : (0,\infty) \to (0,\infty)$, assumed to satisfy (8.4.257). For one thing, taking $\phi(t) := 1$ for each $t > 0$ allows us to recover (see (4.1.313)) the solvability result for the Dirichlet Problem with boundary data in the John-Nirenberg space $[\mathrm{BMO}(\partial\Omega,\sigma)]^M$ formulated in Theorem 8.4.33 (cf. (8.4.230)). For another thing, given any $\alpha \in (0,1)$, taking $\phi(t) := t^\alpha$ for each $t > 0$ yields (see (4.1.314)) the solvability result for the Dirichlet Problem with boundary data in the homogeneous Hölder space $[\dot{\mathscr{C}}^\alpha(\partial\Omega)]^M$ formulated in Theorem 8.4.36 (cf. (8.4.248), (8.4.252)). Furthermore, given any $p \in (1,\infty)$ and $\lambda \in (0, n-1)$, taking $\phi(t) := t^{(\lambda-n+1)/p}$ for each $t > 0$ yields (see (4.1.315)) the solvability result for the Dirichlet Problem with boundary data in the homogeneous Morrey-Campanato space $[\mathring{\mathfrak{L}}^{p,\lambda}(\partial\Omega,\sigma)]^M$ formulated in (8.4.112) (or (8.4.113)).

Next we turn our attention to the task of presenting the proof of Theorem 8.4.38.

Proof of Theorem 8.4.38 Theorem 3.1.5 tells us that if $\|\nu\|_{[\mathrm{BMO}(\partial\Omega,\sigma)]^n} < \delta$ for some sufficiently small $\delta \in (0,1)$ then Ω is a UR domain with an unbounded boundary. Note that (8.4.257) implies that for all $r \in (0,\infty)$ we have

$$\max\left\{ r\int_r^\infty \phi(t)\,\frac{dt}{t^2},\ r\int_r^\infty \phi(t)\,\frac{\ln(t/r)}{t^2}\,dt \right\} \leq \widetilde{C}_*\phi(r), \tag{8.4.262}$$

where $\widetilde{C}_* \in (0,\infty)$ is a (universal) multiple of C_*. Granted this, by further decreasing the value of $\delta \in (0,1)$ matters may also be arranged so that the operator (8.4.259) becomes an isomorphism (see (4.2.37)). As a consequence, given any boundary datum $f \in [\mathrm{BMO}_{\phi,p}(\partial\Omega,\sigma)]^M$ we can find

$$g \in [\mathrm{BMO}_{\phi,p}(\partial\Omega,\sigma)]^M \text{ and } c \in \mathbb{C}^M \text{ such that} \tag{8.4.263}$$
$$\big(\tfrac{1}{2}I + K_{\text{mod}}\big)g + c = f \text{ and } \|g\|_{[\mathrm{BMO}_{\phi,p}(\partial\Omega,\sigma)]^M} \leq C\|f\|_{[\mathrm{BMO}_{\phi,p}(\partial\Omega,\sigma)]^M},$$

for some constant $C = C(\Omega, L, p, C_*) \in (0,\infty)$ independent of f. In particular, the embeddings in (4.1.287) and (4.1.289) (here we use the fact that (8.4.257) implies the Dini condition at infinity for ϕ formulated in (4.1.288)) guarantee that

$$g \in \Big[L^1\Big(\partial\Omega, \frac{\sigma(x)}{1+|x|^n}\Big)\Big]^M. \tag{8.4.264}$$

In view of this, the jump-formula in [132, (1.8.27)], the equality in (8.4.263), and the properties listed in [132, (1.8.7)], it follows that the function u defined in (8.4.260) solves the boundary problem formulated in the statement of the theorem. Moreover,

$$\begin{aligned}
&\sup_{\substack{x\in\partial\Omega \text{ and}\\ r\in(0,\infty)}} \left\{\frac{1}{\phi(r)}\left(\frac{1}{\sigma(\Delta(x,r))}\int_{B(x,r)\cap\Omega}|\nabla u|^p \operatorname{dist}(\cdot,\partial\Omega)^{p-1}\,d\mathcal{L}^n\right)^{1/p}\right\}\\
&\quad\le C\cdot \sup_{\substack{x\in\partial\Omega \text{ and}\\ r\in(0,\infty)}} \left\{\frac{1}{\phi(r)}\int_1^\infty\left(⨍_{\Delta(x,\lambda r)}\Big|g-⨍_{\Delta(x,\lambda r)} g\,d\sigma\Big|^p\,d\sigma\right)^{1/p}\frac{d\lambda}{\lambda^2}\right\}\\
&\quad\le C\cdot \sup_{\substack{x\in\partial\Omega \text{ and}\\ r\in(0,\infty)}} \left\{\frac{1}{\phi(r)}\int_1^\infty \phi(\lambda r)\|g\|_{[\mathrm{BMO}_{\phi,p}(\partial\Omega,\sigma)]^M}\frac{d\lambda}{\lambda^2}\right\}\\
&\quad= C\cdot\|g\|_{[\mathrm{BMO}_{\phi,p}(\partial\Omega,\sigma)]^M}\cdot\sup_{r\in(0,\infty)}\left\{\frac{1}{\phi(r)}\int_1^\infty\phi(\lambda r)\frac{d\lambda}{\lambda^2}\right\}\\
&\quad= C\cdot\|g\|_{[\mathrm{BMO}_{\phi,p}(\partial\Omega,\sigma)]^M}\cdot\sup_{r\in(0,\infty)}\left\{\frac{1}{\phi(r)}\left(r\int_r^\infty\phi(t)\frac{dt}{t^2}\right)\right\}\\
&\quad\le C\cdot C_*\cdot\|g\|_{[\mathrm{BMO}_{\phi,p}(\partial\Omega,\sigma)]^M}\le C\|f\|_{[\mathrm{BMO}_{\phi,p}(\partial\Omega,\sigma)]^M},
\end{aligned}\tag{8.4.265}$$

thanks to [132, Theorem 5.1.8] (recall from [132, Example 5.1.6] that genuine double layers are generalized double layers), (4.1.282) (with $\Sigma := \partial\Omega$ and $w \equiv 1$), the change of variables $t := \lambda r$, the estimate in (8.4.262), and the last property in (8.4.263). Hence, u also satisfies (8.4.261). □

Here are the versions of the Dirichlet Problem with boundary data in the homogeneous and inhomogeneous space of functions of bounded central mean oscillations alluded to earlier, complementing the solvability result from Theorem 8.4.38.

Theorem 8.4.39 *Pick two integers $n, M \in \mathbb{N}$ with $n \ge 2$ and assume $\Omega \subseteq \mathbb{R}^n$ is an Ahlfors regular domain. Abbreviate $\sigma := \mathcal{H}^{n-1}\lfloor\partial\Omega$ and denote by ν the geometric measure theoretic outward unit normal to Ω. In addition, fix a reference point $x_0 \in \partial\Omega$, an aperture parameter $\kappa > 0$, and an integrability exponent $p \in (1,\infty)$. Also, make the assumption that*

$$\begin{aligned}
&\phi:(0,\infty)\to(0,\infty) \textit{ is } \mathcal{L}^1\textit{-measurable and there exists } C_*\in(0,\infty)\\
&\textit{such that } r\int_r^\infty \phi(t)\,\frac{\ln(2t/r)}{t^2}\,dt \le C_*\phi(r) \textit{ for each } r\in(0,\infty).
\end{aligned}\tag{8.4.266}$$

Use the reference point $x_0 \in \partial\Omega$, the integrability exponent $p \in (1,\infty)$, and the above function ϕ to define (cf. (4.1.283)-(4.1.284)) the ϕ-modulated homogeneous *space of functions of bounded L^p-styled x_0-central mean oscillations*

$$\mathrm{BCMO}_{\phi,p}(\partial\Omega,\sigma) := \Big\{\psi \in L^1_{\mathrm{loc}}(\partial\Omega,\sigma) : \|\psi\|_{\mathrm{BCMO}_{\phi,p}(\partial\Omega,\sigma)} < +\infty\Big\}, \tag{8.4.267}$$

where, for each $\psi \in L^1_{\mathrm{loc}}(\partial\Omega,\sigma)$,

$$\|\psi\|_{\mathrm{BCMO}_{\phi,p}(\partial\Omega,\sigma)} := \sup_{r\in(0,\infty)} \left\{ \frac{1}{\phi(r)} \Big(⨍_{\Delta(x_0,r)} \Big|\psi - ⨍_{\Delta(x_0,r)} \psi \, d\sigma\Big|^p d\sigma\Big)^{1/p} \right\}.$$

Lastly, consider a homogeneous, second-order, constant complex coefficient, weakly elliptic $M \times M$ system L in $\mathbb{R}^n$, with the property that $\mathfrak{A}^{\mathrm{dis}}_L \neq \varnothing$.

Then there exists a threshold $\delta \in (0,1)$, depending only on n, p, L, the constant C_* from (8.4.257), *and the* ADR *constants of $\partial\Omega$, with the property that if $\|\nu\|_{[\mathrm{BMO}(\partial\Omega,\sigma)]^n} < \delta$ it follows that Ω is a* UR *domain with an unbounded boundary, and the boundary value problem concerned with finding a function satisfying*

$$\begin{cases} u \in \big[\mathscr{C}^\infty(\Omega)\big]^M, \ Lu = 0 \text{ in } \Omega, \ u\big|^{\kappa-\mathrm{n.t.}}_{\partial\Omega} = f \in \big[\mathrm{BCMO}_{\phi,p}(\partial\Omega,\sigma)\big]^M, \text{ and} \\ \sup\limits_{r\in(0,\infty)} \left\{ \frac{1}{\phi(r)} \left(\frac{1}{\sigma\big(\Delta(x_0,r)\big)} \int_{B(x_0,r)\cap\Omega} |\nabla u|^p \operatorname{dist}(\cdot,\partial\Omega)^{p-1} \, d\mathcal{L}^n \right)^{1/p} \right\} < +\infty, \end{cases}$$

has a solution. Explicitly, with the modified boundary-to-boundary double layer K_{mod} associated with the set Ω and a coefficient tensor $A \in \mathfrak{A}^{\mathrm{dis}}_L$ as in (A.0.117), *the operator* (*cf.* (4.1.311))

$$\tfrac{1}{2}I + \big[K_{\mathrm{mod}}\big] : \big[\mathrm{BCMO}_{\phi,p}(\partial\Omega,\sigma)/\sim\big]^M \longrightarrow \big[\mathrm{BCMO}_{\phi,p}(\partial\Omega,\sigma)/\sim\big]^M$$

is an isomorphism, and the above boundary value problem is solved by

$$u := \mathcal{D}_{\mathrm{mod}} g + c \text{ in } \Omega, \tag{8.4.268}$$

where the function $g \in \big[\mathrm{BCMO}_{\phi,p}(\partial\Omega,\sigma)\big]^M$ is an arbitrarily chosen representative of the equivalence class $\big(\tfrac{1}{2}I + \big[K_{\mathrm{mod}}\big]\big)^{-1}[f] \in \big[\mathrm{BCMO}_{\phi,p}(\partial\Omega,\sigma)/\sim\big]^M$ and $c \in \mathbb{C}^M$ is a suitably chosen constant. Moreover, this solution has the property that

$$\sup_{r\in(0,\infty)} \left\{ \frac{1}{\phi(r)} \left(\frac{1}{\sigma\big(\Delta(x_0,r)\big)} \int_{B(x_0,r)\cap\Omega} |\nabla u|^p \operatorname{dist}(\cdot,\partial\Omega)^{p-1} \, d\mathcal{L}^n \right)^{1/p} \right\} \leq C\|f\|_{[\mathrm{BCMO}_{\phi,p}(\partial\Omega,\sigma)]^M} \tag{8.4.269}$$

for a constant $C = C(\Omega, L, p, C_*) \in (0,\infty)$ independent of the boundary datum f.

A similar solvability result is valid for the Dirichlet Problem with boundary data in the ϕ-modulated inhomogeneous space of functions of bounded L^p-styled x_0-central mean oscillations. Concretely, having defined (*cf.* (4.1.285)-(4.1.286))

$$\dot{\mathrm{B}}\mathrm{CMO}_{\phi,p}(\partial\Omega,\sigma) := \Big\{\psi \in L^1_{\mathrm{loc}}(\partial\Omega,\sigma) : \|\psi\|_{\dot{\mathrm{B}}\mathrm{CMO}_{\phi,p}(\partial\Omega,\sigma)} < +\infty\Big\}, \tag{8.4.270}$$

where, for each $\psi \in L^1_{\mathrm{loc}}(\partial\Omega,\sigma)$,

$$\|\psi\|_{\dot{\mathrm{B}}\mathrm{CMO}_{\phi,p}(\partial\Omega,\sigma)} := \sup_{r\in(1,\infty)} \left\{\frac{1}{\phi(r)}\Big(⨍_{\Delta(x_0,r)}\Big|\psi - ⨍_{\Delta(x_0,r)}\psi\,\mathrm{d}\sigma\Big|^p\,\mathrm{d}\sigma\Big)^{1/p}\right\},$$

the goal now is to find a function satisfying

$$\begin{cases} u \in \big[\mathscr{C}^\infty(\Omega)\big]^M, \ Lu = 0 \text{ in } \Omega, \ u\big|^{\kappa-\mathrm{n.t.}}_{\partial\Omega} = f \in \big[\dot{\mathrm{B}}\mathrm{CMO}_{\phi,p}(\partial\Omega,\sigma)\big]^M, \text{ and} \\ \sup\limits_{r\in(1,\infty)} \left\{\frac{1}{\phi(r)}\left(\frac{1}{\sigma\big(\Delta(x_0,r)\big)}\int_{B(x_0,r)\cap\Omega}|\nabla u|^p \operatorname{dist}(\cdot,\partial\Omega)^{p-1}\,\mathrm{d}\mathcal{L}^n\right)^{1/p}\right\} < +\infty. \end{cases}$$

Formulated as such, this boundary value problem always has a solution with

$$\sup_{r\in(1,\infty)} \left\{\frac{1}{\phi(r)}\left(\frac{1}{\sigma(\Delta(x_0,r))}\int_{B(x_0,r)\cap\Omega}|\nabla u|^p \operatorname{dist}(\cdot,\partial\Omega)^{p-1}\,\mathrm{d}\mathcal{L}^n\right)^{1/p}\right\}$$
$$\leq C\|f\|_{[\mathrm{BCMO}_{\phi,p}(\partial\Omega,\sigma)]^M} \tag{8.4.271}$$

for a constant $C = C(\Omega, L, p, C_*) \in (0,\infty)$ *independent of the boundary datum* f.

Proof All claims may be justified by reasoning as in the proof of Theorem 8.4.38, now making use of the invertibility results from (4.2.38)-(4.2.39). □

In the last portion of this section we shall establish well-posedness results for boundary value problems for weakly elliptic second-order systems with constant complex coefficients in sufficiently flat δ-AR domains $\Omega \subseteq \mathbb{R}^n$, with boundary data in Generalized Banach Function Spaces $\mathbb{X}$ on $\partial\Omega$ (introduced in [130, §5.1]) for which the Hardy-Littlewood maximal operator $\mathcal{M}_{\partial\Omega}$ is bounded both on $\mathbb{X}$ and its associated space $\mathbb{X}'$ (in this regard, see [111, Chapter 8] where "classical" Banach function spaces have been considered). We start by discussing the Dirichlet Problem.

Theorem 8.4.40 *Pick* $n \in \mathbb{N}$ *with* $n \geq 2$ *and let* $\Omega \subseteq \mathbb{R}^n$ *be an Ahlfors regular domain. Denote by* ν *its geometric measure theoretic outward unit normal, set* $\sigma := \mathcal{H}^{n-1}\lfloor\partial\Omega$, *and fix an aperture parameter* $\kappa > 0$. *Let* $\mathbb{X}$ *be a Generalized Banach Function Space on* $(\partial\Omega,\sigma)$, *of the sort introduced in* [130, §5.1], *and denote by* $\mathbb{X}'$ *its associated space (aka Köthe dual; cf.* [130, Definitions 5.1.4, 5.1.11]). *With* $\mathcal{M}_{\partial\Omega}$ *denoting the Hardy-Littlewood maximal operator on* $(\partial\Omega,\sigma)$, *assume*

$$\mathcal{M}_{\partial\Omega} : \mathbb{X} \to \mathbb{X} \text{ and } \mathcal{M}_{\partial\Omega} : \mathbb{X}' \to \mathbb{X}' \text{ are well-defined and bounded.} \tag{8.4.272}$$

Given a homogeneous, second-order, constant complex coefficient, weakly elliptic $M \times M$ *system* L *in* $\mathbb{R}^n$, *consider the Dirichlet Problem*

$$\begin{cases} u \in \big[\mathscr{C}^\infty(\Omega)\big]^M, & Lu = 0 \text{ in } \Omega, \\ \mathcal{N}_\kappa u \in \mathbb{X}, & u\big|_{\partial\Omega}^{\kappa-\mathrm{n.t.}} = f \in \big[\mathbb{X}\big]^M. \end{cases} \tag{8.4.273}$$

Then the following claims are true:

(a) [Existence, Regularity, and Estimates] *If* $\mathfrak{A}_L^{\mathrm{dis}} \neq \emptyset$ *and* $A \in \mathfrak{A}_L^{\mathrm{dis}}$, *then there exists* $\delta \in (0,1)$ *which depends only on n, A, the operator norms of* $\mathcal{M}_{\partial\Omega}$ *on* $\mathbb{X}$, $\mathbb{X}'$, *and the Ahlfors regularity constants of* $\partial\Omega$ *such that if* $\|\nu\|_{[\mathrm{BMO}(\partial\Omega,\sigma)]^n} < \delta$ *(i.e., if* Ω *is a* δ-AR *domain in the sense of Definition 3.1.8) then* $\frac{1}{2}I + K_A$ *is an invertible operator on* $\big[\mathbb{X}\big]^M$ *and the function* $u : \Omega \to \mathbb{C}^M$ *defined as*

$$u(x) := \Big(\mathcal{D}_A\Big(\tfrac{1}{2}I + K_A\Big)^{-1} f\Big)(x) \text{ for all } x \in \Omega, \tag{8.4.274}$$

is a solution of the Dirichlet Problem (8.4.273). *Moreover,*

$$\|\mathcal{N}_\kappa u\|_{\mathbb{X}} \approx \|f\|_{[\mathbb{X}]^M}. \tag{8.4.275}$$

Furthermore, with the $\mathbb{X}$*-styled Sobolev space of order one on* $\partial\Omega$ *defined as*

$$\begin{aligned} &\mathbb{X}_1 := \Big\{\phi \in \mathbb{X} : \partial_{\tau_{jk}}\phi \in \mathbb{X} \text{ for } 1 \leq j,k \leq n\Big\} \text{ equipped with} \\ &\|\phi\|_{\mathbb{X}_1} := \|\phi\|_{\mathbb{X}} + \sum_{j,k=1}^{n} \|\partial_{\tau_{jk}}\phi\|_{\mathbb{X}} \text{ for each } \phi \in \mathbb{X}_1, \end{aligned} \tag{8.4.276}$$

the function u defined in (8.4.274) *satisfies the following regularity result*

$$\mathcal{N}_\kappa(\nabla u) \in \mathbb{X} \Longleftrightarrow f \in \big[\mathbb{X}_1\big]^M, \tag{8.4.277}$$

and if either of these conditions holds then

$$\begin{aligned} &(\nabla u)\big|_{\partial\Omega}^{\kappa-\mathrm{n.t.}} \text{ exists (in } \mathbb{C}^{n\cdot M}) \text{ at } \sigma\text{-a.e. point on } \partial\Omega \text{ and} \\ &\|\mathcal{N}_\kappa u\|_{\mathbb{X}} + \|\mathcal{N}_\kappa(\nabla u)\|_{\mathbb{X}} \approx \|f\|_{[\mathbb{X}_1]^M}. \end{aligned} \tag{8.4.278}$$

(b) [Uniqueness] *If* $\mathfrak{A}_{L^\top}^{\mathrm{dis}} \neq \emptyset$ *there exists* $\delta \in (0,1)$ *depending only on n, L, the operator norms of* $\mathcal{M}_{\partial\Omega}$ *on* $\mathbb{X}$, $\mathbb{X}'$, *and the Ahlfors regularity constants of* $\partial\Omega$ *so that if* $\|\nu\|_{[\mathrm{BMO}(\partial\Omega,\sigma)]^n} < \delta$ *(i.e., if* Ω *is a* δ-AR *domain in the sense of Definition 3.1.8) then the Dirichlet Problem* (8.4.273) *has at most one solution.*

(c) [Well-Posedness] *If* $\mathfrak{A}_L^{\mathrm{dis}} \neq \emptyset$ *and* $\mathfrak{A}_{L^\top}^{\mathrm{dis}} \neq \emptyset$ *then there exists* $\delta \in (0,1)$ *which depends only on n, L, the operator norms of* $\mathcal{M}_{\partial\Omega}$ *on* $\mathbb{X}$, $\mathbb{X}'$, *and the Ahlfors regularity constants of* $\partial\Omega$ *such that if* $\|\nu\|_{[\mathrm{BMO}(\partial\Omega,\sigma)]^n} < \delta$ *(i.e., if* Ω *is a* δ-AR *domain in the sense of Definition 3.1.8) then the Dirichlet Problem* (8.4.273) *is uniquely solvable and the solution satisfies* (8.4.275).

(d) [Sharpness] *If* $\mathfrak{A}_L^{\mathrm{dis}} = \emptyset$ *then the Dirichlet Problem* (8.4.273) *may fail to be solvable (actually for boundary data belonging to an infinite dimensional subspace of the corresponding homogeneous Sobolev space styled after the given*

Generalized Banach Function Space). Also, if $\mathfrak{A}^{\rm dis}_{L^\top} = \varnothing$ *then the Dirichlet Problem* (8.4.273) *may have more than one solution (in fact, the linear space of null-solutions may actually be infinite dimensional).*

(e) [Other Function Spaces] *Similar results to the ones described above are true with* $\mathbb{X}$ *replaced throughout by* $\mathbb{X}'$*, or by* $\mathring{\mathbb{X}} := \overline{L^\infty_{\rm comp}(\partial\Omega,\sigma)}^{\|\cdot\|_{\mathbb{X}}}$.

Proof In view of the extrapolation theory developed in [130, §5.2], the commutator estimates from [131, Theorem 2.7.8], and the operator estimates for the chord-dot-normal singular integral operators in the last part of Theorem 4.1.3, we can follow the blueprint of the proof of [111, Theorem 8.18, pp. 546–547] to obtain all desired conclusions. □

Various choices of the Generalized Banach Function Space $\mathbb{X}$ naturally present themselves. For example, well-posedness results as in Theorem 8.4.40 are valid for the Dirichlet Problem with boundary data (and the size condition for the solution) in variable exponent Lebesgue spaces by choosing $\mathbb{X} := L^{p(\cdot)}(\partial\Omega,\sigma)$ where the exponent function $p(\cdot):\partial\Omega\to[p_-,p_+]$, for some $p_-,p_+\in(1,\infty)$, satisfies a log-Hölder continuity condition both at finite distance and at infinity.

We also wish to note that the assumption made in (8.4.272) has a "structural" character, as far as the integrity of the conclusions in Theorem 8.4.40 is concerned. Indeed, even for $L := \Delta$ in flat domains, say for the upper-half space $\Omega := \mathbb{R}^n_+$, there exists a Generalized Banach Function Space $\mathbb{X}$ over $(\mathbb{R}^{n-1},\mathcal{L}^{n-1})$ such that the corresponding problem (8.4.273) is not solvable. This may be solely attributed to the fact that a portion of (8.4.272) fails in this case. More specifically, making the choice $\mathbb{X} := L^1(\mathbb{R}^{n-1},\mathcal{L}^{n-1})$ leads to a Dirichlet Problem whose space of admissible boundary data $\Big\{u\big|^{\kappa-{\rm n.t.}}_{\partial\mathbb{R}^n_+} : u\in\mathscr{C}^\infty(\mathbb{R}^n_+),\ \Delta u=0,\ \mathcal{N}_\kappa u\in L^1(\mathbb{R}^{n-1},\mathcal{L}^{n-1})\Big\}$ is precisely the Hardy space $H^1(\mathbb{R}^{n-1},\mathcal{L}^{n-1})$ (cf. [60, Theorems 9 and 11]), which is strictly smaller than the full space of boundary data $L^1(\mathbb{R}^{n-1},\mathcal{L}^{n-1})$. As regards the hypothesis (8.4.272), while the Hardy-Littlewood maximal operator is bounded on $\big(L^1(\mathbb{R}^{n-1},\mathcal{L}^{n-1})\big)' = L^\infty(\mathbb{R}^{n-1},\mathcal{L}^{n-1})$, it fails to map $L^1(\mathbb{R}^{n-1},\mathcal{L}^{n-1})$ to itself, and this is the root of the aforementioned failure of solvability.

We continue by discussing the Inhomogeneous Regularity Problem for weakly elliptic systems in δ-AR domains with data in Generalized Banach Function Spaces.

Theorem 8.4.41 *Fix* $n\in\mathbb{N}$ *with* $n\geq 2$ *and let* $\Omega\subseteq\mathbb{R}^n$ *be an Ahlfors regular domain. Abbreviate* $\sigma := \mathcal{H}^{n-1}\lfloor\partial\Omega$*, denote by* ν *the geometric measure theoretic outward unit normal to* Ω*, and fix some aperture parameter* $\kappa\in(0,\infty)$*. Let* $\mathbb{X}$ *be a Generalized Banach Function Space on* $(\partial\Omega,\sigma)$*, of the sort introduced in* [130, §5.1], *and denote by* $\mathbb{X}'$ *its associated space (aka Köthe dual; cf.* [130, Definitions 5.1.4, 5.1.11]). *With* $\mathcal{M}_{\partial\Omega}$ *denoting the Hardy-Littlewood maximal operator on* $(\partial\Omega,\sigma)$*, assume that*

$$\mathcal{M}_{\partial\Omega}:\mathbb{X}\to\mathbb{X} \text{ and } \mathcal{M}_{\partial\Omega}:\mathbb{X}'\to\mathbb{X}' \text{ are well-defined and bounded.} \tag{8.4.279}$$

Recall that $\mathbb{X}_1$ *is the* $\mathbb{X}$*-styled Sobolev space of order one on* $\partial\Omega$ *from* (8.4.276). *Finally, given a homogeneous, second-order, constant complex coefficient, weakly elliptic* $M\times M$ *system* L *in* $\mathbb{R}^n$*, consider the Inhomogeneous Regularity Problem*

$$\begin{cases} u \in \big[\mathscr{C}^\infty(\Omega)\big]^M, & Lu = 0 \text{ in } \Omega, \\ \mathcal{N}_\kappa u,\ \mathcal{N}_\kappa(\nabla u) \in \mathbb{X}, & u\big|_{\partial\Omega}^{\kappa-\mathrm{n.t.}} = f \in \big[\mathbb{X}_1\big]^M. \end{cases} \tag{8.4.280}$$

Then, in relation to this, the following statements are true:

(a) [Existence and Estimates] *If* $\mathfrak{A}_L^{\mathrm{dis}} \neq \varnothing$ *and* $A \in \mathfrak{A}_L^{\mathrm{dis}}$, *then there exists some* $\delta \in (0,1)$ *which depends only on n, A, the operator norms of* $\mathcal{M}_{\partial\Omega}$ *on* $\mathbb{X}$, $\mathbb{X}'$, *and the Ahlfors regularity constants of* $\partial\Omega$ *such that if* $\|\nu\|_{[\mathrm{BMO}(\partial\Omega,\sigma)]^n} < \delta$ *(i.e., if* Ω *is a* δ-AR *domain in the sense of Definition 3.1.8) then* $\frac{1}{2}I + K_A$ *is an invertible operator on* $\big[\mathbb{X}_1\big]^M$ *and the function*

$$u(x) := \Big(\mathcal{D}_A\Big(\tfrac{1}{2}I + K_A\Big)^{-1} f\Big)(x), \qquad \forall\, x \in \Omega, \tag{8.4.281}$$

is a solution of the Inhomogeneous Regularity Problem (8.4.280). *In addition,*

$$\|\mathcal{N}_\kappa u\|_{\mathbb{X}} \approx \|f\|_{[\mathbb{X}]^M} \ \text{ and } \ \|\mathcal{N}_\kappa u\|_{\mathbb{X}} + \|\mathcal{N}_\kappa(\nabla u)\|_{\mathbb{X}} \approx \|f\|_{[\mathbb{X}_1]^M}. \tag{8.4.282}$$

(b) [Uniqueness] *Whenever* $\mathfrak{A}_{L^\top}^{\mathrm{dis}} \neq \varnothing$, *there exists* $\delta \in (0,1)$ *which depends only on n, L, the operator norms of* $\mathcal{M}_{\partial\Omega}$ *on* $\mathbb{X}$, $\mathbb{X}'$, *and the Ahlfors regularity constants of* $\partial\Omega$ *such that if* $\|\nu\|_{[\mathrm{BMO}(\partial\Omega,\sigma)]^n} < \delta$ *(i.e., if* Ω *is a* δ-AR *domain in the sense of Definition 3.1.8) then the Inhomogeneous Regularity Problem* (8.4.280) *has at most one solution.*

(c) [Well-Posedness] *If* $\mathfrak{A}_L^{\mathrm{dis}} \neq \varnothing$ *and* $\mathfrak{A}_{L^\top}^{\mathrm{dis}} \neq \varnothing$ *then there exists* $\delta \in (0,1)$ *which depends only on n, L, the operator norms of* $\mathcal{M}_{\partial\Omega}$ *on* $\mathbb{X}$, $\mathbb{X}'$, *and the Ahlfors regularity constants of* $\partial\Omega$ *such that if* $\|\nu\|_{[\mathrm{BMO}(\partial\Omega,\sigma)]^n} < \delta$ *(i.e., if* Ω *is a* δ-AR *domain in the sense of Definition 3.1.8) then the Inhomogeneous Regularity Problem* (8.4.280) *is uniquely solvable and the solution satisfies* (8.4.282).

(d) [Sharpness] *If* $\mathfrak{A}_L^{\mathrm{dis}} = \varnothing$ *the Inhomogeneous Regularity Problem* (8.4.280) *may fail to be solvable (actually for boundary data belonging to an infinite dimensional subspace of* $[\mathbb{X}_1]^M$*) even when* Ω *is a half-space, and if* $\mathfrak{A}_{L^\top}^{\mathrm{dis}} = \varnothing$ *the Inhomogeneous Regularity Problem* (8.4.280) *may posses more than one solution (in fact, the linear space of null-solutions may actually be infinite dimensional) even when* Ω *is a half-space. In particular, if either* $\mathfrak{A}_L^{\mathrm{dis}} = \varnothing$, *or* $\mathfrak{A}_{L^\top}^{\mathrm{dis}} = \varnothing$, *then the Inhomogeneous Regularity Problem* (8.4.280) *may fail to be well posed even when* Ω *is a half-space.*

(e) [Other Function Spaces] *Analogous results to those described above are valid with* $\mathbb{X}$ *replaced throughout by* $\mathbb{X}'$, *or by* $\mathring{\mathbb{X}} := \overline{L^\infty_{\mathrm{comp}}(\partial\Omega,\sigma)}^{\|\cdot\|_{\mathbb{X}}}$.

Proof All desired conclusions may be justified by reasoning as in the proof of [111, Theorem 8.19, pp. 554–555], keeping in mind the invertibility results from item *(14)* in Theorem 4.2.1, the embeddings in [130, Proposition 5.2.7], the extrapolation theory developed in [130, §5.2], and the mapping properties of boundary layer potentials on Generalized Banach Function Spaces. □

We next proceed to formulate and solve the Homogeneous Regularity Problem with boundary data from homogeneous Sobolev spaces styled after Generalized Banach Function Spaces. This augments solvability results established earlier in Theorem 8.4.40 and Theorem 8.4.41.

Theorem 8.4.42 *Fix $n \in \mathbb{N}$ with $n \geq 2$ and let $\Omega \subseteq \mathbb{R}^n$ be an Ahlfors regular domain. Abbreviate $\sigma := \mathcal{H}^{n-1}\lfloor\partial\Omega$, denote by ν the geometric measure theoretic outward unit normal to Ω, and fix an aperture parameter $\kappa > 0$. Let $\mathbb{X}$ be a Generalized Banach Function Space on $(\partial\Omega, \sigma)$, of the sort introduced in* [130, §5.1], *and denote by $\mathbb{X}'$ its associated space (aka Köthe dual; cf.* [130, Definitions 5.1.4, 5.1.11]). *With $\mathcal{M}_{\partial\Omega}$ denoting the Hardy-Littlewood maximal operator on $(\partial\Omega, \sigma)$, assume that*

$$\mathcal{M}_{\partial\Omega} : \mathbb{X} \to \mathbb{X} \text{ and } \mathcal{M}_{\partial\Omega} : \mathbb{X}' \to \mathbb{X}' \text{ are well-defined and bounded.} \tag{8.4.283}$$

Recall from (4.1.319) *the homogeneous Sobolev space $\dot{\mathbb{X}}_1$, styled after the Generalized Banach Function Space $\mathbb{X}$ as*

$$\dot{\mathbb{X}}_1 := \Big\{\phi \in L^1\big(\partial\Omega, \tfrac{\sigma(x)}{1+|x|^n}\big) \cap \mathbb{X}_{\rm loc} : \partial_{\tau_{jk}}\phi \in \mathbb{X} \text{ for } j,k \in \{1,\dots,n\}\Big\}, \tag{8.4.284}$$

where the membership $\phi \in \mathbb{X}_{\rm loc}$ means that $\phi \cdot \mathbf{1}_K \in \mathbb{X}$ for every compact set $K \subseteq \partial\Omega$. This space is equipped with the semi-norm (cf. (8.4.285))

$$\dot{\mathbb{X}}_1 \ni \phi \longmapsto \|\phi\|_{\dot{\mathbb{X}}_1} := \sum_{j,k=1}^{n} \|\partial_{\tau_{jk}}\phi\|_{\mathbb{X}}. \tag{8.4.285}$$

For a given homogeneous, second-order, constant complex coefficient, weakly elliptic $M \times M$ system L in $\mathbb{R}^n$, consider the Homogeneous Regularity Problem

$$\begin{cases} u \in \big[\mathscr{C}^\infty(\Omega)\big]^M, \quad Lu = 0 \text{ in } \Omega, \\ \mathcal{N}_\kappa(\nabla u) \in \mathbb{X}, \quad u\big|_{\partial\Omega}^{\kappa-\text{n.t.}} = f \in \big[\dot{\mathbb{X}}_1\big]^M. \end{cases} \tag{8.4.286}$$

In relation to this, the following statements are valid:

(a) [Existence, Estimates, and Integral Representations] *If $\mathfrak{A}_L^{\rm dis} \neq \varnothing$ then there exists $\delta \in (0,1)$ which depends only on n, L, the operator norms of $\mathcal{M}_{\partial\Omega}$ on $\mathbb{X}$, $\mathbb{X}'$, and the Ahlfors regularity constants of $\partial\Omega$ with the property that if $\|\nu\|_{[\mathrm{BMO}(\partial\Omega,\sigma)]^n} < \delta$ (a scenario which ensures that Ω is a δ-*AR *domain in the sense of Definition 3.1.8) then the following properties are true. First,*

$$\big[S_{\rm mod}\big] : \big[\mathbb{X}\big]^M \longrightarrow \big[\dot{\mathbb{X}}_1/\sim\big]^M \tag{8.4.287}$$

is a well-defined linear isomorphism and the Homogeneous Regularity Problem (8.4.286) *is solvable. More specifically, with $[f] \in \big[\dot{\mathbb{X}}_1/\sim\big]^M$ denoting the equivalence class (modulo constants) of the boundary datum f, and with*

$$g \in \big[\mathbb{X}\big]^M \text{ selected so that } \big[S_{\rm mod}\big]g = [f], \tag{8.4.288}$$

there exists a constant $c \in \mathbb{C}^M$ *such that the function*

$$u := \mathscr{S}_{\rm mod} g + c \ \text{ in } \ \Omega \tag{8.4.289}$$

is a solution of the Homogeneous Regularity Problem (8.4.286)*. In addition, this solution satisfies (with implicit constants independent of* f*)*

$$\|\mathcal{N}_\kappa(\nabla u)\|_{\mathbb{X}} \approx \|\nabla_{\tan} f\|_{[\mathbb{X}]^{n\cdot M}}. \tag{8.4.290}$$

Second, for each coefficient tensor $A \in \mathfrak{A}_L^{\rm dis}$ *one has the isomorphism*

$$\tfrac{1}{2}I + \big[K_{A,\rm mod}\big] : \big[\dot{\mathbb{X}}_1/\sim\big]^M \longrightarrow \big[\dot{\mathbb{X}}_1/\sim\big]^M, \tag{8.4.291}$$

and the Homogeneous Regularity Problem (8.4.286) *may be solved as*

$$u := \mathcal{D}_{A,\rm mod} h + c \ \text{ in } \ \Omega, \tag{8.4.292}$$

for a suitable constant $c \in \mathbb{C}^M$ *and with*

$$h \in \big[\dot{\mathbb{X}}_1\big]^M \text{ such that } [h] = \Big(\tfrac{1}{2}I + \big[K_{A,\rm mod}\big]\Big)^{-1}[f]. \tag{8.4.293}$$

Moreover, this solution continues to satisfy (8.4.290).

(b) [Uniqueness] *Whenever* $\mathfrak{A}_{L^\top}^{\rm dis} \neq \varnothing$*, there exists* $\delta \in (0,1)$ *which depends only on* n*,* L*, the operator norms of* $\mathcal{M}_{\partial\Omega}$ *on* $\mathbb{X}$*,* $\mathbb{X}'$*, and the Ahlfors regularity constants of* $\partial\Omega$ *with the property that if* $\|\nu\|_{[\mathrm{BMO}(\partial\Omega,\sigma)]^n} < \delta$ *(i.e., if* Ω *is a* δ-AR *domain in the sense of Definition 3.1.8) then the Homogeneous Regularity Problem* (8.4.286) *has at most one solution.*

(c) [Well-Posedness and Additional Integrability/Regularity] *Whenever* $\mathfrak{A}_L^{\rm dis} \neq \varnothing$ *and* $\mathfrak{A}_{L^\top}^{\rm dis} \neq \varnothing$ *it follows that there exists* $\delta \in (0,1)$ *which depends only on* n*,* L*, the operator norms of* $\mathcal{M}_{\partial\Omega}$ *on* $\mathbb{X}$*,* $\mathbb{X}'$*, and the Ahlfors regularity constants of* $\partial\Omega$ *with the property that if* $\|\nu\|_{[\mathrm{BMO}(\partial\Omega,\sigma)]^n} < \delta$ *(i.e., if* Ω *is a* δ-AR *domain in the sense of Definition 3.1.8) then the Homogeneous Regularity Problem* (8.4.286) *is uniquely solvable. Moreover, the unique solution* u *of* (8.4.286) *satisfies (in a quantitative fashion)*

$$\mathcal{N}_\kappa u \in \mathbb{X} \Longleftrightarrow f \in \big[\mathbb{X}_1\big]^M, \tag{8.4.294}$$

where $\mathbb{X}_1$ *stands for the* $\mathbb{X}$*-styled Sobolev space of order one on* $\partial\Omega$ *defined in* (8.4.276)*. In particular, the equivalence in* (8.4.294) *proves that the unique solution of the Homogeneous Regularity Problem* (8.4.286) *for a boundary datum* f *belonging to* $\big[\mathbb{X}_1\big]^M$ *(which is a subspace of* $\big[\dot{\mathbb{X}}_1\big]^M$*) is actually the unique solution of the Inhomogeneous Regularity Problem* (8.4.280) *for the boundary datum* f*.*

(d) [Sharpness] *If* $\mathfrak{A}_L^{\rm dis} = \varnothing$ *the Homogeneous Regularity Problem* (8.4.286) *may fail to be solvable (actually for boundary data belonging to an infinite dimensional*

subspace of the corresponding homogeneous Sobolev space styled after the given Generalized Banach Function Space), and if $\mathfrak{A}^{\text{dis}}_{L^\top} = \varnothing$ *the Homogeneous Regularity Problem* (8.4.286) *may posses more than one solution (in fact, the linear space of null-solutions may actually be infinite dimensional), even in the case when* $\Omega = \mathbb{R}^n_+$. *In particular, if either* $\mathfrak{A}^{\text{dis}}_{L} = \varnothing$, *or* $\mathfrak{A}^{\text{dis}}_{L^\top} = \varnothing$, *then the Homogeneous Regularity Problem* (8.4.286) *may fail to be well posed, again, even in the case when* $\Omega = \mathbb{R}^n_+$.

(e) [Other Function Spaces] *Similar results to the ones above holds with* $\mathbb{X}$ *replaced throughout by* $\mathbb{X}'$, *or by* $\mathring{\mathbb{X}} := \overline{L^\infty_{\text{comp}}(\partial\Omega, \sigma)}^{\|\cdot\|_{\mathbb{X}}}$.

Proof By availing ourselves of the extrapolation theory developed in [130, §5.2], the commutator estimates from [131, Theorem 2.7.8], and the operator estimates for the chord-dot-normal singular integral operators in the last part of Theorem 4.1.3, we can parallel the proof of [111, Theorem 8.20, pp. 555–557] to justify all claims in the statement. □

The next goal is to formulate and solve the Neumann Problem for weakly elliptic systems in δ-AR domains with data in Generalized Banach Function Spaces.

Theorem 8.4.43 *Let* $\Omega \subseteq \mathbb{R}^n$ *be a* UR *domain. Denote by* ν *the geometric measure theoretic outward unit normal to* Ω, *abbreviate* $\sigma := \mathcal{H}^{n-1}\lfloor\partial\Omega$, *and fix an aperture parameter* $\kappa > 0$. *Let* $\mathbb{X}$ *be a Generalized Banach Function Space on* $(\partial\Omega, \sigma)$, *of the sort introduced in* [130, §5.1], *and denote by* $\mathbb{X}'$ *its associated space (aka Köthe dual; cf.* [130, Definitions 5.1.4, 5.1.11]). *With* $\mathcal{M}_{\partial\Omega}$ *denoting the Hardy-Littlewood maximal operator on* $(\partial\Omega, \sigma)$, *assume that*

$$\mathcal{M}_{\partial\Omega} : \mathbb{X} \longrightarrow \mathbb{X} \text{ and } \mathcal{M}_{\partial\Omega} : \mathbb{X}' \longrightarrow \mathbb{X}' \text{ are well-defined and bounded.} \tag{8.4.295}$$

Next, suppose L *is a homogeneous, second-order, constant complex coefficient, weakly elliptic* $M \times M$ *system in* $\mathbb{R}^n$. *Finally, select some coefficient tensor* $A \in \mathfrak{A}_L$ *and consider the Neumann Problem*

$$\begin{cases} u \in \big[\mathscr{C}^\infty(\Omega)\big]^M, & Lu = 0 \text{ in } \Omega, \\ \mathcal{N}_\kappa(\nabla u) \in \mathbb{X}, & \partial^A_\nu u = f \in \big[\mathbb{X}\big]^M. \end{cases} \tag{8.4.296}$$

Then the following statements are valid:

(a) [Existence, Estimates, and Integral Representation] *If* $A^\top \in \mathfrak{A}^{\text{dis}}_{L^\top}$ *then there exists* $\delta \in (0,1)$ *which depends only on n, A, the operator norms of* $\mathcal{M}_{\partial\Omega}$ *on* $\mathbb{X}$, $\mathbb{X}'$, *and the Ahlfors regularity constants of* $\partial\Omega$ *such that if* $\|\nu\|_{[\text{BMO}(\partial\Omega,\sigma)]^n} < \delta$ *(i.e., if* Ω *is a* δ-AR *domain in the sense of Definition 3.1.8) then* $-\frac{1}{2}I + K^\#_{A^\top}$ *is an invertible operator on* $\big[\mathbb{X}\big]^M$ *and the function* $u : \Omega \to \mathbb{C}^M$ *defined as*

$$u(x) := \Big(\mathscr{S}_{\text{mod}}\Big(-\tfrac{1}{2}I + K^\#_{A^\top}\Big)^{-1} f\Big)(x) \text{ for all } x \in \Omega, \tag{8.4.297}$$

is a solution of the Neumann Problem (8.4.296) *which satisfies*

$$\|\mathcal{N}_\kappa(\nabla u)\|_{\mathbb{X}} \leq C\|f\|_{[\mathbb{X}]^M}. \tag{8.4.298}$$

for some constant $C \in (0, \infty)$ independent of f. Also, the operator

$$\begin{aligned} &\partial_\nu^A \mathcal{D}_{A,\text{mod}} : \big[\dot{\mathbb{X}}_1/\sim\big]^M \longrightarrow \big[\mathbb{X}\big]^M \text{ defined as} \\ &(\partial_\nu^A \mathcal{D}_{A,\text{mod}})[f] := \partial_\nu^A(\mathcal{D}_{A,\text{mod}} f) \text{ for each } f \in \big[\dot{\mathbb{X}}_1\big]^M \end{aligned} \tag{8.4.299}$$

is an isomorphism which implies that

$$\begin{aligned} &\text{there exists } g \in [\dot{\mathbb{X}}_1]^M \text{ with } \partial_\nu^A(\mathcal{D}_{A,\text{mod}} g) = f \text{ and such} \\ &\text{that } \|g\|_{[\dot{\mathbb{X}}_1]^M} \leq C\|f\|_{[\mathbb{X}]^M} \text{ for a constant } C \in (0,\infty). \end{aligned} \tag{8.4.300}$$

Consequently, the function $u := \mathcal{D}_{A,\text{mod}} g$ in Ω is a solution of the Neumann Problem (8.4.296) *which continues to satisfy* (8.4.298).

(b) [Uniqueness (modulo constants)] *Whenever $A \in \mathfrak{A}_L^{\text{dis}}$ there exists $\delta \in (0, 1)$ which depends only on n, A, the operator norms of $\mathcal{M}_{\partial\Omega}$ on $\mathbb{X}$, $\mathbb{X}'$, and the Ahlfors regularity constants of $\partial\Omega$ such that if $\|\nu\|_{[\text{BMO}(\partial\Omega,\sigma)]^n} < \delta$ (i.e., if Ω is a δ-*AR *domain in the sense of Definition 3.1.8) then any two solutions of the Neumann Problem* (8.4.296) *differ by a constant from $\mathbb{C}^M$.*

(c) [Well-Posedness] *Whenever $A \in \mathfrak{A}_L^{\text{dis}}$ and $A^\top \in \mathfrak{A}_{L^\top}^{\text{dis}}$ there exists $\delta \in (0, 1)$ which depends only on n, A, the operator norms of $\mathcal{M}_{\partial\Omega}$ on $\mathbb{X}$, $\mathbb{X}'$, and the Ahlfors regularity constants of $\partial\Omega$ such that if $\|\nu\|_{[\text{BMO}(\partial\Omega,\sigma)]^n} < \delta$ (i.e., if Ω is a δ-*AR *domain in the sense of Definition 3.1.8) then the Neumann Problem* (8.4.296) *is solvable, the solution is unique modulo constants from $\mathbb{C}^M$, and each solution satisfies* (8.4.298).

(d) [Other Function Spaces] *Analogous results to those described above are valid with $\mathbb{X}$ replaced throughout by $\mathbb{X}'$, or by $\mathring{\mathbb{X}} := \overline{L^\infty_{\text{comp}}(\partial\Omega,\sigma)}^{\|\cdot\|_{\mathbb{X}}}$.*

Proof We may establish all claims in the statement by reasoning as in the proof of [111, Theorem 8.21, pp. 558–559], while bearing in mind the invertibility results from item it (15) in Theorem 4.2.1, as well as the mapping properties for boundary layer potentials from Theorem 4.1.21 and Theorem 4.1.22. □

Finally, we treat the Transmission Problem for weakly elliptic systems in two-sided NTA domains having unbounded Ahlfors regular boundaries, with boundary data in Generalized Banach Function Spaces. The case of the Transmission Problem with boundary data in Lebesgue/Sobolev spaces for the Laplacian in upper-graph Lipschitz domains in $\mathbb{R}^n$, with $n \geq 2$ arbitrary, has been treated in [48]. For the two-dimensional Laplacian in an infinite sector in the plane, counterexamples to the well-posedness of the aforementioned Transmission Problem for certain values of the integrability exponent (related to the aperture of the sector and the transmission parameter appearing in the formulation of the problem) have been presented in [149].

Theorem 8.4.44 *Fix $n \in \mathbb{N}$ with $n \geq 2$ and assume*

$$\Omega \subseteq \mathbb{R}^n \text{ is a two-sided NTA domain with an unbounded Ahlfors regular boundary.} \tag{8.4.301}$$

Denote by ν the geometric measure theoretic outward unit normal to Ω, abbreviate $\sigma := \mathcal{H}^{n-1} \lfloor \partial\Omega$, and set

$$\Omega_+ := \Omega, \quad \Omega_- := \mathbb{R}^n \setminus \overline{\Omega}. \tag{8.4.302}$$

Let $\mathbb{X}$ be a Generalized Banach Function Space on $(\partial\Omega, \sigma)$, of the sort introduced in [130, §5.1], *and denote by $\mathbb{X}'$ its associated space (aka Köthe dual; cf.* [130, Definitions 5.1.4, 5.1.11]). *With $\mathcal{M}_{\partial\Omega}$ denoting the Hardy-Littlewood maximal operator on $(\partial\Omega, \sigma)$, assume that*

$$\mathcal{M}_{\partial\Omega} : \mathbb{X} \to \mathbb{X} \text{ and } \mathcal{M}_{\partial\Omega} : \mathbb{X}' \to \mathbb{X}' \text{ are well-defined and bounded.} \tag{8.4.303}$$

Also, fix an aperture parameter $\kappa > 0$, and a transmission (or coupling) parameter $\eta \in \mathbb{C}$. Next, assume L is a homogeneous, second-order, constant complex coefficient, weakly elliptic $M \times M$ system in $\mathbb{R}^n$. Having selected some $A \in \mathfrak{A}_L$, consider the Transmission Problem

$$\begin{cases} u^{\pm} \in \left[\mathscr{C}^\infty(\Omega_{\pm})\right]^M, \ Lu^{\pm} = 0 \text{ in } \Omega_{\pm}, \ \mathcal{N}_\kappa(\nabla u^{\pm}) \in \mathbb{X}, \\ u^+\big|_{\partial\Omega}^{\kappa-\mathrm{n.t.}} - u^-\big|_{\partial\Omega}^{\kappa-\mathrm{n.t.}} = f \in \left[\dot{\mathbb{X}}_1\right]^M, \\ \partial_\nu^A u^+ - \eta \cdot \partial_\nu^A u^- = g \in \left[\mathbb{X}\right]^M. \end{cases} \tag{8.4.304}$$

Finally, suppose that either

$$A^\top \in \mathfrak{A}^{\mathrm{dis}}_{L^\top} \text{ and } \eta \in \mathbb{C} \setminus \{-1\}, \tag{8.4.305}$$

or

$$A \in \mathfrak{A}^{\mathrm{dis}}_{L} \text{ and } \eta \in \mathbb{C} \setminus \{0, -1\}. \tag{8.4.306}$$

Then for each $m \in \mathbb{N}$ there exists some small $\delta \in (0, \infty)$, which depends only on n, m, A, the two-sided NTA constants of Ω, the Ahlfors regularity constants of $\partial\Omega$, and the operator norms of $\mathcal{M}_{\partial\Omega}$ on $\mathbb{X}$, $\mathbb{X}'$, such that whenever[105]

$$\left|\tfrac{1-\eta}{1+\eta}\right| \cdot \|\nu\|^{\langle m \rangle}_{[\mathrm{BMO}(\partial\Omega,\sigma)]^n} < \delta \tag{8.4.307}$$

it follows that the Transmission Problem (8.4.304) *is well posed. Specifically, in this scenario* (8.4.304) *is always solvable, any two solutions differ by a constant (from $\mathbb{C}^M$), and any solution of* (8.4.304) *satisfies, for some finite constant $C > 0$ independent of the boundary data f and g,*

$$\|\mathcal{N}_\kappa(\nabla u^{\pm})\|_{\mathbb{X}} \leq C\big(\|f\|_{[\dot{\mathbb{X}}_1]^M} + \|g\|_{[\mathbb{X}]^M}\big). \tag{8.4.308}$$

---

[105] using the piece of notation introduced in (4.1.12)

Similar results to the ones above holds with the space $\mathbb{X}$ *replaced throughout by* $\mathbb{X}'$, *or by* $\mathring{\mathbb{X}} := \overline{L^\infty_{\rm comp}(\partial\Omega, \sigma)}^{\|\cdot\|_{\mathbb{X}}}$.

Before presenting its proof, we make several comments elaborating on the nature of the above theorem. First, in the formulation of the boundary conditions in (8.4.304), the existence of the boundary traces is guaranteed by the Fatou-type result from Theorem 4.1.20. The latter theorem also confirms the fact that the boundary data themselves are selected from spaces which are natural (in view of the first three conditions in (8.4.304)).

Second, the (pointwise) conormal derivative $\partial^A_\nu u^+$ is defined as in [131, Definition 1.7.1] relative to the set Ω_+, while in light of the fact that the geometric measure theoretic outward unit normal to Ω_- is $-\nu$, at σ-a.e. point on $\partial\Omega$, we take $\partial^A_\nu u^-$ to be the opposite of (i.e.,-1 times) the conormal derivative operator from [131, Definition 1.7.1] for the set Ω_-, acting on the function u^- (which is defined in said set). That is, we make the convention that

$$\partial^A_\nu u^- := -\partial^A_{(-\nu)} u^-. \tag{8.4.309}$$

Third, in view of (4.1.16), the demand stipulated in (8.4.307) is satisfied if there exists some $\varepsilon \in (0, 1)$ such that

$$\left|\tfrac{1-\eta}{1+\eta}\right|\left(e^{\varepsilon-1}/\varepsilon\right) \cdot \|\nu\|^{1-\varepsilon}_{[\mathrm{BMO}(\partial\Omega,\sigma)]^n} < \delta. \tag{8.4.310}$$

In particular, for any fixed $\eta \in \mathbb{C} \setminus \{-1\}$, we see that (8.4.310) (hence also (8.4.307)) holds if $\|\nu\|_{[\mathrm{BMO}(\partial\Omega,\sigma)]^n}$ is sufficiently small (in a manner which now also depends on η). Under the latter smallness assumption, in place of (8.4.301) we may merely assume that Ω is just an Ahlfors regular domain (cf. Theorem 3.4.2).

Yet another perspective on the demand made in (8.4.307) comes from the realization that said demand is fulfilled whenever the transmission parameter η is sufficiently close to 1, irrespective of the size of $\|\nu\|_{[\mathrm{BMO}(\partial\Omega,\sigma)]^n}$. Remarkably, the Transmission Problem (8.4.304) is well posed in this case without making any assumptions on the flatness of $\partial\Omega$.

Let us now turn to the task of providing the proof of Theorem 8.4.44.

Proof of Theorem 8.4.44 To prove existence, first work under the assumptions made in (8.4.305) and pick an arbitrary $m \in \mathbb{N}$. Henceforth assume (8.4.307) holds for a sufficiently small $\delta \in (0, \infty)$. In such a scenario, we may rely on estimate (4.1.49) from Theorem 4.1.3, the fact that $\mathbb{X}$ is stable under multiplication by essentially bounded functions (cf. [130, (5.1.13)]), and Proposition 1.3.2 to conclude that

$$\left\|\left(\tfrac{1-\eta}{1+\eta}\right)K^{\#}_{A^\top}\right\|_{[\mathbb{X}]^M \to [\mathbb{X}]^M} < \tfrac{1}{2}. \tag{8.4.311}$$

Hence, a Neumann series argument gives that the operator

$$-\tfrac{1}{2}I + \left(\tfrac{1-\eta}{1+\eta}\right)K^{\#}_{A^\top} : [\mathbb{X}]^M \longrightarrow [\mathbb{X}]^M \text{ is invertible.} \tag{8.4.312}$$

As a consequence, the operator

$$-\tfrac{\eta+1}{2}I+(1-\eta)K^{\#}_{A^\top}:\big[\mathbb{X}\big]^M\longrightarrow\big[\mathbb{X}\big]^M \text{ is invertible.} \tag{8.4.313}$$

In particular, it makes sense to consider the functions $u^\pm:\Omega_\pm\to\mathbb{C}^M$ defined as

$$u^\pm:=\pm\mathcal{D}^\pm_{A,\text{mod}}f+\mathcal{S}^\pm_{\text{mod}}h \text{ in } \Omega_\pm, \tag{8.4.314}$$

where the superscripts $\pm$ indicate that the modified boundary-to-domain layer potentials are associated with the sets $\Omega_\pm$ and where

$$h:=\big(-\tfrac{\eta+1}{2}I+(1-\eta)K^{\#}_{A^\top}\big)^{-1}\big(g-(1-\eta)(\partial^A_\nu\mathcal{D}_{A,\text{mod}})f\big)\in\big[\mathbb{X}\big]^M. \tag{8.4.315}$$

Thanks to the mapping properties for boundary layer potentials from Theorem 4.1.21 and Theorem 4.1.22, it may be readily verified that $u^\pm$ defined as in (8.4.314)-(8.4.315) enjoy (8.4.308), satisfy the first three conditions in (8.4.304), and we have (here the convention made in (8.4.309) is relevant)

$$\begin{aligned} u^+\big|^{\kappa-\text{n.t.}}_{\partial\Omega}-u^-\big|^{\kappa-\text{n.t.}}_{\partial\Omega} &=\big(\tfrac12 I+K_{A,\text{mod}}\big)f+S_{\text{mod}}h-\big(-\tfrac12 I+K_{A,\text{mod}}\big)f-S_{\text{mod}}h\\ &=f \text{ at } \sigma\text{-a.e. point on } \partial\Omega, \end{aligned} \tag{8.4.316}$$

and, at σ-a.e. point on $\partial\Omega$,

$$\begin{aligned} &\partial^A_\nu u^+-\eta\cdot\partial^A_\nu u^-\\ &\quad=(\partial^A_\nu\mathcal{D}_{A,\text{mod}})f+\big(-\tfrac12 I+K^{\#}_{A^\top}\big)h-\eta(\partial^A_\nu\mathcal{D}_{A,\text{mod}})f-\eta\big(\tfrac12 I+K^{\#}_{A^\top}\big)h\\ &\quad=(1-\eta)(\partial^A_\nu\mathcal{D}_{A,\text{mod}})f+\big(-\tfrac{\eta+1}{2}I+(1-\eta)K^{\#}_{A^\top}\big)h=g, \end{aligned} \tag{8.4.317}$$

where the last equality comes from (8.4.315). Consequently, $u^\pm$ from (8.4.320)-(8.4.321) solve the Transmission Problem (8.4.304) and satisfy (8.4.308). This establishes existence and estimates in the current case.

Consider next the task of proving existence for the Transmission Problem (8.4.304) under the assumptions made in (8.4.306). Again, pick an arbitrary $m\in\mathbb{N}$ and assume (8.4.307) holds for a sufficiently small $\delta\in(0,\infty)$. In this case, we may employ estimate (4.1.357) to conclude that

$$\Big\|\big(\tfrac{1-\eta}{1+\eta}\big)\big[K_{A,\text{mod}}\big]\Big\|_{[\dot{\mathbb{X}}_1/\sim]^M\to[\dot{\mathbb{X}}_1/\sim]^M}<\tfrac12. \tag{8.4.318}$$

Then a Neumann series argument shows that the operator

$$\tfrac12(1+\eta^{-1})I+(1-\eta^{-1})\big[K_{A,\text{mod}}\big]:\big[\dot{\mathbb{X}}_1/\sim\big]^M\longrightarrow\big[\dot{\mathbb{X}}_1/\sim\big]^M \tag{8.4.319}$$

is an isomorphism. In turn, this permits us to consider

$$\begin{aligned} u^+&:=\mathcal{D}^+_{A,\text{mod}}h-\mathcal{S}^+_{\text{mod}}g-c \text{ in } \Omega_+,\\ u^-&:=-\eta^{-1}\mathcal{D}^-_{A,\text{mod}}h-\eta^{-1}\mathcal{S}^-_{\text{mod}}g \text{ in } \Omega_-, \end{aligned} \tag{8.4.320}$$

where now the function $h \in \big[\dot{\mathbb{X}}_1\big]^M$ and the constant $c \in \mathbb{C}^M$ are chosen so that

$$\Big(\tfrac{1}{2}\big(1+\eta^{-1}\big)I + \big(1-\eta^{-1}\big)K_{A,\mathrm{mod}}\Big)h = f + \big(1-\eta^{-1}\big)S_{\mathrm{mod}}g + c. \tag{8.4.321}$$

Bearing in mind the mapping properties for boundary layer potentials from Theorem 4.1.21 and Theorem 4.1.22, we may now check that $u^\pm$ defined in (8.4.320)-(8.4.321) satisfy (8.4.308) plus the first three conditions in (8.4.304), and we have (again, here the convention made in (8.4.309) is relevant)

$$\begin{aligned} u^+\Big|_{\partial\Omega}^{\kappa-\mathrm{n.t.}} - u^-\Big|_{\partial\Omega}^{\kappa-\mathrm{n.t.}} &= \big(\tfrac{1}{2}I + K_{A,\mathrm{mod}}\big)h - S_{\mathrm{mod}}g - c - \eta^{-1}\big(-\tfrac{1}{2}I + K_{A,\mathrm{mod}}\big)h + \eta^{-1}S_{\mathrm{mod}}g \\ &= \Big(\tfrac{1}{2}\big(1+\eta^{-1}\big)I + \big(1-\eta^{-1}\big)K_{A,\mathrm{mod}}\Big)h - (1-\eta^{-1})S_{\mathrm{mod}}g - c \\ &= f \text{ at } \sigma\text{-a.e. point on } \partial\Omega, \end{aligned} \tag{8.4.322}$$

thanks to (8.4.321), as well as

$$\begin{aligned} \partial_\nu^A u^+ - \eta\cdot\partial_\nu^A u^- &= (\partial_\nu^A \mathcal{D}_{A,\mathrm{mod}})h - \big(-\tfrac{1}{2}I + K^\#_{A^\top}\big)g - (\partial_\nu^A \mathcal{D}_{A,\mathrm{mod}})h + \big(\tfrac{1}{2}I + K^\#_{A^\top}\big)g \\ &= g \text{ at } \sigma\text{-a.e. point on } \partial\Omega. \end{aligned} \tag{8.4.323}$$

Hence, $u^\pm$ from (8.4.320)-(8.4.321) solve the Transmission Problem (8.4.304) and satisfy (8.4.308). This takes care of existence and estimates.

As far as the issue of uniqueness is concerned, the goal is to show that

if $u^\pm$ solve the homogeneous version of the Transmission Problem (8.4.304) (corresponding to having $f = 0$ and $g = 0$) then there exists a constant $c \in \mathbb{C}^M$ such that $u^\pm = c$ in $\Omega_\pm$. (8.4.324)

For this, we shall present two proofs.

The first proof proceeds as follows. For starters, the first boundary condition in (8.4.304) presently reads

$$u^+\Big|_{\partial\Omega}^{\kappa-\mathrm{n.t.}} = u^-\Big|_{\partial\Omega}^{\kappa-\mathrm{n.t.}} \text{ at } \sigma\text{-a.e. point on } \partial\Omega. \tag{8.4.325}$$

Recall that we are assuming (8.4.301). Bearing this in mind, [129, Proposition 8.4.9], [130, Proposition 11.3.2], the embeddings in [130, Lemma 5.2.5], the Fatou-type result from Theorem 4.1.20, and (8.4.325) allow us to write

$$\begin{aligned} \nu_j\Big((\partial_k u^+)\Big|_{\partial\Omega}^{\kappa-\mathrm{n.t.}}\Big) - \nu_k\Big((\partial_j u^+)\Big|_{\partial\Omega}^{\kappa-\mathrm{n.t.}}\Big) &= \partial_{\tau_{jk}}\Big(u^+\Big|_{\partial\Omega}^{\kappa-\mathrm{n.t.}}\Big) \\ = \partial_{\tau_{jk}}\Big(u^-\Big|_{\partial\Omega}^{\kappa-\mathrm{n.t.}}\Big) &= \nu_j\Big((\partial_k u^-)\Big|_{\partial\Omega}^{\kappa-\mathrm{n.t.}}\Big) - \nu_k\Big((\partial_j u^-)\Big|_{\partial\Omega}^{\kappa-\mathrm{n.t.}}\Big) \end{aligned} \tag{8.4.326}$$

at σ-a.e. point on $\partial\Omega$, for each $j, k \in \{1, \dots, n\}$. For ease of notation, we henceforth agree to abbreviate

$$\begin{aligned}&(\partial_{\tau_{jk}}u_\alpha)(x) := \nu_j(x)\big((\partial_k u_\alpha)\big|^{\kappa-\mathrm{n.t.}}_{\partial\Omega}\big)(x) - \nu_k(x)\big((\partial_j u_\alpha)\big|^{\kappa-\mathrm{n.t.}}_{\partial\Omega}\big)(x)\\ &\text{for each } j,k \in \{1,\dots,n\},\ \alpha \in \{1,\dots,M\}, \text{ and } \sigma\text{-a.e. } x \in \partial\Omega.\end{aligned} \tag{8.4.327}$$

In terms of this abbreviation, we may re-cast (8.4.326) simply as

$$\begin{aligned}&\partial_{\tau_{jk}}u^+_\alpha = \partial_{\tau_{jk}}u^-_\alpha \text{ at } \sigma\text{-a.e. point on } \partial\Omega,\\ &\text{for each } j,k \in \{1,\dots,n\} \text{ and } \alpha \in \{1,\dots,M\}.\end{aligned} \tag{8.4.328}$$

Also, from [131, (1.5.230)] (written for u^+ and Ω_+), (8.4.328), the fact that we are presently assuming

$$\partial^A_\nu u^+ = \eta \cdot \partial^A_\nu u^-, \tag{8.4.329}$$

and [131, (1.5.232)] (written for u^- and Ω_-) we see that for each pair of indices, $\ell \in \{1,\dots,n\}$ and $\gamma \in \{1,\dots,M\}$, and each point $x \in \Omega$ we have

$$\begin{aligned}(\partial_\ell u^+_\gamma)(x) &= \int_{\partial\Omega} a^{\beta\alpha}_{rs}(\partial_r E_{\gamma\beta})(x-y)\big(\partial_{\tau_{\ell s}}u^+_\alpha\big)(y)\,\mathrm{d}\sigma(y)\\ &\quad - \int_{\partial\Omega}(\partial_\ell E_{\gamma\alpha})(x-y)\big(\partial^A_\nu u^+\big)_\alpha(y)\,\mathrm{d}\sigma(y)\\ &= \int_{\partial\Omega} a^{\beta\alpha}_{rs}(\partial_r E_{\gamma\beta})(x-y)\big(\partial_{\tau_{\ell s}}u^-_\alpha\big)(y)\,\mathrm{d}\sigma(y)\\ &\quad - \eta\int_{\partial\Omega}(\partial_\ell E_{\gamma\alpha})(x-y)\big(\partial^A_\nu u^-\big)_\alpha(y)\,\mathrm{d}\sigma(y)\\ &= (1-\eta)\int_{\partial\Omega}(\partial_\ell E_{\gamma\alpha})(x-y)\big(\partial^A_\nu u^-\big)_\alpha(y)\,\mathrm{d}\sigma(y).\end{aligned} \tag{8.4.330}$$

Going nontangentially to the boundary in the most extreme sides of (8.4.330) then proves (on account of the Fatou-type result from Theorem 4.1.20 and the jump-formula in [131, Theorem 2.5.1]) that at σ-a.e. $x \in \partial\Omega$ we have

$$\begin{aligned}\Big((\partial_\ell u^+_\gamma)\Big|^{\kappa-\mathrm{n.t.}}_{\partial\Omega}\Big)(x) &= (1-\eta)\frac{1}{2\sqrt{-1}}\widehat{\partial_\ell E_{\gamma\alpha}}(\nu(x))\big(\partial^A_\nu u^-\big)_\alpha(x)\\ &\quad + (1-\eta)\cdot\lim_{\varepsilon\to0^+}\int\limits_{y\in\partial\Omega,\ |x-y|>\varepsilon}(\partial_\ell E_{\gamma\alpha})(x-y)\big(\partial^A_\nu u^-\big)_\alpha(y)\,\mathrm{d}\sigma(y)\end{aligned} \tag{8.4.331}$$

for each $\ell \in \{1,\dots,n\}$ and $\gamma \in \{1,\dots,M\}$. Based on this and (A.0.203), at σ-a.e. point $x \in \partial\Omega$ we may then write

$$\begin{aligned}\big(\partial^A_\nu u^+\big)_\mu(x) &= \nu_r(x)a^{\mu\beta}_{rs}\Big((\partial_s u^+_\beta)\Big|^{\kappa-\mathrm{n.t.}}_{\partial\Omega}\Big)(x)\\ &= (1-\eta)\frac{1}{2\sqrt{-1}}\widehat{\partial_s E_{\beta\alpha}}(\nu(x))\big(\partial^A_\nu u^-\big)_\alpha(x)\nu_r(x)a^{\mu\beta}_{rs}\\ &\quad + (1-\eta)\cdot\lim_{\varepsilon\to0^+}\int\limits_{y\in\partial\Omega,\ |x-y|>\varepsilon}\nu_r(x)a^{\mu\beta}_{rs}(\partial_s E_{\beta\alpha})(x-y)\big(\partial^A_\nu u^-\big)_\alpha(y)\,\mathrm{d}\sigma(y)\end{aligned} \tag{8.4.332}$$

for each $\mu \in \{1, \dots, M\}$. Note that, thanks to [131, (1.4.27)] and (A.0.158), the first line in the right-hand side above becomes

$$\begin{aligned}(1-\eta)&\frac{1}{2\sqrt{-1}}\widehat{\partial_s E_{\beta\alpha}}\big(\nu(x)\big)\big(\partial_\nu^A u^-\big)_\alpha(x)\nu_r(x)a_{rs}^{\mu\beta}\\ &=(1-\eta)\frac{1}{2}\big(a_{rs}^{\mu\beta}\nu_r(x)\nu_s(x)\big)\widehat{E_{\beta\alpha}}\big(\nu(x)\big)\big(\partial_\nu^A u^-\big)_\alpha(x)\\ &=-(1-\eta)\frac{1}{2}\big[L\big(\nu(x)\big)\big]_{\mu\beta}\big[L\big(\nu(x)\big)\big]^{-1}_{\beta\alpha}\big(\partial_\nu^A u^-\big)_\alpha(x)\\ &=-(1-\eta)\frac{1}{2}\delta_{\mu\alpha}\big(\partial_\nu^A u\big)_\alpha(x)=-(1-\eta)\frac{1}{2}\big(\partial_\nu^A u^-\big)_\mu(x),\end{aligned} \tag{8.4.333}$$

at σ-a.e. point $x \in \partial\Omega$, for each $\mu \in \{1, \dots, M\}$. Also, from (A.0.118) and the first formula in [131, (1.4.32)] we see that the second line in the right-hand side of (8.4.332) becomes

$$\lim_{\varepsilon\to 0^+}\int_{\substack{y\in\partial\Omega,\ |x-y|>\varepsilon}}\nu_r(x)a_{rs}^{\mu\beta}(\partial_s E_{\beta\alpha})(x-y)\big(\partial_\nu^A u^-\big)_\alpha(y)\,d\sigma(y)=\Big(K^{\#}_{A^\top}\big(\partial_\nu^A u^-\big)\Big)_\mu(x)$$

at σ-a.e. point $x \in \partial\Omega$, for each $\mu \in \{1, \dots, M\}$. From this, (8.4.332), and (8.4.333) we conclude that

$$\partial_\nu^A u^+ = -(1-\eta)\frac{1}{2}\partial_\nu^A u^- + (1-\eta)K^{\#}_{A^\top}\big(\partial_\nu^A u^-\big) \text{ at } \sigma\text{-a.e. point on } \partial\Omega. \tag{8.4.334}$$

Upon recalling (8.4.329) we finally arrive at

$$\Big(-\tfrac{\eta+1}{2}I + (1-\eta)K^{\#}_{A^\top}\Big)\big(\partial_\nu^A u^-\big) = 0. \tag{8.4.335}$$

Let us now adopt the assumptions made in (8.4.305). In addition, pick an arbitrary $m \in \mathbb{N}$ and assume (8.4.307) holds a sufficiently small $\delta \in (0, \infty)$ (so that (8.4.311) is valid). As before, this implies (8.4.313). Together with (8.4.335) this forces

$$\partial_\nu^A u^- = 0. \tag{8.4.336}$$

Going back with this in (8.4.330) then yields $\nabla u^+ = 0$ in Ω_+. Since it is implicit in (8.4.301) that $\Omega_\pm$ are connected sets, this goes to show that u^+ is a constant in Ω_+, say $u^+ \equiv c \in \mathbb{C}^M$ in Ω_+. Based on this and (8.4.325) we then conclude that

$$u^-\Big|_{\partial\Omega}^{\kappa-\text{n.t.}} = u^+\Big|_{\partial\Omega}^{\kappa-\text{n.t.}} = c \text{ at } \sigma\text{-a.e. point on } \partial\Omega, \tag{8.4.337}$$

hence also

$$\nu_j\Big((\partial_k u^-)\Big|_{\partial\Omega}^{\kappa-\text{n.t.}}\Big) - \nu_k\Big((\partial_j u^-)\Big|_{\partial\Omega}^{\kappa-\text{n.t.}}\Big) = \partial_{\tau_{jk}}\Big(u^-\Big|_{\partial\Omega}^{\kappa-\text{n.t.}}\Big) = 0 \tag{8.4.338}$$

for each $j, k \in \{1, \dots, n\}$ (cf. (8.4.326)). Keeping (8.4.336) and (8.4.338) in mind and writing [131, (1.5.230)] for u^- and Ω_-, we then see that $\nabla u^- = 0$ in Ω_-. As the set Ω_- is connected, it follows that u^- is a constant in Ω_-. In concert with (8.4.337) this shows that $u^- \equiv c$ in Ω_-, as wanted. This establishes uniqueness (modulo constants) under the assumptions made in (8.4.305).

Going further, once more pick an arbitrary $m \in \mathbb{N}$ and assume (8.4.307) holds a sufficiently small $\delta \in (0, \infty)$. Now the goal is to prove uniqueness, modulo constants, for the Transmission Problem (8.4.304), working under the hypotheses adopted in (8.4.306). To this end, suppose $u^\pm$ solve the homogeneous version of (8.4.304). From [131, (1.5.230)] (written for u^+ and Ω_+), (8.4.328), (8.4.329), and [131, (1.5.232)] (written for u^- and Ω_-) we see that for each pair of indices, $\ell \in \{1, \dots, n\}$ and $\gamma \in \{1, \dots, M\}$, and each point $x \in \Omega$ we have

$$\begin{aligned}
(\partial_\ell u^+_\gamma)(x) &= \int_{\partial\Omega} a^{\beta\alpha}_{rs}(\partial_r E_{\gamma\beta})(x-y)\big(\partial_{\tau_{\ell s}} u^+_\alpha\big)(y)\,d\sigma(y) \\
&\quad - \int_{\partial\Omega} (\partial_\ell E_{\gamma\alpha})(x-y)\big(\partial^A_\nu u^+\big)_\alpha(y)\,d\sigma(y) \\
&= \int_{\partial\Omega} a^{\beta\alpha}_{rs}(\partial_r E_{\gamma\beta})(x-y)\big(\partial_{\tau_{\ell s}} u^-_\alpha\big)(y)\,d\sigma(y) \\
&\quad - \eta \int_{\partial\Omega} (\partial_\ell E_{\gamma\alpha})(x-y)\big(\partial^A_\nu u^-\big)_\alpha(y)\,d\sigma(y) \\
&= (1-\eta)\int_{\partial\Omega} a^{\beta\alpha}_{rs}(\partial_r E_{\gamma\beta})(x-y)\big(\partial_{\tau_{\ell s}} u^-_\alpha\big)(y)\,d\sigma(y) \\
&= (1-\eta)\int_{\partial\Omega} a^{\beta\alpha}_{rs}(\partial_r E_{\gamma\beta})(x-y)\big(\partial_{\tau_{\ell s}} u^+_\alpha\big)(y)\,d\sigma(y).
\end{aligned} \tag{8.4.339}$$

Going nontangentially to the boundary in (8.4.339) then shows, on account of the Fatou-type result from Theorem 4.1.20 and the jump-formula in [131, Theorem 2.5.1], that at σ-a.e. $x \in \partial\Omega$ we have

$$\begin{aligned}
\Big((\partial_\ell u^+_\gamma)\big|^{\kappa-\mathrm{n.t.}}_{\partial\Omega}\Big)(x) &= (1-\eta)\frac{1}{2\sqrt{-1}} a^{\beta\alpha}_{rs}\widehat{\partial_r E_{\gamma\beta}}(\nu(x))\big(\partial_{\tau_{\ell s}} u^+_\alpha\big)(x) \\
&\quad + (1-\eta)\cdot \lim_{\varepsilon\to 0^+} \int_{y\in\partial\Omega,\ |x-y|>\varepsilon} a^{\beta\alpha}_{rs}(\partial_r E_{\gamma\beta})(x-y)\big(\partial_{\tau_{\ell s}} u^+_\alpha\big)(y)\,d\sigma(y).
\end{aligned} \tag{8.4.340}$$

For each $r \in \{1, \dots, n\}$ and $\beta \in \{1, \dots, M\}$ we may rely on [131, (1.4.27)] to write

$$\widehat{\partial_r E_{\gamma\beta}}\big(\nu(x)\big) = \sqrt{-1}\,\nu_r(x)\widehat{E_{\gamma\beta}}\big(\nu(x)\big) = \sqrt{-1}\,\nu_r(x)\big[L\big(\nu(x)\big)\big]^{-1}_{\gamma\beta} \tag{8.4.341}$$

at σ-a.e. $x \in \partial\Omega$. Bring in an additional index $t \in \{1, \dots, n\}$. If we now multiply (8.4.340) by $\nu_t(x)$ then subtract from the resulting formula its version with ℓ and t interchanged we then arrive, bearing in mind (8.4.341), (8.4.327), (1.1.34), at

$$
\begin{aligned}
\big(\partial_{T_{t\ell}} u_\gamma^+\big)(x) &= (1-\eta)\frac{1}{2}\big[L\big(\nu(x)\big)\big]_{\beta\alpha}\big[L\big(\nu(x)\big)\big]^{-1}_{\gamma\beta}\big(\partial_{T_{t\ell}} u_\alpha^+\big)(x) \qquad (8.4.342)\\
&\quad - (1-\eta)\cdot \lim_{\varepsilon\to 0^+} \int\limits_{\substack{y\in\partial\Omega\\ |x-y|>\varepsilon}} \nu_s(y) a_{rs}^{\beta\alpha}(\partial_r E_{\gamma\beta})(x-y)\big(\partial_{T_{t\ell}} u_\alpha^+\big)(y)\,d\sigma(y)\\
&\quad + (1-\eta)\cdot \lim_{\varepsilon\to 0^+} \int\limits_{\substack{y\in\partial\Omega\\ |x-y|>\varepsilon}} \big(\nu_t(x)-\nu_t(y)\big) a_{rs}^{\beta\alpha}(\partial_r E_{\gamma\beta})(x-y)\big(\partial_{T_{\ell s}} u_\alpha^+\big)(y)\,d\sigma(y)\\
&\quad + (1-\eta)\cdot \lim_{\varepsilon\to 0^+} \int\limits_{\substack{y\in\partial\Omega\\ |x-y|>\varepsilon}} \big(\nu_\ell(x)-\nu_\ell(y)\big) a_{rs}^{\beta\alpha}(\partial_r E_{\gamma\beta})(x-y)\big(\partial_{T_{st}} u_\alpha^+\big)(y)\,d\sigma(y),
\end{aligned}
$$

valid for each $t, \ell \in \{1,\dots,n\}$, each $\gamma \in \{1,\dots,M\}$, and σ-a.e. $x \in \partial\Omega$. In relation to (8.4.342), we make several observations. For starters, the first line in the right-hand side of (8.4.342) is

$$
\begin{aligned}
&(1-\eta)\frac{1}{2}\big[L\big(\nu(x)\big)\big]_{\beta\alpha}\big[L\big(\nu(x)\big)\big]^{-1}_{\gamma\beta}\big(\partial_{T_{t\ell}} u_\alpha^+\big)(x) \qquad (8.4.343)\\
&\qquad = (1-\eta)\cdot\frac{1}{2}\delta_{\gamma\alpha}\big(\partial_{T_{t\ell}} u_\alpha^+\big)(x) = (1-\eta)\cdot\frac{1}{2}\big(\partial_{T_{t\ell}} u_\gamma^+\big)(x).
\end{aligned}
$$

This last expression may be absorbed in the left-hand side of (8.4.342), which subsequently becomes $\frac{1+\eta}{2}\big(\partial_{T_{t\ell}} u_\gamma^+\big)(x)$. The second observation is that, as is visible from the very definition of the boundary-to-boundary double layer potential operator K_A associated with the set Ω and the coefficient tensor A as in (A.0.116), the second line in the right-hand side of (8.4.342) is precisely

$$
(1-\eta)\cdot\Big(K_A\big(\partial_{T_{t\ell}} u^+\big)\Big)_\gamma(x), \quad \text{where } \partial_{T_{t\ell}} u^+ := \big(\partial_{T_{t\ell}} u_\alpha^+\big)_{1\le\alpha\le M}. \qquad (8.4.344)
$$

The final observation we make with regard to (8.4.342) is that the third and fourth lines in the right-hand side of (8.4.342) are multiples of commutators of the form

$$
(1-\eta)\cdot\Big(\big[M_\nu, U\big]\big(\partial_T u^+\big)\Big)(x). \qquad (8.4.345)
$$

Above, M_ν denotes the operator of pointwise multiplication by generic scalar components of ν. Also, U stands for the principal-value singular integral operator of formal convolution type with a suitable matrix-valued kernel whose entries are linear combinations with coefficients which are entries from A of generic first-order partial derivatives of generic entries of the matrix E. Finally, $\partial_T u^+$ stands for generic functions of the form $\partial_{T_{t\ell}} u_\alpha^+$ with $1 \le \ell, t \le n$ and $1 \le \alpha \le M$.

In view of these observations (and since$\eta \neq -1$), we may recast (8.4.342) as

$$\frac{1}{2}\partial_{\tau_{t\ell}}u^+ = \Big(\tfrac{1-\eta}{1+\eta}\Big)K_A\big(\partial_{\tau_{t\ell}}u^+\big) + \Big(\tfrac{1-\eta}{1+\eta}\Big)\big[M_\nu, U\big]\big(\partial_\tau u^+\big)$$
$$\text{at } \sigma\text{-a.e. point on } \partial\Omega, \text{ for each } t, \ell \in \{1, \dots, n\}. \tag{8.4.346}$$

Since we are currently assuming that $A \in \mathfrak{A}_L^{\text{dis}}$, from (8.4.346), estimate (4.1.49) from Theorem 4.1.3, the fact that $\mathbb{X}$ is stable under multiplication by essentially bounded functions (cf. [130, (5.1.13)]), Proposition 1.3.2, [131, Corollary 2.7.9], [131, Theorem 1.4.2], (4.1.17), and (8.4.307) we then conclude that for each pair of indices $t, \ell \in \{1, \dots, n\}$ we have

$$\frac{1}{2}\big\|\partial_{\tau_{t\ell}}u^+\big\|_{[\mathbb{X}]^M} \leq C\delta\big\|\partial_{\tau_{t\ell}}u^+\big\|_{[\mathbb{X}]^M} + C\delta\sum_{j,k=1}^{n}\big\|\partial_{\tau_{jk}}u^+\big\|_{[\mathbb{X}]^M}. \tag{8.4.347}$$

After summing up in all $t, \ell \in \{1, \dots, n\}$ we conclude from (8.4.347) that

$$\frac{1}{2}\sum_{t,\ell=1}^{n}\big\|\partial_{\tau_{t\ell}}u^+\big\|_{[\mathbb{X}]^M} \leq C\delta\sum_{t,\ell=1}^{n}\big\|\partial_{\tau_{t\ell}}u^+\big\|_{[\mathbb{X}]^M}, \tag{8.4.348}$$

where $C \in (0, \infty)$ depends only on n, m, A, the two-sided NTA constants of Ω, the Ahlfors regularity constants of $\partial\Omega$, and the operator norms of $\mathcal{M}_{\partial\Omega}$ on $\mathbb{X}$, $\mathbb{X}'$. If $\delta \in (0, \infty)$ is sufficiently small to begin with, relative to C, we see from (8.4.348) that $\sum_{t,\ell=1}^{n}\big\|\partial_{\tau_{t\ell}}u^+\big\|_{[\mathbb{X}]^M} \leq 0$ hence, necessarily,

$$\partial_{\tau_{t\ell}}u_\alpha^+ = 0 \text{ for each } t, \ell \in \{1, \dots, n\} \text{ and } \alpha \in \{1, \dots, M\}. \tag{8.4.349}$$

Feeding this back into (8.4.339) then proves that $\nabla u^+ = 0$ in Ω_+. As is implicit in (8.4.301) that $\Omega_\pm$ are connected sets, this shows that u^+ is a constant in Ω_+, say $u^+ \equiv c^+ \in \mathbb{C}^M$ in Ω_+. Based on this, (8.4.329), (8.4.326), and keeping in mind that $\eta \neq 0$, we then obtain $\partial_\nu^A u^- = \eta^{-1}\cdot\partial_\nu^A u^+ = 0$ at σ-a.e. point on $\partial\Omega$, and $\partial_{\tau_{jk}}u_\alpha^- = 0$ at σ-a.e. point on $\partial\Omega$, for each $j, k \in \{1, \dots, n\}$ and $\alpha \in \{1, \dots, M\}$. With this in hand, the integral representation formula [131, (1.5.230)] written for u^- in Ω_-, then shows that $\nabla u^- = 0$ in Ω_- thus, as before, u^- is a constant in Ω_-, say $u^- = c^- \in \mathbb{C}^M$ in Ω_-. The final step is to invoke (8.4.325) in order to write $c^+ = u^+\big|_{\partial\Omega}^{\kappa-\text{n.t.}} = u^-\big|_{\partial\Omega}^{\kappa-\text{n.t.}} = c^-$, which completes the proof of uniqueness (modulo constants) for the Transmission Problem (8.4.304), working under the hypotheses adopted in (8.4.306).

The second proof of (8.4.324) goes as follows. Suppose $u^\pm$ solve the homogeneous version of (8.4.304). For some constant $c \in \mathbb{C}^M$ (which is allowed to vary from line to line) we may then write

$$\begin{aligned} u^+ &= \mathcal{D}^+_{A,\text{mod}}\Big(u^+\big|_{\partial\Omega}^{\kappa-\text{n.t.}}\Big) - \mathcal{S}^+_{\text{mod}}\big(\partial_\nu^A u^+\big) + c = \mathcal{D}^+_{A,\text{mod}}\Big(u^-\big|_{\partial\Omega}^{\kappa-\text{n.t.}}\Big) - \eta\cdot\mathcal{S}^+_{\text{mod}}\big(\partial_\nu^A u^-\big) + c \\ &= (1-\eta)\cdot\mathcal{S}^+_{\text{mod}}\big(\partial_\nu^A u^-\big) + c \text{ in } \Omega_+. \end{aligned} \tag{8.4.350}$$

Above, the first equality is the integral representation formula (4.1.333) (written for u^+ in Ω_+, bearing in mind that the latter set is connected), the second equality takes into account the homogeneous versions of the boundary conditions in (8.4.304), while the final equality in (8.4.350) uses (4.1.334) (written for u^- in Ω_-, again keeping in mind that this set is connected). In view of the jump-formula (4.1.341), after taking the conormal derivative ∂_ν^A of the most extreme sides in (8.4.350) we arrive at (8.4.335). Under the assumptions made in (8.4.305) this yields, as before, to the conclusion in (8.4.324).

When the hypotheses in (8.4.306) are adopted instead, in place of (8.4.350) we write, for some constant $c \in \mathbb{C}^M$ (which is allowed to vary from line to line),

$$\begin{aligned} u^+ &= \mathcal{D}^+_{A,\mathrm{mod}}\Big(u^+\big|^{\kappa-\mathrm{n.t.}}_{\partial\Omega}\Big) - \mathcal{S}^+_{\mathrm{mod}}\big(\partial_\nu^A u^+\big) + c \\ &= (1-\eta)\cdot \mathcal{D}^+_{A,\mathrm{mod}}\Big(u^+\big|^{\kappa-\mathrm{n.t.}}_{\partial\Omega}\Big) + \eta\Big\{\mathcal{D}^+_{A,\mathrm{mod}}\Big(u^-\big|^{\kappa-\mathrm{n.t.}}_{\partial\Omega}\Big) - \mathcal{S}^+_{\mathrm{mod}}\big(\partial_\nu^A u^-\big)\Big\} + c \\ &= (1-\eta)\cdot \mathcal{D}^+_{A,\mathrm{mod}}\Big(u^+\big|^{\kappa-\mathrm{n.t.}}_{\partial\Omega}\Big) + c \text{ in } \Omega_+. \end{aligned} \tag{8.4.351}$$

Here, the first equality is the integral representation formula (4.1.333) (written for the function u^+ in the connected set Ω_+), the second equality takes into account the homogeneous versions of the boundary conditions in (8.4.304), while the final equality in (8.4.351) uses (4.1.334) (written for the function u^- in the connected set Ω_-). Upon recalling the jump-formula (4.1.344), after going nontangentially to the boundary in the most extreme sides of (8.4.351) we now obtain

$$\Big(-\tfrac{1}{2}(1+\eta)I + (1-\eta)K_{A,\mathrm{mod}}\Big)\Big(u^+\big|^{\kappa-\mathrm{n.t.}}_{\partial\Omega}\Big) = c. \tag{8.4.352}$$

Since under the assumptions made in (8.4.306) we have (8.4.318), a Neumann series argument shows that the operator

$$-\tfrac{1}{2}(1+\eta)I + (1-\eta)\big[K_{A,\mathrm{mod}}\big] : \big[\dot{\mathbb{X}}_1/\sim\big]^M \longrightarrow \big[\dot{\mathbb{X}}_1/\sim\big]^M \tag{8.4.353}$$

is an isomorphism. Together, (8.4.352), (8.4.353), and the Fatou-type result from Theorem 4.1.20 imply that $u^+\big|^{\kappa-\mathrm{n.t.}}_{\partial\Omega}$ is a constant on $\partial\Omega$. Plugging this back into (8.4.351) then further gives (thanks to [132, (1.8.10)]) that u^+ is a constant in Ω_+. In view of the homogeneous versions of the boundary conditions in (8.4.304) we next see that $u^-\big|^{\kappa-\mathrm{n.t.}}_{\partial\Omega}$ is a constant on $\partial\Omega$ and that $\partial_\nu^A u^- = 0$. Using this information in the integral representation formula (4.1.333) (written for the function u^- in the connected set Ω_-) then shows (again, by virtue of [132, (1.8.10)]) that u^- is a constant in Ω_-. Thanks to (8.4.325), we ultimately arrive at the conclusion in (8.4.324). This completes the second proof of (8.4.324). Lastly, the final claim in the statement of the theorem is dealt with similarly, so the proof of Theorem 8.4.44 is finished. □

8.5 Boundary Problems in Domains with Unbounded Boundaries for the Stokes System

Shifting perspectives, in this section we take on the task of proving solvability results for a large spectrum of boundary value problems for the *Stokes system* in sufficiently flat Ahlfors regular domains. While the specific algebraic format of said system prevents it from being part of the general theory developed so far in this chapter (pertaining to homogeneous second-order constant coefficient weakly elliptic systems), the Calderón-Zygmund theory (including invertibility results) we have developed in relation to the boundary layer potentials for the Stokes system together with our Fatou-type theorems for the Stokes system, and the robust function space theory already in place, make this endeavor first possible and, in the end, particularly successful.

We begin by proving the well-posedness of the Homogeneous Regularity Problem for the Stokes system with boundary data in homogeneous Sobolev spaces.

Theorem 8.5.1 *Consider $n \in \mathbb{N}$ with $n \geq 2$ and let $\Omega \subseteq \mathbb{R}^n$ be an Ahlfors regular domain. Abbreviate $\sigma := \mathcal{H}^{n-1} \lfloor \partial\Omega$ and denote by ν the geometric measure theoretic outward unit normal to Ω. Also, fix an aperture parameter $\kappa > 0$ and pick an integrability exponent $p \in (1, \infty)$. In this setting, consider the Homogeneous Regularity Problem for the Stokes system:*

$$\begin{cases} \vec{u} \in \big[\mathscr{C}^\infty(\Omega)\big]^n, \quad \pi \in \mathscr{C}^\infty(\Omega), \\ \Delta\vec{u} - \nabla\pi = 0 \ \textit{and} \ \operatorname{div}\vec{u} = 0 \ \textit{in} \ \Omega, \\ \mathcal{N}_\kappa(\nabla\vec{u}),\ \mathcal{N}_\kappa\pi \in L^p(\partial\Omega, \sigma), \\ \vec{u}\Big|_{\partial\Omega}^{\kappa-\mathrm{n.t.}} = \vec{f} \in \big[\dot{L}^p_1(\partial\Omega, \sigma)\big]^n, \end{cases} \tag{8.5.1}$$

where $\dot{L}^p_1(\partial\Omega, \sigma)$ is the homogeneous boundary Sobolev space introduced in [130, Definition 11.5.3] (*cf.* (A.0.145)-(A.0.146)).

*There exists $\delta \in (0, 1)$ which depends only on n, p, and the Ahlfors regularity constants of $\partial\Omega$ with the property that if $\|\nu\|_{[\mathrm{BMO}(\partial\Omega,\sigma)]^n} < \delta$ (i.e., if Ω is a δ-*AR *domain in the sense of Definition 3.1.8) then Ω is a connected two-sided* NTA *domain with unbounded boundary and the following properties are true:*

(a) [Existence, Uniqueness, Estimates, and Integral Representations] *The Homogeneous Regularity Problem* (8.5.1) *is well-posedness (in the sense that it has a unique solution, which depends continuously on the boundary datum). As far as existence is concerned, the modified boundary-to-boundary single layer operator for the Stokes system S_{mod}, originally introduced in* (4.2.110), *induces a linear isomorphism in the context*

$$\begin{aligned} &\big[S_{\mathrm{mod}}\big] : \big[L^p(\partial\Omega, \sigma)\big]^n \longrightarrow \big[\dot{L}^p_1(\partial\Omega, \sigma)/\sim\big]^n \\ &\big[S_{\mathrm{mod}}\big]\vec{g} := \big[S_{\mathrm{mod}}\vec{g}\big] \ \textit{for each} \ \vec{g} \in \big[L^p(\partial\Omega, \sigma)\big]^n, \end{aligned} \tag{8.5.2}$$

and the pair $(\vec{u}, \pi)$, with

$$\vec{u} := \mathscr{S}_{\rm mod}\vec{g} - c \text{ and } \pi := \mathcal{Q}\vec{g} \text{ in } \Omega, \tag{8.5.3}$$

(*where* $\mathscr{S}_{\rm mod}$ *is the modified boundary-to-domain single layer for the Stokes system recalled in* (A.0.234), *and* $\mathcal{Q}$ *is the associated pressure potential from* (A.0.213)) *constitutes a solution of the Homogeneous Regularity Problem* (8.5.1) *if one takes*

$$\begin{gathered} \vec{g} \in \big[L^p(\partial\Omega, \sigma)\big]^n \text{ to be the unique function} \\ \text{so that } c := S_{\rm mod}\vec{g} - \vec{f} \text{ is a constant in } \mathbb{C}^n. \end{gathered} \tag{8.5.4}$$

In addition, this solution satisfies (*with implicit constants independent of* $\vec{f}$)

$$\big\|\mathcal{N}_\kappa(\nabla\vec{u})\big\|_{L^p(\partial\Omega,\sigma)} \approx \|\nabla_{\tan}\vec{f}\|_{[L^p(\partial\Omega,\sigma)]^{n\cdot n}}, \tag{8.5.5}$$

and

$$\big\|\mathcal{N}_\kappa\pi\big\|_{L^p(\partial\Omega,\sigma)} + \big\|\mathcal{N}_\kappa(\nabla\vec{u})\big\|_{L^p(\partial\Omega,\sigma)} \approx \|\nabla_{\tan}\vec{f}\|_{[L^p(\partial\Omega,\sigma)]^{n\cdot n}}. \tag{8.5.6}$$

Alternatively, if $K^{\rm mod}_{\rm slip}$ *denotes the modified boundary-to-boundary slip double layer operator originally introduced in* (4.2.105), *then*

$$\tfrac{1}{2}I + \big[K^{\rm mod}_{\rm slip}\big] : \big[\dot{L}^p_1(\partial\Omega, \sigma)/\sim\big]^n \longrightarrow \big[\dot{L}^p_1(\partial\Omega, \sigma)/\sim\big]^n \tag{8.5.7}$$

is an isomorphism, and the Homogeneous Regularity Problem (8.5.1) *may be solved as*

$$\vec{u} := \mathcal{D}^{\rm mod}_{\rm slip}\vec{h} - c \text{ and } \pi := \mathcal{P}_{\rm slip}\vec{h} \text{ in } \Omega, \tag{8.5.8}$$

(*where* $\mathcal{D}^{\rm mod}_{\rm slip}$ *stands for the modified boundary-to-domain slip double layer operator introduced in* (4.2.107), *while* $\mathcal{P}_{\rm slip}$ *is the slip pressure potential operator from* (4.2.108)) *and this solution continues to satisfy* (8.5.5), *if one takes*

$$\begin{gathered} \vec{h} \in \big[\dot{L}^p_1(\partial\Omega, \sigma)\big]^n \text{ to be the unique function} \\ \text{so that } c := \big(\tfrac{1}{2}I + K^{\rm mod}_{\rm slip}\big)\vec{h} - \vec{f} \text{ is a constant in } \mathbb{C}^n. \end{gathered} \tag{8.5.9}$$

Finally,

> *if the boundary datum* $\vec{f}$ *actually belongs to the inhomogeneous Sobolev space* $\big[L^p_1(\partial\Omega, \sigma)\big]^n$ *then the unique solution* $(\vec{u}, \pi)$ *of the Homogeneous Regularity Problem* (8.5.1) *enjoys the additional property that* $\mathcal{N}_\kappa\vec{u}$ *belongs to* $L^p(\partial\Omega, \sigma)$. (8.5.10)

(*b*) [Other Spaces of Boundary Data] *Similar well-posedness results to those described in item (a) above are valid with* $\dot{L}^p_1(\partial\Omega, \sigma)$ *replaced by any of the spaces* $\dot{M}^{p,\lambda}_1(\partial\Omega, \sigma)$, $\dot{\mathbb{M}}^{p,\lambda}_1(\partial\Omega, \sigma)$, *and* $\dot{\mathcal{B}}^{q,\lambda}_1(\partial\Omega, \sigma)$ *with* $p, q \in (1, \infty)$ *and* $\lambda \in (0, n-1)$. *More generally, given any Generalized Banach Function Space* $\mathbb{X}$ *on* $(\partial\Omega, \sigma)$ *with the property that*

$$\mathcal{M}_{\partial\Omega} : \mathbb{X} \to \mathbb{X} \text{ and } \mathcal{M}_{\partial\Omega} : \mathbb{X}' \to \mathbb{X}' \text{ are well-defined and bounded} \tag{8.5.11}$$

(with $\mathcal{M}_{\partial\Omega}$ denoting the Hardy-Littlewood maximal operator on $\partial\Omega$, and with $\mathbb{X}'$ denoting the associated space of $\mathbb{X}$; cf. [130, Definitions 5.1.4, 5.1.11]), *there exists $\delta_{\mathbb{X}} \in (0, 1)$, depending only on n, the operator norms of $\mathcal{M}_{\partial\Omega}$ on $\mathbb{X}$ and $\mathbb{X}'$, as well as the Ahlfors regularity constants of $\partial\Omega$, such that if $\|\nu\|_{[\mathrm{BMO}(\partial\Omega,\sigma)]^n} < \delta_{\mathbb{X}}$ then Ω is a connected two-sided* NTA *domain with unbounded boundary, and the following Homogeneous Regularity Problem for the Stokes system is well-posed:*

$$\begin{cases} \vec{u} \in \big[\mathscr{C}^\infty(\Omega)\big]^n, \quad \pi \in \mathscr{C}^\infty(\Omega), \\ \Delta\vec{u} - \nabla\pi = 0 \text{ and } \operatorname{div}\vec{u} = 0 \text{ in } \Omega, \\ \mathcal{N}_\kappa(\nabla\vec{u}),\ \mathcal{N}_\kappa\pi \in \mathbb{X}, \\ \vec{u}\big|_{\partial\Omega}^{\kappa-\mathrm{n.t.}} = \vec{f} \in \big[\dot{\mathbb{X}}_1\big]^n, \end{cases} \tag{8.5.12}$$

where $\dot{\mathbb{X}}_1$ is the homogeneous boundary Sobolev space introduced in (4.2.52)-(4.2.53). *Additionally,*

> *if the boundary datum $\vec{f}$ actually belongs to the inhomogeneous Sobolev space $\big[\mathbb{X}_1\big]^n$ it follows that the unique solution $(\vec{u}, \pi)$ of the Homogeneous Regularity Problem* (8.5.12) *has the additional property that $\mathcal{N}_\kappa\vec{u}$ belongs to $\mathbb{X}$.* (8.5.13)

The proof of Theorem 8.5.1 is slightly postponed and, for now, we make a number of useful observations and record a couple of corollaries. For further use, let us first point out that the Fatou-type result for the Stokes system proved in [132, Theorem 6.3.2] implies that

> if Ω is a UR domain then for any two functions $(\vec{u}, \pi)$ satisfying the properties listed in the first four lines of (8.5.1) the nontangential pointwise traces $\nabla\vec{u}\big|_{\partial\Omega}^{\kappa-\mathrm{n.t.}}$ and $\pi\big|_{\partial\Omega}^{\kappa-\mathrm{n.t.}}$ exist (in $\mathbb{C}^{n\times n}$ and $\mathbb{C}$, respectively) σ-a.e. on $\partial\Omega$, and the slip conormal derivative $\partial_\nu^{\mathrm{slip}}(\vec{u}, \pi)$ (see (4.2.112)) belongs to $\big[L^p(\partial\Omega, \sigma)\big]^n$. (8.5.14)

In addition, [130, Proposition 11.5.12] shows that

> if Ω is actually an NTA domain with an Ahlfors regular boundary then for any two functions $(\vec{u}, \pi)$ satisfying the properties listed in the first four lines of (8.5.1) it follows that the nontangential pointwise trace $\vec{u}\big|_{\partial\Omega}^{\kappa-\mathrm{n.t.}}$ exists (in $\mathbb{C}^n$) at σ-a.e. point on $\partial\Omega$ and, as a function, belongs to the space $\big[\dot{L}_1^p(\partial\Omega, \sigma)\big]^n$. (8.5.15)

In particular, the latter Fatou-type result ensures that the space of boundary data in (8.5.1) is optimal. As a consequence of (8.5.15), Theorem 8.5.1, and [132, Theorem 6.2.6, item *(9)*], we also have the following important structural result:

Corollary 8.5.2 *Fix $n \in \mathbb{N}$ with $n \geq 2$ and let $\Omega \subseteq \mathbb{R}^n$ be an Ahlfors regular domain. Abbreviate $\sigma := \mathcal{H}^{n-1}\lfloor\partial\Omega$ and denote by ν the geometric measure theoretic outward unit normal to Ω. Also, fix an aperture parameter $\kappa > 0$ and pick an integrability exponent $p \in (1, \infty)$. Then there exists some threshold $\delta \in (0, 1)$ which depends only on n, p, and the Ahlfors regularity constants of $\partial\Omega$ with the property that whenever $\|\nu\|_{[\mathrm{BMO}(\partial\Omega,\sigma)]^n} < \delta$ it follows that the set of conditions*

$$\begin{cases} \vec{u} \in \big[\mathscr{C}^\infty(\Omega)\big]^n, \quad \pi \in \mathscr{C}^\infty(\Omega), \\ \Delta\vec{u} - \nabla\pi = 0 \ \text{and} \ \mathrm{div}\,\vec{u} = 0 \ \text{in} \ \Omega, \\ \mathcal{N}_\kappa(\nabla\vec{u}),\ \mathcal{N}_\kappa\pi \in L^p(\partial\Omega, \sigma), \end{cases} \tag{8.5.16}$$

is equivalent to having

$$\vec{u} = \mathscr{S}_{\rm mod}\vec{g} + c \ \text{and} \ \pi = Q\vec{g} \ \text{in} \ \Omega, \ \text{for some} \ \vec{g} \in \big[L^p(\partial\Omega, \sigma)\big]^n \ \text{and} \ c \in \mathbb{C}^n, \tag{8.5.17}$$

where $\mathscr{S}_{\rm mod}$ is the modified boundary-to-domain single layer for the Stokes system introduced in (A.0.234)*, and Q is the (single layer) pressure potential for the Stokes system from* (A.0.213).

Likewise, from the Fatou-type result proved in [132, Theorem 6.3.4] we see that

$$\text{if } \Omega \text{ is a UR domain then for any two functions } (\vec{u}, \pi) \text{ satisfying the properties listed in the first four lines of (8.5.12) the nontangential pointwise traces } \nabla\vec{u}\Big|_{\partial\Omega}^{\kappa-\mathrm{n.t.}} \text{ and } \pi\Big|_{\partial\Omega}^{\kappa-\mathrm{n.t.}} \text{ exist (in } \mathbb{C}^{n\times n} \text{ and } \mathbb{C}, \text{ respectively) } \sigma\text{-a.e. on } \partial\Omega, \text{ and the slip conormal derivative } \partial_\nu^{\rm slip}(\vec{u}, \pi) \text{ (see (4.2.112)) belongs to } \big[\mathbb{X}\big]^n. \tag{8.5.18}$$

Moreover, from [111, Proposition 8.5] we know that

$$\text{if } \Omega \text{ is actually an NTA domain with an Ahlfors regular boundary then for any two functions } (\vec{u}, \pi) \text{ satisfying the properties listed in the first four lines of (8.5.12) it follows that the nontangential pointwise trace } \vec{u}\Big|_{\partial\Omega}^{\kappa-\mathrm{n.t.}} \text{ exists (in } \mathbb{C}^n) \text{ at } \sigma\text{-a.e. point on } \partial\Omega \text{ and, as a function, belongs to the space } \big[\dot{\mathbb{X}}_1\big]^n. \tag{8.5.19}$$

This shows that the space of boundary data in (8.5.12) is optimal. As a consequence of (8.5.19), the last part in Theorem 8.5.1, item *(14)* in Theorem 4.2.3, [132, (6.2.1)-(6.2.2)], [131, Theorem 2.4.1, item *(4)*], the embeddings in [130, Proposition 5.2.7], and [132, (6.2.165), (6.2.166)], we also have the following important structural result (expanding upon Corollary 8.5.2):

Corollary 8.5.3 *Fix $n \in \mathbb{N}$ with $n \geq 2$ and let $\Omega \subseteq \mathbb{R}^n$ be an Ahlfors regular domain. Abbreviate $\sigma := \mathcal{H}^{n-1}\lfloor\partial\Omega$, denote by ν the geometric measure theoretic outward unit normal to Ω, and fix an aperture parameter $\kappa > 0$. Also, consider a Generalized Banach Function Space $\mathbb{X}$ on $(\partial\Omega, \sigma)$ with the property that*

$$\mathcal{M}_{\partial\Omega} : \mathbb{X} \to \mathbb{X} \ \text{and} \ \mathcal{M}_{\partial\Omega} : \mathbb{X}' \to \mathbb{X}' \ \text{are well-defined and bounded} \tag{8.5.20}$$

(with $\mathcal{M}_{\partial\Omega}$ denoting the Hardy-Littlewood maximal operator on $\partial\Omega$, and with $\mathbb{X}'$ denoting the associated space of $\mathbb{X}$; cf. [130, Definitions 5.1.4, 5.1.11]).

Then there exists $\delta_{\mathbb{X}} \in (0,1)$, depending only on n, the operator norms of $\mathcal{M}_{\partial\Omega}$ on $\mathbb{X}$ and $\mathbb{X}'$, as well as the Ahlfors regularity constants of $\partial\Omega$, with the property that if $\|\nu\|_{[\mathrm{BMO}(\partial\Omega,\sigma)]^n} < \delta_{\mathbb{X}}$ it follows that the set of conditions

$$\begin{cases} \vec{u} \in \big[\mathscr{C}^\infty(\Omega)\big]^n, \quad \pi \in \mathscr{C}^\infty(\Omega), \\ \Delta\vec{u} - \nabla\pi = 0 \text{ and } \operatorname{div}\vec{u} = 0 \text{ in } \Omega, \\ \mathcal{N}_\kappa(\nabla\vec{u}),\ \mathcal{N}_\kappa\pi \in \mathbb{X}, \end{cases} \tag{8.5.21}$$

is equivalent to having

$$\vec{u} = \mathscr{S}_{\mathrm{mod}}\vec{g} + c \text{ and } \pi = \mathcal{Q}\vec{g} \text{ in } \Omega, \text{ for some } \vec{g} \in \big[\mathbb{X}\big]^n \text{ and some } c \in \mathbb{C}^n, \tag{8.5.22}$$

where $\mathscr{S}_{\mathrm{mod}}$ is the modified boundary-to-domain single layer for the Stokes system introduced in (A.0.234), *and $\mathcal{Q}$ is the (single layer) pressure potential for the Stokes system from* (A.0.213).

Lastly, we wish to note that, in view of (8.5.10) and (8.5.13), the Homogeneous Regularity Problems (8.5.1) and (8.5.12) contain, as sub-problems, the following well-posedness results for the Inhomogeneous Regularity Problem for the Stokes system:

Corollary 8.5.4 *Fix $n \in \mathbb{N}$ with $n \geq 2$ and let $\Omega \subseteq \mathbb{R}^n$ be an Ahlfors regular domain. Denote by ν its geometric measure theoretic outward unit normal, set $\sigma := \mathcal{H}^{n-1}\lfloor\partial\Omega$, and fix an aperture parameter $\kappa > 0$. Then for each integrability exponent $p \in (1,\infty)$ there exists $\delta \in (0,1)$, depending only on n, p, and the Ahlfors regularity constants of $\partial\Omega$, with the property that if $\|\nu\|_{[\mathrm{BMO}(\partial\Omega,\sigma)]^n} < \delta$ the following the Inhomogeneous Regularity Problem for the Stokes system is well posed:*

$$\begin{cases} \vec{u} \in \big[\mathscr{C}^\infty(\Omega)\big]^n, \quad \pi \in \mathscr{C}^\infty(\Omega), \\ \Delta\vec{u} - \nabla\pi = 0 \text{ and } \operatorname{div}\vec{u} = 0 \text{ in } \Omega, \\ \mathcal{N}_\kappa\vec{u},\ \mathcal{N}_\kappa(\nabla\vec{u}),\ \mathcal{N}_\kappa\pi \in L^p(\partial\Omega,\sigma), \\ \vec{u}\big|_{\partial\Omega}^{\kappa-\mathrm{n.t.}} = \vec{f} \in \big[L^p_1(\partial\Omega,\sigma)\big]^n. \end{cases} \tag{8.5.23}$$

Furthermore, analogous well-posedness results are valid with $L^p_1(\partial\Omega,\sigma)$ replaced by any of the spaces $M^{p,\lambda}_1(\partial\Omega,\sigma)$, $\mathring{M}^{p,\lambda}_1(\partial\Omega,\sigma)$, and $\mathcal{B}^{q,\lambda}_1(\partial\Omega,\sigma)$ with $p,q \in (1,\infty)$ and $\lambda \in (0,n-1)$. More generally, given any Generalized Banach Function Space $\mathbb{X}$ on $(\partial\Omega,\sigma)$ with the property that

$$\mathcal{M}_{\partial\Omega} : \mathbb{X} \to \mathbb{X} \text{ and } \mathcal{M}_{\partial\Omega} : \mathbb{X}' \to \mathbb{X}' \text{ are well-defined and bounded} \tag{8.5.24}$$

(with $\mathcal{M}_{\partial\Omega}$ denoting the Hardy-Littlewood maximal operator on $\partial\Omega$, and with $\mathbb{X}'$ denoting the associated space of $\mathbb{X}$; cf. [130, Definitions 5.1.4, 5.1.11]), *there exists $\delta_{\mathbb{X}} \in (0,1)$, depending only on n, the operator norms of $\mathcal{M}_{\partial\Omega}$ on $\mathbb{X}$ and $\mathbb{X}'$, as well as the Ahlfors regularity constants of $\partial\Omega$, such that if $\|\nu\|_{[\mathrm{BMO}(\partial\Omega,\sigma)]^n} < \delta_{\mathbb{X}}$ then Ω*

is a connected two-sided NTA *domain with unbounded boundary, and the following Inhomogeneous Regularity Problem for the Stokes system is well-posed:*

$$\begin{cases} \vec{u} \in \big[\mathscr{C}^\infty(\Omega)\big]^n, \quad \pi \in \mathscr{C}^\infty(\Omega), \\ \Delta\vec{u} - \nabla\pi = 0 \text{ and } \operatorname{div}\vec{u} = 0 \text{ in } \Omega, \\ \mathcal{N}_\kappa\vec{u},\ \mathcal{N}_\kappa(\nabla\vec{u}),\ \mathcal{N}_\kappa\pi \in \mathbb{X}, \\ \vec{u}\big|_{\partial\Omega}^{\kappa-\text{n.t.}} = \vec{f} \in \big[\mathbb{X}_1\big]^n, \end{cases} \tag{8.5.25}$$

where $\mathbb{X}_1 := \Big\{f \in \mathbb{X} : \partial_{\tau_{jk}} f \in \mathbb{X} \text{ for } 1 \leq j,k \leq n\Big\}$.

We now turn to the task of providing the proof of Theorem 8.5.1.

Proof of Theorem 8.5.1 On account of Theorem 3.1.5, choosing $\delta \in (0,1)$ small enough guarantees that Ω is a connected two-sided NTA domain with unbounded boundary (in particular, Ω is a UR domain). The claims made in relation to the operators (8.5.2) and (8.5.7) are seen from items *(10)-(11)* of Theorem 4.2.3. From these invertibility results and the properties of the boundary layer potentials for the Stokes system described in [132, Theorem 6.2.6, items *(9)-(10)*] all claims regarding existence, estimates, and integral representation formulas for the Homogeneous Regularity Problem for the Stokes system formulated in (8.5.1) follow.

Let us deal with the issue of uniqueness. To this end, assume $(\vec{u},\pi)$ solves the version of the boundary value problem (8.5.1) corresponding to $\vec{f} := 0$. In concert, [129, Proposition 8.4.9] and (8.5.14) show that all hypotheses of [132, Theorem 6.2.7] are presently satisfied. In the current setting, the integral representation formulas established in [132, (6.2.226), (6.2.227)] with $\lambda := 1$ reduce (bearing in mind that $\vec{u}\big|_{\partial\Omega}^{\kappa-\text{n.t.}} = 0$ on $\partial\Omega$, and that the set Ω is connected) to

$$\vec{u} = -\mathscr{S}_{\text{mod}}\big(\partial_\nu^{\text{slip}}(\vec{u},\pi)\big) + c \text{ and } \pi = -\mathcal{Q}\big(\partial_\nu^{\text{slip}}(\vec{u},\pi)\big) \text{ in } \Omega \tag{8.5.26}$$

for some constant vector $c \in \mathbb{C}^n$. In turn, from (8.5.26), the jump-formula in [132, (6.2.210)] written for $\lambda := 1$, and (4.2.112), at σ-a.e. point on $\partial\Omega$ we see that

$$\partial_\nu^{\text{slip}}(\vec{u},\pi) = -\partial_\nu^{\text{slip}}\Big(\mathscr{S}_{\text{mod}}\big(\partial_\nu^{\text{slip}}(\vec{u},\pi)\big),\, \mathcal{Q}\big(\partial_\nu^{\text{slip}}(\vec{u},\pi)\big)\Big) = -\Big(-\tfrac{1}{2}I + K^{\#}_{\text{slip}}\Big)\big(\partial_\nu^{\text{slip}}(\vec{u},\pi)\big),$$

hence

$$\Big(\tfrac{1}{2}I + K^{\#}_{\text{slip}}\Big)\big(\partial_\nu^{\text{slip}}(\vec{u},\pi)\big) = 0. \tag{8.5.27}$$

If $\delta \in (0,1)$ is taken sufficiently small to begin with, we may employ item *(1)* in Theorem 4.2.3 (with $z = 1/2$) to ensure that the operator

$$\tfrac{1}{2}I + K^{\#}_{\text{slip}} : \big[L^p(\partial\Omega,\sigma)\big]^n \longrightarrow \big[L^p(\partial\Omega,\sigma)\big]^n \tag{8.5.28}$$

is an isomorphism (cf. (4.2.119)). In light of this, we may then conclude from (8.5.27) that $\partial_\nu^{\text{slip}}(\vec{u},\pi) = 0$. Using this back in (8.5.26) gives that $\vec{u}$ is a constant and $\pi = 0$ in Ω. Upon recalling that we are currently assuming $\vec{u}\big|_{\partial\Omega}^{\kappa-\text{n.t.}} = 0$, we finally see that both $\vec{u}$ and π vanish identically in Ω. This proves that the Homogeneous Regularity Problem (8.5.1) has at most one solution.

Next, the left-pointing inequality in (8.5.5) is implied by the fact that $\vec{g}$ in (8.5.4) may be chosen so that

$$\|\vec{g}\|_{[L^p(\partial\Omega,\sigma)]^n} \leq C\big\|[\vec{f}]\big\|_{[\dot{L}^p_1(\partial\Omega,\sigma)/\sim]^n} \tag{8.5.29}$$

for some constant $C \in (0, \infty)$ independent of $\vec{f}$ (itself a consequence of (8.5.2) and the completeness results established in [130, Proposition 11.5.14]), together with (8.5.3), [130, (11.4.9)], the first estimate in [132, (6.2.210)], and (A.0.146). The right-pointing inequality in (8.5.5) follows from [130, Proposition 11.3.2] (whose current applicability makes use of the last membership in [132, (6.2.210)]) and (A.0.88). This finishes the proof of (8.5.5). Note that (8.5.6) follows from (8.5.5), (8.5.3), the last estimate in [132, (6.2.210)], and (8.5.29).

To completes the treatment of item *(a)*, there remains justify (8.5.10). In this regard, may rely on Theorem 4.2.3 to conclude (assuming $\delta \in (0, 1)$ is sufficiently small to begin with) that the operator

$$\tfrac{1}{2}I + K_{\text{slip}} : \big[L^p_1(\partial\Omega,\sigma)\big]^n \longrightarrow \big[L^p_1(\partial\Omega,\sigma)\big]^n \tag{8.5.30}$$

is an isomorphism. Then, with the inverse interpreted in this context, we conclude from [132, (1.5.8), (1.5.20)] that the pair $(\vec{u}, \pi)$ with

$$\vec{u} := \mathcal{D}_{\text{slip}}\Big(\big(\tfrac{1}{2}I + K_{\text{slip}}\big)^{-1}\vec{f}\Big) \text{ and } \pi := \mathcal{P}_{\text{slip}}\Big(\big(\tfrac{1}{2}I + K_{\text{slip}}\big)^{-1}\vec{f}\Big) \text{ in } \Omega, \tag{8.5.31}$$

is, thanks to [132, Theorem 6.2.4], a solution of the Homogeneous Regularity Problem (8.5.1) with the additional property that $\mathcal{N}_\kappa\vec{u}$ belongs to $L^p(\partial\Omega,\sigma)$. This concludes the treatment of item *(a)*.

Lastly, item *(b)* is dealt with in a similar fashion, now making use of the invertibility results from item *(14)* in Theorem 4.2.3, the nature of [132, (6.2.1)-(6.2.2)], the nontangential maximal function estimates from item *(4)* in [131, Theorem 2.4.1], the embeddings in [130, Proposition 5.2.7], the smoothness properties and PDE recorded in [132, (6.2.165), (6.2.166)]. □

The next goal is to prove the solvability result for the Dirichlet Problem for the Stokes system with boundary data in Lebesgue spaces, formulated below.

Theorem 8.5.5 *Consider $n \in \mathbb{N}$ with $n \geq 2$ and let $\Omega \subseteq \mathbb{R}^n$ be an Ahlfors regular domain. Abbreviate $\sigma := \mathcal{H}^{n-1}\lfloor\partial\Omega$ and denote by ν the geometric measure theoretic outward unit normal to Ω. Also, fix an aperture parameter $\kappa > 0$ and pick an integrability exponent $p \in (1, \infty)$. In this setting, consider the Dirichlet Problem for the Stokes system:*

$$\begin{cases} \vec{u} \in \big[\mathscr{C}^\infty(\Omega)\big]^n, \quad \pi \in \mathscr{C}^\infty(\Omega), \\ \Delta\vec{u} - \nabla\pi = 0 \textit{ and } \operatorname{div}\vec{u} = 0 \textit{ in } \Omega, \\ \mathcal{N}_\kappa\vec{u} \in L^p(\partial\Omega,\sigma), \\ \vec{u}\big|^{\kappa-\text{n.t.}}_{\partial\Omega} = \vec{f} \in \big[L^p(\partial\Omega,\sigma)\big]^n. \end{cases} \tag{8.5.32}$$

Then there exists $\delta \in (0,1)$ *which depends only on* n, p, *and the Ahlfors regularity constants of* $\partial\Omega$ *such that if* $\|\nu\|_{[\mathrm{BMO}(\partial\Omega,\sigma)]^n} < \delta$ *(i.e., if* Ω *is a* δ-AR *domain in the sense of Definition 3.1.8) it follows that* Ω *is a* UR *domain, the operator*

$$\tfrac{1}{2}I + K_{\text{slip}} : \big[L^p(\partial\Omega,\sigma)\big]^n \longrightarrow \big[L^p(\partial\Omega,\sigma)\big]^n \tag{8.5.33}$$

is an isomorphism (where K_{slip} *denotes the boundary-to-boundary slip double layer operator originally introduced in* (4.2.102)*), and the pair* $(\vec{u},\pi)$ *with*

$$\vec{u} := \mathcal{D}_{\text{slip}}\Big(\big(\tfrac{1}{2}I + K_{\text{slip}}\big)^{-1}\vec{f}\Big), \quad \pi := \mathcal{P}_{\text{slip}}\Big(\big(\tfrac{1}{2}I + K_{\text{slip}}\big)^{-1}\vec{f}\Big) \tag{8.5.34}$$

(where $\mathcal{D}_{\text{slip}}$ *stands for the boundary-to-domain slip double layer operator introduced in* (4.2.106)*, while* $\mathcal{P}_{\text{slip}}$ *is the slip pressure potential operator from* (4.2.108)*) is a solution of the Dirichlet Problem* (8.5.32) *which satisfies*

$$\big\|\mathcal{N}_\kappa\vec{u}\big\|_{L^p(\partial\Omega,\sigma)} \leq C\|\vec{f}\|_{[L^p(\partial\Omega,\sigma)]^n} \tag{8.5.35}$$

for a constant $C \in (0,\infty)$ *independent of* $\vec{f}$.

Similar solvability results are valid with $L^p(\partial\Omega,\sigma)$ *replaced by any of the spaces* $M^{p,\lambda}(\partial\Omega,\sigma)$, $\mathring{M}^{p,\lambda}(\partial\Omega,\sigma)$, *and* $\mathcal{B}^{q,\lambda}(\partial\Omega,\sigma)$ *with* $p, q \in (1,\infty)$ *and* $\lambda \in (0, n-1)$.

More generally, given a Generalized Banach Function Space $\mathbb{X}$ *on* $(\partial\Omega,\sigma)$ *with the property that*

$$\mathcal{M}_{\partial\Omega} : \mathbb{X} \to \mathbb{X} \text{ and } \mathcal{M}_{\partial\Omega} : \mathbb{X}' \to \mathbb{X}' \text{ are well-defined and bounded,} \tag{8.5.36}$$

where $\mathcal{M}_{\partial\Omega}$ *is the Hardy-Littlewood maximal operator on* $\partial\Omega$, *and* $\mathbb{X}'$ *is the associated space of* $\mathbb{X}$ *(cf.* [130, Definitions 5.1.4, 5.1.11]*), there exists* $\delta_{\mathbb{X}} \in (0,1)$ *which depends only on* n, *the operator norms of* $\mathcal{M}_{\partial\Omega}$ *on* $\mathbb{X}$ *and* $\mathbb{X}'$, *as well as the Ahlfors regularity constants of* $\partial\Omega$, *such that if*

$$\|\nu\|_{[\mathrm{BMO}(\partial\Omega,\sigma)]^n} < \delta_{\mathbb{X}} \tag{8.5.37}$$

it follows that Ω *is a* UR *domain, the operator*

$$\tfrac{1}{2}I + K_{\text{slip}} : \big[\mathbb{X}\big]^n \longrightarrow \big[\mathbb{X}\big]^n \text{ is an isomorphism} \tag{8.5.38}$$

and the pair $(\vec{u},\pi)$ *with*

$$\vec{u} := \mathcal{D}_{\text{slip}}\Big(\big(\tfrac{1}{2}I + K_{\text{slip}}\big)^{-1}\vec{f}\Big), \quad \pi := \mathcal{P}_{\text{slip}}\Big(\big(\tfrac{1}{2}I + K_{\text{slip}}\big)^{-1}\vec{f}\Big) \tag{8.5.39}$$

is a solution of the Dirichlet Problem

$$\begin{cases} \vec{u} \in \big[\mathscr{C}^\infty(\Omega)\big]^n, \quad \pi \in \mathscr{C}^\infty(\Omega), \\ \Delta\vec{u} - \nabla\pi = 0 \text{ and } \operatorname{div}\vec{u} = 0 \text{ in } \Omega, \\ \mathcal{N}_\kappa\vec{u} \in \mathbb{X}, \quad \vec{u}\big|_{\partial\Omega}^{\kappa-\text{n.t.}} = \vec{f} \in \big[\mathbb{X}\big]^n. \end{cases} \tag{8.5.40}$$

which satisfies $\|\mathcal{N}_\kappa\vec{u}\|_{\mathbb{X}} \leq C\|\vec{f}\|_{[\mathbb{X}]^n}$ *for a constant* $C \in (0,\infty)$ *independent of* $\vec{f}$. *Moreover, a similar result holds with* $\mathbb{X}$ *replaced by* $\mathring{\mathbb{X}} := \overline{L^\infty_{\rm comp}(\partial\Omega,\sigma)}^{\|\cdot\|_{\mathbb{X}}}$. *As a corollary,*

granted (8.5.37), *it follows that* (8.5.38) *holds and* (8.5.39) *solves the Dirichlet Problem* (8.5.40) *whenever* $\mathbb{X}$ *is a rearrangement invariant Banach function space whose Boyd indices satisfy* $1 < p_{\mathbb{X}} \leq q_{\mathbb{X}} < \infty$. (8.5.41)

In particular,

under the assumption made in (8.5.37), *it follows that* (8.5.38) *holds and* (8.5.39) *solves the Dirichlet Problem* (8.5.40) *whenever* $\mathbb{X} := L^\Phi(\partial\Omega,\sigma)$, *the Orlicz space associated as in* (A.0.133) *with the measure space* $(\partial\Omega,\sigma)$ *and any Young function* Φ *satisfying* $1 < i(\Phi) \leq I(\Phi) < \infty$. (8.5.42)

Proof All claims in the statement are consequences of Theorem 3.1.5, Theorem 4.2.3, [132, Theorem 6.2.4], [130, Proposition 5.2.7], [130, Proposition 5.3.14], [130, Proposition 5.3.15], [130, (5.3.62)], and item *(4)* of [131, Theorem 2.4.1]. □

As and end-point case of the Dirichlet Problem for the Stokes system with boundary data in Lebesgue spaces from Theorem 8.5.5 we have the following solvability result for the Dirichlet Problem for the Stokes system with boundary data in the John-Nirenberg space BMO.

Theorem 8.5.6 *Fix* $n \in \mathbb{N}$ *with* $n \geq 2$ *and suppose* $\Omega \subseteq \mathbb{R}^n$ *is an Ahlfors regular domain. Abbreviate* $\sigma := \mathcal{H}^{n-1}\lfloor\partial\Omega$ *and denote by* ν *the geometric measure theoretic outward unit normal to* Ω. *Also, fix an arbitrary aperture parameter* $\kappa > 0$.

Then there exists some threshold $\delta \in (0,1)$ *which depends only on n and the* ADR *constants of* $\partial\Omega$ *with the property that whenever* $\|\nu\|_{[\mathrm{BMO}(\partial\Omega,\sigma)]^n} < \delta$ *it follows that* Ω *is a* UR *domain with an unbounded boundary, and the boundary value problem*

$$\begin{cases} \vec{u} \in \left[\mathscr{C}^\infty(\Omega)\right]^n, \quad \pi \in \mathscr{C}^\infty(\Omega), \\ \Delta\vec{u} - \nabla\pi = 0 \ \textit{and} \ \mathrm{div}\,\vec{u} = 0 \ \textit{in} \ \Omega, \\ |\nabla\vec{u}|^p\,\mathrm{dist}\,(\cdot,\partial\Omega)^{p-1}\,d\mathcal{L}^n \ \textit{is a Carleson} \\ \textit{measure in } \Omega \textit{ for each } p \in (1,\infty), \\ \vec{u}\big|^{\kappa-\mathrm{n.t.}}_{\partial\Omega} = \vec{f} \in \left[\mathrm{BMO}(\partial\Omega,\sigma)\right]^n, \end{cases} \tag{8.5.43}$$

is solvable. More specifically, if $K^{\rm mod}_{\rm slip}$ *denotes the modified boundary-to-boundary slip double layer operator originally introduced in* (4.2.105), *then the operator*

$$\tfrac{1}{2}I + \left[K^{\rm mod}_{\rm slip}\right] : \left[\mathrm{BMO}(\partial\Omega,\sigma)/\!\sim\right]^n \longrightarrow \left[\mathrm{BMO}(\partial\Omega,\sigma)/\!\sim\right]^n \tag{8.5.44}$$

is an isomorphism, and the boundary value problem (8.5.43) *is solved by the pair* $(\vec{u},\pi)$ *defined as*

$$\vec{u} := \mathcal{D}^{\rm mod}_{\rm slip}\vec{g} - c \ \textit{and} \ \pi := \mathcal{P}_{\rm slip}\vec{g} \ \textit{in} \ \Omega, \tag{8.5.45}$$

if one takes

$$\begin{aligned} &\vec{g} \in \big[\mathrm{BMO}(\partial\Omega,\sigma)\big]^n \ \text{ to be the unique function} \\ &\text{so that } c := \big(\tfrac{1}{2}I + K^{\text{mod}}_{\text{slip}}\big)\vec{g} - \vec{f} \text{ is a constant in } \mathbb{C}^n. \end{aligned} \tag{8.5.46}$$

Furthermore, this solution has the property that for each exponent $p \in (1,\infty)$ there exists some constant $C = C(\Omega,p) \in (0,\infty)$ such that

$$\begin{aligned} \sup_{x\in\partial\Omega,\, r>0} \frac{1}{\sigma\big(B(x,r)\cap\partial\Omega\big)} \int_{B(x,r)\cap\Omega} |\nabla\vec{u}|^p \operatorname{dist}(\cdot,\partial\Omega)^{p-1}\, d\mathcal{L}^n \\ \leq C\|\vec{f}\|^p_{[\mathrm{BMO}(\partial\Omega,\sigma)]^n}. \end{aligned} \tag{8.5.47}$$

Moreover, for each $p \in (1,\infty)$ there exists a constant $C \in (0,\infty)$ with the property that the solution $(\vec{u},\pi)$ given in (8.5.45) *of the boundary value problem* (8.5.43) *satisfies*

$$\begin{aligned} \lim_{R\to 0^+} \sup_{\substack{x\in\partial\Omega \text{ and} \\ r\in(0,R)}} \Big(\frac{1}{\sigma\big(B(x,r)\cap\partial\Omega\big)} \int_{B(x,r)\cap\Omega} |\nabla\vec{u}|^p \operatorname{dist}(\cdot,\partial\Omega)^{p-1}\, d\mathcal{L}^n\Big)^{\frac{1}{p}} \\ \leq C \operatorname{dist}\big(\vec{f},\, [\mathrm{VMO}(\partial\Omega,\sigma)]^n\big), \end{aligned} \tag{8.5.48}$$

where the distance in the right-hand side is considered in the space $\big[\mathrm{BMO}(\partial\Omega,\sigma)\big]^n$.

As a corollary, for each $p \in (1,\infty)$ the solution $(\vec{u},\pi)$ given in (8.5.45) *of the boundary value problem* (8.5.43) *satisfies*

$$\begin{aligned} &|\nabla\vec{u}|^p \operatorname{dist}(\cdot,\partial\Omega)^{p-1}\, d\mathcal{L}^n \ \text{ is a vanishing Carleson measure in } \Omega \\ &\text{whenever the boundary datum } \vec{f} \text{ belongs to } \big[\mathrm{VMO}(\partial\Omega,\sigma)\big]^n. \end{aligned} \tag{8.5.49}$$

Finally, for each $p \in (1,\infty)$ there exists a constant $C \in (0,\infty)$, which depends only on n, p, and the ADR *constants of $\partial\Omega$, with the property that the solution $(\vec{u},\pi)$ from* (8.5.45)-(8.5.46) *of the boundary value problem* (8.5.43) *satisfies*

$$\begin{aligned} \max\Bigg\{ &\lim_{R\to\infty} \sup_{\substack{x\in\partial\Omega,\, r>0 \\ B(x,r)\subseteq\mathbb{R}^n\setminus B(0,R)}} \Big(\tfrac{1}{\sigma(B(x,r)\cap\partial\Omega)} \int_{B(x,r)\cap\Omega} |\nabla\vec{u}|^p \operatorname{dist}(\cdot,\partial\Omega)^{p-1}\, d\mathcal{L}^n\Big)^{1/p}, \\ &\lim_{R\to 0^+} \sup_{x\in\partial\Omega,\, r\in(0,R)} \Big(\tfrac{1}{\sigma(B(x,r)\cap\partial\Omega)} \int_{B(x,r)\cap\Omega} |\nabla\vec{u}|^p \operatorname{dist}(\cdot,\partial\Omega)^{p-1}\, d\mathcal{L}^n\Big)^{1/p}, \\ &\lim_{R\to\infty} \sup_{x\in\partial\Omega,\, r>R} \Big(\tfrac{1}{\sigma(B(x,r)\cap\partial\Omega)} \int_{B(x,r)\cap\Omega} |\nabla\vec{u}|^p \operatorname{dist}(\cdot,\partial\Omega)^{p-1}\, d\mathcal{L}^n\Big)^{1/p} \Bigg\} \\ &\leq C \operatorname{dist}\Big(\vec{f},\, \big[\mathrm{CMO}(\partial\Omega,\sigma)\big]^n\Big), \end{aligned} \tag{8.5.50}$$

with the distance in the right-hand side considered in $[\mathrm{BMO}(\partial\Omega,\sigma)]^n$ (recall that $\mathrm{CMO}(\partial\Omega,\sigma)$ *has been introduced in* (A.0.53)). *Thus, as a consequence of* (8.5.50) *and* [132, Definition 1.8.1], *given any* $p \in (1, \infty)$ *it follows that the solution* $(\vec{u}, \pi)$ *given in* (8.5.45)-(8.5.46) *of the boundary value problem* (8.5.43) *has the property that*

$$|\nabla \vec{u}|^p \operatorname{dist}(\cdot, \partial\Omega)^{p-1} \, d\mathcal{L}^n \text{ is a super vanishing Carleson measure in } \Omega, \\ \text{whenever the boundary datum } \vec{f} \text{ belongs to } \big[\mathrm{CMO}(\partial\Omega, \sigma)\big]^n. \tag{8.5.51}$$

In view of (8.5.49), Theorem 8.5.6 implies the solvability of the Dirichlet Problem for the Stokes system with data in the Sarason space VMO, formulated as

$$\begin{cases} \vec{u} \in \big[\mathscr{C}^\infty(\Omega)\big]^n, \quad \pi \in \mathscr{C}^\infty(\Omega), \\ \Delta\vec{u} - \nabla\pi = 0 \text{ and } \operatorname{div}\vec{u} = 0 \text{ in } \Omega, \\ |\nabla\vec{u}|^p \operatorname{dist}(\cdot, \partial\Omega)^{p-1} \, d\mathcal{L}^n \text{ is a vanishing Carleson} \\ \text{measure in } \Omega \text{ for each } p \in (1, \infty), \\ \vec{u}\big|_{\partial\Omega}^{\kappa-\mathrm{n.t.}} = \vec{f} \in \big[\mathrm{VMO}(\partial\Omega, \sigma)\big]^n, \end{cases} \tag{8.5.52}$$

which is contained in (8.5.43) as a sub-problem. A direct proof of this solvability result may be given based on item *(3)* of Theorem 4.2.3 and [132, (1.8.32)].

Here is the proof of Theorem 8.5.6.

Proof of Theorem 8.5.6 Theorem 3.1.5 ensures that having $\|\nu\|_{[\mathrm{BMO}(\partial\Omega,\sigma)]^n} < \delta$ for some sufficiently small $\delta \in (0, 1)$ makes Ω a UR domain with an unbounded boundary. After possibly further decreasing the value of $\delta \in (0, 1)$ we may rely on (4.2.122) to conclude that the operator (8.5.44) is an isomorphism follows. By virtue of (4.2.123), matters may also be arranged so that the operator

$$\tfrac{1}{2}I + \big[K^{\mathrm{mod}}_{\mathrm{slip}}\big] : \big[\mathrm{VMO}(\partial\Omega, \sigma)/\sim\big]^n \longrightarrow \big[\mathrm{VMO}(\partial\Omega, \sigma)/\sim\big]^n \tag{8.5.53}$$

is an isomorphism. Thanks to the invertibility of (8.5.44) and (8.5.53), the fact that $(\vec{u}, \pi)$ defined as in (8.5.45)-(8.5.46) solves the boundary value problem (8.5.43) and satisfies the properties claimed in (8.5.47)-(8.5.51) is seen from [129, (7.4.118)], [132, Theorem 6.2.6, items *(1)*, *(4)*, *(6)*], and the Open Mapping Theorem. □

Moving on, we shall now treat the Neumann Problem for the Stokes system in the theorem below, in a formulation which involves the slip conormal derivative introduced in (4.2.112).

Theorem 8.5.7 *Consider* $n \in \mathbb{N}$ *with* $n \geq 2$ *and let* $\Omega \subseteq \mathbb{R}^n$ *be an Ahlfors regular domain. Abbreviate* $\sigma := \mathcal{H}^{n-1}\lfloor\partial\Omega$ *and denote by* ν *the geometric measure theoretic outward unit normal to* Ω*. Also, fix an aperture parameter* $\kappa > 0$ *and pick an integrability exponent* $p \in (1, \infty)$*. In this setting, consider the following Neumann Problem for the Stokes system:*

$$\begin{cases} \vec{u} \in \left[\mathscr{C}^\infty(\Omega)\right]^n, \quad \pi \in \mathscr{C}^\infty(\Omega), \\ \Delta\vec{u} - \nabla\pi = 0 \text{ and } \operatorname{div}\vec{u} = 0 \text{ in } \Omega, \\ \mathcal{N}_\kappa(\nabla\vec{u}),\ \mathcal{N}_\kappa\pi \in L^p(\partial\Omega,\sigma), \\ \partial_\nu^{\text{slip}}(\vec{u},\pi) = \left[\nabla\vec{u} + (\nabla\vec{u})^\top\right]\Big|_{\partial\Omega}^{\kappa-\text{n.t.}} \nu - \left(\pi\big|_{\partial\Omega}^{\kappa-\text{n.t.}}\right)\nu = \vec{f} \in \left[L^p(\partial\Omega,\sigma)\right]^n. \end{cases} \tag{8.5.54}$$

There exists $\delta \in (0,1)$ which depends only on n, p, and the Ahlfors regularity constants of $\partial\Omega$ with the property that if

$$\|\nu\|_{[\text{BMO}(\partial\Omega,\sigma)]^n} < \delta \tag{8.5.55}$$

then Ω is a connected two-sided NTA *domain with unbounded boundary and the following properties are true:*

(a) [Existence, Uniqueness, Estimates, and Integral Representations] *The Neumann Problem* (8.5.54) *is well-posedness in the sense that it has a solution which is unique (modulo constants, in the case of the pressure vector), and which depends continuously on the boundary data. Specifically, if $K^{\#}_{\text{slip}}$ denotes the transpose slip double layer operator originally introduced in* (4.2.103)*, then*

$$-\tfrac{1}{2}I + K^{\#}_{\text{slip}} : \left[L^p(\partial\Omega,\sigma)\right]^n \longrightarrow \left[L^p(\partial\Omega,\sigma)\right]^n \tag{8.5.56}$$

is an isomorphism, and any solution $(\vec{u},\pi)$ of the Neumann Problem (8.5.54) *is of the form*

$$\vec{u} := \mathscr{S}_{\text{mod}}\Big(\big(-\tfrac{1}{2}I + K^{\#}_{\text{slip}}\big)^{-1}\vec{f}\Big) + c \text{ and } \pi := Q\Big(\big(-\tfrac{1}{2}I + K^{\#}_{\text{slip}}\big)^{-1}\vec{f}\Big) \tag{8.5.57}$$

(where $\mathscr{S}_{\text{mod}}$ is the modified boundary-to-domain single layer for the Stokes system introduced in (A.0.234)*, and Q is the corresponding pressure potential from* (A.0.213)*) for some constant $c \in \mathbb{C}^n$. In addition, any solution $(\vec{u},\pi)$ of the Neumann Problem* (8.5.54) *satisfies*

$$\left\|\mathcal{N}_\kappa\pi\right\|_{L^p(\partial\Omega,\sigma)} + \left\|\mathcal{N}_\kappa(\nabla\vec{u})\right\|_{L^p(\partial\Omega,\sigma)} \leq C\|\vec{f}\|_{[L^p(\partial\Omega,\sigma)]^n} \tag{8.5.58}$$

for some constant $C \in (0,\infty)$ independent of $\vec{f}$.

(b) [Other Spaces of Boundary Data] *Similar well-posedness results to those described in item (a) above are valid with $L^p(\partial\Omega,\sigma)$ replaced by any of the spaces $M^{p,\lambda}(\partial\Omega,\sigma)$, $\mathring{M}^{p,\lambda}(\partial\Omega,\sigma)$, and $\mathcal{B}^{q,\lambda}(\partial\Omega,\sigma)$ with $p,q \in (1,\infty)$ and $\lambda \in (0,n-1)$. More generally, given any Generalized Banach Function Space $\mathbb{X}$ on $(\partial\Omega,\sigma)$ with the property that*

$$\mathcal{M}_{\partial\Omega} : \mathbb{X} \to \mathbb{X} \text{ and } \mathcal{M}_{\partial\Omega} : \mathbb{X}' \to \mathbb{X}' \text{ are well-defined and bounded,} \tag{8.5.59}$$

where $\mathcal{M}_{\partial\Omega}$ is the Hardy-Littlewood maximal operator on $\partial\Omega$, and $\mathbb{X}'$ is the associated space of $\mathbb{X}$ (cf. [130, Definitions 5.1.4, 5.1.11]*), there exists some $\delta_{\mathbb{X}} \in (0,1)$ which depends only on n, the operator norms of $\mathcal{M}_{\partial\Omega}$ on $\mathbb{X}$ and $\mathbb{X}'$,* *as well as the Ahlfors regularity constants of* $\partial\Omega$, *such that if*

$$\|\nu\|_{[\mathrm{BMO}(\partial\Omega,\sigma)]^n} < \delta_{\mathbb{X}} \tag{8.5.60}$$

it follows that Ω *is a connected two-sided* NTA *domain with unbounded boundary, the operator*

$$-\tfrac{1}{2}I + K^{\#}_{\text{slip}} : [\mathbb{X}]^n \longrightarrow [\mathbb{X}]^n \text{ is an isomorphism} \tag{8.5.61}$$

and any solution of the Neumann Problem

$$\begin{cases} \vec{u} \in \big[\mathscr{C}^\infty(\Omega)\big]^n, \quad \pi \in \mathscr{C}^\infty(\Omega), \\ \Delta\vec{u} - \nabla\pi = 0 \text{ and } \operatorname{div}\vec{u} = 0 \text{ in } \Omega, \\ \mathcal{N}_\kappa(\nabla\vec{u}),\ \mathcal{N}_\kappa\pi \in \mathbb{X}, \\ \partial^{\text{slip}}_\nu(\vec{u},\pi) = \vec{f} \in \big[\mathbb{X}\big]^n, \end{cases} \tag{8.5.62}$$

is of the form

$$\begin{gathered} \vec{u} := \mathscr{S}_{\text{mod}}\Big(\big(-\tfrac{1}{2}I + K^{\#}_{\text{slip}}\big)^{-1}\vec{f}\Big) + c \text{ and } \pi := \mathcal{Q}\Big(\big(-\tfrac{1}{2}I + K^{\#}_{\text{slip}}\big)^{-1}\vec{f}\Big) \\ \text{where } \big(-\tfrac{1}{2}I + K^{\#}_{\text{slip}}\big)^{-1} \text{ is the inverse of the operator (8.5.61),} \end{gathered} \tag{8.5.63}$$

for some constant $c \in \mathbb{C}^n$. *Moreover, any solution* $(\vec{u},\pi)$ *of the Neumann Problem* (8.5.62) *satisfies* $\|\mathcal{N}_\kappa\vec{u}\|_{\mathbb{X}} \le C\|\vec{f}\|_{[\mathbb{X}]^n}$ *for a constant* $C \in (0,\infty)$ *independent of* $\vec{f}$. *In addition, a similar well-posedness result holds with* $\mathbb{X}$ *replaced by* $\mathring{\mathbb{X}} := \overline{L^\infty_{\text{comp}}(\partial\Omega,\sigma)}^{\|\cdot\|_{\mathbb{X}}}$. *As a corollary,*

> *granted* (8.5.60), *it follows that the Neumann Problem* (8.5.62) *is well posed if* $\mathbb{X}$ *is any rearrangement invariant Banach function space with Boyd indices* $1 < p_{\mathbb{X}} \le q_{\mathbb{X}} < \infty$. (8.5.64)

In particular,

> *under the assumption made in* (8.5.60), *it follows that the Neumann Problem* (8.5.62) *for* $\mathbb{X} := L^\Phi(\partial\Omega,\sigma)$, *the Orlicz space associated as in* (A.0.133) *with the measure space* $(\partial\Omega,\sigma)$ *and any Young function* Φ *satisfying* $1 < i(\Phi) \le I(\Phi) < \infty$. (8.5.65)

As in the case of (8.5.14), the Fatou-type result for the Stokes system proved in [132, Theorem 6.3.2] implies that whenever Ω is actually a UR domain (which may be arranged by assuming that (8.5.55) holds for some $\delta \in (0,1)$ sufficiently small; cf. Theorem 3.1.5) it follows that for any two functions $(\vec{u}, \pi)$ satisfying the properties listed in the first four lines of (8.5.54) the nontangential pointwise traces $\nabla\vec{u}\big|^{\kappa-\mathrm{n.t.}}_{\partial\Omega}$ and $\pi\big|^{\kappa-\mathrm{n.t.}}_{\partial\Omega}$ exist (in $\mathbb{C}^{n\times n}$ and $\mathbb{C}$, respectively) σ-a.e. on $\partial\Omega$, and the slip conormal derivative $\partial_\nu^{\mathrm{slip}}(\vec{u}, \pi)$ (as in (4.2.112)) belongs to the space $\big[L^p(\partial\Omega, \sigma)\big]^n$. (8.5.66)

In particular, this implies that whenever Ω is a UR domain, the boundary condition in (8.5.54) is meaningfully formulated, in an intrinsic fashion. It is also implicit in (8.5.66) that the choice of $\big[L^p(\partial\Omega, \sigma)\big]^n$ as the space of boundary data is optimal in the context of said problem.

Here is the proof of Theorem 8.5.7.

Proof of Theorem 8.5.7 That (8.5.55) with $\delta \in (0,1)$ small makes Ω a connected two-sided NTA domain with unbounded boundary follows from Theorem 3.1.5. By eventually further decreasing the value of $\delta \in (0,1)$, we may invoke Theorem 4.2.3 to ensure that the operator (8.5.56) is an isomorphism. Granted this, we conclude that the functions $(\vec{u}, \pi)$ in (8.5.57) are meaningfully defined. Furthermore, [132, Theorem 6.2.6] implies that the functions in question solve (for any choice of $c \in \mathbb{C}^n$) the Neumann Problem (8.5.54) and obey the estimate claimed in (8.5.58).

In the opposite direction, Corollary 8.5.2 tells us that any two given functions $(\vec{u}, \pi)$ satisfying the properties listed in the first four lines of (8.5.54) have the structure described in (8.5.17). In light of the jump-formula in [132, Theorem 6.2.6, item *(9)*], this structural result implies that $\big(-\frac{1}{2}I + K^{\#}_{\mathrm{slip}}\big)\vec{g} = \vec{f}$, from which we see that $\vec{g} = \big(-\frac{1}{2}I + K^{\#}_{\mathrm{slip}}\big)^{-1}\vec{f}$. Thus, any solution $(\vec{u}, \pi)$ of the Neumann Problem (8.5.54) has the form predicted in (8.5.57). In particular, we have uniqueness (modulo constants, for the velocity vector) in the Neumann Problem (8.5.54). This finishes the treatment of item *(a)*, and item *(b)* is handled similarly. □

We next turn our attention to the Neumann Problem with data from the Hardy space H^p with $\frac{n-1}{n} < p \leq 1$, for the Stokes system in sufficiently flat Ahlfors regular domains with unbounded boundaries in $\mathbb{R}^n$. In the formulation of this boundary value problem we shall employ the *weak slip conormal derivative* operator considered in a distributional sense, as $\overset{\bullet}{\partial}{}^{\lambda}_{\nu}$ in [132, Definition 6.2.9] with $\lambda := 1$. In the special case $p = 1$ we also allow the slip conormal derivative operator to be taken in a strong pointwise sense, as in (4.2.112).

Theorem 8.5.8 *Fix $n \in \mathbb{N}$ with $n \geq 2$, and let $\Omega \subseteq \mathbb{R}^n$ be an Ahlfors regular domain. Denote by ν the geometric measure theoretic outward unit normal to Ω, abbreviate $\sigma := \mathcal{H}^{n-1}\lfloor\partial\Omega$, and fix an aperture parameter $\kappa > 0$. In this setting, fix an exponent*

$$p \in \big(\tfrac{n-1}{n}, 1\big] \tag{8.5.67}$$

and consider the Neumann Problem

$$\begin{cases} \vec{u} \in \left[\mathscr{C}^\infty(\Omega)\right]^n, \quad \pi \in \mathscr{C}^\infty(\Omega), \\ \Delta\vec{u} - \nabla\pi = 0 \text{ and } \operatorname{div}\vec{u} = 0 \text{ in } \Omega, \\ \mathcal{N}_\kappa(\nabla\vec{u}),\ \mathcal{N}_\kappa\pi \in L^p(\partial\Omega, \sigma), \\ \overset{\bullet}{\partial}{}_\nu^{\text{slip}}(\vec{u}, \pi) = \vec{f} \in \left[H^p(\partial\Omega, \sigma)\right]^n, \end{cases} \tag{8.5.68}$$

where $\overset{\bullet}{\partial}{}_\nu^{\text{slip}}$ *denotes the weak slip conormal derivative considered as in* [132, Definition 6.2.9] *for the choice* $\lambda := 1$. *In relation this boundary value problem, the following statements are true:*

(a) [Existence, Uniqueness, Estimates, and Integral Representations] *There exists some threshold* $\delta \in (0, 1)$ *which depends only on n, p, and the Ahlfors regularity constants of* $\partial\Omega$ *such that if* $\|\nu\|_{[\mathrm{BMO}(\partial\Omega,\sigma)]^n} < \delta$ *(i.e., if* Ω *is a* δ-AR *domain in the sense of Definition 3.1.8) then* Ω *is a connected two-sided* NTA *domain with unbounded boundary, the operator* $-\frac{1}{2}I + K^{\#}_{\text{slip}}$ *is invertible on the Hardy space* $\left[H^p(\partial\Omega, \sigma)\right]^n$, *and the pair of functions* $(\vec{u}, \pi)$ *defined as*

$$\vec{u} := \mathcal{S}\Big(\big(-\tfrac{1}{2}I + K^{\#}_{\text{slip}}\big)^{-1}\vec{f}\Big) \text{ and } \pi := \mathcal{Q}\Big(\big(-\tfrac{1}{2}I + K^{\#}_{\text{slip}}\big)^{-1}\vec{f}\Big) \tag{8.5.69}$$

is a solution of the Neumann Problem (8.5.68) *satisfying, for some constant* $C \in (0, \infty)$ *independent of* $\vec{f}$,

$$\big\|\mathcal{N}_\kappa(\nabla\vec{u})\big\|_{L^p(\partial\Omega,\sigma)} + \big\|\mathcal{N}_\kappa\pi\big\|_{L^p(\partial\Omega,\sigma)} \leq C\|\vec{f}\|_{[H^p(\partial\Omega,\sigma)]^n}. \tag{8.5.70}$$

Moreover, the solution of the Neumann Problem (8.5.68) *is unique, modulo constants from* $\mathbb{C}^n$ *for the velocity vector. All in all, the Neumann Problem* (8.5.68) *is well posed.*

(b) [The Neumann Problem with a Strong Slip Conormal Derivative in H^1] *In the case* $p = 1$ *similar results to those in item (a) are valid for the Neumann Problem formulated with the strong slip conormal derivative (defined in a pointwise sense), i.e., for the boundary value problem*

$$\begin{cases} \vec{u} \in \left[\mathscr{C}^\infty(\Omega)\right]^n, \quad \pi \in \mathscr{C}^\infty(\Omega), \\ \Delta\vec{u} - \nabla\pi = 0 \text{ and } \operatorname{div}\vec{u} = 0 \text{ in } \Omega, \\ \mathcal{N}_\kappa(\nabla\vec{u}),\ \mathcal{N}_\kappa\pi \in L^1(\partial\Omega, \sigma), \\ \partial_\nu^{\text{slip}}(\vec{u}, \pi) = \Big[\nabla\vec{u} + (\nabla\vec{u})^\top\Big]\Big|_{\partial\Omega}^{\kappa-\text{n.t.}}\nu - \big(\pi\big|_{\partial\Omega}^{\kappa-\text{n.t.}}\big)\nu = \vec{f} \in \left[H^1(\partial\Omega, \sigma)\right]^n \\ \text{at } \sigma\text{-a.e. point on } \partial\Omega. \end{cases} \tag{8.5.71}$$

From [132, Theorem 6.2.10, item *(1)*] we know that the properties in first four lines of (8.5.68) imply that the weak slip conormal derivative $\overset{\bullet}{\partial}{}_\nu^{\text{slip}}(\vec{u}, \pi)$ is a well-defined distribution in the Hardy space $\left[H^p(\partial\Omega, \sigma)\right]^n$. In particular, this explains the choice of the space of boundary data in the Neumann Problem (8.5.68).

Together, [132, Theorem 6.2.10, items *(1)-(2)*] also show that the properties in first four lines of (8.5.71) imply that the strong (i.e., pointwise) slip conormal derivative

$\partial_\nu^{\rm slip}(\vec{u},\pi)$ is a well-defined function in the Hardy space $\big[H^1(\partial\Omega,\sigma)\big]^n$. Thus, the choice of the space of boundary data in the Neumann Problem (8.5.71) is natural.

Proof of Theorem 8.5.8 The claims in the current item *(a)* are justified much as those in item *(a)* of Theorem 8.5.7, now making use of item *(2)* in Theorem 4.2.3, and the Calderón-Zygmund theory for boundary layer operators associated with the Stokes system on Hardy spaces from [132, Theorem 6.2.11]. The claims in the present item *(b)* then follow from what we have proved so far, [132, Theorem 6.2.10, item *(3)*], and the Fatou-type result for null-solutions of the Stokes system from [132, Theorem 6.3.2]. □

We end this section by sampling, in the remark below, other solvability results for the Stokes system in sufficiently flat Ahlfors regular domains, which our tools and techniques can handle.

Remark 8.5.9 *Thanks to the invertibility results in Theorem 4.2.3 and the Calderón-Zygmund theory for boundary layer operators associated with the Stokes system developed in* [132, Theorem 6.2.4], [132, Theorem 6.2.6], *and* [132, Theorem 6.2.11], *similar solvability results (including estimates and integral representation formulas) to those presented in Theorem 8.4.15 for the Dirichlet Problem with boundary data in inhomogeneous Morrey-Campanato spaces* $\mathfrak{L}^{p,\lambda}$, *Theorem 8.4.29 for the Homogeneous Regularity Problem with boundary data in homogeneous Hardy-based Sobolev spaces* $\dot{H}^p_1$ *with* $\frac{n-1}{n} < p \leq 1$, *Theorem 8.4.36 for the Dirichlet Problem with boundary data in homogeneous Hölder spaces* $\dot{\mathscr{C}}^\alpha$ *with* $0 < \alpha < 1$, *Theorem 8.4.37 for the Dirichlet Problem with data in homogeneous vanishing Hölder spaces* $\dot{\mathscr{C}}^\alpha_{\rm van}$ *with* $0 < \alpha < 1$, *Theorem 8.4.38 for the Dirichlet Problem with boundary data in the* ϕ*-modulated space of functions of bounded mean oscillations* $\mathrm{BMO}_{\phi,p}$, *as well as Theorem 8.4.39 for the Dirichlet Problem with boundary data in the homogeneous and inhomogeneous* ϕ*-modulated space of functions of bounded central mean oscillations,* $\mathrm{BCMO}_{\phi,p}$ *and* $\dot{\mathrm{B}}\mathrm{CMO}_{\phi,p}$, *are also valid for the Stokes system, in the same geometric settings.*

Furthermore, much as in Theorem 8.4.32, the Transmission Problem for the Stokes system formulated as

$$\begin{cases} \vec{u}_\pm \in \big[\mathscr{C}^\infty(\Omega_\pm)\big]^n, \quad \pi_\pm \in \mathscr{C}^\infty(\Omega_\pm), \\ \Delta\vec{u}_\pm - \nabla\pi_\pm = 0, \ \operatorname{div}\vec{u}_\pm = 0 \ \textit{in} \ \Omega_\pm, \\ \mathcal{N}_\kappa(\nabla\vec{u}_\pm), \ \mathcal{N}_\kappa\pi_\pm \in L^p(\partial\Omega,\sigma), \\ \vec{u}_+\big|^{\kappa-\mathrm{n.t.}}_{\partial\Omega} - \vec{u}_-\big|^{\kappa-\mathrm{n.t.}}_{\partial\Omega} = \vec{f} \in \big[\dot{H}^p_1(\partial\Omega,\sigma)\big]^n, \\ \dot{\partial}_\nu^{\rm slip}(\vec{u}_+,\pi_+) - \eta\cdot\dot{\partial}_\nu^{\rm slip}(\vec{u}_-,\pi_-) = \vec{g} \in \big[H^p(\partial\Omega,\sigma)\big]^n, \end{cases} \tag{8.5.72}$$

is well posed (with uniqueness understood modulo constants for the velocity vectors) for each given $p \in \big(\frac{n-1}{n},\infty\big)$ *and coupling parameter* $\eta \in \mathbb{C}\setminus\{-1\}$, *assuming* $\Omega \subseteq \mathbb{R}^n$ *is a two-sided* NTA *domain with a sufficiently flat Ahlfors regular boundary. For the* L^p *Transmission Problem for the Stokes system in upper-graph Lipschitz domains (and Lipschitz domains with compact boundaries) the reader is referred to* [160].

8.6 Non-Fredholm Boundary Value Problems for Weakly Elliptic Systems

As already alluded to in §2, a wealth of boundary value problems for the certain types of second-order, homogeneous, symmetric, constant coefficient, weakly elliptic systems fail to be Fredholm solvable. This includes the Dirichlet Problem, the Inhomogeneous Regularity Problem, and the Homogeneous Regularity Problem for the $n \times n$ system $L_D := \Delta - 2\nabla\text{div}$ in the upper half-space $\mathbb{R}^n_+$, with $n \geq 2$ arbitrary, Bitsadze's operator $L_B := \partial^2_{\bar{z}}$ in the unit disk in the complex plane, or the complement of its closure, and the family of weakly elliptic operators $L_\lambda := \partial^2_{\bar{z}} - \lambda^2 \partial^2_z$, indexed by $\lambda \in (-1, 1) \setminus \{0\}$, in the upper half-plane. In all cases, the aforementioned failure of Fredholm solvability manifests itself at all levels, namely both the space of null-solutions is infinite dimensional and the space of admissible boundary data has infinite codimension in the full space of boundary data. This phenomenon is remarkable in and of itself, and in this case may be attributed to the lack of existence of a distinguished coefficient tensor for the aforementioned systems/operators. In spite of this, we can precisely describe both the space of null-solutions and the space of admissible boundary data, by providing intrinsic characterizations in each case.

We shall first carry out this program for the system L_D in $\mathbb{R}^n_+$, staring with the Dirichlet Problem which we treat in Theorem 8.6.1 below. This stands in sharp contrast with the fact that the Dirichlet Problem for any symmetric homogeneous constant coefficient second-order weakly elliptic system possessing a distinguished coefficient tensor is Fredholm solvable in any sufficiently flat Ahlfors regular domain with an unbounded boundary (cf. Theorem 8.4.3).

Theorem 8.6.1 *Pick an integer $n \in \mathbb{N}$ satisfying $n \geq 2$, and choose an integrability exponent $p \in (1, \infty)$ together with an aperture parameter $\kappa \in (0, \infty)$. Formulate the L^p Dirichlet Problem for the system $L_D = \Delta - 2\nabla\text{div}$ in the upper half-space with boundary datum*

$$\vec{f} = (f_1, \dots, f_n) \in \big[L^p(\mathbb{R}^{n-1}, \mathcal{L}^{n-1})\big]^n \tag{8.6.1}$$

as the task of finding a vector-valued function $\vec{u}$ satisfying

$$\begin{cases} \vec{u} \in \big[\mathscr{C}^\infty(\mathbb{R}^n_+)\big]^n, \quad \Delta\vec{u} - 2\nabla\text{div}\vec{u} = 0 \text{ in } \mathbb{R}^n_+, \\ \mathcal{N}_\kappa \vec{u} \in L^p(\mathbb{R}^{n-1}, \mathcal{L}^{n-1}), \\ \vec{u}\big|^{\kappa-\text{n.t.}}_{\partial\mathbb{R}^n_+} = \vec{f} \text{ at } \mathcal{L}^{n-1}\text{-a.e. point on } \mathbb{R}^{n-1}. \end{cases} \tag{8.6.2}$$

Then the boundary value problem (8.6.2) *has a solution if and only if the boundary datum satisfies the compatibility condition*

$$f_n = -\sum_{j=1}^{n-1} R_j f_j \tag{8.6.3}$$

where R_j with $1 \leq j \leq n-1$ are the standard Riesz transforms in $\mathbb{R}^{n-1}$ (defined as in (A.0.215) *with $\Sigma := \mathbb{R}^{n-1}$), acting from the Lebesgue space $L^p(\mathbb{R}^{n-1}, \mathcal{L}^{n-1})$ into* *itself. The space of admissible boundary data for* (8.6.2)*, i.e., the collection of all* $\vec{f}$ *as in* (8.6.1) *satisfying* (8.6.3)*, is a closed infinite codimensional subspace of the full space of boundary data* $\big[L^p(\mathbb{R}^{n-1},\mathcal{L}^{n-1})\big]^n$.

Furthermore, whenever (8.6.3) *holds, any solution* $\vec{u}$ *of* (8.6.2) *has the form*

$$\begin{gathered}\vec{u}(x) = \big(P^{\Delta}_{x_n} * \vec{f}\big)(x') + x_n(\nabla w)(x)\\ \textit{for each } x=(x',x_n)\in\mathbb{R}^{n-1}\times(0,\infty)=\mathbb{R}^n_+,\end{gathered} \tag{8.6.4}$$

where w *is any scalar function satisfying*

$$\begin{gathered}w\in\mathscr{C}^\infty(\mathbb{R}^n_+),\quad \Delta w=0 \textit{ in } \mathbb{R}^n_+,\quad \mathcal{N}_\kappa\big(x\mapsto x_n(\nabla w)(x)\big)\in L^p(\mathbb{R}^{n-1},\mathcal{L}^{n-1}).\\ \textit{and } \big[x\mapsto x_n(\nabla w)(x)\big]\Big|^{\kappa-\mathrm{n.t.}}_{\partial\mathbb{R}^n_+}=0 \textit{ at } \mathcal{L}^{n-1}\textit{-a.e. point on } \mathbb{R}^{n-1}.\end{gathered} \tag{8.6.5}$$

As a consequence, the space of null-solutions for the L^p *Dirichlet Problem* (8.6.2) *is the infinite dimensional space*

$$\Big\{x_n\nabla w : w\in\mathscr{C}^\infty(\mathbb{R}^n_+),\ \Delta w=0 \textit{ in } \mathbb{R}^n_+,\ \big[x\mapsto x_n(\nabla w)(x)\big]\Big|^{\kappa-\mathrm{n.t.}}_{\partial\mathbb{R}^n_+}=0,\\ \textit{and } \mathcal{N}_\kappa\big(x\mapsto x_n(\nabla w)(x)\big)\in L^p(\mathbb{R}^{n-1},\mathcal{L}^{n-1})\Big\}. \tag{8.6.6}$$

Finally, similar results to those above are valid for Muckenhoupt weighted Lebesgue spaces, Morrey spaces, vanishing Morrey spaces, and block spaces.

Proof That the boundary value problem (8.6.2) has a solution if and only if the boundary datum $\vec{f}$ satisfies the compatibility condition (8.6.3) is clear from Theorem 2.6.4. Also, the fact that the space of admissible boundary data for (8.6.2) is a closed subspace of the full space of boundary data $\big[L^p(\mathbb{R}^{n-1},\mathcal{L}^{n-1})\big]^n$, of infinite codimension, is seen from Corollary 2.6.5. In the case that (8.6.3) holds, Theorem 2.5.4 implies (bearing in mind the format of the unique solution of the L^p Dirichlet Problem for the Laplacian in $\mathbb{R}^n_+$) that any solution $\vec{u}$ of (8.6.2) may be expressed as in (8.6.4)-(8.6.5). From (2.6.13) and Corollary 2.6.3 we see the space of null-solutions for the L^p Dirichlet Problem (8.6.2) is the infinite dimensional space described in (8.6.6). Remark 2.6.11 gives the last claim in the statement. □

We continue by discussing the Inhomogeneous Regularity Problem in the upper half-space for the system $L_D=\Delta-2\nabla\mathrm{div}$ in $\mathbb{R}^n_+$. Once again, the focus is on the lack of Fredholm solvability, something that strongly contrasts with the fact that the Inhomogeneous Regularity Problem for any symmetric homogeneous constant coefficient second-order weakly elliptic system possessing a distinguished coefficient tensor is Fredholm solvable in any sufficiently flat Ahlfors regular domain with an unbounded boundary (cf. Theorem 8.4.1, and (8.4.10) in particular).

Theorem 8.6.2 *Fix an integer* $n\in\mathbb{N}$ *satisfying* $n\geq 2$*, and pick an integrability exponent* $p\in(1,\infty)$ *along with an aperture parameter* $\kappa\in(0,\infty)$*. Consider the* L^p_1*-Inhomogeneous Regularity Problem for the system* $L_D=\Delta-2\nabla\mathrm{div}$ *in the upper half-space with boundary datum*

$$\vec{f} = (f_1, \dots, f_n) \in \left[L^p_1(\mathbb{R}^{n-1}, \mathcal{L}^{n-1})\right]^n, \tag{8.6.7}$$

i.e.,

$$\begin{cases} \vec{u} \in \left[\mathscr{C}^\infty(\mathbb{R}^n_+)\right]^n, \quad \Delta\vec{u} - 2\nabla \mathrm{div}\vec{u} = 0 \text{ in } \mathbb{R}^n_+, \\ \mathcal{N}_\kappa \vec{u} \in L^p(\mathbb{R}^{n-1}, \mathcal{L}^{n-1}), \quad \mathcal{N}_\kappa(\nabla\vec{u}) \in L^p(\mathbb{R}^{n-1}, \mathcal{L}^{n-1}), \\ \vec{u}\big|^{\kappa-\mathrm{n.t.}}_{\partial\mathbb{R}^n_+} = \vec{f} \text{ at } \mathcal{L}^{n-1}\text{-a.e. point on } \mathbb{R}^{n-1}. \end{cases} \tag{8.6.8}$$

Then the boundary value problem (8.6.8) *has a solution if and only if the boundary datum satisfies the compatibility condition*

$$f_n = -\sum_{j=1}^{n-1} R_j f_j \tag{8.6.9}$$

where R_j with $1 \le j \le n-1$ are the standard Riesz transforms in $\mathbb{R}^{n-1}$ (defined as in (A.0.215) *with $\Sigma := \mathbb{R}^{n-1}$), acting on Sobolev spaces of order one as in* [132, (5.3.12)]. *The space of admissible boundary data for* (8.6.8)*, i.e., the collection of all $\vec{f}$ as in* (8.6.7) *satisfying* (8.6.9)*, is a closed infinite codimensional subspace of the full space of boundary data $\left[L^p_1(\mathbb{R}^{n-1}, \mathcal{L}^{n-1})\right]^n$.*

In addition, whenever (8.6.9) *holds, any solution $\vec{u}$ of* (8.6.8) *has the form*

$$\vec{u}(x) = \big(P^\Delta_{x_n} * \vec{f}\big)(x') + x_n(\nabla w)(x) \\ \text{for each } x = (x', x_n) \in \mathbb{R}^{n-1} \times (0, \infty) = \mathbb{R}^n_+, \tag{8.6.10}$$

where w is any scalar function satisfying

$$w \in \mathscr{C}^\infty(\mathbb{R}^n_+), \quad \Delta w = 0 \text{ in } \mathbb{R}^n_+, \quad \mathcal{N}_\kappa(\nabla w) \in L^p(\mathbb{R}^{n-1}, \mathcal{L}^{n-1}), \\ \text{and } \mathcal{N}_\kappa\big(x \mapsto x_n(\nabla w)(x)\big) \in L^p(\mathbb{R}^{n-1}, \mathcal{L}^{n-1}). \tag{8.6.11}$$

As corollary, the space of null-solutions for the L^p_1-Inhomogeneous Regularity Problem (8.6.8) *is the infinite dimensional space*

$$\Big\{x_n \nabla w : w \in \mathscr{C}^\infty(\mathbb{R}^n_+), \ \Delta w = 0 \text{ in } \mathbb{R}^n_+, \ \mathcal{N}_\kappa(\nabla w) \in L^p(\mathbb{R}^{n-1}, \mathcal{L}^{n-1}), \\ \text{and } \mathcal{N}_\kappa\big(x \mapsto x_n(\nabla w)(x)\big) \in L^p(\mathbb{R}^{n-1}, \mathcal{L}^{n-1})\Big\}. \tag{8.6.12}$$

Finally, similar results to those described above are valid for Muckenhoupt weighted Sobolev spaces, Morrey-based Sobolev spaces, vanishing Morrey-based Sobolev spaces, and block-based Sobolev spaces (all inhomogeneous, of order one).

Proof From the last part in the statement of Theorem 2.6.7 we see that the boundary value problem (8.6.8) has a solution if and only if the boundary datum $\vec{f}$ satisfies the compatibility condition (8.6.9). Whenever the latter membership materializes, Theorem 2.5.1 implies (keeping in mind the format of the unique solution of the L^p_1 Regularity Problem for the Laplacian in $\mathbb{R}^n_+$) that any solution $\vec{u}$ of (8.6.8) may be written as in (8.6.10)-(8.6.11). That the space of null-solutions for the problem

(8.6.8) is the infinite dimensional space (8.6.12) follows from Corollary 2.6.2. The very last claim in the statement is implied by Remark 2.6.11. □

We can generalize Theorem 8.6.2 by considering higher order Sobolev spaces as the full space of boundary data. This gives rise to what we call the Higher Order Regularity Problem for the system $L_D = \Delta - 2\nabla\mathrm{div}$. Its formulation now demands nontangential maximal function control for higher order derivatives of the solution. The precise statement is given in the theorem below, whose proof proceeds via induction on the order of smoothness. The initial step is contained in Theorem 8.6.2, while the inductive step makes crucial use of the "structure theorems" from §2.5.

Theorem 8.6.3 *Pick two integers, $n \in \mathbb{N}$ satisfying $n \geq 2$, and $k \in \mathbb{N}$ arbitrary. Also, pick an integrability exponent $p \in (1, \infty)$ along with an aperture parameter $\kappa \in (0, \infty)$. In this setting, consider the Higher Order Regularity Problem for the system $L_D = \Delta - 2\nabla\mathrm{div}$ in the upper half-space with boundary datum in the higher order Sobolev space*

$$\vec{f} = (f_1, \dots, f_n) \in \big[L^p_k(\mathbb{R}^{n-1}, \mathcal{L}^{n-1})\big]^n, \tag{8.6.13}$$

i.e.,

$$\begin{cases} \vec{u} \in \big[\mathscr{C}^\infty(\mathbb{R}^n_+)\big]^n, \quad \Delta\vec{u} - 2\nabla\mathrm{div}\vec{u} = 0 \ \text{in} \ \mathbb{R}^n_+, \\ \mathcal{N}_\kappa(\nabla^\ell \vec{u}) \in L^p(\mathbb{R}^{n-1}, \mathcal{L}^{n-1}) \ \text{for} \ \ell = 0, 1, \dots, k, \\ \vec{u}\big|^{\kappa-\mathrm{n.t.}}_{\partial\mathbb{R}^n_+} = \vec{f} \ \text{at} \ \mathcal{L}^{n-1}\text{-a.e. point on} \ \mathbb{R}^{n-1}. \end{cases} \tag{8.6.14}$$

Then the boundary value problem (8.6.14) *has a solution if and only if the boundary datum satisfies the compatibility condition*

$$f_n = -\sum_{j=1}^{n-1} R_j f_j \tag{8.6.15}$$

where R_j with $1 \leq j \leq n-1$ are the standard Riesz transforms in $\mathbb{R}^{n-1}$ (defined as in (A.0.215) *with $\Sigma := \mathbb{R}^{n-1}$), acting on higher order Sobolev spaces. The space of admissible boundary data for* (8.6.14)*, i.e., the collection of all $\vec{f}$ as in* (8.6.13) *satisfying* (8.6.15)*, is a closed infinite codimensional subspace of the full space of boundary data $\big[L^p_k(\mathbb{R}^{n-1}, \mathcal{L}^{n-1})\big]^n$.*

Moreover, whenever (8.6.15) *holds, any solution $\vec{u}$ of* (8.6.14) *has the form*

$$\begin{aligned} &\vec{u}(x) = \big(P^\Delta_{x_n} * \vec{f}\big)(x') + x_n(\nabla w)(x) \\ &\text{for each } x = (x', x_n) \in \mathbb{R}^{n-1} \times (0, \infty) = \mathbb{R}^n_+, \end{aligned} \tag{8.6.16}$$

where w is any scalar function satisfying

$$\begin{aligned} &w \in \mathscr{C}^\infty(\mathbb{R}^n_+), \quad \Delta w = 0 \ \text{in} \ \mathbb{R}^n_+, \quad \mathcal{N}_\kappa\big(x \mapsto x_n(\nabla w)(x)\big) \in L^p(\mathbb{R}^{n-1}, \mathcal{L}^{n-1}), \\ &\text{and } \mathcal{N}_\kappa(\nabla^\ell w) \in L^p(\mathbb{R}^{n-1}, \mathcal{L}^{n-1}) \ \text{for} \ \ell = 1, \dots, k. \end{aligned} \tag{8.6.17}$$

In particular, the space of null-solutions for the Higher Order Regularity Problem (8.6.14) *is the infinite dimensional space*

$$\Big\{x_n\nabla w : w \in \mathscr{C}^\infty(\mathbb{R}^n_+),\ \Delta w = 0 \ \textit{in}\ \mathbb{R}^n_+,\ \mathcal{N}_\kappa(\nabla^\ell w) \in L^p(\mathbb{R}^{n-1},\mathcal{L}^{n-1})\ \textit{for}\ 1 \le \ell \le k,$$
$$\textit{and}\ \mathcal{N}_\kappa\big(x \mapsto x_n(\nabla w)(x)\big) \in L^p(\mathbb{R}^{n-1},\mathcal{L}^{n-1})\Big\}. \tag{8.6.18}$$

Proof For starters, the description of the space of admissible boundary data for the Higher-Order Inhomogeneous Regularity Problem provided in Theorem 2.6.7 shows that (8.6.14) has a solution if and only if the boundary datum $\vec{f}$ satisfies the compatibility condition (8.6.15). Corollary 2.6.8 implies that the space of admissible boundary data for the problem (8.6.14) is a closed subspace of the full space of boundary data $\big[L^p_k(\mathbb{R}^{n-1},\mathcal{L}^{n-1})\big]^n$, of infinite codimension.

The fact that if (8.6.15) holds then for any w as in (8.6.17) the vector-valued function $\vec{u}$ defined as in (8.6.16) is a solution of (8.6.14) is seen from definitions, Corollary 2.4.2, and the last part in [129, Corollary 8.9.13].

Next, assume that (8.6.15) holds and $\vec{u}$ is an arbitrary solution of (8.6.14). The goal is to show that $\vec{u}$ may be expressed as in (8.6.16)-(8.6.17). For starters, Theorem 8.6.2 does guarantee that $\vec{u}$ may be written as in (8.6.16) for some scalar function w satisfying

$$\begin{gathered} w \in \mathscr{C}^\infty(\mathbb{R}^n_+), \quad \Delta w = 0 \text{ in } \mathbb{R}^n_+, \\ \mathcal{N}_\kappa\big(x \mapsto x_n(\nabla w)(x)\big) \in L^p(\mathbb{R}^{n-1},\mathcal{L}^{n-1}), \\ \text{and } \mathcal{N}_\kappa(\nabla w) \in L^p(\mathbb{R}^{n-1},\mathcal{L}^{n-1}). \end{gathered} \tag{8.6.19}$$

We shall prove via induction that $\mathcal{N}_\kappa(\nabla^\ell w)$ belongs to $L^p(\mathbb{R}^{n-1},\mathcal{L}^{n-1})$ for each index $\ell \in \{1,\dots,k\}$. The case $\ell = 1$ is contained in (8.6.19). To justify the inductive step, assume that

$$\mathcal{N}_\kappa(\nabla^\ell w) \in L^p(\mathbb{R}^{n-1},\mathcal{L}^{n-1}) \text{ for some } \ell \in \{1,\dots,k-1\}, \tag{8.6.20}$$

the goal being to show that

$$\mathcal{N}_\kappa(\nabla^{\ell+1} w) \in L^p(\mathbb{R}^{n-1},\mathcal{L}^{n-1}). \tag{8.6.21}$$

To this end, apply ∂_n^ℓ to (8.6.16) to obtain

$$\begin{gathered} (\partial_n^\ell \vec{u})(x) = \partial_n^\ell\big[\big(P^\Delta_{x_n} * \vec{f}\big)(x')\big] + x_n(\nabla\partial_n^\ell w)(x) + \ell(\nabla\partial_n^{\ell-1} w)(x) \\ \text{for each } x = (x',x_n) \in \mathbb{R}^{n-1}\times(0,\infty) = \mathbb{R}^n_+. \end{gathered} \tag{8.6.22}$$

As seen from (8.6.14), we have

$$\mathcal{N}_\kappa(\partial_n^\ell \vec{u}) \in L^p(\mathbb{R}^{n-1},\mathcal{L}^{n-1}) \tag{8.6.23}$$

while the discussion in the first part of §2.2 and Proposition 2.2.1 imply that

$$\text{the nontangential maximal operator of the function}\\ \mathbb{R}^n_+ \ni (x', x_n) \mapsto \partial_n^\ell\big[\big(P^\Delta_{x_n} * \vec{f}\big)(x')\big] \text{ belongs to } L^p(\mathbb{R}^{n-1}, \mathcal{L}^{n-1}). \tag{8.6.24}$$

To proceed, define

$$\vec{U} := \partial_n^\ell \vec{u}, \qquad W := \partial_n^\ell w, \tag{8.6.25}$$

and

$$\vec{V}(x) := \partial_n^\ell\big[\big(P^\Delta_{x_n} * \vec{f}\big)(x')\big] + \ell(\nabla\partial_n^{\ell-1} w)(x)\\ \text{for each } x = (x', x_n) \in \mathbb{R}^{n-1} \times (0, \infty) = \mathbb{R}^n_+. \tag{8.6.26}$$

Then (8.6.25)-(8.6.26) together with (8.6.22) imply

$$\vec{U} = \vec{V} + x_n \nabla W \text{ in } \mathbb{R}^n_+ \tag{8.6.27}$$

while (8.6.14) guarantees (bearing in mind that $\ell \leq k-1$) that

$$\begin{cases} \vec{U} \in \big[\mathscr{C}^\infty(\mathbb{R}^n_+)\big]^n, \quad \Delta\vec{U} - 2\nabla\text{div}\vec{U} = 0 \text{ in } \mathbb{R}^n_+,\\ \mathcal{N}_\kappa \vec{U} \in L^p(\mathbb{R}^{n-1}, \mathcal{L}^{n-1}), \ \mathcal{N}_\kappa(\nabla\vec{U}) \in L^p(\mathbb{R}^{n-1}, \mathcal{L}^{n-1}). \end{cases} \tag{8.6.28}$$

Also, (8.6.26), the first line in (8.6.19), (2.4.33), (8.6.24), and (8.6.20) give

$$\begin{cases} \vec{V} \in \big[\mathscr{C}^\infty(\mathbb{R}^n_+)\big]^n,\\ \Delta\vec{V} = 0, \ \text{div}\vec{V} = 0, \ \text{in } \mathbb{R}^n_+,\\ \mathcal{N}_\kappa \vec{V} \in L^p(\mathbb{R}^{n-1}, \mathcal{L}^{n-1}), \end{cases} \tag{8.6.29}$$

whereas from (8.6.25), the first line in (8.6.19), the first membership in the second line of (8.6.28), and the last line of (8.6.29) we deduce that

$$\begin{cases} W \in \mathscr{C}^\infty(\mathbb{R}^n_+), \quad \Delta W = 0 \text{ in } \mathbb{R}^n_+,\\ \mathcal{N}_\kappa\big(x \mapsto x_n(\nabla W)(x)\big) \in L^p(\mathbb{R}^{n-1}, \mathcal{L}^{n-1}). \end{cases} \tag{8.6.30}$$

In view of (8.6.27)-(8.6.30), Theorem 2.5.4 applies, and (2.5.57) gives that

$$\mathcal{N}_\kappa(\nabla W) \in L^p(\mathbb{R}^{n-1}, \mathcal{L}^{n-1}). \tag{8.6.31}$$

We then conclude from (8.6.31) and (8.6.25) that

$$\mathcal{N}_\kappa(\nabla\partial_n^\ell w) \in L^p(\mathbb{R}^{n-1}, \mathcal{L}^{n-1}). \tag{8.6.32}$$

As in (2.5.91), fix $\varepsilon \in \big(0, \frac{\kappa}{4+2\kappa}\big)$. Based on interior estimates for harmonic functions, much as in (2.5.94) write

$$\begin{aligned} |(\nabla^{\ell+1} w)(z)| &\leq \frac{C}{t}\Big(⨍_{B(z,\varepsilon t)} |(\nabla^\ell w)(x)|^p \,dx\Big)^{\frac{1}{p}} \\ &\leq \frac{C}{t}\Big(⨍_{B_{n-1}(z',\kappa t/2)} \mathcal{N}_\kappa(\nabla^\ell w)^p \,d\mathcal{L}^{n-1}\Big)^{\frac{1}{p}} \leq \frac{C}{t^{1+(n-1)/p}}\big\|\mathcal{N}_\kappa(\nabla^\ell w)\big\|_{L^p(\mathbb{R}^{n-1}, \mathcal{L}^{n-1})}, \end{aligned} \tag{8.6.33}$$

for each point $z = (z', t) \in \mathbb{R}^{n-1} \times (0, \infty) = \mathbb{R}^n_+$. In concert with (8.6.20), this implies

$$\lim_{t\to\infty} (\nabla^{\ell+1} w)(z', t) = 0 \text{ for each fixed } z' \in \mathbb{R}^{n-1}. \tag{8.6.34}$$

In a very similar fashion, for each point $z = (z', t) \in \mathbb{R}^{n-1} \times (0, \infty) = \mathbb{R}^n_+$ we have

$$\begin{aligned} |(\nabla^{\ell+2} w)(z)| &\leq \frac{C}{t^2} \Big(\fint_{B(z,\varepsilon t)} |(\nabla^\ell w)(x)|^p \, dx \Big)^{\frac{1}{p}} \\ &\leq \frac{C}{t^2} \Big(\fint_{B_{n-1}(z',\kappa t/2)} \mathcal{N}_\kappa (\nabla^\ell w)^p \, d\mathcal{L}^{n-1} \Big)^{\frac{1}{p}} \leq \frac{C}{t^{2+(n-1)/p}} \big\| \mathcal{N}_\kappa (\nabla^\ell w) \big\|_{L^p(\mathbb{R}^{n-1}, \mathcal{L}^{n-1})}, \end{aligned} \tag{8.6.35}$$

which, thanks to (8.6.20), also implies

$$\lim_{t\to\infty} (\nabla^{\ell+2} w)(z', t) = 0 \text{ for each fixed } z' \in \mathbb{R}^{n-1}. \tag{8.6.36}$$

In turn, from (8.6.34), (8.6.36), (8.6.32), and repeated applications of Lemma 2.5.3 we conclude that the membership in (8.6.21) holds. This concludes the proof of the fact that any solution $\vec{u}$ of (8.6.14) has the form (8.6.16) for some w as in (8.6.17).

That the space of null-solutions for the Higher Order Regularity Problem (8.6.14) is the space described in (8.6.18) is a consequence of what we have proved above. There remains to show that the space described in (8.6.18) is infinite dimensional. For this, consider the mapping

$$\begin{gathered} L^p_k(\mathbb{R}^{n-1}, \mathcal{L}^{n-1}) \ni f \mapsto \Lambda f \text{ where} \\ (\Lambda f)(x) := x_n \nabla \big[\big(P^\Delta_{x_n} * f \big)(x') \big] \text{ for each } (x', x_n) \in \mathbb{R}^n_+. \end{gathered} \tag{8.6.37}$$

From this and Proposition 2.2.1 we see that for each $f \in L^p_k(\mathbb{R}^{n-1}, \mathcal{L}^{n-1})$ we have

$$\begin{gathered} \Lambda f \in \mathscr{C}^\infty(\mathbb{R}^n_+), \ \Delta(\Lambda f) = 0 \text{ in } \mathbb{R}^n_+, \text{ and} \\ \mathcal{N}_\kappa(\nabla^\ell \Lambda f) \in L^p(\mathbb{R}^{n-1}, \mathcal{L}^{n-1}) \text{ for } 0 \leq \ell \leq k. \end{gathered} \tag{8.6.38}$$

In addition, the last membership above with $\ell = 0$ and the last part in [129, Corollary 8.9.13] ensure that for each $f \in L^p_k(\mathbb{R}^{n-1}, \mathcal{L}^{n-1})$ we also have

$$\mathcal{N}_\kappa\big(x \mapsto x_n (\nabla \Lambda f)(x)\big) \in L^p(\mathbb{R}^{n-1}, \mathcal{L}^{n-1}). \tag{8.6.39}$$

As a consequence of the definition and properties recorded in (8.6.37)-(8.6.39), the assignment sending each $f \in L^p_k(\mathbb{R}^{n-1}, \mathcal{L}^{n-1})$ into Λf is a well-defined linear mapping taking values into the space described in (8.6.18). Since from (8.6.37) it is also clear that the mapping in question is injective, and since $L^p_k(\mathbb{R}^{n-1}, \mathcal{L}^{n-1})$ is infinite dimensional, we ultimately conclude that the space described in (8.6.18) is indeed infinite dimensional. □

To complete the discussion pertaining to the system $L_D = \Delta - 2\nabla\text{div}$ we shall also treat the Homogeneous Regularity Problem in the upper half-space for this system. The principal aim is to elucidate the issues of solvability, uniqueness, and structure of solutions. Much like the other problems discussed earlier, this boundary value problem fails to be Fredholm solvable, in stark contrast with the fact that the Homogeneous Regularity Problem for any symmetric homogeneous constant coefficient second-order weakly elliptic system possessing a distinguished coefficient tensor is Fredholm solvable in any sufficiently flat Ahlfors regular domain with an unbounded boundary (cf. Theorem 8.4.1). We can, nonetheless, prove that this boundary value problem has "index zero," in the sense that we have an isomorphism[106]

$$\frac{\text{Full Space of Boundary Data}}{\text{Space of Admissible Boundary Data}} \cong \text{Space of Null-Solutions.} \tag{8.6.40}$$

Here is a precise statement.

Theorem 8.6.4 *Fix an integer $n \in \mathbb{N}$ satisfying $n \geq 2$, and pick an integrability exponent $p \in (1, \infty)$ along with an aperture parameter $\kappa \in (0, \infty)$. Consider the $\dot{L}^p_1$-Homogeneous Regularity Problem for the system $L_D = \Delta - 2\nabla\mathrm{div}$ in the upper half-space with boundary datum*

$$\vec{f} = (f_1, \dots, f_n) \in \big[\dot{L}^p_1(\mathbb{R}^{n-1}, \mathcal{L}^{n-1})\big]^n, \tag{8.6.41}$$

i.e.,

$$\begin{cases} \vec{u} \in \big[\mathscr{C}^\infty(\mathbb{R}^n_+)\big]^n, \quad \Delta\vec{u} - 2\nabla\mathrm{div}\vec{u} = 0 \ \text{ in } \mathbb{R}^n_+, \\ \mathcal{N}_\kappa(\nabla\vec{u}) \in L^p(\mathbb{R}^{n-1}, \mathcal{L}^{n-1}), \\ \vec{u}\big|^{\kappa-\mathrm{n.t.}}_{\partial\mathbb{R}^n_+} = \vec{f} \ \text{ at } \mathcal{L}^{n-1}\text{-a.e. point on } \mathbb{R}^{n-1}. \end{cases} \tag{8.6.42}$$

Then (8.6.42) has a solution if and only if the boundary datum satisfies the compatibility condition

$$f_n = -\sum_{j=1}^{n-1} R^{\mathrm{mod}}_j f_j + c \ \text{ for some constant } c \in \mathbb{C}, \tag{8.6.43}$$

where R^{mod}_j with $1 \leq j \leq n-1$ are the modified Riesz transforms in $\mathbb{R}^{n-1}$ (defined in (A.0.222)) acting from the homogeneous Sobolev space $\dot{L}^p_1(\mathbb{R}^{n-1}, \mathcal{L}^{n-1})$ into itself (cf. [132, (5.3.56)]). In other words, the space of admissible boundary data for (8.6.42), i.e.,

$$\begin{aligned} \dot{\mathscr{U}}^p_1 := \Big\{ \vec{u}\big|^{\kappa-\mathrm{n.t.}}_{\partial\mathbb{R}^n_+} : \ \vec{u} \in \big[\mathscr{C}^\infty(\mathbb{R}^n_+)\big]^n, \ \Delta\vec{u} - 2\nabla\mathrm{div}\vec{u} = 0 \ \text{ in } \mathbb{R}^n_+, \\ \text{and } \mathcal{N}_\kappa(\nabla\vec{u}) \in L^p(\mathbb{R}^{n-1}, \mathcal{L}^{n-1}) \Big\}, \end{aligned} \tag{8.6.44}$$

is precisely the collection of all vector-valued functions $\vec{f}$ as in (8.6.13) satisfying (8.6.15). Also, if $\dot{\mathscr{U}}^p_1/\!\sim$ denotes the space of equivalence classes, modulo constants,

---

[106] In the case of Fredholm solvable boundary value problems, these would be finite dimensional spaces, a scenario in which the existence of an isomorphism simply reduces to the matter of them having equal (finite) dimensions.

of admissible boundary data for (8.6.42), *then* $\dot{\mathscr{U}}_1^p/\sim$ *is a closed infinite codimensional subspace of* $\big[\dot{L}_1^p(\mathbb{R}^{n-1},\mathcal{L}^{n-1})/\sim\big]^n$, *the full space of boundary data modulo constants. Moreover, when* (8.6.43) *holds, any solution* $\vec{u}$ *of* (8.6.42) *has the form*

$$\begin{gathered}\vec{u}(x) = \big(P^{\Delta}_{x_n} * \vec{f}\,\big)(x') + x_n(\nabla w)(x)\\ \textit{for each } x = (x', x_n) \in \mathbb{R}^{n-1}\times(0,\infty) = \mathbb{R}^n_+,\end{gathered} \tag{8.6.45}$$

where w *is any scalar function satisfying*

$$w \in \mathscr{C}^\infty(\mathbb{R}^n_+), \quad \Delta w = 0 \ \textit{in} \ \mathbb{R}^n_+, \quad \mathcal{N}_\kappa(\nabla w) \in L^p(\mathbb{R}^{n-1},\mathcal{L}^{n-1}). \tag{8.6.46}$$

In particular, the space of null-solutions for the $\dot{L}_1^p$*-Homogeneous Regularity Problem* (8.6.42) *is the infinite dimensional space*

$$\dot{\mathscr{N}}_{p,1} := \Big\{x_n\nabla w : \ w \in \mathscr{C}^\infty(\mathbb{R}^n_+),\ \Delta w = 0 \ \textit{in} \ \mathbb{R}^n_+,\ \mathcal{N}_\kappa(\nabla w) \in L^p(\mathbb{R}^{n-1},\mathcal{L}^{n-1})\Big\}. \tag{8.6.47}$$

The $\dot{L}_1^p$*-Homogeneous Regularity Problem* (8.6.42) *has "index zero," in the sense that for this problem one has the isomorphism (compare with* (8.6.40)*)*

$$\frac{\big[\dot{L}_1^p(\mathbb{R}^{n-1},\mathcal{L}^{n-1})/\sim\big]^n}{\dot{\mathscr{U}}_1^p/\sim} \cong \dot{\mathscr{N}}_{p,1}. \tag{8.6.48}$$

Lastly, similar results to those described above are valid for Muckenhoupt weighted Sobolev spaces, Morrey-based Sobolev spaces, vanishing Morrey-based Sobolev spaces, and block-based Sobolev spaces (all homogeneous, of order one).

Proof For one thing, the description of the space of admissible boundary data for the $\dot{L}_1^p$-Homogeneous Regularity Problem (8.6.42) claimed in the statement has been established in Theorem 2.6.9. For another thing, Corollary 2.6.10 tells us that the space of admissible boundary data, modulo constants, for the $\dot{L}_1^p$-Homogeneous Regularity Problem (8.6.42) (i.e., the space $\dot{\mathscr{U}}_1^p/\sim$) is a closed subspace, of infinite codimension, of $\big[\dot{L}_1^p(\mathbb{R}^{n-1},\mathcal{L}^{n-1})/\sim\big]^n$. In concert, (2.6.3) and either (2.6.53) or (2.6.54) show that we have the isomorphism claimed in (8.6.48). Whenever (8.6.43) holds, Theorem 2.5.1 implies (keeping in mind the format of the unique solution of the $\dot{L}_1^p$-Homogeneous Regularity Problem for the Laplacian in $\mathbb{R}^n_+$) that any solution $\vec{u}$ of (8.6.42) may be written as in (8.6.45)-(8.6.46). Next, Corollary 2.6.1 shows that the space of null-solutions for the $\dot{L}_1^p$-Homogeneous Regularity Problem (8.6.42) is the infinite dimensional space described in (8.6.47). Finally, the very last claim in the statement is a consequence of Remark 2.6.11. □

Changing topics, we shall now discuss boundary value problems for the Bitsadze operator $\partial_{\bar z}^2$ in the complex plane. Concretely, we shall treat the interior/exterior L^p Dirichlet Problem, and the interior/exterior L_1^p Regularity Problem for the unit disk in the plane. A trade-mark feature of all these boundary value problems is the fact that they fail to be Fredholm solvable, at a basic level, namely both the space of null-solutions is infinite dimensional and the space of admissible boundary data has infinite codimension in the full space of boundary data. However, we are able to precisely identify both the space of null-solutions and the space of admissible boundary data, by providing explicit characterizations in each case.

We begin by exploring the lack of Fredholmness for the interior L^p Dirichlet Problem for the Bitsadze operator in $\mathbb{D}$, the unit disk in the complex plane. See Theorem 8.6.5 below which stands in sharp contrast with the fact that the L^p Dirichlet Problem for any symmetric homogeneous constant coefficient second-order weakly elliptic system possessing a distinguished coefficient tensor is Fredholm solvable in any sufficiently flat Ahlfors regular domain with compact boundary (cf. Theorem 8.1.6). The reader is reminded that we agree to denote $\mathbb{D}_+ := \mathbb{D}$ and $\mathbb{D}_- := \mathbb{C} \setminus \overline{\mathbb{D}}$, and that the boundary Hardy spaces $\mathcal{H}^p(\partial\mathbb{D}_\pm, \sigma)$ have been introduced in [132, (8.1.7)].

Theorem 8.6.5 *Let $\mathbb{D}$ be the unit disk in $\mathbb{C}$, and set $\sigma := \mathcal{H}^1 \lfloor \partial\mathbb{D}$. Also, fix an integrability exponent $p \in (1, \infty)$ and some aperture parameter $\kappa \in (0, \infty)$. Consider the L^p Dirichlet Problem for the Bitsadze operator $\partial_{\bar{z}}^2$ in the unit disk, with a given complex-valued function $f \in L^p(\partial\mathbb{D}, \sigma)$ as boundary datum, i.e.,*

$$\begin{cases} u \in \mathscr{C}^\infty(\mathbb{D}), \ \partial_{\bar{z}}^2 u = 0 \ \text{in } \mathbb{D}, \ \mathcal{N}_\kappa u \in L^p(\partial\mathbb{D}, \sigma), \\ u\big|_{\partial\mathbb{D}}^{\kappa-\text{n.t.}} = f \ \text{at } \sigma\text{-a.e. point on } \partial\mathbb{D}. \end{cases} \tag{8.6.49}$$

Then (8.6.49) *has a solution if and only if*

$$f \in \mathcal{H}^p(\partial\mathbb{D}_+, \sigma) \oplus \langle \bar{z} \rangle, \tag{8.6.50}$$

with $\langle \bar{z} \rangle$ the (one-dimensional) linear span over $\mathbb{C}$ of the function $\partial\mathbb{D} \ni z \mapsto \bar{z} \in \mathbb{C}$. Alternatively, the L^p Dirichlet Problem (8.6.49) *has a solution if and only if the boundary datum $f \in L^p(\partial\mathbb{D}, \sigma)$ satisfies the cancellation conditions*

$$\int_{\partial\mathbb{D}} f(z) z^k \, d\sigma(z) = 0 \ \text{for all } k \in \mathbb{N} \text{ with } k \geq 2. \tag{8.6.51}$$

As a result, the space of admissible boundary data for the L^p Dirichlet Problem (8.6.49) *has the following descriptions:*

$$\begin{aligned} &\Big\{ u\big|_{\partial\mathbb{D}}^{\kappa-\text{n.t.}} : u \in \mathscr{C}^\infty(\mathbb{D}), \ \partial_{\bar{z}}^2 u = 0 \text{ in } \mathbb{D}, \ \mathcal{N}_\kappa u \in L^p(\partial\mathbb{D}, \sigma), \ u\big|_{\partial\mathbb{D}}^{\kappa-\text{n.t.}} \text{ exists } \sigma\text{-a.e. on } \partial\mathbb{D} \Big\} \\ &\quad = \Big\{ f \in L^p(\partial\mathbb{D}, \sigma) : \int_{\partial\mathbb{D}} f(z) z^k \, d\sigma(z) = 0 \text{ for all } k \in \mathbb{N} \text{ with } k \geq 2 \Big\} \\ &\quad = \mathcal{H}^p(\partial\mathbb{D}_+, \sigma) \oplus \langle \bar{z} \rangle. \end{aligned} \tag{8.6.52}$$

As a consequence of this, the space of admissible boundary data for the L^p Dirichlet Problem (8.6.49)*, appearing in the left-most side of* (8.6.52)*, is a closed subset of the full space of boundary data $L^p(\partial\mathbb{D}, \sigma)$, whose cokernel is isomorphic to the infinite dimensional space $\mathcal{H}^p(\partial\mathbb{D}_-, \sigma) / \langle \bar{z} \rangle$.*

Moreover, for each given boundary datum

$$f \in \mathcal{H}^p(\partial\mathbb{D}_+, \sigma) \oplus \langle \bar{z} \rangle, \text{ say, } f = g + \lambda\bar{z} \text{ with } g \in \mathcal{H}^p(\partial\mathbb{D}_+, \sigma) \text{ and } \lambda \in \mathbb{C}, \tag{8.6.53}$$

any solution u of (8.6.49) *is of the form*

$$u = \mathscr{C}g + (1 - |z|^2)W + \lambda\bar{z} \text{ in } \mathbb{D}, \tag{8.6.54}$$

where $\mathscr{C}$ is the boundary-to-domain Cauchy operator associated with $\mathbb{D}$, and W is any holomorphic function in $\mathbb{D}$ satisfying[107]

$$\mathcal{N}_\kappa(\delta_{\partial\mathbb{D}} \cdot W) \in L^p(\partial\mathbb{D}, \sigma) \text{ and } \big[\delta_{\partial\mathbb{D}} \cdot W\big]\Big|_{\partial\mathbb{D}}^{\kappa-\text{n.t.}} = 0 \text{ at } \sigma\text{-a.e. point on } \partial\mathbb{D}. \tag{8.6.55}$$

In particular, the space of null-solutions for the L^p Dirichlet Problem for the Bitsadze operator in the unit disk, i.e.,

$$\Big\{u \in \mathscr{C}^\infty(\mathbb{D}) : \partial_{\bar{z}}^2 u = 0 \text{ in } \mathbb{D},\ \mathcal{N}_\kappa u \in L^p(\partial\mathbb{D}, \sigma),\ u\Big|_{\partial\mathbb{D}}^{\kappa-\text{n.t.}} = 0 \text{ at } \sigma\text{-a.e. point on } \partial\mathbb{D}\Big\},$$

coincides with the infinite dimensional space

$$\Big\{(1 - |z|^2)W : \ W \text{ holomorphic in } \mathbb{D},\ \mathcal{N}_\kappa(\delta_{\partial\mathbb{D}} \cdot W) \in L^p(\partial\mathbb{D}, \sigma), \text{ and } \big[\delta_{\partial\mathbb{D}} \cdot W\big]\Big|_{\partial\mathbb{D}}^{\kappa-\text{n.t.}} = 0 \text{ at } \sigma\text{-a.e. point on } \partial\mathbb{D}\Big\}. \tag{8.6.56}$$

Proof All claims are direct consequences of [132, Theorems 8.1.1, 8.1.2]. □

We wish to remark that, thanks to the result recorded in [132, (8.1.43)], it is necessary to actually assume the existence of the nontangential boundary trace $u\big|_{\partial\mathbb{D}}^{\kappa-\text{n.t.}}$ in the description of the space of admissible boundary data for the L^p-Dirichlet Problem (8.6.49), appearing in the left-most side of (8.6.52).

There is a companion result to Theorem 8.6.5, regarding the L^p_1 Regularity Problem for the Bitsadze operator in the unit disk in the complex plane. This is formulated in Theorem 8.6.6 below, which sharply contrasts the fact that the L^p_1 Regularity Problem for any symmetric homogeneous constant coefficient second-order weakly elliptic system possessing a distinguished coefficient tensor is Fredholm solvable in any sufficiently flat Ahlfors regular domain with compact boundary (cf. Theorem 8.1.1). Before stating it, the reader is reminded that the boundary Hardy spaces with regularity, $\mathcal{H}^p_1(\partial\mathbb{D}_\pm, \sigma)$, have been defined in [132, (8.1.22)].

Theorem 8.6.6 *Let $\mathbb{D}$ be the unit disk in $\mathbb{C}$, and set $\sigma := \mathcal{H}^1\lfloor\partial\mathbb{D}$. Also, pick an integrability exponent $p \in (1, \infty)$ and fix an aperture parameter $\kappa \in (0, \infty)$. Consider the L^p_1 Regularity Problem for the Bitsadze operator $\partial_{\bar{z}}^2$ in the unit disk, with a given complex-valued function $f \in L^p_1(\partial\mathbb{D}, \sigma)$ as boundary datum, that is,*

[107] recall that $\delta_{\partial\mathbb{D}} := \text{dist}(\cdot, \partial\mathbb{D})$

$$\begin{cases} u \in \mathscr{C}^\infty(\mathbb{D}),\ \partial_{\bar z}^2 u = 0 \ \textit{in}\ \mathbb{D},\ \mathcal{N}_\kappa(\nabla u) \in L^p(\partial\mathbb{D}, \sigma), \\ u\big|_{\partial\mathbb{D}}^{\kappa-\mathrm{n.t.}} = f \ \textit{at}\ \sigma\textit{-a.e. point on}\ \partial\mathbb{D}. \end{cases} \tag{8.6.57}$$

Then the problem (8.6.57) *has a solution if and only if*

$$f \in \mathcal{H}_1^p(\partial\mathbb{D}_+, \sigma) \oplus \langle \bar z \rangle, \tag{8.6.58}$$

with $\langle \bar z \rangle$ *the (one-dimensional) linear span over* $\mathbb{C}$ *of the function* $\partial\mathbb{D} \ni z \mapsto \bar z \in \mathbb{C}$. *Alternatively, the* L_1^p *Regularity Problem* (8.6.57) *has a solution if and only if the boundary datum* $f \in L_1^p(\partial\mathbb{D}, \sigma)$ *satisfies the cancellation conditions*

$$\int_{\partial\mathbb{D}} f(z) z^k \, d\sigma(z) = 0 \ \textit{for all}\ k \in \mathbb{N}\ \textit{with}\ k \geq 2. \tag{8.6.59}$$

In light of these results, the space of admissible boundary data for the L_1^p *Regularity Problem* (8.6.49) *has the following descriptions:*

$$\begin{aligned} &\Big\{u\big|_{\partial\mathbb{D}}^{\kappa-\mathrm{n.t.}} : u \in \mathscr{C}^\infty(\mathbb{D}),\ \partial_{\bar z}^2 u = 0 \ \textit{in}\ \mathbb{D},\ \mathcal{N}_\kappa(\nabla u) \in L^p(\partial\mathbb{D}, \sigma)\Big\} \\ &\quad = \Big\{f \in L_1^p(\partial\mathbb{D}, \sigma) : \int_{\partial\mathbb{D}} f(z) z^k \, d\sigma(z) = 0 \ \textit{for all}\ k \in \mathbb{N}\ \textit{with}\ k \geq 2\Big\} \\ &\quad = \mathcal{H}_1^p(\partial\mathbb{D}_+, \sigma) \oplus \langle \bar z \rangle. \end{aligned} \tag{8.6.60}$$

As such, the space in the left side of (8.6.60) *is a closed subset of the full space of boundary data* $L_1^p(\partial\mathbb{D}, \sigma)$, *whose cokernel is isomorphic to the infinite dimensional space* $\mathcal{H}_1^p(\partial\mathbb{D}_-, \sigma) / \langle \bar z \rangle$. *In addition, for each given boundary datum*

$$f \in \mathcal{H}_1^p(\partial\mathbb{D}_+, \sigma) \oplus \langle \bar z \rangle,\ \textit{say,}\ f = g + \lambda \bar z \ \textit{with}\ g \in \mathcal{H}_1^p(\partial\mathbb{D}_+, \sigma)\ \textit{and}\ \lambda \in \mathbb{C}, \tag{8.6.61}$$

any solution u *of* (8.6.57) *is of the form*

$$u = \mathscr{C} g + (1 - |z|^2) W + \lambda \bar z \ \textit{in}\ \mathbb{D}, \tag{8.6.62}$$

where $\mathscr{C}$ *is the boundary-to-domain Cauchy operator associated with the set* $\mathbb{D}$, *and* W *is any function in the Hardy space* $\mathcal{H}^p(\mathbb{D})$ *(cf.* [132, (8.1.2)]). *In particular, the space of null-solutions for the* L_1^p *Regularity Problem for the Bitsadze operator in the unit disk, i.e.,*

$$\Big\{u \in \mathscr{C}^\infty(\mathbb{D}) : \partial_{\bar z}^2 u = 0 \ \textit{in}\ \mathbb{D},\ \mathcal{N}_\kappa(\nabla u) \in L^p(\partial\mathbb{D}, \sigma),\ u\big|_{\partial\mathbb{D}}^{\kappa-\mathrm{n.t.}} = 0 \ \textit{on}\ \partial\mathbb{D}\Big\},$$

coincides with the infinite dimensional space

$$\big\{(1 - |z|^2) W : W \in \mathcal{H}^p(\mathbb{D})\big\}. \tag{8.6.63}$$

We wish to note that whenever $u \in \mathscr{C}^\infty(\mathbb{D})$ has $\mathcal{N}_\kappa(\nabla u) \in L^p(\partial\mathbb{D}, \sigma)$, the result established in [129, (8.9.236)] ensures that the nontangential trace $u\big|_{\partial\mathbb{D}}^{\kappa-\mathrm{n.t.}}$ exists at σ-a.e. point on $\partial\mathbb{D}$. This is relevant in the context of (8.6.57) and (8.6.60).

Proof of Theorem 8.6.6 Collectively, [132, Theorem 8.1.3] together with [132, Theorem 8.1.1] yield all claims in the statement. □

We next turn to the topic of *exterior* boundary problems for the Bitsadze operator, and first consider the L^p Dirichlet Problem in the complement of the closed unit disk in the complex plane. Here is a result which should be contrasted with Theorem 8.1.6.

Theorem 8.6.7 *Let $\mathbb{D}$ be the unit disk in the complex plane and set $\sigma := \mathcal{H}^1 \lfloor \partial\mathbb{D}$. Also, select an integrability exponent $p \in (1, \infty)$ and fix some aperture parameter $\kappa \in (0, \infty)$. In this setting, consider the L^p Dirichlet Problem for the Bitsadze operator $\partial_{\bar z}^2$ in the complement of the closed unit disk, with a given complex-valued function $f \in L^p(\partial\mathbb{D}, \sigma)$ as boundary datum, i.e.,*

$$\begin{cases} u \in \mathscr{C}^\infty(\mathbb{C} \setminus \overline{\mathbb{D}}), \quad \partial_{\bar z}^2 u = 0 \ \text{in } \mathbb{C} \setminus \overline{\mathbb{D}}, \\ \mathcal{N}_\kappa u \in L^p(\partial\mathbb{D}, \sigma), \quad u(z) = o(1) \ \text{as } z \to \infty, \\ u\big|_{\partial\mathbb{D}}^{\kappa-\mathrm{n.t.}} = f \ \text{at } \sigma\text{-a.e. point on } \partial\mathbb{D}. \end{cases} \tag{8.6.64}$$

Then the exterior boundary problem (8.6.64) *has a solution if and only if*

$$f \in \mathcal{H}^p(\partial\mathbb{D}_-, \sigma). \tag{8.6.65}$$

Alternatively, the exterior L^p Dirichlet Problem (8.6.64) *has a solution if and only if the boundary datum $f \in L^p(\partial\mathbb{D}, \sigma)$ satisfies the cancellation conditions*

$$\int_{\partial\mathbb{D}} f(z) z^{-k} \, d\sigma(z) = 0 \ \text{for all } k \in \mathbb{N}_0. \tag{8.6.66}$$

In view of these, the space of admissible boundary data for the exterior L^p Dirichlet Problem (8.6.64) *has the following descriptions:*

$$\begin{aligned} &\Big\{ u\big|_{\partial\mathbb{D}}^{\kappa-\mathrm{n.t.}} : u \in \mathscr{C}^\infty(\mathbb{C} \setminus \overline{\mathbb{D}}),\ \partial_{\bar z}^2 u = 0 \ \text{in } \mathbb{C} \setminus \overline{\mathbb{D}},\ \mathcal{N}_\kappa u \in L^p(\partial\mathbb{D}, \sigma),\ u(\infty) = 0, \\ &\qquad \text{and } u\big|_{\partial\mathbb{D}}^{\kappa-\mathrm{n.t.}} \ \text{exists } \sigma\text{-a.e. on } \partial\mathbb{D} \Big\} \\ &= \Big\{ f \in L^p(\partial\mathbb{D}, \sigma) : \int_{\partial\mathbb{D}} f(z) z^{-k} \, d\sigma(z) = 0 \ \text{for all } k \in \mathbb{N}_0 \Big\} \\ &= \mathcal{H}^p(\partial\mathbb{D}_-, \sigma). \end{aligned} \tag{8.6.67}$$

As a consequence, the space of admissible boundary data for the exterior L^p Dirichlet Problem (8.6.64), *appearing in the left-most side of* (8.6.67), *is a closed subset of the full space of boundary data $L^p(\partial\mathbb{D}, \sigma)$, whose cokernel is isomorphic to the infinite dimensional space $\mathcal{H}^p(\partial\mathbb{D}_+, \sigma)$.*

Also, if in place of the vanishing condition at infinity one now assumes mere boundedness for the function u, then in place of (8.6.67) *one obtains*

$$\begin{aligned}\Big\{u\Big|_{\partial\mathbb{D}}^{\kappa-\mathrm{n.t.}} &: u\in\mathscr{C}^\infty(\mathbb{C}\setminus\overline{\mathbb{D}}),\ \partial_{\bar z}^2u=0 \textit{ in } \mathbb{C}\setminus\overline{\mathbb{D}},\ \mathcal{N}_\kappa u\in L^p(\partial\mathbb{D},\sigma),\\ &\qquad u(z)=O(1) \textit{ as } z\to\infty,\ u\Big|_{\partial\mathbb{D}}^{\kappa-\mathrm{n.t.}} \textit{ exists } \sigma\textit{-a.e. on } \partial\mathbb{D}\Big\}\\ &=\Big\{f\in L^p(\partial\mathbb{D},\sigma):\int_{\partial\mathbb{D}}f(z)z^{-k}\,d\sigma(z)=0 \textit{ for all } k\in\mathbb{N}\Big\}\\ &=\mathcal{H}^p(\partial\mathbb{D}_-,\sigma)\oplus\mathbb{C}.\end{aligned}\tag{8.6.68}$$

Moreover, for each $f\in\mathcal{H}^p(\partial\mathbb{D}_-,\sigma)$*, any solution* u *of* (8.6.64) *is of the form*

$$u=\mathscr{C}f+(1-|z|^2)W \textit{ in } \mathbb{C}\setminus\overline{\mathbb{D}},\tag{8.6.69}$$

where $\mathscr{C}$ *is the boundary-to-domain Cauchy operator associated with* $\mathbb{C}\setminus\overline{\mathbb{D}}$*, and* W *is any holomorphic function in* $\mathbb{C}\setminus\overline{\mathbb{D}}$ *satisfying*

$$\begin{aligned}&\mathcal{N}_\kappa\big((1-|z|^2)W\big)\in L^p(\partial\mathbb{D},\sigma),\quad W(z)=o(1/|z|^2) \textit{ as } z\to\infty\\ &\textit{and } \big[(1-|z|^2)W\big]\Big|_{\partial\mathbb{D}}^{\kappa-\mathrm{n.t.}}=0 \textit{ at } \sigma\textit{-a.e. point on } \partial\mathbb{D}.\end{aligned}\tag{8.6.70}$$

In particular, the space of null-solutions for the L^p *Dirichlet Problem for the Bitsadze operator in the complement of the closed unit disk, i.e.,*

$$\begin{aligned}\Big\{u\in\mathscr{C}^\infty(\mathbb{C}\setminus\overline{\mathbb{D}}):\ &\partial_{\bar z}^2u=0 \textit{ in } \mathbb{C}\setminus\overline{\mathbb{D}},\ \mathcal{N}_\kappa u\in L^p(\partial\mathbb{D},\sigma),\ u(z)=o(1) \textit{ as } z\to\infty,\\ &\textit{and } u\Big|_{\partial\mathbb{D}}^{\kappa-\mathrm{n.t.}}=0 \textit{ at } \sigma\textit{-a.e. point on } \partial\mathbb{D}\Big\},\end{aligned}\tag{8.6.71}$$

coincides with the infinite dimensional space

$$\begin{aligned}\Big\{(1-|z|^2)W:\ &W \textit{ holomorphic in } \mathbb{C}\setminus\overline{\mathbb{D}},\ \mathcal{N}_\kappa\big((1-|z|^2)W\big)\in L^p(\partial\mathbb{D},\sigma),\\ &\big[(1-|z|^2)W\big]\Big|_{\partial\mathbb{D}}^{\kappa-\mathrm{n.t.}}=0 \textit{ at } \sigma\textit{-a.e. point on } \partial\mathbb{D}, \textit{ and } W(z)=o(1/|z|^2) \textit{ as } z\to\infty\Big\}.\end{aligned}\tag{8.6.72}$$

Finally, the same characterization remains valid provided little "o" is replaced by big "O" both in (8.6.71) *and in* (8.6.72).

Proof Together, [132, Theorems 8.1.4, 8.1.5] imply all desired conclusions. □

Lastly, here is a companion to Theorem 8.6.7, in which the focus is to precisely characterize the lack of Fredholmness for the L^p_1 Regularity Problem for the Bitsadze operator in the complement of the closed unit disk in the complex plane. The result below should be compared with Theorem 8.1.1.

Theorem 8.6.8 *Denote by* $\mathbb{D}$ *the unit disk in the complex plane, and abbreviate* $\sigma:=\mathcal{H}^1\lfloor\partial\mathbb{D}$*. Also, choose an integrability exponent* $p\in(1,\infty)$ *together with an aperture parameter* $\kappa\in(0,\infty)$*. Consider the* L^p_1 *Regularity Problem for the Bitsadze operator* $\partial_{\bar z}^2$ *in the complement of the closed unit disk, with a given complex-valued function* $f\in L^p_1(\partial\mathbb{D},\sigma)$ *as boundary datum, i.e.,*

$$\begin{cases} u \in \mathscr{C}^\infty(\mathbb{C} \setminus \overline{\mathbb{D}}), \quad \partial_{\bar z}^2 u = 0 \text{ in } \mathbb{C} \setminus \overline{\mathbb{D}}, \\ \mathcal{N}_\kappa(\nabla u) \in L^p(\partial\mathbb{D}, \sigma), \quad u(z) = o(1) \text{ as } z \to \infty, \\ u\big|_{\partial\mathbb{D}}^{\kappa-\mathrm{n.t.}} = f \text{ at } \sigma\text{-a.e. point on } \partial\mathbb{D}. \end{cases} \tag{8.6.73}$$

Then the exterior boundary value problem (8.6.73) *has a solution if and only if*

$$f \in \mathcal{H}_1^p(\partial\mathbb{D}_-, \sigma). \tag{8.6.74}$$

Alternatively, the exterior L_1^p Regularity Problem (8.6.73) *has a solution if and only if the boundary datum $f \in L_1^p(\partial\mathbb{D}, \sigma)$ satisfies the cancellation conditions*

$$\int_{\partial\mathbb{D}} f(z) z^{-k} \, \mathrm{d}\sigma(z) = 0 \text{ for all } k \in \mathbb{N}_0. \tag{8.6.75}$$

Hence the space of admissible boundary data for the exterior L_1^p Regularity Problem (8.6.73) *may be described as:*

$$\begin{aligned} &\Big\{ u\big|_{\partial\mathbb{D}}^{\kappa-\mathrm{n.t.}} : u \in \mathscr{C}^\infty(\mathbb{C} \setminus \overline{\mathbb{D}}), \ \partial_{\bar z}^2 u = 0 \text{ in } \mathbb{C} \setminus \overline{\mathbb{D}}, \ \mathcal{N}_\kappa(\nabla u) \in L^p(\partial\mathbb{D}, \sigma), \ u(\infty) = 0 \Big\} \\ &\quad = \Big\{ f \in L_1^p(\partial\mathbb{D}, \sigma) : \int_{\partial\mathbb{D}} f(z) z^{-k} \, \mathrm{d}\sigma(z) = 0 \text{ for all } k \in \mathbb{N}_0 \Big\} \\ &\quad = \mathcal{H}_1^p(\partial\mathbb{D}_-, \sigma). \end{aligned} \tag{8.6.76}$$

As a consequence, the space of admissible boundary data for the exterior L_1^p Regularity Problem (8.6.73)*, appearing in the left-most side of* (8.6.76)*, is a closed subset of the full space of boundary data $L_1^p(\partial\mathbb{D}, \sigma)$, whose cokernel is isomorphic to the infinite dimensional space $\mathcal{H}_1^p(\partial\mathbb{D}_+, \sigma)$.*

Furthermore, if in place of the vanishing condition at infinity one now assumes mere boundedness for the function u, then in place of (8.6.76) *one obtains*

$$\begin{aligned} &\Big\{ u\big|_{\partial\mathbb{D}}^{\kappa-\mathrm{n.t.}} : u \in \mathscr{C}^\infty(\mathbb{C} \setminus \overline{\mathbb{D}}), \ \partial_{\bar z}^2 u = 0 \text{ in } \mathbb{C} \setminus \overline{\mathbb{D}}, \ \mathcal{N}_\kappa(\nabla u) \in L^p(\partial\mathbb{D}, \sigma), \\ &\qquad\qquad \text{and } u(z) = O(1) \text{ as } z \to \infty \Big\} \\ &\quad = \Big\{ f \in L_1^p(\partial\mathbb{D}, \sigma) : \int_{\partial\mathbb{D}} f(z) z^{-k} \, \mathrm{d}\sigma(z) = 0 \text{ for all } k \in \mathbb{N} \Big\} \\ &\quad = \mathcal{H}_1^p(\partial\mathbb{D}_-, \sigma) \oplus \mathbb{C}. \end{aligned} \tag{8.6.77}$$

In addition, for each given boundary datum $f \in \mathcal{H}_1^p(\partial\mathbb{D}_-, \sigma)$, any solution u of (8.6.73) *is of the form*

$$u = \mathscr{C} f + (1 - |z|^2) W \text{ in } \mathbb{C} \setminus \overline{\mathbb{D}}, \tag{8.6.78}$$

where $\mathscr{C}$ is the boundary-to-domain Cauchy operator associated with $\mathbb{C} \setminus \overline{\mathbb{D}}$, and W is any function in the Hardy space $\mathcal{H}^p(\mathbb{D}_-)$ satisfying $W(z) = o(1/|z|^2)$ as $z \to \infty$.

In particular, the space of null-solutions for the L^p_1 Regularity Problem for the Bitsadze operator in the complement of the closed unit disk, i.e.,

$$\Big\{u \in \mathscr{C}^\infty\big(\mathbb{C}\setminus\overline{\mathbb{D}}\big) : \ \partial_{\bar z}^2 u = 0 \ \textit{in} \ \mathbb{C}\setminus\overline{\mathbb{D}}, \ \mathcal{N}_\kappa(\nabla u) \in L^p(\partial\mathbb{D},\sigma), \tag{8.6.79}$$
$$u(z) = o(1) \ \textit{as} \ z \to \infty, \ \textit{and} \ u\Big|_{\partial\mathbb{D}}^{\kappa-\mathrm{n.t.}} = 0 \ \textit{at} \ \sigma\textit{-a.e. point on} \ \partial\mathbb{D}\Big\},$$

coincides with the infinite dimensional space

$$\Big\{(1-|z|^2)W : \ W \in \mathcal{H}^p(\mathbb{D}_-) \ \textit{and} \ W(z) = o(1/|z|^2) \ \textit{as} \ z \to \infty\Big\}. \tag{8.6.80}$$

Finally, the aforementioned identification continues to hold provided little "o" is replaced by big "O" both in (8.6.79) *and in* (8.6.80).

It is of significance to observe that whenever $u \in \mathscr{C}^\infty\big(\mathbb{C}\setminus\overline{\mathbb{D}}\big)$ has the property that $\mathcal{N}_\kappa(\nabla u) \in L^p(\partial\mathbb{D},\sigma)$, the result proved in [129, (8.9.236)] guarantees that the nontangential trace $u\big|_{\partial\mathbb{D}}^{\kappa-\mathrm{n.t.}}$ exists at σ-a.e. point on $\partial\mathbb{D}$. This is relevant in the context of (8.6.73), (8.6.76), and (8.6.77).

Proof of Theorem 8.6.8 All desired conclusions may be justified based on [132, Theorems 8.1.4, 8.1.7]. □

We wish to note that, thanks to [132, Remark 8.1.8], analogous results hold for a multitude of other spaces of boundary data:

Remark 8.6.9 *Natural versions of Theorem 8.6.5, Theorem 8.6.6, Theorem 8.6.7, and Theorem 8.6.8 are valid for Muckenhoupt weighted Lebesgue spaces, Morrey spaces, vanishing Morrey spaces, block spaces, as well as for their Sobolev space counterparts.*

Moving on, we next discuss boundary value problem in the upper half-plane for the operator $L_\lambda := \partial_{\bar z}^2 - \lambda^2\partial_z^2$, which has been introduced and studied in [132, §8.2]. We begin by considering the Homogeneous Regularity Problem in Theorem 8.6.10 below. This result stands in sharp contrast to the fact that the Homogeneous Regularity Problem for any symmetric, homogeneous, constant (complex) coefficient, second-order weakly elliptic system, possessing a distinguished coefficient tensor is well posed in any sufficiently flat Ahlfors regular domain with unbounded boundary (cf. Theorem 8.4.1).

Theorem 8.6.10 *Fix an aperture parameter $\kappa > 0$ along with an integrability exponent $p \in (1,\infty)$ and define the homogeneous Hardy spaces with regularity (as in* (A.0.108) *for the upper half-plane)*

$$\dot{\mathcal{H}}^p_{1,\pm}(\mathbb{R},\mathcal{L}^1) := \Big\{w\Big|_{\partial\mathbb{C}_\pm}^{\kappa-\mathrm{n.t.}} : \ w \in \mathscr{O}(\mathbb{C}_\pm) \ \textit{with} \ \mathcal{N}_\kappa(\nabla w) \in L^p(\mathbb{R},\mathcal{L}^1)\Big\}. \tag{8.6.81}$$

Also, pick a real number $\lambda \in (-1,1)\setminus\{0\}$ and use this to define $L_\lambda := \partial_{\bar z}^2 - \lambda^2\partial_z^2$, which is a scalar, homogeneous, constant (complex) coefficient, weakly elliptic second-order operator in $\mathbb{C}$. Given a complex-valued function $f \in \dot{L}^p_1(\mathbb{R},\mathcal{L}^1)$, consider the

Homogeneous Regularity Problem for the operator L_λ in the upper half-plane with boundary datum f, i.e.,

$$\begin{cases} u \in \mathscr{C}^\infty(\mathbb{C}_+), \ L_\lambda u = 0 \ \text{in} \ \mathbb{C}_+, \ \mathcal{N}_\kappa(\nabla u) \in L^p(\mathbb{R}, \mathcal{L}^1), \\ u\big|_{\partial\mathbb{C}_+}^{\kappa-\text{n.t.}} = f \ \text{at} \ \mathcal{L}^1\text{-a.e. point on} \ \partial\mathbb{C}_+ \equiv \mathbb{R}. \end{cases} \tag{8.6.82}$$

Then the boundary value problem (8.6.82) *has a solution if and only if*

$$f \in \dot{\mathcal{H}}^p_{1,+}(\mathbb{R}, \mathcal{L}^1). \tag{8.6.83}$$

Hence the space of admissible boundary data for the boundary value problem (8.6.82) *is precisely the "positive" homogeneous Hardy space with regularity* $\dot{\mathcal{H}}^p_{1,+}(\mathbb{R}, \mathcal{L}^1)$. *Alternatively, the Homogeneous Regularity Problem* (8.6.82) *has a solution if and only if there exists a constant $c \in \mathbb{C}$ such that*

$$H_{\text{mod}} f = -\mathrm{i} f + c \ \text{at} \ \mathcal{L}^1\text{-a.e. point on} \ \mathbb{R}, \tag{8.6.84}$$

where H_{mod} is the modified Hilbert transform on the real line (cf. (A.0.109)*). As a consequence, the space of admissible boundary data for the Homogeneous Regularity Problem* (8.6.82) *modulo constants, is a closed subspace, of infinite codimension, in the full space of boundary data $\dot{L}^p_1(\mathbb{R}, \mathcal{L}^1)/\sim$, modulo constants.*

In addition, for each given boundary datum $f \in \dot{\mathcal{H}}^p_{1,+}(\mathbb{R}, \mathcal{L}^1)$, any solution of the Homogeneous Regularity Problem (8.6.82) *is of the form*

$$u(z) = \big(\mathscr{C}_{\text{mod}} f\big)\Big(\frac{z - \lambda\overline{z}}{1-\lambda}\Big) - \frac{\mathrm{i}c}{2} + \phi\big((1+\lambda)(z - \lambda\overline{z})\big) - \phi\big((1-\lambda)(z + \lambda\overline{z})\big)$$

for each $z \in \mathbb{C}_+$, where $\mathscr{C}_{\text{mod}}$ is the modified boundary-to-domain Cauchy integral operator defined as in (A.0.58) *with Ω the upper half-plane, c is the complex number for which* (8.6.84) *holds, and where ϕ is any function satisfying*

$$\phi \in \mathscr{O}(\mathbb{C}_+) \ \text{and} \ \mathcal{N}_\kappa(\nabla\phi) \in L^p(\mathbb{R}, \mathcal{L}^1). \tag{8.6.85}$$

As a consequence, the space on null-solutions for the Homogeneous Regularity Problem (8.6.82)*, i.e.,*

$$\Big\{u \in \mathscr{C}^\infty(\mathbb{C}_+) : \ L_\lambda u = 0 \ \text{in} \ \mathbb{C}_+, \ \mathcal{N}_\kappa(\nabla u) \in L^p(\mathbb{R}, \mathcal{L}^1), \ u\big|_{\partial\mathbb{C}_+}^{\kappa-\text{n.t.}} = 0\Big\}, \tag{8.6.86}$$

coincides with the infinite dimensional space

$$\Big\{\phi\big((1+\lambda)(z - \lambda\overline{z})\big) - \phi\big((1-\lambda)(z + \lambda\overline{z})\big) : \ \phi \in \mathscr{O}(\mathbb{C}_+), \ \mathcal{N}_\kappa(\nabla\phi) \in L^p(\mathbb{R}, \mathcal{L}^1)\Big\}.$$

Proof All claims in the statement follow from [132, Theorem 8.2.3], also taking into account [132, (1.8.288), Example 1.8.4, Remark 1.8.13], the fact that we presently have $C_{\text{mod}} = (\mathrm{i}/2) H_{\text{mod}}$, and reasoning as in [132, (8.2.38)-(8.2.42)]. □

We next discuss the Inhomogeneous Regularity Problem for the operator L_λ in the upper half-plane. Theorem 8.6.11 strongly contrasts (8.4.10) which implies that the Inhomogeneous Regularity Problem for any symmetric, homogeneous, constant (complex) coefficient, second-order weakly elliptic system, possessing a distinguished coefficient tensor is well posed in any sufficiently flat Ahlfors regular domain with unbounded boundary.

Theorem 8.6.11 *Select some aperture parameter $\kappa > 0$ together with some integrability exponent $p \in (1,\infty)$ and define the inhomogeneous Hardy spaces with regularity*

$$\mathcal{H}^p_{1,\pm}(\mathbb{R},\mathcal{L}^1) := \Big\{ w\big|^{\kappa-\mathrm{n.t.}}_{\partial\mathbb{C}_\pm} : w \in \mathscr{O}(\mathbb{C}_\pm) \text{ with } \mathcal{N}_\kappa w,\ \mathcal{N}_\kappa(\nabla w) \in L^p(\mathbb{R},\mathcal{L}^1) \Big\}.$$

Next, fix a real number $\lambda \in (-1,1)\setminus\{0\}$ and use this to define $L_\lambda := \partial_{\bar z}^2 - \lambda^2\partial_z^2$, which is a scalar, homogeneous, constant (complex) coefficient, weakly elliptic second-order operator in $\mathbb{C}$. Given a complex-valued function $f \in L^p_1(\mathbb{R},\mathcal{L}^1)$, consider the Inhomogeneous Regularity Problem for the operator L_λ in the upper half-plane with boundary datum f, namely

$$\begin{cases} u \in \mathscr{C}^\infty(\mathbb{C}_+), \quad L_\lambda u = 0 \text{ in } \mathbb{C}_+, \\ \mathcal{N}_\kappa u,\ \mathcal{N}_\kappa(\nabla u) \in L^p(\mathbb{R},\mathcal{L}^1), \\ u\big|^{\kappa-\mathrm{n.t.}}_{\partial\mathbb{C}_+} = f \text{ at } \mathcal{L}^1\text{-a.e. point on } \partial\mathbb{C}_+ \equiv \mathbb{R}. \end{cases} \tag{8.6.87}$$

Then the boundary value problem (8.6.87) *has a solution if and only if*

$$f \in \mathcal{H}^p_{1,+}(\mathbb{R},\mathcal{L}^1). \tag{8.6.88}$$

As such, the space of admissible boundary data for the boundary value problem (8.6.87) *is precisely the "positive" inhomogeneous Hardy space with regularity $\mathcal{H}^p_{1,+}(\mathbb{R},\mathcal{L}^1)$. Alternatively, the Inhomogeneous Regularity Problem* (8.6.87) *has a solution if and only if*

$$Hf = -\mathrm{i}f \text{ at } \mathcal{L}^1\text{-a.e. point on } \mathbb{R}, \tag{8.6.89}$$

where H is the standard Hilbert transform on the real line (cf. (1.0.6)*). Consequently, the space of admissible boundary data for the Inhomogeneous Regularity Problem* (8.6.87) *is a closed subspace, of infinite codimension, in the full space of boundary data $L^p_1(\mathbb{R},\mathcal{L}^1)$.*

Moreover, for each given boundary datum $f \in \mathcal{H}^p_{1,+}(\mathbb{R},\mathcal{L}^1)$, any solution of the Inhomogeneous Regularity Problem (8.6.87) *is of the form*

$$u(z) = (\mathscr{C}f)\Big(\frac{z-\lambda\bar z}{1-\lambda}\Big) + \phi\big((1+\lambda)(z-\lambda\bar z)\big) - \phi\big((1-\lambda)(z+\lambda\bar z)\big) \text{ for all } z \in \mathbb{C}_+$$

where $\mathscr{C}$ is the boundary-to-domain Cauchy integral operator defined as in (A.0.57) *with $\Omega := \mathbb{C}_+$, and ϕ is any function satisfying*

$$\phi \in \mathscr{O}(\mathbb{C}_+) \text{ and } \mathcal{N}_\kappa \phi,\ \mathcal{N}_\kappa(\nabla\phi) \in L^p(\mathbb{R}, \mathcal{L}^1). \tag{8.6.90}$$

In particular, the space on null-solutions for the Inhomogeneous Regularity Problem (8.6.87), *i.e.*,

$$\Big\{u \in \mathscr{C}^\infty(\mathbb{C}_+) : L_\lambda u = 0 \text{ in } \mathbb{C}_+,\ \mathcal{N}_\kappa u,\ \mathcal{N}_\kappa(\nabla u) \in L^p(\mathbb{R}, \mathcal{L}^1),\ u\Big|_{\partial\mathbb{C}_+}^{\kappa-\mathrm{n.t.}} = 0\Big\},$$

coincides with the infinite dimensional space

$$\Big\{\phi\big((1+\lambda)(z-\lambda\bar{z})\big) - \phi\big((1-\lambda)(z+\lambda\bar{z})\big) : \phi \in \mathscr{O}(\mathbb{C}_+),\ \mathcal{N}_\kappa\phi,\ \mathcal{N}_\kappa(\nabla\phi) \in L^p(\mathbb{R}, \mathcal{L}^1)\Big\}.$$

Proof This is justified much like Theorem 8.6.10, now relying on [132, Theorem 8.2.5] in place of [132, Theorem 8.2.3]. □

We augment Theorems 8.6.10-8.6.11 by also considering the Dirichlet Regularity Problem for the operator L_λ in the upper half-plane. Our main result in this regard, Theorem 8.6.12, is in stark contrast with Theorem 8.4.3 which implies that the Dirichlet Problem for any symmetric, homogeneous, constant (complex) coefficient, second-order weakly elliptic system, possessing a distinguished coefficient tensor is well posed in any sufficiently flat Ahlfors regular domain with unbounded boundary.

Theorem 8.6.12 *Select some aperture parameter* $\kappa > 0$ *together with some integrability exponent* $p \in (1, \infty)$ *and bring in the Hardy spaces*

$$\mathcal{H}^p_\pm(\mathbb{R}, \mathcal{L}^1) := \Big\{w\Big|_{\partial\mathbb{C}_\pm}^{\kappa-\mathrm{n.t.}} : w \in \mathscr{O}(\mathbb{C}_\pm) \text{ with } \mathcal{N}_\kappa w \in L^p(\mathbb{R}, \mathcal{L}^1)\Big\}. \tag{8.6.91}$$

Next, fix a real number $\lambda \in (-1, 1)\setminus\{0\}$ *and use this to define* $L_\lambda := \partial_{\bar{z}}^2 - \lambda^2\partial_z^2$, *which is a scalar, homogeneous, constant (complex) coefficient, weakly elliptic second-order operator in* $\mathbb{C}$. *Given a complex-valued function* $f \in L^p(\mathbb{R}, \mathcal{L}^1)$, *consider the Dirichlet Problem for the operator* L_λ *in the upper half-plane with boundary datum* f, *namely*

$$\begin{cases} u \in \mathscr{C}^\infty(\mathbb{C}_+),\ L_\lambda u = 0 \text{ in } \mathbb{C}_+,\ \mathcal{N}_\kappa u \in L^p(\mathbb{R}, \mathcal{L}^1), \\ u\Big|_{\partial\mathbb{C}_+}^{\kappa-\mathrm{n.t.}} = f \text{ at } \mathcal{L}^1\text{-a.e. point on } \partial\mathbb{C}_+ \equiv \mathbb{R}. \end{cases} \tag{8.6.92}$$

Then the boundary value problem (8.6.92) *has a solution if and only if*

$$f \in \mathcal{H}^p_+(\mathbb{R}, \mathcal{L}^1). \tag{8.6.93}$$

As such, the space of admissible boundary data for the boundary value problem (8.6.92) *is precisely the "positive" Hardy space* $\mathcal{H}^p_+(\mathbb{R}, \mathcal{L}^1)$. *Alternatively, the Dirichlet Problem* (8.6.92) *has a solution if and only if*

$$Hf = -\mathrm{i}f \text{ at } \mathcal{L}^1\text{-a.e. point on } \mathbb{R}, \tag{8.6.94}$$

where H is the standard Hilbert transform on the real line (cf. (1.0.6)). *Consequently, the space of admissible boundary data for the Dirichlet Problem* (8.6.92) *is a closed subspace, of infinite codimension, in the full space of boundary data $L^p(\mathbb{R}, \mathcal{L}^1)$.*

Moreover, the space on null-solutions for the Dirichlet Problem (8.6.92), *i.e.,*

$$\mathcal{N}^p_\lambda := \Big\{u \in \mathscr{C}^\infty(\mathbb{C}_+) : \ L_\lambda u = 0 \ \text{in} \ \mathbb{C}_+, \ \mathcal{N}_\kappa u \in L^p(\mathbb{R}, \mathcal{L}^1), \ u\Big|^{\kappa-\text{n.t.}}_{\partial\mathbb{C}_+} = 0\Big\},$$

is infinite dimensional, and for each given boundary datum $f \in \mathcal{H}^p_+(\mathbb{R}, \mathcal{L}^1)$ any solution of the Dirichlet Problem (8.6.92) *is of the form*

$$u(z) = (\mathscr{C}f)\Big(\frac{z - \lambda\bar{z}}{1-\lambda}\Big) + w(z) \ \text{for all} \ z \in \mathbb{C}_+ \tag{8.6.95}$$

where $\mathscr{C}$ is the boundary-to-domain Cauchy integral operator defined as in (A.0.57) *with $\Omega := \mathbb{C}_+$, and w is any function in $\mathcal{N}^p_\lambda$.*

Proof This is justified arguing much like in the proof of Theorem 8.6.10, now using [132, Theorem 8.2.4] in lieu of [132, Theorem 8.2.3]. □

Here is a comment pointing to further natural generalizations.

Remark 8.6.13 *All our earlier results in this section, regarding the nature of boundary problems for the scalar operators L_B and L_λ, may naturally adapted to genuine systems of the sort considered in* [132, (8.1.136), (8.1.152), (8.2.95)], *in the manner indicated in the last part of* [132, §8.1].

Our last topic in this section elaborates on the failure of Fredholm solvability in the case of the *Neumann Problem* for a Legendre-Hadamard (strongly) elliptic system if the conormal derivative involves a coefficient tensor which is not distinguished. In the remark below we consider the Neumann Problem for the vector Laplacian in the higher-dimensional setting.

Remark 8.6.14 *Fix $n \in \mathbb{N}$ with $n \geq 2$, and consider the system*[108]

$$L := \Delta \cdot I_{n\times n} \ \text{in} \ \mathbb{R}^n. \tag{8.6.96}$$

Then the coefficient tensor (defined as in (1.6.87) *with $\mu := 1$, $\lambda := -1$, and $\zeta := -1$)*

$$\begin{aligned} A = \big(a^{\alpha\beta}_{jk}\big)_{\substack{1\le\alpha,\beta\le n\\ 1\le j,k\le n}} \ \text{with entries given by} \\ a^{\alpha\beta}_{jk} := \delta_{jk}\delta_{\alpha\beta} + \delta_{j\alpha}\delta_{k\beta} - \delta_{j\beta}\delta_{k\alpha}, \quad 1 \le j,k,\alpha,\beta \le n, \end{aligned} \tag{8.6.97}$$

is symmetric (that is, $A^\top = A$), and belongs to $\mathfrak{A}_L$, i.e., we have

$$L = \Big(a^{\alpha\beta}_{jk}\partial_j\partial_k\Big)_{1\le\alpha,\beta\le n}. \tag{8.6.98}$$

[108] i.e., L is the "vector Laplacian"

As such, any given an exponent $p \in (1, \infty)$ along with an aperture parameter $\kappa \in (0, \infty)$, and a bounded UR *domain $\Omega \subseteq \mathbb{R}^n$, if we denote by ν the geometric measure theoretic outward unit normal to Ω and abbreviate $\sigma := \mathcal{H}^{n-1}\lfloor\partial\Omega$, it makes sense to consider the Neumann Problem*

$$\begin{cases} u \in \big[\mathscr{C}^\infty(\Omega)\big]^n, \quad Lu = 0 \text{ in } \Omega, \\ \mathcal{N}_\kappa(\nabla u) \in L^p(\partial\Omega, \sigma), \quad \partial_\nu^A u = f \in \big[L^p(\partial\Omega, \sigma)\big]^n. \end{cases} \tag{8.6.99}$$

We claim that this boundary value problem fails to be Fredholm solvable.

To substantiate this claim, note that for any function $u = (u_\beta)_{1 \le \beta \le n}$ satisfying the first three conditions in (8.6.99) *the Fatou-type result from* [131, Theorem 3.3.4] *enures that $(\nabla u)\big|_{\partial\Omega}^{\kappa-\text{n.t.}}$ exists (in $\mathbb{C}^{n^2}$) at σ-a.e. point on $\partial\Omega$. In particular, $\partial_\nu^A u$ is a well-defined function in $\big[L^p(\partial\Omega, \sigma)\big]^n$. In fact, from* [131, Definition 1.7.1] *and* (8.6.97) *we see (with the summation convention over repeated indices in effect) that*

$$\begin{aligned} \partial_\nu^A u &= \Big(\nu_j a_{jk}^{\alpha\beta}(\partial_k u_\beta)\big|_{\partial\Omega}^{\kappa-\text{n.t.}}\Big)_{1\le\alpha\le n} \\ &= \Big(\nu_k(\partial_k u_\alpha)\big|_{\partial\Omega}^{\kappa-\text{n.t.}} + \nu_\alpha(\partial_k u_k)\big|_{\partial\Omega}^{\kappa-\text{n.t.}} - \nu_k(\partial_\alpha u_k)\big|_{\partial\Omega}^{\kappa-\text{n.t.}}\Big)_{1\le\alpha\le n} \\ &= \Big[\nabla u - (\nabla u)^\top + (\operatorname{div} u)I_{n\times n}\Big]\Big|_{\partial\Omega}^{\kappa-\text{n.t.}} \cdot \nu, \text{ at } \sigma\text{-a.e. point on } \partial\Omega. \end{aligned} \tag{8.6.100}$$

As a consequence, any $u = (u_j)_{1\le j\le n} \in \big[\mathscr{C}^1(\Omega)\big]^n$ with $\mathcal{N}_\kappa(\nabla u) \in L^p(\partial\Omega, \sigma)$ and satisfying the Moisil-Teodorescu system/generalized Cauchy-Riemann equations in the set Ω

$$(\text{MT/GCR}) \begin{cases} \partial_k u_j = \partial_j u_k \text{ in } \Omega \text{ for } 1 \le j, k \le n, \\ \partial_j u_j = 0 \text{ in } \Omega, \end{cases} \tag{8.6.101}$$

is a nullsolution of the Neumann Problem (8.6.99)*. In turn, this shows that the space of nullsolutions for the Neumann Problem* (8.6.99) *is infinite dimensional*[109].

Let K_A, $K_A^\#$ be the boundary-to-boundary double layer potential operator and its formal transpose, associated with the coefficient tensor A and the set Ω as in (A.0.116) *and* (A.0.118)*, respectively. Reasoning as in* (8.1.497)-(8.1.498) *we see that the space of admissible boundary data for the Neumann Problem* (8.6.99) *is*

$$\operatorname{Im}\Big(-\tfrac{1}{2}I + K_{A^\top}^\#;\ \big[L^p(\partial\Omega,\sigma)\big]^n\Big) = \operatorname{Im}\Big(-\tfrac{1}{2}I + K_A^\#;\ \big[L^p(\partial\Omega,\sigma)\big]^n\Big), \tag{8.6.102}$$

where the second equality makes use of the fact that $A^\top = A$. Thanks to [130, (2.1.49)] *and* [132, Theorem 1.5.1, item *(iii)*]*, it follows that the norm-closure of* $\operatorname{Im}\big(-\tfrac{1}{2}I + K_A^\#; \big[L^p(\partial\Omega,\sigma)\big]^n\big)$ *in the space $\big[L^p(\partial\Omega,\sigma)\big]^n$ is the annihilator*

---

[109] for example, since Ω is bounded, said space of nullsolutions contains all functions of the form ∇w with $w \in \mathscr{C}^\infty(\overline{\Omega})$ harmonic in Ω, hence contains the gradients of all harmonic polynomials in $\mathbb{R}^n$

$$^{\perp}\Big[\mathrm{Ker}\,\big(-\tfrac{1}{2}I + K_A;\, \big[L^{p'}(\partial\Omega,\sigma)\big]^n\big)\Big] \ \textit{with } p' := (1-1/p)^{-1} \in (1,\infty). \tag{8.6.103}$$

In addition, for any vector-valued function $u = (u_j)_{1\leq j\leq n} \in \big[\mathscr{C}^\infty(\Omega)\big]^n$ *whose components satisfy* (8.6.101) *and with the property that* $\mathcal{N}_\kappa u \in L^{p'}(\partial\Omega,\sigma)$ *we have*

$$u = \mathcal{D}_A\big(u\big|^{\kappa-\mathrm{n.t.}}_{\partial\Omega}\big) \ \textit{in } \Omega, \tag{8.6.104}$$

where $\mathcal{D}_A$ *is the boundary-to-domain double layer potential operator associated with the coefficient tensor A and the set* Ω *as in* (A.0.72). *Indeed, this is seen from the boundary layer integral representation formula from* [131, Theorem 1.5.1] (*whose applicability is ensured by the choice of A in* (8.6.97) *and Comment 9 following the statement of* [131, Theorem 3.1.6]), *and the fact that*

$$\Big(a^{\alpha\beta}_{jk}\partial_k u_\beta\Big)_{1\leq\alpha,j\leq n} = \nabla u - (\nabla u)^\top + (\mathrm{div}\, u)I_{n\times n} = 0 \ \textit{in } \Omega \tag{8.6.105}$$

(*compare with* (8.6.100)). *From the integral representation formula* (8.6.104) *and the jump-formula from* [132, Theorem 1.5.1, item *(iv)*] *we then conclude that*

$$u \mapsto u\big|^{\kappa-\mathrm{n.t.}}_{\partial\Omega} \in \mathrm{Ker}\,\big(-\tfrac{1}{2}I + K_A;\, \big[L^{p'}(\partial\Omega,\sigma)\big]^n\big) \tag{8.6.106}$$

is well defined, linear, and injective assignment. As a consequence,

$$\dim \mathrm{Ker}\,\big(-\tfrac{1}{2}I + K_A;\, \big[L^{p'}(\partial\Omega,\sigma)\big]^n\big) = +\infty. \tag{8.6.107}$$

Finally, we note that [130, (2.1.26)] *implies that*

for a Banach space X, the annihilator $^{\perp}W$ *of any given linear subspace W of the dual* X^* *has the property that* $\dim\big(X/{}^{\perp}W\big) = \dim W$. (8.6.108)

Ultimately, from (8.6.102)-(8.6.103) *and* (8.6.107)-(8.6.108) *we conclude that the space of admissible boundary data for the Neumann Problem* (8.6.99) *has infinite codimension in the full-space of data,* $\big[L^p(\partial\Omega,\sigma)\big]^n$. *As such, the failure of Fredholm solvability for the Neumann Problem* (8.6.99) *is total, even though L in* (8.6.96) *is a symmetric system with real coefficients satisfying the Legendre-Hadamard strong ellipticity condition* (*hence L is also weakly elliptic*) *and with the property that* $\mathfrak{A}^{\mathrm{dis}}_L \neq \varnothing$. *The source of this failure can be linked to the fact that for the coefficient tensor* $A \in \mathfrak{A}_L$ *chosen in* (8.6.97) *we have* $A \notin \mathfrak{A}^{\mathrm{dis}}_L$.

Of course, similar considerations apply to other function spaces of interest, playing the role of the Lebesgue scale in the formulation of (8.6.99), *including Muckenhoupt weighted Lebesgue spaces, Lorentz spaces, Morrey spaces, vanishing Morrey spaces, and block spaces.*

Appendix A
Terms and notation used in Volume V

A

Ahlfors regular domain (cf. [129, Definition 5.9.15]):

$$\text{a nonempty open subset } \Omega \text{ of } \mathbb{R}^n \text{ such that } \partial\Omega \text{ is} \\ \text{an Ahlfors regular set and } \mathcal{H}^{n-1}(\partial\Omega \setminus \partial_*\Omega) = 0 \tag{A.0.1}$$

$A_\kappa(\partial\Omega)$, the κ-accessible (from within Ω) subset of $\partial\Omega$ (cf. [129, (8.8.2)]):

$$A_\kappa(\partial\Omega) := \big\{x \in \partial\Omega : x \in \overline{\Gamma_\kappa(x)}\big\} \tag{A.0.2}$$

$[w]_{A_p}$, the characteristic of the weight w on a space of homogeneous type (X, ρ, μ) (cf. [129, §7.7]):

$$[w]_{A_p} := \sup_{B\ \rho\text{-ball}} \Big(⨍_B w \,\mathrm{d}\mu\Big)\Big(⨍_B w^{-1/(p-1)} \,\mathrm{d}\mu\Big)^{p-1} \tag{A.0.3}$$

$A_p(X, \rho, \mu)$, the Muckenhoupt A_p-class on a space of homogeneous type (X, ρ, μ) (cf. [129, §7.7]):

$$A_p(X, \rho, \mu) := \big\{w \ \text{ weight function } : [w]_{A_p} < \infty\big\} \tag{A.0.4}$$

$A_\infty(X, \rho, \mu)$, the Muckenhoupt A_∞-class on a space of homogeneous type (X, ρ, μ) (cf. [129, §7.7]):

$$A_\infty(X, \rho, \mu) := \bigcup_{1 \le p < \infty} A_p(X, \mu) \tag{A.0.5}$$

$V^\perp$, the annihilator $V^\perp$ of a subspace V of a Banach space X:

$$V^\perp := \big\{\Lambda \in X^* : \Lambda(x) = 0 \ \text{ for all } \ x \in V\big\} \tag{A.0.6}$$

${}^\perp W$, the annihilator of a subspace W of X^* (where X is Banach):

$${}^\perp W := \big\{x \in X : \Lambda(x) = 0 \ \text{ for all } \ \Lambda \in W\big\} \tag{A.0.7}$$

D. Mitrea et al., *Geometric Harmonic Analysis V*, Developments in Mathematics 76,
https://doi.org/10.1007/978-3-031-31561-9

T^*, the adjoint of the operator $T \in \mathcal{L}(X \to Y)$:

$$T^* : Y^* \longrightarrow X^*, \quad T^*(\Lambda) := \Lambda \circ T \text{ for each } \Lambda \in Y^* \tag{A.0.8}$$

$a \otimes b$, the tensor product of vectors $a = (a_j)_{1 \le j \le N} \in \mathbb{C}^N$ and $b = (b_k)_{1 \le k \le M} \in \mathbb{C}^M$, defined as the $N \times M$ matrix

$$a \otimes b := \big(a_j b_k\big)_{\substack{1 \le j \le N \\ 1 \le k \le M}} \tag{A.0.9}$$

$A^\top$, the transpose of a coefficient tensor $A := \big(a_{rs}^{\alpha\beta}\big)_{\substack{1 \le r,s \le n \\ 1 \le \alpha,\beta \le M}}$ (cf. [131, §2.7]):

$$\big(A^\top\big)_{sr}^{\beta\alpha} := a_{rs}^{\alpha\beta} \text{ for all } \alpha, \beta, r, s \tag{A.0.10}$$

$A\nabla u$, the action of the coefficient tensor A on the Jacobian matrix ∇u, defined if $A = \big(a_{rs}^{\alpha\beta}\big)_{\substack{1 \le r,s \le n \\ 1 \le \alpha,\beta \le M}}$ $u = (u_\beta)_{1 \le \beta \le N} \in \big[\mathcal{D}'(\Omega)\big]^M$ as (cf. [131, §2.7]):

$$A\nabla u := \big(a_{rs}^{\alpha\beta} \partial_s u_\beta\big)_{\substack{1 \le \alpha \le M \\ 1 \le r \le n}} \tag{A.0.11}$$

$\langle A\cdot, \cdot\rangle$, the bilinear form associated with the coefficient tensor $A = \big(a_{rs}^{\alpha\beta}\big)_{\substack{1 \le r,s \le n \\ 1 \le \alpha,\beta \le M}}$ (cf. [131, §2.7]):

$$\langle A\zeta, \eta\rangle := a_{rs}^{\alpha\beta} \zeta_s^\beta \eta_r^\alpha \text{ for all } \zeta := (\zeta_s^\beta)_{\beta,s} \in \mathbb{C}^{N \times n} \text{ and } \eta := (\eta_r^\alpha)_{\alpha,r} \in \mathbb{C}^{M \times n} \tag{A.0.12}$$

$\mathcal{A}_{q,\kappa}$, the L^q-based area-function in $\mathbb{R}^n \setminus \Sigma$, acting on each $u \in W^{1,1}_{\rm loc}(\mathbb{R}^n \setminus \Sigma)$ as (cf. [131]):

$$(\mathcal{A}_{q,\kappa} u)(x) := \Big(\int_{\Gamma_\kappa^{\Sigma^c}(x)} |(\nabla u)(y)|^q |x - y|^{q-n}\, dy\Big)^{1/q}, \quad x \in \Sigma \tag{A.0.13}$$

$\mathfrak{A}_{\rm WE}(n, M)$, the class of weakly elliptic coefficient tensors, defined as the collection of all coefficient tensors $A = \big(a_{rs}^{\alpha\beta}\big)_{\substack{1 \le \alpha,\beta \le M \\ 1 \le r,s \le n}}$ with complex entries, with the property that the $M \times M$ homogeneous second-order system L_A associated with A in $\mathbb{R}^n$ is weakly elliptic.

$A_s^{p,q}(\Omega)$, the Besov/Triebel-Lizorkin space in the open set $\Omega \subseteq \mathbb{R}^n$ (with $A := B$ corresponding to Besov spaces, and with $A := F$ corresponding to Triebel-Lizorkin spaces), with $0 < p, q \le \infty$ and $s \in \mathbb{R}$, defined as (cf. [130, §9.2]):

$$A_s^{p,q}(\Omega) := \big\{u \in \mathcal{D}'(\Omega) : \text{there exists } U \in A_s^{p,q}(\mathbb{R}^n) \text{ such that } U\big|_\Omega = u\big\} \tag{A.0.14}$$

and equipped with the quasi-norm

$$\|u\|_{A_s^{p,q}(\Omega)} := \inf\Big\{\|U\|_{A_s^{p,q}(\mathbb{R}^n)} : U \in A_s^{p,q}(\mathbb{R}^n),\ U\big|_\Omega = u\Big\} \tag{A.0.15}$$

$A^{p,q}_s(\Omega)_{\rm bdd}$, the space of all distributions u in the open set $\Omega \subseteq \mathbb{R}^n$ with the property that $(\psi\big|_\Omega)u \in A^{p,q}_s(\Omega)$ for each cutoff function $\psi \in \mathscr{C}^\infty_c(\mathbb{R}^n)$ (cf. [130, Convention 8.3.7] and (A.0.256)).

B

$B_\rho(x, r)$, the ρ-ball with center at $x \in X$ and radius $r > 0$ in the quasi-metric space (X, ρ) (cf. [129, §7.1]):

$$B_\rho(x, r) := \{y \in X : \rho(x, y) < r\} \tag{A.0.16}$$

$B_{n-1}(x', r) := \big\{y' \in \mathbb{R}^{n-1} : |y' - x'| < r\big\}$, the $(n-1)$-dimensional (open) ball in $\mathbb{R}^{n-1}$ centered at $x' \in \mathbb{R}^{n-1}$ and of radius $r \in (0, \infty)$

BMO_1, the BMO-based Sobolev spaces of order one (locally integrable functions with distributional first-order partial derivatives in BMO)

$\|f\|_*(\Delta)$, the local BMO norm of f on the surface ball Δ (cf. [129, §7.4]):

$$\|f\|_*(\Delta) := \sup_{\Delta' \subseteq \Delta} ⨍_{\Delta'} |f - f_{\Delta'}|\, d\mu \tag{A.0.17}$$

$\|f\|_{\dot{\mathrm{BMO}}(X,\mu)}$, the homogeneous BMO semi-norm of the function f in the context of a space of homogeneous type (X, ρ, μ) (cf. [129, §7.4]):

$$\|f\|_{\dot{\mathrm{BMO}}(X,\mu)} := \sup_{x \in X,\, r>0} ⨍_{B_\rho(x,r)} \big|f - f_{B_\rho(x,r)}\big|\, d\mu \tag{A.0.18}$$

$\|\cdot\|_{\mathrm{BMO}(X,\mu)}$, the inhomogeneous BMO "norm" in the context of a space of homogeneous type (X, ρ, μ) (cf. [129, §7.4]):

$$\|f\|_{\mathrm{BMO}(X,\mu)} := \begin{cases} \|f\|_{\dot{\mathrm{BMO}}(X,\mu)} & \text{if } X \text{ is unbounded} \\ \|f\|_{\dot{\mathrm{BMO}}(X,\mu)} + \Big|\int_X f\, d\mu\Big| & \text{if } X \text{ is bounded} \end{cases} \tag{A.0.19}$$

$\mathrm{BMO}(X, \mu)$, the space of functions of bounded mean oscillations for a space of homogeneous type (X, ρ, μ) (cf. [129, §7.4]):

$$\mathrm{BMO}(X, \mu) := \Big\{f \in L^1_{\rm loc}(X, \mu) : \|f\|_{\mathrm{BMO}(X,\mu)} < +\infty\Big\} \tag{A.0.20}$$

$\widetilde{\mathrm{BMO}}(X, \mu)$, the space BMO modulo constants for a space of homogeneous type (X, ρ, μ) (cf. [129, (7.4.96)]):

$$\widetilde{\mathrm{BMO}}(X, \mu) := \mathrm{BMO}(X, \mu)/\sim\, = \big\{[f] : f \in \mathrm{BMO}(X, \mu)\big\} \tag{A.0.21}$$

∂E, the topological boundary of the set E

$Borel_\tau(X)$, the Borelians of the topological space (X, τ)

$[A; B] := [A, B] := AB - BA$, the commutator of A and B

$\{A;B\} := AB + BA$, the anti-commutator of A and B

$\mathrm{Bd}(X \to Y)$, the space of linear and bounded operators between the quasi-normed vector spaces $(X, \|\cdot\|_X)$ and $(Y, \|\cdot\|_Y)$ (cf. [130, §1.2]):

$$\mathrm{Bd}(X \to Y) := \big\{T : X \to Y : T \text{ linear mapping with } \|T\|_{X\to Y} < +\infty\big\} \tag{A.0.22}$$

$\mathrm{Bd}(X)$, the space of linear and bounded operators from the quasi-normed vector space X into itself:

$$\mathrm{Bd}(X) := \mathrm{Bd}(X \to X) \tag{A.0.23}$$

$\mathscr{B}(X \to Y)$, the space of linear and bounded operators from X to Y, where X, Y are two linear topological spaces (cf. [130, §1.1]):

$$\mathscr{B}(X \to Y) := \big\{T : X \longrightarrow Y : T \ \text{ linear and bounded}\big\} \tag{A.0.24}$$

$\mathscr{B}^{q,\lambda}(\Sigma,\sigma)$, the block space on the Ahlfors regular set $\Sigma \subseteq \mathbb{R}^n$, defined for $q \in (1,\infty)$ and $\lambda \in (0, n-1)$ as (cf. [130, §6.2]):

$$\begin{aligned}\mathscr{B}^{q,\lambda}(\Sigma,\sigma) := \Big\{f \in \big(\mathrm{Lip}_c(\Sigma)\big)' : f = \sum_{j=1}^{\infty}\lambda_j b_j \text{ in } \big(\mathrm{Lip}_c(\Sigma)\big)' \text{ for } \{\lambda_j\}_{j\in\mathbb{N}} \in \ell^1(\mathbb{N})\\ \text{and } \{b_j\}_{j\in\mathbb{N}} \text{ family of } \mathscr{B}^{q,\lambda}\text{-blocks on } \Sigma\Big\}\end{aligned} \tag{A.0.25}$$

and equipped with the norm

$$\begin{aligned}\|f\|_{\mathscr{B}^{q,\lambda}(\Sigma,\sigma)} := \inf\Big\{\sum_{j=1}^{\infty}|\lambda_j| : f = \sum_{j=1}^{\infty}\lambda_j b_j \ \text{ in } \ \big(\mathrm{Lip}_c(\Sigma)\big)' \ \text{ with}\\ \{\lambda_j\}_{j\in\mathbb{N}} \in \ell^1(\mathbb{N}) \ \text{ and each } b_j \text{ a } \mathscr{B}^{q,\lambda}\text{-block on } \Sigma\Big\}\end{aligned} \tag{A.0.26}$$

$\dot{B}^{p,q}_s(\Sigma,\sigma)$, the homogeneous Besov space on the Ahlfors regular set $\Sigma \subseteq \mathbb{R}^n$, defined (cf. [130, Definition 7.1.2]) for

$$\begin{aligned} s \in (-1,1), \qquad \max\big\{\tfrac{n-1}{n}, \tfrac{n-1}{n+s}\big\} < p \le \infty, \qquad 0 < q \le \infty,\\ \max\Big\{(s)_+, -s + (n-1)\Big(\tfrac{1}{p} - 1\Big)_+\Big\} < \beta < 1,\\ \max\Big\{s - \tfrac{n-1}{p}, (n-1)\Big(\tfrac{1}{p} - 1\Big)_+, -s + (n-1)\Big(\tfrac{1}{p} - 1\Big)\Big\} < \gamma < 1\end{aligned} \tag{A.0.27}$$

as the collection of "distributions" f on Σ (i.e., functionals $f \in \big(\mathring{\mathscr{G}}^{\beta,\gamma}_0(\Sigma)\big)^*$) with

$$\|f\|_{\dot{B}^{p,q}_s(\Sigma,\sigma)} := \Big\{\sum_{k\in\mathbb{Z}}\big[2^{ks}\|E_k f\|_{L^p(\Sigma,\sigma)}\big]^q\Big\}^{1/q} < \infty \tag{A.0.28}$$

with the natural alterations when $p = \infty$ or $q = \infty$

$B^{p,q}_s(\Sigma,\sigma)$, the inhomogeneous Besov space on the Ahlfors regular set $\Sigma \subseteq \mathbb{R}^n$, defined (cf. [130, §7.1]) for

$$s \in (-1,1), \qquad \max\left\{\tfrac{n-1}{n}, \tfrac{n-1}{n+s}\right\} < p \leq \infty, \qquad 0 < q \leq \infty, \tag{A.0.29}$$
$$\max\left\{(s)_+,\, -s+(n-1)\Big(\tfrac{1}{p}-1\Big)_+\right\} < \beta < 1, \quad (n-1)\Big(\tfrac{1}{p}-1\Big)_+ < \gamma < 1,$$

as the collection of all "distributions" f on Σ such that, if $\{E_k\}_{k\in\mathbb{Z},\, k\geq\widetilde{\kappa}_\Sigma}$ is the family of conditional expectation operators on the Ahlfors regular set Σ, then

$$\|f\|_{B^{p,q}_s(\Sigma,\sigma)} := \left\{\sum_{\tau\in I_{\widetilde{\kappa}_\Sigma}} \sum_{\nu=1}^{N(\widetilde{\kappa}_\Sigma,\tau)} \sigma(Q^{\widetilde{\kappa}_\Sigma,\nu}_\tau)\Big[m_{Q^{\widetilde{\kappa}_\Sigma,\nu}_\tau}(|E_{\widetilde{\kappa}_\Sigma} f|)\Big]^p\right\}^{1/p}$$
$$+\left\{\sum_{k\in\mathbb{Z},\, k\geq\widetilde{\kappa}_\Sigma+1} \big[2^{ks}\|E_k f\|_{L^p(\Sigma,\sigma)}\big]^q\right\}^{1/q} < \infty \tag{A.0.30}$$

$B^{p,q}_s(\mathbb{R}^n)$, the Besov space in $\mathbb{R}^n$ equipped with the quasi-norm $\|\cdot\|_{B^{p,q}_s(\mathbb{R}^n)}$ for $0 < p,q \leq \infty$ and $s \in \mathbb{R}$ (cf. [130, §9.1])

$B^{p,q}_s(\Omega)$, the Besov space in the open set $\Omega \subseteq \mathbb{R}^n$ with $0 < p,q \leq \infty$ and $s \in \mathbb{R}$, defined as the collection of $u \in \mathcal{D}'(\Omega)$ for which there exists $U \in B^{p,q}_s(\mathbb{R}^n)$ such that $U\big|_\Omega = u$, and equipped with the quasi-norm (cf. [130, (9.2.1)])

$$\|u\|_{B^{p,q}_s(\Omega)} := \inf\left\{\|U\|_{B^{p,q}_s(\mathbb{R}^n)} : U \in B^{p,q}_s(\mathbb{R}^n),\ U\big|_\Omega = u\right\}$$

$\mathrm{BMO}_{-1}(\partial\Omega,\sigma)$, the BMO-based negative Sobolev space of order minus one on $\partial\Omega$ (cf. [130, Definition 11.10.9]):

$$\mathrm{BMO}_{-1}(\partial\Omega,\sigma) := \Big(H_1^{\frac{n-1}{n-2},1}(\partial\Omega,\sigma)\Big)^* \tag{A.0.31}$$

$\mathcal{B}^{q,\lambda}_1(\partial\Omega,\sigma)$, the block-based Sobolev space of order one on $\partial\Omega$ ([130, (11.7.20)]):

$$\mathcal{B}^{q,\lambda}_1(\partial\Omega,\sigma) := \Big\{f \in \mathcal{B}^{q,\lambda}(\partial\Omega,\sigma) : \text{for all } j,k \in \{1,\dots,n\}$$
$$\text{one has } \partial_{\tau_{jk}} f \in \mathcal{B}^{q,\lambda}(\partial\Omega,\sigma)\Big\} \tag{A.0.32}$$

equipped with the norm

$$\mathcal{B}^{q,\lambda}_1(\partial\Omega,\sigma) \ni f \longmapsto \|f\|_{\mathcal{B}^{q,\lambda}(\partial\Omega,\sigma)} + \sum_{j,k=1}^n \big\|\partial_{\tau_{jk}} f\big\|_{\mathcal{B}^{q,\lambda}(\partial\Omega,\sigma)} \tag{A.0.33}$$

$\mathcal{B}^{p,q,\lambda}_1(\partial\Omega,\sigma)$, the off-diagonal block-based Sobolev space of order one on the set $\partial\Omega$ (cf. [130, (11.7.18)-(11.7.19)]):

$$\mathcal{B}^{p,q,\lambda}_1(\partial\Omega,\sigma) := \Big\{ f \in \mathcal{B}^{p,\lambda}(\partial\Omega,\sigma) : \text{ for each } j,k \in \{1,\dots,n\} \text{ one has } \partial_{\tau_{jk}} f \in \mathcal{B}^{q,\lambda}(\partial\Omega,\sigma) \Big\} \tag{A.0.34}$$

equipped with the norm

$$\mathcal{B}^{p,q,\lambda}_1(\partial\Omega,\sigma) \ni f \longmapsto \|f\|_{\mathcal{B}^{p,\lambda}(\partial\Omega,\sigma)} + \sum_{j,k=1}^{n} \big\|\partial_{\tau_{jk}} f\big\|_{\mathcal{B}^{q,\lambda}(\partial\Omega,\sigma)} \tag{A.0.35}$$

$\mathcal{B}^{q,\lambda}_{-1}(\partial\Omega,\sigma)$, the block-based negative Sobolev space of order minus one on the set $\partial\Omega$ (cf. [130, Definition 11.8.9]):

$$\mathcal{B}^{q,\lambda}_{-1}(\partial\Omega,\sigma) := \Big(\mathring{M}^{p,\lambda}_1(\partial\Omega,\sigma)\Big)^* \tag{A.0.36}$$

if $p,q \in (1,\infty)$ are such that $1/p + 1/q = 1$ and $\lambda \in (0, n-1)$

$\dot{\mathcal{B}}^{q,\lambda}_1(\partial\Omega,\sigma)$, the block-based homogeneous Sobolev space of order one on $\partial\Omega$, for $q \in (1,\infty)$ and $\lambda \in (0, n-1)$ (cf. [130, Definition 11.13.11]):

$$\dot{\mathcal{B}}^{q,\lambda}_1(\partial\Omega,\sigma) := \Big\{ f \in L^1\Big(\partial\Omega, \frac{\sigma(x)}{1+|x|^n}\Big) \cap L^{q_\lambda}_{\rm loc}(\partial\Omega,\sigma) \text{ satisfying } \partial_{\tau_{jk}} f \in \mathcal{B}^{q,\lambda}(\partial\Omega,\sigma) \text{ for } 1 \le j,k \le n \Big\} \tag{A.0.37}$$

where

$$q_\lambda := \frac{q(n-1)}{n-1+\lambda(q-1)} \in (1,q) \tag{A.0.38}$$

and equipped with the semi-norm

$$\dot{\mathcal{B}}^{q,\lambda}_1(\partial\Omega,\sigma) \ni f \mapsto \|f\|_{\dot{\mathcal{B}}^{q,\lambda}_1(\partial\Omega,\sigma)} := \sum_{j,k=1}^{n} \|\partial_{\tau_{jk}} f\|_{\mathcal{B}^{q,\lambda}(\partial\Omega,\sigma)} \tag{A.0.39}$$

$\mathscr{B}^{p,q}_\alpha(\Omega; D)$, the Besov-Hardy space of null-solutions of the standard Dirac operator $D = \sum_{j=1}^{n} \mathbf{e}_j \odot \partial_j$ in an open set $\Omega \subseteq \mathbb{R}^n$, defined for any $p \in (0,\infty]$, any $q \in (0, \le \infty]$, and any $\alpha \in \mathbb{R}$ as in [132, Theorem 4.4.10]:

$$\mathscr{B}^{p,q}_\alpha(\Omega; D) := \big\{ u \in B^{p,q}_\alpha(\Omega) \otimes C\ell_n : Du = 0 \text{ in } \Omega \big\} \tag{A.0.40}$$

and equipped with the quasi-norm inherited from $B^{p,q}_\alpha(\Omega) \otimes C\ell_n$

$\mathscr{B}^{p,q}_s(\partial\Omega; D)$, the boundary Besov-Hardy space associated with the Dirac operator $D = \sum_{j=1}^{n} \mathbf{e}_j \odot \partial_j$ in a bounded NTA domain $\Omega \subseteq \mathbb{R}^n$ with an Ahlfors regular boundary, defined for integrability exponents $p \in \big(\frac{n-1}{n}, \infty\big)$ and $q \in (0,\infty]$, and smoothness index $(n-1)\big(\frac{1}{p} - 1\big)_+ < s < 1$ as in [132, Theorem 4.4.10]:

$$\mathscr{B}^{p,q}_s(\partial\Omega; D) := \big\{\mathrm{Tr}_{\Omega\to\partial\Omega}\, u : u \in \mathscr{B}^{p,q}_{s+\frac{1}{p}}(\Omega; D)\big\} \tag{A.0.41}$$

and equipped with the quasi-norm inherited from $B^{p,q}_s(\partial\Omega, \sigma) \otimes C\ell_n$

C

$\overline{U}$, the closure of the set $U \subseteq \mathbb{R}^n$

$\mathscr{C}^k(\overline{\Omega})$, the space of functions of class $\mathscr{C}^k$ in an open neighborhood of $\overline{\Omega}$

$\mathscr{C}^k_c(\Omega)$, the space of functions of class $\mathscr{C}^k$ with compact support in the open set Ω

$\mathscr{C}^k_b(\Omega)$, the space of bounded functions of class $\mathscr{C}^k$ in Ω

$\big(\mathscr{C}^\infty_b(\Omega)\big)^*$, the algebraic dual of $\mathscr{C}^\infty_b(\Omega)$

$\mathrm{CBM}(\Omega)$, the space of complex Borel measures in the open set $\Omega \subseteq \mathbb{R}^n$

$C\ell_n$, the Clifford algebra $(C\ell_n, +, \odot)$ generated by n imaginary units defined as the minimal enlargement of $\mathbb{R}^n$ to a unitary real algebra which is not generated (as an algebra) by any proper subspace of $\mathbb{R}^n$, and such that $x \odot x = -|x|^2$ for each $x \in \mathbb{R}^n \hookrightarrow C\ell_n$ (cf. [129, §6.4])

$\|\cdot\|_{\dot{\mathscr{C}}^\alpha(U,\rho)}$, the homogeneous Hölder space semi-norm of order $\alpha > 0$ in the set $U \subseteq X$, in the context of a quasi-metric space (X, ρ), defined for each function $f : U \to \mathbb{R}$ as (cf. [129, (7.3.2)]):

$$\|f\|_{\dot{\mathscr{C}}^\alpha(U,\rho)} := \sup_{x,y\in U,\, x\neq y} \frac{|f(x) - f(y)|}{\rho(x, y)^\alpha} \tag{A.0.42}$$

$\dot{\mathscr{C}}^\alpha(U, \rho)$, the homogeneous Hölder space of order α in $U \subseteq X$, defined in the setting of a quasi-metric space (X, ρ) as (cf. [129, (7.3.1)]):

$$\dot{\mathscr{C}}^\alpha(U, \rho) := \Big\{f : U \to \mathbb{R} : \|f\|_{\dot{\mathscr{C}}^\alpha(U,\rho)} < +\infty\Big\} \tag{A.0.43}$$

$\dot{\mathscr{C}}^\alpha(U, \rho)/\sim$, the homogeneous Hölder space of order α modulo constants, in $U \subseteq X$, defined in the context of a quasi-metric space (X, ρ) as (cf. [129, (7.3.6)]):

$$\dot{\mathscr{C}}^\alpha(U, \rho)/\sim\, := \big\{[f] : f \in \dot{\mathscr{C}}^\alpha(U, \rho)\big\} \tag{A.0.44}$$

$\dot{\mathscr{C}}^\alpha_{\mathrm{loc}}(U, \rho)$, the local homogeneous Hölder space of order $\alpha > 0$ in the set $U \subseteq X$, defined in the context of a quasi-metric space (X, ρ) as (cf. [129, (7.3.7)]):

$$\dot{\mathscr{C}}^\alpha_{\mathrm{loc}}(U, \rho) := \Big\{f : U \to \mathbb{C} : f\big|_{B_\rho(x,r)\cap U} \in \dot{\mathscr{C}}^\alpha\big(B_\rho(x, r) \cap U, \rho\big) \text{ for each } x \in U \text{ and } r \in (0, \infty)\Big\} \tag{A.0.45}$$

$\|\cdot\|_{\mathscr{C}^\alpha(U,\rho)}$, the inhomogeneous Hölder space norm of order $\alpha > 0$ in the set $U \subseteq X$, in the context of a quasi-metric space (X, ρ), defined for each function $f : U \to \mathbb{R}$ as (cf. [129, (7.3.20)]):

$$\|f\|_{\mathscr{C}^\alpha(U,\rho)} := \sup_U |f| + \|f\|_{\dot{\mathscr{C}}^\alpha(U,\rho)}, \qquad \forall f \in \mathscr{C}^\alpha(U,\rho) \tag{A.0.46}$$

$\mathscr{C}^\alpha(U,\rho)$, the inhomogeneous Hölder space of order $\alpha > 0$ in the set $U \subseteq X$, defined in the context of a quasi-metric space (X,ρ) as (cf. [129, (7.3.19)]):

$$\mathscr{C}^\alpha(U,\rho) := \big\{f \in \dot{\mathscr{C}}^\alpha(U,\rho) : f \text{ is bounded in } U\big\} \tag{A.0.47}$$

$\mathscr{C}^\alpha_c(U,\rho)$, the space of Hölder functions of order $\alpha > 0$ with ρ-bounded support in the set $U \subseteq X$, defined in the context of a quasi-metric space (X,ρ) as (cf. [129, (7.3.26), (7.3.27)]):

$$\mathscr{C}^\alpha_c(U,\rho) := \big\{f \in \dot{\mathscr{C}}^\alpha(U,\rho) : f \text{ vanishes outside of a } \rho\text{-bounded subset of } U\big\} \tag{A.0.48}$$

$\dot{\mathscr{C}}^\gamma_{\mathrm{van}}(\Sigma)$, the homogeneous "vanishing" Hölder space of order γ on the set Σ (cf. [130, (3.2.5)]):

$$\dot{\mathscr{C}}^\gamma_{\mathrm{van}}(\Sigma) := \Big\{f \in \dot{\mathscr{C}}^\gamma(\Sigma) : \lim_{r\to 0^+}\Big(\sup_{x\in\Sigma}\|f\|_{\dot{\mathscr{C}}^\gamma(B(x,r)\cap\Sigma)}\Big) = 0\Big\} \tag{A.0.49}$$

$\mathscr{C}^\gamma_{\mathrm{van}}(\Sigma)$, the inhomogeneous "vanishing" Hölder space of order γ on the set Σ (cf. [130, (3.2.8)]):

$$\mathscr{C}^\gamma_{\mathrm{van}}(\Sigma) := \Big\{f \in \mathscr{C}^\gamma(\Sigma) : \lim_{r\to 0^+}\Big(\sup_{x\in\Sigma}\|f\|_{\dot{\mathscr{C}}^\gamma(B(x,r)\cap\Sigma)}\Big) = 0\Big\} \tag{A.0.50}$$

$\mathrm{Cp}(X \to Y)$, the space of compact linear operators from the topological vector space X into the topological vector space Y:

$$\mathrm{Cp}(X \to Y) := \big\{T : X \to Y : T \text{ linear compact mapping}\big\} \tag{A.0.51}$$

$\mathrm{Cp}(X)$, the space of compact linear operators from the topological vector space X into itself:

$$\mathrm{Cp}(X) := \mathrm{Cp}(X \to X) \tag{A.0.52}$$

$\mathrm{CMO}(\Sigma,\sigma)$, the Coifman-Weiss space (cf. [130, (4.6.11)]):

$\mathrm{CMO}(\Sigma,\sigma)$ is the closure in $\mathrm{BMO}(\Sigma,\sigma)$ of $\mathscr{C}^0_0(\Sigma)$, the space of all continuous functions on Σ which vanish at infinity (A.0.53)

C, the boundary-to-domain Cauchy-Clifford integral operator associated with the set $\Omega \subseteq \mathbb{R}^n$ of locally finite perimeter, acting on any Clifford algebra-valued function $f \in L^1\big(\partial_*\Omega, \frac{\sigma_*(x)}{1+|x|^{n-1}}\big) \otimes \mathcal{C}\ell_n$ at each point $x \in \mathring{\Omega}$ as in [131, (2.5.282)]:

$$Cf(x) := \frac{1}{\omega_{n-1}}\int_{\partial_*\Omega} \frac{x-y}{|x-y|^n} \odot \nu(y) \odot f(y)\,\mathrm{d}\sigma_*(y) \tag{A.0.54}$$

$\mathfrak{C}$, the boundary-to-boundary Cauchy-Clifford integral operator, acting on any Clifford algebra-valued function $f \in L^1\big(\partial_*\Omega, \frac{\sigma_*(x)}{1+|x|^{n-1}}\big) \otimes \mathcal{C}\ell_n$ at σ_*-a.e. point $x \in \partial_*\Omega$ as in [131, (2.5.284)]:

$$\mathfrak{C}f(x) := \lim_{\varepsilon\to 0^+} \frac{1}{\omega_{n-1}} \int_{y\in\partial_*\Omega,\ |x-y|>\varepsilon} \frac{x-y}{|x-y|^n} \odot \nu(y) \odot f(y)\, d\sigma_*(y) \tag{A.0.55}$$

$\mathfrak{C}^{\#}$, the transpose Cauchy-Clifford integral operator associated with some given set $\Omega \subseteq \mathbb{R}^n$ of locally finite perimeter, acting on any Clifford algebra-valued function $f \in L^1\big(\partial\Omega, \frac{\sigma(y)}{1+|y|^{n-1}}\big) \otimes \mathcal{C}\ell_n$ at σ-a.e. point $x \in \partial_*\Omega$ as in [132, (1.6.1)]:

$$\mathfrak{C}^{\#}f(x) := \lim_{\varepsilon\to 0^+} \frac{-1}{\omega_{n-1}} \int_{y\in\partial\Omega,\ |x-y|>\varepsilon} \nu(x) \odot \frac{x-y}{|x-y|^n} \odot f(y)\, d\sigma(y) \tag{A.0.56}$$

$\mathscr{C}$, the boundary-to-domain Cauchy integral operator in the plane, acting on any complex-valued function $f \in L^1\big(\partial_*\Omega, \frac{\sigma(\zeta)}{1+|\zeta|}\big)$ at each $z \in \Omega$ as in [132, (1.6.35)]:

$$\mathscr{C}f(z) := \frac{1}{2\pi i} \int_{\partial_*\Omega} \frac{f(\zeta)}{\zeta - z}\, d\zeta. \tag{A.0.57}$$

$\mathscr{C}_{\rm mod}$, the modified boundary-to-domain Cauchy integral operator in the plane, acting on $f \in L^1\big(\partial_*\Omega, \frac{\sigma(\zeta)}{1+|\zeta|^2}\big)$ at each $z \in \Omega$ as in [132, (1.8.227)]:

$$(\mathscr{C}_{\rm mod})(z) := \frac{1}{2\pi i} \int_{\partial_*\Omega} \Big\{\frac{1}{\zeta - z} - \frac{1}{\zeta}\mathbf{1}_{\mathbb{C}\setminus\overline{B(0,1)}}(\zeta)\Big\} f(\zeta)\, d\zeta \tag{A.0.58}$$

$C_{\rm mod}$, the modified boundary-to-domain Cauchy-Clifford integral operator, acting on any function $f \in L^1\big(\partial_*\Omega, \frac{\sigma(x)}{1+|x|^n}\big) \otimes \mathcal{C}\ell_n$ at each point $x \in \Omega$ as in [132, (1.8.238)]:

$$C_{\rm mod}f(x) := \frac{1}{\omega_{n-1}} \int_{\partial_*\Omega} \Big\{\frac{x-y}{|x-y|^n} + \frac{y}{|y|^n}\mathbf{1}_{\mathbb{R}^n\setminus\overline{B(0,1)}}(y)\Big\} \odot \nu(y) \odot f(y)\, d\sigma(y) \tag{A.0.59}$$

$\dot{C}^p_{q,\eta}(\Sigma,\sigma)$, the homogeneous Calderón space on the closed Ahlfors regular set $\Sigma \subseteq \mathbb{R}^n$ for $p \in [1,\infty]$, $q \in [1,\infty)$, and $\eta \in \mathbb{R}$, defined as in [132, (3.1.10)],

$$\dot{C}^p_{q,\eta}(\Sigma,\sigma) := \Big\{f \in L^1_{\rm loc}(\Sigma,\sigma) : f^{\#}_{q,\eta} \in L^p(\Sigma,\sigma)\Big\} \tag{A.0.60}$$

and equipped with the semi-norm (defined as in [132, (3.1.11)])

$$\|f\|_{\dot{C}^p_{q,\eta}(\Sigma,\sigma)} := \|f^{\#}_{q,\eta}\|_{L^p(\Sigma,\sigma)}, \qquad \forall f \in \dot{C}^p_{q,\eta}(\Sigma,\sigma) \tag{A.0.61}$$

where $f^{\#}_{q,\eta}(x) := \sup_{R>0} \big\{R^{-\eta}\big(⨍_{\Delta(x,R)} |f(y) - f_{\Delta(x,R)}|^q\, d\sigma(y)\big)^{1/q}\big\}$ for each $x \in \Sigma$

$C^p_{q,\eta}(\Sigma,\sigma)$, the inhomogeneous Calderón space on the closed Ahlfors regular set $\Sigma\subseteq\mathbb{R}^n$, defined for $p\in[1,\infty]$, $q\in[1,\infty)$, and $\eta\in\mathbb{R}$ as in [132, (3.1.14)],

$$C^p_{q,\eta}(\Sigma,\sigma):=L^p(\Sigma,\sigma)\cap\dot{C}^p_{q,\eta}(\Sigma,\sigma)=\Big\{f\in L^p(\Sigma,\sigma):\ f^{\#}_{q,\eta}\in L^p(\Sigma,\sigma)\Big\} \tag{A.0.62}$$

and equip this space with the norm (cf. [132, (3.1.15)])

$$\|f\|_{C^p_{q,\eta}(\Sigma,\sigma)}:=\|f\|_{L^p(\Sigma,\sigma)}+\big\|f^{\#}_{q,\eta}\big\|_{L^p(\Sigma,\sigma)},\qquad \forall f\in C^p_{q,\eta}(\Sigma,\sigma) \tag{A.0.63}$$

D

$u\cdot w=\langle u,w\rangle$, the dot product of two vectors $u,w\in\mathbb{R}^n$

$\mathrm{div}\vec{F}$, the divergence of the vector field $\vec{F}$

$\mathcal{D}'(\Omega)$, the space of distributions in the open set Ω

${}_{\mathcal{D}'(\Omega)}\langle\cdot,\cdot\rangle_{\mathcal{D}(\Omega)}$, the distributional pairing in the open set Ω

$\Delta:=\partial_1^2+\cdots+\partial_n^2$, the Laplace operator in $\mathbb{R}^n$

δ_x, the Dirac distribution with mass at x

D, the classical (homogeneous) Dirac operator in $\mathbb{R}^n$ defined as (cf. [129, (6.4.139)]):

$$D=\sum_{j=1}^n \mathbf{e}_j\odot\partial_j \tag{A.0.64}$$

$D=\Big(\sum_{j=1}^n a_j^{\alpha\beta}\partial_j+b^{\alpha\beta}\Big)_{\substack{1\leq\alpha\leq N\\ 1\leq\beta\leq N'}}$, a (generic) $N\times N'$ first-order system

$D^\top$, the (real) transpose of the first-order system D:

$$D^\top:=\Big(-\sum_{j=1}^n a_j^{\alpha\beta}\partial_j+b^{\alpha\beta}\Big)_{\substack{1\leq\beta\leq N'\\ 1\leq\alpha\leq N}} \tag{A.0.65}$$

$\overline{D}$, the complex conjugate of the first-order system D

D^*, the Hermitian adjoint of the first-order system D

d, the exterior derivative operator acting on the differential form $u=\sum_J u_J\mathrm{d}x^J$ according to (cf. [129, (1.11.32)], and also [129, (6.4.140)-(6.4.141)] for the Clifford algebra context):

$$\mathrm{d}u=\sum_{j=1}^n\sum_J\frac{\partial u_J}{\partial x_j}\mathrm{d}x_j\wedge\mathrm{d}x^J \tag{A.0.66}$$

δ, the formal adjoint of the exterior derivative operator d on differential forms (see also [129, (6.4.142)] for the Clifford algebra context)

δ_{jk}, the Kronecker symbol, i.e., $\delta_{jk}:=1$ if $j=k$ and $\delta_{jk}:=0$ if $j\neq k$

$\delta_{\partial\Omega}(\cdot)$, the distance function to the boundary of Ω

$U \Delta V := (U \setminus V) \cup (V \setminus U)$, the symmetric difference of the sets U and V

$\Delta(x,r) := B(x,r) \cap \partial\Omega$, the surface ball with center at $x \in \partial\Omega$ and radius r

$\mathbf{D} := D_L$, the (homogeneous) Dirac operator in $\mathbb{R}^n$ acting from the left (cf. [129, (6.4.48)]):

$$\mathbf{D} := D_L := \sum_{j=1}^{n} \mathbf{e}_j \odot \partial_j \tag{A.0.67}$$

$D_R u := \sum_{j=1}^{n} (\partial_j u) \odot \mathbf{e}_j$, the (homogeneous) Dirac operator acting from the right on the Clifford algebra-valued function u (cf. [129, (6.4.49)])

$\mathbb{D}_k(X)$, the k-th generation of dyadic cubes in the geometrically doubling quasi-metric space X, defined as in [129, Proposition 7.5.4]:

$$\mathbb{D}_k(X) := \{Q_\alpha^k\}_{\alpha \in I_k} \tag{A.0.68}$$

$\mathbb{D}(X)$, the dyadic grid on the geometrically doubling quasi-metric space X, defined as in [129, Proposition 7.5.4]:

$$\mathbb{D}(X) := \bigcup_{k \in \mathbb{Z},\, k \geq \kappa_X} \mathbb{D}_k(X) \tag{A.0.69}$$

$\dim X$, the dimension of a vector space X

$\mathrm{d}\zeta$, the complex arc-length, defined as

$$\mathrm{d}\zeta := \mathrm{i}\nu(\zeta)\,\mathrm{d}\sigma(\zeta) \tag{A.0.70}$$

Def, the deformation tensor, acting on the vector-distribution $\vec{u} = (u_1, \dots, u_n)$ as

$$\mathrm{Def}\,\vec{u} := \tfrac{1}{2}\big(\partial_j u_k + \partial_k u_j\big)_{1 \leq j,k \leq n} \tag{A.0.71}$$

$\Delta + k^2$, the Helmholtz operator with wave number k

$\mathcal{D} = \mathcal{D}_A$, the boundary-to-domain double layer for a second-order, homogeneous, constant coefficient, $M \times M$ weakly elliptic system $L_A := \big(a_{rs}^{\alpha\beta} \partial_r \partial_s\big)_{1 \leq \alpha,\beta \leq M}$ corresponding to a given coefficient tensor $A := \big(a_{rs}^{\alpha\beta}\big)_{\substack{1 \leq \alpha,\beta \leq M \\ 1 \leq r,s \leq n}}$ (with canonical fundamental solution $E = \big(E_{\gamma\beta}\big)_{1 \leq \gamma,\beta \leq M}$), whose action on any vector-valued function $f = (f_\alpha)_{1 \leq \alpha \leq M} \in \big[L^1\big(\partial\Omega, \frac{\sigma(x)}{1+|x|^{n-1}}\big)\big]^M$ at each $x \in \Omega$ as in [132, (1.3.18)] is:

$$\mathcal{D}_A f(x) := \Big(-\int_{\partial_*\Omega} \nu_s(y) a_{rs}^{\beta\alpha} (\partial_r E_{\gamma\beta})(x-y) f_\alpha(y)\,\mathrm{d}\sigma(y)\Big)_{1 \leq \gamma \leq M} \tag{A.0.72}$$

$\mathcal{D}_{\text{mod}} = \mathcal{D}_{A,\text{mod}}$, the boundary-to-domain modified double layer potential operator for an $M \times M$ weakly elliptic system $L_A := \big(a_{rs}^{\alpha\beta} \partial_r \partial_s\big)_{1 \leq \alpha,\beta \leq M}$ corresponding to the coefficient tensor $A := \big(a_{rs}^{\alpha\beta}\big)_{\substack{1 \leq \alpha,\beta \leq M \\ 1 \leq r,s \leq n}}$ (with fundamental solution $E = \big(E_{\gamma\beta}\big)_{1 \leq \gamma,\beta \leq M}$)

acting on $f = (f_\alpha)_{1\leq\alpha\leq M} \in \big[L^1\big(\partial\Omega, \frac{\sigma(x)}{1+|x|^n}\big)\big]^M$ at $x \in \Omega$ as in [132, (1.8.6)]:

$$(\mathcal{D}_{\rm mod} f)(x) := \Big(- \int_{\partial_*\Omega} \nu_s(y) a^{\beta\alpha}_{rs} \big\{ (\partial_r E_{\gamma\beta})(x-y) - k_1^{(r\gamma\beta)}(-y) \big\} f_\alpha(y)\, d\sigma(y) \Big)_{1\leq\gamma\leq M} \tag{A.0.73}$$

where $k_1^{(r\gamma\beta)} := (\partial_r E_{\gamma\beta}) \cdot \mathbf{1}_{\mathbb{R}^n \setminus \overline{B(0,1)}}$ for every r, γ, β

$\mathcal{D}_{\Delta,\rm mod}$, the boundary-to-domain modified harmonic double layer potential operator whose action on $f \in L^1\big(\partial_*\Omega, \frac{\sigma(y)}{1+|y|^n}\big)$ is defined at $x \in \Omega$ as in [132, (1.8.50)]:

$$\mathcal{D}_{\Delta,\rm mod} f(x) := \frac{1}{\omega_{n-1}} \int_{\partial_*\Omega} \Big\{ \frac{\langle \nu(y), y - x\rangle}{|x-y|^n} - \frac{\langle \nu(y), y\rangle}{|y|^n} \cdot \mathbf{1}_{\mathbb{R}^n \setminus \overline{B(0,1)}}(y) \Big\} f(y)\, d\sigma(y) \tag{A.0.74}$$

$\mathcal{D}_\lambda$, the boundary-to-domain double layer for the Stokes system associated with an arbitrary open set $\Omega \subseteq \mathbb{R}^n$ of locally finite perimeter and with the coefficient tensor $A_\lambda := \big(\delta_{jk}\delta_{\alpha\beta} + \lambda\delta_{j\beta}\delta_{k\alpha}\big)_{\substack{1\leq\alpha,\beta\leq n\\ 1\leq j,k\leq n}}$ corresponding to any $\lambda \in \mathbb{C}$, acting on $\vec{f} = (f_j)_{1\leq j\leq n} \in \Big[L^1\Big(\partial_*\Omega, \frac{\sigma(x)}{1+|x|^{n-1}}\Big)\Big]^n$ at $x \in \Omega$ as in [132, (6.2.18)]:

$$\begin{aligned} \mathcal{D}_\lambda \vec{f}(x) = \Bigg(\int_{\partial_*\Omega} \Big\{ &(\lambda - 1)\frac{\delta_{j\gamma}}{2\omega_{n-1}} \frac{\langle x-y, \nu(y)\rangle}{|x-y|^n} \\ &- (\lambda+1)\frac{n}{2\omega_{n-1}} \frac{\langle x-y, \nu(y)\rangle (x_j - y_j)(x_\gamma - y_\gamma)}{|x-y|^{n+2}} \\ &+ (\lambda-1)\frac{1}{2\omega_{n-1}} \frac{\nu_j(y)(x_\gamma - y_\gamma) - \nu_\gamma(y)(x_j - y_j)}{|x-y|^n} \Big\} f_j(y)\, d\sigma(y) \Bigg)_{1\leq\gamma\leq n} \end{aligned} \tag{A.0.75}$$

${}_{\mathcal{D}'(\Omega)}\langle\langle \cdot, \cdot \rangle\rangle_{\mathcal{D}(\Omega)}$, the distributional Hermitian inner product for double forms (cf. [132, (7.3.11)])

E

(ε, δ)-domain: (cf. [129, Definition 5.11.8])

$$\begin{array}{l} \text{a nonempty, open, proper subset } \Omega \text{ of } \mathbb{R}^n \text{ with the property that} \\ \text{for any } x, y \in \Omega \text{ with } |x-y| < \delta \text{ there exists a rectifiable curve} \\ \gamma : [0,1] \to \Omega \text{ such that } \gamma(0) = x,\ \gamma(1) = y, \text{ as well as length}(\gamma) \leq \\ \frac{1}{\varepsilon}|x-y| \text{ and } \frac{|z-x|\,|z-y|}{|x-y|} \leq \frac{1}{\varepsilon}\text{dist}(z, \partial\Omega) \text{ for each } z \in \gamma([0,1]) \end{array} \tag{A.0.76}$$

$\mathbf{e}_j$, the unit vector in the j-th direction in $\mathbb{R}^n$, defined for each $j \in \{1, \dots, n\}$ as $\mathbf{e}_j := (\delta_{jk})_{1\leq k\leq n} \in \mathbb{R}^n$ where δ_{jk} is the Kronecker symbol

$\{\mathbf{e}_j\}_{1\leq j\leq n}$, the standard orthonormal basis in $\mathbb{R}^n$

E_Δ, the standard fundamental solution for the Laplacian, defined for $x \in \mathbb{R}^n \setminus \{0\}$ as

$$E_\Delta(x) := \begin{cases} \dfrac{1}{\omega_{n-1}(2-n)} \dfrac{1}{|x|^{n-2}} & \text{if } n \geq 3, \\ \dfrac{1}{2\pi} \ln |x| & \text{if } n = 2 \end{cases} \tag{A.0.77}$$

$\mathscr{E}'(\Omega)$, the space of distributions compactly supported in the open set $\Omega \subseteq \mathbb{R}^n$

$[x]_{X/Y} := x + Y$, the equivalence class of the vector $x \in X$ in the quotient space X/Y

$\mathcal{E}_p(X)$, the p-envelope of a quasi-normed space X whose dual separates points, defined for $p \in (0, 1]$ as the completion of X in the quasi-norm $|\!|\!| \cdot |\!|\!|_p$ (cf. [130, Definition 7.8.4])

$\mathrm{Ex}_{\partial\Omega\to\Omega}$, the extension operator from $\partial\Omega$ to Ω (cf. [130, Theorem 8.4.1])

$E = E_L = \big(E_{\alpha\beta}\big)_{1\leq\alpha,\beta\leq M}$, the canonical fundamental solution of the weakly elliptic $M \times M$ system L (cf. [131, Theorem 1.4.2])

ε^A_B, the generalized Kronecker symbol for arrays A, B (cf. [129, (6.4.116)]):

$$\varepsilon^A_B := \begin{cases} \det\big((\delta_{ab})_{a\in A, b\in B}\big) & \text{if } |A| = |B|, \\ 0 & \text{otherwise} \end{cases} \tag{A.0.78}$$

$_{\mathcal{E}'(\Omega)}\langle\langle \cdot, \cdot \rangle\rangle_{\mathcal{E}(\Omega)}$, the (compact support) distributional Hermitian inner product for double forms (cf. [132, (7.3.12)])

$E_n(\cdot, \cdot)$, the fundamental solution for the complex Laplacian $\square := \bar\partial\vartheta + \vartheta\bar\partial$ in $\mathbb{C}^n$ (cf. [132, (7.3.18)])

F

$\Phi(X \to Y)$, the collection of Fredholm operators from the linear topological space X into the linear topological space Y (cf. [130, Definition 2.2.1])

$\Phi_+(X \to Y)$, the collection of finite-dim kernel semi-Fredholm operators from the Banach space X into the Banach space Y (cf. [130, §2.1])

$\Phi_-(X \to Y)$, the collection of finite-dim cokernel semi-Fredholm operators from the Banach space X into the Banach space Y (cf. [130, §2.1])

$f^\#_\gamma$, the Fefferman-Stein grand maximal function (with parameter $\gamma \in (0, 1)$), associating to each "distribution" $f \in \big(\mathrm{Lip}_c(\Sigma)\big)'$ on the Ahlfors regular set $\Sigma \subseteq \mathbb{R}^n$ the function defined at each point $x \in \Sigma$ by (cf. [130, (4.1.6)]):

$$f^\#_\gamma(x) := \sup_{\psi\in\mathcal{T}_\gamma(x)} \big|\langle f, \psi\rangle\big| \tag{A.0.79}$$

where $\mathcal{T}_\gamma(x) \subseteq \mathrm{Lip}_c(\Sigma)$ is the collection of all normalized (in $\dot{\mathscr{C}}^\gamma(\Sigma)$) bump functions centered at the point x

$\rho_{\mathrm{Fred}}(T; X)$, the Fredholm (or essential spectral) radius of $T \in \mathrm{Bd}(X)$ (cf. [130, Definition 2.2.5]):

$$\rho_{\text{Fred}}(T;X) := \inf\left\{r > 0 : zI - T \in \Phi(X \to X),\ \ \forall z \in \mathbb{C} \setminus \overline{B(0,r)}\right\} \tag{A.0.80}$$

$[\vec{F}]_\infty$, the contribution of the vector field $\vec{F}$ at infinity (cf. [129, (1.3.2)])

$\dot{F}^{p,q}_s(\Sigma,\sigma)$, the homogeneous Triebel-Lizorkin space on the Ahlfors regular set $\Sigma \subseteq \mathbb{R}^n$, defined (cf. [130, Definition 7.1.2]) for

$$\begin{gathered} s \in (-1,1), \quad \max\left\{\tfrac{n-1}{n}, \tfrac{n-1}{n+s}\right\} < p \leq \infty, \quad \max\left\{\tfrac{n-1}{n}, \tfrac{n-1}{n+s}\right\} < q \leq \infty, \\ \max\left\{(s)_+, -s + (n-1)\left(\tfrac{1}{p} - 1\right)_+\right\} < \beta < 1, \\ \max\left\{s - \tfrac{n-1}{p}, (n-1)\left(\tfrac{1}{p} - 1\right)_+, -s + (n-1)\left(\tfrac{1}{p} - 1\right)\right\} < \gamma < 1, \end{gathered} \tag{A.0.81}$$

as all "distributions" f on Σ (i.e., functionals in $\big(\mathring{\mathscr{G}}^{\beta,\gamma}_0(\Sigma)\big)^*$) such that

$$\|f\|_{\dot{F}^{p,q}_s(\Sigma,\sigma)} := \left\|\left\{\sum_{k\in\mathbb{Z}} \left[2^{ks}|E_k f|\right]^q\right\}^{1/q}\right\|_{L^p(\Sigma,\sigma)} < \infty \tag{A.0.82}$$

if $p < \infty$ (with natural alterations if $q = \infty$) and, for $p = \infty$,

$$\|f\|_{\dot{F}^{\infty,q}_s(\Sigma,\sigma)} := \sup_{\ell\in\mathbb{Z}}\sup_{\tau\in I_\ell}\left[\fint_{Q^\ell_\tau} \sum_{k=\ell}^{\infty} \left[2^{ks}|E_k f|\right]^q \,\mathrm{d}\sigma\right]^{1/q} < \infty \tag{A.0.83}$$

again with natural alterations when $q = \infty$

$F^{p,q}_s(\Sigma,\sigma)$, the inhomogeneous Triebel-Lizorkin space on the Ahlfors regular set $\Sigma \subseteq \mathbb{R}^n$, defined as the collection of all "distributions" f on Σ with the property that $\|f\|_{F^{p,q}_s(\Sigma,\sigma)} < +\infty$ (cf. [130, Definition 7.1.2])

$F^{p,q}_s(\mathbb{R}^n)$, the (inhomogeneous) Triebel-Lizorkin space in $\mathbb{R}^n$ for $0 < p,q \leq \infty$ and $s \in \mathbb{R}$, equipped with the quasi-norm $\|\cdot\|_{F^{p,q}_s(\mathbb{R}^n)}$ (cf. [130, (9.1.4)-(9.1.5)])

$F^{p,q}_s(\Omega) := \big\{u \in \mathcal{D}'(\Omega) :$ there exists $U \in F^{p,q}_s(\mathbb{R}^n)$ such that $U\big|_\Omega = u\big\}$, the Triebel-Lizorkin space in the (arbitrary) open set $\Omega \subseteq \mathbb{R}^n$ for $0 < p,q \leq \infty$ and $s \in \mathbb{R}$, equipped with the quasi-norm (cf. [130, (9.2.1)])

$$\|u\|_{F^{p,q}_s(\Omega)} := \inf\left\{\|U\|_{F^{p,q}_s(\mathbb{R}^n)} : U \in F^{p,q}_s(\mathbb{R}^n),\ U\big|_\Omega = u\right\}$$

$f_\Delta := \fint_\Delta f\,\mathrm{d}\sigma$, the integral average of f on the "surface ball" Δ

$\widehat{\phi}$, the Fourier transform of ϕ in $\mathbb{R}^n$:

$$\widehat{\phi}(\xi) := \int_{\mathbb{R}^n} e^{-\mathrm{i}\langle x,\,\xi\rangle}\phi(x)\,\mathrm{d}x, \qquad \xi \in \mathbb{R}^n \tag{A.0.84}$$

$\mathscr{F}^{p,q}_\alpha(\Omega;D)$, the Triebel-Lizorkin-Hardy space of null-solutions of the Dirac operator $D = \sum_{j=1}^n \mathbf{e}_j \odot \partial_j$ in an open set $\Omega \subseteq \mathbb{R}^n$ defined for $p \in (0,\infty)$, $q \in (0,\infty]$, and $\alpha \in \mathbb{R}$

as in [132, (4.4.130)]:

$$\mathscr{F}^{p,q}_{\alpha}(\Omega; D) := \big\{u \in F^{p,q}_{\alpha}(\Omega) \otimes C\ell_n \,:\, Du = 0 \text{ in } \Omega\big\} \tag{A.0.85}$$

G

∇u, the gradient (Jacobian matrix) of a $\mathbb{C}^M$-valued function $u = (u_\alpha)_{1\leq\alpha\leq M}$ defined in an open subset of $\mathbb{R}^n$, defined as:

$$\nabla u := \big(\partial_j u_\alpha\big)_{\substack{1\leq\alpha\leq M\\ 1\leq j\leq n}} = \begin{bmatrix} \partial_1 u_1 & \cdots & \partial_n u_1 \\ \vdots & \vdots & \vdots \\ \partial_1 u_M & \cdots & \partial_n u_M \end{bmatrix} \tag{A.0.86}$$

∇', the gradient operator in $\mathbb{R}^{n-1}$

$\Gamma_\kappa(x) = \Gamma_{\Omega,\kappa}(x)$, the ($\kappa$-)nontangential approach region with vertex at $x \in \partial\Omega$, defined as (cf. [129, (8.1.2)]):

$$\Gamma_\kappa(x) = \Gamma_{\Omega,\kappa}(x) := \big\{y \in \Omega : |x - y| < (1+\kappa)\delta_{\partial\Omega}(y)\big\}, \quad \forall x \in \partial\Omega \tag{A.0.87}$$

$\mathbb{X}$, a Generalized Banach Function Space on the measure space $(X, \mathfrak{M}, \mu)$ equipped with the norm $\|\cdot\|_{\mathbb{X}}$ (cf. [130, Definition 5.1.4])

$\mathring{\mathbb{X}} := \overline{L^\infty_{\text{comp}}(X,\mu)}^{\|\cdot\|_{\mathbb{X}}}$, the closure of L^∞_{comp} in the Generalized Banach Function Space $(\mathbb{X}, \|\cdot\|_{\mathbb{X}})$ (cf. [130, Definition 5.2.6])

$\mathbb{X}$, a Generalized Banach Function Space on the measure space $(X, \mathfrak{M}, \mu)$ equipped with the norm $\|\cdot\|_{\mathbb{X}}$ (cf. [130, Definition 5.1.4])

$\mathring{\mathbb{X}} := \overline{L^\infty_{\text{comp}}(X,\mu)}^{\|\cdot\|_{\mathbb{X}}}$, the closure of L^∞_{comp} in the Generalized Banach Function Space $(\mathbb{X}, \|\cdot\|_{\mathbb{X}})$ (cf. [130, Definition 5.2.6])

$\nabla_{\tan} f$, the tangential gradient of the function $f \in L^1_{1,\text{loc}}(\partial_*\Omega, \sigma_*)$ (cf. [130, §11.4]):

$$\nabla_{\tan} f := \Big(\sum_{k=1}^{n} \nu_k \partial_{\tau_{kj}} f\Big)_{1\leq j\leq n} \quad \text{at } \sigma_*\text{-a.e. point on } \partial_*\Omega \tag{A.0.88}$$

H

$\mathcal{H}^{n-1}$, the $(n-1)$-dimensional Hausdorff measure in $\mathbb{R}^n$

$\mathcal{H}^s$, the s-dimensional Hausdorff measure in $\mathbb{R}^n$

$H^p(\Sigma,\sigma)$, the Lebesgue-based Hardy space on the Ahlfors regular set $\Sigma \subseteq \mathbb{R}^n$, defined for $p \in \big(\frac{n-1}{n}, \infty\big)$ and $\gamma \in \big((n-1)\big(\frac{1}{p} - 1\big)_+, 1\big)$ as (cf. [130, Definition 4.2.1])

$$H^p(\Sigma,\sigma) := \Big\{f \in \big(\text{Lip}_c(\Sigma)\big)' : f^{\#}_{\gamma} \in L^p(\Sigma,\sigma)\Big\} \tag{A.0.89}$$

and equipped with the quasi-norm

$$\|f\|_{H^p(\Sigma,\sigma)} := \|f^{\#}_{\gamma}\|_{L^p(\Sigma,\sigma)}, \qquad \forall f \in H^p(\Sigma,\sigma) \tag{A.0.90}$$

$\dot{H}^p(\Sigma,\sigma)$, the homogeneous Hardy space on the Ahlfors regular set $\Sigma \subseteq \mathbb{R}^n$, defined as $\dot{H}^p(\Sigma,\sigma) := \{f \in H^p(\Sigma,\sigma) : \langle f, 1\rangle = 0\}$ if Σ is bounded, and simply as $\dot{H}^p(\Sigma,\sigma) := H^p(\Sigma,\sigma)$ if Σ is unbounded (cf. [130, (4.2.12)])

$H^{p,q}(\Sigma,\sigma)$, the Lorentz-based Hardy space on the Ahlfors regular set $\Sigma \subseteq \mathbb{R}^n$, defined for $p \in \big(\frac{n-1}{n},\infty\big)$, $p \in (0,\infty]$, and $\gamma \in \big((n-1)\big(\frac{1}{p}-1\big)_+, 1\big)$ as (cf. [130, Definition 4.2.3])

$$H^{p,q}(\Sigma,\sigma) := \Big\{f \in \big(\mathrm{Lip}_c(\Sigma)\big)' : f^{\#}_{\gamma} \in L^{p,q}(\Sigma,\sigma)\Big\} \tag{A.0.91}$$

and equipped with the quasi-norm

$$\|f\|_{H^{p,q}(\Sigma,\sigma)} := \|f^{\#}_{\gamma}\|_{L^{p,q}(\Sigma,\sigma)}, \qquad \forall f \in H^{p,q}(\Sigma,\sigma) \tag{A.0.92}$$

$H^{p,q}_{\mathrm{fin}}(\Sigma,\sigma)$, the vector space of all finite linear combinations of (p,q)-atoms on Σ equipped with the quasi-norm $\|f\|_{H^{p,q}_{\mathrm{fin}}(\Sigma,\sigma)}$ defined as the infimum of all $\Big(\sum_{j=1}^{N}|\lambda_j|^p\Big)^{1/p}$ such that $f = \sum_{j=1}^{N}\lambda_j a_j$ for some $N \in \mathbb{N}$, $\{\lambda_j\}_{1\leq j\leq N} \subseteq \mathbb{C}$, and (p,q)-atoms $\{a_j\}_{1\leq j\leq N}$ (cf. [130, (4.4.113)])

$\mathfrak{H}$, the L^p-filtering operator, acting on each given distribution $f \in H^p(\Sigma,\sigma)$ with $\frac{n-1}{n} < p < \infty$ at each $x \in \Sigma$ according to (cf. [130, Theorem 4.9.1]):

$$(\mathfrak{H}f)(x) := \lim_{t\to 0^+} {}_{(H^p(\Sigma,\sigma))^*}\big\langle S_t(x,\cdot), f\big\rangle_{H^p(\Sigma,\sigma)} \tag{A.0.93}$$

where $\{S_t(\cdot,\cdot)\}_t$ are the integral kernels of a suitable approximation to the identity

$\mathscr{H}^{q,\lambda}(\Sigma,\sigma)$, the space defined for $q \in (1,\infty)$, $\lambda \in (0, n-1)$ as (cf. [130, §6.1]):

$$\mathscr{H}^{q,\lambda}(\Sigma,\sigma) := \Big\{f = \sum_{j=1}^{\infty}\lambda_j a_j \text{ in } \big(\mathrm{Lip}_c(\Sigma)\big)' : \text{ for } \{\lambda_j\}_{j\in\mathbb{N}} \in \ell^1(\mathbb{N}) \text{ and } \{a_j\}_{j\in\mathbb{N}} \text{ family of } \mathscr{H}^{q,\lambda}\text{-atoms on } \Sigma\Big\} \tag{A.0.94}$$

and equipped with the norm

$$\|f\|_{\mathscr{H}^{q,\lambda}(\Sigma,\sigma)} := \inf\Big\{\sum_{j=1}^{\infty}|\lambda_j| : f = \sum_{j=1}^{\infty}\lambda_j a_j \text{ in } \big(\mathrm{Lip}_c(\Sigma)\big)' \text{ with } \{\lambda_j\}_{j\in\mathbb{N}} \in \ell^1(\mathbb{N}) \text{ and each } a_j \text{ a } \mathscr{H}^{q,\lambda}\text{-atom on } \Sigma\Big\} \tag{A.0.95}$$

$H^s(\Omega)$, the L^2-based fractional Sobolev space of order $s \in \mathbb{R}$ in an open set $\Omega \subseteq \mathbb{R}^n$ (cf. [130, §9.2]):

$$H^s(\Omega) := \Big\{U\big|_\Omega : U \in H^s(\mathbb{R}^n)\Big\} = \Big\{U\big|_\Omega : U \in F^{2,2}_s(\mathbb{R}^n)\Big\}, \qquad \forall s \in \mathbb{R}, \quad \text{(A.0.96)}$$

and equipped with the norm

$$\begin{aligned}\|u\|_{H^s(\Omega)} &:= \inf\Big\{\|U\|_{H^s(\mathbb{R}^n)} : U \in H^s(\mathbb{R}^n) \text{ such that } u = U\big|_\Omega\Big\}\\ &\approx \inf\Big\{\|U\|_{F^{2,2}_s(\mathbb{R}^n)} : U \in F^{2,2}_s(\mathbb{R}^n) \text{ such that } u = U\big|_\Omega\Big\} \qquad \text{(A.0.97)}\end{aligned}$$

$H^p_\kappa(\Omega)$, the Hardy space of harmonic functions u in the open set $\Omega \subseteq \mathbb{R}^n$ with a p-th power integral nontangential maximal function (cf. [131, (5.5.4)]):

$$H^p_\kappa(\Omega) := \big\{u \in \mathscr{C}^\infty(\Omega) : \Delta u = 0 \text{ in } \Omega, \text{ and } \mathcal{N}_\kappa u \in L^p(\partial\Omega, \sigma)\big\} \qquad \text{(A.0.98)}$$

$\dot{H}^p_1(\partial\Omega, \sigma)$, the Hardy-based homogeneous Sobolev space of order one on $\partial\Omega$, defined for each $p \in \big(\frac{n-1}{n}, \infty\big)$ as (cf. [130, Definition 11.10.5]):

$$\begin{aligned}\dot{H}^p_1(\partial\Omega, \sigma) := \Big\{f \in L^1\Big(\partial\Omega, \frac{\sigma(x)}{1+|x|^n}\Big) : \partial_{\tau_{jk}} f \in H^p(\partial\Omega, \sigma) \qquad \text{(A.0.99)}\\ \text{for each } j, k \in \{1, \dots, n\}\Big\}\end{aligned}$$

and equipped with the semi-norm

$$\dot{H}^p_1(\partial\Omega, \sigma) \ni f \mapsto \|f\|_{\dot{H}^p_1(\partial\Omega,\sigma)} := \sum_{j,k=1}^n \|\partial_{\tau_{jk}} f\|_{H^p(\partial\Omega,\sigma)} \qquad \text{(A.0.100)}$$

$\dot{H}^p_1(\partial\Omega, \sigma)/\sim$, the quotient space of classes $[\cdot]$ of equivalence modulo constants of functions in $\dot{H}^p_1(\partial\Omega, \sigma)$ with $p \in \big(\frac{n-1}{n}, \infty\big)$ (cf. [130, (11.10.33)-(11.10.34)]):

$$\dot{H}^p_1(\partial\Omega, \sigma)/\sim := \big\{[f] : f \in \dot{H}^p_1(\partial\Omega, \sigma)\big\} \qquad \text{(A.0.101)}$$

equipped with the semi-quasinorm

$$\dot{H}^p_1(\partial\Omega, \sigma)/\sim \ni [f] \longmapsto \big\|[f]\big\|_{\dot{H}^p_1(\partial\Omega,\sigma)/\sim} := \sum_{j,k=1}^n \big\|\partial_{\tau_{jk}} f\big\|_{H^p(\partial\Omega,\sigma)} \qquad \text{(A.0.102)}$$

$H^{q,p}_1(\partial\Omega, \sigma)$, the Hardy-based inhomogeneous Sobolev space of order one on $\partial\Omega$, defined for $p \in \big(\frac{n-1}{n}, \infty\big)$ and $q \in [1, \infty]$ as (cf. [130, Definition 11.10.6]):

$$\begin{aligned}H^{q,p}_1(\partial\Omega, \sigma) &:= L^q(\partial\Omega, \sigma) \cap \dot{H}^p_1(\partial\Omega, \sigma) \qquad \text{(A.0.103)}\\ &= \Big\{f \in L^q(\partial\Omega, \sigma) : \partial_{\tau_{jk}} f \in H^p(\partial\Omega, \sigma) \text{ for } 1 \leq j, k \leq n\Big\}\end{aligned}$$

and equipped with the quasi-norm

$$\|f\|_{H_1^{q,p}(\partial\Omega,\sigma)} := \|f\|_{L^q(\partial\Omega,\sigma)} + \|f\|_{\dot{H}_1^p(\partial\Omega,\sigma)}, \quad \forall f \in H_1^{q,p}(\partial\Omega,\sigma) \qquad \text{(A.0.104)}$$

$\mathcal{H}^p(\Omega; D)$, the Hardy space in Ω associated with the first-order $N \times M$ system D, defined as in [131, (3.2.2)]:

$$\begin{array}{l}\mathcal{H}^p(\Omega; D) \text{ is the collection of all functions } u \in \left[\mathscr{C}^\infty(\Omega)\right]^M \text{ satisfying} \\ \mathcal{N}_\kappa u \in L^p(\partial\Omega,\sigma) \text{ and } Du = 0 \text{ in } \Omega, \text{ and which also vanish at infinity} \\ \text{(as described in [131, Definition 1.6.3]) when } \Omega \text{ is an exterior domain}\end{array} \qquad \text{(A.0.105)}$$

and equipped with the quasi-norm (cf. [131, (3.2.3)])

$$\|u\|_{\mathcal{H}^p(\Omega;D)} := \|\mathcal{N}_\kappa u\|_{L^p(\partial\Omega,\sigma)}, \qquad \forall u \in \mathcal{H}^p(\Omega; D) \qquad \text{(A.0.106)}$$

$\mathscr{H}^p_\bullet(\partial\Omega; D)$, the "bullet" boundary Hardy space associated with the Dirac operator $D := \sum_{j=1}^n \mathbf{e}_j \odot \partial_j$ in the open set $\Omega \subseteq \mathbb{R}^n$, defined as in (A.0.107):

$$\mathscr{H}^p_\bullet(\partial\Omega; D) := \left\{\nu \bullet u : u \in \mathcal{H}^p(\Omega; D)\right\} \qquad \text{(A.0.107)}$$

and equipped with the quasi-norm inherited from $H^p(\partial\Omega,\sigma) \otimes \mathcal{C}\ell_n$

$\dot{\mathcal{H}}^p_{1,\pm}(\partial\Omega,\sigma)$, the "positive/negative" homogeneous Hardy spaces with regularity on $\partial\Omega$, defined as in [132, (1.8.285)]:

$$\dot{\mathcal{H}}^p_{1,\pm}(\partial\Omega,\sigma) := \Big\{u\big|^{\kappa-\text{n.t.}}_{\partial\Omega} : u \text{ holomorphic in } \Omega_\pm \text{ and } \mathcal{N}_\kappa(\nabla u) \in L^p(\partial\Omega,\sigma)\Big\} \qquad \text{(A.0.108)}$$

where $\Omega_+ := \Omega$ and $\Omega_- := \mathbb{C} \setminus \overline{\Omega}$

$H_{\rm mod}$, the modified Hilbert transform on the real line, whose action on each function $f \in L^1\big(\mathbb{R}, \frac{dx}{1+|x|^2}\big)$ at $\mathcal{L}^1$-a.e. point $x \in \mathbb{R}$ is defined as in [132, (5.3.99)]:

$$(H_{\rm mod} f)(x) := \lim_{\varepsilon\to 0^+} \frac{1}{\pi} \int_{\mathbb{R}} \Big\{\frac{1}{x-y}\mathbf{1}_{\mathbb{R}\setminus[x-\varepsilon,x+\varepsilon]}(y) + \frac{1}{y}\mathbf{1}_{\mathbb{R}\setminus[-1,1]}(y)\Big\} f(y)\,dy \qquad \text{(A.0.109)}$$

$\mathcal{H}^p(\Omega)$, the (several complex variable) holomorphic Hardy space in the UR domain $\Omega \subseteq \mathbb{R}^{2n} \equiv \mathbb{C}^n$ defined for each $p \in \big(\frac{2n-1}{2n}, \infty\big)$ as in [132, (7.6.17)]:

$$\mathcal{H}^p(\Omega) := \Big\{F : \Omega \to \mathbb{C} : F \text{ holomorphic}, \ \mathcal{N}_\kappa F \in L^p(\partial\Omega,\sigma), \text{ and } \lim_{|z|\to\infty} F(z) = 0 \text{ if } \Omega \text{ is an exterior domain}\Big\}, \qquad \text{(A.0.110)}$$

and equipped with the quasi-norm $\mathcal{H}^p(\Omega) \ni F \mapsto \|F\|_{\mathcal{H}^p(\Omega)} := \|\mathcal{N}_\kappa F\|_{L^p(\partial\Omega,\sigma)}$

I

$\langle \cdot, \cdot \rangle$, the (real) inner product in $\mathbb{C}^M$ defined for any vectors $u = (u_k)_{1\leq k\leq M} \in \mathbb{C}^M$ and $w = (w_k)_{1\leq k\leq M} \in \mathbb{C}^M$ as:

$$\langle u, w\rangle := \sum_{k=1}^{M} u_k w_k \tag{A.0.111}$$

$\mathrm{i} := \sqrt{-1} \in \mathbb{C}$, the complex imaginary unit

ι^*, the pull-back map induced by the canonical inclusion ι

$\mathring{U}$, the interior of the set $U \subseteq \mathbb{R}^n$

$⨍_E f\,\mathrm{d}\mu := \frac{1}{\mu(E)}\int_E f\,\mathrm{d}\mu$, the integral average of the function f on the set $E \subseteq X$, in a measure space (X, μ)

$f_{B_\rho(x,r)}$, the integral average of f over the ρ-ball $B_\rho(x,r)$, in the context of a space of homogeneous type (X, ρ, μ), defined as (cf. [129, (7.4.9)]):

$$f_{B_\rho(x,r)} := ⨍_{B_\rho(x,r)} f\,\mathrm{d}\mu := \frac{1}{\mu\big(B_\rho(x,r)\big)}\int_{B_\rho(x,r)} f(y)\,\mathrm{d}\mu(y) \tag{A.0.112}$$

$I_{E,\alpha}$, the fractional integral operator of order α on the set E contained in a metric space (X, ρ) equipped with an upper d-dimensional Borel measure μ on (X, τ_ρ), acting on functions $f \in L^1\big(E, \frac{\mu(x)}{1+\rho(x,x_0)^{d-\alpha}}\big)$ according to (cf. [129, (7.8.3)]):

$$I_{E,\alpha} f(x) := \int_E \frac{f(y)}{\rho(x,y)^{d-\alpha}}\,\mathrm{d}\mu(y) \;\text{ for } \mu\text{-a.e. } x \in E \tag{A.0.113}$$

$\mathrm{Im}\,(T : X \to Y) := \{Tx : x \in X\}$, the image (or range) of the operator $T : X \to Y$

$(X_0, X_1)_{\theta,q} := \big\{x \in X_0 + X_1 : \|x\|_{(X_0,X_1)_{\theta,q}} < +\infty\big\}$, the intermediate space for the real method of interpolation for the compatible pair of quasi-Banach spaces X_0 and X_1, equipped with the real interpolation quasi-norm $\|\cdot\|_{(X_0,X_1)_{\theta,q}}$ (cf. [130, §1.3])

$i(\Phi) := \sup\limits_{0<t<1} \frac{\ln h_\Phi(t)}{\ln t} = \lim\limits_{t\to 0^+} \frac{\ln h_\Phi(t)}{\ln t}$ where $h_\Phi(t) := \sup\limits_{s>0} \frac{\Phi(st)}{\Phi(s)}$ for $t \in (0,\infty)$, the lower dilation index of the Young function Φ (cf. [130, (5.3.14)])

$I(\Phi) := \inf\limits_{1<t<\infty} \frac{\ln h_\Phi(t)}{\ln t} = \lim\limits_{t\to \infty} \frac{\ln h_\Phi(t)}{\ln t}$ where $h_\Phi(t) := \sup\limits_{s>0} \frac{\Phi(st)}{\Phi(s)}$ for $t \in (0,\infty)$, the upper dilation index of the Young function Φ (cf. [130, (5.3.15)])

K

K_Δ, the boundary-to-boundary harmonic double layer potential (cf. [129, (1.1.32)]):

$$K_\Delta f(x) := \lim_{\varepsilon\to 0^+} \frac{1}{\omega_{n-1}} \int_{\partial\Omega\setminus\overline{B(x,\varepsilon)}} \frac{\langle \nu(y), y-x\rangle}{|x-y|^n} f(y)\,\mathrm{d}\mathcal{H}^{n-1}(y), \quad x \in \partial\Omega \tag{A.0.114}$$

$K^{\#}_\Delta$, the transpose harmonic double layer potential, defined as (cf. [129, (1.1.33)]):

$$K^{\#}_\Delta f(x) := \lim_{\varepsilon\to 0^+} \frac{1}{\omega_{n-1}} \int_{\partial\Omega\setminus\overline{B(x,\varepsilon)}} \frac{\langle \nu(x), x-y\rangle}{|x-y|^n} f(y)\,\mathrm{d}\mathcal{H}^{n-1}(y), \quad x \in \partial\Omega \tag{A.0.115}$$

$\operatorname{Ker}(T : X \to Y) := \{x \in X : Tx = 0\}$, the kernel (or null-space) of the operator T

$\operatorname{Ker} L := \left\{u \in \left[\mathscr{C}^\infty(\Omega)\right]^M : Lu = 0 \text{ in } \Omega\right\}$, the null-space of the $M \times M$ system L, in an open set Ω

$K = K_A$, the boundary-to-boundary double layer for a second-order complex coefficient weakly elliptic, $M \times M$ system $L_A := (a_{rs}^{\alpha\beta}\partial_r\partial_s)_{1\leq\alpha,\beta\leq M}$ corresponding to the coefficient tensor $A := \left(a_{rs}^{\alpha\beta}\right)_{\substack{1\leq\alpha,\beta\leq M\\ 1\leq r,s\leq n}}$ (with canonical fundamental solution $E = \left(E_{\gamma\beta}\right)_{1\leq\gamma,\beta\leq M}$) acting on any function $f = (f_\alpha)_{1\leq\alpha\leq M} \in \left[L^1\left(\partial\Omega, \frac{\sigma(x)}{1+|x|^{n-1}}\right)\right]^M$ at σ-a.e. point $x \in \partial_*\Omega$ as in [132, (1.3.68)]:

$$Kf(x) := \left(-\lim_{\varepsilon\to 0^+}\int_{y\in\partial_*\Omega,\ |x-y|>\varepsilon} \nu_s(y)a_{rs}^{\beta\alpha}(\partial_r E_{\gamma\beta})(x-y)f_\alpha(y)\,d\sigma(y)\right)_{1\leq\gamma\leq M} \tag{A.0.116}$$

$K_{\mathrm{mod}} = K_{A,\mathrm{mod}}$, the modified boundary-to-boundary double layer potential operator for a second-order, homogeneous, constant complex coefficient, weakly elliptic $M \times M$ system $L = L_A := \left(a_{rs}^{\alpha\beta}\partial_r\partial_s\right)_{1\leq\alpha,\beta\leq M}$, corresponding to the coefficient tensor $A := \left(a_{rs}^{\alpha\beta}\right)_{\substack{1\leq\alpha,\beta\leq M\\ 1\leq r,s\leq n}}$ (with fundamental solution $E = \left(E_{\gamma\beta}\right)_{1\leq\gamma,\beta\leq M}$) acting on any $f = (f_\alpha)_{1\leq\alpha\leq M} \in \left[L^1\left(\partial_*\Omega, \frac{\sigma(x)}{1+|x|^n}\right)\right]^M$ at σ-a.e. point $x \in \partial\Omega$ as in [132, (1.8.24)]:

$$K_{\mathrm{mod}}f(x) := \left(-\lim_{\varepsilon\to 0^+}\int_{\partial_*\Omega} \nu_s(y)a_{rs}^{\beta\alpha}\{k_\varepsilon^{(r\gamma\beta)}(x-y) - k_1^{(r\gamma\beta)}(-y)\}f_\alpha(y)\,d\sigma(y)\right)_{1\leq\gamma\leq M} \tag{A.0.117}$$

where $k_\varepsilon^{(r\gamma\beta)} := (\partial_r E_{\gamma\beta})\cdot \mathbf{1}_{\mathbb{R}^n\setminus\overline{B(0,\varepsilon)}}$ for each $\varepsilon > 0$

$K^\# = K_A^\#$, the transpose double layer for a second-order, homogeneous, constant complex coefficient, weakly elliptic $M \times M$ system $L = L_A := \left(a_{rs}^{\alpha\beta}\partial_r\partial_s\right)_{1\leq\alpha,\beta\leq M}$ corresponding to the coefficient tensor $A := \left(a_{rs}^{\alpha\beta}\right)_{\substack{1\leq\alpha,\beta\leq M\\ 1\leq r,s\leq n}}$ (with canonical fundamental solution $E = \left(E_{\gamma\beta}\right)_{1\leq\gamma,\beta\leq M}$) acting on any function $f \in \left[L^1\left(\partial\Omega, \frac{\sigma(x)}{1+|x|^{n-1}}\right)\right]^M$ at σ-a.e. point $x \in \partial_*\Omega$ as in [132, (1.3.72)]:

$$K^\# f(x) := \left(\lim_{\varepsilon\to 0^+}\int_{y\in\partial\Omega,\ |x-y|>\varepsilon} \nu_s(x)a_{rs}^{\beta\alpha}(\partial_r E_{\gamma\beta})(x-y)f_\gamma(y)\,d\sigma(y)\right)_{1\leq\alpha\leq M} \tag{A.0.118}$$

$\operatorname{Ker} D$, the null-space of a first-order constant complex coefficient $N \times M$ system $D = \left(\sum_{j=1}^n a_j^{\alpha\beta}\partial_j\right)_{\substack{1\leq\alpha\leq N\\ 1\leq\beta\leq M}}$ in $\mathbb{R}^n$ which is injectively elliptic (i.e., its symbol, given by the formula $\operatorname{Sym}(D;\xi) := \mathrm{i}\left(\sum_{j=1}^n a_j^{\alpha\beta}\xi_j\right)_{\substack{1\leq\alpha\leq N\\ 1\leq\beta\leq M}}$ is an injective mapping from $\mathbb{C}^M$ into $\mathbb{C}^N$, for each vector $\xi \in \mathbb{R}^n \setminus \{0\}$), defined as in [132, (4.4.62)]:

$$\operatorname{Ker} D := \{u \in [\mathscr{C}^\infty(\Omega)]^M : Du = 0 \text{ in } \Omega\} \tag{A.0.119}$$

K_λ, the boundary-to-boundary double layer for the Stokes system associated with the coefficient tensor $A_\lambda := \big(\delta_{jk}\delta_{\alpha\beta} + \lambda\delta_{j\beta}\delta_{k\alpha}\big)_{\substack{1\le\alpha,\beta\le n\\ 1\le j,k\le n}}$ corresponding to any $\lambda \in \mathbb{C}$, acting on $\vec{f} = (f_j)_{1\le j\le n} \in \Big[L^1\Big(\partial_*\Omega, \frac{\sigma(x)}{1+|x|^{n-1}}\Big)\Big]^n$ at σ-a.e. $x \in \partial_*\Omega$ as in [132, (6.2.96)]:

$$\begin{aligned} K_\lambda\vec{f}(x) = \Big(\lim_{\varepsilon\to 0^+} \int_{\substack{y\in\partial_*\Omega\\ |x-y|>\varepsilon}} \Big\{ & (\lambda-1)\frac{\delta_{j\gamma}}{2\omega_{n-1}}\frac{\langle x-y, \nu(y)\rangle}{|x-y|^n} \\ & -(\lambda+1)\frac{n}{2\omega_{n-1}}\frac{\langle x-y,\nu(y)\rangle(x_j-y_j)(x_\gamma-y_\gamma)}{|x-y|^{n+2}} \\ & +(\lambda-1)\frac{1}{2\omega_{n-1}}\frac{\nu_j(y)(x_\gamma-y_\gamma)-\nu_\gamma(y)(x_j-y_j)}{|x-y|^n}\Big\} f_j(y)\,\mathrm{d}\sigma(y)\Big)_{1\le\gamma\le n} \end{aligned} \tag{A.0.120}$$

$K^\#_\lambda$, the transpose double layer for the Stokes system associated with the coefficient tensor $A_\lambda := \big(\delta_{jk}\delta_{\alpha\beta} + \lambda\delta_{j\beta}\delta_{k\alpha}\big)_{\substack{1\le\alpha,\beta\le n\\ 1\le j,k\le n}}$ corresponding to any $\lambda \in \mathbb{C}$, acting on $\vec{f} = (f_j)_{1\le j\le n} \in \Big[L^1\Big(\partial_*\Omega, \frac{\sigma(x)}{1+|x|^{n-1}}\Big)\Big]^n$ at σ-a.e. $x \in \partial_*\Omega$ as in [132, (6.2.99)]:

$$\begin{aligned} K^\#_\lambda\vec{f}(x) = \Big(\lim_{\varepsilon\to 0^+} \int_{\substack{y\in\partial\Omega\\ |x-y|>\varepsilon}} \Big\{ & -(\lambda-1)\frac{\delta_{j\gamma}}{2\omega_{n-1}}\frac{\langle x-y, \nu(x)\rangle}{|x-y|^n} \\ & +(\lambda+1)\frac{n}{2\omega_{n-1}}\frac{\langle x-y,\nu(x)\rangle(x_j-y_j)(x_\gamma-y_\gamma)}{|x-y|^{n+2}} \\ & -(\lambda-1)\frac{1}{2\omega_{n-1}}\frac{\nu_j(x)(x_\gamma-y_\gamma)-\nu_\gamma(x)(x_j-y_j)}{|x-y|^n}\Big\} f_\gamma(y)\,\mathrm{d}\sigma(y)\Big)_{1\le j\le n} \end{aligned} \tag{A.0.121}$$

k^{x_0}, the Poisson kernel for the Laplacian for the domain Ω (cf. [131, (5.3.38)]):

$$k^{x_0} := \frac{\mathrm{d}\omega^{x_0}}{\mathrm{d}\sigma} \tag{A.0.122}$$

i.e., the Radon-Nikodym derivative of the harmonic measure ω^{x_0} with pole at $x_0 \in \Omega$ with respect to the surface measure $\sigma := \mathcal{H}^{n-1}\lfloor\partial\Omega$ on the topological boundary $\partial\Omega$

L

local John condition, satisfied by an open set $\Omega \subseteq \mathbb{R}^n$ (cf. [129, Definition 5.11.7]):

$$\begin{array}{l}\text{there exist } \theta \in (0,1),\ M_o \in (1,\infty), \text{ and } R \in \big(0, \operatorname{diam}\partial\Omega\big] \text{ (the} \\ \text{latter required to be } \infty \text{ if } \partial\Omega \text{ is unbounded) such that for every} \\ x \in \partial\Omega \text{ and } r \in (0,R_*) \text{ one can find a point } x_r \in B(x,r)\cap\Omega \\ \text{with the property that } B(x_r,\theta r) \subseteq \Omega \text{ and for each } y \in \Delta(x,r) \text{ it} \\ \text{is possible to find a rectifiable path } \gamma_y : [0,1] \to \overline{\Omega} \text{ whose length} \\ \text{is } \leq M_o \cdot r, \text{ which satisfies } \gamma_y(0) = y,\ \gamma_y(1) = x_r, \text{ and such that} \\ \operatorname{dist}\big(\gamma_y(t),\partial\Omega\big) > \theta \cdot |\gamma_y(t) - y| \text{ for every } t \in (0,1]\end{array} \tag{A.0.123}$$

$\mathcal{L}^n$, the (n-dimensional) Lebesgue measure in $\mathbb{R}^n$

$L^\infty_{\rm comp}$, the space of essentially bounded functions with compact support

$L^0(X,\mu)$, the space of measurable functions which are pointwise finite μ-a.e. on X

$L^p_{\rm bdd}(\Omega,\mathcal{L}^n)$, the space of functions which are p-th power integrable on bounded subsets of Ω (cf. [129, (4.2.4)]):

$$\begin{array}{l} L^p_{\rm bdd}(\Omega,\mathcal{L}^n) \text{ be the collection of all } \mathcal{L}^n\text{-measurable functions defined} \\ \text{in } \Omega \text{ which are } p\text{-th power absolutely integrable with respect to the} \\ \text{Lebesgue measure on each bounded } \mathcal{L}^n\text{-measurable subset of the set } \Omega. \end{array} \tag{A.0.124}$$

$\|\cdot\|_{{\rm Lip}(X)}$, the natural semi-norm on ${\rm Lip}(X)$, defined in the context of a metric space $(\mathscr{X},d)$ as (cf. [129, (3.7.1)]):

$$\|f\|_{{\rm Lip}(X)} := \sup_{x,y\in X,\ x\neq y} \frac{|f(x)-f(y)|}{d(x,y)} \tag{A.0.125}$$

${\rm Lip}(X)$, the space of Lipschitz functions on the (quasi-)metric space X, defined as (cf. [129, (3.7.2)]):

$${\rm Lip}(X) := \big\{f : X \to \mathbb{C} :\ \|f\|_{{\rm Lip}(X)} < +\infty\big\} \tag{A.0.126}$$

${\rm Lip}_c(X)$, Lipschitz functions with bounded support in the (quasi-)metric space X

$\big({\rm Lip}_c(\Sigma)\big)'$, the space distributions on a given set $\Sigma \subseteq \mathbb{R}^n$ (cf. [129, (4.1.34)]):

$$\text{the topological dual of } \big({\rm Lip}_c(\Sigma), \tau_{\mathscr{D}}\big) \tag{A.0.127}$$

${}_{({\rm Lip}_c(\Sigma))'}\langle\cdot,\cdot\rangle_{{\rm Lip}_c(\Sigma)}$, or simply $\langle\cdot,\cdot\rangle$, the distributional pairing on the set Σ

$\|\cdot\|_{L^{p,q}(X,\mu)}$, the Lorentz space quasi-norm, defined as (cf. [129, (6.2.14)]):

$$\|f\|_{L^{p,q}(X,\mu)} := \begin{cases} \Big(\displaystyle\int_0^\infty \big[t^{1/p} f_X^*(t)\big|^q \frac{dt}{t}\Big)^{1/q} & \text{if } 0<p,q<\infty, \\ \sup_{t>0}\big[t^{1/p} f_X^*(t)\big] & \text{if } 0<p\leq\infty,\ q=\infty, \\ \|f\|_{L^\infty(X,\mu)} & \text{if } p=\infty,\ 0<q\leq\infty \end{cases} \tag{A.0.128}$$

$L^{p,q}(X,\mu)$, the Lorentz space on the measure space (X,μ) (cf. [129, (6.2.13)]):

$$L^{p,q}(X,\mu) := \Big\{ f : X \to \overline{\mathbb{R}} \ \mu\text{-measurable} : \|f\|_{L^{p,q}(X,\mu)} < +\infty \Big\} \tag{A.0.129}$$

$L^{p,q}_{\star}(\Omega,\mu)$, the maximal Lorentz space with respect to the Borel measure μ in the open set $\Omega \subseteq \mathbb{R}^n$, defined as (cf. [129, (6.6.41)]):

$$L^{p,q}_{\star}(\Omega,\mu) := \big\{ u : \Omega \to \mathbb{C} : u \text{ is } \mathcal{L}^n\text{-measurable, } u_{\star,\theta} \in L^{p,q}(\Omega,\mu) \big\} \tag{A.0.130}$$

$L^{p}_{\star}(\Omega,\mu)$, the maximal Lebesgue space with respect to the Borel measure μ in the open set $\Omega \subseteq \mathbb{R}^n$, defined as (cf. [129, (6.6.43)]):

$$\begin{aligned} L^{p}_{\star}(\Omega,\mu) &:= L^{p,p}_{\star}(\Omega,\mu) \qquad\qquad \text{(A.0.131)} \\ &= \big\{ u : \Omega \to \mathbb{C} : u \text{ is } \mathcal{L}^n\text{-measurable and } u_{\star,\theta} \in L^{p}(\Omega,\mu) \big\} \end{aligned}$$

$\log_+$, the positive part of ln, defined for each $t \in [0,\infty)$ as (cf. [129, (7.6.68)]):

$$\log_+ t := \begin{cases} 0 & \text{if } t \in [0,1], \\ \ln t & \text{if } t \in [1,\infty) \end{cases} \tag{A.0.132}$$

$\mathcal{L}(X \to Y)$, the space of linear and continuous operators from X to Y

$L^{\Phi}(X,\mu)$, the Orlicz space on the sigma-finite measure space $(X,\mathfrak{M},\mu)$ associated with a Young function Φ (cf. [130, §5.3]):

$$L^{\Phi}(X,\mu) := \big\{ f \in \mathscr{M}(X,\mu) : \|f\|_{L^{\Phi}(X,\mu)} < \infty \big\} \tag{A.0.133}$$

where the Luxemburg norm $\|\cdot\|_{L^{\Phi}(X,\mu)}$ is defined as

$$\begin{aligned} \|f\|_{L^{\Phi}(X,\mu)} &:= \inf \Big\{ \lambda > 0 : \int_X \Phi(|f(x)|/\lambda)\, d\mu(x) \le 1 \Big\} \qquad \text{(A.0.134)} \\ &= \inf \Big\{ \lambda > 0 : \int_0^{\infty} \mu\Big(\big\{ x \in X : |f(x)| > \lambda \Phi^{-1}(t) \big\} \Big)\, dt \le 1 \Big\} \end{aligned}$$

$L^p(\log L)^{\alpha}(X,\mu) = \Big\{ f \in \mathscr{M}(X,\mu) : \int_X |f(x)|^p \big[\ln(e + |f(x)|) \big]^{\alpha} d\mu(x) < +\infty \Big\}$ for $p \in (1,\infty)$ and $\alpha \in \mathbb{R}$, the Zygmund space (cf. [130, §5.3])

$L^p(\Omega, w\mathcal{L}^n)$, the weighted L^p Lebesgue space on the set $\Omega \subseteq \mathbb{R}^n$, equipped with the natural norm $\|u\|_{L^p(\Omega, w\mathcal{L}^n)} := \Big(\int_{\Omega} |u|^p\, w\, d\mathcal{L}^n \Big)^{1/p}$ (cf. [130, §8.3])

$L^p_s(\Omega) := \Big\{ U\big|_{\Omega} : U \in L^p_s(\mathbb{R}^n) \Big\}$, the Bessel potential space in an open set $\Omega \subseteq \mathbb{R}^n$ for $p \in (1,\infty)$ and $s \in \mathbb{R}$, equipped with the norm (cf. [130, §9.2]):

$$\|u\|_{L^p_s(\Omega)} := \inf \Big\{ \|U\|_{L^p_s(\mathbb{R}^n)} : U \in L^p_s(\mathbb{R}^n),\ u = U\big|_{\Omega} \Big\}$$

$\dot{\mathfrak{L}}^{p,\lambda}(\Sigma,\sigma)$, the homogeneous Morrey-Campanato space on a given Ahlfors regular set $\Sigma \subseteq \mathbb{R}^n$, defined for each integrability exponent $p \in (1,\infty)$ and each parameter

$\lambda \in (0, n-1)$ as (cf. [130, §6.1]):

$$\dot{\mathfrak{L}}^{p,\lambda}(\Sigma,\sigma) := \Big\{ f \in L^1_{\rm loc}(\Sigma,\sigma) : \|f\|_{\dot{\mathfrak{L}}^{p,\lambda}(\Sigma,\sigma)} < +\infty \Big\} \tag{A.0.135}$$

where, for each $f \in L^1_{\rm loc}(\Sigma,\sigma)$,

$$\|f\|_{\dot{\mathfrak{L}}^{p,\lambda}(\Sigma,\sigma)} := \sup_{\substack{x\in\Sigma \text{ and} \\ 0<R<2\,\mathrm{diam}(\Sigma)}} \left\{ R^{\frac{n-1-\lambda}{p}} \Big(⨍_{\Sigma\cap B(x,R)} |f(y) - f_{\Delta(x,R)}|^p \, d\sigma(y) \Big)^{\frac{1}{p}} \right\} \tag{A.0.136}$$

$\mathfrak{L}^{p,\lambda}(\Sigma,\sigma)$, the inhomogeneous Morrey-Campanato space on a given Ahlfors regular set $\Sigma \subseteq \mathbb{R}^n$, defined for each integrability exponent $p \in (1,\infty)$ and each parameter $\lambda \in (0, n-1)$ as (cf. [130, §6.1]):

$$\mathfrak{L}^{p,\lambda}(\Sigma,\sigma) := L^p(\Sigma,\sigma) \cap \dot{\mathfrak{L}}^{p,\lambda}(\Sigma,\sigma) = \Big\{ f \in L^p(\Sigma,\sigma) : \|f\|_{\dot{\mathfrak{L}}^{p,\lambda}(\Sigma,\sigma)} < +\infty \Big\} \tag{A.0.137}$$

where

$$\|f\|_{\mathfrak{L}^{p,\lambda}(\Sigma,\sigma)} := \|f\|_{L^p(\Sigma,\sigma)} + \|f\|_{\dot{\mathfrak{L}}^{p,\lambda}(\Sigma,\sigma)} \tag{A.0.138}$$

$L^p_1(\partial_*\Omega,\sigma_*)$, the L^p-based Sobolev space of order one on $\partial_*\Omega$, defined for each $p \in [1,\infty]$ as (cf. [130, Definition 11.1.2]):

$$L^p_1(\partial_*\Omega,\sigma_*) := \Big\{ f \in L^p(\partial_*\Omega,\sigma_*) : \partial_{\tau_{jk}} f \text{ exists in } L^p(\partial_*\Omega,\sigma_*) \text{ for each } j,k \in \{1,\dots,n\} \Big\} \tag{A.0.139}$$

and equipped with the norm (cf. [130, Proposition 11.1.9]):

$$\|f\|_{L^p_1(\partial_*\Omega,\sigma_*)} := \|f\|_{L^p(\partial_*\Omega,\sigma_*)} + \sum_{1\le j,k\le n} \|\partial_{\tau_{jk}} f\|_{L^p(\partial_*\Omega,\sigma_*)} \tag{A.0.140}$$

$L^{p,q}_1(\partial_*\Omega,\sigma_*)$, the off-diagonal Sobolev space of order one on $\partial_*\Omega$, defined for each $p,q \in [1,\infty]$ as (cf. [130, Definition 11.1.2]):

$$L^{p,q}_1(\partial_*\Omega,\sigma_*) := \Big\{ f \in L^p(\partial_*\Omega,\sigma_*) : \partial_{\tau_{jk}} f \text{ exists in } L^q(\partial_*\Omega,\sigma_*) \text{ for each } j,k \in \{1,\dots,n\} \Big\} \tag{A.0.141}$$

and equipped with the norm (cf. [130, Proposition 11.1.9]):

$$\|f\|_{L^{p,q}_1(\partial_*\Omega,\sigma_*)} := \|f\|_{L^p(\partial_*\Omega,\sigma_*)} + \sum_{j,k=1}^{n} \|\partial_{\tau_{jk}} f\|_{L^q(\partial_*\Omega,\sigma_*)} \tag{A.0.142}$$

$L^{p,q}_{1,\mathrm{loc}}(\partial_*\Omega, \sigma_*)$, the local off-diagonal (boundary) Sobolev space defined for any two integrability exponents $p, q \in [1, \infty]$ as (cf. [130, Definition 11.1.2]):

$$L^{p,q}_{1,\mathrm{loc}}(\partial_*\Omega, \sigma_*) := \Big\{ f \in L^p_{\mathrm{loc}}(\partial_*\Omega, \sigma_*) : \partial_{\tau_{jk}} f \text{ exists in } L^q_{\mathrm{loc}}(\partial_*\Omega, \sigma_*) \quad \text{for each } j, k \in \{1, \dots, n\} \Big\} \tag{A.0.143}$$

$L^p_{1,\mathrm{loc}}(\partial_*\Omega, \sigma_*)$, the (boundary) Sobolev space defined for $p \in [1, \infty]$ as (cf. [130, Definition 11.1.2]):

$$L^p_{1,\mathrm{loc}}(\partial_*\Omega, \sigma_*) := L^{p,p}_{1,\mathrm{loc}}(\partial_*\Omega, \sigma_*) \tag{A.0.144}$$

$\dot{L}^p_1(\partial\Omega, \sigma)$, the L^p-based homogeneous Sobolev space of order one on $\partial\Omega$ as (cf. [130, Definition 11.5.3]):

$$\dot{L}^p_1(\partial\Omega, \sigma) := \Big\{ f \in L^1\Big(\partial\Omega, \frac{\sigma(x)}{1 + |x|^n}\Big) : \partial_{\tau_{jk}} f \in L^p(\partial_*\Omega, \sigma) \quad \text{for each } j, k \in \{1, \dots, n\} \Big\} \tag{A.0.145}$$

and equipped with the semi-norm

$$\|f\|_{\dot{L}^p_1(\partial\Omega,\sigma)} := \sum_{j,k=1}^{n} \|\partial_{\tau_{jk}} f\|_{L^p(\partial_*\Omega,\sigma)} \tag{A.0.146}$$

$L^p_1(\partial_*\Omega, w\sigma_*) = L^p_1(\partial_*\Omega, w)$, the weighted Sobolev space of order one on $\partial_*\Omega$, defined for $p \in [1, \infty]$ and a generic weight w on $\partial_*\Omega$ as (cf. [130, §11.7]):

$$L^p_1(\partial_*\Omega, w\sigma_*) := \big\{ f \in L^p(\partial_*\Omega, w\sigma_*) : \partial_{\tau_{jk}} f \in L^p(\partial_*\Omega, w\sigma_*) \quad \text{for each } j, k \in \{1, \dots, n\} \big\} \tag{A.0.147}$$

equipped with the norm

$$\|f\|_{L^p(\partial_*\Omega, w\sigma_*)} + \sum_{j,k=1}^{n} \|\partial_{\tau_{jk}} f\|_{L^p(\partial_*\Omega, w\sigma_*)} \tag{A.0.148}$$

$L^p_1\big(\partial_*\Omega, \frac{\sigma_*(x)}{1+|x|^a}\big)$, the "power" weighted Sobolev space on $\partial_*\Omega$, defined for each $p \in [1, \infty]$ and $a \in [0, \infty)$ as (cf. [130, (11.7.6)]):

$$L^p_1\big(\partial_*\Omega, \tfrac{\sigma_*(x)}{1+|x|^a}\big) = \Big\{ f \in L^p\big(\partial_*\Omega, \tfrac{\sigma_*(x)}{1+|x|^a}\big) : \partial_{\tau_{jk}} f \in L^p\big(\partial_*\Omega, \tfrac{\sigma_*(x)}{1+|x|^a}\big) \quad \text{for each } j, k \in \{1, \dots, n\} \Big\} \tag{A.0.149}$$

$L^p_1(\partial\Omega, w)$, the Muckenhoupt weighted Sobolev space of order one on $\partial\Omega$, defined for $p \in (1, \infty)$ and $w \in A_p(\partial\Omega, \sigma)$ as (cf. [130, §11.7]):

$$L_1^p(\partial\Omega, w) := \{f \in L^p(\partial\Omega, w) : \partial_{\tau_{jk}} f \in L^p(\partial\Omega, w),\ 1 \leq j,k \leq n\} \tag{A.0.150}$$

and equipped with the norm

$$\|f\|_{L_1^p(\partial\Omega,w)} := \|f\|_{L^p(\partial\Omega,w)} + \sum_{j,k=1}^{n} \|\partial_{\tau_{jk}} f\|_{L^p(\partial\Omega,w)} \tag{A.0.151}$$

$L_1^{p,q}(\partial\Omega, \sigma)$, the Lorentz-based Sobolev space of order one on $\partial\Omega$, defined for $p \in (1,\infty)$ and $q \in (0,\infty]$ as (cf. [130, §11.7]):

$$L_1^{p,q}(\partial\Omega, \sigma) := \{f \in L^{p,q}(\partial\Omega, \sigma) : \partial_{\tau_{jk}} f \in L^{p,q}(\partial\Omega, \sigma),\ 1 \leq j,k \leq n\} \tag{A.0.152}$$

and equipped with the quasi-norm

$$\|f\|_{L_1^{p,q}(\partial\Omega,\sigma)} := \|f\|_{L^{p,q}(\partial\Omega,\sigma)} + \sum_{j,k=1}^{n} \|\partial_{\tau_{jk}} f\|_{L^{p,q}(\partial\Omega,\sigma)} \tag{A.0.153}$$

$L_{-1}^p(\partial_*\Omega, \sigma_*)$, the ($L^p$-based) negative Sobolev space of order minus one on $\partial_*\Omega$ defined for $p, p' \in (1,\infty)$ with $\frac{1}{p} + \frac{1}{p'} = 1$ as (cf. [130, Definition 11.8.1]):

$$L_{-1}^p(\partial_*\Omega, \sigma_*) := \Big(L_1^{p'}(\partial_*\Omega, \sigma_*)\Big)^* \tag{A.0.154}$$

$L_{-1}^{p,q}(\partial_*\Omega, \sigma_*)$, the off-diagonal negative Sobolev space on $\partial_*\Omega$ defined for any two integrability exponents $p, q \in (1,\infty)$ as (cf. [130, (11.8.28)]):

$$L_{-1}^{p,q}(\partial_*\Omega, \sigma_*) := \Big(L_1^{p',q'}(\partial_*\Omega, \sigma_*)\Big)^* \text{ where } \tfrac{1}{p} + \tfrac{1}{p'} = 1 \text{ and } \tfrac{1}{q} + \tfrac{1}{q'} = 1 \tag{A.0.155}$$

$L_{-1}^p(\partial\Omega, w)$, the Muckenhoupt weighted negative Sobolev space on $\partial\Omega$ defined for $p \in (1,\infty)$ and $w \in A_p(\partial\Omega, \sigma)$ as (cf. [130, Definition 11.8.7]):

$$L_{-1}^p(\partial\Omega, w) := \Big(L_1^{p'}(\partial\Omega, w')\Big)^* \tag{A.0.156}$$

where $p' := (1 - 1/p)^{-1} \in (1,\infty)$ is the Hölder conjugate exponent of p and where $w' := w^{1-p'} \in A_{p'}(\partial\Omega, \sigma)$ is the conjugate weight of w

$L = \big(a_{rs}^{\alpha\beta}\partial_r\partial_s\big)_{1\leq\alpha,\beta\leq M}$, a homogeneous, second-order, constant (complex) coefficient, $M \times M$ system in $\mathbb{R}^n$

L_A, the second-order homogeneous constant coefficient $M \times M$ system in $\mathbb{R}^n$ associated with a given coefficient tensor $A := \big(a_{rs}^{\alpha\beta}\big)_{\substack{1\leq r,s\leq n \\ 1\leq\alpha,\beta\leq M}}$ as:

$$L_A := \Big(a_{rs}^{\alpha\beta}\partial_r\partial_s\Big)_{1\leq\alpha,\beta\leq M} \tag{A.0.157}$$

$L(\xi)$, the characteristic matrix of the homogeneous, complex constant coefficient, second-order $M \times M$ system $L = \big(a_{rs}^{\alpha\beta}\partial_r\partial_s\big)_{1\le\alpha,\beta\le M}$ in $\mathbb{R}^n$, defined for each vector $\xi = (\xi_r)_{1\le r\le n} \in \mathbb{R}^n$ as:

$$L(\xi) := -\big(a_{rs}^{\alpha\beta}\xi_r\xi_s\big)_{1\le\alpha,\beta\le M} \tag{A.0.158}$$

$L_{\lambda,\mu}(\xi) := -\mu|\xi|^2 I_{n\times n} - (\lambda+\mu)\xi\otimes\xi$ for each $\xi \in \mathbb{R}^n$, the characteristic matrix of the complex Lamé system $L_{\lambda,\mu} := \mu\Delta + (\lambda+\mu)\nabla\mathrm{div}$

$L^p_{A,z}(\partial\Omega,\sigma)$, the space defined for $p, p' \in (1,\infty)$ with $1/p + 1/p' = 1$ and $z \in \mathbb{C}$ as in [132, (1.7.14)]:

$$L^p_{A,z}(\partial\Omega,\sigma) := \Big\{f \in \big[L^p(\partial\Omega,\sigma)\big]^M : \int_{\partial\Omega}\langle f, g\rangle\,\mathrm{d}\sigma = 0 \ \text{ for each } \ g \in \big[L^{p'}(\partial\Omega,\sigma)\big]^M \ \text{ with } \ (zI + K_A)g = 0\Big\} \tag{A.0.159}$$

L_λ, the Stokes system in the special writing given in [132, (6.1.2)]:

$$L_\lambda := \big((\delta_{jk}\delta_{\alpha\beta} + \lambda\delta_{j\beta}\delta_{k\alpha})\partial_j\partial_k\big)_{1\le\alpha,\beta\le n} \ \text{ for any } \lambda \in \mathbb{C} \tag{A.0.160}$$

L_B, the Bitsadze operator, defined as in [132, (1.4.77)]:

$$L_B := \partial_{\bar z}^2 = \tfrac{1}{4}(\partial_x + \mathrm{i}\partial_y)^2 \tag{A.0.161}$$

M

$\mathcal{M}_{X,s,\alpha}$, the L^s-based fractional Hardy-Littlewood maximal operator of order α in the space of homogeneous type (X,ρ,μ), defined for each μ-measurable function f on X as (cf. [129, (7.6.1)]):

$$\mathcal{M}_{X,s,\alpha}f(x) := \sup_{r>0}\left[\mu(B_\rho(x,r))^\alpha\Big(⨍_{B_\rho(x,r)}|f|^s\,\mathrm{d}\mu\Big)^{\frac{1}{s}}\right], \quad \forall\, x \in X \tag{A.0.162}$$

$\mathcal{M}_{X,s}$, the L^s-based Hardy-Littlewood maximal operator in the space of homogeneous type (X,ρ,μ), defined for μ-measurable functions f as (cf. [129, (7.6.7)]):

$$\mathcal{M}_{X,s}f(x) := \sup_{r>0}\Big(⨍_{B_\rho(x,r)}|f|^s\,\mathrm{d}\mu\Big)^{\frac{1}{s}}, \quad \forall\, x \in X \tag{A.0.163}$$

$\mathcal{M}_X$, the Hardy-Littlewood maximal operator on the space of homogeneous type (X,ρ,μ), defined for μ-measurable functions f as (cf. [129, (7.6.16)]):

$$\mathcal{M}_X f(x) := \sup_{r\in(0,\infty)}\frac{1}{\mu\big(B_\rho(x,r)\big)}\int_{B_\rho(x,r)}|f|\,\mathrm{d}\mu, \quad \forall\, x \in X \tag{A.0.164}$$

$u^{\max}_M$, the tangential maximal function of u (with exponent M), defined at each $x \in \partial\Omega$ as (cf. [129, (8.5.2)]):

$$u_M^{\max} : \partial\Omega \longrightarrow [0,+\infty] \text{ defined by}$$
$$u_M^{\max}(x) := \Big\| u(y)\Big(\frac{\delta_{\partial\Omega}(y)}{|x-y|}\Big)^M \Big\|_{L_y^\infty(\Omega,\mathcal{L}^n)} \quad \text{for each } x \in \partial\Omega \tag{A.0.165}$$

$\mathscr{M}(X,\mu)$, the collection of μ-measurable functions defined on an arbitrary measure space $(X,\mathfrak{M},\mu)$

$\mathscr{M}_+(X,\mu)$, the collection of non-negative μ-measurable functions defined on an arbitrary measure space $(X,\mathfrak{M},\mu)$

$M^{p,\lambda}(\Sigma,\sigma)$, the Morrey space on the Ahlfors regular set $\Sigma \subseteq \mathbb{R}^n$, defined for each $p \in (0,\infty)$ and $\lambda \in (0,n-1)$ as (cf. [130, §6.2]):

$$M^{p,\lambda}(\Sigma,\sigma) := \Big\{ f : \Sigma \to \mathbb{C} : f \text{ is } \sigma\text{-measurable and } \|f\|_{M^{p,\lambda}(\Sigma,\sigma)} < +\infty \Big\} \tag{A.0.166}$$

where, for each σ-measurable function f on Σ,

$$\|f\|_{M^{p,\lambda}(\Sigma,\sigma)} := \sup_{\substack{x\in\Sigma \text{ and} \\ 0<R<2\,\mathrm{diam}(\Sigma)}} \Bigg\{ R^{\frac{n-1-\lambda}{p}} \Big(⨍_{\Sigma\cap B(x,R)} |f|^p \, d\sigma \Big)^{\frac{1}{p}} \Bigg\} \tag{A.0.167}$$

$\mathring{M}^{p,\lambda}(\Sigma,\sigma)$, the vanishing Morrey space on the Ahlfors regular set $\Sigma \subseteq \mathbb{R}^n$ with $p \in (0,\infty)$ and $\lambda \in (0,n-1)$ (cf. [130, §6.2]):

$$\mathring{M}^{p,\lambda}(\Sigma,\sigma) := \text{the closure of } L^s(\Sigma,\sigma) \text{ with } s := \tfrac{p(n-1)}{n-1-\lambda} \text{ in } M^{p,\lambda}(\Sigma,\sigma) \tag{A.0.168}$$

$M_1^{p,\lambda}(\partial\Omega,\sigma)$, the Morrey-based Sobolev space of order one on $\partial\Omega$, defined for $p \in (1,\infty)$ and $\lambda \in (0,n-1)$ as (cf. [130, (11.7.12)-(11.7.13)]):

$$M_1^{p,\lambda}(\partial\Omega,\sigma) := \Big\{ f \in M^{p,\lambda}(\partial\Omega,\sigma) : \text{ for each } j,k \in \{1,\dots,n\} \text{ one has } \partial_{\tau_{jk}} f \in M^{p,\lambda}(\partial\Omega,\sigma) \Big\} \tag{A.0.169}$$

equipped with the norm

$$M_1^{p,\lambda}(\partial\Omega,\sigma) \ni f \longmapsto \|f\|_{M^{p,\lambda}(\partial\Omega,\sigma)} + \sum_{j,k=1}^n \big\|\partial_{\tau_{jk}} f\big\|_{M^{p,\lambda}(\partial\Omega,\sigma)} \tag{A.0.170}$$

$M_1^{p,q,\lambda}(\partial\Omega,\sigma)$, the off-diagonal Morrey-based Sobolev space of order one on $\partial\Omega$, defined for $p,q \in (1,\infty)$ and $\lambda \in (0,n-1)$ as (cf. [130, (11.7.14)-(11.7.15)]):

$$M_1^{p,q,\lambda}(\partial\Omega,\sigma) := \Big\{ f \in M^{p,\lambda}(\partial\Omega,\sigma) : \text{ for each } j,k \in \{1,\dots,n\} \text{ one has } \partial_{\tau_{jk}} f \in M^{q,\lambda}(\partial\Omega,\sigma) \Big\}, \tag{A.0.171}$$

and endowed with the norm

$$\|f\|_{M^{p,\lambda}(\partial\Omega,\sigma)} + \sum_{j,k=1}^{n} \|\partial_{\tau_{jk}} f\|_{M^{q,\lambda}(\partial\Omega,\sigma)} \tag{A.0.172}$$

$\mathring{M}_1^{p,q,\lambda}(\partial\Omega,\sigma)$, the vanishing off-diagonal Morrey-based Sobolev space of order one on $\partial\Omega$, defined for $p, q \in (1,\infty)$ and $\lambda \in (0, n-1)$ as (cf. [130, (11.7.16)]):

$$\mathring{M}_1^{p,q,\lambda}(\partial\Omega,\sigma) := \Big\{ f \in \mathring{M}^{p,\lambda}(\partial\Omega,\sigma) : \text{ for each } j,k \in \{1,\dots,n\} \\ \text{one has } \partial_{\tau_{jk}} f \in \mathring{M}^{q,\lambda}(\partial\Omega,\sigma) \Big\} \tag{A.0.173}$$

$\mathring{M}_1^{p,\lambda}(\partial\Omega,\sigma)$, the vanishing Morrey-based Sobolev space of order one on $\partial\Omega$, defined for $p \in (1,\infty)$ and $\lambda \in (0, n-1)$ as (cf. [130, (11.7.17)]):

$$\mathring{M}_1^{p,\lambda}(\partial\Omega,\sigma) := \Big\{ f \in \mathring{M}^{p,\lambda}(\partial\Omega,\sigma) : \text{ for each } j,k \in \{1,\dots,n\} \\ \text{one has } \partial_{\tau_{jk}} f \in \mathring{M}^{p,\lambda}(\partial\Omega,\sigma) \Big\} \tag{A.0.174}$$

$M_{-1}^{p,\lambda}(\partial\Omega,\sigma)$, the Morrey-based negative Sobolev space on $\partial\Omega$, defined for two integrability exponents $p, q \in (1,\infty)$ with $1/p + 1/q = 1$ and $\lambda \in (0, n-1)$ as (cf. [130, Definition 11.8.9]):

$$M_{-1}^{p,\lambda}(\partial\Omega,\sigma) := \Big(\mathcal{B}_1^{q,\lambda}(\partial\Omega,\sigma) \Big)^{*} \tag{A.0.175}$$

$\dot{M}_1^{p,\lambda}(\partial\Omega,\sigma)$, the homogeneous Morrey-based Sobolev space of order one on $\partial\Omega$ defined for $p \in (1,\infty)$ and $\lambda \in (0, n-1)$ as (cf. [130, Definition 11.13.1]):

$$\dot{M}_1^{p,\lambda}(\partial\Omega,\sigma) := \Big\{ f \in L^1\Big(\partial\Omega, \frac{\sigma(x)}{1+|x|^n}\Big) \cap L^p_{\mathrm{loc}}(\partial\Omega,\sigma) : \\ \partial_{\tau_{jk}} f \in M^{p,\lambda}(\partial\Omega,\sigma) \text{ for } 1 \le j,k \le n \Big\} \tag{A.0.176}$$

and equipped with the semi-norm

$$\|f\|_{\dot{M}_1^{p,\lambda}(\partial\Omega,\sigma)} := \sum_{j,k=1}^{n} \|\partial_{\tau_{jk}} f\|_{M^{p,\lambda}(\partial\Omega,\sigma)} \tag{A.0.177}$$

$\dot{\mathbb{M}}_1^{p,\lambda}(\partial\Omega,\sigma)$, the homogeneous vanishing Morrey-based Sobolev space of order one on $\partial\Omega$, defined for $p \in (1,\infty)$ and $\lambda \in (0, n-1)$ as (cf. [130, Definition 11.13.15]):

$$\dot{\mathbb{M}}_1^{p,\lambda}(\partial\Omega,\sigma) := \Big\{ f \in L^1\Big(\partial\Omega, \frac{\sigma(x)}{1+|x|^n}\Big) \cap L^p_{\mathrm{loc}}(\partial\Omega,\sigma) : \\ \partial_{\tau_{jk}} f \in \mathring{M}^{p,\lambda}(\partial\Omega,\sigma) \text{ for } 1 \le j,k \le n \Big\} \tag{A.0.178}$$

and equipped with the semi-norm

$$\|f\|_{\dot{\mathbb{M}}_1^{p,\lambda}(\partial\Omega,\sigma)} := \sum_{j,k=1}^{n} \|\partial_{\tau_{jk}} f\|_{M^{p,\lambda}(\partial\Omega,\sigma)} \tag{A.0.179}$$

M_b, the operator of pointwise multiplication by the function $b \in L^\infty(\Sigma, \sigma)$

$[M_b, T]$, the commutator of M_b (the pointwise multiplication by b) with the operator T, defined by $[M_b, T]f := b(Tf) - T(bf)$

N

$\mathbb{N}_0 := \mathbb{N} \cup \{0\} = \{0, 1, 2, \dots\}$

NTA domain: a nonempty open subset Ω of $\mathbb{R}^n$ satisfying a two-sided corkscrew condition as well as a Harnack chain condition (cf. [129, Definition 5.11.1])

two-sided NTA domain: a nonempty open subset Ω of $\mathbb{R}^n$ satisfying a two-sided corkscrew condition as well as a two-sided Harnack chain condition (cf. [129, Definition 5.11.1])

one-sided NTA domain (or interior NTA domain): a nonempty open subset Ω of $\mathbb{R}^n$ satisfying an interior corkscrew condition as well as a Harnack chain condition (cf. [129, Definition 5.11.1])

$\mathcal{N}_\kappa u$, the (κ-)nontangential maximal operator acting on the measurable function $u : \Omega \to \mathbb{R}^n$ according to (cf. [129, (8.2.1)]):

$$\mathcal{N}_\kappa u : \partial\Omega \longrightarrow [0, +\infty], \quad (\mathcal{N}_\kappa u)(x) := \|u\|_{L^\infty(\Gamma_\kappa(x), \mathcal{L}^n)} \text{ for all } x \in \partial\Omega \tag{A.0.180}$$

$\mathcal{N}_\kappa^\varepsilon u$, the ($\kappa$-)nontangential maximal operator truncated at height $\varepsilon > 0$, acting on the measurable function $u : \Omega \to \mathbb{R}^n$ according to (cf. [129, (1.5.5)]):

$$\mathcal{N}_\kappa^\varepsilon u := \mathcal{N}_\kappa(u \cdot \mathbf{1}_{O_\varepsilon}) \text{ where } O_\varepsilon := \big\{x \in \Omega : \operatorname{dist}(x, \partial\Omega) < \varepsilon\big\}. \tag{A.0.181}$$

$\mathcal{N}_\kappa^E u$, the restricted nontangential maximal function of $u : \Omega \to \mathbb{R}$, relative to the set E (cf. [129, (8.2.4)]):

$$\begin{aligned} &\mathcal{N}_\kappa^E u : \partial\Omega \longrightarrow [0, +\infty] \text{ defined as} \\ &(\mathcal{N}_\kappa^E u)(x) := \|u\|_{L^\infty(\Gamma_\kappa(x)\cap E, \mathcal{L}^n)} \text{ for each } x \in \partial\Omega \end{aligned} \tag{A.0.182}$$

ν, the geometric measure theoretic outward unit normal to a set $\Omega \subseteq \mathbb{R}^n$ of locally finite perimeter, defined via the requirement that (cf. [129, (5.6.2)-(5.6.3)]):

$$\text{there exist a locally finite Borel-regular measure } \sigma_* \text{ in } \mathbb{R}^n \text{ and a vector-valued function } \nu \in \big[L^\infty(\mathbb{R}^n, \sigma_*)\big]^n \text{ satisfying } |\nu(x)| = 1 \text{ at } \sigma_*\text{-a.e. } x \in \mathbb{R}^n \text{ and with such that } \nabla \mathbf{1}_\Omega = -\nu\sigma_* \text{ in } \big[\mathcal{D}'(\mathbb{R}^n)\big]^n \tag{A.0.183}$$

$\nu \bullet \vec{F}$, the "bullet" product involving a vector field $\vec{F} \in \big[L^1_{\rm bdd}(\Omega, \mathcal{L}^n)\big]^n$ (where Ω is an arbitrary open subset of $\mathbb{R}^n$) whose divergence, considered in the sense of distributions in Ω, satisfies $\operatorname{div}\vec{F} \in L^1_{\rm bdd}(\Omega, \mathcal{L}^n)$, defined as a functional acting on each $\psi \in \operatorname{Lip}_c(\partial\Omega)$ according to (cf. [129, Proposition 4.2.3]):

$$\langle \nu \bullet \vec{F}, \psi \rangle := \int_\Omega \vec{F} \cdot \nabla\Psi \, d\mathcal{L}^n + \int_\Omega (\mathrm{div}\vec{F})\Psi \, d\mathcal{L}^n, \tag{A.0.184}$$

where Ψ is any complex-valued function satisfying

$$\begin{aligned} &\Psi \in \mathrm{Lip}(\overline{\Omega}),\ \Psi\big|_{\partial\Omega} = \psi, \text{ and } \Psi \equiv 0 \text{ outside} \\ &\text{of some compact subset of } \overline{\Omega} \end{aligned} \tag{A.0.185}$$

$\nu \circledbullet u$, the Clifford bullet product of ν with u (cf. [130, (10.2.100)]):

$$\nu \circledbullet u := (-\mathrm{i})\mathrm{Sym}(D; \nu) \bullet u \tag{A.0.186}$$

$N^p_\kappa(\Omega; \mu)$, the space of functions in Ω with a p-th power integrable non-tangential maximal function on $\partial\Omega$ with respect to the measure μ (cf. [129, (8.3.31)]):

$$\begin{aligned} N^p_\kappa(\Omega; \mu) := \big\{ u : \Omega \to \mathbb{C} :\ & u \text{ is } \mathcal{L}^n\text{-measurable, and} \\ & \|u\|_{N^p_\kappa(\Omega;\mu)} := \|\mathcal{N}_\kappa u\|_{L^p(\partial\Omega,\mu)} < +\infty \big\} \end{aligned} \tag{A.0.187}$$

$\big(u\big|^{\kappa-\mathrm{n.t.}}_{\partial\Omega}\big)(x)$, the nontangential trace of the function $u : \Omega \to \mathbb{R}$ at the point $x \in \partial\Omega$ such that $x \in \overline{\Gamma_\kappa(x)}$, defined as (cf. [129, Definition 8.9.1]):

$$\begin{aligned} &\big(u\big|^{\kappa-\mathrm{n.t.}}_{\partial\Omega}\big)(x) \text{ is the number } a \in \mathbb{R} \text{ with the property that for every } \varepsilon > 0 \\ &\text{there exists } r > 0 \text{ so that } |u(y) - a| < \varepsilon \text{ at } \mathcal{L}^n\text{-a.e. } y \in \Gamma_\kappa(x) \cap B(x, r) \end{aligned} \tag{A.0.188}$$

$\|T\|_{X\to Y}$, the operator "norm" of some given positively homogeneous mapping T, acting between the quasi-normed vector spaces $\big(X, \|\cdot\|_X\big)$ and $\big(Y, \|\cdot\|_Y\big)$

$$\|T\|_{X\to Y} := \sup\big\{\|Tu\|_Y : u \in X,\ \|u\|_X = 1\big\} \in [0, +\infty] \tag{A.0.189}$$

$\|T\|_{\mathrm{Bd}(X\to Y)} := \|T\|_{X\to Y}$

$\|T\|_{\mathrm{Bd}(X)}$, the "norm" of a positively homogeneous mapping T acting from the quasi-normed vector space $\big(X, \|\cdot\|_X\big)$ into itself

$$\|T\|_{\mathrm{Bd}(X)} := \|T\|_{\mathrm{Bd}(X\to X)} = \sup_{x\in X,\ \|x\|_X=1} \|Tx\|_X \tag{A.0.190}$$

$\|T\|^{\mathrm{ess}}_{X\to Y}$, the essential norm of the operator $T \in \mathrm{Bd}(X \to Y)$, where X, Y are quasi-normed spaces (cf. [130, §1.2]):

$$\|T\|^{\mathrm{ess}}_{X\to Y} := \mathrm{dist}\big(T, \mathrm{Cp}(X \to Y)\big) = \inf\big\{\|T - K\|_{X\to Y} : K \in \mathrm{Cp}(X \to Y)\big\} \tag{A.0.191}$$

$\|\cdot\|_{(X_0,X_1)_{\theta,q}}$, the real interpolation quasi-norm

O

$\mathbf{1}_E$, the characteristic function of a given set E

$\omega_{n-1} := \mathcal{H}^{n-1}(S^{n-1})$, the surface area of S^{n-1} (the $(n-1)$-dimensional sphere in the Euclidean space $\mathbb{R}^n$)

$O_\varepsilon := \{x \in \Omega : \delta_{\partial\Omega}(x) < \varepsilon\}$, the one-sided collar neighborhood of $\partial\Omega$ of "width" ε

$\mathscr{O}(\Omega)$, the collection of all holomorphic functions in an open set $\Omega \subseteq \mathbb{C}$

P

$\mathfrak{P}$, the $\mathfrak{P}$-maximal operator acting on a Lebesgue measurable function $u : \Omega \to \mathbb{R}^n$ at the point $x \in \partial\Omega$ according to (cf. [130, §10.1]):

$$(\mathfrak{P}u)(x) := \sup_{0<r<2\,\mathrm{diam}(\partial\Omega)} \left\{ \frac{1}{\sigma\big(\partial\Omega \cap B(x,r)\big)} \int_{\Omega\cap B(x,r)} |u|\,\mathrm{d}\mathcal{L}^n \right\} \tag{A.0.192}$$

${}_{\mathscr{E}'(\Omega)}\big\langle\!\big\langle \cdot, \cdot \big\rangle\!\big\rangle_{\mathscr{E}(\Omega)}$, the pairing between a compactly supported distribution u in Ω and a smooth function $f \in \mathscr{C}^\infty(\mathrm{supp}\,u)$, say $f \in \mathscr{C}^\infty(O)$ with $O \subseteq \Omega$ open set containing $\mathrm{supp}\,u$, defined for each $F \in \mathscr{C}^\infty(\Omega)$ with the property that $F = f$ near $\mathrm{supp}\,u$ as (cf. [129, (2.2.33)]):

$${}_{\mathscr{E}'(\Omega)}\big\langle\!\big\langle u, f \big\rangle\!\big\rangle_{\mathscr{E}(\Omega)} := {}_{\mathscr{E}'(\Omega)}\big\langle u, F \big\rangle_{\mathscr{E}(\Omega)} \tag{A.0.193}$$

$\wedge$, the exterior product of differential forms

$\vee$, the interior product of differential forms (cf. [132, (7.1.52)])

$\overset{\bullet}{\partial}_{\tau_{jk}} u$, the weak tangential derivative of a given function $u \in L^1_{\mathrm{loc}}(\Omega, \mathcal{L}^n)$ with the property that $\nabla u \in \big[L^1_{\mathrm{bdd}}(\Omega, \mathcal{L}^n)\big]^n$, defined as (cf. [129, Example 4.2.4]):

$$\overset{\bullet}{\partial}_{\tau_{jk}} u := \nu \bullet \big((\partial_k u)\mathbf{e}_j - (\partial_j u)\mathbf{e}_k\big), \tag{A.0.194}$$

i.e., the "bullet" product $\nu \bullet \vec{F}^u_{jk}$, where

$$\vec{F}^u_{jk} := (\partial_k u)\mathbf{e}_j - (\partial_j u)\mathbf{e}_k \tag{A.0.195}$$

$\partial_* E$, the measure theoretic boundary of a Lebesgue measurable set $E \subseteq \mathbb{R}^n$, defined as (cf. [129, (5.2.1)]):

$$\partial_* E := \Big\{ x \in \mathbb{R}^n : \limsup_{r\to 0^+} \frac{\mathcal{L}^n(B(x,r)\cap E)}{r^n} > 0 \ \text{ and } \ \limsup_{r\to 0^+} \frac{\mathcal{L}^n(B(x,r)\setminus E)}{r^n} > 0 \Big\} \tag{A.0.196}$$

$\partial^* E$, the reduced boundary of a set $E \subseteq \mathbb{R}^n$ of locally finite perimeter, defined as (cf. [129, (5.6.13)]):

$\partial^* E$ consists of all points $x \in \partial E$ satisfying the following three properties: $0 < \sigma_*(B(x,r)) < +\infty$ for each $r \in (0,\infty)$, formula $\lim\limits_{r\to 0^+} ⨍_{B(x,r)} \nu\,\mathrm{d}\sigma_* = \nu(x)$ is valid, and $|\nu(x)| = 1$. (A.0.197)

$\partial_{\text{nta}}\Omega$, the nontangentially accessible boundary of an open proper subset Ω of $\mathbb{R}^n$, defined as (cf. [129, Definition 8.8.5]):

$$\partial_{\text{nta}}\Omega := \bigcap_{\kappa>0} A_\kappa(\partial\Omega) = \big\{x \in \partial\Omega : \; x \in \overline{\Gamma_\kappa(x)} \text{ for each } \kappa > 0\big\} \tag{A.0.198}$$

$(a)_+ := \max\{a, 0\}$, the positive part of the number $a \in \mathbb{R}$

$\pi_\kappa(E) = \pi_{\Omega,\kappa}(E)$, the "shadow" (or projection) of a given set $E \subseteq \Omega$ onto $\partial\Omega$, defined as (cf. [129, (8.1.15)]):

$$\pi_\kappa(E) = \pi_{\Omega,\kappa}(E) := \big\{x \in \partial\Omega : \Gamma_\kappa(x) \cap E \neq \varnothing\big\} \tag{A.0.199}$$

$p_{\mathbb{X}}$, the lower Boyd index of a rearrangement invariant Banach function space $\mathbb{X}$ on a non-atomic sigma-finite measure space $(X, \mathfrak{M}, \mu)$

$\partial_\nu^A(\cdot,\cdot)$, the conormal derivative operator with respect to the coefficient tensor A acting from weighted Sobolev spaces (cf. [130, Proposition 8.5.3]), and from Besov/Triebel-Lizorkin spaces (cf. [130, Proposition 9.5.2])

$\dot{\partial}_\nu^A u$, the weak conormal derivative of u with respect to the coefficient tensor A defined (cf. [130, Definition 10.2.18]) by first introducing

$$\vec{F}^\alpha := (A\nabla u)_\alpha = \big(a_{rs}^{\alpha\beta}\partial_s u_\beta\big)_{1\leq r\leq n} \text{ for each } \alpha \in \{1,\dots,M\} \tag{A.0.200}$$

and then setting

$$\dot{\partial}_\nu^A u := \big(\nu \bullet \vec{F}^\alpha\big)_{1\leq\alpha\leq M} \in \Big[\big(\mathrm{Lip}_c(\partial\Omega)\big)'\Big]^M \tag{A.0.201}$$

$\partial_{\tau_{jk}}\varphi$ with $\varphi \in \mathscr{C}^1(O)$, the pointwise tangential derivative operator (cf. [130, §11.1]):

$$\partial_{\tau_{jk}}\varphi := \nu_j\big[\partial_k\varphi\big]\big|_{O\cap\partial_*\Omega} - \nu_k\big[\partial_j\varphi\big]\big|_{O\cap\partial_*\Omega} \text{ at } \sigma_*\text{-a.e. point on } O \cap \partial_*\Omega \tag{A.0.202}$$

$\partial_{\tau_{jk}}$, the tangential derivative operator defined in a weak, or distributional, sense (cf. [130, (11.1.17) and Definition 11.1.2], and [130, Definition 11.2.1])

$\overline{\partial} := \partial_{\bar{z}} := \frac{1}{2}(\partial_x + i\partial_y)$, the Cauchy-Riemann operator

$\partial := \partial_z := \frac{1}{2}(\partial_x - i\partial_y)$, the (complex) conjugate of the Cauchy-Riemann operator

∂_ν^A, the pointwise conormal derivative operator with respect to the coefficient tensor $A := \big(a_{rs}^{\alpha\beta}\big)_{\substack{1\leq\alpha,\beta\leq M\\ 1\leq r,s\leq n}}$ acting on some $u = (u_\beta)_{1\leq\beta\leq N} \in \big[\mathcal{D}'(\Omega)\big]^N$ (which is of function type near $\partial\Omega$) at $\mathcal{H}^{n-1}$-a.e. point on $\partial_*\Omega$ as (cf. [131, (1.7.9) in Definition 1.7.1]):

$$\partial_\nu^A u := \Big(\nu_r\big(a_{rs}^{\alpha\beta}\partial_s u_\beta\big)\Big|_{\partial\Omega}^{\kappa-\text{n.t.}}\Big)_{1\leq\alpha\leq M} \tag{A.0.203}$$

$\partial_\nu^{\widetilde{D},D}$, the conormal derivative operator associated with the factorization of the second-order system L as $\widetilde{D}D$, where $\widetilde{D}$ and D are homogeneous, constant complex coefficient, first-order systems in $\mathbb{R}^n$, acting on a given $u \in \big[\mathcal{D}'(\Omega)\big]^N$ (which is of function type near $\partial\Omega$) as in [131, (1.7.36)]:

$$\partial_\nu^{\widetilde{D},D} u := (-\mathrm{i})\mathrm{Sym}(\widetilde{D};\nu)(Du)\Big|_{\partial\Omega}^{\kappa-\mathrm{n.t.}} \tag{A.0.204}$$

$\mathrm{P.V.}\big(b\,k(x-\cdot)\big|_\Sigma$, the principal-value distribution associated with the smooth odd kernel k and the bounded function b on the countably rectifiable upper Ahlfors regular set $\Sigma \subseteq \mathbb{R}^n$, acting on each $\phi \in \mathrm{Lip}_c(\Sigma)$ as (cf. [130, Proposition 11.9.1]):

$$\big\langle \mathrm{P.V.}\big(b\,k(x-\cdot)\big|_\Sigma\big), \phi\big\rangle := \lim_{\varepsilon\to 0^+} \int_{y\in\Sigma,\ |y-x|>\varepsilon} b(y)k(x-y)\phi(y)\,\mathrm{d}\sigma(y) \tag{A.0.205}$$

Π_Ω, the Newtonian (volume) potential operator associated with the system L in the set Ω acting on a given function w as in [132, (1.3.4)]:

$$\Pi_\Omega w := \int_\Omega E(\cdot - y)w(y)\,\mathrm{d}y \tag{A.0.206}$$

where E is the canonical fundamental solution of the system L

∂_τ, the tangential partial derivative in two dimensions (cf. [132, (1.6.40)]):

$$\partial_\tau := \partial_{\tau_{12}} = \nu_1\partial_2 - \nu_2\partial_1 \tag{A.0.207}$$

$\partial_\nu^\lambda(\vec{u},\pi)$, the (pointwise) conormal derivative for the Stokes system associated with the coefficient tensor $A_\lambda := \big(\delta_{jk}\delta_{\alpha\beta} + \lambda\delta_{j\beta}\delta_{k\alpha}\big)_{\substack{1\le\alpha,\beta\le n\\ 1\le j,k\le n}}$ for each $\lambda \in \mathbb{C}$ at σ-a.e. point on $\partial_*\Omega$ as in [132, (6..1.6)]:

$$\partial_\nu^\lambda(\vec{u},\pi) := \big[\nabla\vec{u} + \lambda(\nabla\vec{u})^\top\big]\Big|_{\partial\Omega}^{\kappa-\mathrm{n.t.}}\nu - \Big(\pi\Big|_{\partial\Omega}^{\kappa-\mathrm{n.t.}}\Big)\nu \tag{A.0.208}$$

$\overset{\bullet}{\partial}{}_\nu^\lambda(\vec{u},\pi)$, the weak conormal derivative for the Stokes system associated for each $\lambda \in \mathbb{C}$ with the coefficient tensor $A_\lambda := \big(\delta_{jk}\delta_{\alpha\beta} + \lambda\delta_{j\beta}\delta_{k\alpha}\big)_{\substack{1\le\alpha,\beta\le n\\ 1\le j,k\le n}}$ by first introducing the family of vector fields (cf. [132, (6.2.239)])

$$\vec{F}^\alpha := \nabla u_\alpha + \lambda\,\partial_\alpha\vec{u} - \pi\,\mathbf{e}_\alpha \ \text{ for each } \alpha \in \{1,\dots,n\} \tag{A.0.209}$$

and then defining as in [132, Definition 6.2.9]:

$$\overset{\bullet}{\partial}{}_\nu^\lambda(\vec{u},\pi) := \big(\nu \bullet \vec{F}^\alpha\big)_{1\le\alpha\le n} \in \Big[\big(\mathrm{Lip}_c(\partial\Omega)\big)'\Big]^n \tag{A.0.210}$$

$\mathcal{P}_\lambda$, the (double layer) pressure operator for the Stokes system acting on any vector-valued function $\vec{f} \in \big[L^1\big(\partial_*\Omega, \frac{\sigma(x)}{1+|x|^n}\big)\big]^n$ at each point $x \in \Omega$ as in [132, (6.2.21)]:

$$\mathcal{P}_\lambda\vec{f}(x) := -(1+\lambda)\int_{\partial_*\Omega} \nu_j(y)\langle(\partial_j\vec{q})(x-y), \vec{f}(y)\rangle\,\mathrm{d}\sigma(y) \tag{A.0.211}$$

where the pressure vector $\vec{q}$ is defined as in [132, (6.2.2)]:

$$\vec{q}(x) := -\frac{1}{\omega_{n-1}} \frac{x}{|x|^n} \quad \text{for each } x \in \mathbb{R}^n \setminus \{0\} \tag{A.0.212}$$

Q

X/Y, the quotient space of a vector space X and a linear subspace Y of X

$q_{\mathbb{X}}$, the upper Boyd index of a rearrangement invariant Banach function space $\mathbb{X}$ on a non-atomic sigma-finite measure space $(X, \mathfrak{M}, \mu)$

Q, the (single layer) pressure potential for the Stokes system acting on any vector-valued function $\vec{f} \in \big[L^1\big(\partial\Omega, \frac{\sigma(x)}{1+|x|^{n-1}}\big)\big]^n$ at each point $x \in \Omega$ as in [132, (6.2.14)]:

$$Q\vec{f}(x) := \int_{\partial\Omega} \langle \vec{q}(x-y), \vec{f}(y)\rangle \, d\sigma(y) \tag{A.0.213}$$

where the pressure vector $\vec{q}$ is defined as in (A.0.212)

R

$\mathbb{R}^n_+$, the (open) upper half-space in $\mathbb{R}^n$

$\mathbb{R}^n_-$, the (open) lower half-space in $\mathbb{R}^n$

rad(Ω), the number associated with any nonempty open set $\Omega \subseteq \mathbb{R}^n$ as (cf. [129, (5.11.31)]):

$$\begin{aligned} &\mathrm{rad}(\Omega) := \inf_j \inf_{x\in\Omega_j} \sup_{y\in\Omega_j} |x-y|, \quad \text{where} \\ &\{\Omega_j\}_j \ \text{ are the connected components of } \ \Omega \end{aligned} \tag{A.0.214}$$

R_j, the j-th boundary-to-boundary Riesz transform, with $j \in \{1, \dots, n\}$, acting on $f \in L^1\big(\Sigma, \frac{\sigma}{1+|\cdot|^{n-1}}\big)$ at σ-a.e. $x \in \Sigma$ as in [131, (2.5.336)]:

$$R_j f(x) := \lim_{\varepsilon\to 0^+} \frac{2}{\omega_{n-1}} \int\limits_{|x-y|>\varepsilon,\, y\in\Sigma} \frac{x_j - y_j}{|x-y|^n} f(y)\, d\sigma(y) \tag{A.0.215}$$

$\mathcal{R}_j$, the j-th boundary-to-domain Riesz transform, with $j \in \{1, \dots, n\}$, acting on each $f \in L^1\big(\Sigma, \frac{\sigma}{1+|\cdot|^{n-1}}\big)$ at every point $x \in \mathbb{R}^n \setminus \Sigma$ as in [131, (2.5.335)]:

$$\mathcal{R}_j f(x) := \frac{2}{\omega_{n-1}} \int_\Sigma \frac{x_j - y_j}{|x-y|^n} f(y)\, d\sigma(y) \tag{A.0.216}$$

$\mathcal{R}_{jk}$, the boundary-to-domain integral operator, with $j, k \in \{1, \dots, n\}$, acting on each $f \in L^1\big(\partial_*\Omega, \frac{\sigma(x)}{1+|x|^{n-1}}\big)$ at any point $x \in \Omega$ as in [131, (2.5.256)]:

$$\mathcal{R}_{jk} f(x) := \frac{-1}{\omega_{n-1}} \int_{\partial_*\Omega} \frac{\nu_j(y)(x_k - y_k) - \nu_k(y)(x_j - y_j)}{|x-y|^n} f(y)\, d\sigma_*(y) \tag{A.0.217}$$

R_{jk}, the principal-value singular integral operator, with $j,k \in \{1,\dots,n\}$, acting on each $f \in L^1\big(\partial_*\Omega, \frac{\sigma(x)}{1+|x|^{n-1}}\big)$ at σ_*-a.e. point $x \in \partial_*\Omega$ as in [131, (2.5.268)]:

$$R_{jk}f(x) := \lim_{\varepsilon\to 0^+} \int_{\substack{y\in\partial_*\Omega \\ |x-y|>\varepsilon}} \Big[\nu_j(y)(\partial_k E_\Delta)(y-x) - \nu_k(y)(\partial_j E_\Delta)(y-x)\Big] f(y)\,\mathrm{d}\sigma_*(y) \tag{A.0.218}$$

$(R_{jk})_{\max}$, the maximal operator, with $j,k \in \{1,\dots,n\}$, acting on each function $f \in L^1\big(\partial_*\Omega, \frac{\sigma(x)}{1+|x|^{n-1}}\big)$ at any point $x \in \partial_*\Omega$ as in [131, (2.5.270)]:

$$(R_{jk})_{\max} f(x) := \sup_{\varepsilon>0} \Big| \int_{y\in\partial_*\Omega,\ |x-y|>\varepsilon} \big[\nu_j(y)(\partial_k E_\Delta)(y-x) - \nu_k(y)(\partial_j E_\Delta)(y-x)\big] f(y)\,\mathrm{d}\sigma_*(y)\Big| \tag{A.0.219}$$

$R^{\#}_{jk}$, the principal-value singular integral operator on $\partial\Omega$, for $j,k \in \{1,\dots,n\}$, acting on any $f \in L^p(\partial\Omega,\sigma)\otimes \mathcal{C}\ell_n$ at σ_*-a.e. point $x \in \partial_*\Omega$ as in [132, (1.6.16)]:

$$R^{\#}_{jk} f(x) := \lim_{\varepsilon\to 0^+} \frac{-1}{\omega_{n-1}} \int_{y\in\partial\Omega,\ |x-y|>\varepsilon} \frac{\nu_j(x)(x_k-y_k) - \nu_k(x)(x_j-y_j)}{|x-y|^n} f(y)\,\mathrm{d}\sigma(y) \tag{A.0.220}$$

$\mathcal{R}_{\mathcal{C}\ell}$, the boundary-to-domain Clifford-Riesz transform, acting on any Clifford algebra-valued function $f \in L^1\big(\partial\Omega, \frac{\sigma(x)}{1+|x|^{n-1}}\big)\otimes \mathcal{C}\ell_n$ as in [132, (2.1.77)]:

$$\mathcal{R}_{\mathcal{C}\ell} f(x) := \frac{2}{\omega_{n-1}} \int_{\partial\Omega} \frac{x-y}{|x-y|^n} \odot f(y)\,\mathrm{d}\sigma(y), \qquad \forall\, x \in \Omega \tag{A.0.221}$$

R_j^{mod}, the j-th modified Riesz transform acting on any function $f \in L^1\big(\mathbb{R}^{n-1}, \frac{dx'}{1+|x'|^n}\big)$ at $\mathcal{L}^{n-1}$-a.e. point $x' \in \mathbb{R}^{n-1}$ as in [132, (5.3.53)]:

$$\big(R_j^{\mathrm{mod}} f\big)(x) = \lim_{\varepsilon\to 0^+} \frac{2}{\omega_{n-1}} \int_{\mathbb{R}^{n-1}} \Big\{ \tfrac{x_j-y_j}{|x'-y'|^n} \mathbf{1}_{\mathbb{R}^{n-1}\setminus \overline{B_{n-1}(x',\varepsilon)}}(y') - \tfrac{-y_j}{|-y'|^n} \mathbf{1}_{\mathbb{R}^{n-1}\setminus \overline{B_{n-1}(0',1)}}(y') \Big\} f(y')\,\mathrm{d}y' \tag{A.0.222}$$

where $B_{n-1}(z',r)$ is the $(n-1)$-dimensional ball centered at $z' \in \mathbb{R}^{n-1}$ of radius r

S

$\sigma := \mathcal{H}^{n-1}\lfloor\partial\Omega$, the surface measure on $\partial\Omega$

$\sigma_* := \mathcal{H}^{n-1}\lfloor\partial_*\Omega$, the surface measure on $\partial_*\Omega$

$\sigma := \mathcal{H}^{n-1}\lfloor\Sigma$, the surface measure on the closed Ahlfors regular set $\Sigma \subseteq \mathbb{R}^n$

$\mathrm{Sym}(D;\cdot)$, the principal symbol of the first-order system $D = \Big(\sum_{j=1}^{n} a_j^{\alpha\beta}\partial_j\Big)_{\substack{1\le\alpha\le N\\ 1\le\beta\le M}}$ defined at each $\xi \in \mathbb{R}^n$ as:

$$\mathrm{Sym}(D;\xi) := \mathrm{i}\Big(\sum_{j=1}^{n} a_j^{\alpha\beta}\xi_j\Big)_{\substack{1\le\alpha\le N\\ 1\le\beta\le M}} \tag{A.0.223}$$

$*$, the Hodge star operator

$S^{n-1} := \partial B(0,1)$, the unit sphere in $\mathbb{R}^n$

$S^{n-1}_{\pm} := S^{n-1} \cap \mathbb{R}^n_{\pm}$, the upper and lower (open) hemispheres

$\mathrm{supp} f$, the support of the μ-measurable function f defined on a topological space (X,τ) (where μ is a Borel measure measure on (X,τ)), defined as (cf. [129, (3.8.7) in Definition 3.8.3]):

$$\mathrm{supp} f := \Big\{x \in X : \int_O |f|\,\mathrm{d}\mu > 0 \text{ for each open set } O \subseteq X \text{ with } x \in O\Big\} \tag{A.0.224}$$

$\mathcal{S}(\mathbb{R}^n)$, the space of (smooth, rapidly decreasing) Schwartz functions in $\mathbb{R}^n$

$\mathcal{S}'(\mathbb{R}^n)$, the space of tempered distributions in $\mathbb{R}^n$

$u_{\star,\theta}$, the solid maximal function of $u : \Omega \to \mathbb{C}$, defined at each point $x \in \Omega$ according to (cf. [129, (6.6.2)]):

$$u_{\star,\theta}(x) := \|u\|_{L^\infty(B(x,\theta\delta_{\partial\Omega}(x)),\mathcal{L}^n)} \in [0,\infty] \tag{A.0.225}$$

$f_p^{\#}$, the L^p-based Fefferman-Stein sharp maximal function of $f \in L^1_{\mathrm{loc}}(X,\mu)$, defined as (cf. [129, (7.4.110)]):

$$f_p^{\#}(x) := \sup_{r>0}\Big(⨍_{B_\rho(x,r)} |f(y) - f_{B_\rho(x,r)}|^p\,\mathrm{d}\mu(y)\Big)^{1/p}, \quad \forall x \in X \tag{A.0.226}$$

$\rho_{\mathrm{inv}}(T;X)$, the spectral radius of $T \in \mathrm{Bd}(X)$, with X a quasi-Banach space (cf. [130, Definition 2.2.5]):

$$\rho_{\mathrm{inv}}(T;X) := \inf\big\{r > 0 : zI - T : X \to X \text{ homeomorphism} \\ \text{for each choice of } z \in \mathbb{C}\setminus\overline{B(0,r)}\big\} \tag{A.0.227}$$

$\mathcal{S}$, the boundary-to-domain single layer potential operator for an $M\times M$ second-order weakly elliptic system L (with canonical fundamental solution E) acting on any function $f \in \big[L^1\big(\partial\Omega, \frac{\sigma(x)}{1+|x|^{n-2}}\big)\big]^M$ at each point $x \in \Omega$ as in [132, (1.3.6)]:

$$\mathcal{S} f(x) := \int_{\partial\Omega} E(x-y)f(y)\,\mathrm{d}\sigma(y) \tag{A.0.228}$$

$\mathcal{S}_{\mathrm{mod}}$, the boundary-to-domain modified single layer potential operator for an $M\times M$ second-order weakly elliptic system L (with canonical fundamental solution E) acting on any $f \in \big[L^1\big(\partial\Omega, \frac{\sigma(x)}{1+|x|^{n-2}}\big)\big]^M$ at each point $x \in \Omega$ as in [132, (1.5.50)]:

$$\mathscr{S}_{\text{mod}}f(x) := \int_{\partial\Omega} \{E(x-y) - E(-y)\cdot \mathbf{1}_{\mathbb{R}^n\setminus B(0,1)}(-y)\}f(y)\,d\sigma(y) \qquad \text{(A.0.229)}$$

$\mathscr{S}_{\Delta,\text{mod}}$, the boundary-to-domain modified harmonic single layer potential operator acting on any $f \in L^1\big(\partial\Omega, \frac{\sigma(x)}{1+|x|^{n-1}}\big)$ at each point $x \in \Omega$ as in [132, (1.8.53)]:

$$\mathscr{S}_{\Delta,\text{mod}}f(x) := \int_{\partial\Omega} \big\{E_\Delta(x-y) - E_\Delta(-y)\cdot \mathbf{1}_{\mathbb{R}^n\setminus B(0,1)}(-y)\big\}f(y)\,d\sigma(y) \qquad \text{(A.0.230)}$$

S, the boundary-to-boundary single layer for an $M\times M$ second-order weakly elliptic system L (with canonical fundamental solution E) acting on each function f in $\big[L^1\big(\partial\Omega, \frac{\sigma(x)}{1+|x|^{n-2}}\big)\big]^M$ (assuming $n \geq 3$) at σ-a.e. $x \in \partial\Omega$ as in [132, (1.3.62)]:

$$Sf(x) := \int_{\partial\Omega} E(x-y)f(y)\,d\sigma(y) \qquad \text{(A.0.231)}$$

S_{mod}, the boundary-to-boundary modified single layer potential operator for an $M\times M$ second-order weakly elliptic system L (with canonical fundamental solution E) acting on any $f \in \big[L^1\big(\partial\Omega, \frac{\sigma(x)}{1+|x|^{n-1}}\big)\big]^M$ at σ-a.e. point $x \in \partial\Omega$ as in [132, (1.5.73)]:

$$S_{\text{mod}}f(x) := \int_{\partial\Omega} \{E(x-y) - E(-y)\cdot \mathbf{1}_{\mathbb{R}^n\setminus B(0,1)}(-y)\}f(y)\,d\sigma(y) \qquad \text{(A.0.232)}$$

$\mathscr{S}$, the boundary-to-domain single layer for the Stokes system acting on any vector-valued function $\vec{f} \in \big[L^1\big(\partial\Omega, \frac{\sigma(x)}{1+|x|^{n-2}}\big)\big]^n$ at each point $x \in \Omega$ as in [132, (6.2.12)]:

$$\mathscr{S}\vec{f}(x) := \int_{\partial\Omega} E(x-y)\vec{f}(y)\,d\sigma(y) \qquad \text{(A.0.233)}$$

where E is the Kelvin matrix-valued fundamental solution of the Stokes system in $\mathbb{R}^n$ (cf. [132, (6.2.1)]).

$\mathscr{S}_{\text{mod}}$, the modified boundary-to-domain single layer for the Stokes system defined as in [132, (6.2.164)]:

$$\begin{gathered}\mathscr{S}_{\text{mod}}\vec{f}(x) := \int_{\partial\Omega} \{E(x-y) - E_*(-y)\}\vec{f}(y)\,d\sigma(y) \ \text{ for each } \ x \in \Omega,\\ \text{for each } \ \vec{f} \in \big[L^1\big(\partial\Omega, \tfrac{\sigma(x)}{1+|x|^{n-1}}\big)\big]^n, \quad \text{with} \quad E_* := E\cdot \mathbf{1}_{\mathbb{R}^n\setminus B(0,1)},\end{gathered} \qquad \text{(A.0.234)}$$

where E is the Kelvin matrix-valued fundamental solution of the Stokes system in $\mathbb{R}^n$ (cf. [132, (6.2.1)]).

T

$\mathcal{T}_\gamma(x)$, the family of "bump" (i.e., localized, and normalized in the Hölder norm) functions centered at $x \in \Sigma$ (cf. [130, §4.1])

$\mathrm{Tr}_{\mathbb{R}^n\to\Sigma}$, the trace operator from $\mathbb{R}^n$ to Σ defined for each $u \in W^{1,p}\big(\mathbb{R}^n, \delta_\Sigma^{ap}\,\mathcal{L}^n\big)$ as the limit (cf. [130, Theorem 8.1.1]):

$$(\mathrm{Tr}_{\mathbb{R}^n\to\Sigma}\, u)(x) := [u]_{\mathbb{R}^n}(x) := \lim_{r\to 0^+} \fint_{B(x,r)} u(y)\, dy \tag{A.0.235}$$

$\mathrm{Tr}_{\Omega\to\partial\Omega}$, the trace operator from the entire Euclidean space $\mathbb{R}^n$ to the set $\partial\Omega$, defined for each function $u \in W^{1,p}_a(\Omega) := W^{1,p}(\Omega, \delta^{ap}_{\partial\Omega}\mathcal{L}^n)$ as the limit (cf. [130, Theorem 8.3.6]):

$$(\mathrm{Tr}_{\Omega\to\partial\Omega}\, u)(x) := [u]_{\overline{\Omega}}(x) := \lim_{r\to 0^+} \fint_{B(x,r)\cap\Omega} u(y)\, dy \tag{A.0.236}$$

$\mathrm{Tr}_{\Omega\to\partial\Omega}$, the trace operator from the open set $\Omega \subseteq \mathbb{R}^n$ to its boundary $\partial\Omega$, acting on each given $u \in A^{p,q}_\alpha(\Omega)$ (and with $w \in A^{p,q}_\alpha(\mathbb{R}^n)$ such that $w\big|_\Omega = u$) according to (cf. [130, Theorem 9.4.5]):

$$\big(\mathrm{Tr}_{\Omega\to\partial\Omega}\, u\big)(x) := \lim_{r\to 0^+} \fint_{B(x,r)} w\, d\mathcal{L}^n \ \text{ at } \sigma\text{-a.e. } x \in \partial\Omega \tag{A.0.237}$$

T_ε, the truncated singular integral operator acting on each $f \in L^1\big(\partial\Omega, \frac{\sigma(x)}{1+|x|^n}\big)$ at any point $x \in \partial\Omega$ as:

$$(T_\varepsilon f)(x) := \int_{y\in\partial\Omega,\ |x-y|>\varepsilon} k(x-y) f(y)\, d\sigma(y) \tag{A.0.238}$$

$T_{\max}$, the maximal operator acting on each $f \in L^1\big(\partial\Omega, \frac{\sigma(x)}{1+|x|^n}\big)$ at any $x \in \partial\Omega$ as:

$$(T_{\max} f)(x) := \sup_{\varepsilon>0} \big|(T_\varepsilon f)(x)\big| \tag{A.0.239}$$

T_{mod}, the modified principal-value singular integral operator acting on each function $f \in L^1\big(\Sigma, \frac{\sigma(x)}{1+|x|^n}\big)$ at σ-a.e. point in Σ as:

$$T_{\text{mod}} f := \lim_{\varepsilon\to 0^+} \int_\Sigma \big\{k_\varepsilon(\cdot - y) - k_1(-y)\big\} f(y)\, d\sigma(y) \tag{A.0.240}$$

where $k_\varepsilon := k \cdot \mathbf{1}_{\mathbb{R}^n\setminus\overline{B(0,\varepsilon)}}$ for each $\varepsilon > 0$

$\mathcal{T}_{\text{mod}}$, the boundary-to-domain modified integral operator, acting on any given function $f \in L^1\big(\partial\Omega, \frac{\sigma(x)}{1+|x|^n}\big)$ at each point $x \in \Omega$ as:

$$(\mathcal{T}_{\text{mod}} f)(x) := \int_{\partial\Omega} \big\{k(x-y) - k_1(-y)\big\} f(y)\, d\sigma(y) \tag{A.0.241}$$

where $k_1 := k \cdot \mathbf{1}_{\mathbb{R}^n\setminus\overline{B(1,0)}}$

U

UR set (cf. [129, Definition 5.10.1]): a closed set $\Sigma \subset \mathbb{R}^n$ which is (upper) Ahlfors regular and has Big Pieces of Lipschitz Images (in a uniform, quantitative, scale-invariant fashion)

UR domain (cf. [129, Definition 5.10.6]): a nonempty open subset Ω of $\mathbb{R}^n$ such that $\partial\Omega$ is a UR set and

$$\mathcal{H}^{n-1}(\partial\Omega \setminus \partial_*\Omega) = 0 \tag{A.0.242}$$

$\mathrm{UC}(X, \rho)$, the space of uniformly continuous functions on the metric space (X, ρ)

$[u]^A_\infty$, the contribution at infinity of a null-solution u for the vector-Helmholtz operator, associated with the coefficient tensor $A := \big(a^{JI}_{rs}\big)_{\substack{1\le J,I\le M\\ 1\le r,s\le n}}$ in the writing of the vector Laplacian as $\Delta = \mathrm{div} A\nabla$, defined at each $x \in \mathbb{R}^n$ as:

$$[u]^A_\infty := \lim_{R\to\infty} \Big(\int_{|y|=R} \widehat{y}_s \big\{ a^{IJ}_{rs} (\partial_r \Phi_k)(x-y) u_J(y) \\ + a^{IJ}_{sr} \Phi_k(x-y)(\partial_r u_J)(y) \big\}\, \mathrm{d}\mathcal{H}^{n-1}(x) \Big)_{1\le I\le M} \tag{A.0.243}$$

where Φ_k is the radiating fundamental solution of the Helmholtz operator $\Delta + k^2$

$u \otimes w$, the tensor product of differential forms (cf. [132, (7.3.4)])

$U \sqcup V$, the union of two disjoint sets U, V

V

$\mathrm{VMO}(X, \mu)$, the Sarason space of functions of vanishing mean oscillations on the measure metric space (X, ρ, μ) defined (with $\mathrm{UC}(X, \rho)$ denoting the space of uniformly continuous functions on (X, ρ)) as the space (cf. [130, §3.1]):

$$\mathrm{VMO}(X, \mu) := \text{the closure of } \mathrm{UC}(X, \rho) \cap \mathrm{BMO}(X, \mu) \text{ in } \mathrm{BMO}(X, \mu) \tag{A.0.244}$$

$\mathrm{VMO}_{-1}(\partial\Omega, \sigma)$, the VMO-based negative Sobolev space on $\partial\Omega$, defined for $n \ge 3$ as [130, Definition 11.10.9]):

$$\mathrm{VMO}_{-1}(\partial\Omega, \sigma) := \text{ the closure of } L^{n-1}(\partial\Omega, \sigma) \text{ in } \mathrm{BMO}_{-1}(\partial\Omega, \sigma) \tag{A.0.245}$$

$\mathrm{VMO}_{-1}(\partial\Omega, \sigma)$ in the two-dimensional setting defined as the closure of $L^q(\partial\Omega, \sigma)$ in $\mathrm{BMO}_{-1}(\partial\Omega, \sigma)$, for any $q \in (1, \infty)$ (cf. [132, (2.3.134)])

$\vee$, the interior product of differential forms (cf. [132, (7.1.52)])

W

weakly elliptic system: a system L whose characteristic matrix is invertible, i.e., satisfies $\det[L(\xi)] \neq 0$ for each $\xi \in \mathbb{R}^n \setminus \{0\}$

weakly elliptic coefficient tensor: a coefficient tensor A with the property that the canonically associated second-order system L_A is weakly elliptic, in the sense that $\det[L_A(\xi)] \neq 0$ for each $\xi \in \mathbb{R}^n \setminus \{0\}$

$W^{k,p}(\Omega)$, the L^p-based Sobolev space of order k in Ω (intrinsically defined)

$W^{k,p}_{\mathrm{loc}}(\Omega)$, the local L^p-based Sobolev space of order k in Ω

$W^{k,p}_{\mathrm{bdd}}(\Omega)$, the space of Sobolev functions on any bounded measurable subset of Ω (cf. [129, (3.0.4)]):

$$W^{k,p}_{\mathrm{bdd}}(\Omega) \text{ denote the space of functions } u \in W^{k,p}_{\mathrm{loc}}(\Omega) \text{ with the property that } \partial^\alpha u \in L^p(O, \mathcal{L}^n) \text{ for each } \alpha \in \mathbb{N}_0^n \text{ with } |\alpha| \leq k \text{ and each bounded Lebesgue measurable subset } O \text{ of } \Omega. \tag{A.0.246}$$

$W^{k,p}(\mathbb{R}^n, w\mathcal{L}^n)$, the weighted Sobolev space in $\mathbb{R}^n$, defined for $p \in (0, \infty)$, $k \in \mathbb{N}_0$ and a generic weight w in $\mathbb{R}^n$ as (cf. [130, §8.1]):

$$W^{k,p}(\mathbb{R}^n, w\mathcal{L}^n) := \Big\{ f \in W^{k,1}_{\mathrm{loc}}(\mathbb{R}^n) : \|f\|_{W^{k,p}(\mathbb{R}^n, w\mathcal{L}^n)} < +\infty \Big\} \tag{A.0.247}$$

where

$$\|f\|_{W^{k,p}(\mathbb{R}^n, w\mathcal{L}^n)} := \sum_{|\beta|\leq k} \Big(\int_{\mathbb{R}^n} |\partial^\beta f|^p w \, d\mathcal{L}^n \Big)^{1/p} \tag{A.0.248}$$

$W^{k,p}_a(\Omega)$, the weighted Sobolev space in Ω defined as in [130, (8.3.5)] for the weight $w := \delta^{ap}_{\partial\Omega}$, with $k \in \mathbb{N}_0$, $p \in (0, \infty)$, $a \in \mathbb{R}$, and equipped with the quasi-norm (cf. [130, Definition 8.3.4]):

$$\|u\|_{W^{k,p}_a(\Omega)} := \sum_{|\alpha|\leq k} \Big(\int_\Omega |(\partial^\alpha u)(x)|^p \, \delta_{\partial\Omega}(x)^{ap} \, dx \Big)^{1/p} \tag{A.0.249}$$

$\mathring{W}^{1,p}_a(\Omega)$, the weighted Sobolev space in the open set Ω, defined for $p \in (1, \infty)$ and $a \in (-1/p, 1 - 1/p)$ as (cf. [130, (8.3.65)]):

$$\mathring{W}^{1,p}_a(\Omega) := \text{ the closure of } \mathscr{C}^\infty_c(\Omega) \text{ in } W^{1,p}_a(\Omega) \tag{A.0.250}$$

$W^{-1,p}_a(\Omega)$, the negative weighted Sobolev space in Ω, defined for $p \in (1, \infty)$ and $a \in (-1/p, 1 - 1/p)$ as (cf. [130, (8.5.1)-(8.5.2)]):

$$W^{-1,p}_a(\Omega) := \Big\{ f = f_0 + \sum_{j=1}^{n} \partial_j f_j \in \mathcal{D}'(\Omega) : \ f_j \in L^p\big(\Omega, \delta^{ap}_{\partial\Omega}\mathcal{L}^n\big), \ 0 \leq j \leq n \Big\} \tag{A.0.251}$$

equipped with the norm

$$W^{-1,p}_a(\Omega) \ni f \mapsto \inf \Big\{ \sum_{j=0}^{n+1} \|f_j\|_{L^p(\Omega, \delta^{ap}_{\partial\Omega}\mathcal{L}^n)} : \ f = f_0 + \sum_{j=1}^{n} \partial_j f_j, \quad f_j \in L^p\big(\Omega, \delta^{ap}_{\partial\Omega}\mathcal{L}^n\big), \ 0 \leq j \leq n \Big\} \tag{A.0.252}$$

$W^{k,p}_{a,\star}(\Omega)$, the weighted maximal Sobolev space in Ω, defined for $p \in (0, \infty)$, $k \in \mathbb{N}_0$, and $a \in \mathbb{R}$ as (cf. [130, Definition 8.6.1]):

$$W^{k,p}_{a,\star}(\Omega) := \Big\{ u \in W^{k,1}_{\mathrm{loc}}(\Omega) : \ \partial^\alpha u \in L^p_\star\big(\Omega, \delta^{ap}_{\partial\Omega}\mathcal{L}^n\big) \text{ for all } \alpha \in \mathbb{N}_0^n \text{ with } |\alpha| \leq k \Big\} \tag{A.0.253}$$

equipped with the quasi-norm

$$\|u\|_{W^{k,p}_{a,\star}(\Omega)} := \sum_{|\alpha|\le k} \|\partial^\alpha u\|_{L^p_\star(\Omega,\delta^{ap}_{\partial\Omega}\mathcal{L}^n)}$$
$$\approx \sum_{|\alpha|\le k} \Big(\int_\Omega |(\partial^\alpha u)_{\star,\theta}|^p \delta^{ap}_{\partial\Omega}\, d\mathcal{L}^n\Big)^{\frac{1}{p}} \text{ for } \theta \in (0,1) \qquad \text{(A.0.254)}$$

$\mathscr{W}^{1,p}_a(\Omega; D)$, the weighted Sobolev-Hardy space of null-solutions of the Dirac operator $D = \sum_{j=1}^{n} \mathbf{e}_j \odot \partial_j$ in an open set $\Omega \subseteq \mathbb{R}^n$ defined for $p \in (0,\infty)$ and $a \in \big(-\frac{1}{p}, 1-\frac{1}{p}\big)$ as in [132, (4.4.131)]:

$$\mathscr{W}^{1,p}_a(\Omega; D) := \big\{u \in W^{1,p}_a(\Omega) \otimes \mathcal{C}\ell_n : Du = 0 \text{ in } \Omega\big\} \qquad \text{(A.0.255)}$$

X

${}_{X^*}\langle\cdot,\cdot\rangle_X$, the duality pairing between a vector space X and its algebraic dual X^*

$\mathbb{X}'$, the associated space (aka Köthe dual) of the Generalized Banach Function Space $\mathbb{X}$, equipped with the norm $\|\cdot\|_{\mathbb{X}'}$ (cf. [130, Definition 5.1.11])

$\mathscr{X}_{\rm bdd}(\Omega)$, or $\mathscr{X}(\Omega)_{\rm bdd}$, the space of distributions in the open set $\Omega \subseteq \mathbb{R}^n$ defined as (cf. [130, Convention 8.3.7]):

$$\Big\{u \in \mathcal{D}'(\Omega) : \big(\psi\big|_\Omega\big)u \in \mathscr{X}(\Omega) \text{ for each } \psi \in \mathscr{C}^\infty_c(\mathbb{R}^n)\Big\} \qquad \text{(A.0.256)}$$

$\mathscr{X} \otimes \Lambda^{\alpha,\beta}$, the space of differential forms with coefficients from $\mathscr{X}$, where $\mathscr{X}$ is an arbitrary subspace of $\mathcal{D}'(\Omega)$ for some open subset Ω of $\mathbb{C}^n$

References

1. M. Abramowitz and I.A. Stegun, *Handbook of Mathematical Functions*, Dover, New York, 1972.
2. S. Agmon, *Multiple layer potentials and the Dirichlet problem for higher order elliptic equations in the plane. I*, Comm. Pure Appl. Math., 10 (1957), 179–239.
3. S. Agmon, *The L^p approach to the Dirichlet problem. I. Regularity theorems*, Ann. Scuola Norm. Sup. Pisa (3), 13 (1959), 405–448.
4. S. Agmon, A. Douglis, and L. Nirenberg, *Estimates near the boundary for solutions of elliptic partial differential equations satisfying general boundary conditions, I*, Comm. Pure Appl. Math., 12 (1959), 623–727.
5. S. Agmon, A. Douglis, and L. Nirenberg, *Estimates near the boundary for solutions of elliptic partial differential equations satisfying general boundary conditions, II*, Comm. Pure Appl. Math., 17 (1964), 35–92.
6. M.S. Agranovich, *Elliptic Boundary Problems*, pp. 1–144 in "Partial Differential Equations IX," Encyclopaedia of Mathematical Sciences, Vol. 79, M.S. Agranovich, Yu.V. Egorov, and M.A. Shubin eds., Springer-Verlar, Berlin, Heidelberg, 1997.
7. J.F. Ahner, *On the eigenvalues of the electrostatic integral operator, II*, J. Math. Anal. Applications, 181 (1994), no. 2, 328–334.
8. R. Alvarado, D. Brigham, V. Maz'ya, M. Mitrea, and E. Ziadé, *On the regularity of domains satisfying a uniform hour-glass condition and a sharp version of the Hopf-Oleinik Boundary Point Principle*, Journal of Mathematical Sciences, 176 (2011), no. 3, 281–360.
9. R. Alvarado and M. Mitrea, *Hardy Spaces on Ahlfors-Regular Quasi-Metric Spaces. A Sharp Theory*, Lecture Notes in Mathematics, Vol. 2142, Springer, 2015.
10. T. Archibald, *From attraction theory to existence proofs: the evolution of potential-theoretic methods in the study of boundary-value problems, 1860-1890*, Rev. Histoire Math., 2 (1996), no. 1, 67–93.
11. J.M. Ash and R. Brown, *Uniqueness and non uniqueness for harmonic functions with zero nontangential limits*, pp. 30–40 in "Harmonic Analysis", ICM-90 Satellite Conference Proceedings, S. Igari editor, Springer-Verlag, Tokyo, 1991.
12. F.V. Atkinson, *Normal solvability of linear equations in normed spaces*, Mat. Sb., Nov. Ser., 28 (1951), no. 70, 3–14.
13. A. Barton and S. Mayboroda, *Layer Potentials and Boundary-Value Problems for Second Order Elliptic Operators With Data in Besov Spaces*, Memoirs of the American Mathematical Society, Vol. 243, No. 1149, 2016.
14. R.E. Basye, *Simply connected sets*, Trans. Amer. Math. Soc., 38 (1935), 341–356.
15. C. Bennett and R. Sharpley, *Interpolation of Operators*, Pure and Applied Mathematics, Vol. 129, Academic Press, 1988.

D. Mitrea et al., *Geometric Harmonic Analysis V*, Developments in Mathematics 76,
https://doi.org/10.1007/978-3-031-31561-9

16. S. Bortz, M. Engelstein, M. Goering, T. Toro, and Z. Zhao, *Two phase free boundary problem for Poisson kernels*, preprint, (2020).
17. D.W. Boyd, *Spectra of convolution operators*, Acta Scientiarum Mathematicarum, 35 (1973), 31–37.
18. K. Brewster, D. Mitrea, I. Mitrea, and M. Mitrea, *Extending Sobolev functions with partially vanishing traces from locally (ε, δ)-domains and applications to mixed boundary problems*, J. Funct. Anal., 266 (2014), 4314–4421.
19. F.E. Browder, *The Dirichlet problem for linear elliptic equations of arbitrary even order with variable coefficients*, Proc. Nat. Acad. Sci. U. S. A., 38 (1952), 230–235.
20. R.M. Brown, *The Neumann problem on Lipschitz domains in Hardy spaces of order less than one*, Pacific Journal of Mathematics, 171 (1995), no. 2, 389–407.
21. R. Brown, I. Mitrea, M. Mitrea, and M. Wright, *Mixed boundary value problems for the Stokes system*, Trans. Amer. Math. Soc., 362 (2010), no. 3, 1211–1230.
22. Yu.D. Burago and V.G. Maz'ya, *Potential theory and function theory for irregular regions*, Sem. Math. V.A. Steklov Inst. Leningr., 3 (1969), 1–68 (translation from Zap. Nauchn. Semin. Leningr. Otd. Mat. Inst. Steklova 3, (1967), 1–152).
23. Yu.D. Burago, V.G. Maz'ya, and V.D. Sapozhnikova, *On the double layer potential for non-regular regions*, Sov. Math. Dokl., 3 (1962), 1640–1642 (translation from Dokl. Akad. Nauk SSSR 147 (1962), 523–525).
24. Yu.D. Burago, V.G. Maz'ya, and V.D. Sapozhnikova, *On the theory of simple and double layer potentials for domains with irregular boundaries*, Probl. Mat. Anal., 1 (1968), 1–30 (translation from Probl. Mat. Analiza, Kraevye Zadachi, Integral. Uravn. Leningrad (1966), 3–34).
25. D.L. Burkholder and R.F. Gundy, *Distribution function inequalities for the area integral*, Studia Math., 44 (1972), 527–544.
26. A.P. Calderón, *On the behavior of harmonic functions near the boundary*, Trans. Amer. Math. Soc., 68 (1950), 47–54.
27. A. Calderón, *The multipole expansion of radiation fields*, J. Rat. Mech. Anal., 3 (1954), 523–537.
28. A.P. Calderón, *Commutators, singular integrals on Lipschitz curves and applications*, pp. 85–96 in "Proceedings of the International Congress of Mathematicians" (Helsinki, 1978), Acad. Sci. Fennica, Helsinki, 1980.
29. A.P. Calderón, *Boundary value problems for the Laplace equation in Lipschitzian domains*, pp. 33–48 in "Recent Progress in Fourier Analysis," I. Peral and J. Rubio de Francia, eds., Math. Stud. Vol. 111, Elsevier/North-Holland, Amsterdam, 1985.
30. L. Capogna, C.E. Kenig, and L. Lanzani, *Harmonic Measure: Geometric and Analytic Points of View*, University Lecture Series, Vol. 35, Amer. Math. Soc., Providence, Rhode Island, 2005.
31. T. Carleman, *Über das Neumann-Poincarésche Problem für ein Gebiet mit Ecken*, Almqvist & Wiksell, 1916.
32. B.C. Carlson, *Special Functions of Applied Mathematics*, Academic Press, New York, San Francisco, London, 1997.
33. R. Coifman, A. McIntosh, and Y. Meyer, *L'intégrale de Cauchy definit un opérateur borné sur L^2 pour les courbes lipschitziennes*, Ann. Math., 116 (1982), 361–388.
34. D. Colton and R. Kress, *Integral Equation Methods in Scattering Theory*, Wiley Interscience Publications, New York, 1983.
35. D. Colton and R. Kress, *Inverse Acoustic and Electromagnetic Scattering Theory*, Springer-Verlag, Applied Math. Series, No. 93, 1992.
36. M. Costabel, *Some historical remarks on the positivity of boundary integral operators*, pp. 1–27 in "Boundary Element Analysis", M. Schanz and O. Steinbach eds., Lecture Notes in Applied and Computational Mechanics, Vol. 29, Springer, Berlin, Heidelberg, 2007.
37. B.E.J. Dahlberg, D.S. Jerison and C.E. Kenig, *Area integral estimates for elliptic differential operators with non-smooh coefficients*, Ark. Mat., 22 (1984), no. 1-2, 97–108.
38. B.E.J. Dahlberg and C.E. Kenig, *Hardy spaces and the Neumann problem in L^p for Laplace's equation in a Lipschitz domain*, Ann. of Math., 125 (1987), 437–465.

39. B. Dahlberg and C. Kenig, *L^p estimates for the three-dimensional systems of elastostatics on Lipschitz domains*, pp. 621–634 in Lecture Notes Pure Appl. Math., Vol. 122, 1990.
40. B.E.J. Dahlberg, C.E. Kenig, and G.C. Verchota, *Boundary value problems for the systems of elastostatics in Lipschitz domains*, Duke Math. J., 57 (1988), no. 3, 795–818.
41. W.R. Derrick, *Open problems in singular integral theory*, J. Integral Equations Applications, 5 (1993), no. 1, 23–28.
42. E. DiBenedetto, *Partial Differential Equations*, Birkhäuser, second edition, 2010.
43. J. Dieudonné, *History of Functional Analysis*, North Holland, 1981.
44. M. Dindoš and M. Mitrea, *The stationary Navier-Stokes system in nonsmooth manifolds: the Poisson problem in Lipschitz and C^1 domains*, Arch. Ration. Mech. Anal., 174 (2004), no. 1, 1–47.
45. X.T. Duong, S. Hofmann, D. Mitrea, M. Mitrea, and L. Yan, *Hardy spaces and regularity for the inhomogeneous Dirichlet and Neumann problems*, Revista Matemática Iberoamericana, 29 (2013), no. 1, 183–236.
46. J. Elschner, *The double layer potential operator over polyhedral domains. I: Solvability in weighted Sobolev spaces*, Appl. Anal., 45 (1992), 117–134.
47. L. Escauriaza, E.B. Fabes, and G. Verchota, *On a regularity theorem for weak solutions to transmission problems with internal Lipschitz boundaries*, Proc. Amer. Math. Soc., 115 (1992), 1069–1076.
48. L. Escauriaza and M. Mitrea, *Transmission problems and spectral theory for singular integral operators on Lipschitz domains*, J. Funct. Anal., 216 (2004), no. 1, 141–171.
49. L.C. Evans and R.F. Gariepy, *Measure Theory and Fine Properties of Functions*, Studies in Advanced Mathematics, CRC Press, Boca Raton, FL, 1992.
50. E.B. Fabes, M. Jodeit Jr., and J.E. Lewis, *On the spectra of a Hardy kernel*, J. Funct. Anal., 21 (1976), 187–194.
51. E.B. Fabes, M. Jodeit Jr., and J.E. Lewis, *Double layer potentials for domains with corners and edges*, Indiana University Mathematics Journal, 26, (1977), no. 1, 95–114.
52. E.B. Fabes, M. Jodeit Jr., and N.M. Rivière, *Potential techniques for boundary value problems on C^1-domains*, Acta Math., 141 (1978), no. 3–4, 165–186.
53. E.B. Fabes, R.L. Johnson, and U. Neri, *Spaces of harmonic functions representable by Poisson integrals of functions in BMO and $\mathcal{L}_{p,\lambda}$*, Indiana Univ. Math. J., 25 (1976), no. 2, 159–170.
54. E. Fabes and C. Kenig, *On the Hardy space H^1 of a C^1 domain*, Arkiv Math., 19 (1981), 1–22.
55. E. Fabes, C. Kenig, and G.C. Verchota, *The Dirichlet problem for the Stokes system on Lipschitz domains*, Duke Math. J., 57 (1988), no. 3, 769–793.
56. E. Fabes, O. Mendez, and M. Mitrea, *Boundary layers on Sobolev-Besov spaces and Poisson's equation for the Laplacian in Lipschitz domains*, J. Funct. Anal., 159 (1998), no. 2, 323–368.
57. E. Fabes and U. Neri, *Dirichlet problem in Lipschitz domains with bmo data*, Proc. Amer. Math. Soc., 78 (1980), 33–39.
58. E. Fabes, M. Sand, and J.K. Seo, *The spectral radius of the classical layer potentials on convex domains*, pp. 129–137 in "Partial Differential Equations with Minimal Smoothness and Applications," B. Dahlberg et al (eds), The IMA Volumes in Mathematics and its Applications, Vol. 42, Springer, New York, NY, 1992.
59. C. Fefferman, *Characterizations of bounded mean oscillation*, Bull. of AMS, 77 (1971), 587–588.
60. C. Fefferman and E.M. Stein, *H^p spaces of several variables*, Acta Math., 129 (1972), no. 3-4, 137–193.
61. I. Fredholm, *Sur une nouvelle méthode pour la résolution du problème de Dirichlet*, Stockholm Öfver (Öfversigt af Kongl. Svenska Vetenscaps Akademiens Förhandingar), 57 (1900), 39–46.
62. I. Fredholm, *Sur une classe de transformations rationnelles*, Comptes Rendus de l'Académie des Sciences, 134 (1902), 219–222.
63. I. Fredholm, *Sur une classe d'équations fonctionelles*, Acta Math., 27 (1903), 365–390.
64. I. Fredholm, *Solution d'un problème fondamental de la théorie de l'élasticité*, Arkiv för Matematik, Astronomi och Fysik, 2 (1906), no. 28, 1–8.

65. W. Gao, *Layer potentials and boundary value problems for elliptic systems in Lipschitz domains*, J. Funct. Anal., 95 (1991), 377–399.
66. J. García-Cuerva and J.L. Rubio de Francia, *Weighted Norm Inequalities*, North Holland, Mathematics Studies Vol. 116, 1985.
67. L. Gårding, *Dirichlet's problem for linear elliptic partial differential equations*, Math. Scand., 1 (1953), 55–72.
68. J. Garnett and D.E. Marshall, *Harmonic Measure*, Cambridge University Press, 2005.
69. C.F. Gauss, *Allegemeine Lehrsätze*, Collected Works, Vol. V, 1863–1871.
70. J.E. Gilbert and M. Murray, *Clifford Algebras and Dirac Operators in Harmonic Analysis*, Cambridge University Press, 1991.
71. I.C. Gohberg, *Vladimir Maz'ya: Friend and mathematician. Recollections*, pp. 1–5 in "The Maz'ya Anniversary Collection", J. Rossmann et al eds., Vol. I, Operator Theory: Advances and Applications, 109/110, Birkhäuser, Bassel-Boston-Berlin, 1999.
72. I.Ts. Gohberg and M.G. Krein, *The fundamentals on defect numbers, and indices of linear operators*, Usp. Mat. Nauk, 12 (1957), 43–118 (English translation: Transl., II Ser., Amer. Math. Soc., 10 (1960), 185–264).
73. I.Ts. Gohberg and A.S. Marcus, *Some remarks on topologically equivalent norms*, Izv. Mold. Fil. Akad. Nauk SSSR, 10 (1960), no. 76, 91–94.
74. L. Grafakos, *Classical and Modern Fourier Analysis*, Pearson Education, Inc., New Jersey, 2004.
75. G. Green, *An Essay on the Application of Mathematical Analysis to the Theories of Electricity and Magnetism*, Nottingham, England, T. Wheelhouse, 1828.
76. G. Grubb and N.J. Kokholm, *A global calculus of parameter-dependent pseudodifferential boundary roblems in L_p Sobolev spaces*, Acta Math., 171 (1993), 165–229.
77. R.F. Gundy and R.L. Wheeden, *Weighted integral inequalities for the nontangential maximal function, Lusin area integral and Walsh-Paley series*, Studia Math., 49 (1974), 107–124.
78. C. Hermite, *Correspondance d'Hermite et de Stieltjes*, chapter Lettre 374 d'Hermite à Stieljtes du 20 mai 1893, pages 317–319. Gauthier-Villars, 1905.
79. S. Hofmann, E. Marmolejo-Olea, M. Mitrea, S. Perez-Esteva, and M. Taylor, *Hardy spaces, singular integrals and the geometry of Euclidean domains of locally finite perimeter*, Geometric and Functional Analysis (GAFA), 19 (2009), no. 3, 842–882.
80. S. Hofmann, M. Mitrea, and M. Taylor, *Geometric and transformational properties of Lipschitz domains, Semmes-Kenig-Toro domains, and other classes of finite perimeter domains*, J. Geom. Anal., 17 (2007), no. 4, 593–647.
81. S. Hofmann, M. Mitrea, and M. Taylor, *Singular Integrals and Elliptic Boundary Problems on Regular Semmes-Kenig-Toro Domains*, Int. Math. Res. Not. IMRN (2010), no. 14, 2567–2865.
82. S. Hofmann, M. Mitrea, and M. Taylor, *Symbol calculus for operators of layer potential type on Lipschitz surfaces with VMO normals, and related pseudodifferential operator calculus*, Analysis and PDE, 8 (2015), no. 1, 115–181.
83. L. Hörmander, *Linear Partial Differential Operators*, Springer Verlag, Berlin-New York, 1976.
84. B. Jawerth and M. Mitrea, *Higher dimensional scattering theory on C^1 and Lipschitz domains*, Amer. J. Math., 117 (1995), no. 4, 929–963.
85. D.S. Jerison and C.E. Kenig, *Boundary behavior of harmonic functions in nontangentially accessible domains*, Advances in Mathematics, 46 (1982), no. 1, 80–147.
86. D. Jerison and C.E. Kenig, *The inhomogeneous Dirichlet problem in Lipschitz domains*, J. Funct. Anal., 130 (1995), 161–219.
87. N. Kalton, S. Mayboroda, and M. Mitrea, *Interpolation of Hardy-Sobolev-Besov-Triebel-Lizorkin spaces and applications to problems in partial differential equations*, pp. 121–177 in "Interpolation Theory and Applications," AMS Contemporary Mathematics, Vol. 445, 2007.
88. N. Kalton and M. Mitrea, *Stability of Fredholm properties on interpolation scales of quasi-Banach spaces and applications*, Trans. Amer. Math. Soc., 350, (1998), 3837–3901.
89. T. Kato, *Perturbation Theory for Linear Operators*, reprint of the 1980 edition, Springer-Verlag, Berlin, 1995.

90. O.D. Kellogg, *Foundations of Potential Theory*, Dover Publications, Inc., 1953 (republication of the work printed in 1929 by J. Springer, Berlin).
91. C.E. Kenig, *Weighted H^p spaces on Lipschitz domains*, Amer. J. Math., 102 (1980), 129–163.
92. C.E. Kenig, *Elliptic boundary value problems on Lipschitz domains*, pp. 131–183 in "Beijing Lectures in Harmonic Analysis," Ann. of Math. Stud., Vol. 112, 1986.
93. C.E. Kenig, *Harmonic Analysis Techniques for Second Order Elliptic Boundary Value Problems*, CBMS Regional Conference Series in Mathematics, Vol. 83, AMS, Providence, RI, 1994.
94. C. Kenig and T. Toro, *Harmonic measure on locally flat domains*, Duke Math. J., 87 (1997), no. 3, 509–551.
95. C. Kenig and T. Toro, *Free boundary regularity for harmonic measures and Poisson kernels*, Annals of Mathematics, 150 (1999), 369–454.
96. J. Kilty, *The L^p Dirichlet problem for the Stokes system on Lipschitz domains*, Indiana Univ. Math. Journal, 58 (2009), no. 3, 1219–1234.
97. J. Kolar and J. Kristensen, *Gradient ranges of bumps on the plane*, Proc. Amer. Math. Soc., 133 (2005), no. 6, 1699–1706.
98. V.A. Kozlov, V.G. Maz'ya and J. Rossmann, *Elliptic Boundary Value Problems in Domains with Point Singularities*, Mathematical Surveys and Monographs, Vol. 52, Amer. Math. Soc., Providence, RI, 1997.
99. J. Král, *On the logarithmic potential of the double distribution*, Czechoslovak Math. J., 14 (1964), 306–321.
100. J. Král, *The Fredholm radius of an operator in potential theory*, Czechoslovak Math. J., 15 (1965), 454–473 and 565–588.
101. J. Král, *Integral Operators in Potential Theory*, Lecture Notes in Mathematics, Vol. 823, Springer-Verlag, 1980.
102. G. Kresin and V. Maz'ya, *Maximum Principles and Sharp Constants for Solutions of Elliptic and Parabolic Systems*, Mathematical Surveys and Monographs, Vol. 183, American Mathematical Society, Providence, 2012.
103. V.D. Kupradze, *On Sommerfeld's radiation principle*, Doklady Akad. Nauk SSSR, 2 (1934), 1–7.
104. V.D. Kupradze, *Fundamental Problems of the Mathematical Theory of Diffraction*, Moscow, 1935.
105. C. Kuratowski, *Topology*, Vol. II, Academic Press and PWN-Polish Scientific Publishers, 1966.
106. O.A. Ladyzhenskaya, *The Mathematical Theory of Viscous Incompressible Flow*, Gordon and Breach Science Publishers, New York-London, 1963.
107. T. Lalescu, *Introduction à la théorie des équations intégrales, Avec une préface de É. Picard*, Paris: A. Hermann et Fils, VII+152 pp., 1912.
108. H. Lebesgue, *Sur des cas d'impossibilité du problème de Dirichlet*, Comptes Rendus de la Société Mathématique de France, 17 (1913), 48–50.
109. Ya.B. Lopatinskiĭ, *On a method of reducing boundary value problems for systems of differential equations of elliptic type to regular integral equations*, (Russian) Ukrain. Mat. Žurnal, 5 (1953), no. 2, 123–151.
110. F. Maggi, *Sets of Finite Perimeter and Geometric Variational Problems. An introduction to Geometric Measure Theory*, Cambridge Studies in Advance Mathematics, Vol, 135, Cambridge University Press, 2012.
111. J.J. Marín, J. María Martell, D. Mitrea, I. Mitrea, and M. Mitrea, *Singular Integrals, Quantitative Flatness, and Boundary Problems*, Progress in Mathematics, Vol. 344, Birkhäuser, 2022.
112. E. Marmolejo-Olea, I. Mitrea, M. Mitrea, and Q. Shi, *Transmission boundary problems for Dirac operators on Lipschitz domains and applications to Maxwell's and Helmholtz's equations*, Trans. Amer. Math. Soc., 364 (2012), no. 8, 4369–4424.
113. J. Martell, D. Mitrea, I. Mitrea, and M. Mitrea, *The higher order regularity Dirichlet problem for elliptic systems in the upper half-space*, pp. 123–141 in "Harmonic Analysis and Partial Differential Equations," Contemporary Mathematics, Vol. 612, Amer. Math. Soc., Providence, RI, 2014.

114. J.M. Martell, D. Mitrea, I. Mitrea, and M. Mitrea, *The Dirichlet problem for elliptic systems with data in Köthe function spaces*, Revista Matemática Iberoamericana, 32 (2016), no. 3, 913–970.
115. J.M. Martell, D. Mitrea, I. Mitrea, and M. Mitrea, *The* BMO*-Dirichlet problem for elliptic systems in the upper half-space and quantitative characterizations of* VMO, Analysis and PDE, 12 (2019), no. 3, 605–720.
116. J.M. Martell, D. Mitrea, I. Mitrea, and M. Mitrea, *Fatou-type theorems and boundary value problems for elliptic systems in the upper-half space*, Algebra and Analysis (Algebra i Analiz), 31 (2019), no. 2, 3–50.
117. V. Maz'ya, personal communication, (2005).
118. V. Maz'ya, *Seventy five (thousand) unsolved problems in analysis and partial differential equations*, Integr. Equ. Oper. Theory, 90 (2018), no. 2. 1–44.
119. V.G. Maz'ya, *Boundary Integral Equations*, Part II in "Analysis IV: Linear and Boundary Integral Equations", V.G. Mazya and S.M. Nikol'skii (eds), pp. 127–233, Springer-Verlag, 1991.
120. V. Maz'ya, M. Mitrea, and T. Shaposhnikova, *The inhomogeneous Dirichlet problem for the Stokes system in Lipschitz domains with unit normal close to VMO*, Journal of Functional Analysis and Its Applications, 43 (2009), no. 3, 217–235. (translated from Funktsional'nyi Analiz i Ego Prilozheniya, 43 (2009), no. 3, 65–88).
121. V. Maz'ya, M. Mitrea, and T. Shaposhnikova, *The Dirichlet problem in Lipschitz domains for higher order elliptic systems with rough coefficients*, Journal d'Analyse Mathématique, 110 (2010), no. 1, 167–239.
122. V. Maz'ya and T. Shaposhnikova, *Higher regularity in the classical layer potential theory for Lipschitz domains*, Indiana Univ. Math. J., 54 (2005), no. 1, 99–142.
123. V. Maz'ya and A. Soloviev, *Boundary Integral Equations on Contours with Peaks*, Operator Theory: Advances and Applications , Vol. 196, Birkhäuser, Basel, 2009.
124. N. Miller, *Weighted Sobolev spaces and pseudodifferential operators with smooth symbols*, Trans. Amer. Math. Soc., 269, (1982), no. 1, 91–109.
125. D. Mitrea, *Layer potentials and Hodge decompositions in two dimensional Lipschitz domains*, Mathematische Annalen, 322 (2002), no. 1, 75–101.
126. D. Mitrea, *The method of layer potentials for non-smooth domains with arbitrary topology*, Integral Equations and Operator Theory, 29 (1997), 320–338.
127. D. Mitrea, *Distributions, Partial Differential Equations, and Harmonic Analysis*, Second edition, Springer Nature Switzerland, 2018.
128. D. Mitrea, I. Mitrea, and M. Mitrea, *A sharp divergence theorem with nontangential traces*, Notices of the American Mathematical Society, 67 (2020), no. 9, 1295–1305.
129. D. Mitrea, I. Mitrea, and M. Mitrea, *Geometric Harmonic Analysis I: A Sharp Divergence Theorem with Nontangential Pointwise Traces*, Developments in Mathematics Vol. 72, Springer Nature, Switzerland, 2022.
130. D. Mitrea, I. Mitrea, and M. Mitrea, *Geometric Harmonic Analysis II: Function Spaces Measuring Size and Smoothness on Rough Sets*, Developments in Mathematics Vol. 73, Springer Nature, Switzerland, 2023.
131. D. Mitrea, I. Mitrea, and M. Mitrea, *Geometric Harmonic Analysis III: Integral Representations, Calderón-Zygmund Theory, Fatou Theorems, and Applications to Scattering*, Developments in Mathematics Vol. 74, Springer Nature, Switzerland, 2023.
132. D. Mitrea, I. Mitrea, and M. Mitrea, *Geometric Harmonic Analysis IV: Boundary Layer Potentials in Uniformly Rectifiable Domains, and Applications to Complex Analysis*, Developments in Mathematics Vol. 75, Springer Nature, Switzerland, 2023.
133. D. Mitrea, I. Mitrea, and M. Mitrea, *Geometric Harmonic Analysis V: Fredholm Theory and Finer Estimates for Integral Operators, with Applications to Boundary Problems*, Developments in Mathematics Vol. 76, Springer Nature, Switzerland, 2023.
134. D. Mitrea, I. Mitrea, and M. Mitrea, *The Dirichlet problem with* VMO *data in upper-graph Lipschitz domains*, Rendiconti Lincei Matematica e Applicazioni (Atti Accad. Naz. Lincei Rend. Lincei Mat. Appl.), 30 (2019), no. 4, 701–732.

135. D. Mitrea, I. Mitrea, M. Mitrea, and S. Monniaux, *Groupoid Metrization Theory with Applications to Analysis on Quasi-Metric Spaces and Functional Analysis*, Birkhäuser, 2013.
136. D. Mitrea, I. Mitrea, M. Mitrea, and M. Taylor, *The Hodge-Laplacian: Boundary Value Problems on Riemannian Manifolds*, Studies in Mathematics, Vol. 64, De Gruyter, 2016.
137. D. Mitrea and M. Mitrea, *Uniqueness for inverse conductivity and transmission problems in the class of Lipschitz domains*, Comm. Partial Differential Equations, 23 (1998), 1419–1448.
138. D. Mitrea and M. Mitrea, *General second order, strongly elliptic systems in low dimensional nonsmooth manifolds*, pp. 61–86 in "Harmonic Analysis and Boundary Value Problems", L. Capogna and L. Lanzani eds., Contemporary Math., Vol. 277, AMS, 2001.
139. D. Mitrea and M. Mitrea, *Finite energy solutions of Maxwell's equations and constructive Hodge decompositions on nonsmooth Riemannian manifolds*, J. Funct. Anal., 190 (2002), no. 2, 339–417.
140. D. Mitrea, M. Mitrea, and S. Monniaux, *The Poisson problem for the exterior derivative operator with Dirichlet boundary condition on nonsmooth domains*, Communications in Pure and Applied Analysis, 7 (2008), no. 6, 1295–1333.
141. D. Mitrea, M. Mitrea, and J. Pipher, *Vector potential theory on Lipschitz domains in $\mathbb{R}^3$ and applications to electromagnetic scattering*, Journal of Fourier Analysis and Applications, 3 (1997), no. 2, 131–192.
142. D. Mitrea, M. Mitrea, and J. Verdera, *Characterizing regularity of domains via the Riesz transforms on their boundaries*, Analysis and PDE, 9 (2016), no. 4, 955–1018.
143. D. Mitrea, M. Mitrea, and L. Yan, *Boundary value problems for the Laplacian in convex domains and semiconvex domains*, Journal of Functional Analysis, 258 (2010), no. 8, 2507–2585.
144. I. Mitrea, *Spectral radius properties for layer potentials associated with the elastostatics and hydrostatics equations in nonsmooth domains*, J. Fourier Anal. Appl., 5 (1999), no. 4, 385–408.
145. I. Mitrea, *On the spectra of elastostatic and hydrostatic layer potentials on curvilinear polygons*, J. Fourier Anal. Appl., 8 (2002), no. 5, 443–487.
146. I. Mitrea, *On the traction problem for the Lamé system on curvilinear polygons*, J. Integral Equations and Applications, 16 (2004), no. 2, 175–218.
147. I. Mitrea and M. Mitrea, *The Poisson problem with mixed boundary conditions in Sobolev and Besov spaces in non-smooth domains*, Trans. Amer. Math. Soc., 359 (2007), no. 9, 4143–4182.
148. I. Mitrea and M. Mitrea, *Multi-Layer Potentials and Boundary Problems for Higher-Order Elliptic Systems in Lipschitz Domains*, Lecture Notes in Mathematics, Vol. 2063, Springer, Berlin, 2013.
149. I. Mitrea and K. Ott, *Counterexamples to the well-posedness of L^p transmission boundary value problems for the Laplacian*, Proc. Amer. Math Soc., 135 (2007), no. 7, 2037–2043.
150. M. Mitrea, *The method of layer potentials in electro-magnetic scattering theory on non-smooth domains*, Duke Math. J., 77 (1995), no. 1, 111–133.
151. M. Mitrea, *Boundary value problems and Hardy spaces associated to the Helmholtz equation in Lipschitz domains*, J. Math. Anal. Appl., 202 (1996), no. 3, 819–842.
152. M. Mitrea, *Generalized Dirac operators on non-smooth manifolds and Maxwell's equations*, J. Fourier Anal. Appl., 7 (2001), 207–256.
153. M. Mitrea and M. Taylor, *Boundary layer methods for Lipschitz domains in Riemannian manifolds*, J. Funct. Anal., 163 (1999), no. 2, 181–251.
154. M. Mitrea and M. Taylor, *Potential theory on Lipschitz domains in Riemannian manifolds: Hölder continuous metric tensors*, Communications in Partial Differential Equations, 25 (2000), no. 7-8, 1487–1536.
155. M. Mitrea and M. Taylor, *Potential theory on Lipschitz domains in Riemannian manifolds: Sobolev-Besov space results and the Poisson problem*, J. Funct. Anal., 176 (2000), no. 1, 1–79.
156. M. Mitrea and M. Taylor, *Potential theory on Lipschitz domains in Riemannian manifolds: L^p, Hardy, and Hölder space results*, Communications in Analysis and Geometry, 9 (2001), no. 2, 369–421.

157. M. Mitrea and M. Taylor, *Boundary problems for the Navier-Stokes system on non-smooth Riemannian manifolds*, Math. Annalen, 321 (2001), no. 4, 955–987.
158. M. Mitrea and M. Taylor, *Potential theory for Lipschitz subdomains of Riemannian manifolds equipped with Dini continuous metric tensors*, Transactions of the American Mathematical Society, 355 (2003), no. 5, 1961–1985.
159. M. Mitrea and M. Taylor, *The Poisson problem in weighted Sobolev spaces on Lipschitz domains*, Indiana Univ. Math. J., 55 (2006), no. 3, 1063–1089.
160. M. Mitrea and M. Wright, *Boundary Value Problems for the Stokes System in Arbitrary Lipschitz Domains*, Astérisque, Soc. Math. de France, Vol. 344, 2012.
161. R.L. Moore, *A characterization of Jordan regions by properties having no reference to their boundaries*, Proceedings of the National Academy of Sciences of the United States of America, 4 (1918), no. 12, 364–370.
162. M. Mourgoglou and X. Tolsa, *The regularity problem for the Laplace equation in rough domains*, preprint, (2022).
163. B. Muckenhoupt and R.L. Wheeden, *Weighted bounded mean oscillations and the Hilbert transform*, Studia Matematica, 54 (1976), 221–237.
164. C. Müller, *Über die Beugung elektromagnetischer Schwingungen an endlichen homogenen Körpern*, Math. Ann., 123 (1951), 345–378.
165. C. Müller, *Foundations of the Mathematical Theory of Electromagnetic Waves*, Springer-Verlag, Berlin, Heidelberg, New York, 1969.
166. F. Nazarov, X. Tolsa, and A. Volberg, *On the uniform rectifiability of AD-regular measures with bounded Riesz transform operator: the case of codimension* 1, Acta Math., 213 (2014), 237–321.
167. I. Netuka, *Fredholm radius of a potential theoretic operator for convex sets*, Čas. Pěstovani Mat., 100 (1975), no. 4, 374–383.
168. C. Neumann, *Untersuchungen über das Logarithmische und Newtonsche Potential*, B.G. Teubner, Leipzig, 1877.
169. S.M. Nikol'skii *Linear equations in linear normed spaces*, Izv., Akad., Nauk SSSR, Ser. Mat., 7 (1943), no. 3, 147–166.
170. F. Olver, *NIST Digital Library of Mathematical Functions*, Chapter 10, `http://dlmf.nist.gv/10$#$PT1`, Section 10.6, [edited by] Frank W.J. Olver and L.C. Maximon, New York: Cambridge University Press: NIST, 2010.
171. S. V. Piro, *Harmonic functions on Lipschitz domains with* VMO *data*, Ann. Univ. Ferrara-Sez. VII-Sc. Mat., XXVIII (1982), 97–104.
172. J. Plemelj *Potentialtheoretische Untersuchungen*, Preisschriften der Fürstlich Jablonowskischen Gesellschaft zu Leipzig, Teubner-Verlag, Leipzig, 1911.
173. H. Poincaré, *La méthode de Neumann et le problème de Dirichlet*, Acta Mathematica, 20 (1897), 59–142.
174. H. Poincaré, *Théorie du potentiel newtonien*, Paris 1899.
175. H. Poincaré, *Science et Méthode*, Flammarion, 1908 (English translation in Dover Publications, New York, 1952).
176. J. Radon, *Über lineare Funktionaltransformationen und Funktionalgleichungen*, Sitzungsberichte Akad. Wiss., Abt. 2a, Wien, 128 (1919), 1083–1121.
177. J. Radon, *Über die Randwertaufgaben beim logarithmischen Potential*, Sitzungsberichte Akad. Wiss., Abt. 2a, Wien, 128 (1919), 1123–1167.
178. A. Rathsfeld, *The invertibility of the double layer potential operator in the space of continuous functions defined on a polyhedron. The panel method*, Appl. Anal., 45 (1992), 135–177.
179. F. Rellich, *Uber das asymptotische Verhalten der Losungen von* $\Delta u + \lambda u = 0$ *in unendlichen Gebieten*, Jber. Deutsch. Math. Verein., 53 (1943), 57–65.
180. F. Riesz and B. Sz.-Nagy, *Functional Analysis*, Ungar Publishing Co., New York, 1955, French original: Akad. Kiado, Budapest, 1952.
181. W. Rudin, *Functional Analysis*, 2nd edition, McGraw-Hill, Inc., New York, 1991.
182. S. Saks, *Theory of the Integral*, 2nd edition, Dover Publications, New York, 1964.
183. W.K. Saunders, *On solutions of Maxwell equations in an exterior region*, Proc. Nat. Acad. Sci. U.S.A., 38 (1952), 342–348.

184. S. Semmes, *Chord-arc surfaces with small constant. I*, Adv. Math., 85 (1991), no. 2, 198–223.
185. S. Semmes, *Chord-arc surfaces with small constant. II. Good parametrization*, Adv. Math., 88 (1991), no. 2, 170–199.
186. Z.Ya. Shapiro, *The first boundary value problem for an elliptic system of differential equations*, (Russian) Mat. Sb., 28 (1951), no. 1, 55–78.
187. Z.Ya. Shapiro, *On general boundary problems for equations of elliptic type*, (Russian) Izvestiya Akad. Nauk SSSR. Ser. Mat., 17 (1953), 539–562.
188. V.Yu. Shelepov, *On the index of an integral operator of the potential type in the space L_p*, Dokl. Akad. Nauk SSSR 186 (6) (1969), 1266–1268 (English translation: Sov. Math. Dokl. 10/A, 754–757).
189. V.Yu. Shelepov, *On the index and spectrum of integral operators of potential type along Radon curves*, Math. USSR Sbornik, 70 (1991), no. 1, 175–203.
190. V.Yu. Shelepov, *On investigations by Ya.B. Lopatinskiĭ's method of matrix integral equations in the space of continuous functions* (Russian), General Theory of Boundary Value Problems, Collect. Sci. Works, Naukova Dumka, Kiev, 1983, pp. 220–226.
191. Z. Shen, *A note on the Dirichlet problem for the Stokes system in Lipschitz domains*, Proc. Amer. Math. Soc., 123 (1995), no. 3, 801–811.
192. Z. Shen, *Boundary value problems in Morrey spaces for elliptic systems on Lipschitz domains*, Amer. J. Math., 125 (2003), no. 5, 1079–1115.
193. Z. Shen, *Necessary and sufficient conditions for the solvability of the L^p Dirichlet problem on Lipschitz domains*, Mathematische Annalen, 336 (2006), 697–725.
194. Z. Shen, *The L^p boundary value problems on Lipschitz domains*, Advances in Mathematics, 216 (2007), 212–254.
195. V.A. Solonnikov, *General boundary value problems for systems elliptic in the sense of A. Douglis and L. Nirenberg. I*, (Russian) Izv. Akad. Nauk SSSR, Ser. Mat., 28 (1964), 665–706.
196. V.A. Solonnikov, *General boundary value problems for systems elliptic in the sense of A. Douglis and L. Nirenberg. II*, (Russian) Trudy Mat. Inst. Steklov, Vol. 92 (1966), 233–297.
197. S.L. Sobolev, *Partial Differential Equations of Mathematical Physics*, Pergamon Press, Oxford, 1964.
198. E.M. Stein, *Singular Integrals and Differentiability Properties of Functions*, Princeton Mathematical Series, No. 30, Princeton University Press, Princeton, NJ, 1970.
199. E.M. Stein and G. Weiss, *Generalization of the Cauchy-Riemann equations and representations of the rotation group*, Amer. J. Math., 90 (1968), no. 1., 163–196.
200. E.M. Stein and G. Weiss, *Introduction to Fourier Analysis on Euclidean Spaces*, Princeton University Press, Princeton, N.J., 1971.
201. O. Tapiola and X. Tolsa, *Connectivity conditions and boundary Poincaré inequalities*, preprint, (2022).
202. T. Toro, *Potential analysis meets geometric measure theory*, Proceedings of the International Congress of Mathematicians 2010 (ICM 2010), pp. 1485–1497 (2011).
203. G. Verchota, *Layer potentials and regularity for the Dirichlet problem for Laplace's equation in Lipschitz domains*, J. Funct. Anal., 59 (1984), no. 3, 572–611.
204. M. I. Višik, *On strongly elliptic systems of differential equations*, Mat. Sbornik N.S., 29 (1951), 615–676.
205. H. Weyl, *Das asymptotische Verteilungsgesetz der Eigenschwingungen eines beliebig gestalteten elastischen Körper*, Rend. Circ. Mat. Palermo, 34 (1914), 1–49. (The asymptotic distribution law for the eigen-oscillations of an elastic body of arbitrary form; English translation by D.H. Delphenich.)
206. H. Weyl, *Die natürlichen Randwertaufgaben im Aussenraum für Strahlungsfeldern beliebiger Dimensionen und beliebiger Ranges*, Math. Zeit., 56 (1952), 105–119.
207. J.T. Wloka, B. Rowley, and B. Lawruk, *Boundary Value Problems for Elliptic Systems*, Cambridge University Press, 1995.
208. T. Xiangxing, *Noncontinuous data boundary value problems for Schrödinger equation in Lipschitz domains*, Front. Math. China, 4 (2006), 589–603.

209. G. Zecca, *The unsolvability of the Dirichlet problem with* L(log L)$^{\alpha}$ *boundary data*, Rend. Acc. Sc. Fis. Mat. Napoli, 72 (2005), 71–80.
210. G. Zecca, *On the Dirichlet problem with Orlicz boundary data*, Bollettino della Unione Matematica Italiana B, 10 (2007), no. 3, 661–679.

Subject Index

D. Mitrea et al., *Geometric Harmonic Analysis V*, Developments in Mathematics 76,
https://doi.org/10.1007/978-3-031-31561-9

Symbol Index

D. Mitrea et al., *Geometric Harmonic Analysis V*, Developments in Mathematics 76, https://doi.org/10.1007/978-3-031-31561-9